D1552101

The Water Encyclopedia

THIRD EDITION

Hydrologic Data and Internet Resources

The Water Encyclopedia

THIRD EDITION

Hydrologic Data and Internet Resources

Edited by
Pedro Fierro, Jr.
Evan K. Nyer

CRC is an imprint of the Taylor & Francis Group,
an informa business

CRC Press
Taylor & Francis Group
6000 Broken Sound Parkway NW, Suite 300
Boca Raton, FL 33487-2742

Printed in the United States of America on acid-free paper
10 9 8 7 6 5 4 3 2 1

International Standard Book Number-10: 1-56670-645-9 (Hardcover)
International Standard Book Number-13: 978-1-56670-645-2 (Hardcover)

Library of Congress Cataloging-in-Publication Data

The water encyclopedia : hydrologic data and Internet resources / editors, Pedro Fierro, Jr. and Evan K. Nyer. -- 3rd ed.
p. cm.
Includes bibliographical references and index.
ISBN-13: 978-1-56670-645-2 (alk. paper)
ISBN-10: 1-56670-645-9 (alk. paper)
1. Water resources development--Encyclopedias. 2. Hydrology--Encyclopedias. 3. Water resources development--Computer network resources. 4. Hydrology--Computer network resources. 5. Water resources development--Charts, diagrams, etc. 6. Hydrology--Charts, diagrams, etc. I. Fierro, Pedro. II. Nyer, Evan K. III. Title.

TC403.W38 2006
553.703--dc22 2006008689

Visit the Taylor & Francis Web site at
http://www.taylorandfrancis.com

and the CRC Press Web site at
http://www.crcpress.com

Preface

"Just do an Internet search." "It's on the Internet." How often have we said or been told that we could find it on the Internet. This third edition of *The Water Encyclopedia: Hydrologic Data and Internet Resources* started from a premise that most of the information provided within this publication could be found on the Internet. As our team of contributing authors started reviewing each section within each chapter, it soon became apparent that you cannot always find it on the internet. This edition represents many hours of effort to identify the most current information on a wide range of water-related topics whether it can be found on the internet or in other sources.

The *Encyclopedia* has retained many of the elements of the previous editions but has also been expanded to reflect the many changes within the environmental industry as well as the current and topical water-related matters of the last decade. Prepared by scientists and engineers, this publication is intended to serve as a valuable resource to all professionals dealing with water-related issues as well as the general public. The material presented has been footnoted to provide the user with the opportunity to return to the original source material for additional research. Where possible, an Internet URL address is provided to guide the user to the appropriate source.

The third edition of the *Encyclopedia* has been significantly expanded beyond the previous edition. The first two chapters of this edition are new and discuss data management and international data collection. Data management concepts are presented to review the use of databases, geographic information systems (GIS), data reporting and metadata. Data repositories and availability vary around the world and range in ease of access and usability. The international data collection provides some direction on potential data sources in less developed areas as well as case histories of actual project work and Internet sources for international water-related data.

This edition contains more than 1100 tables and 500 figures providing data related to weather, surface water, groundwater, water use, water quality, waste water, pollution, and water resource management. The pollution chapter alone has grown to contain some 450 plus tables and figures. Wastewater, previously included within the pollution chapter, is presented as a stand-alone chapter to facilitate use of this reference. A chapter of useful conversion factors and constants concludes this edition.

Whether you are looking for a specific piece of information or exploring one or more of the many topics related to water, this edition provides its users with a tremendous wealth of data whether on the Internet or not.

Acknowledgments

We want to extend our thanks and appreciation to the many individuals, publishers, and organizations that have made this third edition a reality. Without their time, cooperation, collaboration, this work would not have been possible. Most importantly, the support and access to resources for the management of this compilation provided by ARCADIS G&M was invaluable, and their on-going support and encouragement to undertake these efforts are deeply appreciated.

A number of individual contributors were involved in compiling the relevant information for each of the chapters and they are identified at the start of their chapters. Our thanks and appreciation to you and your families for the time committed to completing this task. Behind the scenes and the backbone of keeping everything organized, we want to extend a special thanks to Chris Worden and Carla Gerstner for their encouragement, patience, and the occasional stern word. Additionally, we want to acknowledge Barbara Kelly and Amanda Fierro for their efforts in preparing materials for the manuscript.

The Editors

Pedro Fierro, Jr. is a hydrogeologist and associate vice president with ARCADIS G&M, Inc., where he is involved with a wide variety of environmental assessments and remediation programs. He has been responsible for the direction of several hundred sites addressing environmental issues. Fierro has addressed various audiences on topics ranging from sampling methodologies, regulatory compliance, site assessment techniques, liability management, and remediation technologies.

Fierro received his bachelor's degree in geology from the University of Rochester, Rochester, New York and his master's degree in geology with an emphasis on groundwater studies from the University of Kentucky. He currently holds geological professional licenses/registrations in Alabama, Florida, Georgia, Kentucky, Pennsylvania, and Tennessee. He is a certified groundwater professional and a certified professional geologist. He was a contributing author to *In Situ Treatment Technology*.

Evan K. Nyer is a senior vice president with ARCADIS G&M, Inc., where he is responsible for maintaining and expanding the company's technical expertise in geology/hydrogeology, engineering, fate and transport, and remediation technologies. He has been active in the development of new treatment technologies for many years. He has been responsible for the strategies, technical designs and installations of more than 400 groundwater and soil remediation systems at contaminated sites throughout the United States. Nyer also lectures, provides expert testimony, and serves as the public spokesperson for one technically complicated site.

Nyer received his graduate degree in environmental engineering from Purdue University and has authored five books: *Practical Techniques for Groundwater and Soil Remediation*, published by Lewis Publishers, Inc.; *Groundwater Treatment Technology*, first and second edition, published by Van Nostrand Reinhold; *Groundwater and Soil Remediation*, and *In Situ Treatment Technology* (now in its second edition) published by CRC Press; and is co-author of *Bioremediation*, published by the American Academy of Environmental Engineers. Nyer is a regular contributor to *Groundwater Monitoring and Remediation* having had his own column "Treatment Technology" in the periodical for the past 20 years.

Contributors

James M. Bedessem
ARCADIS G&M, Inc.
Tampa, Florida

Brian Burke
ARCADIS G&M, Inc.
Tampa, Florida

Pedro Fierro, Jr.
ARCADIS G&M, Inc.
Tampa, Florida

William H. Lynch
ARCADIS G&M, Inc.
West Palm Beach, Florida

Daniel J. McCarthy
ARCADIS G&M, Inc.
Philadelphia, Pennsylvania

Melvin Rivera
ARCADIS G&M, Inc.
Tampa, Florida

Christopher Spooner
ARCADIS G&M, Inc.
Tampa, Florida

Gustavõ Suarez
ARCADIS G&M, Inc.
Tampa, Florida

Katherine L. Thalman
ARCADIS G&M, Inc.
Tampa, Florida

Daniel Zell
Dewberry & Davis, LLC
Fairfax, Virginia

Contents

CHAPTER 1

Data Management

Daniel J. McCarthy

CONTENTS

1.1 INTRODUCTION

Data management encompasses many tasks, priorities, and decisions. Underlying these activities is the need for an accurate data from sampling and monitoring programs designed to measure the effects of operational activities. To understand the value of good data management, it is helpful to understand the nature of how this information is generated and used to support management decisions.

So what is data? The American Heritage Dictionary defines data as factual information, especially information organized for analysis or used to reason or make decisions, or values derived from scientific experiments. Scientific professionals generate huge quantities of data every day. It is estimated that scientists spend 80% of their time managing the data and 20% analyzing and interpreting. By establishing sound data management practices, more time can be spent in data analysis and interpretation.

Throughout this chapter, we will provide examples of data management practices within the context of an investigation of contaminated groundwater and surface water. These practices are directly applicable to managing other types of data, such as those found in this book.

1.2 DATABASE OVERVIEW

A discussion of data management would be incomplete without a general discussion of databases.

A very general definition of a database might be "*A collection of related items of information contained on*

Table 1.1 Groundwater and Surface Water Location Data

Location ID	Area of Concern	*x*	*y*	*z*
SW-01	Upstream	1,75,470.994	16,35,550.124	100.203
SW-02	Upstream	1,77,126.487	16,35,925.814	100.102
SW-03	Outfall 1	1,77,047.029	16,35,676.853	100.00
SW-04	Downstream	1,76,871.093	16,35,674.137	98.97
SW-05	Downstream	1,75,790.418	16,35,597.208	97.96
MW-01	Background area	1,74,345.077	16,32,431.087	96.597
MW-02	Oil storage area	1,74,251.127	16,32,466.059	97.384
MW-03	Oil storage area	1,74,690.942	16,31,435.707	97.384

various media organized in a way that allows easy search and retrieval of subsets of the items of information." Note that, strictly speaking, a database does not have to be electronic: Boxes containing recipes, telephone books, or paper address books are all databases. A database used for environmental purposes might be composed of a combination of paper copies of information along with items of information contained in electronic form, with perhaps some sort of paper or electronic index to or inventory list of all of the data.

Electronic data are nearly always organized into tables. Consider the example shown in Table 1.1.

Each row of this table represents a single data point; in this case the first row provides data about location SW-1, and only SW-1. Each column of this table represents a type of data that is stored for each row. In our example, the Area of Concern column identifies the spatial group that each location belongs to.

The rows in a database table are typically called "records," while the columns are called "fields." Thus, a useful definition of an electronic database is "A collection of related items of data organized into one or more tables." Each field is constrained to a single data type. Table 1.2 lists the most common data types.

Electronic databases are typically either "flat file databases," in which the entire database resides in a single table, or "relational databases," in which the data are distributed into more than one table, which are then linked together by a common key field. The tables will be related to one another according to a one-to-one (each record in one table has a single matching record in the other table) or one-to-many relationship (each record in one table may have one or more matching records in the other table, but not the reverse).

Table 1.2 Common Data Types

Data Type	Description or Example
Integer	Typically stores numbers that relate to counts quantities, or, ID numbers
Decimal	Numbers with fractional parts such as percentages or rates
Floating point	Numbers with a scientific notation that can be calculated approximately, such as distance or weight
Fixed length character	Names, descriptions, addresses
Date/Time	Storage of date and/or time or intervals of dates or times
Boolean	Explicit constraints (Yes/No, True/False) or logical constraints (AND/OR)
Unstructured data	Images, video, audio

One significant advantage of relational databases to flat-file databases is the ability to query the data in different ways. A query is defined as a statement to retrieve database records that match certain criteria. By structuring the query statement a certain way, different information can be returned from the data set.

The most popular general-purpose software for managing data is an electronic spreadsheet program such as Microsoft Excel. Electronic spreadsheets are excellent tools for managing electronic data that fit in a single table. However, spreadsheets are cumbersome or inadequate tools for managing relational data, where more than one table is required. For managing relational data, other more powerful data management programs should be used. There are many popular relational database management systems available, including Microsoft Access. It should be noted that, in contrast to flat-file databases, which typically can be managed by the casual computer user, large relational databases require management by trained individuals, and will usually be beyond the capabilities of the casual user.

The field of database management is continually in flux, and would have changed by the time this book is published. Thus, it is impossible to cover all the facets of data management and database theory here. However, it can be said that the relational database model overwhelmingly dominates large-scale data management and database theory. For further information about relational databases, the reader is directed to any of the numerous references on this subject. A particularly helpful book designed for the casual database user is by Michael J. Hernandez (2003) entitled *Database Design for Mere Mortals: A Hands-On Guide to Relational Database Design*, Addison–Wesley Developers Press.

1.3 GEOGRAPHIC INFORMATION SYSTEMS OVERVIEW

Data are often presented in a tabular format. Sometimes, a visual representation is helpful in drawing conclusions, particularly if the data have a spatial component. A Geographical Information System (GIS) is a way to display information with a spatial component. GIS can be defined as a software package that manages and displays information in a database composed of data that are associated with spatial information. That is to say, there will be both tabular and spatial information in the database, it will be possible to query the database for specific data, and the user will be able to display the data spatially, as a map.

The software package comprising the GIS may be a single program, a set of programs from a single vendor, a combination of programs that together constitute the GIS, a custom-programmed software package, or any combination thereof.

In the groundwater arena, the tabular data could typically be depth to water data, water table elevation data, water chemistry data, and water quality data. In addition, the tabular data will have some spatial component in either two or three dimensions, as *x–y–z* coordinates. Spatial data, in addition to the *x–y–z* coordinates mentioned above, often will include digitally processed air or satellite photographs, computer-aided design (CAD) drawings, or other electronic spatial entities. Clearly, depth to water data and water chemistry data, which can be displayed in both plan view (two-dimensional) and side view (three-dimensional), are well suited to management using a GIS. Table 1.3 is an example of data from a GIS system representing depth to groundwater.

When generated in a GIS system in plan view, the view can look like Figure 1.1.

The primary attraction of a GIS is the ability to manage, query and display a large amount of data spatially, in real time. Using most GISs, the user can view the data on a map, query for a subset of the data while viewing the map, and then see the distribution of the subset data when the map is refreshed. This process can be repeated as many times as necessary to answer a question. For example, the temporary wells in the previous figure were not measured during the August sampling event. By querying the GIS system only for locations that were measured during the August event, we return a subset of the data, shown in Table 1.4.

From this set of data, our map would look like Figure 1.2.

The literature on GISs is voluminous. Because GISs have a wide applicability throughout many disciplines, the reader is directed to the internet, where search engines associated with any popular internet portal (Yahoo! or America Online, for example) may be used to find literally thousands of references on the subject.

Table 1.3 Depth to Groundwater Data from GIS System

Well_Id	Loc_Type	X_Coord	Y_Coord	Date	Gwelev
MW131	Monitor well	25,48,766.73000	3,19,202.76570	8/17/2004	19.21000
MW132	Monitor well	25,48,671.72700	3,19,228.04840	8/17/2004	20.38000
MW133	Monitor well	25,48,677.58100	3,19,285.61010	8/17/2004	19.23000
MW134	Monitor well	25,48,740.41300	3,19,287.73680	8/17/2004	18.13000
MW135	Monitor well	25,48,668.82800	3,19,335.81440	8/17/2004	18.78000
MW136	Monitor well	25,48,747.90500	3,19,346.29610	8/17/2004	16.71000
MW137	Monitor well	25,48,704.97300	3,19,107.44400	8/17/2004	21.40000
TW-161	Temp monitor well	25,48,677.70000	3,18,512.69000	8/17/2004	NM
TW-162	Temp monitor well	25,49,393.00000	3,17,485.00000	8/17/2004	NM
TW-163	Temp monitor well	25,49,412.00000	3,17,501.00000	8/17/2004	NM
TW-164	Temp monitor well	25,48,775.80000	3,19,271.63000	8/17/2004	NM
TW-165	Temp monitor well	25,48,788.62000	3,19,196.71000	8/17/2004	NM
TW-166	Temp monitor well	25,48,656.95000	3,19,215.44000	8/17/2004	NM
TW-167	Temp monitor well	25,48,690.37000	3,19,153.66000	8/17/2004	NM
TW-168	Temp monitor well	25,48,618.73000	3,19,285.76000	8/17/2004	NM
TW-169	Temp monitor well	25,48,672.29000	3,19,291.02000	8/17/2004	NM
TW-170	Temp monitor well	25,48,722.90000	3,19,286.09000	8/17/2004	NM
TW-171	Temp monitor well	25,48,745.57000	3,19,224.97000	8/17/2004	NM
TW-172	Temp monitor well	25,48,758.72000	3,19,164.83000	8/17/2004	NM
TW-173	Temp monitor well	25,48,304.43000	3,19,248.26000	8/17/2004	NM
TW-174	Temp monitor well	25,48,314.10000	3,19,441.75000	8/17/2004	NM
TW-175	Temp monitor well	25,48,247.26000	3,19,666.89000	8/17/2004	NM

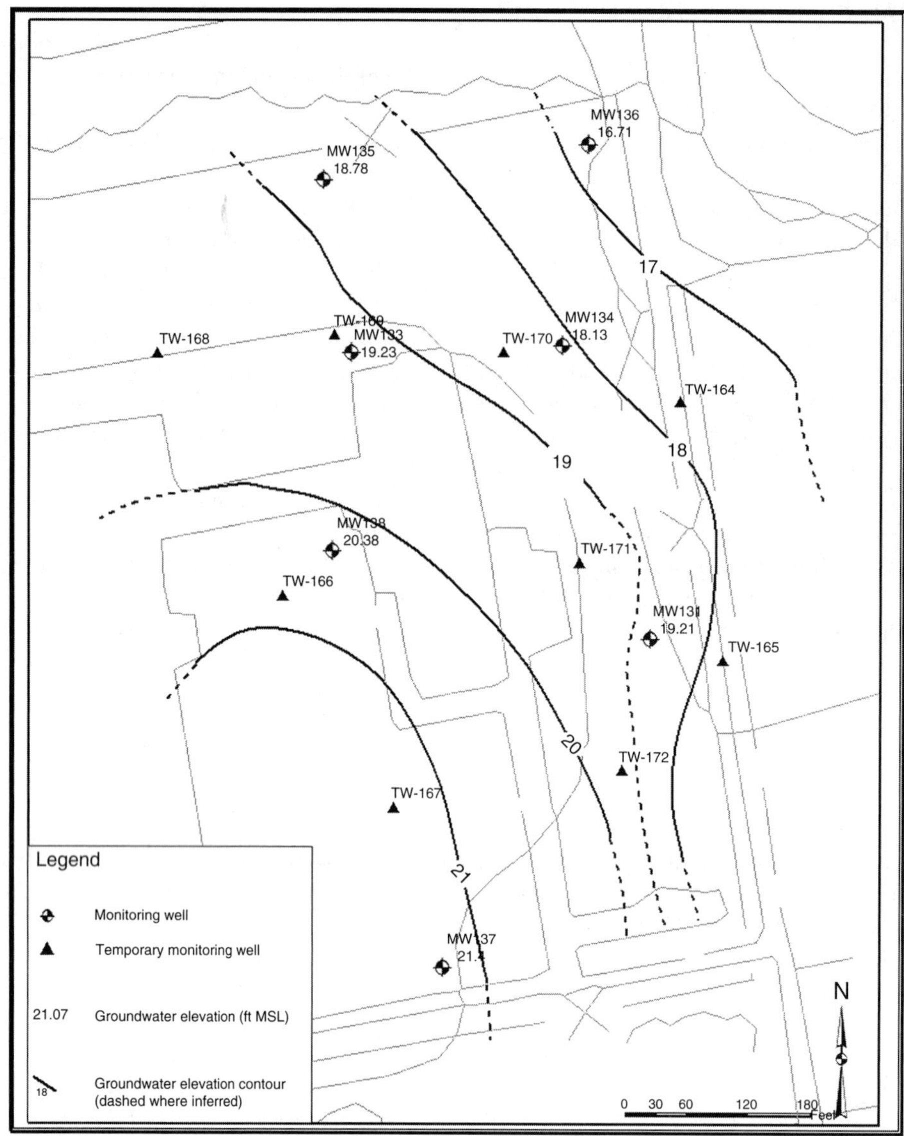

Figure 1.1 Example of groundwater elevation data from GIS System.

1.4 UNDERSTANDING DATA MANAGEMENT NEEDS

Initially designing a data management program for an investigation or experiment requires significant scientific expertise. However, after design and implementation, the processes generally follow a well-defined and straightforward cycle. In our groundwater contamination example, samples are routinely collected and sent to laboratories where they are analyzed with the results reported in the form of a hard-copy analytical results report. From this point on, the data are put to multiple uses to meet a variety of needs. Some portion of the data are collected and reported under the requirements of environmental permits, while additional data are generated voluntarily to further the objectives of sound environmental management. Accountability for effectively managing the collection and utilization of this information according to well-defined processes within

Table 1.4 Subset of Table 1.3 Data for August Events Only

Well_Id	Loc_Type	X_Coord	Y_Coord	Date	Gwelev
MW131	Monitor well	25,48,766.73000	3,19,202.76570	8/17/2004	19.21000
MW132	Monitor well	25,48,671.72700	3,19,228.04840	8/17/2004	20.38000
MW133	Monitor well	25,48,677.58100	3,19,285.61010	8/17/2004	19.23000
MW134	Monitor well	25,48,740.41300	3,19,287.73680	8/17/2004	18.13000
MW135	Monitor well	25,48,668.82800	3,19,335.81440	8/17/2004	18.78000
MW136	Monitor well	25,48,747.90500	3,19,346.29610	8/17/2004	16.71000
MW137	Monitor well	25,48,704.97300	3,19,107.44400	8/17/2004	21.40000

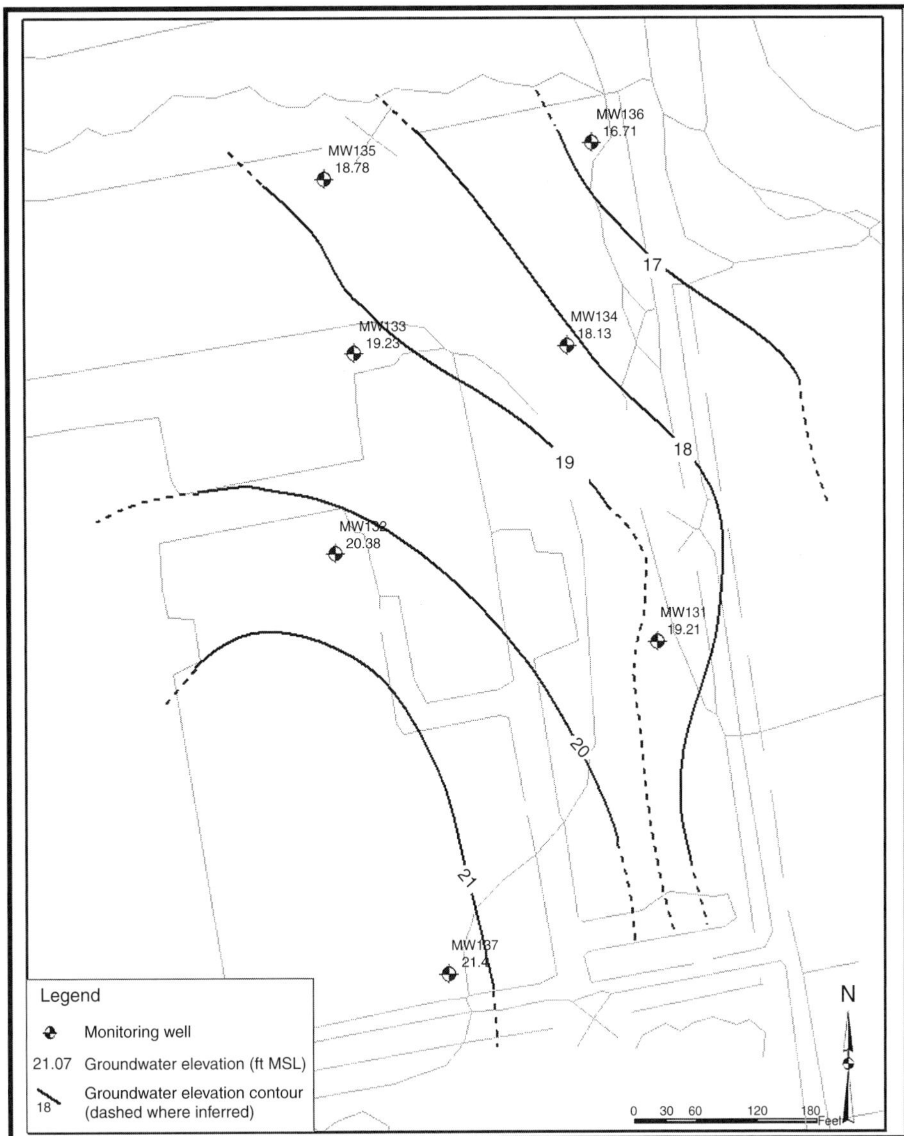

Figure 1.2 Groundwater elevation data from August only.

standardized software environments is essential to effective environmental management.

Problems with data collection can be mitigated with a data management plan. This plan will typically specify how data are to be labeled and categorized, the format that the data are to be stored in, how to handle data collected over a period of time or over a significant geographic area, and procedures to account for changes in the investigation and experiment. The advantage to having a data management plan is that it allows the users to collect, label, and record data in a consistent manner. By doing so, retrieval of that data does not have to take into account bias on the part of the individual collecting the data. Consistent terms, units, methods and procedures, will allow any user to retrieve data accurately and quickly.

1.5 DATA CATEGORIZATION

1.5.1 Spatial Data

Data found in this book are often of two distinct types: data with a spatial representation and that of a temporal representation. Water quality, for example, can be represented as changing over an area based on land use, and can also be represented as changing over time due to urban development changing the drainage pathways.

It is often helpful to have a defined nomenclature when collecting and categorizing data. This nomenclature makes it easy to glean basic information from the raw data, as well as expediting queries from the data management system. As an example, consider the following spatial data, shown in Table 1.5.

Table 1.5 Spatial Data for Groundwater and Surface Water Locations

Location ID	Area of Concern	*x*	*y*	*z*
SW-01	Upstream	1,75,470.994	16,35,550.124	100.203
SW-02	Upstream	1,77,126.487	16,35,925.814	100.102
SW-03	Outfall 1	1,77,047.029	16,35,676.853	100.00
SW-04	Downstream	1,76,871.093	16,35,674.137	98.97
SW-05	Downstream	1,75,790.418	16,35,597.208	97.96
MW-01	Background area	1,74,345.077	16,32,431.087	96.597
MW-02	Oil storage area	1,74,251.127	16,32,466.059	97.384
MW-03	Oil storage area	1,74,690.942	16,31,435.707	97.384

Consistent nomenclature for identifying spatial locations is crucial to maintaining data integrity, particularly where there is sensitivity to geographic, political, and physical boundaries. In addition, each location should be unique in order to maintain referential integrity within the data management system. This unique value in this table is referred to as a *primary key*. The location IDs are coded in such a way to let the reader know at a glance what type of location it is. Acceptable locations types in this example include:

Location Type Identifier	Location Type Description
MW	Monitoring well
SW	Surface water

The Area of Concern column provides a general categorization of where the location occurs within the context of the immediate surroundings. This is useful for queries that would ask for all data found in a particular site-specific area.

The example table also presents spatial data in the form of *x*, *y*, and *z* coordinates. Spatial data for each location should be collected in a form that is consistent with use at the site, but that can also reference features that are located nearby, such as surface water bodies, wetland areas, or other physiographic features. Use of a consistent coordinate system will make sure that all the data can be compared to each other.

Examples of coordinate systems include latitude, longitude and height, Universal Transverse Mercator (UTM), Earth Centered, Earth Fixed Cartesian (ECEF), and State Plane coordinates. In our example, the coordinates are in State Plane Coordinates. In the United States, the State Plane System was developed in the 1930s and was based on the North American Datum 1927 (NAD27), which are based on the foot. A more recent variation is the NAD83 system, which is based on the North American 1983 datum and is based on the meter. The State Plane System was developed to provide local references tied to a national datum. Most USGS 7.5 Minute Quadrangles use several coordinate system grids including latitude and longitude, UTM kilometer tic marks, and applicable State Plane coordinates.

1.6 TEMPORAL DATA

With the establishment of the spatial locations, data collected over time can be collected and referenced. Consider Table 1.6, which summarizes data from samples collected from our groundwater and surface water locations.

Consistent nomenclature for sample identifications is crucial to maintaining data integrity. And sample identifications should be unique in order to maintain referential integrity. Character limits are often in place in database management systems; therefore, care should be taken in minimizing spaces, dashes, or parentheses. In our example, the date of collection is captured in the sample ID in parentheses, which allows each sample to be unique.

By including the Location ID in this table, we establish a relationship to the previous table. This relationship allows us to query the data in different ways. Because the relationship is of one location to many samples, this is referred to as a *one-to-many relationship*. The Location ID in this table is referred to as a *foreign key* because it matches primary key values in our spatial data table presented earlier. Together the primary and foreign keys create a parent/child relationship, which is at the heart of relational database systems.

The Sample Type column provides an identifier to discriminate individual samples from each other based on quality assurance needs. In this example, all of the samples have a normal "N" type. If duplicate samples were required for quality assurance purpose to check on the validity of the data set, one could identify the sample type as a "D" for duplicate.

The Sample Matrix column identifies what the medium collected for each sample was, through

Table 1.6 Sample Information from Groundwater and Surface Water Locations

Sample ID	Location ID	Sample Type	Sample Matrix	Sample Date	Sample Time
SW-1(081602)	SW-1	N	WS	8/16/2002	13:00
SW-2(081602)	SW-2	N	WS	8/16/2002	13:10
SW-3(081602))	SW-3	N	WS	8/16/2002	13:20
SW-4(081602)	SW-4	N	WS	8/16/2002	13:30
SW-5(081602)	SW-5	N	WS	8/16/2002	14:31
MW-01(111301)	MW-01	N	WG	11/13/2001	15:22
MW-01(111401)	MW-01	N	WG	11/14/2001	16:31
MW-01(111501)	MW-01	N	WG	11/15/2001	11:00

an abbreviated two-digit code. Examples of these kinds of matrices are:

Sample Matrix	Matrix Description
WB	Water collected from borehole or during geoprobe investigation
WE	Estuary water
WG	Ground water
WL	Leachate
WO	Ocean water
WP	Drinking water
WQ	Water quality control matrix
WS	Surface water
WW	Waste water

The sample date and time columns allow the data to be sorted from more recent to historical and can provide a context for how the data changes over time.

Continuing with our groundwater contamination example, consider Table 1.7, which summarizes the data obtained from an analytical laboratory.

The primary key (Sample ID) is present again to establish the relationship back to the other tables. The other columns relate information related to the analyses performed on the samples and the results.

Each of these records represents a concentration of a given chemical at a given location for a specific point in time. As chemical concentrations, detection limits and detections change, those data points would be represented as new records in the database. This can be illustrated in the following query result, shown in Table 1.8.

1.7 DATA VALIDATION AND VERIFICATION

Once data are collected into the database, steps must be taken to ensure that it was collected accurately and is representative of the source. Data verification checks the compliance of the collected data against known requirements of the experiment or investigation. For example, using the data set shown above, we would verify with the analytical laboratory that the groundwater samples were analyzed according to specific methods, such as the use of the calibration samples for laboratory equipment. If a verification check fails, then the data may be considered suspect.

Data validation, by contrast, must take into account the suitability of the data, and must take into account how the data were collected, how the data were analyzed, and finally, based on the results of the review of the collection and analysis processes, how the data should be used. If the collection process is found to be flawed, the data might be discarded or used for qualitative purposes only. The United States Environmental Protection Agency has provided guidance documents on validating data, which can be found at www.epa.gov/superfund/programs/clp/guidance.htm.

1.8 DATA REPORTING

Once the data have been collected and validated into the data management system, it is time to retrieve the data and make some conclusions about its meaning. Data retrieval usually takes the form of querying the data and then reporting the data.

1.8.1 Querying Data

A query is a programmatic statement that asks a question about the data. Each query usually specifies a criterion, which is a condition or test that must be met in order for a given record to be selected. Queries produce subsets of data, which are the records that match the conditions of the query. For example, a query might request the locations where a certain chemical, e.g. carbon tetrachloride, exceeds a certain concentration, e.g. 500 μg/L. This query would produce a result like the one presented below in Table 1.9.

Table 1.7 Analytical Data Summary

Sample ID	Matrix	SDG	Lab Method	Chemical	Result	RDL	Detect	Unit
SW-1(081602)	WG	884825	SW8260	Carbon disulfide		2.5	N	μg/L
SW-1(081602)	WG	884825	SW8260	Xylene (total)		2.5	N	μg/L
SW-1(081602)	WG	884825	SW8260	Ethylbenzene		0.2	N	μg/IL
SW-1(081602)	WG	884825	SW8260	Carbon tetrachloride	670	25	Y	μg/L

Table 1.8 Additional Analytical Data

Sample ID	Matrix	SDG	Sample Date	Lab Method	Chemical	Result	RDL	Detect	Unit
SW-1(081602)	WG	884825	8/16/2002	SW8260	Carbon tetrachloride	670	25	Y	μg/L
SW-1(111602)	WG	884825	11/16/2002	SW8260	Carbon tetrachloride	340	25	Y	μg/L
SW-1(011603)	WG	884825	01/16/2003	SW8260	Carbon tetrachloride	100	5	Y	μg/L
SW-1(031603)	WG	884825	03/16/2003	SW8260	Carbon tetrachloride		5	N	μg/L

Table 1.9 Query Results for Carbon Tetrachloride Concentrations over 500 μg/L

Sample ID	Matrix	SDG	Lab Method	Chemical	Result	RDL	Detect	Unit
SW-1(081602)	WG	884825	SW8260	Carbon tetrachloride	670	25	Y	μg/L

Table 1.10 Data Summary for Crosstab Query

Loc ID	AoC	Nitrate	Iron	Sulfate Concentration
SW-01	Upstream	200	599	100.203
SW-02	Upstream	250	342	100.102
SW-03	Background area	300	105	100
SW-04	Downstream	100	20	98.97
SW-05	Downstream	50	40	97.96
MW-01	Background area	75	50	96.597
MW-02	Oil storage area	60	65	97.384
MW-03	Oil storage area	35	10	97.384

Table 1.11 Crosstab Query Results

	Area of Concern			
Data	Background Area	Downstream	Oil Storage Area	Upstream
Average of nitrate	187.5	75	47.5	225
Average of iron	77.5	30	37.5	470.5
Average of sulfate concentration	98.2985	98.465	97.384	100.1525
Max of nitrate	300	100	60	250
Max of iron	105	40	65	599
Max of sulfate concentration	100	98.97	97.384	100.203

This type of query is referred to a selection query; it returns the selection based on the criteria. Other types of queries, available in Microsoft Access for example, include action queries and crosstab queries. Action queries change record information by specifying criteria and changing the values in given fields based on those criteria. An example of an action query would be the application of data qualifiers following a rigorous validation of the collected data. For example, in our surface water example, the data validation process might uncover that the analytical laboratory initially reported the reported detection limit of carbon tetrachloride as 25 μg/L but in fact it should have been 100 μg/L. An action query could be used to specify the criteria (Sample SW-1(081602) and carbon tetrachloride) and correct the reported detection limit (RDL) from 25 to 100.

Crosstab queries perform aggregate calculations on the value of a field, using one or more other fields as rows and one field's data as columns. For example, consider the data set shown in Table 1.10.

An analysis of this data might help refine a remedial course of action for impacted groundwater. Useful calculations to perform would be the average and the maximum values of each of the constituents. A crosstab query of this data would generate the following results, shown in Table 1.11.

1.8.2 Reporting Data

Once data have been retrieved through a query process, it can be reported in tabular format, like those found in this chapter and elsewhere in this book, graphically in the form of charts and graphs, or if there is a spatial representation to the data, as a figure in a GIS system.

1.9 METADATA

The formal definition of metadata is simply "data about data." Metadata is the information about a data source, for example, a book contains information, but there is also information about that book such as the author and publisher—this is the metadata.

Metadata in the context of this book can be used to describe how a particular data table was assembled, who collected the data, what method was used to collect and aggregate the data, the sources of the data. In our groundwater contamination example, metadata might include who did the surveying for the locations, the date of that survey, and the coordinate system specification.

For the sample data, the metadata might consist of the following components, shown in Table 1.12.

Table 1.12 Examples of Metadata Components

Field Name	Field Description
COC^{num}	Chain of custody identifier
Sent_to_lab_date	Date sample was sent to lab
Sample_receipt_date	Date that sample was received at laboratory
Sampler	Name or initials of sampler
Sampling_company	Name or initials of sampling company
Sampling_reason	Reason for sampling
Sampling_technique	Sampling technique
Task_code	Code used to identify the task under which sample was retrieved
Collection_quarter	Quarter of the year sample was collected (e.g., "1Q96")
Composite_yn	Boolean field used to indicate whether a sample is a composite sample
Composite_desc	Description of composite sample (if composite_yn is YES)

Within the context of GISs, metadata almost always refers to "data about digital geospatial data." Throughout this discussion, the term "metadata" is used under this restricted definition to refer to the *content, quality, condition, and other characteristics of digital geospatial data*.

Metadata for electronic images should include, as a minimum:

- How the image was created
- Who created it originally
- What has been done to enhance the image
- Coordinate system to which the image has been rectified
- Projection system to which the image has been rectified
- Other information unique to these particular images

With regard to electronic spatial data, in the United States, efforts are being made to try to develop a universal standard format for metadata for GIS systems. This would make accessing and using metadata much easier than it is today. Toward this end, the Federal Geographic Data Committee (FGDC) has approved a standard for metadata. Development of the standard was a part of the development of the National Spatial Data Infrastructure.

The standard is known as the Content Standard for Digital Geospatial Metadata (Version 2). It may be downloaded as a pdf file from the FGDC web site (www.fgdc.gov/standards/documents/standards/metadata/v2_0698.pdf).

While this standard is intended to facilitate the use of metadata and associated data (particularly images), it appears to not have been universally accepted outside of U.S. government bodies. Accordingly, within this definition metadata may take many forms, ranging from simple (on the paper map, the map legend is the metadata) to separate electronic text files, often multi-page, associated with a specific electronic aerial or satellite photo image.

The primary use of metadata is to correlate spatial image information, particularly aerial photos, satellite photos, and computer aided design (CAD) images, in three-dimensional space, with tabular data obtained in the field. There are, thus, two aspects to this process.

The first aspect is the creation of an electronic base map image on which tabular data may be electronically posted. Typically, the user will need to overlay images together (a CAD image superimposed on an aerial photo, for example) or combine smaller images together to create larger maps, or some combination of both. In this process, it is often possible to obtain access to additional tabular data already associated with (i.e. linked to, or posted on) images.

The second aspect is creating or obtaining the tabular data, and orienting it in two- or three-dimensional space. Tabular data are usually oriented in three-dimensional space according to surveys taken by hand on the ground, either using traditional surveying methods (compass, transit, etc) or global positioning satellite data. Because of the availability of extremely inexpensive receivers, GPS is rapidly becoming the dominate tool to obtain two- and three-dimensional positioning data in the field.

Typically the orientation metadata associated with the spatial data must be used to convert (reproject or recoordinate) the spatial data to conform to the orientation data associated with the tabular data, or vice-versa, or both. This process of correlation is called "rectification." Metadata are thus typically used to rectify the spatial image data to the tabular data.

1.10 CONCLUSIONS

The data management process is intended to reduce the amount of time spent on manipulating data and increase the level of utility of the data to the end users. The goals of the data management process should be as follows:

- Understand your data needs and how you are collecting it;
- Have a plan to accurately categorize all facets of your data;
- Promote accuracy of data through validation and verification;
- Promote consistency of data querying and reporting; and
- Understand the role of metadata in GIS.

The data presented in this book are categorized in a way to facilitate effective data management, and through querying and reporting, would be a verifiable and validated source of data for experimentation and investigation purposes.

CHAPTER 2

International Data Collection

Daniel Zell

CONTENTS

In developing countries, dams, irrigation schemes, watershed management plans, water and wastewater systems, and flood mitigation works have grown in both number and complexity. Because these data intensive approaches, such as river basin planning, are being embraced worldwide and funded by multilateral and bilateral organizations, the need for data has increased. China, for example, has undertaken massive water resources projects on a scale never seen before, and vast irrigation rehabilitation projects are underway in Afghanistan. In many Latin American countries, water management projects are a top priority. With the demand for such types of projects comes the need for data, information, knowledge management, and, in particular, people who fall under the broad category of water resources engineers. This need is made sharper given the imminent retirement of the seasoned professionals of the post-WWII generation who have spent their lives in water resources, leaving a younger, less experienced cohort of engineers to tackle the future.

2.1 INTRODUCTION

When, as often happens, engineers and other technical professionals in the water resources field are asked to render a technical judgment, they usually need a large set of data to analyze the issue. In the United States, general information such as precipitation, topography, streamflow, and other related data is usually readily available from standard sources: previous studies are usually on-hand with the implementing agency, e.g., state government, military, and the private sector. But sometimes, data are harder to find, less reliable, maybe even lost. This latter case is the normal starting point for water resources projects in less developed countries. The engineer will have to invest a good deal of time, effort, and sometimes money to get the sort of data that are, usually, freely available in developed countries. This chapter gives engineers an idea of what to expect, some approaches to gathering data, and international internet data sources.

There has been and always will be a need for rapid and complete data collection for water resources projects in the developing world, a process made even more challenging since data are fragmented among various government and private organizations. Due to both varied organizational arrangements and cultural factors, procedures for data collection vary from region to region and country to country, and it is not possible within the confines of this chapter to give a comprehensive step-by-step method for accomplishing it. Here, we present a brief picture of the challenges of data collection through some selected examples, an outline of a brief process, and a listing of sources of data for international water resources projects. This chapter, then, will be most useful for those with little experience in water resources projects in developing nations.

Now, before getting into the details, a few definitions and explanations.

1. *Less developed countries* is a term of convenience, generally meaning any country except the United States, the former Soviet Union, Canada, Western Europe, Australia, New Zealand, and Japan. But even though such a simple dichotomy does not really exist—other countries undoubtedly deserve to be considered developed and may indeed boast superior data repositories—this term, because of its widespread acceptance, will be used throughout this chapter.
2. *Engineer* is defined here as a person responsible for technical aspects of a water-related project. Although the engineer is normally an outsider to the country or region under study and perhaps initially unfamiliar with its cultural norms and practices, he or she engineer understands not only the data collection need but also its intended application and eventual output. In some cases engineers will be scientists, economists, even policymakers. The term, again, is chosen out of convenience.
3. To make the chapter as useful as possible to the widest audience, it generally refers to data in the generic sense, rather than, for example, average precipitation and uses illustrative examples rather than an analytical examination.

Without digressing too much into the world of development economics, we note that the engineer needs to understand the stage or level of development. In general, there is a positive correlation between wealth, or Gross Domestic Product per capita, and the centralization and quality of water resources data. In Afghanistan, one of the poorest countries in the world, for example, where data sources are widely scattered, the results are incomplete, with much historical data, and studies lost in the warfare over the past few decades. Climate also plays a factor, as natural disasters, humidity, and even rodents often destroy hard copies in countries with limited computerized archives. The overall lesson is that even in the relatively more developed countries of Central and South America, the engineer will need to search for secondary sources of data *outside* of the government, such as previous consultants' reports, private water companies' data, and others (see the list of websites at the end of this chapter). An outside engineer's local contacts will be essential in setting appointments, identifying possible resources, and even collecting data themselves.

But—a word of caution—it is unlikely that this task can be accomplished without the involvement of the engineer. If you are not willing to travel to the country of assignment, then you should probably find somebody else to do the job. Both the credibility and realism of the outputs will suffer from a lack of field presence. Besides, as an outsider, typically seen as immune to local politics and prejudices, the engineer who is willing to travel and spend enough time in country can surmount obstacles that an indigenous assistant could not.

2.2 DATA SOURCES

For general purposes, let us consider a simple dichotomy in terms of data availability:

- Initial Conditions (given data), and
- Needs (data objectives).

Initial data may be given by your client or employer or may be found through some cursory searching of the internet resources listed in this chapter. Data needs or objectives depend on the particular assignment and will likely evolve throughout the problem solving process. In any case, it is well worth taking the time to explore most, if not all, of the websites listed at the end of the chapter with a high-speed internet connection. You will spend much less time than trying to find data from alternate sources, such as government departments, previous and/or current projects, and well-known experts in the field. And the websites listed are generally regarded as reliable.

But website searches are not the only route. I recommend a variety of methods, especially if the country and region are unfamiliar to the engineer. Before arriving in a country, establish primary contacts and introductions through the employer or client. Email, however convenient, is no substitute for phone calls in building relationships. These contacts may, if sufficiently motivated, be in a much better position to collect the required data. Unfortunately, due primarily to the long and frequently inconclusive history of most local water resources projects with international engineers, it will be a difficult task convincing someone unknown to you, that your cause is worthwhile. This is yet another reason why you should use your in-country network, even if it is only your employer or client, as a foundation for collecting data. The social ties and relationships that locals have established are a resource you should not ignore.

A frequent issue when working in an unfamiliar region and culture is payment. How much do things cost? From CAD and GIS operators, to internet usage, to maps, most, if not all, goods and services have a value. In my experience, the issue is not whether or not to pay, but rather, "what is the correct price?" This is not to say that you can not get data for free; indeed, that is how I got most of mine. The engineer will have to rely on his or her in-country network and awareness of local social norms to determine whether or not a payment is required. In general, if work is required beyond normal hours or duties, then a payment is more likely to be required.

Another thing to be prepared for is equipment and procedures that do not work. Copiers, for example, are usually broken because of a lack of basic supplies and/or maintenance. Or it may not exist or be available to anyone except the senior staff. Regulations may not let you take records from the office for off-site duplication. Unless you have a penchant for data entry, this is another case where it may be appropriate to pay the correct fee for the data to be digitally entered. I suggest offering an electronic copy of the records by email to both the department and data entry person. By making them part of the process, rather than a cog in a wheel, you may use these people as resources for a long time to come, as I did with a local hydrologist in Afghanistan months after my return. (It did, however, take much longer than expected and required several phone calls and visits by contacts in Afghanistan.)

2.3 CASE STUDIES

The following case studies will give the engineer a general perspective of the operational aspects of data collection in a few developing nations. Let me note at once, however, the enormous importance of learning about and paying attention to the unique cultures, conditions, and particularly the values of each nation or region. Ignoring them will imperil both the engineer and the project itself.

2.3.1 Latin America

Overall, although collecting water resources data in Latin America is like working in developing countries worldwide,* the relatively higher income in Latin America can make the work easier, as education level generally correlates with income. Because hydropower is a highly developed sector in Latin America, its organizations often have the best data. However, the data are often considered proprietary and, being tightly controlled, take concerted and time-consuming effort by the engineer and local counterparts to get it.

As in other regions, engineers should rely on their counterparts, local staff, or client by insisting that reasonable efforts be expended to gather data in advance. Due to the culture of Latin America, counterparts can be relied on quite heavily, and will likely be highly competent. In this manner, the engineer's time can be best and efficiently used to address the remaining obstacles, like analyzing the data and writing reports.

Still, the engineer should be prepared to complete an assignment with much less data in both quantity and quality than is the norm in developed countries. When no data are found for a specific basin, data for a similar region—in combination with field interviews, maps, and surveys—is used to build a model. The uncertainties resulting from using such data should be plainly and simply stated in the engineer's work.

According to Dr Molina, the process of cleaning data—searching for errors—although time consuming and tedious, has repeatedly proved itself to be worth the trouble. Common errors include converting units, converting gauge readings to flow, and a myriad of other possibilities. These errors, if unchecked and corrected, will skew the results of an analysis. On the other hand, careful advance planning can minimize delays, for example, by contracting a local firm or consultants to carefully screen the data for anomalies and enter it into a useful format.

* According to a June 25, 2004, interview with Dr Medardo Molina, a Peruvian-born water resources expert who has been active in water resources since 1965 as an international consultant and professor. Dr Molina has published numerous papers in both Spanish and English.

2.3.2 Central Asia

In Afghanistan, collecting water resources data were quite complicated, based on my personal experiences in 2003. What little historical data existed was fragmented and incomplete. Thirty years of instability destroyed critical records and the hydrological network. The language barrier, with data sometimes recorded in Dari, Pashto, or Russian, further complicated data collection.

The assigned task—to construct a national water balance and determine water availability for rehabilitating irrigation—necessitated casting a wide net to collect all sources of data. Before visiting Afghanistan, my thorough search of libraries and databases yielded only few results, but one of them, a previous water balance study, was quite useful. Although I made contacts in advance, scheduling appointments from outside of Afghanistan did not work. I now understand that the local custom is to pay a brief introductory visit, without an appointment. Then, a later appointment can be scheduled where the useful work will be accomplished.

This introductory visit may seem full of pleasantries and even a bit useless. To the novice it may seem not along the critical path. Quite to the contrary, the visits are used to evaluate the engineer, to understand what the engineer is doing, and to understand if the engineer is worthy of help. Besides, since data are rarely immediately available, the first meeting gives the local source time to prepare.

The team conducted initial and follow-up visits to the following organizations:

> *Ministries*: Irrigation and Water Resources and Environment, Rehabilitation and Rural Development, Agriculture and Animal Husbandry, and the Central Statistics Office
>
> *United Nations*: Food and Agriculture Organization and World Food Program, in particular the Vulnerability Analysis Mapping Unit for socio-economic data, and the Development Program
>
> *Others*: ICARDA (International Center for Agricultural Research in the Dry Areas), the U.S. Agency for International Development, and various private consultants, firms, and non-government organizations

From these and subsequent visits over a 12-month period, the consultant team gathered the hydrological station data. We hired a local professional consultant to both construct and digitize a map of the monitoring stations. More importantly, the team was able to persuade a wide range of policy makers and water resources professionals of the report's thoroughness and usefulness by conducting follow-up visits to the data repositories, courtesy calls to government officials to present the report and findings, and formal presentations.

The United Nations' organizations provided the most accurate satellite and agricultural data, and previous consultants' reports were essential in constructing the overall water balance. Although we still had to do careful fact checking based on the newly available data, the efforts of independent and knowledgeable local staff helped us get data not only more rapidly but also with a higher degree of reliability than would have otherwise been possible.

It is important to note that the local staff's efforts had to be supplemented with continual phone calls and emails by the consultants as well as periodic visits to Afghanistan. As outsiders, foreign consultants can often bypass the social norms and traditions that hamper local staff. At the same time, continued communication and visits let the team more completely understand the extent of the problem or issues to be addressed and more importantly, convey the engineers' recommended solution to the client as it evolved. This served two purposes:

1. By avoiding the "parachute in" approach where a consultant works rapidly and alone, often behind a closed door, the team won a sense of buy-in and respect from the client, which ensured that the client would actually understand the results of the work.
2. By working hand-in-hand with a client who lacked basic institutional capacity, the consultant team was able to increase the client's capacity to use the team's results.

With so many people interested in the report, the team decided to circulate a draft copy. The resulting informal peer review process increased the credibility of the findings, and although it led to a series of revisions, in the end, it was the only way to accomplish the task.

2.3.3 South Asia

In contrast to the situation in Afghanistan, data availability was relatively high during a 2004 assignment to the Indian State of Orissa. At the state water resources agency, I saw many rooms with neatly bound papers stacked literally from floor to ceiling. More remarkably, when motivated by the department head, almost everything we asked for was quickly found. And because the information requested was relatively recent, we did not encounter the issue of deteriorating paper records in a humid environment without climate control. The main problems were poor database design and data entry, and resistance to improving quality control standards.

To make sure that our primary data on average daily flow from reservoirs were accurate, my team went to one of the more remote locations to look at primary records. This entailed a journey across the small rice paddies that characterize the regions, a courtesy call to the local supervisor, and even more driving to where the road literally ended at the reservoir. The water resources department had told the reservoir about our trip, and the records were ready. However, local customs dictated a long, slow, and quite good meal as well as an exchanging of pleasantries before the ledger was produced. Then, we asked the data recorders how they recorded the gate height and translated that measurement into a daily flow. Spot checking a few calculations for accuracy, we found no mistakes.

A universal issue when dealing with government agencies is territorialism. This issue nearly derailed the entire project in India. An agency insisted that a task was within their realm of authority but showed neither the will nor the capacity to perform the work as required. Rather the agency did the bare minimum and strongly resisted any efforts at improvement. The lesson the engineer should take from this is that relying on someone or some agency out of your control is risky. In hindsight, the solution would have been to accept the agency as a partner, then help them, as cooperatively and diligently as possible, to complete the project.

2.4 GENERAL PROCESS FOR DATA COLLECTION

The following process serves as a guideline for engineers unfamiliar with international water resources engineering. It is highly likely that additional steps will be needed in any actual project depending on the requirements and country.

1. Learn key features of the country's culture norms and values by both talking to someone from the country and reading appropriate articles and even the literature.
2. Identify the problem.
   This step consists of more than simply reading the terms of reference. The engineer should plan on conducting one-to-one discussions with senior and mid-level staff at the local or client agency.
3. Identify data requirements.
4. Identify local staff requirements such as cartographers, hydrologists, statisticians, field surveyors, etc.
5. Contact the client or local agency.
   5.1. Contact made well in advance of the initial visit so that the staff are aware of the planned work and can plan accordingly.
   5.2. Contact at regular intervals, at least weekly. Contact by phone and email is recommended.
6. Make initial visit.
   Prearranged visits with organizations should be conducted. Social outings, while exhausting, may prove useful for particularly difficult data collection situations.
7. Collect the data.
8. Review data collected.
   Determine what steps need to be taken to ensure the data are both accurate and in a usable format.
9. Conduct field validation if required.
10. Make more visits.
    These visits will supplement initial data gatherings and present initial findings.
11. Perform analysis.
12. Prepare and circulate draft report.
13. Incorporate comments as appropriate.
14. Prepare report.

2.5 INTERNET SOURCES

The following addresses give access to a lot of data. The descriptions of the 2006 websites come directly from the web pages themselves.

AQUASTAT is the Food and Agricultural Organization's (FAO) global information system of water and agriculture developed by the Land and Water Development Division of FAO. AQUASTAT is FAO's global information system of water and agriculture developed by the Land and Water Development Division of FAO. The objective of AQUASTAT is to provide users with comprehensive information on the state of agricultural water management across the world, with emphasis on developing countries and countries in transition (www.fao.org/WAICENT/FAOINFO/AGRICULT/AGL/AGLW/aquastat/main/index.stm).

The following is an excerpt from the AQUASTAT country profile of Afghanistan

Water Resources

Based on the hydrographic systems, the country can be divided into four zones:

- the northern basin (24% of the territory) with the Amu Darya and its tributaries (14%), which drain toward the Aral sea, and the rivers of northern Afghanistan (10%), which disappear within the country before joining the Amu Darya;
- the western region (12%) consisting of the Hari Rud river basin (6%) and the Murgab river basin (6%), both rivers disappearing in Turkmenistan;
- the south-western basin (52%) with the Helmand river flowing toward the Sistan swamps, located on the border of Iran and Afghanistan. In 1972, a document was signed between Afghanistan and Iran to allocate a discharge of 26 m/s of Helmand river water to Iran all year round;
- the eastern Kabul basin (12%), which is the only river system having an outlet to the sea, joining the Indus at Attock in Pakistan.

Internal renewable water resources are estimated at 55 $km^3 yr^{-1}$. The Kunar river, which originates in Pakistan, crosses the border with an average annual flow of 10 km^3 and joins the Kabul river at Jalalabad about 180 km further downstream. The Kabul river flows again into Pakistan 80 km further downstream.

Total water withdrawal was estimated at 26.11 km^3 in 1987, of which 99% for agricultural purposes. Recently, there has been a large development of groundwater use in some provinces.

In 1986, there were two dams higher than 15 m. The installed capacity of the hydroelectric plants was 281 MW in 1992, which is about 70% of total installed capacity. There is considerable potential for the generation of hydropower, by both large dams and microhydropower stations.

FAOSTAT is an on-line and multilingual database currently containing over 3 million time-series records covering international statistics in the following areas:

Production	Land use and irrigation	Fertilizer and pesticides
Trade	Forest products	Agricultural machinery
Food balance sheets	Fishery products	Food aid shipments
Producer prices	Population	Exports by destination
Forestry trade flow	Food quality control	

faostat.fao.org/default.jsp?language=EN.

	Year										
	1993	1994	1995	1996	1997	1998	1999	2000	2001	2002	2003
Population-estimates rural (1000)											
Africa developing	428,145	435,774	443,382	450,948	458,471	465,941	473,370	480,738	488,033	495,224	502,287
AGG_COUNTRIES	428,145	435,774	443,382	450,948	458,471	465,941	473,370	480,738	488,033	495,224	502,287
Population-estimates urban (1000)											
Africa developing	205,568	214,269	223,156	232,235	241,525	251,050	260,846	270,932	281,347	292,111	303,245
AGG_COUNTRIES	205,568	214,269	223,156	232,235	241,525	251,050	260,846	270,932	281,347	292,111	303,245

The tables above are an example of the output produced by FAO STAT, in this case rural and urban populations in African Developing Nations from 1993 to 2003. The output is also available in .csv format.

GRDC (Global Runoff Data Center) The GRDC makes the unique offer to the international research and science community of easy and universal access to river flow information on a global scale. On request, data products are developed and specialized databases are assembled for projects on both regional and global scale. The GRDC serves as a communication platform between institutions, advisors, and scientists and also transfers information about other relevant databases with a hydrological content such as the Global Precipitation Climatology Centre (GPCC) and the Programme Office of the Global Environment Monitoring System—Water (GWPO) of UNEP. It also maintains close ties to the UNESCO Water project "Flow Regimes from International Experimental and Network Data" (FRIEND).

The GRDC contains long-term mean annual freshwater surface water fluxes into the world oceans. Estimates are based on 251 discharge stations of major rivers. Data available include mean, minimum, maximum monthly discharges, and time series of mean, minimum, maximum annual discharge for 3035 stations. UNH-GRDC Global Composite Runoff Fields combines observed river discharge information with a climate-driven Water Balance Model (http://grdc.bafg.de).

The following is a sample output:

```
# GRDC-No.:             2903425
# River:                LENA
# Station:              UST-KUT
# Country:              RU
# Lat (dec. degree):      56.7700
# Lon (dec. degree):     105.6500
# Area (km**2):         71400.000
#
#
# General procedure to calculate the statistics:
# Original monthly values are used whenever available.
# In order to extend the time series, monthly data derived from daily values
# is added where applicable. Then the calculation of a monthly value
# allows a maximum to 5 missing daily values per month
#
# -999 indicates missing value or insufficient data
#
# Time series: 1936 - 1948
#
# Missing original monthly values (%): 17.00
# Missing original daily values (%):    not available
#
Number of values used for calc.:      |   127.000
Absolute minimum discharge (m**3/s): |    37.800
Absolute maximum discharge (m**3/s): |  1790.000
Average annual discharge (m**3/s):    |   402.554
Average annual volume (km**3/yr):     |    12.695
Average annual runoff (mm/yr):        |   177.801
#
# Long-term monthly characteristics (m**3/s)
# (Calculation requires at least 5 years of data)
Mo|Min       |Year|Max       |Year|Mean      |S. deviation|No. values
01|    45.800|1944|   171.000|1938|   101.200|    37.64067|  9
02|    39.700|1944|   149.000|1938|    85.222|    32.85977|  9
03|    37.800|1944|   149.000|1938|    81.589|    33.89430|  9
04|    79.600|1936|   693.000|1939|   281.422|   231.32462|  9
05|   594.000|1943|  1790.000|1941|  1141.818|   355.42926| 11
06|   211.000|1943|  1250.000|1942|   641.308|   360.34299| 13
07|   129.000|1943|  1480.000|1948|   607.000|   359.37097| 13
08|    97.000|1943|  1140.000|1948|   468.692|   252.89140| 13
09|   103.000|1943|  1010.000|1948|   457.462|   224.95651| 13
10|    85.800|1943|   383.000|1947|   267.380|   104.19906| 10
11|    56.000|1943|   242.000|1944|   158.633|    65.03188|  9
12|    50.500|1943|   215.000|1937|   138.878|    53.44971|  9
#
# Annual characteristics (m**3/s)
# (Calculation requires at least 9 months per year)
Year|Min       |Mo|Max       |Mo|Mean      |No. values
1936|    71.000|03|  1180.000|06|   410.442| 12
1937|    93.900|03|  1600.000|05|   379.933| 12
1938|   149.000|02|  1250.000|05|   499.833| 12
1939|    87.500|12|   694.000|05|   260.883| 12
1940|    52.400|03|  1010.000|07|   350.800| 12
1941|    79.000|03|  1790.000|05|   408.250| 12
1942|    80.900|12|  1360.000|05|   405.358| 12
```

CGIAR is a global link to research on agriculture, hunger, poverty, and the environment. CGIAR (Consultative Group on International Agricultural Research) is a good starting point for international water issues related to agriculture (www.cigar.org).

To access any of CGIARs publications published by the 15 research centers, go to the CGIAR Library Gateway (www.cgiar.org/publications/library/index.html).

CGIAR Reefbase—ReefBase is the world's premier online information system on coral reefs, and provides information services to coral reef professionals involved in management, research, monitoring, conservation, and education. Its goal is to facilitate sustainable management of coral reefs and related coastal/marine environments, in order to benefit poor people in developing countries whose livelihoods depend on these natural resources. ReefBase's Online Geographic Information System (GIS) allows you to display coral reef related data and information on interactive maps. You can zoom, search, and query datalayers, and save or bookmark the map (www.reefbase.org).

ICARDA (International Center for Research in the Dry Areas) is one of the research centers of CGIAR, whose mission is to improve and integrate the management of soil, water, nutrients, plants, and animals in ways that optimize sustainable agricultural production. There are many relevant articles, publications, and datasets on on-farm water use and water efficiency (www.icarda.cgiar.org).

IWMI's (International Water Management Institute) on-line publication section contains several thousand pages of peer reviewed research on water management. All research outputs and publications produced by IWMI are international public goods, freely available to partners in developing countries and to members of the international development, academic and research communities (www.iwmi.cgiar.org/pubs/mindex.htm).

The Remote Sensing and GIS Unit (RS GIS Unit) of the International Water Management Institute (IWMI), is a centralized facility for all spatial data-related activities of IWMI at the headquarters in Sri Lanka and Regional Offices located in different parts of the world.

Currently, the RS GIS Unit holds over 1 terabyte of data. Although the emphasis is on IWMI benchmark river basins, large volumes of data are also available at National, Regional, and Global levels. These data are catalogued, streamlined, and released to the public through the IWMI Data Storehouse Pathway (DSP). Much of the river basin and other datasets are composed as single mega files of hundreds or sometimes thousands of bands consisting of continuous streams of 8-day or monthly time series data in several wavebands and/or indices.

Large volumes of multitemporal data from multiple satellite sensors are used in several IWMI research projects. These projects include: (a) Global Irrigated Area Mapping (GIAM) at global to local scales, (b) the Wetland project in the Limpopo river basin of four Southern African Nations, (c) the Krishna river basin project in India, (d) the Indo-Gangetic river basin project in India and Pakistan, (e) the Drought Assessment and Mitigation project in Afghanistan, Pakistan, and parts of India, and (f) the biodiversity project in Sri Lanka.

More information on the RS GIS Unit and its activities can be found in several areas of this web site (www.iwmidsp.org/iwmi/info/centerprofile.asp).

Hydrological Processes is a relevant international journal with abstracts freely accessible on the web. Of particular interest are the past articles on mathematical and methodological aspects of hydrological processes and modeling.

Accessed through Wiley Intersciences (www.interscience.wiley.com).

Although traditionally thought of as a repository of United States data sets only, the United States Geological Survey (USGS) has a notably thorough database on selected countries that will likely grow over time. In particular, datasets for Jordan and Israel can be found through their project websites (www.watercare.org and www.exact-me.org).

Also on USGS, a general homepage with linkages to Ukraine, United Arab Emirates, Bangladesh, and Cyprus (international.usgs.gov/disciplines/water.htm).

Winrock Water is both a discussion forum and data clearinghouse that includes an annotated bibliography. Winrock water has selected leading reference materials, research and discussions of major issues in the water resources field through the internet. The links section is of particular interest to those in data collection (www.winrockwater.org).

Sakia.org is an information and communication service in the area of "land and water". Sakia.org hosts several services such as the email discussion list IRRIGATION-L the WWW Virtual Library Irrigation & Hydrology (content filling stage), the WWW Database on Irrigation & Hydrology Software—IRRISOFT (under revision), the e-Journal of Land and Water, an open access and peer reviewed international scientific journal for research and developments and the *Journal of Applied Irrigation Science*. Sakia.org is fostering the open and free access to knowledge in support of the "land and water" community (www.sakia.org).

CHAPTER **3**

Climate and Precipitation

Pedro Fierro, Jr.

CONTENTS

SECTION 3A CLIMATIC DATA — UNITED STATES

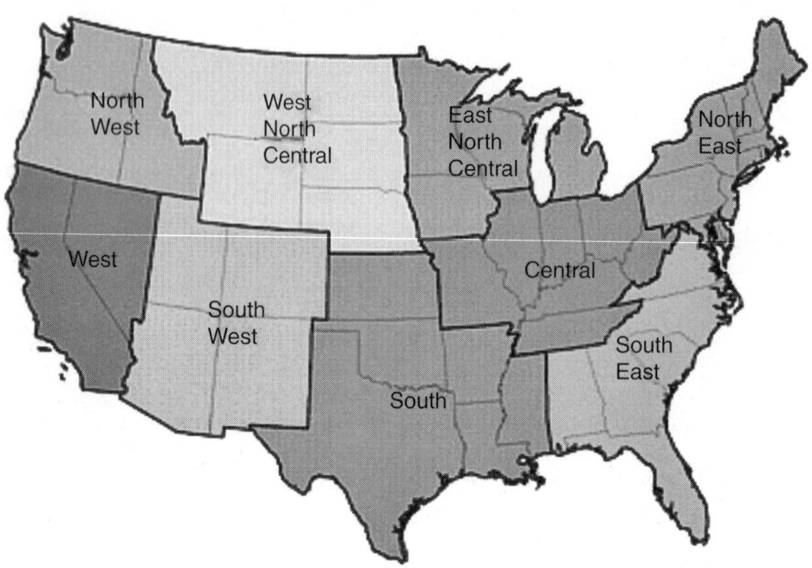

Figure 3A.1 U.S. standard regions for temperature and precipitation. (From U.S. National Oceanic and Atmospheric Administration.)

Table 3A.1 Normal Daily Mean Temperature — Selected Cities of the United States

State	Station	Years	Jan	Feb	Mar	Apr	May	Jun	Jul	Aug	Sep	Oct	Nov	Dec	Annual
AL	Birmingham AP	30	42.6	46.8	54.5	61.3	69.3	76.4	80.2	79.6	73.8	62.9	53.1	45.6	62.2
	Huntsville	30	39.8	44.3	52.3	60.4	68.6	76.0	79.5	78.6	72.4	61.3	51.2	43.1	60.6
	Mobile	30	50.1	53.5	60.2	66.1	73.5	79.3	81.5	81.3	77.2	67.7	58.9	52.3	66.8
	Montgomery	30	46.6	50.5	57.9	64.3	72.3	78.9	81.8	81.2	76.3	65.4	56.1	49.0	65.1
AK	Anchorage	30	15.8	18.7	25.9	36.3	46.9	54.7	58.4	56.4	48.2	34.1	21.8	17.5	36.3
	Annette	30	35.1	37.1	39.5	43.8	49.4	54.3	58.2	58.6	53.8	46.5	39.7	36.4	46.1
	Barrow	30	−13.7	−15.9	−13.7	−0.5	20.1	35.0	40.4	38.7	31.2	14.6	−0.9	−10.6	10.4
	Bethel	30	6.6	7.6	14.5	25.9	41.3	51.4	56.0	53.6	45.4	30.0	17.4	9.4	29.9
	Bettles	30	−11.2	−7.9	4.2	22.4	44.3	57.8	60.2	53.5	41.0	18.7	−0.8	−7.4	22.9
	Big Delta	30	−2.6	2.3	14.2	32.1	47.8	57.5	60.8	55.5	44.4	24.1	6.4	0.1	28.6
	Cold Bay	30	28.2	27.6	30.0	33.5	39.8	45.9	50.6	51.8	47.8	40.0	34.5	31.0	38.4
	Fairbanks	30	−9.7	−3.8	11.1	31.7	48.8	59.7	62.4	56.2	44.5	23.5	2.3	−5.9	26.8
	Gulkana	30	−4.7	3.2	15.3	31.1	43.9	53.1	57.0	53.1	43.1	26.4	5.5	−1.6	27.1
	Homer	30	23.4	24.9	29.4	36.4	43.7	50.0	54.1	53.8	47.9	37.8	29.4	25.8	38.1
	Juneau	30	25.7	28.9	33.7	40.8	47.9	53.9	56.8	55.7	50.0	42.3	33.3	28.7	41.5
	King Salmon	30	15.4	15.6	23.5	33.1	43.5	50.9	55.7	54.8	47.6	33.3	23.2	17.2	34.5
	Kodiak	30	29.7	29.9	32.6	37.3	43.5	49.2	54.1	55.0	49.4	40.3	34.0	30.6	40.5
	Kotzebue	30	−2.5	−3.5	−0.3	11.5	31.6	44.8	54.7	52.1	41.8	23.2	8.3	−0.2	21.8
	Mcgrath	30	−6.7	−0.9	11.8	29.1	46.2	56.7	59.8	54.9	44.7	25.3	5.8	−3.8	26.9
	Nome	30	5.8	5.7	9.4	19.6	37.1	47.3	52.6	50.6	42.9	28.5	16.9	8.4	27.1
	St. Paul Island	30	25.7	23.3	24.2	28.4	35.7	41.9	46.7	48.4	45.0	38.3	33.1	28.8	35.0
	Talkeetna	30	11.0	15.4	22.6	34.3	45.8	55.3	58.9	55.6	46.2	31.4	17.5	13.0	33.9
	Unalakleet	30	3.3	4.3	10.7	22.7	39.5	49.0	55.5	53.1	43.6	26.5	12.6	6.1	27.2
	Valdez	30	21.9	24.8	29.8	37.7	45.8	52.2	55.2	53.6	47.1	38.2	28.3	24.7	38.3
	Yakutat	30	25.8	28.4	31.5	37.2	43.6	49.7	53.6	53.3	48.2	41.1	32.4	28.6	39.5
AZ	Flagstaff	30	29.7	32.2	36.6	42.9	50.8	60.1	66.1	64.4	57.8	47.1	36.5	30.2	46.2
	Phoenix	30	54.2	58.2	62.7	70.2	79.1	88.6	92.8	91.4	86.0	74.6	61.6	54.3	72.9
	Tucson	30	51.7	55.0	59.2	66.0	74.5	84.1	86.5	84.9	80.9	70.5	58.7	51.9	68.7
	Winslow	30	34.2	40.0	46.3	53.4	62.2	72.1	77.5	75.6	68.2	55.9	43.2	34.1	55.2
	Yuma	30	58.1	62.0	66.5	72.7	79.9	88.8	94.1	93.5	88.2	77.2	64.8	57.4	75.3
AR	Fort Smith	30	40.1	45.2	53.4	61.4	70.1	78.4	82.4	81.3	74.4	63.3	51.7	43.2	62.1
	Little Rock	30	40.2	45.6	54.3	63.0	70.9	78.8	83.2	82.1	75.0	64.5	52.5	43.4	62.8
	North Little Rock	30	38.0	43.7	52.6	61.1	69.5	77.5	82.2	81.5	73.9	62.8	50.5	41.0	61.2
CA	Bakersfield	30	47.8	53.3	57.3	62.7	70.3	77.7	83.1	81.9	76.7	67.2	54.8	47.2	65.0
	Bishop	30	38.0	42.4	47.7	54.1	62.5	71.1	76.8	74.8	67.3	56.6	44.8	38.0	56.2
	Eureka	30	47.9	48.9	49.2	50.7	53.6	56.3	58.1	58.7	57.4	54.5	51.0	47.9	52.9
	Fresno	30	46.0	51.4	55.5	61.2	68.8	76.1	81.4	79.9	74.6	65.0	52.7	45.2	63.2
	Long Beach	30	57.0	58.3	59.7	63.0	65.9	69.8	73.8	75.1	73.4	68.6	61.8	57.1	65.3
	Los Angeles AP	30	57.1	58.0	58.3	60.8	63.1	66.4	69.3	70.7	70.1	66.9	61.6	57.6	63.3
	Los Angeles CO	30	58.3	60.0	60.7	63.8	66.2	70.5	74.2	75.2	74.0	69.5	62.9	58.5	66.2
	Mount Shasta	30	35.3	38.2	41.2	46.3	53.2	60.2	66.1	65.1	59.5	50.5	39.9	34.8	49.2
	Redding	30	45.5	49.1	52.5	57.8	66.2	75.2	81.3	78.9	73.4	63.2	51.1	45.3	61.6
	Sacramento	30	46.3	51.2	54.5	58.9	65.5	71.5	75.4	74.8	71.7	64.4	53.3	45.8	61.1

(Continued)

Table 3A.1 (Continued)

State	Station	Years	Jan	Feb	Mar	Apr	May	Jun	Jul	Aug	Sep	Oct	Nov	Dec	Annual
	San Diego	30	57.8	58.9	60.0	62.6	64.6	67.4	70.9	72.5	71.6	67.6	61.8	57.6	64.4
	San Francisco AP	30	49.4	52.4	54.0	56.2	58.7	61.4	62.8	63.6	63.9	61.0	54.7	49.5	57.3
	San Francisco CO	30	52.3	55.0	55.9	57.3	58.4	60.5	61.3	62.4	63.7	62.5	57.5	52.7	58.3
	Santa Barbara	30	53.1	55.2	56.7	58.9	60.9	64.2	67.0	68.6	67.4	63.5	57.5	53.2	60.5
	Santa Maria	30	51.6	53.1	53.8	55.5	57.8	60.9	63.5	64.2	63.9	61.1	55.5	51.6	57.7
	Stockton	30	46.0	51.1	54.9	60.0	66.7	73.2	77.3	76.5	72.8	64.6	53.1	45.3	61.8
CO	Alamosa	30	14.7	22.5	32.7	40.8	50.4	59.4	64.1	62.1	54.5	42.8	28.4	17.1	40.8
	Colorado Springs	30	28.1	31.7	37.8	45.3	54.6	64.4	69.6	67.6	59.8	48.9	36.2	29.0	47.8
	Denver	30	29.2	33.2	39.6	47.6	57.2	67.6	73.4	71.7	62.4	51.0	37.5	30.3	50.1
	Grand Junction	30	26.1	34.1	43.4	50.9	60.5	71.1	76.8	74.7	65.4	52.7	38.1	28.2	51.8
	Pueblo	30	29.3	34.6	41.8	49.9	59.7	69.8	75.4	73.5	64.8	52.4	38.4	30.3	51.7
CT	Bridgeport	30	29.9	31.9	39.5	48.9	59.0	68.0	74.0	73.1	65.7	54.7	45.1	35.1	52.1
	Hartford	30	25.7	28.8	38.0	48.9	59.9	68.5	73.7	71.6	63.2	51.9	41.8	30.8	50.2
DE	Wilmington	30	31.5	34.2	42.7	52.4	62.5	71.5	76.6	75.0	67.7	55.8	45.9	36.4	54.4
DE	Washington Dulles AP	30	31.7	34.8	43.4	53.1	62.3	70.9	75.7	74.4	67.3	55.0	45.2	36.0	54.2
	Washington Nat'l AP	30	34.9	38.1	46.5	56.1	65.6	74.5	79.2	77.4	70.5	58.8	48.7	39.5	57.5
FL	Apalachicola	30	52.7	55.3	60.7	66.8	74.1	80.0	81.9	81.7	79.1	70.2	62.0	55.2	68.3
	Daytona Beach	30	58.4	60.0	64.7	68.9	74.8	79.7	81.7	81.5	79.9	74.0	67.0	60.8	71.0
	Fort Myers	30	64.9	66.0	69.9	73.6	78.8	82.2	83.0	83.1	82.1	77.5	71.7	66.4	74.9
	Gainesville	30	54.3	57.0	62.5	67.6	74.3	79.2	80.9	80.4	77.8	70.1	62.8	56.3	68.6
	Jacksonville	30	53.1	55.8	61.6	66.6	73.4	79.1	81.6	80.8	77.8	69.4	61.7	55.0	68.0
	Key West	30	70.3	70.8	73.8	77.0	80.7	83.4	84.5	84.4	83.4	80.2	76.3	72.0	78.1
	Miami	30	68.1	69.1	72.4	75.7	79.6	82.4	83.7	83.6	82.4	78.8	74.4	69.9	76.7
	Orlando	30	60.9	62.6	67.4	71.5	77.1	81.2	82.4	82.5	81.1	75.3	68.8	63.0	72.8
	Pensacola	30	52.0	54.9	61.0	66.9	74.6	80.6	82.6	82.2	78.7	69.5	60.7	54.1	68.2
	Tallahassee	30	51.8	54.8	61.1	66.4	74.4	80.4	82.4	82.1	78.9	69.1	60.4	53.7	68.0
	Tampa	30	61.3	62.7	67.4	71.5	77.6	81.5	82.5	82.7	81.6	75.8	69.3	63.3	73.1
	Vero Beach	30	63.0	63.9	67.7	71.5	76.2	80.4	81.7	81.6	80.7	76.4	70.5	64.7	73.2
	West Palm Beach	30	66.2	67.2	70.6	73.8	78.2	81.2	82.5	82.8	81.7	78.1	73.1	68.3	75.3
GA	Athens	30	42.2	46.0	53.5	60.9	69.1	76.3	79.8	78.4	72.6	61.8	52.7	44.8	61.5
	Atlanta	30	42.7	46.7	54.3	61.6	69.8	76.8	80.0	78.9	73.3	62.8	53.4	45.4	62.2
	Augusta	30	44.8	48.4	55.9	62.4	70.5	77.5	80.8	79.3	73.8	63.1	54.4	46.9	63.2
	Columbus	30	46.8	50.3	57.6	64.2	72.3	79.2	82.0	81.3	76.2	65.8	56.7	49.1	65.1
	Macon	30	45.5	48.9	56.2	62.7	71.0	78.0	81.1	80.0	74.5	63.9	55.1	47.8	63.8
	Savannah	30	49.2	52.5	59.3	65.3	72.8	78.8	82.1	80.8	76.7	67.1	58.7	51.4	66.2
HI	Hilo	30	71.4	71.5	72.0	72.5	73.7	75.1	75.9	76.3	76.2	75.6	74.0	72.2	73.9
	Honolulu	30	73.0	73.0	74.3	75.6	77.2	79.5	80.8	81.8	81.5	80.2	77.7	74.8	77.5
	Kahului	30	71.8	71.9	73.1	74.2	75.7	77.6	78.8	79.5	79.1	78.1	76.0	73.4	75.8
	Lihue	30	71.7	71.7	72.7	73.9	75.4	77.7	79.0	79.7	79.5	78.2	75.9	73.3	75.7
ID	Boise	30	30.2	36.7	43.8	50.6	58.6	67.2	74.7	73.9	64.2	52.8	39.9	30.6	52.0
	Lewiston	30	33.7	38.4	44.7	51.1	58.5	65.8	73.5	73.4	63.8	51.6	40.4	33.9	52.4
	Pocatello	30	24.4	30.0	37.9	45.6	53.5	62.0	69.2	68.4	58.8	47.7	34.7	25.3	46.5
IL	Chicago	30	22.0	27.0	37.3	47.8	58.7	68.2	73.3	71.7	63.8	52.1	39.3	27.4	49.1
	Moline	30	21.1	26.9	38.7	50.5	61.7	71.2	75.3	73.2	65.0	53.0	39.1	26.4	50.2

	Peoria	30	22.5	28.2	39.8	51.2	61.9	71.1	75.1	73.1	65.4	53.4	40.1	27.8	50.8
	Rockford	30	19.0	24.7	36.1	47.9	59.6	68.8	72.9	70.9	62.8	51.0	37.2	24.4	48.0
	Springfield	30	25.1	30.6	41.8	52.8	63.6	72.6	76.3	74.2	67.0	55.5	42.3	30.3	52.7
IN	Evansville	30	31.0	35.8	45.8	55.5	65.6	74.8	78.6	76.5	69.1	57.3	45.9	35.6	56.0
	Fort Wayne	30	23.6	27.3	38.1	49.0	60.4	69.7	73.4	71.1	64.1	52.4	40.6	29.0	49.9
	Indianapolis	30	26.5	31.2	41.7	52.0	62.6	71.7	75.4	73.5	66.3	54.6	42.9	31.6	52.5
	South Bend	30	23.4	27.3	37.5	48.3	59.6	69.0	73.0	71.0	63.4	52.1	40.1	28.7	49.5
IA	Des Moines	30	20.4	26.6	38.4	50.6	61.9	71.4	76.1	73.9	65.1	52.8	37.9	24.9	50.0
	Dubuque	30	17.0	23.1	34.8	47.5	59.1	68.3	72.3	70.0	61.8	50.4	35.7	22.5	46.9
	Sioux City	30	18.6	25.1	36.5	49.5	61.2	70.5	74.6	72.1	63.1	50.8	34.8	22.3	48.3
	Waterloo	30	16.1	22.6	35.0	47.8	60.2	69.9	73.6	71.2	62.6	50.2	35.1	21.6	47.2
KS	Concordia	30	26.6	32.4	42.5	52.8	63.0	73.4	79.1	77.0	68.0	56.0	40.8	30.2	53.5
	Dodge City	30	30.1	36.0	44.3	53.9	63.8	74.3	79.8	78.2	69.3	57.1	42.4	33.1	55.2
	Goodland	30	27.6	32.4	39.8	48.8	58.7	69.6	75.1	73.2	64.0	51.8	37.4	29.6	50.7
	Topeka	30	27.2	33.4	44.2	54.5	64.4	73.9	78.4	76.7	68.1	56.6	42.6	31.4	54.3
	Wichita	30	30.2	36.3	45.9	55.3	65.0	75.5	81.0	79.8	70.8	58.6	44.2	33.6	56.4
KY	Greater Cincinnati AP	30	29.7	34.1	43.9	53.7	63.7	72.0	76.3	74.5	67.4	55.7	44.7	34.6	54.2
	Jackson	30	33.9	37.9	47.1	56.3	64.1	71.4	75.0	73.8	67.9	57.5	47.7	38.3	55.9
	Lexington	30	32.0	36.4	45.6	54.6	63.8	72.2	76.1	74.8	68.0	56.6	45.9	36.3	55.2
	Louisville	30	33.0	37.6	46.9	56.4	65.8	74.2	78.4	77.0	70.1	58.5	47.6	37.6	57.0
	Paducah	30	32.9	38.1	47.6	57.0	65.9	74.5	78.2	76.2	69.1	58.0	46.8	36.9	56.8
LA	Baton Rouge	30	50.1	53.5	60.3	66.6	74.0	79.7	81.7	81.4	77.5	68.1	59.0	52.4	67.0
	Lake Charles	30	50.9	54.4	61.0	67.3	74.9	80.5	82.6	82.4	78.4	69.5	60.1	53.3	68.0
	New Orleans	30	52.6	55.7	62.4	68.2	75.6	80.7	82.7	82.5	78.9	70.0	61.4	55.1	68.8
	Shreveport	30	46.4	51.2	58.5	65.2	73.0	79.9	83.4	82.9	77.0	66.7	56.1	48.4	65.7
ME	Caribou	30	9.5	13.0	24.6	38.1	51.6	60.8	65.6	63.4	53.8	42.8	30.6	16.4	39.2
	Portland	30	21.7	24.8	33.7	43.7	53.8	62.9	68.7	67.2	58.7	47.7	38.3	27.6	45.8
MD	Baltimore	30	32.3	35.5	43.7	53.2	62.9	71.8	76.5	74.5	67.4	55.4	45.5	36.7	54.6
MA	Blue Hill	30	26.0	28.3	36.3	46.3	57.0	65.7	71.6	69.9	62.1	51.6	41.8	31.2	49.0
	Boston	30	29.3	31.5	38.9	48.3	58.5	68.0	73.9	72.3	64.7	54.1	44.9	34.8	51.6
	Worcester	30	23.6	26.0	34.3	45.0	56.3	64.7	70.1	68.3	60.2	49.6	39.6	28.9	47.2
MI	Alpena	30	17.8	19.0	28.0	40.3	52.2	61.3	66.7	64.5	56.3	45.6	34.6	24.0	42.5
	Detroit	30	24.5	27.2	36.9	48.1	59.8	69.0	73.5	71.8	63.9	51.9	40.7	29.6	49.8
	Flint	30	21.3	23.8	33.7	45.4	57.1	66.2	70.6	68.5	60.7	49.2	38.1	26.7	46.8
	Grand Rapids	30	22.4	25.0	34.6	46.3	58.1	67.1	71.4	69.4	61.3	49.9	38.4	27.6	47.7
	Houghton Lake	30	17.8	19.9	29.3	41.8	53.9	62.2	66.7	64.6	56.8	46.1	34.8	23.7	43.2
	Lansing	30	21.6	24.0	33.9	45.5	57.1	66.2	70.3	68.4	60.5	49.2	38.0	26.9	46.8
	Marquette	30	11.5	14.8	23.7	36.4	50.3	59.3	64.4	62.3	53.5	42.5	28.9	17.2	38.7
	Muskegon	30	23.5	25.4	34.0	44.9	56.1	64.9	69.9	68.5	60.5	49.7	38.7	28.6	47.1
	Sault Ste. Marie	30	13.2	15.6	24.9	38.4	51.3	58.6	63.9	63.3	54.8	44.4	32.4	20.2	40.1
MN	Duluth	30	8.4	14.8	25.4	39.0	51.8	59.9	65.5	63.7	54.7	43.5	28.0	14.0	39.1
	International Falls	30	2.7	10.9	23.6	39.3	53.3	61.6	66.1	63.8	53.2	41.6	24.4	8.5	37.5
	Minneapolis-St. Paul	30	13.1	20.1	32.1	46.6	59.3	68.4	73.2	70.6	61.0	48.7	32.5	18.7	45.4
	Rochester	30	11.8	18.4	30.6	44.7	56.9	66.1	70.1	67.7	58.9	47.0	31.2	17.3	43.4
	Saint Cloud	30	8.8	16.1	28.4	43.6	56.6	65.1	69.8	67.2	57.4	45.3	28.8	14.4	41.8
MS	Jackson	30	45.0	49.2	56.8	63.4	71.5	78.5	81.4	80.9	75.5	64.4	54.8	47.6	64.1

(Continued)

Table 3A.1 (Continued)

State	Station	Years	Jan	Feb	Mar	Apr	May	Jun	Jul	Aug	Sep	Oct	Nov	Dec	Annual
	Meridian	30	46.1	50.2	57.3	63.8	71.7	78.5	81.7	81.4	76.1	64.8	55.7	48.9	64.7
	Tupelo	30	40.4	44.8	53.1	60.9	69.4	76.9	80.6	79.6	73.3	61.9	51.5	43.4	61.3
MO	Columbia	30	27.8	33.7	44.0	54.4	63.7	72.7	77.4	75.7	67.3	56.0	43.2	32.0	54.0
	Kansas City	30	26.9	33.0	43.8	54.4	64.3	73.6	78.5	76.6	68.1	56.8	42.7	31.3	54.2
	St. Louis	30	29.6	35.4	45.8	56.6	66.5	75.6	80.2	78.2	70.2	58.3	45.3	33.9	56.3
	Springfield	30	31.7	37.1	46.3	55.6	64.7	73.4	78.5	77.6	69.3	58.4	45.9	35.7	56.2
MT	Billings	30	24.0	29.8	37.0	46.1	55.7	65.2	72.0	70.9	59.5	48.1	34.1	26.1	47.4
	Glasgow	30	10.8	19.1	30.9	44.5	55.5	64.4	70.2	69.5	57.3	45.0	27.9	15.6	42.6
	Great Falls	30	21.7	26.4	33.4	42.6	51.5	60.0	66.2	65.6	55.4	45.5	32.3	24.3	43.8
	Havre	30	14.6	21.9	32.5	44.3	54.5	62.7	68.3	67.6	56.3	44.6	29.1	19.0	43.0
	Helena	30	20.2	26.4	35.1	44.1	52.9	61.2	67.8	66.7	56.1	44.8	30.9	21.4	44.0
	Kalispell	30	21.4	26.8	34.9	43.4	51.3	57.7	63.5	63.2	53.1	41.9	30.9	23.1	42.6
	Missoula	30	23.5	29.0	37.6	45.2	52.7	60.2	66.9	66.3	56.1	44.4	32.0	23.4	44.8
NE	Grand Island	30	22.4	28.2	38.3	49.9	60.6	71.1	75.8	73.6	64.4	52.0	36.4	25.6	49.9
	Lincoln	30	22.4	28.3	39.4	51.2	62.0	72.7	77.8	75.4	66.0	53.5	38.1	26.5	51.1
	Norfolk	30	20.4	26.4	37.0	49.1	60.3	70.1	74.8	72.7	63.4	51.0	35.1	23.7	48.7
	North Platte	30	23.2	29.4	38.0	48.1	58.3	68.4	74.3	72.6	62.4	49.7	34.6	25.7	48.7
	Omaha Eppley AP	30	21.7	28.0	39.3	51.4	62.2	72.2	76.7	74.5	65.4	53.2	38.0	25.6	50.7
	Omaha (North)	30	22.4	28.5	39.8	52.0	62.3	71.5	75.6	74.0	65.7	53.9	38.4	26.2	50.9
	Scottsbluff	30	24.5	30.0	37.3	46.2	56.8	67.2	73.0	70.9	60.5	47.8	34.0	25.7	47.9
	Valentine	30	20.8	26.6	35.3	46.1	57.5	67.6	73.7	72.1	61.5	48.3	33.0	23.6	47.2
NV	Elko	30	25.6	31.3	38.6	44.6	52.7	61.7	69.1	67.6	58.2	46.7	34.5	26.0	46.4
	Ely	30	25.2	29.8	35.9	42.2	50.4	59.9	67.4	65.8	56.7	45.4	33.5	25.8	44.8
	Las Vegas	30	47.0	52.2	58.3	66.0	75.4	85.6	91.2	89.3	81.3	68.7	55.0	47.0	68.1
	Reno	30	33.6	38.5	43.3	48.6	56.4	64.7	71.3	69.9	62.4	52.0	40.9	33.6	51.3
	Winnemucca	30	30.1	36.1	41.1	46.7	55.2	64.3	72.0	69.9	60.3	48.8	37.4	29.6	49.3
NH	Concord	30	20.1	23.3	33.3	44.6	56.0	64.9	70.0	68.2	59.4	47.8	37.6	25.9	45.9
	Mt. Washington	30	5.2	6.6	13.6	22.9	35.6	44.4	48.7	47.6	40.4	30.2	20.6	10.1	27.2
NJ	Atlantic City AP	30	32.1	34.2	41.8	50.6	60.5	69.7	75.3	73.5	66.3	55.1	45.9	36.8	53.5
	Atlantic City CO	30	35.2	36.9	43.3	51.4	60.5	69.4	75.2	74.8	68.9	58.5	49.0	40.2	55.3
	Newark	30	31.3	33.8	42.2	52.3	62.7	71.9	77.2	75.5	67.8	56.4	46.4	36.4	54.5
NM	Albuquerque	30	35.7	41.4	48.1	55.6	64.7	74.8	78.5	76.1	69.1	57.3	44.4	36.1	56.8
	Clayton	30	33.9	37.6	43.7	51.7	60.5	69.9	73.8	72.2	64.7	54.6	42.2	34.8	53.3
	Roswell	30	40.0	45.7	52.9	60.5	69.6	78.0	80.8	78.9	72.0	61.4	48.9	40.7	60.8
NY	Albany	30	22.2	25.0	35.0	46.6	58.1	66.3	71.1	69.0	60.6	49.3	39.2	28.0	47.6
	Binghamton	30	21.7	23.8	32.7	44.1	55.9	63.9	68.7	66.6	58.8	48.1	37.6	27.1	45.8
	Buffalo	30	24.5	25.9	34.3	45.3	57.0	65.8	70.8	69.1	61.5	50.7	40.2	29.8	48.0
	Islip	30	30.9	32.4	39.8	49.1	59.2	68.5	74.6	73.1	65.8	54.3	44.9	35.7	52.4
	New York C.Park	30	32.1	34.6	42.5	52.5	62.6	71.2	76.5	75.1	67.5	56.6	47.1	37.3	54.6
	New York (JFK AP)	30	31.8	33.5	40.9	50.1	59.7	68.8	74.8	74.1	67.2	56.5	46.8	37.2	53.5
	New York (Laguardia AP)	30	32.6	34.8	42.3	52.2	62.4	71.5	77.1	75.9	68.6	57.7	47.6	37.9	55.1
	Rochester	30	23.9	25.3	33.9	45.3	57.0	65.8	70.7	68.9	61.2	50.4	39.9	29.4	47.7
	Syracuse	30	22.7	24.5	33.6	45.3	57.1	65.8	70.9	69.2	61.3	50.1	39.7	28.6	47.4
NC	Asheville	30	35.8	39.0	46.3	54.1	62.0	69.2	73.0	71.8	65.7	55.2	46.4	39.0	54.8

	Cape Hatteras	30	46.1	46.8	52.4	59.8	67.6	74.8	79.2	78.6	74.8	65.7	57.6	50.0	62.8
	Charlotte	30	41.7	45.2	52.8	60.9	69.0	76.5	80.3	78.9	72.7	61.7	52.3	44.4	61.4
	Greensboro-Wnstn-Slm-HPT	30	37.7	41.2	49.1	57.6	65.8	73.6	77.9	76.2	69.8	58.5	49.2	41.0	58.2
	Raleigh	30	39.7	43.0	50.7	59.1	67.0	74.7	78.8	77.2	71.2	60.0	51.0	43.0	59.6
	Wilmington	30	46.1	48.5	55.0	62.7	70.2	77.0	81.1	79.7	75.0	64.8	56.5	48.9	63.8
ND	Bismarck	30	10.2	18.1	29.7	43.3	56.0	64.7	70.4	69.0	57.7	45.2	28.0	15.2	42.3
	Fargo	30	6.8	14.1	27.2	43.5	57.4	66.0	70.6	69.0	58.0	45.3	27.0	12.5	41.5
	Grand Forks	30	5.3	13.1	25.7	42.3	56.8	65.2	69.4	67.8	57.0	44.3	25.8	11.3	40.3
	Williston	30	8.0	16.8	28.7	42.5	54.6	63.7	69.3	68.3	56.1	43.6	25.6	13.0	40.9
OH	Akron	30	25.2	28.3	37.7	48.1	58.8	67.5	71.8	70.3	63.0	51.6	41.1	30.7	49.5
	Cleveland	30	25.7	28.4	37.5	47.6	58.5	67.5	71.9	70.2	63.3	52.2	41.8	31.1	49.7
	Columbus	30	28.3	32.0	42.0	52.0	62.6	71.2	75.1	73.5	66.5	54.7	43.7	33.5	52.9
	Dayton	30	26.3	30.3	40.2	50.6	61.2	70.2	74.3	72.3	65.1	53.5	42.2	31.4	51.5
	Mansfield	30	24.3	27.3	36.7	47.2	58.0	66.8	71.0	69.3	62.6	51.5	40.5	29.6	48.8
	Toledo	30	23.9	27.0	37.2	48.3	59.6	68.8	73.0	70.8	63.5	51.8	40.5	29.2	49.5
	Youngstown	30	24.9	27.7	36.7	47.4	57.6	65.9	69.9	68.4	61.5	50.8	40.7	30.4	48.5
OK	Oklahoma City	30	36.7	42.3	51.0	59.7	68.4	76.8	82.0	81.2	73.2	62.0	48.9	39.5	60.1
	Tulsa	30	36.4	42.0	51.4	60.8	69.3	78.0	83.5	82.2	73.5	62.6	49.7	39.7	60.8
OR	Astoria	30	42.4	44.2	46.0	48.5	52.7	56.7	60.1	60.8	58.5	52.6	46.6	42.8	51.0
	Burns	30	24.4	30.0	37.0	43.0	50.9	58.1	65.9	64.2	55.0	44.4	32.7	24.9	44.2
	Eugene	30	39.8	42.8	46.3	49.8	54.8	60.2	66.2	66.4	61.7	52.6	44.7	39.5	52.1
	Medford	30	39.1	43.5	47.1	51.6	58.1	65.6	72.7	72.5	65.9	55.1	43.9	38.1	54.5
	Pendleton	30	33.8	38.7	45.1	51.0	58.1	65.4	72.6	72.0	63.4	52.3	41.2	33.9	52.3
	Portland	30	39.9	43.1	47.2	51.2	57.1	62.7	68.1	68.5	63.6	54.3	45.8	40.2	53.5
	Salem	30	40.3	43.0	46.5	50.0	55.6	61.2	66.8	67.0	62.2	52.9	45.2	40.2	52.6
	Sexton Summit	30	37.5	38.6	39.6	43.4	49.9	56.5	63.7	64.2	60.1	51.3	40.5	37.3	48.6
PC	Guam	30	77.6	77.6	78.4	79.5	80.3	80.5	79.8	79.5	79.7	79.5	79.6	78.9	79.2
	Johnston Island	30	77.5	77.7	77.9	78.7	79.8	81.2	82.0	82.6	82.5	81.9	79.9	78.3	80.0
	Koror	30	81.4	81.2	81.8	82.3	82.6	81.8	81.4	81.6	82.0	82.1	82.5	82.0	81.9
	Kwajalein, Marshall IS	30	81.6	81.8	82.3	82.2	82.4	82.2	82.1	82.3	82.3	82.4	82.1	81.9	82.1
	Majuro, Marshall IS	30	80.8	81.1	81.2	81.1	81.3	81.2	81.1	81.4	81.5	81.5	81.4	80.9	81.2
	Pago Pago, Amer Samoa	30	81.5	81.8	82.0	81.6	80.9	80.3	79.7	79.8	80.3	80.7	81.2	81.7	81.0
	Pohnpei, Caroline IS	30	80.8	81.1	81.4	81.2	81.2	81.0	80.7	80.7	80.7	80.7	81.0	80.9	81.0
	Chuuk, E. Caroline IS	30	81.5	81.3	81.6	81.8	81.9	81.6	81.4	80.9	81.3	81.4	82.1	81.5	81.5
	Wake Island	30	77.8	77.3	78.2	79.2	80.8	82.6	83.3	83.3	83.6	82.6	81.0	79.3	80.8
	Yap, W Caroline IS	30	80.1	80.3	80.8	81.5	81.7	81.0	80.6	80.4	80.6	80.7	80.9	80.6	80.8
PA	Allentown	30	27.1	29.9	38.8	49.0	59.6	68.5	73.3	71.2	63.4	52.0	42.0	32.0	50.6
	Erie	30	26.9	28.2	36.5	46.8	58.1	67.4	72.1	70.9	64.0	53.3	42.9	32.7	50.0
	Harrisburg	30	30.3	32.8	41.7	52.1	62.0	70.7	75.9	74.0	66.2	54.5	44.3	34.8	53.3
	Middletown/ Harrisburg AP	30	30.3	32.8	41.7	52.1	62.0	70.7	75.9	74.0	66.2	54.5	44.3	34.8	53.3
	Philadelphia	30	32.3	34.8	43.2	53.1	63.5	72.3	77.6	76.3	68.8	57.2	47.1	37.4	55.3
	Pittsburgh	30	27.5	30.5	39.8	49.9	60.0	68.4	72.6	71.0	64.0	52.5	42.3	32.5	51.0
	Avoca	30	26.3	28.9	37.9	48.7	59.6	67.5	72.1	70.3	62.5	51.5	41.5	31.4	49.9

(Continued)

Table 3A.1 (Continued)

State	Station	Years	Jan	Feb	Mar	Apr	May	Jun	Jul	Aug	Sep	Oct	Nov	Dec	Annual
	Williamsport	30	25.5	28.5	38.0	49.0	59.5	67.8	72.4	70.9	63.1	51.3	40.8	30.7	49.9
RI	Providence	30	28.7	30.9	38.8	48.6	58.7	67.6	73.3	71.9	64.0	53.0	43.8	33.8	51.1
SC	Charleston AP	30	47.9	50.7	57.7	64.2	72.1	78.2	81.7	80.5	76.1	66.2	58.0	50.5	65.3
	Charleston CO	30	49.8	52.4	58.7	65.9	73.5	79.4	82.8	81.6	77.6	68.5	60.5	52.8	67.0
	Columbia	30	44.6	47.9	55.4	63.2	71.6	78.5	82.0	80.3	74.7	63.7	54.7	47.0	63.6
	Greenville-Spartanburg	30	40.8	44.4	51.6	59.0	67.2	74.7	78.8	77.5	71.4	60.5	51.1	43.5	60.1
SD	Aberdeen	30	11.0	18.7	30.7	45.4	57.9	66.8	72.2	70.5	59.8	46.8	29.3	16.0	43.8
	Huron	30	14.2	21.0	32.6	46.1	58.2	67.9	73.4	71.5	61.0	47.9	31.3	18.6	45.4
	Rapid City	30	22.4	27.3	34.9	44.7	55.0	64.6	71.7	71.1	60.6	48.2	33.4	24.7	46.6
	Sioux Falls	30	14.0	20.8	32.6	45.7	57.8	67.5	73.0	70.8	60.9	48.0	31.3	18.3	45.1
TN	Bristol-JhnCty-Kgsprt	30	34.2	38.0	46.5	54.6	63.0	70.7	74.2	72.8	66.6	55.0	45.5	37.3	54.9
	Chattanooga	30	39.4	43.4	51.4	59.6	67.7	75.4	79.6	78.5	72.1	60.4	50.3	42.4	60.0
	Knoxville	30	37.6	41.8	49.7	57.8	66.0	73.8	77.7	76.9	70.8	58.8	49.0	40.9	58.4
	Memphis	30	39.9	44.9	53.5	62.1	70.6	78.7	82.5	81.2	74.8	63.8	52.3	43.3	62.4
	Nashville	30	36.8	41.3	50.1	58.5	67.1	75.1	79.1	77.9	71.3	59.9	49.3	40.5	58.9
	Oak Ridge	30	36.6	40.6	48.8	57.2	65.6	73.3	77.3	76.2	70.0	58.4	47.7	39.4	57.6
TX	Abilene	30	43.5	48.6	56.4	64.6	72.8	79.8	83.5	82.6	75.5	66.0	53.7	45.4	64.4
	Amarillo	30	35.8	40.6	47.9	56.2	65.2	74.3	78.2	76.3	69.1	58.2	45.1	37.0	57.0
	Austin/City	30	50.2	54.6	61.7	68.3	75.1	81.0	84.2	84.5	79.5	70.6	59.7	52.1	68.5
	Austin/Bergstrom	30	48.1	52.6	59.9	67.1	74.7	80.1	82.8	82.1	77.5	68.6	57.8	49.8	66.8
	Brownsville	30	59.6	62.7	68.8	73.8	79.3	82.7	83.9	84.0	81.0	75.0	67.7	61.1	73.3
	Corpus Christi	30	56.1	59.5	66.0	71.5	77.5	81.9	83.8	83.9	80.8	73.8	65.1	58.1	71.5
	Dallas-Fort Worth	30	44.1	49.4	57.4	65.0	73.1	80.9	85.0	84.4	77.5	67.2	55.1	46.7	65.5
	Dallas-Love Field	30	45.9	51.0	58.8	66.3	74.4	82.2	86.5	86.1	78.9	68.4	56.4	48.0	66.9
	Del Rio	30	51.3	56.1	63.8	70.6	77.7	82.9	85.3	85.1	80.0	71.1	60.1	52.4	69.7
	El Paso	30	45.1	50.5	57.0	64.6	73.7	82.1	83.3	81.1	75.4	64.9	52.7	45.4	64.7
	Galveston	30	55.8	58.0	64.1	70.0	76.9	82.2	84.3	84.4	81.1	74.1	65.4	58.1	71.2
	Houston	30	51.8	55.4	62.3	68.5	75.8	81.3	83.6	83.3	78.9	70.4	60.9	53.7	68.8
	Lubbock	30	38.1	43.3	51.2	60.0	69.2	77.1	79.8	78.0	70.9	60.7	48.1	39.7	59.7
	Midland-Odessa	30	43.2	48.6	55.9	63.7	72.8	79.6	81.7	80.4	73.9	64.4	52.3	44.8	63.4
	Port Arthur	30	52.2	55.6	62.2	68.2	75.4	80.9	82.7	82.5	78.7	70.1	60.9	54.2	68.6
	San Angelo	30	44.9	49.7	57.2	65.0	73.1	79.2	82.4	81.3	74.8	65.4	54.0	46.4	64.5
	San Antonio	30	50.3	54.7	62.1	68.6	75.8	81.5	84.3	84.2	79.4	70.7	60.0	52.4	68.7
	Victoria	30	53.2	56.7	63.7	69.7	76.6	81.8	84.2	84.2	80.1	72.3	62.7	55.2	70.0
	Waco	30	46.1	50.8	58.5	65.9	74.1	81.3	85.4	85.2	78.6	68.6	56.8	48.3	66.6
	Wichita Falls	30	40.5	45.7	54.2	62.4	71.4	79.7	84.8	83.5	75.6	64.7	51.9	42.9	63.1
UT	Milford	30	28.1	33.7	41.6	48.3	56.6	66.7	74.2	72.6	63.1	50.6	37.6	28.6	50.1
	Salt Lake City	30	29.2	34.5	43.1	50.0	58.8	69.0	77.0	75.6	65.0	52.5	39.6	30.2	52.0
VT	Burlington	30	18.0	19.9	30.7	43.5	56.5	65.6	70.6	68.2	59.4	47.7	37.1	24.8	45.2
VA	Lynchburg	30	34.5	37.8	46.0	55.3	63.4	71.0	75.1	73.8	67.1	56.1	46.6	38.2	55.4
	Norfolk	30	40.1	42.0	49.0	57.4	66.3	74.5	79.1	77.4	72.1	61.1	52.3	44.2	59.6
	Richmond	30	36.4	39.5	47.7	57.1	65.4	73.5	77.9	76.3	69.8	58.3	49.0	40.4	57.6
	Roanoke	30	35.8	39.1	47.2	56.1	64.1	71.9	76.2	74.7	67.7	56.6	47.3	39.1	56.3

WA	Olympia	30	38.1	40.5	43.6	47.4	53.3	58.2	62.8	63.3	58.3	49.7	42.4	38.0	49.7
	Quillayute	30	40.6	42.2	43.8	46.7	51.2	54.9	58.6	59.3	56.5	50.1	44.2	40.6	49.1
	Seattle CO	30	41.5	43.8	46.9	50.9	56.6	61.1	65.5	66.0	61.3	53.4	46.0	41.3	52.9
	Seattle Sea-Tac AP	30	40.9	43.3	46.2	50.2	55.8	60.7	65.3	65.6	61.1	52.7	45.2	40.7	52.3
	Spokane	30	27.3	32.5	39.5	46.5	54.4	61.6	68.6	68.6	59.2	47.2	34.9	27.2	47.3
	Walla Walla	30	34.7	39.7	46.5	52.7	59.8	67.3	75.3	75.2	66.1	54.7	43.1	35.1	54.2
	Yakima	30	29.1	35.2	42.5	48.7	56.2	62.9	69.1	68.3	60.0	48.6	37.0	28.8	48.9
PR	San Juan	30	76.6	76.9	77.6	79.1	80.6	82.1	82.2	82.4	82.2	81.6	79.6	77.7	79.9
WV	Beckley	30	30.4	33.9	42.1	51.5	59.9	67.0	70.7	69.3	63.1	52.8	43.4	34.8	51.6
	Charleston	30	33.4	36.9	45.3	54.3	62.4	69.9	73.9	72.6	66.2	55.1	45.9	37.5	54.5
	Elkins	30	28.6	31.6	40.0	48.9	57.9	65.6	69.6	68.5	62.1	50.6	41.0	32.7	49.8
	Huntington	30	32.7	36.8	45.9	55.2	63.6	71.3	75.3	73.9	66.9	55.6	45.9	37.1	55.0
WI	Green Bay	30	15.6	20.5	31.3	44.2	56.4	65.4	69.9	67.5	58.8	47.4	34.0	21.2	44.4
	La Crosse	30	15.9	22.6	34.6	48.4	60.6	69.6	74.0	71.6	62.7	50.6	35.5	21.8	47.3
	Madison	30	17.3	22.6	33.7	45.9	57.7	67.0	71.6	69.1	60.7	49.3	35.5	23.0	46.1
	Milwaukee	30	20.7	25.4	34.9	45.2	56.1	66.3	72.0	70.6	63.0	51.4	38.4	26.2	47.5
WY	Casper	30	22.3	26.7	35.0	42.7	52.1	62.7	70.0	68.6	57.6	45.7	32.0	23.8	44.9
	Cheyenne	30	25.9	28.8	34.2	41.6	51.3	61.5	67.7	65.9	56.6	45.4	33.3	27.1	45.0
	Lander	30	20.3	25.6	35.5	43.9	53.4	63.7	70.9	69.4	58.7	46.4	30.3	21.3	45.0
	Sheridan	30	21.3	26.9	35.3	43.9	52.5	61.6	68.8	68.2	57.1	45.1	31.0	22.4	44.6

Note: In Fahrenheit degrees, based on 30-year average values 1971–2000; Temperature data are the normal daily values for each month.

Source: From U.S. National Oceanic and Atmospheric Administration, *Comparative Climatic Data for the United States Through 2000*, www.noaa.gov.

Table 3A.2 Normal Daily Minimum Temperature — Selected Cities of the United States

State	Station	Years	Jan	Feb	Mar	Apr	May	Jun	Jul	Aug	Sep	Oct	Nov	Dec	Annual
AL	Birmingham AP	30	32.3	35.4	42.4	48.4	57.6	65.4	69.7	68.9	63.0	50.9	41.8	35.2	50.9
	Huntsville	30	30.7	34.0	41.2	48.4	57.5	65.4	69.5	68.1	61.7	49.6	40.7	33.8	50.1
	Mobile	30	39.5	42.4	49.2	54.8	62.8	69.2	71.8	71.7	67.6	56.3	47.8	41.6	56.2
	Montgomery	30	35.5	38.6	45.4	51.2	60.1	67.3	70.9	70.1	64.9	52.2	43.5	37.6	53.1
AK	Anchorage	30	9.3	11.7	18.2	28.7	38.9	47.0	51.5	49.4	41.4	28.3	15.9	11.4	29.3
	Annette	30	30.4	32.3	34.2	37.7	43.1	48.3	52.4	52.6	48.0	41.7	35.1	32.1	40.7
	Barrow	30	−19.6	−22.0	−20.0	−7.3	15.3	30.4	34.3	33.8	27.5	9.8	−6.4	−16.4	5.0
	Bethel	30	0.7	1.3	7.2	18.4	33.1	43.3	48.8	47.5	39.1	24.7	11.7	3.2	23.3
	Bettles	30	−19.2	−17.7	−8.0	10.6	33.7	46.9	49.5	43.7	32.8	11.9	−8.0	−15.1	13.4
	Big Delta	30	−9.6	−6.4	3.2	21.7	37.7	47.6	51.1	46.1	35.6	17.0	−0.8	−7.1	19.7
	Cold Bay	30	23.5	22.9	24.9	28.8	34.8	41.1	46.1	47.4	43.0	35.1	29.9	26.5	33.7
	Fairbanks	30	−19.0	−15.6	−2.7	19.8	36.9	48.5	51.9	46.2	34.7	15.6	−6.6	−15.2	16.3
	Gulkana	30	−12.9	−7.4	2.3	19.7	32.2	41.1	45.4	41.7	32.8	18.4	−2.2	−9.5	16.8
	Homer	30	17.5	18.3	22.5	29.3	36.7	43.0	47.2	46.7	41.0	31.4	23.5	20.0	31.4
	Juneau	30	20.7	23.5	27.8	33.4	40.1	46.1	49.2	48.3	43.8	37.7	28.9	24.4	35.3
	King Salmon	30	8.0	7.4	15.1	24.9	34.8	42.2	47.5	47.4	40.3	26.0	15.9	9.3	26.6
	Kodiak	30	24.6	24.3	26.8	31.8	38.2	43.9	48.5	48.6	43.2	34.3	28.9	25.3	34.9
	Kotzebue	30	−8.6	−9.9	−7.7	3.3	25.3	38.8	49.4	47.4	37.2	18.8	3.2	−6.4	15.9
	Mcgrath	30	−15.6	−12.5	−1.8	17.7	35.5	45.7	49.8	45.7	35.9	18.3	−2.2	−12.3	17.0
	Nome	30	−1.8	−2.3	1.0	12.4	31.1	40.6	46.6	45.2	37.2	22.9	10.8	0.9	20.4
	St. Paul Island	30	21.5	18.9	19.5	24.0	31.5	37.6	43.0	45.1	40.7	34.1	29.1	24.7	30.8
	Talkeetna	30	2.3	5.0	11.1	23.9	34.9	45.1	49.9	46.5	37.3	23.6	9.4	4.8	24.5
	Unalakleet	30	−3.9	−4.2	1.8	13.8	32.1	42.7	48.9	46.5	35.8	19.4	5.0	−1.4	19.7
	Valdez	30	17.2	19.6	23.8	30.9	38.6	45.0	48.0	46.4	40.9	33.4	23.9	20.2	32.3
	Yakutat	30	19.4	21.0	23.6	29.2	36.1	42.7	47.1	46.2	40.6	34.8	26.3	22.9	32.5
AZ	Flagstaff	30	16.5	18.8	22.8	27.3	34.0	41.4	49.9	49.1	41.7	31.1	22.1	16.6	30.9
	Phoenix	30	43.4	47.0	51.1	57.5	66.3	75.2	81.4	80.4	74.5	62.9	50.0	43.5	61.1
	Tucson	30	38.9	41.6	45.1	50.5	58.6	68.0	73.4	72.4	67.7	57.0	45.1	39.2	54.8
	Winslow	30	21.3	25.5	31.1	36.9	45.3	54.2	62.0	61.1	52.9	40.1	28.7	21.0	40.0
	Yuma	30	46.2	48.8	52.8	58.1	65.1	73.2	80.8	80.8	75.3	64.0	52.2	45.8	61.9
AR	Fort Smith	30	27.8	32.6	40.9	49.0	58.9	67.2	71.4	70.3	62.9	50.5	39.5	31.1	50.2
	Little Rock	30	30.8	34.8	42.6	50.0	59.2	67.8	72.0	70.5	63.6	51.5	41.5	33.9	51.5
	North Little Rock	30	31.3	36.1	44.5	52.7	61.2	68.9	72.9	71.5	64.9	54.1	43.4	34.9	53.0
CA	Bakersfield	30	39.3	43.0	46.2	49.6	56.8	63.7	69.2	68.4	63.9	54.9	44.2	38.2	53.1
	Bishop	30	22.4	26.4	31.0	36.0	43.7	50.7	55.7	53.7	46.9	37.1	27.1	21.6	37.7
	Eureka	30	40.8	41.8	42.2	44.0	47.6	50.7	52.8	53.4	51.2	47.7	43.9	40.6	46.4
	Fresno	30	38.4	41.4	44.9	48.4	54.9	61.2	66.1	64.9	60.4	51.9	42.3	37.0	51.0
	Long Beach	30	46.0	48.1	50.4	53.2	57.8	61.3	64.6	65.6	63.7	58.3	50.1	45.3	55.4
	Los Angeles AP	30	48.6	50.1	51.3	53.6	56.9	60.1	63.3	64.5	63.6	59.4	52.7	48.5	56.1
	Los Angeles CO	30	48.5	50.3	51.6	54.4	57.9	61.4	64.6	65.6	64.6	59.9	52.6	48.3	56.6
	Mount Shasta	30	26.4	28.7	30.3	33.3	39.0	44.9	48.9	47.5	42.9	36.6	29.9	25.8	36.2
	Redding	30	35.5	38.1	41.1	44.9	51.6	59.6	64.1	60.8	56.5	48.0	39.8	35.0	47.9
	Sacramento	30	38.8	41.9	44.2	46.3	50.9	55.5	58.3	58.1	55.8	50.6	42.8	37.7	48.4
	San Diego	30	49.7	51.5	53.6	56.4	59.8	62.6	65.9	67.4	66.1	61.2	53.6	48.9	58.1

	San Francisco AP	30	42.9	45.5	46.8	48.1	50.5	52.9	54.5	55.5	55.1	52.4	47.5	43.0	49.6
	San Francisco CO	30	46.4	48.5	49.2	50.1	51.4	53.2	54.4	55.6	56.1	54.6	50.8	46.7	51.4
	Santa Barbara	30	40.8	44.0	46.0	47.6	50.5	53.9	57.3	58.4	56.6	51.6	44.0	39.9	49.2
	Santa Maria	30	39.3	41.4	42.7	43.4	46.9	50.4	53.5	54.2	52.9	48.2	41.8	38.2	46.1
	Stockton	30	38.1	41.0	43.6	46.7	52.1	57.5	60.8	60.3	57.4	50.5	42.1	36.7	48.9
CO	Alamosa	30	−3.7	4.7	15.8	22.8	32.4	40.4	46.4	45.2	36.5	23.9	11.1	−0.7	22.9
	Colorado Springs	30	14.5	18.0	23.9	31.4	40.7	49.5	54.8	53.6	45.4	34.3	22.6	15.6	33.7
	Denver	30	15.2	19.1	25.4	34.2	43.8	53.0	58.7	57.4	47.3	35.9	23.5	16.4	35.8
	Grand Junction	30	15.6	22.7	31.0	37.5	46.4	55.3	61.4	59.7	50.4	38.6	26.3	17.5	38.5
	Pueblo	30	14.0	18.8	26.3	34.5	44.8	53.5	59.4	58.1	48.7	35.3	22.5	15.1	35.9
CT	Bridgeport	30	22.9	24.9	32.0	40.7	50.6	59.6	66.0	65.4	57.7	46.3	37.5	28.0	44.3
	Hartford	30	17.2	19.9	28.3	37.9	48.1	57.0	62.4	60.7	52.1	40.6	32.6	22.6	40.0
DE	Wilmington	30	23.7	25.8	33.4	42.1	52.4	61.8	67.3	65.8	58.1	45.6	36.9	28.4	45.1
DC	Washington Dulles AP	30	21.9	24.1	31.8	40.2	49.9	59.0	64.0	62.8	55.6	42.3	33.8	26.0	42.6
	Washington Nat'l AP	30	27.3	29.7	37.3	45.9	55.8	65.0	70.1	68.6	61.8	49.6	40.0	32.0	48.6
FL	Apalachicola	30	43.0	45.8	51.4	57.6	65.1	71.6	73.9	74.0	71.2	60.5	52.0	45.3	59.3
	Daytona Beach	30	47.1	48.8	53.7	58.0	64.5	70.6	72.4	72.8	71.9	65.3	57.0	50.1	61.0
	Fort Myers	30	54.5	55.4	59.3	62.7	68.4	73.1	74.2	74.4	73.9	68.6	62.1	56.2	65.2
	Gainesville	30	42.4	44.7	49.9	54.7	62.0	68.4	70.8	70.6	68.1	59.2	51.1	44.4	57.2
	Jacksonville	30	41.9	44.3	49.8	54.6	62.5	69.4	72.4	72.2	69.4	59.7	50.8	44.1	57.6
	Key West	30	65.2	65.7	68.8	72.1	75.9	78.7	79.6	79.2	78.5	75.7	71.9	67.3	73.2
	Miami	30	59.6	60.5	64.0	67.6	72.0	75.2	76.5	76.5	75.7	72.2	67.5	62.2	69.1
	Orlando	30	49.9	51.3	55.9	59.9	65.9	71.3	72.6	73.0	71.9	65.5	58.7	52.6	62.4
	Pensacola	30	42.7	45.4	51.7	57.6	65.8	72.1	74.5	74.2	70.4	59.6	51.1	44.7	59.2
	Tallahassee	30	39.7	42.1	48.2	52.8	62.3	69.8	72.7	72.7	69.2	56.9	47.9	41.6	56.3
	Tampa	30	52.4	53.8	58.5	62.4	68.9	74.0	75.3	75.4	74.3	67.6	60.7	54.7	64.8
	Vero Beach	30	52.7	53.6	57.8	61.6	67.2	71.8	73.0	72.9	72.7	68.5	61.9	54.7	64.0
	West Palm Beach	30	57.3	58.2	61.9	65.4	70.5	73.8	75.0	75.4	74.7	71.2	65.8	60.1	67.4
GA	Athens	30	32.9	35.4	42.3	48.7	57.6	65.3	69.3	68.5	62.7	50.7	42.2	35.3	50.9
	Atlanta	30	33.5	36.5	43.6	50.4	59.5	67.1	70.6	69.9	64.3	52.8	43.5	36.2	52.3
	Augusta	30	33.1	35.5	42.5	48.1	57.2	65.4	69.6	68.4	62.4	49.6	40.9	34.7	50.6
	Columbus	30	36.6	39.0	45.7	51.8	61.3	68.8	72.3	71.5	66.4	54.5	45.7	39.0	54.4
	Macon	30	34.5	37.0	43.8	49.5	58.6	66.6	70.5	69.5	63.7	51.1	42.5	36.3	52.0
	Savannah	30	38.0	40.9	47.5	52.9	61.3	68.1	71.8	71.3	67.3	56.1	46.9	40.1	55.2
HI	Hilo	30	63.6	63.5	64.7	65.6	66.7	68.0	69.2	69.4	69.0	68.5	67.2	64.9	66.7
	Honolulu	30	65.7	65.4	66.9	68.2	69.6	72.1	73.8	74.7	74.2	73.2	71.1	67.8	70.2
	Kahului	30	63.3	63.1	64.6	66.0	67.0	69.3	70.8	71.0	70.0	69.4	67.9	65.1	67.3
	Lihue	30	65.4	65.5	67.3	68.9	70.3	72.7	74.0	74.5	74.0	72.8	70.8	67.6	70.3
ID	Boise	30	23.6	28.8	34.0	39.4	46.6	54.2	60.3	59.8	51.2	41.3	32.4	24.1	41.3
	Lewiston	30	28.0	31.2	35.6	40.6	47.0	53.6	59.3	59.3	50.9	41.2	34.1	28.5	42.5
	Pocatello	30	16.3	20.9	27.3	32.6	39.2	45.7	50.9	49.9	41.8	33.3	24.9	16.8	33.3
IL	Chicago	30	14.3	19.2	28.5	37.6	47.5	57.2	63.2	62.2	53.7	42.1	31.6	20.4	39.8
	Moline	30	12.3	18.2	29.0	39.3	50.0	59.7	64.5	62.4	53.4	41.6	30.1	18.3	39.9
	Peoria	30	14.3	19.7	30.2	40.3	50.8	60.1	64.6	62.6	54.0	42.3	31.4	20.1	40.9
	Rockford	30	10.8	16.3	26.7	36.8	47.9	57.6	62.6	60.9	51.8	40.1	29.0	16.9	38.1
	Springfield	30	17.1	22.2	32.4	42.2	52.7	61.9	66.0	63.9	55.4	44.4	33.7	22.6	42.9

(Continued)

Table 3A.2 (Continued)

State	Station	Years	Jan	Feb	Mar	Apr	May	Jun	Jul	Aug	Sep	Oct	Nov	Dec	Annual
IN	Evansville	30	22.6	26.2	35.2	43.8	54.0	63.5	67.8	65.1	57.0	44.6	36.0	27.0	45.2
	Fort Wayne	30	16.1	19.2	28.8	38.2	49.1	58.8	62.5	60.4	52.8	41.8	32.7	22.3	40.2
	Indianapolis	30	18.5	22.5	32.0	41.2	51.8	61.3	65.2	63.3	55.2	43.6	34.1	24.0	42.7
	South Bend	30	15.7	19.0	28.2	37.7	48.4	58.3	62.8	61.3	53.3	42.3	32.6	21.7	40.1
IA	Des Moines	30	11.7	17.8	28.7	39.9	51.4	61.0	66.1	63.9	54.3	42.2	29.0	16.7	40.2
	Dubuque	30	9.2	15.4	26.2	37.5	48.8	57.9	62.4	60.2	51.7	40.5	27.8	15.2	37.7
	Sioux City	30	8.5	15.3	25.7	37.3	49.2	58.5	62.9	60.6	50.1	38.0	24.8	12.8	37.0
	Waterloo	30	6.3	13.2	24.9	35.8	48.1	58.1	62.2	59.5	49.8	37.8	25.1	12.5	36.1
KS	Concordia	30	16.9	21.9	31.1	41.2	51.9	61.8	67.4	65.6	56.1	44.0	30.5	20.8	42.4
	Dodge City	30	18.7	23.6	31.2	40.7	51.7	61.6	66.8	65.6	56.5	43.8	30.2	21.7	42.7
	Goodland	30	15.8	19.7	26.4	34.8	45.7	55.5	61.1	59.6	50.0	37.5	25.2	17.8	37.4
	Topeka	30	17.2	23.0	32.9	42.9	53.4	63.2	67.7	65.4	55.9	44.3	32.1	21.8	43.3
	Wichita	30	20.3	25.3	34.4	43.7	54.0	63.9	69.1	67.9	59.3	46.9	33.9	24.0	45.2
KY	Greater Cincinnati AP	30	21.3	25.0	33.8	42.7	52.9	61.6	66.1	64.2	56.8	44.9	35.7	26.4	44.3
	Jackson	30	25.7	28.9	37.4	45.8	54.3	61.9	65.7	64.3	58.4	47.4	38.9	30.2	46.6
	Lexington	30	24.1	27.7	35.9	44.1	53.6	62.2	66.4	64.9	57.9	46.4	37.3	28.4	45.7
	Louisville	30	24.9	28.5	37.1	46.0	56.1	65.1	69.8	68.2	60.9	48.5	39.3	29.9	47.9
	Paducah	30	23.9	28.2	37.1	45.6	55.0	63.8	67.7	64.9	57.1	45.2	36.5	27.5	46.1
LA	Baton Rouge	30	40.2	43.1	49.6	55.8	64.1	70.2	72.7	71.9	67.5	56.4	47.9	42.1	56.8
	Lake Charles	30	41.2	44.3	50.8	57.2	65.7	72.1	74.3	73.6	69.1	58.6	49.7	43.3	58.3
	New Orleans	30	43.4	46.1	52.7	58.4	66.4	72.0	74.2	73.9	70.6	60.2	51.8	45.6	59.6
	Shreveport	30	36.5	40.3	47.2	53.8	62.7	69.9	73.4	72.3	66.4	55.0	45.3	38.3	55.1
ME	Caribou	30	−0.3	2.9	15.2	29.2	40.7	49.9	54.8	52.6	43.6	34.1	23.7	8.0	29.6
	Portland	30	12.5	15.6	25.2	34.7	44.2	52.9	58.6	57.2	48.5	37.4	29.5	18.7	36.3
MD	Baltimore	30	23.5	26.1	33.6	42.0	51.8	60.8	65.8	63.9	56.6	43.7	34.7	27.3	44.2
MA	Blue Hill	30	18.1	20.3	27.8	37.1	47.0	55.9	62.0	60.9	53.2	42.9	34.2	23.8	40.3
	Boston	30	22.1	24.2	31.5	40.5	50.2	59.4	65.5	64.5	56.8	46.4	37.9	27.8	43.9
	Worcester	30	15.8	17.8	25.6	35.5	46.2	55.0	60.8	59.5	51.3	40.7	32.0	21.6	38.5
MI	Alpena	30	9.5	9.7	18.7	30.2	40.0	48.8	54.5	52.9	45.2	35.6	27.0	16.9	32.4
	Detroit	30	17.8	20.0	28.5	38.4	49.4	58.9	63.6	62.2	54.1	42.5	33.5	23.4	41.0
	Flint	30	13.3	15.3	24.3	34.6	45.2	54.6	59.1	57.4	49.4	38.6	29.8	19.1	36.7
	Grand Rapids	30	15.6	17.4	25.9	36.1	46.6	55.8	60.5	59.0	51.0	40.2	31.2	21.4	38.4
	Houghton Lake	30	9.7	10.5	19.2	30.6	40.7	48.9	53.4	52.2	45.3	36.2	27.6	16.8	32.6
	Lansing	30	13.9	15.4	24.3	34.5	44.8	54.3	58.4	57.0	48.9	38.6	30.1	19.7	36.7
	Marquette	30	3.3	5.4	14.3	26.9	39.1	48.3	53.5	52.0	43.8	34.0	22.4	10.2	29.4
	Muskegon	30	17.1	18.3	25.4	35.1	45.1	54.2	59.8	58.8	50.7	40.6	31.8	22.6	38.3
	Sault Ste. Marie	30	4.9	6.6	16.1	28.8	39.3	46.5	52.0	52.4	44.8	36.0	25.9	13.1	30.5
MN	Duluth	30	−1.2	5.1	16.5	28.9	40.2	48.5	54.6	53.5	44.8	34.5	20.7	5.6	29.3
	International Falls	30	−8.4	−0.7	12.3	27.1	40.0	49.1	53.6	51.3	41.6	31.5	16.4	−1.1	26.1
	Minneapolis-St. Paul	30	4.3	11.8	23.5	36.2	48.5	57.8	63.0	60.8	50.8	38.9	24.8	10.9	35.9
	Rochester	30	3.7	10.6	22.6	34.6	46.1	55.6	60.1	58.0	48.7	37.1	23.7	10.1	34.3
	Saint Cloud	30	−1.2	6.4	19.1	32.2	44.1	52.9	57.9	55.5	45.7	34.3	20.4	5.5	31.1
MS	Jackson	30	35.0	38.2	45.4	51.7	61.0	68.1	71.4	70.3	64.6	52.0	43.4	37.3	53.2
	Meridian	30	34.7	37.7	44.3	50.4	59.5	66.8	70.5	69.8	64.2	51.3	42.8	37.2	52.4

	Tupelo	30	30.5	33.5	41.4	48.2	57.7	65.7	69.8	68.2	61.7	48.8	40.0	33.2	49.9
MO	Columbia	30	18.2	23.4	33.0	42.9	52.8	61.8	66.3	64.0	55.4	44.1	33.0	22.5	43.1
	Kansas City	30	17.8	23.3	33.2	43.5	53.9	63.2	68.2	66.1	57.2	45.9	33.4	22.5	44.0
	St. Louis	30	21.2	26.5	36.2	46.5	56.6	65.9	70.6	68.6	60.3	48.2	36.7	25.8	46.9
	Springfield	30	21.8	26.4	34.9	43.6	53.4	62.2	67.1	65.6	57.4	46.1	35.3	25.9	45.0
MT	Billings	30	15.1	20.1	26.4	34.7	44.0	52.5	58.3	57.3	47.1	37.2	25.6	17.7	36.3
	Glasgow	30	1.8	9.9	20.6	32.2	43.0	51.6	56.6	55.7	44.1	33.0	18.5	6.4	31.1
	Great Falls	30	11.3	15.1	21.5	29.7	38.3	46.0	50.4	49.9	41.2	33.0	22.5	14.4	31.1
	Havre	30	3.7	10.4	20.0	30.0	40.2	48.0	52.0	51.3	40.7	29.8	17.3	7.8	29.3
	Helena	30	9.9	15.6	23.5	31.2	39.8	47.5	52.3	50.8	41.2	31.2	20.3	11.3	31.2
	Kalispell	30	13.8	18.4	24.8	30.8	37.9	43.5	46.7	45.8	37.1	28.4	23.2	16.1	30.5
	Missoula	30	16.2	20.5	27.1	32.4	39.3	45.9	50.2	49.3	40.6	31.4	24.0	16.5	32.8
NE	Grand Island	30	12.2	17.7	27.0	37.8	49.3	59.1	64.4	62.3	51.8	39.3	25.9	15.9	38.6
	Lincoln	30	11.5	17.2	27.5	38.8	50.1	60.4	65.9	63.7	53.2	40.4	27.0	16.2	39.3
	Norfolk	30	9.6	15.5	25.4	36.8	48.3	58.0	63.0	61.0	50.4	38.0	24.7	13.7	37.1
	North Platte	30	9.9	15.4	23.8	33.4	44.5	54.2	60.2	58.4	46.7	33.7	20.7	12.1	34.4
	Omaha Eppley AP	30	11.6	18.0	28.1	39.6	50.7	60.6	65.9	63.8	53.5	41.1	28.1	16.4	39.8
	Omaha (North)	30	12.6	19.0	28.8	40.3	51.3	60.5	65.5	64.1	55.0	43.1	29.2	17.2	40.6
	Scottsbluff	30	11.0	15.8	23.0	31.4	42.4	52.1	57.4	54.9	43.7	31.3	19.7	11.6	32.9
	Valentine	30	7.8	13.7	22.1	32.4	43.7	53.2	59.1	57.3	45.8	33.1	20.1	10.5	33.2
NV	Elko	30	14.1	19.7	25.9	29.9	36.8	43.5	48.6	47.0	38.1	28.3	20.9	13.8	30.6
	Ely	30	10.4	15.6	21.9	26.4	33.4	40.6	47.4	46.4	37.5	27.8	18.2	10.6	28.0
	Las Vegas	30	36.8	41.4	47.0	53.9	62.9	72.3	78.2	76.7	68.8	56.5	44.0	36.6	56.3
	Reno	30	21.8	25.4	29.3	33.2	40.2	46.5	51.4	49.9	43.1	34.0	26.4	20.7	35.2
	Winnemucca	30	18.5	23.6	27.0	30.7	38.4	45.8	51.8	49.2	40.2	30.2	23.3	17.0	33.0
NH	Concord	30	9.7	12.6	22.7	32.2	42.4	51.8	57.1	55.6	46.6	35.1	27.6	16.2	34.1
	Mt. Washington	30	−3.7	−1.7	5.9	16.4	29.5	38.5	43.3	42.1	34.6	24.0	13.6	1.7	20.4
NJ	Atlantic City AP	30	22.8	24.5	31.7	39.8	49.8	59.3	65.4	63.7	56.0	43.9	35.7	27.1	43.3
	Atlantic City CO	30	29.0	30.6	37.0	45.2	54.8	63.9	69.8	69.7	63.6	52.5	42.9	34.0	49.4
	Newark	30	24.4	26.6	34.2	43.7	54.1	63.5	69.1	67.7	59.9	48.2	39.1	29.8	46.7
NM	Albuquerque	30	23.8	28.2	33.7	40.5	49.7	59.4	64.7	63.2	56.0	43.8	31.6	24.2	43.2
	Clayton	30	20.3	23.7	29.2	37.2	46.7	55.9	60.2	59.2	51.5	40.6	28.7	21.6	39.6
	Roswell	30	24.4	29.3	35.7	43.3	53.2	62.0	66.7	65.5	58.3	46.3	33.3	25.1	45.3
NY	Albany	30	13.3	15.7	25.4	35.9	46.5	55.0	60.0	58.3	49.9	38.8	30.8	20.1	37.5
	Binghamton	30	15.0	16.7	24.7	35.1	46.2	54.4	59.2	57.4	49.9	39.6	30.9	20.8	37.5
	Buffalo	30	17.8	18.6	26.1	36.4	47.7	56.9	62.1	60.5	52.9	42.6	33.7	23.6	39.9
	Islip	30	22.6	24.3	31.1	40.0	49.4	59.6	65.9	64.5	56.6	44.6	36.1	27.5	43.5
	New York C.Park	30	26.2	28.1	35.1	44.2	54.2	63.3	68.8	67.7	60.3	49.6	41.0	31.6	47.5
	New York (JFK AP)	30	24.7	26.1	32.9	41.6	51.2	60.4	66.7	66.3	59.5	48.7	39.8	30.5	45.7
	New York (Laguardia AP)	30	26.5	28.3	35.1	44.4	54.3	63.7	69.5	68.7	61.6	50.9	41.6	32.0	48.1
	Rochester	30	16.6	17.3	25.2	35.3	46.1	55.0	60.0	58.7	51.3	41.1	32.6	22.7	38.5
	Syracuse	30	14.0	15.5	24.2	34.9	45.8	54.6	60.1	58.8	51.1	40.4	32.0	20.9	37.7
NC	Asheville	30	25.8	28.0	34.9	41.8	50.6	58.3	62.7	61.8	55.4	43.3	35.3	28.8	43.9
	Cape Hatteras	30	38.6	39.0	44.5	51.8	60.2	68.1	72.9	72.3	68.5	58.8	50.3	42.6	55.6
	Charlotte	30	32.1	34.4	41.6	49.1	58.2	66.5	70.6	69.3	63.0	50.9	41.8	34.9	51.0

(Continued)

Table 3A.2 (Continued)

State	Station	Years	Jan	Feb	Mar	Apr	May	Jun	Jul	Aug	Sep	Oct	Nov	Dec	Annual
	Greensboro-Wnstn-Slm-HPT	30	28.2	30.6	37.8	45.5	54.7	63.5	68.1	66.8	60.1	47.5	38.6	31.4	47.7
	Raleigh	30	29.6	31.9	38.9	46.4	55.3	63.8	68.5	67.2	61.0	48.2	39.5	32.6	48.6
	Wilmington	30	35.8	37.5	43.7	51.2	59.8	67.6	72.3	71.0	65.9	53.9	45.1	38.1	53.5
ND	Bismarck	30	−0.6	7.8	19.1	30.6	42.8	51.6	56.4	54.7	43.7	32.1	17.8	4.8	30.1
	Fargo	30	−2.3	5.4	19.0	32.4	45.3	54.5	59.0	57.0	46.1	34.4	18.7	4.2	31.1
	Grand Forks	30	−4.3	3.7	17.1	31.0	43.5	52.8	56.8	54.5	44.3	33.0	17.4	2.5	29.4
	Williston	30	−3.3	5.9	17.2	29.1	40.9	50.1	55.2	53.8	42.2	30.2	14.9	2.1	28.2
OH	Akron	30	17.4	19.8	27.9	37.1	47.8	56.8	61.3	60.2	53.1	42.1	33.4	23.6	40.0
	Cleveland	30	18.8	21.0	28.9	37.9	48.3	57.7	62.3	61.2	54.3	43.7	34.9	24.9	41.2
	Columbus	30	20.3	23.5	32.2	41.2	51.8	60.7	64.9	63.2	55.9	44.0	34.9	25.9	43.2
	Dayton	30	19.0	22.4	31.2	40.4	51.1	60.2	64.4	62.2	54.6	43.5	34.3	24.4	42.3
	Mansfield	30	16.2	18.7	26.8	36.1	46.7	55.8	60.3	58.9	52.1	41.3	32.2	22.0	38.9
	Toledo	30	16.4	18.9	27.9	37.7	48.6	58.2	62.6	60.7	52.9	41.6	32.6	22.3	40.0
	Youngstown	30	17.4	19.3	27.1	36.5	46.2	54.6	58.7	57.5	50.9	40.9	33.0	23.4	38.8
OK	Oklahoma City	30	26.2	31.1	39.4	48.1	57.9	66.4	70.8	69.8	62.2	50.6	38.2	29.2	49.2
	Tulsa	30	26.3	31.1	40.3	49.5	59.0	67.9	73.1	71.2	62.9	51.1	39.3	29.8	50.1
OR	Astoria	30	36.7	37.6	38.6	40.8	45.4	49.8	52.9	53.2	49.5	44.1	40.1	37.1	43.8
	Burns	30	14.0	19.4	24.9	28.6	35.6	41.1	46.4	43.9	35.0	26.4	20.6	14.6	29.2
	Eugene	30	33.0	34.9	36.7	38.9	42.7	47.0	50.8	50.8	46.7	40.5	37.2	33.3	41.0
	Medford	30	30.9	33.1	35.9	39.0	44.0	50.1	55.2	54.9	48.3	40.2	35.0	31.0	41.5
	Pendleton	30	27.4	30.9	35.4	39.7	45.9	52.0	57.5	57.3	49.7	40.7	33.8	27.7	41.5
	Portland	30	34.2	35.9	38.6	41.9	47.5	52.6	56.9	57.3	52.5	45.2	39.8	35.0	44.8
	Salem	30	33.5	34.7	36.6	38.8	43.6	48.4	52.0	52.1	47.7	41.3	37.9	33.9	41.7
	Sexton Summit	30	32.5	33.0	32.5	34.9	39.6	45.4	51.8	52.7	49.8	43.3	35.2	32.5	40.3
PC	Guam	30	71.2	71.1	71.7	72.7	73.2	73.5	72.8	72.8	72.8	72.7	73.6	73.1	72.6
	Johnston Island	30	73.1	73.2	73.4	74.2	75.2	76.6	77.4	78.0	77.9	77.5	75.7	73.9	75.5
	Koror	30	75.1	74.9	75.2	75.8	76.0	75.4	75.3	75.7	76.0	75.8	75.9	75.6	75.6
	Kwajalein, Marshall IS	30	77.5	77.5	77.9	77.8	78.0	77.8	77.6	77.6	77.5	77.8	77.7	77.9	77.7
	Majuro, Marshall IS	30	76.3	76.6	76.5	76.4	76.5	76.3	76.2	76.3	76.4	76.3	76.4	76.3	76.4
	Pago Pago, Amer Samoa	30	76.1	76.3	76.6	76.3	76.2	76.1	75.5	75.5	75.8	76.2	76.5	76.4	76.1
	Pohnpei, Caroline IS	30	74.8	75.1	75.2	74.7	74.6	74.2	73.2	72.8	72.6	72.6	73.3	74.5	74.0
	Chuuk, E. Caroline IS	30	75.9	76.2	76.4	76.4	76.1	76.0	75.0	74.3	74.7	74.5	75.8	75.3	75.6
	Wake Island	30	73.1	72.4	73.1	73.9	75.3	77.1	77.8	77.9	78.4	77.4	76.3	74.7	75.6
	Yap, W Caroline IS	30	73.7	73.8	74.0	74.6	74.9	74.3	74.0	73.7	73.7	73.8	74.0	74.2	74.1
PA	Allentown	30	19.1	21.0	28.9	37.8	48.3	57.7	62.6	60.7	52.7	41.1	32.7	24.0	40.6
	Erie	30	20.3	20.9	28.2	37.9	48.7	58.5	63.7	62.7	55.9	45.5	36.4	26.8	42.1
	Harrisburg	30	23.1	24.7	32.5	41.5	51.4	60.6	66.0	64.2	56.7	44.6	36.1	27.8	44.1
	Middletown/ Harrisburg AP	30	23.1	24.7	32.5	41.5	51.4	60.6	66.0	64.2	56.7	44.6	36.1	27.8	44.1
	Philadelphia	30	25.5	27.5	35.1	44.2	54.8	64.0	69.7	68.5	60.9	48.7	39.5	30.6	47.4
	Pittsburgh	30	19.9	22.3	30.1	39.1	49.2	57.7	62.4	61.0	53.9	42.5	34.2	25.3	41.5
	Avoca	30	18.5	20.4	28.4	38.1	48.4	56.7	61.5	60.1	52.6	41.7	33.7	24.2	40.4

	Williamsport	30	17.9	19.9	28.2	37.8	47.8	56.8	61.7	60.4	52.8	40.9	32.7	23.7	40.1
RI	Providence	30	20.3	22.5	30.0	39.1	48.8	57.9	64.1	62.8	54.5	43.1	35.1	25.6	42.0
SC	Charleston AP	30	36.9	39.1	46.0	52.2	61.3	68.5	72.5	71.6	67.1	55.3	46.4	39.3	54.7
	Charleston CO	30	42.4	44.9	51.5	58.8	67.4	73.8	77.0	76.1	72.2	61.9	53.4	45.5	60.4
	Columbia	30	34.0	36.3	43.5	50.7	60.0	67.9	71.8	70.6	64.6	51.5	42.6	36.1	52.5
	Greenville-Spartanburg AP	30	31.4	33.9	40.5	47.0	56.2	64.3	68.7	67.9	61.7	49.7	41.0	34.3	49.7
SD	Aberdeen	30	0.6	8.8	21.2	33.4	45.6	54.8	59.7	57.4	46.5	34.4	19.7	6.3	32.4
	Huron	30	3.5	10.8	22.3	33.9	45.8	55.4	60.7	58.6	47.3	34.9	21.1	8.4	33.6
	Rapid City	30	11.3	15.9	23.2	32.3	42.7	51.8	57.9	56.6	46.0	34.7	22.1	13.3	34.0
	Sioux Falls	30	2.9	10.1	21.3	32.5	44.6	54.5	60.3	58.4	47.6	34.8	20.7	7.8	33.0
TN	Bristol-JhnCty-Kgsprt	30	24.3	27.0	34.6	42.0	51.0	59.5	63.5	61.7	54.7	41.8	33.6	26.8	43.4
	Chattanooga	30	29.9	32.6	40.0	47.0	56.2	64.6	69.4	68.3	61.7	48.5	39.5	32.7	49.2
	Knoxville	30	28.9	31.8	39.1	46.6	55.6	63.9	68.5	67.3	60.8	47.7	38.9	31.9	48.4
	Memphis	30	31.3	35.5	43.7	51.9	60.8	68.8	72.9	71.2	64.3	52.5	42.6	34.5	52.5
	Nashville	30	27.9	31.2	39.4	47.1	56.7	65.0	69.5	68.0	61.0	48.6	39.5	31.5	48.8
	Oak Ridge	30	27.2	29.5	36.6	43.8	53.4	61.7	66.4	65.2	58.8	45.7	36.4	29.8	46.2
TX	Abilene	30	31.8	36.5	43.8	51.8	61.0	68.5	72.3	71.4	64.4	54.4	42.3	33.9	52.7
	Amarillo	30	22.6	27.0	33.6	41.7	51.7	61.1	65.3	63.8	56.3	44.6	31.8	24.1	43.6
	Austin/City	30	40.0	44.0	50.9	57.6	65.4	71.1	73.4	73.3	68.8	59.8	49.3	41.9	58.0
	Austin/Bergstrom	30	37.3	41.0	48.4	56.3	65.1	70.2	71.5	70.3	65.3	56.3	45.9	38.2	55.5
	Brownsville	30	50.5	53.3	59.5	65.2	71.6	74.9	75.4	75.3	72.6	65.9	58.6	52.0	64.6
	Corpus Christi	30	46.2	49.3	56.2	62.3	69.5	73.5	74.4	74.5	71.6	64.0	55.4	48.1	62.1
	Dallas-Fort Worth	30	34.0	38.7	46.4	54.0	63.0	70.7	74.6	74.0	67.2	56.4	45.1	36.8	55.1
	Dallas-Love Field	30	36.4	41.0	48.5	56.1	64.9	72.7	76.8	76.4	69.2	58.2	46.8	38.6	57.1
	Del Rio	30	39.7	44.1	51.6	58.5	66.7	72.1	74.3	74.1	69.4	60.5	49.2	41.2	58.5
	El Paso	30	32.9	37.5	43.7	51.1	60.6	68.8	72.0	70.2	63.7	51.8	39.8	33.4	52.1
	Galveston	30	49.7	51.5	58.2	64.7	72.3	77.8	79.8	79.5	75.6	68.4	59.4	51.8	65.7
	Houston	30	41.2	44.3	51.3	57.9	66.1	71.8	73.5	73.0	68.4	58.8	49.8	42.8	58.2
	Lubbock	30	24.4	28.9	36.2	45.4	55.6	64.1	67.7	66.0	58.4	47.0	34.5	26.1	46.2
	Midland-Odessa	30	29.6	34.1	40.8	48.6	58.8	66.4	69.1	67.9	61.6	51.3	38.8	31.2	49.9
	Port Arthur	30	42.9	45.9	52.4	58.6	66.4	72.3	73.8	73.2	69.4	59.6	50.8	44.5	59.2
	San Angelo	30	31.8	36.0	43.3	51.0	60.6	67.6	70.4	69.4	63.0	53.0	41.4	33.5	51.8
	San Antonio	30	38.6	42.4	49.9	56.9	65.5	71.6	74.0	73.6	68.8	59.4	48.6	40.8	57.5
	Victoria	30	43.6	46.7	53.9	60.1	68.1	73.3	75.0	74.6	70.3	61.6	52.3	45.2	60.4
	Waco	30	35.1	39.3	46.8	54.2	63.3	70.6	74.1	73.5	67.0	56.7	45.8	37.5	55.3
	Wichita Falls	30	28.9	33.4	41.1	49.3	59.3	67.8	72.4	71.3	63.7	52.4	40.1	31.3	50.9
UT	Milford	30	15.5	20.2	26.4	31.6	38.9	47.1	55.4	54.5	44.9	33.1	23.0	15.0	33.8
	Salt Lake City	30	21.3	25.5	33.4	39.0	46.9	55.8	63.4	62.4	52.4	41.0	30.4	22.4	41.2
VT	Burlington	30	9.3	10.9	21.8	33.6	45.2	54.7	59.8	58.1	49.9	38.9	30.3	17.3	35.8
VA	Lynchburg	30	24.5	26.9	34.4	42.6	51.2	59.5	63.7	62.4	55.9	43.7	35.2	27.9	44.0
	Norfolk	30	32.3	33.6	40.1	47.8	57.6	66.2	71.4	70.1	64.8	52.8	43.7	36.1	51.4
	Richmond	30	27.6	29.7	37.0	45.3	54.6	63.3	68.3	66.8	59.9	47.2	38.4	31.1	47.4
	Roanoke	30	26.6	29.0	36.5	44.2	52.3	60.4	64.9	63.4	56.6	44.6	36.6	29.6	45.4

(Continued)

Table 3A.2 (Continued)

State	Station	Years	Jan	Feb	Mar	Apr	May	Jun	Jul	Aug	Sep	Oct	Nov	Dec	Annual
WA	Olympia	30	31.8	32.6	34.1	36.5	42.0	46.4	49.6	49.5	44.9	38.9	35.3	32.1	39.5
	Quillayute	30	34.6	35.1	35.7	37.6	41.9	46.0	49.0	49.2	45.7	40.9	37.5	34.6	40.7
	Seattle CO	30	36.0	37.1	39.2	42.5	48.2	52.7	56.4	57.1	52.6	46.4	40.4	36.1	45.4
	Seattle Sea-Tac AP	30	35.9	37.2	39.1	42.1	47.2	51.7	55.3	55.7	51.9	45.7	39.9	35.9	44.8
	Spokane	30	21.7	25.7	30.4	35.5	42.6	49.2	54.6	54.5	45.9	35.8	28.7	21.6	37.2
	Walla Walla	30	28.8	32.5	36.9	41.3	47.6	54.3	60.7	61.2	52.9	43.6	36.0	29.3	43.8
	Yakima	30	20.5	24.7	28.9	33.2	40.0	46.2	50.9	50.1	42.3	32.9	26.3	20.5	34.7
PR	San Juan	30	70.8	70.9	71.7	73.2	74.9	76.6	76.9	77.0	76.5	75.6	74.0	72.1	74.2
WV	Beckley	30	22.1	24.9	32.4	40.6	49.2	57.0	61.1	59.8	53.5	42.4	34.4	26.5	42.0
	Charleston	30	24.2	26.7	34.0	41.8	50.3	58.3	62.9	61.7	55.0	43.1	35.3	28.2	43.5
	Elkins	30	18.0	19.7	26.9	34.6	44.1	52.7	57.6	56.7	50.1	37.0	29.3	21.9	37.4
	Huntington	30	24.5	27.5	35.5	43.7	52.6	60.9	65.4	64.1	56.8	44.8	36.6	28.9	45.1
WI	Green Bay	30	7.1	12.1	22.6	33.9	44.7	54.0	58.6	56.5	47.5	36.9	25.6	13.3	34.4
	La Crosse	30	6.3	12.8	24.5	37.1	48.7	57.9	62.8	60.7	51.7	40.1	27.4	13.6	37.0
	Madison	30	9.3	14.3	24.6	35.2	46.0	55.7	61.0	58.7	49.9	38.9	27.7	15.8	36.4
	Milwaukee	30	13.4	18.3	27.3	36.4	46.2	56.3	62.9	62.1	54.1	42.6	31.0	19.4	39.2
WY	Casper	30	12.2	16.4	23.1	29.3	37.9	46.6	53.2	51.8	41.7	31.8	21.3	14.0	31.6
	Cheyenne	30	14.8	17.2	22.0	28.7	38.3	47.5	53.4	52.0	42.9	32.5	22.1	16.1	32.3
	Lander	30	8.7	13.9	23.5	31.3	40.3	48.9	55.4	54.1	44.4	33.2	18.9	9.9	31.9
	Sheridan	30	9.7	14.9	22.5	30.4	38.6	46.8	52.4	51.5	41.0	30.3	18.5	10.4	30.6

Note: In Fahrenheit degrees, based on 30-year average values 1971–2000.

Source: From U.S. National Oceanic and Atmospheric Administration, *Comparative Climatic Data for the United States Through 2000*, www.noaa.gov.

Table 3A.3 Normal Daily Maximun Temperature — Selected Cities of the United States

State	Station	Years	Jan	Feb	Mar	Apr	May	Jun	Jul	Aug	Sep	Oct	Nov	Dec	Annual
AL	Birmingham AP	30	52.8	58.3	66.5	74.1	81.0	87.5	90.6	90.2	84.6	74.9	64.5	56.0	73.4
	Huntsville	30	48.9	54.6	63.4	72.3	79.6	86.5	89.4	89.0	83.0	72.9	61.6	52.4	71.1
	Mobile	30	60.7	64.5	71.2	77.4	84.2	89.4	91.2	90.8	86.8	79.2	70.1	62.9	77.4
	Montgomery	30	57.6	62.4	70.5	77.5	84.6	90.6	92.7	92.2	87.7	78.7	68.7	60.3	77.0
AK	Anchorage	30	22.2	25.8	33.6	43.9	54.9	62.3	65.3	63.3	55.0	40.0	27.7	23.7	43.1
	Annette	30	39.7	41.9	44.7	49.8	55.7	60.3	64.1	64.6	59.6	51.4	44.2	40.7	51.4
	Barrow	30	−7.7	−9.8	−7.4	6.3	24.9	39.5	46.5	43.6	34.8	19.3	4.6	−4.7	15.8
	Bethel	30	12.4	13.9	21.8	33.3	49.4	59.4	63.1	59.7	51.7	35.3	23.1	15.6	36.6
	Bettles	30	−3.1	2.0	16.4	34.1	54.9	68.7	70.8	63.2	49.1	25.4	6.4	0.4	32.4
	Big Delta	30	4.4	10.9	25.1	42.5	57.8	67.3	70.4	64.8	53.2	31.1	13.5	7.2	37.4
	Cold Bay	30	32.8	32.3	35.1	38.2	44.9	50.8	55.1	56.2	52.5	45.0	39.1	35.5	43.1
	Fairbanks	30	−0.3	8.0	25.0	43.6	60.6	70.9	73.0	66.3	54.3	31.4	11.2	3.3	37.3
	Gulkana	30	3.5	13.8	28.2	42.4	55.6	65.0	68.5	64.5	53.4	34.3	13.2	6.4	37.4
	Homer	30	29.3	31.4	36.3	43.4	50.6	57.0	61.0	60.8	54.8	44.1	35.2	31.6	44.6
	Juneau	30	30.6	34.3	39.5	48.1	55.7	61.6	64.3	63.1	56.1	46.9	37.6	33.0	47.6
	King Salmon	30	22.8	23.8	32.0	41.3	52.1	59.5	63.8	62.2	54.9	40.5	30.5	25.1	42.4
	Kodiak	30	34.7	35.5	38.3	42.7	48.8	54.5	59.6	61.4	55.6	46.2	39.0	35.8	46.0
	Kotzebue	30	3.7	3.0	7.2	19.6	37.8	50.8	60.0	56.7	46.4	27.5	13.3	6.0	27.7
	Mcgrath	30	2.3	10.7	25.3	40.5	56.8	67.6	69.7	64.1	53.4	32.2	13.8	4.8	36.8
	Nome	30	13.4	13.6	17.7	26.8	43.0	53.9	58.6	56.0	48.6	34.0	23.0	15.8	33.7
	St. Paul Island	30	29.8	27.6	28.8	32.8	39.8	46.2	50.3	51.6	49.2	42.5	37.1	32.9	39.1
	Talkeetna	30	19.6	25.7	34.0	44.6	56.7	65.4	67.9	64.6	55.1	39.1	25.6	21.2	43.3
	Unalakleet	30	10.5	12.7	19.6	31.5	46.9	55.2	62.0	59.6	51.3	33.6	20.2	13.6	34.7
	Valdez	30	26.6	30.0	35.8	44.4	52.9	59.4	62.3	60.8	53.3	43.0	32.7	29.1	44.2
	Yakutat	30	32.1	35.7	39.3	45.1	51.1	56.6	60.1	60.4	55.7	47.3	38.4	34.3	46.3
AZ	Flagstaff	30	42.9	45.6	50.3	58.4	67.6	78.7	82.2	79.7	73.8	63.1	50.8	43.7	61.4
	Phoenix	30	65.0	69.4	74.3	83.0	91.9	102.0	104.2	102.4	97.4	86.4	73.3	65.0	84.5
	Tucson	30	64.5	68.4	73.3	81.5	90.4	100.2	99.6	97.4	94.0	84.0	72.3	64.6	82.5
	Winslow	30	47.1	54.4	61.5	69.8	79.0	90.0	93.0	90.1	83.5	71.7	57.7	47.1	70.4
	Yuma	30	69.9	75.2	80.1	87.2	94.7	104.4	107.3	106.1	101.0	90.3	77.3	69.0	88.5
AR	Fort Smith	30	48.1	54.8	64.2	73.2	80.0	87.7	92.9	92.6	84.9	75.0	61.4	50.9	72.1
	Little Rock	30	49.5	55.6	64.2	72.9	81.0	89.0	92.8	92.1	85.1	75.1	62.0	52.5	72.7
	North Little Rock	30	49.1	55.1	64.1	73.2	80.5	88.6	93.5	92.6	85.0	74.8	61.6	51.9	72.5
CA	Bakersfield	30	56.3	63.5	68.3	75.7	83.8	91.6	96.9	95.4	89.4	79.5	65.3	56.1	76.8
	Bishop	30	53.6	58.4	64.3	72.1	81.2	91.5	97.9	95.8	87.6	76.0	62.4	54.3	74.6
	Eureka	30	55.0	55.9	56.1	57.4	59.6	61.8	63.3	63.9	63.6	61.3	58.0	55.1	59.3
	Fresno	30	53.6	61.3	66.1	74.0	82.7	90.9	96.6	94.8	88.8	78.1	63.0	53.4	75.3
	Long Beach	30	68.0	68.5	68.9	72.7	74.0	78.3	82.9	84.6	83.1	78.9	73.4	68.8	75.2
	Los Angeles AP	30	65.6	65.8	65.3	68.0	69.3	72.6	75.3	76.8	76.5	74.3	70.4	66.7	70.6
	Los Angeles CO	30	68.1	69.6	69.8	73.1	74.5	79.5	83.8	84.8	83.3	79.0	73.2	68.7	75.6
	Mount Shasta	30	44.2	47.6	52.1	59.2	67.3	75.5	83.2	82.6	76.0	64.4	49.9	43.8	62.2
	Redding	30	55.4	60.1	63.9	70.6	80.7	90.7	98.5	96.9	90.2	78.4	62.4	55.6	75.3
	Sacramento	30	53.8	60.5	64.7	71.4	80.0	87.4	92.4	91.4	87.5	78.2	63.7	53.9	73.7

(Continued)

Table 3A.3 (Continued)

State	Station	Years	Jan	Feb	Mar	Apr	May	Jun	Jul	Aug	Sep	Oct	Nov	Dec	Annual
	San Diego	30	65.8	66.3	66.3	68.7	69.3	72.2	75.8	77.5	77.0	74.0	69.9	66.3	70.8
	San Francisco AP	30	55.9	59.3	61.2	64.3	66.8	69.9	71.1	71.7	72.7	69.7	62.0	56.1	65.1
	San Francisco CO	30	58.1	61.4	62.5	64.5	65.4	67.7	68.2	69.2	71.3	70.4	64.1	58.6	65.1
	Santa Barbara	30	65.4	66.3	67.4	70.1	71.2	74.4	76.7	78.7	78.2	75.4	71.0	66.4	71.8
	Santa Maria	30	63.9	64.8	64.8	67.6	68.6	71.4	73.5	74.2	74.9	74.0	69.2	64.9	69.3
	Stockton	30	53.8	61.2	66.1	73.3	81.3	88.9	93.8	92.6	88.2	78.6	64.0	53.8	74.6
CO	Alamosa	30	33.1	40.2	49.6	58.7	68.3	78.4	81.7	78.9	72.5	61.7	45.7	34.8	58.6
	Colorado Springs	30	41.7	45.4	51.6	59.2	68.4	79.2	84.4	81.6	74.1	63.4	49.8	42.4	61.8
	Denver	30	43.2	47.2	53.7	60.9	70.5	82.1	88.0	86.0	77.4	66.0	51.5	44.1	64.2
	Grand Junction	30	36.6	45.4	55.7	64.3	74.5	86.9	92.1	89.6	80.3	66.7	49.8	38.9	65.1
	Pueblo	30	44.6	50.4	57.3	65.3	74.6	86.1	91.4	88.8	80.8	69.4	54.3	45.4	67.4
CT	Bridgeport	30	36.9	38.8	46.9	57.0	67.4	76.4	81.9	80.7	73.6	63.1	52.6	42.1	59.8
	Hartford	30	34.1	37.7	47.7	59.9	71.7	80.0	84.9	82.5	74.3	63.1	50.9	39.0	60.5
DE	Wilmington	30	39.3	42.5	51.9	62.6	72.5	81.1	86.0	84.1	77.2	65.9	55.0	44.4	63.6
DC	Washington Dulles AP	30	41.4	45.5	55.0	65.9	74.6	82.8	87.4	85.9	78.9	67.7	56.5	45.9	65.6
	Washington Nat'l AP	30	42.5	46.5	55.7	66.3	75.4	83.9	88.3	86.3	79.3	68.0	57.3	47.0	66.4
FL	Apalachicola	30	62.4	64.8	69.9	76.0	83.0	88.3	89.8	89.4	87.0	79.9	72.0	65.0	77.3
	Daytona Beach	30	69.7	71.1	75.6	79.8	85.0	88.8	91.0	90.1	87.9	82.6	76.9	71.4	80.8
	Fort Myers	30	75.3	76.5	80.5	84.5	89.1	91.2	91.7	91.7	90.3	86.3	81.3	76.6	84.6
	Gainesville	30	66.2	69.3	75.1	80.4	86.5	89.9	90.9	90.1	87.4	81.0	74.4	68.1	79.9
	Jacksonville	30	64.2	67.3	73.4	78.6	84.3	88.7	90.8	89.4	86.1	79.1	72.5	65.8	78.4
	Key West	30	75.3	75.9	78.8	81.9	85.4	88.1	89.4	89.5	88.2	84.7	80.6	76.7	82.9
	Miami	30	76.5	77.7	80.7	83.8	87.2	89.5	90.9	90.6	89.0	85.4	81.2	77.5	84.2
	Orlando	30	71.8	73.9	78.8	83.0	88.2	91.0	92.2	92.0	90.3	85.0	78.9	73.3	83.2
	Pensacola	30	61.2	64.4	70.2	76.2	83.4	89.0	90.7	90.1	87.0	79.3	70.3	63.4	77.1
	Tallahassee	30	63.8	67.4	74.0	80.0	86.5	90.9	92.0	91.5	88.5	81.2	72.9	65.8	79.5
	Tampa	30	70.1	71.6	76.3	80.6	86.3	88.9	89.7	90.0	89.0	84.1	78.0	72.0	81.4
	Vero Beach	30	73.3	74.1	77.6	81.4	85.2	89.0	90.4	90.2	88.7	84.3	79.1	74.7	82.3
	West Palm Beach	30	75.1	76.3	79.2	82.1	85.9	88.5	90.1	90.1	88.7	85.0	80.4	76.4	83.2
GA	Athens	30	51.4	56.5	64.7	73.0	80.5	87.2	90.2	88.2	82.5	72.9	63.2	54.2	72.0
	Atlanta	30	51.9	56.8	65.0	72.9	80.0	86.5	89.4	87.9	82.3	72.9	63.3	54.6	72.0
	Augusta	30	56.5	61.3	69.2	76.7	83.9	89.6	92.0	90.2	85.3	76.5	67.8	59.1	75.7
	Columbus	30	56.9	61.6	69.4	76.5	83.2	89.5	91.7	91.0	86.0	77.0	67.6	59.2	75.8
	Macon	30	56.6	60.9	68.5	75.9	83.4	89.5	91.8	90.5	85.4	76.8	67.8	59.2	75.5
	Savannah	30	60.4	64.1	71.0	77.7	84.3	89.5	92.3	90.3	86.0	78.1	70.5	62.6	77.2
HI	Hilo	30	79.2	79.4	79.2	79.3	80.6	82.2	82.5	83.2	83.4	82.7	80.7	79.5	81.0
	Honolulu	30	80.4	80.7	81.7	83.1	84.9	86.9	87.8	88.9	88.9	87.2	84.3	81.7	84.7
	Kahului	30	80.3	80.8	81.5	82.5	84.3	86.0	86.9	87.9	88.1	86.9	84.1	81.7	84.3
	Lihue	30	77.9	77.9	78.1	78.8	80.6	82.7	83.9	84.9	85.0	83.5	81.0	79.0	81.1
ID	Boise	30	36.7	44.5	53.6	61.7	70.7	80.3	89.2	88.0	77.2	64.3	47.5	37.2	62.6
	Lewiston	30	39.4	45.6	53.8	61.6	70.0	78.0	87.6	87.6	76.7	62.0	46.8	39.2	62.4
	Pocatello	30	32.5	39.0	48.5	58.5	67.7	78.3	87.5	86.8	75.7	62.0	44.5	33.8	59.6
IL	Chicago	30	29.6	34.7	46.1	58.0	69.9	79.2	83.5	81.2	73.9	62.1	47.1	34.4	58.3
	Moline	30	29.8	35.6	48.3	61.7	73.3	82.7	86.1	83.9	76.5	64.4	48.0	34.5	60.4

	Peoria	30	30.7	36.6	49.4	62.0	73.0	82.2	85.7	83.6	76.7	64.4	48.8	35.5	60.7
	Rockford	30	27.2	33.0	45.5	59.1	71.2	79.9	83.1	80.9	73.9	61.8	45.5	32.0	57.8
	Springfield	30	33.1	38.9	51.1	63.4	74.4	83.3	86.5	84.5	78.5	66.6	50.9	38.0	62.4
IN	Evansville	30	39.5	45.4	56.4	67.2	77.1	86.1	89.4	87.8	81.3	70.0	55.7	44.1	66.7
	Fort Wayne	30	31.0	35.4	47.4	59.8	71.6	80.6	84.3	81.8	75.4	63.0	48.5	35.8	59.6
	Indianapolis	30	34.5	39.9	51.4	62.9	73.5	82.1	85.6	83.7	77.4	65.6	51.6	39.2	62.3
	South Bend	30	31.0	35.5	46.8	58.9	70.7	79.6	83.1	80.7	73.6	61.8	47.7	35.6	58.8
IA	Des Moines	30	29.1	35.4	48.2	61.3	72.3	81.8	86.0	83.9	75.9	63.5	46.7	33.1	59.8
	Dubuque	30	24.8	30.8	43.3	57.4	69.3	78.6	82.1	79.8	71.9	60.3	43.6	29.7	56.0
	Sioux City	30	28.7	35.0	47.3	61.7	73.2	82.5	86.2	83.7	76.0	63.7	44.8	31.7	59.6
	Waterloo	30	25.8	31.9	45.0	59.7	72.2	81.7	85.0	82.8	75.3	62.5	45.0	30.7	58.1
KS	Concordia	30	36.3	42.9	53.9	64.4	74.0	85.0	90.7	88.4	79.9	67.9	51.0	39.6	64.5
	Dodge City	30	41.4	48.3	57.3	67.1	75.9	86.9	92.8	90.8	82.0	70.4	54.5	44.4	67.7
	Goodland	30	39.4	45.0	53.2	62.7	71.7	83.6	89.1	86.7	78.0	66.0	49.6	41.3	63.9
	Topeka	30	37.2	43.8	55.5	66.1	75.3	84.5	89.1	87.9	80.3	68.9	53.1	40.9	65.2
	Wichita	30	40.1	47.2	57.3	66.9	76.0	87.1	92.9	91.6	82.2	70.2	54.5	43.1	67.4
KY	Greater Cincinnati AP	30	38.0	43.1	53.9	64.7	74.4	82.4	86.4	84.8	78.0	66.4	53.6	42.7	64.0
	Jackson	30	42.0	46.8	56.8	66.8	73.8	80.8	84.2	83.3	77.4	67.5	56.4	46.3	65.2
	Lexington	30	39.9	45.2	55.3	65.1	74.0	82.3	85.9	84.6	78.1	66.9	54.5	44.3	64.7
	Louisville	30	41.0	46.6	56.8	66.8	75.4	83.3	87.0	85.8	79.4	68.4	55.9	45.4	66.0
	Paducah	30	41.9	48.0	58.1	68.4	76.9	85.2	88.6	87.4	81.2	70.8	57.2	46.3	67.5
LA	Baton Rouge	30	60.0	63.9	71.0	77.3	84.0	89.2	90.7	90.9	87.4	79.7	70.1	62.8	77.3
	Lake Charles	30	60.6	64.5	71.3	77.4	84.1	88.9	91.0	91.3	87.7	80.5	70.6	63.3	77.6
	New Orleans	30	61.8	65.3	72.1	78.0	84.8	89.4	91.1	91.0	87.1	79.7	71.0	64.5	78.0
	Shreveport	30	56.2	62.0	69.7	76.6	83.2	89.8	93.3	93.4	87.6	78.3	66.8	58.5	76.3
ME	Caribou	30	19.3	23.2	34.1	47.0	62.6	71.8	76.3	74.2	64.1	51.4	37.4	24.8	48.9
	Portland	30	30.9	34.1	42.2	52.8	63.3	72.8	78.8	77.3	68.9	57.9	47.1	36.4	55.2
MD	Baltimore	30	41.2	44.8	53.9	64.5	73.9	82.7	87.2	85.1	78.2	67.0	56.3	46.0	65.1
MA	Blue Hill	30	33.8	36.3	44.8	55.5	67.0	75.5	81.2	78.9	71.0	60.3	49.3	38.6	57.7
	Boston	30	36.5	38.7	46.3	56.1	66.7	76.6	82.2	80.1	72.5	61.8	51.8	41.7	59.3
	Worcester	30	31.4	34.1	43.0	54.4	66.3	74.4	79.3	77.1	69.0	58.4	47.1	36.2	55.9
MI	Alpena	30	26.1	28.2	37.3	50.3	64.3	73.8	79.0	76.1	67.4	55.6	42.2	31.2	52.6
	Detroit	30	31.1	34.4	45.2	57.8	70.2	79.0	83.4	81.4	73.7	61.2	47.8	35.9	58.4
	Flint	30	29.2	32.3	43.1	56.2	69.0	77.7	82.0	79.5	71.9	59.7	46.3	34.2	56.8
	Grand Rapids	30	29.3	32.6	43.3	56.6	69.6	78.4	82.3	79.7	71.7	59.6	45.5	33.7	56.9
	Houghton Lake	30	25.9	29.3	39.4	53.0	67.2	75.5	80.0	77.1	68.3	56.0	41.9	30.5	53.7
	Lansing	30	29.4	32.6	43.5	56.6	69.4	78.1	82.1	79.7	72.0	59.8	46.0	34.1	56.9
	Marquette	30	19.7	24.2	33.1	45.8	61.5	70.3	75.2	72.6	63.2	50.9	35.4	24.1	48.0
	Muskegon	30	29.8	32.5	42.5	54.6	67.0	75.6	80.0	78.1	70.3	58.7	45.6	34.6	55.8
	Sault Ste. Marie	30	21.5	24.5	33.6	48.0	63.2	70.7	75.7	74.1	64.8	52.8	38.9	27.2	49.6
MN	Duluth	30	17.9	24.4	34.2	49.0	63.4	71.2	76.3	73.9	64.5	52.5	35.2	22.3	48.7
	International Falls	30	13.8	22.4	34.9	51.5	66.6	74.2	78.6	76.3	64.7	51.7	32.5	18.1	48.8
	Minneapolis-St. Paul	30	21.9	28.4	40.6	57.0	70.1	79.0	83.3	80.4	71.1	58.4	40.1	26.4	54.7
	Rochester	30	19.9	26.2	38.7	54.8	67.7	76.6	80.1	77.5	69.2	56.9	38.7	24.5	52.6
	Saint Cloud	30	18.7	25.7	37.7	54.9	69.0	77.3	81.7	78.9	69.0	56.3	37.2	23.2	52.5
MS	Jackson	30	55.1	60.3	68.1	75.0	82.1	88.9	91.4	91.4	86.4	76.8	66.3	57.9	75.0

(Continued)

Table 3A.3 (Continued)

State	Station	Years	Jan	Feb	Mar	Apr	May	Jun	Jul	Aug	Sep	Oct	Nov	Dec	Annual
	Meridian	30	57.5	62.6	70.3	77.1	83.9	90.1	92.9	92.9	88.0	78.3	68.5	60.5	76.9
	Tupelo	30	50.3	56.0	64.8	73.5	81.0	88.0	91.4	90.9	84.9	74.9	63.0	53.6	72.7
MO	Columbia	30	37.4	43.9	55.1	65.9	74.6	83.6	88.6	87.3	79.1	68.0	53.4	41.5	64.9
	Kansas City	30	36.0	42.6	54.4	65.2	74.6	83.9	88.8	87.1	79.0	67.6	52.0	40.0	64.3
	St. Louis	30	37.9	44.3	55.4	66.7	76.5	85.3	89.8	87.9	80.1	68.3	53.8	42.0	65.7
	Springfield	30	41.6	47.7	57.8	67.7	75.9	84.6	89.9	89.5	81.2	70.6	56.4	45.5	67.4
MT	Billings	30	32.8	39.5	47.6	57.5	67.4	78.0	85.8	84.5	71.8	58.9	42.7	34.5	58.4
	Glasgow	30	19.9	28.3	41.3	56.7	67.9	77.1	83.8	83.3	70.4	57.1	37.4	24.8	54.0
	Great Falls	30	32.1	37.7	45.3	55.6	64.7	73.9	82.0	81.2	69.6	58.0	42.1	34.2	56.4
	Havre	30	25.5	33.4	44.9	58.5	68.8	77.4	84.6	83.9	71.9	59.4	40.8	30.1	56.6
	Helena	30	30.5	37.3	46.8	56.9	65.9	75.0	83.4	82.5	71.0	58.4	41.5	31.5	56.7
	Kalispell	30	28.9	35.2	44.9	56.0	64.7	71.9	80.2	80.5	69.0	55.3	38.6	30.1	54.6
	Missoula	30	30.8	37.4	48.1	58.0	66.1	74.5	83.6	83.2	71.5	57.4	40.0	30.3	56.7
NE	Grand Island	30	32.6	38.6	49.5	61.9	71.9	83.0	87.1	84.8	76.9	64.6	46.8	35.3	61.1
	Lincoln	30	33.2	39.3	51.2	63.5	73.8	84.9	89.6	87.1	78.8	66.5	49.1	36.8	62.8
	Norfolk	30	31.2	37.3	48.5	61.3	72.3	82.3	86.5	84.4	76.4	64.0	45.5	33.6	60.3
	North Platte	30	36.5	43.3	52.1	62.7	72.0	82.6	88.4	86.8	78.0	65.6	48.5	39.2	63.0
	Omaha Eppley AP	30	31.7	37.9	50.4	63.2	73.7	83.7	87.4	85.2	77.3	65.2	47.8	34.8	61.5
	Omaha (North)	30	32.1	38.0	50.8	63.6	73.3	82.4	85.6	83.9	76.3	64.6	47.5	35.1	61.1
	Scottsbluff	30	38.0	44.3	51.7	61.0	71.1	82.2	88.7	86.8	77.3	64.4	48.2	39.8	62.8
	Valentine	30	33.8	39.4	48.4	59.8	71.2	81.9	88.3	86.9	77.2	63.5	45.9	36.7	61.1
NV	Elko	30	37.1	42.9	51.2	59.3	68.6	79.9	89.6	88.1	78.2	65.0	48.1	38.2	62.2
	Ely	30	40.0	44.0	49.9	57.9	67.3	79.2	87.3	85.1	75.8	63.0	48.8	41.0	61.6
	Las Vegas	30	57.1	63.0	69.5	78.1	87.8	98.9	104.1	101.8	93.8	80.8	66.0	57.3	79.9
	Reno	30	45.5	51.7	57.2	64.1	72.6	82.8	91.2	89.9	81.7	69.9	55.3	46.4	67.4
	Winnemucca	30	41.6	48.5	55.1	62.6	72.0	82.7	92.2	90.6	80.4	67.3	51.4	42.2	65.6
NH	Concord	30	30.6	34.1	43.8	56.9	69.6	77.9	82.9	80.8	72.1	60.5	47.6	35.6	57.7
	Mt. Washington	30	14.0	14.8	21.3	29.4	41.6	50.3	54.1	53.0	46.1	36.4	27.6	18.5	33.9
NJ	Atlantic City AP	30	41.4	43.9	51.9	61.3	71.1	80.0	85.1	83.3	76.6	66.3	56.0	46.4	63.6
	Atlantic City CO	30	41.4	43.2	49.5	57.5	66.1	74.8	80.6	79.8	74.1	64.5	55.0	46.3	61.1
	Newark	30	38.1	41.1	50.1	60.8	71.4	80.2	85.2	83.2	75.7	64.7	53.7	43.0	62.3
NM	Albuquerque	30	47.6	54.6	62.4	70.6	79.7	90.2	92.3	89.0	82.2	70.7	57.1	47.9	70.4
	Clayton	30	47.4	51.5	58.2	66.1	74.2	83.9	87.4	85.1	77.9	68.5	55.7	48.0	67.0
	Roswell	30	55.6	62.0	70.0	77.7	86.0	94.0	94.8	92.3	85.7	76.5	64.5	56.3	76.3
NY	Albany	30	31.1	34.3	44.5	57.3	69.8	77.5	82.2	79.7	71.3	59.7	47.5	36.0	57.6
	Binghamton	30	28.4	30.9	40.6	53.1	65.6	73.4	78.1	75.8	67.8	56.7	44.3	33.4	54.0
	Buffalo	30	31.1	33.2	42.5	54.1	66.4	74.8	79.6	77.8	70.1	58.9	46.7	36.0	55.9
	Islip	30	39.1	40.5	48.5	58.1	68.9	77.4	83.2	81.7	74.9	64.0	53.7	43.9	61.2
	New York C.Park	30	38.0	41.0	49.8	60.7	70.9	79.0	84.2	82.4	74.7	63.5	53.1	42.9	61.7
	New York (JFK AP)	30	38.8	40.9	48.9	58.6	68.3	77.2	82.9	81.8	74.9	64.3	53.8	44.0	61.2
	New York (Laguardia AP)	30	38.6	41.2	49.6	60.0	70.6	79.3	84.7	83.1	75.6	64.5	53.6	43.7	62.1
	Rochester	30	31.2	33.2	42.7	55.2	67.9	76.6	81.4	79.1	71.1	59.7	47.2	36.1	56.8
	Syracuse	30	31.4	33.5	43.1	55.7	68.5	77.0	81.7	79.6	71.4	59.8	47.4	36.3	57.1

NC	Asheville	30	45.9	50.0	57.7	66.5	73.5	80.0	83.3	81.7	76.0	67.1	57.4	49.3	65.7
	Cape Hatteras	30	53.6	54.6	60.2	67.7	74.9	81.5	85.4	84.8	81.1	72.6	64.8	57.3	69.9
	Charlotte	30	51.3	55.9	64.1	72.8	79.7	86.6	90.1	88.4	82.3	72.6	62.8	54.0	71.7
	Greensboro-Wnstn-Slm-HPT	30	47.2	51.7	60.3	69.7	76.9	83.8	87.6	85.7	79.4	69.6	59.9	50.6	68.5
	Raleigh	30	49.8	54.0	62.5	71.8	78.7	85.5	89.1	87.2	81.3	71.8	62.4	53.3	70.6
	Wilmington	30	56.3	59.5	66.2	74.1	80.6	86.4	89.9	88.3	84.1	75.6	67.8	59.6	74.0
ND	Bismarck	30	21.1	28.5	40.2	55.9	69.1	77.8	84.5	83.3	71.6	58.2	38.2	25.7	54.5
	Fargo	30	15.9	22.8	35.3	54.5	69.5	77.4	82.2	81.0	69.9	56.1	35.2	20.8	51.7
	Grand Forks	30	14.9	22.4	34.3	53.6	70.0	77.6	81.9	81.0	69.7	55.6	34.1	20.1	51.3
	Williston	30	19.4	27.6	40.1	56.0	68.2	77.3	83.4	82.8	70.0	57.0	36.2	24.0	53.5
OH	Akron	30	32.9	36.8	47.5	59.0	69.8	78.2	82.3	80.3	72.8	61.1	48.7	37.7	58.9
	Cleveland	30	32.6	35.8	46.1	57.3	68.6	77.4	81.4	79.2	72.3	60.8	48.7	37.4	58.1
	Columbus	30	36.2	40.5	51.7	62.9	73.3	81.6	85.3	83.8	77.1	65.4	52.4	41.0	62.6
	Dayton	30	33.7	38.2	49.3	60.7	71.2	80.1	84.2	82.3	75.6	63.5	50.1	38.5	60.6
	Mansfield	30	32.4	35.9	46.6	58.4	69.3	77.8	81.8	79.7	73.0	61.7	48.7	37.2	58.6
	Toledo	30	31.4	35.1	46.5	58.9	70.7	79.5	83.4	81.0	74.0	62.1	48.3	36.0	58.9
	Youngstown	30	32.4	36.0	46.3	58.2	69.0	77.1	81.0	79.3	72.1	60.7	48.4	37.3	58.2
OK	Oklahoma City	30	47.1	53.5	62.5	71.2	78.9	87.2	93.1	92.5	84.1	73.4	59.6	49.8	71.1
	Tulsa	30	46.5	52.9	62.4	72.1	79.6	88.0	93.8	93.2	84.1	74.0	60.0	49.6	71.4
OR	Astoria	30	48.1	50.8	53.3	56.1	60.0	63.6	67.2	68.3	67.5	61.0	53.1	48.4	58.1
	Burns	30	34.7	40.5	49.0	57.4	66.1	75.1	85.4	84.5	75.0	62.4	44.8	35.1	59.2
	Eugene	30	46.5	50.7	55.9	60.6	66.8	73.3	81.5	81.9	76.6	64.6	52.1	45.7	63.0
	Medford	30	47.3	53.8	58.3	64.3	72.2	81.2	90.2	90.1	83.5	70.0	52.8	45.2	67.4
	Pendleton	30	40.1	46.5	54.8	62.2	70.2	78.7	87.7	86.6	77.1	63.8	48.5	40.0	63.0
	Portland	30	45.6	50.3	55.7	60.5	66.7	72.7	79.3	79.7	74.6	63.3	51.8	45.4	62.1
	Salem	30	47.0	51.2	56.3	61.1	67.5	74.0	81.5	81.9	76.6	64.5	52.4	46.4	63.4
	Sexton Summit	30	42.5	44.1	46.7	51.9	60.1	67.5	75.5	75.7	70.3	59.3	45.7	42.0	56.8
PC	Guam	30	84.0	84.0	85.0	86.2	87.3	87.4	86.8	86.1	86.5	86.3	85.6	84.7	85.8
	Johnston Island	30	81.9	82.1	82.3	83.1	84.3	85.8	86.5	87.2	87.0	86.2	84.1	82.6	84.4
	Koror	30	87.6	87.5	88.3	88.8	89.1	88.1	87.5	87.5	88.0	88.3	89.0	88.4	88.2
	Kwajalein, Marshall IS	30	85.6	86.1	86.7	86.5	86.7	86.5	86.6	86.9	87.0	86.9	86.5	85.8	86.5
	Majuro, Marshall IS	30	85.2	85.6	85.9	85.7	86.0	86.0	85.9	86.4	86.6	86.6	86.3	85.5	86.0
	Pago Pago, Amer Samoa	30	86.8	87.2	87.3	86.9	85.6	84.5	83.8	84.0	84.8	85.2	85.8	86.9	85.7
	Pohnpei, Caroline IS	30	86.8	87.0	87.5	87.6	87.8	87.8	88.2	88.6	88.8	88.7	88.6	87.3	87.9
	Chuuk, E. Caroline IS	30	87.0	86.4	86.7	87.1	87.6	87.2	87.7	87.4	87.8	88.3	88.3	87.7	87.4
	Wake Island	30	82.4	82.1	83.3	84.5	86.2	88.1	88.8	88.7	88.7	87.8	85.7	83.9	85.9
	Yap, W Caroline IS	30	86.5	86.7	87.5	88.3	88.5	87.7	87.2	87.1	87.4	87.6	87.7	87.0	87.4
PA	Allentown	30	35.0	38.7	48.7	60.1	70.9	79.3	83.9	81.7	74.0	62.9	51.2	40.0	60.5
	Erie	30	33.5	35.4	44.7	55.6	67.4	76.2	80.4	79.0	72.0	61.0	49.3	38.6	57.8
	Harrisburg	30	37.5	40.9	50.9	62.6	72.6	80.8	85.7	83.7	75.7	64.3	52.5	41.7	62.4
	Middletown/ Harrisburg AP	30	37.5	40.9	50.9	62.6	72.6	80.8	85.7	83.7	75.7	64.3	52.5	41.7	62.4
	Philadelphia	30	39.0	42.1	51.3	62.0	72.1	80.6	85.5	84.0	76.7	65.7	54.8	44.2	63.2

(Continued)

Table 3A.3 (Continued)

State	Station	Years	Jan	Feb	Mar	Apr	May	Jun	Jul	Aug	Sep	Oct	Nov	Dec	Annual
	Pittsburgh	30	35.1	38.8	49.5	60.7	70.8	79.1	82.7	81.1	74.2	62.5	50.5	39.8	60.4
	Avoca	30	34.1	37.3	47.3	59.2	70.8	78.2	82.6	80.5	72.4	61.2	49.3	38.6	59.3
	Williamsport	30	33.2	37.1	47.8	60.2	71.3	78.9	83.2	81.4	73.3	61.8	49.0	37.8	59.6
RI	Providence	30	37.1	39.3	47.7	58.1	68.5	77.3	82.6	80.9	73.4	62.9	52.4	42.1	60.2
SC	Charleston AP	30	58.9	62.3	69.3	76.1	82.9	87.9	90.9	89.4	85.0	77.0	69.6	61.6	75.9
	Charleston CO	30	57.1	59.8	65.8	72.9	79.6	84.9	88.5	87.1	83.0	75.1	67.6	60.0	73.5
	Columbia	30	55.1	59.5	67.4	75.7	83.1	89.1	92.1	90.0	84.8	75.8	66.7	57.8	74.8
	Greenville-Spartanburg	30	50.2	54.8	62.7	71.0	78.2	85.1	88.8	87.1	81.1	71.4	61.3	52.7	70.4
SD	Aberdeen	30	21.4	28.5	40.2	57.4	70.2	78.7	84.7	83.5	73.0	59.2	38.8	25.7	55.1
	Huron	30	24.8	31.3	43.0	58.3	70.5	80.3	86.1	84.4	74.7	60.9	41.4	28.8	57.1
	Rapid City	30	33.6	38.6	46.6	57.1	67.2	77.4	85.5	85.5	75.2	61.7	44.8	36.1	59.1
	Sioux Falls	30	25.2	31.6	43.8	58.8	71.0	80.6	85.6	83.2	74.2	61.1	41.9	28.8	57.2
TN	Bristol-JhnCty-Kgsprt	30	44.1	48.9	58.4	67.1	74.9	81.8	84.8	83.9	78.5	68.2	57.4	47.8	66.3
	Chattanooga	30	48.8	54.1	62.8	72.1	79.1	86.2	89.8	88.7	82.5	72.3	61.1	52.0	70.8
	Knoxville	30	46.3	51.7	60.3	69.0	76.3	83.6	86.9	86.4	80.7	69.9	59.0	49.8	68.3
	Memphis	30	48.6	54.4	63.3	72.4	80.4	88.5	92.1	91.2	85.3	75.1	62.1	52.2	72.1
	Nashville	30	45.6	51.4	60.7	69.8	77.5	85.1	88.7	87.8	81.5	71.1	59.0	49.4	69.0
	Oak Ridge	30	45.9	51.6	61.0	70.5	77.8	84.9	88.1	87.2	81.1	71.1	59.0	49.0	68.9
TX	Abilene	30	55.2	60.7	69.1	77.3	84.7	91.1	94.8	93.7	86.7	77.6	65.1	56.9	76.1
	Amarillo	30	48.9	54.1	62.2	70.6	78.6	87.4	91.0	88.7	81.8	71.8	58.4	49.8	70.3
	Austin/City	30	60.3	65.1	72.5	78.9	84.8	90.9	95.0	95.6	90.1	81.4	70.1	62.3	78.9
	Austin/Bergstrom	30	58.9	64.1	71.4	77.8	84.3	89.9	94.0	94.0	89.7	80.8	69.7	61.3	78.0
	Brownsville	30	68.7	72.2	78.0	82.3	86.9	90.5	92.4	92.6	89.4	84.0	76.8	70.2	82.0
	Corpus Christi	30	66.0	69.7	75.8	80.7	85.6	90.2	93.2	93.4	89.9	83.6	74.9	68.0	80.9
	Dallas-Fort Worth	30	54.1	60.1	68.3	75.9	83.2	91.1	95.4	94.8	87.7	77.9	65.1	56.5	75.8
	Dallas-Love Field	30	55.4	61.0	69.1	76.5	83.8	91.6	96.1	95.8	88.5	78.6	66.0	57.4	76.7
	Del Rio	30	62.8	68.0	76.0	82.7	88.7	93.7	96.2	96.0	90.6	81.7	70.9	63.5	80.9
	El Paso	30	57.2	63.4	70.2	78.1	86.7	95.3	94.5	92.0	87.1	77.9	65.5	57.4	77.1
	Galveston	30	61.9	64.4	70.0	75.2	81.4	86.6	88.7	89.3	86.5	79.7	71.3	64.3	76.6
	Houston	30	62.3	66.5	73.3	79.1	85.5	90.7	93.6	93.5	89.3	82.0	72.0	64.6	79.4
	Lubbock	30	51.9	57.8	66.2	74.7	82.8	90.0	91.9	90.0	83.4	74.4	61.6	53.2	73.2
	Midland-Odessa	30	56.8	63.0	70.9	78.8	86.8	92.7	94.3	92.8	86.1	77.4	65.8	58.4	77.0
	Port Arthur	30	61.5	65.3	72.0	77.8	84.3	89.4	91.6	91.7	88.0	80.5	70.9	63.9	78.1
	San Angelo	30	57.9	63.5	71.1	79.0	85.6	90.8	94.4	93.1	86.6	77.8	66.5	59.3	77.2
	San Antonio	30	62.1	67.1	74.3	80.4	86.0	91.4	94.6	94.7	90.0	82.0	71.4	64.0	79.8
	Victoria	30	62.8	66.6	73.4	79.2	85.1	90.3	93.4	93.7	89.9	83.0	73.0	65.2	79.6
	Waco	30	57.0	62.3	70.2	77.6	84.8	92.0	96.7	96.9	90.1	80.4	67.8	59.1	77.9
	Wichita Falls	30	52.1	58.1	67.2	75.5	83.5	91.7	97.2	95.8	87.5	77.1	63.7	54.5	75.3
UT	Milford	30	40.6	47.1	56.7	65.0	74.2	86.2	92.9	90.7	81.3	68.0	52.2	42.2	66.4
	Salt Lake City	30	37.0	43.4	52.8	60.9	70.6	82.2	90.6	88.7	77.6	64.0	48.7	38.0	62.9
VT	Burlington	30	26.7	29.0	39.6	53.3	67.8	76.5	81.4	78.4	68.9	56.4	44.0	32.3	54.5
VA	Lynchburg	30	44.5	48.6	57.6	68.0	75.5	82.5	86.4	85.1	78.3	68.4	58.0	48.4	66.8

	Norfolk	30	47.8	50.3	57.8	67.0	74.9	82.8	86.8	84.7	79.4	69.4	60.9	52.3	67.8
	Richmond	30	45.3	49.3	58.4	68.9	76.2	83.6	87.5	85.7	79.7	69.3	59.7	49.7	67.8
	Roanoke	30	45.0	49.1	57.9	68.0	75.9	83.3	87.5	86.0	78.8	68.6	58.0	48.6	67.2
WA	Olympia	30	44.4	48.3	53.0	58.2	64.6	70.0	76.1	77.0	71.7	60.4	49.6	43.8	59.8
	Quillayute	30	46.6	49.2	51.8	55.7	60.4	63.8	68.2	69.3	67.3	59.2	50.8	46.5	57.4
	Seattle CO	30	46.9	50.5	54.5	59.3	64.9	69.5	74.5	74.9	69.9	60.3	51.5	46.5	60.3
	Seattle Sea-Tac AP	30	45.8	49.5	53.2	58.2	64.4	69.6	75.3	75.6	70.2	59.7	50.5	45.5	59.8
	Spokane	30	32.8	39.3	48.6	57.5	66.2	73.9	82.5	82.6	72.5	58.5	41.1	32.8	57.4
	Walla Walla	30	40.6	46.9	56.0	64.1	72.0	80.3	89.9	89.1	79.3	65.8	50.1	40.8	64.6
	Yakima	30	37.7	45.6	56.0	64.1	72.4	79.6	87.2	86.5	77.6	64.3	47.7	37.1	63.0
PR	San Juan	30	82.4	82.8	83.4	84.9	86.3	87.6	87.4	87.8	87.8	87.5	85.1	83.2	85.5
WV	Beckley	30	38.8	42.8	51.9	62.5	70.6	77.0	80.2	78.9	72.6	63.1	52.4	43.1	61.2
	Charleston	30	42.6	47.0	56.6	66.7	74.6	81.5	84.9	83.5	77.3	67.1	56.4	46.8	65.4
	Elkins	30	39.3	43.5	53.2	63.2	71.7	78.5	81.7	80.4	74.1	64.1	52.8	43.5	62.2
	Huntington	30	41.0	46.1	56.3	66.6	74.6	81.7	85.1	83.7	77.0	66.4	55.1	45.3	64.9
WI	Green Bay	30	24.1	28.9	40.0	54.6	68.0	76.8	81.2	78.5	70.2	57.9	42.4	29.0	54.3
	La Crosse	30	25.5	32.4	44.6	59.7	72.5	81.3	85.2	82.5	73.7	61.1	43.6	29.9	57.7
	Madison	30	25.2	30.8	42.8	56.6	69.4	78.3	82.1	79.4	71.4	59.6	43.3	30.2	55.8
	Milwaukee	30	28.0	32.5	42.6	53.9	66.0	76.3	81.1	79.1	71.9	60.2	45.7	33.1	55.9
WY	Casper	30	32.3	37.0	46.9	56.1	66.4	78.8	86.8	85.3	73.4	59.5	42.6	33.6	58.2
	Cheyenne	30	37.1	40.5	46.4	54.4	64.4	75.4	81.9	79.8	70.3	58.2	44.5	38.1	57.6
	Lander	30	31.9	37.4	47.5	56.5	66.5	78.5	86.3	84.8	73.0	59.5	41.8	32.6	58.0
	Sheridan	30	33.0	39.0	48.2	57.5	66.4	76.4	85.2	84.9	73.1	59.8	43.4	34.4	58.5

Note: In Fahrenheit degrees, based on 30-year average values 1971–2000.

Source: From U.S. National Oceanic and Atmospheric Administration, *Comparative Climatic Data for the United States Through 2000*, www.noaa.gov.

Table 3A.4 Mean Number of Days with Minimum Temperature 32 ° or Less — Selected Cities of the United States

State	Station	Years	Jan	Feb	Mar	Apr	May	Jun	Jul	Aug	Sep	Oct	Nov	Dec	Annual
AL	Birmingham CO	11	18	11	5	1	0	0	0	0	0	—	4	14	52
	Birmingham AP	39	17	12	6	1	0	0	0	0	0	—	6	14	57
	Huntsville	35	18	13	7	1	0	0	0	0	0	—	8	16	63
	Mobile	40	8	5	1	—	0	0	0	0	0	—	1	6	22
	Montgomery	39	13	8	3	—	0	0	0	0	0	—	4	10	39
AK	Anchorage	38	31	27	29	20	3	0	0	—	3	20	28	30	191
	Annette	42	17	13	12	4	—	0	0	0	0	2	10	14	72
	Barrow	82	31	28	31	30	31	24	14	15	25	31	30	31	321
	Barter IS	41	31	28	31	30	31	23	9	11	25	31	30	31	310
	Bethel	44	30	28	31	28	16	1	—	—	6	26	29	30	224
	Bettles	50	31	28	31	29	14	—	—	2	15	30	30	31	240
	Big Delta	57	31	28	30	26	8	—	—	1	10	28	30	31	222
	Cold Bay	59	24	23	25	21	9	—	0	0	—	9	19	24	154
	Fairbanks	39	31	28	31	27	7	0	0	1	9	29	30	31	222
	Gulkana	54	31	28	31	29	15	1	—	3	14	27	30	31	238
	Homer	61	28	25	27	22	9	—	0	—	3	18	25	28	185
	Juneau	58	25	22	23	14	3	—	0	—	1	7	18	23	137
	King Salmon	39	28	26	27	24	11	—	0	—	5	22	25	28	197
	Kodiak	40	22	20	21	13	3	—	0	0	1	12	18	23	134
	Kotzebue	59	31	28	31	30	25	6	—	—	7	28	30	31	248
	Mcgrath	60	31	28	31	28	11	—	0	1	10	28	30	31	229
	Nome	36	31	28	31	29	19	3	—	1	10	25	29	31	238
	St. Paul Island	85	26	26	29	27	18	3	—	—	2	11	19	25	187
	Talkeetna	62	31	28	31	28	13	—	0	1	8	25	29	31	223
	Unalakleet	30	31	28	31	29	18	2	—	1	8	27	30	31	236
	Valdez	30	30	27	29	17	1	—	0	—	1	12	26	30	172
	Yakutat	38	25	23	24	21	8	—	0	—	5	11	22	25	163
AZ	Flagstaff	53	30	28	30	25	13	3	—	—	3	19	28	30	208
	Phoenix	42	3	1	—	0	0	0	0	0	0	0	—	1	5
	Tucson	62	6	4	1	—	0	0	0	0	0	—	1	5	17
	Winslow	42	28	23	19	8	1	0	0	0	—	5	21	28	135
	Yuma	31	1	—	0	0	0	0	0	0	0	0	0	1	1
AR	Fort Smith	38	23	16	7	1	0	0	0	0	0	1	8	19	74
	Little Rock	42	19	12	5	—	0	0	0	0	0	—	5	15	56
	North Little Rock	24	17	10	4	—	0	0	0	0	0	—	4	12	46
CA	Bakersfield	39	5	1	—	0	0	0	0	0	0	—	1	5	11
	Bishop	54	29	24	20	8	1	—	0	0	—	7	24	29	143
	Blue Canyon	54	18	16	18	11	4	—	0	0	—	2	10	15	95
	Eureka	92	2	1	—	—	0	0	0	0	0	—	—	1	4
	Fresno	39	7	3	1	—	0	0	0	0	0	—	2	8	20
	Long Beach	42	—	0	0	0	0	0	0	0	0	0	0	—	0
	Los Angeles AP	43	—	0	0	0	0	0	0	0	0	0	0	—	0
	Los Angeles CO	61	—	0	0	0	0	0	0	0	0	0	0	—	0
	Mount Shasta	42	26	21	22	14	4	—	—	0	1	6	18	25	136

	Redding	16	11	5	2	0	0	0	0	0	0	0	3	12	33
	Sacramento	52	6	2	0	—	0	0	0	0	0	0	1	6	16
	San Diego	42	—	0	0	0	0	0	0	0	0	0	0	0	0
	San Francisco AP	43	1	—	—	0	0	0	0	0	0	0	0	1	2
	San Francisco CO	65	—	—	0	0	0	0	0	0	0	0	0	—	0
	Santa Barbara	66	3	1	0	0	0	0	0	0	0	0	—	3	7
	Santa Maria	39	5	3	1	—	—	0	0	0	0	—	2	6	17
	Stockton	43	8	2	1	—	0	0	0	0	0	0	2	8	21
CO	Alamosa	57	31	28	31	27	14	2	0	—	8	27	30	31	227
	Colorado Springs	42	30	27	26	14	2	0	0	0	1	9	24	29	162
	Denver	35	30	26	24	11	1	0	0	0	1	8	24	29	156
	Grand Junction	39	30	25	17	6	—	0	0	0	—	4	21	30	133
	Pueblo	37	30	27	25	11	1	0	0	0	1	11	26	30	162
CT	Bridgeport	37	26	23	17	3	—	0	0	0	0	1	8	21	98
	Hartford	43	28	25	22	8	1	0	0	0	—	6	16	27	133
DE	Wilmington	55	25	22	15	3	—	0	0	0	0	1	10	22	99
DC	Washington Dulles AP	40	26	22	17	5	—	0	0	0	—	5	14	23	113
	Washington Nat'l AP	42	22	18	8	1	0	0	0	0	0	—	4	15	68
FL	Apalachicola	42	3	1	—	0	0	0	0	0	0	0	—	2	6
	Daytona Beach	59	2	1	—	0	0	0	0	0	0	0	—	1	5
	Fort Myers	59	—	—	0	0	0	0	0	0	0	0	0	—	0
	Gainesville	19	5	3	1	0	0	0	0	0	0	0	1	4	14
	Jacksonville	61	6	3	1	0	0	0	0	0	0	0	1	4	15
	Key West	54	0	0	0	0	0	0	0	0	0	0	0	0	0
	Miami	38	—	0	—	0	0	0	0	0	0	0	0	—	0
	Orlando	39	2	—	—	0	0	0	0	0	0	0	—	1	2
	Pensacola	39	6	3	1	0	0	0	0	0	0	—	1	4	15
	Tallahassee	41	11	7	3	—	0	0	0	0	0	—	4	9	34
	Tampa	39	1	1	—	0	0	0	0	0	0	0	—	1	3
	Vero Beach	19	—	0	—	0	0	0	0	0	0	0	0	—	0
	West Palm Beach	38	—	—	—	0	0	0	0	0	0	0	0	—	0
GA	Athens	59	15	11	5	0	0	0	0	0	0	—	6	13	50
	Atlanta	42	15	11	5	0	0	0	0	0	0	—	5	12	48
	Augusta	38	16	12	5	1	0	0	0	0	0	1	7	14	55
	Columbus	57	13	8	3	—	0	0	0	0	0	—	4	11	39
	Macon	38	14	10	4	—	0	0	0	0	0	—	5	12	44
	Savannah	38	10	6	2	—	0	0	0	0	0	—	2	8	28
HI	Hilo	57	0	0	0	0	0	0	0	0	0	0	0	0	0
	Honolulu	33	0	0	0	0	0	0	0	0	0	0	0	0	0
	Kahului	38	0	0	0	0	0	0	0	0	0	0	0	0	0
	Lihue	53	0	0	0	0	0	0	0	0	0	0	0	0	0
ID	Boise	63	26	21	17	8	2	—	0	0	0	6	17	26	122
	Lewiston	56	21	16	12	3	—	0	0	0	—	3	12	21	87
	Pocatello	39	28	25	25	15	4	—	0	—	3	15	24	28	168
IL	Cairo	45	21	15	7	—	0	0	0	0	0	—	6	17	66
	Chicago	44	29	25	21	7	1	0	0	0	—	5	16	26	130

(Continued)

Table 3A.4 (Continued)

State	Station	Years	Jan	Feb	Mar	Apr	May	Jun	Jul	Aug	Sep	Oct	Nov	Dec	Annual
	Moline	42	29	25	21	7	1	0	0	0	—	6	18	27	134
	Peoria	43	29	25	19	6	—	0	0	0	—	4	17	27	127
	Rockford	39	29	26	23	9	1	0	0	0	0	7	19	28	143
	Springfield	43	28	23	17	4	—	0	0	0	—	4	15	25	115
IN	Evansville	41	25	20	13	3	—	0	0	0	0	3	12	21	97
	Fort Wayne	41	28	24	21	8	1	0	0	0	—	5	16	26	128
	Indianapolis	43	27	23	17	5	—	0	0	0	—	4	14	24	114
	South Bend	39	28	24	20	8	1	0	0	0	—	3	14	25	124
IA	Des Moines	41	30	25	21	6	—	0	0	0	—	5	19	29	134
	Dubuque	35	31	26	23	8	1	0	0	0	0	7	21	30	147
	Sioux City	43	31	27	23	9	1	0	0	0	1	8	23	30	152
	Waterloo	43	31	27	24	10	1	0	0	0	1	9	22	30	155
KS	Concordia	40	29	24	19	5	—	0	0	0	—	3	17	28	125
	Dodge City	39	29	23	18	5	—	0	0	0	—	4	17	28	125
	Goodland	36	30	27	24	12	1	0	0	0	1	8	24	30	157
	Topeka	38	29	22	16	4	—	0	0	0	—	4	16	27	118
	Wichita	49	28	22	15	3	—	0	0	0	—	1	14	26	109
KY	Greater Cincinnati AP	40	26	22	16	5	—	0	0	0	—	3	13	22	106
	Jackson	22	23	16	12	3	—	0	0	0	0	1	9	17	81
	Lexington	39	24	20	14	3	—	0	0	0	0	2	11	20	93
	Louisville	42	24	19	12	2	—	0	0	0	0	1	9	20	86
	Paducah	19	23	16	11	3	0	0	0	0	0	2	11	19	85
LA	Baton Rouge	43	9	5	1	—	0	0	0	0	0	—	2	6	22
	Lake Charles	38	5	3	1	0	0	0	0	0	0	—	1	4	14
	New Orleans	56	5	3	1	—	0	0	0	0	0	0	1	3	12
	Shreveport	50	12	7	3	—	0	0	0	0	0	—	3	10	35
ME	Caribou	63	31	28	29	22	5	—	0	0	3	14	25	30	187
	Portland	62	30	27	26	14	2	0	0	0	1	9	19	28	155
MD	Baltimore	52	25	21	14	3	—	0	0	0	0	2	11	21	97
MA	Blue Hill	117	28	26	23	9	—	0	0	0	—	3	14	26	129
	Boston	38	26	23	16	2	0	0	0	0	0	0	7	21	96
	Worcester	47	29	26	25	10	1	0	0	0	—	5	17	28	141
MI	Alpena	43	30	27	28	19	7	1	0	—	2	11	22	29	176
	Detroit	44	29	25	22	9	1	0	0	0	—	4	16	26	132
	Flint	39	29	26	24	11	2	0	0	0	—	6	17	27	140
	Grand Rapids	39	29	26	24	12	2	0	0	0	—	6	17	28	143
	Houghton Lake	38	31	28	28	17	5	—	0	—	2	9	22	30	171
	Lansing	39	29	26	25	13	3	—	0	0	1	8	18	27	150
	Marquette	24	31	28	29	22	8	0	0	0	2	16	26	30	194
	Muskegon	42	29	26	24	11	2	—	0	0	—	5	16	26	138
	Sault Ste. Marie	61	31	28	29	21	7	1	0	—	2	9	22	30	179
MN	Duluth	41	31	28	29	20	5	—	0	—	2	12	26	31	184
	International Falls	63	31	28	30	22	7	—	0	—	4	16	28	31	197
	Minneapolis-St. Paul	43	31	27	25	11	1	0	0	0	0	7	23	30	155

	Rochester	42	31	27	26	12	2	0	0	0	1	10	24	30	163
	Saint Cloud	62	31	28	28	17	3	—	0	0	2	13	26	31	178
MS	Jackson	39	14	10	4	—	0	0	0	0	0	0	6	12	47
	Meridian	38	15	11	5	—	0	0	0	0	0	1	6	13	51
	Tupelo	19	16	10	6	1	0	0	0	0	0	—	6	14	53
MO	Columbia	33	27	21	15	3	—	0	0	0	—	2	14	25	107
	Kansas City	30	28	21	15	4	—	0	0	0	—	2	14	26	110
	St. Louis	42	26	20	13	2	—	0	0	0	0	1	11	23	96
	Springfield	42	26	20	13	3	—	0	0	0	—	2	12	23	100
MT	Billings	43	28	24	23	13	2	—	0	0	1	8	22	28	147
	Glasgow	38	31	28	28	16	3	0	0	0	2	14	27	31	179
	Great Falls	41	27	24	26	17	4	—	0	—	3	12	22	27	160
	Helena	39	30	27	27	18	4	—	0	—	3	17	27	30	181
	Kalispell	43	29	26	27	18	6	1	—	—	6	21	25	29	189
	Missoula	42	29	26	26	17	5	—	—	—	4	18	26	30	181
NE	Grand Island	41	31	26	23	8	1	0	0	0	0	7	23	30	148
	Lincoln	31	30	26	22	8	1	0	0	0	0	6	22	30	144
	Norfolk	57	31	27	24	10	1	0	0	0	1	7	24	30	155
	North Platte	38	31	28	27	13	3	—	0	0	2	14	28	31	177
	Omaha Eppley AP	38	30	26	22	7	—	0	0	0	—	6	20	29	139
	Omaha (North)	39	30	26	21	7	—	0	0	0	—	4	19	29	137
	Scottsbluff	38	30	27	27	14	2	—	0	0	2	13	27	30	172
	Valentine	47	31	28	27	15	3	—	0	0	2	14	27	31	177
NV	Elko	38	29	26	26	20	8	1	—	—	7	22	26	29	195
	Ely	64	31	27	29	24	13	4	—	0	8	23	28	30	218
	Las Vegas	42	11	4	1	—	0	0	0	0	0	—	2	10	27
	Reno	39	27	24	22	16	5	0	0	—	3	14	24	28	163
	Winnemucca	53	28	24	25	19	8	1	—	—	6	21	25	28	186
NH	Concord	37	30	27	26	17	5	—	0	—	2	14	21	29	171
	Mt. Washington	70	31	28	31	28	20	6	2	3	12	23	28	31	243
NJ	Atlantic City AP	38	25	22	17	6	—	0	0	0	—	3	13	22	109
	Atlantic City CO	40	21	17	10	1	0	0	0	0	0	—	4	15	69
	Newark	37	23	20	12	1	0	0	0	0	0	0	6	19	83
NM	Albuquerque	42	28	21	14	4	—	0	0	0	0	2	15	28	113
	Clayton	54	29	25	22	9	1	0	0	0	—	5	20	28	139
	Roswell	30	25	16	7	2	0	0	0	0	0	1	12	24	88
NY	Albany	37	29	26	24	12	1	0	0	0	0	8	18	27	146
	Binghamton	51	29	26	25	12	1	0	0	0	—	6	18	27	145
	Buffalo	42	28	26	24	10	1	0	0	0	—	3	14	25	130
	Islip	19	25	21	17	3	0	0	0	0	0	1	10	21	99
	New York C.Park	89	22	20	12	1	—	0	0	0	0	—	5	18	78
	New York (JFK AP)	41	23	20	12	1	0	0	0	0	0	—	4	17	77
	New York (Laguardia AP)	40	22	19	10	1	0	0	0	0	0	—	3	15	70
	Rochester	39	28	25	23	11	1	0	0	0	—	4	15	25	133
	Syracuse	39	28	25	24	11	1	0	0	0	—	5	15	26	135

(Continued)

Table 3A.4 (Continued)

State	Station	Years	Jan	Feb	Mar	Apr	May	Jun	Jul	Aug	Sep	Oct	Nov	Dec	Annual
NC	Asheville	38	23	19	13	4	—	0	0	0	—	4	13	21	97
	Cape Hatteras	45	9	8	3	—	0	0	0	0	0	—	1	6	27
	Charlotte	42	18	14	7	1	—	0	0	0	0	1	6	15	63
	Greensboro-Wnstn-Slm-HPT	39	22	18	10	2	—	0	0	0	0	1	9	18	80
	Raleigh	38	20	16	9	2	—	0	0	0	0	1	9	17	75
	Wilmington	39	13	10	4	—	0	0	0	0	0	—	3	10	40
ND	Bismarck	43	31	28	29	18	4	—	0	0	3	15	28	31	186
	Fargo	43	31	28	27	16	4	—	0	0	2	13	27	31	179
	Williston	41	31	28	29	18	5	—	0	0	3	17	29	31	189
OH	Akron	39	28	24	21	9	1	—	0	0	0	4	14	25	125
	Cleveland	42	28	24	21	9	1	—	0	0	0	2	12	24	121
	Columbus	43	27	23	18	6	0	0	0	0	—	3	13	23	115
	Dayton	39	27	23	18	6	—	0	0	0	—	3	14	24	115
	Mansfield	37	28	24	21	9	1	0	0	0	0	4	15	25	126
	Toledo	47	29	25	23	10	1	—	0	0	—	6	17	26	137
	Youngstown	59	28	25	23	11	2	—	0	—	—	4	16	26	134
OK	Oklahoma City	37	23	15	8	1	0	0	0	0	0	1	8	20	75
	Tulsa	42	23	16	8	1	0	0	0	0	0	—	8	19	76
OR	Astoria	49	9	6	6	2	—	0	0	0	0	—	4	7	34
	Burns	18	30	26	27	21	9	3	—	1	8	24	27	30	207
	Eugene	60	14	10	7	2	—	—	0	0	—	2	7	11	54
	Medford	41	19	14	11	5	1	0	0	0	—	4	10	16	80
	Pendleton	67	21	16	10	3	—	0	0	0	—	3	12	20	84
	Portland	62	12	8	4	1	—	0	0	0	0	1	5	9	40
	Salem	40	13	11	9	5	1	—	0	0	—	3	7	13	62
	Sexton Summit	40	19	16	20	15	7	—	0	0	—	3	11	17	107
PC	Guam	43	0	0	0	0	0	0	0	0	0	0	0	0	0
	Johnston Island	28	0	0	0	0	0	0	0	0	0	0	0	0	0
	Koror	51	0	0	0	0	0	0	0	0	0	0	0	0	0
	Kwajalein, Marshall IS	50	0	0	0	0	0	0	0	0	0	0	0	0	0
	Majuro, Marshall IS	48	0	0	0	0	0	0	0	0	0	0	0	0	0
	Pago Pago, Amer Samoa	43	0	0	0	0	0	0	0	0	0	0	0	0	0
	Pohnpei, Caroline IS	51	0	0	0	0	0	0	0	0	0	0	0	0	0
	Chuuk, E. Caroline IS	51	0	0	0	0	0	0	0	0	0	0	0	0	0
	Wake Island	50	0	0	0	0	0	0	0	0	0	0	0	0	0
	Yap, W Caroline IS	54	0	0	0	0	0	0	0	0	0	0	0	0	0
PA	Allentown	59	28	25	20	6	—	0	0	0	—	4	15	26	123
	Erie	37	27	25	23	10	1	—	0	0	0	1	11	24	122
	Harrisburg	37	26	23	17	4	—	0	0	0	—	2	11	23	106
	Middletown/Harrisburg AP	64	26	23	17	3	—	0	0	0	—	2	11	23	105

	Philadelphia	43	25	21	14	2	—	0	0	0	0	1	8	20	92
	Pittsburgh	43	27	24	19	8	1	0	0	0	0	4	14	24	121
	Avoca	47	28	25	22	8	1	0	0	0	—	4	14	25	127
	Williamsport	58	28	25	21	8	1	0	0	0	—	5	15	25	128
RI	Block IS	39	24	21	16	3	0	0	0	0	0	—	5	17	86
	Providence	39	28	24	19	5	—	0	0	0	0	3	13	24	116
SC	Charleston AP	60	10	7	3	—	0	0	0	0	0	—	3	9	32
	Charleston CO	27	4	2	—	0	0	0	0	0	0	0	0	3	9
	Columbia	36	16	12	6	1	0	0	0	0	0	1	7	14	57
	Greenville-Spartanburg AP	40	18	14	7	1	—	0	0	0	0	1	7	15	63
SD	Aberdeen	41	31	28	27	15	3	0	0	—	2	14	27	31	178
	Huron	43	31	27	26	14	2	—	0	0	1	12	26	31	170
	Rapid City	60	30	27	27	15	3	—	0	0	2	11	25	30	169
	Sioux Falls	39	31	27	26	13	2	0	0	0	1	12	26	31	168
TN	Bristol-JhnCty-Kgsprt	41	23	19	13	4	—	0	0	0	0	3	13	21	96
	Chattanooga	62	19	15	8	1	0	0	0	0	0	1	10	17	71
	Knoxville	42	20	16	9	2	—	0	0	0	0	1	8	17	72
	Memphis	61	17	12	5	—	0	0	0	0	0	—	5	14	53
	Nashville	37	21	16	9	2	0	0	0	0	0	1	8	17	74
	Oak Ridge	37	21	17	12	3	—	0	0	0	0	2	12	19	86
TX	Abilene	39	17	10	4	—	0	0	0	0	0	—	5	14	50
	Amarillo	41	27	22	15	4	—	0	0	0	—	2	15	27	111
	Austin	40	8	4	1	0	0	0	0	0	0	—	1	5	18
	Brownsville	36	1	—	—	0	0	0	0	0	0	0	—	1	2
	Corpus Christi	38	3	1	—	0	0	0	0	0	0	—	—	1	5
	Dallas-Fort Worth	39	14	8	3	—	0	0	0	0	0	—	3	10	37
	Del Rio	39	6	3	1	0	0	0	0	0	0	—	1	5	16
	El Paso	42	18	11	5	1	—	0	0	0	0	—	7	18	59
	Galveston	42	2	1	—	0	0	0	0	0	0	0	—	1	3
	Houston	33	7	4	1	—	0	0	0	0	0	—	1	5	18
	Lubbock	55	25	18	11	2	—	0	0	0	0	1	11	24	92
	Midland-Odessa	39	20	13	6	1	0	0	0	0	0	—	7	18	63
	Port Arthur	42	6	3	1	—	0	0	0	0	0	—	1	4	14
	San Angelo	42	17	10	4	1	0	0	0	0	0	—	5	14	51
	San Antonio	60	8	4	2	—	0	0	0	0	0	—	2	6	21
	Victoria	41	5	2	—	0	0	0	0	0	0	—	1	3	10
	Waco	39	12	7	2	—	0	0	0	0	0	—	3	9	33
	Wichita Falls	42	21	13	6	1	0	0	0	0	0	—	6	17	64
UT	Milford	39	30	26	27	18	6	1	0	0	3	15	26	30	180
	Salt Lake City	43	27	22	15	6	1	0	0	0	—	4	18	27	122
VT	Burlington	38	30	26	26	15	2	0	0	0	1	8	18	28	154
VA	Lynchburg	39	23	20	12	3	—	0	0	0	0	3	11	20	92
	Norfolk	54	16	13	6	—	0	0	0	0	0	—	3	13	51
	Richmond	73	21	19	10	2	—	0	0	0	0	2	10	20	83

(Continued)

Table 3A.4 (Continued)

State	Station	Years	Jan	Feb	Mar	Apr	May	Jun	Jul	Aug	Sep	Oct	Nov	Dec	Annual
	Roanoke	38	23	19	12	2	—	0	0	0	0	2	10	19	88
	Wallops Island	28	21	17	10	1	0	0	0	0	0	—	6	17	72
	Olympia	43	15	14	14	8	2	—	0	0	1	5	10	15	84
WA	Quillayute	36	13	10	9	6	—	0	0	0	—	3	7	12	61
	Seattle CO	48	7	4	1	0	0	0	0	0	0	—	2	5	19
	Seattle Sea-Tac AP	43	9	6	3	—	0	0	0	0	0	—	3	8	29
	Spokane	43	26	23	21	10	2	0	0	0	1	10	20	27	139
	Walla Walla	43	19	14	6	0	0	0	0	0	—	2	9	22	73
	Yakima	56	28	24	21	12	3	—	0	0	1	12	22	28	149
PR	San Juan	47	0	0	0	0	0	0	0	0	0	0	0	0	0
WV	Beckley	39	25	21	17	7	1	—	0	0	—	5	14	22	112
	Charleston	55	23	20	15	5	—	0	0	0	0	3	13	21	99
	Elkins	58	26	24	22	12	3	—	—	0	1	11	20	26	144
	Huntington	41	23	20	13	4	—	0	0	0	—	3	11	20	95
WI	Green Bay	41	31	27	27	14	2	0	0	0	1	9	22	29	161
	La Crosse	50	30	27	25	9	1	0	0	0	—	6	21	29	149
	Madison	43	30	27	26	13	3	—	0	0	1	9	21	29	159
	Milwaukee	42	29	26	23	9	1	0	0	0	—	4	17	27	136
WY	Casper	38	29	26	27	19	7	—	—	0	4	16	25	29	181
	Cheyenne	43	29	26	27	18	4	0	0	0	2	13	24	29	172
	Lander	56	31	28	28	18	5	—	0	0	3	14	28	31	184
	Sheridan	38	30	27	27	18	5	—	0	0	3	16	28	30	186

The mean number of days with a minimum temperature of 32°F or lower indicates the frequency of occurrence of days with freezing temperatures.
The annual value is the total of the unrounded monthly values, it may not agree with the sum of the rounded monthly values.

Source: From U.S. National Oceanic and Atmospheric Administration, *Comparative Climatic Data for the United States Through 2000*, www.noaa.gov.

Table 3A.5 Mean Number of Days with Maximum Temperature 90° F or More — Selected Cities of the United States

State	Station	Years	Jan	Feb	Mar	Apr	May	Jun	Jul	Aug	Sep	Oct	Nov	Dec	Annual
AL	Birmingham	11	0	0	0	—	1	10	18	17	5	0	0	0	52
	Birmingham AP	39	0	0	0	—	2	11	18	16	7	—	0	0	54
	Huntsville	35	0	0	0	—	1	9	17	15	6	—	0	0	48
	Mobile	40	0	0	0	—	4	16	22	21	10	1	0	0	74
	Montgomery	39	0	0	0	—	4	17	22	21	12	1	0	0	78
AK	Anchorage	38	0	0	0	0	1	4	6	3	—	0	0	0	14
	Annette	42	0	0	0	—	2	4	6	6	2	—	0	0	19
	Barrow	29	0	0	0	0	0	—	1	—	0	0	0	0	1
	Barter IS	41	0	0	0	0	0	0	—	—	0	0	0	0	0
	Bethel	44	0	0	0	0	1	3	6	2	—	0	0	0	12
	Bettles	50	0	0	0	0	1	13	16	5	—	0	0	0	36
	Big Delta	58	0	0	0	—	2	11	15	8	1	0	0	0	37
	Cold Bay	58	0	0	0	0	0	—	—	—	—	0	0	0	0
	Fairbanks	39	0	0	0	—	4	17	21	10	1	0	0	0	54
	Gulkana	54	0	0	0	0	1	9	13	7	—	0	0	0	30
	Homer	62	0	0	0	0	—	—	1	1	0	0	0	0	2
	Juneau	58	0	0	0	—	1	5	7	6	—	0	0	0	19
	King Salmon	39	0	0	0	0	1	3	6	4	—	0	0	0	13
	Kodiak	40	0	0	0	0	—	1	2	3	—	0	0	0	7
	Kotzebue	59	0	0	0	0	—	1	3	1	0	0	0	0	4
	Mcgrath	60	0	0	0	0	2	11	14	6	—	0	0	0	32
	Nome	36	0	0	0	0	—	2	3	1	—	0	0	0	6
	St. Paul Island	85	0	0	0	0	0	0	0	0	0	0	0	0	0
	Talkeetna	62	0	0	0	0	2	10	13	7	—	0	0	0	32
	Unalakleet	30	0	0	0	0	—	1	3	2	—	0	0	0	6
	Valdez	30	0	0	0	0	—	2	4	3	—	0	0	0	9
	Yakutat	38	0	0	0	—	—	1	1	1	—	0	0	0	3
AZ	Flagstaff	53	0	0	0	0	0	1	2	—	—	0	0	0	3
	Phoenix	42	0	—	2	10	23	29	31	31	28	15	—	0	168
	Tucson	62	0	—	0	5	18	28	29	29	24	9	—	0	144
	Winslow	42	0	0	0	—	3	17	24	19	5	—	0	0	68
	Yuma	31	0	—	3	12	23	29	31	31	28	18	1	0	175
AR	Fort Smith	38	0	0	—	1	2	13	23	22	10	1	0	0	72
	Little Rock	42	0	0	—	—	3	16	23	21	9	1	0	0	73
	North Little Rock	24	0	0	0	—	2	13	22	21	8	—	0	0	67
CA	Bakersfield	39	0	0	—	2	10	19	28	26	17	5	—	0	108
	Bishop	55	0	0	0	—	5	19	29	27	14	1	0	0	95
	Blue Canyon	55	0	0	0	0	0	—	—	0	—	0	0	0	0
	Eureka	92	0	0	0	0	0	0	0	0	0	0	0	0	0
	Fresno	39	0	0	—	2	10	19	28	27	18	4	0	0	108
	Long Beach	42	—	—	—	1	1	2	3	5	6	3	1	0	22
	Los Angeles AP	43	0	—	—	—	—	—	—	—	1	1	—	—	3
	Los Angeles CO	62	—	—	—	1	1	1	4	5	6	3	1	—	22

(Continued)

Table 3A.5 (Continued)

State	Station	Years	Jan	Feb	Mar	Apr	May	Jun	Jul	Aug	Sep	Oct	Nov	Dec	Annual
	Mount Shasta	42	0	0	0	0	—	2	9	7	3	—	0	0	21
	Redding	16	0	0	0	1	7	16	28	25	18	5	0	0	100
	Sacramento	52	0	0	0	—	5	12	22	19	13	3	0	0	74
	San Diego	42	0	—	—	—	—	—	—	—	1	1	—	0	2
	San Francisco AP	43	0	0	0	—	—	1	1	—	1	—	0	0	3
	San Francisco CO	65	0	0	0	—	—	—	—	—	1	—	0	0	1
	Santa Barbara	66	0	0	—	—	—	—	0	—	1	1	—	0	2
	Santa Maria	39	0	0	—	—	—	1	—	—	1	1	—	0	3
	Stockton	43	0	0	0	1	7	14	23	21	14	3	0	0	83
CO	Alamosa	57	0	0	0	0	0	—	1	—	0	0	0	0	1
	Colorado Springs	42	0	0	0	0	—	4	9	3	1	0	0	0	16
	Denver	35	0	0	0	—	—	7	15	10	2	0	0	0	34
	Grand Junction	39	0	0	0	0	1	15	25	19	4	0	0	0	64
	Pueblo	37	0	0	0	—	3	15	23	18	7	—	0	0	65
CT	Bridgeport	37	0	0	0	—	—	1	3	2	—	0	0	0	6
	Hartford	43	0	0	0	—	1	4	7	5	1	—	0	0	18
DE	Wilmington	55	0	0	0	—	1	4	8	5	2	—	0	0	20
DC	Washington Dulles AP	40	0	0	0	—	1	5	11	9	3	—	0	0	29
	Washington Nat'l AP	42	0	0	0	—	2	7	14	10	4	—	0	0	37
FL	Apalachicola	42	0	0	0	—	1	6	8	8	4	—	0	0	26
	Daytona Beach	59	0	0	—	2	6	11	17	14	6	1	0	0	57
	Fort Myers	59	0	—	0	3	14	20	25	25	19	6	—	—	113
	Gainesville	19	0	0	0	2	10	17	22	20	10	1	0	0	81
	Jacksonville	61	0	0	—	1	8	16	24	21	10	1	0	0	81
	Key West	54	0	0	0	—	1	7	16	17	8	1	0	0	48
	Miami	38	0	0	—	2	4	10	17	17	10	2	0	0	62
	Orlando	39	0	0	—	4	12	19	25	25	18	3	0	—	106
	Pensacola	39	0	0	0	—	2	13	18	16	9	—	0	0	59
	Tallahassee	41	0	0	—	1	9	19	23	22	15	2	0	0	92
	Tampa	39	0	0	0	1	9	17	21	22	16	3	—	0	87
	Vero Beach	19	0	0	—	1	4	10	17	18	10	2	—	0	63
	West Palm Beach	38	0	0	—	2	4	10	18	19	10	2	—	0	65
GA	Athens	59	0	0	0	—	3	11	17	14	5	—	0	0	51
	Atlanta	42	0	0	0	—	1	8	13	10	3	0	0	0	36
	Augusta	38	0	0	0	1	6	15	23	19	9	1	0	0	74
	Columbus	57	0	0	0	—	6	17	22	21	10	1	0	0	76
	Macon	38	0	0	—	1	6	17	23	21	10	1	0	0	80
	Savannah	38	0	0	—	1	6	14	22	18	7	1	0	0	70
HI	Hilo	57	—	—	—	0	—	—	0	—	—	—	—	—	0
	Honolulu	33	0	0	0	0	—	2	5	11	10	4	0	0	33
	Kahului	38	0	0	—	—	1	2	3	6	7	5	1	—	25
	Lihue	53	0	0	0	0	0	0	0	—	—	—	0	0	0
ID	Boise	63	0	0	0	—	1	5	19	16	4	—	0	0	45
	Lewiston	56	0	0	0	—	1	4	16	14	4	0	0	0	40

	Pocatello	39	0	0	0	0	0	4	14	13	2	—	0	0	34
IL	Cairo	45	0	0	0	—	2	11	17	13	4	—	0	0	48
	Chicago	44	0	0	0	—	1	4	7	4	2	—	0	0	17
	Moline	42	0	0	0	—	1	6	9	6	2	—	0	0	23
	Peoria	43	0	0	0	—	—	5	8	5	2	—	0	0	20
	Rockford	39	0	0	0	—	0	3	6	3	1	—	0	0	14
	Springfield	43	0	0	0	—	1	7	10	6	3	—	0	0	28
IN	Evansville	41	0	0	0	—	2	9	15	11	4	—	0	0	41
	Fort Wayne	41	0	0	0	0	1	4	6	4	1	0	0	0	15
	Indianapolis	43	0	0	0	0	0	4	7	5	2	0	0	0	18
	South Bend	39	0	0	0	0	1	3	5	3	1	0	0	0	13
IA	Des Moines	41	0	0	—	—	—	4	9	6	2	—	0	0	22
	Dubuque	35	0	0	0	—	—	1	4	2	1	0	0	0	8
	Sioux City	43	0	0	—	0	1	6	9	6	2	—	0	0	24
	Waterloo	43	0	0	0	—	1	4	6	4	1	—	0	0	16
KS	Concordia	40	0	0	0	—	1	9	18	14	5	1	0	0	49
	Dodge City	39	0	0	—	1	3	13	22	19	8	1	—	0	66
	Goodland	36	0	0	0	—	1	9	17	14	6	—	0	0	47
	Topeka	38	0	0	0	—	1	7	16	14	5	—	0	0	43
	Wichita	49	0	0	0	—	2	12	21	19	8	1	0	0	63
KY	Greater Cincinnati AP	40	0	0	0	0	—	4	8	6	2	0	0	0	19
	Jackson	22	0	0	0	—	—	2	6	5	1	0	0	0	14
	Lexington	39	0	0	0	0	—	3	8	7	2	0	0	0	20
	Louisville	42	0	0	0	—	—	6	12	10	3	0	0	0	32
	Paducah	19	0	0	0	—	1	11	18	14	6	0	0	0	50
LA	Baton Rouge	43	0	0	—	—	6	19	24	22	12	2	0	0	84
	Lake Charles	38	0	0	0	—	2	14	23	23	12	1	0	0	75
	New Orleans	56	0	0	0	—	4	16	21	21	9	1	0	0	72
	Shreveport	50	0	0	—	—	4	18	26	25	14	2	0	0	90
ME	Caribou	63	0	0	0	0	—	—	1	—	—	0	0	0	1
	Portland	62	0	0	0	0	—	1	2	1	—	0	0	0	5
MD	Baltimore	52	0	0	0	—	2	6	12	8	3	—	0	0	30
MA	Blue Hill	117	0	0	0	—	—	1	3	2	—	0	0	0	5
	Boston	38	0	0	0	—	—	3	6	3	1	0	0	0	13
	Worcester	47	0	0	0	—	—	1	2	1	—	0	0	0	3
MI	Alpena	43	0	0	0	—	—	2	3	1	—	0	0	0	6
	Detroit	44	0	0	0	0	—	3	5	3	1	—	0	0	12
	Flint	39	0	0	0	0	—	2	3	2	—	0	0	0	7
	Grand Rapids	39	0	0	0	0	0	2	5	2	—	0	0	0	10
	Houghton Lake	38	0	0	0	0	—	1	2	1	—	0	0	0	3
	Lansing	39	0	0	0	0	—	2	4	2	1	0	0	0	10
	Marquette	24	0	0	0	—	—	1	2	1	—	0	0	0	3
	Muskegon	42	0	0	0	0	—	1	1	1	0	0	0	0	2
	Sault Ste. Marie	61	0	0	0	0	0	—	1	1	—	0	0	0	1
MN	Duluth	41	0	0	0	0	—	—	1	1	—	0	0	0	2
	International Falls	63	0	0	0	—	—	1	2	1	—	0	0	0	4

(Continued)

Table 3A.5 (Continued)

State	Station	Years	Jan	Feb	Mar	Apr	May	Jun	Jul	Aug	Sep	Oct	Nov	Dec	Annual
	Minneapolis-St. Paul	43	0	0	0	—	1	3	6	3	1	0	0	0	14
	Rochester	42	0	0	0	—	—	2	3	1	—	0	0	0	6
	Saint Cloud	62	0	0	0	—	—	2	4	3	1	—	0	0	10
MS	Jackson	39	0	0	0	—	5	18	24	23	12	1	0	0	82
	Meridian	38	0	0	—	—	4	16	24	23	12	1	0	0	81
	Tupelo	19	0	0	0	—	2	13	22	20	9	—	0	0	66
MO	Columbia	33	0	0	0	—	—	5	14	12	4	—	0	0	35
	Kansas City	30	0	0	0	—	—	6	15	12	4	—	0	0	36
	St. Louis	42	0	0	0	—	1	8	15	12	4	—	0	0	41
	Springfield	42	0	0	0	—	—	6	16	15	4	—	0	0	40
MT	Billings	43	0	0	0	—	—	4	12	11	2	—	0	0	29
	Glasgow	38	0	0	0	—	1	3	9	9	2	—	0	0	23
	Great Falls	41	0	0	0	0	—	2	8	8	1	—	0	0	19
	Helena	39	0	0	0	0	—	2	8	8	1	0	0	0	19
	Kalispell	43	0	0	0	0	—	1	6	6	—	0	0	0	13
	Missoula	42	0	0	0	0	—	2	9	9	1	0	0	0	22
NE	Grand Island	41	0	0	—	1	2	8	14	11	4	—	0	0	39
	Lincoln	31	0	0	0	0	1	9	15	12	4	—	0	0	42
	Norfolk	57	0	0	0	1	1	7	12	9	3	—	0	0	33
	North Platte	38	0	0	0	—	1	6	13	11	4	—	0	0	35
	Omaha Eppley AP	38	0	0	0	0	2	7	13	8	3	—	0	0	33
	Omaha (North)	39	0	0	0	—	1	5	9	7	2	—	0	0	24
	Scottsbluff	38	0	0	0	—	1	8	16	13	5	—	0	0	43
	Valentine	47	0	0	0	—	1	7	14	13	5	1	0	0	41
NV	Elko	38	0	0	0	0	—	5	19	15	3	0	0	0	42
	Ely	64	0	0	0	0	0	3	11	6	1	0	0	0	20
	Las Vegas	42	0	0	—	3	16	26	30	30	22	6	0	0	133
	Reno	39	0	0	0	0	1	7	20	18	5	—	0	0	52
	Winnemucca	53	0	0	0	—	1	9	22	19	6	—	0	0	57
NH	Concord	37	0	0	0	—	1	2	5	3	—	0	0	0	11
	Mt. Washington	70	0	0	0	0	0	0	0	0	0	0	0	0	0
NJ	Atlantic City AP	38	0	0	0	—	1	4	7	5	1	0	0	0	18
	Atlantic City CO	40	0	0	0	—	—	1	2	1	—	0	0	0	4
	Newark	37	0	0	0	—	2	5	9	7	1	0	0	0	24
NM	Albuquerque	42	0	0	0	0	3	17	23	16	4	—	0	0	62
	Clayton	54	0	0	0	—	1	9	14	9	3	—	0	0	36
	Roswell	30	0	0	—	2	10	22	25	23	11	2	0	0	94
NY	Albany	37	0	0	0	—	—	2	4	2	—	0	0	0	8
	Binghamton	51	0	0	0	0	0	—	1	1	—	0	0	0	2
	Buffalo	42	0	0	0	—	—	1	1	1	—	0	0	0	3
	Islip	19	0	0	0	0	1	2	3	2	—	0	0	0	7
	New York C.Park	89	0	0	0	—	1	3	7	5	1	—	0	0	17
	New York (JFK AP)	41	0	0	0	—	—	2	4	3	1	0	0	0	10

	New York (Laguardia AP)	40	0	0	0	—	1	3	6	4	1	0	0	0	16
	Rochester	39	0	0	0	—	—	2	4	2	1	0	0	0	8
	Syracuse	39	0	0	0	—	—	2	4	2	—	0	0	0	7
NC	Asheville	38	0	0	0	0	—	2	4	2	—	0	0	0	8
	Cape Hatteras	45	0	0	0	0	—	1	2	2	—	0	0	0	5
	Charlotte	42	0	0	0	—	2	8	14	11	4	—	0	0	39
	Greensboro-Wnstn-Slm-HPT	39	0	0	0	—	1	6	12	8	2	—	0	0	29
	Raleigh	38	0	0	—	1	2	8	14	11	3	—	0	0	38
	Wilmington	39	0	0	0	1	2	8	16	12	4	—	0	0	44
ND	Bismarck	43	0	0	0	—	1	3	8	8	2	—	0	0	21
	Fargo	43	0	0	0	—	1	2	5	5	1	—	0	0	14
	Williston	41	0	0	0	—	1	3	8	9	2	—	0	0	23
OH	Akron	39	0	0	0	0	—	2	4	2	—	0	0	0	7
	Cleveland	42	0	0	0	0	—	2	4	2	1	0	0	0	9
	Columbus	43	0	0	0	0	1	4	6	4	1	0	0	0	16
	Dayton	39	0	0	0	0	—	4	7	4	1	0	0	0	16
	Mansfield	37	0	0	0	0	—	1	3	2	—	0	0	0	6
	Toledo	47	0	0	0	0	1	4	6	3	1	—	0	0	15
	Youngstown	59	0	0	0	0	—	1	3	2	0	0	0	0	7
OK	Oklahoma City	37	0	0	—	—	2	11	23	23	9	1	0	0	68
	Tulsa	42	0	0	—	1	2	13	24	22	9	1	0	0	72
OR	Astoria	49	0	0	0	0	0	—	—	—	—	0	0	0	0
	Burns	18	0	0	0	0	—	2	9	9	1	0	0	0	21
	Eugene	60	0	0	0	0	—	1	6	5	2	—	0	0	15
	Medford	41	0	0	0	—	2	7	18	17	9	1	0	0	55
	Pendleton	67	0	0	0	—	1	4	14	11	3	—	0	0	33
	Portland	62	0	0	0	0	—	1	4	4	2	—	0	0	10
	Salem	40	0	0	0	0	—	2	6	6	2	—	0	0	16
	Sexton Summit	40	0	0	0	0	—	—	1	1	—	0	0	0	2
PC	Guam	44	0	0	—	—	3	4	2	1	1	—	—	0	10
	Johnston Island	28	0	0	0	0	0	0	0	0	0	—	0	0	0
	Koror	51	4	3	5	10	14	8	4	4	6	9	12	7	86
	Kwajalein, Marshall IS	50	—	—	1	0	1	1	1	2	2	2	1	—	10
	Majuro, Marshall IS	48	0	0	0	0	—	0	—	—	1	1	—	—	1
	Pago Pago, Amer Samoa	43	4	5	5	3	—	—	0	—	0	—	2	4	22
	Pohnpei, Caroline IS	51	1	1	3	4	6	5	8	10	12	12	9	3	74
	Chuuk, E. Caroline IS	51	—	—	—	1	3	2	3	3	3	3	2	—	21
	Wake Island	50	0	0	—	—	1	3	6	8	7	3	—	—	29
	Yap, W Caroline IS	54	—	—	—	4	7	5	3	4	4	5	4	1	37
PA	Allentown	59	0	0	0	—	1	3	7	4	1	—	0	0	16
	Erie	37	0	0	0	0	—	1	1	1	—	0	0	0	2
	Harrisburg	37	0	0	0	—	1	5	9	6	2	—	0	0	23

(Continued)

Table 3A.5 (Continued)

State	Station	Years	Jan	Feb	Mar	Apr	May	Jun	Jul	Aug	Sep	Oct	Nov	Dec	Annual
	Middletown/ Harrisburg AP	64	0	0	0	—	1	5	9	6	2	—	0	0	23
	Philadelphia	43	0	0	0	—	1	5	9	7	2	0	0	0	24
	Pittsburgh	43	0	0	0	0	—	2	4	2	1	0	0	0	8
	Avoca	47	0	0	0	—	—	2	4	2	—	0	0	0	7
	Williamsport	58	0	0	0	—	1	3	6	3	1	—	0	0	14
RI	Block IS	39	0	0	0	—	0	—	—	—	0	0	0	0	0
	Providence	39	0	0	0	—	1	2	4	3	1	0	0	0	10
SC	Charleston AP	60	0	0	—	1	4	11	17	15	5	—	0	0	53
	Charleston CO	27	0	0	0	—	1	6	13	10	3	—	0	0	34
	Columbia	36	0	0	—	2	6	16	23	18	8	1	0	0	73
	Greenville-Spartanburg	40	0	0	0	—	2	8	14	10	3	—	0	0	37
SD	Aberdeen	41	0	0	0	—	—	3	8	7	2	—	0	0	20
	Huron	43	0	0	0	—	1	4	11	9	3	—	0	0	28
	Rapid City	60	0	0	0	—	0	3	11	12	4	—	0	0	31
	Sioux Falls	39	0	0	0	—	1	4	10	6	2	0	0	0	22
TN	Bristol-JhnCty-Kgsprt	41	0	0	0	0	—	3	6	4	2	0	0	0	14
	Chattanooga	62	0	0	0	—	2	10	17	15	5	—	0	0	49
	Knoxville	42	0	0	0	—	1	5	11	9	3	0	0	0	29
	Memphis	61	0	0	0	—	3	14	22	19	8	1	0	0	67
	Nashville	37	0	0	0	—	1	9	17	13	5	—	0	0	45
	Oak Ridge	37	0	0	0	—	1	7	13	10	4	—	0	0	35
TX	Abilene	39	0	0	1	3	8	19	26	25	12	2	—	0	96
	Amarillo	41	0	0	—	1	5	13	21	17	7	1	0	0	65
	Austin	41	—	—	1	2	7	21	28	28	17	4	0	0	109
	Brownsville	36	—	—	1	4	12	24	28	28	19	6	—	—	122
	Corpus Christi	38	—	—	1	2	5	20	28	27	18	5	—	—	106
	Dallas-Fort Worth	39	0	0	—	1	5	20	28	27	14	3	0	0	97
	Del Rio	39	0	—	2	7	15	25	28	28	20	5	—	—	130
	El Paso	42	0	0	0	2	14	26	27	24	13	2	0	0	108
	Galveston	42	0	0	0	—	—	1	4	5	2	—	0	0	12
	Houston	33	0	—	—	1	6	20	27	26	16	3	0	0	99
	Lubbock	55	0	0	—	2	9	18	23	20	9	1	0	0	81
	Midland-Odessa	39	0	—	—	4	12	21	26	23	11	2	0	0	101
	Port Arthur	42	0	0	0	—	2	17	24	24	13	2	0	0	83
	San Angelo	42	—	—	1	5	12	21	27	26	13	3	—	0	109
	San Antonio	60	0	—	1	2	9	22	28	28	18	4	—	—	113
	Victoria	41	0	—	—	1	6	20	28	28	18	5	—	0	106
	Waco	39	0	—	—	1	7	22	29	28	17	4	—	0	109
	Wichita Falls	42	0	—	1	2	8	20	28	27	14	4	0	0	104
UT	Milford	39	0	0	0	0	1	10	24	18	5	—	0	0	57
	Salt Lake City	43	0	0	0	0	1	9	23	20	4	0	0	0	57

VT	Burlington	38	0	0	0	—	0	1	3	1	—	0	0	0	6
VA	Lynchburg	39	0	0	0	—	1	4	9	7	2	—	0	0	23
	Norfolk	54	0	0	0	0	2	7	12	9	3	—	0	0	33
	Richmond	73	0	0	—	1	3	9	14	11	4	—	0	0	42
	Roanoke	38	0	0	0	—	1	5	11	8	2	0	0	0	27
	Wallops Island	28	0	0	0	—	1	3	7	4	1	0	0	0	15
WA	Olympia	43	0	0	0	0	—	1	2	2	1	—	0	0	6
	Quillayute	36	0	0	0	0	—	—	—	—	—	0	0	0	0
	Seattle CO	48	0	0	0	0	—	—	1	—	—	0	0	0	1
	Seattle Sea-Tac AP	43	0	0	0	0	—	—	1	1	—	0	0	0	2
	Spokane	43	0	0	0	—	—	2	8	7	1	0	0	0	19
	Walla Walla	43	0	0	0	0	2	7	16	15	4	0	0	0	44
	Yakima	56	0	0	0	—	1	4	14	11	2	0	0	0	33
PR	San Juan	47	—	1	2	3	6	9	8	10	11	9	2	1	62
WV	Beckley	39	0	0	0	0	0	—	0	—	0	0	0	0	1
	Charleston	55	0	0	0	—	1	5	8	5	2	—	0	0	21
	Elkins	58	0	0	0	0	—	—	1	1	—	0	0	0	2
	Huntington	41	0	0	0	—	1	4	8	6	2	0	0	0	21
WI	Green Bay	41	0	0	0	0	—	2	3	1	—	0	0	0	6
	La Crosse	49	0	0	0	—	1	4	7	4	1	—	0	0	17
	Madison	43	0	0	0	—	—	3	5	3	1	—	0	0	11
	Milwaukee	42	0	0	0	—	—	2	4	2	1	0	0	0	9
WY	Casper	38	0	0	0	0	—	5	13	11	2	0	0	0	31
	Cheyenne	43	0	0	0	0	—	1	6	3	—	0	0	0	10
	Lander	56	0	0	0	0	—	3	11	7	1	0	0	0	21
	Sheridan	38	0	0	0	0	—	3	11	11	3	—	0	0	28

Note: Through 2002. For Alaska, the reported values are 70° F. The annual value is the total of the unrounded monthly values, it may not agree with the sum of the rounded monthly values.

Source: From U.S. National Oceanic and Atmospheric Administration, *Comparative Climatic Data for the United States Through 2000*, www.noaa.gov.

Table 3A.6 Normal Monthly Heating Degree Days — Selected Cities of the United States

State	Station	Years	Jul	Aug	Sep	Oct	Nov	Dec	Jan	Feb	Mar	Apr	May	Jun	Annual
AL	Birmingham AP	30	0	0	11	133	359	590	691	514	339	154	31	1	2823
	Huntsville	30	0	0	18	165	417	669	780	587	404	180	41	1	3262
	Mobile	30	0	0	2	51	204	387	455	326	182	57	3	0	1667
	Montgomery	30	0	0	3	84	278	487	568	415	250	98	9	0	2192
AK	Anchorage	30	206	268	505	957	1297	1472	1526	1295	1212	861	560	311	10470
	Annette	30	215	206	337	572	760	885	928	781	791	637	484	321	6917
	Barrow	30	763	815	1016	1564	1978	2346	2440	2267	2443	1967	1391	903	19893
	Bethel	30	280	355	587	1085	1428	1724	1813	1608	1566	1175	738	410	12769
	Bettles	30	170	366	721	1437	1975	2245	2365	2041	1888	1280	642	227	15357
	Big Delta	30	144	308	620	1270	1761	2016	2097	1759	1576	987	536	228	13302
	Cold Bay	30	447	409	518	774	915	1054	1142	1048	1085	945	780	573	9690
	Fairbanks	30	121	283	615	1287	1882	2199	2315	1926	1670	999	504	179	13980
	Gulkana	30	250	370	658	1199	1786	2064	2163	1733	1543	1018	655	358	13797
	Homer	30	338	349	514	844	1069	1215	1290	1125	1103	860	663	451	9821
	Juneau	30	257	288	453	704	953	1125	1219	1010	973	728	529	335	8574
	King Salmon	30	290	317	521	984	1254	1481	1538	1384	1286	957	667	425	11104
	Kodiak	30	339	310	468	766	931	1067	1096	983	1007	833	667	474	8941
	Kotzebue	30	327	407	696	1297	1703	2022	2092	1918	2023	1606	1037	607	15735
	Mcgrath	30	174	318	611	1233	1779	2132	2223	1847	1653	1078	583	256	13887
	Nome	30	387	446	664	1134	1444	1756	1836	1663	1727	1361	867	533	13818
	St. Paul Island	30	569	518	603	828	957	1122	1220	1174	1267	1097	911	693	10959
	Talkeetna	30	193	294	563	1043	1425	1613	1676	1391	1317	923	596	293	11327
	Unalakleet	30	297	373	644	1192	1572	1826	1915	1703	1685	1271	791	481	13750
	Valdez	30	306	353	537	832	1101	1251	1336	1126	1091	821	596	383	9733
	Yakutat	30	353	364	506	742	980	1129	1216	1027	1040	837	664	460	9318
AZ	Flagstaff	30	33	56	224	554	850	1085	1099	930	880	668	446	174	6999
	Phoenix	30	0	0	0	11	117	305	304	174	99	29	1	0	1040
	Tucson	30	0	0	0	33	195	397	401	275	194	76	7	0	1578
	Winslow	30	0	1	31	290	649	965	961	713	580	357	133	12	4692
	Yuma	30	0	0	0	5	90	246	228	114	73	23	3	0	782
AR	Fort Smith	30	0	0	23	145	448	745	854	619	410	172	34	1	3451
	Little Rock	30	0	0	13	124	400	666	775	563	369	150	24	0	3084
	North Little Rock	30	0	0	8	99	383	669	770	549	343	128	31	0	2980
CA	Bakersfield	30	0	0	2	51	283	534	521	324	236	119	31	3	2104
	Bishop	30	1	1	46	276	609	847	843	643	545	344	138	21	4314
	Eureka	30	216	198	232	326	423	532	530	451	491	430	354	263	4446
	Fresno	30	0	0	3	70	344	597	578	377	283	140	37	4	2433
	Long Beach	30	0	0	1	16	128	265	268	205	186	99	39	5	1212
	Los Angeles AP	30	1	0	2	21	121	234	252	205	212	141	78	19	1286
	Los Angeles CO	30	0	0	1	11	91	201	206	149	144	83	36	5	927
	Mount Shasta	30	66	65	196	451	753	936	921	753	738	563	374	175	5991
	Redding	30	0	0	13	131	420	611	606	445	390	239	99	7	2961
	Sacramento	30	0	0	11	84	359	595	580	387	335	208	97	10	2666
	San Diego	30	0	0	1	12	109	231	227	176	160	90	47	10	1063

	San Francisco AP	30	77	56	62	131	298	476	482	354	339	266	201	120	2862
	San Francisco CO	30	133	107	95	100	232	383	396	283	288	233	214	150	2614
	Santa Barbara	30	22	23	54	92	234	368	369	277	263	193	151	75	2121
	Santa Maria	30	68	49	70	141	288	422	419	337	350	291	230	135	2800
	Stockton	30	0	0	5	76	348	609	592	391	313	169	54	6	2563
CO	Alamosa	30	47	91	302	675	1082	1475	1551	1189	983	719	451	169	8734
	Colorado Springs	30	11	20	163	471	827	1082	1114	915	816	568	306	76	6369
	Denver	30	1	9	136	436	826	1078	1111	892	788	524	267	60	6128
	Grand Junction	30	1	1	59	342	766	1118	1194	860	643	397	151	20	5552
	Pueblo	30	1	3	88	381	779	1058	1092	843	694	431	172	24	5566
CT	Bridgeport	30	2	4	68	320	591	918	1089	944	803	489	207	32	5467
	Hartford	30	3	12	120	413	697	1054	1218	1024	844	486	195	38	6104
DE	Wilmington	30	1	2	49	297	564	871	1029	864	687	376	132	15	4887
DC	Washington Dulles AP	30	1	4	60	323	589	882	1025	847	670	362	139	21	4923
	Washington Nat'l AP	30	0	1	19	202	467	755	906	741	562	269	72	5	3999
FL	Apalachicola	30	0	0	0	32	152	328	408	285	169	42	3	0	1419
	Daytona Beach	30	0	0	0	6	67	185	245	183	99	29	1	0	815
	Fort Myers	30	0	0	0	1	16	76	103	75	28	3	0	0	302
	Gainesville	30	0	0	0	21	121	268	321	208	123	40	2	0	1104
	Jacksonville	30	0	0	0	30	148	314	374	272	155	55	5	0	1353
	Key West	30	0	0	0	0	0	14	26	18	6	0	0	0	64
	Miami	30	0	0	0	0	4	38	58	39	15	1	0	0	155
	Orlando	30	0	0	0	2	40	142	220	128	57	9	0	0	598
	Pensacola	30	0	0	1	35	175	352	416	299	171	48	1	0	1498
	Tallahassee	30	0	0	1	49	193	369	428	315	185	71	5	0	1616
	Tampa	30	0	0	0	4	46	144	187	136	63	13	0	0	593
	Vero Beach	30	0	0	0	0	18	101	166	114	48	8	0	0	455
	West Palm Beach	30	0	0	0	0	10	58	83	60	27	4	0	0	242
GA	Athens	30	0	0	10	133	353	597	687	522	349	152	28	1	2832
	Atlanta	30	0	0	11	126	352	600	692	523	346	150	26	1	2827
	Augusta	30	0	0	5	112	313	540	617	469	301	129	21	1	2508
	Columbus	30	0	0	3	78	263	481	559	415	252	94	8	0	2153
	Macon	30	0	0	4	91	277	494	570	427	262	104	12	0	2241
	Savannah	30	0	0	1	56	204	403	472	350	202	72	6	0	1766
HI	Hilo	30	0	0	0	0	0	0	0	0	0	0	0	0	0
	Honolulu	30	0	0	0	0	0	0	0	0	0	0	0	0	0
	Kahului	30	0	0	0	0	0	0	0	0	0	0	0	0	0
	Lihue	30	0	0	0	0	0	0	0	0	0	0	0	0	0
ID	Boise	30	12	16	126	408	769	1088	1102	819	675	460	254	80	5809
	Lewiston	30	10	10	102	401	717	951	962	742	616	411	218	71	5211
	Pocatello	30	21	26	201	536	907	1240	1274	1003	842	584	353	129	7116
IL	Chicago	30	5	9	112	401	759	1147	1333	1075	858	513	232	49	6493
	Moline	30	3	8	108	394	782	1191	1374	1090	831	450	172	19	6422
	Peoria	30	2	7	94	368	738	1136	1316	1045	788	423	159	19	6095
	Rockford	30	5	14	136	444	832	1244	1430	1150	912	522	215	36	6940
	Springfield	30	1	4	77	319	674	1060	1239	980	726	376	126	14	5596

(Continued)

Table 3A.6 (Continued)

State	Station	Years	Jul	Aug	Sep	Oct	Nov	Dec	Jan	Feb	Mar	Apr	May	Jun	Annual
IN	Evansville	30	0	1	45	262	565	891	1047	825	591	295	85	5	4612
	Fort Wayne	30	3	11	105	394	722	1094	1275	1063	835	479	188	29	6198
	Indianapolis	30	2	4	77	335	659	1020	1192	957	724	394	141	16	5521
	South Bend	30	6	13	114	392	721	1090	1270	1055	844	498	213	41	6257
IA	Des Moines	30	1	6	103	386	804	1223	1385	1090	826	439	153	16	6432
	Dubuque	30	8	21	158	465	879	1307	1492	1192	949	536	226	40	7273
	Sioux City	30	3	10	128	434	888	1308	1439	1131	885	473	172	25	6896
	Waterloo	30	7	20	155	478	903	1343	1538	1221	948	528	205	29	7375
KS	Concordia	30	1	2	76	307	722	1068	1195	927	702	380	131	13	5524
	Dodge City	30	1	2	65	273	674	978	1087	826	647	351	121	12	5037
	Goodland	30	5	10	117	407	807	1079	1147	916	776	490	224	35	6013
	Topeka	30	1	1	73	287	665	1030	1174	898	647	336	106	7	5225
	Wichita	30	0	0	49	235	620	965	1087	819	594	302	89	5	4765
KY	Greater Cincinnati AP	30	1	3	68	326	626	953	1110	899	684	373	138	19	5200
	Jackson	30	0	4	44	263	522	830	966	761	557	273	128	10	4358
	Lexington	30	1	2	53	284	574	877	1026	819	616	332	119	13	4716
	Louisville	30	0	1	36	240	527	838	992	779	569	280	84	6	4352
	Paducah	30	0	0	38	229	516	833	978	750	529	250	67	2	4192
LA	Baton Rouge	30	0	0	2	49	211	386	456	325	179	55	2	0	1665
	Lake Charles	30	0	0	1	38	191	363	434	304	163	47	1	0	1542
	New Orleans	30	0	0	0	30	169	332	403	288	150	44	1	0	1417
	Shreveport	30	0	0	6	78	296	522	597	416	250	89	8	0	2262
ME	Caribou	30	58	103	344	691	1039	1505	1719	1466	1254	805	417	159	9560
	Portland	30	19	37	199	523	790	1152	1346	1145	988	649	361	116	7325
MD	Baltimore	30	0	1	41	279	549	839	986	816	647	345	119	12	4634
MA	Blue Hill	30	9	22	138	422	698	1040	1207	1034	894	562	271	74	6371
	Boston	30	4	8	84	344	604	932	1104	951	815	503	233	48	5630
	Worcester	30	9	20	158	478	764	1119	1284	1094	952	601	278	74	6831
MI	Alpena	30	46	78	260	583	894	1244	1447	1285	1133	723	394	150	8237
	Detroit	30	5	12	121	429	742	1099	1270	1084	894	527	221	45	6449
	Flint	30	13	28	163	474	781	1149	1329	1147	957	577	267	66	6951
	Grand Rapids	30	10	24	159	471	793	1147	1317	1135	956	571	255	58	6896
	Houghton Lake	30	41	74	250	577	901	1271	1468	1283	1115	685	338	117	8120
	Lansing	30	17	35	179	493	805	1167	1341	1160	970	585	277	69	7098
	Marquette	30	92	134	348	700	1083	1484	1659	1405	1280	859	468	200	9712
	Muskegon	30	15	27	168	476	784	1117	1288	1124	968	602	296	78	6943
	Sault Ste. Marie	30	91	107	320	642	979	1383	1606	1405	1253	798	434	212	9230
MN	Duluth	30	69	106	331	682	1124	1587	1772	1435	1248	787	421	180	9742
	International Falls	30	55	102	360	723	1217	1744	1946	1551	1304	775	378	140	10295
	Minneapolis-St. Paul	30	7	20	178	516	978	1428	1616	1279	1034	560	222	44	7882
	Rochester	30	23	50	208	558	1014	1479	1650	1305	1066	609	281	65	8308
	Saint Cloud	30	19	49	253	604	1077	1551	1742	1381	1135	637	285	79	8812
MS	Jackson	30	0	0	7	100	305	516	607	440	272	110	11	0	2368
	Meridian	30	0	0	6	106	303	506	598	435	274	111	14	0	2353

	Tupelo	30	0	0	14	150	400	639	750	559	368	160	29	1	3070
MO	Columbia	30	1	2	72	291	642	1004	1153	891	656	336	115	10	5173
	Kansas City	30	0	7	58	269	668	1047	1182	897	658	331	124	8	5249
	St. Louis	30	0	1	46	246	583	943	1097	845	613	294	83	6	4757
	Springfield	30	1	1	62	248	578	899	1034	790	581	300	100	8	4602
MT	Billings	30	20	25	205	516	909	1195	1280	1001	876	575	312	90	7004
	Glasgow	30	22	38	253	609	1088	1506	1671	1290	1055	610	308	91	8541
	Great Falls	30	48	64	275	579	952	1237	1323	1063	948	639	389	158	7675
	Havre	30	33	49	270	622	1067	1410	1546	1201	999	613	324	113	8247
	Helena	30	42	56	283	631	1018	1348	1397	1093	932	634	384	157	7975
	Kalispell	30	85	94	341	705	1014	1297	1359	1079	933	640	411	213	8171
	Missoula	30	55	62	276	637	985	1287	1291	1019	852	596	384	178	7622
NE	Grand Island	30	3	7	114	401	835	1192	1310	1031	819	452	175	23	6362
	Lincoln	30	1	5	100	377	806	1188	1328	1043	799	425	154	16	6242
	Norfolk	30	4	11	130	432	879	1266	1388	1099	872	478	180	28	6767
	North Platte	30	6	14	158	481	902	1222	1316	1026	853	519	240	46	6783
	Omaha Eppley AP	30	1	6	105	384	806	1211	1349	1053	805	424	151	17	6312
	Omaha (North)	30	4	15	83	349	800	1204	1323	1022	783	400	154	16	6153
	Scottsbluff	30	7	11	157	489	885	1183	1233	969	837	544	257	53	6625
	Valentine	30	11	19	175	516	952	1285	1386	1101	932	571	260	57	7265
NV	Elko	30	53	57	237	569	916	1208	1222	943	820	612	383	161	7181
	Ely	30	26	48	258	605	938	1220	1240	996	903	690	459	178	7561
	Las Vegas	30	0	0	1	57	320	581	583	380	247	90	16	1	2276
	Reno	30	12	22	130	416	732	987	984	757	683	502	285	91	5601
	Winnemucca	30	12	23	173	509	829	1101	1088	820	742	554	315	105	6271
NH	Concord	30	22	44	212	548	835	1220	1402	1188	999	623	302	90	7485
	Mt. Washington	30	504	542	740	1079	1333	1702	1857	1639	1594	1262	914	619	13785
NJ	Atlantic City AP	30	1	6	69	323	573	868	1019	873	725	437	187	32	5113
	Atlantic City CO	30	0	0	20	228	481	772	924	787	674	409	167	18	4480
	Newark	30	1	2	41	264	541	863	1030	869	697	371	120	13	4812
NM	Albuquerque	30	0	0	29	248	614	898	914	670	525	294	85	4	4281
	Clayton	30	1	7	85	325	685	937	964	767	661	407	179	27	5045
	Roswell	30	0	0	18	144	485	755	775	542	378	182	51	2	3332
NY	Albany	30	10	26	168	484	772	1142	1330	1136	938	553	240	62	6861
	Binghamton	30	22	43	200	514	812	1160	1331	1156	997	617	292	90	7234
	Buffalo	30	8	21	149	442	737	1081	1256	1111	961	594	268	65	6693
	Islip	30	0	1	49	339	604	909	1060	913	782	479	197	24	5357
	New York C.Park	30	1	2	43	261	524	841	1009	853	695	372	127	16	4744
	New York (JFK AP)	30	1	2	42	264	532	838	1007	868	723	420	166	19	4882
	New York (Laguardia AP)	30	1	1	40	249	524	836	1008	861	713	392	136	16	4777
	Rochester	30	10	24	154	447	741	1085	1263	1117	958	582	266	66	6713
	Syracuse	30	10	25	158	460	748	1108	1294	1131	959	572	254	66	6785
NC	Asheville	30	1	2	52	285	531	769	872	708	550	302	116	15	4203
	Cape Hatteras	30	0	0	2	72	244	464	587	518	400	187	44	3	2521
	Charlotte	30	0	0	16	165	404	655	747	585	409	180	44	3	3208

(Continued)

Table 3A.6 (Continued)

State	Station	Years	Jul	Aug	Sep	Oct	Nov	Dec	Jan	Feb	Mar	Apr	May	Jun	Annual
	Greensboro-Wnstn-Slm-HPT	30	0	1	32	232	480	742	851	679	501	245	77	8	3848
	Raleigh	30	0	1	20	194	425	679	783	627	456	214	61	5	3465
	Wilmington	30	0	0	3	95	277	497	589	474	331	134	28	1	2429
ND	Bismarck	30	19	44	256	625	1112	1539	1711	1335	1110	660	305	93	8809
	Fargo	30	17	37	245	614	1137	1610	1808	1446	1185	652	271	73	9095
	Grand Forks	30	27	53	276	655	1186	1660	1860	1484	1233	689	294	88	9505
	Williston	30	22	43	274	645	1146	1567	1734	1336	1104	646	315	90	8922
OH	Akron	30	7	16	120	412	704	1040	1220	1026	836	498	219	50	6148
	Cleveland	30	7	13	110	385	677	1023	1205	1025	847	516	235	54	6097
	Columbus	30	3	7	81	354	656	982	1154	954	742	421	165	27	5546
	Dayton	30	2	7	90	358	669	1016	1185	973	760	427	167	24	5678
	Mansfield	30	8	19	122	407	714	1066	1236	1045	852	509	227	53	6258
	Toledo	30	6	18	129	431	745	1107	1281	1087	878	517	224	45	6468
	Youngstown	30	15	26	148	439	723	1063	1243	1057	879	530	252	71	6446
OK	Oklahoma City	30	0	0	30	152	482	780	884	648	446	197	43	1	3663
	Tulsa	30	0	0	29	152	468	781	898	658	437	179	38	1	3641
OR	Astoria	30	151	130	197	386	542	688	695	583	585	492	375	244	5068
	Burns	30	89	96	313	638	968	1245	1259	982	869	661	439	226	7785
	Eugene	30	40	31	115	370	594	780	769	615	564	443	308	152	4781
	Medford	30	10	7	69	316	632	837	804	610	550	402	233	69	4539
	Pendleton	30	14	15	115	400	711	962	971	747	623	433	247	83	5321
	Portland	30	21	15	78	318	558	756	765	605	529	393	234	94	4366
	Salem	30	39	34	116	376	592	771	765	623	574	452	301	141	4784
	Sexton Summit	30	127	101	211	440	737	860	853	741	788	648	470	267	6243
PC	Guam	30	0	0	0	0	0	0	0	0	0	0	0	0	0
	Johnston Island	30	0	0	0	0	0	0	0	0	0	0	0	0	0
	Koror	30	0	0	0	0	0	0	0	0	0	0	0	0	0
	Kwajalein, Marshall IS	30	0	0	0	0	0	0	0	0	0	0	0	0	0
	Majuro, Marshall IS	30	0	0	0	0	0	0	0	0	0	0	0	0	0
	Pago Pago, Amer Samoa	30	0	0	0	0	0	0	0	0	0	0	0	0	0
	Pohnpei, Caroline IS	30	0	0	0	0	0	0	0	0	0	0	0	0	0
	Chuuk, E. Caroline IS	30	0	0	0	0	0	0	0	0	0	0	0	0	0
	Wake Island	30	0	0	0	0	0	0	0	0	0	0	0	0	0
	Yap, W Caroline IS	30	0	0	0	0	0	0	0	0	0	0	0	0	0
PA	Allentown	30	3	8	95	374	657	985	1147	966	784	450	176	26	5671
	Erie	30	13	18	123	398	684	1016	1209	1074	926	585	288	75	6409
	Harrisburg	30	0	1	52	338	621	937	1076	901	723	390	148	14	5201
	Middletown/ Harrisburg AP	30	0	1	52	338	621	937	1076	901	723	390	148	14	5201
	Philadelphia	30	1	2	39	269	545	857	1020	858	681	362	113	12	4759
	Pittsburgh	30	6	13	105	397	677	996	1163	979	788	462	200	43	5829
	Avoca	30	9	18	138	431	711	1047	1214	1040	866	512	219	53	6258

	Williamsport	30	5	12	116	417	708	1033	1201	1014	824	471	196	38	6035
RI	Providence	30	3	9	101	377	637	961	1125	965	817	494	221	44	5754
SC	Charleston AP	30	0	0	2	68	222	428	510	394	242	95	11	1	1973
	Charleston CO	30	0	0	0	47	183	390	489	362	224	57	3	0	1755
	Columbia	30	0	0	8	121	325	552	628	485	321	131	23	1	2595
	Greenvile-Spartanbrg AP	30	0	0	19	178	417	655	750	586	420	197	47	3	3272
SD	Aberdeen	30	11	27	206	569	1066	1506	1678	1318	1072	591	251	59	8354
	Huron	30	8	21	180	530	996	1417	1572	1242	1004	567	242	49	7828
	Rapid City	30	16	21	190	521	934	1233	1314	1061	925	595	313	88	7211
	Sioux Falls	30	7	20	175	519	986	1417	1563	1236	988	558	231	46	7746
TN	Bristol-JhnCty-Kgsprt	30	1	2	44	279	541	810	919	744	556	303	108	11	4318
	Chattanooga	30	0	0	16	180	442	697	797	618	432	195	48	2	3427
	Knoxville	30	0	0	22	210	470	732	841	652	467	223	65	3	3685
	Memphis	30	0	0	13	121	381	651	770	565	366	144	22	0	3033
	Nashville	30	0	0	24	189	457	730	858	664	462	217	56	1	3658
	Oak Ridge	30	0	0	30	230	518	787	882	696	510	254	80	6	3993
	Abilene	30	0	0	18	93	353	608	678	477	299	113	20	0	2659
TX	Amarillo	30	1	1	56	239	594	874	920	706	542	291	94	7	4325
	Austin/City	30	0	0	2	32	205	406	475	319	163	44	2	0	1648
	Austin/Bergstrom	30	0	0	1	38	248	480	532	360	180	45	5	0	1889
	Brownsville	30	0	0	0	6	64	170	206	123	45	8	0	0	622
	Corpus Christi	30	0	0	0	12	103	246	299	191	77	19	0	0	947
	Dallas-Fort Worth	30	0	0	2	52	312	571	650	448	248	74	13	0	2370
	Dallas-Love Field	30	0	0	7	62	281	527	605	415	238	75	9	0	2219
	Del Rio	30	0	0	2	24	183	384	423	262	111	27	1	0	1417
	El Paso	30	0	0	9	96	388	626	641	435	285	113	11	0	2604
	Galveston	30	0	0	0	6	112	245	316	220	94	15	0	0	1008
	Houston	30	0	0	1	37	189	367	427	298	156	48	2	0	1525
	Lubbock	30	0	0	33	158	472	744	800	588	409	178	38	2	3422
	Midland-Odessa	30	0	0	19	103	380	622	680	472	302	120	18	0	2716
	Port Arthur	30	0	0	1	34	181	349	411	286	143	41	1	0	1447
	San Angelo	30	0	0	12	82	325	558	617	427	258	92	13	0	2384
	San Antonio	30	0	0	2	33	197	390	455	303	149	42	1	0	1572
	Victoria	30	0	0	1	22	145	314	372	249	113	28	1	0	1245
	Waco	30	0	0	6	58	271	512	589	409	235	77	7	0	2164
	Wichita Falls	30	0	0	18	106	395	676	762	550	354	140	23	0	3024
UT	Milford	30	1	4	112	451	821	1129	1144	879	726	503	280	66	6116
	Salt Lake City	30	2	3	87	370	737	1067	1108	857	665	448	215	48	5607
VT	Burlington	30	17	38	203	538	834	1240	1457	1273	1063	642	283	77	7665
VA	Lynchburg	30	1	1	43	268	526	798	918	749	572	288	102	13	4279
	Norfolk	30	0	0	8	150	368	623	758	638	487	240	66	4	3342
	Richmond	30	0	1	27	224	464	736	863	705	526	250	75	7	3878
	Roanoke	30	1	2	48	276	528	798	911	745	569	290	107	13	4288
WA	Olympia	30	90	78	195	465	666	829	819	678	644	511	347	201	5523
	Quillayute	30	193	174	250	460	626	762	758	646	656	548	421	296	5790

(Continued)

Table 3A.6 (Continued)

State	Station	Years	Jul	Aug	Sep	Oct	Nov	Dec	Jan	Feb	Mar	Apr	May	Jun	Annual
	Seattle CO	30	52	50	139	362	571	735	729	593	564	423	266	131	4615
	Seattle Sea-Tac AP	30	55	45	138	383	592	754	747	613	582	447	291	150	4797
	Spokane	30	44	42	196	554	897	1168	1169	916	790	557	338	149	6820
	Walla Walla	30	9	9	99	323	659	928	939	709	574	373	191	69	4882
	Yakima	30	28	28	155	487	813	1100	1090	821	671	463	258	98	6012
PR	San Juan	30	0	0	0	0	0	0	0	0	0	0	0	0	0
WV	Beckley	30	10	16	115	388	647	926	1068	882	714	414	199	48	5427
	Charleston	30	1	3	56	300	558	837	977	794	604	319	122	18	4589
	Elkins	30	15	23	134	456	719	996	1133	959	792	498	251	70	6046
	Huntington	30	1	2	55	292	557	843	991	796	596	308	116	14	4571
WI	Green Bay	30	19	38	208	540	925	1350	1537	1270	1065	638	301	85	7976
	La Crosse	30	6	17	152	467	893	1347	1545	1225	975	526	204	38	7395
	Madison	30	12	33	183	504	892	1298	1490	1209	978	576	261	63	7499
	Milwaukee	30	13	18	134	443	808	1200	1384	1132	949	611	318	86	7096
WY	Casper	30	15	24	229	581	961	1252	1309	1073	921	661	393	115	7534
	Cheyenne	30	28	39	238	574	914	1145	1187	1000	928	686	414	136	7289
	Lander	30	16	24	220	580	1027	1355	1397	1122	921	643	373	116	7794
	Sheridan	30	27	33	238	581	979	1293	1351	1078	925	628	374	129	7636

Note: Based on 30-year average values 1971–2000. Degree-day data are used to estimate amounts of energy required to maintain comfortable indoor temperature levels. Each degree that a day's mean temperature is below 65°F is counted as one heating degree day.

Source: From U.S. National Oceanic and Atmospheric Administration, *Comparative Climatic Data for the United States Through 2000*, www.noaa.gov.

Table 3A.7 Normal Monthly Cooling Degree Days — Selected Cities of the United States

State	Station	Years	Jan	Feb	Mar	Apr	May	Jun	Jul	Aug	Sep	Oct	Nov	Dec	Annual
AL	Birmingham AP	30	1	3	16	51	167	351	476	455	280	69	9	3	1881
	Huntsville	30	0	1	8	40	142	326	446	417	238	47	5	1	1671
	Mobile	30	9	11	45	108	282	450	529	520	384	151	41	18	2548
	Montgomery	30	2	4	24	73	225	415	519	502	350	106	23	9	2252
AK	Anchorage	30	0	0	0	0	0	0	3	0	0	0	0	0	3
	Annette	30	0	0	0	0	0	0	5	8	0	0	0	0	13
	Barrow	30	0	0	0	0	0	0	0	0	0	0	0	0	0
	Bethel	30	0	0	0	0	0	0	0	1	0	0	0	0	1
	Bettles	30	0	0	0	0	0	11	20	7	0	0	0	0	38
	Big Delta	30	0	0	0	0	0	1	12	12	2	0	0	0	27
	Cold Bay	30	0	0	0	0	0	0	0	0	0	0	0	0	0
	Fairbanks	30	0	0	0	0	0	20	42	11	1	0	0	0	74
	Gulkana	30	0	0	0	0	0	0	0	0	0	0	0	0	0
	Homer	30	0	0	0	0	0	0	0	0	0	0	0	0	0
	Juneau	30	0	0	0	0	0	0	0	0	0	0	0	0	0
	King Salmon	30	0	0	0	0	0	0	0	0	0	0	0	0	0
	Kodiak	30	0	0	0	0	0	0	0	0	0	0	0	0	0
	Kotzebue	30	0	0	0	0	0	0	8	4	0	0	0	0	12
	Mcgrath	30	0	0	0	0	0	4	11	6	0	0	0	0	21
	Nome	30	0	0	0	0	0	0	2	0	0	0	0	0	2
	St. Paul Island	30	0	0	0	0	0	0	0	0	0	0	0	0	0
	Talkeetna	30	0	0	0	0	0	1	4	0	0	0	0	0	5
	Unalakleet	30	0	0	0	0	0	0	1	2	0	0	0	0	3
	Valdez	30	0	0	0	0	0	0	0	0	0	0	0	0	0
	Yakutat	30	0	0	0	0	0	0	0	0	0	0	0	0	0
AZ	Flagstaff	30	0	0	0	0	0	23	64	36	3	0	0	0	126
	Phoenix	30	2	15	71	223	473	744	900	859	664	350	53	1	4355
	Tucson	30	0	3	25	107	300	577	672	625	477	211	20	0	3017
	Winslow	30	0	0	0	3	39	221	384	327	124	6	0	0	1104
	Yuma	30	12	29	118	251	466	714	900	882	694	381	82	11	4540
AR	Fort Smith	30	0	0	8	43	155	364	520	491	275	59	5	1	1921
	Little Rock	30	1	1	14	52	188	408	542	502	296	72	8	2	2086
	North Little Rock	30	0	6	10	66	212	412	564	528	307	83	8	0	2196
CA	Bakersfield	30	0	1	7	56	205	392	580	541	374	144	4	0	2304
	Bishop	30	0	0	0	3	45	191	357	293	106	8	0	0	1003
	Blue Canyon	30	0	0	0	1	15	41	129	125	84	28	0	0	423
	Eureka	30	0	0	0	0	0	0	0	2	5	0	0	0	7
	Fresno	30	0	0	3	40	170	351	530	483	316	97	1	0	1991
	Long Beach	30	3	5	10	28	55	135	260	302	244	119	20	2	1183
	Los Angeles AP	30	6	7	6	15	19	58	135	175	154	81	22	4	682
	Los Angeles CO	30	15	23	26	58	84	178	295	325	281	164	44	13	1506
	Mount Shasta	30	0	0	0	0	7	31	100	66	29	2	0	0	235
	Redding	30	0	0	3	22	133	310	504	430	263	74	2	0	1741

(Continued)

Table 3A.7 (Continued)

State	Station	Years	Jan	Feb	Mar	Apr	May	Jun	Jul	Aug	Sep	Oct	Nov	Dec	Annual
	Sacramento	30	0	0	6	24	110	204	320	303	210	66	5	0	1248
	San Diego	30	2	4	5	17	32	81	183	230	199	97	15	1	866
	San Francisco AP	30	0	0	0	4	10	21	23	26	38	20	0	0	142
	San Francisco CO	30	0	2	4	3	9	14	19	26	56	22	5	0	160
	Santa Barbara	30	0	0	5	8	23	50	84	133	125	45	9	0	482
	Santa Maria	30	0	1	1	4	5	10	23	25	33	18	3	0	123
	Stockton	30	0	0	0	18	111	254	390	363	247	73	0	0	1456
CO	Alamosa	30	0	0	0	0	0	7	27	10	0	0	0	0	44
	Colorado Springs	30	0	0	0	1	5	86	184	131	35	1	0	0	443
	Denver	30	0	0	0	2	23	135	261	217	57	0	0	0	695
	Grand Junction	30	0	0	0	3	45	247	414	350	119	4	0	0	1182
	Pueblo	30	0	0	0	2	31	191	343	276	91	2	0	0	936
CT	Bridgeport	30	0	0	0	1	21	125	286	258	91	7	0	0	789
	Hartford	30	0	0	1	5	38	144	277	220	68	5	1	0	759
DE	Wilmington	30	0	0	2	9	62	215	368	317	135	16	1	0	1125
DC	Washington Dulles AP	30	0	0	4	11	60	203	345	302	132	15	3	0	1075
	Washington Nat'l AP	30	0	0	4	21	108	307	464	410	210	32	4	0	1560
FL	Apalachicola	30	25	14	35	95	284	448	522	517	423	193	63	23	2642
	Daytona Beach	30	36	40	86	150	306	441	513	502	436	277	122	52	2961
	Fort Myers	30	97	99	174	260	423	516	564	568	518	394	222	122	3957
	Gainesville	30	17	17	70	131	308	439	507	498	408	207	74	26	2702
	Jacksonville	30	15	21	58	116	277	437	535	509	400	183	64	21	2636
	Key West	30	189	183	277	361	485	552	604	600	550	473	340	233	4847
	Miami	30	155	154	236	315	442	510	568	568	517	433	291	194	4383
	Orlando	30	91	60	129	201	373	485	539	543	484	319	154	79	3457
	Pensacola	30	11	11	42	107	296	466	541	529	412	174	48	19	2656
	Tallahassee	30	9	12	44	96	273	440	515	509	400	162	49	16	2525
	Tampa	30	56	59	124	204	393	501	550	549	489	323	157	76	3481
	Vero Beach	30	104	83	132	202	347	462	517	513	471	354	183	92	3460
	West Palm Beach	30	122	121	195	266	408	485	544	549	499	408	255	160	4012
GA	Athens	30	0	1	8	46	164	351	471	430	257	54	7	1	1790
	Atlanta	30	0	1	11	52	170	354	463	430	262	58	8	1	1810
	Augusta	30	1	2	15	52	191	385	511	468	296	77	15	3	2016
	Columbus	30	1	4	25	77	234	429	533	511	349	107	21	6	2297
	Macon	30	1	3	23	73	227	418	528	495	325	95	19	6	2213
	Savannah	30	6	11	39	101	267	435	547	506	367	141	40	11	2471
HI	Hilo	30	198	180	215	223	268	303	336	350	336	328	268	223	3228
	Honolulu	30	249	225	288	319	379	436	490	521	497	472	382	303	4561

	Kahului	30	210	195	250	277	330	379	429	448	422	407	329	260	3936
	Lihue	30	207	188	239	266	323	382	433	455	435	408	327	257	3920
ID	Boise	30	0	0	0	3	30	116	281	260	75	4	0	0	769
	Lewiston	30	0	0	0	3	29	105	283	285	84	3	0	0	792
	Pocatello	30	0	0	0	0	3	51	167	143	23	0	0	0	387
IL	Chicago	30	0	0	1	9	48	159	283	234	91	10	0	0	835
	Moline	30	0	0	1	12	62	205	322	254	101	12	0	0	969
	Peoria	30	0	0	1	11	64	210	325	263	112	12	0	0	998
	Rockford	30	0	0	1	7	49	162	263	205	74	7	0	0	768
	Springfield	30	0	0	2	17	84	249	358	291	140	23	1	0	1165
IN	Evansville	30	0	0	4	23	108	304	425	356	173	27	2	0	1422
	Fort Wayne	30	0	0	1	7	53	183	278	212	88	8	0	0	830
	Indianapolis	30	0	0	2	10	69	221	331	272	122	14	1	0	1042
	South Bend	30	0	0	1	10	53	172	268	214	86	9	0	0	813
IA	Des Moines	30	0	0	1	12	60	219	353	285	110	12	0	0	1052
	Dubuque	30	0	0	0	6	37	138	233	175	61	6	0	0	656
	Sioux City	30	0	0	0	11	53	198	311	246	87	8	0	0	914
	Waterloo	30	0	0	0	7	47	168	261	198	70	7	0	0	758
KS	Concordia	30	0	0	2	15	63	265	436	366	163	22	1	0	1333
	Dodge City	30	0	0	2	18	79	291	462	407	193	28	1	0	1481
	Goodland	30	0	0	0	4	26	173	320	266	99	6	0	0	894
	Topeka	30	0	0	3	22	85	278	419	357	166	26	1	0	1357
	Wichita	30	0	0	2	19	93	330	503	454	221	35	1	0	1658
KY	Greater Cincinnati AP	30	0	0	3	13	71	209	334	280	126	16	1	0	1053
	Jackson	30	0	0	0	11	100	201	310	277	130	29	1	0	1059
	Lexington	30	0	0	3	16	80	228	350	307	147	21	2	0	1154
	Louisville	30	0	0	6	24	109	287	421	374	189	29	3	1	1443
	Paducah	30	0	0	6	33	122	320	444	377	191	37	3	0	1533
LA	Baton Rouge	30	12	15	55	128	303	462	538	523	389	157	48	22	2652
	Lake Charles	30	9	13	49	126	312	467	544	534	399	178	54	20	2705
	New Orleans	30	15	19	62	136	320	466	538	534	413	182	62	29	2776
	Shreveport	30	5	7	31	87	242	436	553	532	353	119	24	7	2396
ME	Caribou	30	0	0	0	0	7	39	80	56	9	0	0	0	191
	Portland	30	0	0	0	0	7	51	144	120	24	1	0	0	347
MD	Baltimore	30	0	0	4	13	71	236	390	332	153	19	2	0	1220
MA	Blue Hill	30	0	0	1	3	21	93	215	173	49	3	0	0	558
	Boston	30	0	0	1	4	32	139	282	235	76	7	1	0	777
	Worcester	30	0	0	0	0	7	64	166	122	12	0	0	0	371
MI	Alpena	30	0	0	0	3	13	54	117	83	22	1	0	0	293
	Detroit	30	0	0	0	6	41	144	252	204	75	5	0	0	727
	Flint	30	0	0	1	5	33	110	199	151	52	4	0	0	555
	Grand Rapids	30	0	0	1	6	38	124	218	165	56	5	0	0	613
	Houghton Lake	30	0	0	0	3	20	64	127	89	24	1	0	0	328
	Lansing	30	0	0	1	6	34	113	195	151	53	5	0	0	558
	Marquette	30	0	0	0	0	12	30	72	50	3	0	0	0	167

(Continued)

Table 3A.7 (Continued)

State	Station	Years	Jan	Feb	Mar	Apr	May	Jun	Jul	Aug	Sep	Oct	Nov	Dec	Annual
	Muskegon	30	0	0	0	4	24	86	181	145	44	3	0	0	487
	Sault Ste. Marie	30	0	0	0	1	6	20	56	50	12	0	0	0	145
MN	Duluth	30	0	0	0	0	7	28	82	60	12	0	0	0	189
	International Falls	30	0	0	0	1	17	47	91	67	10	0	0	0	233
	Minneapolis-St. Paul	30	0	0	0	4	41	146	259	190	56	3	0	0	699
	Rochester	30	0	0	0	1	30	99	181	135	26	1	0	0	473
	Saint Cloud	30	0	0	0	2	26	90	172	121	31	1	0	0	443
MS	Jackson	30	5	8	32	83	232	424	525	505	340	101	25	10	2290
	Meridian	30	4	6	26	70	213	400	509	495	331	91	20	8	2173
	Tupelo	30	0	1	14	50	171	364	488	453	274	60	8	2	1885
MO	Columbia	30	0	0	3	18	70	245	396	341	152	20	1	0	1246
	Kansas City	30	0	0	0	12	101	264	418	367	151	12	0	0	1325
	St. Louis	30	0	0	7	32	114	316	461	396	196	36	3	0	1561
	Springfield	30	0	0	3	20	83	258	415	379	179	27	1	0	1365
MT	Billings	30	0	0	0	2	13	90	227	204	44	3	0	0	583
	Glasgow	30	0	0	0	1	17	80	185	182	28	1	0	0	494
	Great Falls	30	0	0	0	1	7	47	126	121	22	2	0	0	326
	Havre	30	0	0	0	1	10	59	146	141	19	1	0	0	377
	Helena	30	0	0	0	0	3	39	122	100	13	0	0	0	277
	Kalispell	30	0	0	0	0	3	17	62	63	4	0	0	0	149
	Missoula	30	0	0	0	0	3	33	111	99	10	0	0	0	256
NE	Grand Island	30	0	0	1	11	48	218	349	285	107	8	0	0	1027
	Lincoln	30	0	0	1	13	56	244	390	315	123	12	0	0	1154
	Norfolk	30	0	0	0	11	48	202	324	261	93	7	0	0	946
	North Platte	30	0	0	0	4	22	139	279	230	70	2	0	0	746
	Omaha Eppley AP	30	0	0	1	14	60	233	365	296	114	12	0	0	1095
	Omaha (North)	30	0	0	0	8	71	209	330	295	101	5	0	0	1019
	Scottsbluff	30	0	0	0	2	19	138	279	225	63	1	0	0	727
	Valentine	30	0	0	0	5	27	141	286	242	75	3	0	0	779
NV	Elko	30	0	0	0	0	2	62	181	135	31	1	0	0	412
	Ely	30	0	0	0	0	0	22	98	69	7	0	0	0	196
	Las Vegas	30	0	1	17	93	310	597	792	734	470	150	4	0	3168
	Reno	30	0	0	0	0	11	72	204	164	41	1	0	0	493
	Winnemucca	30	0	0	0	0	11	81	232	174	28	0	0	0	526
NH	Concord	30	0	0	0	2	18	82	173	133	33	1	0	0	442
	Mt. Washington	30	0	0	0	0	0	0	0	0	0	0	0	0	0
NJ	Atlantic City AP	30	0	0	1	5	44	168	322	269	110	15	1	0	935
	Atlantic City CO	30	0	0	0	0	25	147	316	302	136	25	0	0	951
	Newark	30	0	0	2	10	70	240	403	350	146	19	2	0	1242
NM	Albuquerque	30	0	0	0	6	70	297	417	343	148	9	0	0	1290
	Clayton	30	0	0	0	6	37	173	274	228	77	2	0	0	797

	Roswell	30	0	0	1	47	194	391	488	431	228	32	2	0	1814
NY	Albany	30	0	0	1	3	27	102	206	157	46	2	0	0	544
	Binghamton	30	0	0	1	4	23	74	158	115	32	2	0	0	409
	Buffalo	30	0	0	0	4	28	101	203	158	50	4	0	0	548
	Islip	30	0	0	0	0	15	129	296	251	72	7	0	0	770
	New York C. Park	30	0	0	2	10	63	214	379	335	138	17	2	0	1160
	New York (JFK AP)	30	0	0	0	2	31	162	335	306	125	13	1	0	975
	New York (Laguardia AP)	30	0	0	1	6	54	209	377	336	141	17	1	0	1142
	Rochester	30	0	0	1	5	32	109	210	162	54	4	0	0	577
	Syracuse	30	0	0	1	4	29	105	203	158	48	3	0	0	551
NC	Asheville	30	0	0	0	6	47	165	278	243	104	8	0	0	851
	Cape Hatteras	30	1	1	5	29	122	297	440	422	297	96	24	4	1738
	Charlotte	30	0	1	7	40	142	323	451	405	226	43	5	1	1644
	Greensboro-Wnstn-Slm-HPT	30	0	0	4	25	97	263	398	345	172	24	3	1	1332
	Raleigh	30	0	1	9	38	119	293	429	379	206	39	6	2	1521
	Wilmington	30	3	4	17	65	187	361	501	455	304	90	25	5	2017
ND	Bismarck	30	0	0	0	2	18	80	180	161	30	0	0	0	471
	Fargo	30	0	0	0	3	33	104	191	162	38	2	0	0	533
	Grand Forks	30	0	0	0	2	30	85	148	127	27	1	0	0	420
	Williston	30	0	0	0	1	20	82	177	166	24	1	0	0	471
OH	Akron	30	0	0	1	7	41	136	232	189	69	4	0	0	679
	Cleveland	30	0	0	2	7	40	140	239	195	80	8	1	0	712
	Columbus	30	0	0	2	9	61	188	296	251	106	11	1	0	925
	Dayton	30	0	0	2	9	62	194	305	246	105	11	1	0	935
	Mansfield	30	0	0	1	7	39	136	228	181	73	7	0	0	672
	Toledo	30	0	0	1	7	42	148	248	190	73	6	0	0	715
	Youngstown	30	0	0	1	8	33	112	190	154	57	5	1	0	561
OK	Oklahoma City	30	0	1	7	38	145	360	527	497	271	58	3	0	1907
	Tulsa	30	0	1	10	50	163	385	568	524	277	64	6	1	2049
OR	Astoria	30	0	0	0	0	1	2	4	7	7	1	0	0	22
	Burns	30	0	0	0	0	0	19	117	70	12	0	0	0	218
	Eugene	30	0	0	0	0	5	21	95	93	32	1	0	0	247
	Medford	30	0	0	0	2	24	90	253	240	95	7	0	0	711
	Pendleton	30	0	0	0	2	23	86	243	224	63	3	0	0	644
	Portland	30	0	0	0	1	15	44	138	145	53	2	0	0	398
	Salem	30	0	0	0	0	7	25	95	98	31	1	0	0	257
	Sexton Summit	30	0	0	0	0	0	11	84	76	62	15	0	0	248
PC	Guam	30	390	350	414	435	473	463	458	449	439	449	439	431	5190
	Johnston Island	30	389	354	399	409	457	485	526	545	523	522	448	412	5469
	Koror	30	507	454	518	518	545	503	509	515	509	527	524	526	6155
	Kwajalein, Marshall IS	30	512	470	536	515	538	515	529	534	518	537	513	522	6239

(Continued)

Table 3A.7 (Continued)

State	Station	Years	Jan	Feb	Mar	Apr	May	Jun	Jul	Aug	Sep	Oct	Nov	Dec	Annual
	Majuro, Marshall IS	30	487	451	502	482	504	485	498	507	496	510	490	493	5905
	Pago Pago, Amer Samoa	30	511	469	526	498	493	460	454	457	460	486	485	515	5814
	Pohnpei, Caroline IS	30	490	449	506	485	501	480	486	487	471	486	477	493	5811
	Chuuk, E. Caroline IS	30	510	456	513	501	522	497	506	492	488	508	511	511	6015
	Wake Island	30	396	344	409	427	488	527	567	567	556	546	480	443	5750
	Yap, W Caroline IS	30	467	428	488	493	518	479	484	478	466	486	476	484	5747
PA	Allentown	30	0	0	1	6	45	164	292	235	83	7	0	0	833
	Erie	30	0	0	1	5	30	105	197	166	63	6	0	0	573
	Harrisburg	30	0	0	0	1	54	186	337	279	87	11	0	0	955
	Middletown/ Harrisburg AP	30	0	0	0	1	54	186	337	279	87	11	0	0	955
	Philadelphia	30	0	0	2	10	70	234	395	351	152	19	2	0	1235
	Pittsburgh	30	0	0	2	8	41	143	244	203	78	6	1	0	726
	Avoca	30	0	0	1	5	35	113	220	174	57	4	0	0	609
	Williamsport	30	0	0	0	6	39	135	251	206	68	4	0	0	709
RI	Block IS	30	0	0	0	0	1	51	184	177	48	5	0	0	466
	Providence	30	0	0	0	3	25	122	265	223	71	5	0	0	714
SC	Charleston AP	30	3	7	29	85	242	408	532	494	348	122	35	8	2313
	Charleston CO	30	16	7	27	83	266	431	551	514	377	156	47	11	2486
	Columbia	30	2	4	20	69	206	388	515	467	296	76	15	5	2063
	Greenville- Spartanburg	30	0	0	5	30	127	304	430	384	207	35	3	1	1526
SD	Aberdeen	30	0	0	0	3	29	112	235	196	49	2	0	0	626
	Huron	30	0	0	0	4	29	138	273	228	66	3	0	0	741
	Rapid City	30	0	0	0	2	13	86	227	208	59	3	0	0	598
	Sioux Falls	30	0	0	0	5	37	151	278	217	65	4	0	0	757
TN	Bristol-JhnCty- Kgsprt	30	0	0	1	10	61	200	309	274	128	11	1	0	995
	Chattanooga	30	0	0	5	32	124	312	450	418	229	35	2	1	1608
	Knoxville	30	0	1	5	27	110	282	408	381	205	28	3	0	1450
	Memphis	30	1	2	15	72	210	428	554	504	307	84	11	2	2190
	Nashville	30	0	0	9	37	136	321	455	416	230	46	5	1	1656
	Oak Ridge	30	0	0	2	19	95	254	380	347	180	23	1	0	1301
TX	Abilene	30	0	5	28	94	253	442	568	535	327	118	15	1	2386
	Amarillo	30	0	0	2	18	90	285	405	345	173	26	0	0	1344
	Austin/City	30	7	18	59	147	323	495	605	610	439	207	51	13	2974
	Austin/Bergstrom	30	8	11	23	107	307	453	551	532	377	148	32	8	2557
	Brownsville	30	60	76	180	292	463	551	607	608	497	338	170	82	3924
	Corpus Christi	30	32	43	121	229	402	520	594	598	485	300	123	51	3498

	Dallas-Fort Worth	30	2	11	10	72	265	478	621	601	376	118	15	2	2571
	Dallas-Love Field	30	2	9	39	110	290	511	659	646	417	162	28	5	2878
	Del Rio	30	2	18	86	207	401	545	630	622	453	217	41	4	3226
	El Paso	30	0	1	8	65	238	481	535	473	293	69	2	0	2165
	Galveston	30	30	23	65	163	367	515	596	601	482	286	122	30	3280
	Houston	30	15	21	63	147	328	485	573	563	412	196	65	25	2893
	Lubbock	30	0	0	7	48	180	382	472	413	225	49	1	0	1777
	Midland-Odessa	30	0	2	15	77	254	438	512	473	281	83	4	0	2139
	Port Arthur	30	11	16	55	140	324	480	553	546	417	195	63	23	2823
	San Angelo	30	1	4	30	107	277	446	554	519	322	113	16	1	2390
	San Antonio	30	7	19	68	161	344	505	607	601	439	215	57	15	3038
	Victoria	30	18	26	84	181	368	514	601	597	454	248	83	29	3203
	Waco	30	2	8	39	111	292	497	637	628	416	170	34	6	2840
	Wichita Falls	30	0	2	19	66	220	448	618	574	339	99	10	1	2396
UT	Milford	30	0	0	0	1	18	115	286	239	54	2	0	0	715
	Salt Lake City	30	0	0	0	4	34	184	395	355	111	6	0	0	1089
VT	Burlington	30	0	0	0	3	23	96	192	139	35	1	0	0	489
VA	Lynchburg	30	0	0	3	20	72	218	348	308	141	19	2	0	1131
	Norfolk	30	1	2	8	35	119	303	456	403	240	50	11	2	1630
	Richmond	30	0	1	8	33	107	282	428	374	193	33	6	1	1466
	Roanoke	30	0	0	4	20	74	217	355	309	136	17	2	0	1134
WA	Olympia	30	0	0	0	0	2	10	38	40	7	0	0	0	97
	Quillayute	30	0	0	0	0	1	3	7	8	4	0	0	0	23
	Seattle CO	30	0	0	0	0	4	15	67	79	27	0	0	0	192
	Seattle Sea-TAC AP	30	0	0	0	0	5	19	65	65	19	0	0	0	173
	Spokane	30	0	0	0	1	11	46	155	154	26	1	0	0	394
	Walla Walla	30	0	0	0	4	29	138	329	323	131	3	0	0	957
	Yakima	30	0	0	0	1	18	68	187	163	28	0	0	0	465
PR	San Juan	30	360	332	388	421	484	513	533	539	515	513	436	392	5426
WV	Beckley	30	0	0	1	8	29	99	183	149	56	4	0	0	529
	Charleston	30	0	0	7	25	77	204	324	280	125	18	3	1	1064
	Elkins	30	0	0	0	2	16	76	153	126	41	2	0	0	416
	Huntington	30	0	0	8	25	82	218	341	300	135	19	3	1	1132
WI	Green Bay	30	0	0	0	3	24	95	177	126	36	2	0	0	463
	La Crosse	30	0	0	0	8	49	162	272	208	70	6	0	0	775
	Madison	30	0	0	0	6	33	123	214	154	48	4	0	0	582
	Milwaukee	30	0	0	0	5	27	114	222	180	63	5	0	0	616
WY	Casper	30	0	0	0	0	3	64	186	154	28	0	0	0	435
	Cheyenne	30	0	0	0	0	1	41	126	92	20	0	0	0	280
	Lander	30	0	0	0	0	3	70	190	153	29	0	0	0	445
	Sheridan	30	0	0	0	1	4	52	165	154	30	1	0	0	407

Note: Based on 30-year average values 1971–2000. Degree-day data are used to estimate amounts of energy required to maintain comfortable indoor temperature levels. Each degree that a day's mean temperature is above 65°F is counted as one cooling degree day.

Source: From U.S. National Oceanic and Atmospheric Administration, *Comparative Climatic Data for the United States Through 2000*, www.noaa.gov.

Table 3A.8 Average Wind Speed (MPH) — Selected Cities of the United States

State	Station	Years	Jan	Feb	Mar	Apr	May	Jun	Jul	Aug	Sep	Oct	Nov	Dec	Annual
AL	Birmingham AP	59	8.1	8.7	9.0	8.2	6.8	6.0	5.7	5.4	6.3	6.2	7.2	7.7	7.1
	Huntsville	35	9.0	9.4	9.8	9.2	7.9	6.9	6.0	5.8	6.7	7.3	8.1	9.0	7.9
	Mobile	54	10.1	10.3	10.7	10.1	8.7	7.5	6.9	6.7	7.7	8.0	8.9	9.6	8.8
	Montgomery	58	7.7	8.2	8.3	7.3	6.1	5.8	5.7	5.2	5.9	5.7	6.5	7.1	6.6
AK	Anchorage	49	6.4	6.8	7.1	7.3	8.5	8.4	7.3	6.9	6.7	6.7	6.4	6.3	7.1
	Annette	38	11.6	11.8	10.5	10.6	9.0	8.5	7.7	8.1	8.8	11.2	11.7	12.0	10.1
	Barrow	69	11.9	11.3	11.3	11.5	12.0	11.5	11.7	12.4	13.2	13.3	12.5	11.7	12.0
	Barter IS	33	15.1	14.4	13.7	12.0	12.7	11.6	10.9	11.8	13.2	14.8	14.9	13.9	13.2
	Bethel	44	14.5	14.7	13.7	12.9	11.5	11.0	10.6	11.0	11.6	12.3	13.2	13.6	12.6
	Bettles	27	5.8	6.3	7.0	7.4	7.2	6.8	6.5	6.2	6.4	6.4	5.8	5.6	6.5
	Big Delta	26	11.2	10.2	8.8	8.0	8.2	6.9	6.1	6.6	7.6	8.7	10.2	10.0	8.5
	Cold Bay	47	17.5	17.9	17.4	17.5	16.2	15.8	15.6	16.2	16.2	16.6	17.5	17.5	16.8
	Fairbanks	51	3.0	3.9	5.3	6.6	7.7	7.1	6.6	6.1	6.0	5.3	3.8	3.0	5.4
	Gulkana	14	5.0	4.9	6.3	8.5	8.7	8.1	7.7	7.7	7.3	6.0	4.1	3.3	6.5
	Homer	28	7.8	7.7	7.8	8.1	8.2	7.8	7.1	6.6	7.0	7.3	7.7	7.8	7.6
	Juneau	57	8.1	8.2	8.4	8.5	8.3	7.7	7.5	7.4	8.0	9.5	8.4	8.8	8.2
	King Salmon	47	10.5	11.0	11.3	10.9	11.0	10.5	9.9	10.0	10.4	10.3	10.4	10.1	10.5
	Kodiak	49	12.7	12.5	12.5	11.6	10.6	9.3	7.7	8.4	9.7	11.4	12.5	12.6	11.0
	Kotzebue	56	13.9	13.0	11.9	12.0	10.7	11.9	12.7	13.2	13.2	13.5	14.4	12.9	12.8
	Mcgrath	52	3.2	4.2	5.3	6.5	6.7	6.4	5.9	5.8	5.9	5.4	3.7	3.2	5.2
	Nome	55	10.8	11.0	10.1	10.1	9.9	9.7	9.7	10.4	11.0	10.5	11.5	10.3	10.4
	St. Paul Island	28	19.9	20.0	18.8	17.4	14.9	13.6	12.1	13.7	15.4	17.4	20.0	20.1	16.9
	Talkeetna	19	6.0	5.5	5.5	4.7	4.9	5.1	4.2	3.7	3.7	3.8	5.0	4.9	4.8
	Valdez	22	7.5	7.8	6.7	5.2	5.8	5.9	4.9	4.2	4.3	6.3	7.5	7.0	6.1
	Yakutat	54	7.2	7.3	7.0	7.1	7.5	6.9	6.6	6.3	6.9	7.8	7.2	7.8	7.1
AZ	Flagstaff	35	6.5	6.6	7.1	7.6	7.3	7.0	5.5	5.0	5.6	5.8	6.6	6.6	6.4
	Phoenix	57	5.3	5.8	6.6	6.9	7.0	6.7	7.1	6.6	6.3	5.8	5.3	5.1	6.2
	Tucson	57	7.9	8.1	8.6	8.9	8.8	8.7	8.4	7.9	8.3	8.2	8.1	7.8	8.3
	Winslow	42	7.1	8.5	10.3	11.3	10.7	10.6	9.0	8.4	8.1	7.6	7.3	6.7	8.8
	Yuma	28	7.3	7.4	7.9	8.3	8.3	8.5	9.5	8.9	7.3	6.6	6.9	7.2	7.8
AR	Fort Smith	57	8.2	8.5	9.4	8.9	7.7	6.7	6.3	6.3	6.6	6.8	7.8	8.1	7.6
	Little Rock	60	8.4	8.9	9.6	9.0	7.6	7.1	6.7	6.3	6.6	6.8	8.0	8.1	7.8
CA	Bakersfield	50	5.2	5.8	6.5	7.1	7.9	7.9	7.2	6.8	6.2	5.5	5.1	5.0	6.4
	Blue Canyon	50	7.8	7.7	7.4	6.5	6.5	6.3	5.8	5.9	6.4	6.8	6.6	6.7	6.7
	Eureka	54	6.9	7.2	7.6	8.0	7.9	7.4	6.8	5.8	5.5	5.6	6.0	6.4	6.8
	Fresno	53	5.2	5.7	6.7	7.4	8.1	8.3	7.4	6.8	6.1	5.2	4.7	4.9	6.4
	Long Beach	33	5.2	6.0	6.7	7.4	7.1	7.0	6.8	6.6	6.2	5.6	5.2	5.0	6.2
	Los Angeles AP	54	6.7	7.4	8.1	8.5	8.4	8.0	7.9	7.7	7.3	6.9	6.7	6.6	7.5
	Los Angeles CO	28	6.2	6.4	6.5	6.3	6.0	5.4	5.0	4.9	5.1	5.2	5.8	5.8	5.7
	Mount Shasta	3	5.0	5.2	5.8	6.2	5.4	5.4	4.4	4.2	4.6	4.2	5.2	5.4	5.1
	Redding	16	6.2	7.1	7.3	7.0	7.3	7.5	6.6	6.1	6.0	6.0	5.7	6.4	6.6
	Sacramento	52	7.1	7.3	8.4	8.6	9.0	9.6	8.9	8.4	7.4	6.4	6.0	6.4	7.8
	San Diego	62	6.0	6.6	7.5	7.8	7.9	7.8	7.5	7.4	7.1	6.5	5.9	5.6	7.0
	San Francisco AP	75	7.2	8.6	10.5	12.2	13.4	14.0	13.6	12.8	11.1	9.4	7.5	7.1	10.6
	San Francisco CO	28	6.7	7.5	8.5	9.5	10.4	10.9	11.2	10.5	9.1	7.6	6.3	6.5	8.7

	Santa Barbara	31	5.1	5.9	6.7	7.6	7.0	6.7	6.5	6.1	5.5	5.5	5.2	4.4	6.0
	Santa Maria	22	6.4	7.2	8.1	8.0	8.2	7.8	6.5	6.3	5.9	6.0	6.4	6.2	6.9
	Stockton	42	6.7	7.0	7.7	8.4	9.2	9.2	8.2	7.7	7.1	6.4	5.8	6.4	7.5
C0	Alamosa	11	5.6	6.6	8.4	10.3	9.7	9.0	7.0	6.2	6.5	6.4	5.6	4.9	7.2
	Colorado Springs	54	9.4	10.0	11.1	11.6	11.2	10.4	9.3	8.9	9.4	9.6	9.5	9.4	10.0
	Denver	47	8.6	8.7	9.6	10.0	9.3	8.8	8.3	8.0	7.9	7.8	8.2	8.4	8.6
	Grand Junction	56	5.7	6.7	8.4	9.4	9.6	9.8	9.4	9.1	9.0	7.9	6.8	6.0	8.2
	Pueblo	47	7.8	8.5	9.6	10.3	9.7	9.3	8.7	7.9	7.9	7.4	7.5	7.7	8.5
CT	Bridgeport	30	12.5	12.9	13.0	12.4	11.1	9.9	9.4	9.5	10.5	11.3	12.0	12.1	11.4
	Hartford	48	8.9	9.4	9.9	9.8	8.7	8.0	7.3	7.0	7.3	7.8	8.4	8.7	8.4
DE	Wilmington	54	9.8	10.3	11.0	10.4	9.0	8.3	7.8	7.4	7.8	8.1	9.2	9.3	9.0
DC	Washington Dulles AP	40	8.1	8.6	9.0	8.8	7.4	6.8	6.2	5.8	6.2	6.6	7.6	7.7	7.4
	Washington Nat'l AP	54	10.0	10.3	10.9	10.5	9.3	8.9	8.3	8.1	8.3	8.7	9.4	9.6	9.4
FL	Apalachicola	54	8.3	8.7	8.9	8.5	7.7	7.1	6.4	6.4	7.8	8.0	8.0	8.0	7.8
	Daytona Beach	57	8.8	9.3	9.8	9.4	8.9	8.0	7.3	7.0	8.0	8.9	8.3	8.3	8.5
	Fort Myers	57	8.3	8.9	9.3	8.8	8.0	7.2	6.6	6.7	7.4	8.4	8.1	7.9	8.0
	Gainesville	19	6.9	7.4	7.8	7.2	6.9	6.1	5.6	5.4	5.8	6.3	6.2	6.0	6.5
	Jacksonville	53	8.1	8.7	9.1	8.5	7.9	7.7	7.0	6.7	7.4	7.8	7.6	7.6	7.8
	Key West	49	11.8	12.0	12.1	12.2	10.5	9.6	9.4	9.2	9.6	10.8	12.0	11.8	10.9
	Miami	53	9.5	10.0	10.5	10.5	9.5	8.3	7.9	7.9	8.2	9.2	9.7	9.1	9.2
	Orlando	54	9.0	9.6	9.9	9.4	8.8	8.0	7.3	7.2	7.6	8.6	8.6	8.5	8.5
	Pensacola	38	9.0	9.3	9.8	9.5	8.6	7.6	6.9	6.7	7.6	7.9	8.2	8.8	8.3
	Tallahassee	41	6.7	7.1	7.5	6.8	6.2	5.7	5.0	5.0	5.9	6.3	5.9	6.3	6.2
	Tampa	56	8.6	9.1	9.4	9.2	8.6	7.9	7.1	6.9	7.6	8.3	8.2	8.3	8.3
	Vero Beach	19	8.7	9.0	9.9	9.5	9.1	7.7	6.9	6.5	7.3	8.6	8.6	8.0	8.3
	West Palm Beach	60	10.1	10.5	11.0	10.9	9.9	8.3	7.7	7.7	8.7	10.0	10.4	10.0	9.6
GA	Athens	47	8.3	8.6	8.7	8.3	7.1	6.6	6.3	5.8	6.4	6.6	7.3	8.0	7.3
	Atlanta	64	10.4	10.6	10.9	10.1	8.7	8.1	7.7	7.3	8.0	8.5	9.1	9.8	9.1
	Augusta	51	6.9	7.5	7.9	7.4	6.4	6.0	5.8	5.3	5.4	5.6	5.9	6.5	6.4
	Columbus	44	7.2	7.7	8.0	7.2	6.6	6.1	5.8	5.5	6.4	6.3	6.4	6.9	6.7
	Macon	54	8.0	8.4	8.8	8.3	7.4	7.0	6.7	6.2	6.7	6.5	6.9	7.5	7.4
	Savannah	52	8.2	8.6	9.1	8.6	7.6	7.4	6.9	6.7	7.2	7.3	7.2	7.6	7.7
HI	Hilo	53	7.4	7.7	7.7	7.5	7.4	7.1	6.9	6.8	6.8	6.7	6.8	7.2	7.2
	Honolulu	53	9.4	10.1	11.3	11.6	11.6	12.6	13.1	12.8	11.2	10.5	10.7	10.4	11.3
	Kahului	30	10.8	11.2	12.2	13.4	13.0	14.9	15.5	14.7	13.0	12.2	11.8	11.3	12.8
	Lihue	52	11.1	11.7	12.7	13.5	12.8	13.2	13.7	13.1	11.7	11.8	12.4	12.0	12.5
ID	Boise	63	7.9	8.9	9.9	9.9	9.4	9.0	8.4	8.2	8.2	8.3	8.4	8.1	8.7
	Pocatello	50	10.5	10.5	11.1	11.6	10.5	10.1	9.1	8.9	9.0	9.5	10.1	9.9	10.1
IL	Cairo	22	9.8	9.8	10.6	10.2	8.2	7.4	6.5	6.2	7.0	7.3	9.1	9.3	8.5
	Chicago	44	11.6	11.4	11.8	11.9	10.5	9.3	8.4	8.2	8.9	10.1	11.1	11.0	10.3
	Moline	59	10.7	10.6	11.7	11.8	10.1	8.8	7.5	7.1	8.0	9.2	10.6	10.3	9.7
	Peoria	59	10.9	10.9	11.7	11.6	9.9	8.9	7.8	7.3	8.3	9.3	10.6	10.6	9.8
	Rockford	52	10.5	10.6	11.6	11.8	10.4	9.3	8.1	7.7	8.4	9.5	10.6	10.4	9.9
	Springfield	55	12.2	12.2	13.2	13.0	11.1	9.6	8.3	7.9	8.7	10.1	12.2	12.0	10.9
IN	Evansville	62	9.2	9.3	10.0	9.6	7.9	7.1	6.2	5.8	6.4	6.9	8.7	8.9	8.0

(Continued)

Table 3A.8 (Continued)

State	Station	Years	Jan	Feb	Mar	Apr	May	Jun	Jul	Aug	Sep	Oct	Nov	Dec	Annual
	Fort Wayne	56	11.5	11.0	11.7	11.6	10.0	8.9	8.0	7.3	8.2	9.1	10.9	11.1	9.9
	Indianapolis	54	10.9	10.8	11.6	11.2	9.6	8.5	7.5	7.2	7.9	8.9	10.5	10.5	9.6
	South Bend	54	11.9	11.2	11.9	11.6	10.2	9.1	8.1	7.7	8.5	9.5	11.0	11.2	10.2
IA	Des Moines	53	11.4	11.2	12.4	12.6	11.0	10.1	8.9	8.6	9.4	10.3	11.3	11.1	10.7
	Sioux City	61	11.4	11.2	12.3	13.2	11.8	10.7	9.2	9.1	9.9	10.5	11.4	11.0	11.0
	Waterloo	46	11.4	11.4	12.1	12.6	11.1	9.8	8.4	8.3	9.0	10.2	11.0	11.0	10.5
KS	Concordia	40	11.7	12.1	13.4	13.8	12.1	11.7	11.3	10.9	11.2	11.6	11.8	11.5	11.9
	Dodge City	60	13.5	13.9	15.5	15.5	14.6	14.0	13.1	12.5	13.5	13.5	13.7	13.4	13.9
	Goodland	54	12.4	12.4	14.0	14.4	13.5	12.7	11.9	11.5	12.0	11.9	11.9	11.9	12.5
	Topeka	53	9.7	10.2	11.5	11.7	10.2	9.4	8.4	8.0	8.5	9.0	9.7	9.5	9.6
	Wichita	49	12.0	12.5	13.8	14.0	12.3	12.2	11.3	11.1	11.6	11.9	12.1	11.7	12.2
KY	Greater Cincinnati AP	55	10.4	10.4	11.0	10.6	8.7	7.9	7.2	6.8	7.4	8.1	9.7	10.0	9.0
	Jackson	21	7.3	7.4	7.6	7.7	6.0	5.3	5.0	4.6	5.3	6.0	7.0	7.1	6.4
	Lexington	55	10.6	10.6	10.9	10.4	8.6	7.9	7.2	6.8	7.6	8.1	9.9	10.3	9.1
	Louisville	55	9.5	9.5	10.1	9.7	8.0	7.4	6.8	6.4	6.8	7.2	8.9	9.1	8.3
	Paducah	18	8.8	8.9	9.4	8.9	7.4	6.2	5.7	5.1	5.6	6.5	8.2	8.4	7.4
LA	Baton Rouge	51	8.7	9.1	9.1	8.7	7.6	6.5	5.9	5.6	6.4	6.6	7.4	8.1	7.5
	Lake Charles	41	9.8	10.0	10.2	9.8	8.6	7.5	6.1	6.1	7.1	7.8	8.8	9.3	8.4
	New Orleans	54	9.3	9.8	9.9	9.4	8.1	6.8	6.1	5.9	7.3	7.6	8.7	9.0	8.2
	Shreveport	50	9.2	9.6	10.0	9.6	8.3	7.6	7.1	6.6	7.2	7.4	8.3	8.8	8.3
ME	Caribou	22	11.1	10.8	11.7	10.9	10.3	9.3	8.6	8.1	9.2	10.0	10.0	10.4	10.0
	Portland	62	9.0	9.4	10.0	9.9	9.1	8.2	7.6	7.5	7.8	8.4	8.8	9.0	8.7
MD	Baltimore	52	9.4	9.9	10.7	10.2	8.9	8.2	7.6	7.5	7.7	8.1	8.8	8.9	8.8
MA	Blue Hill	61	17.2	17.2	17.2	16.4	14.6	13.8	12.9	12.6	13.5	15.2	16.2	16.7	15.3
	Boston	45	13.7	13.7	13.6	13.1	12.0	11.4	11.0	10.8	11.3	11.9	12.7	13.4	12.4
	Worcester	36	11.7	11.6	11.5	11.0	10.0	8.9	8.4	8.3	8.6	9.4	10.4	10.9	10.1
MI	Alpena	42	8.8	8.4	8.9	9.2	8.3	7.5	7.0	6.7	7.1	7.8	8.5	8.4	8.1
	Detroit	44	11.9	11.4	11.7	11.3	10.1	9.2	8.5	8.1	8.7	9.7	11.2	11.3	10.3
	Flint	61	11.8	11.2	11.8	11.5	10.1	9.0	8.1	7.8	8.8	9.8	11.2	11.3	10.2
	Grand Rapids	39	11.4	10.6	11.1	11.0	9.7	8.9	8.3	7.9	8.3	9.4	10.5	10.7	9.8
	Houghton Lake	21	9.7	9.1	9.1	9.5	8.8	7.8	7.5	7.0	7.8	8.8	9.6	9.3	8.7
	Lansing	43	11.7	10.9	11.2	11.1	9.9	8.8	8.0	7.5	8.2	9.4	10.7	11.0	9.9
	Muskegon	41	12.2	11.6	11.6	11.5	9.9	9.3	8.8	8.6	9.3	10.7	11.9	11.7	10.6
	Sault Ste. Marie	61	9.6	9.3	10.0	10.3	9.7	8.5	7.8	7.7	8.6	9.2	9.7	9.6	9.2
MN	Duluth	53	11.6	11.3	11.8	12.3	11.6	10.4	9.4	9.4	10.3	11.2	11.6	11.2	11.0
	International Falls	50	8.9	8.8	9.4	9.9	9.4	8.5	7.7	7.5	8.5	9.3	9.4	8.8	8.8
	Minneapolis-St. Paul	64	10.5	10.4	11.3	12.2	11.1	10.4	9.4	9.2	10.0	10.6	11.0	10.4	10.5
	Rochester	42	14.2	13.7	14.1	14.3	13.2	12.1	10.8	10.4	11.5	12.7	13.6	13.7	12.9
	Saint Cloud	16	8.4	8.4	9.0	9.8	9.2	8.3	7.1	6.5	7.2	8.4	8.6	8.2	8.3
MS	Jackson	39	8.2	8.4	8.7	8.0	6.8	6.1	5.2	5.3	6.1	6.2	6.9	7.8	7.0
	Meridian	43	7.1	7.5	7.9	7.1	6.0	5.2	4.9	4.6	5.3	5.2	6.1	6.9	6.2
	Tupelo	19	7.6	8.2	8.3	7.8	6.6	5.7	5.3	5.2	6.1	6.0	6.9	7.5	6.8
MO	Columbia	32	10.7	10.8	11.7	11.5	9.1	8.7	8.2	7.9	8.6	9.3	10.6	10.6	9.8

	Kansas City	30	11.1	11.1	12.3	12.3	10.3	9.9	9.2	8.8	9.6	10.5	11.2	10.9	10.6
	St. Louis	53	10.6	10.8	11.6	11.3	9.4	8.8	8.0	7.6	8.2	8.9	10.2	10.3	9.6
	Springfield	57	11.4	11.5	12.5	12.0	10.2	9.3	8.4	8.4	9.1	10.1	11.0	11.2	10.4
MT	Billings	63	13.0	12.3	11.4	11.4	10.7	10.1	9.5	9.5	10.2	11.0	12.2	13.0	11.2
	Glasgow	33	9.8	10.1	11.2	12.3	12.2	11.1	10.5	10.9	10.9	10.6	9.5	9.7	10.7
	Great Falls	61	14.9	13.9	12.8	12.6	11.3	11.1	10.0	10.1	11.2	12.9	14.5	15.2	12.5
	Havre	3	8.8	9.4	10.2	11.6	12.4	10.8	9.8	9.0	9.4	9.1	10.6	10.8	10.2
	Helena	62	6.7	7.3	8.2	9.1	8.8	8.5	7.8	7.4	7.4	7.1	7.1	6.7	7.7
	Kalispell	40	5.6	5.6	6.7	7.7	7.4	6.9	6.3	6.3	6.1	5.1	5.3	5.1	6.2
	Missoula	58	5.1	5.6	6.7	7.6	7.4	7.2	6.9	6.7	6.0	5.1	5.0	4.7	6.2
NE	Grand Island	53	11.7	11.7	13.1	14.0	12.6	11.8	10.5	10.3	11.0	11.2	11.8	11.6	11.8
	Lincoln	30	9.6	10.0	11.3	12.1	10.5	9.8	9.3	9.1	9.5	9.9	9.9	9.6	10.1
	Norfolk	26	11.6	11.5	12.5	13.1	11.5	10.8	9.7	9.5	10.3	11.0	11.4	11.3	11.2
	North Platte	50	9.2	9.8	11.5	12.6	11.5	10.4	9.5	9.2	9.7	9.6	9.5	9.0	10.1
	Omaha Eppley AP	66	10.9	11.1	12.2	12.6	10.9	10.1	8.8	8.8	9.4	9.8	10.9	10.7	10.5
	Omaha (North)	9	9.9	9.2	10.5	10.2	8.9	8.3	7.5	7.6	8.6	9.0	9.7	9.8	9.1
	Scottsbluff	51	10.6	11.1	12.0	12.5	11.8	10.5	9.3	8.9	9.4	9.7	10.2	10.4	10.5
	Valentine	34	9.3	9.4	10.4	11.1	11.0	9.9	9.1	9.2	9.7	9.5	9.7	9.2	9.8
NV	Elko	47	5.2	5.7	6.6	7.2	6.8	6.7	6.2	5.9	5.4	5.0	5.1	5.0	5.9
	Ely	61	10.1	10.3	10.7	10.9	10.7	10.6	10.3	10.4	10.3	10.0	9.9	9.9	10.3
	Las Vegas	54	7.4	8.5	10.1	11.0	11.0	11.0	10.2	9.6	9.0	8.1	7.8	7.3	9.2
	Reno	60	5.6	6.2	7.8	8.2	8.0	7.7	7.2	6.6	5.8	5.4	5.5	5.3	6.6
	Winnemucca	46	7.6	8.0	8.6	8.6	8.6	8.5	8.4	7.8	7.6	7.3	7.2	7.4	8.0
NH	Concord	60	7.2	7.9	8.1	7.8	7.0	6.5	5.7	5.4	5.6	6.0	6.6	7.0	6.7
	Mt. Washington	67	46.1	44.3	41.4	35.8	29.7	27.3	25.3	24.7	28.8	33.8	39.5	44.5	35.1
NJ	Atlantic City AP	44	10.7	11.1	11.8	11.4	10.1	9.1	8.3	7.9	8.3	8.7	9.9	10.3	9.8
	Newark	58	11.2	11.5	11.9	11.2	10.0	9.5	8.9	8.7	9.0	9.4	10.2	10.8	10.2
NM	Albuquerque	63	8.0	8.8	9.9	10.7	10.5	9.8	8.9	8.1	8.4	8.2	7.9	7.6	8.9
	Clayton	10	11.9	12.4	13.1	14.4	13.2	12.9	11.2	10.2	11.3	11.8	11.8	12.1	12.2
	Roswell	29	7.7	8.6	10.1	10.2	9.9	9.7	8.6	7.9	8.0	7.9	7.7	7.6	8.7
NY	Albany	64	9.8	10.1	10.6	10.5	9.0	8.3	7.5	7.0	7.4	8.0	9.1	9.3	8.9
	Binghamton	51	11.3	11.3	11.5	11.2	9.9	9.3	8.4	8.2	8.8	9.7	10.8	11.0	10.1
	Buffalo	63	14.0	13.3	13.1	12.3	11.4	10.8	10.2	9.7	10.2	11.1	12.6	13.1	11.8
	Islip	19	9.6	10.1	10.5	9.8	9.0	8.4	7.7	7.4	7.7	8.3	9.4	9.3	8.9
	New York C. Park	65	10.6	10.7	11.0	10.2	8.8	8.1	7.6	7.5	8.1	8.8	9.8	10.1	9.3
	New York (JFK AP)	44	13.0	13.3	13.5	12.7	11.6	10.7	10.2	10.0	10.4	11.0	12.2	12.7	11.8
	New York (Laguardia AP)	54	13.7	13.8	13.9	12.9	11.6	11.0	10.4	10.3	11.0	11.6	12.8	13.4	12.2
	Rochester	62	11.6	11.1	11.0	10.7	9.3	8.6	8.0	7.7	8.1	8.8	10.2	10.7	

Note: Through 2002. The average wind speed is based on the speed of the wind regardless of direction.

Source: From U.S. National Oceanic and Atmospheric Administration, *Comparative Climatic Data for the United States Through 2000*, www.noaa.gov.

Table 3A.9 Maximum Wind Speed (MPH) — Selected Cities of the United States

			Jan		Feb		Mar		Apr		May		Jun		Jul		Aug		Sep		Oct		Nov		Dec		Annual	
State	Station	Years	DR	SP	DR	SP	DR	SP	DR	SP	DR	SP	DR	SP	DR	SP	DR	SP	DR	SP	DR	SP	DR	SP	DR	SP	DR	SP
AL	Birmingham AP	39	W	49	SE	59	SW	65	SW	56	NW	65	SW	56	SW	57	NW	50	SE	50	W	43	N	52	SE	41	SW	65
	Huntsville	35	26	44	08	43	12	46	18	44	24	46	24	56	10	52	02	63	23	43	01	43	27	40	30	40	02	63
	Mobile	44	18	44	23	46	10	40	01	44	32	51	16	45	21	60	14	63	09	63	36	46	17	38	22	43	14	63
	Montgomery	17	30	32	26	38	28	46	32	39	36	35	29	44	26	35	10	44	70	35	08	52	30	39	30	35	08	52
AK	Anchorage (G)	23	E	64	NE	61	NE	75	SE	43	S	43	SE	46	SE	40	N	44	S	48	S	55	NE	55	SE	55	NE	75
	Annette	50	16	58	16	50	14	48	14	60	14	44	16	44	16	35	16	40	11	51	16	55	13	51	16	58	14	60
	Barrow (G)	22	E	58	SW	74	E	56	26	51	NE	41	W	43	W	55	27	55	SW	66	W	54	E	53	SW	61	SW	74
	Barter IS. (G)	9	W	75	W	69	W	66	E	49	W	56	E	53	W	48	NW	58	E	58	W	69	E	61	W	70	W	75
	Bethel (G)	23	S	61	NE	59	S	56	S	51	S	53	S	59	S	46	NW	56	SE	69	S	77	W	66	S	67	S	77
	Bettles	10	11	30	08	25	08	29	23	28	11	31	01	28	24	30	25	32	24	25	24	25	26	38	24	40	24	40
	Big Delta	29	29	74	18	67	20	63	18	60	20	55	16	51	18	63	16	47	18	66	18	58	11	56	11	63	29	74
	Cold Bay	47	17	71	16	73	15	71	15	61	14	60	11	63	17	54	16	64	17	75	21	60	14	66	11	64	17	75
	Fairbanks	51	25	31	26	36	22	40	24	32	23	32	25	40	27	32	27	34	08	33	25	40	25	35	24	37	22	40
	Gulkana	9	04	52	16	44	15	35	18	46	19	35	15	38	15	40	16	30	16	36	19	46	19	33	16	49	04	52
	Homer	30	28	39	07	39	20	35	90	38	11	44	90	35	16	29	28	32	36	49	80	41	08	44	40	69	40	69
	Juneau	32	12	45	12	46	11	40	11	40	12	40	12	35	12	32	12	38	12	48	12	49	11	58	11	55	11	58
	King Salmon (G)	23	E	69	E	69	E	62	S	59	S	63	E	58	E	47	SW	56	E	71	E	67	E	67	E	66	E	71
	Kodiak (G)	23	NW	75	NW	67	NW	82	E	67	W	59	NE	52	NW	52	NW	67	NW	78	NW	70	W	82	NW	83	NW	83
	Kotzebue (G)	23	E	72	E	63	E	66	80	56	NE	49	SE	46	SE	45	S	53	NE	54	SE	60	SE	63	E	68	E	72
	Mcgrath (G)	23	S	59	SW	47	SE	46	S	46	S	45	NW	62	S	46	S	49	S	49	E	40	S	53	SW	52	NW	62
	Nome	45	09	54	04	51	02	44	05	45	09	44	04	35	24	35	15	41	18	44	20	52	24	55	05	54	24	55
	St. Paul Island (G)	23		63	N	72	N	67	SE	67	SW	74	S	53	SE	47	N	58	N	61	W	70	SW	84	E	79	SW	84
	Talkeetna	35	04	38	03	35	03	39	34	29	18	32	19	29	17	22	40	28	30	35	20	32	02	31	36	35	03	39
	Unalakleet (G)	6	E	63	E	71	E	60	E	43	E	46	E	37	SW	39	S	44	S	45	SE	53	E	62	E	61	E	71
	Valdez (G)	23	N	94	NE	83	NE	82	N	55	N	44	W	38	N	41	N	56	SW	69	NE	66	N	77	N	75	N	94
	Yakutat (G)	23	SE	81	SE	62	SE	64	SE	64	SE	48	SE	45	SE	44	SE	60	SE	63	SE	60	SE	70	SE	63	SE	81
AZ	Flagstaff	18	SW	38	SW	34	21	38	SW	40	SW	46	SW	35	NW	39	SW	30	W	33	40	38	SW	39	NE	38	SW	46
	Phoenix	17	25	36	26	30	24	43	30	51	11	35	40	33	13	43	14	37	15	39	24	36	25	30	29	39	30	51
	Tucson	54	E	40	E	59	SE	41	SW	46	SE	43	SE	50	SE	71	NE	54	SE	54	SE	47	E	55	W	44	SE	71
	Winslow	37	23	56	22	63	22	61	25	56	25	53	20	52	16	59	21	45	31	45	22	49	22	46	22	52	22	63
	Yuma	40	NW	41	W	50	N	43	NW	47	NW	38	SW	42	NE	61	SE	60	E	57	S	47	N	47	W	47	NE	61
AR	Fort Smith	20	30	43	30	39	34	45	26	45	30	49	28	57	09	51	09	46	31	45	29	51	26	40	22	44	28	57
	Little Rock	41	S	44	SW	57	SE	56	NW	65	NW	61	NE	60	NW	56	NW	54	NW	50	SSW	58	SW	49	SW	48	NW	65
	North Little Rock	3	NW	25	SW	25	S	28	NE	30	SW	30	NE	21	W	30	W	28	NE	24	SE	24	SW	27	SW	25	NE	30
CA	Bakersfield	53	19	36	13	49	17	40	29	40	32	40	15	41	29	25	14	33	14	35	08	38	30	35	13	46	13	49
	Bishop (G)	17		60	W	63		58		62		62		54		60		70		47		52		66		68		70
	Blue Canyon	33	20	67	17	76	07	67	20	50	23	37	07	49	09	32	07	30	09	49	05	70	19	54	07	51	17	76
	Eureka	83	S	54	SW	48	SW	48	N	49	NW	40	NW	39	N	35	N	34	N	44	SW	56	S	55	S	56	SW	56
	Fresno	24	14	39	13	36	29	30	29	36	32	32	29	28	32	23	31	28	31	29	31	26	28	30	14	29	14	39
	Long Beach	33	17	37	18	40	11	39	29	44	27	30	29	24	18	23	16	23	10	26	30	37	25	44	32	39	29	44
	Los Angeles AP (G)	52	NE	51	N	57	W	62	N	59	W	49	W	40	SW	31	SE	33	E	39	W	46	W	60	NW	49	W	62
	Los Angeles CO	40	N	49	NW	40	NW	47	NW	40	NW	39	N	32	W	21	E	24	NW	27	N	48	N	42	SE	44	N	49
	Mount Shasta	6	34	23	36	22	35	21	34	24	32	21	31	18	36	20	34	15	35	20	30	17	34	20	32	21	34	24
	Redding (G)	16	S	70	S	64	S	74	S	47	S	54	N	60	N	36	S	46		44	S	66	S	58	S	85	S	85
	Sacramento	53	SE	60	SE	51	S	66	SW	45	SW	74	SW	47	SW	36	SW	38	NW	42	SE	68	SE	70	SE	70	SW	74
	San Diego	58	SE	56	S	45	SW	46	S	37	S	30	S	26	NW	23	NW	23	S	31	N	31	SE	51	NW	39	SE	56
	San Francisco AP	53	16	58	22	55	26	46	18	47	26	46	28	44	28	40	26	37	28	38	25	44	18	51	22	54	16	58
	San Francisco CO	36	SE	47	SW	47	S	44	W	38	W	38	W	40	W	38	W	34	W	32	SE	43	S	41	SE	45	SE	47

	Santa Maria	13	13	36	13	46	16	35	30	44	30	35	29	36	29	32	29	30	29	32	17	37	26	31	16	36	13	46
	Stockton	39	14	46	16	41	33	39	35	37	35	41	29	35	27	31	26	30	34	33	33	37	15	47	15	44	15	47
CO	Alamosa (G)	19	SW	58	23	62	23	60	24	71	22	63	20	69	30	66	34	51	SW	54	W	62	21	63	26	54	24	71
	Colorado Springs	52	29	55	28	61	29	60	28	61	27	52	20	55	22	49	34	45	30	44	27	59	34	52	27	60	28	61
	Denver	14	32	44	30	36	30	41	33	46	36	43	21	38	29	46	33	33	29	36	01	36	36	36	32	38	33	46
	Grand Junction	23	25	36	20	41	34	53	27	49	20	46	30	57	35	45	24	45	28	49	21	44	31	39	26	35	30	57
	Pueblo	18	29	51	29	46	36	52	35	52	35	60	29	48	30	58	22	48	01	46	30	47	04	48	27	47	35	60
CT	Bridgeport	42	34	67	34	65	08	58	32	55	34	50	35	39	29	40	04	58	18	74	09	58	14	58	25	53	18	74
	Hartford	17	24	46	30	46	30	43	05	41	28	39	36	45	27	39	18	40	17	43	29	38	17	44	26	41	24	46
DE	Wilmington	54	15	51	29	46	28	45	33	46	24	48	34	43	27	48	35	46	07	40	20	58	16	46	32	46	20	58
DC	Washington Dulles AP	39	20	39	30	37	28	44	32	46	35	40	31	55	30	48	34	43	25	35	29	38	29	35	30	40	31	55
	Washington Nat'l AP	17	29	41	33	39	33	44	31	39	32	46	31	49	50	47	34	37	32	39	23	39	32	37	34	38	31	49
FL	Apalachicola	48	E	48	E	42	E	54	SE	51	SE	47	E	55	N	63	NE	59	E	67	NW	56	SE	47	SE	42	E	67
	Daytona Beach	54	26	43	20	44	24	58	18	46	28	46	33	40	25	40	11	50	11	58	05	53	50	39	34	40	24	58
	Fort Myers	51	25	40	25	39	35	46	20	39	32	44	31	48	18	45	25	44	05	92	23	45	30	32	33	35	05	92
	Jacksonville	22	30	38	30	39	22	44	32	46	29	34	28	39	26	57	11	38	25	36	21	31	33	38	31	40	26	57
	Key West	26	27	41	12	57	22	54	01	58	13	46	18	40	12	33	19	41	12	43	35	46	12	47	26	39	01	58
	Miami	45	24	46	19	55	04	46	24	35	32	52	13	37	25	43	12	86	06	69	90	59	07	38	32	38	12	86
	Orlando	53	25	42	25	46	24	46	02	50	35	51	32	64	14	46	32	50	24	46	05	48	26	46	20	35	32	64
	Pensacola	30	31	40	13	40	16	39	32	35	12	39	29	46	20	37	20	40	16	54	22	39	21	35	20	36	16	54
	Tallahassee	43	23	46	09	40	27	48	19	35	29	40	03	44	22	39	02	58	08	46	20	32	16	40	28	37	02	58
	Tampa	50	32	44	32	50	29	43	28	44	36	46	31	67	32	58	11	38	34	56	21	40	25	40	36	45	31	67
	West Palm Beach	53	29	48	11	48	27	51	32	55	27	45	09	71	34	46	13	86	36	58	16	74	10	39	36	38	13	86
GA	Athens	47	25	52	20	52	24	50	23	47	31	35	18	43	13	35	35	47	15	37	05	35	24	41	10	48	25	52
	Atlanta	26	23	46	29	52	31	49	30	44	27	54	24	51	30	60	32	41	27	37	09	35	29	39	30	33	30	60
	Augusta	51	25	40	30	40	23	52	32	39	28	48	08	62	33	48	18	45	04	36	18	40	27	40	28	35	08	62
	Columbus	44	29	46	20	52	27	44	28	40	23	39	29	55	36	52	18	47	36	38	12	40	31	37	33	35	29	55
	Macon	20	32	37	27	46	70	35	33	45	31	47	27	44	30	44	12	38	36	37	18	33	30	40	30	35	31	47
	Savannah	22	31	30	09	31	32	46	23	35	22	44	05	43	04	45	34	37	10	40	31	35	23	40	30	29	32	46
HI	Hilo	23	36	35	34	39	10	29	34	26	35	29	11	25	05	25	36	25	04	25	34	29	02	28	36	29	34	39
	Honolulu	22	70	36	07	35	06	32	50	35	13	30	70	30	05	30	70	31	13	38	60	29	20	46	70	35	20	46
	Kahului	28	SW	44	NE	40	N	43	E	36	E	34	40	38	60	38	NE	35	E	33	E	36	SW	41	21	40	SW	44
	Lihue	23	40	38	23	41	06	39	05	36	06	33	90	35	70	31	05	31	15	84	04	33	18	65	40	41	15	84
ID	Boise	61	SE	50	W	56	W	52	W	50	W	50	SW	50	W	61	SE	56	SE	50	SE	56	NW	57	NW	56	W	61
	Lewiston (G)	31		72		64		60		58		54		54	29	59		51		59		59		59		63		72
	Pocatello	51	SE	61	W	57	W	72	S	61	W	61	W	50	W	57	SW	54	W	57	SW	54	W	67	NW	57	W	72
IL	Cairo	45	SW	50	SW	68	NW	60	SW	59	SW	63	SW	60	NW	49	S	45	SW	47	SW	40	SW	53	SW	63	SW	68
	Chicago	44	28	47	25	45	01	54	24	54	34	52	24	41	36	55	32	46	23	58	20	48	23	51	26	46	23	58
	Moline	13	29	40	26	40	23	46	26	49	30	40	24	49	27	37	20	57	26	36	26	40	26	49	22	45	20	57
	Peoria	17	23	38	27	39	23	54	24	45	20	36	30	44	26	40	30	37	29	35	27	44	20	48	06	46	23	54
	Rockford	52	27	40	22	49	25	46	11	54	27	52	20	47	30	53	29	57	20	52	21	40	20	46	06	46	29	57
	Springfield	23	25	39	29	51	24	56	30	61	28	46	16	43	40	52	33	44	18	39	20	40	24	46	25	36	30	61
IN	Evansville	17	26	34	26	43	25	40	28	44	33	46	28	46	32	35	27	37	28	36	23	36	31	41	24	41	28	46
	Fort Wayne	53	SW	59	W	61	S	65	W	63	S	57	SE	65	NW	61	N	55	W	52	SW	46	SW	57	SW	52	S	65
	Indianapolis	23	19	45	23	46	21	54	25	47	30	40	29	46	28	49	29	45	26	46	27	44	23	44	26	41	28	49
	South Bend	53	22	52	20	47	20	51	27	55	27	68	27	50	34	45	32	63	25	36	25	56	22	58	23	43	27	68
IA	Des Moines	47	NW	66	W	56	S	66	W	76	W	70	NW	76	W	73	SSE	60	NW	55	W	56	W	72	SW	61	W	76
	Dubuque (G)	19		58		52		62		68		74		55	NW	74	32	66	NW	58		54		55	E	56		74
	Sioux City	58	NW	56	NW	54	N	61	W	68	W	80	W	91	NW	66	NW	56	S	66	W	70	NW	59	NW	53	W	91

(Continued)

Table 3A.9 (Continued)

State	Station	Years	Jan		Feb		Mar		Apr		May		Jun		Jul		Aug		Sep		Oct		Nov		Dec		Annual	
			DR	SP	DR	SP	DR	SP	DR	SP	DR	SP	DR	SP	DR	SP	DR	SP	DR	SP	DR	SP	DR	SP	DR	SP	DR	SP
	Waterloo	42	29	46	28	44	23	46	25	52	18	52	33	60	35	58	21	46	28	38	29	43	22	53	32	39	33	60
KS	Concordia	21	32	46	21	41	35	46	24	55	30	41	24	54	25	58	06	44	17	44	26	46	34	46	35	40	25	58
	Dodge City	16	34	56	34	47	35	63	21	53	30	52	29	60	36	56	32	63	14	51	34	49	20	44	32	46	32	63
	Goodland	53	34	53	36	51	33	62	29	62	27	61	33	66	30	64	23	60	34	51	27	61	33	52	34	52	33	66
	Topeka	17	31	39	31	36	18	55	08	51	34	47	34	48	34	44	27	38	32	43	23	39	31	45	30	37	18	55
	Wichita	21	35	48	20	44	24	49	23	56	18	61	30	51	34	70	04	52	19	44	31	49	33	48	32	44	34	70
KY	Greater Cincinnati AP	39	28	46	29	40	27	45	25	46	31	37	27	41	32	45	29	41	31	36	29	48	28	43	21	40	29	48
	Jackson	21	23	37	19	39	31	33	27	43	28	39	29	31	33	32	28	37	29	26	25	28	19	30	30	30	27	43
	Lexington	41	18	47	32	46	27	36	32	46	22	35	30	44	29	37	22	39	29	41	17	40	27	45	22	39	18	47
	Louisville	17	16	38	23	44	19	43	22	56	23	40	04	54	32	46	21	47	29	39	29	40	23	44	27	40	22	56
	Paducah	18	30	41	25	40	19	37	24	38	24	51	29	45	34	51	10	33	18	35	24	43	21	44	22	35	34	51
LA	Baton Rouge	40	24	39	17	39	27	38	23	39	17	48	03	40	03	41	14	46	06	58	33	40	22	33	29	60	29	60
	Lake Charles	41	32	58	25	40	18	40	06	44	31	43	19	53	33	36	11	46	36	40	33	38	21	46	33	36	32	58
	New Orleans	43	27	48	26	43	16	38	10	40	36	55	25	49	13	44	33	42	09	69	17	40	22	38	28	46	09	69
	Shreveport	40	30	41	30	43	29	54	28	52	32	63	36	46	29	46	11	40	19	44	25	37	31	46	32	43	32	63
ME	Caribou	13	31	36	25	41	33	37	31	33	33	37	34	35	32	32	33	28	30	26	30	40	32	41	31	39	32	41
	Portland	17	16	38	08	45	11	41	29	40	28	37	28	35	29	37	28	57	29	33	14	37	10	41	30	44	28	57
MD	Baltimore	50	NE	63	W	68	SE	80	W	70	SW	65	SW	80	NW	57	NE	54	W	56	SE	73	E	58	W	57	SE	80
MA	Blue Hill	39	S	76	S	77	ENE	72	NW	66	S	65	NW	61	NW	78	SSW	67	SSE	92	S	62	S	67	SSE	68	SSE	92
	Boston	17	17	46	08	43	06	54	28	43	24	43	28	45	26	46	08	47	23	47	04	47	11	48	05	51	06	54
	Worcester	46	25	60	32	76	29	76	05	54	31	51	25	39	31	46	36	44	32	41	25	43	20	54	23	51	32	76
MI	Alpena	22	80	30	35	37	20	35	16	38	21	35	21	35	14	37	31	35	19	38	33	31	20	38	26	33	19	38
	Detroit	23	22	48	22	51	21	46	22	47	23	43	23	37	28	53	29	35	28	35	24	47	24	45	29	49	28	53
	Flint	47	26	45	28	40	27	58	24	44	32	81	29	52	29	41	27	37	27	46	25	39	23	46	27	40	32	81
	Grand Rapids	23	24	45	24	55	25	47	24	52	26	47	24	39	26	51	31	41	18	40	24	47	23	49	24	39	19	52
	Houghton Lake	25	26	40	24	37	28	36	24	37	26	40	19	51	13	37	25	29	23	32	16	35	27	40	32	33	19	51
	Lansing	41	SW	54	SW	56	W	59	W	61	W	46	SE	63	NE	56	SW	47	N	57	SW	48	W	56	SW	56	SE	63
	Marquette	6	NW	44	NW	31	NW	40	NW	44	N	34	NW	38	NW	35	NW	37	W	35	SE	38	NW	31	SW	35	NW	44
	Muskegon	43	31	44	34	41	25	41	22	48	27	44	24	44	33	40	26	43	20	40	22	40	23	52	23	40	23	52
	Sault Ste. Marie	31	NW	47	W	47	SE	42	SE	42	E	49	S	37	SE	44	NW	35	W	43	NW	42	NW	60	NW	45	NW	60
MN	Duluth	17	30	45	29	44	E	57	20	43	28	44	27	46	31	41	26	48	26	33	32	46	08	44	32	41	E	57
	International Falls	47	30	35	26	36	29	42	23	52	20	52	18	46	29	46	30	43	34	38	30	47	27	35	31	36	23	52
	Minneapolis-St. Paul	23	32	51	34	37	32	37	18	45	22	49	33	48	35	43	20	44	29	39	31	43	25	41	32	38	32	51
	Rochester	40	30	48	28	45	23	58	30	53	25	69	13	53	17	51	34	46	24	44	30	47	29	47	32	47	25	69
	Saint Cloud	13	1	36	30	46	30	43	30	44	27	41	32	62	22	49	22	45	32	34	31	43	30	39	31	39	32	62
MS	Jackson	26	35	46	13	43	16	44	14	44	22	35	35	40	33	44	17	37	06	55	31	30	14	41	15	41	06	55
	Meridian	43	33	41	02	35	34	39	19	48	17	35	22	46	40	51	34	38	02	45	02	35	19	41	80	37	40	51
	Tupelo	19	24	39	25	35	24	39	10	33	31	41	20	41	35	48	01	41	30	32	25	38	23	37	11	38	35	48
MO	Columbia	14	NW	49	24	41	26	46	23	54	36	39	26	49	31	46	19	52	35	38	25	47	19	45	06	37	23	54
	Kansas City	18	32	39	20	40	23	46	20	48	24	46	10	51	20	58	31	40	14	41	21	40	22	37	20	39	20	58
	St. Louis	23	29	40	30	45	27	48	27	49	34	46	27	48	36	46	31	40	25	41	28	52	11	41	29	39	28	52
	Springfield	17	NW	39	15	39	25	39	34	41	10	43	36	47	28	46	36	48	10	40	22	35	10	43	13	37	36	48
MT	Billings	59	W	66	W	72	NW	61	NW	72	NN	68	NW	79	N	73	NW	69	NW	61	NW	68	NW	63	NW	66	NW	79
	Glasgow	34	33	41	29	46	10	41	30	54	32	44	30	54	30	69	23	66	30	46	29	54	27	48	29	54	30	69
	Great Falls	54	SW	65	W	72	W	73	W	70	SW	65	NW	70	W	73	SW	71	NW	73	W	73	SW	73	SW	82	SW	82
	Havre	3	29	46	24	46	27	46	28	58	10	51	27	47	28	55	25	46	33	45	26	69	33	46	29	45	26	69

	Helena	62	SW	73	W	73	SW	61	W	52	SW	56	W	56	SW	65	S	65	NW	54	W	62	SW	56	NW	59	SW	73
	Kalispell	36	04	52	01	40	03	41	23	43	23	40	03	38	31	38	15	43	22	36	32	38	03	35	03	52	04	52
	Missoula	57	S	52	NW	47	SW	50	NW	51	SW	57	S	51	SE	72	SW	58	16	48	SW	51	SW	42	W	56	SE	72
NE	Grand Island	40	34	54	33	53	34	55	21	52	29	59	30	68	29	59	22	52	25	40	33	51	34	51	31	52	30	68
	Lincoln	30	33	48	NW	48	N	54	NW	52	W	51	NE	67	SW	52	NW	65	32	44	30	47	NW	48	34	46	NE	67
	Norfolk	21	33	46	32	44	30	44	32	51	07	44	28	46	21	58	32	60	30	55	18	46	31	51	35	46	32	60
	North Platte	23	31	45	36	52	02	47	28	49	31	48	32	52	24	55	31	56	30	46	31	53	29	46	02	52	31	56
	Omaha Eppley AP	14	33	45	33	39	33	40	20	51	16	57	24	40	28	58	30	46	19	36	11	48	28	45	34	39	28	58
	Omaha (North)	15	NW	41	NW	38	NW	38	NW	46	N	34	SW	39	SW	50	NW	39	NW	35	NW	34	NW	38	NW	37	SW	50
	Scottsbluff	52	34	53	29	60	29	62	31	55	32	80	29	80	05	52	35	52	32	46	31	48	29	56	32	47	32	80
	Valentine	20	32	43	30	52	33	43	30	49	33	49	29	53	26	51	20	53	31	40	29	51	30	45	31	41	20	53
NV	Elko	39	23	40	27	39	29	41	25	48	34	55	27	61	23	45	16	35	27	58	29	35	20	40	27	50	27	61
	Ely	57	SE	66	S	56	SW	65	S	59	S	74	SW	63	S	50	E	57	S	57	S	65	S	51	SE	61	S	74
	Las Vegas	17	23	45	23	50	23	51	22	49	22	56	34	48	30	45	32	43	16	41	31	47	21	43	34	48	22	56
	Reno	13	16	45	15	44	21	49	SW	50	17	46	23	37	18	41	19	40	23	43	15	41	19	52	19	67	19	67
	Winnemucca	51	W	56	W	59	W	66	W	52	N	61	W	57	W	56	W	51	W	57	NE	54	22	48	SW	61	W	66
NH	Concord	57	NW	44	N	42	NE	71	NW	52	NW	48	SW	44	SW	45	E	56	E	42	NW	39	NE	72	NW	52	NE	72
	Mt. Washington (G)	63	NW	173	E	166	W	180	SE	231	W	164	NW	136	NW	110	ENE	142	SE	174	W	161	NW	163	NW	178	SE	231
NJ	Atlantic City AP	43	29	47	27	43	24	46	07	46	70	40	30	41	31	41	12	35	32	60	29	41	28	47	36	55	32	60
	Atlantic City CO	11	4	43	04	43	07	63	19	37	01	37	05	30	19	44	35	30	07	32	04	44	04	36	11	45	07	63
	Newark	54	30	52	23	46	30	45	29	55	32	50	26	58	35	52	09	46	05	51	11	48	09	82	32	55	09	82
NM	Albuquerque	18	09	52	27	47	19	49	11	47	25	48	27	48	36	52	80	51	34	43	26	48	27	48	06	51	09	52
	Roswell	16	NW	47	NW	56	NW	52	25	49	NW	60	NW	73	19	49	NW	44	35	43	22	44	NE	65	SW	58	NW	73
NY	Albany	19	30	40	29	44	30	46	17	35	32	55	23	43	29	41	30	47	32	35	28	39	28	41	29	43	32	55
	Binghamton	17	27	40	25	41	24	39	24	32	28	33	26	40	30	35	32	43	34	33	26	33	27	41	24	43	24	43
	Buffalo	54	SW	91	SW	70	W	68	W	67	SW	63	NW	56	NW	59	SW	56	SW	59	SW	63	SW	66	S	60	SW	91
	New York C. Park	19	7	40	08	34	8	37	80	35	30	29	08	28	35	29	30	33	90	29	33	28	32	30	05	39	7	40
	New York (JFK AP)	39	26	52	25	46	06	46	31	46	16	44	28	43	29	51	30	46	28	47	26	44	30	44	26	49	26	52
	New York (Laguardia AP)	23	5	40	01	43	06	52	29	55	17	41	29	38	35	53	28	46	33	51	10	52	30	55	05	52	29	55
	Rochester	17	24	45	25	59	25	55	25	41	18	45	36	39	20	52	27	40	27	68	25	43	25	46	24	45	27	68
	Syracuse	53	W	60	W	62	SE	56	NW	52	30	52	NW	49	28	54	NW	43	32	59	SE	63	E	59	W	52	SE	63
NC	Asheville	38	34	45	34	60	33	48	22	44	34	40	36	40	35	43	30	41	36	45	33	35	32	40	34	44	34	60
	Cape Hatteras	32	24	44	16	44	19	52	21	40	18	35	29	37	15	52	33	60	11	60	22	47	15	58	16	46	33	60
	Charlotte	23	31	30	32	33	30	32	19	32	35	34	20	32	20	38	28	46	12	46	21	37	24	30	16	38	28	46
	Greensboro-Wnstn-Slm-HPT	22	19	41	07	38	30	49	26	37	27	62	27	43	35	50	19	30	26	46	30	46	20	38	24	34	27	62
	Raleigh	49	27	41	23	45	24	46	25	40	20	54	33	39	23	69	33	46	04	53	29	73	32	35	22	40	29	73
	Wilmington	22	26	38	25	44	29	58	28	47	27	44	23	46	4	53	06	56	07	67	01	35	24	38	26	43	07	67
ND	Bismarck	23	29	44	30	52	36	43	31	55	27	54	34	52	22	64	30	54	15	46	32	44	31	47	33	45	22	64
	Fargo	17	34	49	33	51	34	49	32	49	24	45	14	43	33	74	34	51	31	45	33	49	31	47	30	45	33	74
	Grand Forks	5	17	39	27	47	35	48	31	49	15	53	31	51	12	40	30	62	29	43	31	46	31	54	34	45	30	62
	Williston	23	32	44	30	46	29	46	30	55	27	46	30	56	26	63	29	53	27	40	28	55	29	52	30	48	26	63
OH	Akron	41	22	44	25	51	26	49	30	40	32	46	31	67	23	44	17	51	36	40	24	43	25	41	24	48	31	67
	Cleveland	25	22	53	24	45	27	46	23	44	20	42	28	41	23	43	24	39	29	35	24	46	18	40	25	46	22	53
	Columbus	21	24	40	27	43	26	47	26	47	25	52	31	40	33	47	29	43	23	44	24	40	27	45	26	47	25	52
	Dayton	17	25	43	28	45	27	49	25	49	24	54	24	43	29	61	26	33	22	43	26	43	22	44	25	46	29	61
	Mansfield	36	24	46	24	44	26	44	33	46	25	37	23	40	26	39	17	41	33	34	32	40	23	39	18	46	24	46
	Toledo	45	W	47	SW	56	W	56	SW	72	25	46	W	50	NW	54	W	47	NW	47	24	45	SW	65	30	48	SW	72
	Youngstown	53	25	48	27	58	25	55	33	51	24	46	23	45	27	58	27	44	36	40	23	44	25	52	25	46	27	58
OK	Oklahoma City	21	34	45	32	45	24	52	32	67	23	53	35	48	31	74	05	46	21	52	30	43	19	46	33	44	31	74
	Tulsa	25	18	37	20	41	18	46	34	55	30	41	04	49	19	51	29	38	27	39	25	40	29	44	33	36	34	55

(Continued)

Table 3A.9 (Continued)

State	Station	Years	Jan		Feb		Mar		Apr		May		Jun		Jul		Aug		Sep		Oct		Nov		Dec		Annual	
			DR	SP	DR	SP	DR	SP	DR	SP	DR	SP	DR	SP	DR	SP	DR	SP	DR	SP	DR	SP	DR	SP	DR	SP	DR	SP
OR	Astoria	49	17	55	19	47	18	47	20	52	22	37	18	30	19	29	20	30	17	36	20	44	20	46	25	52	17	55
	Eugene	46	20	58	19	60	18	48	18	44	25	46	27	29	32	37	11	32	20	32	18	63	23	46	18	40	18	63
	Medford	53	23	50	25	46	16	55	14	35	12	38	17	37	07	44	16	48	14	47	20	40	19	40	14	44	16	55
	Pendleton	47	24	52	25	54	29	63	27	77	27	48	29	62	31	49	23	43	27	47	25	49	27	62	29	63	27	77
	Portland	48	S	54	SW	61	S	57	S	60	SW	42	SW	40	SW	33	SW	29	S	61	S	88	SW	56	S	57	S	88
	Salem	53	18	43	18	46	19	40	18	44	20	31	23	28	24	26	18	25	19	34	18	58	17	49	17	46	18	58
	Sexton Summit	15	16	60	19	53	21	60	20	50	20	44	18	42	35	39	20	45	12	45	14	51	19	59	20	63	20	63
PC	Guam	38	W	64	NE	36	70	35	SW	64	NE	76	E	32	22	74	SE	43	E	35	W	44	NE	80	27	106	27	106
	Johnston Island	5	33	31	09	38	08	35	09	35	07	32	08	31	06	33	08	32	07	35	08	32	03	33	06	43	06	43
	Koror	13	20	35	06	23	36	26	10	25	15	46	30	33	26	33	08	28	27	37	29	31	25	52	09	31	25	52
	Kwajalein, Marshall IS	42	22	55	18	35	90	39	08	37	11	44	12	41	09	41	70	44	07	44	20	40	12	60	09	45	12	60
	Majuro, Marshall IS	13	06	36	70	35	07	25	07	26	07	25	09	29	23	30	32	29	22	29	06	33	22	31	09	35	06	36
	Pago Pago, Amer Samoa	23	34	46	36	63	32	37	35	35	08	35	08	43	09	32	17	33	06	38	18	35	09	39	21	81	21	81
	Pohnpei, Caroline IS	14	29	28	10	21	23	20	29	26	10	23	90	18	30	18	25	32	23	21	22	23	24	35	27	23	24	35
	Chuuk, E. Caroline IS	13	34	30	50	28	33	33	8	29	09	29	40	29	24	33	23	39	21	32	22	35	24	35	24	46	24	46
	Wake Island	8	08	39	02	36	06	37	06	38	09	30	08	28	10	41	19	51	16	30	08	33	07	40	03	39	19	51
	Yap, W Caroline IS	13	05	23	04	23	12	39	13	23	08	20	23	23	28	28	22	25	09	23	16	26	09	36	26	41	26	41
PA	Allentown	54	29	55	25	58	29	58	29	60	30	58	27	81	27	55	23	58	25	46	14	49	30	58	29	52	27	81
	Erie	45	20	53	29	52	14	55	21	46	25	37	36	37	32	46	25	36	17	45	24	43	31	41	24	40	14	55
	Harrisburg	10	27	44	30	35	28	37	28	35	29	47	33	58	29	35	35	46	28	31	25	30	31	40	31	46	33	58
	Middletown/ Harrisburg AP	20	34	46	29	51	29	46	33	46	29	47	33	58	16	52	29	49	28	40	32	37	33	40	31	46	33	58
	Philadelphia	60	NE	61	NW	59	NW	56	SW	59	SW	56	NW	73	SW	49	E	67	NE	49	SW	66	SW	60	NW	48	NW	73
	Pittsburgh	50	23	52	26	58	25	48	30	51	30	48	34	53	25	51	29	46	20	38	31	39	29	45	25	48	26	58
	Avoca	45	SW	52	W	60	S	49	NW	47	SW	46	W	43	NW	43	NE	50	SW	47	E	40	NW	49	SW	47	W	60
	Williamsport	49	27	66	14	60	11	58	18	62	18	55	29	62	20	78	29	60	16	59	11	75	09	77	16	58	20	78
RI	Block IS	12	27	36	36	46	05	45	04	29	15	29	18	28	30	28	04	29	16	53	08	37	28	38	25	40	16	53
	Providence	49	20	46	16	46	18	60	20	51	20	42	20	40	14	39	11	90	18	58	14	41	18	52	14	48	11	90
SC	Charleston AP	27	20	40	30	38	21	46	20	38	24	33	17	44	28	40	27	38	21	52	21	39	15	37	24	39	21	52
	Columbia	49	28	46	20	40	27	60	33	44	28	47	27	47	32	40	30	48	30	48	27	29	35	35	26	41	27	60
	Greenville-Spartanburg AP	12	25	36	22	37	25	39	29	39	36	43	30	43	17	49	24	36	23	30	12	31	25	32	35	29	17	49
SD	Aberdeen	33	34	58	06	52	35	52	31	55	16	46	10	47	31	63	33	46	32	41	33	49	34	46	33	43	31	63
	Huron	60	NW	57	NW	56	NW	68	SE	73	NW	70	SE	65	NW	77	NW	72	NW	64	W	72	NW	73	NW	59	NW	77
	Rapid City	19	32	59	33	59	33	54	32	61	32	57	25	54	21	69	32	54	32	52	32	55	33	57	33	52	21	69
	Sioux Falls	54	32	47	31	45	02	60	31	51	11	46	23	70	36	69	31	58	16	50	27	60	36	52	36	46	23	70
TN	Bristol-JhnCty-Kgsprt	47	25	40	25	46	25	40	25	41	32	50	27	39	23	40	34	46	31	29	27	36	26	37	24	40	32	50
	Chattanooga	27	31	30	25	37	32	44	18	32	18	35	24	37	30	44	06	43	29	33	29	35	30	38	22	29	30	44
	Knoxville	28	27	43	25	39	24	43	28	64	20	40	07	35	24	43	30	38	24	29	26	43	25	49	20	39	28	64
	Memphis	23	34	35	24	38	16	40	24	46	34	40	29	51	34	37	30	37	36	39	28	40	23	40	30	36	29	51
	Nashville	27	26	38	20	36	13	41	25	40	36	41	10	38	36	38	02	40	34	33	17	35	15	39	23	41	13	41
TX	Abilene	22	19	38	31	45	27	41	27	55	31	54	21	49	30	48	19	55	35	43	31	46	33	39	32	40	27	55
	Amarillo	28	25	45	25	48	34	58	25	53	13	47	35	60	30	48	02	46	80	41	31	58	31	46	32	51	35	60
	Austin/City	23	35	37	34	39	33	36	27	46	20	52	34	41	28	40	03	35	02	52	35	33	30	36	31	44	20	52

	Austin/Bergstrom	61	29	73	23	59	20	73	25	60	32	78	29	80	11	62	36	66	02	63	36	51	32	62	34	76	29	80
	Brownsville	23	18	40	23	36	28	44	17	41	17	38	30	41	12	36	19	48	30	51	36	35	17	39	17	38	30	51
	Corpus Christi	26	30	43	06	45	17	45	30	45	10	56	01	41	02	46	11	55	30	49	28	49	90	41	18	38	10	56
	Dallas-Fort Worth	49	36	55	36	51	29	55	32	55	14	55	32	52	36	65	36	73	11	53	23	46	34	50	32	53	36	73
	Dallas-Love Field	3	30	41	17	40	29	47	35	43	28	47	16	40	02	35	33	35	10	35	14	43	03	38	33	45	29	47
	Del Rio	23	31	45	31	39	33	52	15	41	32	48	28	51	16	39	14	60	32	45	33	46	32	36	29	43	14	60
	El Paso	27	26	64	24	51	28	52	26	56	25	45	32	51	30	45	26	54	35	41	23	41	23	52	26	44	26	64
	Galveston	124	S	53	N	60	SE	50	NW	68	W	66	SE	62	NW	68	E	91	NE	100	SE	66	NW	72	NW	50	NE	100
	Houston	33	33	32	26	46	25	35	14	45	23	46	30	45	10	46	08	51	05	37	27	41	33	37	14	46	08	51
	Lubbock	54	28	59	25	58	34	69	25	58	36	70	05	63	25	64	30	46	36	45	25	65	25	59	25	58	36	70
	Midland-Odessa	49	28	44	25	67	27	53	30	53	20	52	24	58	05	58	16	55	23	53	14	46	25	48	26	47	25	67
	Port Arthur	23	06	39	16	39	10	43	30	46	29	52	14	55	05	44	01	44	06	38	33	36	34	38	25	41	14	55
	San Angelo	54	26	45	29	48	27	58	28	75	02	60	02	57	17	46	02	44	34	52	29	60	29	66	30	43	28	75
	San Antonio	26	29	39	31	42	33	46	35	39	28	46	90	38	09	48	11	39	25	43	31	35	33	37	33	33	09	48
	Victoria (G)	19	N	49	N	54	N	54	N	62	N	68	N	60	E	54	N	54	S	44	NW	75	N	55	N	45	NW	75
	Waco	54	32	49	36	58	27	65	36	62	36	60	09	69	36	60	05	60	32	60	34	52	29	62	32	52	09	69
	Wichita Falls	54	32	49	29	57	27	59	14	52	20	59	36	69	33	60	34	55	01	53	29	60	32	56	29	55	36	69
UT	Milford	18	SW	44	W	56	SW	52	SW	52	SW	57	SW	57	NW	45	SW	52	SW	47	NW	45	SW	56	SW	53	SW	57
	Salt Lake City	66	NW	59	SE	56	NW	71	NW	57	NW	57	W	63	NW	51	SW	58	W	61	NW	67	NW	63	S	54	NW	71
VT	Burlington	19	16	38	17	39	16	33	33	36	29	35	29	39	30	37	36	35	19	36	13	32	18	35	27	35	29	39
VA	Lynchburg	58	W	45	S	50	S	43	NE	43	N	56	SW	56	NW	43	W	48	NE	40	N	41	NW	43	SE	45	N	56
	Norfolk	30	20	43	36	44	22	46	02	41	70	40	30	46	34	46	35	46	30	46	04	48	21	40	01	39	04	48
	Richmond	17	22	38	23	39	27	41	33	46	23	46	26	45	23	40	36	44	10	40	10	37	23	36	15	40	33	46
	Roanoke	41	30	53	31	40	32	52	32	58	36	46	28	46	34	46	30	44	10	38	30	36	34	52	30	40	32	58
	Wallops Island	16	NNE	70	SSW	66	W	68	NE	69	S	52	W	69	WNW	67	WNW	75	WNW	70	W	60	WNW	60	E	73	WNW	75
WA	Olympia	54	18	55	18	45	23	40	23	46	29	39	25	32	18	29	27	26	18	35	23	58	18	60	16	45	18	60
	Quillayute	34	21	35	SE	46	23	38	SW	32	W	28	SE	23	NE	23	SE	27	SE	33	SE	42	SE	37	SW	39	SE	46
	Seattle CO (G)	11	SSE	51	SSW	40	WSW	54	SW	44	WSW	46	SW	37	SW	39	S	33	S	33	SSW	41	SSW	63	SW	46	SSW	63
	Seattle Sea-Tac AP	34	S	47	S	51	SW	44	SW	38	SW	32	SW	29	SW	26	SW	29	20	36	SW	38	S	66	S	49	S	66
	Spokane	53	SW	59	SW	54	SW	54	SW	52	W	49	SW	44	SW	43	SW	50	SW	38	SW	56	SW	54	SW	51	SW	59
	Yakima	48	25	44	28	48	23	48	29	46	18	46	20	47	24	43	20	37	30	38	31	41	29	45	23	48	28	48
PR	San Juan	17	50	28	06	29	07	32	60	31	60	28	12	35	32	49	04	34	05	79	40	37	36	33	70	30	05	79
WV	Beckley	39	24	46	26	40	27	58	27	44	24	41	27	40	36	46	32	40	30	46	26	32	16	44	28	41	27	58
	Charleston	53	25	45	19	40	32	46	27	45	25	55	32	50	29	46	29	50	20	35	25	45	29	40	25	55	25	55
	Elkins	19	27	46	25	55	32	46	27	50	30	46	27	40	30	37	32	40	26	35	29	46	30	33	27	40	25	55
	Huntington	40	26	43	26	41	25	37	18	44	29	47	24	35	34	35	24	35	34	29	30	38	23	35	26	36	29	47
WI	Green Bay	17	40	39	27	37	29	44	22	41	29	46	16	41	24	36	28	35	32	37	28	36	20	45	26	38	29	46
	La Crosse	21	32	45	34	37	34	40	25	53	09	58	34	63	27	52	32	63	27	40	34	39	18	46	34	43	34	63
	Madison	53	E	68	W	57	SW	70	SW	73	SW	77	W	59	NW	72	W	47	W	52	SW	73	SE	56	SW	65	SW	77
	Milwaukee	20	80	44	27	52	26	41	24	48	30	46	31	47	30	54	02	47	70	41	24	43	23	52	04	40	30	54
WY	Casper	49	20	58	23	58	25	81	25	54	32	58	26	54	25	52	25	50	32	53	25	55	25	51	20	63	25	81
	Cheyenne	21	25	55	27	59	28	52	30	58	26	48	29	71	25	46	25	56	28	48	26	53	27	59	27	63	29	71
	Lander	13	26	44	20	46	23	68	25	62	17	52	19	53	22	47	18	39	20	52	26	55	27	44	24	55	23	68
	Sheridan	17	30	53	24	45	31	49	30	61	31	60	31	47	20	52	32	58	31	46	29	46	32	55	31	48	31	60

Note: Through 2002. This table expresses both a maximum wind speed and where available, the direction (referenced to true North) from which it blew. Short gusts are listed only for stations denoted with a (G). If the direction is expressed as one of the 16 compass points (N, NNE, NE, etc) the maximum speed is calculated from the minimum time during which one mile of wind passed the station. If the direction is expressed numerically, the maximum speed is the highest one-minute average value recorded by the observer. Direction is given in tens of degrees clockwise from true North.

Source: From U.S. National Oceanic and Atmospheric Administration, *Comparative Climatic Data for the United States Through 2000*, www.noaa.gov.

Table 3A.10 Average Percentage of Possible Sunshine — Selected Cities of the United States

State	Station	Years	Jan	Feb	Mar	Apr	May	Jun	Jul	Aug	Sep	Oct	Nov	Dec	Annual
AL	Birmingham CO	10	48	48	62	61	64	63	60	62	57	63	49	52	57
	Birmingham AP	34	42	50	54	63	66	65	59	63	61	66	55	46	58
	Montgomery	45	47	52	59	65	63	62	61	63	62	64	55	49	58
AK	Anchorage	40	34	42	50	50	50	46	43	39	38	36	32	27	41
	Juneau	33	32	32	37	39	39	34	31	32	26	19	23	20	30
	Nome	40	40	55	54	54	50	43	37	32	36	34	31	34	42
AZ	Flagstaff	15	77	73	76	82	88	86	75	76	81	79	75	73	78
	Phoenix	101	78	80	84	89	93	94	85	85	89	88	83	77	85
	Tucson	53	80	82	86	90	92	93	78	80	87	88	84	79	85
	Yuma	42	84	87	90	94	95	97	91	91	93	92	87	82	90
AR	Fort Smith	51	50	55	56	60	62	69	73	72	66	65	54	50	61
	Little Rock	32	46	54	57	62	68	73	71	73	68	69	56	48	62
	North Little Rock	24	61	62	72	77	74	81	82	79	81	75	62	57	72
CA	Eureka	84	43	46	52	57	58	59	55	51	55	50	44	41	51
	Fresno	46	47	65	77	85	90	95	97	96	94	88	66	46	79
	Los Angeles CO	32	69	72	73	70	66	65	82	83	79	73	74	71	73
	Redding	10	72	82	85	90	91	94	97	97	94	92	84	73	88
	Sacramento	46	48	65	74	82	90	94	97	96	93	86	66	49	78
	San Diego	56	72	71	70	68	59	58	68	70	69	68	75	73	68
	San Francisco CO	38	56	62	69	73	72	73	66	65	72	70	62	53	66
CO	Denver	46	71	69	69	67	64	70	71	71	74	72	64	67	69
	Grand Junction	56	61	65	65	70	73	81	79	77	79	74	63	61	71
	Pueblo	61	75	74	74	74	73	78	79	78	80	79	72	71	76
CT	Hartford	42	53	56	57	55	57	60	62	62	59	57	45	47	56
DC	Washington Nat'l AP	50	46	50	55	57	58	64	62	62	61	59	51	46	56
FL	Apalachicola	57	58	61	65	74	77	71	64	64	66	74	67	57	66
	Jacksonville	50	58	62	68	73	70	66	65	64	58	60	60	54	63
	Key West	38	74	77	82	84	82	76	77	76	72	71	71	70	76
	Miami	20	66	68	74	76	72	68	72	71	70	70	67	63	70
	Pensacola	5	48	53	61	63	67	67	57	58	60	71	64	49	60
	Tampa	50	63	65	71	75	75	67	62	61	61	65	64	61	66
GA	Atlanta	65	49	54	58	66	68	67	63	64	62	66	58	50	60
	Macon	48	56	61	65	73	71	70	67	71	67	69	64	57	66
	Savannah	46	54	57	62	71	68	65	64	62	58	63	61	55	62
HI	Hilo	52	46	46	42	37	37	44	41	42	43	39	33	37	41
	Honolulu	46	65	68	72	70	72	74	76	77	77	70	65	63	71
	Kahului	37	64	64	64	63	68	72	71	71	73	68	62	63	67
	Lihue	51	55	57	56	55	61	63	62	65	67	59	49	49	58
ID	Boise	60	40	50	62	68	70	75	87	85	82	69	43	38	64
	Pocatello	53	40	53	61	66	67	75	83	81	80	72	47	40	64
IL	Cairo	45	45	50	56	62	65	72	74	75	69	67	51	44	61
	Chicago	16	44	49	51	50	58	67	66	62	59	55	38	43	54
	Moline	53	48	50	50	53	57	63	68	66	62	58	42	40	55
	Peoria	52	47	50	51	55	60	67	69	67	64	61	43	42	56

	Springfield	48	48	52	51	56	63	68	71	70	68	63	48	44	58
IN	Evansville	56	42	48	55	60	64	71	73	73	69	65	48	42	59
	Fort Wayne	52	46	51	55	60	68	74	75	74	68	62	42	38	59
	Indianapolis	53	40	49	50	54	60	65	66	68	65	61	41	38	55
IA	Des Moines	50	51	54	57	56	61	68	72	70	66	62	49	46	59
	Sioux City	55	57	56	57	59	61	67	73	70	66	63	51	50	61
KS	Concordia	34	64	63	63	65	67	76	78	76	70	68	59	57	67
	Dodge City	59	67	65	65	68	67	74	79	78	74	73	67	65	70
	Topeka	53	56	55	57	58	61	66	71	70	66	64	54	52	61
	Wichita	49	58	61	62	63	64	69	76	75	68	65	58	57	65
KY	Greater Cincinnati AP	13	33	40	48	56	57	61	61	61	61	54	36	31	50
	Louisville	48	41	48	51	56	60	66	67	66	64	61	46	40	56
	Paducah	18	45	48	54	61	63	60	70	70	63	64	50	42	58
LA	Lake Charles	19	62	66	74	71	72	78	83	81	78	75	67	59	72
	New Orleans	22	46	50	56	62	62	63	58	61	61	64	54	48	57
	Shreveport	50	51	57	58	60	63	71	75	74	70	69	60	54	64
ME	Portland	55	56	59	56	54	54	59	63	63	62	58	48	53	57
MD	Baltimore	40	51	55	56	56	56	62	64	62	60	58	51	49	57
MA	Blue Hill	112	46	50	48	49	52	55	57	58	56	55	47	46	52
	Boston	61	53	56	57	56	58	63	65	65	63	60	50	52	58
MI	Alpena	37	36	45	53	52	59	63	65	59	51	42	28	28	48
	Detroit	31	40	46	52	54	61	66	68	67	61	51	35	31	53
	Grand Rapids	36	28	39	46	51	56	62	64	61	54	44	27	23	46
	Lansing	42	36	44	49	52	61	65	69	64	59	50	31	29	51
	Marquette	21	35	41	49	51	61	64	66	62	54	45	35	34	50
	Sault Ste. Marie	55	36	47	55	54	57	58	62	58	45	38	24	27	47
MN	Duluth	48	48	53	55	56	57	58	65	61	52	46	35	39	52
	Minneapolis-St.Paul	58	53	59	57	58	61	66	72	69	62	55	39	42	58
MS	Jackson	33	49	54	60	66	63	70	66	67	65	70	57	49	61
	Tupelo	13	53	53	61	73	72	74	75	73	72	62	51	46	64
MO	Columbia	27	50	49	50	55	57	64	67	64	60	59	47	45	56
	Kansas City	23	58	55	58	62	61	66	72	67	66	60	49	49	60
	St. Louis	37	50	52	54	56	59	66	68	65	63	60	46	43	57
MT	Billings	57	47	53	61	60	61	64	76	75	68	61	46	45	60
	Great Falls	46	49	56	66	62	62	65	79	76	67	61	46	44	61
	Helena	55	46	55	61	59	60	64	78	74	67	60	44	42	59
	Missoula	57	33	44	54	57	59	63	81	77	69	55	34	29	55
NE	Lincoln	40	58	57	57	58	61	69	73	70	66	63	53	52	61
	North Platte	50	63	62	62	64	65	71	77	75	72	70	60	61	67
	Omaha (North)	57	55	53	54	58	61	67	74	70	68	65	51	48	60
	Valentine	29	63	62	59	59	62	69	76	76	71	68	60	60	65
NV	Ely	56	68	68	71	70	72	80	80	81	82	75	67	67	73
	Las Vegas	47	77	81	83	87	88	93	88	88	91	87	81	78	85
	Reno	45	65	68	75	80	81	85	92	92	91	83	70	64	79
	Winnemucca	42	51	56	60	66	72	77	86	85	82	74	54	52	68
NH	Concord	58	52	55	53	53	55	58	62	60	56	53	42	47	54

(Continued)

Table 3A.10 (Continued)

State	Station	Years	Jan	Feb	Mar	Apr	May	Jun	Jul	Aug	Sep	Oct	Nov	Dec	Annual
	Mt. Washington	64	32	35	34	34	36	32	30	31	35	38	29	30	33
NJ	Atlantic City AP	36	50	53	55	56	56	60	61	65	61	59	51	47	56
NM	Albuquerque	63	72	72	73	77	79	83	76	75	79	79	76	71	76
	Roswell	7	60	68	75	77	80	83	77	73	72	77	73	71	74
NY	Albany	61	46	52	54	54	56	60	64	61	57	52	37	39	53
	Binghamton	51	37	42	46	50	56	62	64	61	55	49	32	29	49
	Buffalo	59	31	38	46	51	56	65	67	64	57	50	29	27	48
	New York C.Park	109	51	55	57	58	61	64	65	64	62	61	52	49	58
	Rochester	57	35	41	49	53	59	66	69	66	59	49	31	30	51
	Syracuse	53	33	39	46	49	55	59	63	59	53	44	26	25	46
NC	Asheville	32	55	59	61	66	61	62	60	54	56	61	58	55	59
	Cape Hatteras	33	48	52	60	67	65	65	65	65	64	60	56	48	60
	Charlotte	48	54	58	61	68	67	67	67	65	64	65	58	55	62
	Greensboro-Wnstn-Slm-HPT	68	51	56	60	63	63	64	62	61	62	64	57	53	60
	Raleigh	42	52	56	60	63	59	60	60	58	58	60	57	53	58
	Wilmington	51	56	59	64	70	67	66	64	62	61	64	63	59	63
ND	Bismarck	63	53	53	58	58	61	64	73	72	65	58	43	47	59
	Fargo	54	50	56	58	60	61	62	71	69	60	54	40	43	57
	Williston	39	51	57	61	60	62	66	74	74	67	59	45	50	60
OH	Cleveland	59	30	37	45	52	58	65	67	63	60	52	31	25	49
	Columbus	45	36	42	44	51	56	60	60	60	61	56	37	31	50
	Dayton	53	40	44	48	52	58	66	66	67	65	59	40	36	53
	Toledo	41	41	46	50	52	60	64	65	63	61	54	37	33	52
OK	Oklahoma City	42	60	60	65	68	66	75	79	79	72	70	61	58	68
	Tulsa	55	53	56	58	60	60	66	74	73	66	64	56	53	62
OR	Portland	46	28	38	48	52	57	56	69	66	62	44	28	23	48
PC	Guam	39	47	52	57	57	56	52	40	36	38	38	40	38	46
	Johnston Island	22	70	74	74	70	74	78	79	76	73	65	59	62	71
	Koror	42	54	54	63	61	54	44	45	44	52	47	50	49	51
	Majuro, Marshall IS	42	61	64	66	59	58	55	56	61	59	55	53	53	58
	Pago Pago, Amer Samoa	34	45	45	45	40	34	34	40	43	51	42	46	45	42
	Pohnpei, Caroline IS	44	38	41	45	42	41	40	44	44	44	40	39	36	41
	Chuuk, E. Caroline IS	42	50	55	56	50	47	44	47	48	44	42	44	42	47
	Wake Island	28	68	71	76	75	75	76	72	68	68	68	65	64	70
	Yap, W Caroline IS	43	59	61	68	67	65	52	47	45	50	47	52	49	55
PA	Allentown	12	43	48	53	46	53	62	57	61	58	57	49	45	53
	Harrisburg	53	49	54	58	59	60	65	68	67	62	58	47	44	58
	Middletown/Harrisburg AP	62	48	54	57	58	59	64	67	66	61	58	47	44	57
	Philadelphia	60	49	53	55	56	57	62	61	62	59	60	52	49	56
	Pittsburgh	49	32	36	43	46	50	55	57	56	55	51	36	28	45
	Avoca	41	41	47	50	53	57	61	62	61	55	52	36	34	51
RI	Providence	42	56	58	58	57	58	61	63	62	62	61	50	52	58

SC	Charleston AP	39	56	59	66	72	68	66	67	64	61	63	59	56	63
	Columbia	46	55	59	64	70	68	67	67	66	65	67	63	59	64
	Greenville-Spartanburg	39	54	57	63	66	62	62	60	61	62	66	58	54	60
SD	Huron	55	57	59	59	61	65	70	76	74	69	63	50	49	63
	Rapid City	55	57	60	63	62	60	65	73	74	70	66	55	55	63
TN	Chattanooga	65	43	49	53	61	65	65	62	63	64	63	53	44	57
	Knoxville	57	40	47	53	63	64	65	64	63	61	61	49	40	56
	Memphis	35	50	54	56	64	69	74	74	75	69	70	58	50	64
	Nashville	54	41	47	52	59	60	65	63	63	62	62	50	42	56
TX	Abilene	49	62	64	70	72	70	78	80	78	71	72	67	62	70
	Amarillo	61	69	68	72	74	71	78	79	77	73	75	72	67	73
	Austin/City	58	49	51	55	54	56	69	75	74	66	64	54	49	60
	Austin/Bergstrom	60	49	51	55	54	56	69	75	74	66	63	56	49	60
	Brownsville	59	41	48	53	58	63	73	80	76	68	65	51	42	60
	Corpus Christi	60	44	49	54	56	59	72	80	76	68	67	54	43	60
	Dallas-Fort Worth	17	52	54	58	61	57	67	75	73	67	63	57	52	61
	El Paso	54	78	82	86	89	90	90	82	81	83	84	83	77	84
	Galveston	103	48	51	56	61	67	75	73	71	68	71	59	48	62
	Houston	27	45	50	54	58	62	68	70	68	66	64	52	51	59
	Lubbock	25	65	66	73	74	71	76	77	76	71	75	69	65	72
	Midland-Odessa	22	66	69	73	78	78	81	81	77	77	72	74	65	74
	Port Arthur	26	42	52	52	52	64	69	65	63	62	67	57	47	58
	San Antonio	53	47	50	57	56	56	67	74	74	67	64	54	48	60
UT	Milford	16	58	64	63	69	73	82	77	79	80	76	62	60	70
	Salt Lake City	64	45	54	64	68	72	80	83	82	82	72	53	42	66
VT	Burlington	59	41	48	51	49	56	59	64	60	54	47	31	33	49
VA	Lynchburg	52	52	56	58	62	62	65	62	62	61	62	56	53	59
	Norfolk	32	53	56	60	63	62	67	62	62	61	59	56	54	60
	Richmond	46	54	58	62	66	66	70	68	66	65	63	59	54	63
WA	Quillayute	30	22	30	34	35	37	35	43	44	47	34	21	19	33
	Seattle CO	31	28	34	42	47	52	49	63	56	53	37	28	23	43
	Seattle Sea-Tac AP	30	28	40	50	52	56	56	65	65	62	43	28	23	47
	Spokane	48	28	41	55	61	65	67	80	78	72	55	29	23	54
PR	San Juan	47	69	70	76	72	64	65	69	68	62	64	60	61	67
WV	Elkins	11	29	32	39	46	44	48	44	44	45	46	37	28	40
WI	Green Bay	53	49	52	53	55	61	65	66	63	56	47	38	40	54
	Madison	50	47	51	52	52	58	64	67	64	60	54	39	40	54
	Milwaukee	55	44	47	50	53	60	65	69	66	59	54	39	38	54
WY	Cheyenne	63	64	67	67	63	61	67	69	68	70	69	61	60	66
	Lander	50	65	68	70	66	64	72	75	75	72	67	58	61	68
	Sheridan	55	57	60	63	60	60	65	75	75	68	62	53	55	63

Note: Through 2002. The total time that sunshine reaches the surface of the earth is expressed as the percentage of the maximum amount possible from sunrise to sunset with clear sky conditions.

Source: From U.S. National Oceanic and Atmospheric Administration, *Comparative Climatic Data for the United States Through 2000*, www.noaa.gov.

Table 3A.11 Mean Number of Cloudy Days (Clear, Partly Cloudy, Cloudy) — Selected Cities of the United States (Through 2002)

			Jan			Feb			Mar			Apr			May			Jun			Jul			Aug			Sep			Oct			Nov			Dec			Annual		
State	Station	Years	CL	PC	CD	CL	PC	CD	CL	PC	CD	CL	PC	CD	CL	PC	CD	CL	PC	CD	CL	PC	CD	CL	PC	CD	CL	PC	CD	CL	PC	CD	CL	PC	CD	CL	PC	CD	CL	PC	CD
AL	Birmingham AP	37	7	6	18	7	6	15	7	8	16	9	8	13	8	11	12	7	13	10	5	14	12	7	15	10	9	9	11	14	8	9	10	7	13	8	7	16	99	111	155
	Huntsville	27	7	6	18	7	6	16	7	8	17	9	7	14	8	10	14	8	11	11	7	13	11	9	13	10	9	9	12	12	7	12	9	7	14	8	6	18	100	101	164
	Mobile	47	8	6	17	8	7	14	9	8	14	9	9	12	9	11	11	7	14	9	4	15	12	6	15	10	9	10	11	14	8	9	11	7	12	9	6	16	102	116	147
	Montgomery	51	7	6	17	8	6	14	8	8	15	10	8	12	9	10	12	8	12	10	6	14	11	8	14	9	10	9	11	14	7	10	11	6	13	9	6	16	107	107	151
AK	Anchorage	44	7	5	19	6	4	18	8	5	18	6	6	18	4	7	20	3	7	20	3	6	22	3	6	22	4	5	21	5	5	21	6	5	20	6	4	21	61	65	239
	Annette	41	4	4	23	3	4	21	4	5	23	3	6	21	4	7	21	2	6	21	4	6	21	4	6	21	4	6	21	2	4	25	3	3	24	3	4	24	40	60	265
	Barrow	54	6	2	3	12	6	11	14	7	10	11	7	12	3	5	23	3	6	21	3	7	21	1	4	26	1	3	26	2	4	25	6	3	11	4	0	0	66	52	187
	Barter IS	34	4	3	8	10	6	12	12	8	12	9	8	13	3	6	22	3	7	20	3	9	19	1	6	23	2	4	24	2	5	24	5	4	15	0	0	0	53	66	192
	Bethel	34	8	5	18	10	4	14	9	5	16	6	6	17	4	8	20	2	7	21	2	5	24	2	4	26	2	5	23	4	5	22	6	5	19	8	5	18	62	65	239
	Bettles	5	15	5	11	10	4	14	9	5	16	7	10	13	7	11	13	4	13	13	2	10	19	2	8	21	7	7	16	5	6	21	9	6	14	9	7	15	86	93	187
	Big Delta	3	14	8	9	5	7	17	8	8	15	2	9	20	2	11	18	4	9	17	—	8	22	2	8	21	5	7	18	4	5	23	7	6	17	7	7	17	60	93	212
	Cold Bay	42	2	5	23	2	5	22	2	6	23	1	4	26	—	3	27	—	3	27	—	2	28	—	2	29	—	4	26	1	5	25	1	6	23	2	5	24	10	50	304
	Fairbanks	45	9	6	16	8	6	14	10	7	14	7	8	16	5	11	15	3	10	17	3	9	19	3	7	21	4	6	20	4	5	22	7	5	18	7	6	18	70	86	210
	Gulkana	5	15	4	12	6	4	18	9	6	15	5	10	15	3	10	18	2	10	18	4	9	18	3	9	19	6	8	17	3	7	21	5	5	21	7	7	17	67	89	209
	Homer	15	8	4	19	7	4	17	6	6	19	5	6	18	4	7	20	4	8	18	3	9	19	5	8	18	5	7	18	5	5	21	6	5	19	7	5	18	65	75	225
	Juneau	47	6	3	23	4	3	21	4	3	23	4	4	22	4	4	23	3	4	22	3	5	23	4	5	22	3	3	24	2	2	27	4	2	24	3	2	25	44	41	280
	King Salmon	42	8	5	18	7	5	16	8	6	18	4	7	20	3	6	22	1	5	24	1	5	25	1	5	25	2	6	22	4	7	20	6	5	19	7	6	19	51	69	245
	Kodiak	25	5	6	20	6	5	18	6	7	18	4	6	19	3	5	22	3	6	21	3	6	22	4	7	20	4	7	19	6	8	17	7	7	16	6	6	19	59	74	232
	Kotzebue	52	11	4	16	12	3	13	12	6	13	10	7	13	8	9	14	6	10	14	4	8	19	4	6	21	6	5	19	8	5	19	9	4	17	11	4	16	100	72	193
	McGrath	55	9	5	17	9	4	15	9	6	15	7	7	15	4	8	19	2	9	19	2	8	21	2	5	23	3	5	22	4	5	22	6	4	19	8	5	19	67	72	226
	Nome	51	10	4	17	12	4	13	11	5	14	9	6	14	7	8	16	6	9	15	4	7	20	3	6	22	5	6	20	7	6	18	7	4	18	10	4	17	91	69	205
	St. Paul Island	66	2	6	23	3	6	20	3	7	21	2	6	22	1	4	26	1	3	26	1	2	28	1	3	28	1	5	24	1	7	23	1	6	22	1	5	24	18	61	287
	Talkeetna	10	6	4	21	8	4	16	9	4	18	5	5	18	6	6	19	1	12	17	5	7	19	3	11	17	5	4	19	6	5	20	7	3	19	5	2	22	67	67	225
	Unalakleet	4	8	4	20	10	3	15	10	6	15	9	5	16	5	9	17	5	7	18	2	8	21	6	6	20	4	7	19	6	5	21	8	5	18	11	4	16	83	67	215
	Valdez	23	5	4	22	5	3	20	8	4	19	6	5	19	4	6	21	3	6	21	3	5	23	4	6	21	3	5	22	5	4	23	6	4	20	5	3	23	57	55	253
	Yakutat	48	5	3	22	4	3	21	5	4	22	4	5	21	2	5	24	2	5	24	2	4	25	3	4	24	3	4	24	3	3	25	4	3	23	4	3	24	41	46	278
AZ	Flagstaff	43	12	6	12	11	6	11	12	8	12	12	9	9	15	9	7	18	8	4	9	13	9	10	13	8	16	10	5	17	7	7	15	7	8	14	6	11	162	102	102
	Phoenix	57	14	7	10	13	7	9	14	8	8	17	7	6	21	7	3	23	5	2	17	10	4	18	10	4	22	5	3	20	6	4	18	6	6	15	6	9	211	85	70
	Tucson	55	14	7	10	13	7	9	15	7	9	17	7	6	20	7	4	21	6	3	10	12	9	12	12	7	19	7	4	20	6	5	18	6	6	15	6	10	193	91	81
	Winslow	37	12	7	12	12	7	10	13	9	9	15	9	7	17	9	5	20	7	4	11	12	8	12	13	6	18	8	4	19	7	6	16	6	8	14	7	10	177	99	89
	Yuma	32	15	7	9	15	7	6	17	7	6	21	6	4	24	5	2	25	4	1	20	8	3	22	6	3	24	4	2	23	5	3	19	6	5	17	7	8	242	71	52
AR	Fort Smith	49	9	6	16	9	6	14	9	7	15	9	7	14	8	10	13	10	10	10	12	11	8	12	11	8	12	8	10	14	7	11	11	6	13	10	7	15	123	95	147
	Little Rock	35	9	6	16	9	6	13	9	7	15	9	7	14	8	11	12	9	12	9	9	13	9	12	11	9	11	9	10	14	7	9	11	6	13	9	6	16	119	100	147
CA	Bakersfield	55	7	8	17	7	8	13	10	9	12	12	9	9	18	9	5	23	5	2	26	3	1	26	4	1	24	4	2	19	7	5	12	8	10	7	8	16	191	81	93
	Bishop	29	11	8	12	11	8	10	13	9	9	14	9	7	15	10	6	20	7	3	22	7	2	23	6	2	23	5	2	20	7	5	15	8	7	14	7	9	201	89	75
	Blue Canyon	31	9	5	17	8	5	15	9	6	17	10	7	13	12	8	11	18	5	6	26	4	1	25	4	2	22	4	3	17	6	9	10	5	15	9	5	17	174	64	126
	Eureka	92	6	6	19	5	6	17	6	8	17	6	8	16	6	10	15	7	10	13	6	11	14	6	11	15	9	9	13	8	9	14	6	7	17	6	7	18	77	101	187
	Fresno	46	5	7	19	8	8	13	11	8	12	14	8	8	18	8	5	23	5	2	27	3	1	26	3	1	24	4	2	20	6	5	12	7	11	7	6	18	194	73	98
	Long Beach	38	12	8	11	10	7	11	11	9	11	12	10	8	10	13	8	12	12	6	18	11	2	19	10	2	15	11	4	13	11	7	13	8	8	13	8	10	159	119	87
	Los Angeles AP	60	12	8	11	11	6	11	12	9	11	11	9	9	10	11	10	10	11	9	13	13	5	14	12	5	13	10	6	13	10	8	15	8	8	13	8	10	147	116	103
	Los Angeles CO	34	14	8	9	12	7	9	13	9	9	12	10	8	11	12	8	14	10	6	21	9	1	22	7	1	18	8	3	16	9	6	16	7	6	15	8	8	186	106	73
	Mount Shasta	22	9	6	16	8	7	13	8	8	15	10	10	10	12	8	10	16	7	7	25	5	2	24	4	2	21	5	4	14	7	9	9	7	13	7	8	16	164	83	118
	Redding	9	8	5	18	8	6	14	8	7	17	8	8	13	11	9	10	17	8	5	25	4	1	25	5	2	22	4	3	18	7	6	11	8	11	10	5	16	172	77	116
	Sacramento	49	7	6	19	8	7	13	10	9	12	12	10	8	17	9	5	22	6	2	27	3	1	26	4	1	24	4	2	19	6	6	10	8	12	8	6	17	188	77	100
	San Diego	55	12	7	11	10	8	10	11	9	11	10	10	10	8	11	11	9	12	9	13	13	5	15	12	4	15	10	6	14	10	8	15	8	7	14	8	10	146	117	102
	San Francisco AP	68	9	8	15	8	7	13	10	9	13	11	9	10	14	10	8	16	9	5	21	8	3	19	9	3	18	8	4	16	9	7	11	8	11	9	8	14	160	100	105
	Santa Maria	45	13	7	11	12	6	10	13	8	10	13	9	8	14	10	7	15	11	4	17	12	2	17	13	2	16	10	4	17	9	5	15	7	8	14	7	10	176	110	80
	Stockton	46	5	6	19	7	7	15	10	8	13	12	9	10	17	8	5	22	6	3	27	3	1	26	4	1	23	4	2	19	6	6	9	8	12	7	7	18	184	77	104
CO	Alamosa	33	13	10	8	11	10	7	10	11	9	10	13	8	9	14	8	14	12	4	9	17	6	11	14	6	15	10	5	17	8	6	14	9	7	14	10	7	148	137	81
	Alamosa	2	16	6	6	8	6	7	8	4	3	8	8	8	11	7	6	14	6	5	16	6	1	5	10	9	10	2	2	10	6	3	6	6	4	12	8	2	124	75	54
	Colorado Springs	44	12	8	11	9	8	11	9	9	13	8	10	12	7	12	12	10	12	8	9	15	7	10	13	8	14	8	7	15	8	8	12	8	10	12	8	11	127	120	118
	Colorado Springs	3	5	2	2	8	5	7	8	6	7	5	6	8	1	5	2	3	4	2	7	4	2	6	8	8	11	4	6	10	1	5	6	2	2	11	4	1	82	51	50
	Denver	61	10	9	11	8	9	11	8	10	13	7	11	13	6	12	13	10	12	8	9	16	6	10	14	8	13	9	8	13	9	9	10	10	10	11	10	10	115	130	120
	Grand Junction	49	9	7	15	8	7	13	8	8	15	8	9	13	10	11	11	15	9	6	14	11	6	13	11	6	16	8	5	15	8	8	11	7	12	10	8	13	136	106	122
	Pueblo	50	12	9	10	10	8	11	10	9	12	9	10	11	9	12	11	12	11	7	11	14	5	12	12	7	15	8	6	15	8	8	12	9	9	12	9	10	139	119	107
	Pueblo	3	11	4	3	9	5	6	7	2	5	10	8	5	4	7	4	11	4	2	15	6	2	8	11	5	9	2	2	11	4	5	10	2	2	15	4	2	120	59	44
CT	Bridgeport	46	8	8	15	8	7	13	8	9	14	7	9	14	7	10	14	8	10	12	7	12	12	9	11	12	10	9	12	11	8	12	8	8	14	8	8	15	99	107	159
	Hartford	41	8	8	15	7	8	14	7	8	16	6	9	15	6	10	16	6	10	14	6	12	13	7	11	13	8	9	13	9	9	13	6	8	16	7	7	17	82	108	175
DE	Wilmington	47	7	7	17	7	7	14	8	8	15	7	8	14	7	10	14	8	11	12	8	11	12	9	10	12	10	8	12	11	8	12	8	8	14	8	7	16	97	104	164
DC	Washington Dulles AP	33	7	7	17	7	6	15	7	8	16	7	9	14	7	10	14	6	12	12	8	11	12	8	11	12	9	9	12	11	8	12	7	8	15	7	7	17	92	105	168
	Washington Nat'l AP	48	7	7	16	7	7	15	7	8	15	7	9	14	7	10	14	7	11	11	7	12	12	9	10	12	10	8	12	11	8	12	8	8	14	8	7	16	96	106	164

FL	Apalachicola	62	10	8	5	10	7	12	11	8	12	12	9	9	13	10	8	9	13	8	6	13	11	7	13	11	10	10	10	16	7	7	13	8	9	10	8	13	128	113	116
	Daytona Beach	52	9	9	13	9	8	12	10	10	12	11	11	9	10	11	10	6	13	12	5	14	12	5	15	11	5	13	12	9	10	11	10	10	10	9	9	13	97	132	136
	Fort Myers	54	11	12	9	10	11	7	11	12	8	11	13	6	10	15	6	5	16	10	2	18	11	2	19	10	4	16	11	11	13	8	11	12	7	11	12	8	98	168	99
	Jacksonville	47	9	8	14	9	7	12	9	9	13	10	10	9	9	12	10	5	13	12	5	14	12	5	15	11	5	12	13	10	9	12	10	9	11	9	8	14	94	127	144
	Key West	43	11	11	9	12	10	7	13	11	6	13	11	6	10	13	8	5	14	11	3	16	11	3	17	11	3	15	12	9	13	9	10	11	8	11	11	9	104	155	107
	Miami	46	9	13	9	8	12	8	9	14	8	8	15	7	6	15	9	3	14	13	3	17	11	2	18	11	2	15	12	7	14	10	8	14	9	9	13	9	74	175	115
	Orlando	47	9	10	12	9	9	11	9	10	12	10	11	9	9	14	9	4	14	12	3	17	11	3	17	11	4	15	12	9	11	11	10	10	10	10	9	12	89	147	130
	Pensacola	29	8	7	16	9	7	13	9	8	14	10	9	11	9	11	11	7	15	8	4	17	10	6	15	10	10	11	9	13	9	9	11	8	11	9	7	16	105	123	137
	Tallahassee	34	8	7	15	8	7	13	9	9	13	11	10	9	9	13	10	6	14	10	4	17	10	5	16	10	8	12	10	13	8	10	11	8	11	9	8	14	102	129	134
	Tampa	49	10	10	12	9	9	10	11	10	10	11	11	8	11	13	8	6	14	10	3	16	12	3	17	11	5	14	11	11	10	9	12	10	9	10	10	11	101	143	121
	West Palm Beach	48	8	11	12	8	11	10	8	13	10	9	13	8	8	14	10	4	13	13	4	14	13	3	16	12	3	14	13	6	14	11	7	13	10	9	12	10	75	159	131
	West Palm Beach	1	1	6	6	8	3	2	7	6	4	9	13	8	5	6	8	6	8	9	4	14	13	3	16	12	3	14	13	6	14	11	7	13	10	2	3	10	30	32	29
GA	Athens	52	9	7	15	9	6	13	9	7	14	10	8	11	9	10	12	8	12	10	7	13	11	8	13	10	10	9	11	14	7	10	12	6	12	10	6	15	113	105	147
	Atlanta	61	8	6	16	8	6	14	9	8	15	10	8	12	9	11	11	8	12	10	6	13	12	7	13	10	10	10	10	14	7	10	12	6	12	9	6	15	110	107	149
	Augusta	44	9	6	16	9	6	13	9	8	14	11	8	11	8	11	12	8	11	11	6	13	12	7	14	10	9	9	11	14	7	10	12	6	12	10	6	15	112	106	147
	Columbus	49	8	6	17	8	6	14	9	7	15	10	9	11	9	10	12	8	12	10	5	13	12	8	14	10	10	9	11	14	7	10	12	6	12	9	6	16	110	104	151
	Macon	46	8	7	16	8	7	13	9	8	14	11	8	11	9	11	11	8	11	11	6	13	12	8	12	10	9	10	11	14	7	10	12	6	12	9	7	15	112	107	147
	Savannah	45	9	6	16	8	6	14	9	9	13	11	9	11	9	10	12	7	11	12	6	14	12	6	14	12	7	10	13	12	8	11	11	7	12	9	7	15	104	110	152
HI	Hilo	50	6	11	13	5	10	13	3	10	18	1	9	20	1	11	19	2	11	17	1	12	18	2	12	17	3	12	15	3	12	16	3	10	17	5	11	15	36	132	197
	Honolulu	47	9	13	9	8	12	8	8	14	9	6	14	10	7	16	9	6	17	6	7	18	5	8	17	6	8	16	6	7	15	8	7	14	9	8	14	9	90	181	94
	Kahului	37	13	10	8	11	9	7	11	11	9	8	12	10	10	14	8	11	13	6	11	15	5	12	14	5	12	13	6	11	13	8	11	11	8	12	11	8	131	145	90
	Lihue	46	8	13	10	7	12	9	5	14	12	3	14	13	3	17	11	3	17	9	3	19	10	4	18	9	5	18	7	5	16	10	4	14	12	6	14	11	56	184	125
ID	Boise	56	4	5	22	5	6	17	6	7	18	6	9	15	8	10	13	12	10	8	20	7	3	19	8	4	17	7	6	12	8	11	6	6	18	5	6	20	120	90	155
	Lewiston	42	3	4	24	3	5	21	4	6	21	5	6	19	6	9	16	8	9	13	18	7	6	16	8	7	13	7	9	9	7	15	3	6	22	3	4	24	91	78	196
	Pocatello	46	3	7	22	4	6	18	5	8	18	6	8	16	7	10	14	12	9	9	17	9	5	16	11	5	15	8	7	12	8	11	5	7	18	3	7	21	106	98	162
IL	Cairo	30	8	6	17	8	6	14	9	8	15	8	9	13	8	10	13	8	12	10	9	13	10	12	11	8	12	9	9	15	7	9	9	8	13	8	7	16	113	104	149
	Chicago	37	7	6	18	6	6	16	5	8	18	6	8	16	7	10	14	7	11	11	8	12	10	9	11	11	9	10	11	9	9	14	5	6	18	6	6	19	84	105	176
	Moline	62	8	7	16	7	6	15	6	8	17	7	9	15	8	9	14	7	11	12	10	12	9	10	11	10	12	8	11	12	8	12	7	7	16	7	6	18	101	100	164
	Peoria	52	7	6	18	7	6	16	6	7	18	6	8	16	7	10	14	7	11	12	9	12	10	10	10	10	11	9	10	11	8	12	7	6	17	7	6	19	95	97	172
	Rockford	45	7	7	17	7	6	15	6	8	17	6	8	16	8	9	14	7	11	12	9	12	10	9	11	11	10	9	12	10	7	13	6	6	18	7	6	19	93	98	174
	Springfield	48	7	6	17	7	6	15	6	7	18	7	8	15	8	9	14	8	9	12	10	11	10	11	10	10	12	8	10	12	7	12	8	6	16	7	6	18	104	94	167
IN	Evansville	55	7	5	19	7	6	15	7	8	16	7	8	15	8	9	14	8	11	10	9	12	9	11	11	9	11	8	10	13	7	11	8	7	15	7	6	18	102	100	163
	Fort Wayne	49	5	6	20	5	7	17	5	7	19	6	7	17	7	9	15	6	12	12	8	12	11	9	12	11	9	9	12	10	8	14	5	6	19	4	6	21	78	102	186
	Indianapolis	64	6	6	19	6	6	16	6	7	18	6	7	17	7	9	15	7	11	12	8	12	10	9	12	10	11	9	11	11	8	12	6	7	17	5	6	20	88	99	179
	South Bend	56	3	6	22	4	6	19	5	7	19	6	8	17	7	10	14	7	10	12	8	13	10	9	12	10	9	9	12	9	8	14	4	6	20	3	6	22	73	100	193
IA	Des Moines	46	8	7	16	8	6	15	7	7	17	7	8	15	8	8	15	8	10	11	10	11	9	11	10	10	12	8	11	12	7	12	7	7	16	7	7	17	105	97	164
	Dubuque	21	7	7	17	8	6	15	6	8	17	6	8	16	7	9	16	6	11	13	8	12	10	9	10	11	10	8	12	10	8	13	7	6	17	7	6	18	92	99	175
	Sioux City	55	8	8	15	7	7	14	7	8	17	7	9	15	7	9	15	8	11	11	11	12	8	12	11	9	12	8	11	12	8	12	7	7	16	7	7	16	104	104	157
	Waterloo	35	8	7	16	7	6	15	6	7	18	6	8	16	7	9	14	7	11	12	9	12	11	9	11	11	10	8	12	10	8	14	5	7	18	7	6	18	91	100	174
KS	Concordia	30	11	7	13	9	7	13	9	7	15	9	8	13	8	9	13	10	10	10	14	10	7	13	10	8	13	7	10	13	7	11	10	7	13	10	7	14	131	95	140
	Concordia	3	6	2	0	8	4	0	10	3	0	7	4	0	11	4	0	13	6	0	11	8	0	16	3	0	8	2	0	10	4	0	7	4	0	9	4	0	116	48	2
	Dodge City	50	11	8	12	9	8	12	9	8	14	9	9	12	9	10	12	11	10	8	13	12	7	14	10	7	14	7	9	15	7	9	12	8	11	11	7	12	137	104	124
	Dodge City	3	6	3	9	5	6	6	10	5	5	9	4	8	12	4	5	14	5	3	16	4	3	16	5	3	9	1	2	10	6	4	8	4	4	13	3	5	128	51	57
	Goodland	72	12	9	10	10	9	10	9	10	12	9	10	11	9	11	11	12	11	6	14	12	5	13	12	6	14	9	7	15	8	7	12	8	9	12	9	10	143	117	105
	Goodland	3	6	4	9	8	4	8	7	4	5	8	3	11	8	5	6	14	4	4	13	5	2	14	4	6	9	9	3	12	3	6	8	4	5	14	2	5	121	42	69
	Topeka	46	10	6	15	8	6	14	7	7	16	8	8	14	7	10	14	8	10	11	11	11	9	12	10	9	12	8	10	13	7	11	10	7	14	9	7	15	114	97	154
	Topeka	3	4	4	10	9	3	7	7	5	6	8	4	10	11	4	6	11	6	5	6	9	6	12	7	4	7	1	4	10	2	8	6	2	8	10	3	8	102	51	83
	Wichita	39	11	6	14	8	7	13	9	7	15	9	8	13	8	10	13	9	11	10	13	11	7	13	10	8	13	7	10	13	7	11	11	7	13	10	7	14	128	97	140
	Wichita	3	6	5	8	7	6	8	7	4	8	6	2	9	11	8	6	10	7	3	15	4	3	16	5	2	6	3	2	10	4	5	7	2	8	11	4	8	114	55	70
KY	Greater Cincinnati AP	44	5	6	20	5	6	17	5	7	19	6	8	17	6	10	15	7	10	13	8	12	12	8	12	11	10	9	11	10	7	13	6	6	18	5	6	20	81	98	186
	Jackson	14	4	5	22	4	5	19	6	7	18	6	8	16	6	8	17	6	12	13	7	12	12	7	12	12	8	9	12	9	8	14	6	6	18	5	5	21	73	97	195
	Lexington	51	6	6	20	6	6	17	6	7	18	6	9	15	7	10	14	7	12	11	8	12	11	9	12	10	10	8	11	12	7	12	7	7	17	6	6	19	89	101	174
	Louisville	47	6	6	20	6	6	16	6	7	18	6	9	15	8	9	14	8	11	11	8	13	10	10	11	9	10	9	11	12	8	12	7	6	16	6	6	19	93	102	171
	Paducah	11	7	6	18	6	8	15	7	8	16	8	8	14	7	9	15	7	12	11	10	11	10	11	11	8	11	8	11	11	7	13	9	4	17	7	6	17	101	97	167
LA	Baton Rouge	42	7	6	18	8	6	15	8	8	15	8	9	13	8	11	12	7	14	9	5	15	11	7	15	9	9	11	10	14	8	8	10	8	13	8	7	16	99	119	147
	Baton Rouge	1	10	4	14	10	6	15	8	3	8	7	6	9	13	6	16	7	18	10	9	7	9	10	8	4	3	2	10	5	3	8	6	3	7	2	2	5	90	67	106
	Lake Charles	34	7	6	19	8	6	15	7	8	16	7	9	15	7	11	12	8	13	9	6	14	10	7	15	9	10	11	9	13	9	9	10	8	13	8	7	17	98	115	152
	New Orleans	47	7	7	17	8	6	14	8	8	15	8	10	12	9	11	11	8	13	9	5	15	12	7	14	10	10	11	10	14	8	9	10	8	12	8	7	16	101	118	146
	Shreveport	43	8	5	18	8	5	15	8	6	16	8	8	14	8	10	13	9	13	9	10	12	9	11	12	8	11	9	10	13	7	10	11	6	13	9	6	16	114	100	151
ME	Caribou	54	7	7	17	6	6	16	7	7	17	5	7	18	4	9	18	3	10	17	3	13	15	5	11	15	5	9	15	5	8	18	3	6	21	6	7	19	59	100	206
	Portland	54	10	7	14	9	7	13	9	7	15	7	7	16	6	9	15	7	10	13	7	11	13	9	11	11	10	8	12	10	8	13	7	7	15	9	7	15	101	99	165
	Baltimore	45	8	8	15	8	7	14	8	9	14	8	9	13	8	10	13	8	11	10	9	12	10	9	11	11	11	9	11	12	8	11	8	8	13	8	7	16	105	108	152
MA	Boston	60	9	7	15	8	7	13	8	8	15	7	8	15	6	10	15	7	10	13	7	12	12	9	11	11	10	8	12	11	8	12	8	7	15	9	7	15	98	103	164
	Worcester	40	9	8	14	8	7	14	8	8	16	7	9	15	6	10	15	6	11	13	6	12	13	8	11	12	9	9	12	10	8	13	7	8	15	8	8	16	90	107	168
MI	Alpena	36	3	7	21	4	7	17	6	8	17	6	8	16	7	10	14	8	11	12	8	13	10	9	11	12	6	10	14	5	8	18	2	6	22	2	6	23	67	104	194
	Detroit	37	4	7	20	5	7	17	6	7	18	6	8	16	7	10	14	8	11	11	9	12	10	9	11	11	8	10	12	8	9	15	4	7	19	3	6	22	75	105	185

(Continued)

Table 3A.11 (Continued)

State	Station	Years	Jan			Feb			Mar			Apr			May			Jun			Jul			Aug			Sep			Oct			Nov			Dec			Annual		
			CL	PC	CD	CL	PC	CD	CL	PC	CD	CL	PC	CD	CL	PC	CD	CL	PC	CD	CL	PC	CD	CL	PC	CD	CL	PC	CD	CL	PC	CD	CL	PC	CD	CL	PC	CD	CL	PC	CD
	Flint	54	4	6	21	4	7	17	5	7	19	6	7	17	6	10	15	6	12	12	7	13	11	8	12	11	7	10	13	7	9	15	3	6	21	3	6	22	66	105	195
	Grand Rapids	32	3	5	23	3	6	19	5	7	19	6	7	17	7	9	15	7	11	12	8	12	11	8	11	12	7	9	14	6	8	17	3	5	22	2	4	25	64	96	205
	Houghton Lake	30	3	6	22	4	7	17	6	7	18	6	7	17	7	9	14	7	11	12	7	13	10	8	11	12	6	9	14	5	8	18	2	5	23	2	6	23	65	100	200
	Lansing	41	3	7	21	4	7	17	5	8	18	6	7	17	7	10	14	7	11	11	8	13	10	8	12	11	8	9	13	7	9	15	4	6	21	3	6	23	71	104	191
	Muskegon	53	2	3	26	3	5	21	5	7	18	7	7	16	8	9	14	9	10	12	11	11	9	11	11	10	9	9	12	7	8	15	2	4	23	2	4	26	75	88	202
	Sault Ste. Marie	54	4	6	22	5	6	17	7	7	17	6	7	16	7	9	15	7	10	13	8	11	12	8	11	13	5	9	17	5	7	19	2	4	24	3	4	23	66	90	209
MN	Duluth	47	7	7	17	7	6	15	7	7	17	6	8	16	6	10	15	5	11	14	7	13	11	7	12	11	6	9	15	7	8	17	5	6	20	6	6	19	77	102	187
	International Falls	56	7	7	16	8	6	14	8	8	16	7	8	15	7	9	15	5	11	14	6	13	11	7	12	12	6	9	15	6	7	18	4	5	21	6	6	19	76	101	188
	Minneapolis-St. Paul	57	8	7	15	8	7	14	7	7	17	7	8	15	7	9	15	7	10	12	10	12	9	10	11	10	10	8	12	10	7	14	5	6	18	6	6	18	95	101	169
	Rochester	35	7	7	16	7	6	15	6	7	18	6	7	16	7	9	15	7	11	13	8	12	11	9	11	11	9	7	14	8	8	15	5	6	19	6	6	19	86	97	182
	Saint Cloud	50	8	8	15	8	7	13	7	7	17	7	7	16	7	10	14	7	11	12	10	12	9	10	11	10	10	9	12	10	8	13	6	6	18	7	7	18	97	102	166
MS	Jackson	30	8	6	17	8	6	14	9	7	15	9	7	13	8	10	12	9	12	9	8	13	10	9	13	9	10	9	10	15	7	10	10	7	13	9	6	16	111	104	150
	Jackson	2	8	4	15	12	4	13	10	4	10	12	6	5	16	8	12	14	12	9	9	4	14	9	7	5	5	9	1	6	7	12	5	1	7	3	3	5	110	52	108
	Meridian	50	7	7	17	7	6	15	8	7	15	9	8	13	9	11	12	8	13	9	6	15	10	9	13	9	10	9	11	14	7	10	10	7	13	9	7	16	108	110	148
	Tupelo	10	8	6	17	7	6	15	8	7	16	10	8	12	8	10	13	9	11	9	11	11	8	12	12	8	13	9	8	14	6	11	9	6	15	8	6	18	119	97	150
	Tupelo	2	10	3	14	9	4	14	10	7	8	14	4	5	18	4	8	7	11	6	8	7	11	10	7	3	4	1	1	7	2	9	5	1	9	3	3	7	104	47	96
MO	Columbia	26	8	6	16	7	6	15	7	7	17	8	7	15	7	8	16	7	11	12	11	10	10	11	10	10	11	8	12	12	7	13	8	6	16	8	6	17	104	91	169
	Kansas City	23	10	6	14	8	6	14	8	8	16	9	7	14	8	9	14	10	10	11	13	11	7	12	11	8	12	8	10	13	7	11	9	7	14	10	6	15	120	96	149
	St. Louis	47	7	7	17	7	6	15	6	8	16	7	8	15	7	9	14	7	11	12	9	11	10	10	11	10	11	8	10	12	7	11	9	7	15	7	7	17	101	101	164
	Springfield	50	8	7	16	7	6	15	8	8	16	8	8	14	7	10	14	8	10	11	12	10	9	12	10	9	12	7	11	13	7	11	10	6	14	9	6	16	115	96	155
MT	Billings	56	6	8	18	4	8	16	4	9	18	4	9	17	6	10	15	7	12	11	13	12	6	14	11	7	10	9	10	9	9	13	6	8	17	6	8	17	89	112	164
	Glasgow	39	5	7	18	5	7	16	5	9	17	5	8	16	6	11	15	7	12	11	12	12	7	12	11	8	9	9	11	8	9	15	6	7	17	5	8	18	86	111	168
	Great Falls	57	5	6	19	4	6	18	4	9	18	4	8	18	5	10	17	5	11	14	13	12	6	12	11	8	10	9	11	7	9	15	5	7	18	5	8	18	79	106	180
	Helena	54	5	6	20	4	7	17	4	8	19	4	9	18	5	10	16	6	11	13	14	11	6	13	11	7	11	9	11	8	9	14	5	7	18	4	7	20	82	104	179
	Kalispell	45	2	4	25	2	5	21	3	6	21	4	6	20	5	9	17	6	10	14	14	9	7	12	9	9	10	8	12	6	7	18	2	5	23	2	3	26	70	81	214
	Missoula	51	2	4	24	3	5	21	3	6	22	4	7	20	5	9	17	6	9	14	15	10	6	14	10	8	10	8	11	7	7	17	2	5	23	2	4	25	75	83	208
NE	Grand Island	54	9	8	14	7	7	13	8	8	15	8	9	13	8	9	14	10	10	9	13	11	7	13	10	8	14	7	9	13	8	10	9	7	13	10	7	14	123	102	140
	Grand Island	2	5	3	11	7	4	7	7	4	7	6	6	9	9	3	9	12	4	5	11	8	4	14	4	5	8	2	3	9	3	6	9	4	6	10	2	10	106	48	81
	Lincoln	28	10	7	15	8	7	13	8	8	15	8	8	14	8	10	13	10	10	10	12	10	9	12	10	9	13	8	10	12	8	12	9	7	15	8	7	15	117	99	149
	Lincoln	3	3	2	5	5	1	6	4	3	4	8	8	14	3	2	7	6	4	4	4	2	4	10	5	4	9	5	5	12	2	7	6	2	4	6	6	12	71	35	62
	Norfolk	50	9	8	15	8	7	14	7	8	16	8	8	14	8	10	14	10	10	10	13	11	7	13	10	8	13	8	10	12	8	11	8	7	14	9	8	15	117	103	146
	North Platte	43	9	9	14	7	7	14	7	8	16	7	9	14	7	10	14	10	11	9	12	12	7	12	11	8	13	8	9	12	8	10	9	8	13	10	8	13	115	109	141
	Omaha Eppley AP	49	9	8	14	7	7	14	7	8	16	7	8	14	7	10	14	8	11	11	11	12	8	12	10	9	12	8	10	13	8	10	9	7	14	8	8	15	111	105	149
	Omaha (North)	18	9	8	14	8	6	15	8	8	16	8	8	14	7	10	14	9	11	10	11	10	10	10	11	9	13	8	9	11	8	12	8	7	15	8	7	15	110	102	153
	Scottsbluff	52	8	8	14	7	8	13	6	10	15	7	9	15	6	11	14	10	11	9	13	12	6	13	11	7	14	8	8	13	8	10	8	8	14	9	9	14	115	112	138
	Valentine	40	9	8	14	7	7	13	7	8	15	8	8	14	8	10	13	11	10	9	14	12	6	14	11	6	14	7	9	12	8	10	9	8	13	10	7	14	123	105	136
NV	Elko	56	7	8	17	6	7	15	6	8	17	6	9	15	8	10	13	13	10	7	17	10	4	18	9	4	18	7	5	14	8	9	8	7	15	8	7	16	130	99	136
	Ely	56	9	8	15	7	7	14	7	8	15	7	9	13	7	11	12	13	10	6	15	11	5	15	12	5	17	8	5	14	8	9	10	8	12	9	8	14	131	109	125
	Las Vegas	47	14	6	11	12	7	9	14	8	9	16	8	7	18	8	5	22	5	3	20	7	3	22	7	3	22	5	2	20	6	4	16	7	7	15	7	10	210	82	73
	Reno	53	8	7	16	7	7	14	8	9	14	8	10	12	12	10	10	16	8	6	22	6	3	22	6	3	21	6	4	16	8	8	9	8	13	8	8	15	158	93	114
	Winnemucca	45	7	7	18	6	6	16	7	7	17	6	9	14	9	10	12	14	9	8	20	7	4	19	8	4	18	6	5	14	7	9	8	7	15	8	6	16	138	89	138
NH	Concord	54	9	7	15	8	8	13	8	8	15	7	8	15	6	10	15	6	12	12	7	12	12	8	11	12	9	9	12	9	9	13	6	8	16	8	8	15	90	109	166
	Mt. Washington	70	5	6	21	4	5	19	4	5	21	4	6	20	3	7	20	2	7	21	1	8	22	3	8	20	4	7	19	6	6	19	3	5	22	4	5	22	44	76	245
NJ	Atlantic City AP	37	8	8	15	8	7	14	8	8	15	7	9	14	6	11	14	7	12	12	7	12	13	8	11	12	10	8	12	11	9	12	8	9	13	8	8	15	94	111	160
	Newark	53	8	8	16	7	8	13	8	8	15	7	9	14	6	11	14	7	11	12	7	12	12	8	12	12	10	9	12	11	8	12	8	8	14	8	8	15	93	112	160
NM	Albuquerque	56	13	8	10	11	8	10	11	10	10	13	10	8	14	10	7	17	9	4	12	14	5	13	13	5	17	8	6	17	8	6	15	8	7	14	7	10	167	111	87
	Clayton	28	13	7	12	11	7	10	11	9	11	12	9	10	11	9	10	13	10	6	12	12	7	15	10	6	17	6	7	18	7	7	15	6	9	14	7	10	162	99	104
	Roswell	9	11	9	11	12	8	8	14	9	8	14	9	7	13	12	6	17	10	3	11	15	5	13	12	6	13	8	9	18	6	7	16	7	7	15	9	7	168	113	84
NY	Albany	57	5	8	18	6	7	15	6	8	17	5	8	16	5	9	16	5	11	13	6	13	12	7	12	13	8	10	12	8	9	14	4	8	18	5	7	19	69	111	185
	Binghamton	44	3	7	22	3	6	19	4	7	20	5	7	18	5	9	17	5	11	14	6	13	13	5	12	14	6	10	14	6	8	17	3	6	21	2	6	23	52	102	212
	Buffalo	52	1	6	24	2	6	20	4	8	19	5	8	17	6	10	16	6	12	12	7	13	11	7	12	12	6	10	14	6	8	16	2	5	23	1	6	24	54	103	208
	New York C. Park	42	8	9	14	8	9	11	9	10	12	8	11	12	8	12	11	8	12	10	8	13	10	9	12	10	11	10	9	12	10	9	9	10	12	9	9	13	107	127	132
	New York (JFK AP)	37	8	8	15	8	8	13	8	9	14	8	9	13	7	11	13	7	11	12	7	12	11	8	13	10	10	9	11	11	9	11	8	8	14	8	8	15	98	116	152
	New York (Laguardia AP)	47	8	8	15	8	7	13	8	9	14	7	9	13	7	11	13	7	11	11	7	13	11	8	12	11	10	9	11	11	9	11	7	8	14	8	9	15	96	117	153
	Rochester	55	2	7	22	2	7	19	5	8	18	6	8	16	6	10	15	7	11	12	8	12	11	8	12	12	7	10	13	6	8	16	2	6	22	2	6	23	61	104	200
	Syracuse	44	3	7	22	3	6	19	5	7	19	6	7	17	6	10	15	7	11	12	8	12	11	7	11	13	7	10	13	6	8	17	2	6	22	2	5	24	63	98	205
	Syracuse	0	3	7	3	3	6	5	3	7	11	6	7	17	5	6	9	3	10	6	8	12	11	7	11	13	7	10	13	6	8	17	2	6	22	2	5	24	11	23	34
NC	Asheville	31	9	7	15	9	6	13	9	8	14	10	8	12	7	10	13	6	12	12	5	14	12	5	13	13	7	11	13	12	8	11	11	7	12	10	7	14	99	113	153
	Cape Hatters	38	9	7	16	9	5	14	10	7	14	10	9	11	9	10	13	7	10	12	7	10	14	8	10	13	10	9	11	11	8	12	10	8	12	10	7	14	109	101	156
	Charlotte	49	9	6	16	8	6	14	9	8	14	10	9	12	8	10	13	7	11	12	7	12	12	7	13	11	9	9	11	13	8	10	11	6	12	10	6	15	109	105	152

	Greensboro-Wnstn-Slm-HPT	67	9	7	15	9	6	13	9	8	14	10	8	12	8	11	12	7	12	11	7	13	12	7	12	11	10	9	12	13	7	10	11	7	12	10	6	15	109	107	150
	Raleigh	47	9	7	15	9	6	14	9	7	14	10	9	11	8	10	13	8	12	11	7	12	12	7	12	11	10	9	11	13	7	11	11	7	11	10	7	14	111	106	149
	Wilmington	44	10	6	15	9	6	13	10	8	14	11	8	11	8	11	12	7	11	12	6	12	13	7	12	12	8	10	12	12	7	11	12	7	11	10	7	14	111	104	150
ND	Bismarck	56	7	8	17	6	8	15	6	8	17	6	9	15	6	10	14	7	10	12	11	13	7	12	11	8	10	9	11	9	8	14	6	7	17	7	7	17	93	107	165
	Fargo	53	7	7	17	6	7	15	5	9	17	6	9	15	7	10	14	6	11	13	10	13	8	10	12	9	9	9	12	9	8	14	5	6	18	6	7	18	88	109	168
	Williston	34	6	8	17	6	8	15	7	8	16	6	9	15	6	11	13	8	11	11	12	12	7	11	12	8	10	9	11	9	8	14	6	7	16	7	7	17	93	112	160
OH	Akron	47	3	6	22	3	6	19	5	6	20	5	8	18	6	9	16	6	11	13	7	13	11	8	12	11	8	9	12	9	8	14	4	6	20	3	5	23	68	99	198
	Cleveland	54	3	5	24	3	6	20	5	6	20	5	8	17	6	10	15	7	11	12	8	12	11	9	11	11	8	10	12	8	8	15	3	6	21	3	5	24	66	97	202
	Columbus	46	4	6	21	4	6	18	5	7	19	5	8	17	6	10	15	6	11	13	7	13	11	7	13	11	9	9	12	10	8	13	5	7	19	4	6	21	72	103	190
	Dayton	52	5	6	20	5	6	17	5	7	19	5	7	17	6	9	16	6	10	13	7	12	11	8	12	11	9	9	12	10	8	13	5	7	18	4	6	21	77	100	188
	Mansfield	30	4	6	21	4	6	18	5	7	19	6	8	16	7	10	15	7	12	11	8	13	10	8	11	11	9	9	12	9	8	14	4	6	20	3	5	23	73	101	191
	Toledo	40	5	7	20	5	7	17	5	7	19	6	8	17	7	10	14	7	12	12	8	13	10	8	12	11	8	10	12	8	9	14	4	7	19	3	6	22	73	107	185
	Youngstown	52	3	5	23	3	5	20	5	6	20	5	7	18	6	9	17	6	11	12	7	13	11	7	12	12	7	10	13	8	8	15	3	6	21	3	5	23	63	97	205
OK	Oklahoma City	44	11	6	14	9	7	13	10	8	13	9	8	13	9	10	12	10	11	9	15	10	7	14	10	7	13	8	9	14	7	10	12	7	11	12	6	13	139	96	130
	Oklahoma City	3	4	5	7	10	3	10	8	3	7	10	2	8	9	6	6	13	5	3	16	4	2	15	4	3	7	3	2	12	4	5	6	4	6	8	5	8	119	49	68
	Tulsa	54	10	7	14	9	6	13	9	8	14	8	9	13	8	10	13	9	11	10	13	11	7	13	11	7	12	8	10	14	7	10	11	7	12	10	7	13	127	103	136
	Tulsa	3	5	4	8	8	3	7	8	1	9	10	1	10	10	4	9	11	8	4	14	6	4	15	5	4	8	4	4	10	4	6	7	1	10	9	5	7	113	45	83
OR	Astoria	40	3	4	24	3	3	22	3	5	23	3	6	21	3	8	20	3	7	19	6	10	15	6	10	15	8	7	14	5	7	19	3	5	22	3	4	24	50	76	239
	Astoria	1	5	8	15	6	4	15	3	5	2	3	5	19	8	6	14	2	7	12	5	9	10	6	6	8	1	2	2	2	4	5	1	2	8	3	2	3	38	55	114
	Burns	11	5	8	18	6	6	17	6	7	18	5	10	15	9	10	12	12	9	9	19	7	5	20	8	3	16	8	6	12	7	12	6	7	17	5	7	18	120	94	151
	Eugene	53	2	4	25	3	4	21	3	6	22	4	7	19	6	9	17	8	8	14	16	8	7	14	9	8	12	8	9	5	9	17	2	6	23	1	4	26	75	82	209
	Medford	67	3	5	24	4	6	19	5	7	19	6	8	16	9	9	13	13	8	9	23	5	3	22	6	4	18	6	6	10	8	12	3	6	21	2	4	25	117	79	169
	Pendleton	60	2	5	24	3	6	20	5	8	19	5	10	15	8	11	13	10	10	10	19	8	4	18	8	5	15	8	7	10	8	13	4	6	20	3	5	24	101	91	173
	Portland	47	3	4	24	3	4	21	3	5	23	3	6	21	5	7	19	6	8	16	13	8	10	11	10	10	10	8	12	5	8	18	3	4	23	2	3	26	68	74	222
	Salem	58	3	5	24	3	4	21	4	6	21	4	7	19	6	8	18	7	8	15	15	8	8	14	9	9	11	8	10	6	8	17	3	5	22	2	4	25	77	80	208
	Sexton Summit	30	5	5	21	4	4	20	5	5	21	6	7	17	9	7	14	13	7	10	23	5	3	21	6	4	17	6	7	11	6	14	6	5	20	5	4	22	126	68	172
PC	Guam	20	3	13	15	2	10	16	3	15	13	1	15	14	1	16	15	1	11	19	—	9	22	0	5	26	—	6	23	—	10	21	2	13	15	3	12	16	16	134	215
	Johnston Island	24	13	14	4	11	13	4	11	14	6	7	12	11	6	13	12	7	15	8	7	16	9	6	14	11	6	13	11	7	12	12	7	12	12	11	13	7	99	159	107
	Koror	50	—	7	24	—	6	21	1	9	21	1	9	20	1	9	21	—	5	25	—	5	26	—	5	26	—	6	23	1	7	23	—	8	22	—	7	24	4	83	276
	Kwajalein, Marshall IS	41	2	7	21	3	6	19	2	7	21	2	6	23	1	6	24	1	6	23	1	6	24	1	6	24	1	5	24	1	5	26	1	6	23	2	6	23	18	72	276
	Majuro, Marshall IS	46	1	7	23	1	8	19	1	8	22	1	7	22	1	7	23	—	7	23	0	7	24	1	8	22	1	7	22	1	7	23	0	7	23	1	6	24	8	85	272
	Pago Pago, Amer Samoa	36	1	10	21	0	9	18	1	12	18	1	12	17	1	13	17	2	14	14	2	15	14	2	16	14	2	14	14	1	13	17	1	11	18	1	10	20	13	148	204
	Pohnpei, Caroline IS	50	—	5	26	—	5	24	—	5	26	—	5	25	—	5	25	—	5	25	0	6	24	1	7	24	1	6	24	0	6	24	—	6	24	—	5	26	2	65	296
	Chuuk, E. Caroline IS	51	—	4	27	—	3	25	—	4	27	—	4	26	—	4	27	—	4	26	—	4	27	—	4	27	—	4	26	—	4	27	—	4	26	—	3	28	0	45	318
	Wake Island	43	12	13	5	12	11	5	12	14	5	11	14	6	10	14	6	10	13	7	7	11	13	6	11	14	7	11	12	9	12	10	12	13	5	14	13	4	121	149	94
	Yap, W Caroline IS	53	1	8	22	0	8	20	1	9	21	1	10	19	1	10	20	—	6	24	0	4	26	—	4	27	—	6	24	0	7	23	—	8	21	0	8	22	5	88	271
PA	Allentown	52	7	8	16	7	8	13	8	8	15	7	9	14	7	11	14	8	11	11	8	12	11	9	11	11	9	9	12	10	9	12	7	8	15	7	7	17	94	110	161
	Erie	40	2	4	25	3	6	20	5	7	19	6	8	17	7	10	14	8	11	11	8	13	10	8	12	11	7	10	14	6	8	17	3	5	23	1	4	26	63	97	205
	Harrisburg	53	7	7	17	7	7	14	7	8	16	6	9	15	6	10	15	7	12	12	7	12	12	8	11	12	9	9	12	10	8	13	6	8	15	6	8	17	87	109	169
	Middletown/Harrisburg AP	61	7	8	17	7	7	14	7	8	16	7	9	15	6	11	14	6	12	12	7	12	12	8	12	12	9	10	12	10	9	13	6	8	15	6	8	17	85	112	168
	Philadelpia	55	7	8	16	7	7	14	8	8	15	7	9	14	6	11	14	7	11	12	7	12	12	8	11	11	10	9	11	11	9	12	7	9	14	7	8	15	93	112	160
	Pittsburgh	43	3	6	22	3	6	19	4	7	20	4	8	17	5	9	17	5	12	13	5	13	13	7	12	13	7	10	13	8	9	14	4	6	20	3	6	23	59	103	203
	Avoca	40	4	7	20	5	7	17	5	8	18	6	8	16	6	10	16	7	11	12	6	13	12	7	12	12	7	10	13	8	8	14	4	7	19	4	7	20	70	106	189
	Williamsport	51	5	7	19	5	7	16	6	8	17	6	8	16	5	10	15	6	12	12	6	13	12	6	13	12	6	11	14	7	9	15	4	7	19	4	7	20	67	110	188
RI	Block Is	19	8	9	14	8	8	12	9	9	13	8	8	13	7	11	13	8	10	12	7	11	13	7	11	13	10	9	11	11	8	12	7	10	14	7	10	14	98	113	155
	Providence	42	10	7	15	8	7	13	8	8	15	7	8	14	6	10	15	7	10	13	7	12	12	8	10	12	9	8	12	11	8	12	8	7	15	8	8	15	98	103	164
SC	Charleston AP	47	9	6	16	9	6	13	9	8	14	11	8	11	8	11	12	6	11	13	5	12	13	6	13	13	7	10	13	11	8	11	12	7	12	9	7	14	102	109	155
	Columbia	48	9	6	16	9	6	13	9	8	14	11	8	11	9	10	11	8	11	11	7	13	12	8	12	11	10	9	12	13	7	11	12	6	12	10	6	15	115	103	147
	Greenville-Spartanburg AP	33	10	6	15	10	5	13	10	8	13	11	8	11	8	10	13	8	11	11	7	13	11	8	12	11	10	9	11	14	7	10	12	6	12	11	6	14	121	100	145
SD	Aberdeen	28	8	8	16	7	6	15	6	7	17	7	7	16	8	10	14	9	10	11	11	13	7	12	11	8	11	8	11	10	8	13	6	6	18	7	7	17	101	101	163
	Huron	56	8	8	15	6	8	14	6	8	17	7	9	15	8	10	14	9	11	10	12	12	7	12	11	7	12	8	10	11	8	12	7	7	16	7	7	16	104	107	154
	Rapid City	53	8	8	15	6	8	14	6	9	15	6	10	15	7	11	13	9	11	10	13	13	5	14	12	5	13	9	8	12	8	11	8	8	14	8	8	14	111	115	139
	Sioux Falls	50	8	8	15	7	7	15	6	8	17	7	8	15	7	10	14	9	11	11	12	12	8	12	11	9	12	8	10	11	8	12	7	7	16	8	7	16	105	103	157
TN	Bristol-JhnCty-Kgsprt	58	6	7	18	6	7	16	7	8	17	8	9	14	7	11	14	6	13	11	6	13	12	7	14	11	10	10	11	12	8	11	8	7	14	7	7	17	88	112	165
	Chattanooga	65	7	7	17	7	6	15	8	8	16	9	8	13	9	10	12	8	12	10	7	13	11	8	13	10	10	10	11	13	8	10	10	7	13	8	6	17	104	106	155
	Knoxville	53	6	7	18	7	6	16	7	7	16	8	9	13	8	10	13	7	12	10	7	13	11	8	12	10	10	9	11	12	8	11	9	7	14	7	7	17	97	107	162
	Memphis	43	8	6	17	8	6	15	8	7	17	9	7	14	8	10	13	10	11	9	10	12	9	12	12	7	12	8	10	14	7	10	10	6	14	9	6	16	118	96	151
	Nashville	54	6	6	18	7	6	15	8	7	16	8	8	13	8	10	13	8	12	10	8	13	10	10	12	9	11	9	10	13	8	10	9	7	14	7	7	17	102	106	156
	Oak Ridge	42	7	7	17	8	6	15	8	7	16	10	7	13	9	9	13	9	10	10	8	12	11	9	11	10	10	9	11	14	7	10	9	7	14	8	6	17	109	98	158
TX	Abilene	56	11	6	14	10	6	12	11	7	12	12	8	11	10	10	11	13	11	7	14	10	7	14	10	7	14	8	8	15	7	9	13	6	11	12	6	13	149	95	121
	Amarillo	51	13	7	11	10	7	10	12	9	11	12	9	9	11	10	10	13	11	6	13	12	5	14	10	7	15	7	8	17	7	8	15	7	9	13	7	10	157	104	104
	Amarillo	3	8	5	5	10	4	6	11	3	3	12	4	5	11	5	6	18	5	2	17	3	2	16	2	4	10	1	2	12	2	4	10	3	4	13	5	4	146	43	44
	Austin	54	9	6	16	8	6	14	9	8	15	8	8	15	6	11	13	8	15	7	12	13	6	12	14	5	11	11	9	12	9	9	11	7	12	10	6	15	115	114	136

(Continued)

Table 3A.11 (Continued)

State	Station	Years	Jan			Feb			Mar			Apr			May			Jun			Jul			Aug			Sep			Oct			Nov			Dec			Annual		
			CL	PC	CD	CL	PC	CD	CL	PC	CD	CL	PC	CD	CL	PC	CD	CL	PC	CD	CL	PC	CD	CL	PC	CD	CL	PC	CD	CL	PC	CD	CL	PC	CD	CL	PC	CD	CL	PC	CD
	Brownsville	52	6	7	18	7	6	15	7	8	16	5	10	15	6	14	11	8	15	6	11	14	6	11	14	7	9	13	8	11	12	7	9	9	12	7	7	17	96	131	138
	Corpus Christi	53	7	7	17	7	6	15	7	8	16	6	9	15	5	12	13	9	14	7	11	14	6	11	13	7	10	12	8	12	10	8	9	9	12	8	6	17	102	121	142
	Dallas-Fort Worth	42	10	6	16	10	6	13	10	8	14	9	8	13	8	10	13	11	12	8	15	10	6	15	10	6	13	9	9	14	7	10	12	6	12	11	6	14	135	97	133
	Del Rio	16	10	7	14	10	7	11	11	8	13	8	7	14	6	9	16	8	13	9	12	11	8	11	12	8	9	11	10	12	8	10	12	7	11	11	6	14	121	106	138
	El Paso	53	14	7	10	14	7	7	15	8	7	17	8	5	19	8	4	20	7	3	12	13	5	14	12	5	18	7	5	19	7	5	18	6	6	15	7	9	193	100	72
	Houston	26	7	5	18	7	5	16	7	6	18	7	7	16	6	11	14	7	13	9	7	16	8	6	17	8	9	11	10	11	9	11	9	7	14	7	6	18	90	114	161
	Lubbock	49	12	6	12	11	7	10	12	9	11	12	9	9	11	11	9	14	11	6	14	11	6	15	10	6	14	8	8	17	7	8	15	7	9	13	7	11	160	102	103
	Midland-Odessa	47	12	6	12	11	7	10	13	8	10	13	8	9	13	10	8	15	9	5	13	11	7	14	10	6	14	8	8	17	6	8	15	6	9	14	6	11	165	96	104
	Port Arthur	42	7	6	18	8	6	15	7	7	17	6	8	16	7	12	12	8	14	8	7	15	9	7	15	9	9	11	10	12	10	9	10	7	13	8	6	17	95	117	153
	San Angelo	47	12	6	13	11	6	11	12	8	11	11	8	11	11	10	10	14	10	6	15	10	7	15	10	6	13	8	8	15	7	8	14	6	10	12	7	12	154	97	114
	San Antonio	53	9	6	16	8	6	14	9	7	15	7	8	15	6	11	14	7	15	8	9	15	7	10	15	6	9	12	9	11	10	9	10	7	13	10	6	16	105	119	141
	Victoria	34	7	6	18	7	6	16	7	7	18	5	8	17	5	11	15	6	15	9	7	15	8	7	15	9	8	12	10	11	10	10	9	7	14	7	6	19	86	118	162
	Waco	50	9	6	16	9	6	14	10	7	14	9	7	14	8	10	13	10	12	8	14	10	6	14	11	6	12	9	8	13	8	9	12	6	12	11	6	15	130	98	136
	Waco	3	5	2	6	9	1	8	9	3	7	7	4	4	6	5	5	10	3	2	10	4	2	14	3	2	5	3	2	8	2	4	3	2	6	4	4	6	89	36	53
	Wichita Falls	50	11	6	14	10	6	12	11	8	12	11	8	11	11	9	11	13	11	7	15	9	7	15	9	6	15	7	8	15	7	9	13	6	11	12	6	13	151	93	120
	Wichita Falls	2	12	6	11	15	4	8	9	3	4	10	6	4	18	6	12	24	10	2	19	6	1	16	4	2	3	2	1	12	3	3	1	3	8	1	5	5	140	58	60
UT	Milford	40	9	8	14	8	8	12	9	8	13	10	10	11	11	10	9	17	8	5	16	10	5	16	10	5	18	7	4	17	7	7	11	8	11	10	8	13	151	104	111
	Salt Lake City	69	6	6	19	5	7	16	7	8	16	7	9	14	9	10	12	14	10	6	17	10	4	16	11	4	16	8	5	14	8	9	8	7	15	6	6	18	125	101	139
VT	Burlington	52	5	6	20	4	7	17	6	7	18	5	8	17	5	9	17	5	11	14	5	13	13	6	12	13	6	10	14	6	8	17	3	5	22	3	6	22	58	101	206
VA	Lynchburg	49	9	7	15	8	7	13	9	9	13	9	9	12	8	10	13	8	12	10	8	12	11	9	11	11	10	9	11	13	7	11	10	7	13	10	7	14	112	107	147
	Norfolk	47	9	7	16	8	6	14	9	8	15	9	9	12	8	10	13	7	12	11	7	12	12	8	12	11	9	9	11	12	7	12	10	8	12	9	7	15	106	107	153
	Richmond	50	8	7	16	8	6	14	8	8	15	8	9	13	7	10	14	7	12	12	7	12	12	7	12	12	9	9	12	12	7	12	9	8	13	9	6	15	100	106	160
	Roanoke	48	8	8	15	8	7	14	8	8	14	9	9	13	7	11	13	7	12	11	7	13	11	8	12	11	10	9	12	13	7	11	9	8	13	9	8	14	102	112	151
WA	Olympia	54	2	4	25	2	4	22	3	6	22	3	7	20	4	9	18	5	8	16	10	10	11	9	10	12	8	10	13	3	8	20	2	5	23	2	4	26	52	84	228
	Quillayute	29	3	3	24	4	4	21	4	6	22	3	6	21	3	8	20	4	7	20	6	9	16	6	9	16	8	8	14	5	7	19	3	5	22	4	4	24	51	75	239
	Seattle CO	24	3	5	23	3	6	19	4	8	19	5	9	16	7	10	14	7	8	15	12	10	9	10	10	11	9	8	13	5	8	18	3	6	21	3	5	23	71	93	201
	Seattle Sea-Tac AP	51	3	4	24	3	4	21	3	6	22	3	7	20	4	9	18	5	8	17	10	10	11	9	10	12	8	9	13	4	7	19	2	4	23	2	4	25	58	82	226
	Spokane	48	3	4	24	3	5	20	5	8	19	4	8	17	6	10	15	7	10	12	16	8	6	15	9	7	12	8	9	8	8	15	3	5	22	3	4	24	86	88	191
	Yakima	49	4	5	22	4	6	18	6	8	17	6	9	14	8	11	12	10	10	10	19	8	4	18	8	6	15	8	7	9	8	13	5	6	19	4	5	22	109	92	164
PR	San Juan	40	9	18	4	7	16	4	9	17	4	7	17	6	4	16	12	4	16	10	5	17	9	5	17	8	4	17	9	5	17	9	5	18	7	7	18	6	73	203	89
WV	Beckley	32	5	5	21	4	5	18	5	6	20	5	7	17	5	9	17	3	10	16	3	12	16	4	12	16	6	9	15	9	8	14	5	7	18	5	6	21	60	95	210
	Charleston	47	4	6	21	4	6	18	5	8	19	6	8	16	6	10	15	5	13	12	5	13	13	5	14	12	7	11	12	9	9	13	5	7	18	5	6	20	65	111	189
	Elkins	49	3	6	22	3	6	19	4	6	20	4	8	18	4	9	17	3	11	16	2	13	16	3	14	15	4	11	15	7	8	13	5	6	19	4	5	21	48	103	212
	Huntington	34	4	6	21	4	5	19	5	7	19	5	8	17	6	9	17	5	11	14	5	12	14	5	12	14	7	9	14	9	8	14	5	6	19	5	5	21	63	99	203
WI	Green Bay	46	8	7	17	7	7	15	7	8	16	6	8	16	7	10	14	8	11	12	8	12	11	8	11	12	8	9	13	7	8	15	5	6	19	6	6	19	85	102	178
	La Crosse	17	7	7	17	8	7	13	7	7	17	7	7	16	7	9	15	7	10	12	10	11	10	9	11	11	9	9	12	11	7	13	6	6	18	7	6	18	95	97	173
	Madison	49	7	6	17	7	6	15	6	7	17	6	8	16	7	9	15	7	10	13	9	11	11	9	10	12	9	8	12	9	7	14	5	6	18	6	6	19	89	96	180
	Milwaukee	55	7	6	18	6	6	15	6	8	17	6	8	16	7	10	14	8	10	12	10	11	10	10	11	10	9	9	12	9	8	13	5	6	18	6	6	19	90	100	175
WY	Casper	45	7	8	16	6	8	14	6	9	16	5	9	16	6	11	15	10	11	9	14	11	6	13	11	7	13	9	8	11	8	12	7	8	14	8	8	15	107	111	147
	Cheyenne	60	9	10	13	7	9	12	7	10	14	6	10	14	5	12	14	9	12	9	10	15	7	10	13	8	13	9	9	13	9	9	10	9	11	9	9	13	106	127	133
	Lander	49	8	10	13	7	10	12	7	10	14	6	10	14	6	11	14	10	11	9	14	11	6	13	12	6	14	9	8	12	9	10	8	10	12	10	10	12	114	122	129
	Sheridan	55	6	8	17	5	8	15	5	9	17	5	9	16	6	10	15	8	12	11	13	12	6	14	11	6	12	9	9	10	9	12	6	8	16	7	8	17	95	113	157

Note: This table shows the mean number of days per category of cloudiness. The categories are determined for daylight hours only. Clear denotes zero to 3/10 average sky cover. Partly cloudy denotes 4/10 to 7/10 average sky cover. Cloudy denotes 8/10 to 10/10 average sky cover.

Source: From U.S. National Oceanic and Atmospheric Administration, *Comparative Climatic Data for the United States Through 2000*, www.noaa.gov.

Table 3A.12 Average Relative Humidity: Morning (M), Afternoon (A) — Selected Cities of the United States (Through 2002)

State	Station	Years M	Years A	Jan M	Jan A	Feb M	Feb A	Mar M	Mar A	Apr M	Apr A	May M	May A	Jun M	Jun A	Jul M	Jul A	Aug M	Aug A	Sep M	Sep A	Oct M	Oct A	Nov M	Nov A	Dec M	Dec A	Annual M	Annual A
AL	Birmingham AP	39	39	80	64	79	60	79	57	83	57	86	60	84	59	86	62	86	61	87	62	87	58	83	60	81	63	83	60
	Huntsville	35	35	82	68	80	63	80	60	82	57	86	60	88	61	90	64	91	63	89	63	87	59	84	62	82	67	85	62
	Mobile	40	40	83	65	83	61	85	60	88	58	88	60	89	61	90	66	91	66	90	65	87	59	86	62	85	66	87	62
	Montgomery	39	39	82	64	80	60	82	57	86	57	88	60	88	60	90	64	91	64	89	62	89	58	87	60	84	64	86	61
AK	Anchorage	49	49	75	73	75	68	70	57	66	53	63	49	67	56	74	63	78	65	80	64	78	66	78	74	77	76	73	64
	Annette	37	37	81	77	80	73	77	69	76	66	76	66	79	69	81	71	84	73	86	75	85	78	83	79	83	80	81	73
	Barrow	51	51	69	69	66	67	67	68	75	74	85	82	87	84	88	84	92	86	91	87	85	84	78	79	73	73	80	78
	Barter IS	40	40	70	70	69	67	67	67	73	74	86	84	89	87	89	86	91	88	90	87	84	83	75	75	70	70	79	78
	Bethel	52	52	78	77	76	74	80	75	83	72	80	62	78	60	86	68	91	73	92	71	88	76	84	81	78	78	83	72
	Bettles	52	52	69	68	67	64	67	60	66	60	62	50	60	47	69	53	79	61	79	62	78	72	73	72	70	70	70	62
	Big Delta	43	43	65	64	68	61	64	52	57	44	50	38	57	44	64	49	69	51	70	53	74	66	69	67	68	68	65	55
	Cold Bay	33	33	85	83	85	81	85	78	84	77	84	75	87	77	90	82	91	82	88	79	84	77	84	81	84	83	86	79
	Fairbanks	50	50	70	71	68	65	68	53	60	45	53	37	60	43	70	50	78	55	79	55	80	68	75	74	72	73	69	57
	Gulkana	50	50	72	71	73	67	70	54	64	45	58	40	60	42	68	48	71	49	75	52	79	65	76	75	73	73	70	57
	Homer	53	53	78	75	78	70	75	65	72	63	70	63	71	64	76	69	79	69	80	68	79	66	77	72	78	76	76	68
	Juneau	36	36	78	74	81	71	79	66	77	61	75	61	75	61	79	67	82	70	86	74	84	76	81	77	81	79	80	70
	King Salmon	54	54	79	76	78	71	78	67	77	62	73	56	77	59	83	64	86	67	86	66	85	68	83	77	79	77	80	68
	Kodiak	55	55	79	76	79	73	76	69	74	69	76	72	79	75	82	77	80	74	81	73	78	68	78	72	77	74	78	73
	Kotzebue	39	39	73	73	73	73	73	71	78	75	82	78	81	77	83	77	85	77	83	74	82	77	78	77	75	75	79	75
	McGrath	48	48	73	72	73	64	73	53	68	49	63	44	66	47	75	56	83	62	84	61	84	70	79	76	76	75	75	61
	Nome	39	39	74	74	73	72	73	70	76	74	76	73	76	73	82	78	83	79	81	73	80	71	77	75	74	74	77	74
	St. Paul Island	24	24	84	83	86	84	87	83	86	81	89	81	92	83	95	89	95	89	90	83	83	78	82	80	84	83	88	83
	Talkeetna	51	51	73	69	74	64	71	57	67	52	65	49	70	53	79	61	84	64	84	64	80	67	75	71	73	73	75	62
	Unalakleet	10	10	67	69	71	73	71	70	76	74	76	71	80	74	81	75	82	73	80	69	77	71	74	73	68	69	75	72
	Valdez	30	30	75	74	74	68	72	64	71	61	73	61	78	64	84	71	85	72	87	74	78	70	74	70	76	75	77	69
	Yakutat	38	38	85	82	86	77	83	71	78	69	78	71	81	74	85	78	86	79	89	78	89	79	87	82	86	85	84	77
AZ	Flagstaff	45	45	74	50	74	45	72	41	67	32	63	27	54	21	67	37	77	44	74	37	72	36	70	43	72	51	70	39
	Phoenix	42	42	64	32	59	27	56	24	42	17	34	14	30	12	43	20	50	23	48	23	49	22	56	27	65	33	50	23
	Tucson	62	62	62	32	58	27	53	23	42	16	34	13	32	13	56	28	65	33	55	27	52	25	54	28	62	34	52	25
	Winslow	25	25	76	47	68	33	61	25	52	20	43	16	37	14	58	27	65	30	65	29	60	26	66	33	75	47	60	29
	Yuma	14	14	57	28	56	24	52	21	47	17	44	15	41	13	49	22	55	24	57	24	54	23	56	27	58	32	52	22
AR	Fort Smith	38	38	82	65	80	60	79	57	82	56	88	62	89	62	88	60	88	58	89	60	87	57	84	61	83	65	85	60
	Little Rock	38	38	80	65	79	61	78	59	82	60	87	62	85	58	86	58	85	58	86	59	84	55	81	60	79	63	83	60
CA	Bakersfield	26	26	85	63	80	53	74	44	67	33	57	27	52	24	49	23	53	25	57	28	62	34	77	50	83	61	66	39
	Bishop	7	7	67	32	68	31	58	20	53	17	49	15	39	14	46	14	45	14	50	16	50	16	63	24	64	29	54	20
	Blue Canyon	13	13	56	58	61	61	66	61	61	54	55	47	47	38	41	34	42	34	48	42	50	48	57	60	55	59	53	50
	Fresno	39	39	91	68	89	56	86	47	80	35	71	27	65	24	61	22	66	24	71	28	77	35	87	54	91	68	78	41
	Long Beach	32	32	76	53	78	54	80	55	80	51	81	55	82	56	82	54	82	53	83	54	81	54	79	53	77	52	80	54
	Los Angeles AP	43	43	71	61	75	64	79	66	80	65	83	67	85	68	86	69	85	69	84	68	80	66	72	62	68	60	79	65
	Los Angeles CO	43	43	64	50	70	52	74	52	79	54	80	55	80	54	80	52	79	53	75	52	72	54	59	47	59	48	73	52
	Mount Shasta	13	13	76	66	75	60	75	52	71	43	69	39	67	36	64	28	65	29	66	33	71	46	78	62	77	67	71	47
	Redding	16	16	84	61	82	51	78	46	76	38	72	33	63	25	60	20	60	19	61	23	68	30	82	51	83	59	72	38
	Sacramento	16	16	91	70	89	61	86	53	83	44	82	38	78	32	77	30	78	29	77	31	79	37	87	57	88	67	83	46
	San Diego	42	42	72	58	74	60	76	61	76	60	78	65	81	67	82	67	83	67	81	67	77	65	73	62	71	59	77	63
	San Francisco AP	43	43	86	68	85	66	82	63	82	60	83	60	84	59	86	60	87	61	85	59	82	59	84	64	85	68	84	62
	San Francisco CO	8	8	81	63	83	63	80	61	82	61	88	68	78	63	80	65	80	64	75	58	70	52	71	55	69	55	78	61
	Santa Barbara	6	6	80	57	79	58	83	60	76	58	83	60	88	62	90	64	90	64	86	63	81	62	77	52	81	55	83	60
	Santa Maria	26	26	82	61	85	62	86	65	89	61	91	61	92	61	89	62	90	62	92	63	85	62	80	62	80	59	87	62
	Stockton	26	26	90	71	89	61	84	50	80	41	74	34	71	29	68	28	69	29	71	32	75	38	83	57	91	71	79	45
CO	Alamosa	45	45	78	57	78	48	74	36	71	30	72	28	74	25	83	35	85	38	81	33	76	34	78	47	77	56	77	39
	Colorado Springs	42	42	58	46	59	40	62	39	63	35	68	37	67	35	69	39	71	43	67	38	59	37	61	46	57	49	63	40
	Denver	35	35	63	49	67	44	67	40	67	35	70	38	69	35	68	34	69	35	68	34	65	36	68	49	65	52	67	40
	Grand Junction	39	39	78	62	73	47	64	35	58	28	54	25	44	19	48	22	52	24	53	27	59	33	71	47	77	59	61	36
	Pueblo	23	23	69	49	66	37	68	34	68	31	70	32	70	28	74	32	76	35	72	32	69	33	74	46	69	51	70	37
CI	Bridgeport	36	36	70	59	70	57	70	55	69	53	75	59	78	61	78	60	79	61	82	61	80	59	77	60	74	60	75	59
	Hartford	43	43	72	57	72	53	73	51	70	45	74	48	77	51	79	51	83	53	86	55	84	51	79	56	76	59	77	52
DE	Wilmington	55	55	76	60	75	56	74	53	73	50	76	54	78	54	79	54	83	56	85	56	85	54	80	56	77	59	78	55
DC	Washington Dulles AP	33	33	77	58	78	54	78	52	77	49	83	55	84	56	86	55	88	55	90	56	89	54	83	54	79	58	83	55
	Washington Nat'l AP	42	42	71	56	71	53	70	50	70	49	75	53	76	53	76	53	80	55	82	56	80	54	76	54	72	57	75	54

(Continued)

Table 3A.12 (Continued)

State	Station	Years		Jan		Feb		Mar		Apr		May		Jun		Jul		Aug		Sep		Oct		Nov		Dec		Annual	
		M	A	M	A	M	A	M	A	M	A	M	A	M	A	M	A	M	A	M	A	M	A	M	A	M	A	M	A
FL	Apalachicola	42	42	85	66	86	65	86	65	86	64	85	65	85	67	86	71	88	75	88	69	86	62	85	63	86	67	86	66
	Daytona Beach	58	58	88	59	87	57	87	55	86	54	85	57	87	63	89	64	91	67	90	67	87	63	88	60	88	61	88	61
	Fort Myers	58	58	89	57	88	54	89	52	88	48	88	50	89	58	89	60	91	60	92	61	90	57	90	56	89	57	89	56
	Gainesville	19	19	90	60	90	56	91	53	91	50	91	50	88	56	89	59	91	60	96	64	94	61	93	60	91	61	91	58
	Jacksonville	66	66	88	58	87	54	87	50	87	48	86	51	88	57	89	59	91	61	92	63	91	59	90	57	89	59	89	56
	Key West	54	54	82	69	81	67	80	66	77	64	77	65	78	68	77	67	78	67	81	70	82	69	83	69	83	70	80	68
	Miami	38	38	84	60	84	58	82	56	79	54	80	58	84	65	83	63	85	65	87	67	86	63	85	62	84	60	83	61
	Orlando	39	39	88	57	88	53	89	50	88	47	88	49	90	58	91	59	93	60	92	61	90	57	90	56	89	58	90	55
	Pensacola	39	39	82	65	82	61	84	62	86	60	87	62	84	61	86	65	87	66	85	63	82	57	82	61	81	65	84	62
	Tallahassee	41	41	87	58	87	54	89	51	91	47	90	50	91	56	94	61	95	61	93	58	91	53	90	55	88	57	90	55
	Tampa	39	39	87	60	86	57	87	55	86	52	85	53	86	60	87	64	90	65	91	63	89	58	88	58	88	60	88	59
	Vero Beach	39	39	88	60	88	56	86	56	84	55	84	58	80	60	83	59	85	60	82	59	81	58	79	56	81	56	83	58
	West Palm Beach	38	38	84	59	83	57	82	56	79	55	80	59	84	66	85	64	86	64	87	66	84	63	84	61	84	60	83	61
GA	Athens	47	47	81	58	80	54	81	53	83	50	86	54	88	56	90	58	92	59	92	59	89	55	85	54	82	58	86	56
	Atlanta	42	42	79	59	77	55	77	52	79	50	82	53	84	56	88	59	89	60	88	59	84	54	82	56	80	59	82	56
	Augusta	38	38	84	54	83	49	84	48	86	45	87	49	87	52	89	55	92	57	92	55	91	50	89	50	85	53	87	51
	Columbus	57	57	84	59	83	54	84	51	85	48	85	50	85	53	89	57	90	56	89	56	88	52	87	54	84	58	86	54
	Macon	38	38	84	58	84	54	86	52	88	49	88	50	88	54	90	57	93	57	93	58	90	52	88	53	85	57	88	54
	Savannah	38	38	83	55	82	51	84	49	84	46	86	51	88	56	89	58	91	60	91	60	89	54	87	53	84	55	87	54
HI	Hilo	53	53	79	67	78	66	80	67	81	69	80	68	79	66	81	68	81	69	80	68	80	69	82	71	81	69	80	68
	Honolulu	33	33	81	61	78	59	74	57	70	56	68	54	67	52	68	52	68	52	69	53	71	56	75	59	79	61	72	56
	Kahului	38	38	83	62	81	60	77	59	75	58	71	56	68	53	71	56	72	56	71	55	74	57	77	60	80	61	75	58
	Lihue	53	53	82	67	80	66	78	66	77	67	75	66	75	66	76	66	76	66	77	65	78	68	80	69	81	69	78	67
ID	Boise	63	63	80	70	79	60	73	44	70	36	69	34	66	29	54	21	52	22	58	29	66	38	77	60	81	71	69	43
	Lewiston	44	44	80	70	79	61	76	50	74	43	75	40	73	36	60	25	57	25	66	32	78	49	82	68	81	74	74	48
	Pocatello	39	39	80	71	80	62	76	50	71	38	71	35	71	32	65	25	62	23	66	28	70	37	77	60	80	72	72	44
IL	Cairo	22	22	80	67	79	64	78	59	77	54	82	57	83	57	84	59	87	59	87	57	85	53	80	58	79	66	82	59
	Chicago	44	44	78	70	78	67	79	63	77	58	77	57	79	58	82	60	86	61	85	61	81	59	80	66	80	71	80	63
	Moline	42	42	76	69	77	67	78	63	78	58	80	57	82	58	85	62	89	64	87	62	81	59	80	66	79	71	81	63
	Peoria	43	43	80	71	81	69	81	65	78	59	81	60	82	60	86	63	89	65	87	63	84	61	83	69	83	73	83	65
	Rockford	39	39	81	73	81	69	81	65	80	60	80	58	82	60	86	62	91	64	90	63	85	61	83	69	83	74	83	65
	Springfield	43	43	80	71	80	70	81	66	79	60	81	58	83	59	86	63	89	64	88	61	83	59	82	67	83	73	83	64
IN	Evansville	41	41	79	69	79	66	79	62	78	58	82	59	83	59	86	61	88	61	88	60	84	57	80	64	80	70	82	62
	Fort Wayne	41	41	81	72	81	69	81	63	79	57	80	55	81	55	85	56	89	59	89	57	86	58	84	68	83	75	83	62
	Indianapolis	43	43	81	71	81	67	80	61	79	56	82	57	83	57	87	60	90	60	90	57	87	57	84	66	83	72	84	62
	South Bend	39	39	82	73	81	68	80	61	78	56	79	54	81	55	84	57	89	59	89	59	84	60	83	68	83	75	83	62
IA	Des Moines	41	41	77	70	79	67	78	63	77	58	78	59	80	60	83	61	85	63	84	62	79	59	79	66	80	72	80	63
	Dubuque	35	35	78	72	79	69	79	65	77	60	79	61	83	64	86	65	89	66	87	65	81	62	81	69	82	74	82	66
	Sioux City	43	43	78	71	80	69	80	66	78	57	79	58	82	61	86	63	89	66	86	63	81	58	82	67	82	73	82	64
	Waterloo	43	43	78	72	80	71	82	68	81	59	81	58	84	60	87	63	90	64	89	63	83	61	83	69	82	74	83	65
KS	Concordia	40	40	78	67	79	64	78	59	80	57	84	61	84	59	81	56	83	59	82	58	78	55	80	63	79	67	80	61
	Dodge City	39	39	76	61	76	58	76	54	76	51	81	56	80	53	76	50	79	52	79	53	75	51	77	57	76	60	77	55
	Goodland	36	36	76	61	77	53	78	48	78	43	84	48	82	42	82	40	83	42	79	41	75	45	76	59	75	61	79	48
	Topeka	38	38	78	66	78	64	78	60	80	58	84	62	86	63	85	63	87	62	87	61	83	58	81	64	80	67	82	62
	Wichita	49	49	79	65	78	62	77	57	78	56	83	60	83	56	79	53	79	53	81	57	80	56	79	60	80	65	80	58
KY	Greater Cincinnati	40	40	80	69	78	64	78	59	77	54	81	56	84	57	86	58	89	58	89	57	84	56	80	63	81	69	82	60
	Jackson	21	21	78	64	76	60	73	53	71	48	82	57	86	62	90	63	91	62	89	60	83	55	76	58	79	65	81	59
	Lexington	39	39	81	69	79	64	77	58	76	55	81	58	84	58	86	59	88	59	88	58	85	57	81	63	81	68	82	60
	Louisville	42	42	78	65	77	61	76	57	76	52	82	56	83	57	85	58	87	57	88	57	85	55	80	61	79	66	81	59
	Paducah	18	18	81	71	80	68	77	65	80	62	87	66	88	67	90	69	92	69	91	67	86	65	82	68	82	71	85	67
LA	Baton Rouge	43	43	85	67	84	62	86	61	89	60	91	62	92	63	92	66	92	66	91	64	89	59	89	63	87	66	89	63
	Lake Charles	38	38	87	72	87	67	89	67	90	65	93	67	93	68	94	69	94	68	92	67	91	62	89	66	89	70	91	67
	New Orleans	54	54	85	68	84	65	84	63	87	62	89	62	90	65	91	68	91	68	89	67	87	62	86	64	85	68	87	65
	Shreveport	50	50	83	66	82	62	83	60	86	60	90	63	90	62	90	61	90	59	89	60	88	59	86	62	85	65	87	62
ME	Caribou	58	58	75	67	75	63	76	60	76	56	74	52	78	56	83	58	86	59	88	61	86	62	85	71	80	71	80	61
	Portland	62	62	76	61	76	57	75	58	73	55	75	58	78	60	80	59	83	59	86	60	84	59	82	62	79	61	79	59
MD	Baltimore	49	49	73	57	72	54	72	51	72	49	77	52	77	52	80	53	84	55	85	55	84	54	79	54	75	57	78	54
MA	Blue Hill	49	49	76	62	75	59	75	57	72	52	75	55	78	58	80	56	82	58	83	59	80	56	79	60	75	61	78	58
	Boston	38	38	69	58	68	56	70	57	69	55	72	59	73	58	74	57	77	59	79	61	77	58	74	59	70	59	73	58
	Worcester	47	47	73	60	72	57	71	55	68	50	70	51	75	57	77	58	79	59	82	61	78	56	78	61	75	62	75	57
MI	Alpena	43	43	81	71	80	66	83	61	80	54	78	52	80	52	85	54	90	58	91	61	87	61	84	69	83	73	83	61

	Detroit	44	44	80	70	79	65	79	60	78	54	78	53	79	54	82	54	86	57	87	57	84	57	82	65	81	70	81	60
	Flint	39	39	81	72	80	68	80	61	78	56	78	54	81	56	84	55	89	58	90	59	85	60	83	68	82	74	83	62
	Grand Rapids	39	39	82	73	81	68	81	63	79	57	79	53	82	56	84	56	89	59	89	60	86	62	83	70	83	75	83	63
	Houghton Lake	38	38	84	73	83	68	84	62	80	54	78	50	82	55	86	55	91	60	92	62	88	64	87	73	85	76	85	63
	Lansing	39	39	82	73	81	68	81	62	79	56	78	54	82	57	86	57	91	60	91	61	88	63	85	70	85	76	84	63
	Muskegon	42	42	81	75	81	70	80	63	77	57	76	55	81	58	84	59	89	62	89	63	84	65	81	70	81	75	82	64
	Sault Ste. Marie	61	61	81	74	81	70	82	66	80	59	79	55	85	62	88	62	92	63	92	67	89	67	86	75	84	77	85	66
MN	Duluth	41	41	78	72	77	68	78	66	76	59	76	56	82	63	85	63	88	66	88	67	82	65	81	72	80	75	81	66
	International Falls	60	60	75	70	74	66	76	62	76	55	77	53	83	59	88	60	91	63	90	66	85	65	84	75	80	75	82	64
	Minneapolis-St.Paul	43	43	76	69	76	67	76	64	74	55	75	55	78	58	81	59	84	61	84	63	80	61	80	68	78	71	78	63
	Rochester	42	42	81	76	81	74	82	71	80	63	80	60	83	62	86	65	89	67	88	67	83	64	84	73	84	78	83	68
	Saint Cloud	50	50	78	71	79	68	81	65	80	55	80	54	85	59	88	59	91	62	90	63	85	61	83	70	80	73	83	63
MS	Jackson	39	39	86	68	86	63	86	60	90	58	91	60	91	61	93	64	94	63	93	62	92	59	90	62	87	66	90	62
	Meridian	38	38	86	65	86	61	87	58	90	58	91	61	91	61	92	64	93	63	91	62	91	58	89	60	87	64	89	61
	Tupelo	19	19	82	70	80	67	80	61	84	61	88	65	89	65	90	66	91	66	90	65	88	63	85	67	83	70	86	66
MO	Columbia	33	33	80	69	80	67	78	62	78	59	85	65	87	65	87	63	88	63	87	63	84	62	81	66	81	70	83	64
	Kansas City	30	30	77	67	78	66	77	62	77	59	83	63	85	64	85	64	86	64	85	64	80	60	79	66	79	68	81	64
	St. Louis	42	42	81	68	80	65	79	61	77	58	81	59	82	59	83	60	86	60	86	61	82	59	81	65	82	69	82	62
	Springfield	42	42	78	65	78	63	77	59	79	59	85	63	87	64	87	62	87	60	87	63	82	59	80	63	80	66	82	62
MT	Billings	43	43	65	57	66	51	68	46	69	42	70	42	72	39	64	32	61	30	65	37	64	43	65	54	65	57	66	44
	Glasgow	38	38	77	73	79	70	80	58	75	42	74	40	77	41	74	35	69	32	72	37	75	47	80	65	79	73	76	51
	Great Falls	41	41	67	61	67	55	69	49	69	42	71	41	72	41	68	31	66	30	68	37	65	44	65	56	65	60	68	46
	Helena	37	37	72	64	73	54	73	46	70	39	71	38	73	38	67	30	68	30	72	35	73	43	74	58	73	66	72	45
	Kalispell	38	38	82	76	82	67	81	54	78	43	80	44	85	46	84	36	82	35	84	42	85	53	84	73	83	79	82	54
	Missoula	42	42	85	76	85	66	83	51	80	42	81	42	83	42	78	31	75	30	82	38	85	50	86	71	86	80	82	52
NE	Grand Island	41	41	77	66	78	64	79	60	79	54	82	57	82	56	83	58	85	59	84	57	79	53	80	61	79	66	81	59
	Lincoln	30	30	78	68	80	66	80	63	80	58	83	61	83	59	83	60	86	63	84	61	81	58	81	64	81	68	82	62
	Norfolk	57	57	76	66	79	66	80	63	78	53	80	54	82	56	83	57	86	59	83	56	79	53	79	61	79	67	80	59
	North Platte	38	38	80	66	79	63	80	58	80	53	83	57	83	58	83	57	85	58	83	54	81	52	81	60	80	64	81	58
	Omaha Eppley AP	38	38	79	68	77	64	78	61	77	57	80	59	82	61	85	63	87	65	86	63	81	59	81	66	81	70	81	63
	Omaha (North)	9	9	75	63	77	63	78	59	76	53	79	55	79	56	85	63	86	62	84	60	78	55	79	65	77	67	79	60
	Scottsbluff	37	37	74	58	75	48	77	44	76	41	80	42	80	39	81	37	83	39	80	37	76	42	77	53	75	58	78	45
	Valentine	35	35	76	63	77	64	79	61	77	54	78	54	79	54	78	53	79	53	77	51	75	52	76	59	75	62	77	57
NV	Elko	35	35	78	60	79	52	77	41	72	33	71	30	63	24	54	18	53	19	59	22	64	27	74	47	76	58	68	36
	Ely	50	50	73	55	75	50	72	42	69	34	67	30	58	23	52	21	55	23	59	24	64	31	71	46	72	54	65	36
	Las Vegas	42	42	55	32	51	28	45	23	35	16	32	14	24	11	28	15	33	17	34	17	36	20	45	27	53	32	39	21
	Reno	39	39	79	50	74	40	69	33	65	28	64	25	62	22	60	18	61	19	66	22	70	27	74	40	77	50	68	31
	Winnemucca	53	53	79	58	77	47	73	38	67	30	64	27	58	22	46	16	45	16	53	20	62	28	74	46	79	58	65	34
NH	Concord	37	37	76	59	76	55	77	53	75	46	77	47	81	52	84	51	88	52	90	55	88	52	84	59	80	61	81	53
	Mt. Washington	34	34	83	83	81	83	84	84	85	84	84	81	86	82	88	84	87	85	82	83	80	79	82	82	82	82	84	83
NJ	Atlantic City AP	38	38	79	59	79	55	77	54	77	52	79	55	81	56	83	57	87	58	88	58	88	56	84	57	79	59	82	56
	Newark	37	37	73	59	72	54	70	51	66	48	70	51	71	52	72	51	76	53	78	55	79	53	76	56	74	58	73	53
NM	Albuquerque	42	42	68	39	63	31	55	24	48	19	47	19	46	18	59	27	65	31	61	30	60	30	64	36	69	43	59	29
	Clayton	45	45	64	48	65	42	65	37	64	32	70	37	71	35	76	41	77	42	72	40	65	41	64	47	62	53	68	41
	Roswell	29	29	71	41	65	33	58	26	54	22	58	24	65	27	68	32	73	35	74	38	69	36	69	39	69	42	66	33
NY	Albany	37	37	78	64	77	58	76	54	72	49	75	53	79	56	81	55	86	58	89	59	86	58	82	63	80	65	80	58
	Binghamton	51	51	80	71	79	66	79	62	76	56	78	56	83	59	84	58	89	60	90	63	85	62	82	69	82	73	82	63
	Buffalo	42	42	79	73	79	69	80	65	77	58	76	56	78	57	79	55	83	58	84	60	82	60	80	68	81	72	80	63
	Islip	42	42	76	62	76	58	77	57	76	55	76	56	71	53	74	56	77	56	79	56	78	54	74	55	70	54	75	56
	New York C.Park	68	68	68	60	68	57	68	54	67	51	71	53	74	56	75	55	78	57	79	57	76	55	73	59	69	60	72	56
	New York (JFK AP)	41	41	71	60	71	57	71	56	70	55	73	59	75	60	75	59	78	60	80	60	78	57	76	59	72	60	74	58
	New York (Laguardia AP)	40	40	67	58	65	55	67	53	67	51	71	53	72	54	72	53	75	55	76	57	74	55	71	58	68	59	70	55
	Rochester	39	39	78	69	79	66	79	62	77	55	77	54	80	56	83	55	87	58	88	61	85	60	81	67	81	72	81	61
	Syracuse	39	39	77	69	78	64	78	60	76	53	76	55	79	56	81	56	86	58	87	61	84	61	81	67	81	71	80	61
NC	Asheville	38	38	85	59	83	55	84	53	85	50	91	56	94	60	95	62	97	62	97	62	93	56	88	56	85	59	90	57
	Cape Hatteras	45	45	80	68	80	65	80	63	78	60	81	65	83	68	85	70	86	70	84	68	82	65	81	65	80	67	82	66
	Charlotte	42	42	78	55	76	52	78	49	78	46	82	52	84	55	86	56	88	57	89	57	86	53	83	53	79	55	82	53
	Greensboro-Wnstn-Slm-HPT	39	39	78	56	76	52	77	50	78	48	82	54	84	56	87	59	90	60	90	59	88	54	83	53	79	56	83	55
	Raleigh	38	38	80	55	78	52	80	49	81	46	85	54	87	56	89	58	91	59	92	59	90	53	85	52	81	55	85	54
	Wilmington	39	39	81	56	80	52	82	52	81	48	84	55	85	60	87	63	90	64	90	62	89	56	85	53	82	55	85	56
ND	Bismarck	43	43	76	71	79	70	81	66	79	55	79	53	84	57	84	53	83	52	82	54	79	55	81	67	79	72	81	61
	Fargo	43	43	76	73	78	74	82	72	79	59	76	53	82	59	85	59	86	58	84	61	80	62	82	72	79	75	81	65
	Williston	41	41	79	74	81	73	83	67	79	55	77	52	81	54	80	51	77	51	79	55	79	58	82	71	81	75	80	61
OH	Akron	39	39	80	71	79	66	78	60	76	54	78	56	81	57	84	56	88	59	88	60	84	59	80	66	80	71	81	61

(Continued)

Table 3A.12 (Continued)

State	Station	Years		Jan		Feb		Mar		Apr		May		Jun		Jul		Aug		Sep		Oct		Nov		Dec		Annual	
		M	A	M	A	M	A	M	A	M	A	M	A	M	A	M	A	M	A	M	A	M	A	M	A	M	A	M	A
	Cleveland	42	42	79	70	78	68	79	63	77	58	78	58	80	58	82	57	86	60	85	60	81	60	78	66	78	71	80	62
	Columbus	43	43	78	68	77	64	76	57	76	53	79	55	81	55	84	56	87	57	87	57	83	55	80	63	79	69	81	59
	Dayton	39	39	79	70	79	66	79	61	77	55	78	55	80	55	82	56	87	57	87	56	83	57	81	65	81	71	81	60
	Mansfield	36	36	82	73	81	69	80	63	78	57	79	57	81	58	83	58	88	61	88	60	83	59	81	67	83	74	82	63
	Toledo	47	47	81	70	80	66	81	60	79	54	80	53	82	54	85	55	91	59	91	57	86	57	83	66	83	73	84	60
	Youngstown	55	55	81	72	80	68	80	63	77	56	79	54	82	56	85	56	89	57	89	59	85	58	82	67	82	72	83	61
OK	Oklahoma City	37	37	78	62	77	60	76	57	77	56	83	62	84	61	80	55	80	55	83	59	80	57	80	60	78	62	80	59
	Tulsa	42	42	78	63	76	59	75	56	77	55	85	62	85	63	81	57	82	56	85	62	82	57	80	61	79	63	81	60
OR	Astoria	49	49	87	78	86	74	88	71	89	70	89	70	90	71	90	70	91	71	91	70	90	73	87	78	86	81	89	73
	Eugene	45	45	92	80	92	72	91	65	90	58	91	55	90	50	87	39	88	39	89	43	93	62	93	78	92	84	91	60
	Medford	41	41	90	70	88	57	86	50	84	45	83	39	78	33	74	26	74	26	78	29	86	42	91	67	91	76	84	47
	Pendleton	61	61	80	75	78	65	74	50	71	42	70	38	66	33	55	24	54	26	61	32	72	47	80	70	81	78	70	48
	Portland	62	62	85	76	85	67	85	60	86	55	85	53	83	49	82	45	83	45	86	48	90	62	88	74	86	78	85	59
	Salem	40	40	87	76	88	68	88	61	88	57	88	53	87	49	85	41	85	40	87	45	90	60	90	77	89	80	88	59
	Sexton Summit	7	7	80	72	78	70	84	68	84	63	73	53	73	49	65	40	66	42	65	45	70	58	81	75	79	75	75	59
PC	Guam	9	9	87	73	87	72	86	70	88	70	89	72	88	73	90	77	91	78	92	79	90	77	88	77	87	76	89	74
	Johnston Island	23	23	76	68	77	68	78	70	79	71	78	70	77	69	78	69	79	71	80	72	80	72	79	73	78	72	78	70
	Koror	51	51	80	76	80	75	78	73	78	74	79	77	81	78	81	78	80	77	79	76	79	77	79	77	80	77	80	76
	Kwajalein, Marshall IS	42	42	79	72	78	70	79	71	81	74	83	77	81	75	84	77	83	77	83	76	83	76	83	76	81	75	81	75
	Majuro, Marshall IS	47	47	80	75	79	74	80	75	83	77	83	78	84	78	84	78	83	77	82	76	82	76	83	77	82	77	82	76
	Pago Pago, Amer Samoa	34	34	88	75	88	75	89	75	89	76	87	76	85	76	83	74	84	74	84	74	84	76	84	76	85	75	86	75
	Pohnpei, Caroline IS	32	32	85	78	84	76	85	77	88	79	90	81	92	80	94	79	95	79	95	79	94	79	93	80	87	79	90	79
	Chuuk, E. Caroline IS	32	32	81	76	81	75	82	75	84	77	85	78	86	78	88	78	88	77	88	77	87	78	86	78	83	79	85	77
	Wake Island	45	45	77	66	78	66	80	67	81	68	82	68	82	68	82	70	83	71	83	71	82	71	80	69	78	67	81	69
	Yap, W Caroline IS	54	54	79	76	78	74	77	73	76	73	78	75	80	77	81	77	82	78	81	78	81	78	80	78	80	77	79	76
PA	Allentown	52	52	76	62	76	57	75	53	74	50	78	53	80	54	82	53	86	56	88	57	87	55	82	59	79	62	80	56
	Erie	37	37	78	72	77	70	77	65	75	62	76	62	79	64	80	64	82	65	82	65	77	63	76	68	77	72	78	66
	Harrisburg	49	49	72	58	71	55	72	52	70	49	74	52	77	53	79	52	83	55	85	56	82	54	77	57	73	58	76	54
	Middletown/Harrisburg AP	37	37	73	59	72	55	72	53	71	50	75	52	76	52	78	52	81	54	84	55	81	54	76	56	72	58	76	54
	Philadelphia	43	43	74	60	72	55	72	53	71	50	75	53	77	53	78	54	81	54	83	56	83	54	78	56	74	59	76	55
	Pittsburgh	42	42	77	66	75	62	76	57	74	51	77	52	80	53	83	54	86	56	87	57	82	55	79	62	78	67	79	58
	Avoca	47	47	76	66	75	61	74	57	72	52	76	52	82	56	83	55	86	58	88	60	84	58	79	63	77	67	79	59
	Williamsport	57	57	77	62	76	58	77	53	75	49	80	51	84	54	87	55	90	57	92	59	89	57	82	61	78	63	82	57
RI	Block IS	15	15	73	65	73	65	75	65	79	65	80	66	83	69	87	72	86	71	84	70	80	66	76	65	72	65	79	67
	Providence	39	39	72	57	71	54	72	53	70	49	73	53	76	56	77	56	80	56	82	57	81	54	78	57	74	58	75	55
SC	Charleston AP	60	60	83	56	82	52	83	50	84	49	85	53	86	59	88	62	90	63	90	62	89	56	86	53	84	55	86	56
	Columbia	36	36	83	54	82	49	84	48	84	44	85	48	86	51	88	53	91	56	92	55	91	51	89	51	84	53	86	51
	Greenville-Spartanburg AP	40	40	77	55	76	51	76	50	78	48	83	53	85	54	87	56	89	58	89	59	86	53	82	53	79	55	82	54
SD	Aberdeen	34	34	79	73	80	74	83	70	82	59	81	57	84	61	86	59	87	58	85	58	82	59	83	71	81	74	83	65
	Huron	43	43	77	70	80	71	83	68	82	59	83	59	85	61	86	58	88	59	86	58	81	58	82	66	80	71	83	63
	Rapid City	52	52	69	64	71	61	75	54	73	47	76	48	78	49	74	41	72	37	68	39	67	46	69	60	69	64	72	51
	Sioux Falls	39	39	78	71	80	71	82	67	81	59	81	58	81	60	84	60	86	62	85	61	81	60	83	69	81	73	82	64
TN	Bristol-JhnCty-Kgsprt	41	41	81	62	80	58	80	52	82	50	89	55	90	58	92	61	93	60	93	57	90	53	85	56	82	62	86	57
	Chattanooga	72	72	82	62	81	57	81	53	82	49	86	53	87	55	89	57	91	57	91	56	90	53	85	56	83	61	86	56
	Knoxville	42	42	82	64	80	59	80	55	82	52	87	57	89	59	90	61	92	60	92	59	90	56	85	59	83	64	86	59
	Memphis	63	63	78	65	77	61	76	58	77	56	81	58	82	59	84	60	85	59	85	58	82	54	79	58	78	64	80	59
	Nashville	37	37	79	66	79	62	77	57	79	56	85	60	86	60	88	61	89	61	89	62	86	58	81	62	80	66	83	61
TX	Abilene	39	39	72	56	72	56	70	51	72	50	78	55	78	54	71	49	72	50	76	56	75	54	75	56	73	56	74	54
	Amarillo	41	41	71	53	72	52	69	46	68	42	75	48	77	49	73	46	77	50	79	52	73	48	73	50	71	52	73	49
	Austin	41	41	78	63	78	61	79	59	82	60	88	64	88	61	87	56	85	55	85	59	83	59	82	62	79	63	83	60
	Brownsville	36	36	88	70	89	66	88	63	89	64	90	65	90	64	91	60	91	61	91	65	90	64	87	66	87	69	89	65
	Corpus Christi	38	38	87	70	87	68	87	65	89	66	92	70	93	67	93	62	92	62	90	65	89	63	87	66	86	68	89	66
	Dallas-Fort Worth	39	39	79	63	78	61	79	59	81	59	86	63	85	59	79	53	78	53	82	59	82	58	81	61	80	63	81	59
	Del Rio	23	23	74	60	72	57	70	54	74	57	78	61	78	60	74	57	75	59	80	62	80	63	80	63	75	59	76	60
	El Paso	42	42	65	34	55	27	47	21	39	17	41	17	45	19	61	29	65	33	66	33	63	30	61	33	65	38	56	27
	Galveston	96	96	85	77	84	74	85	74	86	75	84	73	81	70	81	70	81	69	81	68	80	65	83	72	85	76	83	72
	Houston	33	33	85	68	86	65	87	65	89	64	91	66	92	65	92	63	92	63	92	64	91	62	89	65	87	67	89	65
	Lubbock	55	55	72	52	71	52	67	44	67	42	74	47	76	48	73	49	76	52	79	54	77	51	73	49	72	51	73	49
	Midland-Odessa	39	39	71	51	71	48	65	40	66	38	73	43	75	46	70	45	73	47	78	53	78	49	75	49	72	49	72	46
	Port Arthur	42	42	88	71	87	67	88	66	90	66	92	68	93	68	94	70	94	69	92	68	91	63	89	66	89	70	91	68

	San Angelo	42	42	75	56	75	53	71	48	73	47	79	53	80	54	75	48	76	50	82	58	81	56	80	56	78	56	77	53
	San Antonio	60	60	80	61	80	59	79	57	82	59	87	62	87	60	86	55	85	54	85	57	84	57	81	59	80	60	83	58
	Victoria	41	41	87	69	87	66	87	63	89	64	91	66	92	65	92	62	92	62	92	65	90	62	89	64	87	68	90	65
	Waco	39	39	83	66	82	64	82	62	84	62	88	65	86	60	81	53	79	51	84	58	84	59	84	63	83	65	83	61
	Wichita Falls	42	42	79	59	78	58	78	54	80	53	85	57	84	55	76	48	77	49	83	56	82	55	82	57	80	59	80	55
UT	Salt Lake City	43	43	79	69	78	60	70	46	66	39	65	34	59	26	52	22	53	24	61	30	68	41	75	59	79	71	67	43
VI	Burlington	37	37	73	64	74	61	75	58	73	52	73	51	77	54	78	53	83	56	86	60	81	60	78	65	77	67	77	59
VA	Lynchburg	39	39	73	54	73	50	73	48	73	45	81	52	83	55	86	57	86	56	88	57	86	52	80	52	76	54	80	53
	Norfolk	54	54	75	59	75	57	74	54	74	51	77	56	79	57	81	59	84	61	84	61	83	59	79	57				

Note: The relative humidity is expressed as a percentage measure of the amount of moisture in the air compared to the maximum amount of moisture the air can hold at the same temperature and pressure. Average humidity values are given for selected morning and afternoon observations. Maximum relative humidity values usually occur during morning hours. In this publication, the Local Standard Time (LST) of morning and afternoon humidity is shown below. Atlantic, Alaskan (M morning 8 A.M.) (Afternoon 2 P.M.), Eastern, Bering, 165W Meridian (M morning 7 A.M.) (Afternoon 1 P.M.), Central, 180E Meridian (M morning 6 A.M.) (Afternoon NOON), Mountain, 165E Meridian (M morning 5 A.M.) (Afternoon 5 P.M.), Pacific, 150E Meridian (M morning 4 A.M.) (Afternoon 4 P.M.), 135E Meridian (M morning 9 A.M.) (Afternoon 3 P.M.).

Source: From U.S. National Oceanic and Atmospheric Administration, *Comparative Climatic Data for the United States Through 2000*, www.noaa.gov.

SECTION 3B CLIMATIC DATA — WORLD

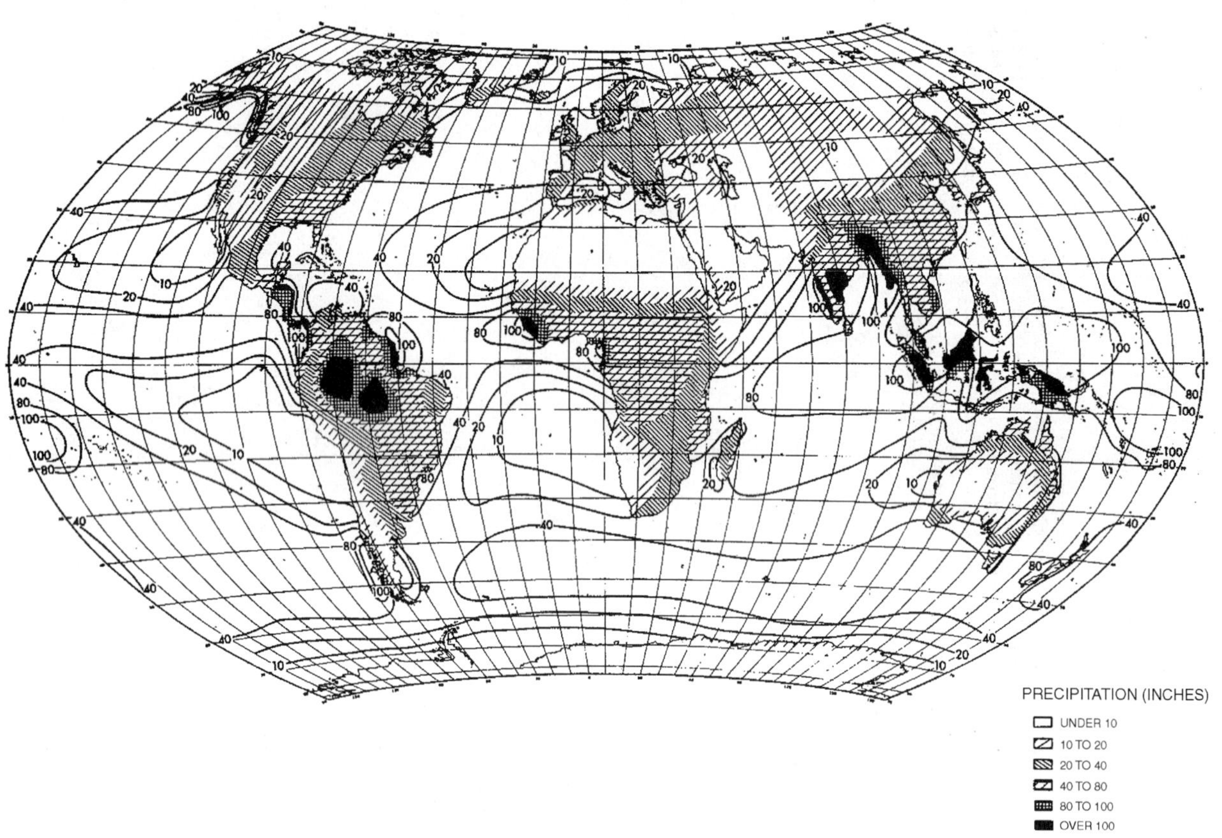

Figure 3B.2 General Pattern of Annual World Precipitation. (From Environmental Science Service: Administration, *Climates of the World*, 1969. www.noaa.gov.)

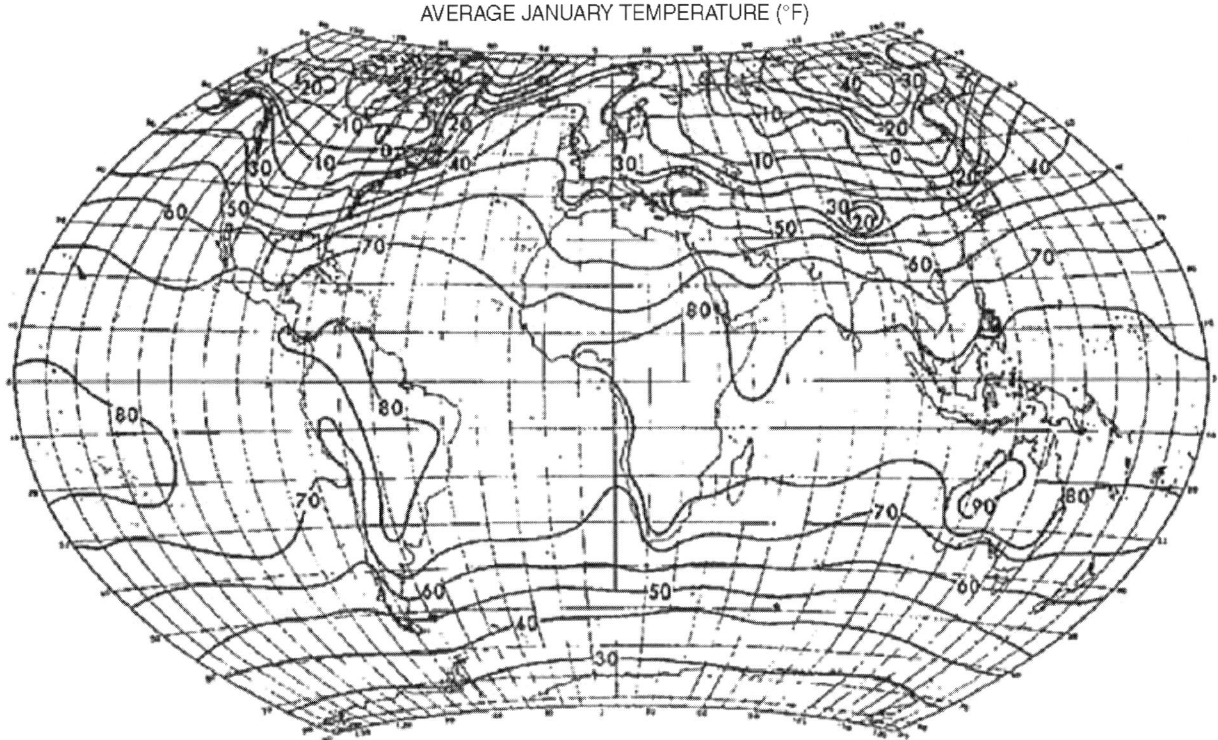

Figure 3B.3 Average January World Temperature. (From *Climates of the World*, Historical Climatology Series 6–14, 1991. www.noaa.gov.)

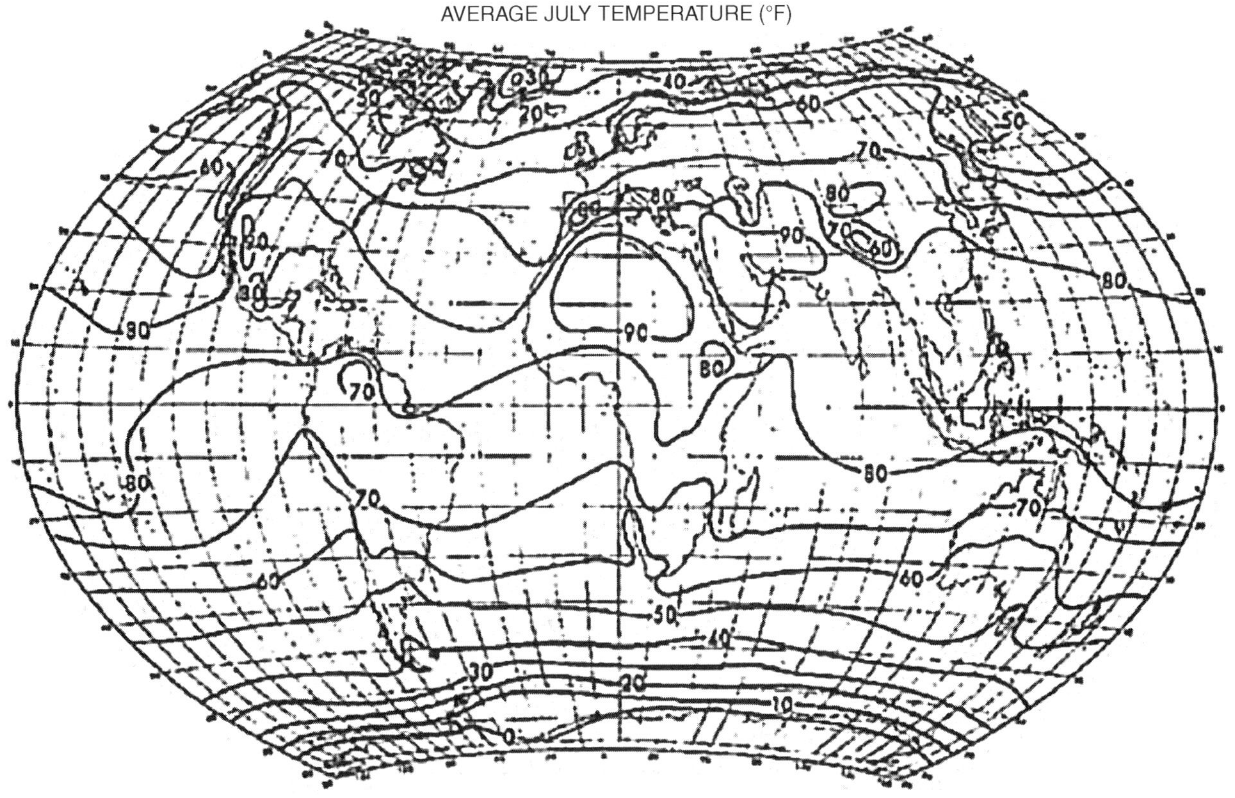

Figure 3B.4 Average July World Temperatures. (From *Climates of the World*, Historical Climatology Series 6–14, 1991. www.noaa.gov.)

Table 3B.13 Temperature and Precipitation Data for Representative World-Wide Stations

Country and Station	Latitude	Longitude	Elevation	Length of Record	Temperature										Average Precipitation													
					Average Daily								Extreme															
					Jan		Apr		July		Oct																	
					Maximum	Minimum	Maximum	Minimum	Maximum	Minimum	Maximum	Minimum	Maximum	Minimum	Length of Record	January	February	March	April	May	June	July	August	September	October	November	December	Year
	°	°	Feet	Year	°F	°F	°F	°F	°F	°F	°F	°F	°F	°F	Year	IN.	IN.	IN.	IN.	IN.	IN.	IN.	IN.	IN.	IN.	IN.	IN.	IN.
North America																												
United States																												
Albuquerque, NM	35 03N	106 37W	5,311	30	46	24	69	42	91	66	71	45	104	−16	30	0.4	0.4	0.5	0.5	0.8	0.6	1.2	1.3	1.0	0.8	0.4	0.5	8.4
Asheville, NC	35 26N	82 32W	2,140	30	48	28	67	42	84	61	68	45	99	−7	30	4.2	4.0	4.8	4.0	3.7	3.5	5.9	4.9	3.6	3.1	2.8	3.6	48.1
Atlanta, GA	33 39N	84 26W	1,010	30	52	37	70	50	87	71	72	52	103	−9	30	4.4	4.5	5.4	4.5	3.2	3.8	4.7	3.6	3.3	2.4	3.0	4.4	47.2
Austin, TX	30 18N	97 42W	597	30	60	41	78	57	95	74	82	60	109	−2	30	2.4	2.6	2.1	3.6	3.7	3.2	2.2	1.9	3.4	2.8	2.1	2.5	32.5
Birmingham, AL	33 34N	86 45W	620	30	57	36	76	50	93	71	79	52	107	−10	30	5.0	5.3	6.0	4.5	3.4	4.0	5.2	4.9	3.3	3.0	3.5	5.0	53.1
Bismark, ND	46 46N	100 45W	1,647	30	20	0	55	32	86	58	59	34	114	−45	30	0.4	0.4	0.8	1.2	2.0	3.4	2.2	1.7	1.2	0.9	0.6	0.4	15.2
Boise, ID	43 34N	116 13W	2,838	30	36	22	63	37	91	59	65	38	112	−28	30	1.3	1.3	1.3	1.2	1.3	0.9	0.2	0.2	0.4	0.8	1.2	1.3	11.4
Brownsville, TX	25 54N	97 26W	16	30	71	52	82	66	93	76	85	67	104	12	30	1.4	1.5	1.0	1.6	2.4	3.0	1.7	2.8	5.0	3.5	1.3	1.7	26.9
Buffalo, NY	42 56N	78 44W	705	30	31	18	53	34	80	59	60	41	99	−21	30	2.8	2.7	3.2	3.0	3.0	2.5	2.6	3.1	3.1	3.0	3.6	3.0	35.6
Cheyenne, WY	41 09N	104 49W	6,126	30	37	14	56	30	85	55	63	32	100	−38	30	0.5	0.6	1.2	1.9	2.5	2.1	1.8	1.4	1.1	0.8	0.6	0.5	15.0
Chicago, IL	41 47N	87 45W	607	30	33	19	57	41	84	67	63	47	105	−23	30	1.9	1.6	2.7	3.0	3.7	4.1	3.4	3.2	2.7	2.8	2.2	1.9	33.2
Des Moines, IO	41 32N	93 39W	938	30	29	11	59	38	87	65	66	43	110	−30	30	1.3	1.1	2.1	2.5	4.1	4.7	3.1	3.7	2.9	2.1	1.8	1.1	30.5
Dodge City, KS	37 46N	99 58W	2,582	30	42	20	66	41	93	68	71	46	109	−26	30	0.6	0.7	1.2	1.8	3.2	3.0	2.3	2.4	1.5	1.4	0.6	0.5	19.2
El Paso, TX	31 48N	106 24W	3,918	30	56	30	78	49	95	69	79	50	109	−8	30	0.5	0.4	0.4	0.3	0.4	0.7	1.3	1.2	1.1	0.9	0.3	0.5	8.0
Indianapolis, IN	39 44N	86 17W	792	30	37	21	61	40	86	64	67	44	107	−25	30	3.1	2.3	3.4	3.7	4.0	4.6	3.5	3.0	3.2	2.6	3.1	2.7	39.2
Jacksonville, FL	30 25N	81 39W	20	30	67	45	80	58	92	73	80	62	105	10	30	2.5	2.9	3.5	3.6	3.5	6.3	7.7	6.9	7.6	5.2	1.7	2.2	53.6
Kansas City, MO	39 07N	94 36W	742	30	40	23	66	46	92	71	72	49	113	−22	30	1.4	1.2	2.5	3.6	4.4	4.6	3.2	3.8	3.3	2.9	1.8	1.5	34.2
Las Vegas, NV	36 05N	115 10W	2,162	30	54	32	78	51	104	76	80	53	117	8	30	0.5	0.4	0.4	0.2	0.1	—	0.5	0.5	0.3	0.2	0.3	0.4	3.8
Los Angeles, CA	33 56N	118 23W	97	30	64	45	67	52	76	62	73	57	110	23	30	2.7	2.9	1.8	1.1	0.1	0.1	—	—	0.2	0.4	1.1	2.4	12.8
Louisville, KY	38 11N	85 44W	477	30	44	27	66	43	89	67	70	46	107	−20	30	4.1	3.3	4.6	3.8	3.9	4.0	3.4	3.0	2.6	2.3	3.2	3.2	41.4
Miami, FL	25 48N	80 16W	7	30	76	58	83	66	89	75	85	71	100	28	30	2.0	1.9	2.3	3.9	6.4	7.4	6.8	7.0	9.5	8.2	2.8	1.7	59.9
Minneapolis, MN	44 53N	93 13W	834	30	22	2	56	33	84	61	61	37	108	−34	30	0.7	0.8	1.5	1.9	3.2	4.0	3.3	3.2	2.4	1.6	1.4	0.9	24.9
Missoula, MT	46 55N	114 05W	3,190	30	28	10	57	31	85	49	58	30	105	−33	30	0.9	0.9	0.7	1.0	1.9	1.9	0.9	0.7	1.0	1.0	0.9	1.1	12.9
Nashville, TN	36 07N	86 41W	590	30	49	31	71	48	91	70	74	49	107	−15	30	5.5	4.5	5.2	3.7	3.7	3.3	3.7	2.9	2.9	2.3	3.3	4.2	45.2
New Orleans, LA	29 59N	90 15W	3	30	64	45	78	58	91	73	80	61	102	7	30	3.8	4.0	5.3	4.6	4.4	4.4	6.7	5.3	5.0	2.8	3.3	4.1	53.7
New York, NY	40 47N	73 58W	132	30	40	27	60	43	85	68	66	50	106	−15	30	3.3	2.8	4.0	3.4	3.7	3.3	3.7	4.4	3.9	3.1	3.4	3.3	42.3
Oklahoma City, OK	35 24N	97 36W	1,285	30	46	28	71	49	93	72	74	52	113	−17	30	1.3	1.4	2.0	3.1	5.2	4.5	2.4	2.5	3.0	2.5	1.6	1.4	30.9
Phoenix, AZ	33 26N	112 01W	1,117	30	64	35	84	50	105	75	87	55	118	16	30	0.7	0.9	0.7	0.3	0.1	0.1	0.8	1.1	0.7	0.5	0.5	0.9	7.3
Pittsburgh, PA	40 27N	80 00W	747	30	40	25	63	42	85	65	65	45	103	−20	30	2.8	2.3	3.5	3.4	3.8	4.0	3.6	3.5	2.7	2.5	2.3	2.5	36.9
Portland, ME	43 39N	70 19W	47	30	32	12	53	32	80	57	60	37	103	−39	30	4.4	3.8	4.3	3.7	3.4	3.2	2.9	2.4	3.5	3.2	4.2	3.9	42.9
Portland, OR	45 36N	122 36W	21	30	44	33	62	42	79	56	63	45	107	−3	30	5.4	4.2	3.8	2.1	2.0	1.7	0.4	0.7	1.6	3.6	5.3	6.4	37.2
Reno, NV	39 30N	119 47W	4,404	30	45	16	65	31	89	46	69	29	106	−19	30	1.2	1.0	0.7	0.5	0.5	0.4	0.3	0.2	0.2	0.5	0.6	1.1	7.2
Salt Lake City, UT	40 46N	111 58W	4,220	30	37	18	63	36	94	60	65	38	107	−30	30	1.4	1.2	1.6	1.8	1.4	1.0	0.6	0.9	0.5	1.2	1.3	1.2	14.1
San Francisco, CA	37 37N	122 23W	8	30	55	42	64	47	72	54	71	51	106	20	30	4.0	3.5	2.7	1.3	0.5	0.1	—	—	0.2	0.7	1.6	4.1	18.7
Sault Ste. Marie, MI	46 28N	84 22W	721	30	23	8	46	30	76	54	55	38	98	−37	30	2.1	1.5	1.8	2.2	2.8	3.3	2.5	2.9	3.8	2.8	3.3	2.3	31.3
Seattle, WA	47 27N	122 18W	400	30	44	33	58	40	76	54	60	44	100	0	30	5.7	4.2	3.8	2.4	1.7	1.6	0.8	1.0	2.1	4.0	5.4	6.3	39.0
Sheridan, WY	44 46N	106 58W	3,964	30	34	9	56	31	87	56	62	33	106	−41	30	0.6	0.7	1.4	2.2	2.6	2.6	1.2	0.9	1.2	1.1	0.8	0.6	15.9
Spokane, WA	47 38N	117 32W	2,356	30	31	19	59	36	86	55	60	38	108	−30	30	2.4	1.9	1.5	0.9	1.2	1.5	0.4	0.4	0.8	1.6	2.2	2.4	17.2
Washington, DC	38 51N	77 03W	14	30	44	30	66	46	87	69	68	50	106	−15	30	3.0	2.5	3.2	3.2	4.1	3.2	4.2	4.9	3.8	3.1	2.8	2.8	40.8
Wilmington, NC	34 16N	77 55W	28	30	58	37	74	51	89	71	76	55	104	5	30	2.9	3.4	4.0	2.9	3.5	4.3	7.7	6.9	6.3	3.0	3.1	3.4	51.4

United States, Alaska																												
Anchorage	61 13N	149 52W	85	30	21	4	44	28	65	50	42	28	86	−38	30	0.8	0.7	0.5	0.4	0.5	1.0	1.9	2.6	2.5	1.9	1.0	0.9	14.7
Annette	55 02N	131 34W	110	30	38	30	50	37	63	51	51	42	90	−4	30	11.4	8.5	9.6	9.1	7.1	5.7	6.0	7.5	9.9	16.9	14.7	12.1	118.5
Barrow	71 18N	156 47W	31	30	−9	−23	7	−7	45	33	21	12	78	−56	30	0.2	0.2	0.1	0.1	0.1	0.4	0.8	0.9	0.6	0.5	0.2	0.2	4.3
Bethel	60 47N	161 48W	125	30	11	−4	34	18	62	48	38	25	90	−52	30	1.1	1.1	1.0	0.6	1.0	1.2	2.0	4.2	2.6	1.5	1.1	1.0	18.4
Cold Bay	55 12N	162 43W	96	30	33	23	38	28	54	45	45	36	78	−9	30	2.3	3.2	1.8	1.5	2.3	2.0	1.8	4.3	4.3	4.6	3.8	2.6	34.5
Fairbanks	64 49N	147 52W	436	30	−1	−21	42	17	72	48	35	17	99	−66	30	0.9	0.5	0.4	0.3	0.7	1.4	1.8	2.2	1.1	0.9	0.6	0.5	11.3
Juneau	58 22N	134 35W	12	30	30	20	45	31	63	48	47	37	89	−21	30	4.0	3.1	3.3	2.9	3.2	3.4	4.5	5.0	6.7	8.3	6.1	4.2	54.7
King Salmon	58 41N	156 39W	49	30	21	6	41	25	63	47	43	29	88	−40	30	1.1	1.0	1.0	0.6	1.0	1.4	2.1	3.4	3.1	2.2	1.5	1.0	19.4
Nome	64 30N	165 26W	13	30	12	−3	28	14	55	44	35	24	84	−47	30	1.0	0.9	0.9	0.8	0.7	0.9	2.3	3.8	2.7	1.7	1.2	1.0	17.9
St. Paul Island	57 09N	170 13W	22	30	30	21	33	24	49	42	41	33	64	−26	30	1.8	1.2	1.1	1.0	1.3	1.2	2.3	3.3	3.1	3.2	2.5	1.8	23.8
Shemya	52 43N	174 06E	122	30	34	29	38	33	49	44	42	38	63	16	30	2.5	2.3	2.6	2.1	2.4	1.3	2.2	2.1	2.3	2.8	2.7	2.1	27.4
Yakutat	59 31N	139 40W	28	30	34	20	45	29	61	48	49	35	86	−24	30	10.9	8.2	8.7	7.2	8.0	5.1	8.4	10.9	16.6	19.6	16.1	12.3	132.0
Canada																												
Aklavik, NWT	68 14N	135 00W	30	22	−10	−26	19	−2	66	47	25	15	93	−62	22	0.5	0.5	0.4	0.5	0.5	0.8	1.4	1.4	0.9	0.9	0.8	0.4	9.0
Alert, NWT	82 31N	62 20W	95	9	−19	−29	−8	−18	44	36	2	−7	67	−53	10	0.2	0.3	0.3	0.3	0.5	0.6	0.5	1.1	1.0	0.9	0.2	0.4	6.3
Calgary, Alta	51 06N	114 01W	3,540	55	24	2	53	27	76	47	54	29	97	−49	55	0.5	0.5	0.8	1.0	2.3	3.1	2.5	2.3	1.5	0.7	0.7	0.6	16.7
Charlottetown, PEI	46 17N	63 08W	181	65	26	10	43	30	73	58	54	41	98	−27	65	3.8	3.0	3.2	2.8	2.7	2.6	3.0	3.4	3.4	4.1	3.8	4.0	39.8
Chatham, NB	47 00N	65 27W	109	50	23	2	47	28	77	56	55	37	102	−43	50	3.4	2.7	3.3	3.0	3.2	3.6	3.9	4.0	3.1	4.0	3.4	3.2	40.8
Churchhill, Man	58 45N	94 04W	94	30	−11	−27	24	4	64	43	34	20	96	−57	30	0.5	0.6	0.9	0.9	0.9	1.9	2.2	2.7	2.3	1.4	1.0	0.7	16.0
Edmonton, Alta	53 34N	113 31W	2,219	71	16	−3	52	28	74	50	51	30	99	−57	71	0.9	0.7	0.7	1.0	1.9	3.2	3.3	2.4	1.3	0.8	0.9	0.9	18.0
Fort Nelson, BC	58 50N	122 35W	1,253	12	1	−15	47	25	74	51	43	25	98	−61	13	0.9	1.2	0.7	0.8	1.4	2.5	2.4	1.5	1.3	1.0	1.4	1.2	16.3
Fort Simpson, NWT	61 45N	121 14W	554	42	−10	−27	38	14	74	50	36	21	97	−70	42	0.7	0.7	0.5	0.7	1.4	1.5	2.0	1.5	1.3	1.1	0.9	0.8	13.1
Frobisher Bay, NWT	63 45N	68 33W	110	18	−9	−23	16	−1	53	39	29	18	76	−49	10	0.7	0.9	0.8	0.8	0.7	0.9	1.5	2.0	1.8	1.1	1.1	1.0	13.3
Gander, Nfld	48 57N	54 34W	496	14	27	13	40	27	71	52	51	37	96	−17	14	2.6	3.3	2.8	2.6	2.6	2.8	3.6	3.6	3.7	4.1	4.2	3.7	39.6
Halifax, NS	44 39N	63 34W	83	75	32	15	47	31	74	55	57	41	99	−21	71	5.4	4.4	4.9	4.5	4.1	4.0	3.8	4.4	4.1	5.4	5.3	5.4	55.7
Kapuskasing, Ont	49 25N	82 28W	743	19	10	−14	43	19	75	50	47	31	101	−53	19	2.0	1.1	1.6	1.8	2.1	2.3	3.4	2.9	3.5	2.5	2.4	1.9	27.5
Knob Lake, Que	54 48N	66 49W	1,712	30	−3	−21	30	12	64	46	37	25	88	−59	10	1.9	1.9	1.4	1.6	1.7	3.3	3.3	4.4	3.4	2.9	2.4	1.5	29.7
Montreal, Que	45 30N	73 34W	187	67	21	6	50	33	78	61	54	40	97	−35	77	3.8	3.0	3.5	2.6	3.1	3.4	3.7	3.5	3.7	3.4	3.5	3.6	40.8
North Bay, Ont	46 21N	79 25W	1,216	17	22	2	48	28	78	56	49	36	99	−46	23	2.0	1.5	1.8	2.2	2.5	3.2	3.2	2.7	3.7	3.2	2.7	2.1	30.8
Ottawa, Ont	45 19N	75 40W	374	65	21	3	51	31	81	58	54	37	102	−38	65	2.9	2.2	2.8	2.7	2.5	3.5	3.4	2.6	3.2	2.9	3.0	2.6	34.3
Penticton, BC	49 28N	119 36W	1,129	32	32	21	61	35	84	53	59	38	105	−16	32	1.0	0.7	0.7	0.7	1.1	1.2	0.8	0.8	1.0	0.8	0.9	1.1	10.8
Port Arthur, Ont	48 22N	89 19W	644	62	17	−4	44	26	74	52	50	34	104	−42	59	0.9	0.8	1.0	1.5	2.1	2.8	3.6	2.8	3.4	2.5	1.5	0.9	23.8
Prince George, BC	53 53N	122 41W	2,218	27	23	3	54	27	75	44	52	30	102	−58	27	1.8	1.2	1.4	0.8	1.3	2.1	1.6	1.9	2.0	2.0	1.9	1.9	19.9
Prince Rupert, BC	54 17N	130 23W	170	26	39	30	50	37	62	49	53	42	90	−3	26	9.8	7.6	8.4	6.7	5.3	4.1	4.8	5.1	7.7	12.2	12.3	11.3	95.3
Quebec, Que	46 48N	71 23W	239	72	18	2	44	29	76	57	51	37	97	−34	72	3.5	2.7	3.0	2.4	3.1	3.7	4.0	4.0	3.6	3.4	3.2	3.2	39.8
Regina, Sask	50 26N	104 40W	1,884	55	10	−11	50	26	79	51	52	27	110	−56	49	0.5	0.3	0.7	0.7	1.8	3.3	2.4	1.8	1.3	0.9	0.6	0.4	14.7
Resolute, NWT	74 43N	94 59W	220	13	−20	−33	−1	−16	45	35	11	0	61	−61	7	0.1	0.1	0.2	0.2	0.5	0.8	0.9	1.1	0.8	0.5	0.2	0.1	5.5
St. John, NB	45 17N	66 04W	119	61	28	11	43	32	69	54	54	41	93	−24	61	4.1	3.1	3.7	3.2	3.1	3.2	3.1	3.6	3.7	4.1	3.9	3.8	42.6
St. Johns, Nfld	47 32N	52 44W	211	68	30	18	41	29	69	51	53	40	93	−21	58	5.3	5.1	4.6	3.8	3.9	3.1	3.1	4.0	3.7	4.8	5.7	6.0	53.1
Saskatoon, Sask	52 08N	106 38W	1,690	38	9	−11	49	26	77	52	51	27	104	−55	38	0.9	0.5	0.7	0.7	1.4	2.6	2.4	1.9	1.5	0.9	0.5	0.6	14.6
The Pas, Man	53 49N	101 15W	890	27	1	−18	45	21	75	54	45	26	100	−54	27	0.6	0.5	0.7	0.8	1.4	2.2	2.2	2.1	2.0	1.2	1.0	0.8	15.5
Toronto, Ont	43 40N	79 24W	379	105	30	16	50	34	79	59	56	40	105	−26	105	2.7	2.4	2.6	2.5	2.9	2.7	3.0	2.7	2.9	2.4	2.8	2.6	32.2
Vancouver, BC	49 17N	123 05W	127	43	41	32	58	40	74	54	57	44	92	2	41	8.6	5.8	5.0	3.3	2.8	2.5	1.2	1.7	3.6	5.8	8.3	8.8	57.4
Whitehorse, YT	60 43N	135 04W	2,303	10	13	−3	41	22	67	45	41	28	91	−62	10	0.6	0.5	0.6	0.4	0.6	1.0	1.6	1.5	1.3	0.7	1.0	0.8	10.6
Winnipeg, Man	49 54N	97 14W	783	66	7	−13	48	27	79	55	51	31	108	−54	66	0.9	0.9	1.2	1.4	2.3	3.1	3.1	2.5	2.3	1.5	1.1	0.9	21.2
Yellow Knife, NWT	62 28N	114 27W	674	13	−8	−23	29	9	69	52	36	26	90	−60	13	0.8	0.6	0.7	0.4	0.7	0.6	1.5	1.4	1.0	1.3	1.0	0.8	10.8
Greenland																												
Angmagssalik	65 36N	37 33W	95	30	23	10	35	16	54	37	35	25	77	−26	38	2.9	2.4	2.6	2.1	2.0	1.8	1.5	2.1	3.3	4.7	3.0	2.7	31.1
Denmarkshaven	76 46N	19 00W	7	2	−1	−15	6	−13	47	34	13	2	63	−42	2	1.2	0.7	0.7	0.1	0.2	0.2	0.5	0.6	0.3	0.3	1.0	0.7	6.0
Elsmitte	70 53N	40 42W	9,843	1	−33	−53	−14	−37	19	1	−23	−42	27	−85	1	0.6	0.2	0.3	0.2	0.1	0.1	0.1	0.4	0.3	0.5	0.5	1.0	4.3
Godthaab	64 10N	51 43W	66	40	19	10	31	20	52	38	35	26	76	−20	45	1.4	1.7	1.6	1.2	1.7	1.4	2.2	3.1	3.3	2.5	1.9	1.5	23.5
Ivigtut	61 12N	48 10W	98	48	24	12	38	24	57	42	40	29	86	−20	50	3.3	2.6	3.4	2.5	3.5	3.2	3.1	3.7	5.9	5.7	4.6	3.1	44.6
Jacobshavn	69 13N	51 02W	104	32	8	−7	24	6	51	40	31	20	71	−46	52	0.4	0.4	0.5	0.5	0.6	0.8	1.2	1.4	1.3	0.9	0.7	0.5	9.2
Nord	81 36N	16 40W	118	8	−15	−28	−5	−18	44	35	3	−6	61	−60	8	0.8	0.8	0.5	0.3	0.1	0.3	1.0	1.4	1.2	0.6	1.4	0.5	8.9
Scoresbysund	70 29N	21 58W	56	12	12	−3	22	6	49	36	25	15	63	−42	12	1.8	1.4	0.9	1.4	0.4	0.8	1.5	0.7	1.7	1.4	1.1	1.9	15.0
Thule	76 31N	68 44W	251	12	−4	−17	10	−7	46	38	19	8	63	−44	12	0.4	0.3	0.2	0.2	0.3	0.2	0.7	0.6	0.6	0.7	0.5	0.2	4.9
Upernivik	72 47N	56 07W	59	40	−1	−13	15	−1	48	35	29	21	69	−44	50	0.4	0.5	0.7	0.6	0.6	0.5	0.9	1.1	1.1	1.1	1.1	0.6	9.2

(Continued)

Table 3B.13 (Continued)

Country and Station	Latitude	Longitude	Elevation	Length of Record	Temperature										Average Precipitation													
					Average Daily								Extreme															
					Jan		Apr		July		Oct																	
					Maximum	Minimum	Maximum	Minimum	Maximum	Minimum	Maximum	Minimum	Maximum	Minimum	Length of Record	January	February	March	April	May	June	July	August	September	October	November	December	Year
	°	°	Feet	Year	°F	°F	°F	°F	°F	°F	°F	°F	°F	°F	Year	IN.	IN.	IN.	IN.	IN.	IN.	IN.	IN.	IN.	IN.	IN.	IN.	IN.
Mexico																												
Acapulco	16 50N	99 56W	10	8	85	70	87	71	89	75	88	74	97	60	40	0.3	—	0.0	—	1.4	12.8	9.1	9.3	13.9	6.7	1.2	0.4	55.1
Chihuahua	28 42N	105 57W	4,429	9	65	36	81	51	89	66	79	51	102	12	22	0.2	0.4	0.3	0.2	0.2	1.7	3.6	3.7	3.3	0.9	0.5	0.4	15.4
Guadalajara	20 41N	103 20W	5,194	26	73	45	85	53	79	60	78	56	101	26	33	0.4	0.2	0.2	0.2	1.1	8.8	9.4	8.5	7.2	2.2	0.8	0.7	39.7
Guaymas	27 57N	110 55W	58	9	74	57	84	65	96	82	91	75	117	41	41	0.5	0.2	0.2	0.1	—	0.1	1.7	2.7	2.1	0.7	0.3	0.8	9.4
La Paz	24 07N	110 17W	85	9	74	54	86	58	96	73	90	68	108	31	12	0.2	0.1	0.0	0.0	0.0	0.2	0.4	1.2	1.4	0.6	0.5	1.1	5.7
Lerdo	25 30N	103 32W	3,740	10	72	45	86	57	90	68	82	58	105	23	14	0.4	0.1	0.2	0.3	0.8	1.5	1.5	1.3	2.0	0.8	0.8	0.5	10.2
Manzanillo	19 04N	104 20W	26	17	86	68	87	67	93	76	91	76	103	54	17	0.1	0.2	—	0.0	0.1	4.7	5.7	6.4	14.5	5.1	0.9	1.8	39.5
Mazatan	23 11N	106 25W	256	10	71	61	76	65	86	77	85	76	93	52	46	0.8	0.5	0.2	0.1	0.1	1.5	5.9	8.3	8.0	2.6	0.9	1.3	30.2
Merida	20 58N	89 38W	72	22	83	62	92	69	92	73	87	71	106	51	40	1.2	0.9	0.7	0.8	3.2	5.6	5.2	5.6	6.8	3.8	1.3	1.3	36.5
Mexico City	19 26N	99 04W	7,340	42	66	42	78	52	74	54	70	50	92	24	48	0.2	0.3	0.5	0.7	1.9	4.1	4.5	4.3	4.1	1.6	0.5	0.3	23.0
Monterrey	25 40N	100 18W	1,732	11	68	48	84	62	90	71	80	64	107	25	33	0.6	0.7	0.8	1.3	1.3	3.0	2.3	2.4	5.2	3.0	1.5	0.8	22.9
Salina Cruz	16 12N	95 12W	184	10	85	72	88	76	89	76	87	75	98	62	22	*	0.4	0.6	0.5	3.3	11.6	4.5	5.5	7.1	4.0	0.9	0.1	38.5
Tampico	22 16N	97 51W	78	12	75	59	83	69	89	75	85	71	104	34	12	1.5	1.2	1.0	1.5	1.9	8.7	4.9	4.8	10.8	5.0	2.0	1.6	44.9
Vera Cruz	19 12N	96 08W	52	10	77	66	83	72	87	74	85	73	98	53	40	0.9	0.6	0.6	0.8	2.6	10.4	4.1	11.1	13.9	6.9	3.0	1.0	65.7
Central America																												
Belize																												
Belize	17 31N	88 11W	17	27	81	67	86	74	87	75	86	72	97	49	33	5.4	2.4	1.5	2.2	4.3	7.7	6.4	6.7	9.6	12.0	8.9	7.3	74.4
Canal Zone																												
Balboa Heights	08 57N	79 33W	118	34	88	71	90	74	87	74	85	73	97	63	46	1.0	0.4	0.7	2.9	8.0	8.4	7.1	7.9	8.2	10.1	10.2	4.8	69.7
Cristobal	09 21N	79 54W	35	36	84	76	86	77	85	76	86	75	97	66	73	3.4	1.5	1.5	4.1	12.5	13.9	15.6	15.3	12.7	15.8	22.3	11.7	130.3
Costa Rica																												
San Jose	09 56N	84 08W	3,760	8	75	58	79	62	77	62	77	60	92	49	34	0.6	0.2	0.8	1.8	9.0	9.5	8.3	9.5	12.0	11.8	5.7	1.6	70.8
El Salvador																												
San Salvador	13 42N	89 13W	2,238	39	90	60	93	65	89	65	87	65	105	45	39	0.3	0.2	0.4	1.7	7.7	12.9	11.5	11.7	12.1	9.5	1.6	0.4	70.0
Guatemala																												
Guatemala City	14 37N	90 31W	4,855	6	73	53	82	58	78	60	76	60	90	41	29	0.3	0.1	0.5	1.2	6.0	10.8	8.0	7.8	9.1	6.8	0.9	0.3	51.8
Honduras																												
Tela	15 46N	87 27W	41	4	82	67	87	72	88	73	86	71	96	58	20	8.9	5.1	2.6	3.3	4.3	5.0	6.4	9.4	7.7	13.5	15.9	14.0	96.1
West Indies																												
Bridgetown, Barbados	13 08N	59 36W	181	35	83	70	86	72	86	74	86	73	95	61	22	2.6	1.1	1.3	1.4	2.3	4.4	5.8	5.8	6.7	7.0	8.1	3.8	50.3
Camp Jacob, Guadaloupe	16 01N	61 42W	1,750	19	77	64	79	65	81	68	81	68	92	54	21	9.2	6.1	8.1	7.3	11.5	14.1	17.6	15.3	16.4	12.4	12.3	10.1	140.4
Fort-de-France, Martinique	14 37N	61 05W	13	22	83	69	86	71	86	74	87	73	96	56	31	4.7	4.3	2.9	3.9	4.7	7.4	9.4	10.3	9.3	9.7	7.9	5.9	80.4
Hamilton, Bermuda	32 17N	64 46W	151	59	68	58	71	59	85	73	79	69	99	40	62	4.4	4.7	4.8	4.1	4.6	4.4	4.5	5.4	5.2	5.8	5.0	4.7	57.6
Havana, Cuba	23 08N	82 21W	80	25	79	65	84	69	89	75	85	73	104	43	72	2.8	1.8	1.8	2.3	4.7	6.5	4.9	5.3	5.9	6.8	3.1	2.3	48.2
Kingston, Jamaica	17 58N	76 48W	110	33	86	67	87	70	90	73	88	73	97	56	59	0.9	0.6	0.9	1.2	4.0	3.5	1.5	3.6	3.9	7.1	2.9	1.4	31.5
La Guerite, St. Christopher (St. Kitts)	17 20N	62 45W	157	19	80	71	83	73	86	76	85	75	91	61	21	4.1	2.0	2.3	2.3	3.8	3.6	4.4	5.2	6.0	5.4	7.3	4.5	50.9

Nassau, Bahamas	25 05N	77 21W	12	35	77	65	81	69	88	75	85	73	94	41	57	1.4	1.5	1.4	2.5	4.6	6.4	5.8	5.3	6.9	6.5	2.8	1.3	46.4
Port-au-Prince, Haiti	18 33N	72 20W	121	42	87	68	89	71	94	74	90	72	101	58	70	1.3	2.3	3.4	6.3	9.1	4.0	2.9	5.7	6.9	6.7	3.4	1.3	53.3
Saint Clair, Trinidad	10 40N	61 31W	67	49	87	69	90	69	88	71	89	71	101	52	97	2.7	1.6	1.8	2.1	3.7	7.6	8.6	9.7	7.6	6.7	7.2	4.9	64.2
Saint Thomas, Virgin Is	18 20N	64 58W	11	9	82	71	85	74	88	77	87	76	92	63	9	2.5	1.9	1.7	2.2	4.6	3.2	3.2	4.1	6.9	5.6	3.9	3.9	43.7
San Juan, Puerto Rico	18 26N	66 00W	13	30	81	67	84	69	87	74	87	73	94	60	30	4.7	2.9	2.2	3.7	7.1	5.7	6.3	7.1	6.8	5.8	6.5	5.4	64.2
Santo Domingo, Dom. Rep	18 29N	69 54W	57	26	84	66	85	69	88	72	87	72	98	59	25	2.4	1.4	1.9	3.9	6.8	6.2	6.4	6.3	7.3	6.0	4.8	2.4	55.8
South America																												
Argentina																												
Bahia Blanca	38 43S	62 16W	95	33	88	62	71	51	57	39	71	48	109	18	46	1.7	2.2	2.5	2.3	1.2	0.9	1.0	1.0	1.6	2.2	2.1	1.9	20.6
Buenos Aires	34 35S	58 29W	89	23	85	63	72	53	57	42	69	50	104	22	70	3.1	2.8	4.3	3.5	3.0	2.4	2.2	2.4	3.1	3.4	3.3	3.9	37.4
Cipolletti	38 57S	67 59W	889	9	89	56	72	40	55	29	72	43	107	9	24	0.4	0.4	0.7	0.4	0.6	0.6	0.5	0.3	0.6	0.9	0.5	0.5	6.4
Corrientes	27 28S	58 50W	177	39	93	71	81	63	71	53	82	60	112	30	40	4.7	4.5	5.3	5.6	3.3	1.9	1.7	1.5	2.8	4.7	5.2	5.2	46.4
La Quiaca	22 06S	65 36W	11,345	23	70	41	69	32	60	16	71	32	95	0	25	3.5	2.6	1.8	0.3	—	0.0	—	—	0.1	0.3	1.0	2.7	12.3
Mendoza	32 53S	68 49W	2,625	23	90	60	73	47	59	35	76	50	109	15	46	0.9	1.2	1.1	0.5	0.4	0.3	0.2	0.3	0.5	0.7	0.7	0.7	7.5
Parana	31 44S	60 31W	210	12	91	67	77	58	62	45	75	54	113	21	23	3.1	3.1	3.9	4.9	2.6	1.2	1.2	1.6	2.1	2.8	3.7	4.5	35.0
Puerto Madryn	42 47S	65 01W	26	50	81	57	70	46	55	36	68	45	104	10	50	0.4	0.6	0.7	0.5	0.9	0.6	0.6	0.4	0.6	0.7	0.4	0.6	7.0
Santa Cruz	50 01S	68 32W	39	12	70	48	57	39	41	28	58	39	94	1	20	0.6	0.3	0.3	0.6	0.4	0.5	0.4	0.5	0.3	0.3	0.4	0.7	5.3
Santiago del Estero	27 46S	64 18W	653	28	97	69	82	59	70	44	87	59	116	19	20	3.4	3.0	3.0	1.3	0.6	0.3	0.2	0.2	0.5	1.4	2.5	4.1	20.4
Ushulaia	54 50S	68 20W	26	16	57	41	48	33	39	25	52	35	85	6	21	2.0	2.6	1.9	2.1	1.5	1.2	1.2	1.1	1.3	1.6	1.5	1.9	19.9
Bolivia																												
Concepcion	16 15S	62 03W	1,607	5	85	66	86	62	81	54	88	62	101	32	16	7.2	4.7	4.4	1.8	2.0	1.5	1.1	0.9	1.2	2.9	5.0	5.9	38.6
La Paz	16 30S	68 08W	12,001	31	63	43	65	40	62	33	66	40	80	26	50	4.5	4.2	2.6	1.3	0.5	0.3	0.4	0.5	1.1	1.6	1.9	3.7	22.6
Sucre	19 03S	65 17W	9,344	5	63	48	63	45	61	37	65	46	88	25	52	7.3	4.9	3.7	1.6	0.2	0.1	0.2	0.3	1.0	1.6	2.6	4.3	27.8
Brazil																												
Barra do Corda	05 35S	45 28W	266	9	89	71	89	71	92	64	94	72	103	45	9	6.7	8.7	8.0	6.1	2.3	1.0	0.7	0.7	1.0	2.5	3.9	5.7	47.2
Bela Vista	22 06S	56 22W	525	13	91	67	85	61	77	49	87	61	108	20	20	6.6	4.9	4.4	4.3	5.0	2.8	1.3	1.8	2.9	5.4	5.8	7.0	52.2
Belem	01 27S	48 29W	42	16	87	72	87	73	88	71	89	71	98	61	20	12.5	14.1	14.1	12.6	10.2	6.7	5.9	4.4	3.5	3.3	2.6	6.1	96.0
Brasilia	15 51S	47 56W	3,481	3	80	65	82	62	78	51	82	64	93	46	3	9.0	7.8	4.8	3.4	1.4	—	0.0	—	1.3	4.9	9.7	11.7	54.0
Conceicao do Araguaia	08 15S	49 12W	53	5	88	70	91	68	95	63	93	68	102	55	5	14.9	12.1	10.8	4.1	1.9	0.4	—	0.5	1.5	6.6	4.9	8.6	66.2
Corumba	19 00S	57 39W	381	8	94	73	92	73	84	64	93	70	106	33	11	7.3	5.9	5.1	4.6	2.9	1.9	0.3	1.2	2.6	4.0	5.6	7.1	48.5
Florianopolis	27 35S	48 33W	96	17	83	72	74	64	68	57	73	63	102	32	25	7.6	5.6	6.3	4.1	3.6	6.5	2.2	3.7	4.3	5.1	3.5	4.3	53.1
Goias	15 58S	50 04W	1,706	11	86	63	91	63	89	56	94	63	104	41	11	12.5	9.9	10.2	4.6	0.4	0.3	0.0	0.3	2.3	5.3	9.4	9.5	64.8
Guarapuava	25 16S	51 30W	3,592	10	79	61	73	55	66	47	74	53	94	23	5	8.7	5.8	5.4	4.5	4.6	6.5	2.7	3.6	4.6	6.9	6.6	6.1	65.8
Manaus	03 08S	60 01W	144	11	88	75	87	75	89	75	92	76	101	63	25	9.8	9.1	10.3	8.7	6.7	3.3	2.3	1.5	1.8	4.2	5.6	8.0	71.3
Natal	05 46S	35 12W	52	18	87	76	86	73	82	69	85	75	100	61	18	1.9	4.8	7.0	9.2	7.1	8.7	7.7	3.8	1.4	0.8	0.7	1.1	54.2
Parana	12 26S	48 06W	853	19	90	58	90	58	91	48	94	58	105	37	19	11.3	9.3	9.4	4.0	0.5	*	0.1	0.2	1.1	5.0	9.1	12.2	62.3
Porto Alegre	30 02S	51.13W	33	22	87	67	78	60	66	49	74	57	105	25	22	3.5	3.2	3.9	4.1	4.5	5.1	4.5	5.0	5.2	3.4	3.1	3.5	49.1
Quixeramobim	05 12S	39 18W	653	9	92	79	86	76	88	74	93	77	100	63	13	0.7	5.0	6.6	5.0	7.0	1.7	0.7	0.6	0.4	0.6	0.7	0.6	29.6
Recife	08 04S	34 53W	97	27	86	77	85	75	80	71	84	75	94	50	56	2.1	3.3	6.3	8.7	10.5	10.9	10.0	6.0	2.5	1.0	1.0	1.1	63.4
Rio de Janeiro	22 55S	43 12W	201	38	84	73	80	69	75	63	77	66	102	46	84	4.9	4.8	5.1	4.2	3.1	2.1	1.6	1.7	2.6	3.1	4.1	5.4	42.6
Salvador (Bahia)	13 00S	38 30W	154	25	86	74	84	74	79	69	83	71	100	50	20	2.6	5.3	6.1	11.2	10.8	9.4	7.2	4.8	3.3	4.0	4.5	5.6	74.8
Santarem	02 30S	54 42W	66	22	86	73	85	73	87	71	91	73	99	65	22	6.8	10.9	13.2	12.9	11.3	6.9	4.1	1.7	1.5	1.9	2.3	4.1	77.9
Sao Paulo	23 37S	46 39W	2,628	44	77	63	73	59	66	53	68	57	100	32	24	8.8	7.8	6.0	2.2	3.0	2.4	1.5	2.1	3.5	4.6	6.0	9.4	57.3
Sena Madureira	09 04S	68 39W	443	12	92	69	91	68	91	63	93	69	100	41	17	11.2	11.3	10.2	9.4	4.1	2.2	1.1	1.5	4.0	7.0	7.5	11.7	81.2
Uaupes	00 08S	67 05W	272	15	88	72	88	72	85	70	89	71	100	52	10	10.3	7.7	10.0	10.6	12.0	9.2	8.8	7.2	5.1	6.9	7.2	10.4	105.4
Uruguaiana	29 46S	57 07W	246	15	91	69	78	59	66	48	77	55	108	27	12	3.6	3.6	5.6	5.1	3.7	4.2	3.2	2.8	3.6	4.1	2.9	4.1	46.6
Chile																												
Ancud	41 47S	73 52W	184	30	62	51	57	47	50	42	55	45	82	30	46	3.1	3.7	5.3	7.4	9.9	11.0	10.3	9.4	6.5	4.2	4.7	4.6	80.1
Antofagasta	26 42S	70 24W	308	22	76	63	70	58	63	51	66	55	86	37	32	0.0	0.0	0.0	—	—	0.1	0.2	0.1	—	0.1	—	0.0	0.5
Arica	18 28S	70 20W	95	15	78	64	74	60	66	54	69	58	93	39	25	—	0.0	0.0	0.0	0.0	0.0	0.0	—	0.0	0.0	0.0	—	—
Cabo Raper	46 50S	75 38W	131	8	58	46	54	44	47	38	51	40	72	28	10	7.8	5.8	7.1	7.7	7.5	7.9	9.5	7.5	5.6	7.0	6.7	7.0	87.1
Los Evangelistas	52 23S	75 07W	190	16	50	44	48	41	43	36	45	39	66	19	27	11.7	10.0	11.3	11.4	9.6	9.4	9.4	8.6	9.2	8.8	9.9	10.1	119.4
Potrerillos	26 30S	69 27W	9,350	7	65	49	63	47	57	40	61	44	75	20	7	—	—	0.3	—	0.7	—	0.5	0.3	0.2	0.2	0.0	—	2.2
Puerto Aisen	42 24S	72 42W	33	8	63	50	55	43	45	37	55	42	93	18	11	7.8	7.8	8.3	7.5	14.7	10.4	11.1	11.1	6.5	7.8	7.0	7.9	107.9

(Continued)

Table 3B.13 (Continued)

Country and Station	Latitude	Longitude	Elevation	Length of Record	Temperature: Average Daily Jan Maximum	Jan Minimum	Apr Maximum	Apr Minimum	July Maximum	July Minimum	Oct Maximum	Oct Minimum	Extreme Maximum	Extreme Minimum	Average Precipitation: Length of Record	January	February	March	April	May	June	July	August	September	October	November	December	Year
	°	°	Feet	Year	°F	°F	°F	°F	°F	°F	°F	°F	°F	°F	Year	IN.	IN.	IN.	IN.	IN.	IN.	IN.	IN.	IN.	IN.	IN.	IN.	IN.
Punta Arenas	53 10S	70 54W	26	15	58	45	50	39	40	31	51	38	86	11	15	1.5	0.9	1.3	1.4	1.3	1.6	1.1	1.2	0.9	1.1	0.7	1.4	14.4
Santiago	33 27S	70 42W	1,706	14	85	53	74	45	59	37	72	45	99	24	58	0.1	0.1	0.2	0.5	2.5	3.3	3.0	2.2	1.2	0.6	0.3	0.2	14.2
Valdivia	39 48S	73 14W	16	29	73	52	62	46	52	41	63	44	97	19	60	2.6	2.9	5.2	9.2	14.2	17.7	15.5	12.9	8.2	5.0	4.9	4.1	102.4
Valparaiso	33 01S	71 38W	135	30	72	56	67	52	60	47	65	50	94	32	41	0.1	—	0.3	0.6	4.1	5.9	3.9	2.9	1.3	0.4	0.2	0.2	19.9
Colombia																												
Andagoya	05 06N	76 40W	197	8	90	75	90	75	89	74	90	74	97	62	15	25.0	21.4	19.5	26.1	25.5	25.8	23.3	25.3	24.6	22.7	22.4	19.5	281.1
Bogota	04 42N	74 08W	8,355	10	67	48	67	51	64	50	66	50	75	30	49	2.3	2.6	4.0	5.8	4.5	2.4	2.0	2.2	2.4	6.3	4.7	2.6	41.8
Cartagena	10 28N	75 30W	39	6	84	73	87	76	88	78	87	77	98	61	10	0.4	0.0	0.4	0.9	3.4	3.4	3.0	0.6	0.5	10.8	8.9	4.5	36.8
Ipiales	00 50N	77 42W	9,680	9	61	50	60	49	57	42	62	49	77	32	13	3.1	2.3	3.5	3.5	2.8	1.9	1.3	1.1	1.4	3.1	3.3	2.6	29.9
Tumaco	01 49N	78 45W	7	10	82	75	84	76	82	75	82	75	90	64	10	16.9	11.7	9.6	14.6	17.4	12.0	7.7	7.3	7.3	5.9	4.9	7.0	122.3
Ecuador																												
Cuenca	02 53S	78 39W	8,301	7	69	50	69	50	65	47	70	49	81	29	10	2.0	1.8	3.2	4.3	4.3	1.7	0.9	1.1	1.6	3.1	1.8	2.5	28.3
Guayaquil	02 10S	79 53W	20	5	87	72	88	72	84	67	86	68	98	52	10	8.3	11.4	11.5	8.1	2.1	0.4	0.2	—	—	—	0.1	1.1	43.2
Quito	00 08S	78 29W	9,222	54	67	46	69	47	71	44	71	46	86	25	33	3.9	4.4	5.6	6.9	5.4	1.7	0.8	1.2	2.7	4.4	3.8	3.1	43.9
French Guiana																												
Cayenne	04 56N	52 27W	20	38	84	74	86	75	88	73	91	74	97	65	51	14.4	12.3	15.8	18.9	21.7	15.5	6.9	2.8	1.2	1.3	4.6	10.7	126.1
Guyana																												
Georgetown	06 50N	58 12W	6	54	84	74	85	76	85	75	87	76	93	68	35	8.0	4.5	6.9	5.5	11.4	11.9	10.0	6.9	3.2	3.0	6.1	11.3	88.7
Lethem	03 24N	59 38W	270	3	91	73	91	74	87	73	92	76	97	63	9	1.2	1.4	1.3	5.7	11.5	11.9	14.8	9.4	3.4	2.3	4.3	1.3	68.5
Paraguay																												
Asuncion	25 17S	57 30W	456	15	95	71	84	65	74	53	86	62	110	29	30	5.5	5.1	4.3	5.2	4.6	2.7	2.2	1.5	3.1	5.5	5.9	6.2	51.8
Bahia Negra	20 14S	58 10W	318	20	92	74	87	68	79	61	90	69	106	35	20	5.4	5.3	4.9	2.9	2.3	1.6	1.5	0.6	2.3	4.2	5.3	4.3	40.6
Peru																												
Arequipa	16 21S	71 34W	8,460	13	67	49	67	48	67	47	68	47	82	25	37	1.3	1.8	0.7	0.2	—	—	—	—	0.0	—	—	0.4	4.4
Cajamarca	07 09S	78 30W	8,662	9	71	48	70	47	70	41	71	47	79	25	9	3.6	4.2	4.6	3.4	1.7	9.5	9.2	9.3	2.3	2.3	1.9	3.2	28.2
Cusco	13 33S	71 59W	10,866	13	68	45	71	40	70	31	72	43	86	16	12	6.4	5.9	4.3	2.0	0.6	0.2	0.2	0.4	1.0	2.6	3.0	5.4	32.0
Iquitos	03 45S	73 13W	384	5	90	71	87	71	88	68	90	70	100	54	5	9.1	10.4	9.4	13.6	10.7	5.7	6.4	5.2	10.5	7.3	9.1	10.3	107.7
Lima	12 05S	77 03W	394	15	82	66	80	63	67	57	71	58	93	49	15	0.1	—	—	—	0.2	0.2	0.3	0.3	0.3	0.1	0.1	—	1.6
Mollendo	17 00S	72 07W	80	10	79	66	76	63	67	57	70	59	90	50	10	—	0.1	—	—	0.1	0.1	—	0.2	0.2	0.1	0.1	—	0.9
Surinam																												
Paramaribo	05 49N	55 09W	12	35	85	72	86	73	87	73	91	73	99	62	75	8.4	6.5	7.9	9.0	12.2	11.9	9.1	6.2	3.1	3.0	4.9	8.8	91.0
Uruguay																												
Artigas	30 24S	56 23W	384	13	91	65	77	55	65	45	75	54	107	24	50	4.3	3.9	4.7	5.1	4.1	4.1	2.8	3.0	4.0	4.7	3.8	4.1	48.6
Montevideo	34 52S	56 12W	72	56	83	62	71	53	58	43	68	49	109	25	56	2.9	2.6	3.9	3.9	3.3	3.2	2.9	3.1	3.0	2.6	2.9	3.1	37.4
Venezuela																												
Caracas	10 30N	66 56W	3,418	30	75	56	81	60	78	61	79	61	91	45	46	0.9	0.4	0.6	1.3	3.1	4.0	4.3	4.3	4.2	4.3	3.7	1.8	32.9
Ciudad Bolivar	08 07N	63 32W	197	10	90	72	93	75	90	75	93	75	100	64	10	1.4	0.8	0.7	1.0	3.8	5.5	6.3	7.1	3.6	4.0	2.8	1.3	38.3
Maracaibo	10 39N	71 36W	20	12	90	73	92	76	94	76	92	76	102	66	36	0.1	—	0.3	0.8	2.7	2.2	1.8	2.2	2.8	5.9	3.3	0.6	22.7
Merida	08 36N	71 10W	5,293	14	73	56	75	60	76	59	75	60	90	48	14	2.5	1.5	3.6	6.7	9.8	7.3	4.7	5.7	6.7	9.5	8.2	3.4	69.7
Santa Elena	04 36N	61 07W	2,976	10	82	61	82	63	81	61	84	61	95	48	10	3.2	3.2	3.2	5.7	9.6	9.5	9.1	7.6	5.3	4.9	4.9	4.5	70.7

Pacific Islands																												
Easter Is. (Isla de Pascua)	27 10S	109 26W	98	4	77	64	78	63	70	58	73	58	88	46	10	4.8	3.7	4.6	4.2	4.6	4.3	3.5	3.0	2.7	3.7	4.6	4.9	48.6
Mas a Tierra (Juan Fernandez)	33 37S	78 52W	20	25	72	60	68	57	60	50	61	51	86	39	29	0.8	1.2	1.6	3.4	5.9	6.4	5.8	4.4	2.9	1.9	1.6	1.0	36.9
Seymour Is. (Galapagos Is.)	00 28S	90 18W	36	3	86	72	87	75	81	69	81	67	93	58	3	0.8	1.4	1.1	0.7	—	—	—	—	—	—	—	—	4.0
Atlantic Islands																												
Fernando de Noronha	03 50S	32 25W	148	32	84	75	82	75	81	73	82	75	93	63	32	1.7	4.7	7.4	10.5	10.5	7.3	5.4	1.9	0.7	0.3	0.4	0.5	51.3
Cumberland Bay, South Georgia	54 16S	36 30W	8	23	48	35	42	29	34	23	41	28	84	−3	24	3.3	4.3	5.3	5.4	5.2	4.9	5.5	5.3	3.5	2.6	3.4	3.0	51.7
Laurie Is., South Orkneys	60 44S	44 44W	13	48	35	29	31	21	20	4	30	19	54	−40	46	1.4	1.5	1.9	1.6	1.2	1.0	1.3	1.3	1.1	1.1	1.3	1.0	15.7
Stanley, Falkland Isles	51 42S	57 51W	6	25	56	42	49	37	40	31	48	35	76	12	41	2.8	2.3	2.5	2.6	2.6	2.1	2.0	2.0	1.5	1.6	2.0	2.8	26.8
Europe																												
Albania																												
Durres	41 19N	19 28E	23	10	51	42	63	55	83	74	68	58	95	21	10	3.0	3.3	3.9	2.2	1.6	1.9	0.5	1.9	1.7	7.1	8.5	7.3	42.9
Andorra																												
Les Escaldes	42 30N	01 31E	3,543	5	43	29	59	39	78	55	61	42	91	0	9	1.5	1.7	2.9	2.4	4.7	3.1	2.2	3.4	3.1	3.5	3.3	2.5	34.3
Austria																												
Innsbruck	47 16N	11 24E	1,909	34	34	20	60	39	78	55	58	40	97	−16	35	2.1	1.8	1.5	2.2	2.9	4.1	5.1	4.5	3.1	2.4	2.2	1.9	33.8
Vienna	48 15N	16 22E	664	50	34	26	57	41	75	59	55	44	98	−14	100	1.5	1.4	1.8	2.0	2.8	2.7	3.0	2.7	2.0	2.0	1.9	1.8	25.6
Bulgaria																												
Sofia	42 42N	23 20E	1,805	30	34	22	62	41	82	57	63	42	99	−17	27	1.3	1.1	1.7	2.3	3.3	3.2	2.4	2.0	2.3	2.1	1.9	1.4	25.0
Varna	43 12N	27 55E	115	30	40	30	59	43	84	63	67	50	107	−12	20	1.5	0.9	1.2	1.2	1.8	2.6	1.9	1.2	1.5	1.9	1.9	2.0	19.6
Cyprus																												
Nicosia	35 09N	33 17E	716	40	58	42	74	50	97	69	81	58	116	23	64	2.9	2.0	1.3	0.8	1.1	0.4	—	—	0.2	0.9	1.7	3.0	14.6
Czechoslovakia																												
Prague	50 05N	14 25E	662	40	34	25	55	40	74	58	54	44	98	−16	70	0.9	0.8	1.1	1.5	2.4	2.8	2.6	2.2	1.7	1.2	1.2	0.9	19.3
Prerov	49 27N	17 27E	702	20	34	25	57	38	77	55	56	40	100	−23	21	1.3	1.1	1.1	2.0	2.4	2.9	3.5	3.2	2.0	2.4	1.5	1.4	24.8
Denmark																												
Copenhagen	55 41N	12 33E	43	30	36	29	50	37	72	55	53	42	91	−3	30	1.6	1.3	1.2	1.7	1.7	2.1	2.2	3.2	1.9	2.1	2.2	2.1	23.3
Aarhus	56 08N	10 12E	161	21	35	27	51	37	70	54	53	42	87	−12	21	2.3	1.5	1.4	1.8	1.2	2.2	2.5	3.3	3.2	2.6	2.5	2.1	26.6
Finland																												
Helsinki	60 10N	24 57E	30	20	27	17	43	31	71	57	45	37	89	−23	50	2.2	1.7	1.7	1.7	1.9	2.0	2.3	3.3	2.8	2.9	2.7	2.4	27.6
Kuusamo	65 57N	29 12E	843	20	17	2	35	18	68	50	36	27	90	−40	20	1.1	1.1	1.1	1.1	1.4	2.3	2.8	3.0	2.1	2.1	1.6	1.1	20.8
Vaasa	63 05N	21 36E	13	18	26	16	41	28	69	55	44	36	89	−29	19	1.1	0.8	0.8	1.0	1.4	1.8	2.4	2.5	2.7	2.3	1.7	1.1	19.6
France																												
Ajaccio (Corsica)	41 52N	08 35E	243	46	56	40	66	48	85	64	72	55	103	23	86	3.0	2.3	2.6	2.2	1.6	0.9	2.8	0.7	1.7	3.8	4.4	3.1	29.1
Bordeaux	44 50N	00 43W	157	51	48	35	63	44	80	58	66	47	102	9	47	2.7	2.8	2.9	2.6	2.5	2.3	2.0	1.9	2.2	3.0	3.9	3.9	32.7
Brest	48 19N	04 47W	56	56	49	40	57	44	70	56	61	49	95	7	56	3.5	3.0	2.5	2.5	1.9	2.0	2.0	2.2	2.3	3.6	4.2	4.4	34.1
Cherbourg	49 39N	01 38W	30	47	47	40	54	43	67	57	59	50	91	14	47	3.3	2.9	2.7	2.0	1.9	1.8	1.9	3.0	2.9	4.6	5.1	5.2	37.3
Lille	50 35N	03 05W	141	40	42	33	58	40	75	55	59	45	96	0	40	2.5	1.9	2.5	2.0	2.4	2.2	2.8	2.3	2.6	3.0	3.0	3.2	30.3
Lyon	45 42N	04 47E	938	70	41	30	61	42	80	58	61	45	105	−13	70	1.4	1.4	1.8	2.1	2.8	2.9	2.8	2.9	3.1	3.1	2.6	1.9	28.8
Marseille	43 18N	05 23E	246	72	53	38	59	41	78	58	76	57	101	9	102	1.9	1.5	1.8	2.0	1.9	1.0	0.6	0.9	2.6	3.7	3.1	2.2	23.2
Paris	48 49N	02 29E	164	66	42	32	60	41	76	55	59	44	105	1	118	1.5	1.3	1.5	1.7	2.0	2.1	2.1	2.0	2.0	2.2	2.0	1.9	22.3
Strasbourg	48 35N	07 46E	465	20	40	31	59	41	78	57	58	43	101	−8	20	1.6	1.4	1.7	2.6	2.6	3.1	3.4	3.4	3.1	2.7	2.0	1.9	29.5
Toulouse	43 33N	01 23E	538	47	47	35	62	43	82	59	66	48	111	1	47	1.9	1.7	2.3	2.7	2.9	2.4	1.5	2.1	2.3	2.2	2.4	2.3	26.7
Germany																												
Berlin	52 27N	13 18E	187	50	35	26	55	38	74	55	55	41	96	−15	40	1.9	1.3	1.5	1.7	1.9	2.3	3.1	2.2	1.9	1.7	1.7	1.9	23.1
Bremen	53 05N	08 47E	52	50	37	30	53	38	71	55	54	43	94	−7	80	1.9	1.6	1.8	1.5	2.1	2.6	3.2	2.8	2.1	2.2	2.0	2.2	26.0
Frankfurt A/M	50 07N	08 40E	338	50	37	29	58	41	75	56	56	43	100	−7	80	1.7	1.3	1.6	1.5	2.0	2.5	2.8	2.6	1.9	2.2	2.0	2.0	24.1
Hamburg	53 33N	09 58E	66	50	35	28	51	39	69	56	53	44	92	−4	80	2.1	1.9	2.0	1.8	2.1	2.7	3.4	3.2	2.5	2.6	2.1	2.5	28.9
Munich	48 09N	11 34E	1,739	50	33	23	54	37	72	54	53	40	92	−14	80	1.7	1.4	1.9	2.7	3.7	4.6	4.7	4.2	3.2	2.2	1.9	1.9	34.1
Munster	51 58N	07 38E	207	50	39	29	56	38	73	54	56	42	96	−17	40	2.6	1.9	2.2	2.0	2.2	2.7	3.3	3.1	2.5	2.7	2.4	2.9	30.5
Numberg	49 27N	11 03E	1,050	50	35	26	56	38	74	55	55	41	99	−18	80	1.5	1.2	1.3	1.7	2.2	2.5	3.1	3.1	2.1	2.1	1.9	1.7	24.4
Gibraltar																												
Windmill Hill	36 06N	05 21W	400	12	58	50	64	55	77	66	70	61	97	35	12	4.6	3.4	3.7	2.5	1.4	0.2	—	0.1	0.8	3.5	4.1	5.4	29.7
Greece																												
Athens	37 58N	23 43E	351	72	54	42	67	52	90	72	74	60	109	20	80	2.2	1.6	1.4	0.8	0.8	0.6	0.2	0.4	0.6	1.7	2.8	2.8	15.8
Iraklion (Crete)	35 20N	25 08E	98	21	60	48	70	54	85	72	77	62	114	32	22	3.7	3.0	1.6	0.9	0.7	0.1	—	0.1	0.7	1.7	2.8	2.8	15.8

(Continued)

Table 3B.13 (Continued)

Country and Station	Latitude	Longitude	Elevation	Length of Record	Temperature, Average Daily, Jan Maximum	Jan Minimum	Apr Maximum	Apr Minimum	July Maximum	July Minimum	Oct Maximum	Oct Minimum	Extreme Maximum	Extreme Minimum	Average Precipitation, Length of Record	January	February	March	April	May	June	July	August	September	October	November	December	Year
	°	°	Feet	Year	°F	°F	°F	°F	°F	°F	°F	°F	°F	°F	Year	IN.	IN.	IN.	IN.	IN.	IN.	IN.	IN.	IN.	IN.	IN.	IN.	IN.
Rhodes	36 26N	28 15E	289	10	59	51	67	59	83	74	76	68	104	30	6	5.7	3.9	2.6	1.7	0.5	0.3	0.0	—	0.4	1.7	5.2	6.7	28.5
Thessaloniki	40 37N	22 57E	78	9	49																							
Hungary																												
Budapest	47 31N	19 02E	394	50	35	26	62	44	82	61	61	45	103	−10	50	1.5	1.5	1.7	2.0	2.7	2.6	2.0	1.9	1.8	2.1	2.4	2.0	24.2
Debrecen	47 36N	21 39E	430	50	33	21	61	39	81	57	60	41	102	−22	80	1.2	1.1	1.4	1.8	2.4	2.8	2.5	2.3	1.8	2.2	2.0	1.6	23.1
Iceland																												
Akureyri	65 41N	18 05W	16	23	34	26	40	30	57	47	43	34	83	−8	26	1.7	1.5	1.7	1.3	0.6	0.9	1.3	1.6	1.9	2.3	1.9	1.9	18.6
Reykjavik	64 09N	21 56W	92	25	36	28	43	33	58	48	44	36	74	4	30	4.0	3.1	3.0	2.1	1.6	1.7	2.0	2.6	3.1	3.4	3.6	3.7	33.9
Ireland																												
Cork	51 54N	08 29W	56	27	48	38	55	41	68	53	58	44	85	15	35	4.9	3.6	3.3	2.6	2.9	2.0	2.9	3.1	2.9	3.9	4.5	4.7	41.3
Dublin	53 22N	06 21W	155	30	47	35	54	38	67	51	57	43	86	8	35	2.7	2.2	2.0	1.9	2.3	2.0	2.8	3.0	2.8	2.7	2.7	2.6	29.7
Shannon Airport	52 41N	08 55W	8	9	46	36	55	41	66	53	58	45	87	12	12	3.8	3.0	2.0	2.2	2.4	2.1	3.1	3.0	3.0	3.4	4.2	4.3	36.5
Italy																												
Ancona	43 37N	13 32E	52	30	46	36	62	50	83	68	67	55	102	18	30	2.6	1.7	1.6	2.3	2.1	1.9	1.5	1.5	3.5	3.7	2.5	3.0	28.0
Cagliari (Sardinia)	39 15N	09 03E	3	30	56	43	66	50	86	67	72	58	102	25	25	2.2	1.5	1.5	1.2	1.5	0.5	0.1	0.4	1.0	3.0	1.8	2.3	17.0
Genoa	44 24N	08 55E	318	10	50	41	65	53	82	70	73	58	100	18	10	3.9	4.0	3.3	3.4	4.6	1.4	1.6	2.3	4.7	6.1	7.2	4.1	46.6
Naples	40 51N	14 15E	82	30	54	40	65	52	84	70	71	60	101	24	30	3.7	3.2	3.0	2.6	1.8	1.8	0.6	0.7	2.8	5.1	4.5	5.4	35.2
Palermo (Sicily)	38 07N	13 19E	354	10	58	47	67	53	86	71	75	62	113	31	30	3.8	3.4	2.4	1.9	1.1	0.6	0.2	0.6	0.2	3.7	4.1	4.5	28.3
Rome	41 48N	12 36E	377	10	54	39	68	46	88	64	73	53	104	20	30	3.3	2.9	2.0	2.0	1.9	0.7	0.4	0.7	2.8	4.3	4.4	4.1	29.5
Taranto	40 28N	17 17E	56	10	55	43	59	50	89	70	73	58	108	26	10	1.6	0.9	1.3	0.8	1.0	0.6	0.4	0.7	1.0	2.2	1.8	1.9	14.2
Venice	45 26N	12 23E	82	10	43	33	63	49	82	67	65	52	97	14	30	2.0	2.1	2.4	2.8	3.2	3.3	2.6	2.6	2.6	3.7	3.5	2.6	33.4
Luxembourg																												
Luxembourg	49 37N	06 03E	1,096	7	36	29	58	40	74	55	56	43	99	−10	100	2.3	2.0	1.9	2.1	2.4	2.5	2.8	2.6	2.4	2.7	2.7	2.8	29.2
Malta																												
Valletta	35 54N	14 31E	233	90	59	51	66	56	84	72	76	66	105	34	90	3.3	2.3	1.5	0.8	0.4	0.4	*	0.2	1.3	2.7	3.6	3.9	20.3
Monaco																												
Monaco	43 44N	07 25E	180	60	54	46	61	53	77	70	67	60	93	27	60	2.4	2.3	3.1	2.2	2.1	1.4	0.7	1.1	2.3	4.7	4.3	3.5	30.1
Netherlands																												
Amsterdam	52 23N	04 55E	5	29	40	34	52	43	69	59	56	48	95	3	29	2.0	1.4	1.3	1.6	1.8	1.8	2.6	2.7	2.8	2.8	2.6	2.2	25.6
Norway																												
Bergen	60 24N	05 19E	141	49	43	27	55	34	72	51	57	38	89	3	75	7.9	6.0	5.4	4.4	3.9	4.2	5.2	7.3	9.2	9.2	8.0	8.1	78.8
Kristiansand	58 10N	07 59E	175	11	32	25	50	35	71	53	53	39	90	−14	56	5.0	3.6	3.6	2.7	2.5	2.8	3.5	5.3	4.7	6.2	5.7	6.4	52.0
Oslo	59 56N	10 44E	308	44	30	20	50	34	73	56	49	37	93	−21	56	1.7	1.3	1.4	1.6	1.8	2.4	2.9	3.8	2.5	2.9	2.3	2.3	26.9
Tromoso	69 39N	18 57E	335	47	30	22	37	27	69	48	40	33	83	−1	75	4.1	3.8	3.3	2.4	2.1	2.1	2.3	2.9	4.7	4.5	4.0	3.9	40.1
Trondheim	63 25N	10 27E	417	44	31	22	45	32	66	51	46	36	95	−22	65	3.1	2.7	2.6	2.0	1.7	1.9	2.4	3.0	3.4	3.7	2.8	2.8	32.1
Vardo	70 22N	31 06E	43	40	27	19	34	26	53	44	38	32	80	−11	56	2.5	2.5	2.3	1.5	1.3	1.3	1.5	1.7	1.9	2.5	2.1	2.4	23.5
Poland																												
Danzig	54 24N	18 40E	36	36	33	25	49	37	70	56	53	42	94	−16	35	1.2	1.0	1.3	1.5	1.8	2.3	2.8	2.6	2.1	1.8	1.8	1.5	21.7
Krakow	50 04N	19 57E	723	35	32	22	55	38	76	57	56	41	97	−28	35	1.1	1.3	1.4	1.8	2.8	4.0	4.5	3.8	2.7	2.2	1.7	1.3	28.6
Warsaw	52 13N	21 02E	294	25	30	21	54	38	75	56	54	41	98	−22	113	1.2	1.1	1.3	1.5	1.9	2.6	3.0	3.0	1.9	1.7	1.4	1.4	22.0
Wroclaw (Breslau)	51 07N	17 05E	482	50	35	25	55	39	74	57	55	42	98	−26	40	1.5	1.1	1.5	1.7	2.4	2.4	3.4	2.7	1.8	1.7	1.5	1.5	23.2
Portugal																												
Braganca	41 49N	06 47W	2,395	11	46	31	59	39	80	54	62	42	103	10	11	11.9	6.9	7.7	3.7	3.0	1.6	0.5	0.6	1.5	3.0	6.3	7.1	53.8
Lagos	37 06N	08 38W	46	21	61	47	67	52	83	64	73	58	107	28	17	3.2	2.6	2.8	1.4	0.8	0.2	*	*	0.4	1.5	2.6	2.8	18.3
Lisbon	38 43N	09 08W	313	75	56	46	64	52	79	63	69	57	103	29	75	3.3	3.2	3.1	2.4	1.7	0.7	0.2	0.2	1.4	3.1	4.2	3.6	27.0
Romania																												
Bucharest	44 25N	26 06E	269	41	33	20	63	41	86	61	65	44	105	−18	41	1.5	1.1	1.7	1.6	2.5	3.8	2.3	1.8	1.5	1.6	1.9	1.5	22.8

Cluj	46 47N	23 40E	1,286	15	31	18	58	38	79	56	60	41	100	−26	16	1.3	1.2	1.0	2.1	3.3	3.3	2.6	3.3	2.0	1.7	1.0	1.2	24.0
Constanta	44 11N	28 39E	13	20	37	25	55	42	79	63	62	49	101	−13	39	1.2	1.2	1.1	1.1	1.3	1.7	1.3	1.1	1.1	1.4	1.2	1.4	15.1
Spain																												
Almeria	36 51N	02 28W	213	20	61	47	69	54	85	69	76	62	108	34	20	0.9	1.0	0.7	0.9	0.7	0.2	*	0.1	0.6	0.9	1.5	1.1	8.6
Barcelona	41 24N	02 09E	312	20	56	42	64	51	81	69	71	58	98	24	30	1.2	2.1	1.9	1.8	1.8	1.3	1.2	1.7	2.6	3.4	2.7	1.8	23.5
Burgos	42 20N	03 42W	2,825	29	42	30	57	38	77	53	61	43	99	0	29	1.5	1.5	2.1	1.9	2.4	1.7	0.8	0.7	1.4	2.0	2.2	2.0	20.2
Madrid	40 25N	03 41W	2,188	30	47	33	64	44	87	62	66	48	102	14	30	1.1	1.7	1.7	1.7	1.5	1.2	0.4	0.3	1.2	1.9	2.2	1.6	16.5
Sevilla	37 29N	05 59W	98	26	59	41	73	51	96	67	78	57	117	27	26	2.2	2.9	3.3	2.3	1.3	0.9	0.1	0.1	1.1	2.6	3.7	2.8	23.3
Valencia	39 28N	00 23W	79	26	58	41	67	51	83	68	73	57	107	20	29	0.9	1.5	0.9	1.2	1.1	1.3	0.4	0.5	2.2	1.6	2.5	1.3	15.4
Sweden																												
Abisko	68 21N	18 49E	1,273	11	20	6	33	19	61	45	35	24	82	−30	11	0.7	0.6	0.5	0.5	0.7	1.8	1.6	1.8	1.2	1.0	0.6	0.6	11.7
Goteberg	57 42N	11 58E	55	39	35	27	48	36	69	56	51	42	88	−13	61	2.5	2.0	2.0	1.7	1.9	2.2	2.8	3.7	3.1	3.1	2.7	2.8	30.5
Haparanda	65 50N	24 09E	30	20	22	10	38	23	71	53	39	30	89	−34	20	2.2	1.6	1.2	1.5	1.4	1.7	2.1	2.8	2.6	2.8	2.5	2.0	24.4
Karlstad	59 23N	13 30E	164	30	30	20	49	32	73	56	49	38	93	−21	30	1.9	1.2	1.2	1.4	1.9	1.9	2.6	3.1	2.9	2.4	2.4	1.9	24.8
Sarna	61 41N	13 07E	1,504	20	19	4	42	23	69	46	42	28	91	−51	20	1.6	0.8	0.9	1.2	1.6	2.8	3.6	3.3	2.6	2.3	1.8	1.8	24.3
Stockholm	59 21N	18 04E	146	30	31	23	45	32	70	55	48	39	97	−26	30	1.5	1.1	1.1	1.5	1.6	1.9	2.8	3.1	2.1	2.1	1.9	1.9	22.4
Visby (Gotland)	57 39N	18 18E	36	30	35	28	44	33	67	55	50	41	88	1	30	1.7	1.1	1.2	1.4	1.1	1.4	2.0	2.7	1.7	1.9	2.1	2.0	20.3
Switzerland																												
Berne	46 57N	07 26E	1,877	30	35	26	56	39	74	56	55	42	96	−9	77	1.9	2.0	2.6	3.0	3.7	4.4	4.4	4.3	3.5	3.5	2.7	2.5	38.5
Geneva	46 12N	06 09E	1,329	30	39	29	58	41	77	58	58	44	101	−1	125	1.9	1.8	2.2	2.5	3.0	3.1	2.9	3.6	3.6	3.8	3.1	2.4	33.9
Zurich	47 23N	08 33E	1,617	23	38	28	57	39	76	55	57	42	98	−12	23	2.3	1.9	2.9	3.4	4.0	4.9	5.0	4.6	3.3	3.2	2.5	2.9	40.9
Turkey																												
Edirne	41 39N	26 34E	154	18	41	28	66	44	88	63	70	49	107	−8	18	2.2	1.9	1.7	1.9	1.7	2.1	1.5	1.1	1.1	2.1	2.9	3.0	23.2
Istanbul	40 58N	28 50E	59	18	45	36	61	45	81	65	67	54	100	17	18	3.7	2.3	2.6	1.9	1.4	1.3	1.7	1.5	2.3	3.8	4.1	4.9	31.5
United Kingdom																												
Belfast	54 35N	05 56W	57	7	42	34	53	38	65	52	55	44	82	14	30	4.2	2.8	2.3	2.4	2.3	2.5	3.5	3.5	3.4	3.8	3.6	3.9	38.2
Birmingham	52 29N	01 56W	535	30	42	35	53	40	69	54	55	45	92	11	30	2.9	2.1	1.7	2.2	2.5	1.8	2.8	2.7	2.3	2.9	3.2	2.6	29.7
Cardiff	51 28N	03 10W	203	30	45	36	55	41	69	54	57	45	91	2	30	4.6	3.0	2.3	2.5	3.0	2.2	3.4	3.9	3.6	4.5	4.6	4.3	41.9
Dublin	53 22N	06 21W	155	30	47	35	54	38	67	51	57	43	86	8	35	2.7	2.2	2.0	1.9	2.3	2.0	2.8	3.0	2.8	2.7	2.7	2.6	29.7
Edinburgh	55 55N	03 11W	441	30	43	35	50	39	65	52	53	44	83	15	30	2.5	1.6	1.6	1.6	2.2	1.9	3.1	3.1	2.6	2.9	2.4	2.1	27.6
London	51 29N	00 00	149	30	44	35	56	40	73	55	58	44	99	9	30	2.0	1.5	1.4	1.8	1.8	1.6	2.0	2.2	1.8	2.3	2.5	2.0	22.9
Liverpool	53 24N	03 04W	198	30	44	36	52	41	66	55	55	46	87	15	30	2.7	1.9	1.5	1.6	2.2	2.0	2.8	3.1	2.6	3.0	3.0	2.5	28.9
Perth	56 24N	03 27W	77	30	43	32	53	38	68	51	55	41	89	0	30	3.1	2.2	1.9	1.7	2.3	2.0	3.1	2.9	2.8	3.3	2.7	2.7	30.7
Plymouth	50 21N	04 07W	87	30	47	40	54	43	66	55	58	49	88	16	30	4.3	3.0	2.6	2.3	2.5	2.0	2.6	2.9	2.8	3.8	4.6	4.4	37.8
Wick	58 26N	03 04W	119	30	42	35	48	38	59	50	52	43	80	8	30	2.9	2.1	1.8	2.1	1.8	2.0	2.6	2.6	2.9	3.2	3.1	2.9	30.0
U.S.S.R.																												
Arkhangelsk	64 33N	40 32E	22	23	9	2	36	23	64	51	36	30	91	−49	25	1.2	1.1	1.1	0.7	1.3	1.9	2.6	2.7	2.2	1.9	1.6	1.3	19.8
Astrakhan	46 21N	48 02E	45	10	23	14	57	40	85	69	56	40	99	−22	25	0.5	0.5	0.4	0.6	0.6	0.7	0.5	0.4	0.6	0.4	0.6	0.6	6.4
Dnepropetrovsk	48 27N	35 04E	259	18	25	16	53	39	80	62	56	40	101	−25	17	1.4	1.1	1.2	1.4	1.8	3.0	1.9	1.6	1.0	1.8	1.6	1.6	19.4
Kaunas	54 54N	23 53E	118	19	26	18	49	34	72	53	50	38	96	−23	19	1.6	1.3	1.3	1.8	2.0	3.2	3.3	3.5	1.9	1.9	1.6	1.6	25.0
Kirov	58 36N	49 41E	594	20	6	−2	41	27	72	55	37	29	92	−43	29	1.2	1.0	0.9	0.9	1.9	2.5	2.1	2.9	2.3	2.0	1.6	1.3	20.6
Kursk	51 45N	36 12E	773	15	19	11	47	35	74	58	48	36	91	−23	20	1.5	1.3	1.2	1.5	2.2	2.5	3.2	2.3	1.6	1.8	1.5	1.7	22.3
Leningrad	59 56N	30 16E	16	26	23	12	45	31	71	57	45	37	91	−36	95	1.0	0.9	0.9	1.0	1.6	2.0	2.5	2.8	2.1	1.8	1.4	1.2	19.2
Lvov	49 50N	24 01E	978	9	31	22	53	38	77	59	55	43	97	−29	35	1.3	1.5	1.8	2.0	2.8	3.7	4.1	3.1	2.4	2.1	0.8	1.6	28.2
Minsk	53 54N	27 33E	738	12	22	13	47	33	70	54	47	36	92	−27	20	1.4	1.5	1.3	1.5	2.0	2.8	3.0	3.1	1.6	1.5	1.5	1.7	22.9
Moscow	55 46N	37 40E	505	15	21	9	47	31	76	55	46	34	96	−27	11	1.5	1.4	1.1	1.9	2.2	2.9	3.0	2.9	1.9	2.7	1.7	1.6	24.8
Odessa	46 29N	30 44E	214	20	28	22	52	41	79	65	57	47	99	−13	15	1.0	0.7	0.7	1.1	1.1	1.9	1.6	1.4	1.1	1.4	1.1	1.1	14.3
Riga	56 57N	24 06E	67	30	29	20	48	35	72	56	49	39	93	−20	57	1.3	1.0	1.1	1.2	1.7	2.4	3.0	3.0	2.1	2.0	1.9	1.5	22.2
Saratov	51 32N	46 03E	197	14	15	7	50	35	82	64	48	36	102	−27	15	1.0	1.0	0.8	1.0	1.3	1.8	1.2	1.3	1.1	1.4	1.4	1.2	14.5
Sevastopol	44 37N	33 31E	75	20	39	30	55	42	79	65	63	50	97	−4	30	1.1	1.1	1.1	0.9	0.6	1.1	0.8	0.6	1.1	1.5	1.2	1.1	12.2
Stalingrad	48 42N	44 31E	136	8	15	4	52	36	84	65	53	37	106	−30	12	0.9	1.0	0.6	0.6	1.0	1.9	0.9	0.8	0.7	1.0	1.5	1.3	12.2
Stavropol	45 02N	41 58E	1,886	18	26	17	50	37	76	60	55	42	95	−22	41	1.4	1.1	1.5	2.4	3.0	4.1	3.0	2.0	2.5	2.3	1.8	1.8	26.9
Tallin	59 26N	24 48E	146	15	27	18	42	31	70	55	47	38	89	−19	63	1.1	1.0	0.9	1.1	1.7	1.9	2.1	2.7	2.3	2.1	1.9	1.5	20.2
Tbilisi	41 43N	44 48E	1,325	10	39	26	61	44	83	65	64	48	95	6	10	0.7	0.8	1.3	1.6	3.6	3.1	2.2	1.7	1.9	1.3	2.0	1.2	21.4
Ust'Shchugor	64 16N	57 34E	279	15	4	−14	35	17	65	49	33	23	90	−67	15	1.1	0.8	0.8	0.7	1.4	2.2	3.0	3.2	2.4	2.2	1.5	1.3	20.6
Ufy	54 43N	55 56E	571	20	6	−3	44	30	75	58	41	31	99	−42	23	1.6	1.3	1.2	0.9	1.6	2.4	2.6	2.2	1.8	2.3	2.2	2.3	22.5
Yugoslavia																												
Belgrade	44 48N	20 28E	453	16	37	27	64	45	84	61	65	47	107	−14	16	1.6	1.3	1.6	2.2	2.6	2.8	1.9	2.5	1.7	2.7	1.8	1.9	24.6
Skopje	41 59N	21 28E	787	10	40	26	67	42	88	60	65	43	105	−11	10	1.5	1.2	1.3	1.5	1.9	1.9	1.3	1.1	1.1	2.6	2.3	1.8	19.5
Split	43 31N	16 26E	420	14	51	29	65	50	87	68	69	55	100	17	51	3.1	2.5	3.2	3.0	2.5	2.1	1.2	1.6	2.9	4.4	4.2	4.4	35.1

(Continued)

Table 3B.13 (Continued)

Country and Station	Latitude	Longitude	Elevation	Length of Record	Temperature: Average Daily: Jan Maximum	Jan Minimum	Apr Maximum	Apr Minimum	July Maximum	July Minimum	Oct Maximum	Oct Minimum	Extreme Maximum	Extreme Minimum	Average Precipitation: Length of Record	January	February	March	April	May	June	July	August	September	October	November	December	Year
	°	°	Feet	Year	°F	°F	°F	°F	°F	°F	°F	°F	°F	°F	Year	IN.	IN.	IN.	IN.	IN.	IN.	IN.	IN.	IN.	IN.	IN.	IN.	IN.
Ocean Islands																												
Bjornoya, Bear Island	74 31N	19 01E	49	10	26	17	27	16	44	36	36	29	71	−25	25	1.6	1.3	1.3	0.9	0.8	0.7	0.8	1.2	1.8	1.7	1.4	1.6	15.1
Gronfjorden, Spitzbergen	78 02N	14 15E	23	19	10	−4	15	−3	46	38	25	17	60	−57	15	1.4	1.3	1.1	0.9	0.5	0.4	0.6	0.9	1.0	1.2	0.9	1.5	11.7
Horta, Azores	38 32N	28 38W	200	30	62	54	64	55	76	65	71	62	88	38	30	4.5	4.1	4.2	3.0	2.9	2.0	1.5	1.9	3.2	4.4	4.1	4.5	40.3
Jan Mayen	77 01N	08 28W	131	5	31	21	31	22	46	38	39	29	60	−18	29	2.1	1.7	1.6	1.4	0.9	0.9	1.4	1.8	2.5	2.5	2.2	2.2	21.2
Lerwick, Shetland Island	60 08N	01 11W	269	30	42	35	46	37	58	49	50	42	71	17	30	4.5	3.4	2.9	2.7	2.2	2.2	2.7	2.9	3.7	4.3	4.5	4.5	40.5
Matochikin Shar, Novaya Zemlya	73 16N	56 24E	61	9	8	−6	13	−1	47	36	30	21	68	−41	9	0.6	0.6	0.6	0.4	0.3	0.4	1.4	1.5	1.5	0.6	0.6	0.4	8.9
Ponta Delgada, Azores	37 45N	25 40W	118	30	62	54	64	55	76	64	71	61	85	37	30	4.0	3.5	3.5	2.5	2.3	1.4	1.0	1.2	2.9	3.6	3.7	3.0	32.6
Stornoway, Hebrides	58 11N	06 21W	34	30	44	37	49	39	61	51	53	44	78	11	15	6.4	3.2	3.2	3.1	2.5	2.4	3.0	4.3	4.7	6.2	4.6	5.5	49.1
Thorshavn, Faeroes	62 02N	06 45W	82	50	42	33	45	36	56	47	58	40	70	8	50	6.6	5.2	4.8	3.6	3.4	2.5	3.1	3.5	4.7	5.9	6.3	6.6	56.2
Africa																												
Algeria																												
Adrar	27 52N	00 17W	948	15	69	39	92	60	115	82	92	63	124	25	15	*	*	0.1	*	*	*	*	*	*	0.2	0.2	*	0.6
Algiers	36 46N	03 03E	194	25	59	49	68	55	83	70	74	63	107	32	25	4.4	3.3	2.9	1.6	1.8	0.6	*	0.2	1.6	3.1	5.1	5.4	30.0
Annaba	36 54N	07 46E	66	26	59	46	67	52	85	69	75	61	115	32	26	5.6	4.1	2.9	2.2	1.5	0.6	0.1	0.3	1.2	3.0	4.3	5.2	31.0
Bordj Omar Driss	28 06N	06 42E	1,224	15	67	38	90	59	110	78	92	63	124	19	15	0.3	0.1	0.1	0.2	*	*	0.0	*	*	*	0.2	0.2	1.1
El Golea	30 35N	02 53E	1,247	15	63	37	84	56	107	79	87	60	120	23	15	0.1	0.3	0.5	*	*	*	*	*	*	0.3	0.4	0.3	1.9
Tamanrasset	22 42N	05 31E	4,593	15	67	39	86	56	95	71	85	59	102	20	15	0.2	*	*	0.2	0.4	0.1	0.1	0.4	0.1	*	*	*	1.5
Touggourt	33 07N	06 04E	226	26	62	38	83	55	107	77	84	59	122	26	26	0.2	0.4	0.5	0.2	0.2	0.2	*	*	0.1	0.3	0.5	0.3	2.9
Angola																												
Cangamba	13 41S	19 52E	4,331	6	84	62	89	58	82	46	87	59	109	20	7	8.9	7.4	6.8	1.8	0.1	0.0	0.0	0.2	0.2	1.6	5.1	8.5	40.6
Huambo	12 48S	15 45E	5,577	14	78	58	78	57	77	47	81	58	90	36	14	8.7	7.8	9.8	5.7	0.4	0.0	*	*	0.6	5.5	9.6	8.9	57.0
Luanda	08 49S	13 13E	194	27	83	74	85	75	74	65	79	71	98	58	59	1.0	1.4	3.0	4.6	0.5	*	*	*	0.1	0.2	1.1	0.8	12.7
Moçâmedes	15 12S	12 09E	10	15	79	65	82	66	68	56	74	61	102	44	21	0.3	0.4	0.7	0.5	*	*	*	*	*	*	0.1	0.1	2.1
Benin																												
Cotonou	06 21N	02 26E	23	5	80	74	83	78	78	74	80	75	95	65	10	1.3	1.3	4.6	4.9	10.0	14.4	3.5	1.5	2.6	5.3	2.3	0.5	52.4
Botswana																												
Francistown	21 13S	27 30E	3,294	20	88	65	83	56	75	41	90	61	107	24	28	4.2	3.1	2.8	0.7	0.2	0.1	*	*	*	0.9	2.3	3.4	17.7
Maun	19 59S	23 25E	3,091	20	90	66	87	58	77	42	95	64	110	24	20	4.3	3.8	3.5	1.1	0.2	*	0.0	0.0	*	0.5	1.9	2.8	18.2
Tsabong	26 03S	22 27E	3,156	10	94	65	83	51	71	34	88	54	107	15	14	2.0	1.9	1.9	1.3	0.4	0.4	0.1	*	0.2	0.7	1.1	1.5	11.5
Burkina Faso																												
Bobo Dioulasso	11 10N	04 15W	1,411	11	92	58	99	71	87	69	90	70	115	46	10	0.1	0.2	1.1	2.1	4.6	4.8	9.8	12.0	8.5	2.5	0.7	0.0	46.4
Ouagadougou	12 22N	01 31W	991	10	92	60	103	79	91	74	95	74	118	48	15	*	0.1	0.5	0.6	3.3	4.8	8.0	10.9	5.7	1.3	*	0.0	35.2
Cameroon																												
Ngaoundere	07 17N	13 19E	3,601	9	87	55	87	64	82	63	82	61	102	46	10	*	*	1.1	5.5	7.0	8.4	10.6	9.6	9.2	5.3	0.5	*	57.2
Yaounde	03 53N	11 32E	2,526	11	85	67	85	66	80	66	81	65	96	57	11	0.9	2.6	5.8	6.7	7.7	6.0	2.9	3.1	8.4	11.6	4.6	0.9	61.2

Central African Republic																											
Bangui	04 22N	18 34E	1,270	5	90	68	91	71	85	69	87	69	101	57	5	1.0	1.7	5.0	5.3	7.4	4.5	8.9	8.1	5.9	7.9	4.9	0.2	60.8
Ndele	08 24N	20 39E	1,939	3	99	67	98	73	86	69	90	68	109	58	3	0.2	1.3	0.6	1.7	8.4	6.1	8.3	10.1	10.7	7.8	0.6	0.0	55.8
Chad																											
Am Timan	11 02N	20 17E	1,430	3	98	56	105	68	89	70	96	67	113	43	3	0.0	0.0	0.1	1.2	4.3	5.0	7.3	12.3	5.8	1.2	0.0	0.0	37.2
Fort Lamy	12 07N	15 02E	968	5	93	57	107	74	92	72	97	70	114	47	5	0.0	0.0	0.0	0.1	1.2	2.6	6.7	12.6	4.7	1.4	0.0	0.0	29.3
Largeau (Faya)	18 00N	19 10E	837	5	84	54	104	69	109	76	103	72	121	37	5	0.0	0.0	0.0	0.0	*	0.0	*	0.7	*	0.0	0.0	0.0	0.7
Congo																											
Brazzaville	04 15S	15 15E	1,043	15	88	69	91	71	82	63	89	70	98	54	18	6.3	4.9	7.4	7.0	4.3	0.6	*	*	2.2	5.4	11.5	8.4	58.0
Ouesso	01 37N	16 04E	1,132	4	88	69	91	71	85	69	87	69	106	60	4	2.4	3.6	6.4	3.2	5.8	4.6	2.9	3.7	7.9	10.0	5.7	2.4	58.6
Pointe Noire (Loango)	04 39S	11 48E	164	7	85	73	87	74	78	66	83	72	93	59	7	5.4	6.7	6.4	8.0	3.9	0.0	0.0	0.0	0.4	4.1	6.6	6.6	48.1
Djibouti																											
Djibouti	11 36N	43 09E	23	16	84	73	90	79	106	87	92	80	117	63	46	0.4	0.5	1.0	0.5	0.2	*	0.1	0.3	0.3	0.4	0.9	0.5	46.0
Egypt																											
Alexandria	31 12N	29 53E	105	45	65	51	74	59	85	73	83	68	111	37	61	1.9	0.9	0.4	0.1	*	*	*	*	*	0.2	1.3	2.2	7.0
Aswan	24 02N	32 53E	366	46	74	50	96	66	106	79	98	71	124	35	11	*	*	*	*	*	*	0.0	0.0	0.0	*	*	*	*
Cairo	29 52N	31 20E	381	42	65	47	83	57	96	70	86	65	117	34	42	0.2	0.2	0.2	0.1	0.1	*	0.0	0.0	*	*	0.1	0.2	1.1
Ethiopia																											
Addis Ababa	09 20N	38 45E	8,038	15	75	43	77	50	69	50	75	45	94	32	37	0.5	1.5	2.6	3.4	3.4	5.4	11.0	11.8	7.5	0.8	0.6	0.2	48.7
Asmara	15 17N	38 55E	7,628	9	74	44	78	51	71	53	72	53	88	31	17	*	*	0.4	1.5	1.5	1.3	6.7	5.0	1.3	0.3	0.4	*	18.4
Diredawa	09 02N	41 45E	3,937	8	81	58	91	69	90	68	89	67	100	49	8	0.8	0.8	3.3	3.0	2.8	1.5	4.3	3.8	2.2	0.5	0.3	0.8	24.1
Gambela	08 15N	34 35E	1,345	26	98	64	98	71	87	69	92	67	111	48	30	0.2	0.4	1.4	3.2	5.9	6.7	8.5	9.5	7.3	3.5	1.8	0.4	48.8
Gabon																											
Libreville	00 23N	09 26E	115	11	87	73	89	73	83	68	86	71	99	62	21	9.8	9.3	13.2	13.4	9.6	0.5	0.1	0.7	4.1	13.6	14.7	9.8	98.8
Mayoumba	03 25S	10 38E	200	8	84	73	86	73	78	68	82	72	91	60	8	6.5	9.3	6.2	10.2	2.3	0.1	0.0	0.2	2.6	9.3	10.7	4.6	62.0
Gambia																											
Banjul	13 21N	16 40W	90	9	88	59	91	65	86	74	89	72	106	45	9	0.1	0.1	*	*	0.4	2.3	11.1	19.7	12.2	4.3	0.7	0.1	51.0
Ghana																											
Accra	05 33N	00 12W	88	17	87	73	88	76	81	73	85	74	100	59	65	0.6	1.3	2.2	3.2	5.6	7.0	1.8	0.6	1.4	2.5	1.4	0.9	28.5
Kumasi	06 40N	01 37W	942	10	88	66	89	71	82	70	86	70	100	51	10	0.8	2.3	5.7	5.1	7.5	7.9	4.3	3.1	6.8	7.1	3.7	0.8	55.2
Guinea																											
Conakry	09 31N	13 43W	23	7	88	72	90	73	83	72	87	73	96	63	10	0.1	0.1	0.4	0.9	6.2	22.0	51.1	41.5	26.9	14.6	4.8	0.4	169.0
Kouroussa	10 39N	09 53W	1,217	9	93	60	99	73	87	69	90	69	109	39	10	0.4	0.3	0.9	2.8	5.3	9.7	11.7	13.6	13.4	6.6	1.3	0.4	66.4
Guinea-Bissau																											
Bolama	11 34N	15 26W	62	31	88	67	91	73	84	74	87	74	106	59	37	*	*	*	*	0.8	7.8	23.1	27.6	16.9	8.0	1.6	0.1	85.9
Ifni (Now in Morocco)																											
Sidi Ifni	29 27N	10 11W	148	14	66	52	71	59	75	64	75	62	124	40	14	1.0	0.6	0.5	0.6	0.1	0.1	*	*	0.4	0.1	0.9	1.8	6.1
Ivory Coast																											
Abidjan	05 19N	04 01W	65	13	88	73	90	75	83	73	85	74	96	59	10	1.6	2.1	3.9	4.9	14.2	19.5	8.4	2.1	2.8	6.6	7.9	3.1	77.1
Bouake	07 42N	05 00W	1,194	12	91	68	92	70	85	68	89	68	104	57	10	0.4	1.5	4.1	5.8	5.3	6.0	3.1	4.6	8.2	5.2	1.5	1.0	46.7
Kenya																											
Mombasa	04 03S	39 39E	52	45	87	75	86	76	81	71	84	74	96	61	54	1.0	0.7	2.5	7.7	12.6	4.7	3.5	2.5	2.5	3.4	3.8	2.4	47.3
Nairobi	01 16S	36 48E	5,971	15	77	54	75	58	69	51	76	55	87	41	17	1.5	2.5	4.9	8.3	6.2	1.8	0.6	0.9	1.2	2.1	4.3	3.4	37.7
Liberia																											
Monrovia	06 18N	10 48W	75	6	89	71	90	72	80	72	86	72	97	62	4	0.2	0.1	4.4	11.7	13.4	36.1	24.2	18.6	29.9	25.2	8.2	2.9	174.9
Libya																											
Benghazi	32 06N	20 04E	82	46	63	50	74	58	84	71	80	66	109	37	46	2.6	1.6	0.8	0.2	0.1	*	*	*	0.1	0.7	1.8	2.6	10.5
Kufra	24 12N	23 21E	1,276	7	69	43	90	62	101	75	90	64	122	26	7	*	0.0	0.0	0.0	*	0.0	0.0	0.0	0.0	0.0	0.0	*	*
Sabhah	27 01N	14 26E	1,457	3	64	41	89	60	102	74	91	64	120	24	10	*	*	*	*	0.1	0.1	0.0	0.0	0.0	*	*	*	0.3
Tripoli	32 54N	13 11E	72	47	61	47	72	57	85	71	80	65	114	33	56	3.2	1.8	1.1	0.4	0.2	0.1	*	*	0.4	1.6	2.6	3.7	15.1
Malagasy Republic																											
Diego Suarez	12 17S	49 17E	100	11	88	75	88	75	84	69	86	72	98	63	31	10.6	9.5	7.6	2.2	0.3	0.2	0.2	0.3	0.3	0.7	1.1	5.8	38.7
Tananarive	18 55S	47 33E	4,500	44	79	61	76	58	68	48	80	54	95	34	62	11.8	11.0	7.0	2.1	0.7	0.3	0.3	0.4	0.7	2.4	5.3	11.3	53.4
Tulear	23 20S	43 41E	20	27	92	72	89	64	81	58	86	65	108	43	15	3.1	3.2	1.4	0.3	0.7	0.4	0.1	0.2	0.3	0.7	1.4	1.7	13.5
Malawi																											
Karonga	09 57S	33 56E	1,596	8	86	71	85	70	81	59	91	66	99	51	8	7.1	7.0	10.8	6.2	1.7	0.1	*	*	0.0	0.3	0.3	4.7	38.3
Zomba	15 23S	35 19E	3,141	27	80	65	78	62	72	53	85	64	95	41	29	12.1	9.9	10.1	2.7	0.7	0.4	0.3	0.3	0.2	1.0	4.3	10.9	52.9
Mali																											
Araouane	18 54N	03 33W	935	8	81	48	110	67	111	79	103	70	130	37	10	*	*	0.0	0.0	0.0	0.2	0.2	0.5	0.6	0.1	0.1	*	1.7
Bamako	12 39N	07 58W	1,116	11	91	61	103	76	89	71	93	71	117	47	10	*	*	0.1	0.6	2.9	5.4	11.0	13.7	8.1	1.7	0.6	*	44.1
Gao	16 16N	00 03W	902	15	83	58	105	77	97	80	100	78	116	44	19	*	0.0	*	0.1	0.4	1.0	2.9	5.4	1.5	0.2	*	0.0	11.5
Mauritania																											
Atar	20 31N	13 04W	761	7	84	54	97	67	106	81	98	72	117	39	10	*	0.0	*	*	*	0.1	0.3	1.2	1.1	0.1	*	*	2.8
Nema	16 36N	07 16W	883	9	86	62	105	79	99	78	101	79	120	47	10	0.1	*	*	*	0.7	1.1	2.3	4.7	2.1	0.7	*	0.1	11.6

(Continued)

Table 3B.13 (Continued)

Country and Station	Latitude	Longitude	Elevation	Length of Record	Temperature: Average Daily, Jan Maximum	Jan Minimum	Apr Maximum	Apr Minimum	July Maximum	July Minimum	Oct Maximum	Oct Minimum	Extreme Maximum	Extreme Minimum	Average Precipitation: Length of Record	January	February	March	April	May	June	July	August	September	October	November	December	Year
	°	°	Feet	Year	°F	°F	°F	°F	°F	°F	°F	°F	°F	°F	Year	IN.	IN.	IN.	IN.	IN.	IN.	IN.	IN.	IN.	IN.	IN.	IN.	IN.
Nouakchott	18 07N	15 36W	69	5	85	57	90	64	89	74	91	71	115	44	10	*	0.1	*	*	*	0.1	0.5	4.1	0.9	0.4	0.1	*	6.2
Morocco																												
Casablanca	33 35N	07 39W	164	48	63	45	69	52	79	65	76	58	110	31	40	2.1	1.9	2.2	1.4	0.9	0.2	0.0	*	0.3	1.5	2.6	2.8	15.9
Marrakech	31 36N	08 01W	1,509	35	65	40	79	52	101	67	83	57	120	27	31	1.0	1.1	1.3	1.2	0.6	0.3	0.1	0.1	0.4	0.9	1.2	1.2	9.4
Rabat	34 00N	06 50W	213	35	63	46	71	52	82	63	77	58	118	32	29	2.6	2.5	2.6	1.7	1.1	0.3	*	*	0.4	1.9	3.3	3.4	19.8
Tangier	35 48N	05 49W	239	35	60	47	65	51	80	64	72	59	106	28	35	4.5	4.2	4.8	3.5	1.7	0.6	*	*	0.9	3.9	5.8	5.4	35.3
Mozambique																												
Beira	19 50S	34 51E	28	37	89	75	86	71	77	61	87	71	109	48	39	10.9	8.4	10.1	4.2	2.2	1.3	1.2	1.1	0.8	5.2	5.3	9.2	59.9
Chicoa	15 36S	32 21E	899	8	96	65	93	63	86	55	101	68	117	32	8	7.8	5.7	4.4	0.6	*	*	*	*	*	1.1	2.6	5.2	27.4
Maputo	25 58S	32 36E	194	42	86	71	83	66	76	55	82	64	114	45	42	5.1	4.9	4.9	2.1	1.1	0.8	0.5	0.5	1.1	1.9	3.2	3.8	29.9
Namibia																												
Keetmanshoop	26 35S	18 08E	3,295	17	95	65	85	57	70	42	87	55	108	26	45	0.8	1.1	1.4	0.6	0.2	*	*	*	0.1	0.2	0.3	0.4	5.2
Windhoek	22 34S	17 06E	5,669	30	85	63	77	55	68	43	84	59	97	25	60	3.0	2.9	3.1	1.6	0.3	*	*	*	0.1	0.4	0.9	1.9	14.3
Niger																												
Agades	16 59N	07 59E	1,706	8	86	50	105	70	104	75	101	68	115	40	10	0.0	0.0	*	*	0.2	0.3	1.9	3.7	0.7	0.0	0.0	0.0	6.8
Bilma	18 41N	12 55E	1,171	9	81	45	101	63	108	75	101	62	116	29	10	0.0	0.0	0.0	*	*	0.0	0.1	0.5	0.3	0.0	0.0	0.0	0.9
Niamey	13 31N	02 06E	709	10	93	58	108	77	94	74	101	74	114	47	10	*	*	0.2	0.3	1.3	3.2	5.2	7.4	3.7	0.5	*	0.0	21.6
Nigeria																												
Enugu	06 27N	07 29E	763	11	90	72	91	74	83	71	87	71	99	55	33	0.7	1.1	2.6	5.9	10.4	11.4	7.6	6.7	12.8	9.8	2.1	0.5	71.5
Kaduna	10 35N	06 26E	2,113	18	89	59	95	72	83	68	89	66	105	46	34	*	0.1	0.5	2.5	5.9	7.1	8.5	11.9	10.6	2.9	0.1	*	50.1
Lagos	06 27N	03 24E	10	32	88	74	89	77	83	74	85	74	104	60	47	1.1	1.8	4.0	5.9	10.6	18.1	11.0	2.5	5.5	8.1	2.7	1.0	72.3
Maiduguri	11 51N	13 05E	1,162	15	90	54	104	72	90	73	96	68	112	43	40	*	*	*	0.3	1.6	2.7	7.1	8.7	4.2	0.7	*	0.0	25.3
Senegal																												
Dakar	14 42N	17 29W	131	25	79	64	81	65	88	76	89	76	109	53	26	*	*	*	*	*	0.7	3.5	10.0	5.2	1.5	0.1	0.3	21.3
Kaolack	14 08N	16 04W	20	9	93	60	103	68	91	75	93	74	114	48	10	*	0.0	*	*	0.3	2.6	6.9	10.7	7.0	2.7	0.1	*	30.3
Sierra Leone																												
Freetown/Lungi	08 37N	13 12W	92	8	87	73	88	76	82	73	85	72	98	62	8	0.4	0.2	1.2	3.1	9.5	14.3	29.2	36.5	22.3	14.2	5.5	1.2	137.6
Somalia																												
Berbera	10 26N	45 02E	45	30	84	68	89	77	107	88	92	76	117	58	30	0.3	0.1	0.2	0.5	0.3	*	*	0.1	*	0.1	0.2	0.2	2.0
Mogadiscio	02 02N	45 21E	39	13	86	73	90	78	83	73	86	76	97	59	21	*	*	*	2.3	2.3	3.8	2.5	1.9	1.0	0.9	1.6	0.5	16.9
South Africa, Republic of																												
Cape Town	33 54S	18 32E	56	19	78	60	72	53	63	45	70	52	103	28	18	0.6	0.3	0.7	1.9	3.1	3.3	3.5	2.6	1.7	1.2	0.7	0.4	20.0
Durban	29 50S	31 02E	16	15	81	69	78	64	72	52	75	62	107	39	78	4.3	4.8	5.1	3.0	2.1	1.3	1.1	1.5	2.8	4.3	4.8	4.7	39.7
Kimberley	28 48S	24 46E	3,927	19	91	64	77	52	65	36	83	54	103	20	57	2.4	2.5	3.1	1.5	0.7	0.2	0.2	0.3	0.6	1.0	1.6	2.0	16.1
Port Elizabeth	33 59S	25 36E	190	14	78	61	73	55	67	45	70	54	104	31	84	1.2	1.3	1.9	1.8	2.4	1.8	1.9	2.0	2.3	2.2	2.2	1.7	22.7
Port Nolloth	29 14S	16 52E	23	20	67	53	66	50	62	45	64	49	107	31	64	0.1	0.1	0.2	0.2	0.3	0.3	0.3	0.3	0.2	0.1	0.1	0.1	2.3
Pretoria	25 45S	28 14E	4,491	13	81	60	75	50	66	37	80	55	96	24	12	5.0	4.3	4.5	1.7	0.9	0.6	0.3	0.2	0.8	2.2	5.2	5.2	30.9
Walvis Bay	22 56S	14 30E	24	20	73	59	75	55	70	47	67	51	104	25	20	*	0.2	0.3	0.1	0.1	*	*	0.1	*	*	*	*	0.9
Sudan																												
El Fasher	13 38N	25 21E	2,395	17	88	50	102	64	96	70	99	64	113	33	17	*	0.0	*	*	0.3	0.7	4.5	5.3	1.2	0.2	0.0	0.0	12.2
Khartoum	15 37N	32 33E	1,279	46	90	59	105	72	101	77	104	75	118	41	46	*	*	*	*	0.1	0.3	2.1	2.8	0.7	0.2	*	0.0	6.2
Port Sudan	19 37N	37 13E	18	30	81	68	89	71	106	83	93	76	117	50	40	0.2	0.1	*	*	*	*	0.3	0.1	*	0.4	1.7	0.9	3.7
Wadi Halfa	21 55N	31 20E	410	39	75	46	98	62	106	74	98	67	127	28	39	*	*	*	*	*	0.0	*	*	*	*	*	0.0	*
Wau	07 42N	28 03E	1,443	38	96	64	99	72	89	69	93	69	115	50	38	*	0.2	0.9	2.6	5.3	6.5	7.5	8.2	6.6	4.9	0.6	*	43.3

Tanzania																												
Dares Salaam	06 50S	39 18E	47	44	83	77	86	73	83	66	85	69	96	59	49	2.6	2.6	5.1	11.4	7.4	1.3	1.2	1.0	1.2	1.6	2.9	3.6	41.9
Iringa	07 47S	35 42E	5,330	14	76	59	75	59	72	52	80	57	90	42	24	6.8	5.1	7.1	3.5	0.5	*	*	*	0.1	0.2	1.5	4.5	29.3
Kigoma	04 53S	29 38E	2,903	26	80	67	81	67	83	63	84	69	100	53	18	4.8	5.0	5.9	5.1	1.7	0.2	0.1	0.2	0.7	1.9	5.6	5.3	36.5
Togo																												
Lome	06 10N	01 15E	72	5	85	72	86	74	80	71	83	72	94	58	15	0.6	0.9	1.9	4.6	5.7	8.8	2.8	0.4	1.4	2.4	1.1	0.4	31.0
Tunisia																												
Gabes	33 53N	10 07E	7	50	61	43	74	54	89	71	81	62	122	27	50	0.9	0.7	0.8	0.4	0.3	*	*	0.1	0.5	1.2	1.2	0.6	6.7
Tunis	36 47N	10 12E	217	50	58	43	70	51	90	68	77	59	118	30	50	2.5	2.0	1.6	1.4	0.7	0.3	0.1	0.3	1.3	2.0	1.9	2.4	16.5
Uganda																												
Kampala	00 20N	32 36E	4,304	15	83	65	79	64	77	62	81	63	97	53	15	1.8	2.4	5.1	6.9	5.8	2.9	1.8	3.4	3.6	3.8	4.8	3.9	46.2
Lira	02 15N	32 54E	3,560	14	91	61	86	64	81	61	86	61	100	50	14	0.7	1.0	3.5	6.9	7.9	4.9	6.4	10.0	8.3	6.1	3.2	1.8	60.7
Western Sahara																												
Semara	26 46N	11 31W	1,509	6	73	47	88	58	99	66	88	61	121	37	6	0.1	*	0.0	*	*	0.0	0.0	*	1.0	*	0.4	0.0	1.5
Villa Cisneros	23 42N	15 52W	35	12	71	56	74	60	78	65	80	65	107	48	14	*	*	*	*	0.1	0.0	*	0.2	1.4	0.1	0.2	1.0	3.0
Zaire																												
Kalemie	05 54S	29 12E	2,493	5	85	66	83	67	82	58	87	67	92	50	20	4.2	4.7	6.3	8.4	3.3	0.3	0.1	0.3	0.8	2.8	7.9	6.3	45.4
Kinshasa	04 20S	15 18E	1,066	8	87	70	89	71	81	64	88	70	97	58	12	5.3	5.7	7.7	7.7	6.2	0.3	0.1	0.1	1.2	4.7	8.7	5.6	53.3
Kisangani	00 26N	25 14E	1,370	8	88	69	88	70	84	67	86	68	97	61	14	2.1	3.3	7.0	6.2	5.4	4.5	5.2	6.5	7.2	8.6	7.8	3.3	67.1
Luluabourg	05 54S	22 25E	2,198	3	85	68	86	68	85	63	85	68	94	57	14	5.4	5.6	7.7	7.6	3.3	0.8	0.5	2.3	4.6	6.5	9.1	8.9	62.3
Zambia																												
Balovale	13 34S	23 06E	3,577	8	82	65	84	61	81	47	91	64	108	38	9	8.5	6.9	5.8	1.2	*	0.0	0.0	*	0.3	2.3	4.4	8.9	38.3
Kasama	10 12S	31 11E	4,544	10	79	61	79	60	76	50	87	62	95	39	10	10.7	9.9	10.9	2.8	0.5	*	*	*	*	0.8	6.4	9.5	51.5
Lusaka	15 25S	28 19E	4,191	10	78	63	79	59	73	49	88	64	100	39	10	9.1	7.5	5.6	0.7	0.1	*	*	0.0	*	0.4	3.6	5.9	32.9
Zimbabwe																												
Bulawayo	20 09S	28 37E	4,405	15	81	61	79	56	70	45	85	59	99	28	50	5.6	4.3	3.3	0.7	0.4	0.1	*	*	0.2	0.8	3.2	4.8	23.4
Salisbury	17 50S	31 08E	4,831	15	78	60	78	55	70	44	83	58	95	32	50	7.7	7.0	4.6	1.1	0.5	0.1	*	0.1	0.2	1.1	3.8	6.4	32.6
Atlantic Islands																												
Funchal, Madeira Island	32 38N	16 55W	82	30	66	56	67	58	75	66	74	65	103	40	30	2.5	2.9	3.1	1.3	0.7	0.2	*	*	1.0	3.0	3.5	3.3	21.5
Georgetown, Ascension Island	07 56S	14 25W	55	29	85	73	88	75	84	72	83	71	95	65	45	0.2	0.4	0.7	1.1	0.5	0.5	0.5	0.4	0.3	0.3	0.2	0.1	5.2
Hutts Gate, St. Helena	15 57S	05 40W	2,062	30	68	60	69	61	62	55	61	54	82	50	30	2.1	3.1	4.2	3.1	2.8	3.2	4.3	2.6	2.2	1.7	1.2	1.6	32.1
Las Palmas, Canary Islands	28 11N	15 28W	20	45	70	58	71	61	77	67	79	67	99	46	48	1.4	0.9	0.9	0.5	0.2	*	*	*	0.2	1.1	2.1	1.6	8.6
Porto da Praia, Cape Verde Is	14 54N	23 31W	112	25	77	68	79	69	83	75	85	76	94	56	25	0.1	*	*	*	0.0	*	0.2	3.8	4.5	1.2	0.3	0.1	10.2
Porto da Praia, Cape Verde Is	14 54N	23 31W	112	25	77	68	79	69	83	75	85	76	94	56	25	0.1	*	*	*	0.0	*	0.2	3.8	4.5	1.2	0.3	0.1	10.2
Santa Isabel, Fernando Po	03 46N	08 46E	—	2	87	67	89	70	84	69	86	70	102	61	16	1.3	2.5	4.2	7.2	9.4	11.1	7.4	6.6	9.6	10.4	3.5	1.7	74.9
Sao Tome, Sao Tome	00 20N	06 43E	16	10	86	73	86	73	82	69	84	71	91	56	10	3.2	4.2	5.9	5.0	5.3	1.1	*	*	0.9	4.3	4.6	3.5	38.0
Tristan da Cunha	37 03S	12 19W	75	5	66	59	64	57	57	50	59	51	75	38	5	3.5	3.5	6.4	4.7	7.1	5.9	6.1	6.9	7.9	5.8	4.3	4.0	66.1
Indian Ocean Islands																												
Agalega Island	10 26S	56 40E	10	3	86	77	87	77	83	75	84	75	91	69	2	5.9	10.1	4.9	6.9	13.2	8.9	8.7	3.2	1.8	4.2	7.0	10.0	84.7
Cocos (Keeling) Island	12 05S	96 53E	15	36	86	77	85	78	82	76	84	76	94	68	38	5.4	7.7	8.5	10.4	7.9	9.0	8.7	4.8	3.7	3.3	4.2	4.6	78.2
Heard Island	53 01S	73 23E	16	5	41	35	39	33	34	27	35	28	58	13	5	5.8	5.8	5.7	6.1	5.8	3.9	3.6	2.2	2.5	3.7	4.0	5.1	54.3
Hellburg, Reunion Island	21 04S	55 22E	3,070	5	74	59	73	56	65	48	69	51	84	40	11	22.4	8.0	16.4	7.2	5.3	4.4	3.1	3.0	2.0	2.3	3.5	12.9	90.5
Port Victoria, Seychelles	04 37S	55 27E	15	60	83	76	86	77	81	75	83	75	92	67	64	15.2	10.5	9.2	7.2	6.7	4.0	3.3	2.7	5.1	6.1	9.1	13.4	92.5
Royal Alfred Observatory, Mauritius	20 06S	57 32E	181	40	86	73	82	70	75	62	80	64	95	50	43	8.5	7.8	8.7	5.0	3.8	2.6	2.3	2.5	1.4	1.6	1.8	4.6	50.6
Asia-Far East																												
China																												
Canton	23 10N	113 20E	59	26	65	49	77	65	91	77	85	67	101	31	36	0.9	1.9	4.2	6.8	10.6	10.6	8.1	8.5	6.5	3.4	1.2	0.9	63.6
Chanasha	28 15N	112 58E	161	14	45	35	70	56	94	78	75	59	109	16	26	1.9	3.7	5.3	5.7	8.2	8.7	4.4	4.3	2.7	3.0	2.7	1.5	52.1
Chungking	29 30N	106 33E	855	27	51	42	73	59	93	76	71	61	111	28	60	0.7	0.8	1.5	3.8	5.7	7.1	5.6	4.7	5.8	4.3	1.9	0.8	42.9

(Continued)

Table 3B.13 (Continued)

					Temperature										Average Precipitation													
					Average Daily								Extreme															
					Jan		Apr		July		Oct																	
Country and Station	Latitude	Longitude	Elevation	Length of Record	Maximum	Minimum	Maximum	Minimum	Maximum	Minimum	Maximum	Minimum	Maximum	Minimum	Length of Record	January	February	March	April	May	June	July	August	September	October	November	December	Year
	°	°	Feet	Year	°F	°F	°F	°F	°F	°F	°F	°F	°F	°F	Year	IN.	IN.	IN.	IN.	IN.	IN.	IN.	IN.	IN.	IN.	IN.	IN.	IN.
Hankow	30 35N	114 17E	75	29	46	34	69	55	93	78	74	60	108	9	55	1.8	1.9	3.6	5.8	7.0	9.0	7.0	4.1	3.0	3.1	1.9	1.2	49.4
Harbin	45 45N	126 38E	476	35	7	−14	54	31	84	65	54	31	102	−43	38	0.2	0.2	0.4	0.9	1.7	3.7	6.6	4.7	2.3	1.2	0.5	0.2	22.6
Kashgar	39 24N	76 07E	4,296	27	33	12	71	48	92	68	71	43	106	−15	18	0.6	0.1	0.5	0.2	0.3	0.2	0.4	0.3	0.1	0.1	0.2	0.3	3.2
Kunming	25 02N	102 43E	6,211	32	61	37	76	51	77	62	70	53	91	22	31	0.4	0.5	0.7	0.8	4.3	6.3	8.8	8.6	5.0	3.0	1.7	0.4	40.5
Lanchow	36 06N	103 55E	5,105	8	33	7	65	40	84	61	62	39	100	−3	4	0.2	0.2	0.2	0.5	0.8	0.7	3.3	5.1	2.2	0.6	0.0	0.3	14.1
Mukden	41 47N	123 24E	138	40	20	−2	60	36	87	69	62	39	103	−28	42	0.2	0.2	0.7	1.2	2.6	3.8	7.0	6.3	2.9	1.7	0.9	0.4	28.2
Shanghai	31 12N	121 26E	16	56	47	32	67	49	91	75	75	56	104	10	81	1.9	2.4	3.3	3.6	3.8	7.0	5.8	5.5	5.2	2.9	2.1	1.5	45.0
Tientsin	39 10N	117 10E	13	24	33	16	68	45	90	73	68	48	109	−3	25	0.2	0.1	0.4	0.5	1.1	2.4	7.6	6.0	1.7	0.6	0.4	0.2	21.0
Urumchi	43 45N	87 40E	2,972	6	13	−7	60	36	82	58	50	31	112	−30	6	0.6	0.3	0.5	1.5	1.1	1.5	0.7	1.0	0.6	1.7	1.6	0.4	11.5
Hong Kong	22 18N	114 10E	109	50	64	56	75	67	87	78	81	73	97	32	50	1.3	1.8	2.9	5.4	11.5	15.5	15.0	14.2	10.1	4.5	1.7	1.2	85.1
Japan																												
Kushiro	43 02N	144 12E	315	41	30	8	44	31	66	55	58	40	87	−19	41	1.8	1.4	2.8	3.6	3.8	4.1	4.4	4.9	6.6	4.0	3.1	2.0	42.9
Miyako	39 38N	141 59E	98	30	43	23	58	37	77	62	66	46	99	1	30	2.9	3.0	3.2	3.5	4.5	5.0	5.0	7.2	9.5	6.8	3.0	2.6	56.2
Nagasaki	32 44N	129 53E	436	59	49	36	66	50	85	73	72	58	98	22	59	2.8	3.3	4.9	7.3	6.7	12.3	10.1	6.9	9.8	4.5	3.7	3.2	75.5
Osaka	34 47N	135 26E	49	60	47	32	65	47	87	73	72	55	102	19	60	1.7	2.3	3.8	5.2	4.9	7.4	5.9	4.4	7.0	5.1	3.0	1.9	52.6
Tokyo	35 41N	139 46E	19	60	47	29	63	46	83	70	69	55	101	17	60	1.9	2.9	4.2	5.3	5.8	6.5	5.6	6.0	9.2	8.2	3.8	2.2	61.6
Korea																												
Pusan	35 10N	129 07E	6	36	43	29	62	47	81	71	70	54	97	7	36	1.7	1.4	2.7	5.5	5.2	7.9	11.6	5.1	6.8	2.9	1.6	1.2	53.6
Pyongyang	39 01N	125 49E	94	43	27	8	61	38	84	69	65	43	100	−19	43	0.6	0.4	1.0	1.8	2.6	3.0	9.3	9.0	4.4	1.8	1.6	0.8	36.4
Seoul	37 31N	126 55E	34	22	32	15	62	41	84	70	67	45	99	−12	22	1.2	0.8	1.5	3.0	3.2	5.1	14.8	10.5	4.7	1.6	1.8	1.0	49.2
Mongolia																												
Ulan Bator	47 54N	106 56E	4,287	13	−2	−27	45	18	71	50	44	17	97	−48	15	*	*	0.1	0.2	0.3	1.0	2.9	1.9	0.8	0.2	0.2	0.1	7.7
Taiwan																												
Tainan	22 57N	120 12E	53	13	72	55	82	67	89	77	86	70	95	39	13	0.7	0.7	1.1	3.2	6.3	15.6	16.0	15.8	8.4	1.2	0.9	0.6	70.5
Taipei	25 04N	121 32E	21	12	66	53	77	64	92	76	80	68	101	32	12	3.8	5.3	4.3	5.3	6.9	8.8	8.8	8.7	8.2	5.5	4.2	2.9	72.7
Union of Soviet Socialist Republics																												
Alma-Ata	43 16N	76 53E	2,543	19	23	7	56	38	81	60	55	35	100	−30	27	1.3	0.9	2.2	4.0	3.7	2.6	1.4	1.2	1.0	2.0	1.9	1.3	23.5
Chita	52 02N	113 30E	2,218	10	−10	−27	42	19	75	51	38	18	99	−52	24	0.1	0.1	0.1	0.4	1.1	1.8	3.3	3.3	1.2	0.5	0.2	0.2	12.3
Dubinka	69 07N	87 00E	141	5	−23	−31	6	−10	59	47	19	11	84	−62	5	0.3	0.4	0.2	0.3	0.6	1.9	1.5	2.1	1.8	0.9	0.4	0.3	10.7
Irkutsk	52 16N	104 19E	1,532	10	3	−15	42	20	70	50	41	21	98	−58	38	0.5	0.4	0.3	0.6	1.3	2.2	3.1	2.8	1.7	0.7	0.6	0.6	14.9
Kazalinsk	45 46N	62 06E	207	10	16	5	58	27	90	65	57	35	108	−27	19	0.4	0.4	0.5	0.5	0.6	0.2	0.2	0.3	0.3	0.4	0.5	0.6	4.9
Khabarovsk	48 28N	135 03E	165	7	−2	−13	41	28	75	63	48	34	91	−46	8	0.3	0.2	0.3	0.7	2.0	3.5	4.1	3.3	3.0	0.7	0.6	0.5	19.2
Kirensk	57 47N	108 07E	938	18	−14	−28	38	15	74	51	10	−4	95	−71	19	0.8	0.5	0.5	0.5	1.0	1.8	2.1	2.1	1.7	1.0	1.0	1.0	14.0
Krasnoyarsk	56 01N	92 52E	498	10	3	−10	34	23	67	55	34	26	103	−47	8	0.1	0.2	0.1	0.2	1.0	1.4	1.2	2.1	1.7	0.9	0.5	0.4	9.8
Markovo	64 45N	170 50E	85	15	−19	−29	5	−8	59	47	16	9	84	−72	16	0.2	0.2	0.3	0.1	0.3	0.8	1.0	1.9	1.1	0.4	0.4	0.3	7.0
Narym	58 50N	81 39E	197	13	−7	−18	35	19	71	56	35	25	94	−61	14	0.8	0.5	0.8	0.5	1.3	2.6	2.4	2.7	1.7	1.4	1.1	0.9	16.8
Okhotsk	59 21N	143 17E	18	19	−6	−17	29	10	57	48	33	21	78	−50	25	0.1	0.1	0.2	0.4	0.9	1.6	2.2	2.6	2.4	1.0	0.2	0.1	11.8
Omsk	54 58N	73 20E	279	19	−1	−14	39	21	74	56	40	27	102	−56	22	0.6	0.3	0.3	0.5	1.2	2.0	2.0	2.0	1.1	1.0	0.7	0.8	12.5
Petropavlovsk	52 53N	158 42E	286	7	23	11	35	25	56	47	46	34	84	−29	35	3.0	2.2	3.4	2.5	2.2	2.0	3.1	3.2	3.8	3.9	3.6	3.0	35.9
Salehkard	66 31N	66 35E	60	18	−13	−21	18	4	61	49	26	20	85	−65	27	0.3	0.3	0.3	0.3	0.7	1.3	1.9	2.0	1.5	0.7	0.5	0.4	10.2
Semipalatinsk	50 24N	80 13E	709	10	8	−7	45	26	81	57	46	30	101	−47	10	0.9	0.5	0.5	0.6	1.2	1.5	1.1	1.3	0.7	1.2	1.1	1.0	11.6
Sverdlovsk	56 49N	60 38E	894	21	6	−5	42	26	70	54	37	28	94	−45	29	0.5	0.4	0.5	0.7	1.9	2.7	2.6	2.7	1.6	1.2	1.1	0.8	16.7
Tashkent	41 20N	69 18E	1,569	19	37	21	65	47	92	64	65	41	106	−19	19	2.1	1.1	2.6	2.3	1.4	0.5	0.2	0.1	0.1	1.2	1.5	1.6	14.7
Verkhoyansk	67 34N	133 51E	328	24	−54	−63	19	−10	66	47	12	−3	98	−90	44	0.2	0.2	0.1	0.2	0.3	0.9	1.1	1.0	0.5	0.3	0.3	0.2	5.3
Vladivostok	43 07N	131 55E	94	14	13	0	46	34	71	60	55	41	92	−22	53	0.3	0.4	0.7	1.2	2.1	2.9	3.3	4.7	4.3	1.9	1.2	0.6	23.6
Yakutsk	62 01N	129 43E	535	19	−45	−53	27	6	73	54	23	11	97	−84	22	0.3	0.2	0.1	0.3	0.4	1.1	1.6	1.3	1.1	0.5	0.4	0.3	7.4

Asia-Southeast

			10	5	85	76	87	77	87	76	86	77	99	70	12	14.6	7.6	7.8	9.8	10.9	9.5	9.0	7.3	11.8	14.5	15.2	13.0	131.0
Brunel																												
Brunel	04 55N	114 55E	10	5	85	76	87	77	87	76	86	77	99	70	12	14.6	7.6	7.8	9.8	10.9	9.5	9.0	7.3	11.8	14.5	15.2	13.0	131.0
Burma																												
Mandalay	21 59N	96 06E	252	20	82	55	1.1	77	93	78	89	73	111	44	20	0.1	0.1	0.2	1.2	5.8	6.3	2.7	4.1	5.4	4.3	2.0	0.4	32.6
Moulmein	16 26N	97 39E	150	43	89	65	95	77	83	74	88	75	103	52	60	0.2	0.2	0.5	3.0	19.9	37.1	47.5	44.2	27.1	8.5	1.7	0.3	190.2
Cambodia																												
Phanom Penh	11 33N	104 51E	39	37	88	71	95	76	90	76	87	76	105	55	49	0.3	0.4	1.4	3.1	5.7	5.8	6.0	6.1	8.9	9.9	5.5	1.7	54.8
Indonesia																												
Jakarta	06 11S	106 50E	26	80	84	74	87	75	87	73	87	74	98	66	78	11.8	11.8	8.3	5.8	4.5	3.8	2.5	1.7	2.6	4.4	5.6	8.0	70.8
Manokwari	00 53S	134 03E	10	5	86	73	86	74	86	74	87	74	93	68	40	12.0	9.4	13.2	11.1	7.8	7.2	5.4	5.6	4.9	4.7	6.5	10.3	98.1
Mapanget	01 32N	124 55E	264	21	85	73	86	73	87	73	89	72	97	65	63	18.6	13.8	12.2	8.0	6.4	6.5	4.8	4.0	3.3	4.9	8.9	14.7	106.1
Penfui	10 10S	123 39E	335	21	87	75	89	72	88	70	92	72	101	58	63	15.2	13.7	9.2	2.6	1.2	0.4	0.2	0.0	0.0	0.7	3.3	9.1	55.7
Pontianak	00 00N	109 20E	13	20	87	74	89	75	89	74	89	75	96	68	63	10.8	8.2	9.5	10.9	11.1	8.7	6.5	8.0	9.0	14.4	15.3	12.7	125.1
Tabing	00 52S	100 21E	19	21	87	74	87	75	87	74	86	74	94	68	63	13.9	10.1	12.2	14.5	12.8	11.7	10.5	13.7	16.2	20.1	20.5	19.2	175.4
Tarakan	03 19N	117 33E	20	19	85	73	86	75	87	74	87	74	94	67	31	10.9	10.2	14.0	13.9	13.5	12.6	10.3	12.4	11.6	14.3	15.2	13.4	152.3
Laos																												
Vientiane	17 58N	102 34E	559	13	83	58	95	73	89	75	88	71	108	32	27	0.2	0.6	1.5	3.9	10.5	11.9	10.5	11.5	11.9	4.3	0.6	0.1	67.5
Malaysia																												
Kuala Lumpur	03 06N	101 42E	111	19	90	72	91	74	90	72	89	73	99	64	19	6.2	7.9	10.2	11.5	8.8	5.1	3.9	6.4	8.6	9.8	10.2	7.5	96.1
North Borneo																												
Sandakan	05 54N	118 03E	38	45	85	74	89	76	89	75	88	75	99	70	46	19.0	10.9	8.6	4.5	6.2	7.4	6.7	7.9	9.3	10.2	14.5	18.5	123.7
Philippine Islands																												
Davao	07 07N	125 38E	88	15	87	72	91	73	88	73	89	73	97	65	34	4.8	4.5	5.2	5.8	9.2	9.1	6.5	6.5	6.7	7.9	5.3	6.1	77.6
Manila	14 31N	121 00E	49	61	86	69	93	73	88	75	88	74	101	58	75	0.9	0.5	0.7	1.3	5.1	10.0	17.0	16.6	14.0	7.6	5.7	2.6	82.0
Sarawak:																												
Kuching	01 29N	110 20E	85	5	85	72	90	73	90	72	89	73	98	64	19	24.0	20.1	12.9	11.0	10.3	7.1	7.7	9.2	8.6	10.5	14.1	18.2	153.7
Singapore																												
Singapore	01 18N	103 50E	33	39	86	73	88	75	88	75	87	74	97	66	64	9.9	6.8	7.6	7.4	6.8	6.8	6.7	7.7	7.0	8.2	10.0	10.1	95.0
Thailand																												
Bangkok	13 44N	100 30E	53	10	89	67	95	78	90	76	88	76	104	50	10	0.2	1.1	1.1	2.3	5.2	6.0	6.9	9.2	14.0	9.9	1.8	0.1	57.8
Viet Nam Hanoi	21 03N	105 52E	20	12	68	58	80	70	92	79	84	72	108	41	12	0.8	1.2	2.5	3.6	4.1	11.2	11.9	15.2	10.0	3.5	2.6	2.8	69.4
Saigon	10 49N	106 39E	33	31	89	70	95	76	88	75	88	74	104	57	33	0.6	0.1	0.5	1.7	8.7	13.0	12.4	10.6	13.2	10.6	4.5	2.2	78.1

Asia-Middle East

Aden																												
Riyan	14 39N	49 19E	83	13	82	67	88	74	92	77	88	72	111	57	13	0.3	0.1	0.6	0.2	*	0.1	0.1	0.1	*	*	0.7	0.3	2.5
Afghanistan																												
Kabul	34 30N	69 13E	5,955	9	36	18	66	43	92	61	73	42	104	−6	45	1.3	1.5	3.6	3.3	0.9	0.2	0.1	0.1	*	0.4	0.6	0.6	12.6
Kandhar	31 36N	65 40E	3,462	7	56	31	83	50	102	66	85	44	111	14	7	3.1	1.7	0.8	0.3	0.2	*	0.1	*	0.0	*	*	0.8	7.0
Bangladesh																												
Dacca	23 46N	90 23E	24	60	77	56	92	74	88	79	88	75	108	43	61	0.3	1.2	2.4	5.4	9.6	12.4	13.0	13.3	9.8	5.3	1.0	0.2	73.9
India																												
Ahmadabad	23 03N	72 37E	180	45	85	58	104	75	93	79	97	73	118	36	45	*	0.1	0.1	*	0.4	3.7	12.2	8.1	4.2	0.4	0.1	*	29.3
Bangalore	12 57N	77 40E	2,937	60	80	57	93	69	81	66	82	65	102	46	60	0.2	0.3	0.4	1.6	4.2	2.9	3.9	5.0	6.7	5.9	2.7	0.4	34.2
Bombay	19 06N	72 51E	27	60	88	62	93	74	88	75	93	73	110	46	60	0.1	0.1	0.1	*	0.7	19.1	24.3	13.4	10.4	2.5	0.5	0.1	71.2
Calcutta	22 32N	88 20E	21	60	80	55	97	76	90	79	89	74	111	44	60	0.4	1.2	1.4	1.7	5.5	11.7	12.8	12.9	9.9	4.5	0.8	0.2	63.0
Cherrapunji	25 15N	91 44E	4,309	35	60	46	71	59	72	65	72	61	87	33	35	0.7	2.1	7.3	26.2	50.4	6.1	96.3	70.1	43.3	19.4	2.7	0.5	425.1
Hyderabad	17 27N	78 28E	1,741	50	85	59	101	75	87	73	88	68	112	43	45	0.3	0.4	0.5	1.2	1.1	4.4	6.0	5.3	6.5	2.5	1.1	0.3	29.6
Jalpaiguri	26 32N	88 43E	272	50	74	50	90	68	89	77	87	70	104	36	55	0.3	0.7	1.3	3.7	11.8	25.9	32.2	25.3	21.2	5.6	0.5	0.2	128.7
Lucknow	26 45N	80 52E	400	60	74	47	101	71	92	80	91	67	119	34	60	0.8	0.7	0.3	0.3	0.8	4.5	12.0	11.5	7.4	1.3	0.2	0.3	40.1
Madras	13 04N	80 15E	51	60	85	67	95	78	96	79	90	75	113	57	60	1.4	0.4	0.3	0.6	1.0	1.9	3.6	4.6	4.7	12.0	14.0	5.5	50.0
Mormugao	15 22N	73 49E	157	10	86	70	88	79	83	75	86	75	98	59	30	*	*	*	0.7	2.6	29.6	31.2	15.9	9.5	3.8	1.3	0.2	94.8
New Delhi	28 35N	77 12E	695	10	71	43	97	68	95	80	93	64	115	31	75	0.9	0.7	0.5	0.3	0.5	2.9	7.1	6.8	4.6	0.4	0.1	0.4	25.2
Silchar	24 49N	92 48E	95	60	78	52	88	69	90	77	88	72	103	41	53	0.8	2.1	7.9	14.3	15.6	21.7	19.7	19.7	14.4	6.5	1.4	0.4	124.5
Indian Ocean Islands																												
Port Blair, Andaman Is	11 40N	92 43E	261	60	84	72	89	75	84	75	84	74	97	62	60	1.8	1.1	1.1	2.4	15.1	21.7	15.4	16.3	17.4	12.5	10.5	7.9	123.2

(Continued)

Table 3B.13 (Continued)

Country and Station	Latitude	Longitude	Elevation	Length of Record	Temperature: Average Daily: Jan Maximum	Jan Minimum	Apr Maximum	Apr Minimum	July Maximum	July Minimum	Oct Maximum	Oct Minimum	Extreme Maximum	Extreme Minimum	Average Precipitation: Length of Record	January	February	March	April	May	June	July	August	September	October	November	December	Year
	°	°	Feet	Year	°F	°F	°F	°F	°F	°F	°F	°F	°F	°F	Year	IN.	IN.	IN.	IN.	IN.	IN.	IN.	IN.	IN.	IN.	IN.	IN.	IN.
Amini Divi, Laccadive Is	11 07N	72 44E	13	29	86	74	92	80	86	77	86	77	99	65	30	0.7	*	*	1.5	3.7	14.3	12.0	7.7	6.3	5.8	2.6	1.3	56.0
Minicoy, Maldive Is	08 18N	73 00E	9	20	85	73	87	80	85	76	85	76	98	63	50	1.8	0.7	0.9	2.3	7.0	11.6	8.9	7.8	6.3	7.3	5.5	3.4	63.5
Car Nicobar, Nicobar Is	09 09N	92 49E	47	13	86	77	90	77	86	77	85	75	95	66	30	3.9	1.2	2.1	3.5	12.5	12.4	9.3	10.2	12.9	11.6	11.4	7.8	98.8
Iran																												
Abadan	30 21N	48 13E	10	12	64	44	90	62	112	81	98	63	127	24	10	1.5	1.7	0.6	0.8	0.1	0.0	0.0	0.0	0.0	0.1	1.0	1.8	7.6
Isfahan	32 37N	51 41E	5,238	45	47	25	72	46	98	67	78	47	108	−4	45	0.7	0.6	0.8	0.6	0.3	*	0.1	*	*	0.1	0.4	0.7	4.4
Kermanshah	34 19N	47 07E	4,331	15	45	23	68	38	99	56	79	38	108	−3	15	2.6	2.3	2.8	2.2	1.6	*	*	*	*	0.4	2.0	2.4	16.4
Rezaiyeh	37 32N	45 05E	4,364	3	32	17	67	45	91	64	67	47	99	−11	3	1.9	2.3	2.0	1.7	1.2	0.5	*	0.1	0.2	1.5	0.8	1.6	13.8
Tehran	35 41N	51 19E	3,937	24	45	27	71	49	99	72	76	53	109	−5	33	1.8	1.5	1.8	1.4	0.5	0.1	0.1	0.1	0.1	0.3	0.8	1.2	9.7
Iraq																												
Baghdad	33 20N	44 24E	111	15	60	39	85	57	110	76	92	61	121	18	15	0.9	1.0	1.1	0.5	0.1	*	*	*	*	0.1	0.8	1.0	5.5
Basra	30 34N	47 47E	8	10	64	45	85	63	104	81	94	64	123	24	10	1.4	1.1	1.2	1.2	0.2	0.0	*	*	*	*	1.4	0.8	7.3
Mosul	36 19N	43 09E	730	26	54	35	77	49	109	72	88	51	124	12	29	2.8	3.1	2.1	1.9	0.7	*	*	*	*	0.2	1.9	2.4	15.2
Israel																												
Haifa	32 48N	35 02E	23	16	65	49	77	58	88	75	85	68	112	27	30	6.9	4.3	1.6	1.0	0.2	*	*	*	0.1	1.0	3.7	7.3	26.2
Jerusalem	31 47N	35 13E	2,654	19	55	41	73	50	87	63	81	59	107	26	50	5.1	4.7	2.9	0.9	0.1	*	0.0	0.0	*	0.3	2.2	3.5	19.7
Tel Aviv	32 06N	34 46E	33	10	64	50	70	57	82	72	79	65	102	34	10	4.9	2.7	2.0	0.7	0.1	0.0	0.0	0.0	0.1	0.4	4.1	6.1	21.1
Jammu/Kashmir																												
Srinagar	33 58N	74 46E	5,458	50	41	24	67	45	88	64	74	41	106	−4	50	2.9	2.8	3.6	3.7	2.4	1.4	2.3	2.4	1.5	1.2	0.4	1.3	25.9
Jordan																												
Amman	31 58N	35 59E	2,547	25	54	39	73	49	89	65	81	57	109	21	25	2.7	2.9	1.2	0.6	0.2	0.0	0.0	0.0	*	0.2	1.3	1.8	10.9
Kuwait																												
Kuwait	29 21N	48 00E	16	14	61	49	83	68	103	86	91	73	119	33	10	0.9	0.9	1.1	0.2	*	0.0	0.0	0.0	0.0	0.1	0.6	1.1	5.1
Lebanon																												
Beirut	33 54N	35 28E	111	62	62	51	72	58	87	73	81	69	107	30	71	7.5	6.2	3.7	2.2	0.7	0.1	*	*	0.2	2.0	5.2	7.3	35.1
Nepal																												
Katmandu	27 42N	85 22E	4,423	27	65	36	84	53	84	69	80	56	99	27	9	0.6	1.6	0.9	2.3	4.8	9.7	14.7	13.6	6.1	1.5	0.3	0.1	56.2
Oman and Muscat																												
Muscat	23 37N	58 35E	15	23	77	66	90	78	97	87	93	80	116	51	38	1.1	0.7	0.4	0.4	*	0.1	*	*	0.0	0.1	0.4	0.7	3.9
Pakistan																												
Karachi	24 48N	66 59E	13	43	77	55	90	73	91	81	91	72	118	39	59	0.5	0.4	0.3	0.1	0.1	0.7	3.2	1.6	0.5	0.1	0.1	0.2	7.8
Multan	30 11N	71 25E	400	60	68	42	95	68	102	86	94	64	122	29	60	0.4	0.4	0.4	0.3	0.3	0.6	2.0	1.8	0.5	0.1	0.1	0.2	7.1
Rawalpindi	33 35N	73 03E	1,676	60	62	38	86	59	98	77	89	57	118	25	60	2.5	2.5	2.7	1.9	1.3	2.3	8.1	9.2	3.9	0.6	0.3	1.2	36.5
Saudi Arabia																												
Dhahran	26 16N	50 10E	78	10	69	54	90	69	107	86	95	73	120	40	10	1.1	0.6	0.4	0.2	0.1	0.0	0.0	0.0	0.0	0.0	0.2	0.9	3.5
Jidda	21 28N	39 10E	20	5	84	66	91	70	99	79	95	73	117	49	5	0.2	*	*	*	*	0.0	*	*	*	*	1.0	1.2	2.5
Riyadh	24 39N	46 42E	1,938	3	70	46	89	64	107	78	94	61	113	19	3	0.1	0.8	0.9	1.0	0.4	*	0.0	*	0.0	0.0	*	*	3.2
Sri Lanka																												
Colombo	06 54N	79 52E	22	25	86	72	88	76	85	77	85	75	99	59	40	3.5	2.7	5.8	9.1	14.6	8.8	5.3	4.3	6.3	13.7	12.4	5.8	92.3

Syria																												
Deir Ez Zor	35 21N	40 09E	699	5	53	35	80	52	105	78	86	56	114	16	8	1.6	0.8	0.3	0.8	0.1	*	0.0	0.0	0.0	0.2	1.5	0.9	6.2
Damascus	33 30N	36 20E	2,362	13	53	36	75	49	96	64	81	54	113	21	7	1.7	1.7	0.3	0.5	0.1	*	*	0.0	0.7	0.4	1.6	1.6	8.6
Aleppo	36 14N	37 08E	1,280	8	50	34	75	48	97	69	81	54	117	9	10	3.5	2.5	1.5	1.1	0.3	0.1	0.0	*	*	1.0	2.2	3.3	15.5
Turkey																												
Adana	36 59N	35 18E	82	21	57	39	74	51	93	71	84	58	109	19	31	4.3	4.0	2.5	1.6	2.0	0.7	0.2	0.2	0.7	1.9	2.4	3.8	24.3
Ankara	39 57N	32 53E	2,825	26	39	24	63	40	86	59	69	44	104	−13	24	1.3	1.2	1.3	1.3	1.9	1.0	0.5	0.4	0.7	0.9	1.2	1.9	13.6
Erzurum	39 54N	41 16E	6,402	16	24	8	50	32	78	53	59	37	93	−22	16	1.4	1.6	2.0	2.5	3.1	2.1	1.3	0.9	1.1	2.3	1.8	1.1	21.2
Izmir	38 27N	27 15E	92	39	55	39	70	49	92	69	76	55	108	12	58	4.4	3.3	3.0	1.7	1.3	0.6	0.2	0.2	0.8	2.1	3.3	4.8	25.5
Samsun	41 17N	36 19E	131	24	50	38	59	45	79	65	69	56	103	20	27	2.9	2.6	2.7	2.3	1.8	1.5	1.5	1.3	2.4	3.2	3.5	2.4	29.1
United Arab Emirates																												
Sharjah	25 20N	55 24E	18	11	74	54	86	65	100	82	92	71	118	37	12	0.9	0.9	0.4	0.2	0.0	0.0	0.0	0.0	0.0	0.0	0.4	1.4	4.2
Yemen																												
Kamaran I	15 20N	42 37E	20	26	82	74	89	79	98	85	93	82	105	66	21	0.2	0.2	0.1	0.1	0.1	*	0.5	0.7	0.1	0.1	0.4	0.9	3.4
Australia & Pacific Islands																												
Australia																												
Adelaide	34 57S	138 32E	20	86	86	61	73	55	59	45	73	51	118	32	104	0.8	0.7	1.0	1.8	2.7	3.0	2.6	2.6	2.1	1.7	1.1	1.0	21.1
Alice Springs	23 48S	133 53E	1,791	62	97	70	81	54	67	39	88	58	111	19	30	1.7	1.3	1.1	0.4	0.6	0.5	0.3	0.3	0.3	0.7	1.2	1.5	9.9
Bourke	30 05S	145 58E	361	63	99	70	82	55	65	40	85	56	125	25	72	1.4	1.5	1.1	1.1	1.0	1.1	0.9	0.8	0.8	0.9	1.2	1.4	13.2
Brisbane	27 25S	153 05E	17	53	85	69	79	61	68	49	80	60	110	35	91	6.4	6.3	5.7	3.7	2.8	2.6	2.2	1.9	1.9	2.5	3.7	5.0	44.7
Broome	17 57S	122 13E	56	41	92	79	93	72	82	58	91	72	113	40	50	6.3	5.8	3.9	1.2	0.6	0.9	0.2	0.1	*	*	0.6	3.3	22.9
Burketown	17 45S	139 33E	30	31	93	77	91	69	82	55	93	70	110	40	53	8.2	6.3	5.2	1.0	0.2	0.3	*	*	*	0.4	1.5	4.4	27.5
Canberra	35 18S	149 11E	1,886	23	82	55	67	44	52	33	68	43	109	14	25	1.9	1.7	2.2	1.6	1.8	2.1	1.8	2.2	1.6	2.2	1.9	2.0	23.0
Carnarvon	24 53S	113 40E	13	43	88	72	84	66	71	51	78	61	118	37	57	0.4	0.7	0.7	0.6	1.5	2.4	1.6	0.7	0.2	0.1	*	0.2	9.1
Cloncurry	20 40S	140 30E	622	32	99	77	90	67	77	51	95	68	127	35	59	4.4	4.2	2.4	0.7	0.5	0.6	0.3	0.1	0.3	0.5	1.3	2.7	18.0
Esperance	33 50S	121 55E	14	44	77	60	72	54	62	45	68	50	117	31	60	0.7	0.7	1.2	1.8	3.3	4.1	4.0	3.8	2.7	2.2	1.0	0.9	26.4
Laverton	28 40S	122 23E	1,510	30	96	69	81	57	64	41	82	55	115	25	30	0.8	0.8	1.6	0.8	0.9	0.7	0.6	0.5	0.2	0.3	0.8	0.8	8.8
Melbourne	37 49S	144 58E	115	88	78	57	68	51	56	42	67	48	114	27	88	1.9	1.8	2.2	2.3	2.1	2.1	1.9	1.9	2.3	2.6	2.3	2.3	25.7
Mundiwindi	23 52S	120 10E	1,840	15	101	64	87	61	70	41	89	58	112	22	15	1.0	1.9	2.0	0.8	0.6	0.9	0.1	0.3	0.3	0.5	0.5	1.2	10.1
Perth	31 56S	115 58E	64	44	85	63	76	57	63	48	70	53	112	31	63	0.3	0.4	0.8	1.7	5.1	7.1	6.7	5.7	3.4	2.2	0.8	0.5	34.7
Port Darwin	12 25S	130 52E	104	58	90	77	92	76	87	67	93	77	105	55	70	15.2	12.3	10.0	3.8	0.6	0.1	*	0.1	0.5	2.0	4.7	9.4	58.7
Sydney	33 52S	151 02E	62	87	78	65	71	58	60	46	71	56	114	35	87	3.5	4.0	5.0	5.3	5.0	4.6	4.6	3.0	2.9	2.8	2.9	2.9	46.5
Thursday Island	10 35S	142 13E	200	31	87	77	86	77	82	73	86	76	98	64	49	18.2	15.8	13.9	8.0	1.6	0.5	0.4	0.2	0.1	0.3	1.5	7.0	67.5
Townsville	19 15S	146 46E	18	31	87	76	84	70	75	59	83	71	110	39	67	10.9	11.2	7.2	3.3	1.3	1.4	0.6	0.5	0.7	1.3	1.9	5.4	45.7
William Creek	28 55S	136 21E	247	39	96	69	80	55	65	41	84	56	119	25	30	0.5	0.6	0.3	0.3	0.3	0.5	0.2	0.3	0.3	0.5	0.5	0.7	5.0
Windorah	25 26S	142 36E	390	29	101	74	86	59	70	43	91	61	116	26	50	1.4	1.6	1.6	0.9	0.8	0.8	0.5	0.4	0.5	0.6	0.9	1.4	11.4
Tasmania																												
Hobart	42 53S	147 20E	177	70	71	53	63	48	52	40	63	46	105	28	100	1.9	1.5	1.8	1.9	1.8	2.2	2.1	1.9	2.1	2.3	2.4	2.1	24.0
New Zealand																												
Auckland	37 00S	174 47E	23	36	73	60	67	56	56	46	63	52	90	33	92	3.1	3.7	3.2	3.8	5.0	5.4	5.7	4.6	4.0	4.0	3.5	3.1	49.1
Christchurch	43 29S	172 32E	118	52	70	53	62	45	50	35	62	44	96	21	64	2.2	1.7	1.9	1.9	2.6	2.6	2.7	1.9	1.8	1.7	1.9	2.2	25.1
Dunedin	45 55S	170 12E	4	77	66	50	59	45	48	37	59	42	94	23	77	3.4	2.8	3.0	2.8	3.2	3.2	3.1	3.0	2.7	3.0	3.2	3.5	36.9
Wellington	41 17S	174 46E	415	66	69	56	63	51	53	42	60	48	88	29	79	3.2	3.2	3.2	3.8	4.6	4.6	5.4	4.6	3.8	4.0	3.5	3.5	47.4
Pacific Islands																												
Canton, Phoenix Is	02 46S	171 43W	9	12	88	78	89	78	89	78	90	78	98	70	30	2.6	2.2	2.5	3.6	4.3	2.6	2.6	2.5	1.2	1.1	1.6	2.6	29.4
Guam, Marianas Is	13 33N	144 50E	361	30	84	72	86	73	87	72	86	73	95	54	30	4.6	3.5	2.6	3.0	4.2	5.9	9.0	12.8	13.4	13.1	10.3	6.1	88.5
Honolulu, Hawaii	21 20N	157 55W	7	30	79	66	80	68	85	73	84	72	93	56	30	3.8	3.3	2.9	1.3	1.0	0.3	0.4	0.9	1.0	1.8	2.2	3.0	21.9
Iwo Jima, Bonin Is	24 47N	141 19E	353	15	71	64	77	69	86	78	84	76	95	46	17	3.2	2.5	2.1	3.7	4.9	4.0	6.4	6.5	4.6	5.9	4.8	4.3	52.8
Madang, New Guinea	05 12S	145 47E	19	12	87	75	88	74	88	74	88	75	98	62	20	12.1	11.9	14.9	16.9	15.1	10.8	7.6	4.8	5.3	10.0	13.3	14.5	137.2
Midway Is	28 13N	177 23W	29	21	69	62	71	64	81	74	79	72	92	46	20	4.6	3.7	3.1	2.5	1.9	1.3	2.9	3.9	3.7	3.7	3.6	4.2	40.7
Naha, Okinawa	26 12N	127 39F	96	30	67	56	76	64	89	77	81	69	96	41	30	5.3	5.4	6.1	6.1	8.9	10.0	7.1	10.0	7.1	6.6	5.9	4.3	82.8
Noumea, New Caledonia	22 16S	166 27E	246	24	86	72	83	70	76	62	80	65	99	52	52	3.7	5.1	5.7	5.2	4.4	3.7	3.6	2.6	2.5	2.0	2.4	2.6	43.5
Pago Pago, Samoa	14 19S	170 43W	29	2	87	75	87	76	83	74	85	75	98	67	41	24.5	20.5	19.2	16.5	15.4	12.3	10.0	8.2	13.1	14.9	19.2	19.8	193.6
Ponape, Caroline Is	06 58N	158 13E	123	30	86	75	86	75	87	73	87	72	96	67	30	11.1	9.7	14.6	20.0	20.3	16.7	16.2	16.3	15.8	16.0	16.9	18.3	191.9

(Continued)

Table 3B.13 (Continued)

					Temperature										Average Precipitation													
					Average Daily								Extreme															
					Jan		Apr		July		Oct																	
Country and Station	Latitude	Longitude	Elevation	Length of Record	Maximum	Minimum	Maximum	Minimum	Maximum	Minimum	Maximum	Minimum	Maximum	Minimum	Length of Record	January	February	March	April	May	June	July	August	September	October	November	December	Year
	°	°	Feet	Year	°F	°F	°F	°F	°F	°F	°F	°F	°F	°F	Year	IN.	IN.	IN.	IN.	IN.	IN.	IN.	IN.	IN.	IN.	IN.	IN.	IN.
Port Moresby, New Guinea	09 29S	147 09E	126	20	89	76	87	75	83	73	86	75	98	64	38	7.0	7.6	6.7	4.2	2.5	1.3	1.1	0.7	1.0	1.4	1.9	4.4	39.8
Rabaul, New Guinea	04 13S	152 11E	28	19	90	73	90	73	89	73	92	73	100	65	24	14.8	10.4	10.2	10.0	5.2	3.3	5.4	3.7	3.5	5.1	7.1	10.1	88.8
Suva, Fiji Is	18 08S	178 26E	20	43	86	74	84	73	79	68	81	70	98	55	43	11.4	10.7	14.5	12.2	10.1	6.7	4.9	8.3	7.7	8.3	9.8	12.5	117.1
Tahiti, Society Is	17 33S	149 36W	7	23	89	72	89	72	86	68	87	70	93	61	27	13.2	11.5	6.5	6.8	4.9	3.2	2.6	1.9	2.3	3.4	6.5	11.9	74.7
Tulagi, Solomon Is	09 05S	160 10E	8	20	88	76	88	76	86	76	87	76	96	68	37	14.3	15.8	15.0	10.8	8.1	6.8	7.6	8.7	8.0	8.7	10.0	10.4	123.4
Wake Is	19 17N	166 39E	11	30	82	73	83	74	87	77	86	77	92	64	30	1.1	1.4	1.5	1.9	2.0	1.9	4.6	7.1	5.2	5.3	3.1	1.8	36.9
Yap, Caroline Is	9 31N	138 08E	62	30	85	76	87	77	88	75	88	75	97	69	30	7.9	4.6	5.4	6.4	9.5	10.7	13.8	14.7	14.0	13.2	11.2	10.2	121.6
Antarctica																												
Byrd Station	80 01S	119 32W	5,095	6	10	−2	−11	−30	−25	−45	−15	−33	31	−82	6	0.4	0.4	0.2	0.3	0.4	0.5	0.7	0.7	0.3	0.7	0.0	0.3	4.9
Elisworth	77 44S	41 07W	139	6	22	12	−10	−25	−21	−35	−2	−15	36	−70	6	0.3	0.2	0.3	0.6	0.2	0.2	0.2	0.2	0.3	0.4	0.5	0.2	3.6
McMurdo Station	77 53S	166 48W	8	10	30	21	−1	−13	−9	−24	2	−12	42	−59	10	0.5	0.7	0.4	0.4	0.4	0.3	0.2	0.3	0.4	0.2	0.2	0.3	4.3
South Pole Station	89 59S	000 00W	9,186	5	−16	−23	−66	−79	−67	−81	−55	−64	6	−107	5	*	0.1	0.0	0.0	0.0	0.0	0.0	0.0	0.0	*	0.0	*	0.1
Wilkes	66 16S	110 31E	31	7	34	28	17	9	8	−3	16	6	46	−35	7	0.5	0.4	1.7	1.1	1.4	1.2	1.3	0.8	1.5	1.2	0.8	0.3	12.2

Note: 1. "Length of Record" refers to average daily maximum and minimum temperatures and precipitation. A standard period of the 30 years from 1931–1960 had been used for locations in the United States and some other countries. The length of record of extreme maximum and minimum temperatures includes all available years of data for a given location and is usually for a longer period. 2. * = Less than 0.05″. 3. Except for Antarctica, amounts of solid precipitation such as snow or hail have been converted to their water equivalent. Because of the frequent occurrence of blowing snow, it has not been possible to determine the precise amount of precipitation actually falling in Antarctica. The values shown are the average amounts of solid snow accumulating in a given period as determined by snow markers. The liquid content of the accumulation is undetermined.

Source: From Environmental Science Services Administration, *Climates of the World*, 1969. Geographic names revised by editors in accordance with 1987 usage.

SECTION 3C WEATHER EXTREMES

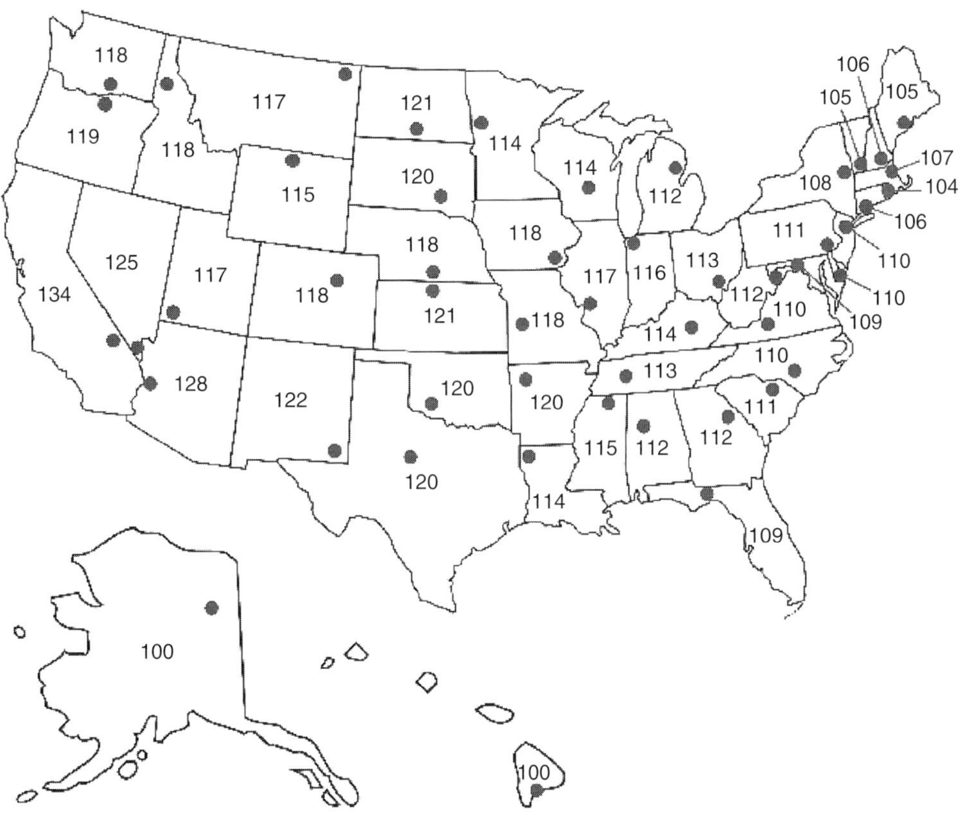

Figure 3C.5 Record highest temperature (°F) (through 2000). (From U.S. National Oceanic and Atmospheric Administration, *Comparative Climatic Data for the United States Through 2000*. www.noaa.gov.)

Table 3C.14 Record Highest Temperatures by State

State	Temperature (°F)	Date	Station	Elevation (ft)
Alabama	112	Sep 5, 1925	Centerville	345
Alaska	100	Jun 27, 1915	Fort Yukon	est. 420
Arizona	128	Jun 29, 1994	Lake Havasu City	505
Arkansas	120	Aug 10, 1936	Ozark	396
California	134	Jul 10, 1913	Greenland Ranch	−178
Colorado	118	Jul 11, 1888	Bennett	5,484
Connecticut	106	Jul 15, 1995	Danbury	450
Delaware	110	Jul 21, 1930	Millsboro	20
Florida	109	Jun 29, 1931	Monticello	207
Georgia	112	Aug 20, 1983	Grenville	860
Hawaii	100	Apr 27, 1931	Pahala	850
Idaho	118	Jul 28, 1934	Orofino	1,027
Illinois	117	Jul 14, 1954	East St. Louis	410
Indiana	116	Jul 14, 1936	Collegeville	672
Iowa	118	Jul 20, 1934	Keokuk	614
Kansas	121	Jul 24, 1936[a]	Alton (Near)	1,651
Kentucky	114	Jul 28, 1930	Greensburg	581
Louisiana	114	Aug 10, 1936	Plain Dealing	268
Maine	105	Jul 10, 1911	North Bridgton	450
Maryland	109	Jul 10, 1936[a]	Cumberland & Frederick	623; 325
Massachusetts	107	Aug 2, 1975	New Bedford & Chester	120; 640
Michigan	112	Jul 13, 1936	Mio	963
Minnesota	114	Jul 6, 1936[a]	Moorhead	904
Mississippi	115	Jul 29, 1930	Holly Springs	600
Missouri	118	Jul 14, 1954[a]	Warsaw & Union	705; 560
Montana	117	Jul 5, 1937	Medicine Lake	1,950
Nebraska	118	Jul 24, 1936[a]	Minden	2,169
Nevada	125	Jun 29, 1994[a]	Laughlin	605
New Hampshire	106	Jul 4, 1911	Nashua	125
New Jersey	110	Jul 10, 1936	Runyon	18
New Mexico	122	Jun 27, 1994	Waste Isolat. Pilot Plt	3,418
New York	108	Jul 22, 1926	Troy	35
North Carolina	110	Aug 21, 1983	Fayetteville	213
North Dakota	121	Jul 6, 1936	Steele	1,857
Ohio	113	Jul 21, 1934[a]	Gallipolis (Near)	673
Oklahoma	120	Jun 27, 1994[a]	Tipton	1,350
Oregon	119	Aug 10, 1898[a]	Pendleton	1,074
Pennsylvania	111	Jul 10, 1936[a]	Phoenixville	100
Rhode Island	104	Aug 2, 1975	Providence	51
South Carolina	111	Jun 28, 1954[a]	Camden	170
South Dakota	120	Jul 5, 1936	Gannvalley	1,750
Tennessee	113	Aug 9, 1930[a]	Perryville	377
Texas	120	Jun 28, 1994[a]	Monahans	2,660
Utah	117	Jul 5, 1985	Saint George	2,880
Vermont	105	Jul 4, 1911	Vernon	310
Virginia	110	Jul 15, 1954	Balcony Falls	725
Washington	118	Aug 5, 1961	Ice Harbor Dam	475
West Virginia	112	Jul 10, 1936[a]	Martinsburg	435
Wisconsin	114	Jul 13, 1936	Wisconsin Dells	900
Wyoming	115	Aug 8, 1983	Basin	3,500

[a] Also on earlier dates at the same or other places.

Source: From U.S. National Oceanic and Atmospheric Administration, *Comparative Climatic Data for the United States Through 2000*, www.noaa.gov.

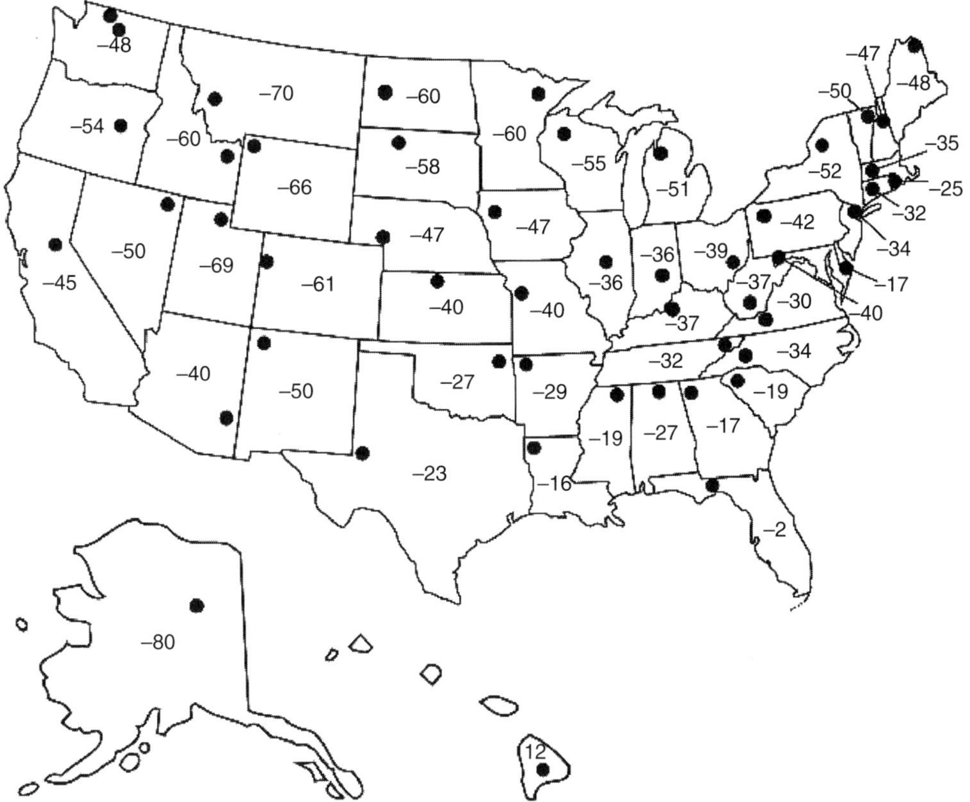

Figure 3C.6 Record lowest temperature (°F) (through 2000). (From U.S. National Oceanic and Atmospheric Administration, *Comparative Climatic Data for the United States Through 2000.* www.noaa.gov.)

Table 3C.15 Record Lowest Temperatures by State

State	Temperature (°F)	Date	Station	Elevation (ft)
Alabama	−27	Jan 30, 1966	New Market	760
Alaska	−80	Jan 23, 1971	Prospect Creek Camp	1,100
Arizona	−40	Jan 7, 1971	Hawley Lake	8,180
Arkansas	−29	Feb 13, 1905	Pond	1,250
California	−45	Jan 20, 1937	Boca	5,532
Colorado	−61	Feb 1, 1985	Maybell	5,920
Connecticut	−32	Jan 22, 1961[a]	Coventry	480
Delaware	−17	Jan 17, 1893	Millsboro	20
Florida	−2	Feb 13, 1899	Tallahassee	193
Georgia	−17	Jan 27, 1940	CCC Camp F-16	est. 1,000
Hawaii	12	May 17, 1979	Mauna Kea Obs 111.2	13,770
Idaho	−60	Jan 18, 1943	Island Park Dam	6,285
Illinois	−36	Jan 5, 1999	Congerville	635
Indiana	−36	Jan 19, 1994	New Whiteland	785
Iowa	−47	Feb 3, 1996[a]	Elkader	770
Kansas	−40	Feb 13, 1905	Lebanon	1,812
Kentucky	−37	Jan 19, 1994	Shelbyville	730
Louisiana	−16	Feb 13, 1899	Minden	194
Maine	−48	Jan 19, 1925	Van Buren	510
Maryland	−40	Jan 13, 1912	Oakland	2,461
Massachusetts	−35	Jan 12, 1981	Chester	640
Michigan	−51	Feb 9, 1934	Vanderbilt	785
Minnesota	−60	Feb 2, 1996	Tower	1,460
Mississippi	−19	Jan 30, 1966	Corinth	420
Missouri	−40	Feb 13, 1905	Warsaw	700
Montana	−70	Jan 20, 1954	Rogers Pass	5,470
Nebraska	−47	Dec 22, 1989[a]	Oshkosh	3,379
Nevada	−50	Jan 8, 1937	San Jacinto	5,200
New Hampshire	−47	Jan 29, 1934	Mt. Washington	6,262
New Jersey	−34	Jan 5, 1904	River Vale	70
New Mexico	−50	Feb 1, 1951	Gavilan	7,350
New York	−52	Feb 18, 1979[a]	Old Forge	1,720
North Carolina	−34	Jan 21, 1985	Mt. Michell	6,525
North Dakota	−60	Feb 15, 1936	Parshall	1,929
Ohio	−39	Feb 10, 1899	Milligan	800
Oklahoma	−27	Jan 18, 1930[a]	Watts	958
Oregon	−54	Feb 10, 1933[a]	Seneco	4,700
Pennsylvania	−42	Jan 5, 1904	Smethport	est. 1,500
Rhode Island	−25	Feb 5, 1996	Greene	425
South Carolina	−19	Jan 21, 1985	Caesars Head	3,115
South Dakota	−58	Feb 17, 1936	McIntosh	2,277
Tennessee	−32	Dec 30, 1917	Mountain City	2,471
Texas	−23	Feb 8, 1933[a]	Seminole	3,275
Utah	−69	Feb 1, 1985	Peter's Sink	8,092
Vermont	−50	Dec 30, 1933	Bloomfield	915
Virginia	−30	Jan 22, 1985	Mtn. Lake Bio. Stn.	3,870
Washington	−48	Dec 30, 1968	Mazama & Winthrop	2,120; 1,755
West Virginia	−37	Dec 30, 1917	Lewisburg	2,200
Wisconsin	−55	Feb 4, 1996	Couderay	1,300
Wyoming	−66	Feb 9, 1933	Riverside R.S.	6,500

[a] Also on earlier dates at the same or other places.

Source: From U.S. National Oceanic and Atmospheric Administration, *Comparative Climatic Data for the United States Through 2000*, www.noaa.gov.

Table 3C.16 Temperature — Highest of Record — Selected Cities of the United States

Data	Through 2002	Years	Jan	Feb	Mar	Apr	May	Jun	Jul	Aug	Sep	Oct	Nov	Dec	Annual
AL	Birmingham CO	11	74	80	87	91	95	99	106	102	99	89	82	76	106
	Birmingham AP	59	81	83	89	92	99	102	106	103	100	94	85	80	106
	Huntsville	35	77	83	85	90	96	101	104	103	101	90	84	79	104
	Mobile	61	84	82	90	94	100	102	104	105	99	93	87	81	105
	Montgomery	58	83	85	89	91	98	105	105	104	101	100	87	85	105
AK	Anchorage	49	50	48	51	65	77	85	82	82	73	61	54	48	85
	Annette	55	61	65	64	82	88	89	89	90	82	71	67	62	90
	Barrow	82	36	36	34	42	47	72	79	76	62	43	39	34	79
	Barter IS	39	39	37	36	43	52	68	78	72	66	46	37	37	78
	Bethel	44	48	46	46	60	80	86	83	84	72	58	51	45	86
	Bettles	52	48	40	49	63	86	92	93	88	79	53	45	38	93
	Big Delta	58	48	51	58	72	90	92	91	90	79	66	52	55	92
	Cold Bay	58	51	50	56	60	67	72	77	78	76	69	59	54	78
	Fairbanks	51	50	47	56	74	89	96	94	93	84	65	49	45	96
	Gulkana	55	46	46	50	67	85	90	91	86	74	65	48	49	91
	Homer	59	51	51	53	63	71	80	81	78	69	64	55	51	81
	Juneau	58	57	57	61	72	82	86	90	83	73	61	56	54	90
	King Salmon	60	53	57	56	65	80	88	86	84	74	67	56	49	88
	Kodiak	54	54	56	57	64	80	86	82	83	73	62	54	56	86
	Kotzebue	60	39	40	39	48	74	85	85	80	69	51	38	37	85
	McGrath	60	54	55	55	67	82	90	89	89	76	61	49	49	90
	Nome	56	43	48	43	51	78	81	86	81	71	59	47	43	86
	St. Paul Island	85	48	44	50	49	59	62	63	66	61	54	50	52	66
	Talkeetna	63	45	51	54	69	85	91	90	89	78	68	51	54	91
	Unalakleet	30	47	44	47	62	78	86	87	85	71	57	48	41	87
	Valdez	31	46	52	53	62	78	86	85	82	74	58	50	52	86
	Yakutat	56	55	54	59	71	79	87	84	86	77	63	55	52	87
AZ	Flagstaff	53	66	71	73	80	89	96	97	92	90	85	74	68	97
	Phoenix	65	88	92	100	105	113	122	121	116	118	107	95	88	122
	Tucson	62	87	92	99	104	108	117	114	112	107	102	93	84	117
	Winslow	71	75	78	85	92	101	106	109	103	99	93	80	74	109
	Yuma	45	88	97	100	107	116	122	124	120	116	112	98	86	124
AR	Fort Smith	57	81	86	94	95	98	105	111	110	109	96	86	82	111
	Little Rock	61	83	85	91	95	98	105	112	109	106	97	86	80	112
	North Little Rock	24	78	83	87	94	97	102	110	111	105	92	83	78	111
CA	Bakersfield	65	82	87	92	101	107	114	115	112	112	103	91	83	115
	Bishop	55	77	81	87	93	101	109	110	107	112	97	84	78	112
	Blue Canyon	55	71	73	72	82	88	92	95	97	93	88	78	75	97
	Eureka	92	78	85	78	80	84	85	76	82	86	87	78	77	87
	Fresno	53	78	80	90	100	107	110	112	112	111	102	89	76	112
	Long Beach	50	91	91	98	105	103	109	107	105	110	111	101	92	111
	Los Angeles AP	67	88	92	95	102	97	104	97	98	110	106	101	94	110
	Los Angeles CO	62	95	95	98	106	102	112	107	105	110	108	100	91	112
	Mount Shasta	42	65	71	80	86	94	98	100	105	103	93	80	72	105
	Redding	16	77	83	88	94	104	111	118	115	116	105	88	78	118
	Sacramento	52	70	76	88	95	105	115	114	110	108	104	87	72	115
	San Diego	62	88	90	93	98	96	101	95	98	111	107	97	88	111
	San Francisco AP	75	72	78	85	92	97	106	105	100	103	99	85	75	106
	San Francisco CO	65	79	81	83	94	101	103	103	98	101	102	86	76	103
	Santa Barbara	66	86	85	90	96	92	109	109	101	102	103	97	83	109
	Santa Maria	60	86	89	95	103	100	102	104	103	103	108	93	90	108
	Stockton	43	71	78	87	100	107	111	114	109	108	101	84	72	114
CO	Alamosa	57	62	66	73	80	90	95	96	91	87	81	71	61	96
	Colorado Springs	54	73	76	81	87	94	100	100	99	94	86	78	77	100
	Denver	61	73	76	84	90	96	104	104	101	97	89	79	75	104
	Grand Junction	56	60	68	81	89	101	105	105	103	100	88	75	64	105
	Pueblo	61	81	81	86	93	102	108	106	104	101	94	84	82	108
CT	Bridgeport	54	68	67	84	91	97	96	103	100	99	86	78	76	103
	Hartford	48	66	73	89	96	99	100	102	102	99	91	81	76	102
DE	Wilmington	55	75	78	86	94	96	100	102	101	100	91	85	75	102
DC	Washington Dulles AP	40	75	79	89	93	97	100	104	104	99	90	84	79	104
	Washington Nat'l AP	61	79	82	89	95	99	101	104	105	101	94	86	79	105
FL	Apalachicola	61	79	80	85	90	98	101	102	99	96	93	87	82	102
	Daytona Beach	59	87	89	92	96	100	102	102	100	99	95	89	88	102
	Fort Myers	59	88	92	93	96	99	103	101	100	96	95	95	90	103

(Continued)

Table 3C.16 (Continued)

Data	Through 2002	Years	Jan	Feb	Mar	Apr	May	Jun	Jul	Aug	Sep	Oct	Nov	Dec	Annual
	Gainesville	19	83	87	89	95	98	102	108	99	97	92	88	84	108
	Jacksonville	61	85	88	91	95	100	103	105	102	100	96	88	84	105
	Key West	50	86	85	88	90	91	94	95	95	94	93	89	86	95
	Miami	60	88	89	92	96	96	98	98	98	97	95	91	87	98
	Orlando	60	87	90	92	96	102	100	101	100	98	95	89	90	102
	Pensacola	39	80	82	86	96	98	101	106	104	98	92	85	81	106
	Tallahassee	42	82	86	90	95	100	103	103	103	99	94	88	84	103
	Tampa	56	86	88	91	93	98	99	97	98	96	94	90	86	99
	Vero Beach	19	88	89	91	94	99	99	99	98	97	94	92	87	99
	West Palm Beach	66	89	90	94	99	96	98	101	98	97	95	91	90	101
GA	Athens	59	80	81	88	93	97	104	104	107	99	98	86	79	107
	Atlanta	54	79	80	89	93	95	101	105	102	98	95	84	79	105
	Augusta	52	82	86	89	96	99	105	107	108	101	97	90	82	108
	Columbus	57	83	83	89	93	97	104	104	104	100	96	86	82	104
	Macon	54	84	85	95	96	99	106	108	105	102	100	88	82	108
	Savannah	52	84	86	91	95	100	104	105	104	98	97	89	83	105
HI	Hilo	56	92	92	93	89	94	90	89	93	92	91	92	93	94
	Honolulu	33	88	88	88	91	93	92	94	93	95	94	93	89	95
	Kahului	38	89	89	90	91	92	94	95	97	96	96	93	90	97
	Lihue	52	86	86	88	88	88	89	89	90	90	90	89	86	90
ID	Boise	63	63	71	81	92	98	109	111	110	102	94	78	65	111
	Lewiston	56	66	72	76	97	100	107	110	115	103	89	77	65	115
	Pocatello	53	57	65	75	86	93	103	104	104	98	91	75	64	104
IL	Cairo	45	75	77	85	91	98	104	104	103	103	92	82	79	104
	Chicago	44	65	72	88	91	93	104	104	101	99	91	78	71	104
	Moline	70	69	71	88	93	104	104	105	106	100	93	80	71	106
	Peoria	63	70	72	86	92	93	105	103	103	100	90	81	71	105
	Rockford	52	63	70	85	91	95	101	103	104	102	90	76	67	104
	Springfield	55	71	74	87	90	95	103	112	103	101	93	83	74	112
IN	Evansville	62	76	79	84	91	95	104	105	102	103	94	83	77	105
	Fort Wayne	56	69	73	82	88	94	106	103	101	100	90	79	71	106
	Indianapolis	63	71	76	85	89	93	102	104	102	100	90	81	74	104
	South Bend	63	68	74	85	91	95	104	102	103	99	92	82	70	104
IA	Des Moines	63	65	73	91	93	98	103	105	108	101	95	81	69	108
	Dubuque	50	60	66	85	93	91	100	101	100	97	90	75	67	101
	Sioux City	62	71	71	91	97	102	108	108	104	103	94	81	70	108
	Waterloo	54	65	66	87	100	94	103	105	105	98	95	80	67	105
	Concordia	41	74	86	88	98	102	109	109	108	109	96	84	82	109
KS	Dodge City	60	80	85	93	100	105	110	109	107	106	96	91	86	110
	Goodland	82	79	81	89	96	104	109	111	110	105	96	87	83	111
	Topeka	56	73	84	89	95	97	107	110	110	109	96	85	73	110
	Wichita	50	75	87	89	96	100	110	113	110	108	95	85	83	113
KY	Greater Cincinnati AP	41	69	75	84	89	93	102	103	102	98	88	81	75	103
	Jackson	22	78	79	87	92	90	99	101	101	95	86	81	79	101
	Lexington	58	76	80	83	88	92	101	103	103	103	91	83	75	103
	Louisville	55	77	77	86	91	95	102	106	101	104	92	84	76	106
	Paducah	19	70	77	84	90	94	103	102	104	100	89	83	74	104
LA	Baton Rouge	52	84	85	91	92	98	103	101	105	104	94	87	85	105
	Lake Charles	38	82	83	86	95	96	99	102	107	105	94	87	82	107
	New Orleans	56	83	85	89	92	96	100	101	102	101	94	87	84	102
	Shreveport	50	84	89	92	94	102	102	107	109	109	97	88	84	109
ME	Caribou	63	53	59	73	86	96	96	95	95	91	79	68	58	96
	Portland	62	64	64	88	85	94	98	99	103	95	88	74	71	103
MD	Baltimore	52	75	79	89	94	98	101	104	105	100	92	83	77	105
	Milton	117	68	68	89	94	93	99	100	101	99	88	81	74	101
MA	Boston	51	66	70	89	94	95	100	102	102	100	90	79	76	102
	Worcester	47	60	67	84	91	92	94	96	96	91	85	78	72	96
MI	Alpena	44	52	65	80	90	94	103	102	102	94	88	76	65	103
	Detroit	44	62	70	81	89	93	104	102	100	98	91	77	69	104
	Flint	46	61	68	80	87	93	101	101	98	94	89	76	70	101
	Grand Rapids	39	62	69	78	88	92	98	100	100	93	87	77	69	100
	Houghton Lake	38	54	59	76	86	90	103	98	96	92	85	70	64	103
	Lansing	44	66	69	79	86	94	99	100	100	97	89	77	69	100

(Continued)

Table 3C.16 (Continued)

Data	Through 2002	Years	Jan	Feb	Mar	Apr	May	Jun	Jul	Aug	Sep	Oct	Nov	Dec	Annual
	Marquette	24	46	61	71	92	93	96	99	96	93	87	73	59	99
	Muskegon	63	63	67	80	86	93	98	96	99	95	83	76	64	99
	Sault Ste. Marie	62	45	49	75	85	89	93	97	98	95	80	67	62	98
MN	Duluth	61	52	55	78	88	90	94	97	97	95	86	71	55	97
	International Falls	63	48	58	76	93	95	99	98	95	95	88	73	57	99
	Minneapolis-St.Paul	64	58	61	83	95	96	102	105	102	98	90	77	68	105
	Rochester	43	55	63	79	91	92	101	102	99	95	93	75	62	102
	Saint Cloud	62	55	55	79	96	97	102	103	103	98	90	75	61	103
MS	Jackson	39	83	85	89	94	99	105	106	107	104	95	88	84	107
	Meridian	57	83	85	90	95	99	104	107	106	105	97	87	84	107
	Tupelo	19	77	84	85	93	94	101	105	106	103	92	86	79	106
MO	Columbia	33	74	82	85	90	92	103	111	110	101	93	83	76	111
	Kansas City	30	69	77	86	93	95	105	107	109	106	92	82	74	109
	St. Louis	45	76	85	89	93	94	102	107	107	104	94	85	76	107
	Springfield	57	76	81	87	93	93	101	113	106	104	93	81	77	113
MT	Billings	68	68	72	79	92	96	105	108	105	103	90	77	69	108
	Glasgow	47	61	71	79	91	102	108	104	108	103	90	79	59	108
	Great Falls	65	67	70	78	89	93	101	105	106	98	91	76	69	106
	Havre	3	62	64	77	91	96	99	103	109	96	82	78	61	109
	Helena	62	72	69	77	86	93	100	105	105	99	87	75	64	105
	Kalispell	53	53	64	72	84	94	96	104	105	99	86	69	57	105
	Missoula	58	59	66	75	87	95	98	105	105	99	85	73	60	105
NE	Grand Island	57	76	80	90	96	101	107	109	110	104	96	82	76	110
	Lincoln	31	73	84	89	97	99	107	108	107	106	93	85	70	108
	Norfolk	57	74	76	88	95	103	106	113	107	101	95	82	71	113
	North Platte	51	73	79	86	98	97	107	112	105	102	94	82	75	112
	Omaha Eppley AP	66	69	78	89	97	99	105	114	110	104	96	83	72	114
	Omaha (North)	39	66	76	88	96	100	104	107	106	103	93	79	66	107
	Scottsbluff	60	74	77	87	93	97	106	109	104	102	92	80	77	109
	Valentine	48	72	78	85	100	99	110	114	108	104	96	86	74	114
NV	Elko	72	64	70	77	86	92	104	107	107	99	88	78	65	107
	Ely	64	68	67	73	82	90	99	101	97	93	84	75	67	101
	Las Vegas	54	77	87	91	99	109	115	116	116	113	103	87	77	116
	Reno	61	70	75	83	89	96	103	108	105	101	91	77	70	108
	Winnemucca	53	68	74	81	90	97	106	109	108	103	91	77	67	109
NH	Concord	61	68	67	89	95	97	98	102	101	98	90	80	73	102
	Mt. Washington	70	47	43	54	60	66	71	71	72	69	59	52	47	72
NJ	Atlantic City AP	59	78	75	87	94	99	106	104	103	99	90	84	77	106
	Atlantic City CO	41	70	72	82	91	94	97	101	102	92	89	78	74	102
	Newark	61	74	76	89	97	99	102	105	105	105	92	85	76	105
MN	Albuquerque	63	69	76	85	89	98	107	105	101	100	91	77	72	107
	Clayton	55	80	81	86	91	99	104	102	102	99	93	85	83	104
	Roswell	30	82	85	93	99	107	114	111	107	103	99	88	81	114
	Albany	56	65	68	89	92	94	99	100	99	100	89	82	71	100
NY	Binghamton	51	63	66	82	88	89	94	98	95	96	82	77	65	98
	Buffalo	59	72	71	81	94	90	96	97	99	98	87	80	74	99
	Islip	19	69	67	82	94	98	96	102	100	92	86	78	77	102
	New York C.Park	134	72	75	86	96	99	101	106	104	102	94	84	75	106
	New York (JFK AP)	42	69	71	85	90	99	99	104	100	98	88	77	75	104
	New York (Laguardia AP)	41	68	73	83	94	97	99	107	104	96	87	80	75	107
	Rochester	62	74	73	84	93	94	100	98	99	99	91	81	72	100
	Syracuse	53	70	69	87	92	96	98	98	101	97	87	81	72	101
NC	Asheville	38	80	78	83	89	93	96	96	100	92	86	81	78	100
	Cape Hatteras	45	75	76	81	89	91	95	96	94	92	89	81	78	96
	Charlotte	63	79	81	90	93	100	103	103	103	104	98	85	78	104
	Greensboro-Wnstn-Slm-HPT	74	78	81	90	94	98	102	102	103	100	95	85	78	103
	Raleigh	58	80	84	92	95	97	104	105	105	104	98	88	80	105
	Wilmington	51	82	85	89	95	98	104	102	103	98	95	87	82	104
ND	Bismarck	63	63	69	81	93	98	111	109	109	105	95	79	65	111
	Fargo	50	52	66	78	100	98	100	106	106	102	93	74	57	106
	Grand Forks	5	47	67	64	87	88	96	93	95	96	77	73	50	96

(Continued)

Table 3C.16 (Continued)

Data	Through 2002	Years	Jan	Feb	Mar	Apr	May	Jun	Jul	Aug	Sep	Oct	Nov	Dec	Annual
	Williston	41	53	66	78	92	106	106	109	107	104	93	76	58	109
OH	Akron	54	70	72	81	88	93	100	101	98	99	86	80	76	101
	Cleveland	61	73	74	83	88	92	104	103	102	101	90	82	77	104
	Columbus	63	74	75	85	89	94	102	100	101	100	90	80	76	102
	Dayton	59	71	73	82	89	93	102	102	102	101	89	79	72	102
	Mansfield	43	65	71	82	86	92	101	100	97	93	85	78	73	101
	Toledo	47	65	71	81	88	95	104	104	99	98	91	78	70	104
	Youngstown	59	71	73	82	88	92	99	100	97	99	87	80	76	100
OK	Oklahoma City	49	80	92	93	100	104	105	110	110	108	96	87	86	110
	Tulsa	64	79	90	96	102	96	103	112	110	109	98	87	80	112
OR	Astoria	49	67	72	73	83	87	93	100	96	95	85	71	64	100
	Burns	18	55	67	71	84	94	98	107	100	97	86	70	57	107
	Eugene	60	67	72	77	86	93	102	105	108	103	94	76	68	108
	Medford	73	71	79	86	93	103	111	115	114	110	99	77	72	115
	Pendleton	67	70	75	79	91	100	108	110	113	102	92	80	67	113
	Portland	62	63	71	80	90	100	100	107	107	105	92	73	65	107
	Salem	65	65	72	80	88	100	105	108	108	104	93	72	68	108
	Sexton Summit	65	66	67	74	82	91	95	100	97	97	87	77	66	100
PC	Guam	44	88	89	90	91	92	93	94	91	95	91	90	89	95
	Johnston Island	29	88	87	85	86	87	88	89	89	89	90	89	86	90
	Koror	53	93	93	94	94	94	95	93	94	92	93	93	94	95
	Kwajalein, Marshall IS	50	90	90	90	90	91	90	91	91	92	92	91	89	92
	Majuro, Marshall IS	47	89	89	90	89	90	90	90	91	90	91	91	90	91
	Pago Pago, Amer Samoa	43	95	96	95	95	93	91	91	92	92	94	95	94	96
	Pohnpei, Caroline IS	52	93	93	95	94	93	94	94	97	96	95	95	95	97
	Chuuk, E. Caroline IS	52	92	91	94	93	94	93	92	92	93	92	91	91	94
	Wake Island	50	89	88	90	91	92	93	94	95	95	92	90	91	95
	Yap, W Caroline IS	54	90	92	90	97	93	94	96	93	93	93	93	94	97
PA	Allentown	59	72	76	87	93	97	100	105	100	99	90	81	72	105
	Erie	49	68	75	82	89	90	100	99	94	94	88	80	75	100
	Harrisburg	49	73	75	86	93	97	100	107	101	102	97	84	75	107
	Middletown/Harrisburg AP	64	73	78	87	93	97	100	107	101	102	97	84	75	107
	Philadelphia	61	74	74	87	95	97	100	104	101	100	96	81	73	104
	Pittsburgh	50	72	76	82	89	91	98	103	100	97	87	82	74	103
PA	Avoca	47	67	71	85	92	93	97	101	98	95	84	80	69	101
	Williamsport	58	69	71	87	92	96	102	103	100	102	91	83	69	103
RI	Block IS	39	58	62	74	92	83	90	91	91	87	80	70	64	92
	Providence	49	69	72	85	98	95	97	102	104	100	86	78	77	104
SC	Charleston AP	60	83	87	90	95	98	103	104	105	99	94	88	83	105
	Charleston CO	17	80	82	88	94	96	104	103	103	98	93	84	81	104
	Columbia	55	84	84	91	94	101	107	107	107	101	101	90	83	107
	Greenville-Spartanburg AP	40	79	81	89	93	97	100	104	103	96	92	85	76	104
SD	Aberdeen	41	60	62	82	98	96	108	110	112	103	96	78	62	112
	Huron	61	63	71	89	97	99	109	112	110	106	102	86	66	112
	Rapid City	60	76	75	82	93	98	109	110	106	104	94	83	75	110
	Sioux Falls	57	66	70	87	94	100	110	108	108	104	94	81	63	110
TN	Bristol-JhnCty-Kgsprt	57	79	80	85	89	92	97	102	101	100	90	81	78	102
	Chattanooga	63	78	79	87	93	99	104	106	105	102	94	84	78	106
	Knoxville	61	77	83	86	92	94	102	103	102	103	91	84	80	103
	Memphis	61	79	81	85	94	99	104	108	107	103	95	86	81	108
	Nashville	63	78	84	86	91	97	106	107	104	105	94	84	79	107
	Oak Ridge	63	75	79	86	92	93	101	105	103	102	90	83	78	105
TX	Abilene	63	89	93	97	99	109	109	110	109	107	103	92	89	110
	Amarillo	62	81	88	94	98	103	108	105	106	103	99	87	81	108
	Austin/City	61	90	99	98	98	102	108	109	107	112	98	91	90	112
	Austin/Bergstrom	61	89	101	98	99	102	109	106	107	112	98	90	91	112
	Brownsville	64	93	94	106	102	102	102	102	102	105	96	97	94	106
	Corpus Christi	64	91	98	102	102	103	106	104	103	109	98	98	91	109

(Continued)

Table 3C.16 (Continued)

Data	Through 2002	Years	Jan	Feb	Mar	Apr	May	Jun	Jul	Aug	Sep	Oct	Nov	Dec	Annual
	Dallas-Fort Worth	49	88	95	96	95	103	113	110	108	111	102	89	88	113
	Dallas-Love Field	3	80	81	86	92	96	97	104	106	110	93	86	80	110
	Del Rio	40	90	99	101	106	109	112	108	109	110	106	96	90	112
	El Paso	63	80	83	89	98	104	114	112	108	104	96	87	80	114
	Galveston	63	78	83	85	92	94	99	101	100	96	94	85	80	101
	Houston	33	84	91	91	95	99	103	104	107	109	96	89	85	109
	Lubbock	56	83	87	95	100	109	114	108	106	103	100	88	81	114
	Midland-Odessa	55	84	90	95	101	108	116	112	107	107	101	90	85	116
	Port Arthur	49	82	85	87	94	97	100	103	108	105	95	88	84	108
	San Angelo	55	90	97	97	103	109	110	111	109	107	100	93	91	111
	San Antonio	61	89	100	100	101	103	107	106	108	111	99	94	90	111
	Victoria	42	88	95	97	98	101	106	104	107	111	99	93	88	111
	Waco	60	88	96	100	101	102	109	109	112	111	101	92	91	112
	Wichita Falls	56	87	93	100	102	110	117	114	113	111	102	89	88	117
UT	Milford	53	66	75	80	87	94	105	104	102	98	90	76	65	105
	Salt Lake City	74	62	69	78	86	96	104	107	106	100	89	75	69	107
VT	Burlington	59	66	62	84	91	93	100	100	101	98	85	75	67	101
VA	Lynchburg	58	80	79	87	94	93	100	103	102	101	93	83	79	103
	Norfolk	54	80	82	88	97	100	101	103	104	99	95	86	80	104
	Richmond	73	81	83	93	96	100	104	105	102	103	99	86	81	105
	Roanoke	55	79	80	87	95	96	100	104	105	101	93	83	80	105
	Wallops Island	28	79	79	86	93	97	96	101	100	96	90	82	77	101
WA	Olympia	61	63	73	76	87	96	101	103	104	98	90	74	64	104
	Quillayute	36	65	73	72	83	92	96	97	99	97	83	69	64	99
	Seattle CO	65	66	74	75	87	92	100	100	97	92	82	73	65	100
	Seattle Sea-Tac AP	58	64	70	75	85	93	96	100	99	98	89	74	64	100
	Spokane	55	59	63	71	90	96	101	103	108	98	86	67	56	108
	Walla Walla	55	67	75	77	87	99	107	112	109	99	87	81	65	112
	Yakima	56	68	69	80	92	102	105	108	110	100	88	73	67	110
PR	San Juan	48	92	96	96	97	96	97	95	97	97	98	96	94	98
WV	Beckley	39	73	74	81	86	89	90	94	96	92	81	78	73	96
	Charleston	55	79	79	89	94	93	98	104	101	102	92	85	80	104
	Elkins	58	76	75	84	89	93	93	99	95	97	86	80	76	99
	Huntington	42	78	79	86	92	93	100	102	100	97	87	82	80	102
WI	Green Bay	53	53	61	78	89	91	98	103	99	95	88	74	64	103
WI	La Crosse	50	57	64	84	93	94	102	108	105	100	93	75	67	108
	Madison	63	56	64	82	94	93	101	104	102	99	90	76	64	104
	Milwaukee	62	62	68	82	91	93	101	103	103	98	89	77	68	103
WY	Casper	52	60	68	74	84	92	102	104	102	97	87	72	63	104
	Cheyenne	67	66	71	74	83	91	100	100	96	95	83	75	69	100
	Lander	56	63	68	76	82	91	100	101	101	94	85	70	64	101
	Sheridan	62	70	76	77	87	95	105	107	106	103	92	81	72	107

Source: From U.S. National Oceanic and Atmospheric Administration, *Comparative Climatic Data for the United States Through 2000*, www.noaa.gov.

Table 3C.17 Temperature — Lowest of Record — Selected Cities of the United States

Data	Through 2002	Years	Jan	Feb	Mar	Apr	May	Jun	Jul	Aug	Sep	Oct	Nov	Dec	Annual
AL	Birmingham CO	11	−6	11	13	29	42	48	60	57	41	32	21	−1	−6
	Birmingham AP	59	−6	3	2	26	35	42	51	51	37	27	5	1	−6
	Huntsville	35	−11	1	6	26	36	45	54	52	38	29	15	−3	−11
	Mobile	61	3	11	21	32	43	49	60	59	42	30	22	8	3
	Montgomery	58	0	10	17	28	40	49	59	56	39	26	13	5	0
AK	Anchorage	49	−34	−28	−24	−4	17	33	38	31	19	−5	−21	−30	−34
	Annette	55	1	2	1	21	31	37	40	40	33	18	−3	1	−3
	Barrow	82	−53	−56	−52	−42	−19	4	22	20	1	−32	−40	−55	−56
	Barter IS	41	−54	−59	−51	−44	−16	13	24	20	4	−26	−51	−51	−59
	Bethel	44	−48	−39	−39	−22	4	28	31	28	18	−6	−24	−37	−48
	Bettles	52	−70	−64	−56	−37	−10	27	29	22	0	−35	−57	−59	−70
	Big Delta	58	−63	−60	−49	−37	−1	30	32	22	−2	−39	−47	−62	−63
	Cold Bay	59	−13	−9	−13	4	18	29	33	32	26	6	1	−1	−13
	Fairbanks	51	−61	−58	−49	−24	−1	31	35	27	3	−27	−46	−62	−62
	Gulkana	55	−60	−65	−48	−42	5	26	29	20	2	−23	−44	−58	−65
	Homer	59	−24	−19	−21	−9	6	29	34	31	20	2	−7	−16	−24
	Juneau	58	−22	−22	−15	6	25	31	36	27	23	11	−5	−21	−22
	King Salmon	60	−48	−41	−42	−15	4	29	33	25	15	−12	−28	−38	−48
	Kodiak	54	−16	−12	−6	7	20	30	37	34	26	10	0	−2	−16
	Kotzebue	60	−49	−52	−48	−44	−18	20	30	29	13	−19	−36	−47	−52
	McGrath	60	−75	−64	−51	−40	−2	30	33	25	2	−28	−53	−67	−75
	Nome	56	−54	−42	−46	−30	−11	23	30	26	9	−10	−39	−41	−54
	St. Paul Island	85	−26	−16	−19	−8	8	16	28	29	22	12	4	−5	−26
	Talkeetna	63	−48	−46	−43	−37	−5	28	33	25	11	−21	−41	−45	−48
	Unalakleet	30	−48	−50	−50	−30	−6	25	32	28	6	−20	−47	−50	−50
	Valdez	31	−20	−10	−6	5	21	31	33	32	25	8	1	−6	−20
	Yakutat	56	−22	−20	−20	3	21	29	35	29	21	6	−6	−24	−24
AZ	Flagstaff	53	−22	−23	−16	−2	14	22	32	24	23	−2	−13	−23	−23
	Phoenix	65	17	22	25	32	40	50	61	60	47	34	25	22	17
	Tucson	62	16	20	20	27	38	47	59	61	44	26	24	16	16
	Winslow	71	−18	−7	−2	16	23	35	44	41	31	13	−1	−12	−18
	Yuma	45	24	28	32	41	46	54	63	63	53	35	30	27	24
AR	Fort Smith	57	−10	−9	7	22	35	47	50	51	33	22	8	−5	−10
	Little Rock	61	−4	−5	11	28	40	46	54	52	37	29	17	−1	−5
	North Little Rock	24	−6	4	14	30	40	52	60	53	41	27	19	−2	−6
CA	Bakersfield	65	20	25	31	33	37	45	52	52	45	29	28	19	19
	Bishop	55	−7	−2	9	15	25	29	34	37	26	16	5	−8	−8
	Blue Canyon	55	5	6	9	17	21	28	36	35	27	17	13	3	3
	Eureka	92	25	27	29	32	35	40	45	44	41	32	29	21	21
	Fresno	53	19	24	26	32	36	44	50	49	37	27	26	18	18
	Long Beach	50	25	33	33	38	40	47	51	52	50	39	34	28	25
	Los Angeles AP	67	23	32	34	39	43	48	49	51	47	41	34	32	23
	Los Angeles CO	62	28	34	26	39	5	49	54	53	51	41	38	30	5
	Mount Shasta	42	−2	1	11	14	21	25	31	34	25	19	9	−5	−5
	Redding	16	19	21	28	28	34	42	53	51	40	33	23	17	17
	Sacramento	52	23	23	26	31	36	41	48	49	43	36	26	18	18
	San Diego	62	29	36	39	41	48	51	55	57	51	43	38	34	29
	San Francisco AP	75	24	25	30	31	36	41	43	42	38	34	25	20	20
	San Francisco CO	65	30	31	38	40	44	47	47	48	48	45	40	28	28
	Santa Barbara	66	26	25	34	36	38	42	49	47	45	36	30	20	20
	Santa Maria	60	20	22	24	31	31	36	43	43	36	26	25	20	20
	Stockton	43	19	22	27	32	38	45	49	50	43	33	25	17	17
CO	Alamosa	57	−50	−35	−20	−6	11	24	30	29	15	−10	−30	−42	−50
	Colorado Springs	54	−26	−27	−11	−3	21	32	42	39	22	5	−8	−24	−27
	Denver	61	−25	−30	−11	−2	22	30	43	41	17	3	−8	−25	−30
	Grand Junction	56	−23	−18	5	11	26	34	44	43	29	18	−2	−17	−23
	Pueblo	61	−29	−31	−20	2	25	36	44	40	21	4	−17	−25	−31
CT	Bridgeport	54	−7	−5	4	18	31	41	49	44	36	26	16	−4	−7
	Hartford	48	−26	−21	−6	9	28	35	44	36	30	17	1	−14	−26
DE	Wilmington	55	−14	−6	2	18	30	41	48	43	36	24	14	−7	−14
DC	Washington Dulles AP	40	−18	−14	−1	17	28	36	41	38	30	15	9	−4	−18
	Washington Nat'l AP	61	−5	4	11	24	34	47	54	49	39	29	16	1	−5
FL	Apalachicola	61	9	21	22	36	47	48	63	62	50	37	24	13	9
	Daytona Beach	59	15	24	26	35	44	52	60	65	52	41	27	19	15
	Fort Myers	59	28	30	33	39	50	60	66	65	63	45	34	26	26
	Gainesville	19	10	18	26	34	42	50	62	61	52	33	25	16	10
	Jacksonville	61	7	19	23	34	45	47	61	59	48	36	21	11	7

(Continued)

Table 3C.17 (Continued)

Data	Through 2002	Years	Jan	Feb	Mar	Apr	May	Jun	Jul	Aug	Sep	Oct	Nov	Dec	Annual
	Key West	50	41	45	47	48	64	68	69	68	69	60	49	44	41
	Miami	60	30	32	32	46	53	60	69	68	68	51	39	30	30
	Orlando	60	19	26	25	38	48	53	64	64	56	43	29	20	19
	Pensacola	39	5	15	22	33	48	56	61	62	43	32	25	11	5
	Tallahassee	42	6	14	20	29	34	46	57	57	40	30	13	10	6
	Tampa	56	21	24	29	40	49	53	63	67	57	40	23	18	18
	Vero Beach	19	21	28	32	42	47	57	67	64	64	46	38	23	21
	West Palm Beach	66	27	32	30	43	51	61	66	65	66	46	36	28	27
GA	Athens	59	−4	5	11	26	37	45	55	54	36	24	7	2	−4
	Atlanta	54	−8	5	10	26	37	46	53	55	36	28	3	0	−8
	Augusta	52	−1	9	12	26	35	47	55	54	36	22	15	5	−1
	Columbus	57	−2	10	16	28	39	44	59	57	38	24	10	4	−2
	Macon	54	−6	9	14	29	40	46	54	55	35	26	10	5	−6
	Savannah	52	3	14	20	32	39	51	61	57	43	28	15	9	3
HI	Hilo	56	54	53	54	56	58	60	62	63	61	62	58	55	53
	Honolulu	33	53	53	55	57	60	65	66	67	66	61	57	54	53
	Kahului	38	48	50	52	54	57	58	58	61	60	58	55	52	48
	Lihue	52	50	52	51	56	58	61	62	66	65	61	57	52	50
ID	Boise	63	−17	−15	6	19	22	31	35	34	23	11	−3	−25	−25
	Lewiston	56	−22	−15	2	20	23	34	41	41	28	15	−3	−22	−22
	Pocatello	53	−30	−33	−12	13	20	28	34	30	19	10	−14	−29	−33
IL	Cairo	45	−12	−5	6	28	38	51	54	50	40	27	5	−4	−12
	Chicago	44	−27	−19	−8	7	24	36	40	41	28	17	1	−25	−27
	Moline	70	−27	−28	−19	7	26	39	46	40	24	16	−9	−24	−28
	Peoria	63	−25	−19	−10	14	25	39	47	41	26	19	−2	−23	−25
	Rockford	52	−27	−24	−11	5	24	37	43	41	27	15	−10	−24	−27
	Springfield	55	−21	−22	−12	19	28	40	48	43	32	17	−3	−21	−22
IN	Evansville	62	−21	−23	−9	23	28	41	47	43	31	21	−3	−15	−23
	Fort Wayne	56	−22	−18	−10	7	27	38	44	38	29	19	−1	−18	−22
	Indianapolis	63	−27	−21	−7	16	28	37	44	41	28	17	−2	−23	−27
	South Bend	63	−22	−17	−13	11	24	35	42	40	29	20	−7	−16	−22
IA	Des Moines	63	−24	−26	−22	9	30	38	47	40	26	14	−4	−22	−26
	Dubuque	50	−28	−27	−20	11	24	36	44	40	28	13	−17	−25	−28
	Sioux City	62	−26	−26	−22	−2	25	38	42	37	24	12	−9	−24	−26
	Waterloo	54	−33	−31	−34	−4	25	38	42	38	22	11	−17	−29	−34
KS	Concordia	41	−17	−15	−7	14	26	41	48	45	29	14	−4	−26	−26
	Dodge City	60	−13	−15	−15	14	26	41	46	47	29	14	0	−21	−21
	Goodland	82	−26	−22	−20	0	21	31	42	38	19	7	−12	−27	−27
	Topeka	56	−20	−23	−7	10	26	43	43	41	29	19	2	−26	−26
	Wichita	50	−12	−21	−2	15	31	43	51	48	31	18	1	−16	−21
KY	Greater Cincinnati AP	41	−25	−11	−11	15	27	39	47	43	31	16	1	−20	−25
	Jackson	22	−18	−8	7	20	32	44	52	45	34	26	13	−13	−18
	Lexington	58	−21	−15	−2	18	26	39	47	42	34	20	−3	−19	−21
	Louisville	55	−22	−19	−1	22	31	42	50	46	33	23	−1	−15	−22
	Paducah	19	−15	−8	11	24	35	44	53	44	35	24	10	−10	−15
LA	Baton Rouge	52	9	15	20	32	44	53	58	59	43	30	21	8	8
	Lake Charles	38	15	17	23	34	49	56	61	59	47	30	23	11	11
	New Orleans	56	14	16	25	32	41	50	60	60	42	35	24	11	11
	Shreveport	50	3	12	17	31	42	52	58	53	42	28	16	5	3
ME	Caribou	63	−33	−41	−28	−2	18	30	36	34	23	14	−8	−31	−41
	Portland	62	−26	−39	−21	8	23	33	40	33	23	15	3	−21	−39
	Baltimore	52	−7	−3	6	20	32	40	50	45	35	25	13	0	−7
MA	Blue Hill	117	−16	−21	−5	6	27	36	44	39	28	21	5	−19	−21
	Boston	51	−12	−4	6	16	34	45	50	47	38	28	15	−7	−12
	Worcester	47	−19	−12	−4	11	28	36	43	38	30	20	6	−13	−19
MI	Alpena	44	−28	−37	−27	0	20	28	34	30	25	16	−6	−18	−37
	Detroit	44	−21	−15	−4	10	25	36	41	38	29	17	9	−10	−21
	Flint	46	−25	−22	−12	6	22	33	40	37	26	19	6	−13	−25
	Grand Rapids	39	−22	−19	−8	3	22	33	41	39	27	18	5	−18	−22
	Houghton Lake	38	−26	−34	−23	3	21	29	33	29	21	16	−5	−21	−34
	Lansing	44	−29	−25	−15	−2	19	30	37	35	22	15	4	−18	−29
	Marquette	24	−27	−34	−23	−5	17	28	36	34	24	14	−5	−28	−34
	Muskegon	63	−13	−19	−10	1	22	31	39	36	27	21	−14	−15	−19
	Sault Ste. Marie	62	−36	−35	−24	−2	18	26	36	29	25	16	−10	−31	−36
MN	Duluth	61	−39	−39	−29	−5	17	27	35	32	22	8	−23	−34	−39
	International Falls	63	−46	−45	−38	−14	11	23	34	30	20	2	−32	−41	−46
	Minneapolis-St.Paul	64	−34	−32	−32	2	18	34	43	39	26	13	−17	−29	−34

(Continued)

Table 3C.17 (Continued)

Data	Through 2002	Years	Jan	Feb	Mar	Apr	May	Jun	Jul	Aug	Sep	Oct	Nov	Dec	Annual
	Rochester	43	−32	−35	−31	5	21	35	42	37	23	11	−20	−33	−35
	Saint Cloud	62	−43	−40	−32	−3	19	32	40	33	18	5	−20	−41	−43
MS	Jackson	39	2	10	15	27	38	47	51	55	35	26	17	4	2
	Meridian	57	0	8	15	28	38	42	55	53	34	24	16	2	0
	Tupelo	19	−6	4	16	29	40	49	58	52	40	29	19	−3	−6
MO	Columbia	33	−19	−15	−5	19	29	40	48	42	32	22	0	−20	−20
	Kansas City	30	−17	−19	−10	12	30	42	51	43	31	17	1	−23	−23
	St. Louis	45	−18	−12	−5	22	31	43	51	47	36	23	1	−16	−18
	Springfield	57	−13	−17	−3	18	30	42	44	44	31	18	4	−16	−17
MT	Billings	68	−30	−38	−19	−5	14	32	41	35	22	−7	−22	−32	−38
	Glasgow	47	−47	−38	−27	−3	20	32	41	37	15	−6	−26	−38	−47
	Great Falls	65	−37	−35	−29	−6	15	31	36	30	16	−11	−25	−43	−43
	Havre	3	−23	−23	−27	4	17	38	39	34	21	4	−6	−30	−30
	Helena	62	−42	−42	−30	1	17	30	36	28	18	−8	−39	−38	−42
	Kalispell	53	−38	−36	−29	10	19	26	30	30	16	−3	−28	−35	−38
	Missoula	58	−33	−27	−13	14	21	30	31	30	20	0	−23	−30	−33
NE	Grand Island	57	−28	−19	−21	7	23	38	42	40	23	9	−11	−26	−28
	Lincoln	31	−33	−24	−19	3	24	39	42	41	26	8	−5	−27	−33
	Norfolk	57	−27	−26	−20	2	24	38	42	40	26	11	−15	−30	−30
	North Platte	51	−23	−22	−22	7	19	29	39	35	17	10	−13	−34	−34
	Omaha Eppley AP	66	−23	−21	−16	5	27	38	44	43	25	13	−9	−23	−23
	Omaha (North)	39	−22	−20	−16	7	25	41	44	44	28	15	−11	−25	−25
	Scottsbluff	60	−32	−28	−27	−8	15	30	40	39	19	−6	−13	−42	−42
	Valentine	48	−30	−31	−29	3	19	30	38	34	17	−1	−22	−39	−39
NV	Elko	72	−43	−37	−9	−2	10	23	30	20	9	1	−12	−38	−43
	Ely	64	−27	−30	−13	−5	7	18	28	24	15	−3	−15	−29	−30
	Las Vegas	54	8	16	23	31	40	48	60	56	46	26	21	11	8
	Reno	61	−16	−16	−2	13	18	25	33	24	20	8	1	−16	−16
	Winnemucca	53	−24	−28	−3	6	10	23	29	28	12	−2	−8	−37	−37
NH	Concord	61	−33	−37	−16	8	21	30	35	29	21	10	−5	−22	−37
	Mt. Washington	70	−47	−46	−38	−20	−2	8	24	20	9	−5	−20	−46	−47
NJ	Atlantic City AP	59	−10	−11	5	12	25	37	42	40	32	20	10	−7	−11
	Atlantic City CO	41	−3	1	2	22	36	45	53	50	42	27	8	4	−3
	Newark	61	−8	−7	6	16	33	43	52	45	35	28	15	−1	−8
NM	Albuquerque	63	−17	−5	8	19	16	40	52	50	37	21	−7	−7	−17
	Clayton	55	−21	−17	−11	9	23	37	45	45	26	12	−10	−14	−21
	Roswell	30	−9	3	9	23	34	47	59	54	40	14	4	−8	−9
NY	Albany	56	−28	−21	−21	10	26	36	40	34	24	16	5	−22	−28
	Binghamton	51	−20	−15	−7	9	25	33	39	37	25	17	3	−18	−20
	Buffalo	59	−16	−20	−7	12	26	35	43	38	32	20	9	−10	−20
	Islip	19	−7	1	8	24	34	43	50	45	38	28	11	7	−7
	New York C.Park	134	−6	−15	3	12	32	44	52	50	39	28	5	−13	−15
	New York (JFK AP)	42	−2	−2	7	20	34	45	55	46	40	25	19	2	−2
	New York (Laguardia AP)	41	−3	−2	8	22	38	46	56	51	44	30	18	−1	−3
	Rochester	62	−17	−19	−7	13	26	35	42	36	28	20	5	−16	−19
	Syracuse	53	−26	−26	−16	9	25	35	45	40	28	19	5	−22	−26
NC	Asheville	38	−16	−2	2	22	28	35	44	42	30	21	8	−7	−16
	Cape hatteras	45	6	14	19	26	39	44	54	56	45	32	22	12	6
	Charlotte	63	−5	5	4	24	32	45	53	53	39	24	11	2	−5
	Greensboro-Wnstn-Slm-HPT	74	−8	−4	5	21	32	42	48	45	35	20	10	0	−8
	Raleigh	58	−9	0	11	23	31	38	48	46	37	19	11	4	−9
	Wilmington	51	5	11	9	30	38	48	55	55	44	27	20	0	0
ND	Bismarck	63	−44	−43	−31	−12	15	30	35	33	11	−10	−30	−43	−44
	Fargo	50	−35	−39	−23	−7	20	30	36	33	19	7	−24	−32	−39
	Grand Forks	5	−29	−22	−14	6	22	33	37	40	28	12	−9	−24	−29
	Williston	41	−40	−41	−28	−15	17	26	34	34	17	−9	−27	−50	−50
OH	Akron	54	−25	−13	−3	10	24	32	43	41	32	20	−1	−16	−25
	Cleveland	61	−20	−15	−5	10	25	31	41	38	32	19	3	−15	−20
	Columbus	63	−22	−13	−6	14	25	35	43	39	31	20	5	−17	−22
	Dayton	59	−25	−16	−7	15	27	40	44	39	32	21	−2	−20	−25
	Mansfield	43	−22	−11	−6	8	25	37	43	40	33	20	2	−17	−22
	Toledo	47	−20	−14	−6	8	25	32	40	34	26	15	2	−19	−20
	Youngstown	59	−22	−14	−10	11	24	30	40	32	29	20	1	−12	−22
OK	Oklahoma City	49	−4	−3	3	20	37	47	53	51	36	16	11	−8	−8
	Tulsa	64	−8	−11	−3	22	35	49	51	52	35	18	10	−8	−11

(Continued)

Table 3C.17 (Continued)

Data	Through 2002	Years	Jan	Feb	Mar	Apr	May	Jun	Jul	Aug	Sep	Oct	Nov	Dec	Annual
OR	Astoria	49	11	9	22	29	30	37	39	39	33	26	15	6	6
	Burns	18	−27	−28	−14	10	15	21	25	22	17	−7	−13	−20	−28
	Eugene	60	−4	−3	20	27	28	32	39	38	31	17	12	−12	−12
	Medford	73	−3	6	16	21	28	31	38	39	29	18	10	−6	−6
	Pendleton	67	−22	−18	1	18	25	35	42	40	30	11	−12	−19	−22
	Portland	62	−2	−3	19	29	29	39	43	44	34	26	13	6	−3
	Salem	65	−10	−4	12	23	25	32	37	36	26	20	9	−12	−12
	Sexton Summit	65	−2	9	15	19	22	27	36	36	26	20	11	2	−2
PC	Guam	44	56	59	54	59	62	63	64	63	61	64	62	61	54
	Johnston Island	29	65	64	67	65	68	69	70	70	71	66	63	62	62
	Koror	53	69	71	69	69	71	70	70	70	70	71	70	71	69
	Kwajalein, Marshall IS	50	68	71	70	71	71	71	70	71	68	71	70	69	68
	Majuro, Marshall IS	47	69	70	70	70	70	70	70	71	70	70	68	70	68
	Pago Pago, Amer Samoa	43	67	67	67	68	66	64	62	64	63	67	67	67	62
	Pohnpei,Caroline IS	52	61	63	65	61	61	67	66	63	62	63	64	65	61
	Chuuk,E.Caroline IS	52	68	67	68	68	67	68	67	69	67	64	66	69	64
	Wake Island	50	65	65	65	65	69	71	69	68	69	68	65	64	64
	Yap,W. Caroline IS	54	67	67	64	66	67	65	66	65	66	66	63	65	63
PA	Allentown	59	−15	−8	−1	16	28	39	46	41	30	21	11	−8	−15
	Erie	49	−18	−17	−9	12	26	32	44	37	33	24	7	−6	−18
	Harrisburg	49	−9	−5	5	19	31	40	49	45	30	23	13	−8	−9
	Middletown/ Harrisburg AP	64	−22	−5	5	19	31	40	49	45	30	23	13	−8	−22
	Philadelphia	61	−7	−4	7	19	28	44	51	44	35	25	15	1	−7
	Pittsburgh	50	−22	−12	−1	14	26	34	42	39	31	16	−1	−12	−22
	Avoca	47	−21	−16	−4	14	27	34	43	38	29	19	9	−9	−21
	Williamsport	58	−20	−13	−2	15	28	36	43	38	28	20	8	−15	−20
RI	Block IS	39	−2	−2	8	18	34	41	51	45	40	30	16	−4	−4
	Providence	49	−13	−7	1	14	29	41	48	40	33	20	6	−10	−13
SC	Charleston A	60	6	12	15	29	36	50	58	56	42	27	15	8	6
	Charleston CO	17	10	22	22	36	13	58	65	59	55	40	35	5	5
	Columbia	55	−1	5	4	26	34	44	54	53	40	23	12	4	−1
	Greenville-Spartanburg AP	40	−6	8	11	25	31	40	54	52	36	25	12	5	−6
SD	Aberdeen	41	−35	−45	−32	−2	19	33	39	32	20	8	−27	−39	−45
	Huron	61	−37	−41	−24	−2	17	32	37	36	19	8	−21	−30	−41
	Rapid City	60	−27	−31	−21	1	18	31	39	38	18	−2	−19	−30	−31
	Sioux Falls	57	−36	−31	−23	5	17	33	38	34	22	9	−17	−28	−36
TN	Bristol-JhnCty -Kgsprt	57	−21	−15	−2	21	30	38	45	43	34	20	5	−9	−21
	Chattanooga	63	−10	1	8	25	34	41	51	50	36	22	4	−2	−10
	Knoxville	61	−24	−8	1	22	32	43	49	49	36	25	5	−6	−24
	Memphis	61	−4	−11	12	29	38	48	52	48	36	25	9	−13	−13
	Nashville	63	−17	−13	2	23	34	42	51	47	36	26	−1	−10	−17
	Oak Ridge	63	−17	−13	1	20	30	39	49	50	33	21	0	−7	−17
TX	Abilene	63	−9	−7	7	25	36	47	55	50	35	23	14	−7	−9
	Amarillo	62	−11	−14	−3	14	28	41	51	49	30	12	0	−8	−14
	Austin City	61	−2	7	18	35	43	53	64	61	41	30	20	4	−2
	Brownsville	64	19	22	32	38	52	60	68	63	55	35	31	16	16
	Corpus Christi	64	14	18	24	33	47	58	64	64	50	28	28	13	13
	Dallas-Fort Worth	49	4	7	15	29	41	51	59	56	43	29	20	−1	−1
	Dallas-Love Field	3	19	21	17	41	50	62	69	70	49	39	29	21	17
	Del Rio	40	15	14	21	33	45	55	64	64	48	28	22	10	10
	El Paso	63	−8	8	14	23	31	46	57	56	41	25	1	5	−8
	Galveston	63	11	8	26	38	52	57	66	67	52	39	26	14	8
	Houston	33	12	20	22	31	44	52	62	60	48	29	19	7	7
	Lubbock	56	−16	−8	2	22	30	44	51	52	33	18	−1	−2	−16
	Midland-Odessa	55	−8	−11	9	20	34	47	53	54	36	24	11	−1	−11
	Port Arthur	49	14	20	23	32	46	56	61	60	45	30	22	12	12
	San Angelo	55	5	−1	8	25	35	48	56	54	37	26	13	−4	−4
	San Antonio	61	0	6	19	31	43	53	62	61	41	27	21	6	0
	Victoria	42	14	19	21	33	49	59	62	62	48	31	24	9	9
	Waco	60	−5	4	15	27	37	52	60	53	40	25	17	−4	−5
	Wichita Falls	56	−5	−8	8	24	36	51	54	53	38	25	14	−7	−8
UT	Milford	53	−28	−29	−14	9	17	24	38	34	23	−2	−13	−32	−32
	Salt Lake City	74	−22	−30	2	14	25	35	40	37	27	16	−14	−21	−30
VT	Burlington	59	−30	−30	−20	2	24	33	39	35	25	15	−2	−26	−30

(Continued)

Table 3C.17 (Continued)

Data	Through 2002	Years	Jan	Feb	Mar	Apr	May	Jun	Jul	Aug	Sep	Oct	Nov	Dec	Annual
VA	Lynchburg	58	−10	−10	7	20	31	40	49	45	35	21	8	−4	−10
	Norfolk	54	−3	8	18	28	36	45	54	49	45	27	20	7	−3
	Richmond	73	−12	−10	11	23	31	40	51	46	35	21	10	−1	−12
	Roanoke	55	−11	−1	9	20	31	39	47	42	34	22	9	−4	−11
	Wallops Island	28	0	10	14	27	37	44	53	48	43	31	20	4	0
WA	Olympia	61	−8	−1	9	23	25	30	35	33	25	14	−1	−7	−8
	Quillayute	36	7	11	19	23	29	33	38	36	28	24	5	7	5
	Seattle CO	65	11	11	22	31	35	42	47	48	40	30	13	9	9
	Seattle Sea-Tac AP	58	0	1	11	29	28	38	43	44	35	28	6	6	0
	Spokane	55	−22	−24	−7	17	24	33	37	35	22	7	−21	−25	−25
	Walla Walla	55	−4	−13	4	29	34	39	46	42	32	15	−11	−14	−14
	Yakima	56	−21	−25	−1	20	25	30	34	35	24	4	−13	−17	−25
PR	San Juan	48	61	62	60	64	66	69	69	70	69	46	66	59	46
WV	Beckley	39	−22	−16	−5	11	23	32	41	36	30	18	4	−18	−22
	Charleston	55	−16	−12	0	19	26	33	46	41	34	17	6	−12	−16
	Elkins	58	−24	−22	−15	3	20	25	32	34	27	11	0	−24	−24
	Huntington	42	−21	−9	−2	20	27	40	46	43	31	16	8	−13	−21
WI	Green Bay	53	−31	−28	−29	7	21	32	40	38	24	15	−9	−27	−31
	La Crosse	51	−37	−36	−28	7	26	37	33	40	28	14	−9	−30	−37
	Madison	63	−37	−29	−29	0	19	31	36	35	25	13	−11	−25	−37
	Milwaukee	62	−26	−26	−10	12	21	33	40	44	28	18	−5	−20	−26
WY	Casper	52	−40	−29	−21	−6	16	28	30	33	16	−3	−21	−41	−41
	Cheyenne	67	−29	−34	−21	−8	16	25	38	36	8	−1	−16	−28	−34
	Lander	56	−37	−28	−16	−2	18	25	39	35	10	−3	−18	−37	−37
	Sheridan	62	−35	−32	−23	−2	13	27	35	32	6	−9	−25	−37	−37

Source: From U.S. National Oceanic and Atmospheric Administration, *Comparative Climatic Data for the United States Through 2000*, www.noaa.gov.

Table 3C.18 World-Wide Extremes of Temperature and Precipitation

Temperature

A. *Highest*

World, 58°C (136°F), El Azizia, Libya, 13 September 1922
Western Hemisphere, 57°C (134°F), Death Valley, California, 10 July 1913
Antarctica, −13.6°C (7.5°F), 27 December 1978
Asia, 54°C (129°F), Tirat Tsvi, Israel, 21 June 1942
Australia, 53°C (128°F), Cloncurry, Queensland, 16 January 1889
Europe, 50°C (122°F), Seville, Spain, 4 August 1881
South America, 49°C (120°F), Rivadavia, Argentina, 11 December 1905
Canada, 45°C (113°F), Midale and Yellow Grass, Saskatchewan, 5 July 1937
Vanda Station, Antarctica, had a 15°C (59°F) maximum, 5 January 1974 (possibly Antarctica's highest)
South Pole, −14°C (7.5°F), 27 December 1978
Persian Gulf had a 36°C (96°F) sea-surface, 5 August 1924
Annual Mean, 34.4°C (94°F), Dallol, Ethiopia

B. *Lowest*

World, −89°C (−129°F), Vostok, Antarctica, 21 July 1983
Northern Hemisphere, −68°C (−90°F), Verkhoyansk, U.S.S.R., 5 and 7 February 1892 and Oimekon, U.S.S.R., 6 February 1933
Greenland, −66°C (−87°F), Northice, 9 January 1954
North America, excluding Greenland, −63°C (−81°F), Snag, Yukon Territory, 3 February 1947
U.S., −62°C (−80°F), Prospect Creek, Endicott Mts., Alaska, 23 January 1971
U.S., excluding Alaska, −56.5°C (−70°F), Rogers Pass, Montana, 20 January 1954
Europe, −55°C (−67°F), Ust 'Shchugor, U.S.S.R., January (date not known, lowest in 15-year period)
South America, −33°C (−27°F), Sarmiento, Argentina, 1 June 1907
Africa, −24°C (−11°F), Ifrane, Morocco, 11 February 1935
Antarctica, annual mean temperature −57°C, (−71°F), Sovietskaya, Antarctica
Australia, −23°C (−9°F), Charlotte Pass, New South Wales, 29 June 1994
Upper Air, −153°C (−243°F) at 93 km (58 mi) above Point Barrow, AK

Precipitation

A. *Greatest Rainfall*

World, 1-minute, 3.1 cm (1.23″), Unionville, Maryland, 4 July 1956
World, 20-minute, 20.5 cm (8″), Curtea-de-Arges, Romania, 7 July 1889
World, 42-minute, 30.5 cm (12″), Holt, Missouri, 22 June 1947
World, 60-minute, 30.5 cm (12″), Holt, Missouri, 22 June 1947 and Kilauea Sugar Plantaion 24–25 January 1956
World, 12-hour, 117 cm (46″), Grand Ilet, La R'eunion Island, 28 January 1980
World, 24-hour, 183 cm (72″), Foc-Foc, La R'eunion Island, 7–8 January 1966
World, 5-day, 430 cm (169″), Commerson, La R'eunion Island, 23–28 January 1980
World, 1-month, 930 cm (366″), Cherrapunji, India, July 1861
World, 12-month, 2,647 cm (1042″), Cherrapunji, India, August 1860–1861
Northern Hemisphere, 24-hour, 125 cm (49″), Paishih, Taiwan, 10–11 September 1963
Australia, 24-hour, 114 cm (44″), Bellenden Ker, Queensland, 4 January 1979
Canada, 24-hour, 49 cm (19″), Ucluelet Brynnor Mines, British Columbia, 6 October 1967
United States, 24-hour, 109 cm (43″), Alvin, Texas, 25–26 July 1979
United States, 12-month, 1878 cm (793″), Kukui, Maui, Hawaii, December 1981–1982

B. *Greatest Average Yearly Precipitation*

World, 1,168 cm (460i″), Mount Waialeale, Kauai, HI (1931–1960), 1187 cm (467 Mawsynram,India (1941–1979), 1330 cm (524″), Lloro, Colombia (1932–1960)
Asia, 1187 cm (467″) during a 38-year period, Mawsynrami, India
Africa, 1029 cm (405″) during a 32-year period, Debundscha, Cameroon
South America, 899 cm (354″) during a 10–16 year period, Quibdo, Colombia
North America, 650 cm (256″) during a 14-year period Henderson Lake, British Columbia
Europe, 465 cm (183″) during a 22-year period, Crkvice, Yugoslavia
Australia, average yearly, 864 cm (340″), Bellenden Ker, Queensland
Bahia Felix, Chile, averages 325 days/year with rain
Canada, highest frequency of days with precipitation, 242 per year average, Langara, Queen Charlotte Islands, British Columbia

C. *Least Precipitation*

Arica, Chile, had no rain for more than 14 consecutive years, October 1903 to January 1918
U.S., longest dry period, 767 days from 3 October 1912 to 8 November 1914, Bagdad, California
Canada, least precipitation during a calendar year, 1.27 cm (0.05″), Arctic Bay, Northwest Territories, 1949 Canada, lowest frequency of days with precipitation, 8 per year average, Rea Point, Northwest Territories

D. *Lowest Average Yearly Precipitation*

World, 0.08 cm (0.03″) during a 59-year period, Arica, Chile

(Continued)

Table 3C.18 (Continued)

Africa, <0.25 cm (<0.1″> during a 39-year period, Wadi Haifa, Sudan
North America, 3.0 cm (1.2″) during a 14-year period, Bataques, Mexico
United States, 4.1 cm (1.63″) during a 42-year period, Death Valley, California
Asia, 4.6 cm (1.8″) during a 50-year period, Aden, South Yemen
Australia, 10 cm (4.05″) during a 42-year period, Troudaninna, South Australia
Europe, 16 cm (6.4″) during a 25-year period, Astrakhan, U.S.S.R.

E. *Hailstones*

U.S., largest hailstone, 44.5 cm (17.5″) circumference, Coffeyville, Kansas, 3 September 1979 Canada, heaviest hailstone, 290 gm (10.23 oz),
Cedoux, Saskatchewan, 27 August 1973 Canada, highest frequency of days with hail, 7 per year average, Edson and Red Deer, Alberta
United States, highest frequency of days with hail, 9.4 per year average, Cheyenne, Wyoming
World, heaviest hailstone, 1.02 kg (2.25 lbs) in the Gopalganj district, Bangladesh, 14 April 1986

F. *Greatest Snowfall*

North America, 24-hour, 192.5 cm (76″), Silver Lake, Colorado, 14–15 April 1921
Bessans, France, had a snowfall of 172 cm (68″) in 19 hours, 5–6 April 1969
Canada, climatological day, 118 cm (46″), Lakelse Lake, British Columbia, 17 January 1974
North America, one storm, 480 cm (189″), Mt. Shasta Ski Bowl, California, 13–19 February 1959
North America, one season, 2850 cm (1122″), Rainier Paradise Ranger Station, Washington, 1971–1972
Canada, one season, 2446.5 cm (964″), Revelstoke Mt. Copeland, British Columbia, 1971–1972
Canada, highest frequency of days with snow, 142 per year average, Old Glory Mountain, British Columbia
North America, greatest depth of snow on the ground, 1145.5 cm (451″), Tamarack, California, 11 March 1911
Canada, greatest depth of snow on the ground, 775 cm (305″), Loch Lomond, British Columbia

Other Elements

A. *Thunderstorms*

Kampala, Uganda, averages 242 days/year with thunderstorms, during a 10-year period
Bogor, Indonesia, averaged 322 days/year with thunderstorms from 1916 to 1920
Canada, highest frequency of days with thunderstorms, 34 per year average, Windsor, Ontario
North America, highest average annual days with thunderstorms, 100 per year average, Tampa, Florida

B. *Fog Frequency*

U.S. West Coast, highest average, 2552 hours per year during a 10-year period or more, Cape Disappointment, Washington U.S. East Coast, highest average, 1580 hours per year during a 10-year period or more, Moose Peak Lighthouse, Mistake Island, Maine
Canada, highest average, 158 days per year, Cape Race, Newfoundland

Source: From Krause, P. and Flood, K., 1997, Weather and climate extremes, U.S. Army Corps of Engineers, Topographic Engineering Center, Alexandria, VA 22315.

Table 3C.19 World Record Point Rainfall

Duration	Units	Rainfall (mm)	Location	Date
1	min	38	Barot, Guadeloupe	26 Nov 1970
8		126	Fussen, Bavaria	25 May 1920
15		198	Plumb Point, Jamaica	12 May 1916
20		206	Curtea-de-Arges, Romania	7 Jul 1889
42		305	Holt, U.S.A.	22 Jun 1947
60		401	Shangdi, Nei Monggol, China	3 Jul 1975
2.17	hours	483	Rockport, U.S.A.	18 Jul 1889
2.75		559	D'Hanis, U.S.A.	31 May 1935
4.5		782	Smethport, U.S.A.	18 Jul 1942
6		840	Muduocaidang, China	1 Aug 1977
9		1,087	Belouve, La Réunion	28 Feb 1964
10		1,400	Muduocaidang, China	1 Aug 1977
18.5		1,689	Belouve, La Réunion	28–89 Feb 1964
24		1,825	Foc Foc, La Réunion	7–8 Jan 1966
2	days	2,467	Aurere, La Réunion	7–9 Apr 1958
3		3,130	Aurere, La Réunion	6–9 Apr 1958
4		3,721	Cherrapunji, India	12–15 Sep 1974
5		4,301	Commerson, La Réunion	23–27 Jan 1980
6		4,653	Commerson, La Réunion	22–27 Jan 1980
7		5,003	Commerson, La Réunion	21–27 Jan 1980
8		5,286	Commerson, La Réunion	20–27 Jan 1980
9		5,692	Commerson, La Réunion	19–27 Jan 1980
10		6,028	Commerson, La Réunion	18–27 Jan 1980
11		6,299	Commerson, La Réunion	17–27 Jan 1980
12		6,401	Commerson, La Réunion	16–27 Jan 1980
13		6,422	Commerson, La Réunion	15–27 Jan 1980
14		6,432	Commerson, La Réunion	15–28 Jan 1980
15		6,433	Commerson, La Réunion	14–28 Jan 1980
31		9,300	Cherrapunji, India	1–31 Jul 1861
2	months	12,767	Cherrapunji, India	Jun–Jul 1861
3		16,369	Cherrapunji, India	May–Jul 1861
4		18,738	Cherrapunji, India	Apr–Jul 1861
5		20,412	Cherrapunji, India	Apr–Aug 1861
6		22,454	Cherrapunji, India	Apr–Sep 1861
11		22,990	Cherrapunji, India	Jan–Nov 1861
12		26,461	Cherrapunji, India	Aug 1860–Jul 1861
2	years	40,768	Cherrapunji, India	1860–1861

Source: From World Meteorological Organization and are published in the *Guide to Hydrological Practices 1994, 5th Edition*, WMO No. 168, www.noaa.gov.

SECTION 3D PRECIPITATION DATA

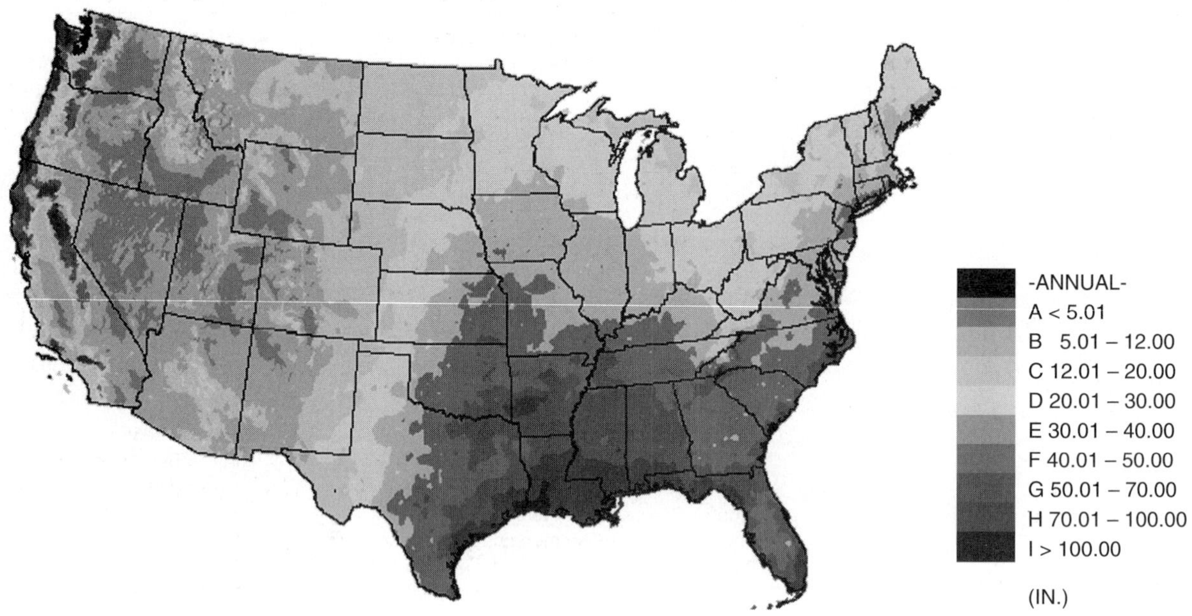

Figure 3D.7 U.S. annual mean total precipitation. (From Climate Atlas of the United States, updated 8/27/02.)

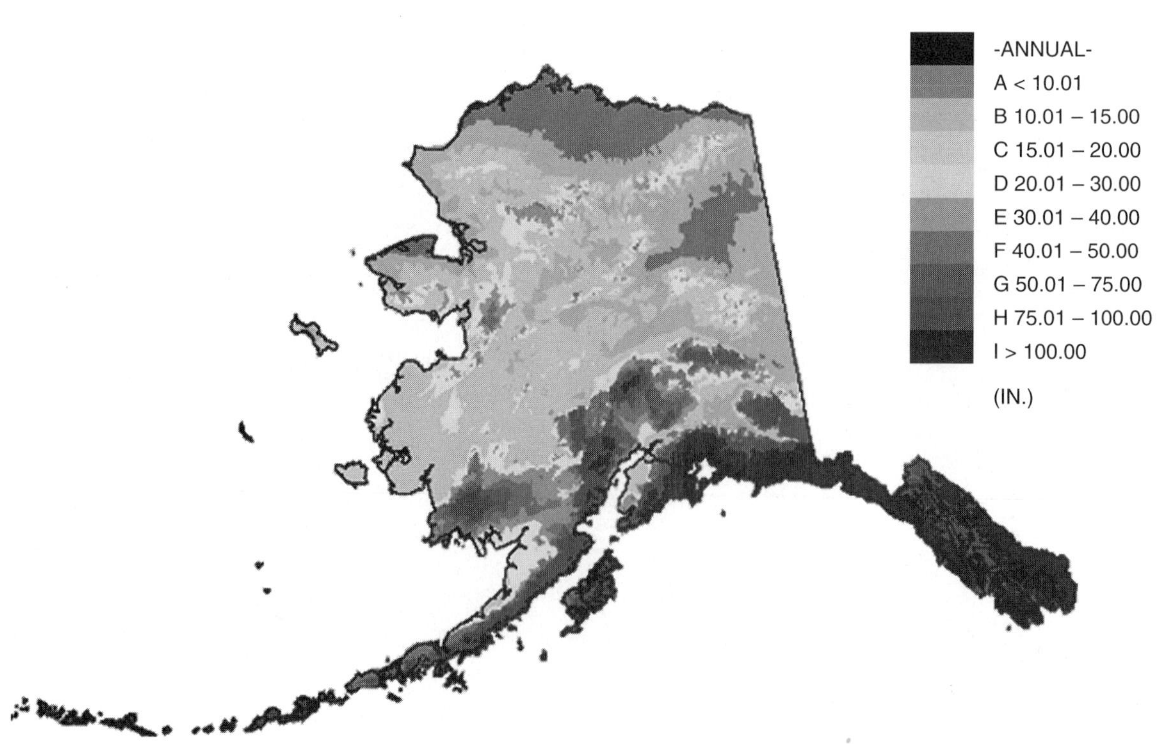

Figure 3D.8 Alaska annual mean total precipitation. (From Climate Atlas of the United States, updated 8/27/02.)

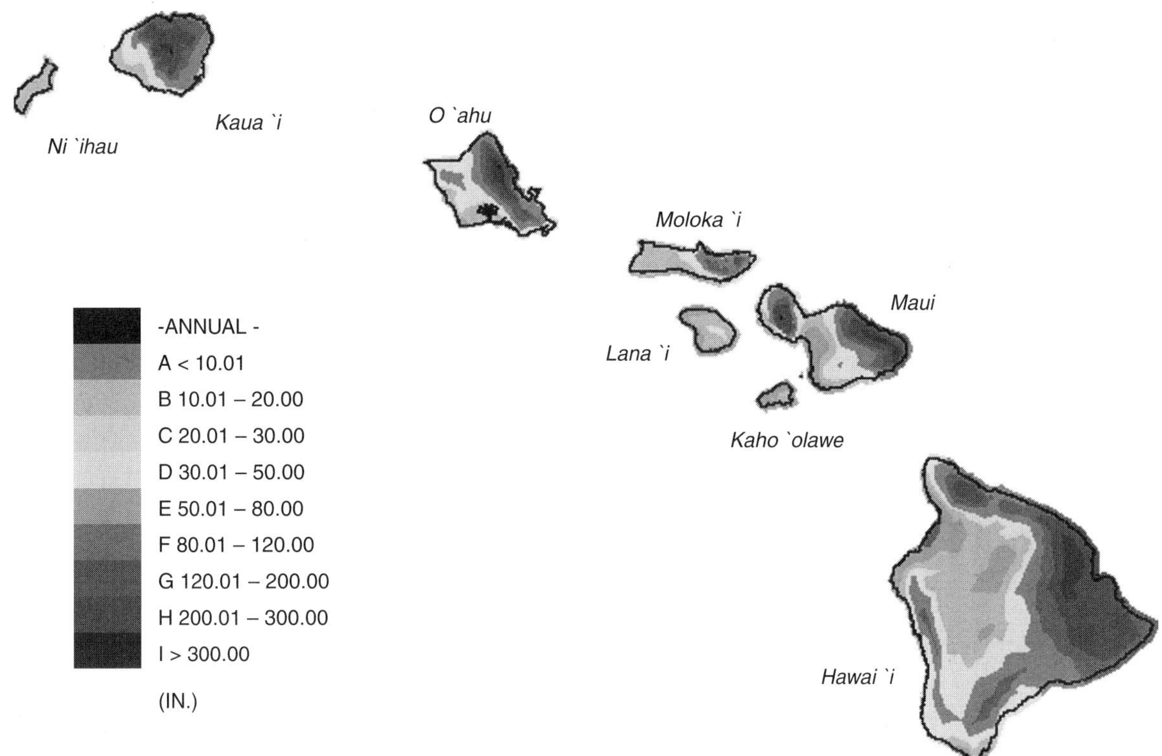

Figure 3D.9 Hawaii annual mean total precipitation. (From Climate Atlas of the United States, updated 8/27/02.)

Table 3D.20 Normal Monthly Precipitation (inches) — Selected Cities of the United States

State	Station	Years	Jan	Feb	Mar	Apr	May	Jun	Jul	Aug	Sep	Oct	Nov	Dec	Annual
AL	Birmingham	30	5.45	4.21	6.10	4.67	4.83	3.78	5.09	3.48	4.05	3.23	4.63	4.47	53.99
	Huntsville	30	5.52	4.95	6.68	4.54	5.24	4.22	4.40	3.32	4.29	3.54	5.22	5.59	57.51
	Mobile	30	5.75	5.10	7.20	5.06	6.10	5.01	6.54	6.20	6.01	3.25	5.41	4.66	66.29
	Montgomery	30	5.04	5.45	6.39	4.38	4.14	4.13	5.31	3.63	4.22	2.58	4.53	4.97	54.77
AK	Anchorage	30	0.68	0.74	0.65	0.52	0.69	1.06	1.70	2.93	2.87	2.08	1.09	1.05	16.08
	Annette	30	9.67	8.05	7.96	7.37	5.73	4.72	4.26	6.12	9.49	13.86	12.21	11.39	100.83
	Barrow	30	0.12	0.12	0.09	0.12	0.12	0.32	0.87	1.04	0.69	0.39	0.16	0.12	4.16
	Bethel	30	0.62	0.51	0.67	0.65	0.85	1.60	2.03	3.02	2.31	1.43	1.37	1.12	16.18
	Bettles	30	0.84	0.61	0.55	0.38	0.85	1.43	2.10	2.54	1.82	1.08	0.90	0.87	13.97
	Big Delta	30	0.34	0.41	0.22	0.20	0.77	2.38	2.77	2.11	1.03	0.73	0.59	0.39	11.94
	Cold Bay	30	3.08	2.59	2.48	2.30	2.65	2.89	2.53	3.59	4.51	4.54	4.79	4.33	40.28
	Fairbanks	30	0.56	0.36	0.28	0.21	0.60	1.40	1.73	1.74	1.12	0.92	0.68	0.74	10.34
	Gulkana	30	0.45	0.52	0.36	0.22	0.59	1.54	1.82	1.80	1.44	1.02	0.67	0.97	11.40
	Homer	30	2.61	2.04	1.82	1.21	1.07	0.96	1.45	2.28	3.37	2.77	2.87	3.00	25.45
	Juneau	30	4.81	4.02	3.51	2.96	3.48	3.36	4.14	5.37	7.54	8.30	5.43	5.41	58.33
	King Salmon	30	1.03	0.72	0.79	0.94	1.35	1.70	2.15	2.89	2.81	2.09	1.54	1.39	19.41
	Kodiak	30	8.17	5.72	5.22	5.48	6.31	5.38	4.12	4.48	7.84	8.36	6.63	7.64	75.35
	Kotzebue	30	0.55	0.42	0.38	0.41	0.33	0.57	1.43	2.00	1.70	0.95	0.71	0.60	10.05
	Mcgrath	30	1.04	0.74	0.81	0.66	1.02	1.45	2.32	2.75	2.36	1.46	1.46	1.44	17.51
	Nome	30	0.92	0.75	0.60	0.65	0.74	1.14	2.15	3.23	2.51	1.58	1.28	1.01	16.56
	St. Paul Island	30	1.74	1.25	1.12	1.12	1.21	1.41	1.91	2.96	2.79	2.70	2.87	2.13	23.21
	Talkeetna	30	1.45	1.28	1.26	1.22	1.64	2.41	3.24	4.53	4.35	3.06	1.78	1.96	28.18
	Unalakleet	30	0.40	0.31	0.39	0.35	0.55	1.25	2.15	2.92	2.10	0.89	0.66	0.47	12.44
	Valdez	30	6.02	5.53	4.49	3.55	3.08	3.01	3.84	6.62	9.59	8.58	5.51	7.59	67.41
	Yakutat	30	13.18	10.99	11.41	10.80	9.78	7.17	7.88	13.27	20.88	24.00	15.17	15.85	160.38
AZ	Flagstaff	30	2.18	2.56	2.62	1.29	0.80	0.43	2.40	2.89	2.12	1.93	1.86	1.83	22.91
	Phoenix	30	0.83	0.77	1.07	0.25	0.16	0.09	0.99	0.94	0.75	0.79	0.73	0.92	8.29
	Tucson	30	0.99	0.88	0.81	0.28	0.24	0.24	2.07	2.30	1.45	1.21	0.67	1.03	12.17
	Winslow	30	0.46	0.53	0.61	0.27	0.36	0.30	1.18	1.31	1.02	0.90	0.55	0.54	8.03
	Yuma	30	0.38	0.28	0.27	0.09	0.05	0.02	0.23	0.61	0.26	0.26	0.14	0.42	3.01
AR	Fort Smith	30	2.37	2.59	3.94	3.91	5.29	4.28	3.19	2.56	3.61	3.94	4.80	3.39	43.87
	Little Rock	30	3.61	3.33	4.88	5.47	5.05	3.95	3.31	2.93	3.71	4.25	5.73	4.71	50.93
	North Little Rock	30	3.37	3.27	4.88	5.03	5.40	3.51	3.15	2.97	3.53	3.81	5.74	4.53	49.19
CA	Bakersfield	30	1.18	1.21	1.41	0.45	0.24	0.12	0.00	0.08	0.15	0.30	0.59	0.76	6.49
	Bishop	30	0.88	0.97	0.62	0.24	0.26	0.21	0.17	0.13	0.28	0.20	0.44	0.62	5.02
	Eureka	30	5.97	5.51	5.55	2.91	1.62	0.65	0.16	0.38	0.86	2.36	5.78	6.35	38.10
	Fresno	30	2.16	2.12	2.20	0.76	0.39	0.23	0.01	0.01	0.26	0.65	1.10	1.34	11.23
	Long Beach	30	2.95	3.01	2.43	0.60	0.23	0.08	0.02	0.10	0.24	0.40	1.12	1.76	12.94
	Los Angeles AP	30	2.98	3.11	2.40	0.63	0.24	0.08	0.03	0.14	0.26	0.36	1.13	1.79	13.15
	Los Angeles CO	30	3.33	3.68	3.14	0.83	0.31	0.06	0.01	0.13	0.32	0.37	1.05	1.91	15.14
	Mount Shasta	30	7.06	6.45	5.81	2.65	1.87	0.99	0.39	0.43	0.87	2.21	5.08	5.35	39.16
	Redding	30	6.50	5.49	5.15	2.40	1.66	0.69	0.05	0.22	0.48	2.18	4.03	4.67	33.52
	Sacramento	30	3.84	3.54	2.80	1.02	0.53	0.20	0.05	0.06	0.36	0.89	2.19	2.45	17.93
	San Diego	30	2.28	2.04	2.26	0.75	0.20	0.09	0.03	0.09	0.21	0.44	1.07	1.31	10.77
	San Francisco AP	30	4.45	4.01	3.26	1.17	0.38	0.11	0.03	0.07	0.20	1.04	2.49	2.89	20.11

	San Francisco CO	30	4.72	4.15	3.40	1.25	0.54	0.13	0.04	0.09	0.28	1.19	3.31	3.18	22.28
	Santa Barbara	30	3.57	4.28	3.51	0.63	0.23	0.05	0.03	0.11	0.42	0.52	1.32	2.26	16.93
	Santa Maria	30	2.64	3.23	2.94	0.91	0.32	0.05	0.03	0.05	0.31	0.45	1.24	1.84	14.01
	Stockton	30	2.71	2.46	2.28	0.96	0.50	0.09	0.05	0.05	0.33	0.82	1.77	1.82	13.84
CO	Alamosa	30	0.25	0.21	0.46	0.54	0.70	0.59	0.94	1.19	0.89	0.67	0.48	0.33	7.25
	Colorado Springs	30	0.28	0.35	1.06	1.62	2.39	2.34	2.85	3.48	1.23	0.86	0.52	0.42	17.40
	Denver	30	0.51	0.49	1.28	1.93	2.32	1.56	2.16	1.82	1.14	0.99	0.98	0.63	15.81
	Grand Junction	30	0.60	0.50	1.00	0.86	0.98	0.41	0.66	0.84	0.91	1.00	0.71	0.52	8.99
	Pueblo	30	0.33	0.26	0.97	1.25	1.49	1.33	2.04	2.27	0.84	0.64	0.58	0.39	12.39
CT	Bridgeport	30	3.73	2.92	4.15	3.99	4.03	3.57	3.77	3.75	3.58	3.54	3.65	3.47	44.15
	Hartford	30	3.84	2.96	3.88	3.86	4.39	3.85	3.67	3.98	4.13	3.94	4.06	3.60	46.16
DE	Wilmington	30	3.43	2.81	3.97	3.39	4.15	3.59	4.28	3.51	4.01	3.08	3.19	3.40	42.81
DC	Washington Dulles AP	30	3.05	2.77	3.55	3.22	4.22	4.07	3.57	3.78	3.82	3.37	3.31	3.07	41.80
	Washington Nat'l AP	30	3.21	2.63	3.60	2.77	3.82	3.13	3.66	3.44	3.79	3.22	3.03	3.05	39.35
FL	Apalachicola	30	4.87	3.76	4.95	3.00	2.62	4.30	7.31	7.29	7.10	4.18	3.62	3.51	56.51
	Daytona Beach	30	3.13	2.74	3.84	2.54	3.26	5.69	5.17	6.09	6.61	4.48	3.03	2.71	49.29
	Fort Myers	30	2.23	2.10	2.74	1.67	3.42	9.77	8.98	9.54	7.86	2.59	1.71	1.58	54.19
	Gainesville	30	3.51	3.39	4.26	2.86	3.23	6.78	6.10	6.63	4.37	2.50	2.17	2.56	48.36
	Jacksonville	30	3.69	3.15	3.93	3.14	3.48	5.37	5.97	6.87	7.90	3.86	2.34	2.64	52.34
	Key West	30	2.22	1.51	1.86	2.06	3.48	4.57	3.27	5.40	5.45	4.34	2.64	2.14	38.94
	Miami	30	1.88	2.07	2.56	3.36	5.52	8.54	5.79	8.63	8.38	6.19	3.43	2.18	58.53
	Orlando	30	2.43	2.35	3.54	2.42	3.74	7.35	7.15	6.25	5.76	2.73	2.32	2.31	48.35
	Pensacola	30	5.34	4.68	6.40	3.89	4.40	6.39	8.02	6.85	5.75	4.13	4.46	3.97	64.28
	Tallahassee	30	5.36	4.63	6.47	3.59	4.95	6.92	8.04	7.03	5.01	3.25	3.86	4.10	63.21
	Tampa	30	2.27	2.67	2.84	1.80	2.85	5.50	6.49	7.60	6.54	2.29	1.62	2.30	44.77
	Vero Beach	30	2.89	2.45	4.20	2.88	3.80	6.03	6.53	6.04	6.84	5.04	3.04	2.19	51.93
	West Palm Beach	30	3.75	2.55	3.68	3.57	5.39	7.58	5.97	6.65	8.10	5.46	5.55	3.14	61.39
GA	Athens	30	4.69	4.39	4.99	3.35	3.86	3.94	4.41	3.78	3.53	3.47	3.71	3.71	47.83
	Atlanta	30	5.02	4.68	5.38	3.62	3.95	3.63	5.12	3.67	4.09	3.11	4.10	3.82	50.20
	Augusta	30	4.50	4.11	4.61	2.94	3.07	4.19	4.07	4.48	3.59	3.20	2.68	3.14	44.58
	Columbus	30	4.78	4.48	5.75	3.84	3.62	3.51	5.04	3.78	3.07	2.33	3.97	4.40	48.57
	Macon	30	5.00	4.55	4.89	3.14	2.98	3.54	4.32	3.79	3.26	2.37	3.22	3.93	45.00
	Savannah	30	3.95	2.92	3.64	3.32	3.61	5.49	6.04	7.20	5.08	3.12	2.40	2.81	49.58
HI	Hilo	30	9.74	8.86	14.35	12.54	8.07	7.36	10.71	9.78	9.14	9.64	15.58	10.50	126.27
	Honolulu	30	2.73	2.35	1.89	1.11	0.78	0.43	0.50	0.46	0.74	2.18	2.26	2.85	18.29
	Kahului	30	3.74	2.36	2.35	1.75	0.66	0.23	0.49	0.53	0.39	1.05	2.17	3.08	18.80
	Lihue	30	4.59	3.26	3.58	3.00	2.87	1.82	2.12	1.91	2.69	4.25	4.70	4.78	39.57
ID	Boise	30	1.39	1.14	1.41	1.27	1.27	0.74	0.39	0.30	0.76	0.76	1.38	1.38	12.19
	Lewiston	30	1.14	0.95	1.12	1.30	1.56	1.16	0.72	0.75	0.80	0.96	1.21	1.05	12.74
	Pocatello	30	1.14	1.01	1.38	1.18	1.51	0.91	0.70	0.66	0.89	0.97	1.13	1.10	12.58
IL	Chicago	30	1.75	1.63	2.65	3.68	3.38	3.63	3.51	4.62	3.27	2.71	3.01	2.43	36.27
	Moline	30	1.58	1.51	2.92	3.82	4.25	4.63	4.03	4.41	3.16	2.80	2.73	2.20	38.04
	Peoria	30	1.50	1.67	2.83	3.56	4.17	3.84	4.02	3.16	3.12	2.76	2.99	2.40	36.03
	Rockford	30	1.41	1.34	2.39	3.62	4.02	4.80	4.10	4.21	3.47	2.57	2.63	2.06	36.63
	Springfield	30	1.62	1.80	3.15	3.36	4.06	3.77	3.53	3.41	2.83	2.62	2.87	2.54	35.56

(Continued)

Table 3D.20 (Continued)

State	Station	Years	Jan	Feb	Mar	Apr	May	Jun	Jul	Aug	Sep	Oct	Nov	Dec	Annual
IN	Evansville	30	2.91	3.10	4.29	4.48	5.01	4.10	3.75	3.14	2.99	2.78	4.18	3.54	44.27
	Fort Wayne	30	2.05	1.94	2.86	3.54	3.75	4.04	3.58	3.60	2.81	2.63	2.98	2.77	36.55
	Indianapolis	30	2.48	2.41	3.44	3.61	4.35	4.13	4.42	3.82	2.88	2.76	3.61	3.03	40.95
	South Bend	30	2.27	1.98	2.89	3.62	3.50	4.19	3.73	3.98	3.79	3.27	3.39	3.09	39.70
IA	Des Moines	30	1.03	1.19	2.21	3.58	4.25	4.57	4.18	4.51	3.15	2.62	2.10	1.33	34.72
	Dubuque	30	1.28	1.42	2.57	3.49	4.12	4.08	3.73	4.59	3.56	2.50	2.49	1.69	35.52
	Sioux City	30	0.59	0.62	2.00	2.75	3.75	3.61	3.30	2.90	2.42	1.99	1.40	0.66	25.99
	Waterloo	30	0.84	1.05	2.13	3.23	4.15	4.82	4.20	4.08	2.95	2.49	2.10	1.11	33.15
KS	Concordia	30	0.66	0.73	2.35	2.45	4.20	3.95	4.20	3.24	2.50	1.84	1.45	0.86	28.43
	Dodge City	30	0.62	0.66	1.84	2.25	3.00	3.15	3.17	2.73	1.70	1.45	1.01	0.77	22.35
	Goodland	30	0.43	0.44	1.20	1.51	3.46	3.30	3.54	2.49	1.12	1.05	0.82	0.40	19.76
	Topeka	30	0.95	1.18	2.56	3.14	4.86	4.88	3.83	3.81	3.71	2.99	2.31	1.42	35.64
	Wichita	30	0.84	1.02	2.71	2.57	4.16	4.25	3.31	2.94	2.96	2.45	1.82	1.35	30.38
KY	Greater Cincinnati AP	30	2.92	2.75	3.90	3.96	4.59	4.42	3.75	3.79	2.82	2.96	3.46	3.28	42.60
	Jackson	30	3.56	3.68	4.38	3.79	5.16	4.67	4.59	4.13	3.77	3.18	4.20	4.27	49.38
	Lexington	30	3.34	3.27	4.41	3.67	4.78	4.58	4.80	3.77	3.11	2.70	3.44	4.03	45.91
	Louisville	30	3.28	3.25	4.41	3.91	4.88	3.76	4.30	3.41	3.05	2.79	3.80	3.69	44.54
	Paducah	30	3.47	3.93	4.27	4.95	4.75	4.51	4.45	2.99	3.56	3.45	4.53	4.38	49.24
LA	Baton Rouge	30	6.19	5.10	5.07	5.56	5.34	5.33	5.96	5.86	4.84	3.81	4.76	5.26	63.08
	Lake Charles	30	5.52	3.28	3.54	3.64	6.06	6.07	5.12	4.85	5.95	3.94	4.61	4.60	57.19
	New Orleans	30	5.87	5.47	5.24	5.02	4.62	6.83	6.20	6.15	5.55	3.05	5.09	5.07	64.16
	Shreveport	30	4.60	4.21	4.18	4.42	5.25	5.05	3.99	2.71	3.21	4.45	4.68	4.55	51.30
ME	Caribou	30	2.97	2.06	2.57	2.64	3.27	3.31	3.89	4.15	3.27	2.99	3.12	3.19	37.44
	Portland	30	4.09	3.14	4.14	4.26	3.82	3.28	3.32	3.05	3.37	4.40	4.72	4.24	45.83
MD	Baltimore	30	3.47	3.02	3.93	3.00	3.89	3.43	3.85	3.74	3.98	3.16	3.12	3.35	41.94
MA	Blue Hill	30	4.78	4.06	4.79	4.32	3.79	3.93	3.74	4.06	4.13	4.42	4.64	4.56	51.22
	Boston	30	3.92	3.30	3.85	3.60	3.24	3.22	3.06	3.37	3.47	3.79	3.98	3.73	42.53
	Worcester	30	4.07	3.10	4.23	3.92	4.35	4.02	4.19	4.09	4.27	4.67	4.34	3.80	49.05
MI	Alpena	30	1.76	1.35	2.13	2.31	2.61	2.53	3.17	3.50	2.80	2.33	2.08	1.83	28.40
	Detroit	30	1.91	1.88	2.52	3.05	3.05	3.55	3.16	3.10	3.27	2.23	2.66	2.51	32.89
	Flint	30	1.57	1.35	2.22	3.13	2.74	3.07	3.17	3.43	3.76	2.34	2.65	2.18	31.61
	Grand Rapids	30	2.03	1.53	2.59	3.48	3.35	3.67	3.56	3.78	4.28	2.80	3.35	2.70	37.13
	Houghton Lake	30	1.61	1.25	2.05	2.29	2.57	2.93	2.75	3.72	3.11	2.26	2.14	1.75	28.43
	Lansing	30	1.61	1.45	2.33	3.09	2.71	3.60	2.68	3.46	3.48	2.29	2.66	2.17	31.53
	Marquette	30	2.60	1.85	3.13	2.79	3.07	3.21	3.01	3.55	3.74	3.66	3.27	2.43	36.31
	Muskegon	30	2.22	1.58	2.36	2.91	2.95	2.58	2.32	3.77	3.52	2.80	3.23	2.64	32.88
	Sault Ste. Marie	30	2.64	1.60	2.41	2.57	2.50	3.00	3.14	3.47	3.71	3.32	3.40	2.91	34.67
MN	Duluth	30	1.12	0.83	1.69	2.09	2.95	4.25	4.20	4.22	4.13	2.46	2.12	0.94	31.00
	International Falls	30	0.84	0.64	0.96	1.38	2.55	3.98	3.37	3.14	3.03	1.98	1.36	0.70	23.93
	Minneapolis-St. Paul	30	1.04	0.79	1.86	2.31	3.24	4.34	4.04	4.05	2.69	2.11	1.94	1.00	29.41
	Rochester	30	0.94	0.75	1.88	3.01	3.53	4.00	4.61	4.33	3.12	2.20	2.01	1.02	31.40
	Saint Cloud	30	0.76	0.59	1.50	2.13	2.97	4.51	3.34	3.93	2.93	2.24	1.54	0.69	27.13
MS	Jackson	30	5.67	4.50	5.74	5.98	4.86	3.82	4.69	3.66	3.23	3.42	5.04	5.34	55.95

	Meridian	30	5.92	5.35	6.93	5.62	4.87	3.99	5.45	3.34	3.64	3.28	4.95	5.31	58.65
	Tupelo	30	5.14	4.68	6.30	4.94	5.80	4.82	3.65	2.67	3.35	3.38	5.01	6.12	55.86
MO	Columbia	30	1.73	2.20	3.21	4.16	4.87	4.02	3.80	3.75	3.42	3.18	3.47	2.47	40.28
	Kansas City	30	1.15	1.31	2.44	3.38	5.39	4.44	4.42	3.54	4.64	3.33	2.30	1.64	37.98
	St. Louis	30	2.14	2.28	3.60	3.69	4.11	3.76	3.90	2.98	2.96	2.76	3.71	2.86	38.75
	Springfield	30	2.11	2.28	3.82	4.31	4.57	5.02	3.56	3.37	4.83	3.47	4.46	3.17	44.97
MT	Billings	30	0.81	0.57	1.12	1.74	2.48	1.89	1.28	0.85	1.34	1.26	0.75	0.67	14.77
	Glasgow	30	0.35	0.26	0.47	0.75	1.72	2.20	1.78	1.25	0.98	0.71	0.39	0.37	11.23
	Great Falls	30	0.68	0.51	1.01	1.40	2.53	2.24	1.45	1.65	1.23	0.93	0.59	0.67	14.89
	Havre	30	0.47	0.36	0.70	0.87	1.84	1.90	1.51	1.20	1.03	0.62	0.45	0.51	11.46
	Helena	30	0.52	0.38	0.63	0.91	1.78	1.82	1.34	1.29	1.05	0.66	0.48	0.46	11.32
	Kalispell	30	1.47	1.15	1.11	1.22	2.04	2.30	1.41	1.25	1.20	0.96	1.45	1.65	17.21
	Missoula	30	1.06	0.77	0.96	1.09	1.95	1.73	1.09	1.15	1.08	0.83	0.96	1.15	13.82
	Grand Island	30	0.54	0.68	2.04	2.61	4.07	3.72	3.14	3.08	2.43	1.51	1.41	0.66	25.89
NE	Lincoln	30	0.67	0.66	2.21	2.90	4.23	3.51	3.54	3.35	2.92	1.94	1.58	0.86	28.37
	Norfolk	30	0.57	0.76	1.97	2.59	3.92	4.25	3.74	2.80	2.25	1.72	1.44	0.65	26.66
	North Platte	30	0.39	0.51	1.24	1.97	3.34	3.17	3.17	2.15	1.32	1.24	0.76	0.40	19.66
	Omaha Eppley AP	30	0.77	0.80	2.13	2.94	4.44	3.95	3.86	3.21	3.17	2.21	1.82	0.92	30.22
	Omaha (North)	30	0.76	0.77	2.25	3.07	4.57	3.84	3.75	2.93	3.03	2.49	1.67	0.95	30.08
	Scottsbluff	30	0.54	0.58	1.16	1.79	2.70	2.65	2.13	1.19	1.22	1.01	0.80	0.56	16.33
	Valentine	30	0.30	0.48	1.11	1.97	3.20	3.01	3.37	2.20	1.61	1.22	0.72	0.33	19.52
NV	Elko	30	1.14	0.88	0.98	0.81	1.08	0.67	0.30	0.36	0.68	0.71	1.05	0.93	9.59
	Ely	30	0.74	0.75	1.05	0.90	1.29	0.66	0.60	0.91	0.94	1.00	0.63	0.50	9.97
	Las Vegas	30	0.59	0.69	0.59	0.15	0.24	0.08	0.44	0.45	0.31	0.24	0.31	0.40	4.49
	Reno	30	1.06	1.06	0.86	0.35	0.62	0.47	0.24	0.27	0.45	0.42	0.80	0.88	7.48
	Winnemucca	30	0.83	0.62	0.86	0.85	1.06	0.69	0.27	0.35	0.53	0.66	0.80	0.81	8.33
	Concord	30	2.97	2.36	3.04	3.07	3.33	3.10	3.37	3.21	3.16	3.46	3.57	2.96	37.60
	Mt. Washington	30	8.52	7.33	9.42	8.43	8.21	8.36	8.02	8.08	8.55	7.66	10.49	8.84	101.91
NJ	Atlantic City AP	30	3.60	2.85	4.06	3.45	3.38	2.66	3.86	4.32	3.14	2.86	3.26	3.15	40.59
	Atlantic City CO	30	3.44	2.88	3.79	3.25	3.16	2.46	3.36	4.16	3.02	2.71	2.96	3.18	38.37
	Newark	30	3.98	2.96	4.21	3.92	4.46	3.40	4.68	4.02	4.01	3.16	3.88	3.57	46.25
NM	Albuquerque	30	0.49	0.44	0.61	0.50	0.60	0.65	1.27	1.73	1.07	1.00	0.62	0.49	9.47
	Clayton	30	0.30	0.27	0.62	0.99	2.08	2.21	2.81	2.69	1.56	0.74	0.54	0.32	15.13
	Roswell	30	0.39	0.41	0.35	0.58	1.30	1.62	1.99	2.31	1.98	1.29	0.53	0.59	13.34
NY	Albany	30	2.71	2.27	3.17	3.25	3.67	3.74	3.50	3.68	3.31	3.23	3.31	2.76	38.60
	Binghamton	30	2.58	2.46	2.97	3.49	3.55	3.80	3.49	3.35	3.59	3.02	3.32	3.03	38.65
	Buffalo	30	3.16	2.42	2.99	3.04	3.35	3.82	3.14	3.87	3.84	3.19	3.92	3.80	40.54
	Islip	30	4.27	3.33	4.76	4.13	3.90	3.71	2.93	4.48	3.39	3.63	3.86	4.13	46.52
	New York C. Park	30	4.13	3.15	4.37	4.28	4.69	3.84	4.62	4.22	4.23	3.85	4.36	3.95	49.69
	New York (JFK AP)	30	3.62	2.70	3.78	3.75	4.13	3.59	3.92	3.64	3.50	3.03	3.48	3.31	42.46
	New York (Laguardia AP)	30	3.56	2.75	3.93	3.68	4.16	3.57	4.41	4.09	3.77	3.26	3.67	3.51	44.36
	Rochester	30	2.34	2.04	2.58	2.75	2.82	3.36	2.93	3.54	3.45	2.60	2.84	2.73	33.98
	Syracuse	30	2.60	2.12	3.02	3.39	3.39	3.71	4.02	3.56	4.15	3.20	3.77	3.12	40.05
NC	Asheville	30	4.06	3.83	4.59	3.50	4.41	4.38	3.87	4.30	3.72	3.17	3.82	3.39	47.07
	Cape Hatteras	30	5.84	3.94	4.95	3.29	3.92	3.82	4.95	6.56	5.68	5.31	4.93	4.56	57.75

(Continued)

Table 3D.20 (Continued)

State	Station	Years	Jan	Feb	Mar	Apr	May	Jun	Jul	Aug	Sep	Oct	Nov	Dec	Annual
	Charlotte	30	4.00	3.55	4.39	2.95	3.66	3.42	3.79	3.72	3.83	3.66	3.36	3.18	43.51
	Greensboro-Wnstn-Slm-HPT	30	3.54	3.10	3.85	3.43	3.95	3.53	4.44	3.71	4.29	3.27	2.96	3.06	43.14
	Raleigh	30	4.02	3.47	4.03	2.80	3.79	3.42	4.29	3.78	4.26	3.18	2.97	3.04	43.05
	Wilmington	30	4.52	3.66	4.22	2.94	4.40	5.36	7.62	7.31	6.79	3.21	3.26	3.78	57.07
ND	Bismarck	30	0.45	0.51	0.85	1.46	2.22	2.59	2.58	2.15	1.61	1.28	0.70	0.44	16.84
	Fargo	30	0.76	0.59	1.17	1.37	2.61	3.51	2.88	2.52	2.18	1.97	1.06	0.57	21.19
	Grand Forks	30	0.68	0.58	0.89	1.23	2.21	3.03	3.06	2.72	1.96	1.70	0.99	0.55	19.60
	Williston	30	0.54	0.39	0.74	1.05	1.88	2.36	2.28	1.48	1.35	0.87	0.65	0.57	14.16
OH	Akron	30	2.49	2.28	3.15	3.39	3.96	3.55	4.02	3.65	3.43	2.53	3.04	2.98	38.47
	Cleveland	30	2.48	2.29	2.94	3.37	3.50	3.89	3.52	3.69	3.77	2.73	3.38	3.14	38.71
	Columbus	30	2.53	2.20	2.89	3.25	3.88	4.07	4.61	3.72	2.92	2.31	3.19	2.93	38.52
	Dayton	30	2.60	2.29	3.29	4.03	4.17	4.21	3.75	3.49	2.65	2.72	3.30	3.08	39.58
	Mansfield	30	2.63	2.17	3.36	4.17	4.42	4.52	4.22	4.60	3.44	2.68	3.76	3.26	43.24
	Toledo	30	1.93	1.88	2.62	3.24	3.14	3.80	2.80	3.19	2.84	2.35	2.78	2.64	33.21
	Youngstown	30	2.34	2.03	3.05	3.33	3.45	3.91	4.10	3.43	3.89	2.46	3.07	2.96	38.02
OK	Oklahoma City	30	1.28	1.56	2.90	3.00	5.44	4.63	2.94	2.48	3.98	3.64	2.11	1.89	35.85
	Tulsa	30	1.60	1.95	3.57	3.95	6.11	4.72	2.96	2.85	4.76	4.05	3.47	2.43	42.42
OR	Astoria	30	9.62	7.87	7.37	4.93	3.28	2.57	1.16	1.21	2.61	5.61	10.50	10.40	67.13
	Burns	30	1.18	1.11	1.24	0.85	1.05	0.66	0.40	0.45	0.50	0.72	1.11	1.30	10.57
	Eugene	30	7.65	6.35	5.80	3.66	2.66	1.53	0.64	0.99	1.54	3.35	8.44	8.29	50.90
	Medford	30	2.47	2.10	1.85	1.31	1.21	0.68	0.31	0.52	0.78	1.31	2.93	2.90	18.37
	Pendleton	30	1.45	1.22	1.26	1.13	1.22	0.78	0.41	0.56	0.63	0.99	1.63	1.48	12.76
	Portland	30	5.07	4.18	3.71	2.64	2.38	1.59	0.72	0.93	1.65	2.88	5.61	5.71	37.07
	Salem	30	5.84	5.09	4.17	2.76	2.13	1.45	0.57	0.68	1.43	3.03	6.39	6.46	40.00
	Sexton Summit	30	4.71	4.29	3.92	2.38	1.35	0.94	0.35	0.61	1.20	2.93	5.32	5.18	33.18
PC	Guam	30	5.58	5.11	4.24	4.16	6.39	6.28	11.66	16.17	13.69	11.88	9.34	6.11	100.61
	Johnston Island	30	1.64	1.29	2.01	1.86	1.14	0.87	1.40	2.07	2.46	2.78	4.78	2.70	25.00
	Koror	30	11.20	9.65	8.79	9.45	11.27	17.54	16.99	14.47	11.65	13.41	11.62	12.33	148.37
	Kwajalein, Marshall IS	30	5.12	3.73	3.82	7.63	8.62	8.86	10.24	10.42	11.82	11.46	10.74	7.94	100.40
	Majuro, Marshall IS	30	8.09	6.86	8.43	11.30	11.53	11.09	12.41	11.95	11.96	13.73	12.81	11.50	131.66
	Pago Pago, Amer Samoa	30	14.02	12.14	11.15	11.16	10.43	5.94	5.76	6.43	7.36	10.03	11.16	13.38	118.96
	Pohnpei, Caroline IS	30	12.52	9.78	13.96	16.94	19.41	17.06	16.72	16.37	14.94	16.30	14.74	15.87	184.61
	Chuuk, E. Caroline IS	30	8.58	8.77	8.15	10.94	11.29	12.82	12.45	15.09	13.12	10.69	11.09	10.98	133.97
	Wake Island	30	1.40	1.89	2.38	2.11	1.70	1.95	3.44	5.62	4.82	4.27	2.78	1.87	34.23
	Yap, W Caroline IS	30	7.24	5.45	6.14	5.58	8.15	13.46	13.25	14.41	13.53	12.25	8.82	9.34	117.62

PA	Allentown	30	3.50	2.75	3.56	3.49	4.47	3.99	4.27	4.35	4.37	3.33	3.70	3.39	45.17
	Erie	30	2.53	2.28	3.13	3.38	3.34	4.28	3.28	4.21	4.73	3.92	3.96	3.73	42.77
	Harrisburg	30	3.18	2.88	3.58	3.31	4.60	3.99	3.21	3.24	3.65	3.06	3.53	3.22	41.45
	Middletown/ Harrisburg AP	30	3.18	2.88	3.58	3.31	4.60	3.99	3.21	3.24	3.65	3.06	3.53	3.22	41.45
	Philadelphia	30	3.52	2.74	3.81	3.49	3.88	3.29	4.39	3.82	3.88	2.75	3.16	3.31	42.05
	Pittsburgh	30	2.70	2.37	3.17	3.01	3.80	4.12	3.96	3.38	3.21	2.25	3.02	2.86	37.85
	Avoca	30	2.46	2.08	2.69	3.28	3.69	3.97	3.74	3.10	3.86	3.02	3.12	2.55	37.56
	Williamsport	30	2.85	2.61	3.21	3.49	3.79	4.45	4.08	3.38	3.98	3.19	3.62	2.94	41.59
RI	Block IS	30	3.68	3.04	3.99	3.72	3.40	2.77	2.62	3.00	3.19	3.04	3.77	3.57	39.79
	Providence	30	4.37	3.45	4.43	4.16	3.66	3.38	3.17	3.90	3.70	3.69	4.40	4.14	46.45
SC	Charleston AP	30	4.08	3.08	4.00	2.77	3.67	5.92	6.13	6.91	5.98	3.09	2.66	3.24	51.53
	Charleston CO	30	3.62	2.62	3.83	2.44	2.77	4.96	5.50	6.54	6.13	3.02	2.18	2.78	46.39
	Columbia	30	4.66	3.84	4.59	2.98	3.17	4.99	5.54	5.41	3.94	2.89	2.88	3.38	48.27
	Greenville- Spartanburg AP	30	4.41	4.24	5.31	3.53	4.59	3.92	4.65	4.08	3.96	3.88	3.79	3.86	50.24
SD	Aberdeen	30	0.48	0.48	1.34	1.83	2.69	3.49	2.92	2.42	1.81	1.63	0.75	0.38	20.22
	Huron	30	0.48	0.57	1.67	2.29	3.00	3.28	2.86	2.07	1.80	1.59	0.89	0.39	20.90
	Rapid City	30	0.37	0.46	1.03	1.86	2.96	2.83	2.03	1.61	1.10	1.37	0.61	0.40	16.64
	Sioux Falls	30	0.51	0.51	1.81	2.65	3.39	3.49	2.93	3.01	2.58	1.93	1.36	0.52	24.69
TN	Bristol-JhnCty- Kgsprt	30	3.52	3.40	3.91	3.23	4.32	3.89	4.21	3.00	3.08	2.30	3.08	3.39	41.33
	Chattanooga	30	5.40	4.85	6.19	4.23	4.28	3.99	4.73	3.59	4.31	3.26	4.88	4.81	54.52
	Knoxville	30	4.57	4.01	5.17	3.99	4.68	4.04	4.71	2.89	3.04	2.65	3.98	4.49	48.22
	Memphis	30	4.24	4.31	5.58	5.79	5.15	4.30	4.22	3.00	3.31	3.31	5.76	5.68	54.65
	Nashville	30	3.97	3.69	4.87	3.93	5.07	4.08	3.77	3.28	3.59	2.87	4.45	4.54	48.11
	Oak Ridge	30	5.13	4.50	5.72	4.32	5.14	4.64	5.16	3.39	3.75	3.02	4.86	5.42	55.05
TX	Abilene	30	0.97	1.13	1.41	1.67	2.83	3.06	1.69	2.63	2.91	2.90	1.30	1.27	23.78
	Amarillo	30	0.63	0.55	1.13	1.33	2.50	3.28	2.68	2.94	1.88	1.50	0.68	0.61	19.71
	Austin/City	30	1.89	1.99	2.14	2.51	5.03	3.81	1.97	2.31	2.91	3.97	2.68	2.44	33.65
	Austin/ Bergstrom	30	2.20	1.73	1.98	2.77	5.87	3.38	1.61	1.48	2.63	2.70	2.61	2.39	31.36
	Brownsville	30	1.36	1.18	0.93	1.96	2.48	2.93	1.77	2.99	5.31	3.78	1.75	1.11	27.55
	Corpus Christi	30	1.62	1.84	1.73	2.05	3.48	3.53	2.00	3.54	5.03	3.94	1.74	1.75	32.26
	Dallas-Fort Worth	30	1.90	2.37	3.06	3.20	5.15	3.23	2.12	2.03	2.42	4.11	2.57	2.57	34.73
	Dallas-Love Field	30	1.89	2.31	3.13	3.46	5.30	3.92	2.43	2.17	2.65	4.65	2.61	2.53	37.05
	Del Rio	30	0.57	0.96	0.96	1.71	2.31	2.34	2.02	1.59	2.06	2.00	0.96	0.75	18.23
	El Paso	30	0.45	0.39	0.26	0.23	0.38	0.87	1.49	1.75	1.61	0.81	0.42	0.77	9.43
	Galveston	30	4.08	2.61	2.76	2.56	3.70	4.04	3.45	4.22	5.76	3.49	3.64	3.53	43.84
	Houston	30	3.68	2.98	3.36	3.60	5.15	5.35	3.18	3.83	4.33	4.50	4.19	3.69	47.84
	Lubbock	30	0.50	0.71	0.76	1.29	2.31	2.98	2.13	2.35	2.57	1.70	0.71	0.67	18.69
	Midland-Odessa	30	0.53	0.58	0.42	0.73	1.79	1.71	1.89	1.77	2.31	1.77	0.65	0.65	14.80
	Port Arthur	30	5.69	3.35	3.75	3.84	5.83	6.58	5.23	4.85	6.10	4.67	4.75	5.25	59.89
	San Angelo	30	0.81	1.18	0.99	1.60	3.09	2.52	1.10	2.05	2.95	2.57	1.10	0.94	20.91

(Continued)

Table 3D.20 (Continued)

State	Station	Years	Jan	Feb	Mar	Apr	May	Jun	Jul	Aug	Sep	Oct	Nov	Dec	Annual
	San Antonio	30	1.66	1.75	1.89	2.60	4.72	4.30	2.03	2.57	3.00	3.86	2.58	1.96	32.92
	Victoria	30	2.44	2.04	2.25	2.97	5.12	4.96	2.90	3.05	5.00	4.26	2.64	2.47	40.10
	Waco	30	1.90	2.43	2.48	2.99	4.46	3.08	2.23	1.85	2.88	3.67	2.61	2.76	33.34
	Wichita Falls	30	1.12	1.57	2.27	2.62	3.92	3.69	1.58	2.38	3.19	3.11	1.68	1.68	28.83
UT	Milford	30	0.73	0.77	1.21	0.99	0.94	0.44	0.76	1.04	0.92	1.12	0.77	0.58	10.27
	Salt Lake City	30	1.37	1.33	1.91	2.02	2.09	0.77	0.72	0.76	1.33	1.57	1.40	1.23	16.50
VT	Burlington	30	2.22	1.67	2.32	2.88	3.32	3.43	3.97	4.01	3.83	3.12	3.06	2.22	36.05
VA	Lynchburg	30	3.54	3.10	3.83	3.46	4.11	3.79	4.39	3.41	3.88	3.39	3.18	3.23	43.31
	Norfolk	30	3.93	3.34	4.08	3.38	3.74	3.77	5.17	4.79	4.06	3.47	2.98	3.03	45.74
	Richmond	30	3.55	2.98	4.09	3.18	3.95	3.54	4.67	4.18	3.98	3.60	3.06	3.12	43.91
	Roanoke	30	3.23	3.08	3.84	3.61	4.24	3.68	4.00	3.74	3.85	3.15	3.21	2.86	42.49
WA	Olympia	30	7.54	6.17	5.29	3.58	2.27	1.78	0.82	1.10	2.03	4.19	8.13	7.89	50.79
	Quillayute	30	13.65	12.35	10.98	7.44	5.51	3.50	2.34	2.67	4.15	9.81	14.82	14.50	101.72
	Seattle CO	30	5.24	4.09	3.92	2.75	2.03	1.55	0.93	1.16	1.61	3.24	5.67	6.06	38.25
	Seattle Sea-Tac AP	30	5.13	4.18	3.75	2.59	1.77	1.49	0.79	1.02	1.63	3.19	5.90	5.62	37.07
	Spokane	30	1.82	1.51	1.53	1.28	1.60	1.18	0.76	0.68	0.76	1.06	2.24	2.25	16.67
	Walla Walla	30	2.25	1.97	2.20	1.83	1.95	1.15	0.73	0.84	0.83	1.77	2.85	2.51	20.88
	Yakima	30	1.17	0.80	0.70	0.53	0.51	0.62	0.22	0.36	0.39	0.53	1.05	1.38	8.26
PR	San Juan	30	3.02	2.30	2.14	3.71	5.29	3.52	4.16	5.22	5.60	5.06	6.17	4.57	50.76
WV	Beckley	30	3.23	2.96	3.63	3.42	4.39	3.92	4.78	3.45	3.23	2.64	2.88	3.09	41.63
	Charleston	30	3.25	3.19	3.90	3.25	4.30	4.09	4.86	4.11	3.45	2.67	3.66	3.32	44.05
	Elkins	30	3.43	3.20	3.92	3.53	4.77	4.61	4.83	4.26	3.82	2.86	3.42	3.44	46.11
	Huntington	30	3.21	3.09	3.83	3.33	4.41	3.88	4.46	3.88	2.80	2.73	3.32	3.37	42.31
WI	Green Bay	30	1.21	1.01	2.06	2.56	2.75	3.43	3.44	3.77	3.11	2.17	2.27	1.41	29.19
	La Crosse	30	1.19	0.99	2.00	3.38	3.38	4.00	4.25	4.28	3.40	2.16	2.10	1.23	32.36
	Madison	30	1.25	1.28	2.28	3.35	3.25	4.05	3.93	4.33	3.08	2.18	2.31	1.66	32.95
	Milwaukee	30	1.85	1.65	2.59	3.78	3.06	3.56	3.58	4.03	3.30	2.49	2.70	2.22	34.81
WY	Casper	30	0.58	0.64	0.90	1.52	2.38	1.43	1.29	0.73	0.98	1.14	0.82	0.62	13.03
	Cheyenne	30	0.45	0.44	1.05	1.55	2.48	2.12	2.26	1.82	1.43	0.75	0.64	0.46	15.45
	Lander	30	0.52	0.54	1.24	2.07	2.38	1.15	0.84	0.57	1.14	1.37	0.99	0.61	13.42
	Sheridan	30	0.77	0.57	1.00	1.77	2.41	2.02	1.11	0.80	1.38	1.41	0.80	0.68	14.72

Note: Based on 30-year average values 1971–2000. The normal precipitation is the arithmetic mean for each month over the 30-year period and includes the liquid water equivalent of snowfall. The annual value is the total of the unrounded monthly values and may not agree with the sum of the rounded monthly values.

Source: From U.S. National Oceanic and Atmospheric Administration, *Comparative Climatic Data for the United States Through 2000*, www.noaa.gov.

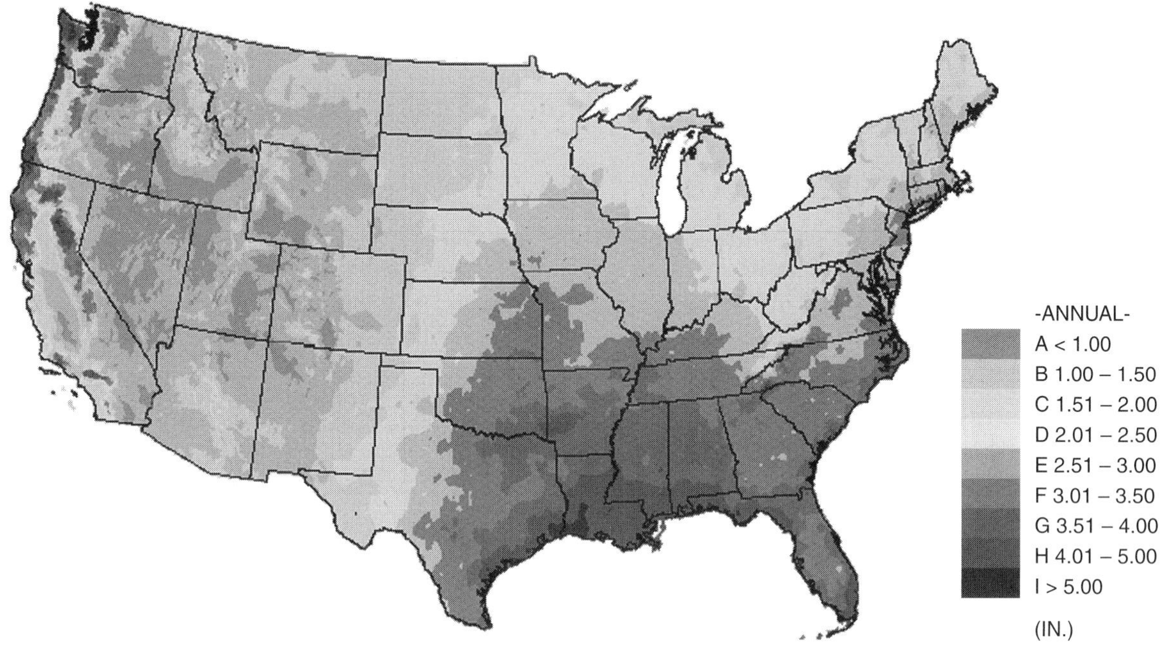

Figure 3D.10 U.S. mean maximum daily precipitation. (From Climate Atlas of the United States, updated 8/27/02.)

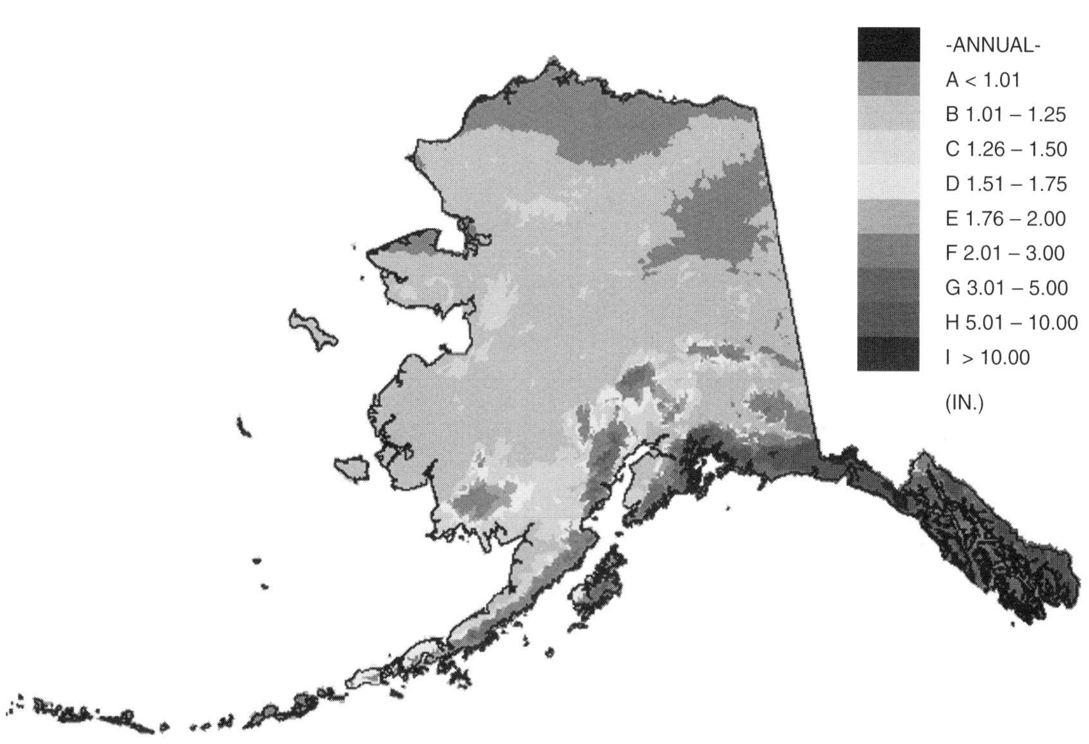

Figure 3D.11 Alaska mean maximum daily precipitation. (From Climate Atlas of the United States, updated 8/27/02.)

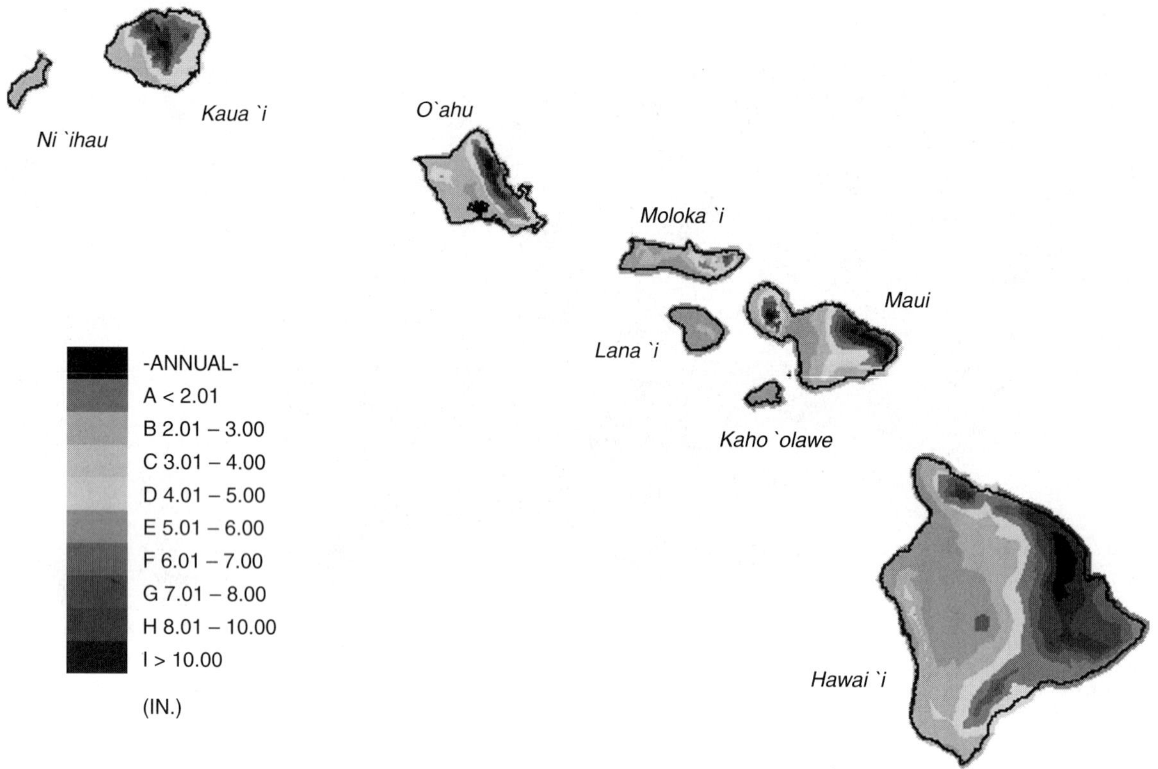

Figure 3D.12 Hawaii mean maximum daily precipitation. (From Climate Atlas of the United States, updated 8/27/02.)

Table 3D.21 Record Maximum Annual Precipitation by State

State	Precipitation (inches)	Date	Station	Elevation (ft)
Alabama	98.22	1961	Citronelle	331
Alaska	332.29	1976	MacLeod Harbor	40
Arizona	58.92	1978	Hawley Lake	8,180
Arkansas	98.55	1957	Newhope	2,420
California	153.54	1909	Monumental	2,420
Colorado	92.84	1897	Ruby	est.10,000
Connecticut	78.53[a]	1955	Burlington	460
Delaware	72.75	1948	Lewes	10
Florida	112.43[a]	1966	Wewahitchka	50
Georgia	122.16	1959	Flat Top	est. 3,600
Hawaii	704.83	1982	Kukui	5,788
Idaho	81.05	1933	Roland	4,150
Illinois	74.58	1950	New Burnside	560
Indiana	97.38	1890	Marengo	570
Iowa	74.50	1851	Muscatine	680
Kansas	68.55[a]	1993	Blaine	1,530
Kentucky	79.68	1950	Russelville	590
Louisiana	113.74[a]	1991	New Orleans (Audubon)	6
Maine	75.64	1845	Brunswick	70
Maryland	76.52	1971	Towson	390
Massachusetts	76.49[a]	1996	New Salem	845
Michigan	64.01	1881	Adrian	770
Minnesota	52.36	1993	Fairmont	1,187
Mississippi	104.36[a]	1991	Waveland	8
Missouri	92.77	1957	Portageville	280
Montana	55.51	1953	Summit	5,210
Nebraska	64.52	1896	Omaha	980
Nevada	59.03	1969	Mt. Rose Resort	est. 7,300
New Hampshire	130.14	1969	Mount Washington	6,260
New Jersey	85.99	1882	Paterson	100
New Mexico	62.45	1941	White Tail	7,450
New York	90.97	1996	Slide Mountain	2,649
North Carolina	129.60	1964	Rosman	2,220
North Dakota	37.98	1944	Milnor	2,600
Ohio	70.82	1870	Little Mountain	1,187
Oklahoma	84.47	1957	Kiamichi Tower	2,350
Oregon	204.04	1996	Laurel Mountain	3,590
Pennsylvania	81.64	1952	Mt. Pocono	1,910
Rhode Island	70.21	1983	Kingston	100
South Carolina	110.79	1994	Jocassee	2,500
South Dakota	48.42	1946	Deadwood	4,550
Tennessee	114.88	1957	Haw Knob	4,900
Texas	109.38	1873	Clarksville	440
Utah	108.54	1983	Alta	8,760
Vermont	100.96[a]	1996	Mt. Mansfield	3,950
Virginia	83.70[a]	1996	Philpott Dam	1,123
Washington	184.56	1931	Wynoochee Oxbow	670
West Virginia	89.01	1926	Bayard	2,381
Wisconsin	62.07	1884	Embarrass	808
Wyoming	55.46	1945	Grassy Lake Dam	7,240

[a] At least one month estimated.

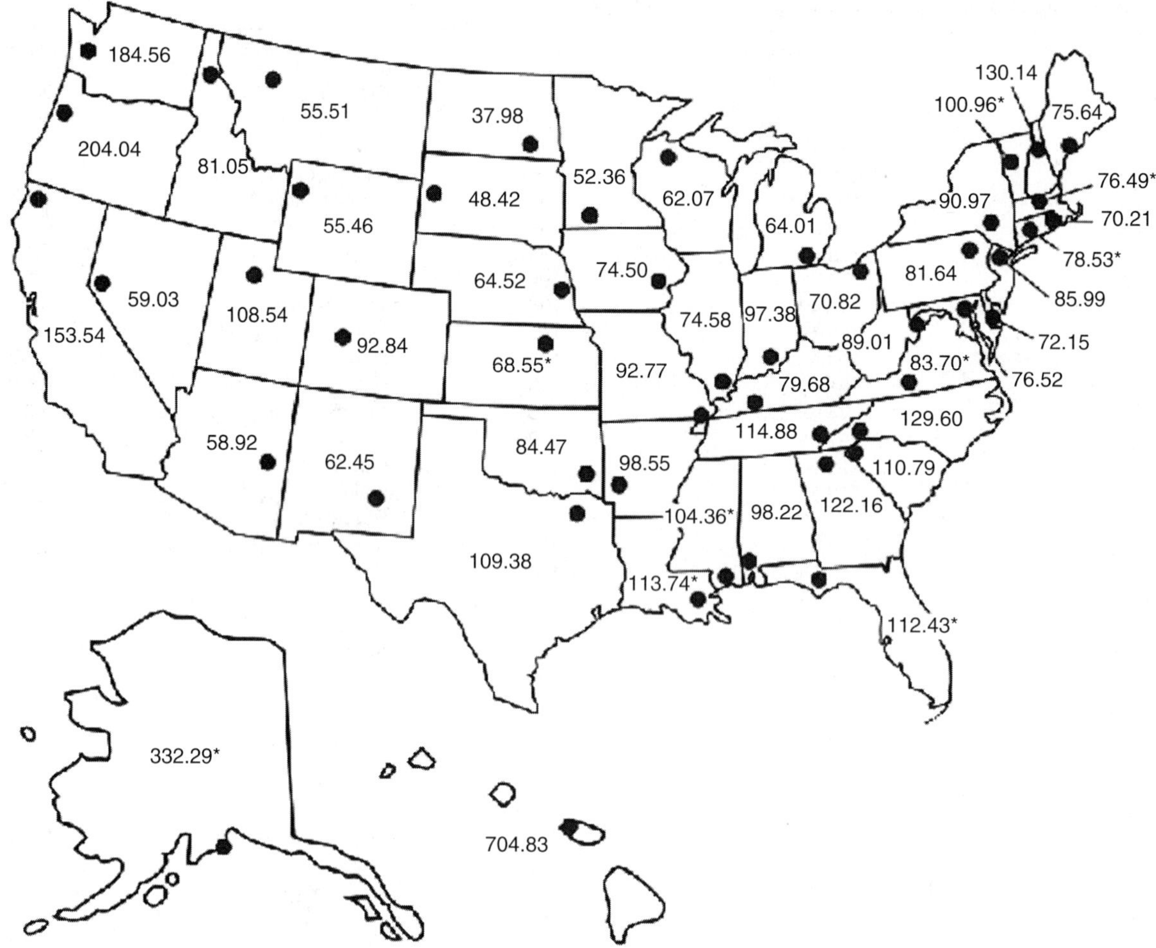

Figure 3D.13 Record maximum annual precipitation (in.) (through 1998) (*at least one month estimate). (From U.S. National Oceanic and Atmospheric Administration, *Comparative Climate Data for the United States Through 2000.* www.noaa.gov.)

Table 3D.22 Record Maximum 24-hour Precipitation by State

State	Precip. (inches)	Date	Station	Elevation (feet)
Alabama	32.52	Jul 19–20, 1997	Dauphin Is Sea Lab	8
Alaska	15.20	Oct 12, 1982	Angoon	15
Arizona	11.40	Sep 4–5, 1970	Workman Creek	6,970
Arkansas	14.06	Dec 3, 1982	Big Fork	1,100
California	26.12	Jan 22–23, 1943	Hoegees Camp	2,760
Colorado	11.08	Jun 17, 1965	Holly	3,390
Connecticut	12.77	Aug 19, 1955	Burlington	460
Delaware	8.50	Jul 13, 1975	Dover	30
Florida	38.70	Sep 5, 1950	Yankeetown	5
Georgia	21.10	Jul 6, 1994	Americus	490
Hawaii	38.00	Jan 24–25, 1956	Kilauea Plantation	180
Idaho	7.17	Nov 23, 1909	Rattlesnake Creek	4,000
Illinois	16.91	Jul 18, 1996	Aurora	640
Indiana	10.50	Aug 6, 1905	Princeton	480
Iowa	16.70	Aug 5–6, 1959	Decatur Co.	1,110
Kansas	12.59	May 31–Jun 1, 1941	Burlington	1,010
Kentucky	10.40	Jun 28, 1960	Dunmor	610
Louisiana	22.00	Aug 28–29, 1962	Hackberry	10
Maine	13.32	Oct 20–21, 1996	Portland	45
Maryland	14.75	Jul 26–27, 1897	Jewell	165
Massachusetts	18.15	Aug 18–19, 1955	Westfield	220
Michigan	9.78	Aug 31–Sep. 1, 1914	Bloomingdale	750
Minnesota	10.84	Jul 21–22, 1972	Fort Ripley	1,140
Mississippi	15.68	Jul 9, 1968	Columbus	190
Missouri	18.18	Jul 20, 1965	Edgarton	856
Montana	11.50	Jun 20, 1921	Circle	2,440
Nebraska	13.15	Jul 8–9, 1950	York	1,610
Nevada	7.13	Jan 31, 1963	Mt. Rose Hwy. Stn.	7,360
New Hampshire	10.38	Feb 10–11, 1970	Mount Washington	6,262
New Jersey	14.81	Aug 19, 1939	Tuckerton	20
New Mexico	11.28	May 18–19, 1955	Lake Maloya	7,400
New York	11.17	Oct 9, 1903	NYC Central Park	130
North Carolina	22.22	Jul 15–16, 1916	Altapass	2,600
North Dakota	8.10	Jun 29, 1975	Litchville	1,470
Ohio	10.75	Aug 7–8, 1995	Lockington Dam	950
Oklahoma	15.68	Oct 11, 1973	Enid	1,245
Oregon	11.65	Nov 19, 1996	Port Orford	150
Pennsylvania	34.50[a]	Jul 17, 1942	Smethport	1,510
Rhode Island	12.13	Sept 16–17, 1932	Westerly	40
South Carolina	17.00	Aug 27, 1995	Antreville	700
South Dakota	8.00	Sep 10, 1900	Elk Point	1,127
Tennessee	11.00	Mar 28, 1902	McMinnville	900
Texas	43.00[a]	Jul 25–26, 1979	Alvin	50
Utah	6.00[a]	Sept 5, 1970	Bug Point	6,600
Vermont	8.77	Nov 3–4, 1927	Somerset	2,080
Virginia	27.00[a]	Aug 20 1969	Nelson Co.	Est 500
Washington	14.26	Nov 23–24, 1986	Mt. Mitchell #2	3,600
West Virginia	19.00[a]	Jul 18, 1889	Rockport	700
Wisconsin	11.72	Jun 24, 1946	Mellen	1,150
Wyoming	6.06	Aug 1, 1985	Cheyenne	6,126

[a] Estimated.

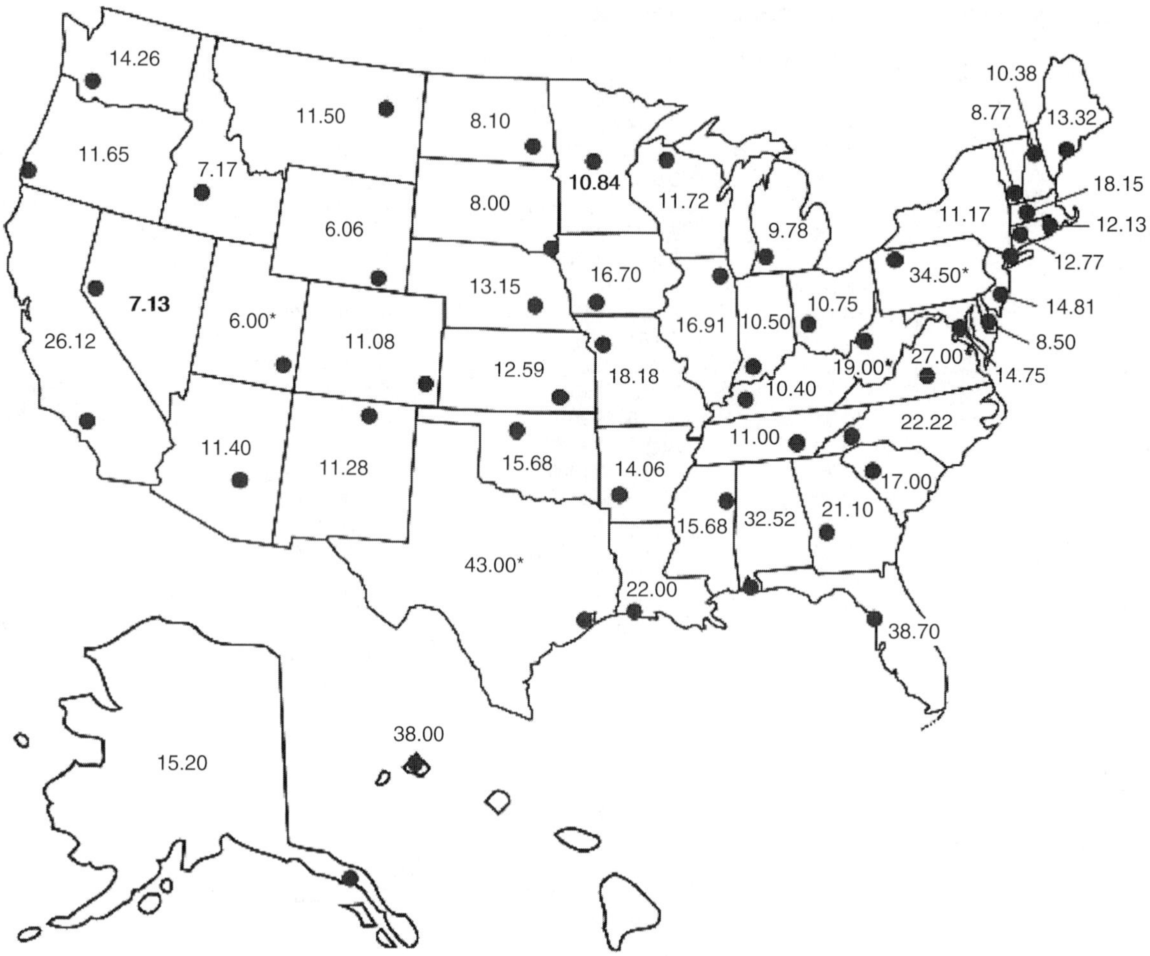

Figure 3D.14 Record maximum 24-h precipitation (in.) (through 1998) (*estimated). (From U.S. National Oceanic and Atmospheric Administration, *Comparative Climate Data for the United States Through 2000*, www.noaa.gov.)

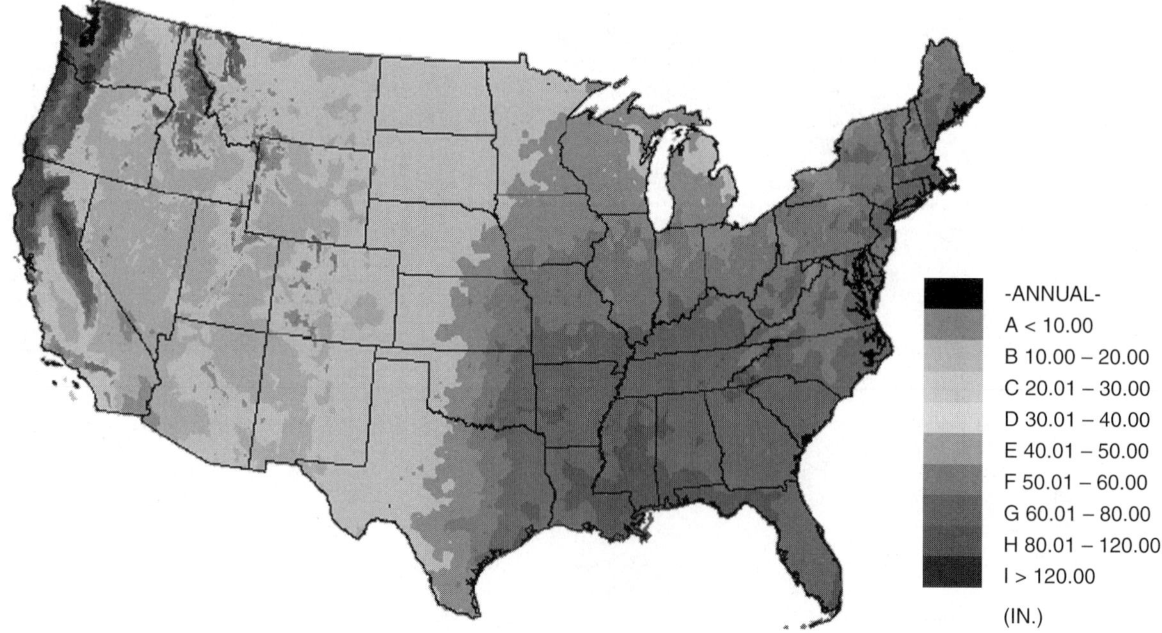

Figure 3D.15 U.S. record total precipitation. (From Climate Atlas of the United States, updated 8/27/02.)

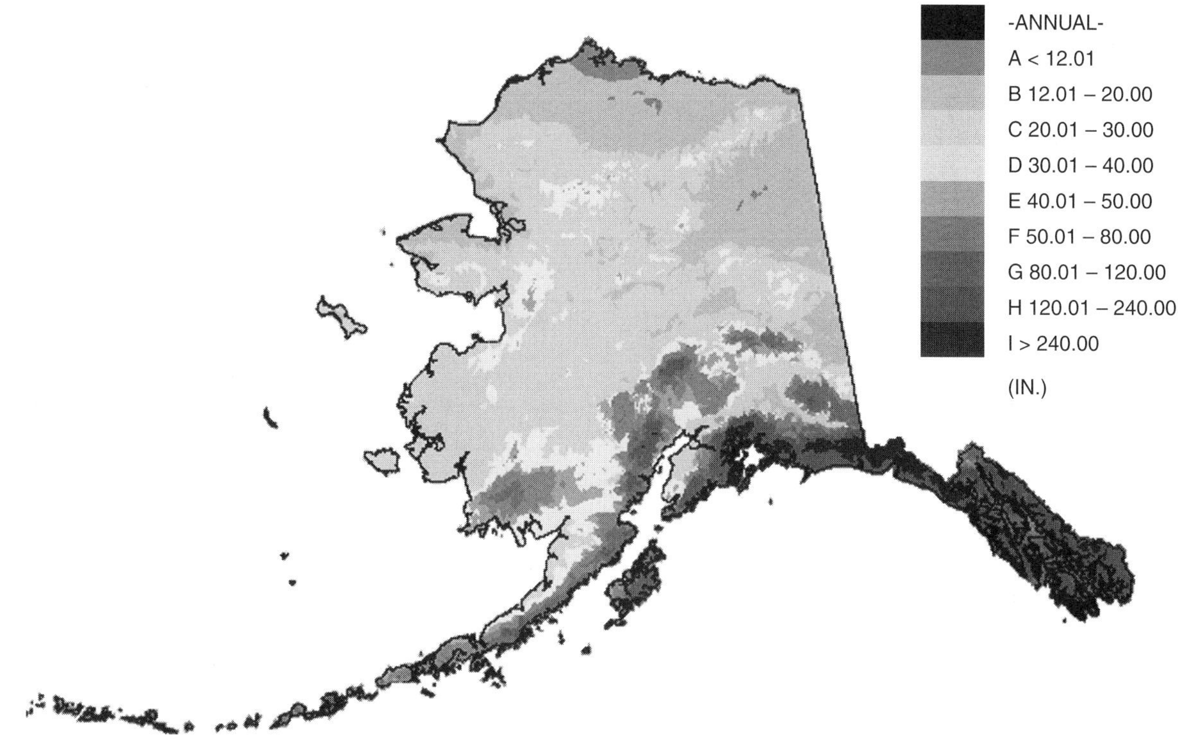

Figure 3D.16 Alaska record total precipitation. (From Climate Atlas of the United States, updated 8/27/02.)

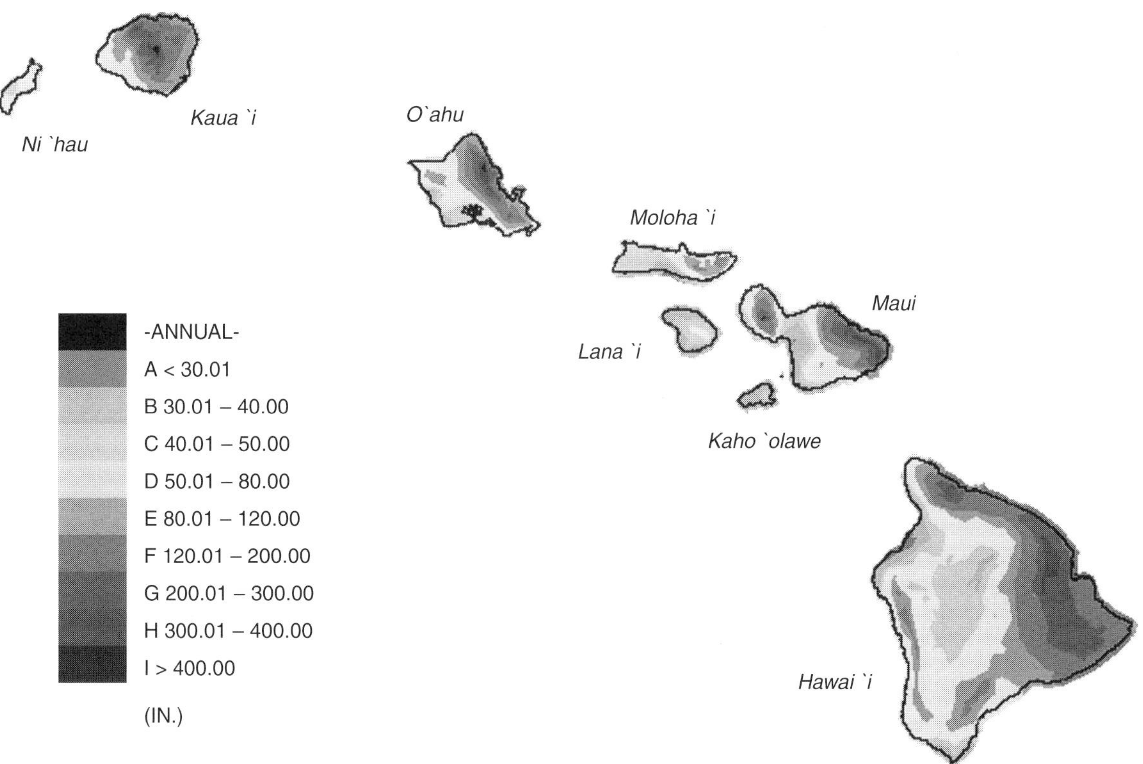

Figure 3D.17 Hawaii record total precipitation. (From Climate Atlas of the United States, updated 8/27/02.)

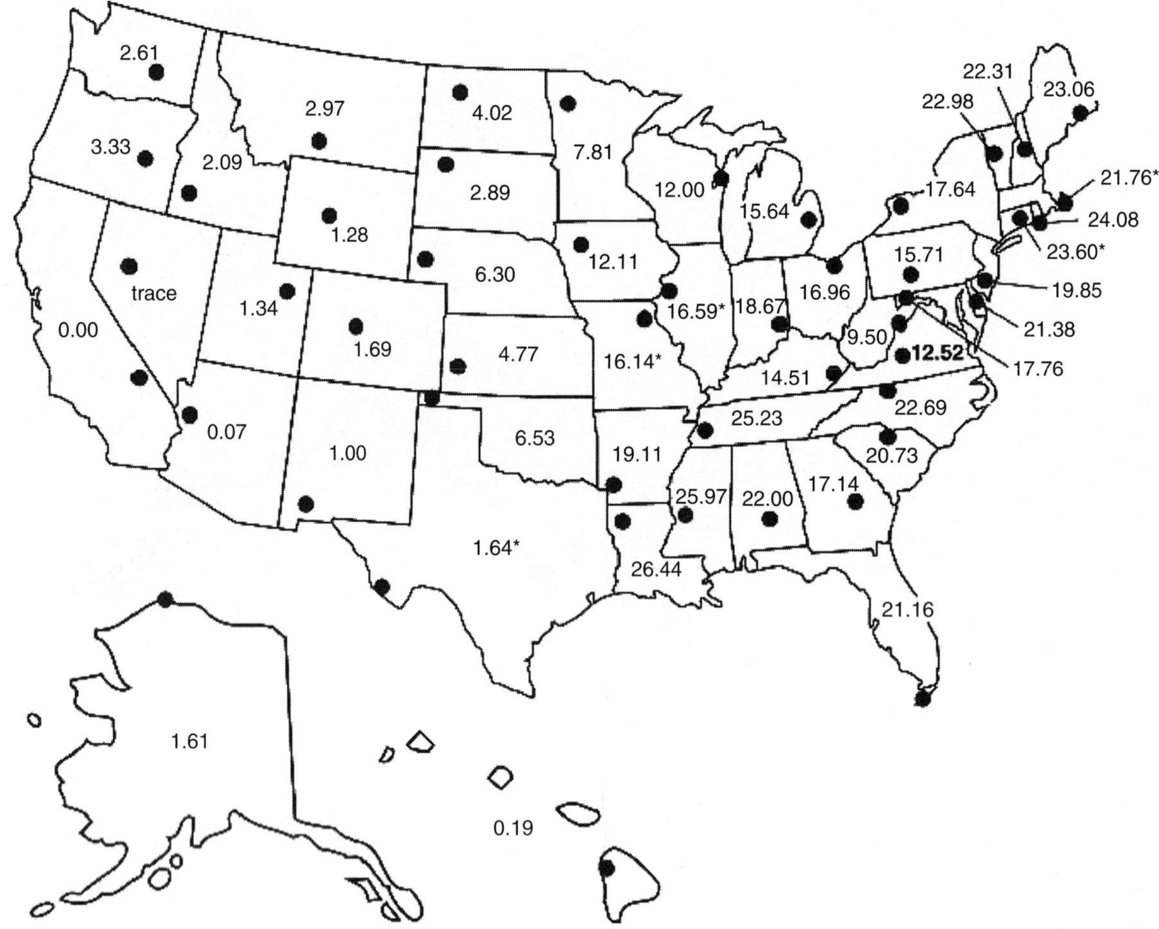

Figure 3D.18 Record minimum annual precipitation (in.) (through 1998) (*at least one month estimate). (From U.S. National Oceanic and Atmospheric Administration, *Comparative Climate Data for the United States Through 2000*, www.noaa.gov.)

Table 3D.23 Mean Number of Days with Precipitation 0.01 in. or More — Selected Cities of the United States

State	Station	Years	Jan	Feb	Mar	Apr	May	Jun	Jul	Aug	Sep	Oct	Nov	Dec	Annual
AL	Birmingham CO	11	10	10	11	9	10	9	12	10	9	8	10	9	117
	Birmingham AP	59	11	10	11	9	10	10	12	10	8	6	9	11	117
	Huntsville	35	12	10	12	10	10	10	11	9	8	7	10	11	118
	Mobile	61	11	9	10	7	8	11	16	14	10	6	8	10	121
	Montgomery	58	11	9	10	8	8	9	12	9	8	6	8	10	108
AK	Anchorage	38	8	8	7	6	7	8	12	14	14	12	9	11	115
	Annette	55	20	19	20	18	17	15	14	15	18	24	22	23	226
	Barrow	82	4	4	4	4	4	5	9	11	11	11	6	5	78
	Barter IS	39	5	5	5	6	6	6	8	12	10	12	8	5	87
	Bethel	44	9	7	9	9	11	13	15	18	16	12	11	10	141
	Bettles	43	8	7	7	5	7	10	12	14	12	11	10	10	114
	Big Delta	42	6	4	4	4	7	12	14	13	9	9	7	6	93
	Cold Bay	57	19	18	18	17	17	16	17	20	21	23	22	21	229
	Fairbanks	51	8	6	6	4	7	11	12	13	10	11	10	9	106
	Gulkana	44	7	6	5	3	5	10	13	12	11	9	7	8	94
	Homer	46	14	12	11	9	10	9	11	13	16	15	13	15	147
	Juneau	58	19	17	18	17	17	16	17	18	21	24	20	21	223
	King Salmon	57	11	9	10	10	12	13	15	17	17	13	12	12	152
	Kodiak	53	17	16	17	16	17	15	15	14	16	17	17	18	195
	Kotzebue	59	8	7	7	7	6	6	11	14	12	10	10	9	108
	McGrath	60	10	8	9	7	9	12	15	17	14	12	12	12	138
	Nome	56	11	9	9	9	8	9	12	16	14	11	12	10	130
	St. Paul Island	84	18	15	15	14	14	12	15	18	20	22	22	20	205
	Talkeetna	48	9	9	9	7	12	13	15	17	16	13	10	11	139
	Unalakleet	17	6	5	6	6	6	8	12	16	12	10	8	8	103
	Valdez	30	17	15	15	14	16	15	17	17	21	19	15	18	200
	Yakutat	56	19	18	19	18	19	17	18	18	21	24	21	22	235
AZ	Flagstaff	53	7	7	8	6	4	3	11	12	7	5	5	6	82
	Phoenix	63	4	4	4	2	1	1	4	5	3	3	2	4	36
	Tucson	62	4	4	4	2	2	2	10	9	5	3	3	4	53
	Winslow	71	4	4	5	3	3	2	7	9	5	4	3	4	54
	Yuma	45	3	2	2	1	—	—	1	2	1	1	1	2	17
AR	Fort Smith	57	8	8	9	10	11	9	7	7	7	7	7	8	97
	Little Rock	60	10	9	10	10	10	8	8	7	7	7	8	9	104
	North Little Rock	24	10	9	10	10	11	9	8	7	8	8	9	10	107
CA	Bakersfield	65	6	6	6	4	2	1	—	—	1	2	4	5	37
	Bishop	55	4	3	3	2	3	2	2	2	2	1	2	3	29
	Blue Canyon	55	12	12	13	10	7	3	1	1	3	6	10	12	90
	Eureka	92	16	14	16	12	9	5	2	3	4	9	13	16	119
	Fresno	53	8	8	7	4	2	1	—	—	1	2	5	7	45
	Long Beach	58	6	5	5	3	1	—	—	—	1	2	3	5	31
	Los Angeles AP	67	6	6	6	3	1	1	1	—	1	2	3	5	35
	Los Angeles CO	62	6	6	6	3	1	1	—	1	1	2	3	5	35

(Continued)

Table 3D.23 (Continued)

State	Station	Years	Jan	Feb	Mar	Apr	May	Jun	Jul	Aug	Sep	Oct	Nov	Dec	Annual
	Mount Shasta	42	12	11	12	9	7	5	2	2	3	6	10	12	93
	Redding	16	14	11	12	8	7	4	1	1	2	4	9	11	83
	Sacramento	63	10	9	9	5	3	1	—	—	1	3	7	9	58
	San Diego	62	7	6	7	5	2	1	—	0	1	2	4	6	42
	San Francisco AP	75	11	10	10	6	3	1	—	—	1	4	7	10	63
	San Francisco CO	65	11	10	11	6	3	1	—	1	2	4	8	10	69
	Santa Barbara	66	6	6	6	2	1	1	1	—	1	2	3	5	33
	Santa Maria	60	8	8	8	5	2	1	—	—	1	2	5	6	46
	Stockton	60	9	9	8	5	3	1	—	—	1	3	7	7	52
CO	Alamosa	56	4	4	5	5	6	5	9	10	6	5	4	4	67
	Colorado Springs	54	5	5	7	8	10	10	13	12	7	5	4	4	90
	Denver	61	6	6	9	9	11	9	9	9	6	5	6	5	89
	Grand Junction	56	7	6	8	7	6	4	5	7	6	6	6	6	73
	Pueblo	60	4	4	6	6	8	7	9	9	5	4	4	4	71
CT	Bridgeport	54	11	10	11	11	11	10	8	9	9	7	10	11	118
	Hartford	48	11	10	12	11	12	11	10	10	10	9	11	12	127
DE	Wilmington	55	11	9	11	11	11	10	9	9	8	8	9	10	117
DC	Washington Dulles AP	39	10	9	10	10	12	10	11	10	9	8	9	10	117
	Washington Nat'l AP	61	10	9	11	10	11	10	10	9	8	7	8	9	112
FL	Apalachicola	61	9	8	8	6	5	10	15	14	11	5	6	8	105
	Daytona Beach	59	7	7	8	6	8	13	13	14	14	11	7	7	115
	Fort Myers	59	6	5	6	4	8	15	18	18	16	8	4	5	112
	Gainesville	19	9	7	8	6	7	15	16	16	12	7	7	7	116
	Jacksonville	61	8	8	8	6	8	13	14	15	13	9	6	8	116
	Key West	54	6	5	5	5	8	12	12	15	16	11	7	7	109
	Miami	60	7	6	6	6	10	15	16	18	17	14	8	7	131
	Orlando	60	6	7	7	6	8	14	17	16	14	9	6	6	116
	Pensacola	39	10	9	9	6	7	10	14	13	9	5	8	9	110
	Tallahassee	41	10	9	9	6	8	13	17	14	9	5	7	8	115
	Tampa	56	7	7	7	5	6	12	16	16	13	7	5	6	106
	Vero Beach	19	9	7	8	6	8	14	14	14	16	13	9	8	126
	West Palm Beach	60	8	7	8	7	11	15	15	16	17	13	9	8	133
GA	Athens	59	11	9	11	9	9	10	11	9	8	7	8	10	111
	Atlanta	68	12	10	11	9	9	10	12	9	8	7	9	10	115
	Augusta	52	10	9	10	8	9	10	11	10	8	6	7	9	107
	Columbus	57	10	9	10	8	8	10	13	10	8	6	8	9	109
	Macon	54	11	9	10	8	9	10	12	10	8	6	7	9	110
	Savannah	52	9	8	9	7	8	12	13	13	10	6	7	8	110
HI	Hilo	60	17	17	23	25	25	25	27	27	24	24	23	21	279
	Honolulu	53	9	9	9	9	7	6	7	6	7	8	9	10	96
	Kahului	44	10	9	10	10	6	5	7	6	5	7	10	11	98
	Lihue	52	15	13	16	17	16	17	20	18	16	18	18	18	200

ID	Boise	63	12	10	10	8	8	6	2	2	4	6	10	11	89
	Lewiston	56	11	9	11	10	10	9	5	4	5	8	11	11	104
	Pocatello	53	12	10	10	8	9	7	4	4	5	5	9	11	95
IL	Cairo	45	10	9	12	11	11	9	9	8	7	7	9	10	112
	Chicago	44	11	9	12	13	11	10	10	9	9	9	11	11	125
	Moline	70	9	8	11	11	12	10	10	9	9	8	9	9	115
	Peoria	63	9	8	11	12	12	10	9	8	8	8	9	10	114
	Rockford	52	9	8	11	12	12	10	10	9	9	9	9	10	117
	Springfield	55	9	8	11	12	11	10	9	8	8	8	9	10	113
IN	Evansville	62	10	9	12	12	11	10	9	7	7	8	9	10	115
	Fort Wayne	56	12	10	13	13	12	10	10	9	9	9	11	12	132
	Indianapolis	63	12	10	13	12	12	10	10	9	8	8	10	12	126
	South Bend	63	16	12	14	13	12	11	9	9	9	10	13	15	144
IA	Des Moines	63	7	7	10	11	12	11	9	9	9	8	7	8	108
	Dubuque	50	9	8	11	12	12	11	10	9	9	9	9	10	117
	Sioux City	62	7	6	9	10	11	11	9	9	8	7	6	7	99
	Waterloo	52	7	7	9	10	12	10	9	9	9	7	7	7	104
KS	Concordia	40	5	5	8	9	11	10	9	9	8	6	5	5	89
	Dodge City	60	5	5	7	7	10	9	9	8	6	5	4	4	78
	Goodland	82	4	5	6	7	10	9	9	8	6	5	4	4	77
	Topeka	56	6	6	9	10	12	11	9	8	8	7	6	6	97
	Wichita	49	5	5	8	8	11	9	8	7	8	6	5	6	86
KY	Greater Cincinnati AP	55	12	11	13	13	12	11	10	9	8	8	11	12	130
	Jackson	22	14	13	14	12	14	12	12	9	9	9	12	14	143
	Lexington	58	12	11	13	12	12	11	11	9	8	8	11	12	130
	Louisville	55	11	10	13	12	12	10	10	8	8	7	10	12	124
	Paducah	19	10	9	11	11	11	9	8	7	7	8	10	10	110
LA	Baton Rouge	51	10	9	9	7	8	10	13	12	9	6	8	10	110
	Lake Charles	41	10	8	8	7	8	9	11	11	9	6	8	9	104
	New Orleans	54	10	9	9	7	8	11	14	13	10	6	8	10	114
	Shreveport	50	9	8	10	8	9	8	8	7	7	7	9	9	99
ME	Caribou	63	15	12	13	13	14	14	14	13	12	13	14	14	160
	Portland	62	11	10	11	12	12	11	10	9	9	9	12	11	129
MD	Baltimore	52	10	9	11	11	11	10	9	9	8	7	9	9	114
MA	Blue Hill	117	12	11	13	12	12	11	11	10	10	10	11	12	134
	Boston	51	12	10	12	11	12	11	9	10	9	9	11	12	126
	Worcester	47	12	11	12	12	13	11	10	10	10	9	12	12	134
MI	Alpena	43	15	11	12	12	12	11	10	11	12	13	14	15	146
	Detroit	44	13	11	13	13	11	10	10	10	10	10	12	13	135
	Flint	61	13	11	12	13	11	10	9	9	10	10	12	13	134
	Grand Rapids	39	16	12	12	13	11	10	9	9	11	11	13	16	144
	Houghton Lake	38	15	11	12	12	11	11	9	10	12	12	13	14	141
	Lansing	48	14	11	13	13	11	11	10	10	10	10	13	14	139
	Marquette	24	18	13	15	12	11	12	11	11	14	15	16	17	165
	Muskegon	62	17	13	13	12	11	9	9	9	10	11	14	16	143

(Continued)

Table 3D.23 (Continued)

State	Station	Years	Jan	Feb	Mar	Apr	May	Jun	Jul	Aug	Sep	Oct	Nov	Dec	Annual
	Sault Ste. Marie	61	19	14	13	11	11	11	10	11	13	14	17	19	165
MN	Duluth	61	12	9	11	11	12	13	12	11	12	10	11	11	134
	International Falls	63	11	9	10	9	12	13	12	11	12	10	11	11	130
	Minneapolis-St.Paul	64	9	7	10	10	11	12	10	10	10	8	8	9	116
	Rochester	42	9	8	10	12	12	11	11	10	10	9	9	9	120
	Saint Cloud	62	9	7	9	9	11	12	10	10	9	8	8	8	109
MS	Jackson	39	11	9	10	8	9	9	11	10	8	7	9	10	110
	Meridian	57	11	9	10	9	9	9	11	9	8	6	8	10	107
	Tupelo	19	11	10	11	9	11	10	8	7	6	7	9	11	111
MO	Columbia	33	8	8	11	11	12	9	8	8	8	9	9	9	112
	Kansas City	30	7	7	10	11	12	10	9	9	8	8	7	7	105
	St. Louis	45	9	8	11	11	11	9	9	8	8	8	9	9	111
	Springfield	57	8	8	10	11	11	10	8	8	8	8	9	9	109
MT	Billings	68	8	7	9	10	11	11	8	6	7	6	6	7	96
	Glasgow	47	8	6	7	7	10	11	8	7	6	5	6	7	90
	Great Falls	65	9	8	9	9	11	12	8	8	7	6	7	7	100
	Helena	62	8	6	8	8	11	12	8	8	6	6	7	8	95
	Kalispell	53	15	12	11	10	12	12	7	8	8	9	13	15	131
	Missoula	58	14	10	11	10	12	12	7	7	8	8	11	13	123
NE	Grand Island	64	5	6	7	9	11	10	9	8	7	5	5	5	87
	Lincoln	31	6	5	8	10	12	9	9	9	8	6	6	6	93
	Norfolk	57	6	6	8	9	11	10	9	8	8	6	5	5	91
	North Platte	50	5	5	7	8	11	10	10	8	7	5	5	4	84
	Omaha Eppley AP	66	6	7	8	10	12	10	9	9	8	6	6	6	99
	Omaha (North)	18	5	6	10	10	12	10	10	9	8	7	6	7	101
	Scottsbluff	59	5	5	7	9	12	11	8	7	7	5	5	5	86
	Valentine	47	4	5	7	9	11	10	9	8	7	5	4	4	83
NV	Elko	72	9	9	9	8	8	6	3	3	4	5	7	9	79
	Ely	64	7	7	8	8	7	5	5	6	4	5	5	6	73
	Las Vegas	54	3	3	3	2	1	1	3	3	2	2	2	3	26
	Reno	60	6	6	6	4	4	3	2	2	3	3	5	6	51
	Winnemucca	53	8	7	8	7	7	5	2	2	3	4	7	8	69
NH	Concord	61	11	9	11	11	12	11	10	10	9	9	11	11	127
	Mt. Washington	70	19	18	19	18	17	16	16	15	15	15	19	20	209
NJ	Atlantic City AP	59	11	10	11	11	10	9	9	9	8	7	9	10	113
	Atlantic City CO	40	10	9	10	10	10	8	9	8	8	7	9	10	109
	Newark	61	11	10	11	11	12	10	10	9	9	8	10	11	122
NM	Albuquerque	63	4	4	5	3	4	4	9	10	6	5	4	4	61
	Clayton	53	3	3	5	5	8	8	10	9	6	4	3	3	68
	Roswell	30	4	3	3	3	4	5	6	8	7	5	3	4	55
NY	Albany	56	13	11	12	12	13	11	10	10	10	9	12	12	135
	Binghamton	51	17	14	15	14	13	12	11	11	11	12	15	17	161
	Buffalo	59	20	17	16	14	13	11	10	10	11	12	16	19	168
	Islip	19	11	9	11	12	11	9	9	9	9	8	10	10	118

	New York C. Park	133	11	10	11	11	11	10	10	10	8	8	9	10	121
	New York (JFK AP)	44	10	10	11	11	11	10	9	9	8	8	10	11	118
	New york (Laguardia AP)	62	11	10	11	11	11	10	9	9	8	8	10	11	118
	Rochester	62	18	16	15	13	12	11	10	10	11	12	15	18	160
	Syracuse	53	19	16	17	14	13	11	11	11	11	13	17	19	171
NC	Asheville	38	11	9	12	10	11	12	12	12	10	8	9	10	125
	Cape Hatteras	45	11	10	11	9	10	9	12	11	9	9	9	10	120
	Charlotte	63	10	9	11	9	9	10	11	10	7	7	8	10	111
	Greensboro-Wnstn-Slm-HPT	74	10	9	11	9	10	10	12	10	8	7	8	9	115
	Raleigh	58	10	10	10	9	10	10	11	10	8	7	8	9	113
	Wilmington	51	11	10	10	8	9	10	13	12	10	7	8	9	117
ND	Bismarck	63	8	7	8	8	10	12	9	8	7	6	6	7	96
	Fargo	60	9	7	8	8	10	11	10	9	8	7	6	8	101
	Williston	41	8	6	8	8	10	11	9	7	7	6	7	8	95
OH	Akron	54	16	14	15	14	13	11	11	10	10	10	14	16	154
	Cleveland	61	16	14	15	15	13	11	10	10	10	11	14	16	155
	Columbus	63	13	11	13	13	13	11	11	9	8	9	11	13	137
	Dayton	59	13	11	13	13	12	11	10	9	8	9	11	12	132
	Mansfield	43	14	12	14	14	13	11	10	10	9	10	13	14	144
	Toledo	47	13	11	13	13	12	10	9	9	10	9	12	14	134
	Youngstown	59	17	15	15	15	13	12	11	10	10	11	15	17	160
OK	Oklahoma City	63	5	6	7	8	10	9	6	6	7	7	5	6	83
	Tulsa	63	6	7	8	9	11	9	6	7	7	7	6	7	91
OR	Astoria	49	22	19	21	18	16	13	8	8	10	16	21	22	193
	Burns	18	11	10	11	9	9	6	3	3	4	6	12	11	95
	Eugene	60	18	15	17	13	10	7	3	4	6	11	17	18	139
	Medford	73	14	11	12	10	8	5	2	2	4	7	13	14	101
	Pendleton	67	12	10	11	9	8	7	3	3	4	7	12	12	98
	Portland	62	18	16	17	15	12	9	4	5	7	12	18	19	152
	Salem	65	18	16	17	14	11	8	3	4	7	12	18	19	147
	Sexton Summit	65	16	15	16	12	9	6	2	3	5	9	16	17	126
PC	Guam	45	20	18	19	20	20	24	26	26	25	25	25	24	271
	Johnston Island	28	11	12	15	14	13	12	12	13	14	16	15	16	162
	Koror	51	23	19	20	19	23	25	24	22	20	23	22	24	265
	Kwajalein, Marshall IS	50	16	13	15	17	21	23	24	23	22	24	23	19	240
	Majuro, Marshall IS	48	18	16	18	20	23	24	24	23	23	23	23	22	258
	Pago Pago, Amer Samoa	36	24	21	23	22	20	19	19	18	17	21	20	23	247
	Pohnpei, Caroline IS	51	22	20	23	24	27	27	27	26	24	25	25	24	294
	Chuuk, E. Caroline IS	51	19	16	19	20	24	24	24	24	22	23	23	23	264
	Wake Island	50	11	10	12	14	15	15	19	19	19	19	15	13	180

(Continued)

Table 3D.23 (Continued)

State	Station	Years	Jan	Feb	Mar	Apr	May	Jun	Jul	Aug	Sep	Oct	Nov	Dec	Annual
	Yap, W Caroline IS	54	21	18	18	18	21	24	24	24	23	24	23	22	258
PA	Allentown	59	11	10	11	12	12	11	10	10	9	8	10	11	124
	Erie	49	19	15	15	14	12	10	10	11	11	13	16	19	163
	Harrisburg	49	11	10	11	13	13	11	10	9	9	9	10	10	126
	Middletown/ Harrisburg AP	24	11	10	11	12	13	10	10	9	9	8	10	10	122
	Philadelphia	62	11	9	11	11	11	10	9	9	8	8	9	10	117
	Pittsburgh	50	16	14	15	14	13	12	11	10	10	10	13	16	152
	Avoca	47	12	11	13	12	13	12	11	11	10	10	12	12	139
	Williamsport	58	12	11	12	13	13	12	11	11	10	10	12	12	139
RI	Block IS	36	10	9	11	10	10	9	7	8	7	8	10	11	110
	Providence	49	11	10	12	11	11	11	9	9	9	9	10	12	124
SC	Charleston AP	60	10	9	10	7	9	11	14	13	10	6	7	9	114
	Charleston CO	27	10	8	8	7	8	10	11	11	10	6	7	9	104
	Columbia	55	10	9	10	8	8	10	12	11	8	6	7	9	109
	Greenville- Spartanburg AP	40	11	9	11	9	10	10	12	10	9	7	9	10	117
SD	Aberdeen	71	6	6	7	8	10	10	9	8	7	5	6	6	89
	Huron	63	6	7	8	9	10	11	9	8	7	6	6	6	92
	Rapid City	60	6	7	8	10	12	12	9	8	6	6	6	6	95
	Sioux Falls	57	6	7	9	10	11	11	10	9	8	6	7	6	98
TN	Bristol- JhnCty- Kgsprt	57	14	12	13	11	12	11	12	10	8	8	10	12	133
	Chattanooga	72	12	10	12	10	10	11	12	10	8	7	9	11	121
	Knoxville	60	12	11	13	11	11	11	11	9	8	8	10	11	127
	Memphis	52	10	9	11	10	9	9	9	7	7	6	9	10	107
	Nashville	61	11	10	12	11	11	10	10	9	8	7	9	11	119
	Oak Ridge	61	12	11	13	11	11	11	12	10	8	8	10	11	128
TX	Abilene	63	5	5	5	6	8	6	5	6	6	6	5	5	67
	Amarillo	61	4	4	5	5	8	8	8	8	6	5	4	4	69
	Austin	61	8	8	8	7	9	7	5	5	7	7	7	8	85
	Brownsville	60	7	6	4	4	5	6	4	7	10	7	6	7	73
	Corpus Christi	63	8	7	5	5	6	6	5	6	9	7	6	7	77
	Dallas-Fort Worth	49	7	7	8	8	9	7	5	5	6	6	6	7	79
	Del Rio	39	5	5	5	5	7	5	4	4	6	5	5	5	62
	El Paso	63	4	3	2	2	2	3	8	8	5	4	3	4	49
	Galveston	63	10	8	8	6	6	7	9	9	9	7	8	10	96
	Houston	33	10	8	9	7	8	10	9	9	9	8	8	9	105
	Lubbock	56	4	4	4	5	7	7	7	7	6	5	3	4	63
	Midland-Odessa	55	4	4	3	3	6	5	5	6	6	5	3	3	51
	Port Arthur	49	10	9	8	7	7	9	11	12	10	7	8	9	105
	San Angelo	55	5	5	4	5	7	5	4	5	6	5	4	4	59
	San Antonio	60	8	7	8	7	8	7	4	5	7	6	7	8	82

	Victoria	41	8	7	7	6	7	8	7	9	10	7	7	8	91
	Waco	59	7	7	8	7	9	6	4	5	6	6	7	6	80
	Wichita Falls	59	5	5	6	7	9	7	5	6	6	6	5	5	71
UT	Milford	56	7	7	8	6	5	3	5	6	4	4	5	6	67
	Salt Lake City	74	10	9	10	10	8	5	4	6	5	6	8	9	91
VT	Burlington	59	15	11	13	12	14	13	12	12	12	12	14	15	154
VA	Lynchburg	58	11	9	11	10	12	10	11	10	8	7	9	9	118
	Norfolk	54	11	10	11	10	10	9	11	10	8	7	8	9	116
	Richmond	65	10	9	11	9	11	9	11	9	8	7	8	9	113
	Roanoke	55	10	10	11	10	12	10	12	10	9	7	9	9	119
	Wallops Island	28	11	10	11	10	10	9	11	8	8	8	9	10	115
WA	Olympia	61	20	17	18	15	11	9	5	6	8	14	19	21	163
	Quillayute	36	23	19	21	19	17	14	11	10	12	18	22	23	209
	Seattle CO	48	18	15	17	14	11	9	5	6	8	11	18	18	151
	Seattle Sea-Tac AP	58	19	16	17	14	11	9	5	6	9	13	18	19	155
	Spokane	55	14	11	11	9	9	8	5	5	6	8	13	14	112
	Walla Walla	55	13	10	12	10	9	7	4	3	4	7	15	12	106
	Yakima	56	10	7	6	5	5	5	2	3	3	5	8	10	69
PR	San Juan	47	17	13	12	13	16	15	19	19	18	17	19	19	197
WV	Beckley	39	16	14	15	15	14	13	13	11	11	10	12	15	159
	Charleston	55	15	14	15	14	13	12	13	11	9	9	12	14	151
	Elkins	58	18	16	17	15	15	13	14	12	11	11	13	17	171
	Huntington	41	13	12	14	13	13	11	11	10	8	9	11	13	139
WI	Green Bay	53	10	8	11	11	11	11	10	11	10	9	10	10	121
	La Crosse	50	9	7	10	11	11	11	10	10	10	8	8	9	114
	Madison	54	10	8	11	12	12	11	10	10	9	9	10	10	120
	Milwaukee	62	11	10	12	12	12	11	10	9	9	9	10	11	125
WY	Casper	52	7	8	9	10	11	9	8	6	7	7	7	7	95
	Cheyenne	67	6	6	9	10	12	11	11	10	8	6	6	6	100
	Lander	56	4	5	7	8	9	6	6	5	6	5	5	4	71
	Sheridan	62	9	8	11	11	12	11	8	6	7	8	8	8	106

Note: Through 2002. This table shows mean number of days per month with at least 0.01 in. of precipitation. This is the smallest amount of precipitation numerically recorded, and includes the liquid water equivalent of frozen precipitation. The frequency of days with precipitation should not be considered as frequency of cloudy days.

Source: From U.S. National Oceanic and Atmospheric Administration, *Comparative Climatic Data for the United States Through 2000*, www.noaa.gov.

Table 3D.24 Record Minimum Annual Precipitation by State

State	Precipitation (in.)	Date	Station	Elevation (ft)
Alabama	22.00	1954	Primrose Farm	180
Alaska	1.61	1935	Barrow	31
Arizona	0.07	1956	Davis Dam	660
Arkansas	19.11	1936	Index	300
California	0.00	1929	Death Valley	−282
Colorado	1.69	1939	Buena Vista	7,980
Connecticut	23.60[a]	1965	Baltic	140
Delaware	21.38	1965	Dover	30
Florida	21.16	1989	Conch Key	6
Georgia	17.14	1954	Swainsboro	320
Hawaii	0.19	1953	Kawaihae	est. 75
Idaho	2.09	1947	Grand View	2,360
Illinois	16.59[a]	1956	Keithsburg	540
Indiana	18.67	1934	Brooksville	630
Iowa	12.11	1958	Cherokee	1,360
Kansas	4.77	1956	Johnson	3,270
Kentucky	14.51	1968	Jeremiah	1,160
Louisiana	26.44	1936	Shreveport	170
Maine	23.06	1930	Machias	30
Maryland	17.76	1930	Picardy	1,030
Massachusetts	21.76[a]	1965	Chatham L.S.	20
Michigan	15.64	1936	Croswell	730
Minnesota	7.81	1936	Angus	870
Mississippi	25.97	1936	Yazoo City	120
Missouri	16.14[a]	1956	La Belle	770
Montana	2.97	1960	Belfry	4,040
Nebraska	6.30	1931	Hull	4,400
Nevada	T	1898	Hot Springs	4,072
New Hampshire	22.31	1930	Bethlehem	1,440
New Jersey	19.85	1965	Canton	20
New Mexico	1.00	1910	Hermanas	4,540
New York	17.64	1941	Lewiston	320
North Carolina	22.69	1930	Mount Airy	1,070
North Dakota	4.02	1934	Parshall	1,930
Ohio	16.96	1963	Elyria	730
Oklahoma	6.53	1956	Regnier	4,280
Oregon	3.33	1939	Warm Springs Reservoir	3,330
Pennsylvania	15.71	1965	Breezewood	1,350
Rhode Island	24.08	1965	Block Island	40
South Carolina	20.73	1954	Rock Hill	667
South Dakota	2.89	1936	Ludlow	2,850
Tennessee	25.23	1941	Halls	310
Texas	1.64[a]	1956	Presidio	2,580
Utah	1.34	1974	Myton	5,080
Vermont	22.98	1941	Burlington	330
Virginia	12.52	1941	Moores Creek Dam	1,950
Washington	2.61	1930	Wahluke	416
West Virginia	9.50	1930	Upper Tract	1,540
Wisconsin	12.00	1937	Plum Is.	590
Wyoming	1.28	1960	Lysite	5,260

[a] At least one month estimated.

Table 3D.25 Velocity of Fall, Number of Drops, and Kinetic Energy for Rainfall of Various Intensities

	Intensity (in./hr)	Median Diameter (mm)	Velocity of Fall (ft/sec)	Drops per Square Foot (no./sec)	Kinetic Energy (ft-lbs.per sq. ft/hr)
Fog	0.005	0.01	0.01	6,264,000	4.043×10^{-8}
Mist	0.002	0.1	0.7	2.510	7.937×10^{-5}
Drizzle	0.01	0.96	13.5	14	0.148
Light Rain	0.04	1.24	15.7	26	0.797
Moderate Rain	0.15	1.60	18.7	46	4.241
Heavy Rain	0.60	2.05	22.0	46	23.47
Excessive Rain	1.60	2.40	24.0	76	74.48
Cloudburst	4.00	2.85	25.9	113	216.9
Do	4.00	4.00	29.2	41	275.8
Do	4.00	6.00	30.5	12	300.7

SECTION 3E SNOW AND SNOW MELT

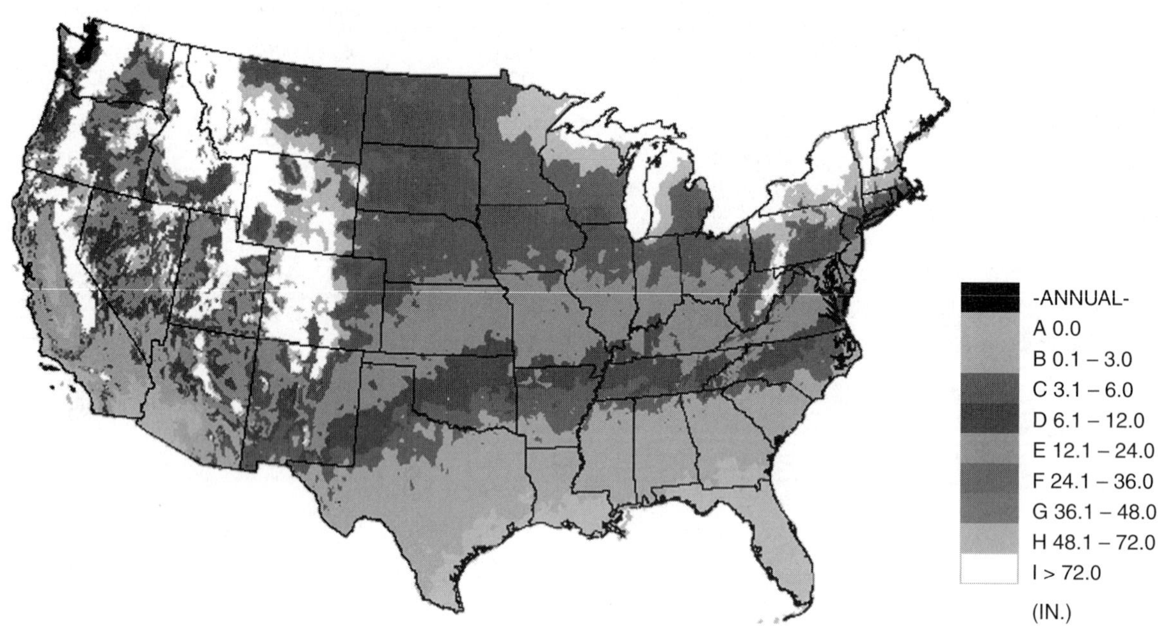

Figure 3E.19 U.S. mean total snow. (From Climate Atlas of the United States, updated 8/27/02.)

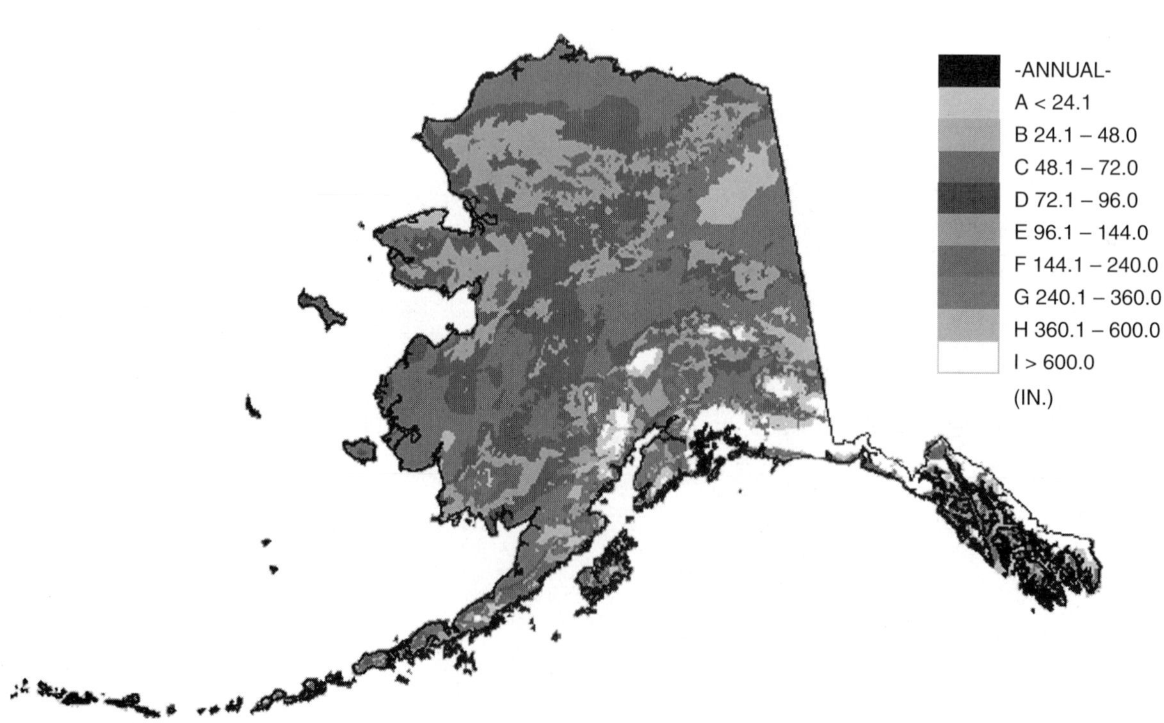

Figure 3E.20 Alaska mean total snow. (From Climate Atlas of the United States, updated 8/27/02.)

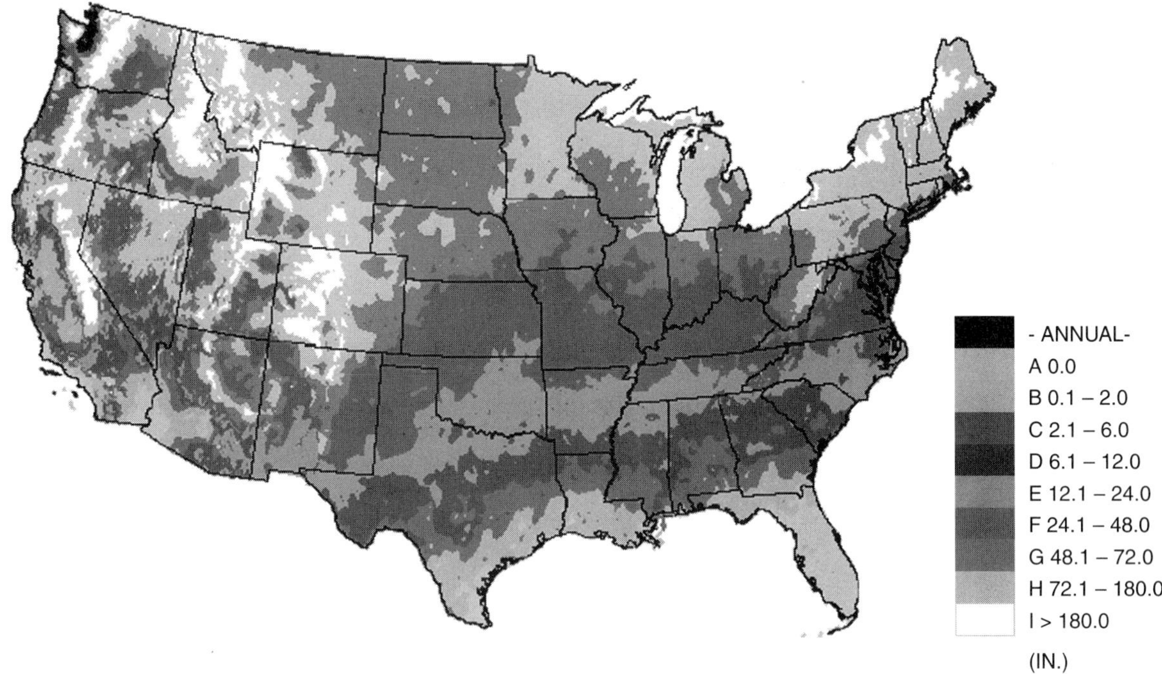

Figure 3E.21 U.S. annual (Aug–Jul) record total snowfall. (From Climate Atlas of the United States, updated 8/27/02.)

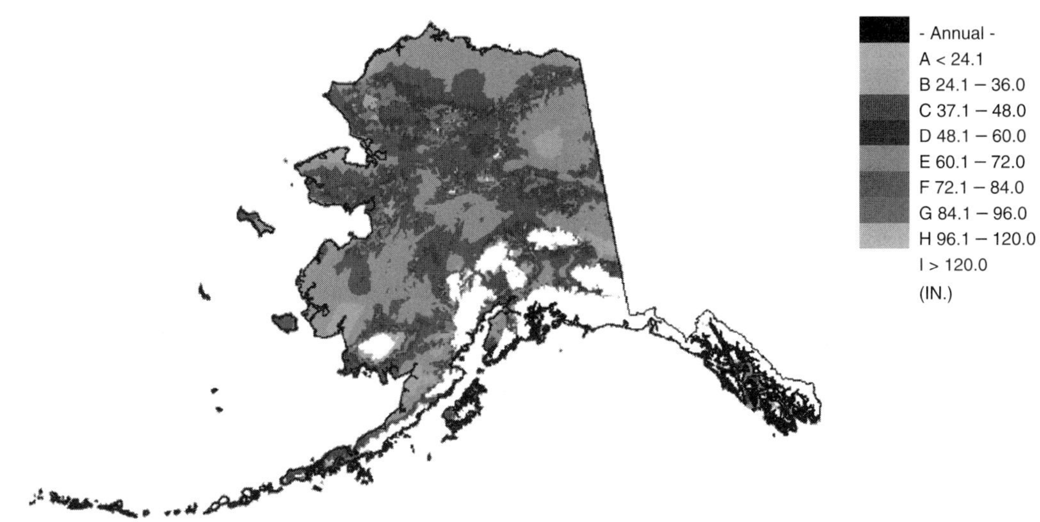

Figure 3E.22 General Pattern of Annual World Precipitation. (Form Environmental Science Service: Administration, *Climates of the World*, 1969, www.noaa.gov.)

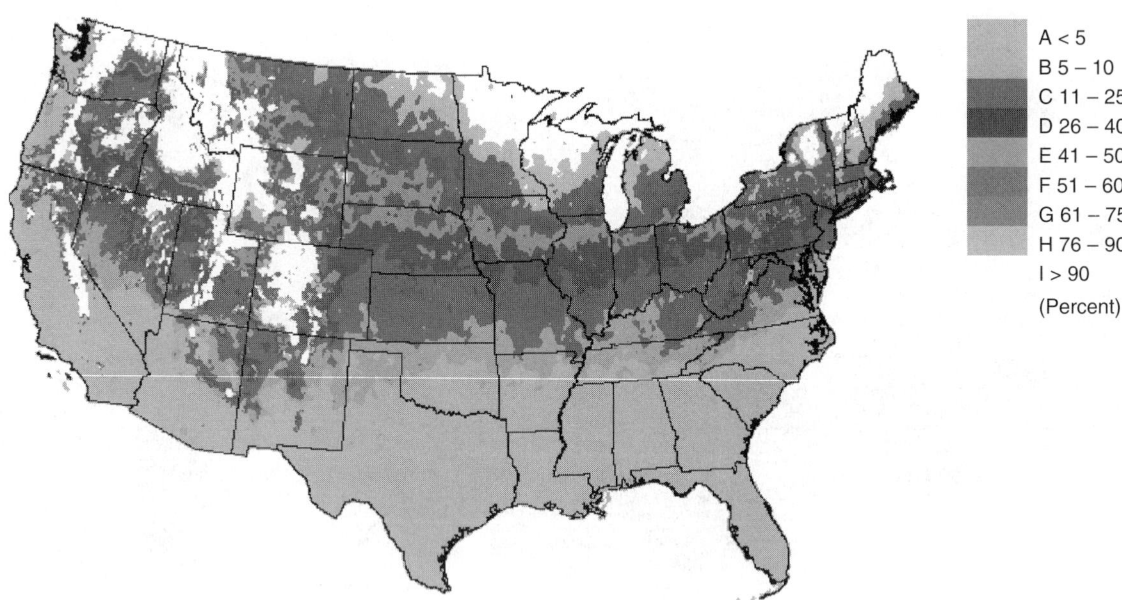

Figure 3E.23 Average January world temperature. (From *Climates of the World*, Historical Climatology Series 6–14, 1991, www.noaa.gov.)

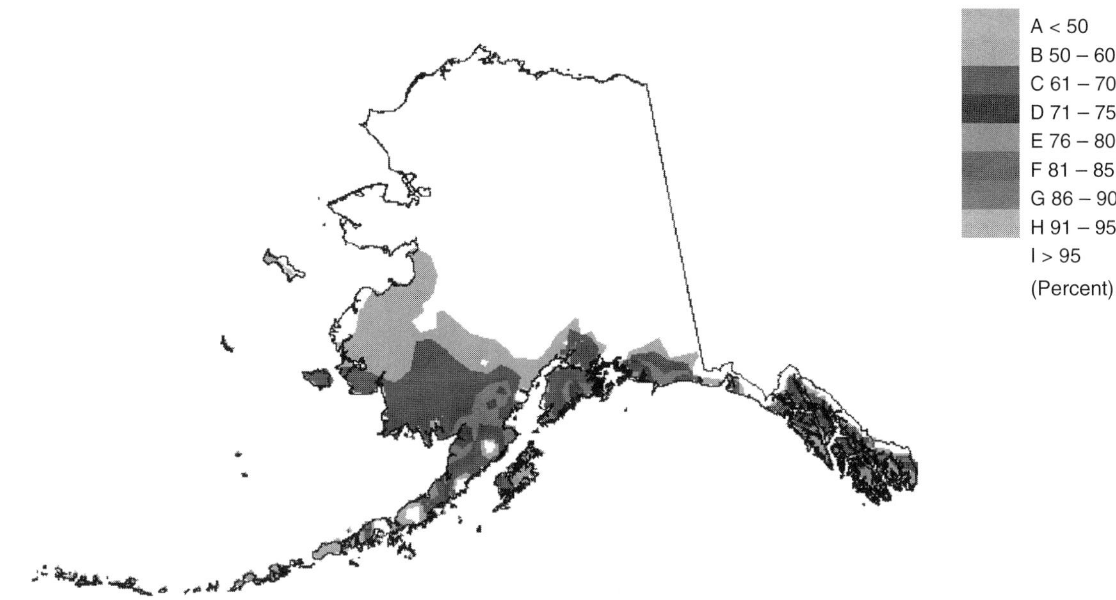

Figure 3E.24 Average July world temperature. (From *Climates of the World*, Historical Climatology Series 6–14, 1991, www.noaa.gov.)

Table 3E.26 Physical Properties of Snow and Ice

	Density (g/cm^3)	Porosity (percent)	Air Permeability (g/cm^3/sec)	Grain Size (mm)
New Snow	0.01–0.3	99–67	>400–40	0.01–5
Old Snow	0.2–0.6	78–35	100–20	0.5–3
Firm[a]	0.4–0.84	56–8	40–0	0.5–5
Glacier Ice	0.84–0.917	8–0	0	1–100

[a] Firm is snow which has been modified into a dense compact material by deformation, refreezing, recrystallization, and other processes.

Table 3E.27 Heat Supplied to Melting Snow by Different Processes

Heat Supply	Extreme Conditions	Approximate Heat Supplied[a]
Convection from turbulent air	70° dry bulb, 20-mile wind	600
Condensation of atmospheric moisture	60° dew point, 20-mile wind	600
Absorption of solar radiation	Very moist air, cloudy at night	200
Warm rain	4 in., 50° wet bulb	100
Conduction from soil	New snow	20

[a] Calories per square centimeter per day.

Table 3E.28 Relation of Snow Melt to Snow Evaporation

Air Temperature (°C)	Relative Humidity (%)	Snow Evaporated (g cm^{-2} day^{1})	Heat Transfer from Air to Snow (cal cm^{-2} day^{1})	Heat Required in Evaporation Process[a] (cal cm^{-2} day^{1})	Heat Available to Melt Snow (cal cm^{-2} day^{1})	Melted Snow (g cm^{-2} day^{1})	Melt/Evaporation
5	20	2.02	900	1,370	0	0	
10	20	1.69	1,790	1,150	640	8.0	4.7
15	20	1.25	2,690	850	1,840	23.0	18.4
20	20	0.67	3,590	460	3,130	39.1	58.4

[a] Heat required in evaporation process is equal to heat transfer from air to snow plus heat obtained by lowering of snow-surface temperature.

Table 3E.29 Melting Constant for Snow

Location	Descriptive Notes	Melting Period	Melting Constant
Albany, NY	Tests of small cylinders	8–12 hours	0.04–0.06
Donner Summit, CA	Observations in 1917	Apr 1–May 6	0.071
Gooseberry Creek, UT	Field measurements	Apr 23–May 9	0.091
Gooseberry Creek, UT	Tests of cores	6–9 hr	0.05–0.07
Finland	All basins, 1934–1937	Apr	0.108
Soda Springs, CA	Average, 1936–1941	Apr	0.051
New England floods	Studies by Boston Soc. C.E.	1–14 days	0.01–0.04
NY and PA basins	Flood runoff studies	Mar or Apr	0.04–0.07
LaGrange Brook, NY	Basin area, 36 acres	Mar 28–Apr 6	0.09
New England floods, 1936	Geol. Survey, average values	Mar 9–22	0.03–0.05
Permigewasset Basin, NH	Flood of March 1936	Mar 17–20	0.16
Crater Lake, OR	Small test plots	Mar 3–Jun 9	0.153
Crater Lake, OR	Small test plots	May 26–Jun 2	0.658

Note: The melting constant is the depth of water in inches melted per degree day. A degree day is a unit of heat resulting from a day with a mean temperature one degree Fahrenheit above 32°F.

Table 3E.30 Snow Survey Reports — Western United States

SNOTEL (SNOWpack Telemetry) is an extensive system operated by the Natural Resources Conservation Services (NRCS). The system supports the Congressional mandate from the mid-1930's "to measure snow pack in the mountains of the West and forecast the water supply." SNOTEL has been in operation since 1980 replacing manual measurements and provides the data necessary to support the data needs of the NRCS and others

Specific products (data) supported include

- Climate Information
- Data Collection Technology
- Snow Survey Information
- Water Supply Forecasting
  - Colorado River Basin
  - Columbia River Basin/Alaska
  - Great Basin/California/Pacific Coastal
  - Missouri River Basin
- Hydraulics and Hydrology
- Irrigation and Water Management
- Water and Wind Erosion
- Water Quality Assessment and Monitoring
- Wetlands and Drainage

Regional

Water and Climate Monitoring
Natural Resources Conservation Service
Unites States Department of Agriculture
101 SW Main, Suite 1600
Portland, OR 97204-3224

Water and Climate Services
Natural Resources Conservation Service
Unites States Department of Agriculture
101 SW Main, Suite 1600
Portland, OR 97204-3224

States

Alaska
Data Collection Officer, Natural Resources Conservation Service, 949 E. 36th Ave., Suite 400, Anchorage, AK 99508-4362, (907) 271-2424

Arizona
Water Supply Specialist, Natural Resources Conservation Service, 3003 N. Central Ave., Suite 800, Phoenix, AZ 85012-2945

California
Water Supply Specialist, 430 G Street, #4164, Davis, CA 95616, (530) 792-5624, (530) 792-5791 (fax)

Colorado
Snow Survey Supervisor, USDA Natural Resources Conservation Service, Snow Survey Office, 655 Parfet St., Rm. E200C, Lakewood, CO 80215-5517, (303)236-2910 Ext. 235

Idaho
Water Supply Specialist, Natural Resources Conservation Service, Snow Survey Office, 9173 West Barnes Drive, Suite C, Boise, Idaho 83709-1574

Montana
Water Supply Specialist, Natural Resources Conservation Service, 10 E. Babcock, Room 443, Bozeman, Montana 59715

Nevada
Water Supply Specialist, Natural Resources Conservation Service, 5301 Longley Lane, Building F, Suite 201, Reno, NV 89511

New Mexico
Water Supply Specialist, USDA—NRCS, 6200 Jefferson NE, Albuquerque, NM

Oregon
Snow Survey Supervisor, USDA, NRCS, 101 SW Main Street, Suite 1300, Portland, OR 97204, 503-414-3266

Utah
Snow Survey Supervisor, USDA–NRCS, Snow Surveys, 245 N. Jimmy Doolittle Road, Salt Lake City, UT 84116

Washington
Water Supply Specialist, Washington Snow Survey Office, 2021 E. College Way, Suite 214, Mount Vernon, WA 98273

Wyoming
Water Supply Specialist, Federal Building, Room 3124, 100 East B Street Casper, Wyoming 82601-1911

http://www.wcc.nrcs.usda.gov

Table 3E.31 Greatest Snowfalls in North America

	Place	Date	In.	Cm
24 hr	Silver Lake, CO	Apr 14–15, 1921	76	195.6
1 month	Tamarack, CA	Jan 1911	390	991
1 storm	Mt. Shasta Ski Bowl, CA	Feb 13–19, 1959	189	480
1 season	Mount Baker, WA	1998–1999	1,140	2,895.6

Table 3E.32 National Snowfall and Snow Depth Extremes

		Location					
Coop Station Number	Snow Amount (in.)	Station Name	State		Ending Date	Number of Years of Non-Missing Data	Data Period Analyzed
Greatest daily snowfall	62.0	509146	Thompson Pass	AK	12/29/1955	19	1952–1973
Greatest 2-day snowfall (snowed both days)	120.6	509146	Thompson Pass	AK	12/30/1955	19	1952–1073
Greatest 3-day snowfall (snowed all 3 days)	147.0	509146	Thompson Pass	AK	12/30/1955	19	1952–1973
Greatest 4-day snowfall (snowed all 4 days)	163.0	509146	Thompson Pass	AK	12/30/1955	19	1952–1973
Greatest 5-day snowfall (snowed all 5 days)	175.4	509146	Thompson Pass	AK	12/31/1955	19	1952–1973
Greatest 6-day snowfall (snowed all 6 days)	172.6	509146	Thompson Pass	AK	2/24/1953	19	1952–1973
Greatest 7-day snowfall (snowed all 7 days)	186.9	509146	Thompson Pass	AK	2/25/1993	19	1952–1973
Greatest monthly snowfall total	297.9	509146	Thompson Pass	AK	2/1953	17	1952–1973
Greatest Aug–July snowfall total	1069.8	456898	Rainier Paradise Rangers	WA	1974	17	1948–2000
Greatest daily snow depth	293.0	456898	Rainier Paradise Rangers	WA	4/12/1974	11	1948–2000

Note: Missing data may cause apparent discrepancies between the daily extreme, monthly total, and seasonal total snowfall values. The monthly and seasonal totals were based on complete data; if any days were missing, then the monthly or seasonal total could not be computed for that year. Daily snowfall extremes were not as susceptible to missing data. Consequently, it may be possible for a 1-day extreme to be greater than a multiple-day extreme, a daily extreme to be greater than a monthly total, and a monthly total to be greater than a seasonal total. Checking the "number of years with non-missing data" parameter is an important part of using this snow climatology.

CHAPTER **4**

Hydrologic Elements

Brian Burke

CONTENTS

SECTION 4A HYDROLOGIC CYCLE

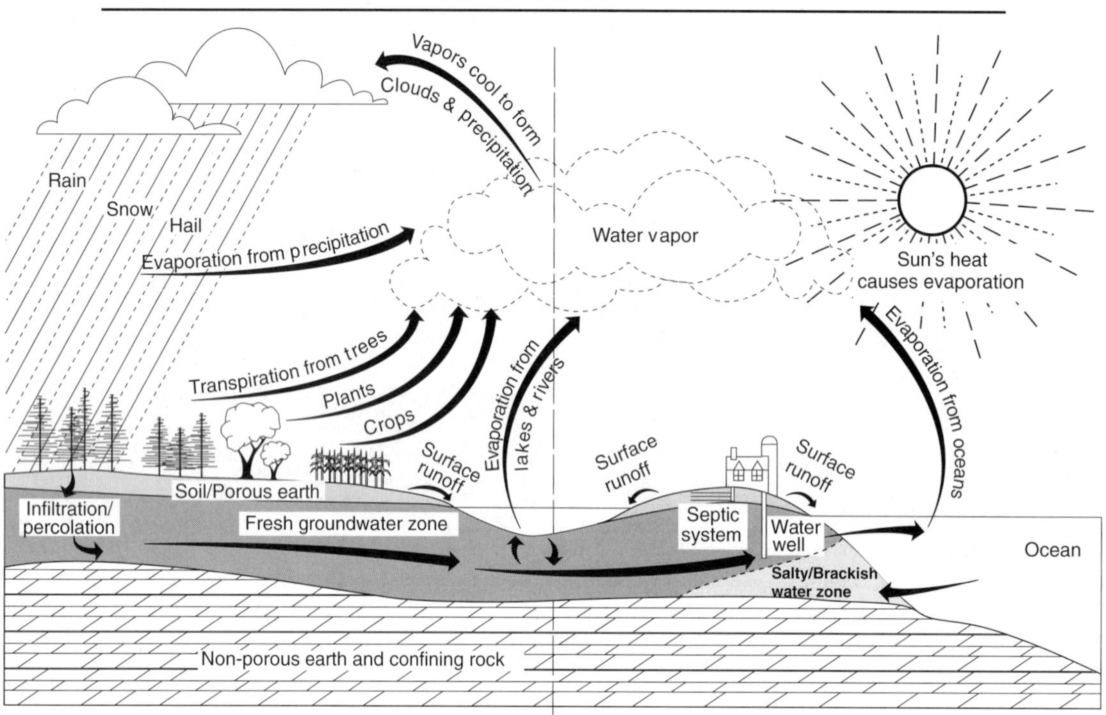

Figure 4A.1 The hydrologic cycle. (From www.dnr.ohio.gov.)

Table 4A.1 Hydrologic Effects of Urbanization

Change in Land or Water Use	Possible Hydrologic Effect
Transition from Pre-Urban to Early-Urban Stage:	
Removal of trees or vegetation	Decrease in transpiration and increase in storm flow
Construction of scattered city-type houses and limited water and sewage facilities	Increased sedimentation of streams
Drilling of wells	Some lowering of water table
Construction of septic tanks and sanitary drains	Some increase in soil moisture and perhaps a rise in water table. Perhaps some waterlogging of land and contamination of nearby wells or streams from overloaded sanitary drain system
Transition from Early-Urban to Middle-Urban State:	
Bulldozing of land for mass housing, some topsoil removed, farm ponds filled in	Accelerated land erosion and stream sedimentation and aggradation. Increased flood flows. Elimination of smallest streams
Mass construction of houses, paving of streets, building of culverts	Decreased infiltration, resulting in increased flood flows and lowered groundwater levels. Occasional flooding at channel constrictions (culverts) on remaining small streams. Occasional overtopping or undermining of banks of artificial channels on small streams
Discontinued use and abandonment of some shallow wells	Rise in water table
Diversion of nearby streams for public water supply	Decrease in runoff between points of diversion and disposal
Untreated or inadequately treated sewage discharged into streams or disposal wells	Pollution of stream or wells. Death of fish and other aquatic life. Inferior quality of water available for supply and recreation at downstream populated areas
Transition from Middle-Urban to Late-Urban Stage:	
Urbanization of area completed by addition of more houses and streets and of public, commercial, and industrial buildings	Reduced infiltration and lowered water table. Streets and gutters act as storm drains, creating higher flood peaks and lower base flow of local streams
Larger quantities of untreated waste discharged into local streams	Increased pollution of streams and concurrent increased loss of aquatic life. Additional degradation of water available to downstream users
Abandonment of remaining shallow wells because of pollution	Rise in water table
Increase in population requires establishment of new water-supply and distribution systems, construction of distant reservoirs diverting water from upstream sources within or outside basin	Increase in local streamflow if supply is from outside basin
Channels of streams restricted at least in part to artificial channels and tunnels	Increased flood damage (higher stage for a given flow). Changes in channel geometry and sediment load. Aggradation
Construction of sanitary drainage system and treatment plant for sewage	Removal of additional water from the area, further reducing infiltration and recharge of aquifer
Improvement of storm drainage system	A definite effect is alleviation or elimination of flooding of basements, streets, and yards, with consequent reduction in damages, particularly with respect to frequency of flooding
Drilling of deeper, large-capacity industrial wells	Lowered water-pressure surface of artesian aquifer; perhaps some local overdrafts (withdrawal from storage) and land subsidence. Overdraft of aquifer may result in salt-water encroachment in coastal areas and in pollution or contamination by inferior or brackish waters
Increased use of water for air conditioning	Overloading of sewers and other drainage facilities. Possibly some recharge to water table, due to leakage of disposal lines
Drilling of recharge wells	Raising of water-pressure surface
Waste-water reclamation and utilization	Recharge to groundwater aquifers. More efficient use of water resources

Note: A selected sequence of changes in land and water use associated with urbanization.

Source: U.S. Geological Survey.

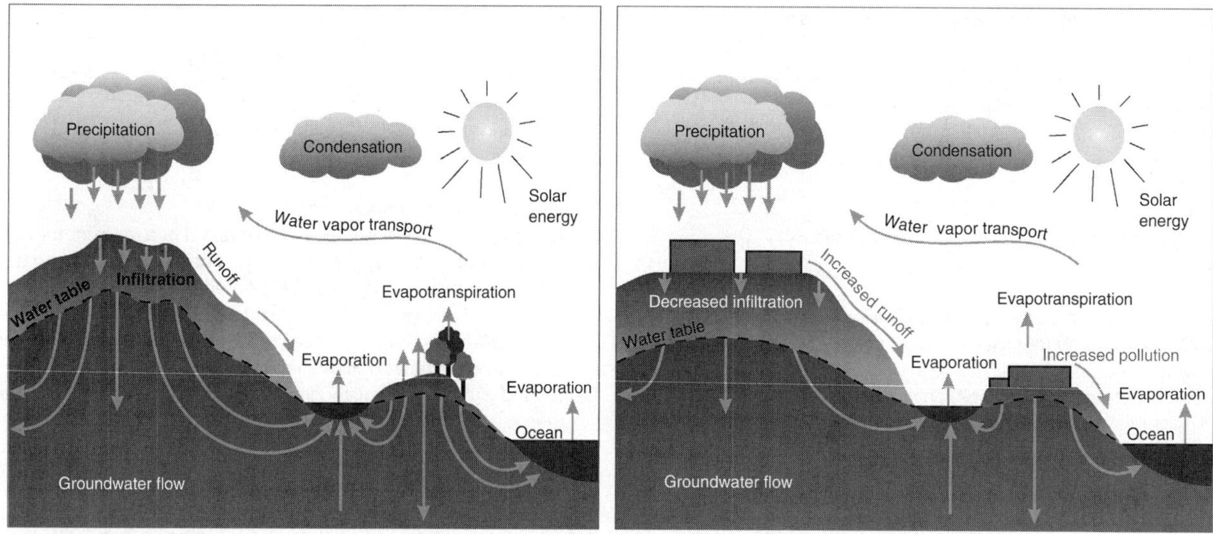

Figure 4A.2 Water cycle before and after urbanization. (From www.unce.unr.edu.)

SECTION 4B WATER RESOURCES — UNITED STATES

Table 4B.2 Distribution of Water in the Continental United States

	Volume		Annual Circulation	Replacement Period
	$\times 10^9$ m^3	%	($\times 10^9$ m^3/yr)	(yr)
Liquid water				
Groundwater				
Shallow (<800 m deep)	63,000	43.2	310	>200
Deep (>800 m deep)	63,000	43.2	6.2	>10,000
Freshwater lakes	19,000	13.0	190	100
Soil moisture (1-m root zone)	630	0.43	3,100	0.2
Salt lakes	58	0.04	5.7	>10
Average in stream channels	50	0.03	1,900	<0.03
Water vapor in atmosphere	190	0.13	6,200	>0.03
Frozen water, glaciers	67	0.05	1.6	>40

Source: From Ad Hoc Panel on Hydrology, *Scientific Hydrology*, Washington, DC: Federal Council for Science and Technology, 1962.

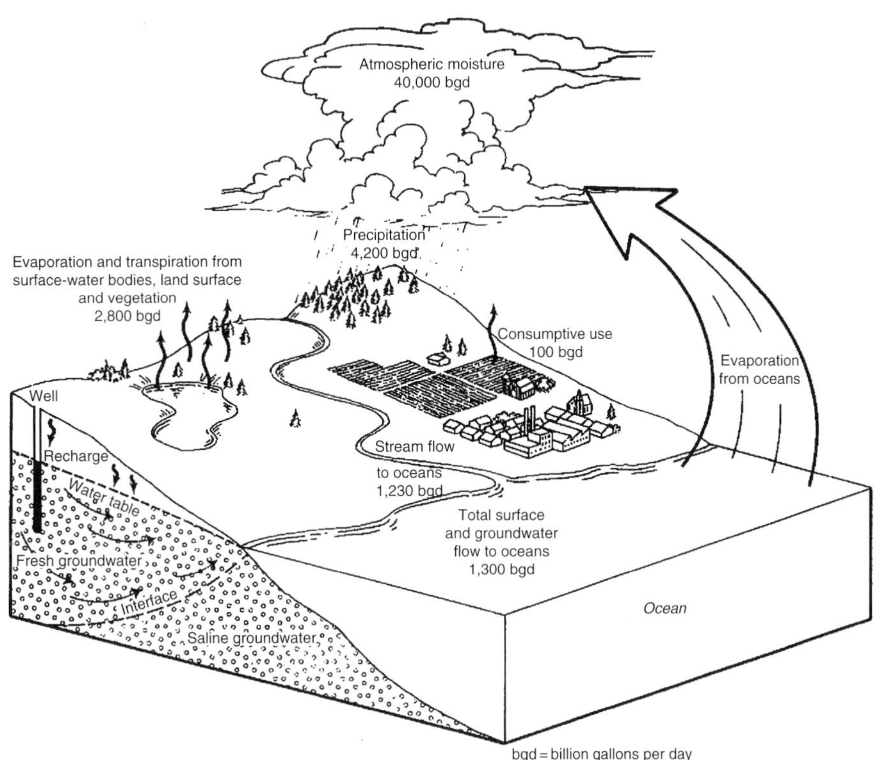

Figure 4B.3 Hydrologic cycle showing the gross water budget of the conterminous United States. (From U.S. Geological Survey, *National Water Summary 1983 — Hydrologic Events and Issues*, Water-Supply Paper 2250, 1984.)

Table 4B.3 Some Purposes of Water-Resources Development

Purpose	Description	Type of Works and Measures
Flood control	Flood-damage abatement or reduction, protection of economic development, conservation storage, river regulation, recharging of groundwater, water supply, development of power, protection of life	Dams, storage reservoirs, levees, floodwalls, channel improvement, floodways, pumping stations, floodplain zoning, flood forecasting
Irrigation	Agricultural production	Dams, reservoirs, walls, canals, pumps and pumping plants, weed-control and desilting works, distribution systems, drainage facilities, farmland grading
Hydroelectricity	Provision of power for economic development and improved living standards	Dams, reservoirs, penstocks, power plants, transmission lines
Navigation	Transportation of goods and passengers	Dams, reservoirs, canals, locks, open-channel improvements, harbor improvements
Domestic and industrial water supply	Provision of water for domestic, industrial, commercial, municipal, and other uses	Dams, reservoirs, walls, conduits, pumping plants, treatment plants, saline-water conversion, distribution systems
Watershed management	Conservation and improvement of the soil, sediment abatement, runoff retardation, forests and grassland improvement, and protection of water supply	Soil-conservation practices, forest and range management practices, headwater-control structures, debris-detention dams, small reservoirs, and farm ponds
Recreational use of water	Increased well-being and health of the people	Reservoirs, facilities for recreational use, works for pollution control, preservation of scenic and wilderness areas
Fish and wildlife	Improvement of habitat for fish and wildlife, reduction or prevention of fish or wildlife losses associated with man's works, enhancement of sports opportunities, provision for expansion of commercial fishing	Wildlife refuges, fish hatcheries, fish ladders and screens, reservoir storage, regulation of streamflows, stocking of streams and reservoirs with fish, pollution control, and land management
Pollution abatement	Protection or improvement of water supplies for municipal, domestic, industrial and agricultural uses and for aquatic life and recreation	Treatment facilities, reservoir storage for augmenting low flows, sewage-collection systems, legal control measures
Insect control	Public health, protection of recreational values, protection of forests and crops	Proper design and operation of reservoirs and associated works, drainage, and extermination measures
Drainage	Agricultural production, urban development and protection of public health	Ditches, tile drains, levees, pumping stations, soil treatment
Sediment control	Reduction or control of slit load in streams and protection of reservoirs	Soil conservation, sound forest practices, proper highway and railroad construction, desilting works, channel and revetment works, bank stabilization, special dam construction and reservoir operations
Salinity control	Abatement or prevention of salt-water contamination of agricultural, industrial, and municipal water supplies	Reservoirs for augmenting low stream-flow, barriers, groundwater recharge, coastal jetties

Source: From Chow, V.T., Water as a World Resource, *Water International*, 4, 6, 1979. With permission.

SECTION 4C WORLD WATER BALANCE

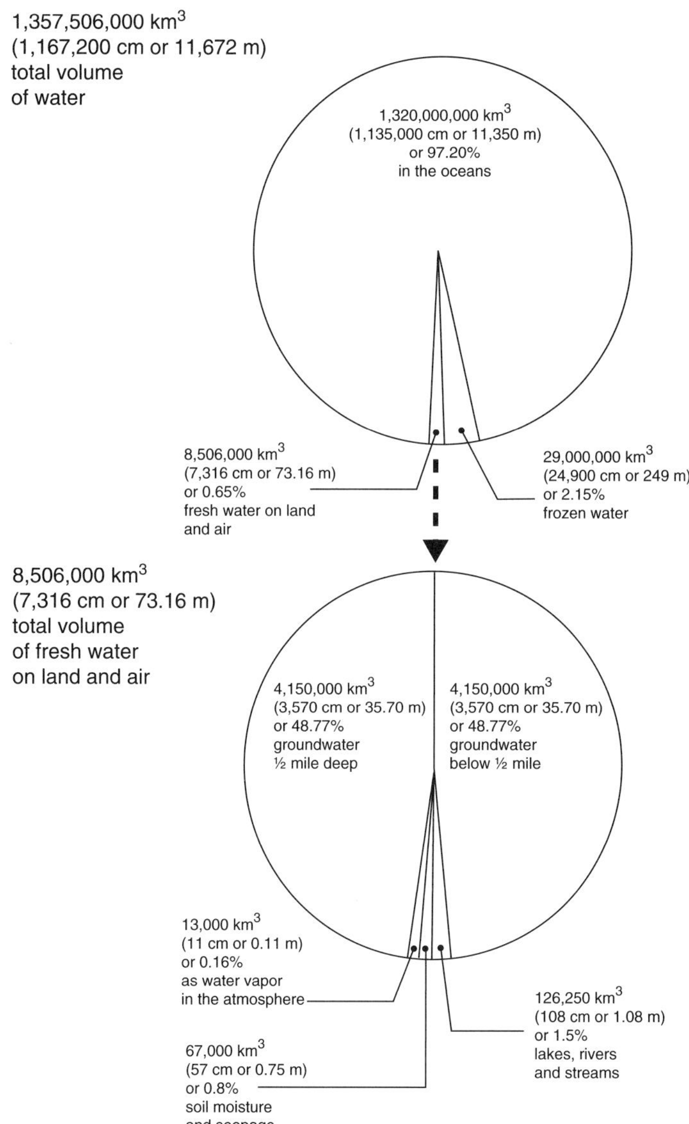

Note: figures in brackets indicate the height that the relevant quantites of water would reach if they were placed on the whole non-frozen land area of the earth which is 116,400,000 km^3

Figure 4C.4 Water availability on earth. (From Doxiadis, C.A., *Water and Environment International Conference on Water for Peace*, Washington, DC, 1967.)

Table 4C.4 Estimated Global Water Cycle

		Volume		
Type of Water	**Location**	**Millions of cu. Miles**	**Millions of cu Kilometer**	**Percent of Total Volume**
Salt Water				97.00
	Oceans	314.2	1308.0 (96.4%)	
	Saline bodies	2.1	8.7 (0.6%)	
Fresh Water				2.90
	Ice & snow	6.9	28.7 (2.1%)	
	Lakes	0.5	2.1 (0.15%)	
	Rivers	0.01	0.04 (0.003%)	
	Accessible groundwater	1.0	4.2 (0.31%)	
Atmospheric				0.10
	Sea evaporation	0.1	0.42 (0.03%)	
	Land evaporation	0.05	0.21 (0.015%)	
	Precipitation over sea	0.09	0.37 (0.03%)	
	Precipitation over land	0.03	0.12 (0.01%)	
	Water vapor	0.005	0.02 (0.002%)	
Rounded Total		326.00	1357.00	100.0

Source: From National Weather Service Northwest River Forecast Center, www.nwrfc.noaa.gov.

Table 4C.5 World Water Balance, by Continent

Water Balance Elements	Europe[a]	Asia	Africa	North America[b]	South America	Australia[c]	Total Land Area[d]
Area, millions of km^2	9.8	45.0	30.3	20.7	17.8	8.7	132.3
in mm							
Precipitation (P)	734	726	686	670	1,648	736	834
Total river runoff (R)	319	293	139	287	583	226	294
Groundwater runoff (U)	109	76	48	84	210	54	90
Surface water runoff (S)	210	217	91	203	373	172	204
Total infiltration and soil moisture (W)	524	509	595	467	1,275	564	630
Evaporation (E)	415	433	547	383	1,065	510	540
in km^3							
Precipitation	7,165	32,690	20,780	13,910	29,355	6,405	110,303
Total river runoff	3,110	13,190	4,225	5,960	10,380	1,965	38,830
Groundwater runoff	1,065	3,410	1,465	1,740	3,740	465	11,885
Surface water runoff	2,045	9,780	2,760	4,220	6,640	1,500	26,945
Total infiltration and soil moisture	5,120	22,910	18,020	9,690	22,715	4,905	83,360
Evaporation	4,055	19,500	16,555	7,950	18,975	4,440	71,475
Relative values							
Groundwater runoff as percent of total runoff	34	26	35	32	36	24	31
Coefficient of groundwater discharge into rivers	0.21	0.15	0.08	0.18	0.16	0.10	0.14
Coefficient of runoff	0.43	0.40	0.23	0.31	0.35	0.31	0.36

[a] Including Iceland.
[b] Excluding the Canadian archipelago and including Central America.
[c] Including Tasmania, New Guinea and New Zealand, only within the limits of the continent: P-440 mm; R-47 mm; U-7 mm; S-40 mm; W-400 mm; E-393 mm.
[d] Excluding Greenland, Canadian archipelago and Antarctica.

Source: From Lvovitch, M.I., *EOS*, 54, 1973. With permission. Copyright by American Geophysical Union.

Table 4C.6 World Water Resources by Region

	Region	Total Area (km^2) (FAOSTATE, 1999) (1)	Total Population (FAOSTAT, 2000) (2)	Average Precipitation 1961–1990 (km^3/yr) (IPCC) (3)	Internal Resources: Total (km^3/yr) (4)	External Resources: Natural (km^3/yr)	External Resources: Actual (km^3/yr)	Total Resources: Natural (km^3/yr)	Total Resources: Natural (km^3/yr)	% of World Resources	IRWR/inhab. (m^3/yr)	TRWR (actual)/inhab. (m^3/yr)
1	Northern America	21,899,600	409,895,363	13,384	6,662	47 (5)	47	6,709	6,709	15.2%	16,253	16,368
2	Central America and Caribbean	749,120	72,430,000	1,506	781	6 (6)	6	787	787	1.8%	10,784	10,867
3	Southern America	17,853,960	345,737,000	28,635	12,380	0	0	12,380	12,380	28.3%	35,808	35,808
4	Western and Central Europe	4,898,416	510,784	4,096	2,170	11	11	2,181	2,181	5.0%	4,249	4,270
5	Eastern Europe	18,095,450	217,051,000	8,452	4,449	244	244	4,693	4,693	10.2%	20,498	21,622
6	Africa	30,044,850	793,288,000	20,415	3,950	0	0	3,950	3,950	9.0%	4,980	4,980
7	Near East	6,347,970	257,114,000	1,378	488	3	3	491	491	1.1%	1,897	1,909
8	Central Asia	4,655,490	78,563,000	1,270	261	28	28	289	289	0.6%	3,321	3,681
9	Southern and Eastern Asia	21,191,290	3,331,938 000	24,017	11,712	8	8	11,720	11,720	26.8%	3,515	3,518
10	Oceania and Pacific	8,058,920	25,388,537	4,772	911	0	0	911	911	2.1%	35,869	35,869
	World	133,795,066	6,042,188,900	107,924	43,764	0	0	43,764	43,764	100.0%	7,243	7,243

Notes: (1) No FAOSTAT data for Spilsbergen (Norway); (2) No FAOSTAT data for West Bank (Palestinian authority); data from Margat and Vallée (2000); (3) No IPCC data on Near East (Saudi Arabia, West Bank (Palestinian Authority); Gaza strip (Palestinian Authority)), South Asia (Taiwan Province of China, east Timor), Caribbean (Aruba). Pacific (Polynesia, Guam) so not included in total. For Europe: no IPCC data for Spilsbergen (Norway), Luxembourg and Belgium; national data source used; (4) No data for various islands in Caribbean (Aruba, Bermuda, Grenada, Guadeloupe, Martinique, St. Lucia. St. Vincent, Dominica) Pacific (French Polynesia, Guam, New Caledonia, Samoa, Tonga), Asia (Macao, Hong Kong); so not included in regional and global totals; (5) 47 km^3/year from Guatemala to Mexico; (6) 6 km^3/year from North America region (Mexico).

Source: www.fao.org.

Table 4C.7 Water Poor Countries

FAO Code	Country	Average Precipitation 1961–1990 (km^3/yr)	Internal Resources Surface (km^3/yr)	Internal Resources Groundwater (km^3/yr)	Internal Resources Overlap (km^3/yr)	Internal Resources Total (km^3/yr)	External Resources Natural (km^3/yr)	External Resources Actual (km^3/yr)	Total Resources Natural (km^3/yr)	Total Resources Actual (km^3/yr)
105	Israel	9.16	0.25	0.50	0.00	0.75	0.92	0.92	1.67	1.67
112	Jordan	9.93	0.40	0.50	0.22	0.68	0.20	0.20	0.88	0.88
124	Libyan Arab Jamahiriya	98.53	0.20	0.50	0.10	0.60	0.00	0.00	0.60	0.60
136	Mauritania	94.66	0.10	0.30	0.00	0.40	11.00	11.00	11.40	11.40
35	Cape Verde	1.70	0.18	0.12	0.00	0.30	0.00	0.00	0.30	0.30
72	Djibouti	5.12	0.30	0.02	0.02	0.30	0.00	0.00	0.30	0.30
225	United Arab Emirates	6.53	0.15	0.12	0.12	0.15	0.00	0.00	0.15	0.15
179	Qatar	0.81	0.00	0.05	0.00	0.05	0.00	0.00	0.05	0.05
134	Malta	0.12	0.00	0.05	0.00	0.05	0.00	0.00	0.05	0.05
76	Gaza Strip (Palestinian Authority)	0.00	0.00	0.05	0.00	0.05	0.01	0.01	0.06	0.06
13	Bahrain	0.06	0.00	0.00	0.00	0.00	0.11	0.11	0.12	0.12
118	Kuwait	2.16	0.00	0.00	0.00	0.00	0.02	0.02	0.02	0.02

Source: From Review of World Water Resources by Country, www.fao.org/documents.

Table 4C.8 Water Rich Countries

FAO Code	Country	Average Precipitation 1961–1990 (km^3/yr)	Internal Resources Surface (km^3/yr)	Internal Resources Groundwater (km^3/yr)	Internal Resources Overlap (km^3/yr)	Internal Resources Total (km^3/yr)	External Resources Natural (km^3/yr)	External Resources Actual (km^3/yr)	Total Resources Natural (km^3/yr)	Total Resources Actual (km^3/yr)	IRWR/inhab. (m^3/yr)
21	Brazil	15,236	5,418	1,874	1,874	5,418	2,815	2,815	8,233	8,233	31,795
185	Russian Federation	7,855	4,037	788	512	4,313	195	195	4,507	4,507	29,642
33	Canada	5,352	2,840	370	360	2,850	52	52	2,902	2,902	92,662
101	Indonesia	5,147	2,793	455	410	2,838	0	0	2,838	2,838	13,381
41	China, Mainland	5,995	2,712	829	728	2,812	17	17	2,830	2,830	2,245
44	Colombia	2,975	2,112	510	510	2,112	20	20	2,132	2,132	50,160
231	United States of America (Cont.)	5,800	1,862	1,300	1,162	2,000	71	71	2,071	2,071	7,153
170	Peru	1,919	1,616	303	303	1,616	297	297	1,913	1,913	62,973
100	India	3,559	1,222	419	380	1,261	647	636	1,908	1,897	1,249

Source: From Review of World Water Resources by Country, www.fao.org/documents.

SECTION 4D HYDROLOGIC DATA

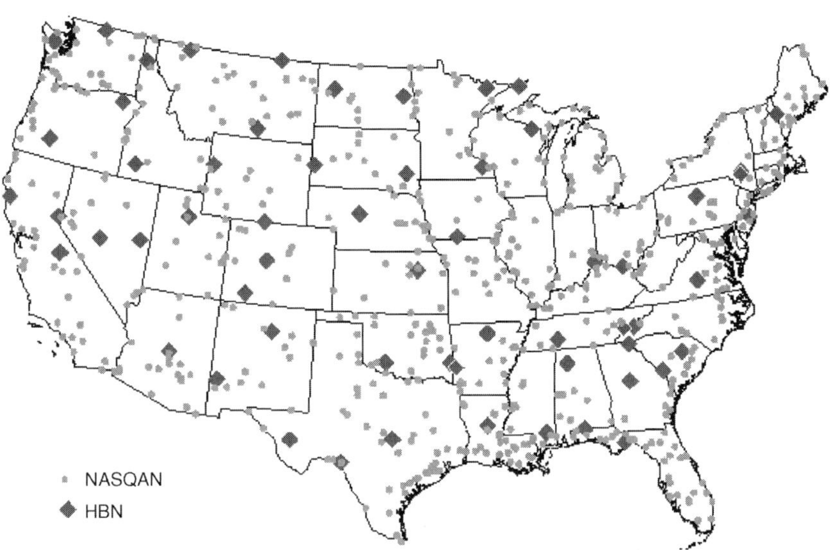

The networks included 63 HBN stations from 1962-95 and 618 NASQAN stations from 1973-95. Stations located outside of the conterminous United States for NASQAN included 13 in Alaska. 8 in Hawaii, 6 in Puerto Rico, and 1 in Guam; HBN stations included 1 each in Alaska and Hawaii.

Figure 4D.5 Locations of NASQAN and national hydrologic bench-mark stations in the United States. (From http://water.usgs.gov.)

Table 4D.9 National Stream Water Quality Accounting Network (NASQAN) — Stations Operated by NASQAN 1996–2000

USGA Station ID	Station Name	USGS Office Sampling Station	Latitude	Longitude	Hydrologic Unit Code	Location of Stream Gage and Sampling Site	Drainage Area	Remarks
Mississippi River Basin								
03216600	Ohio River at Greenup Dam near Greenup, KY	Louisville, KY	38°38′48″	82°51′38″	05090103	At left bank at downstream end of lock guidewall in lower poolat Greenup locks, 1.1 mi upstream from Grays Branch, 4.7 mi downstream from Little Sandy River, 5.0 mi north of Greenup and at mile 341.5	62,000 sq. mi., approximately	
03303280	Ohio River at Cannelton Dam at Cannelton, IN	Louisville, KY	37°53′58″	86°42′20″	05140201	At Cannelton Dam, 0.7 mi upstream from Indian Creek, 3.3 mi upstream from Lead Creek, and at mile 720.8. Water-quality samples are collected 2.0 mi upstream from discharge station	97,000 sq. mi, approximately	
03378500	Wabash River at New Harmony, IN	Paducah, KY	38°07′55″	87°56′25″	05120113	At bridge on U.S. Highway 66 at New Harmony and at mile 51.5	29,234 sq. mi	Water discharge obtained from station Wabash river at Mount Carmel, IL (03377500)
03609750	Tennessee River at Highway 60 near Paducah, KY	Paducah, KY	37°02′16″	88°31′46″	06040006	At auxiliary gaging station at bridge on U.S. Highway 60, 16.3 mi downstream from gagin station, 2.4 mi east of Paducah, and at mile 5.3	40,330 sq. mi., 40,200 sq. mi. at gage	Records of daily discharge are taken from gaging station near Paducah (03609500). Flow completely regulated. Barkley-Kentucky Cannal (03438190) diverts water from and to Lake Barkley in the Cumberland River Basin
03612500	Ohio River at Dam 53 near Grand Chain, IL	Paducah, KY	37°12′11″	89°02′30″	05140206	At auxiliary gaging station, 0.5 mi upstream from Gar Creek, 3.0 mi southwest of Grand Chain, 18.1 mi downstream from gaging station at Metropolis, and at mile 962.2	203,100 sq. mi, approximately	Water discharge obtained from Ohio River at Metropolis, IL (03611500). Flow regulated by many dams and reservoirs
05420500	Mississippi River at Clinton, IA	Iowa City, IA	41°46′50″	90°15′07″	07080101	At river end of 3rd St., at downstream end of ADM repair dock, 10.3 miles upstream from Wapsipinicon River, 4.8 mi upstream from Camanche gage, 5.9 mi downstream from Lock and Dam 13, and at mile 516.6 upstream from Ohio River. Water-quality samples collected at Fulton-Lyons Bridge, 6.4 mi upstream of discharge station	85,600 sq. mi., approximately, at Fulton-Lyons Bridge in Clinton	
05587455	Mississippi River below Grafton, IL	Rolla, MO	38°58′05″	90°25′42″	07110009	On left bank 0.2 mi downstream from the mouth of the Illinois River, 15.3 mi above Lock and Dam 26, 23.0 mi above mouth of Missouri River and at mile 218.6 upstream of the mouth o the Ohio River. Water-quality samples collected 4 mi downstream of discharge station	171,300 sq. mi., approximately	
06185500	Missouri River near Culbertson, MT	Fort Peck, MT	48°07′30″	104°28′20″	10060005	On right bank at upstream side of bridge on State Highway 16, 2.5 mi southeast of Culbertson, 10 mi downstream from Big Muddy Creek and at river mile 1,620.76	91,557 sq. mi	

06329500	Yellowstone River near Sidney, MT	Fort Peck, MT	47°40′42″	104°09′22″	10100004	On left bank at Montana-Dakota Utilities Company powerplant, 0.2 mi downstream from bridge on State Highway 23, 2.5 mi south of Sidney, 3.0 mi downstream from Fox Creek, and at river mile 29.2	69,103 sq. mi;	
06338490	Missouri River at Garrison Dam, ND	Bismark, ND	47°30′08″	101°25′50″	10130101	In control structure of Garrison Dam, 2.5 mi west of Riverdale, 14 mi upstream from Knife River, and at mile 1,389.9	181,400 sq. mi., approximately	
06440000	Missouri River at Pierre SD	Pierre, SD	44°22′23″	100°22′03″	10140101	On left bank downstream from Dakota Minnesota and Eastern Railroad bridge, 1.3 mi upstream from Bad River, 5.8 mi downstream from Oahe Dam, and at mile 1066.5. Water-quality samples collected 0.25 mile below Oahe Dam, about 5.55 mile upstream from gaging station. Inflow between these two locations generally are negligible	243,500 sq. mi., approximately	
06610000	Missouri River at Omaha, NE	Council Bluffs, IA	41°15′32″	95°55′20″	10230006	On right bank on left side of concrete floodwall, at foot of Douglas Street, 275 ft downstream of Interstate 480 Highway bridge in Omaha and at mile 615.9 Water-quality samples are collected at Interstate-80 bridge, 2.0 miles downstream of gaging station	322,800 sq. mi. approximately. The 3,959 sq. mi. in the Great Divide basin are not included	Flow regulated by upstream mainstem reservoirs. US Army Corps of Engineers raingage and satellite data collection platform at station
06805500	Platte River at Louisville, NE	Linclon, NE	41°00′55″	96°09′28″	10200202	On the left bank at the upstream side of bridge on Nebraska Highway 50, 1 mi north of Louisville, and at mile 16.5	85,370 sq. mi., appoximately, of which about 71,000 sq. mi., contributes directly to surface runoff	
06934500	Missouri River at Hermann, MO	Rolla, MO	38°42′36″	91°26′21″	10300300	On downstream side of third pier from right abutment of bridge on State Highway 19 at Hermann, and at mile 97.9	524,200 sq. mi., approximately	
07022000	Mississippi River at Thebes, IL	Rolla, MO	37°13′00″	89°27′50″	07140105	Near center span on downstream side of railroad bridge at Thebes, 5.0 mi downstream from Headwater Diversion Channel and at mile 43.7 above Ohio River	713,200 sq. mi., approximately	
07263620	Arkansas River at David D. Terry Lock & Dam below Little Rock, AR	Little Rock, AR	34°40′07″	92°09′18″	11110207	At upper end of upstream wall at David D. Terry Lock and Dam. 10.7 mi downstream from Main Street bridge at Little Rock, and at mile 124.2	158,288 sq. mi., of which 22,241 sq. mi. is probably non-contributing	Discharge is from station 07263450, 16.8 mi. upstream
07373420	Mississippi River near St. Francisville, LA	Baton Rouge, LA	30°45′30″	91°23′45″	08070100	At State Highway 10 ferry crossing, 2.0 mi southwest of St. Francisville and at mile 266.0	1,125,300 sq. mi. contributing	Discharge is from Mississippi River at Tarbert Landing, MS, station 07295100
07381495	Atchafalaya River at Melville, LA	Baton Rouge, LA	30°41′26″	91°44′10″	08080101	At bridge on Texas and Pacific Railroad in Melville	93,316 sq. mi	Discharge is from station 07381490, Atachafalaya river at Simmesport, LA

(Continued)

Table 4D.9 (Continued)

USGA Station ID	Station Name	USGS Office Sampling Station	Latitude	Longitude	Hydrologic Unit Code	Location of Stream Gage and Sampling Site	Drainage Area	Remarks
					Rio Grande Basin			
08364000	Rio Grand at El Paso, TX	Albuquerque, NM	31°48′10″	106°32′25″	13030102	At gaging stati on the downstream side of the Courchesne Bridge, 5.6 mi upstream from the Santa Fe Street-Juarez Avenue bridge betwen El Paso, Tx, and Cd. Juarez, Chihuahua at mile 1,249 and 1.7 mi upstream from the American Dam	29,267 sq. mi	Discharge measured by International Boundary and Water Commission
08377200	Rio Grande at Foster Ranch, near Langtry, TX	San Angelo, TX	29°46′50″	101°45′20″	13040212	At gaging station 0.1 mi downstream from Terrell-Val Verde Country line, 16.9 mi from Langtry, and 597.2 midownstream from the American Dam at El Paso	80,742 sq. mi	Discharge measured by International Boundary and Water Commission
08447410	Pecos River near Langtry, TX	San Angelo, TX	29°48′10″	101°26′45″	13040212	At gaging station 7.4 mi east of Langtry and 15.0 mi upstream from confluence with Rio Grande	35,179 sq. mi	Discharge measured by International Boundary and Water Commission
08450900	Rio Grande below Amistad Dam near Del Rio, TX	San Angelo, TX	29°25′30″	101°02′27″	13080001	2.2 mi downstream from Amistad Dam and 10 mi northwest of Del Rio	123,143 sq. mi	Discharge measured by International Boundary and Water Commission
08459200	Rio Grande at Pipeline Crossing below Laredo, TX	San Angelo, TX	27°24′01″	99°29′18″	13080002	8.7 Mi (14.0 km) downstream from Texas-Mexican Railway Bridge near Laredo, and at mile 352.69 (567.48 km)	132,578	Discharge measured by International Boundary and Water Commission
08461300	Rio Grande below Falcon Dam, TX	San Angelo, TX	26°33′25″	99°10′05″	13090001	U.S. tailrace at Falcon Dam	159,270 sq. mi	Discharge measured by International Boundary and Water Commission
08470400	Arroyo Colorado at Harlingen, TX map of lower basin	San Angelo, TX	26°10′24″	97°42′01″	13090002	On downstream side of northbound service road on U.S. Highways 83&77, about 18 mi from point of main floodway that divides into North Floodway and Arroyo Colorado	182 sq. mi	Discharge measured by International Boundary and Water Commission
08475000	Rio Grande near Brownsville, TX map of lower basin	San Angelo, TX	25°52′35″	97°27′15″	13090002	At International Boundary and Water Commission gaging station, 1000 ft downstream from El Jardin pumping plant, 6.8 mi below International bridge between Brownsville and Matamoras, Tamps., Mex. And 48.8 miles above the Gulf of Mexico	176,333 sq. mi	Discharge measured by International Boundary and Water Commission
					Colorado River Basin			
09180500	Colorado River near Cisco, UT	Moab, UT	38°48′38″	109°17′34″	14030005	On left bank 1 mi downstream from Dolores River, 11 mi south of Cisco, 36 mi downstream from Colorado-Utah state line, 97 mi upstream from Green River and 235 mi upstream from San Juan River, at mile 1022.3 from Arizona-Sonora	24,100 sq. mi., approximately	

09315000	Green River at Green River, UT	Moab, UT	38°59′10″	110°09′02″	14060008	On right bank, 1,400 ft upstream from railroad bridge, 0.9 mi southeast of town of Green River, 22.7 mi upstream from San Rafael River, at mile 117.6 upstream from mouth	44,850 sq. mi. of which about 4,260 sq. mi. (including 3,959 sq. mi. in Great Divide Basin in southern Wyoming) i noncontributing	Flow regulated by Flaming gorge Reservoir (09234400)
09379500	San Juan River near Bluff, UT	Moab UT	37°08′49″	109°51′52″	14080205	On left bank, 1,600 ft downstream from Gypsum Creek, 1,800 ft upstream from highway bridge, 20 mi southwest of Bluff, at mile 113.5	23,000 sq. mi., approximately	No diversions between station and mouth of river. Flow regulated by Navajo Reservoir, NM (09355100)
09380000	Colorado River at Lees Ferry, AZ	Flagstaff, AZ	36°51′53″	111°35′15″	14070006	In Navajo Indian Reservation, on left bank at head of Marble gorge at lees ferry, just upstream from Paria River, 16 mi downstream of Glen Canyon Dam, 28 mi downstream from UT-AZ state line, and 61.5 mi upstream from Little Colorado River	111,800 sq. mi., approximately, including 3,959 sq. mi. in Great Divide basin in southern Wyoming, which is noontributing	Many diversions above Lake Powell for irrigation, municipal, and industrial use. No diversion or inflow between Lake Powell and the gage
09404200	Colorado River above Diamond Creek near Peach Springs, AZ	Flagstaff, AZ	36°46′25″	113°21′46″	15010002	In Lake Mead NRA, on the right bank, 0.6 mi upstream from Diamond Creek, 138 mi downstream from Phantom Ranch, 25 mi north of Peach Springs, 242 mi downstream from Glen Canyon Dam, and 130 mi upstream from Hoover Dam	149,316 sq. mi., including 3,959 sq. mi. in Great Divide Basin in southern Wyoming nand 697 sq. mi. on the Colorado Plateau	Several unregulated tributaries below Glen Canyon Dam
09421500	Colorado River below Hoover Dam, AZ-NV	Las Vegas, NV	36°00′55″	114°44′16″	15030101	In powerhouse at downstream side of Hoover Dam. Water-quality samples collected at gaging station 0.3 mi downstream from Hoover Dam	171,700 sq. mi., approximately, included 3,959 sq. mi. in Great Divide Basin in southern Wyoming, which is non-contributing	
09429490	Colorado River above Imperial Dam, CA-AZ. schematic map of Lower Colorado River	Yuma, AZ	32°52′59″	114°27′55″	15030104	Imperial Dam is 5 mi upstream from Laguna Dam, 15 mi northeast of Yuma, 90 mi downstream from Palo Verde Dam and 147 mi downstream from Parker Dam. Water-quality samples collected below trash racks at All-American Canal headworks at west end of Imperial Dam	188,500 sq. mi., approximately, including 3,959 sq. mi. in Great Divide basin in southern Wyoming, which is non-contributing	Records show flow of Colorado River reaching Imperial Dam and are synthesized from records of several other stations

(Continued)

Table 4D.9 (Continued)

USGA Station ID	Station Name	USGS Office Sampling Station	Latitude	Longitude	Hydrologic Unit Code	Location of Stream Gage and Sampling Site	Drainage Area	Remarks
09522000	Colorado River at Northerly International Boundary (NIB), above Morelos Dam, near Andrade, CA schematic map of Lower Colorado River	Yuma, AZ	32°43′07″	114°43′05″	15030108	On left bank at northerly international boundary, 0.5 mi east of Andrade. 1.1 mi upstream from Morelos Dam, 1.1 mi downstream from Rockwood Gate, and 6.4 mi downstream from gaging station on Colorado River below Yuma Main Canal wasteway	246,700 sq. mi., approximately, including all closed basins entirely within the drainage boundary, also 3,959 sq. mi. in Great divide basin in southern Wyoming, which is non-contributing	
Columbia River Basin								
12400520	Columbia River at Northport, WA	Spokane, WA	48°55′08″	117°47′11″	17020001	0.4 mi downstream from State Highway 25 bridge at Northport, 10.3 mi downstream from gaging station at boundary, and at mile 735.1	60,200 sq. mi., approximately	Discharge is routed from gaging station at international boundary (12399500)
12472900	Columbia River at Vernita Bridge, near Priest Rapids Dam WA	Pasco, WA	46°38′34″	119°43′54″	17020016	At State Highway 24 Vernita Bridge crossing, 9.0 mi downstream from Priest Rapids Dam and at mile 388.1	96,000 sq. mi., approximately	Discharge determined by routing flows from the gaging station below Priest Rapids Dam (12472800) 6.4 mi upstream
13353200	Snake River at Burbank, WA	Pasco, WA	46°15′00″	118°53′45″	17060110	Approximately 1.0 mi downstream from Ice Harbor Dam	108,800 sq. mi	Discharge is obtained and routed from Ice Harbor Dam, 1.0 mi upstream
14128910	Columbia River at Warrendale, OR	Portland, OR	45°36′45″	122°01′35″	17080001	On left bank 0.1 mi downstream from Tumult Creek, 1.0 mi west of Warrendale, 5.1 mi downstream from Bonneville Dam, and at mile 141.0	240,000 sq. mi., approximately	Stream discharge taken from Columbia River at the Dalles, OR (14105700) at river mile 188.9
14211720	Willamette River at Portland, OR	Portland, OR	45°31′07″	122°40′00″	17090012	In pier at east end of drawspan, on upstream side of Morrison bridge in Portland, and at mile 12.8	11,100 sq. mi., approximately	Water discharge records obtained by flow routing procedures usg sta records
14246900	Columbia river at Beaver Army Terminal, near Quincy, OR	Portland, OR	46°10′55″	123°10′50″	17080003	On left bank, 0.7 mi downstream from Crims Island, 3.0 mi northwest of Qunicy, and at mile 53.8	256,900 sq. mi., approximately	

Source: From http://water.usgs.gov

Figure 4D.6 NASQAN stations, 1996–2000. (From http://water.usgs.gov.)

Table 4D.10 Water Quality Characteristics Are Measured as NASQAN Stations

Code	Parameter	Description
4S	5S	75S
MAJ	00010	Water temperature (degree Celsius)
MAJ	00061	Discharge, instantaneous (cubic feet per second)
MAJ	00076	Turbidity (nephelometric turbidity units, NTU)
MAJ	00095	Specific conconductance (microsiemens per centimeter at 25 Celsius)
MAJ	00300	Oxygen, dissolved (milligrams per liter)
MAJ	00400	Ph, field (standard units)
MAJ	00452	Carbonate, filtered (milligrams per liter as CO_3)
MAJ	00453	Bicarbonate, filtered (milligrams per liter as HCO_3)
MAJ	00608	Ammonia–nitrogen (milligrams per liter as N)
MAJ	00613	Nitrite–nitrogen (milligrams per liter as N)
MAJ	00623	Ammonia-plus-organic-nitrogen, dissolved (milligrams per liter as N)
MAJ	00625	Ammonia-plus-organic-nitrogen (milligrams per liter as N)
MAJ	00631	Nitrite-plus-nitrate-nitrogen, dissolved (milligrams per liter as N)
MAJ	00665	Phosphorus, total (milligrams per liter as P)
MAJ	00666	Phosphorus, dissolved (milligrams per liter as P)
MAJ	00671	Orthophosphate-phosphorus (milligrams per liter as P)
MAJ	00681	Carbon, organic, dissolved (milligrams per liter as C)
MAJ	00689	Carbon, organic, suspended (milligrams per liter as C)
MAJ	00915	Calcium, dissolved (milligrams per liter as Ca)
MAJ	00925	Magnesium, dissolved (milligrams per liter as Mg)
MAJ	00930	Sodium, dissolved milligrams per liter as Na)
MAJ	00935	Potassium, dissolved (milligrams per liter as K)
MAJ	00940	Chloride, dissolved (milligrams per liter as Cl)
MAJ	00945	Sulfate, dissolved (milligrams per liter as SO_4)

(Continued)

Table 4D.10 (Continued)

Code	Parameter	Description
MAJ	00950	Fluoride, dissolved (milligrams per liter as F)
MAJ	00955	Silica, dissolved (milligrams per liter as SiO_2)
MAJ	39086	Alkalinity, filtered (milligrams per liter as $CaCO_3$)
MAJ	70300	Residue on evaporation (180 Celsius) (milligrams per liter)
MAJ	70331	Sediment, finer than 63 microns (percent)
MAJ	80154	Sediment, suspended (milligrams per liter)
SEDCHEM	29816	Antimony, sediment, suspended, total (micrograms per gram)
SEDCHEM	29818	Arsenic, sediment, suspended, total (micrograms per gram)
SEDCHEM	29820	Barium, sediment, suspended, total (micrograms per gram)
SEDCHEM	29822	Beryllium, sediment, suspended, total (micrograms per gram)
SEDCHEM	29826	Cadmium, sediment, suspended, total (micrograms per gram)
SEDCHEM	29829	Chromium, sediment, suspended, total (micrograms per gram)
SEDCHEM	29832	Copper, sediment, suspended, total (micrograms per gram)
SEDCHEM	29836	Lead, sediment, suspended, total (micrograms per gram)
SEDCHEM	29839	Manganese, sediment, suspended, total (micrograms per gram)
SEDCHEM	29841	Mercury, sediment, suspended, total (micrograms per gram)
SEDCHEM	29843	Molybdenum, sediment, suspended, total (micrograms per gram)
SEDCHEM	29846	Nickel, sediment, suspended, total (micrograms per gram)
SEDCHEM	29847	Selenium, sediment, suspended, total (micrograms per gram)
SEDCHEM	29850	Silver, sediment, suspended, total (micrograms per gram)
SEDCHEM	29853	Vanadium, sediment, suspended, total (micrograms per gram)
SEDCHEM	29855	Zinc, sediment, suspended, total (micrograms per gram)
SEDCHEM	30221	Aluminum, sediment, suspended, total (percent)
SEDCHEM	30244	Carbon, sediment, suspended, total (percent)
SEDCHEM	30269	Iron, sediment, suspended, total (percent)
SEDCHEM	30292	Phosphorus, sediment, suspended, total (percent)
SEDCHEM	30308	Sulfur, sediment, suspended (percent)
SEDCHEM	30317	Titanium, sediment, suspended, total (percent)
SEDCHEM	35031	Cobalt, sediment suspended, total (micrograms per gram)
SEDCHEM	35040	Strontium, sediment, suspended, total (micrograms per gram)
SEDCHEM	35046	Uranium, sediment, suspended, total (micrograms per gram)
SEDCHEM	35050	Lithium, sediment, suspended, total (micrograms per gram)
SEDCHEM	49955	Thallium, sediment, suspended, total (micrograms per gram)
SEDCHEM	50279	Sediment, suspended (milligrams per liter)
SEDCHEM	50465	Carbon, organic, suspended, total (percent)
TE	01000	Arsenic, dissolved (milligrams per liter as As)
TE	01005	Barium, dissolved (milligrams per liter as Ba)
TE	01010	Beryllium, dissolved (milligrams per liter as Be)
TE	01020	Boron, dissolved (micrograms per liter as B)
TE	01025	Cadmium, dissolved (micrograms per liter as Cd)
TE	01030	Chromium, dissolved (micrograms per liter as Cr)
TE	01035	Cobalt, dissolved (micrograms per liter as Co)
TE	01040	Copper, dissolved (micrograms per liter as Cu)
TE	01046	Iron, dissolved (micrograms per liter as Fe)
TE	01049	Lead, dissolved (micrograms per liter as Pb)
TE	01056	Manganese, dissolved (micrograms per liter as Mn)
TE	01057	Thallium, dissolved (micrograms per liter as Tl)
TE	01060	Molybdenum, dissolved (micrograms per liter as Mo)
TE	01065	Nickel, dissolved (micrograms per liter as Ni)
TE	01075	Silver, dissolved (micrograms per liter as Ag)
TE	01080	Strontium, dissolved (micrograms per liter as Sr)
TE	01085	Vanadium, dissolved (micrograms per liter as V)
TE	01090	Zinc, dissolved (micrograms per liter as Zn)
TE	01095	Antimony, dissolved (micrograms per liter as Sb)
TE	01106	Aluminum, dissolved (micrograms per liter as Al)
TE	01130	Lithium, dissolved (micrograms per liter as Li)
TE	01145	Selenium, dissolved (micrograms per liter as Se)
PEST	04024	Propachlor, dissolved (micrograms per liter)
PEST	04028	Butylate, dissolved (micrograms per liter)
PEST	04035	Simazine, dissolved (micrograms per liter)

(Continued)

Table 4D.10 (Continued)

Code	Parameter	Description
PEST	04037	Prometon, dissolved (micrograms per liter)
PEST	04040	Desethyl atrazine, dissolved (micrograms per liter)
PEST	04041	Cyanazine, dissolved (micrograms per liter)
PEST	04095	Fonofos, dissolved (micrograms per liter)
PEST	22703	Uranium, natural, dissolved (micrograms per liter)
PEST	34253	Alpha BHC, dissolved (micrograms per liter)
PEST	34653	P, P′ DDE, dissolved (micrograms per liter)
PEST	38933	Chlorpyrifos, dissolved (micrograms per liter)
PEST	39341	Lindane, dissolved (micrograms per liter)
PEST	39381	Dieldrin, dissolved (micrograms per liter)
PEST	39415	Metolachlor, dissolved (micrograms per liter)
PEST	39532	Malathion, dissolved (micrograms per liter)
PEST	39542	Parathion, dissolved (micrograms per liter)
PEST	39572	Diazinon, dissolved (micrograms per liter)
PEST	39632	Atrazine, dissolved (micrograms per liter)
PEST	46342	Alachlor, dissolved (micrograms per liter)
PEST	49260	Acetochlor, dissolved (micrograms per liter)
PEST	82630	Metribuzin, dissolved (micrograms per liter)
PEST	82660	Diethylanilene, dissolved (micrograms per liter)
PEST	82661	Trifluralin, dissolved (micrograms per liter)
PEST	82663	Ethalfluralin, dissolved (micrograms per liter)
PEST	82664	Phorate, dissolved (micrograms per liter)
PEST	82665	Terbacil, dissolved (micrograms per liter)
PEST	82666	Linuron, dissolved (micrograms per liter)
PEST	82667	Methyl parathion, dissolved (micrograms per liter)
PEST	82668	EPTC, dissolved (micrograms per liter)
PEST	82669	Pebulate, dissolved (micrograms per liter)
PEST	82670	Tebuthiuron, dissolved (micrograms per liter)
PEST	82671	Molinate, dissolved (micrograms per liter)
PEST	82672	Ethoprop, dissolved (micrograms per liter)
PEST	82673	Benfluralin, dissolved (micrograms per liter)
PEST	82674	Carbofuran, dissolved (micrograms per liter)
PEST	82675	Terbufos, dissolved (micrograms per liter)
PEST	82676	Pronamide, dissolved (micrograms per liter)
PEST	82677	Disultoton, dissolved (micrograms per liter)
PEST	82678	Triallate, dissolved (micrograms per liter)
PEST	82679	Propanil, dissolved (micrograms per liter)
PEST	82680	Carbaryl, dissolved (micrograms per liter)
PEST	82681	Thiobencarb, dissolved (micrograms per liter)
PEST	82682	DCPA, dissolved (micrograms per liter)
PEST	82683	Pendimethalin, dissolved (micrograms per liter)
PEST	82684	Napropamide, dissolved (micrograms per liter)
PEST	82685	Propargite, dissolved (micrograms per liter)
PEST	82686	Azinphos-methyl, dissolved (micrograms per liter)
PEST	82687	Permethrin, dissolved (micrograms per liter)
PEST	91063	Diazinon, D-10 surrogate (percent)
PEST	91065	HCH, alpha, D-6 surrogate (percent)
PEST	99108	Spike volume (milliters)
PEST	99856	Sample volume (milliters)

Note: ASCII text file containing parameter code definitions for constituents.
Analyzed by the USGS National Stream Quality Accounting Network (1996–2000).
File is tab-delimited.
The first header contains column names.
The second header contains column formats.
Code = Constituent group, defined as follows:
MAJ = instantaneous Q, field parameters, major ions, nutrients, suspended sediment
SEDCHEM = sediment chemistry
TE = trace elements (dissolved)
PEST = pesticides
Parameter = WATSTORE code
Description = Constituent name (units of measure)

Source: http://water.usgs.gov.

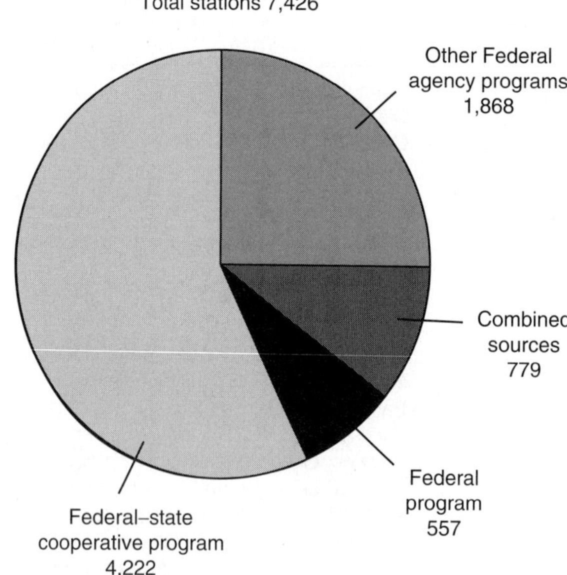

Figure 4D.7 Sources of funds for operation of continuous surface-water discharge stations. (From U.S. Geological Survey Water Data Program, http://water.usgs.gov.)

Table 4D.11 Hydrologic and Related Data Collection Networks in the United States

Type of Network	Number of Stations
Automatic meterological observing stations (full parameter); temperature, dew point, wind, pressure, precipitation)	92
National weather service synoptic and basic observation stations (high quality observations for basic weather program)	67
Cooperative station services (observations by lay persons):	
Temperature and precipitation	5,568
Precipitation only — daigeoly	3,200
Precipitation only — storage	32
Hourly precipitation stations equipped with recording precipitation gages	3,205
Cooperative stations equipped with both recording and nonrecording precipitations gages	1,995
Crop reporting stations	566
River and/or rainfall reporting stations	
River stage reports only	998
Rainfall reports only	3,656
River stage and rainfall reports	1,069
Evaporation storage	448
Reference Climatological Stations	21
Automated Hydrologic Observing System (AHOS) — river and rainfall data for flood forecasting	
AHOS/T[a]	506
AHOS/S[b]	75
Special reporting stations	293
Cooperative station data published	
Temperature and precipitation	8,256
Precipitation only	3,055
Evaporation	431
Soil temperature	308
Miscellaneous (snow density, special meteorological, etc)	473

[a] Data transmitted by telephone.
[b] Data transmitted by satellite.

Source: From National Weather Service, *Operations of the National Weather Service*, 1985.

Table 4D.12 USGS Programs Managed by the Water Resources Discipline

- *Cooperative Water Program* — The Cooperative Program, a partnership between the USGS and state and local agencies, provides information that forms the foundation for many of the Nation's water resources management and planning activities
- *National Streamflow Information Program* (NSIP) — The National Streamflow Information Program (NSIP) is a conceptual plan developed by the USGS for a new approach to the acquisition and delivery of streamflow information
- *National Water Quality Assessment Program* (NAWQA) — Since 1991, USGS scientists with the NAWQA program have been collecting and analyzing data and information in more than 50 major river basins and aquifers across the Nation. The goal is to develop long-term consistent and comparable information on streams, groundwater, and aquatic ecosystems to support sound management and policy decisions. The NAWQA program is designed to answer these questions:
    1. What is the condition of our Nation's streams and groundwater?
    2. How are these conditions changing over time?
    3. How do natural features and human activities affect these conditions?
- *Toxic Substances Hydrology (Toxics) Program* — provides unbiased earth science information on the behavior of toxic substances in the Nation's hydrologic environments. The information is used to avoid human exposure, to develop effective cleanup strategies, and to prevent further contamination
- *Groudwater Resources Program* — The Groundwater Resources Program encompasses regional studies of groundwater systems, multidisciplinary studies of critical groundwater issues, access to groundwater data, and research and methods development. The program provides unbiased scientific information and many of the tools that are used by Federal, State, and local management and regulatory agencies to make important decisions about the Nation's groundwater resources
- *Hydrologic Research and Development* — Hydrologic Research and Development focuses on long-term investigations that integrate hydrological, geological, chemical, climatic, and biological information related to water resources issues. The program provides the primary support for the *National Research Program* (NRP) in the hydrologic sciences and for *Water, Energy, and Biogeochemical Budgets (WEBB) program*
- *State Water Resoruces Research Institute Program* — A matching grant program to support water resources research, education, and information transfer at the 54 university based Water Resources Research Institutes. This program includes the National Institutes for Water Resources USGS *Student Internship Program*

Subprograms:

- *Water Information Coordination Program* (WICP) — ensures the availability of water information required for effective decision making for natural resources management and environmental protection and to do it cost effectively
- *Drinking Water Programs* — The wide range of monitoring, assessment, and research activities conducted by the USGS to help understand the protect the quality of our drinking water resources is described on these pages. These studies are often done in collaboration with other federal, state, tribal, and local agencies
- *National Stream Quality Accounting Network* (NASQAN) — Focus is on monitoring the water quality of four of the Nation's largest river systems — the Mississippi (including the Missouri and Ohio), the Columbia, the Colorado, and the Rio Grande
- *Hydrologic Benchmark Network* (HBN) — was established in 1963 to provide long-term measurements of streamflow and water quality in areas that are minimally affected by human activities. These data were to be used to study time trends and to serve as controls for separating natural from artificial changes in other streams. The network has consisted of as many as 58 drainage basins in 39 State
- *National Atmospheric Deposition Program/National Trends Network* (NADP/NTN) — A nationwide network of precipitation monitoring sites. The first sites in the network were established in 1978. The network currently consists of approximately 200 sites
- *National Research Program* (NRP) — conducts basic and problem-oriented hydrologic research in support of the mission of the U.S. Geological Survey (USGS)
- *National Water Summary Program* — a series of publications designed to increase public understanding of the nature, geographic distribution, magnitude, and trends of the Nation's water resources. It is often referred to as the USGS "encyclopedia of water"
- *National Water-Use Program* — examines the withdrawal, use, and return flow of water on local, state, and national levels
- *USGS Environmental Affairs Program* — provides guidance and information on the National Environmental Policy Act and other environmental issues
- *Water, Energy, and Biogeochemical Budgets* (WEBB) — understands the processes controlling water, energy, and biogeochemical fluxes over a range of temporal and spatial scales and to understand the interactions of these processes, including the effect of atmospheric and climatic variables
- *National Irrigation Water Quality Program* — A Department of Interior program to identify and address irrigation-induced water quality and contamination problems related to Department of Interior water projects in the west

International Programs:

- *Cyprus Water Resources Database Development* — This project met the USGS goal of supporting U.S. foreign policy. It was requested by the U.S. Ambassador to Cyprus and coordinated closely through the U.S. Department of State. It took 5 years of negotiations with senior Cypriot officials, Embassy staff, U.S. Department of State, and selected United Nations offices to design and implement this project. This project enabled water managers on Cyprus to manage their limited water resources, which will directly contribute to enhancement and protection of the quality of life for Cypriot citizens
- *Public Awareness and Water Conservation* — The project, which began in 1996, is part of the Middle East Peace Process and is one of several projects sponsored by the Multilateral Working Group on Water Resources. The U.S. Department of State requested the USGS to undertake this activity and has provided political guidance throughout the project. The project meets the USGS goal of supporting U.S. foreign policy and fostering outreach and public awareness activities

(Continued)

Table 4D.12 (Continued)

- *Regional Water Data Banks* — The Executive Action Team Multilateral Working Group on Water Resoruces, Water Data Banks Project consists of a series of specific actions to be taken by the Israelis, Jordanians, and Palestinians that are designed to foster the adoption of common, standardized data collection and storage techniques among the Parties, improve the quality of the water resources data collected in the region, and to improve communication among the scientific community in the region
- *Ukraine Streamflow Project* — Floods are among the most frequent and costly natural disasters in terms of human hardship and economic loss. In Ukraine, two major floods (one in 1998 and one in 2001) have occurred in the Tisa River Basin in the last 5 years. Both floods caused several fatalities, damaged or destroyed several thousand homes, destroyed bridges and roads, and created severe personal and economic hardship for the residents of Zakarpattia Oblast in western Ukraine. Near real-time streamflow data can be used to forecast and manage floods and improve public safety
- *Groundwater Research Program for the Emirate of Abu Dhabi, United Arab Emirates* — Since 1988 the USGS has been partnering with the National Drilling Company (NDC) of the Abu Dhabi Emirate to collect information on the groundwater resources of the Emirate, to conduct research on the hydrology of the arid environment, to provide training in water resources investigations, and to document the results of the cooperative work in scientific publications
- *Geologic, Hydrologic, and Geochemical Characterization of the Deep Groundwater Aquifer System In the Bengal Delta of Bangladesh* — The USGS is currently conducting research on the deeper aquifer system in Bangladesh in areas associated with high levels of arsenic in the shallow groundwater. This work is an integral step in the characterization of the hydrogeolocial framework needed to define the potential for developing safe and sustainable groundwater sources
- *Botswana–Village Flood Watch* — The Village Flood Watch project, which was completed in 2002, was designed to help establish an early-warning system for potential flooding events by adding or upgrading six gauging stations to near real-time capabilities and providing training on hydrologic runoff modeling
- *Jordan Groundwater Management* — The project objective is to enhance current Jordanian technical capacities for hydrogeologic data and information development, management and analysis; development and use of groundwater management models; and joint design and conduct of outreach workshops and meetings to increase public understanding of the benefits of local efforts in groundwater management and conservation
- *Summary of Palestinian Hydrologic Data 2000* — The project provides a critical tool to the USAID Water Resources Program including several investigative, development, and construction projects, in the West Bank and Gaza, designed to comprehensively develop, manage, and protect water resources. This activity demonstrates the USGS leadership role in the natural sciences and confirms the mission of providing scientific information to manage natural resources to enhance and protect the quality of life

Source: From Water Resources of the United States, http://water.usgs.gov.

Table 4D.13 Number of USGS Data-Collection Stations Operated in 1994, by Source of Funds

Types of Stations	Federal Program	Federal-State Cooperative Program	Other Federal Agency Program	Combined Support	Total
Surface water:					
Discharge	638	6,419	2,219	964	10,240
Stage-only — streams, lakes and reservoirs	47	968	850	183	2,048
Quality	778	1,666	426	228	3,098
Groundwater:					
Water levels	2,344	27,029	2,421	237	32,031
Quality	691	4,602	1,347	216	6,856

Source: From U.S. Geological Survey Water Data Program, http://water.usgs.gov.

Table 4D.14 Increasing Global Data Coverage

Regions	Number of Stations	Number of Data Point	Physical/Chemical	Major Loans	Metals	Nutrients	Organic Contaminants	Microbiology	Date Range
Africa	138	206907	26712	79889	6439	41289	370	832	1977–2004
Americas	682	417994	47198	73210	88124	47284	3583	10401	1965–2004
Asia	332	641940	118868	159329	83005	98796	6794	32018	1971–2004
Europe	318	823323	146747	136392	154742	108815	14539	27260	1978–2003
Oceania	94	206650	31678	12237	2535	46992	1438	1383	1979–2004
Total	1544	2296814	371203	461087	334845	343176	26734	71894	1965–2004

Source: From GEMS Water, *State of the UNEP GEMS/Water Global Network and Annual Report*, United Nations Environment Programme, Global Environment Monitoring System (GEMS) Water Programme, 2004, www.gemswater.org.

SECTION 4E INTERCEPTION

Table 4E.15 Interception by Trees

Type or Species	Age or Size	Place in Succession	Locality	Interception (Percent)
Hemlock	Mature	Climax	Connecticut	48
Douglas fir	25 yr	Climax	Washington	43
Hemlock	Mature	Climax	New Hampshire	38
Spruce-fir	Mature	Climax	Maine	37
Hemlock	Mature	Climax	Adirondacks, New York	34
Douglas fir	Mature	Climax	Washington	34
Hemlock	Mature	Climax	Ithaca, New York	31
Spruce — fir — paper birch	Mature	Climax	Maine	26
White pine — hemlock	Mature	Climax	Massachusetts	24
Western white pine — western hemlock	Overmature	Climax	Idaho	21
Maple — beech	Mature	Climax	New York	43
Mixed	Mature	Climax	New York	40
Maple — hemlock	Mature, cutover	Climax	Wisconsin	25
Beech — birch — maple	Mature	Climax	Ontario	21
Ponderosa pine	Mature	Preclimax	Arizona	40
Lodgepole pine	Mature	Preclimax	Colorado	32
Ponderosa pine	Mature	Preclimax	Idaho	27
Jeffrey pine	Mature	Preclimax	Southern California	26
Lodgepole pine	32 yr	Preclimax	Colorado	23
Ponderosa pine	Mature	Preclimax	Idaho	22
Ponderosa pine	Young	Pioneer	Colorado	18
Calif, scrub oak	6 ft	—	Southern California	31
Mixed brush	Mature	Preclimax	North Fork, California	19
White pine — red pine	40 yr	Preclimax	Ontario	37
Jack pine	50 yr	Pioneer	Wisconsin	21
Shortleaf pine	45 yr	Pioneer	North Carolina	16
Quaking aspen	32 yr	Pioneer	Colorado	16
Chaparral, mixed	6 ft		Southern California	17
Maple — hemlock	Mature (after leaf fall)	Climax (under-stocked)	Wisconsin	16
Hemlock	Mature	Climax	New York	13
Oak-pine	Open, second growth	Preclimax	New Jersey	13
Ponderosa — lodgepole pine	25 ft	Preclimax	Idaho	8
Beech — maple	Mature	Climax	New York	6
Chamise	6 ft	Pioneer	Southern California	3

Note: Interception includes stemflow and is expressed as a percentage of annual precipitation.

Source: From Compilation of data from various references, Kittredge, *Forest Influences*, McGraw-Hill, Copyright 1948. With permission.

Table 4E.16 Interception by Various Forest Types

Forest Type	Gross Interception		Stemflow		Net Interception		Net Snow Interception (%)
	With Leaves (%)	Without Leaves (%)	With Leaves (%)	Without Leaves (%)	With Leaves (%)	Without Leaves (%)	
Northern hardwood	20	17	5	10	15	7	10
Aspen — birch	15	12	5	8	10	4	7
Spruce — spruce-fir	35	—	3	—	32	—	35
White pine	30	—	4	—	26	—	25
Hemlock	30	—	2	—	28	—	25
Red pine	32	—	3	—	29	—	30

Source: U.S. Forest Service.

Table 4E.17 Interception by Various Crops

Description	Alfalfa	Corn	Soybean	Oats
During growing season:				
Rainfall (in.)	10.81	7.12	6.25	6.77
Canopy penetration (in.)	6.18	4.84	4.06	6.30
Stemflow (in.)	0.76	1.18	1.28	
Interception (in.)	3.87	1.10	0.91	0.47
Interception (%)	35.8	15.5	14.6	6.9
During low-vegetation development (%)	21.9	3.4	9.1	3.1

Source: U.S. Department of Agriculture.

SECTION 4F INFILTRATION

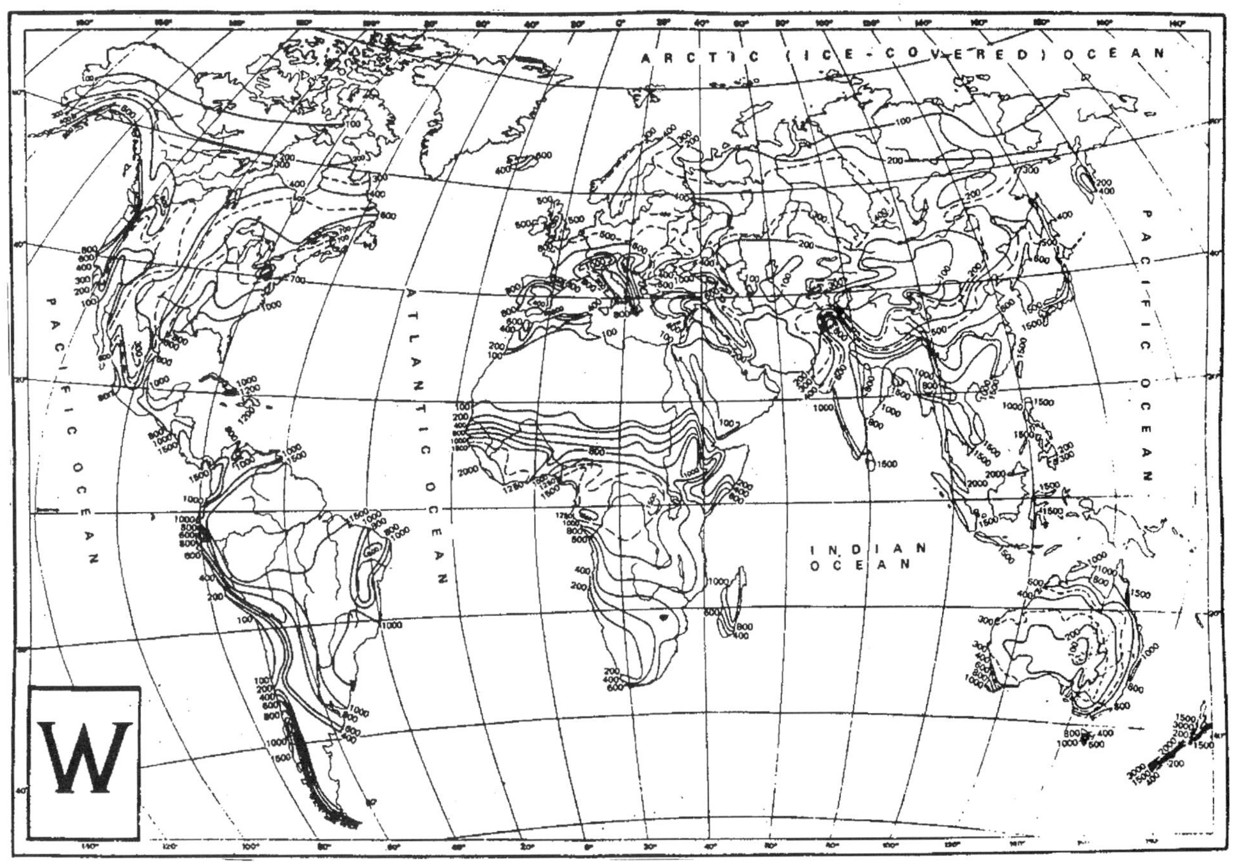

Figure 4F.8 Total annual infiltration and soil moisture in the world (in mm). (From Lvovitch, M.I., *EOS*, 54, 1973, Copyright by American Geophysical Union. With permission.)

Table 4F.18 Seepage Rates for Canals

Canal Soil Material	Seepage (Feet per Day)	Canal Soil Material	Seepage (Feet per Day)
Sandy loam	8.2	Loam and adobe	1.4
Gravelly loam	5.3	Loam	1.1
Fine sandy loam and adobe	3.8	Silty clay	0.9
Sand and sandy loam	3.4	Sand and silty clay	0.4
Loam and sandy loam	3.3	Sand and clay	0.1
Adobe	3.0	Loam and gravelly loam	0.1
Fine sandy loam	2.1		

Note: Values are average maximum rates through the wetted area.

Source: From Rohwer and Stout, Colo, *Agric. Exp. Sta. Bull.*, 1948. With permission.

Table 4F.19 Infiltration Rate and Land Use

1. Fallow	8. Pasture, fair
2. Row crops, poor rotation[a]	9. Woods, poor
3. Row crops, good rotation[b]	10. Pasture, good
4. Pasture, poor	11. Woods, fair
5. Legumes after row crops	12. Meadows
6. Small grains, poor rotation	13. Woods, good
7. Small grains, good rotation	

Notes: Rank of land uses in order of infiltration rate; first use listed has lowest rate

[a] One-fourth or less in hay or sod.
[b] More than one-fourth of rotation in hay or sod.

Source: U.S. Soil Conservation Service.

Table 4F.20 Infiltration Model Classification

Category	Model Selected	Reference
Semi-empirical	SCS model	USDA-SCS (1972)
Homogeneous	Philip's two-term model	Philip (1957)
Nonhomogeneous	Green-Ampt model for layered systems	Flerchinger *et al.* (1988)
Ponding	Green-Ampt explicit model	Salvucci and Entekhabi (1994)
Non-ponding	Constant Flux Green-Ampt model	Swartzendruber (1974)
Wetting and drying	Infiltration/Exfiltration model	Eagleson (1978)

Source: From USEPA *Estimation of Infiltration Rate in the Vadose Zone: Application of Selected Mathematical Models — Volume II*, EPA/600/R-97/128b, 1998.

Table 4F.21 Concentrations of Compounds in Highway Runoff Prior to and after Infiltration through the Second Batch of Medium Formulation Number 9: 90-Percent Sand, 5-Percent Clay, and 5-Percent Mulch

		Effluent After Infiltration		
Analyte	**Influent Prior to Infiltration**	**Halfway Through Experiment**	**End of Experiment**	**Percent Change[a]**
Calcium, dissolved (mg/L)	6.3	5.2	5.1	−19
Magnesium, dissolved (mg/L)	1.2	1.9	1.8	50
Potassium, dissolved (mg/L)	1.2	1.6	1.6	33
Sodium, dissolved (mg/L)	3.2	3.5	3.4	6
Chloride, dissolved (mg/L)	2.8	3.0	3.0	7
Fluoride, dissolved (mg/L)	<0.2	[e]0.1	[e]0.1	—
Silica, dissolved (mg/L)	7.6	8.3	7.9	4
Sulfate, dissolved (mg/L)	11	11	11	0
Nitrate + nitrite (mg/L)	0.83	0.83	0.80	−4
Suspended solids (mg/L)	13	42	65	400
Dissolved solids (mg/L)	48	55	52	8
Arsenic, dissolved (μg/L)	0.3	0.5	0.7	133
Arsenic, total (μg/L)	<2	[e]2	[e]1	—
Cadmium, dissolved (μg/L)	0.1	[e]0.02	[e]0.02	[e]−80
Cadmium, total (μg/L)	0.28	0.17	0.16	−43
Copper, dissolved (μg/L)	6	4	4	−33
Copper, total (μg/L)	10.5	16.9	12.9	23
Iron, dissolved (μg/L)	48	79	19	−60
Lead, dissolved (μg/L)	0.4	0.09	[e]0.06	[e]-85
Lead, total (μg/L)	2.8	8.1	6.4	129
Manganese, dissolved (μg/L)	17	[e]2	<3	>−82
Zinc, dissolved (μg/L)	20	2	2	−90
Zinc, total (μg/L)	55.3	54.0	54.7	−1
Total petroleum hydrocarbons (mg/L)	4	<2	<2	>−50
Calcium, dissolved (mg/L)	6.3	5.2	5.1	−19
Magnesium, dissolved (mg/L)	1.2	1.9	1.8	50
Potassium, dissolved (mg/L)	1.2	1.6	1.6	33
Sodium, dissolved (mg/L)	3.2	3.5	3.4	6

Note: mg/L, milligrams per liter; μg/L, micrograms per liter; <, less than; >, greater than; e, estimated value; —, not computed.

[a] Percent change was calculated using the prior-to-infiltration and end-of-experiment values.

Source: From Kenneth, C. Ames, Emily L., Inkpen, Lonna M., Frans, William R., Bidlake, Technical Report WARD 5122, Wahington State Department of Transportation.

SECTION 4G RUNOFF

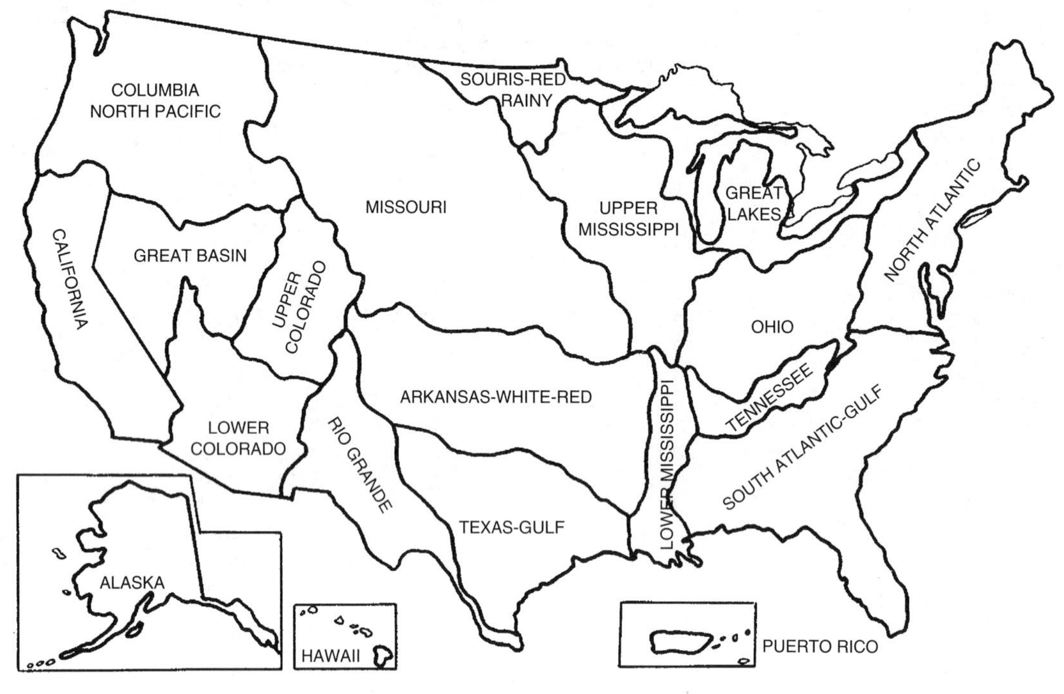

Figure 4G.9 Water resources regions of the United States. (From U.S. Water Resources Council, 1968.)

Table 4G.22 World-Wide Stable Runoff, by Continent

	Stable Runoff[a] (km^2)					
	Of Underground Origin	**Regulated by Lakes**	**Regulated by Water Reservoirs**	**Total**	**Total River Runoff[b]**	**Total Stable Runoff as Present of Total River Runoff**
Europe	1,065	60	200	1,325	3,110	43
Asia	3,410	35	560	4,005	13,190	30
Africa	1,465	40	400	1,905	4,225	45
North America	1,740	150	490	2,380	5,960	40
South America	3,740	—	160	3,900	10,380	38
Australia[c]	465	—	30	495	1,965	25
Total land area except polar zones	11,885	285	1,840	14,010	38,830	36

[a] Excluding flood flows.
[b] Including flood flow.
[c] Including Tasmania, New Guinea, and New Zealand.

Source: From Lvovitch, M.I., *EOS*, 54, 1, 1973.

Figure 4G.10 Annual total river runoff in the world (includes groundwater discharge to rivers; in mm). (From Lvovitch, M.I., *EOS*, 54, 1973. Copyright by American Geophysical Union. With permission.)

Table 4G.23 Runoff in the United States

Region	Mean	50%[a]	90%[a]	95%[a]
North Atlantic[b]	163	163	123	112
South Atlantic-Gulf	197	188	131	116
Great Lakes[b,c]	63.2	61.4	46.3	42.4
Ohio[d]	125	125	80.0	67.5
Tennessee	41.5	41.5	28.2	24.4
Upper Mississippi[d]	64.6	64.6	36.4	28.5
Lower Mississippi[d]	48.4	48.4	29.7	24.6
Souris-Red-Rainy[b]	6.17	5.95	2.60	1.91
Missouri[b]	54.1	53.7	29.9	23.9
Arkansas-White–Red	95.8	93.4	44.3	33.4
Texas-Gulf	39.1	37.5	15.8	11.4
Rio Grande[e]	4.9	4.9	2.6	2.1
Upper Colorado[e]	13.45	13.45	8.82	7.50
Lower Colorado[d,e]	3.19	2.51	1.07	0.85
Great Basin[d]	5.89	5.82	3.12	2.46
Columbia-North Pacific[b]	210	210	154	138
California[f]	65.1	64.1	32.8	25.6
Conterminous United States[g]	1,201			
Alaska[b]	580	—[h]	—[h]	—[h]
Hawaii	13.3	—[h]	—[h]	—[h]
United States[g]	1,794			

Note: Annual natural runoff in billions of gallons per day; regions are shown in Figure 4G.9.

[a] Flow exceeded in indicated percent of years.
[b] Does not include runoff from Canada.
[c] Does not include net precipitation on the lakes.
[d] Does not include runoff from upstream regions.
[e] Does not include runoff from Mexico.
[f] Virgin flow. Mean annual natural runoff estimated to be 13.7 bgd.
[g] Rounded.
[h] Not available.

Source: U.S. Water Resources Council, 1968.

Table 4G.24 Runoff Distribution in the United States

Range in Runoff (Inches per Year)	Area (Square Miles)	Percent of Total Area	Percent of Total Runoff
0–0.25	306,000	10.1	0.1
0.25–0.5	380,000	12.6	.5
0.5–1.0	266,000	8.8	.8
1.0–2.5	413,000	13.7	2.8
2.5–5	247,000	8.2	3.6
5–10	258,000	8.5	7.4
10–20	830,000	27.4	44.8
20–40	290,000	9.6	32.4
40–80	30,000	1.0	6.9
Over 80	2,000	0.1	0.7
Total	3,022,000	100	100

Source: House of Representatives, U.S. Congress.

Table 4G.25 Seasonal Variation of Natural Runoff by Regions of the United States

Region	Months of High Flow	Months of Low Flow
North Atlantic	March, April	August, September
South Atlantic-Gulf	February, March	September, October
Great Lakes	April	January, August, September
Ohio	March	September, October
Tennessee	March	October
Upper Mississippi	March, April	January, September, October
Lower Mississippi	March	October
Souris-Red-Rainy	April	January, February
Missouri	March, June	January
Arkansas-White–Red	April, May, June	January, September
Texas-Gulf	March, May	August, October
Rio Grande	May	June
Upper Colorado	June	January, February
Lower Colorado	March, April	June, November
Great Basin	June	September, January
Columbia-North Pacific	February, April, May	January, February, August, September
California	April, May	September, October, December

Source: U.S. Water Resources Council, 1968.

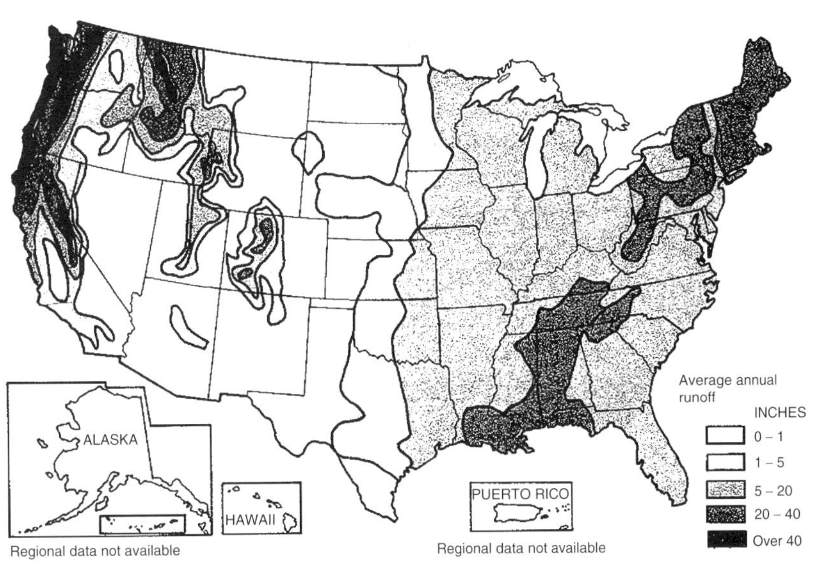

Figure 4G.11 Average annual runoff in the United States. (From U.S. Water Resources Council 1968, The Nation's Water Resources.)

Table 4G.26 Runoff for National Forest and Non-National Forest Areas in Selected Western Drainage Basins

	Area		Average Annual Water Production				
				NF		Outside NF	
Drainage Basin or Area	NF (Percent)	Outside NF (Percent)	Whole Area (Inches)	Inches	Percent of Total Volume	Inches	Percent of Total Volume
Columbia (in U.S.)	37	63	10.4	16.7	59	6.7	41
Colorado (in U.S.)	19	81	2.5	7.2	56	1.3	44
Rio Grande above El Paso	25	75	1.7	3.8	58	0.9	42
Central Valley (California only)	32	68	11.8	23.5	63	6.4	37
Rogue-Umpqua Area	40	60	35.5	37.0	42	34.2	58
Northwest Washington (State less Columbia)	32	68	39.3	51.4	41	33.7	59
Southern California Coast (Los Angeles watershed to Mexican border)	25	75	3.7	6.3	43	2.8	57
North Platte and South Platte	11	89	1.7	6.2	41	1.2	59
Missouri above Fort Randall Dam	9	91	1.7	6.9	37	1.2	63
Arkansas above Dodge City	9	91	1.2	4.7	38	0.7	62

Source: U.S. Geological Survey.

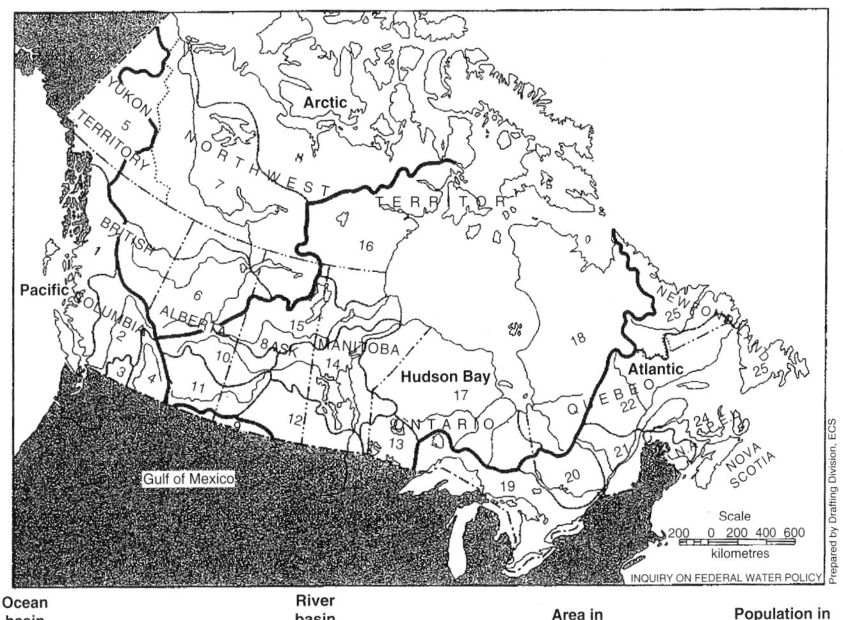

Ocean basin region		River basin region	Area in 000s km^2	Population in 000s 1981
	1	Pacific Coastal	352	616
	2	Fraser-Lower Mainland	234	1 722
Pacific	3	Okanagan-Similkameen[a]	14	189
	4	Columbia[a]	90	161
	5	Yukon[a]	328	23
	6	Peace-Athabasca	487	286
Arctic	7	Lower Mackenzie	1 300	43
	8	Arctic Coast-Islands	2 025	13
Gulf of Mexico	9	Missouri[a]	26	14
	10	North Saskatchewan	146	1 084
	11	South Saskatchewan[a]	170	1 282
	12	Assiniboine-Red[a]	190	1 300
	13	Winnipeg[a]	107	77
Hudson Bay	14	Lower Saskatchewan-Nelson	363	224
	15	Churchill	298	68
	16	Keewatin	689	5
	17	Northern Oontario	694	157
	18	Northern Quebec	950	109
	19	Great Lakes[a]	319	7 579
	20	Ottawa	146	1 270
	21	St. Lawrence[a]	116	5 193
Atlantic	22	North Shore-Gaspé	403	653
	23	St. John-St. Croix[a]	37	393
	24	Maritime Coastal	114	1 314
	25	Newfoundland-Labrador	376	568
CANADA			9 974	24 343

[a]Canadian portion only; area and population on U.S. side of international basin regions are excluded from totals.

Figure 4G.12 Drainage regions of Canada. (From Pearse, P.H., Currents of change, *Final Report Inquiry on Federal Water Policy*, Ottawa, Canada, 1985.)

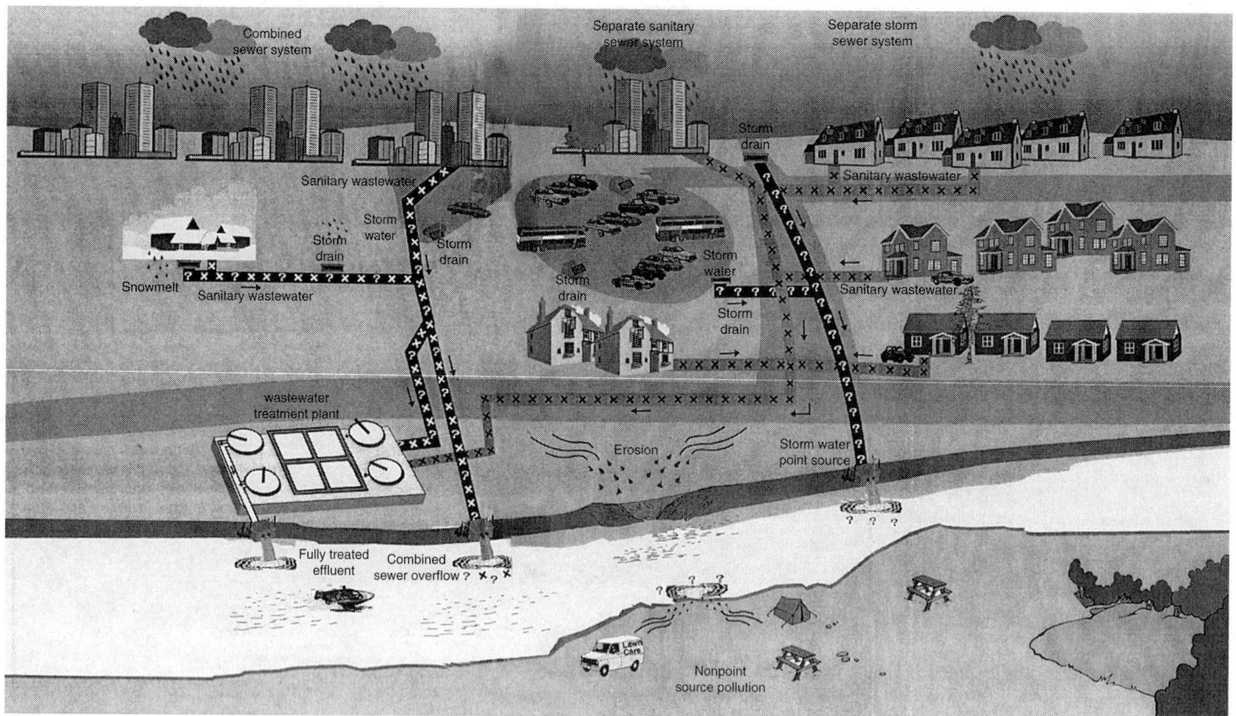

Figure 4G.13 Urban runoff flows in different types of sewer systems. (From www.gao.gov. GAO water quality — better data and evaluation of urban runoff programs needed to assess effectiveness — report to Congressional requesters, U.S. General Accounting Office.)

Table 4G.27 Projections of Average Water Availability in the United States[a]

Region	1965	1980	2000	2020
North Atlantic	163	163	163	163
South Atlantic-Gulf	197	197	197	197
Great Lakes[b]	80.3	80.3	80.3	80.3
Ohio	125	125	125	125
Tennessee	41.5	41.5	41.5	41.5
Upper Mississippi[c]	66.7	66.7	66.7	66.7
Lower Mississippi[d]	408	401	395	390
Souris-Red-Rainy[e]	6.2	6.4	6.8	6.8
Missouri[f]	54.5	54.6	54.8	54.9
Arkansas-White–Red[f]	95.8	95.9	96.0	96.0
Texas-Gulf[g]	39.1	39.2	39.2	39.2
Rio Grande[h]	5.2	5.3	5.3	5.3
Upper Colorado[i]	13.5	13.5	13.5	13.5
Lower Colorado[j]	14.1	12.6	11.9	11.6
Great Basin[k]	6.9	7.0	7.1	7.2
Columbia-North Pacific[l]	258	258	258	258
California[m]	69.7	69.4	69.3	69.2
Alaska[l]	710	710	710	710
Hawaii	13.3	13.3	13.3	13.3

Note: Values in billion gallons per day; for regions, see Figure 4G.9.

[a] Nature runoff adjusted for imports and upstream runoff where appropriate, values rounded.
[b] Includes net precipitation of U.S. portion of Great Lakes.
[c] Includes import from Great Lakes Region.
[d] Includes net upstream runoff and imports.
[e] Includes import from Missouri Region.
[f] Includes imports from Upper Colorado Region.
[g] Includes import from Arkansas-White–Red Region.
[h] Includes imports from Upper Colorado Region and Mexican Treaty deliveries.
[i] Virgin flow at Lee Ferry Compact point.
[j] Includes net upstream runoff.
[k] Includes imports from Upper Colorado Region and nature runoff from California Region.
[l] Includes natural runoff from Canada.
[m] Includes imports from Lower Colorado Region.

Source: U.S. Water Resources Council, 1968.

Table 4G.28 Values of Runoff Coefficient in the Rational Formula

Types of Drainage Area	Runoff Coefficient (*C*)	Type of Drainage Area	Runoff Coefficient (*C*)
Lawns:		Industrial:	
Sandy soil, flat, 2%	0.05–0.10	Light areas	0.50–0.80
Sandy soil, average, 2–7%	0.10–0.15	Heavy areas	0.60–0.90
Sandy soil, steep, 7%	0.15–0.20	Parks, cemeteries	0.10–0.25
Heavy soil, flat, 2%	0.13–0.17	Playgrounds	0.20–0.35
Heavy soil, average, 2–7%	0.18–0.22	Railroad yard areas	0.20–0.40
Heavy soil, steep, 7%	0.25–0.35	Unimproved areas	0.10–0.30
Business:		Streets:	
Downtown areas	0.70–0.95	Asphaltic	0.70–0.95
Neighborhood areas	0.50–0.70	Concrete	0.80–0.95
Residential:		Brick:	0.70–0.85
Single-family areas	0.30–0.50	Drives and walks	0.75–0.85
Multi units, detached	0.40–0.60	Roofs	0.75–0.95
Multi units, attached	0.60–0.75		
Suburban	0.25–0.40		
Apartment dwelling areas	0.50–0.70		

Note: Formula is applicable to drainage areas less than about 5000 acres and has the form $Q = CiA$ where Q is peak discharge in cfs, C is a dimensionless runoff coefficient, i is rainfall intensity for the time of concentration in inches per hour, and A is drainage area in acres.

Source: *Amer. Soc. Civil Engrs.*

Table 4G.29 Watershed Characteristics for Determining Runoff Coefficient in the Rational Formula

Designation of Watershed Characteristics	Runoff-Producing Characteristics			
	100 Extreme	75 High	50 Normal	25 Low
Relief	(40) Steep, rugged terrain, with average slopes generally above 30%	(30) Hilly, with average slopes of 10–30%	(20) Rolling, with average slopes of 5–10%	(10) Relatively flat land, with average slopes of 0–5%
Soil infiltration	(20) No effective soil cover, either rock or thin soil mantle of negligible infiltration capacity	(15) Slow to take up water; clay or other soil of low infiltration capacity, such as heavy gumbo	(10) Normal; deep loam with infiltration about equal to that of typical prairie soil	(5) High; deep sand or other soil that takes up water readily and rapidly
Vegetal cover	(20) No effective plant cover; bare or very sparse cover	(15) Poor to fair; clean-cultivated crops or poor natural cover; less than 10% of drainage area under good cover	(10) Fair to good; about 50% of drainage area in good grassland, woodland, or equivalent cover; not more than 50% of area in clean-cultivated crops	(5) Good to excellent; about 90% of drainage area in good grassland, woodland, or equivalent cover
Surface storage	(20) Negligible; surface depressions few and shallow; drainage-ways steep and small; no ponds or marshes	(15) Low; well-defined system of small drainage-ways; no ponds or marshes	(10) Normal; considerable surface-depression storage; drainage system similar to that of typical prairie lands; lakes, ponds and marshes less than 2% of drainage area	(5) High; surface-depression storage high; drainage system not sharply defined; large flood-plain storage or a large number of lakes, ponds, or marshes

Note: For each watershed characteristic in left column select appropriate descriptive box; add four numerical values given in parentheses to obtain runoff coefficient as a percentage.

Source: U.S. Soil Conservation Service.

Table 4G.30 Time of Concentration of a Watershed

Length (Feet)	Height (Feet)										
	5	10	20	40	60	100	200	400	600	800	1000
200	2	2	1	1	1	1					
400	5	4	3	2	2	1	1				
600	8	6	4	3	3	2	2	1	1		
800	10	8	6	5	4	3	2	2	2	1	
1,000	14	10	8	6	5	4	3	2	2	2	2
2,000	28	21	17	13	10	9	7	5	4	4	3
3,000	48	35	27	20	17	13	10	8	7	6	5
4,000	65	50	36	28	24	19	14	11	9	8	7
5,000	100	65	48	35	28	22	19	14	12	11	9
7,000	150	100	70	52	42	34	27	20	17	15	13
10,000	240	180	120	85	70	54	40	28	25	22	20
15,000	330	270	210	150	120	90	60	46	44	34	30
20,000	430	350	280	210	180	140	95	65	54	48	43
30,000	600	500	450	320	280	230	170	120	100	90	75
40,000	870	660	510	420	360	300	240	180	150	130	120
50,000	1080	840	600	480	430	360	300	230	190	160	150

Note: Values are time in minutes for water to travel from the most distant point in a watershed to the watershed outlet. Length is distance along the main stream from the watershed outlet to the most distant ridge; height is difference in elevation between the watershed outlet and the most distant ridge.

Source: From Kirpich, *Civil Eng*, 1940.

Table 4G.31 Total Impervious Area for Specific Land-Use Categories

	Typical Values of Total Impervious Area (Precent)		
Land-Use Category	**Low**	**Intermediate**	**High**
Single-family residential[a]	16	27	45
Multifamily residential[b]	50	60	70
Commercial[c]	80	88	95
Industrial[d]	50	75	90
Public facilities[e]	50	60	75
Parks and undeveloped land[f]	0	1	3

[a] Single-family residential — Single-family dwellings predominate.
[b] Multifamily residential — Multiple-family units predominate. These include duplexes, apartment buildings, and condominiums.
[c] Commercial — Zone consisting of various types of business.
[d] Industrial — Manufacturing complexes, railroad yards, and large utilities.
[e] Public facilities — School, hospitals, churches, airports, and other public buildings.
[f] Parks and undeveloped land-parks, forests, and open undeveloped land.

Source: From Conger, D.H., Estimating magnitude and frequency of floods for wisconsin urban streams, *U.S. Geological Survey Water-Resources investigations Report 86-4005*, 1986.

SECTION 4H EROSION AND SEDIMENTATION

Table 4H.32 Drainage Area, Water and Suspended Sediment Discharges for Major Rivers of the World

River	Drainage Area ($\times 10^6$ km^2)	Water Discharge (km^3 yr)	Sediment Discharge Millions of Tons per Year: Strakhov (1961) and Lisitzin (1972)	Holeman (1968)	Milliman and Meade (1983)
North America					
St. Lawrence (Canada)	1.03	447	4	4	4
Hudson (U.S.A.)	0.02	12	36	—	1
Mississippi (U.S.A.) (including Atchafalaya)	3.27	580	500	349	210
Brazos (U.S.A.)	0.11	7	32	32	16
Colorado (Mexico)	0.64	20	135	135	0.1
Eel (U.S.A.)	0.008	—	—	16	14
Columbia (U.S.A.)	0.67	251	36	9	8
Fraser (Canada)	0.22	112	—	—	20
Yukon (U.S.A.)	0.84	195	88	—	60
Copper (U.S.A.)	0.06	39	—	—	70
Susitna (U.S.A.)	0.05	40	—	—	25
MacKenzie (Canada)	1.81	306	15	—	100
Total North America	9.57				528
South America					
Chira (Peru)	0.02	5	—	—	4-75
Magdalena (Colombia)	0.24	237	—	—	220
Orinoco (Venezuela)	0.99	1100	86	86	210
Amazon (Brazil)	6.15	6300	498	364	900
Sao Francisco (Brazil)	0.64	97	—	—	6
La Plata (Argentina)	2.83	470	129	82	92
Negro (Argentina)	0.10	30	—	—	13
Total South America	10.85				1420
Europe					
Rhone (France)	0.09	49	31	—	10
Po (Italy)	0.07	46	18	15	15
Danube (Romania)	0.81	206	67	19	67
Semani (Albania)	—	—	—	22	?
Drini (Albania)	0.01	—	—	15	?
Total Europe	0.97				92
Eurasian Arctic					
Yana (U.S.S.R.)	0.22	29	3	—	3
Ob (U.S.S.R.)	2.50	385	16	15	16
Yenisei (U.S.S.R.)	2.58	560	13	—	13
Severnay Dvina (U.S.S.R.)	0.35	106	4.5	—	4.5
Lena (U.S.S.R.)	2.50	514	15	—	12
Kolyma (U.S.S.R.)	0.64	71	6	—	6
Indigirka (U.S.S.R.)	0.36	55	14	—	14
Total Euras. Arctic	9.15				68
Asia					
Amur (U.S.S.R.)	1.85	325	25	—	52
Liaohe (China)	0.17	6	—	—	41
Daling (China)	0.02	1	—	—	36
Haiho (China)	0.05	2	—	—	81
Yellow (Huangho) (China)	0.77	49	1890	1890	1080
Yangtze (China)	1.94	900	500	502	478
Huaihe (China)	0.26	—	—	—	14
Pearl (Zhu Jiang) (China)	0.44	302	—	27	69
Hungho (Vietnam)	0.12	123	130	130	130
Mekong (Vietnam)	0.79	470	170	170	160
Irrawaddy (Burma)	0.43	428	299	300	265
Ganges/Brahmaputra (Bangladesh)	1.48	971	2180	2180	1670
Mehandi (India)	0.13	67	—	62	2

(Continued)

Table 4H.32 (Continued)

River	Drainage Area ($\times 10^6$ km^2)	Water Discharge (km^3 yr)	Sediment Discharge Millions of Tons per Year		
			Strakhov (1961) and Lisitzin (1972)	Holeman (1968)	Milliman and Meade (1983)
Damodar (India)	0.02	10	—	28	?
Godavari (India)	0.31	84	—	—	96
Indus (Pakistan)	0.97	238	435	440	100
Tigris-Euphrates (Iraq)	1.05	46	105	53	?
Total Asia	9.74				4334
Africa					
Nile (Egypt)	2.96	30	110	111	0
Niger (Nigeria)	1.21	192	67	4	40
Zaire (Zaire)	3.82	1250	65	64	43
Orange (S. Africa)	1.02	11	153	—	17
Zambesi (Mozambique)	1.20	223	100	—	20
Limpopo (Mozambique)	0.41	5	—	—	33
Rufiji (Tanzania)	0.18	9	—	—	17
Tana (Kenya)	0.032	—	—	—	32
Total Africa (minus Nile)	7.48				175
Oceania					
Murray (Aust.)	1.06	22	32	32	30
Waiapu (N.Z.)	—	—	—	—	28
Haast (N.Z.)	0.001	6	—	—	13
Fly (New Guinea)	0.061	77			30
Purari (New Guinea)	0.031	77			80
Choshui (Taiwan)	0.003	6	—	—	66
Kaoping (Taiwan)	0.003	9	—	—	39
Tsengwen (Taiwan)	0.001	2	—	—	28
Hualien (Taiwan)	0.002	4	—	—	19
Peinan (Taiwan)	0.002	4	—	—	17
Hsiukuluan (Taiwan)	0.002	4	—	—	16
Total Oceania (excluding Murray)	1.074	39			336

Source: From Milliman, J.D., and Meade, R.H., World-wide Delivery of River sediment to oceans Copyright. *J. Geol.*, 91, 1, 1983. With permission.

Table 4H.33 Losses of Land by Riverbank Erosion in the United States

Region	Areas of Significant Erosion	Average Land Losses
	River-miles	*Acres/yr*
North Atlantic	Not significant	Not significant
South Atlantic-Gulf	5,100	350
Great Lakes	Not significant	Not significant
Ohio	Not significant	Not significant
Tennessee	Not significant	Not significant
Upper Mississippi	Not available	Not available
Lower Mississippi	1,044	4,705
Souris-Red-Rainy	Not available	Not available
Missouri	1,692	5,000
Arkansas-White–Red	2,300	7,300
Texas-Gulf	1,698	1,045
Rio Grande	250	150
Upper Colorado	Not available	Not available
Lower Colorado	Not available	Not available
Great Basin	265	150
Columbia-North Pacific	13,500	1,300
California	2,600	3,837
Alaska	80,000	5,000
Hawaii	Not significant	Not significant
Puerto Rico	Not available	Not available

Note: Estimated average annual losses as of 1966; for regions see Figure 4G.9.

Source: U.S. Water Resources Council, 1968.

Table 4H.34 Losses of Land by Erosion of Beaches and Estuary Shores in the United States

Region	Average Annual Land Losses
	Acres/mile
North Atlantic	0.12
South Atlantic-Gulf	0.11
Great Lakes	0.07
Lower Mississippi	—
Texas-Gulf	0.27
Columbia-North Pacific	0.02
California	0.13
Alaska	0.06
Hawaii	0.24
Puerto Rico	—

Note: Estimated average annual losses as of 1966; for regions see Figure 4G.9.

Source: U.S. Water Resources Council, 1968.

Table 4H.35 Erosion Problems in the Public Domain of the United States

	Extent of Erosion			
Region	Slight	Moderate	Critical	Total
Missouri	7.5	9.0	5.2	21.7
Arkansas-White–Red	0.1	0.4	0.2	0.7
Rio Grande	1.3	6.4	3.6	11.3
Upper Colorado	4.1	14.4	8.3	26.8
Lower Colorado	2.4	13.8	7.9	24.1
Great Basin	15.4	20.1	11.6	47.1
Columbia-North Pacific	14.6	8.5	5.0	28.1
California	6.8	5.4	3.2	15.4
Total	52.2	78.0	45.0	175.2

Note: Values in millions of acres; conterminous United States only; for regions see Figure 4G.9.

Source: U.S. Water Resources Council, 1968.

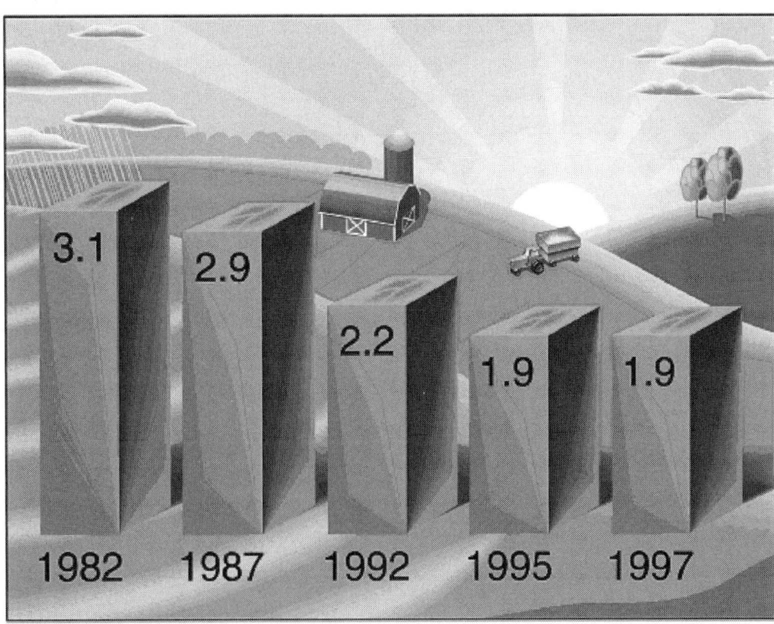

Figure 4H.14 Total erosion on cropland and Conservation Reserve Program Land. (From USDA, Natural Resources Conservation Service, 1997 National Resources Inventory, revised December 2000, www.nrcs.usda.gov.)

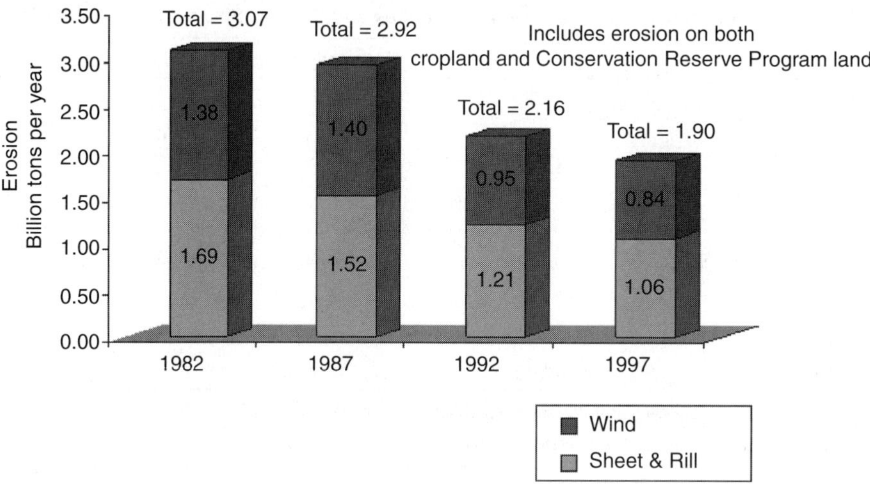

Figure 4H.15 Changes in erosion, 1982–1997. (From www.nrcs.usda.gov.)

Hawaii

Pacific Basin (No Data)

Northern Marianas

Guam

American Samoa

Alaska (No Data)

Each red dot represents 5,000 acres of highly erodible land (57.3 million acres) and each yellow dot represents 5,000 acres of non-highly erodible land (50.5 million acres) with excess erosion above the tolerable soil erosion rate.

95% or more Federal area

108 million acres have excessive erosion with a total of 1.3 billion tons of erosion.

Excess erosion leads to water quality concerns from sediments, nutrients, and pesticides as well as air quality in wind erosion areas of the West, Midwest, Northern Plains, and Southern Plains. Excess erosion is also an indicator of forgone opportunities for improving soil, water and air quality, sequestering carbon dioxide, and helping in goals to reduce greenhouse gases in the atmosphere.

Puerto Rico / U.S. Virgin Islands

USDA NRCS U.S. Department of Agriculture Natural Resources Conservation Service Resource Assessment Division Washington DC December 2000

Map ID: m5083 For proper interpretation, see Explanation of Analyis for this map at our web site. Search for "USDASOTL" to locate our map index.

Data Source: 1997 National Resources Inventory Revised December 2000

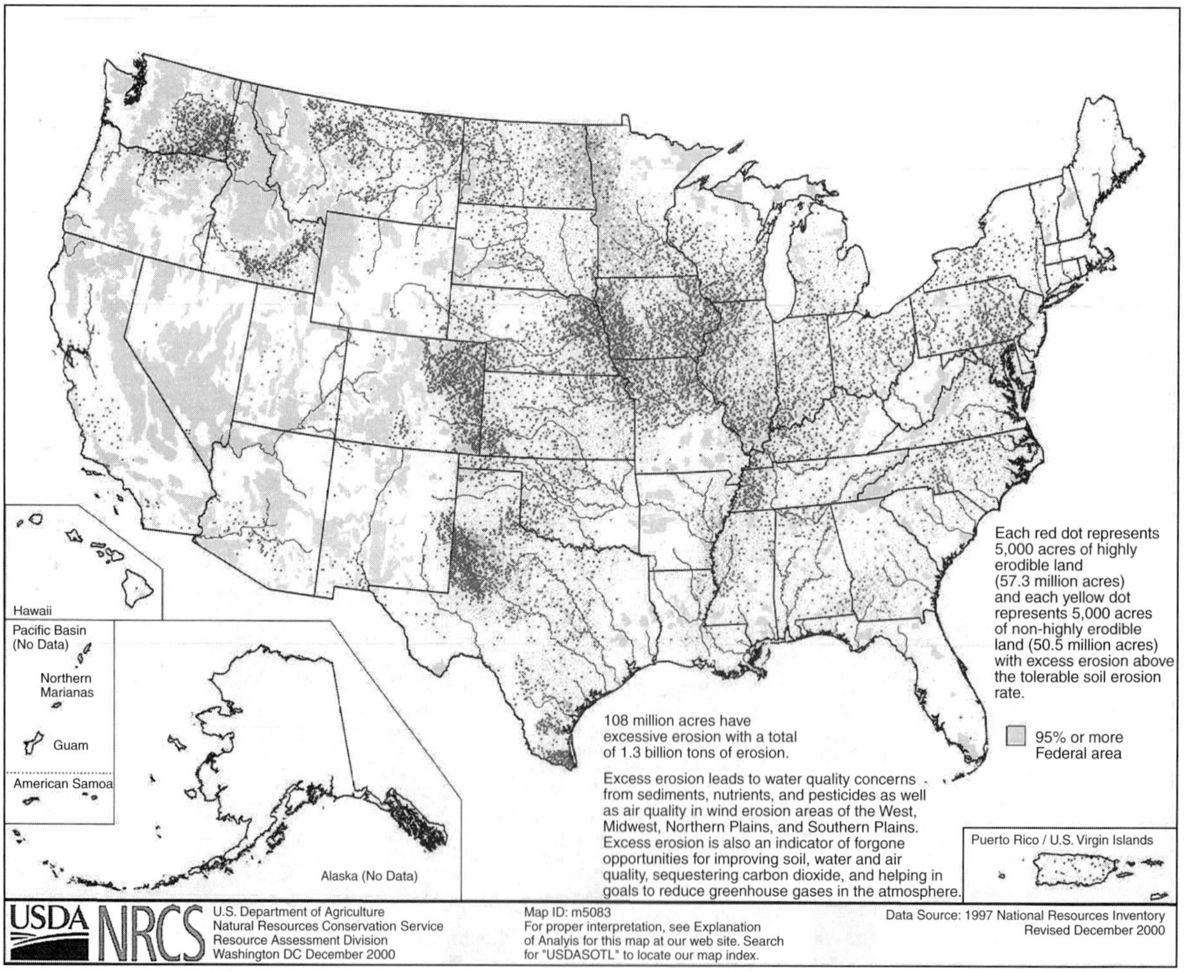

Figure 4H.16 Excessive erosion on cropland, 1997. (From www.nrcs.usda.gov.)

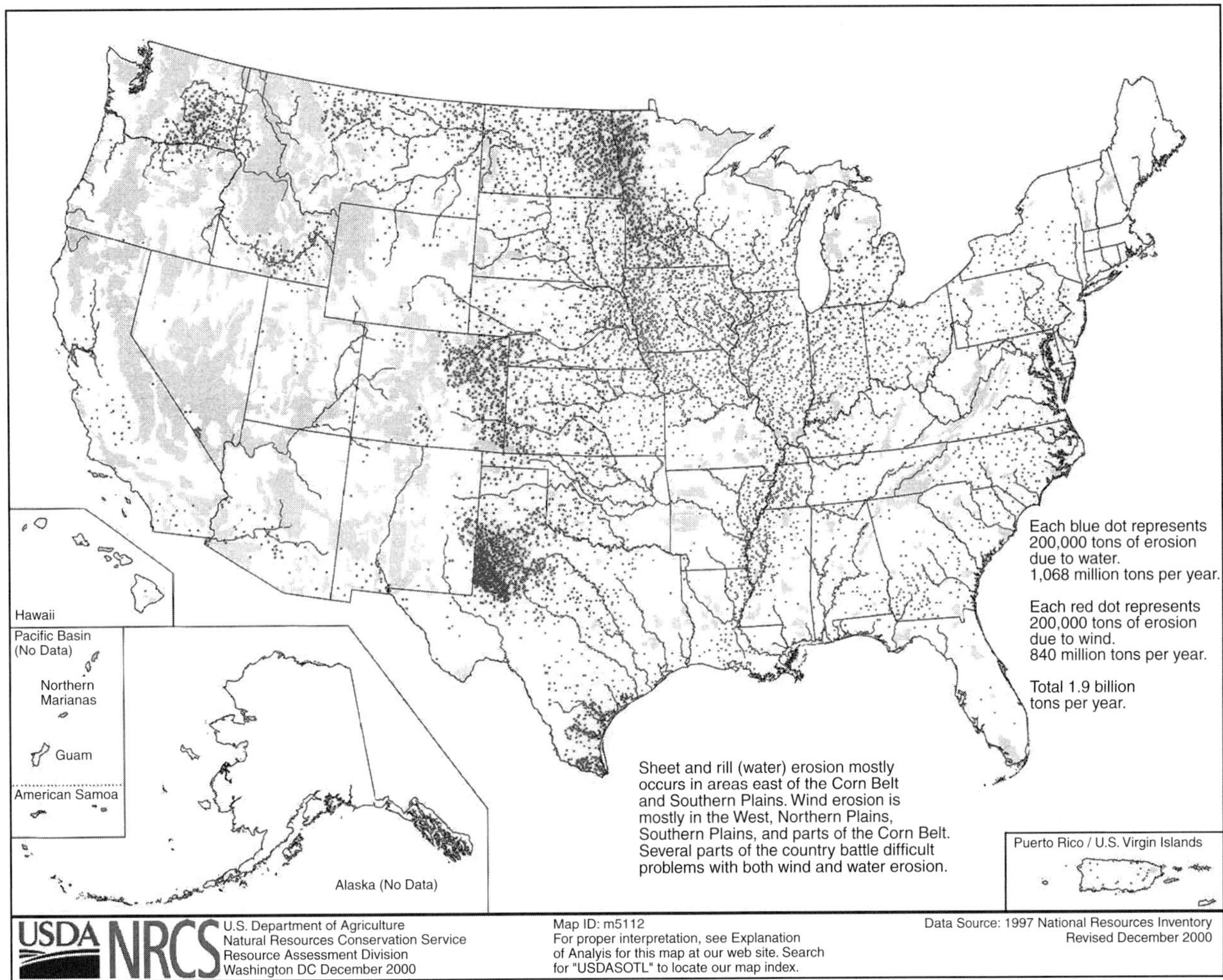

Figure 4H.17 Total wind and water erosion, 1997. (From www.nrcs.usda.gov.)

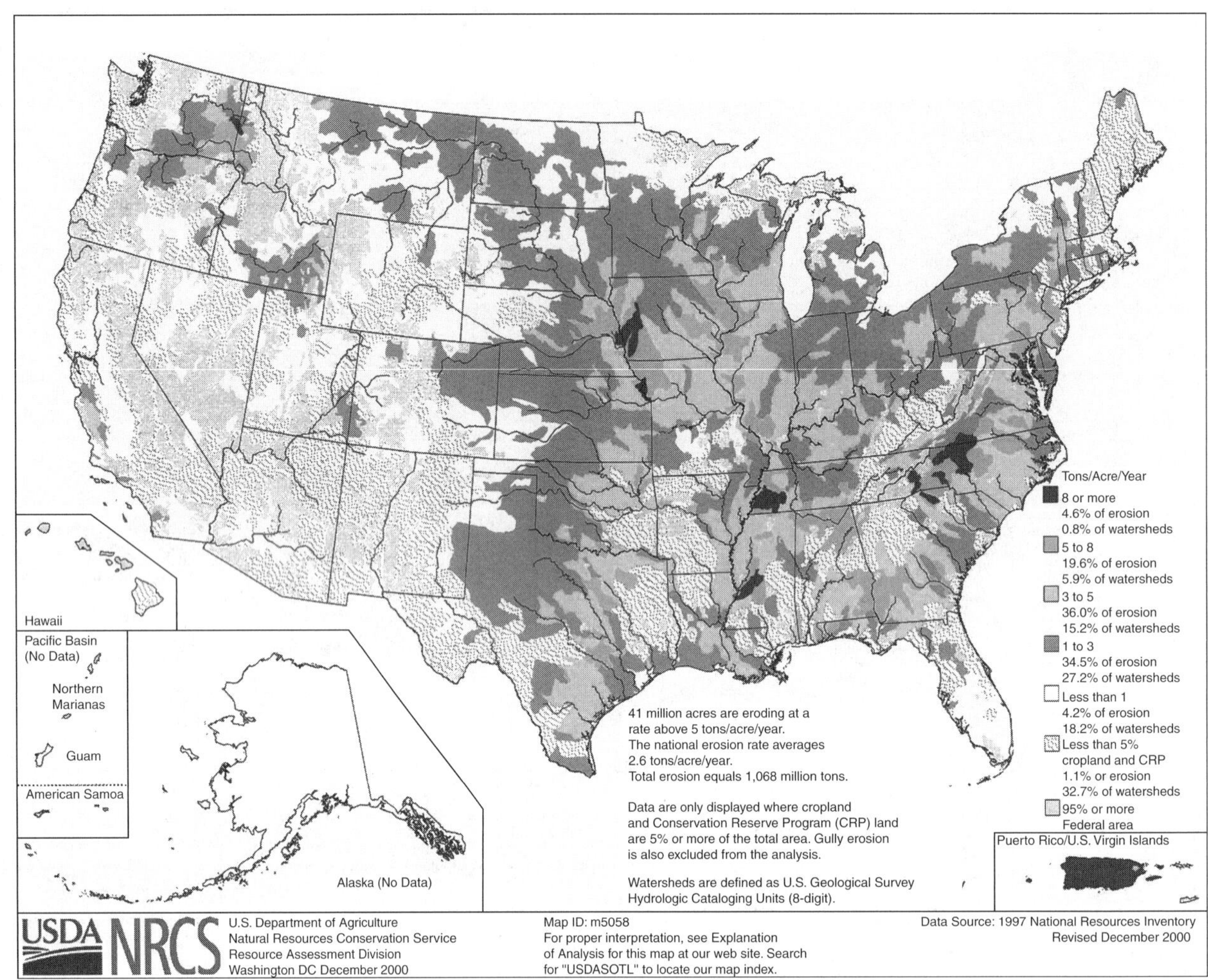
Tons/Acre/Year
8 or more
4.6% of erosion
0.8% of watersheds
5 to 8
19.6% of erosion
5.9% of watersheds
3 to 5
36.0% of erosion
15.2% of watersheds
1 to 3
34.5% of erosion
27.2% of watersheds
Less than 1
4.2% of erosion
18.2% of watersheds
Less than 5% cropland and CRP
1.1% or erosion
32.7% of watersheds
95% or more Federal area
Hawaii
Pacific Basin (No Data)
Northern Marianas
Guam
American Samoa
Alaska (No Data)
Puerto Rico/U.S. Virgin Islands
41 million acres are eroding at a rate above 5 tons/acre/year.
The national erosion rate averages 2.6 tons/acre/year.
Total erosion equals 1,068 million tons.
Data are only displayed where cropland and Conservation Reserve Program (CRP) land are 5% or more of the total area. Gully erosion is also excluded from the analysis.
Watersheds are defined as U.S. Geological Survey Hydrologic Cataloging Units (8-digit).
USDA NRCS
U.S. Department of Agriculture
Natural Resources Conservation Service
Resource Assessment Division
Washington DC December 2000
Map ID: m5058
For proper interpretation, see Explanation of Analysis for this map at our web site. Search for "USDASOTL" to locate our map index.
Data Source: 1997 National Resources Inventory
Revised December 2000

Figure 4H.18 Average annual soil erosion by water on cropland and Conservation Reserve Program land, 1997. (From www.nrcs.usda.gov.)

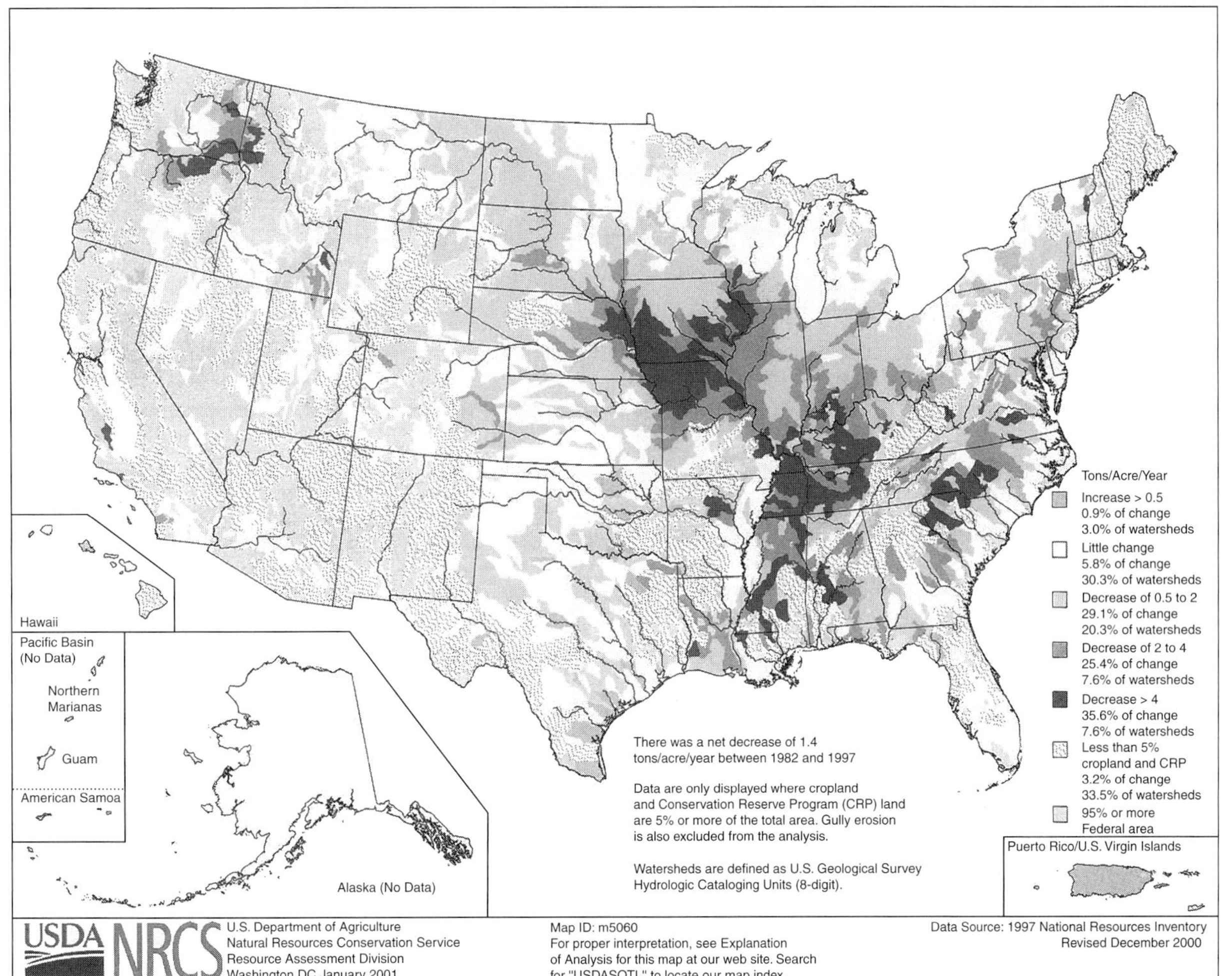

Figure 4H.19 Change in average annual soil erosion by water on cropland and Conservation Reserve Program land, 1982–1997. (From www.nrcs.usda.gov.)

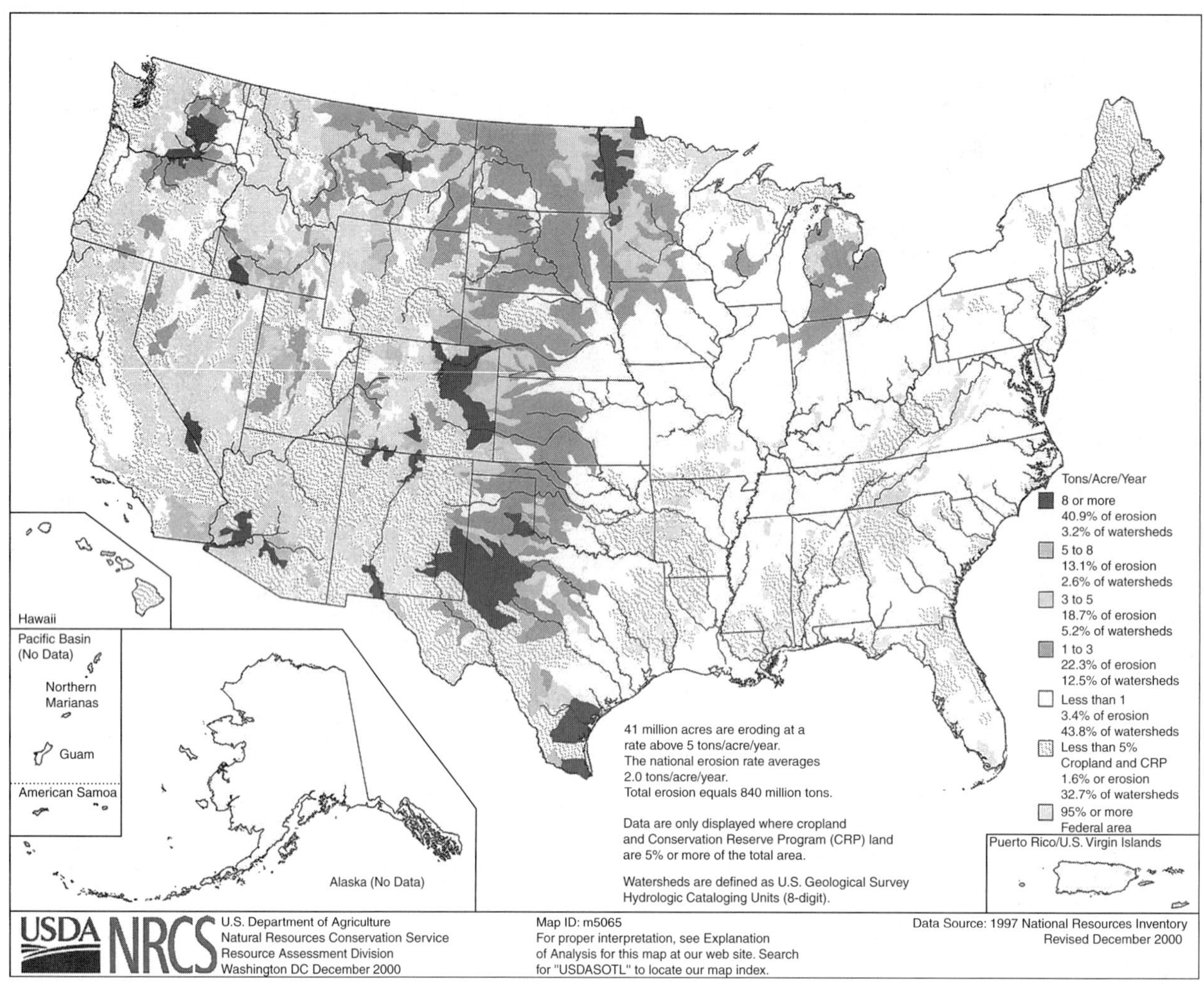

Figure 4H.20 Average annual soil erosion by wind on cropland and Conservation Reserve Program land, 1997. (From www.nrcs.usda.gov.)

Table 4H.36 Selected Quantitative Effects of Man's Activities on Surface Erosion

Initial Status of Land Use	Type of Disturbance	Magnitude of Specific Disturbance[a]
Forestland	Planting row crops	100 to 1,000 Times
Grassland	Planting row crops	20 to 100 Times
Forestland	Building logging roads	220 Times
Forestland	Woodcutting and skidding	1.6 Times
Forestland	Fire	7 to 1,500 Times
Forestland	Mining	1,000 Times
Row crop	Construction	10 Times
Pastureland	Construction	200 Times
Forestland	Construction	2,000 Times

[a] Relative magnitude of surface erosion from disturbed surface, assuming "I" for the initial status. The first row of the table, for example, indicates that transforming a forestland into row crops may increase surface erosion 100 to 1000 times.

Source: U.S. Environmental Protection Agency, Loading Functions for Assessments of Water Pollution from Non-Point Sources, *Environmental Protection Technology Series*, Washington, DC, 1976.

Table 4H.37 Representative Rates of Erosion from Various Land Uses

Land Use	Amount of Erosion (Tons/Square Mile/Year)	Rate of Erosion Relative to Forest = 1
Forest	24	1
Grassland	240	10
Abandoned Surface Mines	2,400	100
Cropland	4,800	200
Harvested Forest	12,000	500
Active Surface Mines	48,000	2,000
Construction	48,000	2,000

Source: U.S. Environmental Protection Agency. *Methods For Identifying and Evaluating the Nature and Extent of Nonpoint Source of Pollutants*, EPA 430/9-73-014, Washington, DC, 1973.

Table 4H.38 Estimated Average Annual Sheet and Rill Erosion on Nonfederal Land, by State and Year (Data in tons/acre/year)

		Cropland				
State	Year	Cultivated	Noncultivated	Total	CRP Land	Pastureland
Alabama	1982	7.6	0.8	7.2	—	0.6
	1987	6.5	0.5	6.0	3.0	0.5
	1992	7.0	0.5	6.3	0.6	0.5
	1997	6.7	0.5	6.0	1.0	0.5
Arizona	1982	0.6	0.3	0.5	—	0.2
	1987	0.6	0.2	0.6	—	0.1
	1992	0.6	0.2	0.6	—	0.1
	1997	0.7	0.2	0.6	—	0.1
Arkansas	1982	3.8	0.8	3.7	—	1.1
	1987	3.8	0.6	3.7	0.7	1.1
	1992	3.5	0.6	3.4	0.7	1.2
	1997	3.5	0.6	3.4	0.6	1.1
California	1982	1.2	0.7	1.1	—	0.2
	1987	1.1	0.8	1.0	2.4	0.2
	1992	1.0	0.5	0.8	1.1	0.1
	1997	0.7	0.5	0.6	0.3	0.1
Colorado	1982	2.2	0.2	1.9	—	0.3
	1987	2.2	0.2	2.0	2.2	0.3
	1992	2.0	0.2	1.8	0.8	0.3
	1997	1.7	0.2	1.5	0.4	0.3
Connecticut	1982	4.8	0.6	2.6	—	0.2
	1987	5.7	1.2	3.1	—	0.2
	1992	6.1	1.4	3.3	—	0.2
	1997	5.6	0.7	2.7	—	0.1
Delaware	1982	2.1	0.2	2.0	—	0.4
	1987	2.0	0.4	2.0	—	0.4
	1992	2.1	0.7	2.1	0.1	0.5
	1997	2.0	0.4	2.0	0.1	0.6
Florida	1982	2.4	0.5	1.8	—	0.1
	1987	2.1	0.4	1.4	0.5	0.1
	1992	1.8	0.4	1.2	0.6	0.1
	1997	1.8	0.5	1.2	0.4	0.1
Georgia	1982	6.2	0.4	5.9	—	0.5
	1987	6.1	1.0	5.7	2.1	0.4
	1992	5.5	0.6	5.0	0.5	0.4
	1997	5.9	0.3	5.2	0.2	0.4
Hawaii	1982	5.3	3.0	5.0	—	0.8
	1987	5.1	2.8	4.8	—	0.7
	1992	4.6	2.8	4.3	—	0.7
	1997	2.5	3.3	2.7	—	0.8
Idaho	1982	5.0	0.4	4.3	—	0.4
	1987	4.4	0.3	3.7	2.9	0.4

(Continued)

Table 4H.38 (Continued)

State	Year	Cropland			CRP Land	Pastureland
		Cultivated	Noncultivated	Total		
	1992	3.5	0.4	2.9	1.5	0.4
	1997	3.4	0.4	2.8	1.3	0.5
Illinois	1982	6.3	1.2	6.2	—	1.6
	1987	5.3	1.5	5.2	4.3	1.3
	1992	4.4	1.6	4.3	1.2	1.0
	1997	4.1	0.6	4.0	0.5	1.0
Indiana	1982	4.8	1.1	4.7	—	1.0
	1987	4.4	0.9	4.2	1.7	0.8
	1992	3.4	1.1	3.3	0.4	0.8
	1997	3.0	0.9	2.9	0.3	0.7
Iowa	1982	7.7	1.8	7.5	—	1.3
	1987	6.5	1.5	6.3	0.8	1.3
	1992	5.6	1.1	5.4	0.5	1.2
	1997	4.9	0.8	4.7	0.5	1.1
Kansas	1982	2.7	0.4	2.5	—	0.8
	1987	2.6	0.5	2.5	2.3	0.8
	1992	2.3	0.4	2.2	0.4	0.7
	1997	2.2	0.4	2.1	0.3	0.7
Kentucky	1982	8.3	1.0	6.9	—	2.4
	1987	8.2	1.1	6.6	4.2	2.4
	1992	5.8	1.2	4.5	0.9	2.5
	1997	4.4	1.2	3.4	0.9	2.0
Louisiana	1982	4.7	0.6	4.6	—	0.2
	1987	4.1	0.3	4.0	0.6	0.2
	1992	3.5	0.6	3.4	0.2	0.2
	1997	3.3	0.6	3.2	0.6	0.2
Maine	1982	3.6	0.2	1.7	—	0.2
	1987	4.0	0.4	1.8	—	0.2
	1992	3.1	0.3	1.3	0.1	0.2
	1997	3.9	0.3	1.7	0.2	0.2
Maryland	1982	5.6	1.3	5.2	—	1.1
	1987	5.3	2.0	5.1	1.0	1.1
	1992	5.0	1.8	4.6	1.3	1.0
	1997	4.4	1.2	4.0	1.0	0.7
Massachusetts	1982	5.7	0.2	1.7	—	0.2
	1987	5.9	0.1	1.7	—	0.1
	1992	4.1	0.2	1.3	—	0.2
	1997	4.5	0.1	1.2	—	0.1
Michigan	1982	2.5	0.6	2.2	—	0.3
	1987	2.5	0.7	2.2	3.8	0.2
	1992	2.3	0.6	1.9	0.6	0.2
	1997	2.0	0.5	1.6	0.2	0.2
Minnesota	1982	2.6	0.6	2.4	—	0.4
	1987	2.6	0.4	2.5	1.3	0.3
	1992	2.3	0.3	2.2	0.3	0.3
	1997	2.1	0.3	2.0	0.2	0.3
Mississippi	1982	7.7	2.8	7.6	—	1.3
	1987	6.6	2.3	6.5	4.4	1.2
	1992	5.7	1.3	5.5	2.6	1.2
	1997	5.3	1.2	5.0	1.1	1.2
Missouri	1982	10.9	1.0	9.6	—	2.0
	1987	8.4	0.7	7.4	6.3	1.7
	1992	6.6	0.7	5.5	1.1	1.6
	1997	5.6	0.7	4.5	0.7	1.3
Montana	1982	2.1	0.2	1.8	—	0.2
	1987	2.3	0.2	2.0	0.8	0.2
	1992	2.0	0.2	1.7	0.2	0.2
	1997	1.9	0.3	1.6	0.2	0.2
Nebraska	1982	4.8	0.7	4.5	—	0.9

(Continued)

Table 4H.38 (Continued)

State	Year	Cropland			CRP Land	Pastureland
		Cultivated	Noncultivated	Total		
	1987	4.2	0.5	3.9	1.5	0.8
	1992	3.5	0.5	3.3	0.7	0.7
	1997	2.9	0.5	2.7	0.5	0.7
Nevada	1982	0.2	0.0	0.1	0.0	0.0
	1987	0.2	0.0	0.1	0.0	0.0
	1992	0.2	0.0	0.1	0.0	0.1
	1997	0.2	0.0	0.1	0.0	0.1
New Hampshire	1982	4.1	0.4	1.4	—	0.5
	1987	4.4	0.4	1.3	—	0.5
	1992	3.7	0.4	0.8	—	0.4
	1997	3.5	0.4	0.9	—	0.5
New Jersey	1982	6.7	1.0	5.5	—	0.5
	1987	6.7	1.1	5.7	—	0.6
	1992	5.5	0.8	4.3	0.3	0.5
	1997	5.6	0.6	4.3	0.3	0.4
New Mexico	1982	1.2	0.1	1.0	—	0.1
	1987	0.9	0.1	0.7	1.0	0.1
	1992	1.0	0.2	0.8	0.4	0.1
	1997	0.9	0.1	0.7	0.2	0.1
New York	1982	4.0	0.7	2.6	—	0.4
	1987	4.1	0.9	2.7	3.8	0.4
	1992	4.0	0.8	2.4	0.5	0.3
	1997	3.9	0.7	2.3	0.3	0.3
North Carolina	1982	6.4	1.5	6.1	—	1.1
	1987	6.3	1.0	6.0	15.7	1.0
	1992	5.6	1.4	5.3	4.5	1.0
	1997	5.0	1.0	4.6	1.2	1.7
North Dakota	1982	1.9	0.4	1.8	—	0.4
	1987	2.0	0.5	1.8	1.1	0.5
	1992	1.5	0.3	1.4	0.3	0.5
	1997	1.4	0.3	1.3	0.2	0.4
Ohio	1982	3.8	1.1	3.6	—	2.3
	1987	3.7	1.1	3.5	3.4	1.7
	1992	3.3	1.2	3.1	0.5	1.7
	1997	2.6	1.4	2.5	0.3	1.7
Oklahoma	1982	2.7	0.6	2.6	—	0.9
	1987	3.0	0.6	2.9	1.1	0.7
	1992	2.9	0.5	2.8	0.4	0.7
	1997	2.8	0.5	2.8	0.3	0.6
Oregon	1982	4.6	0.7	3.8	—	0.6
	1987	3.4	0.5	2.6	2.9	0.5
	1992	3.2	0.4	2.5	0.4	0.5
	1997	3.1	0.4	2.3	0.4	0.5
Pennsylvania	1982	7.0	0.7	4.8	—	1.1
	1987	6.9	1.2	5.0	1.7	1.0
	1992	5.8	1.2	4.2	1.0	1.0
	1997	5.1	1.2	3.8	0.3	0.8
Rhode Island	1982	7.0	1.1	3.0	—	0.1
	1987	5.0	2.2	2.9	—	0.1
	1992	4.8	1.6	2.5	—	0.1
	1997	3.5	1.8	2.2	—	0.1
South Carolina	1982	4.0	1.9	3.9	—	0.4
	1987	3.9	1.4	3.8	3.9	0.4
	1992	3.3	1.0	3.1	1.7	0.4
	1997	3.2	0.7	3.0	0.5	0.4
South Dakota	1982	2.8	0.3	2.5	—	0.3
	1987	2.6	0.3	2.3	2.6	0.3
	1992	2.2	0.3	2.0	0.4	0.3
	1997	2.0	0.2	1.7	0.1	0.2

(Continued)

Table 4H.38 (Continued)

State	Year	Cropland			CRP Land	Pastureland
		Cultivated	Noncultivated	Total		
Tennessee	1982	11.0	0.9	9.4	—	0.8
	1987	10.8	1.0	9.1	9.5	0.7
	1992	9.1	0.9	7.1	0.8	0.7
	1997	7.7	0.6	5.6	0.7	0.8
Texas	1982	2.6	0.9	2.6	—	0.7
	1987	2.6	1.2	2.5	0.6	0.6
	1992	2.6	0.7	2.6	0.3	0.5
	1997	2.6	0.8	2.6	0.2	0.5
Utah	1982	1.4	0.2	0.9	—	0.1
	1987	1.5	0.2	0.8	3.2	0.1
	1992	1.4	0.2	0.8	1.3	0.1
	1997	1.6	0.2	0.8	0.9	0.2
Vermont	1982	4.6	0.2	1.3	—	0.3
	1987	4.2	0.2	1.4	—	0.2
	1992	3.4	0.5	1.2	—	0.1
	1997	3.1	0.7	1.2	—	0.1
Virginia	1982	6.6	1.5	5.3	—	3.4
	1987	6.4	1.6	4.9	0.8	3.4
	1992	6.4	1.4	4.5	0.8	3.4
	1997	5.9	1.5	3.9	0.5	3.3
Washington	1982	6.1	0.5	5.5	—	0.2
	1987	7.0	0.4	6.2	2.4	0.4
	1992	5.0	0.5	4.4	0.5	0.4
	1997	4.7	0.6	4.0	0.6	0.3
West Virginia	1982	7.3	0.7	2.5	—	4.2
	1987	9.2	0.9	2.8	0.7	5.4
	1992	4.7	0.8	1.7	0.3	6.1
	1997	4.3	0.8	1.4	0.0	6.0
Wisconsin	1982	4.7	1.5	4.1	—	0.6
	1987	4.1	2.0	3.7	4.5	0.6
	1992	3.8	0.7	3.2	0.7	0.5
	1997	3.7	1.2	3.3	0.6	0.6
Wyoming	1982	1.5	0.2	0.9	—	0.3
	1987	1.4	0.1	0.8	1.6	0.2
	1992	1.3	0.2	0.7	0.6	0.3
	1997	1.1	0.1	0.6	0.2	0.3
Caribbean	1982	11.1	11.9	11.2	—	7.0
	1987	11.2	13.1	11.5	—	7.3
	1992	12.1	15.4	12.9	—	8.0
	1997	12.2	13.2	12.7	—	6.4
National average	1982	4.4	0.7	4.0	—	1.1
	1987	4.0	0.7	3.7	2.0	1.0
	1992	3.5	0.6	3.1	0.6	1.0
	1997	3.1	0.7	2.8	0.4	0.9

Source: www.nrcs.usda.gov.

Table 4H.39 Dissolved and Suspended Sediment Loads in Selected Rivers of the United States

River and Location	Elevation (ft)	Drainage Area (sq mi)	Average Discharge, *Q* (cfs)	Discharge + Drainage Area (cfs/sq mi)	Years of Record in Sample[a]	Average Suspended Load	Average Dissolved Load (millions of tons/yr)	Total Average Suspended and Dissolved Load	Total Average Load + Drainage Area (tons sq mi/yr)	Dissolved Load as Percent of Total Load (%)
Little Colorado at Woodruff, Arizona	5,129	8,100	63.3	0.0078	6	1.6	0.02	1.62	199	1.2
Canadian River near Amarillo, Texas	2,989	19,445	621	0.032	1	6.41	0.124	6.53	336	1.9
Colorado River near San Saba, Texas	1,096	30,600	1,449	0.047	5	3.02	0.208	3.23	105	6.4
Bighorn River at Kane, Wyoming	3,609	15,900	2,391	0.150	1	1.60	0.217	1.82	114	12
Green River at Green River, Utah	4,040	40,600	6,737	0.166	26–20	19	2.5	21.5	530	12
Colorado River near Cisco, Utah	4,090	24,100	8,457	0.351	25–20	15	4.4	19.4	808	23
Iowa River at Iowa City, Iowa	627	3,271	1,517	0.464	3	1.184	0.485	1.67	510	29
Mississippi River at Red River Landing, Louisiana		1,144,500[b]	569,500[b]	0.497	3	284	101.8	385.8	337	26
Sacramento River at Sacramento, California	0	27,000[c]	25,000[c]	0.926	3	2.85	2.29	5.14	190	44
Flint River near Montezuma, Georgia	256	2,900	3,528	1.22	1	0.400	0.132	0.53	183	25
Juniata River near New Port, Pennsylvania	364	3,354	4,329	1.29	7	0.322	0.566	0.89	265	64
Delaware River at Trenton, New Jersey	8	6,780	11,730	1.73	9–4	1.003	0.830	1.83	270	45

[a] Computation of load, dissolved or suspended, depends on discharge for same period. Years of record pertain to number of years used for related values of discharge and of suspended and dissolved load. Where two figures are shown, the first is for suspended load and the second is for dissolved load.
[b] From USGS records for Vicksburg, Mississippi station.
[c] Estimated.

Source: From Leopold, Wolman, and Miller, *Fluvial Processes in Geomorphology*, W.H. Freeman and Company, 1964. With permission.

Table 4H.40 Discharge of Suspended Sediment to the Coastal Zone by 10 Major rivers of the United States, About 1980

Rivers	Average Annual Sediment Discharge (million ton/yr)
Rivers that discharge the largest sediment loads:	
Mississippi	230[a]
Copper	80
Yukon	65
Susitna	25
Eel	15
Brazos	11
Columbia:	
Before Mount St. Helens eruption	10
Since Mount St. Helens eruption-approximate	40
Rivers with large drainage areas:	
St. Lawrence	1.5
Rio Grande	0.8
Colorado	0.1

[a] Includes Atchafalaya River.

Source: From Meade, R.H., Parker, R.S., Sediment in rivers of the United States, *U.S. Geological Survey Water-Supply Paper 2275*, National Water Summary 1984, 1985.

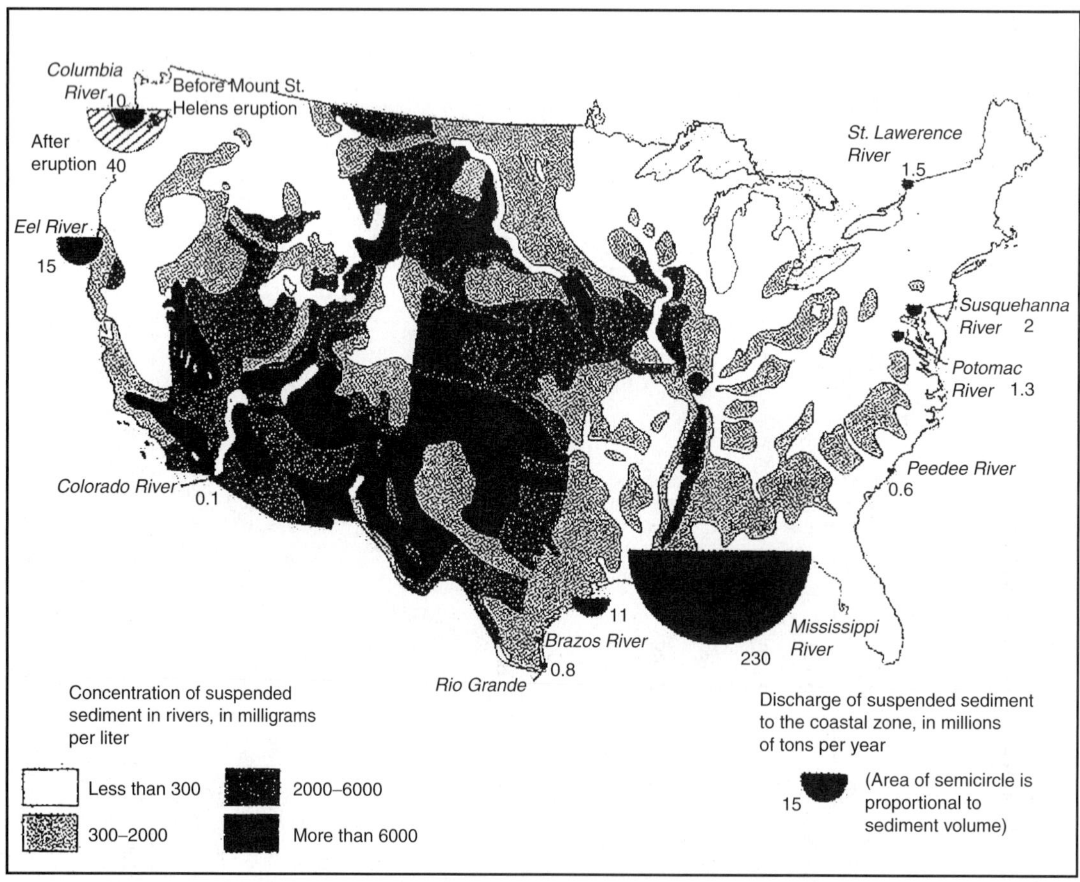

Figure 4H.21 Concentration of suspended sediment in rivers and discharge of suspended sediment to the coastal zone in the conterminous United States. (From Meade, R.H., Parker, R.S., Sediment in the rivers of the United States, National Water Summary 1984, *U.S. Geological Survey Water-Supply Paper 2275*, 1984.)

Table 4H.41 Dimension and Rate of Formation of Modern Deltas

River	Dimension of Subaerial Delta, Statute Mi		Amount of Sediment Discharged		Annual Extension of Subaerial Delta	
	Length	Breadth	River Water by Weight (avg), ppm	Annual Volume of Sediment (mi^3)	Measurement Period (yr)	Approximate Distance (ft)
Mississippi's present bird-foot delta	12	30	550	0.068	1838–1947	250
Hwang-Ho	300	470[a]	50,000[b]		1870–1937	950
Ganges-Brahmaputra	220	200	870	0.043 (Ganges only)		
Rhone into Mediterranean Sea	30	47	400–590	0.005	1737–1870	190
Danube	46	46	310	0.008		40
Nile (prior to barrages)	96	145	1600	0.001	1100–1870	45
Colorado above Hoover Dam	43	0.05–0.6	8300	0.032	1936–1948	3.6 mi (gorge)
Euphrates Tigris	350	90			1793–1853	180

[a] Includes 100 ml of nondeltaic Shantung Peninsula.
[b] Maximum is 400,000 ppm.

Source: From *McGraw-Hill Encyclopedia of the Geological Sciences*. Copyright 1978. With permission.

Table 4H.42 Water Storage Capacity and the Capacity Lost Annually Due to Sedimentation in the Conterminous United States

Farm Production Region	Total Water Storage Capacity (million ac-ft)	Usable Water Storage Capacity (million ac-ft)	Estimated Water Storage Capacity Lost (%)	Estimated Water Storage Capacity Lost (thousand ac-ft)	Stream Sediment Originating on Cropland (%)	Reservoir Sedimentation from Cropland (thousand ac-ft)
Northeast	36.5	25.2	0.08	28.1	29	8.2
Appalachian	59.5	30.6	0.13	75.5	29	21.9
Southeast	73.6	47.6	0.17	127.3	33	42.0
Lake States	29.3	19.5	0.27	79.1	64	50.6
Corn Belt	39.7	15.2	0.26	104.8	63	66.0
Delta States	42.7	20.1	0.21	87.5	41	35.9
Northern Plains	78.9	54.4	0.23	184.6	36	66.5
Southern Plains	110.3	46.6	0.19	207.4	19	39.4
Mountain	167.1	138.1	0.18	302.5	8	24.2
Pacific	90.7	74.7	0.49	441.6	9	39.7
United States (Lower 48)	728.3	472.1	0.22	1,638.5		394.4

Note: Reservoirs with 5000 acre-feet or more total capacity.

Source: From Crowder, B.M., *J Soil Water Conserv., Soil Conserv. Soc. Am.*, 1987.

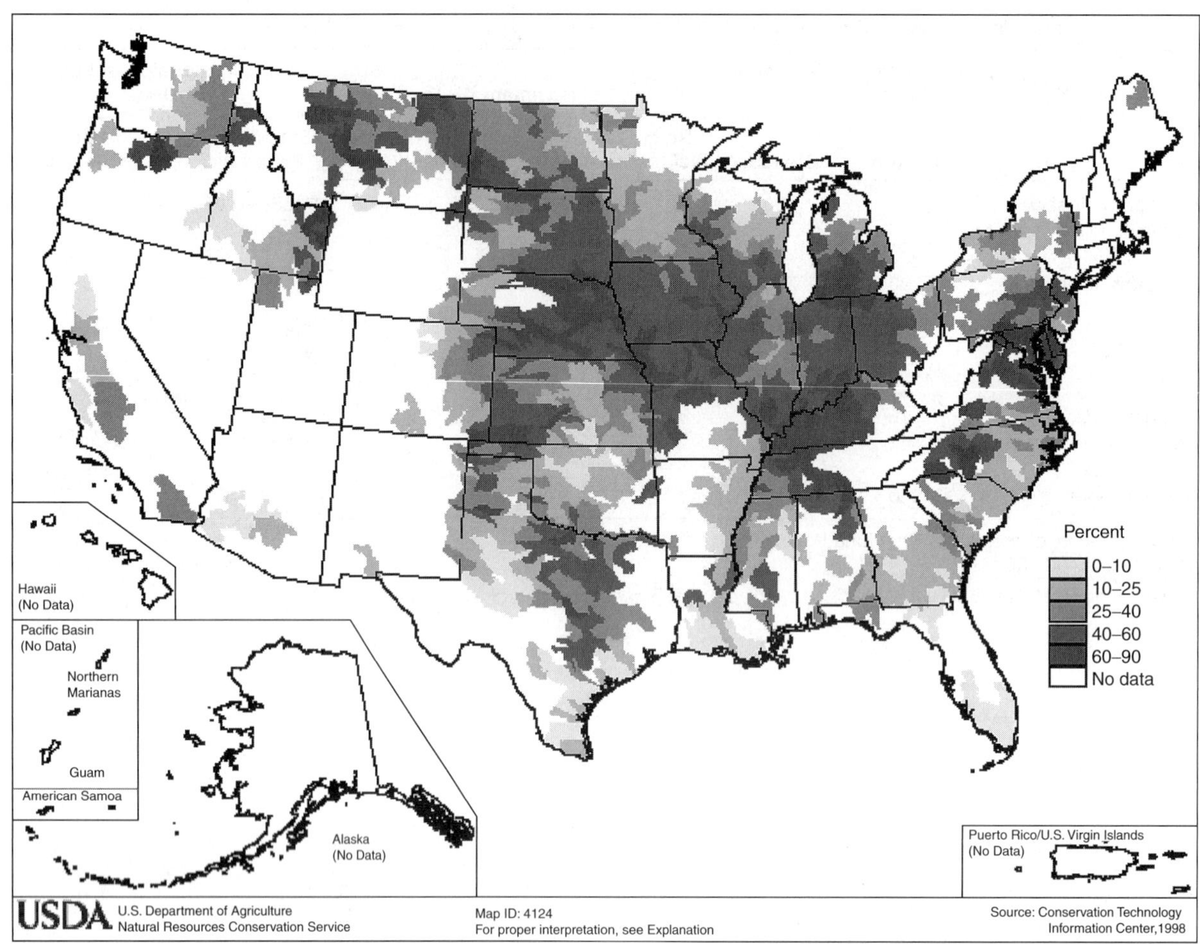

Figure 4H.22 Percent conservation tillage. (From www.epa.gov.)

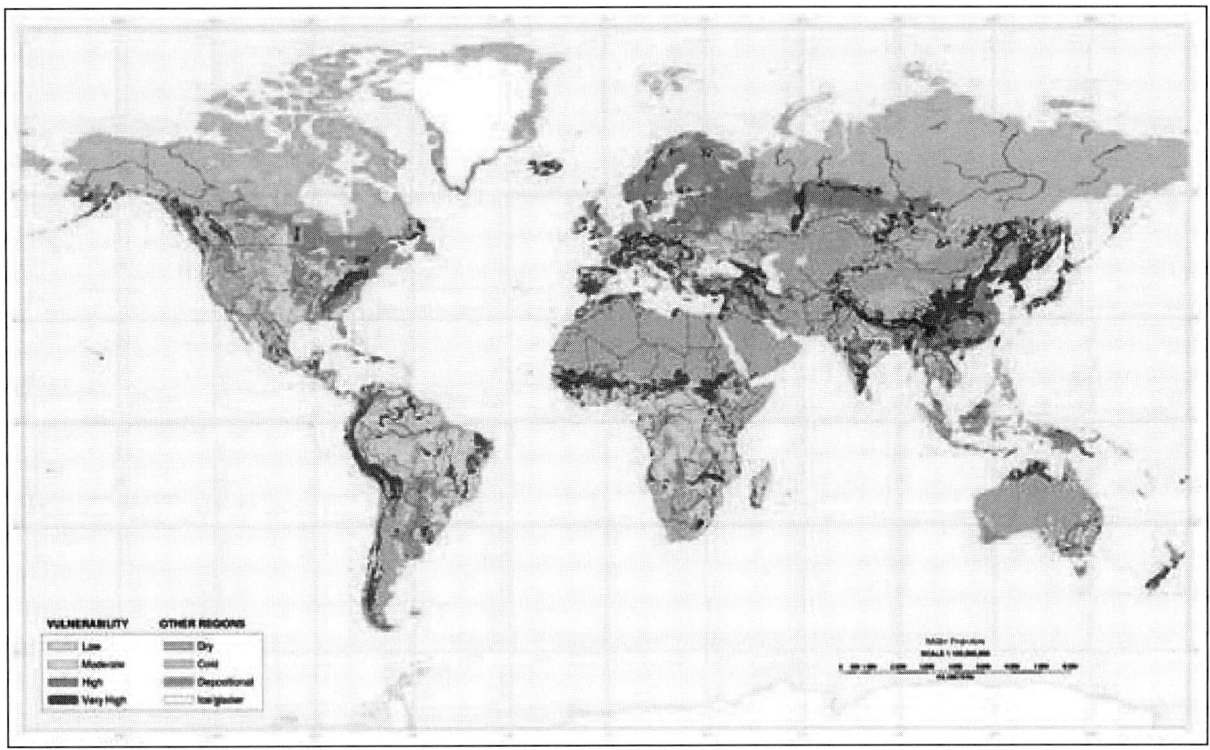

Figure 4H.23 Water erosion vulnerability. (From www.soils.usda.gov.)

Table 4H.43 Matrix of Laws Relating to Beach Nourishment

Law	Provisions	Relevance
P.L. 71-520 of 1930	Authorized the USACE to conduct shoreline erosion control studies (not construction) in cooperation with state governments; the Beach Erosion Board (BEB) was also established	First federal involvement in shoreline protection activities
P.L. 79-727 of 1946	Expanded the use of federal funds to now include one third of construction costs in addition to the studies for projects along publicly owned shores	Expanded federal involvement as a result of major hurricanes
Submerged Lands Act of 1953 (43 USC 1301 and following)	Gave coastal states authority over the resources of submerged lands from the shore out to three miles seaward	Affected the availability of offshore sand for beach nourishment
Outer Continental Shelf Lands Act of 1953 (43 USC 1331 and following)	Provided for the federal government to manage the mineral resources of the OCS lying on or under the seabed that extends seaward from state waters out to the edge of the shelf	Affected the availability of offshore sand for beach nourishment
P.L. 84-826 of 1956	Expanded the authority for federal shore protection to include privately owned shores where substantial public benefits would result; also defined periodic renourishment as construction for the protection of shores for a period of usually ten years	Federal authority now included shore protection on privately owned shores where public benefits result
River and Harbor Act (33 USC 401 and following) of 1962 (P.L. 87-874) and 1968 (P.L. 980-483)	Under Section 103 (33 USC 426g), the Corps was authorized to participate in the cost of protecting the shores of publicly owned property and private property where public benefits result; increased federal aid from one third to 100 percent for shore protection study costs leading to authorization; also increased federal participation in the cost of beach erosion and shore protection to 50 percent of the construction cost when the beaches were publicly owned or used, and 70 percent for seashore parks and conservation areas when certain conditions of ownership and use of the beaches were met	Resulted in a large number of studies and subsequent authorizations in the 1950's and 60's; Required USACE to fund mitigation for downdrift erosion caused by federal navigation works

(Continued)

Table 4H.43 (Continued)

Law	Provisions	Relevance
	Under Section 111 (33 USC 426i), mitigation could be conducted for shoreline erosion that results from federal navigation works	
P.L. 88-172 of 1963	Established the Coastal Engineering Research Board (CERB) and the Coastal Engineering Research Center (CERC), replacing the Beach Erosion Board	Resulted from increased need for additional engineering and study in the area of beach erosion, coupled with increased beach development and more demand for erosion relief from the federal government
Coastal Zone Management Act of 1972 (16 USC 1451 and following) (P.L. 92-583)	Required all federal agencies with activities directly affecting the coastal zone, or with development projects within the zone, to assure that those activities or projects are consistent with the approved state Coastal Zone Management Program	Established a national program to assist the states in comprehensively managing the nation's coastal resources through wise management practices. Encouraged coastal zone management and provided grants (Section 306A) for maintaining coastal areas
Coastal Barrier Resources Act of 1982 (16 USC 3501 and following) (P.L. 97-384)	Established the Coastal Barrier Resources System (CBRS); areas in the CBRS may no longer receive federal financial assistance for new construction or improvements. The CBRS was greatly expanded with the passage of the Coastal Barrier Improvement Act (CBIA) of 1990 (P.L. 101-591)	The intent of the law was to discourage development. The law applied only to areas within the defined CBRS
Water Resources Development Act (33 USC 2201 and following) of 1976 (P.L. 94-587), 1986 (P.L. 99-662), 1988 (P.L. 100-676), 1992 (P.L. 102-580), 1996 (P.L. 104-303), 1999 (P.L. 106-53), and 2000 (P.L. 106-541)	Established a broad congressional policy to encourage conservation efforts among federal, state, and local governments. Authorized the Secretary of the Army to construct, operate, and maintain any water resource development project. The resource development projects over which the USACE currently maintains jurisdiction are navigation, flood control, shore protection, and beach renourishment projects	Authorized beach nourishment projects. Set cost-sharing percentage, with a general trend to reduce federal percentage and increase non-federal percentage (will be 50/50 by 2003)
Shore Protection Act of 1996 (Section 227 of the WRDA of 1996) (33 USC 2601 and following)	Recommended funding for shore protection project studies and construction	Rejected the Administration's position of not authorizing funding for new projects
National Environmental Policy Act of 1969 (42 USC 4321 and following)	Required federal agencies to evaluate the environmental impacts associated with major actions they fund, support, permit, or implement	Required that actions be the least environmentally damaging practicable alternative. Most beach nourishment projects have the potential for adverse impacts and will trigger a required NEPA analysis
Clean Water Act (33 USC 1251 and following)	Under section 404, a permit was required for the discharge of dredged or fill materials into the waters of the U.S. The USACE has the permitting authority for the 404 program	Proponents of beach nourishment projects must obtain a Section 404 permit
Endangered Species Act (16 USC 1531 and following)	Federal agencies must review actions they undertake or support to determine whether they may affect endangered species or their habitats; agency must consult with the USFWS	Significant impacts on beach nourishment projects; limitations on construction typically exclude construction in certain seasons, for example, the nesting season for sea turtles on the Atlantic and Gulf Coasts. There are short-term environmental impacts associated with both removing the sand from the source and depositing it onto the beach
National Historic Preservation Act (16 USC 470 and following)	Federal agencies must consider the effects of their undertakings (including the issuance of permits, the expenditure of federal funding, and the initiation of federal projects) on historic resources that the either eligible for listing or are listed on the National Register of Historic Places	Areas worthy of historic preservation must be avoided in the beach nourishment site selection process

Source: csc.noaa.gov/beachnourishment.

Table 4H.44 Atlantic and Gulf Coast State Beach Nourishment Program Summary

State	Beach Nourishment Policy	Related Policies	State Funding Program
Alabama	Yes	A B C D E	No
Connecticut	Yes	B C D E	Yes
Delaware	Yes	A B C D E	Yes
Florida	Yes	B C D E	Yes
Georgia	Yes	A B C D E	Case-by-Case
Louisiana	Yes	A B C D	Yes
Maine	No	A B C D	No
Maryland	No	A B C	Yes
Massachusetts	Yes	A B C D E	Yes
Mississippi	Yes	A B C	Yes
New Hampshire	Yes	A B C D E	Case-by-Case
New Jersey	Yes	A B C D E	Yes
New York	Yes	A B C D E	Case-by-Case
North Carolina	Yes	A B C D E	Case-by-Case
Rhode Island	Yes	A B C D E	No
South Carolina	Yes	B C D E	Case-by-Case
Texas	Yes	A B D E	Case-by-Case
Virginia	Yes	A B C	Yes

Note: Policies related to beach nourishment. A = near shore sand mining; B = dredge and fill; C = sand scraping/dune reshaping; D = dune creation/restoration; E = public access.

Source: NOAA 2000, csc.noaa.gov.

Table 4H.45 Riverbank Treatment in the United States

Region	Mattress	Jetties	Training Walls	Riprap	Revetment	Dikes	Other	Total Treated
North Atlantic	—	18	18	—	—	—	—	36
South Atlantic-Gulf	35	67	—	—	—	—	—	102
Great Lakes	—	—	—	—	—	—	—	—
Ohio	—	—	—	50	10	—	—	60
Tennessee	—	—	—	—	—	—	—	—
Upper Mississippi	—	—	—	—	—	—	—	—
Lower Mississippi	573	—	—	—	—	75	5836	6484
Souris-Red-Rainy	—	—	—	—	—	—	—	—
Missouri	—	735	—	—	735	—	—	1470
Arkansas-White–Red	250	1	80	—	—	140	—	471
Texas-Gulf	0	0	0	0	0	0	0	0
Bio Grande	—	250	—	—	—	—	—	250
Upper Colorado	—	—	—	—	—	—	—	—
Lower Colorado	—	—	—	—	—	—	—	—
Great Basin	0	0	0	0	0	0	0	0
Columbia-North Pacific	—	—	—	261	—	—	—	261
California	—	—	—	—	—	—	19	19
Alaska	—	—	6	—	—	—	—	6
Hawaii	0	0	0	0	0	0	0	0
Puerto Rico	—	—	—	—	—	—	—	—

Note: Value in miles as of 1966; Corps of Engineers projects only; for regions see Fig. 4G.9.

Source: U.S. Water Resources Council, 1968.

Table 4H.46 Sediment Yield from Drainage Areas of 100 Square Miles or Less of the United States

Region	Estimated Sediment Yield			Region	Estimated Sediment Yield		
	High	Low	Average		High	Low	Average
		tons/sq/mi/yr				tons/sq/mi/yr	
North Atlantic	1210	30	250	Arkansas-White–Red	8210	260	2200
South Atlantic-Gulf	1850	100	800	Texas-Gulf	3180	90	1800
Great Lakes	800	10	100	Rio Grande	3340	150	1300
Ohio	2110	160	850	Upper Colorado	3340	150	1800
Tennessee	1560	460	700	Lower Colorado	1620	150	600
Upper Mississippi	3900	10	800	Great Basin	1780	100	400
Lower Mississippi	8210	1560	5200	Columbia-North Pacific	1100	30	400
Souris-Red-Rainy	470	10	50	California	5570	80	1300
Missouri	6700	10	1500				

Note: For regions see Fig. 4G.9.

Source: U.S. Water Resources Council, 1968.

Table 4H.47 Permissible Velocities for Channels Lined with Vegetation

Cover	Slope Range[a] (Percent)	Permissible Velocity	
		Erosion Resistant Soils (ft/sec)	Easily Eroded Soils (ft/sec)
	0–5	8	6
Bermudagrass	5–10	7	5
	over 10	6	4
Buffalograss	0–5	7	5
Kentucky bluegrass	5–10	6	4
Smooth brome	over 10	5	3
Blue grama			
Grass mixture	0–5[a]	5	4
	5–10	4	3
Lespedeza sericea			
Weeping lovegrass			
Yellow bluestem			
Kudzu	0–5[b]	3.5	2.5
Alfalfa			
Crabgrass			
Common lespedeza[c]			
Sudangrass[c]	0–5[d]	3.5	2.5

Note: Use velocities exceeding 5 feet per second only where good covers and proper maintenance can be obtained. Values apply to average, uniform stands of each type of cover.

[a] Do not use on slopes steeper than 10 percent except for side slopes in a combination channel.
[b] Do not use on slopes steeper than 5 percent except for side slopes in a combination channel.
[c] Annuals — used on mild slopes or as temporary protection until permanent covers are established.
[d] Use on slopes steeper than 5 percent is not recommended.

Source: U.S. Soil Conservation Service.

Table 4H.48 Permissible Velocities for Channels with Linings Other Than Vegetation

Original Material Excavated	Clear Water, No Detritus (ft/sec)	Water Transporting Colloidal Silts (ft/sec)	Water Transporting Noncolloidal Silts, Sands, Gravels, or Rock Fragments (ft/sec)
Fine sand, noncolloidal	1.50	2.50	1.50
Sandy loam, noncolloidal	1.75	2.50	2.00
Silt loam, noncolloidal	2.00	3.00	2.00
Alluvial silts, noncolloidal	2.00	3.50	2.00
Ordinary firm loam	2.50	3.50	2.25
Volcanic ash	2.50	3.50	2.00
Fine gravel	2.50	5.00	3.75
Stiff clay, very colloidal	3.75	5.00	3.00
Graded, loam to cobbles, noncolloidal	3.75	5.00	5.00
Alluvial silts, colloidal	3.75	5.00	3.00
Graded, silt to cobbles, colloidal	4.00	5.50	5.00
Coarse gravel, noncolloidal	4.00	6.00	6.50
Cobbles and shingles	5.00	5.50	6.50
Shales and hardpans	6.00	6.00	5.00

Note: Values apply to aged straight channels with mild bed slopes.

Source: From Fortier and Scobey, *Trans. Am. Soc. Civil Eng.*, 1926. With permission.

Table 4H.49 Settling Velocities of Sand and Silt in Still Water

Diameter of Particle (mm)	Order of Size	Setting Velocity (mm/sec)	Time Required to Settle 1 Foot
10.0	Gravel	1000	0.3 sec
1.0	Coarse sand	100	3.0 sec
0.8	Coarse sand	83	
0.6	Coarse sand	63	
0.5	Coarse sand	53	
0.4	Coarse sand	42	
0.3	Coarse sand	32	
0.2	Coarse sand	21	
0.15	Coarse sand	15	
0.10	Fine sand	8	38.0 sec
0.08	Fine sand	6	
0.06	Fine sand	3.8	
0.05	Fine sand	2.9	
0.04	Fine sand	2.1	
0.03	Fine sand	1.3	
0.02	Fine sand	0.62	
0.015	Fine sand	0.35	
0.010	Silt	0.154	33.0 min
0.008	Silt	0.098	
0.006	Silt	0.065	
0.005	Silt	0.0385	
0.004	Silt	0.0247	
0.003	Silt	0.0138	
0.002	Silt	0.0062	
0.0015	Silt	0.0035	
0.001	Bacteria	0.00154	55.0 h
0.0001	Clay particles	0.000154	230.0 d
0.00001	Colloidal particles	0.000000154	63.0 yr

Note: Temperature 50 °F; all particles assumed to have a specific gravity of 2.65.

Source: Amer. Water Works Assoc.

Table 4H.50 Classification of Alluvial Channels Based on Channel Stability and on Mode of Sediment Transport

Mode of Sediment Transport	Silt-Clay in Channel Sediment, Percent[a]	Proportion of Total Sediment Load		Channel Stability		
		Suspended Load Percent	Bedload, Percent	Stable (Graded Stream)	Depositing (Excess Load)	Eroding (Deficiency of Load)
Suspended load	30–100	85–100	0–15	Stable suspended-load channel. Width-depth ratio less than 7; sinuosity greater than 2.1; gradient relatively gentle	Depositing suspended load channel. Major deposition on banks cause narrowing of channel; streambed deposition minor	Eroding suspended-load channel. Streambed erosion predominant; channel widening minor
Mixed load	8–30	65–85	15–35	Stable mixed-load channel. Width-depth ratio greater than 7 less than 25; sinuosity, less than 2.1 greater than 1.5; gradient moderate	Depositing mixed-load channel. Initial major deposition on banks followed by streambed deposition	Eroding mixed-load channel. Initial streambed erosion followed by channel widening
Bedload	0–8	30–65	35–70	Stable bedload channel. Width-depth ratio greater than 25; sinuosity, less than 1.5; gradient relatively steep	Depositing badload channel. Streambed deposition and island formation	Eroding bedload channel. Little streambed erosion; channel widening predominant

[a] The percentage of sediment finer than 0.074 mm in the perimeter of the channel.

Source: U.S. Geological Survey.

SECTION 4I TRANSPIRATION

Table 4I.51 Transpiration Ratios for Crops

Crop	Varieties Tested, No	Transpiration Ratio	
		Range	Mean
Grains			
Proso	2	531–603	567
Millet	3	863–1117	959
Buckwheat	1	—	969
Sorgo	3	863–1804	1237
Grain sorghum	4	750–1050	868
Barley	4	1128–1464	1241
Corn	2	821–1998	1405
Oats	4	1379–1915	1627
Wheat, emmer	1	—	1167
Wheat, durum	6	1365–1622	1475
Wheat, common	11	1244–3398	1872
Wheat, hybrids	2	1995–2163	2079
Rye	1	—	2142
Flax	6	2010–5162	3252
Legumes			
Clover	2	636–759	698[a]
Clover, sweet	1	—	731[a]
Vetch	5	562–899	708[a]
Alfalfa	10	626–920	844[a]
Cowpeas	1	—	1632
Beans	2	1583–1815	1699
Beans, soy	1	—	1974
Chickpeas	1	—	1685
Peas, Canadian field	1	—	2153
Lupinus albus	1	—	4734
Grassses			
Sudan grass	1	—	380[a]
Wheat grass	1	—	678[a]
Brome grass	1	—	977[a]
Miscellaneous			
Cotton	1	—	568[a]
Sugar beets	1	—	629
Potatoes	2	1325–2877	2101
Cabbage	1	—	518[a]
Rape	1	—	714[a]
Watermelons	1	—	1102
Cantaloupes	1	—	1754
Turnips	1	—	1471
Cucumbers	1	—	1549

Note: Transpiration ratios are weighted values in pounds of water per pound of crop product. Data measured at Akron, Colorado.

[a] Based on total dry matter

Source: From Shantz and Piemeisel, *J. Agric. Res.*, 1927.

Table 4I.52 Transpiration Ratios for Weeds and Native Plants

Weeds or Native Plants	Transpiration Ratio	Weeds or Native Plants	Transpiration Ratio
Weeds		Native weeds (continued)	
Tumbleweed	260	Sunflower	623
Pigweed	305	Mountain sage	654
Russian thistle	314	Verbena	702
Lamb's quarters	658	Fetid marigold	847
Polygonum	678		
Native weeds		Native plants	
Purslane	281	Buffalo grass	296
Cocklebur	415	Buffalo and grama grass	338
Nightshade	487	Clammyweed	483
Buffalo bur	536	Iva	534
Gumweed	585	Western ragweed	912
		Western wheat grass	1035
		Franseria	1131

Note: Transpiration ratios are weighted values in pounds of water per pound of dry matter. Data measured at Akron, Colorado.

Source: From Shantz and Piemeisel, *J. Agric. Res.*, 1927.

Table 4I.53 Transpiration Ratios for Trees

Tree	Scientific Name	Transpiration Ratio
Ash	*Fraxinus excelsior*	981
White birch	*Betula alba*	849
Beech	*Fagus sylvatica*	1043
Hornbeam	*Carpinus betulus*	787
Field elm	*Ulmus campestris*	738
Stiel oak	*Quercus pedunculus*	454
Traubean oak	*Quercus sessilifolia*	790
Zerr oak	*Quercus cerris*	669
Black alder	*Alnus glutinosa*	840
Gray alder	*Alnus incana*	678
Sycamore maple	*Acer platanoides*	520
Mountain maple	*Acer pseudoplat*	635
Field maple	*Acer campestria*	1281
Linden	*Tilia grandifolia*	1038
Aspen	*Populus tremula*	873
Service berry	*Sorbus tormin*	1748
Larch	*Larix europea*	1165
Spruce	*Abies excelsa*	242
Fir	*Abies pectinata*	96
Scotch white pine	*Pinus silvestris*	110
Black Austrian pine	*Pinus larico*	123

Note: Transpiration ratios are expressed as pounds of water per pound of dry-leaf matter. Data measured by Hohnel (1879–1880).

Source: U.S. Weather Bureau.

SECTION 4J EVAPORATION

Table 4J.54 Reservoir Evaporation at Selected Stations in the United States

	Month												
Station	Jan	Feb	Mar	Apr	May	June	July	Aug	Sept	Oct	Nov	Dec	Annual
AZ, Yuma	3.9	4.6	6.5	8.0	9.8	11.5	13.4	12.9	10.7	8.0	6.1	4.5	100
CA, Sacramento	0.8	1.4	2.5	3.6	5.0	7.1	8.9	8.6	7.1	4.8	2.6	1.2	54
CO, Denver	1.6	1.8	2.5	3.7	5.0	7.4	8.8	8.4	6.7	4.6	3.0	1.9	55
FL, Miami	3.0	3.4	4.1	4.9	5.0	4.8	5.3	5.1	4.3	4.1	4.3	2.7	51
GA, Macon	1.7	2.2	3.1	4.3	5.1	6.2	6.3	5.8	5.2	4.2	2.8	1.8	49
ME, Eastport	0.8	0.7	0.9	1.1	1.4	1.7	2.0	2.1	2.0	1.6	1.1	0.7	16
MN, Minneapolis	0.3	0.4	0.9	1.7	3.2	4.4	6.0	5.8	4.6	3.0	1.3	0.4	32
MS, Vicksburg	1.3	1.9	2.9	4.2	5.0	5.7	5.8	5.5	5.2	4.4	2.9	1.6	46
MO, Kansas City	0.9	1.1	1.7	3.1	4.4	6.1	8.0	7.8	6.0	4.5	2.5	1.0	47
MT, Havre	0.5	0.5	1.1	2.5	4.5	6.1	8.2	8.3	5.6	3.3	1.5	0.7	43
NE, North Platte	0.8	1.1	2.2	3.7	5.0	6.5	8.6	8.4	6.9	4.6	2.6	1.1	51
NM, Roswell	2.1	3.2	4.9	6.8	8.3	9.8	9.4	8.3	6.9	5.5	3.5	2.5	71
NY, Albany	0.6	0.7	1.1	2.0	3.2	4.3	5.2	4.7	3.4	2.4	1.4	0.8	30
ND, Bismarck	0.4	0.5	1.0	2.3	4.0	5.3	7.3	7.7	5.8	3.3	1.3	0.5	39
OH, Columbus	0.6	0.8	1.1	2.3	3.5	4.6	5.6	5.1	4.1	3.0	1.6	0.6	33
OK, Oklahoma City	1.5	1.9	3.1	4.7	5.5	7.8	10.2	10.7	8.8	6.3	3.5	2.0	66
OR, Baker	0.5	0.7	1.4	2.5	3.4	4.4	6.9	7.3	4.9	2.9	1.5	0.6	37
SC, Columbia	1.6	2.4	3.2	4.5	5.4	6.3	6.6	6.0	5.5	4.4	3.0	1.9	51
TN, Nashville	0.9	1.3	1.9	3.3	4.1	5.1	5.8	5.4	4.9	3.7	2.1	1.1	39
TX, Galveston	0.9	1.3	1.6	2.6	4.1	5.6	6.2	6.1	5.7	4.6	2.7	1.3	43
TX, San Antonio	2.2	3.1	4.5	5.6	6.5	8.4	9.4	9.4	7.6	5.8	3.7	2.4	69
UT, Salt Lake City	0.8	1.0	2.0	3.5	5.1	7.9	10.6	10.4	7.3	3.9	2.0	1.0	55
VA, Richmond	1.3	1.7	2.2	3.5	4.1	5.0	5.6	4.9	4.1	3.2	2.4	1.5	39
WA, Seattle	0.8	0.8	1.4	2.1	2.7	3.4	3.9	3.4	2.6	1.6	1.1	0.7	24
WI, Milwaukee	0.6	0.7	0.9	1.3	2.1	3.2	5.0	5.4	4.7	3.2	1.6	0.6	29
Gulf off Texas Coast	4.0	4.0	3.5	3.5	4.0	4.5	5.0	5.5	6.5	6.5	6.0	5.0	58
Gulf Stream off Cape Hatteras, NC	9.0	9.5	8.5	7.0	5.5	3.5	3.5	3.5	5.5	9.0	9.5	10.0	84
Ocean off Massachusetts	3.0	2.5	2.0	1.5	1.0	1.5	1.5	2.0	2.5	3.0	3.5	4.0	28

Note: Mean monthly computed values in inches.

Source: Minnesota Resources Commission.

Table 4J.55 Evaporation Equations

For pan evaporation, the expression is

$$E_P = \{\exp[(T_a - 212)(0.1024 - 0.01066 \ln R)] - 0.0001 + 0.025(e_s - e_a)^{0.88}(0.37 + 0.0041U_p)\} \times \{0.025 + (T_a + 398.36)^{-2} 4.7988 \times 10^{10} \exp[-7482.6/(T_a + 398.36)]\}^{-1}$$

For lake evaporation, the expression is

$$E_L = \{\exp[(T_a - 212)(0.1024 - 0.01066 \ln R)] - 0.0001 + 0.0105(e_s - e_a)^{0.88}(0.37 - 0.0041U_P)\} \times \{0.015 + (T_a + 398.36)^{-2} 6.8554 \times 10^{10} \exp[-7482.6/(T_a + 398.36)]\}^{-1}$$

The terms in these expressions are

E_P = pan evaporation, inches
E_L = lake evaporation, inches
T_a = air temperature, degrees Fahrenheit
e_a = vapor pressure, inches of mercury at temperature T_a
e_s = vapor pressure, inches of mercury at temperature T_d
T_d = dew point temperature, degrees Fahrenheit
R = solar radiation, langleys per day
U_P = wind movement, miles per day

Notes: The equations enable pan and lake evaporation to be computed from climatic data at first-order weather stations. Daily values of evaporation are obtained using mean daily temperature and vapor pressure data together with data on solar radiation and wind movement as specified.

Source: U.S. Weather Bureau, 1962.

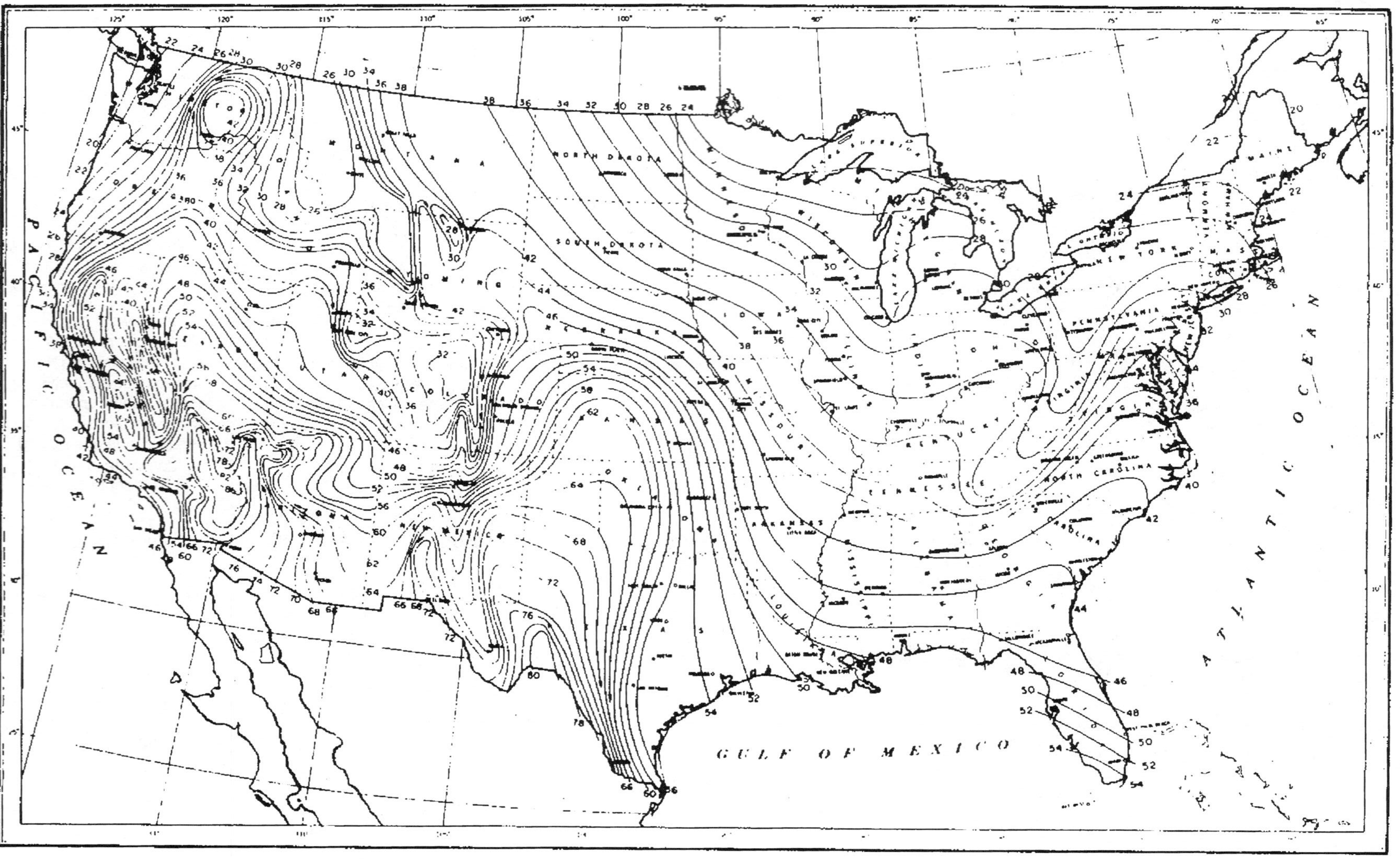

Figure 4J.24 Mean annual lake evaporation in the United States (values in inches for period 1946–1955). (From U.S. Weather Bureau.)

Figure 4J.25 Annual evaporation in the world (in mm). (From Lvovitch, M.I., *EOS*, 54, 1973, Copyright by American Geophysical Union.)

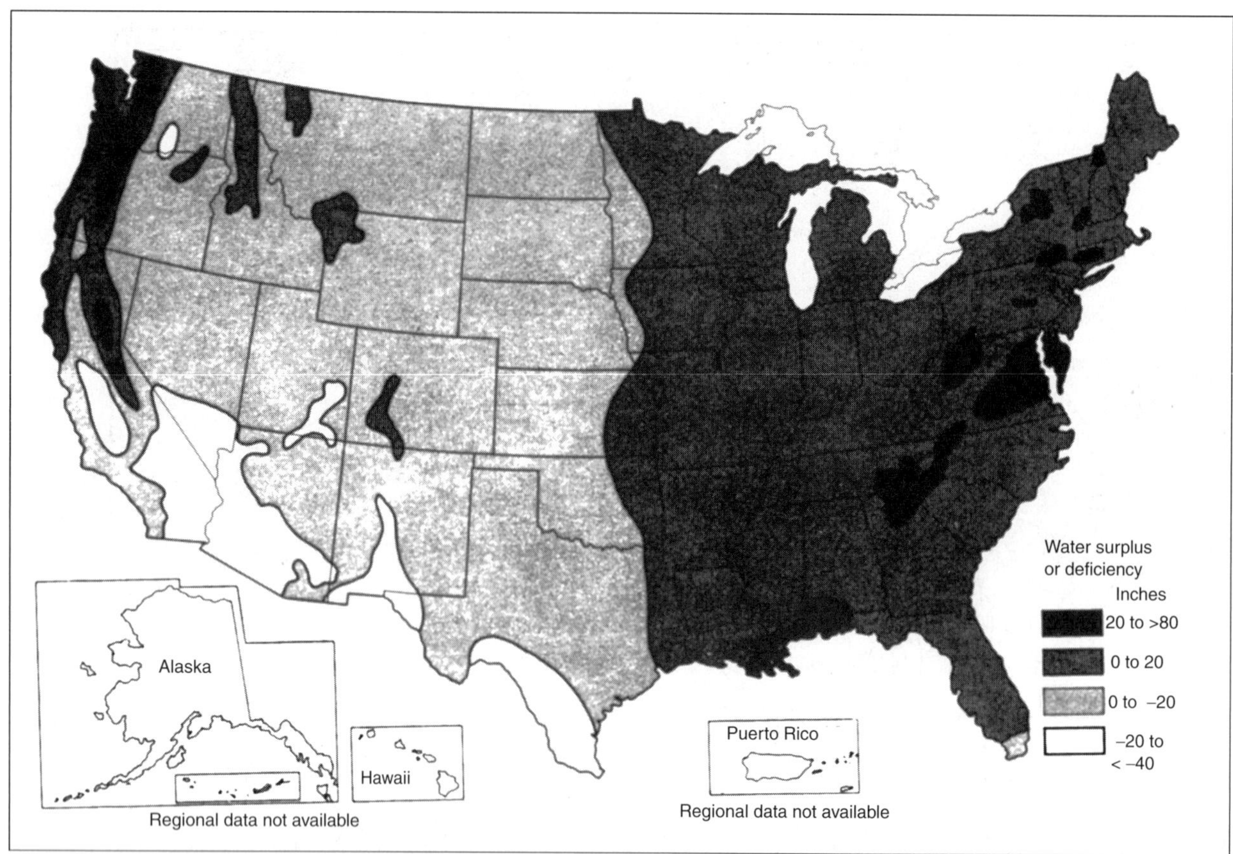

Figure 4J.26 Areas of natural water surplus and natural water deficiency (computed by subtracting values of potential evapotranspiration from average precipitation). (From U.S. Water Resources Council, 1968, The Nation's Water Resources.)

SECTION 4K CONSUMPTIVE USE

Table 4K.56 Consumptive Use by Irrigated Crops in the Western United States

		Consumptive Use (Evapotranspiration), (in.)							
Location	**Crop**	**Apr**	**May**	**June**	**July**	**Aug**	**Sept**	**Oct**	**Total**
Arizona, Mesa	Alfalfa	5.0	6.5	9.0	12.0	10.0	6.0	4.0	52.5
	Dates	6.2	7.6	8.3	9.2	8.4	7.2	5.7	52.6
California									
Los Angeles[a]	Lemons	2.1	2.6	3.3	3.9	3.7	3.4	2.8	21.8
	Oranges	2.2	2.2	3.1	3.4	3.7	3.1	2.9	20.6
	Walnuts	3.8	5.0	5.9	6.1	5.0	2.8	2.0	30.6
	Alfalfa	3.3	6.7	5.4	7.8	4.2	5.6	4.4	37.4
Coastal	Alfalfa	4.9	4.9	4.3	5.2	5.9	5.5	4.7	35.4
Ontario	Peaches	1.0	3.5	6.7	8.0	6.5	2.7	1.4	29.8
Shafter	Cotton	0.5	1.0	4.0	8.5	9.7	5.8	3.2	32.7
Firebaugh	Cotton	—	0.8	1.1	7.3	7.8	3.6	2.0	22.6
	Cotton	—	0.4	0.7	8.4	9.5	3.0	2.5	24.5
Delta[b]	Alfalfa	3.6	4.8	6.0	7.8	6.6	6.0	1.2	36.0
	Potatoes	—	1.8	4.6	6.2	3.6	1.8	—	18.0
	Truck	1.2	3.0	6.0	5.4	5.4	3.6	1.8	26.4
	Sugar beets	1.6	3.8	6.1	7.3	6.4	2.4	—	27.6
	Beans	1.9	2.4	1.7	2.9	6.9	4.4	—	20.2
	Fruit	2.2	3.8	6.0	6.8	4.8	2.8	0.8	27.2
	Onions	1.6	3.2	5.9	5.2	2.4	1.9	—	19.8
Davis	Sugar beets	—	5.2	5.7	7.1	5.8	—	—	23.8
	Tomatoes	—	—	3.2	6.2	4.9	4.7	—	22.3
	Alfalfa	—	6.8	7.9	8.3	7.1	4.3	—	—
	Prunes	—	5.8	6.0	7.6	6.5	5.0	—	—
	Peaches	—	5.4	6.4	7.9	7.2	5.0	—	—
	Walnuts	—	6.6	6.7	8.4	7.2	4.8	—	—
	Grapes	—	4.6	4.9	6.2	5.3	4.3	—	—
Winters	Apricots	—	—	5.6	6.8	6.5	4.9	—	—
Nebraska, Scottsbluff	Alfalfa	1.4	4.0	7.0	7.1	6.4	3.0	—	28.9
	Beets	1.9	3.3	5.2	6.9	5.8	1.1	—	24.2
	Potatoes	—	—	—	3.4	5.8	4.4	—	—
	Oats	—	3.0	6.1	5.1	—	—	—	14.2

Notes: Data for irrigation season only.

[a] In San Fernando Valley, City of Los Angeles, California.
[b] In Sacramento–San Joaquin Delta, California.

Source: U.S. Dept. of Agriculture.

Table 4K.57 Consumptive Use by Principal Crops in the Central Valley, California

Month	Improved Pasture	Alfalfa	Sugar Beets S.V.	Sugar Beets S.J.V.	Cotton a	Cotton b	Deciduous Orchard	Rice S.V.
January	1.0	1.0	0.9	—	—	—	—	(0.8)[c]
February	1.8	1.8	1.2	—	—	—	—	(1.5)
March	3.0	2.8	—	1.6	—	—	1.8	(1.4)
April	4.7	4.2	—	3.6	—	—	3.3	4.8
May	6.1	5.4	1.8	5.6	1.0	0.6	4.9	7.6
June	7.8	7.0	6.0	7.7	6.0	3.6	6.7	9.6
July	8.2	7.6	8.5	8.5	9.7	8.8	7.5	10.0
August	7.1	6.8	7.3	5.1	8.5	7.8	6.4	8.5
September	5.2	5.1	5.6	1.9	5.5	5.3	4.5	6.4
October	3.5	3.5	3.5	—	2.1	1.8	2.7	3.2
November	1.6	1.6	1.6	—	—	—	—	(1.4)
December	0.8	0.8	0.8	—	—	—	—	(0.8)
Total	50.8	47.6	37.2	34.0	32.8	27.9	37.8	56.0

Note: Values in inches; S.V. is Sacramento Valley, and S.J.V. is San Joaquin Valley.

[a] Solid planting or one row skipped in three.
[b] Planting of two rows skipped in four.
[c] Values in parentheses are for nongrowing season. Values may change with differences in rainfall.

Source: Calif. Dept. of Water Resources, 1967.

Table 4K.58 Consumptive Use by Crops in the Sacramento-San Joaquin Delta, California

Month	Alfalfa	Asparagus	Beans	Beets	Celery	Corn	Fruit	Grain and Hay	Onions	Potatoes
January	(0.06)	0.05	(0.06)	(0.06)	(0.04)	(0.04)	(0.04)	(0.04)	(0.04)	(0.06)
February	(0.08)	0.05	(0.08)	(0.08)	(0.04)	(0.04)	(0.04)	(0.04)	(0.04)	(0.08)
March	0.10	0.05	(0.08)	(0.08)	(0.04)	(0.04)	(0.04)	0.07	0.08	(0.08)
April	0.30	0.05	(0.16)	0.13	(0.08)	(0.08)	0.18	0.60	0.13	(0.16)
May	0.40	0.08	(0.20)	0.32	(0.10)	(0.10)	0.32	0.83	0.27	0.15
June	0.50	0.14	0.14	0.51	0.10	0.24	0.50	0.20	0.49	0.38
July	0.65	0.40	0.24	0.61[a]	0.10	0.85	0.57	(0.14)	0.43	0.52
August	0.55	0.68	0.58	0.53[a]	0.20	0.84[a]	0.40	(0.23)	0.20	0.30
September	0.50	0.55	0.37	0.20[a]	0.25	0.40[a]	0.23	(0.21)	(0.16)	0.15
October	0.20	0.42	(0.09)	(0.13)	0.30	0.10	0.07	(0.14)	(0.13)	(0.09)
November	(0.10)	0.12	(0.07)	(0.10)	0.20	(0.10)	(0.07)	(0.07)	(0.10)	(0.07)
December	(0.07)	0.10	(0.05)	(0.07)	0.05	(0.07)	(0.05)	(0.05)	(0.07)	(0.05)
Growing season	3.20	2.69	1.33	2.30	1.20	2.43	2.27	1.70	1.60	1.50
Year	3.51	2.69	2.12	2.82	1.50	2.90	2.51	2.62	2.14	2.09

Note: Depth in feet. Figures in parentheses show estimated losses by soil evaporation and weed transpiration.

[a] Including additional use of water by weeds.

Source: California Department of Public Works.

Table 4K.59 Total Consumptive Use and Peak Daily Use, Western United States

	Southern Coastal				South Pacific Coastal Interior and North Coastal							
	300 Days Plus		250–300 Days		250–300 Days		210–250 Days		180–210 Days		150–180 Days	
Crops	Season Use (in.)	Daily Use (in./d)	Season Use (in.)	Daily Use (in./d)	Season Use (in.)	Daily Use (in./d)	Season Use (in.)	Daily Use (in./d)	Season Use (in.)	Daily Use (in./d)	Season Use (in.)	Daily Use (in./d)
Alfalfa	36.0	0.20	30.0	0.17	37.0	0.27	32.0	0.22	26.0	0.20	22.0	0.18
Pasture	33.5	0.20	28.0	0.17	33.0	0.27	30.0	0.22	24.0	0.20	20.0	0.18
Grain — small	16.0	0.18	14.0	0.16	17.0	0.22	14.5	0.20	12.0	0.20	10.0	0.18
Beets — sugar	29.0	0.20	25.0	0.18	30.0	0.27	26.0	0.22	—	—	—	—
Beans — field	12.0	0.18	10.0	0.16	13.0	0.22	11.0	0.18	—	—	—	—
Corn — field	—	—	—	—	—	—	19.0	0.25	16.0	0.22	14.0	0.20
Potatoes	—	—	—	—	24.0	0.20	20.0	0.18	—	—	—	—
Peas — green	10.0	0.18	8.0	0.16	11.0	0.20	9.0	0.18	8.0	0.16	7.0	0.16
Legume seed	25.0	0.20	22.0	0.18	26.0	0.25	22.0	0.22	20.0	0.20	18.0	0.18
Tomatoes	18.0	0.16	15.0	0.16	19.0	0.20	16.0	0.18	13.0	0.16	11.0	0.16
Vegetable seed	16.0	0.16	14.0	0.16	18.0	0.18	16.0	0.18	14.0	0.16	12.0	0.16
Beans — pole	—	—	—	—	18.0	0.20	16.0	0.20	14.0	0.18	12.0	0.16
Corn — sweet	16.0	0.18	14.0	0.16	16.0	0.20	14.0	0.18	12.0	0.18	11.0	0.16
Apples	—	—	—	—	24.0	0.20	22.0	0.20	20.0	0.18	18.0	0.18
Cherries	—	—	—	—	24.0	0.20	22.0	0.20	—	—	—	—
Peaches	—	—	—	—	24.0	0.20	22.0	0.20	—	—	—	—
Prunes	—	—	—	—	22.0	0.20	20.0	0.20	—	—	—	—
Apricots	—	—	—	—	22.0	0.20	20.0	0.20	—	—	—	—
Oranges	20.0	0.16	18.0	0.16	22.0	0.18	—	—	—	—	—	—
Avocados	18.0	0.16	—	—	—	—	—	—	—	—	—	—
Walnuts	22.0	0.20	18.0	0.18	24.0	0.25	22.0	0.22	20.0	0.20	—	—
Strawberries	22.0	0.20	18.0	0.18	23.0	0.25	18.0	0.22	16.0	0.20	14.0	0.18
Lettuce	4.0	0.16	4.0	0.16	6.0	0.18	5.0	0.18	—	—	—	—
Mint	—	—	—	—	23.0	0.24	21.0	0.22	19.0	0.20	17.0	0.18
Hops	—	—	—	—	20.0	0.22	20.0	0.20	—	—	—	—

	Central Valley—California and Valleys East Side of Cascade Mountains											
	250–300 Days		210–250 Days		180–210 Days		150–180 Days		120–150 Days		90–120 Days	
Alfalfa	40.0	0.30	34.0	0.28	30.0	0.25	26.0	0.22	20.0	0.20	14.0	0.18
Pasture	36.0	0.30	30.0	0.28	28.0	0.25	24.0	0.22	18.0	0.20	13.0	0.18
Grain — small	18.0	0.22	16.0	0.22	15.0	0.20	14.0	0.18	13.0	0.18	12.0	0.16
Beets — sugar	33.0	0.30	28.0	0.25	24.0	0.22	20.0	0.20	18.0	0.18	—	—
Beans — field	17.0	0.22	13.0	0.20	13.0	0.20	12.0	0.18	12.0	0.18	—	—
Corn — field	26.0	0.35	22.0	0.32	22.0	0.30	20.0	0.25	18.0	0.22	17.0	0.20
Potatoes — summer	12.0	0.16	—	—	—	—	—	—	—	—	—	—
Potatoes — fall	—	—	19.0	0.25	18.0	0.22	18.0	0.20	17.0	0.18	16.0	0.16
Peas — green	—	—	8.0	0.18	7.0	0.18	7.0	0.18	7.0	0.16	6.0	0.15
Peas — field	—	—	10.0	0.18	9.0	0.18	9.0	0.18	8.0	0.16	8.0	0.15

Tomatoes	20.0	0.20	18.0	0.18	18.0	0.18	17.0	0.17	16.0	0.16	—	—
Cotton	26.0	0.30	22.0	0.28	—	—	—	—	—	—	—	—
Grain — sorghum	15.0	0.20	13.0	0.18	12.0	0.17	10.0	0.16	9.0	0.15	8.0	0.15
Apples	—	—	26.0	0.20	23.0	0.20	21.0	0.18	—	—	—	—
Cherries	—	—	24.0	0.20	21.0	0.20	19.0	0.18	—	—	—	—
Peaches	22.0	0.22	22.0	0.20	20.0	0.20	18.0	0.18	—	—	—	—
Apricots	20.0	0.22	17.0	0.20	15.0	0.20	—	—	—	—	—	—
Oranges	28.0	0.18	—	—	—	—	—	—	—	—	—	—
Strawberries	24.0	0.20	20.0	0.20	18.0	0.18	—	—	—	—	—	—
Lettuce — winter	4.0	0.20	—	—	—	—	—	—	—	—	—	—
Mint	—	—	20.0	0.22	18.0	0.20	—	—	—	—	—	—
Hops	—	—	18.0	0.20	16.0	0.18	—	—	—	—	—	—
Grapes	30.0	0.25	25.0	0.22	22.0	0.20	—	—	—	—	—	—
Walnuts	24.0	0.22	20.0	0.20	—	—	—	—	—	—	—	—
Almonds	22.0	0.25	20.0	0.22	—	—	—	—	—	—	—	—

	Central Intermountain, Desert, and Western High Plains											
	250–300 Days		210–250 Days		180–210 Days		150–180 Days		120–150 Days		90–120 Days	
Alfalfa	52.0	0.40	44.0	0.32	36.0	0.29	30.0	0.26	24.0	0.22	19.0	0.20
Pasture	48.0	0.40	40.0	0.30	33.0	0.28	28.0	0.25	22.0	0.22	17.0	0.20
Grain — small	21.0	0.25	18.0	0.22	16.0	0.20	16.0	0.20	16.0	0.20	14.0	0.18
Beets — sugar	37.0	0.30	32.0	0.30	30.0	0.28	26.0	0.25	24.0	0.22	18.0	0.20
Beans — field	22.0	0.25	17.0	0.20	14.0	0.20	14.0	0.18	14.0	0.17	12.0	0.15
Corn — field	—	—	30.0	0.35	26.0	0.30	24.0	0.28	22.0	0.24	—	—
Potatoes — fall	—	—	23.0	0.30	21.0	0.28	20.0	0.25	19.0	0.22	17.0	0.20
Peas — field	—	—	—	—	10.0	0.19	10.0	0.18	10.0	0.17	9.0	0.15
Tomatoes	—	—	20.0	0.22	18.0	0.20	17.0	0.18	16.0	0.17	—	—
Cotton	32.0	0.30	30.0	0.28		—	—	—	—	—	—	—
Grain — sorghum	19.0	0.25	18.0	0.20	16.0	0.20	14.0	0.18	12.0	0.17	—	—
Apples	—	—	—	—	28.0	0.22	24.0	0.20	20.0	0.18	—	—
Cherries	—	—	—	—	26.0	0.22	—	—	—	—	—	—
Peaches	—	—	29.0	0.25	27.0	0.22	—	—	—	—	—	—
Apricots	26.0	0.25	24.0	0.25	25.0	0.20	—	—	—	—	—	—
Almonds	22.0	0.25	20.0	0.25	—	—	—	—	—	—	—	—
Vineyards	40.0	0.27	32.0	0.25	26.0	0.22	—	—	—	—	—	—
Legume seed	—	—	—	—	—	—	—	—	16.0	0.18	14.0	0.16
Grass seed	—	—	—	—	—	—	—	—	14.0	0.14	12.0	0.14
Potatoes — seed	—	—	—	—	—	—	—	—	16.0	0.16	14.0	0.15
Grapefruit	45.0	0.20	—	—	—	—	—	—	—	—	—	—
Oranges	36.0	0.18	—	—	—	—	—	—	—	—	—	—
Lettuce — winter	6.0	0.18	—	—	—	—	—	—	—	—	—	—
Melons	22.0	0.25	20.0	0.22	18.0	0.20	16.0	0.18	—	—	—	—
Palm dates	60.0	0.30	—	—	—	—	—	—	—	—	—	—
Truck crops	20.0	0.25	18.0	0.22	14.0	0.20	12.0	0.18	12.0	0.16	10.0	0.15

Source: From Woodward, *Sprinkler Irrigation*, Sprinkler Irrigation Assoc., 1959. With permission.

Table 4K.60 Accumulated Use of Water by Crops with Various Planting to Maturity Periods

		Total Period of Growth									
		2 Month		3 Month		4 Month		5 Month		6 Month	
Percentage of Growing Period[a]	Accumulated Consumptive Use of Water in Percentage of Total Use	Days Since Planting	Accum. Water Use (in.)	Days Since Planting	Accum. Water Use (in.)	Days Since Planting	Accum. Water Use (in.)	Days Since Planting	Accum. Water Use (in.)	Days Since Planting	Accum. Water Use (in.)
10	6.0	6	0.3	9	0.4	12	0.6	15	0.8	18	0.9
20	13.8	12	0.7	18	1.0	24	1.4	30	1.7	36	2.0
30	23.5	18	1.1	27	1.9	36	2.3	45	2.9	54	3.5
40	34.5	24	1.7	36	2.6	48	3.4	60	4.3	72	5.1
50	46.5	30	2.3	45	3.4	60	4.6	75	5.7	90	6.8
60	59.4	36	2.9	54	4.4	72	5.8	90	7.3	108	8.7
70	72.3	42	3.5	63	5.3	84	7.1	105	8.9	126	10.6
80	84.3	48	4.1	72	6.2	96	8.3	120	10.3	144	12.4
90	94.5	54	4.7	81	7.0	108	9.3	135	11.6	162	13.9
100	100.0	60	4.9	90	7.4	120	9.8	150	12.3	180	14.7

[a] Growing Period refers to the entire time from planting to the time the plant dies, which is usually longer than the period from planting to harvesting. Flowering will occur at about 50 to 60 percent of the growing period and fruiting after 60 percent.

Source: From Israelson and Hansen, *Irrigation Principles and Practices*, John Wiley & Sons, 1962. With permission.

Table 4K.61 Variations in Consumptive Use by Crops

Crop	No. of Tests	Range in Water Requirements (ft)
	(A) Farm crops in the southwest	
Alfalfa	369	3.47–5.08
Rhodes grass	12	3.49–4.43
Sudan grass	25	2.88–3.16
Barley	3	1.24–1.83
Oats	2	1.90–2.09
Wheat	46	1.46–2.24
Corn	42	1.44–1.99
Kafir	16	1.32–1.54
Flax	3	1.23–1.59
Broomcorn	9	0.97–1.15
Emmer	6	1.19–1.87
Feterita	8	0.97–1.10
Millet	5	0.91–1.09
Milo	35	0.96–1.67
Sorghum	34	1.69–2.08
Cotton	103	2.35–3.51
Potatoes	12	1.59–2.04
Soybeans	36	1.66–2.81
Sugar beets	5	1.77–2.72
Sugar cane[a]	41	3.48–4.56
	(B) Vegetable crops in the southwest	
Beans, snap	9	0.83–1.44
Beets, table	28	0.87–1.37
Cabbage	21	0.94–1.49
Carrots	6	1.27–1.60
Cauliflower	6	1.43–1.77
Lettuce	49	0.72–1.35
Onions	4	0.73–1.52
Peas	8	1.21–1.56
Melons	3	0.80–1.07
Spinach	12	0.80–1.07
Sweet potatoes	3	1.77–2.25
Tomatoes	17	0.95–1.42
	(C) Crops in the Missouri and Arkansas Basins	
Forage, including alfalfa	648	1.94–2.62
Barley	335	1.33–1.82
Oats	409	1.35–1.81
Wheat	542	1.36–1.80
Corn	70	1.23–1.83
Kafir corn	15	1.43–1.57
Flax	50	1.47–1.85
Millet	14	0.81–0.94
Milo maize	27	1.09–1.70
Sorghum	26	1.06–1.47
Apples	4	2.10–2.60
Beans	4	1.30–1.60
Buckwheat	3	1.05–1.30
Cantaloupes	10	1.50–2.30
Peas	168	1.36–1.94
Potatoes	350	1.38–1.70
Sugar beets	128	1.60–2.50
Sunflowers	16	1.20–1.40
Tomatoes	6	2.10–2.80
Cucumbers	7	1.73–3.75

[a] Not commonly produced in the Southwest.

Source: U.S. Dept. of Agriculture.

Table 4K.62 Consumptive Use of Water by Crops in Florida

Month	Citrus	Pasture[a]	Sugarcane	Rice[b]	Rainfall	Pan Evapotranspiration[c]
January	2.09	2.01	1.42	0	1.97	3.39
February	2.60	2.52	1.10	0	1.97	4.00
March	3.58	3.35	2.52	0	3.21	5.70
April	4.49	4.21	3.39	1.63	2.96	6.54
May	5.31	5.20	4.80	3.07	4.74	7.06
June	4.41	4.25	5.98	5.82	9.08	6.24
July	4.88	4.80	6.50	8.43	8.58	6.36
August	4.80	4.80	6.69	3.05	8.21	6.12
September	4.02	3.86	5.12	(5.00)	8.82	5.31
October	3.59	3.43	5.20	(3.00)	5.65	4.82
November	2.72	2.48	3.19	0	1.74	3.71
December	2.09	1.93	2.59	0	1.80	3.19
Total	44.58	42.84	48.50	22.00	58.76	62.44

Note: Everglades agricultural area; in inches.

[a] Mean monthly values averaged over 5 years and averaged over water table depth of 12, 24, and 36 inches maintained in lysimeters at Ft. Lauderdale. These turfgrass evapotranspiration values are assumed to be valid for pastures adequately supplied with water.

[b] Assuming planting date of April 15 which is approximately the middle of the planting season. Values in parentheses are estimates for a ratoon crop. Rice is not always ratoon cropped.

[c] Pan evapotranspiration is a measure of the capability of the air to evaporate water. A relatively higher indicates relatively high consumptive use of water.

Source: From Bajwa, R.S., *Analysis of Irrigation Potential in the Southeast: Florida, A Special Report*, U.S. Department of Agriculture, 1985.

Table 4K.63 Economic Irrigation Requirements in the Western United States

Div. No	Division Description	State	Irrigation Season, Inclusive	Total Depth (ft)
	(A) Columbia River basin			
1	Snake River Valley	Idaho	Apr–Oct	2.5
2	Upper Snake River Valley	Idaho	Apr–Sept	2.3
3	Jackson Lake and Upper Snake basin	Idaho and Wyoming	May–Sept	1.7
4	Southwest Idaho and north Nevada	Idaho and Nevada	Apr–Sept	1.9
5	Salmon River basin	Idaho	May–Aug	2.0
6	North Idaho	Idaho	May–Sept	1.5
7	Bitterroot and Missoula River basins	Montana	Apr–Nov	2.1
8	Flathead Lake and River basins	Montana	Apr–Sept	1.8
9	Owyhee and Malheur River basins	Oregon	Apr–Sept	2.4
10	Northeast Oregon	Oregon	Apr–Sept	2.0
11	Umatilla, John Day, Deschutes, and Hood basins	Oregon	Apr–Oct	2.5
12	Central Oregon	Oregon	May–Aug	2.4
13	Yakima and Wenatchee river basins	Washington	Apr–Nov	2.6
14	Southeast Washington	Washington	Apr–Oct	2.1
15	Northeast Washington	Washington	Apr–Oct	2.2
16	Okanogan River basin	Washington	Apr–Nov	2.3
17	Lower Columbia River basin	Washington	May–Sept	1.3
18	Willamette River basin	Oregon	May–Sept	1.2
19	Puget Sound region[a]	Washington	May–Sept	1.4
	(B) Pacific Slope basins			
1	Umpqua, Coquill, and Lower Rogue basins	Oregon	Apr–Sept	0.85
2	Upper Rogue River basin	Oregon	Mar–Sept	1.50
3	Klamath Lake and River basins	Oregon and California	Apr–Sept	2.00
4	Northwest California	California	Apr–Oct	1.40
5	Pit River basin	California	Apr–Sept	1.60
6	Feather, Yuba, and American River basins	California	Mar–Nov	1.50
7	Sacramento Valley	California	Mar–Oct	2.10
8	Sacramento-San Joaquin Delta	California	May–Sept	2.00
9	San Francisco Bay basin	California	Mar–Nov	1.50
10	Salinas River basin	California	Mar–Oct	1.70
11	Santa Maria, Santa Inez, and Santa Clara basins	California	Jan–Dec	1.60
12	San Joaquin Valley	California	Feb–Oct	2.30
13	West slope of Sierras	California	Feb–Nov	1.70
14	East slope of Coast Range	California	Feb–Oct	1.80
15	Antelope and Victor Valleys	California	Mar–Oct	1.90
16	Los Angeles, San Gabriel, and Santa Ana basins	California	Jan–Dec	1.70
17	Upper Santa Ana River Valley	California	Jan–Dec	1.80
18	San Diego County	California	Jan–Dec	1.40
	(C) Southwest			
1	Imperial Valley	California	Jan–Dec	3.10
2	South Nevada	Nevada	Jan–Dec	2.90
3	Southwest Arizona	Arizona	Jan–Dec	3.00
4	Northwest Arizona	Arizona	Mar–Oct	2.30
5	Navajo country	Arizona	Mar–Oct	2.30
6	Southeast Arizona	Arizona	Feb–Nov	2.60
7	San Juan basin	New Mexico	Apr–Sept	2.20
8	West New Mexico	New Mexico	Apr–Oct	1.70
9	Rio Grande basin	New Mexico	Jan–Dec	2.60
10	Pecos River basin	New Mexico	Jan–Dec	2.40
11	Northeast New Mexico	New Mexico	Feb–Nov	1.60
12	Central Rio Grande basin	Texas	Jan–Nov	2.40
13	Pecos River basin	Texas	Jan–Nov	2.25
14	West Central Texas	Texas	Jan–Dec	1.60
15	Lower Rio Grande basin	Texas	Jan–Dec	1.75
16	Upper Nueces and Colorado River basins	Texas	Jan–Dec	1.30
17	Upper Brazos and Red River basins	Texas	Jan–Dec	1.10
18	Eastern Panhandle	Texas	Mar–Oct	1.35

(Continued)

Table 4K.63 (Continued)

Div. No	Division Description	State	Irrigation Season, Inclusive	Total Depth (ft)
19	Western Panhandle	Texas	Mar–Oct	1.65
20	Panhandle	Oklahoma	Apr–Oct	1.25
21	West Oklahoma	Oklahoma	Apr–Oct	1.00
22	San Luis basin	Colorado	May–Sept	1.80
23	San Juan basin	Colorado	Apr–Sept	1.90
24	Yampa and White River basins	Colorado	May–Aug	1.35
25	Upper Colorado River basin	Colorado	Apr–Sept	1.70
26	Virgin River basin	Utah	Feb–Nov	2.25
27	San Juan basin	Utah	Apr–Sept	2.10
28	Green River basin	Utah	Apr–Oct	2.00
29	Uintah basin	Utah	Apr–Sept	1.75
30	Green River basin	Wyoming	May–Aug	1.60
	(D) Great basin			
1	Bear River basin	Idaho and Utah	May–Oct	2.0
2	Utah Lake and Great Salt Lake Valleys[b]	Utah	May–Oct	2.2
3	Sevier River basin	Utah	May–Oct	2.1
4	Irrigable lands, southwest Utah	Utah	Apr–Oct	1.8
5	Irrigable lands, southern Nevada	Nevada	Apr–Oct	1.7
6	Antelope Valley and Mohave River areas	California and Nevada	Mar–Oct	1.8
7	Mono, Owens, and Inyo-Kern valleys	California	Mar–Oct	2.1
8	Walker River basin	California and Nevada	Apr–Oct	2.0
9	Truckee and Carson River basins	California and Nevada	May–Oct	2.1
10	Humboldt, Quinn, and White River basins	Nevada	May–Sept	2.0
11	Honey Lake basin	California and Nevada	Apr–Sept	1.7
12	Malheur Lake, Harney Lake, and other basins	Oregon	Apr–Sept	1.5
	(E) Missouri and Arkansas basins			
1	Northeast Montana	Montana	May–Aug	1.40
2	North central Montana	Montana	May–Aug	1.50
3	Central Montana	Montana	May–Aug	1.70
4	Upper Missouri River basin	Montana	May–Aug	1.60
5	Upper Yellowstone River basin	Montana	May–Aug	1.90
6	Southeast Montana	Montana	May–Sept	1.95
7	Big Horn River basin	Wyoming	May–Sept	1.65
8	Yellowstone and Missouri River basins	Wyoming	May–Sept	1.70
9	Upper Platte River basin	Wyoming	May–Sept	1.60
10	Northeast Colorado	Colorado	Apr–Sept	2.05
11	North central Colorado	Colorado	Apr–Sept	2.20
12	South central Colorado	Colorado	Apr–Sept	2.10
13	Southeast Colorado	Colorado	Apr–Oct	2.30
14	West Kansas	Kansas	Apr–Oct	1.75
15	Central Nebraska	Nebraska	Apr–Oct	1.25
16	West Nebraska	Nebraska	Apr–Oct	2.00
17	Western South Dakota	South Dakota	May–Sept	1.50
18	Western North Dakota	North Dakota	May–Sept	1.35

Note: Estimated minimum quantities of irrigation water required, assuming optimal irrigation practices.

[a] Not in the Columbia River Basin.
[b] South of Weber River basin.

Source: U.S. Department of Agriculture.

Table 4K.64 Estimated Evapotranspiration for Types of Vegetation in the Western United States

Vegetation Type	Annual Evapotranspiration (in.)
Forest	
Lodgepole pine	19
Engelmann spruce-fir	15
White pine-larch-fir	22
Mixed conifer	22
True fir	24
Aspen	23
Pacific Douglas-fir-hemlock-redwood	30
Interior ponderosa pine	17
Interior Douglas-fir	21
Chaparral and woodland	
Southern California chaparral	20
California woodland-grass	18
Arizona chaparral	17.5
Pinon-juniper	14.5
Semiarid grass and shrub	10.6
Alpine	20

Source: Select Committee on National Water Resources, U.S. Senate, 1960.

Table 4K.65 Consumptive Use in a Municipal Area

Cultural Classification	Consumptive Use (ft)	Cultural Classification	Consumptive Use (ft)
Estates	2.07	Reservoir sites	1.34
Class A residential	1.92	Park	2.40
Class B residential	1.88	Schools	1.63
Rural residential	1.78	River wash	0.99
Semicommercial	1.32		

Note: Data for Raymond Basin, Los Angeles County, California.

Source: California Department of Water Resources.

Blaney-Criddle Consumptive Use Formula

The consumptive use of an irrigated crop in which ample water supply is available can be estimated by the Blaney-Criddle formula. For a given month the consumptive use is given by

$$u = k(tp) \tag{4.1}$$

where u is the monthly consumptive use measured in inches, k is a monthly consumptive use coefficient dependent on the crop and location, t is the mean monthly temperature in degrees Fahrenheit, and p is the monthly percentage of daytime hours of the year.

Because t and p can be found from climatic data at a given location, they are often combined into a monthly consumptive use factor f, so that Equation 4.1 becomes simply

$$u = kf \tag{4.2}$$

For an entire growing season the consumptive use is given by

$$U = KF \tag{4.3}$$

where U is the seasonal consumptive use measured in inches, K is a seasonal consumptive use coefficient, and F is the sum of the monthly consumptive use factors.

Values for the Western United States of k are given in Table 4K.66, K in Table 4K.67, p in Table 4K.69, and f in Table 4K.70.

Table 4K.66 Monthly Consumptive Use Coefficients (*k*) for Irrigated Crops in the Western United States

Crop	Location	Mar	Apr	May	June	July	Aug	Sept	Oct	Nov
Alfalfa	California, coastal	0.60	0.65	0.70	0.80	0.85	0.85	0.80	0.70	0.60
	California, interior	0.65	0.70	0.80	0.90	1.10	1.00	0.85	0.80	0.70
	North Dakota	—	0.84	0.89	1.00	0.86	0.78	0.72		
	Utah, St. George	—	0.88	1.15	1.24	0.97	0.87	0.81		
Corn (maize)	North Dakota	—	—	0.47	0.63	0.78	0.79	0.70		
Cotton	Arizona	—	0.27	0.30	0.49	0.86	1.04	1.03	0.81	
	Texas	0.24	0.22	0.61	0.42	0.50				
Orchard, citrus	Arizona	0.57	0.60	0.60	0.64	0.64	0.68	0.68	0.65	0.62
	California, coastal	—	0.40	0.42	0.52	0.55	0.55	0.55	0.50	0.45
Pasture	California, Murrieta	—	—	0.84	0.84	0.77	0.82	1.09	0.70	
Potatoes	North Dakota	—	—	0.45	0.74	0.87	0.75	0.54		
	South Dakota	—	—	0.69	0.60	0.80	0.89	0.39		
Small grain	North Dakota	—	0.19	0.55	1.13	0.77	0.30			
Wheat	Texas	0.64	1.16	1.26	0.87					
Sorghum	Arizona	—	—	—	—	0.34	0.72	0.97	0.62	0.60
	Kansas	—	—	—	0.80	0.94	1.17	0.86	0.47	
	Texas	—	—	—	0.26	0.73	1.20	0.85	0.49	
Soy beans	Arizona	—	—	—	0.26	0.58	0.92	0.92	0.55	
Sugar beets	California, coastal	—	0.39	0.38	0.36	0.37	0.35	0.38		
	California, interior	—	0.30	0.60	0.86	0.96	0.91	0.41		
	Montana	—	—	—	—	0.83	1.05	1.02		
Truck crops	California, interior	0.19	0.26	0.38	0.55	0.71	0.82	0.69	0.37	0.35

Note: For use with the Blaney-Criddle consumptive use formula.

Source: U.S. Department of Agriculture.

Table 4K.67 Seasonal Consumptive Use Coefficients (*k*) for Irrigated Crops in the Western United States

Crop	Length of Normal Growing Season or Period[a]	Consumptive Use Coefficient (*K*)[b]
Alfalfa	Between frosts	0.80 to 0.90
Bananas	Full year	0.80 to 1.00
Beans	3 months	0.60 to 0.70
Cocoa	Full year	0.70 to 0.80
Coffee	Full year	0.70 to 0.80
Corn (Maize)	4 months	0.75 to 0.85
Cotton	7 months	0.60 to 0.70
Dates	Full year	0.65 to 0.80
Flax	7 to 8 months	0.70 to 0.80
Grains, small	3 months	0.75 to 0.85
Grain, sorghums	4 to 5 months	0.70 to 0.80
Oilseeds	3 to 5 months	0.65 to 0.75
Orchard crops		
Avocado	Full year	0.50 to 0.55
Grapefruit	Full year	0.55 to 0.65
Orange and lemon	Full year	0.45 to 0.55
Walnuts	Between frosts	0.60 to 0.70
Deciduous	Between frosts	0.60 to 0.70
Pasture crops		
Grass	Between frosts	0.75 to 0.85
Ladino whiteclover	Between frosts	0.80 to 0.85
Potatoes	3 to 5 months	0.65 to 0.75
Rice	3 to 5 months	1.00 to 1.10
Sisal	Full year	0.65 to 0.70
Sugar beets	6 months	0.65 to 0.75
Sugarcane	Full year	0.80 to 0.90
Tobacco	4 months	0.70 to 0.80
Tomatoes	4 months	0.65 to 0.70
Truck crops, small	2 to 4 months	0.60 to 0.70
Vineyard	5 to 7 months	0.50 to 0.60

Note: For use with the Blaney-Criddle consumptive use formula.

[a] Length of season depends largely on variety and time of year when the crop is grown. Annual crops grown during the winter period may take much longer than if grown in the summertime.

[b] The lower values of K for use in the Blaney-Criddle formula, $U=KF$, are for the more humid areas, and the higher values are for the more arid climates.

Source: U.S. Department of Agriculture.

Table 4K.68 Monthly and Seasonal Consumptive Use Coefficients (*k*) for Mature Urban Landscape Plantings

Planting Type	Monthly (Maximum K)	Seasonal (Average)
Warm-season grass (Bermuda, buffalo), Mesa and Tempe, Arizona (elevation 1200 ft)	1.17	0.97
Warm-season grass (Bermuda, buffalo), Laramie and Wheatland, Colorado (elevation 7200 and 4700 ft)	1.08	0.79
Cool-season grass (Kentucky bluegrass), Mesa and Tempe, Arizona (elevation 1200 ft)	1.41	1.20
Cool-season grass (Kentucky bluegrass), Laramie and Wheatland, Colorado (elevation 7200 and 4700 ft)	1.30	0.95
Deciduous fruit or nut tree in bare soil, Mulch, or Paved Area	0.90	0.70
Deciduous fruit or nut tree with cover crop (turf, ground cover, etc)	1.25	1.00
Grapefruit in bare soil, mulch, or paved area	0.75	0.65
Lemon, orange	0.65	0.55
Other trees	0.90	0.80
Grapevine, Mesa and Tempe, Arizona (elevation 1200 ft)	0.80	0.70
Grapevine, elsewhere	0.75	0.60
Other vines	0.80	0.70
Shrubs over 4 ft diameter	0.80	0.70
Shrubs under 4 ft diameter	1.00	0.90
Ground cover plants, other small plants	1.00	0.90
Arid climate native plants	0.35	0.25
Pavements, mulches, nonliving soil covers	0.00	0.00

Note: For use with the Blaney-Criddle consumptive use formula; the higher the monthly or seasonal factor, the greater the planting's water demand in that time period.

Source: From Ferguson, B.K., Water conservation methods in urban landscape irrigation: an exploratory overview, *Water Res. Bull.*, 23, 147, 1987. With permission.

Table 4K.69 Monthly Percentage of Daytime Hours of the Year (p)

Latitude	January	February	March	April	May	June	July	August	September	October	November	December
North												
0	8.50	7.66	8.49	8.21	8.50	8.22	8.50	8.49	8.21	8.50	8.22	8.50
5	8.32	7.57	8.47	8.29	8.65	8.41	8.67	8.60	8.23	8.42	8.07	8.30
10	8.13	7.47	8.45	8.37	8.81	8.60	8.86	8.71	8.25	8.34	7.91	8.10
15	7.94	7.36	8.43	8.44	8.98	8.80	9.05	8.83	8.28	8.26	7.75	7.88
20	7.74	7.25	8.41	8.52	9.15	9.00	8.25	8.96	8.30	8.18	7.58	7.66
25	7.53	7.14	8.39	8.61	9.33	9.23	9.45	9.09	8.32	8.09	7.40	7.42
30	7.30	7.03	8.38	8.72	9.53	9.49	9.67	9.22	8.33	7.99	7.19	7.15
32	7.20	6.97	8.37	8.76	9.62	9.59	9.77	9.27	8.34	7.95	7.11	7.05
34	7.10	6.91	8.36	8.80	9.72	9.70	9.88	9.33	8.36	7.90	7.02	6.92
36	6.99	6.85	8.35	8.85	9.82	9.82	9.99	9.40	8.37	7.85	6.92	6.79
38	6.87	6.79	8.34	8.90	9.92	9.95	10.10	9.47	8.38	7.80	6.82	6.66
40	6.76	6.72	8.33	8.95	10.02	10.08	10.22	9.54	8.39	/7.75	6.72	7.52
42	6.63	6.65	8.31	9.00	10.14	10.22	10.35	9.62	8.40	7.69	6.62	6.37
44	6.49	6.58	8.30	9.06	10.26	10.38	10.49	9.70	8.41	7.63	6.49	6.21
46	6.34	6.50	8.29	9.12	10.39	10.54	10.64	9.79	8.42	7.57	6.36	6.04
48	6.17	6.41	8.27	9.18	10.53	10.71	10.80	9.89	8.44	7.51	6.23	5.86
50	5.98	6.30	8.24	9.24	10.68	10.91	10.99	10.00	8.46	7.45	6.10	5.65
52	5.77	6.19	8.21	9.29	10.85	11.13	11.20	10.12	8.49	7.39	5.93	5.43
54	5.55	6.08	8.18	9.36	11.03	11.38	11.43	10.26	8.51	7.30	5.74	5.18
56	5.30	5.95	8.15	9.45	11.22	11.67	11.69	10.40	8.53	7.21	5.54	4.89
58	5.01	5.81	8.12	9.55	11.46	12.00	11.98	10.55	8.55	7.10	5.31	4.56
60	4.67	5.65	8.08	9.65	11.74	12.39	12.31	10.70	8.57	6.98	5.04	4.22
South												
0	8.50	7.66	8.49	8.21	8.50	8.22	8.50	8.49	8.21	8.50	8.22	8.50
5	8.68	7.76	8.51	8.15	8.34	8.05	8.33	8.38	8.19	8.56	8.37	8.68
10	8.86	7.87	8.53	8.09	8.18	7.86	8.14	8.27	8.17	8.62	8.53	8.88
15	9.05	7.98	8.55	8.02	8.02	7.65	7.95	8.15	8.15	8.68	8.70	9.10
20	9.24	8.09	8.57	7.94	7.85	7.43	7.76	8.03	8.13	8.76	8.87	9.33
25	9.46	8.21	8.60	7.84	7.66	7.20	7.54	7.90	8.11	8.86	9.04	9.58
30	9.70	8.33	8.62	7.73	7.45	6.96	7.31	7.76	8.07	8.97	9.24	9.85
32	9.81	8.39	8.63	7.69	7.36	6.85	7.21	7.70	8.06	9.01	9.33	9.96
34	9.92	8.45	8.64	7.64	7.27	6.74	7.10	7.63	8.05	9.06	9.42	10.08
36	10.03	8.51	8.65	7.59	7.18	6.62	6.99	7.56	8.04	9.11	9.51	10.21
38	10.15	8.57	8.66	7.54	7.08	6.50	6.87	7.49	8.03	9.16	9.61	10.34
40	10.27	8.63	8.67	7.49	6.97	6.37	6.76	7.41	8.02	9.21	9.71	10.49
42	10.40	8.70	8.68	7.44	6.85	6.23	6.64	7.33	8.01	9/9.26	9.82	10.64
44	10.54	8.78	8.69	7.38	6.73	6.08	6.51	7.25	7.99	9.31	9.94	10.80
46	10.69	8.86	8.70	7.32	6.61	5.92	6.37	7.16	7.96	9.37	10.07	10.97

Note: Latitudes 60°N to 46°S. For use with the Blaney-Criddle consumptive use formula.

Source: U.S. Dept. of Agriculture.

Table 4K.70 Monthly Consumptive Use Factors (*f*) for Locations in the Western United States and Hawaii

	Arizona			California
Month	Phoenix *f*	Safford *f*	Yuma *f*	Bakersfield *f*
January	3.64	3.14	3.90	3.33
February	3.82	3.40	4.07	3.58
March	5.07	4.51	5.36	4.74
April	5.89	5.35	6.10	5.55
May	7.28	6.73	7.36	6.87
June	8.17	7.61	8.16	7.61
July	8.85	8.20	8.91	8.38
August	8.25	7.52	8.41	7.69
September	6.91	6.26	6.98	6.20
October	5.58	5.04	5.81	5.13
November	4.20	3.67	4.42	3.68
December	3.62	3.15	3.87	3.38
Total	71.28	64.58	73.35	65.84
Frost-free period	2/5 to 12/6	4/5 to 11/4	1/12 to 12/26	2/21 to 11/25

	California			
Month	El Centro *f*	Escondido *f*	Merced *f*	Red Bluff *f*
January	3.88	3.70	3.16	3.06
February	4.00	3.70	3.38	3.35
March	5.37	4.72	4.45	4.53
April	6.17	5.26	5.27	5.33
May	7.58	6.20	6.57	6.70
June	8.20	6.49	7.35	7.58
July	9.07	7.22	8.08	8.34
August	8.56	6.93	7.39	7.61
September	7.19	5.95	6.06	6.14
October	5.95	5.14	4.95	4.98
November	4.44	4.06	3.67	3.60
December	3.93	3.73	3.14	3.20
Total	74.34	63.10	53.47	64.24
Frost-free period	1/29 to 12/9	3/9 to 11/25	3/9 to 11/20	3/5 to 12/5

	California		Colorado	
Month	Sacramento *f*	Santa Ana *f*	Fort Collins *f*	Grand Junction *f*
January	3.13	3.77	1.76	1.72
February	3.39	3.78	1.89	2.29
March	4.52	4.77	3.02	3.57
April	5.18	5.27	4.10	4.67
May	6.30	6.17	5.49	6.20
June	6.93	6.50	6.45	7.22
July	7.42	7.05	7.11	7.98
August	6.92	6.71	6.50	7.19
September	5.81	5.81	4.98	5.60
October	4.89	5.11	3.73	4.22
November	3.64	4.15	2.42	2.71
December	3.06	3.80	1.81	1.92
Total	61.19	62.89	49.26	55.29
Frost-free period	2/6 to 12/10	2/7 to 12/7	5/7 to 9/29	4/13 to 10/25

	Colorado	Idaho		
Month	Montrose *f*	Boise *f*	Idaho Falls *f*	Lewiston *f*
January	1.68	1.82	1.26	2.09
February	2.15	2.22	1.55	2.42
March	3.32	3.44	2.79	3.77

(Continued)

Table 4K.70 (Continued)

April	4.32	4.44	4.05	4.84
May	5.70	5.74	5.45	6.25
June	6.64	6.68	6.26	7.12
July	7.31	7.58	7.19	8.13
August	6.62	6.87	6.45	8.01
September	5.20	5.15	4.79	5.40
October	3.89	3.83	3.60	4.02
November	2.56	2.58	2.18	2.64
December	1.77	1.90	1.45	2.12
Total	51.16	52.25	47.02	56.81
Frost-free period	5/6 to 10/9	4/23 to 10/17	5/15 to 9/19	4/5 to 10/26

	Idaho	**Kansas**		**Montana**
	Twin Falls	**Garden City**	**Wichita**	**Agricultural College**
Month	***f***	***f***	***f***	***f***
January	1.77	2.12	2.21	1.30
February	2.16	2.32	2.39	1.48
March	3.35	3.65	3.80	2.50
April	4.39	4.81	4.98	3.76
May	5.75	6.32	6.45	5.15
June	6.53	7.30	7.45	6.03
July	7.57	7.97	8.10	6.87
August	6.67	7.37	7.53	6.21
September	5.00	5.83	5.99	4.53
October	3.93	4.41	4.65	3.35
November	2.48	2.93	3.13	2.03
December	9.84	2.17	2.32	1.41
Total	51.44	57.20	59.00	44.63
Frost-free period	5/18 to 9/26	4/25 to 10/16	4/10 to 10/27	5/24 to 9/16

	Montana	**Nebraska**		**New Mexico**
	Missoula	**McCook**	**Scottsbluff**	**Albuquerque**
Month	***f***	***f***	***f***	***f***
January	1.16	1.85	1.72	2.40
February	1.54	2.11	1.88	2.69
March	2.77	3.34	3.02	3.85
April	4.01	4.59	4.20	4.87
May	5.46	6.14	5.71	6.23
June	6.34	7.17	6.79	7.09
July	7.17	7.97	7.58	7.65
August	6.32	7.25	6.87	6.98
September	4.51	5.56	5.19	5.65
October	3.26	4.20	3.83	4.46
November	1.98	2.64	2.45	3.01
December	1.27	1.91	1.81	2.54
Total	45.79	54.73	51.05	57.39
Frost-free period	5/18 to 9/23	5/3 to 10/6	5/11 to 9/26	4/13 to 10/28

	New Mexico		**Nevada**	
	Carlsbad	**State College**	**Carson City**	**Yerington**
Month	***f***	***f***	***f***	***f***
January	3.18	2.96	2.20	2.06
February	3.37	3.12	2.40	2.46
March	4.64	4.29	3.45	3.56
April	5.56	5.15	4.25	4.40
May	6.89	6.29	5.49	5.65
June	7.61	7.29	6.23	6.37
July	7.93	7.72	7.04	7.17
August	7.55	7.17	6.42	6.63

(Continued)

Table 4K.70 (Continued)

September	6.15	5.94	5.00	5.09
October	5.04	4.77	3.87	3.95
November	3.70	3.45	2.69	2.67
December	3.06	2.85	2.24	2.11
Total	64.68	61.00	51.28	52.12
Frost-free period	3/29 to 11/4	4/6 to 10/31	5/23 to 9/19	5/23 to 9/18

	Oklahoma	Oregon		
	Altus	Baker	Hood River	Medford
Month	*f*	*f*	*f*	*f*
January	2.75	1.60	2.09	2.50
February	3.09	1.90	2.40	2.81
March	4.43	3.12	3.61	3.90
April	5.49	4.10	4.57	4.69
May	6.86	5.33	5.83	5.91
June	7.78	6.11	6.47	6.74
July	8.33	6.92	7.15	7.54
August	7.82	6.26	6.50	6.85
September	6.34	4.72	5.04	5.33
October	5.07	3.55	3.92	4.12
November	3.65	2.32	2.65	2.90
December	2.88	1.68	2.15	2.42
Total	64.48	47.64	52.38	55.71
Frost-free period	3/28 to 11/9	5/12 to 10/3	4/20 to 10/20	5/6 to 10/4

	Texas			Utah
	Amarillo	Fort Stockton	Lubbock	Logan
	f	*f*	*f*	*f*
January	2.33	3.45	2.85	1.59
February	2.48	3.61	3.09	1.86
March	3.78	4.87	4.28	3.05
April	4.75	5.74	5.25	4.28
May	6.07	7.08	6.57	5.63
June	6.98	7.69	7.37	6.53
July	7.54	7.94	7.80	7.53
August	6.98	7.47	7.28	6.86
September	5.67	6.25	5.95	5.19
October	4.39	5.25	4.83	3.87
November	2.87	3.93	3.47	2.46
December	2.43	3.39	2.84	1.69
Total	56.27	66.67	61.58	50.54
Frost-free period	4/11 to 11/2	4/1 to 11/3	4/12 to 11/3	5/7 to 10/11

	Utah	Washington		Wyoming
	Salt Lake City	Prosser	Yakima	Cheyenne
Month	*f*	*f*	*f*	*f*
January	1.96	1.95	1.72	1.70
February	2.26	2.36	2.22	1.82
March	3.47	3.80	3.71	2.75
April	4.45	4.82	4.64	3.67
May	5.77	6.21	6.06	5.08
June	6.83	7.03	6.88	6.13
July	7.77	7.74	7.63	6.87

(Continued)

Table 4K.70 (Continued)

August	7.13	6.92	6.79	6.29
September	5.40	5.24	5.21	4.78
October	4.06	3.97	3.91	3.46
November	2.75	2.55	2.41	2.32
December	2.06	2.01	1.92	1.84
Total	53.91	54.55	53.11	46.71
Frost-free period	4/13 to 10/22	4/28 to 10/4	4/15 to 10/22	5/14 to 10/2

	Wyoming	Hawaii	
	Worland	Honolulu W.B. Airport	Waianai
Month	*f*	*f*	*f*
January	0.97	5.62	5.57
February	1.40	5.15	5.16
March	2.79	6.04	6.08
April	4.08	6.26	6.45
May	5.57	6.87	7.05
June	6.58	6.98	7.23
July	7.41	7.20	7.49
August	6.60	7.02	7.35
September	4.87	6.49	6.65
October	3.57	6.27	6.43
November	2.07	5.67	5.74
December	1.19	5.54	5.61
Total	47.10	75.07	76.81
Frost-free period	5/10–9/27	—	—

Note: For use with the Blaney-Criddle consumptive use formula.

Source: U.S. Department of Agriculture.

Table 4K.71 Estimated Oat Yields[a] Based on Stored Soil Water and Growing Season Precipitation[a]

Consumptive Use Area	Stored Soil Water + Growing Season Precipitation (in.)														
	4	5	6	7	8	9	10	11	12	13	14	15	16	17	18
	Bushels per acre[b,c]														
1 High	0	2	11	21	30	39	49	58	68	77	86	96	105	115	124
2 Moderate high	0	2	12	23	33	43	54	64	74	84	95	105	115	126	136
3 Moderate	0	2	14	26	37	49	61	72	84	96	108	119	131	143	154
4 Moderate low	0	2	15	28	40	52	65	78	90	102	115	128	140	152	165

[a] Estimated yields reflect consumptive use data from Huntley, Havre, Sidney, Conrad, Kalispell, Bozeman, and Moccasin.
[b] Yields may vary from estimates due to climatic conditions, weeds, disease, insects, lodging, or stand density.
[c] When rooting depths are limited by rocks, gravel, or impermeable layers such as shale, yields may vary.

Source: From NRCS, MT July 2002, Specification MT590-5.

Table 4K.72 Estimated Safflower Yields[a] Based on Stored Soil Water and Growing Season Precipitation[a]

Consumptive Use Area	Stored Soil Water + Growing Season Precipitation (in.)														
	8	9	10	11	12	13	14	15	16	17	18	19	20	21	228
	Pounds per acre[b,c]														
2 Moderate high	115	279	443	607	771	935	1,099	1,263	1,427	1,591	1,755	1,919	2,083	2,247	2,411

[a] Estimated yields reflect consumptive use data from Huntley, Havre, Sidney, Conrad, Kalispell, Bozeman, and Moccasin.
[b] Yields may vary from estimates due to climatic conditions, weeds, disease, insects, lodging, or stand density.
[c] When rooting depths are limited by rocks, gravel, or impermeable layers such as shale, yields may vary.

Source: From NRCS, MT July 2002, Specification MT590-5.

Table 4K.73 Estimated Spring Wheat Yields[a] Based on Stored Soil Water and Growing Season Precipitation[a]

Consumptive Use Area	Stored Soil Water + Growing Season Precipitation (in.)														
	4	5	6	7	8	9	10	11	12	13	14	15	16	17	18
	Bushels per acre[b,c]														
1 High	0	6	10	15	20	24	29	34	39	43	48	53	57	62	67
2 Moderate high	0	6	11	16	21	27	32	37	42	47	52	57	62	67	72
3 Moderate	0	7	13	19	24	30	36	42	48	53	59	65	71	77	82
4 Moderate low	0	7	13	20	26	32	38	44	50	56	62	68	74	80	87

[a] Estimated yields reflect consumptive use data from Huntley, Havre, Sidney, Conrad, Kalispell, Bozeman, and Moccasin.
[b] Yields may vary from estimates due to climatic conditions, weeds, disease, insects, lodging, or stand density.
[c] When rooting depths are limited by rocks, gravel, or impermeable layers such as shale, yields may vary.

Source: From NRCS, MT July 2002, Specification MT590-5.

Table 4K.74 Estimated Barley Yields[a] Based on Stored Soil Water and Growing Season Precipitation[a]

Consumptive Use Area	Stored Soil Water + Growing Season Precipitation (in.)														
	4	5	6	7	8	9	10	11	12	13	14	15	16	17	18
	Bushels per acre[b,c]														
1 High	6	13	20	27	34	41	48	55	62	69	76	83	90	97	104
2 Moderate high	7	14	22	30	37	45	52	60	68	75	83	90	98	106	113
3 Moderate	8	16	25	33	42	50	59	67	76	84	93	101	110	118	127
4 Moderate low	8	17	26	35	44	53	62	71	80	89	98	107	116	125	134

[a] Estimated yields reflect consumptive use data from Huntley, Havre, Sidney, Conrad, Kalispell, Bozeman, and Moccasin.
[b] Yields may vary from estimates due to climatic conditions, weeds, disease, insects, lodging, or stand density.
[c] When rooting depths are limited by rocks, gravel, or impermeable layers such as shale, yields may vary.

Source: From NRCS, MT July 2002, Specification MT590-5.

Table 4K.75 Estimated Winter Wheat Yields[a] Based on Stored Soil Water and Growing Season Precipitation[a]

Consumptive Use Area	Stored Soil Water + Growing Season Precipitation (in.)														
	4	5	6	7	8	9	10	11	12	13	14	15	16	17	18
	Bushels per acre[b,c]														
1 High	0	6	11	17	22	28	33	38	44	49	55	60	65	71	76
2 Moderate high	0	6	12	18	24	30	35	41	47	53	59	64	70	76	82
3 Moderate	0	7	14	20	27	34	40	47	53	60	67	73	80	86	93
4 Moderate low	0	8	15	22	29	36	43	50	57	64	71	78	85	92	99

[a] Estimated yields reflect consumptive use data from Huntley, Havre, Sidney, Conrad, Kalispell, Bozeman, and Moccasin.
[b] Yields may vary from estimates due to climatic conditions, weeds, disease, insects, lodging, or stand density.
[c] When rooting depths are limited by rocks, gravel, or impermeable layers such as shale, yields may vary.

Source: From NRCS, MT July 2002, Specification MT590-5.

Table 4K.76 Irrigated Land in Farms, 1889–1994, by Region and Crop

Region	1889[a,b]	1949[a]	1969[a]	1974[a]	1978[a]	1982[a]	1987[a]	1992[c]	1993[c]	1994[d]
	1000 acres									
USDA production region										
Atlantic[e]	—	500	1,800	2,000	2,900	2,700	3,000	3,500	3,600	3,700
North Central[f]	—	—	500	600	1,400	1,700	2,000	2,600	2,400	2,700
Northern Plains	—	1,100	4,600	6,200	8,800	9,300	8,700	10,200	9,800	10,300
Delta States	—	1,000	1,900	1,800	2,700	3,100	3,700	5,400	5,200	5,300
Southern Plains	—	3,200	7,400	7,100	7,500	6,100	4,700	5,300	5,500	5,300
Mountain States	2,300	11,600	12,800	12,700	14,800	14,100	13,300	13,600	13,900	14,300
Pacific Coast	1,200	8,300	10,000	10,600	12,000	11,900	10,800	10,500	10,100	10,300
United States[g]	3,600	25,800	39,100	41,200	50,400	49,000	46,400	51,300	50,600	52,000
Crop										
Corn for grain	NA	NA	3,300	5,600	8,700	8,500	8,000	10,300	9,800	10,600
Wheat	NA	NA	2,000	3,300	3,000	4,600	3,700	4,100	3,900	3,900
Rice	NA	NA	2,200	2,600	3,000	3,200	2,400	3,500	3,100	3,400
Soybeans	NA	NA	700	500	1,300	2,300	2,600	3,100	3,200	3,200
Cotton	NA	NA	3,100	3,700	4,700	3,400	3,500	3,700	4,400	4,300
Hay	NA	NA	7,900	8,000	8,900	8,500	8,600	8,400	8,600	8,900

Note: — Indicated > 50,0000 acres. NA = not applicable.

[a] Census of agriculture.
[b] Excludes rice, which was grown on 342,000 acres in South Atlantic and Gulf States in 1899.
[c] Preliminary estimates constructed from unpublished USDA sources and the Census. Partial returns from 1992 Census were incorporated.
[d] Forecast based on March Planting Intentions (NASS).
[e] Northeast, Appalachian, and Southeast farm production regions.
[f] Lake States and Corn Belt production regions. Remaining regions correspond to single farm production regions.
[g] Includes Alaska and Hawaii.

Source: USDA, ERS data.

Table 4K.77 Average Depth of Irrigation Water Applied Per Season, 1969–94, by Region and Crop

Item	1969[a]	1974[a]	1979[b]	1984[b]	1988[b]	1990[c]	1991[c]	1992[c]	1993[c]	1994[d]
	Inches[e]									
Region										
Atlantic[f]	8.5	11.5	15.0	16.5	15.5	15.5	16.0	16.0	16.5	16.5
North Central[g]	7.5	8.0	9.5	9.5	10.5	9.0	9.5	10.0	8.0	10.0
Northern Plains	16.0	17.0	15.5	13.5	14.5	14.0	14.0	13.5	11.5	14.5
Delta States	15.5	17.5	26.0	17.5	18.0	16.5	15.5	16.5	15.5	15.5
Southern Plains	18.0	18.5	17.0	16.5	17.0	16.5	15.0	16.0	16.0	16.0
Mountain States	30.5	28.5	24.0	24.5	24.5	24.0	24.0	24.0	23.0	24.0
Pacific Coast	33.0	34.0	32.0	34.0	34.5	34.5	34.5	34.5	33.0	35.0
United States[h]	25.5	25.0	22.5	22.5	22.5	22.5	21.5	21.5	20.0	21.5
Crop										
Corn for grain	18.5	19.5	17.0	16.0	16.0	15.5	15.0	15.0	13.5	15.5
Wheat	23.0	24.0	20.5	16.5	16.0	15.5	14.5	14.5	14.0	14.5
Rice	28.0	28.5	33.5	33.5	32.5	31.5	30.5	30.0	30.5	30.5
Soybeans	12.0	11.5	14.0	9.5	10.0	8.5	7.0	8.0	6.5	7.5
Cotton	23.0	25.5	24.0	25.0	24.5	23.0	21.0	23.0	20.0	21.5
Alfalfa hay	32.5	30.5	26.0	28.0	29.0	28.5	27.5	27.5	26.5	27.5

[a] Census of Agriculture.
[b] Estimates constructed by State, by crop, from Farm and Ranch Irrigation Surveys (FRIS) (USDC, 1990, 1986, and 1982a) and ERS estimates of irrigated area.
[c] Aggregated from FRIS State/crop application rates adjusted to reflect annual changes in precipitation. Sensitivity to precipitation is estimated as a function of average precipitation and soil hydrologic group.
[d] Forecast using precipitation records through May 1994.
[e] Values rounded to nearest 0.5 inches.
[f] Northeast, Appalachian, and Southeast production regions.
[g] Lake States and Corn Belt farm production regions.
[h] Includes Alaska and Hawaii.

Source: USDA, ERS data.

Table 4K.78 Irrigated Area, 1992 Census of Agriculture

State	1987	1992	Change
	1,000 acres		Percent
Colorado	3,014	3,170	5
Connecticut	7	6	−19
Delaware	61	62	2
Florida	1,623	1,783	10
Idaho	3,219	3,260	1
Illinois	208	328	58
Indiana	170	241	42
Iowa	92	116	25
Kansas	2,463	2,680	9
Maine	6	10	69
Maryland	51	57	12
Massachusetts	20	20	−1
Michigan	315	368	17
Missouri	535	709	33
New Hampshire	3	2	−41
New Jersey	91	80	−12
Ohio	32	29	−9
Oregon	1,648	1,622	−2
Rhode Island	3	3	−15
Vermont	2	2	16
Virginia	79	62	−22
Washington	1,519	1,641	8
West Virginia	3	3	−12
Wisconsin	285	331	16
Wyoming	1,518	1,465	−4
Total, 25 states	16,967	18,050	6

Note: The table includes 1992 data on States released by the Bureau of the Census by June 20, 1994.

Source: USDC, 1994.

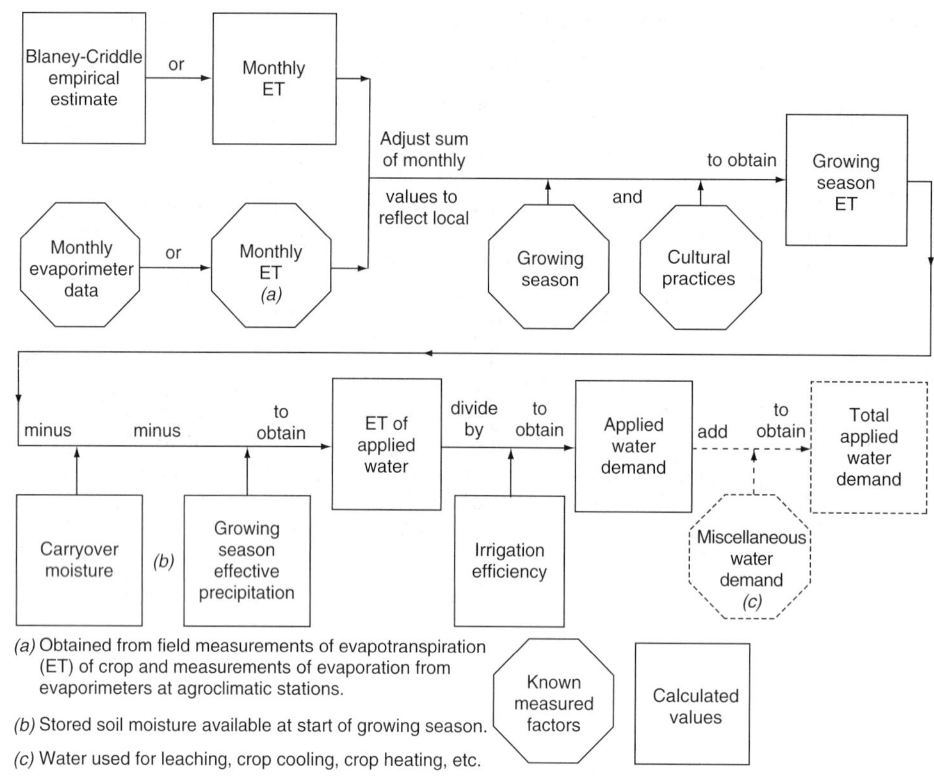

Figure 4K.27 Steps in determining agricultural applied water demand. (From American Society of Civil Engineers, 1972, Copyright, Groundwater Management.)

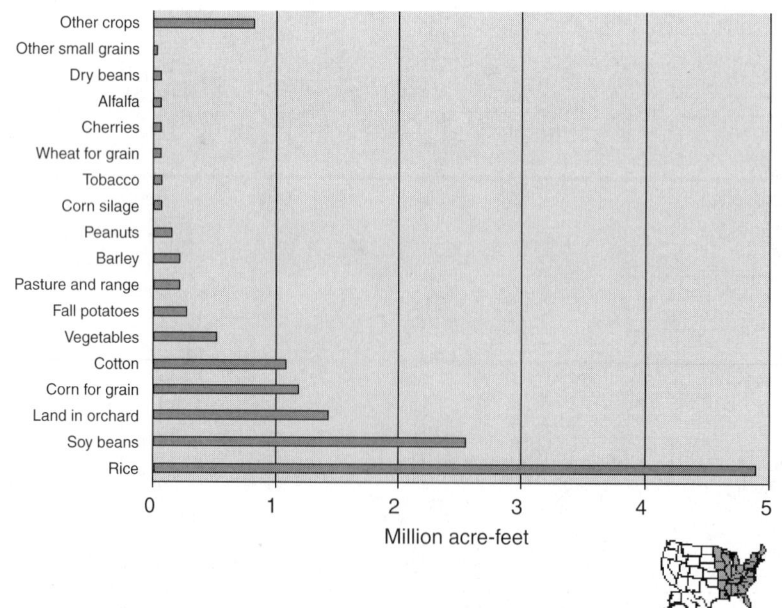

Figure 4K.28 Eastern water applications by crop, 1998. (From ERS, 1998 Farm and Ranch Irrigation Survey, USDA, www.ers.usda.gov.)

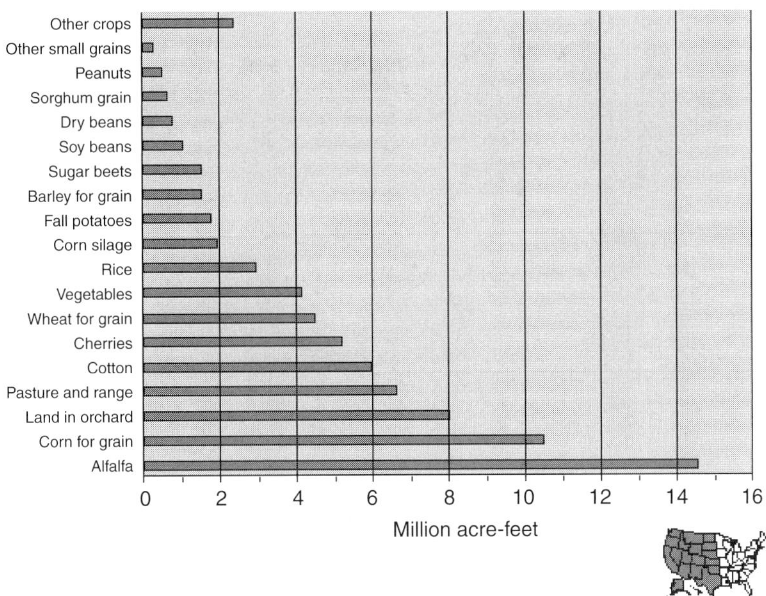

Figure 4K.29 Western water applications by crop, 1998. (From ERS, 1998 Farm and Ranch Irrigation Survey, USDA, http://ers.usda.gov.)

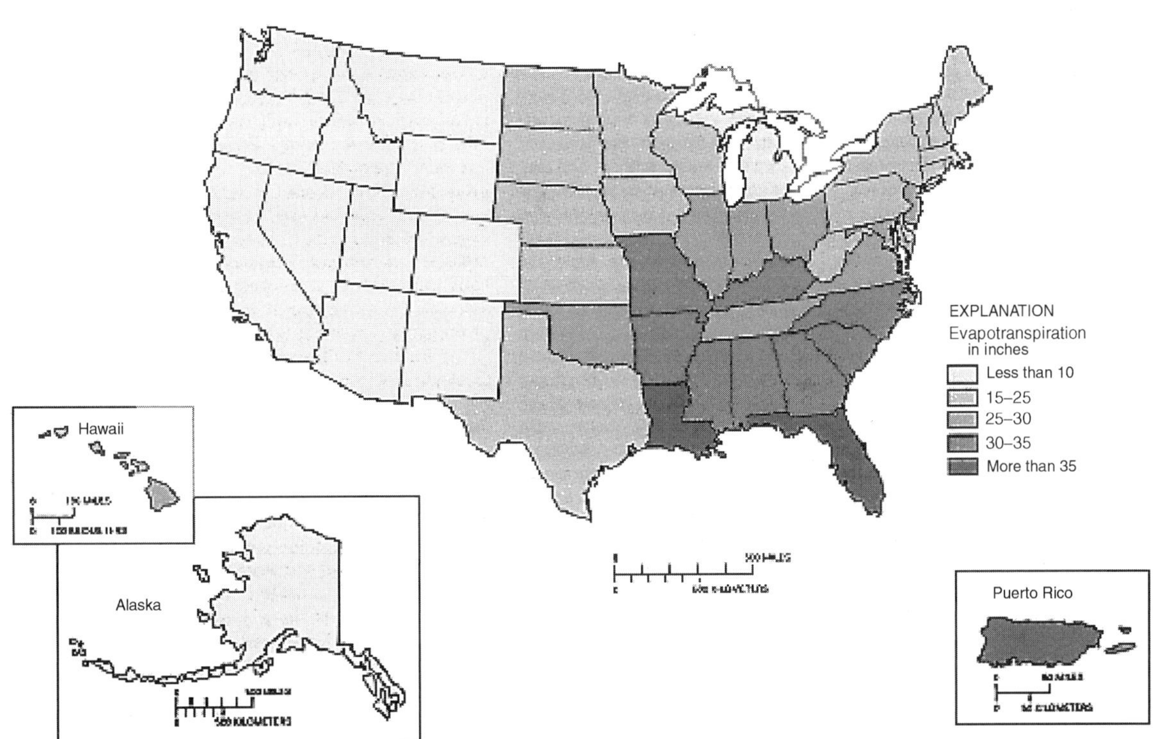

Figure 4K.30 Estimated mean annual evapotranspiration in the United States and Puerto Rico. (From U.S. Geological Survey, 1990, http://geochange.er.usgs.gov.)

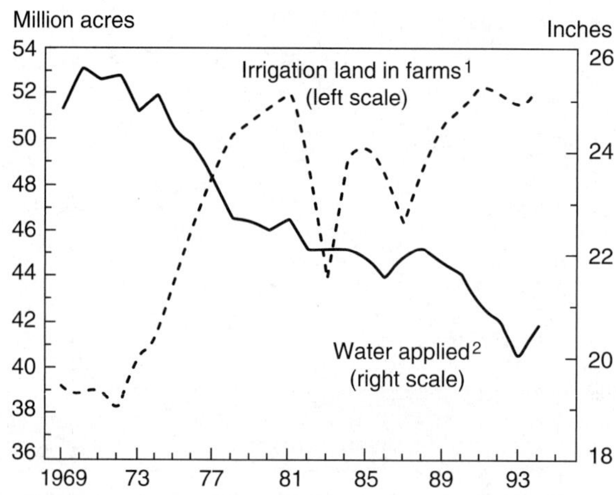

Figure 4K.31 Irrigation trends, 1969–1994. (From USDA, ERS data, www.ers.usda.gov.)

SECTION 4L PHREATOPHYTES

Table 4L.79 Phreatophyte Areas and Their Consumptive Use in Selected Western States (Estimates as of 1953)

State	Area (Acres)	Annual Use (Acre-Feet)
Arizona	405,000	1,280,000
California[a]	317,000	1,150,000
Colorado[a]	737,000	1,056,000
Idaho	500,000	1,000,000
Montana	1,600,000	3,200,000
Nebraska[a]	515,000	709,000
Nevada	2,801,000	1,500,000
New Mexico	300,000	900,000
North Dakota	1,035,000	1,660,000
Oregon[a]	40,800	21,200
South Dakota	850,000	1,240,000
Texas[a]	262,000	436,500
Utah	1,200,000	1,500,000
Wyoming	527,000	1,100,000
Total (approximate)[b]	11,090,000	16,750,000

[a] Partial data, from published reports on areas within the state.
[b] Partial data.

Source: U.S. Geological Survey.

Table 4L.80 Consumptive Use by Some Common Phreatophytes in the Western United States

Plant	Annual Rate Including Precipitation	Volume Density	Depth to Water	Locality
	Acre-feet per acre	Percent	Feet	
Alder	5.3	—	—	Santa Ana River, CA
Batamote or seepwillow	4.7	100	6	Safford Valley, AZ
Cottonwood	7.6–5.2	100	3–4	San Luis Rey River, CA
Do	6.0	100	6	Safford Valley, AZ
Greasewood	2.5–0.08	—	—	Escalante Valley, UT
Mesquaite	3.3	100	10	Safford Valley, AZ
Sacaton	4.0–3.5	—	—	Pecos River Valley, NM
Saltcedar	5.5–4.7	—	—	Do
Do	9.2–7.3	100	4–7	Safford Valley, AZ
Saltgrass	4.1–1.1	—	1.5–5	Owens Valley, CA
Do	2.9–1.1	—	2–4	Santa Ana, CA
Do	2.3–1.1	—	0.3–2.1	San Luis Valley, CO
Do	4.5	—	2.0	Carlsbad, NM
Do	2.6	—	0.65	Isleta, NM
Do	4.0–0.8	—	0.4–3.1	Los Griegos, NM
Do	1.9	—	2.2	Mesilla Dam, NM
Do	2.3–1.6	—	1.9–2.6	Escalante Valley, UT
Do	2.0	—	2.0	Vernal, UT
Willow	4.4	—	2.0	Santa Ana, CA
Do	2.5	—	1.1	Isleta, NM

Source: From Select Committee on National Water Resources U.S. Senate, 1960.

Table 4L.81 Phreatophyte Areas and their Consumptive Use in River Basins of the Western United States

State and River or Valley	Area in Acres	Estimated Annual Water Use in Acre-Feet	Year of Survey or Estimate	Principal Species of Phreatophytes
Arizona				
Little Colorado River				
Main stem only St. Johns to mouth	49,560	64,720	1951	Saltcedar, cottonwood, willow, brush
Total Little Colorado including Zuni and Puerco Rivers	65,310	74,360	1951	Saltcedar, cottonwood, willow, brush
Gila River				
Virden, NM to Clifton, AZ	5,600	6,310	1951	Cottonwood, willow, seepwillow
Head of Safford Valley to Coolidge Dam	25,520	61,500	1951	Saltcedar, mesquite, willow, seepwillow, cotton wood, arrowweed
Coolidge Dam to Kelvin	3,970	7,460	1951	Saltcedar, mesquite, willow, seepwillow, cotton wood, arrowweed
Kelvin to Gillespie Dam downstream from Granite Reef Dam on Salt River, Rillito on Santa Cruz River and Lake Pleasant on Agua Fria River	142,880	282,770	1951	Saltcedar, mesquite, willow, seepwillow, cotton wood, arrowweed
Gillespie Dam to Dome	51,270	69,220	1951	Saltcedar, mesquite, seepwillow
Total Gila River including San Pedro River, Palominas to mouth, Verde River, Bartlett Dam to mouth, Salt River, Stewart Mountain Dam to Granite Reef Dam	300,710	522,510	1951	
Salt River: Stewart Mountain Dam to Granite Reef Dam	2,710	2,070	1951	Saltcedar, mesquite
San Pedro River: Palominas to Gila at mouth	42,510	69,420	1951	Saltcedar, mesquite
Verde River: Bartlett to Gila at mouth	3,790	5,720	1951	Saltcedar, mesquite, seepwillow
California				
San Luis Rey River: IN San Diego County	6,390	17,800	1945	Cottonwood, willow
Santa Anna River: Below Prado Dam	1,071	3,000	1946	Cottonwood, willow
Kings River: Kings River Soil Conservation District	1,080	6,500	1958	Cottonwood, willow
Idaho: Malad Valley	10,900	32,400	1953	Grasses and rushes
New Mexico				
Pecos River: Alamogordo Dam to Texas State line	42,500	117,000	1956	Saltcedar
Rio Grande: Bernardo to San Marcial	52,000	77,800	1955	Saltcedar
Nevada				
Little Humboldt River: Paradise Valley	36,500	23,000	1947	Greasewood, saltgrass, willows
Muddy River: To mouth	4,900	10,330	1951	Saltcedar
Nevada-Utah-Arizona-California				
Virgin River: Littlefield to mouth	7,360	27,160	1951	Saltcedar
Lower Colorado River:				
Main stem				
Lee Ferry to Hoover Dam	1,480	5,920	1951	Saltcedar, willow
Hoover Dam to Davis Dam	4,500	11,450	1951	Saltcedar, willow
Davis Dam to Parker Dam	57,490	209,290	1950	Saltcedar, willow
Parker Dam to Laguna Dam	191,890	314,360	1951	Saltcedar, willow
Laguna Dam to International Boundary	33,510	54,210	1951	Saltcedar, willow
Total main stem	288,920	595,230	1951	
Total Lower Colorado River Basin, including Little Colorado River, Virgin River, and Gila River Basins	667,200	1,230,090	1951	

Source: Select Committee on National Water Resources, U.S. Senate, 1960.

Table 4L.82 Phreatophytes in the Western United States

Scientific Name	Common Name	Occurrence as a Phreatophyte	Relation to Groundwater: Depth to Water Below Land Surface (ft)	Relation to Groundwater: Quality	Relation to Groundwater: Use (acre feet/acre)	Remarks
Acacia greggii A. Gray	Catclaw, devilsclaw, una de gato	Southern California to western Texas	—	1	—	Uses more water than mesquite. Forms thickets along streams and washes
Acer negundo Linnacus	Boxelder	Canada to Oklahoma and Arizona	—	—	—	Occurs in moist places and along streams, chiefly in mountains. Observed in the flood plain of the North Canadian River near Oklahoma City, Okla. Widely used as a shade tree
Alhagi camelorum Fischer	Camelthorn	Arizona	—	—	—	Introduced into Southwestern United States from Asia Minor. Poor browse plant. Observed growing as a phreatophyte along Little Colorado River between Holbrook and Winslow, AZ, in localities where the depth to water ranged from 4 to 6 ft. Aggressive and thicket forming, root system deep and extensive
Allenrolfea occidentalis (S. Watson) Kuntze	Pickleweed, iodinebush	California to western Texas	1–20	3	—	
Alnus	Alder	—	—	—	5.3	Occurs along streams, river bottom land, and other wet sites. The use of 5.3 ft was for the period May to October 1932 in Cold-water Canyon, altitude 2400 ft, San Bernardino Mountains, CA, where alder constituted 82 percent of the vegetation
Anemopsis californica (Nuttall) Hooker and Arnott	Yerba mansa	So. California, So. Nevada to Utah and Texas	Shallow	3	—	Used by Pima Indians as a herbal remedy. Common in saline and wet lowlands
Aster spinosus Bentham	Spiny aster	Arizona	—	—	—	Identified as phreatophyte in bottom land of lower Safford Valley, AZ
Atriplex canescens (Pursh) Nuttall	Fourwing saltbush, Chamiso, chamiza	South Dakota to Oregon, south to Mexico	8–62	1–2	—	Tolerates alkali. Valuable browse plant. Useful in erosion control. Taproots 30–40 ft deep. May not always occur as a phreatophyte
hastata Linnaeus	—	Oregon and California to Kansas and New Mexico	—	3	—	Occurs in saline soils, especially around alkaline lakes, in salt marshes, and in other water-soaked soils
lentiformis (Torrey) Watson	Quailbrush, lenscale Nevada saltbush	Southern Utah and Nevada to California and Sonora, Mexico	6–15	3	—	High tolerance for alkali and saline soil. Fair browse plant. Reaches height of 10 ft where water table is shallow

(Continued)

Table 4L.82 (Continued)

Scientific Name	Common Name	Occurrence as a Phreatophyte	Relation to Groundwater: Depth to Water Below Land Surface (ft)	Relation to Groundwater: Quality	Relation to Groundwater: Use (acre feet/acre)	Remarks
Baccharis emoryi A. Gray	Emory baccharis	Texas to southern California and southern Utah	—	2		
glutinosa Persoon	Batamote, seepwillow, water motie, waterwillow	Colorado and Texas to California and Mexico	2–15	2	4.7	Evapotranspiration for plants grown in tanks ranged from 10.3 ft with water level at 2 ft to 4.6 ft with water level of 6 ft. Safford Valley, AZ
sarothroides A. Gray	Broom baccharis, desertbroom, rosinbrush	Southern California, Arizona, southwestern New Mexico	—	—	—	Occurs along streams in draws, in canyon bottoms and wet alkaline sites
sergiloides A. Gray	Squaw baccharis, waterweed	Arizona, southern California, southern Nevada, southwestern Utah	—	2	—	Occurs as a phreatophyte in lower Safford Valley, AZ
viminea Crandolle	Mulefat	Southwestern Utah, southern California, Nevada, Arizona	—	—	—	Useful in erosion control
Bigelovia hartwegii, probably *Aplopappus heterophyllus* A. Gray	Rayless goldenrod		—	—	—	Will grow in dry places but thrives where groundwater is within reach
Celtis reticulata Torrey	Hackberry, cumaru, kom	Arizona	—	—	—	A large tree that may reach 3 ft in diameter and 50 ft in height. Usually occurs along streams
Cercidium floridum Bentham	Blue palo verde	Southwestern Arizona, southeastern California	—	—	—	Common along washes, canyons, valleys, alluvial plains, grassland at sites where groundwater is plentiful
Chilopsis linearis Sweet	Desertwillow	Western Texas to southern Nevada, Arizona, southern California	To 50	—	—	May not always occur as a phreatophyte
Chrysothamnus pumilus (Nuttall)	Rabbitbrush	Mud Lake, Idaho	—	—	—	

nauseosus consimilis (Greene)	Rubber rabbitbrush	Nevada, Utah, Idaho, Wyoming	—	2–3	—	
nauseosus graveolens (Nuttall)	Rubber rabbitbrush	Montana, Idaho, Utah, Nevada, New Mexico	2.5–15	2–3	—	
nauseosus mohavensis (Greene)	Rubber rabbitbrush	Northern California, Nevada	—	2–3	—	
nauseosus oreophilus (Al Nelson)	Rubber rabbitbrush	Wyoming, Colorado, Utah	—	—	—	
nauseosus viridulus	Rubber rabbitbrush	Colorado to Oregon; Nevada, New Mexico	—	—	—	
Cowania stansburiana (Torrey)	Vanadium bush	Arizona, Idaho, Utah	—	—	—	Used as an indicator of vanadium–uranium deposits by prospectors in the Colorado Plateau. Able to grow in highly mineralized ground and to absorb large amounts of uranium
Dalea spinosa Gray	Smoketree, smokethorn	Southeastern California, southwestern Arizona	—	—	—	Its persistent occurrence in gravelly and sandy washes suggests it depends upon groundwater underflow and occurs as a phreatophyte
Dasiphora fruticosa Linnaeus	Bush or shrubby cinquefoil	Locally in Idaho but widespread in Oregon, Washington, Utah, Nevada and Arizona	Shallow	—	—	Occurs as a phreatophyte in Pahsimeroi Valley, Idaho. Grows on subalpine meadows, along streams, about cold springs in peaty, sandy, or clayey loams
Distichlis spicata (Linnaeus) Greene	Seashore saltgrass	Western United States	—	—	—	
stricta (Torrey) Rydberg	Saltgrass, or desert saltgrass	All Western States	2–14	1–3	—	
Elymus condensatus Presl	Giant wildrye	All Western States except New Mexico	1–12	1–2	—	Fair forage, Killed by overgrazing. Extensive root system
triticoides Buckley	Creeping wildrye	Western United States	—	1–2	—	Good forage. Frequently cut for hay. Associated with giant wild-rye along Humboldt River, Nev
Eragrostis obtusiflora (Fournier) Scribner	Mexican saltgrass, alkali lovegrass	Southeastern Arizona, southwestern New Mexico	4–15	2–3	—	Commonly locally in saline soil near Wilcox, AZ. Observed growing in Sulphur Springs Valley, AZ, where depth to water table was from 4–15 ft

(Continued)

Table 4L.82 (Continued)

Scientific Name	Common Name	Occurrence as a Phreatophyte	Relation to Groundwater: Depth to Water Below Land Surface (ft)	Relation to Groundwater: Quality	Relation to Groundwater: Use (acre feet/acre)	Remarks
Fraxinus velutina Torrey	Velvet ash, Arizona ash	Southwestern Utah, southern Nevada, California, Arizona, New Mexico, and western Texas	—	1	—	Prominent stream-bank and canyon tree; restricted to areas with a permanent groundwater supply. Popular as a shade tree in Arizona and California
Hedysarum boreale Nuttall	Sweet vetch	Colorado, Utah	—	—	—	Deep tap root. Identified as a phreatophyte in Colorado
Heliotropium curassavicum Linnaeus	Heliotrope, Chinese pusley	Southwestern Utah to southern California	Shallow	2–3	—	High tolerance for alkali. Occurs on moist saline soil
Hymenoclea monogyra Torrey and Gray	Burrobush	Western Texas to southern California	Shallow	1–2	—	Occurs largely along streams, washes, and in bottom lands; aggressive. Often forming thickets. Unpalatable to livestock
salsola Torrey and Gray	White burrobush	Utah to Arizona and California	—	—	—	Occurs in sandy desert
Juglans microcarpa Berlandier	Walnut, nogal, butternut	Arizona, New Mexico	2–20	1	—	Occurs along watercourses and washes; intolerant of shade. Deep tap root
Juncus balticus Willdenow	Wirerush, wiregrass	Western United States	—	—	7.8	Grows in wet sites where groundwater is shallow, also in shallow ponds. Appears to occur both as phreatophyte and hydrophyte. Deep root system. Fair to good forage
Juncus cooperi Engelmann	Desertrush	Southern Utah to California	—	2–3	—	Occurs on the margins of salt marshes and alkaline meadows, common in Death Valley, CA, along the edge of the playa often associated with saltgrass
Juniperus scopulorum?	Rocky Mountain juniper; locally "swampcedar."	Nevada	10	1–2	—	Occurs locally as a phreatophyte in White River and Spring Valleys, Nev. May be a hybrid between *J. scopulorum* and *J. utahensis.*
Leptochola fascicularis (Lamarck) A. Gray	Sprangletop	Western United States	—	1–3	—	Occurs along ditches and in moist waste places, often in brackish marshes; most places in alkali plains. Often invades rice fields
Medicago sativa Linnaeus	Alfalfa	Western United States	4+	1–2	—	
Phragmites communis Trinius	Reed, giant reed-grass, carrizo	Western United States	0–8±	1–2	—	Occurs also as a hydrophyte in the shallow water of streams, lakes, ponds, and marshes
Picea engelmanni Parry	Engelmann spruce	Mountain areas of western United States	—	1	—	Requires a good water supply and depends upon groundwater in many localities. Shallow root system

Platanus wrightil Watson	Arizona sycamore	Southern Arizona, southeastern and southwestern California, New Mexico	—	1	—	Common along stream and rocky canyons, in foothills and mountains, upper desert, desert grassland, and oak woodland zones. Valuable in erosion control
Pluchea sericea Coville	Arrowweed	Texas to southern Utah and southern California	0–10±	1–2	—	Occurs along streams and flood plains. Abundant along lower reaches of Colorado River and tributaries. Arrowweed may grow where depth to water is 25 ft
Populus	Cottonwood	Western United States	—	1–2	—	
tremuloides aurea Tidestorm	Quaking aspen	Mountainous areas of Western United States	—	1	—	Considered a phreatophyte when it grows along streams, around springs, and in other wet areas. Shallow root system
Prosopis juliflora (Swartz.)	Mesquite, honey mesquite	Southern Kansas to southeastern California and Mexico	—	1–2	—	Extensive root development. Reported to penetrate 60 ft below surface
velutina Wooton	Velvet mesquite	Southern Arizona	10+	1–2	3.3	Occurs in bottom lands. Extensive root development
pubescens Bentham	Screwbean mesquite, tornillo	Western Texas to southern Nevada and southern California	—	—	—	Characteristic of bottom lands along desert streams and water holes of Mojave and Colorado Deserts
Quercus agrifolia Nee	California live oak	California	35±	1	—	Occurrence related to depth to water table
lobata Nee	Roble oak	California	10–20	—	—	
Salicornia europaea Linnaeus	Glasswort	—	—	3	—	Frequently occurs in salt flats with salts approximately 1.0 percent of weight of soil
rubra Linnaeus	Glasswort	Colorado, New Mexico, Nevada, Utah	—	3	—	Some value as waterfowl feed. In Nevada occurs along edges of channels draining into playas
utahensis Tidestrom	Glasswort	Utah	—	3	—	Occurs on borders of salt lakes and alkaline places
Salix	Willow	Western United States	—	—	—	
Sambucus	Elder, elderberry	Western United States	—	1	—	Eleven species reported to grow in Western United States. Grows along streams, in canyons and in moist sites
Sarcobatus vermiculatus (Hook). Torrey	Big greasewood	Western United States	60±	—	—	
Sequoia gigantea (Lindley)	Giant or bigtree sequoia	California	—	1	—	Appears to prefer localities where a water table is within reach of its roots but will grow elsewhere
Sesuvium portulocostrum	Lowland purslane	Southern Nevada and California	—	3	—	Reported as a plant that grows on moist alkaline soils. Indicative of groundwater but usually of poor quality. Alkali resistant

(Continued)

Table 4L.82 (Continued)

Scientific Name	Common Name	Occurrence as a Phreatophyte	Relation to Groundwater			Remarks
			Depth to Water Below Land Surface (ft)	Quality	Use (acre feet/acre)	
verrucosum Rafinesque	Warty sesuvium, sea purslane	Southern Arizona, California, and New Mexico	Shallow	3	—	
Shepherdia	Buffalo berry	Arizona, New Mexico, Nevada, Oregon, Black Hills	—	1	—	Fruit edible. Grows in moist sites and along streams and river bottoms. One species reported growing as a phreatophyte in Big Smoky Valley, Nev. Occurs also as a phreatophyte in Mason Valley, Nev.
Sporobolus airoides Torrey	Alkali sacaton	Western United States	5–25±	1–3	3.7	Most common in the Southwest where it is important as forage; deep, coarse root system. Prefers moist alkali flats. Grows in very saline or saline-alkali soils. Soil salinity may range from 0.3 to 3.0 percent. Grows best in range 0.3 to 0.5 percent
wrightii Munro	Sacaton	Arizona to western Texas	—	1–2	—	Occurs in alluvial flats and bottom lands. Will not grow on highly alkaline soils
Suaeda depressa Watson	Seepweed, saltwort	Southwest	—	3	—	Browsed when other forage is scarce. Occurs on saline or saline-alkali soils with salt content in first foot as much as 3.2 percent
Suaeda suffrutescens Watson	Desert seepweed	Western Texas, New Mexico, Arizona	—	3	—	Browsed when other forage is scarce. Occurs on saline or saline-alkali soils with salt content in first foot as much as 3.2 percent
torreyana Watson	Torrey seepweed, iodineweed, inkweed	Eastern Oregon to New Mexico and California	4–15	3	—	Browsed when other forage is scarce. Occurs on saline or saline-alkali soils with salt content in first foot as much as 3.2 percent
Tamarix aphylla Linnaeus	Athel tree	Southwest	—	1–3	—	
Gallica Linnaeus	Saltcedar, French tamarisk	Southwest	—	1–3	—	
Washingtonia filifera Wendland	Fan palm, California palm	Southern Arizona, California, SE New Mexico	—	1–3	—	Highly tolerant to alkali. Generally grows where groundwater is at shallow depth

Note: The quality of the groundwater with respect to its suitability for crop growth is indicated by numerals as follows: 1 = excellent to good; 2 = good to poor; 3 = poor to unsatisfactory. The use of groundwater, including precipitation, unless otherwise stated is presumed to be for a plant growth of 100-percent volume density.

Source: U.S. Geological Survey.

CHAPTER **5**

Surface Water

Christopher Spooner

CONTENTS

SECTION 5A RIVERS

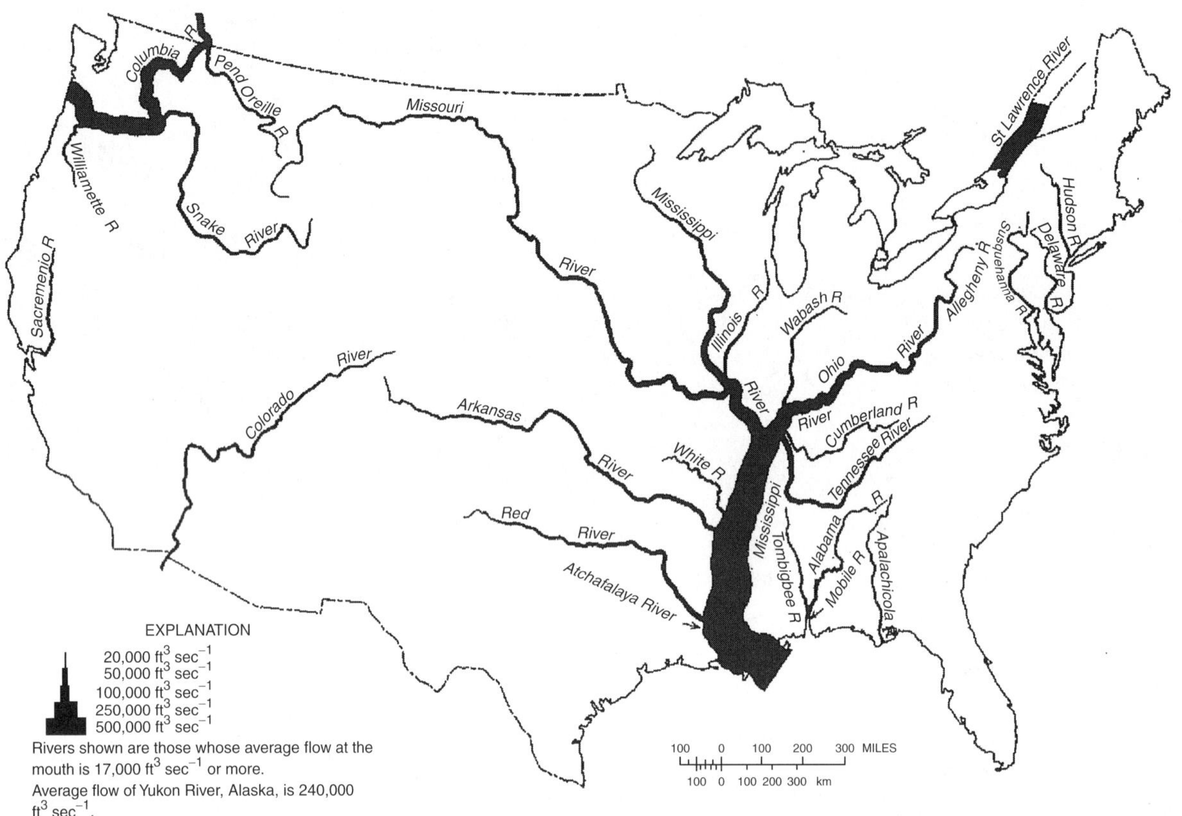

Figure 5A.1 Large rivers in the United States. (From Iseri, K.T., and W.B. Langbein, *Large Rivers of the United States*, U.S. Geol. Survey Circular 686, 1974.)

Table 5A.1 Average Discharge at Downstream Gaging Stations on Large Rivers of the United States, 1931–1960, and 1941–1970

River	Gaging-Station Location	Drainage Area (Square Miles)	Average Discharge (1931–1960) (ft^3/sec)	Average Discharge (1941–1970) (ft^3/sec)
Alabama	At Claiborne, AL	22,000	31,140	31,510
Allegheny	At Natrona, PA	11,410	19,200	18,810
Apalachicola	At Chattahoochee, FL	17,200	20,700	21,700
Arkansas	At Little Rock, AR	158,000	41,300	42,130
Atchafalaya[a]	At Krotz Springs, LA	93,320	160,800[b]	180,800[b]
Colorado	Below Hoover Dam, AZ–NV[c]	167,800	14,580[d]	14,530
Columbia	At The Dalles, OR	237,000	183,000	189,000
Cumberland	Near Grand Rivers, KY	17,598	26,900	28,030[e]
Delaware	At Trenton, NJ[f]	9,397	16,100	14,500[g]
Hudson	At Green Island, NY	8,090	—	12,520[h]
Illinois	At Merdosia, IL	25,300	20,500	20,670
Mississippi	At Alton, IL	171,500	91,300	98,300
Mississippi	At Vicksburg, MS	1,144,500	554,000	565,300
Missouri	At Hermann, MO	528,200	69,200	76,200
Ohio	At Metropolis, IL	203,000	257,000	257,200
Pend Oreille	At international boundary	25,200	26,900	28,420
Red	At Alexandria, LA	67,412	32,470	32,100
Sacramento	At Verona, CA[i]	—	25,700	27,200
St. Lawrence	At Cornwall, Ontario-near Massena, NY[j]	299,000	233,000[k]	239,000[k]
Snake	Near Clarkston, WA	103,200	48,600	48,960
Susquehanna	At Marietta, PA	25,990	36,100	35,060
Tennessee	Near Paducah, KY	40,200	63,400	64,050[f]
Tombigbee	At Jackson Lock and Dam near Coffeeville, AL	18,500	25,200	25,130
Wabash	At Mount Carmel, IL	28,600	26,400	26,600
White	At Clarendon, AR	25,497	29,490	29,360
Willamette	At Salem, OR	7,280	23,870	24,780
Yukon	At Ruby, AK	259,000	—	170,000[l]

[a] Continuation of Red River.
[b] Includes diversion from Mississippi River through Old River or Old River diversion channel.
[c] Very little tributary flow downstream. Downstream station located at Yuma, AZ., drainage area 242,900 square miles. The greater part of the natural flow is diverted for irrigation and other uses in the basin above Yuma. Average flow at Yuma, 1963–1970, is less than 1,000 ft^3sec^{-1}.
[d] For the period 1934–1960.
[e] Interbasin diversion beginning June 1966 between Lake Barkley on Cumberland River and Lake Kentucky on Tennessee River through Barkley–Kentucky Canal.
[f] Five tributaries below Trenton have been added.
[g] Unadjusted for diversion by New York City reservoirs since 1954.
[h] October 1946 to September 1970 (24 years).
[i] American River and Yolo bypass have been added.
[j] Formerly at Ogdensburg, NY
[k] Furnished by the U.S. Army Corps of Engineers through International St. Lawrence River Board of Control.
[l] Average is for 1957–1970; station operated only since 1956.

Source: From Iseri, K.T., and W.B. Langbein, 1974, *Large Rivers of the United States*, U.S. Geol. Survey Circular 686.

Table 5A.2 Flow of Selected Streams in the United States

	Gaging Station			Streamflow Characteristics		
	Name	Drainage Area (mL^2)	Period of Analysis	7-Day, 10-Year Low Flow (ft^3/sec)	Average Discharge (ft^3/sec)	100-Year Flood (ft^3/sec)
	Alabama					
	South Atlantic–Gulf Region Choctawhatcheee-Wseambia Subregion					
1.	Choctawhatchee River, Newton	686	1923–1926	88	983	40,900
			1937–1983			
2.	Conecuh River, Brantley	500	1937–1983	31	680	27,300
	Alabama Subregion					
3.	Coosa River, Childersburg	8,392	1915–1968	2,000	13,860	157,600
			1969–1978	1,330	13,860	144,900
4.	Tallapoosa River, Wadley	1,675	1923–1983	140	2,594	73,800
5.	Alabama River, Montgomery	15,087	1927–1968	5,240	24,260	317,000
			1969–1983	3,860	24,260	219,500
	Mobile–Tombigbee Subregion					
6.	Cahaba River, Centreville	1,027	1902–1907	143	1,633	117,000
			1931			
			1937–1983			
7.	Mulberry Fork, Garden City	365	1928–1983	4.9	681	51,300
8.	Black Warrior River, Northport	4,820	1895–1902	90	8,041	221,000
			1929–1960			
			1961–1983	504	8,041	305,400
9.	Tombigbee River, Coatopa	15,385	1928–1983	685	23,500	
	Tennessee Region					
	Middle Tennessee-Elk Subregion					
10.	Flint River, Chase	342	1930–1983	66	554	75,200
11.	Tennessee River, Florence	30,810	1984–1983	7,490	51,900	—
	Alaska					
	Alaska Region					
	Southeast Alaska Subregion					
1.	Stikine River, Wrangell	19,920	1976–1983	4,500[a]	56,674	299,600[a]
2.	Fish Creek, Ketchikan	32.1	1915–1936[b]	31	421	5,420
			1938–1983			
	South-Central Alaska Subregion					
3.	Copper River, Chitina	20,600	1955–1983	3,040	37,670	321,000
4.	Susitna River, Gold Creek	6,160	1949–1983	723	9,724	115,000
5.	Susitna River, Susitna Station	19,400	1974–1983	5,000[a]	49,940	230,000[a]
	Southwest Alaska Subregion					
6.	Kvichak River, Igiugig	6,500	1967–1983	7,380	18,060	66,500
7.	Nuyakuk River, Dillingham	1,490	1953–1983	1,100	6,156	36,200
8.	Nushagak River, Ekwok	9,850	1977–1983	6,000[a]	23,840	89,200[a]
9.	Kuskokwim River, Crooked Creek	31,100	1951–1983	7,850	41,220	445,000
	Yukon Subregion					
10.	Yukon, River, Eagle	113,500	1911–1913[b]	10,500	82,660	605,000
			1950–1983			
11.	Porcupine River, Fort Yukon	29,500	1964–1979	6[b]	14,230	476,000
12.	Chena River, Fairbanks	1,980	1948–1983	150	1,384	38,800[c]
13.	Tanana River, Nenana	25,600	1962–1983	4,740	23,550	153,000[d]
14.	Koyukuk River, Hughes	18,700	1960–1982	267	14,540	332,000[e]
15.	Yukon River, Pilot Station	321,000	1975–1983	37,000[a]	219,600	751,000[a]
	Northwest Alaska Subregion					
16.	Kobuk River, Kiana	9,520	1976–1983	1,300[a]	15,270	152,000[a]
	Arctic Subregion					
17.	Kaparuk River, Deadhorse	3,130	1971–1983	No flow	1,367	218,000
	Arizona					
	Lower Colorado River Basin					
1.	Colorado River, Lees Ferry	111,800	1912–1962	1,670	17,850	189,500
			1965–1984	—	—	—
2.	Colorado River, below Hoover Dam	171,700	1935–1984	2,550	13,590	—

(Continued)

Table 5A.2 (Continued)

	Gaging Station			Streamflow Characteristics		
	Name	Drainage Area (mL^2)	Period of Analysis	7-Day, 10-Year Low Flow (ft^3/sec)	Average Discharge (ft^3/sec)	100-Year Flood (ft^3/sec)
3.	Bill Williams River, below Alamo Dam	4,730	1940–1968	0.72	92.3	325,000
4.	Colorado River, above Morelos Dam	246,700	1950–1984	541	—	—
	Little Colorado Subregion					
5.	Little Colorado River, Cameron	26,500	1974–1984	—	244	32,800
	Upper Gila Subregion					
6.	Gila River, Clifton	4,010	1928–1984	8.15	192	30,600
7.	Gila River, Solomon	7,896	1914–1984	22.0	468	86,800
	Middle Gila Subregion					
8.	San Pedro River, Palominas	741	1950–1981	0.03	32.1	21,800
9.	San Pedro River, Winkelman	4,471	1966–1979	—	57.1	—
10.	Gila River, Kelvin	18,011	1912–1984	0.82	494	244,000
11.	Santa Cruz River, Tucson	2,222	1915–1981	—	22.7	20,300
	Salt Subregion					
12.	Black River, Fort Apache	1,232	1958–1984	16.7	412	56,100
13.	White River, Fort Apache	632	1958–1984	4.80	201	11,900
14.	Salt River, Roosevelt	4,306	1925–1984	81.9	888	164,000
15.	Verde River, above Horseshoe Dam	5,872	1945–1984	72.5	564	158,000
	Arkansas					
	Lower Mississippi Region					
	Mississippi River Main Stem					
1.	Mississippi River, Memphis, TN	932,800	1933–1981	99,000	474,200	1,860,000
	Lower Mississippi–St. Francis Subregion					
	St. Francis River Basin					
2.	St. Francis Bay, Riverfront	—	1936–1975	57	5,274	—
			1978–1981			
			1944–1975	83	—	—
			1978–1981			
	Lower Red–Ouachita Subregion					
3.	Ouachita River, Malvern	1,585	1928–1984	105	2,380	194,000
			1954–1984	244		
4.	Ouachita River, Camden	5,357	1928–1984	236	7,490	299,000
			1954–1984	548	—	—
5.	Smackover Creek, Smackover	385	1961–1983	0.35	374	39,700
6.	Saline River, Rye	2,102	1937–1983	12.6	2,590	102,000
7.	Bayou Bartholomew, McGehee	576	1939–1942	6.5	676	6,930
			1946–1984			
	Arkansas–White–Red Region					
	Upper White Subregion					
	White River Basin					
8.	Buffalo River, St. Joe	829	1939–1984	16.5	1,027	176,000
9.	White River, Calico Rock	9,978	1939–1983	894	9,830	352,000
			1945–1983	973	—	—
			1958–1983	1,120	—	—
10.	Spring River, Imboden	1,183	1936–1983	279	1,360	163,000
11.	Black River, Black Rock	7,369	1929–1931	1,980	8,410	176,000
			1939–1983			
			1950–1983	1,990	—	—
12.	Middle Fork Little Red River, Shirley	302	1939–1983	<0.19	467	140,000
13.	White River, Clarendon	25,555	1928–1981	4,090	29,510	291,000
			1945–1981	5,050	—	—
			1958–1981	6,020	—	—
	Lower Arkansas Subregion					
	Arkansas River Basin					
14.	Poteau River, Cauthron	203	1939–1983	<0.1	215	47,100
			1950–1983	<0.1	—	—

(Continued)

Table 5A.2 (Continued)

	Gaging Station			Streamflow Characteristics		
	Name	Drainage Area (mL^2)	Period of Analysis	7-Day, 10-Year Low Flow (ft^3/sec)	Average Discharge (ft^3/sec)	100-Year Flood (ft^3/sec)
15.	Mulberry River, Mulberry	373	1938–1983	<0.16	534	82,400
16.	Big Piney Creek, Dover	274	1950–1983	0.15	398	112,000
17.	Petit Jean River, Danville	764	1916–1984	0.74	809	91,900
			1949–1984	1.9	—	—
18.	Arkansas River, Murray Dam	158,030	1927–1984	1,230	40,290	588,000
		Red–Sulphur Subregion				
		Red River Basin				
19.	Red River, Index	48,030	1936–1984	812	11,170	190,000
			1945–1984	934	—	—
			1969–1984	1,110	—	—
		California				
		California Region				
		Sacramento Subregion				
1.	Feather River, Nicolaus	5,921	1944–1969	169	7,957	521,000
			1970–1983	1,061	9,424	332,000
2.	Sacramento River, Verona	21,251	1930–1969	1,618	18,240	77,700[f]
			1970–1983	5,732	22,680	94,700[f]
3.	American River Fair Oaks	1,888	1906–1955	64	3,735	257,000
			1956–1983	426	3,942	150,000
		Tulare–Buena Vista Lakes and San Joaquin Subregions				
4.	Kern River, Kernville	846	1912–1984	104	762	45,800
5.	Kings River, Trimmer	1,342	1953–1983	111	2,177	135,000
6.	Merced River, Stevinson	1,273	1941–1983	52	733	14,400
7.	San Joaquin River, Vernalis	13,536	1930–1983	241	4,783	99,900
		Southern California Coastal Subregion				
8.	San Diego River, Santee	377	1914–1943	0.1	42.3	54,900
			1944–1982	1.0	13.7	5,400
9.	Santa Margarita River, Ysidora	740	1924–1948	0	43.3	46,000
			1949–1983	0	31.0	32,000
10.	Santa Ana River, Santa Ana	1,700	1942–1984	0	52.8	33,800
11.	Los Angeles River, at Long Beach	827	1930–1940	0.1	110	192,000
			1941–1982	3.8	222	118,000
12.	Santa Clara River, Los Angeles–Ventura County Line	625	1953–1971	0.1	36.2	161,000
			1972–1984	2.9	67.8	58,500
		Central California Coastal Subregion				
13.	Salinas River, Spreckels	4,156	1930–1941	0.1	659	145,000[g]
			1942–1965	0.5	262	145,000[g]
			1966–1984	0.6	590	145,000[g]
14.	San Lorenzo River, Big Trees	106	1937–1984	9.2	140	39,600
		Klamath–Northern California Coastal Subregion				
15.	Russian River, Guerneville	1,338	1940–1958	77	2,230	108,000
			1959–1983	40	2,435	93,400
16.	Eel River, Scotia	3,113	1911–1984	43	7,412	608,000
17.	Klamath River, Klamath	12,100	1911–1984	1,859	18,110	556,000
18.	Smith River, Crescent City	609	1932–1984	191	3,891	231,000
		Great Basin Region				
		Central Lahontan Subregion				
19.	Truckee River, Tahoe City	507	1910–1984	2.4	240	2,830
		Colorado				
		Missouri Region				
		North and South Platte Subregions				
1.	North Platte River, Northgate	1,431	1915–1984	35	440	7,870
2.	South Platte River, Hartsel	880	1933–1984	3.3	79.1	2,410
3.	South Platte River, Kersey	9,598	1901–1984	51	834	40,400
4.	South Platte River, Julesburg	23,138	1902–1984	7.6	524	62,300

(Continued)

Table 5A.2 (Continued)

	Gaging Station			Streamflow Characteristics		
	Name	Drainage Area (mL^2)	Period of Analysis	7-Day, 10-Year Low Flow (ft^3/sec)	Average Discharge (ft^3/sec)	100-Year Flood (ft^3/sec)
		Arkansas–White–Red Region				
		Upper Arkansas Subregion				
5.	Arkansas River, Canon City	3,117	1888–1981	129	715	14,300
6.	Arkansas River, La Junta	12,210	1912–1973	4.8	244	96,300
			1974–1984	3.8	233	19,300
7.	Purgatoire River, Trinidad	795	1895–1976	2.7	83.3	34,400
			1977–1981	—	64.3	—
8.	Purgatoire River, Las Animas	3,503	1922–1931	0.34	116	94,000
			1948–1976			
			1977–1984	—	81.0	—
9.	Arkansas River Lamar	19,780	1913–1942	1.1	301	131,000
			1948–1984	0.63	93.6	35,500
		Rio Grande Region				
		Rio Grande Headwaters Subregion				
10.	Rio Grande, Del Norte	1,320	1889–1984	107	901	13,400
11.	Rio Grande, Lobatos	7,700	1899–1984	7.1	575	19,900
		Upper Colorado Region				
		Colorado Headwaters Subregion				
12.	Colorado River, near Dotsero	4,394	1940–1984	536	2,136	23,800
13.	Colorado River, Cameo	8,050	1933–1984	997	3,900	41,900
		Gunnison Subregion				
14.	Gunnison River, Gunnison	1,012	1940–1928	148	888	11,500
			1944–1984	115	709	9,000
15.	Gunnison River Grand Junction	7,928	1896–1965	265	2,611	38,100
			1968–1984	495	2,659	30,500
		White–Yampa Subregion				
16.	Yampa River, Maybell	3,410	1916–1984	39	1,573	19,900
17.	White River, Meeker	755	1901–1984	179	626	6,570
		San Juan Subregion				
18.	Animas River, Durango	692	1912–1984	128	819	15,500
		Connecticut				
		New England Region				
		Connecticut Subregion				
		Connecticut River Basin				
1.	Connecticut River, Thompsonville	9,661	1928–1983	2,200	16,400	209,000
2.	Burlington Brook, Burlington	4.10	1931–1983	0.7	8.3	1,250
3.	Farmington River, Rainbow	590	1928–1960	144	1,030	44,000
			1961–1983	101	1,040	24,000
4.	Salmon Silver, East Hampton	100	1928–1983	5.2	184	16,600
		Connecticut Coastal Subregion				
		Thames River Basin				
5.	Mount Hope River, Warrenville	28.6	1940–1983	0.9	51.2	5,620
6.	Shetucket River, Willimantic	404	1928–1952	46.5	667	25,000
			1953–1983	44.2	734	22,500
7.	Quinebaug River, Jewett City	713	1918–1958	119	1,250	29,500
			1959–1983	90.0	1,330	26,500
8.	Yantic River, Yantic	89.3	1930–1983	5.2	165	10,800
		Quinnipiac River Basin				
9.	Quinnipiac River, Wallingford	115	1930–1983	32.6	211	6,340
		Housatonic River Basin				
10.	Housatonic River, Falls Village	634	1912–1983	119	1,090	24,000
11.	Shepaug River, Roxbury	132	1930–1971	6.2	236	24,000
12.	Pomperaug River, Southbury	75.1	1932–1983	6.0	128	19,900
13.	Housatonic River, Stevenson	1,544	1928–1983	160	2,600	95,100
14.	Naugatuck River, Beacon Falls	260	1928–1959	61.2	484	46,000
			1960–1983	59.4	557	23,000

(Continued)

Table 5A.2 (Continued)

	Gaging Station			Streamflow Characteristics		
	Name	Drainage Area (mL^2)	Period of Analysis	7-Day, 10-Year Low Flow (ft^3/sec)	Average Discharge (ft^3/sec)	100-Year Flood (ft^3/sec)
		Saugatuck River Basin				
15.	Saugatuck River, Westport	79.8	1932–1967	2.25	119	13,400
		Delaware				
		Mid-Atlantic Region				
		Delaware Subregion				
		Christina River Basin				
1.	Christina River, Coochs Bridge	20.5	1943–1984	1.5	28.8	4,840
2.	Brandywine Creek, Wilmington	314	1946–1984	75	488	34,300
		Upper Chesapeake Subregion				
		Indian River Basin				
3.	Stockley Branch, Stockley	5.24	1943–1984	0.66	7.04	200
		Nanticoke River Basin				
4.	Nanticoke River, Bridgeville	75.4	1943–1984	15	92.8	3,570
		Florida				
		South Atlantic–Gulf Region				
		Altamaha–St. Marys Subregion				
1.	St. Marys River, Macclenny	700	1927–1983	18	672	40,500
		St. Johns Subregion				
2.	St. Johns River, Christmas	1,539	1934–1983	24	1,310	18,500
3.	St. Johns River, DeLand	3,066	1934–1983	0	3,120	21,900
4.	Oklawaha River,	2,747	1944–1968	788	2,020	12,900
	Rodman Dam		1969–1983	—	1,550	—
		Southern Florida Subregion				
5.	Fisheating Creek, Palmdale	311	1932–1983	0	257	21,400
6.	Kissimmee River,	2,886	1929–1962	809	2,190	29,800
	S-65E near Okeechobee		1964–1983	36	1,390	
		Peace–Tampa Bay Subregion				
7.	Peace River, Acradia	1,367	1932–1983	57	1,150	34,400
8.	Hillsborough River, Zephyrhills	220	1940–1983	53	259	10,300
9.	Withlacoochee River, Holder	1,825	1932–1983	158	1,090	9,750
		Suwannee Subregion				
10.	Suwannee River, Branford	7,880	1932–1983	1,790	6,940	68,000
11.	Santa Fe River, Fort White	1,017	1928–1929	730	1,610	16,400
			1933–1983			
12.	Suwannee River, Wilcox	9,640	1931	4,020	10,400	66,400
			1942–1983			
		Ochlockonee Subregion				
13.	Ochlockonee River, Havana	1,140	1927–1983	30	1,030	41,200
		Apalachicola Subregion				
14.	Apalachicola River, Chattahoochee	17,200	1929–1983	7,000	22,400	264,000
		Choctawhatchee–Escambia Subregion				
15.	Choctawhatchee River, Bruce	4,384	1931–1983	1,630	7,140	128,000
16.	Yellow River, Milligan	624	1939–1983	184	1,170	45,900
17.	Shoal River, Crestview	474	1939–1983	291	1,100	33,600
18.	Escambia River, Century	3,817	1935–1983	777	6,360	179,000
19.	Perdido River, Barrineau Park	394	1942–1983	221	766	34,200
		Georgia				
		South Atlantic–Gulf Region				
		Ogeechee–Savannah Subregion				
1.	Broad River, Bell	1,430	1927–1932	200	1,809	60,400
			1937			—
2.	Savannah River, Augusta	7,508	1960–1981	5,500	10,200	—
		Altamaha–St. Marys Subregion				
3.	Oconee River, Greensboro	1,090	1903–1932	150	1,446	50,700
			1936–1978			
4.	Altamaha River, Doctortown	13,600	1931–1983	2,250	13,770	225,000

(Continued)

Table 5A.2 (Continued)

	Gaging Station			Streamflow Characteristics		
	Name	Drainage Area (mL^2)	Period of Analysis	7-Day, 10-Year Low Flow (ft^3/sec)	Average Discharge (ft^3/sec)	100-Year Flood (ft^3/sec)
5.	Penholoway Creek, Jesup	210	1958–1983	0	201	7,180
		Suwannee Subregion				
6.	Alapaha River, Statenville	1,400	1931–1983	25	1,044	24,200
		Apalachicola Subregion				
7.	Chattahoochee River, Atlanta	1,450	1959–1981	860	2,840	—
8.	Flint River, Culloden	1,850	1911–1923	180	2,402	99,100
			1928–1931			
			1937–1983			
9.	Flint River, Albany	5,310	1901–1921	1,000	6,303	94,600
			1929–1983			
		Alabama Subregion				
10.	Etowah River, Allatoona Dam	1,120	1950–1981	240	1,944	—
		Tennessee Region				
		Middle Tennessee–Hiawassee Subregion				
11.	Toccoa River, Dial	177	1912–1983	125	498	16,600
		Hawaii				
		Hawaii Region				
		Kauai Subregion				
1.	East Branch of North Fork Wailua River near Lihue	6.27	1916–1983	10.4	48.6	10,400
		Oahu Subregion				
2.	Kalihi Stream, near Honolulu	2.61	1917–1983	0.29	6.74	10,400
		Maui Subregion				
3.	Honopou Stream near Huolo	0.64	1911–1983	0.26	4.69	4,410
		Hawaii Subregion				
4.	Waiakea Stream near Mountain View	17.4	1931–1983	0.10	11.8	1,140
		Idaho				
		Great Basin Region				
		Bear Subregion				
		Bear River Basin				
1.	Bear River, Preston	4,545	1944–1984	80	937	8,190
		Pacific Northwest Region				
		Kootenai–Pond Oreille–Spokane Subregion				
		Pend Oreille River Basin				
2.	Priest River, Priest River	902	1904	200	1,686	11,500
			1930–1984			
		Spokane River Basin				
3.	Spokane River, Post Falls	3,340	1913–1984	180	6,297	46,000
		Upper Snake Subregion				
4.	Snake River, Irwin	5,225	1950–1984	560	6,691	31,700
5.	Henrys Fork, Rexburg	2,920	1910–1984	400	2,088	12,100
6.	Portneuf River, Pocatello	1,250	1913–1916	14	280	2,650
			1918–1984			
7.	Snake River, Milner	17,180	1910–1926	5	2,711	42,400
			1927–1984	—	—	28,300
8.	Big Lost River, below Mackay Reservoir, Mackay	813	1905	36	314	3,280
			1913–1914			
			1920–1984			
9.	Big Wood River below Magic Dam, Richfield	1,600	1913–1984	2	480	10,400
		Middle Snake Subregion				
10.	Snake River, King Hill	35,800	1910–1926	6,000	10,910	54,600
			1927–1984	—	—	39,100
11.	Bruneau River, Hot Spring	2,630	1904–1914	47	409	7,500
			1943–1984			

(Continued)

Table 5A.2 (Continued)

	Gaging Station			Streamflow Characteristics		
	Name	Drainage Area (mL^2)	Period of Analysis	7-Day, 10-Year Low Flow (ft^3/sec)	Average Discharge (ft^3/sec)	100-Year Flood (ft^3/sec)
12.	Boise River, Boise	2,680	1953–1984	1	2,951	10,000
13.	Payette River, Payette	3,240	1936–1984	400	3,183	10,000
14.	Weiser River, Weiser	1,460	1953–1984	54	1,132	26,000
15.	Snake River, Weiser	69,200	1911–1984	6,600	18,490	10,000
	Lower Snake Subregion					
16.	Salmon River, White Bird	13,550	1911–1917	2,400	11,420	126,000
			1920–1984			
17.	Clearwater River, Spalding	9,570	1910–1913	1,500	15,550	188,000
			1925–1984			
	Illinois					
	Upper Mississippi Region					
	Upper and Lower Illinois Subregions					
	Illinois River Main Stem					
1.	Illinois River, Marseilles	8,259	1919–1983	—	—	91,100
			1940–1983	3,180	9,791	—
2.	Illinois River, Meredosia	26,028	1921–1983	—	—	132,300
			1940–1983	3,630	21,976	—
	Illinois River Basin–Tributaries					
3.	Kankakee River, Wilmington	5,150	1915–1983	463	4,233	68,100
4.	Des Plaines, River Riverside	630	1914–1983	—	—	7,830
			1943–1983	6.0	471	—
			1974–1983	48	—	—
5.	Fox River, Dayton	2,642	1915–1983	176	1,703	37,400
			1974–1983	366	—	—
6.	Vermillion River, Leonore	1,251	1931–1983	—	822	40,700
			1973–1983	9.6	—	—
7.	Mackinaw River, Congerville	767	1945–1983	1.3	511	43,900
8.	Spoon River, Seville	1,636	1914–1983	20	1,054	37,600
9.	Sangamon River, Oakford	5,093	1910–1983	147	3,335	82,800
			1974–1983	263	—	—
10.	La Moine River, Ripley	1,293	1921–1983	10	802	27,500
	Rock Subregion					
	Rock River Basin					
11.	Pecatonica River, Freeport	1,326	1914–1983	191	900	21,300
12.	Kishwaukee River, Perryville	1,099	1940–1983	68	713	25,000
13.	Rock River, Joslin	9,549	1940–1983	1,270	6,020	58,800
14.	Green River, Geneseo	1,003	1936–1983	40	610	13,000
	Upper Mississippi–Kaskaskia–Meramec Subregion					
	Kaskaskia and Big Muddy River Basins					
15.	Kaskaskia River, Vandalia	1,940	1908–1969	14	1,412	33,000
			1970–1983	34	1,769	30,400
16.	Big Muddy River, Murphysboro	2,169	1916–1970	—	—	39,300
			1931–1970	2.3	1,788	—
			1971–1983	47	1,888	41,000
	Ohio Region					
	Wabash and Lower Ohio Subregions					
	Embarras and Little Wabash River Basins					
17.	Embarras River, Ste. Marie	1,516	1910–1983	14	1,224	53,700
18.	Little Wabash, River, Carmi	3,102	1940–1983	6.2	2,529	45,300
	Indiana					
	Ohio Region					
	Great Miami Subregion					
	Whitewater River Basin					
1.	Whitewater River, Alpine	522	1928–1983	48	551	49,000
	Wabash Subregion					
	Wabash River Main Stem–White River Basin–Patoka River Basin					
2.	Muscatatuck River, Deputy	293	1947–1983	0.0	348	41,200

(Continued)

Table 5A.2 (Continued)

	Gaging Station			Streamflow Characteristics		
	Name	Drainage Area (mL^2)	Period of Analysis	7-Day, 10-Year Low Flow (ft^3/sec)	Average Discharge (ft^3/sec)	100-Year Flood (ft^3/sec)
3.	South Fork Patoka River, Spurgeon	42.8	1964–1983	2.2	51.9	5,990
4.	Eagle Creek, Indianapolis	174	1938–1968	0.5	148	18,400
			1969–1983	6.0	168	11,800
5.	Driftwood River, Edinburgh	1,060	1940–1983	91	1,144	49,500
6.	Wabash River, Peru	2,686	1943–1967	92	2,290	74,300
			1970–1983	155	2,500	31,000
7.	Wabash River, Mount Carmel, IL	28,635	1927–1983	2,280	27,440	315,000
	Upper Mississippi Region					
	Upper Illinois Subregion					
	Kankakee River Basin					
8.	Kankakee River, Shelby	1,779	1922–1983	417	1,619	6,950
9.	Iroquois River, Foresman	449	1948–1983	11	383	5,660
	Great Lakes Region					
	Southestern Lake Michigan Subregion					
	St. Joseph River Basin					
10.	St. Joseph River, Elkhart	3,370	1947–1983	818	3,177	21,500
11.	Pigeon Creek Angola	106	1947–1983	5.8	78.5	843
	Western Lake Erie Subregion					
	Maumee River Basin					
12.	Muamee River, New Haven	1,967	1956–1983	72	1,645	25,600
	Iowa					
	Upper Mississippi Region					
	Mississippi River Main Stem					
1.	Mississippi River, Clinton	85,600	1873–1983	10,050	47,390	295,000[h]
	Northeast Iowa River Basin[i]					
2.	Upper Iowa River, Decorah	511	1951–1983	32	327	22,400
3.	Turkey River, Garber	1,545	1913–1916	81	949	33,100
			1919–1927			
			1929–1930			
			1932–1983			
4.	Maquoketa River, Maquoketa	1,553	1913–1983	160	1,027	47,700
5.	Wapsipinicon River, De Witt	2,330	1934–1983	98	1,537	31,600
	Iowa–Cedar River Basin[j]					
6.	Iowa River, Iowa City	3,271	1903–1958	60	1,470	43,700
			1959–1983	93	2,180	17,400
7.	English River, Kalona	573	1939–1983	2.3	370	25,300
8.	Shell Rock River, Shell Rock	1,746	1953–1983	64	974	42,800
9.	Cedar River, Waterloo	5,146	1940–1983	284	2,984	98,900
10.	Cedar River, Cedar Rapids	6,510	1902–1983	347	3,414	83,500
11.	Iowa River, Wapello	12,499	1914–1958	555	5,950	102,000
			1959–1983	893	8,650	116,000
	Skunk River Basin[j]					
12.	South Skunk River, Oskaloosa	1,635	1945–1983	10	916	25,800
13.	North Skunk River, Sigourney	730	1945–1983	2.3	436	27,400
14.	Skunk River, Augusta	4,303	1914–1983	31	2,407	55,200
	Des Moines River Basin[k]					
15.	Des Moines River, Stratford	5,452	1920–1983	40	1,882	54,600
16.	North Raccoon River, Jefferson	1,619	1940–1983	8.3	708	27,200
17.	South Raccoon River, Redfield	988	1940–1983	26	449	32,200
18.	Raccoon River, Van Meter	3,441	1915–1983	34	1,346	46,500
19.	Des Moines River, Keosauqua	14,038	1903–1906	143	5,160	123,000
			1911–1968			
			1969–1983	245	7,860	90,600
	Missouri Region					
	Missouri River Main Stem[l]					
20.	Missouri River, Sioux City	314,600	1897–1956[m]	3,810	30,000	437,000
			1957–1983	6,570	28,700	144,500[n]

(Continued)

Table 5A.2 (Continued)

	Gaging Station			Streamflow Characteristics		
	Name	Drainage Area (mL^2)	Period of Analysis	7-Day, 10-Year Low Flow (ft^3/sec)	Average Discharge (ft^3/sec)	100-Year Flood (ft^3/sec)
	Western Iowa River Basin[o]					
21.	Big Sioux River, Akron	9,030	1928–1983	19	901	71,000
22.	Floyd River, James	886	1934–1983	2.7	197	34,300
23.	Little Sioux River, Correctionville	2,500	1918–1925	14	766	32,600
			1928–1932			
			1936–1983			
24.	Boyer River, Logan	871	1918–1925	6.5	315	31,800
			1937–1983			
	Southern Iowa River Basin[p]					
25.	Nishnabotna River, Hamburg	2,806	1922–1923	28	1,057	40,700
			1928–1983			
26.	Nodaway River, Clarinda	762	1918–1925	5.8	338	37,900
			1936–1983			
27.	Thompson River, Davis City	701	1918–1925	1.6	370	25,500
			1941–1983			
28.	Chariton River, Rathbun	549	1956–1969	0.25	303	40,327
			1970–1983	4.0	382	2,130
	Kansas					
	Missouri Region					
	Republican and Smoky Hill Subregions					
	Republican and Smoky Hill River Basins					
1.	Republic River, Clay Center	24,542	1917–1983	75[q]	990	76,000[q]
2.	Smoky Hill River, Elkader	3,555	1940–1983	0.0	30	70,000
3.	Solomon River, Niles	6,770	1897–1903	33[q]	550	51,000[r]
			1917–1983			
4.	Smoky Hill River Enterprise	19,260	1935–1983	120[q]	1,600	85,000[r]
	Kansas, Gasconade–Osage, and Missouri–Nishnabotna Subregions					
	Kansas, Osage, and Missouri River Basins					
5.	Kansas River, Fort Riley	44,870	1964–1983	240[q]	2,600	140,000[r]
6.	Big Blue River, Manhattan	9,640	1955–1983	18[q]	2,000	50,000[r]
7.	Kansas River, De Soto	59,756	1917–1983	800[q]	7,000	300,000[r]
8.	Marais des Cygnes River, Kansas–Missouri State line	3,230	1959–1983	2.5[q]	2,000	67,000[q]
9.	Missouri River, St. Joseph, MO	420,300	1929–1983	6,100[q]	42,000	—
	Arkansas–White–Red Regions					
	Middle Arkansas, Upper Cimarron, and Arkansas–Keystone Subregions					
	Arkansas River Basin					
10.	Arkansas River, Syracuse	25,763	1902–1906	0.3[q]	310	130,000[q]
			1921–1983			
11.	Little Arkansas River, Valley Center	1,327	1922–1983	10	280	43,000
12.	Arkansas River, Arkansas City	43,713	1902–1906	170[q]	1,800	99,000[r]
			1922–1983			
	Middle Arkansas and Neosho–Verdigris Subregions					
	Walnut, Verdigris, and Neosho River Basins					
13.	Verdigris River, Independence	2,892	1895–1904	9.0[q]	1,700	72,000[r]
			1921–1983			
14.	Neosho River, Parsons	4,905	1922–1983	7.5[q]	2,500	56,000[q]
	Kentucky					
	Ohio Region					
	Middle and Lower Ohio Subregions					
	Ohio River Main Stem					
1.	Ohio River, Greenup Dam	62,000	1968–1983	7,400	92,530	699,000
2.	Ohio River, Louisville	91,170	1928–1983	8,200	115,700	862,000
3.	Ohio River, Metropolis, IL	203,000	1928–1983	46,000	271,000	1,580,000
	Salt River Basin					
4.	Salt River, Shepherdsville	1,197	1938–1983	0.22	1,572	61,900
5.	Rolling Fork, Boston	1,299	1938–1983	2.3	1,801	65,600

(Continued)

Table 5A.2 (Continued)

	Gaging Station			Streamflow Characteristics		
	Name	Drainage Area (mL^2)	Period of Analysis	7-Day, 10-Year Low Flow (ft^3/sec)	Average Discharge (ft^3/sec)	100-Year Flood (ft^3/sec)
	Big Sandy–Guyandotte Subregion					
6.	Levisa Fork, Pikeville	1,232	1937–1983	5.8	1,474	76,400
	Kentucky–Licking Subregion					
	Licking River Basin					
7.	Licking River, Catawba	3,300	1914–1983	13	4,143	84,900
	Kentucky River Basin					
8.	Middle Fork Kentucky River, Tallega	537	1930–1983	0.64	730	51,400
9.	Kentucky River, Salvisa	5,102	1925–1983	136	6,737	125,000
	Green Subregion					
	Green River Basin					
10.	Green River, Munfordville	1,673	1915–1983	73	2,722	70,300
11.	Pond River, Apex	194	1940–1983	0	267	25,800
	Cumberland Subregion					
	Cumberland River Basin					
12.	Cumberland River, Williamsburg	1,607	1959–1983	22	2,736	54,000
13.	Little River, Cadiz	244	1940–1983	11	349	18,200
	Tennessee Region					
	Lower Tennessee Subregion					
14.	Tennessee River, Paducah	40,200	1889–1983	8,190	64,060[s]	—
				—	65,450[t]	—
	Louisiana					
	South Atlantic–Gulf Region					
	Pearl Subregion					
	Pearl River Basin					
1.	Pearl River, Bogalusa	6,573	1939–1983	1,320	9,887	129,000
2.	Bogue Chitto, Bush	1,213	1938–1983	460	1,915	93,200
	Lower Mississippi Region					
	Mississippi River Main Stem[u]					
3.	Mississippi River, Vicksburg, MO	1,118,160	1929–1983	127,000	578,800	2,203,000
4.	Mississippi River, Tarbert Landing, MO	1,124,900	1939–1983	142,000	514,200	—
	Lower Red–Ouachita Subregion					
	Ouachita River Basin					
5.	Big Creek, Pollock	51	1943–1983	7.4	61.4	37,200
	Lower Mississippi–Lake Maurepas Subregion					
6.	Amite River, Denham Springs	1,280	1939–1983	304	2,021	136,000
7.	Tangipahoa River, Robert	646	1939–1983	284	1,154	81,900
	Louisiana Coastal Subregion					
	Atchafalaya–Teche–Vermillion and Calcasieu–Mermentau River Basin					
8.	Atchafalaya River, Simmesport	87,570	1939–1983	26,000	196,700	—
9.	Calcasieu River, Oberlin	753	1923–1924 1939–1983	37	1,147	58,900
10.	Calcasieu River, Kinder	1,700	1923–1924 1939–1957 1962–1983	202	2,568	121,000
	Arkansas–White–Red Region					
	Red–Sulphur Subregion					
	Red River Basin					
11.	Red River, Shreveport	60,613	1929–1983	1,150	24,030	297,000
12.	Red River, Alexandria	67,500	1929–1983	1,650	30,870	251,000
13.	Saline Bayou, Lucky	154	1941–1983	4.5	162	17,200
	Texas–Gulf Region					
	Sabine Subregion					
	Sabine River Basin					
14.	Sabine River, Ruliff, TX	9,329	1925–1983	432	7,491	90,700

(Continued)

Table 5A.2 (Continued)

	Gaging Station			Streamflow Characteristics		
	Name	Drainage Area (mL²)	Period of Analysis	7-Day, 10-Year Low Flow (ft³/sec)	Average Discharge (ft³/sec)	100-Year Flood (ft³/sec)
		Maine				
		New England Region				
		St. John Subregion				
1.	St. John River, Ninemile Bridge	1,341	1950–1985	96	2,330	47,900
2.	St. John River, Fort Kent	5,665	1926–1985	747	9,730	167,000
3.	Aroostook River, Washburn	1,654	1930–1985	143	2,670	51,500
		Maine Coastal Subregion				
4.	St. Croix River, Baring	1,374	1958–1985	484	2,760	31,000
5.	Narraguagus River, Cherryfield	227	1948–1985	29	503	11,300
6.	Sheepscot River, North Whitefield	148	1938–1985	8.8	249	7,080
		Penobscot Subregion				
7.	Penobscot River, Dover–Foxcroft	298	1902–1985	19	603	25,400
8.	Penobscot River, West Enfield	6,71	1901–1985	2,970	11,960	150,000
		Kennebec Subregion				
9.	Kennebec River, Bingham	2,715	1907–1910 1930–1985	1,310	4,450	59,200
10.	Carrabassett River, North Anson	353	1902–1907 1925–1985	45	717	39,500
		Androscoggin Subregion				
11.	Swift River, Roxbury	96.9	1929–1985	6.9	199	21,100
12.	Little Androscoggin River, South Paris	75.8	1913–1924 1931–1985	2.6	139	6,700
13.	Androscoggin River, Auburn	3,263	1928–1985	1,690	6,140	99,700
		Saco Subregion				
14.	Royal River, Yarmouth	141	1949–1985	24	275	11,000
15.	Saco River, Cornish	1,293	1916–1985	386	2,710	36,800
		Maryland (and the District of Columbia)				
		Mid-Atlantic Region				
		Potomac Subregion				
1.	Conoccocheague Creek, Fairview	494	1928–1983	53	590	26,800
2.	Antietam Creek, Sharpsburg	281	1899–1983	66	275	14,400
3.	Monocacy River, Frederick	817	1929–1983	50	926	65,900
		Upper Chesapeake Subregion				
4.	Pocomoke River, Willards	60.5	1949–1983	3.4	71	1,830
5.	Choptank River, Greensboro	113	1948–1983	5.4	132	9,360
6.	Patuxent River, Unity	34.8	1944–1983	2.8	39	26,900
		Susquehanna Subregion				
7.	Susquehanna River, Conowingo	27,100	1968–1983	—	42,180	—
		Ohio Region				
		Monongahela Subregion				
8.	Youghiogheny River, Oakland	134	1941–1983	5.9	297	12,800
		Massachusetts				
		New England Region				
		Connecticut Subregion				
1.	Millers River, Erving	372	1915–1983	47	630	—
2.	North River, Shattuckville	89.0	1940–1983	8.1	183	17,000
3.	Deerfield River, West Deerfield	557	1941–1983	97	1,285	56,000
4.	Connecticut River, Montague City	7,860	1905–1983	1,700	13,760	—
5.	Ware River, Barre	96.3	1929–1983	6.4	167	—
6.	East Branch Swift River, Hardwick	43.7	1938–1983	0.2	69.5	3,100
7.	Chicopee River, Indian Orchard	689	1929–1983	130	903	—
8.	West Branch Westfield River, Huntington	94.0	1936–1983	5.7	190	29,000
9.	Westfield River, Westfield	497	1915–1983	84	921	—
		Merrimack Subregion				
10.	Nashua River, East Pepperell	316	1936–1983	46	568	16,000
11.	Concord River, Lowell	307	1937–1983	33	630	6,000

(Continued)

Table 5A.2 (Continued)

	Gaging Station			Streamflow Characteristics		
	Name	Drainage Area (mL^2)	Period of Analysis	7-Day, 10-Year Low Flow (ft^3/sec)	Average Discharge (ft^3/sec)	100-Year Flood (ft^3/sec)
12.	Merrimack River, Lowell	4,423	1924–1983	937	7,530	—
	Massachusetts–Rhode Island Coastal Subregion					
13.	Parker River, Byfield	21.3	1946–1983	0.2	36.7	610
14.	Ipswich River, Ipswich	125	1931–1983	2.0	187	3,120
15.	Charles River, Dover	183	1938–1983	13	302	3,800
16.	Indian Head River, Hanover	30.2	1967–1983	1.4	62.4	1,800
17.	Wading River, Norton	43.3	1926–1983	2.3	73.3	1,500
	Connecticut Coastal Subregion					
18.	Housatonic River, Great Barrington	280	1914–1983	69	526	11,000
	Michigan					
	Great Lakes Region					
	Northwestern Lake Michigan and Southeastern Lake Michigan Subregions					
1.	St. Joseph River, Niles	3,666	1931–1984	945	3,260	20,400
2.	Kalamazoo River, Fennville	1,600	1930–1936 1938–1984	335	1,420	12,300
3.	Red Cedar River, East Lansing	355	1903 1932–1984	9.79	205	6,890
4.	Grand River, Lansing	1,230	1902–1906 1935–1984	80.2	833	8,800
5.	Grand River, Grand Rapids	4,900	1902–1905 1931–1984	721	3,570	53,000
6.	Escanaba River, Cornell	870	1904–1912 1951–1984	168	892	13,000
	Northeastern Lake Michigan–Lake Michigan Subregion					
7.	Muskegon River, Evart	1,450	1931 1934–1984	314	998	9,060
8.	Muskegon River, Newaygo	2,350	1910–1914 1917–1919 1931–1984	672	1,970	14,100
9.	Manistee River, Manistee	1,780	1952–1984	1,210	2,000	8,240
	Southwestern Lake Huron–Lake Huron Subregion					
10.	Shiawassee River, Fergus	637	1940–1984	42.1	420	9,330
11.	Flint River, Fosters	1,188	1940–1984	66.4	743	16,700
12.	Cass River, Frankenmuth	841	1936 1940–1984	20.4	490	20,800
13.	Tittabawassee River, Midland	2,400	1937–1984	187	1,680	44,600
	Southern Lake Superior–Lake Superior and St. Clair–Detroit Subregions					
14.	Ontonagon River, Rockland	1,340	1943–1984	308	1,430	32,400
15.	Sturgeon River, Sidnaw	171	1913–1915 1944–1984	8.19	216	4,830
16.	Tahquamenon River, Paradise	790	1954–1984	196	936	7,660
17.	Clinton River, Mt. Clemens	734	1935–1984	61.4	531	23,200
18.	Huron River, Ann Arbor	729	1905–1984	43.6	456	5,940
	Minnesota					
	Upper Mississippi Region					
	Mississippi River Basin[v]					
1.	Mississippi River, Anoka	19,000	1932–1983	1,194	7,655	98,000
2.	Crow Wing River, Pillager	3,300	1968–1983	173	1,264	15,300
3.	Sauk River, St. Cloud	925	1910–1912 1931 1935–1981	13.1	276	10,000
4.	Crow River, Rockford	2,520	1910–1917 1931 1935–1983	14.7	664	19,000
5.	Rum River, St. Francis	1,360	1931 1934–1983	64.4	602	14,000
6.	Cannon River, Welch	1,320	1911–1913 1931–1971	61.6	501	34,000

(Continued)

Table 5A.2 (Continued)

	Gaging Station			Streamflow Characteristics		
	Name	Drainage Area (mL^2)	Period of Analysis	7-Day, 10-Year Low Flow (ft^3/sec)	Average Discharge (ft^3/sec)	100-Year Flood (ft^3/sec)
7.	Zumbro River, Zumbro Falls	1,130	1910–1917 1931–1980	77.7	517	40,200
8.	Root River, Houston	1,270	1910–1917 1931–1983	178	696	51,500
	Minnesota Subregion					
9.	Minnesota River, Jordan	16,200	1935–1983	171	3,520	115,000
10.	Lac qui Parle River, Lac qui Parle	983	1913 1932 1934–1983	0.20	120	19,300
11.	Chippewa River, Milan	1,870	1938–1983	2.90	269	12,400
12.	Cottonwood River, New Ulm	1,280	1912–1913 1936–1937 1939–1983	2.77	289	33,000
13.	Blue Earth River, Rapidan	2,430	1940–1945 1950–1983	14.9	895	34,600
	St. Croix Subregion					
14.	St. Croix River, St. Croix Falls	6,240	1903–1983	1,099	4,235	61,000
	Souris–Red–Rainy Region					
	Red Subregion Red Lake River Basin					
15.	Otter Tail River, Orwell Dam Fergus Falls	1,830	1931–1983	12.3	304	4,800
16.	Red River of the North, Grand Forks	30,100	1883–1983	71.4	2,558	89,000
17.	Red Lake River, Crookston	5,280	1902–1983	31.6	1,130	31,000
	Rainy Subregion					
	Little Fork and Big Fork River Basins					
18.	Rainy River, Manitou Rapids	19,400	1929–1983	3,597	12,830	80,000
19.	Little Fork River, Littlefork	1,730	1912–1916 1929–1983	40.3	1,053	27,400
20.	Big Fork River, Big Falls	1,460	1929–1979 1983	33.7	715	21,800
	Great Lakes Region					
	Western Lake Superior Subregion					
21.	Pigeon River, Grand Portage	600	1924–1983	44.5	506	13,600
22.	Baptism River, Beaver Bay	140	1928–1983	3.45	169	8,820
23	St. Louis River, Scanlon	3,430	1909–1983	316	2,313	38,000
	Mississippi					
	Lower Mississippi Region					
	Lower Mississippi–Yazoo Subregion					
	Yazoo River Basin					
1.	Yazoo River, Greenwood[w]	7,450	1907–1912 1927–1939 1940–1984	831 741	9,330 10,900	— 45,000
2.	Big Sunflower River, Sunflower[w]	767	1935–1984	87	1,070	14,100
	Lower Mississippi–Big Black Subregion					
	Big Black River Basin					
3.	Big Black River, Bovina	2,810	1936–1984	84	3,800	73,400
	South Atlantic–Gulf Region					
	Pearl Subregion					
	Pearl River Basin					
4.	Pearl River, Monticello	4,993	1938–1960 1961–1984	324 365	6,110 7,530	— 97,100
	Mobile–Tombigbee Subregion					
	Tombigbee River Basin					
5.	Tombigbee River, Columbus	4,463	1899–1912 1928–1982	233	6,520	223,000
	Pascagoula Subregion					
	Pascagoula River Basin					
6.	Pascagoula River, Merrill	6,590	1930–1968	865	9,350	—

(Continued)

Table 5A.2 (Continued)

	Gaging Station			Streamflow Characteristics		
	Name	Drainage Area (mL^2)	Period of Analysis	7-Day, 10-Year Low Flow (ft^3/sec)	Average Discharge (ft^3/sec)	100-Year Flood (ft^3/sec)
			1969–1984	1,080	11,800	221,000
		Missouri				
		Upper Mississippi Region				
		Upper Mississippi–Kaskaskia–Meramec Subregion				
1.	Salt River, New London	2,480	1922–1983	1.7	1,700	87,000
2.	Mississippi River, St. Louis	697,000	1951–1983	43,000	183,000	1,000,000
3.	Meramec River, Eureka	3,788	1921–1983	280	3,100	144,000
4.	Mississippi River, Thebes, IL	713,200	1951–1983	47,100	198,000	1,100,000
		Lower Mississippi Region				
		Lower Mississippi–St. Francis Subregion				
		St. Francis River Basin				
5.	St. Francis River, Patterson	956	1920–1983	15	1,100	89,000
6.	Little River, Morehouse	450	1945–1983	33	530	11,000
		Missouri Region				
		Gasconade–Osage and Chariton–Grand Subregions				
		Osage and Grand River Basins				
7.	Missouri River, Kansas City	485,200	1955–1983	6,400	51,000	—
8.	Grand River, Gallatin	2,250	1921–1983	4.0	1,200	72,000
9.	Osage River, St. Thomas	14,500	1931–1983	480	9,900	—
10.	Gasconade River, Jerome	2,840	1923–1983	320	2,500	106,000
11.	Missouri River, Hermann	524,200	1955–1983	11,000	72,000	—
		Arkansas–White–Red Region				
		Upper White Subregion				
		White River Basin				
12.	James River, Galena	987	1921–1983	38	940	69,000
13.	White River, Branson	4,020	1956–1983	78	3,500	—
14.	Current River, Doniphan	2,038	1918–1983	940	2,700	104,000
15.	Spring River, Waco	1,164	1924–1983	18	840	80,000
		Montana				
		Missouri Region				
		Missouri River Basin[x]				
1.	Beaverhead River, Barretts	2,737	1907–1983	122	430	3,040
2.	Missouri River, Fort Benton	24,749	1890–1983	2,230	7,827	96,000
3.	Marias River, Shelby	3,242	1902–1904	65	940	—
			1905–1906			
			1907–1908			
			1911–1983			
4.	Musselshell River, Mosby	7,846	1929	0.00	301	34,600
			1932–1983			
			1934–1930			
5.	Milk River, Nashua	22,332	1939–1983	13	710	36,600
6.	Missouri River, Culbertson	91,557	1941–1951	1,520	11,000	54,900
			1958–1983			
		Yellowstone River Basin[y]				
7.	Yellowstone River, Billings	11,795	1928–1983	1,090	7,074	80,000
8.	Bighorn River, Bighorn	22,885	1945–1983	767	3,939	41,700
9.	Tongue River, Miles City	5,379	1938–1942	3.3	440	17,800
			1946–1983			
10.	Powder River, Locate	13,194	1938–1983	1.6	612	48,100
11.	Yellowstone River, Sidney	69,103	1910–1931	1,410	13,080	156,000
			1933–1983			
		Pacific Northwest Region				
		Clark Fork Basin[z]				
12.	Clark Fork, St. Regis	10,709	1910–1983	1,440	7,583	79,800
13.	Bitterroot River, Darby	1,049	1937–1983	123	931	13,200
14.	Flathead River, Columbia Falls	4,464	1928–1983	1,090	9,737	84,000
15.	Clark Fork, Plains	19,958	1910–1983	4,440	20,010	145,000

(Continued)

Table 5A.2 (Continued)

	Gaging Station			Streamflow Characteristics		
	Name	Drainage Area (mL^2)	Period of Analysis	7-Day, 10-Year Low Flow (ft^3/sec)	Average Discharge (ft^3/sec)	100-Year Flood (ft^3/sec)
	Kootenai River Basin[z]					
16.	Kootenai River, Libby	10,240	1911–1970	1,610	12,100	116,000
			1973–1983	2,560	11,740	76,300
	Nebraska					
	Missouri Region					
	Missouri River Main Stem[aa]					
1.	Missouri River, Fort Randall Dam, SD	263,500	1947–1983	1,450[bb]	25,230[bb]	99,000[bb]
2.	Missouri River, Rulo	414,900	1950–1983	6,210[bb]	40,190[bb]	241,000[bb]
	Niobrara Subregion					
3.	Niobrara River, Norden	8,390	1953–1963	516	952	10,900
			1964–1983	398	810	
	North Platte Subregion					
4.	Pumpkin Creek, Bridgeport	1,020	1932–1983	0.35	28.3	3,320
5.	North Platte River, North Platte	30,900	1896–1940	135	2,720	36,700
			1941–1983		713	10,700
	South Plate Subregion					
6.	South Platte River, North Platte	24,300	1918–1946	78	435	77,300
			1947–1983		402	57,300
	Platte Subregion					
7.	Platte River, Overton	57,700	1915–1940	46	2,860	60,700
			1941–1983		1,470	32,800
8.	Platte River, Louisville	85,800	1954–1983	430	5,980	169,000
	Loup Subregion					
9.	Middle Loup River, Dunning	1,850	1946–1983	260	401	1,100
10.	Loup River, Genoa	14,400	1943–1983	0.96	574	130,000
	Elkhorn Subregion					
11.	Elkhorn River, Waterloo	6,900	1929–1983	119	1,120	83,500
	Missouri–Nishnabotna Subregion					
12.	Big Nemaha River, Falls City	1,340	1945–1983	11.4	587	80,700
	Republican Subregion					
13.	Medicine River, Harry Strunk Lake	770	1951–1983	16.0	65.9	23,700
14.	Republican River, Cambridge	14,520	1950–1983	18.0[cc]	279[cc]	16,800[cc]
	Kansas Subregion					
	Blue River Basin					
15.	Big Blue River, Barneston	4,447	1933–1983	35.1	787	50,100
16.	Little Blue River, Fairbury	2,350	1911–1915	48.3	369	48,500
			1930–1983			
	Nevada					
	Lower Colorado Regions					
	Lower Colorado–Lake Mead Subregion					
1.	Virgin River, Littlefield, AZ	5,090	1929–1983	48	243	35,300
2.	Muddy River, Moapa	3,820	1913–1915	31	41.5	5,000
			1916–1918			
			1928–1931			
			1944–1983			
3.	Lee Canyon, Charleston Park	9.20	1963–1983	0	0.025	4,300
4.	Las Vegas Wash, Henderson	2,125	1957–1983	—	46.6	6,500
	Great Basin Region					
	Black Rock Desert–Humboldt Subregion					
	Humboldt River Basin					
5.	Humboldt River, Palisade	5,010	1902–1906	8.9	385	7,700
			1911–1983			
6.	Humboldt River, Imlay	15,700	1911–1983	0.3	235	5,700

(Continued)

Table 5A.2 (Continued)

	Gaging Station			Streamflow Characteristics		
	Name	Drainage Area (mL^2)	Period of Analysis	7-Day, 10-Year Low Flow (ft^3/sec)	Average Discharge (ft^3/sec)	100-Year Flood (ft^3/sec)
		Central Lahontan Subregion				
		Walker Lake Basin				
7.	Walker River, Wabuska	2,600	1902–1904	3.8	170	6,700
			1920–1924			
			1925–1935			
			1939–1941			
			1942–1943			
			1944–1983			
		Carson River Basin				
8.	Carson River, Carson City	886	1939–1983	4.8	418	28,300
		Truckee River Basin				
9.	Truckee River, Nixon	1,827	1957–1983	14	538	28,300
		Central Nevada Desert Basins Subregions				
10.	Newark Valley tributary, Hamilton	157	1962–1983	0	0.325	1,100
11.	South Twin River, Round Mountain	20	1965–1983	0.78	7.06	260
		New Hampshire				
		New England Region				
		Androscoggin Subregion				
1.	Androscoggin River, Gorham	1,361	1913–1983[cc]	1,280	2,465	20,900
		Saco Subregion				
2.	Saco River, Conway	385	1903–1909	93	933	53,800
			1929–1983			
3.	Lamprey River, Newmarket	183	1934–1983	4.9	282	6,310
		Merrimack Subregion				
4.	Pemigewasset River, Plymouth	622	1903–1983	115	1,358	60,800
5.	Blackwater River, Webster	129	1918–1920	13	213	—
			1927–1983			
6.	Soucook River, Concord	76.8	1951–1983	3.7	112	4,080
7.	Merrimack River, Goffs Falls Manchester	3,092	1936–1983	663	5,280	—
		Connecticut Subregion				
8.	Connecticut River, Pittsburg	254	1956–1983	35	571	—
9.	Upper Ammonoosuc River, Groveton	232	1940–1980	49	473	10,800
			1982–1983			
10.	Ammonoosuc River, Bethlehem Junction	87.6	1939–1983	27	208	13,600
11.	Connecticut River, Wells River, VT	2,644	1949–1983	632	4,731	—
12.	Sugar River, West Claremont	269	1928–1983	40	404	13,800
13.	Connecticut River, North Walpole	5,493	1942–1983	993	9,380	—
14.	Ashuelot River, Hinsdale	420	1907–1911	46	671	—
			1914–1983			
		New Jersey				
		Lower Hudson–Long Island Subregion				
		Hackensack and Passaic River Basins				
1.	Hackensack River, New Milford	113	1922–1984	0	103	5,570
2.	Passaic River, Chatham	100	1904–1984[dd]	3.7	172	3,730
3.	Passaic River, Little Falls	762	1898–1984	32	1,168	22,500
4.	Saddle River, Lodi	54.6	1924–1984	13	102	5,750
		Raritan River Basin				
5.	South Branch Raritan River, High Bridge	65.3	1919–1984	22	123	6,600
6.	Stony Brook, Princeton	44.5	1954–1984	0.1	65.1	8,390
7.	Raritan River below Calco Dam Bound Brook	785	1904–1984[dd]	72	1,293	40,800
8.	Green Brook, Plainfield	9.75	1939–1984	0	12.9	3,280

(Continued)

Table 5A.2 (Continued)

	Gaging Station			Streamflow Characteristics		
	Name	Drainage Area (mL^2)	Period of Analysis	7-Day, 10-Year Low Flow (ft^3/sec)	Average Discharge (ft^3/sec)	100-Year Flood (ft^3/sec)
	Delaware and Lower Hudson–Long Island Subregions					
	Atlantic Coastal Basins					
9.	Swimming River, Red Bank	49.2	1923–1984	0	80.8[ee]	11,000
10.	Manasquan River, Squankum	44.0	1932–1984	18	75.9	2,870
11.	Oyster Creek, Brookville	7.43	1968–1984	13	28.7	514
12.	Great Egg Harbor River, Folsom	57.1	1926–1984	22	86.8	1,230
	Delaware Subregion					
	Delaware River Basin and Streams Tributary to Delaware Bay					
13.	Maurice River, Norma	112	1933–1984	37	168	2,880
14.	Flat Brook, Flatbrookville	64.0	1924–1984	7.8	110	7,070
15.	Delaware River, Trenton	6,780	1914–1984	1,800[ff]	11,740	217,000[ff]
16.	Crosswicks Creek, Extonville	81.5	1941–1984[gg]	24	136	5,800
17.	McDonalds Branch, Lebanon State Forest	2.35	1954–1984	0.9	2.32	49
18.	Cooper River, Haddonfield	17.0	1964–1984	8.6	36.3	3,840
	New Mexico					
	Arkansas–White–Red Region[hh]					
	Upper Canadian Subregion					
	Canadian River Basin					
1.	Canadian River, Logan	11,141	1904–1983	0.0	392[ii]	333,000
	Rio Grande Region					
	Upper and Lower Pecos Subregions					
	Pecos River Basin					
2.	Pecos River, Pecos	189	1919–1983	12.0	98.1	3,070
3.	Delaware River, Red Bluff, TX	689	1912–83	0.0	13.0	82,500
	Rio Grande River Basin (main stem)[ll]					
4.	Rio Grande, Albuquerque	117,440	1941–1983	0.3	1,068[mm]	22,000
				1,232[nn]	—	—
	Upper Colorado Region					
	San Juan Subregion					
5.	San Juan River, Shiprock	12,900	1927–1983	53.6	2,181	67,200
	Lower Colorado Region[oo]					
6.	Gila River, Gila	1,864	1927–1983	19.7	141	24,900
	New York					
	Mid-Atlantic Region					
	Upper Hudson Subregion					
1.	Sacandaga River, Stewarts Bridge	1,055	1907–1929	106	2,230	30,400
			1931–1984	757	2,090	14,800
2.	Mohawk River, Cohoes	3,456	1917–1984	772	5,750	128,000
3.	Hudson River, Green Island	8,090	1946–1984	2,810	13,700	191,000
4.	Wappinger Creek, Wappingers Falls	181	1928–1984	6.5	253	18,500
	Delaware Subregion					
5.	East Branch Delaware River, Fishs Eddy	784	1912–1954	89	1,670	73,700
			1955–1984	111	1,100	40,100
6.	Delaware River, Port Jervis	3,070	1904–1954	416	5,570	184,000
			1963–1984	832	4,750	170,000
	Susquehanna Subregion					
7.	Susquehanna River, Waverly	4,773	1937–1984	385	7,580	139,000
8.	Chemung River, Chemung	2,506	1903–1984	104	2,530	143,000
	Great Lakes Region					
	Southwestern and Southeastern Lake Ontario Subregions					
9.	Genesee River, Rochester	2,467	1919–1951	511	2,780	44,100
			1952–1984	311	2,880	30,600
10.	Oswego River, Oswego	5,100	1933–1984	980	6,690	38,600

(Continued)

Table 5A.2 (Continued)

	Gaging Station			Streamflow Characteristics		
	Name	Drainage Area (mL^2)	Period of Analysis	7-Day, 10-Year Low Flow (ft^3/sec)	Average Discharge (ft^3/sec)	100-Year Flood (ft^3/sec)
	Northeastern Lake Ontario–Lake Ontario–St. Lawrence Subregion					
11.	Black River, Watertown	1,874	1920–1984	825	4,020	41,000
12.	West Branch Oswegatchie River, Harrisville	244	1916–1984	43	515	7,290
13.	St. Lawrence River, Massena	298,000	1860–1984	179,000	243,000	358,000
	North Carolina					
	South Atlantic–Gulf Region					
	Chowan–Roanoke Subregion					
1.	Roanoke River, Roanoke Rapids	8,386	1911–1949	1,010	8,085	215,000
			1950–1984	1,310	7,700	66,800
	Neuse–Pamlico Subregion					
2.	Tar River, Tarboro	2,183	1896–1900	90	2,234	45,500
			1931–1984			
3.	Neuse River, Kinston	2,692	1930–1981	210	2,892	43,100
			1981–1984	—	—	33,000
	Cape Fear Subregion					
4.	Cape Fear River, Lillington	3,464	1923–1975	75	3,300	117,000
			1975–1981	600	3,300	80,000
	Pee Dee Subregion					
5.	South Yadkin River, Mocksville	306	1939–1984	61	340	15,700
	Tennessee Region					
	Tennessee Subregion					
6.	French Broad River, Asheville	945	1896–1984	455	2,093	49,100
	North Dakota					
	Souris–Red–Rainy Region					
	Souris and Red Subregions					
	Souris River and Red River of the North Basins					
1.	Souris River, Minot	10,600[pp]	1904–1983	<0.1	171	11,500
		6,700[qq]				
2.	Red River of the North, Wahpeton	4,010[pp]	1943–1983	12.3	519	11,000
3.	Big Coulee, Churchs Ferry	2,510[pp]	1951–1979	0	44.1	3,370
		690[qq]				
4.	Sheyenne River, West Fargo	8,870[pp]	1904–1905	12.9	176	5,280
		5,780[qq]	1930–1983			
5.	Red River of the North Grand	30,100[pp]	1883–1983	69.1	2,558	89,000
		3,800[qq]				
	Missouri Region					
	Missouri–Little Missouri and Missouri–Oahe Subregions					
	Missouri River Main Stem and Tributary River Basins					
6.	Little Missouri River, Watford City	8,310[pp]	1935–1983	<0.1	593	78,700
7.	Knife River, Hazen	2,240[pp]	1930–1933	2.7	181	36,200
			1938–1983			
8.	Missouri River, Bismarck	186,400[pp]	1921–1983	6,570	22,740	63,700
9.	Heart River, Mandan	3,310[pp]	1929–1932	<0.3	268	49,500
			1938–1983			
10.	Cannonball River, Breien	4,100[pp]	1935–1983	<0.1	256	60,000
11.	James River, Jamestown	2,820[pp]	1929–1934	1.1	62.2	4,800
		1,650[qq]				
	Ohio					
	Ohio Region					
	Muskingum Subregion					
1.	Tuscarawas River, Massillon	518	1937–1984	71	441	8,670
2.	Tuscarawas River, Newcomerstown	2,443	1921–1937	253	2,541	66,900
			1938–1984			23,000
3.	Muskingum River, McConnelsville	7,422	1921–1937	641	7,596	183,000
			1938–1984			97,100

(Continued)

Table 5A.2 (Continued)

	Gaging Station			Streamflow Characteristics		
	Name	Drainage Area (mL^2)	Period of Analysis	7-Day, 10-Year Low Flow (ft^3/sec)	Average Discharge (ft^3/sec)	100-Year Flood (ft^3/sec)
		Scioto Subregion				
4.	Scioto River, Prospect	567	1925–1932	9.3	454	13,900
			1939–1984			
5.	Olentangy River, Delaware	393	1923–1934	5.2	351	22,000
			1938–1951			6,280
			1951–1984			
6.	Scioto River, Higby	5,131	1930–1984	296	4,579	184,000
		Great Miami Subregion				
7.	Great Miami River, Sidney	541	1914–1984	21	477	27,300
8.	Stillwater River, Englewood	650	1925–1984	15	579	10,500
9.	Mad River, Dayton	635	1914–1984	131	629	21,100
10.	Great Miami River, Hamilton	3,630	1907–1918	284	3,279	140,000
			1927–1984			95,900
		Great Lakes Region				
		Western Lake Erie Subregion				
		Maumee River Basin				
11.	Blanchard River, Findlay	346	1923–1935	2.3	251	14,300
			1940–1984			
12.	Auglaize River, Defiance	2,318	1915–1984	11	1,718	62,700
13.	Maumee River, Waterville	6,330	1898–1901	95	4,926	97,800
			1921–1935			
			1939–1984			
		Southern Lake Erie Subregion				
		Cuyahoga River Basin				
14.	Cuyahoga River, Hiram Rapids	151	1927–1935	16	207	4,410
			1944–1984			3,690
15.	Cuyahoga River, Independence	707	1921–1923	63	817	18,000
			1927–1935			
			1940–1984			
		Oklahoma				
		Arkansas–White–Red Region				
		Arkansas River Basin, Salt Fork Arkansas River and Cimarron River Basin, Verdigris River and Grand (Neosho) River Basins, and Canadian River Basins[rr]				
1.	Salt Fork Arkansas River, Tonkawa	4,528	1942–1982	6.68	730	69,400
2.	Cimarron River Perkins	17,852	1940–1982	8.73	1,180	174,000
3.	Arkansas River, Tulsa	74,615	1926–1964	155	6,550	324,000
			1965–1982	346	6,940	165,000
4.	Verdigris River, Claremore	6,534	1936–1962	3.06	3,720	178,000
			1965–1982	15.4	3,720	46,400
5.	Illinois River, Tahlequah	959	1936–1982	16.8	867	141,000
6.	Little River, Sasakawa	865	1943–1965	0.69	398	66,800
			1966–1982	0.08	242	23,300
7.	Beaver River, Beaver	7,955	1938–1982	0.03	95.9	68,100
8.	Fourche Maline, Red Oak	122	1939–1963	0.10	126	51,400
			1966–1982	0.12	133	17,900
		Red River Basin[ss], Washita River Basin				
9.	North Fork Red River, Headrick	4,244	1946–1982	0.39	266	54,200
10.	Red River, Gainesville	30,782	1947–1982	97.6	2,750	180,000
11.	Washita River, Dickson	7,202	1929–1960	33.7	1,540	117,000
			1962–1982	4.87	1,140	64,200
12.	Muddy Boggy Creek, Farris	1,087	1938–1982	0.14	880	61,200
13.	Red River, Arthur City, TX	44,531	1945–1982	375	7,890	174,000
		Oregon				
		Pacific Northwest Region				
		Oregon Closed Basins Subregion				
1.	Silvies River, Burns	934	1928–1983	1.5	175	4,900

(Continued)

Table 5A.2 (Continued)

	Gaging Station			Streamflow Characteristics		
	Name	Drainage Area (mL^2)	Period of Analysis	7-Day, 10-Year Low Flow (ft^3/sec)	Average Discharge (ft^3/sec)	100-Year Flood (ft^3/sec)
2.	Donner und Blitzen River, Burns	200	1912–1913	20	125	4,200
			1915–1916			
			1918–1921			
			1939–1983			
	Middle Snake Subregion					
3.	Owyhee River, Owyhee Reservoir	11,160	1930–1983	1.7	380	—
	Middle Columbia Subregion					
4.	Umatilla River, Umatilla	2,290	1928–1983	1.3	456	19,700
5.	John Day River, McDonald Ferry	7,580	1906–1983	28	2,036	37,800
6.	Deschutes River, Moody	10,500	1897–1999	3,610	5,846	—
			1907–1983			
	Willamette Subregion					
7.	Santiam River, Jefferson	1,790	1909–1953	323	7,821	—
			1967–1982	1,150	7,821	—
8.	Willamette River, Salem	7,280	1911–1941	2,720	23,650	
			1969–1982	5,160		
	Oregon–Washington Coastal Subregion					
	Rogue River Basin					
9.	Wilson River, Tillamook	161	1932–1983	51	1,205	36,700
10.	Umpqua River, Elkton	3,683	1906–1983	797	7,517	276,000
11.	Rogue River, Raygold	2,053	1905–1983	870	2,978	139,000
	California Region					
	Klamath–Northern California Coastal Subregion					
	Klamath River Basin					
12.	Sprague River, Chiloquin	1,580	1922–1983	127	584	13,300
13.	Williamson River, Chiloquin	3,000	1918–1982	414	1,049	14,100
14.	Klamath River, Keno	3,920	1905–1912	165	1,684	13,000
			1930–1983			
	Pennsylvania					
	Mid-Atlantic Region					
	Delaware Subregion					
	Delaware River Main Stem					
1.	Bush Kill Shoemakers	117	1908–1983	7.6	235	10,800
2.	Delaware River, Belvidere, NJ	4,535	1922–1983	920	7,890	220,000
3.	Delaware River Trenton, NJ	6,780	1913–1983	—	11,685	270,000
	Schuylkill River Basin					
4.	Schuylkill River, Landingville	133	1947–1983	28	292	14,600
5.	Schuylkill River, Pottstown	1,147	1926–1983	260	1,891	74,000
	Susquehanna Subregion					
	Susquehanna River Main Stem					
6.	Susquehanna River, Towanda	7,797	1913–1983	550	10,600	105,000
7.	Susquehanna River, Danville	1,220	1899–1983	980	15,320	260,000
8.	Susquehanna River, Sunbury	18,300	1937–1983	1,600	26,520	530,000
9.	Susquehanna River, Harrisburg	24,100	1890–1983	2,556	34,350	750,000
	West Branch Susquehanna River Basin					
10.	West Branch Susquehanna River, Lewisburg	6,847	1939–1983	655	10,810	280,000
	Juniata River Basin					
11.	Juniata River, Newport	3,354	1899–1983	380	4,295	145,000
	Potomac Subregion					
12.	Tonoloway Creek, Needmore	10.7	1965–1983	0.27	12.4	1,590
	Ohio Region					
	Allegheny Subregion					
13.	Allegheny River, Port Allegany	248	1974–1983	24	476	9,300
14.	Oil Creek Rouseville	300	1932–1983	29	535	19,800
15.	Allegheny River, Franklin	5,982	1914–1983	511	10,470	125,000

(Continued)

Table 5A.2 (Continued)

	Gaging Station			Streamflow Characteristics		
	Name	Drainage Area (mL^2)	Period of Analysis	7-Day, 10-Year Low Flow (ft^3/sec)	Average Discharge (ft^3/sec)	100-Year Flood (ft^3/sec)
		Monongahela Subregion				
16.	Monongahela River, Elizabeth	5,340	1933–1983	698	9,109	170,000
17.	Monongahela River, Braddock	7,337	1938–1983	1,150	12,460	230,000
		Upper Ohio Subregion				
		Ohio River Main Stem				
18.	Connoquenessing Creek, Zelienople	356	1919–1983	11	464	19,450
		Puerto Rico				
		Caribbean Region				
		Puerto Rico Subregion				
		North Coast Area				
1.	Rio Culebrinas, Moca	71.2	1967–1985	20.0	299	111,000
2.	Rio Grande de Arecibo, Central Cambalache	200[tt]	1969–1984	90.0	510	—
3.	Rio Grande de Manati, Manati	197[uu]	1970–1985	60.0	375	255,000
4.	Rio Cibuco, Vega Baja	99.1[vv]	1973–1985	8.2[tt]	125	45,800
5.	Rio de la Plata, Toa Alta	200[ww]	1960–1985	7.8	276	202,000
6.	Rio Grande de Loiza, Caguas	89.8	1960–1985	14.0	219	131,000
7.	Rio Herrera, Colonia Dolores	2.75	1966–1973	1.4[tt]	9.47	5,430[tt]
8.	Rio Espiritu Santo, Rio Grande	8.62	1966–1985	5.0	57.0	22,400
		East Coast Area				
9.	Rio Fajardo, Fajardo	14.9	1961–1985	3.5	68.9	45,400
		South Coast Area				
10.	Rio Grande de Patillas, Patillas	18.3	1966–1985	5.4	60.9	40,500
11.	Rio Inabon, Real Abajo	9.70	1964–1970 1971–1985	1.3	18.6	15,100
12.	Rio Cerrillos, Ponce	17.8	1964–1985	3.1	35.8	22,200
13.	Rio Portugues, Ponce	8.82	1964–1985	1.5	18.2	21,800
		West Coast Area				
14.	Rio Guanajibo, Hormigueros	120	1973–1985	5.9[tt]	220	160,000[tt]
15.	Rio Grande de Anasco, San Sebastian	134[xx]	1963–1985	38.0	304	83,600
		Rhode Island				
		New England Region				
		Massachusetts–Rhode Island Coastal Subregion				
		Blackstone River Basin				
1.	Branch River, Forestdale	91.2	1941–1983	13	171	7,110
2.	Blackstone River, Woonsocket	416	1930–1983	100	763	19,200
3.	South Branch Pawtuxet River, Washington	63.8	1942–1983	16	130	2,890
4.	Pawtuxet River, Cranston	200	1941–1983	73	345	5,220
		Pawcatuck River Basin				
5.	Pawcatuck River, Wood River Jct.	100	1942–1983	28	194	2,090
6.	Wood River, Hope Valley	72.4	1942–1983	20	156	2,630
7.	Pawcatuck River, Westerly	295	1942–1983	67	576	6,850
		South Carolina				
		South Atlantic–Gulf Region				
		Pee Dee Subregion				
		Lower Pee Dee River Basin				
1.	Pee Dee River, Pee Dee	8,830	1938–1983	1,500[yy]	9,850	160,000[yy]
2.	Lynches River, Effingham	1,030	1925–1983	132	1,035	22,100
3.	Little Pee Dee River, Galivants Ferry	2,790	1943–1983	315	3,243	31,300
4.	Black River, Kingstree	1,252	1920–1983	5.7	942	39,100
5.	Waccamaw River, Longs	1,110	1950–1983	0.99	1,223	17,300
		Edisto–Santee Subregion				
		Santee River Basin				
6.	North Pacolet River, Fingerville	116	1931–1983	43	215	13,100

(Continued)

Table 5A.2 (Continued)

	Gaging Station			Streamflow Characteristics		
	Name	Drainage Area (mL^2)	Period of Analysis	7-Day, 10-Year Low Flow (ft^3/sec)	Average Discharge (ft^3/sec)	100-Year Flood (ft^3/sec)
7.	Broad River, Richtex	4,850	1925–1983	970[yy]	6,250	210,000[yy]
8.	Saluda River, Columbia	2,520	1925–1983	260[yy]	2,929	70,000[yy]
9.	Wateree River, Carnden	5,070	1904–1910 1925–1983	490[yy]	6,444	225,000[yy]
10.	Congaree River, Columbia	7,850	1939–1983	1,800[yy]	9,425	220,000[yy]
11.	Lake Marlon–Moultrie Diversion Canal	—	1943–1983	2,320	15,125	—
	Edisto–South Carolina Coastal Basin					
12.	Edisto River, Givhans	2,730	1939–1983	442	2,711	29,200
13.	Salkehatchie River, Miley	341	1951–1983	33	356	4,390
	Ogeechee–Savannah Subregion					
	Savannah River Basin					
14.	Savannah River, Augusta, GA	7,508	1883–1891 1896–1906 1925–1983	4,700[yy]	10,300	—
	South Dakota					
	Misouri Region					
	Missouri River Main Stem[zz]					
1.	Missouri River, Mobridge	208,700	1928–1962[aaa]	3,500	21,560	471,000
2.	Missouri River, Pierre	243,500	1929–1965	2,100[bbb]	21,860	97,500[bbb]
3.	Missouri River, Fort Randall Dam	263,500	1947–1983	1,450[bbb]	25,230	99,000[bbb]
4.	Missouri River, Yankton	279,500	1930–1983	5,980[bbb]	26,430	92,400[bbb]
5.	Missouri River, Sioux City, IA	314,600	1929–1983	6,380[bbb]	29,360	115,000[bbb]
	Western Tributaries[ccc]					
6.	Little Missouri River, Camp Crook	1,970	1903–1983[ddd]	0.2	136	13,300
7.	Grand River, Little Eagle	5,370	1958–1983	0.3[bbb]	238	24,400[bbb]
8.	Moreau River, Whitehorse	4,880	1954–1983	0.0	202	44,900
9.	Cheyenne River, Cherry Creek	23,900	1960–1983	26.1[bbb]	827	84,600[bbb]
10.	Bad River, Fort Pierre	3,107	1928–1983	0.0	147	47,000
11.	White River, Oacoma	10,200	1928–1983	0.5	531	49,200
12.	Keya Paha River, Wewela	1,070	1937–1983[ddd]	3.6	68.9	8,680
	Eastern Tributaries[eee]					
13.	James River, Scotland	20,300	1928–1983	1.5	372	23,600
14.	Vermillion River, Wakonda	1,680	1945–1983	0.9	125	6,050
15.	Big Sioux River, Akron, IA	8,360	1928–1983	18.8	901	73,200
	Tennessee					
	Ohio Region					
	Cumberland Subregion					
	Cumberland Basin					
1.	New River, New River	382	1934–1985	0.47	741	63,500
2.	Wolf River, Byrdstown	106	1942–1985	5.19	192	31,400
3.	Cumberland River, Celina	7,307	1924–1985	850	11,830	78,000
4.	West Fork Stones River, Smyrna	237	1965–1985	9.0	440	57,400
5.	Harpeth River, Kingston Springs	681	1924–1985	25.4	986	69,700
6.	Red River, Port Royal	935	1961–1985	66.6	1,351	18,000
	Tennessee Region					
	Upper Tennessee, Middle Tennessee–Hiwassee, Middle Tennessee–Elk, and Lower Tennessee Subregions					
	Tennessee Basin					
7.	Nolichucky River, Embreeville	805	1919–1985	224	1,370	72,600
8.	Little River, Maryville	269	1951–1985	54.8	535	37,200
9.	Obed River, Lancing	518	1958–1968 1974–1985	1.3	1,062	84,400
10.	South Chickamauga Creek, Chickamauga	428	1928–1978 1980–1985	88.3	698	35,100

(Continued)

Table 5A.2 (Continued)

	Gaging Station			Streamflow Characteristics		
	Name	Drainage Area (mL^2)	Period of Analysis	7-Day, 10-Year Low Flow (ft^3/sec)	Average Discharge (ft^3/sec)	100-Year Flood (ft^3/sec)
11.	Tennessee River, Chattanooga	21,400	1874–1985	10,000	37,100	257,000
12.	Elk River, Prospect	1,784	1905–1907	330	3,076	128,000
			1920–1985			
13.	Duck River, Hurricane Mills	2,557	1925–1985	303	4,121	114,000
14.	Buffalo River, Lobelville	707	1927–1985	174	1,196	88,900
15.	Big Sandy River, Bruceton	205	1929–1985	35.5	294	18,900
	Lower Mississippi Region					
	Lower Mississippi–Hatchie Subregion					
	Lower Mississippi Basin					
16.	Obion River, Obion	1,852	1929–1958	266	2,702	92,800
			1966–1985			
17.	Hatchie River, Bolivar	1,480	1929–1985	126	2,428	68,000
18.	Loosahatchie River, Arlington	262	1969–1985	71	364	24,000
19.	Wolf River, Germantown	699	1969–1985	200	1,040	42,100
	Arkansas–White–Red Region					
	Canadian–Red River Basin[fff]					
1.	Canadian River, Amarillo	15,376	1939–1983	0.3	331	135,000
2.	Red River Terral, OK	22,787	1939–1983	76.4	2,117	—
	Texas–Gulf Region					
	Sabine–Neches–Trinity–San Jacinto River Basin[ggg]					
3.	Trinity River, Dallas	6,106	1903–1983	20.5	1,530	—
4.	Trinity River, Romayor	17,186	1969–1983	64	7,417	—
5.	Neches River, Rockland	3,636	1962–1983	27.6	1,974	68,400
	Brazos–Colorado River Basin[hhh]					
6.	Salt Fork Brazos River, Aspermont	2,496	1940–1983	0.0	108	52,900
7.	Brazos River, South Bend	13,107	1939–1983	0.0	836	—
8.	North Bosque River, Clifton	968	1968–1983	0.0	167	73,200
9.	Colorado River, Colorado City	1,585	1953–1983	0.0	38.9	—
10.	Llano River, Junction	1,849	1916–1983	17.8	194	363,000
11.	Colorado River, Wharton	30,600	1939–1983	224	2,685	—
	Lavaca–Guadalupe–Nueces River Basin[iii]					
12.	Guadalupe River, Spring Branch	1,315	1923–1983	0.1	311	158,000
13.	Nueces River, Laguna	737	1924–1983	9.6	148	408,000
14.	Nueces River, Three Rivers	15,427	1916–1983	0.0	848	116,000
	Rio Grande Region					
	Rio Grande Basin[jjj]					
15.	Pecos River, Girvin	29,560	1940–1983	3.3	84.2	23,300
	U.S. Virgin Islands					
	Caribbean Region					
	U.S. Virgin Islands Subregion					
	St. Thomas					
1.	Bonne Resolution Gut, Bonne Resolution	0.49	1963–1968	0	0.24	[kkk]1,650
			1979–1981			
			1982			
2.	Turpentine Run, Mariendal	2.97	1963–1969	0	1.07	[kkk]9,710
			1979–1980			
			1982			
	St. John					
3.	Guinea Gut, Bethany	0.37	1963–1967	0	0.08	[kkk]946
			1983			
	St. Croix					
4.	Jolly Hill Gut, Jolly Hill	2.10	1963–1969	0	0.02	[lll]223
			1983			

(Continued)

Table 5A.2 (Continued)

	Gaging Station			Streamflow Characteristics		
	Name	Drainage Area (mL[2])	Period of Analysis	7-Day, 10-Year Low Flow (ft^3/sec)	Average Discharge (ft^3/sec)	100-Year Flood (ft^3/sec)
	Guam, American Samoa, and the Trust Territory of the Pacific Islands					
	Aipan					
1.	S. F. Talofofo Stream	0.64	1968–1984	0.0	1.35	—
	Guam					
2.	Ugum River	5.76	1977–1984	3.6	23.3	—
3.	Ylig River	6.48	1952–1984	0.2	28.0	5,980
4.	Pago River	5.67	1951–1982	0.2	26.3	12,300
	Palau					
5.	Diongradid River	4.45	1969–1984	3.2	32.4	2,870
6.	Tabecheding River	6.07	1970–1984	1.7	48.4	4,910
	Yap					
7.	Oaringeel Stream	0.24	1968–1984	0.1	1.07	696
	Truk					
8.	Wichen River	0.57	1968–1983	0.02	3.05	1,060
	Pohnpei					
9.	Nanpil River	3.00	1970–1984	1.8	44.6	10,000
	Kosrae					
10.	Malem River	0.76	1971–1981 1982–1984	0.3	6.71	2,760
	American Samoa					
11.	Aasu Stream	1.03	1958–1984	0.4	6.05	586
12.	Afuelo Stream	0.25	1958–1984	0.03	1.45	683
	Utah					
	Upper Colorado Region					
	Colorado River Main Stem[mmm]					
1.	Colorado River, Cisco	24,100[nnn]	1895–1984	1,100	7,563	87,600
	Upper–Coloradodolores Subregion					
	Dolores River Basin					
2.	Dolores River, Cisco	[nnn]4,580	1951–1984	19	785	23,600
	Great Divide–Upper Green and Lower Green Subregions					
	Green River Basin					
3.	Green River, Jensen	[nnn]29,660	[ooo]1904–1984	480[ppp] 743	4,396[ppp] 4,456	38,200

Gaging station: Period of analysis is for the water years used to compute average discharge and may differ from that used to compute other streamflow characteristics. Streamflow characteristics: The 7-day, 10-year low flow is a discharge statistic; the lowest average discharge during 7 consecutive days of a year will be equal to or less than this value, on the average, once every 10 years. The average discharge is the arithmetic average annual discharges during the period of analysis. The 100-year flood is the peak flow that has a 1-percent chance of being equaled or exceeded in a given year. The degree of regulation is the effect of dams on the natural flow of the river. Abbreviations: Do = ditto; mi^2 = square miles; ft^3/sec = cubic feet per second; … = insufficient data or not applicable.

[a] Less than 10 years of record. Minimum discharge and maximum instantaneous discharge for period of record are shown.
[b] Record interrupted.
[c] Adjusted for no-flow periods.
[d] Adjusted for high-outlier in period of record. Did not use 1981 peak because it was regulated.
[e] Adjusted for high-outlier in period of record.
[f] Sutter and Yolo Bypasses Carry Much of Floodflow Past Verona gage.
[g] Regulation has Little Effect on High Floodflows.
[h] From Upper Mississippi River Basin Commission, 1978.
[i] Within the Upper Mississippi–Black–Root, Upper Mississippi–Maquoketa–Plum, and Upper Mississippi–Iowa–Skunk–Wapsipinicon Subregion (Seaber and Others, 1984).
[j] Within the Upper Mississippi–Iowa–Skunk–Wapsipinicon Subregions (Seaber and Others, 1984).
[k] Within the Minnesota Des Moines Subregions(Seaber and Others, 1984).
[l] Within the Missouri–Big Sioux, Missouri–Little Sioux, and Missouri–Nishnabotna Subregions (Seaber and Others, 1984).
[m] Flow Parameters Based Only on 1929–1931 and 1939–1956 Water Years.
[n] From U.S. Army Corps of Engineers, February 1978.
[o] Within the Missouri–Big Sioux, Missouri–Little Sioux, and Missouri–Nishnabotna Subregions (Seaber and Others, 1984).
[p] Within the Missouri–Nishnabotna, Chariton–Grand, and Upper Mississippi–Salt Subregions (Seaber and Others, 1984).
[q] Based on period of analysis since regulation began. These values are not based on detailed analyses, are approximate estimates, and are for information purposes only.

(Continued)

Table 5A.2 (Continued)

[r] From flood-insurance hydrology study. Based on detailed analyses of regulated-flow conditions.
[s] Prior to opening of Barkley–Kentucky Canal (1889–1965).
[t] Since the opening of Barkley–Kentucky Canal (1965–1983).
[u] Includes all or parts of the Lower Mississippi–Yazoo, Lower Mississippi–Big Black, Lower Mississippi–Lake Maurepas, and the Lower Mississippi Subregions (Seaber, Kapinos, and Knapp, 1984).
[v] Includes the Mississippi Headwaters and the Upper Mississippi Black-Roof Subregions.
[w] Data furnished by U.S. Army Corps of Engineers.
[x] Includes the Saskatchewan, the Missouri Headwaters, the Missouri–Marias, the Missouri–Musselshell, the Milk, and the Missouri–Polar Subregions.
[y] Includes the upper Yellowstone, the Big Horn, the Powder–Tongue, the lower Yellowstone, and the Missouri–Little Missouri Subregions.
[z] Contained within the Kootenai–Pend Oreille–Spokane Subregion.
[aa] Within the Missouri–Big Sioux, Missouri–Little Sioux, and Missouri–Nishnabotna Subregions.
[bb] Analysee based on period of record since regulation began.
[cc] Based on record to 1981.
[dd] Period of record not continuous.
[ee] Adjusted for diversion and change in reservoir contents.
[ff] Analysis based no regulated period 1955–1984.
[gg] Period of record not continuous.
[hh] Also includes parts of the Upper Arkansas, Upper Cimarron, Lower Canadian, North Canadian, and Red Headwaters Subregions.
[ii] Fifteen years, prior to completion of Conchas Dam.
[jj] Twenty-four years, prior to completion of Ute dam.
[kk] Twenty-one years (1963–1983), subsequent to completion of Ute Dam.
[ll] Includes all or parts of Rio Grande Headwaters, Rio Grande–Elephant–Butte, Rio Grande–Mimbres, and Rio Grande Closed basins Subregions.
[mm] Thirty-two years, prior to closure of Cochiti Dam.
[nn] Ten years (1974–1983), subsequent to closure of Cochiti Dam.
[oo] Includes parts of the Little Colorado, Upper Gila and Sonora Subregions.
[pp] Approximate.
[qq] Noncontributing.
[rr] Includes parts or all of the Upper Cimarron, Arkansas–Keystone, Lower Cimarron, Lower Arkansas, Neosho–Verdigris, Lower Canadian and North Canadian Subregions.
[ss] Includes parts or all of the Red Headwaters, Red–Washita, and Red–Sulphur Subregions.
[tt] Estimated.
[uu] Drainage area includes 38 mi^2 which are partly or entirely noncontributing and excludes 6.0 mi^2 upstream from Lago El Guinea and Lago de Matrullas.
[vv] Drainage area includes 25.4 mi^2 which do not contribute directly to surface runoff.
[ww] Drainage area excludes 8.2 mi^2 upstream form Lago Carite, flow from which is diverted to the Rio Guamani.
[xx] Drainage area includes 39.7 mi^2 from headwaters of Lago Yahuecas (17.05 mi^2), Lago Guayo (9.67 mi^2), Lago prieto (9.50 mi^2), and Lago Toro (3.5 mi^3) which does not contribute to surface runoff except at high stages.
[yy] Analysis based on records collected since regulation began.
[zz] Within the Missouri–Oahe, Missouri–White, and Missouri–Big Sioux Subregions.
[aaa] Station discontinued subsequent to construction of Oahe Dam in 1962.
[bbb] Analysis based on period of record after regulation began.
[ccc] Within the Missouri–Oahe, Missouri–Little Missouri, Cheyenne, Missouri–White, and Niobrara Subregions.
[ddd] Period of record not continuous.
[eee] Within the James and Missouri–Big Sioux Subregions.
[fff] Within the Upper Canadian, Lower Canadian, North Canadian, Red Headwaters, Red–Washita, and Red–Sulphur Subregions.
[ggg] Within the Sabine, neches, Triniy, and Galveston Bay–San Subregions.
[hhh] Within the Brazos Headwaters, Middle Brazos, Lower Brazos, Upper Colorado, and Lower Colorado–San Bernard Coastal Subregions.
[iii] Within the Central Texas Coastal and Nueces–Southwestern Texas Coastal Subregions.
[jjj] Within the Rio Grande–Mimbres, Rio Grande Amistad, Rio Grande Closed Basins, Upper Pecos, Lower Pecos, Rio Grande–Falcon, and Lower Rio Grande Subregions.
[kkk] Discharge represents highest recorded. Data available are not adequate to determine a discharge–frequency relation, but it is estimated to have exceeded the 100-year flood.
[lll] Discharge represents highest recorded.
[mmm] Within the Upper Colorado–Dolores and Upper Colorado–Dirty Devil Subregions.
[nnn] Approximate.
[ooo] Period of analysis not continuous.
[ppp] Since completion of Flaming Gorge Reservoir in 1963.
[qqq] Based on record to 1981.

Source: From U.S. Geological Survey, *National Water Summary 1985—Hydrologic Events and Surface-Water Resources, Water-Supply Paper 2300.*

Original Source: Reports of the U.S. Geological Survey and State agencies.

Table 5A.3 Monthly Discharge of Principal Rivers in the United States

River and Station	Basin Area (km²)	Mean Monthly Discharge (m³/sec) Jan	Feb	Mar	Apr	May	Jun	Jul	Aug	Sep	Oct	Nov	Dec	Year	Period of Record
Penobscot, West Enfield, ME	17,000	214	186	290	809	685	345	221	181	176	224	328	282	327	1902–65
Kennebec, Bingham, ME	7,040	94.4	99.5	116	202	259	136	94.8	89.6	89.0	88.2	93.3	92.9	121	1907–10
Androscoggin, Auburn, ME	8,436	123	111	194	428	336	161	93.5	81.7	93.4	107	149	137	168	1929–65
Merrimack, Lowell, MA	12,000	187	188	348	542	320	165	86.6	71.3	90.8	98.4	171	193	205	1923–65
Connecticut, Thompsonville, CT	25,020	385	341	677	1,300	763	365	194	172	202	237	390	393	451[1]	1928–65
Delaware, Trenton, NJ	17,600	351	353	611	653	387	235	197	170	157	179	292	334	327	1912–65
Susquehanna, Harrisburg, PA	62,400	1,070	1,100	2,210	2,090	1,290	703	422	332	306	457	696	877	963[1]	1890–1965
Potomac, Washington, DC	29,940	369	461	642	552	379	221	132	143	110	150	161	245	296	1930–65
James, Richmond, VA	17,500	260	300	358	316	192	126	83.5	106	85.1	95.3	105	185	184	1934–65
Roanoke, Roanoke Rapids, NC	21,800	304	329	362	315	233	182	175	183	158	155	153	221	230	1911–65
Cape Fear, Lillington, NC	8,910	143	181	169	140	66.6	49.7	64.9	61.3	72.7	59.0	57.6	87.7	95.9	1923–65
Pee Dee, Pee Dee, SC	22,900	316	431	471	396	241	179	190	183	210	182	171	224	265	1938–65
Santee, Pineville, SC	38,100	58.6	123	161	139	47.8	24.8	22.2	18.8	63.6	72.9	31.6	43.0	67.0	1942–65
Savannah, Clyo, GA	25,500	418	438	519	520	338	240	244	253	222	277	249	313	336	1929–33, 37–65
Altamaha, Doctortown, GA	35,200	477	636	795	782	388	239	231	214	177	177	162	295	385	1931–65
St. Johns, De Land, FL	8,080	78.8	69.0	73.4	70.8	46.2	46.8	80.2	100	124	168	144	103	92.2	1933–65
Suwannee, Branford, FL	20,000	169	203	287	316	207	145	148	180	185	175	141	142	191	1931–65
Apalachicola, Chattahoochee, FL	44,300	752	876	1,150	1,040	652	465	477	429	341	361	359	550	620[1]	1928–65
Escambia, Century, FL	9,886	223	246	339	337	166	108	120	123	95.6	74.6	90.5	160	173	1934–65
Alabama, Claiborne, AL	57,000	1,280	1,550	1,830	1,770	899	531	519	484	369	367	433	834	902[1]	1930–65
Tombigbee, Leroy, AL	49,500	1,190	1,580	1,750	1,530	714	318	330	208	175	162	347	661	742[1]	1928–60
Pascagoula, Merrill, MS	17,000	381	492	561	515	279	140	170	124	100	78.5	133	269	269	1930–65
Pearl, Bogalusa, LA	17,200	332	483	554	502	298	138	144	103	74.8	70.0	101	222	250	1938–65
Ohio, Louisville, KY	236,100	4,880	5,470	7,210	6,020	3,730	2,280	1,610	1,190	743	738	1,500	2,780	3,120[1]	1928–65
Wabash, Mount Carmel, IL	74,100	1,130	1,110	1,370	1,380	1,130	742	528	270	194	217	332	537	744[1]	1928–65
Cumberland, Smithland, KY	46,395	1,210	1,580	1,740	1,390	718	446	348	272	217	188	383	809	779	1939–65
Tennessee, Paducah, KY	104,000	2,890	3,590	3,160	2,100	1,520	1,170	1,140	1,080	1,050	987	1,370	2,130	1,820[1]	1939–65
Ohio, Metropolis, IL	526,000	10,900	13,400	14,900	13,400	8,650	5,480	4,070	3,040	2,300	2,180	3,640	6,200	7,300[1]	1928–65
Fox, Wrightstown, WS	15,900	109	112	132	199	171	143	95.4	76.2	73.1	86.4	99.8	105	117	1896–1965
Grand, Grand Rapids, MI	12,700	90.7	111	213	185	128	88.9	54.4	41.0	46.0	54.0	64.8	70.9	95.2	1904–65
Maumee, Waterville, OH	16,350	197	203	317	268	157	92.4	47.4	22.7	22.4	31.5	58.7	108	127	1899–1901
St. Lawrence, Ogdensburg, NY	764,600	6,230	6,150	6,430	6,970	7,230	7,340	7,290	7,100	6,860	6,630	6,510	6,450	6,760	1860–1965
Red of the North, Grand Forks, ND	78,000	19.5	18.0	48.1	238	138	108	80.3	44.1	36.0	35.2	31.4	24.8	68.5	1882–1965
Mississippi, Clinton, IA	222,000	677	730	1,330	2,420	2,280	1,990	1,560	1,030	1,040	1,100	1,030	708	1,330[1]	1874–1965
Mississippi, Alton, IL	444,200	1,790	2,120	3,540	4,870	4,290	3,680	2,840	1,740	1,650	1,660	1,790	1,530	2,620[1]	1927–65
Missouri, Culbertson, MT	237,130	187	171	212	259	205	184	236	363	389	377	249	197	252	1941–51; 58–65
Yellowstone, Sidney, MT	178,220	143	173	307	310	527	1,200	690	257	197	228	197	153	365[1]	1910–31; 33–65
Missouri, Yankton, SD	723,900	267	299	600	1,120	829	1,250	1,060	737	696	663	485	263	689[1]	1930–65
Platte, South Bend	221,000	84.7	148	249	233	218	267	134	88.6	93.7	87.6	109	93.4	150	1953–65
Missouri, Nebraska City, NE	1,073,000	374	532	1,060	1,540	1,200	1,690	1,360	918	848	807	674	384	948[1]	1929–65
Kansas, Bonner Springs,	155,100	67.0	106	161	221	261	426	332	169	180	136	99.1	67.5	185	1917–65

(Continued)

Table 5A.3 (Continued)

River and Station	Basin Area (km²)	Mean Monthly Discharge (m³/sec) Jan	Feb	Mar	Apr	May	Jun	Jul	Aug	Sep	Oct	Nov	Dec	Year	Period of Record
Missouri, Herman, MO	1,368,000	1,080	1,420	2,430	3,360	3,020	4,160	3,390	1,930	1,820	1,600	1,440	1,050	2,220[1]	1897–1965
White, DeValls Bluff, AR	60,686	850	943	1,140	1,270	1,230	829	505	339	291	305	402	542	720[1]	1928–45; 50–65
Arkansas, Tulsa, OK	193,250	73.6	90.9	118	261	375	368	277	151	164	185	116	82.6	188	1925–65
Canadian, Whitefield, OK	123,220	61.5	102	147	255	430	291	164	69.0	88.6	117	87.0	76.9	157	1938–65
Arkansas, Little Rock, AR	409,741	987	1,160	1,330	1,810	2,320	1,780	1,070	545	555	762	695	721	1,140[1]	1927–65
Mississippi, Vicksburg, MS	2,964,300	16,200	21,100	24,800	27,900	23,600	18,500	14,600	9,340	7,320	7,090	8,220	11,100	15,800[1]	1928–65
Red, Alexandria, LA	175,000	1,100	1,330	1,380	1,350	1,600	1,150	555	305	260	342	432	673	880[1]	1928–65
Ouachita, Monroe, LA	39,622	627	879	1,010	1,039	940	524	272	103	90.1	109	174	316	507[1]	1932–65
Mississippi, Tarbert Landing	3,923,800	17,900	23,000	27,800	30,800	27,900	21,400	16,500	10,500	8,100	7,930	8,750	12,100	17,700[1]	1928–65
Sabine, Ruliff, TX	24,160	399	416	406	366	381	252	120	82.9	57.5	55.5	116	221	239	1927–65
Neches, Evadale, TX	20,590	273	297	308	308	325	188	73.4	37.3	28.2	36.6	81.1	164	176	1904–06; 21–65
Trinity, Romayor, TX	44,512	247	262	291	278	439	295	118	41.4	53.6	86.4	122	185	202	1924–65
Brazos, Richmond, TX	114,000	200	234	231	257	459	303	145	68.9	99.2	153	140	188	207	1903–06; 22–65
Colorado, Wharton, TX	107,200	67.6	79.5	67.5	101	133	131	82.8	47.0	72.2	77.2	80.6	69.8	80.7	1919–25; 38–65
Nueces, Mathis, TX	43,150	8.6	12.0	11.4	15.0	43.5	38.9	29.5	6.9	42.7	38.2	10.8	4.2	21.9	1939–65
Pecos, Shumla, TX	91,069	5.42	5.28	4.82	6.57	13.4	9.84	7.23	5.28	16.6	12.7	6.31	5.44	8.24	1954–65
Rio Grande, Laredo, TX	352,178	73.7	72.2	61.9	69.7	121	158	136	143	269	188	96.8	82.5	123	1900–14; 22–65
Green, Green River, UT	105,000	49.3	64.8	124	222	485	600	250	108	76.8	77.4	68.5	50.6	181	1895–99; 1905–65
Colorado, Lees Ferry, AZ	279,500	150	192	274	561	1,280	1,600	657	313	236	245	210	163	489[1]	1911–65
Colorado, Yuma, AZ	629,100	222	262	293	367	665	1,040	586	306	230	214	200	213	383	1902–65
Sevier, Juab, UT	13,300	0.67	0.35	1.28	6.86	20.4	13.6	14.2	9.13	5.98	2.36	1.05	0.38	6.36	1911–65
Humboldt, Imlay, NV	40,700	2.09	3.53	6.60	9.83	14.2	13.9	10.8	2.54	1.05	0.71	0.92	1.71	5.66	1935–41; 45–65
San Joaquin, Vernalis, CA	35,070	126	181	187	199	239	212	64.9	28.1	33.8	46.8	58.6	99.2	124	1922–65
Sacramento, Sacramento, CA	60,940	940	1,180	1,020	960	850	501	299	289	323	320	419	681	647[1]	1948–65
Eel, Scotia, CA	8,063	508	566	349	262	108	35.3	9.42	4.08	3.57	22.1	121	370	200	1910–65
Klamath, Klamath, CA	31,300	806	1,050	749	793	648	381	162	94.2	93.6	148	348	629	486	1910–26; 50–65
Chehalis, Porter, WA	3,351	279	251	182	131	64.5	31.0	15.8	10.9	12.4	38.9	176	239	119	1952–65
Pend Oreille, Newport, WA	62,700	377	382	439	723	1,540	2,060	1,210	505	339	340	402	391	726[1]	1903–41; 52–65
Columbia International Bdry.	155,000	1,080	1,110	1,200	1,930	5,040	8,140	5,920	3,130	1,940	1,600	1,430	1,190	2,810	1937–65
Snake, Clarkston, WA	267,300	866	1,010	1,320	2,300	3,540	3,140	1,130	555	549	666	777	887	1,390[1]	1915–65
Columbia, Dalles, OR	614,000	2,750	3,020	3,590	5,700	10,300	14,000	9,630	5,270	3,470	2,810	2,800	2,820	5,520[1]	1878–1965
Willamette, Salem, OR	18,900	1,300	1,250	971	824	624	412	205	134	147	253	772	1,130	665[1]	1909–16; 23–65
Cowlitz, Castle Rock, WA	5,796	371	354	297	323	345	291	150	76.3	65.2	122	302	415	259	1927–65
Umpqua,Elkton, OR	9,539	443	450	350	278	191	113	50.9	33.6	33.4	55.8	194	361	212	1905–65
Rogue, Agness, OR	10,200	312	352	260	229	181	94.2	46.7	37.1	35.9	64.1	154	427	182	1960–65
Copper, Chitina, AK	53,300	160	130	124	147	848	2,320	3,250	2,920	1,460	615	303	200	1,040[1]	1955–65
Kuskokwim, Crooked Creek, AK	80,500	385	328	284	315	2,080	2,740	2,000	2,420	2,470	1,300	599	457	1,280	1951–65
Yukon, Eagle, AK	294,000	475	433	394	408	3,220	6,160	5,130	3,980	2,980	1,930	980	598	2,220[1]	1911–14; 50–65
Yukon, Rampart, AK	516,400	674	591	502	517	5,420	11,500	7,950	6,480	5,090	2,780	1,220	799	3,630[1]	1955–65
Yukon, Kaltag, AK	767,000	1,320	1,100	915	930	7,980	18,300	13,100	11,800	9,820	5,400	2,350	1,490	6,210[1]	1956–65

[1] Monthly and yearly averages rounded to three significant figures.

Source: From UNESCO, 1971.

Table 5A.4 Length of Principal Rivers in the United States and Canada

River/Outflow	Length	
	Miles	km
Alabama/Mobile River	735	1,183
Albany/James Bay	610	981
Arkansas/Mississippi River	1,459	2,348
Black/Chantrey Inlet	600	965
Brazos/Gulf of Mexico	870	1,400
Canadian/Arkansas River	906	1,458
Churchill/Hudson Bay	1,000	1,609
Cimarron/Arkansas River	600	965
Colorado (U.S.-Mex.)/Gulf of California	1,450	2,333
Colorado (Texas)/Matagorda Bay	840	1,352
Columbia/Pacific Ocean	1,243	2,000
Columbia, Upper Mouth of Snake River	890	1,432
Cumberland/Ohio River	720	1,158
Fraser/Strait of Georgia	850	1,368
Gila/Colorado River	630	1,014
Green (UT–WY)/Colorado River	730	1,175
Hamilton/Atlantic Ocean	600	965
James (ND–SD)/Missouri River	710	1,142
Kuskokwim/Kuskokwim Bay	680	1,094
Liard/Mackenzie River	693	1,115
Mackenzie/Arctic Ocean	900	1,448
Milk/Missouri River	625	1,006
Mississippi/Mouth of SW Pass	2,348	3,778
Mississippi, Upper/to Mouth of Missouri River	1,171	1,884
Mississippi–Missouri–Red Rock/Mouth of SW Pass	3,710	5,969
Missouri/Mississippi River	2,315	3,725
Missouri–Red Rock/Mississippi River	2,533	4,076
Mobile-Alabama–Coosa/Mobile Bay	780	1,255
North Canadian/Canadian River	760	1,223
North Platte/Platte River	618	994
Ohio/Mississippi River	981	1,578
Ohio–Allegheny/Mississippi River	1,306	2,101
Ottawa/St. Lawrence	790	1,271
Ouachita/Red River	605	973
Peace/Slave River	1,195	1,923
Pecos/Rio Grande	735	1,183
Red (OK–TX–LA)/Mississippi River	1,270	2,043
Rio Grande/Gulf of Mexico	1,885	3,033
St. Lawrence/Lake Ontario	800	1,287
Saskatchewan N./Lake Winnipeg	1,100	1,770
Saskatchewan S./Lake Winnipeg	1,205	1,939
Severn (Ontario)/Hudson Bay	610	981
Snake/Columbia River	1,038	1,670
Tanana/Yukon River	620	998
Tennessee/Ohio River	652	1,049
Tennessee–French Broad/Ohio River	900	1,448
White (AR–MO)/Mississippi River	720	1,158
Yellowstone/Missouri River	671	1,080
Yukon/Bering Sea	1,979	3,185

Note: Comprises rivers 600 miles or more in length. Length represents distance to designated outflow from (a) original headwater of named river where name applies to entire length of channel, or (b) upper limit of channel so named, usually the junction of two tributaries or headwater streams.

Source: From Statistical Abstract of the United States 1986.

Table 5A.5 Flowing Water Resources of the United States

Stream Order[a]	Number Streams	Average Length Miles	Total Length Miles (L)	Drainage Area Sq. Miles (A_d)	Mean Flow For Area Drained (CFS)	Mean Width Feet (W)	Mean Depth Feet (D)	Mean Velocity ft/sec (V)	Calculated Discharge CFS = WDV	Total Surface Area, A_sSq. Miles (thousands)	Total Channel Storage Acre Feet (millions)
1	1,570,000	1	1,570,000	1	0.65	4	0.15	1.0	0.60	1.2	0.11
2	350,000	2.1	810,000	4.7	3.1	10	0.29	1.3	3.7	1.5	0.29
3	80,000	5.3	420,000	23	15.0	18	0.58	1.5	15.6	1.4	0.53
4	18,000	12	220,000	109	71.0	37	1.10	1.8	73	1.5	1.1
5	4,200	28	116,000	518	340	75	2.20	2.3	380	1.6	2.4
6	950	64	61,000	2,500	1,600	160	4.1	2.7	1,800	1.8	4.9
7	200	147	30,000	12,000	7,600	320	8.0	3.3	8,500	1.8	9.3
8	41	338	14,000	56,000	36,000	650	15.0	3.9	38,000	1.7	16.5
9	8	777	6,200	260,000	171,000	1,300	29.0	5.6	211,000	1.5	28.3
10	1	1,800	1,800	1,250,000	810,000	2,800	55.0	5.9	900,000	1.0	34.3
Total	2,023,000		3,249,000							15.0	97.0

Note: Based on stream order and channel morphology.

[a] Stream order classification based on river characteristics. A first order stream has no tributary channels; a second order stream is formed when two first order streams merge. When two second order streams merge, a third order stream is formed, and so on, downstream in the drainage basin until the water is discharged to the sea.

Source: From Keup, L.E., Flowing Water Resources, *Water Resources Bulletin*, V.21, no.2, 1985. Reprinted with permission.

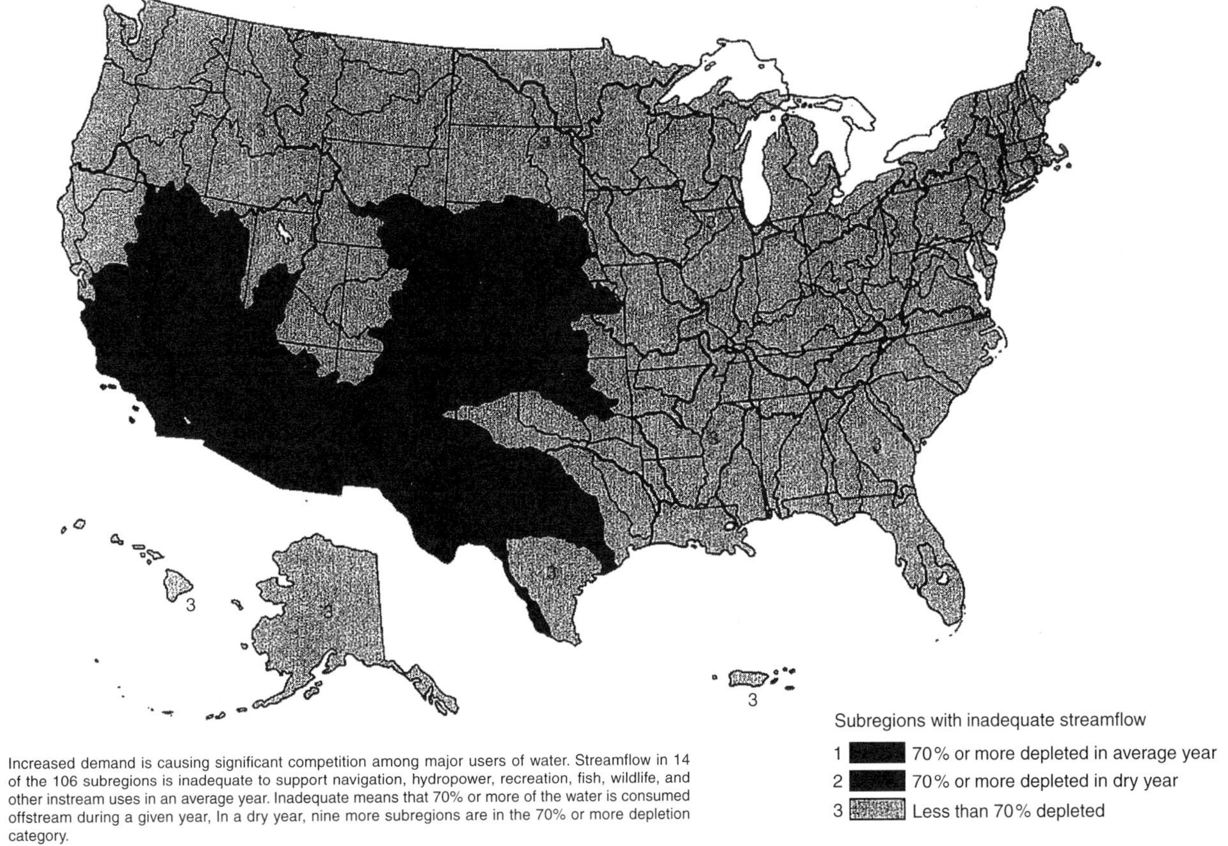

3
3
Subregions with inadequate streamflow
1 70% or more depleted in average year
2 70% or more depleted in dry year
3 Less than 70% depleted
Increased demand is causing significant competition among major users of water. Streamflow in 14 of the 106 subregions is inadequate to support navigation, hydropower, recreation, fish, wildlife, and other instream uses in an average year. Inadequate means that 70% or more of the water is consumed offstream during a given year, In a dry year, nine more subregions are in the 70% or more depletion category.

Figure 5A.2 Inadequate surface water supply for instream use in the United States. (From COUNCIL on Environmental Quality, 1981, Environmental Trends.)

Table 5A.6 Velocity of Low Flows and Average Length of Streams in the United States

Water Resource Region	Mean Velocity (mi/hour)	Average Stream Length (miles)	Mean Depth at Velocities Given in Col. 1 (ft)	Mean Flow at Velocity Given in Col. 1 (ft³/sec)
New England	1½	100	4.0	1,500
Delaware-Hudson	1½	75	3.0	500
Eastern Great Lakes	1½	50	3.0	800
Western Great Lakes	1½	50	3.0	800
Chesapeake Bay	1½	75	5.0	2,500
Ohio	1	100	4.0	1,500
Cumberland	1½	100	3.0	500
Tennessee	1½	75	4.0	1,200
Southeast	1½	150	5.0	2,500
Upper Mississippi	1	150	3.0	500
Lower Mississippi	1½	150	4.0	1,200
Upper Missouri	1½	250	3.0	500
Lower Missouri	1	125	3.5	900
Upper Arkansas, Red	1½	200	2.0	200
Lower Arkansas, Red, and White	1½	175	3.5	1,000
Western Gulf	1	300	4.0	1,500
Rio Grande and Pecos	1½	150	1.5	100
Colorado	1	150	2.0	300
Great Basin	1	100	1.0	50
Pacific Northwest	1	150	4.0	1,500
Central Pacific	1½	100	4.0	1,500
South Pacific		50	0.5	20

Note: Velocities are estimated for discharges which are exceeded 95 percent of the time. Stream lengths are estimated for representative streams in each region. For location of river basins see Figure 2.6.

Source: From U.S. Geological Survey.

Table 5A.7 Annual River Flow Rates in Canada

Ocean Basin Region	River Basin Region	Annual Flow Rates[a](m³/sec)		
		Reliable[b](Low)	Mean	High[c]
Pacific	1. Pacific Coastal	12,570	16,390	20,200
	2. Fraser-Lower Mainland[d]	3,044	3,972	4,900
	3. Okanagan-Similkameen[d]	31	74	116
	4. Columbia[e]	1,644	2,009	2,373
	5. Yukon[e]	1,806	2,506	3,206
Arctic	6. Peace-Athabasca	1,862	2,903	3,946
	7. Lower Mackenzie[f]	6,114	7,337	8,561
	8. Arctic Coast-Islands	5,920	10,251	14,582
Gulf of Mexico	9. Missouri[e]	3	12	41
Hudson Bay	10. North Saskatchewan	160	234	373
	11. South Saskatchewan	147	239	418
	12. Assiniboine-Red[e]	16	50	188
	13. Winnipeg[d,e]	382	758	1,137
	14. Lower Saskatchewan-Nelsone[f]	1,108	1,911	2,714
	15. Churchill[d]	323	701	1,070
	16. Keewatin	2,945	3,876	4,806
	17. Northern Ontario[d]	3,733	5,995	8,258
	18. Northern Quebec[d]	12,820	16,830	20,830
Atlantic	19. Great Lakes	2,403	3,067	3,733
	20. Ottawa	1,390	1,990	2,590
	21. St. Lawrence[e,f]	1,504	2,140	2,777
	22. North Shore-Gaspé	6,437	8,706	10,980
	23. St. John-St. Croix[e]	507	779	1,050
	24. Maritime Coastal	2,079	3,081	4,085
	25. Newfoundland-Labrador	6,908	9,324	11,739
Canada		75,856	105,135	134,674

Note: For map of river basin regions see Figure 2.8.

[a] From recorded flows except in Prairie basins where natural flows have been estimated.
[b] Flow equalled or exceeded in 19 years out of 20.
[c] Flow equalled or exceeded in 1 year out of 20.
[d] Excludes flow transferred into neighboring basin region; because this flow is recorded in importing basin, transfers have little effect on national total.
[e] Excludes inflow from United States portion of basin region.
[f] Excludes inflow from upper basin region.

Source: From Pearse, P.H., Currents of change, *Final Report Inquiry on Federal Water Policy*, Ottawa, Canada, 1985.

Table 5A.8 Longest Rivers of the World

Name	Outflow	Length[a] Miles	km	Rank in World (first 100)
World				
Nile	Mediterranean Sea	4,132	6,650	1
Amazon	South Atlantic Ocean	4,000	6,400	2
Yantze	East China Sea	3,915	6,300	3
Mississippi	Gulf of Mexico	3,710	5,971	4
Yenisey	Kara Sea	3,442	5,540	5
Huang Ho (Yellow)	Gulf of Cihli	3,395	5,464	6
Ob	Gulf of Ob	3,362	5,410	7
Paraná	Rio de la Plata	3,032	4,880	8
Congo	South Atlantic Ocean	2,900	4,700	9
Amur	Sea of Okhotak	2,761	4,444	10
Africa				
Nile	Mediterranean Sea	4,132	6,650	1
Congo	South Atlantic Ocean	2,900	4,700	9
Niger	Bight of Blafra	2,600	4,200	16
Zambezi	Mozambique Channel	2,200	3,500	24
Kasai	Congo River	1,338	2,153	63
Orange	South Atlantic Ocean	1,300	2,100	66
White Nile (al-Bahr al-Abyad)	Nile River	1,295	2,084	67
Lualaba	Congo River	1,100	1,800	87
Limpopo	Mozambique Channel	1,100	1,800	87
Jubba (Juba)	Indian Ocean	1,030	1,658	94
Sénégal	South Atlantic Ocean	1,020	1,641	95
Okavango (Kubango)	Okavango Swamp	1,000	1,600	100
Lomami	Congo River	830	1,500	
Blue Nile (al-Bahr al-Azraq)	White Nile River	907	1,460	
Chari (Shari)	Lake Chad	870	1,400	
Ubangi-Uele	Congo River	870	1,400	
Awash	Lake Abe	750	1,200	
America, North				
Mississippi-Missouri-Red Rock	Gulf of Mexico	3,710	5,971	4
Mackenzie-Slave-Peace	Beaufort Sea	2,635	4,241	15
Missouri-Red Rock	Mississippi River	2,533	4,076	17
St. Lawrence-Great Lakes	Gulf of Saint Lawrence	2,500	4,000	19
Mississippi	Gulf of Mexico	2,348	3,779	20
Missouri	Mississippi River	2,315	3,726	21
Yukon–Nisutlin	Bering Sea	1,979	3,185	28
Rio Grande	Gulf of Mexico	1,885	3,034	29
Yukon	Bering Sea	1,875	3,018	31
Nelson Saskatchewan	Hudson Bay	1,600	2,575	46
Arkansas	Mississippi River	1,459	2,348	55
Colorado	Gulf of California	1,450	2,333	56
Ohio–Allegheny	Mississippi River	1,306	2,102	65
Red	Mississippi River	1,270	2,044	68
Columbia	North Pacific Ocean	1,243	2,000	71
Saskatchewan	Lake Winnipeg	1,205	1,939	76
Peace	Slave River	1,195	1,923	79
Snake	Columbia River	1,038	1,670	93
Churchill	Hudson Bay	1,000	1,609	99
Ohio	Mississippi River	981	1,579	
Canadian	Arkansas River	906	1,456	
Tennessee–French Broad	Ohio River	900	1,448	
Upper Columbia	Columbia River	890	1,432	
Brazos	Gulf of Mexico	870	1,400	
South Saskatchewan	Saskatchewan River	865	1,392	
Fraser	Strait of Georgia	850	1,368	
Colorado (of Texas)	Matagorda Bay	840	1,352	
St. Lawrence	Lake Ontario	800	1,287	
North Saskatchewan	Saskatchewan River	800	1,287	
Ottawa	St. Lawrence	790	1,271	

(Continued)

Table 5A.8 (Continued)

Name	Outflow	Length[a] Miles	km	Rank in World (first 100)
North Canadian	Canadian River	760	1,223	
Pecos	Rio Grande River	735	1,183	
Kuskokwim	Bering Sea	680	1,094	
America, South				
Amazon-Ucayali-Apurimac	South Atlantic Ocean	4,000	6,400	2
Paraná	Rio de la Plata	3,032	4,880	8
Madeira-Mamoré-Guaporé	Amazon River	2,082	3,350	25
Jurua	Amazon River	2,040	3,283	26
Purus	Amazon River	1,995	3,211	27
Sáo Francisco	South Atlantic Ocean	1,811	2,914	33
Japurá (Caquetá)	Amazon River	1,750	2,816	37
Ucayali-Apurimac	Amazon River	1,701	2,738	40
Orinoco	South Atlantic Ocean	1,700	2,736	41
Tocantins	Pará River	1,677	2,699	42
Araguaia	Tocantins River	1,632	2,627	44
Paraguay	Paraná River	1,584	2,550	47
Pilcomayo	Paraguay River	1,550	2,500	51
Negro (Guainia)	Amazon River	1,400	2,253	59
Xingu	Amazon River	1,300	2,100	66
Tapajos-Teles Pires	Amazon River	1,238	1,992	72
Mamoré	Guaporé River	1,200	1,931	77
Marañón-Hualiaga	Amazon River	1,184	1,905	80
Guaporé (Iténez)	Mamore River	1,087	1,749	90
Parnaiba	South Atlantic Ocean	1,056	1,700	92
Madre de Dios	Beni River	1,056	1,700	92
Putumayo (Iça)	Amazon River	1,000	1,609	99
Solimões	Amazon River	1,000	1,609	99
Uruguay	Río de la Plata	990	1,593	
Magdalena	Caribbean Sea	930	1,497	
Guaviare	Orinoco River	930	1,497	
Ucayali	Marañón River	910	1,465	
Teles Pires	Tapajós River	870	1,400	
Grande	Mamoré River	845	1,360	
Cauca	Magdalena River	838	1,349	
Iguaçu	Paraná River	808	1,300	
Asia				
Yangtze	East China Sea	3,915	6,300	3
Yenisey-Balkal-Selenga	Kara Sea	3,442	5,540	5
Huang Ho (Yellow)	Gulf of Chihli	3,395	5,464	6
Oh-Irtysh	Gulf of Ob	3,362	5,410	7
Amur-Argun	Sea of Okhotsk	2,761	4,444	10
Lena	Laptev Sea	2,734	4,400	11
Mekong	South China Sea	2,700	4,350	12
Ob-Katun	Gulf of Ob	2,696	4,338	13
Irtysh–Chorny Irtysh	Ob River	2,640	4,248	14
Yenisey	Kara Sea	2,549	4,102	18
Ob	Gulf of Ob	2,268	3,650	22
Syrdarya–Arabelsu	Aral Sea	1,876	3,019	30
Nizhnyaya Tunguska	Yenisey River	1,857	2,989	32
Brahmaputra	Jamuna River	1,800	2,900	34
Indus	Arabian Sea	1,800	2,900	34
Amur	Sea of Okhotak	1,755	2,824	36
Euphrates	Shatt-al-Arab	1,740	2,800	38
Vilyuy	Lena River	1,647	2,650	43
Amu Darya–Pyandzh	Aral Sea	1,578	2,540	48
Kolyma–Kulu	East Siberian Sea	1,562	2,513	49
Ganges	Padma River	1,560	2,510	50
Ishim	Irtysh River	1,522	2,450	52
Salween	Gulf of Martaban	1,500	2,400	54
Olenyok	Laptev Sea	1,424	2,292	57
Aldan	Lena River	1,412	2,273	58
Syrdarya	Aral Sea	1,374	2,212	60
Chu Chiang (Pearl)-Hsi	South China Sea	1,365	2,197	62

(Continued)

Table 5A.8 (Continued)

Name	Outflow	Length[a] Miles	km	Rank in World (first 100)
Kolyma (Kolima)	East Siberian Sea	1,323	2,129	64
Tarim	Lop Nor Basin	1,261	2,030	69
Chulym-Bely Iyus	Ob River	1,257	2,023	70
Irrawaddy	Andaman Sea	1,238	1,992	72
Vitim-Vitimkan	Lena River	1,229	1,978	73
Indigirka-Khastakh	East Siberian Sea	1,228	1,977	74
Hsi	South China Sea	1,216	1,957	75
Sungari	Amur River	1,197	1,927	78
Tigris	Shatt-al-Arab	1,180	1,900	81
Podkamennaya Tunguska	Yenisey River	1,159	1,865	83
Vitim	Lena River	1,141	1,837	84
Chulym	Ob River	1,118	1,799	88
Angara	Yanisey River	1,105	1,779	89
Indigirka	East Siberian Sea	1,072	1,726	91
Khatanga-Kotuy	Laptev Sea	1,017	1,636	96
Ket	Ob River	1,007	1,621	97
Argun	Amur River	1,007	1,620	98
Shilka-Onon	Amur River	989	1,592	
Tobol-Kokpektysay	Irtysh River	989	1,591	
Alazeya–Kadylchan	East Siberian Sea	988	1,590	
Han Shui	Yangtze River	952	1,532	
Yana-Sartang	Laptev Sea	927	1,492	
Godavari	Bay of Bengal	910	1,465	
Amga	Aldan River	908	1,462	
Sutleg	Indus River	900	1,450	
Ili-Tekes	Lake Balkhash	894	1,439	
Olyokma	Lena River	892	1,436	
Amu Darya	Aral Sea	879	1,415	
Taz	Gulf of Taz	871	1,401	
Yamuna	Ganges River	855	1,376	
Kura	Caspian Sea	848	1,364	
Tavda-Lozva	Tobol River	843	1,356	
Liao	Gulf of Liaotung	836	1,345	
Taseyeva-Chuna	Angara River	820	1,319	
Vyatka	Kama River	817	1,314	
Krishna	Bay of Bengal	800	1,290	
Narmada	Gulf of Cambay	800	1,290	
Zeya	Amur River	772	1,242	
Chu	Betpak Dala Plateau	663	1,067	
Europe				
Volga	Caspian Sea	2,193	3,530	23
Danube	Black Sea	1,770	2,850	35
Ural	Caspian Sea	1,509	2,428	53
Dnepr	Black Sea	1,367	2,200	61
Don	Sea of Azov	1,162	1,870	82
Pechora	Barents Sea	1,124	1,809	85
Kama	Volga River	1,122	1,805	86
Oka	Vola River	932	1,500	
Belaya	Kama River	889	1,430	
Rhine	North Sea	865	1,392	
Dnestr	Black Sea	840	1,352	
Northern Dvina-Sukhona	White Sea	809	1,302	
Oceania				
Darling	Murray River	1,702	2,739	39
Murray	Great Australian Bight	1,609	2,589	45
Murrumbidgee	Murray River	981	1,579	
Lachlan	Murrumidgee River	992	1,484	

[a] Conversions of rounded figures are rounded to nearest hundred miles or kilometres.

Source: From Encyclopaedia Britannica, 15th edition, Copyright 1988 by Encyclopaedia Britannica, Inc. Reprinted with permission.

Table 5A.9 Large Rivers of the World

River	Country	Drainage Area (Thousands of sq mi)	Average Discharge at Mouth (Thousands of cfs)	Rank
		North America		
Mississippi[a]	U.S.A. and Canada	1,244	611	7
St. Lawrence	U.S.A. and Canada	498	500	11
Mackenzie	Canada	697	280	17
Columbia	U.S.A. and Canada	258	256	19
Yukon	Canada	360	180	24
Frazer	Canada	92	113	32
Nelson	Canada	414	80	37
Mobile	U.S.A.	42	58	43
Susquehanna	U.S.A.	28	38	48
		South America		
Amazon	Brazil	2,231	7,500[b]	1
Orinoco	Venezuela	340	600	8
Parana	Argentina	890	526	10
Tocantins	Brazil	350	360	16
Magdalena	Colombia	93	265	18
Uruguay	[c]	90	136	26
Sao Francisco	Brazil	260	100	34
		Africa		
Congo	Congo	1,550	1,400	2
Zambezi	Mozambique	500	250	20
Niger	Nigeria	430	215	22
Nile	Egypt	1,150	100	33
		Asia		
Yangtze	China	750	770	3
Brahmaputra	Bangladesh	361	700	4
Ganges	India	409	660	5
Yenisel	U.S.S.R.	1,000	614	6
Lena	U.S.S.R.	936	547	9
Irrawaddy	Burma	166	479	12
Ob	U.S.S.R.	959	441	13
Mekong	Thailand	310	390	14
Amur	U.S.S.R.	712	388	15
Indus	Pakistan	358	196	23
Kolyma	U.S.S.R.	249	134	27
Sankai (Si)	China	46	127	28
Godavari	India	115	127	29
Hwang Ho (Yellow)	China	260	116	31
Pyasina	U.S.S.R.	74	90	36
Krishna	India	119	69	39
Indigirka	U.S.S.R.	139	64	40
Salween	Burma	108	53	44
Shatt-al Arab[d]	Iraq	209	51	45
Yana	U.S.S.R.	95	35	49
		Europe		
Danube	Romania	315	218	21
Pechora	U.S.S.R.	126	144	25
Dvina (Northern)	U.S.S.R.	139	124	30
Neva	U.S.S.R.	109	92	36
Rhine	Netherlands and Germany	56	78	38
Dnepr	U.S.S.R.	194	59	41
Rhone	France	37	59	42
Po	Italy	27	51	46
Vistula	Poland	76	38	47

[a] Includes Atchafalaya River.
[b] Department of Interior News Release, Feb. 24, 1964.
[c] Argentina and Uruguay.
[d] Tigris, Euphrates and Karun.

Source: From Young, L.L., U.S. Geological Survey, 1964.

SECTION 5B LAKES

Table 5B.10 Natural Fresh-Water Lakes of the United States of 10 sq. mi or More

Name	Latitude	Longitude	Area (sq. mi)
Alaska			
Iliamna	59°35	155°00	1,000
Becharof	57°50	156°25	458
Teshekpuk	70°35	153°30	315
Naknek	58°35	156°00	242
Tustumena	60°25	150°20	117
Clark	60°10	154°00	110
Dall	60°15	163°45	100
Inland[a]	66°30	159°50	95
Imuruk Basin[a]	65°05	165°40	80
Upper Ugashik	57°50	156°25	75
Kukaklek	59°35	155°00	72
Lower Ugashik	57°30	156°55	72
Nerka	59°20	158°45	69
Nuyakuk	59°50	158°50	64
Aropuk	61°10	163°45	57
Tazlina	61°50	146°30	57
Nanwhyenuk or Nonvianuk	59°00	155°25	56
Nunavakpak	60°45	162°40	53
Kaghasuk[a]	60°55	163°40	52
Skilak	60°25	150°20	38
Chauekuktuli	60°05	158°50	34
Chikuminuk	60°15	158°55	34
Beverly	59°40	158°45	33
Whitefish	61°20	160°00	33
Aleknagik	59°20	158°45	31
Brooks	58°30	155°55	31
Kgun	61°35	163°50	31
Nonvianuk	59°00	155°30	31
Takslesluk	61°05	162°55	31
George	61°15	148°35	29
Nunavak Anukslak	61°05	162°30	29
Unnamed	60°55	164°00	28
Grosvenor	58°40	155°15	27
Tetlin	63°05	142°45	27
Chakachamna	61°10	152°30	26
Imuruk	65°35	163°10	26
Nunavakanuk	62°05	164°40	25
Louise	62°20	146°30	23
Minchumina	63°55	152°15	23
Klutina	61°40	145°30	22
Unnamed	61°30	164°30	22
Unnamed	71°05	156°30	21
Beluga	61°25	151°30	20
Unnamed	60°05	164°00	20
Unnamed	61°40	160°25	20
Kenai	60°25	149°35	19
Kyigayalik	61°00	162°30	19
Tikchik	59°55	158°20	19
Bering	60°20	144°20	17
Kulik	59°50	158°50	17
Upnuk	60°05	158°55	17
Unnamed slough	62°40	163°30	17
Teloquana	60°55	153°55	16
Unnamed	60°30	161°40	16
Unnamed	61°00	163°45	16
Unnamed	61°30	164°55	16
Five Day Slough	62°05	162°00	15
Togiak	59°35	159°35	15
Unnamed	59°55	163°15	15
Black	56°25	159°00	14
Ualik	59°05	159°30	14

(Continued)

Table 5B.10 (Continued)

Name	Latitude	Longitude	Area (sq. mi)
Walker	67°05	154°25	14
Unnamed	60°20	164°25	14
Unnamed	60°50	163°30	14
Unnamed	59°50	163°30	14
Amanka	59°05	159°10	13
Whitefish	60°55	154°55	13
Unnamed	71°00	156°00	13
Crosswind	62°20	146°00	12
Kakhomak	59°30	154°10	12
Karluk	57°20	154°05	12
Mother Goose	57°10	157°20	12
Unnamed	60°25	164°10	12
Unnamed	59°50	163°25	12
Unnamed	62°15	162°20	12
Unnamed	70°50	153°30	12
Coleville	58°45	155°40	11
Harlequin	59°25	138°55	11
Unnamed	60°25	164°10	11
Unnamed	60°55	162°20	11
Bear	56°00	160°15	10
Chignik	56°15	158°50	10
Ewan	62°25	145°50	10
Kontrashibuna	60°10	154°00	10
Kukaklik	61°40	160°30	10
Kulik	58°55	155°00	10
Unnamed	61°45	160°40	10
Miles	60°40	144°45	10
Susitna	62°25	146°40	10
Unnamed	60°20	162°00	10
Unnamed	60°25	162°00	10
Unnamed	60°55	162°10	10
Unnamed	59°55	163°15	10
Unnamed	62°00	162°00	10

Name	County	Area (sq. mi)
California		
Tahoe[b]	Placer, Eldorado	193
Clear	Lake	65
Eagle[c]	Lassen	41
Florida		
Okeechobee	Hendry, Glades, Okeechobee, Martin, Palm Beach	700
George	Putnam, Marion, Volusia, Lake	70
Kissimmee	Osceola, Polk	55
Apopka	Orange	48
Istokpoga	Highlands	43
Tsala Apopka	Citrus	30
Tohopekaliga	Osceola	29
Harris	Lake	27
Orange	Alachua, Marion	26
East Tohopekaliga	Osceola	19
Griffin	Lake	14
Monroe	Seminole, Volusia	14
Jessup	Seminole	13
Weohyakapka	Polk	12
Talquin	Gadsden, Leon	11
Eustis	Lake	11
Blue Cypress	Osceola, Indian River	10
Hatchineha	Polk, Osceola	10
Lochloosa	Alachua	10
Idaho		
Pend Oreille	Bonner	148
Bear[d]	Bear Lake	110[d]
Coeur d'Alene	Kootenai	50
Priest	Bonner	37
Grays[f]	Bonneville, Caribou	34
Henrys	Fremont	10

(Continued)

Table 5B.10 (Continued)

Name	County	Area (sq. mi)
Iowa		
Spirit	Dickinson	12
Louisiana		
White[g]	Vermilion	83
Grand	Iberia, St. Mary, St. Martin	64
Caddo[h]	DeSoto	60
Catahoula[i]	LaSalle	32
Grand	Cameron	32
Six Mile	St. Martin, St. Mary	30
Fausse Pointe	St. Mary, Iberia	24
Lac des Allemands	St. John the Baptist	23
Verret	Assumption	22
Polourde	St. Martin, St. Mary, Assumption	18
Maine		
Moosehead	Piscataquis, Somerset	117
Sebago	Cumberland	45
Chesuncook[j]	Piscataquis	43
West Grand	Washington	37
Flagstaff	Somerset, Franklin	28
Spedni[k]	Washington	28
Grand Falls[k]	Washington	27
East Grand[k]	Washington, Aroostook	26
Mooselookmeguntic	Oxford, Franklin	26
Twin	Penobscot, Piscataquis	25
Chamberlain and Telos	Piscataquis	22
Graham	Hancock	19
Churchill and Eagle	Piscataquis	17
Baskahegan	Washington	16
Umbagog[l]	Oxford	16
Brassua	Somerset	15
Square	Aroostook	14
Millinocket	Penobscot, Piscataquis	14
Great	Kennebec	13
Richardson	Oxford	13
Schoodic	Piscataquis	11
Sebec	Piscataquis	11
Aziscohos	Oxford	10
Canada Falls	Somerset	10
Rangeley	Oxford	10
Michigan		
St. Clair[m]		460
Houghton	Roscommon	31
Torch	Antrim, Kalkaska	29
Charlevoix[n]	Charlevoix	27
Burt	Cheboygan	27
Mullet	Cheboygan	26
Gogebic	Ontonagon, Gogebic	21
Manistique	Mackinac, Luce	16
Black	Cheboygan, Presque Isle	16
Crystal	Benzie	15
Portage	Houghton	15
Higgins	Crawford, Roscommon	15
Hubbard	Alcona	14
Leelanau	Leelanau	13
Indian	Schoolcraft	12
Elk	Antrim, Grand Traverse	12
Glen	Leelanau	10
Minnesota		
Lake of the Woods[o]	Lake of the Woods	1,485
Upper and Lower Red	Beltrami	451
Rainy[o]	Koochiching, St. Louis	345
Mille Lacs	Aitken, Crow Wing, Mille Lacs	207
Leech	Cass	176
Winnibigoshish	Itasca, Cass	109
Vermilion	St. Louis	77
Lac La Croix[o]	St. Louis	53
Cass	Cass, Beltrami	46

(Continued)

Table 5B.10 (Continued)

Name	County	Area (sq. mi)
Basswood[o]	Lake	46
Namakan[o]	St. Louis	44
Kabetogama	Itasca	40
Pepin[p]	Goodhue, Wabasha	39
Mud	Marshall	37
Saganaga[o]	Cook	32
Pokegama	Itasca	24
Minnetonka	Hennepin, Carver	22
Otter Tail	Otter Tail	22
Gull	Cass, Crow Wing	20
Pelican	St. Louis	19
Traverse[q]	Traverse	18
Big Stone[q]	Big Stone	17
Crooked[o]	St. Louis, Lake	17
Sandy	Aitkin	15
Swan	Nicollet	15
Island	St. Louis	14
Bowstring	Itaska	14
Burntside	St. Louis	14
Sand Point[o]	St. Louis	14
Trout	St. Louis	14
St. Croix[p]	Washington	13
Lac qui Parle	Chippewa, Lac qui Parle	13
Pelican	Crow Wing	13
Dead	Otter Tail	12
Minnewaska	Pope	12
Thief	Marshall	12
Nett	St. Louis, Koochiching	12
Osakis	Douglas, Todd	10
Bemidji	Beltrami	10
Lida	Otter Tail	10
Montana		
Flathead	Lake, Flathead	197[r]
Medicine	Sheridan	15[s]
McDonald	Flathead	10
Nevada		
Tahoe[b]	Ormsby, Douglas	193
New Hampshire		
Winnipesaukee	Belknap, Carroll	72
Umbagog[l]	Coos	16
Squam	Gafton, Carroll	11
New York		
Champlain[t]	Clinton, Essex	490[u]
Oneida	Oswego, Oneida	80
Seneca	Seneca, Schuyler	67
Cayuga	Cayuga, Seneca, Tompkins	66
George	Warren	44
Chautauqua	Chautauqua	21
Black	St. Lawrence	17
Canandaigua	Ontario, Yates	17
Skaneateles	Onondaga, Cayuga	14
Owasco	Cayuga	10
North Carolina		
Mattamuskeet[v]	Hyde	67
Phelps	Washington	25
Waccamaw	Columbus	14
Oregon		
Upper Klamath	Klamath	142[w]
Crater	Klamath	21
South Dakota		
Traverse[q]	Roberts	18
Big Stone[q]	Roberts	17
Tennessee		
Reelfoot	Lake, Obion	22
Texas		
Caddo[h]	Marion	60

(Continued)

Table 5B.10 (Continued)

Name	County	Area (sq. mi)
Utah		
Utah	Utah	140
Bear[d]	Rich	110
Vermont		
Champlain[t]	Chittenden, Franklin	490[u]
Washington		
Chelan	Chelan	55
Washington	King	35
Ozette	Clallam	12
Wisconsin		
Winnebago	Winnebago, Calumet, Fond du Lac	215
Pepin[p]	Pierce, Pepin	39
Poygan	Winnebago	17
Koshkonong	Jefferson	16
Mendota	Dane	15
St. Croix[p]	St. Croix	12
Green	Green Lake	11
Wyoming		
Yellowstone	Yellowstone National Park	37[x]
Jackson	Teton	39[y]
Shoshone	Yellowstone National Park	11

Note: Lakes are arranged by states; the Great Lakes are excluded.

[a] May be salt water.
[b] California and Nevada.
[c] Mildly saline, less than 1,000 ppm.
[d] Idaho and Utah.
[e] 136 sq. mi including Mud Lake.
[f] Submerged marsh.
[g] Originally brackish; now kept fresh by controls on salt water intrusion.
[h] Louisiana and Texas.
[i] Shrinks to small area at extremely low stages.
[j] Includes Ripogenus and Caribou.
[k] Maine and Quebec.
[l] Maine and New Hampshire.
[m] Michigan and Ontario.
[n] Formerly called Pine.
[o] Minnesota and Ontario.
[p] Minnesota and Wisconsin.
[q] Minnesota and South Dakota.
[r] At normal high water; 188 sq. mi at medium low water; lake regulated for power between these limits.
[s] Includes 4 islands having area of about 1 sq. mi.
[t] New York, Vermont, and Quebec.
[u] Includes islands totaling about 55 sq. mi.
[v] The lake originally landlocked, was drained and provided with outlet and is fresh water; level regulated to some extent by control works on canals draining the area.
[w] At upper level; dam at outlet allows regulation so that area varies between 93 and 142 sq. mi.
[x] Includes islands totaling 3 sq. mi.
[y] Enlarged by dam; original area, 30 sq. mi.

Source: From U.S. Geological Survey.

Table 5B.11 Natural Fresh-Water Lakes of the United States of 100 sq. mi or More

Name	Location	Area (sq. mi)
Lake of the Woods	Minnesota and Ontario	1,485
Iliamna	Alaska	1,000
Okeechobee	Florida	700
Champlain	New York, Vermont, and Quebec	490
St. Clair	Michigan and Ontario	460
Becharof	Alaska	458
Upper and Lower Red	Minnesota	451
Rainy	Minnesota and Ontario	345
Teshekpuk	Alaska	315
Naknek	Alaska	242
Winnebago	Wisconsin	215
Mille Lacs	Minnesota	207
Flathead	Montana	197
Tahoe	California and Nevada	193
Leech	Minnesota	176
Pend Oreille	Idaho	148
Upper Klamath	Oregon	142
Utah	Utah	140
Yellowstone	Wyoming	137
Tustumena	Alaska	117
Moosehead	Maine	117
Clark	Alaska	110
Bear	Idaho and Utah	110
Winnibigoshish	Minnesota	109
Dall	Alaska	100

Note: The Great Lakes are excluded.

Source: From U.S. Geological Survey.

Table 5B.12 Natural Fresh-Water Lakes of the United States, 250 ft Deep or More

Name	Location	Depth (ft)
Crater	Oregon	1,932
Tahoe	California and Nevada	1,645
Chelan	Washington	1,605
Pend Oreille	Idaho	1,200
Nuyakuk	Alaska	930
Deer	Alaska	877
Chauekuktuli	Alaska	700
Crescent	Washington	624
Seneca	New York	618
Clark	Alaska	606
Beverley	Alaska	500
Nerka	Alaska	475
Tokatz	Alaska	474
Long	Alaska	470
Lower Sweetheart	Alaska	459
Cayuga	New York	435
Crater	Alaska	414
Cooper	Alaska	>400
Champlain	New York, Vermont, and Quebec	400
Kasnyku	Alaska	393

(Continued)

Table 5B.12 (Continued)

Name	Location	Depth (ft)
Chakachamna	Alaska	380
Ozette	Washington	331
Aleknagik	Alaska	330
Sebago	Maine	316
Swan	Alaska	>314
Baranoff	Alaska	303
Payette	Idaho	>300
Quinault	Washington	About 300
Crescent	Alaska	291
Wallowa	Oregon	283
Chilkoot	Alaska	282
Odell	Oregon	279
Silver	Alaska	278
Grant	Alaska	>250

Note: The Great Lakes are excluded.

Source: From U.S. Geological Survey.

Table 5B.13 Largest Lake in Each State of the United States

State	Largest Entirely Within State	Largest Partly in Another State	Shared With	Origin	Area sq. mi	Feet Above Sea Level	Maximum Depth (ft)	Shoreline Length (miles)
AL	Wheeler			Man-made	104.84	556	58	1,063
		Guntersville	TN	Man-made	107.97	595	60	962
AK	Illamna			Natural	1,033	50	—	188
AZ	San Carlos[a]			Man-made	30.6	2,523	249	—
		Powell	UT	Man-made	252	3,700	580	—
AR	Ouachita			Man-made	62.65	578	207	690
		Bull Shoals	MO	Man-made	111.31	695	243	1,050
CA	Salton Sea			Natural	360	−231	46	—
		Tahoe	NV	Natural	192	6,229	1,685	71
CO	John Martin[a]			Man-made	28.72	3,765	118	86
CT	Candlewood			Man-made	8.46	429	85	65
DE	Lum's Pond			Man-made	0.31	44	12	3.5
FL	Okeechobee			Natural	700	18.7	—	110
GA	Sidney Lanier			Man-made	57.96	1,035	180	540
		Clark Hill[a]	SC	Man-made	111.09	330	190	1,057
HI	Koloa[a]			Man-made	0.66	233	22.5	3.3
ID	Pend Oreille			Natural	133	2,063	1,150	111.3
		Bear	UT	Natural	136	5,930	30	51.5
IL	Crab Orchard			Man-made	10.96	405	33	103
		Michigan	WI, IN, MI	Natural	22,400	578.8	923	1,660
IN	Wawasee			Natural	4.09	859	68	18
		Michigan	WI, IL, MI	Natural	22,400	578.8	923	1,660
IA	Spirit			Natural	8.84	1,402	—	—
KS	Tuttle Creek[a]			Man-made	24.68	1,075	56	112
KY	Cumberland			Man-made	78.51	723	—	1,255
		Kentucky	TN	Man-made	247.34	375	145	2,380
LA	Pontchartrain			Natural	630	S.L	15	113
ME	Moosehead			Natural	117	1,028	246	—
MD	Deep Creek			Man-made	7.03	2,462	—	—
MA	Quabbin[a]			Man-made	38.6	530	150	118
MI	Houghton			Natural	31.3	1,139	20	30
		Superior	WI, Ont, MN	Natural	31,800	600	1,333	2,980
MN	Red			Natural	451	1,175	31	123
		Superior	WI, Ont, MI	Natural	31,800	600	1,333	2,980
MS	Sardis			Man-made	15.31	234	—	60
MO	Lake of the Ozarks			Man-made	93.75	—	125	1,375
		Bull Shoals	AR	Man-made	111.31	695	243	1,050
MT	Fort Peck[a]			Man-made	382.81	2,250	220	1,600
NE	McConaughy			Man-made	55	3,276	150	50
NV	Pyramid			Natural	187.5	3,800	330	70
		Mead	AZ	Man-made	247	1,221	589	550
NH	Winnipesaukee			Natural	71.55	504	120	128
NJ	Hopatcong			Man-made	4.19	924	58	35

NM	Elephant Butte[a]			Man-made	58.85	4,450	193	250
NY	Oneida			Natural	80	370	50	52
		Erie	MI, PA, OH, Ont	Natural	9,910	570	210	856
NC	Norman			Man-made	50.78	760	115	520
		John H. Kerr[a]	VA	Man-made	76.4	300	100	800
ND	Garrison[a]			Man-made	609.38	1,850	200	1,600
OH	Grand			Man-made	20	869	12	60
		Erie	MI, PA, NY, Ont	Natural	9,910	570	210	856
OK	Eufaula[a]			Man-made	160.16	585	87	600
		Texoma	TX	Man-made	149.06	617	94	540
OR	Upper Klamath (incl. Agency Lake)			Natural	140.63	4,139	40	105
PA	Wallenpaupack			Man-made	9	1,182	50	45
		Erie	MI, NY, OH, Ont	Natural	9,910	570	210	856
RI	Scituate[a]			Man-made	5.68	284	80	38
SC	Marion			Man-made	157.03	75	35	299
		Clark Hill[a]	GA	Man-made	111.09	330	190	1,057
SD	Francis Case			Man-made	160.31	1,375	140	540
TN	Watts Bar			Man-made	60.31	745	80	783
		Kentucky	KY	Man-made	247.34	375	145	2,380
TX	Texarkana			Man-made	46.56	225	39	141
		Texoma	OK	Man-made	149.06	617	94	540
UT	Great Salt			Natural	1,500	4,200	48	350
		Powell	AZ	Man-made	252	3,700	580	—
VT	Bomoseen			Natural	3.69	411	—	—
		Champlain	NY Que	Natural	430	100	399	—
VA	Smith Mountain			Man-made	31.25	795	200	500
		John H. Kerr[a]	NC	Man-made	76.4	300	100	800
WA	F.D. Roosevelt			Man-made	123.44	1,288	375	302
WV	Tygart			Man-made	5.37	1,010		106
		Bluestone[a]	VA	Man-made	3.07	1,409	42	33
WI	Winnebago			Natural	215.26	—	21.6	91.96
		Superior	MN, MI, Ont	Natural	31,800	600	1,333	2,980
WY	Yellowstone			Natural	137	7,735	—	—

[a] Reservoir.

Source: From National Geographic Society.

Figure 5B.3 The Great Lakes. (From U.S. Geological Survey, *National Water Summary 1985—Hydrologic Events and Surface-Water Resources, Water-Supply Paper 2300.*)

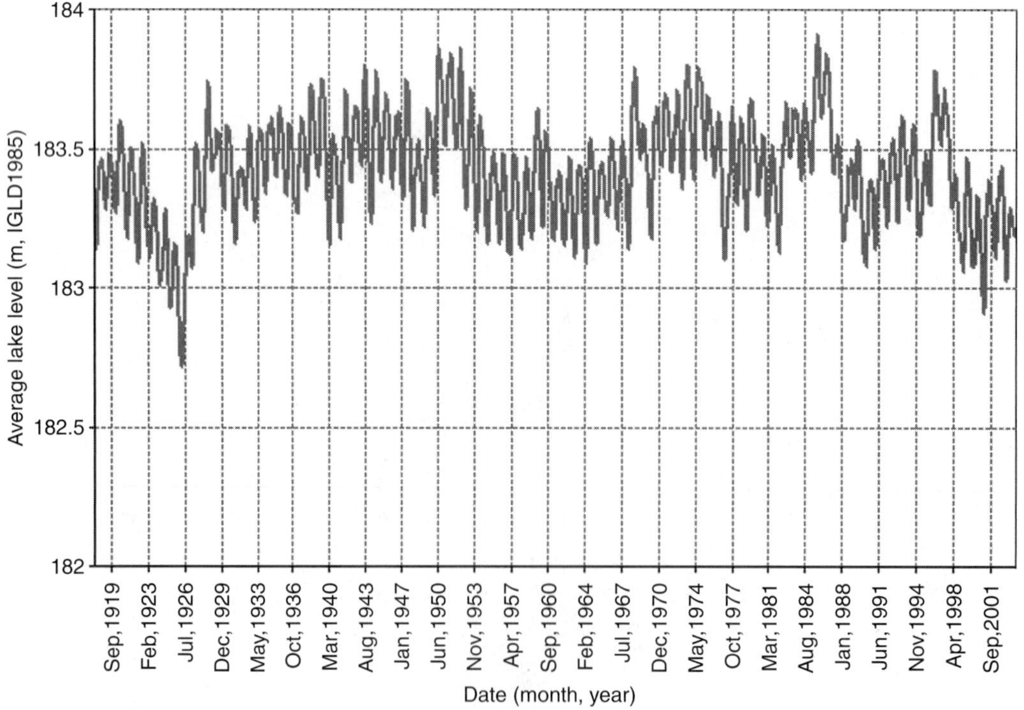

Figure 5B.4 Great Lakes water levels for Lake Superior graphed for consecutive years, Lake Superior: 1918–2003. (From www.usace.army.mil.)

Table 5B.14 Selected Facts about the Great Lakes System

Outlet	Remarks
St. Marys River to Lake Huron	Largest surface area of all the freshwater lakes in the world. Outflow controlled by St. Marys River Compensating works
Straits of Mackinac to Lake Huron	Sixth largest surface area of world's freshwater lakes
St. Clair River to Lake St. Clair	Fifth largest surface area of world's freshwater lakes
Detroit River to Lake Erie	Shallowest lake in the Great Lakes system
Niagara River and Falls to Lake Ontario	Eleventh largest surface area of world's freshwater lakes
St. Lawrence River to Atlantic Ocean	Outflow controlled by St. Lawrence Seaway and Power Project

Source: From U.S. Geological Survey, *National Water Summary 1985—Hydrologic Events and Surface-Water Resources, Water-Supply Paper 2300*, 1985.

Table 5B.15 Great Lakes Physical Features and Population

		Superior	Michigan	Huron	Erie	Ontario	Totals
Elevation[a]	(ft)[f]	600	577	577	569	243	
	(m)	183	176	176	173	74	
Length	(mi)[e]	350	307	206	241	193	
	(km)	563	494	332	388	311	
Breadth	(mi)[e]	160	118	183	57	53	
	(km)	257	190	245	92	85	
Average Depth[a]	(ft)[f]	483	279	195	62	283	
	(m)	147	85	59	19	86	
Maximum Depth[a]	(ft)[e]	1,332	925	750	210	802	
	(m)	406	282	229	64	244	
Volume[a]	(cu. mi.)[e]	2,900	1,180	850	116	393	5,439
	(km^3)	12,100	4,920	3,540	484	1,640	22,684
Water Area	(sq. mi)[e]	31,700	22,300	23,000	9,910	7,340	94,250
	(km^2)	82,100	57,800	59,600	25,700	18,960	244,160
Land Drainage Area[b]	(sq. mi)[e]	49,300	45,600	51,700	30,140	24,720	201,460
	(km^2)	127,700	118,000	134,100	78,000	64,030	521,830
Total Area	(sq. mi)[e]	81,000	67,900	74,700	40,050	32,060	295,710
	(km^2)	209,800	175,800	193,700	103,700	82,990	765,990
Shoreline Length[c]	(mi)[e]	2,726	1,638	3,827	871	712	10,210d
	(km)	4,385	2,633	6,157	1,402	1,146	17,017d
Retention Time	(yr)[f]	191	99	22	2.6	6	
Population:U.S	(1990)[g]	425,548	10,057,026	1,502,687	10,017,530	2,704,284	24,707,075
	Canada (1991)	181,573		1,191,467	1,664,639	5,446,611	8,484,290
	Totals	607,121	10,057,026	2,694,154	11,682,169	8,150,895	33,191,365
	Outlet	St. Marys River	Straits of Mackinac	St. Clair River	Niagara River/ Welland Canal	St. Lawrence River	

Note: [a]Measured at low water datum; [b]Land Drainage Area for Lake Huron includes St. Marys River. Lake Erie includes the St. Clair-Detroit system. Lake Ontario includes the Niagara River; [c]Including islands; [d]These totals are greater than the sum of the shoreline length for the lakes because they include the connecting channels (excluding the St. Lawrence River).

Source: From www.epa.gov.

Original Source: [e]Coordinating Committee on Great Lakes Basic Hydraulic and Hydrologic Data, Coordinated Great Lakes Physical Data. May, 1992; [f]Extension Bulletins E-1866-70, Michigan Sea Grant College Program, Cooperative Extension Service, Michigan State University, E. Lansing, Michigan, 1985; [g]1990–1991 population census data were collected on different watershed boundaries and are not directly comparable to previous years.

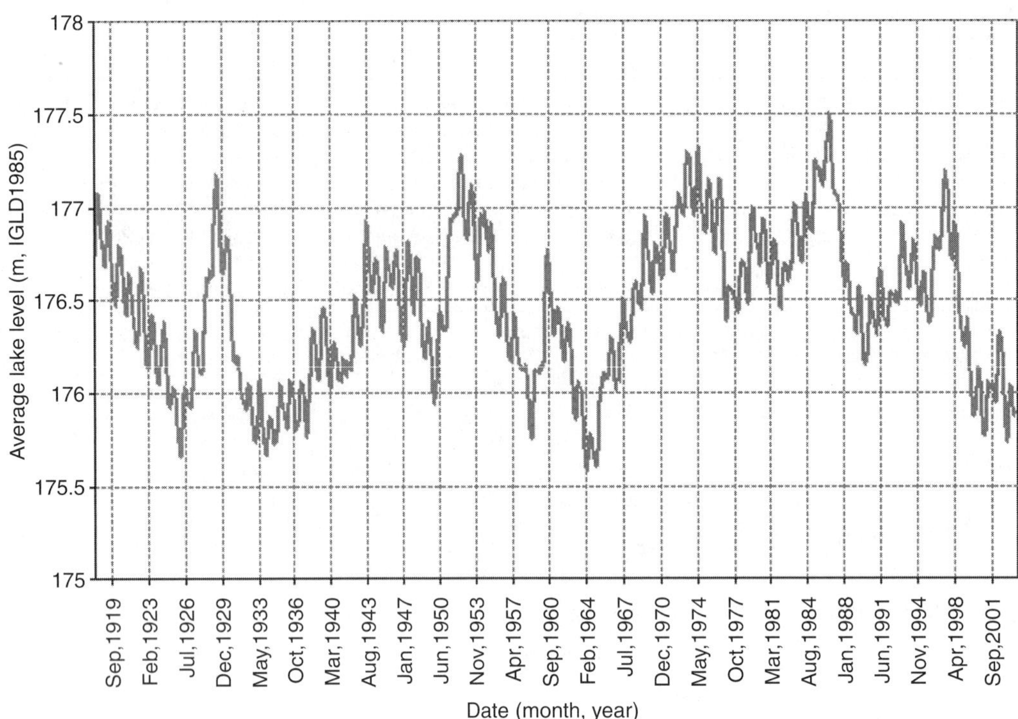

Figure 5B.5 Great Lakes water levels for Lake Michigan/Huron graphed for consecutive years, Lake Michigan/Huron: 1918–2003. (From www.usace.army.mil.)

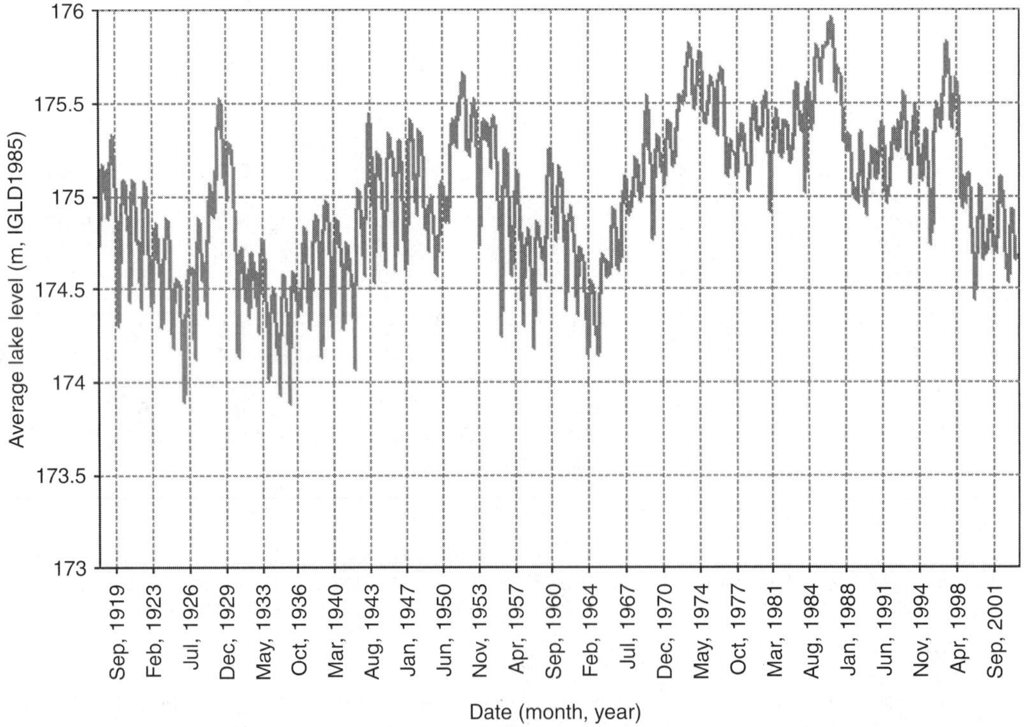

Figure 5B.6 Great Lakes water levels for Lake St. Clair graphed for consecutive years, Lake St. Clair: 1918–2003. (From www.usace.army.mil.)

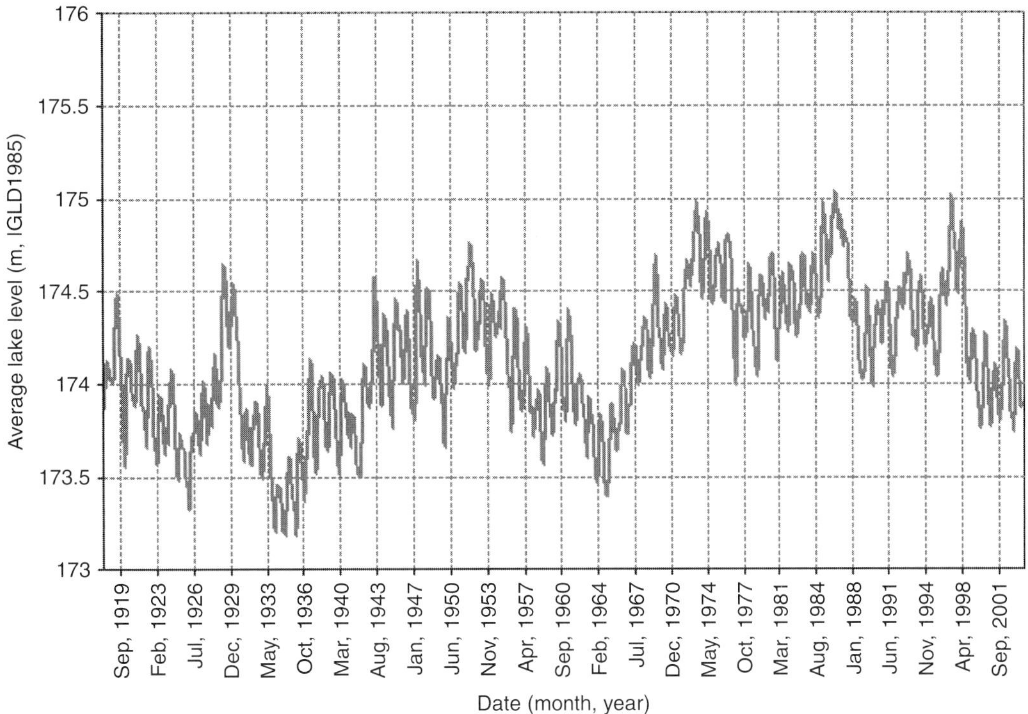

Figure 5B.7 Great Lakes water levels for Lake Erie graphed for consecutive years, Lake Erie: 1918–2003. (From www.usace.army.mil.)

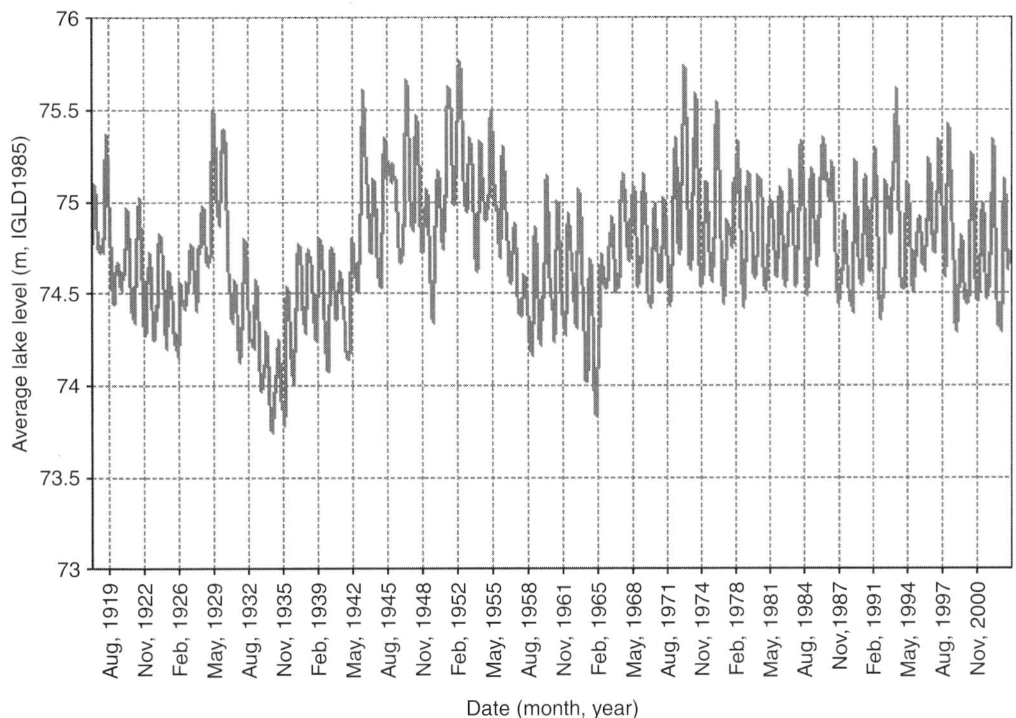

Figure 5B.8 Great Lakes water levels for Lake Ontario graphed for consecutive years, Lake Ontario: 1918–2003. (From www.usace.army.mil.)

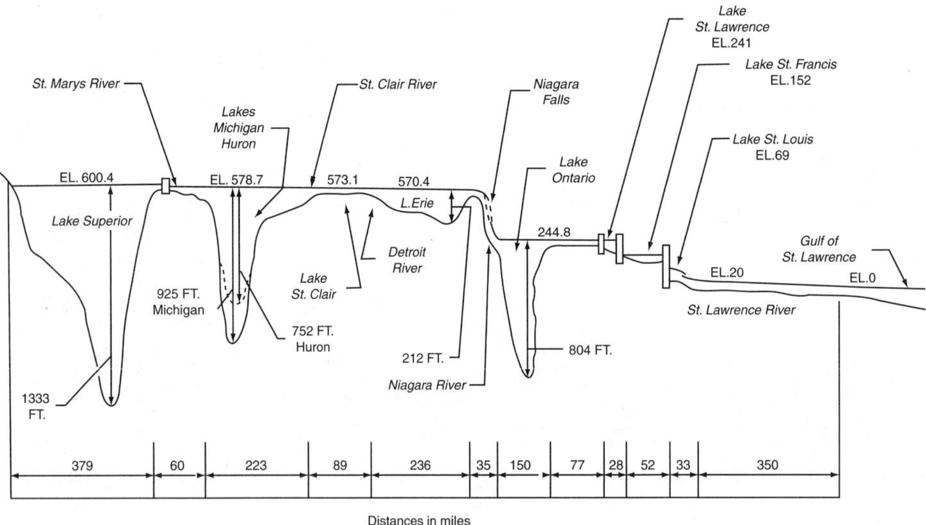

Figure 5B.9 Profile of the Great Lakes—St. Lawrence River Drainage System. (From International great lakes levels board. Regulation of Great Lakes water levels, Report to the international joint commission, Washington, 1973. With permission.)

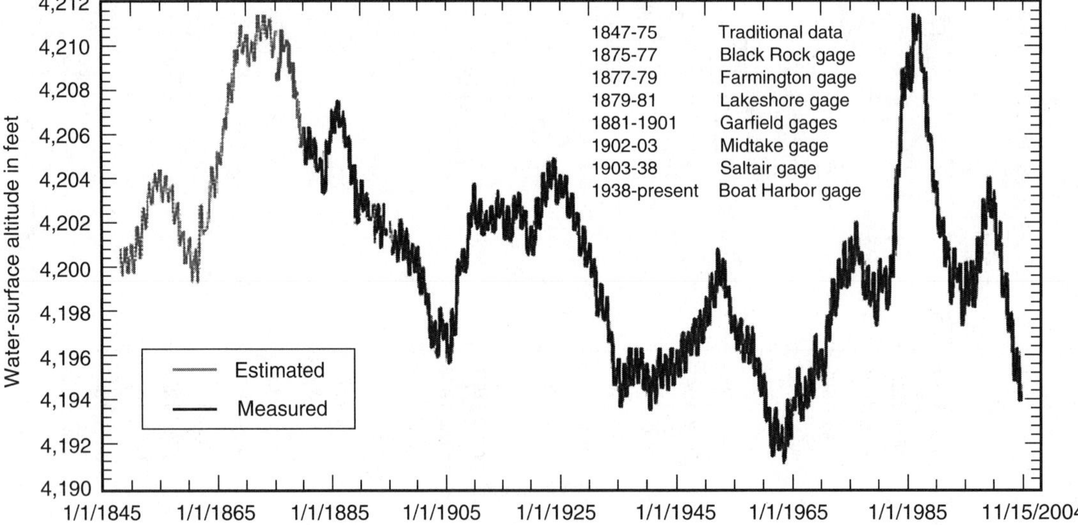

Figure 5B.10 Fluctuation in water-surface altitude of Gilbert Bay (south part), Great Salt Lake, 1847 to present. (From www.usgs.gov.)

Table 5B.16 Hydrologic Characteristics of the Great Lakes

	Lake Surface Elevation, in Feet, 1900–1984							
	Monthly Mean			Monthly Range (From Winter Low to Summer High)			Total Dissolved Solids, 1986 ppm	Mean Discharge (m^2/sec)
Lake	Average	Maximum	Minimum	Average	Maximum	Minimum		
Superior	600.59	602.02	598.23	1.2	2.1	0.4	52	2,076
Michigan–Huron	578.27	581.04	575.35	1.2	2.1	0.4	—	—
Michigan	—	—	—	—	—	—	150	1,558
Huron	—	—	—	—	—	—	118	5,038
St. Clair	573.34	576.23	569.86	1.7	3.3	0.6	—	—
Erie	570.44	573.51	567.49	1.6	2.8	0.9	198	5,545
Ontario	244.71	248.06	241.45	2.0	3.6	0.7	194	6,624

Note: Levels referenced to international Great Lakes datum 1955.

Source: From U.S. Geological Survey, *National Water Summary 1985—Hydrologic Events and Surface-Water Resources, Water-Supply Paper 2300* and U.S. Army Corps of Engineers data.

Table 5B.17 Principal Saline Lakes of the United States

Lake	Present Area (sq. mi)	Remarks
California		
Salton Sea	350	About 650 sq. mi at highest stage in 1905–07
Owens	Dry at times each year since 1943	110 sq. mi in 1872, prior to diversions from Owens River; 35 sq. mi in 1943
Mono	76	Maximum, 89 sq. mi in 1919
Goose (in California and Oregon)	About 100	Maximum, 186 sq. mi, 125 in California and 61 in Oregon; overflowed into Pitt River in 1869 and 1881; dry in 1930; 150 sq. mi in 1958
Eagle	41	Some question as to whether Eagle Lake should be considered saline or fresh water. It has no surface outlet (Martin, 1962), but since 1924 it has been tapped by tunnel to Willow Creek. Salinity is considerably less than 1,000 ppm, according to California Dept. of Water Resources. During period 1895–1925 lake rose to highest level since at least 1650 (Harding, 1935); rise believed due to closing of subterranean outlet by earthquake in 1890 (Antevs, 1938)
Honey	Dry	90 sq. mi in 1867, possibly higher in 1890; dry in 1903; high in 1904; dry in 1924. Contained some water April 1958 to September 1960, and early in 1962
Louisiana		
Pontchartrain	625	These lakes are connected with the Gulf of Mexico, and are subject to tidal fluctuation
Sabine (Louisiana and Texas)	95	As above
Calcasieu	90	As above
Maurepas	90	As above
Salvador	70	As above
Nevada		
Pyramid	180	Maximum size, 220 sq. mi. Low until 1860; reached extreme high level in 1862 and 1868 or 1869; nearly as high in 1890; began to drop in 1917 (Hardman and Venstrom, 1941)
Walker	107	Maximum size, 125 sq. mi
Winnemucca	Dry	Maximum size, 180 sq. mi. Dry in 1840, but began to fill shortly thereafter (Zones, 1961). According to Russell (1885) the lake rose more than 50 ft and approximately doubled its area between 1867 and 1882. Was 87 ft deep in 1882. Dry since 1945
Carson	Nearly dry	Maximum size, 41 sq. mi. A few water-filled pot holes remain. Once called South Carson Lake; received flow of Carson River before Lahontan Reservoir was built

(Continued)

Table 5B.17 (Continued)

Lake	Present Area (sq. mi)	Remarks
Carson Sink	Dry	A shallow playa some 250 sq. mi in area shown on some maps as a body of water. Russell (1885) called it North Carson Lake. Dry in 1882, but probably has had some water at times since. Once received water from both Carson and Humboldt Rivers
Ruby		Maximum size, 37 sq. mi. Shown as swamp on recent maps of Army Map Service and Nevada Dept. of Highways
Franklin		Maximum size, 32 sq. mi. Shown as swamp on recent maps of the Army Map Service and Nevada Dept. of Highways
North Dakota		
Devils	24	140 sq. mi in 1867; 70 sq. mi in 1883; 45 sq. mi in 1900; 10 sq. mi in 1940. Since 1940 lake has been rising
Oregon		
Malheur and Harney	Probably dry	Malheur, the larger of the two lakes, overflows into Harney, which has no outlet. Maximum combined size, 125 sq. mi. Reported dry in 1931; high in late 1950s; about 1 sq. mi in 1961, and expected to go dry in 1962
Goose (see California)		
Abert	52	Maximum size, 60 sq. mi. Dry in 1930 or thereabouts, but fairly high in 1958
Summer	Probably dry	Maximum size, 70 sq. mi. Nearly dry in 1961
Silver	Dry	Maximum size, 15 sq. mi. Dry in 1961. Because of the transient nature of the lake, the water—whenever there is any—is relatively fresh; hay is raised on the dry lake bed
Warner	Probably less than 10	A series of shallow lakes; combined area about 30 sq. mi in 1953, a wet year, estimated from Army Map Service map based on aerial photograph taken in 1953. Present lakes are all that is left of Pleistocene Warner Lake, which covered about 300 sq. mi and was about 270 ft deep
Utah		
Great Salt	About 1,000	Maximum size since 1851, 2,400 sq. mi in 1870s; minimum, 950 sq. mi in October 1961; seasonal high in 1962 was 1,050 sq. mi in June
Sevier	Dry	Maximum size, 125 sq. mi; has been dry for several years

Source: From U.S. Geological Survey, 1963.

Table 5B.18 Hydrologic Data for Great Salt Lake and West Desert Pumping Project

Great Salt Lake	
Dimensions	80×35 mi
Average depth	22 ft
Maximum depth	42 ft
Contents in dissolved minerals (mainly chloride, sodium, sulfate, magnesium, potassium with lesser amounts of calcium, lithium, bromium and boron[a])	4–5 billion tons
West Desert Pumping Project	
Flood control project by State of Utah to lower water level of Great Salt Lake	
Start of construction July 1986	
Cost of construction and 1st year of operation	71.7 million
Projected volume of diversion	2.73 million acre ft total
Water is lifted by 3 large pumps (capacity 1,000 cfs each) through 4.1 mi long outlet canal to evaporation pond (west pond)	
Surface area of west pond	500 sq. mi
Rate of evaporation from west pond	825,000 acre-ft/yr
Salinity of water in west pond	350 g/L

Rock and earthfill Southern Pacific Transportation Co. railroad causeway separates the lake into two parts, Southern part of lake (60% of total area) receives 90% of lake's freshwater inflow, Total annual inflow 1931–76 averaged 2.9 million acre-ft, Northern part of lake receives most of its water as brine flowing through culverts and causeway from southern part of lake, Lake salinity varies with lake level, Northern lake is 16% salt and about 3 times saltier than southern lake (June 1987).

[a] For chemical analysis of brine see Chapter 6 Table 6.8.

Source: From Compiled from information provided by Utah Division of Water-Resources, 1987 and U.S. Geological Survey Circular 913.

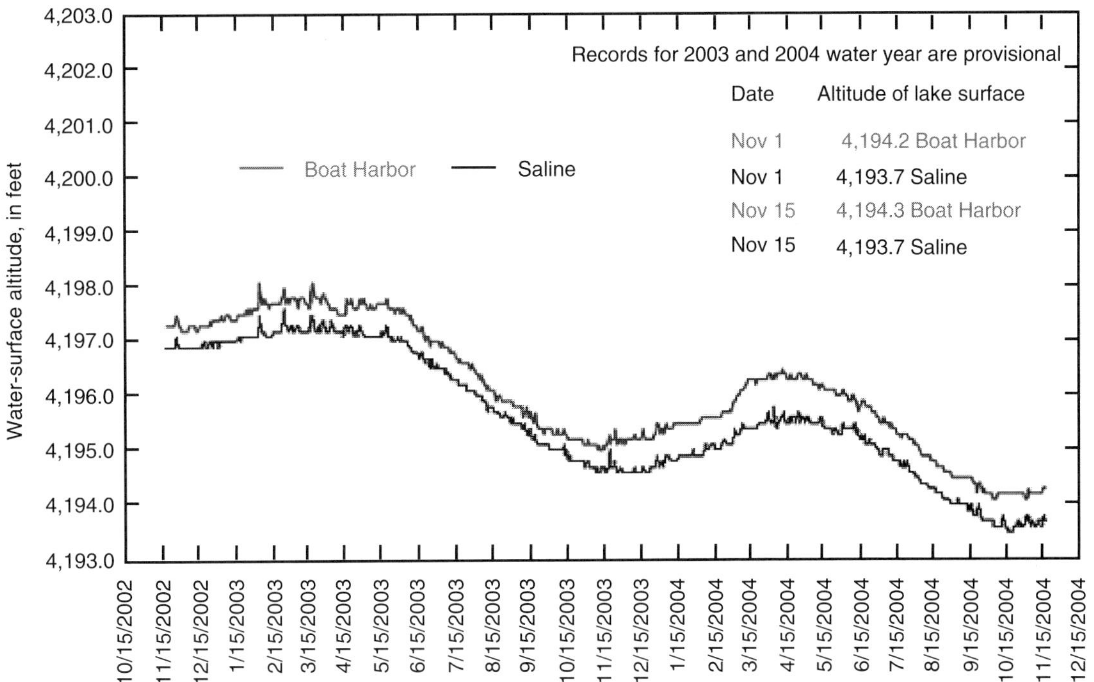

Figure 5B.11 Fluctuation in water-surface altitude of both parts of Great Salt Lake during last 2 years. (From www.usgs.gov.)

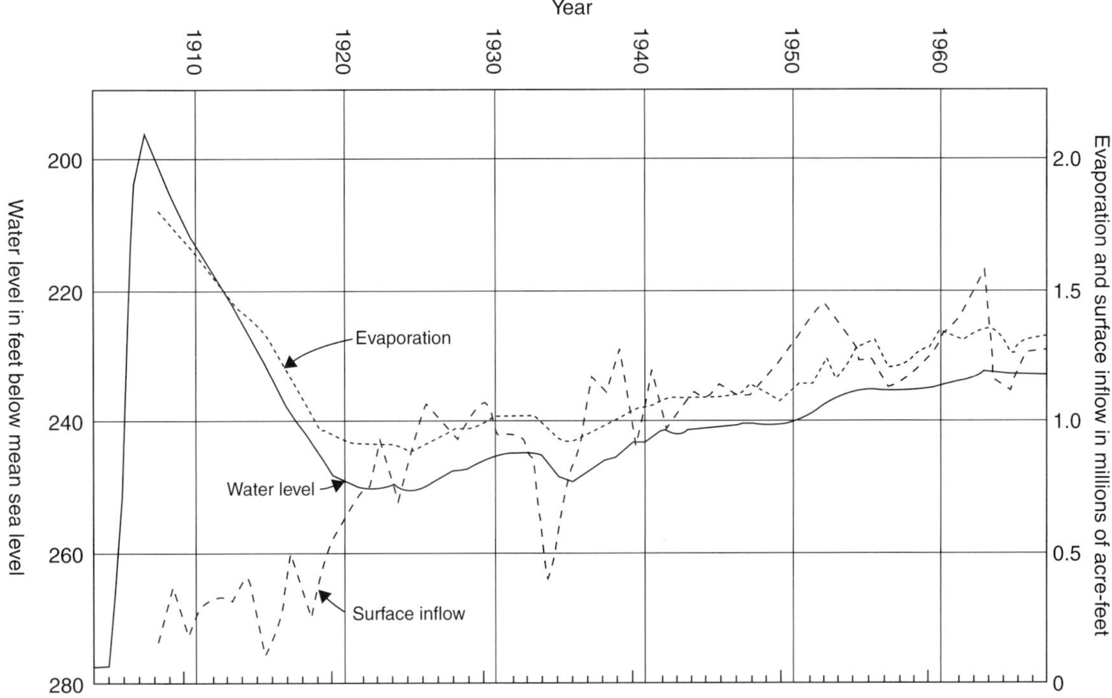

Figure 5B.12 Water levels, evaporation and surface inflow of the Salton Sea, California. (From U.S. Geological survey, professional paper 486-C and California department of water resources bulletin 143-7, geothermal wastes and the water resources of the salton sea area, 1966. With permission.)

Table 5B.19 Long Term Average Min-Max Water Levels

	Jan	Feb	Mar	Apr	May	Jun	Jul	Aug	Sep	Oct	Nov	Dec
Lake Superior[a]												
Mean	601.51	601.31	601.21	601.31	601.64	601.90	602.13	602.20	602.23	602.13	602.00	601.77
Max	602.69 1986	602.46 1986	602.40 1986	602.62 1986	602.82 1986	602.89 1986	603.08 1950	603.22 1952	603.22 1985	603.38 1985	603.31 1985	603.05 1985
Min	599.84 1926	599.61 1926	599.54 1926	599.48 1926	599.61 1926	599.90 1926	600.26 1926	600.46 1926	600.79 1926	600.72 1925	600.43 1925	600.13 1925
Lakes Michigan–Huron[a]												
Mean	578.54	578.48	578.54	578.81	579.13	579.33	579.43	579.36	579.20	579.00	578.81	578.64
Max	581.30 1987	581.07 1986	581.10 1986	581.46 1986	581.63 1986	581.79 1986	581.99 1986	581.99 1986	581.99 1986	582.35 1986	581.96 1986	581.56 1986
Min	576.12 1965	576.08 1964	576.05 1964	576.15 1964	576.57 1964	576.64 1964	576.71 1964	576.67 1964	576.64 1964	576.44 1964	576.28 1964	576.18 1964
Lake St. Clair[a]												
Mean	573.62	573.43	573.82	574.31	574.57	574.70	574.77	574.67	574.44	574.15	573.88	573.88
Max	576.77 1986	576.77 1986	576.77 1986	576.84 1986	576.87 1986	577.17 1986	577.20 1986	577.10 1986	576.90 1986	577.30 1986	576.84 1986	576.77 1986
Min	570.47 1936	570.51 1926	571.03 1934	571.92 1926	572.24 1934	572.34 1934	572.51 1934	572.21 1934	571.98 1934	571.75 1934	571.46 1934	571.65 1964
Lake Erie[a]												
Mean	570.83	570.80	571.10	571.59	571.85	571.95	571.92	571.69	571.39	571.06	570.83	570.83
Max	573.69 1987	573.43 1987	573.75 1986	574.08 1985	574.05 1986	574.28 1986	574.25 1986	573.95 1986	573.59 1986	573.95 1986	573.65 1986	573.82 1986
Min	568.27 1935	568.18 1936	568.24 1934	568.83 1934	569.03 1934	569.06 1934	569.06 1934	569.00 1934	568.83 1934	568.57 1934	568.24 1934	568.21 1934
Lake Ontario[a]												
Mean	244.59	244.69	244.98	245.64	246.10	246.19	246.03	245.67	245.18	244.78	244.55	244.49
Max	246.59 1946	246.95 1952	247.28 1952	248.20 1973	248.46 1973	248.56 1952	248.23 1947	247.97 1947	247.41 1947	246.78 1945	246.65 1945	246.72 1945
Min	246.16 1935	242.06 1936	242.59 1935	242.88 1935	243.14 1935	243.41 1935	243.24 1934	242.78 1934	242.49 1934	242.19 1934	241.96 1934	241.93 1934
Lake Superior[b]												
Mean	183.34	183.28	183.25	183.28	183.38	183.46	183.53	183.55	186.56	183.53	183.49	183.42
Max	183.70 1986	183.63 1986	183.61 1986	183.68 1986	183.74 1986	183.76 1986	183.82 1950	183.86 1952	183.86 1985	183.91 1985	183.89 1985	183.81 1985
Min	182.83 1926	182.76 1926	182.74 1926	182.72 1926	182.76 1926	182.85 1926	182.96 1926	183.02 1926	183.12 1926	183.10 1925	183.01 1925	182.92 1925

Lakes Michigan–Huron[b]												
Mean	176.34	176.32	176.34	176.42	176.52	176.58	176.61	176.59	176.54	176.48	176.42	176.37
Max	177.18 1987	177.11 1986	177.12 1986	177.23 1986	177.28 1986	177.33 1986	177.39 1986	177.39 1986	177.38 1986	177.50 1986	177.38 1986	177.26 1986
Min	175.60 1965	175.59 1964	175.58 1964	175.61 1964	175.74 1964	175.76 1964	175.78 1964	175.77 1964	175.76 1964	175.70 1964	175.65 1964	175.62 1964
Lake St. Clair[b]												
Mean	174.84	174.78	174.80	175.05	175.13	175.17	175.19	175.16	175.09	175.00	174.92	174.92
Max	175.80 1986	175.80 1986	175.80 1986	175.82 1986	175.83 1986	175.92 1986	175.93 1986	175.90 1986	175.84 1986	175.96 1986	175.82 1986	175.80 1986
Min	173.88 1936	173.89 1926	174.05 1934	174.32 1926	174.42 1934	174.45 1934	174.50 1934	174.41 1934	174.34 1934	174.27 1934	174.18 1934	174.24 1964
Lake Erie[b]												
Mean	173.99	173.98	174.07	174.22	174.30	174.33	174.32	174.25	174.16	174.06	173.99	173.99
Max	174.86 1987	174.78 1987	174.88 1986	174.98 1985	174.97 1986	175.04 1986	175.03 1986	174.94 1986	174.83 1986	174.94 1986	174.85 1986	174.90 1986
Min	173.21 1935	173.18 1936	173.20 1934	173.38 1934	173.44 1934	173.45 1934	173.45 1934	173.43 1934	173.38 1934	173.30 1934	173.20 1934	173.19 1934
Lake Ontario[b]												
Mean	74.55	74.58	74.67	74.87	75.01	75.04	74.99	74.88	74.73	74.61	74.54	75.52
Max	75.16 1946	75.27 1952	75.37 1952	75.65 1973	75.73 1973	75.76 1952	75.66 1947	75.58 1947	75.41 1947	75.22 1945	75.18 1945	75.20 1945
Min	73.81 1935	73.78 1936	73.94 1935	74.03 1935	74.11 1935	74.19 1935	74.14 1934	74.00 1934	73.91 1934	73.82 1934	73.75 1934	73.74 1934

Note: Period of Records: 1918–2003. All levels in this table are referenced to the International Great Lakes Datum of 1985 (IGLD85).

[a] English Units (ft).
[b] Metric Units (m).

Source: From www.usace.army.mil.

Table 5B.20 Hydrologic Data for Closed Lakes

Lake	Drainage Area (sq. mi)	Evaporation (ft/yr)		Coeff-cient of Variation of Lake Area[a]	Response Time[b] (yr)	Overflow Expressed as Depth Over Tri-butary Area (ft)	Salinity		Mean Depth (ft)	Lake Area (sq. mi)
		Gross	Net[c]				Date	ppm		
Devils Lake, ND	3,000	2.5	1.2	0.40	14	2.5	1899	8,470	13	45
							1923	15,210	10	26
							1948	25,000	4.5	14
							1952	8,680	10	20
Basin, Lake, Saskatchewan	105	2.25	1.0	0.07	25	—	1938–41	11,900	20	16
Quill Lakes, Saskatchewan	2,700	2.0	0.75	0.15	20	—	1938–41	25,000	10	230
Redberry Lake, Saskatchewan	120	2.25	1.0	0.038	50	—	1938–41	14,000	43	27
Great Salt Lake, UT	21,000	3.3	2.7	0.125	9	1.0	1877	138,000	18	2,200
							1932	276,000	13	1,300
Sevier Lake, UT	16,000	3.7	3.2	0.35	3	4	1872	86,400	8	188
Pyramid Lake, NV	2,650	4.2	3.7	0.04	65	9	1882	3,486	167	200
Walker Lake, NV	3,500	4.2	3.8	0.075	45	13	1882	2,500	120	110
Mono Lake, CA	600	4.1	3.3	0.043	35	200	1882	51,170	61	85
Elsinore Lake, CA	717	4.5	3.2	0.68	3.0	0.3	1949	8,880	5	5
Owens Lake, CA	2,900	5.5	5.0	0.10	10	12	1876	60,000	24	105
							1905	213,700	11	76
Omak Lake, WA	100	3.2	2.2	0.067	30	19	[d]	5,704	50	5.5
Lake Abert, OR	900	3.5	2.5	0.5	6	11	1902	76,000	5	50
							1912	30,000		–
							1956—59	20,000	10	60
Summer Lake, OR	330	3.5	2.5	1.0	2	61	1901	36,000	3	30
							1912	18,000		–
Harney Lake, OR	5,300	3.3	2.5	0.8	2	1.0	1912	22,380	4.8	47
Lake Eyre, Australia	550,000	7.5	7.0	2.5	1.5	10	1950	[e]40,000	8.5	3,100
							1951	[e]240,000	2.8	740
Lake Corangamite, Australia	1,300	4.0	2.0	0.30	10	3.8	1933	105,000	3.5	74
							1950	50,000	6.0	88
							1956	12,000	12	140
Aral Sea, U.S.S.R.	625,000	3.0	2.6	0.10	35	10	—	10,700	52	25,000
Caspian Sea, Asia	1,400,000	3.3	2.8	0.015	300	22	—	11,000	600	170,000
Dead Sea, Palestine	12,000	5.1	4.8	0.03	40	175	—	220,000	460	390
Lake of Urmia, Iran	20,000	3.0	2.5	0.11	9		—	148,000	16	1,800

Lake Van, Turkey	6,000	3.3	2.0	0.02	150	53	1944	[e]22,400	175	1,450
Tuz Golu, Turkey	4,400	3.4	2.4	0.5	1	86	1959	250,000	2	650
Elton Lake, U.S.S.R.	—	—	3.0	1.0	1.0	—	—	300,000	2.3	110
Baskuntschak Lake, U.S.S.R.	—	—	3.0	2.0	0.5	—	—	260,000	1.15	50

Note: These lakes occupy topographic sinks with no discharges by surface streams or seepage and with a groundwater gradient toward the lake.

[a] Coefficient of variation of lake area is equal to the standard deviation of lake volume divided by the area of the lake.
[b] Response time is the ratio of a change in lake volume to the corresponding change in rate of discharge.
[c] Net evaporation is gross evaporation minus precipitation.
[d] Before 1924.
[e] Milligrams per liter.

Source: From U.S. Geological Survey.

Table 5B.21 Water Balance of the Major Lakes of the World

Lake	Observation Period	Volume (km^3)	Inflow (km^3/yr)	Precipitation (mm/yr)	Outflow (km^3/yr)	Evaporation (mm/yr)	Inflow Factor[a]	Outflow Factor[b]	Retentions Time (yr)
Caspian Sea	1940–1966	78200	289	246	10.8	994	0.76	0.03	204
Michigan and Huron	1959–1966	8200	142	780	161	750	0.61	0.65	33
Superior	1959–1966	11600	47.6	760	69.7	470	0.43	0.64	107
Victoria	1925–1959	2700	17.9	1630	21.9	1570	0.14	0.17	21
Aral Sea	1959–1969	1020	49.5	173	0	1050	0.82	0	15
Tanganyika	Long-term	18900	25.8	1000	3.0	1690	0.44	0.05	322
Baikai	1901–1970	23000	60.3	405	59.5	416	0.82	0.82	317
Nyasa	Long-term	7720	34.2	1220	6.3	2130	0.48	0.09	107
Great Slave	Long-term	1070	136	350	141	166	0.93	0.97	7.4
Erie	1959–1966	545	190	860	182	920	0.90	0.88	2.6
Ontario	1959–1966	1710	210	900	210	800	0.93	0.94	7.6
Balkhash	1911–1966	112	15.7	154	0	1020	0.85	0	6.0
Ladoga	1932–1958	908	69.1	606	73.7	344	0.87	0.92	11
Chad	1954–1962	44	45.8	378	0	2260	0.85	0	0.9
Eyre	Long-term	—	4.2	150	0	Dries up	0.77	0	—
Maracaibo	Long-term	—	19.6	977	4.9	2080	0.60	0.15	—
Onega	Long-term	295	15.9	575	18.0	350	0.74	0.84	14
Rudolf	Long-term	—	16.0	750	0.0	2610	0.71	0	—
Titicaca	Long-term	710	7.7	625	0.6	1500	0.60	0.05	55

[a] Percentage of inflow of the sum of inflow and lake precipitation.
[b] Percentage of outflow of the sum of outflow and lake evaporation.

Source: From Kuusisto, E.E., *Lakes Their Physical Aspects, in Facets of Hydrology II*, John C. Rodda, Editor, John Wiley and Sons 1985. Reproduced with permission.

Table 5B.22 Major Lakes in the World

Lake	Country	Surface Area (km^2)	Maximum Depth (m)	Volume (km^3)
		Europe		
Caspian Sea[a]	U.S.S.R. Iran	374,000	1,025	78,200
Ladozskoje	U.S.S.R.	17,700	230	908
Onezskoje	U.S.S.R.	9,630	127	295
Vänern	Sweden	5,550	100	180
Cudskoje with Pskovskoje	U.S.S.R.	3,550	15	25
Vättern	Sweden	1,900	119	72
Saimaa	Finland	1,800	58	36
Beloje	U.S.S.R.	1,290	20	5.2
Vygozero	U.S.S.R.	1,140	18	7.1
Mälaren	Sweden	1,140	64	10
Il'men'	U.S.S.R.	1,100	10	12
Päyänne	Finland	1,065	93	—
Inari	Finland	1,000	80	28
Imandra	U.S.S.R.	900	67	11
Balaton	Hungary	596	12	1.9
Lac de Geneve	Switzerland, France	581	310	90
Bodensee	German Federal Republic, Switzerland, Austria	538	252	48
Hälmaren	Sweden	484	22	—
Stor Sjön	Sweden	464	74	8.0
Kubenskoje	U.S.S.R.	407	13	1.7
Loch Ness	Great Britain	396	31	—
Garda	Italy	370	346	50
Mjøsa	Norway	363	434	56
Skadarsko	Albania, Yugoslavia	362	10	2.2
Ohridsko	Albania, Yugoslavia	350	256	61
Sniardwy	Poland	331	47	2.8
Torne Träsk	Sweden	330	168	17
Neusieler See	Austria, Hungary	323	2	—
Prespansko	Greece, Albania, Yugoslavia	288	54	4.0
Neuchâtel	Switzerland	216	152	—
Lago Maggiore	Italy, Switzerland	214	372	—
Femund	Norway	202	131	6.0
Como	Italy	146	410	—
		Asia		
Aral'skoje More[a]	U.S.S.R.	64,100	68	1,020
Bajkali	U.S.S.R.	31,500	1,741	23,000
Balchas	U.S.S.R.	18,200	26	112
Tonle Sap	Cambodia	10,000[b]	12	40
Issyk-Kul'	U.S.S.R.	6,200	702	1,730
Dongtinghu	China	6,000[c]	10	—
Rizaiyeh (Urumiyeh)[a]	Iran	5,800	16	45
Zajsan	U.S.S.R.	5,510	8.5	53
Tajmyr	U.S.S.R.	4,560	26	13
Kukunor[a]	China	4,220	38	—
Chanka	U.S.S.R. China	4,190	10.6	18.5
Van[a]	Turkey	3,760	145	—
Lob Nor[a]	China	3,500	5	(5)
Ubsa Nor[a]	Mongolia	3,350	—	—
Poyanghu	China	2,700	20	—
Alakol'	U.S.S.R.	2,650	54	58.6
Chövsgöl Nuur	Mongolia	2,620	270	480
Cany	U.S.S.R.	2,500	10	4.5
Tuz[a]	Turkey	2,500	—	—
Namru Tso[a]	China	2,460	—	—
Taihu	China	2,210	—	—
Char Us Nuur	Mongolia	1,760	—	—
Tengiz[a]	U.S.S.R.	1,590	8	—

(Continued)

Table 5B.22 (Continued)

Lake	Country	Surface Area (km^2)	Maximum Depth (m)	Volume (km^3)
Ebi Nor[a]	China	1,420	—	—
Chirgis Nuur	Mongolia	1,480	—	—
Sevan	U.S.S.R.	1,230	86	38
Dalai Nur	China	1,100	—	—
Ulyunger Nor	China	1,000	—	—
Dead Sea[a]	Israel, Jordan	940	400	188
Seletyteniz	U.S.S.R.	777	3.2	1.5
Sasykkol'	U.S.S.R.	736	—	—
P'asino	U.S.S.R.	735	10	—
Kulundinskoje[a]	U.S.S.R.	728	4.9	—
Biwa ko	Japan	688	103	27.5
Gandhi	India	663	64	39.2
Karnaphuli	Bangladesh, India	656	33	13.8
Buir Nuur	Mongolia	610	11	—
Markakol'	U.S.S.R.	449	30	—
Ubinskoje	U.S.S.R.	440	3	—
Karakul'[a]	U.S.S.R.	380	238	—
Tungabharda	India	378	47	12.4
Fumibhal	Thailand	300	123	29.7
Kronockoje	U.S.S.R.	245	128	—
Teleckoje	U.S.S.R.	223	325	40
		Africa		
Victoria	Tanzania, Kenya, Uganda	69,000	92	2,700
Tanganyika	Tanazania, Zaire, Zambia, Burundi, Rwanda	32,900	1,435	18,900
Nyasa	Malawi, Mozambique, Tanzania	30,900	706	7,725
Chad	Chad, Niger, Nigeria	16,600[d]	12	44.4
Rudolf	Kenya	8,660	73	—
Mobutu Sese Seko	Uganda, Zaire	5,300	57	64.0
Mweru	Zambia, Zaire	5,100	15	32.0
Bangweulu	Zambia	4,920[e]	5	5.00
Rukwa	Tanzania	4,500	—	—
Tana	Ethiopia	3,150	14	28.0
Idi Amin Dada	Zaire, Uganda	2,500	131	78.2
Kivu	Zaire, Rwanda	2,370	496	569
Mai Ndombe	Zaire	2,325	6	—
Kamnit	Nigeria	1,270	60	14.0
Abaya	Ethiopia	1,160	13	8.20
Shirwa	Malawi	1,040	2.6	45.0
Tumba	Zaire	765	—	—
Faguibine	Mali	620	14	3.72
Gab el Aulia	Sudan	600	12	—
Chamo	Ethiopia	551	12.7	—
Upemba	Zaire	530	3.5	0.90
Zwai	Ethiopia	434	7	1.10
Shala	Ethiopia	409	266	37.0
Langana	Ethiopia	230	46.2	3.82
L. de Guiers	Senegal	213	7	0.64
Hora Abyata	Ethiopia	205	14.2	1.56
Naivasha	Kenya	140	—	—
Awusa	Ethiopia	130	21	1.34
		North America		
Superior	Canada, U.S.A.	82,680	406	11,600
Huron	Canada, U.S.A.	59,800	229	3,580
Michigan	U.S.A.	58,100	281	4,680
Great Bear	Canada	30,200	137	1,010
Great Slave	Canada	27,200	156	1,070
Erie	Canada, U.S.A.	25,700	64	545

(Continued)

Table 5B.22 (Continued)

Lake	Country	Surface Area (km^2)	Maximum Depth (m)	Volume (km^3)
Winnipeg	Canada	24,600	19	127
Ontario	Canada, U.S.A.	19,000	236	1,710
Nicaragua	Nicaragua	8,430	70	108
Athabaska	Canada	7,900	60	110
Dear Lake	Canada	6,300	—	—
Winnipegosis	Canada	5,470	12	16
Nipigon	Canada	4,800	162	
Manitoba	Canada	4,720	28	17
Great Salt[a]	U.S.A.	4,660	14	19
Forest	Canada, U.S.A.	4,410	21	—
Dubawant	Canada	4,160	—	—
Mistassini	Canada	2,190	120	—
Managua	Nicaragua	1,490	80	—
Saint Clair	Canada	1,200	7.2	5.3
Lesser Slave	Canada	1,190	3	—
Chapala	Mexico	1,080	10	10.2
Winnibago	U.S.A.	818	6	4.1
Marion	U.S.A.	465	—	2.8
Winnipesaukee	U.S.A.	181	55	3.8
		South America		
Maracaibo	Venezuela	13,300	35	—
Titicaca	Peru, Bolivia	8,110	230	710
Poopó[a]	Bolivia	2,530	3	2
Buenos-Aires	Chile, Argentina	2,400	—	—
Lago Argentina	Argentina	1,400	300	—
Valencia	Venezuela	350	—	—
		Australia		
Eyre[a]		up to 15,000	20	—
Amadeus[a]		8,000	—	—
Torrens[a]		5,800	—	—
Gairdner[a]		4,780	—	—
George		145	3	0.3
		New Zealand		
Taupo		611	159	—
Te Anau		352	276	—
Wakatipu		293	378	—
Wanaka		194	—	—
Manapouri		130	—	—
Hawea		119	—	—

[a] Salt lakes.
[b] At low levels 3000 km^2, at high levels 30,000 km^2.
[c] At low levels 4000 km^2, at high levels 12,000 km^2.
[d] At low levels 7000–10,000 km^2, at high levels 18,000–22,000 km^2.
[e] At low levels 4000 km^2, at high levels 15,000 km^2.

Source: From U.S.S.R. National Committee for the International Hydrological Decade, Atlas of World Balance UNESCO, 1977.

SECTION 5C WATERFALLS

Table 5C.23 Major Waterfalls of the World

Name and Location	ft
Africa	
Angola	
Duque de Braganca, Lucala R	344
Ruacana, Cunene R	406
Ethiopia	
Baratieri, Ganale Dorya R	459
Dal Verme, Ganale Dorya R	98
Fincha	508
Tesissat, Blue Nile R[b]	140
Lesotho	
Maletsunyane	630
Rhodesia-Zambia	
Victoria, Zambezi R[b]	355
South Africa	
Aughrabies, Orange R[b]	400
Howick, Umgeni R	311
Tugela (5 falls)[a]	3,110
Highest fall	1,350
Tanzania-Zambia	
Kalambo[b]	726
Uganda	
Murchison, Victoria Nile R	130
Zambia	
Chirombo, Leisa R	880
Asia	
India	
Cauvery[c]	330
Gersoppa (Jog), Sharavati R[a,c]	830
Japan	
Kegon, L. Chuzenji[c]	330
Yudaki, L. Yuno	335
Australasia	
Australia	
New South Wales	
Wentworth[a]	518
Highest fall	360
Wollomombi	1,100
Queensland	
Coomera	210
Tully	450
New Zealand	
Bowen (from Glaciers)[b]	540
Helena	890
Sterling	505
Sutherland, Arthur R[a]	1,904
Europe	
Austria	
Upper Gastein	207
Lower Gastein (Both on Ache R.)	280
Golling, Schwarzbach R[a]	200
Krimml (Krimmler)	1,250
France	1,385
Gavarnie (C)[a]	
Great Britain–Wales	
Pistyll Cain, Afon Gain R	150
Pistyll Rhaiadr	240
Scotland Glomach	370

(Continued)

Table 5C.23 (Continued)

Name and Location	ft
Iceland	
Detti, Jokul R	144
Gull, Hvita R	101
Italy	
Toce (C)	470
Norway	
Eastern Mardalsfoss[a]	1,696
Highest fall	974
Western Mardalsfoss (Both on L. Eikesdal)	1,535
Skjeggedal	525
Skykkje, Skykkjua R	820
Vettis, Morkedöla R	1,214
Highest fall	889
Vöring, Bjoreia R	597
Sweden	
Handöl, Handöl Cr[a]	345
Stora Sjöfallet, Lule R[a,b]	130
Tannforsen, Are R	120
Switzerland	
Giétroz (Glacier) (C)[a]	1,640
Diesbach[a]	394
Giessbach[a]	1,312
Handegg, Aare R	151
Iffigen	394
Pissevache, La Salanfe R	213
Reichenbach[a]	656
Rhine	65
Simmen, Simme R[a]	459
Stäuber	590
Staubbach	984
Trümmelbach[a]	1,312
North America	
Canada	
British Columbia	
Takakkaw (Daly Glacier)[a]	1,650
Highest fall	1,200
Panther, Nigel Cr	600
Labrador	
Churchill Falls, Churchill R	245
Mackenzie District	
Virginia, S. Nahanni R	315
Quebec	
Montmorency	251
Canada-United States	
Ontario-New York	
Niagara: American	193
Horseshoe	186
United States	
Arizona	
Mooney, Havasu Cr	220
California	
Feather, Fall R	640
Illilouette	370
Nevada, Merced R	594
Ribbon[c]	1,612
Silver Strand	1,170
Vernal, Merced R	317
Yosemite[a]	2,425
Bridalveil	620
Yosemite (upper)[b]	1,430

(Continued)

Table 5C.23 (Continued)

Name and Location	ft
Yosemite (lower)[b]	320
Colorado	
Seven	266
Georgia	
Tallulah[a]	251
Idaho	
Henry's Fork (upper)	96
Henry's Fork (lower)	70
Shoshone, Snake R[c]	195
Twin, Snake R[c]	125
Kentucky	
Cumberland	68
Maryland	
Great Potomac R (C)	90
Minnesota	
Minnehaha[c]	54
Montana	
Missouri	75
New Jersey	
Passaic[c]	70
New York	
Taughannock	215
Oregon	
Multnomah[a]	620
Highest fall	542
Tennessee	
Fall Creek	256
Rock House Creek	125
Washington	
Fairy Falls	700
Mt. Rainer National Pk	
Narada, Paradise R	168
Sluiskin, Paradise R	300
Palouse	198
Snoqualmie	270
Wisconsin	
Manitou, Black R	165
Wyoming	
Yellowstone National Pk	
Tower	132
Yellowstone (upper)	109
Yellowstone (lower)	308
Mexico	
El Salto	
Juanacatlán, Rio Grande de Santiago[c]	66
South America	
Argentina-Brazil	
Iguazú[a]	230
Brazil	
Glass	1,325
Herval	400
Paulo Afonso, São Francisco R	275
Patos-Maribondo, Rio Grande	115
Urubupunga, Alto Paraná R	40
Brazil-Paraguay	
Sete Quedas, or Guaira Alto Paraná R	130
Colombia	
Tequendama	
Bogotá R	427

(Continued)

Table 5C.23 (Continued)

Name and Location	ft
Catarata de Candelas, Cusiana R	984
Ecuador	
Agoyan, Pastaza R	200
Guyana	
Kaieteur, Potaro R	741
King Edward VIII, Semang R	840
King George VI, Utshi R	1,600
Marina, Ipobe R[a]	500
Highest fall	300
Venezuela	
Angel	3,212
Highest fall	2,648
Cuquenán	2,000

Note: Height-total drop in one or more leaps. If river names not shown, they are same as the falls. R—river; L—lake; (C)—Cascade-type.

[a] Falls of more than one leap.
[b] Falls that diminish greatly seasonally.
[c] Falls that reduce to a trickle or are dry for part of each year.

Source: From National Geographic Society.

SECTION 5D GLACIERS AND ICE

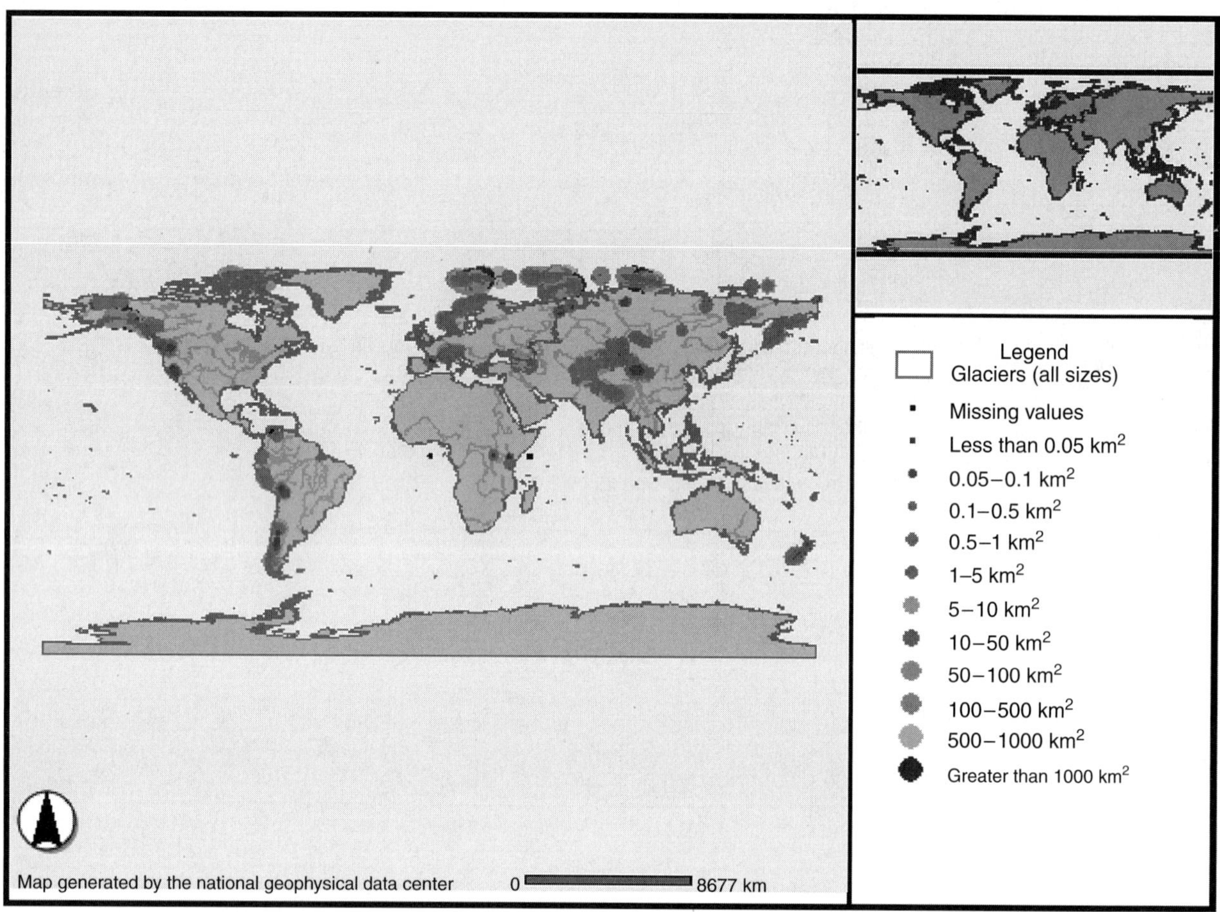

Figure 5D.13 World glacier inventory. (From www.ngdc.noaa.gov.)

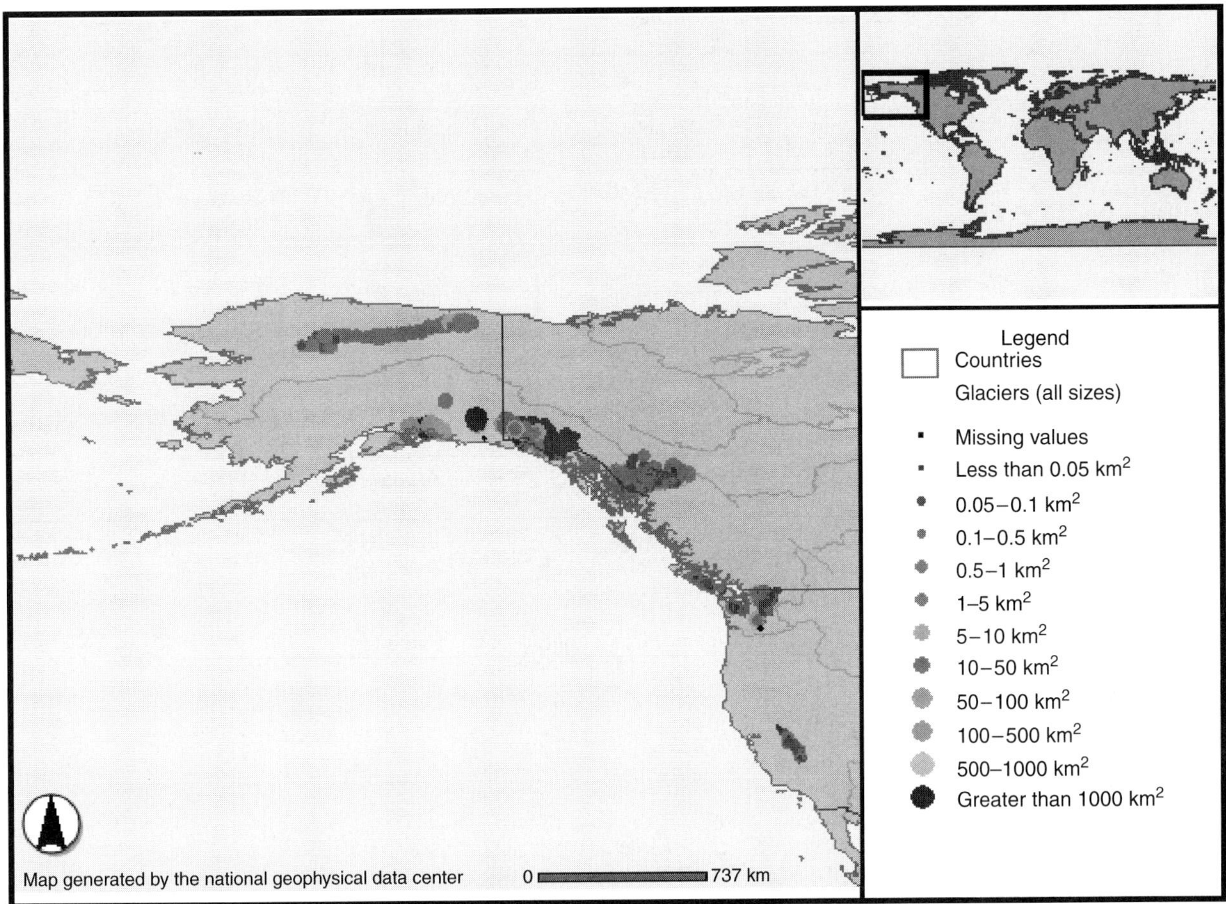

Figure 5D.14 Glacier inventory of western North America. (From www.ngdc.noaa.gov.)

Table 5D.24 Glacial Ice Coverage of the World

Land Area	sq. mi
Continental Europe	3,880
Continental Asia	43,270
Continental North America	30,900
Continental South America	9,600
South polar regions	5,020,450
North polar regions	721,150
Africa	8
New Zealand	386
New Guinea	6
Total	5,829,650

Source: From Huberty and Flock, Natural Resources, McGraw-Hill, Copyright 1959. Reproduced with permission.

Table 5D.25 Glaciers in the United States

State	Approximate Number of Glaciers	Total Glaciated Area (sq. mi)	Glacier Contribution to July–August Streamflow (Estimated)	
			Thousand Acre-feet	Million Gallons
Alaska	(unknown)	29,000	150,000	49,000,000
Washington	950	160	870	280,000
California	290	19	65	21,000
Wyoming	100	19	80	26,000
Montana	200	16	65	21,000
Oregon	60	8	40	13,000
Colorado	25	0.6	2	650
Idaho	20	0.6	2	650
Nevada	5	0.1	0.4	130
Utah	1	0.04	0.1	33

Source: From U.S. Geological Survey, *National Water Summary 1985—Hydrologic Perspectives. Water-Supply Paper 2300.*

Table 5D.26 Areas of Glaciers in the Western Conterminous United States

Location	Area (sq. km)	
	Meier (1961a)	More Recent Source, Where Available
Washington		
1. North Cascades (northern Cascade Range)[a]	251.7	267.0[b]
2. Olympic Mountains	33.0	45.9[c]
3. Mount Rainier	87.8	92.1[d]
4. Goat Rocks area	1.5	1.5
5. Mount Adams	16.1*	16.1*
6. Mount St. Helens	7.3	5.92/2.16[e]
Total	397.4	428.5/424.8
Oregon		
7. Mount Hood	9.9	13.5[d]
8. Mount Jefferson	3.2	3.2
9. Three Sisters area	7.6	8.3[d]
10. Wallowa Mountains	—	*0.1
Total	20.7	25.1
California		
11. Mount Shasta	5.5	6.9[d]
12. Salmon-Trinity Mountains	0.3	0.3
13. Sierra Nevada	13.1	50.0/63.0[f]
Total	18.9	57.2/70.2
Montana		
14. Glacier National Park	13.8	28.4[g]
15. Cabinet Range	0.5	0.5[g]
16. Flathead-Mission-Swan Ranges	*1.2	*2.3[g]
17. Crazy Mountains	*0.5	*0.4[g]
18. Beartooth Mountains	10.8	10.9[g]
Total	26.8	42.5
Wyoming		
19. Big Horn Mountains	0.3	1.0[g]
20. Absaroka Range	0.7*	.7*[g]
21. Teton Range	2.0	1.7[g]
22. Wind River Range	44.5	31.6[g]
Total	47.5	37.5
Colorado		
23. Rocky Mountain Park-Front Range, others	1.7	1.5[g]

(Continued)

Table 5D.26 (Continued)

Location	Area (sq. km) Meier (1961a)	More Recent Source, Where Available
Idaho		
24. Sawtooth Mountains	—	1.0[h*]
Utah		
25. Wasatch Mountains	—	0.2[i]
Nevada		
26. Wheeler Peak	0.2	0.2

Note: Glacier areas in the first column are taken from Meier (1961a); dashes mean not determined by Meier. Glacier areas in the second column are from Meier (1961a) where a more recent source is not available. The change in area between 1961 and the more recent source is normally due to a more complete data set rather than a true change. An asterisk indicates that the value is estimated.

[a] The region bounded by the Canadian border on the north, Snoqualmie Pass on the south, the Puget Lowlands on the west, and the Columbia and Okanogan Rivers on the east.
[b] Post and Others, 1971.
[c] Spicer, 1986.
[d] Driedger and Kennard, 1986.
[e] Brugman and Meier, 1981. Before/after eruption of 18 May 1980.
[f] Raub and Others, 1980; unpub. data. The 50 km^2area includes glaciers plus moraine-covered ice; the 63 km^2 area includes glaciers, moraine-covered ice, and small ice bodies not large enough to be considered glaciers.
[g] Graf, 1977.
[h] Estimated; various observers have reported numerous small glaciers.
[i] Timpanogos Cave, Utah, USGS 1:24,000-scale topographic map.

Source: From www.usgs.gov.

Table 5D.27 Variations in the Position of Glacier Fronts Addenda from Earlier Years

NR	Glacier Name	PSFG NR	Method	1st Survey DMY	2nd Survey DMY	Variations (m)
	Colombia					
1	Alfombralese	CO0013B	A	13.1.1945	10.2.1959	−50.0
				10.2.1959	11.1.1975	−50.0
				11.1.1975	10.12.1985	−80.0
				10.12.1985	19.1.1987	−20.0
2	Azufradoe	CO0005B	A	13.1.1945	10.2.1959	60.0
				10.2.1959	11.1.1975	−20.0
				11.1.1975	10.12.1985	−130.0
				10.12.1985	19.1.1987	ST
3	Azufradow	CO0005A	A	13.1.1945	10.2.1959	−70.0
				10.2.1959	11.1.1975	−20.0
				11.1.1975	10.12.1985	−80.0
				10.12.1985	19.1.1987	ST
4	Lacabana	CO00007	A	10.2.1959	11.1.1975	−200.0
				11.1.1975	10.12.1985	−200.0
				10.12.1985	19.1.1987	−20.0
5	Laplazuela	CO00006	A	13.1.1945	10.2.1959	−20.0
				10.2.1959	11.1.1975	−30.0
				11.1.1975	19.1.1987	−220.0
6	Lagunillas	CO00008	A	13.1.1945	10.2.1959	0.0
				10.2.1959	11.1.1975	0.0
				11.1.1975	10.12.1985	−50.0
				10.12.1985	19.1.1987	−10.0
7	Leonera alta	CO00009	A	13.1.1945	10.2.1959	−330.0
				10.2.1959	11.1.1975	−180.0
				11.1.1975	10.12.1985	−100.0
				10.12.1985	19.1.1987	−30.0

(Continued)

Table 5D.27 (Continued)

NR	Glacier Name	PSFG NR	Method	1st Survey DMY	2nd Survey DMY	Variations (m)
8	Nereidas	CO00014	C	1958	6.3.1986	−644.5
				6.3.1986	5.5.1987	−40.0
				5.5.1987	18.3.1988	−50.0
				18.3.1988	27.12.1990	−150.0
	Chile					
9	Amalia	RC00056	A	1945	1986	−6000.0
10	Asia	RC00055	A	1945	1984	−195.0
				1984	1986	−96.0
11	Balmaceda	RC00060	A	1945	1984	−2496.0
				1984	1986	−80.0
12	Bernardo	RC00037	A	1945	1976	−837.0
				1976	1984	−304.0
13	Calvo	RC00053	A	1945	1984	0.0
				1984	1986	0.0
14	Dickson	RC00063	A	1945	1984	−3120.0
15	EUROPA	RC00049	A	1945	1981	−504.0
				1981	1986	−234.0
16	GREVE	RC00040	A	1945	1976	−3317.0
				1976	1981	−215.0
				1981	1984	−102.0
				1984	1986	−22.0
				1986	1987	−80.0
17	GREY	RC00062	A	1945	1967	−550.0
				1967	1975	− 350.0
18	HPS12	RC00043	A	1981	1984	−180.0
				1984	1986	0.0
19	HPS13	RC00045	A	1945	1984	0.0
				1984	1986	0.0
20	HPS15	RC00046	A	1945	1984	0.0
				1984	1986	0.0
21	HPS19	RC00047	A	1981	1986	−400.0
22	HPS28	RC00051	A	1945	1984	−351.0
				1984	1986	−1028.0
23	HPS29	RC00052	A	1945	1984	−234.0
				1984	1986	−120.0
24	HPS31	RC00050	A	1945	1970	−975.0
				1970	1984	−252.0
25	HPS34	RC00054	A	1945	1984	−39.0
				1984	1986	0.0
26	HPS38	RC00057	A	1945	1984	− 468.0
				1984	1986	240.0
27	HPS41	RC00058	A	1945	1984	−360.0
				1984	1986	0.0
28	HPS8	RC00041	A	1945	1976	−1240.0
				1976	1979	−60.0
				1979	1984	−265.0
				1984	1986	66.0
29	HPS9	RC00042	A	1976	1979	−30.0
				1979	1984	−35.0
				1984	1986	−134.0
30	OCCIDENTAL	RC00039	A	1945	1976	−93.0
				1976	1984	−592.0
				1984	1987	− 462.0
31	FHIDRO	RC00036	A	1945	1976	−1643.0
				1976	1984	−216.0
				1984	1986	134.0
32	PENGUIN	RC00048	A	1981	1986	−60.0
33	PINGO	RC00061	A	1945	1984	−1326.0
				1984	1986	0.0

(Continued)

Table 5D.27 (Continued)

NR	Glacier Name	PSFG NR	Method	1st Survey DMY	2nd Survey DMY	Variations (m)
34	Pio XI	RC00044	A	1925	9.1926	1000.0
				9.1926	12.1928	400.0
				12.1928	1945	−2500.0
				1945	1951	5400.0
				1951	1963	600.0
				1963	1969	500.0
				1969	1976	2400.0
				1976	1981	310.0
				1981	1986	−400.0
35	Snowy	RC00059	A	1945	1984	−936.0
				1984	1986	0.0
36	Tempano	RC00038	A	1945	1976	−1178.0
				1976	1984	−1264.0
				1984	1986	−694.0
	Argentina					
37	Frias	RA00064	A	1984	1986	0.0
	Sweden					
38	Hyllglaciaeren	S00780	C	16.8.1984	15.9.1988	−38.0
	Kenya					
39	Cesar	KN00004	A	1899	21.2.1947	−120.0
				3.9.1947	3.9.1987	−95.0
40	Darwin	KN00006	A	21.2.1947	3.9.1987	−60.0
41	Diamond	KN00010	A	21.2.1947	3.9.1987	−40.0
42	Forel	KN00011	A	21.2.1947	3.9.1987	−9.0
43	Gregory	KN00009	A	21.2.1947	3.9.1987	−120.0
				13.3.1986	1.9.1990	−25.0
44	Heim	KN00012	A	21.2.1947	3.9.1987	−9.0
45	Joseph	KN00003	A	21.2.1947	3.9.1987	−250.0
46	Krapf	KN00001	A	1899	21.2.1947	−150.0
47	Lewis	KN00008	A	1.5.1934	21.2.1947	−130.0
				1.1.1974	13.2.1978	−25.0
				21.2.1947	3.9.1987	−245.0
48	Melhuish	KN00014	A	21.2.1947	3.9.1987	−220.0
49	Northey	KN00013	A	21.2.1947	3.9.1987	−230.0
50	Tyndall	KN00005	A	1899	21.2.1947	−250.0
				21.2.1947	3.9.1987	−70.0
	Poland					
51	Pod Bula	PL00111	C	9.9.1980	26.9.1981	−32.6
				26.9.1981	25.9.1982	−7.0
				25.9.1982	10.10.1983	13.0
				10.10.1983	30.9.1984	−.3
				30.9.1984	30.9.1985	−20.7
				30.9.1985	29.9.1985	24.3
				29.9.1986	10.10.1987	−10.3
				10.10.1987	25.9.1988	17.7
				25.9.1988	8.10.1989	3.3
				8.10.1989	27.9.1990	−33.0
	C.I.S					
52	Dzhelo	SU07106	C	6.9.1985	1986	−X
				1986	1987	−X
				1987	3.9.1988	−50.7
				3.9.1988	6.9.1989	−19.6
				6.9.1989	3.9.1990	−8.7
53	Leviy Karagemsk	SU07107	C	6.9.1985	2.9.1986	−19.0
				2.9.1986	5.9.1987	1.1
				5.9.1987	3.9.1988	−12.6
				3.9.1988	6.9.1989	−11.0
				6.9.1989	5.9.1990	−8.7
54	Mizhirgichiran	SU03043	C	6.9.1989	7.9.1990	12.9

(Continued)

Table 5D.27 (Continued)

NR	Glacier Name	PSFG NR	Method	1st Survey DMY	2nd Survey DMY	Variations (m)
55	Muravlev	SU06002	C	3.9.1989	26.8.1990	−3.3
56	NO. 122 (UNIV.)	SU07108	C	6.9.1985	2.9.1986	−15.6
				2.9.1986	5.9.1987	−6.4
				5.9.1987	3.9.1988	−4.9
				3.9.1988	6.9.1989	−5.5
				6.9.1989	5.9.1990	−13.1
57	Praviy Karagems	SU07109	C	6.9.1985	2.9.1986	−18.6
				2.9.1986	5.9.1987	−.5
				5.9.1987	3.9.1988	2.3
				3.9.1988	5.9.1990	−7.5
58	Shumskiy	SU06001	C	8.9.1989	27.8.1990	−9.4
	Pakistan					
59	Aling	PK00035		1970	1989	−X
60	Bualtar	PK00004		1939	1988	+X
	Nepal					
61	Thulagi	NP00013	A	1958	1.11.1972	−50.0
				1.11.1972	1.11.1977	−50.0
				1.11.1977	1.11.1984	−150.0
				1.11.1984	1.11.1988	−850.0

Note:

NR	Record number
Glacier Name	15 alphabetic or numeric digits
PSFG Number	5 digits identifying glacier with alphabetic prefix denoting country
Method	A = aerial photogrammetry B = terrestrial photogrammetry C = geodetic ground survey (theodolite, tape etc.) D = combination of a, b, or c E = other methods or no information
1ST Survey:	Day, month, and year of survey
2ND Survey:	Day, month, and year of following survey
Variation in Meters:	Variation in the position of the glacier front in horizontal projection expressed as the change in length between the surveys
Key to Symbols:	+X: Glacier in advance −X: Glacier in retreat ST: Glacier stationary SN: Glacier front covered by snow

Source: From World Glacier Monitoring Service by Wilfried Haeberli, Martin Hoelzle, Stephan Suter and Regula Frauenfelder. www.wgms.ch.

Table 5D.28 Variations in the Postion of Glacier Fronts 1990–1995

NR	Glacier Name	PSFG NR	First Survey	Last Survey	Method	Variations (m)				
						1991	1992	1993	1994	1995
	Canada									
1	Overlord	CD01590	1928	1990	C		−27.3	0.8		−50.0
2	Wedgemount	CD02333	1928	1990	A				−39.0	−14.5
	U.S.A.									
3	Blue Glacier	US02126	1938	1990	C	−14.0	−42.0	−6.0	−25.0	8.0
4	Cantwell	US00320	1950	1950	C			−98.0		
5	Mccall	US00001	1957	1971	C			−285.0	−5.4	−19.0
6	Middle Toklat	US00315	1954	1954	C		−769.0			
7	South Cascade	US02013	1957	1990	A	−30.0	−38.0	−22.0	−29.0	−23.0
	Colombia									
8	Nereidas	CO00014	1958	1990	C	−50.0	−80.0	−X	−X	−260.0
	Peru									
9	Broggi	PE00003	1968	1990	C	−36.7	−54.5	−22.2	−7.8	−17.4
10	Uruashraju	PE00005	1968	1990	C	−25.8	−32.7	−24.2	−24.1	−34.7
11	Yanamarey	PE00004	1972	1990	C	−7.2	−4.0	−22.0	−21.7	−35.8
	Bolivia									
12	Chacaltaya	RB05180		1991	C		−5.2	−4.7	−4.6	−17.6
13	Zongo	RB05150		1991	C		−12.3	1.1	−10.2	−6.4
	Chile									
14	Bernardo	RC00037	1945	1984	A		226.0			
15	Grey	RC00062	1945	1975	A	−352.0				−80.0
16	PIO XI	RC00044	1925	1986	A		500.0	400.0	700.0	−600.0
	Iceland									
17	Breidamjok EA	IS1126A	1932	1990	C	−X				
18	Briedamjok EB	IS1126B	1932	1990	C	−40.0	−24.0	−25.0		
19	Breidamjok WA	IS1125A	1932	1990	C	−62.0	−57.0	−40.0	−18.0	0.0
20	Breidamjok WC	IS1125C	1932	1990	C	−70.0	−69.0	−16.0	−20.0	−30.0
21	Brokarjokull	IS01427		1990	C	−12.0		−18.0	11.0	
22	Falljokull	IS01021	1932	1990	C	−3.0	−14.0	−3.0	−14.0	−1.0
23	Fjallsfitjar	IS1024B	1948	1990	C	−9.0	−39.0	−67.0	10.0	5.0
24	Fjallsj. Brmfj	IS1024A	1948	1987	C	−21.0	−22.0	−15.0	−20.0	−5.0
25	Fjallsj.G-Sel	IS1024C	1948	1990	C	−25.0	−25.0	−30.0	−X	25.0
26	Flaajokull	IS1930A		1972	C	5.0	6.0	−7.0	−5.0	
27	Gigjokull	IS00112	1930	1990	C			7.0	22.0	
28	Gljufurarjokull	IS00103	1939	1989	C				−X	−4.0
29	Hagafellsjok E	IS00306	1934	1990	C			−149.0		
30	Hagafellsjok W	IS00204	1934	1990	C			−86.0		
31	Halsjokull	IS00117	1972	1990	C	3.0	5.0	SN		
32	Hoffellsj W	IS02031	1930	1990	C		−2.0			
33	Hrutarjokull	IS00923	1948	1990	C	−5.0	−X	−X	−28.0	5.0
34	Hyrningsjokull	IS00100	1931	1990	C	2.0	9.0	20.0	16.0	28.0
35	Jokulkrokur	IS00007	1965	1985	C	3.0				
36	Kaldalonsjokull	IS00102	1931	1988	C	SN	−4.0	−6.0	−3.0	12.0
37	Kverkjokull	IS02500	1963	1989	C			13.0		
38	Kviarjokull	IS00822	1934	1990	C	−14.0	−X	−X	−11.0	6.0
39	Leirufjjokull	IS00200	1966	1990	C	−22.0	0.0	−37.0	−32.0	39.0
40	Morsarjokull	IS00318	1932	1990	C	−3.0	−52.0	31.0	0.0	0.0
41	Mulajokulls	IS0311A	1933	1990	C	−91.0	49.0	153.0	−15.0	−52.0
42	Nauthagajokull	IS00210	1932	1990	C	−8.0	1.0	3.0	−9.3	−2.0
43	Oldufellsjokull	IS00114	1967	1989	C			57.0		
44	Reykjafjardarj	IS00300	1931	1990	C	−47.0	−10.0	SN	−25.0	−8.0
45	Satujokull	IS00530		1990	C	−11.0	−8.0	0.0	−X	
46	Sidujok EM177	IS0015B	1964	1990	C	−95.0	−27.0	−63.0	1117.0	−3.0
47	Skaftafellsj.	IS00419	1932	1990	C	−12.0	0.0	−6.0	−46.0	−56.0
48	Skalafellsjokul	IS1728A		1990	C		2.0	5.0		2.0
49	Skeidararj E1	IS0117A	1932	1990	C	147.0	−28.0	−9.0	−33.0	−31.0
50	Skeidararj E2	IS0117B	1932	1990	C	64.0	−10.0	−16.0	−11.0	−7.0
51	Skeidararj E3	IS0117C	1932	1990	C	0.0	−3.0	−1.0	−4.0	−3.0
52	Skeidararj W	IS00116	1932	1990	C	429.0	86.0	−X	−X	−77.0

(Continued)

Table 5D.28 (Continued)

NR	Glacier Name	PSFG NR	First Survey	Last Survey	Method	Variations (m)				
						1991	1992	1993	1994	1995
53	Solheimajok W	IS0113A	1930	1990	C	2.0	13.0	12.0	−3.0	13.0
54	Svinafellsj	IS0520A	1932	1990	C	11.0	7.0	−11.0	−7.0	5.0
55	Tungnaarjokull	IS02214	1955	1990	C	−81.0	−9.0	−31.0	−17.0	1175.0
56	Virkisjokull	IS00721	1932	1990	C	−X	−X	−X	−22.0	−X
	Norway									
57	Austerdalsbreen	N31220	1906	1990	C	0.0	5.0	7.0	15.0	15.0
58	Brigsdalsbreen	N37110	1901	1990	C	10.0	35.0	75.0	80.0	65.0
59	Engabreen	N67011	1903	1990	C	−24.0		18.0		115.0
60	Faabergstoelsb.	N31015	1907	1990	C	−18.0	−6.0	10.0	34.0	44.0
61	Hansbreen	N12419	1936	1990	B	−309.0	56.0	−64.0	45.0	42.0
62	Hellstugubreen	N00511	1902	1990	C	−7.0	−9.0	−3.0	−9.0	−6.0
60	Leirbreen	N00548	1910	1990	C			−16.5		−14.5
64	Nigardsbreen	N31014	1907	1990	C	10.0	21.0	14.0	36.0	50.0
65	Stegholtbreen	N31021	1907	1990	C	−7.0	−5.0	−3.0	−10.0	−5.0
66	Storbreen	N00541	1904	1990	C				−1.5	0.0
67	Styggedalsbreen	N30720	1903	1992	C			0.0	4.0	2.0
	Sweden									
68	Hyllglaciaeren	S00780	1965	1990	C	0.0	0.0	0.0	−4.0	0.0
69	Isfallsglac	S00787	1897	1990	C	0.0	5.0	3.0	2.0	5.0
70	Karsojietna	S00798	1905	1990	C		0.0	0.0	−2.0	
71	Mikkajekna	S00766	1896	1990	C	−8.0	−10.0	−15.0	−12.5	−13.5
72	Partejekna	S00763	1965	1990	C	0.0	−4.0	−5.0	−12.0	−9.0
73	Passusjietna E	S00797	1968	1990	C		0.0	0.0		
74	Passusjietna W	S00796	1968	1990	C		−2.0	−4.0	−2.0	0.0
75	Rabots Glaciaer	S00785	1946	1990	C	−11.0	−14.0	−15.0		−18.6
76	Riukojietna	S00790	1963	1990	C		0.0	0.0	0.0	0.0
77	Ruopsokjekna	S00764	1965	1989	C		−10.0	−11.0	−3.9	−3.4
78	Ruotesjekna	S00767	1965	1990	C	−14.0	−7.0	−4.0	−10.0	
79	Salajekna	S00759	1897	1990	C		−10.0		−14.0	−12.5
80	SE Kaskasatj GL	S00789	1910	1989	C	0.0	3.0	1.0	2.0	5.0
81	Storglaciaeren	S00788	1897	1990	C	0.0	0.0	0.0	0.0	0.0
82	Stour Raetta GL	S00784	1963	1990	C		0.0	−5.0	−3.0	−1.0
83	Suottasjekna	S00768	1964	1990	C	−2.0	0.0	0.0	0.0	0.0
84	Unnaraeita GL	S00783	1963	1990	C		0.0	0.0	0.0	0.0
85	Vartasjekna	S00765	1964	1990	C	0.0	0.0	0.0	0.0	0.0
	France									
86	Argentiere	F00002	1878	1990	C	−11.0	−13.0	−25.0	−10.0	−22.0
87	Blanc	F00031	1871	1990	C	−15.5	−18.0	−26.0	−27.0	−23.5
88	Bossons	F00004	1861	1990	C	−110.0	−66.0	−81.0	−74.0	−18.0
89	Gebroulaz	F00009	1730	1990	C	−6.0	−21.0	ST	ST	ST
90	Mer De Glace	F00003	1879	1990	C	5.0	ST	ST	ST	ST
91	Saint Sorlin	F00015	1904	1990	C	−2.7	−6.2	−1.8	2.6	−3.4
	Switzerland									
92	Allalin	CH00011	1880	1990	A	−35.0	−63.0	−20.0	−11.0	−24.0
93	Alpetli (Kander)	CH00109	1893	1990	C	−2.0	−9.0	−6.0	−3.0	−3.0
94	Ammerten	CH00111	1969	1990	C	−6.0	−4.0	−1.0	−4.0	−2.0
95	Arolla (BAS)	CH00027	1884	1990	C	−26.0	−10.0	−15.0	−6.0	−11.0
96	Basodino	CH00104	1893	1990	C	−2.0	−3.0	+X	6.0	−25.0
97	Bella Tola	CH00021	1945	1990	C	−13.0	−1.0	−5.0	−9.0	−39.0
98	Biferten	CH00077	1893	1990	C	−4.0			−39.0	−7.0
99	BIS	CH00107	1883	1990	E	−X	−X		−X	−X
100	Bluemlisalp	CH00064	1893	1990	C	−11.0	−6.0	−17.0	−7.0	−3.0
101	Boveyre	CH00041	1889	1990	C	−25.0	−37.0		−13.0	
102	Breney	CH00036	1892	1990	C	−11.0	−24.0	−27.0	−8.0	−2.0
103	Bresciana	CH00103	1896	1990	C	−12.0	SN	−17.0	−8.0	−14.0
104	Brunegg	CH00020	1941	1990	C	−5.0	−3.0	−4.0	−8.0	−5.0
105	Brunni	CH00072	1882	1990	C				−X	
106	Calderas	CH00095	1920	1990	C	−17.0	−9.0	−8.0	−4.0	−7.0
107	Cambrena	CH00099	1889	1990	C	−5.0	−12.0	−X	−17.0	−9.0

(Continued)

Table 5D.28 (Continued)

NR	Glacier Name	PSFG NR	First Survey	Last Survey	Method	Variations (m) 1991	1992	1993	1994	1995
108	Cavagnoli	CH00119	1893	1990	C	−16.0	−21.0	−X	−12.0	0.0
109	Cheillon	CH00029	1919	1990	C	−79.0	−60.0		11.0	−7.0
110	Corbassiere	CH00038	1889	1990	C	−15.0	−14.0	9.0		
111	Corno	CH00120	1893	1990	C	−7.0	−13.0	−X	−16.0	1.0
112	Damma	CH00070	1920	1990	C	2.0	−9.0	−X	−9.0	−16.0
113	Eiger	CH00059	1893	1990	C	−28.0	−16.0	−17.0	−13.0	−8.0
114	EN Darrey	CH00030	1929	1990	C		−8.0		−65.0	−2.0
115	FEE North	CH00013	1879	1990	C	−10.0	−5.0		−55.0	−38.0
116	Ferpecle	CH00025	1891	1990	C	−19.0	−8.0	−7.0	−23.0	−12.0
117	Fiescher	CH00004	1891	1990	C	−6.0	7.0	−16.0	−21.0	−16.0
118	Findelen	CH00016	1892	1990	D	−61.0	−22.0		−32.0	−35.0
119	Firnalpeli	CH00075	1894	1990	C		−15.0	−X	−8.0	−5.0
120	Forno	CH00102	1894	1990	C	−19.0	−22.0	−16.0	−24.0	−21.0
121	Gamchi	CH00061	1893	1990	C	−3.0	−6.0	−12.0	−9.0	−1.0
122	Gauli	CH00052	1886	1990	C	−11.0	−21.0	−9.0	−30.0	−12.0
123	Gietro	CH00037	1889	1990	A	−X	−X	−X	−X	−X
124	Glaernisch	CH00080	1923	1990	C	−5.0		−22.0	−9.0	−35.0
125	Gorner	CH00014	1883	1990	C	−4.0	−52.0	−12.0	−34.0	−10.0
126	Grand Desert	CH00031	1892	1989	C	−73.0	−7.0		−11.0	−4.0
127	Grand Plan Neve	CH00045	1893	1990	C	−10.0	0.0		4.0	6.0
128	Gries (Aegina)	CH00003	1961	1990	A	−9.0	−18.0	−X	−17.0	−11.0
129	Griess (Klausen)	CH00074	1929	1990	C	−34.0	−2.0		−18.0	0.0
130	Griessen (OBWA)	CH00076	1894	1990	C		−9.0	−2.0	−5.0	ST
131	Grosser Aletsch	CH00005	1886	1990	A	−9.0	−18.0	−26.0	−37.0	−60.0
132	Huefi	CH00073	1882	1990	C		−26.0		−33.0	−33.0
133	Kaltwasser	CH00007	1891	1990	C	4.0	1.0	−X	−8.0	−4.0
134	Kehlen	CH00068	1893	1990	C	−14.0	−12.0	−X	−41.0	−29.0
135	Kessjen	CH00012	1928	1990	A	−28.0	12.0			+X
136	Laemmern	CH00063	1917	1990	C	−10.0	−8.0	−6.0	−6.0	−5.0
137	Lang	CH00018	1888	1990	C	−7.0	−3.0	−7.0	−1.0	−8.0
138	Lavaz	CH00082	1899	1989	C			−386.0		+X
139	Lenta	CH00084	1895	1990	C		−92.0		−11.0	−11.0
140	Limmern	CH00078	1964	1990	C	−5.0	−2.0		−6.0	−2.0
141	Lischana	CH00098	1895	1990	C	−5.0	−7.0	−2.0	−5.0	−6.0
142	Martinets	CH00046	1894	1986	E		−X			
143	Mittelaletsch	CH00106	1970	1990	C	−16.0	−8.0	−X	−X	−X
144	Moiry	CH00024	1891	1990	C	−6.0	−4.0	−2.0	−1.0	0.0
145	Moming	CH00023	1911	1990	C	−10.0	−30.0	−81.0	−32.0	−16.0
146	Mont Durand	CH00035	1885	1990	C	23.0	6.0	20.0	8.0	8.0
147	Mont Fort	CH00032	1892	1988	C	−221.0	−8.0		5.0	23.0
148	Mont Mine	CH00026	1956	1990	C	−25.0	−5.0	−7.0	−58.0	−18.0
149	Morteratsch	CH00094	1879	1990	C	−5.0	−6.0	−24.0	−12.0	−24.0
150	Mutt	CH00002	1918	1990	D	−5.0	−7.0	−X	0.0	
151	OB. Grindelwald	CH00057	1880	1990	C	−53.0	−50.0	−60.0	−62.0	−7.0
152	Oberaar	CH00050	1920	1990	A	−4.0	−18.0	−3.0	−20.0	−10.0
153	Ofental	CH00009	1922	1990	D		−42.0			+X
154	Otemma	CH00034	1887	1990	C	−13.0	−13.0	−40.0	−32.0	−13.0
155	Palue	CH00100	1894	1990	C	−12.0	−7.0	−7.0	−9.0	−8.0
156	Paneyrosse	CH00044	1893	1990	C	−7.0	0.0		−4.0	6.0
157	Paradies	CH00086	1898	1990	C	−21.0	−81.0	−1.0	−10.0	12.0
158	Paradisino	CH00101	1955	1990	C	−12.0	3.0	4.0	−3.0	0.0
159	Pierredar	CH00049	1921	1990	E	−X	−X		−X	−X
160	Pizol	CH00081	1894	1990	C	−22.0	5.0		−14.0	SN
161	Plattalva	CH00114	1969	1990	C	−10.0	1.0		−8.0	−4.0
162	Porchabella	CH00088	1893	1990	C	−9.0	−10.0		−15.0	−6.0
163	Prapio	CH00048	1898	1990	C	−15.0	−10.0	−5.0	0.0	ST
164	Punteglias	CH00083	1895	1989	C	−22.0	−X		−38.0	−22.0
165	Raetzli	CH00065	1924	1990	C	−20.0	−X	−56.0	0.0	−16.0
166	Rhone	CH00001	1970	1990	A	−31.0	−10.0	−12.0	−7.0	1.0

(Continued)

Table 5D.28 (Continued)

NR	Glacier Name	PSFG NR	First Survey	Last Survey	Method	Variations (m)				
						1991	1992	1993	1994	1995
167	Ried	CH00017	1895	1990	C	−6.0	−12.0	−X	−X	−X
168	Roseg	CH00092	1894	1990	C	0.0	−12.0	−47.0	−30.0	−84.0
169	Rosenlaui	CH00056	1880	1990	E	−X	−X	−X	−X	−X
170	Rossboden	CH00105	1891	1990	C	2.0	10.0	2.0	3.0	−9.0
171	Rotfirn Nord	CH00069	1956	1990	C	−5.0	−12.0	−X	−9.0	−11.0
172	Saleina	CH00042	1888	1990	C	−24.0	−10.0	−6.0	−26.0	−18.0
172	Sankt Anna	CH00067	1926	1989	C	−12.0	−5.0		−7.0	−1.0
174	Sardona	CH00091	1895	1990	C	−23.0	−8.0		−16.0	27.0
175	Schwarz	CH00062	1924	1990	C	−8.0	−3.0	−11.0	−43.0	−1.0
176	Schwarzberg	CH00010	1915	1990	A	6.0	−5.0	−X	−X	−10.0
177	Sesvenna	CH00097	1956	1990	C	−7.0	−5.0	−6.0	−9.0	−2.0
178	Sex Rouge	CH00047	1898	1990	C	−22.0	−X	−X	ST	ST
179	Silvretta	CH00090	1956	1990	A	−9.0	−13.0	−14.0	−6.0	−15.0
180	Stein	CH00053	1894	1990	C	−5.0	−7.0	−6.0	−15.0	−12.0
181	Steinlimmi	CH00054	1961	1990	C	−12.0	−11.0	−10.0	−18.0	−3.0
182	Sulz	CH00079	1912	1989	C	−10.0	−5.0			−22.0
183	Suretta	CH00087	1921	1990	C	48.0	−43.0	−38.0	70.0	−14.0
184	Taelliboden	CH00008	1922	1990	D	−8.0	−2.0			+X
185	Tiatscha	CH00096	1926	1990	C	−10.0	−13.0	−2.0	−1.0	ST
186	Tiefen	CH00066	1925	1990	C	−14.0	−7.0		−20.0	3.0
187	Trient	CH00043	1878	1990	C	−19.0	−15.0	−X	−58.0	−40.0
188	Trift (Gadmen)	CH00055	1921	1990	E	−X	−X		−X	−X
189	Tsanfleuron	CH00033	1892	1990	C	−15.0		−6.0	−2.0	−18.0
190	Tschierva	CH00093	1943	1990	C	−9.0	−19.0	−21.0	−45.0	−24.0
191	Tschingel	CH00060	1893	1990	C	−9.0	−4.0	−2.0	−5.0	−0.0
192	Tseudet	CH00040	1890	1990	C	−12.0	−22.0	16.0	−10.0	−27.0
193	Tsidjiore Nouve	CH00028	1880	1990	C	−4.0	−8.0	−6.0	14.0	6.0
194	Turtmann (West)	CH00019	1883	1990	C	7.0	3.0	−12.0		2.0
195	UNT.Grindelwald	CH00058	1880	1990	E	−X	−X	−X	ST	−X
196	Unteraar	CH00051	1893	1990	A	−17.0	−30.0	−37.0	−43.0	−26.0
197	Val Torta	CH00118	1970	1990	C	−9.0	SN	−2.0	9.0	−5.0
198	Valleggia	CH00117	1971	1990	C	−X	−10.0	−6.0	−2.0	−7.0
199	Valsorey	CH00039	1889	1990	C	−9.0	−19.0	−9.0	−25.0	−43.0
200	Verstankla	CH00089	1926	1990	C	−6.0	−10.0	−X	−30.0	−3.0
201	Vorab	CH00085	1893	1990	C	−29.0	−19.0		−22.0	SN
202	Wallenbur	CH00071	1893	1990	C	−6.0		−X	−13.0	−3.0
203	Zinal	CH00022	1891	1990	C	−15.0	−30.0	−7.0	−4.0	−6.0
204	Zmutt	CH00015	1892	1990	C	3.0	−2.0	2.0	0.0	
	Austria									
205	Aeu.Pirchi.Kar	A00229	1981	1990	C	3.5	1.8	−1.5	−3.0	−.8
206	Alp.Kraeul F	A00321	1975	1990	C	−2.6	−8.6	−2.8	−10.6	
207	Alpeiner F	A00307	1881	1990	C	−8.5	−14.3	−10.6	−7.7	−24.9
208	Bachfallen F	A00304	1922	1990	C	−5.5	−11.3	−8.4	−7.7	−11.2
209	Baerenkopf K	A00702	1924	1990	C	−3.4	−27.5	−3.2	−5.3	4.4
210	Berglas F	A00308	1891	1990	C	−1.7	−10.0	−5.6	−10.5	−8.0
211	Bieltal F	A0105A	1924	1990	C	−8.9	−8.8	−5.0	−12.6	−7.1
212	Bockkogel F	A00302	1922	1990	E	−X	−X	−X	−X	
213	Brennkogl K	A00727	1987	1990	C	−2.3	−9.3	−11.5	−14.4	−4.0
214	Daunkogel F	A0310A	1891	1990	C	−2.6	−10.2	−4.3	−6.5	−4.8
215	Diem F	A00220	1893	1990	C	−3.7	−1.6	−3.8	−6.7	−5.1
216	Dorfer K	A00509	1896	1990	C	−6.3	−16.6	−9.8	−13.6	−13.2
217	E.Gruebl F	A00317	1891	1990	C	−1.3	−13.2	−6.4	−7.2	
218	Eiskar G	A01301	1992	1992	C			−2.0	−.4	−5.0
219	Fernau F	A00312	1890	1990	C	−1.5	−4.0	−2.9	−7.2	−3.0
220	Freiger F	A00320	1974	1990	C	−2.9	−7.5	−6.9	−4.2	−10.3
221	Freiwand K	A00706	1950	1990	C	−7.0	−4.4	−1.1	−10.9	SN
222	Frosnitz K	A00507	1923	1990	C	−3.0	−5.5	−5.5	−6.7	−1.6
223	Fruschnitz K	A00722	1974	1986	E	−X				
224	Furtschagl K	A00406	1978	1990	E	−X	−X	−X	−X	−X

(Continued)

Table 5D.28 (Continued)

NR	Glacier Name	PSFG NR	First Survey	Last Survey	Method	Variations (m)				
						1991	1992	1993	1994	1995
225	Gaiskar F	A00325	1983	1990	C	−5.6	−11.5	−X	−5.9	−2.5
226	Gaissberg F	A00225	1891	1990	C	−6.5	−2.1	−17.2	−5.8	−7.5
227	Gepatsch F	A00202	1896	1990	C	−4.4	−6.2	−6.0	−10.8	−8.4
228	Goessnitz K	A01201	1982	1990	C	−3.3	−7.3	−10.9	−21.1	−6.6
229	GR Goldberg Kee	A0802B		1975	C	−2.5	−12.1	−5.5	−15.0	−4.2
230	GR.Gosau G	A01101	1933	1990	C	−2.2	−14.4	−4.0	−8.2	−10.5
231	Grosselend K	A01001	1898	1990	C	0.9	−6.7	−7.8	−29.2	−6.7
232	Gruenau F	A00315	1891	1990	C	6.0	−6.0	−17.2	−12.7	−15.3
233	Gurgler F	A00222	1895	1990	C	−2.5	−6.3	−1.2	−7.3	−3.4
234	Guslar F	A00210	1894	1990	C	−11.5	−12.8	−12.8	−13.8	−10.5
235	Habach Kees	A00504		1990	E	−X				
236	Hallstaetter G	A01102	1843	1990	C	0.4	−7.7		−8.7	−4.9
237	Hinteraeis F	A00209	1891	1990	C	−17.2	−23.4	−22.4	−21.8	−13.1
238	Hochalm K	A01005	1898	1990	C	0.3	−15.0	−12.4	−6.1	−5.1
239	Hochjoch F	A00208	1890	1990	C	−25.7	−25.5	−32.2	−27.4	−31.4
240	Hochmoos F	A00309	1946	1990	C	−4.0	−1.7	−2.0	−2.0	
241	Hofmanns K	A00724	1977	1987	E	−X				
242	Horn K. (Schob.)	A01202	1983	1990	C	−2.8	−5.8	−2.8	−4.1	−4.3
243	Horn K. (Ziller)	A00402	1881	1990	C	−2.7	−8.7	−4.7	−7.7	−9.3
244	INN.Pirchlkar	A00228	1982	1990	C	1.4	0.4	0.1	−11.4	1.4
245	Jamtal F	A00106	1892	1990	C	−14.6	−9.0	−9.2	−11.1	−7.1
246	KA.Tauern K.S	A0602B	1961	1990	C	−2.1	−27.1		−18.8	
247	Kaelberspitz K	A01003	1927	1990	C	−4.6	−14.7	−7.1	−7.1	1.4
248	Karles F	A00207	1950	1990	C	−9.2	−13.0	−7.9	−4.5	−2.0
249	Karlinger K	A00701	1896	1990	C	−12.0	−10.0	−84.0	−37.0	−X
250	Kesselwand F	A00226	1965	1990	C	−17.5	−22.3	−28.4	−36.1	−33.9
251	KL.Fleiss K	A00801	1896	1990	C	−2.6	−20.8	−13.0	−10.8	−2.2
252	Kleineiser K	A00717	1961	1990	C	−.2	−4.7		−8.7	
253	Kleinelend K	A01002	1898	1990	C	−4.3	−19.4	−4.8	−6.0	1.4
254	Klostertaler M	A0102B	1964	1990	C	−9.1	−7.4	−2.9	−11.0	−4.1
255	Klostertaler N	A0102A	1968	1990	C	−4.1	−5.1	−1.0	−4.2	−7.6
256	Klostertaler S	A0102C	1924	1990	C	−10.9	−11.4	−3.8	−12.7	
257	Krimmler K. East	A0501B		1992	C			−7.8	−10.8	−6.3
258	Krimmler K	A0501A	1895	1990	C	−2.5	−7.2	2.5	−4.4	−16.6
259	Kruml K	A00806	1985	1990	C	−1.5	−7.0	−6.5		SN
260	Laengentaler F	A00305	1922	1990	C	−3.2	−5.3	−4.2	−4.9	SN
261	Landeck K	A00604	1979	1990	C	−.7	−1.4		−12.5	
262	Langtaler F	A00223	1891	1990	C	−18.0	−27.5	−6.0	−25.0	−15.8
263	Laperwitz K	A00721	1974	1986	E	−X				
264	Larain F	A00107	1928	1990	C	−24.5	−23.7			
265	Liesenser F	A00306	1922	1990	C	−7.7	−16.6	−6.4	−16.1	−4.9
266	Litzner GL	A00101	1932	1990	C	4.5	−1.7	−1.7	0.9	−3.3
267	Marzell F	A00218	1891	1990	C	−4.0	−6.8	1.7	−6.8	−3.6
268	Maurer K (GLO.)	A00714	1961	1990	C	0.2	−7.6	−1.0	−3.9	SN
269	Maurer K (VEN.)	A00510	1896	1990	C		−61.8	−8.8	−6.5	SN
270	Mittelberg F	A00206	1924	1990	C	−12.1	−6.3	−X	−6.5	0.9
271	Mitterkar F	A00214	1891	1990	C	−22.5	−21.8	−3.0	−10.6	SN
272	Mutmal F	A00227	1969	1990	C	−8.1	−21.4	−16.1	−24.2	−7.4
273	Niederjoch F	A00217	1891	1990	C	−19.7	−14.4	−9.2	−14.2	−6.4
274	Obersulzbach K	A00502	1880	1990	C	−6.1	−13.6	−14.8	−18.5	−8.2
275	Ochsentaler GL	A00103	1901	1990	C	−7.2	−18.7	−15.2	−25.1	−9.2
276	Oedenwinkel K	A00712	1960	1990	C	−4.7	−1.4	0.3	−3.2	0.3
277	Pasterzen K	A00704	1879	1990	C	−33.5	−7.6	−17.5	−17.1	−13.6
278	Pfaffen F	A00324	1981	1990	C	−6.1	−11.5	−2.5	−15.0	−6.5
279	Pfandlscharten	A00707	1931	1990	C	−9.5	−13.1			
280	Praegrat K	A00603	1961	1990	C	−4.5	−9.2		−14.4	
281	Rettenbach F	A00212	1952	1990	C	−5.4	−5.9	−5.9	−9.8	−3.4
282	Riffl K.N	A00718	1961	1990	C	2.7	−8.6	1.1	−4.7	SN
283	Rifflkar Kees	A0713A	1961	1990	E	−X				

(Continued)

Table 5D.28 (Continued)

NR	Glacier Name	PSFG NR	First Survey	Last Survey	Method	Variations (m)				
						1991	1992	1993	1994	1995
284	Rofenkar F	A00215	1891	1990	C	-6.0	-10.6	-3.8	-9.8	-7.3
285	Rotmoos F	A00224	1891	1990	C	-1.0	-8.8	-8.0	-8.0	-12.1
286	Schalf F	A00219	1924	1990	C	-2.9	-9.6	-15.5		-18.1
287	Schattenspitz	A00108	1973	1990	C	-2.7	-15.0	-5.5	-5.2	
288	Schaufel F	A00311	1922	1990	C	-1.9	-4.1	-5.6	-11.1	-3.3
289	Schladminger G	A01103	1933	1990	C	-1.4	-3.0		-1.4	-.1
290	Schlappereben K	A00805	1983	1990	C	-2.0	-2.8	-2.9	-.5	-3.5
291	Schlaten K	A00506	1891	1990	C	-4.6	-7.9	-6.8	-10.5	-7.9
292	Schlegeis K	A00405	1978	1990	E	-X	-X	-X	-X	-X
293	Schmiedinger K	A00726	1981	1990	C	-0.1	-3.6	-2.3	-2.8	
294	Schneeglocken	A00109	1973	1990	C	-5.8	-8.2	-4.3	-7.7	-5.8
295	Schneeloch G	A01104	1969	1990	C	-1.3	-5.2		-2.2	-6.7
296	Schwarzenberg F	A00303	1905	1990	C	-6.8	-13.8	-10.8	-15.3	-6.8
297	Schwarzenstein	A00403	1881	1990	C	5.5	-15.0	-17.0	-16.0	-12.0
298	Schwarzkarl K	A00716	1961	1990	C	-4.4	-11.1	-6.0	-19.7	-2.3
299	Schwarzkoepfl K	A00710	1954	1990	C	-7.0	-11.5	-6.3	-17.9	-14.8
300	Sexegerten F	A00204	1919	1990	C	-6.0	-28.5	-10.7	-39.0	-5.3
301	Simming F	A00318	1922	1990	C	-1.9	-13.4	-17.5	-43.4	-22.6
302	Simony K	A00511	1896	1990	C	-4.8	-9.8	-8.1	-11.5	-15.2
303	Sonnblick K	A0601A	1963	1990	C	-2.1	-7.9	0.4	-8.7	0.7
304	Spiegel F	A00221	1891	1990	C	-6.8	-2.2	-4.4	-10.4	-9.8
305	Sulzenau F	A0314A	1891	1990	C	-4.0	-20.3	-29.9	-34.2	-104.4
306	Sulztal F	A00301	1922	1990	C	-4.3	-18.5	-35.8	-21.6	-7.8
307	Taschach F	A00205	1924	1990	C	-4.9	-12.8	-12.9	-10.5	-13.5
308	Taufkar F	A00216	1891	1990	C	-8.9	-12.4	-3.4	-6.2	-6.2
309	Teischnitz K	A00723	1975	1986	E	-X				
310	Totenfeld	A00110	1976	1990	C	-2.7	-2.1	-3.2	-5.1	-.3
311	Triebenkarlas F	A00323	1978	1990	C	0.1	-12.0	-14.3	-16.3	-12.3
312	Uebergoss.ALM	A00901	1892	1990	C	-5.8	-3.0			
313	Umbal K	A00512	1896	1990	C	-18.2		-24.3	-24.5	-68.5
314	UNT.Riffel Kees	A0713B	1960	1990	C	-3.0	-4.7	-5.8	-4.4	-4.1
315	Untersulzbach K	A00503	1896	1990	C	-5.0	-7.5	-6.1	-11.2	0.7
316	VD.Kasten K	A00719	1961	1980	E	-X				
317	Verborgenberg F	A00322	1977	1990	C	-2.7	-11.3	-2.3	-7.4	-7.3
318	Vermunt GL	A00104	1913	1909	C	-7.4	-12.0	-5.3	-15.2	-4.3
319	Vernagt F	A00211	1888	1990	C	-13.7	-11.0	-15.9	-16.5	-13.3
320	Viltragen K	A00505	1891	1990	C	-3.9	-10.7	-10.4	-12.0	-12.0
321	W.Tripp K	A01004	1925	1990	C	0.4	-6.4		-5.5	SN
322	Wasserfallwinkl	A00705	1943	1990	C	-5.4	-15.6	-1.7	-7.7	3.6
323	Waxegg K	A00401	1895	1990	C	-6.2	-3.0	-12.0	-8.0	-38.0
324	Weissee F	A00201	1891	1990	C	-8.3	-10.6	-25.6	-4.0	-2.9
325	Wielinger K	A00725	1980	1990	C	-47.0	-49.0	-X	-X	-X
326	Wildgerlos	A00404	1973	1990	C	6.1	-30.8	-12.2	-20.0	-8.4
327	Winkl K	A01006	1928	1990	C	-0.7	-26.3	-1.2	-2.4	-5.0
328	Wurfer K	A00715	1961	1990	C	-5.0	-16.4	SN	-16.8	
329	Wurten K	A00804	1896	1990	C	-10.3	-12.5	-5.5	-14.8	-2.6
330	Zettalunitz K	A00508	1896	1990	C	-3.4	-21.0	-13.7	-21.1	-18.0
	Italy									
331	Agnello	I00029	1928	1990	C	-2.5			SN	-X
332	Alta (Vedretta)	I00730	1923	1990	C		-8.5	-7.0	-10.0	-17.0
333	Amola	I00644	1942	1990	C	-1.0	-9.0		-14.0	-16.0
334	Andollanord	I00336	1927	1990	C	-8.0	-24.0	1.5		-2.0
335	Antelao Inf	I00967	1939	1990	C	-2.0	-7.0	-2.0	-3.0	-3.0
336	Antelao Sup	I00966	1934	1990	C	-3.5	-2.5	0.0	-4.5	-3.0
337	Aurona	I00338	1956	1990	C				-X	-X
338	Barbadorso D	I00778	1935	1990	C		-65.0	-29.0		
339	Basei	I00064	1925	1990	C	-3.0	-1.0	0.0	0.0	
340	Belvedere	I00325	1927	1990	C	4.0	-3.0	-4.0	2.0	-2.0
341	Bessanese	I00040	1928	1990	C	-1.0	-3.0	-1.5	-2.0	0.0

(Continued)

Table 5D.28 (Continued)

NR	Glacier Name	PSFG NR	First Survey	Last Survey	Method	Variations (m)				
						1991	1992	1993	1994	1995
342	Brenva	I00219	1929	1990	C	8.0	−9.0	0.0		
343	Caspoggio	I00435	1928	1990	C	−2.0	−X	−X	−43.0	−12.0
344	Cevedale	I00732	1923	1990	C		−X	−17.0	−16.0	−19.0
345	Chavannes	I00204	1930	1989	C	−15.0	−5.0	1.5	−1.0	
346	Ciardoney	I00081		1990	C	−2.5	−2.0	−3.0	−2.0	−.5
347	Collalto	I00927	1932	1990	C	−2.0	−7.5	−8.0	−4.5	−5.0
348	Cristallo	I00937	1949	1990	C	0.0	−12.0	−8.0	−6.0	−17.5
349	Croda Rossa	I00828		1990	C		−4.0		−4.0	
350	Dosde OR	I00473	1932	1990	C	−28.0	−15.5	−32.0	−37.0	−9.0
351	Dosegu	I00512	1925	1990	C		−60.0	−X	−17.0	−22.0
352	Fellaria OCC	I00439	1915	1990	C	−13.0	−12.0	−9.5	−9.5	−15.0
353	Fontana OCC	I00780	1925	1990	C			−14.5		
354	Forcola	I00731	1923	1990	C		−43.0	−13.0	−21.0	−18.0
355	Forni	I00507	1880	1989	C	−124.0	−30.0	−22.0	−95.0	−19.5
356	Gigante Centr	I00929	1930	1990	C		−X	−92.0		−21.5
357	Gigante OCC	I00930	1930	1990	C	−2.0	−7.0	−5.0	−1.5	−2.5
358	Goletta	I00148	1929	1990	C		−19.5		−6.5	−4.0
359	Gran Pilastro	I00893	1932	1990	C	−10.5		−26.0		
360	Hosand Sett	I00357	1926	1990	C	−5.0	−7.5	0.5	0.0	6.5
361	LA Mare	I00699	1899	1990	C	−33.0	−68.0	−23.5	−59.0	−5.5
362	Lana	I00913	1930	1990	C	−3.0	−8.0	−6.5	−9.0	−3.5
363	LEX Blanche	I00209	1929	1990	C		−40.0	−X	−X	
364	Lunga (Vedretta)	I00733	1923	1990	C		−26.0	−5.0	−18.0	−47.0
365	LYS	I00304	1927	1990	C	3.0	−11.0	−7.0	−12.0	−12.0
366	M. Nevoso OCC	I0931X	1930	1990	C	−2.5	−6.0	−14.0	−1.0	
367	Malavalle	I00875	1928	1990	C	−9.0	−8.0	−2.0	−8.0	−7.5
368	Mandrone	I00639	1896	1990	C	−5.0	−14.0	−2.0	−3.0	−2.5
369	Marmolada	I00941	1925	1990	C	−11.5	−13.5		−10.0	
370	Moncorve	I00131	1924	1990	C	−5.0	−5.0	−1.0	−3.0	
371	Nardis OCC	I00640	1925	1990	C	−10.0	−4.5	2.0		−10.0
372	Neves OR.	I00902	1932	1990	C	−13.0	−19.0	−13.0	−12.5	−12.0
373	Niscli	I00633	1919	1990	C	−5.5	−3.5	0.0		−7.5
374	Pendente	I00876	1933	1990	C	−17.0	−9.0	−3.0	−5.0	−15.0
375	Piode	I00312	1924	1990	C	−19.0		−34.0	−51.0	−5.0
376	Pisgana OCC.	I00577	1920	1990	C	−X	−59.0	−X		−X
377	Pizzo Scalino	I00443	1911	1990	C	−8.5	−8.0	−2.0	−14.0	−10.0
378	Pre De Bar	I00235	1929	1990	C	−6.5	−32.0	7.0	−16.0	−16.0
379	Presanella	I00678	1951	1990	C	0.0	−4.5	−12.5	−10.0	−8.5
380	Quaira Bianca	I00889	1931	1990	C	−7.0	−6.0	−11.0		
381	Rosim	I00754	1926	1990	C		−6.5	−7.0	−7.5	0.0
382	Rossa (Vedr.)	I00697	1923	1990	C	−11.0	−4.0	−6.0	−19.0	−2.5
383	Rosso Destro	I00920	1952	1990	C	−9.5	−8.5	−17.0	−10.0	
384	Rutor	I00189	1927	1990	C	−10.5	−7.0	−1.5	−20.5	−6.5
385	Sassolungo OCC	I00926	1930	1989	C	−11.0	−11.0	−2.5	−10.5	
386	Serana (Vedr.)	I00728	1925	1990	C			−8.0	−X	−16.0
387	Sforzellina	I00516	1925	1990	C	4.0	−4.0	−X	−4.5	−5.0
388	Solda	I00762	1922	1990	C		3.0	−12.0	−8.0	−8.0
389	Tessa	I00829	1926	1990	C		0.0		−4.0	
390	Toules	I00221	1929	1990	C	−12.0	−12.0	−2.0		−86.0
391	Travignolo	I00947	1925	1981	C	−59.0	−3.5		−3.0	−3.5
392	Tresero	I00511	1925	1990	C		−10.0		−X	−6.0
393	Tza De Tzan	I00259	1927	1990	C	+X	−X	−15.0	−15.0	−11.5
394	Ultima (Vedr.)	I00729	1925	1990	C		−13.0	−8.0	−5.5	−4.5
395	Valle Del Vento	I00919	1932	1990	C	−9.0	−10.0	−13.0	−2.5	
396	Vallelunga	I00777	1922	1990	C		−15.0	ST		
397	Valtournenche	I00289	1927	1989	C	−.5	−1.5	−9.0	−8.0	
398	Venerocolo	I00581	1919	1990	C	−6.0	−6.0		−14.5	−19.0
399	Venezia (Vedr.)	I00698	1925	1990	C	−9.5	−38.5	−15.0	−16.5	−X
400	Ventina	I00416	1890	1990	C	1.0	−11.0	−15.0	−10.0	−14.0

(Continued)

Table 5D.28 (Continued)

NR	Glacier Name	PSFG NR	First Survey	Last Survey	Method	Variations (m) 1991	1992	1993	1994	1995
401	Vitelli	I00483	1921	1990	C		-42.0	-7.0	0.0	-X
402	Zai Di Dentro	I00749	1924	1979	C		-61.0	-2.5	-5.0	-4.0
403	Zai Di Mezzo	I00750	1930	1979	C		-58.0	-6.0	-5.5	-3.0
	Kenya									
404	Cesar	KN00004	1899	1987	A			-25.0		
405	Darwin	KN00006	1919	1987	A			-15.0		
406	Diamond	KN00010	1899	1987	A			0.0		
407	Forel	KN00011	1899	1987	A			0.0		
408	Gregory	KN00009	1899	1990	A			-30.0		
409	Heim	KN00012	1899	1987	A			0.0		
410	Joseph	KN00003	1899	1987	A			0.0		
411	Krapf	KN00001	1899	1987	A			-25.0		
412	Lewis	KN00008	1899	1990	A			-25.0		
413	Northey	KN00013	1899	1987	A			0.0		
414	Tyndall	KN00005	1899	1987	A			-50.0		
	Poland									
415	Mieguszowieckie	PL00140	1958	1988	C	1.0	-2.0		SN	2.0
416	Pod Bula	PL00111	1978	1990	C	34.0	-14.1	2.4	-1.0	13.1
417	POD Cubryna	PL00180	1978	1988	C	-1.0	1.0	-2.0	2.0	1.0
	C.I.S									
418	Abramov	SU04101	1967	1990	C	-9.1	-12.9	-22.4	-25.3	ST
419	Alibekskiy	SU03002	1954	1990	C				9.0	-8.0
420	Bezengi	SU03006	1956	1990	C	-7.0				-14.0
421	Bolshoy Azau	SU03004	1887	1987	B		-69.0			
422	Djankuat	SU03010	1965	1990	B		-6.0			-10.0
423	Dzhelo	SU07106	1952	1990	C	-19.0	-18.0	-3.0		-14.3
424	Garabashi	SU03031	1887	1987	B		0.0			
425	Kara-Batkak	SU05080	1957	1990	C	-2.5	-5.5	-3.3	-6.8	-7.2
426	Khakel	SU03003	1957	1990	C				-16.0	8.0
427	Korumdu	SU07103	1952	1990	C	-3.0	-12.0	0.0	-3.5	-1.0
428	Kozelskiy	SU08005	1948	1990	C					ST
429	Leviy Artru	SU07102	1952	1990	D	-16.0	-11.0	-1.0	-22.0	-3.0
430	Levivy Karagemsk	SU07107	1952	1990	C	-19.0	0.0	ST	ST	-3.8
431	Maliy Aktru	SU07100	1952	1990	D	-5.0	-10.0	0.5	-9.0	-12.0
432	Maliy Azau	SU03032	1887	1987	B		0.0			
433	Mizhirgichiran	SU03043	1956	1990	C	15.9				61.0
434	Muravlev	SU06002	1966	1990	C	-3.5				
435	NO. 122 (UNIV.)	SU07108	1952	1990	C	-14.0	-8.0	ST		-6.6
436	NO. 125 (VODOP.)	SU07105	1956	1990	D	-1.0	-4.8	0.0	-1.7	-1.5
437	NO. 462V (KULN.)	SU03005	1934	1990	C	-7.5				
438	Praviy Karagems	SU07109	1952	1990	C	-12.0	-7.0	ST		-5.0
439	Shumskiy	SU06001	1966	1990	C	-7.2				
440	TS.Tuyuksuyskiy	SU05075	1956	1990	C	-17.0	-8.4	-7.3	-13.8	-20.8
441	Tseya	SU03007	1927	1990	C	-11.8	-1.7	-16.3		-5.0
442	Yugo-Vostochniy	SU03018	1957	1991	C		-11.4	-1.5	4.2	17.0
443	Yuzhniy	SU03017	1957	1991	C		-4.1	-2.7	3.5	-3.7
	China									
444	Urumqihe S.NO.1	CN00010	1960	1990	C	-6.5	-3.4	-3.8	-6.8	-6.2
	Pakistan									
445	Aling	PK00035		1989				+X		
446	Bualtar	PK00004		1988			1800.0			-X
447	Karambar	PK00028		1993				-X	+X	
	Nepal									
448	AX010	NP00005	1978	1989	C	-30.0				25.0
449	DX080	NP00007	1976	1989	C					-29.6
450	Gyajo	NP00011	1970	1978	C					-75.6
451	Kongma	NP00010	1970	1989	C					-13.0
452	Kongma Tikpe	NP00009	1974	1989	C					-53.7
453	Rikha Samba	NP00012	1974	1974	C				-205.7	

(Continued)

Table 5D.28 (Continued)

NR	Glacier Name	PSFG NR	First Survey	Last Survey	Method	Variations (m)				
						1991	1992	1993	1994	1995
454	Yala	NP00004	1982	1989	C				−30.9	
	New Zealand									
455	Abel	NZ893A3		1989	A			+X	ST	ST
456	Adams	NZ08974		1987	A		−X	−X		+X
457	Almer	NZ888B1		1989	A			ST	+X	+X
458	Andy	NZ863C1		1987	A			ST	−X	+X
459	Ashburton	NZ688A1		1989	A			+X		+X
460	Balfour	NZ882B1		1985	A					+X
461	Barlow	NZ893A2		1989	A		−X	−X		+X
462	Blair	NZ711D1		1989	A			−X	+X	+X
463	Bonar	NZ863A1		1987	A					+X
464	Brewster	NZ868C1		1989	A		−X	+X	−X	ST
465	Burton	NZ888A1		1989	A			ST		ST
466	Cameron	NZ685B2		1988	A			−X	+X	ST
467	Classen	NZ711M1		1989	A				−X	ST
468	Colin Campbell	NZ693C1		1988	A					+X
469	Crow	NZ664C2		1988	A					+X
470	Dart	NZ752C2	1980	1989	A		−X	+X		−X
471	Donne	NZ851B2		1987	A					−X
472	Douglas (KAR.)	NZ880B2		1987	A			ST		ST
473	Douglas (RAK.)	NZ685B1		1989	A		SN	SN	SN	SN
474	Evans	NZ08972		1988	A		ST			ST
475	Fitzgerald	NZ880B3		1984	A			−X		+X
476	Fox	NZ882A1		1989	A	+X	+X	+X	+X	+X
477	Franz Josef	NZ888B2	1867	1989	A	+X	+X	+X	+X	+X
478	Glenmary	NZ711F1		1989	A				+X	ST
479	Godley	NZ711M3		1989	A					−X
480	Grey and Maud	NZ711M2		1989	A				−X	ST
482	Hooker	NZ711H2		1985	A		−X	−X		−X
482	Horace Walker	NZ880B1		1987	A					+X
483	Ivory	NZ09011	1989	1989	A		−X	−X	−X	ST
484	Jack	NZ08751		1989	A			+X	+X	SN
485	Jackson	NZ868B5		1989	A				ST	
486	Jalf	NZ08861		1989	A			SN	−X	SN
487	Kahutea	NZ685E1		1989	A					+X
488	KEA	NZ08971		1989	A			SN	SN	SN
489	LA Perouse	NZ882B2		1985	A					ST
490	Lambert	NZ08973		1989	A		ST	+X	−X	
491	LE Blanc	NZ868B3		1985	A				−X	+X
492	Lindsay	NZ08671		1989	A			SN	SN	SN
493	Lyell	NZ685C2		1989	A					−X
494	Marchant	NZ880A1		1986	A					ST
495	Marion	NZ863B4		1989	A			−X	ST	+X
496	Marmaduke Dixon	NZ664C1		1989	A			+X	SN	
497	MC COY	NZ693C2		1985	A					+X
498	Mueller	NZ711H1		1989	A	−X	−X			−X
499	Murchison	NZ711J1		1989	A			−X		ST
500	Park Pass 1	NZ752B1		1989	A				+X	+X
501	Poet	NZ868B2		1986	A					+X
502	Ramsay	NZ685C3		1983	A					−X
503	Reischek	NZ685C1		1989	A					−X
504	Retreat	NZ906A1		1989	A			SN	SN	SN
505	Richardson	NZ711E1		1987	A			+X		+X
506	Ridge	NZ711L1		1989	A				ST	SN
507	Rolleston	NZ911A2		1989	A			SN	SN	SN
508	Sale	NZ906B1		1993	A					+X
509	Siege	NZ893A1		1989	A		−X	SN	ST	SN
510	Sinclair	NZ693C3		1985	A					+X
511	Snow White	NZ863B2		1987	A			−X		−X

(Continued)

Table 5D.28 (Continued)

NR	Glacier Name	PSFG NR	First Survey	Last Survey	Method	Variations (m)				
						1991	1992	1993	1994	1995
512	Snowball	NZ863B3		1987	A			ST	ST	ST
513	Spencer	NZ888A2		1989	A		+X	+X		+X
514	Strauchon	NZ880A2		1986	A				−X	ST
515	Tasman	NZ71111		1989	A	−X	−X	−X	−X	−X
516	Therma	NZ08641		1987	A					+X
517	Thurneyson	NZ711B1		1989	A		+X		+X	
518	Tornado	NZ863C2		1986	A					−X
519	Unnamed NZ664C	NZ664C1		1989	A			SN		SN
520	Unnamed NZ685C	NZ685C4		1989	A		SN	SN	SN	+X
521	Unnamed NZ685F	NZ685F1		1989	A			SN	SN	
522	Unnamed NZ752E	NZ752E1		1989	A			+X	SN	SN
523	Unnamed NZ752I	NZ75211		1989	A		SN	SN	SN	SN
524	Unnamed NZ797G	NZ797G1		1989	A			SN	SN	SN
525	Unnamed NZ846	NZ08461		1989	A			+X	+X	+X
526	Unnamed NZ851B	NZ851B1		1989	A			ST	ST	ST
527	Unnamed NZ863B	NZ863B1		1989	A			SN	SN	SN
528	Unnsmrf NX868B	NZ868B4		1980	A					ST
529	Unnamed NZ911A	NZ911A1		1989	A			SN	SN	SN
530	Victoria	NZ882A1		1989	A					+X
531	Whitbourne	NZ752C1		1988	A					−X
532	White	NZ664C1		1989	A			+X		+X
533	Whymper	NZ893B1		1980	A					−X
534	Wigley	NZ873B2		1989	A		−X			−X
535	Wilkinson	NZ906B2		1989	A					ST
536	Zora	NZ868B1		1986	A					+X
	Antarctica									
537	Bartley	AN00016	1983	1990	C		−1.8			0.2
538	Clark CPI	AN00012	1982	1990	C					2.6
539	Hart	AN00019	1985	1990	C		0.0			−0.5
540	Heimdall	AN00003	1970	1991	C		3.5			
541	Meserve MPII	AN00017	1983	1990	C		−3.3			
542	Victoria Upper	AN00013	1984	1990	C		1.8			
543	Wright Lower	AN00018	1975	1990	C		−1.8		1.3	0.6
544	Wright Upper B	AN00011	1984	1990	C		−1.2			

Note: NR — Record number
Glacier Name — 15 alphabetic or numeric digits
PSFG Number — 5 digits identifying glacier with alphabetic prefix denoting country
Method — A = aerial photogrammetry
B = terrestrial photogrammetry
C = geodetic ground survey (theodolite, tape etc.)
D = combination of a, b or c
E = other methods or no information
1ST Survey: — Year when glacier was first surveyed
Last Survey: — Last survey before reported period
Variation in Meters: — Variation in the position of the glacier front in horizontal projection expressed as the change in length between the surveys
Key to Symbols: — +X: Glacier in advance
−X: Glacier in retreat
ST: Glacier stationary
SN: Glacier front covered by snow

Source: From World Glacier Monitoring Service by Wilfried Haeberli, Martin Hoelzle, Stephan Suter and Regula Frauenfelder www.wgms.ch.

Table 5D.29 Changes in Area, Volume and Thickness

NR	Glacier Name	Period From	Period To	Altitude From	Altitude To	Area Mean	Area Change	Volume Change	Thickness Change
	U.S.A								
1.1	McCall	1972	1993	2400	2500	110		−248	−2257
	US00001			2300	2400	720		−1383	−1921
				2200	2300	1160		−4284	−3693
				2100	2200	1360		−6126	−4504
				2000	2100	890		−5019	−5639
				1900	2000	810		−5110	−6308
				1800	1900	530		−4286	−8086
				1700	1800	450		−5070	−11267
				1600	1700	600		−11774	−19624
				1500	1600	360		−9733	−27037
				1400	1500	240		−8730	−36374
	Summary Data			1400	2500	7230		−61762	−8543
	Switzerland								
2.1	Gries (Aegina)	1923	1961	3300	3400	16	−7	−146	−9120
	CH00003			3200	3300	201	−68	−2062	−10260
				3100	3200	772	−239	−11148	−14440
				3000	3100	1572	2	−28673	−18240
				2900	3000	1040	−25	−19760	−19000
				2800	2900	813	−61	−18227	−22420
				2700	2800	699	−94	−21250	−30400
				2600	2700	1303	−221	−45553	−34960
				2500	2600	613	−50	−28419	−46360
				2400	2500	716	−419	−56048	−78280
				2300	2400	112	−15	−22387	−199880
	Summary Data			2300	3400	7857	−1167	−247810	−31540
2.2	Gries (Aegina)	1961	1979	3300	3400	9	1	3	360
	CH00003			3200	3300	133	3	72	540
				3100	3200	533	14	1055	1980
				3000	3100	1574	23	4533	2880
				2900	3000	1015	−11	4385	4320
				2800	2900	752	−26	3384	4500
				2700	2800	605	−62	436	720
				2600	2700	1082	−98	−7596	−7020
				2500	2600	563	45	−5472	−9720
				2400	2500	297	−112	−8874	−29880
				2300	2400	127	−123	−9304	−73260
	Summary Data			2300	3400	6690	−353	−12042	−1800
2.3	Gries (Aegina)	1979	1986	3300	3400	10	0	−40	−3990
	CH00003			3200	3300	130	−40	−355	−2730
				3100	3200	547	−117	−3853	−7043
				3000	3100	1597	69	−1006	−630
				2900	3000	1004	57	−2038	−2030
				2800	2900	726	1	−2185	−3010
				2700	2800	543	30	−1368	−2520
				2600	2700	984	−134	−4890	−4970
				2500	2600	608	70	−2724	−4480
				2400	2500	184	−20	−1468	−7980
				2300	2400	4	−4		
	Summary Data			2300	3400	6337	−88	−16413	−2590
2.4	Gries (Aegina)	1986	1991	3300	3400	10	0	14	1350
	CH00003			3200	3300	90	116	−266	−2950
				3100	3200	430	262	−1699	−3950
				3000	3100	1666	−66	−8580	−5150
				2900	3000	1061	−67	−6154	−5800
				2800	2900	727	−69	−4435	−6100
				2700	2800	573	−116	−3495	−6100
				2600	2700	850	−231	−7098	−8350
				2500	2600	678	127	−7831	−11550

(Continued)

Table 5D.29 (Continued)

NR	Glacier Name	Period From	Period To	Altitude From	Altitude To	Area Mean	Area Change	Volume Change	Thickness Change
				2400	2500	164	−11	−1796	−10950
	Summary Data			2400	3400	6249	−55	−41556	−6650
	Austria								
3.1	Hintereis F.	1979	1991	3700	3800	3	−1	−2	−667
	A00209			3600	3700	52	−4	−82	−1576
				3500	3600	48	−16	−58	−1208
				3400	3500	229	−50	−944	−4122
				3300	3400	701	−66	−2694	−3843
				3200	3300	975	−110	−5155	−5287
				3100	3200	1535	−260	−9390	−6117
				3000	3100	1469	−181	−11802	−8034
				2900	3000	1221	−112	−12738	−10432
				2800	2900	909	−64	−12846	−14132
				2700	2800	968	−70	−11516	−11897
				2600	2700	662	−19	−9636	−14556
				2500	2600	280	19	−6046	−21593
				2400	2500	156	−39	−2529	−16212
				2300	2400	9	−18	−48	−5333
	Summary Data			2300	3800	9217	−991	−85486	−9275
	C.I.S								
4.1	Djankuat	1984	1992	3600	3990	207	−41	303	1465
	SU03010			3500	3600	532	81	−363	−683
				3400	3500	358	−30	−165	−460
				3300	3400	370	−10	404	1092
				3200	3300	427	−16	439	1027
				3100	3200	361	−2	−466	−1290
				3000	3100	292	−8	109	374
				2900	3000	286	−1	−589	−2059
				2800	2900	183	−6	−754	−4121
				2698	2800	97	7	−510	−5258
	Summary Data			2698	3990	3113	−26	−1592	−511
5.1	Muravlev	1981	1982	3610	3620	27		1	40
	SU06002			3600	3610	22		4	170
				3590	3600	11		3	290
				3580	3590	9		3	380
				3570	3580	10		4	370
				3560	3570	15		5	320
				3550	3560	15		3	190
				3540	3550	15		1	60
				3530	3540	17		−2	−120
				3520	3530	15		−4	−280
				3510	3520	15		−6	−420
				3500	3510	14		−7	−520
				3490	3500	17		−10	−560
				3480	3490	11		−6	−550
				3470	3480	11		−6	−550
				3460	3470	9		−5	−560
				3450	3460	12		−7	−580
				3440	3450	14		−8	−550
				3430	3440	13		−7	−520
				3420	3430	11		−6	−520
				3410	3420	8		−4	−530
				3400	3410	11		−6	−550
	Summary Data			3400	3620	302		−60	−150
5.2	Muravlev	1982	1983	3610	3620	27		−21	−770
	SU06002			3600	3610	22		−17	−770
				3590	3600	11		−8	−760
				3580	3590	9		−7	−750
				3570	3580	10		−8	−750

(Continued)

Table 5D.29 (Continued)

NR	Glacier Name	Period From	Period To	Altitude From	Altitude To	Area Mean	Area Change	Volume Change	Thickness Change
				3560	3570	15		−11	−720
				3550	3560	15		−10	−690
				3540	3550	15		−10	−680
				3530	3540	17		−11	−660
				3520	3530	15		−10	−650
				3510	3520	15		−9	−630
				3500	3510	14		−9	−650
				3490	3500	17		−12	−710
				3480	3490	11		−8	−740
				3470	3480	11		−8	−770
				3460	3470	9		−7	−800
				3450	3460	12		−10	−840
				3440	3450	14		−12	−870
				3430	3440	13		−11	−880
				3420	3430	11		−10	−880
				3410	3420	8		−7	−890
				3400	3410	10	−1	−9	−880
	Summary Data			3400	3620	301	−1	−225	−740
5.3	Muravlev	1983	1984	3610	3620	27		−16	−580
	SU06002			3600	3610	22		−14	−620
				3590	3600	11		−7	−650
				3580	3590	9		−6	−660
				3570	3580	10		−6	−640
				3560	3570	15		−10	−660
				3550	3560	15		−10	−660
				3540	3550	15		−10	−650
				3530	3540	17		−11	−630
				3520	3530	15		−9	−610
				3510	3520	15		−9	−600
				3500	3510	14		−8	−580
				3490	3500	17		−10	−570
				3480	3490	11		−6	−570
				3470	3480	11		−6	−590
				3460	3470	9		−5	−610
				3450	3460	12		−8	−650
				3440	3450	14		−8	−600
				3430	3440	13		−7	−570
				3420	3430	11		−6	−570
				3410	3420	8		−5	−590
				3400	3410	10		−6	−630
	Summary Data			3400	3620	301		−183	−620
5.4	Muravlev	1984	1985	3610	3620	27		−21	−770
	SU06002			3600	3610	22		−16	−720
				3590	3600	11		−8	−700
				3580	3590	9		−6	−650
				3570	3580	10		−6	−630
				3560	3570	15		−10	−690
				3550	3560	15		−11	−730
				3540	3550	15		−11	−740
				3530	3540	17		−13	−760
				3520	3530	15		−12	−800
				3510	3520	15		−13	−840
				3500	3510	14		−11	−780
				3490	3500	17		−14	−840
				3480	3490	11		−11	−970
				3470	3480	11		−11	−1040
				3460	3470	9		−10	−1070
				3450	3460	12		−13	−1090
				3440	3450	14		−15	−1060

(Continued)

Table 5D.29 (Continued)

NR	Glacier Name	Period From	Period To	Altitude From	Altitude To	Area Mean	Area Change	Volume Change	Thickness Change
				3430	3440	13		−14	−1040
				3420	3430	11		−11	−1040
				3410	3420	8		−8	−1030
				3400	3410	9	−1	−9	−1030
	Summary Data			3400	3620	300	−1	−254	−830
5.5	Muravlev	1985	1986	3610	3620	27		−17	−620
	SU06002			3600	3610	22		−14	−640
				3590	3600	11		−7	−630
				3580	3590	9		−5	−610
				3570	3580	10		−6	−600
				3560	3570	15		−9	−620
				3550	3560	15		−9	−620
				3540	3550	15		−9	−620
				3530	3540	17		−11	−630
				3520	3530	15		−9	−600
				3510	3520	15		−9	−600
				3500	3510	14		−9	−640
				3490	3500	17		−12	−720
				3480	3490	11		−9	−820
				3470	3480	11		−9	−860
				3460	3470	9		−9	−960
				3450	3460	12		−12	−1020
				3440	3450	14		−15	−1070
				3430	3440	13		−15	−1130
				3420	3430	11		−12	−1140
				3410	3420	8		−9	−1150
				3400	3410	8	−1	−9	−1160
	Summary Data			3400	3620	299	−1	−225	−740
5.6	Muravlev	1986	1987	3610	3620	27		−18	−680
	SU06002			3600	3610	22		−14	−630
				3590	3600	11		−7	−600
				3580	3590	9		−5	−560
				3570	3580	10		−5	−530
				3560	3570	15		−9	−580
				3550	3560	15		−9	−620
				3540	3550	15		−9	−630
				3530	3540	17		−11	−630
				3520	3530	15		−9	−600
				3510	3520	15		−9	−570
				3500	3510	14		−8	−550
				3490	3500	17		−10	−610
				3480	3490	11		−8	−740
				3470	3480	11		−10	−950
				3460	3470	9		−10	−1060
				3450	3460	12		−12	−1000
				3440	3450	14		−13	−930
				3430	3440	13		−12	−900
				3420	3430	11		−10	−890
				3410	3420	8		−7	−890
				3400	3410	8		−7	−880
	Summary Data			3400	3620	299		−212	−690
5.7	Muravlev	1987	1988	3610	3620	27		3	100
	SU06002			3600	3610	22		3	120
				3590	3600	11		1	60
				3580	3590	9		1	70
				3570	3580	10		1	140
				3560	3570	15		2	110
				3550	3560	15		2	110
				3540	3550	15		2	130

(Continued)

Table 5D.29 (Continued)

NR	Glacier Name	Period		Altitude		Area Mean	Area Change	Volume Change	Thickness Change
		From	To	From	To				
				3530	3540	17		2	150
				3520	3530	15		2	110
				3510	3520	15		1	40
				3500	3510	14		−2	−150
				3490	3500	17		−4	−230
				3480	3490	11		−3	−240
				3470	3480	11		−3	−310
				3460	3470	9		−3	−330
				3450	3460	12		−5	−450
				3440	3450	14		−7	−510
				3430	3440	13		−6	−500
				3420	3430	11		−5	−490
				3410	3420	8		−4	−490
				3400	3410	7	−1	−3	−480
	Summary Data			3400	3620	298	−1	−25	−80
5.8	Muravlev	1988	1989	3610	3620	27		−3	−120
	SU06002			3600	3610	22		−4	−190
				3590	3600	11		−2	−190
				3580	3590	9		0	−30
				3570	3580	10		0	20
				3560	3570	15		−2	−120
				3550	3560	15		−3	−200
				3540	3550	15		−4	−240
				3530	3540	17		−4	−260
				3520	3530	15		−5	−330
				3510	3520	15		−6	−370
				3500	3510	14		−5	−390
				3490	3500	17		−7	−390
				3480	3490	11		−5	−450
				3470	3480	11		−6	−520
				3460	3470	9		−5	−570
				3450	3460	12		−7	−600
				3440	3450	14		−10	−730
				3430	3440	13		−10	−790
				3420	3430	11		−9	−820
				3410	3420	8		−7	−840
				3400	3410	7		−6	−860
	Summary Data			3400	3620	298		−110	−340
5.9	Muravlev	1989	1990	3610	3620	27		0	−10
	SU06002			3600	3610	22		2	110
				3590	3600	11		1	90
				3580	3590	9		−1	−90
				3570	3580	10		−1	−110
				3560	3570	15		−6	−370
				3550	3560	15		−7	−510
				3540	3550	15		−7	−490
				3530	3540	17		−9	−590
				3520	3530	15		−10	−690
				3510	3520	15		−10	−680
				3500	3510	14		−11	−760
				3490	3500	17		−15	−880
				3480	3490	11		−9	−860
				3470	3480	11		−9	−820
				3460	3470	9		−8	−840
				3450	3460	12		−12	−960
				3440	3450	14		−14	−1020
				3430	3440	13		−12	−950
				3420	3430	11		−10	−930
				3410	3420	8		−7	−920

(Continued)

Table 5D.29 (Continued)

NR	Glacier Name	Period From	Period To	Altitude From	Altitude To	Area Mean	Area Change	Volume Change	Thickness Change
				3400	3410	6	−1	−5	−910
	Summary Data			3400	3620	297	−1	−165	−530
5.1	Muravlev	1990	1991	3610	3620	27		−33	−1220
	SU06002			3600	3610	22		−25	−1160
				3590	3600	11		−13	−1180
				3580	3590	9		−12	−1300
				3570	3580	10		−13	−1350
				3560	3570	15		−22	−1430
				3550	3560	15		−22	−1430
				3540	3550	15		−22	−1430
				3530	3540	17		−26	−1500
				3520	3530	15		−24	−1600
				3510	3520	15		−25	−1650
				3500	3510	14		−22	−1540
				3490	3500	17		−25	−1490
				3480	3490	11		−19	−1690
				3470	3480	11		−22	−1990
				3460	3470	9		−19	−2140
				3450	3460	12		−26	−2180
				3440	3450	14		−31	−2200
				3430	3440	13		−27	−2080
				3420	3430	11		−22	−2030
				3410	3420	8		−16	−2020
				3400	3410	6		−12	−1990
	Summary Data			3400	3620	297		−478	−1590
6.1	Shumskiy	1989	1990	3720	3740	96		−42	−440
	SU06001			3700	3720	114		−42	−370
				3680	3700	145		−46	−320
				3660	3680	111		−32	−290
				3640	3660	77		−30	−390
				3620	3640	89		−34	−380
				3600	3620	76		−29	−380
				3580	3600	76		−20	−260
				3560	3580	92		−28	−300
				3540	3560	104		6	60
				3520	3540	113		−14	−120
				3500	3520	42		−24	−580
				3480	3500	35		−26	−750
				3460	3480	43		−36	−840
				3440	3460	58		−57	−990
				3420	3440	92		−87	−950
				3400	3420	52		−49	−940
				3380	3400	38		−27	−700
				3360	3380	31		−18	−570
				3340	3360	34		−27	−800
				3320	3340	53		−45	−850
				3300	3320	46		−45	−980
				3280	3300	26		−28	−1060
				3260	3280	20		−20	−1020
				3240	3260	23		−28	−1200
				3220	3240	26		−36	−1400
				3200	3220	27		−59	2190
				3180	3200	23		−54	−2370
				3160	3180	13	−1	−35	−2690
				3140	3160	4	−1	−11	−2740
	Summary Data			3140	3740	1779	−2	−1023	−600
6.2	Shumskiy	1990	1991	3720	3740	96		−76	−790
	SU06001			3700	3720	114		−89	−780
				3680	3700	145		−115	−790

(Continued)

Table 5D.29 (Continued)

NR	Glacier Name	Period From	Period To	Altitude From	Altitude To	Area Mean	Area Change	Volume Change	Thickness Change
				3660	3680	111		−84	−760
				3640	3660	77		−59	−770
				3620	3640	89		−82	−920
				3600	3620	76		−71	−940
				3580	3600	76		−65	−850
				3560	3580	92		−114	−1240
				3540	3560	104		−177	−1700
				3520	3540	113		−209	−1850
				3500	3520	42		−69	−1650
				3480	3500	35		−54	−1530
				3460	3480	43		−67	−1550
				3440	3460	58		−103	−1780
				3420	3440	92		−187	−2030
				3400	3420	52		−115	−2210
				3380	3400	38		−90	−2370
				3360	3380	31		−74	−2400
				3340	3360	34		−67	−1960
				3320	3340	53		−99	−1870
				3300	3320	46		−97	−2110
				3280	3300	26		−56	−2170
				3260	3280	20		−44	−2220
				3240	3260	23		−47	−2060
				3220	3240	26		−51	−1980
				3200	3220	27		−63	−2340
				3180	3200	22	−1	−56	−2560
				3160	3180	12	−1	−26	−2210
				3140	3160	3	−1	−7	−2240
	Summary Data			3140	3740	1776	−3	−2513	−1420
7.1	TS.Tuyuksuysk	1990	1991	3800	3820	117		−117	−1000
	SU05075			3780	3800	150		−159	−1060
				3760	3780	184		−206	−1120
				3740	3760	200		−234	−1170
				3720	3740	166		−184	−1110
				3700	3720	138		−135	−980
				3680	3700	73		−88	−1200
				3660	3680	51		−72	−1410
				3640	3660	54		−62	−1150
				3620	3640	48		−55	−1140
				3600	3620	70		−73	−1040
				3580	3600	55		−70	−1280
				3560	3580	99		−144	−1460
				3540	3560	56		−96	−1710
				3520	3540	54		−111	−2060
				3500	3520	53		−118	−2220
				3480	3500	50		−129	−2590
				3460	3480	37		−120	−3230
	Summary Data			3460	3820	1655		−2173	−1330
7.2	TS.Tuyuksuysk	1991	1992	3800	3820	117		−35	−300
	SU05075			3780	3800	150		−39	−260
				3760	3780	184		−52	−280
				3740	3760	200		−74	−370
				3720	3740	166		−70	−420
				3700	3720	138		−63	−460
				3680	3700	73		−48	−660
				3660	3680	51		−38	−740
				3640	3660	54		−42	−770
				3620	3640	48		−37	−770
				3600	3620	70		−57	−820
				3580	3600	55		−51	−930

(Continued)

Table 5D.29 (Continued)

NR	Glacier Name	Period From	Period To	Altitude From	Altitude To	Area Mean	Area Change	Volume Change	Thickness Change
				3560	3580	99		−130	−1310
				3540	3560	56		−88	−1570
				3520	3540	54		−89	−1650
				3500	3520	48	−5	−96	−1990
				3480	3500	40	−10	−91	−2270
				3460	3480	17	−20	−38	−2230
	Summary Data			3460	3820	1620	−35	−1138	−790
7.3	TS.Tuyuksuysk	1992	1993	3800	3820	117			
	SU05075			3780	3800	150		134	891
				3760	3780	184		213	1160
				3740	3760	200		252	1259
				3720	3740	166		196	1180
				3700	3720	138		183	1324
				3680	3700	73		104	1431
				3660	3680	51		58	1141
				3640	3660	54		46	846
				3620	3640	48		40	835
				3600	3620	70		61	866
				3580	3600	55		40	734
				3560	3580	99		69	696
				3540	3560	56		32	567
				3520	3540	54		22	400
				3500	3520	48		−2	−38
				3480	3500	40		−27	−681
				3460	3480	15	−2	−15	−973
	Summary Data			3460	3820	1618	2	1406	936
	Nepal								
8.1	AX010	1978	1991	5340	5360	6	−2		
	NP00005			5320	5340	12	−2		
				5300	5320	12	−2		
				5280	5300	29	−8		
				5260	5280	65	−23	−97	−1500
				5240	5260	101	−9	−324	−3200
				5220	5240	62	7	−353	−5700
				5200	5220	60	4	−576	−9600
				5180	5200	35	3	−378	−109000
				5160	5180	24	−3	−278	−11400
				5140	5160	19	−3	−229	−11800
				5120	5140	15	3	−232	−15900
				5100	5120	16	−6	−271	−16700
				5080	5100	20	−7	−354	−17900
				5060	5080	31	−16	−564	−18500
				5040	5060	21	7	−387	−18800
				5020	5040	15	−2	−321	−21000
				5000	5020	15	−4	−355	−23200
				4980	5000	7	2	−145	−21700
				4960	4980	3	1	−51	−17500
				4952	4960	1	0	−18	−17500
	Summary Data			4952	5360	568	−58	−4934	−8689

Note:

NR	Record number
Glacier Name	15 alphabetic or numeric digits
Period From To	Period in which the changes take place
Altitude	Altitude interval in meters above sea level
Area Mean	Mean area of altitude interval for period of change (thousand square meters)
Area Change	Change in area of altitude interval for period of change (thousand square meters)
Volume Change	Change in volume of altitude interval for period of change (thousand cubic meters)
Thick Change	Change in thickness of altitude interval for period of change (millimeters)

Source: From World Glacier Monitoring Service by Wilfried Haeberli, Martin Hoelzle, Stephan Suter and Regula Frauenfelder www.wgms.ch.

SECTION 5E FLOODS

[In thousands of cfs; for a typical 300 mi^2 (780 km^2) drainage basin; a mean annual flood is one that will be exceeded in about half the years; the probability of a mean annual flood in any given year is about 50%]

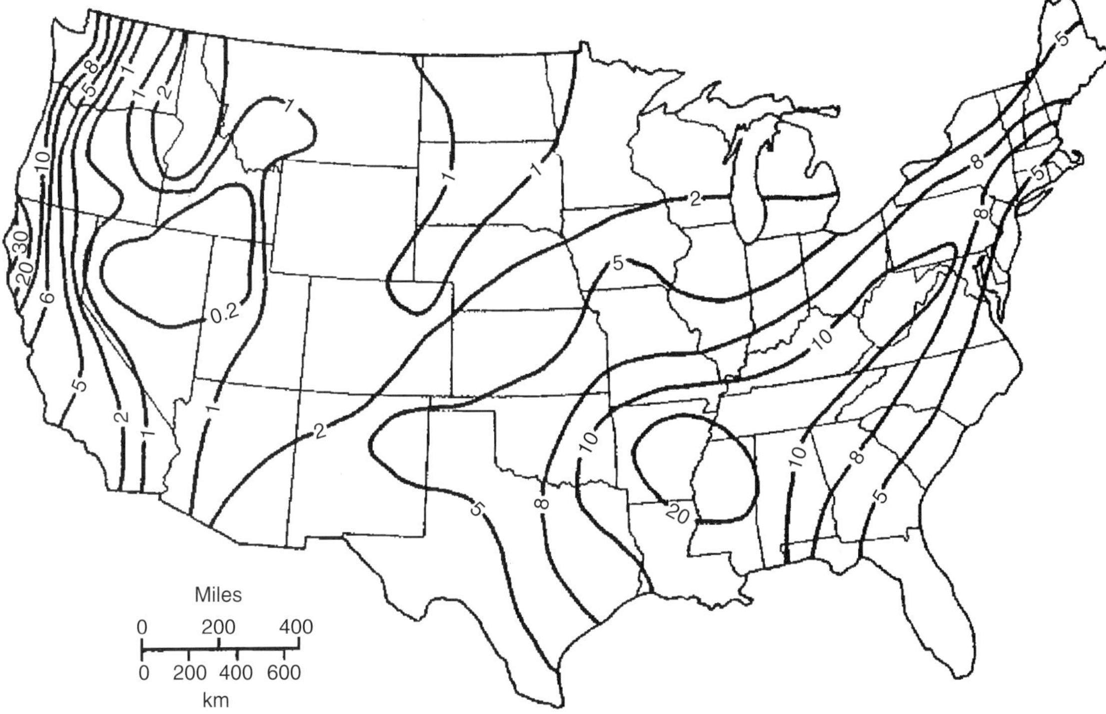

Figure 5E.15 Mean annual flood potential in the United States. (From U.S. Geological Survey, National Atlas Map 121, 1965.)

[In thousands of cfs; for a typical 300 mi^2 (780 km^2) drainage basin; a 10-year flood will be exceeded at irregular intervals that average 10 years; the probability of a 10-year flood in any given year is 10%]

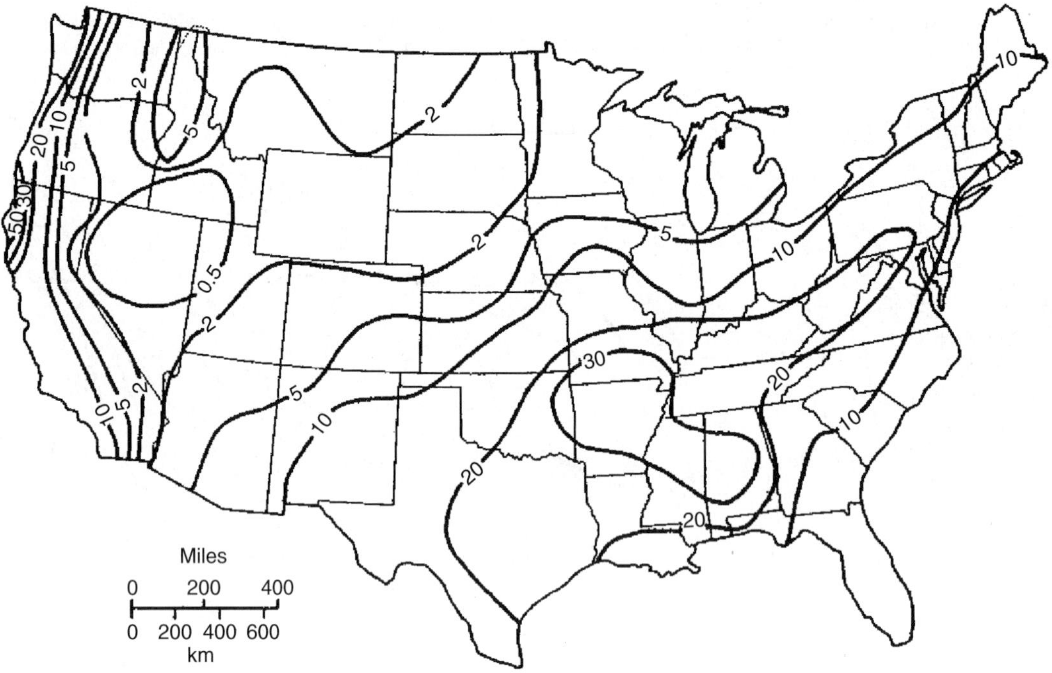

Figure 5E.16 Ten-year flood potential in the United States. (From U.S. Geological Survey, National Atlas Map 121, 1965.)

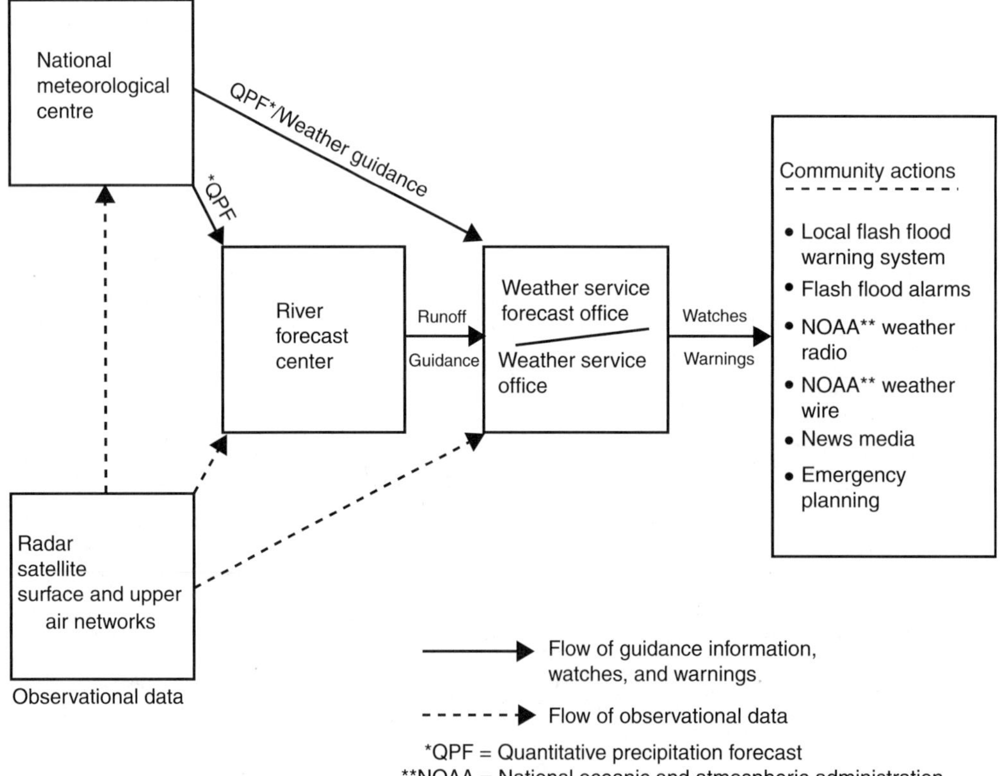

Figure 5E.17 Flash flood warning system in the United States. (From National weather service, Operations of the national weather service. U.S. Department of Commerce, 1985.)

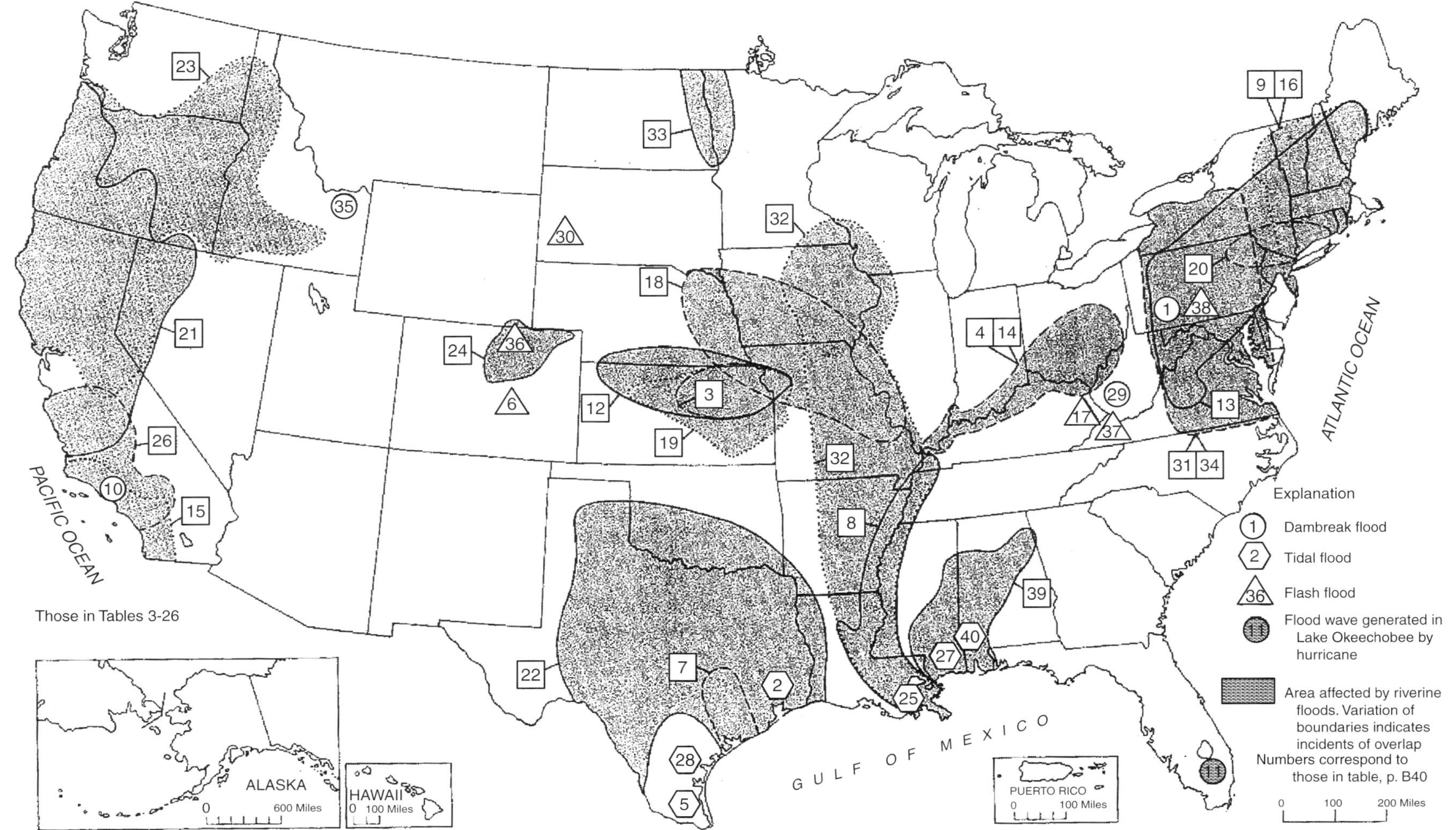

Figure 5E.18 Distribution of great floods in the conterminous United States since 1889. (From Hays, W.W., Facing geologic and hydrologic hazards, Earth-Science considerations, U.S. geological survey professional paper 1240-B, 1981.)

Table 5E.30 Estimated U.S. Flood Damage, by Fiscal Year (Oct–Sep)

Fiscal Year	Damage (Millions Current Dollars)	Implicit Price Deflator	Damage (Millions 1995 Dollars)
1926	9.243	—	—
1927	315.187	—	—
1928	88.155	—	—
1929	61.700	0.12854	480
1930	25.832	0.12385	209
1931	2.070	0.11091	19
1932	10.365	0.09796	106
1933	27.366	0.09541	287
1934	18.903	0.10071	188
1935	123.327	0.10265	1,201
1936	287.137	0.10377	2,767
1937	433.339	0.10815	4,007
1938	108.970	0.10499	1,038
1939	13.861	0.10387	133
1940	40.067	0.10530	381
1941	26.092	0.11244	232
1942	91.548	0.12120	755
1943	220.553	0.12773	1,727
1944	99.789	0.13058	764
1945	159.251	0.13425	1,186
1946	68.930	0.15056	458
1947	281.321	0.16667	1,688
1948	213.716	0.17615	1,213
1949	108.586	0.17594	617
1950	129.903	0.17788	730
1951	1,076.687	0.19072	5,645
1952	254.190	0.19368	1,312
1953	121.752	0.19623	620
1954	74.170	0.19817	374
1955	784.672	0.20163	3,892
1956	305.573	0.20846	1,466
1957	352.145	0.21539	1,635
1958	224.939	0.22059	1,020
1959	121.281	0.22304	544
1960	111.168	0.22620	491
1961	147.680	0.22875	646
1962	86.574	0.23180	373
1963	179.496	0.23445	766
1964	194.512	0.23792	818
1965	1,221.903	0.24241	5,041
1966	116.645	0.24934	468
1967	291.823	0.25698	1,136
1968	443.251	0.26809	1,653
1969	889.135	0.28124	3,161
1970	173.803	0.29623	587
1971	323.427	0.31111	1,040
1972	4,442.992	0.32436	13,698
1973	1,805.284	0.34251	5,271
1974	692.832	0.37329	1,856
1975	1,348.834	0.40805	3,306
1976	1,054.790	0.43119	2,446
1977	988.350	0.45892	2,154
1978	1,028.970	0.49164	2,093
1979	3,626.030	0.53262	6,808
1980	—	0.58145	—
1981	—	0.63578	—
1982	—	0.67533	—
1983	3,693.572	0.70214	5,260

(Continued)

Table 5E.30 (Continued)

Fiscal Year	Damage (Millions Current Dollars)	Implicit Price Deflator	Damage (Millions 1995 Dollars)
1984	3,540.770	0.72824	4,862
1985	379.303	0.75117	505
1986	5,939.994	0.76769	7,737
1987	1,442.349	0.79083	1,824
1988	214.297	0.81764	262
1989	1,080.814	0.84883	1,273
1990	1,636.366	0.88186	1,856
1991	1,698.765	0.91397	1,859
1992	672.635	0.93619	718
1993	16,364.710	0.95872	17,069
1994	1,120.149	0.97870	1,145
1995	5,110.714	1.00000	5,111
1996	6,121.753	1.01937	6,005
1997	8,934.923	1.03925	8,597
1998	2,465.048	1.05199	2,343
1999	5,450.375	1.06677	5,109
2000	1,336.744	1.09113	1,225

Note: — Data unavailable.

Source: From U.S. Bureau of Economic Analysis, 2001.

Table 5E.31 Comparison of Damage Estimates by State, 1955–1978 and 1983–1999.

State	Region	Median Damage (All Years)	Maximum Damage[a]	Years with No Estimate	Years with 0<Est 1.0	Years with Est >100
Rhode Island	New England	0.00	143	33	5	1
Delaware		0.00	7	32	7	0
Massachusetts	New England	0.00	774	25	5	2
New Hampshire	New England	0.00	56	23	6	0
Hawaii		0.00	44	23	2	0
Connecticut	New England	0.00	1,881	21	6	2
Vermont	New England	0.00	194	20	9	1
Wyoming	Arid West	0.05	53	17	14	0
Maine	New England	0.06	77	20	3	0
New Jersey		0.06	749	18	5	8
Alaska (29 yr)		0.07	383	14	4	1
Maryland & DC		0.14	681	15	14	1
Nevada	Arid West	0.16	616	13	12	1
Michigan		0.21	528	17	11	3
N. Dakota	N. Central	0.41	3,280	14	9	4
S. Dakota	N. Central	0.51	796	10	13	4
Colorado	Arid West	0.57	1,866	11	10	4
S. Carolina		0.66	40	5	18	0
New Mexico	Arid West	0.73	34	16	6	0
Utah	Arid West	0.84	712	7	14	2
Montana	Arid West	1.04	229	10	10	1
Idaho		1.21	1,507	9	10	2
Wisconsin		1.61	943	11	8	4
Georgia	Southeast	1.86	307	5	7	3
Virginia		1.91	1,042	9	9	6
Arizona	Arid West	2.27	306	7	9	4
Minnesota		2.40	1,006	4	12	7
Florida	Southeast	2.48	410	6	9	5
N. Carolina		3.99	2,919	5	5	3
Oregon	Pacific NW	4.06	3,143	2	6	4
Washington	Pacific NW	4.32	363	5	7	3

(Continued)

Table 5E.31 (Continued)

State	Region	Median Damage (All Years)	Maximum Damage[a]	Years with No Estimate	Years with 0<Est 1.0	Years with Est >100
Louisiana	Lower Miss.	5.60	3,097	7	7	10
Tennessee	Southeast	6.01	193	2	8	1
Alabama	Southeast	6.10	351	4	4	3
Arkansas	Lower Miss.	6.87	712	2	6	4
Mississippi	Lower Miss.	8.07	1,157	1	3	4
W. Virginia	Ohio R.	8.60	782	1	7	5
Kansas	Central	8.61	575	3	4	6
Oklahoma	Central	8.97	1,045	4	8	5
Pennsylvania		10.39	8,590	3	7	6
Nebraska	Upper Miss.	13.89	307	4	4	4
New York		14.60	2,305	7	3	6
Illinois	Upper Miss.	15.31	2,754	1	3	8
Iowa	Upper Miss.	17.18	5,987	4	6	9
Kentucky	Ohio R.	17.67	453	1	7	7
Indiana	Ohio R.	19.29	310	0	3	3
Ohio	Ohio R.	22.06	313	3	5	4
Missouri	Upper Miss.	25.42	3,577	0	7	12
California		45.64	2,007	3	4	13
Texas		77.44	691	1	1	16

[a] Estimates of maximum damage can be misleading. For example, in Idaho the maximum was caused by failure of the Teton Dam in 1976; the worst damage directly from precipitation and streamflow is estimated at $120 million. In Texas, the maximum appears small but much greater damage occurred in a year not covered by this table ($3.76 billion in 1979). States are ordered by increasing median damage. Missing estimates are treated as zero; all estimates are in millions of 1995 dollars.

Source: From Pielke, Jr., R.A., M.W. Downton, and J.Z. Barnard Miller, 2002: Flood Damage in the United States, 1926–2000: A Reanalysis of National Weather Service Estimates. Boulder, CO: UCAR, www.flooddamagedata.org.

Table 5E.32 Damage in Thousands of Current Dollars

	Deflator	AL	AK	AZ	AR	CA	CO	CT	DE
1955	0.20163	3,379		226	61	165,767	2,567	379,360	117
1956	0.20846	720		0	255	8,745	5,135	0	0
1957	0.21539	2,324		0	27,938	13	2,901	0	0
1958	0.22059	872		0	6,202	33,063	240	0	0
1959	0.22304	0		100	3,090	4	0	0	0
1960	0.22620	670		0	580	516	0	750	0
1961	0.22875	12,625		325	3,503	95	0	0	0
1962	0.23180	3,529		1,000	91	2,780	80	0	0
1963	0.23445	1,280		0	2,500	11,834	50	0	0
1964	0.23792	5,343		55	598	229,168	0	0	0
1965	0.24241	723		11,330	143	11,321	452,293	0	0
1966	0.24934	2,366		3,050	5,055	24,347	707	0	0
1967	0.25698	1,695	98,550	3,576	1,497	1,370	0	0	0
1968	0.26809	408	0	188	21,099	0	0	100	0
1969	0.28124	88	0	0	3,411	423,296	66	528	0
1970	0.29623	10,891	0	5,000	639	47,798	2,040	0	0
1971	0.31111	2,170	8,631	3,476	2,549	3,522	0	0	50
1972	0.32436	2,278	1,090	20,868	1,780	1,132	15	15,414	0
1973	0.34251	5,439	1,500	0	129,579	9,480	121,383	1,950	0
1974	0.37329	1,731	0	2,605	8,746	27,124	0	0	5
1975	0.40805	91,815	0	927	21,387	1,845	0	9,360	0
1976	0.43119	4,710	0	6,000	0	120,100	35,540	7,100	0
1977	0.45892	4,760	200	15,590	130	28,500	1,250	1,570	0
1978	0.49164	3,000	0	131,360	23,900	124,230	70	0	0
1979	0.53262	[a]	0	0	2,620	25,900	50	[a]	0

(Continued)

Table 5E.32 (Continued)

	Deflator	AL	AK	AZ	AR	CA	CO	CT	DE
1980	0.58145								
1981	0.63578								
1982	0.67533								
1983	0.70214	29,431	0	179,938	500,000	673,000	100	0	0
1984	0.72824	23,000	7,150	223,000	5,000	0	107,050	81,700	5,000
1985	0.75117	1,700	50	1,350	19,823	0	7,000	0	50
1986	0.76769	0	0	3,000	2,240	402,000	166	0	0
1987	0.79083	755	20,000	7	15,045	1,015	0	5,000	0
1988	0.81764	1,721	500	71	12,612	52,353	0	0	0
1989	0.84883	178	6,000	33,636	2,320	38,738	481	800	1,600
1990	0.88186	120,000	0	3,220	143,056	570	130	10	0
1991	0.91397	15,055	0	258	12,006	3,376	2,820	16	0
1992	0.93619	320	7,302	5,189	909	93,152	1,602	10,366	2
1993	0.95872	0	0	228,900	2,680	165,920	100	0	0
1994	0.97870	112,696	74,000	1,616	2,024	1,792	1,242	1,316	741
1995	1.00000	0	10,025	6,618	0	1,495,960	18,240	0	0
1996	1.01937	1,649	0	701	205	13,205	4,058	2,092	300
1997	1.03925	1,354	1,271	85	12,874	2,086,125	358,890	52	0
1998	1.05199	368,938	314	66	2,045	621,588	2,550	40	0
1999	1.06677	4,663	0	12,796	1,777	14,176	50,675	1,112	0
2000	1.09113	3,087	110	90	2,773	9,238	297	6,010	0

	Deflator	FL	GA	HI	ID	IL	IN	IA	KS
1955	0.20163	105	1	0	1,371	102	1,003	35	474
1956	0.20846	1,891	212	0	6,222	1,026	4,021	51	33
1957	0.21539	0	1,068	0	20,896	1,206	66,748	1,543	9,164
1958	0.22059	0	323	400	3	17,970	52,302	7,508	4,606
1959	0.22304	150	0	0	500	1,506	12,958	128	4,061
1960	0.22620	12,047	392	0	0	7,503	2,649	7,612	1,947
1961	0.22875	317	5,236	0	939	11,553	13,306	9,389	13,397
1962	0.23180	1,481	0	0	8,112	891	670	6,778	1,826
1963	0.23445	0	445	2,300	2,766	513	8,266	70	168
1964	0.23792	426	3,641	0	11,704	3,044	12,327	240	370
1965	0.24241	144	397	0	4,184	30,564	20	32,462	29,792
1966	0.24934	548	1,628	0	0	577	3,098	904	97
1967	0.25698	95	23	1,029	792	2,629	4,618	4,416	15,093
1968	0.26809	46	133	2,500	0	2,576	22,463	1,650	2,304
1969	0.28124	2,858	79	0	111	9,095	6,672	6,233	10,991
1970	0.29623	145	348	0	38	9,124	2,300	977	4,138
1971	0.31111	476	243	500	1,187	462	1,690	684	1,644
1972	0.32436	41,206	328	0	355	5,927	4,700	13,262	1,646
1973	0.34251	2,282	5,143	0	0	258,704	6,326	12,724	53,772
1974	0.37329	23,050	405	3,869	36,118	75,068	15,805	56,367	3,700
1975	0.40805	15,839	3,002	0	378	20,598	12,317	7,300	3,255
1976	0.43119	0	8,130	270	650,000	3,370	3,680	160	1,330
1977	0.45892	140	4,160	0	0	7,190	8,160	0	46,350
1978	0.49164	3,720	0	0	60	50	38,960	0	0
1979	0.53262	21,990	0	11,000	0	32,250	16,000	2,000	7,000
1980	0.58145								
1981	0.63578								
1982	0.67533								
1983	0.70214	0	0	0	2,200	202,500	20,000	0	0
1984	0.72824	200,000	5,050	6,055	1,000	7,992	22,194	600,550	50,050
1985	0.75117	30,000	0	3,100	0	11,500	50,000	50	5,000
1986	0.76769	7,275	2,000	0	2,005	104,705	2,500	45,307	181,700
1987	0.79083	645	1,470	2,050	17	150,000	1,906	16,755	152,000
1988	0.81764	50,350	230	35,647	0	102	89	0	0
1989	0.84883	2,109	1,792	3,392	178	1,600	716	7,286	3,394
1990	0.88186	500	30,658	665	113	71,045	105,550	351,401	2,048
1991	0.91397	0	106,158	23,715	2,574	19,834	89,504	195,703	16,551

(Continued)

Table 5E.32 (Continued)

	Deflator	FL	GA	HI	ID	IL	IN	IA	KS
1992	0.93619	41,938	1,156	9,260	224	189	45,424	50,800	10,127
1993	0.95872	2,080	7,340	2,910	0	2,640,140	9,550	5,740,000	551,070
1994	0.97870	182,605	300,000	3,700	0	32,606	2,852	9,124	10,437
1995	1.00000	18,536	8,845	0	2,096	27,240	6,789	3,498	8,874
1996	1.01937	158,001	2,581	1,935	49,400	107,585	21,575	165,265	3,969
1997	1.03925	49,707	464	0	125,060	4,295	68,598	3,680	102
1998	1.05199	431,311	166,291	0	1,005	2,380	19,611	168,101	4,888
1999	1.06677	60,080	8,520	0	1,297	3,666	50,124	111,221	60,030
2000	1.09113	499,080	2,101	400	85	3,113	819	14,877	250
	Deflator	**KY**	**LA**	**ME**	**MD**	**MA**	**MI**	**MN**	**MS**
1955	0.20163	6,629	30	0	5,450	155,982	0	0	3,132
1956	0.20846	568	0	0	888	0	1,278	11	1,270
1957	0.21539	55,233	4,147	0	0	0	0	9,128	2,693
1958	0.22059	3,817	2,842	0	100	0	0	17	13,826
1959	0.22304	2,480	0	61	0	0	0	50	280
1960	0.22620	3	112	0	0	6,400	1,181	212	744
1961	0.22875	12,969	6,074	800	0	0	0	552	15,918
1962	0.23180	16,885	1,908	0	0	0	0	1,290	1,982
1963	0.23445	36,917	0	0	0	0	0	26	19
1964	0.23792	35,476	30	0	0	0	0	0	3,152
1965	0.24241	1,044	0	0	53	0	0	97,603	1,931
1966	0.24934	1,671	250	528	0	0	0	4,300	2,706
1967	0.25698	17,583	0	0	125	0	0	0	1,192
1968	0.26809	6,036	2,810	0	0	35,000	100	1,197	6,269
1969	0.28124	8,075	251	300	200	0	13	67,168	1,900
1970	0.29623	707	1,000	0	15	0	0	4,350	3,586
1971	0.31111	6,099	0	0	8,600	0	0	15	12,431
1972	0.32436	15,841	100	0	220,739	10	10	64,318	10,248
1973	0.34251	10,491	334,904	11,200	0	0	530	242	226,885
1974	0.37329	5,218	10,343	3,000	0	0	240	16,939	27,827
1975	0.40805	26,302	90,204	0	27,200	0	54,358	139,726	70,990
1976	0.43119	0	0	3,360	4,900	1,000	790	0	2,840
1977	0.45892	101,000	48,040	4,190	0	0	0	7,870	2,780
1978	0.49164	100,000	145,000	0	150	0	0	65,000	0
1979	0.53262	0	[a]	0	69,000	[a]	0	13,140	[a]
1980	0.58145								
1981	0.63578								
1982	0.67533								
1983	0.70214	100	651,000	375	100	0	0	310	812,600
1984	0.72824	180,236	6,550	10,050	10,015	50,560	0	5,000	6,050
1985	0.75117	460	8,050	45	50	0	80,000	500	2,000
1986	0.76769	25	1,515,250	5,000	0	21,500	405,000	1,501	651
1987	0.79083	68	1,175	61,250	51	47,480	15	27,800	6,380
1988	0.81764	250	8,708	0	0	0	206	555	39,420
1989	0.84883	27,445	322,118	3,200	1,600	0	180	17,600	3,635
1990	0.88186	5,664	115,901	0	23	50	627	3,032	21,805
1991	0.91397	9,034	221,720	16,336	48	9,716	6,133	1,280	313,359
1992	0.93619	46,870	4,191	2,179	339	176	355	1,760	1,010
1993	0.95872	4,980	4,020	3,040	0	160	1,600	964,050	4,480
1994	0.97870	2,544	675	9,323	4,524	0	6,236	1,867	1,352
1995	1.00000	17,673	3,097,250	0	1,620	0	2,900	3,750	1,092
1996	1.01937	21,323	121	4,916	90,481	2,663	26,690	460	200
1997	1.03925	470,915	4,359	26,845	198	75,024	325	743,218	32,774
1998	1.05199	16,639	17,845	0	334	13,510	18,190	2,529	3,498
1999	1.06677	506	5,979	1,580	9,715	250	325	466	1,769
2000	1.09113	17,631	153	2,814	2,452	206	25,430	43,112	408
	Deflator	**MO**	**MT**	**NE**	**NV**	**NH**	**NJ**	**NM**	**NY**
1955	0.20163	666	63	1,500	7,398	0	23,102	1,066	30,072
1956	0.20846	167	317	865	237	0	0	0	1,089

(Continued)

Table 5E.32 (Continued)

	Deflator	MO	MT	NE	NV	NH	NJ	NM	NY
1957	0.21539	9,618	33	5,983	0	0	0	0	166
1958	0.22059	38,718	1	3,064	0	0	3	0	42
1959	0.22304	6,018	82	3,753	0	4,500	0	0	5,667
1960	0.22620	13,506	57	8,884	0	100	0	0	7,229
1961	0.22875	27,375	0	674	891	0	0	0	608
1962	0.23180	557	147	2,630	762	0	0	0	0
1963	0.23445	152	148	13,394	2,858	0	0	620	33,102
1964	0.23792	6,591	54,389	5,146	2,454	0	0	1,235	3,275
1965	0.24241	33,976	253	1,368	4	0	0	4,833	0
1966	0.24934	2,781	0	11,628	307	0	0	1,048	0
1967	0.25698	39,080	2,947	40,644	45	0	1,438	0	777
1968	0.26809	890	0	6,029	1	800	166,690	0	0
1969	0.28124	36,601	388	1,826	0	400	580	0	3,383
1970	0.29623	14,926	581	0	138	0	0	0	3,953
1971	0.31111	191	412	5,941	0	0	138,700	0	1,000
1972	0.32436	5,783	595	73	0	0	15,050	6,613	747,674
1973	0.34251	231,438	0	10,388	0	19,100	50,868	251	5,000
1974	0.37329	62,594	4,217	126	1,000	0	0	0	0
1975	0.40805	7,611	24,123	0	6,200	0	60,687	577	60,064
1976	0.43119	810	50	0	200	0	0	500	38,020
1977	0.45892	52,500	0	1,590	0	610	95,880	0	10,600
1978	0.49164	2,000	19,060	67,000	0	900	14,720	14,450	0
1979	0.53262	0	0	0	0	0	[a]	3,210	[a]
1980	0.58145								
1981	0.63578								
1982	0.67533								
1983	0.70214	50,000	0	0	1,000	75	0	6,000	0
1984	0.72824	96,293	663	100,550	0	6,000	334,200	23,000	217,500
1985	0.75117	100	0	500	0	50	0	24,000	24,700
1986	0.76769	155,000	38,674	28,482	20,650	5,962	0	0	30,820
1987	0.79083	100,550	0	25,890	13	19,100	17,050	10	75,275
1988	0.81764	69	0	61	12	0	50	0	230
1989	0.84883	16,067	2,194	29,772	23	0	1,600	3,378	38,271
1990	0.88186	1,842	1,758	36,536	51	1,200	1	1,187	6,530
1991	0.91397	1,960	10,743	53,615	2	0	16,002	1,567	19,603
1992	0.93619	2,044	1,403	6,683	1,621	0	500	32,264	1,862
1993	0.95872	3,429,630	6,720	294,500	0	0	0	210	55,480
1994	0.97870	37,864	3,392	2,710	160	0	3,520	2,000	25,707
1995	1.00000	25,415	510	5,129	11,970	110	0	954	1,485
1996	1.01937	871	2,243	31,233	370	4,000	36,720	1,285	220,011
1997	1.03925	692	2,874	10,273	640,110	10,952	38,700	380	55,909
1998	1.05199	10,227	3,001	1,483	1,300	700	750	713	38,627
1999	1.06677	36,862	184	22,765	25,009	1,002	800,000	3,980	18,715
2000	1.09113	109,760	30	23,456	221	515	179,100	160	18,498

	Deflator	NC	ND	OH	OK	OR	PA	RI	SC
1955	0.20163	625	2	753	977	9,515	141,381	28,830	74
1956	0.20846	831	0	1,056	0	6,376	7,199	0	0
1957	0.21539	788	100	7	35,665	310	1,048	0	60
1958	0.22059	3,201	0	4,867	169	363	3,582	0	680
1959	0.22304	506	28	54,840	8,907	20	21,109	0	122
1960	0.22620	100	136	191	2,638	360	3,072	0	72
1961	0.22875	1,400	0	1,217	2,483	757	612	0	369
1962	0.23180	0	0	6,512	792	1,550	15	0	97
1963	0.23445	0	0	22,359	413	299	5,397	0	89
1964	0.23792	15,816	0	28,039	798	187,101	16,938	0	1,809
1965	0.24241	88	5,192	0	2,508	5,679	0	0	268
1966	0.24934	198	9,700	1,893	12	2,283	705	0	140
1967	0.25698	1,168	0	6,622	3	1,044	7,251	588	579
1968	0.26809	0	0	20,074	3,021	538	421	9,000	0

(Continued)

Table 5E.32 (Continued)

	Deflator	NC	ND	OH	OK	OR	PA	RI	SC
1969	0.28124	1,338	37,436	87,916	762	938	3,310	0	625
1970	0.29623	2,326	13,832	2,478	5,212	2,518	365	0	52
1971	0.31111	965	1,266	782	23,166	4,350	20,899	0	295
1972	0.32436	10,772	537	12,929	12,006	12,977	2,786,294	0	69
1973	0.34251	39,004	0	8,317	38,119	2,699	5,935	0	7,674
1974	0.37329	1,028	8,291	1,500	29,083	64,017	0	0	78
1975	0.40805	7,932	154,715	15,513	300	7,898	270,600	0	1,477
1976	0.43119	9,120	2,420	40	52,640	1,170	440	0	11,210
1977	0.45892	52,500	80	370	12,720	10,690	330,020	0	260
1978	0.49164	0	13,000	1,520	0	0	6,630	0	60
1979	0.53262	0	20,100	60,000	0	0	0	[a]	0
1980	0.58145								
1981	0.63578								
1982	0.67533								
1983	0.70214	470	0	0	0	7,300	0	0	0
1984	0.72824	40,000	5	10,122	268,000	52,900	75,500	5	1,110
1985	0.75117	50	0	10,000	15,030	50	100	0	100
1986	0.76769	1,990	315	10,000	802,250	33,900	71,540	0	3,070
1987	0.79083	20,461	4,943	20,518	22,250	900	28	550	31,771
1988	0.81764	0	0	2	3,437	125	62	0	0
1989	0.84883	21,072	16,000	52,240	2,121	98	7,106	0	370
1990	0.88186	1,075	0	40,846	40,650	1,070	792	50	677
1991	0.91397	2,694	32	55,165	90	9,010	8,342	174	11,871
1992	0.93619	12,927	0	20,078	10,871	32	1,805	16	0
1993	0.95872	1,400	413,600	25,800	44,720	1,760	440	0	17,920
1994	0.97870	2,032	58,552	39,913	166	0	16,194	0	6,228
1995	1.00000	26,596	44,366	28,511	3,275	11,320	10,385	0	28,169
1996	1.01937	42,119	220	22,721	0	3,203,500	494,862	0	668
1997	1.03925	17,994	3,408,298	66,666	155	173,200	3,136	0	1,105
1998	1.05199	16,135	2,583	181,409	262	10	1,103	0	4,044
1999	1.06677	3,117,160	100,355	963	9,578	2,100	27,642	0	75
2000	1.09113	7,605	191,177	8,839	11,691	5,734	27,476	0	2,885
	Deflator	**SD**	**TN**	**TX**	**UT**	**VT**	**VA**	**WA**	**WV**
1955	0.20163	11	977	5,165	226	0	10,695	1,165	5,187
1956	0.20846	10	279	3,715	210	0	0	6,472	3,185
1957	0.21539	3,969	5,118	78,881	169	3	139	1,664	11,052
1958	0.22059	0	128	18,101	10	0	0	50	1,170
1959	0.22304	0	0	2,886	4	0	28	4,914	709
1960	0.22620	3,417	226	8,093	0	0	211	0	370
1961	0.22875	1	2,263	2,846	281	0	231	130	3,455
1962	0.23180	3,030	651	1,948	1,272	0	0	0	5,914
1963	0.23445	0	6,262	20	64	0	5,937	1,013	17,624
1964	0.23792	0	156	5,435	70	692	0	11,817	4,169
1965	0.24241	740	2,472	39,395	1,746	0	2	1,012	49
1966	0.24934	470	1,608	28,001	1,577	0	0	592	1,868
1967	0.25698	1,125	1,090	98,259	453	0	581	1,910	14,235
1968	0.26809	123	648	24,267	1,260	100	0	611	47
1969	0.28124	31,898	1,090	12,878	237	680	123,552	2,722	5,996
1970	0.29623	19	13,260	3,150	222	0	148	380	297
1971	0.31111	0	86	26,538	1,033	0	1,158	3,908	1,653
1972	0.32436	165,086	6,634	20,605	358	40	180,770	21,029	37,974
1973	0.34251	0	66,273	136,758	2,270	66,466	1,615	0	3,359
1974	0.37329	268	2,243	41,707	0	0	100	21,318	10,375
1975	0.40805	0	12,700	23,074	212	200	18,340	42,289	5,913
1976	0.43119	5,500	200	33,390	0	0	0	2,500	3,260
1977	0.45892	0	21,000	2,450	300	2,710	268,700	5,630	50,500
1978	0.49164	250	0	132,730	0	0	10,000	0	2,900
1979	0.53262	49,000	0	2,000,000	130	0	24,800	3,100	2,000
1980	0.58145								

(Continued)

Table 5E.32 (Continued)

	Deflator	SD	TN	TX	UT	VT	VA	WA	WV
1981	0.63578								
1982	0.67533								
1983	0.70214	0	40,100	0	500,000	0	30	16,943	0
1984	0.72824	206,015	50,500	51,500	50,500	51,600	55,055	1,500	229,000
1985	0.75117	55	1,550	38,650	0	0	290	0	1,050
1986	0.76769	6,665	15,150	34,100	479,000	0	800,000	20,351	600,000
1987	0.79083	3	95	546,515	250	10,500	1,510	30,150	125
1988	0.81764	0	5,165	2,226	0	0	0	11	1
1989	0.84883	16	11,482	341,098	15,403	50	39,363	320	1,010
1990	0.88186	3,000	18,059	386,886	56	15,657	3,472	58,770	8,930
1991	0.91397	2,934	13,109	188,766	6,005	19	984	227,634	908
1992	0.93619	3,460	204	199,356	24	2	7,371	176	5,791
1993	0.95872	763,380	5,070	56,990	160	7,550	0	2,080	620
1994	0.97870	20,399	51,039	1,721	0	1,502	16,169	160	5,397
1995	1.00000	12,270	1,264	85,050	1,500	5,150	66,759	250	8,595
1996	1.01937	360	2,740	407,066	312	5,123	153,516	370,060	224,172
1997	1.03925	100,541	23,479	136,472	10,100	170	898	54,675	18,391
1998	1.05199	50	25,427	163,407	4,485	23,805	2,381	3,120	35,506
1999	1.06677	619	554	612,634	1,314	1,036	255,062	2,371	363
2000	1.09113	0	230	25,130	679	1,845	1,368	488	11,003

	Deflator	WI	WY
1955	0.20163	50	200
1956	0.20846	335	11
1957	0.21539	0	526
1958	0.22059	0	3
1959	0.22304	1,791	0
1960	0.22620	996	0
1961	0.22875	1,442	0
1962	0.23180	57	0
1963	0.23445	142	899
1964	0.23792	0	138
1965	0.24241	14,067	390
1966	0.24934	361	0
1967	0.25698	0	1,096
1968	0.26809	0	0
1969	0.28124	4,763	0
1970	0.29623	0	500
1971	0.31111	0	503
1972	0.32436	0	0
1973	0.34251	6,121	304
1974	0.37329	50	48
1975	0.40805	3,041	0
1976	0.43119	0	100
1977	0.45892	0	100
1978	0.49164	71,000	16,320
1979	0.53262	0	0
1980	0.58145		
1981	0.63578		
1982	0.67533		
1983	0.70214	0	0
1984	0.72824	6,000	0
1985	0.75117	2,300	40,000
1986	0.76769	80,000	250
1987	0.79083	2,992	16
1988	0.81764	32	0
1989	0.84883	160	1,602
1990	0.88186	31,159	44
1991	0.91397	180	2,160
1992	0.93619	29,305	0
1993	0.95872	903,660	0

(Continued)

Table 5E.32 (Continued)

	Deflator	WI	WY
1994	0.97870	62,052	0
1995	1.00000	675	0
1996	1.01937	218,025	181
1997	1.03925	93,346	192
1998	1.05199	82,825	22
1999	1.06677	9,305	0
2000	1.09113	74,298	20

[a] Damage estimate available for large region, but not for individual state.

Source: From Pielke, Jr., R.A., M.W. Downton, and J.Z. Barnard Miller, 2002: Flood damage in the United States, 1926–2000: *A Reanalysis of National Weather Service Estimates*. Boulder, CO: UCAR, www.flooddamagedata.org.

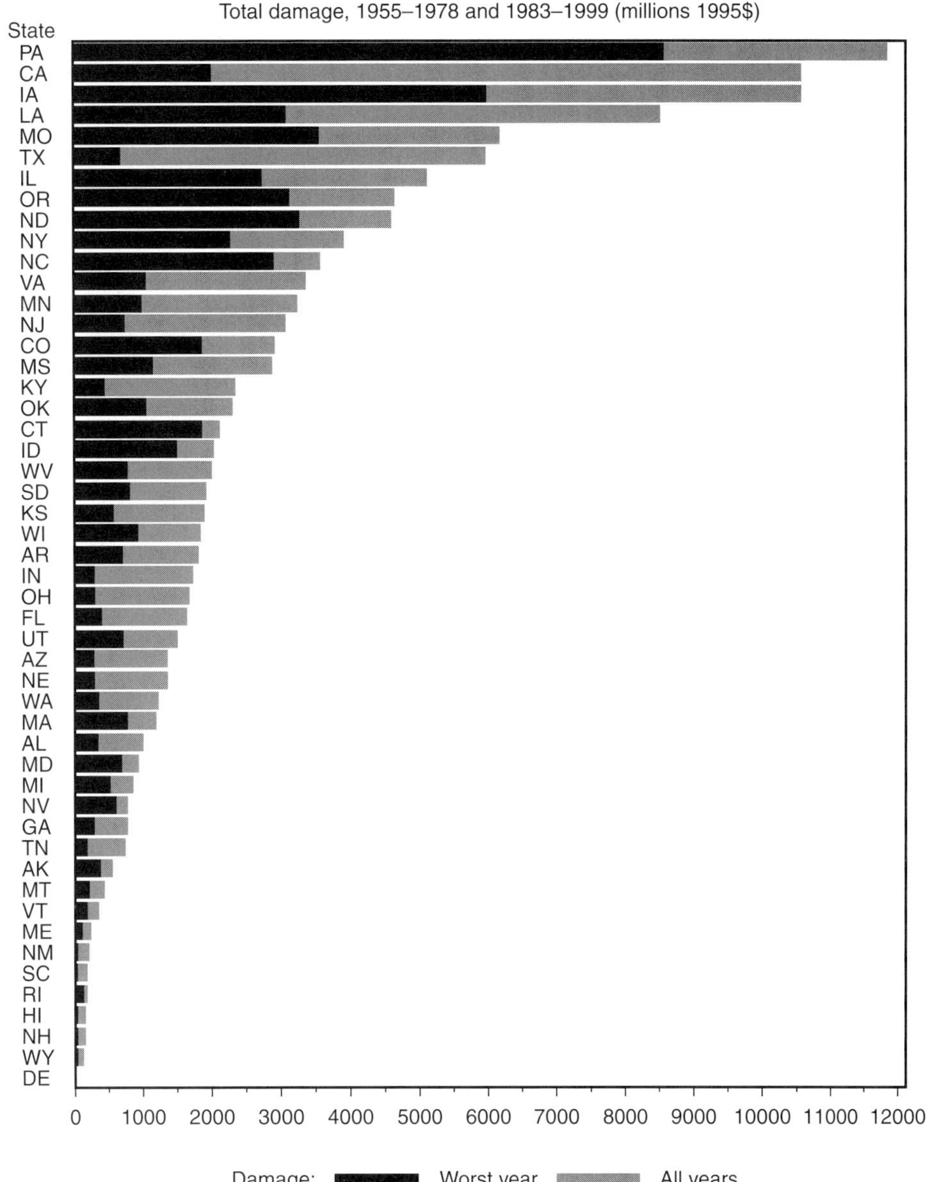

Figure 5E.19 States ranked by estimated total damage during 1955–1978 and 1983–1999. (From Pielke, Jr., R.A., M.W. Downton, and J.Z. Barnard Miller; Flood damage in the United States, 1926–2000: *A Reanalysis of National Weather Service Estimates*. Boulder, CO:UCAR, 2002, www.flooddamagedata.org.)

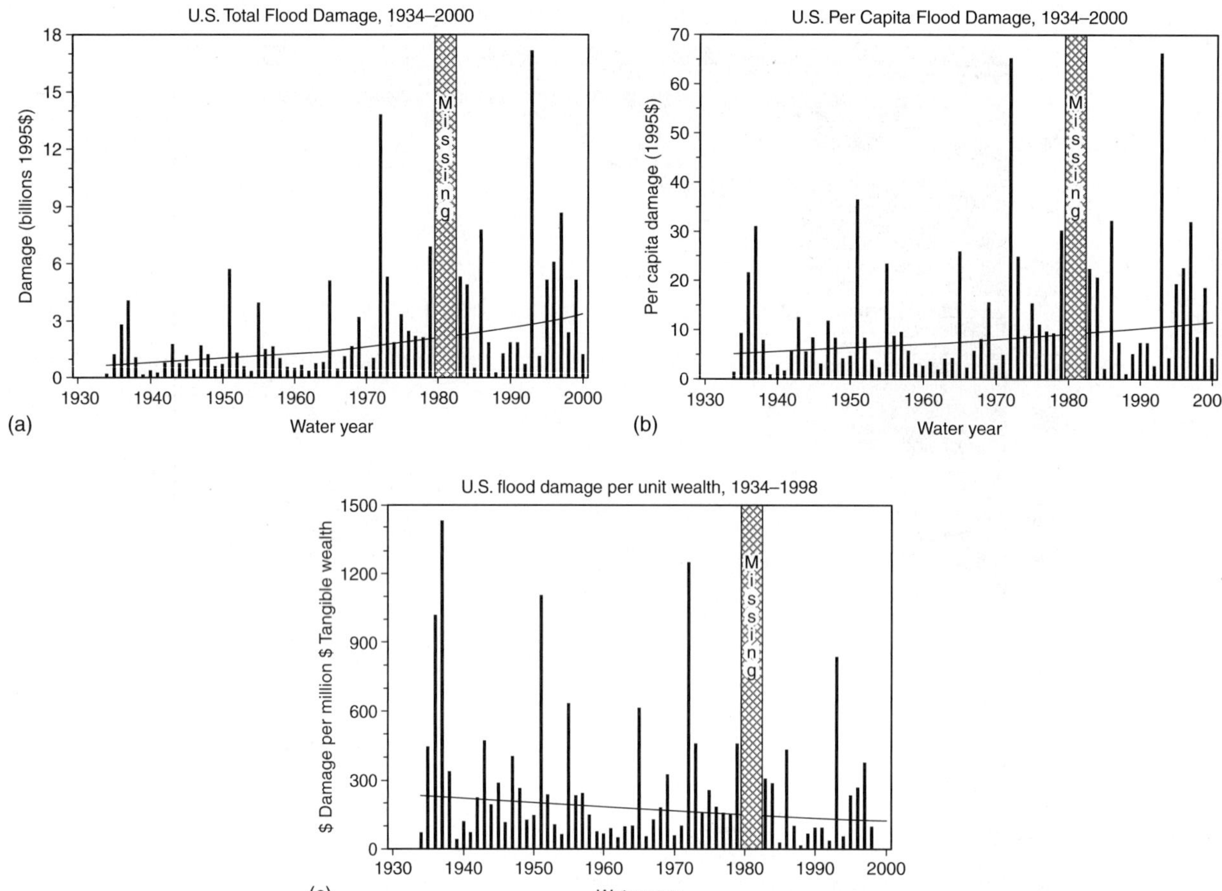

Figure 5E.20 Estimated annual flood damage in the United States, 1934–1999: (a) Total flood damage, (b) Flood damage per capita, (c) Flood damage per million dollars of tangible wealth. (From Pielke, Jr., R.A., M.W. Downton, and J.Z. Barnard Miller; Flood damage in the United States, 1926–2000: *A Reanalysis of National Weather Service Estimates*. Boulder, CO:UCAR, 2002, www.flooddamagedata.org.)

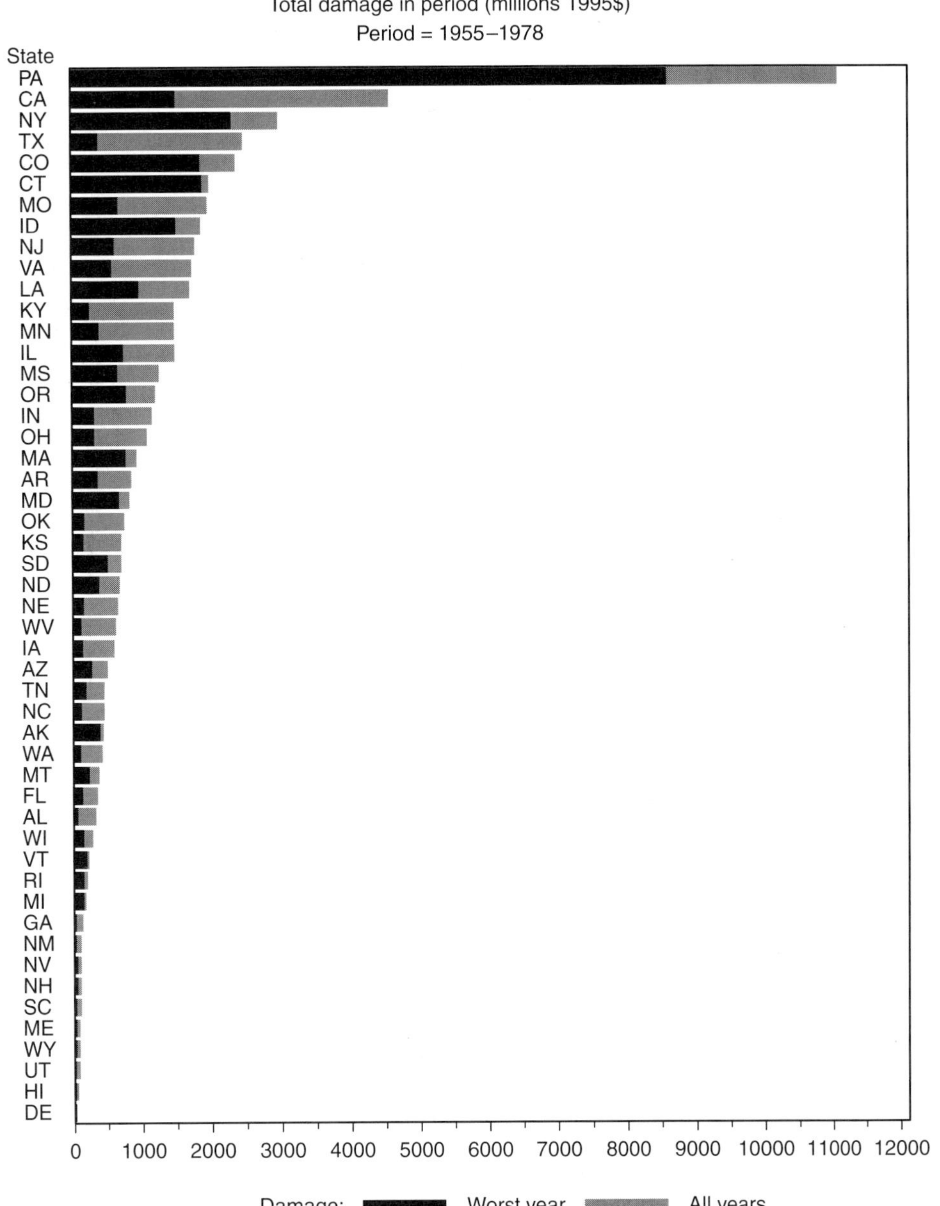

Figure 5E.21A States ranked based on total flood damage during 1955–1978. (From Pielke, Jr., R.A., M.W. Downton, and J.Z. Barnard Miller; Flood damage in the United States, 1926–2000: *A Reanalysis of National Weather Service Estimates*. Boulder, CO:UCAR, 2002, www.flooddamagedata.org.)

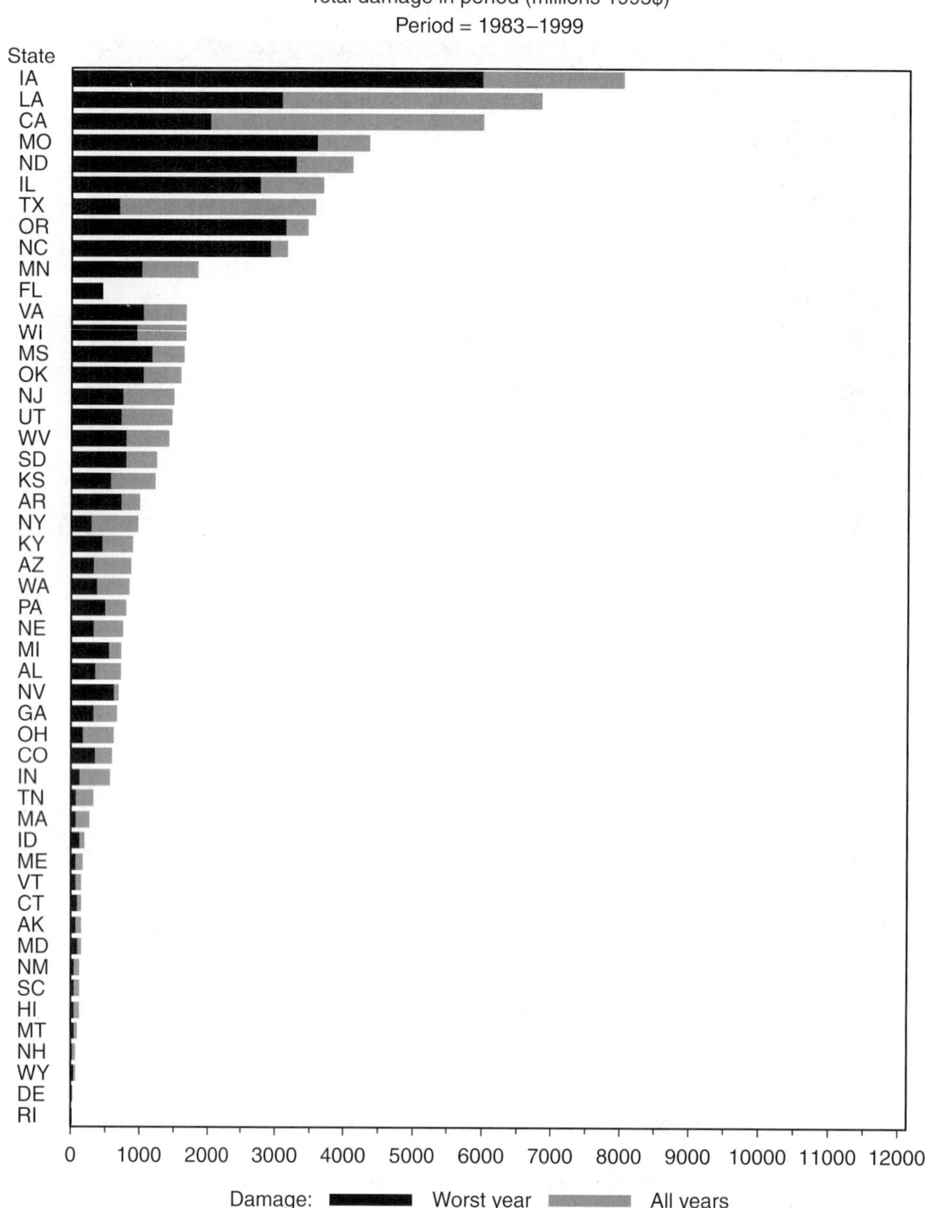

Figure 5E.21B States ranked based on total flood damage during 1983–1999.

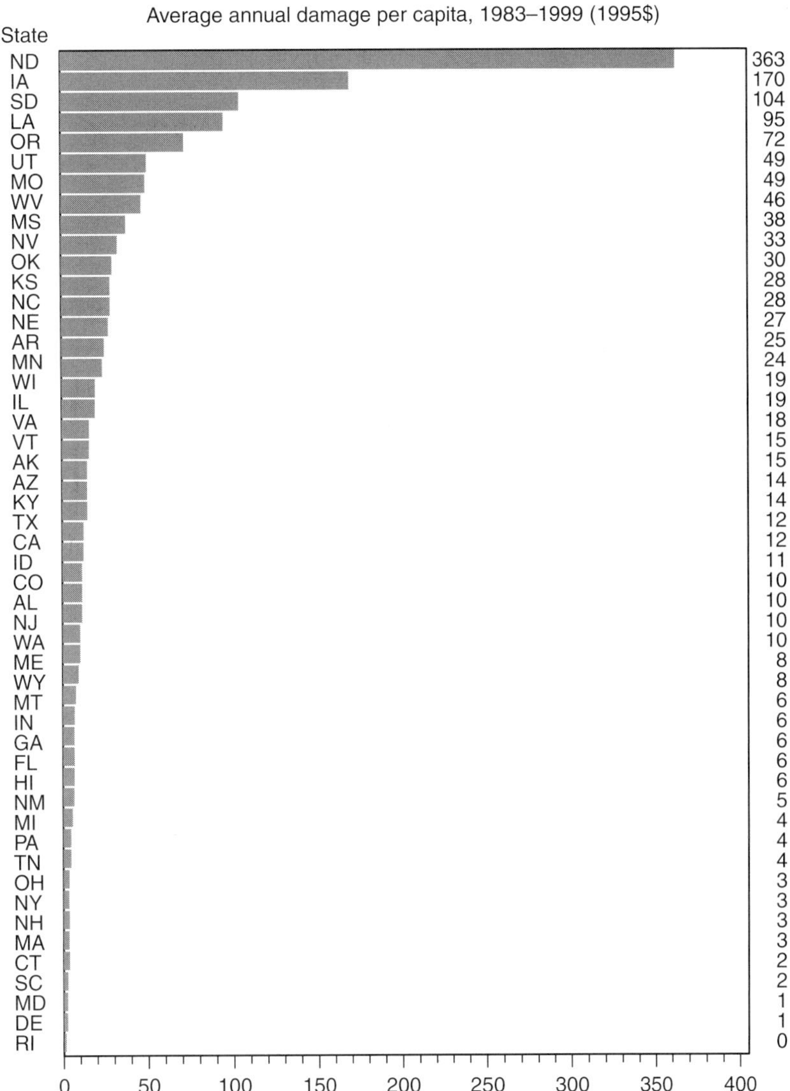

Figure 5E.22 States ranked based on average annual flood damage per capita, 1983–1999. (From Pielke, Jr., R.A., M.W. Downton, and J.Z. Barnard Miller; Flood damage in the United States, 1926–2000: *A Reanalysis of National Weather Service Estimates.* Boulder, CO:UCAR, 2002, www.flooddamagedata.org.)

Table 5E.33 California 1998 El Niño Disaster: Estimated and Actual Public Assistance Costs, in Thousands of Current Dollars

County	Actual Cost (By 6/1/01)	IDE		PDA	
		Estimate	Prop. of Actual	Estimate	Prop. of Actual
State agencies	30,091	7,129	0.24	14,497	0.48
Alameda	18,471	12,971	0.70	8,176	0.44
Amador	258	235	0.91	176	0.68
Butte	1,726	665	0.39	706	0.41
Calaveras	131	—	—	162	1.24
Colusa	4,652	25,000	5.37	1,829	0.39
Contra Costa	5,631	3,885	0.69	4,760	0.85
Del Norte	271	—	—	461	1.70
Fresno	1,701	820	0.48	1,052	0.62
Glenn	3,802	21,250	5.59	9,884	2.60
Humboldt	7,748	1,049	0.14	1,753	0.23
Kern	12,312	—	—	10,306	0.84
Lake	1,889	1,395	0.74	3,044	1.61
Los Angeles	31,229	5,660	0.18	35,516	1.14
Marin	6,449	3,319	0.51	5,447	0.84
Mendocino	2,836	4,259	1.50	3,846	1.36
Merced	2,327	490	0.21	734	0.32
Monterey	26,182	20,181	0.77	11,822	0.45
Napa	468	720	1.54	448	0.96
Orange	12,617	3,992	0.32	16,720	1.33
Riverside	3,130	—	—	5,964	1.91
Sacramento	2,366	—	—	3,066	1.30
San Benito	6,455	26,870	4.16	10,595	1.64
San Bernardino	7,525	—	—	30,429	4.04
San Diego	6,977	—	—	9,180	1.32
San Francisco	3,859	12,300	3.19	3,703	0.96
San Joaquin	2,657	655	0.25	3,155	1.19
San Luis Obispo	4,006	772	0.19	4,915	1.23
San Mateo	21,951	16,110	0.73	26,328	1.20
Santa Barbara	15,816	75	0.00	12,954	0.82
Santa Clara	13,638	9,846	0.72	13,310	0.98
Santa Cruz	12,459	13,673	1.10	6,320	0.51
Solano	3,346	3,628	1.08	8,564	2.56
Sonoma	11,779	11,180	0.95	4,127	0.35
Stanislaus	2,122	—	—	909	0.43
Sutter	1,039	1,582	1.52	758	0.73
Tehama	881	20,000	22.70	616	0.70
Trinity	1,091	1,970	1.81	975	0.89
Tulare	2,149	—	—	919	0.43
Ventura	20,391	3,302	0.16	14,350	0.70
Yolo	909	4,321	4.75	4,484	4.93
Yuba	592	196	0.33	249	0.42
Total	315,929	239,500	0.86[a]	297,204	0.94

[a] Proportion of actual cost ($279 million) of cases with an IDE.

Source: From Pielke, Jr., R.A., M.W. Downton, and J.Z. Barnard Miller, 2002: Flood Damage in the United States, 1926–2000: *A Reanalysis of National Weather Service Estimates*. Boulder, CO: UCAR, www.flooddamagedata.org.

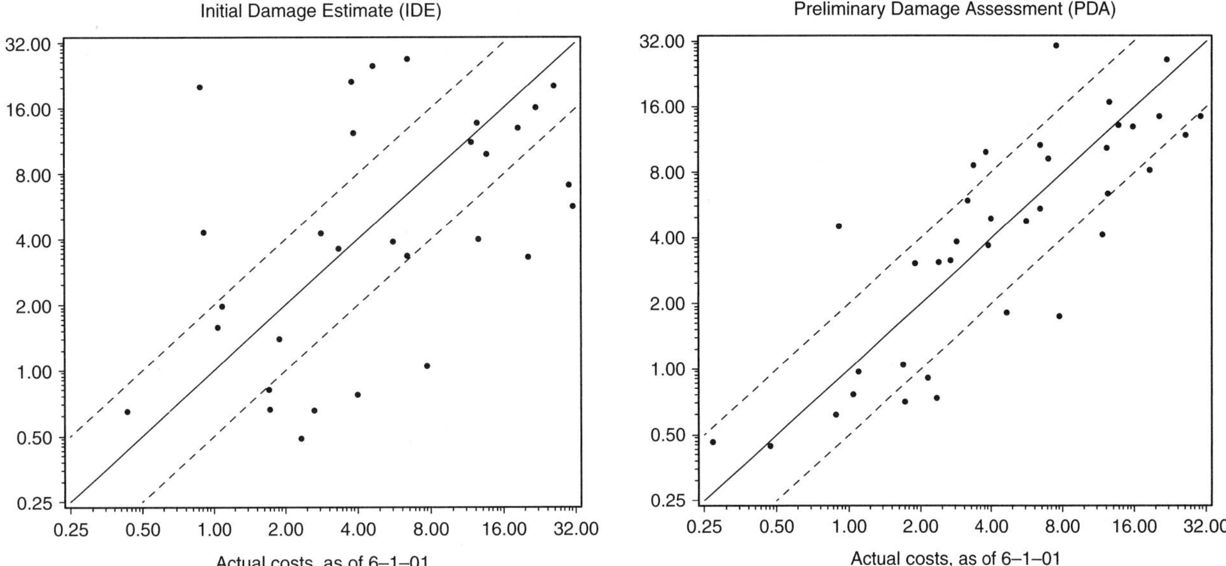

Figure 5E.23 Flood damage estimates in California 1998 El Nino disaster (millions of dollars) estimated flood damage in California counties in the 1998 El Nino disaster, compared with actual cost as of June 1, 2001: (A) Initial damage estimate, (B) Preliminary damage assessment. (From Pielke, Jr., R.A., M.W. Downton, and J.Z. Barnard Miller; Flood damage in the United States, 1926–2000: *A Reanalysis of National Weather Service Estimates*. Boulder, CO:UCAR, 2002, www.flooddamagedata.org.)

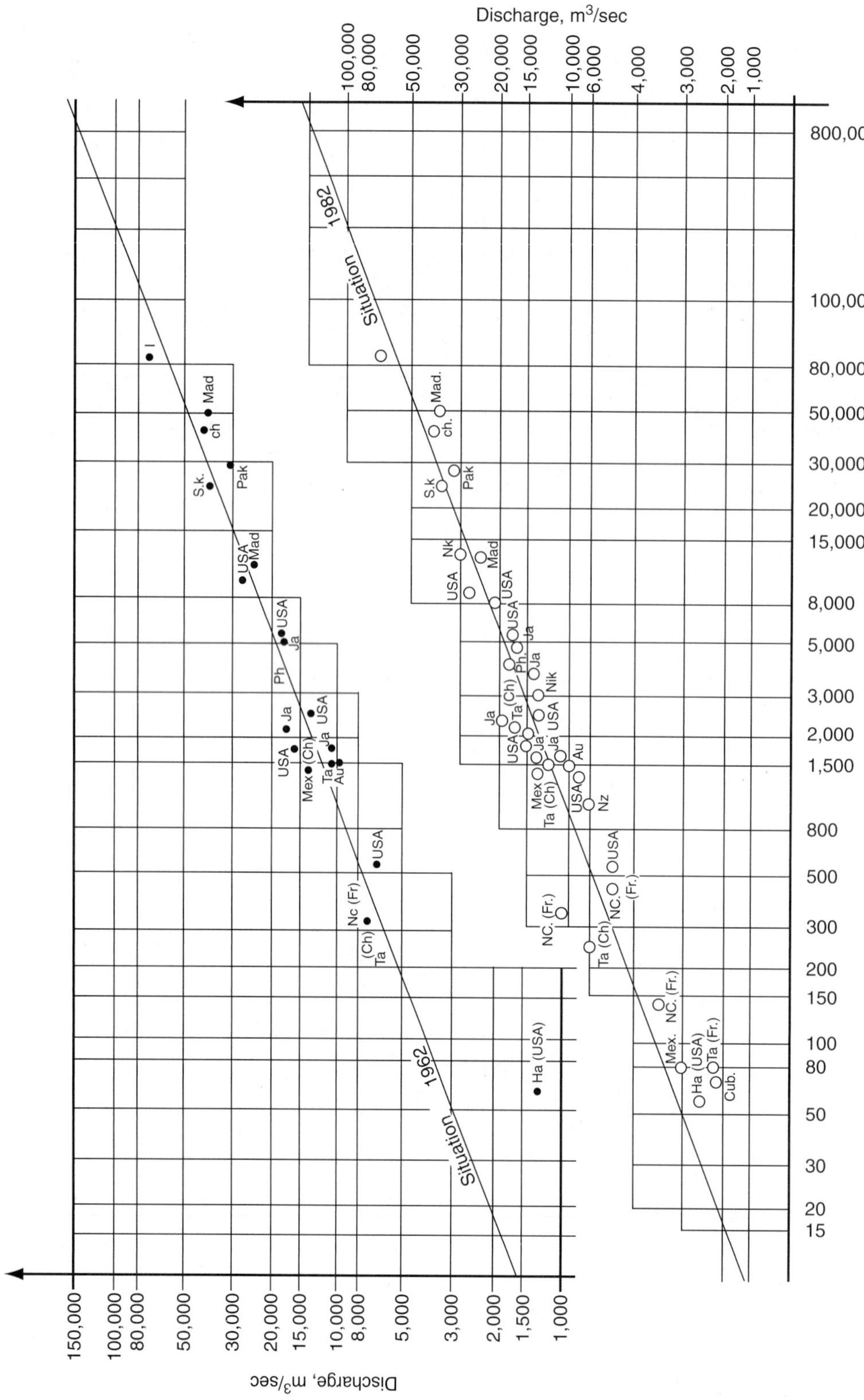

Figure 5E.24 Relationship between maximum flood flows in the world and size of drainage basin. (From Rodier, J.A., and Roche, M. World Catalogue of Maximum Observed Floods, International Assoc. Hydrological Sciences Publ. No. 143, 1984.)

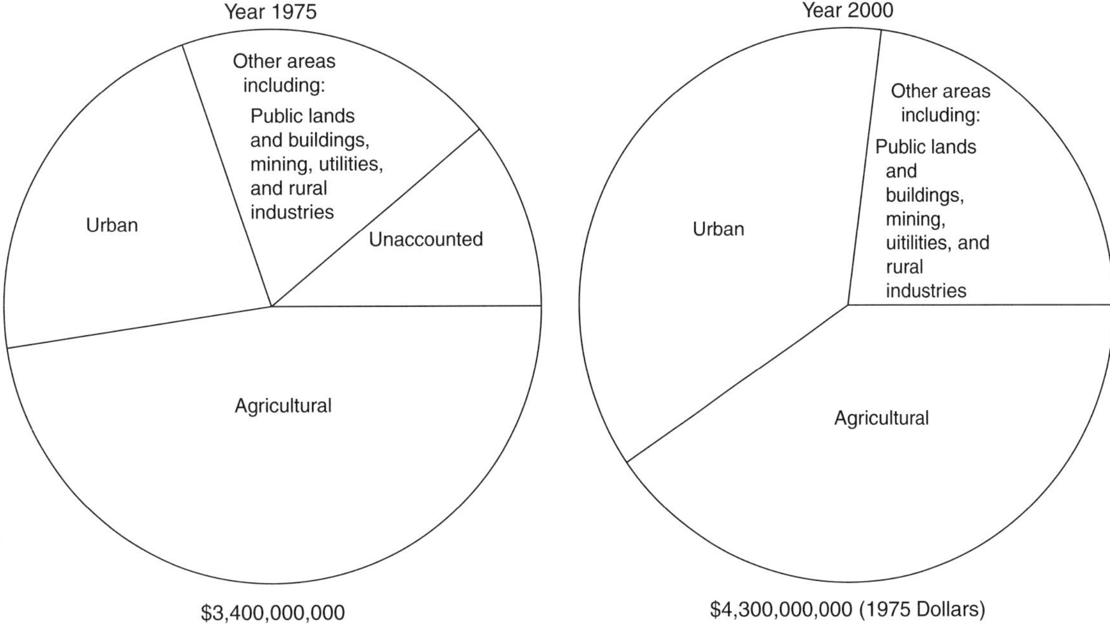

Figure 5E.25 Trends in distribution of annual flood losses in the United States, 1975–2000. (From Hays, W.W., Facing Geologic and Hydrologic Hazards, Earth-Science Considerations, U.S. Geological Survey Professional Paper 1240-B, 1981.)

Table 5E.34 Great Floods in the United States Since 1889

Number[a]	Type of Flood	Date	Location	Lives Lost	Estimated Damages (Millions of Dollars)
1	b	May 1889	Johnstown, Pennsylvania, Dam failure	3,000	—
2	c	September 8, 1900	Hurricane–Galveston, Texas	6,000	30
3	d	May–June 1903	Kansas, Lower Missouri, and Upper Mississippi River	100	40
4	d	March 1913	Ohio River and Tributaries	467	147
5	c	September 14, 1919	Hurricane–South of Corpus Christi, Texas	600–900	22
6	b,e	June 1921	Arkansas River, Colorado	120	25
7	d	September 1921	Texas Rivers	215	19
8	d	Spring 1927	Mississippi River Valley	313	284
9	d	November 1927	New England Rivers	88	46
10	b	March 12–13, 1928	St. Francis Dam failure, Southern California	450	14
11	f	September 13, 1928	Lake Okeechobee, Florida	1,836	26
12	d	May–June 1935	Republican and Kansas Rivers	110	18
13	d	March–April 1936	Rivers in Eastern United States	107	270
14	d	January–February 1937	Ohio and Lower Mississippi River Basins	137	418
15	d	March 1938	Streams in Southern California	79	25
16	d	September 21, 1938	New England	600	306
17	e	July 1939	Licking and Kentucky Rivers	78	2
18	d	May–July 1947	Lower Missouri and Middle Mississippi River Basins	29	235
19	d	June–July 1951	Kansas and Missouri	28	923
20	d	August 1955	Hurricane Diane floods–Northeastern United States	187	714
21	d	December 1955	West coast rivers	61	155
22	d	June 27–30, 1957	Hurricane Audrey–Texas and Louisiana	390	150
23	d	December 1964	California and Oregon	40	416
24	d	June 1965	South Platte River Basin, Colorado	16	415
25	c	September 10, 1965	Hurricane Betsy–Florida and Louisiana	75	1,420
26	d	January–February 1969	Floods in California	60	399

(Continued)

Table 5E.34 (Continued)

Number[a]	Type of Flood	Date	Location	Lives Lost	Estimated Damages (Millions of Dollars)
27	c,d	August 17–18, 1969	Hurricane Camille–Mississippi, Louisiana, and Alabama	256	1,421
28	c	July 30–August 5, 1970	Hurricane Celia–Texas	11	453
29	b	February 1972	Buffalo Creek, West Virginia	125	10
30	e	June 1972	Black Hills, South Dakota	237	165
31	c,d	June 1972	Hurricane Agnes floods–Eastern United States	105	4,020
32	d	Spring 1973	Mississippi River Basin	33	1,155
33	d	June–July 1975	Red River of the North Basin	<10	273
34	c,d	September 1975	Hurricane Eloise floods-Puerto Rico and Northeastern United States	50	470
35	b	June 1976	Teton Dam failure, Southeast Idaho	11	1,000
36	e	July 1976	Big Thompson River, Colorado	139	30
37	e	April 1977	Southern Appalachian Mountains area	22	424
38	b,e	July 1977	Johnstown–Western Pennsylvania	78	330
39	d	April 1979	Mississippi and Alabama	<10	500
40	c	September 12–13, 1979	Hurricane Frederic floods–Mississippi, Alabama, and Florida	13	2,000

[a] Number corresponds to those shown on Figure 3.12.
[b] Dam break flood.
[c] Tidal flood.
[d] Riverine flood.
[e] Flash flood.
[f] Flood wave generated in Lake Okeechobee by hurricane.

Source: From Hays, W.W., 1981, Facing Geologic and Hydrologic Hazards, Earth–Science Considerations, U.S. Geological Survey Professional Paper 1240-B.

Table 5E.35 Sources of Flood Damage Estimates

Source	Timespan	Spatial Scale	Scope
National weather service flood damage data sets	1925–present	Nation State Basin	Estimates of direct physical damage from significant flooding events that result from rainfall or snowmelt
Insurance records (National flood insurance program, private insurers)	1969–present	Nation Community	Personal property claims made by individuals holding flood insurance
Disaster assistance records (Federal emergency management agency)	1992–present	Nation State	Federal and state outlays for public assistance, individual assistance, and temporary housing in presidentially declared disasters
State and local government records	Varies	State	Varies
Newspaper archives	Varies	Community	Varies

Table 5E.36 Major Flood Disasters of the World

Date	Location	Deaths	Date	Location	Deaths
1228	Holland	100,000	1969 Mar 17	Mundau Valley, Alagoas, Brazil	218
1642	China	300,000	1969 Aug 20–22	Western Virginia	189
1887	Huang He River, China	900,000	1969 Sep 15	South Korea	250
1889 May 31	Johnstown, PA	2,200	1969 Oct 1–8	Tunisia	500
1900 Sep 8	Galveston, TX	5,000	1970 May 20	Central Romania	160
1903 June 15	Heppner, OR	325	1970 July 22	Himalayas, India	500
1911	Chang Jiang River, China	100,000	1971 Feb 26	Rio de Janeiro, Brazil	130
1913 Mar 25–27	Ohio, IN	732	1972 Feb 26	Buffalo Creek, WV	118
1915 Aug 17	Galveston, TX	275	1972 June 9	Rapid City, SD	236
1928 Mar 13	Collapse of St. Francis Dam, Santa Paula, CA	450	1972 Aug 7	Luzon Is., Philippines	454
1928 Sep 13	Lake Okeechobee, FL	2,000	1973 Aug 19–31	Pakistan	1,500
1931 Aug	Huang He River, China	3,700,000	1974 Mar 29	Tubaro, Brazil	1,000
1937 Jan 22	Ohio, Miss. Valleys	250	1974 Aug 12	Monty-Long, Bangladesh	2,500
1939	Northern China	200,000	1976 June 5	Teton Dam collapse, ID	11
1946 Apr 1	Hawaii, Alaska	159	1976 July 31	Big Thompson Canyon, CO	139
1947	Honshu Island, Japan	1,900	1976 Nov 17	East Java, Indonesia	136
1951 Aug	Manchuria	1,800	1977 July 19–20	Johnstown, PA	68
1953 Jan 31	Western Europe	2,000	1978 June–Sep	Northern India	1,200
1954 Aug 17	Farahzad, Iran	2,000	1979 Jan–Feb	Brazil	204
1955 Oct 7–12	India, Pakistan	1,700	1979 July 17	Lomblem Is., Indonesia	539
1959 Nov 1	Western Mexico	2,000	1979 Aug 11	Morvi, India	5,000–15,000
1959 Dec 2	Frejus, France	412	1980 Feb 13–22	So. CA., AR	26
1960 Oct 10	Bangladesh	6,000	1981 Apr	Northern China	550
1960 Oct 31	Bangladesh	4,000	1981 July	Sichuan, Hubei Prov., China	1,300
1962 Feb 17	German North Sea coast	343	1982 Jan 23	Nr. Lima, Perui	600
1962 Sep 27	Barcelona, Spain	445	1982 May 12	Guangdong, China	430
1963 Oct 9	Dam collapse, Vaiont, Italy	1,800	1982June 6	So. Conn	12
1966 Nov 3–4	Florence, Venice, Italy	113	1982 Sep 17–21	El Salvador, Guatemala	1,300+
1967 Jan 18–24	Eastern Brazil	894	1982 Dec 2–9	IL., MO., AR	22
1967 Mar 19	Rio de Janeiro, Brazil	436	1983 Feb–Mar	CA coast	13
1967 Nov 26	Lisbon, Portugal	464	1983 Apr 6–12	AL., LA, MS., TN	15
1968 Aug 7–14	Gujarat State, India	1,000	1984 May 27	Tulsa, OK	13
1968 Oct 7	Northeastern India	780	1984 Aug–Sep	S. Korea	200+
1969 Jan 18–26	So. CA	100	1985 July 19	Northern Italy, dam burst	361

Source: From The World Almanac and Book of Facts 1988. *Copyright Pharos Books*, A Scripps Howard Co., New York. Reproduced with permission.

Table 5E.37 Maximum Flood Flows in the World

Country	Station	Basin Area (km^2)	Maximum Discharge (m^3/sec)	K[a]Value	Year
U.S.A. (California)	San Rafael San Rafael	3.2	250	5.194	1973
U.S.A. (California)	L. San Gorgonio Beaumont	4.5	311	5.226	1969
U.S.A. (Hawaii)	Halawa	12	762	5.494	1965
U.S.A. (Hawaii)	Waäilua Lihue	58	2,470	5.819	1963
Cuba	Buey San Miguel	73	2,060	5.623	1963
Tahiti	Papenoo	78	2,200	5.650	1983
Mexico	San Bartolo	81	3,000	5.859	1976
New Caledonia	Ouinne Embouchure	143	4,000	5.845	1975
Taiwan	Cho Shui	259	7,780	6.225	1979
New Caledonia	Ouaälème derniers rapides	330	10,400	6.389	1981

(Continued)

Table 5E.37 (Continued)

Country	Station	Basin Area (km²)	Maximum Discharge (m³/sec)	K[a]Value	Year
New Caledonia	Yaté	435	5,700	5.810	1981
U.S.A. (New York)	Little Nemaha Syracuse	549	6,370	5.826	1950
New Zealand	Haast Roaring Billy	1,020	7,690	5.765	1979
U.S.A. (California)	M.F. American	1,360	8,780	5.770	1964
Mexico	Cithuatian Paso del Mojo	1,370	13,500	6.156	1959
Australia	Pioneer Pleystowe	1,490	9,840	5.840	1918
Taiwan	Hualien Hualien Bridge	1,500	11,900	6.011	1973
Japan	Nyodo Ino	1,560	13,510	6.111	1963
Japan	Kiso Imujama	1,680	11,150	5.910	1961
U.S.A. (Texas)	W. Nueces Bracketville	1,800	15,600	6.156	1959
India	Macchu	1,900	14,000	6.060	1979
Taiwan	Tam Shui Taipei Bridge	2,110	16,700	6.199	1963
Japan	Shingu Oga	2,350	19,025	6.290	1959
U.S.A. (Texas)	Pedernales Johnson City	2,450	12,500	5.873	1952
North Korea	Daeryong Gang	3,020	13,500	5.830	1975
Japan	Yoshino Iwazu	3,750	14,470	5.844	1974
Philippines	Cagayan Echague Isabella	4,244	17,550	5.980	1959
Japan	Tone Yattajima	5,110	16,900	5.871	1947
U.S.A. (Texas)	Nueces Uvalde	5,504	17,400	5.870	1935
U.S.A. (California)	Eel Scotia	8,060	21,300	5.917	1964
U.S.A. (Texas)	Pecos Comstock	(9,100)	26,800	6.110	1954
Madagascar	Betsiboka Ambodiroka	11,800	22,000	5.780	1927
North Korea	Toedong Gang Mirim	12,175	29,000	6.060	1967
South Korea	Han Koan	23,880	37,000	6.047	1925
Pakistan	Jhelum Mangla	29,000	31,100	5.739	1929
China	Hanjiang Hankang	41,400	40,000	5.868	1583
Madagascar	Mangoky Banyan	50,000	38,000	5.698	1933
India	Narmada Garudeshwar	88,000	69,400	6.210	1970
China	Chang Jiang Yitchang	1,010,000	110,000	5.197	1870
U.S.S.R.	Lena Kusur	2,430,000	189,000	5.520	1967
Brazil	Amazonas Obidos	4,640,000	370,000	6.760	1953

Note: Arranged by size of drainage basin.

[a] Flood coefficient $K = 10\,[(1 - (\log(Q) - 6)/(\log(A) - 8)]$ where Q is the largest flood in m³/sec; A is the basin area in km².

Source: From Rodier, J.A., and Roche, M., 1984, World Catalogue of Maximum Observed Floods, International Assoc. Hydrological Sciences Publ. No. 143. Reproduced with permission.

Table 5E.38 Tornadoes, Floods, and Tropical Cyclones in the United States, 1931–1984

Item	1931–1935	1936–1945	1946–1955	1956–1965	1966–1975	1976	1977	1978	1979	1980	1981	1982	1983	1984
Tornadoes, number[a]	830	1,514	2,969	6,572	8,030	835	852	788	852	866	783	1,046	931	907
Lives lost, total	909	1,896	1,751	924	1,172	44	43	53	84	28	24	64	34	122
Most in a single tornado	37	216	169	44	58	5	22	16	42	5	5	10	3	16
Property loss of $500,000 and over	15	56	130	191	428	46	46	59	73	92	55	92	95	125
Floods: Lives lost	368	953	808	557	1,528	187	212	125	103	97	90	155	200	126
Property loss (mil. dot.)	187	1,484	3,350	2,721	10,225	1,000	1,393	1,000	4,000	1,500	1,000	3,500	4,100	4,000
North Atlantic tropical cyclones and hurricanes:[b]														
Number reaching U.S. coast	21	41	40	33	25	2	1	2	5	2	2	1	2	2
Hurricanes only	12	19	21	14	13	1	1	—	3	1	—	—	1	1
Lives lost in U.S.	494	768	495	692	504	9	—	35	11	2	—	—	22	4

Note: — Represents zero.

[a] A violent, rotating column of air descending from a cumulonimbus cloud in the form of a tubular- or funnel-shaped cloud, usually characterized by movements along a narrow path and wind speeds from 100 to over 300 miles per hour. Also known as a "twister" or "waterspout."

[b] Tropical cyclones have maximum winds of 39–73 miles per hour; hurricanes have maximum winds of 74 miles per hour or higher.

Source: From Bureau of the Census, Statistical Abstract of the United States 1987 and data from the U.S. National Oceanic and Atmospheric Administration.

Table 5E.39 Tornadoes: Floods, Tropical Storms, and Lightning: 1993–2003

Weather type	1993	1994	1995	1996	1997	1998	1999	2000	2001	2002	2003, prel.
Tornadoes[a]											
Lives lost	33	69	30	26	67	130	94	41	40	55	54
Injuries	(NA)	(NA)	650	705	1,033	1,868	1,842	882	743	968	1,087
Property loss (mil. dol.)	(NA)	(NA)	410.8	719.6	730.7	1,714.2	1,989.9	423.6	630.1	801.3	1,263.2
Floods and flash floods:											
Lives lost	103	91	80	131	118	136	68	38	48	49	85
Injuries	(NA)	(NA)	57	95	525	6,440	301	47	277	88	65
Property loss (mil. dol.)	(NA)	(NA)	1,250.5	2,120.7	6,910.6	2,324.8	1,420.7	1,255.1	1,220.3	655.0	2,540.9
North Atlantic tropical storms and hurricanes[b]	8	7	19	13	7	14	12	15	15	12	(NA)
Number of hurricanes reaching U.S. mainland	1	—	2	2	1	3	3	—	—	1	(NA)
Direct deaths on U.S. mainland	2	9	17	37	1	9	19	—	24	51	14
Property loss in U.S. (mil. dol.)	57.0	973.0	5,932.3	1,436.1	667.6	3,546.6	4,190.1	8.1	5,187.8	1,104.4	1,879.5
Lightning:											
Deaths	43	69	85	52	42	44	46	51	44	51	44
Injuries	295	577	433	309	306	283	243	364	371	256	237

Note: — Represents zero. NA Not available.

[a] U.S. National Weather Service, Internet site <www.spc.noaa.gov/climo/torn/monthlytornstats.html> (accessed 14 April 2004). A violent, rotating column of air descending from a cumulonimbus cloud in the form of a tubular- or funnel-shaped cloud, usually characterized by movements along a narrow path and wind speeds from 100 to over 300 miles per hour. Also known as a "twister" or "waterspout."

[b] National Hurricane Center (NHC), Coral Gables, FL, unpublished data. For data on individual hurricanes, see the NHC web site at www.nhc.noaa.gov/. Tropical storms have winds of 39–73 miles per hour, hurricanes have winds of 74 miles per hour or higher.

Source: From, except as noted, U.S. National Oceanic and Atmospheric Administration (NOAA), *Storm Data*, monthly. See also NOAA website at www.nws.noaa.gov/om/hazstats.shtml and www.nws.noaa.gov/om/severeweather/sum03.pdf (released 03 March 2004).

Table 5E.40 Deaths, Injuries, and Damage Caused by Floods in the United States, 1965–1985

Fiscal Year	No. of Events	Persons Killed	Persons Injured	Dwellings Destroyed	Dwellings Damaged	Dwellings Destroyed & Damaged
1965–66	67	22	102	91	9,131	9,222
1966–67	NA	16	161	108	22,353	22,461
1967–68	NA	38	824	84	14,224	14,308
1968–69	NA	24	284	71	17,674	17,745
1969–70	NA	51	783	83	33,769	33,852
1970–71	49	22	58	105	6,993	7,098
1971–72	77	519	16,587	7,346	133,805	141,151
1972–73	78	105	1,559	3,229	81,467	84,696
1973–74	83	71	366	1,417	31,309	32,726
1974–75	90	48	500	803	25,008	25,811
1975–76	70	55	2,071	1,377	26,179	27,556
1976–77	68	165	1,469	3,581	35,942	39,523
1977–78	106	196	3,712	1,489	48,508	49,997
1978–79	148	143	3,842	2,659	56,646	59,305
1979–80	122	79	1,121	887	37,439	38,326
1980–81	115	NA	NA	NA	NA	19,578
1981–82	133	70	2,561	NA	NA	46,256
1982–83	149	69	1,988	NA	NA	48,874
1983–84	121	65	1,478	NA	NA	41,578
1984–85	48	9	29	NA	NA	2,308
Totals	—	—	—	—	—	762,371

Note: Based on American National Red Cross data which are by fiscal year (July 1–June 30).

Source: From Rubin, C.B., Yezer, A. M., Hussain, Q, and Webb, A., 1986, Summary of Major Natural Disaster Incidents in the U.S. 1965–85, Natural Hazards Research and Applications Information Center, George Washington University Spec. Publ. 17.

Table 5E.41 Deaths Caused by Floods in the United States in 1987

State	Boat	Open	Other	Outside	Perm. Home	Playing	Auto	All
AL	0	0	0	0	0	0	2	2
AR	0	0	0	0	0	0	4	4
GA	0	0	0	0	0	0	4	4
HI	0	1	0	2	0	0	0	3
IL	0	0	0	0	0	0	1	1
IN	0	0	0	1	0	0	1	2
KY	0	0	0	1	0	1	0	2
MA	0	0	0	1	0	0	0	1
MI	0	0	0	0	0	0	1	1
MN	0	0	0	0	1	0	1	2
NY	0	0	0	1	0	0	10	11
OK	0	0	0	1	0	0	2	3
OR	1	0	0	0	0	0	0	1
PA	0	1	0	0	0	0	0	1
PR	0	0	0	0	0	0	7	7
SC	0	0	0	0	0	1	0	1
TN	0	0	0	0	0	0	3	3
TX	0	1	1	1	0	0	14	17
VA	0	0	0	2	0	0	1	3
VT	0	0	0	0	0	0	1	1
Total	1	3	1	10	1	2	52	70
Percent	1%	4%	1%	14%	1%	3%	74%	99%[a]

Note: By location; flash floods and floods.

[a] Rounding to the nearest percent causes the column to sum to less than 100 percent.

Source: From Peters, B.E., 1988, *National Weather Service*, Fort Worth, TX.

Table 5E.42 Flood Fatalities

Flood Fatalities													
Year	Annual Total	Jan	Feb	Mar	Apr	May	Jun	Jul	Aug	Sep	Oct	Nov	Dec
2004	79/51	3/3	2/0	3/3	9/9	13/13	2/2	5/4	9/8	25/5	2/1	3/2	311
2003	99/47	0/0	7/5	0/0	8/2	14/9	14/12	9/3	14/9	1/1	3/3	12/3	17/0
2002	50/31	7/4	0/0	4/2	2/1	15/8	0/0	10/6	2/1	1/1	3/3	1/0	5/5
2001	66/31	0/0	5/4	3/1	1/1	6/2	21/8	12/5	7/2	0/0	0/0	9/7	2/1
2000	41/20	0/0	5/2	2/2	0/0	0/0	6/5	3/0	6/3	4/2	9/1	4/4	2/1
1999	77/40	12/11	0/0	0/0	7/2	2/1	7/4	6/2	2/0	38/17	3/3	0/0	0/0
1998	136/86	18/14	10/6	16/13	6/4	3/0	16/9	8/4	16/4	3/0	38/31	2/1	0/0
1997	98										3/0	4/0	2/2

Flood Fatalities									
Year	Year Annual	Year	Annual Total	Year	Annual Total	Year	Annual Total	Year	Annual Total
1996	131	1979	121	1959	25	1939	83	1919	2
1995	103	1978	143	1958	47	1938	180	1918	0
1994	70	1977	210	1957	82	1937	142	1917	80
1993	109	1976	193	1956	42	1936	142	1916	118
1992	87	1975	107	1955	302	1935	236	1915	49
1991	58	1974	121	1954	55	1934	88	1914	180
1990	125	1973	148	1953	40	1933	33	1913	527
1989	74	1972	554	1952	54	1932	11	1912	2
1988	37	1971	74	1951	51	1931	0	1911	0
1987	88	1970	135	1950	93	1930	14	1910	0
1986	208	1969	297	1949	48	1929	89	1909	5
1985	70	1968	31	1948	82	1928	15	1908	11
1984	125	1967	34	1947	55	1927	423	1907	7
1983	200	1966	31	1946	28	1926	16	1906	1
1982	155	1965	119	1945	91	1925	36	1905	2
1981	84	1964	100	1944	33	1924	27	1904	0
1980	82	1963	39	1943	107	1923	42	1903	178
		1962	19	1942	68	1922	215		
		1961	52	1941	47	1921	143		
		1960	32	1940	60	1920	42		

Table 5E.43 Major Natural Disasters in the United States, 1965–1985

Type of Disaster	Number	Federal Outlay (thousands of current dollars)	Federal Outlay (thousands of 1982 dollars)
Ice and snow events	19	151,427	205,511
Hurricanes/tropical storms	39	1,173,141	1,947,939
Earthquakes	7	203,881	405,706
Dam and levee failures	7	55,764	80,806
Rains, storms & flooding[a]	337	1,684,702	2,439,852
High winds & waves	2	125,313	120,536
Coastal storms & flooding	7	158,261	205,357
Tornadoes	109	441,685	648,352
Drought/water shortage	4	1,134	5,344
Totals	531	3,995,308	6,059,403

Note: Federally-declared disasters, by type.

[a] Includes land, mud, and debris flows and slides.

Source: From Rubin, C.B., Yezer, A.M., Hussain, Q, and Webb, A., 1986, Summary of Major Natural Disaster Incidents in the U.S. 1965–85, Natural Hazards Research and Applications Information Center, George Washington University Spec. Publ. 17.

Table 5E.44 No. 375. Major U.S. Weather Disasters: 1990–2003

Event	Description	Estimated cost		
		Time period	(bill. dot.)	Deaths
Southern California wildfires	Dry weather, high winds, and resulting wildfires in southern CA burned 743,000 acres & destroyed 3,700 homes	Oct–early Nov 2003	2.5	22
Hurricane Isabel	Category 2 hurricane makes landfall in eastern NC, causing damage along coasts of NC, VA, and MD with wind damage and flooding in NC, VA, MD, DE, WV, NJ, NY and PA	September 2003	over 4	47
Midwest severe storms and tornadoes	Numerous tornadoes over the Midwest, MS River valley, and OH/TN River valleys with record 400 tornadoes in one week	Early May 2003	over 3.1	41
Storms and hail	Sever storms and large hail over southern plains, lower MS River valley, and TX	Early April 2003	over 1.6	—
Widespread drought	Moderate to extreme drought over large portions of 30 states	Spring to fall 2002	over 10	—
Western fire season	Major fires over 11 western states from Rockies to west coast	Spring to fall 2002	over 2	21
Tropical Storm Allison	Tropical storm produced rainfall and severe flooding in coastal portions of TX & LA & damage in MS, FL, VA. and PA	June 2001	5.0	43
Midwest and Ohio Valley hail and tornadoes	Storms, tornadoes, and hail in TX, OK, KS, NE, IA, MO, IL, IN, WI, MI, OH, KY, and PA	April 2001	1.7	3
Southern drought/heat wave	Severe drought and heat over south-central and southeast states cause significant losses in agriculture and related industries	Springer-summer 2000	over 4.0	140
Western fire season	Severe fire season in western states	Spring-summer 2000	over 2.0	—
Hurricane Floyd	Category 2 hurricane in NC, causing severe flooding in NC and some flooding in SC, VA, MD, PA, NY, NJ, DE, RI, CT, MA, and VT	September 1999	6.0	75
Drought/heat wave	Drought/heatwave over eastern U.S.	Summer 1999	1.0	256
Oklahoma-Kansas tornadoes	Category F4–F5 tornadoes hit OK, KS, TX, and TN	May 1999	1.0	55
Arkansas-Tennessee tornadoes	Two outbreaks of tornadoes in 6-day period	January 1999	1.3	31
Texas flooding	Severe flooding in southeast Texas from 2 heavy rain events with 10–20 in. totals	Oct–Nov 1998	1.0	31
Hurricane Georges	Category 2 hurricane in Puerto Rico, Florida Keys, and Gulf coasts fo LA, MS, AL, and FL	September 1998	3–4	16
Hurricane Bonnie	Category 3 hurricane in eastern NC and VA	August 1998	1.0	2
Southern drought/heat wave	Severe drought and heat wave from TX/OK to the Carolinas	Summer 1998	6.0	200
Minnesota severe storms/hail	Very damaging severe thunderstorms with large hail over wide areas of Minnesota	May 1998	1.5	1
Southeast severe weather	Tornadoes and flooding related to strong El Nino in the southeast	Winter/Spring 1998	1.0	Over 130
Northeast ice storm	Intense ice storm hits ME, NH, VT, and NY	January 1998	1.4	16
Northern plains flooding	Severe flooding in Dakotas and Minnesota due to heavy spring snowmelt	April–May 1997	2.0	11
MS and OH valleys flooding and tornadoes	Tornadoes and severe flooding hit the slates of AR, MO, MS, TN, IL, IN, KY, OH, and WV	March 1997	1.0	67
West Coast flooding	Flooding from rains and snowmelt in CA, WA, OR, ID, NV, & MT	Dec 1996–Jan 1997	2–3	36
Hurricane Fran	Category 3 hurricane in NC and VA	Sep 1996	5.0	37
Southern Plains severe drought	Drought in agricultural areas of TX & OK	Fall 1995-summer 1996	Over 4	(NA)
Pacific Northwest severe flooding	Flooding from heavy rain & snowmelt in OR, WA, ID, and MT.	Feb 1996	1.0	9
Blizzard of '96 followed by flooding	Heavy snowstorm followed by severe flooding in Appalachians, Mid-Atlantic, and Northeast	Jan 1996	3.0	187
Hurricane Opal	Category 3 hurricane in FL, AL, parts of GA, TN, & Carolinas.	Oct 1995	Over 3	27
Hurricane Marilyn	Category 2 hurricane in Virgin Islands	Sep 1995	2.1	13
TX/OK/LA/MS severe weather and flooding	Flooding, hail, & tornadoes across TX, OK, parts of LA, MS, Dallas & New Orleans hardest hit	May 1995	5–6	32

California flooding	Flooding from frequent winter storms across much of CA	Jan–Mar 1995	3.0	27
Western Fire Season	Severe fire season in western states due to dry weather	Summer-Fall 1994	1.0	(NA)
Texas flooding	Flooding from torrential rain & thunderstorms across southeast TX	Oct 1994	1.0	19
Tropical Storm Alberto	Flooding due to 10–25 inch rain across GA, AL, part of FL intense ice storm in pts of TX, OK, AR, LA, MS, AL, TN, GA,	July 1994	1.0	32
Southeast ice storm	SC, NC, & VA	Feb 1994	3.0	9
California wildfires	Out-of-control wildfires over southern CA	Fall 1993	1.0	4
Midwest flooding	Extreme flooding across central U.S.	Summer 1993	15–20	48
Drought/heat wave	Extreme drought/heatwave across southeastern U.S.	Summer 1993	1.0	(NA)
Storm/blizzard	"Storm of the Century" hits entire eastern seaboard	Mar 1993	3–6	270
Nor'easter of 1992	Slow-moving storm batters northeast U.S. coast, New England.	Dec 1992	1–2	19
Hurricane Iniki	Category 4 hurricane hit Hawaiian Island of Kauai	Sep 1992	1.8	7
Hurricane Andrew	Category 4 hurricane hit FL & LA	Aug 1992	27.0	58
Oakland Firestorm	Oakland, CA firestorm due to low humidity & high winds	Oct 1991	1.5	25
Hurricane Bob	Category 2 hurricane-mainly coastal NC, Long Island, & New England	Aug 1991	1.5	18
TX/OK/LA/AR Flooding	Torrential rains cause flooding along Trinity, Red, and Arkansas rivers	May 1990	1.0	13

Note: 5.0 represents $5,000,000,000. Covers only weather related disasters costing $1 billion or more. — Represents zero. NA not available or not reported. $del.

Source: From U.S. National Oceanic and Atmospheric Administration. National Climatic Data Center, "Billion Dollar U.S. Weather Disasters, 1980–2003" (release date: March 2, 2004). See also <www.ncdc.noaa.gov/oa/reports/billionz.html#TOP> (released 02 March 2004).

Table 5E.45 Deaths, Injuries, and Damage Caused by Hurricanes in the United States, 1965–1985

Fiscal Year	No. of Events	Persons Killed	Persons Injured	Dwellings Destroyed	Dwellings Damaged	Dwellings Destroyed & Damaged
1965–66	5	72	25,202	2,059	148,607	150,666
1966–67	NA	0	13	6	316	322
1967–68	NA	19	11,396	388	29,405	29,793
1968–69	NA	2	45	1	705	706
1969–70	NA	272	9,062	6,046	48,734	54,780
1970–71	5	9	4,498	1,887	34,442	36,329
1971–72	4	2	235	36	24,258	24,294
1972–73	0	0	0	0	0	0
1973–74	0	0	0	0	0	0
1974–75	2	3	8	45	2,514	2,559
1975–76	3	32	4,409	4,642	31,670	36,312
1976–77	1	2	23	15	498	513
1977–78	3	0	8	6	142	148
1978–79	1	0	0	1	3	4
1979–80	6	20	6,765	6,897	65,033	71,930
1980–81	2	NA	NA	NA	NA	14,865
1981–82	1	0	0	NA	NA	3
1982–83	2	2	961	NA	NA	7,454
1983–84	4	16	3,094	NA	NA	18,663
1984–85	0	0	0	0	0	0
Totals	—	—	—	—	—	449,341

Note: Based on American National Red Cross data which are by fiscal year (July 1–June 30).

Source: From Rubin, C.B., Yezer, A.M., Hussain, Q, and Webb, A., 1986, Summary of Major Natural Disaster Incidents in the U.S. 1965–85, Natural Hazards Research and Applications Information Center, George Washington University Spec. Publ. 17.

Table 5E.46 Public and Private Outlays for Hurricane Damage in the United States, 1965–1985

State	Year	Federal Outlay (in Thousands of Dollars)	Insurance Payment (in Thousands of Dollars)	States Affected
LA	1965	38,543	500,000	LA,FL,MS
FL	1965	1,706		
MS	1965	1,783		
		42,032		
TX	1967	9,925	34,800	TX
FL	1968	640	2,580	FL
MS	1969	74,524	165,300	MS,LA,AL,FL
LA	1969	15,167		
AL	1969	918		
		90,609		
TX	1970	35,808	309,950	TX
LA	1971	1,160	4,730	LA,MS
FL	1972	3,361	97,853	FL,NY,VA,PA MD,WV,OH,GA SC,NC,MI,DE DC,NJ,CT,RI MA,VT,ME
NY	1972	98,098		
VA	1972	16,815		
PA	1972	351,531		
MD	1972	23,309		
WV	1972	1,294		
OH	1972	1,453		
		495,861		
LA	1974	4,565	14,721	LA

(Continued)

Table 5E.46 (Continued)

State	Year	Federal Outlay (in Thousands of Dollars)	Insurance Payment (in Thousands of Dollars)	States Affected
NY	1976	6,773	22,697	NY,NJ,CT,MA
CA	1976	8,507	NA	—
AL	1979	189,893	752,510	AL,MS,FL,GA SC,NC,VA,MD DC,DE,PA,NJ NY,CT,MA
MS	1979	33,684		
FL	1979	3,691		
		227,268		
TX	1980	31,817	57,911	TX
TX	1980	386	NA	—
HI	1982	11,920	137,000	HI
TX	1983	40,038	675,520	TX
NC	1984	3,460	36,000	NC,SC
MS	1985	18,929	543,304	MS,AL,FL,LA
AL	1985	4,647		
FL	1985	13,933		
		37,509		
PA	1985	9,233	418,750	PA,CT,RI,NJ NY,MA,NC,VA MD,DE,NH,VT ME
CT	1985	21,359		
RI	1985	5,846		
NJ	1985	4,613		
NY	1985	38,750		
MA	1985	13,862		
		93,663		
LA	1985	23,962	44,000	LA,MS,AL,FL
FL	1985	7,238	77,600	FL,GA
TOTAL NO=39		1,173,141	3,895,226	

Source: From Rubin, C.B., Yezer, A.M., Hussain, Q, and Webb, A., 1986, Summary of Major Natural Disaster Incidents in the U.S. 1965–85, Natural Hazards Research and Applications Information Center, George Washington University Spec. Publ. 17.

Table 5E.47 U.S. Hurricane Strikes by Decade Number of Hurricanes by Saffir-Simpson Category to Strike the Mainland U.S. Each Decade.

Decade	Saffir-Simpson Category[a] 1	2	3	4	5	All 1–5	Major 3–5
1900–1909	5	5	4	2	0	16	6
1910–1919	8	3	5	3	0	19	8
1920–1929	6	4	3	2	0	15	5
1930–1939	4	5	6	1	1	17	8
1940–1949	7	8	7	1	0	23	8
1950–1959	8	1	7	2	0	18	9
1960–1969	4	5	3	2	1	15	6
1970–1979	6	2	4	0	0	12	4
1980–1989	9	1	5	1	0	16	6
1990–1999	3	6	4	0	1[b]	14	5
2000–2009	1	0	0	0	0	1	0
1900–1999	61	39	48	14	3	165	65

[a] Only the highest Saffir-Simpson Category to affect the U.S. has been used.
[b] This reflects the Hurricane Andrew reanalysis and upgrade from Category 4 to Category 5

Source: From www.nhc.noaa.gov.

SECTION 5F FLOOD PREVENTION

Table 5F.48 Strategies and Tools for Achieving Flood Hazard Reduction

Nonstructural

A. Modify susceptibility to flood damage and disruption
  1. Floodplain regulations
    a. State regulations for flood hazard areas
    b. Local regulations for flood hazard areas
      (1) Zoning
      (2) Subdivision regulations
      (3) Building codes
      (4) Housing codes
      (5) Sanitary and well codes
      (6) Other regulatory tools
  2. Development and redevelopment policies
    a. Design and location of services and utilities
    b. Land-right acquisition and open-space use
    c. Redevelopment and renewal
    d. Permanent evacuation
  3. Disaster preparedness and response planning
  4. Floodproofing
  5. Flood forecasting and warning systems and emergency plans

B. Modify the impact of flooding on individuals and the community
  1. Information and education
  2. Flood insurance
  3. Tax adjustments
  4. Flood emergency measures
  5. Postflood recovery

Structural

C. Modify flooding
  1. Dams and reservoirs
  2. Dikes, levees, and floodwalls
  3. Channel alterations
  4. High-flow diversions and spillways
  5. Land treatment measures
  6. On-site detention measures

Source: From U.S. Water Resources Council, 1981.

Table 5F.49 Structural Adjustments as Floodproofing Measures

Measure	Material Protected	Class of Measure	Prerequisites	
			Structural	Hydrologic
Seepage control	St–Co	P–C	Well constructed	None
Sewer adjustment	St–Co	P–C	None	H–W
Permanent closure	St–Co	P	Impervious walls	H–S
Openings protected	St–Co	C–E	Impervious walls	H–S–W
Interiors protected	St	P–C	None	S–W
Protective coverings	St–Co	P–C–E	None	H–W–F
Fire protection	St–Co	P	None	None
Appliance protection	Co	E	None	W
Utilities service	Co	P–C–E	None	S–W–V
Roadbed protection	St	P–E	Sound structure	H–W–V–D
Elevation	St–Co	P–C–E	Sound structure	S–W–V–F
Temporary removal	Co	E	None	W–F
Rescheduling	Co	E	Alternatives	W
Proper salvage	Co	—	None	None
Watertight caps	Co	P–C	None	W
Proper anchorage	St–Co	P–C	Sound structure	S–W–V–D
Underpinning	St	P	Sound structure	V
Timber treatment	St	P	None	None
Deliberate flooding	St–Co	E	None	None
Structural design	St–Co	P	Design	H–S
Reorganized use	Co	P	Alternatives	None

St = structure; P = permanent; H = hydrostatic pressure; F = flood-to-peak interval; Co = content; C = contingent; S = stage of flood; V = velocity of flow; E = emergency; W = warning; D = duration of flood.

Source: From Schaeffer, Univ. Chicago, Dept. Geography Research Paper, 1960.

Table 5F.50 Flood Safety Rules

Before the Flood

1. Keep on hand materials like sandbags, plywood, plastic sheeting, and lumber
2. Install check valves in building sewer traps, to prevent flood water from backing up in sewer drains
3. Arrange for auxiliary electrical supplies for hospitals and other operations which are critically affected by power failure
4. Keep first aid supplies at hand
5. Keep your automobile fueled; if electric power is cut off, filling stations may not be able to operate pumps for several days
6. Keep a stock of food which requires little cooking and no refrigeration; electric power may be interrupted
7. Keep a portable radio, emergency cooking equipment, lights and flashlights in working order

When You Receive a Flood Warning

8. Store drinking water in clean bathtubs, and in various containers. Water service may be interrupted
9. If forced to leave your home and time permits, move essential items to safe ground; fill tanks to keep them from floating away; grease immovable machinery
10. Move to a safe area before access is cut off by flood water

During the Flood

11. Avoid areas subject to sudden flooding
12. Do not attempt to cross a flowing stream where water is above your knees
13. Do not attempt to drive over a flooded road—you can be stranded, and trapped

After the Flood

14. Do not use fresh food that has come in contact with flood waters
15. Test drinking water for potability; wells should be pumped out and the water tested before drinking
16. Seek necessary medical care at nearest hospital. Food, clothing, shelter, and first aid are available at Red Cross shelters
17. Do not visit disaster area; your presence might hamper rescue and other emergency operations
18. Do not handle live electrical equipment in wet areas; electrical equipment should be checked and dried before returning to service
19. Use flashlights, not lanterns or torches, to examine buildings; flammables may be inside
20. Report broken utility lines to appropriate authorities

During any flood emergency, stay tuned to your radio or television station. Information from NOAA and civil emergency forces may save your life.

Source: From Environmental Science Services Administration, 1966.

Table 5F.51 Methods of Flood Control and Organization

Solutions for the flood problem fall into two distinct classes. The first includes those aimed at preventing the overflow of valley lands. The second embraces measures for human adjustment to the flood hazard

The overflow of the valley lands may be prevented, or reduced in frequency and extent, by

1. Providing an additional or an alternative channel to carry flood flows;
2. Increasing the capacity of the existing channel, so that the same flood may be passed downstream at lesser heights, thus reducing flood damages—a solution commonly known as channel improvement;
3. Reducing flood heights and damages by holding back a part of the floodwaters by means of reservoirs;
4. Constructing levees and flood walls to prevent the spread of floodwaters, or
5. Any combination of the above.

Measures of the second class, aimed at adjustment to the hazard include

1. Zoning of the flood plain to inhibit the development of high damageable values in hazardous areas;
2. Abandonment of efforts to use parts of the flood plain;
3. Use of flood forecasting so that damage may be minimized by removal of people and movable property;
4. Use of flood insurance, not to reduce flood damages, but to spread out the cost of floods over a period of years and thus minimize economic shock;
5. Flood relief in the event of disasters.

Source: From Task Force on Water Resources and Power, 1955.

SECTION 5G FLOOD CONTROLS WORKS

Table 5G.52 Upstream Flood Control Works in the United States

Region	Projects	Watershed Area	Flood Prevention Cost: Land Treatment	Reservoirs	Channel Improvement	Total
	No.	*1,000 acres*	*Mil. dol.*	*Mil. dol.*	*Mil. dol*	*Mil. dol.*
North Atlantic	120	4,906	27.0	108.8	52.5	181.3
South Atlantic-Gulf	163	8,897	93.9	88.6	75.5	258.0
Great Lakes	21	879	11.3	6.4	7.8	25.5
Ohio	89	4,442	45.5	78.8	30.7	155.0
Tennessee	21	1,440	11.8	18.8	10.1	40.7
Upper Mississippi	49	2,052	10.9	23.8	8.4	43.1
Lower Mississippi	104	6,895	80.6	65.9	65.9	212.4
Souris-Red-Rainy	21	2,882	3.8	4.0	15.8	23.6
Missouri	184	4,938	118.9	176.4	31.1	326.4
Arkansas-White-Red	153	12,438	94.4	274.1	27.5	369.0
Texas-Gulf	83	14,616	65.2	113.6	24.9	203.7
Rio Grande	29	1,512	4.7	14.3	2.0	21.0
Upper Colorado	6	517	3.2	9.7	4.3	17.2
Lower Colorado	10	884	5.6	17.0	7.7	30.3
Great Basin	13	871	5.6	8.8	1.2	15.6
Columbia-North Pacific	24	768	5.7	15.0	6.8	27.5
California	31	1,402	11.2	25.7	77.2	114.1
Alaska	0	0	0	0	0	0
Hawaii	5	278	0.5	0.9	10.3	11.7
Puerto Rico-Virgin Islands	3	252	2.7	3.2	7.6	13.5
Total	1,129	70,869	602.5	1,046.8	467.3	2,089.6

Note: Data covers the existing and approved program of the Soil Conservation Service only. In addition, there have been many projects in upstream areas constructed by the Corps of Engineers, the Bureau of Reclamation, and the Bureau of Land Management. Neither the totals nor regional data are available, but this construction amounted to over 1,000 projects, with an estimated cost of about $1 billion.

Source: From U.S. Water Resources Council, 1968.

Table 5G.53 Downstream Flood Control Works in the United States

Region	Reservoirs			Levees and Floodwalls			Channel Improvement		
	Projects	Storage	Cost[a]	Projects	Structures	Cost	Projects	Improvement	Cost
	No.	*1,000 af*	*Mil. dol.*	*No.*	*Miles*	*Mil. dol.*	*No.*	*Miles*	*Mil. dol.*
North Atlantic	5	492	36.0	36	132	144.4	25	54	27.3
South Atlantic-Gulf	7	3,090	49.8	[b]	876	154.4	23	185	5.1
Great Lakes	1	377	23.4	6	7	1.3	9	28	7.1
Ohio	36	12,500	600.0	65	252	202.0	[b]	138	[b]
Tennessee	17	11,590	180.2	0	0	0	0	0	0
Upper Mississippi	14	3,020	54.0	65	861	242.0	19	70	8.7
Lower Mississippi	5	4,400	76.7	[b]	1,621	841.0	[b]	3,348	980.0
Souris-Red-Rainy	5	1,030	1.8	2	2	2.7	7	224	2.9
Missouri	56	20,700	656.0	50	1,130	193.6	7	75	22.2
Arkansas-White-Red	43	24,800	635.5	56	1,023	52.4	39	563	54.0
Texas-Gulf	20	8,600	234.2	5	128	121.4	8	106	94.4
Rio Grande[c]	4	795	31.1	5	205	7.6	7	114	10.5
Upper Colorado	3	1,500	5.5	8	5	0.2	1	1	20.0
Lower Colorado	6	12,100	59.6	2	7	3.2	1	4	0.5
Great Basin	6	386	7.4	4	33	1.6	3	23	1.4
Columbia-North Pacific	24	15,210	320.7	103	546	28.1	27	55	27.7
California	11	3,720	186.5	13	1,515	260.6	12	84	91.4
Alaska	0	0	0	2	3	0.6	1	1	0.1
Hawaii	0	0	0	4	6	0.2	4	3	0.2
Puerto Rico-Virgin Islands	0	0	0	0	0	0	0	0	0
Total[d]	263	124,310	3,158	426+	8,352	2,257	193+	5,076	1,354+

[a] Does not include cost of flood control storage in Bureau of Reclamation projects.
[b] Not reported.
[c] Does not include some facilities constructed by the International Boundary and Water Commission.
[d] Rounded.

Source: From U.S. Water Resources Council, 1968.

SECTION 5H WATER AREAS — UNITED STATES

Table 5H.54 Coastline of the United States (by State)

State	General Coastline[a]		Tidal Shoreline[b]	
	Statute Miles	Kilometers	Statute Miles	Kilometers
U.S.	12,383	19,924	88,633	142,610
Atlantic coast	2,069	3,329	28,673	46,135
Connecticut	—	—	618	994
Delaware	28	45	381	613
Florida	580	933	3,331	5,360
Georgia	100	161	2,344	3,771
Maine	228	367	3,487	5,596
Maryland	31	50	3,190	5,133
Massachusetts	192	309	1,519	2,444
New Hampshire	13	21	131	211
New Jersey	130	209	1,792	2,883
New York	127	204	1,850	2,977
North Carolina	301	484	3,375	5,430
Pennsylvania	—	—	89	143
Rhode Island	40	64	384	618
South Carolina	187	301	2,876	4,627
Virginia	112	180	3,315	5,334
Gulf coast	1,631	2,624	17,141	27,580
Alabama	53	85	607	977
Florida	770	1,239	5,095	8,198
Louisiana	397	639	7,721	12,423
Mississippi	44	71	359	578
Texas	367	591	3,359	5,405
Pacific coast	7,623	12,265	40,298	64,839
Alaska	5,580	8,978	31,383	50,495
California	840	1,352	3,427	5,514
Hawaii	750	1,207	1,052	1,693
Oregon	296	476	1,410	2,269
Washington	157	253	3,026	4,869
Arctic coast				
Alaska	1,060	1,706	2,521	4,056

Note: — Represents zero.

[a] Figures are lengths of general outline of seacoast. Measurements were made with a unit measure of 30 min of latitude on charts as near the scale of 1:1,200,000 as possible. Coastline of sounds and bays is included to a point where they narrow to width of unit measure, and includes the distance across at such point.

[b] Figures obtained in 1939–1940 with a recording instrument on the largest-scale charts and maps then available. Shoreline of outer coast, offshore islands, sounds, bays, rivers, and creeks is included to the head of tidewater or to a point where tidal waters narrow to a width of 100 ft.

Source: From Statistical Abstract of the United States, 1987.

Table 5H.55 Land and Water Area of States and Other Entities: 2000

State and Other Area	Total Area		Land Area		Water Area: Total		Water Area			
	Sq. mi.	Sq. km.	Sq. mi.	Sq. km.	Sq. mi.	Sq. km.	Inland (sq. mi.)	Coastal (sq. mi.)	Great Lakes (sq. mi.)	Territorial (sq. mi.)
Total	**3,800,286**	**9,842,696**	**3,540,999**	**9,171,146**	**259,287**	**671,550**	**79,018**	**42,241**	**60,251**	**77,777**
United States	**3,794,083**	**9,826,630**	**3,537,439**	**9,161,923**	**256,645**	**664,707**	**78,797**	**42,225**	**60,251**	**75,372**
Alabama	52,419	135,765	50,744	131,426	1,675	4,338	956	519	—	200
Alaska	663,267	1,717,854	571,951	1,481,347	91,316	236,507	17,243	27,049	—	47,024
Arizona	113,998	295,254	113,635	294,312	364	942	364	—	—	—
Arkansas	53,179	137,732	52,068	134,856	1,110	2,876	1,110	—	—	—
California	163,696	423,970	155,959	403,933	7,736	20,037	2,674	222	—	4,841
Colorado	104,094	269,601	103,718	268,627	376	974	376	—	—	—
Connecticut	5,543	14,357	4,845	12,548	699	1,809	161	538	—	—
Delaware	2,489	6,447	1,954	5,060	536	1,388	72	371	—	93
District of Columbia	68	177	61	159	7	18	7	—	—	—
Florida	65,755	170,304	53,927	139,670	11,828	30,634	4,672	1,311	—	5,845
Georgia	59,425	153,909	57,906	149,976	1,519	3,933	1,016	48	—	455
Hawaii	10,931	28,311	6,423	16,635	4,508	11,677	38	—	—	4,470
Idaho	83,570	216,446	82,747	214,314	823	2,131	823	—	—	—
Illinois	57,914	149,998	55,584	143,961	2,331	6,037	756	—	1,575	—
Indiana	36,418	94,321	35,867	92,895	551	1,427	316	—	235	—
Iowa	56,272	145,743	55,869	144,701	402	1,042	402	—	—	—
Kansas	82,277	213,096	81,815	211,900	462	1,197	462	—	—	—
Kentucky	40,409	104,659	39,728	102,896	681	1,763	681	—	—	—
Louisiana	51,840	134,264	43,562	112,825	8,278	21,440	4,154	1,935	—	2,189
Maine	35,385	91,646	30,862	79,931	4,523	11,715	2,264	613	—	1,647
Maryland	12,407	32,133	9,774	25,314	2,633	6,819	680	1,843	—	110
Massachusetts	10,555	27,336	7,840	20,306	2,715	7,031	423	977	—	1,314
Michigan	96,716	250,494	56,804	147,121	39,912	103,372	1,611	—	38,301	—
Minnesota	86,939	225,171	79,610	206,189	7,329	18,982	4,783	—	2,546	—
Mississippi	48,430	125,434	46,907	121,489	1,523	3,945	785	590	—	148
Missouri	69,704	180,533	68,886	178,414	818	2,120	818	—	—	—
Montana	147,042	380,838	145,552	376,979	1,490	3,859	1,490	—	—	—
Nebraska	77,354	200,345	76,872	199,099	481	1,247	481	—	—	—
Nevada	110,561	286,351	109,826	284,448	735	1,903	735	—	—	—
New Hampshire	9,350	24,216	8,968	23,227	382	989	314	—	—	68
New Jersey	8,721	22,588	7,417	19,211	1,304	3,377	396	401	—	507
New Mexico	121,590	314,915	121,356	314,309	234	606	234	—	—	—
New York	54,556	141,299	47,214	122,283	7,342	19,016	1,895	981	3,988	479
North Carolina	53,819	139,389	48,711	126,161	5,108	13,229	3,960	—	—	1,148
North Dakota	70,700	183,112	68,976	178,647	1,724	4,465	1,724	—	—	—
Ohio	44,825	116,096	40,948	106,056	3,877	10,040	378	—	3,499	—
Oklahoma	69,898	181,036	68,667	177,847	1,231	3,189	1,231	—	—	—
Oregon	98,381	254,805	95,997	248,631	2,384	6,174	1,050	80	—	1,254
Pennsylvania	46,055	119,283	44,817	116,075	1,239	3,208	490	—	749	—
Rhode Island	1,545	4,002	1,045	2,706	500	1,295	178	9	—	314
South Carolina	32,020	82,932	30,110	77,983	1,911	4,949	1,008	72	—	831
South Dakota	77,117	199,731	75,885	196,540	1,232	3,191	1,232	—	—	—
Tennessee	42,143	109,151	41,217	106,752	926	2,399	926	—	—	—
Texas	268,581	695,621	261,797	678,051	6,784	17,570	5,056	404	—	1,324
Utah	84,899	219,887	82,144	212,751	2,755	7,136	2,755	—	—	—
Vermont	9,614	24,901	9,250	23,956	365	945	365	—	—	—
Virginia	42,774	110,785	39,594	102,548	3,180	8,237	1,006	1,728	—	446
Washington	71,300	184,665	66,544	172,348	4,756	12,317	1,553	2,537	—	666
West Virginia	24,230	62,755	24,078	62,361	152	394	152	—	—	—
Wisconsin	65,498	169,639	54,310	140,663	11,188	28,976	1,830	—	9,358	—
Wyoming	97,814	253,336	97,100	251,489	713	1,847	713	—	—	—
Other areas										
Puerto Rico	5,325	13,790	3,425	8,870	1,900	4,921	67	16	—	1,817
U.S. Minor Outlying Islands	141	365	3	7	138	359	138	—	—	—
Virgin Islands of the U.S.	737	1,910	134	346	604	1,564	16	—	—	588

Note: One square mile = 2.59 sq. km. Area is calculated from the specific boundary recorded for each entity in the U.S. Census Bureau's geographic TIGER database; —, Represents or rounds to zero.

Source: From U.S. Census Bureau, 2000 Census of Population and Housing, Summary Population and Housing Characteristics, Series PHC-1; and unpublished data from the Census TIGER™ data base.

Table 5H.56 U.S. Wetland Resources and Deepwater Habitats by Type: 1986 and 1997

Wetland or Deepwater Category	1986	1997	Change, 1986 to 1997
All wetlands and deepwater habitats, total	**144,673.3**	**144,136.8**	**−536.5**
All deepwater habitats, total	38,537.6	38,645.1	107.5
Lacustrine[a]	14,608.9	14,725.3	116.4
Riverine[b]	6,291.1	6,255.9	−35.2
Estuarine Subtidal[c]	17,637.6	17,663.9	26.3
All wetlands, total	106,135.7	105,491.7	−644
Intertidal wetlands[d]	5,336.6	5,326.2	−10.4
Marine intertidal	133.1	130.9	−2.2
Estuarine intertidal non-vegetated	580.4	580.1	−0.3
Estuarine intertidal vegetated	4,623.1	4,615.2	−7.9
Freshwater wetlands	100,799.1	100,165.5	−633.6
Freshwater non-vegetated	5,251.0	5,914.3	663.3
Freshwater vegetated	95,548.1	94,251.2	−1,296.9
Freshwater emergent[e]	26,383.3	25,157.1	−1,226.2
Freshwater forested[f]	51,929.6	50,728.5	−1,201.1
Freshwater shrub[g]	17,235.2	18,365.6	1,130.4

Note: In thousands of acres (144,673.3 represents 144,677,300). Wetlands and deepwater habitats are defined separately because the term wetland does not included permanent water bodies. Deepwater habitats are permanently flooded land lying below the deepwater boundary of wetlands. Deepwater habitats include environments where surface water is permanent and often deep, so that water, rather than air, is the principal medium within which the dominant organisms live, whether or not they are attached to the substrate. As in wetlands, the dominant plants are hydrophytes; however, the substrates are considered nonsoil because the water is too deep to support emergent vegetation. In general terms, wetlands are lands where saturation with water is the dominant factor determining the nature of soil development and the types of plant and animal communities living in the soil and on its surface. The single feature that most wetlands share is soil or substrate that is at least periodically saturated with or covered by water. Wetlands are lands transitional between terrestrial and aquatic systems where the water table is usually at or near the surface or the land is covered by shallow water.

[a] The lacustrine system includes deepwater habitats with all of the following characteristics: (1) situated in a topographic depression or a dammed river channel: (2) lacking trees, shrubs, persistent emergents, emergent mosses or lichens with greater than 30 percent coverage; (3) total area exceeds 20 acres.

[b] The riverine system includes deepwater habitats contained within a channel, with the exception of habitats with water containing ocean derived salts in excess of 0.5 parts per thousand.

[c] The estuarine system consists of deepwater tidal habitats and adjacent tidal wetland that are usually semi-enclosed by land but have open, partly obstructed, or sporadic access to the open ocean, and in which ocean water is at least occasionally diluted by freshwater runoff from the land. Subtidal is where the substrate is continuously submerged by marine or estuarine waters.

[d] Intertidal is where the substrate is exposed and flooded by tides. Intertidal includes the splash zone of coastal waters.

[e] Emergent wetlands are characterized by erect, rooted herbaceous hydrophytes, excluding mosses and lichens. This vegetation is present for most of the growing season in most years. These wetlands are usually dominated by perennial plants.

[f] Forested wetlands are characterized by woody vegetation that is 20 ft tall or taller.

[g] Shrub wetlands include areas dominated by woody vegetation less than 20 ft tall. The species include true shrubs, young trees, and trees or shrubs that are small or stunted because of environmental conditions.

Source: From U.S. Fish and Wildlife Service, Status and Trends of Wetlands In the Conterminous United States, 1986 to 1997, January 2001.

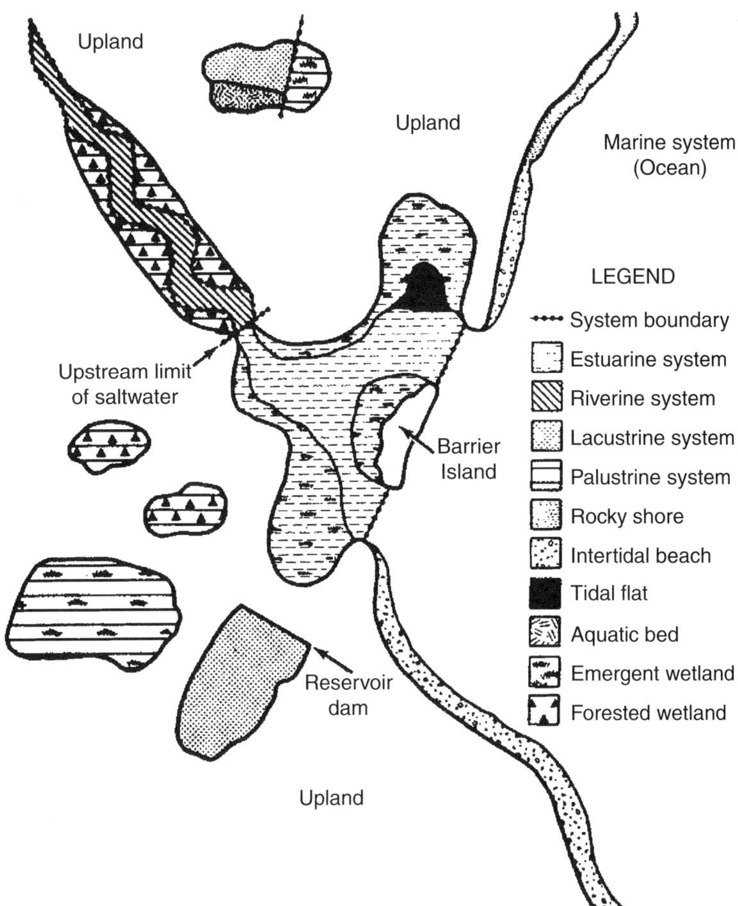

Wetlands occur in every state of the country and due to regional differences in climate, vegetation, soil and hydrologic conditions, they exist in a variety of sizes, shapes and types. Although more abundant in other areas, wetlands even exist in deserts.

Wetlands and deepwater habitats are divided into five ecological systems: (1) Marine, (2) Estuarine, (3) Riverine, (4) Lacustrine, and (5) Palustrine. The Marine System generally consists of the open ocean and its associated coastline. It is mostly a deepwater habitat system, with marine wetlands limited to intertidal areas like beaches, rocky shores and some coral reefs. The Estuarine System includes coastal wetlands like salt and brackish tidal marshes, mangrove swamps, and intertidal flats, as well as deepwater bays, sounds and coastal rivers. The Riverine System is limited to freshwater river and stream channels and is mainly a deepwater habitat system. The Lacustrine System is also a deep water dominated system, but includes standing waterbodies like lakes, reservoirs and deep ponds. The Palustrine System encompasses the vast majority of the country's inland marshes, bogs and swamps and does not include any deepwater habitat.

Figure 5H.26 Major wetland types in the United States. (From Tiner, R.W., Jr., Wetlands of the United States: Current Status and Recent Trends, U.S. Department of the Interior, Fish and Wildlife Service, 1984.)

Table 5H.57 Water Areas for Selected Major Bodies of Water: 2000

Body of Water and State	Area	
	Sq. mi.	Sq. km.
Atlantic Coast water bodies		
Chesapeake Bay (MD-VA)	2747	7115
Pamlico Sound (NC)	1622	4200
Long Island Sound (CT-NY)	914	2368
Delaware Bay (DE-NJ)	614	1591
Cape Cod Bay (MA)	598	1548
Albemarle Sound (NC)	492	1274
Biscayne Bay (FL)	218	565
Buzzards Bay (MA)	215	558
Tangier Sound (MD-VA)	172	445
Currituck Sound (NC)	116	301
Pocomoke Sound (MD-VA)	111	286
Chincoteague Bay (MD-VA)	105	272
Gulf Coast water bodies		
Mississippi Sound (AL-LA-MS)	813	2105
Laguna Madre (TX)	733	1897
Lake Pontchartrain (LA)	631	1635
Florida Bay (FL)	616	1596
Breton Sound (LA)	511	1323
Mobile Bay (AL)	310	802
Lake Borgne (LA-MS)	271	702
Matagorda Bay (TX)	253	656
Atchafalaya Bay (LA)	245	635
Galveston Bay (TX)	236	611
Tampa Bay (FL)	212	549
Pacific Coast water bodies		
Puget Sound (WA)	808	2092
San Francisco Bay (CA)	264	684
Willapa Bay (WA)	125	325
Hood Canal (WA)	117	303
Interior water bodies		
Lake Michigan (IL-IN-MI-WI)	22,342	57,866
Lake Superior (MI-MN-WI)[a]	20,557	53,243
Lake Huron (MI)[a]	8,800	22,792
Lake Erie (MI-NY-OH-PA)[a]	5,033	13,036
Lake Ontario (NH)[a]	3,446	8,926
Great Salt Lake (UT)	1,836	4,756
Green Bay (MI-WI)	1,396	3,617
Lake Okeechobee (FL)	663	1,717
Lake Sakakawea (ND)	563	1,459
Lake Oahe (ND-SD)	538	1,394
Lake of the Woods (MN)[a]	462	1,196
Lake Champlain (NY-VT)[a]	414	1,072
Alaska water bodies		
Chatham Strait	1,559	4,039
Prince William Sound	1,382	3,579
Clarence Strait	1,199	3,107
Iliamna Lake	1,022	2,646
Frederick Sound	792	2,051
Sumner Strait	791	2,048
Stephens Passage	702	1,819
Kvichak Bay	640	1,659
Montague Strait	463	1,198
Becharof Lake	447	1,158
Icy Strait	436	1,130

Note: Includes only that portion of body of water under the jurisdiction of the United States, excluding Hawaii. One square mile = 2.59 sq. km.

[a] Area measurements for Lake Champlain, Lake Erie, Lake Huron, Lake Ontario, Lake St. Clair, Lake Superior, and Lake of the Woods include only those portions under the jurisdiction of the United States.

Source: From U.S. Census Bureau, unpublished data from the Census TIGER™ data base.

Table 5H.58 Wetlands Lost in the United States

State or Region	Original Wetlands (acres)	Wetlands in 1984 (acres)	Percentage of Wetlands Lost
Iowa's natural marshes	2,333,000	26,470	99
California	5,000,000	450,000	91
Nebraska's rainwater basin	94,000	8,460	91
Mississippi alluvial plain	24,000,000	5,200,000	78
Michigan	11,200,000	3,200,000	71
North Dakota	5,000,000	2,000,000	60
Minnesota	18,400,000	8,700,000	53
Louisiana's forested wetlands	11,300,000	5,635,000	50
Connecticut's coastal marshes	30,000	15,000	50
North Carolina's pocosins	2,500,000	1,503,000[a]	40
South Dakota	2,000,000	1,300,000	35
Wisconsin	10,000,000	6,750,000	32

[a] Only 695,000 acres of pocosins remain undisturbed; the rest are partially drained, developed or planned for development.

Source: From Tiner, R.W., Jr., 1984, Wetlands of the United States: Current Status and Recent Trends, U.S. Department of the Interior, Fish and Wildlife Service.

Table 5H.59 Change in Wetland Area for Selected Wetland and Deepwater Categories, 1986 to 1997

	Area in Thousands of Acres			
Wetland/Deepwater Category	Estimated Area, 1986	Estimated Area, 1997	Change, 1986–97	Change (in Percent)
Marine intertidal	133.1	130.9	−2.2	−1.7
	(19.6)	(19.9)	(88.5)	
Estuarine intertidal non-vegetated[a]	580.4	580.1	−0.3	−0.1
	(10.7)	(10.6)	(*)	
Estuarine intertidal vegetated[b]	4,623.1	4,615.2	−7.9	−0.2
	(4.0)	(4.0)	(75.1)	
All intertidal wetlands	5,336.6	5,326.2	−10.4	−0.2
	(3.8)	(3.8)	(73.0)	
Freshwater non-vegetated[c]	5,251.0	5,914.3	663.3	12.6
	(4.1)	(3.9)	(13.4)	
Freshwater vegetated[d]	95,548.1	94,251.2	−1,296.9	−1.4
	(3.0)	(3.0)	(17.1)	
Freshwater emergent	26,383.3	25,157.1	−1,226.2	−4.6
	(8.1)	(8.4)	(18.2)	
Freshwater forested	51,929.6	50,728.5	1,201.1	−2.3
	(2.8)	(2.8)	(23.8)	
Freshwater shrub	17,235.2	18,365.6	1,130.4	6.6
	(4.2)	(4.1)	(25.7)	
All freshwater wetlands	100,799.1	100,165.5	633.6	−0.6
	(2.9)	(2.9)	(36.5)	
All wetlands	106,135.7	105,491.7	−644.0	0.6
	(2.8)	(2.8)	(36.0)	
Deepwater habitats				
Lacustrine[e]	14,608.9	14,725.3	116.4	0.8
	(10.6)	(10.5)	(*)	
Riverine	6,291.1	6,225.9	−35.2	−0.6
	(9.6)	(9.4)	(*)	
Estuarine subtidal	17,637.6	17,663.9	26.3	0.1
	(2.2)	(2.2)	(95.6)	
All deepwater habitats	38,537.6	38,645.1	107.5	0.3
	(4.4)	(4.4)	(*)	
All wetlands and deepwater habitats[a,b]	144,673.3	144,136.8	−536.5	−0.4
	(2.4)	(2.4)	(30.7)	

Note: The coefficient of variation (CV) for each entry (expressed as a percentage) is given in parentheses.
*Statistically unreliable.

[a] Includes the categories: estuarine intertidal aquatic bed and estuarine intertidal unconsolidated shore.
[b] Includes the categories: estuarine intertidal emergent and estuarine intertidal shrub.
[c] Includes the categories: palustrine aquatic bed, palustrine unconsolidated bottom and palustrine unconsolidated shore.
[d] Includes the categories: palustrine emergent, palustrine forested and palustrine shrub.
[e] Does not include the great lakes.

Source: From Dahl, T.E. 2000. Status and trends of wetlands in the conterminous United States 1986 to 1997. U.S. Department of the Interior, Fish and Wildlife Service, Washington, DC, 82pp. http://training.fws.gov.

Table 5H.60 Estuarine and Marine Intertidal Wetland Area and Change, 1986 to 1997

	Area in Thousands of Acres			
Wetland Category	**Estimated Area, 1986**	**Estimated Area, 1997**	**Gain or Loss, 1986–1997**	**Area (as Percent) of All Intertidal Wetland, 1997**
Marine Intertidal	133.1 (19.6)	130.9 (19.9)	−2.2 (88.5)	2.5
Estuarine Unconsolidated shore	551.3 (10.9)	550.8 (10.8)	−0.5 (*)	10.3
Estuarine aquatic bed	29.1 (27.1)	29.3 (26.9)	0.2 (*)	0.6
Marine and estuarine intertidal Non-vegetated[a]	580.4 (10.7)	580.1 (10.6)	−0.3 (*)	13.4
Estuarine emergent	3,956.9 (4.1)	3,942.4 (4.1)	−14.5 (49.2)	74.0
Estuarine shrub	666.2 (12.6)	672.8 (12.6)	6.6 (76.5)	12.6
Estuarine intertidal Vegetated[b]	4,623.1 (4.0)	4,615.2 (4.0)	−7.9 (75.1)	86.6
	Changes in coastal deepwater area, 1986–1997			
Estuarine Subtidal	17,637.6 (2.2)	17,663.9 (2.2)	26.3 (95.6)	—

Note: The coefficient of variation (CV) for each entry (expressed as a percentage) is given in parentheses.
*Statistically unreliable.

[a] Includes the categories: estuarine unconsolidated shore and estuarine aquatic bed.
[b] Includes the categories: estuarine emergent and estuarine shrub.

Source: From Dahl, T.E. 2000. Status and trends of wetlands in the conterminous United States 1986 to 1997. U.S. Department of the Interior, Fish and Wildlife Service, Washington, DC, 82pp. http://training.fws.gov.

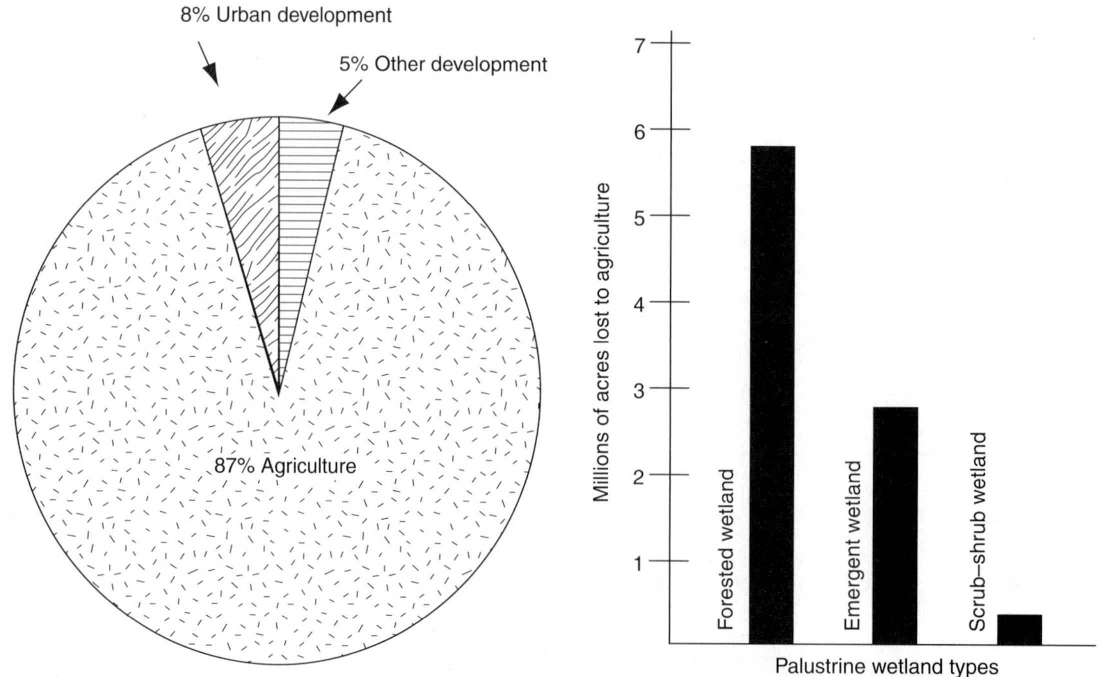

Figure 5H.27 Causes of recent wetland losses in the conterminous United States. (From Tiner, R.W., Jr., Wetlands of the United States: Current Status and Recent Trends, U.S. Department of the Interior, Fish and Wildlife Service, 1984.)

Table 5H.61 Major Causes of Wetland Loss and Degradation

Human Threats

Direct

1. Drainage for crop production, timber production and mosquito control
2. Dredging and stream channelization for navigation channels, flood protection, coastal housing developments, and reservoir maintenance
3. Filling for dredged spoil and other solid waste disposal, roads and highways, and commercial, residential and industrial development
4. Construction of dikes, dams, levees and seawalls for flood control, water supply, irrigation and storm protection
5. Discharges of materials (e.g., pesticides, herbicides, other pollutants, nutrient loading from domestic sewage and agricultural runoff, and sediments from dredging and filling, agricultural and other land development) into waters and wetlands
6. Mining of wetland soils for peat, coal, sand, gravel, phosphate and other materials

Indirect

1. Sediment diversion by dams, deep channels and other structures
2. Hydrologic alterations by canals, spoil banks, roads and other structures
3. Subsidence due to extraction of groundwater, oil, gas, sulphur, and other minerals

Natural Threats

1. Subsidence (including natural rise of sea level)
2. Droughts
3. Hurricanes and other storms
4. Erosion
5. Biotic effects, e.g., muskrat, nutria and goose "eat-outs"

Source: From Tiner, R.W. Jr., 1984, Wetlands of the United States: Current Status and Trends, U.S. Fish and Wildlife Service.

Table 5H.62 Major Wetland Values

Fish and Wildlife Values

Fish and shellfish habitat
Waterfowl and other bird habitat
Furbearer and other wildlife habitat

Environmental Quality Values

Water quality maintenance
Pollution filter
Sediment removal
Oxygen production
Nutrient recycling
Chemical and nutrient absorption
Aquatic productivity
Microclimate regulator
World climate (ozone layer)

Socio-economic Values

Flood control
Wave damage protection
Erosion control
Groundwater recharge and water supply
Timber and other natural products
Energy source (peat)
Livestock grazing
Fishing and shellfishing
Hunting and trapping
Recreation
Aesthetics
Education and scientific research

Source: From Tiner, R.W., Jr., 1984, Wetlands of the United States: Current Status and Trends, U.S. Fish and Wildlife Service.

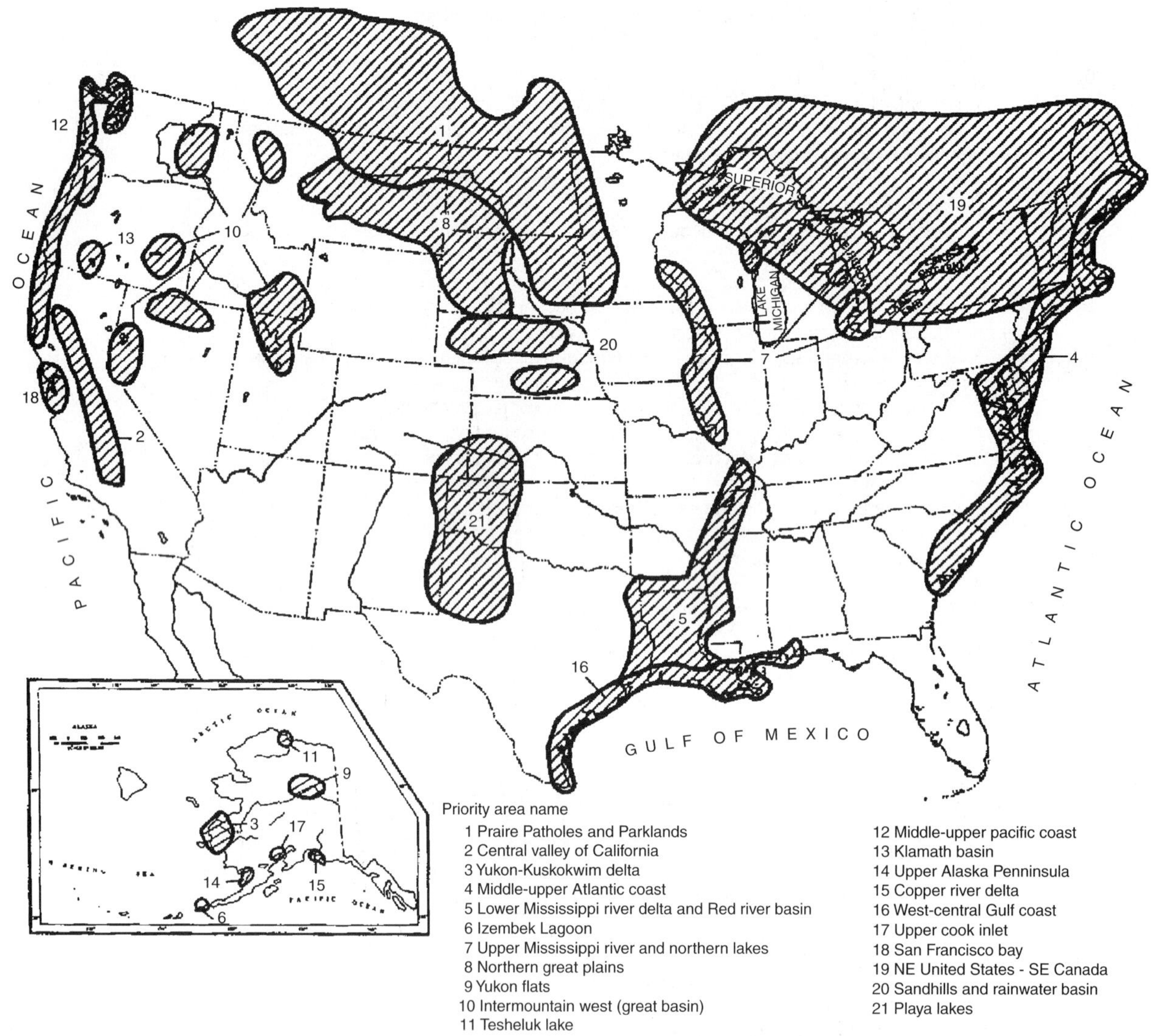

PACIFIC OCEAN
ATLANTIC OCEAN
GULF OF MEXICO
SUPERIOR
LAKE MICHIGAN
Priority area name
1 Praire Patholes and Parklands
2 Central valley of California
3 Yukon-Kuskokwim delta
4 Middle-upper Atlantic coast
5 Lower Mississippi river delta and Red river basin
6 Izembek Lagoon
7 Upper Mississippi river and northern lakes
8 Northern great plains
9 Yukon flats
10 Intermountain west (great basin)
11 Tesheluk lake
12 Middle-upper pacific coast
13 Klamath basin
14 Upper Alaska Penninsula
15 Copper river delta
16 West-central Gulf coast
17 Upper cook inlet
18 San Francisco bay
19 NE United States - SE Canada
20 Sandhills and rainwater basin
21 Playa lakes

Figure 5H.28 Principal waterfowl habitat areas in the United States. (From Tiner, R.W. Jr., Wetlands of the United States: Current Status and Recent Trends, U.S. Department of the Interior, Fish and Wildlife Service, 1984.)

Freshwater wetlands 14%
Urban development 24%
Deepwater 12%
Agriculture 1%
Rural development 19%
Other uplands 30%

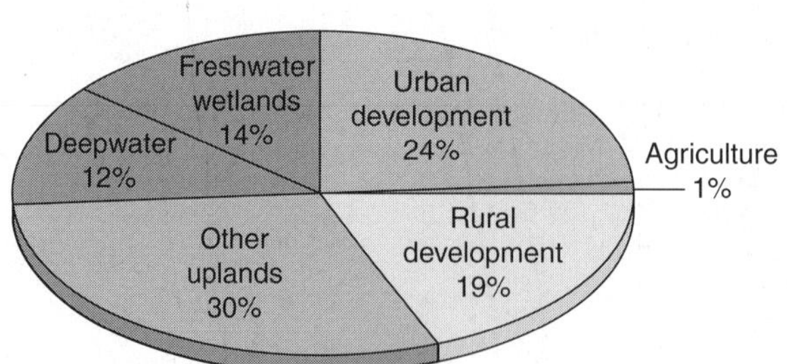

Figure 5H.29 Percent of estuarine and marine wetlands lost to freshwater wetlands, deepwater, or upland categories between 1986 and 1997. (From Dahl, T.E. Status and trends of wetlands in the conterminous United States 1986 to 1997. U.S. Department of the Interior, Fish and Wildlife Service, Washington, DC, 82pp, 2000.) (http://training.fws.gov.)

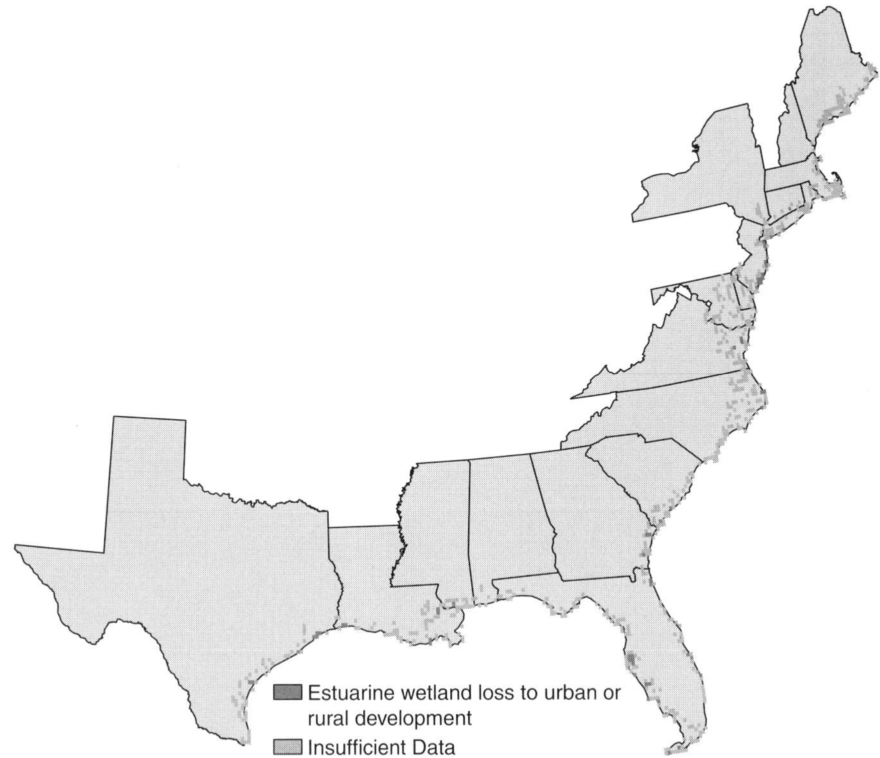

Figure 5H.30 Areas along the Gulf and Atlantic coasts where estuarine wetlands were lost to urban or rural development (shown in orange) between 1986 and 1997. (From Dahl, T.E. Status and trends of wetlands in the conterminous United States 1986 to 1997. U.S. Department of the Interior, Fish and Wildlife Service, Washington, DC, 82pp, 2000.) (http://training.fws.gov.)

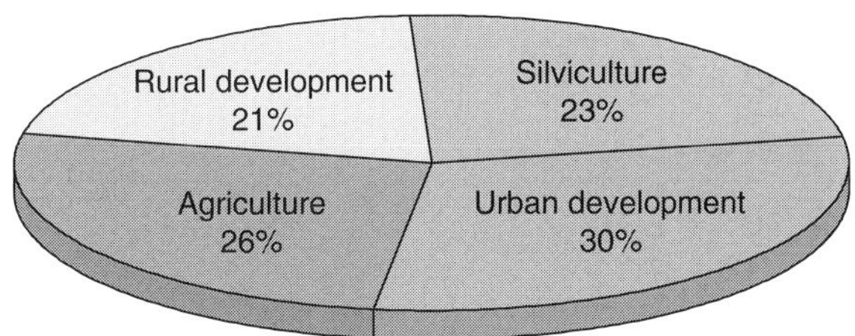

Figure 5H.31 Change in wetlands converted to various land uses between 1986 and 1997. (From Dahl, T.E. Status and trends of wetlands in the conterminous United States 1986 to 1997. U.S. Department of the Interior, Fish and Wildlife Service, Washington, DC, 82pp, 2000.) (http://training.fws.gov.)

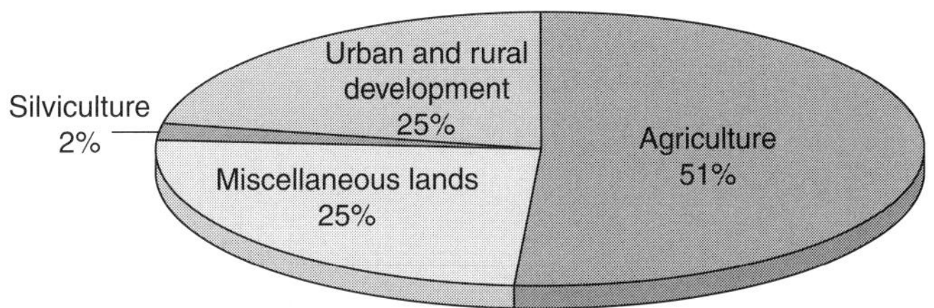

Figure 5H.32 Current upland classification of areas where emergent wetlands were lost between 1986 and 1997. (From Dahl, T.E. Status and trends of wetlands in the conterminous United States 1986 to 1997. U.S. Department of the Interior, Fish and Wildlife Service, Washington, DC, 82pp, 2000.) (http://training.fws.gov.)

Table 5H.63 Effect of Wetlands on Flood Peak Reduction in Wisconsin

Wetland Present in a Basin (percent)	Storm Recurrence Interval (years) 2	25	50	100
	Percentage of Flood Peak Reduction			
1	19	22	26	27
2	28	33	38	39
3	34	39	45	46
5	42	48	54	55
10	51	58	63	64
15	56	63	70	71
30	64	71	77	79

Source: From Conger, D.H., 1971, Estimating Magnitude and Frequency of Floods in Wisconsin, U.S. Geological Survey Open-File Rept.

Table 5H.64 Terms Descriptive of Water Landscapes

Fluvial Types	Fluvial lakes	Rill
Bayou	Freshet	River
Braided stream	Inlet	Slough
Brook	Influent	Spring
Canal	Intermittent stream	Stream
Connecting stream	Interrupted stream	Torrent
Creek	Misfit river	Vigorously meandering stream
Disappearing stream	Outlet	Watercourse
Feeder stream	Raft	Waterway
Lacustrine Types	Fluvial lake	Pit lake
Aestival ponds	Fluviatile lake	Playa
Alkali lakes	Fosse lake	Pond
Alluvial dam lakes	Glacial lake	Pool
Alpine lakes	Grass lake	Pothole
Bar lake	Headwaters lake	Puddle
Barrier lake	Holding pond	Quarry pond
Bayou	Holm lake	Raft lake
Blind lake	Hot springs	Reflection basin
Blowout pond	Impoundment	Rejuvenated lake
Bog lake	Intermittent lake	Reservoir
Borrow pit pond	Kettle lake	Ria lakes
Caldera lake	Lagoon	Rift lakes
Chain of lakes	Lake	Riverine lakes
Charco	Lakelet	Rock lakes
Cirque lake	Landslide lake	Sag pond
Clear lake	Laguna	Salt lakes
Closed lake	Lateral lake	Satellite lakes
Crater lake	Marl lake	Scour lakes
Dead lake	Marsh lake	Seepage lakes
Deflation lake	Meadow lake	Senescent lake
Delta lake	Mesotrophic lake	Sink lakes
Doline lake	Mill pond	Slough
Drainage lake	Mirror lake	Snag lake
Dry lake	Moat lake	Strath lake
Dugout pond	Morainal lake	Swarm of lakes
Dune lake	Nova lake	Tailing pond
Dystrophic lake	Oligotrophic lake	Tanks
Effluent lake	Open lake	Tarn
Evanescent lake	Oriented lake	Thaw lakes
Extinct lake	Oxbow	Tundra lakes
Farm pond	Palodolac	Vernal & autumnal ponds
Finger lake	Perched lake	Walled lakes
Fission lake	Perennial lake	

Source: From Litton, R.B., Tetlow, R.J., Sorensen, J., and Beatty, R.A., 1974, Water and Landscape, Water Information Center, Inc.

Table 5H.65 Aesthetic Evaluation of Rivers

Aesthetic factors
Landscape
Views and vistas
Diversity of flora and geologic features
Color
Form and contrasts
Sensual stimuli
Temperature regime
Winds and other aerial features
Sounds
Odors
Visual patterns
Intellectual interests
Opportunities for interpretive programs
Ecology
Geology
Wildlife
Range and diversity of subjects available for study
Emotional interest
Physical stimuli
Intellectual potentials
Possibility for adventure
Interaction of flora, fauna, and people
Access
Climatic factors
Obstacles or discomforts
Troublesome flora and fauna
Access
Climatic factors
Culture
Quality of land use management construction
Scenic pollution
Historic artifacts
Subjective analysis of most important factors in viewing river — most significant in deriving-pleasurable feelings
Vista
Color
Vegetation (amount and variety)
Spaciousness
Serenity
Naturalness
Riffles in water
Turbidity
Lack of pollution

Source: From Morisawa, M., and Murie, M., 1969, *Evaluation of Natural Rivers.* Antioch College, Water Resources Research, Yellow Springs, OH, in Litton, R.B. and others, 1974, Water and Landscape, Water Information Center, Inc. Reprinted with permission.

SECTION 5I OCEANS AND SEAS

Table 5I.66 Dimensions of the Oceans

Ocean	Area ($10^9 m^2$)	Mean Depth (meters)	Volume ($10^{15} m^3$)
Arctic	14,090	1205	17.0
North Pacific	83,462	3858	322.0
South Pacific	65,521	3891	254.9
North Atlantic	46,772	3285	153.6
South Atlantic	37,364	4091	152.8
Indian	81,602	4284	349.6
Antarctic	32,249	3730	120.3

Source: From U.S. Naval oceanographic office, 1966.

Table 5I.67 Maximum Depths of the Oceans

Name of Area	Location		Depth m	Depth Fathoms	Depth ft
	Pacific Ocean				
Mariana trench	11°20′N	142°12′E	10,924	5973	35,840
Tonga trench	23°16′S	174°44′W	10,800	5906	35,433
Philippine trench	10°38′N	126°36′E	10,057	5499	32,995
Kermadec trench	31°53′S	177°21′W	10,047	5494	32,963
Bonin trench	24°30′N	143°24′E	9,994	5464	32,788
Kuril trench	44°15′N	150°34′E	9,750	5331	31,988
Izu trench	31°05′N	142°10′E	9,695	5301	31,808
New Britain trench	06°19′S	153°45′E	8,940	4888	29,331
Yap trench	08°33′N	138°02′E	8,527	4663	27,976
Japan trench	36°08′N	142°43′E	8,412	4600	27,599
Peru-Chile trench	23°18′S	71°14′W	8,064	4409	26,457
Palau trench	07°52′N	134°56′E	8,054	4404	26,424
Aleutian trench	50°51′N	177°11′E	7,679	4199	25,194
New Hebrides trench	20°36′S	168°37′E	7,570	4139	24,836
North Ryukyu trench	24°00′N	126°48′E	7,181	3927	23,560
Mid America trench	14°02′N	93°39′W	6,662	3643	21,857
	Atlantic Ocean				
Puerto Rico trench	19°55′N	65°27′N	8,605	4705	28,232
So. Sandwich trench	55°42′S	25°56′E	8,325	4552	27,313
Romanche gap	0°13′S	18°26′W	7,728	4226	25,354
Cayman trench	19°12′N	80°00′W	7,535	4120	24,721
Brazil basin	09°10′S	23°02′W	6,119	3346	20,076
	Indian Ocean				
Java trench	10°19′S	109°58′E	7,125	3896	23,376
Ob trench	09°45′S	67°18′E	6,874	3759	22,553
Diamantina trench	35°50′S	105°14′E	6,602	3610	21,660
Vema trench	09°08′S	67°15′E	6,402	3501	21,004
Agulhas basin	45°20′S	26°50′E	6,195	3387	20,325
	Arctic Ocean				
Eurasia basin	82°23′N	19°31′E	5,450	2980	17,881
	Mediterranean Sea				
Ionian basin	36°32′N	21°06′E	5,150	2816	16,896

Source: From The World Almanac and Book of Facts 1987. Copyright Pharos Books, a Scripps Howard Co., New York.

Table 5I.68 Water Residence Times of the Oceans

		North Polar	Atlantic	Pacific	Indian	Total
A.	Ocean volume (km^3×10^4)	8.85	350	695	295	1,349
B.	Surface ocean volume (200 m) (km^3×10^6)	1.7	19.6	35.4	15.5	72.2
C.	Ocean flow compensation (km^3/yr)	3,000	−17,100	+28,000	−13,900	0
D.	Total compensation time: A/C (yr)	2,950	20,500	25,000	21,200	
E.	Surface-only compensation time: B/C (yr)	570	1,150	1,250	1,100	
F.	Stream runoff to oceans (km^3/yr)	2,600	19,400	21,100	5,600	39,700
G.	Runoff residence time: A/F (yr)	3,400	18,000	57,500	52,700	34,000
H.	Surface runoff residence time: B/F (yr)	650	1,000	2,900	2,800	1,800
I.	Atmospheric cycling (precipitation minus evaporation) (km^3/yr)	400	−36,500	15,900	−19,500	−39,700
J.	Whole-ocean atmospheric-cycling residence time: A/I (yr)	22,125	9,600	43,700	15,100	34,000
K.	Atmospheric-cycling surface ocean residence time: B/I (yr)	4,250	500	2,200	1,000	1,800

Source: From Speidel, D.H., and Agnew, A.F., 1979, The Natural Geochemistry of our Environment, in An Overview of Research in Biogeochemistry and Environmental Health, Committee Print 825, Committee on Science and Technology. U.S. House of 77–239.

Table 5I.69 Water Balance of the Oceans

Ocean	Area	Precipitation		Evaporation		Inflow of Water from Continents		Water Balance	
	1000 km^3	mm	1000 km^3	mm	1000 km^3	mm	1000 km^3	mm	1000 km^3
Pacific	178,700	1460	260	1510	269.7	83	14.8	30	5.1
Southern sector	25,300	1140	28.9	684	17.3	72	1.8	530	13.4
Without southern sector	153,400	1510	231.1	1640	252.4	85	13.0	−50	−8.4
Atlantic	91,700	1010	92.7	1360	124.4	226	20.8	−120	−10.9
Southern sector	15,500	1190	18.4	466	7.2	37	0.6.	760	11.8
Without southern sector	76,200	975	74.3	1540	117.2	265	20.6	−300	−22.7
Indian	76,200	1320	100.4	1420	108.0	81	6.1	−20	−1.5
Southern sector	28,500	1240	35.4	688	19.6	30	0.8	580	16.6
Without southern sector	47,700	1360	65.0	1850	88.4	111	5.3	−380	−18.1
Arctic	14,700	361	5.3	220	8.2	355	5.2	500	7.3
World	361,300	1270	458.0	1400	505.0	130	47.0	0	0

Source: From UNESCO, 1977, Atlas of World Water Balance. Reproduced with permission.

Table 5I.70 Dimensions of Individual Seas

Sea	Area ($10^9 m^2$)	Mean Depth (m)	Volume ($10^{12} m^3$)
Tributary to Arctic Ocean			
Norwegian Sea	1383	1742	2408
Greenland Sea	1205	1444	1740
Barents Sea	1405	229	322
White Sea	90	89	8
Kara Sea	883	118	104
Laptev Sea	650	519	338
East Siberian Sea	901	58	53
Chukchi Sea	582	88	51
Beaufort Sea	476	1004	478
Baffin Bay	689	861	593
Tributary to North Atlantic			
North Sea	600	91	55
Baltic Sea	386	86	33
Mediterranean Sea	2516	1494	3758
Black Sea	461	1166	537
Marmara	11	357	4
Azov	40	9	0.4
Caribbean Sea	2754	2491	6860
Gulf of Mexico	1543	1512	2332
Gulf of St. Lawrence	238	127	30
Hudson Bay	1232	128	158
Tributary to South Atlantic			
Gulf of Guinea	1533	2996	4592
Tributary to Indian Ocean			
Red Sea	450	558	251
Persian Gulf	241	40	10
Arabian Sea	3863	2734	10561
Bay of Bengal	2172	2586	5616
Andaman Sea	602	1096	660
Great Australian Bight	484	950	459
Tributary to North Pacific			
Gulf of California	177	818	145
Gulf of Alaska	1327	2431	3226
Bearing Sea	2304	1598	3683
Okhotsk Sea	1590	859	1365
Japan Sea	978	1752	1713
Yellow Sea	417	40	17
East China Sea	752	349	263
Sulu Sea	420	1139	478
Celebes Sea	472	3291	1553
In both North and South Pacific			
South China Sea	3685	1060	3907
Makassar Strait	194	967	188
Molukka Sea	307	1880	578
Ceram Sea	187	1209	227
Tributary to South Pacific			
Java Sea	433	46	20
Bali Sea	119	411	49
Flores Sea	121	1829	222
Savu Sea	105	1701	178
Banda Sea	695	3064	2129
Ceram Sea	187	1209	227
Timor Sea	615	406	250
Arafura Sea	1037	197	204
Coral Sea	4791	2394	11470

Source: From U.S. Naval Oceanographic Office, 1966; amended.

Table 5I.71 Average Rise and Fall of Tides in the United States and Canada

Location	Feet
East Coast	
Quebec	13.7
Halifax, NS	4.4
St. John, NB	20.8
Eastport, ME	18.4
Portland, ME	9.1
Boston, MA	9.5
Newport, RI	3.5
New London, CT	2.6
Bridgeport, CT	6.7
New York (The Battery)	4.5
Port Jefferson, NY	6.6
Albany, NY	4.6
Newark, NJ	5.1
Sandy Hook, NJ	4.7
Philadelphia, PA	6.2
Cape May, NJ	4.6
Washington, DC	2.8
Cape Hatteras, NC	3.6
Wilmington, NC	4.2
Charlotte, SC	5.2
Savannah, GA	7.9
Miami, FL	2.4
Key West, FL	1.3
Mobile, AL	1.5[a]
Galveston, TX	1.0
San Juan, PR	1.1
West Coast	
Vancouver, BC	10.6[a]
Seattle, WA	7.7
San Francisco, CA	4.1
Los Angeles, CA	3.8
San Diego, CA	4.0

[a] Diurnal range.

Source: From National Oceanic and Atmospheric Administration, National Ocean Survey Tide Tables.

Table 5I.72 Temperatures and Salinities of the Oceans

	Temperature	Salinity
North Atlantic		
1. North Polar water	−1 to +2	34.9
2. Subarctic water	+3 to +5	34.7 to 34.9
3. North Atlantic central water	+4 to +17	35.1 to 36.2
4. North Atlantic water	+3 to +4	34.9 to 35.0
5. North Atlantic bottom water	+1 to +3	34.8 to 34.9
6. Mediterranean water	+6 to +10	35.3 to 36.4
South Atlantic		
1. South Atlantic central water	+5 to +16	34.3 to 35.6
2. Antarctic intermediate water	+3 to +5	34.1 to 34.6
3. Subantarctic water	+3 to +9	33.8 to 34.5
4. Antarctic circumpolar water	+0.5 to +2.5	34.7 to 34.8
5. South Atlantic deep and bottom water	0 to +2	34.5 to 34.9
6. Antarctic bottom water	−0.4	34 to 36
Indian Ocean		
1. Equatorial water	4 to 16	34.8 to 35.2
2. Indian central water	6 to 15	34.5 to 35.4
3. Antarctic intermediate water	2 to 6	34.4 to 34.7
4. Subantarctic water	2 to 8	34.1 to 34.6
5. Indian Ocean deep and Antarctic circumpolar water	0.5 to 2	34.7 to 34.75
6. Red Sea water	9	35.5
South Pacific		
1. Eastern South Pacific water	9 to 16	34.3 to 35.1
2. Western South Pacific water	7 to 16	34.5 to 35.5
3. Antarctic intermediate water	4 to 7	34.3 to 34.5
4. Subantarctic water	3 to 7	34.1 to 34.6
5. Pacific deep water and Antarctic circumpolar water	(−1) to 3	34.6 to 34.7
North Pacific		
1. Subarctic water	2 to 10	33.5 to 34.4
2. Pacific equatorial water	6 to 16	34.5 to 35.2
3. Eastern North Pacific water	10 to 16	34.0 to 34.6
4. Western North Pacific water	7 to 16	34.1 to 34.6
5. Arctic intermediate water	6 to 10	34.0 to 34.1
6. Pacific deep water and Arctic circumpolar water	(−1) to 3	34.6 to 34.7

Source: From U.S. Oceanographic Office, 1966.

Table 5I.73 Composition of Sea Water

Constituent	Concentration (ppm)
Chloride	18,980
Sodium	10,560
Sulfate	2,560
Magnesium	1,272
Calcium	400
Potassium	380
Bicarbonate	142
Bromide	65
Strontium	13
Boron	4.6
Fluoride	1.4
Rubidium	0.2
Aluminum	0.16–1.9
Lithium	0.1
Barium	0.05
Iodide	0.05
Silicate	0.04–8.6
Nitrogen	0.03–0.9

(Continued)

Table 5I.73 (Continued)

Constituent	Concentration (ppm)
Zinc	0.005–0.014
Lead	0.004–0.005
Selenium	0.004
Arsenic	0.003–0.024
Copper	0.001–0.09
Tin	0.003
Iron	0.002–0.02
Cesium	$\sim$0.002
Manganese	0.001–0.01
Phosphorous	0.001–0.10
Thorium	$\leq$0.0005
Mercury	0.0003
Uranium	0.00015–0.0016
Cobalt	0.0001
Nickel	0.0001–0.0005
Radium	8×10^{-11}
Beryllium	—
Cadmium	—
Chromium	—
Titanium	Trace

Source: From U.S. Geological survey.

Table 5I.74 Approximate Mineral Content of One Cubic Mile of Sea Water

Mineral	Weight (in tons)	Mineral	Weight (in tons)
Sodium chloride	120,000,000	Fluorine	6,400
Magnesium chloride	18,000,000	Barium	900
Magnesium sulfate	8,000,000	Iodine	100 to 12,000
Calcium sulfate	6,000,000	Arsenic	50 to 350
Potassium sulfate	4,000,000	Rubidium	200
Calcium carbonate	550,000	Silver	up to 45
Magnesium bromide	350,000	Copper, Manganese, Zinc, Lead	10 to 30
Bromine	300,000	Gold	up to 25
Strontium	60,000	Radium	about 1/6 (ounce)
Boron	21,000	Uranium	7

Source: From Smith, The Sun, the Sea, and Tomorrow; Potential Sources of Food, Energy and Minerals from the Sea, Charles Scribners, 1954. With permission.

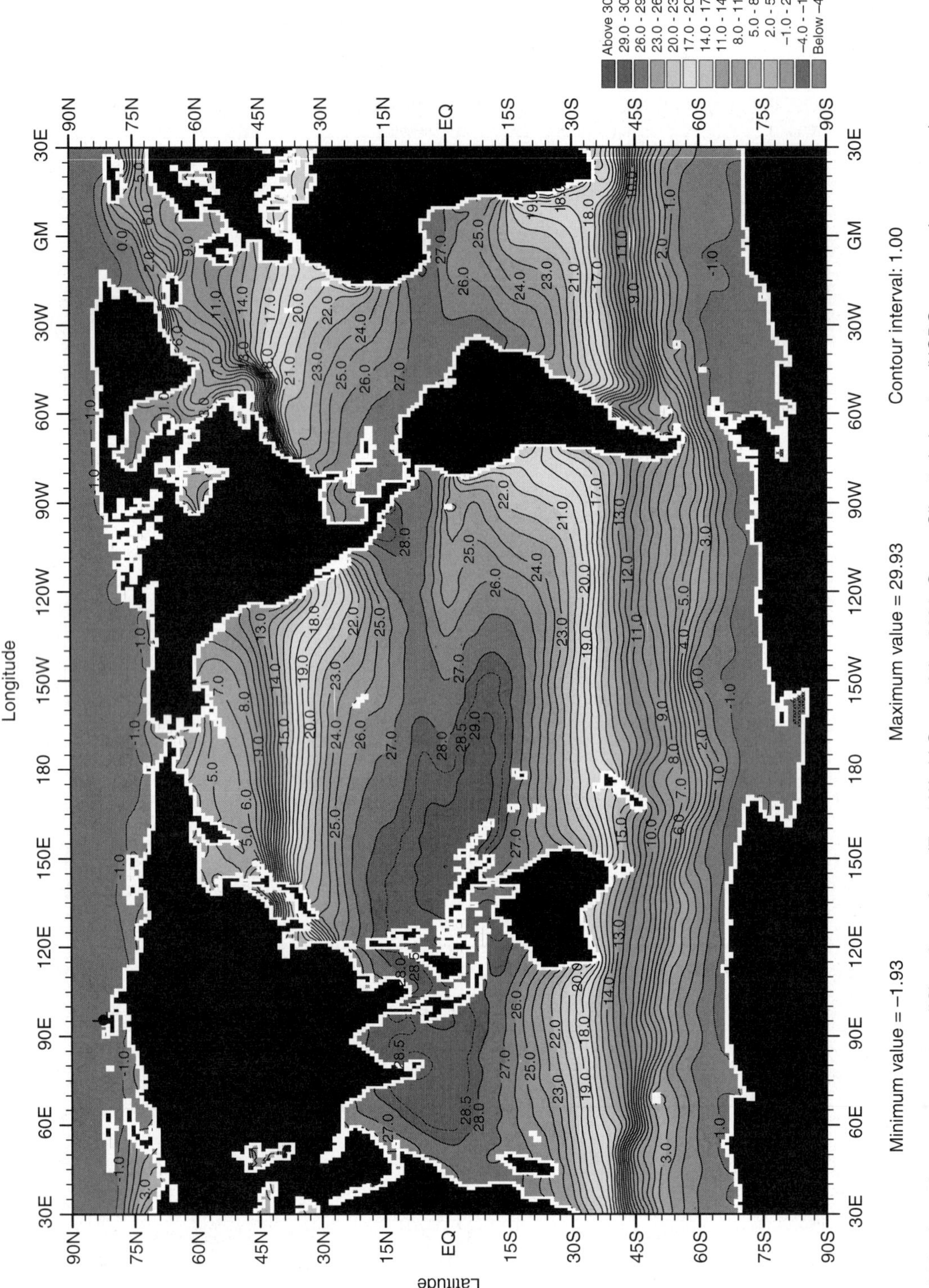

Figure 5I.33 Annual mean temperature (°C) at the surface. (From World Ocean Atlas 2001, Ocean Climate Laboratory/NODC, www.nodc.noaa.gov.)

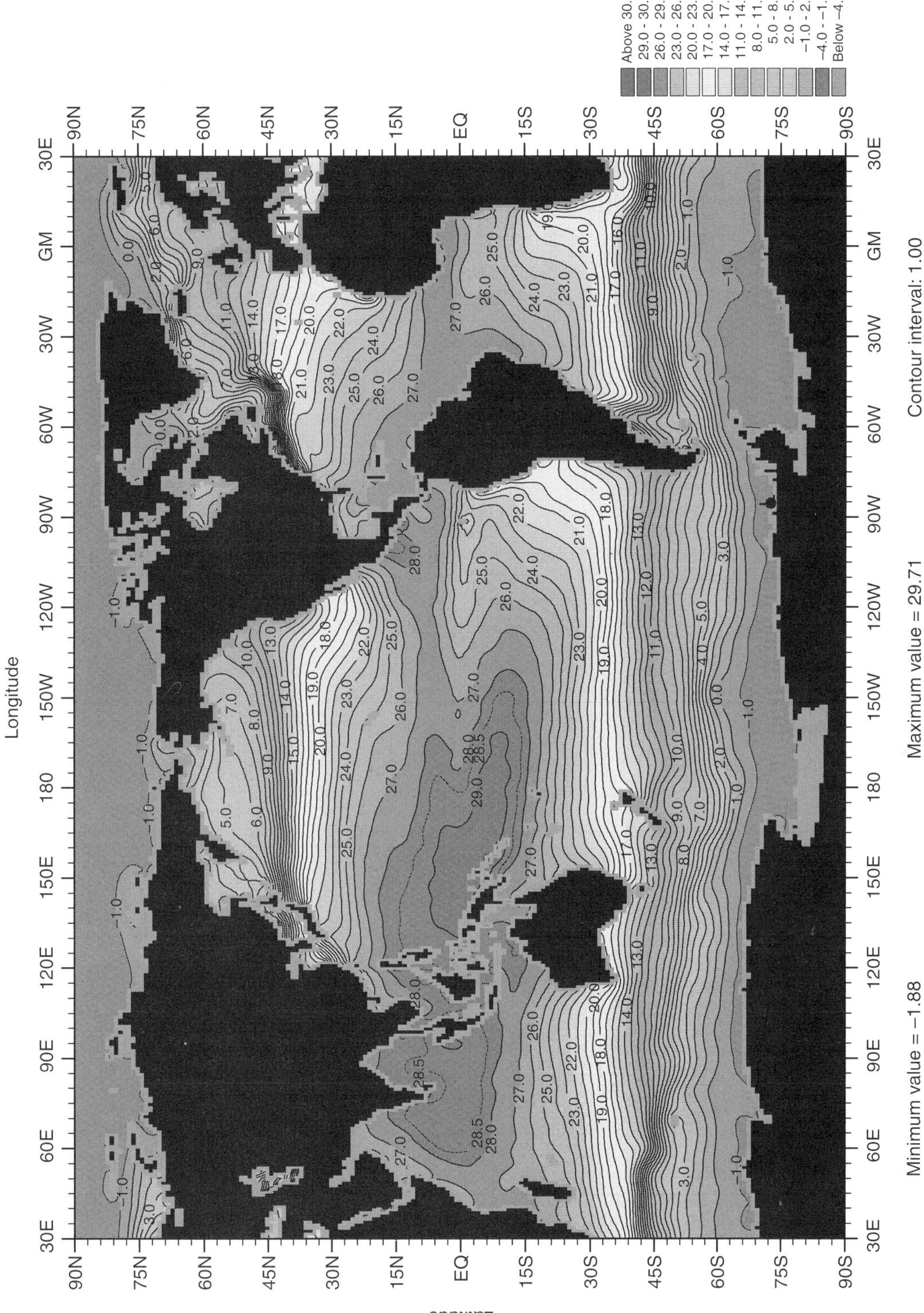

Figure 5I.34 Annual mean temperature (°C) at 10 m depth. (From World Ocean Atlas 2001, Ocean Climate Laboratory/NODC, www.nodc.noaa.gov.)

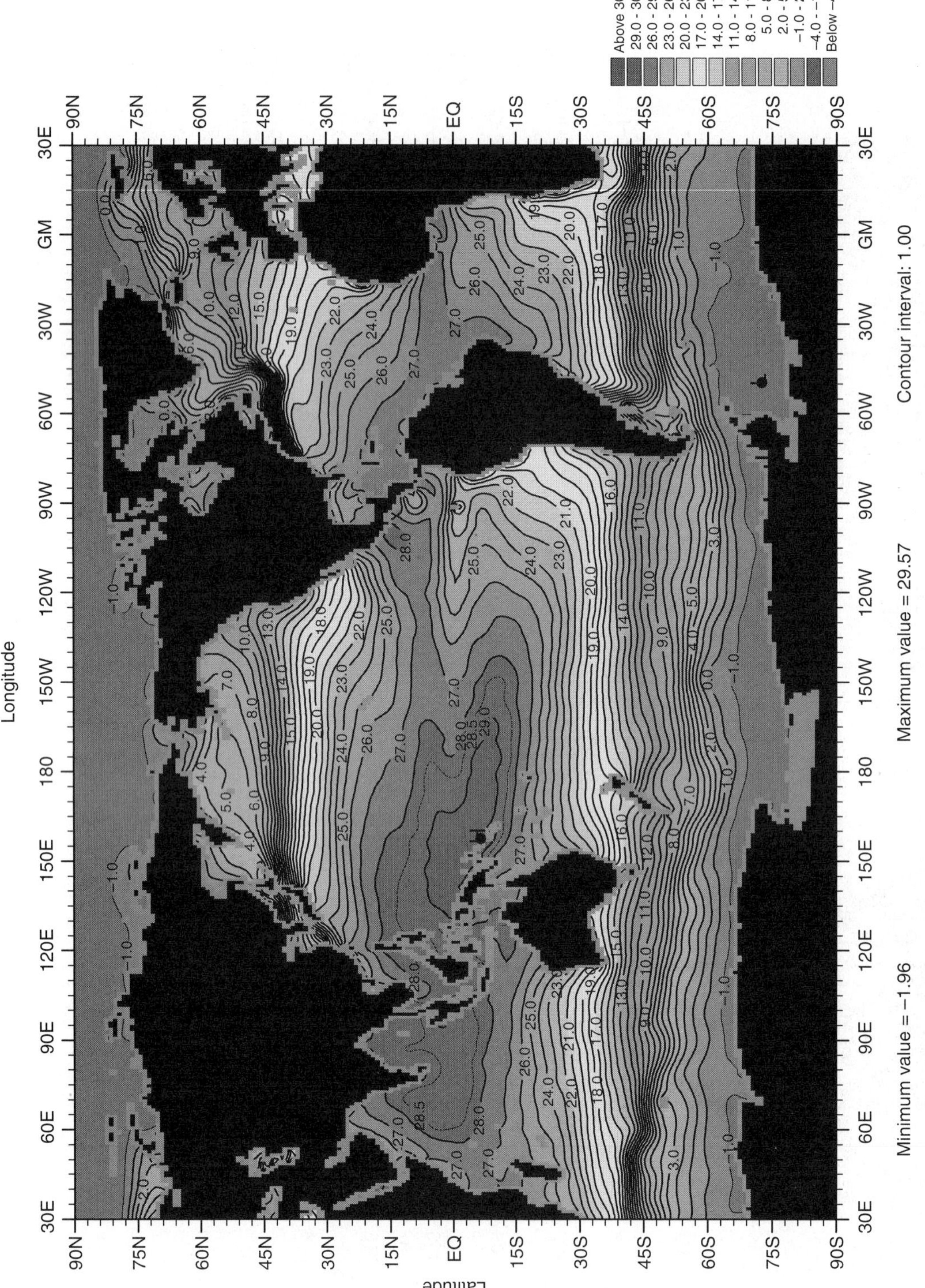

Figure 5I.35 Annual mean temperature (°C) at 20 m depth. (From World Ocean Atlas 2001, Ocean Climate Laboratory/NODC, www.nodc.noaa.gov.)

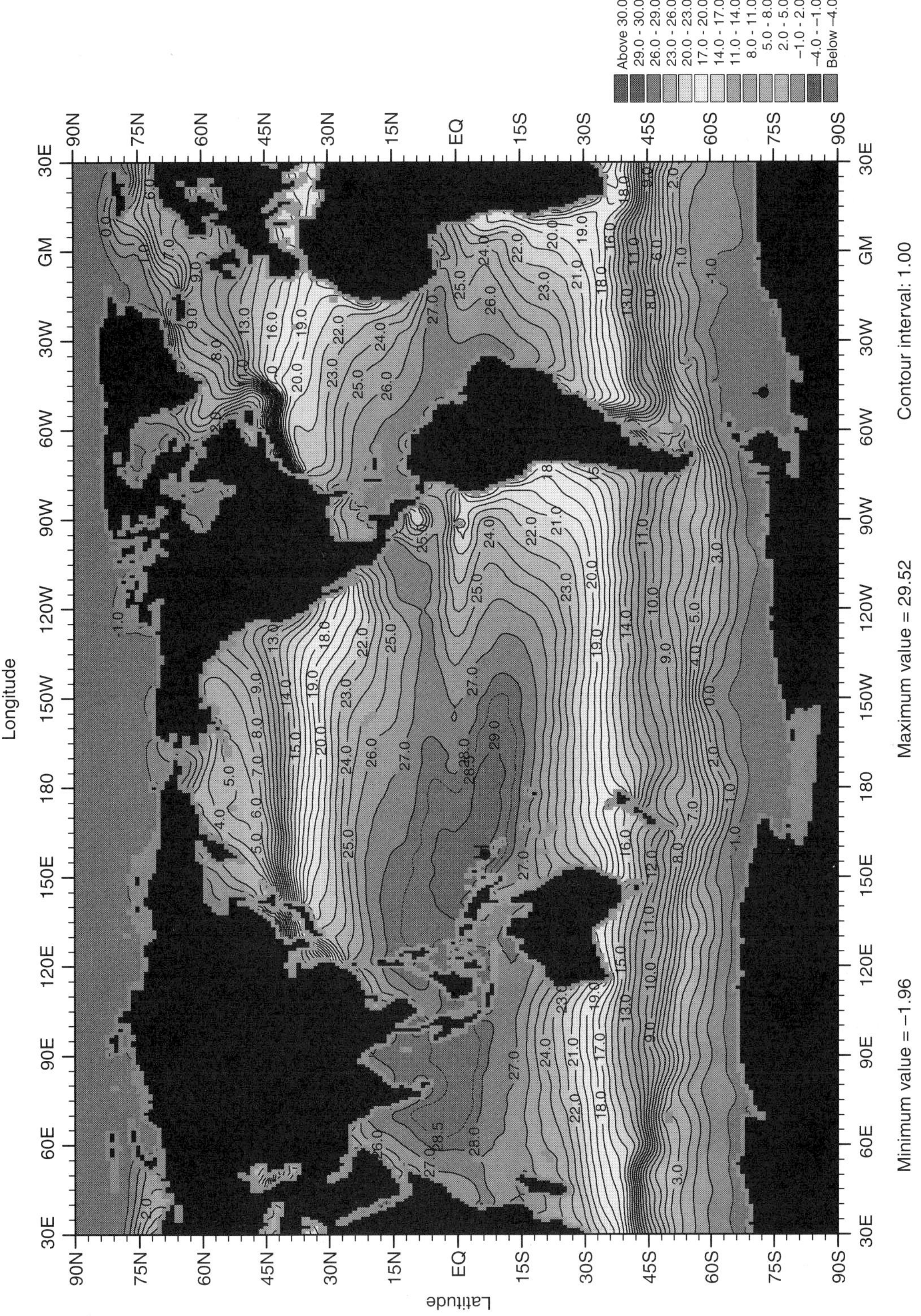

Figure 5I.36 Annual mean temperature (°C) at 30 m depth. (From World Ocean Atlas 2001, Ocean Climate Laboratory/NODC, www.nodc.noaa.gov.)

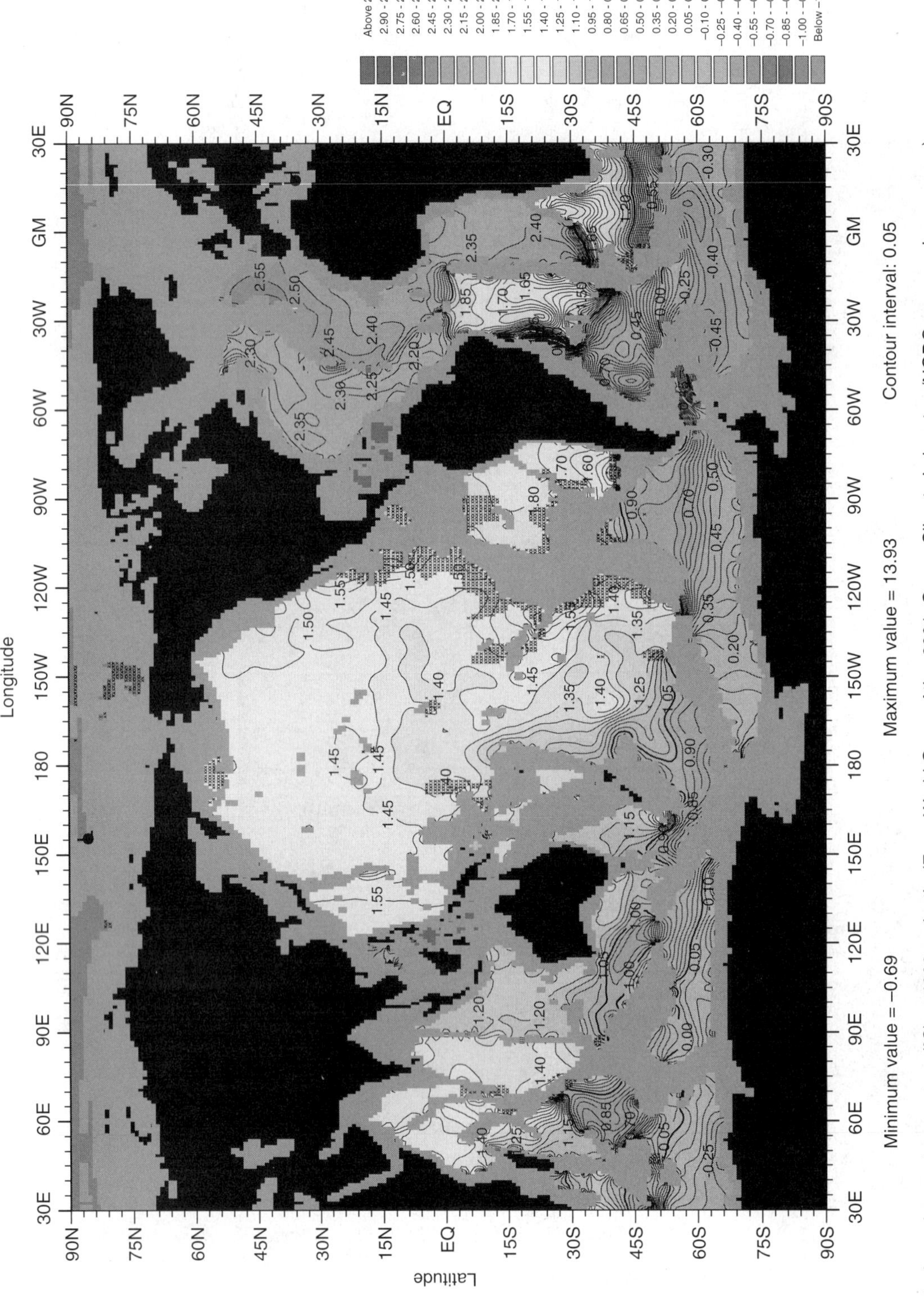

Figure 5I.37 Annual mean temperature (°C) at 4000 m depth. (From World Ocean Atlas 2001, Ocean Climate Laboratory/NODC, www.nodc.noaa.gov.)

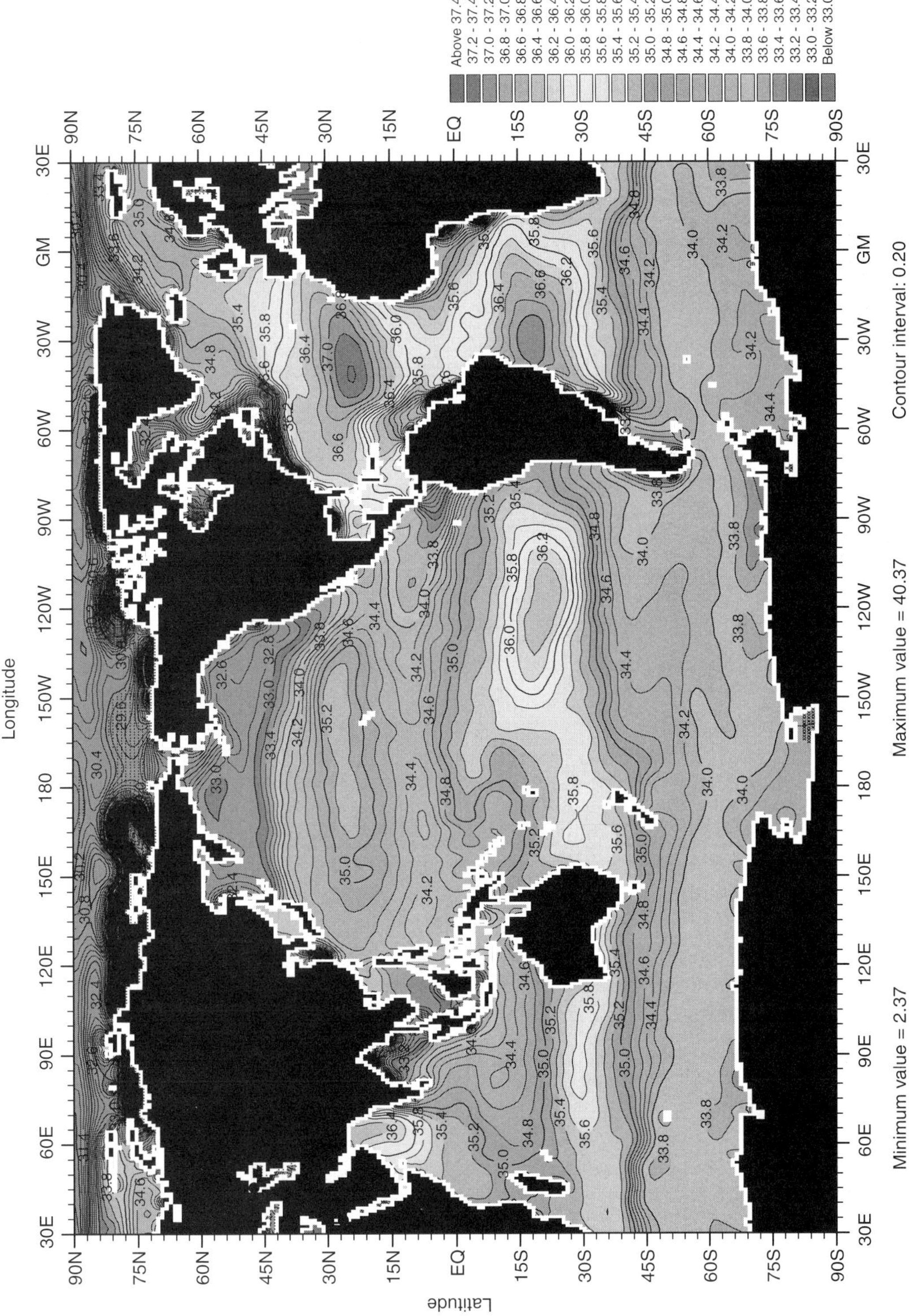

Figure 5I.38 Annual mean salinity (PSS) at the surface. (From World Ocean Atlas 2001, Ocean Climate Laboratory/NODC, www.nodc.noaa.gov.)

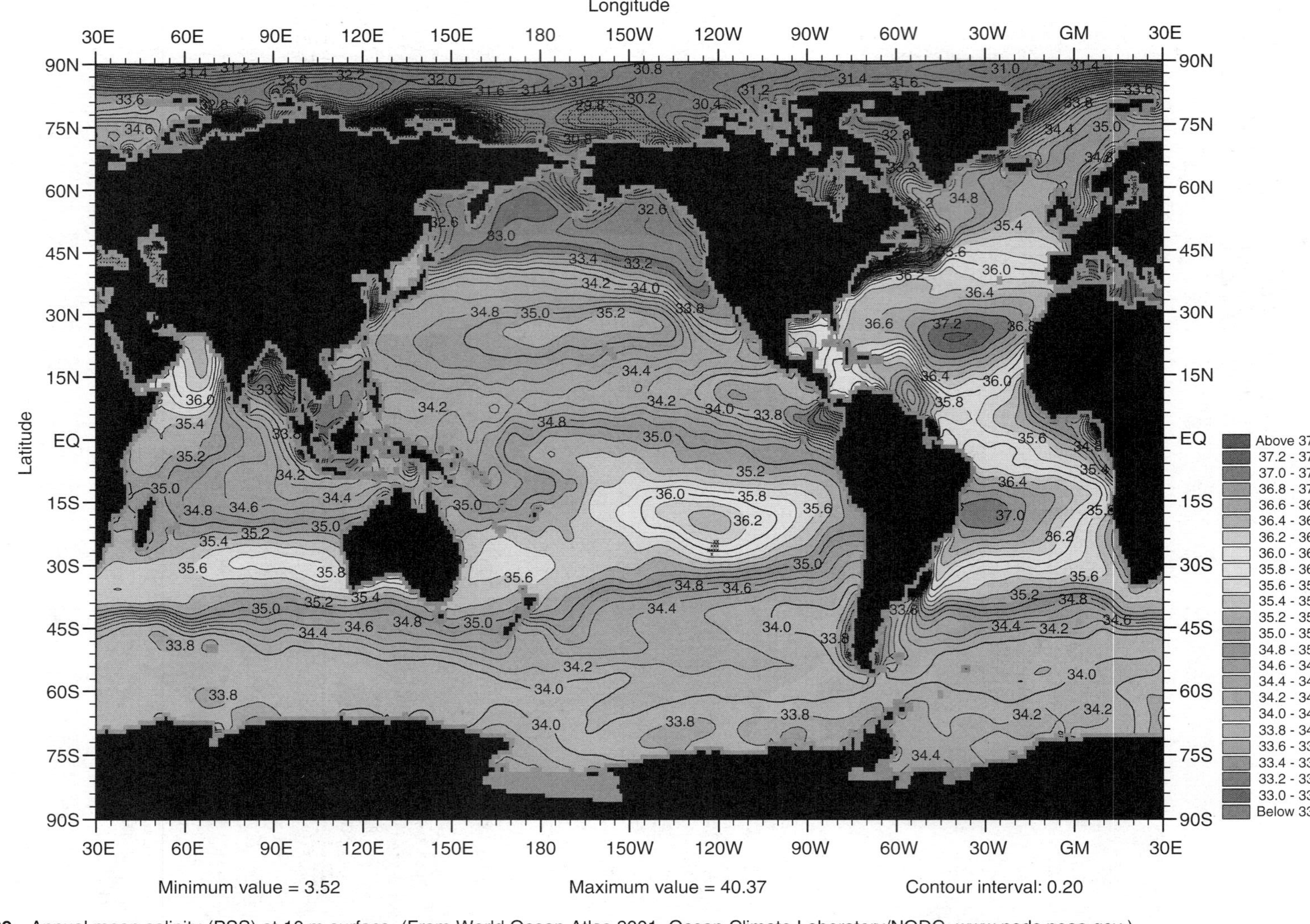

Figure 5I.39 Annual mean salinity (PSS) at 10 m surface. (From World Ocean Atlas 2001, Ocean Climate Laboratory/NODC, www.nodc.noaa.gov.)

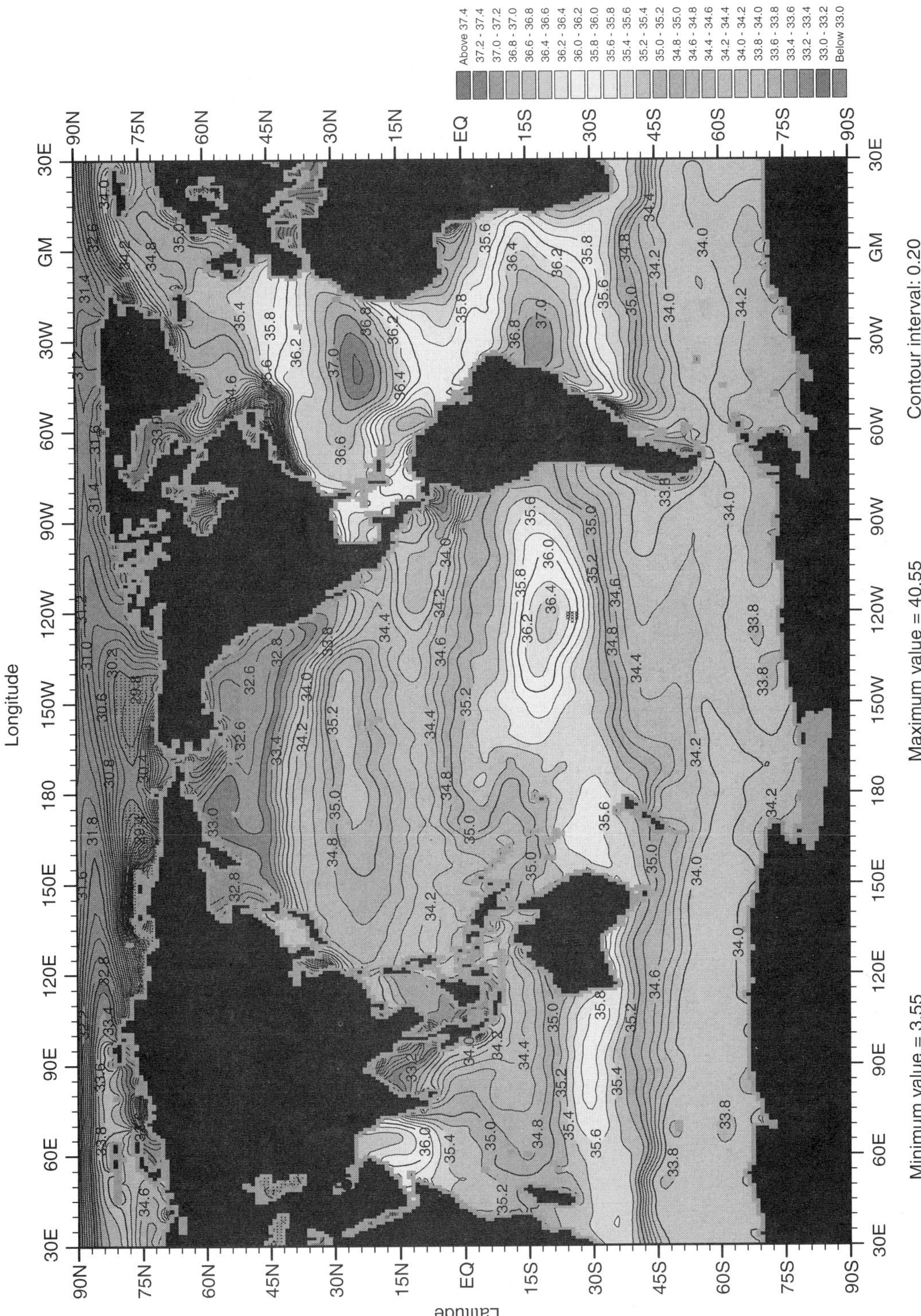

Figure 5I.40 Annual mean salinity (PSS) at 20 m surface. (From World Ocean Atlas 2001, Ocean Climate Laboratory/NODC, www.nodc.noaa.gov.)

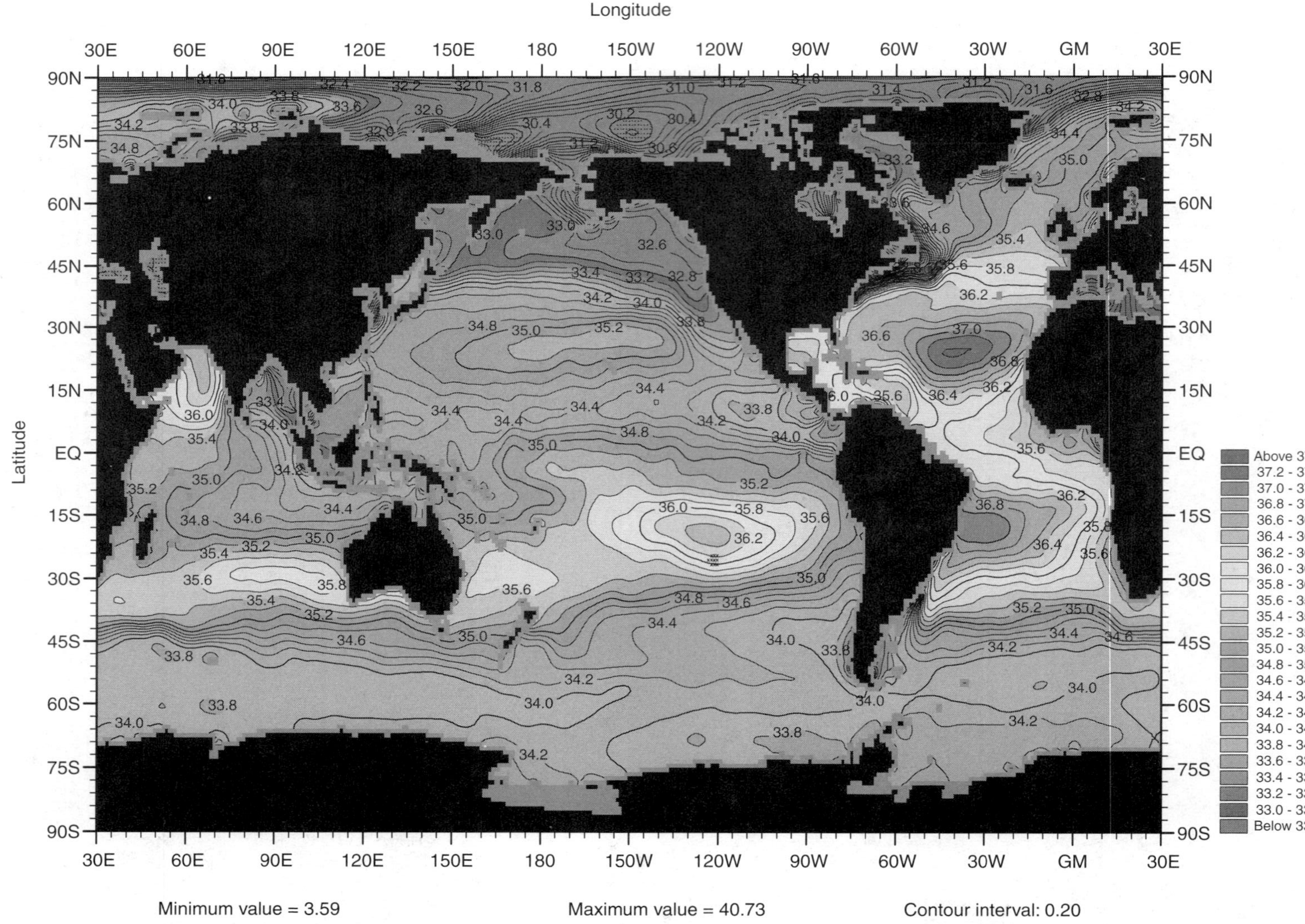

Figure 5I.41 Annual mean salinity (PSS) at 30 m surface. (From World Ocean Atlas 2001, Ocean Climate Laboratory/NODC, www.nodc.noaa.gov.)

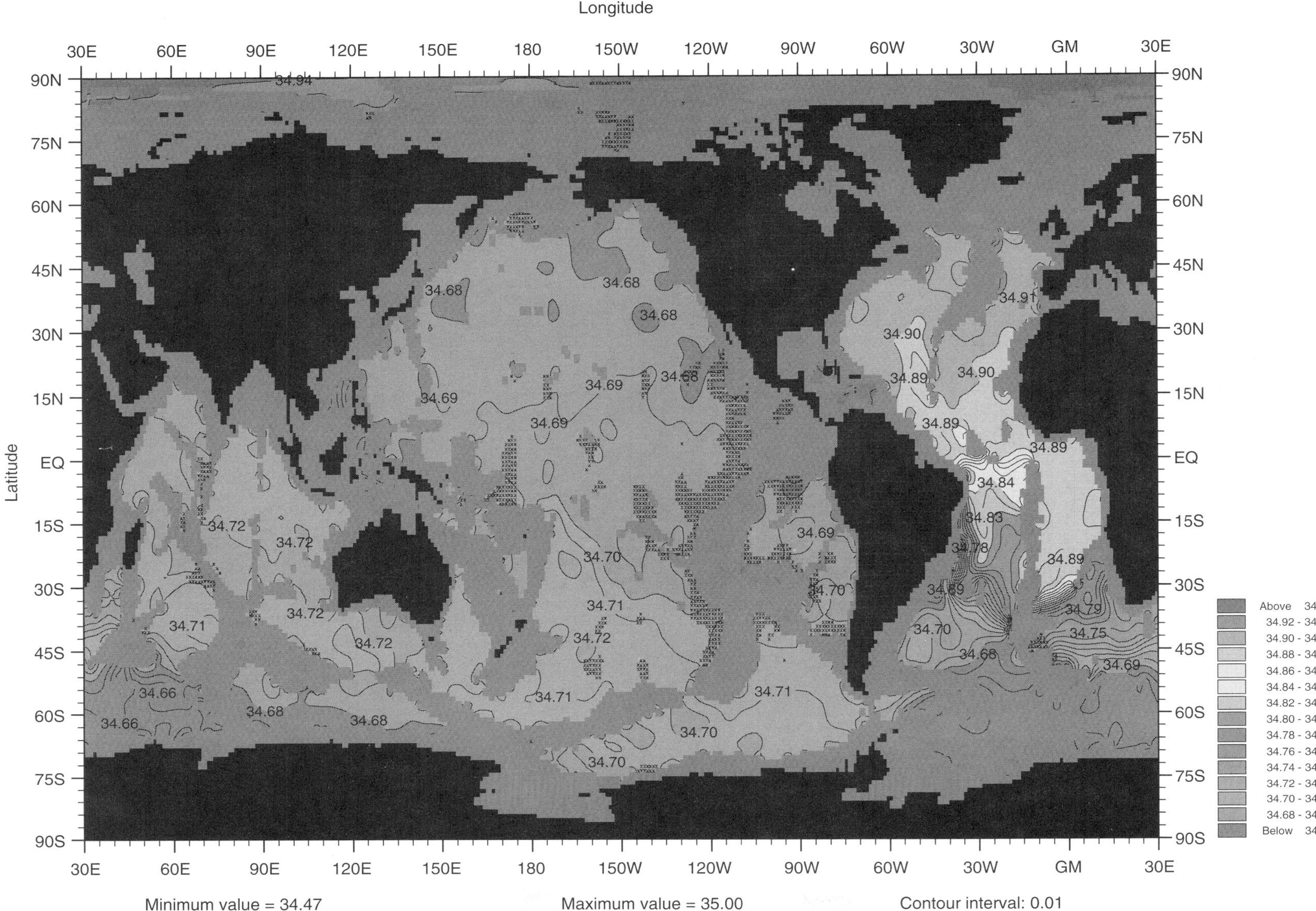

Figure 5I.42 Annual mean salinity (PSS) at 4000 m surface. (From World Ocean Atlas 2001, Ocean Climate Laboratory/NODC, www.nodc.noaa.gov.)

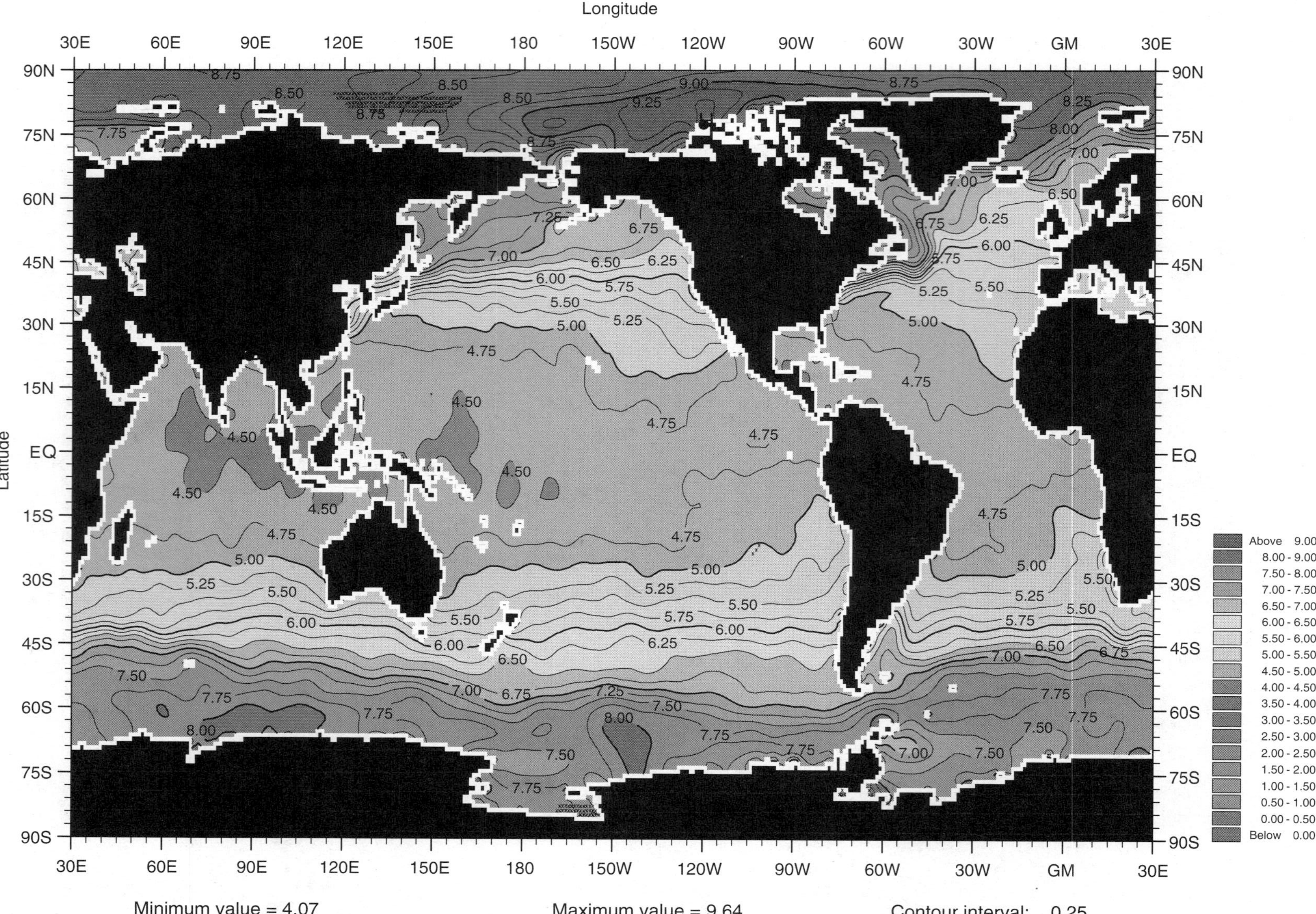

Figure 5I.43 Annual mean oxygen (mL/l) at the surface. (From World Ocean Atlas 2001, Ocean Climate Laboratory/NODC, www.nodc.noaa.gov.)

CHAPTER 6

Groundwater

Melvin Rivera

CONTENTS

SECTION 6A GROUNDWATER — UNITED STATES

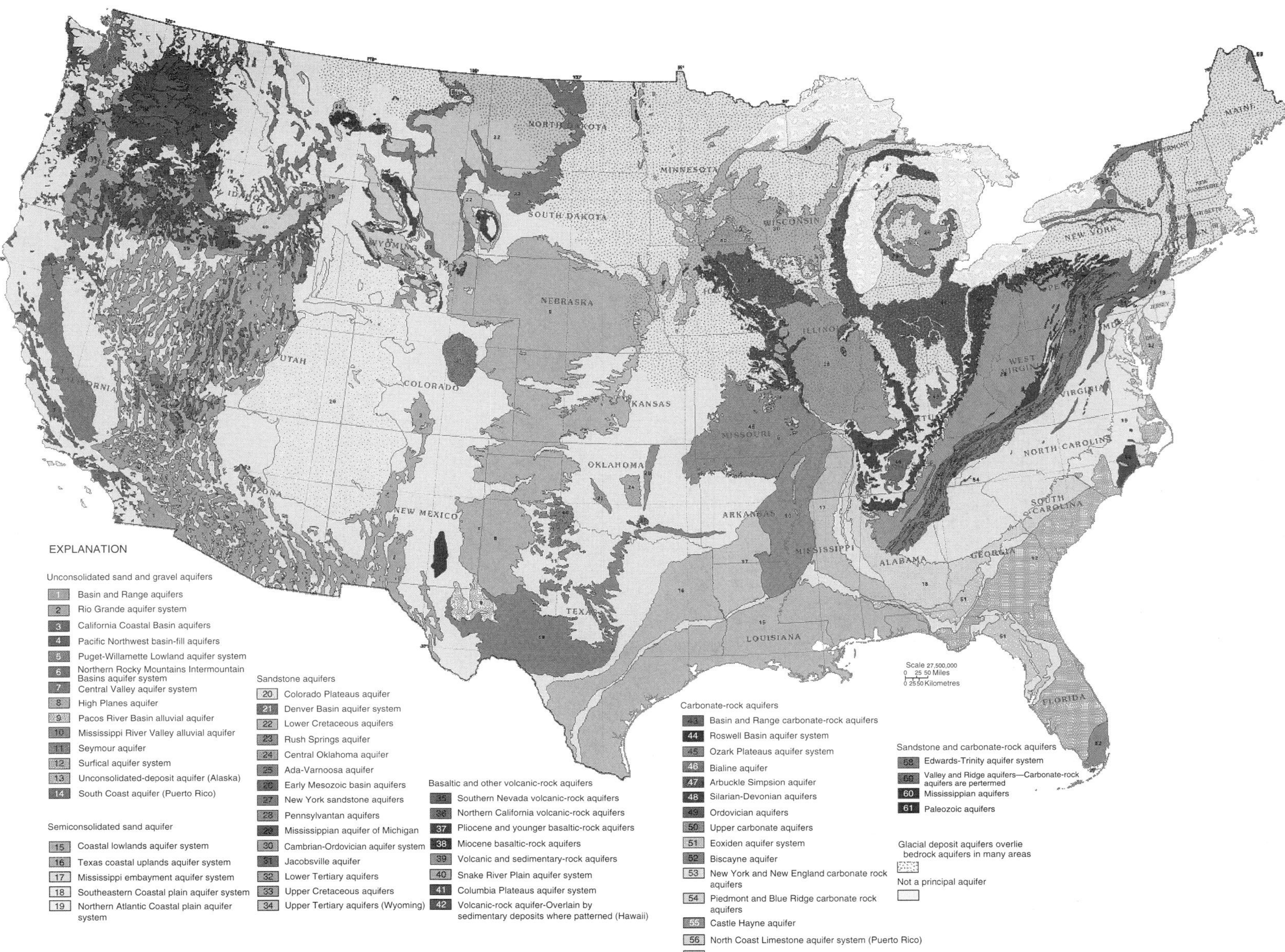
EXPLANATION
Unconsolidated sand and gravel aquifers
1 Basin and Range aquifers
2 Rio Grande aquifer system
3 California Coastal Basin aquifers
4 Pacific Northwest basin-fill aquifers
5 Puget-Willamette Lowland aquifer system
6 Northern Rocky Mountains Intermountain Basins aquifer system
7 Central Valley aquifer system
8 High Planes aquifer
9 Pacos River Basin alluvial aquifer
10 Mississippi River Valley alluvial aquifer
11 Seymour aquifer
12 Surfical aquifer system
13 Unconsolidated-deposit aquifer (Alaska)
14 South Coast aquifer (Puerto Rico)
Semiconsolidated sand aquifer
15 Coastal lowlands aquifer system
16 Texas coastal uplands aquifer system
17 Mississippi embayment aquifer system
18 Southeastern Coastal plain aquifer system
19 Northern Atlantic Coastal plain aquifer system
Sandstone aquifers
20 Colorado Plateaus aquifer
21 Denver Basin aquifer system
22 Lower Cretaceous aquifers
23 Rush Springs aquifer
24 Central Oklahoma aquifer
25 Ada-Varnoosa aquifer
26 Early Mesozoic basin aquifers
27 New York sandstone aquifers
28 Pennsylvantan aquifers
29 Mississippian aquifer of Michigan
30 Cambrian-Ordovician aquifer system
31 Jacobsville aquifer
32 Lower Tertiary aquifers
33 Upper Cretaceous aquifers
34 Upper Tertiary aquifers (Wyoming)
Basaltic and other volcanic-rock aquifers
35 Southern Nevada volcanic-rock aquifers
36 Northern California volcanic-rock aquifers
37 Pliocene and younger basaltic-rock aquifers
38 Miocene basaltic-rock aquifers
39 Volcanic and sedimentary-rock aquifers
40 Snake River Plain aquifer system
41 Columbia Plateaus aquifer system
42 Volcanic-rock aquifer-Overlain by sedimentary deposits where patterned (Hawaii)
Carbonate-rock aquifers
43 Basin and Range carbonate-rock aquifers
44 Roswell Basin aquifer system
45 Ozark Plateaus aquifer system
46 Bialine aquifer
47 Arbuckle Simpsion aquifer
48 Silarian-Devonian aquifers
49 Ordovician aquifers
50 Upper carbonate aquifers
51 Eoxiden aquifer system
52 Biscayne aquifer
53 New York and New England carbonate rock aquifers
54 Piedmont and Blue Ridge carbonate rock aquifers
55 Castle Hayne aquifer
56 North Coast Limestone aquifer system (Puerto Rico)
57 Kingshill aquifer (St. Croix)
Sandstone and carbonate-rock aquifers
58 Edwards-Trinity aquifer system
59 Valley and Ridge aquifers—Carbonate-rock aquifers are pertermed
60 Mississippian aquifers
61 Paleozoic aquifers
Glacial deposit aquifers overlie bedrock aquifers in many areas
Not a principal aquifer
Scale 27,500,000
0 25 50 Miles
0 25 50 Kilometres

Figure 6A.1 Principal aquifers of the United States. (From http://capp.water.usgs.gov.)

Figure 6A.2 River valley aquifers in the United States. (From Water Information Center, 1973, Water Atlas of the United States. H.E. Thomas, *The Conservation of Ground Water*, McGraw-Hill, 1951. With permission.)

Table 6A.1 Occurrence of Aquifers in the United States

(1) Geologic Age and Rock Type	(2) Western Mountain Ranges	(3) Arid Basin	(4) Columbi Lava Plateau	(5) Colorado Plateau	(6) High Plains	(7) Unglaciated Central Region	(8) Glaciated Central Region	(9) Unglaciated Appalachian Region	(10) Glaciated Appalachian Region	(11) Atlantic and Gulf Coastal Plain	(12) Special Comments
Cenozoic											
Quaternary											
Aluvium and related deposits (primarily. Recent and Pleistocene sediments and may include some of Pliocene age)	S and G deposits in valleys and along stream courses. Highly productive but not greatly developed — P to M	S and G deposits in valleys and along stream courses. Highly developed with local depletion. Storage large but perennial recharge limited-P	S and G deposits along streams, interbedded with basalt — I to M	U	S and G along water courses. Sand dune deposits — P (in part)	S and G along water courses and in terrace deposits — I (limited)	S and G along water courses — M	S and G along water courses and in terrace deposits. Not developed		S and G along water courses and in terrace and littoral deposits, especially in the Mississippi and tributary valleys. Not highly developed in East and South. Some depletion in Gulf Coast — I	The most widespread and important aquifers in the United States. Well over one-half of all groundwater pumped in the United States is withdrawn from these aquifers. Many are easily available for artificial recharge and induced infiltration. Subject to saltwater contamination in coastal areas
Glacial drift, especially outwash (Pleisto-cene)	S and G deposits in northern part of region — I	S and G deposits especially in northern part of region and in some valleys — I	S and G outwash, especially in Spokane area — I	U	S and G outwash, much of it reworked (see above) — I	S and G outwash especially along northern boundary of region — I	S and G outwash, terrace deposits and lenses in till throughout region — P (in part)	S and G outwash in northern part. Not highly developed — M	S and G outwash, terrace deposits and lenses in till. Locally highly deve-loped — I	S and G outwash in Mississippi Valley (see above) — I	
Other Pleisto-cene sediments	Alluvial Fm and other basin deposits in the southern part — M to P (see Alluvium above)		U	U	Alluviated plains and valley fills — M to I		U	U	U	Coquina, limestone, sand, and marl Fms in Florida — M	

(Continued)

Table 6A.1 (Continued)

(1) Geologic Age and Rock Type	(2) Western Mountain Ranges	(3) Arid Basin	(4) Columbi Lava Plateau	(5) Colorado Plateau	(6) High Plains	(7) Unglaciated Central Region	(8) Glaciated Central Region	(9) Unglaciated Appalachian Region	(10) Glaciated Appalachian Region	(11) Atlantic and Gulf Coastal Plain	(12) Special Comments
Tertiary Sediments, Pliocene	S and G in valley fill and terrace deposits. Not highly developed — M	Some S and G in valley fill — M	U	U	Ogalalla Fm in High Plains. Extensive S and G with huge storage but little recharge locally. Much depletion — P (in part)	U	U	Absent	Absent	Dewitt Ss in Texas. Citronelle and LaFayette Fms in Gulf States — I	
Miocene	Ellensburg Fm in Washington — I; elsewhere — U	U	Ellensburg Fm in Washington — I; elsewhere — U	U	Arikaree Fm — M	Arikaree Fm — M	Flaxville and other terrace deposits, S and G in north western part — M	Absent	Absent	New Jersey, Maryland, Delaware, Virginia — Cohansey and Calvert Fms — I Delaware to North Carolina — St. Marys and Calvert Fms — I Georgia and Florida — Tampa Ls, Alluvium Bluff Gp, and Tamiami Fm — I Eastern Texas — Oakville and Catahoula Ss — I	Aquifers in coastal areas subject to saltwater encroachment and contamination
Oligocene	U	U	U	U	Brule, clay, locally — I; else where — U	U	U	Absent	Absent	Suwannee Fm, Byram Ls, and Vicksburg Gp — I	

Eocene	Knight and Almy Fm in south-west Wyoming — M	U	U	Knight and Almy Fm in southwest Wyoming, Chuska Ss, and Tohatchi Sh in northwest Arizona and north-east New Mexico — M	U	Claibourne and Wilcox Gp in southern Illinois (?), Kentucky, and Missouri — M; else-where — U	Absent	Absent	Absent	New Jersey, Maryland, Delaware, Virginia — Pamunkey Gp — I. North Carolina to Florida — Ocal a Ls and Castle Hayne Marl — P (in part) Florida — Avon Park Ls, South Carolina to Mexican border, Claibourne Gp, Wilcox Gp — I	Includes the principal formations (Ocala Ls, especially) of the great Floridan aquifer. Subject to saltwater contamination in coastal areas but source of largest groundwater supply in southeastern United States
Paleocene	U	U	U	U	Feet Union Gp — M	Feet Union Gp — M	Feet Union Gp — M	Absent	Absent	Clayton Fm in Georgia — I	
Volcanic rocks, primarily basalt	U	Local flows — M	Many interbeded basalt flows from Eocene to Plio-cene — P	Local flows — M	Absent	Absent	Absent	Absent	Absent	Absent	
Mesozoic upper creta-ceous	U	Ss lenses in southern California — M; elsewhere — U	U	Dakota Ss and other not clearly distinguishable Ss a notable source of water from Minnesota and Iowa to the Rocky Mountains and south into New Mexico; also in Utah and Arizona — I In northwestern part of region Fox Hills and related Ss (Lennep, Colgate, etc.) locally valuable as water sources — M Ss of Montana Gp — M Ss members of Mesaverde Gp in Wyoming, Colorado, Utah, New Mexico, and Arizona — M In Texas aquifers listed under col. 11 — I				U	U	New Jersey, Maryland, Delaware — Magothy and Raritan Fm — I North and South Carolina — Peedee and Black Creek Fms — I Alabama and Georgia — Ripley and Eutah Fms — I	In coastal areas subject to saltwater encroachment and contamination. Ss aquifers of the central regions and the west primarily valuable when water from other sources is unavailable

(Continued)

Table 6A.1 (Continued)

(1) Geologic Age and Rock Type	(2) Western Mountain Ranges	(3) Arid Basin	(4) Columbi Lava Plateau	(5) Colorado Plateau	(6) High Plains	(7) Unglaciated Central Region	(8) Glaciated Central Region	(9) Unglaciated Appalachian Region	(10) Glaciated Appalachian Region	(11) Atlantic and Gulf Coastal Plain	(12) Special Comments
										Tennessee, Kentucky, Illinois — McNairy Ss — I Arkansas to Texas — Navarro Gp and Taylor Fm — I	
Lower creta-ceous	U	U	U	In northern part of these regions Lakota, Cloverly, and Kootenai Ss — M In southern part Purgatoire and Dakota Ss—M. Texas aquifers listed in col. 11—I			U	U	U	Texas — Woodbine Ss — I. New Jersey, Maryland, Delaware — Patapsco and Patuxent Fms — I West of Mississippi River, especially in Texas — Edwards Ls and Ss in Trinity Gp — I	
Jurassic	Locally — Ss Fm — M	Locally — Ss Fm — M	U	Ss Fms. Some may not be develo-ped — I	U	U	Absent	Absent	Absent	U	
Triassic	Locally — Ss and C Fms — M		U	Ss and C Fms used locally. Shinarump C and corre-latives give rise to springs — I	U	U	Absent	Ss, C, jointed shale, and basalt beds of Newark Gp in Massachusetts, Connecticut, New Jersey, Pennsylvania, Maryland, Virginia, and North Carolina — M			Water from Ss, C, and Ls Fms west of Mississippi river, especially valuable when water from other sources is unavailable

Paleozoic Permian	U	U	U	DeChelly Ss — I Kaibab Ls — M	U	San Andres Ls in Roswell Basin — P Quartermaster Gp gives rise to many springs — M Other Ss and Ls in Kansas, Oklahoma, and Texas — M	U	U	Absent	U	
Pennsylvanian	Tensleep Ss in Wyoming and other Ss elsewhere — M	U	U	U	U	Ss and C beds from the Appalachians to Iowa and eastern Kansas — M to I			Jointed and weathered Sh, Ss, and C in Rhode Island and Massachusetts — M	U	
Mississippian	Ls locally but little developed; springs arise from Ls in Rocky Mountains — M	A few springs arise from Ls locally — U	U	Some springs arise from Ls locally — U	U	In Illinois, Iowa, Missouri, and Kentucky the Burlington, Keokuk, and St. Louis Ls — I Some Ss (primarily Chester) — M In Alabama and Tennessee — the Feet Payne chert, Gaspar Fm, and St. Genevieve and Tuscumbia Ls — I In Kentucky many springs arise in Ls			U	U	Do
Devonian	U	U	U	U	U	U, except locally in Michigan (Traverse Fm), Illinois, Missouri, Ohio (Columbia Ls), and Kentucky — M		Jointed Ls, Ss, and Sh, some highly metamorphosed M locally and little used		U	
Silurian	U	U	U	U	U	Ls and dolomite Fms in New York, Kentucky, Tennessee, Ohio, Illinois, and Iowa Better-known aquifers include Monroe dolomite and related carbonate Fms in Ohio — I; "Niagaran" dolomite in Illinois — P (in part)		U	U	U	

(Continued)

Table 6A.1 (Continued)

(1) Geologic Age and Rock Type	(2) Western Mountain Ranges	(3) Arid Basin	(4) Columbi Lava Plateau	(5) Colorado Plateau	(6) High Plains	(7) Unglaciated Central Region	(8) Glaciated Central Region	(9) Unglaciated Appalachian Region	(10) Glaciated Appalachian Region	(11) Atlantic and Gulf Coastal Plain	(12) Special Comments
Ordovician	U	U	U	U	U	In Arkansas, Missouri, Iowa, Illinois, eastern Indiana, southern Wisconsin, south-eastern Minnesota, the St. Peter Ss — I Overlying and subjacent Ls and Ss where present in above states and in Kansas, Oklahoma, and New York — M to I In Kentucky and Tennessee — Ls Fm — M to I		Locally Ls and Ss Fms; not highly developed — M		U	
Cambrian	U	U	U	U	U	Ss beds in Wisconsin, Minnesota, Iowa, and Illinois include Jordan Ss, "Dresbach Fm" (Galesville Ss, Eau Claire Fm, Mt. Simon Ss) — P (in part) Ls and Ss Fms in Missouri and Arkansas give rise to many large springs and yield water to many wells — P		Ls Fms give rise to large springs in southern Appal-achians. Otherwise — U	Eastern New York and New England Ss Fms — M; other-wise — U	U	
Precambrian (including crystalline rocks which may be younger)	Weathered and jointed rocks locally — M		U	U	U	U	Weathered and jointed rocks locally in Minnesota, Wisconsin, northern Michigan, Piedmont Plateau, New England — M to I. Some Ss in North Central States			U	Do

Note: Abbreviations: (1) Aquifers: P, principal aquifer in region; I, important aquifer in region; M, minor aquifer in region; U, unimportant as an aquifer in region. (2) Rock terms; S, sand, Ss, sandstone; G, gravel; C, conglomerate; Sh, shale; Ls, limestone; Fm, formation; Gp, group.

Source: From Maxey, In Chow, *Handbook of Applied Hydrology*, McGraw-Hill, Copyright 1964. With permission.

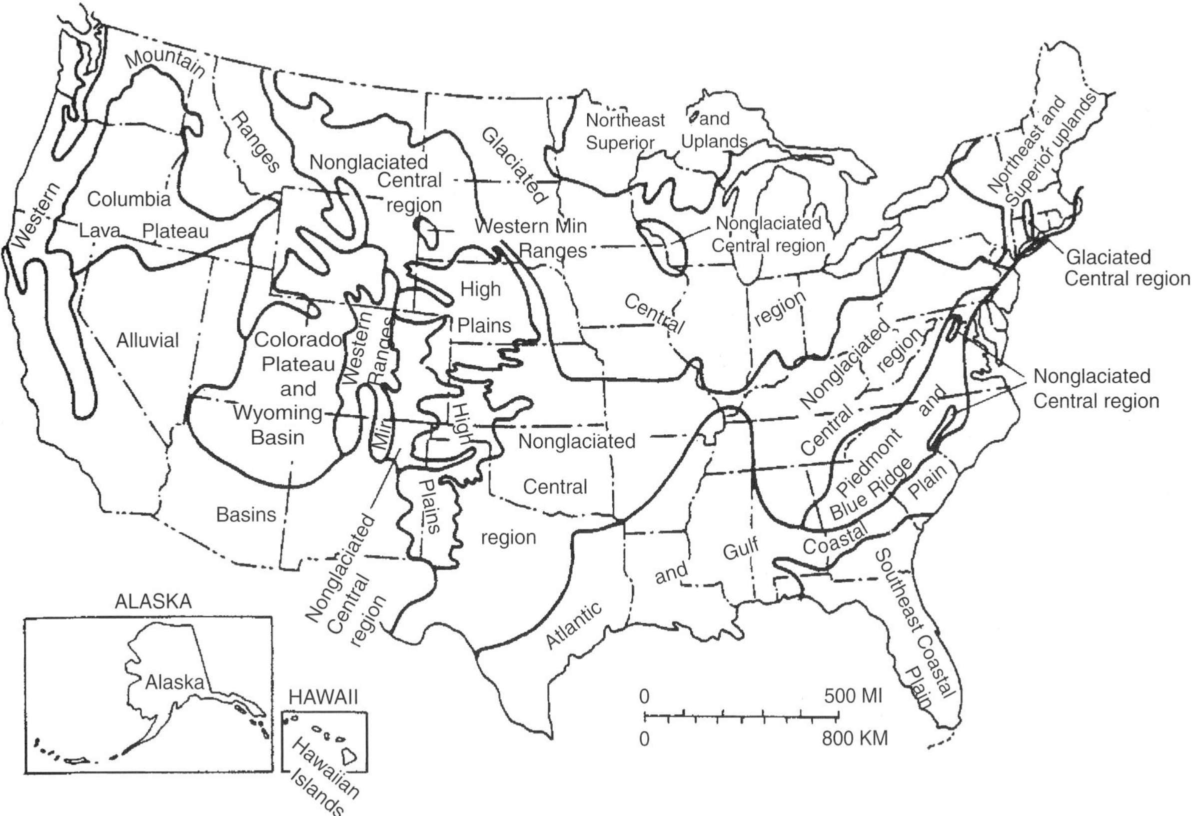

Figure 6A.3 Groundwater regions of the United States. (From Heath, R.C., Classification of ground-water regions of the United States, *Groundwater*, 20, 4, 1982.)

Table 6A.2 Principal Physical and Hydrologic Characteristics of Groundwater Regions in the United States

	Components of the System															Characteristics of the Dominant Aquifers																				
																Water-Bearing Openings					Compo-sition			Storage and Transmission Properties							Recharge and Discharge Conditions					
	Unconfined Aquifer			Confining Beds				Confined Aquifers				Presence and Arrangement				Primary			Secon-dary		Degree of Solubility			Porosity			Trans-missivity				Recharge			Discharge		
Name	Hydrologically Insignificant	Minor Aquifer	Dominant Aquifer	Hydrologically Insignificant	Thin, Discontinuous, or v. Leaky	Thick, Impermeable	Interlayered with Aquifers	Hydrologically Insignificant	Not Highly Productive	Multiple Aquifers	A Single, Dominant Aquifer	Single Dominant Unconfined Aquifer	Two Interconnected Aquifers	Unconfined Aquifer, Confining Bed, Confined Aquifer	Complex Interbedded Sequence	Pores in Unconsolidated Deposits	Pores in Semiconsolidated Rocks	Tubes and Cooling Cracks in Lava	Fractures and Faults	Solution Enlarged Openings	Insoluble	Mixed Soluble and Insoluble	Soluble	Large (>0.2)	Moderate (0.01–0.2)	Small (<0.01)	Large (>2500 m² day⁻¹)	Moderate (250 – 2500 m² day⁻¹)	Small (25 – 250 m² day⁻¹)	Very Small (<25 m² day⁻¹)	Uplands between Streams	Losing Streams	Leakage through Confining Beds	Springs and Surface Seepage	Evaporation and Basin Sinks	Into Other Aquifers
Western Mountain Ranges		X		X					X				X			X			X		X				X				X		X	X		X		
Alluvial Basins		X					X			X					X	X					X			X			X					X			X	X
Columbia Lava Plateau		X					X			X					X	X		X			X					X	X					X		X		
Colorado Plateau and Wyoming Basin	X						X			X					X		X		X		X					X			X		X		X	X		
High Plains			X	X				X				X				X					X			X			X					X		X		
Nonglaciated Central Region		X					X			X					X		X		X	X		X				X		X			X		X	X		X
Glaciated Central Region		X			X					X					X	X	X		X	X		X			X			X					X			X
Piedmont and Blue Ridge		X		X					X				X			X			X		X					X				X	X			X		
Northeast and Superior Uplands		X			X				X				X			X			X		X					X			X		X			X		
Atlantic and Gulf Coastal Plain		X					X			X					X	X				X		X			X			X			X		X	X		X
Southeast Coastal Plain		X				X					X			X		X				X			X	X			X				X		X	X		
Alluvial Valleys			X		X			X				X				X					X			X			X				X			X		
Hawaii			X		X			X				X						X			X			X			X				X	X		X		
Alaska			X		X				X						X	X			X			X			X			X				X		X		

Note: For map of regions, see Figure 6A.3.

Source: From Heath, R.C., Classification of ground-water regions of the United States, *Groundwater*, 20, 4, 1982.

Table 6A.3 Basic Data Required for Groundwater Studies

A. Maps, Cross Sections, and Fence Diagrams

1. Planimetric
2. Topographic
3. Geologic
   a. Structure
   b. Stratigraphy
   c. Lithology
4. Hydrologic
   a. Location of wells, observation wells, and springs
   b. Groundwater table and potentiometric contours
   c. Depth to water
   d. Quality of water
   e. Recharge, discharge, and contributing areas
5. Vegetative cover, location of wetlands
6. Soils
7. Aerial photographs

B. Data on Wells and Springs

1. Location, depth, diameter, types of well, and logs
2. Static and pumping water level, hydrographs, yield, specific capacity, quality of water
3. Present and projected groundwater development and use
4. Corrosion, incrustation, well interference, and similar operation and maintenance problems
5. Location, type, geologic setting, and hydrographs of springs
6. Observation well networks
7. Water sampling sites

C. Aquifer Data

1. Type, such as unconfined, artesian, or perched
2. Thickness, depths, and formational designation
3. Boundaries
4. Transmissivity, storativity, and permeability
5. Specific retention
6. Discharge and recharge
7. Ground and surface water relationships
8. Aquifer models

D. Climatic Data

1. Precipitation
2. Temperature
3. Evapotranspiration

E. Surface Water

1. Use
2. Quality
3. Runoff distribution, reservoir capacities, inflow and outflow data
4. Return flows, section gain or loss
5. Recording stations
6. Low flow data

F. Environment

1. Location of hazardous waste sites or other potential sources of pollution
2. Use of herbicides, pesticides, fertilizers, and road salt
3. Site history

G. Local Drilling Facilities and Practices

1. Size and types of drilling rigs locally available
2. Logging services locally available
3. Locally used materials, well designs, and drilling practices
4. State or local rules and regulations

Source: From U.S. Bureau of Reclamation, *Groundwater Manual*; Amended, 1977.

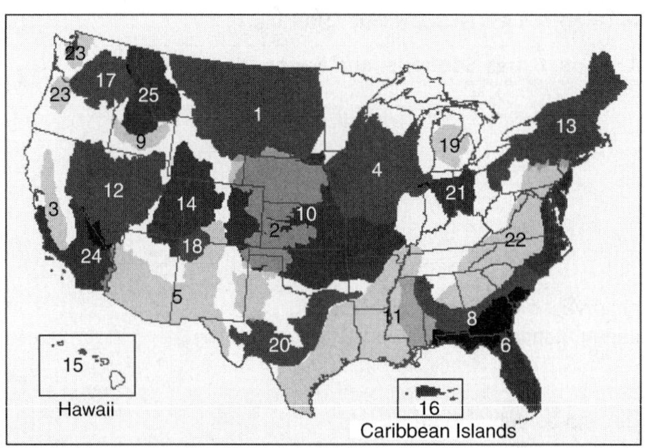

1 Northern Great Plains
2 High Plains
3 Central Valley, California
4 Northern Midwest
5 Southwest Alluvial Basins
6 Floridan
7 Northern Atlantic Coastal Plain
8 Southeastern Coastal Plain
9 Snake River Plain
10 Central Midwest
11 Gulf Coastal Plains
12 Great Basin
13 Northeast Glacial Aquifers
14 Upper Colorado River basin
15 Oahu, Hawaii
16 Caribbean Islands
17 Columbia Plateau
18 San Juan Basin
19 Michigan Basin
20 Edwards-Trinity
21 Midwestern Basins and Arches
22 Appalachian Valleys and Piedmont
23 Puget-Willamette Lowland
24 Southern California Alluvial Basins
25 Northern Rocky Mountain Intermontane

The U.S. Geological Survey initiated the Regional Aquifer-System Analysis (RASA) Program in 1978 in response to Federal and State needs for information to improve management of the Nation's groundwater resources. The objective of the RASA Program is to define the regional geohydrology and establish a framework of background information—geologic, hydrologic, and geochemical—that can be used for regional assessment of groundwater resources and in support of detailed local studies. The program was completed in 1995.

A total of 25 aquifer systems were studied under the RASA Program.

Figure 6A.4 Regional aquifer study areas. (From http://water.usgs.gov.)

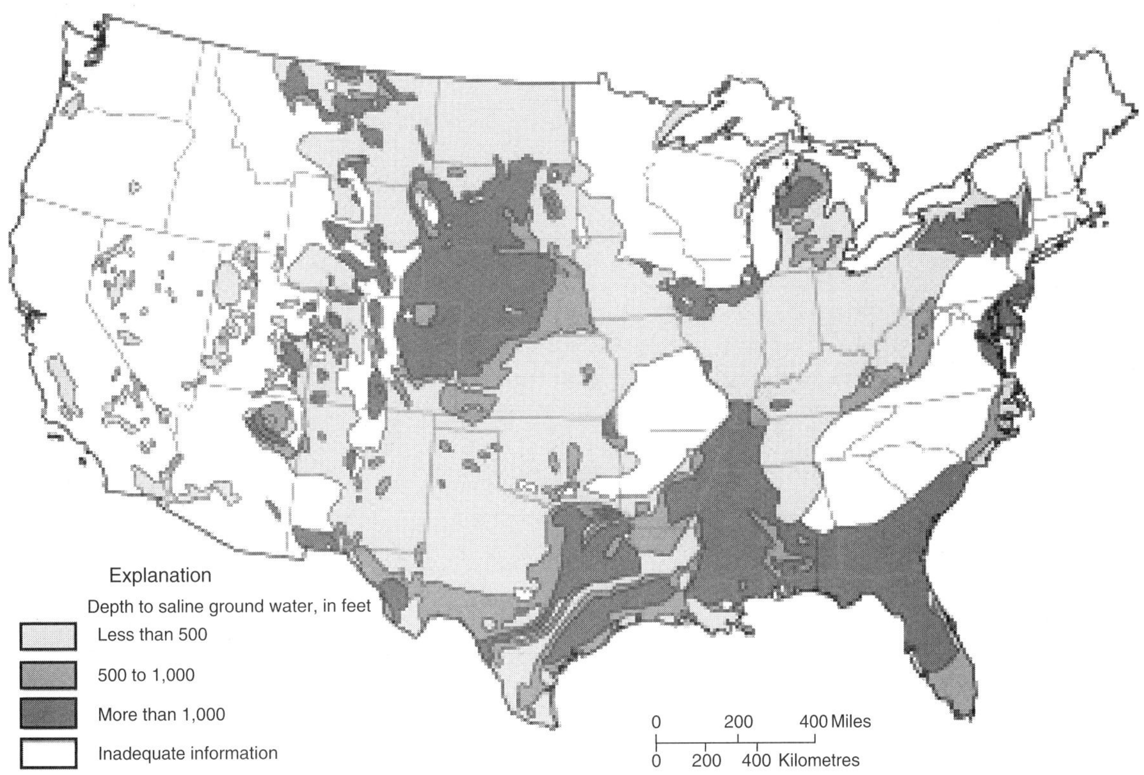

Figure 6A.5 Depth to saline groundwater in the United States (generalized from Feth and others, 1965). (From USGS fact sheet 075-03, October 2003.)

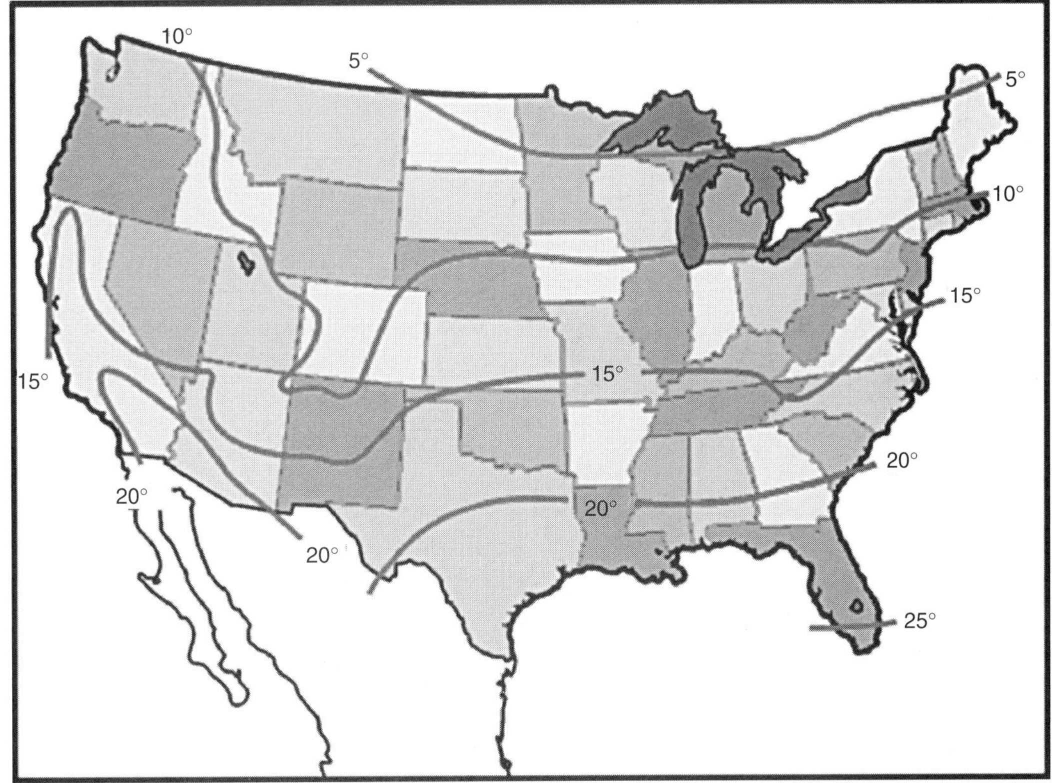

Figure 6A.6 Average shallow groundwater temperatures in the United States developed by Collins. (From www.epa.gov.)

Table 6A.4 Estimated Groundwater in Storage, by Continent

Continent	0–100 m	100–200 m	200–2000 m	Total
Europe	0.2	0.3	1.1	1.6
Asia	1.3	2.1	4.4	7.8
Africa	1	1.5	3.0	5.5
North America	0.7	1.2	2.4	4.3
South America	0.3	0.9	1.8	3
Australia	0.1	0.2	0.9	1.2
Total	3.6	6.2	13.6	23.4

Note: In millions of km^3; based on publications by soviet hydrologists.

Source: From Castany, G., Hydrogeology of deep aquifers, *Episodes*, 1981, 3, 1981.

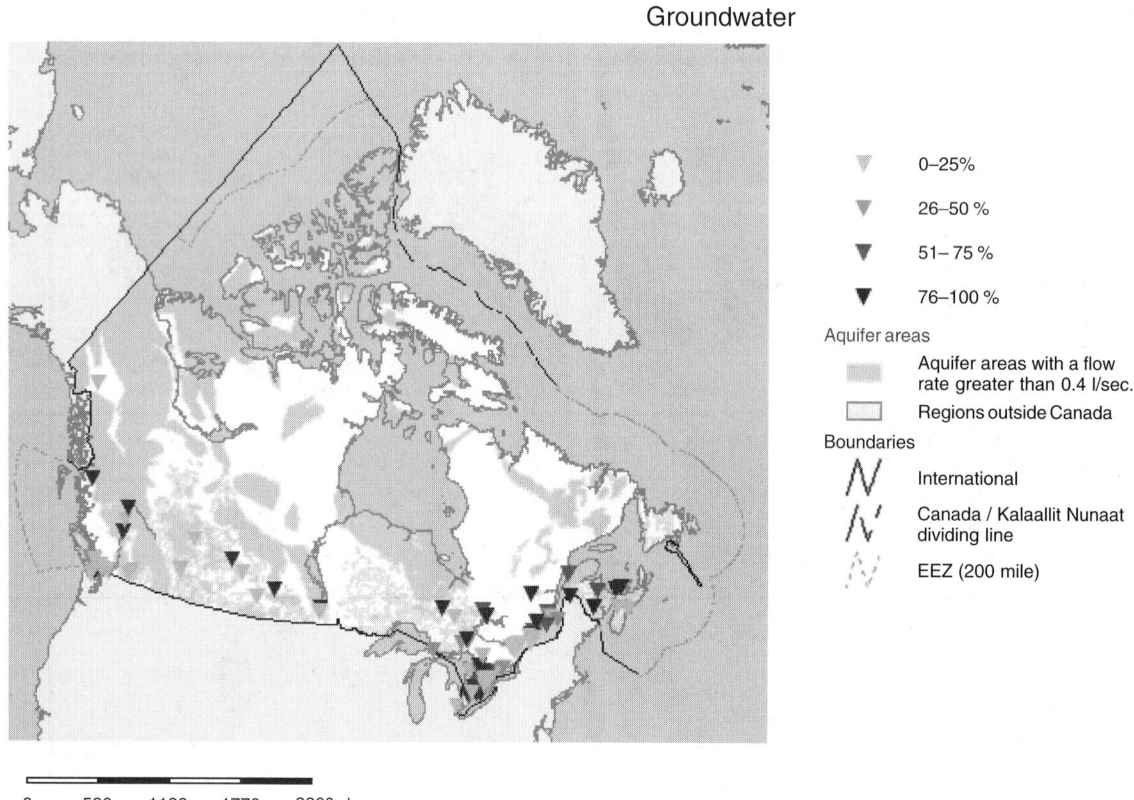

Figure 6A.7 Groundwater potential in Canada and a percentage of people using groundwater resources in Canadian municipalities over 10,000 people. (From www.atlas.gc.ca.)

SECTION 6B WATER WELLS — UNITED STATES

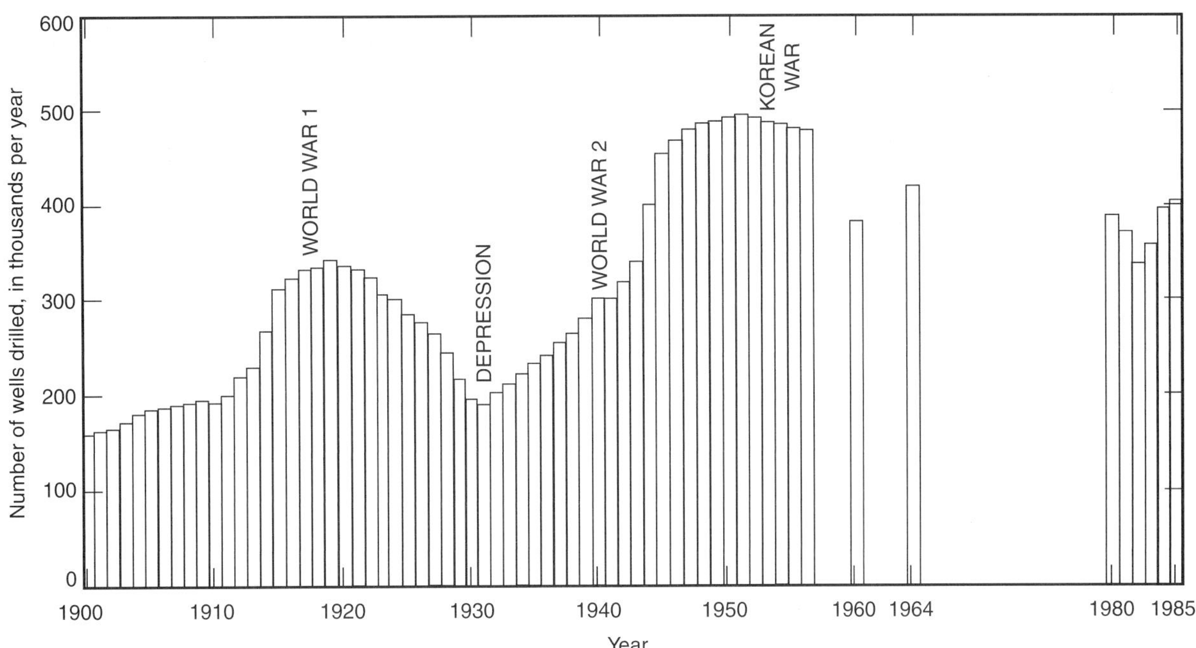

Figure 6B.8 Number of water wells drilled in the United States and relation to major events in the United Stated history. (From Hindall, S.M., Eberle, Michael,1987, National and regional trends in water-well drilling in the United States 1964–1984, *U.S. Geological Survey, Open File Report 87–247*; 1985 data from National Water Well Association.)

Table 6B.5 Number and Type of Water Wells and Boreholes Constructed in the United States in 1985

Application/type	1985
Commercial/industrial	49,379
Heat pump supply/return	18,029
Agricultural irrigation	21,583
Private household	488,918
Public supply	20,010
Monitoring	121,294
Livestock watering[a]	29,343
Lawn/turf irrigation[a]	27,036
Other	34,482
Total	**810,074**

PRIVATE HOUSEHOLD WELLS 60.35%
HEAT PUMP 2.23%
PUBLIC 2.47%
IRRIGATION 2.66%
COMM/IND 6.1%
OTHER 11.22%
MONITORING 14.97%

Note: Based on Water Well Journal Survey of 8,043 firms.

[a] Included in "Other" category on pie chart.

Source: From McCray, Kevin. Copyright Water Well Journal September 1986. Reprinted with permission.

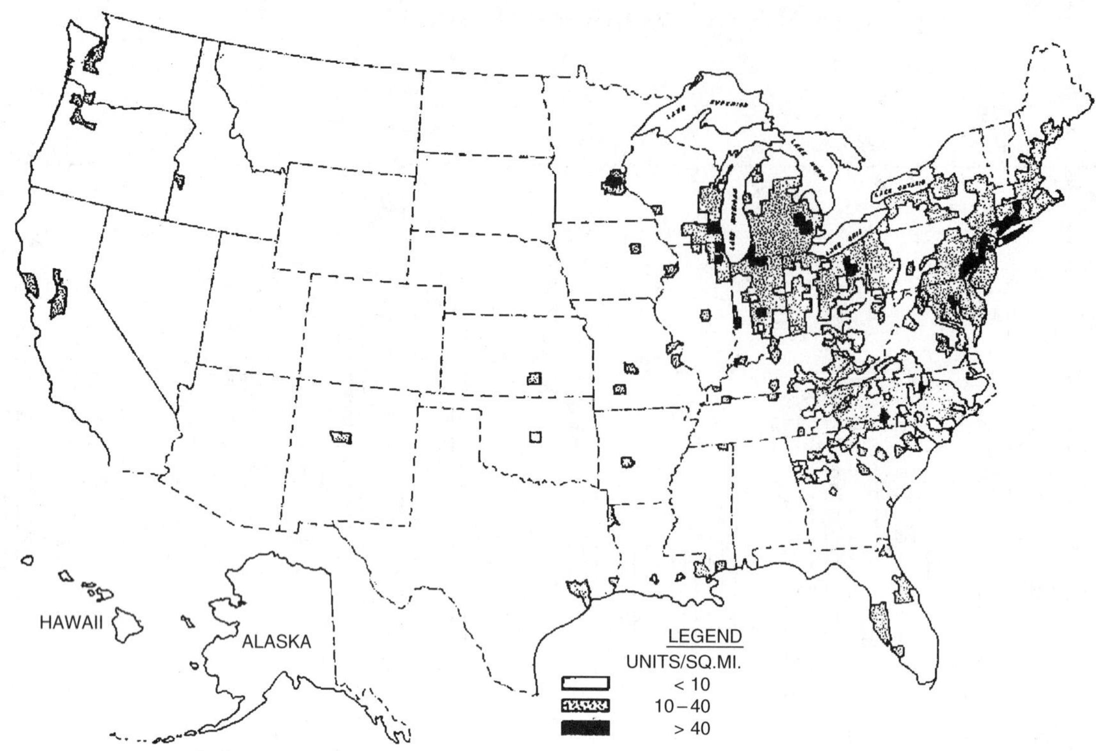

Figure 6B.9 Density of housing units using on site domestic water supply systems in the United States [By county]. (From U.S. Environmental Protection Agency, Office of Water Supply, Office of Solid Waste Management Programs, 1977, *The Report to Congress: Waste Disposal Practices and Their Effects on Groundwater.*)

Table 6B.6 Number and Type of Water Wells in the United States, 1988

State	Irrigation	Public Supply[a]	Community Supply	Household
Alabama	N/A	1,706	1,013	267,202
Alaska	10	1,283	454	32,391
Arizona	6,125	2,533	1,650	43,226
Arkansas	21,078	3,220	1,073	213,672
California	67,770	8,143	3,320	359,584
Colorado	17,809	3,116	1,465	84,459
Connecticut	N/A	5,373	1,073	241,130
Delaware	559	1,141	497	52,701
Florida	29,017	11,337	4,650	573,059
Georgia	4,492	4,139	2,460	405,078
Hawaii	N/A	219	218	536
Idaho	7,371	3,114	1,312	88,853
Illinois	1,107	10,018	1,492	443,681
Indiana	N/A	16,100	1,805	546,381
Iowa	2,210	5,052	3,674	236,709
Kansas	18,658	2,456	2,405	116,567
Kentucky	N/A	1,144	386	247,506
Louisiana	4,558	3,162	1,996	200,446
Maine	N/A	3,433	595	149,331
Maryland	N/A	4,955	953	252,142
Massachussets	N/A	2,894	1,384	132,119
Michigan	N/A	12,188	1,138	934,184
Minnesota	4,250	13,163	2,467	382,572
Missisippi	N/A	3,109	2,257	150,816
Missouri	3,700	3,996	1,633	305,853
Montana	997	2,504	999	80,817
Nebraska	61,361	1,558	1,501	112,740
Nevada	2,332	1,156	697	24,142
New Hampshire	N/A	2,009	733	110,712
New Jersey	N/A	7,765	2,256	227,326
New Mexico	8,031	2,478	1,545	70,157
New York	651	18,068	5,381	659,973
North Carolina	530	15,972	5,094	821,995
North Dakota	808	1,261	718	54,008
Ohio	N/A	13,306	3,508	692,062
Oklahoma	4,351	3,265	2,181	164,506
Oregon	9,241	3,330	1,267	178,407
Pennsylvania	N/A	17,477	5,578	800,292
Rhode Island	N/A	916	169	33,987
South Carolina	185	3,376	2,174	297,435
South Dakota	1,266	1,244	872	56,512
Tennessee	320	3,125	1,006	258,997
Texas	59,636	13,297	9,207	490,453
Utah	2,295	1,961	1,164	14,511
Vermont	N/A	1,739	653	58,380
Virginia	N/A	6,771	2,869	455,556
Washington	5,853	6,130	3,810	195,132
West Virginia	N/A	2,771	834	181,069
Wisconsin	N/A	22,982	2,239	521,579
Wyoming	1,409	1,373	647	30,900
Totals	**348,116**	**282,827**	**98,472**	**13,101,846**
Grand total				**13,732,680**

Note: N/A, Not available.

[a] Includes community supply (systems with at least 15 service connections used by year-round residents or regularly serving at least 25 year-round residents).

Source: From National Water Well Association, 1988.

Table 6B.7 Number of Water Wells Drilled in the United States, 1960–1984

State	Estimated Number of Wells Drilled[a]							Percent of Total Drilled in 1984	Percentage Change between Annual Totals[b]		
	1960	1964	1980	1981	1982	1983	1984		1960 and 1964	1964 and 1984	1980 and 1984
Alabama	4,000	4,500	5,960	6,420	5,920	5,570	6,260	1.6	+13	+39	+5
Alaska	726	1,000	2,440	2,400	2,400	2,800	2,700	0.68	+38	+170	+11
Arizona	1,400	1,520	2,190	2,220	2,380	2,710	2,760	0.7	+8.6	+82	+26
Arkansas	5,000	5,000	4,010	5,910	2,750	3,320	4,200	1.1	NC	−16	+5
California	9,100	10,000	17,100	15,900	11,300	11,000	14,300	3.6	+10	+43	−16
Colorado	3,100	5,910	4,910	4,570	4,390	4,360	4,060	1.0	+91	−31	−17
Connecticut	6,500	6,500	5,470	5,410	4,500	5,140	5,780	1.5	NC	−11	+6
Delaware	3,800	3,440	2,000	2,680	2,290	2,700	3,100	0.78	−9.5	−10	+55
District of Columbia	12	12	ND	0	0	0	ND	ND	NC	ND	ND
Florida	33,900	55,000	40,200	40,500	38,900	43,200	45,600	11	+62	−17	+13
Georgia	10,500	10,000	11,000	13,400	10,100	10,800	12,200	3.1	−4.8	+22	+11
Hawaii	17	21	11	11	7	7	2	<0.1	+24	−90	−82
Idaho	1,400	1,400	2,880	1,470	2,400	1,590	1,630	0.35	NC	+16	−43
Illinois	21,000	19,500	14,000	12,200	13,400	13,600	15,300	3.9	−7.1	−22	+9.3
Indiana	17,700	15,000	9,670	8,180	9,700	9,180	10,300	2.6	−15	−31	+6.5
Iowa	9,000	15,000	5,890	6,850	4,120	3,780	3,140	0.79	+67	−79	−47
Kansas	4,700	5,500	4,530	5,050	3,380	3,420	3,910	1.0	+17	−29	−14
Kentucky	9,880	9,620	5,060	5,100	4,800	5,440	5,740	1.4	−2.6	−40	+13
Louisiana	974	2,620	6,050	6,830	6,580	5,180	5,560	1.4	+170	+110	−8.1
Maine	1,500	1,700	2,860	2,570	2,440	3,470	3,900	1.0	+13	+130	+36
Maryland	4,020	6,900	7,200	8,000	6,700	8,800	8,300	2.1	+72	+20	+15
Massachusetts	8,000	9,000	6,330	6,270	5,370	6,820	7,670	1.9	+12	−15	+21
Michigan	25,000	25,000	24,000	20,000	16,000	17,000	18,500	4.7	NC	−26	−23
Minnesota	13,000	9,000	14,400	10,500	10,800	11,100	12,500	3.1	−31	+39	−13
Mississippi	5,300	5,900	2,670	3,550	2,540	2,400	2,640	0.66	+11	−55	−1
Missouri	6,380	9,990	10,900	8,530	7,830	10,200	11,500	2.9	+57	+15	+5.5
Montana	1,900	2,000	3,580	6,410	6,260	2,360	2,560	0.64	+5.3	+28	−28
Nebraska	5,510	6,000	4,500	5,940	3,470	3,260	3,660	0.92	+8.9	−39	−19
Nevada	824	825	775	765	503	639	718	0.18	NC	−13	−7.4
New Hampshire	3,600	4,400	3,050	4,190	2,630	5,210	5,860	1.5	+22	+33	+92
New Jersey	3,800	3,440	8,620	8,540	8,580	10,900	13,100	3.3	−9.5	+280	−52
New Mexico	2,290	3,150	2,750	2,880	3,370	3,430	3,110	0.78	+38	−1.3	+13
New York	25,000	25,000	16,800	17,000	15,600	17,800	20,000	5.0	NC	−20	+19
North Carolina	20,000	25,000	10,500	12,000	13,500	15,900	17,100	4.3	+25	−32	+63
North Dakota	4,200	3,760	1,710	2,190	1,450	1,480	1,450	0.36	−10	−61	−15
Ohio	17,100	18,600	16,700	14,300	14,200	14,000	15,700	4.0	+8.8	−160	−6.0
Oklahoma	4,400	5,000	7,980	7,630	6,500	5,870	6,590	1.8	+14	+32	−17
Oregon	3,500	4,500	7,500	6,620	3,800	3,550	3,530	0.89	+29	−22	−53
Pennsylvania	13,500	16,200	15,600	12,400	9,620	8,140	10,800	2.7	+20	−33	−31
Rhode Island	200	250	319	240	206	387	548	0.14	+25	+120	+72
South Carolina	5,300	5,400	11,400	5,340	4,640	7,780	8,740	2.2	+1.9	+62	−23
South Dakota	6,080	5,430	2,210	1,820	1,590	1,330	1,500	0.38	−11	−72	−32
Tennessee	10,000	8,000	7,080	7,130	6,710	7,600	8,020	2.0	−20	NC	+13
Texas	19,000	25,000	16,200	17,700	21,700	17,700	21,200	5.3	+32	−15	+31
Utah	630	650	630	547	507	488	548	0.14	+3.2	−16	−13
Vermont	1,240	1,460	3,100	2,280	1,900	2,330	3,050	0.77	+18	+110	−1.6
Virginia	8,500	10,000	10,900	8,830	9,060	15,300	16,900	4.3	+18	+69	+55
Washington	1,400	1,700	5,040	4,290	3,550	4,320	4,030	1.0	+21	+137	−20
West Virginia	5,500	5,900	3,280	3,510	2,730	2,580	2,900	0.73	+7.3	−51	−12
Wisconsin	11,000	12,000	11,600	9,900	9,590	10,400	11,700	2.9	+9.1	−2.5	+0.86

(Continued)

Table 6B.7 (Continued)

State	Estimated Number of Wells Drilled[a] 1960	1964	1980	1981	1982	1983	1984	Percent of Total Drilled in 1984	Percentage Change between Annual Totals[b] 1960 and 1964	1964 and 1984	1980 and 1984
Wyoming	1,000	1,000	3,010	3,680	2,970	2,500	2,520	0.6	NC	+152	−16
Totals	**381,000**	**434,000**	**387,000**	**371,000**	**336,000**	**359,000**	**397,000**	**100**	**+14**	**+8.5**	**+2.6**

[a] Numbers rounded to three significant figures.
[b] Numbers rounded to two significant figures.

Source: From Hindall, S.M., Eberle, Michael, national and regional trends in water-well drilling in the United States 1964–1984, *U.S. Geological Survey, Open File Report*, 87–247, 1987.

Table 6B.8 Regional Trends in Water-Well Construction in the United States, 1960–1984

Region	Number of Wells Drilled 1960	1964	1980	1981	1982	1983	1984	Average Annual Total 1980 Through 1984
Northeast (includes DC)	63,300	68,000	62,100	60,600	50,800	60,200	70,800	60,900
Southeast	95,300	126,000	102,000	106,000	93,600	112,000	121,000	107,000
Great Lakes and Central Appalachians	130,000	123,000	106,000	95,300	87,900	90,900	101,000	96,200
South-Central	54,800	77,200	63,200	62,100	60,600	57,300	63,300	61,300
Northern Rockies and Northern Great Plains	20,100	19,600	17,900	16,700	18,100	12,500	13,300	15,700
Southwest (includes Hawaii)	12,000	13,000	20,700	17,700	15,100	14,800	18,300	17,300
Pacific Northwest (includes Alaska)	5,620	7,200	15,000	11,400	9,750	8,670	10,300	11,000
Totals	**381,000**	**434,000**	**387,000**	**370,000**	**336,000**	**359,000**	**397,000**	**370,000**

Region	Percentage of Total Wells Drilled 1964	1984	Percentage Change between Annual Totals 1960 and 1964	1964 and 1984	1980 and 1984
Northeast (includes DC)	16	18	+7.4	+4.1	+14
Southeast	29	30	+32	−3.9	+19
Great Lakes and Central Appalachians	28	25	−5.4	−18	−4.7
South-Central	18	16	+41	−18	NC
Northern Rockies and Northern Great Plains	4.5	3.3	−2.5	−32	−26
Southwest (includes Hawaii)	3.0	4.6	+8.3	+41	−12
Pacific Northwest (includes Alaska)	1.7	2.6	+28	+43	−31
Totals	**100**	**100**	**+14**	**−8.5**	**+2.6**

Source: From Hindall, S.M., Eberle, Michael, national and regional trends in water-well drilling in the United States 1964–1984, *U.S. Geological Survey, Open File Report*, 87–247, 1987.

SECTION 6C WATER WELLS

Table 6C.9 Water Well Construction Methods and Applications

Method	Materials for Which Best Suited	Water Table Depth for Which Best Suited (m)	Usual Maximum Depth (m)	Usual Diameter Range (cm)	Usual Casing Material	Customary Use	Yield m^3/day[a]	Remarks
Augering Hand auger	Clay, silt sand, gravel less than 2 cm	2–9	10	5–20	Sheet metal	Domestic, drainage	15–250	Most effective for penetrating and removing clay. Limited by gravel over 2 cm. Casing required if material is loose
Power auger	Clay, silt sand, gravel less than 5 cm	2–15	25	15–90	Concrete, steel or wrought-iron pipe	Domestic, irrigation, drainage	15–500	Limited by gravel over 5 cm, otherwise same as for hand auger
Driven Wells Hand, air hammer	Silt, sand, gravel less than 5 cm	2–5	15	3–10	Standard weight pipe	Domestic, drainage	15–200	Limited to shallow water table, no large gravel
Jetted Wells Light, portable rig	Silt, sand, gravel less than 2 cm	2–5	15	4–8	Standard weight pipe	Domestic, drainage	15–150	Limited to shallow water table, no large gravel
Drilled Wells Cable tool	Unconsolidated and consolidated medium hard and hard rock	Any depth	450[b]	8–60	Steel or wrought-iron pipe	All uses	15–15,000	Effective for water exploration. Requires casing in loose materials. Mudscow and hollow rod bits developed for drilling unconsolidated fine to medium sediments
Rotary	Silt, sand, gravel less than 2 cm; soft to hard consolidated rock	Any depth	450[b]	8–45	Steel or wrought-iron pipe	All uses	15–15,000	Fastest method for all except hardest rock. Casing usually not required during drilling. Effective for gravel envelope wells
Reserve-circulation rotary	Silt, sand, gravel, cobble	2–30	60	40–120	Steel or wrought-iron pipe	Irrigation, industrial, municipal	2500–20,000	Effective for large-diameter holes in unconsolidated and partially consolidated deposits. Requires large volume of water for drilling. Effective for gravel envelope wells
Rotary-percussion	Silt, sand, gravel less than 5 cm; soft to hard consolidated rock	Any depth	600[b]	30–50	Steel or wrought–iron pipe	Irrigation, industrial, municipal	2500–15,000	Now used in oil exploration. Very fast drilling. Combines rotary and percussion methods (air drilling) cuttings removed by air. Would be economical for deep water wells

[a] Yield influenced primarily by geology and availability of groundwater.
[b] Greater depths reached with heavier equipment.

Source: From U.S. Soil Conservation Service, *Engineering Field Manual for Conservation Practices*, 1969.

Table 6C.10 Relative Performance of Different Drilling Methods in Various Types of Geologic Formations

Type of Formation	Cable Tool	Direct Rotary (with Fuids)	Direct Rotary (with Air)	Direct Rotary (Down-the-Hole Air Hammer)	Direct Rotary (Drill-Through Casing Hammer)	Reverse Rotary (with fluids)	Reverse Rotary (Dual Wall)	Hydraulic Percussion	Jetting	Driven	Auger
Dune sand	2	5	Not recommended		6	5[a]	6	5	5	3	1
Loose sand and gravel	2	5			6	5[a]	6	5	5	3	1
Quicksand	2	5		Not recommended	6	5[a]	6	5	5	Not recommended	1
Loose boulders in alluvial fans or glacial drift	3–2	2–1			5	2–1	4	1	1		1
Clay and slit	3	5			5	5	5	3	3		3
Firm shale	5	5			5	5	5	3	Not recommended		2
Sticky shale	3	5			5	3	5	3			2
Brittle shale	5	5			5	5	5	3			
Sandstone — poorly cemented	3	4			Not applicable	4	5	4			Not applicable
Sandstone — well cemented	3	3	5			3	5	3			
Chert nodules	5	3	3			3	3	5			
Limestone	5	5	5	6		5	5	5			
Limestone with chert nodules	5	3	5	6		3	3	5			
Limestone with small cracks or fractures	5	3	5	6		2	5	5			
Limestone, cavernous	5	3–1	2	5		1	5	1			
Dolomite	5	5	5	6		5	5	5			
Basalts, thin layers in sedimentary rocks	5	3	5	6		3	5	5			
Basalts — thick layers	3	3	4	5		3	4	3			
Basalts — highly fractured (lost circulation zones)	3	1	3	3		1	4	1			
Metamorphic rocks	3	3	4	5		3	4	3			
Granite	3	3	5	5		3	4	3			

Note: Rate of Penetration: 1, Impossible; 2, Difficult; 3, Slow; 4, Medium; 5, Rapid; 6, Very rapid.

[a] Assuming sufficient hydrostatic pressure is available to contain active sand (under high confining pressures).

Source: From Driscoll, F.G., 1986, *Groundwater and Wells*. Copyright Johnson Division.

Table 6C.11 Description of Drilling Methods

Methods Without Drilling Fluids

Displacement Boring

Pros:
- Does not require heavy equipment (by hand or lightweight equipment)
- Clean method for shallow well installation

Cons:
- Method limited to shallow depths
- Method limited to soft soils and boulder, cobble-free zones
- Not efficient if necessary to install several wells
- Practical limitation up to $\sim 2''$ diameter sampler

Similar to the above method is "Direct Push Technology" or DPT. A common trade name is GeoProbe. DPT does not require heavy equipment, most units are pickup mounted or ATV mounted for easy accessibility

Driven Wells

Pros:
- Cost effective
- Easy access in most conditions

Cons:
- Limited to shallow depths ($<$ 50 ft)
- Limited to unconsolidated, soft formations relatively free of cobbles or boulders
- May require pre-drilling a hole of slightly greater diameter that the well point

Solid-Stem Auger

Pros:
- Rapid and low-cost drilling in clayey formations
- Clean method, does not require circulation fluids
- No casing necessary where the formation is stable
- Allows collection of representative sample in semi-consolidated formations

Cons:
- Practical limitation to $24''$ diameter
- Inefficient in loose, sandy material (depends on the depth)
- Inefficient below the water table (depends on the depth)

Hollow-Stem Auger (HSA)

Pros:
- Allows collection of uncontaminated sample in unconsolidated formation
- Can be used as temporary casing to prevent caving
- Relatively rapid, especially in clayey formations

Cons:
- Ineffective through boulders
- Limited drilling in loose, granular soils, particularly below the water table where sample recovery can be compromised
- Difficult to retrieve a sample in loose, granular soil because cuttings do not always want to come to the surface. Samples must be collected with a split spoon or a continuous corer, either of which can provide excellent samples if done correctly
- Limited to rather shallow depths

Sonic Drilling

Pros:
- Drilling can proceed with or without the use of drilling fluids
- Method can be utilized in unconsolidated and some consolidated formations
- Minimal disturbance to soil samples
- Good recovery of quasi-continuous samples
- Conventional air rotary or down-hole hammer methods can be employed through the outer drive casing
- The rig can also be operated as a fluid rotary machine

Cons:
- A relatively new method that is not available everywhere
- Relatively expensive compared to other drilling methods

(Continued)

Table 6C.11 (Continued)

- Dry casing advancement generates heat that can affect the sample integrity
- Maximum nominal diameter of less than 12 in.
- Practical depth limitation of less than 500 ft

Methods that Use Drilling Fluids

Rotary (Direct) Drilling

Pros:

- High penetration rate
- Drilling operation requires a minimum amount of casing
- Rapid mobilization and demobilization

Cons:

- Use of a drilling fluid, both in terms of sample contamination and water management (in the case of water-based fluids and air injected by gasoline compressors)
- Circulation of drilling fluid may be lost in loose/coarse formations, hence making difficult to transport drill cuttings
- Difficult to collect accurate samples, i.e. a sample from a discrete zone since the cuttings accumulate at surface around the rim of the borehole

Reverse Circulation Rotary Drilling (RC)

Pros:

- Applicable to a wide variety of formations
- Possible to drill large-diameter holes, both quickly and economically
- Minimal disturbance to the formation due to the pressure being applied inside and outside the pipe string
- Easier recovery of cuttings since the up-hole velocity is controlled by the size of the drill pipe and less subject to lost-circulation
- No casing required during drilling and advantageous when high risks of caving inches. If there is a risk of caving, mud should be used as a stabilizer. In the case of air drilling, it presents the same risk than regular air rotary, since the flow is down the annular space

Cons:

- High water requirements (not for air drilling)
- Collection of a representative sample is difficult due to potential material mixing
- Rig size can render access difficult
- Need for drilling mud management (not for air drilling)

Dual-Wall Reverse Circulation Drilling

Pros:

- Good sample recovery due to controlled up-hole fluid velocity
- Fast penetration in coarse alluvial or broken, fissured rock
- Possible to obtain continuous representative samples of the formation and groundwater
- Easy estimate of aquifer yield at many depths in the formation
- Reduction of lost-circulation problems

Cons:

- Practical borehole diameter limited to 10 in.
- Maximum depth of ~1,400 ft, although greater depths can be achieved in hard rock
- Possible to dry out or to not detect a thin of low-yield aquifer
- Possible sample contamination due to the oil used in the air-compressor unless quality air filters are used (this is true for all air methods, unless the contractor uses filters)

Cable-Tool Percussion

Pros:

- In situations where the aquifer is thin and yield is low, the method permits identification of zones that might be overlooked by other drilling methods
- Recovery of representative soil samples at every depth, although samples are disturbed due to the impact of the blow which can affect material several feet below the bottom of the hole
- Allows well construction with low chance of contamination
- Borehole can be bailed at any time to determine approximate yield of the formation at a given depth
- Easy access to rough terrain

Cons:

- Slow penetration rate
- Due to the constant mixing of water, it is not possible to obtain groundwater samples during drilling

(Continued)

Table 6C.11 (Continued)

- Expensive casing for larger diameters
- Difficult to pull back casing in some geologic conditions

Air Percussion Down-the-Hole Hammer

Pros:

- Rapid removal of cuttings
- No use of drilling mud
- High penetration rate, especially in resistant rock formation (e.g. basalt)
- Easy soil and groundwater sampling during drilling
- Possible to measure yield estimate at selected depth in the formation

Cons:

- Restricted to semi-consolidated to consolidated formations

Air Percussion Casing Hammer

Pros:

- Wells can be drilled in unconsolidated materials that could be difficult to drill with cable-tool or direct rotary method
- No water-based fluid (drilling mud) is required in unconsolidated materials
- Representative formation and groundwater samples can be collected
- Borehole is fully stabilized during drilling operations through the use of casing
- Rapid penetration rates even in difficult drilling conditions
- Lost circulation problem is rarely a concern, except in very loose materials (e.g. mine waste rock)
- Operates well in cold weather

Cons:

- Method does not permit yield measurements during drilling
- When groundwater static levels are low, the high air pressure in the hole can prevent water from entering the borehole; a "rest" period is necessary to assess the true static level
- Relatively expensive method (increased cost of driving casing in)
- Very noisy (driving of casing)
- Borehole diameter limited to 12 in.

ODEX Percussion Down-the-Hole Hammer (Odex, Stratex, and Tubex are Trade Names)

Pros:

- Rapid removal of cuttings
- No use of drilling mud
- High penetration rate, especially in resistant rock formation (e.g. basalt)
- Easy soil and groundwater sampling during drilling
- Possible to measure yield estimate at selected depth in the formation
- Advantageous in unconsolidated formations with a high risk of caving (this is the probably the most important feature)

Cons:

- Practically restricted to unconsolidated formations
- Relatively a more expensive method

Source: From technologyinfomine.com. With permission.

Table 6C.12 Data on Standard and Line Pipe Commonly Used for Water Well Casing

Nominal Size (in.)	Outside Diameter (in.)	Outside Diameter Couplings (in.)	Schedule or Class[a]	Wall Thickness (in.)	Weight per Foot-Plain End (Pounds)	Inside Diameter (in.)	Suggested Maximum Setting (ft)[b]
4	4.500	5.200	—	0.219	10.10	4.062	1,190
			40	0.237	10.79	4.026	1,060
6	6.625	7.390	—	0.250	17.02	6.125	705
			40(S)	0.280	18.97	6.065	850
8	8.625	9.625	20	0.250	22.36	8.125	420
			30	0.277	24.70	8.071	525
			40(S)	0.322	28.55	7.981	695
10	10.750	11.750	20	0.250	28.04	10.250	235
			30	0.307	34.24	10.136	410
			40(S)	0.365	40.48	10.020	580
12	12.750	14.000	20	0.250	33.38	12.250	140
			30	0.330	43.77	12.090	320
			S	0.375	49.56	12.000	435
			40[c]	0.406	53.56	11.938	515
14	14.000	15.000	10	0.250	36.71	13.500	105
			20	0.312	45.68	13.376	195
			30(S)	0.375	54.57	13.250	350
			40	0.438	63.37	13.124	495
16	16.000	17.000	10	0.250	42.05	15.500	70
			20	0.312	52.36	15.376	140
			30(S)	0.375	62.58	15.250	240
			40	0.500	82.77	15.000	495
18	18.000	19.000	10	0.250	47.39	17.500	50
			20	0.312	59.03	17.376	100
			S	0.375	70.59	17.250	170
			30	0.438	82.06	17.124	270
			40	0.562	104.76	16.876	495
20	20.000	21.000	10	0.250	52.73	19.500	35
			20(S)	0.375	78.60	19.250	125
			30	0.500	104.13	19.000	295
			40[c]	0.594	123.06	18.802	445
22	22.000	—	10	0.250	58.07	21.500	30
			20(S)	0.375	86.61	21.250	95
			30	0.500	114.81	21.000	220
24	24.000	—	10	0.250	63.41	23.500	20
			20(S)	0.375	94.62	23.250	70
			30	0.562	140.80	22.876	240
			40	0.688	171.17	22.624	410
26	26.000	—	10	0.312	85.73	25.376	30
			S	0.375	102.63	25.250	55
			20	0.500	136.17	25.000	135
28	28.000	—	10[c]	0.312	92.41	27.376	25
			(S)	0.375	110.41	27.250	45
			20	0.500	146.85	27.000	105
			30	0.625	182.73	26.750	210
30	30.000	—	10[c]	0.312	99.08	29.376	20
			(S)	0.375	118.65	29.250	35
			20	0.500	157.53	29.000	85
			30	0.625	196.08	28.750	170

(Continued)

Table 6C.12 (Continued)

Nominal Size (in.)	Outside Diameter (in.)	Outside Diameter Couplings (in.)	Schedule or Class[a]	Wall Thickness (in.)	Weight per Foot-Plain End (Pounds)	Inside Diameter (in.)	Suggested Maximum Setting (ft)[b]
32	32.000	—	10[c]	0.312	105.76	31.376	20
			(S)	0.375	126.66	31.250	30
			20	0.500	168.21	33.000	70
			30	0.625	209.43	32.750	140
34	34.000	—	10[c]	0.312	112.43	33.376	15
			S	0.375	134.67	33.250	25
			20	0.500	178.89	33.000	60
			30	0.625	222.78	32.750	115
36	36.000	—	10[c]	0.312	119.11	35.376	10
			(S)	0.375	142.68	35.250	20
			20	0.500	189.57	35.000	50
			30	0.625	236.13	34.750	100

[a] ASA Standard B36.10 schedule numbers (S) indicates standard weight pipe.
[b] Maximum settings were estimated for the worst possible conditions in unconsolidated formation. A design factor of approximately 1.5 was used for steel with yield strength less than 40,000 lb/in^2. A 50-percent increase in depth of setting beyond those given is considered safe under favorable conditions.
[c] Indicates a non-API standard.

Source: From Bureau of Reclamation, *Groundwater Manual*, 1977.

Table 6C.13 Recommended Casing Diameters for Water Wells

Yield, Gallons per Minute	Recommended Casing Size (in.)	
Less than 100	6	I.D.[a]
75–175	8	I.D.
150–400	10	I.D.
350–600	12	I.D.
600–1,300	16	O.D.[b]
1,300–1,800	20	O.D.
1,800–3,000	24	O.D.
3,000–4,500	30	O.D.
Over 4,500	30	O.D.

Note: For line shaft vertical turbine pumps 1800 rpm.

[a] I.D., Inside Diameter.
[b] O.D., Outside Diameter.

Source: From U.S. Environmental Protection Agency, *Manual of Water Well Construction Practices*, EPA-570/ 9–75–001.

Table 6C.14 Recommended Casing Sizes for Domestic Water Wells

Yield at 50″ Drawdown	Recommended Casing Diameter (in.)	Pump Type		
		Jet	Double Jet	Submersible
Less than 8 gpm	2	X	X	
	3	X	X	X
	4	X	X	X
	5		X	X
	6			X
8–16.5 gpm	2	X	X	
	3	X	X	X
	4	X	X	X
	5		X	X
	6			X
Greater than 16.5 gpm	3	X		
	4	X	X	X
	5		X	X
	6			X

Source: From U.S. Environmental Protection Agency, *Manual of Water Well Construction Practices*, EPA-570/9–75–001.

Table 6C.15 Recommended Maximum Depth of Setting for California Stovepipe Casing

Diameter (in.)	Gauge[a]						Thickness (in.)			
	12		10		8	6				
	D	S	D	S	D	D	3/16	1/4	5/16	3/8
8	340	125	750	260	X	X	X	X	X	X
10	150	60	390	135	X	X	320	750	X	X
12	100	35	225	75	390	X	180	435	875	X
14	60	20	140	45	250	X	115	270	530	X
16	40	15	90	30	165	275	75	180	360	630
18	30		65	20	115	190	55	125	260	445
20	20		45		85	140	35	90	180	320
22			35		60	105	X	X	X	X
24			25		45	80	20	50	100	185
26			20		35	60	X	X	X	X
30			10		25	40	10	25	50	95

Note: D, Telescoping; S, Single thickness; X, Not commonly made in these sizes. Includes similar sheet steel and steel-plate fabricated casing; in feet.

[a] U.S. Standard Gauge.

Source: From Bureau of Reclamation, *Groundwater Manual*, 1977.

Table 6C.16 Recommended Diameter and Thickness of PVC Casing for Water Wells

Nominal Size	Outside Diameter	Inside Diameter	Minimum Wall Thickness
Well Diameters 1.5 in. Through 4 in.-ASTMD 2241-73[a] SDR 21 (Type 1120–1220)			
1.5	1.900	1.720	0.090
2	2.375	2.149	0.113
2.5	2.875	2.601	0.137
3	3.500	3.166	0.177
4	4.500	4.072	0.214
Well Diameters 5 in. Through 12 in.-ASTMD 1785-73[a] Schedule 40 (Type 1120–1220)			
5	5.563	5.047	0.258
6	6.625	6.065	0.280
8	8.625	7.981	0.322
10	10.750	10.020	0.365
12	12.750	11.938	0.406

[a] New ASTM Standards are currently under view.

Source: From U.S. Environmental Protection Agency, *Manual of Water Well Construction Practices*, EPA-570/9–75–001.

Table 6C.17 Well Screen Selection Chart for Small-Capacity Wells

Gradation of Sand	Average Slot Size Thousandths of an inch	Minimum Suggested Length for Corresponding Screen Diameter[a] and Desired Well Yield					
		1¼″ Screen	2″ Screen	3″ Screen	4″ Screen	5″ Screen	6″ Screen
Very Fine Sand 6–7–8 Slot About the finest material that can be utilized for a water supply. A line composed of 12 grains would measure about 1/16″	7	300 gph– 5 ft 450 gph– 8 ft 600 gph–12 ft	450 gph– 6 ft 600 gph– 9 ft 900 gph–13 ft	600 gph– 8 ft 900 gph–11 ft 1200 gph–14 ft	600 gph– 6 ft 1200 gph–10 ft 1800 gph–14 ft	900 gph– 6 ft 1200 gph– 9 ft 1800 gph–11 ft	1200 gph– 8 ft 2000 gph– 2 ft 2400 gph–15 ft
Fine Sand 9–10–12 Slot Often called "sugar sand." Line of 6 or 7 average grains measures 1/16″	10	300 gph–4 ft 450 gph–6 ft 600 gph–9 ft	450 gph–4 ft 600 gph–6 ft 900 gph–9 ft	600 gph–6 ft 900 gph–8 ft 1200 gph–10 ft	600 gph–4 ft 1200 gph–7 ft 1800 gph–10 ft	900 gph–5 ft 1200 gph–7 ft 1800 gph–8 ft	1200 gph–6 ft 1200 gph–9 ft 2400 gph–11 ft
Medium Sand 16–18–20 Slot Average grain size is about 4 grains 1/16″	18	300 gph–4 ft 450 gph–5 ft 600 gph–7 ft	600 gph–5 ft 900 gph–7 ft 1200 gph–9 ft	600 gph–4 ft 1200 gph–9 ft 1800 gph–13 ft	600 gph–3 ft 1200 gph–6 ft 1800 gph–9 ft	900 gph–4 ft 1200 gph–6 ft 1800 gph–7 ft	1200 gph–5 ft 2000 gph–8 ft 2400 gph–10 ft
Medium and Coarse Sand Mixed Average grain size a little less than 1/32″, or between 2 and 3 grains to 1/16″	25	300 gph–3 ft 450 gph–5 ft 600 gph–6 ft	600 gph–5 ft 900 gph–6 ft 1200 gph–8 ft	600 gph–4 ft 1200 gph–7 ft 1800 gph–11 ft	600 gph–3 ft 1200 gph–5 ft 1800 gph–8 ft	900 gph–4 ft 1200 gph–5 ft 1800 gph–6 ft	1200 gph–5 ft 2000 gph–7 ft 2400 gph–91 ft
Coarse Sand Average grain size a little over 1/30 (2 grains to 1/16″)	35	450 gph–4 ft 600 gph–5 ft 900 gph–7 ft	600 gph–4 ft 900 gph–5 ft 1200 gph–7 ft	900 gph–4 ft 1200 gph–6 ft 1800 gph–10 ft	900 gph–3 ft 1200 gph–4 ft 1800 gph–7 ft	1200 gph–4 ft 1800 gph–6 ft 2000 gph–8 ft	1200 gph–4 ft 2000 gph–7 ft 2400 gph–8 ft
Coarse Sand and Fine Gravel Mixed Average grain size about 1/16″. In coarser gravels, No. 80 and No. 100 slot are often used	50	450 gph–4 ft 600 gph–5 ft 900 gph–7 ft	600 gph–4 ft 900 gph–5 ft 1200 gph–6 ft	900 gph–4 ft 1200 gph–6 ft 1800 gph–10 ft	900 gph–3 ft 1200 gph–4 ft 1800 gph–7 ft	1200 gph–4 ft 1800 gph–6 ft 2000 gph–7 ft	1200 gph–4 ft 2000 gph–6 ft 2400 gph–8 ft

[a] Nominal size of screen.

Source: From Edward E. Johnson, Inc. With permission.

Table 6C.18 Recommended Minimum Screen Assembly Diameters

Discharge (gal/min)	Minimum Nominal Screen Assembly Diameter (in.)
Up to 50	2
50 to 125	4
125 to 350	6
350 to 800	8
800 to 1,400	10
1,400 to 2,500	12
2,500 to 3,500	14
3,500 to 5,000	16
5,000 to 7,000	18
7,000 to 9,000	20

Source: From U.S. Bureau of Reclamation, *Groundwater Manual*, 1977.

Table 6C.19 Cost of Water Well Screens

304SS Drive Points

1-1/4″ PS 1.71″OD×1″ ID					2″PS 2.38″OD × 1.75″ID					3″PS All Drive 3.7″OD × 3″ID			4″ PS All Drive 4.69″OD × 4″ID		
Model Number	Screen Length	Ship Wt/pc	Price W60	Price W90	Model Number	Screen Length	Ship Wt/pc	Price W60	Price W90	Screen Length	Ship Wt/pc	Price W60	Screen Length	Ship Wt/pc	Price W60
824	24″	5	$106	$96	924	24″	8	$129	$117	24″	10	$176	24″	13	$217
836	36″	7	$133	$120	936	36″	11	$159	$144	36″	14	$224	36″	17	$279
848	48″	9	$160	$144	948	48″	14	$190	$172	48″	17	$272	48″	21	$340
860	60″	11	$186	$169	960	60″	16	$220	$200	60″	22	$320	60″	25	$404

Fittings are 304 SS MIP X Carbon Steel Point w/guardian plate or 304 SS MIP X MIP for open end extensions. For other screen lengths add $30/ft for 1 ¼″, $34/ft for 2″, $38/ft for 3″, and $49/ft for 4″. All Drive screens are MIPXFIP only, if cast iron point is required add; 3″PS-$80.00, 4″PS-$85.00 to the list price above.

304 SS Small Diameter Waterwell Screens

Nom Dia Wt/ft	Dimensions[a] OD Dim[a] W60	Dimensions[a] OD Dim[a] W90	ID	Screen Price per foot 60	Screen Price per foot 90	Weld Ring[c]	NPT Thd	Fig K Pkr w/WR	Plate Btm w/WR	Flush Threads Sch40	Flush Threads Sch80	Misc Attach Clean and Bag	Misc Attach Ball Loop[d]	Misc Attach SS Point[e]	Type-Max Depth Based on Collapse Strength[a]
2P/3T 3.0#	2.60	2.50	2.02	$34	$31	$41	$39	n/a	$61	$37	$39	$13	$41	$79	W60–1000 W90–500
3P/4T 4.0#	3.74	3.64	3.16	$38	$34	$43	$41	n/a	$64	$39	$41	$21	$48	$92	W60–900 W90–300
4P/5T 5.0#	4.74	4.64	4.16	$49	$45	$49	$49	n/a	$72	$47	$49	$21	$50	$151	W60–500 W90–200
5P/6T 6.0#	5.63	5.53	5.05	$57	$52	$51	$60	$161	$86	$58	$60	$32	$61	n/a	W60–300 W90–100[b]

304 SS Small Diameter Environmental Screens

Nom Dia Wt/ft	Dimensions[a] OD Dim[a] W60	Dimensions[a] OD Dim[a] W90	ID	Screen Price per foot 60	Screen Price per foot 90	Weld Ring[c]	NPT Thd	Fig K Pkr w/WR	Plate Btm w/WR	Flush Threads Sch40	Flush Threads Sch80	Misc Attach Clean and Bag	Misc Attach Ball Loop[d]	Misc Attach SS Point[e]	Type-Max Depth Based on Collapse Strength[a]
1/2″ PS[b] 1.0#	0.89	0.83	0.53	$30	$27	$20	$19	n/a	$30	$73	$80	$13	n/a	n/a	W60-2000 W90-1500
3/4″ PS[b] 1.5#	1.09	1.06	0.73	$31	$28	$21	$21	n/a	$32	$73	$80	$13	n/a	n/a	W60-2000 W90-1000

(Continued)

Table 6C.19 (Continued)

304 SS Small Diameter Environmental Screens

Nom Dia Wt/ft	Dimensions[a] OD Dim[a] W60	W90	ID	Screen Price per foot 60	90	Direct Attached Standard Fittings Per End Weld Ring[c]	NPT Thd	Fig K Pkr w/WR	Plate Btm w/WR	Flush Threads Sch40	Sch80	Misc Attach Clean and Bag	Ball Loop[d]	SS Point[e]	Type-Max Depth Based on Collapse Strength[a]
1″ PS[b] 2.0#	1.33	1.27	0.88	$31	$28	$22	$28	n/a	$35	$66	$72	$13	n/a	$53	W60-2000 W90-1000
1-1/4″PS 1.5#	1.67	1.60	1.19	$30	$29	$24	$36	n/a	$36	$33	$36	$13	$38	$58	W60-1000 W90-600
1-1/2″ PS 3.0#	2.01	1.95	1.54	$33	$31	$25	$38	n/a	$37	$36	$38	$13	$38	$62	W60-1000 W90-600
2″ PS 3.0#	2.47	2.41	1.99	$34	$31	$41	$39	n/a	$61	$37	$39	$13	$41	$79	E60-1000 E90-600
2-1/2″ PS 4.0#	3.03	2.93	2.45	$39	$35	$41	$39	n/a	$61	$37	$39	$21	$44	$81	W60-1000 W90-600
3″ PS 4.0#	3.56	3.46	2.98	$38	$34	$43	$41	n/a	$64	$39	$41	$21	$48	$92	E60-1000 E90-300
4″ PS 5.0#	4.56	4.46	3.98	$49	$45	$49	$49	n/a	$72	$47	$49	$21	$50	$151	E60-500 E90-200
6″ PS 7.0#	6.64	6.54	6.02	$81	$80	$79	$106	n/a	$126	$104	$106	$32	$66	n/a	E60-250 E90-50

Minimum billing length is 3 ft for all diameters. For Tightwind or Super Construction Screen: Add 25% to Screen Price/Feet. Please specify water well or environmental when ordering. Must add "Clean & Bag" charge for environmental screen. Environmental screens supplied in "Shoulder to Shoulder" lengths. For loose plate bottom deduct (1) weld righ from plate bottom assembly.

[a] Dimensions, weights and collapse strength are approximate (based on an average slot and depth).

[b] Minimum order requirement for ½″ through 1″PS is $300 (net).

[c] Standard weld ring length 2PS″ through 6PS″ is 1-1/2″.

[d] Bail Loop prices Do Not include plate.

[e] SS point is weld on. For threaded point add appropriate thread price.

Large Diameter "Free Flow" 304 Stainless Steel

Nom Diam	Standard Dimensions[f] OD	ID	Screen Price Per Foot Max Depth Based on Collapse Strength[f] 100 ft	250 ft	600 ft	1000 ft	Direct Attached Standard Fittings Per End Weld Rings Std	4″	6″	Threads Npt	Jws	Flush Sch40	Flush Sch80	Fig "K" Pkr w/WR	Plate Btm w/WR	Misc Attachments Bail Loop[g]	Lift Lugs[h]	4″WR W/4″ Collar
6″PS	6.70	6.00 Wt/Foot 6#	$81	$83	$85	$93	$79	$107	$131	$106	n/a	$104	$106	n/a	$126	$66	$84	$279
8″Tel	7.50	6.75 Wt/Foot 10#	$88	$88	$96	$96	$108	$141	$173	n/a	$154	n/a	n/a	$435	$173	$70	$93	
8″PS	8.63	7.88 Wt/Foot 12#	$95	$95	$99	$99	$126	$148	$183	$196	n/a	$193	$290	n/a	$202	$70	$93	$338
10″Tel	9.50	8.68 Wt/Foot 13#	$119	$119	$129	$129	$144	$187	$230	n/a	$300	n/a	n/a	$460	$245	$70	$105	
10″PS	10.75	9.88 Wt/Foot 14#	$135	$135	$135	$162	$149	$194	$238	$254	n/a	$292	$440	n/a	$253	$70	$105	$410
12″Tel	11.25	10.40 Wt/Foot 18#	$143	$143	$143	$170	$169	$223	$276	n/a	n/a	n/a	N/a	$605	$287	$70	$116	
12″PS	12.75	11.80 Wt/Foot 23#	$146	$146	$169	$192	$196	$249	$304	$387	n/a	$623	$924	n/a	$333	$70	$116	$512
14″Tel	12.50	11.60 Wt/Foot 25#	$145	$145	$164	$189	$215	$280	$344	n/a	n/a	n/a	N/a	$809	$366	$76	$121	

(Continued)

Table 6C.19 (Continued)

Large Diameter "Free Flow" 304 Stainless Steel

Nom Diam	Standard Dimensions[f] OD	ID	Screen Price Per Foot Max Depth Based on Collapse Strength[f] 100 ft	250 ft	600 ft	1000 ft	Direct Attached Standard Fittings Per End: Weld Rings Std	4″	6″	Threads Npt	Jws	Flush Sch40	Flush Sch80	Fig "K" Pkr w/WR	Plate Btm w/WR	Misc Attachments Bail Loop[g]	Lift Lugs[h]	4″WR W/4″ Collar
14P/16T	14.00 Wt/Foot 30#	13.00	$175	$175	$192	$209	$252	$329	$404	n/a	n/a	na	na	$923	$441	$76	$121	$601
16P/18T	16.00 Wt/Foot 35#	15.00	$185	$202	$215	$235	$302	$377	$464	n/a	n/a	n/a	N/a	$1,014	$529	$76	$128	$670
18P/20T	18.00 Wt/Foot 40#	16.70	$214	$224	$255	$304	$360	$468	$576	n/a	n/a	n/a	N/a	$1,045	$630	$76	$128	$755
20″ PS	20.00 Wt/Foot 42#	18.80	$252	$272	$295	$320	$382	$497	$611	n/a	n/a	n/a	N/a	n/a	$669	$76	$139	$845
24″ Tel	22.00 Wt/Foot 46#	20.70	$342	$348	$391	$407	$404	$525	$646	n/a	na	n/a	N/a	n/a	$707	$87	$139	
24P/26T	24.00 Wt/Foot 58#	22.75	$369	$415	$493	$493	$462	$601	$739	n/a	n/a	n/a	N/a	n/a	$809	$87	$151	$998
26″ PS	26.00 Wt/Foot 62#	24.40	$436	$436	$527	$627	$539	$674	$830	n/a	n/a	n/a	n/a	n/a	$943	$87	$151	$1,102
30″ Tel	27.25 Wt/Foot 65#	25.75	$450	$490	$570	Call	$595	$774	$952	n/a	n/a	n/a	n/a	n/a	$1,041	$105	$169	
30P/36T	30.00 Wt/Foot 74#	28.30	$530	$593	$633	Call	$671	$872	$1,074	n/a	n/a	n/a	n/a	n/a	$1,174	$105	$169	
36″ PS	36.00 Wt/Foot 90#	34.30	$610	$696	$696	Call	$773	$1,005	$1,237	n/a	n/a	n/a	n/a	n/a	$1,353	$105	$169	

Call for pricing on screens deeper than 1000 feet. Standard weld ring lengths: 6″PS-16P/18T are 1-1/2″ long; 18P/20T and larger are 2″ long. To price Weld Rings longer than 6″, combine prices shown above. For Tightwind: Add 25% to Screen Price/Ft (Min TW slot < =0.010 Inch). Minimum billing length for LG Diam SS is 3 feet. Furnished in Full Screen Lengths, unless specified otherwise at the time of order.

[f] Dimensions, weights and collapse strength are approximate (based on an average slot and depth).

[g] Standard weld ring length 2PS″ through 6PS″ is 1-1/2″.

[h] Bail Loop prices Do Not include plate.

Large Diameter 304 Stainless Steel High Flow (HIQ) and Remediation[i]

Nom Diam	Standard Dimensions[i] OD	ID	Screen Price Per Foot Max Depth Based on Collapse Strength[i] 100 ft	250 ft	600 ft	1000 ft	Direct Attached Standard Fittings Per End: Weld Rings STD	4″	6″	Threads NPT	JWS	Flush Sch40	Flush Sch80	Fig "K" Pkr w/WR	Plate Btm w/WR	Misc Attachments Bail Loop[j]	Lift Lugs[k]	4″WR W/4″ Collar
6″PS	6.70 Wt/Foot 6#	6.00	$81	$89	$96	$96	$79	$107	$131	$106	n/a	$104	$106	n/a	$126	$66	$84	$279
8″Tel	7.50 Wt/Foot 10#	6.75	$96	$105	$105	$110	$108	$141	$173	n/a	$153	n/a	n/a	$435	$173	$70	$93	
8″PS	8.63 Wt/Foot 12#	7.88	$116	$116	$125	$125	$126	$148	$183	$196	n/a	$193	$290	n/a	$202	$70	$93	$338
10″Tel	9.50 Wt/Foot 13#	8.68	$131	$131	$141	$141	$144	$187	$230	n/a	$301	n/a	n/a	$460	$245	$70	$105	
10″PS	10.75 Wt/Foot 14#	9.88	$155	$155	$186	$186	$149	$194	$238	$255	n/a	$292	$440	n/a	$253	$70	$105	$410
12″Tel	*11.25* Wt/Foot 18#	*10.40*	$163	$163	$194	$194	$179	$233	$276	n/a	n/a	n/a	n/a	$605	$304	$70	$116	

(Continued)

Table 6C.19 (Continued)

	Standard Dimensions[i]		Screen Price Per Foot Max Depth Based on Collapse Strength[i]				Direct Attached Standard Fittings Per End: Weld Rings			Threads				Fig "K"	Plate	Misc Attachments		
Nom Diam	OD	ID	100 ft	250 ft	600 ft	1000 ft	STD	4″	6″	NPT	JWS	Flush Sch40	Flush Sch80	Fig "K″ Pkr w/WR	Plate Btm w/WR	Bail Loop[j]	Lift Lugs[k]	4″WR W/4″ Collar
Large Diameter 304 Stainless Steel High Flow (HIQ) and Remediation[i]																		
12″PS	12.75 Wt/Foot 23#	11.80	$178	$178	$217	$217	$196	$249	$304	$388	n/a	$623	$924	n/a	$333	$80	$116	$512
14″Tel	12.50 Wt/Foot 25#	11.60	$176	$176	$215	$215	$215	$280	$344	n/a	n/a	n/a	n/a	$809	$366	$76	$121	
14P/16T	14.00 Wt/Foot 30#	13.00	$206	$206	$226	$226	$252	$329	$404	n/a	n/a	na	na	$923	$441	$76	$121	$601
Large Diameter 304 Stainless Steel High Flow (HIQ)[k]																		
16P/18T	16.00 Wt/Foot 35#	15.00	$210	$230	$230	$235	$302	$377	$464	n/a	n/a	n/a	n/a	$1,014	$529	$76	$128	$670
18P/20T	18.00 Wt/Foot 40#	16.70	$285	$285	$285	$330	$360	$468	$576	n/a	n/a	n/a	n/a	$1,045	$630	$76	$128	$755
20″ PS	20.00 Wt/Foot 42#	18.80	$340	$340	$355	$400	$382	$497	$611	n/a	n/a	n/a	n/a	n/a	$669	$76	$139	$845
24″ Tel	22.00 Wt/Foot 46#	20.70	$380	$380	$421	$446	$404	$525	$646	n/a	n/a	n/a	n/a	n/a	$707	$87	$139	
24P/26T	24.00 Wt/Foot 58#	22.75	$420	$420	$488	$488	$462	$601	$739	n/a	n/a	n/a	n/a	n/a	$809	$87	$151	$998
26″ PS	26.00 Wt/Foot 62#	24.40	$436	$436	$516	$527	$539	$674	$830	n/a	n/a	n/a	n/a	n/a	$943	$87	$151	$1,102
30″ Tel	*27.25* Wt/Foot 65#	*25.75*	$525	$525	$570	Call	$595	$774	$952	n/a	n/a	n/a	n/a	n/a	$1,041	$105	$169	
30P/36T	30.00 Wt/Foot 74	28.30	$575	$580	$610	Call	$671	$872	$1,074	n/a	n/a	n/a	n/a	n/a	$1,174	$105	$169	
36″ PS	36.00 Wt/Foot 90#	34.30	$610	$680	$696	Call	$773	$1,005	$1,237	n/a	n/a	n/a	n/a	n/a	$1,353	$105	$169	

Call for pricing on screens deeper than 1000 feet. Standard weld ring lengths: 6″PS-16P/18T are 1-1/2″ long; 18P/20T and larger are 2″ long. To price Weld Rings longer than 6″, combine prices shown above. For Tightwind: Add 25% to Screen Price/Ft (Min TW slot < =0.010 Inch). Minimum billing length for LG Diam SS is 3 feet. Furnished in Full Screen Lengths, unless specified otherwise at the time of order.

[i] Dimensions, weights and collapse strength are approximate (based on an average slot and depth).
[j] Standard weld ring length 2PS″ through 6PS″ is 1–1/2″.
[k] Bail Loop prices do not include plate.

304 Stainless Steel Casing[l]

	Dimensions					Direct Attached Std Fittings Per End: Flush Threads					Loose Fittings: Threaded			
Nom Diam	Sch	OD	ID	Wt Per ft	List Price per Foot	Sch 40	Sch 80	Npt Thread	Weld Ring	Plate Bottom	Point	Cap/Plug	Locking Cap	Add/Joint Cln and Bag
1″PS	5	1.315	1.185	0.88	$12	$66	$72	$28	$22	$28	$96	$94	$31	$13
	10		1.097	1.42	$15									
	40		1.049	1.70	$16									
1-1/4″PS	5	1.660	1.530	1.12	$15	$33	$36	$36	$24	$29	$56	$49	$34	$13
	10		1.442	1.82	$16									
	40		1.380	2.29	$18									
1-1/2″PS	5	1.90	1.770	1.29	$14	$36	$38	$38	$25	$30	$60	$52	$35	$13
	10		1.682	2.10	$16									
	40		1.610	2.74	$19									

(Continued)

Table 6C.19 (Continued)

304 Stainless Steel Casing[1]														
						Direct Attached Std Fittings Per End					Loose Fittings			
	Dimensions				List	Flush Threads		Npt	Weld	Plate	Threaded		Locking	Add/Joint
Nom Diam	Sch	OD	ID	Wt Per ft	Price per Foot	Sch 40	Sch 80	Thread	Ring	Bottom	Point	Cap/Plug	Cap	Cln and Bag
2″PS	5	2.375	2.245	1.62	$16	$37	$39	$39	$41	$34	$67	$52	$36	$13
	10		2.157	2.66	$21									
	40		2.067	3.69	$25									
3″PS	5	3.50	3.334	3.06	$29	$39	$41	$41	$43	$44	$90	$64	$43	$21
	10		3.260	4.37	$34									
	40		3.068	7.65	$40									
4″PS	5	4.50	4.334	3.95	$31	$47	$49	$49	$49	$48	$108	$65	$56	$21
	10		4.260	5.67	$38									
	40		4.026	10.90	$60									
5″PS	5	5.563	5.345	6.41	$49	$58	$60	$60	$54	$70	n/a	$92	$73	$32
	10		2.295	7.84	$59									
	40		5.047	14.75	$92									
6″PS	5	6.625	6.407	7.66	$56	$104	$106	$106	$79	$82	n/a	$101	$90	$32
	10		6.357	9.38	$65									
	40		6.065	19.15	$109									
8″PS	10	8.625	8.329	13.53	$104	$193	$290	$196	$126	$88	n/a	$188	$186	n/a
	40		7.981	28.82	$190									
10″PS	10	10.75	10.420	18.83	$130	$292	$440	$254	$149	$148	n/a	$316	$281	n/a
	40		10.020	40.86	$255									
12″PS	10	12.75	12.390	24.39	$172	$623	$924	$387	$196	$218	n/a	$448	$372	n/a
	40		12.000	50.03	$340									

Minimum billing length is 3 ft. Sumps: Add Weld Ring and Plate Bottom price to the sump length needed (minimum billing is 3 ft). Sch40 & Sch80 Threads: 1″ = 8 TPI; 1.25 & 1.5 = 4 TPI; 2″ > 2 TPI. Locking Cap Lugs are $7.00 and shipped loose. (Part Number 248242). 1 For Slip Cap deduct $7.00 from Locking Cap Price. * Price per foot includes beveling

[1] Must add "Clean & Bag" charge for environmental casing.

Table 6C.20 Intake Areas of Well Screens

Wire-Wound Telescopic Screens Intake Areas (sq. in. per ft of Screen)								
	Slot Opening Size							
Nom Diam (in.)	10-Slot	20-Slot	40-Slot	60-Slot	80-Slot	100-Slot	150-Slot	250-Slot
3	15	26	41	52	59	65	73	82
4	20	35	57	71	81	88	101	115
5	26	45	72	90	102	112	112	132
6	30	53	85	106	100	112	132	156
8	28	51	87	113	133	149	160	194
10	36	65	108	141	166	186	200	243
12	42	77	130	143	171	195	237	265
14	37	68	97	132	161	185	232	292
16	42	60	108	148	180	208	261	327
18	36	69	124	169	206	237	298	375
20	41	77	139	189	229	264	280	366
24	61	113	131	182	226	265	343	449
26	63	118	138	191	237	278	360	471

(Continued)

Table 6C.20 (Continued)

Wire-Wound Telescopic Screens
Intake Areas (sq. in. per ft of Screen)

	Slot Opening Size							
Nom Diam (in.)	10-Slot	20-Slot	40-Slot	60-Slot	80-Slot	100-Slot	150-Slot	250-Slot
30	75	138	161	224	278	325	422	552
36	84	157	184	255	317	371	481	629

PVC Plastic Screens
Intake areas (sq. in. per ft of screen)

	Slot Opening Size									
Size (in.)	6-Slot	8-Slot	10-Slot	12-Slot	15-Slot	20-Slot	25-Slot	30-Slot	35-Slot	40-Slot
1¼	3.0	3.4	4.8	3.0	7.0	8.9	10.8	12.5	14.1	15.6
1½	3.4	4.5	5.5	6.5	8.1	10.2	12.3	14.2	16.2	17.9
2	4.3	5.5	6.8	8.1	10.0	12.8	15.4	17.9	20.3	22.4
3	5.4	7.1	8.8	10.4	12.8	16.5	20.0	23.2	26.5	29.3
4	7.0	9.0	11.3	13.5	16.5	21.2	25.8	30.0	3.9	37.7
5	8.1	10.6	13.1	15.5	19.1	24.7	30.0	34.9	39.7	44.2
6	8.1	10.6	13.2	15.6	19.2	25.0	30.5	35.8	40.7	45.4
8	13.4	17.6	21.7	25.7	31.5	40.6	49.3	57.4	65.0	72.3

Note: The maximum transmitting capacity of the screen can be derived from these figures. To determine GPM per feet of screen, multiply the intake area in square inches by 0.31. It must be remembered that this is the maximum capacity of the screen under ideal conditions with an entrance velocity of 0.1 ft/sec.

Source: From Johnson Division of Signal Environmental Systems, Inc., St. Paul, MN.

Table 6C.21 Optimum Well Screen Entrance Velocities

Coefficient of Permeability (gallons per day per square foot)	Optimum Screen Entrance Velocities (ft/min)
>6000	12
6000	11
5000	10
4000	9
3000	8
2500	7
2000	6
1500	5
1000	4
500	3
<500	2

Source: From Illinois State Water Survey, 1962.

Table 6C.22 Chlorinated Lime Required to Disinfect a Well or Spring

Capacity of Well or Spring in Gallons	Chlorinated Lime Required (25% Available Chlorine)		Approximate Volume of Water, in Gallons to be Used in Preparing Chlorine Solution
	Pounds and Ounces		
50	—	1.5	5
100	—	3.0	5
200	—	6.0	5
300	—	9.0	5
400	—	12.0	5
500	—	15.0	5
1,000	1	14.0	10
2,000	3	12.0	15
3,000	5	10.0	20

Note: Values provide a dosage of approximately 50 parts per million of available chlorine.

Source: From U.S. Public Health Service.

Table 6C.23 Volume of Water in Well Per Foot of Depth

Nominal Casing Size (in.)	Schedule No.	Volume (Gallons per Foot of Depth)
4	40	0.66
5	40	1.04
6	40	1.50
8	30	2.66
10	30	4.19
12	30	5.80
14	30	7.16
16	30	9.49
18	30	11.96
20	30	14.73
22	30	17.99
24	30	21.58

Source: From U.S. Bureau of Reclamation, *Groundwater Manual* 1977.

SECTION 6D INJECTION WELLS

Table 6D.24 Statistical Analysis of Injection Well Data

Distribution of Injection Wells by Industry Type

Type of Industry	Percentage of Existing Wells (%)
Refineries and natural gas plants	20
Chemical, petrochemical, and pharmaceutical companies	55
Metal products companies	7
Other	18

Total Depth of Injection Wells

Total Well Depth	Percentage of Wells (%)
0–1,000 ft	5
1,001–2,000 ft	32
2,001–4,000 ft	27
4,001–6,000 ft	28
6,001–12,000 ft	6
over 12,000 ft	2

Type of Rock Used for Injection

Rock Type	Percentage of Wells (%)
Sand	33
Sandstone	41
Limestone and dolomite	22
Other	4

Rate of Injection

Injection Rate	Percentage of Wells (%)
0–50 gpm	23
51–100 gpm	11
101–200 gpm	25
201–400 gpm	19
401–800 gpm	3
over 800 gpm	1
Unknown	18

Pressure at Which Waste Is Injected

Injection Pressure	Percentage of Wells (%)
Gravity flow	11
Gravity–150 psi	19
151–300 psi	15
301–600 psi	6
601–1,500 psi	13
over 1,500 psi	2
Unknown	34

Source: From Water Well Journal, 1968.

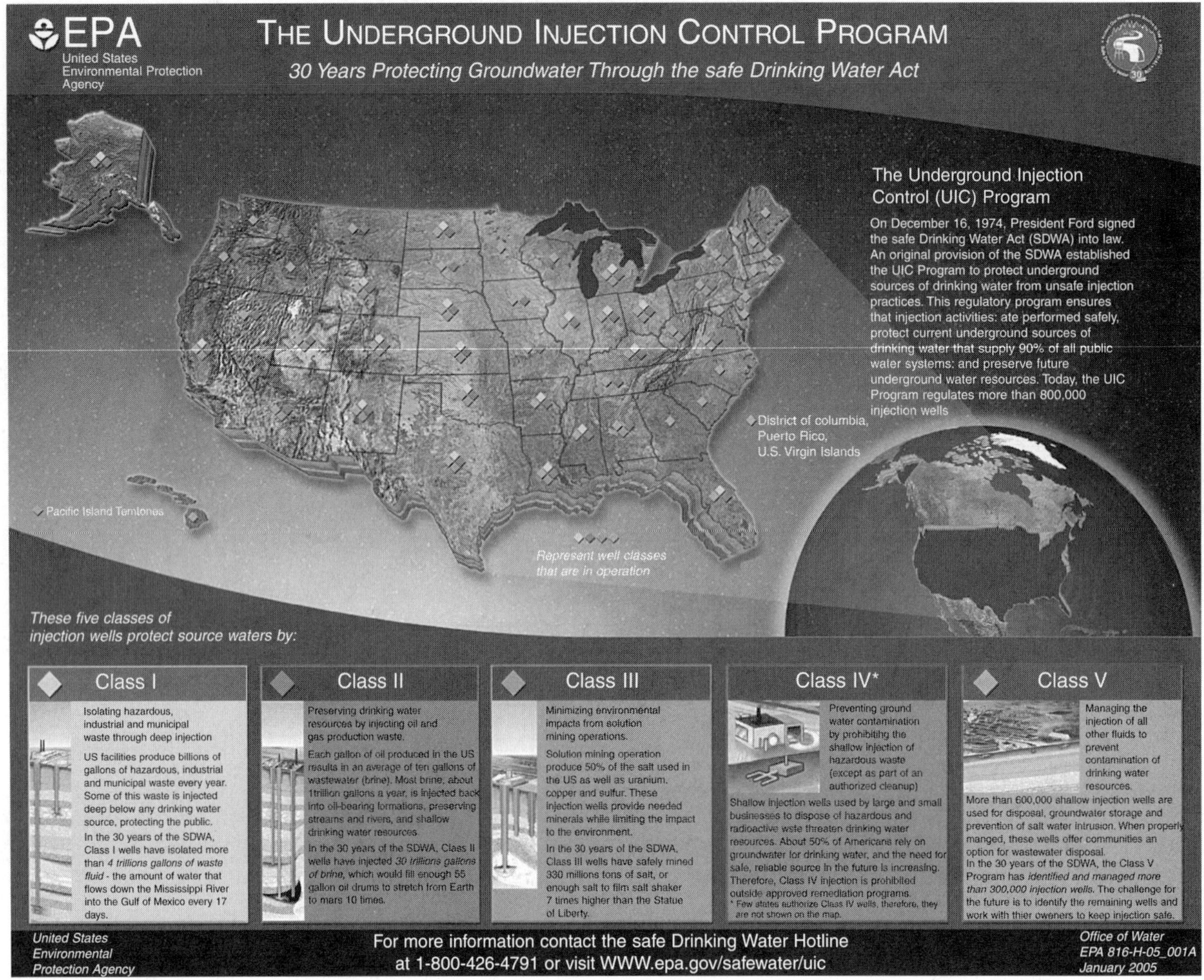

Figure 6D.10 The underground injection control program. (From www.epa.gov.)

Table 6D.25 Classes of Injection Wells

Regulatory Definitions of Injection Wells (§144.6)

The UIC Program provides standards, technical assistance and grants to State governments to regulate injection wells in order to prevent them from contaminating drinking water resources. EPA defines the five classes of wells according to the type of fluid they inject and where the fluid is injected. EPA has published regulations related to the sitting, drilling, construction and operation of many types of injection wells

Class I wells are technologically sophisticated and inject hazardous and non-hazardous wastes below the lowermost underground source of drinking water (*USDW*). Injection occurs into deep, isolated rock formations that are separated from the lowermost USDW by layers of impermeable clay and rock	In general, owners and operators of most new Classes I, II and III injection wells are required to: • Site the wells in a location that is free of faults and other adverse geological features
Class I wells are oil and gas production brine disposal and other related wells. Operators of these wells inject fluids associated with oil and natural gas production. Most of the injected fluid is brine that is produced when oil and gas are extracted from the earth (about 10 barrels of brine for every barrel of oil)	• Drill to a depth athat allows the injection into formations that do not contain water that can potentially be used as a source of drinking water. These injection zones are confined from any formation that may contain water that may potentially be used as a source of drinking water
Class III wells are wells that inject superheated steam, water, or other fluids into formations in order to extract minerals. The injected fluids are then pumped to the surface and the minerals in solution are extracted. Generally, the fluid is treated and re-injected into the same formation. More than 50 percent of the salt and 80 percent of the uranium extraction in the U.S. is produced this way	• Build to inject through an internal pipe (tubing) that is located inside another pipe (casing). This outer pipe has cement on the outside to fill any voids occurring between the outside pipe and the hole that was bored for the well (borehole). This allows for multiple layers of containment of the potentially contaminating injection fluids
Class IV wells inject hazardous or radioactive wastes into or above underground sources of drinking water. These wells are banned under the UIC program because they directly threaten public health	• Test for integrity at the time of completion and every five years thereafter (more frequently for hazardous waste wells, §146.68(d)) • Monitor continuously to assure the integrity of the well
Class V wells are injection wells that are not included in the other classes. Some Class V wells are technologically advanced wastewater disposal systems used by industry, but most are "low-tech" wells, such as septic systems and cesspools. Generally, they are shallow and depend upon gravity to drain or "inject" liquid waste into the ground above or into underground sources of drinking water. Their simple construction provides little or no protection against possible groundwater contamination, so it is important to control what goes into them	Operators of Class I wells injecting hazardous waste are required to demonstrate that the waste will never return to the surface or impact an underground source of drinking water (for 10,000 years). These wells inject at 4,000 ft below the surface or more. Over 9 billion gallons of hazardous waste is injected into wells each year in the US The largest number of injection wells are shallow wells that inject non-hazardous fluids into very shallow aquifers that are or can be used as sources of drinking water. Some of the wells in this category are: • Drainage wells in industrial setting that can receive surface runoff contaminated with a variety of pollutants; • Septic tank systems and dry-wells used in automotive shops that receive fluids from repair and maintenance bays; • Cesspools that receive sewage from a community; • Agricultural drainage wells that may receive water contaminated with pesticides and fertilizers

Source: From www.epa.gov.

Table 6D.26 Underground Injection Control Program — Inventory of Wells

	Class I Hazardous Waste Wells	Class I non-Hazardous Waste Wells	Class II Wells	Class III Facilities (Which May Contain Multiple Wells)	Class III Wells	Class IV Facilities (Which May Contain Multiple Wells)
AK		6	925			
AL			347	1	3	
AR	6	10	1,078			
AZ				3	15	
CA		6	24,955	1	73	
CO		5	787	2	35	
FL	1	136	64			
IA			1			
IL	3	1	8,949			
IN	4	7	1,340			
KS	6	44	16,371	5	157	
KY		1	3,429			
LA	17	28	3,824	15	67	
MI	9	15	1,459	3	22	
MO			211			
MS	6	1	881			
MT			897			
ND		2	377			
NE		1	704	1	1,879	
NM		5	5,577	19	104	
NV			12			
NY		1	503	6	126	
OH	10	2	2,890	3	47	1
OK		11	11,448	1	1	
OR			2			
PA		10	1,897	1	10	
SD			63			
TN			14			
TX	60	46	51,998	86	5,789	
UT			256	1	4	2
VA		3	252	1	3	1
WA			1			
WV			700	3	31	
WY		23	4,666	12	8,211	1
EPA R2 tribal			12			
EPA R6 tribal	1		2,674			
EPA R8 tribal			723			
EPA R9 tribal			542			
	122	364	146,878	164	16,577	5

Note: Types of wells: Class I, Deep industrial waste disposal; Class II, For oil and natural gas production; Class III, Related to mineral recovery; Class IV, Banned, inject hazardous or radioactive waste above USDWs; Class V, Generally shallow disposal wells. There are an estimated 650,000 Class V wells nationally. Class I through IV data from annual reporting by states and EPA Regions. The Class V estimate is based on state estimates and modeling of storm water and large capacity septics.

Source: From www.epa.gov.

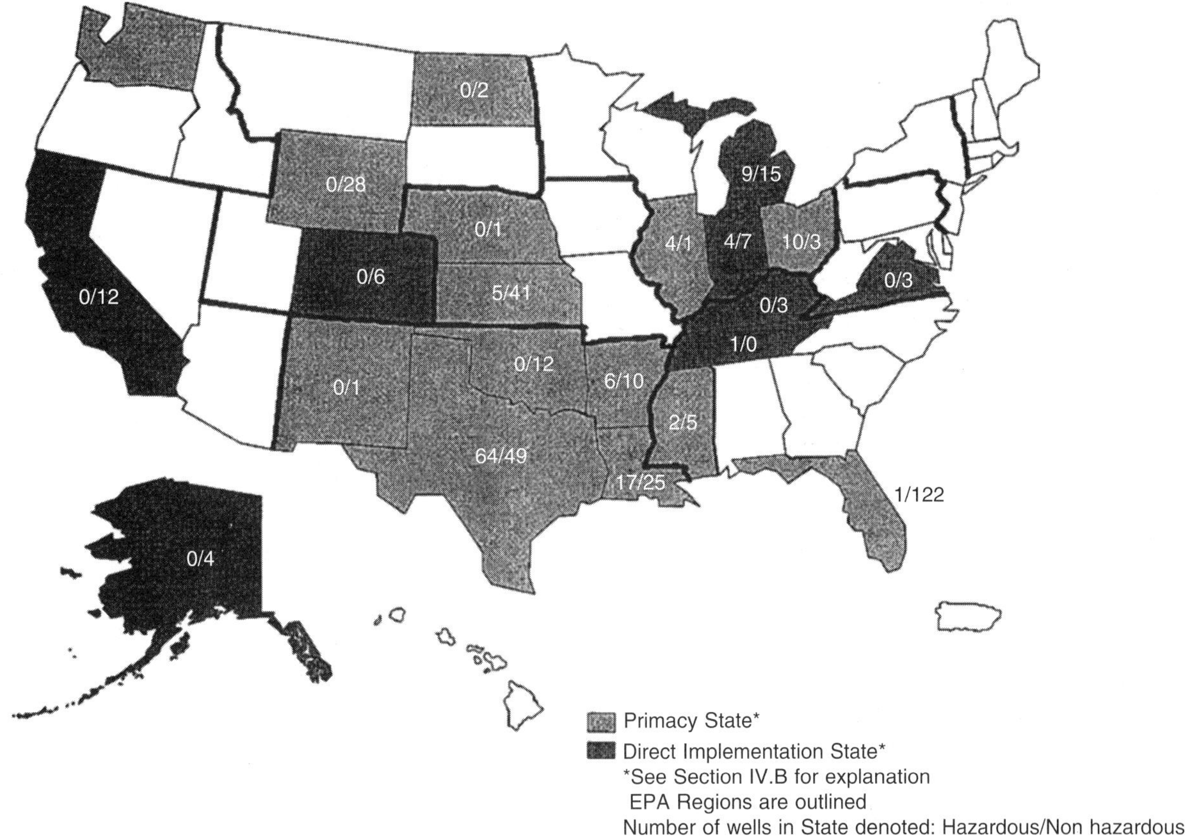

Figure 6D.11 Number of class I wells by state. (From www.epa.gov EPA's class I well inventory, 1999.)

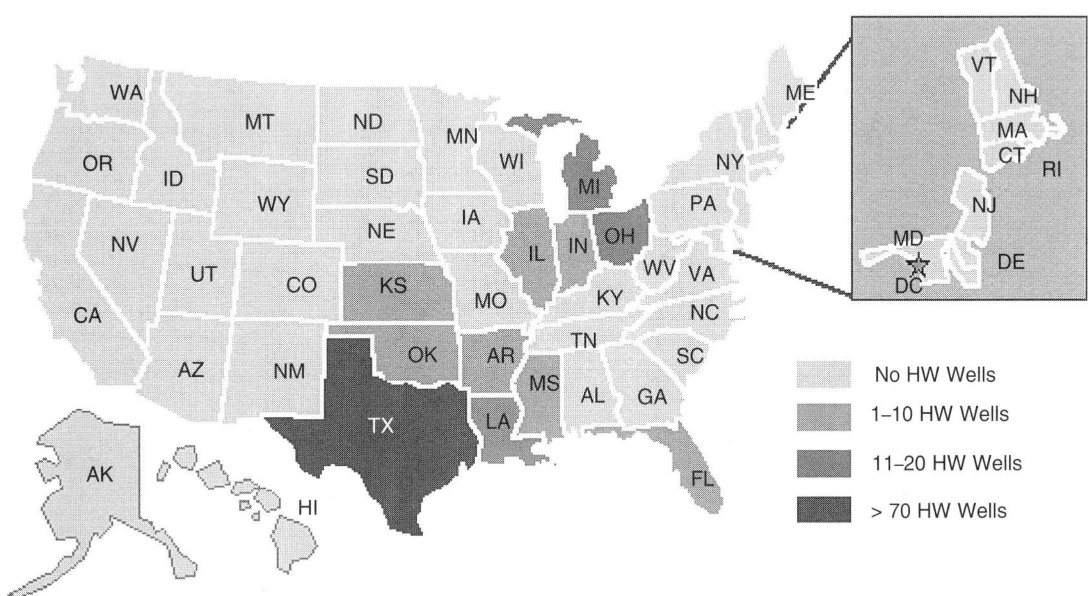

Figure 6D.12 UIC class I deep/high technology hazardous waste wells. (From www.epa.gov.)

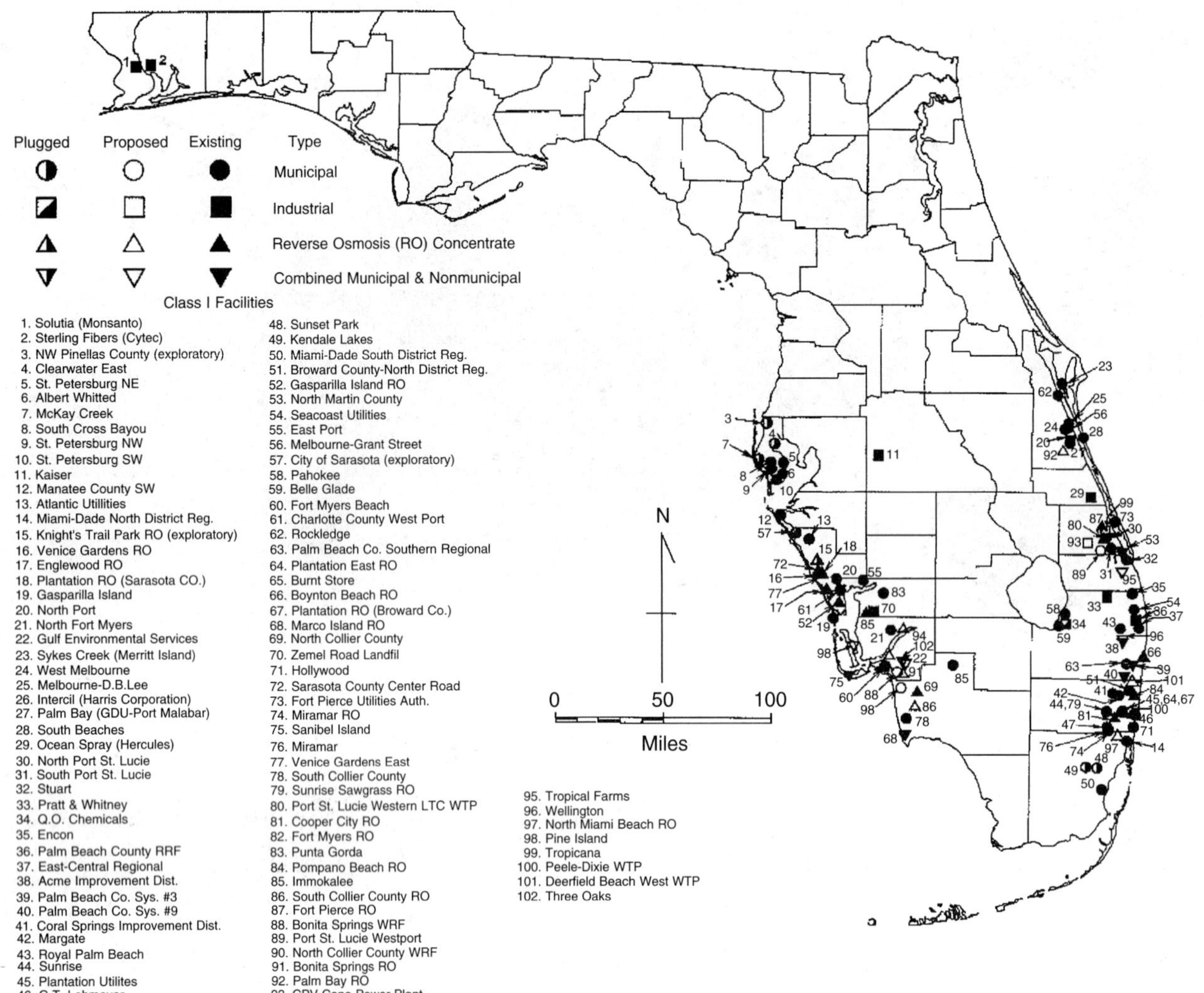

Plugged
Proposed
Existing
Type
Municipal
Industrial
Reverse Osmosis (RO) Concentrate
Combined Municipal & Nonmunicipal
Class I Facilities
1. Solutia (Monsanto)
2. Sterling Fibers (Cytec)
3. NW Pinellas County (exploratory)
4. Clearwater East
5. St. Petersburg NE
6. Albert Whitted
7. McKay Creek
8. South Cross Bayou
9. St. Petersburg NW
10. St. Petersburg SW
11. Kaiser
12. Manatee County SW
13. Atlantic Utillities
14. Miami-Dade North District Reg.
15. Knight's Trail Park RO (exploratory)
16. Venice Gardens RO
17. Englewood RO
18. Plantation RO (Sarasota CO.)
19. Gasparilla Island
20. North Port
21. North Fort Myers
22. Gulf Environmental Services
23. Sykes Creek (Merritt Island)
24. West Melbourne
25. Melbourne-D.B.Lee
26. Intercil (Harris Corporation)
27. Palm Bay (GDU-Port Malabar)
28. South Beaches
29. Ocean Spray (Hercules)
30. North Port St. Lucie
31. South Port St. Lucie
32. Stuart
33. Pratt & Whitney
34. Q.O. Chemicals
35. Encon
36. Palm Beach County RRF
37. East-Central Regional
38. Acme Improvement Dist.
39. Palm Beach Co. Sys. #3
40. Palm Beach Co. Sys. #9
41. Coral Springs Improvement Dist.
42. Margate
43. Royal Palm Beach
44. Sunrise
45. Plantation Utilites
46. G.T. Lohmeyer
47. Pembroke Pines (Century Village)
48. Sunset Park
49. Kendale Lakes
50. Miami-Dade South District Reg.
51. Broward County-North District Reg.
52. Gasparilla Island RO
53. North Martin County
54. Seacoast Utilities
55. East Port
56. Melbourne-Grant Street
57. City of Sarasota (exploratory)
58. Pahokee
59. Belle Glade
60. Fort Myers Beach
61. Charlotte County West Port
62. Rockledge
63. Palm Beach Co. Southern Regional
64. Plantation East RO
65. Burnt Store
66. Boynton Beach RO
67. Plantation RO (Broward Co.)
68. Marco Island RO
69. North Collier County
70. Zemel Road Landfil
71. Hollywood
72. Sarasota County Center Road
73. Fort Pierce Utilities Auth.
74. Miramar RO
75. Sanibel Island
76. Miramar
77. Venice Gardens East
78. South Collier County
79. Sunrise Sawgrass RO
80. Port St. Lucie Western LTC WTP
81. Cooper City RO
82. Fort Myers RO
83. Punta Gorda
84. Pompano Beach RO
85. Immokalee
86. South Collier County RO
87. Fort Pierce RO
88. Bonita Springs WRF
89. Port St. Lucie Westport
90. North Collier County WRF
91. Bonita Springs RO
92. Palm Bay RO
93. CPV Cana Power Plant
94. North Lee County WTP
95. Tropical Farms
96. Wellington
97. North Miami Beach RO
98. Pine Island
99. Tropicana
100. Peele-Dixie WTP
101. Deerfield Beach West WTP
102. Three Oaks
N
0
50
100
Miles

Figure 6D.13 Class I injection facilities. (From www.dep.state.fl.us.)

Table 6D.27 Class I Injection Well Status

Map	Facility	Status — November 2003 Proposed	Active	Other	Total Wells
1	Solutia (Monsanto) (I)	0	3	0	3
2	Sterling Fibers (Cytec) (I)	0	1	1SB	2
3	NW Pinellas County (EX)	0	0	3PA	3
4	Clearwater East	0	0	1PA	1
5	St. Petersburg NE	0	3	0	3
6	Albert Whitted	0	2	0	2
7	McKay Creek	0	0	2PA	2
8	South Cross Bayou	0	3	1PA	4
9	St. Petersburg NW	0	2	0	2
10	St. Petersburg SW	0	3	0	3
11	K.C. Industries (Kaiser) (HW)	0	1	0	1
12	Manatee County SW Subregional	0	1	1EXM	2
13	Atlantic Utilities	0	1	1EXM	2
14	MDW&SA North District Regional	0	2	2IA	4
15	Knight's Trail (EX)	0	0	1EXM	1
16	Venice Gardens (RO)	0	1	0	1
17	Englewood (RO)	0	1	1UC	2
18	Plantation RO (Sarasota Co.)	0	0	1IA	1
19	Gasparilla Island	0	1	0	1
20	North Port	0	1	0	1
21	North Fort Myers Utilities	0	1	1EXM	2
22	Gulf Environmental Services (MN)	1	0	0	1
23	Sykes Creek (Merritt Island)	0	2	1EXM	3
24	West Melbourne	0	1	0	1
25	Melbourne-D.B. Lee	0	0	1TA	1
26	Intercil (Harris Corporation) (I)	0	2	0	2
27	Palm Bay (GDU-Port Malabar)	0	1	0	1
28	South Beaches	0	1	0	1
29	Ocean Spray (Hercules) (I)	0	1	0	1
30	North Port St. Lucie (MN)	0	1	0	1
31	South Port St. Lucie	0	1	0	1
32	Stuart	0	2	0	2
33	Pratt and Whitney (I)	0	1	0	1
34	QO Chemicals (I)	0	0	4PA	4
35	Encon	0	1	0	1
36	Palm Beach County RRF (I)	0	2	0	2
37	East-Central Regional	0	6	1UC	7
38	Acme Improvement District (MN)	0	1	0	1
39	Palm Beach County System #3 (MN)	0	1	0	1
40	Palm Beach County System #9 (MN)	0	1	0	1
41	Coral Springs Improvement District	0	2	0	2
42	Margate	0	2	0	2
43	Royal Palm Beach	0	1	0	1
44	Sunrise	0	3	0	3
45	Plantation Regional (Broward Co.)	0	2	0	2
46	G. T. Lohmeyer	0	5	0	5
47	Pembroke Pines	0	2	0	2
48	Sunset Park	0	0	1PA	1
49	Kendale Lakes	0	0	1PA	1
50	MDW&SA South District Regional	0	13	4IA	17
51	Broward County-North District Regional	2	6	0	8
52	Gasparilla Island RO	0	1	0	1
53	North Martin County (MN)	0	2	0	2
54	Seacoast Utilities	0	1	0	1
55	East Port (Charlotte)	0	2	0	2
56	Melbourne-Grant St.	0	1	0	1
57	City of Sarasota (EX)	0	0	1EXM	1
58	Pahokee	0	1	0	1
59	Belle Glade	0	1	0	1

(Continued)

Table 6D.27 (Continued)

Map	Facility	Status — November 2003			Total Wells
		Proposed	Active	Other	
60	Fort Myers Beach	0	1	0	1
61	West Port (Charlotte)	0	1	0	1
62	Rockledge	0	1	0	1
63	Palm Beach County Southern Regional	0	2	0	2
64	Plantation East RO (Broward County)	0	1	0	1
65	Burnt Store (RO)	0	1	0	1
66	Boynton Beach (RO)	0	1	0	1
67	Plantation RO (Broward Co.)	0	1	0	1
68	Marco Island (MN)	0	1	0	1
69	North Collier County (RO)	0	2	0	2
70	Zemel Road Landfill (I)	0	1	0	1
71	Hollywood	0	0	2UC	2
72	Sarasota County Center Road	0	1	0	1
73	Fort Pierce Utility Authority	0	1	0	1
74	Miramar RO	0	2	0	2
75	Sanibel Island (MN)	0	1	0	1
76	Miramar	0	2	0	2
77	Venice Gardens East RO	0	1	0	1
78	South Collier County	1	1	0	2
79	Sunrise Sawgrass RO	0	1	0	1
80	Port St. Lucie Western LTC WTP	0	1	0	1
81	Cooper City RO	0	1	0	1
82	Fort Myers RO	0	1	0	1
83	Punta Gorda	0	1	1 EXM	2
84	Pompano Beach RO	0	1	0	1
85	Immokalee	0	1	0	1
86	South Collier County RO	0	0	2 UC	2
87	Fort Pierce RO	0	1	0	1
88	Bonita Springs RO	0	0	1UC	1
89	Port St. Lucie Westport	0	0	1UC	1
90	North Collier County WRF	0	0	2UC	2
91	Bonita Springs WRF	0	0	1UC	1
92	Palm Bay RO	1	0	1 EX	2
93	CPV Cana Power Plant	1	0	0	1
94	North Lee County WTP	0	0	1UC	1
95	Tropical Farms	0	0	2UC	2
96	Wellington (MN)	1	0	0	1
97	North Miami Beach RO	1	0	0	1
98	Pine Island (MN)	1	0	0	1
99	Tropicana	1	0	0	1
100	Peele-Dixie WTP	1	0	0	1
101	Deerfield Beach West WTP	1	0	0	1
102	Three Oaks	1	0	0	1
	TOTAL	**13**	**126**	**45**	**184**

Key to Abbreviations

Facility
EX — Exploratory well only
HW — Hazardous waste
I — Industrial (non-hazardous)
RO — Reverse osmosis concentrate
RRF — Resource recovery facility.

Status
TA — Temporarily abandoned well
PA — Permanently abandoned well
IA — Inactive well
SB — Standby well
UC — Under Construction/testing

EX — Exploratory well
EXI — Inactive exploratory well
EXW — Exploratory well converted to an injection well
EXM — Exploratory well converted to a monitor well
EXP — Exploratory well plugged and abandoned.

Source: From www.dep.state.fl.us.

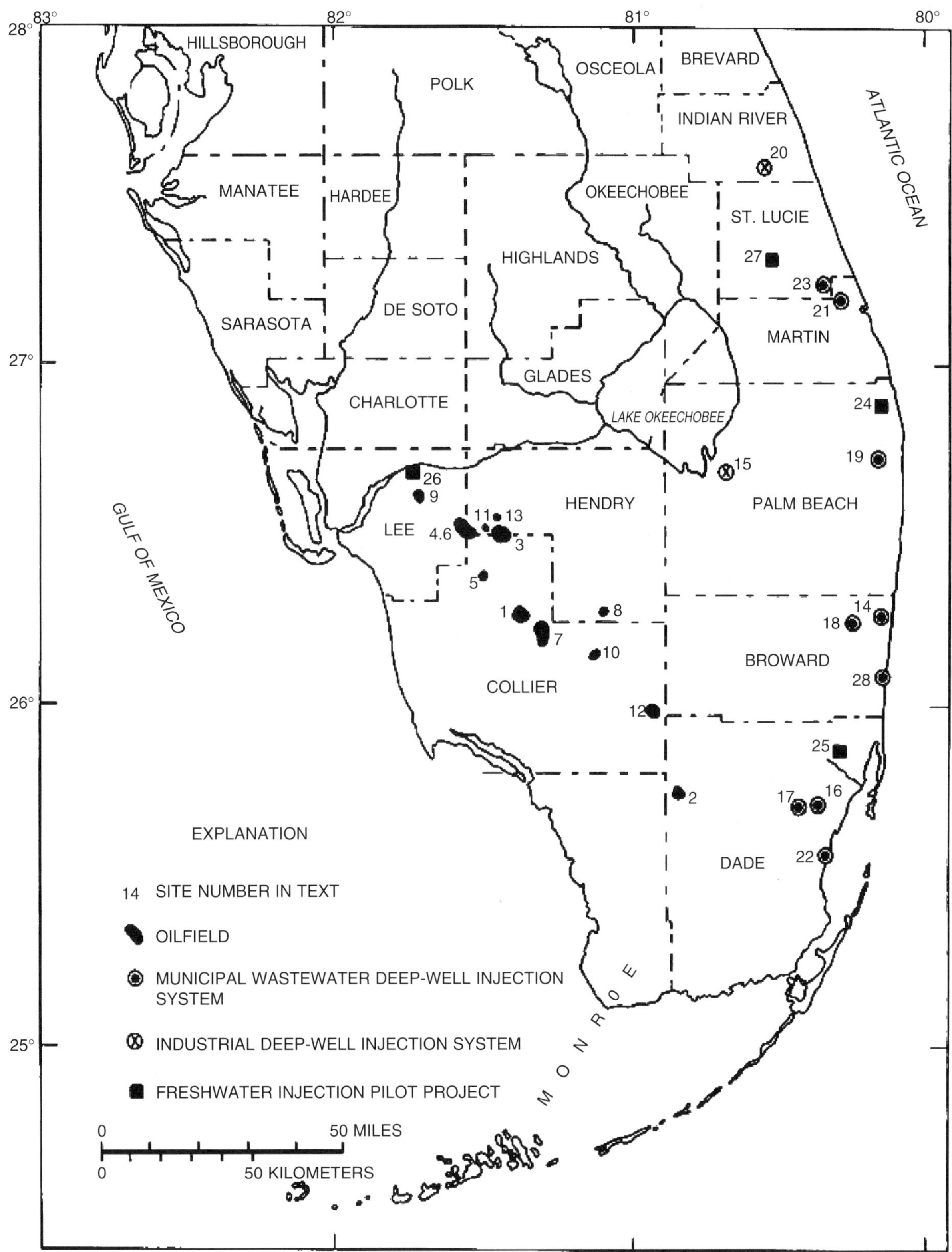

Figure 6D.14 Deep well injection systems. (From www.sofia.usgs.gov.)

Table 6D.28 Summary of Deep-Well Injection Systems in Florida

	Injected Effluent			Design Capacity		Injection Wells			Well Casings		
							Depth			Depth	
Location	Type	Pretreatment	First Year of Operation	ML/d	mgd	Number of Wells	m	ft[a]	Number of Casings	m	ft[a]
Belle Glade, Palm Beach County	Industrial	Cooling	1966	5.6	1.5	2	975	3200	4	883	2900
Sunset Park, South Miami	Municipal	Secondary	1971	22.7	6.0	1	914	3000	3	563	1850
Mulberry, Polk County	Industrial	None	1971	1.3	0.35	1	1371	4500	3	1219	4000
Kendale Lakes, South Miami	Municipal	Secondary	1973	22.7	6.0	1	975	3200	3	670	2200
Margate, Broward County	Municipal	Secondary	1975	56.7	15.0	1	975	3200	3	731	2400
St. Petersburg, Southwest plant	Municipal	Tertiary	1976	75.7	20.0	3	304	1000	3	274	900
Gainesville, Kanapaha plant	Municipal	Advanced wastewater treatment	1976	28.3	7.5	3	304	1000	3	152	500
West Palm Beach	Municipal	Secondary	1978	302.8	80.0	5	1097	3600	4	914	3000
Vero Beach, Indian River County	Industrial	Neutralization	1979	1.1	0.3	1	914	3000	4	731	2400
Miami-Dade Water and Sewer Authority	Municipal	Secondary	1983	423.9	112.0	9	944	3100	4	792	2600

[a] Rounded to nearest 100 ft.

Source: From Garcia-Bengochea, J.I., Protecting water supply aquifers in areas using deep-well wastewater disposal, *J. Am. Water Works Assoc.*, 75, 6, 1983. Copyright AWWA. Reprinted with permission.

SECTION 6E PUMPING OF WATER

Table 6E.29 Useful Factors in Preliminary Planning of Small Pumping Plants

Pump or Pipe Size (in.)	Gallons/min	Acre-Inches per 24 hr	Pipe Velocity (ft/sec)	Velocity Head ($V^2/2g$ ft)	Friction in Feet Per 100 Feet of Pipe	Horsepower Required for 10 ft Total Head. Pump and Transmission Efficiency=70 Percent
6	400	21.2	4.54	0.32	2.21	1.4
6	600	31.8	6.72	0.70	4.7	2.2
6	800	42.4	9.08	1.28	8.0	2.9
6	1,000	53.0	11.32	1.99	12.0	3.6
8	900	47.7	5.75	0.52	2.46	3.2
8	1,100	58.3	7.03	0.77	3.51	4.0
8	1,300	68.9	8.32	1.07	4.72	4.7
8	1,500	79.5	9.60	1.43	6.27	5.4
10	1,200	63.6	4.91	0.38	1.46	4.3
10	1,600	84.8	6.56	0.67	2.35	5.8
10	2,000	108.1	8.10	1.02	3.65	7.2
10	2,400	127.3	9.73	1.47	5.04	8.7
12	2,000	106.1	5.60	0.48	1.43	7.2
12	2,500	132.6	7.00	0.77	2.28	9.0
12	3,000	159.1	8.40	1.10	3.15	10.8
12	3,500	185.6	9.80	1.49	4.10	12.6
14	2,000	106.1	4.20	0.27	0.66	7.2
14	3,000	159.1	6.30	0.61	1.47	10.8
14	4,000	212.1	8.40	1.09	2.47	14.4
14	5,000	265.2	10.50	1.71	3.92	18.0
16	3,600	190.9	5.74	0.51	1.10	13.0
16	4,400	233.3	7.01	0.76	1.58	15.9
16	5,200	275.8	8.29	1.06	2.16	18.8
16	6,000	318.2	9.56	1.42	2.60	21.6
18	4,500	238.6	5.70	0.50	0.93	16.2
18	5,500	291.7	6.96	0.75	1.32	19.8
18	6,500	344.7	8.22	1.05	1.82	23.4
18	8,000	424.2	10.02	1.56	2.65	28.9
20	5,000	265.2	5.13	0.41	0.68	18.0
20	6,500	344.7	6.66	0.69	1.06	23.4
20	8,000	424.2	8.17	1.03	1.63	28.9
20	10,000	530.3	10.40	1.68	2.53	36.1
24	8,000	424.2	5.68	0.50	0.66	28.9
24	10,000	630.3	7.07	0.78	0.98	36.1
24	12,000	636.4	8.50	1.12	1.40	43.3
24	14,000	742.4	9.95	1.54	1.87	50.5
30	12,000	636.4	5.44	0.46	0.47	43.3
30	16,000	848.5	7.36	0.84	0.83	57.7
30	20,000	1061.0	9.09	1.29	1.22	72.2
30	24,000	1273.0	10.90	1.86	1.71	86.6

Source: From U.S. Department of Agriculture.

Table 6E.30 Characteristics of Pumps Frequently Employed in Wells

Type of Pump	Practical Suction Lift[a]	Usual Well-Pumping Depth	Usual Pressure Heads	Advantages	Disadvantages
Reciprocating					
Shallow well	6–7 m	6–7 m	30–60 m	Positive action; discharge against variable heads; pumps water containing sand and silt; especially adapted to low capacity and high lifts	Pulsating discharge; subject to vibration and noise; maintenance cost may be high; may cause destructive pressure if operated against closed valve
Deep well	6–7 m	Up to 180 m	Up to 180 m above cylinder		
Centrifugal					
Shallow well straight centrifugal (single stage)	6 m max	3–6 m	30–45 m	Smooth, even flow; pumps water containing sand and silt; pressure on system is even and free from shock; low-starting torque; usually reliable and good service life	Loses prime easily; efficiency depends on operating under design heads and speed
Regenerative vane turbine type (single impeller)	8 m max	8 m	30–60 m	Same as straight centrifugal except not suitable for pumping water containing sand or silt; self-priming	Same as straight centrifugal except maintains priming easily
Deep well					
Vertical line shaft turbine (multistage)	Impellers submerged	15–90 m	30–250 m	Same as shallow well turbine	Efficiency depends on operating under design head and speed; requires straight well large enough for turbine bowls and housing; lubrication and alignment of shaft critical; abrasion from sand
Submersible turbine (multistage)	Pump and motor submerged	15–120 m	15–120 m	Same as shallow well turbine; easy to frostproof installation; short pump shaft to motor	Repair to motor or pump requires pulling from well; sealing of electrical equipment from water vapor critical; abrasion from sand
Jet					
Shallow well	4–6 m below ejector	Up to 4–6 m below ejector	25–45 m	High capacity at low heads; simple in operation; does not have to be installed over well; no moving parts in well	Capacity reduces as lift increases; air in suction or return line will stop pumping
Deep well	4–6 m below ejector	7–35 m 60 m max.	25–45 m	Same as shallow well jet	Same as shallow well jet
Rotary					
Shallow well (gear type)	7 m	7 m	15–75 m	Positive action; discharge constant under variable heads; efficient operation	Subject to rapid wear if water contains sand or silt; wear of gears reduces efficiency
Deep well (helical rotary type)	Usually submerged	15–150 m	30–150 m	Same as shallow well/rotary; only one moving pump device in well	Same as shallow well rotary except no gear wear

[a] Practical suction lift at sea level. Reduce lift 0.3 m for each 300 m above sea level.

Source: From U.S. Public Health Service, *Manual of Individual Water-Supply Systems*, Publication. 24, 1962.

Table 6E.31 Selection of Pump Size and Diameter of Wells

Anticipated Well Yield			Nominal Size of Pump Bowls (in.)	Optimum Well Diameter (in.)
In gal min^{-1}	In ft^3 min^{-1}	In m^3 min^{-1}		
Less than 100	Less than 13	Less than 0.38	4	6 ID
75–175	10–23	0.28–0.66	5	8 ID
150–400	20–53	0.57–1.52	6	10 ID
350–650	47–87	1.33–2.46	8	12 ID
600–900	80–120	2.27–3.41	10	14 OD
850–1,300	113–173	3.22–4.93	12	16 OD
1,200–1,800	160–240	4.55–6.82	14	20 OD
1,600–3,000	213–400	6.06–11.37	16	24 OD

Note: ID, inside diameter; OD, outside diameter.

Source: From Health, R.C., *Basic Ground-Water Hydrology*, U.S. Geological Survey Water-Supply Paper 2220, 1983.

Table 6E.32 Pumping Plant Performance Standards

Type of Power Unit	Standard Consumption of Fuel or Energy per Water Horsepower[a]
Diesel engine	0.091 gal hr^{-1}
Gasoline engine	0.116 gal hr^{-1}
Propane engine	0.145 gal hr^{-1}
Natural gas	160 ft^3 hr^{-1}
Electric motor	0.885 kw-hr hr^{-1}

[a] Based on pump efficiency of 75 percent.

Source: From College of Agriculture, University of Nebraska.

Table 6E.33 Standard Fuel Requirements for Good Pumping Plants

Pumping Rate (in gpm)	Head (in ft)	Water Horsepower	Fuel or Energy Required: Diesel (gal hr^{-1})	Gasoline (gal hr^{-1})	Propane (gal hr^{-1})	Natural Gas (ft hr^{-1})	Electricity (kwh hr^{-1})
500	100	13	1¼	1½	2	190	14
	150	19	1¾	2¼	2¾	280	21
	200	25	2¼	3	3¾	380	29
700	100	18	1¾	2	2¾	270	20
	150	27	2½	3¼	4	400	30
	200	35	3¼	4¼	5¼	530	40
800	100	20	1¾	2½	3	300	23
	150	30	2¾	3½	4½	450	34
	200	40	3¾	4¾	6	610	46
1000	100	25	2¼	3	3¾	380	29
	150	38	3½	4½	5¾	570	43
	200	50	4½	6	7½	760	57

Note: Based on performance standards in Table 6E.32.

Source: From College of Agriculture, University of Nebraska.

Table 6E.34 Pumping Costs in the Texas High Plains (THP) and in South/Central Texas (SCT) Per Acre-inch of Water at 100 ft Total Head from Irrigation Pumping Plant Efficiency Tests Conducted by the Texas Agricultural Extension Service

		Cost ($) Per Acre-Inches per 100 ft Head		
Type and Price[a]	**Region[b]**	**Lowest**	**Highest**	**Average**
Natural Gas	THP	0.40	3.93	0.81
@ $3.00 MCF	SCT	0.31	1.96	0.76
Electricity	THP	0.49	3.10	1.35
@ $0.07/KWH	SCT	0.29	20.20	1.49
Diesel	THP	0.57	1.91	0.77
@ $0.65/gal	SCT	0.36	3.43	0.83

[a] Assumed price — actual prices varied in each region.
[b] THP (Texas High Plains) results are from more than 240 efficiency tests. SCT (South/Central Texas) results are from 240 efficiency tests.

Source: From Texas Agricultural Extension Service, the Texas A&M University.

Table 6E.35 Irrigation Pumping Equipment Efficiency

Equipment	Attainable Efficiency (Percent)
Pumps (centrifugal, turbine)	75–82
Right-angle pump drives (gear head)	95
Automotive-type engines	20–26
Industrial engines	
Diesel	25–37
Natural gas	24–27
Electric motors	
Small	75–85
Large	85–92

Source: From New L.L. Pumping plant efficiency and irrigation cost L-2218, Texas Agricultural Extension Service.

Table 6E.36 Typical Values of Overall Efficiency for Representative Pumping Plants, Expressed as Percent

Power Source	Recommended as Acceptable	Average Values from Field Tests[a]
Electric	72–77	45–55
Diesel	20–25	13–15
Natural gas	18–24	9–13
Butane, propane	18–24	9–13
Gasoline	18–23	9–12

Ranges are given because of the variation in efficiencies of both pumps and power units. The difference in efficiency for high and low compression engines used for natural gas, propane and gasoline must be considered especially. The higher value of efficiency can be used for higher compression engines.

[a] Typical average observed values reported by pump efficiency test teams.

Source: From New L.L. Pumping plant efficiency and irrigation cost L-2218, Texas Agricultural Extension Service.

Table 6E.37 Nebraska Performance Criteria for Pumping Plants. Fuel Use by New or Reconditioned Plants Should Equal or Exceed These Rates

Energy Source	Water Horsepower-hours[a] Per Unit of Energy	Energy Units
Diesel	12.5	gal
Gasoline[b]	8.7	gal
Natural gas	66.7[c]	1000 ft[c]
Electricity	0.885[d]	Kwh

[a] Based on 75 percent efficiency.
[b] Includes drive losses and assumes no cooling fan.
[c] Assumes natural gas content of 1000 btu ft^{-3}
[d] Direct connection-no drive.

Source: From College of Agriculture, University of Nebraska.

Table 6E.38 Approximate Maximum Flow Rate in Different Pipe Sizes to Keep Velocity ≤ 5 ft sec^{-1}

Pipe Diameter	Flow Rate (gpm)
½	6
¾	10
1	15
1¼	25
1½	35
2	50
3	110
4	200
5	310
6	440
8	780
10	1225
12	1760
16	3140

Source: From Texas Agricultural Extension Service, the Texas A&M University.

Table 6E.39 Friction Losses in Feet of Head Per 100 ft of Pipe (for Pipes with Internal Diameters Shown)

Pipe Size	4-inch			6-inch			8-inch			10-inch			12-inch		
Flow rate (gpm)	Steel	Alum.	PVC	Steel	Alum.	PVC	Steel	Alum.	PVC	Steel	Alum.	PVC	Steel	Alum.	PVC
100	1.2	0.9	0.6	—	—	—	—	—	—	—	—				
150	2.5	1.8	1.2	0.3	0.2	0.2	—	—	—	—	—	—			
200	4.3	3.0	2.1	0.6	0.4	0.3	0.1	0.1	0.1	—	—	—			
250	6.7	4.8	3.2	0.9	0.6	0.4	0.2	0.1	0.1	0.1	0.1	—	—	—	
300	9.5	6.2	4.3	1.3	0.8	0.6	0.3	0.2	0.1	0.1	0.1	—	—	—	—
400	16.0	10.6	7.2	2.2	1.5	1.0	0.5	0.3	0.2	0.2	0.1	0.1	0.1	—	—
500	24.1	17.1	11.4	3.4	2.4	1.6	0.8	0.6	0.4	0.3	0.2	0.1	0.1	0.1	0.1
750	51.1	36.3	24.1	7.1	5.0	3.4	1.8	1.3	0.8	0.6	0.4	0.3	0.2	0.1	0.1
1000	87.0	61.8	41.1	12.1	8.6	5.7	3.0	2.1	1.4	1.0	0.7	0.5	0.4	0.3	0.2
1250	131.4	93.3	62.1	18.3	13.0	8.6	4.5	3.2	2.1	1.5	1.1	0.7	0.6	0.4	0.3
1500	184.1	130.7	87.0	25.6	18.2	12.1	6.3	4.5	3.0	2.1	1.5	1.0	0.9	0.6	0.4
1750	244.9	173.9	115.7	34.1	24.2	16.1	8.4	6.0	4.0	2.8	2.0	1.3	1.2	0.9	0.6
2000	313.4	222.5	148.1	43.6	31.0	20.6	10.8	7.7	5.1	3.6	2.6	1.7	1.5	1.1	0.7

Note: Flow rates below horizontal line for each pipe size exceed the recommended 5-feet-per-second velocity.

Source: From Texas Agricultural Extension Service, the Texas A&M University.

Table 6E.40 Friction Loss in Fittings. Friction Loss in Terms of Equivalent Length of Pipe (ft) of Same Diameter

	Inside Pipe Diameter (in.)					
Type of Fitting	**4**	**5**	**6**	**8**	**10**	**12**
45-degree elbow	5	6	7	10	12.5	15
Long-sweep elbow	7	9	11	14	17	20
Standard elbow	11	13	16	20	25	32
Close return bend	24	30	36	50	61	72
Gate value (open)	2	3	3.5	4.5	5.5	7
Gate value (1/2 open)	65	81	100	130	160	195
Check valve	100	110	30	40	45	35

Source: From Texas Agricultural Extension Service, the Texas A&M University.

Table 6E.41 Pumping Capacities of Aermotor Windmills Shown in the Table Below are Approximate, Based on the Mill Set on the Long Stroke, Operating in a 15 to 20 Mile Per Hour Wind

	[a]Capacity Per Hour Gallons			Elevation in Feet to Which Water Can Be Raised				
				Size of Aermotor Windmill				
Size of Cylinder (in.)	**6 ft**	**8–16 ft**	**6 ft**	**8 ft**	**10 ft**	**12 ft**	**14 ft**	**16 ft**
1–7/8	125	180	120	175	260	390	560	920
2	130	190	95	140	215	320	460	750
2–1/4	180	260	77	112	170	250	360	590
2–1/2	225	325	65	94	140	210	300	490
2–3/4	265	385	56	80	120	180	260	425
3	320	470	47	68	100	155	220	360
3–1/2	440	640	35	50	76	115	160	265
3–3/4		730			65	98	143	230
4	570	830	27	37	58	86	125	200
5	900	1300	17	25	37	55	80	130
6		1875		17	25	38	55	85

The short stroke increases elevation by one-third, and reduces pumping capacities by one-fourth.

[a] Approximate Capacity.

Source: From Aermotor company. With permission.

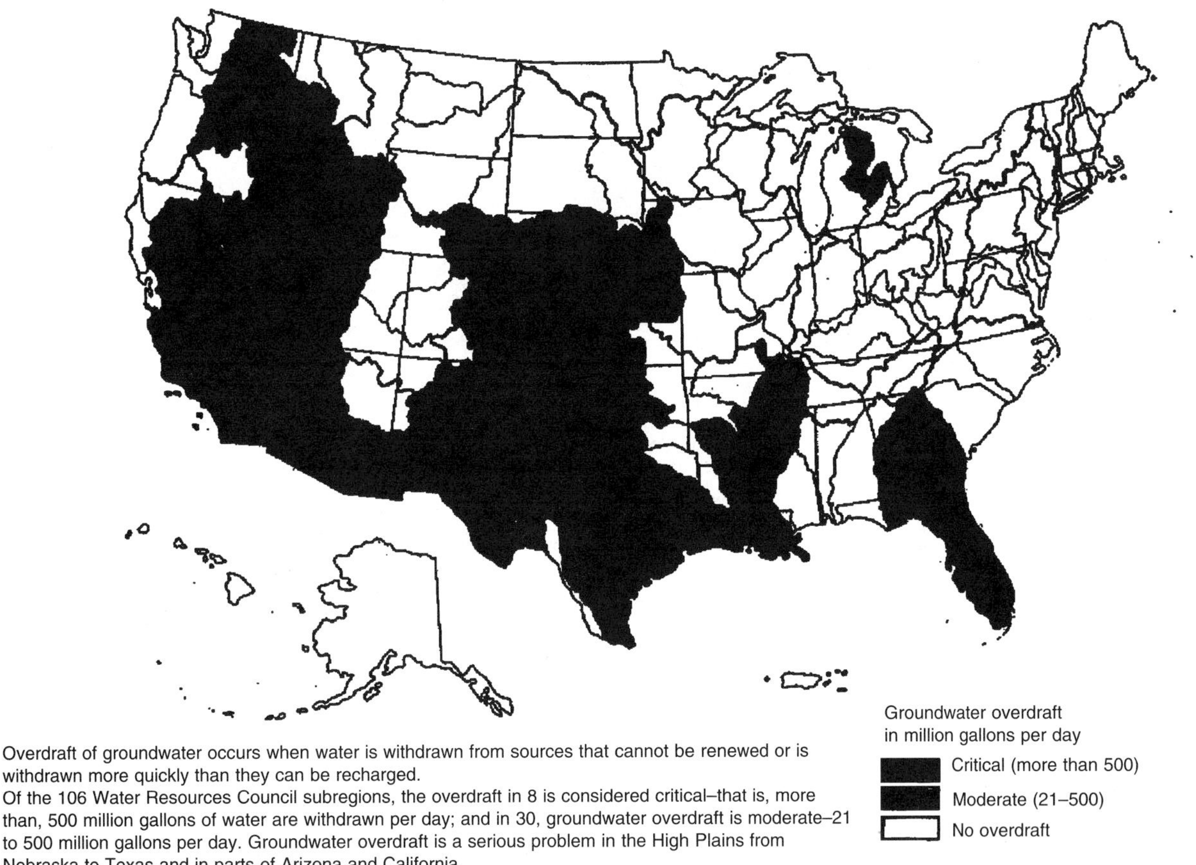

Overdraft of groundwater occurs when water is withdrawn from sources that cannot be renewed or is withdrawn more quickly than they can be recharged.
Of the 106 Water Resources Council subregions, the overdraft in 8 is considered critical–that is, more than, 500 million gallons of water are withdrawn per day; and in 30, groundwater overdraft is moderate–21 to 500 million gallons per day. Groundwater overdraft is a serious problem in the High Plains from Nebraska to Texas and in parts of Arizona and California.
Groundwater overdraft
in million gallons per day
Critical (more than 500)
Moderate (21–500)
No overdraft

Figure 6E.15 Groundwater overdraft in the United States [Data as of 1975; by water resource subregion]. (From council on environmental quality, 1981, *Environmental trends.*)

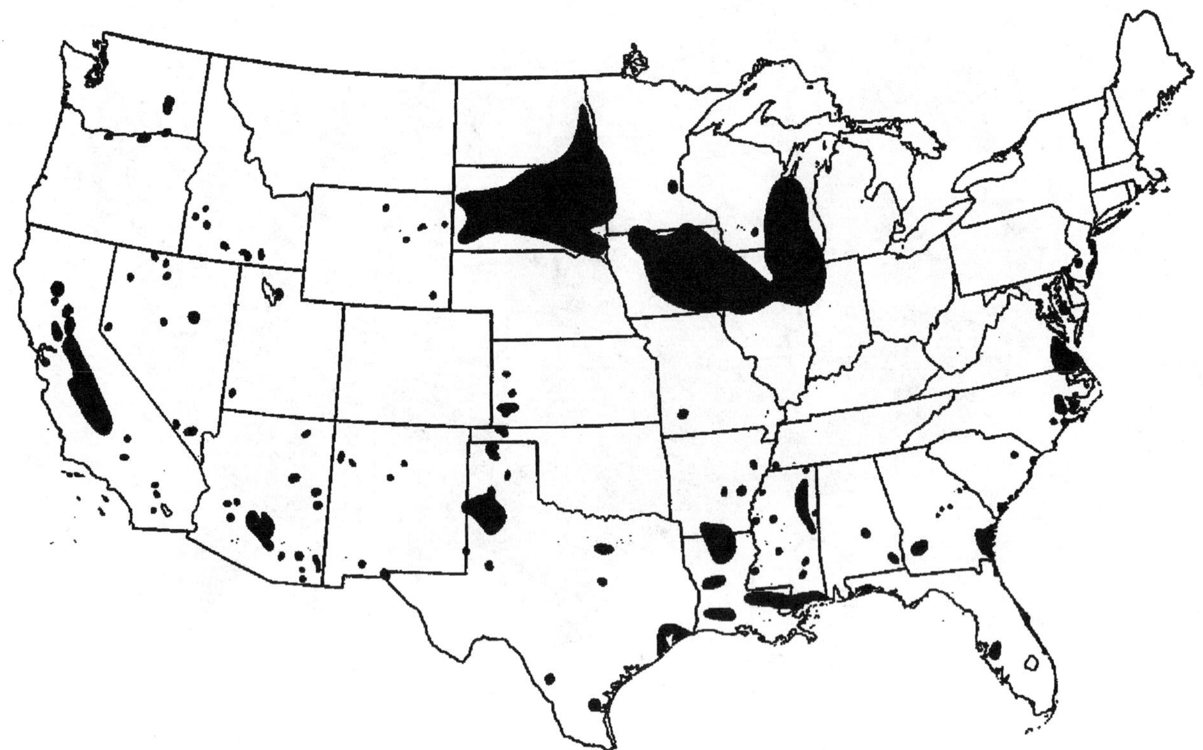

Figure 6E.16 Areas of water-table or Artesian water-level decline in excess of 40 ft in the United States [Decline in at least one aquifer since predevelopment]. (From U.S. Geological Survey, 1984, *National Water Summary 1983—Hydrologic Events and Issues*, Water-Supply Paper 2250.)

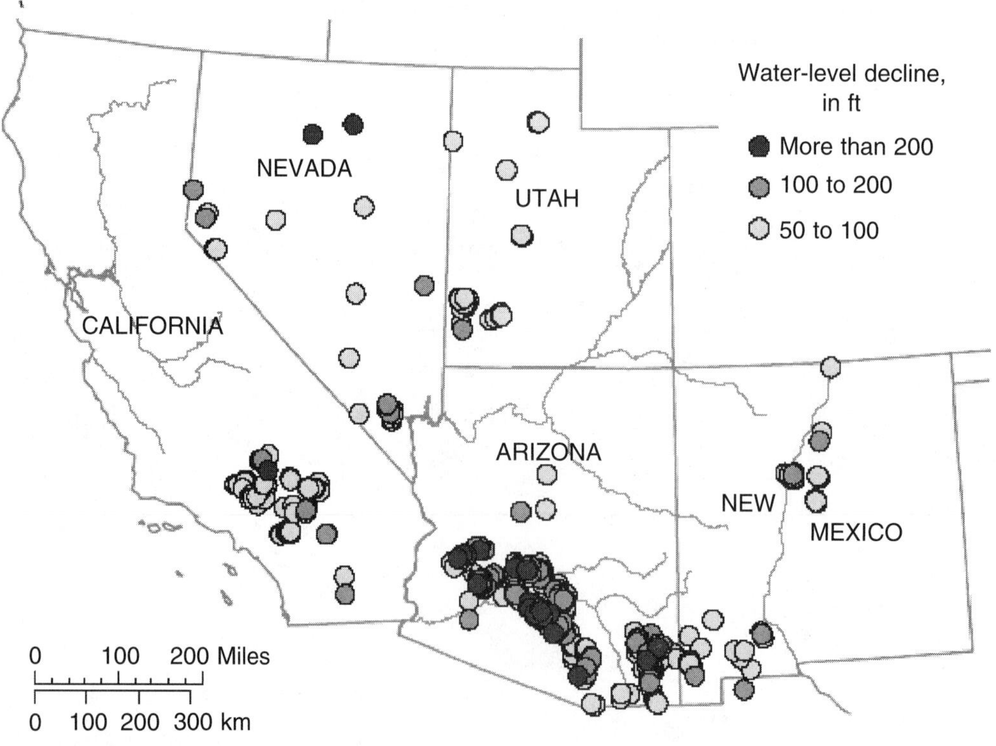

Figure 6E.17 Locations in the basins of Southern California, Nevada, Utah, Arizona, and New Mexico where substantial groundwater level declines have been measured. (From http://water.usgs.gov.)

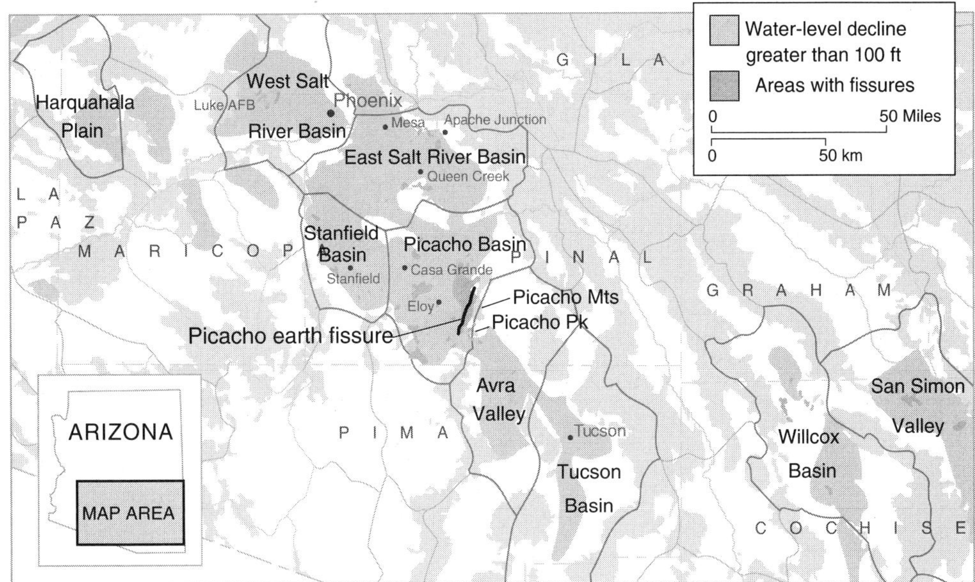

Figure 6E.18 Water-level decline greater than 100 ft and areas with fissures in Picacho, Arizona. (From http://water.usgs.gov.)

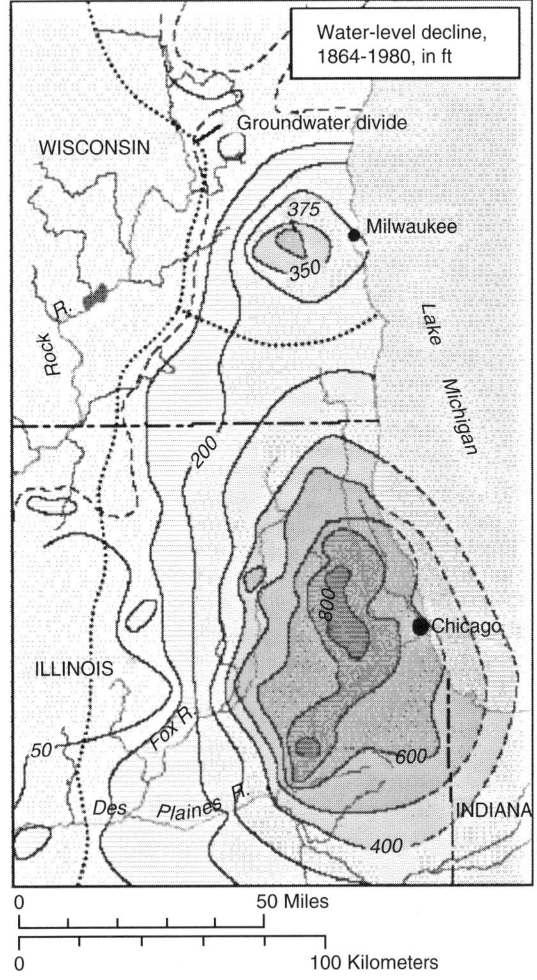

Figure 6E.19 Chicago, Illinois water-level decline, 1864–1980, in ft. (From http://water.usgs.gov.)

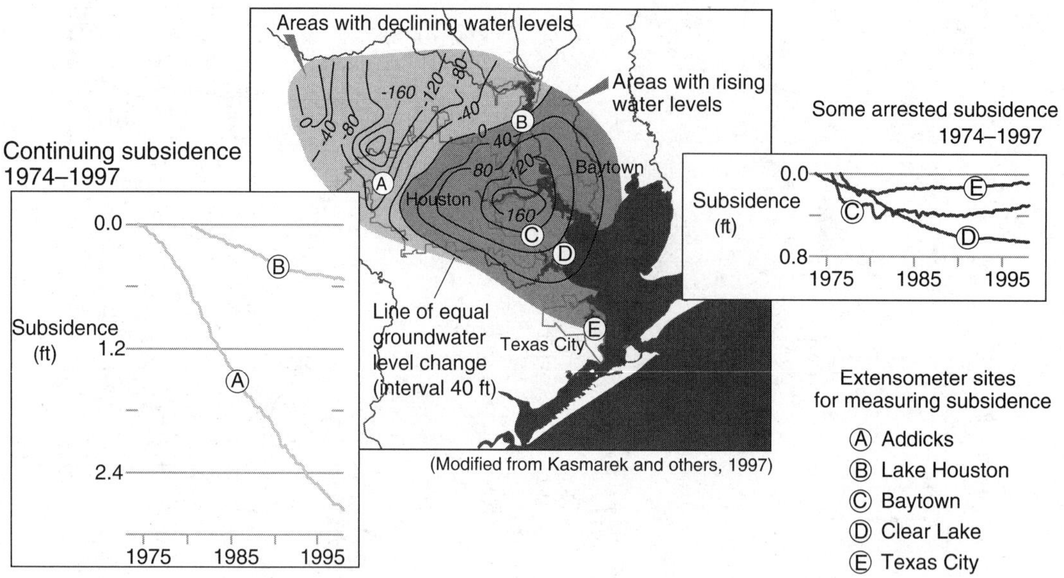

Figure 6E.20 Change in groundwater levels in wells in the Evangeline aquifer, 1977–1997. (From http://water.usgs.gov.)

SECTION 6F SUBSIDENCE

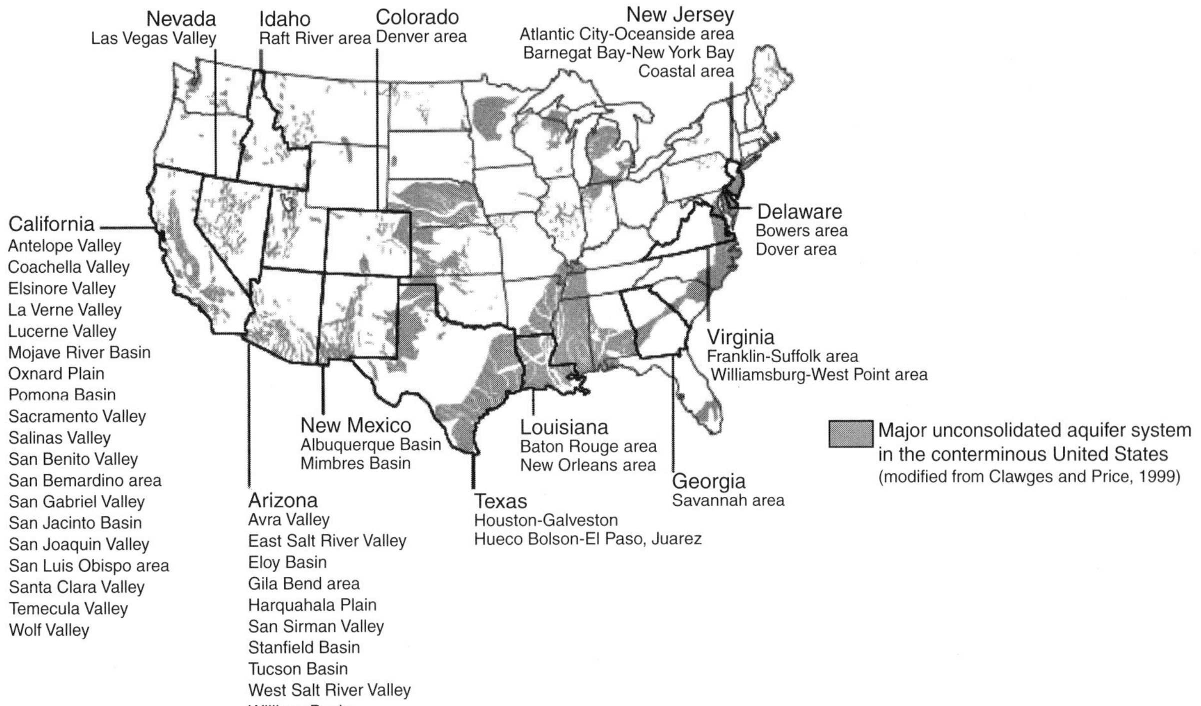

Figure 6F.21 Areas where subsidence has been attributed to the compaction of aquifer system due to groundwater pumpage (modified from Clawges and Price, 1999). (From http://water.usgs.gov.)

Table 6F.42 Areas of Major Land Subsidence Due to Groundwater Overdraft

Location	Depositional Environment and Age	Depth Range of Compacting Beds (m)	Maximum Subsidence (m)	Areas of Subsidence (sq km)	Time of Principal Occurrence
Japan					
Osaka	Alluvial and shallow marine; Quaternary	10–400	3	190	1928–1968
Tokyo	As above	10–400	4	190	1920–1970+
Mexico					
Mexico City	Alluvial and lacustrine; late Cenozoic	10–50	9	130	1938–1970+
Taiwan					
Taipei basin	Alluvial and lacustrine; Quaternary	10–240	1.3	130	1961–1969+
United States					
Arizona, central	Alluvial and lacustrine; late Cenozoic	100–550	2.3	650	1948–1967
California					
Santa Clara Valley	Alluvial and shallow marine; late Cenozoic	55–300	4	650	1920–1970
San Joaquin Valley (three subareas)	Alluvial and lacustrine; late Cenozoic	60–1000	2.9–9	11,000	1935–1970+
				(>0.3 ml)	
Lancaster area	Alluvial and lacustrine; late Cenozoic	60–300(?)	1	400	1955–1967+
Nevada					
Las Vegas	Alluvial; late Cenozoic	60–300	1	500	1935–1963
Texas					
Houston-Galveston area	Fluvial and shallow marine; late Cenozoic	60–600(?)	1–1.5	6,860	1943–1964+
				(>0.15 m)	
Louisiana					
Baton Rouge	Fluvial and shallow marine; Miocene to Holocene	50–600(?)	0.3	650	1934–1965+

Source: From Poland, J.F., Subsidence and its control, in underground waste management and environmental implications, *Amer Assoc. Petr. Geologists*, Memoir 18, 1972.

Table 6F.43 Amounts of Subsidence in Selected Areas in the Southwest

Arizona		Nevada		California		Texas	
Eloy	15 ft	Las Vegas	6 ft	Lancaster	6 ft	El Paso	1 ft
West of Phoenix	18 ft	**New Mexico**		South west of Mendota	29 ft	Houston	9 ft
Tucson	<1 ft	Albuquerque	"<" 1 ft	Davis	4 ft		
		Mimbres Basin	2 ft	Santa Clara Valley	12 ft		
				Ventura	2 ft		

Source: From http://geochange.er.usgs.gov.

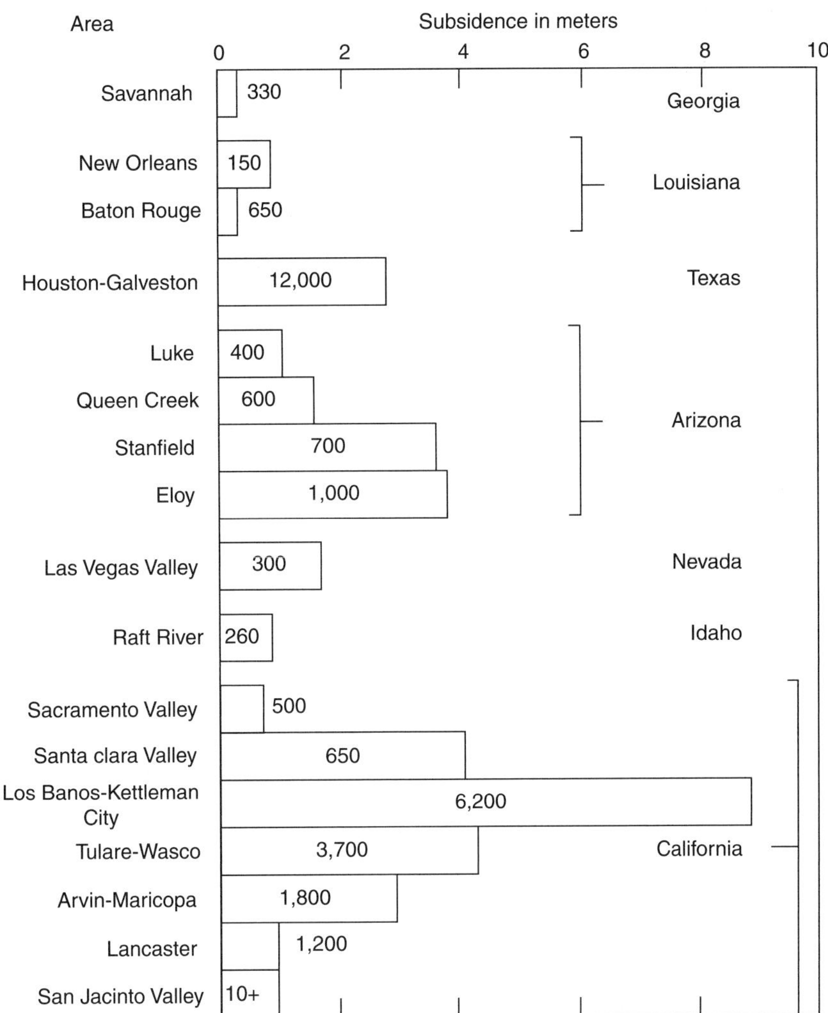

Figure 6F.22 Magnitude of land subsidence from groundwater withdrawal in the United States [numbers in columns represent area in square kilometers]. (From Poland, J.F., *Irrigation and Drainage division ASCE*, 107,1981, IR2. Copyright American Society of Civil Engineers. Reprinted with permission.)

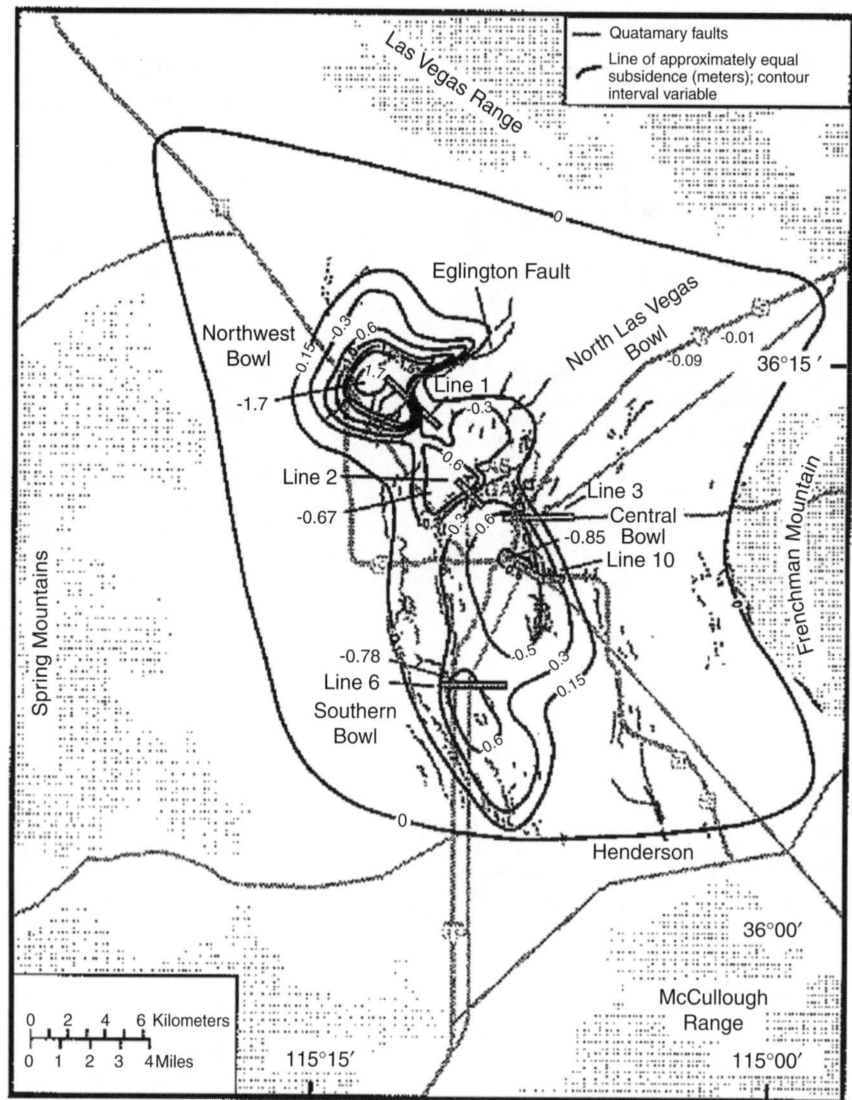

Figure 6F.23 Revised subsidence contour map for Las Vegas Valley, 1963–2000, showing maximum subsidence measured in four localized bowls and location of level lines. (From http://water.usgs.gov.)

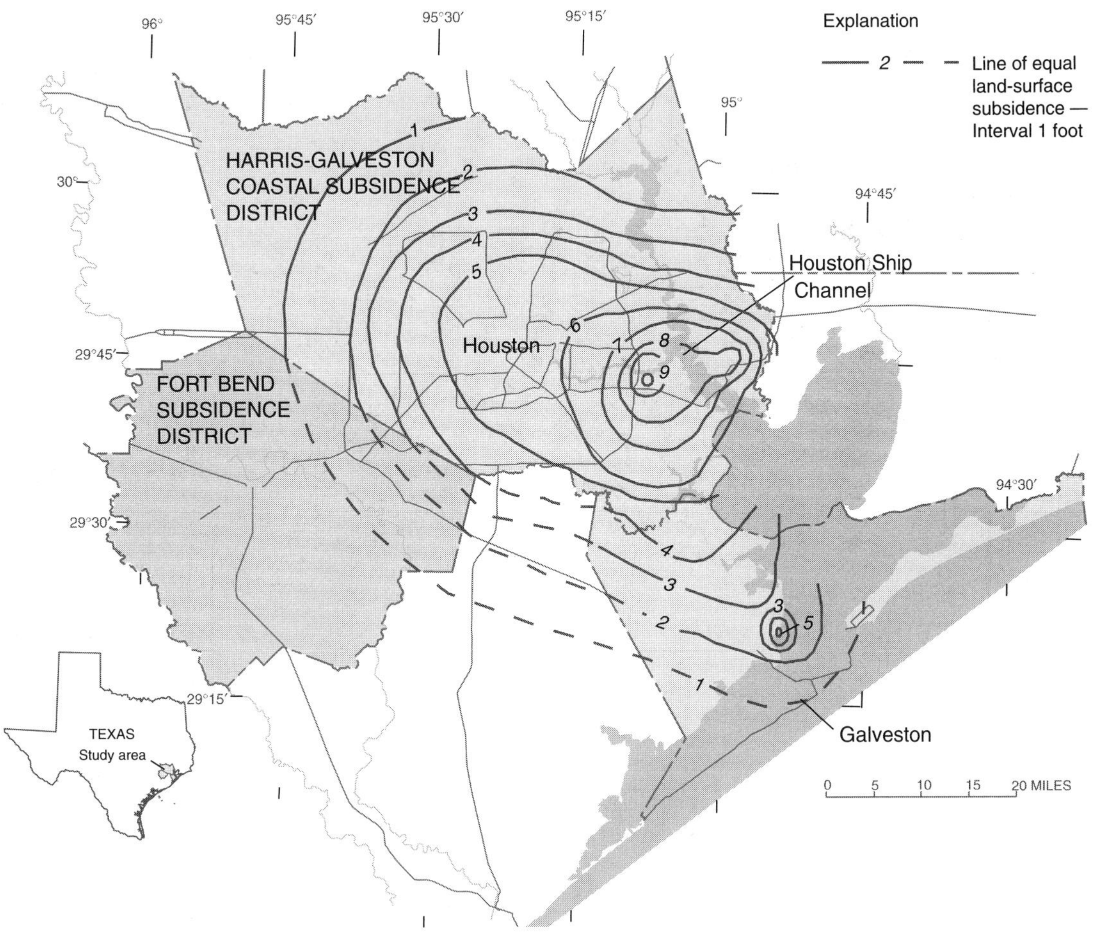

Figure 6F.24 Subsidence occurring between 1906 and 1987 in the Houston-Galveston region, Texas. (From http://water.usgs.gov.)

SECTION 6G AQUIFER CHARACTERISTICS

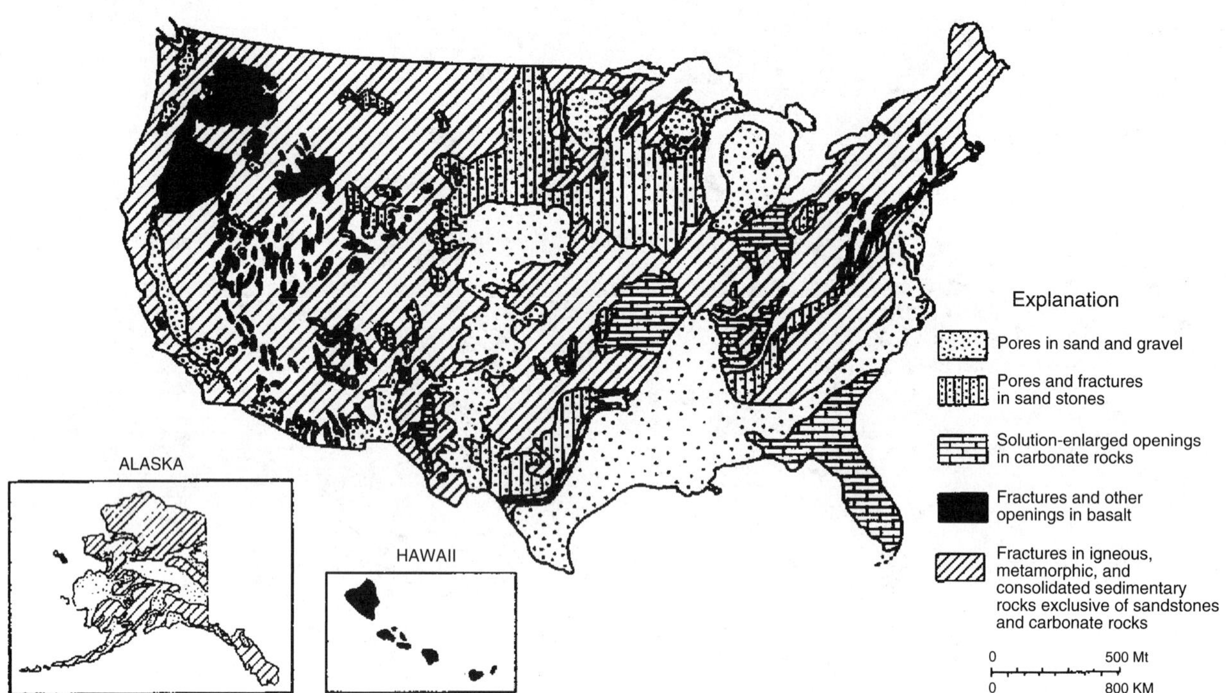

Figure 6G.25 Types of water-bearing openings in dominant aquifers of the United States. (From Heath, R.C., Classification of groundwater regions of the United States, *Groundwater*, 20, 4, 1982, Reprinted with permission.)

Table 6G.44 Features of Groundwater Systems Useful in the Classification of Groundwater Regions

Feature	Aspect	Range in Conditions	Significance of Feature
Component of the system	Unconfined aquifer	Thin, discontinuous, hydrologically insignificant Minor aquifer, serves primarily as a storage reservoir and recharge conduit for underlying aquifer The dominant aquifer	Affects response of the system to pumpage and other stresses. Affects recharge and discharge conditions for the system. Determines susceptibility of the system to pollution
	Confining beds	Not present, or hydrologically insignificant Thin, markedly discontinuous, or very leaky Thick, extensive, and impermeable Complexly interbedded with aquifers or productive zones	
	Confined aquifers	Not present, or hydrologically insignificant Thin or not highly productive Multiple thin aquifers interbedded with non productive zones The dominant aquifer — thick and productive	
	Presence and arrangement of components	A single, hydrologically-dominant, unconfined aquifer Two interconnected aquifers of essentially equal hydrologic importance A three-unit system consisting of an unconfined aquifer, a confining bed, and a confined aquifer A complexly interbedded sequence of aquifers and confining beds	
Water-bearing openings of dominant aquifer	Primary openings	Pores in unconsolidated deposits Pores in semiconsolidated rocks Pores, tubes, and cooling fractures in volcanic (extrusive-igneous) rocks	Controls water-storage and transmission characteristics. Affects dispersion and dilution of wastes
	Secondary openings	Fractures and faults in crystalline and consolidated sedimentary rocks Solution-enlarged openings in limestones and other soluble rocks	
Composition of rock matrix of dominant aquifer	Insoluble Soluble	Essentially insoluble Both relatively insoluble and soluble constituents Relatively soluble	Affects water-storage and transmission characteristics. Has major influence on water quality
Storage and transmission characteristics of dominant aquifer	Porosity	Large, as in well-sorted, unconsolidated deposits Moderate, as in poorly-sorted unconsolidated deposits and semiconsolidated rocks Small, as in fractures crystalline and consolidated sedimentary rocks	Controls response to pumpage and other stresses. Determines yield of wells. Affects long-term yield of system. Affects rate at which pollutants move

(Continued)

Table 6G.44 (Continued)

Feature	Aspect	Range in Conditions	Significance of Feature
	Transmissivity	↑ Large, as in cavernous limestones, lava flows with flow tubes, and clean gravels Moderate, as in well-sorted, coarse-grained sands, and semiconsolidated limestones ↓ Small, as in poorly-sorted, fine-grained deposits and fractures rocks Very small, as in confining beds, which are commonly clay-rich	
Recharge and discharge conditions of dominant aquifer	Recharge	↑ In upland areas between streams Through channels of losing streams ↓ Largely or entirely by leakage across confining beds from adjacent aquifers	Affects (a) response to stress and (b) long-term yields. Determines susceptibility to pollution. Affects water quality
	Discharge	↑ Through springs or by seepage to stream channels, lakes, estuaries, or the ocean By evaporation on flood plains and in basin "sinks" ↓ By seepage across confining beds into adjacent aquifers	

Source: From Heath, R.C., Classification of ground-water systems of the United States, *Ground-Water*, 20, 4, 1982.

Table 6G.45 Geologic Origin of Aquifers Based on Type of Porosity and Rock Type

	Sedimentary				Volcanic	
Type of Porosity	Consolidated	Unconsolidated	Igneous and Metamorphic		Consolidated	Unconsolidated
Intergranular		Gravelly sand Clayey sand Sandy clay	Weathered zone of granite-gneiss		Weathered zone of basalt	Volcanic ejecta, blocks, and fragments Ash
Intergranular and fracture	Breccia Conglomerate Sandstone Slate		Zoogenic limestone Oolitic limestone Calcareous grit		Volcanic tuff Cinder Volcanic breccia Pumice	
Fracture			Limestone Dolomite Dolomitic limestone	Granite Gneiss Gabbro Quartzite Diorite Schist Mica schist	Basalt Andesite Rhyolite	

Source: From United Nations Department of Economic and Social Affairs, 1975, Ground-water storage and artificial recharge, Natural Resources, *Water Series No.2*.

Table 6G.46 Rocks of Greatest Importance in Groundwater Hydrology

Sedimentary Rocks			Igneous Rocks	
Unconsolidated (Pores)	**Consolidated (Pores, Fractures, and Solution Openings)**	**Metamorphic Rocks (Fractures)**	**Intrusive (Fractures)**	**Extrusive (Pores, Tubes, Rubble Zones, and Fractures)**
GRAVEL[a]	Conglomerate[b]	Gneiss	Granite and other coarse-grained igneous rocks	BASALT and other fine-grained igneous rocks
SAND	SANDSTONE	Quartzite-schist		
Silt	Siltstone	Schist		
Clay[c]	*Shale*	Slate-schist		
Till	Tillite (rare)			
Marl	LIMESTONE-DOLOMITE	Marble		
Coquina				

[a] Capitalized names indicate rocks that are major sources of large groundwater supplies.
[b] Lowercase names indicate rocks of relatively wide extent that are sources of small to moderate groundwater supplies.
[c] Italic names indicate rocks that function primarily as confining beds.

Source: From Heath, R.C., Ground-water, in *Perspectives on Water, Uses and Abuses*, D. Speidel, editor, Copyright Oxford University Press, 1988. Reprinted With permission.

Table 6G.47 Common Ranges in Hydraulic Characteristics of Groundwater Regions in the United States

			Common Ranges in Hydraulic Characteristics of the Dominant Aquifers							
			Transmissivity		Hydraulic Conductivity		Recharge Rate		Well Yield	
Region No	Region	Geologic Situation	m^2 day^{-1}	ft^2 day^{-1}	m day^{-1}	ft day^{-1}	mm yr^{-1}	in yr^{-1}	m^3 min^{-1}	gal min^{-1}
1	Western mountain ranges	Mountains with thin soils over fractured rocks, alternating with narrow alluvial and, in part, glaciated valleys	0.5–100	5–1,000	0.0003–15	0.001–50	3–50	0.1–2	0.04–0.4	10–100
2	Alluvial basins	Thick[a] alluvial (locally glacial) deposits in basins and valleys bordered by mountains	20–20,000	2,000–200,000	30–600	100–2,000	0.03–30	0.001–1	0.4–20	100–5,000
3	Columbia lava plateau	Thick lava sequence interbedded with unconsolidated deposits and overlain by thin soils	2,000–500,000	20,000–5,000,000	200–3000	500–10,000	5–300	0.2–10	0.4–80	100–20,000
4	Colorado plateau and Wyoming basin	Thin[a] soils over fractured sedimentary rocks	0.5–100	5–1,000	0.003–2	0.01–5	0.3–50	0.01–2	0.04–2	10–1,000
5	High plains	Thick alluvial deposits over fractured sedimentary rocks	1,000–10,000	10,000–100,000	30–300	100–1,000	5–80	0.2–3	0.4–10	100–3,000
6	Nonglaciated central region	Thin regolith over fractured sedimentary rocks	300–10,000	3,000–100,000	3–300	10–1,000	5–500	0.2–20	0.4–20	100–5,000
7	Glaciated central region	Thick glacial deposits over fractured sedimentary rocks	100–2,000	1,000–20,000	2–300	5–1,000	5–300	0.2–10	0.2–2	50–500
8	Piedmont and Blue Ridge	Thick regolith over fractured crystalline and metamorphosed	9–200	100–2,000	0.001–1	0.003–3	30–300	1–10	0.2–2	50–500
9	Northeast and superior uplands	Thick glacial deposits over fractured crystalline rocks	50–500	500–5,000	2–30	5–100	30–300	1–10	0.1–1	20–200
10	Atlantic and gulf coastal plain	Complexly interbedded sands, silts and clays	500–10,000	5,000–100,000	3–100	10–400	50–500	2–20	0.4–20	100–5,000
11	Southeast coastal plain	Thick layers of sand and clay over semi-consolidated carbonate rocks	1,000–100,000	10,000–1,000,000	30–3000	100–10,000	30–500	1–20	4–80	1,000–20,000
12	Alluvial valleys	Thick and gravel deposits beneath flood-plains and terraces of streams	200–50,000	2,000–50,000	30–2000	100–5,000	50–500	2–20	0.4–20	100–5,000
13	Hawaiian islands	Lava flows segmented by dikes, interbedded with ash deposits, and partly overlain by alluvium	10,000–100,000	100,000–1,000,000	200–3000	500–10,000	30–1,000	1–40	0.4–20	100–5,000
14	Alaska	Galcial and alluvial deposits in part perennially frozen and overlying crystalline, metamorphic, and sedimentary rocks	100–10,000	1,000–100,000	30–600	100–2,000	3–300	0.1–10	0.04–4	10–1,000

Note: All values rounded to one significant figure; for map of regions, see Figure 6A.3.

[a] An average thickness of about 5 m was used as the break point between thick and thin.

Source: From Heath, R.C., Classification of groundwater regions of the United States, *Groundwater*, 20, 4, 1982.

Table 6G.48 Representative Values of Porosity

Material	Porosity, Percent
Gravel, coarse	28[a]
Gravel, medium	32[a]
Gravel, fine	34[a]
Sand, coarse	39
Sand, medium	39
Sand, fine	43
Silt	46
Clay	42
Sandstone, fine-grained	33
Sandstone, medium-grained	37
Limestone	30
Dolomite	26
Dune sand	45
Loess	49
Peat	92
Schist	38
Siltstone	35
Claystone	43
Shale	6
Till, predominantly silt	34
Till, predominantly sand	31
Tuff	41
Basalt	17
Gabbro, weathered	43
Granite, weathered	45

[a] These values are for repacked samples; all others are undisturbed.

Source: From Johnson, A. I., *Specific Yield-Compilation of Specific Yields for Various Materials*, U.S. Geological Survey Water-Supply Paper 1662-D, 1967.

Table 6G.49 Representative Values of Specific Yield

Material	Specific Yield, Percent
Gravel, coarse	23
Gravel, medium	24
Gravel, fine	25
Sand, coarse	27
Sand, medium	28
Sand, fine	23
Silt	8
Clay	3
Sandstone, fine-grained	21
Sandstone, medium-grained	27
Limestone	14
Dune sand	38
Loess	18
Peat	44
Schist	26
Siltstone	12
Till, predominantly silt	6
Till, predominantly sand	16
Till, predominantly gravel	16
Tuff	21

Source: From Johnson, A.I., *Specific Yield-Compilation of Specific Yields for Various Materials*, U.S. Geological Survey Water-Supply Paper 1662-D, 1967.

Table 6G.50 Drillers' Terms Used in Estimating Specific Yield

Crystalline Bedrock (fresh)
Specific yield zero

Granite	Hard rock
Hard boulders	Graphite and rocks
Hard granite	Rock (if in area of known crystalline rocks)

Clay and Related Materials
Specific yield 3 percent

Adobe	Lava
Brittle clay	Loose shale
Caving clay	Muck
Cement	Mud
Cement ledge	Packed clay
Choppy clay	Poor clay
Clay	Shale
Clay, occasional rock	Shell
Crumbly clay	Slush
Cube clay	Soapstone
Decomposed granite	Soapstone float
Dirt	Soft clay
Good clay	Squeeze clay
Gumbo clay	Sticky
Hard clay	Sticky clay
Hardpan (H.P.)	Tiger clay
Hardpan shale	Tule mud
Hard shell	Variable clay
Joint clay	Volcanic rock

Clay and Gravel, Sandy Clay, and Similar Materials
Specific yield 5 percent

Cemented gravel (cobbles)	Clay and sandy clay
Cemented gravel and clay	Cay and silt
Cemented gravel, hard	Clay, cemented sand
Cement and rocks (cobbles)	Clay, compact loam and sand
Clay and gravel (rock)	Clay to coarse sand
Clay and boulders (cobbles)	Clay, streaks of hard packed
Clay, pack sand, and gravel	Sand
Cobbles in clay	Clay, streaks of sandy clay
Conglomerate	Clay, water
Dry gravel (below water table)	Clay with sandy pocket
Gravel and clay	Clay with small streaks of sand
Gravel (cement)	Clay with some sand
Gravel and sandy clay	Clay with streaks of fine sand
Gravel and tough shale	Clay with thin streaks of sand
Gravelly clay	Porphyry clay
Rocks in clay	Quicksandy clay
Rotten cement	Sand — clay
Rotten concrete mixture	Sand shell
Sandstone and float rock	Shale and sand
Silt and gravel	Solid clay with strata of cemented sand
Soil and boulders	Sticky sand clay
	Tight muddy sand
	Very fine tight muddy sand
Cemented sand	
Cemented sand and clay	Dry sandy silt
Clay sand	Fine sandy loam
Dry hard packed sand	Fine sandy silt
Dry and (below water table)	Ground surface
	Loam
Dry sand and dirt	Loam and clay
Fine muddy sand	Sandy clay loam

(Continued)

Table 6G.50 (Continued)

Clay and Gravel, Sandy Clay, and Similar Materials Specific yield 5 percent	
Fine sand, streaks of clay	Sediment
Fine tight muddy sand	Silt
Hard packed sand, streaks of clay	Silty and clay
	Silty clay loam
Hard sand and clay	Silty loam
Hard set sand and clay	Soft loam
Muddy sand and clay	Soil
Packed sand and clay	Soil and clay
Packed sand and shale	Soil and mud
Sand and clay mix	Soil and sandy shale
Sand and tough shale	Surface formation
Sand rock	Top hardpan soil
Sandstone	Topsoil
Sandstone and lava	Topsoil and sandy silt
Set sand and clay	Topsoil — silt
Set sand, streaks of clay	
Cemented sandy clay	
Hard sandy clay (tight)	Decomposed hardpan
Sandy clay	Hardpan and sandstone
Sandy clay with small sand streaks, very fine	Hardpan and sandy clay
	Hardpan and sandy shale
Sandy shale	Hardpan and sandy stratas
Set sandy clay	Hard rock (alluvial)
Silty clay	Sandy hardpan
Soft sandy clay	Semi-hardpan
Clay and fine sand	Washboard
Clay and pumice streaks	
Ash	Hard pumice
Caliche	Porphyry
Chalk	Seepage soft clay
Hard lava formation	Volcanic ash
Fine Sand, Tight Sand, Tight Gravel, and Similar Materials Specific yield 10 percent	
Sand and clay	Sandy loam
Sand and clay strata (traces)	Sandy loam, sand, and clay
Sand and dirt	Sandy silt
Sand and hardpan	Sandy soil
Sand and hard sand	Surface and fine sand
Sand and lava	
Sand and pack sand	Cloggy sand
Sand and sandy clay	Coarse pack sand
Sand and soapstone	Compacted sand and silt
Sand and soil	Dead sand
Sand and some clay	Dirty sand
Sand, clay, and water	Fine pack sand
Sand crust	Fine quicksand with alkali streak
Sand-little water	Fine sand
Sand, mud, and water	Fine sand, loose
Sand (some water)	Hard pack sand
Sand streaks, balance clay	Hard sand
Sand, streaks of clay	Hard sand and streaks of sandy clay
Sand with cemented streaks	
Sand with thin streaks of clay	Hard sand rock and some water sand
	Hard sand, soft streaks
Coarse, and sandy	Loamy fine sand
Loose sandy clay	Medium muddy sand
Medium sandy	Milk sand

(Continued)

Table 6G.50 (Continued)

Fine Sand, Tight Sand, Tight Gravel, and Similar Materials
Specific yield 10 percent

Sandy
Sandy and sandy clay
Sandy clay, sand, and clay
Sandy clay — water bearing
Sandy clay with streaks of sand
Sandy formation
Sandy muck
Sandy sediment
Very sandy clay

Boulders, cemented sand
Cement, gravel, sand, and rocks
Clay and gravel, water bearing
Clay & rock, some loose rock
Clay, sand and gravel
Clay, silt, sand, and gravel
Conglomerate, gravel, and boulders
Conglomerate, sticky clay, sand and gravel
Dirty gravel
Fine gravel, hard
Gravel and hardpan strata
Gravel, cemented sand
Gravel with streaks of clay
Hard gravel
Hard sand and gravel
Packed gravel
Packed sand and gravel
Quicksand and cobbles
Rock sand and clay
Sand and gravel, cemented Streaks
Sand and silt, many gravel
Sand, clay, streaks of gravel
Sandy clay and gravel
Set gravel
Silty sand and gravel (cobbles)
Tight gravel

More of less sand
Muddy sand
Pack sand
Poor water sand
Powder sand
Pumice sand
Quicksand
Sand, mucky or dirty
Set sand
Silty sand
Sloppy sand
Sticky sand
Streaks fine and coarse sand
Surface sand and clay
Tight sand

Brittle clay and sand
Clay and sand
Clay, sand, and water
Clay with sand
Clay with sand streaks
More or less clay, hard sand and boulders
Mud and sand
Mud, sand, and water
Sand and mud with chunks of clay
Silt and fine sand
Silt and sand
Soil, sand, and clay
Topsoil and light sand
Water sand sprinkled with clay

Float rock (stone)
Laminated
Pumice
Seep water
Soft sandstone
Strong seepage

Gravel, Sand, Sand and Gravel, and Similar Materials
Specific yield 25 percent

Boulders
Coarse gravel
Coarse sand
Cobbles
Cobbles stones
Dry gravel (if above water table
Float rocks
Free sand
Gravel
Loose gravel
Loose sand
Rocks

Gravel and sand
Gravel sand sandrock
Medium sand
Rock and gravel
Running sand
Sand
Sand, water
Sand and boulders
Sand and cobbles
Sand and fine gravel
Sand and gravel
Sand gravel
Water gravel

Source: From U.S. Geological Survey.

Table 6G.51 Representative Values of Hydraulic Conductivity

Material	Hydraulic Conductivity ft day^{-1}	Hydraulic Conductivity m day^{-1}	Type of Measurement[a]
Gravel, coarse	490	150	R
Gravel, medium	890	270	R
Gravel, fine	1,500	450	R
Sand, coarse	150	45	R
Sand, medium	40	12	R
Sand, fine	8.2	2.5	R
Silt	0.62	0.08	H
Clay	0.00068	0.0002	H
Sandstone, fine-grained	0.66	0.2	V
Sandstone, medium-grained	10	3.1	V
Limestone	3	0.94	V
Dolomite	0.0033	0.001	V
Dune sand	66	20	V
Loess	0.26	0.08	V
Peat	19	5.7	V
Schist	0.66	0.2	V
Slate	0.00026	0.0008	V
Till, predominantly sand	1.6	0.49	R
Till, predominantly gravel	100	30	R
Tuff	0.66	0.2	V
Basalt	0.033	0.01	V
Gabbro, weathered	0.66	0.2	V
Granite, weathered	4.6	1.4	V

[a] H is horizontal hydraulic conductivity, R is repacked sample, and V is vertical hydraulic conductivity.

Source: From Morris, D.A. and Johnson, A.I., *U.S. Geological Survey Water-Supply Paper 1839-D*, 1967.

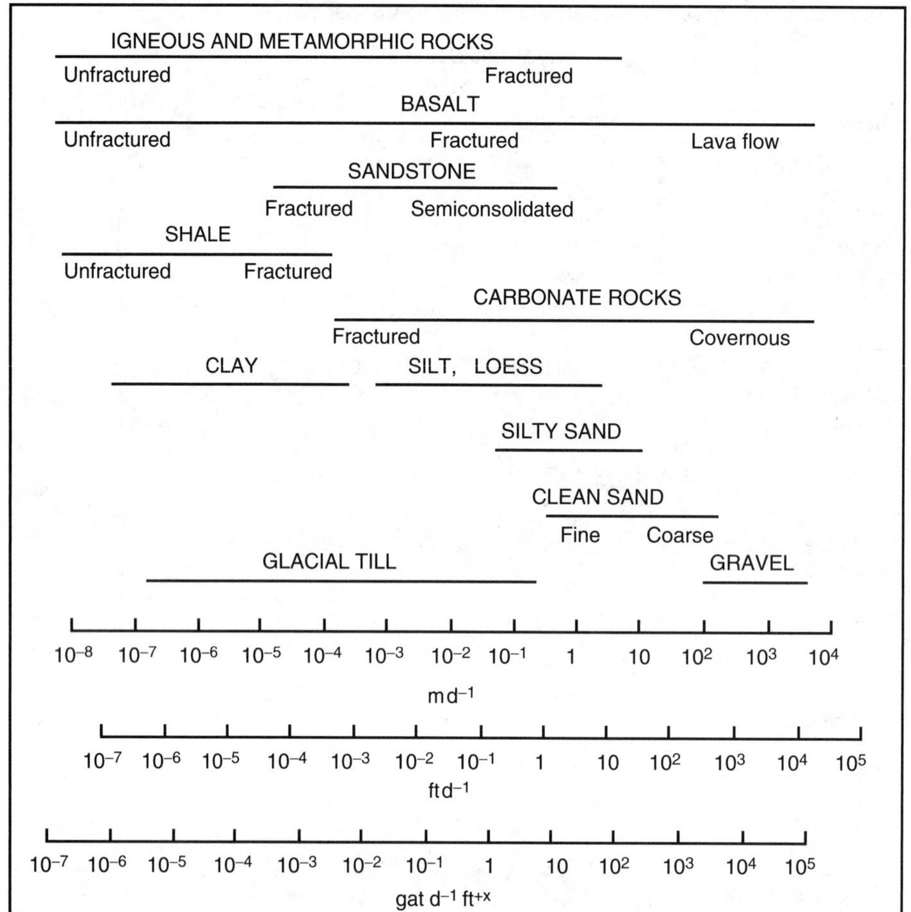

Figure 6G.26 Hydraulic conductivity of selected consolidated and unconsolidated aquifers. (From Heath, R.C., *Basic Ground-Water Hydrology*, U.S. Geological Survey Water-Supply Paper 2220, 1983.)

Table 6G.52 Representative Permeability Ranges for Sedimentary Materials

Material	Permeability, gal day^{-1} ft^{-2}	Material	Permeability, gal day^{-1} ft^{-2}
Clay	10^{-5}–10^{-3}	Very fine sand	1–10^{2}
Silty clay	10^{-5}–10^{-3}	Fine sand	10^{1}–10^{3}
Sandy clay	10^{-4}–10^{-2}	Medium sand	10^{2}–10^{3}
Silty clay loam	10^{-3}–10^{-1}	Coarse sand	10^{2}–10^{4}
Sandy clay loam	10^{-2}–1	Gravel and sand	10^{2}–10^{4}
Silt	10^{-2}–1	Gravel	10^{2}–10^{4}
Silt loam	10^{-2}–1	Sandstone	10^{1}–10^{3}
Loam	10^{-2}–1	Limestone[a]	1–10^{2}
Sandy loam	10^{-1}–10^{1}	Shale	1–10^{2}

[a] Excluding cavernous limestone.

Table 6G.53 Temperature Correction for Permeability

°F	Tc	°F	Tc	°F	Tc	°F	Tc
40	1.37	53	1.11	66	0.92	79	0.78
41	1.35	54	1.09	67	0.91	80	0.77
42	1.33	55	1.08	68	0.89	81	0.76
43	1.31	56	1.06	69	0.88	82	0.75
44	1.28	57	1.04	70	0.87	83	0.74
45	1.26	58	1.03	71	0.86	84	0.73
46	1.24	59	1.01	72	0.85	85	0.72
47	1.22	60	1.00	73	0.84	86	0.71
48	1.20	61	0.99	74	0.83	87	0.70
49	1.18	62	0.97	75	0.82	88	0.69
50	1.16	63	0.96	76	0.81	89	0.68
51	1.15	64	0.95	77	0.80	90	0.67
52	1.13	65	0.93	78	0.79		

Note: To convert coefficient of permeability computed at water temperature shown in table to coefficient of permeability at 60°F, multiply by appropriate factor Tc.

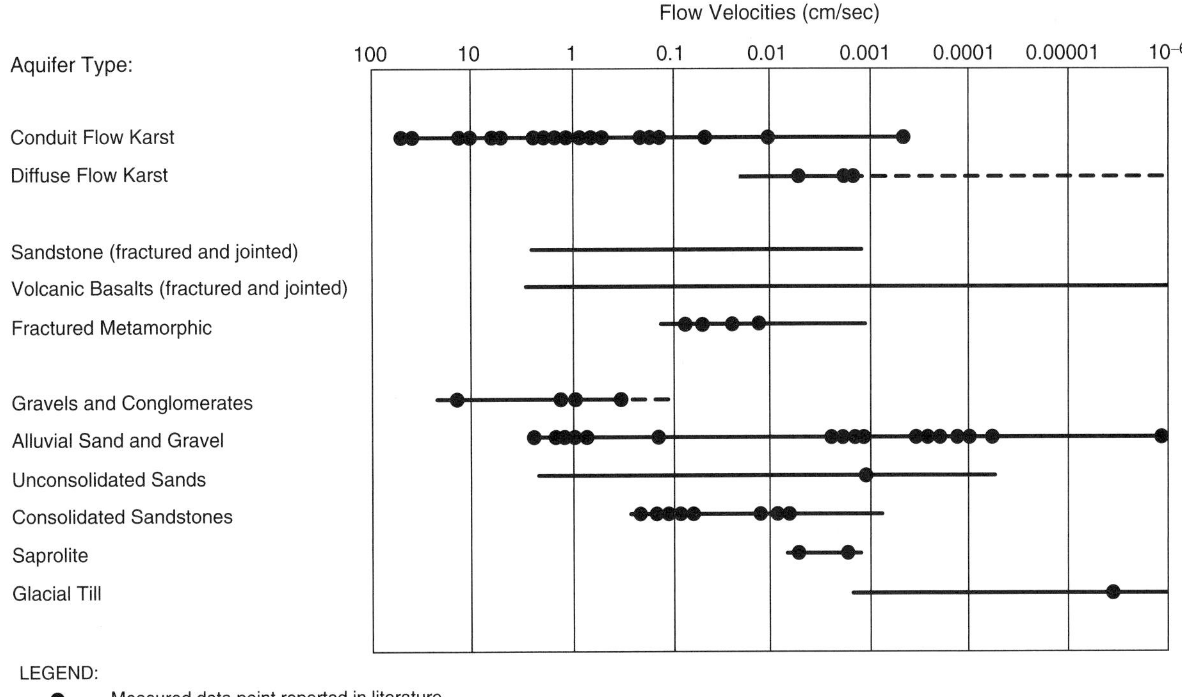

Figure 6G.27 Groundwater flow velocity ranges. (From U.S. Environmental Protection Agency, 1987, *Guidelines for Delineation of Wellhead Protection Areas*, PB88-111430. Original Everett, A.G., 1987.)

SECTION 6H SOIL MOISTURE

Table 6H.54 Guide for Judging How Much of the Available Moisture Has Been Removed from Soil

	Feel or Appearance of Soil and Moisture Deficiency in Inches of Water Per Feet of Soil			
Soil Moisture Deficiency	**Coarse Texture**	**Moderately Coarse Texture**	**Medium Texture**	**Fine and Very Fine Texture**
0% (Field capacity)	Upon squeezing, no free water appears on soil but wet outline of ball is left on hand 0.0	Upon squeezing, no free water appears on soil but wet outline of ball is left on hand 0.0	Upon squeezing, no free water appears on soil but wet outline of ball is left on hand 0.0	Upon squeezing, no free water appears on soil but wet outline of ball is left on hand 0.0
0–25%	Tend to stick together slightly, sometimes forms a very weak ball under pressure 0.0–0.2	Forms weak ball, breaks easily, will not slick 0.0–0.4	Forms a ball, is very pliable, slicks readily if relatively high in clay 0.0–0.5	Easily ribbons out between fingers, has slick feeling 0.0–0.6
25–50%	Appears to be dry, will not form a ball with pressure 0.2–0.5	Tends to ball under pressure but seldom holds together 0.4–0.8	Forms a ball somewhat plastic, will sometimes slick slightly with pressure. 0.5–1.0	Forms a ball, ribbons, out between thumb and forefinger 0.6–1.2
50–75%	Appears to be dry, will not form a ball with pressure[a] 0.5–0.8	Appears to be dry, will not form a ball[a] 0.8–1.2	Somewhat crumbly but holds together from pressure 1.0–1.5	Somewhat pliable, will ball under pressure[a] 1.2–1.9
75–100% (100% is permanent wilting)	Dry, loose, single-grained, flows through fingers 0.8–1.0	Dry, loose, flows through fingers 1.2–1.5	Powdery, dry, sometimes slightly crusted but easily broken down into powdery condition 1.6–2.0	Hard, baked, cracked, sometimes has loose crumbs on surface 1.9–2.5

[a] Ball is formed by squeezing a handful of soil very firmly.

Source: From Israelson and Hansen, *Irrigation Principles and Practices*, John Wiley & Sons, Copyright 1962. With permission.

Table 6H.55 Approximate Limits of Moisture Conditions in Most Irrigation Soils

		Approximate Limits, percent by Weight	
Item	**Soil-Moisture Condition**	**Lower**	**Upper**
1	Hygroscopic moisture content	1–	15
2	Hygroscopic coefficient	1–	15
3	Saturation capacity	15	60
4	Field capacity	7	40
5	Moisture equivalent	5	50
6	Permanent wilting point		30
7	Ultimate wilting point	1	25
8	Moisture in wilting range	1–	5
9	Available moisture capacity	5	20
10	Maximum available storage	1[a]	3
11	Gravity water in saturated soils	8	40

[a] Inches per feet of soil depth.

Source: From Houk, *Irrigation Engineering*, 1, John Wiley & Sons, Copyright 1951.

Table 6H.56 Water-Holding Characteristics of Various Soils

Soil Type	Approximate Depth of Water Per Feet Depth of Soil in Plant Root Zone (in.)		
	Field Capacity	Irrigation Desirable	Wilting Point
Sand	1.2	0.6	0.3
Fine sand	1.5	0.7	0.4
Sandy loam	1.9	1.0	0.6
Fine sandy loam	2.5	1.3	0.8
Loam	3.2	1.7	1.2
Silt loam	3.5	2.0	1.4
Light clay loam	3.7	2.2	1.6
Clay loam	3.8	2.4	1.8
Heavy clay loam	3.8	2.6	2.1
Clay	3.9	2.8	2.4

Source: From U.S. department of agriculture.

Table 6H.57 Representative Physical Properties of Soils

Soil Texture	Infiltration[a] and Permeability (in./hr)	Total Pore Space (%)	Apparent Specific Gravity	Field Capacity (%)	Permanent Wilting (%)	Total Available Moisture[b]		
						Dry Weight Basis (%)	Volume Basis (%)	Inches Per Foot
Sandy	2 (1–10)	38 (32–42)	1.65 (1.55–1.80)	9 (6–12)	4 (2–6)	5 (4–6)	8 (6–10)	1.0 (0.8–1.2)
Sandy loam	1 (0.5–3)	43 (40–47)	1.50 (1.40–1.60)	14 (10–18)	6 (4–8)	8 (6–10)	12 (9–15)	1.4 (1.1–1.8)
Loam	0.5 (0.3–0.8)	47 (43–49)	1.40 (1.35–1.50)	22 (18–26)	10 (8–12)	12 (10–14)	17 (14–20)	2.0 (1.7–2.3)
Clay loam	0.3 (0.1–0.6)	49 (47–51)	1.35 (1.30–1.40)	27 (23–31)	13 (11–15)	14 (12–16)	19 (16–22)	2.3 (2.0–2.6)
Silty clay	0.1 (0.01–0.2)	51 (49–53)	1.30 (1.25–1.35)	31 (27–35)	15 (13–17)	16 (14–18)	21 (18–23)	2.5 (2.2–2.8)
Clay	0.2 (0.05–0.4)	53 (51–55)	1.25 (1.20–1.30)	35 (31–39)	17 (15–19)	18 (16–20)	23 (20–25)	2.7 (2.4–3.0)

Note: Normal ranges are shown in parentheses.

[a] Intake rates vary greatly with soil structure and structural stability, even beyond the normal ranges shown above.

[b] Readily available moisture is approximately 75% of the total available moisture.

Source: From Israelson and Hansen, *Irrigation Principles and Practices*, John Wiley & Sons, Copyright 1962. With permission.

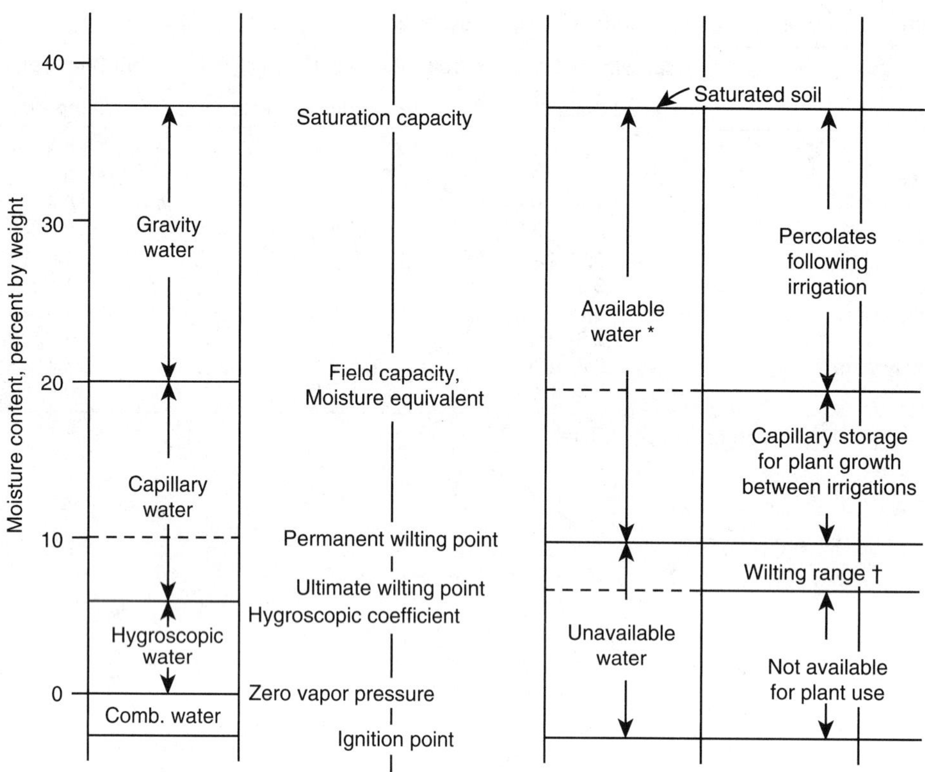

Figure 6H.28 Soil-moisture forms and properties for an assumed fine sandy loam soil. (From Houk, *Irrigation Engineering*, 1, John Wiley & Sons, Copyright 1951. With permission.)

SECTION 6I SPRINGS

Table 6I.58 Classification of Springs According to Magnitude of Discharge

Magnitude	Old System English Units	New System Metric Units
First	Greater than 100 ft^3 sec	$>$10 m^3/sec
Second	10 to 100 ft^3 sec	1–10 m^3/sec
Third	1 to 10 ft^3 sec	0.1–1 m^3/sec
Fourth	100 gal/min to 1 ft^3/sec	10–100 L/sec
Fifth	10 to 100 gal/min	1–10 L/sec
Sixth	1 to 10 gal/min	0.1–1 L/sec
Seventh	1 pt/min to 1 gal/min	10–100 mL/sec
Eighth	Less than 1 pt/min	$<$10 mL/sec

Source: From U.S. Geological Survey.

Table 6I.59 Number of First Magnitude Springs in the United States, by State

State	Number	Rock type	References[a]
Florida	27	Limestone	—
Idaho	14	Limestone and basalt	3, 5
Oregon	15	Balsalt	3
Missouri	8	Limestone	3, 8
California	4	Basalt	3
Hawaii	3	do.	2, 6, 7
Montana	3	Sandstone	3, 4
Texas	2	Limestone	1, 3
Arkansas	1	Basalt	3

[a] References are listed at back of book. 1, Brune (1975); 2, Hirashima (1967); 3, Meinzer (1927); 4, Moore, L. Grady, U.S. Geological Survey, written commun., January 1977; 5, Ray, Herman A., U.S. Geological Survey, written commun., January 1977; 6, Stearns (1966); 7, Stearns and Macdonald (1946); 8, Vineyard and Feder (1974).

Source: From www.flmnh.ufl.edu.

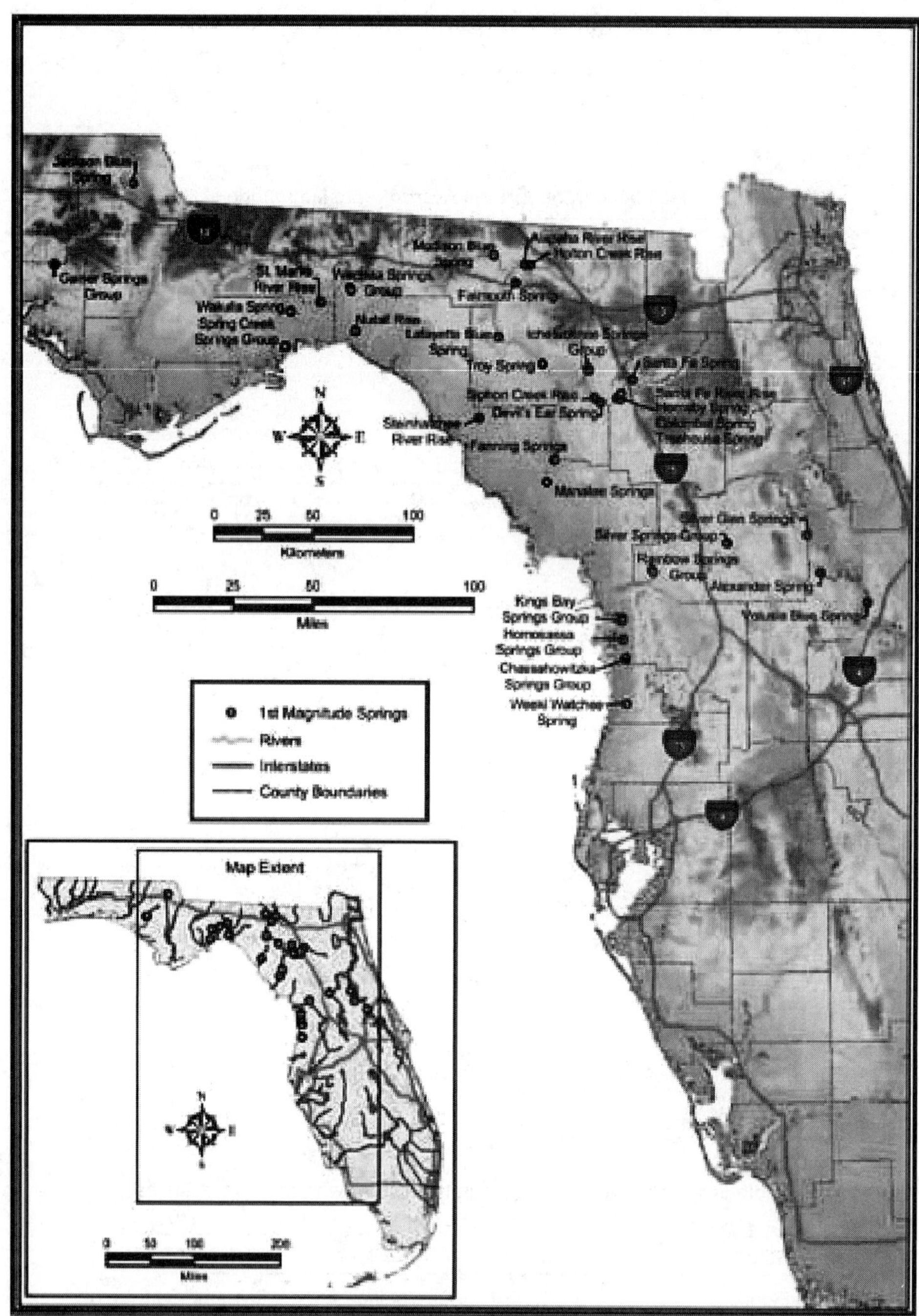

Figure 6I.29 First magnitude springs of Florida. (From www.dep.state.fl.us.)

Table 6I.60 The 27 First-Magnitude Springs and Spring Groups of Florida — with Period of Record, Discharge and Representative Temperatures and Dissolved Solids — Known through December 1976

Spring and Number by County (Refer to Figures 11–15 and Figure 17)		Period of Record	Average (ft^3/sec)	Discharge Range (ft^3/sec)	Number of Measurements	Average Water Temperature C	Average Water Temperature F	Average Dissolved Solids (mg/L)
	Alachua County							
9.	Hornsby Spring	1972–1975	163	76–250	2	22.5	73	230
	Bay County							
1.	Gainer Springs	1941–1972	159	131–185	7	22.0	72	60
	Citrus County							
2.	Chassahowitzka Springs	1930–1972	139	32–197	81	23.5	74	740
4.	Crystal River Springs	1964–1975	916	[a]	[b]	25.0	75	144
5.	Homosassa Springs	1932–1974	175	125–257	90	23.0	73	1800
	Columbia County							
4.	Ichetucknee Springs	1917–1974	361	241–578	375	22.5	73	170
	Hamilton County							
3.	Alapaha Rise	1975–1976	608	508–699	4	19.0	66	130
4.	Holton Spring	1976	288	69–482	3	—	—	—
	Hernando County							
19.	Weeki Wachee Springs	1917–1974	176	101–275	364	23.5	74	150
	Jackson County							
3.	Blue Springs	1929–1973	190	56–287	10	21.0	70	116
	Jefferson County							
1.	Wacissa Springs Group	1971–1974	389	280–605	20	20.5	69	150
	Lafayette County							
11.	Troy Spring	1942–1973	156	148–205	4	22.0	72	171
	Lake County							
1.	Alexander Springs	1931–1972	120	74–162	13	23.5	74	512
	Leon County							
2.	Natural Bridge Spring	1942–1973	106	79–132	5	20.0	68	138
4.	St. Marks Spring	1956–1973	519	310–950	130	20.5	69	154
	Levy County							
3.	Fannin Springs	1930–1973	103	64–139	8	22.0	72	194
5.	Manatee Spring	1932–1973	181	110–238	9	22.0	72	215
	Madison County							
1.	Blue Spring	1932–1973	115	75–145	6	21.0	70	146
	Marion County							
5.	Rainbow Springs	1898–1974	763	487–1,230	402	23.0	73	93
7.	Silver Glen Springs	1931–1972	112	90–129	11	23.0	73	1,200
8.	Silver Springs	1906–1974	820	539–1,290	155	23.0	73	245
	Suwannee County							
8.	Falmouth Spring	1908–1973	158	60–220[c]	8	21.0	70	190
	Volusia County							
1.	Blue Spring	1932–1974	162	63–214	360	23.0	73	826
	Wakulla County							
2.	Kini Spring	1972	176	—	1	20.0	68	110
5.	River Sink Spring	1942–1973	164	102–215	6	20.0	68	110
6.	Wakulla Springs	1907–1974	390	25–1,920	276	21.0	70	153
13.	Spring Creek Springs[a,d]	1972–1974	2,003	[a]	1	19.5	67	2,400

[a] Tidal affected.
[b] Continuous record, vane gage.
[c] Reverse flow of 365 ft^3/sec measured on 02-10-33.
[d] See Figure 11.17.

Source: From www.flmnh.ufl.edu.

Twenty-three states in the United States have Thermal Springs that are included in the database. This table shows the number of Thermal Springs in each state. The first column shows the state abbreviation, the second shows the number of thermal spring in the state, and the third shows a graph of those counts. Nevada has 312 thermal springs, the largest number in any state.

State	Number of Springs	Graph
AK	108	
AR	6	
AZ	60	
CA	304	
CO	47	
FL	2	
GA	7	
HI	11	
ID	232	
MA	1	
MT	61	
NC	1	
NM	77	
NV	312	
NY	1	
OR	126	
SD	2	
TX	9	
UT	116	
VA	11	
WA	30	
WV	5	
WY	132	

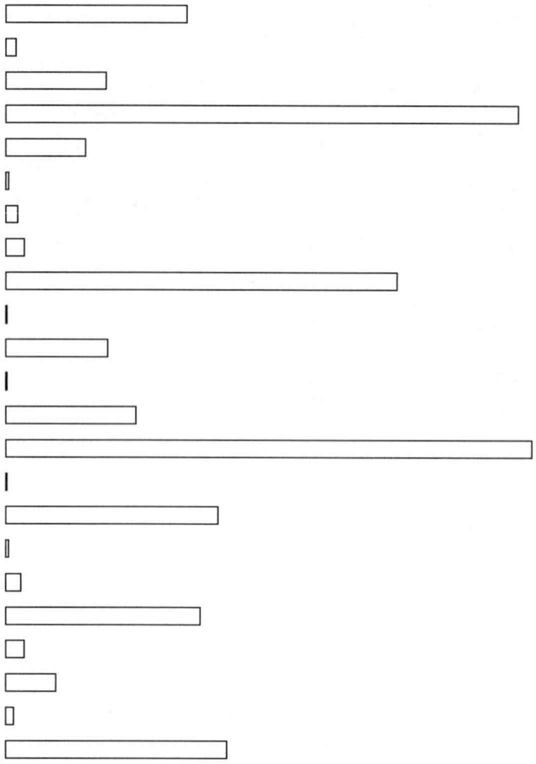

Figure 6I.30 Thermal springs in the United States. (From www.ngdc.noaa.gov.)

Table 6I.61 Major Springs Reported in Other Countries

Spring	Country	Average Discharge (ft^3/sec)	Rock Type	Reference[a]
Ras-El-Ain	Syria	1,370	Limestone	1,2
Stella Spring	Italy	1,290	do.	6
Rio Maule Spring	Chile	1,000	Basalt	2
Fontaine de Vaucluse	France	800	Limestone	3, 6
Timaso Spring	Italy	800	do.	5
Komishimigawa	Japan	700	Basalt	5
Ain Zarka	Syria	490	Limestone	5
Sinn River	Syria	430	do.	5
El Gato	Mexico	185	Basalt	4, 7
Lanza	Bolivia	135	do.	5

[a] References are listed at back of book. 1, Burdon and Safadi (1963); 2, Davis and DeWiest (1966, p. 63, 367–369); 3, Meinzer (1927, p. 91–92, 94); 4, Thomas (1975); 5, Thomas, H. E. (written commun., October 1974); 6, Vineyard and Feder (1974, p. 14); 7, Waring (1965, p. 61);

Source: From www.flmnh.ufl.edu.

Table 6I.62 Natural Heat Flows of Some Hot Spring Areas of the World

Area	Approximate Size[a] (km^2)	Maximum Recorded Temperature[b] (°C)	Total Heat Flow[c] (10^6 cal/sec)
British West Indies			
Qualibou, St. Lucia	~0.1	(S) 185	8.6
St. Vincent	~1	(S)>27	18
Dominica	~1	(S) 90	17
Montserrat	~0.1	(S) 97	1.6
El Salvador			
Total of country	—	—	200
Northern belt, total	—	—	50
Southern belt, total	—	—	>150
Ahuachapán group	80	(D) 174	80
El Playón de Ahuachapán	~0.25	(S) boiling	0.46
Agua Shuca	~0.25	(S) boiling	0.32
Fiji Islands			
Savusavu	~1	(S) 100	2
Iceland			
Steam fields, total heat flow	—	—	630
Hengill, total	50	(D) 230	55–80
Do	—	—	25–125
Hengill, southern part only	—	(D) 230	28
Torfajökull	100	(S) boiling	500
Reykjanes	1	?	5–25
Trölladyngja	5	?	5–25
Krysuvik	10	(D) 230	5–25
Kerlingafjöll	5	(S) boiling	25–125
Vonarskard	?	?	5–125
Grimsvötn	12	?	125–750
Kverkfjöll	10	?	25–125
Askja	25	?	5–25
Námafjall	2.5	?	25–125
Krafla	0.5	(S) boiling	5–25
Theistareykir	2.5	(S) boiling	5–25
Low temperature areas; about 250 areas	—	(D) 146	100
Six lines of thermal springs, each	—	(S) 100	5–25
Reykjavik	~5	(D) 146	1.7
Reykir	~5	(D) 98	11
Deilartunga line, total	—	(S) 100	25–125
Deliartunga spring	—	(S) 100	24
Italy			
(Larderello)	(~50)	((D) 240)	(5)
Ischia and Flegreian Fields	~10	(D) 296	?
(Monta Amiata)	(~3)	((D) 165)	(?)
Vulcano	~1	(D) 194	?
Japan			
Otaki, Kyushu	—	(D) 185	?
Atami, Shizuoka-ken	5	(D) 180	16
Do	—	—	22
Ito, Shizuoka-ken	—	—	44
Obama, Nagasaki-ken	1.5	(D) 180	57
Beppu, Oita-ken	~10	(D) 150	19
Kawayu, Hokkaido	0.7	(S) 65	8
Yunokawa, Hokkaido	~1	(S) 66	4.0
Yachigashira, Hokkaido	?	(S) 69	0.5
Shikabe, Hokkaido	~0.5	(D) 113	1.2
Toyako, Hokkaido	~3	(S) 55	2.2
Noboribetsu, Hokkaido:			
Hot Lake area, total	~0.2	(D) 112	14
Jigokudani Valley (variable)	~0.3	(D) 160	~6–11.2
Matsukawa, N. Honshu	—	(D) 189	?

(Continued)

Table 6I.62 (Continued)

Area	Approximate Size[a] (km^2)	Maximum Recorded Temperature[b] (°C)	Total Heat Flow[c] (10^6 cal/sec)
Onikobe, N. Honshu	~80	(D) 185	?
Narugo, N. Honshu	—	(D) 175	?
Mexico			
Pathé, Hidalgo	~2	(D) 155	?
Ixtlan, Michoacan	—	(D) 150	?
New Zealand			
Wairakei, 1951, 1952	7	(D) 266	133
1954	7	(D) 266	82
1956?	7	(D) 266	143
1958, 1959	7	(D) 266	163
1958	7	(D) 266	101
Waiotapu	~15	(D) 295	272
Orakei Korako	~5	(S) boiling	130
Tikitere	5	(S) boiling	40
Tokopia	—	—	30
Waikiti	—	(S) 91	20
Ngatamariki	~1	(S) hot	12.6
Rotokaua	~5	(S) boiling	52
Ohaki	~1	(S) boiling	12.8
Taupo Spa	~3	(S) boiling?	36
Kawerau (Onepu) 1959?	?	(D) 277	25
1962	?	(D) 285	18
Rotorua	?	(D) > 160	—
Union of South Africa			
Seven scalding springs	?	(S) 64	1.7
United States			
California			
The Geysers	~1	(D) 208	0.4
Sulphur Bank	~2	(D) 136	0.2
Wilbur Springs area	~5	(S) 69	0.4
Casa Diablo — Hot Creek	> 25(?)	(D) 180	70
Alkali Lakes area	—	—	—
(Salton Sea)	(~50)	(D) > 270	(4)
Nevada			
Steamboat Springs	5	(D) 187	7
Bradys Springs	~2	(D) 168	?
Beowawe	~3	(D) 207	?
Wyoming			
Yellowstone Park, Wyoming	9,000	—	—
Total, discharging water	~70	(S) 138	207
Total, calculated	~70	(D) 205	500
Norris Geyser Basin	~3	(D) 205	8
Upper Geyser Basin	~10	(D) 180	90
Mammoth-Hot River	~8	(S) 73	34
U.S.S.R.			
Pauzhetsk, Kamchatka	~1	(D) 195	18
Total			~2700

[a] The limits of a hydrothermal area are very difficult to define and meaningful criteria are difficult to apply. Depending upon the definition, the "limits" of an area can vary by at least an order of magnitude. The definition used here is: "The rather broad boundaries containing specific areas with some surface evidence for abnormally high temperatures at depth. The evidence can consist of one or more of the following: hot springs, fumaroles, active hydrothermal alteration, and abnormally high near-surface geothermal gradient. Closely spaced 'hot spots' not separated by areas of approximately 'normal' gradient for the region are included in a single thermal area".

[b] (S) indicates temperatures measured at the surface; (D) temperatures from drill holes.

[c] Most heat flows are relative to mean annual surface temperatures but a few are relative to 0° or 4°C; such differences are small compared to the uncertainties and have not been modified. 1×10^6 cal/cm^2/sec approximates the "normal" heat flow from 60 to 70 km^2.

Source: From U.S. Geological Survey, 1965.

SECTION 6J ARTIFICIAL RECHARGE

Table 6J.63 Artificial Recharge Projects in the United States and Other Countries

United States			
Alabama	2	Mississippi	1
Arizona	3	Missouri	1
Arkansas	2	Montana	2
California	42	Nebraska	4
Colorado	3	Nevada	1
Connecticut	1	New Jersey	3
Florida	1	New Mexico	1
Georgia	2	New York	1
Hawaii	1	North Carolina	3
Illinois	1	Ohio	3
Indiana	2	Oklahoma	4
Iowa	1	Pennsylvania	3
Kansas	2	South Carolina	2
Kentucky	1	Tennessee	1
Louisiana	3	Texas	9
Maine	1	Utah	1
Maryland	5	Virginia	4
Massachusetts	3	Washington	4
Michigan	3	Wisconsin	1
Minnesota	2		
Other Countries			
Denmark	1	Morocco	1
Fed. Rep. of Germany	10	Namibia	1
Finland	1	Netherlands	20
Greece	4	New Zealand	2
India	3	Oman	1
Israel	2	Qatar	1
Italy	5	Switzerland	1
Jamaica	1	Thailand	2
Japan	3		

Note: Number of projects reported based on return of American Society of Civil Engineers' questionnaires, 1988.

Source: From Johnson, A.I., Personal communication, October 30, 1988.

Table 6J.64 Operation and Maintenance Problems of Artificial Recharge Projects

Problem	Manifestation	Corrective Actions to be Considered
Silt	Lodging of particles within interstices of soil near the surface area, reducing the infiltration rate	(1) Desilt in retention reservoir and/or in uppermost series of basins. Flocculating agent such as "Separan" has been used with success. (2) Bypass water until concentration of silt will not be detrimental, with concentration depending upon soil condition. Ditches and furrows generally can accept waters containing higher concentrations of silt if sufficient velocity is maintained through the project to carry silt back to the main canal. (3) Scrape, harrow, and/or disc after proper drying. Period of drying usually ranges from one to seven days depending upon soil and weather conditions. (4) Remove silt after drying. Silt may be used to build up levees of basins or bridges of ditches or furrows. (5) Sustain vegetative growth. (6) Sluice the silt out of ditches and furrows, and from channels, with due regard to erosion problems. (7) Pump injection well to loosen silt from interstices and remove silt from the well.
Weeds	Increases percolation rate and shortens drying period required for working an area or removing silt from the basin. There is a disadvantage in that vegetative growth may be a fire hazard	(1) Control by chemical means and/or remove when weeds become a fire hazard, especially around structures. Consider use of hand labor instead of mechanical means in order to maintain infiltration rates. However, if possible, leave vegetation undisturbed in wetted area. (2) Prolonged deep submergence will kill vegetation. (3) The control of weeds is generally not considered a problem in the operation of pits and shafts, or injection wells.
Rodents	Leaks and failures of dikes and levees. Public nuisance near urban area	(1) Set out poison about twice a year. (2) Use of traps.
Public health and safety	Rodents and mosquito problem and possible injury to individuals. Potential problems of injury is greatest when depth of water is large in basins and pits	(1) Enclose area with fence and gates with locks. (2) Patrol area with particular attention to children and structural failures, before and during oper-ation, especially near inhabited area. (3) Vector control by use of mosquito fish, chemicals, and/or drying. (4) Rodent control by poisoning or traps. (5) Proper posting of signs when using chemical which is poisonous.

(Continued)

Table 6J.64 (Continued)

Problem	Manifestation	Corrective Actions to be Considered
Maintaining of percolation rates	Reduction of percolation rates will decrease efficiency of system, increasing the unit cost of the amount of water actually recharged	(1) Proper treatment of water. Desilt water to concentration desired. Use chlorine or copper sulfate for control of bacterial slime and algae. Use of chemicals to reduce the possibility of chemical incrustation, which is usually deposition of calcium carbonate. (2) Schedule intermittent drying periods to prevent problems due to swelling of soil particles. Permit growth of vegetation to decrease the drying period by removing the water in the root zone and loosening the soil. Studies have shown that bermuda grass has been successfully used to maintain rates, even under prolonged periods of deep submergence. (3) Prevent aeration of water, especially when operating recharge wells, pits, and shafts. (4) Increase head of water generally by increasing depth of water. (5) Use hand labor whenever possible to decrease the possibility of using heavy equipment which will cause surface compaction especially when soil is wet. (6) Scrape, harrow, and/or disc after proper drying. (7) Remove silt, chemical incrustation, and/or any material decreasing infiltration rates after proper drying period. (8) Maintain the design velocity to reduce silting to a minimum in use of ditches and furrows. (9) Recondition injection wells by use of dry ice, hydrochloric acid, and/or sulfuric acid. (10) Prevent freezing of water during winter by continuous spreading. (11) Check the possibility of base exchange reactions. (12) Soil can be reconditioned by using organic material such as cotton gin trash, or chemical agents, such as krilium.
Maintenance to diversion structures	Breakdown of spreading operations	(1) Systematic and routine maintenance check as well as patrolling when in operation. Attention should be given to wooden structures since they deteriorate faster due to frequent wetting and drying cycles. Also attention should be given to settlement of structures thus changing flow condition. (2) Attention should be given to undercutting of structure particularly on the downstream end with preventive maintenance primarily in the form of riprapping. (3) Sluicing of channel to remove silt and debris which have accumulated near and at diversion structure.

Source: From Richter and Chun, *Proc. Am. Soc. Civil Eng.*, 1959. With permission.

SECTION 6K GEOPHYSICAL LOGGING

Table 6K.65 Borehole Geophysical Logging Methods and Their Uses in Hydrologic Studies

Method	Uses	Recommended Conditions
Electric logging: Single-electrode resistance	Determining depth and thickness of thin beds. Identification of rocks, provided general lithologic information is available, and correlation of formations. Determining casing depths	Fluid-filled hole. Fresh mud required. Hole diameter less than 8 to 10 in. Log only in uncased holes
Short normal (electrodes spacing of 16 in.)	Picking tops of resistive beds. Determining resistivity of the invaded zone. Estimating porosity of formations (deeply invaded and thick interval). Correlation and identification, provided general lithologic information is available	Fluid-filled hole. Fresh mud. Ratio of mud resistivity to formation-water resistivity should be 0.2 to 4. Log only in uncased part of hole
Long normal (electrode spacing of 64 in.)	Determining true resistivity in thick beds where mud invasion is not too deep. Obtaining data for calculation of formation-water resistivity	Fluid-filled hole. Ratio of mud resistivity to formation-water resistivity should be 0.2 to 4. Log only in uncased part of hole
Deep lateral (electrode spacing approximately 19 ft)	Determining true resistivity where mud invasion is relatively deep. Locating thin beds	Fluid-filled uncased hole. Fresh mud. Formations should be of thickness different from electrode spacing and should be free of thin limestone beds
Limestone sonde (electrode spacing of 32 in.)	Detecting permeable zones and determining porosity in hard rock. Determining formation factor *in situ*	Fluid-filled uncased hole. May be salty mud. Uniform hole size. Beds thicker than 5 ft
Laterolog	Investigating true resistivity of thin beds. Used in hard formations drilled with very salty muds. Correlation of formations, especially in hard-rock regions	Fluid-filled uncased hole. Salty mud satisfactory. Mud invasion not too deep
Microlog	Determining permeable beds in hard or well-consolidated formations. Detailing beds in moderately consolidated formations. Correlation in hard-rock country. Determining formation factor in situ in soft or moderately consolidated formations. Detailing very thin beds	Fluid required in hole. Log only in uncased part of hole. Bit-size hole (caved sections may be logged, provided hole enlargements are not too great)
Microlaterolog	Determining detailed resistivity of flushed formation at wall of hole when mudcake thickness is less than three-eighths inch in all formations. Determining formation factor and porosity. Correlation of very thin beds	Fluid-filled uncased hole. Thin mud cake. Salty mud permitted
Spontaneous potential	Helps delineate boundaries of many formations and the nature of these formations. Indicating approximate chemical quality of water. Indicate zones of water entry in borehole. Locating cased interval. Detecting and correlating permeable beds	Fluid-filled uncased hole. Fresh mud

(Continued)

Table 6K.65 (Continued)

Method	Uses	Recommended Conditions
Radiation logging: Gamma Ray	Differentiating shale, clay, and marl from other formations. Correlations of formations. Measurement of inherent radioactivity in formations. Checking formation depths and thicknesses with reference to casing collars before perforating casing. For shale differentiation when holes contain very salty mud. Radioactive tracer studies. Logging dry or cased holes. Locating cemented and cased intervals. Logging in oil-base muds. Locating radioactive ores. In combination with electric logs for locating coal or lignite beds	Fluid-filled or dry cased or uncased hole. Should have appreciable contrast in radioactivity between adjacent formations
Neutron	Delineating formations and correlation in dry or cased holes. Qualitative determination of shales, tight formations, and porous sections in cased wells. Determining porosity and water content of formations, especially those of low porosity. Distinguishing between water- or oil-filled and gas-filled reservoirs. Combining with gamma-ray log for better identification of lithology and correlation of formations. Indicating cased intervals. Logging in oil-base muds	Fluid-filled or dry cased or uncased hole. Formations relatively free from shaly material. Diameter less than 6 inches for dry holes. Hole diameter similar throughout
Induction logging	Determining true resistivity, particularly for thin beds (down to about 2 feet thick) in wells drilled with comparatively fresh mud. Determining resistivity of formations in dry holes. Logging in oil-base muds. Defining lithology and bed boundaries in hard formations. Detection of water-bearing beds	Fluid-filled or dry uncased hole. Fluid should not be too salty
Sonic logging	Logging acoustic velocity for seismic interpretation. Correlation and identification of lithology. Reliable indication of porosity in moderate to hard formations; in soft formations of high porosity it is more responsive to the nature rather than quantity of fluid contained in pores	Not affected materially by type of fluid, hole size, or mud invasion
Temperature logging	Locating approximate position of cement behind casing. Determining thermal gradient. Locating depth of lost circulation. Locating active gas flow. Used in checking depths and thickness of aquifers. Locating fissures and solution openings in open holes and leaks or perforated sections in cased holes. Reciprocal-gradient temperature log may be more useful in correlation work	Cased or uncased hole. Can be used in empty hole if logged at very slow speed, but fluid preferred in hole. Fluid should be undisturbed (no circulation) for 6 to 12 hr minimum before logging; possibly several days may be required to reach thermal equilibrium
Fluid-conductivity logging	Locating point of entry of different quality water through leaks or perforations in casing or opening in rock hole. (Usually resistivity is determined and must be converted to conductivity.) Determining quality of fluid in hole for improved interpretation of electric logs. Determining fresh-water-salt-water interface	Fluid required in cased or uncased hole. Temperature log required for quantitative information

(Continued)

Table 6K.65 (Continued)

Method	Uses	Recommended Conditions
Fluid-velocity logging	Locating zones of water entry into hole. Determining relative quantities of water flow into or out of these zones. Determine direction of flow up or down in sections of hole. Locating leaks in casing. Determine approximate permeability of lithologic sections penetrated by hole, or perforated section of casing	Fluid-filled cased or uncased hole. Injection pumping, flowing, or static (at surface) conditions. Flange or packer units required in large diameter holes. Caliper (section gage) logs required for quantitative interpretation
Casing-collar locator	Locating position of casing collars and shoes for depth control during perforating. Determining accurate depth references for use with other types of logs	Cased hole
Caliper (section gage) survey	Determining hole or casing diameter. Indicates lithologic character of formations and coherency of rocks penetrated. Locating fractures, solution opening, and other cavities. Correlation of formations. Selection of zone to set a packer. Useful in quantitative interpretation of electric, temperature, and radiation logs. Used with fluid-velocity logs to determine quantities of flow. Determining diameter of underreamed section before placement of gravel pack. Determining diameter of hole for use in computing volume of cement to seal annular space. Evaluating the efficiency of explosive development of rock wells. Determining construction information on abandoned wells	Fluid-filled or dry cased or uncased hole. (In cased holes does not give information on beds behind casing.)
Dipmeter survey	Determining dip angle and dip direction (from magnetic north) in relation to well axis in the study of geologic structure. Correlation of formations	Fluid-filled uncased hole. Carefully picked zones needing survey, because of expense and time required. Directional survey required for determination of true dip and strike (generally obtained simultaneously with dipmeter curves)
Directional (inclinometer) survey	Locating points in a hole to determine deviation from the vertical. Determining true depth. Determining possible mechanical difficulty for casing installation or pump operation. Determining true dip and strike from dipmeter survey	Fluid-filled or dry uncased hole
Magnetic logging	Determining magnetic field intensity in bore-hole and magnetic susceptibility of rocks surrounding hole. Studying lithology and correlation, especially in igneous rocks	Fluid-filled or dry uncased hole

Source: From U.S. Geological Survey, 1968.

Table 6K.66 Geophysical Logging Quick Reference Guide

Most Digital Logging Tools from Wherever Include a Gamma Ray Detector for Correlation between Logging Runs

Tool Size	Min/Max Hole Size	TYPE OF LOG	Digital	Analog	PROPERTIES MEASURED
(D = Digital) (A = Analog) (O = Open Hole) (C = Cased Hole) (S = Steel) (P = Plastic) Diameter (in.)		**RESISTIVITY / CONDUCTIVITY**			
D1.7 A2.25	D 2.0/240 (D) A 3.0/36.0 (D)	**ELECTRIC LOG**	X	X	Determination of water quality Bed boundary positions Lithology information Indication of permeable zones are porosity
D1.5 A3.0	D 2.0/36.0 (O) A 5.0/36.0 (O)	**GUARD LOG**	X	X	Lithology correlation between boreholes Determination of formation water quality Indications of fractures and permeable zones Bed boundary and thickness measurements Thin bed resolution
D1.7	D 2.0/24 (O) (C, P)	**DUAL INDUCTION LOG** *Includes New Features*	X		Conductivity measurements Indications of permeable zones and porosity Determination of formation water quality Long term well monitoring Lithology measurements in 2 in. diameter PVC boreholes Magnetic susceptibility (qualitative only)
D2.28	D 2.0/28.7 (O)	**MICRO-RESISTIVITY LOG** *New Logging Tool*	X		High precision bed boundary and thickness determination Invasion profile information Extreme thin bed resolution in environmental wells Porosity values when combined with deep resistivity readings Flushed zone resistivity
D1.7	D 3.0/24 (O)	**INDUCED POLARIZATION LOG** *New Logging Tool*	X		Indicator of mineralization Qualitative permeability studies Lithology information
RADIOACTIVITY					
Various	Various All	**NATURAL GAMMA RAY**	X	X	Correlation between boreholes Lithology information Bed thickness Clay/Shale content
D2.36	D 3.0/11.8 All	**SPECTRAL GAMMA RAY**	X		Lithology determination Mineral detection Sedimentology Improved clay content computation Correlation
A1.7	A 2.0/24 All	**NEUTRON**		X	Lithology identification Porosity/Hydrogen index indicator
Various	A 1.5/24 All	**DENSITY (GAMMA/GAMMA)**		X	Bulk density calculation Lithology identification Correlation
HOLE SIZE					
D1.5 D2.36 A1.25 A2.125	D 20/50.0 All D 25/50.0 All A 1.5/50 All A 2.5/50 All	**3 ARM CALIPER LOG**	X	X	Average borehole diameter Borehole volume Identification of hard and soft formations Correction of other logs affected by borehole diameter
D2.4	D 3.0/27.5 All	**BOREHOLE GEOMETRY LOG** *New Logging Tool*	X		Borehole diameter in two orthogonal directions Borehole break out for stress analysis More accurate borehole volume calculation Borehole verticality (from magnetic deviation section)

Source: From www.welenco.com.

CHAPTER 7

Water Use

Katherine L. Thalman

CONTENTS

SECTION 7A WATER USE — UNITED STATES

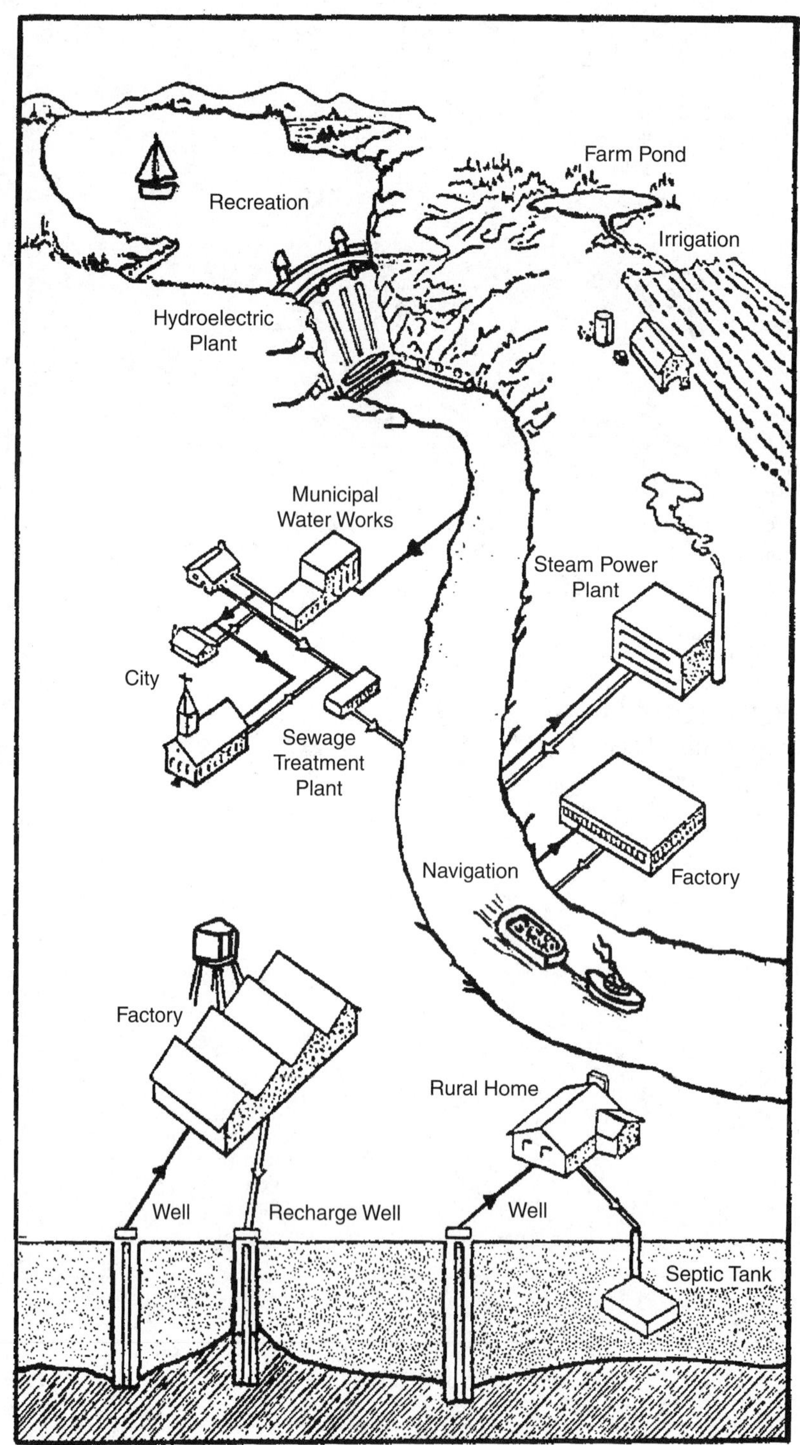

Figure 7A.1 The many uses of water. (From Laas and Beicos, *The Water in Your Life*, Popular Library In 2nd Edition, 1967.)

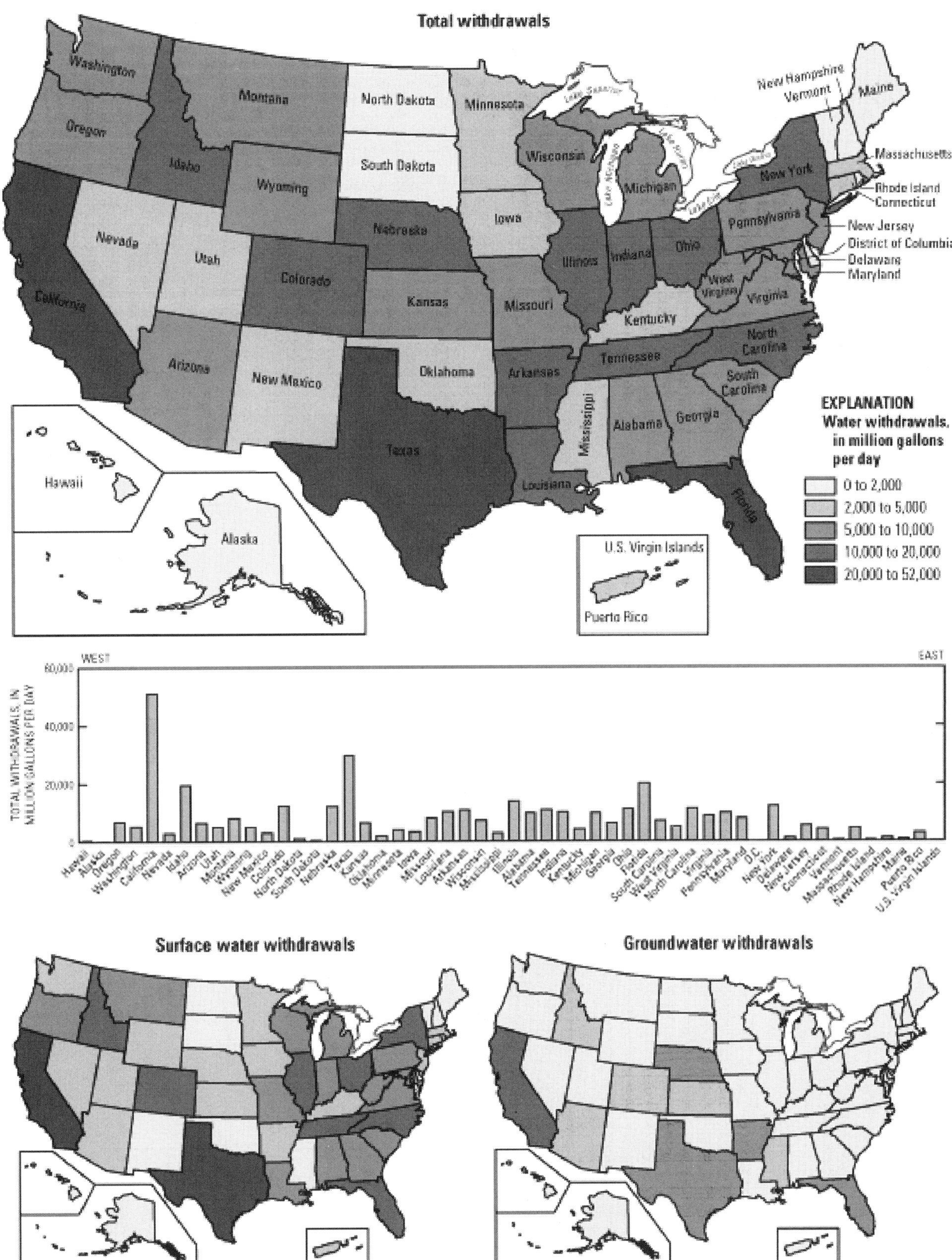

Figure 7A.2 Total water withdrawals for all off-stream water-use categories in the United States, 2000. (From Hutson, S.S. et al., 2004, Estimated use of water in the United States in 2000, *U.S. Geological Survey Circular 1268*, www.usgs.gov/pubs/circ/2004/circ1268.)

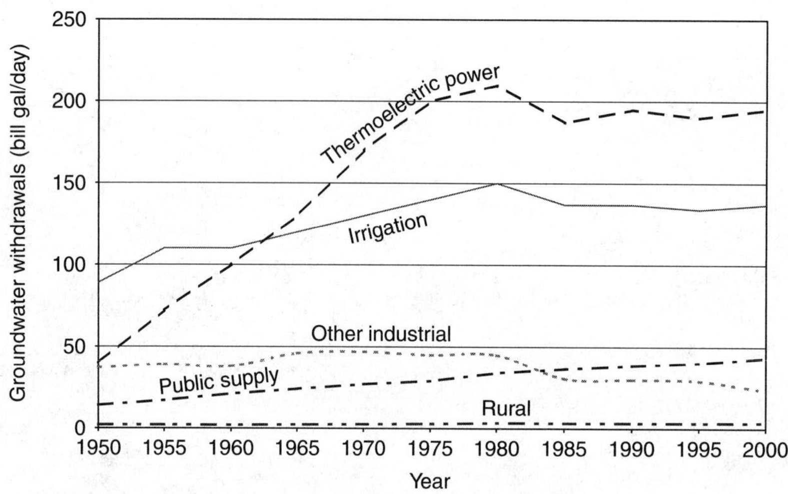

Note: The 2000 data for rural domestic and livestock and other industrial are partial totals.

Figure 7A.3 Total water withdrawal for public supply, rural, irrigation, thermoelectric power, and other industries in the United States, 1950–2000. (Based on data from Hutson, S.S. et al., 2004, Estimated use of water in the United States in 2000, *U.S. Geological Survey Circular 1268*, www.usgs.gov.)

Table 7A.1 Trends in Estimated Water Use in the United States, 1950–2000

	Year											Percentage Change
	1950[a]	1955[b]	1960[c]	1965[d]	1970[d]	1975[c]	1980[c]	1985[c]	1990[c]	1995[c]	2000[c]	1995–2000
Population (million)	150.7	164	179.3	193.8	205.9	216.4	229.6	242.4	252.3	267.1	285.3	7
Offstream use												
Total withdrawals	180	240	270	310	370	420	440	399	408	402	408	2
Public supply	14	17	21	24	27	29	34	36.5	38.5	40.2	43.3	8
Rural domestic and livestock												
Self-supplied domestic	2.1	2.1	2	2.3	2.6	2.8	3.4	3.32	3.39	3.39	3.59	6
Livestock and aquaculture	1.5	1.5	1.6	1.7	1.9	2.1	2.2	4.47[e]	4.5	5.49	[f]	—
Irrigation	89	110	110	120	130	140	150	137	137	134	137	2
Industrial												
Thermoelectric power use	40	72	100	130	170	200	210	187	195	190	195	3
Other industrial use	37	39	38	46	47	45	45	30.5	29.9	29.1	[g]	—
Source of water												
Ground												
Fresh	34	47	50	60	68	82	83	73.2	79.4	76.4	83.3	9
Saline	[h]	0.6	0.4	0.5	1	1	0.9	0.65	1.22	1.11	1.26	14
Surface												
Fresh	140	180	190	210	250	260	290	265	259	264	262	−1
Saline	10	18	31	43	53	69	71	59.6	68.2	59.7	61	2

Note: Data for 1995 and earlier from Solley and others (1998). The water-use data are in billion gallons per day (thousand million gallons per day) and are rounded to two significant figures for 1950–80, and to three significant figures for 1985–2000; percentage change is calculated from unrounded numbers. —, not available.

[a] 48 States and district of Columbia, and Hawaii.
[b] 48 States and district of Columbia.
[c] 50 States and district of Columbia, Puerto Rico, and U.S. Virgin Islands.
[d] 50 States and district of Columbia, and Puerto Rico.
[e] From 1985 to present this category includes water use for fish farms.
[f] Data not available for all States; partial total was 5.46.
[g] Commercial use not available; industrial and mining use totaled 23.2.
[h] Data not available.

Source: From Hutson, S.S. et al., 2004, Estimated use of water in the United States in 2000, *U.S. Geological Survey Circular 1268*, http://water.usgs.gov.

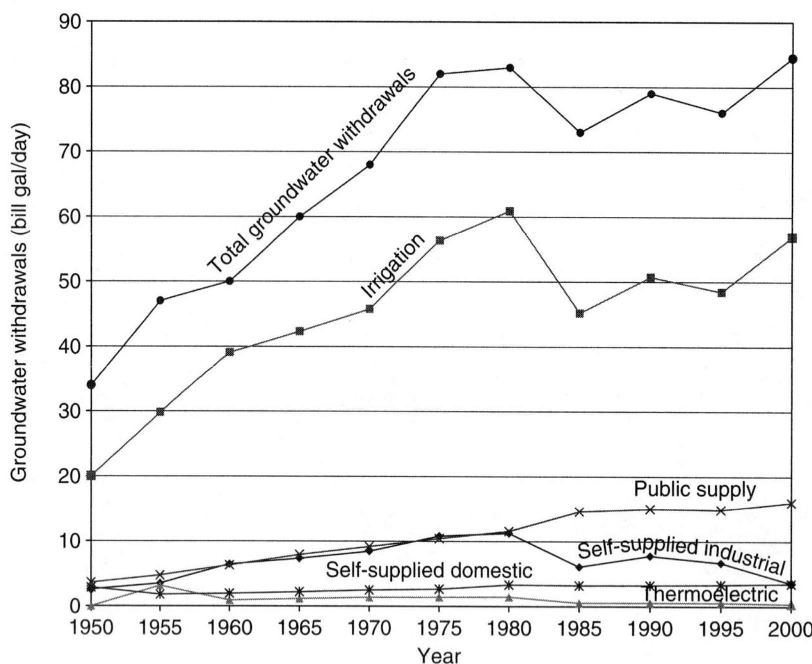

Figure 7A.4 Trends in groundwater use in the United States, 1950–2000. (Based on information from Dziegielewski, B. et al., 2002, Analysis of Water Use Trends in the United States: 1950–1995 University of Illinois at Urbana-Champaign, Illinois Water Resources Center, Special Report 28, February 2002, environ.uivc.edu/iwrc and Hutson, S.S. et al., 2004, Estimated use of water in the United States in 2000, *U.S. Geological Survey Circular 1268*, www.usgs.gov.)

Table 7A.2 Changes in Sectoral Withdrawals in the United States, 1950–1995

	Withdrawals (bgd)			Change (bgd)		Percent Change	
Sector	**1950**	**1980**	**1995**	**1980–1995**	**1980–1995**	**1950–1980**	**1980–1995**
Domestic	17.6	39.6	49.1	22.0	9.5	+125.0	+24.0
Irrigation	89.0	150.0	134.0	61.0	−16.0	+68.5	−10.7
Thermoelectric	40.0	210.0	190.0	170.0	−20.0	+425.0	−9.5
Industrial	37.0	45.0	29.1	8.0	−15.9	+21.6	−35.3
Total	183.6	444.6	402.2	261.0	−42.4	+142.2	−9.5

Note: This sectoral breakdown of the 1980–1995 change in total withdrawals shows a consistent decline in off-stream withdrawals for all sectors, with the exception of public supply and domestic use.

Source: From Dziegielewski, B. et al., 2002, *Analysis of Water Use Trends in the United States: 1950–1995*. Southern Illinois University at Carbondale, Carbondale, IL, February 28, 2002. Reprinted with permission, http://info.geography.siu.edu/geography_info/research/.

Table 7A.3 Percent of United States Population Relying on Groundwater as a Source of Drinking Water, 1995

Alabama	40
Alaska	42
Arizona	57
Arkansas	42
California	43
Colorado	14
Connecticut	41
Delaware	57
Florida	92
Georgia	28
Hawaii	96
Idaho	94
Illinois	24
Indiana	51
Iowa	71
Kansas	45
Kentucky	14
Louisiana	56
Maine	31
Maryland	16
Massachusetts	41
Michigan	25
Minnesota	72
Mississippi	91
Missouri	43
Montana	37
Nebraska	84
Nevada	26
New Hampshire	37
New Jersey	46
New Mexico	88
New York	27
North Carolina	24
North Dakota	44
Ohio	35
Oklahoma	26
Oregon	17
Pennsylvania	22
Rhode Island	17
South Carolina	26
South Dakota	63
Tennessee	37
Texas	42
Utah	55
Vermont	35
Virginia	12
Washington	52
West Virginia	21
Wisconsin	57
Wyoming	42
Total United States Including P.R. and V.I.	41

Source: Abstracted from Solley, W.B. et al., 1998, Estimated use of water in the United States in 1995, *U.S. Geological Survey Circular 1200*.

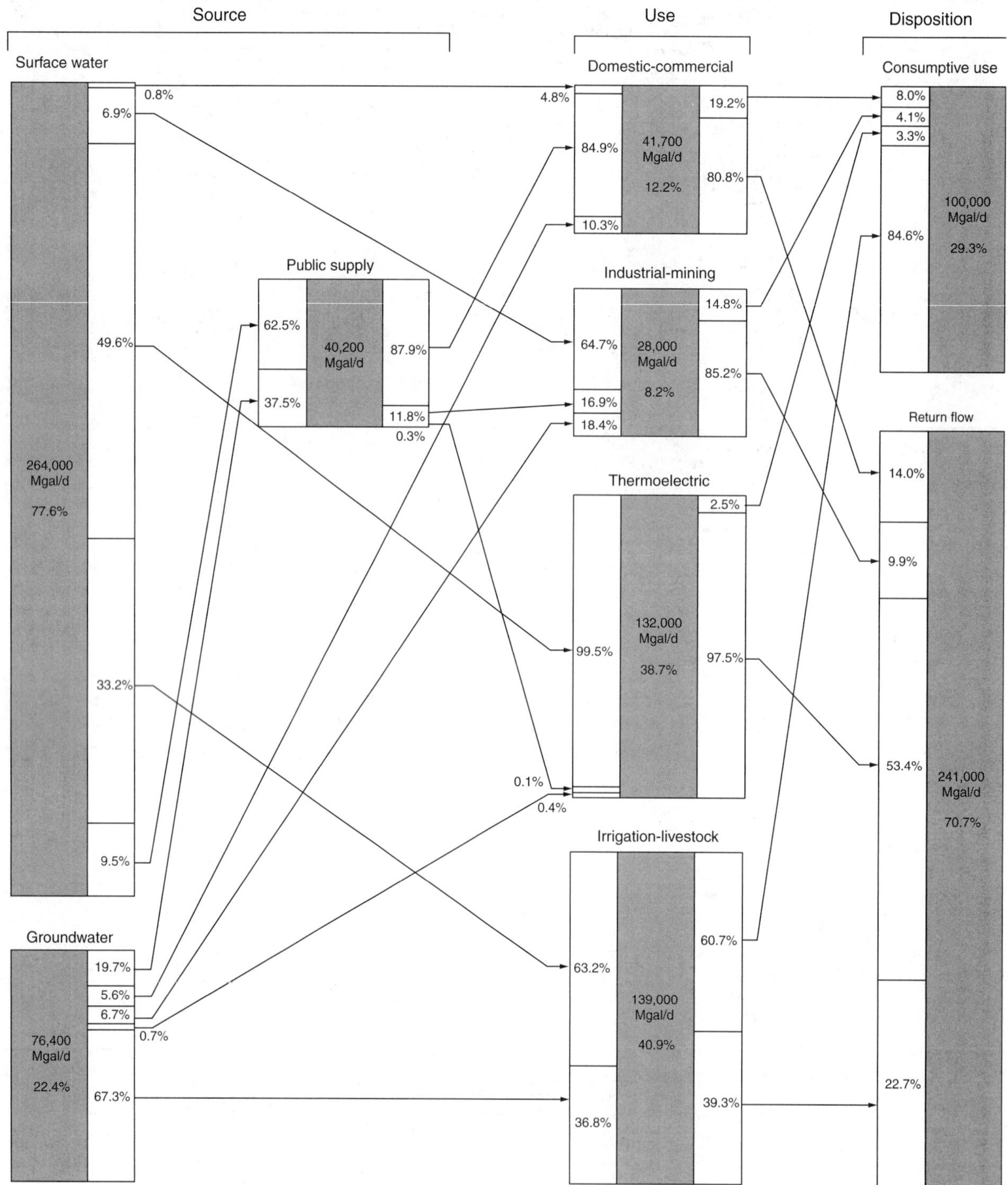

Figure 7A.5 Source, use, and disposition of freshwater in the United States, 1995. For each water-use category, this diagram shows the relative proportion of water source and disposition and the general distribtion of water from source to disposition. The lines and arrows indicate the distribution of water from source to disposition for each category; for example, surface water was 77.6 percent of total freshwater withdrawn, and going from "Source" to "Use" columns, the line from the surface-water block to the domestic and commercial block indicates that 0.8 percent of all surface water withdrawn was the source for 4.8 percent of total water (self-supplied withdrawals, public-supply deliveries) for domestic and commercial purposes. In addition, going from the "Use" to "Disposition" columns, the line from the domestic and commercial block to the consumptive use block indicates that 19.2 percent of the water for domestic and commercial purposes was consumptive use; this represents 8.0 percent of total consumptive use by all water-use categories. (From Solley, W.B. et al., 1998, Estimated use of water in the United States in 1995, *U.S. Geological Survey Circular 1200*, www.usgs.gov.)

Table 7A.4 Surface Water Withdrawals by Water-Use Category, 2000

State	Public Supply Fresh	Domestic Fresh	Irrigation Fresh	Livestock Fresh	Aquaculture Fresh	Industrial Fresh	Industrial Saline	Mining Fresh	Mining Saline	Thermoelectric Power Fresh	Thermoelectric Power Saline	Total Fresh	Total Saline	Total
Alabama	553	0	28.7	—	1.44	777	0	—	—	8,190	0	9,550	0	9,550
Alaska	50.7	0.25	0.02	—	—	3.8	3.86	27.4	49.5	28.9	0	111	53.4	164
Arizona	613	0	2,660	—	—	0	0	4.43	0	26.2	0	3,300	0	3,300
Arkansas	289	0	1,410	—	10.4	66.8	0	2.57	0	2,170	0	3,950	0	3,950
California	3,320	28.6	18,900	227	380	5.65	13.6	2.71	0.46	349	12,600	23,200	12,600	35,800
Colorado	846	0	9,260	—	—	96.4	0	—	—	122	0	10,300	0	10,300
Connecticut	358	0	13.4	—	—	6.61	0	—	—	186	3,440	565	3,440	4,010
Delaware	49.8	0	7.89	0.22	0	42.5	3.25	—	—	366	738	466	741	1,210
District of Columbia	0	0	0.18	—	—	0	0	—	—	9.69	0	9.87	0	9.87
Florida	237	0	2,110	1.51	0.21	74.7	1.18	57.8	0	629	12,000	3,110	12,000	15,100
Georgia	968	0	392	17.7	7.72	333	30	2.05	0	3,240	61.7	4,960	91.7	5,060
Hawaii	7.6	7.22	193	—	—	0	0	—	—	0	0	208	0	208
Idaho	25.3	0	13,300	7.2	1,920	19.7	0	—	—	0	0	15,300	0	15,300
Illinois	1,410	0	4.25	0	—	259	0	—	—	11,300	0	12,900	0	12,900
Indiana	326	0	45.4	14.6	—	2,300	0	78.3	0	6,700	0	9,460	0	9,460
Iowa	79.8	0	1.08	27.1	—	11.7	0	30.3	0	2,530	0	2,680	0	2,680
Kansas	244	0	288	23.5	2.27	6.74	0	17.4	0	2,240	0	2,820	0	2,820
Kentucky	455	8	28.2	—	—	222	0	—	—	3,250	0	3,970	0	3,970
Louisiana	404	0	232	3.31	115	2,400	0	—	—	5,580	0	8,730	0	8,730
Maine	72.5	0	5.23	—	—	237	0	—	—	108	295	423	295	718
Maryland	740	0	12.6	3.18	14.8	49.9	227	4.1	0.02	377	6,260	1,200	6,490	7,690
Massachusetts	542	0	106	—	—	26.2	0	—	—	108	3,610	783	3,610	4,390
Michigan	896	0	73.2	1.15	—	589	0	—	—	7,710	0	9,260	0	9,260
Minnesota	171	0	36.6	0	—	97.8	0	581	0	2,260	0	3,150	0	3,150
Mississippi	40.4	0	99.1	—	49.8	124	0	—	—	318	148	632	148	781
Missouri	594	0	48.1	54.1	81.3	33.5	0	12.8	0	5,620	0	6,450	0	6,450
Montana	92.4	1.29	7,870	—	—	29.3	0	—	—	110	0	8,100	0	8,100
Nebraska	63.8	0	1,370	17.4	—	2.6	0	122	0	2,810	0	4,390	0	4,390
Nevada	478	0	1,540	—	—	5	0	—	—	24.7	0	2,050	0	2,050
New Hampshire	64.1	0.16	4.25	—	13.1	37.9	0	6.72	0	235	761	362	761	1,120
New Jersey	650	0	117	0	0	66.2	0	104	0	648	3,390	1,590	3,390	4,980
New Mexico	33.8	0	1,630	—	—	1.67	0	—	—	45	0	1,710	0	1,710
New York	1,980	0	12.1	—	—	152	0	—	—	4,040	5,010	6,190	5,010	11,200
North Carolina	779	0	221	32.3	0	267	0	—	0	7,850	1,620	9,150	1,620	10,800
North Dakota	31.2	0	73.2	—	—	10.7	0	—	—	902	0	1,020	0	1,020
Ohio	966	2.71	17.8	17.1	0	645	0	35.5	0	8,590	0	10,300	0	10,300
Oklahoma	562	0	151	97.2	16.1	19.1	0	0.23	0	143	0	990	0	990
Oregon	447	7.97	5,290	—	—	183	0	—	—	12.8	0	5,940	0	5,940
Pennsylvania	1,250	0	12.5	—	—	1,030	0	20.9	0	6,970	0	9,290	0	9,290

(Continued)

Table 7A.4 (Continued)

State	Public Supply Fresh	Domestic Fresh	Irrigation Fresh	Livestock Fresh	Aquaculture Fresh	Industrial Fresh	Industrial Saline	Mining Fresh	Mining Saline	Thermoelectric Power Fresh	Thermoelectric Power Saline	Total Fresh	Total Saline	Total
Rhode Island	102	0	2.99	—	—	2.09	0	—	—	2.4	290	110	290	400
South Carolina	462	0	162		—	514	0	—	—	5,700	0	6,840	0	6,840
South Dakota	39.1	0.01	236	25.2	—	1.96	0	—	—	4.01	0	306	0	306
Tennessee	569	0	15.1		—	785	0	—	—	9,040	0	10,400	0	10,400
Texas	2,970	0	2,130	172		1,200	906	91.5	0	9,760	3,440	16,300	4,350	20,700
Utah	274	0	3,390	—	0	8.38	0	17.7	177	49.2	0	3,740	177	3,920
Vermont	40.6	0.25	3.45	—	—	4.86	0	—	—	355	0	404	0	404
Virginia	650	0	22.8	—	—	365	53.3	—	—	3,850	3,580	4,880	3,640	8,520
Washington	552	0.02	2,290	—	—	439	39.9	—	—	518	0	3,800	39.9	3,840
West Virginia	149	0.81	0.02	—	—	958	0	—	—	3,950	0	5,060	0	5,060
Wisconsin	293	0	1.57	6.02	30.4	364	0	—	—	6,090	0	6,780	0	6,780
Wyoming	49.4	0	4,090	—	—	1.47	0	20.7	0	242	0	4,400	0	4,400
Puerto Rico	425	0	57.5	—	—	0	0	—	—	0	2,190	483	2,190	2,670
U.S. Virgin Islands	5.57	1.69	0.21	—	—	3.12	0	—	—	0	136	10.6	136	147
Total	27,300	58.9	80,000	747	2,640	14,900	1,280	1,240	227	135,000	59,500	262,000	61,000	323,000

Note: Figures may not sum of totals because of independent rounding. All values are in million gallons per day. —, data not collected.

Source: From Hutson, S.S. et al., Estimated use of water in the United States in 2000, *U.S. Geological Survey Circular 1268*, 2004, http://water.usgs.gov/pubs/circ/2004/circ1268/.

Table 7A.5 Groundwater Withdrawals for Off-Stream Water-Use Categories in the United States, 2000

State	Public Supply Fresh	Domestic Fresh	Irrigation Fresh	Livestock Fresh	Aquaculture Fresh	Industrial Fresh	Industrial Saline	Mining Fresh	Mining Saline	Thermo-Electric Power Fresh	Total Fresh	Total Saline	Total
Alabama	281	78.9	14.5	—	8.93	56	0	—	—	0	440	0	440
Alaska	29.3	10.9	0.99	—	—	4.32	0	0.01	90.4	4.65	50.2	90.4	141
Arizona	469	28.9	2,750	—	—	19.8	0	81.2	8.17	74.3	3,420	8.17	3,430
Arkansas	132	28.5	6,510	—	187	67	0.08	0.21	0	2.92	6,920	0.08	6,920
California	2,800	257	11,600	182	158	183	0	21	152	3.23	15,200	152	15,400
Colorado	53.7	66.8	2,160	—	—	23.6	0	—	—	16.1	2,320	0	2,320
Connecticut	66	56.2	17	—	—	4.13	0	—	—	0.08	143	0	143
Delaware	45	13.3	35.6	3.7	0.07	17	0	—	—	0.47	115	0	115
District of Columbia	0	0	0	—	—	0	0	—	—	0	0	0	0
Florida	2,200	199	2,180	31	7.81	216	0	160	0	29.5	5,020	0	5,020
Georgia	278	110	750	1.66	7.7	290	0	7.75	0	1.03	1,450	0	1,450
Hawaii	243	4.82	171	—	—	14.5	0.85	—	—	0	433	0.85	434
Idaho	219	85.2	3,720	27.7	51.5	35.8	0	—	—	0	4,140	0	4,140
Illinois	353	135	150	37.6	—	132	0	—	—	5.75	813	0	813
Indiana	345	122	55.5	27.3	—	99.7	0	4.2	0	2.58	656	0	656
Iowa	303	33.2	20.4	81.8	—	226	0	2.49	0	11.9	679	0	679
Kansas	172	21.6	3,430	87.2	3.33	46.6	0	14	0	14.9	3,790	0	3,790
Kentucky	71	19.5	1.14	—	—	95.2	0	—	—	2.71	189	0	189
Louisiana	349	41.2	791	4.03	128	285	0	—	—	28.4	1,630	0	1,630
Maine	29.6	35.7	0.61	—	—	9.9	0	—	—	4.92	80.8	0	80.8
Maryland	84.6	77.1	29.8	7.18	4.81	15.9	0	4.21	0	1.8	225	0	225
Massachusetts	197	42.2	19.7	—	—	10.7	0	—	—	0	269	0	269
Michigan	247	239	128	10.2	—	110	0	—	—	0	734	0	734
Minnesota	329	80.8	190	52.8	—	56.3	0	6.9	0	4.17	720	0	720
Mississippi	319	69.3	1,310	—	321	118	0	—	—	43.5	2,180	0	2,180
Missouri	278	53.6	1,380	18.3	2.01	29.2	0	4.1	0	12.2	1,780	0	1,780
Montana	56.1	17.3	83	—	—	31.9	0	—	—	0	188	0	188
Nebraska	266	48.4	7,420	76	—	35.5	0	5.64	4.55	6.87	7,860	4.55	7,860
Nevada	151	22.4	567	—	—	5.29	0	—	—	12	757	0	757
New Hampshire	33	40.9	0.5	—	3.12	6.95	0	0.08	0	0.71	85.2	0	85.2
New Jersey	400	79.7	22.8	1.68	6.46	65.3	0	6.12	0	2.24	584	0	584
New Mexico	262	31.4	1,230	—	—	8.8	0	—	—	11.4	1,540	0	1,540
New York	583	142	23.3	—	—	145	0	—	—	0	893	0	893
North Carolina	166	189	65.8	89.1	7.88	25.6	0	36.4	0	0.09	580	0	580
North Dakota	32.4	11.9	72.2	—	—	6.88	0	—	—	0	123	0	123
Ohio	500	132	13.9	8.2	1.36	162	0	53.1	0	7.57	878	0	878
Oklahoma	113	25.5	566	53.6	0.29	6.83	0	2.25	256	3.27	771	256	1,030
Oregon	118	68.3	792	—	—	12.1	0	—	—	2.47	993	0	993
Pennsylvania	212	132	1.38	—	—	155	0	162	0	3.98	666	0	666

(Continued)

Table 7A.5 (Continued)

State	Public Supply Fresh	Domestic Fresh	Irrigation Fresh	Livestock Fresh	Aquaculture Fresh	Industrial Fresh	Industrial Saline	Mining Fresh	Mining Saline	Thermo-Electric Power Fresh	Total Fresh	Total Saline	Total
Rhode Island	16.9	8.99	0.46	—	—	2.19	0	—	—	0	28.6	0	28.6
South Carolina	105	63.5	106	—	—	50.9	0	—	—	5.83	330	0	330
South Dakota	54.2	9.52	137	16.9	—	3.16	0	—	—	1.23	222	0	222
Tennessee	321	32.6	7.33	—	—	56.3	0	—	—	0	417	0	417
Texas	1,260	131	6,500	137	—	244	0.5	129	504	60.2	8,470	504	8,970
Utah	364	16.1	469	—	116	34.3	5.08	8.6	21.5	13.1	1,020	26.5	1,050
Vermont	19.5	20.7	0.33	—	—	2.05	0	—	—	0.66	43.2	0	43.2
Virginia	70.7	133	3.57	—	—	104	0	—	—	1.5	314	0	314
Washington	464	125	747	—	—	138	0	—	—	0.92	1,470	0	1,470
West Virginia	41.6	39.6	0.02	—	—	9.7	0	—	—	0	90.9	0	90.9
Wisconsin	330	96.3	195	60.3	39.8	83	0	—	—	8.99	813	0	813
Wyoming	57.2	6.57	413	—	—	4.31	0	58.8	222	1.13	541	222	763
Puerto Rico	88.5	0.88	36.9	—	—	11.2	0	—	—	0	137	0	137
U.S. Virgin Islands	0.52	0	0.29	—	—	0.22	0	—	—	0	1.03	0	1.03
Total	16,000	3,530	56,900	1,010	1,060	3,570	6.51	767	1,260	409	83,300	1,260	84,500

Note: Figures may not sum to totals because of independent rounding. All values are in million gallons per day. —, data not collected.

Source: From Hutson, S.S. et al., 2004, Estimated use of water in the United States in 2000, *U.S. Geological Survey Circular 1268*, www.usgs.gov.

Table 7A.6 Total Off-Stream Water Withdrawals by Source and State in the United States, 2000

State	Population (thousands)	Withdrawals (mil gal/day)									Withdrawals (thousand acre-feet/yr)		
		By Source and Type											
		Groundwater			Surface Water			Total			Total		
		Fresh	Saline	Total	Fresh	Saline	Total	Fresh	Saline	Total	Fresh	Saline	Total
Alabama	4,450	440	0	440	9,550	0	9,550	9,990	0	9,990	11,200	0	11,200
Alaska	627	50.2	90.4	141	111	53.4	164	161	144	305	181	161	342
Arizona	5,130	3,420	8.17	3,430	3,300	0	3,300	6,720	8.17	6,730	7,530	9.16	7,540
Arkansas	2,670	6,920	0.08	6,920	3,950	0	3,950	10,900	0.08	10,900	12,200	0.09	12,200
California	33,900	15,200	152	15,400	23,200	12,600	35,800	38,400	12,800	51,200	43,100	14,300	57,400
Colorado	4,300	2,320	0	2,320	10,300	0	10,300	12,600	0	12,600	14,200	0	14,200
Connecticut	3,410	143	0	143	565	3,440	4,010	708	3,440	4,150	794	3,860	4,650
Delaware	784	115	0	115	466	741	1,210	582	741	1,320	652	831	1,480
District of Columbia	572	0	0	0	9.87	0	9.87	9.87	0	9.87	11.1	0	11.1
Florida	16,000	5,020	0	5,020	3,110	12,000	15,100	8,140	12,000	20,100	9,120	13,400	22,500
Georgia	8,190	1,450	0	1,450	4,960	91.7	5,060	6,410	91.7	6,500	7,190	103	7,290
Hawaii	1,210	433	0.85	434	208	0	208	640	0.85	641	718	0.95	719
Idaho	1,290	4,140	0	4,140	15,300	0	15,300	19,500	0	19,500	21,800	0	21,800
Illinois	12,400	813	0	813	12,900	0	12,900	13,700	0	13,700	15,400	0	15,400
Indiana	6,080	656	0	656	9,460	0	9,460	10,100	0	10,100	11,300	0	11,300
Iowa	2,930	679	0	679	2,680	0	2,680	3,360	0	3,360	3,770	0	3,770
Kansas	2,690	3,790	0	3,790	2,820	0	2,820	6,610	0	6,610	7,410	0	7,410
Kentucky	4,040	189	0	189	3,970	0	3,970	4,160	0	4,160	4,660	0	4,660
Louisiana	4,470	1,630	0	1,630	8,730	0	8,730	10,400	0	10,400	11,600	0	11,600
Maine	1,270	80.8	0	80.8	423	295	718	504	295	799	565	330	895
Maryland	5,300	225	0	225	1,200	6,490	7,690	1,430	6,490	7,910	1,600	7,270	8,870
Massachusetts	6,350	269	0	269	783	3,610	4,390	1,050	3,610	4,660	1,180	4,050	5,220
Michigan	9,940	734	0	734	9,260	0	9,260	10,000	0	10,000	11,200	0	11,200
Minnesota	4,920	720	0	720	3,150	0	3,150	3,870	0	3,870	4,340	0	4,340
Mississippi	2,840	2,180	0	2,180	632	148	781	2,810	148	2,960	3,150	166	3,320
Missouri	5,600	1,780	0	1,780	6,450	0	6,450	8,230	0	8,230	9,220	0	9,220
Montana	902	188	0	188	8,100	0	8,100	8,290	0	8,290	9,300	0	9,300
Nebraska	1,710	7,860	4.55	7,860	4,390	0	4,390	12,200	4.55	12,300	13,700	5.1	13,700
Nevada	2,000	757	0	757	2,050	0	2,050	2,810	0	2,810	3,140	0	3,140
New Hampshire	1,240	85.2	0	85.2	362	761	1,120	447	761	1,210	501	854	1,350
New Jersey	8,410	584	0	584	1,590	3,390	4,980	2,170	3,390	5,560	2,430	3,800	6,230
New Mexico	1,820	1,540	0	1,540	1,710	0	1,710	3,260	0	3,260	3,650	0	3,650
New York	19,000	893	0	893	6,190	5,010	11,200	7,080	5,010	12,100	7,940	5,610	13,600
North Carolina	8,050	580	0	580	9,150	1,620	10,800	9,730	1,620	11,400	10,900	1,810	12,700
North Dakota	642	123	0	123	1,020	0	1,020	1,140	0	1,140	1,280	0	1,280
Ohio	11,400	878	0	878	10,300	0	10,300	11,100	0	11,100	12,500	0	12,500

(Continued)

Table 7A.6 (Continued)

		Withdrawals (mil gal/day)									Withdrawals (thousand acre-feet/yr)		
		By Source and Type											
		Groundwater			Surface Water			Total			Total		
State	Population (thousands)	Fresh	Saline	Total	Fresh	Saline	Total	Fresh	Saline	Total	Fresh	Saline	Total
Oklahoma	3,450	771	256	1,030	990	0	990	1,760	256	2,020	1,970	287	2,260
Oregon	3,420	993	0	993	5,940	0	5,940	6,930	0	6,930	7,770	0	7,770
Pennsylvania	12,300	666	0	666	9,290	0	9,290	9,950	0	9,950	11,200	0	11,200
Rhode Island	1,050	28.6	0	28.6	110	290	400	138	290	429	155	326	481
South Carolina	4,010	330	0	330	6,840	0	6,840	7,170	0	7,170	8,040	0	8,040
South Dakota	755	222	0	222	306	0	306	528	0	528	592	0	592
Tennessee	5,690	417	0	417	10,400	0	10,400	10,800	0	10,800	12,100	0	12,100
Texas	20,900	8,470	504	8,970	16,300	4,350	20,700	24,800	4,850	29,600	27,800	5,440	33,200
Utah	2,230	1,020	26.5	1,050	3,740	177	3,920	4,760	203	4,970	5,340	228	5,570
Vermont	609	43.2	0	43.2	404	0	404	447	0	447	501	0	501
Virginia	7,080	314	0	314	4,880	3,640	8,520	5,200	3,640	8,830	5,830	4,080	9,900
Washington	5,890	1,470	0	1,470	3,800	39.9	3,840	5,270	39.9	5,310	5,910	44.7	5,960
West Virginia	1,810	90.9	0	90.9	5,060	0	5,060	5,150	0	5,150	5,770	0	5,770
Wisconsin	5,360	813	0	813	6,780	0	6,780	7,590	0	7,590	8,510	0	8,510
Wyoming	494	541	222	763	4,400	0	4,400	4,940	222	5,170	5,540	248	5,790
Puerto Rico	3,810	137	0	137	483	2,190	2,670	620	2,190	2,810	695	2,460	3,150
U.S. Virgin Islands	109	1.03	0	1.03	10.6	136	147	11.6	136	148	13	153	166
Total	285,000	83,300	1,260	34,500	262,000	61,000	323,000	345,000	62,300	408,000	387,000	69,800	457,000

Note: Figures may not sum to totals because of independent rounding.

Source: From Hutson, S.S. et al., 2004, Estimated use of water in the United States in 2000, *U.S. Geological Survey Circular 1268*, usgs.gov/pubs/circ/2004/circ1268.

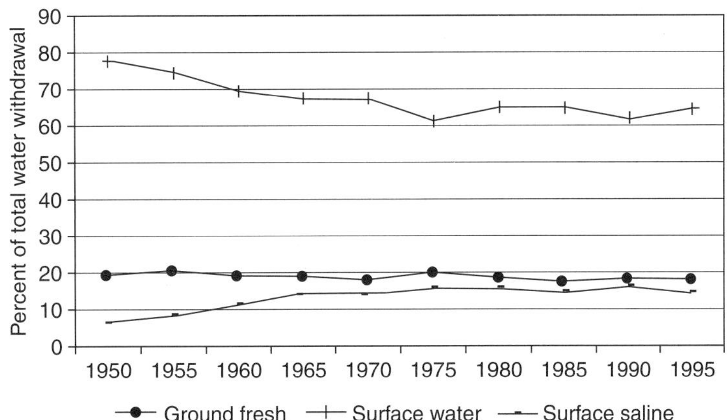

Figure 7A.6 Total off-stream withdrawals by source in the United States, 1950–1995. (From Dziegielewski, B. et al., 2002, *Analysis of Water Use Trends in the United States: 1950–1995*, Southern Illinois University at Carbondale, Carbondale, IL, February 28, 2002. Reprinted with permission. http://info.geography.siu.edu/geography_info/research/.)

Table 7A.7 Total and Percent of Withdrawals by Source in the United States, 1950–1995

Year	Total Withdrawals (bgd)	Groundwater Fresh (bgd)	Surface Water Fresh (bgd)	Surface Water Saline (bgd)	Groundwater Fresh (%)	Surface Water Fresh (%)	Surface Water Saline (%)
1950	180	34	140	10	19	78	6
1955	240	47	180	18	20	75	8
1960	270	50	190	31	19	70	11
1965	310	60	210	43	19	68	14
1970	370	68	250	53	18	68	14
1975	420	82	260	69	20	62	16
1980	440	83	290	71	19	66	16
1985	399	73	265	60	18	66	15
1990	408	79	259	68	19	63	17
1995	402	76	264	60	19	66	15

Source: From Dziegielewski, B. et al., 2002, *Analysis of Water Use Trends in the United States: 1950–1995*, Southern Illinois University at Carbondale, Carbondale, IL, February 28, 2002. Reprinted with Permission. http://info.geography.siu.edu/geography_info/research/.

Table 7A.8 Common Uses of Water in Relation to Consumptive and Nonconsumptive Uses

Common Uses of Water	Consumptive	Nonconsumptive
Steam generation (locomotive or stationary)	Steam released to atmosphere	[a]
Air conditioning		
Evaporative	Cooling achieved by evaporation	[a]
Recirculating	Some water evaporated with each use	[a]
Other cooling (recirculating)	Some water evaporated with each use	[a]
Storage in surface reservoirs	Evaporation from water surface	Seepage underground
Irrigation by sprinkling	Evaporation and transpiration	Little seepage
Irrigation by flooding	Evaporation from ponds, transpiration	Seepage varies
Cooking	Steam to atmosphere	Contributes to sewage
Processing foods, beverages, plastics	Some water goes into manufactured products	Carries organic compounds
Processing		
Petroleum products	Proportion of consumptive use is increased by reuse of the nonconsumptive water	Carries chemicals
Paper and pulp	Proportion of consumptive use is increased by reuse of the nonconsumptive water	Carries pulp and chemicals
Chemicals	Proportion of consumptive use is increased by reuse of the nonconsumptive water	Carries toxic or other chemicals
Metal products	Proportion of consumptive use is increased by reuse of the nonconsumptive water	Carries sludge and soluble chemicals
Atomic fission	Evaporation from tanks	Carries radioactive materials
Stock watering	Evaporation from tanks and ponds	Organic wastes into ground
Drinking	Perspiration	Organic wastes into sewage
Irrigation by furrow	Evapotranspiration	Dissolves chemicals from soil
Washing	Evaporation in drying	Carries sediment and soluble matter
Mining (metals, coal, oil)		Carries natural brines and acids, sediments
Cooling (once-through)	[b]	Water temperature increased by use
Air conditioning (once-through)	[b]	Water temperature increased by use
Fish culture	[c]	
Steam heating	[d]	Steam condenses and is reused
Year-round heat exchange	[d]	Requires storage of water and heat from one season to another
Sanitation (bath, toilet, dishwasher)	[d]	Sewage carries chiefly organic wastes
Hydroelectric power	[e]	Takes water toward oceans
Navigation	[e]	Inland waterways require maintenance of flow

[a] With efficient operation, nonconsumptive use is limited to that required for cooling and/or cleaning equipment.
[b] Increase in water temperature may cause increased evaporation.
[c] Consumptive use by evaporation from water surfaces; may be increased by aerators.
[d] Consumptive use is limited to losses through leaking pipes, valves, etc.
[e] Consumptive use is limited to evaporation from lakes, reservoirs, etc. that are required for continuous operation.

Source: From House of Representatives, U.S. Congress.

Table 7A.9 United States Water Withdrawals and Consumptive Use Per Day by End Use, 1940–1995

Year	Total (bil gal)	Per Capita[b] (gal)	Irrigation (bil gal)	Public supply[a] Total (bil gal)	Public supply[a] Per Capita[c] (gal)	Rural[d] (bil gal)	Industrial and Misc[e] (bil gal)	Steam Electric Utilities (bil gal)
Withdrawals								
1940	140	1,027	71	10	75	3.1	29	23
1950	180	1,185	89	14	145	3.6	37	40
1955	240	1,454	110	17	148	3.6	39	72
1960	270	1,500	110	21	151	3.6	38	100
1965	310	1,602	120	24	155	4.0	46	130
1970	370	1,815	130	27	166	4.5	47	170
1975	420	1,972	140	29	168	4.9	45	200
1980	440	1,953	150	34	183	5.6	45	210
1985	399	1,650	137	38	189	7.8	31	187
1990	408	1,620	137	41	195	7.9	30	195
1995	402	1,500	134	43	192	8.9	26	190
Consumptive Use								
1960	61	339	52	3.5	25	2.8	3.0	0.2
1965	77	403	66	5.2	34	3.2	3.4	0.4
1970	87	427	73	5.9	36	3.4	4.1	0.8
1975	96	451	80	6.7	38	3.4	4.2	1.9
1980	100	440	83	7.1	38	3.9	5.0	3.2
1985	92	380	74	[f]	[f]	9.2	6.1	6.2
1990	94	370	76	[f]	[f]	8.9	6.7	4.0
1995	100	374	81	[f]	[f]	9.9	4.8	3.7

[a] Includes commercial water withdrawals.
[b] Based on U.S. Census Bureau resident population as of July 1.
[c] Based on population served.
[d] Rural farm and nonfarm household and garden use, and water for farm stock and dairies.
[e] For 1940 to 1960, includes manufacturing and mineral industries, rural commercial industries, air-conditioning, resorts, hotels, motels, military and other state and Federal agencies, and miscellaneous; thereafter, includes manufacturing, mining and mineral processing, ordnance, construction, and miscellaneous.
[f] Public supply consumptive use included in end-use categories.

Source: From U.S. Census Bureau, Statistical Abstracts of the United States: 2000, www.census.gov.

Original Source: From 1940–1960, U.S. Bureau of Domestic Business Development, based principally on committee prints, Water Resources Activities in the United States, for the Senate Committee on National Water Resources, U.S. Senate, thereafter, U.S. Geological Survey, Estimated Use of Water in the United States in 1995, circular 1200, and previous quinquennial issues.

SECTION 7B WATER USE — WORLD

Table 7B.10 Major Withdrawal Uses of Water in Canada, 1981, 1986, 1991, and 1996

Sector/industry	Year	Total Intake[a] Quantity (mill m^3)	Total Intake[a] Change from Previous Period (Percent)	Recirculation[b] Quantity (mill m^3)	Recirculation[b] Change from Previous Period (Percent)	Gross Water Use[c] Quantity (mill m^3)	Gross Water Use[c] Change from Previous Period (Percent)	Total Discharge[d] Quantity (mill m^3)	Total Discharge[d] Change from Previous Period (Percent)	Consumption[e] Quantity (mill m^3)	Consumption[e] Change from Previous Period (Percent)
Business sector											
Primary resource industries											
Agriculture	1981	3,125	—	0	—	3,125	—	713	—	2,412	—
	1986	3,559	13.9	0	—	3,559	13.9	807	13.2	2,752	14.1
	1991	3,991	12.1	0	—	3,991	12.1	902	11.8	3,089	12.2
	1996	4,098	2.7	0	—	4,098	2.7	1,062	17.7	3,036	−1.7
Mining	1981	624	—	1,742	—	2,366	—	621	—	3	—
	1986	544	−12.8	1,159	−33.5	1,703	−28.0	542	−12.7	2	−33.3
	1991	489	−10.1	1,221	5.3	1,710	0.4	489	−9.8	1	−50.0
	1996	681	39.3	1,196	−2.0	1,878	9.8	672	37.4	9	800.0
Other primary resource industries	1981	251	—	1,050	—	1,302	—	188	—	63	—
	1986	180	−28.3	873	−16.9	1,054	−19.0	118	−37.2	62	−1.6
	1991	183	1.7	735	−15.8	918	−12.9	111	−5.9	71	14.5
	1996	231	26.2	1,013	37.8	1,244	35.5	138	24.3	92	29.6
Manufacturing industries											
Paper and allied products	1981	3,170	—	4,612	—	7,782	—	2,989	—	181	—
	1986	3,082	−2.8	3,121	−32.3	6,203	−20.3	2,876	−3.8	206	13.8
	1991	2,943	−4.5	2,206	−29.3	5,149	−17.0	2,758	−4.1	185	−10.2
	1996	2,505	−14.9	3,141	42.4	5,646	9.7	2,277	−17.4	228	23.2
Primary metal	1981	2,074	—	1,325	—	3,399	—	2,003	—	71	—
	1986	2,057	−0.8	1,945	46.8	4,002	17.7	2,014	0.5	43	−39.4
	1991	1,610	−21.7	1,689	−13.2	3,298	−17.6	1,518	−24.6	92	114.0
	1996	1,428	−11.3	1,416	−16.2	2,845	−13.7	1,308	−13.8	120	30.4
Chemical and chemical products	1981	3,188	—	1,285	—	4,473	—	2,963	—	225	—
	1986	1,694	−46.9	1,494	16.3	3,189	−28.7	1,630	−45.0	64	−71.6
	1991	1,326	−21.7	979	−34.5	2,305	−27.7	1,231	−24.5	95	48.4
	1996	1,182	−10.9	1,357	38.6	2,539	10.2	1,083	−12.0	99	4.2
Other manufacturing industries	1981	1,721	—	2,286	—	4,007	—	1,588	—	133	—
	1986	1,548	−10.1	1,880	−17.8	3,427	−14.5	1,422	−10.5	126	−5.3
	1991	1,532	−1.0	1,808	−3.8	3, 340	−2.5	1,357	−4.6	175	38.9
	1996	1,282	−16.3	1,067	−41.0	2,349	−29.7	1,131	−16.7	151	−13.7
Electric power and other utilities	1981	18,166	—	1,868	—	20,034	—	18,084	—	82	—
	1986	24,963	37.4	3,776	102.1	28,740	43.5	24,702	36.6	261	218.3
	1991	28,288	13.3	3,374	−10.6	31,662	10.2	28,183	14.1	105	−59.8
	1996	28,664	1.3	11,617	244.3	40,281	27.2	28,183	0.0	481	358.1

(Continued)

Table 7B.10 (Continued)

Sector/industry	Year	Total Intake[a]		Recirculation[b]		Gross Water Use[c]		Total Discharge[d]		Consumption[e]	
		Quantity (mill m^3)	Change from Previous Period (Percent)	Quantity (mill m^3)	Change from Previous Period (Percent)	Quantity (mill m^3)	Change from Previous Period (Percent)	Quantity (mill m^3)	Change from Previous Period (Percent)	Quantity (mill m^3)	Change from Previous Period (Percent)
Other industries	1981	638	—	0	—	638	—	575	—	63	—
	1986	736	15.4	0	—	736	15.4	660	14.8	76	20.6
	1991	816	10.9	0	—	816	10.9	737	11.7	79	3.9
	1996	880	7.8	0	—	880	7.8	796	8.0	84	6.3
Subtotal, business sector	1981	32,957	—	14,168	—	47,126	—	29,724	—	3,233	—
	1986	38,363	16.4	14,248	0.6	52,613	11.6	34,771	17.0	3,592	11.1
	1991	41,178	7.3	12,012	−15.7	53,189	1.1	37,286	7.2	3,892	8.4
	1996	40,951	−0.6	20,807	73.2	61,760	16.1	36,650	−1.7	4,300	10.5
Personal and government sectors	1981	3,760	—	0	—	3,760	—	3,363	—	397	—
	1986	3,719	−1.1	0	—	3,719	−1.1	3,338	−0.7	381	−4.0
	1991	3,802	2.2	0	—	3,802	2.2	3,374	1.1	428	12.4
	1996	3,922	3.2	0	—	3,922	3.2	3,482	3.2	440	2.8
Total, whole economy	1981	36,717	—	14,169	—	50,886	—	33,087	—	3,630	—
	1986	42,083	14.6	14,248	0.6	56,330	10.7	38,109	15.2	3,973	9.4
	1991	44,979	6.9	12,012	−15.7	56,991	1.2	40,659	6.7	4,320	8.7
	1996	44,873	−0.2	20,807	73.2	65,682	15.2	40,132	−1.3	4,740	9.7

Note: Figures may not add up to totals due to rounding.

[a] The quantity of water withdrawn from a water source.
[b] The amount of water used more than once in an industrial application.
[c] Gross water use equals total water intake plus recirculation.
[d] The quantity of water returned to the water source.
[e] Consumption is that part of water intake that is evaporated, incorporated into products or crops, consumed by humans or livestock, or otherwise removed from the local hydrologic environment.

Source: From Statistics Canada, "Human Activity and the Environment", Annual Statistics 2003, Catalogue 16-201-XPE released December 3, 2003, page 12. Statistics Canada information is used with the permission of Statistics Canada. Users are forbidden to copy this material and/or redisseminate the data, in an original or modified form, for commercial purposes, without the expressed permission of Statistics Canada. Information on the availability of the wide range of data from Statistics Canada can be obtained from Statistics Canada's Regional Offices, its World Wide Web site at www.statcan.ca, and its toll-free access number 1-800-263-1136. With permission.

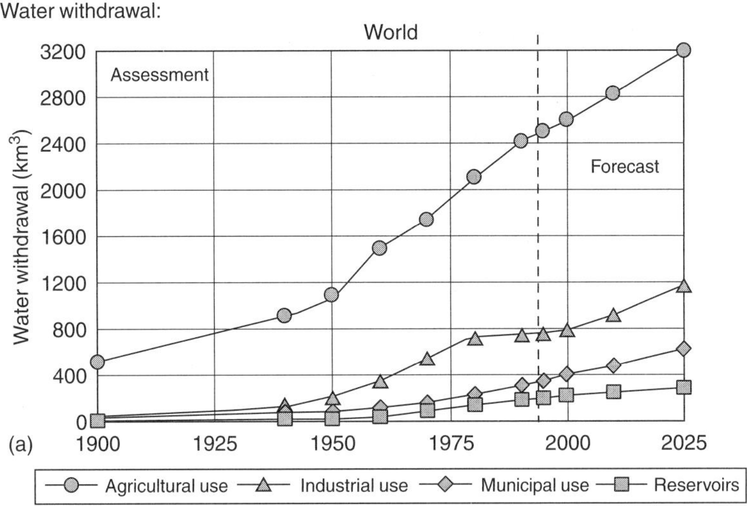

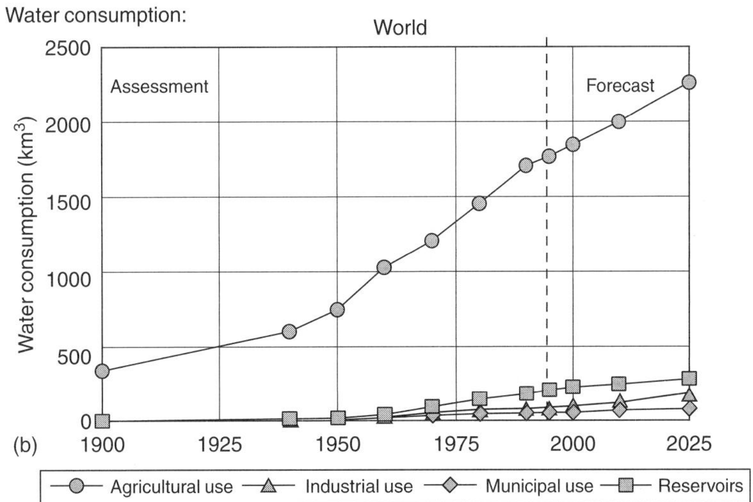

Figure 7B.7 Dynamics of water use in the world by kind of economic activity. (From Shiklomanov, I.A., 1999, Summary of the Monograph "World Water Resources at the Beginning of the 21st Century" Prepared in the Framework of IHP UNESCO, International Hydrological Programme, UNESCO's Intergovernmental Scientific Programme in Water Resources, World Water Resources and Their Use a Joint State Hydrological Institute (SHI)/UNESCO Product, http://webworld.unesco.org/water/ihp/db/shiklomanov/. Copyright © UNESCO 1999. Reproduced by permission of UNESCO.)

Table 7B.11 World Wide Freshwater Resources Availability and Use

	Renewable Water Resources (annual)[a]						Water Withdrawals (annual)							
	Internal Renewable Water Resources (IRWR)				Natural Renewable Water Resources[b]						Sectoral Share (Percent)[c]			
	Groundwater Recharge (km³)[d]	Surface Water (km³)[d]	Overlap (km³)	Total[e] (km³)	Total (km³)	Per Capita (m² person)[f]	Year	Total (mill m³)	Per Capita (m³ person)	as a % of Renewable Water Resources	Agriculture	Domestic	Industry	Desalinated Water Production (mill m³)g[g]
World	**11,358**	**40,594**	**10,067**	**43,219**	—	—	**1990**	**3,414,000**	**650**	—	**71**	**9**	**20**	—
Asia (Excl. Middle East)	**2,472**	**10,985**	**2,136**	**11,321**	—	—	—	—	—	—	—	—	—	—
Armenia	4.2	6.3	1.4	9.1	11	2,778	1994	2,925	784	28	66	30	4	0
Azerbaijan	6.5	6.0	4.4	8.1	30	3,716	1995	16,533	2,151	58	70	5	25	0
Bangladesh	21	84	0	105	1,211	8,444	1990	14,636	133	2	86	12	2	0
Bhutan	—	95	—	95	95	43,214	1987	20	13	0	54	36	10	0
Cambodia	18	116	13	121	476	34,561	1987	520	60	0	94	5	1	0
China	829	2,712	728	2,812	2,830	2,186	1993	525,489	439	20	78	5	18	0
Georgia	17	57	16	58	63	12,149	1990	3,468	635	5	59	21	20	0
India	419	1,222	380	1,261	1,897[h]	1,822[h]	1990	500,000	592	32	92	5	3	0
Indonesia	455	2,793	410	2,838	2,838	13,046	1990	74,346	407	3	93	6	1	0
Japan	27	420	17	430	430	3,372	1992	91,400	735	22	64	19	17	0
Kazakhstan	6.1	69	0	75	110	6,839	1993	33,674	2,010	29	81	2	17	1,328
Korea, Dem People's Rep	13	66	12	67	77	3,415	1987	14,160	742	22	73	11	16	0
Korea, Rep	13	62	11	65	70	1,471	1994	23,668	531	36	63	26	11	0
Kyrgyzstan	14	44	11	46	21[h]	4,078[h]	1994	10,086	2,231	55	94	3	3	0
Lao People's Dem Rep	38	190	38	190	334	60,318	1987	990	259	0	82	8	10	0
Malaysia	64	566	50	580	580	25,178	1995	12,733	636	3	77	11	13	0
Mongolia	6.1	33	4.0	35	35	13,451	1993	428	182	1	53	20	27	0
Myanmar	156	875	150	881	1,046	21,358	1987	3,960	103	0	90	7	3	0
Nepal	20	198	20	198	210	8,703	1994	28,953	1,451	17	99	1	0	0
Pakistan	55	47	50	52	223[h]	2,812[h]	1991	155,600	1,382	100	97	2	2	0
Philippines	180	444	145	479	479	6,093	1995	55,422	811	13	88	8	4	0
Singapore	—	—	—	—	—	—	1975	—	—	—	4	45	51	—
Sri Lanka	7.8	49	7.0	50	50	2,592	1990	9,770	574	22	96	2	2	0
Tajikistan	6.0	63	3.0	66	16[h]	2,587[h]	1994	11,874	2,096	81	92	3	4	0
Thailand	42	199	31	210	410	6,371	1990	33,132	605	10	91	5	4	0
Turkmenistan	0.4	1.0	0	1.4	25[h]	5,015[h]	1994	23,779	5,801	116	98	1	1	0
Uzbekistan	8.8	9.5	2	16	50[h]	1,968[h]	1994	58, 051	2,598	132	94	4	2	0

Vietnam	48	354	35	367	891	11,109	1990	54,330	822	7	87	4	10	0
Europe	**1,318**	**6,223**	**986**	**6,590**	—	—	—	—	—	—	—	—	—	—
Albania	6.2	23	2.4	27	42	13,178	1995	1,400	440	3	71	29	0	—
Austria	6.0	55	6.0	55	78	9,629	1991	2,360	303	3	9	33	58	—
Belarus	18	37	18	37	58	5,739	1990	2,734	266	5	35	22	43	0
Belgium	0.9	12	0.9	12	18	1,781	—	—	—	—	—	—	—	—
Bosnia and Herzegovina	—	—	—	36	38	9,088	1995	1,000	292	3	60	30	10	—
Bulgaria	6.4	20	5.5	21	21	2,734	1988	13,900	1,573	58	22	3	75	—
Croatia	11	27	0.5	38	105	22,654	1996	764	164	1	0	50	50	—
Czech Rep	1.4	13	1.4	13	13	1,283	1991	2,740	266	21	2	41	57	—
Denmark	4.3	3.7	2.0	6.0	6	1,123.0	1990	1,200	233	21	43	30	27	—
Estonia	4.0	12	3.0	13	13	9,413	1995	158	106	1	5	56	39	0
Finland	2.2	107	2.0	107	110	21,223	1991	2,200	439	2	3	12	85	—
France	100	177	98	179	204	3,414	1999	32,300	547	16	10	18	72	—
Germany	46	106	45	107	154	1,878	1991	46,270	579	31	20	11	69	—
Greece	10	56	7.8	58	74	6,984	1997	8,700	826	12	87	10	3	—
Hungary	6.0	6.0	6.0	6.0	104	10,541	1991	6,810	659	6	36	9	55	—
Iceland	24	166	20	170	170	599,944	1991	160	622	0	6	31	63	—
Ireland	11	48	10	49	52	13,408	1980	790	232	2	10	16	74	—
Italy	43	171	31	183	191	3,330	1998	42,000	730	22	48	19	34	—
Latvia	2.2	17	2.0	17	35	14,820	1994	285	112	1	13	55	32	0
Lithuania	1.2	15	1.0	16	25	6,763	1995	254	68	1	3	81	16	0
Macedonia, FYR	—	5.4	—	5.4	6	3,120.6	1996	1,850	936	30	74	12	15	—
Moldova, Rep	0.4	1.0	0.4	1.0	12	2,726	1992	2,963	678	25	26	9	65	0
Netherlands	4.5	11	4.5	11	91	5,691	1991	7,810	519	9	34	5	61	—
Norway	96	376	90	382	382	84,787	1985	2,030	489	1	8	20	72	—
Poland	13	53	12	54	62	1,598	1991	12,280	321	20	11	13	76	—
Portugal	4.0	38	4.0	38	69[h]	6,837[h]	1990	7,290	736	11	48	15	37	—
Romania	8.3	42	8.0	42	212	9,486	1994	26,000	1,141	12	59	8	33	—
Russian Federation	788	4,037[i]	512	4,313[i]	4,507[i]	31,354[i]	1994	77,100	519	2	20	19	62	0
Serbia and Montenegro	3.0	42	1.4	44	209	19,815	1995	13,000	1,233	6	8	6	86	—
Slovakia	1.7	13	1.7	13	50	9,265	1991	1,780	337	4	—	—	—	—
Slovenia	14	19	13	19	32	16,070	1996	1,280	642	4	1	20	80	—
Spain	30	110	28	111	112	2,793	1997	35,210	884	32	68	13	19	—
Sweden	20	170	19	171	174	19,721	1991	2,930	340	2	9	36	55	—

(Continued)

Table 7B.11 (Continued)

	Renewable Water Resources (annual)[a]						Water Withdrawals (annual)							
	Internal Renewable Water Resources (IRWR)				Natural Renewable Water Resources[b]						Sectoral Share (Percent)[c]			
	Groundwater Recharge (km^3)[d]	Surface Water (km^3)[d]	Overlap (km^3)	Total[e] (km^3)	Total (km^3)	Per Capita (m^2 person)[f]	Year	Total (mill m^3)	Per Capita (m^3 person)	as a % of Renewable Water Resources	Agriculture	Domestic	Industry	Desalinated Water Production (mill m^3)g[g]
Switzerland	2.5	40	2.5	40	54	7,464	1991	1,190	172	2	4	23	73	—
Ukraine	20	50	17	53	140	2,868	1992	25,991	500	17	30	18	52	0
United Kingdom	9.8	144	9.0	145	147	2,464	1991	11,790	204	8	3	20	77	—
Middle East & N. Africa	**149**	**374**	**60**	**518**	—	—	—	—	—	—	—	—	—	—
Afghanistan	—	—	—	55	65	2,790	1987	26,110	2,007	72	99	1	0	0
Algeria	1.7	13	1.0	14	14	460	1995	5,000	181	39	52	34	14	64
Egypt	1.3	0.5	0	1.8	58[h]	830[h]	1996	66,000	1,055	127	82	7	11	25
Iran, Islamic Rep	49	97	18	129	138	1,900	1993	70,034	1,122	59	92	6	2	2.9
Iraq	1.2	34	0	35	75[h]	3,111[h]	1990	42,800	2,478	80	92	3	5	0
Israel	0.5	0.3	0	0.8	2	265.0	1997	1,620	287	108	54	39	7	—
Jordan	0.5	0.4	0.2	0.7	1	169.4	1993	984	255	151	75	22	3	2.0
Kuwait	0	0	0	0	0.02	9.9	1994	538	306	3,097	60	37	2	231
Lebanon	3.2	4.1	2.5	4.8	4[h]	1,219.5[h]	1996	1,300	400	33	68	27	6	0
Libyan Arab Jamahiriya	0.5	0.2	0.1	0.6	1	108.5	1999	4,500	870	801	84	13	3	70
Moroco	10	22	3.0	29	29	936	1998	11,480	399	43	89	10	2	3.4
Oman	1.0	0.9	0.9	1.0	1	363.6	1991	1,223	658	181	94	5	2	34
Saudi Arabia	2.2	2.2	2.0	2.4	2	110.6	1992	17,018	1,056	955	90	9	1	714
Syrian Arab Rep	4.2	4.8	2.0	7.0	26[h]	1,541[h]	1995	12,000	844	55	90	8	2	0
Tunisia	1.5	3.1	0.4	4.2	5	576.5	1996	2,830	312	54	86	13	1	8.3
Turkey	69	186	28	227	229[h]	3,344[h]	1997	35,500	558	17	73	16	12	0.5
United Arab Emirates	0.1	0.2	0.1	0.2	0	55.5	1995	2,108	896	1,614	67	24	9	385
Yemen	1.5	4.0	1.4	4.1	4	205.9	1990	2,932	253	123	92	7	1	10
Sub-Saharab Africa	**1,549**	**3,812**	**1,468**	**3,901**	—	—	—	—	—	—	—	—	—	—
Angola	72	182	70	184	184	1,3203	1987	480	54	0	76	14	10	0
Benin	1.8	10	1.5	10	25	3,741	1994	145	27	1	67	23	10	0
Botswana	1.7	1.7	0.5	2.9	14	9,209	1992	113	86	1	48	32	20	0

Burkina Faso	9.5	8.0	5.0	13	13	1,024	1992	376	40	4	81	19	0	0
Burundi	2.1	3.5	2.0	3.6	4	538.3	1987	100	19	4	64	36	0	0
Cameroon	100	268	95	273	286	18,378	1987	400	38	0	35	46	19	0
Central African Rep	56	141	56	141	144	37,565	1987	70	25	0	74	21	5	0
Chad	12	14	10	15	43	5,125	1987	180	34	1	82	16	2	0
Congo	198	222	198	222	832	259,547	1987	40	20	0	11	62	27	0
Congo, Dem Rep	421	899	420	900	1,283	23,639	1990	357	10	0	23	61	16	0
Côte d'Ivoire	38	74	35	77	81	4,853	1987	709	62	1	67	22	11	0
Equatorial Guinea	10	25	9.0	26	26	53,841	1987	10	30	0	6	81	13	0
Eritrea	—	—	—	2.8	6	1,577.7	—	—	—	—	—	—	—	0
Ethiopia	40	110	40	110	110	1,666	1987	2,200	51	3	86	11	3	0
Gabon	62	162	60	164	164	126,789	1987	60	70	0	6	72	22	0
Gambia	0.5	3.0	0.5	3.0	8	5,836.0	1982	20	29	1	91	7	2	0
Ghana	26	29	25	30	53	2,637	1970	300	35	1	52	35	13	0
Guinea	38	226	38	226	226	26,964	1987	740	132	0	87	10	3	0
Guinea-Bissau	14	12	10	16	31	24,670	1991	17	17	0	36	60	4	0
Kenya	3.0	17	0	20	30	947	1990	2,050	87	9	76	20	4	0
Lesotho	0.5	5.2	0.5	5.2	3[h]	1,455.6[h]	1987	50	32	2	56	22	22	0
Liberia	60	200	60	200	232	70,348	1987	130	59	0	60	27	13	0
Madagascar	55	332	50	337	337	19,925	1984	16,300	1,611	8	99	1	—	0
Malawi	1.4	16	1.4	16	17	1,461	1994	936	95	6	86	10	3	0
Mali	20	50	10	60	100	8,320	1987	1,360	167	2	97	2	1	0
Mauritania	0.3	0.1	0	0.4	11	4,029	1985	1,630	923	23	92	6	2	1.7
Mozambique	17	97	15	99	216	11,382	1992	605	42	0	89	9	2	0
Namibia	2.1	4.1	0.04	6.2	18[h]	9,865[h]	1991	249	175	2	68	29	3	0
Niger	2.5	1.0	0	3.5	34	2,891	1988	500	69	2	82	16	2	0
Nigeria	87	214	80	221	286	2,384	1987	3,630	46	2	54	31	15	0
Rwanda	3.6	5.2	3.6	5.2	5	638.2	1993	768	141	22	94	5	2	0
Senegal	7.6	24	5.0	26	39	3,977	1987	1,360	202	5	92	5	3	0
Sierra Leone	50	150	40	160	160	33,237	1987	370	98	0	89	7	4	0
Somalia	3.3	5.7	3.0	6.0	14	1,413	1987	810	119	8	97	3	0	0.1
South Africa	4.8	43	3.0	45	50	1,131	1990	13,309	366	32	72	17	11	0
Sudan	7.0	28	5.0	30	65[h]	1,981[h]	1995	17,800	637	32	94	4	1	0.4
Tanzania, United Rep	30	80	28	82	91	2,472	1994	1,165	39	2	89	9	2	0
Togo	5.7	11	5.0	12	15	3,076	1987	91	29	1	25	62	13	0
Uganda	29	39	29	39	66	2,663	1970	200	21	1	60	32	8	0

(Continued)

Table 7B.11 (Continued)

	Renewable Water Resources (annual)[a]						Water Withdrawals (annual)							
	Internal Renewable Water Resources (IRWR)				Natural Renewable Water Resources[b]						Sectoral Share (Percent)[c]			
	Groundwater Recharge (km^3)[d]	Surface Water (km^3)[d]	Overlap (km^3)	Total[e] (km^3)	Total (km^3)	Per Capita (m^2 person)[f]	Year	Total (mill m^3)	Per Capita (m^3 person)	as a % of Renewable Water Resources	Agriculture	Domestic	Industry	Desalinated Water Production (mill m^3)g[g]
Zambia	47	80	47	80	105	9,676	1994	1,706	190	2	77	16	7	0
Zimbabwe	5.0	13	4.0	14	20	1,530	1987	1,220	131	9	79	14	7	0
North America	**1,670**	**4,702**	**1,522**	**4,850**	—	—	—	—	—	—	—	—	—	—
Canada	370	2,840	360	2,850	2,902	92,810	1991	45,100	1,607	2	12	18	70	—
United States	1,300[i]	1,862[i]	1,162[i]	2,800	3,051	10,574	1990	467,340	1,834	26	42	13	45	—
C. America & Caribean	**359**	**1,050**	**231**	**1,186**	—	—	—	—	—	—	—	—	—	—
Belize	—	—	—	16	19	78,763	1993	95	485	1	0	12	88	0
Costa Rica	37	75	0	112	112	26,764	1997	5,772	1,540	6	80	13	7	0
Cuba	6.5	32	0	38	38	3,382	1995	5,211	475	14	51	49	0	0
Dominican Rep	12	21	12	21	21	2,430	1994	8,339	1,102	45	89	11	0	0
El Salvador	6.2	18	6	18	25	3,872	1992	729	137	4	46	34	20	0
Guatemala	34	101	25	109	111	9,277	1992	1,158	126	1	74	9	17	0
Haiti	2.2	11	—	13	14	1,670	1991	980	139	8	94	5	1	0
Honduras	39	87	30	96	96	14,250	1992	1,520	294	2	91	4	5	0
Jamaica	3.9	5.5	0	9.4	9	3,587.5	1993	900	371	10	77	15	7	0
Mexico	139	361	91	409	457	4,490	1998	77,812	812	18	78	17	5	0
Nicaragua	59	186	55	190	197	36,784	1998	1,285	267	1	84	14	2	0
Panama	21	144	18	147	148	50,299	1990	1,643	685	1	70	28	2	0
Trinidad and Tobago	—	—	—	3.8	4	2,940.4	1997	297	233	8	6	68	26	0
South America	**3,693**	**12,198**	**3,645**	**12,246**	—	—	—	—	—	—	—	—	—	—
Argentina	128	276	128	276	814	21,453	1995	28,583	822	4	75	16	9	0
Bolivia	130	277	104	304	623	71,511	1987	1,210	197	0	87	10	3	0
Brazil	1,874	5,418	1,874	5,418	8,233	47,125	1992	54,870	359	1	61	21	18	0
Chile	140	884	140	884	922	59,143	1987	20,289	1,629	3	84	5	11	0
Colombia	510	2,112	510	2,112	2,132	49,017	1996	8,938	228	0	37	59	4	0
Ecuador	134	432	134	432	432	32,948	1997	16,985	1,423	4	82	12	6	0
Guyana	103	241	103	241	241	314,963	1992	1,460	1,993	1	99	1	1	0
Paraguay	41	94	41	94	336	58,148	1987	430	112	0	78	15	7	0
Peru	303	1,616	303	1,616	1,913	72,127	1992	18,973	849	1	86	7	7	0

Suriname	80	88	80	88	122	289,848	1987	460	1,171	0	89	6	5	0
Uruguay	23	59	23	59	139	41,065	1965	650	—	—	91	6	3	0
Venezuela	227	700	205	722	1,233	49,144	1970	4,100	382	1	46	44	10	0
Oceania	**—**	**1,241**	**20**	**1,693**	—	—	—	—	—	—	—	—	—	—
Australia	72	440	20	492	492	25,185	1985	14,600	933	4	33	65	2	—
Fiji	—	—	—	29	29	34,330	1987	30	42	0	60	20	20	—
New Zealand	—	—	—	327	327	85,221	1991	2,000	588	1	44	46	10	—
Papua New Guinea	—	801	—	801	801	159,171	1987	100	29	0	49	29	22	0
Solomon Islands	—	—	—	45	45	93,405	1987	—	—	—	40	40	20	—
Developed	**3,153**	**12,084**	**2,584**	**13,016**	—	—	—	—	—	—	—	—	—	—
Developing	**8,128**	**28,500**	**7,483**	**29,289**	—	—	—	—	—	—	—	—	—	—

Variable Definitions and Methodology: *Internal Renewable Water Resources* (IRWR) include the average annual flow of rivers and the recharge of groundwater (aquifers) generated from endogenous precipitation — precipitation occurring within a country's borders. IRWR are measured in cubic kilometers per year (km^3/year).

Groundwater recharge is the total volume of water entering aquifers within a country's borders from endogenous precipitation and surface water flow. Groundwater resources are estimated by measuring rainfall in arid areas where rainfall is assumed to infiltrate into aquifers. Where data are available, groundwater resources in humid areas have been considered as equivalent to the base flow of rivers.

Surface water produced internally includes the average annual flow of rivers generated from endogenous precipitation and base flow generated by aquifers. Surface water resources are usually computed by measuring or assessing total river flow occurring in a country on a yearly basis.

Overlap is the volume of water resources common to both surface and groundwater. It is subtracted when calculating IRWR to avoid double counting. Two types of exchanges create overlap: contribution of aquifers to surface flow, and recharge of aquifers by surface run-off. In humid temperate or tropical regions, the entire volume of groundwater recharge typically contributes to surface water flow. In karstic domains (regions with porous limestone rock formations), a portion of groundwater resources are assumed to contribute to surface water flow. In arid and semi-arid countries, surface water flows recharge groundwater by infiltrating through the soil during floods. This recharge is either directly measured or inferred by characteristics of the aquifers and piezometric levels.

Total internal renewable water resources is the sum of surface and groundwater resources minus overlap; in other words, IRWR = Surface Water Resources + Groundwater Recharge — Overlap.

Natural Renewable Water Resources, measured in cubic kilometers per year (km^3/year), is the sum of internal renewable water resources and natural flow originating outside of the country. Natural Renewable Water Resources are computed by adding together both internal renewable water resources (IRWR — see above) and natural flows (flow to and from other countries). Natural incoming flow is the average amount of water which would flow into the country without human influence. In some arid and semi-arid countries, actual water resources are presented instead of natural renewable water resources. These actual totals, labeled with a footnote in the freshwater data table, include the quantity of flows reserved to upstream and downstream countries through formal and informal agreements or treaties. The actual flows are often much lower than natural flow due to water scarcity in arid and semi-arid regions.

Per Capita Natural Renewable Water Resources are measured in cubic meters per person per year (m^3/person/year). Per capita values were calculated by using national population data for 2002.

Water Withdrawals (annual), measured in million cubic meters, refers to total water removed for human uses in a single year, not counting evaporative losses from storage basins. Water withdrawals also include water from nonrenewable groundwater sources, river flows from other countries, and desalination plants.

Per Capita Annual Withdrawals were calculated using national population data for the year the withdrawal data were collected.

Water Withdrawals as a Percent of Renewable Water Resources is the proportion of renewable water resources withdrawn on a per capita basis, expressed in cubic meters per person per year (m^3/person/year). The value is calculated by dividing water withdrawals per capita by actual renewable water resources per capita.

Sector Share of water withdrawals, expressed as a percentage, refers to the proportion of water used for one of three purposes: agriculture, industry, and domestic uses. All water withdrawals are allocated to one of these three categories.

Agricultural uses of water primarily include irrigation and, to a lesser extent, livestock maintenance.

Domestic uses include drinking water plus water withdrawn for homes, municipalities, commercial establishments, and public services (e.g. hospitals).

Industrial uses include cooling machinery and equipment, producing energy, cleaning and washing goods produced as ingredients in manufactured items, and as a solvent.

(Continued)

Table 7B.11 (Continued)

Desalinated Water Production, expressed in million cubic meters, refers to the amount of water produced by the removal of salt from saline waters — usually seawater — using a variety of techniques including reverse osmosis. Most desalinated water is used for domestic purposes.

Most Freshwater resources data were provided by AQUASTAT, a global database of water statistics maintained by the Food and Agriculture Organization of the United Nations. AQUASTAT collects its information from a number of sources — national water resources and irrigation master plans; national yearbooks, statistics and reports; FAO reports and project documents; international surveys; and, results from surveys done by national or international research centers. In most cases, a critical analysis of the information was necessary to ensure consistency among the different data collected for a given country.

When possible, cross-checking of information among countries was used to improve assessment in countries where information was limited. When several sources gave different or contradictory figures, preference was always given to information collected at the national or sub-national level. This preference is based on the assumption by FAO that no regional information can be more accurate than studies carried out at the country level. Unless proven to be wrong, official rather than unofficial sources were used. In the case of shared water resources, a comparison among countries was made to ensure consistency at river-basin level.

For more information on the methodology used to collect these data, please refer to the original source or: Food and Agriculture Organization of the United Nations (FAO): Water Resources, Development and Management Service. October, 2001. Statistics on Water Resources by Country in FAO's AQUASTAT Programme (available on-line at fao.org/ag/agl/aglw/aquastat/water_res/index.stm). Rome: FAO.

Frequency of Update by Data Providers: AQUASTAT was developed by the Food and Agriculture Organization of the United Nations in 1993; data have been available on-line since 2001. Most freshwater data are not available in a time series, and the global data set contains data collected over a time span of up to 30 years. AQUASTAT updates their website as new data become available, or when FAO conducts special regional studies. Studies were conducted in Africa in 1994, the Near East in 1995–96, the former Soviet republics in 1997, selected Asian countries in 1998–99, and Latin America & the Caribbean in 2000. Data from the Blue Plan on Mediterranean water withdrawals were last updated in 2002. Most data updates include revisions of past data.

Data Reliability and Cautionary Notes: While AQUASTAT represents the most complete and careful compilation of country-level water resources statistics to date, freshwater data are generally of poor quality. Information sources are various but rarely complete. Some governments will keep internal water resources information confidential because they are competing for water resources with bordering countries. Many instances of water scarcity are highly localized and are not reflected in national statistics. In addition, the accuracy and reliability of information vary greatly among regions, countries, and categories of information, as does the year in which the information was gathered. As a result, no consistency can be ensured among countries on the duration and dates of the period of reference. All data should be considered order-of-magnitude estimates.

Groundwater Recharge tends to be overestimated in arid areas and underestimated in humid areas.

Natural Renewable Water Resources vary with time. Exchanges between countries are complicated when a river crosses the same border several times. Part of the incoming water flow may thus originate from the same country in which it enters, making it necessary to calculate a "net" inflow to avoid double counting of resources. In addition, the water that is actually accessible to humans for consumption is often much smaller than the total renewable water resources indicated in the data table.

Renewable Water Resources Per Capita contains water resources data from a different set of years than the population data used in the calculation. While the water resources data are usually long-term averages, inconsistencies may arise when combining it with 2002 population data.

Water Withdrawals as a Percentage of Actual Water Resources are also calculated using per capita data from two different years. While this ratio can indicate that some countries are depleting their water resources, it does not accurately reflect localized over-extraction from aquifers and streams. In addition, the calculation does not distinguish ground and surface water.

Sectoral Withdrawal Data may not add to 100 because of rounding. Evaporative losses from storage basins are not considered; users should keep in mind, however, that in some parts of the world up to 25 percent of water that is withdrawn and placed in reservoirs evaporates before it is used by an sector.

Desalinated Water Production may exist in some countries where the volume of production is indicated to be zero, since AQUASTAT assumes that production is zero if no value has been given for those countries where information on water use is available.

[a] Although data were obtained from FAC in 2002, they are long-term averages originating from multiple sources and years. For more information, please consult the original source at fao.org/waicent/faoinfo/agricult/agl/aglw/aquastat/water_res/index.stm.
[b] Natural renewable water resources include internal renewable water resources plus or minus the flows of surface and groundwater entering or leaving the country.
[c] Sectoral withdrawal data may not add up to 100 because of rounding.
[d] Groundwater and surface water cannot be added together to calculate total available water resources because of overlap — water that is counted in both the groundwater and surface water totals.
[e] At the country level, total internal renewable water resources = surface water + groundwater — overlap. Regional and global totals represent a sum of available country-level data.
[f] Calculation is based on withdrawals from various years, and population data from 2002.
[g] Data on desalinated water originate from FAO country surveys conducted in various regions between 1992 and 2000.
[h] Data account for the portion of flow secured through treaties or agreements to other countries.
[i] River discharges in Siberia are not well documented and highly uncertain.
[j] Data are for the continental United States.

Source: From World Resources Institute, Earth trends environmental information, *Water Resources and Freshwater Ecosystems*, Data Tables, Freshwater Resources, www.earthtrends.wri.org. With permission.

Original Source: Renewable Water Resources: Food and Agriculture Organization of the United Nations (FAO): Water Resources, Development and Management Service. 2002. AQUASTAT Information System on Water in Agriculture: Review of Water Resource Statistics by Country. Rome: FAO. Available on-line at fao.org/waicent/faoinfo/agl/aglw/aquastat/water_res/index.htm.

Water Withdrawals: Food and Agriculture Organisation of the United Nations (FAO): Water Resources, Development and Management Service.2002. AQUASTAT Information System on Water in Agriculture. Rome: FAO. Avaliable on-line at www.fao.org/waicent/faoinfo/agricult/agl/aglw/aquastat/dbase/index.htm. Data for Mediterranean countries were provided directly to WRI from: J. Margat, 2002. Present Water Withdrawals in Mediterranean Countries. Paris: Blue Plan.

Population Data (for per capita calculations): Population Division of the Department of Economic and Social Affairs of the United Nations Secretariat. 2002.World Population prospects: The 2000 Revision. New York: United Nations. Data set on CD-ROM.

Table 7B.12 Worldwide Freshwater Withdrawal, by Country and Sector

Region and Country	Year	Total Freshwater Withdrawal (km^3/yr)	Per-Capita Withdrawal (m^3/p/yr)	USE Domestic (%)	USE Industrial (%)	USE Agricultural (%)	USE Domestic (m^3/p/yr)	USE Industrial (m^3/p/yr)	USE Agricultural (m^3/p/yr)	Source	2000 Population (millions)
Africa											
Algeria	2000	6.07	192	22	13	65	42	25	125	a	31.60
Angola	2000	0.34	27	22	16	61	6	4	16	a	12.80
Benin	2000	0.25	40	15	11	74	6	4	30	a	6.20
Botswana	2000	0.14	86	38	19	43	33	16	37	a	1.62
Burkina Faso	2000	0.78	65	11	0	88	7	0	57	a	12.06
Burundi	2000	0.23	33	17	1	82	6	0	27	a	6.97
Cameroon	2000	0.73	48	18	8	74	9	4	36	a	15.13
Cape Verde	2000	0.03	68	15	3	83	10	2	56	a	0.44
Central African Republic	2000	0.02	5	77	19	4	4	1	0	a	3.64
Chad	2000	0.23	32	19	1	80	6	0	25	a	7.27
Comoros	1987	0.01	14	48	5	47	7	1	7	b	0.71
Congo, Democratic Republic (formerly Zaire)	2000	0.36	7	52	16	31	4	1	2	a	2.98
Congo, Republic of	2000	0.04	13	59	30	10	8	4	1	a	51.75
Côte d'Ivoire	2000	0.93	61	23	12	65	14	7	40	a	15.14
Djibouti	2000	0.01	15	11	0	89	2	0	13	a	0.69
Egypt	2000	68.65	1,008	8	14	78	77	140	791	a	68.12
Equatorial Guinea	2000	0.11	243	83	16	1	202	39	2	a	0.45
Eritrea	2000	0.30		4	1	95				a	
Ethiopia	2000	2.65	38	1	6	93	0	2	35	a	69.99
Gabon	2000	0.13	105	48	11	40	51	12	43	a	1.24
Gambia	2000	0.03	24	22	11	67	5	3	16	a	1.24
Ghana	2000	0.52	26	37	15	48	10	4	13	a	19.93
Guinea	2000	1.52	193	8	2	90	15	4	174	a	7.86
Guinea-Bissau	2000	0.11	93	9	1	91	8	1	84	a	1.18
Kenya	2000	1.58	52	30	6	64	16	3	33	a	30.34
Lesotho	2000	0.05	22	40	41	19	9	9	4	a	2.29
Liberia	2000	0.11	34	28	15	56	10	5	19	a	3.26
Libya	2000	4.81	753	8	3	89	62	23	668	a	6.39
Madagascar	2000	14.97	861	3	2	96	24	13	823	a	17.40
Malawi	2000	1.01	92	15	5	81	14	4	74	a	10.98
Mali	2000	6.93	552	1	0	99	4	1	547	a	12.56
Mauritania	2000	1.70	659	9	3	88	58	19	582	a	2.58
Mauritius	2000	0.61	517	25	14	60	132	74	312	a	1.18
Morocco	2000	12.76	440	8	2	90	37	7	396	a	28.98
Mozambique	2000	0.64	33	11	2	87	4	1	28	a	19.56
Namibia	2000	0.27	156	33	5	63	51	7	98	a	1.733
Niger	2000	2.19	203	4	1	95	9	1	193	a	10.81
Nigeria	2000	8.00	62	21	10	69	13	6	43	a	128.79

Rwanda	2000	0.08	10	48	14	39	5	1	4	a	7.67
Senegal	2000	1.59	167	6	4	90	10	6	151	a	9.50
Sierra Leone	2000	0.38	78	5	2	93	4	1	73	a	4.87
Somalia	2000	3.30	286	0	0	100	1	0	285	a	11.53
South Africa	2000	15.31	331	17	10	73	56	35	241	a	46.26
Sudan	2000	37.31	1,251	3	1	97	33	9	1,209	a	29.82
Swaziland	2000	0.83	843	3	6	92	23	48	773	a	0.98
Tanzania	2000	2.00	59	6	1	93	4	1	55	a	33.69
Togo	2000	0.17	36	45	8	47	16	3	17	a	4.68
Tunisia	2000	2.73	278	16	2	82	44	7	227	a	9.84
Uganda	2000	0.30	13	45	15	39	6	2	5	a	22.46
Zambia	2000	1.74	191	16	8	76	31	14	145	a	9.13
Zimbabwe	2000	2.61	210	10	5	86	20	10	180	a	12.42
North and Central America											
Antigua and Barbuda	1990	0.005	75	60	20	20	45	15	15	g	0.07
Barbados	2000	0.08	308	33	44	23	103	136	69	a	0.26
Belize	2000	0.12	500	11	89	0	56	443	1	a	0.24
Canada	1990	43.89	1,431	20	69	12	280	982	168	d	30.68
Costa Rica	2000	2.68	706	29	17	53	208	120	377	a	3.80
Cuba	2000	8.20	732	19	12	69	139	89	504	a	11.20
Dominica	1996	0.02	239	—	—	—				h	0.07
Dominican Republic	2000	3.39	399	32	2	66	128	7	264	a	8.50
El Salvador	2000	1.27	201	25	16	59	50	32	119	a	6.32
Guatemala	2000	2.00	164	6	13	80	11	22	131	a	12.22
Haiti	2000	0.98	125	5	1	94	6	1	118	a	7.82
Honduras	2000	0.86	133	8	11	81	11	15	107	a	6.49
Jamaica	2000	0.41	158	34	17	49	54	27	77	a	2.59
Mexico	2000	78.22	791	17	5	77	137	43	610	a	98.88
Nicaragua	2000	1.30	277	14	3	83	40	7	230	a	4.69
Panama	2000	0.82	287	66	5	28	191	15	81	a	2.86
St. Lucia	1997	0.01	89	—	—	—				g	0.15
St. Vincent and the Grenadines	1995	0.01	88	—	—	—				g	0.11
Trinidad and Tobago	2000	0.30	221	67	27	6	149	60	12	a	1.34
United States of America	2000	563.00	2,026	13	46	41	257	933	836	m	277.83
South America											
Argentina	2000	29.07	785	16	9	74	129	74	581	a	37.03
Bolivia	2000	1.39	167	13	3	83	22	6	139	a	8.33
Brazil	2000	59.30	350	20	18	62	71	63	216	a	169.20
Chile	2000	12.54	824	11	25	64	93	208	524	a	15.21
Colombia	2000	10.71	275	50	4	46	138	10	126	a	38.91
Ecuador	2000	16.98	1,343	12	5	82	167	71	1,104	a	12.65
Guyana	2000	1.64	1,876	2	1	97	32	17	1,828	a	0.87
Paraguay	2000	0.49	89	20	9	72	18	8	64	a	5.50

(Continued)

Table 7B.12 (Continued)

Region and Country	Year	Total Freshwater Withdrawal (km³/yr)	Per-Capita Withdrawal (m³/p/yr)	USE Domestic (%)	USE Industrial (%)	USE Agricultural (%)	USE Domestic (m³/p/yr)	USE Industrial (m³/p/yr)	USE Agricultural (m³/p/yr)	Source	2000 Population (millions)
Peru	2000	20.13	784	8	10	82	66	79	640	a	25.66
Suriname	2000	0.67	1,482	4	3	93	67	43	1,373	a	0.45
Uruguay	2000	3.15	962	2	1	96	23	11	928	a	3.27
Venezuela	2000	8.37	346	45	7	47	158	24	164	a	24.17
Asia											
Afghanistan	2000	23.26	909	2	0	98	16	0	893	a	25.59
Armenia	2000	2.95	806	30	4	66	241	35	529	a	3.66
Azerbaijan	2000	17.25	2,204	5	28	68	106	609	1,488	a	7.83
Bahrain	2000	0.30	485	40	4	57	192	17	276	a	0.62
Bangladesh	2000	76.39	595	3	1	96	19	4	572	a	128.31
Bhutan	2000	0.42	207	4	1	95	8	2	197	a	2.03
Brunei	1994	0.92	2,788	nd	nd	nd				e	0.33
Cambodia	2000	4.09	365	2	1	98	6	2	357	a	11.21
China	2000	549.76	431	7	26	68	28	111	292	k	1,276.30
Cyprus	2000	0.18	228	27	1	71	62	3	163	p	0.79
Georgia	2000	3.61	666	20	21	59	133	140	393	a	5.42
India	2000	645.84	641	8	5	86	52	35	555	a	1,006.77
Indonesia	2000	82.77	389	8	1	91	31	3	356	a	212.57
Iran	2000	72.88	954	7	2	91	65	22	867	a	76.43
Iraq	2000	42.70	1 848	3	5	92	58	85	1,704	a	23.11
Israel	2000	2.04	336	31	7	63	103	23	210	a	6.08
Japan	2000	88.43	699	20	18	62	138	125	437	a	126.43
Jordan	2000	1.02	161	21	4	75	33	7	121	a	6.33
Kazakhstan	2000	35.01	2 068	2	17	82	35	342	1,691	a	16.93
Korea Democratic People's Republic	2000	9.02	377	20	25	55	75	95	207	a	23.91
Korea Rep	2000	18.59	397	36	16	48	141	65	190	a	46.88
Kuwait	2000	0.45	229	45	3	52	102	7	120	a	1.97
Kyrgyz Republic	2000	10.08	2 219	3	3	94	70	69	2,081	a	4.54
Lao People's Dem Rep	2000	2.99	525	4	6	90	22	30	473	a	5.69
Lebanon	2000	1.37	417	33	1	67	136	2	278	a	3.29
Malaysia	2000	9.02	405	17	21	62	68	85	251	a	22.30
Maldives	1987	0.003	10	98	2	0	10	0	0	e	0.29
Mongolia	2000	0.44	161	20	28	52	33	45	83	a	2.74
Myanmar	2000	33.22	673	1	1	98	8	4	661	a	49.34
Nepal	2000	10.18	418	3	1	96	12	3	403	a	24.35
Oman	2000	1.35	497	7	2	91	35	10	452	a	2.72
Pakistan	2000	169.38	1,086	2	2	96	21	22	1,043	a	156.01
Philippines	2000	28.52	380	17	9	74	63	36	281	a	75.04
Qatar	2000	0.29	484	25	3	72	122	14	348	a	0.60
Saudi Arabia	2000	17.32	800	10	1	89	78	9	712	a	21.66

Singapore	1975	0.19	53	45	51	4	24	27	2	c	3.59
Sri Lanka	2000	12.60	669	2	2	95	16	16	637	a	18.82
Syria	2000	19.95	1,237	3	2	95	41	23	1,174	a	16.13
Tajikistan	2000	11.96	1,869	4	5	92	69	87	1,713	a	6.40
Thailand	2000	87.07	1,439	2	2	95	36	35	1,368	a	60.50
Turkey	2001	39.78	605	15	11	74	90	66	449	p	65.73
Turkmenistan	2000	24.64	5,501	2	1	98	93	42	5,366	a	4.48
United Arab Emirates	2000	2.31	945	23	9	68	218	82	645	a	2.44
Uzbekistan	2000	58.33	2,332	5	2	93	111	48	2,173	a	25.02
Vietnam	2000	71.39	886	8	24	68	69	214	604	a	80.55
Yemen	2000	6.63	366	4	1	95	15	2	349	a	18.12
Europe											
Albania	2000	1.70	487	27	11	62	130	54	302	a	3.49
Austria	1997	3.56	429	35	64	1	151	274	4	p	8.29
Belarus	2000	2.79	271	23	46	30	64	126	82	a	10.28
Belgium	1998	7.44	725	13	85	1	96	620	9	p	10.26
Bulgaria	2002	6.59	793	3	78	19	24	621	149	p	8.31
Czech Republic	2002	1.91	187	41	57	2	76	107	4	p	10.20
Denmark	2002	0.67	127	32	26	42	41	32	54	p	5.27
Estonia	2002	1.41	994	56	39	5	554	392	49	p	1.42
Finland	1999	2.33	450	14	84	3	61	377	12	p	5.18
France	2000	30.90	523	16	74	10	82	390	51	p	59.06
Germany	2001	38.00	460	12	68	20	57	312	91	p	82.69
Greece	1997	8.70	821	16	3	81	134	26	661	p	10.60
Hungary	2001	4.55	464	9	59	32	43	272	149	p	9.81
Iceland	2002	0.16	567	34	66	0	193	374	1	p	0.28
Ireland	1980	1.07	299	23	77	0	68	232	0	p	3.57
Italy	1998	42.00	734	18	37	45	134	270	331	p	57.19
Latvia	2001	0.26	108	55	33	12	60	36	13	p	2.40
Lithuania	2002	3.13	848	78	15	7	662	131	56	p	3.69
Luxembourg	1999	0.06	133	42	45	13	56	60	17	p	0.43
Malta	2000	0.02	53	74	1	25	39	0	13	p	0.38
Moldova	2000	2.31	518	9	58	33	49	298	171	a	4.46
Netherlands	2001	8.80	554	6	60	34	34	333	188	p	15.87
Norway	1996	2.40	545	23	67	10	123	364	57	p	4.41
Poland	2002	11.73	303	13	79	8	39	238	25	p	38.73
Portugal	1990	7.29	745	10	12	78	72	90	583	p	9.79
Romania	2002	7.24	322	9	34	57	28	111	183	p	22.51
Russian Federation	2000	76.69	525	19	63	18	98	333	93	a	146.20
Slovak Republic	2002	1.09	203							p	5.37
Slovenia	2001	0.30								p	
Spain	2001	38.60	670	13	19	68	130	180	660	p	39.80
Sweden	2002	2.69	302	37	54	9	111	164	27	p	8.90
Switzerland	2001	2.54	343	24	74	2	83	253	7	p	7.41
Ukraine	2000	37.52	739	12	35	52	90	261	388	a	50.80
United Kingdom	1994	11.75	201	22	75	3	44	152	6	d	58.34
Yugoslavia,[q] former	1980	8.77	368	16	72	12	59	265	44	c	23.81

(Continued)

Table 7B.12 (Continued)

Region and Country	Year	Total Freshwater Withdrawal (km^3/yr)	Per-Capita Withdrawal (m^3/p/yr)	USE Domestic (%)	Industrial (%)	Agricultural (%)	Domestic (m^3/p/yr)	Industrial (m^3/p/yr)	Agricultural (m^3/p/yr)	Source	2000 Population (millions)
Oceania											
Australia	1995	17.80	945	15	10	75	139	95	711	f	18.84
Fiji	2000	0.07	83	11	11	78	9	9	65	a	0.85
New Zealand	2000	2.11	561	49	9	42	272	53	236	a	3.76
Papua New Guinea	1987	0.10	21	56	43	1	12	9	0	c	4.81
Solomon Islands	1987			40	20	40	0	0	0	c	0.44

Note: Figures may not add to totals due to independent rounding. 2000 Population numbers: medium United Nations variant.

Limitations: Extreme care should be used when applying these data — they are often the least reliable and most inconsistent of all water resources information. They come from a wide variety of sources and are collected using different approaches with few formal standards. Consistent data collection is needed in this area, using standard methods and assumptions. As a result, this table includes data that are actually measured, estimated, modeled using different assumptions, or derived from other data. The data also come from different years, making direct comparisons difficult, though the effort of FAO to standardize water-use data for 2000 has somewhat reduced this problem. Industrial withdrawals for Panama, St. Lucia, St. Vincent, and the Grenadines are included in the domestic category. Another major limitation of these data is that they do not include the use of rainfall in agriculture. Many countries use a significant fraction of the rain falling on their territory for agricultural production, but this water use is neither accurately measured nor reported.

[a] New FAO Aquastat estimates from www.fao.org. March 2004.
[b] World Resources Institute, 1990, *World resources* 1990–1991, New York: Oxford University Press.
[c] World Resources Institute, 1994, *World resources* 1994–1995, in collaboration with the United Nations Environment Programme and the United Nations Development Programme, New York: Oxford University Press.
[d] Eurostat Yearbook, 1997. *Statistics of the European Union*, EC/C/6/Ser.26GT, Luxembourg.
[e] UNFAO 1999. Irrigation in Asia in figures. Food and Agriculture Organization, United Nations, Rome.
[f] Nix, H. 1995. Water/Land/Life, Water Research Foundation of Australia, Canberra.
[g] UNFAO. 2000. *Irrigation in Latin America and the Caribbean.* Food and Agricultural Organization, United Nations, Rome.
[h] AQUASTAT web site January 2002. www.fao.org.
[k] Ministry of Water Resources, China. 2001. *Water resources bulletin of China, 2000.* People's Republic of China, Beijing, September.
[m] Hutson, S. S., Barber, N. L., Kenny, J. E., Linsey, K. S., Lumia, D. S., and Maupin, M. A. 2004. *Estimated use of water in the United States in 2000.* United States Geological Survey, Circular 1268. Reston, Virginia.
[p] See Wieland, U. 2003. *Water use and waste water treatment in the European Union and in candidate countries.* Eurostat Statistics in Focus, Theme 8. European Communities. And Eurostat. 2004. Statistics in Focus. europa.eu.int/comm/eurostat europa.eu.int/comm/eurostat/newcronos/queen/display.do?screen=detail&language=en&product=THEME8&root=THEME8_ copy_151979619462/yearlies_copy_1067300085946/dd_copy_251110364103/dda_copy_649289610368/dda10512_copy_729379227605.
[q] Includes Bosnia and Herzegovina, Macedonia, Croatia.

Source: From *World's Water 2004–2005* by Peter H. Gleick.

Table 7B.13 Freshwater Abstractions by Major Use in Selected Countries, 1980–1999

	Public Water Supply (b)				Irrigation (b)				Manufacturing Industry No Cooling (b)				Electrical Cooling (b) Percent			
	1980	1985	1990	1999 (c)	1980	1985	1990	1999 (c)	1980	1985	1990	1999 (c)	1980	1985	1990	1999 (c)
Canada	11.3	11.1	11.3	—	7.4	7.0	7.1	—	8.6	8.9	7.9	—	39.9	56.9	59.7	—
Mexico	7.5	—	—	12.6	82.1	—	—	82.7	10.4	—	—	4.5	0.1	—	—	0.2
U.S.A.	9.1	10.8	11.4	—	38.7	40.5	40.2	—	10.4	6.6	5.7	—	40.4	38.7	38.6	—
Japan	14.9	16.5	17.7	18.5	67.4	67.1	65.9	66.1	17.6	16.6	16.3	15.4	—	—	—	—
Korea	10.9	14.8	21.8	27.0	80.5	76.4	68.4	60.1	8.6	8.8	9.7	12.9	—	—	—	—
Australia	—	12.3	—	—	74.3	69.9	—	74.6	—	5.5	—	—	—	—	—	—
New Zealand	—	—	—	—	—	—	—	—	—	—	—	—	—	—	—	—
Austria	16.7	17.8	16.4	17.0	1.5	1.6	1.3	1.5	46.7	45.7	39.0	36.1	33.9	33.7	41.9	44.1
Belgium	—	—	—	9.8	—	—	—	0.1	—	—	—	5.1	—	—	—	57.0
Czech Rep.	28.0	32.1	35.0	42.0	1.1	1.4	2.7	0.4	28.4	26.7	24.5	21.7	33.6	31.3	29.3	27.6
Denmark	58.1	—	45.3	58.4	7.5	—	15.4	18.6	3.7	—	17.8	8.4	—	—	—	—
Finland	10.5	10.2	18.1	17.4	—	0.5	0.9	1.7	—	—	69.0	67.4	—	—	10.6	11.0
France	17.5	16.9	16.2	19.4	14.1	12.8	13.0	11.0	17.9	14.7	11.8	12.8	50.4	55.5	59.1	56.7
Germany	12.0	12.4	13.6	13.7	—	—	3.3	0.4	6.8	5.8	15.8	14.3	60.4	62.0	60.1	65.0
Greece	12.2	11.9	—	9.9	82.5	83.7	—	87.4	2.7	2.4	—	1.3	1.8	2.0	—	1.4
Hungary	16.7	14.8	16.0	12.7	7.0	4.6	8.5	1.2	2.2	1.4	1.4	0.2	49.7	60.0	62.2	75.4
Iceland	85.2	85.7	50.9	47.4	—	—	—	—	9.3	8.9	6.0	6.4	—	—	—	0.0
Ireland	34.0	—	—	—	12.1	—	—	—	23.4	—	—	—	25.9	—	—	—
Italy	14.2	—	14.1	18.0	57.3	—	—	46.0	14.2	—	—	17.0	12.5	—	—	19.0
Luxembourg	—	—	95.0	62.6	—	—	0.3	0.3	—	—	4.7	23.2	—	—	—	—
The Netherlands	11.2	11.9	16.4	28.6	—	—	—	—	2.1	2.0	2.5	4.3	65.4	70.4	65.9	54.5
Norway	—	30.1	—	—	—	3.4	—	—	—	64.6	—	—	—	—	—	—
Poland	19.2	18.9	21.1	21.2	2.0	3.4	3.3	0.8	12.9	10.6	9.7	4.7	50.7	53.1	51.3	60.1
Portugal	—	—	4.7	6.8	—	—	59.3	79.1	—	—	12.8	3.4	—	—	23.3	11.2
Slovak Rep.	—	28.3	30.4	37.5	—	5.9	12.6	0.8	—	63.7	53.5	58.4	—	—	—	—
Spain	11.8	11.6	11.9	13.2	65.7	65.7	64.2	68.2	—	—	5.1	4.7	22.5	22.7	12.2	13.9
Sweden	23.3	32.8	32.9	35.1	1.6	3.2	3.2	3.7	43.3	37.4	37.5	28.3	—	0.9	0.9	1.0
Switzerland	42.6	43.2	43.6	41.4	—	—	—	—	—	—	—	—	57.4	56.8	56.4	58.6
Turkey	—	14.4	14.2	14.7	—	73.7	53.9	75.1	—	11.9	7.6	10.3	—	—	—	0.1
UK	43.7	52.8	55.5	53.1	0.6	0.7	1.5	1.1	13.8	9.1	7.4	5.4	28.1	21.6	20.0	11.3
Russian Fed.	9.7	—	12.0	18.7	—	—	—	—	—	—	12.9	12.0	—	—	24.5	27.9

Note: a) In general, the data for the four sectors will not sum to 100%, since "agricultural uses other than irrigation", "agricultural uses other than irrigation", "industrial cooling", and "other uses" are not covered here. Exceptions occur when the % are based on partial totals or the categories presented include other uses.

b) "Public water supply" refers to water supply by waterworks, and may include other uses besides the domestic sector. "Irrigation", "Industry no cooling" and "Electrical cooling" refer to self supply (abstraction for own final use).

c) Data refer to 1999 or latest available year. Data prior to 1995 have not been considered.

Note: CAN, 1980, 1985 and 1990: 1981, 1986 and 1991 data; MEX, 1980: % based on totals excluding agricultural uses other than irrigation. Industry no cooling: includes cooling. Electrical cooling 1980: data include Secretariat estimates; U.S.A., Industry no cooling: includes cooling; JPN, Industry no cooling: includes industrial and electrical cooling; KOR, % based on partial totals: electrical cooling excluded. Public supply: data refer to domestic sector. Irrigation: includes other agric, uses. Industry no cooling includes cooling, 1999: 1997 data; AUS, 1980: 1977 data adjusted for average climatic year. 1985: data refer to fiscal year 1983/84 and to both waterworks and self-supply; public supply: data refer to domestic sector; industry no cooling: may include industrial and electrical cooling, 1999: 1996/97 data; NZL, % based on partial totals; AUT, % based on partial totals. Irrigation and industry no cooling: groundwater only. Electrical cooling (includes all industrial cooling): surface water only. 1990:1992 data; irrigation includes other agricultural abstractions. 1999: 1997 data; BEL, 1999: 1998 data including Secretariat estimates; CZE, Industry no cooling: includes cooling; DNK, 1980: 1977 data. 1990; % based on totals referring to groundwater abstractions only, which represent the majority of total freshwater abstractions (e.g. 95–99% for 1995). Industry no cooling: includes some industrial and electrical cooling (self-supply). 1999: 1998 data, Irrigation: 1995 data; FIN, % based on partial total: 1985 and 1990 exclude agricultural uses besides irrigation (1985 data); industry no cooling: includes cooling. Irrigation: 1999 data is country estimate; FRA, 1980 and 1999: 1981 and 1997 data. Irrigation: includes other agricultural uses, but irrigation is the main use. Industry no cooling: includes cooling. 1997: break in time series; DEU, % based on totals excluding agricultural uses other than irrigation. Industry no cooling: includes cooling. 1980 and 1985: 1979 and 1983 data, for western Germany only. 1990 and 1999: 1991 and 1998 data for total Germany; GRC, % based on partial totals excluding agricultural uses other than irrigation, 1999: 1997 data. Public water supply: supply by 42 out of 75 great water distribution enterprises; HUN, 1999: 1998 data; ISL, Public supply: includes the domestic use of geothermal water. Industry no cooling: includes cooling. After 1985, fish farming is a major user of abstracted water, explaining the change in the relative contribution of other sectors. 1990: 1992 data; IRL, Industry no cooling: includes cooling. Irrigation: includes other agricultural uses (e.g. rural domestic use). % based on totals including 1980 data for electrical cooling; ITA, % based on totals excluding agricultural uses besides irrigation. 1990 and 1999: 1989 and 1998 data; LUX, Industry no cooling: includes cooling. 1990: 1989 data, except for industry and electrical cooling (1983 data); irrigation; estimated data; NLD, % based on partial totals excluding all agricultural uses. 1980, 1985, 1990 and 1999: 1981, 1986, 1991 and 1996 data; NOR, Data include 1978 data for industry. 1985: 1983 data. Industry no cooling: includes industrial cooling; POL, % based on totals including abstractions for agriculture, which include aquaculture (areas over 10 ha) and irrigation (Arabic land and forest areas greater than 20 ha); animal production and domestic needs of rural inhabitants are not covered (selfsupply); PRT, % based on totals excluding agricultural uses besides irrigation. 1990 and 1999: 1991 and 1998 data; SLO, Irrigation: Secretariat estimates. Industry no cooling: includes cooling; ESP, % based on totals excluding agricultural uses other than irrigation. Industry no cooling: surface water only; includes industrial cooling. Electrical cooling: until 1990 data include total industrial use. 1990 and 1999: 1991 and 1997 data; SWE, Irrigation: 1980 and refer to 1976; since 1985 data are estimates for dry year. Industry no cooling: 1980 data refer to 1974 and include mining and quarrying and electrical cooling: 1985 and 1990 data refer to 1983. Electrical cooling: 1985 and 1990 data refer to 1983. 1999: 1995 data; CHE, % based on partial totals excluding all agricultural uses. Public supply: includes industry (total industry—ISIC 10–45 rev. 3), which totals 215 million m^3 (1994), and other activities (101 million m^3 (1994)). 1999: 1998 data; TUR, % based on totals excluding agricultural uses other than irrigation. 1985: % based on partial totals excluding electrical cooling. Industry no cooling: includes cooling. 1990: 1991 data. Electrical cooling 1999: estimation (1995 data); UKD, England and Wales only. Data include miscellaneous uses for power generation, but exclude hydroelectric power water use; RUS, 1990 and 1999: 1991 and 1996 data.

Source: From Table 3.1C, OECD Environmental Data Compendium 2002, © OCED 2002, www.oecd.org.

Table 7B.14 Worldwide Fresh Water Utilization by Purpose, 2000

World/ Continent	Total Volume of Fresh Water Utilization (km^3/yr)	Fresh Water Utilization by Purpose: Domestic Use (km^3/yr)	Domestic Use %	Industrial Use (km^3/yr)	Industrial Use %	Agricultural Use (km^3/yr)	Agricultural Use %	Utilization[a] as % of Resources
World	3820.3	377.4	9.9	790.7	20.7	2 652.2	69.4	8.9
Developed countries	1239.2	165.2	13.3	509.1	41.1	564.9	45.6	9.0
Industrialized countries	893.5	131.3	14.7	395.6	44.3	366.6	41.0	10.2
Transition economies	345.7	33.9	9.8	113.5	32.8	198.3	57.4	7.1
Developing countries	2581.1	212.2	8.2	281.6	10.9	2 087.3	80.9	8.9
Latin America and the Caribbean	265.1	50.4	19.0	27.4	10.3	187.3	70.7	2.0
Near East and North Africa	322.6	25.1	7.8	19.5	6.0	278.0	86.2	62.5
Sub-Saharan Africa	98.1	6.9	7.0	2.8	2.9	88.3	90.1	2.5
East and Southeast Asia	977.4	71.2	7.3	192.3	19.7	714.0	73.0	11.2
South Asia	917.8	58.7	6.4	39.6	4.3	819.6	89.3	52.1
Oceania developing	0.1	0.0	34.2	0.0	27.6	0.1	38.2	0.0
North America developing	—	—	—	—	—	—	—	—
Continental groupings								
Africa	208.4	17.9	8.6	15.2	7.3	175.2	84.1	5.3
Asia	2377.1	171.5	7.2	270.2	11.4	1 935.5	81.4	20.4
Latin America	251.7	47.3	18.8	26.1	10.4	178.3	70.8	1.9
Caribbean	13.4	3.1	22.9	1.3	9.4	9.1	67.7	15.7
North America	525.3	69.8	13.3	252.3	48.0	203.2	38.7	9.3
Oceania	26.2	4.6	17.6	2.6	10.1	19.0	72.4	1.6
Europe	418.2	63.2	15.1	223.0	53.3	132.0	31.6	6.5

Source: From FAO, 2003, *Summary of Food and Agricultural Statistics*, www.fao.org. With permission.

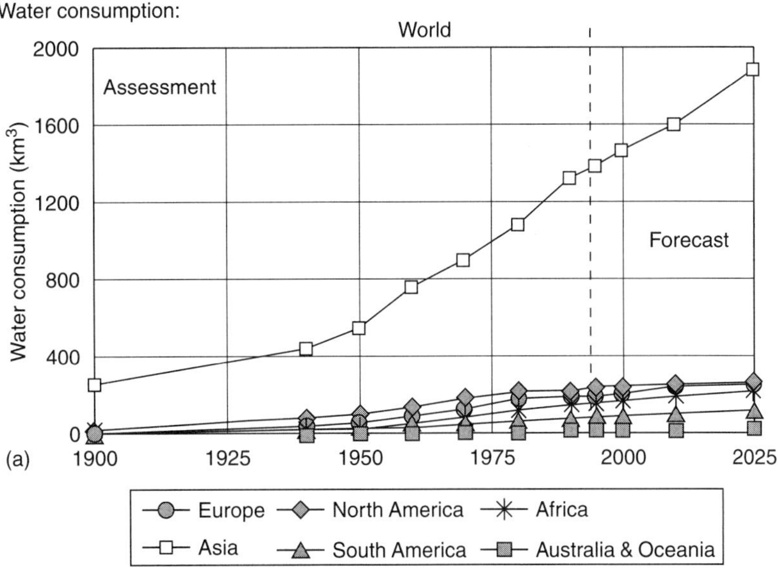

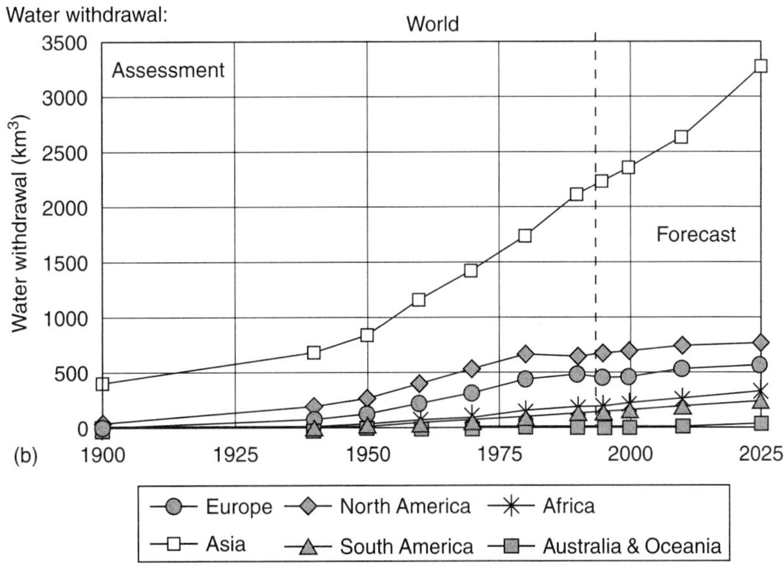

Figure 7B.8 Dynamics of water use in the world by continents. (From Shiklomanov, I.A., 1999, Summary of the Monograph "World Water Resources at the Beginning of the 21st Century" Prepared in the Framework of IHP UNESCO, International Hydrological Programme, UNESCO'S, Intergovernmental Scientific Programme in Water Resources, World Water Resources and Their Use a Joint State Hydrological Institute (SHI)/UNESCO Product, http://webworld.unesco.org/water/ihp/db/shiklomanov/. Copyright © UNESCO 1999. Reproduced by permission of UNESCO.)

Table 7B.15 Freshwater Abstractions by Source, in Selected Countries 1980–1999

	Per Capita/ (m^3/Capita) (b)	As % of Resources/ (a) (b)	Total Abstractions (mill m^3)				Surface Water (mill m^3)				Groundwater (mill m^3)			
			1980	1985	1990	1999 (c)	1980	1985	1990	1999 (c)	1980	1985	1990	1999 (c)
Canada	1610	1.7	37594	42383	45096	47250	36733	41486	44059	—	861	897	1037	—
Mexico	800	16.2	56003	—	—	78402	39374	—	—	53017	16629	23500	24453	25385
U.S.A.	1870	19.9	517720	467335	468620	492260	402750	366095	358790	—	114970	101420	109830	—
Japan	710	21.2	86000	87200	88900	89100	74100	75300	76600	77300	11900	11900	12200	11900
Korea	540	34.3	17510	18580	20600	24800	—	—	—	22200	—	—	—	2600
Australia	1300	6.8	10900	14600	—	24071	—	12360	—	19109	—	2240	—	4962
N. Zealand	570	0.6	1200	1900	—	2000	—	—	—	1200	—	—	—	800
Austria	440	4.2	3342	3363	3734	3561	2207	2195	2561	2496	—	1168	1135	1065
Belgium	730	45.1	—	—	—	7442	—	—	—	6802	—	—	—	641
Czech Rep.	190	12.4	3622	3679	3623	1976	2820	2873	2787	1419	802	806	836	557
Denmark	140	12.3	1205	—	1261	754	45	—	—	20	1160	—	1261	734
Finland	450	2.1	3700	4000	2347	2328	3510	3680	2107	2043	190	320	240	285
France	520	15.9	30972	34887	37687	30341	25268	28714	31486	24240	5704	6173	6201	6101
Germany	490	22.3	42206	41216	47873	40591	35344	34225	—	33880	6862	6991	—	6710
Greece	830	12.1	5040	5496	—	8695	3470	—	—	5023	1570	—	—	3563
Hungary	560	4.7	4805	6267	6293	5653	3551	4880	5266	4822	1254	1386	1026	831
Iceland	560	0.1	108	112	167	156	5	8	7	4	103	104	160	152
Italy	980	32.1	56200	52000	56200	56200	—	4000	—	—	—	12000	—	—
Luxembourg	140	3.7	—	67	59	60	—	22	32	29	—	45	27	32
The Netherlands	290	4.9	9198	9302	7806	4425	8190	8231	6757	3427	1008	1071	1049	998
Norway	600	0.7	—	2025	—	2588	—	1620	—	—	—	405	—	—
Poland	290	17.9	14184	15453	14248	11275	11899	13076	11928	9339	2285	2377	2320	1936
Portugal	1110	15.2	—	—	8600	11090	—	—	—	4800	—	—	—	6290
Slovak Rep.	210	1.4	2232	2061	2116	1149	1575	1390	1388	684	657	671	728	465
Spain	1040	36.8	39920	46250	36900	40855	34800	40840	31400	35323	5120	5410	5500	5532
Sweden	300	1.5	4106	2970	2968	2668	3511	2348	2360	2026	595	622	608	642
Switzerland	360	4.8	2589	2646	2665	2566	1667	1693	1724	1689	922	953	941	877
Turkey	590	16.6	16200	19400	32200	38900	11800	14100	25600	29552	4400	5300	6600	6000
UK	210	17.4	13514	11533	12052	11162	11024	9012	9344	8658	2490	2521	2709	2504
Russian Fed.	560	1.9	113178	117273	126083	82064	101799	103407	111297	72420	11379	13866	14786	11624
N. America	1600	11.0	611300	572000	581700	617900	—	—	—	—	—	—	—	—
OECD/Europe	560	13.9	271400	282700	299100	285600	—	—	—	—	—	—	—	—
EU-15	600	19.4	225600	231000	235400	221300	—	—	—	—	—	—	—	—
OECD	950	11.8	998300	977000	1006800	1043500	—	—	—	—	—	—	—	—

Source: From Table 3.1C, OECD Environmental Data Compendium 2002, © OECD 2002, www.oecd.org.

Note: CAN, 1980, 1985, and 1990: 1981, 1986, and 1991 data. 1999: WWF estimate for 1995; MEX, 1980: excluding agricultural uses besides irrigation. Data include Secretariat estimates for electrical cooling based on electricity generated in power stations in 1980. 1999: Total gross abstraction excluding 143 km^3 used in hydroelectric energy generation; U.S.A., 1999: WWF estimate for 1995; JPN, 1999: 1997 data; KOR, Partial totals excluding electrical cooling. 1999: 1997 data; AUS, 1980: 1977 data adjusted for an average climatic year. 1985: fiscal year 1983/84. 1999: 1996/97 data; NZL, Partial totals excluding industrial and electrical cooling. 1980: composite total based on data for various years. 1999: 1993 estimates; AUT, Partial totals. Surface water: excluding agriculture, irrigation and industry except cooling. Groundwater: excluding industry and electrical cooling. 1999: 1997 data; BEL, 1999: 1998 data; data include Secretariat estimates; DNK, 1980 and 1999: 1977 and 1998 data. 1990 refer only to groundwater abstractions, which represent the majority of total freshwater abstractions (e.g. 95–99% for 1995); FIN, Partial totals. 1985 and 1990: exclude agricultural uses besides irrigations. 1999: includes country estimate for agriculture; FRA, 1980 and 1999: 1981 and 1997 data. 1997: Break in time series; DEU, Excluding agricultural uses besides irrigation. 1980 and 1985: 1979 and 1983 data for western Germany only. 1990 and 1999: 1991 and 1998 data for total Germany. Data include electrical cooling; GRC, Partial totals excluding agricultural uses besides irrigation. 1999: 1997 data including, for public water supply, data from 42 out of 75 great water distribution enterprises; HUN, 1999: 1998 data; ISL, Totals include the domestic use of geothermal water. 1990: 1992 data; IRL, 1999: 1994 data; totals include 1980 data for electrical cooling; ITA, Excluding agricultural uses besides irrigation. 1980: including 1973 estimates for industrial cooling. 1990 and 1999: 1989 and 1998 data; LUX, 1990: 1989 data, including 1983 data; NLD, Partial totals excluding all agricultural uses. 1980, 1985, 1990 and 1999: 1981, 1986, 1991 and 1996 data; NOR, Data include 1978 data for industry. 1985: 1983 data. 1999: data are estimates for 1994; POL, Totals include abstractions for agriculture, which refer to aquaculture (areas over 10 ha) and irrigation (arable land and forest areas greater than 20 ha); animal production and domestic needs of rural inhabitants are not covered (selfsupply); PRT, Excluding agricultural uses besides irrigation. 1990: 1991 data. 1999: 1998 data; ESP, Excluding agricultural uses besides irrigation. Groundwater: excluding industry. 1990 and 1999: 1991 and 1997 data; SWE, 1980, 1985 and 1990: include data from different years. 1999: 1995 data; CHE, Partial totals excluding all agricultural uses. 1999: 1998 data; TUR, Partial totals. Excluding agricultural uses besides irrigation. 1980 and 1985: excluding electrical cooling. 1990: 1991 data. 1999: total: country estimates; surface and groundwater: 1997 data; UKD, Partial totals. England Wales only. Data include miscellaneous uses for power generation, but exclude hydroelectric power water use; RUS, 1990: 1991 data; Totals, Rounded figures, including Secretarial estimates. OECD and EU until 1985: western Germany only. % of renewable resources: calculated using the estimated totals for internal resources (not total resources as for countries), and considering England and Wales only.

[a] Data refer to total abstraction divided by total renewable resources, except for regional totals, where the internal resource estimates were used to avoid double counting. Total renewable resources represent the maximum quantity of water available on average.
[b] Data refer to 1999 or latest available year. Data prior to 1994 have not been considered.
[c] Data refer to 1999 or latest available year.

Source: From Table 3.1B, OECD Environmental Data Compendium 2002, © OCED 2002, www.oecd.org.

Table 7B.16 Worldwide Annual Groundwater Withdrawals and Desalinization

	Average Annual Groundwater Recharge		Annual Groundwater Withdrawals							Desalinated Water Production (mill m³) 1990
							Sectoral Share (percent) [a]			
	Total (km³) Years Vary	Per Capita (m³) Year 2000	Year	Total (km³)	Percentage of Annual Recharge	Per Capita (m³)	Domestics	Industry	Agriculture	
World	**X**	**X**	**1995**	**600–700**	**X**	**106–124**	**65**	**15**	**20**	**X**
Asia (Excl Middle East	**X**	**X**	**X**	**X**	**X**	**X**	**X**	**X**	**X**	**X**
Armenia	4.2	1,193	X	X	X	X	X	X	X	X
Azerbaijan	6.5	842	X	X	X	X	X	X	X	X
Bangladesh	21.0	163	1990	10.7	50.9	97.6	13	1	88[l]	X
Bhutan	X	X	X	X	X	X	X	X	X	X
Cambodia	17.6	1,576	X	X	X	X	X	X	X	X
China	828.8	649	1988	52.9	6.4	47.1	X	X	54	X
Georgia	17.2	3,469	1990	3.0	17.4	549.5	X	X	X	X
India	418.5	413	1990	190.0	45.4	223.3	9	2	89[m]	X
Indonesia	455.0	2,145	X	X	X	X	X	X	X	X
Japan	27.0[n]	213	1995[h]	13.6	50.3	108.2	29	41	30[o]	40.0[p]
Kazakhstan	35.9	2,211	1993[q]	2.4	6.7	143.9	21	71	8[q]	1,328.0[r]
Korea, Dem People's Rep	21.0[b]	874	X	X	X	X	X	X	X	X
Korea, Rep	13.3[b]	284	1995[h]	2.5	18.6	55.1	X	X	17[s]	X
Kyrgyzstan	13.6	2,894	1994[t]	0.6	4.4	132.0	50	25	25[t]	X
Lao People's Dem Rep	38.0	6,994	X	X	X	X	X	X	X	X
Malaysia	64.0	2,877	1995	0.4	0.6	19.0	62	33	5	X
Mongolia	6.1	2,291	1993	0.4	5.8	149.1	X	X	X	X
Maynmar	156.0[b]	3,420	X	X	X	X	X	X	X	X
Nepal	X	X	X	X	X	X	X	X	X	X
Pakistan	55.0	351	1991	60.0	109.1	489.5	X	X	90[u]	X
Philippines	180.0[b]	2,369	1980	4.0	2.2	82.8	50	50	X	X
Singapore	X	X	X	X	X	X	X	X	X	X
Sri Lanka	7.8	414	X	X	X	X	X	X	X	X
Tajikistan	6.0	970	1994	2.3	37.7	398.7	X	X	X	X
Thailand	41.9[b]	682	1980	0.7	1.7	15.0	60	26	14	X
Turkmenistan	3.4	753	1994	0.4	11.9	100.3	53	9	38	X
Uzbekistan	19.7	809	1994	7.4	37.6	334.3	33	11	57[v]	X
Vietnam	48.0	601	1990	0.8	1.7	11.9	X	X	X	X
Europe	**X**	**X**	**X**	**X**	**X**	**X**	**X**	**X**	**X**	**X**
Albania	7.0	2,248	1989	0.6	9.0	193.6	48	X	52	X
Austria	22.3[b]	2,716	1995[h]	1.4	6.2	172.5	52	43	5[w]	X
Belarus	18.0	1,758	1989	1.2	6.6	115.7	52	13	28[x]	X
Belgium	0.9[c]	89	1980	0.8	86.4	79.0	55	22	4	X
Bosnia and Herzegovina	X	X	X	X	X	X	X	X	X	X
Bulgaria	13.4[y]	1,629	1988	5.0	37.3	566.1	X	X	X	X
Croatia	11.0	2,459	X	X	X	X	X	X	X	X
Czech Rep	X	X	1995[h]	0.5	X	48.0	X	X	X	X

Denmark	30.0[z]	5,668	1995[h]	0.9	3.0	169.8	40	22	38[aa]	X
Estonia	4.0	2,865	X	X	X	X	X	X	X	X
Finland	1.9[bb]	367	1995[h]	0.2	12.8	47.8	65	11	24[cc]	X
France	100.0[dd]	1,693	1994	6.0	6.0	103.8	56	27	17	X
Germany	45.7[b]	556	1990	7.1	15.5	89.4	48	47	4[ee]	X
Greece	10.3	968	1990	2.0	19.4	195.7	37	5	58	X
Hungry	6.8[c]	678	1995[h]	1.0	14.5	96.5	35	48	18[ff]	X
Iceland	24.0[b]	85,419	1995[h]	0.2	0.6	558.9	X	X	X	X
Ireland	3.5[c]	928	1995	0.2	6.5	62.3	35	38	29[gg]	X
Italy	43.0	750	1992	13.9	32.3	243.2	39	4	58	X
Latvia	2.2	934	X	X	X	X	X	X	X	X
Lithuania	1.2	327	1995	0.2	17.1	55.1	X	X	X	X
Macedonia, FYR	X	X	X	X	X	X	X	X	X	X
Moldova, Rep	0.4	91	X	X	X	X	X	X	X	X
Netherlands	4.5[hh]	285	1990	1.0	23.3	70.2	32	45	23[ii]	X
Norway	96.0[b]	21,502	1985	0.4	0.4	97.5	27	73	X[jj]	X
Poland	36.0	929	1995[h]	2.0	5.5	51.5	70	30	X[kk]	X
Portugal	5.1[c]	516	1995	3.1	60.1	311.0	39	23	39[ll]	X
Romania	8.3[hh]	372	1993	3.6	43.7	158.0	61	38	1[mm]	X
Russian Federation	788.0	5,363	1988	12.6	1.6	85.5	X	X	X	X
Slovakia	X	X	1995[h]	0.6	X	113.0	X	X	X	X
Slovenia	X	X	1994	0.2	X	88.9	X	X	X	X
Spain	28.9	729	1995[h]	5.4	18.8	137.2	18	2	80	X
Sweden	20.0[c]	2,245	1995[h]	0.6	3.2	72.8	92	8	X[nn]	X
Switzerland	2.7	366	1995[h]	0.9	33.4	126.3	72	40	X[oo]	X
Ukraine	20.0	396	1989	4.0	20.1	77.5	30	18	52[pp]	X
United Kingdom	9.8	167	1995[h]	2.5	25.2	42.4	51	47	2[qq]	X
Yugoslavia	3.0	282	X	X	X	X	X	X	X	X
Middle East & N. Africa	**X**	**X**	**X**	**X**	**X**	**X**	**X**	**X**	**X**	**X**
Afghanistan	29.0[b]	1,276	X	X	X	X	X	X	X	X
Algeria	1.7[c]	54	1989	2.9	167.6	117.1	46	5	49	64.0
Egypt	1.3[c]	19	1995	5.3	407.7	85.1	58	0	42[rr]	25.0
Iran, Islamic Rep	42.0[c]	620	1980	29.0	69.0	738.8	X	X	X	2.9[ss]
Iraq	13.0[c]	562	1985	0.2	1.5	13.1	50	40	X	X
Israel	0.5	80	1996	1.2	234.0	204.5	18	2	80[tt]	20.0[uu]
Jordan	0.6[c]	87	1993	0.5	91.4	100.7	30	4	66[vv]	2.0[r]
Kuwait	X	X	1994	0.3	X	142.7	0	0	100[ww]	231.0[r]
Lebanon	4.8	1,463	1991	0.4	8.3	153.2	13	9	78	X
Libyan Arab Jamahiriya	0.7	116	1995	3.7	561.5	734.9	9	4	87[yy]	70.0[xx]
Morocco	9.0	317	1998	2.7	29.8	97.9	16	X	84[zz]	3.4[aaa]
Oman	1.0	376	1985	0.4	41.9	280.7	X	X	X	34.0[bbb]
Saudi Arabia	1.0[c]	44	1990	14.4	1518.9	899.3	10	X	90	714.0[bbb]
Syrian Arab Rep	6.6	409	1993	1.8	27.3	133.5	13	4	83[ddd]	X
Tunisia	4.2	433	1995	1.6	39.2	181.8	10	4	86	8.3
Turkey	20.0	300	1995[h]	7.6	38.0	124.0	31	9	60[ddd]	0.5
United Arab Emirates	0.1	49	1995	1.6	1333.3	724.1	X	19[j]	81[ccc]	385.0[bbb]

(Continued)

Table 7B.16 (Continued)

	Average Annual Groundwater Recharge		Annual Groundwater Withdrawals							Desalinated Water Production (mill m³) 1990
	Total (km³) Years Vary	Per Capita (m³) Year 2000	Year	Total (km³)	Percentage of Annual Recharge	Per Capita (m³)	Sectoral Share (percent) [a] Domestics	Industry	Agriculture	
Yemen	1.5	84	1985[eee]	1.4	88.5	139.2	X	X	X	10.0[fff]
Sub-Saharan Africa	**X**	**X**	**X**	**X**	**X**	**X**	**X**	**X**	**X**	**X**
Angola	72.0[b]	5,591	X	X	X	X	X	X	X	X
Benin	1.8[c]	295	X	X	X	X	X	X	X	X
Botswana	1.7[c]	1,048	X	X	X	X	X	X	X	X
Burkina Faso	9.5[c]	796	X	X	X	X	X	X	X	X
Burundi	2.1[b]	314	X	X	X	X	X	X	X	X
Cameron	100.0[b]	6,629	X	X	X	X	X	X	X	X
Central African Rep	56.0[b]	15,490	X	X	X	X	X	X	X	X
Chad	11.5[b]	1,503	1990	0.1	0.8	15.7	29	X	71	X
Congo	198.0[b]	67,268	X	X	X	X	X	X	X	0.2
Congo, Dem Rep	421.0[b]	8,150	X	X	X	X	X	X	X	X
Côte d'Ivoire	37.7[c]	2,550	X	X	X	X	X	X	X	X
Equatorial Guinea	10.0[b]	22,097	X	X	X	X	X	X	X	X
Eritrea	X	X	X	X	X	X	X	X	X	X
Ethiopia	44.0[b]	703	X	X	X	X	X	X	X	X
Gabon	62.0[b]	50,566	1989	0.0	0.0	0.6	100	0	0	X
Gambia	0.5[b]	383	X	X	X	X	X	X	X	X
Ghana	26.3[c]	1,301	X	X	X	X	X	X	X	X
Guinea	38.0[b]	5,114	X	X	X	X	X	X	X	X
Guinea-Bissau	14.0[b]	11,541	X	X	X	X	X	X	X	X
Kenya	3.0[c]	100	X	X	X	X	X	X	X	X
Lesotho	0.5[b]	232	X	X	X	X	X	X	X	X
Liberia	60.0[b]	19,023	X	X	X	X	X	X	X	X
Madagascar	55.0[b]	3,450	1984	4.8	8.7	482.9	0	X	X[d]	X
Malawi	1.4[b]	128	X	X	X	X	X	X	X	X
Mali	20.0[c]	1,780	1989	0.1	0.5	11.6	X	X	X	X
Mauritania	0.3[c]	112	1985	0.9	293.3	498.3	X	X	X	1.7
Mozambique	17.0[b]	864	X	X	X	X	X	X	X	0.1
Namibia	2.1[b]	1,217	X	X	X	X	X	X	X	3.0
Niger	2.5[c]	233	1988	0.1	5.2	17.9	58	4	39	X
Nigeria	87.0[b]	780	X	X	X	X	X	X	X	3.0
Rwanda	3.6[b]	466	X	X	X	X	X	X	X	X
Senegal	7.6[b]	802	1985	0.3	3.3	39.2	24	X	72	0.1
Sierra Leone	50.0[b]	10,300	X	X	X	X	X	X	X	X
Somalia	3.3[b]	327	1985	0.3	9.1	45.8	X	X	X	0.1
South Africa	4.8	119	1980	1.8	37.3	64.9	11	6	84	17.5
Sudan	7.0	237	1985	0.3	4.0	13.0	X	X	X	0.4
Tanzania, United Rep	30.0[b]	895	X	X	X	X	X	X	X	X
Togo	5.7[c]	1,231	X	X	X	X	X	X	X	X
Uganda	29.0[b]	1,332	X	X	X	X	X	X	X	X

Zambia	47.1	5,137	X	X	X	X	X	X	X	X
Zimbabwe	5.0[b]	428	X	X	X	X	X	X	X	X
North America	**XX**	**X**	**X**	**X**	**X**	**X**	**X**	**X**	**X**	**X**
Canada	370.0[b]	11,878	1889	1.0	0.3	37.3	34	11	34[e]	X
United States	1,514.0[f]	5,439	1990	109.8	7.3	432.3	20	5	62[g]	X
C. America & Caribbean	**X**	**X**	**X**	**X**	**X**	**X**	**X**	**X**	**X**	**X**
Belize	X	X	X	X	X	X	X	X	X	X
Costa Rica	21.0[b]	5,219	X	X	X	X	X	X	X	X
Cuba	8.0[b]	714	1975	3.8	47.5	408.3	X	X	X	X
Dominican Rep	3.0[b]	353	X	X	X	X	X	X	X	X
El Salvador	X	X	X	X	X	X	X	X	X	X
Guatemala	31.0[b]	2.723	X	X	X	X	X	X	X	X
Haiti	2.5[b]	304	X	X	X	X	X	X	X	X
Honduras	39.0[b]	6,013	X	X	X	X	X	X	X	X
Jamaica	X	X	X	X	X	X	X	X	X	X
Mexico	139.0[b]	1,406	1995[h]	25.1	18.1	275.4	13	23	64[i]	X
Nicaragua	59.0[b]	11,627	X	X	X	X	X	X	X	X
Panama	42.0[b]	14,708	X	X	X	X	X	X	X	X
Trinidad and Tobago	X	X	X	X	X	X	X	X	X	X
South America	**X**	**X**	**X**	**X**	**X**	**X**	**X**	**X**	**X**	**X**
Argentina	128.0[b]	3,456	1975	4.7	3.7	180.4	11	19	70	X
Bolivia	130.0[b]	15,609	X	X	X	X	X	X	X	X
Brazil	1,874.0[b]	11,016	1987	8.0	0.4	57.0	38	25	38	X
Chile	140.0[b]	9,204	X	X	X	X	X	X	X	X
Colombia	510.0[b]	12,051	X	X	X	X	X	X	X	X
Ecuador	134.0[b]	10,596	X	X	X	X	X	X	X	X
Guyana	103.0[b]	119,582	X	X	X	X	X	X	X	X
Paraguay	41.0[b]	7,459	X	X	X	X	X	X	X	X
Peru	303.0[b]	11,807	1973	2.0	0.7	139.4	25	15	60	X
Suriname	80.0[b]	191,787	X	X	X	X	X	X	X	X
Uruguay	23.0[b]	6,892	X	X	X	X	X	X	X	X
Venezuela	227.0[b]	9,392	X	X	X	X	X	X	X	X
Oceania	X	X	X	X	X	X	X	X	X	X
Australia	72.0[c]	3,812	1985	2.2	3.1	143.2	X	20[j]	57[k]	X
Fiji	X	X	X	X	X	X	X	X	X	X
New Zealand	198.0[b]	51,270	X	X	X	X	X	X	X	X
Papua New Guinea	X	X	X	X	X	X	X	X	X	X
Solomon Islands	X	X	X	X	X	X	X	X	X	X

a Estimates are typically approximate and therefore the sum of the sectoral data may not add to 100 percent.
b Sum of all groundwater flows, including base flow (as a constituent of surface water flows).
c Sum of all aquifer recharge flows.
d Sectoral data for Madagascar equal 0.32 percent for domestic for 1984 as reported by Margat 1990.
e Sectoral data for Canada are calculated using a groundwater withdrawal value of 1.6 km^3 from 1985 as reported by Margat 1990.
f Data for the United States are from Economic Commission for Europe (1992) without a specific date. Data reported by Margat (1990) are 660 km^3 from a source dated 1974 and refer to the U.S. Including the 50 states and Puerto Rico.
g Sectoral data for the U.S. are calculated using a groundwater withdrawal value of 101.3 km^3 from 1985 as reported by Margat 1990.
h Data refer to 1995 or latest available year (generally from 1991, 1992, 1993, or 1994).

(Continued)

Table 7B.16 (Continued)

[i] Sectoral data are calculated using a groundwater withdrawal value of 23.5 km^3 from around 1985 as reported by Margat 1990.
[j] Domestic and industrial withdrawals have been combined.
[k] Sectoral data for Australia are calculated using a groundwater withdrawal value of 2.46 km^3 from 1983 as reported by Margat 1990.
[l] Sectoral data for Bangladesh are calculated using a groundwater withdrawal value of 3.4 km^3 from 1979 as reported by Margat 1990.
[m] Sectoral data are from around 1990 as provided by Shiklomonov; total withdrawal data also are from 1990 but are from FAO, Irrigation in Asia in Figures, p. 95.
[n] Data are from FAO, Irrigation in Asia in Figures (1999) which states "The renewable potential of groundwater resources is estimated at about 27 km^3/yr—." A value of 185 km^3/yr is provided by Margat (1990) cited from L'vovich 1974.
[o] Sectoral data for Japan are from 1987 as provided by Shiklomonov 1997 based on groundwater withdrawal of 12.88 km^3.
[p] Data are from 1996.
[q] Both withdrawal and sectoral data are estimated from a bar graph from FAO Report: Irrigation in the Former Soviet Union Countries in Figures, p. 116.
[r] Data are from 1993.
[s] Sectoral data for the Republic of Korea are calculated using a groundwater withdrawal value of 1.2 km^3 from around 1985 as provided by Margat 1990.
[t] Kyrgystan data: FAO Irrigation in the Former Soviet Union Countries in Figures, p. 129, "In 1994, more than 0.6 km^3 of water was withdrawn from groundwater." We have entered a value of 0.6 but the figure may be higher; we have calculated the sectoral data from the figure in this report on page 129, using the 0.6 figure for total withdrawal.
[u] Sectoral data for Pakistan are from Shiklomonov who reports "approximately 90 percent" for agriculture share; total withdrawal also is approximately 60 km^3/yr for around 990 (table p. 57).
[v] Sectoral data for Uzbekistan are from 1994 FAO Irrigation in the Former Sovet Union Countries in Figures estimated from a bar graph, p. 217.
[w] Sectoral data for Austria are calculated using a groundwater withdrawal value of 1.17 km^3 from 1980 as reported by Margat 1990.
[x] Sectoral data Belarus are calculated using a groundwater withdrawal value of 1.06 km^3 from 1985 as reported by Margot 1990.
[y] Data for Bulgaria are from ECE (1992) and refer to the year 1988; Margat (1990) reports data from a 1989 source (Anonyme 1989) as 3.1 km^3.
[z] Data for Denmark are from ECE (1992) and refer to the year 1985; Margat (1990) reports data from a 1981 source (Anonyme 1981) as 4.3 km^3.
[aa] Sectoral data are calculated using a groundwater withdrawal value of 1.32 km^3 from 1977 as reported by Margat 1990.
[bb] Data for Finland are from ECE (1992) without a year specified; Margat (1990) reports data from a 1989 source (Anonyme 1989) as 2.2 km^3.
[cc] Sectoral data are calculated using a groundwater withdrawal value of 0.37 km^3 from 1980 as reported by Margat 1990.
[dd] Data are from Margat (1990) and refer to a source dated 1989 (Margat 1989); data reported from ECE (1992) is 26.0 km^3 for 1981.
[ee] Sectoral data are from Margat (1990) and combine his data for both Germanys. Margat's total withdrawal data are from different dates for the two Germanys. Germany—RFA is 7.77 km^3 from 1981 and Germany—ex DDR is from 1975; the combined total is 9.55 which is used to calculate the sectoral percentage. The sectoral data also are from 1981 (Germany RFA) and 1975 (Germany—ex DDR).
[ff] Sectoral data are calculated using a groundwater withdrawal value of 1.6 km^3 from 1972 as reported by Margat 1990.
[gg] Sectoral data for Ireland are calculated using a groundwater withdrawal value of 0.17 km^3 from 1980 as reported by Margat 1990.
[hh] Sum of the total groundwater flow that is exploitable.
[ii] Sectoral data are calculated using a groundwater withdrawal value of 1.28 km^3 from 1981 as reported by Margat 1990.
[jj] Sectoral data for Norway are calculated using a groundwater withdrawal value of 0.11 km^3 from 1985 as reported by Margat 1990.
[kk] Sectoral data are calculated using a groundwater withdrawal value of 2.0 km^3 from 1980–81 as reported by Margat 1990.
[ll] Sectoral data are calculated using a groundwater withdrawal value of 2.0 km^3 from 1980 as reported by Margat 1990.
[mm] Sectoral data are calculated using a groundwater withdrawal value of 1.18 km^3 from 1975 as reported by Margat 1990.
[nn] Sectoral data for Sweden are calculated using a groundwater withdrawal value of 0.48 km^3 from 1985 as reported by Margat 1990.
[oo] Sectoral data for Switzerland are calculated using a groundwater withdrawal value of 1.0 km^3 from 1983 as reported by Margat 1990.
[pp] Sectoral data are calculated using a groundwater withdrawal value of 4.22 km^3 from 1985 at reported by Margat 1990.
[qq] Sectoral data are calculated using a groundwater withdrawal value of 2.38 km^3 from 1975 as reported by Margat 1990.
[rr] Sectoral data are from 1992, Margat Blue Plan.
[ss] Data are from 1991.
[tt] Sectoral data are from 1994, Margat Blue Plan.
[uu] Data are from Margat, personal communication February 2000.
[vv] Groundwater withdrawal and sectoral data are estimated from a bar graph for 1993 from FAD water Report: Irrigation in the Near East Region in Figures, Rome, 1997, p. 115.
[ww] Groundwater withdrawal and sectoral data are estimated from a bar graph for 1994 from FAD Water Report: Irrigation in the Near East Region in Figures, Rome, 1997, p. 124.
[xx] Data are from 1994; from FAD irrigation in the Near East Region in Figures, p. 29.
[yy] Sectoral percentages are calculated using groundwater withdrawal of 3.81 km^3 which is an estimate provided with sectoral data for 1995 in Margat Blue Plan.
[zz] Sectoral data are from 1991, Margat Blue Plan.
[aaa] Data are from 1992.
[bbb] Data are from 1995.
[ccc] Sectoral percentage for UAE are a combination of data from text and a bar graph for 1995 from FAO Water Report: Irrigation in the Near East Region in Figures, Rome 1997, p. 266.

[ddd] Sectoral data are from 1990, Margat Blue Plan.
[eee] Groundwater withdrawal data are from Margat 1990, presented as two separate values, one for Yemen du Nord for around 1985 equal to 1 bill m^3/yr; the other for Yemen du Sud for 1975 equal to 0.35 billion m3 per year. These two figures have been added.
[fff] Data are from 1989.
[ggg] Sectoral data are from Margat Blue Plan for 1989.
[hhh] Sectoral data refer only to Mauritius Island only.

Technical Notes

Source: From World Resources Institute, Earth Trends Environmental Information, Water Resources and Freshwater Ecosystems, Data Tables, Table FW.2 Groundwater and Desalinization, www.earthtrends.wri.org.

Original Source: From Groundwater resources and withdrawal data: J. Margat, Les eau — souterraines sand le monde (Bureau de recherches géologiques et minières [BRGM], Département eau, Oriéans, France, December 1990); J. Margat and D. Vallée, Water Resources and Uses in the Mediterranean Countries (Blue Plan, Sophia Antipolis, 1999); I.A. Shiklomanov, Comprehensive Assessment of the Freshwater Resources of the World (Stockholm Environment Institute, Stockholm, 1997); Organisation for Economic Co-Operation and Development (OECD), OECD Environmental Data Compendium 1997 (OECD, Paris, 1997); and Economic Commission for Europe, The Environment in Europe and North America (United Nations, New York, 1992).

Groundwater resources and desalinization activities: J. Margat, Le — Eau — Souterraines Dans Le Bassin Mediterraneen. Resources et Utilisations Plan Bleu, Doc. BRGM 282 (Ed. BRGM, Orléans, France, 1998); Food and Agriculture Organization of the United Nations (FAO), Irrigation in Africa in Figures, Water Reports No. 7 (FAO, Rome, 1995); FAO, Irrigation in the Near East Region in Figures, Water Reports No. 9 (FAO, Rome, 1997); FAO, Irrigation in the Former Soviet Union in Figures, Water Report No. 15 (FAO, Rome, 1997); FAO, Irrigation in Asia in Figures, Water Reports No. 18 (FAO, Rome, 1999); and FAO, Irrigation in Latin America in Figures, Water Reports (FAO, Rome, in preparation). Population data: United Nations (U.N.) Population Division, World Population Prospects, 1950–2050 (The 1998 Revision), on diskette (U.N., New York, 1999).

Average annual groundwater recharge is the amount of water that is estimated to annually infiltrate soils, including water from rivers and streams that lose it to underlying strata. In general, this figure would represent the maximum amount of water that could be withdrawn annually without ultimately depleting the groundwater resource. These data are estimated in a variety of ways and caution should be used in comparing values for different countries.

Per capita recharge is the amount of water that annually infiltrates soils on a per person basis, using 2000 population estimates from the U.N. Population Division.

Annual total groundwater withdrawals refers to abstractions from all groundwater sources — even nonrenewable sources. The percentage of annual recharge refers to total groundwater withdrawals. Per capita annual withdrawals were calculated using national population data for the year of data shown.

Sectoral share of withdrawals of groundwater is classified as domestic (drinking water, homes, commercial establishments, public services, and municipal use), industry including water withdrawn to cool thermoelectric plants), and agriculture (irrigation and livestock).

Desalinated water production refers to the removal of salt from saline waters — usually seawater — using a variety of techniques including reverse osmosis. Most desalinated water is used for domestic purposes.

Totals may not add due to rounding.

SECTION 7C PUBLIC WATER SUPPLY — UNITED STATES

Table 7C.17 Public Water Supply Supreme in the United States, 2003

		System Size by Population Served					Water Source Type		
		Very Small 500 or Less	Small 501–3,300	Medium 3,301–10,000	Large 10,001–100,000	Very Large >100,000	Groundwater	Surface Water	Total
CWS	# systems	30,417	14,394	4,686	3,505	361	41,499	11,864	53,363
	Pop. served	5,010,834	20,261,508	27,201,137	98,706,485	122,149,436	86,348,074	186,981,326	273,329,400
	% of systems	57%	27%	9%	7%	1%	78%	22%	100%
	% of pop	2%	7%	10%	36%	45%	32%	68%	100%
NTNCWS	# systems	16,785	2,786	97	16	2	18,908	778	19,686
	Pop. served	2,327,575	2,772,334	506,124	412,463	279,846	5,568,192	730,150	6,298,342
	% of systems	85%	14%	0%	0%	0%	96%	4%	100%
	% of pop	37%	44%	8%	7%	4%	88%	12%	100%
TNCWS	# systems	85,366	2,657	96	29	4	86,061	2,091	88,152
	Pop. served	7,315,647	2,602,706	528,624	619,248	12,269,000	10,527,089	12,808,136	23,335,225
	% of systems	97%	3%	0%	0%	0%	98%	2%	100%
	% of pop	31%	11%	2%	3%	53%	45%	55%	100%
	Total # systems	**132,568**	**19,837**	**4,879**	**3,550**	**367**	**146,468**	**14,733**	**161,201**
	Total Population Served						**102,443,355**	**200,519,612**	

Note: Active, current systems, from Safe Drinking Water Information System/Federal version (SDWIS/FED) 03Q4 frozen inventory table. CWS, Community Water System: A public water system that supplies water to the same population year-round; NTNCWS, Non-Transient Non-Community Water System: A public water system that regularly supplies water to at least 25 of the same people at least six months per year, but not year-round. Some examples are schools, factories, office buildings, and hospitals which have their own water systems; TNCWS, Transient Non-Community Water System: A public water system that provides water in a place such as a gas station or compground where people do not remain for long periods of time. Groundwater systems, groundwater (GW), purchased groundwater (GWP); Surface water systems, surface water (SW), purchased surface water (SWP), groundwater under the direct influence of surface water (GU), purchased groundwater under the direct influence of surface water (GUP).

Source: From USEPA, *FACTOIDS: Drinking Water and Groundwater Statistics for 2003*, www.epa.gov.

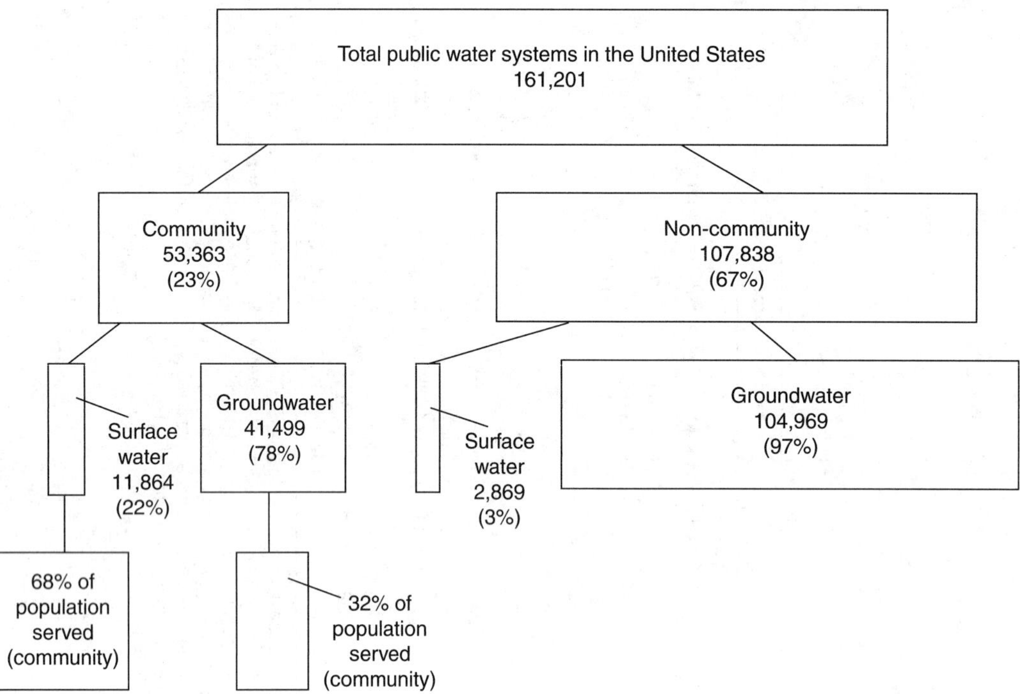

Figure 7C.9 Number and type of public water systems in the United States in 2003. (From American Water Works Association, 1988, *New Dimensions in Safe Drinking Water*, Copyright AWWA modified with data from USEPA, FY2003, Drinking Water Factors, www.epa.gov.)

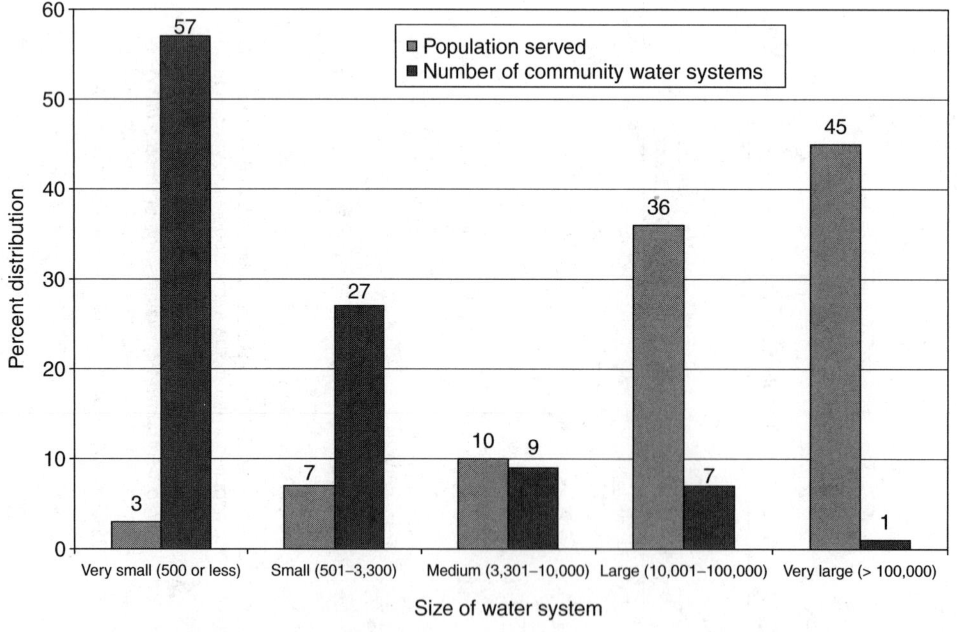

Figure 7C.10 Size distribution of community water systems in the United States in 2003. (From American Water Works Association, 1988, *New Dimensions in Safe Drinking Water*, modified with data from USEPA, *FACTOIDS: Drinking Water and Groundwater Statistics for 2003*, www.epa.gov.)

Table 7C.18 Average Daily Production (MGD) by Primary Water Source in Public Supply Systems in the United States in 2000 (Based on a Survey of 1246 Systems)

	System Service Population Category								
Primary Source of Water	**100 or Less**	**101–500**	**501–3,300**	**3,301–10,000**	**10,001–50,000**	**50,001–100,000**	**100,001–500,000**	**Over 500,000**	**All Sizes**
Primarily Groundwater Systems									
100% Groundwater									
Average Daily Production	0.011	0.034	0.158	1.058	2.628	11.892	17.631	125.642	0.313
Confidence Interval	±0.009	±0.009	±0.034	±0.193	±0.382	±1.643	±13.186	±23.115	±0.052
Mostly Groundwater									
Average Daily Production	0.003	0.019	0.524	0.872	3.718	11.225	29.778	144.963	3.587
Confidence Interval	±0.004	±0.000	±0.552	±0.365	±1.154	±1.490	±4.390	±32.023	±1.236
% Groundwater	66.7	83.3	58.1	78.0	84.5	74.1	73.0	83.5	78.6
Confidence Interval	±0.0	±0.0	±4.9	±11.1	±9.3	±3.6	±5.5	±14.8	±5.9
% Surface Water	0.0	0.0	41.9	7.4	2.1	12.4	6.1	0.0	6.0
Confidence Interval	±0.0	±0.0	±4.9	±7.3	±3.9	±6.1	±3.1	±0.0	±3.9
% Purchased Water	33.3	16.7	0.0	14.7	13.4	13.5	20.9	16.5	15.4
Confidence Interval	±0.0	±0.0	±0.0	±9.6	±8.8	±6.4	±5.8	±14.8	±5.3
Observations	86	88	88	53	35	20	20	5	395
Primarily Surface Water Systems									
100% Surface Water									
Average Daily Production	0.010	0.071	0.279	0.935	4.766	11.041	31.875	271.287	4.280
Confidence Interval	±0.007	±0.031	±0.059	±0.153	±1.430	±2.681	±2.733	±79.561	±0.995
Mostly Surface Water									
Average Daily Production	0.014	0.013	0.269	0.779	4.253	11.537	42.431	219.892	8.636
Confidence Interval	±0.012	±0.010	±0.141	±0.318	±1.325	±1.682	±5.065	±96.067	±4.026
% Groundwater	35.2	24.0	13.5	18.9	14.3	10.4	14.0	8.9	15.6
Confidence Interval	±19.6	±6.5	±10.3	±10.6	±6.5	±4.7	±2.9	±4.1	±4.1
% Surface Water	64.8	76.0	69.4	76.7	81.5	85.4	82.8	86.4	77.7
Confidence Interval	±19.6	±6.5	±8.0	±8.5	±5.8	±5.0	±3.0	±4.1	±3.6
% Purchased Water	0.0	0.0	17.0	4.4	4.1	4.3	3.3	4.7	6.6
Confidence Interval	±0.0	±0.0	±13.0	±4.4	±4.3	±3.4	±1.6	±2.3	±3.7
Observations	41	49	56	63	64	35	70	26	404
Primarily Purchased Water Systems									
100% Purchased Water									
Average Daily Production	0.008	0.017	0.150	0.886	2.302	9.595	23.975	129.602	0.866
Confidence Interval	±0.008	±0.004	±0.046	±0.445	±0.918	±1.417	±3.277	±69.896	±0.309
Mostly Purchased Water									

(Continued)

Table 7C.18 (Continued)

	System Service Population Category								
Primary Source of Water	**100 or Less**	**101–500**	**501–3,300**	**3,301–10,000**	**10,001–50,000**	**50,001–100,000**	**100,001–500,000**	**Over 500,000**	**All Sizes**
Average Daily Production	*	0.016	0.103	1.407	5.595	16.498	32.220	280.185	5.538
Confidence Interval	*	±0.000	±0.093	±0.490	±1.969	±8.407	±4.606	±68.668	±3.383
% Groundwater	*	33.3	42.8	28.4	12.8	22.4	19.5	11.2	31.0
Confidence Interval	*	±0.0	±4.6	±10.9	±7.3	±10.8	±4.4	±5.8	±7.3
% Surface Water	*	0.0	0.0	3.9	6.9	8.1	8.3	13.1	2.8
Confidence Interval	*	±0.0	±0.0	±6.1	±8.0	±9.2	±3.6	±6.0	±2.5
% Purchased Water	*	66.7	57.2	67.7	80.3	69.5	72.2	75.7	66.2
Confidence Interval	*	±0.0	±4.6	±9.7	±10.3	±7.2	±4.7	±8.2	±6.1
Observations	6	22	47	19	17	22	23	7	163
All Systems									
Average Daily Production	0.011	0.033	0.166	0.995	3.589	11.305	28.811	222.672	1.103
Confidence Interval	±0.008	±0.007	±0.024	±0.131	±0.518	±1.093	±5.427	±41.741	±0.124
Observations	138	166	208	170	159	116	201	58	1,216

Note: *No purchased water systems of this size in sample. Definitions: Production is the amount of water drawn from each source. It includes water delivered to customers and system losses.

The tabulations presented in the Community Water System Survery 2000 are based on data collected from a sample of U.S. Water Systems, not from a census of every water system in the United States. A confidence interval is one way to gauge how precisely a given tabulation of survey data can be generalized to the entirety of U.S. Systems represented by the surveyed systems. Any result presented in the table must be viewed as the center of a range that would encompass the precise number that would be found if every U.S. water system could have been included in the tabulation, and not only those who were sampled and responded to the 2000 Community Water System Survey. The confidence interval expresses the range as a "plus/minus," that is, an amount to be added to and subtracted from the calculated data point actually presented in the table. The confidence interval is designed to include the true value in the stated rage 95 perecent of the time.

Source: From USEPA, 2002, *Community Water System Survery 2000*, EPA 815-R-02-005A, www.epa.gov.

Table 7C.19 Number and Percentage of Systems by Primary Source of Water in Public Supply Systems in the United States in 2000 (Based on a Survey of 1246 September)

Primary Source of Water	System Service Population Category								
	100 or Less	101–500	501–3,300	3,301–10,000	10,001–50,000	50,001–100,000	100,001–500,000	Over 500,000	All Sizes
Primary Groundwater Systems									
100% Groundwater									
Number	10,358	12,521	8,687	2,576	971	80	108	7	35,308
Percent	82	76	62	51	32	17	25	8	68
Mostly Groundwater									
Number	1,398	624	283	495	368	56	53	3	3,280
Percent	11	4	2	10	12	12	12	4	6
Primarily Surface Water Systems									
100% Surface Water									
Number	790	897	1,015	835	769	140	113	36	4,595
Percent	6	5	7	17	26	30	26	43	9
Mostly Surface Water									
Number	43	239	197	173	220	70	65	17	1,024
Percent	0	1	1	3	7	15	15	20	2
Primarily Purchased Water Systems									
100% Purchased Water									
Number	69	2,050	3,412	773	476	94	46	13	6,933
Percent	1	12	24	15	16	20	11	15	13
Mostly Purchased Water									
Number	[a]	130	423	200	209	31	45	8	1,046
Percent	[a]	1	3	4	7	7	10	10	2
All									
Number	12,658	16,461	14,017	5,052	3,013	471	430	84	52,186
Percent	100	100	100	100	100	100	100	100	100

Note: The tabulations presented in the Community Water System Survey 2000 are based on data collected from a sample of U.S. Water Systems, not from a census of every water system in the United States.

[a] No purchased water systems of this size in sample.

Source: From USEPA, 2002, *Community Water Systems Survey 2000*, EPA 815-R-02-005A, www.epa.gov.

Table 7C.20 Public Water Supply Systems in United States by State, 2003

	Water System Type				Water Source Type	
	CWS	NTNCWS	TNCWS	Total	Groundwater	Surface Water
AK	434	213	936	1,583	1,337	246
	461,264	43,412	91,513	596,189	294,193	301,996
AL	607	32	72	711	417	294
	5,167,519	19,448	7,399	5,194,366	1,476,656	3,717,710
AR	730	61	329	1,120	705	415
	2,482,867	12,622	17,365	2,512,854	920,088	1,592,766
AZ	799	210	644	1,653	1,442	211
	4,766,531	132,626	132,193	5,031,350	1,440,327	3,591,023
CA	3,140	1,413	3,017	7,570	6,488	1,082
	36,207,653	495,441	11,031,005	47,734,099	10,688,260	37,045,839
CO	836	170	1,004	2,010	1,569	441
	4,923,405	91,311	264,435	5,279,151	782,198	4,496,953
CT	573	667	1,754	2,994	2,924	70
	2,660,903	129,913	60,447	2,851,263	501,499	2,349,764
DC	2			2		2
	595,000			595,000		595,000
DE	226	102	176	504	501	3
	790,178	25,247	57,545	872,970	596,840	276,130
FL	1,925	1,045	3,433	6,403	6,328	75
	16,281,037	263,623	298,672	16,843,332	14,222,761	2,620,571
GA	1,678	251	559	2,488	2,268	220
	7,119,376	72,646	88,806	7,280,828	1,599,694	5,681,134
HI	116	9	4	129	113	16
	1,285,174	6,385	500	1,292,059	1,169,290	122,769
IA	1,140	146	688	1,974	1,814	160
	2,526,314	45,068	88,128	2,659,510	1,473,634	1,185,876
ID	748	245	1,048	2,041	1,969	72
	924,721	52,667	114,320	1,091,708	839,838	251,870
IL	1,797	416	3,370	5,583	4,860	723
	11,300,507	150,302	374,357	11,825,166	3,297,942	8,527,224
IN	853	715	2,928	4,496	4,364	132
	4,499,532	210,497	417,480	5,127,509	2,822,028	2,305,481
KS	912	58	102	1,072	745	327
	2,581,726	23,945	4,054	2,609,725	830,695	1,779,030
KY	424	57	138	619	260	359
	4,588,463	21,968	11,453	4,621,884	364,993	4,256,891
LA	1,131	185	323	1,639	1,556	83
	4,878,378	72,684	74,731	5,025,793	3,022,291	2,003,502
MA	517	250	920	1,687	1,504	183
	8,955,179	71,673	142,832	9,169,684	2,157,850	7,011,834
MD	501	569	2,687	3,757	3,661	96
	4,502,328	162,742	167,496	4,832,566	874,747	3,957,819
ME	395	373	1,211	1,979	1,900	79
	615,264	72,052	197,391	884,707	454,721	429,986
MI	1,447	1,689	8,986	12,122	11,815	307
	7,197,342	471,246	1,063,724	8,732,312	3,203,122	5,529,190
MN	963	565	6,257	7,785	7,675	110
	4,040,822	84,158	490,353	4,615,333	3,184,810	1,430,523
MO	1,456	254	1,001	2,711	2,475	236
	4,864,516	79,159	133,877	5,077,552	1,785,200	3,292,352
MS	1,168	94	116	1,378	1,374	4
	3,007,045	81,400	24,648	3,113,093	2,789,142	323,951
MT	667	225	1,129	2,021	1,803	218
	659,667	59,886	176,587	896,140	502,724	393,416
NC	2,430	576	4,492	77,498	6,830	668
	6,431,967	159,401	377,666	6,969,034	1,963,876	5,005,158
ND	321	28	179	528	444	84
	551,709	4,102	15,523	571,334	264,665	306,669
NE	610	181	586	1,377	1,302	75

(Continued)

Table 7C.20 (Continued)

	Water System Type				Water Source Type	
	CWS	NTNCWS	TNCWS	Total	Groundwater	Surface Water
	1,424,246	42,530	113,001	1,579,777	803,980	775,797
NH	688	441	1,054	2,183	2,127	56
	794,419	91,054	220,026	1,105,499	582,034	523,465
NJ	607	870	2,644	4,121	4,019	102
	7,891,239	217,056	381,246	8,489,541	3,128,541	5,361,000
NM	657	152	509	1,318	1,242	76
	1,602,714	40,246	187,878	1,830,838	1,595,947	234,891
NV	256	107	260	623	590	33
	1,941,253	34,981	28,731	2,004,965	303,722	1,701,243
NY	2,849	759	6,564	10,172	8,983	1,189
	17,784,525	337,668	2,917,223	21,039,416	5,566,889	15,472,527
OH	1,334	1,018	3,221	5,573	5,148	425
	10,025,325	249,361	490,709	10,765,395	3,430,287	7,335,108
OK	1,151	110	376	1,637	912	725
	3,429,060	18,197	28,498	3,475,755	675,542	2,800,213
OR	883	332	1,454	2,669	2,365	304
	2,867,570	76,404	222,081	3,166,055	828,976	2,337,079
PA	2,162	1,224	6,627	10,013	9,433	580
	10,399,770	522,640	843,251	11,765,661	2,825,638	8,940,023
RI	83	74	322	479	454	25
	1,273,301	26,827	56,259	1,356,387	212,598	1,143,789
SC	655	185	575	1,415	1,202	213
	3,343,194	61,982	35,559	3,440,735	600,670	2,840,065
SD	468	30	195	693	560	133
	653,374	12,166	26,706	692,246	295,907	396,339
TN	675	50	435	1,160	602	558
	5,308,927	30,945	59,727	5,399,599	1,450,421	3,949,178
TX	4,516	788	1,245	6,549	5,379	1,170
	22,148,772	322,457	248,086	22,719,315	6,728,199	15,991,116
UT	454	61	438	953	828	125
	3,688,604	42,442	74,740	3,805,786	774,271	3,031,515
VA	1,309	618	1,309	3,236	2,846	390
	5,946,179	274,867	218,251	6,439,297	847,878	5,591,419
VT	438	227	695	1,360	1,223	137
	497,717	41,013	132,407	671,137	310,621	360,516
WA	2,282	309	1,559	4,150	3,884	266
	5,389,076	253,867	261,297	5,904,240	2,876,867	3,027,373
WI	1,096	937	9,389	11,422	11,373	49
	3,824,455	196,102	694,934	4,715,491	2,969,519	1,745,972
WV	547	169	560	1,276	897	379
	1,425,835	55,375	35,435	1,516,645	310,377	1,206,268
WY	279	87	381	747	634	113
	437,627	19,379	75,824	532,830	189,996	342,834
	51,935	19,347	87,901	159,183	145,144	14,039
Total	266,963,569	6,083,183	23,276,369		100,832,964	195,490,157

Note: First row is number of systems; Second row is population served.
CWS, Community Water System; NTNCWS, Non-Transient Non-Community Water System; TNCWS, Transient Non-Community Water System.

Source: From USEPA, *FACTOIDS: Drinking Water and Groundwater Statistics for 2003*, www.epa.gov.

Table 7C.21 Public Supply Freshwater Use in the United States, 2000 (Figures May Not Sum to Totals Because of Independent Rounding)

	Population (thousands)			Withdrawals (mill gall/day)			Withdrawals (thousand Acre-Feet/yr)		
		Served by Public Supply		By Source			By Source		
State	Total	Population	Population (Percent)	Groundwater	Surface Water	Total	Groundwater	Surface Water	Total
Alabama	4,450	3,580	80	281	553	834	315	620	935
Alaska	627	421	67	29.3	50.7	80	32.9	56.9	89.7
Arizona	5,130	4,870	95	469	613	1,080	526	688	1,210
Arkansas	2,670	2,320	87	132	289	421	148	324	472
California	33,900	30,100	89	2,800	3,320	6,120	3,140	3,730	6,860
Colorado	4,300	3,750	87	53.7	846	899	60.2	948	1,010
Connecticut	3,410	2,660	78	66	358	424	74	402	476
Delaware	784	617	79	45	49.8	94.9	50.5	55.9	106
District of Columbia	572	572	100	0	0	0	0	0	0
Florida	16,000	14,000	88	2,200	237	2,440	2,470	266	2,730
Georgia	8,190	6,730	82	278	968	1,250	311	1,090	1,400
Hawaii	1,210	1,140	94	243	7.6	250	272	8.52	281
Idaho	1,290	928	72	219	25.3	244	245	28.3	274
Illinois	12,400	10,900	88	353	1,410	1,760	396	1,580	1,970
Indiana	6,080	4,480	74	345	326	670	386	365	751
Iowa	2,930	2,410	83	303	79.8	383	340	89.5	429
Kansas	2,690	2,500	93	172	244	416	193	273	466
Kentucky	4,040	3,490	86	71	455	525	79.5	510	589
Louisiana	4,470	3,950	88	349	404	753	392	453	844
Maine	1,270	726	57	29.6	72.5	102	33.2	81.3	115
Maryland	5,300	4,360	82	84.6	740	824	94.8	829	924
Massachusetts	6,350	5,880	93	197	542	739	220	608	828
Michigan	9,940	7,170	72	247	896	1,140	277	1,000	1,280
Minnesota	4,920	3,770	77	329	171	500	369	192	561
Mississippi	2,840	2,190	77	319	40.4	359	357	45.3	402
Missouri	5,600	4,770	85	278	594	872	311	666	978
Montana	902	664	74	56.1	92.4	149	62.9	104	167
Nebraska	1,710	1,390	81	266	63.8	330	299	71.6	370
Nevada	2,000	1,870	94	151	478	629	169	536	705
New Hampshire	1,240	756	61	33	64.1	97.1	37	71.9	109
New Jersey	8,410	7,460	89	400	650	1,050	449	729	1,180
New Mexico	1,820	1,460	80	262	33.8	296	294	37.9	332
New York	19,000	17,100	90	583	1,980	2,570	653	2,220	2,880
North Carolina	8,050	5,350	66	166	779	945	186	873	1,060
North Dakota	642	493	77	32.4	31.2	63.6	36.3	35	71.3
Ohio	11,400	9,570	84	500	966	1,470	560	1,080	1,640
Oklahoma	3,450	3,150	91	113	562	675	127	631	757
Oregon	3,420	2,730	80	118	447	566	133	501	634
Pennsylvania	12,300	10,100	82	212	1,250	1,460	237	1,400	1,640
Rhode Island	1,050	922	88	16.9	102	119	19	115	134

South Carolina	4,010	3,160	79	105	462	566	117	517	635
South Dakota	755	625	83	54.2	39.1	93.3	60.7	43.9	105
Tennessee	5,690	5,240	92	321	569	890	360	638	997
Texas	20,900	19,700	94	1,260	2,970	4,230	1,420	3,330	4,740
Utah	2,230	2,180	97	364	274	638	408	307	715
Vermont	609	362	59	19.5	40.6	60.1	21.8	45.6	67.4
Virginia	7,080	5,310	75	70.7	650	720	79.3	728	808
Washington	5,890	4,900	83	464	552	1,020	520	619	1,140
West Virginia	1,810	1,300	72	41.6	149	190	46.6	167	213
Wisconsin	5,360	3,620	67	330	293	623	370	329	699
Wyoming	494	406	82	57.2	49.4	107	64.1	55.3	119
Puerto Rico	3,810	3,800	100	88.5	425	513	99.2	476	576
U.S. Virgin Islands	109	53.4	49	0.52	5.57	6.09	0.58	6.24	6.83
Total	285,000	242,000	85	16,000	27,300	43,300	17,900	30,600	48,500

Source: From Hutson, S.S. et al., 2004, *Estimated Use of Water in the United States in 2000*, U.S. Geological Survey Circular 1268, www.usgs.gov.

Table 7C.22 Water Produced by Selected Public Systems in the United States

City/County[a]	Utility	AWWA Study Group[c]	Service Pop.[b](000)	Total Number of Accounts	System Ownership	Daily Gallons Sold	Daily Capacity (MGD)	Max-Day Prod. (MGD)
Alaska								
Anchorage, AK	Anchorage Water and Wastewater Utility	B	217	53,432	City	21.75	79	49
Juneau, AK	City & Borough of Juneau Water Utility	C	30	8,430	City	3.29	10	6
Fairbanks, AK	Golden Heart Utilities	C	48	6,146	Private	3.04	8	6
Alabama								
Birmingham, AL	Birmingham Water Works Board	A	700	195,321	Dist./Auth.	77.53	188	140
Mobile Area, AL	Mobile Area Water and Sewer System	B	259	89,168	Dist/Auth.	60.78	70	57
Decatur, AL	Decatur Utilities	B	103	23,871	City	27.78	48	37
Huntsville, AL	Huntsville Utilities	B	280	73,707	City	27.58	90	60
Madison County, AL	Madison County Water Department	C	67	22,481	County	5.16	12	10
Sheffield, AL	Sheffield Utilities	C	14	4,792	City	1.78	3	3
Arkansas								
Little Rock, AR	Central Arkansas Water	B	366	132,267	Other	53.91	174	106
North Little Rock, AR	Central Arkansas Water, North Little Rock	C	N/R	N/R	N/R	N/R	N/R	N/R
Jonesboro, AR	City Water and Light	C	56	24,958	Other	10.88	31	22
Russellville, AR	City Corporation	C	25	11,036	City	6.13	20	11
Arizona								
Phoenix, AZ	City of Phoenix —Water Services Department	A	1,591	355,202	City	284.00	670	421
Tucson, AZ	Tucson Water	A	680	211,337	City	98.40	N/R	160
Scottsdale, AZ	City of Scottsdale Water Resources	B	219	81,598	City	61.25	137	92
Tempe, AZ	City of Tempe—Water Utilities Department	B	164	41,681	City	49.31	120	74
Peoria, AZ	City of Peoria	B	106	37,300	City	20.27	54	28
Tuscon Area, AZ	Metropolitan Domestic Water Improvement District	C	45	16,974	Dist./Auth.	8.59	20	14
Flagstaff, AZ	City Of Flagstaff	C	60	16,948	City	7.50	22	12
California								
Los Angeles, CA	Los Angeles Dept. of Water and Power	A	3,890	673,542	City	539.38	600	417
San Francisco, CA	San Francisco Public Utilities Commission	A	2,390	170,653	City	246.99	400	381
Oakland, CA	East Bay Municipal Utility District	A	1,300	377,094	Dist./Auth.	208.99	211	312

San Diego, CA	City of San Diego Water Department	A	1,357	261,348	City	207.36	294	274
Contra Costa, CA	Contra Costa Water District	A	430	60,019	Dist./Auth.	170.14	75	64
Sacramento, CA	City of Sacramento, Department of Utilities	A	435	125,780	City	123.42	275	185
Santa Clara, CA	Santa Clara Valley Water District	A	1,700	8	Dist./Auth.	118.63	220	N/A
Riverside County, CA	Eastern Municipal Water District	A	501	89,576	Dist./Auth.	96.48	43	43
Long Beach, CA	Long Beach Water Department	B	461	89,139	City	62.23	N/R	N/R
Riverside, CA	Riverside Public Utilities	B	319	59,178	City	60.29	72	102
Alameda County, CA	Alameda County Water District	B	323	78,274	Dist/Auth.	45.29	91	75
San Bernardino,CA	San Bernardino Municipal Water Department	B	160	41,218	City	42.27	70	70
Santa Ana, CA	City of Santa Ana, Public Works Agency	B	344	47,468	City	40.38	132	57
Palm Springs, CA	Desert Water Agency	B	65	19,805	Dist/Auth.	37.02	N/R	N/R
Glendale, CA	Glendale Water and Power	B	200	32,478	City	27.61	9	38
Oceanside, CA	City of Oceanside	B	167	40,257	City	27.15	28	28
Escondido, CA	City of Escondido	B	130	24,285	City	27.08	67	67
Covina Area, CA	Suburban Water Systems	B	170	34,297	Private	23.65	7	N/A
Covina, CA	Suburban Water Systems	B	128	32,601	Private	21.49	17	33
Chula Vista, CA	Sweetwater Authority	B	176	33,785	Dist./Auth.	21.04	36	31
Azusa, CA	Azusa Light & Water	C	110	N/R	N/R	19.44	8	N/R
Burbank, CA	Burbank Water and Power	C	106	26,240	City	19.02	9	38
Mesa, CA	Mesa Consolidated Water District	C	111	23,530	Dist./Auth.	18.48	25	25
Manteca, CA	City of Manteca	C	55	14,723	City	11.33	34	26
Mllpitas, CA	City of Milpitas	C	64	15,164	City	10.42	N/R	20
Santa Cruz, CA	City of Santa Cruz Water Department	C	90	23,590	City	9.91	26	15
Marin County, CA	North Marin Water District	C	58	18,500	Dist./Auth.	9.66	25	19
Colorado								
Denver, CO	Denver Water	A	1,081	217,607	City	197.86	645	419
Colorado Springs, CO	Colorado Springs Utilities	B	400	115,000	City	64.98	232	182
Arvada, CO	City of Arvada	C	105	31,855	City	15.70	52	38

(Continued)

Table 7C.22 (Continued)

City/County[a]	Utility	AWWA Study Group[c]	Service Pop.[b](000)	Total Number of Accounts	System Ownership	Daily Gallons Sold	Daily Capacity (MGD)	Max-Day Prod. (MGD)
Longmont, CO	City of Longmont Water/Wastewater Utilities	C	79	23,717	City	12.22	30	32
Grand Junction, CO	City of Grand Junction Water	C	26	9,193	City	5.25	16	12
Connecticut								
Bridgeport, CT	Aquarion Water Company of CT	A	575	172,000	Private	96.27	99	151
New Haven, CT	South Central Connecticut Regional Water Authority	B	391	106,467	Dist/Auth.	45.41	119	84
Waterbury, CT	City of Waterbury Bureau of Water	C	120	26,873	City	11.86	19	19
Hartford, CT	The Metropolitan District	C	359	93,520	Dist./Auth.	4.98	105	88
Manchester, CT	Manchester Water Department	C	56	15,301	City	4.77	12	9
District of Columbia								
Greater Washington, DC	Washington Aqueduct	A	1,000	3	Federal Gov.	178.42	350	229
Delaware								
New Castle County, DE	United Water Delaware	C	106	35,185	Private	18.10	36	27
Florida								
Miami, FL	Miami-Dade Water and Sewer Department	A	2,343	408,187	County	271.89	434	363
Orlando, FL	Orlando Utilities Commission	A	395	123,464	City	78.13	183	107
Pinellas County, FL	Pinellas County Utilities	B	636	109,602	County	67.50	86	80
Orange County, FL	Orange County Utilities	B	596	98,001	County	46.69	110	80
Jacksonville, FL	JEA	B	860	240,469	City	42.99	60	46
Fort Lauderdale, FL	City of Fort Lauderdale	B	168	55,553	City	41.15	90	64
Palm Beach County, FL	Palm Beach County Water Utilities Department	B	440	140,188	County	39.40	73	61
Manatee County, FL	Manatee County Utility Operations	B	340	79,840	County	36.76	51	61
Escambia County, FL	Escambia County Utilities Authority	B	216	91,298	Dist./Auth.	34.22	75	58
Pompano Beach, FL	Office of Environmental Services	B	207	51,359	Dist/Auth.	27.17	46	32
Tallahassee, FL	City of Tallahassee, Water Utility	B	252	70,688	City	26.87	59	49
Hollywood, FL	City of Hollywood	B	N/R	38,400	City	23.38	41	28
Cocoa, FL	City of Cocoa Utilities	B	195	70,264	City	23.37	60	34

Gainesville, FL	Gainesville Regional Utilities	B	168	60,348	City	22.77	54	37
City of Lakeland, FL	City of Lakeland, Department of Water Utilities	B	156	50,116	City	22.67	51	34
Osceola County, FL	Tohopekaliga Water Authority	C	151	40,931	Dist./Auth.	16.71	44	27
Jupiter, FL	Town of Jupiter Utilities	C	62	23,397	City	15.96	27	20
Cape Coral, FL	City of Cape Coral	C	100	40,000	City	6.96	15	11
Martin County, FL	Martin County Utilities	C	67	20,626	County	6.45	14	8
Winter Springs, FL	City of Winter Springs	C	30	14,500	City	4.30	12	6
Georgia								
Gwinnett County, GA	Gwinnett County DPU	B	700	202,198	County	71.45	150	123
DeKalb County, GA	DeKalb County Water & Sewer	B	659	180,673	County	70.03	128	111
Augusta, GA	Augusta Utilities Department	B	201	66,820	City	40.50	84	58
Columbus, GA	Columbus Water Works	B	187	63,665	City	N/R	90	49
Macon, GA	Macon Water Authority	B	130	54,262	Dist./Auth.	26.08	76	43
Savannah, GA	City of Savannah, Georgia	B	248	81,062	City	22.18	113	87
Douglas County, GA	Douglasville-Douglas County Water and Sewer Authority	C	76	31,277	Dist./Auth.	8.48	16	14
Hawaii								
Honolulu, HI	Honolulu Board of Water Supply	A	878	158,141	City	143.57	30	N/R
Maui, HI	County of Maui—Department of Water Supply	B	128	31,021	County	33.73	27	20
Hawaii County, HI	Department of Water Supply	B	111	37,000	County	24.45	25	37
Iowa								
Council Bluffs, IA	Council Bluffs Water Works	C	61	20,035	City	9.37	20	18
Cedar Rapids, IA	Cedar Rapids Water Department	B	126	44,266	City	33.02	65	50
Des Moines, IA	Des Moines Water Works	B	390	72,428	City	40.96	125	82
Waterloo, IA	Waterloo Water Works	C	100	24,497	City	12.02	50	29
Ames, IA	City of Ames Water and Pollution Control	C	51	14,035	City	5.33	11	8
Idaho								
Boise, ID	United Water Idaho	B	190	73,862	Private	41.14	100	91

(Continued)

Table 7C.22 (Continued)

City/County[a]	Utility	AWWA Study Group[c]	Service Pop.[b](000)	Total Number of Accounts	System Ownership	Daily Gallons Sold	Daily Capacity (MGD)	Max-Day Prod. (MGD)
Idaho Falls, ID	City of Idaho Falls	B	52	23,500	City	22.93	82	57
Twin Falls, ID	City of Twin Falls Water Department	C	36	13,097	City	12.51	40	24
Illinois								
Chicago, IL	Chicago Department of Water	A	5,306	490,689	N/R	N/R	N/R	1,346
Evanston, IL	City of Evanston	B	357	14,327	City	N/R	N/R	N/R
Rockford, IL	City of Rockford Water Division	B	155	52,100	City	21.01	65	46
Decatur, IL	City of Decatur	C	90	31,515	City	19.39	36	31
Lake Bluff Village, IL	Central Lake County Joint Action Water Agency	C	180	9	Dist./Auth.	19.17	38	32
Naperville, IL	City of Naperville	C	139	39,906	City	15.48	33	N/A
Indiana								
Evansville, IN	Evansville Water & Sewer Utility	B	155	58,068	City	22.58	50	38
South Bend, IN	South Bend Water Works	B	103	43,441	City	21.53	64	47
Columbus, IN	Columbus City Utilities	C	36	15,455	City	N/R	28	19
Kansas								
Wichita, KS	Wichita Water & Sewer Utility	B	414	135,552	City	55.53	160	115
Johnson County, KS	Water District No. 1 of Johnson County, KS	B	369	123,386	Dist./Auth.	53.41	165	130
Kansas City, KS	Kansas City Board of Public Utilities	B	157	51,793	City	22.54	45	42
Leavenworth, KS	Leavenworth Water Department	C	44	10,000	City	4.81	12	7
Olathe, KS	City of Olathe	C	108	31,412	City	11.44	21	25
Kentucky								
Campbell County, KY	Northern Kentucky Water District	B	N/R	70,966	N/R	N/R	N/R	N/R
Owensboro, KY	Owensboro Municipal Utilities	C	92	23,861	City	11.51	30	17
Paducah, KY	Paducah Waterworks	C	70	22,180	City	7.07	12	11
Winchester, KY	Winchester Municipal Utilities	C	33	10,974	City	2.77	6	5
Campbellsville, KY	Campbellsville Water Co.	C	9	8,600	City	2.23	9	4
Louisiana								
Jefferson Parish, LA	Jefferson Parish Water Department	B	427	142,172	N/R	54.95	148	95

Lafayette, LA	Lafayette Utilities System	C	191	44,448	City	19.18	47	27
Terrebonne Parish, LA	Consolidated Waterworks District No. 1, Terrebonne Parish	C	104	39,138	Dist./Auth.	10.94	24	20
Lafourche Parish, LA	Lafourche Parish Water District No. 1	C	78	28,087	County	9.11	18	12
Baton Rouge Area, LA	Louisiana Water Company	C	88	31,794	Private	8.71	18	13
Massachusetts								
Greater Boston Area, MA	Massachusetts Water Resources Authority	A	2,200	47	Dist/Auth.	231.40	405	310
Boston, MA	Boston Water and Sewer Commission	B	589	87,160	City	N/R	N/R	N/R
Cambridge, MA	Cambridge Water Department	C	101	15,057	City	14.50	24	24
Maryland								
Laurel, MD	Washington Suburban Sanitary Commission	A	1,557	433,555	Dist./Auth.	138.73	315	253
Anne Arundel County, MD	Anne Arundel County	B	324	98,469	County	34.27	42	N/R
Frederick, MD	City of Frederick	C	55	16,582	City	4.92	10	8
Ocean City, MD	Ocean City, MD Water Department	C	30	7,030	City	3.99	18	12
Maine								
Portland, ME	Portland Water District	C	200	48,421	Dist./Auth.	19.81	52	43
Kennebec ME	Kennebec Water District	C	25	8,853	Dist./Auth.	2.63	12	5
Michigan								
Lansing, MI	Lansing Board of Water S Light	B	235	56,088	City	22.12	50	39
Kalamazoo, MI	City of Kalamazoo Dept. of Public Utilities	C	120	36,601	City	18.51	67	41
Ann Arbor, MI	City of Ann Arbor Water Utilities	C	150	27,502	City	17.05	50	26
Port Huron, MI	City of Port Huron	C	57	12,774	City	7.67	20	15
Grand Rapids, MI	City of Grand Rapids Water System	B	280	76,540	City	34.74	135	80
Waterford, MI	Waterford Township Water and Sewer Dept.	C	75	23,430	City	N/R	9	21
Minnesota								
Saint Paul, MN	Saint Paul Regional Water Services	B	417	92,959	City	40.95	150	80
Dututh, MN	City of Duluth	C	99	27,555	N/R	14.74	32	28
Rochester, MN	Rochester Public Utilities	C	90	32,400	City	11.70	31	28

(Continued)

Table 7C.22 (Continued)

City/County[a]	Utility	AWWA Study Group[c]	Service Pop.[b](000)	Total Number of Accounts	System Ownership	Daily Gallons Sold	Daily Capacity (MGD)	Max-Day Prod. (MGD)
Missouri								
Louisville, MO	Louisville Water Company	A	820	258,548	City	110.15	240	191
Kansas City, MO	City of Kansas City, Missouri, Water Services Department	A	600	167,035	City	86.30	240	200
Springfield, MO	City Utilities of Springfield, MO	B	195	74,946	City	26.02	61	50
Columbia, MO	City of Columbia Water Department	C	90	37,702	City	10.87	24	19
Cape Girardeau, MO	City of Cape Girardeau	C	35	15,000	City	N/R	8	7
Montana								
Great Falls, MT	City of Great Falls Utilities	C	57	19,413	City	9.99	66	32
Missoula, MT	Mountain Water Company	C	60	20,935	Private	7.79	60	38
Kalispell, MT	City of Kafispell	C	22	6,427	City	1.89	9	8
North Carolina								
Charlotte, NC	Charlotte-Mecklenburg Utilities	A	700	212,503	City	93.97	242	156
Greensboro, NC	City of Greensboro Water Resources Department	B	250	93,215	City	26.11	54	41
Durham, NC	City of Durham, Env. Res. Dept.	B	181	69,794	City	25.52	61	38
Fayetteville, NC	Public Works Commission	B	138	73,783	City	20.01	50	40
Asheville, NC	City of Asheville—Water Resources Dept.	C	126	46,666	Dist./Auth.	14.50	36	26
Wilmington, NC	City of Wilmington	C	103	40,400	City	11.99	29	21
Cary, NC	Town of Cary	C	114	33,900	City	11.75	40	26
Rocky Mount, NC	City of Rocky Mount	C	68	24,006	City	10.45	26	20
Greenville, NC	Greenville Utilities Commission	C	68	28,381	City	8.98	24	15
Chapel Hill, NC	Orange Water and Sewer Authority	C	70	18,844	Dist./Auth.	7.27	20	14
Dare County, NC	Dare County Water Department	C	32	13,706	County	4.62	13	12
North Dakota								
Bismarck, ND	City of Bismarck	C	61	16,251	City	9.60	30	29
Fargo, ND	Fargo Water Treatment Plant	C	96	24,208	City	9.33	30	24
Grand Forks, ND	Grand Forks Water	C	50	12,754	City	6.80	17	11

Nebraska								
Omaha, NE	Metropolitan Utilities District of Omaha	A	550	180,540	Dist./Auth.	91.62	234	205
Lincoln, NE	Lincoln Water System	B	236	72,260	City	35.08	110	79
New Hampshire								
Concord, NH	City of Concord	B	41	11,000	City	23.46	10	8
Manchester, NH	Manchester Water Works	C	150	27,114	City	15.82	40	31
New Jersey								
Morris County, NJ	Southeast Morris County Municipal Utilities Authority	C	63	17,333	Dist./Auth.	6.90	21	21
New Mexico								
Albuquerque, NM	City of Albuquerque Public Works Department	A	509	155,213	City	87.81	294	160
Nevada								
Reno, NV	Truckee Meadows Water Authority	B	290	82,904	Dist./Auth.	35.56	159	150
Washoe County, NV	Washoe County Department of Water Resources	C	44	16,323	County	7.60	N/R	15
New York								
New York, NY	New York City Water Board	A	9,000	827,072	Dist./Auth.	1,236.16	1,661	1,650
Suffolk County, NY	Suffolk County Water Authority	A	1,200	357,614	Dist./Auth.	185.21	700	470
New Rochelle, NY	United Water New Rochelle	A	137	30,941	Private	150.79	50	31
Rockland County, NY	United Water New York	B	260	68,638	Private	N/R	N/R	N/R
Onondaga County, NY	Onondaga County Water Authority	B	340	80,957	Dist/Auth.	38.94	70	59
Niagara Falls, NY	Niagara Falls Water Board	C	55	19,503	Dist./Auth.	17.57	35	21
Nassau County, NY	New York Water Service Corporation	C	170	45,800	Private	15.62	50	33
Mohawk Valley, NY	Mohawk Valley Water Authority	C	120	37,800	Dist/Auth.	11.45	32	20
Elmira, NY	Elmira Water Board	C	65	17,575	City	4.88	17	9
Canton, NY	Village of Canton	C	8	1,400	City	0.90	1	1
Ohio								
Cleveland, OH	Cleveland Division of Water	A	1,500	414,012	City	167.50	545	395
Columbus, OH	City of Columbus, Division of Water	A	1,070	257,697	City	122.12	265	216

(Continued)

Table 7C.22 (Continued)

City/County[a]	Utility	AWWA Study Group[c]	Service Pop.[b](000)	Total Number of Accounts	System Ownership	Daily Gallons Sold	Daily Capacity (MGD)	Max-Day Prod. (MGD)
Cincinnati, OH	Greater Cincinnati Water Works	A	1,100	236,017	City	112.26	260	220
Akron, OH	City of Akron, Public Utilities Bureau	B	281	82,939	City	29.90	67	54
Canton, OH	City of Canton Water Department	B	140	N/R	City	22.23	41	28
Dayton, OH	City of Dayton, Department of Water	C	420	57,420	City	7.04	196	114
Newark, OH	City of Newark, OH Division of Water and Wastewater	C	47	18,600	City	5.88	15	12
Oklahoma								
Tulsa, OK	Tulsa Metropolitan Utility Authority	A	655	142,753	City	94.03	220	176
Oklahoma City, OK	Oklahoma City Water Utility	A	1,200	177,016	City	76.71	190	185
Broken Arrow, OK	Broken Arrow	C	88	30,286	City	10.98	6	0
Midwest City, OK	City of Midwest City	C	55	18,060	City	5.81	13	12
Oregon								
Portland, OR	Portland Bureau of Water Works	A	787	163,819	City	105.21	325	187
Eugene, OR	Eugene Water & Electric Board	B	140	55,629	City	27.32	72	63
Salem, OR	Salem Public Works	B	170	40,760	City	25.70	85	55
Medford, OR	Medford Water Commission	B	114	25,385	City	24.58	66	54
Washington County, OR	Tualatin Valley Water District	B	222	51,963	Dist./Auth.	23.13	N/A	N/A
Corvallis, OR	City of Corvallis	C	52	15,217	City	5.59	25	14
Pennsylvania								
Philadelphia, PA	Philadelphia Water Department	A	1,665	476,931	City	177.60	540	334
Delaware County, PA	Aqua Pennsylvania, Inc.	A	1,200	340,013	Private	95.79	176	150
Westmoreland County, PA	Municipal Authority of Westmoreland County	B	400	114,219	N/R	N/R	84	58
Chester, PA	Chester Water Authority	B	200	38,617	Dist./Auth.	29.45	60	41
Lancaster, PA	City of Lancaster, Bureau of Water	C	115	42,886	City	15.63	34	26
Bucks County, PA	Bucks County Water and Sewer Authority	C	194	15,671	County	15.40	35	21
Allentown, PA	City of Allentown	C	123	32,670	City	12.28	39	26

Harrisburg, PA	City of Harrisburg, Bureau of Water	C	66	19,997	Dist./Auth.	8.92	20	11
North Wales, PA	North Wales Water Authority	C	90	26,923	Dist./Auth.	7.94	17	14
Williamsport, PA	Williamsport Municipal Water Authority	C	54	18,726	Dist./Auth.	6.91	13	8
Lebanon, PA	City of Lebanon Authority	C	57	17,152	City	4.79	8	9
State College, PA	State College Water Authority	C	65	12,341	Dist/Auth.	4.44	9	7
Carlisle, PA	Borough of Carlisle Municipal Authority	C	20	6,205	Dist./Auth.	1.75	7	5
Rhode Island								
Providence, RI	Providence Water	B	600	73,493	City	57.91	144	120
South Carolina								
Greenville, SC	Greenville Water System	B	400	144,015	City	50.71	142	106
Charleston, SC	Commissioners of Public Works, Charleston, SC	B	409	90,121	Dist/Auth.	43.86	118	67
Spartanburg, SC	Spartan burg Water System	B	125	45,780	Dist/Auth.	23.55	N/R	N/R
Horry County, SC	Grand Strand Water & Sewer Authority	C	160	41,552	Dist./Auth.	19.81	45	34
Mount Pleasant, SC	Mount Pleasant Waterworks	C	66	26,395	Dist./Auth.	5.90	7	11
Georgetown County, SC	Georgetown County Water & Sewer District	C	46	14,455	Dist./Auth.	3.87	8	7
South Dakota								
Sioux—Falls, SD	City of Sioux —Falls	C	139	38,853	City	19.53	55	52
Tennessee								
Nashville, TN	Metro Water Services	A	458	155,712	City	613.70	180	104
Erwin, TN	Erwin Utilities	C	13	4,858	City	1.83	3	3
Seliner, TN	Setmer Utility Division	C	26	6,884	City	1.55	5	2
Texas								
Dallas, TX	Dallas Water Utilities	A	1,979	314,722	City	348.39	815	641
Houston, TX	City of Houston, Public Works and Engineering	A	1,954	412,000	City	309.96	884	440
San Antonio, TX	San Antonio Water System	A	1,145	304,340	City	142.06	964	230
Fort Worth, TX	Fort Worth Water Department	A	838	171,828	City	133.39	380	274
Austin, TX	Austin Water Utility	A	770	184,608	City	120.56	260	214

(Continued)

Table 7C.22 (Continued)

City/County[a]	Utility	AWWA Study Group[c]	Service Pop.[b](000)	Total Number of Accounts	System Ownership	Daily Gallons Sold	Daily Capacity (MGD)	Max-Day Prod. (MGD)
El Paso, TX	El Paso Water Utilities	A	696	173,653	City	96.91	235	174
Arlington, TX	City of Arlington Water Utilities	B	346	92,755	City	52.17	174	113
Plano, TX	City of Plano—Utility Operations	B	250	75,100	City	51.97	N/A	109
Amarillo, TX	City of Amarillo TX	B	175	63,360	City	45.60	N/R	N/R
Irving, TX	City of Irving Water Utilities	B	198	43,584	City	35.66	73	69
Garland, TX	City of Garland	B	221	62,068	City	28.60	44	N/A
Laredo, TX	City of Laredo	B	178	56,203	City	27.07	65	58
Midland, TX	City of Midland	B	96	32,494	City	23.65	57	37
Waco, TX	City of Waco Water Utilities	B	150	37,206	City	23.64	66	53
Tyler, TX	City of Tyler Water Utilities	C	175	27,726	City	17.61	40	37
Denton, TX	Denton Municipal Utilities	C	92	24,053	City	12.99	50	26
Sugar Land, TX	City of Sugar Land	C	65	20,940	City	11.52	43	23
College Station, TX	College Station Utilities	C	79	29,360	City	8.51	23	21
Bell County, TX	Central Texas WSC	C	50	15	N/R	5.43	14	12
Utah								
Salt Lake City, UT	Salt Lake City Public Utilities	A	325	91,283	City	82.54	282	206
Salt Lake City, UT	Jordan Valley Water Conservancy District	B	845	8,541	Dist./Auth.	62.64	200	159
Provo, UT	Provo City Water Resources	B	112	17,372	City	21.64	73	53
Virginia								
Richmond, VA	City of Richmond	A	502	59,747	City	83.70	132	127
Virginia Beach, VA	City of Virginia Beach Public Utilities	B	421	127,314	City	31.90	60	31
Chesterfield County, VA	Chesterfield County Department of Utilities	B	250	88,001	County	29.24	12	11
Chesapeake, VA	City of Chesapeake Public Utilities	C	167	57,613	City	16.02	10	12
Charlottesville, VA	City of Charlottesville	C	40	13,153	City	5.48	N/A	N/A
Vermont								
Burlington, VT	Burlington Public Works	C	44	9,800	City	4.59	12	7
South Burlington, VT	South Burlington Water Department	C	15	5,300	City	N/R	N/R	N/R
Washington								
Seattle, WA	Seattle Public Utilities	A	1,330	176,359	City	126.69	405	223

Tacoma, WA	Tacoma Public Utilities, Water Division	B	300	91,376	City	52.66	134	93
Lakehaven, WA	Lakehaven Utility District	C	112	27	Dist./Auth.	11.13	34	21
Olympia, WA	City of Olympia	C	51	17,064	City	9.00	25	13
Kennewick, WA	City of Kennewick	C	62	18,881	City	8.86	22	18
Renton, WA	City of Renton	C	54	15,042	City	6.40	22	14
Kent, WA	Highline Water District	C	66	18,078	Dist./Auth.	6.33	2	2
Snohomish County, WA	Snohomish County PUD #1	C	53	15,162	Dist./Auth.	3.95	20	8
Wisconsin								
Green Bay, WI	Green Bay Water Utility	C	103	35,466	City	15.95	28	31
Oak Creek, WI	Oak Creek Water and Sewer Utility	C	60	9,503	City	6.61	20	15
Milwaukee, WI	Milwaukee Water Works	A	831	160,976	City	109.35	380	170
Kenosha, WI	Kenosha Water Utility	C	118	28,319	City	11.67	42	25
Janesville, WI	Janesville Water Utility	C	61	22,299	City	11.10	29	22
River Falls, WI	River Falls Municipal Utility	C	13	3,755	City	0.95	7	3
West Virginia								
Lubeck, WV	Lubeck Public Service District	C	11	3,794	Dist./Auth.	0.73	3	1
Wyoming								
Cheyenne, WY	Cheyenne Board of Public Utilities	C	53	19,998	City	13.71	32	N/R
Casper, WY	City of Casper Public Utilities	C	54	18,785	City	9.21	52	28

Note: Group A >75 mgd sold; Group B 20–75 mgd sold; Group C <20 mgd sold.

[a] The primary city, county, or area served by the responding utility is listed.

[b] Includes retail and wholesale population.

Source: Adapted from *2004 Water and Wastewater Rate Survey* by permission. Copyright © 2004, American Water Works Association. (Updated information available electronically from AWWA at 800-926-7337), www.awwa.org.

Table 7C.23 Summary of 2004 Residential and Industrial Water Charges by Public Water Systems (Based on a Survey of 266 Systems)

		Monthly Water Charges								
		5/8-inch Meter						2-inch Meter	4-inch Meter	8-inch Meter
Summary Statistics[a] (11,220,000 gal)	Effective Date	Residential 0 cf (0 gal)	Residential 500 cf (3,740 gal)	Residential 1,000 cf (7,480 gal)	Residential 1,500 cf (11,220 gal)	Residential 3,000 cf (22,440 gal)	Nonmanuf./ Commercial 3,000 cf (22,440 gal)	Commercial/ Light Industrial 50,000 cf (374,000 gal)	Industrial 1,000,000 cf (7,480,000 gal)	Industrial 1,500,000 cf
All Systems (excluding Canadian systems)										
Average	11/26/01	$6.37	$12.29	$19.85	$28.12	$53.35	$52.81	$759.42	$13.734.00	$20,611.53
Median	1/1/03	$5.55	$11.67	$19.08	$27.15	$51.37	$49.85	$732.40	$12,954.04	$19,265.80
Number of Systems	263	257	255	256	255	255	254	253	251	240
Group A Systems (75-MGD Sold)										
Average	8/16/02	$5.76	$12.00	$19.31	$27.63	$54.77	$52.57	$772.28	$13,943.78	$20,993,93
Median	3/3/03	$4.75	$10.72	$18.66	$26.21	$49.45	$49.86	$755.55	$13,516,66	$20,503.95
Number of Systems	47	43	42	43	42	42	42	42	42	42
Group B Systems (20–75 MGD Sold)										
Average	11/24/01	$6.21	$11.47	$18.11	$25.66	$48.42	$45.84	$689.73	$12,681.52	$18,248.64
Median	1/1/03	$5.60	$10.76	$16.37	$25.07	$46.17	$42.95	$672.88	$11,747.97	$17,137.98
Number of Systems	93	93	93	93	93	93	92	92	91	89
Group C Systems (20 MGD Sold)										
Average	8/18/01	$6.71	$13.03	$21.40	$30.19	$56.68	$58.24	$808.76	$14,470.99	$22,393.52
Median	1/1/03	$5.85	$12.60	$21.21	$29.85	$54.53	$54.69	$758.86	$13,069.17	$20,134.83
Number of Systems	123	121	120	120	120	120	120	119	118	109
Canadian Systems[b]										
Average	12/28/02	$7.76	$15.10	$23.74	$62.05	$104.96	$104,96	$623.62	$13,995.92	$22,312.50
Median	1/1/03	$6.71	$15.60	$24.00	$69.64	$119.89	$119.89	$736.26	$15,062,03	$24,033.88
Number of Systems	11	8	10	10	10	9	9	10	10	10

[a] Seasonal and zonal water charges are not included in average and median charge calculations. The number of systems indicates the size of sample for which data was provided.
[b] Water charges for Canadian systems are presented in Canadian dollars. The exchange rate on July 21, 2004, was 1.3242 Canadian dollars to 1 US dollar.

Source: Adapted from *2004 Water and Wastewater Rate Survey* by permission. Copyright © 2004, American Water Works Association. (Updated information available electronically from AWWA at 800-926-7337), www.awwa.org.

Table 7C.24 Unaccounted for Water by Ownership in Public Supply Systems in the United States in 2000 (Based on a Survey of 1246 Systems)

Ownership Type	System Service Population Category								
	100 or Less	101–500	501–3,300	3,301–10,000	10,001–50,000	50,001–100,000	100,001–500,000	Over 500,000	All Sizes
Public Systems									
Average Unaccounted for Water	0.003	0.001	0.017	0.113	0.296	1.166	2.152	18.272	0.188
Confidence Interval	±0.006	±0.000	±0.006	±0.032	±0.074	±0.268	±0.567	±4.405	±0.032
% of Total Water Produced	1.140	5.990	9.935	11.562	8.528	9.693	6.306	7.959	8.949
Confidents Interval	±1.419	±7.067	±3.763	±3.161	±1.861	±2.044	±1.556	±1.027	±2.211
Private Systems									
Average Unaccounted for Water	0.000	0.001	0.012	0.100	0.287	1.055	3.050	7.347	0.019
Confidence Interval	±0.000	±0.000	±0.009	±0.061	±0.110	±0.395	±0.835	±2.972	±0.006
% of Total Water Produced	0.187	2.188	6.237	10.686	12.330	11.093	10.636	7.744	2.323
Confidence Interval	±0.269	±2,660	±3.940	±4.252	±4.511	±4.317	±2.582	±4.017	±1.237
All Systems									
Average Unaccounted for Water	0.000	0.001	0.016	0.111	0.294	1.152	2.227	17.258	0.098
Confidence Interval	±0.000	±0.000	±0.006	±0.028	±0.064	±0.238	±0.536	±4.031	±0.012
% of Total Water Produced	0.258	3.355	9.136	11.435	9.135	9.867	6.670	7.939	5.561
Confidence interval	±0.269	±2.859	±3.218	±2.724	±1.747	±1.864	±1.508	±1.001	±1.305

Note: Unaccounted for water includes system losses, water for fire suppression, and water used in the treatment process.
The tabulations presented in the Community Water System Survery 2000 are based on data collected from a sample of U.S. Water Systems, not from a census of every water system in the United States. A confidence interval is one way to gauge how precisely a given tabulation of survey data can be generalized to the entirety of U.S. Systems represented by the surveyed systems. Any result presented in the table must be viewed as the center of a range that would encompass the precise number that would be found if every U.S. water system could have been included in the tabulation, and not only those who were sampled and responded to the 2000 Community Water System Survey. The confidence interval expresses the range as a "plus/minus," that is, an amount to be added to and subtracted from the calculated data point actually presented in the table. The confidence interval is designed to include the true value in the stated range 95 percent of the time.

Source: From USEPA, 2002, Community Water System Survery 2000, EPA 815-R-02-005A, www.epa.gov.

SECTION 7D PUBLIC WATER SUPPLY—WORLD

Table 7D.25 Groundwater Use in Canada, 1996

Province/Territory	Population Reliant on Groundwater[a]		Municipal Water Systems Reliant on Groundwater[b]	
	Number	Percent	Number	Percent
Newfoundland and Labrador	189,921	33.9	19	23.5
Prince Edward Island	136,188	100.0	5	100.0
Nova Scotia	426,433	45.8	15	41.7
New Brunswick	501,075	66.5	40	72.7
Quebec	2,013,340	27.7	142	36.7
Ontario	3,166,662	28.5	132	42.7
Manitoba	342,601	30.2	22	50.0
Saskatchewan	435,941	42.8	44	65.7
Alberta	641,350	23.1	36	29.0
British Columbia	1,105,803	28.5	63	45.3
Yukon Territory	15,294	47.9	4	100.0
Northwest Territories[c]	18,971	28.1	0	0.0
Canada	8,993,579	30.3	522	41.2

[a] It is assumed that the population not covered by the Municipal Water Use Database is rural and that 90% of this population is groundwater reliant (except in Prince Edward Island, where 100% of the population is known to be groundwater reliant).

[b] Includes population and municipal water systems that are reliant on groundwater only, as well as those that are reliant on groundwater and surface water.

[c] Includes Nunavut.

Source: Statistics Canada, Human Activity and the Environment Annual Statistics 2003, Catalogue 16-201-XPE, released December 3, 2003, page 25. Statistics Canada information is used with the permission of Statistics Canada. Users are forbidden to copy this material and/or redisseminate the data, in an original or modified form, for commercial purposes, without the expressed permission of Statistics Canada. Information on the availability of the wide range of data from Statistics Canada can be obtained from Statistics Canada's Regional Offices, its World Wide Web site at www.statcan.ca, and its toll-free access number 1-800-263-1136.

Original Source: From Statistics Canada, Environment Accounts and Statistics Division, special compilation using data from Environment Canada, Municipal Water Use Database; Statistics Canada, 1996, *Quarterly Estimates of the Population of Canada, the Provinces and the Territories*, 11:3, Catalogue no. 91–001, Ottawa.

Table 7D.26 Water Flows and Metering Rates in Canada, by Province/Territory and Municipal Population

Province/Territory	Percentage of Flow from Surface Water	Total Average Daily Flow (Litres Per Capita)	Average Daily Residential Flow (Litres Per Capita)	Percentage of Residential Clients That Are Metered	Percentage of Business Clients That Are Metered
Newfoundland	95.1	971	664	0.0	47.4
P.E.I.	0.0	529	218	13.4	100.0
Nova Scotia	91.4	667	351	89.1	99.4
New Brunswick	79.7	1,314	416	49.6	89.5
Quebec	93.4	777	395	16.2	32.8
Ontario	88.4	533	285	89.9	98.4
Manitoba	81.6	410	223	96.6	98.6
Saskatchewan	87.5	517	236	98.5	99.6
Alberta	93.9	519	282	82.3	98.9
British Columbia	84.6	651	425	26.5	93.9
Yukon	69.0	803	556	52.8	100.0
N.W.T.	100.0	424	204	97.3	n/a
Nunavut	100.0	105	88	76.7	20.0

(Continued)

Table 7D.26 (Continued)

Province/Territory	Percentage of Flow from Surface Water	Total Average Daily Flow (Litres Per Capita)	Average Daily Residential Flow (Litres Per Capita)	Percentage of Residential Clients That Are Metered	Percentage of Business Clients That Are Metered
Municipal Population					
Under 2,000	61.2	715	446	42.4	53.2
2,000–5,000	57.3	732	466	35.4	55.5
5,000–50,000	78.9	665	397	47.5	75.0
50,000–500,000	88.9	596	326	61.7	91.3
More than 500,000	99.6	614	300	69.0	81.8
Total	89.2	622	335	60.6	83.1
Responding Population	21,634,144	23,822,869	23,822,869	24,235,565	16,075,854

Note: n/a = not applicable.

Source: From *Environment Canada, 2005, 2004 Municipal Water Use Report, Municipal Water Use 2001 Statistics*, www.ec.gc.ca/water/en/manage/use/e_data.htm, Environment Canada, 2005. Reproduced with the permission of the Minister of Public Works and Government Services, 2006.

Original Source: From values derived from the *2001 Municipal Water Use Database, Sustainable Water Use Branch*, Environment Canada.

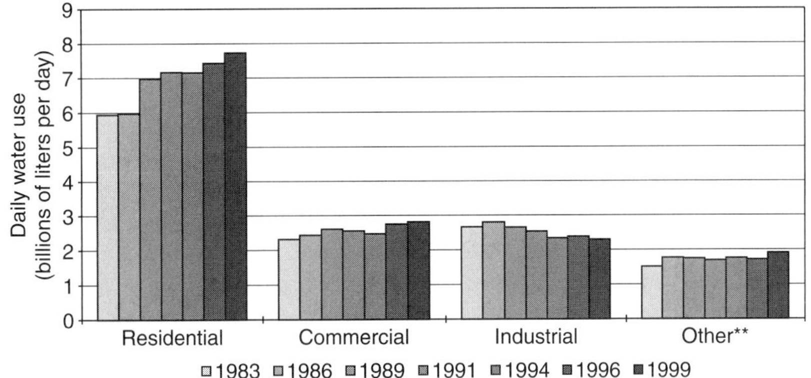

Note: *Water use values are based on (1) municipalities that responded in a given year, and (2) a national level estimate for all municipalities that did not respond, or in earlier years, were not surveyed.

**The "Other" category includes: water lost through leakage; unaccounted water uses, such as water used in firefighting or to flush out pipes; and water that a municipality was unable to one of the other three sectoral categories.

Figure 7D.11 Canada—total daily municipal water use* by sector, 1983–1999. (From www.ec.gc.ca. Environment Canada, 2001, Urban Waler Indicators: Municipal Water Use and Wastewater Treatment, SOE Bulletin No. 2001-1, Environment Canada, 2001. Reproduced with the permission of the Minister of Public Works and Government Services, 2006.)

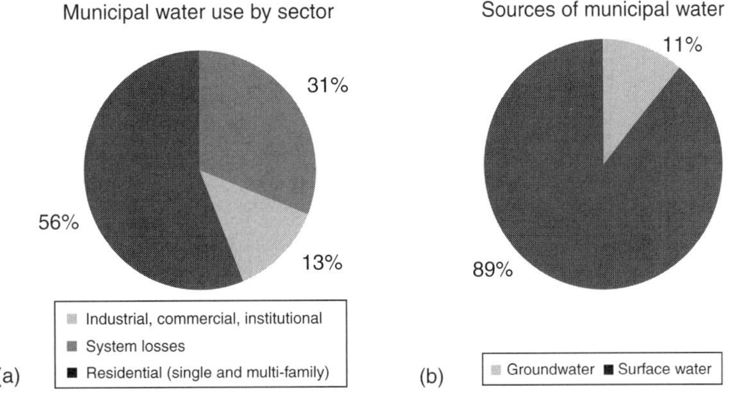

Figure 7D.12 Canada—municipal water use by sector and sources of municipal water, 2001. (From Environment Canada, 2005, 2004, *Municipal Water Use Report, Municipal Water Use 2001 Statistics*, www.ec.gc.ca/water/en/manage/use/e_data.htm, Environment Canada, 2005. Reproduced with the permission of the Minister of Public Works and Government Services, 2006.)

Table 7D.27 Water Use (as a Percentage of Water Served) in Canada, by Province/Territory, by Sector, and Responding Population

Province/Territory	Residential (%)	Commercial/Industrial (%)	System Losses (%)	Responding Population
Newfoundland	73	21	6	279,376
P.E.I.	42	41	17	43,037
Nova Scotia	59	25	16	462,020
New Brunswick	50	41	9	309,203
Quebec	56	25	19	5,892,601
Ontario	53	35	12	8,157,365
Manitoba	55	36	9	775,398
Saskatchewan	46	44	10	613,659
Alberta	56	35	9	2,327,245
British Columbia	65	28	6	2,986,953
Yukon	68	32	n/a	17,635
N.W.T.	45	30	25	23,135
Nunavut	78	16	6	6,204
Municipal Population				
Under 2,000	70	24	5	243,218
2,000–5,000	68	27	6	662,738
5,000–50,000	61	31	9	4,035,190
50,000–500,000	57	33	10	8,344,616
More than 500,000	51	31	18	8,608,069
Total, 2001	56	31	13	21,893,832
Total, 1999	52	35	13	

Note: n/a = not applicable.

Source: From *Environment Canada 2005, 2004 Municipal Water Use Report, Municipal Water Use 2001 Statistics*, www.ec.gc.ca/water/en/manage/use/e_data.htm, Environment Canada, 2005. Reproduced with the permission of the Minister of Public Works and Government Services, 2006.

Original Source: From values derived from the *2001 Municipal Water Use Database, Sustainable Water Use Branch*, Environment Canada.

Table 7D.28 Public Water System Characteristics of Selected Water Systems in Canada

City/County (a)	Utility	Service Pop. (b) (000)	Total Number of Accounts	System Ownership	Daily Gallons Sold	Daily Capacity (MGD)	Max-Day Prod. (MGD)
Canadian Water Systems (d)							
Montreal, PQ	Service de la gestion des infrastructures et de l'environnement	1,845	335,255	City	314.84	766	608
Vancouver, BC	Greater Vancouver Water District	2,021	17	Dist/Auth.	306.34	N/R	406
Edmonton, AB	EPCOR Water Services Inc	872	198,707	Private	91.63	114	133
Winnipeg, MB	City of Winnipeg—Water and Waste Department	631	183,803	City	51.86	N/R	166
Regina, SK	City of Regina	190	59,045	City	17.37	53	39
Barrie, ON	City of Barrie Water Section	135	36,000	City	9.26	26	26
Coquitlam, BC	City of Coquitlam	114	33,647	City	15.19	N/R	N/R
Victoria, BC	City of Victoria	90	18,843	City	12.93	N/R	N/R
Sault Ste Marie, ON	City of Sault Ste Marie Public Utilities Commission	75	25,366	City	7.73	21	17
Strathcona County, AB	Strathcona County	76	18,077	City	7.87	N/R	N/R
Prince Albert, SK	City of Prince Albert	40	10,500	City	N/R	12	10

Note: N/R, Not Reported

(a) The primary city, county, or area served by the responding utility is listed.

(b) Includes retail and wholesale population.

(d) Daily gallons sold, daily capacity, and max-day production for the Canadian systems have been converted to U.S. gallons (1 U.S. gallon = 0.8327 imperial gallons s 0.003785 m^3).

N/R = Not Reported.

Source: Adapted from *2004 Water and Wastewater Rate Survey* by permission. Copyright © 2004, American Water Works Association. (Updated information available electronically from AWWA at 800-926-7337), www.awwa.org.

Table 7D.29 International Public Water System Characteristics, 2004

City/Service Area	Utility	Service Pop.[a](000)	System Ownership	Gallons Sold (MGD)[b]	Active Water Accounts	Water Source[c](%) Groundwater	Surface	Purchased
Osaka City, Japan	Osaka Municipal Waterworks Bureau	2,619,494	City	53,814.94	887,735	N/R	100	N/R
Taichung, Chinese Taiwan	Taiwan Water Supply Corporation	16,548,877	Dist/Auth.	1,421.59	5,424,515	19	79	2
Sao Paulo, Brazil	Companhia de Saneamento Basico do Estado de Sao Pauto—SABESP	24,972,000	State Owned	1,281.07	5,305,883	6	94	2
Rijswijk, Netherlands	VEWIN, The Netherlands Waterworks Association	16,109,000	Dist/Auth.	831.31	7,231,025	61	39	0
Seoul, South Korea	Office of Waterworks Seoul Metropolitan Government	10,280,183	Dist./Auth.	794.48	1,939,847	0	89	11
Wan Chai, Hong Kong	Water Supplies Department	6,809,000	Dist./Auth.	N/R	2,545,717	0	25	75
Sydney, Australia	Sydney Water	4,198,000	Dist./Auth.	376.57	1,638,000	0	0	100
Bangkok, Thailand	Provincial Waterworks Authority (PWA)	10,000,000	Dist./Auth.	372.74	1,897,758	14	60	26
Alexandria, Egypt	Alexandria Water General Authority (AWGA)	4,500,000	Dist./Auth.	364.63	1,119,023	0	0	100
Yokohama, Japan	Yokohama Waterworks Bureau	3,506,966	City	317.25	1,200,908	0	44	57
Curitiba, Brazil	SANEPAR—Companhia de Saneamento do Parana'	7,761,171	Dist/Auth.	300.51	2,020,030	17	83	0
Denmark	DANVA—Danish Water and Waste Water Association	5,370,000	Private	280.91	1,287,000	98	2	0
Busan, South Korea	Busan Metropolitan Waterworks Headquarters	3,701,000	Dist/Auth.	254.33	331,350	0	100	0
Nagoya, Japan	Nagoya Waterworks & Sewerage Bureau	2,283,381	City	203.1	1,129,695	0	100	0
Leederville, Australia	Water Corporation of Western Australia	1,790,953	Dist/Auth.	180.11	754,810	62	38	0
Leederville, Australia	Water Corporation	1,817,750	Private	171.43	805,988	61	39	0
Adelaide, Australia	South Australia Water Corporation	1,492,000	Dist./Auth.	169.84	645,431	17	83	0
Lisbon, Portugal	EPAL—Empressa Portuguesa das Aguas Livres	2,203,386	Dist./Auth.	158.76	334,544	13	87	0
Barcelona, Spain	Aguas de Barcelona	2,848,869	City	138.17	208,021	38	11	51
Sapporo, Japan	Sapporo Waterworks Bureau	1,836,629	City	129.9	809,456	0	100	0
Turin, Italy	Societa Metropolitana Acque Torino Spa	1,437,536	City	121.26	144,627	81	19	1
Concepcion, Chile	Empresas de servicios sanrtarios del Bio–Bio	1,962,726	Private	84.36	545,202	53	47	0
Dublin, Ireland	Dublin City Council	1,033,000	City	65.78	221,005	0	21	79
Bristol, United Kingdom	Bristol Water pic	1,110,000	Private	56.25	473,001	20	80	0
Helsinki, Finland	Helsinki Water	882,000	City	52.26	25,809	0	100	0

Oslo, Norway	Oslo Water and Sewage Works	521,000	City	39.28	51,300	0	100	0
Addis Ababa, Ethiopia	Addis Ababa Water and Sewerage Authority	2,600,000	Dist./Auth.	36.92	210,000	11	89	0
Angered, Sweden	Gothenburg Water and Sewage Works	471,500	City	34.31	39,500	0	100	0
Cluj, Romania	R.A.J.A.C. Cluj	410,000	County	28.51	26,058	0	0	100
Yucatan, Mexico	Junta de Agua Potable y Alcantarillado de Yucatan	652,000	State Owned	26.20	215,679	100	0	0
Cebu City, Philippines	Metropolitan Cebu Water District	785,760	Dist./Auth.	24.81	88,271	93	4	3
Uijeongbu, South Korea	Uijeongbu City	364,530	Dist./Auth.	23.11	36,672	0	5	95
Iasi, Romania	Regia Aittonoma Judeteana Apa Canal	348,951	County	21.75	18,660	0	0	100
Manukau City, New Zealand	Manukau Water	305,000	City	21.12	91,594	0	0	100
Timisoara, Romania	Reg1a Autonoma Apa Si Canal Aquatim Timisoara	320,000	City	18.82	21,643	31	69	0
Davao, Philippines	Davao City Water District	1,271,436	Dist./Auth.	18.79	133,537	98	2	0
Wellington, New Zealand	Wellington City Council	177,348	City	18.00	66,010	0	0	100
Ljubljana, Slovenia	Javno podjetje Vodovod-Kanalizacija	322,800	City	17.08	44,417	99	1	0
Zamboanga, Philippines	Zamboanga City Water District	350,000	Dist./Auth.	12.33	39,288	16	84	0
Nicosia, Cyprus	Water Board of Nicosia —	200,000	State Owned	7.98	80,760	0	100	0
Lemesos, Cyprus	Water Board of Lemesos	150,000	Dist./Auth.	7.06	57,742	30	0	70
Bistrita, Romania	Raja Aquabis	89,132	County	5.97	10,911	2	98	0
Larnaca, Cyprus	Water Board of Larnaca	55,000	Dist./Auth.	2.60	27,970	25	0	75
Kiev, Ukraine	Ukrainian Water Association	N/R	Dist./Auth.	N/R	N/R	N/R	N/R	N/R

Note: N/R, not reported.

[a] Includes retail and wholesale population.
[b] Gallons sold has been converted to U.S. gallons.
[c] Due to rounding, water source percentages may not total 100.

Source: Adapted from 2004 Water and Wastewater Rate Survey by permission. Copyright © 2004, American Water Works Association. (Updated information available electronically from AWWA at 800-926-7337), www.awwa.org.

Table 7D.30 International Public Water System Charges, 2004

City/Service Area	Effective Date	Rate Structure	Monthly Water Charges (U.S. Dollars[a])							
			0 m^3 (0 gal)	15 m^3 (3,960 gal)	30 m^3 (7,920 gal)	85 m^3 (22,400 gal)	150 m^3 (39,600 gal)	1,000 m^3 (264,000 gal)	25,000 m^3 (6,600,000 gal)	40,000 m^3 (10,570,000 gal)
Osaka City, Japan	6/1/97	IB-7	$8.98	$13.57	$29.89	$137.82	$309.04	$3,035.73	$86,581.68	$138,797.89
Taichung, Chinese Taiwan	7/1/94	IB-4	0.50	1.00	1.85	5.50	10.00	53.52	295.09	817.74
Sao Paulo, Brazil	8/28/03	IB-4	3.24	5.77	20.94	87.99	345.75	2,500.48	40,873.08	60,812.41
Rijswijk, Netherlands	1/1/03	N/R	N/R	N/R	N/R	N/R	N/R	N/R	N/R	N/R
Seoul, South Korea	5/20/02	Uniform	N/R	4.90	15.25	60.29	149.17	1,280.83	26,562.50	42,387.50
Wan Chai, Hong Kong	2/1/95	IB-4	N/R	7.71	25.02	88.83	88.08	587.18	14,679.49	23,487.18
Sydney, Australia	7/1/03	Uniform	5.05	16.69	28.33	71.01	121.44	783.82	19,682.38	31,321.33
Bangkok, Thailand	6/9/98	IB-11	N/R	3.74	N/R	N/R	N/R	N/R	25,794.07	N/R
Alexandria, Egypt	1/1/03	IB-2	0.18	0.21	0.30	0.73	2.60	15.87	416.55	666.23
Yokohama, Japan	4/1/01	IB-7	7.31	15.43	43.67	188.44	377.28	3,213.39	94,102.28	150,907.83
Curitiba, Brazil	12/13/03	IB-3	5.06	8.86	20.26	91.39	152.83	1,025.50	25,665.50	41,065.50
Danva, Denmark	N/R	Uniform	127.01	N/R	N/R	N/R	N/R	N/R	N/R	N/R
Busan, South Korea	7/1/01	Uniform	N/R	7.75	18.08	54.54	135.83	1,295.83	30,326.67	48,494.17
Nagoya, Japan	4/1/97	Flat	6.65	17.92	49.62	241.93	459.43	3,115.09	7,864.15	123,770.75
Leederville, Australia	7/1/01	IB-9	8.43	13.11	20.10	58.59	232.33	679.76	13,781.67	21,677.54
Leederville, Australia	7/1/03	IB-7	29.91	N/R	N/R	N/R	N/R	N/R	N/R	N/R
Adelaide, Australia	7/1/03	IB-2	8.65	15.54	27.08	69.38	110.73	764.58	19,240.31	30,778.77
Lisbon, Portugal	12/31/02	IB-4	0.87	5.02	17.15	N/R	N/R	N/R	N/R	N/R
Barcelona, Spain	12/31/03	IB-3	9.25	20.60	44.90	120.54	182.39	1,214.10	30,292.45	48,371.20
Sapporo, Japan	4/1/97	IB-7	12.13	21.31	63.75	185.51	N/R	N/R	N/R	N/R
Turin, Italy	1/1/03	IB-5	N/R	53.28	170.81	270.64	824.18	5,472.69	129,604.52	207,354.88
Concepcion, Chile	11/14/03	Flat	1.02	6.49	11.97	32.05	55.78	366.11	9,128.35	14,604.76
Dublin, Ireland	12/31/03	Flat	N/R	N/R	N/R	N/R	183.06	1,175.08	28,914.13	46,290.60
Bristol, United Kingdom	4/1/03	Uniform	2.43	18.29	34.14	92.71	161.71	1,064.00	20,404.29	29,607.14
Helsinki, Finland	12/12/03	DB-3	2.50	29.03	58.00	163.76	287.83	1,817.02	43,428.81	69,486.09
Oslo, Norway	1/1/03	Uniform	6.40	16.85	27.30	65.60	112.23	708.29	17,438.15	27,897.72
Addis Ababa, Ethiopia	7/7/03	IB-3	0.15	0.24	0.34	N/R	0.34	0.34	0.34	N/R
Angered, Sweden	11/21/03	Uniform	7.68	45.37	78.44	140.29	330.34	934.64	15,068.47	24,171.01
Cluj, Romania	7/1/03	Uniform	N/R	N/R	N/R	N/R	N/R	N/R	N/R	N/R
Yucatan, Mexico	9/19/99	IB-11	1.79	1.79	6.43	23.85	71.17	N/R	N/R	N/R
Cebu City, Philippines	7/1/01	IB-4	1.94	3.01	6.59	44.51	90.53	712.45	17,360.93	27,772.94
Uijeongbu, South Korea	1/1/02	Uniform	N/R	6.75	18.00	96.33	162.50	1,666.67	41,666.67	66,666.67
Iasi, Romania	8/1/03	Uniform	N/R	6.91	13.82	39.14	69.08	460.50	11,512.50	18,420.00
Manukau City, New Zealand	N/R	Uniform	0.74	N/R	N/R	N/R	N/R	N/R	N/R	N/R
Timisoara, Romania	11/1/03	Uniform	N/R	4.50	8.99	25.48	44.96	299.74	7,493.38	11,989.41
Davao, Philippines	6/1/00	IB-5	1.42	2.17	4.85	24.22	48.50	366.00	9,330.78	14,933.77
Wellington, New Zealand	6/30/03	Uniform	N/R	N/R	N/R	N/R	N/R	N/R	N/R	N/R

Ljubljana, Slovenia	12/31/02	Uniform	$6.01	$8.32	$23.33	$34.87	$8.61	$35.16	$92.87	$346.81
Zambonga, Philippines	11/1/02	IB-6	1.81	2.78	5.34	13.02	3.63	10.66	64.48	115.73
Nicosia, Cyprus	1/1/02	IB-7	2.75	14.84	30.22	162.09	208.79	1,483.52	50,054.95	81,373.63
Lemesos, Cyprus	1/1/02	IB-4	6.00	13.00	28.88	131.04	106.60	1,644.60	32,853.92	492,780.28
Bistrita, Romania	11/1/03	Uniform	N/R	5.28	10.57	29.94	52.84	352.24	8,805.88	14,089.41
Larnaca, Cyprus	1/1/02	IB-5	2.16	7.80	21.99	107.68	105.73	964.08	23,580.61	N/R
Kiev, Ukraine	N/R	N/R	N/R	N/R	N/R	N/R	N/R	N/R	N/R	N/R

Note: N/R, not reported.

[a] All charges have been converted to U.S. dollars based on conversion data provided by the utility.

Source: Adapted from 2004 Water and Wastewater Rate Survey by permission. Copyright © 2004, American Water Works Association. (Updated information available electronically from AWWA at 800-926-7337), www.awwa.org.

Table 7D.31 Water Prices in Capitals or Major International Cities[a], 1996, 1998

		Current Exchange Rates (USD/m³)		Current PPPs (USD/m³)				Current Exchange Rates (USD/m³)		Current PPPs (USD/m³)	
		1996	1998	1996	1998			1996	1998	1996	1998
Canada	Ottawa	1.70	0.34	1.95	0.43	Germany	(national average)	1.70	—	1.47	—
	Toronto	0.63	0.31	0.73	0.39		Berlin	—	1.94	—	1.70
	Winnipeg	0.75	0.73	0.87	0.92		Dosseldorf	—	1.92	—	1.68
	Vancouver	—	0.35	—	0.45		Gelsenkirchen	—	1.47	—	1.29
	London	1.02	0.72	1.18	0.92		Hamburg	—	1.74	—	1.53
	Edmonton	0.93	0.90	1.07	1.14		Monchen	—	1.35	—	1.19
U.S.A.	Washington	0.80	—	0.80	—		Stuttgart	—	1.46	—	1.28
	New York	0.88	0.43	0.88	0.43	Greece	Athens	0.77	0.86	0.87	1.05
	Los Angeles	0.60	0.58	0.60	0.58		Thessaloniki	0.82	0.55	0.94	0.68
	Orlando	0.29	—	0.29	—		Chanea	—	1.02	—	1.25
	Miami	—	0.36	—	0.36		Patras	—	0.77	—	0.94
	Indianapolis	—	0.88	—	0.88	Hungary	Budapest	0.28	0.32	0.58	0.71
	Detroit	—	0.35	—	0.35		Debrecen	0.55	0.37	1.16	0.83
Japan	Tokyo	1.16	0.92	0.76	0.74		Pecs	0.69	0.61	1.45	1.35
	Osaka	0.70	0.68	0.46	0.54		Miskolc	0.55	0.44	1.16	0.98
	Sapporo	1.29	1.13	0.85	0.90	Iceland	Reykjavik	0.61	—	0.53	—
	Yokohama	0.95	0.74	0.63	0.59		Hafnarflorour	0.51	—	0.44	—
	Nagoya	1.13	0.72	0.74	0.58	Italy	Rome	0.33	0.28	0.32	0.29
Korea	(national average)	0.36	—	0.46	—		Bologna	0.65	0.61	0.63	0.64
	Seoul	—	0.18	—	0.38		Milan	0.13	0.13	0.13	0.13
	Daegu	—	0.19	—	0.41		Naples	0.65	0.57	0.64	0.59
	Daeieon	—	0.18	—	0.37		Turin	0.28	0.28	0.27	0.29
	Inchon	—	0.14	—	0.29	Luxembourg	Luxembourg	1.64	1.60	1.28	1.40
	Pusan	—	0.22	—	0.46	Netherlands	Amsterdam	1.20	1.02	0.99	0.99
Australia	Sydney	0.93	0.73	0.91	0.89		The Hague	1.92	1.91	1.59	1.85
	Brisbane	0.80	0.68	0.79	0.82		Utrecht	0.94	0.94	0.78	0.92
	Melbourne	0.80	0.59	0.79	0.72	Norway	Oslo	0.32	0.47	0.22	0.39
	Canberra	—	0.63	—	0.76		Bergen	1.14	1.30	0.81	1.07
	Perth	0.76	0.64	0.75	0.78		Trondheim	1.05	0.80	0.74	0.65
N.Zealand	Wellington	0.98	0.63	1.01	0.80	Ponugal	Lisbon	0.99	0.97	1.24	1.39
	Auckland	—	0.46	—	0.58		Coimbra	1.02	0.72	1.28	1.04
	North Shore City	—	0.59	—	0.75		Porto	0.98	1.02	1.23	1.46
Austria	Vienna	1.75	1.48	1.36	1.35	Spain	Madrid	0.85	0.81	0.87	0.94
	Salzburg	1.59	1.43	1.24	1.30		Barcelona	0.81	0.78	0.83	0.91
	Linz	1.11	1.12	0.87	1.01		Bilbao	0.48	0.41	0.49	0.48
Belgium	Brussels	1.80	1.51	1.51	1.45		Seville	0.69	0.57	0.70	0.67
	Antwerp	0.97	0.88	0.82	0.84	Sweden	Stockholm	0.86	0.76	0.60	0.62
	Liège	1.50	1.48	1.26	1.42		Goteborg	0.58	0.59	0.40	0.48
Czech R.	Praha	0.37	0.45	0.86	1.06		Malmo	0.99	0.54	0.69	0.44
	Brno	0.29	0.37	0.67	0.88	Switzerland	Beme	1.22	1.33	0.74	0.97
	Ostrava	0.38	0.44	0.88	1.05		Geneva	2.25	2.14	1.35	1.56
Denmark	Copenhagen	1.34	1.68	0.93	1.32		Zurich	2.26	1.88	1.36	1.37

	Aarhus	0.89	1.26	0.62	0.98	Turkey	Ankara	0.18	—	0.37	—
	Odense	0.98	1.32	0.68	1.03		Canakkale	0.20	—	0.41	—
Finland	Helsinki	0.85	0.76	0.66	0.66		Eskisehir	0.19	—	0.40	—
	Tampere	0.90	0.86	0.70	0.75	UK	London	0.78	0.62	0.78	0.57
	Vaasa	1.32	—	1.03	—		Bristol	0.78	0.57	0.78	0.52
	Turku	1.36	1.19	1.06	1.04		Manchester	0.93	0.55	0.93	0.51
	Espoo	—	1.35	—	1.17		Newcastle	0.74	0.76	0.74	0.69
France	Paris	0.93	0.87	0.73	0.76		Cardiff	1.08	0.56	1.08	0.52
	Banlieue Paris	1.73	1.46	1.35	1.28						
	Bordeaux	1.39	1.16	1.08	1.02						
	Lille	1.33	1.06	1.03	0.93						
	Lyon	1.78	1.45	1.38	1.27						

Note: [a]Prices calculated on the basis of a family of four (two adults and two children) living in a house with garden rather than an apartment. Where there are water meters, the price is based on annual consumption of 200 m^3. Where supply is normally unmeasured the average price has been used (Norway and UK). VAT is not included.
KOR, 1996: national data.
NZL, 1996: Secretariat estimates based on country data for water meter charges for the 1997/98 fiscal year, and considering an annual consumption of 200 m^3.
DEU, 1996: country data which refer to 1997 and are provisional.
GRC, 1996 data refer to 1995; source: Ministry of Development.
NOR, Unmeasured data: refer to the average price.
TUR, 1996 data refer to 1995.
UKD, Unmeasured data: refer to the average price.

Source: Table 3.2, *OECD Environmental Data Compendium 2002*, © OCED 2002, www.oecd.org.

Table 7D.32 Worldwide Unccounted for Water

Location		Percent	Source
Africa (large city average)	1990	39	e
Algiers, Algeria	1990	51	a
Amman, Jordan	1990	52	a
Asia (large city average)	1990	35–42	e, f
Bahrain	1993	36	o
Bahrain	2000	24	o
Barbados	1996	43	j
Buenos Aires, Argentina	1993	43	n
Buenos Aires, Argentina	1996	31	n
Canada (average)	1990	15	b
Casablanca, Morocco	1990	34	a, b
Damascus, Syria	1995	64	a
Dubai, United Arab Emirates	1990	15	a
Gaza	1995	47	a
Gaza	1999	31	a
Haiphong, Vietnam	1998	70	m
Hanoi, Vietnam	1995	63	f
Hebron	1990	48	a
Johor Bahru, Malaysia	1995	21	f
Kansas, United States (average)	1997	15	h
Kansas, United States (range)	1997	3–65	h
Lae, Papua New Guinea	1995	61	f
Latin America and Caribbean (large city average)		42	e
Lebanon (average)		40	a
Male, Maldives	1995	10	f
Mandalay, Myanmar	1995	60	f
Mexico City, Mexico	1997	37	l
Mexico City, Mexico	1999	32	l
Nairobi, Kenya	2000	50	g
Nicosia, Cyprus	1990	16	b
North America (large city average)	1990	15	e
Oran, Algeria	1990	42	a
Penang, Malaysia	1995	20	f
Phnom Penh, Vietnam	1995	61	f
Poland (medium utility range)	1990	19–51	d
Rabat, Morocco	1990	18	b
Ramallah	1990	25	a
Rarotonga, Cook Islands	1995	70	f
Sana'a, Yemen	1990	50	a
Seoul, South Korea	1996	35	p
Singapore	1990	11	b
Singapore	1995	6	f
Sydney, Australia	1990	13.4	c
Tamir, Yemen	1990	28	b
Teheran, Iran	1990	35	a
Tunisia (large utility range)	1990	8–21	a, d
United Kingdom (small utility range)	1990	14–30	d

(Continued)

Table 7D.32 (Continued)

Location		Percent	Source
United States (average)	1990s	12	b
Vietnam (average)	1998	50	m
Washington, DC, area suppliers	1999	10–28	k

[a] Saghir, J., Sehiffler, M., and Woldu, M., 1999, *World Bank Urban Water and Sanitation in the Middle East and North Africa Region: The Way Forward*, the World Bank, Middle East and North Africa Region Infrastructure Development Group, December, available at worldbank.org/wbi/mdf/mdf3/papers/fnance/Saghir.pdf.

[b] Saghir, J., World Bank compilation.

[c] www.sydneywater.com.au/html/tsr/performanceindicators/esd2.html.

[d] World Bank benchmarking data, www.worldbank.org/watsan/topics/bench/bench_network_iup.html.

[e] World Health Organization, 2000, *Global Water Supply and Sanitation Assessment 2000 Report*, available at www.who.int/water_sanitation_health/Globassessment/Global4-4.htm.

[f] Asian Development Bank, 1997, *Second Water Utilities Data Book*, Manila, Philippines, available at www.adb.org/Documents/News/1997/nr1997111.asp.

[g] Makuro, M., 2000, *Nairobi's Response to the Water Crisis*, UNCHS (Habitat) United Nations Centre for Human Settlements, Vol. 6, No. 3, available at unchs.org/unchs/english/hdv6n3/nairobi_response.htm.

[h] Kenny, J.F., 2000, *Public Water-Supply Use in Kansas, 1987–97*, USGS Fact Sheet 187-99, January, United States Geological Survey, available at ks.water.usgs.gov/Kansas/pubs/fact-sheets/fs.187-99.htm1#HDR2.

[i] AWWA recommends the use of audits to reduce unaccounted-for water, awwa.org/govtaff/watcopap.htm.

[j] Barbados Water, 1999, *Managing Water Resources in an Integrated and Participatory Way Report of the First Stakeholder Meeting in Barbados*, September 29–30, 1999, available at commonwealthknowledge.net/Thanni/wwevh.htm.

[k] League of Women Voters, 1999, *Drinking Water Supply in the Washington DC Metropolitan Area: Prospects and Options for the 21st Century*, available at www.dcwatch.com/lwvdc/lwv9903b.htm.

[l] Adelson, N., 2000, *Water Woes: Private Investment Plugs Leaks in Water Sector, Business Mexico*, available at mexconnect.com/mex_/travel/bzm/bzmwaterwoes.html.

[m] Trung, D.Q., Snow, R., Doukas, L., Thanh, N., and Trung, N., 1998, Water-loss reduction program in Vietnam, 24th WECD Conference, Water and Sanitation for All, Islamabad, Pakistan, available at lboro.ac.uk/departments/cv/wedc/papers/24/S/trung.pdf.

[n] Esmay, J., 1998, *Roundtable on Municipal Water*, Vancouver, Canada, available at idrc.ca/industry/canada_e7.html.

[o] Qamber, M., 2000, *Water Demand Management in State of Bahrain, Bahrain Ministry of Electricity & Water*, available at emro.who.int/ceha/AmmanConferenceWaterDemandManagement.pdf.

[p] metro.seoul.kr/eng/smg/agenda/2-3.html.

Source: From *World's Water 2002–2003*, by Peter H. Gleick.

Table 7D.33 Population Supplied with Safe Drinking Water, by Country 1970–2000

Region and Country	Fraction of Population with Access to Drinking Water																				
	Urban							Rural							Total						
	1970	1975	1980	1985	1990	1994	2000	1970	1975	1980	1985	1990	1994	2000	1970	1975	1980	1985	1990	1994	2000
Data Source	a,b	a,b	a,b	a,b	c	d	e	a,b	a,b	a,b	a,b	c	d	e	a,b	a,b	a,b	a,b	c	d	e
Africa	66	68	69	77																	
Algeria	84	100		85			88		61		55			94		77		68			94
Angola			85	87	73	69	34			10	15	20	15	40			26	33	35	32	38
Benin	83	100	26	80	73	41	74	20	20	15	34	43	53	55	29	34	18	50	54	50	63
Botswana	71	95		84	100		100	26	39		46	88			29	45		53	91		
Burkina Faso	35	50	27	43			84	10	26	31	69	70			12	25	31	67		78	
Burundi	77		90	98	92	92	96			20	21	43	49				23	25	45	52	
Cameroon	77			43	42		82	21			24	45		42	32			32	44		62
Cape Verde			100	83		70	64			21	50		34	89			25	52		51	74
Central African Republic				13	19	18	80					26	18	43					23	18	60
Chad	47	43				48	31	24	23				17	26	27	26				24	27
Comoros							98							95							96
Congo	63	81	42				71	6	9	7				17	27	38	20				51
Congo, Democratic Rep.	33	38		52	68	37	89	4	12		21	24	23	26	11	19		32	36	27	45
Côte d'Ivoire	98				57	59	90	29				80	81	65	44				71	72	77
Djibouti			50	50		77	100			20	20		100	100			43	45		90	100
Egypt	94		88		95	82	96	93		64		86	50	94	93		84		90	64	95
Equatorial Guinea			47		65	88	45					18	100	42					32	95	43
Eritrea							63							42							46
Ethiopia	61	58		69			77		1		9			13	6	8		16			24
Gabon							73							55							70
Gambia	97		85	97	100		80	3			50	48		53	12			59	60	76	62
Ghana	86	86	72	93	63	70	87		14	33	39		49	49	35	35	45	56	21	56	64
Guinea	68	69	69	41	100	61	72			2	12	37	62	36		14	15	18	53	62	48
Guinea-Bissau			18	17		38	29			8	22		57	55			10	21		53	49
Kenya	100	100	85			67	87	2	4	15			49	31	15	17	26			53	49
Lesotho	100	65	37	65		14	98	1	14	11	30		64	88	3	17	15	36		52	91
Liberia	100			100		58		6			23		8		15			53		30	
Libya	100	100	100				72	42	82	90				68	58	87	96				72
Madagascar	67	76	80	81		83	85	1	14	7	17		10	31	11	25	21	31		29	47
Malawi			77	97		52	95			37	50		44	44			41	56		45	57
Mali	29		37	46	41	36	74				10	4	38	61				16	13	37	65
Mauritania	98		80	73		84	34	10		85			69	40	17		84			76	37
Mauritius	100	100	100	100	100	95	100	29	22	98	100	100	100	100	61	60	99	100	100	98	100
Morocco	92		100	100	100	98	100	28			25	18	14	58	51			59	56	52	82
Mozambique				38		17	86				9		40	43				15		32	60

Namibia					90	87	100					37	42	67					52	57	77
Niger	37	36	41	35	98	46	70	19	26	32	49	45	55	56	20	27	33	47	56	53	59
Nigeria				100	100	63	81				20	22	26	39				38	49	39	57
Reunion																					
Rwanda	81	84	48	79	84		60	66	68	55	48	67		40	67	68	55	50	68		41
Sao Tome and Principe											45							45			
Senegal	87	56	77	79	65	82	92			25	38	26	28	65			43	53	42	50	78
Seychelles											95							95			
Sierra Leone	75		50	68	80	58	23	1		2	7	20	21	31	12		14	24	39	34	28
Somalia	17	77		58				14	22		22				15	38		34			
South Africa							92							80						70	86
Sudan	61	96	100			66	86	13	43	31			45	69	19	50	51			50	75
Swaziland		83		100		41			29		7		44			37		31		43	
Tanzania	61	88		90			80	9	36		42			42	13	39		53			54
Togo	100	49	70	100		74	85	5	10	31	41		58	38	17	16	38	54		63	54
Tunisia	92	93	100	100		100		17		17	31		89		49		60	70		99	
Uganda	88	100		37	60	47	72	17	29		18	30	32	46	22	35		20	33	34	50
Zambia	70	86		76		64	88	22	16		41		27	48	37	42		58		43	64
Zimbabwe					95		100				32	80		77					84		85
North & Central America & Caribbean																					
Anguilla																					
Antigua and Barbuda																					
Aruba																					
Bahamas	100	100	100	100	98		99	12	13			75		86	65	65	100	100	90		96
Barbados	95	100	99	100	100		100	100	100	98	99	100		100	98	100	99	99	100		100
Belize			99	100	95	96	83			36	26	53	82	69			68	64	74	89	76
British Virgin Islands					100							100									
Canada							100							99							100
Cayman Islands			100	98																	
Costa Rica	98	100	100	100		85	98	59	56	82	83		99	98	74	72	90	91		92	98
Cuba	82	96			100	96	99	15				91	85	82	56				98	93	95
Dominican Republic	72	88	85	85	82	74	83	14	27	34	33	45	67	70	37	55	60	62	67	71	79
Dominica							100							100							100
El Salvador	71	89	67	68	87	78	88	20	28	40	40	15	37	61	40	53	50	51	47	55	74
Grenada	100	100					97	47	77					93							94
Guadeloupe							94							94							94
Guatemala	88	85	90	72	92		97	12	14	18	14	43		88	38	39	46	37	62		92
Haiti		46	51	59	56	37	46		3	8	30	35	23	45		12	19	38	41	28	46
Honduras	99	99	93	56	85	81	97	10	13	40	45	48	53	82	34	41	59	49	64	65	90
Jamaica	100	100	55	99			81	48	79	46	93			59	62	86	51	96			71
Martinique																					
Mexico	71	70	90	99	94	91	94	29	49	40	47		62	63	54	62	73	83	69	83	86

(Continued)

Table 7D.33 (Continued)

Region and Country	Fraction of Population with Access to Drinking Water																				
	Urban							Rural							Total						
	1970	1975	1980	1985	1990	1994	2000	1970	1975	1980	1985	1990	1994	2000	1970	1975	1980	1985	1990	1994	2000
Montserrat																					
The Netherlands Antilles																					
Nicaragua	58	100	67	76		81	95	16	14	6	11		27	59	35	56	39	48		61	79
Panama	100	100	100	100			88	41	54	62	64			86	69	77	81	82		83	87
Puerto Rico																					
St. Kitts																					
St. Lucia																					
St. Vincent																					
Trinidad/ Tobago	100	79	100	100	100			95	100	93	95	88			96	93	97	98	96		86
Turks/Caicos Islands				87							68							77			
United States of America							100							100							100
United States Virgin Islands																					
South America																					
Argentina	69	76	61	63			85	12	26	17	17			30	56	66	54	56			79
Bolivia	92	81	69	75	76	78	93	2	6	10	13	30	22	55	33	34	36	43	53	55	79
Brazil	78	87	83	85	95	85	95	28		51	56	61	31	54	55		72	77	87	72	87
Chile	67	78	100	98		94	99	13	28	17	29		37	66	56	70	84	87		85	94
Colombia	88	86	93	100	87	88	98	28	33	73	76	82	48	73	63	64	86		86	76	91
Ecuador	76	67	79	81	63	82	81	7	8	20	31	44	55	51	34	36	50	57	55	70	71
Falkland Islands (Malvinas)																					
French Guiana							88							71							84
Guyana	100	100	100	100	100	90	98	63	75	60	65	71	45	91	75	84	72	76	81	61	94
Paraguay	22	25	39	53	61		95	5	5	9	8	9		58	11	13	21	28	34		79
Peru	58	72	68	73	68	74	87	8	15	18	17	24	24	51	35	47	50	55	55	60	77
Suriname			100	71			94			79	94			96			88	83			95
Uruguay	100	100	96	95	100		98	59	87	2	27			93	92	98	81	85	89		98
Venezuela	92		93	93		80	88	38		53	65	36	75	58	75		86	89		79	84
Asia																					
Afghanistan	18	40	28	38	40	39	19	1	5	8	17	19	5	11	3	9	8	17	23	12	13
Armenia																					
Azerbaijan																					
Bahrain	100	100		100	100			94	100		100	0			99	100		100			

Bangladesh	13	22	26	24	39	100	99	47	61	40	49	89	97	97	45	56	39	46	81	97	97
Bhutan			50		60	75	86			5	19	30	54	60			7		32	64	62
Brunei Darus			100							95											
Cambodia							53							25							30
China					87	93	94					68	89	66					73	90	75
Cyprus	100	94		100	100		100	92	96		100	100		100	95	95		100	100		100
East Timor																					
Gaza Strip																					
Georgia																					
Hong Kong			100		100					95		96							100		
India	60	80	77	76	86	85	92	6	18	31	50	69	79	86	17	31	42	56	73	81	88
Indonesia	10	41	35	43	35	78	91	1	4	19	36	33	54	65	3	11	23	38	34	62	76
Iran	68	76	82		100	89	99	11	30	50		75	77	89	35	51	66		89	83	95
Iraq	83	100		100	93		96	7	11	54	54	41		48	51	66		86	78	44	85
Israel																					
Japan																					
Jordan	98		100	100	100		100	59		65	88	97		84	77		86	96	99	89	96
Kazakhstan							98							82							91
Korea DPR							100							100							100
Korea Rep	84	95	86	90	100		97	38	33	61	48	76		71	58	66	75	75	93		92
Kuwait	60	100	86	97						100					51	89	87				
Kyrgyzstan							98							66							77
Laos	97	100	28		47	40	59	39	32	20		25	39	100	48	41	21		29	39	90
Lebanon						100	100						100	100						100	100
Macau																					
Malaysia	100	100	90	96	96			1	6	49	76	66		94	29	34	63	84	79		
Maldives			11	58	77	98	100			3	12	68	86	100			2	21		89	100
Mongolia					100		77					58		30					82		60
Myanmar	35	31	38	36	79	36	88	13	14	15	24	72	39	60	18	17	21	27	74	38	68
(Burma)																					
Nepal	53	85	83	70	66	66	85		5	7	25	34	41	80	2	8	11	28	38	44	81
Oman		100		90			41		48		49			30		52		53		63	39
Pakistan	77	75	72	83	82	77	96	4	5	20	27	42	52	84	21	25	35	44	55	60	88
Philippines	67	82	49	49	93	93	92	20	31	43	54	72	77	80	36	50	45	52	81	85	87
Qatar	100	100	76		100			75	83	43					95	97	71				
Saudi Arabia	100	97	92	100			100	37	56	87	88			64	49	64	90	94			95
Singapore			100	100	100		100										100	100	100		100
Sri Lanka	46	36	65	82	80	43	91	14	13	18	29	55	47	80	21	19	28	40	60	46	83
Syria	98		98			92	94	50		54			78	64	71		74			85	80
Tajikistan																					
Thailand	60	69	65	56			89	10	16	63	66	85		77	17	25	63	64			80
Turkey			95				82			62				84			76				83

(Continued)

Table 7D.33 (Continued)

Region and Country	Fraction of Population with Access to Drinking Water																				
	Urban							Rural							Total						
	1970	1975	1980	1985	1990	1994	2000	1970	1975	1980	1985	1990	1994	2000	1970	1975	1980	1985	1990	1994	2000
Turkmenistan																					
United Arab Emirates			95							81							92				
Uzbekistan							96							78							85
Vietnam				70	47	53	81			32	39	33	32	50				45	36	36	56
Yemen A R	45		100	100			85	2		18	25			64	4		31	40			69
Yemen Dem	88		85				85	43		25				64	57		52				69
Oceania																					
American Samoa							100							100							100
Australia							100							100							100
Cook Islands			100	99	100		100				88	100		100				92			100
Fiji	78	89	94		96	100	43	15	56	66		69	100	51	37	69	77		80	100	47
French Polyneisa					100		100					18		100							100
Guam																					
Kiribati			93		91		82			25		63		25							47
Marshall Islands					100							45									
Micronesia						100						38	100							100	
Nauru																					
New Caledonia																					
New Zealand							100														
Niue					0	100	100				100	100	100	100						100	100
Northern Mariana Islands					100							0									
Palau					100		100					97		20							79
Papua New Guinea	44	30	55	95	94	84	88	72	19	10	15	20	17	32	70	20	16	26	32	28	42
Pitcairn																					
Samoa	86	100	97		100		95		23	94		77		100	17	43					99
Solomon Islands			96		82		94			45		58		65					62		71
Tokelau						100					100		100							100	
Tonga	100	100	86	99	92	100	100	53	71	70	99	98	100	100	63	83	17	99	96	100	100
Tuvalu				100		100	100				100		95	100						98	100
Vanuatu			65	95			63			53	54			94				64			88

Wallis and Futuna Islands					
Western Samoa	97	75	94	67	69

[a] United Nations Environment Programme, 1989, *Environmental Data Report*, GEMS Monitoring and Assessment Research Centre, Basil Blackwell, Oxford.

[b] WRI 1988, World Health Organization data, cited by the *World Resources Institute, 1988, World Resources 1988–89*, World Resources Institute and the International Institute for Environment and Development in collaboration with the United Nations Environment Programme, Basic Books, New York.

[c] United Nations Environment Programme, 1993–94, *Environmental Data Report*, GEMS Monitoring and Assessment Research Centre in cooperation with the World Resources Institute and the UK Department of the Environment, Basil Blackwell, Oxford.

[d] WHO 1996, *Water Supply and Sanitation Sector Monitoring Report: 1996* (Sector Status as of 1994), in collaboration with the Water Supply and Sanitation Collaborative Council and the United Nations Children's Fund, UNICEF, New York.

[e] WHO 2000, *Global Water Supply and Sanitation Assessment 2000 Report*, available in full at who.int/water_sanitation_health/Globassessment/GlobalTOC.htm.

Source: From *World's Water 2002–2003*, by Peter H. Gleick. Copyright © 2002 Island Press. Reproduced by permission of Island Press, Washington, DC.

Table 7D.34 Population Supplied with Access to Sanitation, by Country, 1970–2000

Region and Country	Fraction of Population with Access to Sanitation																				
	Urban							Rural							Total						
	1970	1975	1980	1985	1990	1994	2000	1970	1975	1980	1985	1990	1994	2000	1970	1975	1980	1985	1990	1994	2000
Data Source	a,b	a,b	a,b	a,b	c	d	e	a,b	a,b	a,b	a,b	c	d	e	a,b	a,b	a,b	a,b	c	d	e
Africa	47	75	57	75																	
Algeria	13	100	80			90	6	50		40			47	9	67		57			73	
Angola			40	29	25	34	70			15	16	20	8	30			20	19	21	16	44
Benin	83		48	58	60	54	46	1		4	20	35	6	6	14		16	33	45	20	23
Botswana				93	100						28	85						40	89		
Burkina Faso	49	47	38	44		42	88			5	6		11	16	4	4	7	9		18	29
Burundi	96		40	84	64	60	79			35	56	16	50				35	58	18	51	
Cameroon				100			99				1			85				43			92
Cape Verde			34	32		40	95			10	9		10	32			11	10		24	71
Central African Republic	64	100			45		43	96	100			46		23	72	100			46	46	31
Chad	7	9				73	81		1				7	13	1	1				21	29
Comoros							98							98							98
Congo	8	10					14	6	9						6	9					
Congo, Democratic Republic	5	65			46	23	53	5	6		9	11	4	6	5	22			21	9	20
Cote D'Ivoire	23				81	59						100	51		5				92	54	
Djibouti			43	78		77	99			20	17		100	50			39	64		90	91
Egypt					80	20	98			10		26	5	91					50	11	94
Equatorial Guinea					54	61	60					24	48	46					33	54	53
Eritrea							66							1							13
Ethiopia	67	56		96			58	8	8		96			6	14	14					15
Gabon							25							4							21
Gambia					100	83	41					27	23	35					44	37	37
Ghana	92	95	47	51	63	53	62	40	40	17	16	60	36	64	55	56	26	30	61	42	63
Guinea	70		54				94	2		1		0		41	13		11			70	58
Guinea-Bissau			21	29		32	88			13	18		17	34			15	21		20	47
Kenya	85	98	89			69	96	45	48	19			81	81	50	55	30			77	86
Lesotho	44	51	13	22		1	93	10	12	14	14		7	96	11	13	14	15		6	92
Liberia	100			6		38		9			2		2		19					18	
Libya	100	100	100				97	54	69	72				30	67	79	88				97
Madagascar	88		9	55		50	70		9				3	70						15	42
Malawi			100			70	96			81			51	98			83			53	77
Mali	63		79	90	81	58	93				3	10	21	100	8			19	27	31	69
Mauritania	100		5	8			44							19	7						33
Mauritius	51	63	100	100	100	100	100	99	100	90	86	100	100	99	77	82	94	92	100	100	99
Morocco	75			62	100	69	100	4			16		18	42	29					40	75

Mozambique				53		70	69				12		70	26				20			43
Namibia					24		96					11		17					15		41
Niger	10	30	36		71	71	79		1	3		4	4	5	1	3	7		17	15	20
Nigeria					80	61	85				5	11	21	45					35	36	63
Reunion																					
Rwanda	83	87	60	77	88		12	52	56	50	55	17		8	53	57	51	56	21		8
Sao Tome and Principe											15							15			
Senegal			100	87	57	83	94			2		38	40	48			36		46	58	70
Seychelles																					
Sierra Leone			31	60	55	17	23			6	10	31	8	31			12	24	39	11	28
Somalia		77		44					35		5					47		18			
South Africa						79	99						12	73						46	86
Sudan	100	100	73	73		79	87	4	10				4	48	16	22				22	62
Swaziland		99		100		36			25		25		37			36		45		36	
Tanzania		88		93			98		14		58			86		17	66				90
Togo	4	36	24	31		57	69		12	10	9		13	17	1	15	13	14		26	34
Tunisia	100		100	84		100		34			16		85		62			55		96	
Uganda	84	82		32	32	75	96	76	95		30	60	55	72	76	94		30	57	57	75
Zambia	12	87		76		40	99	18	46		34		10	64	16	42		55		23	78
Zimbabwe					95		99				15	22		51					43		68
North & Central America & Caribbean																					
Anguilla																					
Antigua and Barbuda																					
Aruba																					
Bahamas	100	100	88	100	98		93	13	13			2		94	66	65	88	100	63		93
Barbados	100	100		100	100		100	100	100			100		100	100	100			100		100
Belize			62	87	76	23	59			75	45	22	87	21			69	66	50	57	42
British Virgin Islands					100							100									
Canada							100							99							100
Cayman Islands			94	96							94				96						
Costa Rica	66	94	99	99		85	98	43	93	84	89		99	95	52	93	91	95		92	96
Cuba	57	100			100	71	96					68	51	91					92	66	95
Dominican Republic	63	74	25	41	95	76	75	54	16	4	10	75	83	64	58	42	15	23	87	78	71
Dominica																					
El Salvador	66	71	48	82	85	78	88	18	17	26	43	38	59	78	37	39	35	58	59	68	83
Grenada							96							97							97
Guadeloupe							61							61							61
Guatemala			45	41	72		98	11	16	20	12	52		76			30	24	60		85
Haiti			42	42	44	42	50	43	1	10	13	17	16	16			19	21	25	24	28

(Continued)

Table 7D.34 (Continued)

Region and Country	Fraction of Population with Access to Sanitation																				
	Urban							Rural							Total						
	1970	1975	1980	1985	1990	1994	2000	1970	1975	1980	1985	1990	1994	2000	1970	1975	1980	1985	1990	1994	2000
Honduras	64	53	49	24	89	81	94	9	13	26	34	42	53	57	24	26	35	30	63	65	77
Jamaica	100	100	12	92			98	92	91	2	90			66	94	94	7	91			84
Martinique																					
Mexico			77	77	85	81	87	13	14	12	13		26	32			55	58		66	73
Montserrat																					
The Netherlands Antilles																					
Nicaragua			34	35		34	96	8	24		16		27	68				27		31	84
Panama	87	78	83	99			87	69	76	59	61			94	78	77	71	81		86	99
Puerto Rico																					
St. Kitts																					
St. Lucia																					
St. Vincent																					
Trinidad/ Tobago					100							92							97		88
Turks/Caicos Islands	51	83	96	100				96	97	88	95				81	92	93	98			
United States of America							100							100							100
United States Virgin Islands																					
South America																					
Argentina	87	100	80	75			89	79	83	35	35			48	85	97		69			85
Bolivia	25		37	33	38	58	82	4	9	4	10	14	16	38	12		18	21	26	41	66
Brazil	85			86	84	55	85	24		1	1	32	3	40	58			63	71	44	77
Chile	33	36	100	100		82	98	10	11	10	4			93	29	32	83	84			97
Colombia	75	73	93	96	84	76	97	8	13	4	13	18	33	51	47	48	61		64	63	85
Ecuador			73	98	56	87	70		7	17	29	38	34	37			43	65	48	64	59
Falkland Islands (Malvinas)																					
French Guiana							85							57							79
Guyana	95	99	73	100	97		97	92	94	80	79	81		81	93	96	78	86	86		87
Paraguay	16	28	95	89	31		95			80	83	60		95	6	10	86	85	46		95
Peru	52		57	67	76	62	90	16		0	12	20	10	40	36		36	49	59	44	76
Suriname			100	78			100			79	48			34			88	62			83
Uruguay	97	97	59	59			96	13	17	6	59			89	82	83	51	59			95
Venezuela			60	57		64	86	45		12	5	72	30	69			52	50		58	74
Asia																					
Afghanistan	69	63		5	13	38	25	16	15				1	8	21	21				8	12

Armenia																					
Azerbaijan																					
Bahrain				100	100						100	0						100			
Bangladesh	87	40	21	24	40	77	82			1	3	4	30	44	6	5	3	5	10	35	53
Bhutan					80	66	65					3	18	70					7	41	69
Brunei Darus																					
Cambodia			100				58			76				10							18
China					100	58	68					81	7	24					86	21	38
Cyprus	100	94		100	96		100	92	95		100	100		100	95	95		100	98		100
East Timor																					
Gaza Strip																					
Georgia																					
Hong Kong					90							50							88		
India	85	87	27	31	44	70	73	1	2	1	2	3	14	14	18	20	7	9	14	29	31
Indonesia	50	60	29	33	79	73	87	4	5	21	38	30	40	52	12	15	23	37	44	51	66
Iran	100	100	96		100	89	86	48	59	43		35	37	74	70	78	69		72	67	81
Iraq	82	75		100	96		93		1		11			31	47	47		74		36	79
Israel																					
Japan																					
Jordan			94	92	100		100			34		100		98			70		100	95	99
Kazakhstan							100							98							99
Korea DPR							99							100							99
Korea Rep	59	80	100	100	67		76		50	100	100	12		4	25	64	100	100	52		63
Kuwait			100	100						100							100				
Kyrgyzstan							100							100							100
Lao People's Dem Rep		10	13		30	70	84		2	4		8	13	34		3	5		12	24	46
Lebanon						100	100						100	92						100	99
Macau																					
Malaysia	100	100	100	100	94			43	43	55	60	94		41	59	60	70	75	94		
Maldives		21	60	100	95	95	100			1	2	4	26	58		3	13	22		44	56
Mongolia					100		46					47		2					78		30
Myanmar (Burma)	45	38	38	33	50	42	65					13	40	39	35	33	20	24	22	41	46
Nepal	14	14	16	17	34	51	75			1	1	3	16	20	1	1	1		6	20	27
Oman	100	100		88			98		5		25			61		12		31		76	92
Pakistan	12	21	42	51	53	53	94			2	6	12	19	42	3	6	13	19	25	30	61
Philippines	90	76	81	83	79		92	40	44	67	56	63		71	57	56	75	67	70		83
Qatar	100	100			100			16	100			85			83	100					
Saudi Arabia	67	91	81	100			100	11	35	50	33			100	21	47	70	82			100
Singapore			80	99	99		100										80	99			100
Sri Lanka	76	68	80	65	68	33	91	61	55	63	39	45	58	80	64	59	67	44	50	52	83
Syria			74			77	98			28			35	81			50			56	90
Tajikistan																					
Thailand	65	58	64	78			97	8	36	41	46	86		96	17	40	45	52			96
Turkey			56				98							70							91

(Continued)

Table 7D.34 (Continued)

Region and Country	Fraction of Population with Access to Sanitation																				
	Urban							Rural							Total						
	1970	1975	1980	1985	1990	1994	2000	1970	1975	1980	1985	1990	1994	2000	1970	1975	1980	1985	1990	1994	2000
Turkmenistan																					
United Arab Emirates			93							22							80				
Uzbekistan							100							100							100
Vietnam	100				23	43	87		2	55		10	15	70	26				13	21	73
Yemen A R			60	83			99							31							45
Yemen Dem			70				99			15				31			35				45
Oceania																					
American Samoa																					
Australia							100							100							100
Cook Islands			100	100	100		100			76	99	100		100				99			100
Fiji	100	100	85		91	100	75	87	93	60		65	85	12	91	96	70		75	92	43
French Polyneisa					98		99					95		97							98
Guam																					
Kiribati					91	100	54					49	100	44						100	48
Marshall Islands					100							45									
Micronesia					99	100						46	100							100	
Nauru																					
New Caledonia																					
New Zealand																					
Niue					0	100	100				100	100	100	100						100	100
Northern Mariana Islands					100							71		92							
Palau					95		100					100		100							100
Papua New Guinea	100	100	96	99	57	82	92	5	5	3	35		11	80	14	18	15	44		22	82
Pitcairn																					
Samoa	100	100	86		100		95	80	99	83		92	17	100	84	99					99
Solomon Islands			80		73		98			21		2		18					13		34
Tokelau						100				41			100							100	
Tonga	100	100	97	99	88	100		100	100	94	40	78	100		100	100	19	52	82	100	
Tuvalu			100	81		90	100			80	73		85	100						87	100
Vanuatu			95	86			100			68	25			100				40			100

Wallis and Futuna Islands					
Western Samoa	86	88	83	83	84

[a] United Nations Environment Programme, 1989, *Environmental Data Report*, GEMS Monitoring and Assessment Research Center, Basil Blackwell, Oxford.

[b] WRI 1988, World Health Organization data, cited by the World Resources Institute, 1988, *World Resources 1988–89*, World Resources Institute and the International Institute for Environment and Development in collaboration with the United Nations Environment Programme, Basic Books, New York.

[c] United Nations Environment Programme, 1993–94, *Environmental Data Report*, GEMS Monitoring and Assessment Research Centre in cooperation with the World Resources Institute and the UK Department of the Environment, Basil Blackwell, Oxford.

[d] WHO 1996, World Health Organization, 1996, *Water Supply and Sanitation Sector Monitoring Report: 1996* (Sector Status as of 1994), in collaboration with the Water Supply and Sanitation Collaborative Council and the United Nations Children's Fund, UNICEF, New York.

[e] WHO 2000, World Health Organization, 2000, *Global Water Supply and Sanitation Assessment 2000 Report*, available in full at www.who.int/water_sanitation_health/Globassessment/GlobalTOC.htm.

Source: From *World's Water 2002–2003*, by Peter H. Gleick.

Table 7D.35 Water Supply and Sanitation Coverage by Region, 1990 and 2000

Region	1990 Population (millions)				2000 Population (millions)			
	Total Population	Population Served	Population Unserved	% Served[a]	Total Population	Population Served	Population Unserved	% Served[a]
Global	(76% of regional population represented)				(89% of regional population represented)			
Urban water supply	2,292	2,179	113	95	2,845	2,672	173	94
Rural water supply	2,974	1,961	1,013	66	3,210	2,284	926	71
Total water supply	5,266	4,140	1,126	79	6,055	4,956	1,099	82
Urban sanitation	2,292	1,877	415	82	2,845	2,442	403	86
Rural sanitation	2,974	1,028	1,946	35	3,210	1,210	2,000	38
Total sanitation	5,266	2,905	2,361	55	6,055	3,652	2,403	60
Africa	(72% of regional population represented)				(96% of regional population represented)			
Urban water supply	197	166	31	84	297	253	44	85
Rural water supply	418	183	235	44	487	231	256	47
Total water supply	615	349	266	57	784	484	300	62
Urban sanitation	197	167	30	85	297	251	46	84
Rural sanitation	418	206	212	49	487	220	267	45
Total sanitation	615	373	242	61	784	471	313	60
Asia	(88% of regional population represented)				(94% of regional population represented)			
Urban water supply	1,029	972	57	94	1,352	1,254	98	93
Rural water supply	2,151	1,433	718	67	2,331	1,736	595	75
Total water supply	3,180	2,405	775	76	3,683	2,990	693	81
Urban sanitation	1,029	690	339	67	1,352	1,055	297	78
Rural sanitation	2,151	496	1,655	23	2,331	712	1,619	31
Total sanitation	3,180	1,186	1,994	37	3,683	1,767	1,916	48
Latin American and the Caribbean	(77% of regional population represented)				(99% of regional population represented)			
Urban water supply	313	287	26	92	391	362	29	93
Rural water supply	128	72	56	56	128	79	49	62
Total water supply	441	359	82	82	519	441	78	85
Urban sanitation	313	267	46	85	391	340	51	87
Rural sanitation	128	50	78	39	128	62	66	49
Total sanitation	441	317	124	72	519	402	117	78
Oceania	(64% of regional population represented)				(85% of regional population represented)			
Urban water supply	18	18	0	100	21	21	0	98
Rural water supply	8	5	3	62	9	6	3	63
Total water supply	26	23	3	88	30	27	3	88
Urban sanitation	18	18	0	99	21	21	0	99
Rural sanitation	8	7	1	89	9	7	2	81

Total sanitation	26	25	1	96	30	28	2	93
Europe	(15% of regional population represented)				(44% of regional population represented)			
Urban water supply	522	522	0	100	545	542	3	100
Rural water supply	200	199	1	100	184	161	23	87
Total water supply	722	721	1	100	729	703	26	96
Urban sanitation	522	522	0	100	545	537	8	99
Rural sanitation	200	199	1	100	184	137	47	74
Total sanitation	722	721	1	100	729	674	55	92
Northern America	(99.9% of regional population represented)				(99.9% of regional population represented)			
Urban water supply	213	213	0	100	239	239	0	100
Rural water supply	69	69	0	100	71	71	0	100
Total water supply	282	282	0	100	310	310	0	100
Urban sanitation	213	213	0	100	239	239	0	100
Rural sanitation	69	69	0	100	71	71	0	100
Total sanitation	282	282	0	100	310	310	0	100

[a] Due to rounding, coverage figures might not total 100% even if the population unserved is shown as 0.

Source: From *World Health Organization and United Nations Children's Fund Global Water Supply and Sanitation Assesment Report 2000*, Copyright © 2000 World' Health Organization and United Nations Children's Fund, www.wssinfo.org.

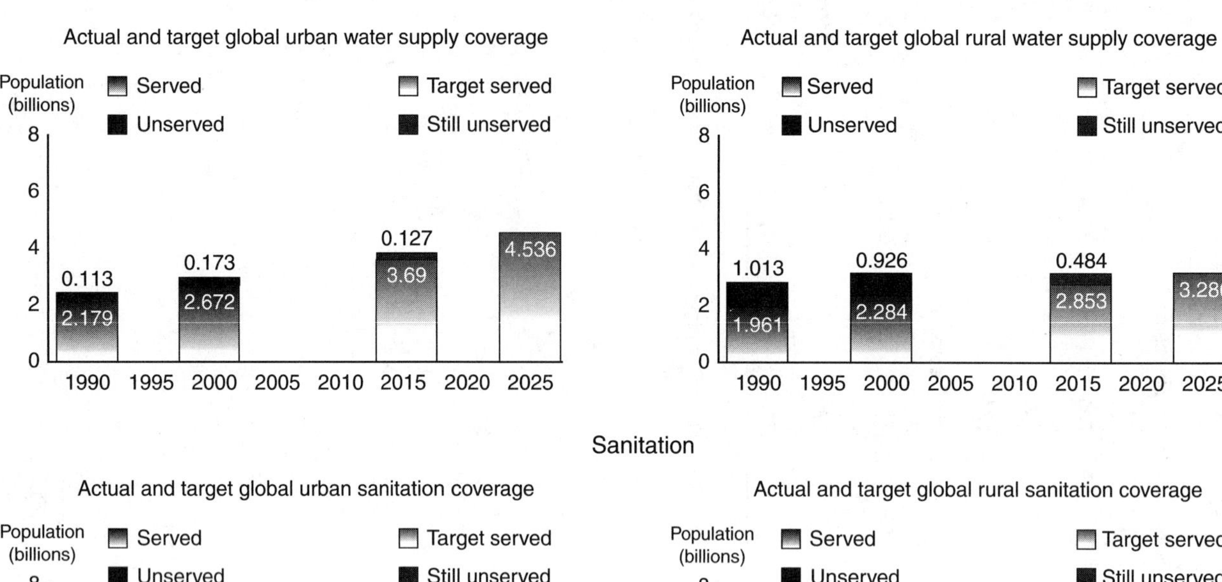

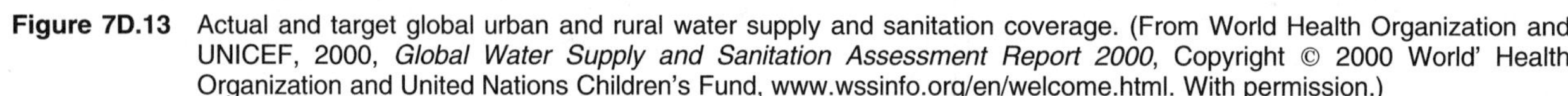

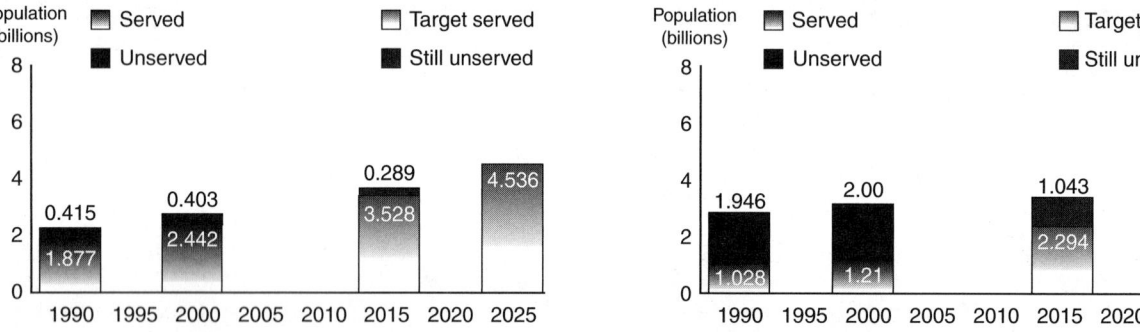

Figure 7D.13 Actual and target global urban and rural water supply and sanitation coverage. (From World Health Organization and UNICEF, 2000, *Global Water Supply and Sanitation Assessment Report 2000*, Copyright © 2000 World' Health Organization and United Nations Children's Fund, www.wssinfo.org/en/welcome.html. With permission.)

Table 7D.36 Constraints in Improving Water Supply and Sanitation Services in Developing Countries

Constraints	Number of Countries Indicting Constraint			Ranking Index[a]
	Very Severe	Severe	Moderate	
Insufficiency of trained personnel (professional)	16	40	27	155
Funding limitations	21	31	30	155
Insufficiency of trained personnel (sub-professional)	16	38	29	153
Operation and maintenance[b]	16	36	23	143
Logistics[b]	11	35	23	126
Inadequate cost-recovery framework	11	34	22	123
Inappropriate institutional framework	6	30	35	113
Insufficient health education efforts	7	24	43	112
Intermittent water service	10	19	32	100
Lack of planning and design criteria	6	17	41	93
Noninvolvement of communities	6	15	44	92
Inadequate or outmoded legal framework	10	14	34	92
Inappropriate technology	5	18	33	84
Insufficient knowledge of water resources	1	20	39	82
Inadequate water resources	5	11	40	77
Lack of definite government policy for sector	4	10	44	76
Import restrictions	5	12	21	60

Note: Ranking and frequency of constraints as reported by 87 countries.

[a] Ranking index = (No. very severe × 3) + (No. severe × 2) + (No. moderate × 1).

[b] "Logistics" is ranked ahead of "Operation and maintenance" in the group of Least Developed Countries.

Source: World Health Organization, 1984. *The International Drinking Water Supply and Sanitation Decade*, WHO Publ. 85. Reprinted with permission.

SECTION 7E DOMESTIC WATER CONSUMPTION

Table 7E.37 Domestic Freshwater Use in the United States, 2000

State	Population (thousands)				Withdrawals (mil gal/day)			Withdrawals (thousand acre-feet/yr)		
			Self-Supplied Domestic		By Source			By Source		
	Total	Served by Public Supply Total	Population	Population (Percent)	Groundwater	Surface Water	Total	Groundwater	Surface Water	Total
Alabama	4,450	3,580	868	20	78.9	0	78.9	88.4	0	88.4
Alaska	627	421	206	33	10.9	0.25	11.2	12.2	0.28	12.5
Arizona	5,130	4,870	265	5	28.9	0	28.9	32.4	0	32.4
Arkansas	2,670	2,320	351	13	28.5	0	28.5	31.9	0	31.9
California	33,900	30,100	3,810	11	257	28.6	286	288	32	320
Colorado	4,300	3,750	555	13	66.8	0	66.8	74.9	0	74.9
Connecticut	3,410	2,660	749	22	56.2	0	56.2	63	0	63
Delaware	784	617	166	21	13.3	0	13.3	14.9	0	14.9
District of Columbia	572	572	0	0	0	0	0	0	0	0
Florida	16,000	14,000	1,950	12	199	0	199	223	0	223
Georgia	8,190	6,730	1,450	18	110	0	110	123	0	123
Hawaii	1,210	1,140	72.9	6	4.82	7.22	12	5.4	8.09	13.5
Idaho	1,290	928	366	28	85.2	0	85.2	95.6	0	95.6
Illinois	12,400	10,900	1,500	12	135	0	135	152	0	152
Indiana	6,080	4,480	1,600	26	122	0	122	137	0	137
Iowa	2,930	2,410	511	17	33.2	0	33.2	37.2	0	37.2
Kansas	2,690	2,500	193	7	21.6	0	21.6	24.2	0	24.2
Kentucky	4,040	3,490	552	14	19.5	8	27.5	21.9	8.97	30.8
Louisiana	4,470	3,950	523	12	41.2	0	41.2	46.2	0	46.2
Maine	1,270	726	549	43	35.7	0	35.7	40	0	40
Maryland	5,300	4,360	932	18	77.1	0	77.1	86.4	0	86.4
Massachusetts	6,350	5,880	473	7	42.2	0	42.2	47.2	0	47.2
Michigan	9,940	7,170	2,770	28	239	0	239	268	0	268
Minnesota	4,920	3,770	1,150	23	80.8	0	80.8	90.6	0	90.6
Mississippi	2,840	2,190	654	23	69.3	0	69.3	77.7	0	77.7
Missouri	5,600	4,770	824	15	53.6	0	53.6	60.1	0	60.1
Montana	902	664	238	26	17.3	1.29	18.6	19.4	1.45	20.8
Nebraska	1,710	1,390	324	19	48.4	0	48.4	54.3	0	54.3
Nevada	2,000	1,870	124	6	22.4	0	22.4	25.2	0	25.2
New Hampshire	1,240	756	479	39	40.9	0.16	41	45.8	0.18	46
New Jersey	8,410	7,460	952	11	79.7	0	79.7	89.3	0	89.3
New Mexico	1,820	1,460	360	20	31.4	0	31.4	35.2	0	35.2
New York	19,000	17,100	1,890	10	142	0	142	159	0	159
North Carolina	8,050	5,350	2,700	34	189	0	189	212	0	212
North Dakota	642	493	149	23	11.9	0	11.9	13.3	0	13.3
Ohio	11,400	9,570	1,790	16	132	2.71	134	148	3.04	151
Oklahoma	3,450	3,150	299	9	25.5	0	25.5	28.5	0	28.5
Oregon	3,420	2,730	692	20	68.3	7.97	76.2	76.5	8.93	85.5
Pennsylvania	12,300	10,100	2,190	18	132	0	132	148	0	148

(Continued)

Table 7E.37 (Continued)

	Population (thousands)				Withdrawals (mil gal/day)			Withdrawals (thousand acre-feet/yr)		
			Self-Supplied Domestic		By Source			By Source		
State	Total	Served by Public Supply Total	Population	Population (Percent)	Groundwater	Surface Water	Total	Groundwater	Surface Water	Total
Rhode Island	1,050	922	127	12	8.99	0	8.99	10.1	0	10.1
South Carolina	4,010	3,160	847	21	63.5	0	63.5	71.2	0	71.2
South Dakota	755	625	129	17	9.52	0.01	9.53	10.7	0.01	10.7
Tennessee	5,690	5,240	453	8	32.6	0	32.6	36.6	0	36.6
Texas	20,900	19,700	1,190	6	131	0	131	147	0	147
Utah	2,230	2,180	56.2	3	16.1	0	16.1	18	0	18
Vermont	609	362	247	41	20.7	0.25	21	23.2	0.28	23.5
Virginia	7,080	5,310	1,770	25	133	0	133	150	0	150
Washington	5,890	4,900	993	17	125	0.02	125	140	0.02	140
West Virginia	1,810	1,300	505	28	39.6	0.81	40.4	44.4	0.91	45.3
Wisconsin	5,360	3,620	1,750	33	96.3	0	96.3	108	0	108
Wyoming	494	406	87.5	18	6.57	0	6.57	7.36	0	7.36
Puerto Rico	3,810	3,800	12.8	0	0.88	0	0.88	0.99	0	0.99
U.S. Virgin Islands	109	53.4	55.2	51	0	1.69	1.69	0	1.89	1.89
Total	**285,000**	**242,000**	**43,500**	**15**	**3,530**	**58.9**	**3,590**	**3,960**	**66.1**	**4,030**

Note: Figures may not sum to totals because of independent rounding.

Source: From Hutson, S.S. et al., 2004, Estimated Use of Water in the United States in 2000, U.S. Geological Survey Circular 1268, www.usgs.gov.

Table 7E.38 Groundwater and Surface Water Withdrawals in Public Supply and Rural Domestic Sectors, 1950–1995

	Public Supply			Self-Supplied Domestic		
Year	Total Withdrawals (bgd)	Groundwater (%)	Surface Water (%)	Total Withdrawals (bgd)	Groundwater (%)	Surface Water (%)
1950	14.0	26	74	3.6	80	20
1955	17.0	28	72	2.5	72	28
1960	21.0	30	70	2.1	93	7
1965	24.0	33	67	2.3	95	5
1970	27.0	34	66	2.6	96	4
1975	29.0	36	64	2.8	95	5
1980	34.0	34	66	3.5	95	5
1985	36.5	40	60	3.3	98	2
1990	38.5	39	61	3.4	96	4
1995	40.2	37	63	3.4	99	1

Source: From Dziegielewski, B. et al., 2002, *Analysis of Water Use Trends in the United States: 1950–1995*. Southern Illinois University at Carbondale, Carbondale, IL, February 28, 2002. Reprinted with permission. http://info.geography.siu.edu/geography_info/research/.

Table 7E.39 Population and Domestic Withdrawals, 1950–1995

Year	Domestic Withdrawals (bgd)	Population (millions)	Per Capita Withdrawals (gpcd)
1950	17.6	150.7	116.8
1955	20.6	164.0	125.6
1960	24.6	179.3	137.2
1965	28.0	193.8	144.5
1970	31.5	205.9	153.0
1975	33.9	216.4	156.7
1980	39.6	229.6	172.5
1985	44.3	242.4	182.7
1990	46.4	252.3	183.9
1995	49.1	267.1	183.8

Source: From Dziegielewski, B. et al., 2002, *Analysis of Water Use Trends in the United States: 1950–1995*. Southern Illinois University at Carbondale, Carbondale, IL, February 28, 2002. Reprinted with permission. http://info.geography.siu.edu/geography_info/research/.

Table 7E.40 Typical Urban Water Use by a Family of Four

	Daily Use		
	Per Family		
	Gal/day	%	Per Capita gal/day
Drinking and water used in kitchen	8	2	2.00
Dishwasher (3 loads per day)	15	4	3.75
Toilet (16 flushes per day)	96	28	24.00
Bathing (4 baths or showers per day)	80	23	20.00
Laundering (6 loads per week)	34	10	8.50
Automobile washing (2 car washes per month)	10	3	2.50
Lawn watering and swimming pools (180 hr/yr)	100	29	25.00
Garbage disposal unit (1 percent of all other uses)	3	1	0.75
Total	346	100	86.50

Note: Assuming no water delivery losses.

Source: From U.S. Water Resources Council, Second National Water Assessment, The Nation's Water Resources 1975–2000; percentage added.

Table 7E.41 Flow Rates for Certain Plumbing, Household, and Farm Fixtures

Location	Flow Pressure[a] (psi)	Flow Rate (gpm)
Ordinary basin faucet	8	2.0
Self-closing basin faucet	8	2.5
Sink faucet, 3/8 in.	8	4.5
Sink faucet, 1/2 in.	8	4.5
Bathtub faucet	8	6.0
Laundry tub faucet, 1/2 in.	8	5.0
Shower	8	5.0
Ball-cock for closet	8	3.0
Flush valve for closet (toilet)	15	15–40[b]
Flushometer valve for closet (toilet)	15	15.0
Garden hose (50 ft, 3/4 in. sill cock)	30	5.0
Garden hose (50 ft, 5/8 in. outlet)	15	3.33
Drinking fountains	15	0.75
Fire hose 1 1/2 in., 1/2 in. nozzle	30	40.0

[a] Flow pressure is the pressure in the supply near the faucet or water outlet while the faucet or water outlet is wide open and flowing.
[b] Wide Range due to variation in design and type of closet (toilet) flush valves.

Source: From U.S. Public Health Service, 1962, *Manual of Individual Water Supply Systems.*

Table 7E.42 Summary of California and Federal Plumbing Fixture Requirements

Plumbing Device	California (Covers Sale and Installation)	Effective Date	Energy Policy Act of 1992 (Covers Only Manufacture)
Showerheads	2.5 gpm	CA 3/20/92	
		US 1/1/94	2.5 gpm[c]
Lavatory Faucets[a]	2.75 gpm	CA 12/22/78	
	2.2 gpm	CA 3/20/92	
		US 1/1/94	2.5 gpm[c]
Sink Faucets[a]	2.2 gpm	CA 3/20/92	
		US 1/1/94	2.5 gpm[c]
Metering (self-closing) Faucets[b] (public restrooms)	Hot water maximum flow rates range from 0.25 to 0.75 gal/cycle and/or from 0.5 gpm to 2.5 gpm, depending on controls and hot water system	CA 7/1/92 US 1/1/94	0.25 gal/cycle (maximum water delivery per cycle)[c]
Tub Spout Diverter[a]	0.1 (new), to 0.3 gpm (after 15,000 cycles of diverting)	CA 3/20/92	(does not appear to be included in EPA)
Toilets (residential)	1.6 gpf	CA 1/1/92 (new construction)	
		CA 1/1/94 (all toilets for sale or installation)	
		US 1/1/94 (non-commercial)	1.6 gpf
Flushometer valves[a]	1.6 gpf	CA 1/1/92 (new construction)	
		CA 1/1/94 (all toilets)	
		US 1/1/94 (commercial)	3.5 gpf
		US 1/1/97 (commercial)	1.6 gpf
Toilets (Commercial)[a]	1.6 gpf	CA 1/1/94 (all toilets for sale or installation)	
		US 1/1/97	1.6 gpf
Urinals	1.0 gpf	CA 1/1/92 (new)	
		CA 1/1/94 (all)	
		US 1/1/94	1.0 gpf

[a] California requirements are pre-existing and more stringent than federal law; therefore California requirements prevail in California.
[b] Federal law is more stringent than California requirements.
[c] Measured at a flowing water pressure of 80 pounds/sq.in.

Source: From California Department of Water Resources, 1998, "Urban, Agricultural and Environmental Water Use," Californaia Water Plan Update Bulletin-160-98, California Dept. of Water Resources, Sacremento, CA, November 1998, www.water.ca.gov.

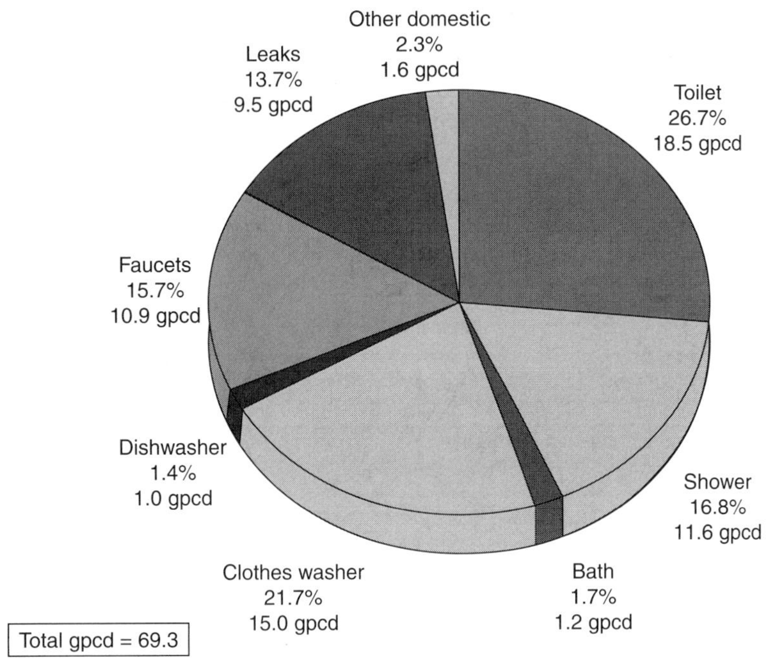

Figure 7E.14 From United States Environmental Protection Agency, 2002. Onsite Wastewater Treatment Systems Manual, EPA/625-R-00/008, February 2002. Original Soure: Mayer, P.W., DeOreo, W.B. et al., 1999. *Residential End Uses of Water*, American Water Works Association Research Foundation, Denver, Colorado.

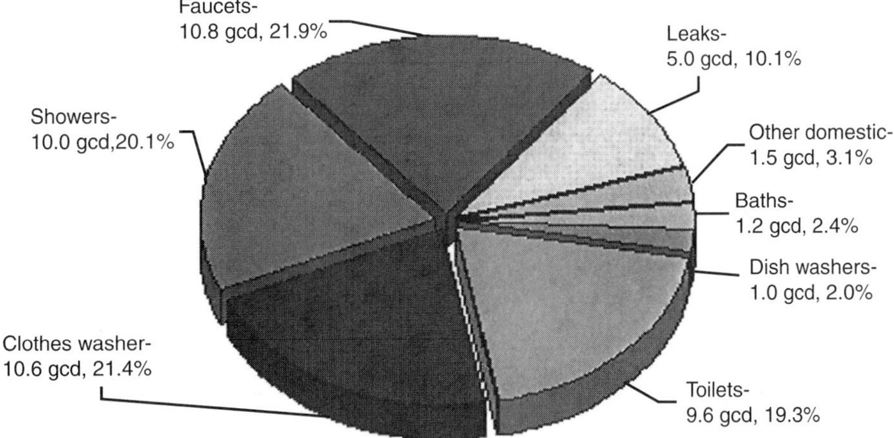

Figure 7E.15 Typical single family home water use—with conservation. (Reprinted from Waterwiser www.waterwiser.org, by permission. Copyright © 1999, American Water Works Association.)

Table 7E.43 Indoor Residential Water Use by Fixture or Appliance

Fixture/Use	Gal/Use: Average Range	Uses/Person/Day: Average Range	Gal/Person/Day: Average Range[a]	% Total: Average Range
Toilet	3.5	5.05	18.5	26.7
	2.9–3.9	4.5–5.6	15.7–22.9	22.6–30.6
Shower	17.2[b]	0.75[b]	11.6	16.8
	14.9–18.6	0.6–0.9	8.3–15.1	11.8–20.2
Bath	See shower	See shower	1.2	1.7
			0.5–1.9	0.9–2.7
Clothes washer	40.5	0.37	15.0	21.7
	—	0.30–0.42	12.0–17.1	17.8–28.0
Dishwasher	10.0	0.10	1.0	1.4
	9.3–10.6	0.06–0.13	0.6–1.4	0.9–2.2
Faucets	1.4[c]	8.1[d]	10.9	15.7
	—	6.7–9.4	8.7–12.3	12.4–18.5
Leaks	NA	NA	9.5	13.7
			3.4–17.6	5.3–21.6
Other domestic	NA	NA	1.6	2.3
			0.0–6.0	0.0–8.5
Total	NA	NA	69.3	100
			57.1–83.5	

Note: Results from AWWARF REUWS at 1,188 homes in 12 metropolitan areas. Homes surveyed were served by public water supplies, which operate at higher pressure than private water sources. Leakage rates might be lower for homes on private water supplies. Results are averages over range. Range is the lowest to highest average for 12 metropolitan areas.

[a] Gal/person/day might not equal gal/use multiplied by uses/person/day because of differences in the number of data points used to calculate means.
[b] Includes shower and bath.
[c] Gallons per minute.
[d] Minutes of use per person per day.

Source: United States Environmental Protection Agency, 2002, Onsite Wastewater Treatment Systems Manual, EPA/625-R-00/008, February 2002.

Original Source: Adapted from Residential End Uses of Water, by permission. Copyright © 1999 American Water Works Association and AWWA Research Foundation, www.awwa.org.

Table 7E.44 Household End Use of Water with and without Conservation

	Without Conservation		With Conservation		
Type of Use	Amount (gpcd)	Percent of Total (%)	Amount (gpcd)	Percent of Total (%)	Savings (%)
Toilets	18.3	28.4	10.4	23.2	44
Clothes washers	14.9	23.1	10.5	23.4	30
Showers	12.2	18.8	10.0	22.4	18
Faucets	10.3	16.0	10.0	22.5	2
Leaks	6.6	10.2	1.5	3.4	77
Baths	1.2	1.9	1.2	2.7	0
Dish washers	1.1	1.6	1.1	2.4	0
Total indoor water use	64.6	100	44.7	100	31

Note: These data are provided for illustrative purposes only and may not be applicable to a given situation. To the extent practical, planners use system-specific assumptions and estimates. gpcd, gallons per capita per day.

Source: From USEPA, 1998 Water Conservation Plan Guidelines, August 6, 1998.

Original Source: From AWWA Waterwiser, "Household End Use of Water without and with Conservation," 1997 Residential Water Use Summary — Typical Single Family Home. With permission (waterwiser.org/wateruse/tables.htm.)

Table 7E.45 Potential Water Savings from Efficient Fixtures

Fixture[a]	Fixture Capacity[b]	Water Use (gpd)		Water Savings (gpd)	
		Per Capita	2.7-Person Household	Per Capita	2.7-Person Household
Toilets[c]					
Efficient	1.5 gallons/flush	6.0	16.2	na	na
Low-flow	3.5 gallons/flush	14.0	37.8	8.0	21.6
Conventional	5.5 gallons/flush	22.0	59.4	16.0	43.2
Conventional	7.0 gallons/flush	28.0	75.6	22.0	59.4
Showerheads[d]					
Efficient	2.5 [1.71 gal/min	8.2	22.1	na	na
Low-flow	3.0 to 5.0 f2.61 gal/min	12.5	33.8	4.3	11.7
Conventional	5.0 to 8.0 [3.4] gal/min	16.3	44.0	8.1	22.0
Faucets[e]					
Efficient	2.5 [1.7] gal/min	6.8	18.4	na	na
Low-flow	3.0 [2.0] gal/min	8.0	21.6	1.2	3.2
Conventional	3.0 to 7.0 [3.3] gal/min	13.2	36.6	6.4	17.2
Toilets, Showerheads, and Faucets Combined					
Efficient	na	21.0	56.7	na	na
Low-flow	na	34.5	93.2	13.4	36.4
Conventional	na	54.5	147.2	33.5	90.4

Note: na = not applicable

[a] Efficient = post-1994
Low-flow = post-1980
Conventional = pre-1980

[b] For showerheads and faucets: maximum rated fixture capacity (measured fixture capacity). Measured fixture capacity equals about two-thirds the maximum.

[c] Assumes four flushes per person per day; does not include losses through leakage.

[d] Assumes 4.8 shower-use-minutes per person per day.

[e] Assumes 4.0 faucet-use-minutes per person per day.

Source: From United States Environmental Protection Agency, 1998, Water Conservation Plan Guidelines, August 6, 1998, www.epa.gov.

Original Source: From Amy Vickers, "Water Use Efficiency Standards for Plumbing Fixtures: Benefits of National Legislation," *American Water Works Association Journal.* Vol. 82 (May 1990): 53.

Table 7E.46 Various Water Requirements in Developing Countries

Category	Typical Water Use
Schools	
Day schools	15–30 L/day per pupil
Boarding schools	90–140 L/day per pupil
Hospitals (with laundry facilities)	220–300 L/day per bed
Hostels	80–120 L/day per resident
Restaurants	65–90 L/day per seat
Cinema houses, concert halls	10–15 L/day per seat
Offices	25–40 L/day per person
Railway and bus stations	15–20 L/day per user
Livestock	
Cattle	25–35 L/day per head
Horses and mules	20–25 L/day per head
Sheep	15–25 L/day per head
Pigs	10–15 L/day per head
Poultry	
Chicken	15–25 L/day per 100

Source: From Vigneswaran, S., 1995, *Water Treatment Processes: Simple Options*, CRC Press, Inc., Boca Raton.

Original Source: IRC, 1981.

Table 7E.47 Water Requirements for Various Types of Establishments in the United States

Types of Establishments	Gal/day
Airports (per passenger)	3–5
Apartments, multiple family (per resident)	60
Bath houses (per bather)	10
Camps: Construction, semipermanant (per worker)	50
Day with no meals served (per camper)	15
Luxury (per camper)	100–150
Resorts, day and night, with limited plumbing (per camper)	50
Tourist with central bath and toilet facilities (per person)	35
Cottages with seasonal occupancy (per resident)	50
Courts, tourist with individual bath units (per person)	50
Clubs: Country (per resident member)	100
Country (per nonresident member)	25
Dwelling: Boardinghouse (per boarder)	50
Luxury (per person)	100–150
Multiple-family apartments (per resident)	40
Rooming houses (per resident)	60
Single family (per resident)	50–75
Estates (per resident)	100–150
Factories (gallons per shift)	15–30
Highway rest area (per person)	5
Hotels with private baths (2 persons per room)	60
Hotels without private baths (per person)	50
Institutions other than hospitals (per person)	75–125
Hospitals (per bed)	250–400
Laundaries, self-serviced (gallons per washing, i.e., per customer)	50
Motels with bath, toilet, and kitchen facilities (per bed space)	50
With bed and toilet (per bed space)	40
Parks: Overnight with flush toilets (per camper)	25
Trailers with individual bath units, no sewer connections (per trailer)	25
Trailers with individual baths, connected to sewer (per person)	50
Picnic: With bathhouses, showers, and flush toilets (per picnicker)	20
With toilet facilities only (gallons per picnicker)	10
Resturants with toilets facilities (per patron)	7–10
Without toilet facilities (per patron)	2 1/2–3
With bars and cocktail lounge (additional quantity per patron)	2
Schools: Boarding (per pupil)	75–100
Day with cafeteria, gymnasiums, and showers (per pupil)	25
Day with cafeteria but no gymnasium or showers (per pupil)	20
Day without cafeteria, gymnasiums, or showers (per pupil)	15
Service stations (per vehicle)	10
Stores (per toilet room)	400
Swimming pools (per swimmer)	10
Theaters (per seat)	5
Workers: Construction (per person per shift)	50
Day (school or offices per person per shift)	15

Source: From USEPA, 1991, Manual of Individual and Non-Public Water Supply-Systems (EPA 570991004).

SECTION 7F BOTTLED WATER

Table 7F.48 Consumption of Bottled Water in the United States Ranked by Leading States, 1994–1999

		Millions of Gallons					
State	**1999 Rank**	**1994**	**1995**	**1996**	**1997**	**1998**	**1999**
California	1	882.2	928.1	1,003.70	1,038.30	1,048.90	1,130.70
Texas	2	290.2	326.3	362.2	400.2	451.9	519.2
Florida	3	179.9	202.7	231.1	256.8	298.5	339.4
New York	4	179.9	193.2	210.4	237.9	269.5	303.2
Arizona	5	107.4	117.2	131.1	143.5	153.4	176.1
Massachusetts	6	95.8	104.5	110.4	124.6	141	158.9
Illinois	7	110.3	117.2	124.2	128.4	136.8	156.8
Pennsylvania	8	87.1	95	103.5	113.3	124.8	141
Louisiana	9	84.2	95	103.5	113.3	120.2	136.5
Maryland/DC	10	81.3	91.9	96.6	101.9	111.9	125.4
Total Top 10		**2,098.10**	**2,271.10**	**2,476.60**	**2,658.20**	**2,857.00**	**3,187.20**
New Jersey	11	66.7	72.9	79.3	86.8	99.5	111.1
Ohio	12	58	63.4	69	71.7	78.8	88.9
Connecticut	13	40.6	44.3	48.3	52.9	62.2	70.5
Michigan	14	34.8	38	41.4	45.3	49.8	57.1
Colorado	15	20.3	25.3	24.1	26.4	33.2	40.1
Total Top 15		**2,318.60**	**2,515.00**	**2,738.70**	**2,941.30**	**3,180.40**	**3,554.90**
All Others		583.3	652.5	710.6	834.5	965.6	1,091.20
Total U.S.		**2,901.90**	**3,167.50**	**3,449.30**	**3,775.80**	**4,146.00**	**4,646.10**

Note: (r) Revised.

Source: From International Bottled Water Association, www.bottledwater.org.

Original Source: From Beverage Marketing Corp., www.beveragemarketing.com. Reprinted with permission.

Table 7F.49 Per Capita Consumption of Bottled Water in the United States by Region, 1994–1999

	Gallons Per Person					
Region	**1994**	**1995**	**1996**	**1997**	**1998**	**1999**
Pacific	23.5	23.5	25.3	25.9	27.4	29.2
Southwest	17.1	18.9	20.9	22.8	24.7	27.7
Northeast	12.4	13.3	14.1	16.1	18.4	20.6
South	6.7	7.6	8.5	9.4	10.5	11.8
West	5.9	6.8	7.7	8	8	9.6
West Central	4.8	5.2	5.8	6.1	6.7	7.7
East Central	5	5.3	5.6	5.8	6.5	7.4
Total	**11.3**	**12**	**13**	**14**	**15.3**	**17**

Note: (r) Revised.

Source: From International Bottled Water Association, www.bottledwater.org. copyright 1999–2004 IBWA.

Original Source: From Beverage Marketing Corp., U.S. Bureau of Census, www.beveragemarketing.com. Reprinted with permission.

Table 7F.50 Bottled Water Market Volume, Growth, and Consumption in the United States, 1976–1999

Year	Millions of Gallons	Volume Change (%)	Gallons Per Capita	Annual Change (%)
1976	354.3	—	1.6	—
1977	388.6	9.70	1.8	—
1978	468.1	20.50	2.1	—
1979	547.2	16.90	2.4	—
1980	605	10.60	2.7	—
1981	691.1	14.20	3	—
1982	782.6	13.20	3.4	—
1983	910.1	16.30	3.9	—
1984	1,058.50	16.30	4.5	—
1985	1,214.20	14.70	5.1	—
1986	1,365.70	12.50	5.7	—
1987	1,554.00	13.80	6.4	—
1988	1,777.20	14.40	7.2	—
1989	2,029.40	14.20	8.2	—
1990	2,237.60	10.30	9	9.30
1991	2,286.50	2.20	9.1	1.30
1992	2,422.00	5.90	9.5	5.10
1993	2,623.90	8.30	10.3	7.50
1994	2,901.90	10.60	11.3	9.70
1995	3,167.50	9.20	12	6.80
1996	3,449.30	8.90	13	8.00
1997	3,775.80	9.50	14	8.30
1998	4,146.00	9.80	15.3	9.20
1999	4,646.10	12.10	17	11.10

Note: (r) Revised.

Source: From International Bottled Water Association, www.bottledwater.org. Copyright 2002 International Bottled Water Association.

Original Source: From Beverage Marketing Corp., U.S. Bureau of Census, www.beveragemarketing.com. Reprinted with permission.

Table 7F.51 Consumption of Bottled Water in the United States by Type of Water, 1993–2003(P)

	Nonsparkling		Sparkling		Imports		Total	
Year	Volume[a]	Change (%)	Volume[a]	Change (%)	Volume[a]	Change (%)	Volume[a]	Change (%)
1993	2,422.20	8.70	174.7	1.40	92.5	7.20	2,689.40	8.20
1994	2,687.60	11.00	174.8	0.10	104	12.40	2,966.40	10.30
1995	2,965.60	10.30	164.2	−6.10	97.1	−6.60	3,226.90	8.80
1996	3,224.30	8.70	159	−3.20	111.8	15.10	3,495.10	8.30
1997	3,491.40	8.30	153.8	−3.30	149.1	33.40	3,794.30	8.60
1998	3,823.80	9.50	146.1	−5.00	160.8	7.80	4,130.70	8.90
1999	4,286.30	12.10	146	−0.10	151.1	−6.00	4,583.40	11.00
2000	4,622.40	7.80	144.2	−1.20	137.8	−8.80	4,904.40	7.00
2001	5,104.20	10.40	144	−0.10	123.9	−10.10	5,372.10	9.50
2002	5,677.50	11.20	149.5	3.80	123.7	−0.20	5,950.70	10.80
2003(P)	6,115.80	7.70	149.5	0.00	129.7	4.90	6,395.00	7.50

Note: (P) Preliminary.

[a] Millions of gallons.

Source: From International Bottled Water Association, www.bottledwater.org.

Original Source: From Beverage Marketing Corp., www.beveragemarketing.com. Reprinted with permission.

Table 7F.52 Bottled Water Volume Producer Revenues and Per Capita Consumption in the United States, 1993–2003(P)

Year	Millions of Gallons	Annual Change (%)	Millions of Dollars	Annual Change (%)	Gallons Per Capita	Annual Change (%)
1993	2,689.40	8.20	$2,876.70	8.20	10.5	—
1994	2,966.40	10.30	$3,164.30	10.00	11.5	9.40
1995	3,226.90	8.80	$3,521.90	11.30	12.2	6.40
1996	3,495.10	8.30	$3,835.40	8.90	13.1	7.40
1997	3,794.30	8.60	$4,222.70	10.10	14.1	7.40
1998	4,130.70	8.90	$4,666.10	10.50	15.3	8.30
1999	4,583.40	11.00	$5,314.70	13.90	16.8	10.00
2000	4,904.40	7.00	$5,809.00	9.30	17.8	6.00
2001	5,372.10	9.50	$6,880.00	18.40	19.3	8.50
2002	5,950.70	10.80	$7,725.00	12.30	21.2	9.80
2003(P)	6,395.00	7.50	$8,277.20	7.10	22.6	6.30

Note: (P) Preliminary.

Source: From International Bottled Water Association, www.bottledwater.org.

Original Source: From Beverage Marketing Corp., www.beveragemarketing.com. Reprinted with permission.

Table 7F.53 Global Bottled Water Market, Leading Countries' Consumption and Compound Annual Growth Rates 1998–2003(P)

2003 Rank	Countries	Millions of Gallons 1998	Millions of Gallons 2003(P)	CAGR 1998/03(P) (%)
1	United States	4,130.70	6,395.00	9.10
2	Mexico	2,873.00	4,354.70	8.70
3	Brazil	1,251.80	2,840.10	17.80
4	China	934.6	2,805.80	24.60
5	Italy	2,038.70	2,791.00	6.50
6	Germany	2,169.10	2,727.20	4.70
7	France	1,733.20	2,351.40	6.30
8	Indonesia	722.2	1,963.30	22.10
9	Thailand	1,014.40	1,302.50	5.10
10	Spain	981.1	1,213.90	4.40
	Top 10 Subtotal	**17,848.80**	**28,744.90**	**10.00**
	All Others	**5,340.70**	**9,278.40**	**11.70**
	World Total	**23,189.50**	**38,023.30**	**10.40**

Note: (P) Preliminary. CAGR, Compound annual growth rate.

Source: From International Bottled Water Association, www.bottledwater.org.

Original Source: From Beverage Marketing Corp., www.beveragemarketing.com. Reprinted with permission.

Table 7F.54 Global Bottled Water Market, Per Capita Consumption by Leading Countries 1998–2003(P)

2003 Rank	Countries	Gallons Per Capita 1998	2003(P)
1	Italy	35.9	48.1
2	Mexico	29.2	41.5
3	France	29.5	39.1
4	United Arab Emirates	28.1	38.1
5	Belgium-Luxembourg	30.7	35.1
6	Germany	26.4	33.1
7	Spain	25.1	30.2
8	Switzerland	23.8	25.4
9	Lebanon	16.2	25.3
10	Saudi Arabia	18.9	23.3
11	Cyprus	17.2	22.8
12	Austria	19.8	22.7
13	United States	15.3	22.6
14	Czech Republic	15.4	22.2
15	Portugal	17.2	20.6
	Global Average	**3.9**	**6**

Note: (P) Preliminary.

Source: From International Bottled Water Association, www.bottledwater.org.

Original Source: From Beverage Marketing Corp., www.beveragemarketing.com. Reprinted with permission.

Table 7F.55 Bottled Water Consumption by Country in North America and South America, 1997–2002

Region	Country	Year (1000 m^3) 1997	1998	1999	2000	2001	2002(P)
North America	United States	14,361.6	15,634.8	17,348.2	18,563.2	20,534.8	22,893.4
North America	Mexico	10,484.1	10,882.5	11,579.0	12,424.3	13,244.3	14,767.4
South America	Brazil	3,931.7	4,741.7	5,658.3	6,816.6	8,166.3	9,628.1
North America	Canada	541.2	650.5	754.6	847.9	938.6	1,027.3
South America	Argentina	568.9	575.2	594.2	598.9	600.1	603.1
South America	Colombia	562.8	579.0	560.0	549.0	548.4	556.7
South America	Venezuela, Republic of Bolivia	201.6	220.9	230.3	247.9	263.3	289.6
South America	Peru	75.6	80.9	89.1	103.5	118.0	132.1
South America	Chile	66.8	80.6	88.6	95.8	103.8	113.0
South America	Paraguay	46.2	50.2	53.5	56.9	60.6	64.5
South America	Uruguay	17.3	18.7	20.5	21.9	23.2	24.5
South America	Nicaragua	13.2	14.5	15.7	17.2	19.0	20.4
North America	Cuba	10.7	12.0	13.2	14.9	16.8	18.7

Note: (P) Preliminary. Not all of the water put into a water system reaches customers or is paid for by water users. This water is typically called "unaccounted-for water," but it is measured and define in a variety of ways. Many in the water industry consider all water that is not metered and billed to customer accounts to be unaccounted-for water. High rates of unaccounted-for water result in financial losses and poor performance of the water agency.

Source: From *World's Water 2004–2005*, by Peter H. Gleick. Copyright © 2004 Island Press. Reproduced by permission of Island Press, Washington, DC.

Original Source: Data were provided by the Beverage Marketing Corporation (BMC) to the author in 2003 and were used with permission, www.worldwater.org.

Table 7F.56 Bottled Water Consumption by Country in Europe, 1997–2002

Region	Country	Year (1000 m^3)					
		1997	1998	1999	2000	2001	2002(P)
Europe	Italy	7,558.5	7,722.5	8,924.6	9,221.5	9,479.9	9,690.1
Europe	Germany	8,207.2	8,216.2	8,602.9	8,693.7	8,850.2	8,983.0
Europe	France	6,053.1	6,565.2	6,947.3	7,462.2	7,820.4	8,430.4
Europe	Spain	3,542.6	3,716.1	3,879.7	4,003.8	4,133.9	4,294.3
Europe	Turkey	931.7	1,185.4	1,368.8	1,667.2	1,870.6	2,007.2
Europe	Poland	828.6	943.7	1,106.3	1,279.0	1,460.6	1,722.7
Europe	Russian Federation	524.5	610.6	790.7	967.8	1,162.4	1,406.4
Europe	United Kingdom	724.1	812.8	967.5	1,071.6	1,195.9	1,339.4
Europe	Belgium-Luxembourg[a]	1,213.0	1,233.6	1,299.0	1,262.4	1,264.9	1,329.7
Europe	Czech Republic	555.1	598.1	639.0	701.5	763.2	820.5
Europe	Portugal	647.1	645.8	705.9	719.3	735.8	761.1
Europe	Romania	406.2	447.6	504.4	553.9	620.9	698.5
Europe	Switzerland	622.7	653.4	651.8	653.6	656.9	668.1
Europe	Austria	569.3	610.0	605.1	609.8	631.5	645.4
Europe	Hungary	202.1	245.0	300.4	397.6	467.2	514.8
Europe	Greece	392.8	410.1	436.3	450.3	463.3	483.3
Europe	Ukraine	241.8	274.2	315.6	362.0	420.6	479.1
Europe	Netherlands	248.8	240.7	273.3	286.1	296.1	316.8
Europe	Croatia	135.8	157.7	176.9	199.9	223.7	247.2
Europe	Slovakia	158.9	161.8	168.8	170.2	173.5	178.2
Europe	Bulgaria	53.9	67.2	88.1	112.2	142.1	177.8
Europe	Sweden	126.8	127.1	143.7	150.8	158.1	164.2
Europe	Slovenia	67.9	80.9	93.2	108.5	124.3	137.9
Europe	Ireland	50.5	61.5	75.1	83.5	92.4	99.4
Europe	Norway	67.3	76.3	76.9	77.2	80.5	82.8
Europe	Denmark	71.6	72.0	71.8	72.0	72.3	74.1
Europe	Finland	45.2	51.0	58.3	62.0	65.6	68.4
Europe	Cyprus	48.4	48.7	50.8	54.9	58.3	62.3
Europe	Lithuania	15.0	17.6	20.4	23.8	28.6	34.6
Europe	Estonia	15.2	18.5	20.9	23.9	26.8	30.2
Europe	Latvia	2.4	3.0	3.7	4.6	5.6	6.9

Note: (P) Preliminary. Not all of the water put into a water system reaches customers or is paid for by water users. This water is typically called "unaccounted-for water," but it is measured and define in a variety of ways. Many in the water industry consider all water that is not metered and billed to customer accounts to be unaccounted-for water. High rates of unaccounted-for water result in financial losses and poor performance of the water agency.

Source: From *World's Water 2004–2005*, by Peter H. Gleick. Copyright © 2004 Island Press. Reproduced by permission of Island Press, Washington, DC.

Original Source: Data were provided by the Beverage Marketing Corporation (BMC) to the author in 2003 and were used with permission, www.worldwater.org.

Table 7F.57 Bottled Water Consumption by Country in Asia, 1997–2002

Region	Country	Year (1000 m^3)					
		1997	1998	1999	2000	2001	2002(P)
Asia	China	2,750.0	3,540.1	4,610.0	5,993.0	7,605.1	9,886.7
Asia	Indonesia	2,261.5	2,735.7	3,435.9	4,300.3	5,121.6	6,145.9
Asia	Thailand	3,567.1	3,842.4	4,063.7	4,286.2	4,539.0	4,837.0
Asia	India	1,047.0	1,364.2	1,681.8	2,149.7	2,667.8	3,361.4
Asia	Japan	646.6	789.5	922.7	1,149.0	1,230.6	1,461.3
Asia	Korea, Republic of	892.5	1,008.8	1,110.3	1,191.5	1,273.7	1,359.0
Asia	Philippines	727.8	837.2	999.1	1,119.0	1,213.0	1,291.8
Asia	Pakistan	69.3	108.2	157.6	242.3	360.3	547.7
Asia	China, Hong Kong SAR	191.0	222.1	245.4	271.1	298.2	331.1
Asia	Malaysia	137.5	157.9	179.6	199.1	217.9	236.8
Asia	Viet Nam	114.9	139.5	159.3	179.6	199.9	219.4
Asia	Singapore	57.2	63.5	69.7	75.6	81.8	88.2
Asia	Brunei Darussalam	9.9	11.1	12.2	13.6	14.9	16.3
Asia	Bangladesh[a]	0.0	0.0	0.0	0.0	0.0	0.0
Asia	Fiji Islands[b]	0.0	0.0	0.0	0.0	0.0	0.0

Note: (P) Preliminary. Not all of the water put into a water system reaches customers or is paid for by water users. This water is typically called "unaccounted-for water," but it is measured and define in a variety of ways. Many in the water industry consider all water that is not metered and billed to customer accounts to be unaccounted-for water. High rates of unaccounted-for water result in financial losses and poor performance of the water agency.

[a] Commercial bottled water essentially does not exist in Bangladesh.

[b] Consumption in Fiji is virtually nil, immeasurable in terms of thousands of cubic meters.

Source: From *World's Water 2004–2005*, by Peter H. Gleick. Copyright © 2004 Island Press. Reproduced by permission of Island Press, Washington, DC.

Original Source: Data were provided by the Beverage Marketing Corporation (BMC) to the author in 2003 and were used with permission, www.worldwater.org.

Table 7F.58 Bottled Water Consumption by Country in Africa/Mideast/Oceania, 1997–2002

Region	Country	Year (1000 m^3)					
		1997	1998	1999	2000	2001	2002(P)
Mideast	Saudi Arabia	1,297.7	1,490.1	1,610.0	1,769.9	1,938.0	2,116.3
Oceania	Australia	304.2	354.9	389.6	443.3	488.0	566.5
Mideast	Lebanon	180.7	214.8	239.8	275.3	309.7	346.2
Mideast	United Arab Emirates	229.9	245.3	256.1	269.9	285.2	326.4
Mideast	Israel	100.2	111.7	132.6	170.3	224.9	283.4
Mideast	Egypt	133.7	145.7	167.7	188.4	208.6	234.6
Mideast	Kuwait	68.2	77.5	95.6	112.5	128.4	144.0
Africa	South Africa	30.9	41.1	57.9	68.8	81.2	96.3
Mideast	Qatar	33.0	38.1	42.8	47.2	51.6	56.2
Mideast	Jordan	27.9	31.1	35.7	40.3	44.4	48.5
Mideast	Bahrain	26.7	28.4	31.4	34.1	36.7	39.5
Mideast	Oman	16.1	18.2	20.3	23.2	26.2	29.5
Oceania	Pacific Islands[a]	10.2	10.9	12.0	12.9	14.0	15.1
All Others		507.9	595.3	737.1	891.0	1,032.7	1,233.6

Note: (P) Preliminary. Not all of the water put into a water system reaches customers or is paid for by water users. This water is typically called "unaccounted-for water," but it is measured and define in a variety of ways. Many in the water industry consider all water that is not metered and billed to customer accounts to be unaccounted-for water. High rates of unaccounted-for water result in financial losses and poor performance of the water agency.

[a] Includes the Caroline Islands (Micronesia excluding Palau), the Marshall Islands, and the Northern Marianas (excluding Guam).

Source: From *World's Water 2004–2005*, by Peter H. Gleick. Copyright © 2004 Island Press. Reproduced by permission of Island Press, Washington, DC.

Original Source: Data were provided by the Beverage Marketing Corporation (BMC) to the author in 2003 and were used with permission, www.worldwater.org.

Table 7F.59 Global Bottled Water Production Volume and Value and Per Capita Consumption, 2000–2003

Region	Production Volume (mil L)				Production Value (mil USD)				Per Capita Consumption (L)			
	2003	2002	2001	2000	2003	2002	2001	2000	2003	2002	2001	2000
Africa & Middle East	12,400	11,220	9,200	8,720	2,110	1,825	1,450	1,250	11	10	9	9
Asia	33,465	30,100	24,030	19,990	7,395	6,490	4,500	3,650	10	9	7	6
Australia	695	650	850	740	440	340	400	350	35	33	37	33
Canada	1,490	1,310	920	820	650	525	350	310	47	41	29	26
East Europe	9,500	8,330	6,770	6,010	2,630	2,250	1,500	1,400	24	21	17	15
Latin America	27,050	26,060	26,950	25,150	3,970	3,800	5,050	5,809	51	50	53	50
U.S.A.[a]	24,463	23,803	24,414	22,020	8,277	7,724	14,500	13,600	90	85	74	67
West Europe	44,020	39,970	38,210	36,350	20,300	15,200	14,500	14,600	112	102	97	93
Total	**153,083**	**141,443**	**131,344**	**119,800**	**45,772**	**38,154**	**34,227**	**30,819**				

[a] Beverage Marketing Corporation.

Source: From International Council of Bottled Water Associations, Bottled Water Statistics.

Original Source: From Zenith International and Beverage Marketing Corporation, www.icbwa.org. Reprinted with permission.

Table 7F.60 Global Projected Bottled Water Sales

Country/Region	1996 Sales (mil L)	Projected 2006 Sales (mil L)	Annual Percentage of Growth (%)
Australia	500	1,000	11
Africa	500	800	4
CIS	600	1,500	13
Asia	1,000	5,000	12
East Europe	1,200	8,500	14
Middle East	1,500	3,000	3
South America	1,700	4,000	7
Pacific Rim	4,000	37,000	18
Central America	6,000	25,000	11
North America	13,000	25,000	4.5
West Europe	27,000	33,000	2.5
Total	**57,000**	**143,800**	

Source: From *World's Water 2002–2003*, by Peter H. Gleick. Copyright © 2002 Island Press. Reproduced by permission of Island Press, Washington, DC.

Original Source: Modified from soc.duke.edu/-s142tm16/World%20Markets.htm.

SECTION 7G INDUSTRIAL AND COMMERCIAL WATER USE — UNITED STATES

Table 7G.61 Industrial Self-Supplied Water Withdrawals in the United States, 2000

	Withdrawals (mill gal/day)									Withdrawals (thousand acre-feet/yr)		
	By Source and Type						Total			By Type		
	Groundwater			Surface Water								
State	Fresh	Saline	Total	Fresh	Saline	Total	Fresh	Saline	Total	Fresh	Saline	Total
Alabama	56	0	56	777	0	777	833	0	833	934	0	934
Alaska	4.32	0	4.32	3.8	3.86	7.66	8.12	3.86	12	9.1	4.33	13.4
Arizona	19.8	0	19.8	0	0	0	19.8	0	19.8	22.2	0	22.2
Arkansas	67	0.08	67.1	66.8	0	66.8	134	0.08	134	150	0.09	150
California	183	0	183	5.65	13.6	19.3	188	13.6	202	211	15.3	226
Colorado	23.6	0	23.6	96.4	0	96.4	120	0	120	135	0	135
Connecticut	4.13	0	4.13	6.61	0	6.61	10.7	0	10.7	12	0	12
Delaware	17	0	17	42.5	3.25	45.7	59.4	3.25	62.7	66.6	3.64	70.3
District of Columbia	0	0	0	0	0	0	0	0	0	0	0	0
Florida	216	0	216	74.7	1.18	75.9	291	1.18	292	326	1.32	328
Georgia	290	0	290	333	30	363	622	30	652	698	33.6	731
Hawaii	14.5	0.85	15.4	0	0	0	14.5	0.85	15.4	16.2	0.95	17.2
Idaho	35.8	0	35.8	19.7	0	19.7	55.5	0	55.5	62.2	0	62.2
Illinois	132	0	132	259	0	259	391	0	391	438	0	438
Indiana	99.7	0	99.7	2,300	0	2,300	2,400	0	2,400	2,690	0	2,690
Iowa	226	0	226	11.7	0	11.7	237	0	237	266	0	266
Kansas	46.6	0	46.6	6.74	0	6.74	53.3	0	53.3	59.8	0	59.8
Kentucky	95.2	0	95.2	222	0	222	317	0	317	356	0	356
Louisiana	285	0	285	2,400	0	2,400	2,680	0	2,680	3,010	0	3,010
Maine	9.9	0	9.9	237	0	237	247	0	247	277	0	277
Maryland	15.9	0	15.9	49.9	227	277	65.8	227	292	73.8	254	328
Massachusetts	10.7	0	10.7	26.2	0	26.2	36.8	0	36.8	41.3	0	41.3
Michigan	110	0	110	589	0	589	698	0	698	782	0	782
Minnesota	56.3	0	56.3	97.8	0	97.8	154	0	154	173	0	173
Mississippi	118	0	118	124	0	124	242	0	242	271	0	271
Missouri	29.2	0	29.2	33.5	0	33.5	62.7	0	62.7	70.3	0	70.3
Montana	31.9	0	31.9	29.3	0	29.3	61.3	0	61.3	68.7	0	68.7
Nebraska	35.5	0	35.5	2.6	0	2.6	38.1	0	38.1	42.7	0	42.7
Nevada	5.29	0	5.29	5	0	5	10.3	0	10.3	11.5	0	11.5
New Hampshire	6.95	0	6.95	37.9	0	37.9	44.9	0	44.9	50.3	0	50.3
New Jersey	65.3	0	65.3	66.2	0	66.2	132	0	132	147	0	147
New Mexico	8.8	0	8.8	1.67	0	1.67	10.5	0	10.5	11.7	0	11.7
New York	145	0	145	152	0	152	297	0	297	333	0	333
North Carolina	25.6	0	25.6	267	0	267	293	0	293	329	0	329
North Dakota	6.88	0	6.88	10.7	0	10.7	17.6	0	17.6	19.7	0	19.7
Ohio	162	0	162	645	0	645	807	0	807	905	0	905
Oklahoma	6.83	0	6.83	19.1	0	19.1	25.9	0	25.9	29.1	0	29.1
Oregon	12.1	0	12.1	183	0	183	195	0	195	218	0	218
Pennsylvania	155	0	155	1,030	0	1,030	1,190	0	1,190	1,330	0	1,330

Rhode Island	2.19	0	2.19	2.09	0	2.09	4.28	0	4.28	4.8	0	4.8
South Carolina	50.9	0	50.9	514	0	514	565	0	565	633	0	633
South Dakota	3.16	0	3.16	1.96	0	1.96	5.12	0	5.12	5.74	0	5.74
Tennessee	56.3	0	56.3	785	0	785	842	0	842	944	0	944
Texas	244	0.5	244	1,200	906	2,110	1,450	907	2,350	1,620	1,020	2,640
Utah	34.3	5.08	39.4	8.38	0	8.38	42.7	5.08	47.8	47.8	5.69	53.5
Vermont	2.05	0	2.05	4.86	0	4.86	6.91	0	6.91	7.75	0	7.75
Virginia	104	0	104	365	53.3	419	470	53.3	523	526	59.7	586
Washington	138	0	138	439	39.9	479	577	39.9	617	647	44.7	692
West Virginia	9.7	0	9.7	958	0	958	968	0	968	1,090	0	1,090
Wisconsin	83	0	83	364	0	364	447	0	447	501	0	501
Wyoming	4.31	0	4.31	1.47	0	1.47	5.78	0	5.78	6.48	0	6.48
Puerto Rico	11.2	0	11.2	0	0	0	11.2	0	11.2	12.5	0	12.5
U.S. Virgin Islands	0.22	0	0.22	3.12	0	3.12	3.34	0	3.34	3.74	0	3.74
Total	3,570	6.51	3,580	14,900	1,280	16,200	18,500	1,280	19,700	20,700	1,440	22,100

Note: Figures may not sum to totals because of independent rounding.

Source: From Hutson, S.S. et al., 2004, Estimated use of water in the United States in 2000, *U.S. Geological Survey Circular 1268*, www.usgs.gov.

Table 7G.62 Self-Supplied Industrial Withdrawals by Source, 1950–1995

Year	Groundwater Fresh (%)	Groundwater Saline (%)	Groundwater Total (%)	Surface Water Fresh (%)	Surface Water Saline (%)	Surface Water Total (%)	Reclaimed Sewage (%)
1950	—	—	7	—	—	93	—
1955	93	7	9	82	18	91	0.1
1960	94	6	17	85	15	83	0.2
1965	93	7	16	78	22	84	0.3
1970	88	12	18	83	17	82	0.3
1975	91	9	24	84	16	76	0.4
1980	92	8	25	87	13	75	0.4
1985	89	11	20	84	16	80	0.5
1990	83	17	26	83	17	74	0.3
1995	84	16	23	91	9	77	0.5

Source: From Dziegielewski, B. et al., 2002, *Analysis of Water Use Trends in the United States: 1950–1995*, Southern Illinois University at Carbondale, Carbondale, IL February 28, 2002. Reprinted with permission. http://info.geography.siu.edu/geography_info/research/.

Table 7G.63 Self-Supplied Industrial Withdrawals and Manufacturing Employment, 1950–1995

Year	Self-Supplied Industrial Withdrawals (bgd)	Industrial Withdrawals Per Capita (gpcd)	Total Manufacturing Employment (1,000s)	Per-Employee Withdrawals (gped)	Employment in Primary Metal Industries (1,000s)
1950	37.0	246	15,241	2,428	1,276
1955	39.0	238	16,882	2,310	1,321
1960	38.0	212	16,796	2,262	1,073
1965	46.0	237	18,062	2,547	1,243
1970	47.0	228	19,367	2,427	1,206
1975	45.0	208	18,323	2,456	1,115
1980	45.0	196	20,285	2,218	1,140
1985	30.5	126	19,248	1,585	788
1990	29.9	119	19,076	1,567	746
1995	29.1	109	18,469	1,576	713

Source: From Dziegielewski, B. et al., 2002, *Analysis of Water Use Trends in the United States: 1950–1995*, Southern Illinois University at Carbondale, Carbondale, IL February 28, 2002. Reprinted with permission. http://info.geography.siu.edu/geography_info/research/.

Table 7G.64 Commercial Freshwater Use in the United States, 1995

State	Self-Supplied Withdrawals: Source: Groundwater	Self-Supplied Withdrawals: Source: Surface Water	Self-Supplied Withdrawals: Total	Public-Supply Deliveries	Total Use: Withdrawals and Deliveries	Total Use: Consumptive Use
Alabama	4.9	0	4.9	122	127	28
Alaska	11	0.1	11	23	34	5.1
Arizona	21	0	21	135	155	78
Arkansas	0.4	100	100	58	158	12
California	77	309	385	994	1,380	259
Colorado	7.7	0.9	8.6	101	109	16
Connecticut	25	1.5	27	89	116	12
Delaware	2.8	0	2.8	20	22	2.2
DC	0	0	0	50	50	5.0
Florida	50	0.2	50	386	436	54
Georgia	33	13	46	168	215	39
Hawaii	45	0.4	46	47	92	43
Idaho	9.8	297	306	18	324	1.4
Illinois	16	88	104	440	544	44
Indiana	45	48	93	119	212	32
Iowa	18	25	43	65	108	14
Kansas	4.9	0.3	5.2	67	72	38
Kentucky	8.0	14	22	23	45	1.6
Louisiana	10	0.7	11	55	66	8.8
Maine	9.8	1.7	11	25	37	3.7
Maryland	19	14	33	85	118	11
Massachusetts	12	0	12	188	200	25
Michigan	16	25	41	253	294	31
Minnesota	46	20	66	103	169	18
Mississippi	18	0	18	33	51	8.6
Missouri	13	0.5	14	59	73	5.3
Montana	0	0	0	26	26	9.6
Nebraska	0.3	0	3	79	79	30
Nevada	7.1	14	21	116	137	24
New Hampshire	12	18	30	21	51	3.5
New Jersey	17	1.2	18	179	197	7.5
New Mexico	18	1.6	20	78	97	56
New York	136	65	200	409	609	61
North Carolina	7.3	0.3	7.6	138	146	7.2
North Dakota	0.1	0.2	0.2	15	15	2.3
Ohio	28	41	68	355	424	66
Oklahoma	6.6	16	23	170	193	18
Oregon	4.4	752	756	79	835	0.7
Pennsylvania	16	14	30	218	247	11
Rhode Island	1.5	0	1.5	20	21	2.1
South Carolina	1.7	0	1.7	50	52	7.8
South Dakota	6.1	4.1	10	21	31	3.1
Tennessee	2.0	18	20	214	234	21
Texas	33	11	44	130	174	35
Utah	3.8	0	3.8	115	119	35
Vermont	9.6	16	26	7.7	33	2.4
Virginia	28	13	41	152	193	23
Washington	24	0.4	24	161	185	37
West Virginia	36	9.2	46	23	68	10
Wisconsin	17	0	17	111	128	26
Wyoming	0.9	0.6	1.6	16	18	2.7
Puerto Rico	1.2	1.5	2.7	61	64	19
Virgin Islands	0.1	0.6	0.8	3.3	4.1	0.6
Total	**939**	**1,950**	**2,890**	**6,690**	**9,590**	**1,310**

Note: Figures may not add to totals because of independent rounding. All values in million gallons per day.

Source: From Salley, W.B. et al., 1998, Estimated use of water in the United States in 1995, *U.S. Geological Survey Circular 1200*, www.usgs.gov.

Table 7G.65 Average Rates of Nonresidential Water Use from Establishment Level Data

Category	SIC Code	Use Rate (gal/employee day)	Sample Size
Construction	—	31	246
General building contractors	15	118	66
Heavy construction	16	20	30
Special trade contractors	17	25	150
Manufacturing	—	164	2790
Food and kindred products	20	469	252
Textile mill products	22	784	20
Apparel and other textile products	23	26	91
Lumber and wood products	24	49	62
Furniture and fixtures	25	36	83
Paper and allied products	26	2614	93
Printing and publishing	27	37	174
Chemicals and allied products	28	267	211
Petroleum and coal products	29	1045	23
Rubber and miscellaneous plastics products	30	119	116
Leather and leather products	31	148	10
Stone, clay, and glass products	32	202	83
Primary metal industries	33	178	80
Fabricated metal products	34	194	395
Industrial machinery and equipment	35	68	304
Electronic and other electrical equipment	36	95	409
Transportation equipment	37	84	182
Instruments and related products	38	66	147
Miscellaneous manufacturing industries	39	36	55
Transportation and public utilities	—	50	226
Railroad transportation	40	68	3
Local and interurban passenger transit	41	26	32
Trucking and warehousing	42	85	100
U.S. postal service	43	5	1
Water transportation	44	353	10
Transportation by air	45	171	17
Transportation services	47	40	13
Communications	48	55	31
Electric, gas, and sanitary services	49	51	19
Wholesale trade	—	53	751
Wholesale trade—durable goods	50	46	518
Wholesale trade—nondurable goods	51	87	233
Retail trade	—	93	1044
Building materials and garden supplies	52	35	56
General merchandise stores	53	45	50
Food stores	54	100	90
Automotive dealers and service stations	55	49	198
Apparel and accessory stores	56	68	48
Furniture and homefurnishings stores	57	42	100
Eating and drinking places	58	156	341
Miscellaneous retail	59	132	161
Finance, insurance, and real estate	—	192	238
Depository institutions	60	62	77
Nondepository institutions	61	361	36
Security and commodity brokers	62	1240	2
Insurance carriers	63	136	9
Insurance agents, brokers, and service	64	89	24
Real Estate	65	609	84
Holding and other investment offices	67	290	5
Services	—	137	1878
Hotels and other lodging places	70	230	197
Personal services	72	462	300
Business services	73	73	243
Auto repair, services, and parking	75	217	108

(Continued)

Table 7G.65 (Continued)

Category	SIC Code	Use Rate (gal/employee day)	Sample Size
Miscellaneous repair services	76	69	42
Motion pictures	78	110	40
Amusement and recreation services	79	429	105
Health services	80	91	353
Legal services	81	821	15
Educational services	82	117	300
Social services	83	106	55
Museums, botanical, zoological gardens	84	208	9
Membership organizations	86	212	45
Engineering and management services	87	58	5
Services, NEC	89	73	60
Public administration	—	106	25
Executive, legislative, and general	91	155	2
Justice, public order, and safety	92	18	4
Administration of human resources	94	87	6
Environmental quality and housing	95	101	6
Administration of economic programs	96	274	5
National security and international affairs	97	112	2

Source: From Mays, L.W. ed., 1996, *Water Resources Handbook*, Copyright © The McGraw-Hill, Companies, Inc., NY. Reprinted with permission.

Original Source: From Planning and Management Consultants, Ltd. (1994; unpublished data).

Table 7G.66 Selected Commercial and Institutional Unit Use Coefficients

CI Category	Unit	Gal/Unit/Day
Barber shops	Chairs	54.60
Beauty shops	Station	269.00
Bus/rail depots	Square foot	3.33
Car washes	Inside square foot	4.78
Churches	Member	0.14
Golf/swim clubs	Member	22.20
Bowling alleys	Alley	133.00
Residential colleges	Student	106.00
Hospitals	Bed	346.00
Hotels	Square foot	0.26
Laundromats	Square foot	2.17
Laundry	Square foot	0.25
Medical offices	Square foot	0.62
Motels	Square foot	0.22
Drive-in movies	Car stall	5.33
Nursing homes	Bed	133.00
New office buildings	Square foot	0.19
Old office buildings	Square foot	0.14
Jails and prisons	Person	133.00
Restaurants	Seat	24.20
Drive-in restaurants	Car stall	109.00
Night clubs	Person served	1.33
Retail space	Sale square foot	0.11
Elementary schools	Student	3.83
High schools	Student	8.02
YMCA/YWCA	Person	33.30
Service stations	Inside square foot	0.25
Theaters	Seat	3.33

Source: From Dziegielewski, B. et al., 2000, Commercial and Institutional End Uses of Water, AWWA.

Original Source: From Crews, J.E. and Miller, M.A., 1983. Forecasting Municipal and Industrial Water Use. IWR Research Report 83R-3. U.S. Army Corps of Engineers, Fort Belvoir, Virginia, www.awwa.org.

Table 7G.67 Distribution of Commercial Water Use by Category in Selected Cities (Percent of CI Sector Use)

Commercial Users Reporting Year	Austin TX 1992	Buffalo NY 1995	Burbank CA 1995	EB- MUD CA 1994	Glendale CA 1995	Miami FL 1995	Orlando FL 1995	Portland OR 1995	San Diego CA 1995	Santa Monica CA 1995	St. Paul MN 1994–95	Santa Rosa CA 1994	Weighted Average 1992–1995
Hospitality[a]	13.26	20.94	11.75	7.94	13.45	17.53	34.86	5.45	34.28	38.55	15.96	28.12	14.80
Warehousing	1.79	10.83		30.77	0.45	6.73	30.94	2.78	0.03		16.87	0.25	12.40
Offices[b]	13.97	15.81	11.37	7.09	12.78	12.29	9.7	5.69	7.59		13.03	15.4	9.20
Irrigation[c]	2.18	5.13		21.94	5.12		0.8	1.57	4.25	10.32	3.12	0.3	6.15
Miscellaneous commercial[d]						31.05	0.45		0.06		0.46		5.72
Sales[e]	6.82	18.15	9.36	3.91	3.54	8.29	2.32	2.99	7.23	6.59	11.97	7.54	5.48
Services[f]	5.64	0.22	0.59	2.61	4.97		0.45	0.75	13.07		0.21	0.43	2.36
Laundries		3.41	3.52	2.53		2.89	2.13	1.10		3.91		5.88	1.73
Vehicle dealers and services	0.90	3.39	0.24	0.59	4.17	0.95	2.11	0.50	2.63	0.57	3.37	4.83	1.15
Meeting and recreation[g]	0.96		2.48	2.13	9.59	0.26	0.53	0.01	2.17	3.14	4.98	0.44	1.11
Communication and research	0.11	0.06	27.84	0.15	7.77		1.04		2.97	1.43	0	0.26	0.72
Landscape[h]	0.05	2.26	1.01	0.42			0.15	1.63				0.3	0.58
Transportation and fuels		1.15		1.40	0.58		0.74	0			0.61	1.12	0.43
Car wash		2.15	1.17	0.38	0.40		0.20		0.77	2.54	1.24	1.23	0.28
Passenger terminals	0.45	1.17	2.31		0.05		0.01	0.30	0.22	0.33	0.16		0.20
Share of Reported CI Use	46.13	84.67	76.64	81.86	62.87	79.99	86.43	22.77	75.27	67.38	71.98	66.9	62.28

Note: Tabular Valves are in percentages.

[a] Hospitality includes restaurant/bar, overnight accommodations, and other group shelter.
[b] Office includes finance, insurance, real estate, and government.
[c] Irrigation includes parks, gardens, botanical, zoological, cemeteries, and open land.
[d] Miscellaneous commercial includes warehousing, warehouse-cold storage, and boat dock.
[e] Sales include grocery stores, convenience stores, and dry goods.
[f] Services include miscellaneous repair services, crematories, funeral homes, laboratories, and printing.
[g] Meeting and recreation include convention center, recreation and theaters, and amusement parks.
[h] Landscape includes landscape horticultural service, agriculture, soil preparation, crop services, veterinary, equestrian, livestock, poultry, and game propagation.

Source: From Dziegielewski, B. et al., 2002, *Analysis of Water Use Trends in the United States: 1950–1995*, Southern Illinois University at Carbondale, Carbondale, IL February 28, 2002. Reprinted with permission. http://info.geography.siu.edu/geography_info/research/.

Original Source: Derived from U.S. Environmental Protection Agency (1997). Table 2.

Table 7G.68 Characteristics of Significant Commercial and Institutional Categories in Five Participating Agencies

Customer Category Description	Average Annual Daily Use (gpdc)[a]	Coefficient of Variation in Daily Use (gpdc)[b]	Percent of Total CI Use (%)	Percent of CI Customers (%)[c]	Percent Seasonal Use (%)[d]	Scaled Average Daily Use (gpdc)[e]
Urban irrigation	2,596	8.73	28.48	30.22	86.90	739.0
Schools and colleges	2,117	12.13	8.84	4.79	57.99	187.0
Hotels and motels	7,113	5.41	5.82	1.92	23.07	414.0
Laundries and laundromats	3,290	8.85	3.95	1.38	13.35	130.0
Office buildings	1,204	6.29	10.19	11.67	29.04	123.0
Hospitals and medical offices	1,236	78.50	3.90	4.19	23.16	48.0
Restaurant	906	7.69	8.83	11.18	16.13	80.0
Food stores	729	16.29	2.86	5.20	19.37	21.0
Auto shops	687	7.96	1.97	6.74	27.16	14.0
Membership organizations	629	6.42	1.95	5.60	46.18	12.0
Car washes	3,031	3.12	0.82	0.36	14.22	25.0

[a] gpdc: gallons per day per customer.
[b] Coefficient of variation in daily use: The ratio of standard deviation of daily use to average of daily use.
[c] Percent of CI customers pertains to CI customers in agencies that have respective category only.
[d] Percent seasonal use = [(total annual use —12 × minimum month use]/total annual use.
[e] Scaled average daily use = average annual daily use in category × percent of total CI use attributed to the category.

Source: From Dziegielewski, B. et al., 2002, *Analysis of Water Use Trends in the United States: 1950–1995*, Southern Illinois University at Carbondale, Carbondale, IL February 28, 2002. Reprinted with permission. http://info.geography.siu.edu/geography_info/research/.

Table 7G.69 Water Use in Mining and Manufacturing in the United States, 1968–1983, and by Industry Group, 1983

Industry	Establishments Reporting[a]	Gross Water Used: Total: Quantity (bil gal)	Gross Water Used: Total: Average per Establishment (mil gal)	Gross Water Used: Water Intake (bil gal)	Gross Water Used: Water Recycled[b] (bil gal)	Water Discharged: Quantity (bil gal)	Water Discharged: Percent Untreated	Water Pollutants Abatement: Capital Expenditures (mil dol)	Water Pollutants Abatement: Operating Cost (mil dol)
Mining									
1968	1,801	3,694	2,051	1,408	(NA)	1,365	78.8	(NA)	(NA)
1973	1,687	3,965	2,350	1,665	2,300	1,605	54.3	38	124
1978	1,056	3,554	3,366	1,473	2,430	1,592	67.3	244	201
1983, total	1,534	3,328	2,169	1,197	2,131	1,037	31.9	189	499
Metal mining	135	735	5,444	170	564	133	39.8	22	65
Anthracite mining	16	5	313	2	3	8	12.5	(Z)	1
Bituminous coal, lignite mining	275	119	433	45	73	116	26.7	14	69
Oil and gas extraction	555	1,452	2,616	602	850	476	31.1	131	318
Nonmetallic minerals, exc. fuels	553	1,018	1,841	378	640	304	32.6	22	46
Manufacturing									
1968	9,402	35,701	3,797	15,467	(NA)	14,276	69.5	(NA)	(NA)
1973	10,668	43,413	4,069	15,024	[c]	14,144	56.5	511	866
1978	9,605	44,494	4,632	12,992	34,199	11,682	59.7	1,249	2,119
1983, total	10,262	33,835	3,297	10,039	23,796	8,914	54.9	819	3,259
Food and kindred products	2,656	1,406	529	648	759	552	64.5	105	187
Tobacco products	20	34	1,700	5	29	4	(D)	(D)	5
Textile mill products	761	333	438	133	200	116	52.6	(D)	25
Lumber and wood products	223	218	978	86	132	71	63.4	4	23
Furniture and fixtures	66	7	106	3	3	3	100.0	2	4
Paper and allied products	600	7,436	12,393	1,899	5,537	1,768	27.1	66	438
Chemicals and allied products	1,315	9,630	7,323	3,401	6,229	2,980	67.0	187	1,013
Petroleum and coal products	260	6,177	23,758	818	5,359	699	46.2	165	543
Rubber, misc. plastic products	375	328	875	76	252	63	63.5	4	37
Leather and leather products	69	7	101	6	1	6	(D)	(S)	6
Stone, clay, and glass products	602	337	560	155	182	133	75.2	10	38
Primary metal products	776	5,885	7,584	2,363	3,523	2,112	58.1	100	421
Fabricated metal products	724	258	356	65	193	61	49.2	33	100
Machinery, exc. electrical	523	307	587	120	186	105	67.6	19	76
Electric and electronic equipment	678	335	494	74	261	70	61.4	45	108
Transportation equipment	380	1,011	2,661	153	859	139	67.6	55	171
Instruments and related products	154	112	727	30	82	28	50.0	10	45
Miscellaneous manufacturing	80	15	188	4	11	4	(D)	2	7

Note: Based on establishments reporting water intake of 20 mil gal. This represented 95 percent and 96 percent of the total water use estimated for mining and manufacturing industries. Water intake refers to that which is used/consumed in the production and processing operations and for sanitary services. D, Withheld to avoid disclosing individual company data; NA, Not available; S, Figure does not meet publication standards; Z, Less than $500,000.

[a] Establishments reporting water intake of 20 million gallons or more. These counts do not apply to water pollutants abatement columns for manufacturing in 1983.
[b] Refers to water recirculated and water reused.
[c] Data estimated; not strictly comparable to other years.

Source: From U.S. Department of Commerce, Statistical Abstract of the United States, 1987.

Table 7G.70 Mining Water Use in the United States, 2000

	Withdrawals (mil gal/day)									Withdrawals, (thousand acre-feet/yr)		
	By Source and Type									By Type		
	Groundwater			Surface Water			Total					
State	Fresh	Saline	Total	Fresh	Saline	Total	Fresh	Saline	Total	Fresh	Saline	Total
Alabama	—	—	—	—	—	—	—	—	—	—	—	—
Alaska	0.01	90.4	90.4	27.4	49.5	76.9	27.4	140	167	30.7	157	188
Arizona	81.2	8.17	89.4	4.43	0	4.43	85.7	8.17	93.8	96	9.16	105
Arkansas	0.21	0	0.21	2.57	0	2.57	2.78	0	2.78	3.12	0	3.12
California	21	152	173	2.71	0.46	3.17	23.7	153	177	26.6	171	198
Colorado	—	—	—	—	—	—	—	—	—	—	—	—
Connecticut	—	—	—	—	—	—	—	—	—	—	—	—
Delaware	—	—	—	—	—	—	—	—	—	—	—	—
District of Columbia	—	—	—	—	—	—	—	—	—	—	—	—
Florida	160	0	160	57.8	0	57.8	217	0	217	244	0	244
Georgia	7.75	0	7.75	2.05	0	2.05	9.8	0	9.8	11	0	11
Hawaii	—	—	—	—	—	—	—	—	—	—	—	—
Idaho	—	—	—	—	—	—	—	—	—	—	—	—
Illinois	—	—	—	—	—	—	—	—	—	—	—	—
Indiana	4.2	0	4.2	78.3	0	78.3	82.5	0	82.5	92.5	0	92.5
Iowa	2.49	0	2.49	30.3	0	30.3	32.8	0	32.8	36.8	0	36.8
Kansas	14	0	14	17.4	0	17.4	31.4	0	31.4	35.2	0	35.2
Kentucky	—	—	—	—	—	—	—	—	—	—	—	—
Louisiana	—	—	—	—	—	—	—	—	—	—	—	—
Maine	—	—	—	—	—	—	—	—	—	—	—	—
Maryland	4.21	0	4.21	4.1	0.02	4.12	8.31	0.02	8.33	9.32	0.02	9.34
Massachusetts	—	—	—	—	—	—	—	—	—	—	—	—
Michigan	—	—	—	—	—	—	—	—	—	—	—	—
Minnesota	6.9	0	6.9	581	0	581	588	0	588	659	0	659
Mississippi	—	—	—	—	—	—	—	—	—	—	—	—
Missouri	4.1	0	4.1	12.8	0	12.8	16.9	0	16.9	19	0	19
Montana	—	—	—	—	—	—	—	—	—	—	—	—
Nebraska	5.64	4.55	10.2	122	0	122	128	4.55	132	143	5.1	148
Nevada	—	—	—	—	—	—	—	—	—	—	—	—
New Hampshire	0.08	0	0.08	6.72	0	6.72	6.8	0	6.8	7.62	0	7.62
New Jersey	6.12	0	6.12	104	0	104	110	0	110	124	0	124
New Mexico	—	—	—	—	—	—	—	—	—	—	—	—
New York	—	—	—	—	—	—	—	—	—	—	—	—
North Carolina	36.4	0	36.4	0	0	0	36.4	0	36.4	40.8	0	40.8
North Dakota	—	—	—	—	—	—	—	—	—	—	—	—
Ohio	53.1	0	53.1	35.5	0	35.5	88.5	0	88.5	99.2	0	99.2

(Continued)

Table 7G.70 (Continued)

	Withdrawals (mil gal/day)									Withdrawals, (thousand acre-feet/yr)		
	By Source and Type									By Type		
	Groundwater			Surface Water			Total					
State	**Fresh**	**Saline**	**Total**	**Fresh**	**Saline**	**Total**	**Fresh**	**Saline**	**Total**	**Fresh**	**Saline**	**Total**
Oklahoma	2.25	256	258	0.23	0	0.23	2.48	256	258	2.78	287	290
Oregon	—	—	—	—	—	—	—	—	—	—	—	—
Pennsylvania	162	0	162	20.9	0	20.9	182	0	182	205	0	205
Rhode Island	—	—	—	—	—	—	—	—	—	—	—	—
South Carolina	—	—	—	—	—	—	—	—	—	—	—	—
South Dakota	—	—	—	—	—	—	—	—	—	—	—	—
Tennessee	—	—	—	—	—	—	—	—	—	—	—	—
Texas	129	504	633	91.5	0	91.5	220	504	724	247	565	812
Utah	8.6	21.5	30.1	17.7	177	194	26.3	198	225	29.4	222	252
Vermont	—	—	—	—	—	—	—	—	—	—	—	—
Virginia	—	—	—	—	—	—	—	—	—	—	—	—
Washington	—	—	—	—	—	—	—	—	—	—	—	—
West Virginia	—	—	—	—	—	—	—	—	—	—	—	—
Wisconsin	—	—	—	—	—	—	—	—	—	—	—	—
Wyoming	58.8	222	280	20.7	0	20.7	79.5	222	301	89.1	248	338
Puerto Rico	—	—	—	—	—	—	—	—	—	—	—	—
U.S. Virgin Islands	—	—	—	—	—	—	—	—	—	—	—	—
Total	767	1,260	2,030	1,240	227	1,470	2,010	1,490	3,490	2,250	1,660	3,920

Note: Figures may not sum to totals because of independent rounding; —, data not collected.

Source: From Hutson, S.S. et al., 2004, Estimated use of water in the United States in 2000, *U.S. Geological Survey Circular 1268*, www.usgs.gov.

Table 7G.71 Changes in Manufacturing and Mining Employment and Industrial Withdrawals (Selected States), 1980–1995

State	Change in Industrial Withdrawals (mgd)	Change in Manufacturing Employment (employees)
Decreasing Withdrawals/Decreasing Employment		
New York	−711	−578,907
Pennsylvania	−1,687	−465,730
Illinois	−1,217	−305,198
New Jersey	−1,247	−262,443
California	−88	−265,131
Massachusetts	−277	−239,587
Ohio	−1,252	−248,244
Connecticut	−240	−177,407
Michigan	−128	−112,866
Increasing Withdrawals/Decreasing Employment		
Virginia	112	−24,057
South Carolina	218	−26,298
West Virginia	543	−68,941
Oregon	635	9,667
Texas	1,131	−111,709

Source: From Dziegielewski, B. et al., 2002, *Analysis of Water Use Trends in the United States: 1950–1995*, Southern Illinois University at Carbondale, Carbondale, IL February 28, 2002. Reprinted with permission. http://info.geography.siu.edu/geography_info/research/.

Table 7G.72 Water Use by Mineral Industries in the United States, 1983

		Gross Water Used[a]					Water Discharged[b]		
		Total		Water Intake					
Industry Group and Industry	Establishments Reporting Water Intake of 20 bil gal or More During 1982 (Number)	Quantity (bil gal)	Percent of All Mineral Industries	Quantity (bil gal)	Percent of All Mineral Industries	Water Re-Circulated and Reused (bil gal)	Total (bil gal)	Untreated (bil gal)	Treated (bil gal)
All mineral industries	1,534	3,328.3	100	1,197.1	100	2,131.1	1,036.7	331.2	705.5
Metal mining	135	734.5	22	170.2	14	564.3	133.0	52.8	80.2
Iron ores	15	421.6	13	45.7	4	375.9	51.2	40.3	11.0
Copper ores	29	223.4	7	89.7	7	133.7	17.7	(D)	(D)
Lead and zinc ores	14	(D)	(D)	(D)	(D)	(D)	26.6	0.7	25.9
Gold and silver ores	28	12.4	(Z)	7.3	1	5.0	5.5	1.2	4.3
Gold ores	20	(D)	(D)	5.8	(Z)	(D)	4.0	(D)	(D)
Silver ores	8	(D)	(D)	1.6	(Z)	(D)	1.5	(D)	(D)
Miscellaneous metal ores	42	46.0	1	14.4	1	31.6	16.2	6.6	9.7
Uranium-radium-vanadium ores	34	15.5	(Z)	10.5	1	5.0	11.3	6.5	4.8
Metallic ores, n.e.c.	8	30.5	1	3.9	(Z)	26.6	4.9	(Z)	4.9
Anthracite mining	16	5.4	(Z)	2.2	(Z)	3.2	7.5	0.9	6.5
Anthracite	16	5.4	(Z)	2.2	(Z)	3.2	7.5	0.9	6.5
Bituminous coal and lignite mining	275	118.6	4	45.3	4	73.3	116.2	30.5	85.7
Bituminous coal and lignite mining	275	118.6	4	45.3	4	73.3	116.2	30.5	85.7
Bituminous coal and lignite	275	118.6	4	45.3	4	73.3	116.2	30.5	85.7
Oil and gas extraction	555	1,451.8	44	601.6	50	850.1	475.6	147.5	328.1
Crude petroleum and natural gas	312	672.8	20	436.6	36	236.2	318.8	84.7	234.1
Natural gas liquids	161	750.7	23	146.7	12	604.1	(D)	(D)	(D)
Oil and gas field services	82	28.2	1	18.4	2	9.8	(D)	(D)	(D)
Drilling oil and gas wells	67	27.3	1	17.5	1	9.7	16.7	8.1	8.5
Oil and gas field services, n.e.c.	14	0.9	(Z)	0.8	(Z)	0.1	(D)	(D)	(D)
Nonmetallic minerals, except fuels	553	1,018.1	31	377.8	32	640.2	304.3	99.3	205.0
Dimension stone	3	(D)	(D)	1.3	(Z)	(D)	1.0	(D)	(D)
Crushed and broken stone, including riprap	178	63.1	2	46.5	4	16.6	45.6	19.2	26.4
Crushed and broken limestone	97	52.9	2	39.0	3	13.9	39.7	15.7	24.1
Crushed and broken granite	46	5.1	(Z)	3.3	(Z)	1.8	2.6	(D)	(D)
Crushed and broken stone, n.e.c.	35	5.1	(Z)	4.2	(Z)	0.9	3.3	(D)	(D)
Sand and gravel	248	201.4	6	123.1	10	78.3	67.0	25.6	41.4
Construction sand and gravel	215	123.5	4	85.1	7	38.4	57.3	23.3	34.1
Industrial sand	33	77.8	2	38.0	3	39.8	9.7	2.3	7.4
Clay, ceramic, and refractory minerals	35	(D)	(D)	(D)	(D)	(D)	(D)	(D)	(D)
Kaolin and ball clay	11	16.2	(Z)	15.0	1	1.2	13.4	(D)	(D)
Clay, ceramic, and refractory minerals, n.e.c.	16	11.4	(Z)	4.4	(Z)	7.0	2.6	(D)	(D)
Chemical and fertilizer mineral mining	71	684.8	21	149.6	12	535.1	141.3	(D)	(D)
Barite	10	2.7	(Z)	(D)	(D)	(D)	(D)	(D)	(D)
Potash, soda, and borate minerals	20	125.5	4	26.9	2	98.6	12.4	(D)	(D)
Phosphate rock	19	489.5	15	60.2	5	429.3	63.4	0.3	63.1

Miscellaneous nonmetallic minerals	17	(D)	(D)	(D)	(D)	1.4	(D)	(D)	0.7
Talc, soapstone, and pyrophyllite	5	(D)	(D)	0.4	(Z)	(D)	0.3	(D)	(D)
Miscellaneous nonmetallic minerals, n.e.c.	12	(D)	(D)	(D)	(D)	(D)	(D)	(D)	(D)

Note: D, Withheld to avoid disclosing data for individual companies; Z, Less than half the unit shown; n.e.c., not elsewhere classified.

[a] Total gross water used is equal to sum of water intake plus water reciruclated and reused without regard to evaporation.
[b] Volume of water discharged may be greater than water intake due to mine water that is drained and discharged.

Source: From 1982 Census of Mineral Industries, U.S. Dept. of Commerce Bureau of the Census, 1985.

Table 7G.73 Water Use by Manufacturing Industries in the United States, 1983 (Bil Gal)

Industry Group and Industry	Establishments Reporting Water Intake of 20 Mil Gal or More During 1982	Gross Water Used[a]						
			Water Intake			Water Discharged		
		Total	Total	From Public Water System	Water Re-Circulated and Reused	Total	Untreated	Treated
All manufacturing industries[b]	10,262	33,835.2	10,038.9	1,310.7	23,796.3	8,913.7	4,889.8	4,023.9
Food and kindred productions	2,656	1,406.2	647.7	219.0	758.6	552.0	355.8	196.1
Meat products	552	119.5	92.7	49.5	26.8	85.5	28.7	56.8
Meat packing plants	205	56.5	44.7	16.9	11.8	41.9	14.7	27.2
Sausages and other prepared meats	104	13.9	11.3	7.8	2.6	9.8	4.1	5.6
Poultry dressing plants	217	47.3	35.2	24.0	12.1	32.4	9.0	23.4
Poultry and egg processing	26	1.8	1.5	0.8	0.3	1.4	0.9	0.6
Dairy products	498	69.8	38.8	17.9	30.9	35.9	29.4	6.4
Creamery butter	17	1.3	1.0	0.3	0.2	0.9	0.8	0.1
Cheese, natural and processed	111	15.2	10.2	3.9	5.1	9.3	6.6	2.7
Condensed and evaporated milk	67	14.2	9.5	2.4	4.8	8.6	7.1	1.5
Ice cream and frozen desserts	40	2.0	1.4	1.1	0.6	1.1	1.1	0.1
Fluid milk	263	37.0	16.7	10.3	20.3	16.0	13.9	2.1
Preserved fruits and vegetables	489	201.4	100.1	38.6	101.3	88.6	46.7	41.9
Canned specialties	52	21.5	17.4	10.3	4.1	13.4	7.0	6.5
Canned fruits and vegetables	224	81.3	30.6	13.9	50.7	26.3	13.8	12.5
Dehydrated fruits, vegetables, and soups	26	(D)	5.6	1.8	(D)	4.9	1.8	3.0
Pickles, sauces, and salad dressings	36	7.2	2.4	2.1	4.9	1.9	0.8	1.1
Frozen fruits and vegetables	117	70.4	40.0	8.1	30.4	38.8	21.5	17.3
Frozen specialties	34	(D)	4.1	2.5	(D)	3.4	1.8	1.6
Grain mill products	119	147.0	79.3	19.7	67.6	74.3	65.7	8.6
Flour and other grain mill products	19	0.9	0.8	0.5	0.1	0.5	(D)	(D)
Cereal breakfast foods	19	12.7	5.9	3.6	6.9	3.8	(D)	(D)
Rice milling	10	(D)	0.6	0.6	(D)	0.4	(D)	(D)
Blended and prepared flour	4	(D)	0.1	0.1	(D)	0.1	0.1	–
Wet corn milling	21	126.9	68.3	12.9	58.6	66.6	59.7	7.0
Dog, cat, and other pet food	32	(D)	2.9	1.9	(D)	2.1	1.1	1.1
Prepared feeds, n.e.c.	14	1.1	0.8	0.3	0.3	0.7	0.6	0.2
Bakery products	76	7.6	3.0	2.3	4.6	2.0	1.7	0.3
Bread, cake, and related products	61	2.8	1.9	(D)	0.9	1.1	1.0	0.1
Cookies and crackers	15	4.8	1.1	(D)	3.7	0.9	0.8	0.1
Sugar and confectionery products	151	377.2	178.7	12.7	198.5	142.0	93.8	48.3
Raw cane sugar	35	198.9	83.1	(D)	115.8	63.0	40.1	23.0
Cane sugar refining	20	76.0	62.8	(D)	13.3	54.9	39.1	15.9
Beet sugar	37	65.3	14.6	0.9	50.7	9.6	(D)	(D)
Confectionery products	44	(D)	(D)	2.1	15.2	(D)	(D)	(D)
Chocolate and cocoa products	10	(D)	(D)	1.0	3.4	(D)	(D)	0.3
Fats and oils	154	103.3	34.1	10.4	69.2	28.9	18.4	10.6

Cottonseed oil mills	18	2.8	1.2	(D)	1.5	(D)	(D)	(D)
Soybean oil mills	60	70.4	20.1	(D)	50.4	18.2	13.5	4.7
Vegetable oil mills, n.e.c.	5	0.8	0.5	(D)	0.3	(D)	(D)	(D)
Animal and marine fats and oils	31	6.3	2.8	0.5	3.6	1.5	0.8	8
Shortening and cooking oils	40	23.0	9.5	3.0	13.5	7.8	3.3	4.5
Beverages	406	308.7	88.5	46.9	220.3	68.2	50.9	17.3
Malt beverages	56	(D)	53.3	32.1	(D)	41.4	30.8	10.7
Malt	20	19.5	7.3	0.8	12.1	6.5	4.7	1.8
Wines, brandy, and brandy spirits	27	2.7	2.6	0.3	0.1	2.3	0.8	1.5
Distilled liquor, except brandy	28	17.2	10.3	1.5	6.9	9.0	(D)	(D)
Bottled and canned soft drinks	247	14.3	12.3	10.9	2.0	6.8	5.1	1.7
Flavoring extracts and syrups, n.e.c.	28	(D)	2.7	1.4	(D)	2.3	(D)	(D)
Miscellaneous foods and kindred products	211	71.9	32.6	20.9	39.3	26.5	20.6	5.9
Canned and cured seafoods	10	(D)	3.0	(D)	(D)	(D)	(D)	(D)
Fresh or frozen packaged fish	43	4.5	4.4	(D)	0.1	4.3	3.0	1.3
Roasted coffee	9	(D)	(D)	—	30.4	(D)	(D)	(D)
Manufactured ice	24	(D)	1.2	0.4	(D)	0.9	0.8	0.1
Macaroni and spaghetti	7	(D)	0.1	01	(D)	(D)	(D)	—
Food preparations, n.e.c.	118	(D)	(D)	8.7	8.5	11.3	7.4	3.8
Tobacco products	20	33.9	5.3	(D)	28.6	4.0	(D)	(D)
Cigarettes	11	(D)	(D)	(D)	(D)	3.1	(D)	(D)
Textile mill products	761	332.9	132.6	70.2	200.3	115.6	61.2	54.4
Weaving mills, cotton	69	70.9	20.7	(D)	50.2	17.9	13.2	4.7
Weaving mills, manmade fiber and silk	135	67.7	18.1	11.0	49.6	14.5	9.4	5.1
Weaving and finishing mills, wool	23	2.9	2.7	0.7	0.2	2.6	(D)	(D)
Narrow fabric mills	12	0.5	0.5	(D)	0.1	0.4	(D)	(D)
Knitting mills	191	48.2	27.2	20.6	21.0	25.3	15.9	9.3
Women's hosiery, except socks	11	(D)	0.7	(D)	(D)	0.7	0.7	(Z)
Hosiery, n.e.c.	20	2.8	0.9	08	1.9	0.8	(D)	(D)
Knit outerwear mills	23	2.6	2.2	(D)	0.4	2.1	1.5	0.7
Knit underwear mills	12	(D)	0.6	(D)	(D)	0.5	(D)	(D)
Circular knit fabric mills	89	26.2	17.2	13.0	9.0	16.2	9.9	6.3
Warp knit fabric mills	36	(D)	5.6	4.1	(D)	4.9	2.8	2.1
Textile finishing, except wool	125	62.3	39.0	15.4	23.3	33.6	11.5	22.1
Finishing plants, cotton	27	8.9	6.8	1.9	2.1	6.2	1.0	5.2
Finishing plants, manmade	69	45.4	26.9	9.2	18.5	22.5	7.6	14.9
Finishing plants, n.e.c.	29	8.1	5.4	4.2	2.7	4.8	2.8	2.0
Floor covering mills	65	14.1	11.3	8.8	2.8	9.8	4.7	5.1
Tufted carpets and rugs	61	(D)	10.4	(D)	(D)	(D)	(D)	(D)
Carpets are rugs, n.e.c.	4	(D)	1.0	(D)	(D)	(D)	(D)	(D)
Yarn and thread mills	77	40.5	8.1	5.7	32.4	7.4	4.0	3.3
Yarn mills, except wool	49	30.2	4.6	3.2	25.6	4.2	2.8	1.4
Throwing and winding mills	15	(D)	1.8	1.4	(D)	1.6	0.9	0.7
Thread mills	11	(D)	(D)	1.1	(D)	1.5	0.4	1.1
Miscellaneous textile goods	64	25.7	5.0	2.3	20.7	4.3	1.6	2.7
Felt goods, except woven felts and hats	10	(D)	1.4	0.2	(D)	1.4	0.1	1.2
Processed textile waste	4	0.2	0.2	0.2	—	(D)	(D)	—

(Continued)

Table 7G.73 (Continued)

Industry Group and Industry	Establishments Reporting Water Intake of 20 Mil Gal or More During 1982	Gross Water Used[a]				Water Discharged		
			Water Intake					
		Total	Total	From Public Water System	Water Re-Circulated and Reused	Total	Untreated	Treated
Coated fabrics, not rubberized	9	(D)	0.6	0.3	(D)	0.6	(D)	(D)
Tire cord and fabric	8	(D)	0.3	(D)	(D)	0.2	(D)	(D)
Nonwoven fabrics	22	(D)	1.5	1.1	(D)	1.0	0.6	0.5
Textile goods, n.e.c.	7	(D)	(D)	0.1	(D)	(D)	(Z)	(D)
Lumber and wood products	223	218.2	86.0	(D)	132.2	71.0	44.9	26.1
Logging camps and logging contractors	9	(D)	0.2	0.1	(D)	0.3	(D)	(D)
Sawmills and planning mills	89	168.5	68.4	(D)	100.1	58.9	39.4	19.5
Sawmills and planning mills, general	89	168.5	68.4	(D)	100.1	58.9	39.4	19.5
Millwork, plywood, and structural members	78	25.9	7.6	1.9	18.3	4.4	(D)	(D)
Millwork	3	(D)	0.4	(Z)	(D)	0.4	(Z)	0.3
Hardwood veneer and plywood	10	0.4	0.3	0.1	0.1	0.1	(D)	(D)
Softwood veneer and plywood	64	(D)	6.9	1.7	(D)	3.9	(D)	(D)
Miscellaneous wood products	42	(D)	9.6	(D)	(D)	7.3	(D)	(D)
Wood preserving	7	0.4	(D)	—	(D)	(D)	(D)	(D)
Particleboard	8	(D)	(D)	(D)	(D)	(D)	(D)	(D)
Wood products, n.e.c.	27	21.2	8.5	1.3	12.7	6.6	(D)	(D)
Furniture and fixtures	66	6.8	3.4	2.2	3.4	3.3	2.9	0.4
Household furniture	26	3.6	1.9	(D)	1.7	1.8	(D)	(D)
Wood household furniture	15	1.7	1.5	(D)	0.2	1.5	(D)	(D)
Metal household furniture	6	(D)	0.2	0.2	(D)	0.2	(D)	(D)
Office furniture	11	(D)	0.6	(D)	(D)	0.6	0.4	0.2
Metal office furniture	10	(D)	0.6	(D)	(D)	0.6	0.4	0.2
Partitions and fixtures	17	(D)	0.5	0.5	(D)	0.5	(D)	(D)
Metal partitions and fixtures	13	(D)	(D)	(D)	(D)	(D)	(D)	(D)
Miscellaneous furniture and fixtures	8	(D)	(D)	(D)	(Z)	(D)	(D)	(D)
Drapery hardware and blinds and shades	7	(D)	(D)	(D)	(Z)	(D)	(D)	(D)
Paper and allied products	600	7,435.8	1,899.3	257.3	5,536.5	1,768.1	479.0	1,289.1
Pulp mills	36	1,020.0	283.2	49.1	736.8	282.7	45.7	237.0
Paper mills, except building paper	234	3,908.6	1,009.5	92.1	2,899.0	958.2	304.3	653.9
Paperboard mills	151	2,353.9	538.7	107.2	1,815.2	462.3	85.4	377.0
Miscellaneous converted paper products	104	125.2	56.5	(D)	68.7	55.0	36.9	18.1
Paper coating and glazing	41	16.0	6.5	(D)	9.4	6.2	3.7	2.4
Bags, except textile bags	16	26.2	5.6	(D)	20.6	5.5	(D)	(D)
Pressed and molded pulp goods	11	(D)	32.1	(D)	(D)	(D)	(D)	(D)
Sanitary paper products	20	(D)	8.9	(D)	(D)	9.4	(D)	(D)
Converted paper products, n.e.c.	12	(D)	3.3	(D)	(D)	3.2	(D)	(D)
Paperboard containers and boxes	57	16.8	6.6	(D)	10.3	5.9	4.5	1.4

Folding paperboard boxes	17	6.2	4.1	04	2.1	(D)	(D)	(D)
Corrugated and solid fiber boxes	21	0.9	0.4	0.4	0.5	0.4	0.3	0.1
Sanitary food containers	14	(D)	0.9	0.4	(D)	0.9	0.8	(Z)
Fiber cans, drums, and similar products	5	(D)	1.1	(D)	(D)	(D)	(D)	(D)
Building paper and board mills	18	11.3	4.8	(D)	6.5	3.9	2.2	1.7
Chemicals and allied products	1,315	9,630.1	3,400.7	210.5	6,229.4	2,979.8	1,996.3	983.6
Industrial inorganic chemicals	301	2,164.0	885.0	57.0	1,279.6	758.4	556.3	202.1
Alkalies and chlorine	32	287.0	157.4	(D)	129.6	142.9	40.8	102.1
Industrial gases	76	492.4	18.6	8.9	473.8	11.9	7.3	4.6
Inorganic pigments	22	96.0	48.9	(D)	47.1	49.5	26.0	23.5
Industrial inorganic chemicals, n.e.c.	171	1,289.1	660.1	28.6	629.0	554.0	482.2	71.9
Plastics materials and synthetics	206	1,435.7	427.1	23.4	1,008.6	391.7	277.0	114.6
Plastics materials and resins	143	580.5	132.7	16.0	447.9	108.0	62.3	45.8
Synthetic rubber	22	236.6	62.9	(D)	173.7	58.5	(D)	(D)
Cellulosic manmade fibers	7	133.4	71.7	(D)	61.7	67.6	(D)	(D)
Organic fibers, noncellulosic	34	485.2	159.9	4.1	325.3	157.5	123.1	34.3
Drugs	112	240.1	90.5	19.9	149.6	87.1	55.8	31.3
Biological products	17	3.3	0.8	0.8	2.5	0.5	0.3	0.2
Medicinals and botanicals	23	122.9	55.3	5.3	67.7	54.9	(D)	(D)
Pharmaceutical preparations	72	113.8	34.4	13.9	79.4	31.6	(D)	(D)
Soaps, cleaners, and toilet goods	108	103.6	64.8	8.5	38.8	61.2	53.7	7.5
Soap and other detergents	32	42.0	16.6	4.3	25.4	14.3	(D)	(D)
Polishes and sanitation goods	22	(D)	(D)	(D)	(D)	3.8	(D)	(D)
Surface active agents	25	(D)	(D)	1.3	8.0	(D)	(D)	(D)
Toilet preparations	29	(D)	2.3	(D)	(D)	(D)	(D)	(D)
Paints and allied products	41	3.5	2.1	1.5	1.4	2.1	1.7	0.3
Industrial organic chemicals	296	4,122.3	1,515.9	78.6	2,606.4	1,381.0	815.2	565.8
Gum and wood chemicals	10	(D)	(D)	(D)	21.3	6.5	(D)	(D)
Cyclic crudes and intermediates	75	(D)	(D)	7.0	287.2	30.9	(D)	(D)
Industrial organic chemicals, n.e.c.	211	3,765.5	1,467.6	(D)	2,297.9	1,343.5	798.5	545.0
Agricultural chemicals	116	1,381.4	305.0	14.5	1,076.4	202.7	154.6	48.1
Nitrogenous fertilizers	59	757.4	70.5	12.1	686.9	45.2	(D)	(D)
Phosphatic fertilizers	34	559.0	216.1	(D)	342.9	142.2	(D)	(D)
Agricultural chemicals, n.e.c.	18	64.9	18.4	(D)	46.5	15.3	(D)	(D)
Miscellaneous chemical products	135	178.9	110.3	7.1	68.6	95.7	82.0	13.7
Adhesives and sealants	31	(D)	(D)	0.6	(D)	(D)	(D)	0.9
Explosives	13	53.4	(D)	—	(D)	(D)	(D)	(D)
Carbon black	14	(D)	1.6	(D)	(D)	0.3	(D)	(D)
Chemical preparations, n.e.c.	73	93.6	63.3	5.8	30.3	52.4	(D)	(D)
Petroleum and coal products	260	6,177.3	818.4	137.7	5,358.9	699.3	323.4	375.9
Petroleum refining	202	6,170.3	814.4	135.9	5,355.9	695.1	321.0	374.1
Paving and roofing materials	39	4.9	2.6	1.0	2.3	3.3	(D)	(D)
Paving mixtures and blocks	15	1.8	0.8	(D)	1.0	1.9	(D)	(D)
Asphalt felts and coatings	24	3.0	1.8	(D)	1.3	1.4	(D)	(D)
Miscellaneous petroleum and coal products	19	2.2	1.4	0.8	0.7	0.9	(D)	(D)
Lubricating oils and greases	7	(D)	0.2	0.2	(D)	0.1	(D)	(D)
Petroleum and coal products, n.e.c.	12	(D)	1.3	0.6	(D)	07	(D)	(D)

(Continued)

Table 7G.73 (Continued)

Industry Group and Industry	Establishments Reporting Water Intake of 20 Mil Gal or More During 1982	Gross Water Used[a]	Water Intake		Water Re-Circulated and Reused	Water Discharged		
		Total	Total	From Public Water System		Total	Untreated	Treated
Rubber and miscellaneous plastics products	375	327.8	76.0	27.4	251.8	62.6	39.8	22.8
Tires and inner tubes	47	121.5	(D)	4.2	(D)	16.5	8.2	8.3
Rubber and plastics footwear	3	0.1	0.1	(Z)	(Z)	0.1	0.1	(Z)
Rubber and plastics hose and belting	24	58.0	(D)	3.0	(D)	5.0	(D)	(D)
Fabricated rubber products, n.e.c.	82	26.8	8.3	4.6	18.4	7.4	(D)	(D)
Miscellaneous plastics products	219	121.4	41.9	15.6	79.5	33.6	22.9	10.8
Leather and leather products	69	6.5	6.1	2.7	0.4	5.7	(D)	(D)
Leather tanning and finishing	65	(D)	(D)	(D)	0.4	(D)	(D)	(D)
Stone, clay, and glass products	602	336.7	154.7	24.0	181.9	132.8	99.5	33.3
Flat glass	16	20.9	4.8	(D)	16.0	4.7	(D)	(D)
Glass and glassware, pressed or blown	120	92.2	13.3	8.7	78.9	11.4	7.1	4.3
Glass containers	75	38.3	7.1	4.0	31.2	5.9	3.8	2.1
Pressed and blown glass, n.e.c.	45	53.9	6.2	4.7	47.7	5.5	3.3	2.2
Products of purchased glass	26	(D)	7.2	(D)	(D)	7.1	(D)	(D)
Cement, hydraulic	100	115.2	80.0	2.4	35.2	68.7	58.6	10.1
Structural clay products	10	(D)	1.4	0.8	(D)	0.7	0.5	0.2
Ceramic wall and floor tile	4	(D)	(D)	—	(D)	(D)	(D)	(D)
Clay refractories	5	(D)	(D)	—	(D)	(D)	(D)	(D)/
Pottery and related products	30	2.1	1.5	1.1	0.6	1.1	0.5	0.6
Vitreous plumbing fixtures	9	(D)	0.5	0.3	(D)	0.4	0.1	0.3
Vitreous china food utensils	5	0.4	0.3	0.3	0.1	0.3	0.1	0.2
Porcelain electrical supplies	11	0.9	0.6	0.4	0.3	0.4	0.3	0.2
Concrete, gypsum, and plaster products	160	16.7	10.8	2.7	5.8	6.5	3.9	2.7
Concrete products, n.e.c.	20	(D)	(D)	0.1	(D)	0.3	(D)	(D)
Ready-mixed concrete	65	4.5	2.1	0.7	2.4	1.5	1.0	0.4
Lime	23	7.0	4.3	(D)	2.8	3.2	(D)	(D)
Gypsum products	51	4.3	3.7	(D)	0.6	1.5	1.0	0.5
Cut stone and stone products	15	(D)	1.1	0.5	(D)	1.1	(D)	(D)
Miscellaneous nonmetallic mineral products	125	70.4	34.5	(D)	35.9	31.5	22.7	8.7
Abrasive products	17	(D)	3.5	1.0	(D)	3.5	2.6	0.9
Asbestos products	13	8.2	1.8	0.8	6.3	1.4	1.0	0.3
Gaskets, packing, and sealing devices	12	3.6	1.8	0.5	1.7	0.9	0.6	0.2
Minerals, ground or treated	22	22.2	18.7	0.8	3.5	18.8	(D)	(D)
Mineral wool	42	27.4	6.2	(D)	21.2	4.4	2.7	1.7
Nonclay refractories	12	2.4	2.0	0.3	0.3	2.0	(D)	(D)
Nonmetallic mineral products, n.e.c.	7	(D)	0.5	(D)	(D)	0.5	(D)	(D)
Primary metal industries	776	5,885.2	2,362.5	108.7	3,522.8	2,112.0	1,227.9	884.1

Blast furnace and basic steel products	259	4,990.5	2,077.6	62.2	2,912.9	1,868.1	1,113.7	754.3
Blast furnaces and steel mills	137	4,908.4	2,038.9	53.9	2,869.5	1,.829.8	1,090.9	739.0
Electrometallurgical products	13	11.3	1.2	0.4	10.1	1.5	(D)	(D)
Steel wire and related products	34	4.5	2.1	1.4	2.5	2.1	(D)	(D)
Cold finishing of steel shapes	32	18.0	11.3	1.4	6.7	11.0	5.8	5.2
Steel pipe and tubes	43	48.3	24.1	5.2	24.1	23.7	(D)	(D)
Iron and steel foundries	154	218.1	69.1	22.6	149.0	51.5	24.2	27.3
Gray iron foundries	93	200.8	63.0	18.8	137.8	45.7	18.7	27.0
Malleable iron foundries	7	(D)	(D)	1.4	(D)	1.4	1.4	—
Steel investment foundries	10	(D)	(D)	0.4	(D)	0.5	(D)	(D)
Steel foundries, n.e.c.	44	7.4	4.2	2.0	3.2	3.9	(D)	(D)
Primary nonferrous metals	71	417.1	125.6	3.6	291.4	111.6	51.6	60.0
Primary copper	20	77.0	(D)	(D)	(D)	10.6	8.0	2.5
Primary lead	5	5.1	0.7	(D)	4.4	0.3	(D)	(D)
Primary zinc	5	18.7	7.2	0.5	11.5	(D)	(D)	(D)
Primary aluminum	24	(D)	67.8	1.1	(D)	62.9	(D)	(D)
Primary nonferrous metals, n.e.c.	17	(D)	(D)	0.5	0.8	(D)	(D)	(D)
Secondary nonferrous metals	34	13.5	3.7	1.3	9.8	3.1	2.2	0.9
Nonferrous rolling and drawing	170	213.2	79.9	15.7	133.3	71.4	32.9	38.6
Copper rolling and drawing	29	43.8	18.6	5.2	25.2	14.8	8.0	6.8
Aluminum sheet, plate, and foil	23	93.5	37.0	2.1	56.4	34.6	12.4	22.2
Aluminum extruded products	40	9.1	3.9	2.6	5.3	3.7	2.1	1.6
Aluminum rolling and drawing, n.e.c.	6	12.0	7.1	0.5	4.9	6.9	3.6	3.3
Nonferrous rolling and drawing, n.e.c.	26	30.6	6.6	3.0	24.0	4.9	1.5	3.4
Nonferrous wire drawing and insulating	46	24.3	6.8	2.5	17.5	6.5	5.3	1.2
Nonferrous foundries	46	15.5	2.4	1.6	13.1	2.2	1.1	1.1
Aluminum foundries	29	14.2	1.7	1.2	12.5	(D)	(D)	(D)
Brass, bronze, and copper foundries	6	(D)	0.2	(D)	(D)	(D)	(D)	(D)
Nonferrous foundries, n.e.c.	11	(D)	0.5	(D)	(D)	0.5	0.1	0.4
Miscellaneous primary metal products	42	17.3	4.1	1.8	13.2	4.0	2.1	1.9
Metal heat treating	28	1.9	1.5	1.3	0.4	1.5	0.5	1.0
Primary metal products, n.e.c.	14	15.4	2.6	0.4	12.8	2.6	1.6	1.0
Fabricated metal products	724	257.9	65.4	38.1	192.5	61.4	29.7	31.7
Metal cans and shipping containers	89	62.1	6.6	5.6	55.5	6.0	1.5	4.4
Metal cans	80	(D)	6.2	5.2	(D)	5.6	(D)	(D)
Metal barrels, drums, and pails	9	(D)	0.4	0.4	(D)	0.4	(D)	(D)
Cutlery, hand tools, and hardware	72	30.8	14.9	4.0	15.8	14.5	(D)	(D)
Cutlery	4	(D)	(D)	0.2	(D)	(D)	(D)	(D)
Hand and edge tools, n.e.c.	28	(D)	1.0	0.8	(D)	0.9	(D)	(D)
Hand saws and saw blades	4	(D)	0.4	(D)	(D)	(D)	(D)	(D)
Hardware, n.e.c.	36	(D)	(D)	(D)	(D)	2.9	1.2	1.7
Plumbing and heating, except electric	25	3.7	3.0	2.5	0.7	3.0	(D)	(D)
Metal sanitary ware	6	(D)	0.6	(D)	(D)	0.7	(D)	(D)
Plumbing fittings and brass goods	12	(D)	2.1	(D)	(D)	2.1	(D)	(D)
Heating equipment, except electric	7	(D)	0.3	0.3	(D)	0.2	0.2	—
Fabricated structural metal product	76	28.3	7.0	4.4	21.4	6.0	3.8	2.2
Fabricated structural metal	11	(D)	1.2	(D)	(D)	1.1	(D)	(D)

(Continued)

Table 7G.73 (Continued)

Industry Group and Industry	Establishments Reporting Water Intake of 20 Mil Gal or More During 1982	Gross Water Used[a]				Water Discharged		
			Water Intake					
		Total	Total	From Public Water System	Water Re-Circulated and Reused	Total	Untreated	Treated
Metal doors, sash, and trim	10	(D)	0.6	0.6	(D)	0.6	0.2	0.3
Fabricated plate work (boiler shops)	30	22.2	4.0	2.6	18.2	3.3	1.7	1.6
Sheet metal work	11	0.7	0.3	0.2	0.4	0.3	0.2	0.1
Prefabricated metal buildings	5	(D)	(D)	(D)	—	(D)	(D)	(D)
Miscellaneous metal work	8	1.6	0.6	—	1.0	0.6	(D)	(D)
Screw machine products, bolts, etc.	49	8.4	2.8	1.8	5.6	2.7	1.6	1.1
Screw machine products	5	(D)	0.1	0.1	(D)	0.1	(D)	(D)
Bolts, nuts, rivets, and washers	44	(D)	2.7	1.7	(D)	2.6	(D)	(D)
Metal forgings and stampings	122	81.5	10.1	8.0	71.5	9.7	7.7	2.1
Iron and steel forgings	36	12.0	2.8	2.3	9.2	2.8	1.9	0.9
Nonferrous forgings	6	3.7	1.4	1.4	2.3	1.4	1.1	0.3
Automotive stampings	42	62.4	3.3	2.8	59.2	3.0	(D)	(D)
Crowns and closures	4	0.3	0.3	(D)	—	0.3	(D)	(D)
Metal stampings, n.e.c.	34	3.1	2.3	(D)	0.8	2.2	1.7	0.5
Metal services, n.e.c.	138	7.8	6.9	4.8	1.0	6.7	2.9	3.8
Plating and polishing	126	(D)	6.0	(D)	(D)	5.8	2.6	3.2
Metal coating and allied services	12	(D)	0.9	(D)	(D)	0.9	0.3	0.5
Ordnance and accessories, n.e.c.	35	9.5	6.7	2.7	2.8	6.0	3.5	2.6
Small arms ammunition	7	(D)	1.4	(D)	(D)	1.2	(D)	(D)
Ammunition, except for small arms, n.e.c.	14	(D)	2.2	(D)	(D)	2.0	0.7	1.3
Small arms	6	(D)	0.8	(D)	(D)	0.8	(D)	(D)
Ordnance and accessories, n.e.c.	8	(D)	2.3	1.7	(D)	2.0	1.6	0.4
Miscellaneous fabricated metal products	118	25.8	7.5	4.3	18.3	6.8	(D)	(D)
Steel springs, except wire	4	(D)	0.2	0.2	(D)	(D)	(D)	(D)
Valves and pipe fittings	54	10.4	2.9	2.0	7.4	2.8	2.1	0.7
Wire springs	4	0.9	0.1	0.1	0.8	0.1	(D)	(D)
Miscellaneous fabricated wire products	8	(D)	0.2	0.2	(D)	0.2	(D)	(D)
Metal foil and leaf	11	(D)	1.7	(D)	(D)	1.4	1.0	0.4
Fabricated pipe and fittings	9	0.2	0.2	0.1	(Z)	(D)	(D)	(D)
Fabricated metal products, n.e.c.	28	(D)	2.2	(D)	(D)	2.1	(D)	(D)
Machinery, except electrical	523	306.5	120.0	52.5	186.4	104.9	71.2	33.8
Engines and turbines	47	66.2	32.0	(D)	34.1	20.7	12.8	7.9
Internal combustion engines, n.e.c.	35	(D)	(D)	3.8	(D)	(D)	(D)	(D)
Farm and garden machinery	41	40.1	32.8	(D)	7.3	32.0	(D)	(D)
Farm machinery and equipment	28	39.2	32.1	(D)	7.1	31.3	(D)	(D)
Lawn and garden equipment	13	0.9	0.7	(D)	0.2	0.7	(D)	(D)
Construction and related machinery	69	51.2	11.4	3.5	39.8	11.8	6.8	5.1
Construction machinery	31	42.6	10.3	(D)	32.3	10.8	6.2	4.6
Mining machinery	8	0.3	(D)	(D)	(D)	0.1	0.1	(Z)
Oil field machinery	21	(D)	0.8	(D)	(D)	(D)	(D)	(D)

Metalworking machinery	57	(D)	4.0	1.8	(D)	3.7	3.6	0.2
Machine tools, metal cutting types	22	3.2	3.0	(D)	0.2	2.7	(D)	(D)
Machine tools, metal forming types	3	(D)	(D)	—	—	(D)	(D)	—
Special dies, tools, jigs, and fixtures	6	(D)	0.2	0.2	(D)	(D)	(D)	—
Machine tool accessories	14	(D)	(D)	—	(D)	0.4	(D)	(D)
Power driven hand tools	10	(D)	0.3	0.2	(D)	0.3	0.2	0.1
Special industry machinery	36	(D)	6.4	(D)	(D)	6.2	(D)	(D)
Food products machinery	9	(D)	(D)	—	(Z)	0.3	0.2	(Z)
Paper industries machinery	5	(D)	(D)	(D)	0.2	0.1	0.1	(Z)
Printing trades machinery	4	(D)	(D)	—	(D)	0.1	(D)	(D)
Special industry machinery, n.e.c.	13	(D)	(D)	0.3	(D)	(D)	(D)	(D)
General industrial machinery	101	38.5	7.1	4.1	31.4	6.1	4.7	1.4
Pumps and pumping equipment	26	(D)	1.6	0.8	(D)	1.5	1.2	0.3
Ball and roller bearings	22	26.5	2.6	1.4	23.9	2.4	1.7	0.7
Air and gas compressors	15	(D)	1.4	1.0	(D)	0.7	(D)	(D)
Blowers and fans	9	(D)	(D)	—	0.1	(D)	(D)	(Z)
Speed changers, drives, and gears	5	(D)	0.4	(D)	(D)	0.4	(D)	(D)
Power transmission equipment, n.e.c.	15	1.1	0.7	0.3	0.4	0.7	(D)	(D)
General industrial machinery, n.e.c.	7	(D)	0.2	0.2	(D)	0.2	(D)	(D)
Office and computing machines	71	39.7	15.0	6.0	24.7	13.3	12.0	1.3
Electronic computing equipment	57	(D)	(D)	—	(D)	12.3	(D)	(D)
Office machines, n.e.c. and typewriters	12	(D)	(D)	—	(D)	(D)	(D)	(D)
Refrigeration and service machinery	70	32.5	9.6	8.7	23.0	9.6	(D)	(D)
Refrigeration and heating equipment	61	(D)	9.2	8.4	(D)	9.3	(D)	(D)
Measuring and dispensing pumps	4	(D)	0.2	0.2	(D)	0.2	0.2	—
Service industry machinery, n.e.c.	3	(D)	(D)	—	—	0.1	0.1	—
Miscellaneous machinery, except electrical	31	18.3	1.7	(D)	16.5	1.5	0.8	0.7
Carburetors, pistons, rings, valves	18	17.8	1.5	(D)	16.3	1.3	0.6	0.7
Machinery, except electrical, n.e.c.	13	0.5	0.2	(D)	0.2	0.2	0.2	(Z)
Electric and electronic equipment	678	334.8	74.1	55.1	260.7	70.3	42.5	27.8
Electric distributing equipment	43	(D)	3.4	3.2	(D)	4.7	(D)	(D)
Transformers	14	(D)	1.6	(D)	(D)	2.9	(D)	(D)
Switchgear and switchboard apparatus	29	3.3	1.9	(D)	1.4	1.8	(D)	(D)
Electrical industrial apparatus	76	18.5	9.5	6.6	9.1	9.4	7.4	1.9
Motors and generators	30	6.7	4.8	(D)	2.0	5.0	4.5	0.5
Industrial controls	18	(D)	(D)	—	(D)	0.6	0.5	0.1
Welding apparatus, electric	6	0.4	0.2	0.2	0.2	0.2	0.1	(Z)
Carbon and graphite products	19	9.8	3.8	3.6	6.1	3.5	2.2	1.3
Electrical industrial apparatus, n.e.c.	3	(D)	(D)	(D)	(D)	0.1	0.1	—
Household appliances	75	24.1	9.4	8.8	14.7	9.1	5.7	3.4
Household cooking equipment	15	2.8	1.8	1.5	1.0	1.7	(D)	(D)
Household refrigerators and freezers	11	(D)	2.6	(D)	(D)	2.5	(D)	(D)
Household laundry equipment	10	4.5	2.1	2.1	2.3	2.2	1.4	0.9
Electric housewares and fans	17	1.6	1.0	(D)	0.6	0.9	0.8	0.2
Household vacuum cleaners	4	(D)	(D)	—	(D)	0.2	0.1	0.1
Household appliances, n.e.c.	15	(D)	1.5	1.5	(D)	1.5	1.1	0.4
Electric lighting and wiring equipment	75	17.8	6.9	4.2	10.9	6.0	4.3	1.7

(Continued)

Table 7G.73 (Continued)

Industry Group and Industry	Establishments Reporting Water Intake of 20 Mil Gal or More During 1982	Gross Water Used[a]				Water Discharged		
		Total	Water Intake		Water Re-Circulated and Reused			
			Total	From Public Water System		Total	Untreated	Treated
Electric lamps	15	1.2	(D)	—	(D)	0.8	0.6	0.1
Current-carrying wiring devices	15	2.7	2.1	1.4	0.5	2.1	1.7	0.4
Noncurrent-carrying wiring devices	20	2.8	1.7	1.2	1.1	1.6	1.0	0.6
Residential lighting fixtures	10	0.4	0.3	0.3	0.1	0.3	0.2	0.1
Commercial lighting fixtures	5	(D)	0.1	0.1	(D)	0.1	(D)	(D)
Lighting equipment, n.e.c.	4	(D)	0.2	0.2	(D)	(D)	(Z)	(D)
Radio and TV receiving equipment	22	(D)	2.3	(D)	(D)	2.4	(D)	(D)
Radio and TV receiving sets	12	2.6	1.7	(D)	0.8	1.8	(D)	(D)
Phonograph records and prerecorded tape	10	(D)	0.6	(D)	(D)	0.5	(D)	(D)
Communication equipment	122	61.4	13.3	9.1	48.1	11.7	8.3	3.5
Telephone and telegraph apparatus	26	9.2	3.5	2.3	5.7	3.2	2.0	1.3
Radio and TV communication equipment	96	52.2	9.8	6.8	42.4	8.5	6.3	2.2
Electronic components and accessories	193	166.0	23.6	17.6	142.4	21.9	10.3	11.6
Electron tubes, all types	19	(D)	2.8	(D)	(D)	2.6	1.8	0.8
Semiconductors and related devices	65	78.0	11.6	9.6	66.4	10.7	3.3	7.3
Electronic capacitors	12	(D)	3.0	0.5	(D)	2.9	2.4	0.6
Electronic resistors	8	(D)	0.4	(D)	(D)	0.4	(D)	(D)
Electronic connectors	21	1.5	1.4	0.7	0.2	1.2	(D)	(D)
Electronic components, n.e.c.	68	63.3	4.6	4.1	58.7	4.2	1.8	2.4
Miscellaneous electrical equipment and supplies	72	26.7	5.6	(D)	21.0	5.1	(D)	(D)
Storage batteries	36	(D)	1.9	(D)	(D)	1.5	0.7	0.8
X-ray electromedical, and electrotherapeutic apparatus	8	0.6	0.4	0.4	0.3	(D)	(D)	(D)
Engine electrical equipment	13	17.0	2.7	(D)	14.2	2.7	(D)	(D)
Electrical equipment and supplies, n.e.c.	7	(D)	(D)	—	(D)	0.2	0.2	(Z)
Transportation equipment	380	1,011.3	152.8	82.3	858.6	139.2	94.0	45.2
Motor vehicles and equipment	194	(D)	66.4	42.8	(D)	59.6	30.7	28.9
Motor vehicles and car bodies	55	(D)	22.6	(D)	(D)	21.0	(D)	(D)
Truck and bus bodies	6	(D)	(D)	—	(D)	0.2	(D)	(D)
Motor vehicle parts and accessories	130	229.7	43.5	22.0	186.2	38.4	19.2	19.2
Aircraft and parts	93	(D)	58.4	(D)	(D)	54.1	43.6	10.5
Aircraft	26	54.2	18.0	(D)	36.2	(D)	(D)	2.8
Aircraft engines and engine parts	34	(D)	(D)	6.9	(D)	(D)	(D)	(D)
Aircraft equipment, n.e.c.	33	(D)	(D)	4.2	21.5	4.5	(D)	(D)
Ship and boat building and repairing	30	27.6	16.3	(D)	11.3	15.8	12.6	3.2
Ship building and repairing	30	27.6	16.3	(D)	11.3	15.8	12.6	3.2
Railroad equipment	16	8.1	3.1	2.4	5.0	2.8	1.8	1.0
Motorcycles, bicycles, and parts	3	5.0	1.0	0.9	4.0	0.9	0.6	0.3

Guided missiles and space vehicles, parts	34	30.2	6.5	6.0	23.8	4.9	4.0	0.9
Guided missiles and space vehicles	17	19.7	2.8	(D)	16.9	2.2	1.8	0.4
Space propulsion units and parts	12	5.2	3.3	(D)	1.9	2.3	1.9	0.5
Space vehicle equipment, n.e.c.	5	5.4	0.4	0.4	4.9	0.4	0.3	(Z)
Miscellaneous transportation equipment	10	5.3	1.2	0.9	4.1	1.1	0.7	0.3
Tanks and tank components	5	5.2	1.1	0.8	4.1	1.0	0.7	0.3
Transportation equipment, n.e.c.	4	(D)	(D)	—	—	(Z)	(Z)	(Z)
Instruments and related products	154	112.0	29.8	9.8	82.3	27.6	13.6	13.9
Engineering and scientific instruments	5	(D)	0.2	(D)	(D)	0.2	(D)	(D)
Measuring and controlling devices	57	21.6	4.2	3.2	17.4	3.8	2.6	1.2
Environmental controls	10	(D)	1.0	(D)	(D)	1.0	(D)	(D)
Process control instruments	12	(D)	0.8	(D)	(D)	0.7	(D)	(D)
Fluid meters and counting devices	11	(D)	0.4	0.3	(D)	0.3	0.3	(Z)
Instruments to measure electricity	14	(D)	1.7	1.3	(D)	1.5	0.7	0.9
Measuring and controlling devices, n.e.c.	10	(D)	0.3	(D)	(D)	0.3	0.2	0.1
Optical instruments and lenses	15	2.1	(D)	—	(D)	0.4	(D)	(D)
Medical instruments and supplies	40	4.2	2.2	(D)	2.0	2.1	1.5	0.5
Surgical and medical instruments	16	(D)	(D)	—	(D)	0.8	(D)	(D)
Surgical appliances and supplies	23	(D)	1.4	1.4	(D)	1.3	(D)	(D)
Photographic equipment and supplies	25	77.7	17.3	(D)	60.4	15.7	(D)	(D)
Watches, clocks, and watchcases	5	(D)	(D)	—	(D)	(D)	(D)	(D)
Miscellaneous manufacturing industries	80	15.4	4.3	3.4	11.1	4.0	(D)	(D)
Toys and sporting goods	28	(D)	1.5	1.3	(D)	1.4	1.2	0.2
Games, toys, and children's vehicles	13	(D)	(D)	—	(D)	0.4	(D)	(D)
Sporting and athletic goods, n.e.c.	14	(D)	1.0	1.0	(D)	0.9	(D)	(D)
Pens, pencils, and office and art supplies	8	(D)	0.6	0.3	(D)	0.6	0.6	(Z)
Pens and mechanical pencils	3	(D)	(D)	—	(D)	(D)	(D)	(D)
Costume jewelry and notions	22	(D)	0.8	(D)	(D)	0.8	(D)	(D)
Costume jewelry	8	(D)	(D)	—	—	(D)	(D)	(D)
Needles, pins, and fasteners	13	(D)	0.6	(D)	(D)	(D)	(D)	(D)
Miscellaneous manufactures	15	(D)	1.2	(D)	(D)	1.1	(D)	(D)
Manufacturing industries, n.e.c.	7	(D)	0.6	(D)	(D)	(D)	(D)	(D)

Note: Billion gallons. D., Withheld to avoid disclosing data for individual companies; Z, Less than half the unit shown; n.e.c., not elsewhere classified.

[a] Total gross water used is equal to sum of water intake plus water recirculated and reused without regard to evaporation.

[b] Excludes data for establishments classified as Apparel and Other Textile Products; Printing and Publishing; and Publishing; and administrative and auxiliary establishments for all major groups.

Source: From 1982 Census of Manufacturers, U.S. Dept. of Commerce Bureau of the Census, 1986. National Aluminate Corp. With permission.

Table 7G.74 Make-Up Water Required in Industrial Cooling Systems

	Temperature Drop				
Cycles of Concentration	**15°F (gal)**	**20°F (gal)**	**25°F (gal)**	**30°F (gal)**	**35°F (gal)**
1.5	45	60	75	90	105
2.0	30	40	50	60	70
2.5	25	33	42	50	58.5
3.0	22.5	30	37.5	45	52.5
3.5	21	28	35	42	49.1
4.0	20	26.7	33.2	40	46.9
4.5	19.2	25.7	32.0	38.7	45.1
5.0	18.7	25.0	31.8	37.5	44
5.5	18.3	24.5	30.7	36.8	43
6.0	18	24	30	36.1	42.1
6.5	17.7	23.7	29.5	35.5	41.5
7.0	17.5	23.3	29.1	35	40.9

Note: Estimated amounts in gallons per 1,000 gal/min recirculation.

Source: From National Aluminate Corp.

Table 7G.75 Geographic Distribution of Water-Intensive Manufacturing Industries in the United States

	Paper		Chemicals		Petroleum Refining		Primary Metals	
Water Resource Region	**ML/day[a]**	**mgd**	**ML/day**	**mgd**	**ML/day**	**mgd**	**ML/day**	**mgd**
New England	9,401	2,541	1,598	432			643	174
Mid-Atlantic	7,104	1,920	15,736	4,253	10,704	2,893	10,341	2,795
South Atlantic Gulf	31,627	8,548	13,590	3,673	1,417	383	4,329	1,170
Great Lakes	8,156	2,196	10,859	2,935	6,389	1,727	38,298	10,351
Ohio	2,471	668	21,108	5,705	3,193	863	33,562	9,071
Tennessee	4,151	1,122	1,342	2,525	973	263	395	107
Upper Mississippi	2,101	568	2,882	779	2,638	713	2,752	744
Lower Mississippi	6,471	1,749	19,920	5,384	10,744	2,904	2,186	591
Souris-Red-Rainy	510	138						
Missouri	99	27	2,527	683	1,986	537	381	103
Arkansas-White-Red	1,809	489	3,455	934	7,425	2,007	1,653	447
Texas Gulf	3,814	1,031	41,425	11,196	32,996	8,918	4,495	1,215
Rio Grande			1,010	273			103	28
Upper Colorado					44	12		
Lower Colorado	432	117	399	108	40	11	136	37
Great Basin			48	13	614	166	1,949	527
Columbia-North Pacific	13,819	3,735	1,798	486	1,017	275	1,646	445
California	4,310	1,165	2,682	725	8,029	2,170	795	215
Alaska	1,306	353	162	44	51	14		
Hawaii					310	84		
Total	**96,658**	**26,124**	**148,543**	**40,147**	**87,601**	**23,676**	**104,251**	**28,176**

Note: 1975 Data.

[a] Million liters per day.

Source: From Kollar, K.L. and MacAuley, P., 1980, Water requirements for industrial development, *J. Am. Water Works Assoc.*, vol. 72, no. 1. Copyright AWWA. Reprinted with permission.

Table 7G.76 Percentage of Gross Industrial Water Use by Purpose in the United States

Industry	Parameters of Water Use	Gross Water Use by Unit of Production	Percentage Noncontact Cooling	Percentage Process and Related	Percentage Sanitary and Miscellaneous
Meatpacking	gal/lb carcass weight	3.6 gal/lb	42	46	12
Poultry dressing	gal/bird poultry slaughter	11.6 gal/bird	12	77	12
Dairy products	gal/lb milk processed	0.85 gal/lb	53	27	19
Canned fruits and vegetables	gal/case 24–303 cans eq	225 gal/case	19	67	13
Frozen fruits and vegetables	gal/lb frozen product	11.2 /gal/lb	19	72	8
Wet corn milling	gal/lb corn grind	416 gal/bu	36	63	1
Cane sugar	gal/ton cane sugar	28,100 gal/ton	30	69	1
Beet sugar	gal/ton beet sugar	33,100 gal/ton	31	67	2
Malt beverages	gal/barrel malt beverage	1,500 gal/bbl	72	13	15
Textile mills	gal/lb fiber consumption	34 gal/lb	57	37	6
Sawmills	gal/bd ft lumber	5.4 gal/bd ft	58	36	6
Pulp and paper mills	gal/ton pulp and paper	130,000 gal/ton	18	80	1
Paper converting	gal/ton paper converted	6,600 gal/ton	18	77	5
Alkalis and chlorine	gal/ton chlorine	29,800 gal/ton	85	14	1
Industrial gases	gal/1,000 cu ft industrial gases	636 gal/mcf	86	13	1
Inorganic pigments	gal/ton inorganic pigments	97,800 gal/ton	41	58	1
Industrial inorganic chemicals	gal/ton chemicals 100 percent basic	14,500 gal/ton	83	16	1
Plastic materials and resins	gal/lb plastic	24 gal/lb	93	7	
Synthetic rubber	gal/lb synthetic rubber	55 gal/lb	83	17	Z
Cellulosic man-made fibers	gal/lb fibers	231 gal/lb	69	30	1
Organic fibers, noncellulosic	gal/lb fibers	101 gal/lb	94	6	1
Paints and pigments	gal/gal paint	13 gal/gal	79	17	4
Industrial organic chemicals	gal/ton chemical building blocks	125,000 gal/ton	91	9	1
Nitrogenous fertilizers	gal/ton fertilizer	28,506 gal/ton	92	8	Z
Phosphatic fertilizers	gal/ton fertilizer	35,602 gal/ton	71	28	1
Carbon black	gal/lb carbon black	4.6 gal/lb	57	38	6
Petroleum refining	gal/barrel crude oil input	1,851 gal/bbl	95	5	Z
Tires and inner tubes	gal/tire car and truck tires	518 gal/tire	81	16	3
Hydraulic cement	gal/ton cement	1,360 gal/ton	82	17	1
Steel	gal/ton steel net production	62,600 gal/ton	56	43	1
Iron and steel foundries	gal/ton ferrous castings	12,400 gal/ton	34	58	8
Primary copper	gal/lb copper	53 gal/lb	52	46	2
Primary aluminum	gal/lb aluminum	49 gal/lb	72	26	2
Automobiles	gal/car domestic automobiles	36,500 gal/car	28	69	3

Note: Z = less than 0.5 percent of gross water use; percentages may not add evenly due to rounding.

Source: From Kollar, K.L. and MacAuley, P., 1980, Water requirements for industrial development, *J. Am. Water Works Assoc.*, vol. 72, no. 1. Copyright AWWA. Reprinted with permission.

Table 7G.77 Industrial Water Use Per Employee in the United States

	Gross Water Use Per Employee		Intake Per Employee	
Industry Group	**L/day**	**gal/d**	**L/day**	**gal/d**
Food and kindred products	15,540	4,200	10,360	2,800
Tobacco manufacturers	22,570	6,100	1,480	400
Textile mill products	6,660	1,800	2,960	800
Apparel and related products	370	100	370	100
Lumber and wood products	5,920	1,600	3,700	1,000
Furniture and fixtures	440	120	370	100
Paper and allied products	43,930	38,900	42,920	11,600
Printing and publishing	370	100	370	100
Chemicals and allied products	149,110	40,300	56,240	15,200
Petroleum and coal products	603,100	163,000	94,350	25,500
Rubber and plastic products	10,730	2,900	3,700	1,000
Leather and leather products	740	200	703	190
Stone, clay, and glass products	11,470	3,100	5,550	1,500
Primary metal industries	78,440	21,200	44,030	11,900
Fabricated metal products	2,960	800	1,110	300
Machinery, except electrical	3,700	1,000	1,480	400
Electrical machinery	9,250	2,500	1,110	300
Transportation equipment	17,020	4,600	2,220	600
Instruments and related products	4,440	1,200	1,110	300
Miscellaneous manufacturing	1,110	300	2,960	200

Source: From Kollar, K.L. and MacAuley, P., 1980, Water requirements for industrial development, *J. Am. Water Works Assoc.*, vol. 72, no. 1. Copyright AWWA. Reprinted with permission.

Table 7G.78 Water Use Versus Industrial Units of Production in the United States

Industry	Parameters of Water Use	Gross Water Used by Unit of Production	Intake by Unit of Production	Consumption by Unit of Production	Discharge by Unit of Production
Meatpacking	gal/lb carcass weight	3.6 gal/lb	2.2 gal/lb	0.1 gal/lb	2.1 gal/lb
Poultry dressing	gal/bird poultry slaughter	11.6 gal/bird	10.3 gal/bird	0.5 gal/bird	9.8 gal/bird
Dairy products	gal/lb milk processed	0.85 gal/lb	0.52 gal/lb	0.03 gal/lb	0.48 gal/lb
Canned fruits and vegetables	gal/case 24–303 cans eq	225 gal/case	107 gal/case	10 gal/case	98 cal/case
Frozen fruits and vegetables	gal/lb frozen product	11.2 gal/lb	7.1 gal/lb	0.2 gal/lb	6.9/gal/lb
Wet corn milling	gal/lb corn grind	416 gal/bu	223 gal/bu	18 gal/bu	205 gal/bu
Cane sugar	gal/ton cane sugar	28,100 gal/ton	18,250 gal/ton	950 gal/ton	17,300 gal/ton
Beet sugar	gal/ton beet sugar	33,100 gal/ton	11,100 gal/ton	390 gal/ton	10,700 gal/ton
Malt beverages	gal/barrel malt beverage	1,500 gal/bbl	420 gal/bbl	90 gal/bbl	330 gal/bbl
Textile mills	gal/lb fiber consumption	34 gal/lb	14 gal/lb	1.4 gal/lb	12.8 gal/lb
Sawmills	gal/bd ft lumber	5.4 gal/bd ft	3.3 gal/bd ft	0.6 gal/bd ft	2.7 gal/bd ft
Pulp and paper mills	gal/ton pulp and paper	130,000 gal/ton	38,000 gal/ton	1,800 gal/ton	36,200 gal/ton
Paper converting	gal/ton paper converted	6,600 gal/ton	3,900 gal/ton	270 gal/ton	3,600 gal/ton
Alkalis and chlorine	gal/ton chlorine	29,800 gal/ton	22,200 gal/ton	700 gal/ton	21,600 gal/ton
Industrial gases	gal/1,000 cu ft industrial gases	636 gal/mcf	226 gal/mcf	31 gal/mcf	193 gal/mcf
Inorganic pigments	gal/ton inorganic pigments	97,800 gal/ton	49,400 gal/ton	1,600 gal/ton	47,800 gal/ton
Industrial inorganic chemicals	gal/ton chemicals 100 percent basic	14,500 gal/ton	4,750 gal/ton	470 gal/ton	4,300 gal/ton
Plastic materials and resins	gal/lb plastic	24 gal/lb	6.7 gal/lb	0.6 gal/lb	6.1 gal/lb
Synthetic rubber	gal/lb synthetic rubber	55 gal/lb	6.5 gal/lb	1.4 gal/lb	5.1 gal/lb
Cellulosic man-made fibers	gal/lb fibers	231 gal/lb	68 gal/lb	4.6 gal/lb	63 gal/lb
Organic fibers, noncellulosic	gal/lb fibers	101 gal/lb	38 gal/lb	1.1 gal/lb	37 gal/lb
Paints and pigments	gal/gal paint	13 gal/gal	7.8 gal/gal	0.4 gal/gal	7.4 gal/gal
Industrial organic chemicals	gal/ton chemical building blocks	125,000 gal/ton	54,500 gal/ton	2,800 gal/ton	51,700 gal/ton
Nitrogenous fertilizers	gal/ton fertilizer	28,506 gal/ton	4,001 gal/ton	701 gal/ton	3,299 gal/ton
Phosphatic fertilizers	gal/ton fertilizer	35,602 gal/ton	8,461 gal/ton	1,277 gal/ton	7,184 gal/ton
Carbon black	gal/lb carbon black	4.6 gal/lb	3.9 gal/lb	0.9 gal/lb	3.1 gal/lb
Petroleum refining	gal/barrel crude oil input	1,851 gal/bbl	289 gal/bbl	28 gal/bbl	261 gal/bbl
Tires and inner tubes	gal/tire car and truck tires	518 gal/tire	153 gal/tire	14 gal/tire	139 gal/tire
Hydraulic cement	gal/ton cement	1,360 gal/ton	830 gal/ton	150 gal/ton	680 gal/ton
Steel	gal/ton steel net production	62,600 gal/ton	38,200 gal/ton	1,400 gal/ton	36,800 gal/ton
Iron and steel foundries	gal/ton ferrous castings	12,400 gal/ton	3,030 gal/ton	260 gal/ton	2,760 gal/ton
Primary copper	gal/lb copper	53 gal/lb	17 gal/lb	4.1 gal/lb	13 gal/lb
Primary aluminum	gal/lb aluminum	49 gal/lb	12 gal/lb	0.2 gal/lb	11.8 gal/lb
Automobiles	gal/car domestic automobiles	36,500 gal/car	11,464 gal/car	649 gal/car	10,814 gal/car

Source: From Kollar, K.L. and MacAuley, P., 1980, Water requirements for industrial development, *J. Am. Water Works Assoc.*, vol. 72, no. 1.

Table 7G.79 Water Use Versus Standardized Units of Production in the United States

Industry	Parameters of Water Use	Gross Water Used by Unit of Production	Intake by Unit of Production	Consumption by Unit of Production	Discharge by Unit of Production
Meatpacking	gal/lb carcass weight	7,194 gal/ton	4,331 gal/ton	78 gal/ton	4,253 gal/ton
Poultry dressing	gal/ton ready-to-cook weight	7,389 gal/ton	6,542 gal/ton	296 gal/ton	6,246 gal/ton
Dairy products	gal/lb milk processed	1,692 gal/ton	1,035 gal/ton	63 gal/ton	964 gal/ton
Canned fruits and vegetables	gal/ton vegetables canned	19,700 gal/ton	9,400 gal/ton	850 gal/ton	8,550 gal/ton
Frozen fruits and vegetables	gal/ton vegetables frozen	22,500 gal/ton	14,100 gal/ton	300 gal/ton	13,800 gal/ton
Wet corn milling	gal/lb corn grind	14,869 gal/ton	7,988 gal/ton	643 gal/ton	7,345 gal/ton
Cane sugar	gal/ton cane sugar	28,102 gal/ton	18,256 gal/ton	944 gal/ton	17,312 gal/ton
Beet sugar	gal/ton beet sugar	33,145 gal/ton	11,118 gal/ton	386 gal/ton	10,731 gal/ton
Malt beverages	gal/beer and malt liquor	49 gal/gal	14 gal/gal	3 gal/gal	11 gal/gal
Textile mills	gal/ton textile fiber input	69,808 gal/ton	30,016 gal/ton	3,008 gal/ton	27,008 gal/ton
Sawmills	gal/bd ft lumber	5.4 gal/bd ft	3.3 gal/bd ft	0.63 gal/bd ft	2.7 gal/bd ft
Pulp and paper mills	gal/ton paper	130,047 gal/ton	37,971 gal/ton	1,178 gal/ton	36,193 gal/ton
Paper converting	gal/ton paper converted	6,584 gal/ton	3,861 gal/ton	273 gal/ton	3,588 gal/ton
Alkalis and chlorine	gal/ton chlorine	29,840 gal/ton	22,302 gal/ton	676 gal/ton	21,626 gal/ton
Industrial gases	gal/ton weight of gas	16,080 gal/ton	5,700 gal/ton	780 gal/ton	4,900 gal/ton
Inorganic pigments	gal/ton pigments	97,800 gal/ton	49,400 gal/ton	1,600 gal/ton	47,800 gal/ton
Industrial inorganic chemicals	gal/ton chemical products	14,500 gal/ton	4,700 gal/ton	470 gal/ton	4,300 gal/ton
Plastic materials and resins	gal/ton plastics	47,061 gal/ton	13,338 gal/ton	1,078 gal/ton	12,278 gal/ton
Synthetic rubber	gal/ton synthetic rubber	110,600 gal/ton	13,200 gal/ton	2,800 gal/ton	10,373 gal/ton
Cellulosic man-made fibers	gal/ton fibers	462,230 gal/ton	135,100 gal/ton	9,200 gal/ton	125,846 gal/ton
Organic fibers, noncellulosic	gal/ton fibers	202,123 gal/ton	76,523 gal/ton	2,153 gal/ton	74,369 gal/ton
Paints and pigments	gal/gal paint	13.2 gal/gal	7.8 gal/gal	0.4 gal/gal	7.4 gal/gal
Industrial organic chemicals	gal/ton chemical building blocks	124,700 gal/ton	54,500 gal/ton	2,800 gal/ton	51,700 gal/ton
Nitrogenous fertilizers	gal/ton fertilizer	28,506 gal/ton	4,001 gal/ton	701 gal/ton	3,299 gal/ton
Phosphatic fertilizers	gal/ton fertilizer	35,602 gal/ton	8,461 gal/ton	1,277 gal/ton	7,184 gal/ton
Carbon black	gal/ton carbon black	9,200 gal/ton	7,885 gal/ton	1,771 gal/ton	6,114 gal/ton
Petroleum refining	gal/gal crude petroleum input	44 gal/gal	6.9 gal/gal	0.7 gal/gal	6.2 gal/gal
Tires and inner tubes	gal/tire car and truck tires	518 gal/tire	153 gal/tire	14 gal/tire	139 gal/tire
Hydraulic cement	gal/ton cement	1,355 gal/ton	831 gal/ton	146 gal/ton	685 gal/ton
Steel	gal/ton steel net tons	62,601 gal/ton	38,200 gal/ton	1,400 gal/ton	36,800 gal/ton
Iron and steel foundries	gal/ton ferrous castings	12,407 gal/ton	3,024 gal/ton	260 gal/ton	2,764 gal/ton
Primary copper	gal/ton copper	106,000 gal/ton	34,000 gal/ton	8,200 gal/ton	26,000 gal/ton
Primary aluminum	gal/ton aluminum	96,300 gal/ton	23,900 gal/ton	381 gal/ton	23,500 gal/ton
Automobiles	gal/car automobiles	36,500 gal/car	11,464 gal/car	649 gal/car	10,814 gal/car

Source: From Kollar, K.L. and MacAuley, P., 1980, Water requirements for industrial development, *J. Am. Water Works Assoc.*, vol. 72, no. 1. Copyright AWWA. Reprinted with permission.

Table 7G.80 Typical Water Uses in Paper Mills

Purpose	Gross Water Use ML/day	Gross Water Use mgd	Intake Requirement (Low Reuse) ML/day	Intake Requirement (Low Reuse) mgd	Intake Requirement (High Reuse) ML/day	Intake Requirement (High Reuse) mgd
Kraft pulping (process use)	118	32	51	14	22	6
Kraft pulping (cooling system)	44	12	44	12	1.5	0.4
Bleaching	140	38	70	19	18	5
Paper forming (process system)	129	35	44	12	22	6
Paper forming (cooling system)	14	4	14	4	0.7	0.2
Electric power cooling[a]	51	14	51	14	1.8	0.5
Net totals[b]	499	135	225	61	44	12.1

Note: 1,000 ton per day integrated bleached kraft paper mill.

[a] Condenser cooling requirements for a steam electric plant producing half of the total electric power needs.

[b] Intake net totals are less than the sum of the individual components because much of the wastewater from high quality uses is cascaded to lower quality uses.

Source: From Kollar, K.L. and MacAuley, P., 1980, Water requirements for industrial development, *J. Am. Water Works Assoc.*, vol. 72, no. 1. Copyright AWWA. Reprinted with permission.

Table 7G.81 Water Intake Requirements in the United States — Average Plants Versus High Recycling Plants

Industry	Parameters of Water Use	Intake		Recycling Rate[a]	
		1973 Industry Average	BAT[b] with Maximum Feasible Recycling	1973 Industry Average	BAT[b] with Maximum Feasible Recycling
Meatpacking	gal/lb carcass weight	2.2 gal/lb	0.5 gal/lb	1.66	6.67
Poultry dressing	gal/bird poultry slaughter	10.3 gal/bird	1.7 gal/bird	1.13	6.71
Dairy products	gal/lb milk processed	0.52 gal/lb	0.13 gal/lb	1.64	6.67
Canned fruits and vegetables	gal/case 24–303 cans eq	107 gal/case	29 gal/case	2.10	7.75
Frozen fruits and vegetables	gal/lb frozen product	7.1 gal/lb	1.6 gal/lb	1.60	7.25
Wet corn milling	gal/bu corn grind	223 gal/bu	46 gal/bu	1.86	9.09
Cane sugar	gal/ton cane sugar	18,250 gal/ton	5,300 gal/ton	1.54	5.26
Beet sugar	gal/ton beet sugar	11,100 gal/ton	6,200 gal/ton	2.98	5.38
Malt beverages	gal/barrel malt beverage	420 gal/bbl	105 gal/bbl	3.50	14.3
Textile mills	gal/lb fiber consumption	14 gal/lb	1.8 gal/lb	2.23	18.2
Sawmills	gal/bd ft lumber	3.3 gal/ft	0.8 gal/ft	1.64	6.85
Pulp and paper mills	gal/ton pulp and paper	38,000 gal/ton	10,700 gal/ton	3.42	12.2
Paper converting	gal/ton paper converted	3,900 gal/ton	750 gal/ton	1.70	8.93
Alkalis and chlorine	gal/ton chlorine	22,200 gal/ton	860 gal/ton	1.34	34.5
Industrial gases	gal/1,000 cu ft industrial gases	226 gal/mcf	18 gal/mcf	2.82	34.5
Inorganic pigments	gal/ton inorganic pigments	49,400 gal/ton	6,100 gal/ton	1.98	16.1
Industrial inorganic chemicals	gal/ton chemicals 100 percent basis	4,750 gal/ton	470 gal/ton	3.08	31.2
Plastic materials and resins	gal/lb plastic	6.7 gal/lb	0.7 gal/lb	3.53	33.3
Synthetic rubber	gal/lb synthetic rubber	6.5 gal/lb	1.6 gal/lb	8.38	33.3
Cellulosic man-made fibers	gal/lb fibers	68 gal/lb	8.4 gal/lb	3.42	27.8
Organic fibers, noncellulosic	gal/lb fibers	38 gal/lb	5.0 gal/lb	2.64	20.0
Paints and pigments	gal/gal paint	7.8 gal/gal	0.8 gal/gal	1.69	16.1
Industrial organic chemicals	gal/ton chemical building blocks	54,500 gal/ton	4,000 gal/ton	2.29	31.2
Nitrogenous fertilizers	gal/ton fertilizer	4,000 gal/ton	900 gal/ton	7.12	31.2
Phosphatic fertilizers	gal/ton fertilizer	8,500 gal/ton	2,400 gal/ton	4.21	14.7
Carbon black	gal/lb carbon black	3.9 gal/lb	0.3 gal/lb	1.17	16.1
Petroleum refining	gal/barrel crude oil input	289 gal/bbl	55 gal/bbl	6.38	33.3
Tires and inner tubes	gal/tire car and truck tires	153 gal/tire	18 gal/tire	3.39	29.4
Hydraulic cement	gal/ton cement	830 gal/ton	180 gal/ton	1.63	7.41
Steel	gal/ton steel net production	38,200 gal/ton	5,300 gal/ton	1.64	11.9
Iron and steel foundries	gal/ton ferrous castings	3,030 gal/ton	1,080 gal/ton	4.10	11.5
Primary copper	gal/lb copper	17 gal/lb	4.5 gal/lb	3.12	11.9
Primary aluminum	gal/lb aluminum	12 gal/lb	2.9 gal/lb	4.11	16.9
Automobiles	gal/car domestic automobiles	11,500 gal/car	2,200 gal/car	3.18	16.3

[a] The recycling rate is obtained by dividing gross water use by intake.
[b] Best available technology economically achievable as defined by Water Pollution Control Act amendments of 1972.

Source: From Kollar, K.L. and MacAuley, P., 1980, Water requirements for industrial development, *J. Am. Water Works Assoc.*, vol. 72, no. 1. Copyright AWWA. Reprinted with permission.

Table 7G.82 Water Recycling in the 20 Plants with the Highest Rates in 34 Major Water-Using Industries in the United States, 1970

Industry	Gross Water Use[a]	Intake[a]	Mean Recycling Rate[b]	Highest Recycling Rate[b]	Tenth Highest Recycling Rate[b]	Twentieth Highest Recycling Rate[b]
Meat packing plants	49.732	20.335	2.45	7.05	2.41	1.85
Poultry dressing	3.473	1.990	1.75	4.28	1.30	1.14
Fluid milk	8.118	0.859	9.45	71.71	7.92	3.96
Canned fruit and vegetables	10.673	3.419	3.12	18.24	2.50	1.76
Frozen fruit and vegetables	17.353	9.259	1.87	7.13	1.97	1.39
Wet corn milling	53.986	32.109	1.68	11.91	2.31	1.11
Beet sugar	58.949	16.829	3.50	22.24	2.97	1.84
Malt liquors	64.350	12.675	5.08	10.00	2.85	1.11
Shortening and cooking oils	48.106	5.425	8.87	113.53	8.23	1.30
Cigarettes	60.765	2.292	26.51	33.39	15.31	1.11
Weaving mills, cotton	74.289	1.186	62.64	285.31	64.25	27.99
Weaving mills, synthetics	88.114	0.717	122.89	558.25	111.27	48.53
Weaving and finishing, wool	19.163	2.637	7.27	93.44	24.19	1.18
Pulp mills	713.440	208.179	3.43	7.57	3.84	1.41
Papermills, except building paper	723.008	71.057	10.18	76.54	8.96	6.06
Paperboard mills	272.670	14.515	18.79	50.00	14.68	8.22
Alkalis and chlorine	198.798	87.167	2.28	25.11	1.79	1.12
Industrial gases	141.450	1.490	94.93	157.80	84.83	46.23
Cyclic intermediate and crudes	327.354	55.446	5.90	160.00	13.45	2.24
Inorganic pigments	120.387	50.222	2.40	15.22	1.53	1.11
Industrial organic chemicals	962.830	35.142	27.40	48.18	23.20	15.80
Industrial inorganic chemicals	505.919	16.670	30.35	70.95	30.10	23.81
Plastic materials and resins	704.229	5.131	137.25	613.60	27.37	13.81
Cellulosic man-made fibers	209.801	48.088	4.36	20.83	4.30	1.37
Organic fibers, noncellulosic	392.335	151.969	2.58	28.06	2.82	1.16
Pharmaceutical preparations	70.621	15.385	4.59	104.73	7.36	1.11
Fertilizers	282.251	23.373	12.08	90.60	9.72	2.45
Petroleum refining	2,026.521	30.221	67.06	251.05	44.08	34.36
Cement, hydraulic	20.868	4.320	4.83	97.35	2.58	1.77
Blast furnaces and steel mills	394.549	29.050	13.58	95.13	18.66	6.76
Electrometallurgical products	22.732	1.827	12.44	65.81	25.64	5.07
Gray iron foundaries	35.396	10.254	3.45	15.23	2.86	1.82
Primary copper	78.473	33.218	2.36	9.85	2.23	1.18
Primary aluminum	65.519	15.723	4.17	10.10	3.50	1.66

[a] Billions of gallons per year: 1 bil gal=3.7 GL.
[b] The recycling rate is obtained by dividing gross water use by intake.

Source: From Kollar, K.L. and MacAuley, P., 1980, Water requirements for industrial development, *J. Am. Water Works Assoc.*, vol. 72, no. 1. Copyright AWWA. Reprinted with permission.

Table 7G.83 Typical Unit Water Requirements for Energy Production in the United States

Fuel and Process	Standard Unit	Gallons per Standard Unit	Gallons per Million Btu	Major Use
Coal				
Western coal mining	ton	6.0–14.7	0.25–0.61	Dust control and washing
Eastern coal mining	Ton	15.8–18.0	0.66–0.75	Dust control and washing
Coal gasification	MSCF[a]	72.0–158	72–158	Process and cooling
Coal liquefaction	Barrel	1,134.0–1,750	31–200	Process and cooling
Petroleum				
Oil and gas production	Barrel	1.7–3.0	3.05	Well drilling and recovery
Oil refining	Barrel	43.0	7.58	Process and cooling
Oil shale production	Barrel	145.4	30.1	Mining, cooling, processing, and waste disposal
Gas processing	MSCF[a]	1.67	1.67	Cooling
Nuclear fuels	—	—	14.3	Mining and processing
Power generation				
Fossil fuels	kWh	0.41	120.16	Cooling
Nuclear fuels	kWh	0.80	234.46	Cooling
Geothermal	—	—	527	Cooling and extraction

[a] Million standard cubic feet.

Source: From United States Federal Energy Administration, 1974, "Project Independence," Project Independence Report, p. 304: U.S. Government Printing Office, Washington, DC, 20402.

Table 7G.84 Thermoelectric Power Water Use by Energy Source and State, 1995

State	Fossil Fuel						Nuclear					
	Withdrawals, by Source and Type				Consumptive Use		Withdrawals, by Source and Type				Consumptive Use	
	Groundwater	Surface Water					Groundwater	Surface Water				
	Fresh	Fresh	Saline	Total	Fresh	Saline	Fresh	Fresh	Saline	Total	Fresh	Saline
Alabama	6.0	4,330	0	4,330	30	0	0	862	0	862	1.7	0
Alaska	4.2	26	0	26	3.1	0	0	0	0	0	0	0
Arizona	42	20	0	20	54	0	0	0	0	0	0	0
Arkansas	5.2	798	0	798	27	0	0	967	0	967	1.2	0
California	3.5	190	4,730	4,920	9.4	2.8	0.1	12	4,690	4,710	0.3	1.3
Colorado	22	93	0	93	41	0	0	0	0	0	0	0
Connecticut	0.1	276	882	1,160	5.9	74	0.1	484	2,300	2,780	0	0
Delaware	0.2	534	740	1,270	0.2	2.9	0	0	0	0	0	0
District of Columbia	0	9.7	0	9.7	0.8	0	0	0	0	0	0	0
Florida	21	615	9,140	9,760	54	0	0.3	0	1,810	1,810	1.2	0
Georgia	3.9	2,910	33	2,950	52	0	1.0	122	0	122	93	0
Hawaii	67	0	903	903	0.7	9.0	0	0	0	0	0	0
Idaho	0	0	0	0	0	0	0	0	0	0	0	0
Illinois	9.5	9,570	0	9,570	144	0	1.3	7,520	0	7,520	263	0
Indiana	11	5,680	0	5,680	114	0	0	0	0	0	0	0
Iowa	13	2,100	0	2,100	7.8	0	2.0	8.1	0	8.1	2.6	0
Kansas	14	1,230	0	1,230	45	0	0	22	0	22	13	0
Kentucky	38	3,410	0	3,410	203	0	0	0	0	0	0	0
Louisiana	31	4,430	0	4,430	212	0	0.1	1,020	0	1,020	10	0
Maine	0.7	30	105	135	3.5	1.7	0	0	0	0	0	0
Maryland	1.6	358	2,780	3,140	3.7	32	0.2	0	3,220	3,220	0	16
Massachusetts	46	150	3,910	4,060	0	0	0	0	454	454	0	6.0
Michigan	3.0	6,030	0	6,030	50	0	0.1	2,340	0	2,340	76	0
Minnesota	1.8	1,210	0	1,210	28	0	0.1	886	0	886	20	0
Mississippi	10	220	112	333	8.0	3.6	32	0	0	0	19	0
Missouri	9.1	5,520	0	5,520	40	0	0.4	21	0	21	11	0
Montana	0	22	0	22	22	0	0	0	0	0	0	0
Nebraska	4.4	1,290	0	1,290	12	0	0	1,060	0	1,060	0	0
Nevada	6.2	21	0	21	28	0	0	0	0	0	0	0
New Hampshire	0.8	228	292	521	4.3	0	0	0	585	585	0	0
New Jersey	1.2	578	980	1,560	3.7	9.9	0.7	0	2,800	2,800	0.7	22
New Mexico	9.3	46	0	46	48	0	0	0	0	0	0	0
New York	0	5,140	5,470	10,600	103	109	0	1,420	1,010	2,440	68	20
North Carolina	0.1	3,210	0	3,210	56	0	0	2,660	1,550	4,210	1.5	17
North Dakota	0	879	0	879	25	0	0	0	0	0	0	0
Ohio	19	8,040	0	8,040	309	0	0	137	0	137	27	0
Oklahoma	3.5	121	0	121	60	0	0	0	0	0	0	0

(Continued)

Table 7G.84 (Continued)

	Fossil Fuel						Nuclear					
	Withdrawals, by Source and Type						Withdrawals, by Source and Type					
	Groundwater	Surface Water			Consumptive Use		Groundwater	Surface Water			Consumptive Use	
State	**Fresh**	**Fresh**	**Saline**	**Total**	**Fresh**	**Saline**	**Fresh**	**Fresh**	**Saline**	**Total**	**Fresh**	**Saline**
Oregon	0	9.0	0	9.0	7.8	0	0	0	0	0	0	0
Pennsylvania	6.2	3,870	0	3,870	120	0	0	2,050	0	2,050	119	0
Rhode Island	0	0	275	275	0	5.5	0	0	0	0	0	0
South Carolina	0.4	1,290	0	1,290	23	0	39	3,470	0	3,470	28	0
South Dakota	2.6	1.9	0	1.9	0.1	0	0	0	0	0	0	0
Tennessee	0	6,830	0	6,830	0.5	0	0	1,470	0	1,470	0	0
Texas	58	6,710	3,870	10,600	271	12	0.8	2,820	0	2,820	26	0
Utah	0	48	0	48	47	0	0	0	0	0	0	0
Vermont	0.4	0.5	0	0.5	0.7	0	0	452	0	452	3.2	0
Virginia	0.1	1,820	973	2,790	8.8	0	0.3	2,080	1,760	3,830	0	0
Washington	0.4	17	0	17	0.4	0	0.1	358	0	358	9.8	0
West Virginia	0.5	3,010	0	3,010	122	0	0	0	0	0	0	0
Wisconsin	5.6	3,860	0	3,860	39	0	0.1	1,970	0	1.970	20	0
Wyoming	1.0	219	0	219	50	0	0	0	0	0	0	0
Puerto Rico	2.2	0	2,260	2,260	0.7	0	0	0	0	0	0	0
Virgin Islands	0	0	173	173	0.2	0	0	0	0	0	0	0
Total	**486**	**97,000**	**37,600**	**135,000**	**2,500**	**263**	**78**	**34,300**	**20,200**	**54,500**	**815**	**82**

Note: Figures may not add to totals because of independent rounding. All values in million gallons per day.

Source: From Solley, W.B. et al., 1998. Esimated Use of Water in the United States in 1995, *U.S. Geological Survey Circular 1200*, www.usgs.gov.

Table 7G.85 Thermoelectric Generation from Different Sources, 1960–1995

Year	Generation by Coal (%)	Generation by Petroleum (%)	Generation by Gas (%)	Generation by Nuclear (%)	Generation by Conventional Steam (%)	Generation by Internal Combustion (%)
1960	65.88	7.37	26.49	0.26	99.31	0.72
1965	66.50	7.25	25.82	0.43	99.04	0.58
1970	55.25	13.84	29.20	1.71	97.90	0.46
1975	51.20	16.94	17.93	10.33	85.96	0.32
1980	58.23	11.93	17.26	12.59	87.53	0.15
1985	64.67	4.29	13.34	17.70	82.74	0.05
1990	62.21	4.33	10.43	23.03	77.40	0.04
1995	61.56	2.01	11.35	25.08	75.12	0.03

Source: From Dziegielewski, B. et al., 2002. *Analysis of Water Use Trends in the United States: 1950–1995*, Southern Illinois University at Carbondale, Carbondale, IL February 28, 2002. Reprinted with permission. http://info.geography.siu.edu/geography_info/research/.

Table 7G.86 Thermoelectric Power Water Withdrawals by Cooling Type in the United States, 2000

	Withdrawals for Once-Through Cooling				Withdrawals for Closed-Loop Cooling by Source and Type					
	Surface Water			Groundwater	Surface Water			Total		
State	Fresh	Saline	Total	Fresh	Fresh	Saline	Total	Fresh	Saline	Total
Alabama	8,020	0	8,020	0	167	0	167	167	0	167
Alaska	28.9	0	28.9	4.65	0	0	0	4.65	0	4.65
Arizona	0	0	0	74.3	26.2	0	26.2	100	0	100
Arkansas	1,690	0	1,690	2.92	478	0	478	481	0	481
California	344	12,600	12,900	3.23	5.41	0	5.41	8.64	0	8.64
Colorado	90.2	0	90.2	16.1	31.8	0	31.8	48	0	48
Connecticut	115	3,440	3,560	0.08	71.7	0	71.7	71.8	0	71.8
Delaware	0	0	0	0.47	366	738	1,100	366	738	1,100
District of Columbia	0	0	0	0	9.69	0	9.69	9.69	0	9.69
Florida	559	11,800	12,400	29.5	69.8	150	219	99.3	150	249
Georgia	2,800	61.7	2,860	1.03	444	0	444	445	0	445
Hawaii	0	0	0	0	0	0	0	0	0	0
Idaho	0	0	0	0	0	0	0	0	0	0
Illinois	11,000	0	11,000	5.75	239	0	239	245	0	245
Indiana	6,450	0	6,450	2.58	252	0	252	254	0	254
Iowa	2,510	0	2,510	11.9	15.6	0	15.6	27.6	0	27.6
Kansas	2,210	0	2,210	14.9	29.1	0	29.1	44	0	44
Kentucky	824	0	824	2.71	2,430	0	2,430	2,430	0	2,430
Louisiana	4,500	0	4,500	28.4	1,080	0	1,080	1,110	0	1,110
Maine	90.8	295	385	4.92	17.2	0	17.2	22.2	0	22.2
Maryland	377	5,670	6,050	1.8	0	589	589	1.8	589	591
Massachusetts	108	3,610	3,720	0	0.45	0	0.45	0.45	0	0.45
Michigan	7,710	0	7,710	0	0	0	0	0	0	0
Minnesota	1,330	0	1,330	4.17	935	0	935	939	0	939
Mississippi	307	148	456	43.5	11.2	0	11.2	54.7	0	54.7
Missouri	5,200	0	5,200	12.2	422	0	422	434	0	434
Montana	84.4	0	84.4	0	25.6	0	25.6	25.6	0	25.6
Nebraska	2,390	0	2,390	6.87	418	0	418	424	0	424
Nevada	0	0	0	12	24.7	0	24.7	36.7	0	36.7
New Hampshire	234	761	995	0.71	1.37	0	1.37	2.08	0	2.08
New Jersey	648	3,330	3,980	2.24	0	57.6	57.6	2.24	57.6	59.9
New Mexico	0	0	0	11.4	45	0	45	56.4	0	56.4
New York	4,040	5,010	9,050	0	0	0	0	0	0	0
North Carolina	7,850	1,620	9,470	0.09	0	0	0	0.09	0	0.09
North Dakota	887	0	887	0	14.5	0	14.5	14.5	0	14.5
Ohio	7,790	0	7,790	7.57	799	0	799	806	0	806
Oklahoma	37.9	0	37.9	3.27	105	0	105	109	0	109
Oregon	0	0	0	2.47	12.8	0	12.8	15.3	0	15.3
Pennsylvania	4,330	0	4,330	3.98	2,650	0	2,650	2,650	0	2,650

Rhode Island	0	290	290	0	2.4	0	2.4	2.4	0	2.4
South Carolina	3,860	0	3,860	5.83	1,850	0	1,850	1,850	0	1,850
South Dakota	0	0	0	1.23	4.01	0	4.01	5.24	0	5.24
Tennessee	8,860	0	8,860	0	174	0	174	174	0	174
Texas	6,990	3,440	10,400	60.2	2,770	0	2,770	2,830	0	2,830
Utah	0	0	0	13.1	49.2	0	49.2	62.2	0	62.2
Vermont	354	0	354	0.66	0.55	0	0.55	1.21	0	1.21
Virginia	3,850	3,580	7,430	1.5	0	0	0	1.5	0	1.5
Washington	444	0	444	0.92	74.2	0	74.2	75.1	0	75.1
West Virginia	3,790	0	3,790	0	163	0	163	163	0	163
Wisconsin	6,090	0	6,090	8.99	0	0	0	8.99	0	8.99
Wyoming	179	0	179	1.13	63.4	0	63.4	64.6	0	64.6
Puerto Rico	0	2,190	2,190	0	0	0	0	0	0	0
U.S. Virgin Islands	0	136	136	0	0	0	0	0	0	0
Total	**119,000**	**58,000**	**177,000**	**409**	**16,300**	**1,530**	**17,900**	**16,800**	**1,530**	**18,300**

Note: Figures may not sum to totals because of independent rounding. All values are in million gallons per day.

Source: From Hutson, S.S. et al., 2004, Estimated use of water in the United States in 2000, *U.S. Geological Survey Circular 1268*, www.usgs.gov.

Table 7G.87 Thermoelectric Power Water Use in the United States, 2000

	Withdrawals (mil gal/day)							Withdrawals (thousand acre-feet/yr)		
	By Source and Type							By Type		
	Groundwater	Surface Water			Total					
State	Fresh	Fresh	Saline	Total	Fresh	Saline	Total	Fresh	Saline	Total
Alabama	0	8,190	0	8,190	8,190	0	8,190	9,180	0	9,180
Alaska	4.65	28.9	0	28.9	33.6	0	33.6	37.6	0	37.6
Arizona	74.3	26.2	0	26.2	100	0	100	113	0	113
Arkansas	2.92	2,170	0	2,170	2,180	0	2,180	2,440	0	2,440
California	3.23	349	12,600	12,900	352	12,600	12,900	395	14,100	14,500
Colorado	16.1	122	0	122	138	0	138	155	0	155
Connecticut	0.08	186	3,440	3,630	187	3,440	3,630	209	3,860	4,070
Delaware	0.47	366	738	1,100	366	738	1,100	411	827	1,240
District of Columbia	0	9.69	0	9.69	9.69	0	9.69	10.9	0	10.9
Florida	29.5	629	12,000	12,600	658	12,000	12,600	738	13,400	14,100
Georgia	1.03	3,240	61.7	3,310	3,250	61.7	3,310	3,640	69.2	3,710
Hawaii	0	0	0	0	0	0	0	0	0	0
Idaho	0	0	0	0	0	0	0	0	0	0
Illinois	5.75	11,300	0	11,300	11,300	0	11,300	12,600	0	12,600
Indiana	2.58	6,700	0	6,700	6,700	0	6,700	7,510	0	7,510
Iowa	11.9	2,530	0	2,530	2,540	0	2,540	2,850	0	2,850
Kansas	14.9	2,240	0	2,240	2,260	0	2,260	2,530	0	2,530
Kentucky	2.71	3,250	0	3,250	3,260	0	3,260	3,650	0	3,650
Louisiana	28.4	5,580	0	5,580	5,610	0	5,610	6,290	0	6,290
Maine	4.92	108	295	403	113	295	408	127	330	457
Maryland	1.8	377	6,260	6,640	379	6,260	6,640	425	7,020	7,440
Massachusetts	0	108	3,610	3,720	108	3,610	3,720	121	4,050	4,170
Michigan	0	7,710	0	7,710	7,710	0	7,710	8,640	0	8,640
Minnesota	4.17	2,260	0	2,260	2,270	0	2,270	2,540	0	2,540
Mississippi	43.5	318	148	467	362	148	510	406	166	572
Missouri	12.2	5,620	0	5,620	5,640	0	5,640	6,320	0	6,320
Montana	0	110	0	110	110	0	110	123	0	123
Nebraska	6.87	2,810	0	2,810	2,820	0	2,820	3,160	0	3,160
Nevada	12	24.7	0	24.7	36.7	0	36.7	41.1	0	41.1
New Hampshire	0.71	235	761	997	236	761	997	265	854	1,120
New Jersey	2.24	648	3,390	4,040	650	3,390	4,040	729	3,800	4,530
New Mexico	11.4	45	0	45	56.4	0	56.4	63.2	0	63.2
New York	0	4,040	5,010	9,050	4,040	5,010	9,050	4,530	5,610	10,100
North Carolina	0.09	7,850	1,620	9,470	7,850	1,620	9,470	8,800	1,810	10,600
North Dakota	0	902	0	902	902	0	902	1,010	0	1,010
Ohio	7.57	8,590	0	8,590	8,590	0	8,590	9,630	0	9,630
Oklahoma	3.27	143	0	143	146	0	146	164	0	164
Oregon	2.47	12.8	0	12.8	15.3	0	15.3	17.2	0	17.2

Pennsylvania	3.98	6,970	0	6,970	6,980	0	6,980	7,820	0	7,820
Rhode Island	0	2.4	290	293	2.4	290	293	2.69	326	328
South Carolina	5.83	5,700	0	5,700	5,710	0	5,710	6,400	0	6,400
South Dakota	1.23	4.01	0	4.01	5.24	0	5.24	5.87	0	5.87
Tennessee	0	9,040	0	9,040	9,040	0	9,040	10,100	0	10,100
Texas	60.2	9,760	3,440	13,200	9,820	3,440	13,300	11,000	3,860	14,900
Utah	13.1	49.2	0	49.2	62.2	0	62.2	69.8	0	69.8
Vermont	0.66	355	0	355	355	0	355	398	0	398
Virginia	1.5	3,850	3,580	7,430	3,850	3,580	7,430	4,310	4,020	8,330
Washington	0.92	518	0	518	519	0	519	582	0	582
West Virginia	0	3,950	0	3,950	3,950	0	3,950	4,430	0	4,430
Wisconsin	8.99	6,090	0	6,090	6,090	0	6,090	6,830	0	6,830
Wyoming	1.13	242	0	242	243	0	243	273	0	273
Puerto Rico	0	0	2,190	2,190	0	2,190	2,190	0	2,460	2,460
U.S. Virgin Islands	0	0	136	136	0	136	136	0	153	153
Total	409	135,000	59,500	195,000	136,000	59,500	195,000	152,000	66,700	219,000

Note: Figures may not sum to totals because of independent rounding.

Source: From Hutson, S.S. et al., 2004, Estimated use of water in the United States in 2000, *U.S. Geological Survey Circular 1268*, www.usgs.gov.

Table 7G.88 Self-Supplied Thermoelectric Withdrawals and Production, 1950–1995

Year	Self-supplied Thermoelectric Withdrawals (bgd)	Withdrawals Per Capita (gpcd)	Thermoelectric Production (bil kWh)	Annual Growth Rate in Production (%)	Energy Production Per Capita (kWh/Capita/d)	Withdrawals Per Unit Production (gal/kWh)
1950	40	264.5	232.8	—	4.2	62.7
1955	72	438.2	433.8	13.26	7.2	60.6
1960	100	560.4	609.6	7.04	9.4	59.9
1965	130	674.3	861.0	7.15	12.2	55.1
1970	170	841.1	1,283.3	8.31	17.4	48.4
1975	200	938.8	1,614.2	4.69	20.8	45.2
1980	210	934.5	2,004.9	4.43	24.4	38.2
1985	187	788.8	2,177.9	1.67	25.2	31.3
1990	195	789.3	2,517.6	2.94	27.9	28.3
1995	190	726.8	2,694.4	1.37	28.2	25.7

Source: From Dziegielewski, B. et al., 2002, *Analysis of Water Use Trends in the United States: 1950–1995*, Southern Illinois University at Carbondale, Carbondale, IL February 28, 2002, Reprinted with permission. http://info.geography.siu.edu/geography_info/research/.

SECTION 7H INDUSTRIAL WATER USE — WORLD

Table 7H.89 Water Intake in Manufacturing (MCM/yr) by Purpose of Initial Use and Industry in Canada, 1996

Industry Group	Number of Plants	Processing	Cooling, Condensing, and Steam	Sanitary Services	Other	Total Intake	%
Food	1,264	128.6	107.3	27.8	5.9	269.5	4.5
Beverages	121	38.4	29.0	4.6	1.1	73.1	1.2
Rubber products	96	3.6	7.7	0.9	0.1	12.3	0.2
Plastic products	486	5.9	5.9	1.3	0.2	13.3	0.2
Primary textiles	87	15.5	64.6	6.5	0.0	86.7	1.4
Textile products	47	12.8	1.8	0.4	0.1	15.0	0.2
Wood products	454	9.7	24.4	2.2	8.8	45.1	0.7
Paper + allied products	292	1,847.5	508.3	49.1	16.4	2,421.3	40.1
Primary metals	217	557.6	830.1	21.5	13.8	1,423.0	23.6
Fabricated metals	543	11.3	6.4	1.6	0.1	19.4	0.3
Transportation equipment	547	28.5	25.0	11.1	0.4	65.0	1.1
Nonmetallic mineral products	726	21.6	44.9	3.5	32.1	102.1	1.7
Petroleum + coal products	27	34.4	324.6	4.9	6.6	370.5	6.1
Chemicals + chemical products	599	220.9	879.8	10.9	9.7	1,121.3	18.6
Total	5,506	2,936.3	2,859.6	146.3	95.3	6,037.5	100.0
%		**48.6**	**47.4**	**2.4**	**1.6**	**100.0**	

Source: From Scharf, D., Burke, D., Villeneuve, and Leigh, L., 1996. Industrial Water Use 1996, Scharf, D., Burke, D.W., Villeneuve, M., and Leigh, L., Environmental Economics Branch, Environment Canada, 2002. Reproduced with the permission of the Minister of Public Works and Government Services, 2006.

Table 7H.90 Water Intake in Manufacturing (MCM/yr) by Source and Industry Group in Canada, 1996

		Fresh Water				Brackish Water			
		Public Supplied	Self-Supplied			Self-Supplied			
Industry Group	Number of Plants	Municipal	Surface	Ground	Other	Ground	Tidewater	Other	Total Intake
Food	1,254	118.7	61.8	44.6	3.4	1.9	38.7	0.2	269.3
Beverages	121	49.0	16.1	8.1	0.0	0.0	0.0	0.0	73.1
Rubber products	96	8.2	1.3	2.4	0.5	0.0	0.0	0.0	12.3
Plastic products	482	7.0	4.8	1.2	0.1	0.1	0.0	0.0	13.2
Primary textiles	87	34.6	51.4	0.1	0.0	0.1	0.0	0.5	86.7
Textile products	47	13.1	0.0	2.0	0.0	0.0	0.0	0.0	15.0
Wood products	454	18.8	16.4	9.5	0.2	0.1	0.1	0.0	45.1
Paper + allied products	292	70.4	2,240.0	65.8	45.3	0.0	0.0	0.0	2,421.3
Primary metals	217	61.2	1,314.0	22.9	12.8	0.0	12.1	0.0	1,423.0
Fabricated metals	543	12.1	6.8	0.5	0.0	0.0	0.0	0.0	19.4
Transportation equipment	547	59.5	4.7	0.7	0.0	0.0	0.0	0.0	65.0
Non-metallic mineral products	725	19.5	36.3	9.9	36.0	0.0	0.4	0.0	102.1
Petroleum + coal products	27	11.4	249.0	2.5	1.3	0.0	102.1	4.2	370.5
Chemicals + chemical products	599	66.1	940.1	7.2	67.2	0.1	40.5	0.1	1,121.3
Total	5,491	549.6	4,942.5	177.3	166.8	2.3	193.9	5.0	6,037.4
%		**9.1**	**81.9**	**2.9**	**2.8**	**0.0**	**3.2**	**0.1**	**100.0**

Source: From Scharf, D., Burke, D., Villeneuve, and Leigh, L., 1996. Industrial Water Use 1996, Scharf, D., Burke, D.W., Villeneuve, M., and Leigh, L., Environmental Economics Branch, Environment Canada, 2002. Reproduced with the permission of the Minister of Public Works and Government Services, 2006.

Table 7H.91 Competing Water Uses for Main Income Groups of Countries

	Agricultural Use (%)	Industrial Use (%)	Domestic Use (%)
World	70	22	8
Low income	87	8	5
Middle income	74	13	12
Lower middle income	75	15	10
Upper middle income	73	10	17
Low and middle income	82	10	8
East Asia and Pacific	80	14	6
Europe and central Asia	63	26	11
Latin America and Caribbean	74	9	18
Middle East and North Africa	89	4	6
South Asia	93	2	4
Sub-Saharan Africa	87	4	9
High income	30	59	11
Europe Economic and Monetary Union (EMU)	21	63	16

Source: From *Water for People Water for Life*, The United Nations World Water Development Report. Copyright © United Nations Educational, Scientific, and Cultural Organization - World Water Assessment Programme (UNESCO-WWAP), 2003. Reproduced by permission of UNESCO. www.unesco.org.

Original Source: From World Bank, 2001, World Development Indicators (WDI), Washington, DC, Available on CD-ROM, Copyright © International Bank for Reconstruction and Development/The World Bank, www.worldbank.org. Reprinted with permission.

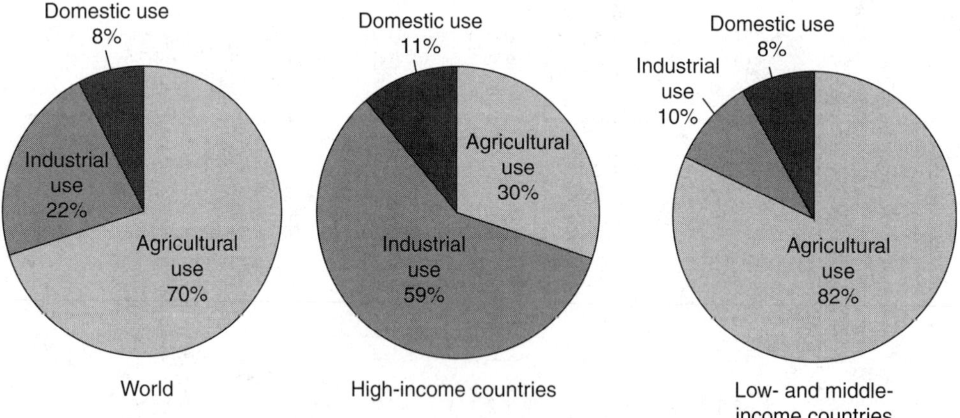

Figure 7H.16 Industrial use of water increases with country income. (From *Water for People Water for Life*, The United Nations World Water Development Report. Copyright © United Nations Educational, Scientific, and Cultural Organization - World Water Assessment Programme (UNESCO-WWAP), 2003. Reproduced by permission of UNESCO. www.unesco.org. Original Source: World Bank, 2001, World Development Indicators (WDI), Washington, DC, Available on CD-ROM, Copyright © International Bank for Reconstruction and development/The World Bank, www.worldbank.org. Reprinted with permission.

Table 7H.92 World-Industrial Water Efficiency

Country	Total Annual Freshwater Withdrawal (bill m)[3a]	% for Industry[b]	Industrial Value Added (IVA) (mi US$)[c]	Population (mi)[d]	IVA/Industrial Annual Withdrawal (US$/m³/Capita)[e]
Algeria	4.5	15	22,618	30	1.11
Angola	0.5	10	4,182	12	7.26
Argentina	28.6	9	77,171	37	0.84
Armenia	2.9	4	1,029	4	2.14
Austria	2.2	60	76,386	8	7.14
Azerbaijan	16.5	25	1,213	8	0.04
Bangladesh	14.6	2	11,507	128	0.31
Belarus	2.7	43	9,543	10	0.81
Benin	0.2	10	333	6	3.59
Bolivia	1.4	20	1,529	8	0.68
Botswana	0.1	20	2,593	2	58.94
Brazil	54.9	18	231,442	168	0.14
Cameroon	0.4	19	2,360	15	2.07
Central African Republic	0.1	6	211	4	13.46
Chad	0.2	2	233	7	8.75
Chile	21.4	11	24,385	15	0.70
China	525.5	18	498,292	1,254	0.00
Colombia	8.9	4	23,120	42	1.40
Congo Dem Rep.	0.0	27	852	3	26.29
Costa Rica	5.8	7	4,456	4	2.88
Côte d'Ivoire	0.7	11	3,039	16	2.47
Croatia	0.1	50	4,995	4	31.22
Czech Republic	2.5	57	20,512	10	1.42
Denmark	0.9	9	40,142	5	100.23
Dominican Republic	8.3	1	5,530	8	16.58
Ecuador	17.0	6	6,535	12	0.57
Egypt, Arab Rep.	55.1	8	22,221	63	0.08
El Salvador	0.7	20	3,158	6	3.57
Estonia	0.2	39	1,494	1	23.88
Ethiopia	2.2	3	726	63	0.17
Finland	2.4	82	48,807	5	4.89
Gabon	0.1	22	2,752	1	208.45
Gambia	0.0	2	50	1	83.74
Georgia	3.5	20	378	5	0.11
Germany	46.3	86	760,536	82	0.23
Ghana	0.3	13	1,927	19	2.60
Guatemala	1.2	17	3,468	11	1.60
Guinea	0.7	3	1,431	7	9.21
Guinea Bissau	0.0	4	46	1	63.33
Haiti	1.0	1	641	8	13.62
Honduras	1.5	5	1,234	6	2.71
India	500.0	3	113,041	998	0.01
Indonesia	74.3	1	85,633	207	0.56
Iran, Islamic Rep.	70.0	2	34,204	63	0.37
Italy	57.5	37	323,494	58	0.27
Jamaica	0.9	7	1,619	3	8.33
Jordan	1.0	3	1,738	5	10.43
Kenya	2.0	4	1,325	29	0.57
Korea, Rep.	23.7	11	249,268	23	4.16
Kyrgyztan	10.1	3	699	6	0.46
Latvia	0.3	32	1,627	2	8.71
Lithuania	0.3	16	2,156	4	13.56
Malawi	0.9	3	288	11	0.82
Malaysia	12.7	13	43,503	23	1.14
Mali	1.4	1	580	11	3.88
Mauritania	16.3	2	284	3	0.32
Mauritius	0.4	7	1,419	1	57.13
Mexico	77.8	5	96,949	97	0.25

(Continued)

Table 7H.92 (Continued)

Country	Total Annual Freshwater Withdrawal (bill m)[3a]	% for Industry[b]	Industrial Value Added (IVA) (mi US$)[c]	Population (mi)[d]	IVA/Industrial Annual Withdrawal (US$/m^3/Capita)[e]
Moldova	3.0	65	508	4	0.07
Mongolia	0.4	27	362	2	1.56
Morocco	11.1	3	12,558	28	1.40
Mozambique	0.6	2	1,020	17	4.92
Namibia	0.3	3	971	2	57.12
The Netherlands	7.8	68	116,700	16	1.37
New Zealand	2.0	13	15,683	4	15.08
Nicaragua	1.3	2	538	5	3.97
Niger	0.5	2	376	10	3.76
Nigeria	4.0	15	14,918	124	0.20
Norway	2.0	68	47,599	4	8.61
Pakistan	155.6	2	14,685	135	0.04
Panama	1.6	2	1,561	3	19.84
Papua New Guinea	0.1	22	1,779	5	16.17
Paraguay	0.4	7	2,334	5	15.51
Peru	19.0	7	20,714	25	0.61
Philippines	55.4	4	26,364	74	0.16
Poland	12.1	67	47,846	39	0.15
Russian Federation	77.1	62	97,800	146	0.01
Rwanda	0.8	1	356	8	4.13
Senegal	1.5	3	1,235	9	3.05
Sierra Leone	0.4	4	170	5	2.30
Slovak Republic	1.4	50	7,036	5	2.01
Slovenia	0.5	50	7,337	2	14.67
South Africa	13.3	11	49,363	42	0.81
Sri Lanka	9.8	2	3,862	19	1.04
Sweden	2.7	30	74,703	9	10.07
Tanzania, United Republic of	1.2	2	928	33	1.15
Thailand	33.1	4	64,800	60	0.81
Togo	0.1	13	309	5	5.21
Tunisia	2.8	2	6,297	9	13.01
Turkey	35.5	11	51,575	64	0.20
Turkmenistan	23.8	1	2,957	5	2.49
Uganda	0.2	8	1,191	21	3.55
Ukraine	26.0	52	17,854	50	0.03
United Kingdom	9.3	8	330,097	60	7.10
Uruguay	4.2	3	5,703	3	15.61
Uzbekistan	58.0	2	4,340	24	0.16
Venezuela	4.1	10	30,083	24	3.12
Vietnam	54.3	10	9,052	78	0.02
Yemen	2.9	1	1,683	17	3.07
Zambia	1.7	7	996	10	0.84
Zimbabwe	1.2	7	2,005	12	1.96

Note: The industrial water productivity shows the economic value (US$) obtained annually by industry per cubic meter of water used. Very high differences can be noted, between high-income countries such as the United Kingdom, showing a per capita industrial water efficiency of US$ 7.10/m^3, and many low-income countries, such as Moldova, with only US$ 0.07/m^3. Observe, however, that countries having small populations or highly specialized industries (high-value gems tourism)—such as Gabon, Namibia, or Mauritius—have also achieved high productivity.

[a] Data refer to any year from 1980 to 1999.
[b] Withdrawal shares are mostly estimated for 1987.
[c] US constant dollar 1995, data for 1999.
[d] Data estimated for 1999.
[e] Population is expressed in millions, U.S. constant dollar 1995.

Source: From *Water for People Water for Life*, The United Nations World Water Development Report. Copyright © United Nations Educational, Scientific, and Cultural Organization - World Water Assessment Programme (UNESCO-WWAP), 2003. Reproduced by permission of UNESCO. www.unesco.org.

Original Source: From World Bank, 2001, World Development Indicators (WDI), Washington, DC, Available on CD-ROM, Copyright © International Bank for Reconstruction and Development/The World Bank, www.worldbank.org. Reprinted with permission.

Table 7H.93 Water Requirements for Selected Industries in the World

Industry, Product, and Country	Unit of Product (Ton, Except as Specified)	Water Required per Unit (L)
Food Products		
Bread or pastry, Belgium		1,100
Bread, United States		2,100–4,200
Bread, Cyprus[a]		600
Canned food		
Belgium		
Fish, canned		400
Fish, preserved		1,500
Fruit		15,000
Vegetables		8,000–80,000
Canada		
Fruits and vegetables[a]		10,000–50,000
Cyprus		
Citrus/tomato juice[a]		2,800
Grapefruit sections[a]		16,000
Peaches/pears[a]		10,000–15,000
Grapes[a]		30,000
Tomatoes, whole[a]		2,000
Tomato paste[a]		21,000
Peas[a]		10,000
Carrots[a]		16,000
Spinach[a]		30,000
Israel		
Citrus fruits[a]	Ton of raw citrus	4,000
Vegetables[a]		10,000–15,000
United States		
Apricots		21,200
Asparagus		20,500
Beans, green		9,300
Beans, lima		69,800
Beets, corn and peas		7,000
Grapefruit juice		2,800
Grapefruit sections		15,600
Peaches and pears		18,100
Pork and beans		9,300
Pumpkin and squash		7,000
Sauerkraut		950
Spinach		49,400
Succotash		34,800
Tomato products		20,500
Tomatoes, whole		2,200
Industry average, fruits, vegetables, and juices (1965)[a]		24,000
Meat		
Meat freezing, Cyprus[a]	Ton of carcass	500
Meat freezing, New Zealand		3,000–8,600
Meat packing, United States[a]	Ton of prepared meat	23,000
Meat packing, Canada[a]	Ton of carcass	8,800–34,000
Meat products, Belgium	Ton of prepared meat	200
Sausage factory, Finland		20,000–35,000
Sausage factory, Cyprus[a]		25,000
Slaughtering, Finland	Ton, live weight	4,000–9,000
Slaughtering, Cyprus[a]	Ton of carcass	10,000
Meat preserving, Israel[a]	Ton of prepared meat	10,000
Fish		
Fresh and frozen fish, Canada[a]		30,000–300,000
Canned fish, Canada[a]		58,000
Canning and preserving fish, Israel[a]	Ton of raw fish	16,000–20,000
Poultry		
Poultry, Canada[a]	Ton	6,000–43,000
Chickens, Israel[a]	Ton of dressed chicken	33,000
Chickens, United States[a]	Per bird	25
Turkeys, United States[a]	Per bird	75
Milk and Milk Products		
Butter		
New Zealand[a]		20,000

(Continued)

Table 7H.93 (Continued)

Industry, Product, and Country	Unit of Product (Ton, Except as Specified)	Water Required per Unit (L)
Cheese		
Cyprus[a]		10,000
New Zealand[a]		2,000
United States[a]		27,500
Milk		
Belgium	1,000 L	7,000
Finland		2,000–5,000
Israel[a]		2,700
Sweden		2,000–4,000
United States[a]		3,000
Milk powder		
New Zealand[a]		45,000
South Africa		200,000
Whey, United States[a]		10,000
Dairy products, general, Canada[a]		12,200
Ice cream, United States[a]		10,000
Yogurt, Cyprus[a]		20,000
Sugar		
Denmark[a]	Ton of sugar beets	4,800–15,800
Finland	Ton of sugar beets	10,000–20,000
France[a]	Ton of sugar beets	10,900
Germany, Federal[a]	Ton of sugar beets	10,400–14,000
Great Britain[a]	Ton of sugar beets	14,900
Israel[a]	Ton of sugar beets	1,800
Italy[a]	Ton of sugar beets	10,500–12,500
Republic of China[a]	Ton of sugar cane	15,000
United States[a]	Ton of sugar beets (range)	3,200–8,300
United States[a]	Ton of sugar beets (average)	6,000
Beverages		
Beer		
Belgium	Kiloliter	7,000–20,000
Canada[a]	Kiloliter	10,000–20,000
Cyprus[a]	Kiloliter (incl. cleaning bottles)	22,000–30,000
Finland	Kiloliter	10,000–20,000
France[a]	Kiloliter	14,500
Israel[a]	Kiloliter	13,500
United Kingdom[a]	Kiloliter	6,000–10,000
United States	Kiloliter	15,200
Whiskey, United States[a]	Kiloliter of proof spirit	2,600–76,000
Distilled spirits, Israel[a]	Kiloliter	30,000
Wine, France[a]	Kiloliter	2,900
Wine, Israel[a]	Kiloliter	500
Miscellaneous Food Products		
Chocolate confectionery, Belgium		15,000–17,000
Gelatin (edible), United States		55,100–83,500
Maize (wet milling), United States	Liter of maize	15.0–25.5
Maize syrup, United States	Liter of maize	3.8–4.3
Wheat milling, Cyprus[a]		2,000
Wheat milling, Israel[a]		700–1,300
Potato flour, Finland	Ton of potatoes	10,000–20,000
Potato starch, Canada[a]	Ton of starch	80,000–150,000
Macaroni, Cyprus[a]		1,200
Molasses, Belgium	Hectoliter of raw material	1,000–12,000
Molasses, United States	Hectoliter of 100 proof	840
Pulp and Paper		
Groundwood pulp		
Finland	Ton of wood-pulp	30,000–40,000
Sulphate pulp		
China, Republic of[a]	Ton of bleached pulp	340,000
China, Republic of[a]	Ton of unbleached pulp	230,000
Finland	Ton of pulp	250,000–350,000
Sweden[a]	Ton of unbleached pulp	75,000–300,000
Sweden[a]	Ton of bleached pulp	170,000–500,000
Sulphite pulp		

(Continued)

Table 7H.93 (Continued)

Industry, Product, and Country	Unit of Product (Ton, Except as Specified)	Water Required per Unit (L)
Finland	Ton of bleached pulp	450,000–500,000
Finland	Ton of unbleached pulp	250,000–300,000
Sweden[a]	Ton of bleached pulp	300,000–700,000
Sweden[a]	Ton of unbleached pulp	140,000–500,000
Wood pulp		
Sweden[a]	Ton of dry pulp	50,000–100,000
South Africa		150,000
Blotting paper, Sweden		350,000–400,000
Craft, printing and fine paper, Finland		375,000
Printing paper, Republic of China[a]		340,000
Newsprint, Republic of China[a]		190,000
Newsprint, Canada[a]		165,000–200,000
Fine paper, Republic of China[a]		800,000
Fine paper, Sweden		900,000–1,000,000
Newsprint paper, Sweden		200,000
Packing and cartridge paper, Sweden		125,000
Press paper, Finland		200,000
Printing paper, Sweden		500,000
Cardboard, Finland		125,000
Paperboard, United States		62,000–376,000
Paper and cardboard, Belgium		180,000
Strawboard, United States		109,000
Wallboard, Finland		125,000
Wallboard, Sweden[a]		50,000
Industry average, United States[a]	Ton of pulp and paper	236,000
Industry average, United Kingdom[a]	Ton of paper and board	90,000[b]
Industry average, France[a]	Ton of pulp and paper	150,000
Petroleum and Synthetic Fuels		
Aviation gasoline, United States	Kiloliter	25,000
Aviation gasoline, Republic of China[a]	Kiloliter	25,000
Gasoline, United States	Kiloliter	7,000–10,000
Gasoline, Republic of China[a]	Kiloliter	8,000
Gasoline, polymerization, United States	Kiloliter	34,000
Kerosene, Belgium	Ton	40,000
Synthetic gasoline, United States	Kiloliter	377,000
Oilfields, United States	Kiloliter of crude petroleum	4,000
Oil refineries		
Belgium[a]		
China, Republic of[a]	Ton of crude petroleum	30,500
Sweden	Ton of crude petroleum	10,000
United States[a]		
Synthetic fuel		
From coal		
South Africa		50,100
United States	Kiloliter	265,500
From natural gas, United States	Kiloliter	88,900
From shale, United States	Kiloliter	20,800
Chemicals		
Acetic acid, United States		417,000–1,000,000
Alcohol, 100 proof, United States	Liter	138
Alcohol, 190 proof, United States	Liter	52–100
Alumina (Bayer process), United States		26,300
Ammonia, synthetic, United States	Ton of liquid NH_3	129,000
Ammonia (Naphtha, reforming), Japan[a]		255,000
Ammonium nitrate, Belgium		52,000
Ammonium sulphate, United States		835,000
Calcium carbide, United States		125,000
Calcium metaphosphate, United States		16,700
Carbon dioxide, United States		83,500
Caustic soda and chlorine, Canada[a]		125,000
Caustic soda (Solvay process), United States[a]		60,500
Caustic soda (Dual process), Federal Republic of Germany[a]		160,000
Caustic soda (Dual process), Republic of China[a]		200,000

(Continued)

Table 7H.93 (Continued)

Industry, Product, and Country	Unit of Product (Ton, Except as Specified)	Water Required per Unit (L)
Caustic soda (Solvay process), Republic of China[a]		150,000
Cellulose nitrate, United States		41,700
Charcoal and wood chemicals, United States	Ton of crude $CaAc_2$	271,000
Chlorine, Federal Republic of Germany[a]		12,600
Ethylene, Israel[a]		16,000
Gases, compressed and liquified, Canada[a]	Cubic meter	60–70
Glycerine, United States		4,600
Gunpowder, United States		401,000–835,000
Hydrochloric acid (salt process), United States	Ton of 20 Be HCl	12,100
Hydrochloric acid (synthetic process), United States	Ton of 20 Be HCl	2,000–4,200
Hydrogen, United States		2,750,000
Lactose, United States		835,000–918,000
Magnesium carbonate, basic	Ton of basic $MgCO_3$	18,000
United States	Ton of $MgCO_3$	163,000
Oxygen, United States	Cubic meter of O_2	243
Polyethylene, Federal Republic of Germany[a]		231,000 (incl. 225,000 cooling water)
Polyethylene, Israel[a]		8,400
Potassium chloride (sylvinite), United States		167,000–209,000
Smokeless powder, United States		209,000
Soap, Belgium		37,000
Soap, Cyprus[a]		4,500
Soap (laundry), United States		960–2,100
Soda ash (ammonia soda process), 58 percent, United States		62,600–75,100
Sodium chlorate, United States		250,000
Sodium silicate, United States	Ton of 40 Be water-glass	670
Stearine, soap and washing agents, Sweden	Ton of fat	70,000–200,000
Sulfuric acid, Belgium		20,000–25,000
Sulfuric acid (chamber process), United States	Ton of 100 percent H_2SO_4	10,400
Sulfuric acid (contact process), United States	Ton of 100 percent H_2SO_4	2,700–20,300
Sulfuric acid, Federal Republic of Germany[a]	Ton of SO_3	83,500
Textiles		
Steeping, dressing, scouring and bleaching		
Steeping flax, Belgium		30,000–40,000
Dressing flax, Sweden		30,000–40,000
Scouring wool, Belgium		240,000–250,000
Washing wool, Sweden		10,000
Bleaching textiles, Belgium		180,000
Dyeing		
Textiles, Belgium		200,000
Textiles, France (range)[a]		52,000–560,000
Textiles, France (average)[b]		180,000
Finishing		
Wet finishing of textiles, Belgium		100,000–150,000
Dyeing and finishing		
Cotton yarn, Israel[a]		60,000–180,000
Synthetic yarn, Israel[a]		90,000–180,000
Wool yarn, Israel		70,000–140,000
Fabrics, Israel[a]		60,000–100,000
Mills		
Cotton		
Finland		50,000–150,000
Sweden		10,000–250,000
Canada[a]	Square yard	1.0
Wool		
Finland	Ton of cloth or yarn	150,000–350,000
Sweden	Ton of wool	400,000
Synthetic fibers		
Artificial silk, Sweden		2,000,000
Rayon		
Belgium		2,000,000
Finland		1,000,000–2,000,000
Rayon staple, Belgium		550,000

(Continued)

Table 7H.93 (Continued)

Industry, Product, and Country	Unit of Product (Ton, Except as Specified)	Water Required per Unit (L)
Industrial duck products, Canada[a]		22,000
Carpets, Canada[a]	Square yard	20
Mining and Quarrying		
Gold, South Africa	Ton of ore	1,000
Iron ore (brown), United States		4,200
Bauxite, United States[a]	Ton of ore	300
Sulfur, United States		12,500
Copper, Finland		3,750
Copper, Israel[a]		3,100
Gravel, Israel[a]		400
Limestone and by-products, Belgium		200–6,500
Iron and Steel Products		
Belgium		
Blast furnace, no recycling		58,000–73,000
Blast furnace, with recycling		50,000
Finished and semi-finished steel, no recycling[a]		61,000
Finished and semi-finished steel, with recycling		27,000
Canada		
Pig iron[b]		130,000
Open hearth steel[a]		22,000
France		
Smelting[a]		46,000
Martin process (Open hearth)[a]		15,000
Thomas process (Bessemer converter)[a]		10,000
Electric furnace steel[a]		40,000
Rolling mills[a]		30,000
Germany, Federal Republic		
Steel works[a]		8,000–12,000
South Africa		
Steel		12,500
Sweden		
Iron and steel works		10,000–30,000
United States (average)		
Fully integrated mills[a]		86,000
Rolling and drawing mills[a]		14,700
Blast furnace smelting[a]		103,000
Electrometallurgical ferroalloys[a]		72,000
Industry, consumptive use (est.)[a]		3,800
Miscellaneous Products		
Automobiles, United States[a]	Vehicle	38,000
Boilers, steam, United States	Horsepower-hour	15
Casein, New Zealand[a]		55,000
Cement, Portland		
Belgium[a]		1,900
Cyprus (dry process)[a]		550
Finland		2,500
United States (wet process)[a]		900
Ceramics and tiles, Belgium		1,800–2,000
Coal:[c]		
Ruhr (Fed. Rep. Of Germany)[a]		1,000 (min.)–1,750 (avg.)
Great Britain[a]		<3,000
Netherlands[a]		2,650
Coal, Belgium		5,000–6,000
Coal, coke and by-product coke, United States		6,300–15,000
Coal washing, United States		840
Condensers, surface, United States	Pound of condensed steam	9.1–27.3
Distilling, grain		
Belgium	Hectoliter of grain treated	6,000–7,000
United States	Hectoliter of grain treated	6,450
Distilling, Sweden	Kiloliter of 100 percent alcohol	15,000–100,000
Electric power (conventional thermal)		
Sweden	Ton of coal	200,000–400,000
South Africa	Kilowatt-hour (consumptive use)	5
United States[a]	Kilowatt-hour	200

(Continued)

Table 7H.93 (Continued)

Industry, Product, and Country	Unit of Product (Ton, Except as Specified)	Water Required per Unit (L)
Republic of China[a]	Kilowatt-hour	230
Explosives		
Sweden		800,000
United States		835,000
Fertilizer plant, Finland	Ton of saltpeter (25 percent nitrogen)	270,000
Glass, Belgium		68,000
Laundry		
Cyprus[a]	Ton of washed goods	45,000
Finland	Ton of washed goods	20,000
Sweden	Ton of washed goods	30,000–50,000
Leather, South Africa		50,100
Leather factory, Finland	Ton of hides	50,000–125,000
Leather tanning, United States[a]	Sq. meter of hide	20–2,550 (range)
Leather tanning, United States[a]	Sq. meter of hide	440 (average)
Leather tanning, Cyprus[a]	Sq. meter of small animal skins	110
Non-ferrous metals, raw and semi-finished, Belgium		80,000
Rock wool, United States		16,700–20,900
Rubber, synthetic, United States		
Butadiene		83,500–2,750,000
Buna S		125,000–2,630,000
Grade GR-S		117,000–2,800,000
Starch		
Belgium	Ton of maize	13,000–18,000
Sweden	Ton of potatoes	10,000

Note: Water requirements for unit of product produced. Other figures based on older data (pre-1950).

[a] Figures based on newer data (post-1960).
[b] Does not include cooling water for power generating plants.
[c] Includes generation of electricity. If this is not included, the quantities above are reduced by about one-half.

Source: From Dept. of Economic and Social Affairs, United Nations, 1969.

SECTION 7I IRRIGATION — UNITED STATES

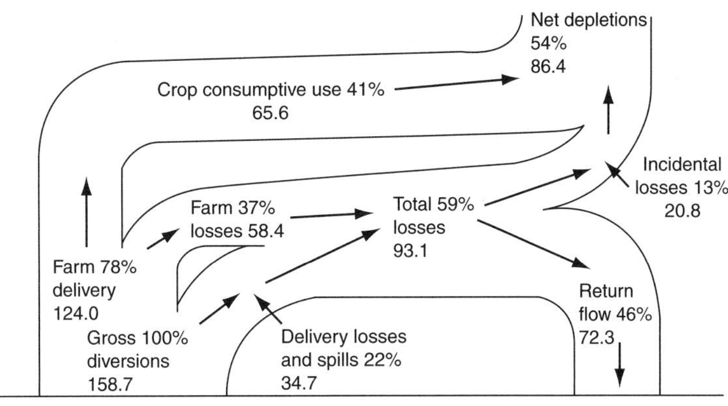

Figure 7I.17 Irrigation water budget of the United States. (From Soil Conservation Service, *America's Soil and Water: Condition and Trends*, 1981.)

Table 7I.94 Acreage Irrigated with On-Farm and Off-Farm Pumped Water in the United States, 1994, 1998, and 2003

	Groundwater			On-Farm Surface Water			Off-Farm Surface Water			Total		
	1994	1998	2003	1994	1998	2003	1994	1998	2003	1994	1998	2003
Northeast												
Connecticut	—	172	290	—	1,699	1,523	—	55	409	—	1,913	2,213
Delaware	—	67,262	64,840	—	11,207	4,803	—	(D)	375	—	77,382	69,088
Maine	—	700	1,480	—	17,331	15,278	—	(D)	1	—	18,324	18,163
Maryland	—	33,427	40,131	—	22,006	13,124	—	401	891	—	55,150	53,734
Massachusetts	—	779	1,289	—	15,696	13,972	—	251	954	—	16,367	16,151
New Hampshire	—	117	162	—	542	592	—	62	64	—	718	818
New Jersey	—	37,406	26,487	—	26,853	21,219	—	996	289	—	63,508	46,679
New York	—	9,762	12,289	—	19,024	33,598	—	2,355	2,712	—	29,176	48,545
Pennsylvania	—	3,469	5,230	—	14,418	13,103	—	199	1,594	—	17,916	19,633
Rhode Island	—	33	52	—	317	567	—	58	29	—	408	648
Vermont	—	42	208	—	537	509	—	—	108	—	579	825
Lake States												
Michigan	184,943	228,137	286,450	116,109	134,858	140,338	7,827	5,598	6,533	305,481	367,992	432,665
Minnesota	292,643	301,550	400,925	24,929	16,382	23,335	9,729	5,448	10,240	326,781	322,346	434,500
Wisconsin	270,362	315,241	345,408	35,196	37,668	45,436	1,928	1,834	1,279	306,096	351,023	391,763
Corn Belt												
Illinois	251,551	280,060	360,740	18,428	12,347	12,905	1,760	289	1,274	271,725	290,825	374,919
Indiana	—	176,427	227,750	—	42,329	48,533	—	960	129	—	217,197	276,294
Iowa	—	62,718	127,128	—	4,109	4,909	—	1,138	2,147	—	67,852	134,164
Missouri	650,580	783,174	961,231	56,363	49,155	55,341	1,744	2,702	4,156	702,183	832,591	1,020,728
Ohio	—	4,718	6,995	—	5,428	5,126	—	2,037	2,355	—	12,037	14,476
Northern Plains												
Kansas	2,446,630	2,577,611	2,430,557	18,945	46,457	75,062	59,251	34,520	60,901	2,501,925	2,650,486	2,543,950
Nebraska	5,408,543	5,069,036	6,974,942	133,254	183,012	121,860	544,246	534,208	443,354	5,979,661	5,692,215	7,516,171
North Dakota	93,983	103,714	152,714	27,321	14,846	25,381	37,272	46,355	30,400	157,426	164,741	207,772
South Dakota	178,827	137,788	233,600	54,117	52,406	39,400	72,576	110,822	117,406	304,454	297,205	390,406
Appalachia												
Kentucky	—	11,005	10,264	—	14,032	10,386	—	526	280	—	25,454	20,685
North Carolina	—	13,209	14,559	—	120,610	85,237	—	1,713	3,109	—	134,468	101,055
Tennessee	—	16,387	25,636	—	6,256	9,119	—	98	574	—	22,741	34,429
Virginia	—	6,411	6,619	—	60,221	28,077	—	692	496	—	65,734	33,635
West Virginia	—	207	247	—	939	532	—	65	22	—	1,211	801
Southeast												
Alabama	—	14,079	24,451	—	33,087	26,994	—	465	1,333	—	46,811	52,722
Florida	746,071	1,032,746	772,907	246,527	274,388	418,190	447,724	340,507	319,598	1,416,019	1,613,719	1,497,653
Georgia	450,029	435,770	518,239	169,433	218,532	194,926	1,000	1,871	56	619,536	647,749	711
South Carolina	—	28,515	31,157	—	32,306	22,017	—	584	232	—	61,015	52,046
Delta States												
Arkansas	2,581,693	3,722,635	3,421,365	271,512	349,574	509,914	13,196	23,485	26,793	2,853,929	4,043,382	3,944,867
Louisiana	557,431	709,207	633,455	240,593	219,653	186,544	34,880	9,810	23,596	820,816	920,823	838,717
Mississippi	624,182	1,078,511	1,115,676	25,361	39,012	46,834	378	1,245	9,213	646,761	1,109,079	1,169,793

Southern Plains												
Oklahoma	392,063	355,414	413,067	34,105	33,189	62,572	48,579	66,743	36,513	474,201	451,788	508,842
Texas	4,319,337	4,576,456	4,174,840	203,619	246,693	184,173	606,841	488,571	588,932	5,100,979	5,237,584	4,947,745
Mountain												
Arizona	340,306	243,313	291,025	26,012	6,412	72,213	492,398	683,637	516,422	752,019	873,589	836,587
Colorado	1,357,765	1,331,615	1,095,601	517,976	559,887	694,262	1,383,255	1,211,561	810,713	2,998,888	2,942,230	2,562,329
Idaho	1,307,688	1,226,924	1,202,870	352,423	388,335	352,666	1,643,919	1,656,908	1,609,148	3,183,733	3,188,406	3,126,857
Montana	136,969	46,606	96,469	691,982	695,480	724,542	1,134,537	1,015,467	1,313,763	1,936,292	1,740,873	2,131,955
Nevada	202,566	223,747	259,721	184,808	296,811	241,713	155,591	191,692	145,684	519,507	694,930	639,310
New Mexico	381,244	425,530	502,021	56,803	49,881	89,790	275,091	281,300	206,658	685,695	720,319	769,787
Utah	196,191	139,693	191,842	162,912	170,750	104,527	741,809	791,945	790,434	1,085,083	1,076,346	1,082,213
Wyoming	78,842	90,730	114,676	556,782	700,511	512,153	774,364	771,299	794,498	1,374,447	1,533,468	1,415,037
Pacific												
California	3,876,870	3,071,740	3,823,115	628,817	745,056	1,045,025	3,910,823	5,120,793	4.166,854	7,245,487	8,139,834	8,471,936
Oregon	501,297	304,579	421,642	564,577	622,453	629,850	682,894	650,475	720,621	1,587,152	1,534,961	1,731,660
Washington	501,662	445,927	484,392	163,723	190,423	292	827,817	963,804	1,041,882	1,434,800	1,554,813	1,806,782
Alaska	—	2,424	2,128	—	191	113	—	3	11	—	2,618	2,252
Hawaii	—	43,996	33,938	—	34,850	17,036	—	46,893	50,999	—	96,543	78,538
Total	**28,816,442**	**32,222,665**	**32,342,820**	**5,926,902**	**7,402,653**	**7,277,527**	**13,919,132**	**16,408,547**	**13,867,438**	**46,418,380**	**54,249,965**	**52,583,431**

Source: Abstracted from: USDA, 2004, Farm and Ranch Irrigation Survey (2003), vol. 3, Special Studies, Part 1,—2002; Census of Agriculture; USDA, 1999, 1998 Farm and Ranch Irrigation Survey, Vol. 3, Special Studies, Part 1—1997 Census of Agriculture and USDA, 1994 Farm and Ranch Irrigation Survey—1992, Census of Agriculture; (D) Withheld to avoid disclosing data for individual farms, www.nass.usda.gov.

Table 7I.95 Growth of Irrigated Farmland in the United States, 1889–2002

Year	Land in Farms[a] (mil acres)	Irrigated Land in Farms (mil acres)	Share of Farmland Irrigated (percent)	Acreage Irrigated per Farm Irrigated (acres)	Change In Irrigated Acreage (mil acres)	Average Annual Growth in Irrigated Acreage (percent)
1889	623	3.6	0.6	67	NA	NA
1900	839	7.5	0.9	70	3.9	6.9
1910	879	14.4	1.6	89	6.9	6.7
1920	956	19.2	2.0	83	4.8	2.9
1930	987	19.5	2	74	0.3	0.2
1939	1,065	18	1.7	60	−1.5	−0.9
1949	1,161	25.8	2.2	84	7.8	3.7
1959	1,123	33	3	108	7.2	2.5
1969	1,063	39.1	3.7	152	6.0	1.7
1974	1017	41.2	4.1	174	2.1	1.1
1978	1,015	50.3	5	179	9.1	2.8
1982	987	49	4.9	176	−1.3	−0.7
1987	964	46.4	4.8	159	−2.6	−1.1
1992	946	49.4	5.2	177	3.0	1.3
1997	932	55.1	5.9	197	5.7	2.2
Adjusted for coverage[b]						
1997	955	56.3	5.9	182	NA	NA
2002	938	55.3	5.9	184	−1.0	−0.4

Note: NA, Not applicable.

[a] Land in Farms includes agricultural land used for crops, pasture, or grazing. It also includes woodland and wasteland not actually under cultivation or used for pasture or grazing, provided it is part of the farm operator's total operation.

[b] The 2002 Census of Agriculture estimates include an area-frame adjustment for incompleteness of the list frame. Similar estimates were calculated for 1997 for the purpose of comparison with the 2002 Census estimates.

Source: From U.S. Dept. of Commerce and U.S. Dept of Agriculture, Census of Agriculture, www.usda.gov.

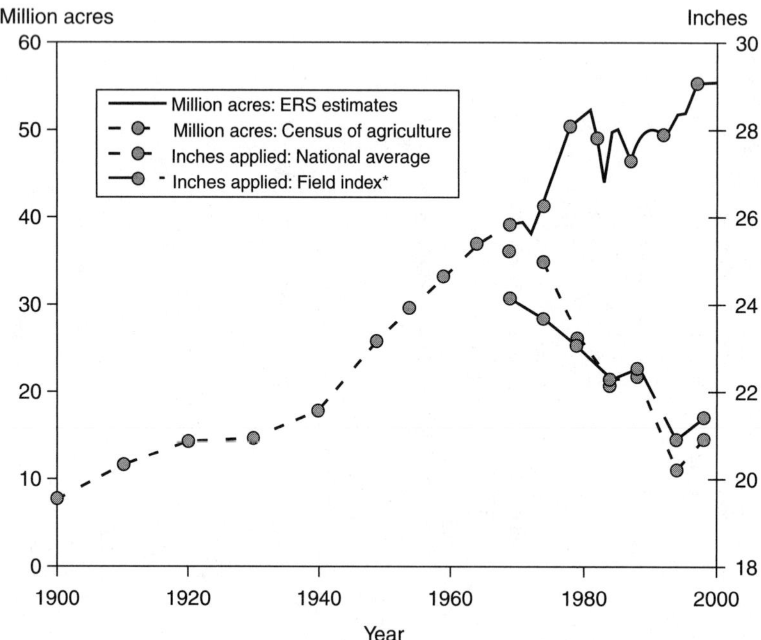

Figure 7I.18 Irrigation trends-in the United States, 1900–2000. (From Heimlich, R., 2003, *Agricultural Resources and Environmental Indicators 2003*, Agriculture Handbook No. (AH722), February 2003.) www.ers.usda.gov.

Table 7I.96 Irrigated Land in Farms, by Region and Crop, 1900–2000

Region	1900[a]	1949[a]	1969[a]	1978[a]	1982[a]	1987[a]	1992[a]	1997[a]	1998[b]	1999[b]	2000[b]
USDA Production Region							1,000 acres				
Atlantic Regions[c]	—	500	1,800	2,900	2,700	3,000	3,200	3,600	3,600	3,500	3,500
North Central[d]	—	—	500	1,400	1,700	2,000	2,500	2,800	2,800	2,900	3,000
Northern Plains	200	1,100	4,600	8,800	9,300	8,700	9,600	10,200	10,100	10,200	10,500
Delta States	200	1,000	1,900	2,700	3,100	3,700	4,500	5,700	6,300	6,000	5,900
Southern Plains	100	3,200	7,400	7,500	6,100	4,700	5,400	6,000	5,900	5,900	5,800
Mountain States	5,300	11,600	12,800	14,800	14,100	13,300	13,300	14,400	14,300	14,400	14,200
Pacific Coast	2,000	8,300	10,000	12,000	11,900	10,800	10,800	12,400	12,200	12,300	12,400
United States[e]	7,800	25,800	39,100	50,300	49,000	46,400	49,400	55,100	55,200	55,300	55,300
Irrigated Crop											
Corn for Grain			3,200	8,700	8,500	8,000	9,700	10,600	10,700	9,900	10,200
Sorghum for grain			3,500	2,000	2,200	1,300	1,600	900	600	800	600
Barley			1,600	2,000	1,900	1,300	1,100	1,100	1,000	1,000	1,000
Wheat			1,900	3,000	4,600	3,700	4,100	4,000	3,700	3,400	3,300
Rice			2,200	3,000	3,200	2,400	3,100	3,100	3,400	3,500	3,100
Soybeans			700	1,300	2,300	2,600	2,500	4,200	4,400	4,800	5,200
Cotton			3,100	4,700	3,400	3,500	3,700	4,900	4,600	4,800	5,300
Alfalfa hay			5,000	5,900	5,500	5,500	5,700	6,000	6,300	6,400	6,300
Other hay			2,900	3,000	3,000	3,100	2,900	3,600	3,400	3,500	3,300
Vegetables			1,500	1,900	1,900	2,000	2,200	2,400	2,500	2,600	2,700
Land in orchards			2,400	3,000	3,300	3,400	3,600	4,100	4,100	4,200	4,300
Other irrigated land in farms			11,100	11,800	9,200	9,500	9,100	10,300	10,500	10,400	10,300

Note: Indicates none or fewer than 5,000 acres.

[a] Census of Agriculture.

[b] Estimates constructed from the Census of Agriculture and other USDA sources.

[c] Northeast, Appalachian, and Southeast farm production regions.

[d] Lake States and Corn Belt production regions.

[e] Includes Alaska and Hawaii.

Source: From Heimlich, R., 2003, *Agricultural Resources and Environmental Indicators 2003*, Agriculture Handbook No. (AH722), February 2003, www.ers.usda.gov.

Original Source: USDA, ERS, based on Census of Agriculture, various years (USDA, 1999a; USDC, 1994; and previous versions); and USDA, ERS data.

Table 7I.97 Acreage and Value of Irrigated Cropland in the United States, 1982

	Acreage			Value				
	Irrigated (1,000 acres)	Share of Crop Irrigated (percent)	Share of Total Cropland[a] (percent)	Irrigated (million dollars)	Irrigated Share of Crop Value (percent)	Irrigated Share of Total Crop Value[b] (percent)	Irrigated Value per Acre (dollars)	Rainfed Value per Acre (dollars)
Corn	9,604	12.3	2.8	3,440	17.2	4.5	358	235
Sorghum	2,295	17.0	0.7	901	53.2	1.2	393	66
Wheat	4,650	6.6	1.4	1,144	16.7	1.4	246	84
Barley and oats	2,098	11.8	0.6	375	20.0	0.5	179	93
Rice	3,233	100.0	1.0	1,226	100.0	1.7	379	NA
Cotton	3,424	35.0	1.0	1,883	58.4	2.5	549	200
Soybeans	2,321	3.6	0.7	491	4.4	0.6	211	17
Irish potatoes	812	64.0	0.2	1,261	81.7	1.7	1,553	546
Hay	8,507	15.0	2.5	2,275	27.7	3.0	267	120
Vegetables and melons	2,029	60.7	0.6	3,375	79.7	4.5	1,663	591
Orchard crops	3,343	70.4	1.0	4,732	85.1	6.2	1,415	502
Sugar beets	550	53.2	0.2	491	77.7	0.6	893	262
Other crops[c]	2,428	17.9	0.7	2,424	25.4	3.1	998	638
Total[d]	45,289	NA	13.4	24,047	NA	31.8	531[e]	176[e]

Note: NA, Not applicable. By crop; contiguous United States.

[a] This is the share each irrigated crop represents of the total acreage of crops produced in the 48 States in 1982.
[b] This is the share each irrigated crop represents of the total value of crops produced in the 48 States in 1982.
[c] Includes peanuts, tobacco, dry edible beans, and the minor acreage crops rye, flax, sugarcane, and dry edible pears.
[d] Includes about 932,000 acres of double-cropped land. Figures might not add to totals due to rounding.
[e] Average weighted by acreage.

Source: From Day, J.C. and Horner, G.L., 1987, U.S. Irrigation, extent and economic importance, *U.S. Department of Agriculture Information Bulletin 523.*

Table 7I.98 Irrigation Water Use in the United States, 2000

State	Irrigated Land (1000 acres)				Withdrawals (1000 mil gal/day)			Withdrawals (1000 acre-feet/yr)			Application Rate (acre-feet/acre)
	By Type of Irrigation				By Source			By Source			
	Sprinkler	Micro-Irrigation	Surface	Total	Ground-water	Surface Water	Total	Ground-water	Surface Water	Total	
Alabama	68.7	1.3	0	70	14.5	28.7	43.1	16.2	32.2	48.4	0.69
Alaska	2.43	0	0.07	2.5	0.99	0.02	1.01	1.11	0.02	1.13	0.45
Arizona	183	14	779	976	2,750	2,660	5,400	3,080	2,980	6,060	6.21
Arkansas	631	0	3,880	4,510	6,510	1,410	7,910	7,290	1,580	8,870	1.97
California	1,660	3,010	5,470	10,100	11,600	18,900	30,500	13,100	21,100	34,200	3.37
Colorado	1,190	1.16	2,220	3,400	2,160	9,260	11,400	2,420	10,400	12,800	3.76
Connecticut	20.6	0.39	0	21	17	13.4	30.4	19	15	34	1.62
Delaware	81.1	0.71	0	81.8	35.6	7.89	43.5	39.9	8.84	48.7	0.6
District of Columbia	0.32	0	0	0.32	0	0.18	0.18	0	0.2	0.2	0.63
Florida	515	704	839	2,060	2,180	2,110	4,290	2,450	2,370	4,810	2.34
Georgia	1,470	73.8	0	1,540	750	392	1,140	841	439	1,280	0.83
Hawaii	16.7	105	0	122	171	193	364	191	216	407	3.35
Idaho	2,440	4.7	1,300	3,750	3,720	13,300	17,100	4,170	15,000	19,100	5.1
Illinois	365	0	0	365	150	4.25	154	168	4.76	173	0.47
Indiana	250	0	0	250	55.5	45.4	101	62.2	51	113	0.45
Iowa	84.5	0	0	84.5	20.4	1.08	21.5	22.9	1.21	24.1	0.28
Kansas	2,660	2.14	647	3,310	3,430	288	3,710	3,840	323	4,160	1.26
Kentucky	66.6	0	0	66.6	1.14	28.2	29.3	1.28	31.6	32.9	0.49
Louisiana	110	0	830	940	791	232	1,020	887	261	1,150	1.22
Maine	35	0.95	0.03	36	0.61	5.23	5.84	0.68	5.86	6.55	0.18
Maryland	57.3	3.32	0	60.6	29.8	12.6	42.4	33.4	14.1	47.6	0.78
Massachusetts	26.6	2.35	0	29	19.7	106	126	22.1	119	141	4.88
Michigan	401	8.67	4.87	415	128	73.2	201	144	82	226	0.54
Minnesota	546	0	26.9	573	190	36.6	227	213	41.1	254	0.44
Mississippi	455	0	966	1,420	1,310	99.1	1,410	1,470	111	1,580	1.11
Missouri	532	1.43	792	1,330	1,380	48.1	1,430	1,550	53.9	1,600	1.21
Montana	506	0	1,220	1,720	83	7,870	7,950	93	8,820	8,920	5.18
Nebraska	4,110	0	3,710	7,820	7,420	1,370	8,790	8,320	1,540	9,860	1.26
Nevada	192	0	456	647	567	1,540	2,110	635	1,730	2,360	3.65
New Hampshire	6.08	0	0	6.08	0.5	4.25	4.75	0.56	4.76	5.32	0.88
New Jersey	109	15.7	3.7	128	22.8	117	140	25.5	131	156	1.22
New Mexico	461	7.17	530	998	1,230	1,630	2,860	1,380	1,830	3,210	3.22
New York	70	8.73	1.84	80.6	23.3	12.1	35.5	26.1	13.6	39.8	0.49
North Carolina	193	3.7	0	196	65.8	221	287	73.8	248	322	1.64
North Dakota	200	0	26.7	227	72.2	73.2	145	80.9	82.1	163	0.72
Ohio	61	0	0	61	13.9	17.8	31.7	15.6	19.9	35.5	0.58
Oklahoma	392	1.5	113	507	566	151	718	635	170	804	1.59
Oregon	1,160	4.02	1,000	2,170	792	5,290	6,080	887	5,920	6,810	3.14

(Continued)

Table 7I.98 (Continued)

	Irrigated Land (1000 acres)				Withdrawals (1000 mil gal/day)			Withdrawals (1000 acre-feet/yr)			
	By Type of Irrigation				By Source			By Source			
State	**Sprinkler**	**Micro-Irrigation**	**Surface**	**Total**	**Ground-water**	**Surface Water**	**Total**	**Ground-water**	**Surface Water**	**Total**	**Application Rate (acre-feet/acre)**
Pennsylvania	28.9	7.17	0	36	1.38	12.5	13.9	1.55	14	15.6	0.43
Rhode Island	4.48	0.29	0.05	4.82	0.46	2.99	3.45	0.52	3.35	3.87	0.8
South Carolina	166	3.66	17.5	187	106	162	267	118	181	300	1.6
South Dakota	276	0	78.3	354	137	236	373	153	264	418	1.18
Tennessee	51.2	5.35	3.96	60.5	7.33	15.1	22.4	8.22	16.9	25.1	0.41
Texas	4,010	89.4	2,390	6,490	6,500	2,130	8,630	7,290	2,390	9,680	1.49
Utah	526	1.68	880	1,410	469	3,390	3,860	526	3,800	4,330	3.08
Vermont	4.95	0	0	4.95	0.33	3.45	3.78	0.37	3.87	4.24	0.86
Virginia	64.3	13.9	0	78.2	3.57	22.8	26.4	4	25.6	29.6	0.38
Washington	1,270	49.9	252	1,570	747	2,290	3,040	837	2,570	3,400	2.16
West Virginia	2.21	0	0.98	3.19	0.02	0.02	0.04	0.02	0.02	0.04	0.01
Wisconsin	355	0	0	355	195	1.57	196	218	1.76	220	0.62
Wyoming	190	4.73	964	1,160	413	4,090	4,500	463	4,580	5,050	4.36
Puerto Rico	15.5	33	5.35	53.8	36.9	57.5	94.5	41.4	64.5	106	1.97
U.S. Virgin Islands	0.2	0	0	0.2	0.29	0.21	0.5	0.33	0.24	0.56	2.8
Total	**28,300**	**4,180**	**29,400**	61,900	**56,900**	**80,000**	**137,000**	**63,800**	**89,700**	**153,000**	**2.48**

Note: Figures may not sum to totals because of independent rounding.

Source: From Hutson, S.S. et al., 2004, Estimated use of water in the United States in 2000, *U.S. Geological Survey Circular 1268*, www.usgs.gov.

Table 7I.99 Irrigation Withdrawals by Source in the United States, 1950–1995

Year	Irrigation Withdrawals (bgd)	Percent from Surface Water Sources (%)	Percent of Total from Reclaimed (%)
1950	89	77.5	—
1955	110	72.8	0.10
1960	110	63.8	0.68
1965	120	64.3	0.44
1970	130	64.5	0.28
1975	140	59.5	0.26
1980	150	59.2	0.19
1985	137	66.7	0.32
1990	137	62.6	0.38
1995	134	63.3	0.54

Note: bgd, billion gallons per day.

Source: From Dziegielewski, B. et al., 2002, *Analysis of Water Use Trends in the United States: 1950–1995*, Southern Illinois University at Carbondale, Carbondale, IL, February 28, 2002, http://info.geography.siu.edu/geography_info/research/. Reprinted with permission.

Table 7I.100 Total Irrigation Withdrawals and Irrigated Acres in the United States, 1950–1995

Year	Total Irrigation Withdrawals (bgd)	USGS Estimated Total Irrigated Acres (thousands)	Average Depth of Water applied (ft/yr) (USGS Acreage Estimate)	Interpolated US Census Bureau Estimated Total Irrigated Acres (thousands)	Average Depth of Water Applied (ft/yr) (USCB Acreage Estimate)
1950	89	25,000	3.20	26,634	3.00
1955	110	34,000	2.68	30,274	3.01
1960	110	39,000	2.41	33,942	2.77
1965	120	44,000	2.95	37,470	3.47
1970	130	50,000	2.80	39,546	3.54
1975	140	54,000	2.96	43,519	3.68
1980	150	58,000	2.93	49,676	3.42
1985	137	57,275	2.68	47,432	3.24
1990	137	57,400	2.67	48,196	3.17
1995	134	57,900	2.59	52,796	2.84

Source: From Dziegielewski, B. et al., 2002, *Analysis of Water Use Trends in the United States: 1950–1995*, Southern Illinois University at Carbondale, Carbondale, IL, February 28, 2002, http://info.geography.siu.edu/geography_info/research/. Reprinted with permission.

Table 7I.101 Changes in Irrigation Withdrawals in Selected States, 1960–1995 and 1980–1995

State	1960–1995 Change (mgd)	1980–1995 Change (mgd)
California	10,894	−8,106
Nebraska	5,350	−1,750
Arkansas	5,016	836
Colorado	3,735	−1,265
Wyoming	3,495	1,695
Montana	3,446	−2,454
Florida	2,809	469
Washington	2,769	69
Idaho	2,048	−2,952
Kansas	1,583	−2,217
Oregon	1,368	268
Mississippi	1,232	762
New Mexico	1,193	−607
Arizona	972	−1,428
Lower 48 States	**49,960**	**−17,589**

Note: mgd, million gallons per day.

Source: From Dziegielewski, B. et al., 2002, *Analysis of Water Use Trends in the United States: 1950–1995*, Southern Illinois University at Carbondale, Carbondale, IL, February 28, 2002, http://info.geography.siu.edu/geography_info/research/. Reprinted with permission.

Table 7I.102 Number of Farms Supplied with Off-Farm Water from the Bureau of Reclamation and Other Federal Agencies, 2003

	Farms Using Any Off-Farm Water						Farms Using Only Off-Farm Water					
				Supplier of Off-Farm Water[a] (Number of Farms)						Supplier of Off-Farm Water[a] (Number of Farms)		
Geographic Area	Farms	Acres Irrigated with Off-Farm Water	Quantity of Off-Farm Water (acre-ft)	Bureau of Reclamation	Other Federal Agencies	All Other Suppliers	Farms	Acres Irrigated with Off-Farm Water	Quantity of Off-Farm Water (acre-ft)	Bureau of Reclamation	Other Federal Agencies	All Other Suppliers
Corn Belt												
Ohio	203	2,355	1,234	16	1	36	198	637	258	16	—	33
Northern Plains												
Kansas	315	60,901	56,658	114	16	43	204	30,554	26,836	62		
Nebraska	1,736	443,354	488,295	1,268	19	451	493	197,822	193,616	277	—	173
North Dakota	90	30,400	46,896	46	22	16	79	26,632	43,352	36	20	14
South Dakota	491	117,406	740,110	302	40	90	465	109,780	732,712	282	38	84
Southern Plains												
Oklahoma	64	36,513	40,668	46	2	9	55	27,481	29,431	44		5
Texas	4,208	588,932	916,727	1,474	33	1,342	3,911	501,965	711,220	1,423	33	1,190
Mountain												
Arizona	1,726	516,422	2,332,795	619	91	568	1,601	397,087	1,835,249	551	66	532
Colorado	6,476	810,713	1,365,054	1,684	99	3,004	6,034	676,181	1,172,001	1,567	61	2,752
Idaho	11,059	1,609,148	3,360,210	4,884	520	3,193	10,123	1,159,032	2,510,827	4,425	394	2,924
Montana	4,861	1,313,763	1,855,914	1,886	419	821	4,637	1,211,250	1,696,914	1,772	406	762
Nevada	623	145,684	366,463	249	—	168	576	97,238	256,905	233	—	146
New Mexico	2,944	206,658	492,319	1,692	4	776	2,670	178,546	433,484	1,490	4	707
Utah	8,429	790,434	1,604,167	1,381	549	4,596	7,856	704,474	1,410,309	1,339	393	4,379
Wyoming	2,999	794,498	1,329,559	1,075	121	1,031	2,876	743,246	1,248,919	1,041	117	971
Pacific												
California	22,558	4,166,854	12,189,829	6,737	1,541	10,491	17,810	2,383,051	7,563,592	4,802	1,215	8,625
Oregon	5,615	720,621	1,450,562	2,552	173	1,602	5,306	548,155	1,074,139	2,350	151	1,510
Washington	8,271	1,041,882	2,464,111	5,185	139	1,692	7,517	842,209	2,024,760	4,584	69	1,601
Other States	3,737	470,900	536,895	—	43	1,326	3,365	335,129	318,623	—	31	2,371
U.S. Total (all states)	86,405	13,867,438	31,638,466	31,210	3,832	31,255	75,776	10,170,469	23,283,147	26,294	2,998	27,589

[a] Counts only include those reporting some or all of their water from a given source. Those reporting an unknown water supplier are excluded.

Source: Abstracted from USDA, 2004, *Farm and Ranch Irrigation Survey (2003)*, vol. 3, Special Studies, Part 1, 2002; Census of Agriculture, www.nass.usda.gov.

Table 7I.103 Irrigation in Areas of the United States with Declining Groundwater Supplies

State	Total Groundwater Irrigation		Irrigated Area with Declining Water[a]		Percent of Irrigated Land Experiencing Declining Water Levels in 1983
	1977	1983	1977	1983	
		1,000 acres			
Arizona	940	938	—[b]	606	65
Arkansas	1,400	2,337	—[b]	425	18
California	4,388	4,265	—[b]	2,068	48
Colorado	1,650	1,660	570	590	36
Florida	1,076	1,610	—[b]	250	16
Idaho	1,149	1,450	—[b]	223	15
Kansas	3,083	3,504	1,950	2,180	62
Nebraska	5,855	7,025	1,842	2,029[c]	29
New Mexico	760	805	560	560	70
Oklahoma	730	645	507	523	81
Texas	7,846	6,685	6,425	4,565[c]	73
Total	**28,877**	**30,924**	—[a]	**14,029**	**45**

Note: In 11 states with major groundwater irrigation. Total groundwater area irrigated was estimated for 1977 and 1983. Decline area irrigated was estimated from data for the latest year available.

[a] Only areas experiencing at least a six-inch average annual decline are included in these estimates.
[b] Data insufficient to make time comparisons.
[c] Data for 1984.

Source: From Ogg, C.W., Hostetter, J.E., and Lee, D.J., 1988, Expanding the conservation reserve to achieve multiple environmental goals, *J. of Soil and Water Conservation*. Copyright 1988. Soil and Water Conservation Society. Reprinted with permission.

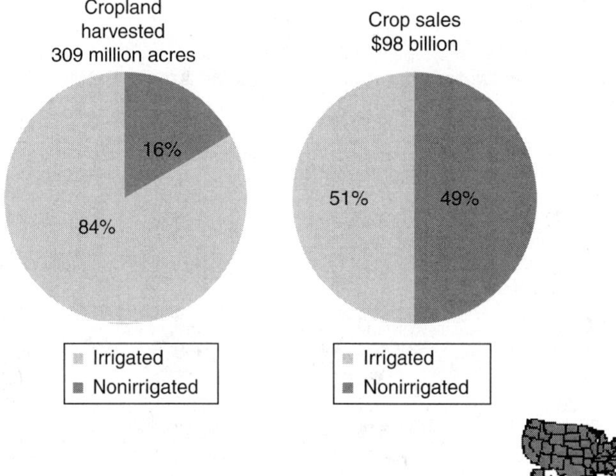

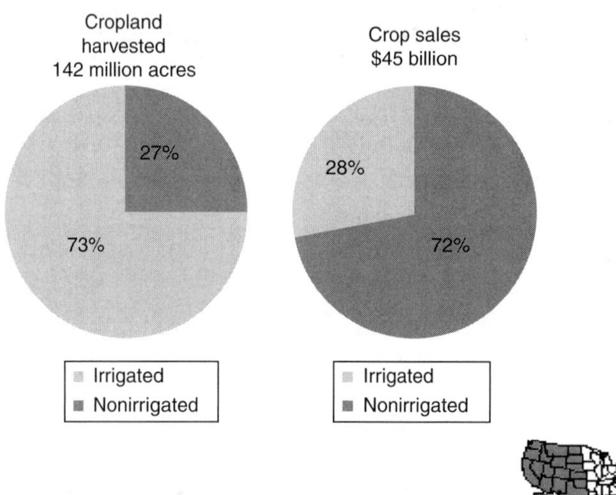

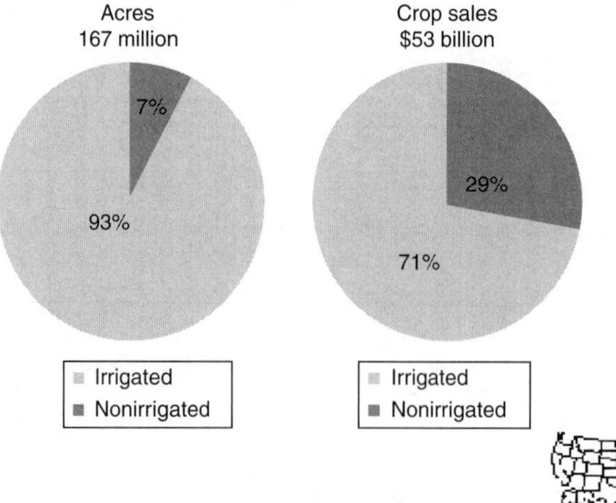

Figure 7I.19 Percent of United States harvested acres and crop sales irrigated, 1997. (From United States Department of Agriculture, Economic Research Service, www.ers.usda.gov, from 1997 census of Agriculture Data.)

Table 7I.104 Top 10 States in Irrigated Agriculture in the United States

Item	Unit	Ranking by State									
		1	2	3	4	5	6	7	8	9	10
Number of farms in 2002											
All farms	1000	TX	MO	IA	TN	KY	OK	CA	MN	OH	WI
		230	107	92.5	90	89	87	84	79	78	77
Irrigated farms 2003	do	CA	NE	TX	ID	OR	WA	CO	UT	MT	FL
		46.8	16.3	15.4	14.3	14.2	12.9	11.6	10.1	8.6	8.3
Percentage of farms with irrigated land in 2003	Percent	UT	ID	NV	CA	WY	NM	CO	AZ	OR	WA
		67.1	59.6	57.4	55.8	52.4	44.9	38.6	38.0	34.6	33.0
Irrigated land in 2003	Mil acre	CA	NE	TX	AR	ID	CO	KS	MT	WA	OR
		8.5	7.5	4.9	3.9	3.1	2.6	2.5	2.1	1.8	1.7
Percentage of harvested cropland irrigated in 2003	Percent	AZ	NV	CA	NM	FL	ID	WY	WA	UT	HI
		100	96.8	96.6	94.3	93.1	92.1	90.3	88.8	88.0	85.3
Market value of crops sold in 2002											
All farms	$bil	CA	IA	IL	FL	MN	TX	WA	NE	IN	ND
		19.2	6.1	5.9	5.0	4.6	3.7	3.6	3.4	3.0	2.5
Major irrigated crops in 2003											
Corn (all)	1000 acres	NE	KS	TX	CA	CO	MO	IL	MI	AR	MN
		4684	1298	831	665	633	307	223	200	195	187
Wheat	do	TX	ID	KS	CA	WA	CO	OR	OK	AZ	MT
		657	501	415	407	278	156	143	132	122	119
Hay (all)	do	CA	CO	MT	ID	WY	OR	UT	WA	NE	NV
		1353	1102	1098	862	833	741	623	510	483	455
Energy expense for on-farm pumping of irrigation water in 2003											
Total	$mil	CA	TX	NE	KS	ID	WA	AR	CO	NM	OR
		378.0	246.8	219.2	105.4	102.8	70.7	68.7	58.8	34.8	32.8
Electricity	do	CA	TX	ID	NE	WA	CO	OR	NM	AR	AZ
		304.4	105.0	100.3	83.1	69.9	52.4	31.6	29.0	24.6	22.7
Natural gas	do	TX	KS	NE	CA	OK	AZ	NM	CO	AR	LA
		123.5	70.0	47.8	15.3	8.2	5.3	3.8	3.7	2.1	0.4
Acres irrigated by source of water in 2003											
Wells	Thous. acre	NE	TX	CA	AR	KS	ID	MS	CO	MO	FL
		6975	4175	3823	3421	2431	1203	1116	1096	961	773
On-farm surface	do	CA	MT	CO	OR	WY	AR	FL	ID	WA	NV
		1045	725	694	630	512	510	418	353	292	242
Off-farm suppliers	do	CA	ID	MT	WA	CO	WY	UT	OR	TX	AZ
		4167	1609	1314	1042	811	794	790	721	589	516
Water applied in irrigation by source in 2003											
Total	Mil acre-feet	CA	NE	ID	TX	AR	CO	WA	AZ	OR	KS
		24.8	8.5	6.1	6.0	4.3	4.0	3.9	3.8	3.3	3.1

(Continued)

Table 7I.104 (Continued)

Item	Unit	Ranking by State 1	2	3	4	5	6	7	8	9	10
Wells	do	CA	NE	TX	AR	KS	ID	CO	NM	AZ	MS
		9.7	7.9	4.8	3.5	3.0	2.1	1.5	1.2	1.2	0.9
On-farm sources	do	CA	OR	CO	MT	AR	WY	FL	ID	WA	NV
		3.0	1.1	1.1	0.8	0.7	0.7	0.7	0.6	0.6	0.5
Off-farm suppliers	do	CA	ID	WA	AZ	MT	UT	OR	CO	WY	TX
		12.2	3.4	2.5	2.3	1.9	1.6	1.5	1.4	1.3	0.9
Irrigation water applied per acre in 2003	Acre-feet	MA	AZ	HI	CA	NV	SD	NM	WA	UT	ID & OR
		5.6	4.5	4.0	2.9	2.5	2.5	2.4	2.2	2.1	1.9
Average land value per acre in 2003											
Irrigated land	$ per acre	CA	AZ	FL	UT	WA	NM	OR	ID	MO	CO
		6300	6000	4180	3500	3200	2650	2250	2200	2150	2000
Nonirrigated cropland	do	FL	GA	CA	MO	LA	OR	MS	AR	NE	WA
		2670	2200	2000	1540	1280	1200	1140	1100	980	950

Note: In selected categories.

Source: Abstracted from: USDA, 2004, *Farm and Ranch Irrigation Survey (2003)*, vol. 3, Special Studies, Part 1, 2002; Census of Agriculture; USDA, Farms and Land in Farms, February 2003; USDA, Census of Agriculture State Data, State Summary Highlights 2002, USDA-NASS, Agricultural Statistics, 2003, www.nass.usda.gov.

Table 7I.105 Standards for Classification of Lands as Irrigable

Land Characteristics	Minimum Requirements
Soils	
Texture	Loamy sand to permeably clay
Depth	
To sand, gravel, or cobble	18 in. of good free-working soil of the fine sandy loam or heavier, or from 24–30 in. of lighter textured soil
To bedrock	At least 18 inches over shattered bedrock or tilted shale bedrock; or 24 in. over massive bedrock or hardpan
Topography	
Slopes	Smooth slopes up to no more than 30 percent in general gradient in reasonably large-sized bodies sloping in the same plane; or undulating slopes which are less than 20 percent in general gradient
Rock cover	No more than enough loose rock and rock outcroppings to moderately reduce productivity and interfere with cultural practices. Varies with soil depth and topographic conditions
Erosion	No more than moderate erosion, with very few gullies which are not crossable by tillage implements
Drainage	
Soil and topography	Such that moderate farm drainage may be required, but without excessive cost
Salinity	Total salts in the soil solution do not exceed 0.5 percent, except in readily drained soils where reclamation appears feasible
Alkalinity	The pH value is 9.0 or less, unless the soil is calcareous in which case higher values may be allowed. If there is evidence of black alkali a lower pH value may be limiting

Source: From California State Water Resources Board, 1955.

Table 7I.106 Summary of Losses and Waste of Irrigation Water

During Delivery	
Average field evaporation before topsoil dries	0.5 in. per irrigation
Surface waste, allowance for large projects	10 per cent of diversions
Seasonal percolation losses, except on porous soils	0–1.5 acre-ft per acre
Losses of flow in farm ditches	5–50 per cent per mile
Deliveries to farms	1–7 acre-ft per acre
Consumptive use, diversified crops	1–3.5 acre-ft per acre
Irrigation efficiencies, common farm crops	20–50 per cent
Irrigation efficiencies, fruit, and special crops	35–70 per cent
Average irrigation efficiencies on large projects	30–50 per cent
During Conveyance	
Evaporation from canal surfaces	Negligible
Evapotranspiration at canal banks	Negligible
Canal seepage, large projects, mostly unlined canals	15–45 per cent of diversions
Seepage losses, most canals lined	5–15 per cent of diversions
Waste on large projects, ample water supplies	5–30 per cent of diversions
Waste on large projects, limited water supplies	1–10 per cent of diversions
Over-all efficiencies, large projects	20–35 per cent
Diversions for large projects	2–10 acre-ft per acre

Source: From Houk, *Irrigation Engineering*, vol.1, John Wiley & Sons, Copyright 1951.

Table 7I.107 Typical Water-Application Losses and Irrigation Efficiencies for Different Soil Conditions

	General Soil Type		
Item	**Open, Porous (%)**	**Medium Loam (%)**	**Heavy Clay (%)**
Farm-lateral loss[a]	15	10	5
Surface runoff loss	5	10	25
Deep percolation loss	35	15	10
Field-irrigation efficiency[b]	60	75	65
Farm-irrigation efficiency[c]	45	65	60

[a] Unlined ditches (loss in new-lined ditches and pipelines is usually about one percent).
[b] For water measured at the field.
[c] For water measured at the farm headgate.

Source: From U.S. Department of Agriculture.

Table 7I.108 Adaptations and Limitations of Common Irrigation Methods

Irrigation Method	Suitabilities and Conditions of Use				Remarks
	Crops	Topography	Water Supply	Soils	
Small rectangular basins	Grain, field crops, orchards, rice	Relatively flat land; area within each basin should be levelled	Can be adapted to streams of various sizes	Suitable for soils of high or low intake rates; should not be used on soils that tend to puddle	High installation costs. Considerable labor required for irrigating. When used for close-spaced crops, a high percentage of land is used for levees and distribution ditches. High efficiencies of water use possible
Large rectangular basins	Grain, field crops, rice	Flat land; must be graded to uniform plane	Large flows of water	Soils of fine texture with low intake rates	Lower installation costs and less labor required for irrigation than with small basins. Substantial levees needed
Contour checks	Orchards, grain, rice, forage crops	Irregular land; slopes less than 2 percent	Flows greater than 30 L (1 cubic foot) per second	Soils of medium to heavy texture which do not crack on drying	Little land grading required. Checks can be continuously flooded as for rice, water ponded as for orchards, or intermittently flooded as for pastures
Narrow borders up to 5 m (16 ft) wide	Pasture, grain, lucerne, vineyards, orchards	Uniform slopes less than 7 percent	Moderately large flows	Soils of medium to heavy texture	Borders should be in direction of maximum slope. Accurate cross-levelling required between guide levees
Wide borders up to 30 m (100 ft) wide	Grain, lucerne, orchards	Land graded to uniform plane with maximum slope less than 0.5 percent	Large flows, up to 600 L (20 cubic feet) per second	Deep soils of medium to fine texture	Very careful land grading necessary. Minimum of labor required for irrigation. Little interference with use of farm machinery
Wild flooding	Pasture, grain	Irregular surfaces with slopes up to 20 percent	Can utilize small continuous flows on steeper land or large flows or large flows on flatter land	Soils of medium to fine texture with stable aggregate which do not crack on drying	Little land grading required. Low initial cost for system. Best adapted to shallow soils since percolation losses may be high on deep permeable soils
Benched terraces	Grain, field crops, forage crops, orchards, vineyards	Slopes up to 20 percent	Streams of small to medium size	Soils must be sufficiently deep that grading operations will not impair crop growth	Care must be taken in constructing benches and providing adequate drainage channels for excess water. Irrigation water must be properly managed. Misuse of water can result in serious soil erosion

(Continued)

Table 7I.108 (Continued)

Irrigation Method	Suitabilities and Conditions of Use				Remarks
	Crops	Topography	Water Supply	Soils	
Straight furrows	Vegetables, row crops, orchards, vineyards	Uniform slopes not exceeding 2 percent for cultivated crops	Flows up to 350 L (12 cubic ft) per second	Can be used on all soils if length of furrows is adjusted to type of soil	Best suited for crops which cannot be flooded. High irrigation efficiency possible. Well adapted to mechanized farming
Graded contour furrows	Vegetables, field crops, orchards, vineyards	Undulating land with slopes up to 8 percent	Flows up to 100 L (3 cubic ft) per second	Soils of medium to fine texture which do not crack on drying	Rodent control is essential. Erosion hazard from heavy rains or water breaking out of furrows. High labor requirement for irrigation
Corrugations	Close-spaced crops such as grain, pasture, lucerne	Uniform slopes of up to 10 percent	Flows up to 30 L (1 cubic foot) per second	Best on soils of medium to fine texture	High water losses possible from deep percolation or surface run-off. Care must be used in limiting size of flow in corrugations to reduce soil erosion. Little land grading required
Basin furrows	Vegetables, cotton, maize and other row crops	Relatively flat land	Flows up to 150 L (5 cubic ft) per second	Can be used with most soil types	Similar to small rectangular basins, except crops are planted on ridges
Zigzag furrows	Vineyards, bush berries, orchards	Land graded to uniform slopes of less than 1 percent	Flows required are usually less than for straight furrows	Used on soils with low intake rates	This method is used to slow the flow of water in furrows to increase water penetration into soil

Source: From Bouher, L.J., 1974, Surface irrigation, *FAO Agricultural Development Paper 95.* Reprinted With permission.

Table 7I.109 Water Application Efficiencies of Irrigation Systems

Type of System	Attainable Efficiencies (%)
Surface Irrigation	
Basin	80–90
Border	70–85
Furrow	60–75
Sprinkler Irrigation	
Hand move or portable	65–75
Traveling gun	60–70
Center pivot & linear move	75–90
Solid set or permanent	70–80
Trickle Irrigation	
With point source emitters	75–90
With line source products	70–85

Source: From Solomon, K.H., 1998, *Irrigation Systems and Water Application Efficiencies, Center of Irrigation Technology*, Irrigation Notes, California State University, January 1998, www.wateright.org. Reprinted with permission.

Table 7I.110 Irrigation Efficiencies

A. Field Efficiencies by Method of Irrigation	
Method of Irrigation	**Range of Efficiency (percent)**
Graded borders	60–75
Basins and level borders	60–80
Contour ditch	50–55
Furrows	55–70
Corrugations	50–70
Subsurface	Up to 80
B. Average Efficiencies for Selected Crops in California	
Crop	**Average Efficiency (percent)**
Alfalfa and irrigated pasture	85
Citrus	80
Deciduous	85
Truck	70
Vineyard	80
Walnuts	85
C. Sprinkler Efficiencies	
Climate	**Efficiency (percent)**
Hot dry	60
Moderate	70
Humid or cool	80

Source: From U.S. Soil Conservation Service and California State Water Rights Board.

Table 7I.111 Crop Irrigation Depths

Crop	Humid Areas	Semiarid to Arid Areas[a]
Alfalfa	36–42	60–120
Beans	—	36–48
Beets (sugar)	—	48–72
Broccoli	—	24
Cabbage	—	24
Clover (ladino)	—	24
Corn (maize)	24–36	48–60
Cotton	24–36	48–72
Grapes	24–30	48–72
Orchards		
Citrus	—	48–72
Deciduous	36–60	72–96
Pasture	18–36	36–48
Peas	—	36–48
Potatoes (white)	12–24	26–48
Small grain	18–30	48
Sorghum	20–30	—
Soybeans	18–36	—
Tobacco	15–24	—
Tomatoes	—	72–120
Truck crops		
Shallow-rooted	9–12	—
Medium-rooted	12–24	—
Deep-rooted	24–30	—

Note: Soil depth in inches.

[a] Larger figure applies to arid areas.

Source: From U.S. Soil Conservation Service.

Table 7I.112 Water Requirements for Various Irrigation and Soil Types

		Coarse Sandy Soils		Light Sandy Loam		Medium Silt Loam		Clay Loam Soils		Very Heavy Clay Soils	
Irrigation Type	**Slope Land Percent**	**Q per Unit**	**Length of Run**	**Q per Unit**	**Length of Run**	**Q per Unit**	**Length of Run**	**Q per Unit**	**Length of Run**	**Q per Unit**	**Length or Run**
Basins[a]	0–2	20 cfs per acre		7.5 cfs per acre		5 cfs per acre		3 cfs per acre		2 cfs per acre	
Border[a] or checks	0–2	1.5 cfs per 10′ width	220′	0.75 cfs per 10′ width	440′	0.5 cfs per 10′ width	550′–880′	0.33 cfs per 10′ width	660′–880′	3 cfs per 10′ width	1,000′
Furrows	0–2	0.2 cfs per each	220′	0.01 cfs per each	330′	0.01 cfs per each	440′–660′	0.008 cfs each	660′	0.005 cfs per each	880′
	2–5	—		0.005 each	220′	0.005 per each	220′–440′	0.003 per each	440′	0.003 per each	550′
	5–8	—		—		0.002 per each	110′–220′	0.001 per each	330′	0.001 per each	330′
Sprinkling	0–2	2″ per hour		0.75″ per hour		0.5″ per hour		0.2″ per hour			
	2–5	2″ per hour		0.75″ per hour		0.5″ per hour		0.2″ per hour			
	5–8	1.5″ per hour		0.5″ per hour		0.4″ per hour		0.15″ per hour			
	8–12	1.0″ per hour		0.4″ per hour		0.3″ per hour					

[a] The range in slope 0–2 per cent is in itself a very rough picture of field practices where the actual slopes, particularly with borders, tend to be closer to 0.2 or 0.3 per cent rather than this higher limit 2 per cent.

Source: From Calif. Agric. Exp. Station.

Table 7I.113 Border Irrigation Relationships for Various Soils, Slopes, and Depths of Application

Soil Texture	Slope of Land (percent)	Depth of Application (in.)	Suggested Borderstrip Size		Size of Irrigation Stream (ft^3/s)
			Width (ft)	Length (ft)	
Coarse	0.25	2	50	500	8.0
		4	50	800	7.0
		6	50	1,320	6.0
	1.00	2	40	300	2.75
		4	40	500	2.50
		6	40	900	2.50
	2.00	2	30	200	1.25
		4	30	300	1.00
		6	30	600	1.00
Medium	0.25	2	50	800	7.0
		4	50	1,320	6.0
		6	50	1,320	3.5
	1.00	2	40	500	2.5
		4	40	1,000	2.5
		6	40	1,320	2.5
	2.00	2	30	300	1.0
		4	30	600	1.0
		6	30	1,000	1.0
Fine	0.25	2	50	1,320	4.0
		4	50	1,320	2.5
		6	50	1,320	1.5
	1.00	2	40	1,320	2.5
		4	40	1,320	1.25
		6	40	1,320	0.75
	2.00	2	30	660	1.0
		4	30	1,320	1.0
		6	30	1,320	0.67

Source: From U.S. Dept. of Agriculture.

Table 7I.114 Irrigation Frequency in Relation to Soil Texture and Depth of Root Zone Wetted

Soil Depth Irrigated (in.)	Soil Type		
	Sands	Loams	Clays
6	3	5	8
12	5	10	17
18	8	16	25
24	10	21	35
30	13	26	—
36	15	31	—
42	18	36	—
48	20	—	—

Note: Approximate number of days between irrigations assuming water use to be 1 in./wk.

Source: From U.S. Dept. of Agriculture.

Table 7I.115 Furrow Irrigation Relationships for Various Soils, Slopes, and Depths of Application

Soil Texture		Coarse				Medium				Fine			
Furrow Slope (percent)	Maximum Allowable Nonerosive Furrow Stream (gal/min)	Depth or Irrigation Application (in.)											
		2	4	6	8	2	4	6	8	2	4	6	8
		Maximum Allowable Length of Run (ft)											
0.25	40	500	720	875	1,000	820	1,150	1,450	1,650	1,050	1,500	1,750	2,140
0.50	20	345	480	600	680	560	800	975	1,120	730	1,020	1,250	1,460
0.75	13	270	380	480	550	450	630	775	900	580	820	1,000	1,150
1.00	10	235	330	400	470	380	540	650	760	500	750	850	990
1.50	7	190	265	330	375	310	430	530	620	400	570	700	800
2.00	5	160	225	275	320	260	370	450	530	345	480	600	675
3.00	3	125	180	220	250	210	295	360	420	270	385	470	550
5.00	2	95	135	165	190	160	225	270	320	210	290	350	410

Source: From U.S. Dept. of Agriculture.

Table 7I.116 Types of Microirrigation Systems

Type	Description	Discharge Rate
Point-source emitters (drip/trickle/bubbler)	Water is applied to the soil surface as discrete or continuous drops, tiny streams, or low volume fountain through small openings	Discharge is in units of gallons per hour (gph) or gallons per minute (gpm) over a specified pressure range. Discharge rates typically range from 0.5 gph to nearly 0.5 gpm for individual drip emitters. Discharge rates are adjusted by varying the length of the tubing. The longer the tube, the greater the friction loss, which decreases the discharge rate. Flows for bubblers are generally less than 1 gal/min
	Microtubes (spaghetti tubing) are classed as point-source emitters even though they are actually tubes rather than emitters. Microtubes consist of various lengths of flexible tubing that is small in diameter (0.020–0.040 in.). Typically, no other water control device is used. Because discharge orifices are small, complete filtration of water is required	
Surface or subsurface line-source emitter systems	This type of irrigation uses surface or buried flexible tubing with uniformly spaced emitter points (or porous tubing). The tubing comes as layflat tubing, flexible tubing or as a semirigid tubing that retains its shape. Because discharge orifices are small, complete filtration of water is required	Surface or subsurface line-source emitter systems have a uniform discharge units of gallons per hour per foot (gph/ft) or gallons per minute per 100 feet (gpm/100 ft) over a specified pressure range
Basin bubblers	The basin bubbler microirrigation system applies water to the soil surface in small fountain type streams. The discharge rate normally exceeds the infiltration rate of the soil, so small basins are used to contain the water until infiltration occurs. Discharge is generally from a small diameter (3/8–1/2 in.) flexible tube that is attached to a buried or surface lateral and located at each plant vine or tree. The typical emitter device is not used, and discharge pressures are very low (<5 lb/in^2) The discharge orifice is larger than that of the other systems, so little or no water filtration is required. Generally, screening of coarse debris and small creatures is sufficient. Drains must be provided to allow discharge of any collected sediment. Bubbler basins apply water to a larger soil volume than do drip emitters; therefore, only one outlet device is needed per plant or tree	The streams have a point discharge rate greater than that for a typical drip or line source system, but generally less than 1 gal/min
Spray or mini sprinkler	With spray or mini sprinkler micro irrigation systems, water is applied to the soil surface as spray droplets from small, low-pressure heads. The typical wetted diameter is 2–7 ft. The wetted pattern is larger than that of typical drip emitter devices, and generally fewer application devices are needed per plant. Spray and mini sprinklers also have less plugging problems and less filtration required than point-source emitters (drippers). Many spray heads only require the replacement of the orifice to change discharge rate. If an orifice becomes plugged, it is easily removed and cleaned or replaced. Spray or mini sprinkler head application patterns can be full, half circle, or partial circle (both sides)	Discharge rates are generally less than 30 gallons per hour (0.5 gph)

Source: From USDA, National Resources Conservation Service (NRCS) Irrigation—Handbooks and Manuals—*National Engineering Handbook Part 652—Irrigation Guide*, www.info.usda.gov.

Table 7I.117 Advantages and Disadvantages of Microirrigation

Advantages	Disadvantages
Little if any run-off and little evaporation occur, and deep percolation can be controlled with good water management. Water is applied at the point of use (plant transpiration)	Microirrigation is considered expensive to install and maintain. In general, the cost of micro systems is greater than that for sprinkle or surface systems
Systems are easily automated with soil moisture sensors and computer controlled for low labor requirements	Frequent maintenance is essential, and a high level of management is required to obtain optimum application efficiencies
Soil moisture levels can be maintained at predetermined levels for start-stop operation	Clogging is a major problem in all micro systems. Emitter outlets are very small, and can be easily clogged with chemical precipitates, soil particles, or organic materials. Clogging can reduce or stop water emission. Chemical treatment of the water is often necessary, and filters are almost always required. Filtration and treatment can be costly, especially where water is taken from surface sources containing sediment and debris. During installation, care should be taken to clean all construction debris from the inside of pipelines as this material can cause plugging
Fertilizer can be efficiently added to irrigation water. With proper water management, there is minimum waste caused by deep percolation, and less opportunity for groundwater pollution	Animals, especially rodents, can damage surface (and shallow subsurface) installed plastic pipe less than 4 in. in diameter
Much of the soil surface remains dry, reducing weed growth and soil surface evaporation	With low operating pressures, poor distribution uniformity can result because of elevation differences on undulating ground. Pressure regulators or pressure compensated emitters are then necessary. However, they require about 2 pounds per square inch for operation
The soil surface remains firm for use by farm workers and equipment	On steep terrain, automatic gravity draining of laterals to a low point within the field can cause low distribution uniformity, especially in low pressure, high volume systems. This problem is aggravated by frequent on-off cycles, but can be overcome by installing air-vacuum valves in a raised pipe arch (i.e., dog leg) at one or more locations in the lateral. Drains are installed just upstream of each pipe arch. This increases the number of sites affected by lateral pipe drainage, thus decreasing effects on distribution uniformity because each drain discharges less water
Frequent irrigations can be used to keep salts in the soil water more diluted and moved away from plant roots. Irrigation with water of higher salinity is possible (requires a high level of management). Where salts are present, soil-water movement must always be toward the edges of the wetted bulb (away from roots). A common mistake is to shut the system down when precipitation occurs, often creating soil-water movement into the plant root zone	When soil water is reduced in the plant root zone, light rains can move salts in surrounding soil into the plant root zone, which can constitute a potential hazard. Salts also concentrate below the soil surface at the perimeter of the soil volume wetted by each emitter. If the soil dries between irrigations, reverse movement of soil water can carry salts from the perimeter back into the root zone. To avoid salt damage to roots, water movement must always be away from the emitter and from the plant root zone. In high soil salinity areas or when using high saline or sodic water for irrigation, one may need to irrigate when it rains
Microirrigation can be used on all terrain and most agricultural crops and soils and is often used on steep, rocky ground that is unsuitable for other forms of irrigation	A smaller volume of soil is wetted at each plant. Plants can be quickly stressed if the system fails (i.e., pump failure, water source cutoff, pipeline or valve failure). Daily checking of the system is necessary even when all or part is automated. Storing a 3-day plant-water supply in the soil is recommended along with daily replacement of water used
Low tension water availability to plants enhances growth and improves crop yield and quality	Multiple emitters at each plant are recommended to decrease effects of manufacturer variability, to increase area of root development, and to reduce risk of plant damage should an emitter become plugged

Source: From USDA, National Resources Conservation Service (NRCS) Irrigation—Handbooks and Manuals—*National Engineering Handbook Part 652—Irrigation Guide*, www.info.usda.gov.

Table 7I.118 Percentage of Water Obtained by Plants from Various Depths in the Soil

Item	Crop	Soil	Water Obtained from					
			First Foot (Percent)	Second Foot (Percent)	Third Foot (Percent)	Fourth Foot (Percent)	Fifth Foot (Percent)	Sixth Foot (Percent)
1	Cotton	Heavy clay	66.0	15.9	6.1	5.2	5.7	1.1
2	...do	Sandy loam	34.6	27.2	20.7	12.5	4.9	—
3	...do	Clay loam	33.2	25.6	17.3	12.6	6.7	4.5
4	Alfalfa	Fine sandy loam	47.0	15.0	15.0	12.0	8.0	3.0
5	Orange trees, mature	Sandy loam	37.0	30.0	15.0	11.0	4.0	3.0

Source: From Univ. of California Agric. Extension and U.S. Dept. of Agriculture.

Table 7I.119 Irrigation Frequency and Amount of Water to Be Applied by Sprinkling Irrigation When Varying Amounts of Available Water Remain on the Soil

Time Since Planting in Days	Consumptive Use of Water (in./day)	Average Depth of Rooting (in.)	Maximum Depth of Available Root Zone (in.)	Interval in Days between Irrigations and Depth in Inches to Be Applied When Different Amounts of Available Water Remain in the Soil							
				75%		50%		25%		0%	
				Days	Depth Inches	Days	Depth Inches	Days	Depth Inches	Days	Depth Inches
0–9	0.05	2	0.2	1	0.3	2	0.4	3	0.5	4	0.6
9–18	0.07	6	0.6	2	0.5	5	0.7	8	1.0	10	1.3
18–27	0.08	11	1.1	3	0.6	7	1.1	10	1.5	13	1.9
27–36	0.09	16	1.6	4	0.8	8	1.4	13	2.0	17	2.5
36–45	0.10	21	2.1	5	1.0	10	1.7	15	2.5	20	3.1
45–54	0.11	26	2.6	6	1.1	12	2.1	17	2.9	24	3.7
54–63	0.11	31	3.1	7	1.3	14	2.4	21	3.3	28	4.2
63–72	0.10	35	3.5	8	1.5	17	2.7	25	3.7	34	4.7
72–81	0.09	38	3.8	11	1.6	21	2.9	33	4.0	44	5.1
81–90	0.05	41	4.2	20	1.7	41	3.1	61	4.2	82	5.5

Note: Length of growing period is 3 months; soil texture is coarse.

Source: From Israelson and Hansen, *Irrigation Principles and Practices*, John Wiley & Sons, Copyright 1962. With permission.

Table 7I.120 Acres Irrigated by Method of Irrigation and Irrigation Water Used by Source in the United States, 1998 and 2003

	1998	2003	Percent Change
Irrigation Method (acres irrigated)			
Sprinklers	24,865,142	26,937,835	+8.3
Center pivot—low pressure	9,292,022	9,696,930	+4.4
Center pivot—medium pressure	7,419,409	9,657,353	30.2
Center pivot—high pressure	1,983,869	1,938,808	−2.3
Linear move towers	284,756	344,162	20.9
Solid set and permanent	1,222,683	1,177,953	−3.7
Side roll	2,033,825	1,825,901	−10.3
Big gun or traveler	765,794	633,188	−17.3
Hand move	1,862,784	1,663,540	−10.7
Gravity flow	27,273,419	23,124,131	−15.2
Down rows or furrows	14,025,125	11,723,084	−16.4
Controlled flooding	8,472,646	8,847,392	+4.4
Uncontrolled flooding	3,273,796	2,297,956	−29.8
Other gravity	1,501,852	255,699	−83
Drip, trickle, or low-flow	2,259,176	2,988,101	+32.3
Subirrigation	549,655	279,522	−49.2
Irrigation Water Used by Source			
Total: Acre-feet (million)	97.3	86.9	
Wells: Acre-feet (million)	43.8	43.5	
Percent	45	50	
On-farm: Acre-feet (million)	11.9	11.8	
Percent	12	14	
Off-farm: Acre-feet (million)	41.5	31.6	
Percent	43	36	
Average Acre-Feet of Water Applied	1.79	1.65	

Source: From USDA, 2004, *Farm and Ranch Irrigation Survey (2003)* Vol. 3, Special Studies, Part 1, 2002; Census of Agriculture, www.nass.usda.gov.

Table 7I.121 Sprinkler Irrigation in the United States, 1998 and 2003

Region and State	1998 (Acres)	2003 (Acres)	Share of Total in 2003 (Percent)	Acreage Change, 1998–2003 (Percent)
Eastern states				
Arkansas	579,218	550,123	2.0	−5.0
Florida	301,735	176,384	0.7	−41.5
Georgia	613,379	596,133	2.2	−2.8
Illinois	288,513	372,212	1.4	29.0
Indiana	212,606	271,231	1.0	27.6
Iowa	65,879	128,824	0.5	95.5
Louisiana	105,525	99,285	0.4	−5.9
Michigan	354,350	418,778	1.6	18.2
Minnesota	311,627	416,901	1.5	33.8
Mississippi	350,358	377,268	1.4	7.7
Missouri	342,009	412,214	1.5	20.5
Wisconsin	352,699	389,434	1.4	10.4
All other states	602,638	500,499	1.9	−16.9
Subtotal	4,480,536	4,709,286	17.5	5.1
Western states				
Arizona	81,385	152,754	0.6	87.7
California	1,528,038	1,723,040	6.4	12.8
Colorado	1,290,045	1,246,601	4.6	−3.4
Idaho	2,186,806	2,202,917	8.2	0.7
Kansas	2,054,238	2,283,103	8.5	11.1
Montana	570,550	773,008	2.9	35.5
Nebraska	3,686,495	5,605,283	20.8	52.0
Nevada	179,859	211,890	0.8	17.8
New Mexico	346,535	389,774	1.4	12.5
North Dakota	121,389	171,673	0.6	41.4

(Continued)

Table 7I.121 (Continued)

Region and State	1998 (Acres)	2003 (Acres)	Share of Total in 2003 (Percent)	Acreage Change, 1998–2003 (Percent)
Oklahoma	335,014	433,099	1.6	29.3
Oregon	769,310	1,048,211	3.9	36.3
South Dakota	199,855	287,014	1.1	43.6
Texas	3,195,982	3,506,636	13.0	9.7
Utah	427,319	472,180	1.8	10.5
Washington	1,263,572	1,450,274	5.4	14.8
Wyoming	234,847	262,994	1.0	12.0
Subtotal	18,471,239	22,220,451	82.5	20.3
U.S. Total (excluding Hawaii and Alaska)	22,951,775	26,929,737	100.0	17.3

Source: From USDA, 1999, 1998 Farm and Ranch Irrigation Survey, vol. 3, Special Studies, Part 1 1997 Census of Agriculture, USDA, 2004, *Farm and Ranch Irrigation Survey (2003)*, vol. 3, Special Studies, Part 1, 2002 Census of Agriculture, www.nass.usda.gov.

Table 7I.122 Water Applications for Selected Crops in the United States, 2003

Crop	Average Application Rate AF/Acre	Irrigated Acreage Harvested 1000 Acres	Estimated Water Use Million AF	Share of Total Water Used Percent (%)
Corn for grain or seed	1.2	9,750	11.7	15
Corn for silage or greenchop	2.2	1,313	2.9	4
Sorghum for grain or seed	1.0	1,108	1.1	1
Wheat for grain or seed	1.5	3,269	4.9	6
Barley for grain or seed	1.5	991	1.5	2
Rice	2.3	2,995	6.9	9
Cotton	1.4	4,080	5.7	7
Soybeans for beans	0.8	5,347	4.3	6
Beans, dry edible	1.6	458	0.7	1
Alfalfa and alfalfa mixtures (dry hay, haylage and greenchop)	2.3	6,222	14.3	19
All other hay (dry hay, haylage, grass silage, and greenchop)	1.7	3,254	5.5	7
Sugarbeets for sugar	2.7	562	1.5	2
Potatoes	1.8	1,033	1.9	2
Vegetables	2.1	2,081	4.4	6
Orchards	2.2	4,105	9.0	12
Total		**46,567**	**76**	**100**

Source: Abstracted from *Farm and Ranch Irrigation Survey (2003)*, vol. 3, Special Studies, Part 1, 2002; Census of Agriculture issued November 2004, www.nass.usda.gov.

Table 7I.123 Depth of Irrigation Water Applied, by Region and Crop, 1969–1998

	1969[a]	1974[a]	1979[b]	1984[b]	1988[b]	1994[b]	1998[b]
			Inches[c]				
Region							
Atlantic[d]	8.0	11.0	13.0	14.0	12.5	10.5	13.0
North Central[e]	7.5	8.0	8.5	9.0	10.0	7.5	8.0
Northern Plains	16.0	17.5	15.0	13.5	14.5	12.0	12.0
Delta States	15.5	17.0	17.5	17.5	18.0	13.0	16.5
Southern Plains	18.0	18.5	17.5	16.5	17.0	17.0	17.0
Mountain States	30.5	28.5	26.5	24.5	24.5	24.0	24.5
Pacific Coast	32.5	33.5	33.5	33.5	34.5	32.5	33.0
United States[e]	25.2	25.0	23.2	22.1	22.3	20.2	20.9
Crop							
Corn for grain	18.5	19.5	16.5	16.0	16.0	13.5	14.5
Sorghum for grain	19.0	19.0	16.5	14.5	14.5	13.5	12.5
Barley	30.0	26.5	23.0	18.5	18.0	19.0	19.5
Wheat	23.5	24.0	21.0	16.5	16.0	17.0	17.0
Rice	28.0	28.5	30.0	33.5	32.5	27.0	28.5
Soybeans	12.0	11.0	10.5	9.5	10.0	8.5	10.0
Cotton	23.0	25.5	26.0	24.5	24.0	20.0	19.0
Alfalfa	32.5	30.5	28.0	28.0	29.0	26.5	29.0
Other hays	22.0	21.0	20.0	21.0	19.5	20.5	24.5
Vegetables	25.0	25.5	25.5	27.0	26.5	24.0	24.0
Land in orchards	29.0	30.0	30.0	31.0	31.5	27.0	28.0

[a] Census of Agriculture.
[b] Estimates constructed by State/crop from the Farm and Ranch Irrigation Survey and ERS estimates of irrigated area.
[c] Includes Alaska and Hawaii.
[d] Northeast, Appalachian, and Southeast production regions.
[e] Lake States and Corn Belt farm production regions.

Source: From Heimlict. R., 2003, *Agricultural Resources and Environmental Indicators 2003*, Agriculture Handbook No. (AH722), February 2003, www.ers.usda.gov.

Original Source: USDA, ERS, based on USDC Census of Agriculture, various years; Farm and Ranch Irrigation Surveys (USDA, 1999b; USDC, 1996; USDC, 1990, and previous versions).

Table 7I.124 Irrigation System versus Crops Grown

	Crop Category			
Irrigation System	1[a]	2[b]	3[c]	4[d]
Surface				
Basins, borders		X	X	X
Furrows, corrugations	X	X		X
Contour levee-rice		X	X	
Sprinkler				
Side (wheel) roll lateral	X	X		
Hand move lateral	X	X		X
Fixed (solid) set		X		X
Center pivot, linear move	X	X		
Big guns-traveling, stationary	X	X		
Micro				
Point source				X
Line source	X			X
Basin bubbler				X
Mini sprinklers & spray heads				X
Subirrigation	X	X	X	X

[a] Row or bedded crops: sugar beets, sugarcane, potatoes, pineapple, cotton, soybeans, corn, sorghum, milo, vegetables, vegetable and flower seed, melons, tomatoes, and strawberries.
[b] Close-growing crops (sown, drilled, or sodded): small grain, alfalfa, pasture, and turf.
[c] Water flooded crops: rice and taro.
[d] Permanent crops: orchards of fruit and nuts, citrus groves, grapes, cane berries, blueberries, cranberries, bananas and papaya plantations, hops, and trees and shrubs for windbreaks, wildlife, landscape, and ornamentals.

Source: From USDA, National Resources Conservation Service (NRCS) Irrigation-Handbook and Manuals – *National Engineering Handbook Part 652-Irrigation Guide*, www.info.usda.gov.

Table 7I.125 Typical Life and Annual Maintenance Cost Percentage for Irrigation System Components

System and Components	Life (yr)	Annual Maint. (% of Cost)	System and Components	Life (yr)	Annual Maint. (% of Cost)
Sprinkler systems	10–15	2–6	**Surface & subsurface systems**	15	5
Handmove	15+	2			
Side or wheel roll	15+	2	**Related components**		
End tow	10+	3	Pipelines		
Side move w/drag lines	15+	4	buried thermoplastic	25+	1
Stationary gun type	15+	2	buried steel	25	1
Center pivot—standard	15+	5	surface aluminum	20+	2
Linear move	15+	6	surface thermoplastic	5+	4
Cable tow	10+	6	buried nonreinforced concrete	25+	1
Hose pull	15+	6	buried galv. steel	25+	1
Traveling gun type	10+	6	buried corrugated metal	25+	1
Fixed or solid set			buried reinforced PMP	25+	1
Permanent	20+	1	gated pipe, rigid, surface	10+	2
Portable	15+	2	surge valves	10+	6
Sprinkler gear driven, impact & spray heads	5–10	6			
			Pumps		
Valves	10–25	3	pump only	15+	3
			w/electric motors	10+	3
Micro system[a]	1–20	2–10	w/internal combustion engine	10+	6
Drip	5–10	3			
Spray	5–10	3	Wells	25+	1
Bubbler	15+	2	Linings		
Semi-rigid, buried	10–20	2	nonreinforced concrete	15+	5
Semi-rigid, surface	10	2	flexible membrane	10	5
Flexible, thin wall, buried	10	2	reinforced concrete	20+	1
Flexible, thin wall, surface	1–5	10			
Emitters & heads	5–10	6	Land grading, leveling	[b]	
Filters, injectors, valves	10+	7	Reservoirs	[c]	

[a] With no disturbance from tillage and harvest equipment.
[b] Indefinite with adequate maintenance.
[c] Indefinite with adequate maintenance of structures, watershed.

Source: From USDA, National Resources Conservation Service (NRCS) *Irrigation-Handbook and Manuals – National Engineering Handbook Part 652-Irrigation Guide*, wcc.nrcs.usda.

Table 7I.126 Energy Requirement and Energy Costs for Various Irrigation Systems in the United States

	Total Lift (Feet)	Field Efficiency (Percent)	Potential Efficiency	Water Per Acre (Inches)	Fuel Per Acre (Gallons)	Cost of Energy (Dollars) 1973	1979	1980
Traveling big guns	389	70	75	17	91.1	$13.67	$74.70	$113.88
Center-Pivot								
(High) 75	273	80	85	15	56.4	8.46	46.25	70.50
(Low) 50	215	80	85	15	44.4	6.66	36.41	55.50
Skid-Tow	215	75	80	16	47.4	7.11	38.87	59.25
Gated pipe without reuse	120	60	70	20	33.1	4.97	27.14	41.38
Gated pipe with reuse	120	70	85	17	28.1	4.22	23.04	35.13
Auto-surface	120	85	92	14	23.1	3.47	18.94	28.88
Drip trickle	150	85	92	14	28.9	4.34	23.70	36.13

Note: Diesel fuel only. Based on 100 foot of lift from groundwater reservoir; 12 in. net of water; pumping plant operating at 75% of Nebraska performance standard of 10.94 water horsepower hours per gallon of diesel fuel or 14.6 brake horsepower hours per gallon of diesel fuel.
Fuel cost per gallon: $0.15–1973, $0.82–1979; $1.25–1980.

Source: From Fischbach, P.E., 1980, *Energy Requirements of Auto-Surface Irrigation*, Copyright, The Irrigation Assoc., 1980 Technical Conference Proceedings. Reprinted with permission.

Table 7I.127 On-Farm Energy Expense in the United States, 1998 and 2003

	1998		2003		
		Expenses Per Acre Irrigated (Dollars)		Expenses Per Acre Irrigated (Dollars)	
On-Farm Energy Pumping Expenses	**Expenses ($1,000)**	**Wells and Surface Water**	**Expenses ($1,000)**	**Water from Wells**	**Surface Water**
Total energy expenses for pump	1,223,106	31.92	1,551,847	39.50	26.39
Electricity	801,184	39.75	953,247	42.64	29.84
Natural gas	206,900	33.99	281,029	57.25	33.67
LP gas, propane, and butane	27,716	16.99	34,053	27.21	22.68
Diesel fuel	182,832	17.78	281,490	25.09	16.27
Gasoline and gasohol	4475	28.14	2,027	11 60	18.05

Source: Abstracted from USDA, 1999, 1998 *Farm and Ranch Irrigation Survey*, vol. 3, Special Studies, Part 1, 1997 Census of Agriculture. USDA, 2004, Farm and Ranch Irrigation Survey (2003), vol. 3, Special Studies, Part 1, 2002; Census of Agriculture, nass.usda.gov.

Table 7I.128 Supply Sources and Variables Costs of Irrigation Water, 1998[a]

Water Source	Acre Irrigated (Million)	Share of Acres Irrigated[b] (Percent)	Average cost[b] ($/Acre)	Cost Range[b] ($/Acre)	Comments
Groundwater			32[c]	7–69[d]	Pumping cost varies with energy prices, depth to water, and efficiency of pumping system
Only source[e]	23.5	47			
Combined sources	6.3	13			
Onfarm surface water			n/a	0–15[f]	Costs are very low in most cases. Some water is pumped from surface sources at higher costs, since energy is required
Only source	4.2	8			
Combined sources	2.7	5			
Off-farm surface water[g]			41[h]	10–85[i]	Most actress relying on off-farm sources are located in West
Only source	10.3	21			
Combined sources	4.8	10			
Total			n/a	n/a	The sum of acres is greater than the irrigated total in the Farm and Ranch Irrigation Survey due to double counting of combined water sources
Only source	37.9	76			
Combined sources	13.8	27			

Note: n/a indicates data not available.

[a] These values include only energy costs for pumping or purchased water costs. Management and labor costs associated with irrigation decisions, system maintenance, and water distribution are not included.
[b] Available data are from the 1998 Farm and Ranch Irrigation Survey.
[c] Reported national average energy expense for the on-farm pumping of irrigation water.
[d] Range in State energy expenses for the onfarm pumping of irrigation water.
[e] Only source means that farms used no other irrigation water source.
[f] Cost estimates based on engineering formulas with an efficient electric system.
[g] Included a minor amount of groundwater supplied from off-farm suppliers.
[h] Reported average cost for off-farm supplies.
[i] Range in reported State average cost of water from off-farm suppliers for State irrigating 50,000 or more acres from off-farm sources. If all States are included, the range expands to $2–$175 per acre.

Source: From Helmich, R., 2003, *Agricultural Resource and Environmental Indicators 2003*, Agricultural Handbook No (AH 777) February 2003, ers.200usda.gov.

Original Source: USDA, ERS based on USDA (1999), 1998 Farm and Ranch Irrigation Survey.

Table 7I.129 Distribution of Irrigation Withdrawals, Acres, Annual Application and Estimated Pumping Costs: 1995

State	Irrigation Withdrawals (mgd)	Irrigated Acres (1,000)	Irrigation Depth (ft/yr)	Surface Irrigation (% area)	Estimated Pumping Cost ($1995/ac/yr)
California	28,894	9,484	3.42	47.4	76.38
Idaho	13,048	3,011	4.86	33.2	39.02
Colorado	12,735	3,307	4.32	75.9	34.74
Texas	9,451	6,313	1.68	55.7	40.48
Montana	8,546	1,810	5.29	71.0	16.24
Nebraska	7,550	7,449	1.14	47.1	18.99
Wyoming	6,595	1,991	3.71	85.3	20.83
Washington	6,469	2,120	3.42	24.1	41.83
Oregon	6,168	1,844	3.75	41.6	30.83
Arkansas	5,936	3,511	1.90	85.0	20.10
Arizona	5,672	1,088	5.84	73.5	71.62
Utah	3,533	1,143	3.47	63.2	28.91
Florida	3,469	2,134	1.82	48.9	19.81
Kansas	3,383	3,086	1.23	32.0	24.55
New Mexico	2,993	959	3.50	56.7	53.55
Mississippi	1,742	1,374	1.42	71.7	17.43
Nevada	1,644	560	3.29	75.7	49.96
48-State Total/Average	132,969	58,066	2.57	54.9	34.63

Source: From Dziegielewski, B. et al., 2002, *Analysis of Water Use Trends in the United States: 1950–1995*, Southern Illinois University at Carbondale, Carbondale, IL, February 28, 2002, http://info.geography.siu.edu/geography_info/research/. Reprinted with permission.

Table 7I.130 Fuel Energy Requirements for Pumping One Acre-Foot of Water at One Pound Per Square Inch

Energy	Horsepower Hours[a]	Percentage of Water Pump Efficiency (Unit Fuel Per Acre-Foot Per psi)		
		65	60	55
Electricity	1.206 per kWh	4.0503	4.3876	4.7866
Diesel	12.35 per gal	0.4000	0.4330	0.4659
Gasoline	9.875 per gal	0.5004	0.5417	0.5830
Natural gas	79 per MCF[b]	0.0625	0.0677	0.0729
LPG	7.9 per gal	0.6254	0.6771	0.7287

[a] This column refers to the assumed number of horsepower hours produced per unit of fuel.
[b] MCF equals 1,000 ft^3.

Source: From Sloggett, G., 1985, *Energy and U.S. Agriculture: Irrigation Pumping, 1974–83*, U.S. Dept. of Agriculture Economic Report 545.

Table 7I.131 Natural Gas, Electricity, and Petroleum Prices by Sector in the United States, 1995–2003

	Natural Gas (cents/ft^3)				Electricity (cents/kwh)				Petroleum (dollars/gal)			
	Industrial[a] Prices				Industrial[b] Prices							
Year	Nominal		Real[c]		Nominal		Real[c]		All Grades Nominal	Real[c]		On-Highway Diesel Fuel[d]
1995	0.271		0.294	[R]	4.66		5.06	[R]	1.21	1.31	[R]	1.11
1996	0.342		0.364	[R]	4.60		4.90	[R]	1.29	1.37	[R]	1.24
1997	0.359		0.376	[R]	4.53		4.75	[R]	1.29	1.35	[R]	1.20
1998	0.314		0.325	[R]	4.48		4.64	[R]	1.12	1.16	[R]	1.04
1999	0.312		0.319	[R]	4.43		4.53	[R]	1.22	1.25	[R]	1.12
2000	0.445		0.445	[R]	4.64		4.64	[R]	1.56	1.56	[R]	1.49
2001	0.524	[R]	0.512	[R]	5.04		4.92	[R]	1.53	1.50	[R]	1.40
2002	0.402	[R]	0.387	[R]	4.88	[R]	4.69	[R]	1.44	1.39	[R]	1.32
2003	0.578	[P]	0.547	[P]	4.95		4.68		1.64	1.55		1.51

Note: R = Revised, P = Preliminary.

[a] Residential, commercial, and industrial prices do not include the price of natural gas delivered to consumers on behalf of third parties.

[b] Retail customers are classified as "Commercial" or "Industrial" based on NAICS (North American Industry Classification System) codes or usage falling within specified limits by rate schedule.

[c] In chained (2000) dollars, calculated by using gross domestic product implicit price deflators. Corrected for inflation.

[d] Nominal dollars.

Source: DOE, *Annual Energy Review 2003*, www.eia.doe.gov.

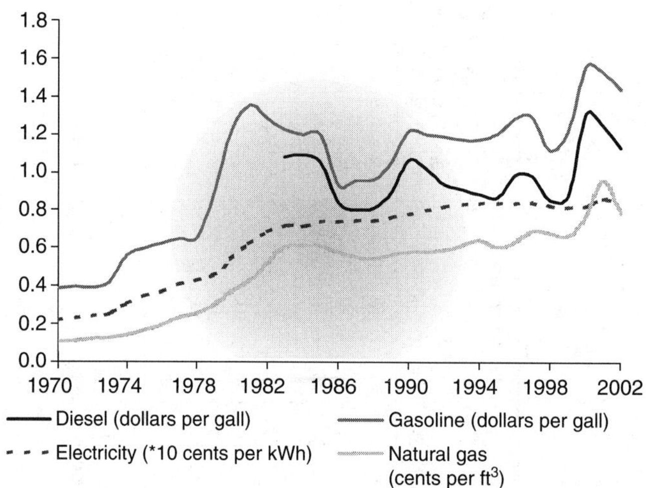

Figure 7I.20 Nominal prices of major fuel sources: 1970–2002. (From Miranowski, J.A., *Energy Demand and Capacity to Adjust in Agricultural Production*, U.S. Department of Agriculture, Agricultural Outlook Forum 2005 on Feb. 24 and 25, 2005.) www.usda.gov.

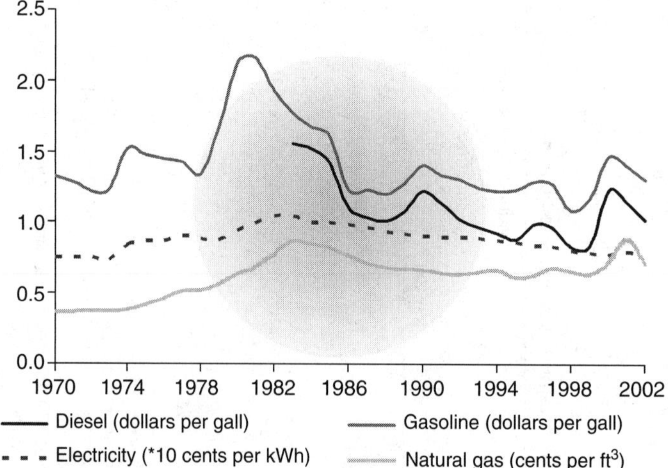

Figure 7I.21 Real prices of major fuel sources: 1970–2002 (1996 dollars). (From Miranowski, J.A., *Energy Demand and Capacity to Adjust in Agricultural Production*, U.S. Department of Agriculture, Agricultural Outlook Forum 2005 on Feb. 24 and 25, 2005.) www.usda.gov.

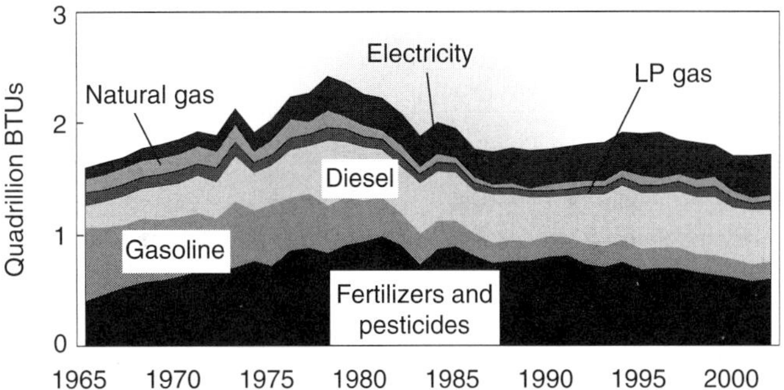

Figure 7I.22 Total energy used on farms, 1965–2002. (From Miranowski, J.A., *Energy Demand and Capacity to Adjust in Agricultural Production*, U.S. Department of Agriculture, Agricultural Outlook Forum 2005 on Feb. 24 and 25, 2005.) www.usda.gov.

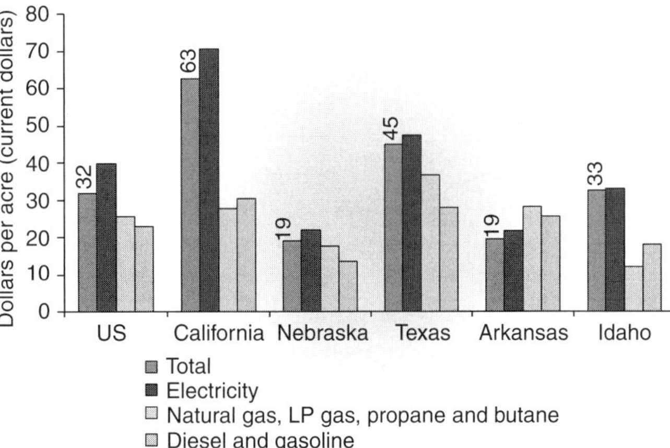

Figure 7I.23 Average irrigation costs per acre—by energy source. (From Miranowski, J.A., *Energy Demand and Capacity to Adjust in Agricultural Production*, U.S. Department of Agriculture, Agricultural Outlook Forum 2005 on Feb. 24 and 25, 2005.) usda.gov/oce/forum/Archives/pastyears.htm.

Table 7I.132 Irrigation Technology and Water Management: Conventional Methods and Improved Practices

System and Aspect	Conventional Technology or Management Practice	Improved Technology or Management Practice
Onfarm conveyance	Open earthen ditches	Concrete or other ditch linings; aboveground pipe; belowground pipe
Gravity application systems		
Release of water	Dirt or canvas checks with siphon tubes	Ditch portals or gates; gated pipe; gated pipe with surge flow or cablegation
Length of irrigation run	Length of field, often one-half mile or more	Shorter runs, one-quarter mile or less
Field gradient	Natural field slope, often substantial; uneven field surface	Land leveled to reduce and smooth field surface gradient
Field runoff	Water allowed to move off field	Applications controlled to avoid runoff; tailwater return systems
Furrow management	Full furrow wetting; furrow bottoms uneven	Alternate furrow wetting; furrow bottoms smooth and consistent
Pressurized application systems		
Pressure requirements	High pressure, typically above 60 pounds per square inche (psi)	Reduced pressure requirements, often 10–30 psi
Water distribution	Large water dispersal pattern	More narrow water dispersal through sprinkler droptubes, improved emitter spacing, and low-flow systems
Automation	Handmove systems; manually operated systems	Self-propelled systems; computer control of water applications
Versatility	Limited to specific crops; used only to apply irrigation	Multiple crops; various uses—irrigation, chemigation, manure applications, frost protection, crop cooling
Water management		
Assessing crop needs	Judgment estimates	Soil moisture monitoring; plant tissue monitoring; weather-based computations
Timing of applied water	Fixed calendar schedule	Water applied at needed by crop; managed for profit (not yield); managed for improved effectiveness of rainfall
Measurement of water	Not metered	Measured using canal flumes, weirs, and meters; external and in-pipe flow meters
Drainage	Runoff to surface-water system or Evaporation ponds; percolation to aquifers	Applications managed to limit drainage; reuse through tailwater pumpback; dual-use systems with subirrigation

Source: From USDA, ERS.

Table 7I.133 Efficient Water Management Practices for Agricultural Water Suppliers in California

List A-Generally Applicable EWMPs

Prepare and adopt and water management plan

Designate a water conservation coordinator

Support the availability of water management services to water users

Improve communication and among water suppliers, water users, and other agencies

Evaluate the need, if any for change in institutional policies to which the water supplier is subject

Evaluate and improve efficiencies of the water supplier's pumps

List B-Conditionally Applicable EWMPs

Facilitate alternative land use

Facilitate using available recycled water that otherwise would not be used beneficially, meets all health and safety criteria and does not cause harm to crops or soil

Facilitate financing capital improvements for on-farm irrigation systems

Facilitate voluntary water transfers that do not unreasonably affect the water user, water supplier, the environment, or third parties

Line or pipe ditches and canals

Increase flexibility in water ordering by, and delivery to, water users within operational limits

Construct and operate water supplier spill and tailwater recovery systems

Optimize conjunctive use of surface and groundwater

Automate canal structures

List C-Other EWMPs

Water measurement and water use reporting

Pricing or other incentives

Source: From California Department of Water Resources 1998, "Urban, Agricultural and Environmental Water Use," California Water Plan Update Bulletin-160-98, California Dept. of Water Resources, Sacremento, Calif., November 1998, water.ca.gov.

SECTION 7J IRRIGATION — WORLD

Table 7J.134 Water Use in the Agriculture Industry by Province, in Canada, 1996

Province	Livestock Watering (1000 m^3)	Irrigation (1000 m^3)	Total (1000 m^3)
Newfoundland and Labrador	483	144	627
Prince Edward Island	1,904	1,715	3,618
Nova Scotia	3,199	2,272	5,471
New Brunswick	2,369	1,443	3,812
Quebec	45,001	58,394	103,395
Ontario	59,233	114,000	173,233
Manitoba	23,843	24,670	48,513
Saskatchewan	39,890	271,370	311,260
Alberta	61,468	2,609,000	2,670,468
British Columbia	14,682	763,110	777,791
Canada	252,071	3,846,117	4,098,188

Note: There is no significant agricultural activity in the Territories.

Source: From Statistics Canada "Human Activity and the Environment," Annual Statistics 2003, page 14.Statistics Canada information is used with the permission of Statistics Canada. Users are forbidden to copy this material and/or redisseminate the data, in an original or modified form, for commercial purposes, without the expressed permissionof Statistics Canada. Information on the availability of the wide range of data from Statictics Canada can be obtained from Statistics Canada's Regional Offices, its World Wide Web site at statcan.ca, and its toll-free access number 1-800-263-1136.

Table 7J.135 Land Areas under Irrigation in Various Countries of the World, 1961–1997

Country and Region	1961	1965	1970	1975	1980	1985	1990	1995	1997
Africa									
Algeria	229	233	238	244	253	338	384	555	560
Angola	75	75	75	75	75	75	75	75	75
Benin	0	2	2	4	5	6	6	10	20
Botswana	1	2	1	1	2	2	2	1	1
Burkina Faso	2	2	4	8	10	12	20	25	25
Burundi	3	5	5	5	10	14	14	14	14
Cameroon	2	4	7	10	14	21	21	21	21
Cape Verde	2	2	2	2	2	2	3	3	3
Chad	5	5	5	6	6	10	14	17	20
Congo	0	0	1	2	1	1	1	1	1
Congo, Dem. Rep. (formerly Zaire)				0	7	9	10	11	11
Côte d'Ivoire	4	6	20	34	44	54	66	73	73
Djibouti	1	1	1	1	1	1	1	1	1
Egypt	2,568	2,672	2,843	2,825	2,445	2,497	2,648	3,283	3,300
Eritrea								28	28
Ethiopia	150	150	155	158	160	162	162	190	190
Gabon	4	4	4	4	4	4	4	7	7
Gambia	1	1	1	1	1	1	1	2	2
Ghana	0	0	7	7	7	7	6	11	11
Guinea	20	20	50	50	90	90	90	93	95
Guinea Bissau	17	17	17	17	17	17	17	17	17
Kenya	14	14	29	40	40	42	54	67	67
Lesotho	3	3	3	3	3	3	3	3	3
Liberia	0	0	2	2	2	2	2	2	2
Libya	121	130	175	200	225	300	470	470	470
Madagascar	300	330	330	465	645	826	1,000	1,087	1,090
Malawi	1	1	4	13	18	18	20	28	28
Mali	60	60	61	60	60	60	78	85	86
Mauritania	20	20	30	30	49	49	49	49	49
Mauritius	8	12	15	15	16	17	17	18	18
Morocco	875	895	920	1,060	1,217	1,245	1,258	1,258	1,251
Mozambique	8	16	26	40	65	93	105	107	107
Namibia	4	4	4	4	4	4	4	7	7
Niger	16	16	18	18	23	30	66	66	66
Nigeria	200	200	200	200	200	200	230	235	233
Reunion	3	5	5	5	5	8	11	12	12
Rwanda	4	4	4	4	4	4	4	4	4
Sao Tome and Principe	10	10	10	10	10	10	10	10	10
Senegal	70	85	78	78	62	90	94	71	71
Sierra Leone	1	2	6	13	20	28	28	29	29
Somalia	90	90	95	100	125	180	180	200	200
South Africa	808	890	1,000	1,017	1,128	1,128	1,290	1,270	1,270
Sudan	1,480	1,550	1,625	1,700	1,800	1,946	1,946	1,946	1,950
Swaziland	36	40	47	56	58	62	67	69	69
Tanzania	20	28	38	52	120	127	144	150	155
Togo	2	2	4	6	6	7	7	7	7
Tunisia	100	100	200	200	243	300	300	361	380
Uganda	2	3	4	4	6	9	9	9	9
Zambia	2	2	9	18	19	28	30	46	46
Zimbabwe	22	34	46	70	80	90	100	150	150
North and Central America									
Barbados	1	1	1	1	1	1	1	1	1
Belize	0	0	1	1	1	2	2	3	3
Canada	350	380	421	500	596	748	718	720	720
Costa Rica	26	26	26	36	61	110	118	126	126
Cuba	230	330	450	580	762	861	900	910	910

(Continued)

Table 7J.135 (Continued)

Country and Region	1961	1965	1970	1975	1980	1985	1990	1995	1997
Dominican Republic	110	115	125	140	165	198	225	259	259
El Salvador	18	20	20	33	110	110	120	120	120
Guadeloupe	1	1	2	1	2	2	2	2	2
Guatemala	32	43	56	72	87	102	117	125	125
Haiti	35	40	60	70	70	70	75	90	90
Honduras	50	66	66	70	72	72	74	74	74
Jamaica	22	24	24	32	33	33	33	33	33
Martinique	1	1	1	2	5	4	4	3	3
Mexico	3,000	3,200	3,583	4,479	4,980	5,285	5,600	6,100	6,500
Nicaragua	18	18	40	67	80	83	85	88	88
Panama	14	18	20	23	28	30	31	32	32
Puerto Rico	39	39	39	39	39	39	39	40	40
Saint Lucia	1	1	1	1	1	1	2	3	3
St. Vincent	0	1	1	1	1	1	1	1	1
Trindad and Tobago	11	11	15	18	21	22	22	22	22
United States	14,000	15,200	16,000	16,690	20,582	19,831	20,900	21,400	21,400
South America									
Argentina	980	1,110	1,280	1,440	1,580	1,620	1,680	1,700	1,700
Bolivia	72	75	80	120	140	125	110	78	88
Brazil	490	610	796	1,100	1,600	2,100	2,700	3,169	3,169
Chile	1,075	1,100	1,180	1,242	1,255	1,257	1,265	1,265	1,270
Colombia	226	235	250	300	400	465	680	1,037	1,061
Ecuador	440	450	470	506	500	300	290	240	250
French Guiana	1	1	1	1	1	2	2	2	2
Guyana	90	109	115	120	125	127	130	130	130
Paraguay	30	30	40	55	60	65	67	67	67
Peru	1,016	1,060	1,106	1,130	1,160	1,210	1,450	1,753	1,760
Suriname	14	15	28	33	42	55	59	60	60
Uruguay	27	35	52	57	79	97	120	140	140
Venezuela	60	62	70	90	137	171	180	200	205
Asia									
Afghanistan	2,160	2,260	2,340	2,430	2,505	2,586	3,000	2,800	2,800
Armenia	Formerly included in Soviet Union							290	290
Azerbaijan	Formerly included in Soviet Union							1,453	1,455
Bahrain	1	1	1	1	1	1	2	4	5
Banglaesh	426	572	1,058	1,441	1,569	2,073	2,936	3,429	3,693
Bhutan	8	10	18	22	26	30	39	39	40
Brunei Darsm				0	1	1	1	1	1
Cambodia	62	100	89	89	100	130	160	270	270
China	30,402	33,579	38,113	42,776	45,467	44,581	47,965	49,857	51,819
Cyprus	30	30	30	30	30	30	36	40	40
Gaza Strip	8	8	9	10	10	11	11	12	12
Georgia	Formerly included in Soviet Union							469	470
Hong Kong	9	8	8	6	3	3	2	2	2
India	24,685	26,510	30,440	33,730	38,478	41,779	45,144	53,000	57,000
Indonesia	3,900	3,900	3,900	3,900	4,301	4,300	4,410	4,687	4,815
Iran	4,700	4,900	5,200	5,900	4,948	6,800	7,000	7,264	7,265
Iraq	1,250	1,350	1,480	1,567	1,750	1,750	3,525	3,525	3,525
Israel	136	151	172	180	203	233	206	199	199
Japan	2,940	2,943	3,415	3,171	3,055	2,952	2,846	2,745	2,701
Jordan	31	32	34	36	37	48	63	75	75
Kazakstan	Formerly included in Soviet Union							2,380	2,149
Korea, DPR	500	500	500	900	1,120	1,270	1,420	1,460	1,460
Korea, Rep.	1,150	1,199	1,184	1,277	1,307	1,325	1,345	1,206	1,163
Kuwait	0	0	1	1	1	2	3	5	5
Kyrgyzstan	Formerly included in Soviet Union							1,077	1,074
Laos	12	13	17	40	115	119	130	155	164

(Continued)

Table 7J.135 (Continued)

Country and Region	1961	1965	1970	1975	1980	1985	1990	1995	1997
Lebanon	41	61	68	86	86	86	86	105	117
Malaysia	228	236	262	308	320	334	335	340	340
Mongolia	5	5	10	23	35	60	77	84	84
Myanmar (Burma)	536	753	839	976	999	1,085	1,005	1,555	1,556
Nepal	70	86	117	230	520	760	900	1,134	1,135
Oman	20	23	29	34	38	41	58	62	62
Pakistan	10,751	11,472	12,950	13,630	14,680	15,760	16,940	17,200	17,580
Philippines	690	730	826	1,040	1,219	1,440	1,560	1,550	1,550
Qatar	1	1	1	1	3	5	6	13	13
Saudi Arabia	343	353	365	375	600	800	900	1,620	1,620
Sri Lanka	335	341	465	480	525	583	520	570	600
Syria	558	522	451	516	539	652	693	1,089	1,168
Tajikistan	Formerly included in Soviet Union							719	720
Thailand	1,621	1,768	1,960	2,419	3,015	3,822	4,238	4,642	5,010
Turkey	1,310	1,400	1,800	2,200	2,700	3,200	3,800	4,186	4,200
Turkmenistan	Formerly included in Soviet Union							1,750	1,800
United Arab Emirates	30	35	45	50	53	58	63	68	72
Uzbekistan	Formerly included in Soviet Union							4,281	4,281
Vietnam	1,000	980	980	1,000	1,542	1,770	1,840	2,000	2,300
West Bank	10	10	9	8	9	10	10	9	9
Yemen	207	231	260	282	289	302	348	485	485
Europe									
Albania	156	205	284	331	371	399	423	340	340
Austria	4	4	4	4	4	4	4	4	4
Bel-Lux	1	1	1	1	1	1	1	24	35
Belarus	Formerly included in Soviet Union							115	115
Bulgaria	720	945	1,001	1,128	1,197	1,229	1,263	800	800
Bosnia Herzegovina	Formerly included in Yugoslavia							2	2
Croatia	Formerly included in Yugoslavia							3	3
Czechoslovakia	108	116	126	136	123	187	282		
Czech Republic	Formerly included in Czechoslovakia							24	24
Denmark	40	65	90	180	391	410	430	481	476
Estonia	Formerly included in Soviet Union							4	4
Finland	2	7	16	40	60	62	64	64	64
France	360	440	539	680	870	1,050	1,300	1,630	1,670
Germany	321	390	419	448	460	470	482	475	475
Greece	430	576	730	875	961	1,099	1,195	1,325	1,385
Hungary	133	100	109	156	134	138	204	210	210
Italy	2,400	2,400	2,400	2,400	2,400	2,425	2,711	2,698	2,698
Latvia	Formerly included in Soviet Union							20	20
Lithuania	Formerly included in Soviet Union							9	9
Macedonia	Formerly included in Yugoslavia							61	55
Malta	1	1	1	1	1	1	1	1	2
Moldova Rep.	Formerly included in Soviet Union							309	309
Netherlands	290	330	380	430	480	530	555	565	565
Norway	18	25	30	40	74	90	97	127	127
Poland	295	275	213	231	100	100	100	100	100
Portugal	620	621	622	625	630	630	630	632	632
Romania	206	230	731	1,474	2,301	2,956	3,109	3,110	3,089
Russia	Formerly included in Soviet Union							5,362	4,990
Slovakia	Formerly included in Czechoslovakia							217	190
Slovenia	Formerly included in Yugoslavia							2	2
Spain	1,950	2,226	2,379	2,818	3,029	3,217	3,402	3,527	3,603
Sweden	20	22	33	45	70	99	114	115	115
Switzerland	20	23	25	25	25	25	25	25	25
Ukraine	Formerly included in Soviet Union							2,585	2,466
United Kingdom	108	105	88	86	140	152	164	108	108

(Continued)

Table 7J.135 (Continued)

Country and Region	1961	1965	1970	1975	1980	1985	1990	1995	1997
Yugoslavia	121	118	130	133	145	164	170		
Yugoslav SFR	Formerly included in Yugoslavia							65	65
Former Soviet Union	9,400	9,900	11,100	14,500	17,200	19,689	20,800		
Oceania									
Australia	1,001	1,274	1,476	1,469	1,500	1,700	1,832	2,500	2,700
Fiji	1	1	1	1	1	1	1	3	3
New Zealand	77	93	111	150	183	256	280	285	285
Total Irrigated Area (000 ha)	**138,813**	**149,740**	**167,331**	**187,559**	**209,233**	**223,304**	**242,185**	**260,083**	**267,727**
Rate of Change over Period		0.08	0.12	0.12	0.12	0.07	0.08	0.07	0.03
Average Annual Change (%)		1.97	2.35	2.42	2.31	1.35	1.69	1.48	1.47

Note: Data for the former Soviet Union after 1990 are split among the separate independent states, now included in Asia and Europe. Data from Yugoslavia and Czechoslovakia after 1990 are now split among several independent states. Original Source Food and Agriculture Organization, 1999. Web site at www.fao.org.

Source: From *World's Water 2000–2001*, by Peter H. Gleick. Copyright © 2000 Island Press. Reproduced by permission of Island Press, Washington, DC.

Table 7J.136 Irrigated Areas for Selected Countries of the World, 1980–2000

	1980	1985	1990	1991	1992	1993	1994	1995	1996	1997	1998	1999	2000
Canada	5,960	7,480	7,180	7,200	7,200	7,200	7,200	7,200	7,200	7,200	7,200	7,200	7,200
Mexico	49,800	52,850	56,000	58,000	61,000	62,000	63,000	64,000	65,000	65,000	65,000	65,000	65,000
U.S.A.	205,820	198,310	209,000	209,000	214,000	215,000	217,000	218,000	220,000	222,820	223,000	224,000	224,000
Japan	30,550	29,520	28,460	28,250	28,020	27,820	27,640	27,450	27,240	27,010	26,790	26,590	26,410
Korea	13,070	13,250	13,450	13,350	13,000	12,700	12,350	12,060	11,760	11,630	11,590	11,530	11,490
Australia	15,000	17,000	18,320	20,120	20,690	21,070	24,080	24,000	23,900	23,800	23,650	22,510	23,850
N.Zealand	1,830	2,560	2,800	2,830	2,850	2,850	2,850	2,850	2,850	2,850	2,850	2,850	2,850
Austria	40	40	40	40	40	40	40	40	40	40	40	40	40
Belgium	130	170	180	180	180	180	180	240	300	350	350	350	350
Czech R.	—	—	—	—	—	240	240	240	240	240	240	240	240
Denmark	3,910	4,100	4,300	4,350	4,350	4,550	4,650	4810	4,810	4,760	4,600	4,470	4470
Finland	600	620	640	640	640	640	6,40	640	640	640	640	640	640
France	8,700	10,500	13,000	13,500	14,000	14,770	15,000	16,300	17,500	1,9070	20,000	21,000	22,000
Germany	4,600	4,700	4,820	4,820	4,820	4,850	4,850	4,850	4,850	4,850	4,850	4,850	4,850
Greece	9,610	11,240	11,950	12,020	12,890	13,300	13,520	13,830	14,140	1,4820	14,220	14,410	14,510
Hungary	1,340	1,380	2,040	2,100	2,240	2,060	2,100	2,100	2,100	2,100	2,100	2,100	2,100
Iceland	—	—	—	—	—	—	—	—	—	—	—	—	—
Ireland	—	—	—	—	—	—	—	—	—	—	—	—	—
Italy	24,000	24,250	27,110	27,100	27,000	27,000	27,000	26,980	26,980	26,980	26,980	26,980	27,000
The Netherl.	4,800	5,300	5,550	5,570	5,600	5,600	5,650	5,650	5,650	5,650	5,650	5,650	5,650
Norway	740	900	970	970	970	1,000	1,100	1,270	1,270	1,270	1,270	1,270	1,270
Poland	1,000	1,000	1,000	1,000	1,000	1,000	1,000	1,000	1,000	1,000	1,000	1,000	1,000
Portugal	6,300	6,300	6,310	6,310	6,310	6,310	6,310	6,320	6,500	6,500	6,500	6,500	6,500
Slov.R.	—	—	—	—	—	2,990	2,350	2,170	1,940	1,710	1,740	1,780	1,830
Spain	30,290	32,170	34,020	33,880	34,030	34,530	36,570	35,270	36,030	36,340	36,520	36,550	36,550
Sweden	700	990	1,140	1,160	1,150	1,150	1,150	1,150	1,150	1,150	1,150	1,150	1,150
Switzerland	250	250	250	250	250	250	250	250	250	250	250	250	250
Turkey	27,000	32,000	38,000	40,000	40,000	40,000	41,860	41,860	42,000	42,000	43,800	45,000	45,000
UK	1,400	1,520	1,640	1,360	1,080	1,080	1,080	1,080	1,080	1,080	1,080	1,080	1,080
Russ.F.	49,940	58,050	61,220	60,540	55,530	52,980	51,580	53,620	51,080	49,900	46,630	46,000	46,000
N.America	261,580	258,640	272,180	274,200	282,200	284,200	287,200	289,200	292,200	295,020	295,200	296,200	296,200
Australia	16,830	19,560	21,120	22,950	23,540	23,920	26,930	26,850	26,750	26,650	26,500	25,360	26,700
OECD	126,640	139,300	155,780	156,950	157,900	161,540	165,540	166,050	168,470	170,800	172,980	175,310	176,480
EU	95,080	101,900	110,700	110,930	112,090	114,000	116,640	117,160	119,670	122,230	122,580	123,670	124,790
OECD	448,670	460,270	490,990	495,700	504,660	510,180	519,660	521,610	526,420	531,110	533,060	534,990	537,280
World	2,097,160	2,251,380	2,443,060	2,479,660	2,514,540	2,555,470	2,579,820	2,614,280	2,638,870	2,673,220	2,686,880	2,699,340	2,716,890

Note: Units are given in km^2.

a. Areas equipped to provide water to the crops. These include areas equipped for full and partial control irrigation, spate irrigation areas, and equipped wetland or inland valley bottoms. All figures are rounded to the nearest 10 km^2.

- JPN, Rice irrigation only;
- KOR, Rice irrigation only;
- BEL, Data for Belgium include Luxembourg.

Source: Table 10.2, *OECD Environment Data Compendium 2002*, © OECD 2002, www.oecd.org.

Table 7J.137 Irrigation Water Use Per Country in the Year 2000

	Total Renewable Water Resources (km^3)	Irrigation Water Requirements (km^3)	Water Requirement Ratio in Percentages (%)	Water Withdrawal for Agriculture (km^3)	Water Withdrawal as Percentage of Renewable Water Resources (%)
Afghanistan	65	8.78	38	22.84	35
Algeria	14.32	1.45	37	3.94	27
Angola	184	0.04	20	0.21	0
Argentina	814	3.43	16	21.52	3
Bangladesh	1210.644	19.09	25	76.35	6
Benin	24.8	0.06	30	0.19	1
Bolivia	622.531	0.26	23	1.16	0
Botswana	14.4	0.02	30	0.06	0
Brazil	8233	6.21	17	36.63	0
Burkina Faso	12.5	0.21	30	0.69	5
Burundi	3.6	0.06	30	0.19	5
Cambodia	476.11	1.2	30	4	1
Cameroon	285.5	0.22	30	0.73	0
Chad	43	0.07	35	0.19	0
Chile	922	1.59	20	7.97	1
China	2829.569	153.9	36	426.85	15
Colombia	2132	1.23	25	4.92	0
Congo, Republic of	832	0	30	0	0
Congo, Dem Republic of	1283	0.03	30	0.11	0
Costa Rica	112.4	0.36	25	1.43	1
Côte d'Ivoire	81	0.17	28	0.6	1
Cuba	38.12	1.41	25	5.64	15
Dominican Republic	20.995	0.56	25	2.24	11
Ecuador	432	2.67	19	13.96	3
Egypt	58.3	28.43	53	53.85	92
El Salvador	25.23	0.19	25	0.76	3
Eritrea	6.3	0.09	32	0.29	5
Ethiopia	110	0.56	22	2.47	2
Gabon	164	0.02	30	0.05	0
Gambia	8	0.01	30	0.02	0
Ghana	53.2	0.06	26	0.25	0
Guatemala	111.27	0.4	25	1.61	1
Guinea	226	0.41	30	1.36	1
Guyana	241	0.45	28	1.6	1
Haiti	14.025	0.18	20	0.93	7
Honduras	95.929	0.17	25	0.69	1
India	1896.66	303.24	54	558.39	29
Indonesia	2838	21.49	28	75.6	3
Iran, Islamic Rep of	137.51	21.06	32	66.23	48
Iraq	75.42	11.2	28	39.38	52
Jamaica	9.404	0.01	25	0.02	0
Jordan	0.88	0.29	39	0.76	86
Kenya	30.2	0.3	30	1.01	3
Korea, Dem People's Rep	77.135	1.49	30	4.96	6
Korea, Republic of	69.7	2.67	30	8.92	13
Laos	333.55	0.81	30	2.7	1
Lebanon	4.407	0.37	40	0.92	21
Libyan Arab Jamahiriya	0.6	2.56	60	4.27	712
Madagascar	337	3.58	25	14.31	4
Malawi	17.28	0.2	25	0.81	5
Malaysia	580	1.68	30	5.6	1
Mali	100	2.06	30	6.87	7
Mauritania	11.4	0.44	29	1.5	13
Mexico	457.222	18.53	31	60.34	13
Morocco	29	4.28	37	11.48	40
Mozambique	216.11	0.22	39	0.55	0
Myanmar	1045.601	9.79	30	32.64	3
Namibia	17.94	0.07	40	0.17	1

(Continued)

Table 7J.137 (Continued)

	Total Renewable Water Resources (km^3)	Irrigation Water Requirements (km^3)	Water Requirement Ratio in Percentages (%)	Water Withdrawal for Agriculture (km^3)	Water Withdrawal as Percentage of Renewable Water Resources (%)
Nepal	210.2	2.45	25	9.82	5
Nicaragua	196.69	0.3	27	1.08	1
Niger	33.65	0.62	30	2.08	6
Nigeria	286.2	1.65	30	5.51	2
Pakistan	222.67	72.14	44	162.65	73
Panama	147.98	0.05	20	0.23	0
Paraguay	336	0.08	23	0.35	0
Peru	1913	5.07	31	16.42	1
Philippines	479	6.33	30	21.1	4
Rwanda	5.2	0.01	30	0.03	1
Saudi Arabia	2.4	6.68	43	15.42	643
Senegal	39.4	0.43	30	1.43	4
Sierra Leone	160	0.12	33	0.35	0
Somalia	13.5	0.98	30	3.28	24
South Africa	50	2.34	21	11.12	22
Sri Lanka	50	2.92	24	12	24
Sudan	64.5	14.43	40	36.07	56
Suriname	122	0.18	30	0.62	1
Swaziland	4.51	0.12	16	0.76	17
Syrian Arab Republic	26.26	8.52	45	18.93	72
Tanzania, United Rep of	91	0.56	30	1.85	2
Thailand	409.944	24.83	30	82.75	20
Togo	14.7	0.02	30	0.08	1
Tunisia	4.56	1.21	54	2.23	49
Turkey	229.3	11.27	40	27.86	12
Uganda	66	0.03	30	0.12	0
Uruguay	139	0.66	22	3.03	2
Venezuela, Boliv Rep of	1233.17	1.24	31	3.97	0
Vietnam	891.21	15.18	31	48.62	5
Yemen	4.1	2.53	40	6.32	154
Zambia	105.2	0.26	19	1.32	1
Zimbabwe	20	0.67	30	2.24	11

Source: From Food and Agriculture Organization of the United Nations (FAO). AQUASTAT - FAO's Information System on Water and Agriculture, www.fao.org. Reprinted with permission.

Table 7J.138 Summary Results of Agricultural Water Use and Comparison with Water Resources, 2000

	Total Renewable Water Resources (km^3)	Irrigation Water Requirements (km^3)	Water Requirement Ratio (%)	Water Withdrawal for Agriculture (km^3)	Water Withdrawal as Percentage of Renewable Water Resources (%)
Latin America	13,409	45	24	187	1
Near East and North Africa	541	109	40	274	51
Sub-Saharan Africa	3,518	31	32	97	3
East Asia	8,609	232	34	693	8
South Asia	2,469	397	44	895	36
90 developing countries	28,545	814	38	2146	8

Source: From Food and Agriculture Organization of the United Nations (FAO), AQUASTAT - FAO's Information System on Water and Agriculture, www.fao.org. Reprinted with permission.

Table 7J.139 Irrigated Area, by World Region, 1961–2001

	Thousand Hectares				
Region	**1961**	**1965**	**1970**	**1975**	**1980**
Africa	7,410	7,795	8,483	9,010	9,491
Asia	90,166	97,093	109,666	121,565	132,377
Europe	8,468	9,401	10,583	12,704	14,479
North & Central America	17,950	19,526	20,939	22,833	27,597
Oceania	1,079	1,368	1,588	1,620	1,684
South America	4,661	5,070	5,673	6,403	7,392
U.S.S.R.	9,400	9,900	11,100	14,500	17,200
World	139,134	150,153	168,032	188,635	210,220
Region	**1985**	**1990**	**1995**	**2000**	**2001**
Africa	10,331	11,235	12,383	12,700	12,813
Asia	141,922	155,009	180,508	190,014	190,385
Europe	16,018	17,414	26,104	25,382	25,347
North & Central America	27,471	28,913	30,473	31,223	31,344
Oceania	1,957	2,114	2,689	2,674	2,674
South America	8,296	9,499	10,086	10,489	10,489
U.S.S.R.	19,689	20,800			
World	225,684	244,984	262,243	272,482	273,052

Note: After 1990, all irrigated area in the former U.S.S.R. is split among Europe and Asia.

Source: From *World's Water 2004–2005*, by Peter H. Gleick. Copyright © 2004 Island Press. Reproduced by permission of Island Press, Washington, DC.

Table 7J.140 Agricultural Land Use for Developing Countries

	Arable Land (mi ha)			Harvested Land (mi ha)			Cropping Intensity (%)		
	Total	**Rainfed**	**Irrigated**	**Total**	**Rainfed**	**Irrigated**	**Total**	**Rainfed**	**Irrigated**
Developing countries									
1997–99	956	754	202	885	628	257	93	83	127
2015	1,017	796	221	977	671	306	96	84	138
2030	1,076	834	242	1,063	722	341	99	87	141
Sub-Saharan Africa									
1997–99	228	223	5.3	154	150	4.5	68	67	86
2015	262	256	6	185	179	5.7	71	70	95
2030	288	281	6.8	217	210	7	76	75	102
Near East and North Africa									
1997–99	86	60	26	70	43	27	81	72	102
2015	89	60	29	77	45	32	86	75	110
2030	93	60	33	84	46	37	90	78	112
Latin America and Caribbean									
1997–99	203	185	18	127	112	16	63	60	86
2015	223	203	20	150	131	19	67	64	95
2030	244	222	22	172	150	22	71	68	100
South Asia									
1997–99	207	126	81	230	131	100	111	103	124
2015	210	123	87	248	131	117	118	106	134
2030	216	121	95	262	131	131	121	109	137
East Asia									
1997–99	232	161	71	303	193	110	130	120	154
2015	233	155	78	317	186	131	136	120	168
2030	237	151	85	328	184	144	139	122	169

Source: Food and Agriculture Organization of the United Nations (FAO), *World Agricultrure: towards 2015/2030 Summary Report*, www.fao.org. Reprinted with permission.

Table 7J.141 Irrigated Land and Percentage of Arable Land That Is Irrigated

	Irrigated Land					
	Area (1000 ha)			As % of Arable Land		
World/Continent	**1980**	**1990**	**2001**	**1980**	**1990**	**2001**
World	210,220	244,984	273,052	15.7	17.6	19.5
Developed countries	58,926	66,286	67,988	9.0	10.2	11.1
Industralized countries	37,355	39,935	43,226	9.9	10.5	11.8
Transition economies	21,571	26,351	24,762	7.9	9.8	10.0
Developing countries	151,294	178,698	205,064	21.9	24.1	26.0
Latin America and the Caribbean	13,811	16,794	18,613	10.8	12.5	12.5
Near East and North Africa	17,982	24,864	27,808	21.8	28.8	32.3
Sub-Saharan Africa	3,980	4,885	5,221	3.2	3.7	3.6
East and Southeast Asia	59,722	65,624	74,605	37.0	33.9	35.0
South Asia	55,798	66,529	78,813	28.6	33.9	40.3
Oceania developing	1	2	4	0.2	0.4	0.7
North America developing	—	—	—	—	—	—
Continental groupings						
Africa	9,491	11,235	12,813	6.0	6.7	7.0
Asia	132,377	155,009	190,385	31.3	33.8	37.4
Caribbean	1,074	1,269	1,308	22.0	23.3	22.2
Latin America	12,737	15,525	17,305	10.4	12.0	12.1
North America	21,178	21,618	23,220	9.0	9.3	10.5
Oceania	1,684	2,114	2,674	3.6	4.2	5.1
Europe	14,479	17,414	25,347	11.5	14.0	8.8

Source: From Food and Agriculture Organization of the United Nations (FAO), 2003, *Summary of Food and Agricultural Statistics*, www.fao.org. Reprinted with permission.

Table 7J.142 Irrigated (Arable) Land: Past and Projected

	Irrigated Land in Use (mi ha)					Annual Growth (% p.a.)		Land in Use as % of Potential (%)		Balance (mi ha)	
	1961/63 (1)	**1979/81 (2)**	**1997/99 (3)**	**2015 (5)**	**2030 (6)**	**1961–1999 (7)**	**1997/99–2030 (8)**	**1997/99 (9)**	**2030 (10)**	**1997/99 (11)**	**2030 (12)**
Sub-Saharan Africa	3	4	5	6	7	2.0	0.9	14	19	32	30
Near East/North Africa	15	18	26	29	33	2.3	0.6	62	75	17	11
Latin America and the Caribbean	8	14	18	20	22	1.9	0.5	27	32	50	46
South Asia	37	56	81	87	95	2.2	0.5	57	67	61	47
excl. India	12	17	23	24	25	1.9	0.2	84	89	4	3
East Asia	40	59	71	78	85	1.5	0.6	64	76	41	27
excl. China	10	14	19	22	25	2.1	0.9	40	53	29	23
All above	103	151	202	221	242	1.9	0.6	50	60	200	161
excl. China	73	106	150	165	182	2.1	0.6	44	54	188	157
excl. China/India	48	67	93	102	112	2.0	0.6	41	50	132	114
Industrial countries	27	37	42			1.3					
Transition countries	11	22	25			2.6					
World	142	210	271			1.8					

Source: From Food and Agriculture Organization, of the United Nations (FAO), World Agriculture Towards 2015/2030, www.fao.org. Reprinted with permission.

Original Source: From Columns (1)–(3): FAOSTAT, November 2001.

Table 7J.143 Irrigation Lending by Continent and Decade

	Irrigation Lending (% of global amount)				Projects (% of global number)				Average Irrigation Lending Amount (1991 US$M)				
Decade	Asia	America	Africa	Europe	Asia	America	Africa	Europe	Asia	America	Africa	Europe	World
1950	75	25			67	33			70	47			62
1960	75	12	13		71	20	10		89	51	112		84
1970	66	12	13	9	47	14	31	7	61	37	18	55	44
1980	73	12	11	4	47	13	34	5	76	44	16	39	49
1990–91	61	23	16		60	17	23		69	93	47		68
Total	69	13	12	5	50	14	30	5	71	47	21	48	51

Source: From Jones, W.I., 1995, *The World Bank and Irrigation, The World Bank*, Washington, DC, Copyright © International Bank for Reconstruction and Development/The World Bank, www.worldbank.org. Reprinted with permission.

Table 7J.144 Average Unit Costs for World Bank Irrigation Projects

	Unit Cost ($/ha)	Number	Adjusted Unit Cost ($/ha)	Number	Adjusted/ Unadjusted Ratio
All	4,837	191	7,950	184	1.64
Satisfactory	2,643	128	2,906	125	1.10
Unsatisfactory	9,294	63	18,637	59	2.01
Gravity	5,584	113	10,355	112	1.85
Pump (mostly from groundwater)	3,766	52	4,415	46	1.17
Mixed	3,727	26	3,846	26	1.03
New construction	7,740	86	12,915	81	1.67
Rehabilitation	1,633	34	5,258	34	3.22
Rehabilitation and extension	3,171	55	3,834	54	1.21
Paddy	6,374	73	11,063	72	1.74
Nonpaddy	3,886	118	5,950	112	1.53
Selected areas					
East and South Asia	2,831	112	4,694	107	1.66
East Asia	4,291	56	7,379	56	1.72
South Asia	1,370	56	1,746	51	1.27
India	1,421	30	1,596	27	1.12
Europe	4,743	17	4,759	17	1.00
Middle East	5,062	9	4,663	7	0.92
Africa	12,925	30	20,833	30	1.61
North Africa	4,911	12	5,226	12	1.06
Sub-Saharan Africa	18,269	18	31,238	18	1.71
Latin America and Caribbean	3,923	20	10,283	20	2.62

Note: Two measures of unit cost are used. "Unit cost" is defined as the actual project cost ($US) measured at evaluation divided by the completion command area (hectares) measured at evaluation. "Adjusted unit cost" is "unit cost" with the denominator adjusted by the completion achievement of construction target (%) measured at evaluation. Two projects, Lake Chad Polders (Chad) and Black Bush Irrigation (Guyana), had completion command areas measured at evaluation equal to zero. This resulted in infinite measures of unit cost, and thus they were not included in the calculation of averages.

Source: From Jones, W.I., 1995, *The World Bank and Irrigation, The World Bank*, Washington, DC, Copyright © International Bank for Reconstruction and Development/The World Bank, www.worldbank.org. Reprinted with permission.

SECTION 7K LIVESTOCK

Table 7K.145 Water Requirements for Farm Animals and Poultry

Horse, work	12
Mule	12
Cattle	
Holstein calves (liquid milk or dried milk and water supplied)	
4 weeks of age	1.2–1.4
8 weeks of age	1.6
12 weeks of age	2.2–2.4
16 weeks of age	3.0–3.4
20 weeks of age	3.8–4.3
26 weeks of age	4.0–5.8
Dairy heifers—Pregnant	7.2–8.4
Steers	
Maintenance ration	4.2
Fattening ration	8.4
Range cattle	4.2–8.4
Jersey cows[a]	
Milk production 5–30 lbs/day	7.2–12
Holstein cows[a]	
Milk production 20–50 lbs/day	7.8–22
Milk production 80 lbs/day	23
Dry	11
Pigs	
Body weight—30 lbs	0.6–1.2
Body weight—60–80 lbs	0.8
Body weight—75–125 lbs	1.9
Body weight—200–380 lbs	1.4–3.6
Pregnant sows	3.6–4.6
Lactating sows	4.8–6.0
Sheep	
On range of dry pasture	0.6–1.6
On range (salty feeds)	2.0
On rations of hay and grain or hay, roots and grain	0–0.7
On good pasture	Little, if any
Chickens (100 birds)	
1–3 weeks of age	0.4–2.0
3–6 weeks of age	1.4–3.0
6–10 weeks of age	3.0–4.0
9–13 weeks of age	4.0–5.0
Pullets	3.0–4.0
Nonlaying hens	5.0
Laying hens (moderate temperatures)	5.0–7.5
Laying hens (temperature 90 °F)	9.0
Turkeys (100 birds)	
1–3 weeks of age	1.1–2.6
7–4 weeks of age	3.7–8.4
9–13 weeks of age	9–14
15–19 weeks of age	17
21–26 weeks of age	14–15

Note: Gallons per day.

[a] Allow 15–20 additional gallons per day for each cow for flushing stables and washing dairy utensils.

Source: From U.S. Dept. of Agriculture.

Table 7K.146 Livestock Freshwater Use in the United States, 2000

	Withdrawals (mi/gal/day)			Withdrawals (thousand acre-feet/yr)		
	By Source			By Source		
	Groundwater	Surface Water	Total	Groundwater	Surface Water	Total
Alabama	—	—	—	—	—	—
Alaska	—	—	—	—	—	—
Arizona	—	—	—	—	—	—
Arkansas	—	—	—	—	—	—
California	182	227	409	204	255	458
Colorado	—	—	—	—	—	—
Connecticut	—	—	—	—	—	—
Delaware	3.7	0.22	3.92	4.15	0.25	4.39
District of Columbia	—	—	—	—	—	—
Florida	31	1.51	32.5	34.7	1.69	36.4
Georgia	1.66	17.7	19.4	1.86	19.9	21.7
Hawaii	—	—	—	—	—	—
Idaho	27.7	7.2	34.9	31	8.07	39.1
Illinois	37.6	0	37.6	42.1	0	42.1
Indiana	27.3	14.6	41.9	30.6	16.4	47
Iowa	81.8	27.1	109	91.8	30.4	122
Kansas	87.2	23.5	111	97.7	26.3	124
Kentucky	—	—	—	—	—	—
Louisiana	4.03	3.31	7.34	4.52	3.71	8.23
Maine	—	—	—	—	—	—
Maryland	7.18	3.18	10.4	8.05	3.56	11.6
Massachusetts	—	—	—	—	—	—
Michigan	10.2	1.15	11.3	11.4	1.29	12.7
Minnesota	52.8	0	52.8	59.2	0	59.2
Mississippi	—	—	—	—	—	—
Missouri	18.3	54.1	72.4	20.5	60.6	81.1
Montana	—	—	—	—	—	—
Nebraska	76	17.4	93.4	85.2	19.5	105
Nevada	—	—	—	—	—	—
New Hampshire	—	—	—	—	—	—
New Jersey	1.68	0	1.68	1.88	0	1.88
New Mexico	—	—	—	—	—	—
New York	—	—	—	—	—	—
North Carolina	89.1	32.3	121	99.9	36.2	136
North Dakota	—	—	—	—	—	—
Ohio	8.2	17.1	25.3	9.19	19.2	28.4
Oklahoma	53.6	97.2	151	60	109	169
Oregon	—	—	—	—	—	—
Pennsylvania	—	—	—	—	—	—
Rhode Island	—	—	—	—	—	—
South Carolina	—	—	—	—	—	—
South Dakota	16.9	25.2	42	18.9	28.2	47.1
Tennessee	—	—	—	—	—	—
Texas	137	172	308	153	192	346
Utah	—	—	—	—	—	—
Vermont	—	—	—	—	—	—
Virginia	—	—	—	—	—	—
Washington	—	—	—	—	—	—
West Virginia	—	—	—	—	—	—
Wisconsin	60.3	6.02	66.3	67.6	6.75	74.4
Wyoming	—	—	—	—	—	—
Puerto Rico	—	—	—	—	—	—
U.S. Virgin Islands	—	—	—	—	—	—
Total	**1,010**	**747**	**1,760**	**1,140**	**838**	**1,980**

Note: Figures may not sum to totals because of independent rounding. —, data not collected.

Source: From Hutson, S.S. et al., 2004, Estimated Use of Water in the United States in 2000, *U.S. Geological Survey Circular 1268*, www.usgs.gov.

SECTION 7L NAVIGATION AND WATERWAYS

Table 7L.147 Projected Flows Required for Efficient Navigation on Inland Waterways of the United States

Waterway	Critical Flow[a] (ft^3/s)			Comments
	1,959	1,980	2,000	
New England, existing waterways: all	0	0	0	All waterways are in tidal reaches
Middle Atlantic, existing waterways				
Great Lakes to Hudson River and Champlain Canals	2,000	2,000	2,000	Available flow 13,000 c.f.s.
Delaware	b	b	b	If depth increased to 45 ft minimum flow will have to be increased by 2,900, c.f.s. to repel salinity
South Atlantic				
Existing waterways				
Cape Fear River above Wilmington, NC	50	100	c	
Savannah River below Augusta, GA	5,000	5,800	c	
Altamaha River below junction of Ocmulgee and Oconee Rivers	1,000	5,000	c	
Okeechobee Waterway	50	50	c	
Apalachicola below Jim Woodruff lock and dam	9,300	9,300	c	
Alabama River below Selma, AL	3,000	3,000	c	
Warrior system	540	540	c	
Possible future waterways				
Santee-Congaree	—	5,200	c	Project might be found feasible if river developed for power
Ocmulgee River below Macon, GA	—	300	c	
Cross-Florida Barge Canal	—	700	c	Authorized project
Chattahoochee River below Atlanta, GA	—	300	c	
Flint River below Albany, GA	—	200	c	
Coosa River	—	300	c	Do
Tennessee-Tombigbee	—	1,246	c	Completed project. (1,246 c.f.s. to be diverted from the Tennessee River.)
Arkansas-White-Red				
Existing waterways				
White River (to mile 168.7)	6,500	10,000	10,000	
Ouachita-Black	100	100	100	
Waterways under construction				
Verdigris River (at Catoosa, OK)	—	115	115	Provision of 9-ft waterway as part of multiple-purpose development of the Arkansas River
Arkansas (at Webber Falls)	—	300	300	Do
Arkansas (at Short Mountain)	—	530	530	Do
Arkansas (at Dardanelle)	—	505	505	Do
Arkansas (Dardenelle to mouth)	—	1,000	1,000	Do
Possible future waterways: Overton-Red Waterway	—	300		Authorized project
Gulf-Southwest				
Waterways under construction: Guadalupe to Victoria	—	0–1,800	0–1,800	For periodic flushing

(Continued)

Table 7L.147 (Continued)

Waterway	Critical Flow[a] (ft³/s)			Comments
	1,959	1,980	2,000	
Possible future waterways				
Trinity River (below Dallas)	—	400	800	Authorized project
Trinity River (above Dallas)	—	150	325	Do
Missouri River Basin, existing waterways				
Missouri at Sioux City[d]	—	30,000	30,000	Could be reduced to 25,000 by dredging
Missouri at Omaha[d]	28,000	28,000	28,000	
Missouri at Nebraska City	31,000	31,000	31,000	
Missouri at Kansas City[d]	35,000	32,500	32,500	Could be reduced to 27,500 by dredging
Missouri at mouth[d]	—	35,000	35,000	
Upper Mississippi Basin				
Existing waterways				
Main stem, pools 1–10	375	750	940	Not required in winter
Main stem, pools 11–22	2,000	2,000	2,000	Do
Main stem, pools 24–26	70,000	25,000	25,000	Required all year
Main stem, Missouri to Ohio	54,000	75,000	75,000	Do
Minnesota	0	0	0	Mississippi backwater
St. Croix	0	0	0	Do
Illinois Waterway	720[e]	1,826[e,f]	1,826[e,f]	
Fox River	300	300	300	Not required in winter
Possible future waterways				
Kaskaskia River	0	200	300	
Big Muddy River	0	200	400	
Tittabawassee River	0	200	200	
Ohio River Basin				
Existing waterways				
Ohio River (main stem)	[c]	[c]	[c]	River canalized
Allegheny River	100	140	150	
Monongahela River	340	440	670	
Kanawha River	300	400	400	
Kentucky River	100	150	200	
Green River	150	250	400	
Cumberland River	100	150	250	
Possible future waterways				
Big Sandy River (main stem)	0	329	329	
Big Sandy River (Levisa Fork)	0	176	176	
Lake Erie-Ohio River Canal	0	430	430	
Lower Mississippi River				
Existing waterways				
Main stem at Cairo	100,000	120,000	120,000	
Main stem below Arkansas	140,000	150,000	150,000	
Ouachita-Black	100	150	150	
Possible future waterways				
Yazoo below Greenwood	—	100	100	
Columbia River Basin				
Existing waterways				
Main stem at Bonneville	40,000	40,000	40,000	
Main stem at The Dallas	700	1,500	2,500	
Willamette, mouth to Salem	6,000	6,000	6,000	
Willamette, Salem to Corvallis	5,000	5,000	5,000	
Snake, mouth to Ice Harbor	300	1,000	2,000	
Waterways under construction				
Main stem John Day to McNary	110,000	1,500	2,500	
Snake, Ice Harbor to Lewiston	—	1,000	2,000	
Possible future waterways				
Main stem head McNary Pool to Rock Island	—	36,000	—	Under study
Snake, Lewiston to mile 174	—	500	1,000	Do

(Continued)

Table 7L.147 (Continued)

Waterway	Critical Flow[a] (ft³/s) 1,959	1,980	2,000	Comments
Snake, mile 174 to mile 188	—	500	1,000	
Snake, mile 188 to mile 232	—	350	700	
North Pacific Coast, Possible future waterways	—	10,000	10,000	
Skagit, mouth to Concrete				
Central Valley, existing waterways				
Sacramento	5,000	5,000	5,000	Shallow draft
San Joaquin above Mossdale	100	100	100	Do

[a] Rate below which streamflow cannot drop without reducing the efficiency of navigation.
[b] Anticipated flows will meet the needs of navigation.
[c] Estimates not available.
[d] Flows required from April through November, assuming continuation of open river navigation. Canalization now being studied. If Missouri canalized these flows would be greatly reduced.
[e] Average annual flow required at Lockport, Ill. Includes 250 c.f.s. to prevent reversal of flow into Lake Michigan when storm runoff occurs. Excludes 120 c.f.s. of present industrial usage which bypasses Lockport lock.
[f] Requirements for recommended duplicate lock system.

Source: From Select Committee on National Water Resources, U.S. Senate, 1960; amended.

Table 7L.148 Navigation Locks and Dams in the United States (Operable as of May 2005)

Project	River Mile	Type of Structure[a]	Status[b]	Community in Vicinity	Locks: Width of Chamber	Locks: Chamber Useable Length	Locks: Lift at Normal Pool Level	Locks: Depth of Miter Sill Upper	Locks: Depth of Miter Sill Lower	Type	Dam: Length (ft)	Dam: Year Opened	Length (miles)	Depth (ft)	Authorized Channel: Width (above) (ft)	Authorized Channel: Width (below) (ft)
Alabama-Coosa Rivers																
Claiborne Lock & Dam	117.5	L	1	Claiborne, AL	84	600	30	16	13	Moveable	1,603.0	1973	61	9	200	200
Millers Ferry Lock & Dam	178	L	1	Camden, AL	84	600	45	16	13	Moveable	9,900.0	1969	103	9	200	200
Jones Bluff Lock & Dam	281.2	L	1	Benton, AL	84	600	45	16	13	Moveable	1,496.0	1974	88	9	200	200
Allegheny River, PA and NY																
Lock & Dam 2	6.7	L	1	Aspinwall, PA	56	360	11	11	12	Miter	1,393.0	1934	8	9	200	200
Lock & Dam 3	14.5	L	1	Cheswick, PA	56	360	14	12	11	Miter	1,436.0	1934	10	9	200	200
Lock & Dam 4	24.2	L	1	Natrona, PA	56	360	11	9	10	Miter	876.0	1927	6	9	200	200
Lock & Dam 5	30.4	L	2	Freeport, PA	56	360	12	10	11	Miter	780.0	1927	6	9	200	200
Lock & Dam 6	36.3	L	2	Clinton, PA	56	360	12	11	11	Miter	1,140.0	1928	9	9	200	200
Lock & Dam 7	45.7	L	2	Kittanning, PA	56	360	13	11	10	Miter	916.0	1930	7	9	200	200
Lock & Dam 8	52.6	L	2	Templeton, PA	56	360	18	14	10	Miter	984.0	1931	10	9	200	200
Lock & Dam 9	62.2	L	3	Rimer, PA	56	360	22	11	11	Miter	950.0	1938	10	9	200	200
Apalachicola, Chattahoochee, & Flint Rivers, GA, AL and FL																
Jim Woodruff Lock & Dam	106.3	L	1	Chattahoochee, FL	82	450	33	14	14	Moveable	6,359.0	1957	47	6	100	100
Gordon, GA (no name given)	46.7	L	1	Gordon, GA	82	450	25	19	13	Moveable	750.0	1963	29	9	100	100
Walter F George Lock & Dam	75.1	L	1	Fort Gaines, GA	82	450	88	18	13	Moveable	1,325.0	1963	85	9	100	100
Atlantic Intracoastal Waterway between Norfolk, VA & St. Johns River, FL																
Great Bridge Guard Lock	12.2	L	1	Chesapeake, VA	72	530	3	16	16	Miter	NA	1932	6	12	200	200
Deep Creek Lock	10.6	L	1	Chesapeake, VA	52	300	12	12	12	Miter	NA	1940	22	50	50	50
South Mills Lock	33.2	L	1	South Mills, NC	52	300	12	12	12	Miter	NA	1941	22	6	50	50
Bayou Teche, LA																
Bayou Teche, LA (FCMR&T)	4	F	2	Berwick, LA	45	90	0	0	0	Sector	NA		80	8	80	80
Keystone Lock	72	L	1	New Iberia, LA	36	160	8	9	8	Miter	175.0	1913	35	6	80	80
Bayou Teche, LA (FCMR&T)–Grand Lake	35.7	F	2	Charenton, LA	45	0	0	0	0	Sector	NA		80	8	80	80
Berwick Lock	1.5	L	1	Berwick, LA	45	300	14	9	9	Sector	NA	1950	125	8	80	80
Black Rock Channel & Tonawanda Harbor, NY																
Black Rock Lock	4	L	1	Buffalo, NY	70	625	5	22	22	Fixed	NA	1914	6	21	200	200
Black Warrior & Tombigbee Rivers, AL																
Coffeeville Lock	116.6	L	1	Coffeeville, AL	110	600	34	13	13	Moveable	1,185.0	1965	97	9	200	200
Demopolis Lock & Dam	213.2	L	1	Demopolis, AL	110	600	40	13	13	Moveable	1,485.0	1962	48	9	200	200
Warrior Lock & Dam	262	L	1	Eutaw, AL	110	600	22	13	13	Moveable	1,832.0	1962	77	9	200	200
William Bacon Oliver Lock & Dam	337.6	L	1	Tuscaloosa, AL	110	600	28	18	18	Moveable	800.0	1991	43	9	200	200
Holt Lock and Dam	347	L	1	Holt, AL	110	600	64	19	13	Moveable	1,138.0	1969	18	9	200	200
John Hollis Bankhead Lock & Dam	366	L	1	Adger, AL	110	600	68	13	13	Moveable	1,170.0	1975	43	9	200	200
Calcasieu River And Pass, LA																
Calcasieu Salt Wtr Barrier	38.9	B	1	West Lake, LA	56	575	0	0	0	Sector	450.0	1968	15	13	56	56
Canaveral Harbor, FL																
Canaveral Lock	3	L	1	Cape Canaveral, FL	90	600	3	13	13	Sector	NA	1965	6	12	100	100
Cape Fear River, NC																
Lock & Dam 1	39	L	1	Kings Bluff, NC	40	200	11	9	9	Miter	275.0	1915	32	8	100	100
Lock & Dam 2	71	L	1	Browns Landing, NC	40	200	9	12	12	Miter	229.0	1917	24	8	100	100
William O. Huske Lock & Dam	95	L	1	Tolars Landing, NC	40	300	9	9	9	Miter	220.0	1935	20	8	100	100
Chicago Harbor																
Chicago Lock	327.2	L	1	Chicago, IL	80	600	4	27	23	Sector	NA	1939	1	21	470	190
Colorado River, TX																
Colorado River East Lock	441.1	L	1	Matagorda, TX	75	1,180	12	15	15	Sector	520.0	1944	3,700	12	125	125
Colorado River West Lock	441.8	L	1	Matagorda, TX	75	1,180	12	15	15	Fixed	520.0	1944	3,700	12	125	125
Columbia River, OR and WA																
Bonneville Lock & Dam	145.3	L	1	Cascade Locks, OR	86	650	65	19	24	Miter	2,680.0	1938	1,080	14	300	300
The Dalles Dam	191.7	L	1	The Dalles, WA	86	650	88	15	15	Vertical	8,735.0	1957	1,000	14	250	250
John Day Lock & Dam	216.5	L	1	Rufus, WA	86	650	110	15	15	Vertical	5,900.0	1968	1,600	14	250	250
McNary Lock & Dam	292	L	1	Plymouth, WA	86	650	103	15	21	Miter	7,365.0	1953	64	14	250	250

Cross Florida Barge Canal																
Henry Holland Buckman Lock	20	L	1	Palatka, FL	84	600	20	14	15	Miter		1968	21	12	0	0
Inglis Lock Dam And and Spillway	92	L	3	Inglis, FL	84	600	28	18	15	Miter	5,100.0	1968	11	12	0	0
Cumberland River, TN & KY																
Barkley Lock & Dam	30.6	L	1	Grand Rivers, KY	110	800	57	24	13	Miter	9,959.0	1964	118	9	300	300
Cheatham Lock & Dam	148.7	L	1	Ashland City, TN	110	798	26	14	12	Miter	801.0	1956	68	9	200	200
Cordull Hull Lock & Dam	313.5	L	1	Carthage, TN	84	400	59	14	13	Miter	1,138.0	1973	72	9	150	150
Old Hickory Lock & Dam	216.2	L	1	Old Hickory, TN	84	397	60	14	13	Miter	3,605.0	1957	97	9	150	150
Four Rivers Basins																
Oklawaha River [no name]	66.9	L	1	Ocklawaha, FL	30	125	22	8	8	Moveable	50.0	1969	0	8	0	0
Freshwater Bayou, LA																
Freshwater Bayou Lock	1.2	L	1	Intracoastal City, LA	84	590	4	16	16	Sector	401.0	1968	250	12	120	120
Gulf Intracoastal Waterway between Apalachee Bay, FL & Mexican Border																
Algiers Lock	0	L	1	Algiers, LA	75	797	18	13	13	Sector	NA	1956	384	16	150	150
Harvey Lock	0	L	1	Harvey, LA	75	415	20	12	12	Miter	NA	1935	384	12	120	120
Inner Harbor Nav Canal Lock	7	L	1	New Orleans, LA	75	626	17	31	31	Miter	NA	1923	384	12	150	150
Bayou Sorrel Lock	37.5	L	1	Plaquemine, LA	56	790	21	14	14	Sector	NA	1952	64	12	120	120
Port Allen Lock	64.1	L	1	Port Allen, LA	84	1,188	45	13	14	Miter	NA	1961	64	12	120	120
Bayou Boeuf Lock	93.3	L	1	Morgan City, LA	75	1,148	11	13	13	Sector	NA	1956	384	16	150	150
Leland Bowman Lock	162.7	L	1	Abbeville, LA	110	1,190	5	11	11	Sector	NA	1934	384	16	200	200
Calcasieu Lock	238.5	L	1	Lake Charles, LA	75	1,194	4	13	13	Sector	NA	1950	384	16	200	200
Brazos East Gate	400.8	F	1	Freeport, TX	75	750	0	15	15	Sector	520.0	1943	3,700	12	125	125
Brazos West Gate	401.1	F	1	Freeport, TX	75	750	0	15	15	Sector	520.0	1944	3,700	12	125	125
Green & Barren Rivers, KY																
Green River Lock & Dam 1	9.1	L	1	Spottsville, KY	84	600	8	12	15	Miter	760.0	1956	54	9	150	150
Green River lock Lock & Dam 2	63.1	L	1	Calhoun, KY	84	600	14	12	12	Miter	512.0	1956	45	9	150	150
Hudson River, NY																
Troy Lock & Dam	153.8	L	1	Troy, NY	45	493	17	16	13	Fixed	1,495.0	1916	2	14	200	200
Illinois Waterway, IL																
Lagrange Lock & Dam	80.2	L	1	Beardstown, IL	110	600	10	16	13	Miter	1,066.0	1939	78	9	300	300
Peoria Lock & Dam	157.7	L	1	Creve Coeur, IL	110	600	11	15	12	Miter	3,446.0	1938	73	9	300	300
Starved Rock Lock & Dam	231	L	1	Utica, IL	110	600	19	17	14	Miter	1,280.0	1933	14	9	300	300
Marseilles Lock & Dam	244.6	L	1	Marseilles, IL	110	600	24	19	14	Miter	778.5	1933	27	9	300	300
Dresden Island Lock & Dam	271.5	L	1	Morris, IL	110	600	22	17	12	Miter	1,615.5	1933	15	9	300	300
Lockport Lock	291.1	L	1	Lockport, IL	110	600	39	20	15	Miter	500.0	1933	35	9	300	300
Thomas J. O'Brien Lock & Dam–Calumet River																
Thomas J. O'Brien Lock & Dam	326.5	L	1	Chicago, IL	110	1,000	4	18	18	Sector	256.75	1960	7	9	300	300
Brandon Road Lock &Dam–Illinois River																
Brandon Road Lock & Dam	286	L	1	Joliet, IL	110	600	34	18	14	Miter	2,373.0	1933	5	9	160	160
Kanawha Lock & Dam, WV																
Winfield Locks & Dam Main 800	31.1	L	1	Winfield, WV	110	800	28	18	18	Miter	700.0	1936	37	9	300	300
Winfield Locks & Dam Main 1	31.1	L	1	Winfield, WV	56	360	28	18	12	Miter	700.0	1,935	37	9	300	300
Winfield Locks & Dam Main 2	31.1	L	1	Winfield, WV	56	360	28	18	12	Miter	700.0	1935	37	9	300	300
Marmet Locks & Dam Main 1	67.7	L	1	Marmet, WV	56	360	24	18	12	Miter	557.0	1934	16	9	150	150
Marmet Locks & Dam Main 2	67.7	L	1	Marmet, WV	56	360	24	18	12	Miter	557.0	1934	16	9	150	150
London Locks & Dam Main 1	82.8	L	1	London, WV	56	360	24	18	12	Miter	557.0	1934	8	9	150	150
London Locks & Dam Main 2	82.8	L	1	London, WV	56	360	24	18	12	Miter	557.0	1934	8	9	150	150
Kaskaskia River, IL																
Kaskaskia River Navigation Lock	0.8	L	1	Modoc, IL	84	600	29	19	11	Miter	120.0	1973	35	9	225	225
Kentucky River, KY																
Lock & Dam 1	4	L	1	Carrollton, KY	38	145	8	8	15	Miter	424.0	1839	27	6	150	150
Lock & Dam 2	31	L	1	Lockport, KY	38	145	14	8	6	Fixed	400.0	1839	11	6	150	150
Lock & Dam 3	42	L	1	Gest, KY	38	145	13	9	6	Miter	465.0	1844	23	6	150	150
Lock & Dam 4	65	L	1	Frankfort, KY	38	145	13	6	7	Fixed	534.0	1844	17	6	150	150
Lock & Dam 5	82.2	L	2	Tyrone, KY	38	145	15	10	6	Fixed	556.0	1844	14	6	100	100
Lock & Dam 6	96.2	L	2	High Bridge, KY	52	147	14	9	6	Fixed	413.0	1891	21	6	100	100
Lock & Dam 7	117	L	2	High Bridge, KY	52	147	15	9	7	Fixed	350.0	1897	23	6	100	100
Lock & Dam 8	139.9	L	2	Camp Nelson, KY	52	146	19	11	6	Fixed	257.0	1900	18	6	100	100
Lock & Dam 9	157.5	L	2	Valley View, KY	52	148	17	11	7	Fixed	362.0	1907	19	6	100	100
Lock & Dam 10	176.4	L	2	Ford, KY	52	148	17	9	6	Fixed	472.0	1907	25	6	100	100
Lock & Dam 11	201	L	2	Irvine, KY	52	148	18	10	6	Fixed	208.0	1906	20	6	100	100

(Continued)

Table 7L.148 (Continued)

Project	River Mile	Type of Structure[a]	Status[b]	Community in Vicinity	Locks: Width of Chamber	Locks: Chamber Useable Length	Locks: Lift at Normal Pool Level	Locks: Depth of Miter Sill: Upper	Locks: Depth of Miter Sill: Lower	Locks: Type	Dam: Length (ft)	Dam: Year Opened	Dam: Length (miles)	Dam: Depth (ft)	Authorized Channel: Width (above) (ft)	Authorized Channel: Width (below) (ft)
Lock & Dam 12	220.9	L	2	Ravenna, KY	52	148	17	10	6	Fixed	240.0	1910	19	6	100	100
Lock & Dam 13	239.9	L	2	Willow, KY	52	148	18	10	6	Fixed	248.0	1915	9	6	100	100
Lock & Dam 14	249	L	2	Heidelberg, KY	52	148	17	9	6	Fixed	248.0	1917	26	6	100	100
Lake Washington Ship Canal																
Hiram M. Chittenden Lock AUX —1 [Ballard Locks]	0	L	1	Seattle, WA	28	123	26	16	16	Miter	235.0	1916	0	30	0	0
Hiram M. Chittenden Lock [Ballard Locks]	0	L	1	Seattle, WA	80	760	26	36	29	Miter	235.0	1916	8	30	100	150
McClellan-Kerr Arkansas River Navigation System, AR and OK																
Norrell Lock & Dam [Lock & Dam 1]	10.3	L	1	Arkansas Post, AR	110	600	30	16	15	Fixed	277.0	1967	3	9	300	300
Lock & Dam 2	13.3	L	1	Arkansas Post, AR	110	600	20	18	14	Tainter	1,120.0	1967	37	9	300	300
Joe Hardin L & D [Lock & Dam 3]	50.2	L	1	Grady, AR	110	600	20	18	14	Tainter	1,260.0	1968	16	9	250	250
Emmett Sanders Lock & Dam [Lock & Dam 4]	66	L	1	Pine Bluff, AR	110	600	14	18	14	Tainter	1,190.0	1968	20	9	250	250
Lock & Dam 5	86.3	L	1	Redfield, AR	110	600	17	18	14	Tainter	1,050.0	1968	22	9	250	250
David D. Terry Lock & Dam [Lock & Dam 6]	108.1	L	1	Little Rock, AR	110	600	18	18	14	Tainter	1,190.0	1968	17	9	250	250
Murray Lock & Dam [Lock & Dam 7]	125.4	L	1	Little Rock, AR	110	600	18	18	14	Tainter	980.0	1969	31	9	250	250
Toad Suck Ferry Lock & Dam [Lock & Dam 8]	155.9	L	1	Conway, AR	110	600	16	18	14	Tainter	1,200.0	1969	21	9	250	250
Arthur V. Ormond L & D [Lock & Dam 9]	176.9	L	1	Morrilton, AR	110	600	19	18	14	Tainter	1,797.0	1969	29	9	250	250
Dardanelle Lock & Dam [Lock & Dam 10]	205.5	L	1	Russellville, AR	110	600	55	18	14	Tainter	1,210.0	1969	51	9	250	250
Ozark Lock & Dam [Lock & Dam 12]	256.8	L	1	Ozark, AR	110	600	34	18	15	Tainter	900.0	1969	36	9	250	250
James W. Trimble Lock & Dam [Lock & Dam 13]	292.8	L	1	Fort Smith, AR	110	600	20	18	14	Tainter	1,050.0	1969	27	9	250	250
W.D. Mayo Lock & Dam [Lock & Dam 14]	319.6	L	1	Spiro, OK	110	600	21	14	14	Moveable	840.0	1970	17	9	250	250
Robert S. Kerr Lock & Dam & Res [Lock & Dam 15]	336.2	L	1	Salisaw, OK	110	600	48	16	14	Moveable	1,090.0	1970	30	9	250	250
Webbers Falls Lock & Dam [Lock & Dam 16]	366.6	L	1	Webber Falls, OK	110	600	30	16	14	Moveable	720.0	1970	35	9	250	250
Chouteau Lock & Dam [Lock & Dam 17]	5	L	1	Muskogee, OK	110	600	21	15	14	Moveable	210.0	1970	20	9	150	150
Newt Graham Lock & Dam [Lock & Dam 18]	26	L	1	Inola, OK	110	600	21	15	14	Moveable	210.0	1970	23	9	150	150
Montgomery Point Lock & Dam	0.5	L	1	Tichnor, AR	110	600	20	18	15	Miter	300	2004	10	9	300	300
Mermentau River, LA																
Schooner Bayou Control Structure	3.4	C	1	Abbeville, LA	75	525	0	0	0	Sector	NA		11	6	60	60
Catfish Point Control Structure	25	C	1	Creole, LA	56	500	0	0	0	Sector	NA		5	15	200	200
Mississippi River between Missouri River & Minneapolis, MN																
Chains Of of Rocks L/D 27	185.5	L	1	Granite City, IL	110	1,200	21	15	15	Vertical	3,000.0	1962	15	9	350	350
Chains Of of Rocks L/D 27 AUX —1	185.5	L	1	Granite City, IL	110	600	21	15	15	Miter	3,000.0	1962	15	9	350	350
Mel Price Lock & Dam [Locks No 26 Main)	200.8	L	1	East Alton, IL	110	1,200	24	23	18	Vertical	990.0	1989	40	9	300	300
Mel Price Lock & Dam [Locks No 26 AUX)	200.8	L	1	East Alton, IL	110	600	24	42	18	Miter	990.0	1989	40	9	300	300
Lock & Dam 25	241.4	L	1	Winfield, MO	110	600	15	19	12	Miter	1,140.0	1939	32	9	300	300
Lock & Dam 24	273.4	L	1	Clarksville, MO	110	600	15	19	12	Miter	1,200.0	1940	27	9	300	300
Lock & Dam 22	301.2	L	1	Saverton, MO	110	600	10	18	14	Miter	3,084.0	1938	24	9	300	300
Lock & Dam 21	324.9	L	1	Quincy, IL	110	600	10	17	12	Miter	2,955.0	1938	18	9	300	300
Lock & Dam 20	343.2	L	1	Canton, MO	110	600	10	15	12	Miter	2,144.0	1936	21	9	300	300
Lock & Dam 19	364.3	L	1	Keokuk, IA	110	1,200	38	15	13	Vertical	8,809.0	1913	46	9	300	300
Lock & Dam 18	410.5	L	1	Gladstone, IL	110	600	10	17	14	Miter	6,960.0	1937	27	9	300	300
Lock & Dam 17	437.1	L	1	New Boston, IL	110	600	8	16	13	Miter	3,196.0	1939	20	9	300	300
Lock & Dam 16	457.2	L	1	Muscatine, IL	110	600	9	17	12	Miter	3,555.0	1937	26	9	300	300
Lock & Dam 15	482.9	L	1	Rock Island, IL	110	600	16	27	11	Miter	1,203.0	1934	10	9	300	300
Lock & Dam 15 AUX —1	482.9	L	1	Rock Island, IL	110	360	16	27	11	Miter	1,203.0	1934	0	9	300	300
Lock & Dam 14	493	L	1	LeClaire, IA	110	600	11	20	13	Miter	2,703.0	1922	29	9	300	300
Lock & Dam 14 AUX —1	493	L	2	LeClaire, IA	80	320	11	21	14	Miter	2,703.0	1939	29	9	300	300
Lock & Dam 13	522.5	L	1	Clinton, IL	110	600	11	19	13	Miter	1,407.0	1939	34	9	300	300
Lock &Dam 12	556.7	L	1	Bellevue, IA	110	600	9	17	13	Miter	8,369.0	1938	26	9	300	300
Lock & Dam 11	583	L	1	Dubuque, IA	110	600	12	19	13	Miter	4,784.0	1937	32	9	300	300
Lock & Dam 10	615.1	L	2	Guttenburg, IA	110	600	8	15	12	Miter	763.0	1937	33	9	300	300
Lock & Dam 9	647.9	L	2	Lynxville, WI	110	600	9	16	13	Miter	811.0	1937	31	9	300	300
Lock & Dam 8	679.2	L	2	Genoa, WI	110	600	11	22	14	Miter	935.0	1937	23	9	300	300
Lock & Dam 7	702.5	L	2	Dresbach, MN	110	600	8	18	12	Miter	940.0	1937	12	9	300	300

Lock & Dam 6	714.3	L	2	Trempealeau, WI	110	600	6	17	13	Miter	893.0	1936	14	9	300	300
Lock & Dam 5A	728.5	L	2	Winona,MN, MN	110	600	5	18	13	Miter	682.0	1936	10	9	300	300
Lock & Dam 5	738.1	L	2	Minneiska, MN	110	600	9	18	12	Miter	1,619.0	1935	15	9	300	300
Lock & Dam 4	752.8	L	2	Alma, WI	110	600	7	17	13	Miter	1,367.0	1935	44	9	300	300
Lock & Dam 3	796.9	L	2	Red Wing, MN	110	600	8	17	14	Miter	365.0	1938	18	9	300	300
Lock & Dam 2	815.2	L	2	Hastings, MN	110	500	12	22	13	Miter	822.0	1948	32	9	300	300
Lock & Dam 1 Main Chamber 1	847.6	L	2	Minn. St. Paul, MN	56	400	38	13	8	Miter	574.0	1932	6	9	300	300
	847.6	L	2	Minn. St. Paul, MN	56	400	38	13	10	Miter	579.0	1917	6	9	300	300
St. Anthony Falls Lower Lock & Dam	853.3	L	2	Minneapolis, MN	56	400	25	14	10	Miter	188.0	1956	1	9	200	200
St. Anthony Falls Upper Lock & Dam	853.9	L	2	Minneapolis, MN	56	400	49	16	14	Miter	NA	1963	4	9	100	100
Monongahela River, PA and WV																
Lock & Dam 2	11.2	L	1	Braddock, PA	110	720	9	15	16	Miter	748.0	1951	13	9	300	300
Lock & Dam 2 AUX	11.2	L	1	Braddock, PA	56	360	9	15	16	Miter	748.0	1951	13	9	300	300
Lock & Dam 3	23.8	L	1	Elizabeth, PA	56	720	8	11	11	Miter	670.0	1907	18	9	300	300
Lock & Dam 3 AUX	23.8	L	1	Elizabeth, PA	56	360	8	11	11	Miter	670.0	1907	18	9	300	300
Lock & Dam 4	41.5	L	1	Monessen, PA	56	720	17	20	10	Miter	535.0	1932	20	9	300	300
Lock & Dam 4 AUX	41.5	L	1	Monessen, PA	56	360	17	20	10	Miter	535.0	1932	20	9	300	300
Maxwell Lock & dam	61.2	L	1	Maxwell, PA	84	720	20	20	14	Miter	460.0	1964	24	9	300	300
Maxwell Lock & dam AUX	61.2	L	1	Maxwell, PA	84	720	20	20	14	Miter	460.0	1964	24	9	300	300
Grays Landing Lock & Dam	82	L	1	Grays Landing, PA	84	720	15	27	18	Miter	576.0	1995	21	9	300	300
L & D 8	90.8	L	1	Point Marion, PA	84	720	19	16	35	Miter	682.0	1925	11	9	300	300
Morgantown Lock & Dam	102	L	1	Morgantown, WV	84	600	17	17	14	Miter	410.0	1950	6	9	300	300
Hildebrand Lock & Dam	108	L	1	Morgantown, WV	84	600	21	14	15	Miter	530.0	1959	7	9	300	300
Opekiska Lock & Dam	115.4	L	1	Opekiska, WV	84	600	22	18	14	Miter	366.0	1964	7	9	300	300
Ohio River																
Emsworth Lock & Dam	6.2	L	1	Emsworth, PA	110	600	18	17	13	Miter	1,717.0	1921	6	9	300	300
Emsworth Lock & Dam AUX	6.2	L	1	Emsworth, PA	56	360	18	16	13	Miter	1,717.0	1921	6	9	300	300
Dashields Lock & Dam	13.3	L	1	Glenwillard, PA	110	600	10	13	18	Miter	1,585.0	1929	7	9	300	300
Dashields Lock & Dam AUX	13.3	L	1	Glenwillard, PA	56	360	10	13	18	Miter	1,585.0	1929	7	9	300	300
Montgomery Lock & Dam	31.7	L	1	Monaca, PA	110	600	18	16	15	Miter	1,379.0	1936	18	9	300	300
Montgomery Lock & Dam AUX	31.7	L	1	Monaca, PA	56	360	18	16	15	Miter	1,379.0	1936	18	9	300	300
New Cumberland Lock & Dam	54.4	L	1	Stratton, OH	110	1200	21	17	15	Miter	1,315.0	1961	23	9	300	300
New Cumberland Lock & Dam AUX	54.4	L	1	Stratton, OH	110	600	21	17	15	Miter	1,315.0	1961	23	9	300	300
Pike Island Lock & Dam	84.2	L	1	Warwood, WV	110	1200	18	17	18	Miter	1,315.0	1965	30	9	300	300
Pike Island Lock & Dam	84.2	L	1	Warwood, WV	110	600	18	17	18	Miter	1,315.0	1965	30	9	300	300
Hannibal Locks & Dam	126.4	L	1	Hannibal, OH	110	1200	21	38	17	Miter	1,098.0	1972	36	9	300	300
Hannibal Locks & Dam AUX	126.4	L	1	Hannibal, OH	110	600	21	38	17	Miter	1,098.0	1972	36	9	300	300
Willow Island Locks & Dam	161.7	L	1	Newport, OH	110	1200	20	35	15	Miter	1,128.0	1973	35	9	300	300
Willow Island Locks & Dam AUX	161.7	L	1	Newport, OH	110	600	20	35	15	Miter	1,128.0	1973	35	9	300	300
Belleville Locks & Dam	203.9	L	1	Reedsville, OH	110	1200	22	37	15	Miter	1,206.0	1969	36	9	300	300
Belleville Locks & Dam AUX	203.9	L	1	Reedsville, OH	110	600	22	37	15	Miter	1,206.0	1969	36	9	300	300
Racine Locks & Dam	237.5	L	1	Letart, WV	110	1200	22	37	15	Miter	1,173.0	1970	34	9	300	300
Racine Locks & Dam AUX	237.5	L	1	Letart, WV	110	600	22	37	15	Miter	1,173.0	1970	34	9	300	300
Robert C. Byrd	279.2	L	1	Hogsett, WV	110	1200	23	41	18	Miter	1,132.0	1937	42	9	300	300
Robert C. Byrd AUX	279.2	L	1	Hogsett, WV	110	600	23	41	18	Miter	1,132.0	1937	42	9	300	300
Greenup Locks & Dam	341	L	1	Greenup, KY	110	1200	30	45	15	Miter	1,287.0	1962	62	9	300	300
Greenup Locks & Dam AUX	341	L	1	Greenup, KY	110	600	30	45	15	Miter	1,287.0	1962	62	9	300	300
Capt Anthony Meldahl Lock & Dam	436.2	L	1	Chilo, OH	110	1200	30	45	15	Miter	1,756.0	1962	95	9	300	300
Capt Anthony Meldahl Lock & Dam AUX	436.2	L	1	Chilo, OH	110	600	30	45	15	Miter	1,756.0	1962	95	9	300	300
Markland Locks & Dam	531.5	L	1	Warsaw, KY	110	1200	35	50	15	Miter	1,395.0	1963	95	9	300	300
Markland Locks & Dam AUX	531.5	L	1	Markland, KY	110	600	35	50	15	Miter	1,395.0	1963	75	9	300	300
Mcalpine Locks & Dam	606.8	L	1	Louisville, KY	110	1200	37	49	12	Miter	8,725.0	1963	75	9	300	300
Mcalpine Locks & Dam AUX	606.8	L	1	Louisville, KY	110	600	37	19	11	Miter	8,725.0	1963	75	9	300	300
Cannelton Lock & Dam	720.7	L	1	Cannelton, IN	110	1200	25	40	15	Miter	2,054.0	1971	114	9	300	300
Cannelton Lock & Dam AUX	720.7	L	1	Cannelton, IN	110	600	25	40	15	Miter	2,054.0	1971	114	9	300	300
Newburgh Lock & Dam	776.1	L	1	Newburgh, IN	110	1200	16	31	15	Miter	2,275.6	1975	55	9	300	300
Newburgh Lock & Dam AUX	776.1	L	1	Newburgh, IN	110	600	16	31	15	Miter	2,275.6	1975	55	9	300	300
John T. Myers Lock & Dam	846	L	1	Mount Vernon, IN	110	1200	18	34	16	Miter	3,504.0	1975	70	9	300	300
John T. Myers Lock & Dam AUX	846	L	1	Mount Vernon, IN	110	600	18	34	16	Miter	3,504.0	1975	70	9	300	300
Smithland Lock & Dam	918.5	L	1	Hamletsburg, IL	110	1200	22	34	12	Miter	2,962.0	1980	73	9	300	300
Smithland Lock & Dam AUX	918.5	L	1	Hamletsburg, IL	110	1200	22	34	12	Miter	2,962.0	1980	73	9	300	300

(Continued)

Table 7L.148 (Continued)

					Locks						Dam				Authorized Channel	
								Depth of Miter Sill								
Project	River Mile	Type of Structure[a]	Status[b]	Community in Vicinity	Width of Chamber	Chamber Useable Length	Lift at Normal Pool Level	Upper	Lower	Type	Length (ft)	Year Opened	Length (miles)	Depth (ft)	Width (above) (ft)	Width (below) (ft)
Lock & Dam 52	938.9	L	1	Brookport, IL	110	1200	12	15	11	Miter	2,998.0	1928	20	9	300	300
Lock & Dam 52 AUX	938.9	L	1	Brookport, IL	110	600	12	15	11	Miter	2,978.0	1928	20	9	300	300
Lock & Dam 53	962.6	L	1	Mound City, IL	110	1200	12	15	10	Miter	3,560.0	1929	24	9	300	300
Lock &Dam 53 AUX	962.6	L	1	Mound City, IL	110	600	12	15	10	Miter	3,560.0	1929	24	9	300	300
Okeechobee Waterway, FL																
S-78 Waterway (Ortona Lock & Dam)	93.6	L	1	LaBelle, FL	50	225	11	12	15	Sector	104.0	1937	16	8	90	90
W.P.Franklin Lock/Cntl Strct [Olga Lock]	122	L	1	Fort Myers, FL	56	400	3	13	13	Sector	1,150.0	1965	8	8	90	90
S-308-B	38.5	L	1	Port Mayaca, FL	56	400	2	17	17	Sector	116.0	1977	5	8	100	100
St. Lucie Lock & Dam																
St. Lucie Lock & Dam	15.3	L	1	Stuart, FL	50	225	13	15	13	Sector	170.0	1944	15	8	80	80
Moore Haven Lock																
Moore Haven Lock	78	L	1	Moore Haven, FL	50	250	2	10	11	Sector	89.8	1953	16	8	90	90
Old River, LA (MR&T)																
Old River Lock	1	L	1	Simmesport, LA	75	1190	35	11	11	Miter	1,100.0	1963	7	12	120	120
Ouachita & Black Rivers Below Camden, AR & LA																
Columbia Lock & Dam	117.2	L	1	Columbia, LA	84	600	18	18	18	Miter	400.0	1972	109	9	100	100
Felsenthal Lock & Dam [Lock & Dam No 3]	226.8	L	1	Felsenthal, AR	84	600	18	18	13	Miter	350.0	1984	55	9	100	100
H. K. Thatcher Lock & Dam [Lock & Dam No 4]	281.7	L	1	Calion, AR	84	600	12	18	13	Miter	350.0	1984	50	7	100	100
Jonesville Lock & Dam	25	L	1	Jonesville, LA	84	600	30	18	15	Miter	450.0	1972	92	9	100	100
Pearl River Lateral Canal																
Lock 1	29.7	L	1	Pearl River, LA	65	274	27	10	10	Miter	NA	1951	11	7	80	80
Lock 2	40.8	L	1	Bush, LA	65	274	15	10	10	Miter	NA	1951	3	7	80	80
Lock 3	43.9	L	1	Sun, LA	65	274	11	10	10	Miter	NA	1951	5	7	80	80
Red River WW-Mississippi R to Shreveport, LA																
Lindy Claiborne Boggs [Red River Lock & Dam No. 1]	44	L	1	Larto, LA	84	685	36	22	13	Miter	630.0	1984	30	9	200	200
John H.Overton [Red River Lock & Dam No. 2]	74	L	1	Ruby, LA	84	685	24	23	23	Miter	348.0	1986	42	9	200	200
Red River Lock & Dam No. 3	116.4	L	1	Colfax, LA	84	685	31	25	18	Miter	432.0	1992	40	9	100	100
Red River Lock & Dam No. 4	169	L	1	Coushatta, LA	84	685	25	25	18	Miter	690.0	1994	31	9	200	200
Joe D. Waggonner Lock & Dam [5]	200	L	1	Caspiana, LA	84	685	25	22	23	Miter	663.0	1994	28	9	200	200
Savannah River, GA																
New Savannah Bluff Lock & Dam	187.2	L	1	Augusta, GA	56	360	15	14	12	Miter	360.0	1937	16	9	90	90
Snake River, WA																
Ice Harbor Lock & Dam	9.7	L	1	Pasco, WA	86	650	103	15	14	Vertical	2,790.0	1962	32	14	250	250
Lower Monumental Lock & Dam	41.6	L	1	Kahlotus, WA	86	650	103	15	15	Vertical	3,800.0	1969	29	14	250	250
Little Goose Lock & Dam	70.3	L	1	Starbuck, WA	86	650	101	15	15	Miter	2,655.0	1970	37	14	250	250
Lower Granite Lock & Dam	107.5	L	1	Pomeroy, WA	86	650	105	15	15	Miter	3,200.0	1975	38	14	250	250
St. Marys River, MI																
Macarthur Lock	47	L	1	Sault Ste. Marie, MI	80	780	22	31	31	Leaf	1,300.0	1914	0	27	300	300
Davis Lock	47	L	2	Sault Ste. Marie, MI	80	1320	22	24	23	Leaf	1,300.0	1914	0	0	300	300
New Poe Lock	47	L	1	Sault Ste. Marie, MI	110	1200	22	32	32	Leaf	1,300.0	1914	0	0	300	300
Tennessee River, TN, AL, & KY																
Kentucky Lock	22.4	L	1	Grand Rivers, KY	110	600	57	24	13	Miter	7,976.0	1944	184	9	300	300
Pickwick Landing Lock	206.7	L	1	Pickwick Dam, TN	110	1000	55	19	17	Miter	7,385.0	1984	53	9	300	300
Pickwick Landing Lock AUX	206.7	L	1	Pickwick Dam, TN	110	600	55	16	17	Miter	7,385.0	1937	53	9	300	300
Wilson Lock	259.4	L	1	Florence, AL	110	600	94	11	11	Miter	3,728.0	1959	16	9	300	300
Wilson Lock	259.4	L	1	Florence, AL	60	300	49	13	13	Miter	3,728.0	1927	16	9	300	300
	259.4	L	1	Florence, AL	60	300	45	11	11	Miter	3,728.0	1927	16	9	300	300
Gen Jos Wheeler Lock	274.9	L	1	Rogersville, AL	110	600	48	15	13	Miter	5,738.0	1933	74	9	300	300
Gen Jos Wheeler Lock AUX	274.9	L	1	Rogersville, AL	60	400	48	15	13	Miter	5,738.0	1933	74	9	300	300
Guntersville Lock [Lock 5, CH 1, Main Channel]	349	L	1	Guntersville, AL	110	600	39	17	18	Miter	3,837.0	1939	76	9	300	300
Guntersville Lock [Lock 5, CH 1, AUX]	349	L	1	Guntersville, AL	60	360	39	17	18	Miter	3,837.0	1939	76	9	300	300
Nickajac Lock	424.7	L	1	Jasper, TN	110	600	39	13	11	Miter	3,763.0	1967	46	9	300	300
Chickamauga Lock	471	L	1	Chattanooga, TN	60	360	49	10	14	Miter	5,654.0	1940	59	9	300	300
Watts Bar Lock	529.9	L	1	Breendenton, TN	60	360	58	12	12	Miter	2,646.0	1942	72	9	300	300

Fort Loudon Lock	602.3	L	1	Lenoir City, TN	60	360	72	12	12	Miter	3,687.0	1943	50	9	300	300
Melton Hill Lock & Dam (Clinch River)	23.1	L	1	Kingston, TN	75	400	58	13	13	Miter	1,072.0	1963	38	9	300	300
Tennessee-Tombigbee Waterway AL, MS																
Howell Heflin Lock & Dam [Gainesville Lock & Dam]	266.1	L	1	Gainesville, AL	110	600	36	15	15	Moveable	817.0	1978	49	9	300	300
Tom Bevill Lock & Dam [Aliceville Lock & Dam]	306.8	L	1	Aliceville, AL	110	600	27	15	15	Moveable	647.0	1979	41	9	300	300
John C. Stennis Lock & Dam [Columbus Lock & Dam]	334.7	L	1	Columbus, MS	110	600	27	15	15	Moveable	573.0	1981	28	9	300	300
Aberdeen Lock & Dam	357.5	L	1	Aberdeen, MS	110	600	27	15	15	Moveable	641.0	1984	23	9	300	300
Amory Lock [Lock A]	371.1	L	1	Amory, MS	110	600	30	15	15	Moveable	284.0	1985	14	9	300	300
Glover Wilkins Lock [Lock B]	376.3	L	1	Smithville, AL	110	600	25	18	18	Moveable	779.0	1985	15	9	300	300
Fulton Lock [Lock C]	391	L	1	Fulton, MS	110	600	25	18	18	Moveable	396.0	1985	7	9	300	300
John Rankin Lock [Lock D]	398.4	L	1	Fulton, MS	110	600	30	18	18	Moveable	282.0	1985	8	9	300	300
G.V. "Sonny" Montgomery Lock [Lock E]	406.7	L	1	Belmont, MS	110	600	30	18	18	Moveable	449.0	1985	5	9	300	300
Jamie Whitten Lock & Dam [Bay Springs Lock & Dam]	411.9	L	1	Tupelo, MS	110	600	84	18	18	Moveable	2,750.0	1985	65	9	300	300
Willamette River at Willamette Falls																
2	26	L	1	West Linn, OR	40	198	10	6	8	Leaf	NA	1872	0	6	150	150
3	26	L	1	West Linn, OR	40	198	10	6	8	Leaf	NA	1872	0	6	150	150
4	26	L	1	West Linn, OR	40	198	10	6	8	Leaf	NA	1872	0	6	150	150
Willamette River At at Willamette Falls	26	L	1	West Linn, OR	40	198	20	6	8	Leaf	NA	1872	0	6	150	150
Willamette Falls Guard Lock	26	L	1	West Linn, OR	40	198	10	6	8	Leaf	NA	1872	24	6	150	150

[a] L, Lock & Dam; F-Flood Gate; B-Barrier; C-Control Structure.

[b] 1, Operational; 2, Seasonal; 3, Operational Weekend/Summer.

Source: From USACE, Navigation Data Center, U.S. Waterway Data Lock Characteristics, www.iwr.usace.army.mil.

Table 7L.149 Lock Characteristics Operational Statistics—1999

						Average Tow Size			Lock Closures	
River/ Lock	Chamber	River Mile	Year Open	Average Processing Time[a]	Total Tonnage	Barges/ Tow	Tons/ Tow	Percent Chamber Available	Total Number	Avg Time[b]
Alabama										
Claiborne	Main	117.5	1969	23	236,895	4	2,096	100.00	0	0.00
Albemarle & Chesapeake Canal										
Great Bridge	Main	11.5	1932	18	545,630	1	610	99.72	8	3.05
Allegheny										
2	Main	6.7	1934	30	2,611,580	3	1,687	93.47	9	63.53
3	Main	14.5	1934	32	2,567,580	3	1,691	97.63	2	103.93
4	Main	24.2	1927	28	1,612,797	2	1,122	99.76	9	2.35
5	Main	30.4	1927	24	766,237	4	1,593	94.25	4	125.85
6	Main	36.3	1928	20	161,199	2	844	87.77	3	357.13
7	Main	45.7	1930	19	133,662	2	941	70.99	6	423.52
8	Main	52.6	1931	22	673,120	1	784	68.80	5	546.67
9	Main	62.2	1938	15	0	0	0	48.22	7	647.97
Apalachicola										
Jim Woodruff	Main	106.3	1954	29	337,305	2	799	100.00	0	0.00
Arkansas										
Norrell	Main	10.3	1967	37	8,797,103	6	5,215	99.63	13	2.47
2	Main	13.3	1967	44	8,826,997	6	5,239	98.59	188	0.67
Joe Hardin	Main	50.2	1968	45	8,048,954	7	6,211	99.95	1	4.33
Emmett Sanders	Main	66.0	1968	43	8,043,052	7	6,274	100.00	0	0.00
5	Main	86.3	1968	40	7,431,516	7	6,219	100.00	0	0.00
David D. Terry	Main	108.1	1968	39	7,387,876	7	6,224	99.99	1	0.58
Murray	Main	125.4	1969	41	6,490,453	7	6,193	99.75	14	1.53
Toad Suck Ferry	Main	155.9	1969	51	6,364,044	7	6,137	100.00	0	0.00
Arthur V. Ormond	Main	176.9	1969	53	5,973,962	7	6,322	99.70	9	2.93
Dardanelle	Main	205.5	1969	50	5,993,626	7	6,336	99.94	2	2.72
Ozark—Jeta Taylor	Main	256.8	1969	46	4,361,324	6	6,500	99.95	1	4.00
James W. Trimble	Main	292.8	1969	38	4,608,165	3	3,278	99.94	4	1.22
W.D. Mayo	Main	319.6	1970	57	4,132,431	5	6,477	100.00	0	0.00
Robert S. Kerr	Main	336.2	1970	53	4,155,658	5	6,544	100.00	0	0.00
Webbers Falls	Main	366.6	1970	54	4,007,128	5	6,232	100.00	0	0.00
Atchafalaya										
Berwick	Main	1.5	1950	10	169,881	1	504	99.96	1	3.65
Black										
Jonesville	Main	25.0	1972	30	1,573,714	2	2,377	100.00	0	0.00
Black Rock Channel										
Black Rock	Main	4.0	1914	14	285,595	1	4,327	39.50	251	21.12
Black Warrior										
Armistead I. Seldon	Main	262.0	1957	51	9,458,220	5	5,753	100.00	0	0.00

Calcasieu River										
Calcasieu Barrier	Cntrl	38.9	1968	3	627,857	2	1,618	100.00	0	0.00
Caloosahatchee										
Moore Haven	Main	78.0	1953	15	21,326	1	117	64.70	384	8.05
W. P. Franklin	Main	122.0	1965	15	14,285	1	121	65.71	367	8.18
Canaveral Barge Canal										
Canaveral	Main	3.0	1965	14	873,290	1	1,078	66.99	391	7.40
Cape Fear										
1	Main	39.0	1915	20	0	0	0	100.00	0	0.00
2	Main	71.0	1917	19	0	0	0	85.96	2	614.98
William O. Huske	Main	95.0	1935	20	0	0	0	85.48	3	423.98
Chicago Harbor Channel										
Chicago	Main	327.2	1939	17	315,530	1	1,005	75.31	19	113.85
Columbia										
Bonneville	Main	145.3	1993	37	9,678,824	3	4,228	97.77	8	24.38
The Dalles	Main	191.7	1957	31	9,271,751	3	4,259	97.93	3	60.52
John Day	Main	216.5	1968	39	8,600,616	3	4,379	97.85	3	62.67
Mcnary	Main	292.0	1953	26	7,604,466	3	4,282	96.95	4	66.70
Cumberland										
Barkley	Main	30.6	1964	56	9,072,278	8	5,305	99.07	33	2.47
Cheatham	Main	148.7	1952	61	9,542,172	8	6,101	95.94	112	3.17
Old Hickory	Main	216.2	1954	55	3,933,051	5	4,733	94.45	27	18.00
Dismal Swamp Canal										
Deep Creek	Main	10.6	1940	20	300	0	300	78.31	5	379.95
South Mills	Main	33.2	1941	23	0	0	0	80.60	6	283.18
Freshwater Bayou										
Freshwater Bayou	Main	1.2	1968	26	4,482,515	38	8,379	100.00	0	0.00
GIWW										
Harvey	Main	0.0	1935	35	4,215,364	2	880	98.98	32	2.78
Inner Harbor Nvg Chnl	Main	5.6	1923	46	19,416,663	3	2,143	90.23	479	1.78
Bayou Boeuf	Main	93.3	1954	38	22,193,235	2	1,689	98.53	23	5.62
Leland Bowman	Main	162.7	1985	30	40,563,247	3	2,895	95.65	40	9.53
Calcasieu	Main	238.5	1950	36	39,525,354	3	2,948	96.07	59	5.83
Giww Algiers Canal										
Algiers	Main	0.0	1956	48	19,222,359	3	2,878	86.51	132	8.95
Giww Port Allen–Morgan Cty Alt. Rte										
Bayou Sorrel	Main	37.5	1952	55	23,157,835	3	3,742	95.03	68	6.40
Port Allen	Main	64.1	1961	81	24,167,141	3	3,436	98.02	59	2.93
Giww Texas										
Brazos East	Main	400.4	1943	0	21,131,151	2	1,938	99.98	1	1.50
Brazos West	Main	400.9	1943	0	21,184,202	2	1,981	100.00	0	0.00
Colorado River East	Main	441.1	1944	13	21,110,876	2	2,097	97.84	21	9.02
Colorado River West	Main	441.8	1944	3	20,783,851	2	2,168	97.32	26	9.03
Green										
1	Main	9.1	1956	18	4,353,350	3	3,049	96.00	4	87.67
2	Main	63.1	1956	22	1,491,500	3	2,302	100.00	0	0.00

(Continued)

Table 7L.149 (Continued)

River/ Lock	Chamber	River Mile	Year Open	Average Processing Time[a]	Total Tonnage	Average Tow Size: Barges/ Tow	Average Tow Size: Tons/ Tow	Percent Chamber Available	Lock Closures: Total Number	Lock Closures: Avg Time[b]
Hudson										
Troy	Main	153.8	1916	23	13,819	1	494	58.56	6	604.97
Illinois										
Lagrange	Main	80.2	1939	64	35,597,851	10	11,435	97.73	47	4.23
Peoria	Main	157.7	1938	73	31,128,998	9	9,427	98.70	43	2.65
Lockport	Main	291.1	1933	82	16,039,564	6	5,443	98.36	32	4.48
Thomas J. O'Brien	Main	326.5	1960	22	7,371,509	3	2,913	99.65	17	1.82
Kanawha										
Winfield	Main8	31.1	1997	56	19,521,262	9	6,711	99.01	50	1.73
Winfield	Main2	31.1	1937	0	0	0	0	100.00	0	0.00
Winfield	Main1	31.1	1937	0	0	0	0	100.00	0	0.00
Marmet	Main2	67.7	1934	161	7,801,139	6	3,807	98.19	60	2.65
Marmet	Main1	67.7	1934	172	6,928,584	6	3,739	95.06	47	9.20
London	Main2	82.8	1934	85	2,315,757	4	1,722	96.99	211	1.25
London	Main1	82.8	1933	119	4,184,898	5	3,864	97.44	216	1.03
Kaskaskia										
Kaskaskia	Main	0.8	1973	14	583,183	2	1,122	99.46	4	11.80
Lake Washington Ship Canal										
Hiram M. Chittenden	Main	1.3	1916	28	1,816,245	2	1,408	96.19	3	111.38
Hiram M. Chittenden	Aux 1	1.3	1916	13	1,931	2	149	96.88	1	273.00
Mermentau										
Catfish Point	Cntrl	25.0	1951	7	84,226	1	354	99.93	2	3.27
Mississippi										
27	Main	185.5	1953	40	79,858,929	11	11,810	96.89	29	9.40
27	Aux 1	185.5	1953	30	3,519,785	4	3,520	91.82	31	23.10
Melvin Price	Main	200.8	1990	44	69,623,228	13	13,236	91.06	49	15.98
Melvin Price	Aux	200.8	1994	30	7,957,608	6	6,140	85.47	10	127.28
25	Main	241.4	1939	79	39,536,830	13	12,870	81.55	61	26.50
24	Main	273.4	1940	85	39,296,994	13	12,846	82.07	60	26.17
22	Main	301.2	1938	92	38,074,304	13	12,500	98.78	49	2.18
21	Main	324.9	1938	80	37,863,139	12	12,047	97.69	25	8.08
20	Main	343.2	1936	80	36,512,515	12	12,058	82.57	58	26.32
19	Main	364.2	1957	55	35,803,139	12	12,088	83.02	33	45.08
18	Main	410.5	1937	71	35,707,505	12	11,883	99.21	34	2.02
17	Main	437.1	1939	89	34,170,210	12	11,591	98.67	39	3.00
16	Main	457.2	1937	68	33,139,184	11	10,534	98.87	52	1.90
15	Main	482.9	1934	98	30,582,032	11	10,645	98.55	150	0.85
15	Aux 1	482.9	1934	20	627,728	1	890	99.91	19	0.43
14	Main	493.0	1939	78	30,839,734	11	9,803	97.78	85	2.30
14	Aux 1	493.0	1922	17	0	0	0	100.00	0	0.00
13	Main	523.0	1938	58	24,803,042	12	11,552	99.65	13	2.35

12	Main	556.0	1938	56	24,426,919	12	11,555	86.96	22	51.95
11	Main	583.0	1937	52	22,495,873	11	10,621	87.19	18	62.35
10	Main	615.0	1936	48	22,005,796	11	11,182	77.79	4	486.32
9	Main	647.0	1938	47	18,820,219	11	11,800	78.06	14	137.28
8	Main	679.0	1937	47	16,826,021	11	11,195	78.01	7	275.15
7	Main	702.0	1937	38	15,856,894	11	10,592	78.53	7	268.70
6	Main	714.0	1936	42	15,793,578	11	10,593	78.02	6	320.88
5A	Main	728.5	1936	33	12,760,903	10	10,266	77.56	6	327.68
5	Main	738.1	1935	39	12,770,886	10	10,274	74.76	4	552.83
4	Main	752.8	1935	35	12,340,409	10	10,033	73.85	5	458.12
3	Main	796.9	1938	30	11,549,156	10	9,616	74.13	5	453.30
2	Main	815.0	1930	36	11,539,256	9	9,490	73.86	6	381.63
1	Main	847.6	1930	20	2,071,780	2	1,734	69.07	5	541.97
Lower Saint Anthony Falls	Main	853.3	1959	24	2,066,980	2	1,733	69.05	5	542.27
Upper Saint Anthony Falls	Main	853.9	1963	20	2,064,130	2	1,727	69.04	5	542.40
Monongahela										
2	Main	11.2	1905	57	20,725,032	7	6,032	99.92	1	6.73
2	Aux 1	11.2	1905	18	399,555	1	249	98.59	2	61.63
3	Main	23.8	1907	37	15,810,950	4	2,800	91.13	7	111.02
3	Aux 1	23.8	1907	25	2,180,600	2	1,458	98.93	35	2.68
4	Main	41.5	1932	45	12,824,375	5	2,933	99.15	6	12.42
4	Aux 1	41.5	1932	25	387,050	2	732	99.96	3	1.20
Maxwell	Main	61.2	1963	33	4,114,250	5	2,033	95.76	1	371.33
Maxwell	Aux 1	61.2	1963	37	9,265,800	5	4,520	99.86	3	4.12
Grays Landing	Main	82.0	1993	30	5,772,720	5	3,090	99.97	5	0.50
Pt. Marion	Main	90.8	1994	30	5,211,735	5	3,192	99.86	3	4.13
Morgantown	Main	102.0	1950	26	383,435	4	2,062	100.00	0	0.00
Hildebrand	Main	108.0	1959	24	26,235	3	1,093	100.00	0	0.00
Opekiska	Main	115.4	1964	23	27,910	2	1,469	100.00	0	0.00
Ohio										
Emsworth	Main	6.2	1921	70	22,529,859	7	5,726	98.02	13	13.33
Emsworth	Aux 1	6.2	1921	24	1,021,740	1	866	99.48	2	22.93
Dashields	Main	13.3	1929	68	24,019,078	8	6,157	99.72	21	1.18
Dashields	Aux 1	13.3	1929	23	494,215	1	619	97.29	3	79.27
Montgomery	Main	31.7	1936	73	25,975,596	8	6,550	97.82	68	2.82
Montgomery	Aux 1	31.7	1936	26	568,956	2	791	84.04	11	127.13
New Cumberland	Main	54.4	1959	61	31,846,533	11	9,796	99.46	18	2.62
New Cumberland	Aux 1	54.4	1959	36	2,051,242	3	1,784	99.49	15	2.95
Pike Island	Main	84.2	1965	56	39,404,927	11	10,381	99.56	21	1.83
Pike Island	Aux 1	84.2	1965	36	1,866,363	2	1,637	99.78	5	3.82
Hannibal	Main	126.4	1973	52	46,249,423	12	12,104	99.96	2	1.75
Hannibal	Aux 1	126.4	1973	38	1,001,031	3	1,738	97.02	7	37.33
Willow Island	Main	161.7	1972	57	38,901,384	12	12,452	89.87	4	221.90
Willow Island	Aux	161.7	1972	54	5,413,991	6	6,166	99.70	7	3.77
Belleville	Main	203.9	1969	57	47,457,409	12	12,525	99.83	8	1.90
Belleville	Aux 1	203.9	1969	36	521,700	3	1,793	100.00	0	0.00

(Continued)

Table 7L.149 (Continued)

						Average Tow Size			Lock Closures	
River/ Lock	Chamber	River Mile	Year Open	Average Processing Time[a]	Total Tonnage	Barges/ Tow	Tons/ Tow	Percent Chamber Available	Total Number	Avg Time[b]
Racine	Main	237.5	1969	58	47,763,667	12	12,599	99.98	1	2.00
Racine	Aux 1	237.5	1969	34	878,229	3	1,807	100.00	0	0.00
Robert C. Byrd	Main	279.2	1993	58	55,188,585	12	12,302	99.25	33	1.98
Robert C. Byrd	Aux	279.2	1993	37	784,300	2	1,491	99.98	1	1.58
Greenup	Main	341.0	1959	51	63,998,888	12	11,420	89.24	153	6.17
Greenup	Aux 1	341.0	1959	49	6,040,145	6	5,541	93.62	80	6.98
Captain Anthony Meldahl	Main	436.2	1962	58	62,006,780	12	12,033	98.92	25	3.80
Captain Anthony Meldahl	Aux 1	436.2	1962	27	772,018	3	2,347	98.31	6	24.60
Markland	Main	531.5	1959	57	52,796,082	11	11,972	96.53	23	13.22
Markland	Aux 1	531.5	1959	35	2,053,048	4	3,733	99.90	4	2.20
Mcalpine	Main	606.8	1961	58	52,983,711	10	11,451	94.58	56	8.48
Mcalpine	Aux 1	606.8	1961	145	1,846,382	10	12,734	99.96	3	1.18
Cannelton	Main	720.7	1971	59	55,135,133	11	12,588	96.20	74	4.50
Cannelton	Aux	720.7	1971	45	1,516,795	6	5,767	66.04	12	247.88
Newburgh	Main	776.1	1975	53	63,546,772	12	12,303	99.89	6	1.58
Newburgh	Aux 1	776.1	1975	30	963,038	2	1,351	100.00	0	0.00
John T. Myers	Main	846.0	1975	52	70,689,316	12	12,790	99.48	60	0.77
John T. Myers	Aux 1	846.0	1975	21	704,872	2	1,500	99.87	16	0.70
Smithland	Main	918.5	1980	43	36,106,180	10	10,387	99.00	130	0.67
Smithland	Aux 1	918.5	1980	43	46,005,193	11	11,912	99.56	16	2.40
52	Main	938.9	1969	35	87,088,317	11	11,533	98.62	48	2.52
52	Aux 1	938.9	1928	56	8,034,188	4	3,881	98.19	22	7.20
53[c]	Main	962.6	1980	19	84,347,119	11	11,705	99.29	34	1.83
53[c]	Aux	962.6	1929	46	3,421,135	3	3,719	99.84	10	1.37
Okeechobee										
St. Lucie	Main	15.3	1941	22	54,561	1	134	64.39	368	8.48
Port Mayaca	Main	38.5	1977	11	20,922	1	176	77.36	240	8.27
Ortona	Main	93.6	1937	21	19,920	1	150	64.31	384	8.15
Old River										
Old River	Main	304.0	1963	35	8,348,430	3	2,882	99.92	2	3.62
Ouachita										
Columbia	Main	117.2	1972	20	1,064,862	2	2,223	100.00	0	0.00
Felsenthal	Main	226.8	1984	11	225,329	1	1,587	100.00	0	0.00
H. K. Thatcher	Main	281.7	1984	11	214,300	1	1,520	100.00	0	0.00
Red										
Lindy Claiborne Boggs	Main	44.0	1984	23	2,119,661	4	2,423	100.00	0	0.00
John H. Overton	Main	74.0	1987	31	2,005,959	4	2,452	99.98	2	0.95
3	Main	116.4	1992	22	1,168,038	3	2,364	100.00	0	0.00
4	Main	169.0	1994	27	753,684	3	2,054	100.00	0	0.00
Joe D. Waggonner	Main	200.0	1994	17	643,066	3	1,991	100.00	0	0.00

Schooner Bayou										
Schooner Bayou	Cntrl	3.4	1950	4	51,265	4	276	86.24	6	200.85
Snake										
Ice Harbor	Main	9.7	1962	33	4,067,370	2	3,345	95.75	20	18.62
Lower Monumental	Main	41.6	1969	37	3,495,982	3	3,800	96.13	10	33.88
Little Goose	Main	70.3	1970	25	3,127,538	3	3,612	95.39	3	134.60
Lower Granite	Main	107.5	1975	28	1,987,253	2	2,843	93.94	2	265.48
St. Marys										
Davis	Main	47.0	1914	41	0	1	0	54.16	7	573.68
Macarthur	Main	47.0	1943	40	16,043,560	5	42,783	68.64	15	183.17
New Poe	Main	47.0	1963	54	66,372,865	11	214,106	81.13	18	91.82
Tennessee										
Ft. Loudon	Main	602.3	1943	63	636,791	3	2,367	95.58	34	11.38
Tenn–Tombigbee										
Howell Heflin	Main	266.1	1978	39	8,274,930	5	3,927	99.86	4	2.98
Jamie Whitten	Main	411.9	1985	41	5,775,215	6	4,170	99.87	8	1.42
Tombigbee										
Demopolis	Main	213.2	1954	46	15,472,123	5	4,670	100.00	0	0.00
Verdigris										
Chouteau	Main	401.4	1970	42	3,508,854	5	5,316	100.00	0	0.00
Newt Graham Lock	Main	421.6	1970	44	3,350,217	4	4,956	100.00	0	0.00
West Pearl										
1	Main	29.7	1949	0	0	0	0	100.00	0	0.00
2	Main	40.8	1950	0	0	0	0	100.00	0	0.00
3	Main	43.9	1950	0	0	0	0	100.00	0	0.00

[a] Average time from start of lockage to end of lockage; expressed in minutes.
[b] Average time lock is closed (unavailable) expressed in hours.
[c] LPMS data not collected at this site.

Source: From www.iwr.usace.army.mil.

Table 7L.150 Important Ship Canals of the World

Canal	Location	Length		Minimum Width		Minimum Depth		Number of Locks	Year Opened
		mi	km	ft	m	ft	m		
Albert	Belgium	80.8	130	335	102	15	4.5	6	1939
Amsterdam-Rhine	The Netherlands	45	72.4	246	75	7.2	2.2	4	1952
Cape Cod	Massachusetts	17.5	28.2	500	152	32	10	0	1914
Chesapeake and Delaware	Delaware, Maryland	46	74	450	137	35	10.7	0	1829
Chicago Sanitary and Ship	Illinois	30.6	49.2	175	53.3	9	2.7	1	1900
Corinth	Greece	3.9	6.3	81	24.6	26	8	0	1893
Erie Canal	New York	363	584	120	36.6	12	3.7	57	1825
Grand Canal	China	1000	1609	100	30.5	2	0.6	24	610
Kiel (Nord-Ostsee)	Germany	61.3	98.6	336.3	102.5	36	11	8	1895
Manchester Ship	England	36	58	120	36.6	22	6.7	5	1894
McClellan-Kerr Arkansas River System	Arkansas, Oklahoma	445	716	150	45.7	9	2.7	17	1970
Moscow	Russia	80	128	98	30	18	5.5	7	1937
North Sea	Netherlands	15.3	24.7	525	160	49.5	15.1	4	1876
Panama	Panama	50.7	81.6	550	168	45	13.7	12	1914
Sabine-Neches Waterway	Texas	93.7	150.8	200	61	30	9.1	0	1916
Saint Lawrence Seaway[a]	Canada, New York	182	293	200	61	27	8.2	7	1959
Soo (Sault Sainte Marie)	Canada	1.4	2.2	61	18.6	19	5.8	1	1895
Soo (St. Marys Falls Canal and Locks)	Michigan	1.8	2.9	300	91.4	25.5	7.8	4	1855
Suez[b]	Egypt	117.9	189.7	741	226	64	19.5	0	1869
Tennessee-Tombigbee Waterway	Alabama, Mississippi	234	377	300	91.4	9	2.7	10	1985
Volga-Baltic	Russia	528	850	70	21.4	11	3.5	7	1964
Welland Ship	Canada	26	42	200	61	27	8.2	8	1932
White Sea-Baltic	Russia	138	222	46	14	10	3.2	19	1933

[a] Excludes passage through Lake Ontario and Welland Ship Canal.
[b] Includes entrance channels at both ends.

Source: From The World Book Encyclopedia © 2006 World Book, Inc. By permission of the publisher, www.worldbook.com.

SECTION 7M WATERBORNE COMMERCE

Table 7M.151 Waterborne Commerce of the United States, 1984–2003

Year	Grand Total	Foreign			Domestic				
		Total	Great Lakes	Coastal	Total	Coastwise	Lakewise	Internal	Intraport
Tons. (millions)									
1984	1,832.6	803.3	58.8	744.6	1,029.3	307.7	98.0	542.5	81.1
1985	1,785.0	774.3	51.3	723.0	1,010.7	309.8	92.0	534.7	74.3
1986	1,870.5	837.2	45.8	791.4	1,033.2	308.0	87.4	560.5	77.4
1987	1,962.8	891.0	45.9	845.1	1,071.8	323.5	96.5	569.8	82.0
1988	2,082.9	976.2	52.5	923.7	1,106.6	325.2	109.7	588.1	83.7
1989	2,135.2	1,037.9	54.8	983.1	1,097.3	302.0	109.1	606.0	80.2
1990	2,159.3	1,041.6	50.5	991.1	1,117.8	298.6	110.2	622.5	86.4
1991	2,087.6	1,013.6	41.8	971.8	1,074.0	294.5	103.4	600.4	75.6
1992	2,127.9	1,037.5	45.5	992.0	1,090.4	285.1	107.4	621.0	76.8
1993	2,123.3	1,060.0	43.6	1,016.4	1,063.2	271.7	109.8	607.3	74.4
1994	2,208.8	1,115.7	50.1	1,065.6	1,093.1	277.0	114.8	618.4	82.9
1995	2,233.5	1,147.4	51.9	1,095.5	1,086.2	266.6	116.1	620.3	83.1
1996	2,276.7	1,183.4	56.4	1,127.0	1,093.4	267.4	114.9	622.1[a]	89.0[a]
1997	2,326.9	1,220.6	57.7	1,162.9	1,106.3	263.1	122.7	630.6	89.8
1998	2,332.3	1,245.4	62.4	1,183.0	1,086.9	249.6	122.2	625.0	90.1
1999	2,316.7	1,260.8	62.4	1,198.3	1,055.9	228.8	113.9	624.6	88.60
2000	2,419.1	1,354.8	64.0	1,290.7	1,064.3	226.9	114.4	628.4	94.6
2001	2,387.4	1,350.8	62.5	1,288.3	1,036.6	223.6	100.0	619.8	93.2
2002	2,335.2	1,319.3	59.6	1,259.7	1,015.9	216.4	101.5	608.0	90.0
2003	2,387.9	1,378.1	56.3	1,321.8	1,009.7	223.5	89.8	609.6	86.9

[a] Beginning in 1996, fish was excluded.

Source: From U.S. Army Corps of Engineers, *2003 Waterborne Commerce of the United States (WCUS) Waterways and Harbors on the: Part V National Summaries of Domestic and Foreign Waterborne Commerce*, www.iwr.usace.army.mil.

Table 7M.152 Freight Carried on Inland Waterways of the United States, 2001–2003

Waterway	Length (miles)	Tons		Tons		Tons	
		2001	%[a]	2002	%[b]	2003	%[c]
Atlantic Coast							
Atlantic Intracoastal Waterway, VA-FL	793	2.5	−20.1	1.9	−26.1	1.9	3.5
Intracoastal Wtwy, Jacksonville to Miami, FL	349	1.0	18.5	0.5	−43.2	0.9	69.5
Gulf Coast							
Bayou Teche, LA	107	1.7	10.4	1.6	−9.3	1.4	−10.0
Black Warrior and Tombigbee Rivers, AL	449	18.9	−19.4	19	−0.5	21.0	10.3
Chocolate Bayou, TX	13	3.4	−11.4	2.9	−14.0	3.3	13.8
Gulf Intracoastal Waterway, TX-FL	1,109	112.2	−1.4	107.7	−4.0	117.8	9.5
GIWW: Morgan City-Port Allen Route, LA	64	23.3	1.0	20.8	−10.6	24.3	16.6
Petit Anse, Tigre, Carlin Bayous, LA	16	3.1	19.7	2.2	−27.8	2.5	13.1
Tennessee-Tombigbee Waterway, AL and MS	234	6.8	−3.6	6.2	−8.1	6.2	−0.2
Mississippi River System							
Allegheny River, PA	72	3.0	−21.9	2.8	−4.6	3.3	17.7
Atchafalaya River, LA	121	11.6	−14.0	10.7	−7.3	9.8	−8.8
Big Sandy River, KY and WV	27	24.2	3.0	25.1	3.7	22.6	−9.9
Cumberland River, KY and TN	381	23.2	2.2	22.6	−2.5	20.6	−8.7
Green and Barren Rivers, KY	109	7.8	88.7	10.4	33.9	7.9	−24.2
Illinois Waterway, IL	357	43.5	−1.7	43.0	−1.0	45.0	4.6
Kanawha River, WV	91	22.2	−1.4	19.2	−13.2	19.4	0.8
McClellan-Kerr Arkansas R. Nav. Sys., AR/OK	462	11.2	4.4	11.9	6.2	13.0	9.1
Mississippi River Mpls, MN to Mouth of Passes	1,814	316.5	−3.3	316.2	−0.1	308.2	−2.5
Minneapolis, MN to Mouth of Missouri River	663	78.8	−5.4	84.1	6.7	77.8	−7.4

(Continued)

Table 7M.152 (Continued)

Waterway	Length (miles)	Tons 2001	%[a]	Tons 2002	%[b]	Tons 2003	%[c]
Mouth of Missouri R. to Mouth of Ohio R.	195	119.1	−2.1	121.5	2.0	111.5	−8.2
Mouth of Ohio River up to Baton Rouge, LA	720	200.7	−1.8	198.3	−1.2	185.5	−6.5
Baton Rouge up to New Orleans, LA[d]	130	220.0	−2.7	222.4	1.1	212.9	−4.3
New Orleans, LA to Mouth of Passes[d]	106	116.7	−3.8	114.8	−1.6	115.8	−0.8
Missouri R. (MO, KS, NE & IA) to Sioux City, IA	732	9.7	11.4	8.3	−15.1	8.1	−2.6
Monongahela River, PA and WV	129	38.1	2.1	38.2	0.4	27.6	−27.8
Ohio River, PA, WV, OH, KY, IN, and IL	981	242.5	2.6	248.2	0.3	228.3	−5.9
Ouachita and Black Rivers, AR and LA	332	1.6	6.7	1.4	−11.5	2.2	57.7
Red River, LA	212	3.7	−2.1	3.7	0.6	4.2	12.6
Tennessee River, TN, KY, MS and AL	652	47.9	−3.3	43.9	−8.3	49.8	13.4
Pacific Coast							
Columbia River System, OR, WA, and ID	596	20.2	−12.2	16.5	18.5	16.5	0.5
Columbia River and Willamette River below Vancouver, WA and Portland, OR[d]	113	19.8	−11.6	15.9	−19.4	16.2	1.5
Vancouver, WA to The Dalles, OR	85	9.8	−8.2	8.0	−18.4	9.4	18.0
The Dalles Dam to McNary Lock and Dam	100	8.9	−9.2	7.3	−18.2	8.5	17.5
Above McNary L & D to Kennewick, WA	39	6.7	−14.6	5.1	−24.5	6.5	27.0
Snake River (WA and ID) to Lewiston, ID	141	5.6	−15.9	4.3	−24.0	5.3	24.6
Willamette River above Portland, OR	118	1.6	83.7	1.6	1.4	1.3	−22.3

[a] Percent change from 2000
[b] Percent change from 2001
[c] Percent change from 2002
[d] Includes deep draft waterways (2001, 2002). Includes coastwise entrance channel miles.

Source: From U.S., Army Corps of Engineers, *2003 Waterborne Commerce of United States (WCUS) Waterways and Harbors on the: Part V - Natinal summaries of Domestic and Foreign Waterbome Commerce:* U.S. Army Corps of Engineers, *2002 Waterborne Commerce of the United States (WCUS) Waterways and Harbors on the: Part V - National Summaries of Domestic and Foreign Waterborne Commerce:* U.S. Army Corps of Engineers, *2001 Waterborne Commerce of the United States (WCUS) Waterways and Harbors on the: Part V - national Summaries of Domestic and Foreign Waterborne commer commerce,* www.iwr.usace.army.mil.

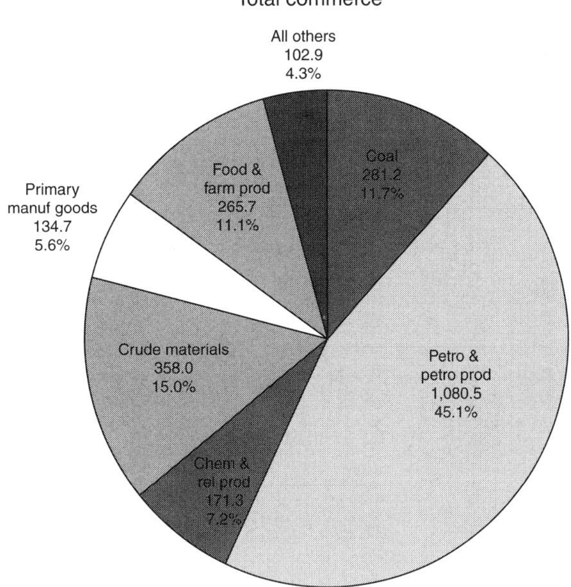

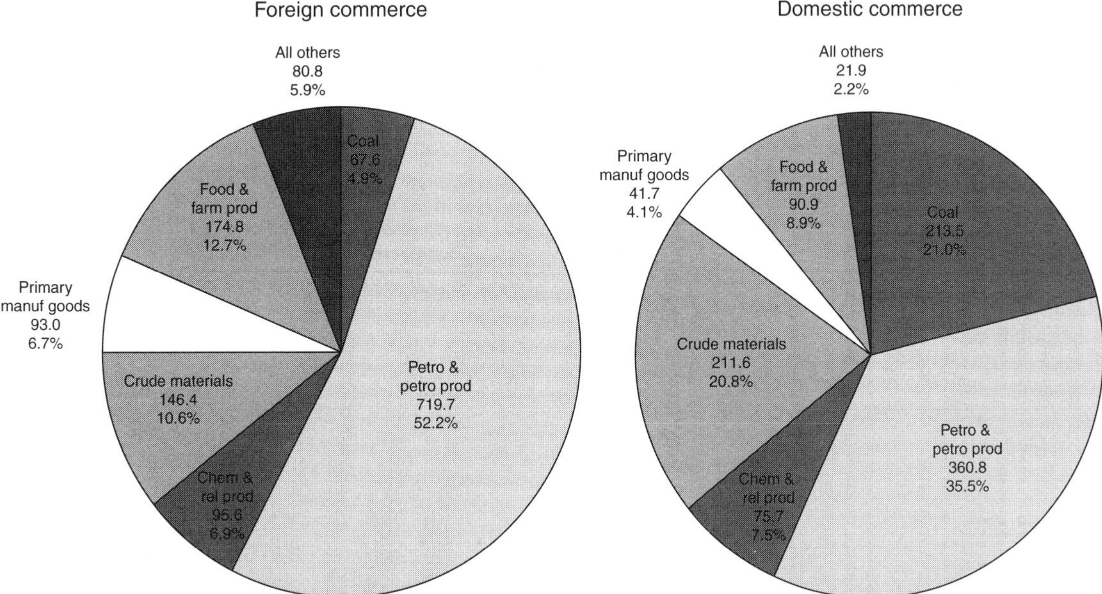

Note: Million short tons and percentage of short tons.

Figure 7M.24 Principal commodities carried on waterways of the United States in 2003. (From U.S. Army Corps of Engineers, *2003 Waterborne Commerce of the United States (WCUS) Waterways and Harbors on the: Part V National Summaries of Domestic and Foreign Waterborne Commerce*, www.iwr.usace.army.mil.)

Table 7M.153 Domestic and Foreign Waterborne Commerce by Type of Commodity in the United States, 2000–2003

	2000			2001			2002			2003		
		Foreign			Foreign			Foreign			Foreign	
Commodity	Domestic	Imports	Exports	Domestic	Imports	Exports	Domestic	Imports	Exports	Domestic	Imports	Exports
Total, all commodoties	1,069,798	976,784	415,042	1,042,472	945,075	399,011	1,021,001	934,941	384,350	1,016,136	1,004,791	373,324
Coal and lignite	220,679	15,462	60,879	227,718	19,857	55,764	227,011	16,671	43,252	213,517	25,015	42,624
Petroleum and petroleum products[a]	370,603	651,650	58,696	369,745	620,556	58,292	348,671	609,289	59,929	360,809	661,540	58,199
Crude Petroleum	83,806	521,619	3,064	85,971	486,249	1,344	85,506	479,318	1,208	87,491	515,747	1,248
Gasoline	94,658	24,157	6,434	93,098	27,732	6,906	86,602	29,282	6,726	87,524	32,294	6,630
Distillate fuel oil	65,695	21,111	4,953	65,591	20,589	4,982	62,651	19,936	5,861	65,569	29,115	7,046
Residual fuel oil	78,534	40,361	12,693	78,742	40,891	14,032	69,420	35,411	12,129	75,958	31,330	9,420
Chemicals and Related Products	75,996	38,479	57,888	71,080	43,830	54,741	73,063	39,572	54,962	75,718	42,007	53,575
Crude Materials, inedible except fuels[a,b]	234,210	97,350	48,753	213,673	95,041	45,283	214,745	92,225	44,983	211,563	102,215	44,235
Forest products, wood and chips	13,475	5,505	14,131	10,222	6,119	10,725	9,243	7,024	8,853	8,090	7,415	8,090
Soil, sand, gravel, rock and stone[b]	132,275	29,061	3,635	129,267	29,793	3,534	128,777	33,691	3,657	130,829	37,453	2,539
Limestone	54,185	10,703	2,476	56,034	10,439	2,422	54,418	11,706	2,709	50,340	12,483	1,838
Phosphate rock	3,418	—	7	1,736	—	—	3,497	2,670	1	3,141	2,547	6
Sand and gravel	71,447	6,611	850	68,510	8,144	864	67,474	7,929	639	73,651	11,103	519
Iron ore and scrap	69,215	19,412	9,281	54,006	13,431	9,319	58,471	15,513	11,818	52,918	16,886	11,107
Nonferrous ores and scrap	6,730	19,370	3,135	6,159	17,117	2,430	6,594	15,505	2,243	6,869	16,870	2,443
Primary manufactured goods	45,788	91,959	15,256	40,336	82,854	13,892	42,426	83,922	14,502	41,695	76,428	16,537
Food and farm products[a,c]	96,940	30,026	156,333	96,514	30,766	154,577	97,556	32,150	150,323	90,923	32,793	142,003
Wheat	12,885	147	30,378	12,074	155	28,039	9,677	248	26,066	10,758	230	26,358
Corn	37,960	163	50,048	38,999	99	48,772	41,720	95	47,936	36,673	83	41,447
Soybeans	20,242	30	27,013	20,034	31	29,352	21,538	39	28,175	20,378	33	29,270
Other agricultural products[d]	7,489	15,920	22,724	7,302	22,905	15,887	7,322	24,213	15,697	7,073	24,985	15,454
All manufactured equipment, machinery and products	21,245	48,598	13,762	19,870	47,726	12,753	14,576	54,865	12,419	18,714	58,502	12,809
Waste and scrap nec	4,267	—	—	3,469	—	—	2,745	—	—	3,108	—	—
Unknown or not elsewhere classified	71	3,259	3,474	67	4,445	3,708	208	6,247	3,981	89	6,293	3,342

Note: Thousand of short tons.

[a] Includes commodities not shown seperately.
[b] Dose not include waterways improvement material, 2000-5.3 million short tons, 2001-6.5 million short tons, 2002-6.4 million short tons, 2003-4.4 million short tons.
[c] Does not include fish landings.
[d] Food, alcoholic beverages, tobacco products, cotton, etc.

Source: From U.S. Army Corps of Engineers, *2003 Waterborns Commerce of the United Sates (WCUS) Waterways and Harbors on the: Part V-National Summaries of Domestic and Foreign Waterborne Commerce*; U.S. Army Corps of Engineers, *2002 Waterborne Commerce of the United States Domestic and Foreign Waterborne Commerce*; U.S. Army Corps of Engineers, *2001 Waterborne Commerce of the United States (WCUS) Waterways and Harborne Commerce*; U.S. Army Corps of Engineers, *2000 Waterborne Commerce of the United States (WCUS) Waterways and Harbors on the: Part V-National Summaries of Domestic and Foreign Waterborne Commerce*, www.iwr.usace.army.mil.

Table 7M.154 Freight Carried on the Mississippi River System by Type of Traffic, 1984–2003

	1984	1985	1986	1987	1988	1989	1990	1991	1992	1993	1994	1995	1996	1997	1998	1999	2000	2001	2002	2003
Total	543.5	527.8	556.4	583.4	601.6	626.4	659.1	646.6	673.1	660.4	693.3	707.2	701.8	707.1	707.4	716.9	715.5	714.8	712.8	676.8
Foreign																				
Inbound	34.1	27.0	35.1	38.1	45.3	59.9	63.1	60.1	63.0	76.9	89.8	81.5	77.3	83.5	85.2	83.5	87.7	87.8	86.3	78.4
Outbound	85.9	81.0	81.1	93.7	97.5	104.0	106.0	109.9	112.3	100.0	92.4	115.8	108.7	98.5	94.7	99.2	100.5	99.8	99.2	91.3
Domestic																				
Coastwise	36.7	40.9	37.4	37.4	35.8	31.1	32.6	31.3	32.3	31.8	32.4	30.6	33.0	36.3	36.7	34.6	34.1	30.9	33.0	31.8
Internal	386.6	378.9	402.8	414.2	423.0	431.5	457.5	445.1	465.4	451.7	478.7	479.4	482.8	488.9	490.7	499.6	493.1	496.3	494.3	475.2

Note: Million short tons. Foreign Inbound includes Upbound Thru Traffic. Foreign Outbound includes Downbound Thru Traffic.

Source: From U.S. Army Corps of Engineers, 2003 *Waterborne Commerce of the United States (WCUS) Waterways and Harbors on the: Part V National Summaries of Domestic and Foreign Waterborne Commerce*, www.iwr.usace.army.mil.

Table 7M.155 U.S. Freight Carried on the Great Lakes by Type of Traffic, 1984–2003

	1984	1985	1986	1987	1988	1989	1990	1991	1992	1993	1994	1995	1996	1997	1998	1999	2000	2001	2002	2003
Total	**162.8**	**148.1**	**137.9**	**148.1**	**168.8**	**168.9**	**167.1**	**151.1**	**159.9**	**159.6**	**175.3**	**177.8**	**181.8**	**188.6**	**192.2**	**182.9**	**187.5**	**171.4**	**167.2**	**156.5**
Foreign																				
Inbound	17.8	17.1	13.8	13.9	15.9	17.8	17.6	14.2	15.4	18.1	23.1	18.9	24.5	24.5	25.6	22.2	23.9	22.0	21.5	23.3
Outbound	40.9	34.2	32.0	32.1	36.6	37.1	32.9	27.5	30.4	25.8	27.8	33.5	32.8	33.7	37.9	40.8	42.9	43.6	38.4	33.1
Domestic																				
Coastwise	0.0	0.0	—	0.0	—	0.0	0.0	—	0.0	0.0	0.0	—	—	—	—	—	0.0	—	—	0.0
Lakewise	98.0	92.0	87.4	96.5	109.7	109.1	110.2	103.4	107.4	109.9	114.8	116.2	115.0	122.8	122.2	113.4	114.4	100.1	101.6	89.8
Internal	1.8	2.2	1.7	1.4	1.6	1.7	3.1	3.2	3.4	3.1	4.6	3.6	3.3	2.3	2.3	1.5	2.1	1.7	2.5	2.8
Intraport	4.2	2.7	3.0	4.3	5.0	3.3	3.4	2.7	3.3	2.8	5.0	5.5	6.2	5.2	4.3	4.4	4.2	4.0	3.2	7.4

Note: Million Short Tons. Foreign Outbound includes Great Lakes Thru Foreign Traffic.

Source: From U.S. Army Corps of Engineers, 2003 *Waterborne Commerce of the United States (WCUS) Waterways and Harbors on the: Part V. National Summaries of Domestic and Foreign Waterborne Commerce*, www.iwr.usace.army.mil.

SECTION 7N WATER–BASED RECREATION

Table 7N.156 Recreational Boats in Use by Boat Type 1997–2003 and Estimates of Retail Expenditures on Recreational Boats

Estimates of Retail Expenditures on Recreational Boating 1997–2003	1997	1998	1999	2000	2001	2002	2003
New Boats	610,100	575,300	584,900	576,900	882,300[a]	846,000	840,800[b]
Total New Boat Retail dollars	$6,636,856,000	$6,308,685,000	$7,711,369,000	$9,515,192,800	$10,223,862,700	$10,898,635,200	$10,603,725,600
Average New Boat Retail Price	$10,878	$10,966	$12,176	$16,494	$11,588	$12,883	$12,611
Used Boats	1,038,819	979,565	995,911	982,289	1,502,295	1,440,486	1,431,632
Total Used Boat Retail Dollars	$4,067,750,452	$3,866,613,387	$4,365,102,290	$5,831,892,361	$6,266,238,429	$6,679,808,671	$6,499,057,626
Average Used Boat Retail Price	$3,916	$3,947	$4,383	$5,937	$4,171	$4,637	$4,540
New Outboard Motors	302,000	314,000	331,900	348,700	299,100	302,100	305,400
Total New Outboard Motor Retail Dollars	$2,006,186,000	$2,155,610,000	$2,602,096,000	$2,901,861,400	$2,411,045,100	$2,479,938,900	$2,554,533,570
Average New Outboard Motor Retail Price	$6,643	$6,865	$7,840	$8,322	$8,061	$8,209	$8,365
Used Outboard Motors	514,216	534,649	565,127	393,732	509,278	514,386	520,005
Total Used Outboard Motor Retail Dollars	$1,229,597.871	$1,321,180,323	$1,594,833,032	$1,778,572,471	$1,477,737,319	$1,519,962,552	$1,565,681,865
Average Used Outboard Motor Retail Price	$2,391	$2,471	$2,822	$2,996	$2,902	$2,955	$3,011
New Boat Trailers	181,000	174,000	168,000	158,500	135,900	141,200	130,600
Total New Boat Retail Dollars	$190,050,000	$189,660,000	$190,008,000	$184,494,000	$181,698,300	$200,645,200	$202,012,080
Average New Boat Trailer Retail Price	$1,050	$1,090	$1,131	$1,164	$1,337	$1,421	$1,547
Estimated Boat/Motor/Trailer Dollars	$14,130,440,323	$13,841,748,710	$16,463,408,323	$20,212.033,032	$20,560,581,848	$21,778,990,523	$21,425,010,741
Estimated Accessory Aftermarket Sales	1,214,057,000	1,650,000,000	1,848,000,000	2,032,800,000	1,937,268,400	2,028,309,545	2,123,640,093
Sub Total	$15,344,497,323	$15,491,748,710	$17,722,048,323	$22,244,833,032	$22,497,840,248	$23,807,300,067	$23,548,650,835
Estimated Other (fuel, finance, insurance, docking, maintenance etc)	$3,573,140,718	$3,500,139,754	$4,014,043.945	$5,625,013,577	$6,058,281,452	$6,554,969,366	$6,448,429,689
Total Expenditures	$18,917,638,040	$18,991,888,464	$21,736,092,268	$27,869,846,610	$28,556,121,701	$30,362,269,433	$29,997,080,523
Percent Change		0.39%	14.45%	28.22%	2.46%	6.32%	–1.20%
New Boat/Motor Expenditures	$8,643,042,000	$8,464,295,000	$10,313,465,000	$12,417,074,200	$12,634,907,800	$13,378,574,100	$13,158,259,170
Percent Change		–2.07%	21.85%	20.40%	1.75%	5.89%	–1.65%
Recreational Boats in Use by Boat Type 1997–2003							
Outboard boats (millions)	8.13	8.19	8.21	8.29	8.34	8.38	8.42
Inboard boats (millions)	1.59	1.61	1.63	1.66	1.69	1.71	1.74
Sterndrive boats (millions)	1.58	1.62	1.67	1.71	1.74	1.77	1.79
Personal Watercraft (millions)	1.00	1.10	1.18	1.24	1.29	1.35	1.42
Sailboats (millions)	1.65	1.67	1.65	1.64	1.63	1.61	1.60
Others (millions)	2.29	2.45	2.49	2.50	2.51	2.53	2.53
Total (millions)	16.23	16.65	16.82	17.03	17.20	17.36	17.49

[a] Includes 357,100 kayaks not previously reported.
[b] Includes 30,580 inflatables not previously reported.

Source: From National Marine Manufactures Association, 2003 Recreational Statistical Abstract, www.nmma.org. Reprinted with permission.

Original Source: From Recreational Boats in Use by Type 1997 to 2003 data USCG/NMMA.

Table 7N.157 The Retail Boating Market 1997–2003 Total Units Sold, Total Retail Value, and Average Retail Unit Cost

	1997	1998	1999	2000	2001	2002	2003
Outboard Boats							
Total units sold	20,000	213,700	230,200	241,200	217,800	212,000	207,100
Retail value	$1,421,400,000	$1,547,188,000	$1,988,928,000	$2,306,577,000	$2,195,859,600	$2,280,908,000	$2,742,825,960
Average unit cost	$7,107	$7,240	$8,640	$9,563	$10,082	$10,759	$13,244
Outboard Motors							
Total units sold	302,000	314,000	331,900	348,700	299,100	302,100	305,400
Retail value	$2,006,186,000	$2,155,610,000	$2,602,096,000	$2,901,881,400	$2,411,045,100	$2,478,838,900	$2,554,533,570
Average unit cost	$6,643	$6,865	$7,840	$8,322	$8,061	$8,205	$8,365
Boat Traillers							
Total units sold	181,000	174,000	168,000	158,500	135,900	141,200	130,600
Retail value	$190,050,000	$189,660,000	$190,008,000	$184,494,000	$181,698,300	$200,645,200	$202,012,080
Average unit cost	$1,050	$1,090	$1,131	$1,164	$1,337	$1,421	$1,547
Inboard Boats-Ski/Wakeboard Boats							
Total units sold	6,100	10,900	12,100	13,600	11,100	10,500	11,100
Retail value	$136,408,200	253,348,700	$308,429,000	$366,438,400	$352,569,300	$398,811,000	$403,289,640
Average unit cost	$22,362	23,243	$25,490	$26,944	$31,763	$37,982	$36,332
Inboard Boats-Cruisers							
Total units sold	6,300	6,700	7,000	10,300	10,800	11,800	9,300
Retail value	$1,669,103,100	$1,704,245,500	$1,799,420,000	$2,925,756,200	$3,758,475,600	$4,336,559,000	$3,467,322,720
Average unit cost	$264,937	$254,365	$257,060	$284,054	$348,007	$367,505	$372,830
Stemdrive Boats							
Total units sold	92,000	77,700	79,600	78,400	72,000	69,300	69,200
Retail value	$2,068,528,000	$1,746,696,000	$2,054,476,000	$2,253,843,200	$2,216,448,000	$2,192,929,200	$2,221,116,840
Average unit cost	$22,484	$22,480	$25,810	$28,748	$30,784	$31,644	$32,097
Canoes							
Total units sold	103,600	107,800	121,000	111,800	105,800	100,000	86,700
Retail value	$61,124,000	$64,033,200	$67,034,000	$64,508,600	$57,449,400	$56,900,000	$49,644,420
Average unit cost	$590	$594	$554	$577	$543	$569	$573
Kayaks							
Total units sold	N/A	N/A	N/A	N/A	357,100	340,300	324,000
Retail value	N/A	N/A	N/A	N/A	$176,764,500	$157,558,900	$151,048,800
Average unit cost	N/A	N/A	N/A	N/A	$495	$463	$465
Inflatables							
Total units sold	N/A	N/A	N/A	N/A	N/A	N/A	30,500
Retail value	N/A	N/A	N/A	N/A	N/A	N/A	$67,417,200
Average unit cost	N/A	N/A	N/A	N/A	N/A	N/A	$2,210
Personal Water Craft							
Total units sold	176,000	130,000	106,000	92,000	80,900	79,300	80,600
Retail value	$1,135,904,000	$868,530,000	$771,044,000	$720,176,000	$641,456,100	$697,681,400	$716,501,760
Average unit cost	$6,454	$6,681	$7,274	$7,828	$7,929	$8,798	$8,890
Jet Boats							
Total units sold	11,700	10,100	7,800	7,000	6,200	5,100	5,600
Retail value	$144,389,700	$167,033,800	$132,678,000	$123,641,000	$118,692,800	$107,997,600	$115,268,160
Average unit cost	$12,341	$16,538	$17,010	$17,663	$19,144	$21,176	$20,584
Sailboats[a]							
Total units sold	14,400	18,400	21,600	22,600	20,600	17,700	16,700
Retail value	N/A	N/A	N/A	$754,252,400	$706,139,303	$669,290,100	$669,290,100
Average unit cost	N/A	N/A	N/A	$33,374	$34,279	$37,813	$40,077

Note: [a]The Sailing Company's Annual Sailing Business Review.

Source: From National Marine Manufactures Association, 2003 Recreational Statistical Abstract, www.nmma.org. Reprinted with permission.

Original Source: From The Sailing Company's Annual Sailing Business Review.

Table 7N.158 Boat Registration Data by State, 2003

	Boat Registration Data by State[a]									
	Powered					Nonpowered			Other	Total
	Inboard	Outboard	Sterndrive	Auxiliary Sail	PWC	Rowboat	Canoe or Kayak	Sail Only	Other Boats	Total
Totals	**1,457,376**	**8,003,686**	**1,601,156**	**139,885**	**744,473**	**118,295**	**282,612**	**146,084**	**301,049**	**12,794,616**
Alabama	17,582	205,393	20,256	994	14,141	534	375	2,631	343	262,249
Alaska	4,195	31,284	4,337	607	1,673	8,767	0	193	360	51,416
Arizona	45,865	65,061	0	1,572	28,081	0	0	0	6,634	147,213
Arkansas	29,704	148,573	0	0	0	0	0	472	17,466	196,215
California	115,697	356,955	209,788	20,414	184,105	8,691	6,863	35,853	25,013	963,379
Colorado	21,496	49,595	7,715	0	17,136	0	0	3,937	696	100,575
Connecticut	7,746	68,915	16,998	4,997	8,433	302	56	207	253	107,907
Delaware	6,193	32,037	11,190	0	0	0	0	0	515	49,935
District of Columbia	543	614	337	150	39	262	13	150	44	2,152
Florida	66,694	624,342	92,408	9,932	106,783	4,224	3,511	7,657	24,417	939,968
Georgia	17,839	218,634	36,247	0	37,673	0	0	4,414	11,911	326,718
Hawaii	7,860	3,267	2,217	1,246	0	0	0	690	320	15,600
Idaho	17,579	40,559	16,160	838	4,634	0	0	800	2,106	82,676
Illinois	30,268	225,527	50,920	3,032	13,995	16,012	0	8,078	12,420	360,252
Indiana	25,293	129,966	35,494	713	0	0	0	1,308	23,371	216,145
Iowa	19,569	131,236	23,592	471	0	956	23,379	4,324	7,309	210,836
Kansas	7,177	65,842	9,816	406	12,731	1,305	272	2,618	296	100,463
Kentucky	15,750	119,450	17,224	342	8,298	0	0	0	12,354	173,418
Louisiana	21,225	262,311	10,583	0	12,932	0	0	0	0	307,051
Maine	7,021	73,542	9,570	0	0	0	0	0	471	90,604
Maryland	15,493	110,726	36,046	10,318	16,753	0	0	483	8,576	198,395
Massachusetts	8,361	98,952	27,392	6,565	8,503	0	0	0	6,348	156,121
Michigan	268,629	577,579	41,288	13,569	0	26,408	0	26,081	0	953,554
Minnesota	23,735	510,654	56,709	2,864	40,394	12,791	169,933	15,622	12,677	845,379
Mississippi	17,109	168,113	12,305	3,930	0	0	0	0	0	201,457
Missouri	9,326	227,118	43,244	155	0	500	502	1,948	43,360	326,153
Montana	19,125	33,596	0	152	0	147	18	346	0	53,384
Nebraska	5,939	47,034	9,968	24	8,819	75	191	155	3,558	75,763
Nevada	3,574	21,223	18,681	483	13,524	233	0	165	697	58,580
New Hampshire	15,887	53,193	16,282	2,324	0	0	0	4,183	8,966	100,835
New Jersey	18,623	112,269	38,683	7,174	22,589	6,011	0	1,793	446	207,588
New Mexico	3,284	21,246	7,035	182	6,821	0	0	1,291	435	40,294
New York	99,062	280,721	132,021	6,687	0	0	0	0	9,603	528,094
North Carolina	16,929	245,772	44,575	3,583	39,759	0	0	1,788	7,451	359,857
North Dakota	2,909	36,549	4,295	130	4,457	0	513	67	329	49,249
Ohio	31,626	189,359	66,634	1,923	42,849	11,721	52,339	9,418	7,179	413,048
Oklahoma	40,787	164,219	22,950	1,822	0	0	0	0	0	229,778

(Continued)

Table 7N.158 (Continued)

Boat Registration Data by State[a]

	Powered					Nonpowered			Other	Total
	Inboard	Outboard	Sterndrive	Auxiliary Sail	PWC	Rowboat	Canoe or Kayak	Sail Only	Other Boats	Total
Oregon	64,352	125,047	0	4,971	0	0	0	0	3,221	197,591
Pennsylvania	30,445	237,196	47,972	344	0	2,082	24,337	1,801	11,058	355,235
Rhode Island	4,824	22,502	9,983	3,372	2,326	0	0	0	0	43,007
South Carolina	13,464	278,142	36,498	5,345	25,421	17,226	180	1,782	2,256	380,314
South Dakota	1,877	36,069	10,739	244	0	0	0	0	4,540	53,469
Tennessee	41,701	188,816	28,864	525	0	0	0	1,730	0	261,636
Texas	117,514	404,827	87,257	0	0	0	0	2,398	7,092	619,088
Utah	11,527	29,363	19,329	0	14,645	0	0	1,314	0	76,178
Vermont	8,517	24,743	0	0	0	0	0	0	0	33,260
Virginia	6,261	156,813	43,019	4,602	25,968	0	0	234	5,096	241,993
Washington	0	156,447	98,586	10,740	0	0	0	0	0	265,773
West Virginia	4,419	39,239	5,566	0	2,405	0	0	0	7,088	58,717
Wisconsin	42,696	514,895	52,881	0	0	0	0	0	328	610,800
Wyoming	14,977	1,962	5,599	0	2,555	0	89	100	443	25,725
Guam[b]	0	0	0	0	0	0	0	0	4,000	4,000
Puerto Rico	8,195	33,993	1,804	967	15,949	0	0	0	3	60,911
Virgin Islands	809	1,834	96	1,168	14	46	41	53	0	4,061
Amer Samoa	19	75	0	8	0	0	0	0	0	102
No. Marianas	85	297	3	0	68	2	0	0	0	455

[a] The figures in this table are derived from reports from the State and jurisdictions. There are a total of 12,794,616 registered recreational vessels. This table classifies registered motorboats and registered nonpowered boats for each State and jurisdiction. Please note that the scope of the boat registration system for each State and jurisdiction is not the same. This explains why some States report the number of non-powered vessels such as rowboats, canoes, and non-powered sailboats and others do not. Also notice that some State and jurisdictions report Personal Watercraft (PWC) as a separate vessel category and others report PWC as an inboard motorboat. An accurate figure on the number of PWC will be provided when all States and jurisdictions classify and report PWC as a separate as a separate vessel category.

[b] Estimate.

Source: From U.S. Department of Homeland Security, United States Coast Guard, 2004, Boating Statistics 2003, COMDTPUB P16754017, www.uscgboating.org.

Table 7N.159 Fatality Rate in Recreational Boating in the United States, 1991–2003

Year	Fatalities	Number of Registered Boats	Fatalities Per 100,000 Registered Boats
1991	924	11,068,440	8.3
1992	816	11,132,386	7.3
1993	800	11,282,736	7.1
1994	784	11,429,585	6.9
1995	829	11,734,710	7.1
1996	709	11,877,938	5.9
1997	821	12,312,982	6.7
1998	815	12,565,930	6.5
1999	734	12,738,271	5.8
2000	701	12,782,143	5.5
2001	681	12,876,346	5.3
2002	750	12,854,054	5.8
2003	703	12,794,616	5.5

Source: From U.S. Department of Homeland Security, United States Coast Guard, 2004, Boating Statistics 2003, COMDTPUB P16754.17, www.uscgboating.org.

Table 7N.160 Types of Boating Accidents in the United States, 2003

	Types of Boating Accidents				
	Accidents	Vessels Involved	Drowning Deaths	Other Deaths	Total Fatalities
Total	**5,438**	**7,363**	**481**	**22**	**703**
Capsizing	514	576	170	36	206
Collision with fixed object	558	629	19	31	50
Collision with floating object	152	215	1	2	3
Collision with another vessel	1,469	2,972	9	61	70
Falls within boat	233	249	3	3	6
Falls overboard	509	530	169	32	201
Fire/explosion (fuel)	142	163	4	3	7
Fire/explosion (other than fuel)	68	79	0	2	2
Flooding/swamping	274	293	36	5	41
Grounding	291	297	2	6	8
Sinking	128	135	6	2	8
Skier mishap	451	477	1	5	6
Struck by boat	89	128	1	8	9
Struck by motor or propeller	107	120	1	5	6
Struck submerged object	128	129	2	2	4
Other (not specified)	80	88	1	3	4
Carbon monoxide poisoning	20	20	0	7	7
Departed vessel (swimming)	34	34	28	1	29
Departed vessel (other)	11	11	10	0	10
Ejected from vessel	7	7	3	2	5
Falls on PWC	15	18	0	1	1
Not reported	158	193	15	5	20

Source: From U.S. Department of Homeland Security, United States Coast Guard, 2004, Boating Statistics 2003, COMDTPUB P16754.17, www.uscgboating.org.

Table 7N.161 Boating Accidents in the United States, 2003

	Time Period	Accidents	Fatalities
Time of Day	Midnight to 2:30 am	174	44
	2:31 am to 4:30 am	54	9
	4:31 am to 6:30 am	65	11
	6:31 am to 8:30 am	131	22
	8:31 am to 10:30 am	305	33
	10:31 am to 12:30 pm	579	65
	12:31 pm to 2:30 pm	878	90
	2:31 pm to 4:30 pm	1,163	124
	4:31 pm to 6:30 pm	973	104
	6:31 pm to 8:30 pm	575	85
	8:31 pm to 10:30 pm	297	61
	10:31 pm to midnight	132	21
	Unknown	112	34
Month of Year	January	70	24
	February	98	18
	March	180	34
	April	225	47
	May	638	87
	June	849	95
	July	1,480	112
	August	1,075	109
	September	383	67
	October	235	50
	November	127	35
	December	78	25
Day of Week	Monday	485	76
	Tuesday	393	62
	Wednesday	395	66
	Thursday	397	62
	Friday	768	113
	Saturday	1,560	159
	Sunday	1,440	165
Totals		**5,438**	**703**

Source: U.S. Department of Homeland Security, United States Coast Guard, 2004, Boating Statistics 2003, COMDTPUB P16754.17, www.uscgboating.org.

Table 7N.162 Life Expectancy in Water

	Water Temperature				
Duration, Hours	**30°F**	**40°F**	**50°F**	**60°F**	**70°F**
1	M	M	M	S	S
2	L	L	M	M	S
3	L	L	M	M	S
4	L	L	L	M	S

Note: Exposure by immersion in low temperature water can have serious consequences. Life expectancies for various durations of exposure are indicated by: L — Lethal, 100 percent expectancy of death; M — Marginal, 50 percent expectancy of unconsciousness which will probably result in drowning; S — Safe. It should also be noted that sudden immersion in ice cold water can cause temporary paralysis with resulting helplessness and loss of buoyancy, causing the victim to sink to the bottom.

Source: Pan American Airways and Calif. Dept. of Harbors and Watercraft. With permission.

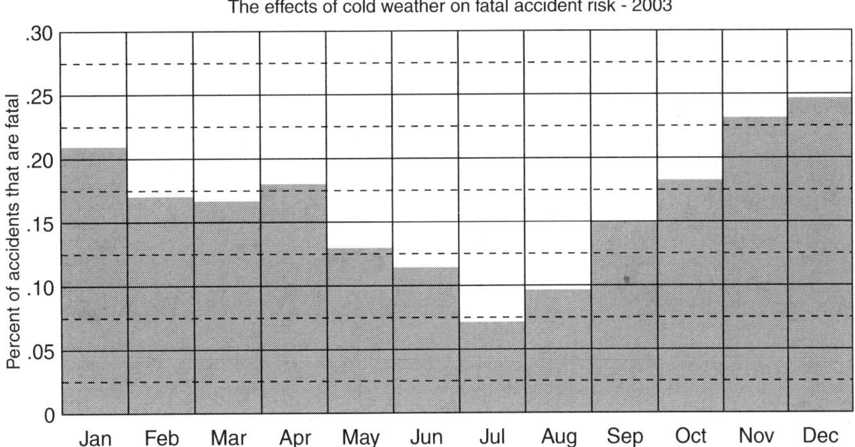

Boaters are more likely to perish if they are involved in a reported accident during the fall & winter months

Month	Fatal accidents	Non-fatal accidents	Total accidents	Fatal accident risk	Total fatalities
January	15	55	70	21%	24
February	17	81	98	17%	18
March	29	151	180	16%	34
April	41	184	225	18%	47
May	81	557	638	13%	87
June	90	759	849	11%	95
July	106	1,374	1,480	7%	112
August	96	979	1,075	9%	109
September	56	327	383	15%	67
October	42	193	235	18%	50
November	29	98	127	23%	35
December	19	59	78	24%	25
Total	621	4,817	5,438		703

Figure 7N.25 The effects of cold water on fatal accidents, 2002. U.S. Department of Homeland Security, United States Coast Guard, 2004, Boating Statistics 2003, COMDTPUB P16754.17, uscgboating.org/statistics/Boating_Statistics_2003.pdf. http://library.fws.gov/nat_survey2001_trends.pdf.

Table 7N.163 Number of Fishing Participants, Days, and Expenditures of Sport Fishermen in the United States, 1991, 1996, and 2001

	1991		1996		2001		1991–2001	1996–2001
	Number	Percent	Number	Percent	Number	Percent	%change	%change
Participants								
Anglers, Total	**35,578**	**100**	**35,246**	**100**	**34,067**	**100**	**−4**	**−3**
All freshwater	31,041	87	29,734	84	28,439	83	−8	−4
Freshwater, except Great Lakes	30,186	85	28,921	82	27,913	82	−8	−3[a]
Great Lakes	2,552	7	2,039	6	1,847	5	−28	−9[a]
Saltwater	8,885	25	9,438	27	9,051	26	2[a]	−4[a]
Days								
Total	**511,329**	**100**	**625,893**	**100**	**557,394**	**100**	**9**	**−11**
All freshwater	439,536	86	515,115	82	466,984	84	6	−9
Freshwater, except Great Lakes	430,922	84	485,474	78	443,247	80	3[a]	−9
Great Lakes	25,335	5	20,095	3	23,138	4	−9[a]	15[a]
Saltwater	74,696	15	103,034	17	90.838	16	22	−12[a]
Expenditures (in 2001 dollars)								
Fishing, Total	**$31,175,168**	**100**	**$42,710,679**	**100**	**$35,632,132**	**100**	**14**	**−17**
Trips	15,396,151	49	17,380,775	41	14,656,001	41	−5[a]	−16
Equipment	12,170,062	39	21,666,341	51	16,963,398	48	39	−22
Fishing equipment	4,860,266	16	5,998,802	14	4,617,488	13	−5[a]	−23
Auxiliary equipment	804,953	3	1,171,540	3	721,048	2	−10[a]	−38
Special equipment	6,504,844	21	14,495,999	34	11,624,862	33	79	−20[a]
Other	3,608,953	12	3,663,563	9	4,012,733	11	11	10[a]

Note: U.S. Population 16 Years Old and Older, Numbers in Thousands.

[a] Not different from zero at the 5 percent level.

Source: From U.S. Department of Interior, Fish and Wildlife Service and U.S. Department of Commerce, U.S. Census Bureau, 2002,2001 National Survey of Fishing, Hunting and Wildlife-Associated Recreation, cenovs.gov.

Table 7N.164 Number and Cost of Sport Fishing Licenses in the United States, 1991, 1996, and 2001

Anglers	1991 Number (Thousands)	1991 Percent	1996 Number (Thousands)	1996 Percent	2001 Number (Thousands)	2001 Percent
Total anglers	35,578	100	35,246	100	34,071	100
Total license purchasers[a]	23,302	65	23,203	66	21,396	63
Anglers purchasing licenses: In state of residence	21,445	60	21,437	61	20,004	59
In other states	3,653	10	4,356	12	3,781	11
Total exempt from purchasing licenses	3,037	9	3,281	9	4,284	13
Anglers exempt from license purchase: In state of residence	2,596	7	2,365	7	3,959	12
In other states	375	1	427	1	608	2
Other[b]	6,586	19	9,143	26	10,268	30
Not reported	3,329	9	558	2	448	1

Expenditures	1991 Thousands of Dollars	1996 Thousands of Dollars	2001 Thousands of Dollars
Total, all licenses, stamps, tags and permits	486,700	579,753	639876
Licenses	443,287	519,061	597,210
Stamps, tags, permits	43,414	60,692	42,666

Note: Detail does not add to total because of multiple responses and nonresponse. Respondents could have been licensed in one state and exempt in another.

[a] Includes persons who had license bought for them. Does not include persons who purchase license and did not fish in 1991, 1996, and 2001.

[b] Includes persons engaged in activities requiring to license or exemptions and those who failed to buy a license for activities requiring a license.

Source: From U.S. Department of Interior, Fish and Wildlife Service and U.S. Department of Commerce, U.S. Census Bureau, 2002, 2001 National Survey of Fishing, Hunting and Wildlife-Associated Recreation; U.S. Department of Interior, Fish and Wildlife Service and U.S. Department of Commerce, U.S. Census Bureau, 1998, 1996 National Survey of Fishing, Hunting, and Wildlife-Associated Recreation; U.S. Department of Interior, Fish and Wildlife Service and U.S. Department of Commerce, U.S. Census Bureau, 1993, 1991 National Survey of Fishing, Hunting, and Wildlife-Associated Recreation, www.census.gov.

Table 7N.165 Fish Trips and Fish Harvested: Estimated Number of Fishing Trips Taken by Marine Recreational Fisherman by Subregion and Year, Atlantic and Pacific Coasts, 1998–2001

	1998		1999		2000		2001	
Subregion	**Fish Trips**	**Fish Harvested**	**Fish Trips**	**Fish Harvested**	**Fish Trips**	**Fish Harvested**	**Fish Trips**	**Fish Harvested**
North Atlantic	6,796	6783	6,478	8841	8,765	17185	9,035	12153
Mid-Atlantic	14,453	29447	14,105	24756	19,451	50,652	21,206	34704
South Atlantic[b]	16,837	24704	14,435	33914	20,075	40,414	21,596	43824
Gulf[b]	16,703	60561	15,894	55525	21,018	67385	22,890	76571
Total	54,789	121495	50,912	123036	69,309	175,616	74,727	167252
Pacific:[c]								
Southern California	2,973	5827	2,437	5098	3,787	7494	4,052	7726
Northern California	1,932	6123	1,713	3909	2,158	3787	2,208	4799
Oregon	649	1712	554	1153	930	1848	1,170	21283
Washington	1,463	5220	1,256	2486	1,643	3288	2,191	4796
Total	7,017	18882	5,960	12646	8,518	16,367	9,621	19,446

Note: "Harvested" includes dead discards and fish used for bait but does not include fish released alive. Numbers in thousands.

[a] 1998–2000 were revised. Data do not include recreational catch in Texas.

[b] Does not include catch from headboats (party boats) in the South Atlantic and Gulf of Mexico.

[c] Data do not include recreational catch in Hawaii or Alaska. Pacific state estimates do not include salmon data collected by recreational surveys.

Source: From USDA–NASS, Agricultural Statistics 2003, Miscellaneous, Agricultural Statistics, www.usda.gov.

Original Source: U.S. Department of Commerce, NOAA, NMFS, Fisheries Statistics and Economics Division, www.nass.usda.gov.

SECTION 7O FISHERIES

Table 7O.166 United States Fisheries—Quantity and Value of Catch, by Region and by State, 1999–2003

Regions and States	1999		2000		2001		2002		2003	
	Thousand Pounds	Thousand Dollars	Thousand Pounds	Thousand Dollars	Thousand Pounds	Thousand Dollars	Thousand Pounds	Thousand Dollars	Thousand Pounds	Thousand Dollars
New England	583,863	655,377	570,728	681,092	635,162	646,447	583,915	685,428	666,179	683,395
Maine	229,633	265,236	226,849	275,107	239,868	251,441	197,057	279,396	232,284	283,802
New Hampshire	11,258	12,542	17,160	13,951	18,584	17,865	23,201	16,691	27,410	15,125
Massachusetts	198,336	260,239	187,861	288,263	242,066	281,059	243,824	297,312	294,477	291,596
Rhode Island	126,206	79,270	119,295	72,544	115,957	65,457	103,656	64,250	95,727	63,054
Connecticut	18,430	38,090	19,563	31,227	18,687	30,625	16,177	27,779	16,281	29,818
Middle Atlantic	225,278	180,673	219,661	173,296	217,975	172,503	206,697	170,134	214,454	177,404
New York	48,175	76,049	41,181	59,426	42,422	55,038	38,665	51,334	39,409	51,628
New Jersey	168,676	97,731	171,804	107,163	168,430	109,820	162,175	112,733	170,017	120,556
Delaware	8,427	6,893	6,676	6,707	7,123	7,645	5,857	6,067	5,018	5,204
Chesapeake	527,407	172,012	492,110	172,210	617,244	174,968	495,675	172,320	10	16
Maryland	67,118	63,759	48,913	53,874	55,536	55,586	53,185	49,013	496,178	179,701
Virginia	460,289	108,253	443,197	118,336	561,708	119,382	442,490	123,307	49,350	49,038
South Atlantic	230,971	198,347	221,350	204,480	199,554	176,488	214,799	173,429	446,828	130,663
North Carolina	154,869	97,304	155,214	95,305	139,277	90,202	159,557	98,723	203,566	161,445
South Carolina	17,773	29,265	15,835	30,344	14,111	23,398	13,458	20,760	139,215	82,960
Georgia	11,234	21,100	9,694	21,331	9,036	14,752	9,563	15,068	22,043	29,075
Florida, East Coast	47,095	50,678	40,607	57,500	37,130	48,136	32,221	38,878	7,453	13,510
Gulf	1,945,063	757,857	1,759,993	910,685	1,605,564	798,319	1,716,140	692,717	34,855	35,900
Florida, West Coast	83,792	146,976	79,415	155,200	78,105	143,810	78,975	138,968	1,600,481	683,276
Alabama	27,399	50,415	29,931	63,275	24,740	43,170	23,380	35,102	76,448	135,912
Mississippi	267,546	48,526	217,744	58,715	213,889	50,561	217,053	46,093	25,344	39,521
Louisiana	1,480,045	302,735	1,344,913	401,095	1,191,460	342,748	1,308,531	305,534	213,116	45,508
Texas	86,281	209,205	87,990	232,400	97,370	218,030	88,201	167,020	1,189,448	294,011
Pacific Coast	5,765,700	1,422,258	5,750,364	1,320,763	6,173,671	1,187,106	6,138,249	1,130,633	96,125	168,324
Alaska	4,492,649	1,105,946	4,465,987	956,990	5,036,338	869,885	5,066,263	811,545	6,277,566	1,375,763
Washington	392,555	98,471	380,223	145,311	337,231	134,454	362,049	142,521	5,305,960	989,781
Oregon	233,177	67,590	262,917	79,351	234,097	72,516	211,183	68,431	379,732	170,158
California	647,319	150,251	641,237	139,111	526,005	110,251	498,754	108,136	225,528	85,549
Great Lakes	23,843	16,009	22,245	18,508	18,818	17,844	17,848	15,544	366,346	130,275
Illinois	86	50	49	35	16	14	–	–	17,471	13,174
Indiana	—	—	—	—	—	—	—	—	—	—
Michigan	13,546	9,339	12,704	8,963	10,322	9,235	9,459	7,362	8,690	5,702
Minnesota	443	197	377	172	501	202	449	180	435	228
New York	1	2	49	75	71	113	47	81	43	50
Ohio	3,932	2,186	3,497	2,442	3,535	3,287	3,427	3,093	3,994	3,037
Pennsylvania	32	43	20	29	25	44	15	37	11	23
Wisconsin	5,803	4,192	5,549	6,792	4,348	4,949	4,451	4,791	4,298	4,134
Hawaii	36,907	64,557	32,531	68,447	23,870	54,561	23,841	52,113	23,556	52,433
Utah	—	—	—	—	—	—	—	—	5,997	15,593
Total, United States	9,339,032	3,467,090	9,068,982	3,549,481	9,491,858	3,228,236	9,397,164	3,092,318	9,505,448	3,342,184

Note: Landings are reported in round (live) weight for all items except univalve and bivalve mollusks such as clams, oysters, scallops, which are reported in weight of meats (excluding the shell). Landings for Mississippi River drainage area States are not available. Data are preliminary. Landings of Alaska pollock, Pacific whiting, and other Pacific groundfish that are caught in waters off Washington, Oregon, and Alaska and are processed at-sea aboard U.S. vessels are credited to the State nearest to the area of capture. Totals may not add due to roundings. Data do not include landings by U.S.-flag vessels at Puerto Rico and other ports outside the 50 States. Data do not include aquaculture products, except oysters and clams.

Source: From U.S. Department of Commerce, NOAA, and National Marine Fisheries Services, 2004, Fisheries of the United States 2003; U.S. Department of Commerce, NOAA, and National Marine Fisheries Services, 2003, Fisheries of the United States 2002; U.S. Department of Commerce, NOAA, and National Marine Fisheries Services, 2002, Fisheries of the United States 2001; U.S. Department of Commerce, NOAA, and National Marine Fisheries Services, 2001, Fisheries of the United States 2000, www.st.nmfs.gov.

Table 7O.167 Major U.S. Domestic Species Landed in 2003 Ranked by Quantity and Value

Rank	Species	Pounds	Rank	Species	Dollars
1	Pollock	3,372,338	1	Crabs	483,586
2	Menhaden	1,599,444	2	Shrimp	424,027
3	Salmon	674,096	3	Lobsters	308,005
4	Cod	591,130	4	Flatfish	266,618
5	Flatfish	444,075	5	Scallops	229,240
6	Hakes	339,944	6	Pollock	208,581
7	Crabs	338,854	7	Salmon	200,838
8	Shrimp	313,624	8	Cod	187,113
9	Herring (sea)	286,050	9	Clams	162,838
10	Sardines	159,493	10	Oysters	103,045

Note: Numbers in thousands.

Source: From U.S. Department of Commerce, NOAA, and National Marine Fisheries Services, 2004, Fisheries of the United States 2003, www.st.nmfs.gov.

Table 7O.168 United States Domestic Fish and Shellfish Catch and Value by Major Species Caught, 1990–2002

	Quantity (1,000 lb)				Value ($1,000)			
Species	1990	2000	2001	2002	1990	2000	2001	2002
Cod: Atlantic	95,881	25,060	33,211	29,841	61,329	26,384	32,086	30,715
Pacific	526,396	530,505	471,711	512,827	91,384	142,330	118,071	96,206
Flounder	254,519	412,723	352,363	372,697	112,921	109,910	105,240	102,370
Halibut	70,454	75,190	77,978	82,044	96,700	143,826	115,169	135,603
Herring, sea; Atlantic	113,095	160,269	209,191	135,871	5,746	9,972	12,717	9,106
Herring, sea; Pacific	108,120	74,835	91,297	78,408	32,178	12,043	13,213	11,534
Menhaden	1,962,160	1,760,498	1,741,430	1,750,609	93,896	112,403	102,690	105,102
Pollock, Alaska	3,108,031	2,606,802	3,179,407	3,341,105	268,344	160,525	230,723	203,696
Salmon	733,146	628,638	722,832	567,179	612,367	270,213	208,926	155,010
Tuna	62,393	50,779	51,854	49,358	105,040	95,176	93,497	84,116
Whiting (Atlantic, silver)	44,500	26,855	28,479	17,622	11,281	11,370	13,232	7,454
Whiting (Pacific, hake)	21,232	452,718	379,304	285,714	1,229	18,809	16,147	13,584
Shellfish, total[a]	1,312,503	1,379,324	1,138,512	1,178,590	1,621,898	1,954,666	1,711,391	1,706,426
Clams	139,198	118,482	122,764	130,076	130,194	153,973	161,992	167,215
Crabs	499,416	299,006	272,246	307,601	483,837	405,006	381,667	397,695
Lobsters: American	61,017	83,180	73,637	82,252	154,677	301,300	254,334	293,329
Oysters	29,193	41,146	32,673	34,397	93,718	90,667	80,946	89,071
Scallops, sea	39,917	32,747	46,958	53,056	153,696	164,609	175,349	203,707
Shrimp	346,494	332,486	324,481	316,727	491,433	690,453	568,547	460,878
Squid, Pacific	36,082	259,508	191,532	160,677	2,636	27,077	17,834	18,262
Total	9,403,571	9,068,985	9,491,836	9,397,164	3,521,995	3,549,481	3,228,285	3,092,318
Fish, total[a]	8,091,068	7,689,661	8,242,490	8,089,987	1,900,097	1,594,815	1,479,988	1,359,392

[a] Includes other types of fish and shellfish, not shown separately.

Source: From U.S. Census Bureau, *Statistical Abstract of the United States.* 2004–2005, www.census.gov.

Original Source: National Oceanic and Atmospheric Administration, National Marine Fisheries Service, Fisheries of the United States, annual.

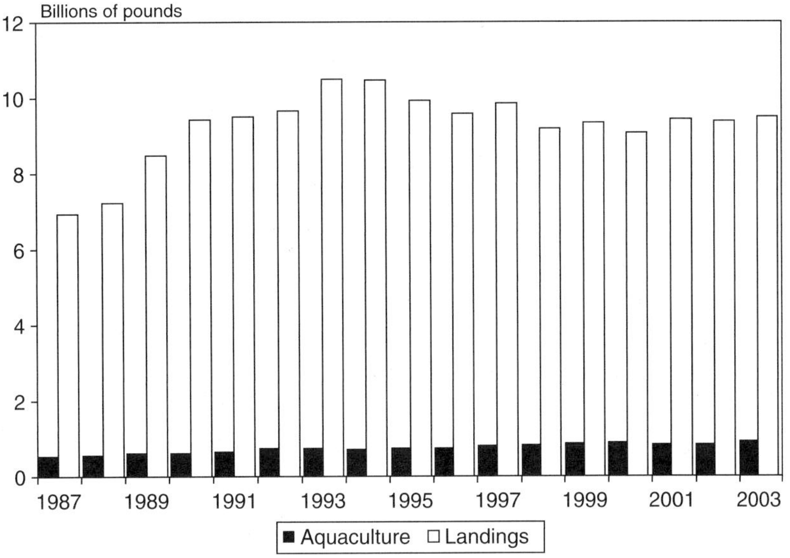

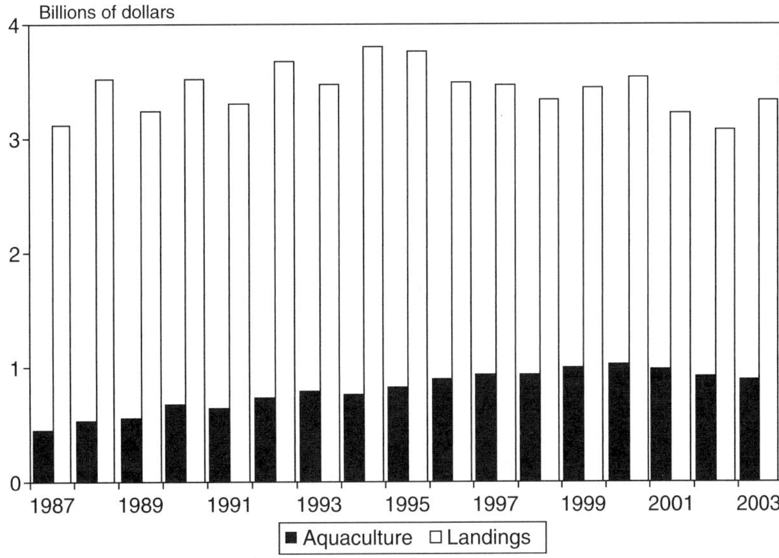

Figure 7O.26 Quantity and volume of domestic commercial landings and aquaculture production in the United States, 2003. (From U.S. Department of Commerce, NOAA, and National Marine Fisheries Service, 2004, Fisheries of the United States 2003, www.st.nmfs.gov.)

Table 7O.169 United States Fisheries Estimated Number of Commercial Fishing Vessels and Fishing Boats by Region and State, 1998–2002

	1998			1999			2000			2001			2002		
Area and State	**Vessels**	**Boats**	**Total**	**Vessels**	**Boats**	**Total**	**Vessels**	**Boats**	**Total**	**Vessels**	**Boats**	**Total**	**Vessels**	**Boats**	**Total**
Northeast															
Connecticut	245	318	563	232	281	513	182	243	425	182	243	425	NA	NA	NA
Delaware	175	NA	175	178	NA	NA	184	NA	NA	184	NA	NA	NA	NA	NA
Maine	1,642	5,799	7,441	1,653	5,821	7,474	1,656	5,836	7,492	1,656	5,836	7,492	NA	NA	NA
Maryland[a]	33	NA	33	34	NA	NA	32	NA	NA	32	NA	NA	NA	NA	NA
Massachusetts	715	4,500	5,215	700	4,520	5,220	695	4,540	5,235	695	4,540	5,235	NA	NA	NA
New Hampshire	115	461	576	121	468	589	109	471	580	109	471	580	NA	NA	NA
New Jersey	387	1,147	1,534	421	NA	NA	397	NA	NA	397	NA	NA	NA	NA	NA
New York[b]	689	2,931	3,620	678	2,825	3,503	NA	NA	NA	NA	NA	NA	NA	NA	NA
Rhode Island	312	2,401	2,713	330	2,239	2,569	344	2,920	3,264	344	2,920	3,264	NA	NA	NA
Virginia[a]	122	NA	122	241	NA	NA	261	NA	NA	261	NA	NA	NA	NA	NA
South Atlantic and Gulf															
North Carolina	891	NA	891	667	NA	NA	773	NA	NA	773	NA	NA	763	NA	NA
South Carolina	569	NA	569	577	NA	NA	520	NA	NA	520	NA	NA	556	NA	NA
Georgia	350	NA	350	350	NA	NA	265	NA	NA	265	NA	NA	226	NA	NA
Florida	2,384	6,157	8,541	2,214	5,602	7,816	2,136	5,502	7,638	2,136	5,502	7,638	1,934	4,438	6,372
Alabama	398	1,338	1,736	454	1,231	1,685	443	1,328	1,771	443	1,328	1,771	425	1,350	1,775
Mississippi	454	688	1,142	502	707	1,209	504	743	1,247	504	743	1,247	522	843	1,365
Louisiana	2,535	11,637	14,172	2,450	11,414	13,864	2,393	11,830	14,223	2,393	11,830	14,223	2,084	8,874	10,958
Texas	NA	NA	NA	NA	NA	NA	NA	NA	NA	NA	NA	NA	NA	NA	NA
West Coast															
Alaska	6,384	9,445	15,829	6,232	9,374	15,606	6,169	9,461	15,630	6,126	9,062	15,188	5,494	8,541	14,035
Washington	904	428	1,332	783	343	1,126	726	355	1,081	726	355	1,081	695	329	1,024
Oregon	668	296	964	643	308	951	721	376	1,097	721	376	1,097	639	359	998
California	1,392	1,191	2,583	1,438	1,142	2,580	1,307	1,132	2,439	1,307	1,132	2,439	1,201	997	2,198
Hawaii	2,855	NA	2,855	NA	NA	NA	347	2,467	2,901	347	2,467	2,814	NA	NA	NA
Great Lakes[c]															
Illinois	5	NA	5	5	NA	NA	5	NA	NA	5	NA	NA	NA	NA	NA
Indiana	NA	NA	NA	NA	NA	NA	NA	NA	NA	NA	NA	NA	NA	NA	NA
Michigan	61	74	135	NA	NA	NA	NA	NA	NA	NA	NA	NA	NA	NA	NA
Minnesota	1	22	23	1	24	25	1	24	25	1	24	25	NA	NA	NA
New York	2	NA	2	2	NA	NA	1	NA	NA	1	NA	NA	NA	NA	NA
Ohio	40	19	59	34	21	55	31	19	50	31	19	50	NA	NA	NA
Pennsylvania	2	1	3	2	1	3	2	1	3	2	1	3	NA	NA	NA
Wisconsin	89	16	105	68	18	86	78	18	96	78	18	96	NA	NA	NA

Note: NA, Data not available or provided seperately. Vessels are documented craft greater than 5 net registered tons. Boats are craft less than 5 net registered ton.

[a] Only federally collected data are available. Inshore data are not available.

[b] Excludes vessels and boats in the Great Lakes.

[c] Commercial fishing fleet size of the Great Lakes states represent only the number of licenses issued by the State; therefore, may not be an accurate total. Tribal data are not included in this table.

Source: From U.S. Department of Commerce, NOAA, and National Marine Fisheries Services, 2004, Fisheries of the United States 2003; U.S. Department of Commerce, NOAA, and National Marine Fisheries Services, 2003, Fisheries of the United States 2002; U.S. Department of Commerce, NOAA, and National Marine Fisheries Services, 2002, Fisheries of the United States 2001; U.S. Department of Commerce, NOAA, and National Marine Fisheries Services, 2001, Fisheries of the United States 2000, www.st.nmfs.gov.

Table 7O.170 Aquaculture Water Withdrawals in the United States, 2000

State	Withdrawals (mil gal/day)			Withdrawals (thousand acre-feet yr)		
	By Source			By Source		
	Groundwater	Surface Water	Total	Groundwater	Surface Water	Total
Alabama	8.93	1.44	10.4	10	1.61	11.6
Alaska	—	—	—	—	—	—
Arizona	—	—	—	—	—	—
Arkansas	187	10.4	198	210	11.6	222
California	158	380	537	177	426	603
Colorado	—	—	—	—	—	—
Connecticut	—	—	—	—	—	—
Delaware	0.07	0	0.07	0.08	0	0.08
District of Columbia	—	—	—	—	—	—
Florida	7.81	0.21	8.02	8.76	0.24	8.99
Georgia	7.7	7.72	15.4	8.63	8.65	17.3
Hawaii	—	—	—	—	—	—
Idaho	51.5	1,920	1,970	57.7	2,150	2,210
Illinois	—	—	—	—	—	—
Indiana	—	—	—	—	—	—
Iowa	—	—	—	—	—	—
Kansas	3.33	2.27	5.6	3.73	2.54	6.28
Kentucky	—	—	—	—	—	—
Louisiana	128	115	243	144	129	273
Maine	—	—	—	—	—	—
Maryland	4.81	14.8	19.6	5.39	16.6	22
Massachusetts	—	—	—	—	—	—
Michigan	—	—	—	—	—	—
Minnesota	—	—	—	—	—	—
Mississippi	321	49.8	371	360	55.9	416
Missouri	2.01	81.3	83.3	2.25	91.2	93.4
Montana	—	—	—	—	—	—
Nebraska	—	—	—	—	—	—
Nevada	—	—	—	—	—	—
New Hampshire	3.12	13.1	16.3	3.5	14.7	18.2
New Jersey	6.46	0	6.46	7.24	0	7.24
New Mexico	—	—	—	—	—	—
New York	—	—	—	—	—	—
North Carolina	7.88	0	7.88	8.83	0	8.83
North Dakota	—	—	—	—	—	—
Ohio	1.36	0	1.36	1.52	0	1.52
Oklahoma	0.29	16.1	16.4	0.33	18.1	18.4
Oregon	—	—	—	—	—	—
Pennsylvania	—	—	—	—	—	—
Rhode Island	—	—	—	—	—	—
South Carolina	—	—	—	—	—	—
South Dakota	—	—	—	—	—	—
Tennessee	—	—	—	—	—	—
Texas	—	—	—	—	—	—
Utah	116	0	116	130	0	130
Vermont	—	—	—	—	—	—
Virginia	—	—	—	—	—	—
Washington	—	—	—	—	—	—
West Virginia	—	—	—	—	—	—
Wisconsin	39.8	30.4	70.2	44.6	34.1	78.7
Wyoming	—	—	—	—	—	—
Puerto Rico	—	—	—	—	—	—
U.S. Virgin Islands	—	—	—	—	—	—
Total	**1,060**	**2,640**	**3,700**	**1,180**	**2,960**	**4,150**

Note: Figures may not sum to totals because of independent rounding; —, data not collected.

Source: From Hutson, S.S. and others, 2004, Estimated Use of Water in the United States in 2003. *U.S. Geological Survey Circular 1268*, www.usgs.gov.

Table 7O.171 U.S. Private Aquaculture — Trout and Catfish Production and Value: 1990–2002

Item	Unit	1990	1995	1997	1998	1999	2000	2001	2002
Trout Foodsize									
Number sold	Millions	67.8	60.2	59.3	57.6	61.0	58.5	54.5	50.2
Total weight	Mil lb	56.8	55.6	56.9	57.9	60.2	59.2	56.9	54.5
Total value of sales	Mil dol	64.6	60.8	60.7	60.3	64.7	63.7	64.4	58.3
Average price received	Dol lb	1.14	1.09	1.07	1.04	1.07	1.08	1.13	1.07
Percent sold to processors	Percent	58	68	63	62	68	70	68	69
Catfish Foodsize									
Number sold	Millions	272.9	321.8	391.8	409.8	424.5	420.1	406.9	405.8
Total weight	Mil lb	392.4	481.5	569.6	601.4	635.2	633.8	647.5	673.7
Total value of sales	Mil dol	305.1	378.1	406.8	445.4	464.7	468.8	410.7	378.5
Average price received	Dol /lb	0.78	0.79	0.71	0.74	0.73	0.74	0.63	0.56
Fish sold to processors	Mil lb	360.4	446.9	524.9	564.4	596.6	593.6	597.1	630.6
Avg. price paid by processors	Cents/lb	75.8	78.6	71.2	74.3	73.7	75.1	64.7	56.8
Processor sales	Mil lb	183.1	227.0	261.8	281.4	292.7	297.2	296.4	317.6
Avg. price received by processors	Cents/lb	224.1	240.3	226.0	229.0	234.0	236.0	226.0	207.0
Inventory (Jan 1)	Mil lb	9.4	10.9	11.9	10.8	12.6	13.6	15.0	12.3

Note: 67.8 represents 67,800,000. Periods are from Sept 1 of the previous year to Aug 31 of stated year. Data are for foodsize fish, those over 12 in. long.

Source: From U.S. Census Bureau, Statistical Abstract of the United States. 2004–2005, www.census.gov.

Original Source: From U.S. Dept. of Agriculture, National Agricultural Statistics Service, *Trout Production* released February; *Catfish Production* released February; and *Catfish Processing* released February. Also in annual.

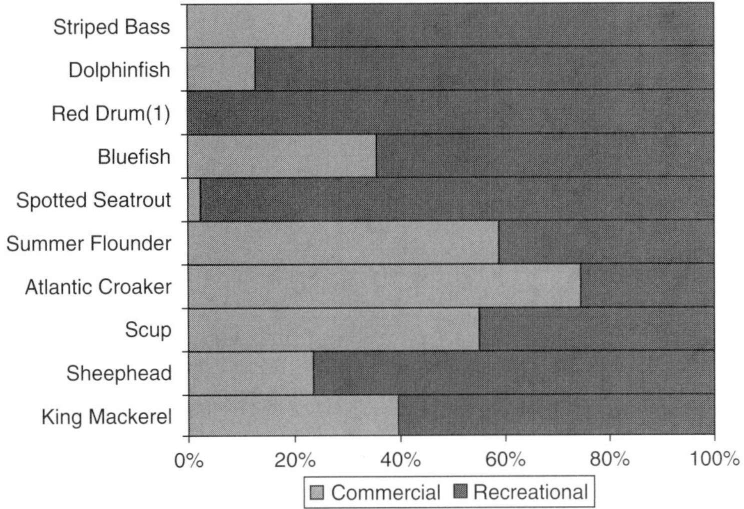

Top Ten Commercial Species
Versus Recreational Harvest - 2003

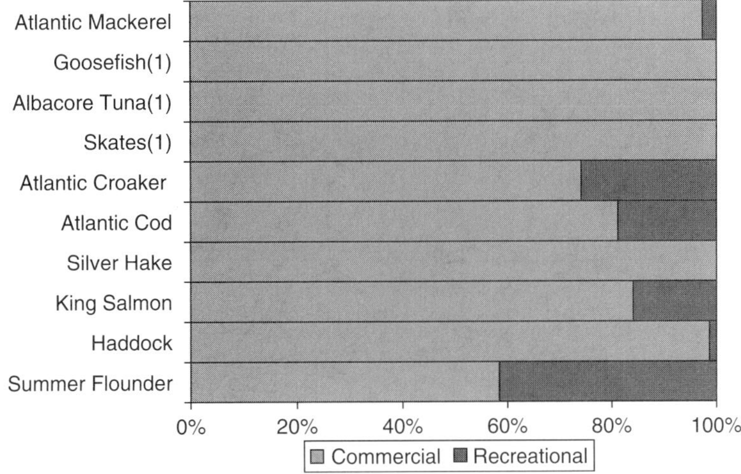

Figure 7O.27 Top ten recreational species versus commercial harvest/top ten commercial species versus recreational harvest, 2003. (From U.S. Department of Commerce, NOAA, and (1) less than 1 percent National Marine Fisheries Service, 2004, Fisheries of the United States 2003, www.st.nmfs.gov.)

Table 7O.172 World Aquaculture and Commercial Catches by Area of Fish, Crustaceans, and Mollusks, 2001–2002

Country	2001			2002		
	Aquaculture	Catch	Total	Aquaculture	Catch	Total
Marine Areas	Metric tons Live-Weight			Metric tons Live-Weight		
Atlantic Ocean						
Northeast	1,315,707	11,143,204	12,458,911	1,307,923	11,048,962	12,356,885
Northwest	108,149	2,240,365	2,348,514	104,761	2,245,008	2,349,769
Eastern central	251	3,929,630	3,929,881	342	3,373,623	3,373,965
Western central	85,094	1,686,404	1,771,498	99,919	1,764,352	1,864,271
Southeast	2,680	1,648,084	1,650,764	2,675	1,701,440	1,704,115
Southwest	52,877	2,287,502	2,340,379	71,793	2,089,660	2,161,453
Mediterranean and Black Sea	367,777	1,570,335	1,938,112	339,264	1,550,099	1,889,363
Indian Ocean						
Eastern	432,253	4,877,380	5,309,633	432,048	5,100,261	5,532,309
Western	30,563	3,981,292	4,011,855	44,074	4,243,330	4,287,404
Pacific Ocean						
Northeast	134,724	2,759,090	2,893,814	141,812	2,702,885	2,844,697
Northwest	11,286,336	22,550,874	33,837,210	12,063,628	21,436,229	33,499,857
Eastern central	60,875	1,860,373	1,921,248	63,540	2,037,267	2,100,807
Western central	640,227	10,103,215	10,743,442	538,639	10,510,202	11,048,841
Southeast	633,595	12,653,427	13,287,022	611,092	13,765,143	14,376,235
Southwest	93,343	752,661	846,004	106,053	739,868	845,921
Antarctic	—	120,159	120,159	—	144,158	144,158
Inland Areas						
Africa	366,787	2,051,183	2,417,970	405,320	2,092,924	2,498,244
Asia	21,053,159	5,734,686	26,787,845	22,295,148	5,722,141	28,017,289
Europe	479,242	347,242	826,484	467,769	354,270	822,039
North America	414,512	174,959	589,471	448,661	170,614	619,275
South America	227,141	368,803	595,944	250,864	377,313	628,177
Oceania	3,803	21,219	25,022	3,246	20,905	24,151
Total	**37,789,095**	**92,862,087**	**130,651,182**	**39,798,571**	**93,190,654**	**132,989,225**

Note: Data for marine mammals and aquatic plants are excluded.

Source: From U.S. Census Bureau, *Statistical Abstract of the United States. 2004–2005*, www.census.gov.

Original Source: From Food and Agrculture Organizaiton of the United Nations (FAO).

Table 7O.173 Capture and Aquaculture Production Totals for Marine and Inland Fisheries

	Total Capture and Aquaculture Production for All Species (thousand metric tons)			Marine Aquaculture Production (thousand metric tons)			Inland Aquaculture Production (thousand metric tons)			Marine Capture Production (thousand metric tons)			Inland Capture Production (thousand metric tons)		
	1979–1981	1989–1991	1999–2001	1979–1981	1989–1991	1999–2001	1979–1981	1989–1991	1999–2001	1979–1981	1989–1991	1999–2001	1979–1981	1989–1991	1999–2001
World	**72,900**	**99,449**	**129,429**	**2,136**	**5,020**	**14,240**	**2,632**	**8,025**	**21,372**	**63,018**	**80,101**	**85,153**	**5,114**	**6,303**	**8,665**
Asia (Excl. Middle East)	**29,517**	**44,556**	74,156	**1,404**	**3,700**	**11,481**	**1,973**	**6,774**	**19,663**	**23,602**	**30,786**	**37,521**	**2,539**	**3,296**	**5,491**
Armenia	—	7	2	—	—	—	—	4	1	—	—	—	—	2	1
Azerbaijan	—	45	17	—	—	—	—	2	0.1	—	—	—	—	44	17
Bangladesh	648	862	1,643	6	19	66	86	177	588	116	255	325	441	412	663
Bhutan	0.3	0.3	0.3	—	—	—	—	0.03	0.03	—	—	—	0.3	0.3	0.3
Cambodia	34	104	332	—	0	1	0.1	6	14	3	34	38	30	63	279
China	4,437	13,200	41,392	422	2,051	9,361	910	4,418	15,113	2,743	5,833	14,696	362	898	2,223
Georgia	—	105	2	—	—	—	—	1	0.1	—	103	2	—	0.1	0.02
India	2,412	3,830	5,752	4	34	94	363	1,048	1,999	1,493	2,258	2,784	552	490	875
Indonesia	1,812	3,072	4,889	25	115	147	164	373	654	1,366	2,291	3,775	256	292	313
Japan	10,377	10,279	5,738	461	700	714	94	97	60	9,691	9,374	4,896	131	107	68
Kazakhstan	—	86	35	—	—	—	—	11	1	—	—	—	—	76	34
Korea, Dem People's Rep	1,410	1,394	270	6	48	63	3	6	4	1,332	1,290	184	69	50	20
Korea, Rep	2,207	2,733	2,274	299	361	285	1	14	12	1,868	2,334	1,970	39	24	6
Kyrgyzstan	—	1	0.1	—	—	—	—	1	0.1	—	—	—	—	0.3	0.1
Lao People's Dem Rep	24	28	71	—	—	—	1	10	41	—	—	—	23	19	30
Malaysia	750	977	1,418	85	47	109	4	10	46	655	919	1,259	6	1	3
Mongolia	0.4	0.2	0.4	—	—	—	—	—	—	—	—	—	0.4	0.2	0.4
Myanmar	580	738	1,156	—	0	5	3	8	99	429	593	857	148	137	195
Nepal	4	14	30	—	—	—	1	9	15	—	—	—	2	5	15
Pakistan	299	480	643	0	0.04	0.1	5	11	17	251	369	446	43	101	179
Philippines	1,586	2,209	2,300	13	81	82	195	302	312	1,153	1,595	1,761	224	231	145
Singapore	16	13	10	0.2	2	4	0	0	1	16	11	5	1	0.1	0
Sri Lanka	186	191	294	0.01	1	5	1	5	4	164	158	252	21	27	33
Tajikistan	—	4	0.2	—	—	—	—	4	0.1	—	—	—	—	0.3	0.1
Thailand	1,194	2,822	3,631	74	198	445	44	104	271	1,691	2,389	2,709	105	131	206
Turkmenistan	—	47	11	—	—	—	—	2	0.1	—	—	—	—	45	11
Uzbekistan	—	27	8	—	—	—	—	22	5	—	—	—	—	4	3

(Continued)

Table 7O.173 (Continued)

	Total Capture and Aquaculture Production for All Species (thousand metric tons)			Marine Aquaculture Production (thousand metric tons)			Inland Aquaculture Production (thousand metric tons)			Marine Capture Production (thousand metric tons)			Inland Capture Production (thousand metric tons)		
	1979–1981	1989–1991	1999–2001	1979–1981	1989–1991	1999–2001	1979–1981	1989–1991	1999–2001	1979–1981	1989–1991	1999–2001	1979–1981	1989–1991	1999–2001
Vietnam	585	963	1,941	7	39	97	91	124	402	403	665	1,273	84	135	170
Europe	**12,622**	**21,829**	**18,138**	**546**	**871**	**1,608**	**198**	**601**	**459**	**11,713**	**19,798**	**15,661**	**165**	**559**	**410**
Albania	9	10	3	0.1	2	0.2	0.2	0.5	0.1	6	6	2	3	2	1
Austria	4	4	3	—	—	—	3	4	3	—	—	—	1	1	0.4
Belarus	—	19	6	—	—	—	—	16	6	—	—	—	—	2	1
Belgium	47	41	32	—	—	—	0.1	1	2	47	40	29	0.00	1	1
Bosnia and Herzegovina	—	—	3	—	—	—	—	—	—	—	—	0	—	—	3
Bulgaria	108	73	12	0	0	0.04	11	9	4	95	62	6	1	2	2
Croatia	—	—	29	—	—	4	—	—	4	—	—	21	—	—	0.02
Czech Rep	—	—	24	—	—	—	—	—	19	—	—	—	—	—	4
Denmark	1,878	1,747	1,526	0.4	6	7	19	33	36	1,858	1,707	1,483	0.2	0.4	0.2
Estonia	—	377	110	—	0.1	0	—	1	0.3	—	372	107	—	4	3
Finland	157	159	171	2	14	13	3	5	3	99	91	120	53	49	35
France	787	854	867	174	197	207	24	45	54	586	608	603	2	4	2
Germany	579	371	285	9	23	25	25	45	42	526	293	196	21	9	23
Greece	105	143	197	0.02	7	89	2	2	3	94	131	101	9	3	3
Hungary	35	31	20	—	—	—	24	18	13	—	—	—	12	13	7
Iceland	1,534	1,354	1,904	0.03	2	3	0.02	0.2	1	1,533	1,350	1,900	0.5	1	0.2
Ireland	145	237	358	5	24	51	0.5	1	1	140	208	305	0.1	4	1
Italy	498	548	512	52	109	164	24	42	49	413	385	293	8	13	5
Latvia	—	479	129	—	—	—	—	4	0	—	474	128	—	0.3	1
Lithuania	—	416	103	—	—	—	—	5	2	—	407	99	—	4	2
Macedonia, FYR	—	—	2	—	—	—	—	—	1	—	—	—	—	—	0.2
Moldova, Rep	—	8	1	—	—	—	—	6	1	—	—	—	—	1	0.3
Netherlands	381	499	588	95	86	73	0.1	1	6	285	408	507	2	3	2
Norway	2,540	1,945	3,166	9	142	493	—	—	—	2,530	1,803	2,672	0.4	1	1
Poland	625	498	261	—	—	—	11	27	35	603	456	210	10	15	16
Portugal	260	327	203	0.1	5	6	0.5	2	1	259	321	196	0	0	0
Romania	182	159	17	—	—	—	36	37	10	129	107	2	16	15	5

Russian Federation	—	7,733	3,992	—	0.3	1	—	179	77	—	7,196	3,646	—	358	269
Serbia and Montenegro	—	—	4	—	—	0.01	—	—	3	—	—	0.4	—	—	1
Slovakia	—	—	2	—	—	—	—	—	1	—	—	—	—	—	1
Slovenia	—	—	3	—	—	0.1	—	—	1	—	—	2	—	—	0.2
Spain	1,371	1,378	1,415	198	196	282	11	21	33	1,144	1,151	1,091	18	10	9
Sweden	230	254	340	0.2	5	3	1	4	3	226	244	332	2	2	1
Switzerland	4	5	3	—	—	—	0.2	1	1	—	—	—	4	3	2
Ukraine	0	1,016	146	0	0.5	0.04	0	77	32	—	901	379	—	38	5
United Kingdom	867	856	936	1	38	146	2	15	13	862	800	773	1	3	4
Middle East & N. Africa	**1,407**	**2,148**	**3,432**	**0.1**	**4**	**76**	**45**	**111**	**385**	**1,220**	**1,765**	**2,535**	**141**	**268**	**437**
Afghanistan	1	1	1	—	—	—	—	—	—	—	—	—	1	1	1
Algeria	48	90	101		0.03	0.04	—	0.3	0.2	48	90	101	0	0	0
Egypt	140	302	701	0	0.5	30	19	61	273	34	79	140	86	162	258
Iran, Islamic Rep	60	268	414	—	—	4	9	24	40	47	201	256	4	42	114
Iraq	45	21	23	—	—	—	4	2	2	27	2	13	14	17	9
Israel	29	24	25	—	0.1	3	12	15	17	13	7	4	4	2	2
Jordan	0.04	0.4	1	—	—	—	—	0.05	1	0.04	0.01	0.2	0	0.4	0.4
Kuwait	3	5	6	—	0	0.2	—	0	0.04	3	5	6	0	0	0
Lebanon	2	2	4	—	—	—	0.05	0.1	0.3	2	2	4	0	0.01	0.02
Libyan Arab Jamahiriya	10	24	33	—	—	—	—	0.1	0.1	10	24	33	0	0	0
Morocco	335	562	910	0.1	0.4	1	0.01	0.05	1	335	560	907	1	2	2
Oman	80	118	121	—	0	2	—	—	—	80	118	119	0	0	0
Saudi Arabia	26	44	55	0	0.1	3	—	2	4	26	43	48	0	0	0
Syrian Arab Rep	4	6	14	—	—	—	1	3	6	1	2	3	2	2	5
Tunisia	60	90	97	0.1	1	1	0	0.1	1	60	89	95	0	0	1
Turkey	417	401	605	0	2	30	1	4	40	387	356	490	29	40	45
United Arab Emirates	66	93	113	0	0	0	—	0	0	66	93	113	0	0	0
Yemen	73	77	127	—	—	—	—	—	—	73	77	127	0	0	0
Sub-Saharan Africa	**3,173**	**3,929**	**5,122**	**0.4**	**3**	**8**	**7**	**23**	**44**	**1,999**	**2,309**	**3,296**	**1,166**	**1,594**	**1,775**
Angola	110	129	222	—	—	—	—	—	—	102	122	216	7	8	6
Benin	37	38	37	—	—	—	—	—	—	4	8	8	33	30	29
Botswana	1	1	0.1	—	—	—	—	—	—	—	—	—	1	1	0.1
Burkina Faso	7	7	8	—	—	—	0	0.01	0.01	—	—	—	7	7	8

(Continued)

Table 7O.173 (Continued)

	Total Capture and Aquaculture Production for All Species (thousand metric tons)			Marine Aquaculture Production (thousand metric tons)			Inland Aquaculture Production (thousand metric tons)			Marine Capture Production (thousand metric tons)			Inland Capture Production (thousand metric tons)		
	1979–1981	1989–1991	1999–2001	1979–1981	1989–1991	1999–2001	1979–1981	1989–1991	1999–2001	1979–1981	1989–1991	1999–2001	1979–1981	1989–1991	1999–2001
Burundi	13	16	12	—	—	—	—	0.04	0.1	—	—	—	13	16	12
Cameroon	81	70	111	—	—	—	0.04	0.1	0.1	61	48	59	20	22	53
Central African Rep	13	13	15	—	—	—	0.1	0.1	0.1	—	—	—	13	13	15
Chad	62	65	84	—	—	—	—	—	—	—	—	—	62	65	84
Congo	30	47	43	—	—	—	0	0.2	0.2	19	21	18	11	26	25
Congo, Dem Rep	107	165	209	—	—	—	—	1	0.4	1	3	4	106	162	205
Côte d'Ivoire	88	92	76	—	—	—	—	0.2	1	71	68	64	18	24	11
Equatorial Guinea	3	4	5	—	—	—	—	—	—	3	3	4	0	0.4	1
Eritrea	—	—	10	—	—	—	—	—	—	—	—	10	—	—	0
Ethiopia	3	5	16	—	—	—	—	0.04	0	0.2	1	0	3	4	16
Gabon	20	21	47	—	—	—	0	0	0.4	19	19	36	2	2	10
Gambia	13	22	31	—	0.1	—	—	0	0	10	19	29	3	3	3
Ghana	237	374	468	—	—	—	0.3	0.4	5	197	316	389	40	58	75
Guinea	20	45	90	—	—	—	—	0	0	19	42	86	1	3	4
Guinea-Bissau	4	5	5	—	—	—	—	—	—	4	5	5	0.01	0.2	0.2
Kenya	53	183	196	0.0	0.1	0	0.2	1	1	5	9	7	47	173	189
Lesotho	0.02	0.03	0.04	—	—	—	0.02	0.02	0.01	—	—	—	0	0.01	0.03
Liberia	13	10	13	—	—	—	0	0	0.02	9	6	9	4	4	4
Madagascar	56	100	139	—	0.02	5	0.1	0.2	2	17	71	102	39	29	30
Malawi	59	70	44	—	—	—	0.1	0.2	1	—	—	—	59	69	43
Mali	82	70	103	—	—	—	—	0.02	0.05	—	—	—	82	70	103
Mauritania	35	70	80	—	—	—	—	—	—	29	64	75	6	6	5
Mozambique	34	32	35	—	—	—	—	0.02	0	31	29	25	4	4	10
Namibia	10	163	573	—	0.02	0.04	—	—	0.01	10	162	571	0.1	1	2
Niger	9	4	16	—	—	—	—	0.03	0.02	—	—	—	9	4	16
Nigeria	263	295	474	0.2	1	—	6	15	24	149	192	308	108	86	142
Rwanda	1	3	7	—	—	—	0.02	0.1	0.3	—	—	—	1	2	7
Senegal	228	307	407	0.1	—	—	0	0.01	0.2	213	290	381	15	18	25
Sierra Leone	52	58	70	—	—	—	0	0.02	0.03	38	43	56	14	15	14
Somalia	12	23	20	—	—	—	—	—	—	12	22	20	0.1	0.4	0.2

South Africa	905	639	666	0.01	2	3	—	2	2	904	635	661	1	1	1
Sudan	28	32	55	—	—	—	—	0.2	1	1	1	5	27	30	48
Tanzania, United Rep	213	373	327	—	—	—	—	0.4	0.2	37	54	52	176	319	274
Togo	9	15	23	—	—	—	—	0.02	0.1	6	10	18	4	5	5
Uganda	171	224	223	—	—	—	—	0.1	1	—	—	—	171	224	222
Zambia	47	68	71	—	—	—	0.03	2	4	—	—	—	47	66	66
Zimbabwe	13	24	13	—	—	—	0.1	0.2	0.2	—	—	—	13	24	13
North America	**5,225**	**7,485**	**6,596**	**100**	**142**	**242**	**83**	**248**	**345**	**4,926**	**7,005**	**5,939**	**116**	**91**	**69**
Canada	1,395	1,608	1,160	3	35	118	1	5	13	1,340	1,519	990	51	48	39
United States	3,718	5,713	5,270	97	107	123	82	242	333	3,474	5,321	4,784	65	42	30
C. America & Caribbean	**1,742**	**1,912**	**1,972**	**1**	**16**	**63**	**10**	**28**	**92**	**1,676**	**1,721**	**1,703**	**54**	**147**	**115**
Belize	2	2	44	—	0.3	4	—	0	—	2	2	40	0.03	0	0
Costa Rica	18	18	43	—	0.3	2	0.03	1	8	17	17	32	0.3	0.4	1
Cuba	168	184	114	1	2	1	1	6	53	160	161	55	6	14	5
Dominican Rep	10	20	13	—	0.1	0.3	0	0.3	2	8	18	10	2	1	1
El Salvador	16	10	15	—	0.5	0.2	—	0.01	0.1	14	6	12	2	4	2
Guatemala	4	6	27	—	1	2	0.01	0.2	3	4	4	16	0.5	2	7
Haiti	5	5	5	—	—	—	—	—	—	5	5	5	0.3	0.4	0.5
Honduras	7	19	19	0.1	3	8	0.01	0.2	1	6	16	11	0.1	0.04	0.1
Jamaica	9	45	44	—	0	0	0.03	3	4	9	8	6	0.2	0.4	0.5
Mexico	1,280	1,416	1,367	0	6	39	9	16	21	1,228	1,269	1,210	43	125	97
Nicaragua	7	5	30	0	0.1	5	0	0	0.01	7	4	24	0.1	0.2	1
Panama	175	162	196	0.2	3	2	0	0.3	1	175	158	193	0	0.02	0.02
Trinidad and Tobago	4	9	10	—	—	—	—	0	0.01	4	9	10	0	0	0
South America	**8,237**	**15,274**	**17,226**	**10**	**123**	**549**	**5**	**32**	**208**	**7,960**	**14,810**	**16,124**	**262**	**309**	**345**
Argentina	438	561	973	—	0	0.02	0.1	0.3	1	427	550	944	11	10	27
Bolivia	5	6	6	—	—	—	—	0.4	0.4	—	—	—	5	6	6
Brazil	801	814	923	0.2	2	37	4	19	139	630	601	552	167	193	195
Chile	2,948	5,885	4,793	1	28	409	0.1	4	2	2,947	5,853	4,383	0	0.02	0.00
Colombia	78	113	184	0.03	5	11	0.2	4	49	30	74	98	48	30	26
Ecuador	595	496	645	9	84	79	0.4	1	7	585	410	559	0.2	1	0.4
Guyana	33	38	53	—	0.02	0.2	0	0.03	0.4	32	37	51	1	1	1
Paraguay	3	12	25	—	—	—	—	0.1	0.1	—	—	—	3	12	25
Peru	3,027	6,878	9,033	1	4	4	0.5	2	4	3,011	6,837	8,990	15	35	35

(Continued)

Table 7O.173 (Continued)

	Total Capture and Aquaculture Production for All Species (thousand metric tons)			Marine Aquaculture Production (thousand metric tons)			Inland Aquaculture Production (thousand metric tons)			Marine Capture Production (thousand metric tons)			Inland Capture Production (thousand metric tons)		
	1979–1981	1989–1991	1999–2001	1979–1981	1989–1991	1999–2001	1979–1981	1989–1991	1999–2001	1979–1981	1989–1991	1999–2001	1979–1981	1989–1991	1999–2001
Suriname	3	7	18	—	—	0.2	—	0	0.1	3	6	17	0.1	0.4	0.2
Uruguay	125	119	108	—	—	0	—	0	0.04	125	118	106	0.4	0.3	2
Venezuela	179	335	405	0	0.3	8	0.3	0.5	5	167	313	364	12	21	28
Oceania	**419**	**788**	**1,246**	**11**	**47**	**124**	**1**	**2**	**4**	**396**	**718**	**1,095**	**11**	**22**	**22**
Australia	136	215	240	8	12	37	1	1	3	126	199	199	1	2	2
Fiji	21	30	43	—	0.01	1	0	0.02	0.3	21	26	35	1	4	6
New Zealand	149	379	654	3	35	84	—	0.2	0.5	146	344	568	1	1	1
Papua New Guinea	42	26	95	—	0	0	—	0.01	0.01	34	12	81	8	14	14
Solomon Islands	35	51	38	—	0	0.01	—	0	0	35	51	38	0	0	0
Developed	**29,330**	**41,007**	**31,968**	**1,118**	**1,762**	**2,689**	**387**	**1,011**	**894**	**27,407**	**37,300**	**27,765**	**418**	**935**	**620**
Developing	**33,012**	**56,915**	**95,919**	**954**	**3,145**	**11,459**	**1,935**	**6,808**	**20,306**	**26,087**	**41,612**	**56,110**	**4,035**	**5,350**	**8,044**

Note: All figures are three year averages for the range of years specified. Numbers in this table are reported with varying precision: values greater than 0.5 thousand metric tons are rounded to the nearest whole number, values between 0.05 and 0.5 thousand metric tons are rounded to the nearest tenth (one decimal place), value between 0.005 and 0.05 are rounded to the nearest hundredth (two decimal places), and "0" represents a value of less than 0.005 thousand metric tons.

Variable Definitions and Methodology.

Total Capture and Aquaculture Production for All Species includes both capture and aquaculture production of fish, crustaceans, molluscs, aquatic mammals, and other aquatic animals, taken for commercial, industrial, recreational and subsistence purposes from inland, brackish and marine waters. For each country, the four sub-categories of production (Marine Aquaculture, Inland Aquaculture, Marine Capture, and Inland Capture) listed in this table should sum to *Total Capture and Aquaculture Production for All Species.*

Marine Aquaculture Production data include fish, crustaceans, molluscs, aquatic mammals, and other aquatic animals, taken for commercial, industrial, recreational and subsistence purposes from marine and brackish environments. The harvest from inland waters is not included. Aquaculture production is defined by FAO as "the farming of aquatic organisms, including fish, molluscs, crustaceans, and aquatic plants. Farming implies some form of intervention in the rearing process to enhance production, such as regular stocking, feeding, protection from predators, etc. [It] also implies ownership of the stock being cultivated." Aquatic organisms that are exploitable by the public as a common property resource are not included in the aquaculture production.

Inland Aquaculture production data include fish, molluscs, crustaceans, aquatic mammals, and other aquatic animals, taken for commercial, industrial, recreational and subsistence purposes from inland waters, such as lakes and rivers. The harvest from marine areas is not included. Some inland lakes are saltwater, thus the data include not only freshwater species but also saltwater species. Aquaculture production is defined by FAO as "the farming of aquatic organisms, including fish, molluscs, crustaceans, and aquatic plants. Farming implies some form of intervention in the rearing process to enhance production, such as regular stocking, feeding, protection from predators, etc. [It] also implies ownership of the stock being cultivated." Aquatic organisms that are exploitable by the public as a common property resource are not included in the aquaculture production.

Marine Capture Production data include fish, molluscs, crustaceans, aquatic mammals, and other aquatic animals, taken for commercial, industrial, recreational and subsistence purposes from marine and brackish environments. The harvest from inland waters is not included. Capture production data refer to the nominal catch of fish, crustaceans, molluscs, aquatic mammals, and other aquatic animals, taken for commercial, industrial, recreational and subsistence purposes from marine, brackish, and inland waters. The harvest from aquaculture and other kind of farming are excluded.

Inland Capture Production data include fish, molluscs, crustaceans, aquatic mammals, and other aquatic animals, taken for commercial, industrial, recreational and subsistence purposes from inland waters, such as lakes and rivers. The harvest from marine areas is not included. Some inland lakes are saltwater, thus the data include not only freshwater species but also saltwater species. Capture production data refer to the nominal catch of fish, crustaceans, molluscs, aquatic mammals, and other aquatic animals, taken for commercial, industrial, recreational and subsistence purposes from marine, brackish, and inland waters. The harvest from aquaculture and other kind of farming are excluded.

Statistics for aquatic plants are excluded from country totals. For a more detailed listing of species groups represented in the table, please refer to the original source at fao.org/waicent/faostat/agricult/fishitems-e-e.html.

Data include all quantities caught for both food and feed purposes but exclude discards. The harvest of fish, crustaceans and molluscs are expressed in live weight, that is the nominal weight of the aquatic organisms at the time of capture.

Aquatic organisms included in the FAO FISHSTAT capture production database have been classified according to approximately 1290 commercial species items, further arranged within the 50 groups of species constituting the nine divisions of the FAO International Standard Statistical Classification of Aquatic Animals and Plants (ISSCAAP). Most fisheries statistics are collected by FAO from official national reports. When these data are missing or considered unreliable, FAO estimates fishery production based on regional fishery organizations, project documents, industry magazines, or statistical interpolations. Fishery production statistics were revised completely by FAO in the 1990s. At this time, FAO estimated missing data points, updated taxonomical classifications, and discriminated more clearly between aquaculture and capture fisheries production.

Fisheries production statistics typically refer to the calendar years (January 1–December 31) with exception of data from Antarctic waters, which use a split year (July 1–June 30).

Frequency of Update by Data Providers. FAO publishes updated data sets annually for FishStat. This table represents data through the year 2001; data were acquired by WRI in May, 2003.

Data Reliability and Cautionary Notes. While the FAO data set provides the most extensive, global time series of fishery statistics since 1970, there are some problems associated with the data. Funding for the development and maintenance of fisheries statistics at the national level has been decreasing in real terms since 1992, while the demand is growing for a variety of global statistics on discards, fish inventories, aquaculture, and illegal activites. Country-level data are often submitted with a 1–2 year delay, and countries are declaring an increasing percentage of their catch as "unidentified fish." As a result, species-item totals frequently underestimate the real catch of individual species. Stock assessment working groups can more accurately estimate the composition of a catch; however, due to financial constraints, these groups are rare, especially in developing countries. Statistics from smaller artisanal and subsistence fisheries are particularly sparse.

FAO states that "general trends are probably reliably reflected by the available statistics...but the annual figures and the assessments involve a certain degree of uncertainty and small changes from year to year are probably not statistically significant." These statistics provide a good overview of regional fisheries trends. However, when reviewing the state of fisheries stocks, evaluating food security, etc., these data should be used with caution and supplemented with estimates from regional organizations, academic literature, expert consultations, and trade data.

For more information, please consult Fishery Statistics: Reliability and Policy Implications, published by the FAO Fisheries Department and available on-line at fao.org/fi/statist/naturechina/30jan02.asp.

Source: From World Resources Institute, Earth Trends Environmental Information, Coastal and Marine Ecosystems, www.earthtrends.wri.org.

Original Source: Fishery Information, Data and Statistics Unit, Food and Agriculture Organization of the United Nations (FAO). 2003. *FISHSTAT Plus: Universal software for fishery statistical time series*, Version 2.3 (available on-line at fao.org/fi/statist/FISOFT/FISHPLUS.asp); Total production dataset. Rome: FAO.

Table 7O.174 World-Fisheries and Aquaculture

State	Marine Catch[a] (annual average) Metric Tons (000) 1998–2000	Marine Catch Percent Change since 1988–1990	Freshwater Catch[b] (annual average) Metric Tons (000) 1998–2000	Freshwater Catch Percent Change since 1988–1990	Total Aquaculture Production (annual average) Metric Tons (000) 1998–00	Total Aquaculture Percent Change since 1988–1990	Trade in Fish and Fish Products[c] (annual average mil US$) Exports 1998–2000	Trade Imports 1998–2000	Food Supply from Fish and Fish Products (kg/person/yr)[d] 1997–1999	Fish Protein as a Percent of All Animal Protein 1997–1999	Number of Fishers 2000	Number of Decked Fishery Vessels 1995[e]	Population within 100 km of the Coast (Percent) 1995
World	**81,601.9**	**2**	**9,550.7**	**31**	**33,179.7**	**63**	**52,548.9**	**57,624.7**	**16.0**	**16**	**34,501,411**	**1,256,841**	**39**
Asia	**36,527.8**	**20**	**5,751.2**	**61**	**26,625.3**	**62**	**15,235.0**	**20,418.1**	**18.0**	**28**	**28,890,352**	**1,057,966**	**38**
(Excl. Middle East)													
Armenia	—	—	1.0	(63)	0.7	—	0.3	1.0	0.5	1	244[f]	6	0
Azerbaijan	0.0	(67)	14.8	(70)	0.2	—	1.7	1.3	0.9	1	1,500[f]	—	56
Bangladesh	179.6	27	754.6	47	597.4	69	313.6	2.5	10.2	47	1,320,480	61	55
Bhutan	—	—	0.3	—	0.0	67	—	—	—	—	450[f]	—	0
Cambodia	36.4	25	184.1	212	14.5	62	35.3	3.1	12.0	35	73,425[g]	—	24
China	14,395.9	170	2,367.1	188	22,722.0	73	3,081.3	1,315.0	24.5	21	12,233,128[h]	432,674	24
Georgia	2.2	(99)	0.2	(60)	0.1	—	0.3	1.6	1.3	2	1,900[f]	82	39
India	2,726.5	33	753.5	48	2,039.2	52	1,221.4	24.0	4.7	14	5,958,744[f]	56,600	26
Indonesia	3,624.7	69	375.3	18	722.5	37	1,582.2	69.7	19.0	56	5,118,571	67,325	96
Japan	4,836.3	(52)	285.1	(3)	763.0	(5)	756.2	14,406.3	65.4	45	260,200	360,747	96
Kazakhstan	0.0	(83)	23.3	(70)	1.2	—	13.2	13.3	1.9	2	16,000[f]	1,970	4
Korea, Dem People's Rep	190.2	(87)	20.0	(66)	67.9	23	69.6	5.6	9.4	36	129,000[f]	2,900	93
Korea, Rep	1,968.3	(16)	16.4	(59)	317.9	(30)	1,346.6	1,037.6	47.3	39	176,928[j]	76,801	100
Kyrgyzstan	—	—	0.1	(79)	0.1	—	—	2.0	0.7	1	154[f]	—	0
Lao People's Dem Rep	—	—	26.3	34	31.2	73	0.0	1.4	10.0	31	15,000[f]	—	6
Malaysia	1,201.8	42	20.4	74	146.8	65	189.6	262.6	57.0	35	100,666[f]	17,965	98
Mongolia	—	—	0.4	91	—	—	0.2	0.0	0.1	0	0	—	0
Myanmar	772.7	33	166.2	21	90.7	93	162.0	1.0	16.0	45	610,000[f]	140	49
Nepal	—	—	13.8	131	13.6	48	0.2	0.3	1.1	3	50,000[f]	—	0
Pakistan	448.3	28	173.9	78	17.6	50	141.9	0.4	2.5	3	272,273	5,064	9
Philippines	1,719.0	14	146.4	(37)	342.7	(5)	408.7	109.2	29.6	42	990,872[f]	3,220	100
Singapore	6.5	(44)	0.04	(68)	4.3	55	413.6	483.9	—	—	364	110	100
Sri Lanka	255.3	67	32.7	4	10.2	45	103.5	66.5	21.2	54	146,188	2,990	100
Tajikistan	—	—	0.1	(81)	0.1	—	—	0.2	0.1	0	200[f]	—	0
Thailand	2,654.6	14	206.5	77	664.5	61	4,180.5	841.3	28.2	37	354,495	17,600	39
Turkmenistan	0.0	(93)	9.4	(79)	0.6	(338)	0.4	0.1	1.7	2	611[i]	45	8
Uzbekistan	—	—	3.0	(40)	5.9	(257)	0.0	2.0	0.5	1	4,800	—	3
Vietnam	1,217.6	92	156.3	19	463.6	66	1,080.4	12.1	18.1	37	1,000,000	140	83
Europe	**15,710.1**	**(24)**	**674.7**	**(18)**	**1,726.0**	**13**	**19,063.8**	**22,875.8**	**20.6**	**10**	**855,333**	**105,324**	**40**
Albania	2.1	(73)	0.8	(64)	0.2	—	6.5	4.5	2.4	1	1,590[j]	2	97
Austria	—	—	0.6	4	2.9	(37)	9.0	189.7	14.3	4	2,300	—	2
Belarus	—	—	0.5	(84)	5.6	(203)	16.4	72.8	8.5	4	5,000[f]	—	0
Belgium	29.7	(27)	0.5	4	1.4	49	471.9	1,059.0	—	—	544[j]	156	83
Bosnia and Herzegovina	0.0	—	2.5	—	—	—	—	8.3	1.9	2	3,500[f]	—	47
Bulgaria	10.2	(87)	1.9	37	5.2	(102)	7.2	15.5	4.2	3	1,483[f]	30	29
Croatia	20.6	—	0.4	—	6.3	—	40.9	38.1	5.2	5	65,151[i]	305	38
Czech Rep	—	—	4.3	—	18.5	—	27.7	79.4	12.7	5	2,243	—	0
Denmark	1,497.3	(15)	1.5	(71)	42.9	19	2,856.3	1,804.9	26.0	10	6,711[j]	4,285	100

Estonia	110.2	(72)	4.4	(36)	0.2	(286)	86.6	37.6	19.7	12	13,346	186	86
Finland	108.7	28	56.4	(16)	15.6	(14)	19.1	127.5	33.6	14	5,879	3,838	73
France	573.2	(10)	3.3	(40)	266.8	11	1,104.7	3,275.8	31.3	9	26,113[g]	6,586	40
Germany	212.6	(33)	24.5	96	66.8	(1)	1,044.5	2,403.1	14.9	7	4,358	2,406	15
Greece	104.4	(16)	4.4	15	73.1	92	248.5	301.0	26.0	11	19,847	18,375	99
Hungary	—	—	7.3	(58)	11.7	(60)	8.1	45.7	4.3	2	4,900	—	0
Iceland	1,799.9	13	0.3	(57)	3.8	50	1,352.0	80.3	93.1[n]	30	6,100	826	100
Ireland	290.2	38	2.6	(52)	45.8	52	356.4	113.5	16.0	6	8,478[i]	1,353	100
Italy	298.7	(24)	5.2	(61)	208.8	34	370.5	2,705.7	24.2	11	48,770	16,000	79
Latvia	120.2	(77)	1.2	(43)	0.4	(853)	54.4	35.9	15.4	11	6,571	351	75
Lithuania	57.8	(85)	1.9	(63)	1.7	(152)	40.0	56.1	22.0	15	4,700	131	23
Macedonia, FYR	—	—	0.2	—	1.5	—	0.6	8.8	5.1	5	8,472	—	14
Moldova, Rep	—	—	0.3	(86)	1.1	(537)	2.0	4.6	3.3	4	40[f]	—	9
Netherlands	513.6	27	2.1	(47)	101.4	5	1,490.2	1,237.0	19.7	9	3,743	1,008	93
Norway	2,726.8	59	1.4	(56)	458.2	74	3,668.3	635.4	52.2	26	23,552	8,664	95
Poland	211.1	(60)	19.5	23	33.1	21	266.8	293.7	12.8	11	8,640[i]	445	14
Portugal	206.6	(37)	0.04	(23)	7.1	(12)	276.6	936.1	65.7	23	25,021	9,265	93
Romania	3.0	(98)	5.1	(75)	9.4	(369)	5.4	38.8	2.5	2	8,519	33	6
Russian Federation	3,700.0	(50)	488.3	(10)	68.6	(179)	1,269.1	230.6	21.7	15	316,300	3,584	15
Serbia and Montenegro	0.4	—	1.2	—	4.3	—	0.8	44.0	2.9	1	1,429[f]	5	8
Slovakia	—	—	1.7	—	0.8	—	2.2	36.3	8.3	5	215	—	0
Slovenia	1.8	—	0.2	—	1.1	—	6.4	28.5	6.9	3	231	11	61
Spain	1,133.8	(8)	8.9	(6)	316.3	26	1,582.1	3,399.6	44.4	18	75,434[f]	15,243	68
Sweden	363.2	51	3.6	(46)	5.5	(55)	472.0	688.9	30.4	14	2,783	1,240	88
Switzerland	—	—	1.8	(49)	1.1	26	3.1	374.1	18.3	7	522	—	0
Ukraine	409.3	(57)	11.6	(81)	31.0	(193)	56.4	109.6	11.4	10	120,000[f]	444	21
United Kingdom	830.6	(1)	4.2	81	148.2	69	1,421.8	2,294.9	21.8	10	17,847	9,562	99
Middle East & N. Africa	**2,348.0**	**24**	**411.0**	**74**	**355.9**	**62**	**—**	**756.3**	**7.2**	**9**	**746,955**	**21,990**	**47**
Afghanistan	—	—	1.1	10	—	—	—	—	—	—	1,500[f]	—	0
Algeria	98.2	(1)	0.0	—	0.3	(35)	2.7	11.2	3.5	6	26,151[i]	2,184	69
Egypt	156.0	81	219.8	44	235.3	75	1.6	157.3	11.2	19	250,000	—	53
Iran, Islamic Rep	248.3	23	140.3	424	35.2	25	48.2	56.1	4.4	7	138,965	900	24
Iraq	12.5	204	10.1	(43)	3.8	16	—	0.6	1.5	8	12,000[f]	8	6
Israel	4.2	(57)	1.8	8	19.1	23	8.2	133.2	23.4	9	1,535[f]	384	97
Jordan	0.1	—	0.4	10	0.5	87	—	23.4	5.1	5	721	—	29
Kuwait	5.8	(19)	1.0	127	0.3	—	5.3	16.4	12.1	5	670[j]	917	100
Lebanon	3.6	122	0.0	—	0.4	75	—	24.2	8.0	7	9,825	5	100
Libyan Arab Jamahiriya	33.0	45	0.0	—	0.1	50	35.0	11.3	6.1	7	9,500[f]	93	79
Morocco	782.3	43	1.8	16	2.2	89	815.3	11.3	8.4	17	106,096[f]	3052	65
Oman	110.1	(18)	0.0	—	5.1	22	46.6	5.3	—	—	28,003[j]	390	88
Saudi Arabia	49.1	10	0.1	—	5.4	78	8.6	108.5	7.6	6	25,360	23	30
Syrian Arab Rep	2.6	81	4.6	282	6.7	58	—	48.9	1.8	2	11,292	5	34
Tunisia	90.9	(4)	1.0	291	1.5	37	94.7	13.1	9.4	12	50,815	17	84
Turkey	491.3	5	28.9	(12)	66.2	93	96.4	62.7	8.0	10	33,614[f]	9,710	58
United Arab Emirates	112.5	22	0.1	82	0.0	—	36.8	27.5	25.9	12	15,543	4,050	85
Yemen	122.3	64	0.0	—	—	—	26.1	4.8	6.8	22	12,200	71	63
Sub-Saharan Africa	**2806.5**	**15**	**1808.0**	**13**	**37.2**	**25**	**1,691.4**	**845.5**	**7.6**	**25**	**1,995,694[i]**	**71**	**21**
Angola	186.4	50	6.0	(25)	—	—	10.8	14.3	10.4	28	30,364[f]	580	29

(Continued)

Table 7O.174 (Continued)

State	Marine Catch[a] (annual average) Metric Tons (000) 1998–2000	Marine Catch[a] Percent Change since 1988–1990	Freshwater Catch[b] (annual average) Metric Tons (000) 1998–2000	Freshwater Catch[b] Percent Change since 1988–1990	Total Aquaculture Production (annual average) Metric Tons (000) 1998–00	Total Aquaculture Production Percent Change since 1988–1990	Trade in Fish and Fish Products[c] (annual average mil US$) Exports 1998–2000	Trade in Fish and Fish Products[c] Imports 1998–2000	Food Supply from Fish and Fish Products (kg/person/yr)[d] 1997–1999	Fish Protein as a Percent of All Animal Protein 1997–1999	Number of Fishers 2000	Number of Decked Fishery Vessels 1995[e]	Population within 100 km of the Coast (Percent) 1995
Benin	13.8	6	24.5	(6)	0.0	—	2.2	4.7	8.7	26	61,793	5	62
Botswana	—	—	0.2	(89)	—	—	0.1	5.3	6.1	5	2,620[f]	—	0
Burkina Faso	—	—	8.1	7	0.0	—	0.0	1.4	1.9	6	8,300	—	0
Burundi	—	—	10.9	(18)	0.1	60	0.2	0.1	2.3	23	7,030[j]	—	0
Cameroon	59.6	21	50.0	138	0.1	(117)	2.7	30.2	12.3	31	24,500	25	22
Central African Rep	—	—	14.8	14	0.1	20	—	0.4	4.2	9	5,410	—	0
Chad	—	—	84.0	31	—	—	—	—	6.9	14	300,000[g]	—	0
Congo	20.6	(6)	25.5	10	0.2	(29)	2.4	19.7	21.4	46	10,500	26	25
Congo, Dem Rep	3.9	97	194.4	21	0.4	(66)	0.5	42.5	6.7	34	108,400	23	3
Côte d'Ivoire	65.5	(2)	11.5	(59)	1.0	86	171.0	171.7	14.2	42	19,707[f]	63	40
Equatorial Guinea	4.5	34	1.0	162	—	—	2.8	2.1	—	—	9,218	5	72
Eritrea	7.0	—	0.0	—	—	—	1.0	0.1	0.9	3	14,500[f]	—	73
Ethiopia[k]	—	—	15.2	365	0.0	—	—	—	0.2	1	6,272	—	1
Gabon	40.4	114	10.1	421	0.4	—	14.0	7.1	49.6	37	8,259[g]	39	63
Gambia	26.5	69	2.5	(7)	0.0	—	4.9	1.3	24.1	64	2,000[f]	—	91
Ghana	384.6	24	77.9	24	0.5	18	81.1	97.3	28.1	66	230,000[f]	500	42
Guinea	78.9	108	4.0	33	0.0	—	23.1	14.3	11.2	51	10,707[f]	15	41
Guinea-Bissau	5.1	3	—	—	—	—	3.1	0.4	4.4	14	2,500[f]	8	95
Kenya	6.0	(29)	191.7	25	0.3	(188)	36.8	6.3	5.4	10	59,565	32	8
Lesotho	—	—	0.0	494	0.00	—	—	—	0.0	0	60[f]	—	0
Liberia	8.5	3	4.1	(1)	0.0	—	0.0	1.9	5.9	26	5,143	14	58
Madagascar	98.9	47	30.0	(10)	5.9	96	77.0	6.4	7.5	16	83,310[j]	65	55
Malawi	—	—	43.8	(41)	0.4	55	0.2	0.3	4.5	34	42,922[j]	57	0
Mali	—	—	102.1	55	0.1	80	0.4	2.2	8.8	15	70,000[i]	—	0
Mauritania	32.9	(51)	5.0	(17)	—	—	70.1	0.5	10.6	11	7,944[g]	126	40
Mozambique	25.8	(16)	10.8	215	0.0	—	84.0	8.8	2.7	21	20,000[f]	291	59
Namibia	305.0	191	1.5	49	0.0	50	266.1	—	11.6	20	2,700[f]	218	5
Niger	—	—	11.4	226	0.0	(100)	0.7	0.6	0.9	3	7,983[f]	—	0
Nigeria	316.4	66	136.9	46	22.6	35	4.8	231.6	8.8	32	481,264[g]	318	26
Rwanda	—	—	6.6	287	0.2	65	—	0.1	1.0	7	5,690	—	0
Senegal	378.8	42	27.3	47	0.1	82	287.6	7.0	32.1	45	55,547[j]	180	83
Sierra Leone	49.5	32	16.3	(0)	0.03	33	14.6	3.3	13.6	61	17,990[f]	27	55
Somalia	20.7	(2)	0.2	(50)	—	—	3.7	—	2.9	2	18,900[f]	12	55
South Africa	596.4	(34)	0.9	10	4.4	47	259.0	64.1	6.9	8	10,500[f]	600	39
Sudan	5.7	336	44.0	52	1.0	88	0.4	0.4	1.7	2	27,700[j]	—	3
Tanzania, United Rep	49.6	(4)	280.0	(18)	0.2	(30)	66.8	0.4	8.9	32	92,529	30	21
Togo	15.4	34	5.2	20	0.1	89	1.8	14.2	13.4	51	14,120	3	45
Uganda	—	—	267.5	19	0.2	80	33.8	0.1	8.9	28	57,862[j]	—	0
Zambia	—	—	68.0	6	4.2	70	0.4	0.9	7.4	25	23,833[f]	235	0
Zimbabwe	—	—	14.0	(41)	0.2	11	2.2	9.3	2.7	10	1,804[i]	—	0

North America	**5457.1**	**(19)**	**419.4**	**(19)**	**559.8**	**32**	**5,682.6**	**10,840.9**	**21.5**	**12**	**303,784**	**45,480**	**41**
Canada	933.5	(37)	68.9	(56)	109.1	70	2,575.9	1,318.9	23.8	10	8,696	18,280	24
United States	4365.8	(15)	350.5	(3)	450.7	23	2,847.5	9,511.3	21.3	7	290,000[f]	27,200	43
C. America & Caribbean	**1582.5**	**(7)**	**117.0**	**(25)**	**132.8**	**(697)**	**1,529.2**	**423.0**	**8.8**	**14**	**446,390**	**7,161**	**55**
Belize	37.8	—	0.0	(50)	2.5	92	27.7	2.4	13.0	13	1,872	12	100
Costa Rica	23.2	40	1.0	233	9.0	95	177.4	27.4	5.9	5	6,510[j]	1,003	100
Cuba	58.4	(68)	5.0	(61)	51.5	86	90.1	26.9	13.1	16	11,865[f]	1,250	100
Dominican Rep	9.2	(44)	0.6	(57)	1.2	80	0.9	52.4	12.6	10	9,286	—	100
El Salvador	7.5	2	2.6	(5)	0.3	(119)	31.9	6.9	2.9	4	24,534	80	99
Guatemala	13.7	324	6.9	477	4.0	79	29.9	7.6	1.6	3	17,275	85	61
Haiti	4.6	(9)	0.5	58	—	—	3.4	7.4	3.1	11	4,700[f]	1	100
Honduras	10.8	(25)	0.1	115	8.3	62	40.2	14.8	2.9	3	21,000[i]	280	65
Jamaica	6.5	(18)	0.5	1	4.0	23	13.0	56.0	25.5	20	23,465	5	100
Mexico	1130.8	(9)	98.5	(27)	47.7	60	694.0	125.2	9.6	8	262,401	3,100	29
Nicaragua	21.7	444	1.2	813	4.8	99	87.1	6.5	3.3	7	14,502	280	72
Panama	182.2	27	0.0	21	5.3	28	232.8	15.4	11.0	8	13,062	695	100
Trinidad and Tobago	9.1	12	0.0	—	0.02	—	11.8	7.7	14.2	14	7,297	19	100
South America	**14649.6**	**1**	**345.7**	**6**	**318.2**	**61**	**4,980.1**	**687.7**	**8.9**	**12**	**784,051**	**13,106**	**49**
Argentina	1006.7	101	24.7	133	1.3	77	824.7	86.5	8.5	4	12,320	800	45
Bolivia	0.9	(64)	5.2	59	0.4	21	0.1	5.6	1.7	2	7,754[f]	—	0
Brazil	520.5	(16)	180.9	(6)	132.7	86	168.7	357.0	6.5	4	290,000[f]	1,450	49
Chile	4150.8	(26)	0.0	(97)	319.6	94	1,694.4	54.1	17.6	10	50,873	563	82
Colombia	101.7	71	25.1	(37)	53.6	88	195.1	86.5	4.5	5	129,410[i]	167	30
Ecuador	466.4	(18)	0.4	(32)	112.0	33	915.7	16.1	7.0	9	162,870[g]	515	61
Guyana	51.2	44	0.7	(16)	0.5	92	38.6	0.7	59.9	47	6,571	55	77
Paraguay	—	—	25.0	124	0.1	44	0.1	2.3	5.5	4	4,469[g]	—	0
Peru	7773.0	15	34.6	2	7.6	34	852.2	15.4	20.3	21	66,361	7,710	57
Suriname	16.0	209	0.2	(27)	0.2	—	6.9	4.1	24.6	24	3,628[f]	22	87
Uruguay	117.9	11	2.2	878	0.0	—	115.0	13.4	8.6	4	4,023	958	78
Venezuela	389.9	37	46.7	41	11.1	94	126.9	45.5	18.3	19	44,302[i]	866	73
Oceania	**1110.1**	**75**	**23.0**	**(1)**	**127.6**	**62**	**1,681.7**	**629.9**	**22.7**	**25**	**85,324**	**1,917**	**87**
Australia	214.6	13	4.1	9	33.9	62	885.7	518.8	21.3	7	13,800[i]	246	90
Fiji	27.9	17	5.5	18	1.3	99	28.5	16.8	32.1	21	8,985[j]	—	100
New Zealand	594.9	97	1.6	(20)	90.4	69	682.2	55.9	30.3	13	1,928	1,375	100
Papua New Guinea	47.1	271	11.7	(7)	0.0	—	31.7	11.5	15.1	31	16,000[f]	35	61
Solomon Islands	46.8	(0)	0.0	—	0.0	—	10.2	0.2	52.5	82	11,000[m]	130	100
Developed	**27258.0**	**(30)**	**1439.3**	**(21)**	**3180.4**	**12**	**27,094.4**	**48,905.7**	**23.7**	**10**	**1,467,401**	**516,259**	**45**
Developing	**53010.2**	**32**	**8110.6**	**49**	**26702.3**	**60**	**24,010.7**	**8,571.6**	**13.8**	**20**	**32,640,482**	**740,322**	

Note: Negative values are shown in parentheses.

Variable Definitions and Methodology

Marine and Freshwater Catch data refer to marine and freshwater fish caught or trapped for commercial, industrial, and subsistence use (catches from recreational activities are included where available); they include fish caught by a country's fleet anywhere in the world. Statistics for mariculture, aquaculture, and other kinds of fish or shellfish farming are not included in the country totals. Marine fish includes demersal fish (flounders, halibuts, soles, etc; cods, hakes, haddocks, etc; redfishes, basses, congers, etc; and sharks, rays, chimeras, etc.), pelagic fish (jacks, mullets, sauries, etc; herrings, sardines, anchovies, etc; tunas, bonitos, billfishes, etc; and mackerels, snooks, cutlassfishes, etc.), and diadromous fish caught in marine areas (i.e. sturgeons, paddlefishes, river eels, salmons, trouts, smelt, shads, and miscellaneous diadromous fishes), marine molluscs (squids, cuttlefishes, octopuses, etc; abalones, winkles, conchs, etc; oysters; mussels; scallops, pectens, etc; clams, cockles, arkshells, etc; and miscellaneous marine molluscs) and marine crustaceans (sea-spiders, crabs, etc; lobsters, spiny-rock lobsters, etc; squat lobsters; shrimps, prawns, etc; krill, planktonic crustaceans, etc; and miscellaneous marine crustaceans).

Freshwater fish includes fish caught in inland waters (i.e. carps, barbels, and other cyprinids; tilapias and other cichlids; and miscellaneous and freshwater fishes), and diadromous fish caught in inland waters, as well as freshwater molluscs and crustaceans. Catch figures are the national totals averaged over a 3-year period. Data are represented as nominal catches, which are the landings converted to a live-weight basis, that is, the weight when caught. Fish catch does not include discards. Landings for some countries are identical to catches. Catch data are provided annually to the FAO Fisheries Department by national fishery offices and regional fishery commissions. Some recent data are provisional. If no data are submitted, FAO uses the previous year's figures or makes estimates based on other information.

Aquaculture is defined by FAO as "the farming of aquatic organisms, including fish, molluscs, and crustaceans. Farming implies some form of intervention in the rearing process to enhance production, such as regular stocking, feeding, and protection from predators, etc. [It] also implies ownership of the stock being cultivated¥" Aquatic organisms that are exploitable by the public as a common property resource are included in the harvest of fisheries.

FAO's global collection of aquaculture statistics from questionnaires to national fishery offices was begun in 1984. FAO's aquaculture database has 337 "species items" that are grouped into six categories. **Total Aquaculture Production** includes marine, freshwater, and diadromous fishes, molluscs and crustaceans cultivated in marine, inland, or brackish environments. For a detailed listing of species, please refer to the original source. Aquaculture production is expressed as an annual average over a 3-year period.

Trade in Fish and Fish Products expresses the value associated with imports and exports of fish that are live, fresh, chilled, frozen, dried, salted, smoked, or canned, and other derived products and preparations. Trade includes freshwater and marine fish, aquaculture, molluscs and crustaceans, meals, and solubles. Aquatic plants are not included. Figures are the national totals averaged over a 3-year period in millions of U.S. dollars. Exports are generally on a free-on-board basis (i.e. not including insurance or freight costs). Imports are usually on a cost, insurance, and freight basis (i.e. insurance and freight costs added in). Regional totals are calculated by adding up imports or exports of each country included in that region.

Therefore, the regional totals should not be taken as a net trade for that region, since there may also be trade occurring within a region. To collate national data, FAO uses its International Standard Statistical Classification of Fishery Commodities. Commodities produced by aquaculture and other kinds of fish farming are also included.

Food Supply from Fish and Fish Products is defined as the quantity of both freshwater and marine fish, seafood and derived products available for human consumption. Data were calculated by taking a country's fish production plus imports of fish and fishery products, minus exports, minus the amount of fishery production destined to non-food uses (i.e. reduction to meal, etc.), and plus or minus variations in stocks. The quantity of fish and fish products consumed include the bones and all parts of the fish.

Fish Protein as a Percent of Animal Protein Supply is defined as the quantity of protein from both freshwater and marine fish, seafood and derived products available for human consumption as a percentage of all available animal protein. FAO calculates food supply for all products, including fish, in its food balance sheets. FAOSTAT maintains statistics on apparent consumption of fish and fishery products, in live weight, for 220 countries in a collection of Supply/Utilization Accounts (SUAs). For each product, the SUA traces supplies from production, imports, and stocks to its utilization in different forms–addition to stocks; exports; animal feed; seed; processing for food and non-food purposes; waste (or losses); and lastly; as food available for human consumption, where appropriate. For more detailed information, please refer to the following article: "Supply Utilization Accounts and Food Balance Sheets in the Context of a National Statistical System", maintained on-line by FAO at fao.org/es/ESS/Suafbs.htm.

Number of Fishers includes the number of people employed in commercial and subsistence fishing (both personnel on fishing vessels and on shore), operating in freshwater, brackish and marine areas, and in aquaculture production activities. Data on people employed in fishing and aquaculture are collected by the FAO through annual questionnaires submitted to the national reporting offices of the member countries. When possible, other national and/or regional published sources are also used to estimate figures. Please refer to the original source for further information on collection methodologies (available online at fao.org/fi/statist/fisoft/fishers.asp) or to the following publication: *Numbers of Fishers 1970–1997, FAO Fisheries Circular N.929 Revision 2, Fishery Information, Data and Statistics Unit (FAO, Rome, 1999).*

Decked Fishery Vessels include trawlers, purse seiners, gill netters, long liners, trap setters, other seiners and liners, multipurpose vessels, dredgers, and other fishing vessels. Data on undecked vessels are being collected by FAO, but are not yet available. Fleet data are collected by the FAO through questionnaires submitted to the national reporting offices of the member countries. Other national or regional published sources, such as the registry of fishing vessels, are also used to estimate fleet size. The flag of the vessel is used to assign its nationality. However, in many cases vessels are flagged in one country, while the ownership, landings, and trade resides with another nation. This approach is referred to as a "flag of convenience," and fishers or corporations use this method to facilitate registration of a vessel (i.e. some countries have fewer registration restrictions), to gain access to fish in different Exclusive Economic Zones, or to avoid having to follow set fishing quotas in their own nation, among other reasons.

Population within 100 km of the Coast refers to estimates of the percentage of the population living within the coastal area based on 1995 population figures. These estimates were calculated using a data set that provides information on the spatial distribution of the world's human population on a 2.5-minute grid. Populations are distributed according to administrative districts, which vary in scale, level, and size from country to country. A 100-km coastal buffer was used to calculate the number of people in the coastal zone for each country. The percentage of the population in the coastal zone was calculated from 1995 United Nations Population Division totals for each country.

Frequency of Update by Data Providers

The Food and Agriculture Organization updates the FishStat database annually. Updates can be found on the FishStat website at fao.org/fi/statist/FISOFT/FISHPLUS.asp. The FAO updates the data on Food Supply variables annually; the most recent updates incorporated in these tables are from July 2002. Data on the number of fishers and decked fishery vessels are updated by the Fishery Information, Data and Statistics Unit (FIDI) of FAO.

Data Reliability and Cautionary Notes

Marine Catch, Freshwater Catch, Total Aquaculture Production, and Trade in Fish and Fishery Products. While the FAO data set provides the most extensive, global time series of fishery statistics since 1 970, there are some problems associated with the data. Funding for the development and maintenance of fisheries statistics at the national level has been decreasing in real terms since 1992, while the demand is growing for a variety of global statistics on discards, fish inventories, aquaculture, and illegal activities. Country-level data are often submitted with a 1–2 year delay, and countries are declaring an increasing percentage of their catch as "unidentified fish." Stock assessment working groups can more accurately estimate the composition of a catch; however, due to financial constraints, these groups are rare, especially in developing countries. Statistics from smaller artisanal and subsistence fisheries are particularly sparse. In addition, fishers sometimes underreport their catches because they have not kept within harvest limits established to manage the fishery. In some cases, catch statistics are inflated to increase the importance of the fishing industry to the national economy. FAO states that "general trends are probably reliably reflected by the available statistics¥ but the annual figures and the assessments involve a certain degree of uncertainty and small changes from year to year are probably not statistically significant." The quality of the aquaculture production estimates varies because many countries lack the resources to adequately monitor landings within their borders.

These statistics provide a good overview of regional fisheries trends. However, when reviewing the state of fisheries stocks, evaluating food security, etc., these data should be used with caution and supplemented with estimates from regional organizations, academic literature, expert consultations, and trade data. For more information, please consult Fishery Statistics: Reliability and Policy Implications, published by the FAO Fisheries Department and available on-line at fao.org/fi/statist/nature-china/30janO2.asp.

Food Supply from Fish and Fishery Products and Fish Protein as a Percent of Total Protein: Food supply is different from actual consumption. Figures do not account for discards (including bones) and losses during storage and preparation. Supply data should only be used to assess food security if it is combined with an analysis of food availability and accessibility. Per capita supply averages can also mask disparate food availability within a particular country. Nonetheless, the data are subject to "vigorous consistency checks." According to FAO, the food supply statistics, "while often far from satisfactory in the proper statistical sense, do provide an approximate picture of the overall food situation in a country and can be useful for economic and nutritional studies, for preparing development plans and for formulating related projects." For more information see Food Balance Sheets: A Handbook, maintained on-line by FAO at fao.org/DOCREP/003/X9892E/X9892E00.htm.

Number of Fishers: Numbers presented in this table are gross estimates. Many countries do not submit data on fishers, or submit incomplete information; therefore the quality of these data is poor. Apart from the gaps and the heavy presence of estimates due to non-reporting, the information provided by national statistical offices may not be strictly comparable since different definitions and methods are used in assessing the number of people engaged in fishing and aquaculture.

FAO recognizes that these statistics are incomplete and may not accurately reflect the current level of employment in the fishing sector. Specifically, it is aware that some countries failed to report for several years, that those which reported regularly have occasionally omitted fish farmers from the total or included subsistence and sport fishers as well as family members living on fishing.

Decked Fishery Vessels: As with the number of fishers, FAO recognizes that these fleet statistics are incomplete and may not accurately reflect current world fishing capacity. These data may include vessels that are no longer in operation. The quality of the estimates varies because many countries lack the resources to adequately monitor and report on fleet size. For further information, please refer to the original source or to *"Fishery Fleet Statistics, 1970 1975, 1980 1985, 1989–95"* Bulletin of Fishery Statistics No. 35 (FAO, Rome, 1998).

[a] Includes marine fish and diadromous fish caught in marine areas, as well as molluscs and crustaceans.
[b] Includes freshwater fish and diadromous fish caught in inland waters or low-salinity marine areas, as well as molluscs and crustaceans.
[c] Includes trade of all marine and freshwater catch, and total aquaculture production, excluding aquatic plants.
[d] Per capita values are expressed on a live-weight equivalent basis, which means that all parts of the fish, including bones, are taken into account when estimating consumption of fish and fishery products.
[e] Includes fishing vessels such as trawlers, long liners, etc., and nonfishing vessels such as motherships, fish carriers, etc.
[f] Data were collected between 1991 and 1996.
[g] Data are for 1997.
[h] Does not include Taiwan or Hong Kong.
[i] Data are for 1998.
[j] Data are for 1999.
[k] Data for Ethiopia before 1993 include Eritrea.
[l] Since independence, data include a substantial but unquantifiable number of sport fishers.
[m] Data are for 1980.
[n] Per capita fish consumption in Iceland includes quantities of fish and fish products destined for the export market.

Source: From World Resources Institute, Earth Trends Environmental Information, *Coastal and Marine Ecosystems*, www.earthtrends.wri.org.

Original Source: **Catch, Aquaculture Production, and Trade in Fish and Fishery Products** data produced by the Fishery Information, Data and Statistics Unit, Food and Agriculture Organization of the United Nations (FAO). 2002. FISHSTAT Plus: *Universal software for fishery statistical time series, Version 2.3* Rome: FAO. Available on-line at: fao.org/fi/statist/FISOFT/FISH PLUS. asp **Food Supply Variables** are from the Food and Agriculture Organization of the United Nations (FAO), *FAOSTAT on-line statistical service.* 2002. Rome: FAO. Available on-line at: http://apps.fao.org. **Data on the Number of Fishers** are from the Food and Agriculture Organization of the United Nations (FAO), Fishery Information, Data and Statistics Unit (FIDI) December, 1999. **Number of People within 100 km of the Coast** is derived from the Center for International Earth Science Information Network (CIESIN), World Resources Institute, and International Food Policy Research Institute. 2000. *Gridded Population of the World, Version 2 alpha* Columbia University, Palisades, NY. Available online at: sedac.ciesin.org/plue/gwp. **Population** (used to calculate per capita values): Population Division of the Department of Economic and Social Affairs of the United Nations Secretariat. 2002. World Population Prospects: The 2000 Revision. Data set on CD-ROM New York: United Nations.

Table 7O.175 World Aquaculture

World/Region	Production (thousand MT) 1980	1990	2001	Average Annual Rate of Growth (%) 1980–1990	1990–2001
World	**4,707**	**13,080**	**37,851**	**10.8**	**10.1**
Developed countries	**1,685**	**2,867**	**3,652**	**5.5**	**2.2**
Industrialized countries	1,419	2,327	3,419	5.1	3.6
Transition economies	266	540	233	7.3	−7.3
Developing countries	**3,022**	**10,212**	**34,199**	**12.9**	**11.6**
Latin America and the Caribbean	27	193	1,112	21.7	17.3
Near East and North Africa	34	104	498	11.8	15.4
Sub-Saharan Africa	7	13	50	6.6	12.8
East and Southeast Asia	2,490	8,668	29,603	13.3	11.8
South Asia	464	1,235	2,931	10.3	8.2
Oceania developing	0	1	4	0.0	13.8
North America developing	0	0	0	0.0	0.0
Continental groupings					
Africa	26	81	402	12.0	15.7
Asia	3,553	10,806	33,514	11.8	10.8
Latin America	25	179	1,051	22.0	17.4
Caribbean	2	13	62	18.9	15.1
North America	172	357	613	7.6	5.1
Oceania	12	42	124	13.1	10.3
Europe	916	1,602	2,086	5.7	2.4

Note: Aquaculture excludes aquatic plants.

Source: From Food and Agriculture Organization of the United Nations (FAO), 2003, *Summary of Food and Agricultural Statistics*, www.fao.org. Reprinted with permission.

Table 7O.176 World Capture Fisheries

World/Region	Production (thousand MT) 1980	1990	2001	Average Annual Rate of Growth (%) 1980–1990	1990–2001
World	**67,706**	**85,507**	**92,356**	**2.4**	**0.7**
Developed countries	**37,476**	**38,138**	**28,428**	**0.2**	**−2.6**
Industralized countries	26,873	27,414	23,546	0.2	−1.4
Transition economies	10,603	10,725	4,881	0.1	−6.9
Developing countries	**30,229**	**47,368**	**63,928**	**4.6**	**2.8**
Latin America and the Caribbean	9,530	16,082	16,800	5.4	0.4
Near East and North Africa	1,360	2,027	3,174	4.1	4.2
Sub-Saharan Africa	2,233	3,517	4,391	4.6	2.0
East and Southeast Asia	13,720	21,235	33,254	4.5	4.2
South Asia	3,136	4,157	5,792	2.9	3.1
Oceania developing	133	184	355	3	6.1
North America developing	116	167	163	4	−0.2
Continental groupings					
Africa	3,663	5,075	6,890	3.3	2.8
Asia	27,519	36,226	45,262	2.8	2.0
Latin America	9,279	15,818	16,620	5.5	0.5
Caribbean	251	264	180	0.5	−3.4
North America	5,001	7,359	6,157	3.9	−1.6
Oceania	411	741	1,108	6.1	3.7
Europe	21,284	19,947	15,963	−0.6	−2.0

Source: From Food and Agriculture Organization of the United Nations (FAO), 2003, *Summary of Food and Agricultural Statistics*, www.fao.org. Reprinted with permission.

Table 7O.177 Capture Fisheries and Aquaculture — Leading Species in 2001

Species	Production (thousand MT)	% of World Production
Capture Fisheries		
Anchoveta (Peruvian anchovy)	7,213	7.8
Alaska pollock (Walleye poll.)	3,136	3.4
Chilean jack mackerel	2,509	2.7
Atlantic herring	1,953	2.1
Japanese anchovy	1,837	2.0
Skipjack tuna	1,836	2.0
Blue whiting (Poutassou)	1,823	2.0
Chub mackerel	1,799	1.9
Capelin	1,671	1.8
Largehead hairtail	1,472	1.6
Other species	67,107	72.7
Aquaculture		
Pacific cupped oyster	4,110	10.9
Grass carp (White amur)	3,636	9.6
Silver carp	3,546	9.4
Common carp	2,849	7.5
Japanese carpet shell	2,091	5.5
Bighead carp	1,663	4.4
Crucian carp	1,527	4.0
Yesso scallop	1,196	3.2
Nile tilapia	1,109	2.9
Atlantic salmon	1,025	2.7
Other species	15,097	39.9

Source: From Food and Agriculture Organization of the United Nations (FAO), 2003, *Summary of Food and Agricultural Statistics*, www.fao.org. With permission.

SECTION 7P WATER IN FOODS

Table 7P.178 Water Content in Various Foods

Tortilla chips	1
Potato chips	2
Peanut butter	2
Popcorn	4
Margarine	14
Butter	16
Croissant	23
Jam	29
Bagel	29
Parmesan cheese	30
Angel food cake	32
Maple syrup	33
Swiss cheese	38
Whole wheat bread	38
English muffin	42
Cheese cake	46
Pizza	48
Brie	49
Apple pie	51
Frankfurter	54
Cream cheese	54
Hamburger	55
Whipping cream	58
Flounder, baked	58
Veal, chuck	59
Chicken, roasted	60
Ice cream	61
Turkey, roasted	62
Tuna, in water	62
Salmon, broiled	63
Ham, smoked, cooked	66
Halibut, broiled	67
Liver, beef, raw	70
Sour cream	71
Lima beans, cooked	71
Avocado	73
Corn, cooked	74
Ricotta cheese	74
Banana	74
Egg, boiled	75
Cottage cheese	79
Potato, raw	80
Clams	81
Grapes	81
Pear	84
Oysters	85
Orange	87
Beets, raw	87
Apple juice	88
Milk, whole	88
Yoghurt, whole milk	88
Carrots, raw	88
Broccoli, raw	89
Mushroom, raw	90
Cantaloupe	90
Milk, skim	91
Beer	92
Asparagus, cooked	94
Tomato, raw	94
Squash, boiled	96
Lettuce, raw	96

Note: Percentage by Weight.

Source: From Calculated from weight and water content values given in *Bowes and Church's Food Values of Portions Commonly Used*, 14th ed., Harper & Row.

CHAPTER 8

Water Quality

Katherine L. Thalman and James M. Bedessem

CONTENTS

SECTION 8A WATER QUALITY

Table 8A.1 Summary of Quality Inputs to Surface and Groundwaters

Contributing Factor	Principal Quality Input to Surface Waters
Meteorological water	Dissolved gases native to atmosphere Soluble gases from man's industrial activities Particulate matter from industrial stacks, dust, and radioactive particles Material washed from surface of earth, e.g., Organic matter such as leaves, grass, and other vegetation in all stages of biodegradation Bacteria associated with surface debris (including intestinal organisms) Clay, silt, and other mineral particles Organic extractives from decaying vegetation Insecticide and herbicide residues
Domestic use (exclusive of industrial)	Undecomposed organic matter, such as garbage ground to sewer, grease, etc. Partially degraded organic matter such as raw wastes from human bodies Combination of above two after biodegradation to various degrees of sewage treatment Bacteria (including pathogens), viruses, worm eggs Grit from soil washings, eggshells, ground bone, etc. Miscellaneous organic solids, e.g., paper, rags, plastics, and synthetic materials Detergents
Industrial use	Biodegradable organic matter having a wide range of oxygen demand Inorganic solids, mineral residues Chemical residues ranging from simple acids and alkalis to those of highly complex molecular structure Metal ions
Agricultural use	Increased concentration of salts and ions Fertilizer residues Insecticide and herbicide residues Silt and soil particles Organic debris, e.g., crop residue
Consumptive use (all sources)	Increased concentration of suspended and dissolved solids by loss of water to atmosphere
Contributing Factor	**Principal Quality Input to Groundwater**
Meteorological water	Gases, including O_2 and CO_2, N_2, H_2S, and H Dissolved minerals, e.g.: Bicarbonates and sulfates of Ca and Mg dissolved from earth minerals Nitrates and chlorides of Ca, Mg, Na and K dissolved from soil and organic decay residues Soluble iron, Mn, and F salts
Domestic use (principally via septic tank systems and seepage from polluted surface waters)	Detergents Nitrates, sulfates, and other residues of organic decay Salts and ions dissolved in the public water supply Soluble organic compounds
Industrial use (not much direct disposal to soil)	Soluble salts from seepage of surface waters containing industrial wastes
Agriculture use	Concentrated salts normal to water applied to land Other materials as per meteorological waters
Land disposal of solid wastes (not properly installed)	Hardness-producing leaching from ashes Soluble chemical and gaseous products or organic decay

Note: This list includes the types of things that may come from any contributing factor. Not all are present in each specific instance.

Source: From McGauhey, *Engineering Management of Water Quality*, McGraw-Hill, Copyright 1968.

Table 8A.2 Conditions That May Cause Variations in Water Quality

Climatic conditions	Runoff from snowmelt—muddy, soft, high bacterial count Runoff during drought—high mineral content, hard, groundwater characteristics Runoff during floods—less bacteria than snowmelt, may be muddy (depending upon other factors listed below)
Geographic conditions	Steep headwater runoff differs from lower valley areas in ground cover, gradients, transporting power, etc.
Geologic conditions	Clay soils produce mud Organic soils or swamps produce color Cultivated land yields silt, fertilizers, herbicides, and insecticides Fractured or fissured rocks may permit silt, bacteria, etc., to move with groundwater Mineral content dependent upon geologic formations
Season of year	Fall runoff carries dead vegetation—color, taste, organic extractives, bacteria Dry season yields dissolved salts Irrigation return water, in growing season only Cannery wastes seasonal Aquatic organisms seasonal Overturn of lakes and reservoirs seasonal Floods generally seasonal Dry period, low flows, seasonal
Resource management practices	Agricultural soils and other denuded soils are productive of sediments, etc. (See third item under Geologic conditions.) Forested land and swampland yield organic debris Overgrazed or denuded land subject to erosion Continuous or batch discharge of industrial wastes alters shock loads Inplant management of waste streams governs nature of waste
Diurnal variation	Production of oxygen by planktonic algae varies from day to night Dissolved oxygen in water varies in some fashion Raw sewage flow variable within 24-hr period; treated sewage variation less pronounced Industrial wastes variable—process wastes during productive shift; different material during washdown and cleanup

Source: From McGauhey, *Engineering Management of Water Quality*, McGraw-Hill, Copyright 1968.

Table 8A.3 Principal Chemical Constituents in Water — Their Sources, Concentrations, and Effects upon Usability

Constituent	Major Sources	Concentration in Natural Water	Effect upon Usability of Water
Silica (SiO_2)	Feldspars, ferromagnesium and clay minerals, amorphous silicachert, opal	Ranges generally from 1.0 to 30 mg/L, although as much as 100 mg/L is fairly common; as much as 4,000 mg/L is found in brines	In the presence of calcium and magnesium, silica forms a scale in boilers and on steam turbines that retards heat; the scale is difficult to remove. Silica may be added to soft water to inhibit corrosion of iron pipes
Iron (Fe)	1. Natural sources *Igneous rocks:* Amphiboles, ferromagnesian micas, ferrous sulfide (FeS), ferric sulfide or iron pyrite (FeS_2), magnetite (Fe_3O_4) *Sandstone rocks:* Oxides, carbonates, and sulfides or iron clay minerals 2. Man-made sources: Well casing, piping, pump parts, storage tanks, and other objects of cast iron and steel which may be in contact with the water Industrial wastes	Generally less than 0.50 mg/L in fully aerated water. Groundwater having a pH less than 8.0 may contain 10 mg/L; rarely as much as 50 mg/L may occur. Acid water from thermal springs, mine wastes and industrial may contain more than 6,000 mg/L	More than 0.1 mg/L precipitates after exposure to air; causes turbidity, stains plumbing fixtures, laundry and cooking utensils, and imparts objectionable tastes and colors to foods and drinks. More than 0.2 mg/L is objectionable for most industrial uses
Manganese (Mn)	Manganese in natural water probably comes most often from soils and sediments. Metamorphic and sedimentary rocks and mica biotite and amphibole hornblende minerals contain large amounts of manganese	Generally 0.20 mg/L or less. Groundwater and acid mine water may contain more than 10 mg/L. Reservoir water that has "turned over" may contain more than 150 mg/L	More than 0.2 mg/L precipitates upon oxidation; causes undesirable tastes, deposits on foods during cooking, stains plumbing fixtures and laundry and fosters growths in reservoirs, filters, and distribution systems. Most industrial users object to water containing more than 0.2 mg/L
Calcium (Ca)	Amphiboles, feldspars, gypsum, pyroxenes, aragonite, calcite, dolomite, clay minerals	As much as 600 mg/L in some western streams; brines may contain as much as 75,000 mg/L	Calcium and magnesium combine with bicarbonate, carbonate, sulfate, and silica to form heat-retarding, pipe-clogging scale in boilers and in other heat-exchange equipment. Calcium and magnesium combine with ions of fatty acid in soaps to form soap suds; the more calcium and magnesium, the more soap required to form suds. A high concentration of magnesium has a laxative effect, especially on new users of the supply
Magnesium (Mg)	Amphiboles, olivine, pyroxenes, dolomite, magnesite, clay minerals	As much as several hundred mg/L in some western streams; ocean water contains more than 1,000 mg/L and brines may contain as much as 57,000 mg/L	
Sodium (Na)	Feldspars (albite), clay minerals, evaporates, such as halite (NaCl) and mirabilite ($Na_2SO_410H_2O$), industrial wastes	As much as 1,000 mg/L in some western streams; about 10,000 mg/L in sea water; about 25,000 mg/L in brines	More than 50 mg/L sodium and potassium in the presence of suspended matter causes foaming, which accelerates scale formation

Potassium (K)	Feldspars (orthoclase and microcline), feldspathoids, some micas, clay minerals	Generally less than about 10 mg/L; as much as 100 mg/L in hot springs; as much as 25,000 mg/L in brines	and corrosion in boilers. Sodium and potassium carbonate in recirculating cooling water can cause deterioration of wood in cooling towers. More than 65 mg/L of sodium can cause problems in ice manufacture
Carbonate (CO_3)		Commonly 0 mg/L in surface water; commonly less than 10 mg/L in groundwater. Water high in sodium may contain as much as 50 mg/L of carbonate	Upon heating, bicarbonate is changed into steam, carbon dioxide, and carbonate. The carbonate combines with alkaline earths—principally calcium and magnesium—to form a crustlike scale of calcium carbonate that retards flow of heat through pipe walls and restricts flow of fluids in pipes. Water containing large amounts of biocarbonate and alkalinity are undesirable in many industries
Bicarbonate (HCO_3)	Limestone, dolomite	Commonly less than 500 mg/L; may exceed 1,000 mg/L in water highly charged with carbon dioxide	
Sulfate (SO_4)	Oxidation of sulfide ores; gypsum; anhydrite; industrial wastes	Commonly less than 1,000 mg/L except in streams and wells influenced by acid mine drainage. As much as 200,000 mg/L in some brines	Sulfate combines with calcium to form an adherent, heat-retarding scale. More than 250 mg/L is objectionable in water in some industries. Water containing about 500 mg/L of sulfate tastes bitter; water containing about 1,000 mg/L may be cathartic
Chloride (Cl)	Chief source is sedimentary rock (evaporates); minor sources are igneous rocks. Ocean tides force salty water upstream in tidal estuaries	Commonly less than 10 mg/L in humid regions; tidal streams contain increasing amounts of chloride (as much as 19,000 mg/L) as the bay or ocean is approached. About 19,300 mg/L in seawater, and as much as 200,000 mg/L in brines	Chloride in excess of 100 mg/L imparts a salty taste. Concentrations greatly in excess of 100 mg/L may cause physiological damage. Food processing industries usually require less than 250 mg/L. Some industries—textile processing, paper manufacturing, and synthetic rubber manufacturing—desire less than 100 mg/L
Fluoride (F)	Amphiboles (hornblende), apatite, fluorite, mica	Concentrations generally do not exceed 10 mg/L in groundwater or 1.0 mg/L in surface water. Concentrations may be as much as 1,600 mg/L in brines	Fluoride concentration between 0.6 and 1.7 mg/L in drinking water has a beneficial effect on the structure and resistance to decay of children's teeth. Fluoride in excess of 1.5 mg/L in some areas causes "mottled enamel" in children's teeth. Fluoride in excess of 6.0 mg/L causes pronounced mottling and disfiguration of teeth

(Continued)

Table 8A.3 (Continued)

Constituent	Major Sources	Concentration in Natural Water	Effect upon Usability of Water
Nitrate (NO_3)	Atmosphere; legumes, plant debris, animal excrement, nitrogenous fertilizer in soil and sewage	In surface water not subjected to pollution, concentration of nitrate may be as much as 5.0 mg/L but is commonly less than 1.0 mg/L. In groundwater the concentration of nitrate may be as much as 1,000 mg/L	Water containing large amount of nitrate (more than 100 mg/L) is bitter tasting and may cause physiological distress. Water from shallow wells containing more than 45 mg/L has been reported to cause methemoglobinemia in infants. Small amounts of nitrate help reduce cracking of high-pressure boiler steel
Dissolved solids	The mineral constituents dissolved in water constitute the dissolved solids	Surface water commonly contains less than 3,000 mg/L; streams draining salt beds in arid regions may contain in excess of 15,000 mg/L. Groundwater commonly contains less than 5,000 mg/L; some brines contain as much as 300,000 mg/L	More than 500 mg/L is undesirable for drinking and many industrial uses. Less than 300 mg/L is desirable for dyeing of textiles and the manufacture of plastics, pulp paper, rayon. Dissolved solids cause foaming in steam boilers; the maximum permissible content decreases with increases in operating pressure

Source: From U.S. Geological Survey, 1962; amended.

Table 8A.4 Relative Abundance of Dissolved Solids in Potable Water

Major Constituents (1.0 to 1000 mg/L)	Secondary Constituents (0.01 to 10.0 mg/L)	Minor Constituents (0.0001 to 0.1 mg/L)	Trace Constituents (generally less than 0.001 mg/L)
Sodium	Iron	Antimony[a]	Beryllium
Calcium	Strontium	Aluminum	Bismuth
Magnesium	Potassium	Arsenic	Cerium[a]
Bicarbonate	Carbonate	Barium	Cesium
Sulfate	Nitrate	Bromide	Gallium
Chloride	Fluoride	Cadmium[a]	Gold
Silica	Boron	Chromium[a]	Indium
		Cobalt	Lanthanum
		Copper	Niobium[a]
		Germanium[a]	Platinum
		Iodide	Radium
		Lead	Ruthenium[a]
		Lithium	Scandium[a]
		Manganese	Silver
		Molybdenum	Thallium[a]
		Nickel	Thorium[a]
		Phosphate	Tin
		Rubidium[a]	Tungsten[a]
		Selenium	Ytterbium
		Titanium[a]	Yttrium[a]
		Uranium	Zirconium
		Vanadium	
		Zinc	

[a] These elements occupy an uncertain position in the list.

Source: From Davis and DeWiest, *Hydrogeology*, John Wiley & Sons, Copyright 1966.

Table 8A.5 Characteristics of Water That Affect Water Quality

Characteristic	Principal Cause	Significance	Remarks
Hardness	Calcium and magnesium dissolved in the water	Calcium and magnesium combine with soap to form an insoluble precipitate (curd) and thus hamper the formation of a lather. Hardness also affects the suitability of water for use in the textile and paper industries and certain others and in steam boilers and water heating	USGS classification of hardness (mg/L as $CaCO_3$) 0–60: Soft 61–120: Moderately hard 121–180: Hard More than 180: Very hard
pH (or hydrogen-ion activity)	Dissociation of water molecules and of acids and bases dissolved in water	The pH of water is a measure of its reactive characteristics. Low values of pH, particularly below pH 4, indicate a corrosive water that will tend to dissolve metals and other substances that it contacts. High values of pH, particularly above pH 8.5, indicate an alkaline water that, on heating, will tend to form scale. The pH significantly affects the treatment and use of water	pH values: less than 7, water is acidic; value of 7, water is neutral; more than 7, water is basic

(Continued)

Table 8A.5 (Continued)

Characteristic	Principal Cause	Significance	Remarks
Specific electrical conductance	Substances that form ions when dissolved in water	Most substances dissolved in water dissociate into ions that can conduct an electrical current. Consequently, specific electrical conductance is a valuable indicator of the amount of material dissolved in water. The larger the conductance, the more mineralized the water	Conductance values indicate the electrical conductivity, in micromhos, of 1 cm^3 of water at a temperature of 25°C
Total dissolved solids	Mineral substances dissolved in water	Total dissolved solids is a measure of the total amount of minerals dissolved in water and is, therefore, a very useful parameter in the evaluation of water quality. Water containing less than 500 mg/L is preferred for domestic use and for many industrial processes	USGS classification of water based on dissolved solids (mg/L) Less than 1,000: Fresh 1,000–3,000: Slightly saline 3,000–10,000: Moderately saline 10,000–35,000: Very saline More than 35,000: Briny

Source: From Heath, R.C., 1984, Basic groundwater hydrology, *U.S. Geological Survey Water-Supply Paper 2220.*

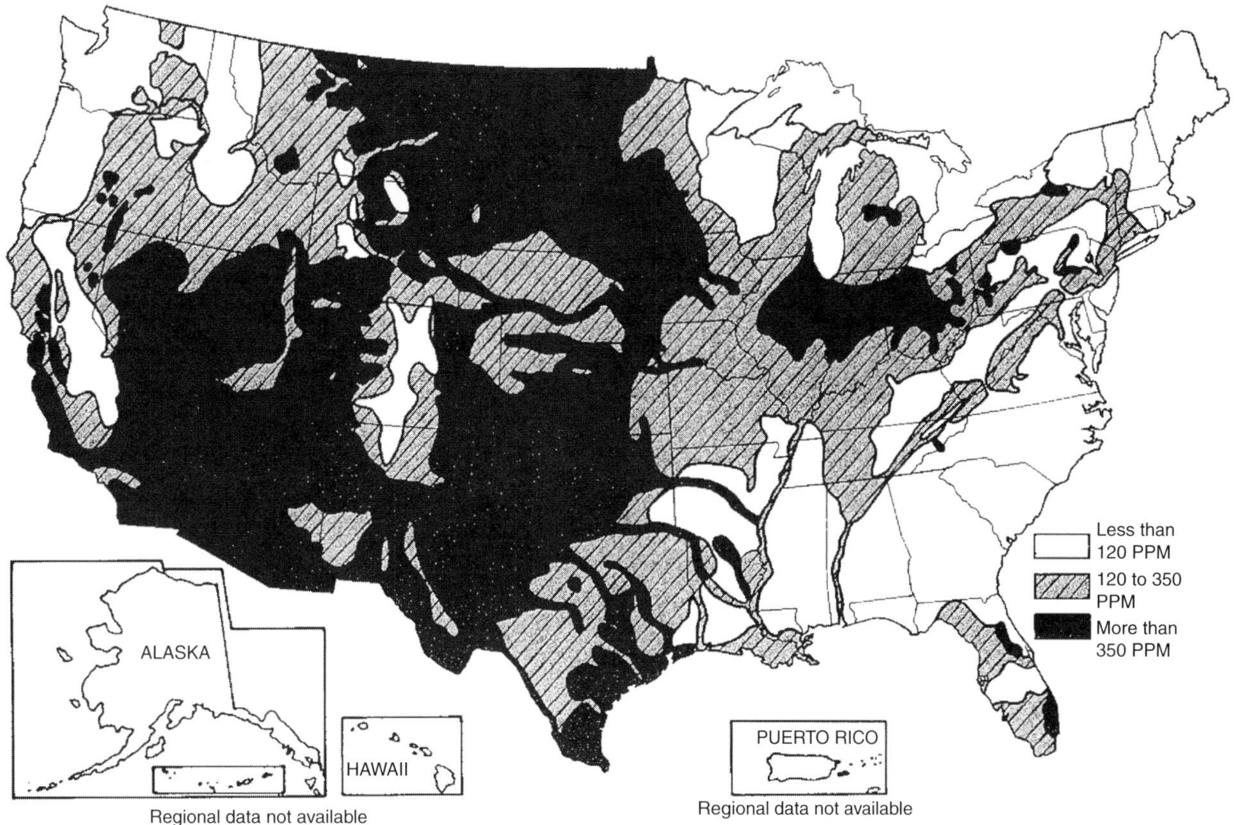

Figure 8A.1 Dissolved solids in surface water. (From U.S. Water Resources Council, 1968.)

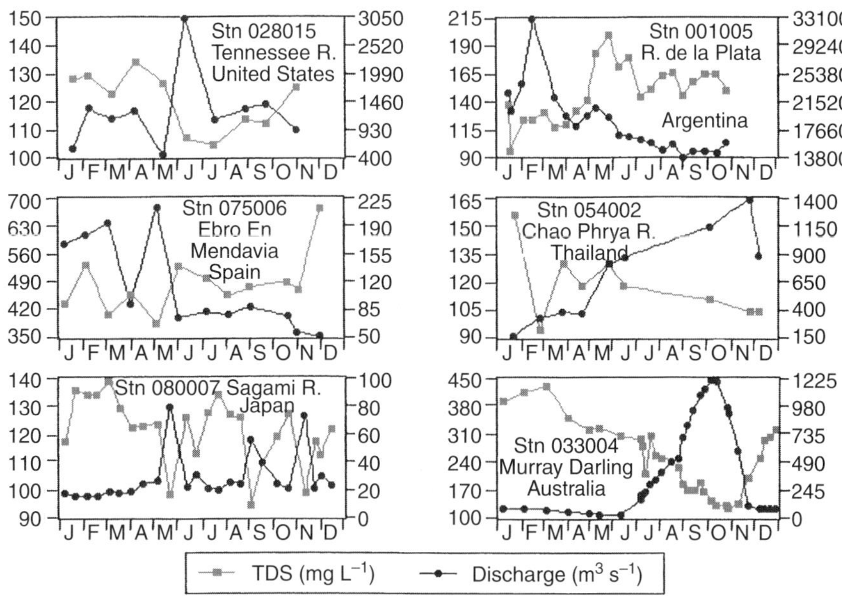

Figure 8A.2 Seasonal variation of total dissolved solids (TDS) and water discharge at selected world river stations for selected years. (From United Nations Environment Programme, Global Environment Monitoring System Water Programme (GEMS/WATER), The annotated digital atlas of global water quality, www.gemswater.org. Reprinted with permission.)

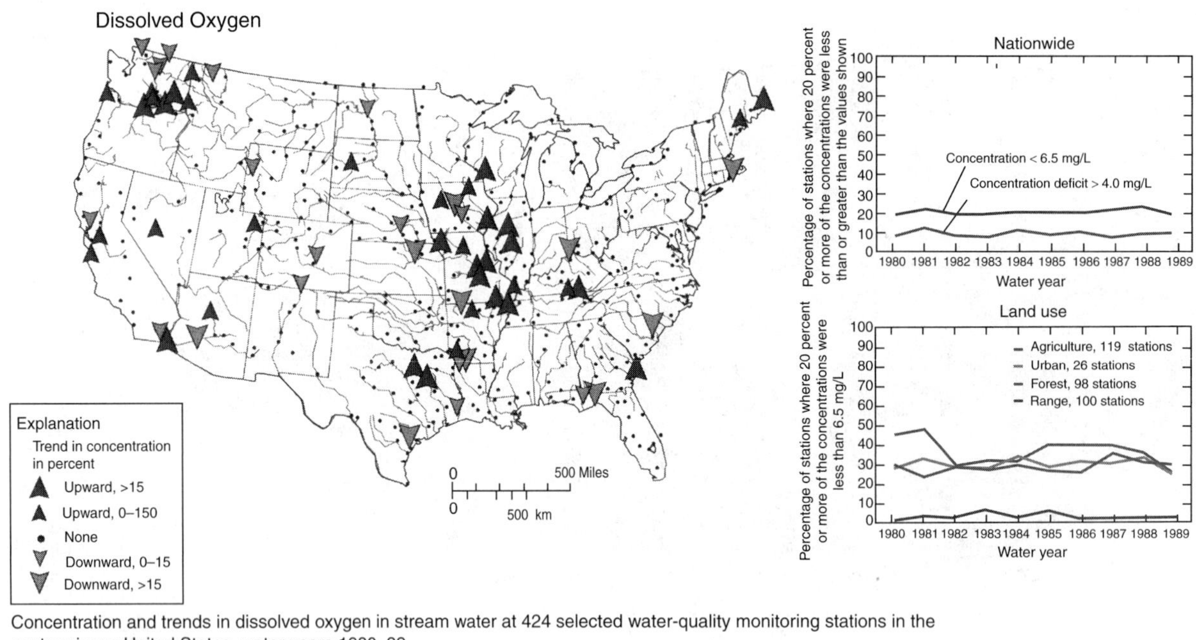

Concentration and trends in dissolved oxygen in stream water at 424 selected water-quality monitoring stations in the conterminous United States, water years 1980–89.

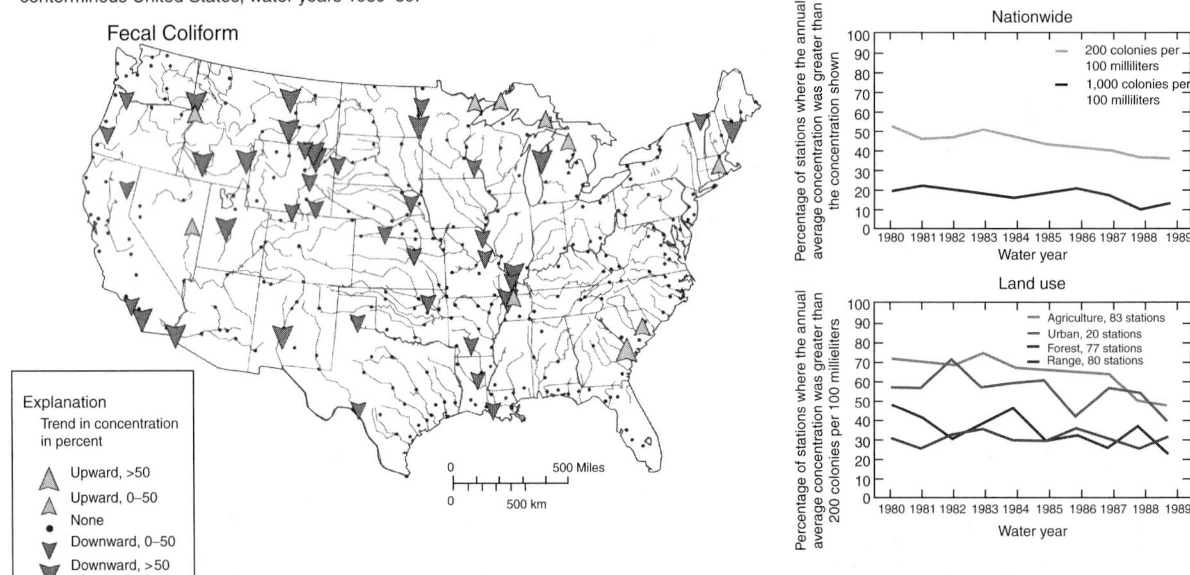

Concentration and trends in fecal coliform bacteria in stream water at 313 selected water-quality monitoring stations in the conterminous United States, water years 1980–89.

Figure 8A.3 Concentration trends in dissolved oxygen and fecal coliform bacteria in United States rivers, 1980–1989. (From USDA, Natural Resources Conservation Services, 1997, Water Quality and Agriculture, Status, Conditions, and Trends, www.nrcs.usda.gov. *Original Source*: Smith, R.A., Alexander, R.B., and Lanfear, K.J., 1993, Stream water quality in the conterminous United States – status and trends of selected indicators during the 1980's in National Water Summary 1990–91 – Stream water quality, *U.S. Geological Survey Water-Supply Paper 2400*, www.usgs.gov.)

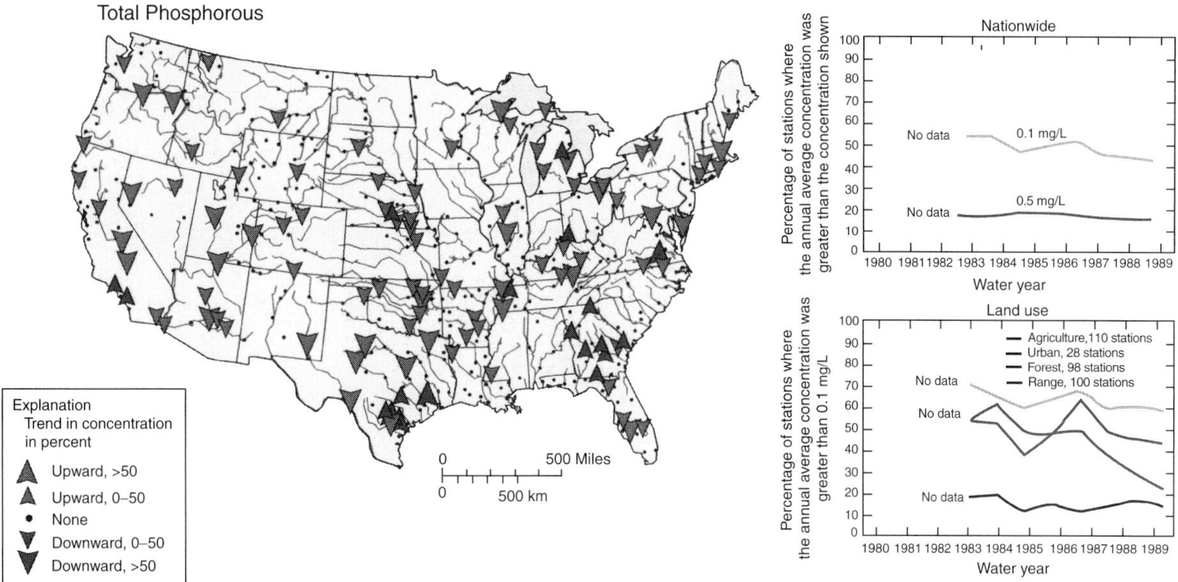

Concentration and trends total phosphorus in stream water at 410 selected water-quality monitoring stations in the conterminous United States, water years 1982–1989.

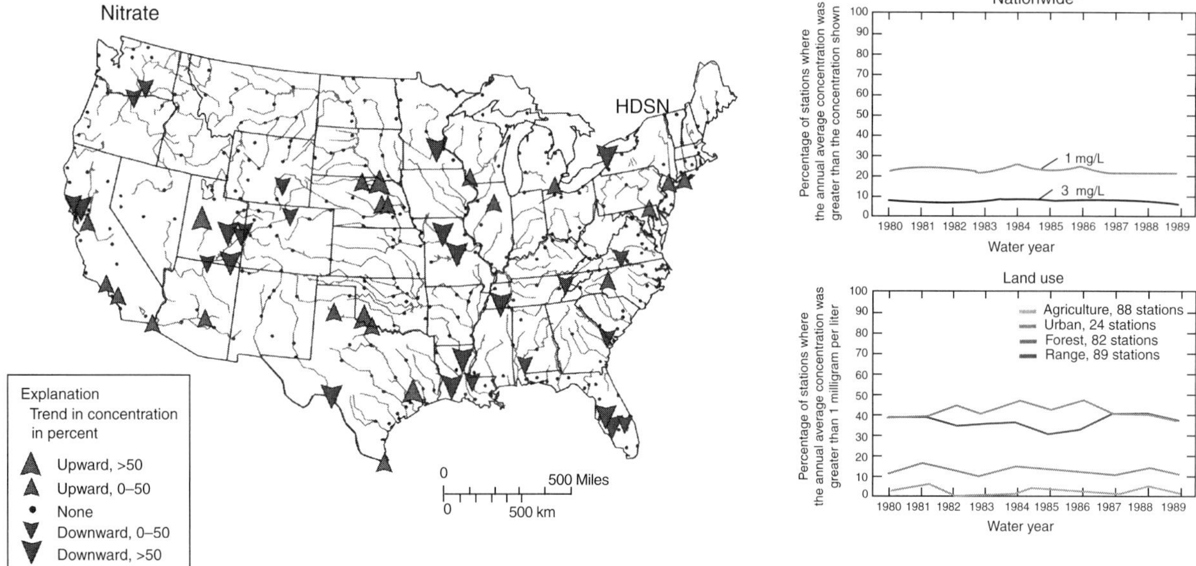

Concentration and trends in nitrate in stream water at 344 selected water-quality monitoring stations in the conterminous United States, water years 1980–1989.

Figure 8A.4 Concentration trends in phosphorous, nitrate, and suspended solids in United States rivers, 1980 to 1989. (From USDA, Natural Resources Conservation Services, 1997, Water quality and agriculture, status, conditions, and trends, www.nrcs.usda.gov. *Original Source*: Smith, R.A., Alexander, R.B., and Lanfear, K.J., 1993, Stream water quality in the conterminous United States – status and trends of selected indicators during the 1980's in National Water Summary 1990–91–Stream water quality, *U.S. Geological Survey Water-Supply Paper 2400*, www.usgs.gov.)

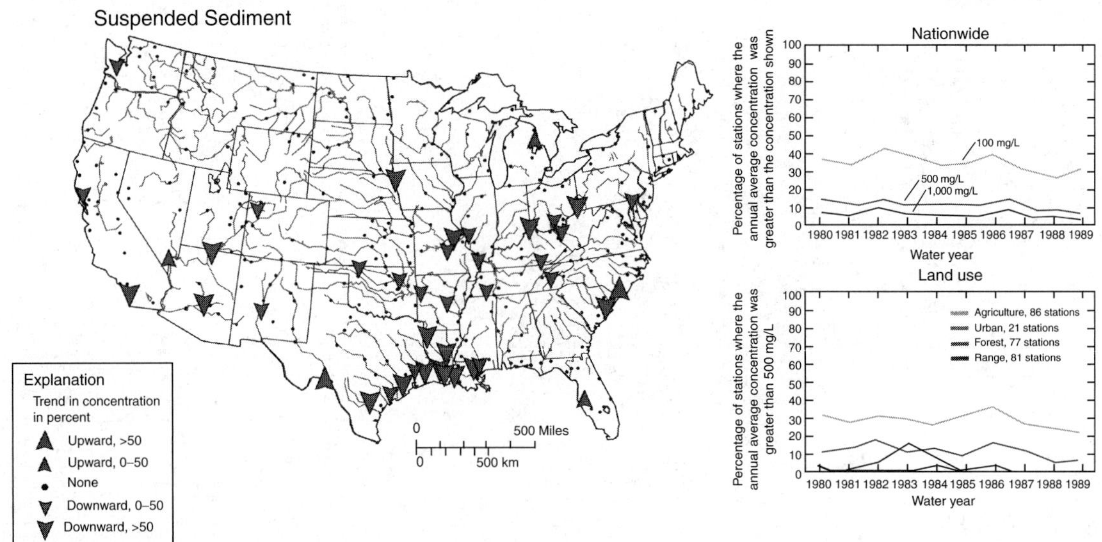

Concentration and trends in suspended sediment in stream water at 324 selected water-quality monitoring stations in the conterminous United States, water years 1980–1989.

Figure 8A.4 *(Continued)*

Table 8A.6 Trends of Surface-Water Quality in the United States, 1974–1981

Constituents and Properties	Number of Stations with— Increasing Trends	No Change	Decreasing Trends	Total Stations
Temperature	39	218	46	303
pH	74	174	56	304
Alkalinity	18	207	79	304
Sulfate	82	182	40	304
Nitrate-nitrite	76	203	25	304
Ammonia	31	221	30	282
Total organic carbon	36	230	13	279
Phosphorus	39	232	30	301
Calcium	23	198	83	304
Magnesium	50	208	46	304
Sodium	103	173	28	304
Potassium	69	193	42	304
Chloride	104	164	36	304
Silica	48	213	41	302
Dissolved solids	68	183	51	302
Suspended sediment	44	204	41	289
Conductivity	69	193	43	305
Turbidity	42	199	18	259
Fecal coliform bacteria	19	216	34	269
Fecal streptococcus bacteria	2	190	78	270
Phytoplankton	22	234	44	300
Dissolved trace metals				
Arsenic	68	228	11	307
Barium	4	81	1	86
Boron	2	15	3	20
Cadmium	32	264	7	303
Chromium	12	152	2	166
Copper	6	83	6	95
Iron	28	258	21	307
Lead	5	232	76	313
Manganese	30	250	19	299
Mercury	8	194	2	204
Selenium	2	201	21	224
Silver	1	32	0	33
Zinc	19	251	32	302

Note: Selected water-quality constituents and properties at NASQAN stations.

Source: From *U.S. Geological Survey Water-Supply Paper 2250.*

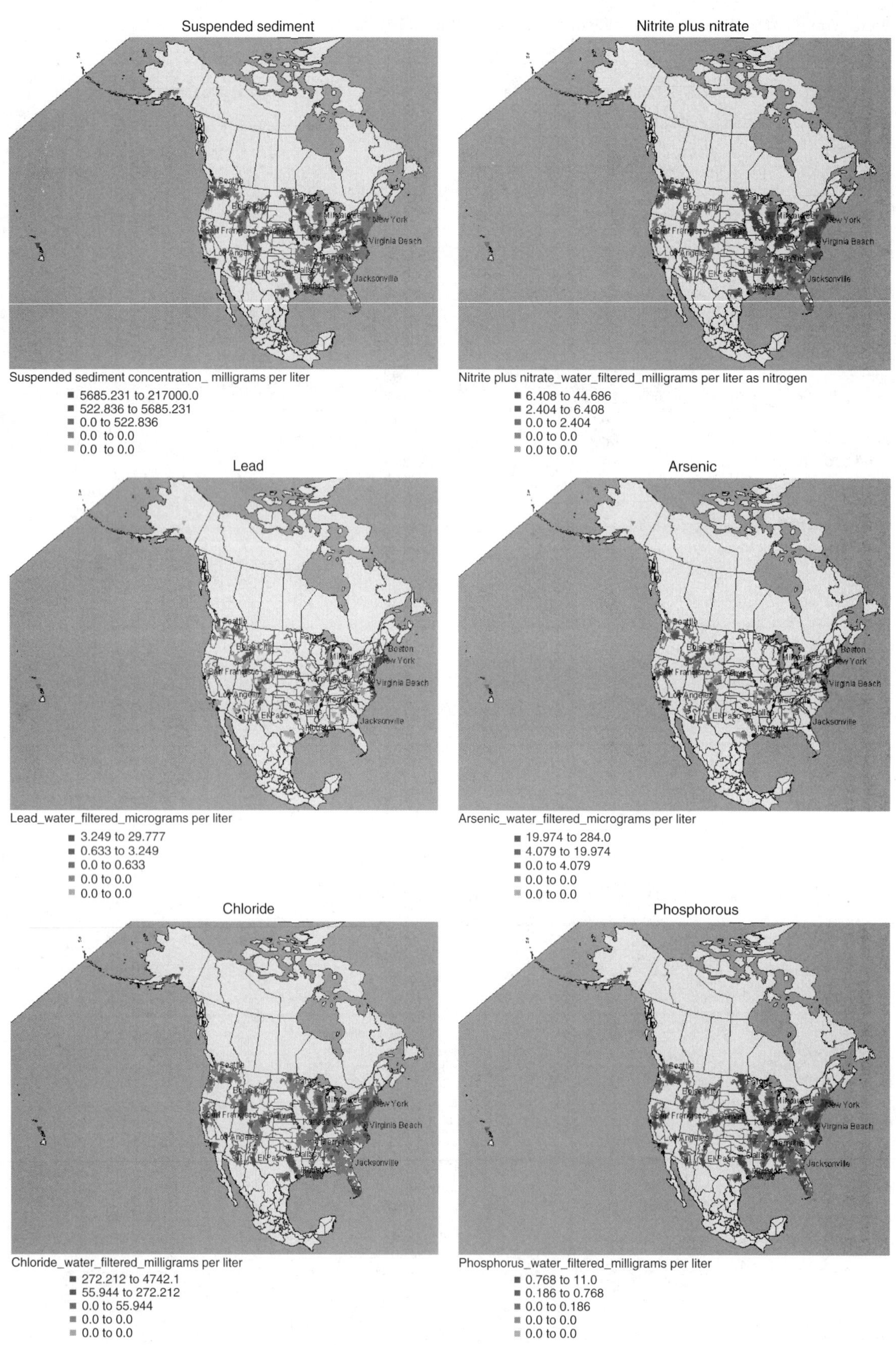

Figure 8A.5 United States Geological Survey NAWQA water quality thematic maps showing maximum concentrations of suspended sediment, nitrite plus nitrate, lead, arsenic, chloride, and phosphorous detected in rivers of the United States. (From United States Geological Survey, NAWQA Date Warehouse Mapper, www.maptrek.er.usgs.gov/NAWQAMapTheme/index.jsp, Maps generated in May 2005.)

Table 8A.7 Estimates of National Background Nutrient Concentrations in the United States

Nutrient	Background Concentration (mg/L)
Total nitrogen in streams (Data from 28 watersheds in first 20 study units)	1.0
Nitrate in streams[(26)]	0.6
Ammonia in streams[(26)]	0.1
Nitrate in shallow groundwater[(27)]	2.0
Total phosphorus in streams[(26)]	0.1
Orthophosphate in shallow groundwater (Data from 47 wells in first 20 study units)	0.02

Source: From U.S. Geological Survey, 1999, The quality of our nation's waters, nutrients and pesticides, *U.S. Geological Survey Circular 1225*, http://usgs.gov.

Table 8A.8 Water Quality of Great Salt Lake, Utah, 1850–1998

	Silica (SiO_2)	Calcium (Ca)	Magnesium (Mg)	Sodium (Na)	Potassium (K)	Lithium (Li)	Bicarbonate ($AsCO_3$)	Sulfate (SO_4)	Chloride (Cl)	Fluoride (F)	Boron (B)	Bromlum (Br)	Total Percent
							Precauseway						
1850	—		0.27	38.29	—	—	—	5.57	55.87	—	—	—	100
1869	—	0.17	2.52	33.15	1.60	—	—	6.57	55.99	—	—	—	100
August 1892	—	1.05	1.23	33.22	1.71	—	—	6.57	56.22	—	—	—	100
October 1913	—	0.16	2.76	33.17	1.66	—	0.09	6.68	55.48	—	—	—	100
March 1930	—	0.17	2.75	32.90	1.61	—	0.05	5.47	57.05	—	—	—	100
							South of causeway						
April 1960	0.00	0.12	2.91	32.71	1.71	—	0.06	6.60	55.88		0.01	—	100
December 1963	0.00	0.09	3.29	31.02	1.86	—	0.07	9.02	54.64	—	0.01	—	100
May 1966	0.00	0.09	3.80	30.56	2.22	0.02	0.10	7.99	65.21	0.00	0.01	—	100
June 1976	—	0.17	3.47	31.29	2.66	0.02	—	7.22	55.11	—	0.01	0.04	100
July 1998	—	0.23	3.52	31.67	2.16	—	—	6.36	56.07	—	—	—	100
							North of causeway						
December 1963	0.00	0.09	4.66	29.08	2.75	—	0.09	7.28	56.04	—	0.01	—	100
May 1966	—	0.05	4.38	29.67	2.61	0.02	0.09	8.58	54.59	0.00	0.01	—	100
June 1976	—	0.13	3.17	32.04	2.58	0.02	—	6.62	55.39	—	0.01	0.04	100
July 1998	—	0.11	3.09	32.59	1.53	—	—	6.40	56.29	—	—	—	100

Note: Composition, in percentage by weight, of dissolved ions in brine.

Source: From Modified from Arnow, Ted, 1984, Water-level and water-quality changes in Great Salt Lake, Utah, 1847–1983, *U.S. Geological Survey Circ. 913*; 1998 Data Utah Geological Survey met.utah.edu/jhorel/homepages/jhorel/saltlake/chemistry.html.

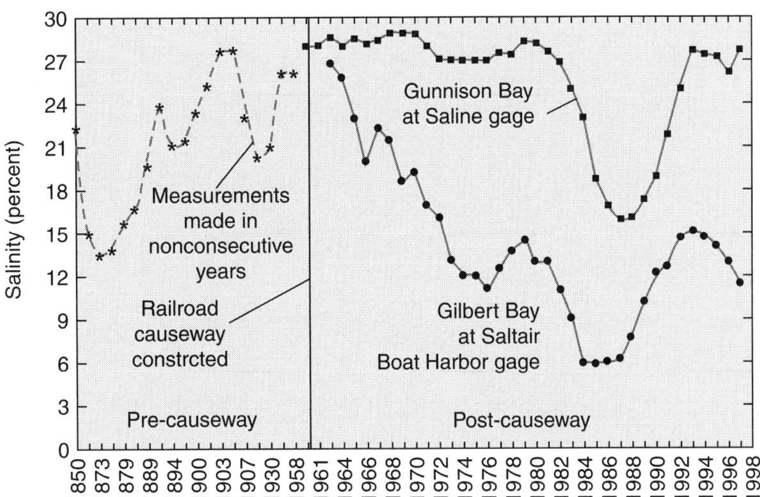

Figure 8A.6 Salinity in the Great Salt Lake, Utah 1950–1998. The Salinity of Great Salt Lake is determined by the amount of inflow (and its salt content) and the amount of evaporation. When there is a lot of inflow, the lake elevation increases and the salinity of the water decreases. When there is less inflow or the evaporation rate is high, the lake elevation declines and the water becomes saltier. In 1959, a solid-fill railroad causeway was constructed across the middle of the lake. The causeway divides the lake into two parts: the north part (Gunnison Bay), which receives little freshwater inflow, and the south part (Gilbert Bay), which receives almost all the inflow. For any given lake elevation, the salinity of Gunnison Bay is always greater than the salinity of Gilbert Bay. The USGS measures salinity periodically at Saltair Boat Harbor and at Promontory (Gilbert Bay) and at Saline (Gunnison Bay). (From U.S. Geological Survey, http://ut.water.usgs.gov/salinity/index.html.)

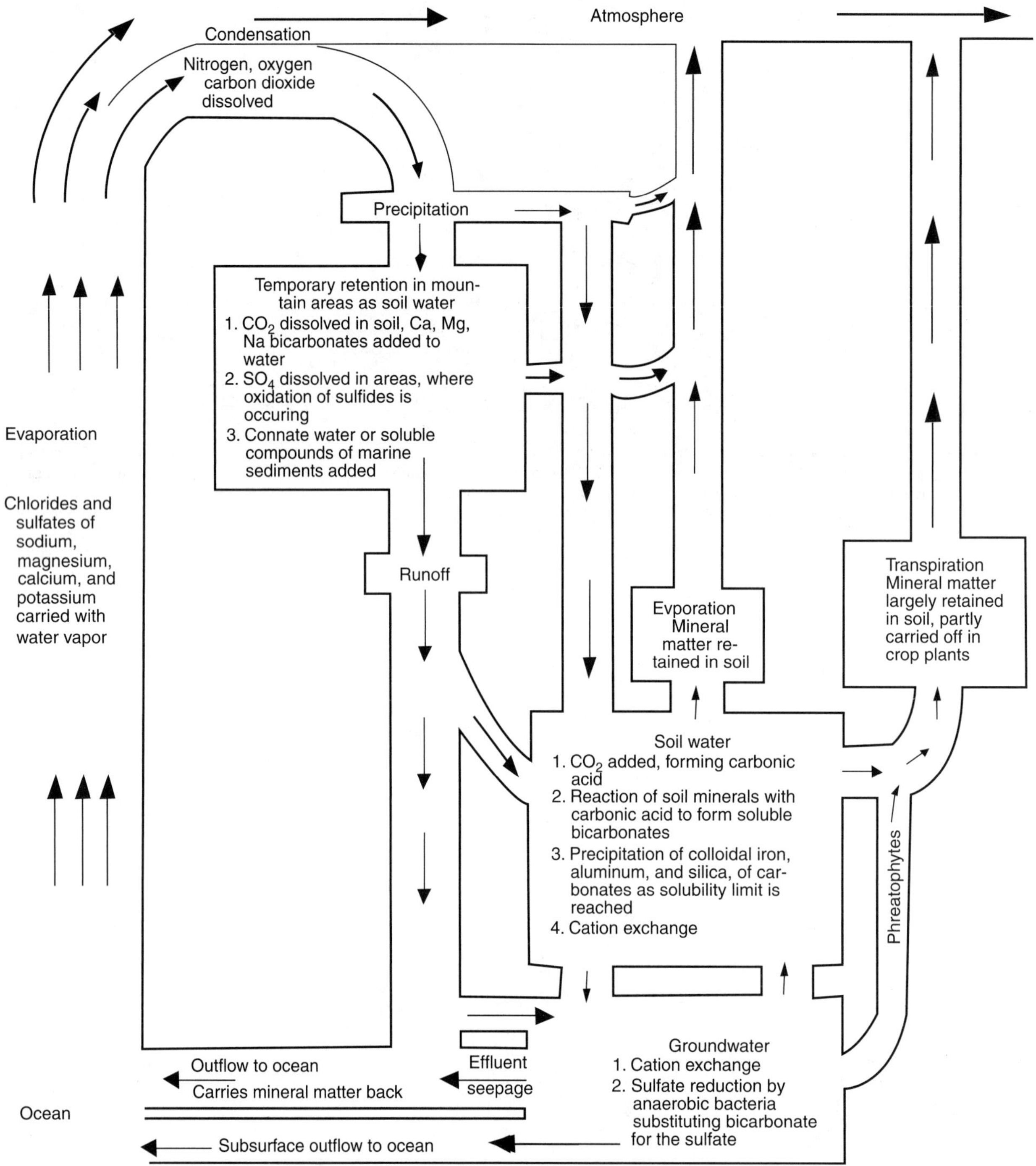

Figure 8A.7 Geochemical cycle of surface and groundwater. (From U.S Geological Survey.)

Table 8A.9 Natural Inorganic Constituents Commonly Dissolved in Groundwater That Are Most Likely to Affect Use of the Water

Substance	Major Natural Sources	Effect on Water Use	Concentrations of Significance (mg/L)[a]
Bicarbonate (HCO_3) and carbonate (CO_3)	Products of the solution of carbonate rocks, mainly limestone ($CaCO_3$) and dolomite ($CaMgCO_3$), by water containing carbon dioxide	Control the capacity of water to neutralize strong acids. Bicarbonates of calcium and magnesium decompose in steam boilers and water heaters to form scale and release corrosive carbon dioxide gas. In combination with calcium and magnesium, cause carbonate hardness	150–200
Calcium (Ca) and magnesium (Mg)	Soils and rocks containing limestone, dolomite, and gypsum ($CaSO_4$). Small amounts from igneous and metamorphic rocks	Principal cause of hardness and of boiler scale and deposits in hot-water heaters	25–50
Chloride (Cl)	In inland areas, primarily from seawater trapped in sediments at time of deposition; in coastal areas, from seawater in contact with freshwater in productive aquifers	In large amounts, increase corrosiveness of water and, in combination with sodium, gives water a salty taste	250
Fluoride (F)	Both sedimentary and igneous rocks. Not widespread in occurrence	In certain concentrations, reduces tooth decay; at higher concentrations, causes mottling of tooth enamel	0.7–1.2[b]
Iron (Fe) and manganese (Mn)	Iron present in most soils and rocks; manganese less widely distributed	Stain laundry and are objectionable in food processing, dyeing, bleaching, ice manufacturing, brewing, and certain other industrial processes	Fe $>$ 0.3, Mn $>$ 0.05
Sodium (Na)	Same as for chloride. In some sedimentary rocks, a few hundred milligrams per liter may occur in freshwater as a result of exchange of dissolved calcium and magnesium for sodium in the aquifer materials	See chloride. In large concentrations, may affect persons with cardiac difficulties, hypertension, and certain other medical conditions. Depending on the concentrations of calcium and magnesium also present in the water, sodium may be detrimental to certain irrigated crops	69 (irrigation), 20–170 (health)[c]
Sulfate (SO_4)	Gypsum, pyrite (FeS), and other rocks containing sulfur (S) compounds	In certain concentrations, gives water a bitter taste and, at higher concentrations, has a laxative effect. In combination with calcium, forms a hard calcium carbonate scale in steam boilers	300–400 (taste), 600–1,000 (laxative)

[a] A range in concentration is intended to indicate the general level at which the effect on water use might become significant.
[b] Optimum range determined by the U.S. Public Health Service, depending on water intake.
[c] Lower concentration applies to drinking water for persons on a strict diet; higher concentration is for those on a moderate diet.

Source: From Heath, R.C., 1982, Basic groundwater hydrology, *U.S. Geological Survey Water-Supply Paper 2220.*

Table 8A.10 Inorganic Substances Found in Groundwater

	Concentration (mg/L)
Aluminum	0.1–1,200
Ammonia	1.0–900
Antimony	—
Arsenic	0.01–2,100
Barium	2.8–3.8
Beryllium	less than 0.01
Boron	—
Cadmium	0.01–180
Calcium	0.5–225
Chlorides	1.0–49,500
Chromium	0.06–2,740
Cobalt	0.01–0.18
Copper	0.01–2.8
Cyanides	1.05–14
Fluorides	0.1–250
Iron	0.04–6,200
Lead	0.01–5.6
Lithium	—
Magnesium	0.2–70
Manganese	0.1–110
Mercury	0.003–0.01
Molybdenum	0.4–40
Nickel	0.05–0.5
Nitrates	1.4–433
Nitrites	—
Palladium	—
Potassium	0.5–2.4
Phosphates	0.4–33
Selenium	0.6–20
Silver	9.0–330
Sodium	3.1–211
Sulfates	0.2–32,318
Sulfites	—
Thallium	—
Titanium	—
Vanadium	243.0
Zinc	0.1–240

Source: From Office of Technology Assessment 1984, Protecting the nation's groundwater from contamination, U.S. Congress, Washington DC.

Table 8A.11 Summary of Inorganic Elements Found in Rural Water Supplies

		In % of Rural Households				
Element	**Level Exceeded (mg/L)**	**Nationwide**	**West**	**North-Central**	**Northeast**	**South**
Mercury	0.002	24.1	10.4	31.8	22.0	25.0
Iron	0.3	18.7	7.0	28.2	16.0	17.0
Cadmium	0.01	16.8	27.1	20.7	1.6	17.3
Lead	0.05	16.6	16.9[a]	10.8[a]	9.6[a]	23.1[a]
Manganese	0.05	14.2	4.7	19.9	16.9	12.3
Sodium	100	14.2	15.0	19.2	6.0	14.1
Selenium	0.01	13.7	41.3	25.7	0.0	2.1
Silver	0.05	4.7	2.1	3.7	4.8	4.8
Sulfates	250.0	4.0	11.7	7.4	0.5	0.7
Nitrate-N	10.0	2.7	4.0	5.8	0.3	1.3
Fluoride	1.4	2.5	6.2	1.8	0.0	2.7
Arsenic	0.05	0.8	2.1	1.8	0.0	0.0
Barium	1.0	0.3	0.0	0.0	0.0	0.7
Magnesium	125.0	0.1	0.5	0.1	0.0	0.0
Chromium	0.05	[b]	0.0	0.0	0.0	0.0
Boron	[c]					

Note: According to survey conducted by United States Environmental Protection Agency.

[a] May be distorted upwards.
[b] Not detected.
[c] Not tested.

Source: From U.S Environmental Protection Agency, 1984, *National Statistical Assessment of Rural Conditions, Executive Summary.* Office of Drinking Water.

Table 8A.12 Water Quality in Selected Rivers in the World, 1996–1999

	Dissolved Oxygen (DO) (mg/L)					Biological Oxygen Demand (BOD) (mg/L)					Nitrates (c) (mg/L)				
	1996	1997	1998	1999	Average Last 3 yrs (b)	1996	1997	1998	1999	Average Last 3 yrs (b)	1996	1997	1998	1999	Average Last 3 yrs (b)
Canada															
Mackenzie	X	X	X	X	X	X	X	X	X	X	X	X	X	X	X
Saskatchewan	8.8	9.2	—	—	9.1	X	X	X	X	X	0.14	0.15	—	—	0.14
Columbia	X	X	X	X	X	X	X	X	X	X	0.11	0.10	0.08	—	0.10
Saint John	X	X	X	X	X	X	X	X	X	X	—	—	—	—	—
Mexico															
Bravo	7.5	8.0	7.7	—	7.7	3.1	2.0	2.4	—	2.5	0.15	0.18	0.16	—	0.16
Lerma	5.8	3.5	0.7	—	3.3	17.0	12.0	92.3	—	40.4	0.78	0.30	0.82	—	0.63
Pánuco	6.6	7.6	6.5	—	6.9	1.7	1.6	1.1	—	1.4	0.13	1.06	0.19	—	0.46
Grijalva	5.1	6.4	6.2	—	5.9	4.4	4.3	5.4	—	4.7	—	0.10	0.14	—	0.10
U.S.A															
Delaware	10.6	11.8	11.9	11.1	11.6	1.6	1.9	1.3	2.6	1.9	—	—	—	—	—
Mississippi	8.4	8.5	8.2	8.8	8.5	0.9	1.1	1.2	1.4	1.2	—	—	—	—	—
Japan															
Ishikari	11.0	11.0	11.0	11.0	11.0	1.2	1.2	1.2	0.9	1.1	X	X	X	X	X
Yodo	9.6	9.4	9.4	9.3	9.4	2.0	1.6	1.7	1.6	1.6	X	X	X	X	X
Tone {Sakae-hashi}	9.5	9.4	9.4	9.5	9.4	2.3	1.5	1.5	2.0	1.7	X	X	X	X	X
Chikugo	9.9	9.8	9.8	10.0	9.9	1.4	1.2	1.3	1.4	1.3	X	X	X	X	X
Korea															
Keum	8.7	8.9	9.8	10.0	9.6	3.9	2.7	2.6	3.0	2.8	2.15	2.09	2.05	2.36	2.17
NakDong	9.3	9.7	9.8	10.5	10.0	3.6	3.4	2.2	2.9	2.8	3.01	2.88	2.78	2.91	2.86
YoungSan	9.4	9.7	9.8	9.5	9.7	2.1	2.1	2.2	2.0	2.1	1.89	2.87	3.23	2.89	3.00
Han	8.7	9.1	10.0	8.3	9.1	3.9	4.1	3.6	3.3	3.7	1.39	2.05	2.36	2.44	2.29
Austria															
Donau	11.0	11.5	10.8	—	11.1	3.7	2.5	2.8	—	3.0	2.29	2.21	0.85	—	1.78
Inn	11.5	11.3	11.0	—	11.3	2.4	2.2	2.4	—	2.3	1.48	1.29	1.26	—	1.35
Grossache	11.7	11.2	—	—	11.3	1.2	1.1	—	—	1.4	0.73	0.68	—	—	0.72
Belgium															
Meuse	11.0	—	—	—	—	2.5	—	—	—	—	2.00	—	—	—	—
Escaut	7.4	—	—	—	—	5.7	—	—	—	—	4.67	—	—	—	—
Czech R.															
Labe	10.3	10.3	9.9	10.3	10.1	3.5	3.9	3.7	3.7	3.8	4.57	4.31	3.89	4.02	4.07
Odra	9.7	9.5	9.9	10.2	9.9	4.9	4.2	3.9	4.8	4.3	3.77	2.35	3.24	3.03	2.87
Morava	11.2	11.0	11.0	10.9	11.0	5.0	5.4	4.4	5.5	5.1	3.98	3.27	3.29	3.46	3.34
Dyje	10.3	10.4	10.5	11.2	10.7	4.5	5.1	6.4	4.5	5.4	5.43	4.44	3.63	3.55	3.87
Denmark															
Gudená	X	X	X	X	X	2.4	3.0	2.3	—	2.6	0.98	1.36	1.50	—	1.28
Skjerná	10.5	9.7	8.7	—	9.6	1.3	1.6	1.3	—	1.4	2.29	2.33	2.90	—	2.50
Susá	X	X	X	X	X	1.9	2.7	—	—	2.1	1.01	1.18	—	—	1.83
Odense	X	X	X	X	X	2.0	1.9	—	—	1.9	4.22	3.69	5.98	—	4.63

Finland															
Torniojoki	11.8	11.7	—	—	11.6	X	X	X	X	X	0.07	0.06	—	—	0.06
Kymijoki	8.6	12.1	—	—	10.6	X	X	X	X	X	0.32	0.19	—	—	0.25
Kokemäenjoki	10.8	10.6	—	—	10.7	X	X	X	X	X	0.53	0.71	—	—	0.58
France															
Loire	10.1	10.3	10.7	9.9	10.3	5.8	5.9	5.9	4.6	5.4	2.32	2.53	2.65	3.03	2.74
Seine	4.2	4.6	5.9	7.1	5.8	4.7	5.3	4.4	3.4	4.4	6.62	6.27	5.90	5.77	5.98
Garonne	9.5	9.6	9.3	8.5	9.1	1.1	1.4	1.4	1.7	1.5	1.92	2.14	1.85	2.22	2.07
Rhöne	9.1	9.9	9.7	10.5	10.0	2.1	1.3	1.3	1.5	1.4	1.44	1.50	1.59	1.41	1.50
German															
Rhein	10.0	9.9	9.7	10.3	10.0	X	X	X	X	X	3.49	3.15	3.15	2.59	2.96
Elbe	11.6	11.2	11.3	11.6	11.4	X	X	X	X	X	4.29	3.92	3.57	3.41	3.63
Weser	9.7	10.8	10.1	9.9	10.3	3.8	2.9	2.1	2.4	2.5	4.52	4.52	4.20	4.43	4.38
Donau	11.1	11.1	10.9	11.1	11.0	2.2	2.4	2.2	2.1	2.2	2.34	2.22	2.07	2.10	2.13
Donar	11.1	11.1	10.9	11.1	11.9	X	X	X	X	X	X	X	X	X	X
Greece															
Strimonas	—	—	9.8	11.1	10.8	X	X	X	X	X	0.87	2.29	1.04	1.52	1.62
Axios	—	—	8.3	8.5	9.6	X	X	X	X	X	1.13	2.31	0.66	1.24	1.40
Akeloos	—	—	—	—	—	X	X	X	X	X	X	X	X	X	X
Nestos	—	—	—	—	—	X	X	X	X	X	X	X	X	X	X
Hungary															
Maros	9.9	9.7	9.9	10.1	9.9	4.3	2.9	3.3	3.7	3.3	1.95	2.06	2.16	2.34	2.19
Duna	9.7	10.8	9.5	10.1	10.1	2.6	2.7	2.6	2.1	2.5	2.59	2.59	1.90	2.12	2.20
Dráva	10.6	10.1	9.8	9.8	9.9	3.1	3.3	3.0	2.9	3.1	1.64	1.37	1.63	1.47	1.49
Tisza	11.1	12.6	12.1	12.7	12.5	1.5	2.0	2.6	3.6	2.7	0.77	0.66	0.77	0.64	0.69
Ireland															
Boyne	10.9	10.9	11.0	10.7	10.9	2.0	1.5	1.9	1.6	1.7	3.69	3.25	3.14	2.48	2.96
Clare	11.6	10.9	10.3	10.7	10.6	1.6	1.1	1.5	2.5	1.7	1.82	1.97	1.80	1.40	1.72
Barrow	11.4	10.8	10.9	11.0	10.9	2.4	1.6	1.8	1.5	1.6	5.22	4.58	4.95	4.14	4.55
Blackwater	10.7	10.2	10.8	10.5	10.5	1.9	2.0	2.1	2.7	2.2	2.89	2.76	2.74	2.20	2.57
Italy															
Po	X	X	X	X	X	X	X	X	X	X	2.89	2.21	2.05	2.10	2.12
Adige	10.4	—	10.7	8.3	9.8	X	X	X	X	X	1.26	1.37	1.22	1.15	1.25
Arno	9.2	13.0	7.2	8.1	9.4	X	X	X	X	X	X	X	X	X	X
Metauro	9.9	10.2	9.9	10.0	10.0	X	X	X	X	X	—	—	—	1.53	—
Luxembourg															
Moselle	9.4	9.1	9.1	9.2	9.1	2.6	2.6	2.2	2.6	2.5	3.07	2.88	2.92	2.62	2.81
Süre	10.0	10.2	10.4	10.6	10.4	3.1	2.9	2.7	3.1	2.9	5.0	5.45	5.74	4.90	5.36
The Netherlands															
Maas-Keizersveer	9.8	9.2	9.2	—	9.4	X	X	X	X	X	X	X	X	X	X
Maas-Eysden	X	X	X	X	X	3.0	2.0	3.6	—	2.9	X	X	X	X	X
Rijn/Maas Delta	11.9	11.4	10.5	—	11.3	X	X	X	X	X	3.64	3.03	3.55	2.55	3.04
Rijn-Lobith	10.2	10.0	10.3	—	10.2	3.0	5.0	—	—	3.3	3.64	3.55	3.55	2.53	3.14
Ijssel-Kampen	10.0	9.6	9.5	—	9.7	X	X	X	X	X	4.02	3.55	3.95	2.64	3.38
Norway															
Skienselva	X	X	X	X	X	X	X	X	X	X	0.21	0.20	0.21	0.20	0.20
Glomma	X	X	X	X	X	X	X	X	X	X	0.40	0.39	0.38	0.36	0.38
Drammenselva	X	X	X	X	X	X	X	X	X	X	0.28	0.29	0.27	0.26	0.27

(Continued)

Table 8A.12 (Continued)

	Dissolved Oxygen (DO) (mg/L)					Biological Oxygen Demand (BOD) (mg/L)					Nitrates (c) (mg/L)				
	1996	1997	1998	1999	Average Last 3 yrs (b)	1996	1997	1998	1999	Average Last 3 yrs (b)	1996	1997	1998	1999	Average Last 3 yrs (b)
Otra	X	X	X	X	X	X	X	X	X	X	0.16	0.14	0.14	0.14	0.14
Poland															
Wisia	10.2	10.6	10.8	10.5	10.6	3.9	4.6	3.4	3.4	3.8	1.62	1.37	1.84	1.42	1.54
Odra	9.8	9.6	11.6	10.9	10.7	3.7	5.1	5.5	4.0	4.9	2.20	1.89	2.63	2.51	2.34
Slovak R.															
Maly Dunaj	9.1	10.5	9.5	9.6	9.8	2.6	3.7	2.9	3.1	3.2	2.55	2.27	2.31	2.48	2.35
Vah	10.1	9.7	9.9	9.3	9.6	4.5	3.9	2.6	2.1	2.9	2.41	2.12	1.71	2.06	1.97
Hron	10.4	10.9	10.5	—	10.6	3.9	3.3	3.2	—	3.5	1.86	1.96	1.89	—	1.90
Hornad	9.6	9.9	9.5	9.6	9.7	6.8	5.4	3.1	2.7	3.7	2.85	3.06	2.64	2.62	2.77
Spain															
Guadalquivir	7.0	5.0	6.0	4.0	5.0	14.5	2.6	3.4	6.6	4.2	3.67	6.55	6.08	5.65	6.09
Duero	7.0	7.0	9.0	10.0	8.7	3.8	2.5	2.1	3.7	2.8	1.79	1.81	2.37	1.31	1.83
Ebro	9.6	9.5	10.0	10.0	9.8	5.2	5.5	4.3	—	5.0	2.26	2.75	2.42	—	2.48
Guadiana	9.9	9.3	8.0	—	9.1	2.9	1.8	2.9	—	2.6	2.10	1.97	1.82	—	1.96
Sweden															
Delàlven	X	X	X	X	X	X	X	X	X	X	0.12	0.10	0.12	0.12	0.11
Ràne alv	X	X	X	X	X	X	X	X	X	X	0.04	0.04	0.03	0.03	0.03
Mònumsán	X	X	X	X	X	X	X	X	X	X	0.16	0.12	0.16	0.19	0.16
Rònneàn	X	X	X	X	X	X	X	X	X	X	1.35	1.74	1.59	1.30	1.54
Switzerland															
Rhin	10.6	10.4	10.4	11.1	10.6	X	X	X	X	X	1.56	1.40	1.37	1.34	1.37
Aare	10.1	10.5	10.6	10.4	10.5	X	X	X	X	X	1.99	1.77	1.78	1.49	1.68
Rhöne	11.2	11.5	11.5	11.6	11.5	X	X	X	X	X	X	X	X	X	X
Turkey															
Porsuk	9.2	9.5	8.7	8.2	8.8	1.3	1.1	1.3	1.3	1.2	1.45	1.20	1.15	1.21	1.19
Sakarya	8.8	9.1	9.3	8.6	9.0	3.7	3.2	3.5	3.4	3.4	1.37	1.42	1.43	1.50	1.45
Yesilirmak	10.2	9.6	9.6	8.8	9.3	2.0	2.7	2.4	2.2	2.4	2.90	1.70	7.53	5.05	4.76
Gediz	3.9	4.9	6.3	5.5	5.6	2.2	2.0	5.5	3.3	3.6	1.15	0.57	0.29	0.06	0.31
UK															
Thames	10.2	11.0	10.8	10.5	10.8	3.0	2.7	1.7	1.7	2.0	8.13	7.85	7.68	6.79	7.44
Severn	10.7	10.8	10.3	—	10.6	2.8	2.0	1.9	7.9	3.9	6.95	6.64	6.70	6.20	6.51
Clyde	8.5	8.0	9.6	8.7	8.8	3.9	2.1	2.5	2.3	2.3	1.95	1.88	2.06	1.70	1.88
Mersey	8.1	7.7	8.2	8.2	8.0	3.9	3.6	3.1	2.8	3.2	4.83	4.43	4.73	5.64	4.94
Lower Bann (N. Ireland)	9.8	9.1	—	—	9.2	3.8	2.8	—	—	3.4	0.80	1.09	—	—	1.01

	Total Phosphorus (c) (mg/L)					Ammonium (c) (mg/L)					Lead (c) (mg/L)				
	1996	1997	1998	1999	Average Last 3 yrs (b)	1996	1997	1998	1999	Average Last 3 yrs (b)	1996	1997	1998	1999	Average Last2 3 yrs (b)
Canada															
Mackenzie	0.07	—	—	—	—	X	X	X	X	X	1.63	—	—	—	—
Saskatchewan	0.04	0.05	—	—	0.044	0.041	0.041	—	—	0.039	0.87	2.57	—	—	1.39
Columbia	0.01	0.01	0.01	0.01	0.009	X	X	X	X	X	0.96	0.76	0.42	0.34	0.51
Saint John	—	—	—	—	—	X	X	X	X	X	—	—	—	—	—
Mexico															
Bravo	0.11	0.13	0.10	—	0.113	0.030	0.030	0.030	—	0.030	X	X	X	X	X
Lerma	—	2.60	5.63	—	—	—	1.350	18.450	—	—	X	X	X	X	X
Panuco	—	—	0.03	—	—	0.020	0.060	0.030	—	0.037	X	X	X	X	X
Grijalya	0.19	0.03	0.04	—	0.085	—	0.080	0.120	—	0.067	X	X	X	X	X
USA															
Delaware	0.05	0.06	0.05	0.09	0.067	0.030	0.040	0.030	0.040	0.037	—	—	—	—	—
Mississippi	0.19	0.15	0.18	0.24	0.190	0.020	0.020	0.060	0.020	0.033	—	—	—	—	—
Japan															
Ishikari	X	X	X	X	X	X	X	X	X	X	X	X	X	X	X
Yodo	X	X	X	X	X	X	X	X	X	X	X	X	X	X	X
Tone {Sakae-hashi}	X	X	X	X	X	X	X	X	X	X	X	X	X	X	X
Chikugo	X	X	X	X	X	X	X	X	X	X	X	X	X	X	X
Korea															
Keum	0.13	0.10	0.05	0.04	0.064	0.838	0.922	0.530	0.411	0.621	—	—	—	—	0
NakDong	0.07	0.14	0.08	0.04	0.086	0.901	0.516	0.283	0.124	0.308	—	—	—	—	0
YoungSan	0.15	0.07	0.14	0.10	0.105	0.510	0.230	0.236	0.317	0.261	—	—	—	—	0
Han	0.28	0.37	0.21	0.21	0.264	2.368	2.416	1.624	1.540	1.860	—	—	—	—	0
Austria															
Donau	0.04	0.11	0.16	—	0.104	0.160	0.124	0.150	—	0.145	0.64	0.54	5.00	—	2.06
Inn	0.04	0.18	0.14	—	0.117	0.098	0.101	0.080	—	0.093	0.58	1.34	1.60	—	1.17
Grossache	0.08	0.07	—	—	0.070	0.015	0.030	—	—	0.025	3.20	4.00	—	—	2.40
Belgium															
Meuse	0.15	—	—	—	—	0.050	—	—	—	—	3.90	—	—	—	—
Escaut	0.80	—	—	—	—	4.450	—	—	—	—	8.50	—	—	—	—
Czech R.															
Labe	—	0.23	0.23	0.21	0.223	0.505	0.429	0.300	0.250	0.326	2.00	1.20	0.85	1.10	1.05
Odra	0.44	0.45	0.39	0.33	0.389	1.912	1.336	0.590	0.500	0.809	17.50	6.38	1.90	1.30	3.19
Morava	0.31	0.25	0.20	0.28	0.243	0.688	0.660	0.390	0.440	0.497	2.67	2.42	1.18	2.20	1.93
Dyje	0.32	0.38	0.51	0.37	0.420	0.462	0.475	0.340	0.390	0.402	X	X	X	X	X
Denmark															
Gudená	0.10	0.11	0.10	—	0.101	0.089	0.095	0.047	—	0.077	X	X	X	X	X
Skjerná	0.05	0.05	0.06	—	0.053	0.119	0.106	0.094	—	0.106	X	X	X	X	X
Susá	0.32	0.36	—	—	0.278	0.096	0.103	—	—	0.097	X	X	X	X	X
Odense	0.14	0.14	0.16	—	0.145	0.154	0.076	0.076	—	0.102	X	X	X	X	X
Finland															
Torniojoki	0.01	0.02	—	—	0.017	0.017	—	—	—	0.017	0.12	0.23	—	—	0.15

(Continued)

Table 8A.12 (Continued)

	Total Phosphorus (c) (mg/L)					Ammonium (c) (mg/L)					Lead (c) (mg/L)				
	1996	1997	1998	1999	Average Last 3 yrs (b)	1996	1997	1998	1999	Average Last 3 yrs (b)	1996	1997	1998	1999	Average Last2 3 yrs (b)
Kymijoki	0.03	0.01	—	—	0.019	0.035	0.004	—	—	0.022	0.19	0.15	—	—	0.20
Kokemäenjoki	0.04	0.05	—	—	0.048	0.079	0.075	—	—	0.074	1.00	0.95	—	—	1.00
France															
Loire	0.25	0.27	0.17	0.22	0.220	0.060	0.160	0.043	0.062	0.088	X	X	X	X	X
Seine	1.23	1.22	1.06	0.82	1.031	0.278	0.484	0.275	0.164	0.308	X	X	X	X	X
Garonne	0.11	0.15	0.13	0.34	0.208	0.130	0.152	0.071	0.080	0.101	1.08	0.11	—	—	1.03
Rhöne	0.13	0.11	0.08	0.06	0.084	0.102	0.126	0.122	0.095	0.114	X	X	X	X	X
German															
Rhein	0.16	0.16	0.16	0.14	0.153	0.160	0.130	0.110	0.079	0.106	3.00	3.70	3.90	4.70	4.10
Elbe	0.23	0.22	0.21	0.19	0.207	0.300	0.210	0.140	0.130	0.160	—	—	3.50	2.63	2.38
Weser	0.24	0.21	0.24	0.16	0.203	0.400	0.240	0.100	0.100	0.147	4.10	4.60	6.10	3.86	4.85
Donau	0.09	0.08	0.08	0.08	0.079	0.080	0.067	0.069	0.078	0.071	1.00	1.00	1.00	1.00	1.00
Greece															
Strimonas	—	—	0.06	0.06	0.078	—	—	0.141	0.093	0.091	X	X	X	X	X
Axios	—	—	0.90	0.52	0.700	—	—	0.404	0.179	0.229	X	X	X	X	X
Akeloos	X	X	X	X	X	X	X	X	X	X	X	X	X	X	X
Nestos	X	X	X	X	X	—	—	0.233	0.095	0.118	X	X	X	X	X
Hungary															
Maros	0.26	0.30	0.23	0.26	0.263	0.240	0.130	0.080	0.140	0.117	0.50	0.90	0.70	0.50	0.70
Duna	0.09	0.11	0.10	0.12	0.108	0.086	0.078	0.074	0.072	0.075	3.20	2.50	2.80	1.30	2.20
Drava	0.16	0.14	0.12	0.14	0.133	0.053	0.067	0.055	0.047	0.056	1.30	1.30	1.70	1.10	1.37
Tisza	X	X	X	X	X	0.032	0.080	0.060	0.068	0.069	X	X	X	X	X
Ireland															
Boyne	0.09	—	0.08	0.06	0.046	0.053	0.043	0.068	0.047	0.053	X	X	X	X	X
Clare	0.07	0.06	0.05	0.14	0.084	0.042	0.022	0.039	0.078	0.046	X	X	X	X	X
Barrow	0.11	0.08	0.10	0.08	0.087	0.057	0.039	0.031	0.031	0.034	X	X	X	X	X
Blackwater	0.17	0.10	0.09	0.09	0.092	0.039	0.048	0.086	0.109	0.081	X	X	X	X	X
Italy															
Po	0.18	0.14	0.15	0.18	0.157	0.140	0.086	0.163	0.130	0.126	X	X	X	X	X
Adige	—	0.05	0.05	—	0.048	0.062	0.062	0.054	0.110	0.076	X	X	X	X	X
Amo	—	0.16	0.16	0.23	0.183	X	X	X	X	X	X	X	X	X	X
Metauro	0.13	0.22	0.11	0.18	0.170	—	—	—	—	—	X	X	X	X	X
Luxembourg															
Moselle	0.61	0.55	0.57	0.17	0.429	0.184	0.165	0.284	0.174	0.208	3.90	2.00	2.10	2.20	2.10
Süre	0.52	0.45	0.56	0.18	0.394	0.224	0.220	0.211	0.162	0.198	4.10	2.00	2.00	2.30	2.10
The Netherlands															
Maas-Kezersveer	0.28	0.39	0.24	0.22	0.283	X	X	X	X	X	4.90	2.30	4.11	3.83	3.41
Maas-Eysden	X	X	X	X	X	X	X	X	X	X	X	X	X	X	X
Rijn/Maas Delta	0.14	0.16	0.14	0.19	0.163	X	X	X	X	X	1.10	1.50	1.90	2.20	1.87
Rijn-Lobith	0.22	0.21	0.22	0.18	0.203	0.220	0.100	0.110	—	0.143	4.50	3.90	4.60	3.60	4.03

Ijssel-Kampen	0.20	0.20	0.19	0.17	0.187	X	X	X	X	X	3.10	3.80	2.60	3.30	3.23
Norway															
Skienselva	0.00	0.01	0.00	0.01	0.005	0.017	0.015	0.014	0.015	0.015	0.11	0.07	0.19	0.07	0.11
Glomma	0.02	0.02	0.01	0.03	0.021	0.045	0.033	0.032	0.035	0.033	0.50	0.55	3.70	0.40	1.55
Drammenselva	0.01	0.01	0.00	0.01	0.007	0.015	0.017	0.016	0.019	0.017	0.17	0.12	0.13	0.15	0.13
Otra	0.00	0.00	0.00	0.01	0.003	0.016	0.015	0.013	0.018	0.015	0.34	0.27	0.30	0.30	0.29
Poland															
Wisia	0.14	0.21	0.20	0.23	0.212	0.449	0.225	0.210	0.368	0.268	0.45	0.60	1.00	0.40	0.67
Odra	0.34	0.31	0.30	0.26	0.291	0.605	0.330	0.120	0.091	0.180	1.37	0.50	0.20	0.10	0.27
Slovak R.															
Maly Dunaj	0.28	0.29	0.27	0.24	0.265	0.423	0.274	0.246	0.109	0.210	1.40	—	2.29	0.81	1.50
Vah	0.24	0.15	0.12	0.15	0.140	0.626	0.539	0.371	0.378	0.429	0.58	1.15	1.10	0.85	1.03
Hron	0.23	0.21	0.18	—	0.205	0.456	0.303	0.313	—	0.357	X	X	X	X	X
Hornad	0.26	0.18	0.18	0.14	0.170	0.840	0.528	0.637	0.534	0.566	8.42	13.62	4.45	3.40	7.16
Spain															
Guadalquivir	0.67	—	—	—	—	0.414	0.143	—	—	0.285	3.00	—	—	—	0.00
Duero	0.19	0.20	0.15	0.15	0.165	0.288	0.266	0.266	—	0.273	—	1.00	1.00	—	0.67
Ebro	0.28	0.19	0.12	—	0.199	0.167	0.023	0.041	—	0.077	—	1.67	—	—	0.56
Guadiana	0.25	0.19	—	—	0.316	0.151	0.100	0.106	—	0.119	—	—	—	—	0.00
Sweden															
Dalàlven	0.02	0.02	0.01	0.01	0.015	0.020	0.017	0.018	0.020	0.018	0.39	0.42	0.49	0.44	0.45
Ràne alv	0.01	0.02	0.01	0.01	0.015	0.013	0.016	0.012	0.012	0.013	0.12	0.26	0.15	0.09	0.17
Mònumsán	0.02	0.02	0.02	0.02	0.022	0.026	0.019	0.017	0.027	0.021	0.35	0.28	0.38	0.47	0.38
Rònneàn	0.04	0.04	0.45	0.71	0.402	0.060	0.076	0.061	0.050	0.062	0.23	0.37	0.44	0.57	0.46
Switzerland															
Rhin	0.05	0.05	0.05	0.04	0.047	X	X	X	X	X	0.60	0.60	0.80	0.70	0.70
Aare	0.07	0.07	0.06	0.04	0.055	X	X	X	X	X	X	X	X	X	X
Rhöne	0.12	0.14	0.14	0.14	0.138	X	X	X	X	X	3.00	4.50	4.60	4.10	4.40
Turkey															
Porsuk	—	—	0.04	0.09	0.061	0.045	0.068	0.100	0.170	0.113	5.00	—	8.00	5.00	4.33
Sakarya	0.50	0.49	0.16	0.27	0.307	0.340	0.340	0.260	0.300	0.300	13.00	—	12.00	6.00	6.00
Yesilinmak	0.27	0.42	0.10	0.07	0.197	0.190	0.750	0.290	0.210	0.417	X	X	X	X	X
Gediz	0.32	0.42	0.05	0.02	0.163	0.013	—	—	—	0.004	—	—	—	—	—
UK															
Thames	1.69	1.88	1.03	1.16	1.357	0.313	0.198	0.176	0.187	0.187	3.50	2.70	3.10	4.10	3.30
Severn	0.86	1.08	0.72	0.47	0.753	0.198	0.174	0.127	0.114	0.138	6.80	4.30	4.10	4.50	4.30
Clyde	0.59	0.56	0.40	0.41	0.458	1.404	0.867	0.657	0.727	0.750	7.20	3.00	2.40	3.50	2.97
Mersey	1.54	1.16	1.02	1.09	1.089	4.342	3.705	2.073	1.445	2.408	4.80	5.00	6.60	5.60	5.73
Lower Bann (N. Ireland)	0.16	0.16	—	—	0.161	0.105	0.077	—	—	0.108	0.40	0.41	—	—	0.41

(Continued)

Table 8A.12 (Continued)

	Cadmium (c) (mg/L)					Chromium (c) (mg/L)					Copper (c) (mg/L)				
	1996	1997	1998	1999	Average Last 3 yrs (b)	1996	1997	1998	1999	Average Last 3 yrs (b)	1996	1997	1998	1999	Average Last 3 yrs (b)
Canada															
Mackenzie	0.250	—	—	—	—	1.58	—	—	—	—	3.52	—	—	—	—
Saskatchewan	0.325	0.730	—	—	0.412	1.38	1.86	—	—	1.49	3.05	2.61	—	—	2.69
Columbia	0.075	0.079	0.102	0.100	0.094	0.36	0.22	0.21	0.21	0.21	1.21	1.03	0.86	0.86	0.92
Saint John	—	—	—	—	—	X	X	X	X	X	—	—	—	—	—
Mexico															
Bravo	X	X	X	X	X	X	X	X	X	X	X	X	X	X	X
Lerma	X	X	X	X	X	X	X	X	X	X	X	X	X	X	X
Pánuco	X	X	X	X	X	X	X	X	X	X	X	X	X	X	X
Grijalva	X	X	X	X	X	X	X	X	X	X	X	X	X	X	X
USA															
Delaware	—	—	—	—	—	—	—	—	—	—	—	—	—	—	—
Mississippi	1.000	1.000	1.000	1.000	1.000	—	—	—	—	—	—	—	—	—	—
Japan															
Ishikari	X	X	X	X	X	X	X	X	X	X	X	X	X	X	X
Yodo	X	X	X	X	X	X	X	X	X	X	X	X	X	X	X
Tone {Sakae-hashi}	X	X	X	X	X	X	X	X	X	X	X	X	X	X	X
Chikugo	X	X	X	X	X	X	X	X	X	X	X	X	X	X	X
Korea															
Keum	—	—	—	—	0.000	—	—	—	—	0.00	X	X	X	X	X
NakDong	—	—	—	—	0.000	—	—	—	—	0.00	X	X	X	X	X
YoungSan	—	—	—	—	0.000	—	—	—	—	0.00	X	X	X	X	X
Han	—	—	—	—	0.000	—	—	—	—	0.00	X	X	X	X	X
Austria															
Donau	—	—	<0.20	—	0.000	0.27	—	1.00	—	0.42	2.00	2.35	2.40	—	2.25
Inn	—	—	0.100	—	0.033	0.25	0.91	0.80	—	0.65	2.33	3.10	2.80	—	2.74
Grossache	0.200	0.200	—	—	0.133	2.00	4.00	—	—	2.00	3.80	5.00	—	—	2.93
Belgium															
Meuse	0.100	—	—	—	—	1.00	—	—	—	—	1.80	—	—	—	—
Escaut	0.300	—	—	—	—	2.80	—	—	—	—	16.20	—	—	—	—
Czech R															
Labe	0.312	0.154	0.230	0.160	0.181	1.80	2.42	2.27	2.19	2.29	9.79	9.23	12.77	12.15	11.38
Odra	2.583	0.090	0.080	0.060	0.077	—	1.92	2.25	1.79	1.99	8.83	3.58	5.42	6.75	5.25
Morava	0.150	0.108	0.130	0.120	0.119	1.92	2.82	1.10	0.50	1.47	3.79	2.80	2.32	3.30	2.81
Dyje	X	X	X	X	X	—	0.94	0.80	0.30	0.68	X	X	X	X	X
Denmark															
Gudená	X	X	X	X	X	X	X	X	X	X	X	X	X	X	X
Skjerná	X	X	X	X	X	X	X	X	X	X	X	X	X	X	X
Susá	X	X	X	X	X	X	X	X	X	X	X	X	X	X	X

Odense	X	X	X	X	X	X	X	X	X	X	X	X	X	X	X
Finland															
Torniojoki	0.030	0.030	—	—	0.033	0.34	0.55	—	—	0.45	0.84	0.71	—	—	0.79
Kymijoki	0.034	0.030	—	—	0.032	0.61	0.53	—	—	0.53	1.34	1.13	—	—	1.18
Kokemäenjoki	0.074	0.138	—	—	0.096	1.78	1.61	—	—	1.82	15.53	20.60	—	—	28.38
France															
Loire	—	—	—	—	—	X	X	X	X	X	X	X	X	X	X
Seine	X	X	X	X	X	X	X	X	X	X	X	X	X	X	X
Garonne	0.389	0.000	—	—	0.184	3.21	1.17	—	—	2.33	5.62	3.56	—	—	4.71
Rhöne	X	X	X	X	X	X	X	X	X	X	X	X	X	X	X
German															
Rhein	0.200	0.200	0.200	0.200	0.200	2.70	3.10	3.30	3.87	3.42	6.30	10.40	12.00	8.55	10.32
Elbe	—	—	0.320	0.243	0.248	—	—	2.10	1.93	1.68	—	—	6.40	6.44	4.98
Weser	0.200	0.200	0.200	0.200	0.200	2.00	2.00	2.10	2.00	2.03	4.60	4.80	5.10	4.33	4.74
Donau	0.100	0.100	0.100	0.100	0.100	1.00	1.00	1.00	1.00	1.00	3.30	2.70	3.00	3.07	2.92
Greece															
Strimonas	0.140	0.140	0.040	1.600	0.593	X	X	X	X	X	X	X	X	X	X
Axios	0.110	0.270	1.200	0.400	0.623	X	X	X	X	X	X	X	X	X	X
Akeloos	0.200	—	—	—	—	X	X	X	X	X	X	X	X	X	X
Nestos	0.200	0.200	1.800	0.470	0.823	X	X	X	X	X	X	X	X	X	X
Hungary															
Maros	0.140	0.170	0.150	0.100	0.140	15.00	15.60	16.90	26.80	19.77	5.30	7.30	10.10	4.80	7.40
Duna	0.630	0.700	0.980	0.580	0.753	1.60	4.00	2.10	1.20	2.43	13.40	8.40	9.40	10.10	9.30
Drava	0.090	0.060	0.070	0.070	0.067	0.50	0.50	0.60	0.70	0.60	1.50	1.80	2.40	2.80	2.33
Tisza	X	X	X	X	X	1.40	1.20	0.70	1.00	0.97	X	X	X	X	X
Ireland															
Boyne	0.200	0.200	0.200	0.200	0.200	X	X	X	X	X	X	X	X	X	X
Clare	0.200	0.200	0.200	0.200	0.200	X	X	X	X	X	X	X	X	X	X
Barrow	0.200	0.200	0.200	0.200	0.200	X	X	X	X	X	X	X	X	X	X
Blackwater	0.200	0.200	0.200	0.200	0.200	X	X	X	X	X	X	X	X	X	X
Italy															
Po	X	X	X	X	X	X	X	X	X	X	X	X	X	X	X
Adige	X	X	X	X	X	X	X	X	X	X	X	X	X	X	X
Arno	X	X	X	X	X	X	X	X	X	X	X	X	X	X	X
Metauro	<0.0002	<0.0002	<0.0002	—	0.000	X	X	X	X	X	X	X	X	X	X
Luxembourg															
Moselle	0.100	0.100	0.100	0.100	0.100	1.50	1.10	1.10	1.10	1.10	4.30	3.40	4.00	3.70	3.70
Süre	0.100	0.100	0.100	0.100	0.100	1.50	1.00	1.00	1.00	1.00	2.70	2.40	2.60	2.50	2.50
The Netherlands															
Maas-Keizersveer	0.290	0.140	0.300	0.200	0.213	2.98	2.06	2.16	1.95	2.06	4.15	3.35	4.72	4.67	4.25
Maas-Eysden	X	X	X	X	X	X	X	X	X	X	X	X	X	X	X
Rijn/Maas Delta	0.500	0.040	0.060	0.140	0.080	—	1.30	1.81	1.40	1.50	3.30	3.10	3.60	2.80	3.17
Rijn-Lobith	0.070	0.060	0.080	0.080	0.073	2.90	3.57	3.76	2.40	3.24	5.20	5.00	5.30	4.70	5.00
Ijssel-Kampen	0.080	0.100	0.050	0.090	0.080	—	3.26	2.42	2.00	2.56	5.30	5.50	4.20	11.10	6.93
Norway															
Skienselva	0.060	0.020	0.020	0.010	0.017	0.50	0.50	0.50	0.13	0.38	1.79	0.90	0.70	0.54	0.71
Glomma	0.080	0.030	0.160	0.020	0.070	0.50	0.50	0.50	0.62	0.54	2.20	2.10	3.20	2.01	2.44

(Continued)

Table 8A.12 (Continued)

	Cadmium (c) (mg/L)					Chromium (c) (mg/L)					Copper (c) (mg/L)				
	1996	1997	1998	1999	Average Last 3 yrs (b)	1996	1997	1998	1999	Average Last 3 yrs (b)	1996	1997	1998	1999	Average Last 3 yrs (b)
Drammenselva	0.020	0.020	0.020	0.020	0.020	0.50	0.50	0.50	0.46	0.49	0.91	1.00	0.80	1.03	0.94
Otra	0.030	0.030	0.030	0.030	0.030	0.50	0.50	0.50	0.19	0.40	0.50	0.50	0.50	0.62	0.54
Poland															
Wisla	0.070	0.100	0.100	0.100	0.100	0.22	0.00	1.00	0.10	0.37	1.47	1.30	2.00	2.00	1.77
Odra	0.096	0.040	0.020	0.000	0.020	0.30	0.00	0.10	0.00	0.03	2.77	2.30	2.00	3.00	2.43
Slovak															
Maly Dunaj	0.032	0.024	0.080	0.070	0.058	0.52	—	0.26	0.17	0.32	17.45	3.32	3.37	3.51	3.40
Vah	0.055	0.092	0.040	0.060	0.064	0.65	1.02	1.65	1.53	1.40	2.02	1.73	2.33	2.68	2.25
Hron	X	X	X	X	X	X	X	X	X	X	X	X	X	X	X
Hornad	1.267	1.333	0.820	0.670	0.941	0.14	4.79	2.34	2.19	3.11	X	X	X	X	X
Spain															
Guadalquivir	1.500	—	—	—	0.000	—	—	—	—	0.00	5.61	—	—	—	0.00
Duero	0.680	0.240	0.010	0.030	0.093	—	0.10	0.50	2.00	0.87	—	0.70	6.00	—	2.23
Ebro	0.030	0.040	—	—	0.023	—	1.67	0.25	—	0.64	0.83	0.83	1.00	3.00	1.61
Guadiana	—	—	—	—	0.000	—	—	—	—	0.00	—	—	—	—	0.00
Sweden															
Dalàlven	0.016	0.018	0.017	0.016	0.017	0.30	0.29	0.35	0.38	0.34	1.30	1.20	1.40	1.40	1.33
Ràne alv	0.008	0.012	0.024	0.006	0.014	0.11	0.13	0.19	0.17	0.16	0.80	1.70	0.80	0.40	0.97
Mòrrumsán	0.011	0.013	0.009	0.012	0.011	0.35	0.41	0.36	0.45	0.41	1.10	1.40	1.30	1.60	1.43
Rònneàn	0.022	0.027	0.040	0.032	0.033	0.55	0.62	0.59	0.61	0.61	1.20	1.40	1.70	1.70	1.60
Switzerland															
Guadiana	X	X	X	X	X	X	X	X	X	X	X	X	X	X	X
Rhin	X	X	X	X	X	X	X	X	X	X	1.60	1.50	1.70	1.70	1.63
Aare	X	X	X	X	X	X	X	X	X	X	X	X	X	X	X
Rhöne	X	X	X	X	X	X	X	X	X	X	X	X	X	X	X
Turkey															
Porsuk	0.000	5.000	5.000	5.000	5.000	0.00	9.00	5.00	5.00	6.33	3.00	5.00	5.00	5.00	5.00
Sakarya	—	—	5.000	—	1.667	—	—	6.00	—	2.00	14.00	—	8.00	15.00	7.67
Yesilirmak	—	—	—	—	—	X	X	X	X	X	—	—	—	—	—
Gediz	—	—	—	—	—	—	—	—	—	—	X	X	X	X	X
UK															
Thames	0.100	0.100	0.100	0.100	0.100	1.30	1.20	1.30	1.30	1.27	6.30	7.10	6.30	6.50	6.63
Severn	0.100	0.100	0.100	0.100	0.100	2.80	1.90	1.80	2.30	2.00	5.50	5.10	4.90	5.50	5.17
Clyde	0.000	0.000	0.100	0.000	0.033	17.00	13.00	14.70	12.30	13.33	4.90	4.30	3.60	3.70	3.87
Mersey	0.100	0.100	0.100	0.100	0.100	5.70	4.90	5.30	4.40	4.87	7.60	6.90	7.50	7.40	7.27
Lower Bann (N. Ireland)	0.100	0.100	—	—	0.100	1.00	1.00	—	—	1.00	3.53	3.37	—	—	3.28

Note: a) Measured at the mouth or downstream frontier of river; b) Average over the last 3 years available: data prior to 1993 have not been taken into account; c) Data refer to total concentrations unless otherwise specified. DO: Concentrations are annual mean conventration; x=Data not available; JPN) Data refer to fiscal year (April to March); KOR) Han: samples were taken at 26 km upstream from the mouth of the river due to the tidal influence; AUT) 1985: 1984 data; FRA) Data refer to hydrological year (September-August). Seine: station under marine influence. Rhône since 1987 data refer to another station; DEU) Elbe: Measuring station—Elbel/Geestacht; 1988 Elbe/Brunsbuttel; since 1989 Elbel/Zollenspieker; GRC) 1980: 1982 data; ESP) Guadalquivir: from 1990 onwards data refer to another station closer to the mouth and farther away from Sevilla influence; TUR) 1980: 1982 data.

BOD: MEX) 1985: 1984 data. 1998: the data's variations can be explained by fluctuations of meteorological conditions and the CNA's actions on control of residual water discharges; JPN) Data refer to fiscal year (April to March); KOR) Han: samples were taken at 26 km upstream from the mouth of the river due to the tidal influence; FRA) Data refer to hydrological year (September–August). Seine: station under marine influence. Rhône since 1987 data refer to another station; DEU) Weser: 1990–1997—BOD7 (20°); NLD) Mass-Eijsden 1990 and 1993–1994; Rijn-Lobith 1993–1996: average include limit of detection values; ESP) Guadalquivir: from 1990 onwards data refer to another station closer to the mouth and further away from Sevilla influence; TUR) 1980: 1982 data; UKD) When the parameter is unmeasurable (quantity is too small), the limit of detection values are used when calculating annual averages. Actual averages may therefore to lower. Clyde 1980: 1982 data.

Nitrates: CAN) Saskatchewan: N02+N03; U.S.A.) Delaware 1985: 1984 data; KOR) Han: samples were taken at 26 km upstream from the mouth of the rivers due to the tidal influence; AUT) 1985: 1984 data; DNK) Data refer to N02+N03; FRA) Data refer to hydrological year (September–August). Loire and Seine: dissolved concentrations. Seine: station under marine influence. Rhône: since 1987 data refer to another station; DEU) Dissolved concentrations; ITA) Po: until 1986 data refer to Ponte Polesella (76 km far from the mouth); since 1989 data refer to Pontelagoscuro (91 km far from the mouth). Metaure 1985: 1984 data; NLD) Rijn-Lobith: dissolved concentrations; NOR) Skienselva and Drammenselva: until 1990 data refer to stations which may have marine influence; from 1990 onwards, data refer to new stations further away from the outlet Skienselva and Glomma 1985: 1983 data. Drammenselva 1985: 1984 data; SPAIN) Dissolved concentrations. Guadalquivir: from 1990 onwards data refer to another station closer to the mouth and farther away from Sevilla influence. Ebro 1980: 1981 data; UKD) When the parameter is unreasonable (quantity too small) the limit of detection values are used when calculating annual averages. Actual averages may therefore be lower.

Phosphorus: CAN) Columbia 1980: 1981 data; MEX) Orthophosphate concentrations; U.S.A.) Mississippi 1985 and 1990 and Delaware 1998 and 1999: annual averages include estimated values; KOR) Han samples were taken at 26 km upstream from the mouth of the river due to the tidal influence; AUT) 1985: 1984 data; FRA) Data refer to hydrological year (September–August). Loire—1980: 1982 data: since 1982 data refer to another station. Seine: station under marine influence. Rhône: sicne 1987 data refer to another station; GRC) Strimonas: 1998 and 1999 data refer to ortho-phosphate; IRL) Boyne: Data refers to ortho-phosphate; ITA) Po: Data until 1988 refer to Ponte Polesella (76 km from the mouth); since 1989 data refer to Pontelagoscuro (91 km from the mouth); Metauro 1985: 1984 data; NOR) Skienselva and Drammenselva: until 1990 data refer to stations which may have marine influence; from 1990 onwards, data refer to new stations further away from the outlet; Skienselva and Glomma 1985: 1983 data; SLO) Maly Dunaj: orthophosphate concentrations; 1980: 1981 data; SPAIN) Guadalquivir: from 1990 onwards data refer to another station closer to the mouth and farther away from Sevilla influence; TUR) Orthophosphate concentrations; Yesilirmak 1980 and Gediz 1980: 1982 and 1981 data; UKD) Orthophosphate concentrations. When a parameter is unmeasurable (quantity too small), limit of detection values are used when calculating annual averages. Actual averages may therefore be lower.

Ammorium: CAN) Dissolved concentrations. 1980: 1981 data; U.S.A.) Delaware and Mississippi: dissolved concentrations; Mississippi: 1980, 1985 and 1999 data include limit of detection values; Delaware: 1982, 1983, 1985, 1988, 1992–1999 include limit of detection values; KOR) Han: samples were taken at 26 km upstream from the mouth of the river due to the tidal influence; AUT) 1985: 1984 data; FRA) Data refer to hydrological year (September–August). Loire and Seine: data refer to dissolved concentrations. Seine: station under marine influence. Rhône: since 1987 data refer to another station; DEU) Dissolved concentrations; GRC) 1980: 1982 data; ITA) Po: until 1988 data refer to Ponte Polesella (76 km from the mouth): since 1989 data refer to Pontelagoscuro (91 km from the mouth). Adige 1988 and Metauro 1995: averages represent upper limits. Adige 1985: 1984 data; LUX) Moselle 91, 96 to 99: upper limits; Süre-Wasserbillig: 1985, 1990–1992, 1994–1999: upper detection limits; NLD) Rhine-Lobith: dissolved concentrations; NOR) Skienselva: until 1990 data refer to a station which may have marine influence; from 1990 onwards, data refer to a different station further away from the outlet; ESP) Dissolved concentrations. Guadalquivir: from 1990 onwards data refer to another station closer to the mouth and farther away from Sevilla influence; TUR) Excepted for 1990–1991 data refer to NH3. Yesihrmak 1980: 1982 data; UKD) When the parameter is unmeasurable (quantity too small) the limit of detection values are used when calculating annual averages. Actual averages may therefore be lower.

Lead: U.S.A.) Delaware: 1988 data represent upper limits: dissolved concentrations. Mississippi: dissolved concentrations; KOR) Han: samples were taken at 26 km upstream from the mouth of the river due to the tidal influence; AUT) 1985: 1984 data. Donau 1980, 82, 86, Inn 1982, 84 and Grossache 1980, 86: limit of detection values; CZE) Labe: from 1988 to 1993 data are upper limit values. Morava: 1995 data is an upper limit value; FIN) Tornionjoki and Kymijoki: include limit of detection values; Kokemäenjoki 1980: 1981 data; FRA) Data refer to hydrological year (September–August); DEU) Elbe: dissolved concentrations: 1988–1989, 1991–1993 and 1995: include limit of detection values. Rhein 1994, 95, Weser 1988–1991 and Donau 1996, 97: include limit of detection values; HUN) Until 1994: total concentrations: 1994–1999: dissolved concentrations; LUX) Moselle and Süre all years: include limit of detection values; Both analysis methods and limit of detection have changed over the years; NLD) Rijn-Maas Delta 1992 and 1996, and Rijn-Lobith 1995: include limit of detection values; NOR) Glomma 1985: 1983 data. Drammenselva: until 1990 data refer to a station which may have marine influence; from 1990 onwards, data refer to a new station further away from the outlet. All rivers: from 1991 heavy metal concentrations have been determined by a different analysing method; SPAIN) Dissolved concentration. Guadalquivir: from 1990 onwards data refer to another station closer to the mouth and farther away from Sevilla influence; SWE) Dissolved concentrations based on analysis of unfiltered samples; TUR) Porsuk 1999: upper limit; UKD) When the parameter is unmeasurable (quantity too small), the limit of detection values are used when calculating annual averages; actual averages may therefore be lower.

(Continued)

Table 8A.12 (Continued)

Cadmium: U.S.A.) Delaware and Mississippi dissolved concentrations. Delaware 1982–1989, 1992–1993, and Mississippi 1980, 1989–1999; include limit of detection values; KOR) Han: samples were taken at 26 km upstream from the mouth of the river due to the tidal influence; AUT) Donau 1980: figure is approximate: Donau 1982, 86–87, 91 and 93, Inn 1984, 86, 88–90, 94 and Grossache 1980, 82 and 84: upper limits. 1985: 1984 data; BEL) Meuse (Agimont): 1994–1996 are upper limits; CZE Labe:from 1990 to 1993 data are upper limit values. Morava: 1993 figure is an upper limit value; FIN) Tornionjoki and Kymijoki: upper limits; 1985: 1984 data; FRA) Data refer to hydrological year (September–August). Loire: since 1988 data refer to another station; 1980 and 1985; 1982 and 1984 data; DEU) Rhein 1984–1989 and 95–97: upper limits: Elbe: data refer to dissolved concentrations; 1990–1991: upper limits. Weser 1988–1997, and Donau: upper limits; GRC) Strimonas 1986–1987, 92–94, Axios 1986–1987, Axeloos 1990, 92–96 and Nestos 1986, 92–97: include limit of detection values. Akeloos and Nestlos 1980: 1982 data; HUN) Until 1994. total concentrations; 1994–1999: dissolved concentrations; Duna: until 1996 total concentrations, 1996–1999 dissolved concentrations; IRL) Data represent upper limits; ITA) Metauro 1996: upper limits; LUX) Moselle and Süre 90 to 99 and Süre 1980, 1985, 1989: upper limits; NLD) Rijn/Maas, Delta 1993–1996, Rijn-Lobith 1993–1996 and Ijssel-Kampen 1993, 95–96: upper limits; NOR) Skienselva and Drammenselva: until 1990 data refer to stations which may have marine influence; from 1990 onwards, data refer to new stations further away from the outlet. Drammenselva 1980 (1981 data): refers to median values, and represents an upper limit; 1986 figure is a time-weighted average. All rivers: from 1991 heavy metal concentrations have been determined by a different analysing method; SPAIN) Dissolved concentrations. Guadalquivir: from 1990 onwards data refer to another station closer to the mouth and farther away from Sevilla influence; Guadiana 1980: 1981 data; SWE) Dalàven and Morrumsàn: dissolved concentrations based on analysis of unifiltered samples; TUR) Porsak 1991–1993, 1995, 1997–1999. Sakarya 1989, 1991–1992, 1995, 1998 and Gediz 1995: upper limits; UKD) When the parameter is unmeasurable (quantity is too small), limit of detection values are used when calculating annual averages. Actual averages may therefore be lower.

Chromium: U.S.A.) Dissolved concentrations. Delaware 1980–1982, 1986–1988 and Mississippi 1985, 1988–1989: included limit of detection values; KOR) Han: samples were taken at 26 km upstream from the mouth of the river due to the tidal influence; AUT) Donau 1982, Inn 1994 and Grossache 1980, 82: include limit of detection values. 1985: 1984 data; BEL) Meuse (Agimont): 1994–1995 are upper limits; CZE) Labe 1988–1993: upper limits. Odra 1991–1992, 94–95: upper limits. Morava 1991–1992 and 1995: upper limits; FIN) Tornionjoki: include limit of detection values. Kymijoki 1985: 1984 data; FRA) Data refer to hydrological year (September–August); DEU) Elbe: dissolved concentrations. Elbe 1988, 90, 92, Weser 1987–1997, and Donau 1989–1997: include limit of detection values; HUN) Until 1994: total concentrations; 1994–1999; dissolved concentrations; Duna: until 1996 total concentrations, 1996–1999 dissolved concentrations; LUX) Moselle 91, 92, 95 to 99 and Süre 1991, 93, 95 to 99: include limit of detection values; NOR) Glomma 1985: 1983 data Drammenselva—1980: 1982 data; until 1990 data refer to a station which may have marine influence; from 1990 onwards, data refer to new station further away from the outlet. All rivers: since 1991 heavy metal concentrations have been determined by a different analysing method. Average of last 3 years represent or include the detection limit value (including 1998: the detection limit for Cr was 0.5); SPAIN) Dissolved concentrations. Guadalquivir: from 1990 onwards data refer to another station closer to the mouth and farther away from Sevilla influence; Guadiana 1985: 1983 data; TUR) Porsuk 1998–1999: upper limits; UKD) When the parameter is unmeasarable (quantity is too small), limit of detection values are used when calculating annual averages. Actual averages may therefore be lower. Lower Bann: dissolved concentrations.

Copper: U.S.A.) Delaware and Mississippi dissolved concentrations; AUT) 1985: 1984 data. Grossache 1980: includes limit of detection values; CZE) Morava 1995: upper limit; FRA) Data refer to hydrological year (September–August); DEU) Elbe: dissolved concentrations; HUN) Until 1994: total concentrations; 1994–1999: dissolved concentrations; Duna: until 1996 total concentrations, 1995–1999 dissolved concentrations; LUX) Moselle 91 to 94, 96 to 98 and Süre 1990–1991, 93, 95, 99: upper limits; NOR) Skienselva and Drammenselva: until 1990 data refer to stations which may have marine influence: from 1990 onwards, data refer to new stations further away from the outlets. Glomma 1985: 1983 data. Drammenselva 1980 (1981data): include limits of detection values and represent a median value. All rivers: from 1991 heavy metal concentrations have been determined by a different analysing method; SPAIN) Dissolved concentrations. Guadalquivir: from 1990 onwards data refer to another station closer to the mouth and farther away from Sevilla influence; SWE) Data refer to dissolved concentrations based on analysis of unfiltered samples; TUR) Porsuk 1988–1998: upper limits; UKD) When the parameter is unreasonable (quantity is too small), limit of detection values are used when calculating annual averages. Actual averages may therefore be lower.

Source: From Tables 3.4A through 3.4I (data from 1996, 1997, 1998, 1999, and average last 3 years), *OECD Environmental Data Compendium 2002*, © OECD 2002, www.oecd.org.

SECTION 8B DRINKING WATER QUALITY STANDARDS UNITED STATES

The U.S. Environmental Protection Agency's National Primary Drinking-Water Regulations and National Secondary Drinking-Water Regulations are summarized in the following tables. The primary regulations specify maximum contaminant levels (MCLs), and health advisories. The MCLs, which are the maximum permissible level of a contaminant in water at the tap, are health related and are legally enforceable. If these concentrations are exceeded or if required monitoring is not performed the public must be notified. The secondary drinking-water regulations specify the secondary maximum contaminant levels (SMCL). The SMCLs are for contaminants in drinking water that primarily affect the esthetic qualities related to public acceptance of drinking water; they are intended to be guidelines for the States and are not federally enforceable. Health advisories are guidance contaminant levels that would not result in adverse health effects over specified short-time periods for most people.

As provided by the Safe Drinking Water Act of 1974, the U.S. Environmental Protection Agency has the primary responsibility for establishing and enforcing regulations. However, States may assume primacy if they adopt regulations that are at least as stringent as the Federal regulations in levels specified for protection of public health and in provision of surveillance and enforcement. The States may adopt more stringent regulations and may establish regulations for other constituents.

Table 8B.13 Contaminants Regulated under the Safe Drinking Water Act

	Contaminants Regulated under the Safe Drinking Water Act								
	1976	1979	1986	1987	1989		1991		1992
Final Regulations	NPD WRs 12/75; 7/76	TTHMs 11/79	Fluoride 4/86	Phase 1 (VOCs) 7/87	TCR 6/89	SWTR 6/89	Phase II 1/91; 7/91	LCR 6/91	Phase V 7/92
Summary of Final Actions	New regs	New reg	*Revision*	New regs	*Revision*	1 *Revision* 4 New regs	38 SOCs & IOCs; 11 *Revisions* 27 New regs	1 *Revision* 1 New reg	23 SOCs & IOCs; 1 *Revision* 22 New regs
# in Regulation	22	1	1	8	1	5	39	2	23
Cumulative # of regulated contaminants	22	23	23	31	31	35	61	62	84
Contaminants regulated	2,4-D 2,4,5-TP (Silvex) arsenic barium cadmium chromium coliform bacterial cadrin fluoride gross alpha gross beta lead lindane mercury methoxychlor nitrate radium-226[1] radium-228[1] selenium silver toxaphene turbidity	total THMs[2]	*fluoride*	benzone carbon tetrachloride 1,2-dichloroethane p-dichlorobenzene 1,1-dichloroethylene 1,1,1-trichloroethlane trichloroethylene vinyl chloride[3]	*total coliforms*[2]	Giardia[4] *turbidity*[4] HPC bacteria[4] Legionella[4] viruses[4]	2-4-D 2,4,5-*TP* acrylamide[4] alachlor aldicarb[5] aldicarb sulfone[5] aldicarb sulfoxide[5] asbestos atrazinc *barium* *cadmium* carbofuran chlordane (mono) chlorobenzene *chromium* dibromochloropropane o-dichlorobenzene cis-1,2-dichloroethylene trans-1,2-dichloroethylene 1,2-dichloropropane epichlorohydrin[4] ethylbenzene ethylene dibromide heptachlor heptachlor epoxide *lindane* *mercury (inorganic)* *methoxychlor* *nitrate* nitrite total nitrate/nitrite PCBs pentachlorophenol *selenium* styrene tetrachlorethylene toluene *toxophene* xylenes	copper[4] *lead*[4]	Adupate di(2-ethylhexyl) antimony beryilium eyanide dalapon dichloromethane[4] dinoseb dioxin 2,3,7,8-TCDD) diqual endothall *endrin* glyphosate hexachlorobenzene hexachlorocyclopentadiene nickel oxamyl (vydate) PAHs (benzo(a) pyrene) phthalate, di(2-ethylhexyl) picloram simazine thalium 1,2,4-trichlorobenzene 1,1,2-trichloroethane

	1995	1998		2000	2001
Final Regulations		Stage 1 DBPR 12/98	Interim ERSWTR 12/98	Radionuclides 12/00	Arsenic 1/01
Summary of Final Actions	*Remand*	1 *Revision* 6 New regs	2 *Revisions* 1 New reg	4 *Revisions* 1 new reg	1 *Revision*
# in Regulation	1	1	3	5	1
Cumulative # of regulated contaminants	83	89	90	91	91
Contaminants regulated	*nickel*	bromate chloramine chlorine chlorine dioxide chlorine haloacetic acids (HAAS)[2] *TTHMs*	Cryptosporidium *Glardia* *turbidity*	grass alpha gross-beta radium-226[1] radium-228[1] uranium	arsenic

Notes:
1. Radium-226 and radium-228 arecontrol as two contaminants althoughtheir standard is combined.
2. Total THMs, haloacetic acids, and total coliforms are counted as one contaminant although both are combined standards: THMs (chloroform, bromodichloromethane, dibromochloromethane, bromoform), TC (total coliform bacteria including fecal coliform and E coli); HAAS (monochloroacetic acid, dichloroacetic acid, trichloroacetic acid, bromoacetic acid, and dibromoacetic acid).
3. Vinyl chloride is also known aschloroethylene & monochloroethylene.
4. These nine contaminants have a treatment technique instead of aMCL.
5. Aldicarb, aldicarb sulfone, and aldicarb sulfoxide are considered regulated contaminants although their MCLs are strayed.
6. Dichloromethane is also known as methylene chloride.

Updated 13 February 2001
www.epa.gov/safowater/mcl.html.

Table 8B.14 National Primary Drinking Water Standards

	MCLG[1] (mg/L)[2]	MCL or TT[1] 3(mg/L)[2]
Microorganisms		
Cryptosporidium	zero	TT[3]
Giardia lamblia	zero	TT[3]
Heterotrophic plate count	n/a	TT[3]
Legionella	zero	TT[3]
Total Coliforms (including fecal coliform and *E. Coli*)	zero	5.0%[4]
Turbidity	n/a	TT[3]
Viruses (enteric)	zero	TT[3]

Disinfection Byproducts

Constituent	MCLG[1] (mg/L)[2]	MCL or TT[1] (mg/L)[2]
Bromate	zero	0.01
Chlorite	0.8	1
Haloacetic acids (HAA5)	n/a[6]	0.06
Total Trihalomethanes (TTHMs)	none[7]	0.1
	—	—
	n/a[6]	0.08

Disinfectants

Constituent	MCLG[1] (mg/L)[2]	MCL or TT[1] (mg/L)[2]
Chloramines (as Cl_2)	MRDLG=41	MRDL=4.01
Chlorine (as Cl_2)	MRDLG=41	MRDL=4.01
Chlorine dioxide (as ClO_2)	MRDLG=0.81	MRDL=0.81

Inorganic Chemicals

Constituent	MCLG[1] (mg/L)[2]	MCL or TT[1] (mg/L)[2]
Antimony	0.006	0.006
Arsenic	7	0.01 as of 01/23/06
Asbestos (fiber $>10\ \mu m$)	7 million fibers/L	7 MFL
Barium	2	2
Beryllium	0.004	0.004
Cadmium	0.005	0.005
Chromium (total)	0.1	0.1
Copper	1.3	TT[8]; Action Level=1.3
Cyanide (as free cyanide)	0.2	0.2
Fluoride	4	4
Lead	zero	TT[8]; Action Level=0.015
Mercury (inorganic)	0.002	0.002
Nitrate (measured as Nitrogen)	10	10
Nitrite (measured as Nitrogen)	1	1
Selenium	0.05	0.05
Thallium	0.0005	0.002

(Continued)

Table 8B.14 (Continued)

Constituent	MCLG[1] (mg/L)[2]	MCL or TT[1] (mg/L)[2]
Organic Chemicals		
Acrylamide	zero	TT[9]
Alachlor	zero	0.002
Atrazine	0.003	0.003
Benzene	zero	0.005
Benzo(a)pyrene (PAHs)	zero	0.0002
Carbofuran	0.04	0.04
Carbon tetrachloride	zero	0.005
Chlordane	zero	0.002
Chlorobenzene	0.1	0.1
2,4-D	0.07	0.07
Dalapon	0.2	0.2
1,2-Dibromo-3-chloropropane (DBCP)	zero	0.0002
o-Dichlorobenzene	0.6	0.6
p-Dichlorobenzene	0.075	0.075
1,2-Dichloroethane	zero	0.005
1,1-Dichloroethylene	0.007	0.007
cis-1,2-Dichloroethylene	0.07	0.07
trans-1,2-Dichloroethylene	0.1	0.1
Dichloromethane	zero	0.005
1,2-Dichloropropane	zero	0.005
Di(2-ethylhexyl) adipate	0.4	0.4
Di(2-ethylhexyl) phthalate	zero	0.006
Dinoseb	0.007	0.007
Dioxin (2,3,7,8-TCDD)	zero	0.00000003
Diquat	0.02	0.02
Endothall	0.1	0.1
Endrin	0.002	0.002
Epichlorohydrin	zero	TT[9]
Ethylbenzene	0.7	0.7
Ethylene dibromide	zero	0.00005
Glyphosate	0.7	0.7
Heptachlor	zero	0.0004
Heptachlor epoxide	zero	0.0002
Hexachlorobenzene	zero	0.001
Hexachlorocyclopentadiene	0.05	0.05
Lindane	0.0002	0.0002
Methoxychlor	0.04	0.04
Oxamyl (Vydate)	0.2	0.2
Polychlorinated biphenyls (PCBs)	zero	0.0005
Pentachlorophenol	zero	0.001
Picloram	0.5	0.5
Simazine	0.004	0.004
Styrene	0.1	0.1
Tetrachloroethylene	zero	0.005
Toluene	1	1
Toxaphene	zero	0.003
2,4,5-TP (Silvex)	0.05	0.05
1,2,4-Trichlorobenzene	0.07	0.07
1,1,1-Trichloroethane	0.2	0.2
1,1,2-Trichloroethane	0.003	0.005
Trichloroethylene	zero	0.005
Vinyl chloride	zero	0.002
Xylenes (total)	10	10

(Continued)

Table 8B.14 (Continued)

Constituent	MCLG[1] (mg/L)[2]	MCL or TT[1] (mg/L)[2]
Radionuclides		
Alpha particles	none[7] — zero	15 picocuries per Liter (pCi/L)
Beta particles and photon emitters	none[7] — zero	4 millirems/yr
Radium 226 and Radium 228 (combined)	none[7] — zero	5 pCi/L
Uranium	zero	30 ug/L as of 12/08/03

Note:

[1]Definitions:

Maximum Contaminant Level (MCL)—The highest level of a contaminant that is allowed in drinking water. MCLs are set as close to MCLGs as feasible using the best available treatment technology and taking cost into consideration. MCLs are enforceable standards.

Maximum Contaminant Level Goal (MCLG)—The level of a contaminant in drinking water below which there is no known or expected risk to health. MCLGs allow for a margin of safety and are non-enforceable public health goals.

Maximum Residual Disinfectant Level (MRDL)—The highest level of a disinfectant allowed in drinking water. There is convincing evidence that addition of a disinfectant is necessary for control of microbial contaminants.

Maximum Residual Disinfectant Level Goal (MRDLG)—The level of a drinking water disinfectant below which there is no known or expected risk to health. MRDLGs do not reflect the benefits of the use of disinfectants to control microbial contaminants.

Treatment Technique—A required process intended to reduce the level of a contaminant in drinking water.

[2]Units are in milligrams per liter (mg/L) unless otherwise noted. Milligrams per liter are equivalent to parts per million.

[3]EPA's surface water treatment rules require systems using surface water or groundwater under the direct influence of surface water to (1) disinfect their water, and (2) filter their water or meet criteria for avoiding filtration so that the following contaminants are controlled at the following levels:

Cryptosporidium: (as of 1/1/02 for systems serving >10,000 and 1/14/05 for systems serving <10,000) 99% removal;
Giardia lamblia: 99.9% removal/inactivation;

Viruses: 99.99% removal/inactivation;

Legionella: No limit, but EPA believes that if *Giardia* and viruses are
Turbidity: At no time can turbidity (cloudiness of water) go above 5 nephelolometric
HPC: No more than 500 bacterial colonies per milliliter.

Long Term 1 Enhanced Surface Water Treatment (Effective Date: January 14, 2005); Surface water systems or (GWUDI) systems serving fewer than 10,000 people must comply with the applicable Long Term 1 Enhanced Surface Water Treatment Rule provisions (e.g. turbidity standards, individual filter monitoring, *Cryptosporidium* removal requirements, updated watershed control requirements for unfiltered systems).

Filter Backwash Recycling; The Filter Backwash Recycling Rule requires systems that recycle to return specific recycle flows through all processes of the system's existing conventional or direct filtration system or at an alternate location approved by the state.

[4]More than 5.0% samples total coliform-positive in a month. (For water systems that collect fewer than 40 routine samples per month, no more than one sample can be total coliform-positive per month.) Every sample that has total coliform must be analyzed for either fecal coliforms or *E. coli* if two consecutive TC-positive samples, and one is also positive for *E.coli* fecal coliforms, system has an acute MCL violation.

(Continued)

Table 8A.14 (Continued)

[5]Fecal coliform and *E. coli* are bacteria whose presence indicates that the water may be contaminated with human or animal wastes. Disease-causing microbes (pathogens) in these wastes can cause diarrhea, cramps, nausea, headaches, or other symptoms. These pathogens may pose a special health risk for infants, young children, and people with severely compromised immune systems.
[6]Although there is no collective MCLG for this contaminant group, there are individual MCLGs for some of the individual contaminants:

Trihalomethanes: bromodichloromethane (zero); bromoform (zero); dibromochloromethane (0.06 mg/L). Chloroform is regulated with this group but has no MCLG.

Haloacetic acids: dichloroacetic acid (zero); trichloroacetic acid (0.3 mg/L). Monochloroacetic acid, bromoacetic acid, and dibromoacetic acid are regulated with this group but have no MCLGs.

[7]MCLGs were not established before the 1986 Amendments to the Safe Drinking Water Act. Therefore, there is no MCLG for this contaminant.

[8]Lead and copper are regulated by a Treatment Technique that requires systems to control the corrosiveness of their water. If more than 10% of tap water samples exceed the action level, water systems must take additional steps. For copper, the action level is 1.3 mg/L, and for lead is 0.015 mg/L.

[9]Each water system must certify, in writing, to the state (using third-party or manufacturer's certification) that when acrylamide and epichlorohydrin are used in drinking water systems, the combination (or product) of dose and monomer level does not exceed the levels specified, as follows:

Acrylamide = 0.05% dosed at 1 mg/L (or equivalent);
Epichlorohydrin = 0.01% dosed at 20 mg/L (or equivalent)

Source: From United States Environmental Protection Agency, www.epa.gov.

Table 8B.15 National Secondary Drinking Water Standards

Constituent	SMCL Level
Aluminum	0.05 to 0.2 mg/L
Chloride	250 mg/L
Color	15 (color units)
Copper	1.0 mg/L
Corrosivity	Noncorrosive
Fluoride	2.0 mg/L
Foaming Agents	0.5 mg/L
Iron	0.3 mg/L
Manganese	0.05 mg/L
Odor	3 threshold odor number
pH	6.5–8.5
Silver	0.10 mg/L
Sulfate	250 mg/L
Total Dissolved Solids	500 mg/L
Zinc	5 mg/L

Note: National Secondary Drinking Water Regulations (NSDWRs or secondary standards) are nonenforceable guidelines regulating contaminants that may cause cosmetic effects (such as skin or tooth discoloration) or aesthetic effects (such as taste, odor, or color) in drinking water. EPA recommends secondary standards to water systems but does not require systems to comply. However, states may choose to adopt them as enforceable standards.

Source: From United States Environmental Protection Agency, www.epa.gov.

Table 8B.16 National Proposed MRDLGs, MRDLs, MCLGs, MCLs, AND AMCLs for Radon, Disinfectanct Residuals, and Disinfection Byproducts

Radon	MCLG	MCL	AMCL
Radon	zero	300 pCi/L	4000 pCi/L
Stage 1 Disinfectants and Disinfection Byproducts Rule			
Disinfectant Residual	**MRDLG (mg/L)**	**MRDL (mg/L)**	**Compliance Based on**
Chlorine	4 (as Cl_2)	4.0 (as Cl_2)	Annual Average
Chloramine	4 (as Cl_2)	4.0 (as Cl_2)	Annual Average
Chlorine Dioxide	0.8 (as ClO_2)	0.8 (as ClO_2)	Daily Samples
Disinfection Byproducts	**MCLG (mg/L)**	**MCL (mg/L)**	**Compliance Based on**
Total trihalomethanes (TTHM)[a]	N/A	0.080	Annual Average
Chloroform	***		
Bromodichloromethane	0		
Dibromochloromethane	0.06		
Bromoform	0		
Haloacetic acids (five) (HAA5)[b]	N/A	0.060	Annual Average
Dichloroacetic acid	0		
Trichloroacetic acid	0.3		
Chlorite	0.8	1.0	Monthly Average
Bromate	0	0.010	Annual Average

Notes: N/A-Not applicable because there are individual MCLGs for TTHMs or HAAs; MRDLGs, Maximum residual disinfectant level goals; MRDLs, Maximum residual disinfectant level; MCLGs, Maximum contaminant level goal; MCLs, Maximum contaminant level; AMCL, Alternate Maximum Contaminant Level; pCi/L, picoCuries per liter; mg/L, milligrams per liter.

[a] Total trihalomethanes is the sum of the concentrations of chloroform, bromodichloromethane, dibromochloromethane, and bromoform.

[b] Haloacetic acids (five) is the sum of the concentrations of mono-, di-, and trichloroacetic acids and mono—and dibromoacetic acids.

Source: From United States Environmental Protection Agency, www.epa.gov.

Table 8B.17 Summary of State Drinking Water Quality Standards That Differ from USEPA Standards

Compound (mg/L)	Alabama	California	Connecticut	Delaware	Florida	Hawaii	Illinois	Massachusetts	New Hampshire	New Jersey	New York	North Carolina	Pennsylvania	Utah	Wisconsin
aldicarb	0.003	—	—	—	—	—	0.002	—	—	—	0.003	0.002	—	—	—
aldicarb sulfone	0.002	—	—	—	—	—	0.002	—	—	—	0.002	0.003	—	—	—
aldicarb sulfoxide	0.004	—	—	—	—	—	0.004	—	—	—	0.004	0.004	—	—	—
aldrin	—	—	—	—	—	—	0.001	—	—	—	—	—	—	—	—
aluminum	—	1	—	—	—	—		—	—	—	—	—	—	—	—
atrazine	—	0.003	—	—	—	—	—	—	—	—	—	—	—	—	—
barium	—	1	—	—	—	—	—	—	—	—	—	—	—	—	—
bentazon	—	0.018	—	—	—	—	—	—	—	—	—	—	—	—	—
benzene	—	0.001	—	—	0.001	—	—	—	—	0.001	—	—	—	—	—
bis(2-ethylhexyl)phthalate	—	0.004	—	—	—	—	—	—	—	—	—	—	—	—	—
carbofuran	—	0.018	—	—	—	—	—	—	—	—	—	—	—	—	—
carbon tetrachloride	—	0.0005	—	—	0.003	—	—	—	—	0.002	—	—	—	—	—
chlordane	—	0.0001	—	—	—	—	—	—	—	0.0005	—	—	—	—	—
chlorobenzene	—	0.07	—	—	—	—	—	—	—	0.05	—	—	—	—	—
chromium	—	0.05	—	—	—	—	—	—	—	—	—	—	—	—	—
cis-1, 2-dichloroethylene	—	0.006	—	—	—	—	—	—	—	—	—	—	—	—	—
2,4-D	—	—	—	—	—	—	0.01	—	—	—	0.05	—	—	—	—
DDT	—	—	—	—	—	—	0.05	—	—	—	—	—	—	—	—
1, 2-dibromo-3-chloropropane	—	0.0002	—	—	—	0.00004	—	—	—	—	—	—	—	—	—
1,2-dichlorobenzene	—	—	—	—	—	—	—	—	—	0.6	—	—	—	—	—
1,3-dichlorobenzene	—	—	—	—	—	—	—	—	—	0.6	—	—	—	—	—
1,4-dichlorobenzene	—	0.005	—	—	—	—	—	0.005	—	—	—	—	—	—	—
1,1-dichloroethane	—	0.005	—	—	—	—	—	—	—	0.05	—	—	—	—	—
1,2-dichloroethane	—	0.0005	—	—	0.003	—	—	—	—	0.002	—	—	—	—	—
1,1-dichloroethylene	—	0.006	—	—	—	—	—	—	—	0.002	—	—	—	—	—
dichloromethane	—	—	—	—	—	—	—	—	—	0.002	—	—	—	—	—
1,2-dichloropropane	—	0.005	—	—	—	—	—	—	—	—	—	—	—	—	—
1,3-dichloropropene	—	0.0005	—	—	—	—	—	—	—	—	—	—	—	—	—
dieldrin	—	—	—	—	—	—	0.001	—	—	—	—	—	—	—	—
endrin	—	—	—	—	—	—	—	—	—	—	0.0002	—	—	—	—
ethion	—	—	—	—	—	—	—	—	—	—	—	—	—	—	—
ethylene dibromide	—	—	—	—	0.00002	0.00004	—	0.00002	—	—	—	—	—	—	—
fluoride	—	2	—	1.8	—	1.4–2.4	—	—	—	—	—	4	2	—	—

formaldehyde	—	0.03	—	—	—	—	—	—	—	—	—	—	—	—	—
heptachlor	—	0.00001	—	—	—	—	0.0001	—	—	—	—	—	—	—	—
heptachlor epoxide	—	0.00001	—	—	—	—	0.0001	—	—	—	—	—	—	—	—
iron	—	—	—	—	—	—	1	—	—	—	—	0.3	—	—	—
isopropanol	—	—	—	—	—	—	—	—	—	—	—	—	—	—	—
manganese	—	—	—	—	—	—	0.15	—	—	—	0.3	0.05	—	—	—
methyl t-butyl ether	—	0.005	—	—	—	—	—	—	—	0.07	—	—	—	—	—
molinate	—	0.02	—	—	—	—	—	—	—	—	—	—	—	—	—
naphthalene	—	—	—	—	—	—	—	—	—	0.3	—	—	—	—	—
polychlorinated biphenyls	—	—	—	—	—	—	—	—	—	0.0005	—	—	—	—	—
silver	—	—	0.05	—	—	—	—	—	—	—	—	—	—	—	—
sodium	—	—	—	—	0.16	—	—	—	—	—	—	—	—	—	—
strontium 90	0.008	0.008	0.008	—	—	—	0.008	—	0.008	—	—	—	—	—	0.008
sulfate	500	—	—	—	—	—	—	—	—	—	—	—	—	1,000	—
tetrachloroethylene	—	—	—	—	0.003	—	—	—	—	0.001	—	—	—	—	—
1,1,2,2-tetrachloroethane	—	0.001	—	—	—	—	—	—	—	0.001	—	—	—	—	—
thiobencarb	—	0.07	—	—	—	—	—	—	—	—	—	—	—	—	—
2,4,5-TP	—	—	—	—	—	—	—	—	—	—	0.01				
toluene	—	0.15	—	—	—	—	—	—	—	—	—	—	—	—	1
1,2,4-trichloro-benzene	—	—	—	—	—	—	—	—	—	0.009	—	—	—	—	—
1,1,1-trichloroethane	—	—	—	—	—	—	—	—	—	0.03	—	—	—	—	—
1,1,2-trichloroethane	—	—	—	—	—	—	—	—	—	0.003	—	—	—	—	—
trans-1,2-dichloroethylene	—	0.01	—	—	—	—	—	—	—	—	—	—	—	—	—
trichloroethylene	—	—	—	—	0.003	—	—	—	—	0.001	—	—	—	—	—
trichlorofluoro-methane	—	0.15	—	—	—	—	—	—	—	—	—	—	—	—	—
1,2,3-trichloropropane	—	—	—	—	—	0.0008	—	—	—	—	—	—	—	—	—
1,1,2-trichloro-1,2,2~trifluoro-ethane	—	1.2	—	—	—	—	—	—	—	—	—	—	—	—	—
tritium	20	20	20	—	—	—	20	—	20	—	—	—	—	—	20
uranium	—	0.02	—	—	—	—	—	—	—	—	—	—	—	—	—
vinyl chloride	—	0.0005	—	—	0.001	—	—	—	—	0.002	—	—	—	—	0.0002

(Continued)

Table 8B.17 (Continued)

Compound (mg/L)	Alabama	California	Connecticut	Delaware	Florida	Hawaii	Iillinois	Massachusetts	New Hampshire	New Jersey	New York	North Carolina	Pennsylvania	Utah	Wisconsin
2-xylene	—	1.75	—	—	—	—	—	—	—	—	—	—	—	—	—
3-xylene	—	1.75	—	—	—	—	—	—	—	—	—	—	—	—	—
4-xylene	—	1.75	—	—	—	—	—	—	—	—	—	—	—	—	—
xylene	—	—	—	—	—	—	—	—	—	1	—	—	—	—	10
zinc, elemental	—	—	—	—	—	—	5	—	—	—	—	—	—	—	—

Note: Concentrations are in milligram per liter (mg/L).

Source: From Data Bank Update Committee, Federal-State Toxicology and Risk Analysis Committee (FSTRAC), *Summary of State and Federal Drinking Water Standards and Guidelines 1998–1999*, www.sis.nlm.nih.gov.

Table 8B.18 Public Water Supply Standard

	IOCs, SOCs, VOCs	Second Cycle									Third Cycle								
		1st Period			2nd Period			3rd Period			1st Period			2nd Period			3rd Period		
		2002	2003	2004	2005	2006	2007	2008	2009	2010	2011	2012	2013	2014	2015	2016	2017	2018	2019
Inorganic Contaminants (IOCs)[1]	**Groundwater (Below MCL)**																		
	Waiver[2]					*									*				
	No Waiver		*			*			*			*			*			*	
	Surface Water (Below MCL)																		
	Waiver[2]					*									*				
	No Waiver	*	*	*	*	*	*	*	*	*	*	*	*	*	*	*	*	*	*
	Groundwater and Surface Water (Above MCL)[3]																		
	Reliably and Consistently ≤ MCL for Groundwater Systems		*			*			*			*			*			*	
	Reliably and Consistently ≤ MCL for Surface Water Systems	*	*	*	*	*	*	*	*	*	*	*	*	*	*	*	*	*	*
	> MCL or Not Reliably and Consistently ≤ MCL	****	****	****	****	****	****	****	****	****	****	****	****	****	****	****	****	****	****
Synthetic Organic Contaminants (SOCs)	**Population >3,300 (Below Detection Limit)**	**02**	**03**	**04**	**05**	**06**	**07**	**08**	**09**	**10**	**11**	**12**	**13**	**14**	**15**	**16**	**17**	**18**	**19**
	Waiver		×			×			×			×			×			×	
	< Detect and No Waiver		**			**			**			**			**			**	
	Population 3,300 (Below Detection Limit)																		
	Waiver		×			×			×			×			×			×	
	< Detect and No Waiver		*			*			*			*			*			*	
	Above Detection Limit																		
	Reliably and Consistently ≤ MCL[4]	*	*	*	*	*	*	*	*	*	*	*	*	*	*	*	*	*	*
	≥ Detect or Not Reliably and Consistently ≤ MCL	****	****	****	****	****	****	****	****	****	****	****	****	****	****	****	****	****	****
		02	**03**	**04**	**05**	**06**	**07**	**08**	**09**	**10**	**11**	**12**	**13**	**14**	**15**	**16**	**17**	**18**	**19**
Volatile Organic Contaminants (VOCs)	**Groundwater (Below Detection Limit)**																		
	< Detect, Vulnerability Assessment, and Waiver[5]				*						*						*		
	No Waiver[6]	*	*	*	*	*	*	*	*	*	*	*	*	*	*	*	*	*	*
	Surface Water (Below Detection Limit)																		
	< Detect, Vulnerability Assessment, and Waiver[7]		×			×			×			×			×			×	
	No Waiver[8]	*	*	*	*	*	*	*	*	*	*	*	*	*	*	*	*	*	*
	Above Detection Limit																		
	Reliably and Consistently < MCL[4]	*	*	*	*	*	*	*	*	*	*	*	*	*	*	*	*	*	*
	≥ Detect or Not Reliably and Consistently ≤ MCL	****	****	****	****	****	****	****	****	****	****	****	****	****	****	****	****	****	****

Table 8B.18 (Continued)

	EXCEPTIONS	Second Cycle									Third Cycle								
		1st Period			2nd Period			3rd Period			1st Period			2nd Period			3rd Period		
Nitrate		2002	2003	2004	2005	2006	2007	2008	2009	2010	2011	2012	2013	2014	2015	2016	2017	2018	2019
	CWSs & NTNCWSs																		
	Surface Water with 4 Quarters of Results < 1/2 MCL[9]	*	*	*	*	*	*	*	*	*	*	*	*	*	*	*	*	*	*
	Groundwater Reliably and Consistently < MCL[9]	*	*	*	*	*	*	*	*	*	*	*	*	*	*	*	*	*	*
	≥ 1/2 MCL	****	****	****	****	****	****	****	****	****	****	****	****	****	****	****	****	****	****
	TNCWSs																		
	Standard Monitoring	*	*	*	*	*	*	*	*	*	*	*	*	*	*	*	*	*	*
Nitrite		02	03	04	05	06	07	08	09	10	11	12	13	14	15	16	17	18	19
	< 1/2 MCL					#									#				
	Reliably and Consistently < MCL[9]	*	*	*	*	*	*	*	*	*	*	*	*	*	*	*	*	*	*
	≥ 1/2 MCL or Not Reliably and Consistently < MCL	****	****	****	****	****	****	****	****	****	****	****	****	****	****	****	****	****	****
Radio nuclides		02	03	04	05	06	07	08	09	10	11	12	13	14	15	16	17	18	19
	< Detection Limit					****						*						*	
	≥ Detection Limit but ≤ 1/2 MCL		!			****				*							*		
	> 1/2 MCL but ≤ MCL					****			*			*			*			*	
	> MCL			****	****	****	****	****	****	****	****	****	****	****	****	****	*** *	****	****
Asbestos		02	03	04	05	06	07	08	09	10	11	12	13	14	15	16	17	18	19
	Waiver		×			×			×			×			×			×	
	No Waiver, Reliably and Consistently ≤ MCL, or Vulnerable to Asbestos Contamination[10]					*									*				
	> MCL	****	****	****	****	****	****	****	****	****	****	****	****	****	****	****	****	****	****

Legend
* = 1 sample at each entry point to distribution system (EPTDS).
** = 2 quarterly samples at each EPTDS. Samples must be taken during 1 calendar year during each 3-year compliance period.
**** = 4 quarterly samples at each EPTDS within time frame designated by the primacy agency.
× = No sampling required unless required by the primacy agency.
= Systems must monitor at a frequency specified by the primacy agency.
! = When allowed by the primacy agency, data collected between June 2000 and December 8, 2003 may be grandfathered to satisfy the initial monitoring requirements due in 2004 for gross alpha, radium 226/228, and uranium.

[1]Until January 22, 2006 the maximum contaminant level (MCL) for arsenic is 50 μg/L; on January 23, 2006 the MCL for arsenic becomes 10 μg/L.

[2]Based on 3 rounds of monitoring at each EPTDS with all analytical results below the MCL. Waivers are not permitted under the current arsenic requirements, however systems are eligible for arsenic waivers after January 23, 2006.

[3]A system with a sampling point result above the MCL must collect quarterly samples, at that sampling point, until the system is determined by the primacy agency to be reliably and consistently below the MCL.

[4]Samples must be taken during the quarter which previously resulted in the highest analytical result. Systems can apply for a waiver after 3 consecutive annual sampling results are below the detection limit.

[5]Groundwater systems must update their vulnerability assessments during the time the waiver is effective. Primacy agencies must re-confirm that the system is nonvulnerable within 3 years of the initial determination or the system must return to annual sampling.

[6]If all monitoring results during initial quarterly monitoring are less than the detection limit, the system can take annual samples. If after a minimum of 3 years of annual sampling with all analytical results less than the detection limit, the primacy agency can allow a system to take 1 sample during each compliance period. Systems are also eligible for a waiver.

[7]Primacy agencies must determine that a surface water system is nonvulnerable based on a vulnerability assessment during each compliance period or the system must return to annual sampling.

[8]If all monitoring results during initial quarterly monitoring are less than the detection limit, the system can take annual samples. Systems are also eligible for a waiver.

[9]Samples must be taken during the quarter which previously resulted in the highest analytical result.

[10]Systems are required to monitor for asbestos during the first 3-year compliance period of each 9-year compliance cycle. A system vulnerable to asbestos contamination due solely to corrosion of asbestos-cement pipe must take 1 sample at a tap served by that pipe. A system vulnerable to asbestos contamination at the source must sample at each EPTDS.

Note: The Standardized Monitoring Framework (SMF), was promulgated in the Phase II Rule on January 30, 1991 (56 FR 3526). The purpose of the SMF is to standardize, simplify, and consolidate monitoring requirements across contaminant groups. The SMF increases public health protection by simplifying monitoring plans and synchronizing monitoring schedules leading to increased compliance with monitoring requirements. The SMF reduces the variability within monitoring requirements for chemical and radiological contaminants across system sizes and types. The SMF summarizes existing systems' ongoing federal monitoring requirements only. Primacy agencies have the flexibility to issue waivers, with EPA approval, which take into account regional and state specific characteristics and concerns. To determine exact monitoring requirements, the SMF must be used in conjunction with any EPA approved waiver and additional requirements as determined by the primacy agency. New water systems may have different and additional requirements as determined by the primacy agency.

Regulated Contaminants:
Inorganic Contaminants (IOCs) — Fifteen (15) (Nitrate, Nitrite, total Nitrate/Nitrite, and Asbestos are exceptions to SMF)
Synthetic Organic Contaminants (SOCs) & Volatile Organic Contaminants (VOCs) — Fifty-One (51)
Radionuclides — Four (4)

Utilities Covered:
All PWS must monitor for Nitrate and Nitrite
CWSs must monitor for IOCs, SOCs, VOCs, and Radionuclides
NTNCWSs must monitor for IOCs SOCs, and VOCs.

Source: From United States Environmental Protection Agency, www.epa.gov.

Table 8B.19 Drinking Water Priority Rulemaking: Microbial and Disinfection Byproduct Rules Summary

Existing M–DBP Regulations

- Microbial Contaminants: The Surface Water Treatment Rule (SWTR), promulgated in 1989, applies to all public water systems using surface water sources or groundwater sources under the direct influence of surface water. It establishes maximum contaminant level goals (MCLGs) for viruses, bacteria and *Giardia lamblia.* It also includes treatment technique requirements for filtered and unfiltered systems that are specifically designed to protect against the adverse health effects of exposure to these microbial pathogens. The Total Coliform Rule, revised in 1989, applies to all PWSs and establishes a maximum contaminant level (MCL) for total coliforms
- Disinfection Byproducts: In 1979, EPA set an interim MCL for total trihalomethanes of 0.10 mg/L as an annual average. This applies to any community water system serving at least 10,000 people that adds a disinfectant to the drinking water during any part of the treatment process

Information Collection Rule (ICR): To support the M–DBP rulemaking process, the ICR required large public water systems serving at least 100,000 people to monitor and collect data on microbial contaminants, disinfectants and disinfection byproducts for 18 months. The data provide EPA with information about disinfection byproducts, disease-causing microorganisms, including *Cryptosporidium*, and engineering data to control these contaminants

Interim Enhanced Surface Water Treatment Rule (IESWTR):

The IESWTR applies to systems using surface water, or groundwater under the direct influence of surface water, that serve 10,000 or more persons. The rule also includes provisions for states to conduct sanitary surveys for surface water systems regardless of system size. The rule builds upon the treatment technique requirements of the SWTR with the following key additions and modifications

- Maximum contaminant level goal (MCLG) of zero for *Cryptosporidium*
- 2-log *Cryptosporidium* removal requirements for systems that filter
- Strengthened combined filter effluent turbidity performance standards
- Individual filter turbidity monitoring provisions
- Disinfection profiling and benchmarking provisions
- Systems using groundwater under the direct influence of surface water now subject to the new rules dealing with *Crypdosporidium*
- Inclusion of *Cryptosporidium* in the watershed control requirements for unfiltered public water systems
- Requirements for covers on new finished water reservoirs
- Sanitary surveys, conducted by states, for all surface water systems regardless of size

The IESWTR, with tightened turbidity performance criteria and required individual filter monitoring, is designed to optimize treatment reliability and to enhance physical removal efficiencies to minimize the *Cryptosporidium* levels in finished water. In addition, the rule includes disinfection benchmark provisions to assure continued levels of microbial protection while facilities take the necessary steps to comply with new DBP standards

Stage 1 Disinfectants and Disinfection Byproducts Rule (DBPR):

The final Stage 1 DBPR applies to community water systems and nontransient noncommunity systems, including those serving fewer than 10,000 people, that add a disinfectant to the drinking water during any part of the treatment process

The final Stage 1 DBPR includes the following key provisions

- Maximum residual disinfectant level goals (MRDLGs) for chlorine (4 mg/L), chloramines (4 mg/L), and chlorine dioxide (0.8 mg/L)
- Maximum contaminant level goals (MCLGs) for four trihalomethanes (chloroform (zero), bromodichloromethane (zero), dibromochloromethane (0.06 mg/L), and bromoform (zero)), two haloacetic acids (dichloroacetic acid (zero) and trichloroacetic acid (0.3 mg/L)), bromate (zero), and chlorite (0.8 mg/L); EPA subsequently removed the zero MCLG for chloroform from its National Primary Drinking Water Regulations, effective May 30, 2000, in accordance with an order of the U.S. Court of Appeals for the District of Columbia Circuit
- MRDLs for three disinfectants (chlorine (4.0 mg/L), chloramines (4.0 mg/L), and chlorine dioxide (0.8 mg/L))
- MCLs for total trihalomethanes—a sum of the four listed above (0.080 mg/L), haloacetic acids (HAA5) (0.060 mg/L)—a sum of the two listed above plus monochloroacetic acid and mono—and dibromoacetic acids), and two inorganic disinfection byproducts (chlorite (1.0 mg/L)) and bromate (0.010 mg/L)); and
- A treatment technique for removal of DBP precursor material

The terms MRDLG and MRDL, which are not included in the SDWA, were created during the negotiations to distinguish disinfectants (because of their beneficial use) from contaminants. The final rule includes monitoring, reporting, and public notification requirements for these compounds. This final rule also describes the best available technology (BAT) upon which the MRDLs and MCLs are based

(Continued)

Table 8B.19 (Continued)

Filter Backwash Recycling Rule (FBRR)

The FBRR requires public water systems (PWSs) to review their backwash water recycling practices to ensure that they do not compromise microbial control. Under the FBRR, recycled filter backwash water, sludge thickener supernatant, and liquids from dewatering processes must be returned to a location such that all processes of a system's conventional or direct filtration including coagulation, flocculation, sedimentation (conventional filtration only) and filtration, are employed. Systems may apply to the State for approval to recycle at an alternate location. The Filter Backwash Rule applies to all public water systems, regardless of size

Long-Term 1 Enhanced Surface Water Treatment Rule (Long-Term 1 ESWTR)

While the Stage 1 DBPR applies to systems of all sizes, the IESWTR only applies to systems serving 10,000 or more people. The Long Term 1 ESWTR, promulgated in January 2002, will strengthen microbial controls for small systems i.e. those systems serving fewer than 10,000 people. The rule will also prevent significant increase in microbial risk where small systems take steps to implement the Stage 1 DBPR

EPA believes that the rule will generally track the approaches in the IESWTR for improved turbidity control, including individual filter monitoring and reporting. The rule will also address disinfection profiling and benchmarking. The Agency is considering what modifications of some large system requirements may be appropriate for small systems

Future M–DBP Rules

Groundwater Rule

EPA has proposed a Groundwater Rule that specifies the appropriate use of disinfection while addressing other components of groundwater systems to ensure public health protection. There are more than 158,000 public groundwater systems. Almost 89 million people are served by community groundwater systems, and 20 million people are served by noncommunity groundwater systems. Ninety-nine percent (157,000) of groundwater systems serve fewer than 10,000 people. However, systems serving more than 10,000 people serve 55% (more than 60 million) of all people who get their drinking water from public groundwater systems. The Groundwater Rule will be promulgated summer 2001

Long-Term 2 Enhanced Surface Water Treatment Rule (Long-Term 2 ESWTR)

EPA is proposing the Long-Term 2 ESWTR to reduce disease incidence associated with Cryptosporidium and other pathogenic microorganisms in drinking water. The Long-Term 2 ESWTR will supplement existing regulations by targeting additional Cryptosporidium treatment requirements to higher risk systems. This proposed regulation also contains provisions to mitigate risks from uncovered finished water storage facilities and to ensure that systems maintain microbial protection as they take steps to reduce the formation of disinfection byproducts. The Long-Term 2 ESWTR will apply to all systems that use surface water or groundwater under the direct influence of surface water

Stage 2 Disinfectants and Disinfection Byproducts Rule (Stage 2 DBPR)

The Stage 2 DBPR is one part of the Stage 2 Microbial and Disinfection Byproducts Rules (M–DBP), which are a set of interrelated regulations that address risks from microbial pathogens and disinfectants/disinfection byproducts (D/DBPs). The Stage 2 M–DBP Rules are the final phase in the M–DBP rulemaking strategy, affirmed by Congress as part of the 1996 Amendments to the SDWA. The Stage 2 DBPR focuses on public health protection by limiting exposure to DBPs, specifically total trihalomethanes (TTHM) and five haloacetic acids (HAA5), which can form in water through disinfectants used to control microbial pathogens. This rule will apply to all community water systems (CWSs) and nontransient noncommunity water systems (NTNCWSs) that add a primary or residual disinfectant other than ultraviolet (UV) light or deliver water that has been disinfected by a primary or residual disinfectant other than UV

Source: From United States Environmental Protection Agency, www.epa.gov.

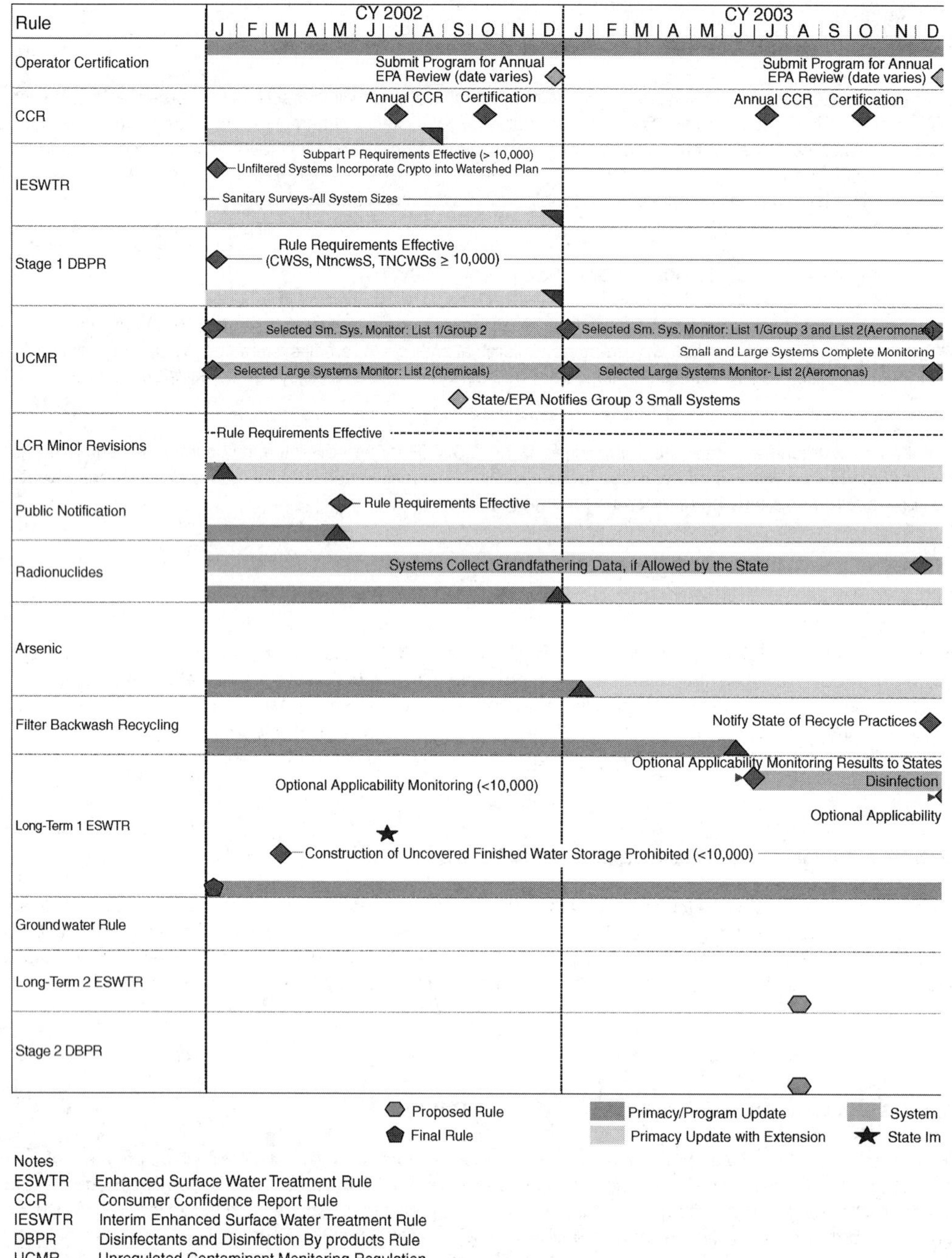

Notes
ESWTR Enhanced Surface Water Treatment Rule
CCR Consumer Confidence Report Rule
IESWTR Interim Enhanced Surface Water Treatment Rule
DBPR Disinfectants and Disinfection By products Rule
UCMR Unregulated Contaminant Monitoring Regulation

Figure 8B.8 Public water supply rule implementation and milestone timeline. (From United States Environmental Protection Agency, www.epa.gov.)

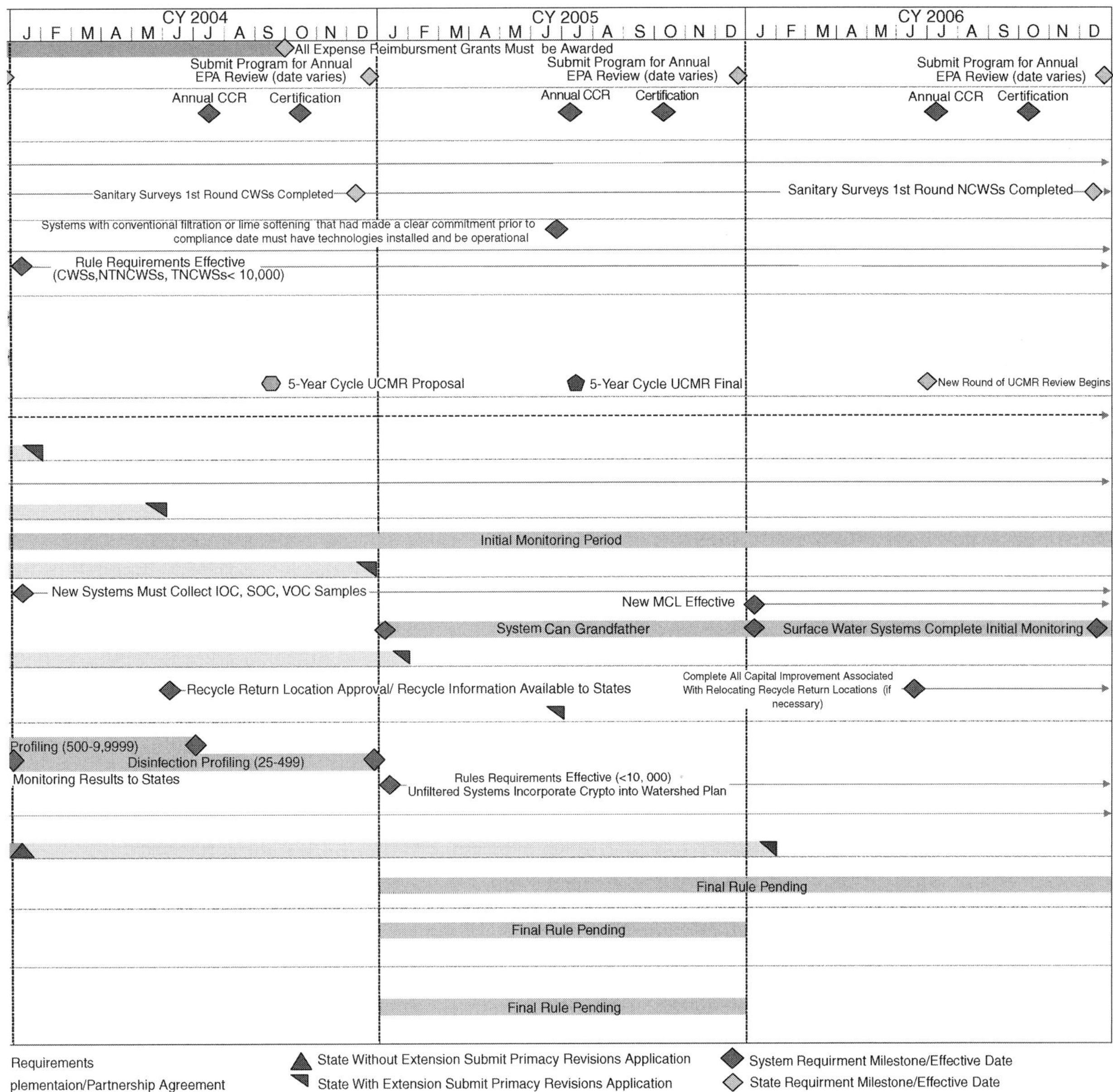
CY 2004
CY 2005
CY 2006
J F M A M J J A S O N D
All Expense Reimbursment Grants Must be Awarded
Submit Program for Annual EPA Review (date varies)
Annual CCR
Certification
Sanitary Surveys 1st Round CWSs Completed
Sanitary Surveys 1st Round NCWSs Completed
Systems with conventional filtration or lime softening that had made a clear commitment prior to compliance date must have technologies installed and be operational
Rule Requirements Effective (CWSs,NTNCWSs, TNCWSs< 10,000)
5-Year Cycle UCMR Proposal
5-Year Cycle UCMR Final
New Round of UCMR Review Begins
Initial Monitoring Period
New Systems Must Collect IOC, SOC, VOC Samples
New MCL Effective
System Can Grandfather
Surface Water Systems Complete Initial Monitoring
Recycle Return Location Approval/ Recycle Information Available to States
Complete All Capital Improvement Associated With Relocating Recycle Return Locations (if necessary)
Profiling (500-9,9999)
Disinfection Profiling (25-499)
Monitoring Results to States
Rules Requirements Effective (<10, 000)
Unfiltered Systems Incorporate Crypto into Watershed Plan
Final Rule Pending
Final Rule Pending
Final Rule Pending
Requirements
plementaion/Partnership Agreement
State Without Extension Submit Primacy Revisions Application
State With Extension Submit Primacy Revisions Application
System Requirment Milestone/Effective Date
State Requirment Milestone/Effective Date

Table 8B.20 Unregulated Chemical Contanminents (1999 List) to Be Monitored in Public Supply Systems in the United States

List 1 Contaminants—Assessment Monitoring		
2,4-dinitrotoluene	DCPA mono-acid degradate	MTBE
2,6-dinitrotoluene	DDE	Nitrobenzene
Acetochlor	EPTC	Perchlorate
DCPA di-acid degradate	Molinate	Terbacil
List 2 Contaminants—Screening Surveys		
1,2-diphenylhydrazine	Diazinon	Prometon
2-methyl-phenol	Disulfoton	RDX
2,4-dichlorophenol	Diuron	Terbufos
2,4-dinitrophenol	Fonofos	*Aeromonas*
2,4,6-trichlorophenol	Linuron	
Alachlor ESA	Nitrobenzene (low-level)[a]	
List 3 Contaminants—Pre-Screen Testing		
Lead-210	Cyanobacteria (blue–green algae), other fresh water algae, and their toxins	Echoviruses
Polonium-210	Caliciviruses	*Helicobacter pylori*
Adenoviruses	Coxsackieviruses	Microsporidia

[a] Nitrobenzene has been added to List 2 from the original UCMR (1999) List to track its occurrence at a concentration lower than the List 1 nitrobenzene minimum reporting level.

Source: From USEPA, 2001, *Unregulated Contaminant Monitoring Regulation Reporting Guidance*, (EPA 815-R-01-029, November 2001), www.epa.gov.

Table 8B.21 Monitoring Requirements for Unregulated Contaminant Program

Contaminant Type	Water Source Type	Timeframe	Frequency
Chemical	Surface water	Twelve (12) months	Four (4) quarterly samples taken as follows: Select either the first, second, or third month of a quarter and sample in that same month of each of four (4) consecutive quarters[a] to ensure that one of those sampling events occurs during the vulnerable time[b]
	Groundwater	Twelve (12) months	Two (2) times in a year taken as follows: Sample during one (1) month of the vulnerable time[b] and during one (1) month five (5) to seven (7) months earlier or later[c]
Microbiological	Surface and groundwater	Twelve (12) months	Six (6) times in a year taken as follows: Select either the first, second, or third month of a quarter and sample in that same month of each of four (4) quarters, and sample an additional two (2) months during the warmest (vulnerable) quarter of the year[d]

[a] "Select either the first, second, or third month of a quarter and sample in that same month of each of four consecutive quarters" means that the system must monitor during each of the 4 months of either: January, April, July, October; or February, May, August, November; or March, June, September, December.

[b] "Vulnerable time" means May 1 through July 31, unless the State or EPA informs the system that it has selected a different time period for sampling as its vulnerable time.

[c] "Sample during one month of the vulnerable time and during one month five to seven months earlier or later" means, for example, that if the system selects May as its "vulnerable time" month to sample, then one month five to seven months earlier would be either October, November, or December of the preceding year, and one month five to seven months later would be either, October, November, or December of the same year.

[d] This means that the system must monitor during each of the six months of either: January, April, July, August, September, October, or February, May, July, August, September, November, or March, June, July, August, September, December; unless the State or EPA informs the system that a different vulnerable quarter has been selected for it.

Source: From USEPA, 2001, *Reference Guide for the Unregulated Contaminant Monitoring Program*, (EPA 815-R-01-023, October 2001), www.epa.gov.

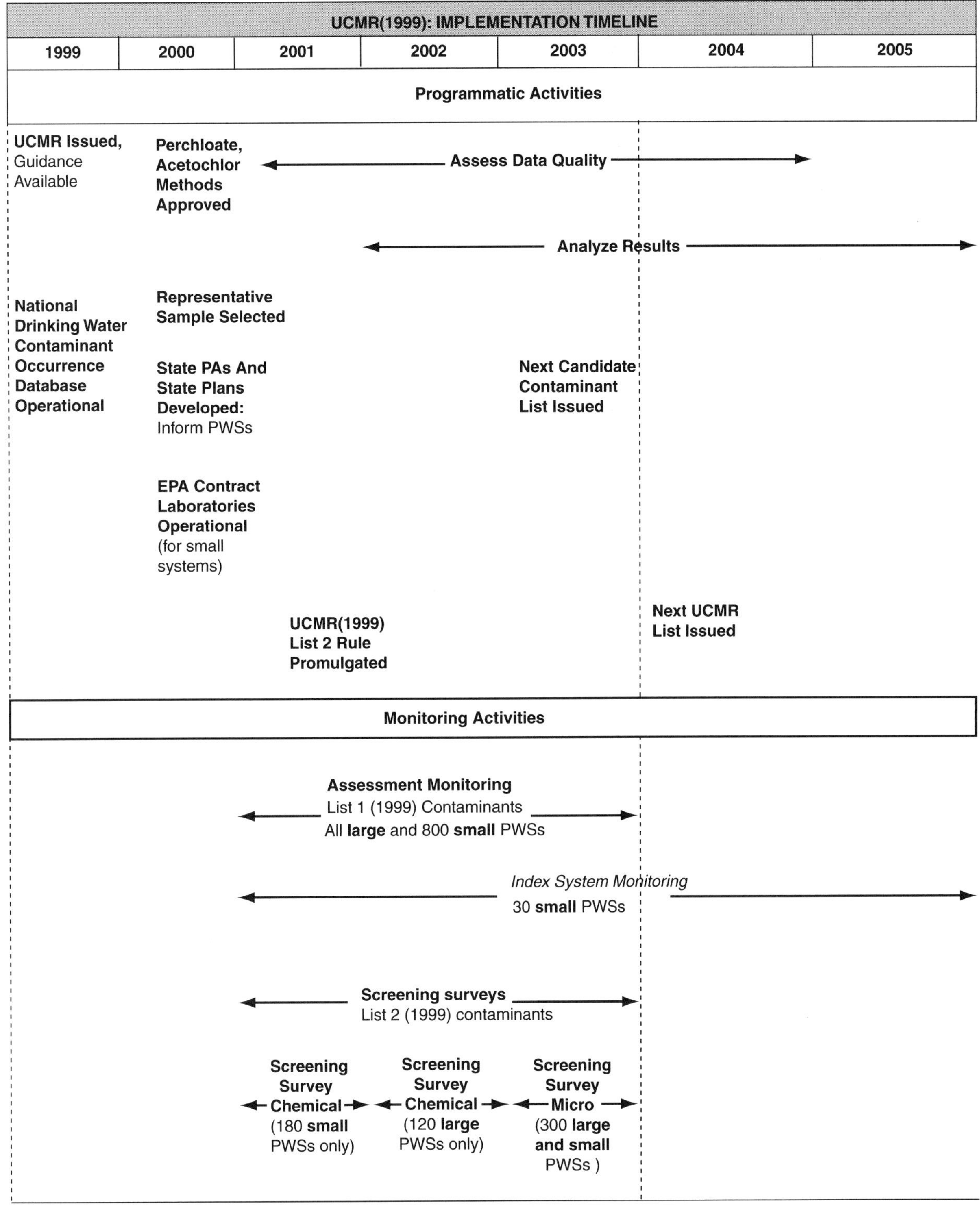

Figure 8B.9 Unregulated chemical contaminant monitoring regulation (1999) timeline and related activities. (From USEPA, 2001, *Reference Guide for the Unregulated Contaminant Monitoring Program*, (EPA 815-R-01-023, October 2001), www.epa.gov.

Table 8B.22 Drinking Water Contaminant Candidate List 2

Microbial Contaminant Candidates

Adenoviruses
Aeromonas hydrophila
Caliciviruses
Coxsackieviruses
Cyanobacteria (blue–green algae), other freshwater algae, and their toxins
Echoviruses
Helicobacter pylori
Microsporidia (Enterocytozoon & Septata)
Mycobacterium avium intracellulare (MAC)

Chemical Contaminant Candidates

1,1,2,2-tetrachloroethane
1,2,4-trimethylbenzene
1,1-dichloroethane
1,1-dichloropropene
1,2-diphenylhydrazine
1,3-dichloropropane
1,3-dichloropropene
2,4,6-trichlorophenol
2,2-dichloropropane
2,4-dichlorophenol
2,4-dinitrophenol
2,4-dinitrotoluene
2,6-dinitrotoluene
2-methyl-Phenol (o-cresol)
Acetochlor
Alachlor ESA & other acetanilide pesticide degradation products
Aluminum
Boron
Bromobenzene
DCPA mono-acid degradate
DCPA di-acid degradate
DDE
Diazinon
Disulfoton
Diuron
EPTC (s-ethyl-dipropylthiocarbamate)
Fonofos
p-Isopropyltoluene (p-cymene)
Linuron
Methyl bromide
Methyl-t-butyl ether (MTBE)
Metolachlor
Molinate
Nitrobenzene
Organotins
Perchlorate
Prometon
RDX
Terbacil
Terbufos
Triazines & degradation products of triazines including, but not limited to Cyanazine and atrazine-desethyl
Vanadium

Source: From United States Environmental Protection Agency, Office of Water (4607m), EPA 815-F-05-001, February 2005, www.epa.gov.

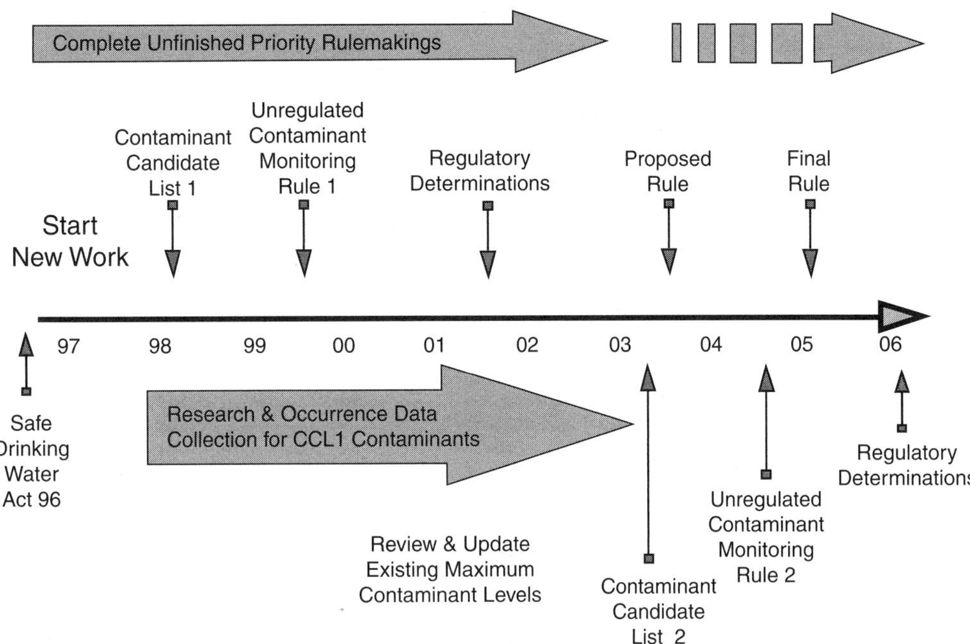

Figure 8B.10 Contaminant candidate identification and selection cycle. (From United States Environmental Protection Agency), www.epa.gov.)

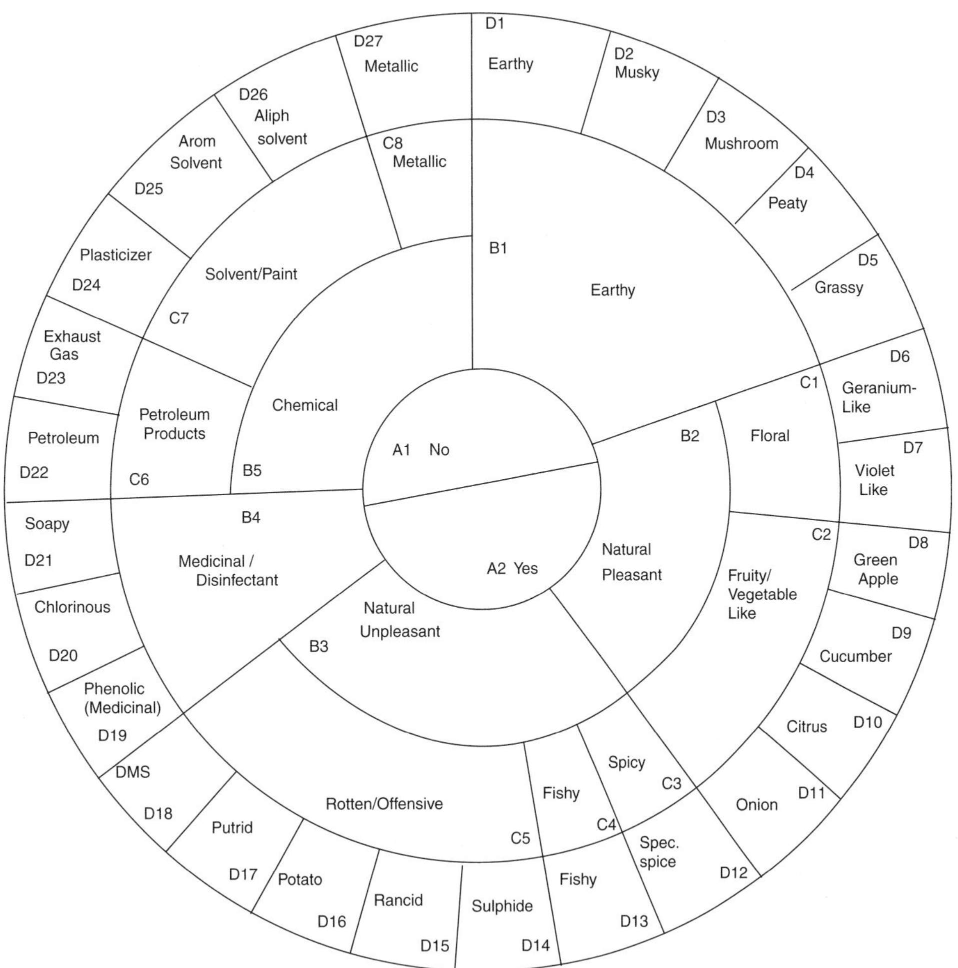

Figure 8B.11 Flavor wheel for drinking water. (From International Association of Water Pollution Research and Control; *Water Quality Bulletin*, Vol. 13, no. 2–3, April–July 1998.)

SECTION 8C DRINKING WATER STANDARDS — WORLD

Table 8C.23 World Health Organization Drinking Water Guideline Values for Chemicals That Are of Health Significance

Chemical	Guideline Value[a] (mg/L)	Remarks
Acrylamide	0.0005[b]	
Alachlor	0.02[b]	
Aldicarb	0.01	Applies to aldicarb sulfoxide and aldicarb sulfone
Aldrin and dieldrin	0.00003	For combined aldrin plus dieldrin
Antimony	0.02	
Arsenic	0.01 (P)	
Atrazine	0.002	
Barium	0.7	
Benzene	0.01[b]	
Benzo[a]pyrene	0.0007[b]	
Boron	0.5 (T)	
Bromate	0.01[b] (A, T)	
Bromodichloromethane	0.06[b]	
Bromoform	0.1	
Cadmium	0.003	
Carbofuran	0.007	
Carbon tetrachloride	0.004	
Chloral hydrate (trichloroacetaldehyde)	0.01 (P)	
Chlorate	0.7 (D)	
Chlordane	0.0002	
Chlorine	5 (C)	For effective disinfection, there should be a residual concentration of free chlorine of ≥ 0.5 mg /L after at least 30 min contact time at pH <8.0
Chlorite	0.7 (D)	
Chloroform	0.2	
Chlorotoluron	0.03	
Chlorpyrifos	0.03	
Chromium	0.05 (P)	For total chromium
Copper	2	Staining of laundry and sanitary ware may occur below guideline value
Cyanazine	0.0006	
Cyanide	0.07	
Cyanogen chloride	0.07	For cyanide as total cyanogenic compounds
2,4-D (2,4-dichlorophenoxyacetic acid)	0.03	Applies to free acid
2,4-DB	0.09	
DDT and metabolites	0.001	
Di(2-ethylhexyl)phthalate	0.008	
Dibromoacetonitrile	0.07	
Dibromochloromethane	0.1	
1,2-Dibromo-3-chloropropane	0.001[b]	
1,2-Dibromoethane	0.0004[b] (P)	
Dichloroacetate	0.05 (T, D)	
Dichloroacetonitrile	0.02 (P)	
Dichlorobenzene, 1,2−	1 (C)	
Dichlorobenzene, 1,4−	0.3 (C)	
Dichloroethane, 1,2−	0.03[b]	
Dichloroethene, 1,1−	0.03	
Dichloroethene, 1,2−	0.05	
Dichloromethane	0.02	
1,2-Dichloropropane (1,2-DCP)	0.04 (P)	
1,3-Dichloropropene	0.02[b]	
Dichlorprop	0.1	
Dimethoate	0.006	
Edetic acid (EDTA)	0.6	Applies to the free acid
Endrin	0.0006	

(Continued)

Table 8C.23 (Continued)

Chemical	Guideline Value[a] (mg/L)	Remarks
Epichlorohydrin	0.0004 (P)	
Ethylbenzene	0.3 (C)	
Fenoprop	0.009	
Fluoride	1.5	Volume of water consumed and intake from other sources should be considered when setting national standards
Formaldehyde	0.9	
Hexachlorobutadiene	0.0006	
Isoproturon	0.009	
Lead	0.01	
Lindane	0.002	
Manganese	0.4 (C)	
MCPA	0.002	
Mecoprop	0.01	
Mercury	0.001	For total mercury (inorganic plus organic)
Methoxychlor	0.02	
Metolachlor	0.01	
Microcystin-LR	0.001 (P)	For total microcystin-LR (free plus cell-bound)
Molinate	0.006	
Molybdenum	0.07	
Monochloramine	3	
Monochloroacetate	0.02	
Nickel	0.02 (P)	
Nitrate (as NO_3^-)	50	Short-term exposure
Nitrilotriacetic acid (NTA)	0.2	
Nitrite (as NO_2^-)	3	Short-term exposure
	0.2 (P)	Long-term exposure
Pendimethalin	0.02	
Pentachlorophenol	0.009[b] (P)	
Pyriproxyfen	0.3	
Selenium	0.01	
Simazine	0.002	
Styrene	0.02 (C)	
2,4,5-T	0.009	
Terbuthylazine	0.007	
Tetrachloroethene	0.04	
Toluene	0.7 (C)	
Trichloroacetate	0.2	
Trichloroethene	0.07 (P)	
Trichlorophenol, 2,4,6−	0.2[b] (C)	
Trifluralin	0.02	
Trihalomethanes		The sum of the ratio of the concentration of each to its respective guideline value should not exceed 1
Uranium	0.015 (P, T)	Only chemical aspects of uranium addressed
Vinyl chloride	0.0003[b]	
Xylenes	0.5 (C)	

[a] P, provisional guideline value, as there is evidence of a hazard, but the available information on health effects is limited; T, provisional guideline value because calculated guideline value is below the level that can be achieved through practical treatment methods, source protection, etc.; A, provisional guideline value because calculated guideline value is below the achievable quantification level; D, provisional guideline value because disinfection is likely to result in the guideline value being exceeded; C, concentrations of the substance at or below the health-based guideline value may affect the appearance, taste or odour of the water, leading to consumer complaints.

[b] For substances that are considered to be carcinogenic, the guideline value is the concentration in drinking water associated with an upper-bound excess lifetime cancer risk of 10^{-5} (one additional cancer per 100000 of the population ingesting drinking-water containing the substance at the guideline value for 70 years). Concentrations associated with upper-bound estimated excess lifetime cancer risks of 10^{-4} and 10^{-6} can be calculated by multiplying and dividing, respectively, the guideline value by 10.

Source: From World Health Organization, 2004, *Guidelines for Drinking-Water Quality*, Third Edition, Volume 1, Recommendations.

Table 8C.24 World Health Organization Drinking Water Guideline Levels for Radionuclides

Radionuclides	Guidance Level (Bq/L)[a]	Radionuclides	Guidance Level (Bq/L)[a]	Radionuclides	Guidance Level (Bq/L)[a]
^{3}H	10000	^{93}Mo	100	^{140}La	100
^{7}Be	10000	^{99}Mo	100	^{139}Ce	1000
^{14}C	100	^{96}Tc	100	^{141}Ce	100
^{22}Na	100	^{97}Tc	1000	^{143}Ce	100
^{32}P	100	^{97m}Tc	100	^{144}Ce	10
^{33}P	1000	^{99}Tc	100	^{143}Pr	100
^{35}S	100	^{97}Ru	1000	^{147}Nd	100
^{36}Cl	100	^{103}Ru	100	^{147}Pm	1000
^{45}Ca	100	^{106}Ru	10	^{149}Pm	100
^{47}Ca	100	^{105}Rh	1000	^{151}Sm	1000
^{46}Sc	100	^{103}Pd	1000	^{153}Sm	100
^{47}Sc	100	^{105}Ag	100	^{152}Eu	100
^{48}Sc	100	^{110m}Ag	100	^{154}Eu	100
^{48}V	100	^{111}Ag	100	^{155}Eu	1000
^{51}Cr	10000	^{109}Cd	100	^{153}Gd	1000
^{52}Mn	100	^{115}Cd	100	^{160}Tb	100
^{53}Mn	10000	^{115m}Cd	100	^{169}Er	1000
^{54}Mn	100	^{111}In	1000	^{171}Tm	1000
^{55}Fe	1000	^{114m}In	100	^{175}Yb	1000
^{59}Fe	100	^{113}Sn	100	^{182}Ta	100
^{56}Co	100	^{125}Sn	100	^{181}W	1000
^{57}Co	1000	^{122}Sb	100	^{185}W	1000
^{58}Co	100	^{124}Sb	100	^{186}Re	100
^{60}Co	100	^{125}Sb	100	^{185}Os	100
^{59}Ni	1000	^{123m}Te	100	^{191}Os	100
^{63}Ni	1000	^{127}Te	1000	^{193}Os	100
^{65}Zn	100	^{127m}Te	100	^{190}Ir	100
^{71}Ge	10000	^{129}Te	1000	^{192}Ir	100
^{73}As	1000	^{129m}Te	100	^{191}Pt	1000
^{74}As	100	^{131}Te	1000	^{193}M	1000
^{76}As	100	^{131m}Te	100	^{198}Au	100
^{77}As	1000	^{132}Te	100	^{199}Au	1000
^{75}Se	100	^{125}I	10	^{197}Hg	1000
^{82}Br	100	^{126}I	10	^{203}Hg	100
^{86}Rb	100	^{129}I	1000	^{200}Tl	1000
^{85}Sr	100	^{131}I	10	^{201}Tl	1000
^{89}Sr	100	^{129}Cs	1000	^{202}Tl	1000
^{90}Sr	10	^{131}Cs	1000	^{204}Tl	100
^{90}Y	100	^{132}Cs	100	^{203}Pb	1000
^{91}Y	100	^{134}Cs	10	^{206}Bi	100
^{93}Zr	100	^{135}Cs	100	^{207}Bi	100
^{95}Zr	100	^{136}Cs	100	^{210}Bi[b]	100
^{93m}Nb	1000	^{137}Cs	10	^{210}Pb[b]	0.1
^{94}Nb	100	^{131}Ba	1000	^{210}Po[b]	0.1
^{95}Nb	100	^{140}Ba	100	^{223}Ra[b]	1
^{224}Ra[b]	1	^{235}U[b]	1	^{242}Cm	10
^{225}Ra	1	^{236}U[b]	1	^{243}Cm	1
^{226}Ra[b]	1	^{237}U	100	^{244}Cm	1
^{228}Ra[b]	0.1	^{238}U[b,c]	10	^{245}Cm	1
^{227}Th[b]	10	^{237}Np	1	^{246}Cm	1
^{228}Th[b]	1	^{239}Np	100	^{247}Cm	1
^{229}Th	0.1	^{236}Pu	1	^{248}Cm	0.1
^{230}Th[b]	1	^{237}Pu	1000	^{249}Bk	100
^{231}Th[b]	1000	^{238}Pu	1	^{246}Cf	100
^{232}Th[b]	1	^{239}Pu	1	^{248}Cf	10
^{234}Th[b]	100	^{240}Pu	1	^{249}Cf	1
^{230}Pa	100	^{241}Pu	10	^{250}Cf	1
^{231}Pa[b]	0.1	^{242}Pu	1	^{251}Cf	1
^{233}Pa	100	^{244}Pu	1	^{252}Cf	1
^{230}U	1	^{241}Am	1	^{253}Cf	100

(Continued)

Table 8C.24 (Continued)

Radionuclides	Guidance Level (Bq/L)[a]	Radionuclides	Guidance Level (Bq/L)[a]	Radionuclides	Guidance Level (Bq/L)[a]
^{231}U	1000	^{242}Am	1000	^{254}Cf	1
^{232}U	1	^{242m}Am	1	^{253}Es	10
^{233}U	1	^{243}Am	1	^{254}Es	10
^{234}U[b]	10			^{254m}Es	100

[a] Guidance levels are rounded according to averaging the log scale values (to 10^n if the calculated value was below 3×10^n and above $3\times10^{n-1}$).
[b] Natural radionuclides.
[c] The provisional guideline value for uranium in drinking-water is 15 μg/litre based on its chemical toxicity for the kidney (www.ec.gc.ca).
Source: From World Health Organization, 2004, *Guidelines for Drinking-Water Quality*, Third Edition, Volume 1, Recommendations. Copyright © World Health Organization 2004, www.who.int.

Table 8C.25 Canadian Guidelines for Drinking Water Quality — Chemical and Physical Parameters

Parameter	Maximum Acceptable Concentration (mg/L)	Aesthetic Objectives (mg/L)
Aldicarb	0.009	
Aldrin + dieldrin	0.0007	
Aluminum[a]		
Antimony	0.006[b]	
Arsenic	0.025	
Atrazine + metabolites	0.005	
Azinphos-methyl	0.02	
Barium	1.0	
Bendiocarb	0.04	
Benzene	0.005	
Benzo[a]pyrene	0.00001	
Boron	5	
Bromate	0.01	
Bromoxynil	0.005	
Cadmium	0.005	
Carbaryl	0.09	
Carbofuran	0.09	
Carbon tetrachloride	0.005	
Chloramines (total)	3.0	
Chloride		≤250
Chlorpyrifos	0.09	
Chromium	0.05	
Colour		≤15 TCU[d]
Copper[b]		≤1.0
Cyanazine	0.01	
Cyanide	0.2	
Cyanobacterial toxins (as microcystin-LR)[c]	0.0015	
Diazinon	0.02	
Dicamba	0.12	
Dichlorobenzene, 1,2–[e]	0.20	≤0.003
Dichlorobenzene, 1,4–[e]	0.005	≤0.001
Dichloroethane, 1,2–	0.005	
Dichloroethylene, 1,1–	0.014	
Dichloromethane	0.05	
Dichlorophenol, 2,4–	0.9	≤0.0003
Dichlorophenoxyacetic acid, 2,4– (2,4–D)	0.1	
Diclofop-methyl	0.009	
Dimethoate	0.02	
Dinoseb	0.01	
Diquat	0.07	
Diuron	0.15	
Ethylbenzene		≤0.0024
Fluoride[f]	1.5	
Glyphosate	0.28	
Iron		≤0.3
Lead[b]	0.010	
Malathion	0.19	

(Continued)

Table 8C.25 (Continued)

Parameter	Maximum Acceptable Concentration (mg/L)	Aesthetic Objectives (mg/L)
Manganese		≤0.05
Mercury	0.001	
Methoxychlor	0.9	
Metolachlor	0.05	
Metribuzin	0.08	
Monochlorobenzene	0.08	≤0.03
Nitrate[g]	45	
Nitrilotriacetic acid (NTA)	0.4	
Odour		Inoffensive
Paraquat (as dichloride)	0.01[h]	
Parathion	0.05	
Pentachlorophenol	0.06	≤0.030
pH		6.5–8.5[i]
Phorate	0.002	
Picloram	0.19	
Selenium	0.01	
Simazine	0.01	
Sodium[j]		≤200
Sulphate[k]		≤500
Sulphide (as H_2S)		≤0.05
Taste		Inoffensive
Temperature		≤15°C
Terbufos	0.001	
Tetrachloroethylene	0.03	
Tetrachlorophenol, 2,3,4,6–	0.1	≤0.001
Toluene		≤0.024
Total dissolved solids (TDS)		≤500
Trichloroethylene	0.05	
Trichlorophenol, 2,4,6–	0.005	≤0.002
Trifluralin	0.045	
Trihalomethanes (total)[l]	0.1	
Turbidity	1 NTU[m]	≤5 NTU[m,n]
Uranium	0.02	
Vinyl chloride	0.002	
Xylenes (total)		≤0.3
Zinc[b]		≤5.0

[a] A health-based guideline for aluminum in drinking water has not been established. However, water treatment plants using aluminum-based coagulants should optimize their operations to reduce residual aluminum levels in treated water to the lowest extent possible as a precautionary measure. *Operational guidance values* of less than 100 μg/L total aluminum for conventional treatment plants and less than 200 μg/L total aluminum for other types of treatment systems are recommended. Any attempt to minimize aluminum residuals must not compromise the effectiveness of disinfection processes or interfere with the removal of disinfection by-product precursors.

[b] Because first-drawn water may contain higher concentrations of metals than are found in running water after flushing, faucets should be thoroughly flushed before water is taken for consumption or analysis.

[c] The guideline is considered protective of human health against exposure to other microcystins (total microcystins) that may also be present.

[d] TCU = true colour unit.

[e] In cases where total dichlorobenzenes are measured and concentrations exceed the most stringent value (0.005 mg/L), the concentrations of the individual isomers should be established.

[f] It is recommended, however, that the concentration of fluoride be adjusted to 0.8–1.0 mg/L, which is the optimum range for the control of dental caries.

[g] Equivalent to 10 mg/L as nitrate-nitrogen. Where nitrate and nitrite are determined separately, levels of nitrite should not exceed 3.2 mg/L.

[h] Equivalent to 0.007 mg/L for paraquation.

[i] No units.

[j] It is recommended that sodium be included in routine monitoring programmes, as levels may be of interest to authorities who wish to prescribe sodium-restricted diets for their patients.

[k] There may be a laxative effect in some individuals when sulphate levels exceed 500 mg/L.

[l] The IMAC for trihalomethanes is expressed as a running annual average. It is based on the risk associated with chloroform, the trihalomethane most often present and in greatest concentration in drinking water. The guideline is designated as interim until such time as the risks from other disinfection by-products are ascertained. The preferred method of controlling disinfection byproducts is precursor removal; however, any method of control employed must not compromise the effectiveness of water disinfection.

[m] NTU = Nephelometric turbidity unit.

[n] At the point of consumption.

Source: From Health Canada, 2004, *Summary of Guidelines for Canadian Drinking Water Quality*, Prepared by the Federal-Provincial-Territorial Committee on Drinking Water of the Federal-Provincial-Territorial Committee on Health and the Environment, April 2004, ec.gc.ga.

Table 8C.26 Canadian Guidelines for Drinking Water Quality—Radionuclides

Radionuclide		Half-life $t_{1/2}$	DCF (Sv/Bq)	MAC (Bq/L)
Primary List of Radionuclides—Maximum Acceptable Concentrations (MACs)				
Natural Radionuclides				
Lead-210	^{210}Pb	22.3 yrs	1.3×10^{-6}	0.1
Radium-224	^{224}Ra	3.66 d	8.0×10^{-8}	2
Radium-226	^{226}Ra	1600 yrs	2.2×10^{-7}	0.6
Radium-228	^{228}Ra	5.76 yrs	2.7×10^{-7}	0.5
Thorium-228	^{228}Th	1.91 yrs	6.7×10^{-8}	2
Thorium-230	^{230}Th	7.54×10^{4} yrs	3.5×10^{-7}	0.4
Thorium-232	^{232}Th	1.40×10^{10} yrs	1.8×10^{-6}	0.1
Thorium-234	^{234}Th	24.1 d	5.7×10^{-9}	20
Uranium-234	^{234}U	2.45×10^{5} yrs	3.9×10^{-8}	4[a]
Uranium-235	^{235}U	7.04×10^{8} yrs	3.8×10^{-8}	4[a]
Uranium-238	^{238}U	4.47×10^{9} yrs	3.6×10^{-8}	4[a]
Artificial Radionuclides				
Cesium-134	^{134}Cs	2.07 yrs	1.9×10^{-8}	7
Cesium-137	^{137}Cs	30.2 yrs	1.3×10^{-8}	10
Iodine-125	^{125}I	59.9 d	1.5×10^{-8}	10
Iodine-131	^{131}I	8.04 d	2.2×10^{-8}	6
Molybdenum-99	^{99}Mo	65.9 hr	1.9×10^{-9}	70
Strontium-90	^{90}Sr	29 yrs	2.8×10^{-8}	5
Tritium[b]	^{3}H	12.3 yrs	1.8×10^{-11}	7000
Secondary List of Radionuclides—Maximum Acceptable Concentrations (MACs)				
Natural Radionuclides[c]				
Beryllium-7	^{7}Be	53.3 d	3.3×10^{-11}	4000
Bismurh-210	^{210}Bi	5.01 d	2.1×10^{-9}	70
Polonium-210	^{210}Po	138.4 d	6.2×10^{-7}	0.2
Artificial Radionuclides				
Americium-241	^{241}Am	432 yrs	5.7×10^{-7}	0.2
Antimony-122	^{122}Sb	2.71 d	2.8×10^{-9}	50
Antimony-124	^{124}Sb	60.2 d	3.6×10^{-9}	40
Antimony-125	^{125}Sb	2.76 yrs	9.8×10^{-10}	100
Barium-140	^{140}Ba	12.8 d	3.7×10^{-9}	40
Bromine-82	^{82}Br	35.3 hr	4.8×10^{-10}	300
Calcium-45	^{45}Ca	165 d	8.9×10^{-10}	200
Calcium-47	^{47}Ca	4.54 d	2.2×10^{-9}	60
Carbon-14[b]	^{14}C	5730 yrs	5.6×10^{-10}	200[b]
Cerium-141	^{141}Ce	32.5 d	1.2×10^{-9}	100
Cerium-144	^{144}Ce	284.4 d	8.8×10^{-9}	20
Cesium-131	^{131}Cs	9.69 d	6.6×10^{-11}	2000
Cesium-136	^{136}Cs	13.1 d	3.0×10^{-9}	50
Chromium-51	^{51}Cr	27.7 d	5.3×10^{-11}	3000
Cobalt-57	^{57}Co	271.8 d	3.5×10^{-9}	40
Cobalt-58	^{58}Co	70.9 d	6.8×10^{-9}	20
Cobalt-60	^{60}Co	5.27 yrs	9.2×10^{-8}	2
Gallium-67	^{67}Ga	78.3 hr	2.6×10^{-10}	500
Gold-198	^{198}Au	2.69 d	1.6×10^{-9}	90
Indium-111	^{111}In	2.81 d	3.9×10^{-10}	400
Iodine-129	^{129}I	1.60×10^{7} yrs	1.1×10^{-7}	1
Iron-55	^{55}Fe	2.68 yrs	4.0×10^{-10}	300
Iron-59	^{59}Fe	44.5 d	3.1×10^{-9}	40
Manganese-54	^{54}Mn	312.2 d	7.3×10^{-10}	200
Mercury-197	^{197}Hg	64.1 hr	3.3×10^{-10}	400
Mercury-203	^{203}Hg	46.6 d	1.8×10^{-9}	80
Neptunium-239	^{239}Np	2.35 d	1.2×10^{-9}	100
Niobium-95	^{95}Nb	35.0 d	7.7×10^{-10}	200
Phosphorus-32	^{32}P	14.3 d	2.6×10^{-9}	50
Plutonium-238	^{238}Pu	87.7 yrs	5.1×10^{-7}	0.3
Plutonium-239	^{239}Pu	2.41×10^{4} yrs	5.6×10^{-7}	0.2
Plutonium-240	^{240}Pu	6560 yrs	5.6×10^{-7}	0.2
Plutonium-241	^{241}Pu	14.4 yrs	1.1×10^{-8}	10

(Continued)

Table 8C.26 (Continued)

Radionuclide		Half-life $t_{1/2}$	DCF (Sv/Bq)	MAC (Bq/L)
Rhodium-105	^{105}Rh	35.4 hr	5.4×10^{-10}	300
Rubidium-81	^{81}Rb	4.58 hr	5.3×10^{-11}	3000
Rubidium-86	^{86}Rb	18.6 d	2.5×10^{-9}	50
Ruthenium-103	^{103}Ru	39.2 d	1.1×10^{-9}	100
Ruthenium-106	^{106}Ru	372.6 d	1.1×10^{-8}	10
Selenium-75	^{75}Se	119.8 d	2.1×10^{-9}	70
Silver-108m	^{108m}Ag	127 yrs	2.1×10^{-9}	70
Silver-110m	^{110m}Ag	249.8 d	3.0×10^{-9}	50
Silver-111	^{111}Ag	7.47 d	2.0×10^{-9}	70
Sodium-22	^{22}Na	2.61 yrs	3.0×10^{-9}	50
Strontium-85	^{85}Sr	64.8 d	5.3×10^{-10}	300
Strontium-89	^{89}Sr	50.5 d	3.8×10^{-9}	40
Sulphur-35	^{35}S	87.2 d	3.0×10^{-10}	500
Technetium-99	^{99}Tc	2.13×10^{5} yrs	6.7×10^{-10}	200
Technetium-99m	^{99m}Tc	6.01 hr	2.1×10^{-11}	7000
Tellurium-129m	^{129m}Te	33.4 d	3.9×10^{-9}	40
Tellurium-131m	^{131m}Te	32.4 hr	3.4×10^{-9}	40
Tellurium-132	^{132}Te	78.2 hr	3.5×10^{-9}	40
Thallium-201	^{201}Tl	3.04 d	7.4×10^{-11}	2000
Ytterbium-169	^{169}Yb	32.0 d	1.1×10^{-9}	100
Yttrium-90	^{90}Y	64 hr	4.2×10^{-9}	30
Yttrium-91	^{91}Y	58.5 d	4.0×10^{-9}	30
Zinc-65	^{65}Zn	243.8 d	3.8×10^{-9}	40
Zirconium-95	^{95}Zr	64.0 d	1.3×10^{-9}	100

[a] The activity concentrations of natural corresponding to the guideline of 0.02 mg/L is about 0.5 Bg/L.
[b] Tritium is also produced naturally in the atmosphere in significant quanties.
[c] The activity concentration of natural uranium corresponding to the chemical guideline of 0.1 mg/L (see separate criteria summary on uranium in the Supporting Documentation) is about 2.6 Bq/L.
[d] Tritium and ^{14}C are also produced naturally in the atmosphere in significant quantities.

Source: From Health Canada, 2004, *Summary of Guidelines for Canadian Drinking Water Quality*, Prepared by the Federal-Provincial-Territorial Committee on Drinking Water of the Federal-Provincial-Territorial Committee on Health and the Environment, April 2004, ec.gc.ga.

Table 8C.27 Australian Drinking Water Guideline Values for Physical and Chemical Characteristics

	Guideline Values[a]	
Characteristic	Health	Aesthetic[b]
Acrylamide	0.0002	
Aluminum (acid-soluble)	c	0.2
Ammonia (as NH_3)	c	0.5
Antimony	0.003	
Arsenic	0.007	
Asbestos	c	
Barium	0.7	
Benzene	0.001	
Beryllium	c	
Boron	4	
Bromate	0.02	
Cadmium	0.002	
Carbon tetrachloride	0.003	
Chloramine—see monochloramine		
Chlorate	c	
Chloride	e	250
Chlorinated furanones (MX)	c	
Chlorine	5	0.6
Chlorine dioxide	1	0.4
Chlorite	0.3	
Chloroacetic acids		
Chloroacetic acid	0.15	
Dichloroacetic acid	0.1	

(Continued)

Table 8C.27 (Continued)

Characteristic	Guideline Values[a] Health	Aesthetic[b]
Trichloroacetic acid	0.1	
Chlorobenzene	0.3	0.01
Chloroketones		
1,1-Dichloropropanone	c	
1,3-Dichloropropanone	c	
1,1,1-Trichloropropanone	c	
1,1,3-Trichloropropanone	c	
Chlorophenols		
2-Chlorophenol	0.3	0.0001
2,4-Dichlorophenol	0.2	0.0003
2,4,6-Trichlorophenol	0.02	0.002
Chloropicrin	c	
Chromium (as Cr(VI))	0.05	
Copper	2	1
Cyanide	0.08	
Cyanogen chloride (as cyanide)	0.08	
Dichlorobenzenes		
1,2-Dichlorobenzene	1.5	0.001
1,3-Dichlorobenzene	c	0.02
1,4-Dichlorobenzene	0.04	0.003
Dichloroethanes		
1,1-Dichloroethane	c	
1,2-Dichloroethane	0.003	
Dichloroethenes		
1,1-Dichloroethene	0.03	
1,2-Dichloroethene	0.06	
Dichloromethane (methylene chloride)	0.004	
Dissolved oxygen	Not necessary	> 85%
Epichlorohydrin	0.0005[d]	
Ethylbenzene	0.3	0.003
Ethylenediamine tetraacetic acid (EDTA)	0.25	
Fluoride	1.5	
Formaldehyde	0.5	
Haloacetonitriles		
Dichloroacetonitrile	c	
Trichloroacetonitrile	c	
Dibromoacetonitrile	c	
Bromochloroacetonitrile	c	
Hardness (as $CaCO_3$)	Not necessary	200
Hexachlorobutadiene	0.0007	
Hydrogen sulfide	c	0.05
Iodine	c	
Iodide	0.1	
Iron	c	0.3
Lead	0.01	
Manganese	0.5	0.1
Mercury	0.001	
Molybdenum	0.05	
Monochloramine	3	0.5
Nickel	0.02	
Nitrate (as nitrate)	50	
Nitrite (as nitrite)	3	
Nitrilotriacetic acid	0.2	
Organotins		
Dialkyltins	c	
Tributyltin oxide	0.001	
Ozone		
pH	c	pH 6.5–8.5
Plasticisers		
Di(2-ethylhexyl) phthalate	0.01	
Di(2-ethylhexyl) adipate	c	
Polycyclic aromatic hydrocarbons (PAHs)		
Benzo-(a)-pyrene	0.00001 (10 ng/L)	
Selenium	0.01	
Silver	0.1	

(Continued)

Table 8C.27 (Continued)

Characteristic	Guideline Values[a]	
	Health	Aesthetic[b]
Sodium	e	180
Styene (vinylbenzene)	0.03	0.004
Sulfate	500	250
Taste and odor	Not necessary	Acceptable to most people
Temperature	Not necessary	No value set
Tetrachloroethene	0.05	
Tin	e	
Toluene	0.8	0.025
Total dissolved solids	Not necessary	500
Trichloroacetaldehyde (chloral hydrate)	0.02	
Trichlorobenzenes (total)	0.03	0.005
1,1,1-Trichloroethane	c	
Trichloroethylene	c	
Trihalomethanes (THMs) (Total)	0.25	
True Color	Not necessary	15 HU
Turbidity	c	5 NTU
Uranium	0.02	
Vinyl chloride	0.0003	
Xylene	0.6	0.02
Zinc	c	3

Note: All values are as "total" unless otherwise stated; Routine monitoring for these compounds is not required unless there is potential for contamination of water supplies (e.g. accidental spillage); The concentration of all chlorination byproducts can be minimized by removing naturally occurring organic matter from the source water, reducing the amount of chlorine added, or using an alternative disinfectant (which may produce other byproducts). Action to reduce trihalomethanes and other byproducts is encouraged, but must not compromise disinfection.
HU, Hazen units; NTU, Nephelometric turbidity units; THMs, trihalomethanes.

[a] All values mg/L unless otherwise stated.
[b] Aesthetic values are not listed if the compound does not cause aesthetic problems, or if the value determined from health considerations is the same or lower.
[c] Insufficient data to set a guideline value based on health considerations.
[d] The guideline value is below the limit of determination. Improved analytical procedures are required for this compound.
[e] No health-based guideline value is considered necessary.

Source: From Australian Government, National Health and Medical Research Council and Natural Resource Management Ministerial Council, National Water Quality Management Strategy, *Australian Drinking Water Guidelines, 2004*, www.waterquality.crc.org.au. With permission.

Table 8C.28 Australian Drinking Water Guideline Values for Pesticides

Pesticide	Guideline Value[a] (mg/L)	Health Value[b] (mg/L)
Acephate		0.01
Aldicarb	0.001	0.001
Aldrin[c] (and dieldrin)	0.00001	0.0003
Ametryn	0.005	0.05
Amitrole[c]	0.001	0.01
Asulam		0.05
Atrazine[c]	0.0001	0.04
Azinphos-methyl	0.002	0.003
Benomyl		0.1
Bentazone		0.03
Bioresmethrin		0.1
Bromacil	0.01	0.3
Bromophos-ethyl		0.01
Bromoxynil		0.03
Carbaryl	0.005	0.03
Carbendazim		0.1
Carbofuran	0.005	0.01
Carbophenothion		0.0005
Carboxin	0.002	0.3
Chlordane[c]	0.00001	0.001
Chlorfenvinphos		0.005

(Continued)

Table 8C.28 (Continued)

Pesticide	Guideline Value[a] (mg/L)	Health Value[b] (mg/L)
Chlorothalonil	0.0001	0.03
Chloroxuron		0.01
Chlorpyrifos[c]		0.01
Chlorsulfuron		0.1
Clopyralid[c]	1	1
2,4-D[c]	0.0001	0.03
DDT[c]	0.00006	0.02
Diazinon	0.001	0.003
Dicamba		0.1
Dichlobenil		0.01
Dichlorvos	0.001	0.001
Diclofop-methyl		0.005
Dicofol		0.003
Dieldrin[c](see aldrin)	0.00001	0.0003
Difenzoquat		0.1
Dimethoate		0.05
Diphenamid	0.002	0.3
Diquat[c]	0.0005	0.005
Disulfoton	0.001	0.003
Diuron[c]		0.03
DPA (2,2-DPA)		0.5
EDB	0.001	0.001
Endosulfan[c]	0.00005	0.03
Endothal	0.01	0.1
EPTC	0.001	0.03
Ethion		0.003
Ethoprophos	0.001	0.001
Etridiazole	0.0001	0.1
Fenamiphos		0.0003
Fenarimol	0.001	0.03
Fenchlorphos		0.03
Fenitrothion		0.01
Fenoprop		0.01
Fensulfothion	0.01	0.01
Fenvalerate		0.05
Flamprop-methyl		0.003
Fluometuron		0.05
Formothion		0.05
Fosamine[c]		0.03
Glyphosate	0.01	1
Heptachlor[c](including its epoxide)	0.00005	0.0003
Hexaflurate		0.03
Hexazinone[c]	0.002	0.3
Lindane[c]	0.00005	0.02
Maldison		0.05
Methidathion		0.03
Methiocarb	0.005	0.005
Methomyl	0.005	0.03
Methoxychlor	0.0002	0.3
Metolachlor	0.002	0.3
Metribuzin	0.001	0.05
Metsulfuron-methyl		0.03
Mevinphos	0.005	0.005
Molinate[c]	0.0005	0.005
Monocrotophos		0.001
Napropamide	0.001	1
Nitralin		0.5
Norflurazon	0.002	0.05
Oryzalin		0.3
Oxamyl	0.005	0.1
Paraquat[c]	0.001	0.03
Parathion		0.01
Parathion methyl	0.0003	0.1
Pebulate	0.0005	0.03
Pendimethalin		0.3
Pentachlorophenol	0.00001	0.01
Permethrin	0.001	0.1
Picloram[c]		0.3

(Continued)

Table 8C.28 (Continued)

Pesticide	Guideline Value[a] (mg/L)	Health Value[b] (mg/L)
Piperonyl butoxide		0.1
Pirimicarb		0.005
Pirimiphos-ethyl		0.0005
Pirimiphos-methyl		0.05
Profenofos		0.0003
Promecarb		0.03
Propachlor	0.001	0.05
Propanil	0.0001	0.5
Propargite		0.05
Propazine	0.0005	0.05
Propiconazole[c]	0.0001	0.1
Propyzamide	0.002	0.3
Pyrazophos		0.03
Quintozene		0.03
Simazine	0.0005	0.02
Sulprofos		0.01
Silvex (see Fenoprop)		
2,4,5-T	0.00005	0.1
Temephos[c]	0.3	0.3
Terbacil	0.01	0.03
Terbufos	0.0005	0.0005
Terbutryn	0.001	0.3
Tetrachlovinphos	0.002	0.1
Thiobencarb		0.03
Thiometon		0.003
Thiophanate		0.005
Thiram		0.003
Triadimefon	0.1	0.002
Trichlorfon		0.005
Triclopyr[c]		0.01
Trifluralin	0.0001	0.05
Vernolate	0.0005	0.03

Note: Routine monitoring for pesticides is not required unless potential exists for contamination of water supplies.

[a] These are generally based on the analytical limit of determination (the level at which the pesticide can be reliably detected using practicable, readily available and validated analytical methods). If a pesticide is detected at or above this value the source should be identified and action taken to prevent further contamination.

[b] Based on 10% of acceptable daily intake (ADI).

[c] These pesticides have either been detected on occasions in Australian drinking water or their likely use would indicate that they may occasionally be detected.

Source: From Australian Government, National Health and Medical Research Council and Natural Resource Management Ministerial Council, National Water Quality Management Strategy, Australian Drinking Water Guidelines, 2004, www.waterquality.crc.org.au. With permission.

Table 8C.29 Australian Drinking Water Guideline Values for Radiological Quality

Guideline value

The total estimated dose per year from all radionuclides in drinking water, excluding the dose from potassium-40, should not exceed 1.0 mSv

If this guideline value of exceeded, the water provider, in conjunction with the relevant health authority, should evaluate possible remedial actions on a cost-benefit basis of assess what action can be justified to reduce the annual exposure

Screening of water supplies

Compliance with the guideline for radiological quality of drinking water should be assessed, initially, by screening for gross alpha and gross beta activity concentration. The recommended screening level for gross alpha activity is 0.5 Bq/L. The recommended screening level for gross beta activity is 0.5 Bq/L after subtraction of the contribution form potassium-40

If either of these activity concentrations is exceeded, specific radionuclides should be identified and their activity concentrations determined. The concentration of both radium-226 and radium-228 should always be determined, as these are the most significant naturally occurring radionuclides in Australian water supplies. Other radionuclides should be identified in necessary of ensure all gross alpha and beta activity is accounted for, after taking into account the counting and other analytical uncertainties involved in the determination

Source: From Australian Government, National Health and Medical Research Council and Natural Resource Management Ministerial Council, National Water Quality Management Strategy, Australian Drinking Water Guidelines, 2004, www.waterquality.crc.org.au. With permission.

Table 8C.30 Comparison of Inorganic Drinking Water Guidelines Recommended by WHO and Standards for Several Developed and Developing Countries

Inorganics (mg/L)	WHO (2004)	U.S.A. (2005)	EEC (1998)	Canada (2004)	Australia (2000)	Thailand (1978)	Vietnam (2002)	Korea (2000)	Indonesia (1995)	Japan (2004)	China (1985)	South Africa (1996)	Chile (1984)	Brazil (2004)
Antimony	0.02	0.006	0.005	0.006	0.003		0.005			0.002				0.005
Arsenic	0.01(P)	7; 0.01 as of 01/23/06	0.01	0.025	0.007	0.05	0.01	0.05	0.05	0.01	0.05	0.01	0.05	0.01
Barium	0.7	2		1	0.7	1	0.7		1					0.7
Boron	0.5 (T)		1	5	4		0.3	0.3		1				
Cadmium	0.003	0.005	0.005	0.005	0.002	0.01	0.003	0.01	0.005	0.01	0.01	5	0.01	0.005
Chromium	0.05 (P)	0.1	0.05	0.05	0.05 (VI)	0.05 (VI)	0.05	0.05 (VI)	0.05 (VI)	0.05 (VI)	0.05 (VI)	0.05 (VI)	0.05 (VI)	0.05
Copper	2	TT, Action Level = 1.3	2.0		2		2		1.0	1.0	1.0	1.0	1.0	2
Cyanide	0.07	0.2	0.05	0.2	0.08	0.2	0.07	0.01	0.1	0.01	0.05		0.2	0.07
Fluoride	1.5	4	1.5	1.5	1.5	0.7	0.7–1.5	1.5	0.5	0.8	1.0	1.0	1.5	1.5
Lead	0.01	TT, Action Level = 0.015	0.025, 0.01[a]	0.01	0.01	0.05	0.01	0.05	0.05	0.01	0.05	0.01	0.05	0.01
Manganese	0.4 (C)				0.5					0.05			0.1	
Mercury (total)	0.001	0.002	0.001	0.001	0.001	0.002	0.001	0.001	0.001	0.0005	0.001	0.001	0.001	0.001
Molybdenum	0.07				0.05		0.07							
Nickel	0.02 (P)		0.02		0.02		0.02			0.01				
Nitrate	50 (acute) as NO_3	10 as N	50 as NO_3	10 as N	50 as NO_3	45 as NO_3	50 as NO_3	10 as N	10 as N	10 as N	20 as N	6 as N	10 as N	10 as N
Nitrite	3 (acute), 0.2 (P) (chronic) as NO_2	1 as N	0.5 as NO_2 (tap)	3.2 as N	3 as NO_2		3 as NO_2		1 as N	10 as N		6 as N	1 as N	1 as N
Selenium	0.01	0.05	0.01	0.01	0.01	0.01	0.01	0.01	0.01	0.01	0.01	0.02	0.01	0.01
Uranium	0.015 (P,T)			0.02	0.02					0.002				

Note: P, provisional guideline value, as there is evidence of a hazard, but the available information on health effects is limited; T, provisional guideline value because calculated guideline value is below the level that can be achieved through practical treatment methods, source protection, etc; A, provisional guideline value because calculated guideline value is below the achievable quantification level; D, provisional guideline value because disinfection is likely to result in the guideline value being exceeded; C, concentrations of the substance at or below the health-based guideline value may affect the appearance, taste or odour of the water, leading to consumer complaints; TT, Treatment Technique—A required process intended to reduce the level of a contaminant in drinking water.

[a] 0.025 (12/25/03–12/25/13), 0.01 (2/25/13+).

Source: Modified from Vigneswaran, Saravanamuthu, 1995, *Water Treatment Processes: Simple Options*, CRC Press, Inc., Boca Raton. **WHO**, World Health Organization, 2004, *Guidelines for Drinking-Water Quality, Third Edition*, Volume 1, Recommendations. www.who.int/water_sanitation_health/dwq/gdwq3/en/index.html. **U.S.A.**, United States Environmental Protection Agency, www.epa.gov/OGWDW/mcl.html. **Australia**, Australian Government, National Health and Medical Research Council and Natural Resource Management Ministerial Council, National Water Quality Management Strategy, Austrlian Drinking Water Guidelines, 2004. www.waterquality.crc.org.au. EEC, Lenntech Drinking Water Standards, www.lenntech.com. **Canada**, Health Canada, 2004, Summary of Guidelines for Canadian Drinking Water Quality, Prepared by the Federal-Provincial-Territorial Committee on Drinking Water of the Federal-Provincial-Territorial Committee on Health and the Environment, April 2004. ec.gc.ga. **Thailand**, Notification of the Ministry of Industry, No. 322, B.E. 2521 (1978), issued under the Industrial Products Standards Act B.E. 2511 (1968), published in the Royal Gazette, Vol. 95, Part 68, dated July 4, B.E. 2521 (1978). pcdv1.pcd.go.th/Information/Regulations/WaterQuality/WaterQualityStandards.cfm. **Vietnam**, Hue, N.D. and Viet, P.H., 2003, Environmental Quality Standards in Vietnam, in United Nations University, Capacity Development Training for Monitoring of POPs in the East Asia Hydrosphere, 1–2 September 2003, Tokyo. **Korea**, Oh, J.R, 2003, Environmental Standards of Korea in United Nations University, Capacity Development Training for Monitoring of POPs in the East Asia Hydrosphere, 1–2 September 2003, Tokyo. **Indonesia**, Vigneswaran, Saravanamuthu, 1995, *Water Treatment Processes: Simple Options*, CRC Press, Inc., Boca Raton (Criteria of Water Quality, Category A). **Japan**, Japan Ministry of Health, Labour and Welfare jwwa.or.jp/water-e07.html. **China**, Chinese Drinking Water Standards (GB 5749–85). **South Africa**, Republic of South Africa, Department of Water Affairs and Forestry, 1996, South African Water Quality Guidelines, Volume 8, Field Guide, First Edition 1996, dwaf.pwv.gov.za/IWQS/wq_guide/field.pdf#search='Recreation%20Water%20Quality%20Guidelines'. **Chile**, Normas oficiales para la calidad del agua Chile, NORMA CHILENA OFICIAL 409/1.Of.84, Agua Potable Parte 1: Requisitos cepis.ops-oms.org/bvsacg/e/normas2/Norma-Chi.pdf#search="NORMA%20CHILENA%20OFICIAL%20409/1.Of.84". **Brazil**, Ministerio da Saude, Portaria N.o 518, DE 25 De Marco De 2004 www.saneago.com.br/novasan/leis/port518.pdf#search="Ministerio%20da%20Saude%2C%20Portaria%20N.%20518".

Table 8C.31 Comparison of Organic Compound Drinking Water Guidelines Recommended by WHO and Standards for Several Developed and Developing Countries

	WHO (2004)	U.S.A. (2005)	EEC (1998)	Canada (2004)	Australia (2000)	Thailand (1978)	Vietnam (2002)	Korea (2000)	Indonesia (1995)	Japan (2004)	China (1985)	South Africa (1996)	Chile (1984)	Brazil (2004)
Organic Compounds (mg/L)														
Acrylamide	0.0005	TT9	0.0001		0.0002		0.0005							0.0005
Benzene	0.01	0.005	0.001	0.005	0.001		0.01	0.01	0.01	0.01				0.005
Benzo[a]pyrene	0.0007	0.0002	0.00001	0.00001	0.00001		0.0007		0.00001		0.00001			0.0007
Carbon tetrachloride	0.004	0.005		0.005	0.003		0.002	0.002		0.002	0.003			0.002
1,2-Dichlorobenzene	1(C)	0.6		0.2	1.5		1							
1,4-Dichlorobenzene	0.3(C)	0.075		0.005	0.04		0.3							
1,2-Dichloroethane	0.03	0.005	0.003	0.005	0.003		0.03		0.01					0.01
1,1-Dichloroethene	0.03	0.007		0.014	0.03		0.03	0.03		0.02				0.03
1,2-Dichloroethene	0.05	0.07 (cis), 0.1 (trans)			0.06		0.05			0.04 (cis)				
Dichloromethane	0.02	0.005		0.05	0.004		0.02	0.02		0.02				0.02
Di(2-ethylhexyl)phthalate	0.008	0.006			0.01		0.008	0.08						
Edetic acid (EDTA)	0.6				0.25		0.2							
Epichlorohydrin	0.0004 (P)	TT9	0.0001		0.0005		0.0004							
Ethylbenzene	0.3	0.7			0.3		0.3	0.3						
Hexachlorobutadiene	0.0006				0.0007		0.0006							
Microcystin-LR	0.001 (P)			0.0015										0.001
Monochlorobenzene		0.1		0.08	0.3		0.3							
Nitrilotriacetic acid	0.2			0.4	0.2		0.2							
Styrene	0.02	0.1			0.03		0.02							0.02
Tetrachloroethene	0.04	0.005	0.01[a]	0.03	0.05		0.04	0.01		0.01				0.04
Toluene	0.7	1			0.8		0.7	0.7						
Trichloroethene	0.07 (P)	0.005	0.01[a]	0.05			0.07	0.03		0.03				0.07
Vinyl chloride	0.0003	0.002	0.0005	0.002	0.0003		0.005							0.005
Xylenes	0.5	10			0.6		0.5	0.5						
Pesticides (mg/L)														
Alachlor	0.02	0.002					0.02							0.02
Aldicarb	0.01			0.009	0.001		0.01							
Aldrin/dieldrin	0.00003		0.0003	0.0007	0.0003		0.0003		0.0007				0.00003	0.00003
Atrazine	0.002	0.003		0.005	0.04		0.002					0.002		0.002
Carbofuran	0.007	0.04		0.09	0.01		0.005							
Chlordane	0.0002	0.002			0.001		0.0002		0.0003				0.0003	0.0002
Chlorpyrifos	0.03			0.09	0.01									
Chlorotoluron	0.03						0.03							
Cyanazine	0.0006			0.01										
DDT and metabolites	0.001				0.02		0.002		0.03		0.001		0.001	0.002
1,2-Dibromo-3-chloropropane	0.001	0.0002					0.001							
1,2-Dibromoethane	0.0004 (P)													
2,4-Dichlorophenoxyacetic acid (2,4-D)	0.03	0.07		0.1	0.03		0.03		0.1				0.1	0.03
1,2-Dichloropropane (1,2-DCP)	0.04 (P)	0.005					0.02							
1,3-Dichloropropane	0.04													
1,3-Dichloropropene	0.02													

Isoproturon	0.009						0.009							
Lindane	0.002	0.0002			0.02		0.002		0.004				0.003	0.002
MCPA	0.002						0.002							
Methoxychlor	0.02	0.04		0.9	0.3		0.02		0.03				0.03	0.02
Metolachlor	0.01			0.05	0.3		0.01							0.01
Molinate	0.006				0.0005		0.006							0.006
Pendimethalin	0.02				0.3		0.02							0.02
Pentachlorophenol	0.009 (P)	0.001		0.06	0.01		0.009		0.01					0.009
Phenol						0.001		0.005		0.005	0.002	1	0.002	
Pyriproxyfen	0.3													
Simazine	0.002	0.004		0.01	0.02		0.02							0.002
Terbuthylazine (TBA)	0.007													
Trifluralin	0.02			0.045	0.05		0.02							0.02
Chlorophenoxy herbicides other than 2,4-D and CPA (mg/L)														
2,4-DB	0.09						0.09							
Dichlorprop	0.1						0.1							
Dimethoate	0.006			0.02	0.05									
Endrin	0.0006	0.002											0.0002	0.0006
Fenoprop	0.009				0.01		0.009						0.01	
Mecoprop	0.01						0.01							
2,4,5-T	0.009				0.1		0.009							
Disinfectants and disinfectant by-products (mg/L)														
Monochloramine	3	4.01		3 (total)	3		3							3
Chlorine	5 (C)	4.01			5		0.003–0.005 (active)						0.2 (free)	5
Disinfectant by-products														
Bromate	0.01 (A,T)	0.01	0.01	0.01	0.02		0.025			0.01				0.025
Chlorate	0.7 (D)													0.2
Chlorite	0.7 (D)	1			0.3		0.2							
2,4,6-Trichlorophenol	0.2 (C)			0.005	0.02		0.2		0.01					0.2
Formaldehyde	0.9				0.5		0.9			0.08				
Trihalomethanes	0.001	0.1	0.1	0.1				0.1		0.01				0.1
Bromoform	0.1				0.25		0.1			0.09				
Dibromochloromethane	0.1						0.1			0.1				
Bromodichloromethane	0.06						0.06			0.03				
Chloroform	0.2						0.2	0.08	0.03	0.06	0.06			
Chlorinated acetic acids														
Monochloroacetic acid	0.02				0.15					0.02				
Dichloroacetic acid	0.05 (T,D)	0.06			0.1		0.05			0.04				
Trichloroacetic acid	0.2	0.06			0.1		0.1			0.02				
Chloral hydrate (trichloroacetaldehyde)	0.01 (P)				0.02		0.01							

(Continued)

Table 8C.31 (Continued)

	WHO (2004)	U.S.A. (2005)	EEC (1998)	Canada (2004)	Australia (2000)	Thailand (1978)	Vietnam (2002)	Korea (2000)	Indonesia (1995)	Japan (2004)	China (1985)	South Africa (1996)	Chile (1984)	Brazil (2004)
Halogenated acetonitriles														
Dichloroacetonitrile	20						0.09							
Dibromoacetonitrile	0.07						0.1							
Cyanogen chloride (as CN)	0.07				0.08		0.01	0.01						

Note: P = provisional guideline value, as there is evidence of a hazard, but the available information on health effects is limited; T = provisional guideline value because calculated guideline value is below the level that can be achieved through practical treatment methods, source protection, etc; A = provisional guideline value because calculated guideline value is below the achievable quantification level; D = provisional guideline value because disinfection is likely to result in the guideline value being exceeded; C = concentrations of the substance at or below the health-based guideline value may affect the appearance, taste or odour of the water, leading to consumer complaints; TT = Treatment Technique—A required process intended to reduce the level of a contaminant in drinking water.

[a] Sum of trichloroethene and tretrachloroethene.

Source: From **WHO**, World Health Organization, 2004, *Guidelines for Drinking-Water Quality, Third Edition*, Volume 1, Recommendations. who.int/water_sanitation_health/dwq/gdwq3/en/index.html. **U.S.A.**, United States Environmental Protection Agency, www.epa.gov/OGWDW/mcl.html. **Australia**, Australian Government, National Health and Medical Research Council and Natural Resource Management Ministerial Council, National Water Quality Management Strategy, Australian Drinking Water Guidelines, 2004, www.waterquality.crc.org.au. **EEC**, Lenntech Drinking Water Standards, www.lenntech.com. **Canada**, Health Canada, 2004, *Summary of Guidelines for Canadian Drinking Water Quality*, Prepared by the Federal-Provincial-Territorial Committee on Drinking Water of the Federal-Provincial-Territorial committee on Health and the Environment, April 2004, ec.gc.ga. **Thailand**, Notification of the Ministry of Industry, No. 322, B.E. 2521 (1978), issued under the Industrial Products Standards Act B.E. 2511 (1968), published in the Royal Gazette, Vol. 95, Part 68, dated July 4, B.E. 2521 (1978). pcdv1.pcd.go.th/Information/Regulations/WaterQuality/WaterQualityStandards.cfm. **Vietnam**, Hue, N.D. and Viet, P.H., 2003, Environmental Quality Standards in Vietnam, in United Nations University, Capacity Development Training for Monitoring of POPs in the East Asia Hydrosphere, 1–2 September 2003, Tokyo. **Korea**, Oh, J.R, 2003, Environmental Standards of Korea in United Nations University, Capacity Development Training for Monitoring of POPs in the East Asia Hydrosphere, 1–2 September 2003, Tokyo. **Indonesia**, Vigneswaran, Saravanamuthu, 1995, Water Treatment Processes: Simple Options, CRC Press, Inc., Boca Raton (Criteria of Water Quality, Category A). **Japan**, Japan Ministry of Health, Labour and Welfare jwwa.or.jp/water-e07.html. **China**, Chinese Drinking Water Standards (GB 5749–85). **South Africa**, Republic of South Africa, Department of Water Affairs and Foresty, 1996, South African Water Quality Guidelines, Volume 8, Field Guide, First Edition 1996, dwaf.pwv.gov.za/IWQS/wq_guide/field.pdf#search = 'Recreation%20Water%20Quality%20Guidelines'. **Chile**, Normas oficiales para la calidad del agua Chile, NORMA CHILENA OFICIAL 409/1.Of.84;, Agua Potable Parte 1: Requisitos cepis.ops-oms.org/bvsacg/e/normas2/Norma-Chi.pdf#search = "NORMA%20CHILENA%20OFICIAL%20409/1.Of.84". **Brazil**, Ministerio da Saude, Portaria N.o 518, DE 25 De Marco De 2004 saneago.com.br/novasan/leis/port518.pdf#search = "Ministerio%20da%20Saude%2C%20Portaria%20N.%20518".

Table 8C.32 Comparison of Aesthetic Drinking Water Guidelines Recommended by WHO and Standards for Several Developed and Developing Countries

	WHO (2004)	U.S.A. (2005)	EEC (1998)	Canada (2004)	Australia (2000)	Thailand (1978)	Vietnam (2002)	Korea (2000)	Indonesia (1995)	Japan (2004)	China (1985)	South Africa (1996)	Chile (1984)	Brazil (2004)
Physical parameters														
Color	15 TCU	15 (color units)	Acceptable to consumers and no abnormal change	15 TCU	15 HU	5 Pt–Co	15 TCU	5	15 TCU	5 Degrees	<15°		20 Pt–Co	15
Taste and odor	Should be acceptable	3 threshold odor number	Acceptable to consumers and no abnormal change	Inoffensive	Acceptable to most people	Non objectionable	Nil	Not abnormal	Odorless, tasteless	Not Abnormal		1 (odour)	Tasteless, odorless	Not objectionable
Turbidity	5 NTU	5 NTU	Acceptable to consumers and no abnormal change	1 NTU (MAC) 5 NTU (AO)	5 NTU	5 SSU	2 NTU	1 NTU, 0.5 NTU for Tap Water	5 NTU	2 Degrees	Less than 3° not to exceed 5	1 NTU	5	5
pH (standard units)	<8.0	6.5–8.5	6.5–9.5	6.5–8.5	6.5–8.5	6.5–8.5	6.5–8.5	5.8–8.5	6.5–8.5	5.8–8.6	6.5–8.5	6–9		
Inorganic constituents (mg/L)														
Aluminum	0.1–0.2	0.05 to 0.2	0.2	0.1	0.2		0.2	0.2	0.2	0.2		0.15		0.2
Ammonia	1.5 (odor), 35 (taste)		0.5		0.5		1.5	0.5				1	0.25	1.5
Chloride	250	250	250	250	250	250	250	250	250	200	250	100	250	250
Copper	1	1	2	1	1	1		1	1	1	1			
Hydrogen sulphide	0.05			0.3	0.05		0.05		0.05					0.05
Iron	0.3	0.3	0.2	0.005	0.3	0.5	0.5	0.3	0.3	0.3	0.3	0.1	0.3	0.3
Manganese	0.1	0.05	0.05	0.05	0.1	0.3	0.5	0.3	0.1	0.05	0.1	0.05	0.1	0.1
Dissolved oxygen	Narrative				>85%									
Sodium	200		200	200	180		200		200	200		100		200
Sulfate	250	250	250	500	250	200	250	200	400	200	250	200	250	250
Total dissolved solids	1000	500		500	500	500	1000	500	1000	500	1000	450	1000	1000
Zinc	3	5		5	3	5	3	1	5	1	1	3	5	5
Organic constituents (mg/L)[a]														
Toluene	0.024–0.17			0.024	0.025									0.17
Xylene	0.020–1.80			0.3	0.02									0.3
Ethylbenzene	0.002–0.20			0.0024	0.003									0.2
Styrene	0.004–2.6				0.004									
Monochlorobenzene	0.010–0.120			0.3	0.01									0.12
1,2-dichlorobenzene	0.001–0.01			0.003	0.001									
1,4-dichlorobenzene	0.0003–0.03			0.001	0.003									
Trichlorobenzenes (total)	0.005–0.05				0.005		0.020							0.02
Synthetic detergents	no foam or taste problems													
Disinfectants and disinfectant by-products (mg/L)														
Chlorine	0.6–1.0				0.6									0.2
Chlorophenols														5
2-chlorophenol	0.0001–0.010				0.0001									
2,4-dichlorophenol	0.0003–0.040			0.0003	0.0003									
Monochloramine	0.3				0.0002									3
2,4,6-trichlorophenol	0.002–0.3			0.002	0.002									0.2

(Continued)

Table 8C.32 (Continued)

[a] The levels indicated are not precise numbers. Problems may occur at lower or higher values according to local circumstances. A range of taste and odour threshold concentrations is given for organic constituents.

Source: From **WHO**, World Health Organization, 2004, *Guidelines for Drinking-Water Quality, Third Edition*, Volume 1, Recommendations. www.who.int/water_sanitation_health/dwq/gdwq3/en/index.html. **U.S.A.**, United States Environmental Protection Agency, www.epa.gov/OGWDW/mcl.html. **Australia**, Australian Government, National Health and Medical Research Council and Natural Resource Management Ministerial Council, National Water Quality Management Strategy, Austrailian Drinking Water Guidelines, 2004. www.waterquality.crc.org.au. **EEC**, Lenntech Drinking Water Standards, www.lenntech.com. **Canada**, Health Canada, 2004, Summary of Guidelines for Canadian Drinking Water Quality, Prepared by the Federal-Provincial-Territorial Committee on Drinking Water of the Federal-Provincial-Territorial Committee on Health and the Environment, April 2004, ec.gc.ga. **Thailand**, Notification of the Ministry of Industry, No. 322, B.E. 2521 (1978), issued under the Industrial Products Standards Act B.E. 2511 (1968), published in the Royal Gazette, Vol. 95, Part 68, dated July 4, B.E. 2521 (1978). pcdv1.pcd.go.th/Information/Regulations/WaterQuality/WaterQualityStandards.cfm. **Vietnam**, Hue, N.D. and Viet, P.H., 2003, Environmental Quality Standards in Vietnam, in United Nations University, Capacity Development Training for Monitoring of POPs in the East Asia Hydrosphere, 1–2 September 2003, Tokyo. **Korea**, Oh, J.R, 2003, Environmental Standards of Korea in United Nations University, Capacity Development Training for Monitoring of POPs in the East Asia Hydrosphere, 1–2 September 2003, Tokyo. **Indonesia**, Vigneswaran, Saravanamuthu, 1995, Water Treatment Processes: Simple Options, CRC Press, Inc., Boca Raton (Criteria of Water Quality, Category A). **Japan**, Japan Ministry of Health, Labour and Welfare jwwa.or.jp/water-e07.html. **China**, Chinese Drinking Water Standards (GB 5749–85). **South Africa**, Republic of South Africa, Department of Water Affairs and Foresty, 1996, South African Water Quality Guidelines, Volume 8, Field Guide, First Edition 1996, dwaf.pwv.gov.za/IWQS/wq_guide/field.pdf#search="Recreation%20Water%20Quality%20Guidelines". **Chile**, Normas oficiales para la calidad del agua Chile, NORMA CHILENA OFICIAL 409/1.Of.84, Agua Potable Parte 1: Requisitos cepis.ops-oms.org/bvsacg/e/normas2/Norma-Chi.pdf#search="NORMA%20CHILENA%20OFICIAL%20409/1.Of.84". **Brazil**, Ministerio da Saude, Portaria N.o 518, DE 25 De Marco De 2004 saneago.com.br/novasan/leis/port518.pdf#search="Ministerio%20da%20Saude%2C%20Portaria%20N.%20518".

SECTION 8D MUNICIPAL WATER QUALITY

Table 8D.33 Range in Quality of Finished Water in Public Water Supplies of the 100 Largest Cities in the United States

Constituent or Property	Maximum	Median	Minimum
Chemical Analyses (parts per million)			
Silica (SiO_2)	72	7.1	0.0
Iron (Fe)	1.30	0.02	0.00
Manganese (Mn)	2.50	0.00	0.00
Calcium (Ca)	145	26	0.0
Magnesium (Mg)	120	6.25	0.0
Sodium (Na)	198	12	1.1
Potassium (K)	30	1.6	0.0
Bicarbonate (HCO_3)	380	46	0
Carbonate (CO_3)	26	0	0
Sulfate (SO_4)	572	26	0.0
Chloride (Cl)	540	13	0.0
Fluoride (F)	7.0	0.4	0.0
Nitrate (NO_3)	23	0.7	0.0
Dissolved solids	1,580	186	22
Hardness as $CaCO_3$	738	90	0
Noncarbonate hardness as $CaCO_3$	446	34	0
Specific conductance micromhos at 25°C	1,660	308	18
pH, pH units	10.5	7.5	5.0
Color, color units	24	2	0
Turbidity	13	0	0
Spectrographic Analyses (μg/L)			
Silver (Ag)	7.0	0.23	ND
Aluminum (Al)	1,500	54	3.3
Boron (B)	590	31	2.5
Barium (Ba)	380	43	1.7
Chromium (Cr)	35	0.43	ND
Copper (Cu)	250	8.3	<0.61
Iron (Fe)	1,700	43	1.9
Lithium (Li)	170	2.0	ND
Manganese (Mn)	1,100	5.0	ND
Molybdenum (Mo)	68	1.4	ND
Nickel (Ni)	34	<2.7	ND
Lead (Pb)	62	3.7	ND
Rubidium (Rb)	67	1.05	ND
Strontium (Sr)	1,200	110	2.2
Titanium (Tl)	49	<1.5	ND
Vanadium (V)	70	<4.3	ND
Radiochemical Analyses			
Beta activity picocuries per liter	130	7.2	<1.1
Radium (Ra) picocuries per liter	2.5	<0.1	<0.1
Uranium (U) micrograms per liter	250	0.15	<0.1

Note: Maximum, median, and minimum values of as of 1962; ND means not detected.

Source: From U.S. Geological Survey.

Table 8D.34 Quality Limits of Finished Water in Public Water Supplies of the 100 Largest Cities in the United States

	Water Supplies Having Less Than Stated Concentration	
Constituent or Property	**Concentration**	**Percent of Water Supplies**
Chemical Analyses (parts per million)		
Silica (SiO_2)	30	94
Iron (Fe)	0.25	98
Manganese (Mn)	0.10	95
Calcium (Ca)	50	93
Magnesium (Mg)	20	96
Sodium (Na)	50	93
Potassium (K)	5.0	93
Bicarbonate (HCO_3)	150	91
Carbonate (CO_3)	1.0	86
Sulfate (SO_4)	100	93
Chloride (Cl)	50	93
Fluoride (F)	1.0	92
Nitrate (NO_3)	5.0	93
Dissolved solids	500	97
	250	86
Hardness as $CaCO_3$	200	94
Noncarbonate hardness as $CaCO_3$	75	94
Specific conductance (micro mhos at 25°C)	500	93
pH, pH units	9.0	90
Color, color units	10	96
Turbidity	3	94
Spectrographic Analyses (μg/L)		
Silver (Ag)	0.50	95
Aluminum (Al)	500	87
Boron (B)	100	94
Barium (Ba)	100	94
Chromium (Cr)	5.0	95
Copper (Cu)	100	94
Iron (Fe)	150	94
Lithium (Li)	50	96
Manganese (Mn)	100	97
Molybdenum (Mo)	10	96
Nickel (Ni)	10	95
Phosphorus (P)	ND	92
Lead (Pb)	10	95
Rubidium (Rb)	5.0	91
Strontium (Sr)	500	96
Titanium (Ti)	5.0	96
Vanadium	10	91
Radiochemical Analyses		
Beta activity picocuries per liter	20	92
Radium (Ra) picocuries per liter	0.2	91
Uranium (U) micrograms per liter	2.0	93

Note: Data as of 1962; ND means not detected.

Source: From U.S. Geological Survey.

Table 8D.35 Quality of Raw and Treated Water in Public Water Supplies of the 100 Largest Cities in the United States

	Raw-Water Supplies[a]		Treated-Water Supplies	
	Population Served (millions)	Number of Cities	Population Served (millions)	Number of Cities
Hardness (ppm)				
Less than 61	21	29	23	30
61–120	15	16	22	41
21–180	16	22	11	16
More than 180	8	27	3.7	13
Dissolved solids (ppm)				
Less than 100	21	27	21	27
101–250	23	38	28	48
251–500	11	29	8	22
More than 500	1.5	6	1	3
pH				
Less than 7.0	16	18	14	9
7.0–9.0	42	80	38	74
More than 9.0			7	17

Note: Data as of 1962.

[a] A few cities are not included because data are lacking.

Source: From U.S. Geological Survey.

Table 8D.36 Standards for Raw Water Used as Sources of Domestic Water Supply

Constituents		Excellent Source of Water Supply, Requiring Disinfection Only, as Treatment	Good Source of Water Supply Requiring Usual Treatment Such as Filtration and Disinfection	Poor Source of Water Supply, Requiring Special or Auxiliary Treatment and Disinfection
B.O.D. (5-day) ppm	Monthly Average	0.75	1.5–2.5	2.0–5.5
	Maximum Day, or sample	1.0	3.0–3.5	4.0–7.5
Coliform MPN per 100 mL	Monthly Average	50–100	240–5,000	10,000–20,000
	Maximum Day, or sample	—	$<20\%>5{,}000$ $<5\%>20{,}000$	
Dissolved oxygen	ppm. average	4.0–7.5	2.5–7.0	2.5–6.5
	% saturation	50–75	25–75	—
pH	Average	6.0–8.5	5.0–9.0	3.8–10.5
Chlorides, max	ppm.	50	250	500
Iron and manganese together	Max. ppm.	0.3	1.0	15
Fluorides	ppm.	1.0	1.0	1.0
Phenolic compounds	Max. ppm	none	0.005	0.025
Color	ppm.	0–20	20–70	150
Turbidity	ppm.	0–10	40–250	—

Source: From Calif. State Water Pollution Control Board, 1952.

Table 8D.37 Summer Temperatures of Selected Municipal Water Supplies in the United States

Location	Temperature at Main Outlet, °F			
	June	July	August	September
Surface Water Sources				
Atlanta, GA	78.1	83.5	79.5	77.8
Baltimore, MD	61.0	66.0	70.0	64.0
Birmingham, AL	78.0	82.0	81.0	79.0
Boston, MA	68.3	74.3	73.4	69.4
Buffalo, NY	62.0	71.0	73.0	66.0
Chicago, IL	55.4	68.0	69.4	62.5
Cincinnati, OH	76.0	82.0	81.0	77.0
Cleveland, OH	58.0	68.0	73.5	71.0
Detroit, MI	64.0	75.0	74.0	68.0
Kansas City, MO	84.0	93.0	91.0	85.0
Louisville, KY	77.0	82.0	82.0	77.0
Nashville, TN	84.0	88.0	88.0	84.0
New Orleans, LA	86.0	89.0	90.0	90.0
Oakland, CA	59.0	62.0	64.0	64.0
Philadelphia, PA	71.0	79.0	77.0	72.0
Pittsburgh, PA	75.2	80.6	80.6	75.2
Sacramento, CA	70.7	70.7	80.6	77.0
St. Louis, MO	77.0	85.0	83.0	75.0
Washington, DC	43.0	67.0	73.0	75.0
Groundwater Sources				
Albuquerque, NM	72.0	72.0	72.0	72.0
Aurora, CO	60.0	60.0	60.0	60.0
Camden, NJ	58.0	58.0	58.0	58.0
El Paso, TX	84.0	85.0	85.0	84.0
Fresno, CA	72.0	72.0	72.0	72.0
Houston, TX	84.0	84.0	84.0	84.0
Jacksonville, FL	84.8	86.3	86.7	82.4
Kalamazoo, MI	52.0	52.0	52.0	52.0
Lafayette, LA	53.0	53.0	53.0	53.0
Lansing, MI	57.5	58.0	59.0	59.0
Lincoln, NE	58.0	59.0	59.0	59.0
Lowell, MA	50.0	50.0	50.0	50.0
Madison, WI	53.0	52.0	52.0	53.0
Marion, FL	54.0	54.0	55.0	55.0
Montgomery, AL	70.0	70.0	71.0	71.0
Pensacola, FL	70.0	70.0	70.0	70.0
Peoria, IL	56.0	56.0	56.0	54.0
Pontiac, MI	55.0	55.0	55.0	55.0
San Antonio, TX	76.0	76.0	76.0	76.0
Sioux Falls, SD	55.0	55.0	55.0	55.0

Source: From U.S. Dept. of Commerce.

Table 8D.38 Quality of Water Supplied by Municipal Water Systems in the United States–1984

State and Water Utility	Avg Temp Raw (°F)	Hardness ($CaCO_3$) Raw (mg/L)	Hardness ($CaCO_3$) Finished (mg/L)	Alkalinity ($CaCO_3$) Raw (mg/L)	Alkalinity ($CaCO_3$) Finished (mg/L)	pH Raw	pH Finished
Alabama							
Birmingham	60.8	75	104	73	77	7.9	8.2
Montgomery	54.0	18	30	15	32	6.7	8.6
Alaska							
Anchorage	37.4	63	61	52	42	7.6	7.2
Arizona							
Phoenix/Phoenix Wtr & Swr Dept	60.1	190	190	125	105	8.2	7.6
Tucson	75.2	141	141	150	150	7.5	7.5
Arkansas							
Fort Smith	56.5	39	18	14	28	6.6	9.0
Little Rock	64.2	9	19	7	11	6.7	7.7
California							
Anaheim	64.0	331	345	123	186	8.3	7.7
Burbank	60.0	165		159		7.6	
Concord/Contra Costa Water Dept	63.0	80	68	64	59	8.0	8.5
Corte Madera/Marin Munic Wtr Dist	58.1	57	67	58	71	7.4	8.7
Fremont/Alameda Cnty Wtr Dist	62.1	97	97	65	65	7.6	8.7
Fresno	70.0	119	119	123	123	7.8	7.8
Glendale	69.8	190	296	156	108	7.9	8.0
La Mesa/Helix Wtr District	71.6	300	300	115	115	7.7	8.0
La Puente/Southwest Suburban Wtr	68.0	220	220	180	180	7.4	7.4
Los Angeles/Los Angeles Wtr & Power	58.1	69	69	89	89	7.9	7.9
Oakland/East Bay Munic Util Dist	60.4	26	27	22	22	8.9	8.9
Palm Springs/Desert Wtr Agency	64.9	150	150	120	120	7.0	7.0
San Diego	66.2	224	235	131	135	8.0	8.2
San Jose/California Wtr Serv Co	68.0		181		134		7.9
Santa Barbara/Santa Barbara Pub	66.2	550	550	210	200	8.1	7.9
Santa Monica	68.0	400	180	220	220	7.8	8.1
Sunnyvale	60.1	300	140	200		8.0	7.3
Colorado							
Boulder	54.5	13	23			7.0	7.2
Colorado Spgs	42.3	34	36	22	29	7.5	7.7
Denver	50.0	102	99	72	63	8.0	7.5
Greeley	50.5	34	36	42	41	7.4	7.1
Connecticut							
Bridgeport	55.8	28	45	14	19	6.7	7.1
Hartford	55.4		20		8	6.5	7.1
New Britain	50.0	50	35	12	14	6.5	8.3
New Haven/Regional Wtr Authority	53.6	50	69	46	53	7.1	6.9
Delaware							
Newark/Artesian Wtr Co	55.4	77	77	40	45	5.8	7.4
District of Columbia							
Washington/Washington Aqueduct	62.6	127	137	77	70	8.0	7.7
Florida							
Clearwater/Pinellas Cnty Wtr Sys	75.2	215	215	200	210	7.2	7.8
Fort Lauderdale	78.8	259	90	231	59	7.6	9.6
Jacksonville	79.0	248	248	139	138	7.5	7.5
Miami	77.0	250	66	224	37	7.3	9.0
Orlando	75.2	125	125	120	120	7.8	7.4
Pompano Beach	68.0	240	63	198	30	7.1	8.8
Tallahassee	68.0		143				7.8
Tampa	72.5	107	151	90	85	7.5	7.7
W. Palm Beach/City of West Palm	68.2	71	102	58	58	8.1	8.4
Winter Haven	74.5	140	140	125	125	8.1	7.8
Georgia							
Acworth/Wyckoff Treatment Div	61.3	17	30	15	22	6.7	7.5
Marietta/Quarles Treatment Div	53.6	15	25	10	18	6.8	8.7
Stone Mountain/Dekalb Cnty Wtr	64.4	14	28	11	17	6.8	8.9
Hawaii							
Honolulu			60		60		8.0
Pearl Harbor	72.0	121		69		6.9	
Idaho							
Boise			117		223		7.6

(Continued)

Table 8D.38 (Continued)

State and Water Utility	Avg Temp Raw (°F)	Hardness ($CaCO_3$) Raw (mg/L)	Hardness ($CaCO_3$) Finished (mg/L)	Alkalinity ($CaCO_3$) Raw (mg/L)	Alkalinity ($CaCO_3$) Finished (mg/L)	pH Raw	pH Finished
Illinois							
Champaign/Northern Illinois Wtr	53.6	258	80	336	110	7.7	9.0
East St. Louis/Interurban Dist	60.8	228	228	159	144	7.7	7.2
Peoria/Peoria Dist/IL-Amer Wtr	53.6	250	250	170	161	8.0	7.6
Springfield	60.8	195	105	140	25	8.2	9.8
Indiana							
Bloomington	59.0	64	80	28	34	7.3	8.0
Fort Wayne	55.4	272	97	200	25	7.9	9.7
Indianapolis	57.2		252	170	159	8.1	7.6
South Bend	51.8	340	340	260	260	8.5	7.8
Iowa							
Cedar Rapids	53.6	276	129	217	67	7.7	9.5
Davenport/Davenport Water Co	55.9	200	200	155	129	8.2	7.1
Des Moines	51.1	304	145	239	63	8.3	9.3
Sioux City	54.0	445	445	267	267	7.2	7.5
Kansas							
Kansas City	62.6	294	272	205	215	8.1	8.0
Mission/Johnson Cnty Wtr Dist 1	59.0	245	121	191	54	7.8	9.1
Topeka	57.2	250	114	195	79	8.2	9.4
Kentucky							
Lexington/Kentucky–American Wtr	59.9	165	148	66	72	7.7	8.1
Louisville	57.2	138	148	66	65	7.5	8.5
Louisiana							
Baton Rouge/Baton Rouge Wtr Wks		5	5		175	8.7	8.5
Jefferson	64.4	154	152	104	99	7.5	7.4
Lake Charles	69.8	110	110	160	175	7.1	8.0
New Orleans	64.0	161	117	106	61	8.0	10.1
Maine							
Portland	50.0	9	9	4		6.9	6.7
Maryland							
Annapolis/City of Annapolis Wtr	64.4	25	70	20	50	6.0	8.7
Baltimore		47	70	40	46	7.3	8.0
Glen Burnie/Anne Arundel Cnty	53.6	41	56			5.1	8.5
Massachusetts							
Fall River	46.0	2	3	1	5	5.7	8.5
Springfield	45.7	11	11	8	11	6.4	6.9
Weymouth	55.0	21	47	7	11	6.5	8.4
Michigan							
Ann Arbor	55.8	269	142	210	60	8.0	9.3
Grand Rapids	45.9	140	141	115	106	8.4	7.6
Lansing	51.8	412	88	318	39	7.0	9.4
Saginaw		96	106	82	84	8.0	8.2
Minnesota							
Bloomington	50.0	310	90	315	92	7.6	8.3
Duluth	39.6	45	45	43	36	7.9	7.1
St. Paul	51.8	175	92	166	62	8.1	8.4
Mississippi							
Jackson	65.7	19	52	15	17	6.5	8.8
Missouri							
Independence/Missouri Wtr Co	59.5	297	119	227	48	7.3	9.6
Kansas City/Kansas City, Mo Wtr	55.0	261	173	182	85	8.2	9.5
Springfield	61.5	150	152	132	123	7.6	7.3
St. Louis/St. Louis Cnty Wtr Co	57.2	221	124	155	48	8.3	9.5
Lincoln	54.0	240	240	180	180	7.8	7.8
Omaha	53.2	286	185	177	68	8.2	9.0
Nevada							
Las Vegas	59.4	288	287	128	130	8.0	7.9
Reno/Sierra Pacific Power Co	53.6	40	40	40	27	7.7	6.9
New Hampshire							
Manchester	59.0	11	11	3	4	6.3	7.2
Nashua/Pennichuck Wtr Wks	59.0	30	50	9	16	6.4	7.3
New Jersey							
Clifton/Passaic Valley Wtr Comm	54.0	86	85	58	55	7.2	7.1
East Orange	50.0	280	280	158	158	7.7	7.7

(Continued)

Table 8D.38 (Continued)

State and Water Utility	Avg Temp Raw (°F)	Hardness ($CaCO_3$) Raw (mg/L)	Hardness ($CaCO_3$) Finished (mg/L)	Alkalinity ($CaCO_3$) Raw (mg/L)	Alkalinity ($CaCO_3$) Finished (mg/L)	pH Raw	pH Finished
Elizabeth/Elizabethtown Wtr Co	51.8	73	86	47	40	7.5	7.2
Harrington Park/Hackensack Wtr	55.4	120	120	85	80	7.4	7.9
Parsippany	52.0	177	177	139	139	7.1	7.1
Short Hills/Commonwealth Wtr Co	55.0	72	135	42	77	7.8	7.4
Toms River	65.3	12	60	38	70	6.2	7.4
Wanaque/North NJ Dist Wtr Supply	50.0	28	36	16	20	6.7	7.4
New Mexico							
Albuquerque	69.8	120	120	138	138	7.3	7.3
Santa Fe/Sangre De Cristo Wtr Co	59.0	150	150	200	200	7.8	7.8
New York							
Albany	48.4	43	54	38	48	7.3	8.9
Buffalo/Tonawanda Wtr Dept	55.4	140	135	95	90	8.2	7.8
East Meadow/Hempstead Wtr Dept		30	30	5	30	5.6	8.6
Lake Success/Jamaica Water Co.		50	50	30	50	6.3	7.2
Massapequa	62.1	8	13	6	30	5.8	7.2
New York/New Bur Wtr Supply			65	37		7.3	
Rochester/Monroe Cnty Wtr Auth	55.9	130	130	95	90	7.8	7.4
Syracuse	52.0	120	120	100	100	8.2	8.1
Yonkers	59.0	110	100	80	65	7.1	6.5
North Carolina							
Charlotte	64.9	13	28	12	19	7.2	9.2
Greensboro	64.4	27	44	24	26	7.0	7.7
Raleigh	68.0	20	40	29	30	6.8	7.4
Winston-Salem/Cnty Util Com	60.8	16	30	16	20	7.0	7.4
North Dakota							
Fargo	44.6	289	123	203	84	8.1	9.1
Ohio							
Akron	54.3	112	112	78	74	7.6	7.3
Cincinnati	60.1	112	130	44	54	7.6	8.5
Cleveland	50.9	125	125	92	84	8.0	7.5
Dayton		362	149	278	62	7.5	8.6
Toledo		127	74	93	38	8.1	9.2
Oklahoma							
Oklahoma City	63.9	154	101	97	40	8.2	10.3
Tulsa	66.2	140	136	97	102	8.0	8.2
Oregon							
Eugene	55.4	24	22	23	22	7.5	7.3
Medford	44.6	33	33	35	35	6.8	6.8
Portland	50.0	12	12	12	10	7.1	6.8
Pennsylvania							
Allentown	55.9	176	199	128	144	7.7	7.6
Bryn Mawr/Philadelphia Suburban	58.3	109	208	35	38	7.0	7.4
Hershey/Riverton Consol Wtr Co	58.6	145	145	121	110	8.0	7.4
Lancaster	57.2	195	208	135	126	7.8	7.6
Philadelphia	55.4	148	132	64	53	7.6	7.0
Pittsburgh/Western Pennsylvania	56.8	111	122	21	31	7.3	7.3
Rhode Island							
Newport	60.8	60	70	22	26	6.5	7.5
West Warwick/Kent Cnty Wtr Auth		32	32	13	13	5.8	6.8
South Carolina							
Anderson	66.0	6	6	9	10	7.0	7.1
Charleston	68.0	29	29	24	30	6.8	8.2
Spartanburg	62.6	11	20	11	11	6.9	7.1
South Dakota							
Rapid City	55.0	283	283	180	180	7.7	7.5
Tennessee							
Chattanooga/Tennessee-American	64.9	68	77	53	50	7.1	7.2
Knoxville/Knoxville Utils Brd	60.8	81	83	78	70	7.6	7.5
Memphis	63.3	47	47	55	55	6.4	7.2
Nashville/Metro Dept Wtr & Swr	60.8	90	90	65	65	7.6	8.0
Texas							
Arlington	69.8	109	110	90	90	8.0	8.2
Dallas	64.4	130	86	110	50	8.0	8.9
El Paso	68.0	250	150	192	80	8.2	8.7

(Continued)

Table 8D.38 (Continued)

State and Water Utility	Avg Temp Raw (°F)	Hardness ($CaCO_3$) Raw (mg/L)	Hardness ($CaCO_3$) Finished (mg/L)	Alkalinity ($CaCO_3$) Raw (mg/L)	Alkalinity ($CaCO_3$) Finished (mg/L)	pH Raw	pH Finished
Fort Worth	66.2	105	107	87	93	8.0	8.5
Lubbock	60.1	223	222	177	166	8.3	7.8
San Antonio	75.7	250		215		7.2	
Wichita Falls	69.8	125	70	125	50	8.1	9.4
Utah							
Ogden	53.6	170	125	131	118	7.4	7.1
Salt Lake City/Salt Lake Cnty Co	46.0	159	156	137	132	7.4	7.3
Vermont							
Burlington	50.0	69	59	47	47	7.6	8.0
Virginia							
Merrifield/Fairfax Cnty Wtr Auth	59.0	60	90	33	44	7.3	7.5
Newport News	60.8	80	80	55	50	7.4	7.1
Norfolk	64.9	45	69	40	37	6.8	7.0
Richmond/City of Richmond	57.2	70	70	45	35	7.6	7.5
Washington							
Everett	51.1	12	12	11	20	6.5	7.2
Seattle/Seattle Wtr Dept	47.3	15	17	11	16	7.2	7.7
Tacoma	50.0	15	15	20	20	6.9	6.9
Vancouver	53.0	98	98			6.7	6.7
West Virginia							
Charleston/W VA Wtr Co - Kanawha	57.9	47	62	19	26	6.8	8.7
Huntington	69.0	109	121	39	38	7.4	7.3
Wheeling	56.5	109	129	32	40	7.4	8.9
Wisconsin							
Green Bay	46.4		130	132	118	8.2	7.6
Madison	50.0	350	350	300	300	7.4	7.4
Milwaukee		93	93	112	104	8.2	7.5
Racine	44.3	140	140	112	107	8.3	7.7
Wyoming							
Casper	48.9	175	208		124	8.1	7.4
Cheyenne	61.9	110	110	63	65	7.2	7.1

Note: Average temperature, hardness, alkalinity, and pH; selected systems only. The quality of water supplied by municipal water systems in 1996 is available from the American Water Works Association, American Water Works Association WaterStats2ev2, Water:\STATS 1996 Survey, Water Quality, www.awwa.org.

Source: From American water works association 1984 Water Utility Operating Data. Copyright AWWA.

Table 8D.39 Radium-226, Radium-228, Radon-222, and Uranium in Public Drinking Water Supplies in the United States (Population-Weighted Average Activates)

State	Radium-226 (pCi/L)	Radium-228 (pCi/L)	Radon-222 (pCi/L)	Uranium (μg/L)
Alabama	0.202	1.00	420.1	0.30
Alaska	0.180	1.00	128.5	0.16
Arizona	0.180	1.03	1435.1	3.65
Arkansas	0.285	1.01	100.0	0.15
California	0.332	1.02	228.4	1.54
Colorado	0.199	1.00	329.9	6.81
Connecticut	0.206	1.00	1208.9	1.24
Delaware	0.180	1.62	123.3	0.10
Florida	0.341	1.01	127.3	0.22
Georgia	1.274	1.01	563.4	1.31
Hawaii	—	—	—	—
Idaho	0.181	1.00	437.4	2.60
Illinois	5.290	4.24	193.2	0.36
Indiana	0.347	1.22	187.4	1.05
Iowa	0.390	1.02	136.4	0.96
Kansas	0.195	1.00	396.1	3.39
Kentucky	0.184	1.00	205.5	0.38
Louisiana	0.218	1.00	108.2	0.12
Maine	0.180	1.00	1228.4	0.42
Maryland	0.486	1.00	266.1	0.08
Massachusetts	0.180	1.00	587.8	0.48
Michigan	0.357	1.10	185.2	0.23
Minnesota	1.899	1.82	388.7	0.87
Mississippi	0.390	1.10	104.3	0.10
Missouri	1.283	1.13	143.7	1.27
Montana	0.244	1.00	344.6	1.95
Nebraska	0.281	1.04	351.6	3.56
Nevada	0.180	1.00	743.2	2.85
New Hampshire	0.255	1.00	2673.5	1.70
New Jersey	0.181	1.00	137.1	0.09
New Mexico	0.396	1.15	309.1	7.99
New York	0.185	1.01	223.7	0.24
North Carolina	0.389	1.14	2277.7	1.13
North Dakota	0.194	1.02	114.0	0.90
Ohio	0.373	1.02	175.2	0.91
Oklahoma	0.383	1.00	158.0	4.02
Oregon	0.180	1.00	118.2	0.14
Pennsylvania	0.227	1.00	507.8	0.83
Rhode Island	0.180	1.00	1170.0	0.10
South Carolina	0.250	1.06	557.7	0.52
South Dakota	0.185	1.00	281.6	3.84
Tennessee	0.228	1.01	113.7	0.16
Texas	0.327	1.03	150.5	0.86
Utah	0.252	1.00	226.8	2.94
Vermont	0.189	1.00	997.1	0.42
Virginia	0.215	1.00	485.4	0.61
Washington	0.180	1.00	432.5	1.52
West Virginia	0.189	1.00	263.6	0.22
Wisconsin	2.688	3.32	367.2	0.95
Wyoming	0.770	1.41	558.0	1.32
United States	0.905	1.41	249.0	0.82

Note: Minimum Reporting Limit (MRL) was used in the average for those cases in which the activity or concentration was les than the MRL. The international unit of activity is the Becquerel (Bq), which is approximately equal to 27 pCi.

Source: From Longtin, J., 1990, Chapter 8, Occurrence of radionuclides in drinking water, a national study In Cothern, R. and Rebers, P (Editors), 1990, *Radon, Radium, and Uranium in Drinking Water*, Lewis Publishers, Inc, Chelsea, Michigan.

Original Source: From Longtin, J., 1988.

Table 8D.40 Occurrence of Selected Radionuclides in Groundwater Used for Drinking Water in the United States

	Concentration (Picocuries/L)				
Radionuclide	**Mean**	**Median**	**Standard Deviation**	**Maximum**	**Number of Samples**
Ra-224	3.2	0.3	10.1	73.6	99
Ra-226	1.6	0.4	2.8	16.9	99
Ra-228	2.1	0.5	7.9	72.3	99
Pb-210	0.6	0.5	0.5	4.1	96
Po-210	0.1	0.01	0.5	4.9	96

Source: From Focazio, M.J., et al., 2001, *Occurrence of Selected Radionuclides in Groundwater Used for Drinking Water in the United States: A Reconnaissance Survey*, 1998, USGS Water-Resources Investigations Report 00-4273, www.usgs.gov.

Table 8D.41 Concentrations of Uranium and Thorium Series Radionuclides in Drinking Water in the United States, Asia, and Europe

	Concentration (mBq kg^{-1})								
Region/country	^{238}U	^{230}Th	^{226}Ra	^{210}Pb	^{210}Po	^{232}Th	^{228}Ra	^{228}Th	^{235}U
North America	0.3–77	0.1	0.4–1.8	0.1–1.5		0.05	0–0.5		0.04
United States									
Asia									
China	0.1–700		0.2–120			0.04–12			
India	0.09–1.5								
Europe									
Finland	0.5–150,000		10–49,000	0.2–21,000	0.2–7,600		18–570		
France	4.4–930		7–700			0–4.2			
Germany	0.4–600		1–1,800	0.2–200	0.1–200				
Italy	0.5–130		0.2–1,200						
Poland	7.3	1.4	1.7–4.5	1.6	0.5	0.06			
Romania	0.4–37		0.7–21	7–44	7–44	0.04–9.3			
Switzerland	0–1,000		0–1,500				0–200		0–50
Spain	3.7–4.4		<20–4,000						
UK			0–180	40–200					
Reference value	1	0.1	0.5	10	5	0.05	0.5	0.05	0.04

Source: From The United Nations is the author of the original material. United Nations Scientific Committee on the Effects of Atomic Radiation, 2000, *UNSCEAR 2000 Report Vol. I, Sources and Effects of Ionizing Radiation.*

Table 8D.42 Radon in Public Drinking Water Supplies in the United States (Population-Weighted Average Activities (pCi/L))

	Sites with ≤ 1000 People		Sites with > 1000 People	
State	**Cothern**	**NIRS**	**Cothern**	**NIRS**
Alabama	160 (40)	2,025 (5)	160 (35)	171 (26)
Alaska	100 (47)	129 (44)	100 (47)	—
Arizona	120 (44)	1,302 (7)	320 (17)	1,610 (1)
Arkansas	75 (51)	—	75 (50)	100 (42)
California	500 (18)	538 (18)	500 (10)	161 (28)
Colorado	380 (23)	336 (29)	380 (14)	317 (12)
Connecticut	1,500 (3)	3,328 (1)	770 (4)	646 (2)
Delaware	100 (48)	116 (48)	126 (42)	126 (33)
Florida	1,000 (9)	393 (25)	148 (40)	118 (35)
Georgia	1,100 (6)	419 (24)	150 (37)	583 (4)
Hawaii	50 (52)	—	50 (51)	—
Idaho	256 (30)	431 (22)	256 (25)	438 (9)
Illinois	100 (49)	136 (40)	167 (34)	198 (20)
Indiana	105 (45)	136 (41)	105 (45)	195 (22)
Iowa	250 (31)	166 (35)	200 (29)	130 (32)
Kansas	250 (32)	365 (27)	106 (44)	370 (11)
Kentucky	250 (33)	148 (39)	110 (43)	220 (19)
Louisiana	180 (39)	116 (49)	180 (31)	107 (41)
Maine	10,000 (1)	1,228 (9)	2,000 (1)	—
Maryland	700 (15)	2,161 (4)	450 (11)	112 (36)
Massachusetts	1,500 (4)	253 (33)	770 (5)	596 (3)
Michigan	105 (46)	370 (26)	105 (46)	164 (27)
Minnesota	210 (36)	342 (28)	210 (28)	397 (10)
Mississippi	150 (41)	133 (42)	82 (49)	100 (43)
Missouri	300 (24)	125 (46)	100 (48)	148 (33)
Montana	500 (19)	535 (19)	328 (16)	112 (37)
Nebraska	300 (25)	291 (31)	290 (19)	444 (8)
Nevada	550 (17)	743 (12)	550 (9)	—
New Hampshire	1,400 (5)	2,674 (3)	1,183 (2)	—
New Jersey	150 (42)	737 (13)	300 (18)	125 (34)
New Mexico	200 (37)	423 (23)	180 (32)	250 (16)
New York	500 (20)	647 (14)	132 (41)	173 (25)
North Carolina	1,100 (7)	2,876 (2)	278 (21)	100 (44)
North Dakota	300 (26)	125 (47)	150 (38)	109 (40)
Ohio	200 (38)	164 (36)	169 (33)	177 (24)
Oklahoma	250 (34)	164 (37)	160 (36)	158 (29)
Oregon	300 (27)	130 (43)	264 (23)	112 (38)
Pennsylvania	1,000 (10)	467 (20)	720 (6)	535 (5)
Rhode Island	3,400 (2)	1,170 (10)	1,151 (3)	—
South Carolina	1,100 (8)	1,260 (8)	276 (22)	196 (21)
South Dakota	300 (28)	334 (30)	290 (20)	273 (15)
Tennessee	100 (50)	128 (45)	24 (52)	112 (39)
Texas	150 (43)	264 (32)	150 (39)	138 (31)
Utah	500 (21)	157 (38)	360 (15)	238 (18)
Vermont	250 (35)	1,533 (6)	656 (8)	497 (7)
Virginia	700 (16)	952 (11)	450 (12)	313 (13)
Washington	300 (29)	238 (34)	264 (24)	520 (6)
West Virginia	1,000 (11)	459 (21)	720 (7)	240 (17)
Wisconsin	750 (14)	540 (17)	234 (27)	300 (14)
Wyoming	880 (12)	558 (16)	415 (13)	—
Puerto Rico	500 (22)	—	200 (30)	—
United States	780 (13)	602 (15)	240 (26)	194 (23)

Note: Numbers in parentheses are relative rankings; The international unit of activity is the Becquerel (Bq), which is approximately equal to 27 pCi.

Source: From Longtin, J., 1990, Chapter 8, Occurrence of radionuclides in drinking water, a national study In Cothern, R. and Rebers, P (Editors), 1990, *Radon, Radium, and Uranium in Drinking Water*, Lewis Publishers, Inc, Chelsea, Michigan.

Original Source: Longtin, J., 1988.

Table 8D.43 Number of Public Drinking Water Supplies in the United States Exceeding Various Levels of Radon

		Estimated Numbers That Exceed the Concentration in Column 2	
Lifetime Risk Level	Radon Concentration (pCi/L)	Public Drinking Water Supplies	Population thousands
10^{-3}	10,000	500–4000	20–300
10^{-4}	1,000	1000–10,000	200–4000
10^{-5}	100	5000–30,000	10,000–100,000
10^{-6}	10	10,000–40,000	50,000–100,000

Note: Rounded off to one significant figure.

Source: From Cothern, C.R., 1987, Estimating the health risk of radon in drinking water, *J. Am. Water Works Assoc.*, vol. 79, no. 4. Copyright AWWA. Reprinted with permission.

Table 8D.44 Occurrence of Radon in Well Water in the United States

State	Number of Wells Sampled	Range of Detected Radon Levels (pCi/L)	Mean Radon Concentration (pCi/L)	Associated Error ± pCi/L
Arizona	5	434–681	582	105
California	44	<100–2,003	589	91
Connecticut	3	757–984	841	98
Iowa	6	All <100	12	75
Illinois	16	182–714	449	115
Indiana	28	<100–624	324	106
Massachusetts	28	<100–3,288	1,145	101
New Hampshire	12	880–4,609	1,716	134
New Jersey	113	<100–3,805	394	79
New Mexico	36	<100–678	253	266
Ohio	10	<100–343	148	116
Pennsylvania	64	<100–4,622	1,570	89
Rhode Island	3	640–787	702	61
Virginia	2	465–468	467	53
West Virginia	7	<100–281	93	42
Cumulative	377	<100–4,622	686	104

Note: Range of detected radon concentrations and corresponding mean radon levels.

Source: From Dixon, K.L., and Lee, R.G., 1988, Occurrence of radon in well supplies, *J. Am. Water Works Assoc.*, vol. 80, no. 7. Copyright AWWA. Reprinted with permission.

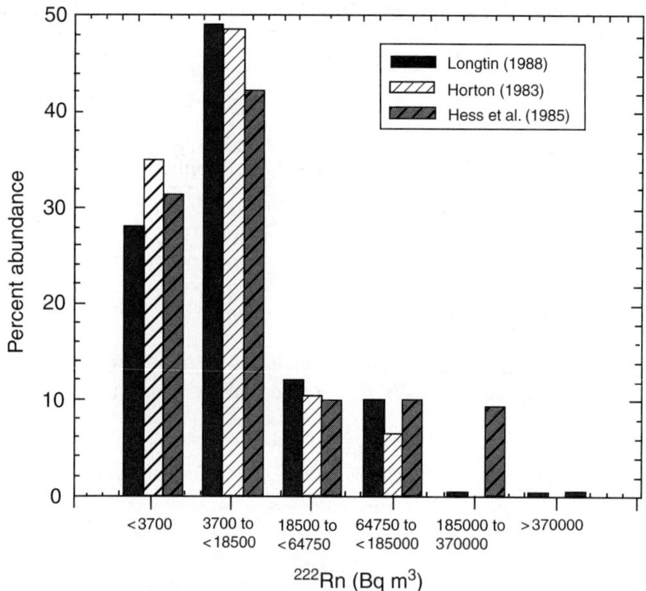

Figure 8D.12 Distributions of radon in drinking water in several studies in the United States. (Reprinted with permission from (*Risk Assessment of Radon in Drinking Water*) © (1999) by the National Academy of Sciences, Courtesy of the National Academies Press, Washington, DC)

Table 8D.45 Aluminum in Public Drinking Water Supplies in the United States

Category	Samples	Samples With Concentrations >0.05 mg/L percent	Samples with Concentrations >0.014 mg/L			Overall Median (mg/L)
			Percent	Median (mg/L)	Maximum (mg/L)	
Region						
I	46	13	33	0.043	0.179	<0.014
II	71	25	44	0.066	0.249	<0.014
III	123	37	54	0.070	2.670	0.022
IV	80	54	68	0.161	0.449	0.060
V	100	51	64	0.082	2.160	0.051
VI	35	29	60	0.040	0.889	0.029
VII	53	2	13	0.026	0.051	<0.014
VIII	105	30	47	0.083	2.580	<0.014
IX	89	29	52	0.053	1.167	0.020
X	14	0	7			<0.014
Population served						
25–9999	286	15	28	0.051	1.167	<0.014
10,000–99,999	92	39	61	0.087	2.580	0.023
100,000–999,999	222	48	66	0.094	2.670	0.045
≥1,000,000	116	38	62	0.058	0.402	0.033

Note: Finished water; by USEPA region and population category.

Source: From Miller, R.G. and others, 1984. The occurrence of aluminum in drinking water, *J. Am. Water Works Assoc.*, vol. 76, no. 1. Copyright Am. Water Works Assoc. Reprinted with permission.

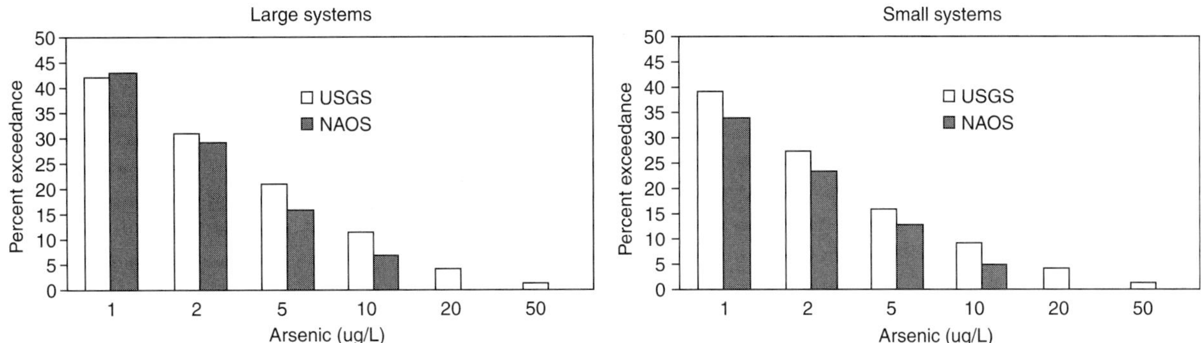

Figure 8D.13 Exceedance frequency of arsenic concentrations in small and large regulated water supply systems in the United States. (From Welch, A.H., et al. 1999, Arsenic in groundwater supplies of the United States, In: *Arsenic Exposure and Health Effects*, W.R. Chappell, C.O. Abernathy and R.L. Calderon, Eds., Elsevier Science, New York, pp. 9–17, http://water.usgs.gov.)

Table 8D.46 Statistical Results of Differences in Arsenic Concentrations in Water Collected from Public Water-Supply Wells and Other Types of Wells, by Physiographic Provinces in the United States

Physiographic Province	Number of Samples Public Water-Supply Wells / All Other Wells	Mean (μg/L) Public Water-Supply wells / All Other Wells	Median (μg/L) Public Water-Supply Wells / All Other Wells	95th Percentile (μg/L) Public Water-Supply Wells / All Other Wells	99th Percentile (μg/L) Public Water-Wupply Wells / All Other Wells	Wilcoxon Test Statistic[a] (p>/z/)
1. Appalachian Highlands	376 2,212	1 3	≤1 ≤1	5 8	10 25	0.6552
2. Atlantic Coast Plain	646 2,047	1 2	≤1 ≤1	2 6	7 21	0.0067
3. Interior Highlands, Interior Plains, and Laurentian Upland	342 3,947	5 5	≤1 ≤1	19 16	75 48	0.3289
4. Intermontane Plateaus	458 4,640	9 15	3 3	39 44	100 200	0.1389
5. Pacific Mountain System	303 2,401	6 9	2 2	21 27	92 82	0.7159
6. Rocky Mountain System	74 1,028	2 7	≤1 ≤1	6 20	30 100	0.6444

Note: μg/L, micrograms per liter; ≥, greater than or equal to; <, less than.

[a] A value < 0.05 indicates the two data sets are different.

Source: From Focazio, M.J., et al., 2000, A Retrospective Analysis on the Occurrence of Arsenic in Groundwater Resources of the United States and Limitations in Drinking-Water-Supply Characterizations, *USGS, Water Resources Investigations Report 99-4279*, www.usgs.gov. With permission.

Locations of major physiographic provinces of the United States

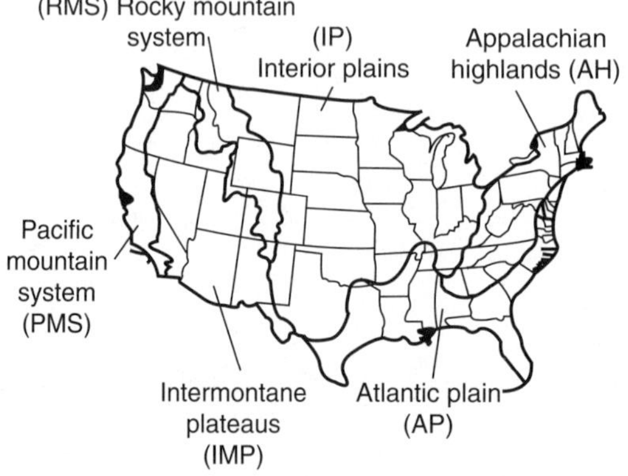

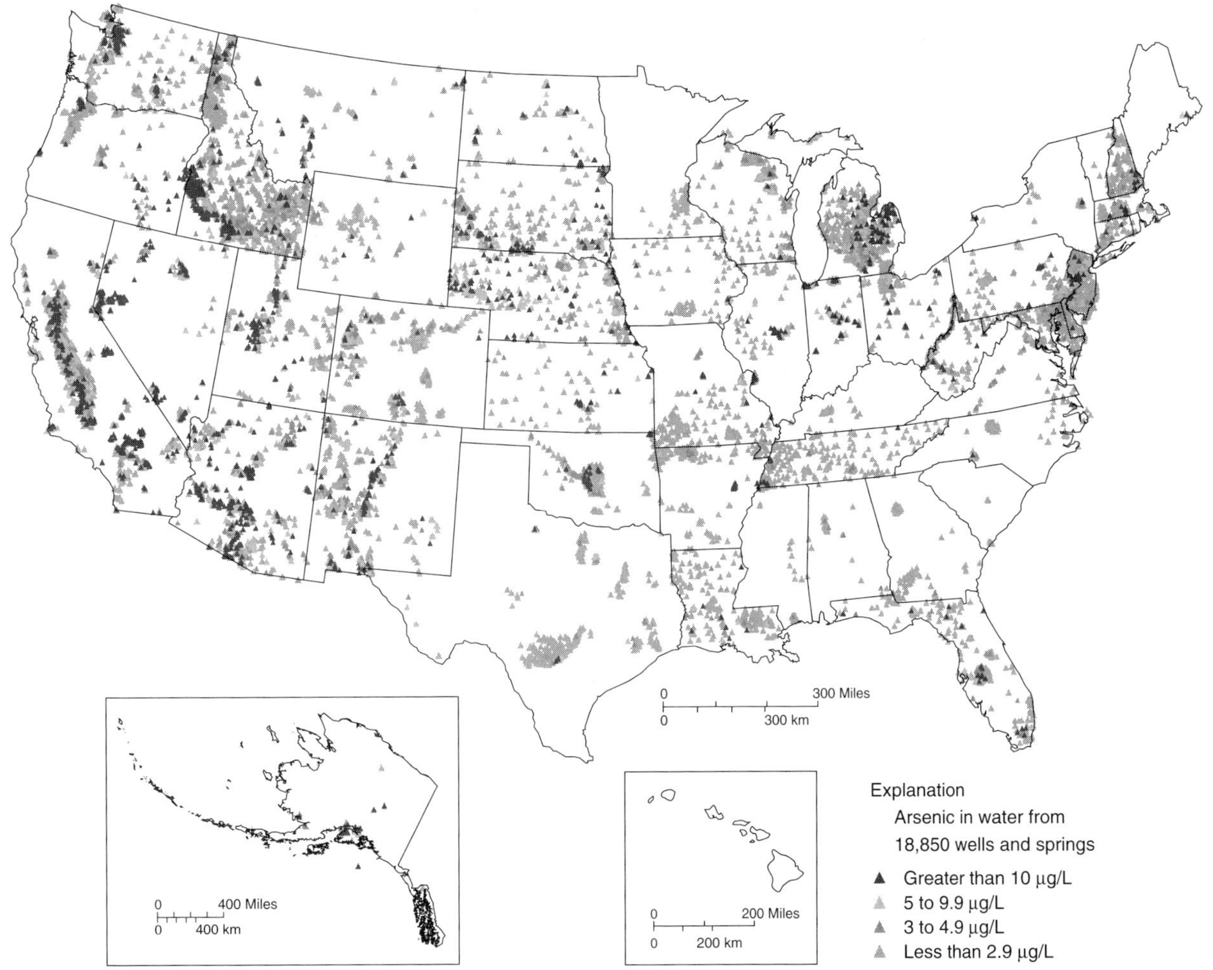

Figure 8D.14 Locations and concentration ranges of samples in the USGS arsenic point data base. (From Focazio, M.J., et al. A Retrospective Analysis on the Occurrence of Arsenic in Groundwater Resources of the United States and Limitations in Drinking-Water-Supply Characterizations, USGS, *Water Resources Investigations Report 99–4279*, 2000, http://water.usgs.gov.)

Table 8D.47 Estimated Arsenic Occurrence in United States Groundwater Community Water Systems

System Size (Population Served)	Total Number of Systems[a]	Number of Systems with Arsenic Concentrations (μg/L)									
		>2	>3	>5	>10	>15	>20	>25	>30	>40	>50
<25	178	49	35	22	9	5	4	3	2	1	1
25–100	14,025	3,833	2,788	1,696	743	429	281	199	147	90	60
101–500	14,991	4,097	2,980	1,812	795	459	300	213	157	96	64
501–1,000	4,671	1,277	929	565	248	143	93	66	49	30	20
1,001–3,300	5,710	1,561	1,135	690	303	175	114	81	60	37	25
3,301–10,000	2,459	672	489	297	130	75	49	35	26	16	11
10,001–50,000	1,215	332	242	147	64	37	24	17	13	8	5
50,001–100,000	131	36	26	16	7	4	3	2	1	1	1
100,001–1,000,000	61	17	12	7	3	2	1	1	1	0	0
> 1,000,000	2	1	0	0	0	0	0	0	0	0	0
Total Systems[b]	434,43	11,873	8,636	5,252	2,302	1,329	869	617	456	278	187
Lower 95% CI		11,543	8,363	5,100	2,250	1,269	821	573	421	252	165
Upper 95% CI		13,007	9,501	5,665	2,567	1,499	995	712	534	335	226

Note: CI, confidence interval.

[a] Based on 1998 Baseline SDWIS data for purchased and non-purchased systems. Systems characterized as GW under the influence of SW are considered to be surface water systems.

[b] Totals may not add up due to rounding of the number of systems to the nearest whole number.

Source: From USEPA, 2000, *Arsenic Occurrence in Public Drinking Water Supplies*, EPA-815-R-00-023, December 2000, www.epa.gov.

Table 8D.48 Estimated Arsenic Occurrence in United States Surface Water Community Water Systems

System Size (Population Served)	Total Number of Systems[a]	Number of Systems with Arsenic Concentrations (μg/L)									
		>2	>3	>5	>10	>15	>20	>25	>30	>40	>50
<25	74	7	4	2	1	0	0	0	0	0	0
25–100	1,001	98	56	30	8	5	3	2	2	1	1
101–500	1,983	195	110	60	16	9	6	5	4	3	2
501–1,000	1,219	120	68	37	10	6	4	3	2	2	1
1,001–3,300	2,420	238	135	73	19	11	8	6	5	3	2
3,301–10,000	1,844	181	103	56	15	9	6	4	3	2	2
10,001–50,000	1,606	158	89	49	13	7	5	4	3	2	2
50,001–100,000	300	29	17	9	2	1	1	1	1	0	0
100,001–1,000,000	261	26	15	8	2	1	1	1	0	0	0
> 1,000,000	13	1	1	0	0	0	0	0	0	0	0
Total Systems	10,721	1,052	597	325	86	50	34	26	20	14	10
Lower 95% CI		973	514	193	56	25	14	9	6	3	2
Upper 95% CI[b]		2,730	2,212	1,036	167	107	88	77	71	65	63

Note: CI, confidence interval.

[a] Based on 1998 Baseline SDWIS data for purchased and non-purchased systems. Systems characterized as GW under the influence of SW are considered to be surface water systems.

[b] Totals may not add up due to rounding of the number of systems to the nearest whole number.

Source: From USEPA, 2000, *Arsenic Occurence in Public Drinking Water Supplies*, EPA-815-R-00-023, December 2000, www.epa.gov.

Table 8D.49 Summary of Documented Cases of Naturally-Occurring Arsenic Problems in World Groundwaters

Country/Region	Area (km^2)	Population Exposed[a]	Concentration Ranges (μg L)	Aquifer Type	Groundwater Conditions	Reference
Bangladesh	150,000	ca. 3×10^7	<0.5–2,500	Holocene alluvial/deltaic sediments. Abundance of solid organic matter	Strongly reducing, neutral pH, high alkalinity, slow groundwater flow rates	DPHE/BGS/MML (1999)
West Bengal	23,000	6×10^6	<10–3,200	As Bangladesh	As Bangladesh	CGWB (1999); POA (1999)
China		5.6×10^6				Sun et al. (2000)
Taiwan	4000	10^5 (formerly)	10–1820	Sediments, including black shale	Strongly reducing, artesian conditions, some groundwaters contain humic acid	Kuo (1968), Tseng et al. (1968)
Inner Mongolia (Huhhot Basin (HB), Bayingao, Hexi, Ba Meng, Tumet Plain)	4300 (HB) 30,000 total	ca. 10^5 in HB	<1–2400	Holocene alluvial and lacustrine sediments	Strongly reducing conditions, neutral pH, high alkalinity. Deep groundwaters often artesian, some have high concentrations of humic acid	Luo et al. (1997), Zhai et al. (1998), Ma et al. (1999), Sun et al. (1999), Smedley et al. (2000b, 2001b)
Xinjiang (Tianshan Plain)	38,000	(500 diagnosed)	40–750	Holocene alluvial plain	Reducing, deep wells (up to 660 m) are artesian	Wang and Huang (1994)
Shanxi				Alluvial plain	Reducing	Sun et al. (1999)
Hungary, Romania (Danube Basin)	110,000	29,000	<2–176	Quaternary alluvial plain	Reducing groundwater, some artesian. Some high in humic acid	Varsányi et al. (1991); Gurzau (2000)
Argentina (Chaco-Pampean Plain	10^6	2×10^6	<1–5300 (7800 in some porewaters)	Holocene and earlier loess with rhyolitic volcanic ash	Oxidizing, neutral to high pH, high alkalinity. Groundwaters often saline. As(V), accompanied by high B, V, Mo, U. Also high As concentrations in some river waters	Nicolli et al., 1989; Nicolli and Merino (2001); Smedley et al. (2001a); Sancha and Castro (2000)
Northern Chile (Antofagasta)	125,000	500,000	100–1000	Quaternary volcanogenic sediment	Generally oxidizing. Arid conditions, high salinity, high B. Also high-As river waters	Cáceres et al. (1992), Karcher et al. (1999); Sancha and Castro (2000)
Southwest U.S.A.		3.5×10^5 (tot)				Smith et al. (1992)
Basin & Range, Arizona	200,000		up to 1300	Alluvial basins, some evaporites	Oxidizing, high pH. As (mainly As (V)) correlates positively with Mo, Se, V, F	Robertson (1989)

Tulare Basin, San Joaquin Valley, California	5000		<1–2600	Holocene and older basin-fill sediments	Internally-drained basin. Mixed redox conditions. Proportion of As(III) increases with well depth. High salinity in some shallow groundwaters. High Se, U, B, Mo	Fujii and Swain (1995)
Southern Carson Desert, Nevada	1300		up to 2600	Holocene mixed aeolian, alluvial, lacustrine sediments, some thin volcanic ash bands	Largely reducing, some high pH. Some with high salinity due to evaporation. Associated high, U, P, Mn, DOC (Fe to a lesser extent)	Welch and Lico (1998)
Salton Sea Basin					Some saline groundwaters, with high U	Welch and Lico (1998)
Mexico (Lagunera)	32,000	4×10^5	8–620	Volcanic sediments	Oxidising, neutral to high pH, As mainly as As(V)	Del Razo et al. (1990)
Some problem areas related to mining activity and mineralized areas						
Thailand (Ron Phibun)	100	15,000	1–5000	Dredged quatemary alluvium (some problems in limestone), tailings	Oxidation of disseminated arsenopyrite due to former tin mining, subsequent groundwater rebound	Williams et al. (1996), Williams (1997)
Greece (Lavrion)				Mine tailings	Mining	
Fairbanks, Alaska, U.S.A.			up to 10,000	Schist, alluvium, mine tailings	Gold mining, arsenopyite, possibly scorodite	Wilson and Hawkins (1978); Welch et al. (1988)
Moira Lake, Ontario, Canda	100		50–3000	Mine tailings	Ore mining (gold, hematite, magnetite, lead, cobalt)	Azcue and Nriagu (1995)
Coeur d'Alene, Idaho, U.S.A.			up to 1400	Valley-fill deposits	River water and groundwater affected by lead-zinc-silver mining	Welch et al. (1988). Mok and Wai (1990)
Lake Oahe, South Datoka, U.S.A.			up to 2000	Lake sediments	As in sediment porewaters from gold mining in the Black Hills	Ficklin and Callender (1989)
Bowen Island, British Colombia	50		0.5–580	Sulphide mineral veins in volcanic country rocks	Neutral to high-pH groundwaters (up to 8.9), As correlated with B, F	Boyle et al. (1998)

[a] Exposed refers to population drinking water with As $> 50\ \mu g\ L^{-1}$ (drinking-water standard of most countries).

Source: From Smedley, P.L. and Kinniburgh, D.G., Chapter 1, Source and Behavior of Arsenic in Natural Waters in WHO, 2001, *United Nations Synthesis Report on Arsenic in Drinking Water Developed on Behalf of the United Nations Administrative Committee on Cooperation Sub-Committee on Water Resources*, with active participation of UNICEF, UNIDO, IAEA and the World Bank, April 20, 2001, www.who.int.

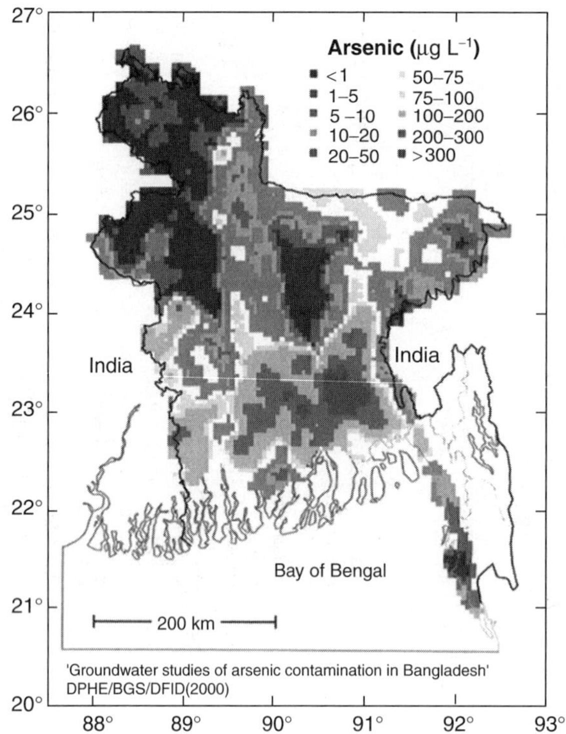

Figure 8D.15 Smoothed map showing the regional trends in groundwater arsenic concentrations in shallow wells in Bangladesh. (From British Geological Survey (BGS) and Government of the People's Republic of Bangladesh, Ministry of Local Government, Rural Development Co-operatives, Department of Public Health Engineering, 2001, *Arsenic Contamination of Groundwater in Bengladesh*, Kinniburgh, D.G. and Smedley, P.L. (eds), vol. 1: Summary, British Geological Survey (BGS) technical Report WC/00/19, British Geological Survey, Keyworth, www.bgs.ac.uk.)

[Total number of arsenic affected districts 9 and blocks 85]

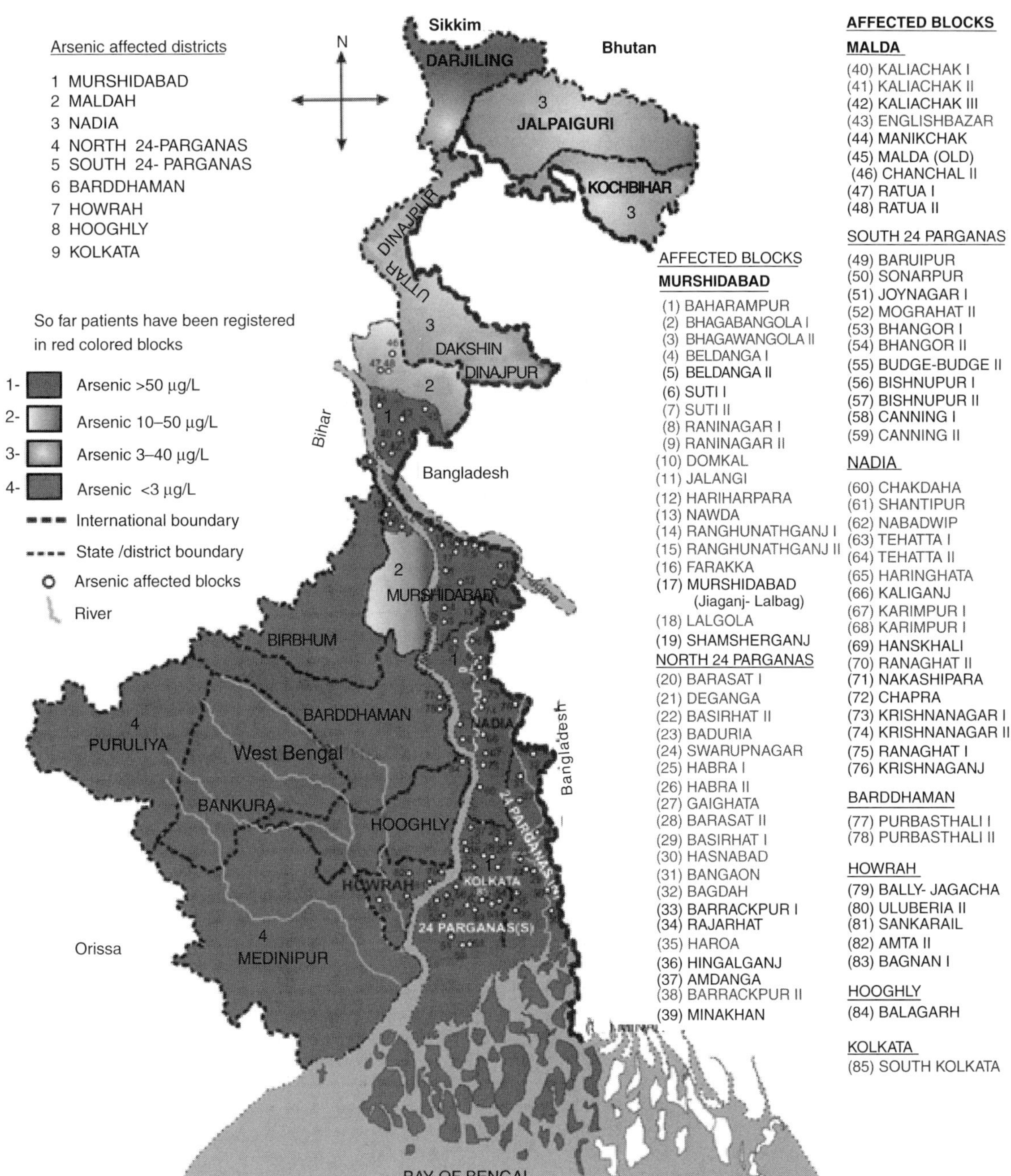

Figure 8D.16 Groundwater arsenic contamination status in West Bengal-India. (From School of Environment Studies, Jadavpur University, Kolkata 700 032, India, Groundwater and Arsenic Contamination Stutus in West Bengal-India, www.soesju.org.)

Table 8D.50 The Effects of Too Little — and Too Much — Fluoride

Level in Water	Effects
0.8–1.2 mg/L	Prevention of tooth decay, strengthening of skeleton
Above 1.5 mg/L	Fluorosis: pitting of tooth enamel and deposits in bones
Above about 10 mg/L	Crippling skeletal fluorosis

Note: Fluoride is a desirable substance: it can prevent or reduce dental decay and strengthen bones, thus preventing bone fractures in older people. Where the fluoride level is naturally low, studies have shown higher levels of both dental caries (tooth decay) and fractures. Because of its positive effect, fluoride is added to water during treatment in some areas with low levels. But you can have too much of a good thing; and in the case of fluoride, water levels above 1.5 mg/L may have long-term undesirable effects. Much depends on whether other sources, such as vegetables, also have high levels. The risk of toxic effect rises with the concentration. It only becomes obvious at much higher levels than 1.5 mg/L.

Source: From World Health Organization, Water Sanitation and Health (WSH), *Naturally Occuring Hazards—Fluoride*, www.who.int/water_sanitation_health/naturalhazards/en/index2.html. Downloaded July 17, 2005. Copyright © World Health Organization.

Table 8D.51 National Surveys of Fluoride in Drinking Water

Location	Fluoride Concentration (mg/L)	Comment	References
Canada	0.05–0.21	Range of mean concentration in nonfluoridated[a] samples collected between 1984 and 1989 from 67 communities in 5 provinces	Health Canada (1993)
	0.73–1.25	Range of mean concentrations in fluoridated[b] samples collected between 1986 and 1989 from 320 communities in 8 provinces	Health Canada (1993)
Czech Republic	0.05–3.0	Range of concentrations in more than 4000 samples of drinking water collected between 1994 and 1996 from 36 districts within the Czech Republic	NIPH (1996)
Finland	<0.1–3.0	Range of concentrations in 5900 groundwater samples	Lahermo et al. (1990)
	<0.1–5.8	Range of concentrations in 1421 well-water samples	Korkka-Niemi et al. (1993)
	<0.1–6.0	Range of concentrations in groundwater database of Geological Survey of Finland, 7229 shallow wells	Lahermo & Backman (2000)
	<0.1–9.3	Range of concentrations in groundwater database of Geological Survey of Finland, 571 drilled wells	Lahermo & Backman (2000)
Germany	0.02–0.17	Range of concentrations of drinking water collected from various facilities in Germany between 1975 and 1986	Bergmann (1995)
The Netherlands	0.04–0.23	Public drinking water samples collected in 1985 from water treatment plants in 12 provinces	Sloof et al. (1989)
Poland	0.02–3.0	Range of mean concentrations in samples of drinking water collected from 94 localities in central and northern Poland between 1993 and 1995	Czarnowski et al. (1996)
U.S.A.	<0.1–1.0	Fluoride levels in drinking water of approximately 62% of the U.S. population served by public supplies range from <0.1 to 1.2 mg/L; levels of fluoride in drinking water of approximately 14% of the US population served by public supplies range from 1 to 2 mg/L	US EPA (1985); US DHHS (1991)

[a] Drinking water in which inorganic fluoride was not intentionally added for the prevention of dental caries.
[b] Drinking water in which inorganic fluoride was intentionally added for the prevention.

Source: From WHO, 2002, *Fluorides, Environmental Health Criteria 227*, Copyright © World Health Organization 2002, www.who.int.

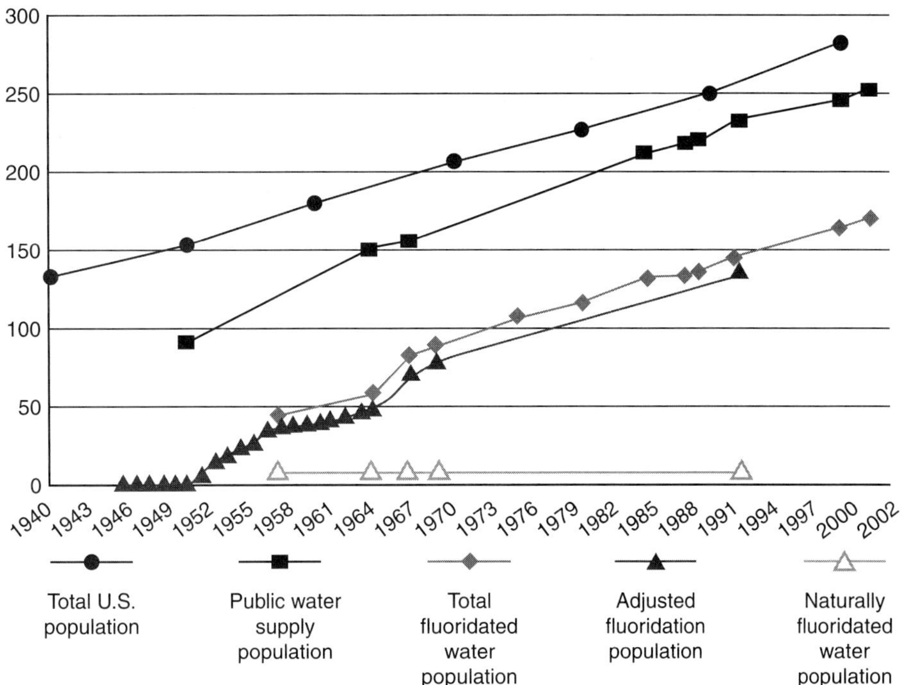

Figure 8D.17 Fluroidation growth, by population, in the United States 1940–2002. (From Center of Disease Control, www.cdc.gov.)

Table 8D.52 Number of Persons and Percentage of the Population Receiving Optimally Fluoridated Water through Public Water Systems (PWS), by State in the United States, 1992 and 2000

State	2000 Fluoridated Population	2000 Total PWS Population	2000 Percentage Fluoridated (%)	1992 Percentage Fluoridated (%)	Change in Percentage 1992–2000
Alabama[a]	3,967,059	4,447,100	89.2	82.6	6.6
Alaska	270,099	489,371	55.2	61.2	−6.0
Arizona	2,700,354	4,869,065	55.5	49.9	5.6
Arkansas[b]	1,455,767	2,431,477	59.9	58.7	1.2
California	9,551,961	33,238,057	28.7	15.7	13.0
Colorado[b]	2,852,386	3,708,061	76.9	81.7	−4.8
Connecticut	2,398,227	2,701,178	88.8	85.9	2.9
Delaware	505,747	624,923	80.9	67.4	13.5
District of Columbia	595,000	595,000	100.0	100.0	0.0
Florida	9,407,494	15,033,574	62.6	58.3	4.3
Georgia	6,161,139	6,634,635	92.9	92.1	0.8
Hawaii[a]	109,147	1,211,537	9.0	13.0	−4.0
Idaho	383,720	845,780	45.4	48.3	−2.9
Illinois	10,453,837	11,192,286	93.4	95.2	−1.8
Indiana	4,232,907	4,441,502	95.3	98.6	−3.3
Iowa	2,181,649	2,390,661	91.3	91.4	−0.1
Kansas	1,513,306	2,421,274	62.5	58.4	4.1
Kentucky	3,235,053	3,367,812	96.1	100.0	−3.9
Louisiana[a]	2,375,702	4,468,976	53.2	55.7	−2.5
Maine	466,208	618,033	75.4	55.8	19.6
Maryland[b]	4,124,953	4,547,908	90.7	85.8	4.9
Massachusetts[a,b]	3,546,099	6,349,097	55.8	57.0	−1.2
Michigan	6,568,151	7,242,531	90.7	88.5	2.2
Minnesota	3,714,465	3,780,942	98.2	93.4	4.8
Mississippi	1,227,268	2,665,075	46.0	48.4	−2.4
Missouri[a]	4,502,722	5,595,211	80.5	71.4	9.1
Montana	143,092	645,452	22.2	25.9	−3.7
Nebraska[b]	966,262	1,243,713	77.7	62.1	15.6
Nevada[b]	1,078,479	1,637,105	65.9	2.1	63.8
New Hampshire	347,007	807,438	43.0	24.0	19.0
Naw Jersey	1,120,410	7,208,514	15.5	16.2	−0.7
New Mexico	1,187,404	1,548,084	76.7	66.2	10.5
New York[b]	12,000,000	17,690,198	67.8	69.7	−1.9
North Carolina	4,862,220	5,837,936	83.3	78.5	4.8
North Dakota	531,738	557,595	95.4	96.4	−1.0
Ohio	8,355,002	9,535,188	87.6	87.9	−0.3
Oklahoma[b]	2,164,330	2,900,000	74.6	58.0	16.6
Oregon[b]	612,485	2,700,000	22.7	24.8	−2.1
Pennsylvania	5,825,328	10,750,095	54.2	50.9	3.3
Rhode Island	842,797	989,786	85.1	100.0	−14.0
South Carolina	3,086,974	3,383,434	91.2	90.0	1.2
South Dakota[b]	553,503	626,221	88.4	100.0	−11.6
Tennessee	4,749,493	5,025,998	94.5	92.0	2.5
Texas	11,868,046	18,072,680	65.7	64.0	1.7
Utah[a,b]	43,816	2,233,169	2.0	3.1	−1.1
Vermont	240,579	443,901	54.2	57.4	−3.2
Virginia	5,677,551	6,085,436	93.3	72.1	21.2
Washington[b]	2,844,893	4,925,540	57.8	53.2	4.6
West Virginia[a]	1,207,000	1,387,000	87.0	82.1	4.9
Wisconsin	3,108,738	3,481,285	89.3	93.0	−3.7
Wyoming[a]	149,774	493,782	30.3	35.7	−5.4
Total	162,067,341	246,120,616	65.8	62.1	3.7

[a] Reported PWS population exceeded total state population; PWS population was set to the 2000 U.S. census of state populations.
[b] Complete data were not available from Water Flouridation Reporting System: additional information was obtained from states.

Source: From CDC, 2002, *Populations Receiving Optimally Fluoridated Public Drinking Water - United States, 2000*, MMWR Weekly, February 22, 2002/51(07); 144-7, 2002, www.cdc.gov.

Table 8D.53 Percentage of United States Population on Public Water Supply Systems Receiving Fluoridated Water, in 2002

Location	2002[a]
United States	67.3
Alabama	82.0[b]
Alaska	57.3
Arizona	55.4
Arkansas	62.1
California	27.7
Colorado	75.4
Connecticut	87.5[b]
Delaware	80.9
District of Columbia	100.0[c,d]
Florida	68.6[c]
Georgia	93.0
Hawaii	8.8[c]
Idaho	47.5
Illinois	99.1
Indiana	95.5
Iowa	91.3
Kansas	62.1
Kentucky	99.7
Louisiana	46.1[c]
Maine	74.4
Maryland	93.7
Massachusetts	60.7[b]
Michigan	86.2
Minnesota	98.4
Mississippi	48.4[c]
Missouri	82.0[c]
Montana	23.8
Nebraska	69.5
Nevada	69.4
New Hampshire	42.7
New Jersey	20.8
New Mexico	76.5
New York	72.9
North Carolina	84.6
North Dakota	95.6
Ohio	90.6[b]
Oklahoma	74.6[b]
Oregon	19.4
Pennsylvania	54.0
Rhode Island	89.2[b]
South Carolina	91.4[b]
South Dakota	78.0[b]
Tennessee	96.0[b]
Texas	68.6[b]
Utah	2.3[c]
Vermont	55.7[b]
Virginia	93.8
Washington	58.9[b]
West Virginia	91.5
Wisconsin	89.4
Wyoming	36.7

[a] Data Source: CDC Water Fluoridation Reporting System (WFRS).
[b] Complete data not available from WFRS, additional information obtained from States.
[c] Reported PWS population exceeded U.S. Census 2000 state population, thus PWS population set to U.S. Census 2000 state population.
[d] Reported fluoridated population exceeded U.S. Census 2000 state population, thus fluoridated population was set to U.S. Census 2000 state population.

Source: From Center of Disease Control, cdc.gov/nohss/FluoridationMapV.asp?Year=2000.

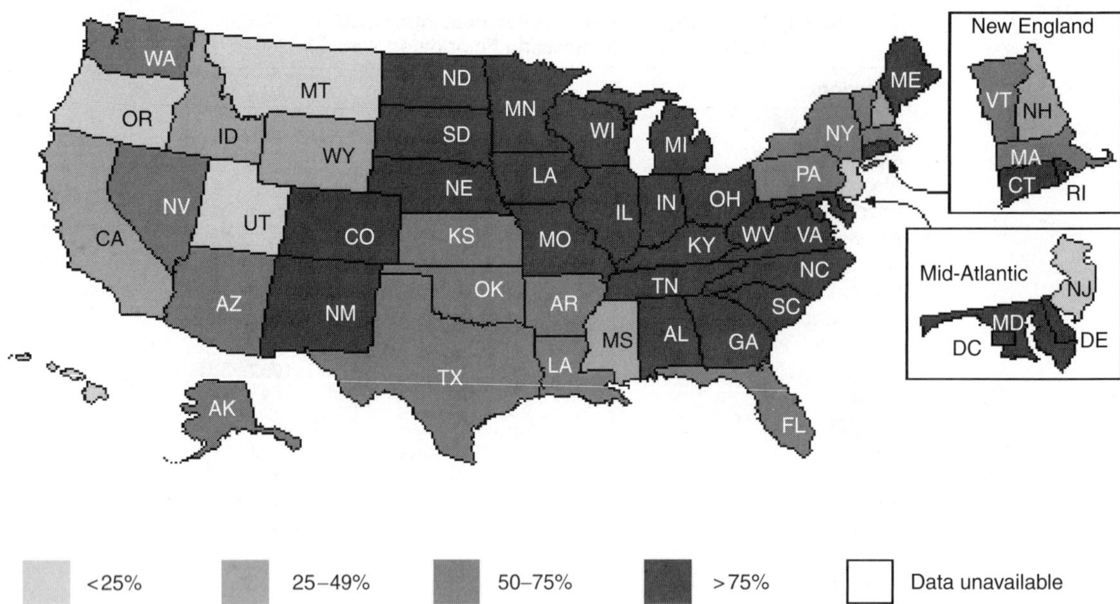

Figure 8D.18 Percent of United States population on fluoridated water, 2002. (From Center of Disease Control, cdc.gov/nohss/FluoridationMapV.asp?Year=2002.)

Geographical belts of high fluoride concentrations in groundwater extend from Syria through Jordan, Egypt, Libya, Algeria, Morocco and the Rift Valley of Western Africa through the Sudan and Kenya. Another belt stretches from Turkey through Iraq, Iran and Afghanistan to India, Northern Thailand and China. The highest natural fluoride concentration ever found in water (2,800 mg/L) was recorded in Lake Nakuru in the Rift Valley in Kenya. High groundwater fluoride concentrations are associated with igneous and metamorphic rocks such as granites and gneisses, volcanic rocks, and salt deposits of marine origin.

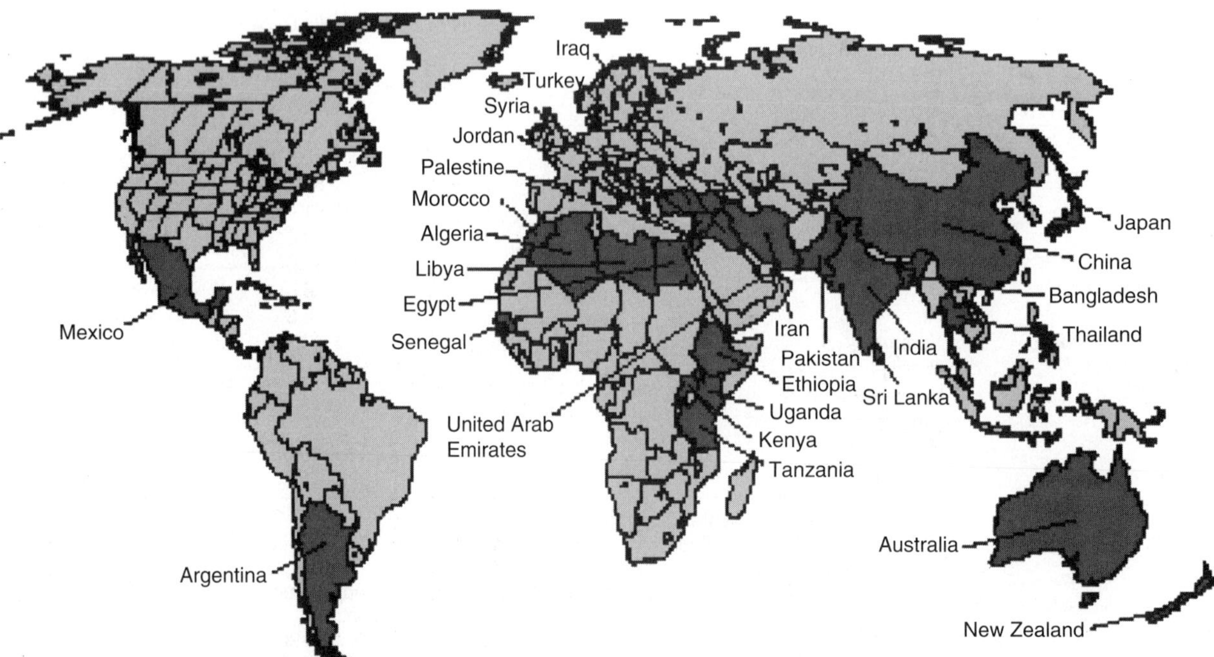

Figure 8D.19 Countries with endemic fluorosis due to excess fluoride in drinking water. (From UNICEF, *Fluoride and Fluoridation*, UNICEF Questions Benefits and Safety, rvi.net/~Flouride/000133.htm.)

Table 8D.54 Summary of Concentrations of Naturally-Occurring Fluoride Detected in Selected Countries with Endemic Fluorosis

Country	Fluoride Concentrations in Groundwater	Population Affected	Reference
Africa			
Ghana	up to 3.8 mg/L		BGS, 2000
Ethiopia	0.8–24.5 mg/L		Alem, 1998
Ethiopia (Lake Ziway)	0.1–23.3 mg/L		Haile, 1999
Kenya	up to 11 mg/L		Nyanchaga and Bailey, 2003
Tanzania	30% of drinking water > 1.5 mg/L, up to 8.0 mg/L		WHO 2004a, EHC 227
South Africa (Port Elizabeth)	0.05–14.00 mg/L		Maclear et al., 2003
Australia	50% groundwater boreholes > 1.5 mg/L with several in range of 3 to 9 mg/L		WHO, 2004a
China		Over 26 million-dental fluorosis	WHO, 2004b
		1 million-skeletal fluorosis	WHO, 2004b
Inner Mongolia	0.4–6.9 mg/L		Xu et al., 1997
India		66 million consume drinking water with elevated Fluoride-15 of India's 32 States	UNICEF
Rajasthan	0.2–5.1 mg/L		Choubisa, 2001
Punjab	0.5–16.2 mg/L		Jolly, 1968
Karnataka State	0.97–7.40 mg/L		Latha et al., 1998
Mexico		5 million (about 6% of population) affected by fluoride in groundwater	UNICEF, WHO, 2004b
Los Altos de Jalisco	0.14–12.97 mg/L		Hurtado et al., 2000
Durango	1–5.67 mg/L		Oritz et al., 1998
Argentina			
La Pampa	0.3–29 mg/L		Smedley et al., 2000

Source: From Alem, Getachew, 1998, Groundwater for rural water supply in the Rift Valley, 24th WEDC Conference, *Sanitation and Water for All. British Geological Survey, 2000,* Water Aid Country Information Sheet: Ghana.

Choubisa, SL., 2001, *Endemic Fluorosis in Southern Rajasthan*, India, Fluoride 34: 61–70.

Haile, G., 1999, Hydrogeochemistry of the Waters in the Lake Ziway Area, 25th WEDC Conference, Integrated Development for Water Supply and Sanitation.

Hurtado, R., Gardea-Torresdey, J, and Tiemann, K.J., 2000, *Fluoride Occurrence in Tap Water at "Los Altos de Jalisco" in the Central Mexico Region*, Proceedings of the 2000 Conference on Hazardous Waste Research.

Jolly SS, 1968, Fluoride 1: 65–75.

Latha, S.S, Ambika, S.R., and Prasad, S.J, 1998, *Fluoride Contamination Status of Groundwater in Karnataka* iisc. ernet.in/currsci/mar25/articles13.htm.

Nyanchaga, E.N. and Bailey, T., 2003, Flouride contamination in drinking water in the Rift Valley, Kenya and evaluation of the efficiency of locally manufactured defluoridation filter, *Journal of Civil Engineering*, JKUAT Vol 8, pp. 79–88.

Maclear, LGA, Adlem, M, and Libala, M.B., 2003, *COEGA Water Quality Monitoring, Trend Analysis of Fluoride Concentrations in Surface Water and Groundwater: 2000–2003*, SRK Consulting.

Ortiz, D., Castro, L., Turrubiartes, F., Milan, J., Diaz-Barriga, F., 1998, Assessment of the exposure to fluoride from drinking water in durango, Mexico, using a geographic information system, *Fluoride* 31 (4), pp 183–187.

Smedley, P, Nicolli, H, and MacDonald, D., 2000, Hydrogeochemisty of arsenic and other problem constituents in groundwaters from La Pampa, Argentina, *Journal of Conference Abstracts*, Volume 5(2), 936, Cambridge Publications.

UNICEF, Fluorides and Fluoridation, UNICEF Questions Benefits and Safety, Fluoride in Water: An Overview, rvi.net/~Flouride/000133.htm Printed 7/17/05.

WHO, 2004a, *Fluoride in Drinking Water Background Document for the Development of WHO Guidelines for Drinking-water Quality*, WHO/SDE/WSH/03.04/96.

WHO, 2004b, *WHO Issues Revised Drinking Water Guidelines to Help Prevent Water-Related Outbreaks and Disease, Press Release WHO/67*, September 21, 2004.

WHO, 2002, Fluorides, *Environmental Health Criteria*, 227.

Xu RQ, Wu DQ, Xu RY, 1997, Water Fluoride and Skeletal Fluorosis-Inner Mongolia, *Fluoride* 30: 26–28.

Table 8D.55 Detections of Giardia Cysts in Source Waters of Public Drinking Water Supplies in the United States

Classification	Samples	No. of Sites	No. of Positive Samples	No. of Positive Sites	Percent Positive of Samples	Percent Positive of Sites
Creeks	444	75	181	38	41	51
Rivers	449	74	163	38	36	51
Lakes	829	49	138	19	17	39
Springs[a]	84	6	16	2	19	33
Wells[a]	63	40	2	2	3	5

[a] Samples represent finished water. Most water from springs and wells is unfiltered and may or may not be disinfected before consumption.

Source: From U.S. Environmental Protection Agency, 1987; Hibler, 1987.

Table 8D.56 Detections of Giardia Cysts in Finished Drinking Waters Supplies of the United States

Classification	Samples	No. of Sites	No. of Positive Samples	No. of Positive Sites	Percent Positive of Samples	Percent Positive of Sites
Unfiltered, chlorinated	1,214	94	80	16	6.6	17
Direct filtration[a]	615	92	148	17	24.0	18.5
Conventional treatment	357	86	12	5	3.4	5.8
Slow sand and diatomaceous earth filtration	18	3	0	0	0	0
Commercial filters and/or pressure filters	33	12	4	2	12.1	16.7
Cartridge filters	51	13	11	7	21.6	53.8
Infiltration galleries	37	16	7	5	18.9	31.3
Filter type unknown	83	24	15	6	18.0	25.0

[a] May or may not include coagulation or disinfection. Number of systems applying coagulant and/or polymer, or whether disinfection was interrupted, could not be determined.

Source: From U.S. Environmental Protection Agency, 1987; Based on data collected from 1979–1986, Hibler, 1987.

Table 8D.57 United States National Source Water (Untreated) Levels of Giardia, January to December 1998

Level of Total Giardia (#cysts/100 mL)	Jan	Feb	Mar	Apr	May	Jun	Jul	Aug	Sep	Oct	Nov	Dec
	Number of Water Treatment Plants											
0	220	232	248	268	262	281	276	280	290	265	264	233
0.1–0.9		1	1	1		1	1	2		1		1
1–9	18	18	10	13	14	9	14	6	6	7	8	11
10–99	45	35	39	25	20	25	24	20	19	33	32	26
100–999	37	38	28	23	24	15	10	21	13	21	16	28
1,000–9,999		2	3		2	1		1		1	3	

Note: Data Purpose, The Information Collection Rule (ICR) data were collected as part of a national research project to support development of national drinking water standards. They should NOT be used to determine local water systems compliance with drinking water standards, nor should they be used to make personal judgements about health risks.

Results of Zero: A measurement of zero means that no *Giardia* was found in the sample volume that was analyzed. Zero results do not indicate the absence of Giardia in the source water because the method recovery is low and the amount of sample analyzed is small. In other words, Giardia may be present in source water even if no oocysts are counted for the sample volume analyzed. The presence of Giardia may also be mistakenly identified as other organic material such as algae. The current ICR method for detecting *Giardia* and *Cryptosporidium* has significant technical limitations: Better detection methods are currently being developed to detect and count protozoa. It is difficult to accurately estimate the numbers of protozoan cysts without testing large quantities of water, and this is not always feasible. The actual levels of these pathogens in source water may be much higher than those found by the tests. However, the current ICR detection method does not distinguish between species of *Giardia* and *Cryptosporidium* that may cause illness and those that do not. The method may also misidentify algae as a Cryptosporidium. With the ICR detection method, both false positive (microbe is counted when it is not actually present) and false negative (microbe is not counted when it is present) results are possible. The ICR detection method cannot determine whether the microbes are alive or whether they are able to cause illness.

Source: From United States Environmental Protection Agency, www.epa.gov.

Table 8D.58 Occurrence of *Cryptosporidium* in Surface and Groundwaters

Water Type	Samples (*n*)	% Positive	Range of Oocyst Concentration (oocysts/L)	Mean Concentration (oocysts/L)	Reference
Stream/river[a] N Amer	6	100	0.8–5,800	1920	Madore et al., 1987[c]
River[a] (2 sites) UK	375	4.5	0.07–4	0.95(g)	The National Cryptosporidium Survey Group, 1992[c]
River[a] (4 sites) UK	691	55.2	0.04–3.0	0.38(g)	The National Cryptosporidium Survey Group, 1992[c]
River[a] (4 sites) UK	430	4.4	0.007–2.75	0.5(g)	The National Cryptosporidium Survey Group, 1992[c]
6 Rivers[a] N Amer	11	100	2–112	25(adj)	Ongerth and Stibbs, 1987[c]
Stream N Amer	19	73.7	0–240	1.09(g)	Rose et al., 1988[c]
Stream/river N Amer	58	77.6	0.04–18	0.94(g)	Rose, 1988[c]
Stream/river N Amer	38	73.7	<0.001–44	0.66(g)	Rose et al., 1991[c]
River/lake N Amer	85	87.1	0.07–484	2.7(g)	LeChevallier et al., 1991a[c]
River/lake N Amer	262	51.5	0.065–65.1	2.4(g)	LeChevallier and Norton, 1995
River N Amer	22	31.8	0.01–75.7	0.58(g)	Stetzenbach et al., 1988[c]
3 Rivers[a] Ottawa, Canada	41	78.8	<0.02–2.25	0.26	Chauret et al., 1995[b]
Surface waters Canada	1173	4.5	NR	NR	Wallis et al., 1996
Surface waters South Africa	NR	18(?)	0–25/L	0.6	Kfir et al., 1995
River/lake N Amer	18	NR	7.1–28.5	17.8	Rose et al., 1988[c]
Stream/lake (impact if any, NR) UK	84	40.5	0.006–2.3	NR	Smith et al., 1991[c]
Lake N Amer	20	70.7	0–22	0.58(g)	Rose et al., 1988[c]
Lake/reservoir N Amer	32	75	1.1–8.9	0.91(g)	Rose, 1988[c]
Lake N Amer	24	58.3	<0.001–3.8	1.03(g)	Rose et al., 1991[c]
Lake N Amer	44	27.3	0.11–251.7	4.74	Stetzenbach et al., 1988[c]

(Continued)

Table 8D.58 (Continued)

Water Type	Samples (*n*)	% Positive	Range of Oocyst Concentration (oocysts/L)	Mean Concentration (oocysts/L)	Reference
Pristine river N Amer	3	NR	NR	0.08(g)	Rose et al., 1988[c]
Pristine river N Amer	59	32.2	NR	0.29	Rose et al., 1991[c]
Pristine lake N Amer	34	52.9	NR	0.093(g)	Rose et al., 1991[c]
Pristine spring N Amer	7	28.6	<0.003–0.13	0.04(g)	Rose et al., 1991[c]
Pristine lake Yukon, Canada	11	9.1	0–0.003	0.003	Roach et al., 1993[c]
Groundwater well N Amer	18	5.6	NR	0.003 (single value)	Rose et al., 1991[c]
Deep pristine groundwater well UK	120	0	—	—	The National Cryptosporidium Survey Group, 1992[c]
Groundwater well[b] UK	138	5.8	0.004–0.922	0.23(g)	The National Cryptosporidium Survey Group, 1992[c]

Note: Pristine = little or no human activity in the watershed or water, restricted access, no agricultural activity within the watershed and no sewage treatment facility discharges impacting the water upstream from the sampling site (Lisle and Rose, 1995); NR = not recorded; (g) = geometric mean; (adj) = data adjusted for recovery efficiencies; (?) legend is missing from the relevant figure in Kfir et al.

[a] Affected by domestic or agricultural waste.
[b] History of coliform contamination.
[c] As cited by Lisle and Rose (1995).

Source: From Butler, B.J. and Mayfield, C.I., 1996, *Cryptospordium* spp. A review of the organism. *Disease and Implications of Managing Water Resources*, Prepared for Waterloo Centre of Groundwater Research, Waterloo, Ontario, Canada, www.inweh.unu.edu. Reprinted with permission.

Table 8D.59 United States National Source Water (Untreated) Levels of Cryptosporidium, January to December 1998

Level of Total Cryptosporidium (oocysts/100 L)	Jan	Feb	Mar	Apr	May	Jun	Jul	Aug	Sep	Oct	Nov	Dec
	Number of Water Treatment Plants											
0	293	292	304	312	305	307	310	309	301	306	302	283
0.01–0.9	1											1
1–9	4	6	4	6	3	2	1	1	2	7	2	4
10–99	17	21	16	6	10	17	12	12	15	11	17	9
100–999	5	7	5	6	4	6	2	8	10	4	2	2
1,000–9,999												

Note: Data Purpose, The Information Collectin Rule (ICR) data were collected as part of a national research project to support development of national drinking water standards. They should NOT be be used to determine local water systems compliance with drinking water standards, nor should they be used to make personal judgements about health risks. Results of Zero: A measurement of zero means that no Cryptosporidium was found in the sample volume that was analyzed. Zero results do not indicate the absence of Cryptosporidium in the source water because the method recovery is low and the amount of sample analyzed is small. In other words, Cryptosporidium may be present in source water even if no oocysts are counted for the sample volume analyzed. The presence of Cryptosporidium may also be mistakenly identified as other organic material such as algae. The current ICR method for detecting *Giardia* and *Cryptosporidium* has significant technical limitations:

It is difficult to accurately estimate the numbers of protozoan cysts without testing large quantities of water, and this is not always feasible. The actual levels of these pathogens in source water may be much higher than those found by the tests

However, the current ICR detection method does not distinguish between species of *Giardia* and *Cryptosporidium* that may cause illness and those that do not. The method may also misidentify algae as a Cryptosporidium. With the ICR detection method, both false positive (microbe is counted when it is not actually present) and false negative (microbe is not counted when it is present) results are possible. The ICR detection method cannot determine whether the microbes are alive or whether they are able to cause illness. Better detection methods are currently being developed to detect and count protozoa.

Source: From United States Environmental Protection Agency, www.epa.gov.

Table 8D.60 Occurrence of Cryptosporidium in Treated Drinking Water

Study Site	Filtration	Samples (*n*)	% Positive	Range of Oocyst Concentration (oocysts/L)	Mean Concentration (oocysts/L)	Reference
U.S.A.	yes	28	14.3	0.005–0.007	0.001(g)	Rose et al., 1991[a]
U.S.A.	yes	82	26.8	0.001–0.48	0.015(g)	LeChevallier et al., 1991b[a]
U.S.A.	no	6	33.3	0.001–0.017	0.002(g)	Rose et al., 1991[a]
Yukon, Canada	no	42	3.8	0.002–0.005	NR	Roach et al., 1993[a]
Scotland	NR	142	40.1	0.007–0.72	NR	Smith et al., 1991[a]
South Africa	NR	NR	1.1	0–1	0	Kfir et al., 1995

Note: NR = not recorded; (g) = geometric mean.

[a] As cited by Lisle and Rose, 1995.

Source: From Butler, B.J. and Mayfield, C.I., 1996, *Cryptospordium* spp. A review of the organism. *Disease and Implications of Managing Water Resources*, Waterloo Centre of Groundwater Research, Waterloo, Ontario, Canada, www.inweh.unu.edu. Reprinted with permission.

SECTION 8E INDUSTRIAL WATER QUALITY

Table 8E.61 Water Quality Tolerance for Certain Industrial Applications

Industry	Turbidity	Color	Color $+O_2$ Consumed	D.O.[a] (mL/L)	Odor	Hardness	Alkalinity	pH	Total Solids	Fe	Mn	Fe+ Mn	Al_2O_3	SiO_2	Cl	F	CO_3	HCO_3	OH	Na_2SO_4 to Na_2SO_3 (ratio)	General[b]
Air																					
Conditioning[c]	—	—	—	—	—	—	—	—	—	0.5	0.5	0.5	—	—	—	—	—	—	—	—	A,B
Baking	10	10	—	—	Low	—[d]	—	—	—	0.2	0.2	0.2	—	—	—	—	—	—	—	—	C
Boiler Feed (pounds per sq. in.)																					
0–150	20	80	100	2	—	80	—	8.0+	3000–1000	—	—	—	5	40	—	—	200	50	50	1–1	—
150–250	10	40	50	0.2	—	40	—	8.5+	2500–500	—	—	—	0.5	20	—	—	100	30	40	2–1	—
250–400	5	5	10	0.0	—	10	—	9.0+	1500–100	—	—	—	0.05	5	—	—	40	5	30	3–1	—
400–over	1	2		0.0	—	2	—	9.6+	50	—	—	—	0.01	1	—	—	20	0	15	3–1	—
Brewing[e]																					
Light	10	10	—	—	Low	—	75	6.5–7.0	500	0.1	0.1	0.1	—	50	100	1.0	50	—	—	—	C,D,G
Dark	10	10	—	—	Low	—	150	7.0+	1000	0.1	0.1	0.1	—	50	100	1.0	50	—	—	—	C.D,H
Canning																					
Legumes	10	—	—	—	Low	25–75	—	7.5+	850	0.2	0.2	0.3	—	—	—	1.0	—	—	—	—	C
General	10	—	—	—	Low	50–400	—	7.5+	850	0.2	0.2	0.3	—	—	—	1.0	—	—	—	—	C
Carbonated																					
Beverages[f]	2	10	10	—	Low	250	125	—	850	0.2	0.2	0.3	—	—	250	0.2–1.0	—	—	—	—	C
Confectionery	—	—	—	—	Low	—	—	—[g]	100	0.2	0.2	0.2	—	—	—	—	—	—	—	—	—
Cooling[h]	50	—	—	—	—	50	—	—	—	0.5	0.5	0.5	—	—	—	—	—	—	—	—	A,B
Food																					
general	10	5–10	—	—	Low	10–250	30–250	—	850	0.2	0.2	0.2	—	—	—	1.0	—	—	—	—	C
Ice (raw water)[i]	1–5	5	—	—	—	—	30–50	—	300	0.2	0.2	0.2	—	10	—	—	—	—	—	—	C
Laundering	—	—	—	—	—	50	60	6.0–6.8	—	0.2	0.2	0.2	—	—	—	—	—	—	—	—	—
Plastics, clear																					
uncolored	2	2	—	—	—	—	—	—	200	0.02	0.02	0.02	—	—	—	—	—	—	—	—	—
Paper and Pulp[j]																					
Groundwood	50	30	—	—	—	200	150	—	500	0.3	0.1	0.3	—	50	75	—	—	—	—	—	E
Kraft, paper, bleached	40	25	—	—	—	100	75	—	300	0.2	0.1	0.2	—	50	200	—	—	—	—	—	E
Soda and sulfite pulps	25	5	—	—	—	100	75	—	250	.01	0.05	0.1	—	20	75	—	—	—	—	—	E
Fine paper	10	5	—	—	—	100	75	—	200	.01	0.05	0.1	—	20	—	—	—	—	—	—	E
Rayon (viscose)																					
Pulp: Production	5	5	—	—	—	8	50	—	100	0.05	0.03	0.05	8.0	25	5	—	—	—	—	—	F
Manufacture	0.3	—	—	—	—	55	—	7.8–8.3	—	—	—	—	—	—	—	—	—	—	—	—	—
Tanning[k]	20	10–100	—	—	—	50–135	135	6.0–8.0	—	0.2	0.2	0.2	—	—	—	—	—	—	—	—	—
Textiles																					
General	5	20	—	—	—	20	—	—	—	0.25	0.25	—	—	—	100	—	—	—	—	—	—
Dyeing[l]	5	5–20	—	—	—	20	—	—	—	0.25	0.25	0.25	—	—	—	—	—	—	—	—	—

Wool scouring[m]		70	—	—	—	20	—	—		1.0	1.0	1.0	—	—	—	—	—	—	—	—	—
Cotton bandage[m]	5	5	—	—	Low	20	—	—	—	0.2	0.2	0.2	—	—	—	—	—	—	—	—	—

Note: Milligrams per liter, except as indicated.

[a] Abbreviations as follows: D.O., dissolved oxygen; ppm, parts per million; pH, hydrogen-ion concentration.

[b] A—no corrosiveness; B—no slime formation; C—conformity with federal drinking water standards necessary; D—NaCl, 275 ppm; E—free CO_2 less than 10 mg/L; F—copper less than 5 mg/L; G—calcium 100–200 mg/L; H—calcium 200–500 mg/L

[c] Water with algae, or hydrogen sulphide odors, is most unsuitable for air conditioning.

[d] Some hardness desirable.

[e] Water for distilling must meet the same general requirements as for brewing (gin and spirits mashing water of light-beer quality, whiskey mashing water of dark-beer quality).

[f] Clear, odorless, sterile water for syrup and carbonization. Water consistent in character. Most high quality filtered municipal water not satisfactory for beverages.

[g] Hard candy requires pH of 7.0 or greater, as low value favors inversion of sucrose, causing sticky products.

[h] Control of corrosiveness is necessary, as is also control of organisms, such as sulphur and iron bacteria, which tend to form slimes.

[i] Ca $(HCO_3)_2$ particularly troublesome. $Mg(HCO_3)_2$ tends to greenish color. CO_2 assists in preventing cracking. Sulphates and chlorides of Ca, Mg, Na should each be less than 300 ppm (white butts).

[j] Uniformity of composition and temperature desirable. Iron objectionable since cellulose absorbs iron from dilute solutions. Manganese very objectionable, clogs pipelines and is oxidized to permanganates by chlorine, causing reddish color.

[k] Excessive iron, manganese, or turbidity creates spots and discoloration in tanning of hides and leather goods.

[l] Constant composition; residual alumina <0.5 ppm.

[m] Calcium, magnesium, iron, manganese, suspended matter, and soluble organic matter may be objectionable.

Source: From American Water Works Association, *Water Quality and Treatment, second edition (New York, 1950). Water Quality Criteria*, California State Water Quality Control Board, second edition (Sacramento, 1963.)

Table 8E.62 Water Quality Guidelines for the Pulp and Paper Industry

	Concentration (mg/L)					
			Kraft		Chem. Pulp & Paper	
Parameter	Fine Paper	Ground Wood	Bleached	Unbleached	Bleached	Unbleached
pH	—	6–8	—	—	6–8	6–8
Color (HU)	<40	<100	<25	<100	<50	<100
Turbidity (NTU)	<10	<20	<40	<100	<10	<20
Calcium	<20	<20	—	—	<20	<20
Magnesium	<12	<12	—	—	<12	<12
Iron	<0.1	<0.1	<0.2	<1.0	<0.1	<1.0
Manganese	<0.3	<0.1	<0.1	<0.5	<0.5	<0.5
Chloride	—	25–75	<200	<200	<200	<200
Silica	<20	<100	<50	<100	<50	<50
Hardness	<100	<100	<100	<100	<100	<100
Alkalinity	40–75	<150	<75	<150	—	—
Dissolved solids	<200	<250	<300	<500	<200	<250
Suspended solids	<10	—	—	—	<10	<10
Temperature (°C)	—	—	—	—	<36	—
CO_2	<10	<10	<10	<10	—	—
Corrosion tendency	NIL	NIL	NIL	NIL	NIL	NIL
Residual chloride	<2.0	—	—	—	—	—

Source: From Canadian Council of Resource and Environment Ministers, *Canadian Water Quality Guidelines*, March 1987.

Table 8E.63 Water Quality Guidelines for the Iron and Steel Industry

	Concentration (mg/L)				
			Rinse Water		
Parameter	Hot-Rolling, Quenching, Gas Cleaning	Cold-Rolling	Softened	Demineralized	Steel Manufacturing
pH	5.0–9.0	5.0–9.0	6.0–9.0	—	6.8–7.0
Suspended solids	<25	<10	ND[a]	ND	—
Dissolved solids	<1000	<1000	ND	ND	—
Settleable solids	<100	<5.0	ND	ND	—
Dissolved oxygen	minimum for aerobic conditions				
Temperature (°C)	<38	<38	<38	<38	<38
Hardness	NS[b,c]	NS[b]	<100	<0.1	<50
Alkalinity	NS[c]	NS[c]	NS[c]	<0.5	—
Sulfate	<200	<200	<200	—	<175
Chloride	<150	<150	<150	ND	<150
Oil	NS	ND	ND	ND	ND
Floating material	NS	ND	ND	ND	ND

[a] ND = not detected.
[b] Controlled by other treatments.
[c] NS = not specified; the parameter has never been a problem at concentrations encountered.

Source: From Canadian Council of Resource and Environment Ministers, *Canadian Water Quality Guidelines*, March 1987; U.S. Environmental Protection Agency, 1973.

Table 8E.64 Water Quality Guidelines for the Petroleum Industry

Parameter	Concentration (mg/L)[a]
pH units	6.0–9.0
Color	NS[b]
Calcium	<75
Magnesium	<25
Iron	<1
Bicarbonate	NS
Sulphate	NS
Chloride	<200
Nitrate	NS
Fluoride	NS
Silica	NS
Hardness (as $CaCO_3$)	<350
Dissolved solids	<750
Suspended solids	<10

[a] Unless otherwise indicated.
[b] NS = not specified. The parameter has never been a problem at concentrations encountered.

Source: From *Canadian Water Quality Guidelines* 1987; Federal Water Pollution Control Administration 1968; Ontario Ministry of the Environment 1974.

Table 8E.65 Water Quality Guidelines for Power Generation Stations

	Concentration (mg/L)			
	Cooling Once-Through			
Parameter	Fresh	Brackish[a]	Boiler Feedwater (10.35–34.48 MPa)	Miscellaneous Uses
Silica	<50	<25	<0.01	—
Aluminum	NS[b]	NS	<0.01	—
Iron	NS	NS	<0.01	<1.0
Manganese	NS	NS	<0.01	—
Calcium	<200	<420	<0.01	—
Magnesium	NS	NS	<0.01	—
Ammonia	NS	NS	<0.07	—
Bicarbonate	<600	<140	<0.5	—
Sulphate	<680	<2,700	NS[c]	—
Chloride	<600	<19,000	NS[c]	—
Dissolved solids	<1000	<35,000	<0.5	<1000
Copper	NS	NS	<0.01	—
Hardness	<850	<6250	<0.07	—
Zinc	NS	NS	<0.01	—
Alkalinity (as $CaCO_3$)	<500	<115	<1	—
pH units	5.0–8.3	6.0–8.3	8.8–9.4	5.0–9.0
Organic material				
Methylene blue active substances	NS	NS	<0.1	<10
Carbon tetrachloride extract	NS[d]	NS[d]	NS	<10
Chemical oxygen demand (COD)	<75	<75	<1.0	—
Dissolved oxygen	—	—	<0.007	—
Suspended solids	<5000	<2500	<0.05	<5

[a] Brackish water—dissolved solids more than 1000 mg/L.
[b] NS = not specified; the parameter has never been a problem at concentrations encountered.
[c] Controlled by treatment for other constituents.
[d] No floating oil.

Source: From Canadian Council of Resource and Environment Ministers, *Canadian Water Quality Guidelines*, March 1987; Krisher, A.S., 1978, Raw water treatment in the CPI. *Chem. Eng.* (N.Y.), vol. 85, pp. 78–98. *Chemical Engineering*, Aug. 28, 1978. © McGraw-Hill, Inc.

Table 8E.66 Water Quality Guidelines for the Food and Beverage Industry

	Concentration (mg/L)							
Parameter	Baking	Brewing	Carbonate Beverages	Confectionary	Dairy	Food Canning, Freezing, Dried, Frozen Fruits, Vegetables	Food Process (General)	Sugar Manufacturing
pH	—	6.5–7.0	<6.9	>7.0	—	6.5–8.5	—	—
Color (HU)	<10	<5	<10	—	ND	<5	5–10	—
Turbidity (NTU)	<10	<10	1–2	—	—	<5	1–10	—
Taste, odor (units)	low	low	ND[a]	low	ND	ND	low	—
Suspended solids	—	—	—	50–100	<500	<10	—	ND
Dissolved solids	—	<800	<850	50–100	<500	<500	<850	—
Calcium	NS[b,c]	<100	—	—	—	<100	—	<20
Magnesium	—	<30	—	—	—	—	—	<10
Iron	<0.2	0.1–1.0	<0.1	<0.2	0.1–0.3	<0.2	<0.2	<1
Manganese	<0.2[d]	<0.1[d]	<0.05	<0.2[d]	0.03–0.1	<0.2[d]	<0.2	<0.1
Copper	—	—	—	—	ND	—	—	—
Ammonium	—	—	—	—	Trace	<0.5	—	—
Bicarbonate	—	ND	—	—	—	—	—	<100
Carbonate	—	<50	<5	—	—	—	—	—
Sulphate	—	<100	<200	—	<60	<250	—	<20
Chloride	—	20–60	<250	<250	<30	<250	—	<20
Nitrate	—	<10	—	—	<20	<10	—	—
Fluoride	—	<1	0.2–1.0	—	—	<1	<1	—
Silica	—	<50	ND	—	—	<50	—	—
Hardness	NS[b]	<70	200–250	—	<180	<250	10–250	<100
Alkalinity	—	<85	50–128	—	—	30–250	30–250	—
Hydrogen sulphide	<0.2	<0.2	<0.2	<0.2	—	—	—	—
Oxygen consumed	—	—	<15	—	—	<1	—	—
Carbon tetrachloride extract	—	—	Slight	—	<10	<0.2	—	—
Chloroform extract	—	—	<0.2	—	—	—	—	—
Acidity	—	—	—	—	—	ND	—	—
Phenol	—	ND	ND	—	—	ND	—	ND
Nitrite	—	—	—	—	—	ND	—	—
Organic matter	—	trace	trace	—	—	—	—	trace

[a] ND = not detected.
[b] Some required for yeast action; excess retards fermentation.
[c] NS = not specified.
[d] Total Fe and Mn.

Source: From Canadian Council of Resource and Environment Ministers, *Canadian Water Quality Guidelines*, March, 1987.

Table 8E.67 International Council of Bottled Water Associations Standard of Quality

Monitoring Parameter Group	Standard
Inorganic Chemicals (IOCs)	Standard
Antimony	0.005
Arsenic	0.01
Barium	0.7
Boron	0.3
Bromate	0.010
Cadmium	0.003
Chlorine	5.0
Chloramine	3.0
Chlorite	0.2
Chromium	0.05
Cyanide	0.07
Fluoride	1.5
Lead	0.01
Mercury	0.001
Molybdenum	0.07
Nickel	0.02
Nitrate-N	50
Nitrite-N	3
Selenium	0.01
Secondary Inorganic Parameters	
Copper	2
Manganese	0.05
Volatile Organic Chemicals (VOCs)	
1,1,1-Trichloroethane	2
1,1-Dichloroethylene	0.03
1,2,4-Trichlorobenzene	0.02
1,2-Dichloroethane	0.03
1,2-Dichloropropane	0.02
1,3-Dichloropropene	0.02
Benzene	0.01
Carbon tetrachloride	0.002
1,2-Dichloroethylene	0.05
Ethylbenzene	0.3
Methylene chloride (Dichloromethane)	0.02
Monochlorobenzene	0.3
o-Dichlorobenzene	1
p-Dichlorobenzene	0.3
Styrene	0.02
Tetrachloroethylene	0.04
Toluene	0.7
Trichloroethylene	0.07
Vinyl chloride	0.005
Xylenes (total)	0.5
Bromodichloromethane	0.06
Chlorodibromomethane	0.1
Chloroform	0.2
Bromoform	0.1
Semivolatile Organic Chemicals (SVOCs)	
Benzo(a)pyrene	0.0007
Di(2-ethyhexyl)adipate	0.08
Di(2-ethyhexyl)phthalate	0.008
Hexachlorobenzene	0.001
Synthetic Organic Chemical (SOCs)	
Alachlor	0.02
Aldicarb	0.01
Aldrin/Dieldrin	0.00003
Atrazine	0.002
Bentazone	0.03
Carbofuran	0.005

(Continued)

Table 8E.67 (Continued)

Monitoring Parameter Group	Standard
Chlordane	0.0002
Chlorotoluron	0.03
DDT	0.002
Dibromochloropropane (DBCP)	0.001
2,4-D	0.03
2,4-DB	0.09
Dichlorprop	0.1
Fenoprop	0.009
Heptachlor	0.03
Heptachlor epoxide	0.03
Isoproturon	0.009
Lindane	0.002
Methoxychlor	0.02
MCPA	0.002
Mecoprop	0.01
Metolachlor	0.01
Molinate	0.006
Pendimethalin	0.02
Pentachlorophenol	0.009
Permethrin	0.02
Propanil	0.02
Pyridate	0.1
Simazine	0.002
2,4,5-T	0.009
Trifluralin	0.02
Additional Regulated Conlaminants	
Acrylamide	0.0005
Cyanogen chloride	0.07
Epichlorohydrin	0.0004
Hexachlorobutadiene	0.0006
Edetic acid (EDTA)	0.2
Nitrilotriacetic acid	0.2
2,4,6-Trichlorophenol (DBP)	0.2
Dichloroacetic acid	0.05
Trichloroacetic acid	0.1
Formaldehyde (DBP)	0.9
Chloral hydrate (Trichloroacetaldehyde)	0.01
Dichloroacetonitrile	0.09
Dibromoacetonitrile	0.1
Trichloroacetonitrile	0.07
Tributyltin oxide	0.002
Microbiological Contaminants	
Total coliform / *E. coli*	0
Radiological Contaminants	Standard
Gross alpha	0.1 Bq/l
Gross beta	1.0 Bq/l
Water Properties	Standard
Color	15 TCU
Turbidity	5 NTU
PH	6.5–8.0
Odor	Not Offensive

Note: All Standards are in mg/L (ppm) except as noted.

Source: From International Council of Bottled Water Associations, ICBWA Model Code, September 27, 2004, www.icbwa.org. With permission.

Table 8E.68 United States Bottled Water Standards

Physical	Maximum
Color (units)	15
Odor	Threshold Odor No.3
Turbidity (units)	5
Total Dissolved Solids	500[a]
Inorganic Substances	**Maximum (mg/L)**
Aluminum	0.2
Antimony	0.006
Arsenic	0.05
Barium	2
Beryllium	0.004
Cadmium	0.005
Chloride	250[a]
Chromium	0.1
Copper	1
Cyanide	0.2
Fluoride	[b]
Iron	0.3[a]
Lead	0.005
Manganese	0.05[a]
Mercury	0.002
Nickel	0.1
Nitrate (as nitrogen)	10
Nitrite (as nitrogen)	1
Total Nitrate & Nitrite (sum as nitrogen)	10
Phenols	0.001
Selenium	0.05
Silver	0.1
Sulfate	250[a]
Thallium	0.002
Zinc	5[a]
Volatile Organic Chemicals or VOCs	**Maximum (mg/L)**
Benzene	0.005
Carbon Tetrachloride	0.005
o-Dichlorobenzene	0.6
p-Dichlorobenzene	0.075
1,2-Dichloroethane	0.005
1,1-Dichloroethylene	0.007
cis-1,2-Dichloroethylene	0.07
trans-1,2-Dichloroethylene	0.1
Dichloromethane	0.005
1,2-Dichloropropane	0.005
Ethylbenzene	0.7
Monochlorobenzene	0.1
Styrene	0.1
Tetrachloroethylene	0.005
Toluene	1
1,2,4-Trichlorobenzene	0.07
1,1,1-Trichloroethane	0.2
1,1,2-Trichloroethane	0.005
Trichloroethylene	0.005
Vinyl chloride	0.002
Xylenes	10
Pesticides and Other Synthetice Organic Chemicals	**Maximum (mg/L)**
Alachlor	0.002
Atrazine	0.003
Benzo(a)pyrene	0.0002
Carbofuran	0.04
Chlordane	0.002
Dalapon	0.2
1,2-Dibromochloropropane (DBCP)	0.0002
2, 4-D	0.07
Di(2-ethylhexyl)adipate	0.4
Dinoseb	0.007
Diquat	0.02

(Continued)

Table 8E.68 (Continued)

Pesticides and Other Synthetice Organic Chemicals	Maximum (mg/L)
Endothall	0.1
Endrin	0.002
Ethyiene Dibromide (EDB)	0.00005
Glyphosate	0.7
Heptachlor	0.0004
Heptachlor Epoxide	0.0002
Hexachlorobenzene	0.001
Hexachlorocyclopentadiene	0.05
Lindane	0.0002
Methoxychlor	0.04
Oxamyl	0.2
Pentachlorophenol	0.001
PCB's (as decachiorobiphenyls)	0.0005
Picloram	0.5
Simazine	0.004
2,3,7,8-TCDD (Dioxin)	3×10^{-8}
Toxaphene	0.003
2,4,5-TP(Si!vex)	0.05
Radioactivity	**Maximum**
Combined Radium-226 and Radium-228	5 pCi/L
Gross alpha particle activity (including Radium-226 but excluding Radon and Uranium)	15 pCi/L
Gross beta particle activity	50 pCi/L
Uranium	30 ug/L
Bacteriological	**Maximum**
Coliforms: Multiple Tube Fermentation Method	2.2MPN/100mL
Membrane Filter Method	1c/100mL
Disinfection Byproducts (DBp's)	**Maximum (mg/L)**
Bromate	0.01
Chlorite	1
Haloacetic acids (five)(HAA5)	0.06
Total Trihalomethanes (THMs)[b]	0.08
Residual Disinfectants	**Maximum (mg/L)**
Chloramine (as Cl_2)	4
Chlorine (as Cl_2)	4
Chlorine dioxide (as ClO_2)	0.8

Annual Average of Maximum Daily Air Temperatures (°F) at the Location Where the Bottled Water Is Sold at Retail	Fluoride Maximum (mg/L) No Fluoride Added	Fluoride Maximum (mg/L) Fluoride Added
53.7 and below	2.4	1.7
53.8–58.3	2.2	1.5
58.4–63.8	2	1.3
63.9–70.6	1.8	1
70.7–79.2	1.6	1
79.3–90.5	1.4	0.8

Imported bottled water with no fluoride added: 1.4 mg/L Fluoride

[a] Mineral water is exempt from allowable level. The exemptions are aesthetically based allowable levels and do not relate to a health concern.

[b] Fluoride standards: Bottled water packaged in the United States.

[c] Total Trihalomethanes (TTHM): Sum of chloroform, bromodichloromethane, chiorodibromomethane, and bromoform. 10 ppb Pursuant to H&SC 111080(b).

Source: From United States Food and Drug Administration 21 CFR 165 Beverages: Title 21—Food and Drugs, Chapter 1, Food and Drug Adminstration Department of Health and Human Services, Subchapter B, Food for Human Consuption, Subpart B, Requirements for Specifics Standardized Beverages, Section 165.110 Bottled Water, www.accessdata.fda.gov.

SECTION 8F IRRIGATION WATER QUALITY

Table 8F.69 Relative Tolerance of Crop Plants to Salt

	Field Crops	
$EC \times 10^3 = 16$	$EC \times 10^3 = 10$	$EC \times 10^3 = 4$
Barley (grain)	Rye (grain)	Field beans
Sugar beet	Wheat (grain)	
Rape	Oats (grain)	
Cotton	Rice	
	Sorghum (grain)	
	Corn (field)	
	Flax	
	Sunflower	
	Castorbeans	
$EC \times 10^3 = 10$	$EC \times 10^3 = 6$	
	Vegetable Crops	
$EC \times 10^3 = 12$	$EC \times 10^3 = 10$	$EC \times 10^3 = 4$
Garden beets	Tomato	Radish
Kale	Broccoli	Celery
Asparagus	Cabbage	Green beans
Spinach	Bell pepper	
	Cauliflower	
	Lettuce	
	Sweet corn	
	Potatoes (white rose)	
	Carrot	
	Onion	
	Peas	
	Squash	
	Cucumber	
$EC \times 10^3 = 10$	$EC \times 10^3 = 4$	$EC \times 10^3 = 3$
	Fruit and Nut Crops	
High Salt Tolerance	**Medium Salt Tolerance**	**Low Salt Tolerance**
Date palm	Cantaloupe	Almond
	Fig	Apple
	Grape	Apricot
	Jujube	Avocado
	Olive	Blackberry
	Papaya	Boysenberry
	Pineapple	Cherimoya
	Pomegranate	Cherry, sweet
		Cherry, sand
		Currant
		Gooseberry
		Grapefruit
		Lemon
		Lime
		Loquat
		Mango
		Orange
		Passion fruit
		Peach
		Pear
		Persimmon
		Plum: prune

(Continued)

Table 8F.69 (Continued)

High Salt Tolerance	Medium Salt Tolerance	Low Salt Tolerance
		Pummelo
		Raspberry
		Rose, apple
		Sapote, white
		Strawberry
		Tangerine
	Forage Crops	
$EC \times 10^3 = 18$	$EC \times 10^3 = 12$	$EC \times 10^3 = 4$
Alkali sacaton	White sweetclover	White Dutch clover
Saltgrass	Yellow sweetclover	Meadow foxtail
Nuttall alkaligrass	Perennial ryegrass	Alsike clover
Bermuda grass	Mountain brome	Red clover
Rhodes grass	Strawberry clover	Ladino clover
Fescue grass	Dallis grass	Burnet
Canada wild rye	Sudan grass	
Western wheat-grass	Hubam clover	
	Alfalfa (California common)	
Barley (hay)	Tall fescue	
Bridsfoot trefoil	Rye (hay)	
	Wheat (hay)	
	Oats (hay)	
	Orchardgrass	
	Blue grama	
	Meadow fescue	
	Reed canary	
	Big trefoil	
	Smooth brome	
	Tall meadow oat-grass	
	Cicer milkvetch	
	Sourclover	
	Sickle milkvetch	
$EC \times 10^3 = 12$	$EC \times 10^3 = 4$	$EC \times 10^3 = 2$

Note: The numbers following $EC \times 10^3$ are the electrical conductivity values of the saturation extract in millimhos per centimeter at 25°C associated with a 50-percent decrease in yields. The saturation extract is the solution extracted from a soil at its saturation percentage.

Source: From U.S. Department of Agriculture, 1954; Kandiah, A., FAO, 1987.

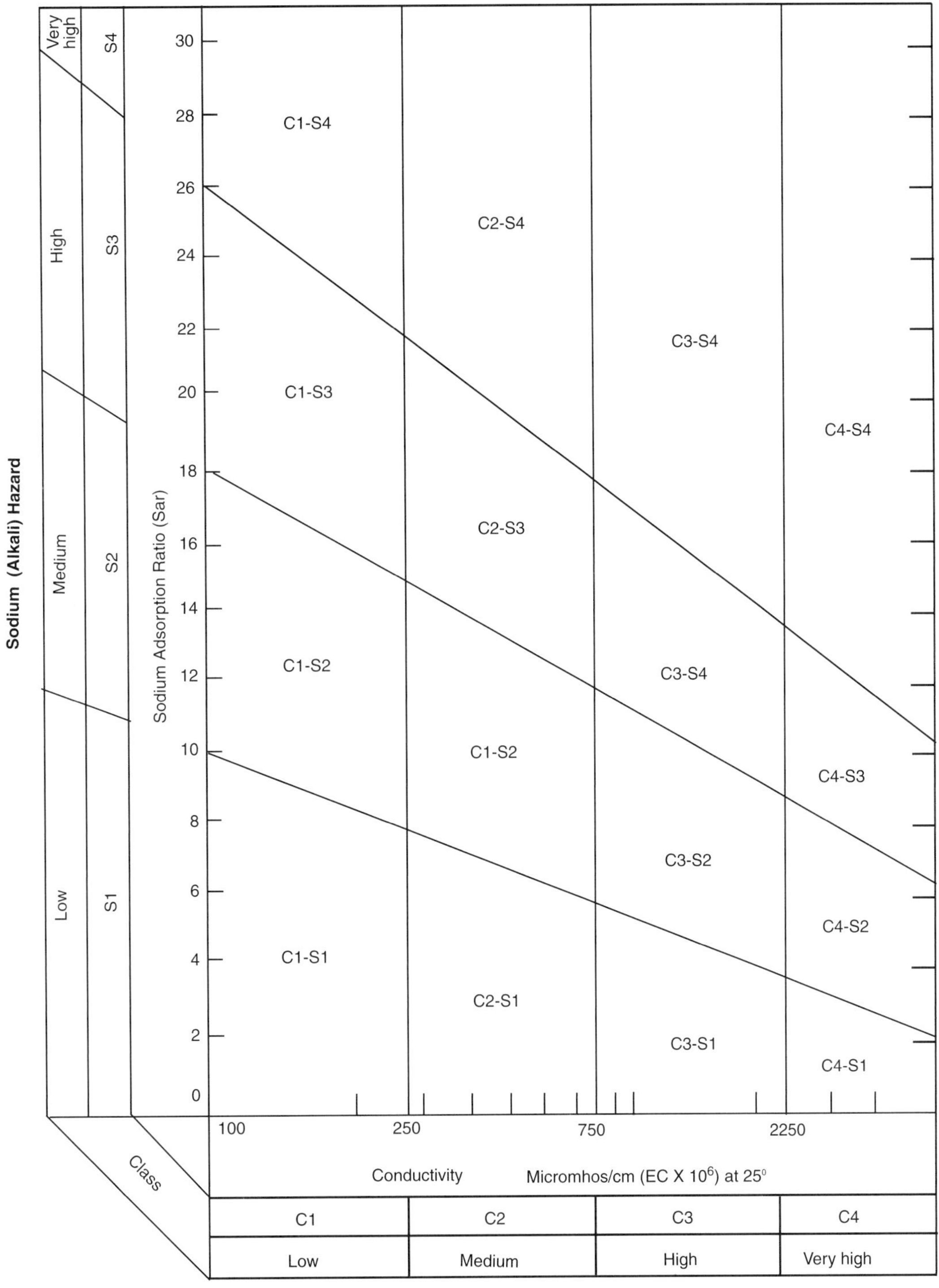

Figure 8F.20 Quality criteria for irrigation water. Note: Sodium Adsorption Ratio SAR = Na/(Ca + Mg)/2 where concentrations are expressed in millequivalents per liter.

Conductivity

Low-salinity water (C1) can be used for irrigation with most crops on most soils with little likelihood that soil salinity will develop. Some leaching is required, but this occurs under normal irrigation practices except in soils of extremely low permeability.

Medium-salinity water (C2) can be used if a moderate amount of leaching occurs. Plants with moderate salt tolerance can be grown in most cases without special practices for salinity control.

High-salinity water (C3) cannot be used on soils with restricted drainage. Even with adequate drainage, special management for salinity control may be required and plants with good salt tolerance should be selected.

Very high-salinity water (C4) is not suitable for irrigation under ordinary conditions, but may be used occasionally under very special circumstances. The soils must be permeable, drainage must be adequate, irrigation water must be applied in excess to provide considerable leaching, and very salt-tolerant crops should be selected.

Sodium

Low-sodium water (S1) can be used for irrigation on almost all soils with little danger of the development of harmful levels of exchangeable sodium. However, sodium-sensitive crops such as stone-fruit trees and avocados may accumulate injurious concentrations of sodium.

Medium-sodium water (S2) will present an appreciable sodium hazard in fine-textured soils having high cation-exchange-capacity, especially under low-leaching conditions, unless gypsum is present in the soil. This water may be used on coarse-textured or organic soils with good permeability.

High-sodium water (S3) may produce harmful levels of exchangeable sodium in most soils and will require special soil management-good drainage, high leaching, and organic matter additions. Gypsiferous soils may not develop harmful levels of exchangeable sodium from such waters. Chemical amendments may be required for replacement of exchangeable sodium, except that amendments may not be feasible with waters of very high salinity.

Very high sodium water (S4) is generally unsatisfactory for irrigation purposes except at low and perhaps medium salinity, where the solution of calcium from the soil or use of gypsum or other amendments may make the use of these waters feasible.

Another criterion for the evaluation of irrigation water is:

Residual Sodium Carbonate (RSC) = (CO3 + HCO3) − (Ca + Mg) where concentrations are expressed in meq/liter.

When RSC > 2.5 Probably not suitable for irrigation

1.25-2.5 Marginal

< 1.25 Probably safe for irrigation. (From U.S. Department of Agriculture.)

Table 8F.70 Guides for Evaluating the Quality of Water Used for Irrigation

Quality Factor	Threshold Concentration[a]	Limiting Concentration[b]
Coliform organisms, MPN per 100 mL	1000[c]	—[d]
Total dissolved solids (TDS), mg/L	500[c]	1500[c]
Electrical conductivity, μmhos/cm	750[c]	2250[c]
Range of pH	7.0–8.5	6.0–9.0
Sodium adsorption ratio (SAR)	6.0[c]	15
Residual sodium carbonate (RSC), meq	1.25[c]	2.5
Arsenic, mg/L	1.0	5.0
Boron, mg/L	0.5[c]	2.0
Chloride, mg/L	100[c]	350
Sulfate, mg/L	200[c]	1000
Copper, mg/L	0.1[c]	1.0

Note: MPN is most probable number. Sodium absorption ratio is defined by the formula $SAR = Na/(Ca + Mg)/2$ where the concentrations are expressed in milliequivalents per liter. Residual sodium carbonate is the sum of the equivalents of normal carbonate and bicarbonate minus the sum of the equivalents of calcium and magnesium.

[a] Threshold values at which irrigator might become concerned about water quality and might consider using additional water for leaching. Below these values, water should be satisfactory for almost all crops and almost any arable soil.

[b] Limiting values at which the yield of high-value crops might be reduced drastically, or at which an irrigator might be forced to less valuable crops.

[c] Values not to be exceeded more than 20 percent of any 20 consecutive samples, nor in any 3 consecutive samples. The frequency of sampling should be specified.

[d] Aside from fruits and vegetables which are likely to be eaten raw, no limits can be specified. For such crops, the threshold concentration would be limiting.

Source: From Calif. State Water Quality Control Board, 1963.

Table 8F.71 FAO Guidelines for Evaluating the Quality of Water for Irrigation

		Degree of Restriction on Use		
Potential Irrigation Problem	**Units**	**None**	**Slight to Moderate**	**Severe**
Salinity (affects crop water availability)[a]				
EC_W	dS/m	<0.7	0.7–3.0	>3.0
or TDS	mg/L	<450	450–2000	>2000
Infiltration (affects infiltration rate of water into the soil. Evaluate using EC_W and SAR together)[b]				
SAR =0–3 and EC_W =		>0.7	0.7–0.2	<0.2
=3–6 =		>1.2	1.2–0.3	<0.3
=6–12=		>1.9	1.9–0.5	<0.5
=12–20=		>2.9	2.9–1.3	<1.3
=20–40=		>5.0	5.0–2.9	<2.9
Specific Ion Toxicity (affects sensitive crops)				
Sodium (Na)				
Surface irrigation	SAR	<3	3–9	>9
Sprinkler irrigation	me/L	<3	>3	
Chloride (Cl)				
Surface irrigation	me/L	<4	4–10	>10
Sprinkler irrigation	me/L	<3	<3	
Boron (B)	mg/L	<0.7	0.7–3.0	>3.0
Trace elements (see Table 8F–77)				
Miscellaneous Effects (affects susceptible crops)				
Nitrogen (NO_3-N)	mg/L	<5	5–30	>30
Bicarbonate (HCO_3)(overhead sprinkling only)	me/L	<1.5	1.5–8.5	>8.5
pH			Normal range 6.5–8.4	

[a] EC_W means electrical conductivity, a measure of the water salinity, reported in deciSiemens per meter at 25°C (dS/m) or in units millimhos per centimeter (mmho/cm). Both are equivalent. TDS means total dissolved solids, reported in milligrams per liter (mg/L).

[b] SAR means sodium absorption ratio.

Source: From Food and Agriculture Organization of the United Nations, 1985, Water quality for agriculture, irrigation, and drainage paper no. 29. Kandiah, A., *Water Quality in Food Production, Water Quality Bulletin*, vol. 12, no. 1, Jan. 1987.

Table 8F.72 Chloride Concentrations (mg/L) Causing Foliar Injury in Crops of Varying Sensitivity

Sensitivity <175	Moderately Sensitive 175–350	Moderately Tolerant 350–700	Tolerant >700
Almond	Pepper	Barley	Cauliflower
Apricot	Potato	Maize	Cotton
Citrus	Tomato	Cucumber	Sugar beet
Plum		Luceme	Sunflower
Grape		Safflower	
		Sorghum	

Source: From Australian and New Zealand Environment and Conservation Council and Agriculture and Resource Management Council of Australia and New Zealand, 2000, *National Water Quality Management Strategy*, Paper No. 4, *Australian and New Zealand Guidelines for Fresh and Marine Water Quality*, vol. I, The Guidelines, October 2000, www.deh.gov.au.

Original Source: From Mass (1990).

Table 8F.73 Risks of Increasing Cadmium Concentrations in Crops Due to Chloride in Irrigation Waters

Irrigation Water Chloride Concentration (mg/L)	Risk of Increasing Crop Cadmium Concentrations
0–350	Low
350–750	Medium
>750	High

Note: If high chloride concentrations are present in irrigation water, it is recommended that produce is tested for cadmium concentration in the edible portions (e.g., tubers for potatoes, leaves for leafy vegeatables, grain for cereals, etc).

Source: From Australian and New Zealand Environment and Conservation Council and Agriculture and Resource Management Council of Australia and New Zealand, 2000, *National Water Quality Management Strategy*, Paper No. 4, *Australian and New Zealand Guidelines for Fresh and Marine Water Quality*, vol. I, The Guidelines, October 2000, www.deh.gov.au.

Orginal Source: From McLaughlin et al. (1999).

Table 8F.74 Sodium Concentration (mg/L) Causing Foliar Injury in Crops of Varying Sensitivity

Sensitivity <115	Moderately Sensitive 115–230	Moderately Tolerant 230–460	Tolerant >460
Almond	Pepper	Barley	Cauliflower
Apricot	Potato	Maize	Cotton
Citrus	Tomato	Cucumber	Sugar beet
Plum		Lucerne	Sunflower
Grape		Safflower	
		Sesame	
		Sorghum	

Source: From Australian and New Zealand Environment and Conservation Council and Agriculture and Resource Management Council of Australia and New Zealand, 2000, *National Water Quality Management Strategy*, Paper No. 4, *Australian and New Zealand Guidelines for Fresh and Marine Water Quality*, vol. I, The Guidelines, October 2000, www.deh.gov.au.

Original Source: From Mass (1990).

Table 8F.75 Effect of Sodium Expressed as Sodium Adsorption Ratio (SAR) on Crop Yield and Quality under Nonsaline Conditions

Tolerance to SAR and Range at Which Affected	Crop	Growth Response under Field Conditions
Extremely sensitive SAR = 2–8	Avocado	Leaf tip burn, leaf scorch
	Deciduous fruits	
	Nuts	
	Citrus	
Sensitive SAR = 8–18	Beans	Stunted growth
Medium SAR = 18–46	Clover	Stunted growth, possible sodium toxicity, possible calcium or magnesium deficiency
	Oats	
	Tall fescue	
	Rice	
	Dallis grass	
High SAR = 46–102	Wheat	Stunted growth
	Cotton	
	Lucerne	
	Barley	
	Beets	
	Rhodes grass	

Note: SAR = Sodium Adsorption Ratio.

Source: From Australian and New Zealand Environment and Conservation Council and Agriculture and Resource Management Council of Australia and New Zealand, 2000, *National Water Quality Management Strategy*, Paper No. 4, *Australian and New Zealand Guidelines for Fresh and Marine Water Quality*, vol. I, The Guidelines, October 2000, www.deh.gov.au.

Original Source: From Pearson (1960).

Table 8F.76 Limits of Boron in Irrigation Water

A. Permissible Limits (Boron in parts per million)

	Crop Group		
Class of Water	**Sensitive**	**Semitolerant**	**Tolerant**
			Excellent
<0.33	<0.67	<1.00	Good
0.33–0.67	0.67–1.33	1.00–2.00	Permissible
0.67–1.00	1.33–2.00	2.00–3.00	Doubtful
1.00–1.25	2.00–2.50	3.00–3.75	Unsuitable
>1.25	>2.50	>3.75	

B. Crop Groups of Boron Tolerance (In each group, the plants first named are considered as being more tolerant; the last named, more sensitive.)

Sensitive	Semitolerant	Tolerant
Pecan	Sunflower (native)	Athel (*Tamarix aphylla*)
Walnut (Black; and Persian, or English)	Potato	Asparagus
	Cotton (Acala and Pima)	Palm (*Phoenix canariensis*)
Jerusalem-artichoke	Tomato	Date palm (*P. dactylifera*)
Navy bean	Sweetpea	Sugar beet
American elm	Radish	Mangel
Plum	Field pea	Garden beet
Pear	Ragged robin rose	Alfalfa
Apple	Olive	Gladiolus
Grape (Sultanina and Malaga)	Barley	Broadbean
Kadota fig	Wheat	Onion
Persimmon	Corn	Turnip
Cherry	Milo	Cabbage
Peach	Oat	Lettuce
Apricot	Zinnia	Carrot
Thornless blackberry	Pumpkin	
Orange	Bell pepper	
Avocado	Sweet potato	
Grapefruit	Lima bean	
Lemon		

Source: From U.S. Dept. of Agriculture.

Table 8F.77 FAO Recommended Maximum Concentrations of Trace Elements in Irrigation Water

Element	Recommended Maximum Concentration[a] (mg/L)	Remarks
Al	5.0	Can cause nonproductivity in acid soils (pH <5.5), but more alkaline soils at >pH 7.0 will precipitate the Ion and eliminate any toxicity
As	0.10	Toxicity to plants varies widely, ranging from 12 mg/L for Sudan grass to less than 0.05 mg/L for rice
Be	0.10	Toxicity to plants varies widely, ranging from 5 mg/L for kale to 0.5 mg/L for bush beans
Cd	0.01	Toxic to beans, beets and turnips at concentrations as low as 0.1 mg/L in nutrient solutions. Conservative limits recommended due to its potential for accumulation in plants and soils to concentrations that may be harmful to humans
Co	0.05	Toxic to tomato plants at 0.1 mg/L in nutrient solution. Tends to be inactivated by neutral and alkaline soils
Cr	0.10	Not generally recognized as an essential growth element. Conservative limits recommended due to lack of knowledge on its toxicity to plants
Cu	0.20	Toxic to a number of plants at 0.1 to 1.0 mg/L in nutrient solutions
F	1.0	Inactivated by neutral and alkaline soils

(Continued)

Table 8F.77 (Continued)

Element	Recommended Maximum Concentration[a] (mg/L)	Remarks
Fe	5.0	Not toxic to plants in aerated soils, but can contribute to soil acidification and loss of availability of essential phosphorus and molybdenum. Overhead sprinkling may result in unsightly deposits on plants, equipment and buildings
Li	2.5	Tolerated by most crops up to mg/L; mobile in soil. Toxic to citrus at low concentrations (0.075 mg/L). Acts similarly to boron
Mn	0.20	Toxic to a number of crops at a few-tenths to a few mg/L, but usually only in acid soils
Mo	0.01	Not toxic to plants at normal concentrations in soil and water. Can be toxic to livestock if forage is grown in soils with high concentrations of available molybdenum
Ni	0.20	Toxic to a number of plants at 0.5 mg/L to 1.0 mg/L; reduced toxicity at neutral or alkaline pH
Pd	5.0	Can inhibit plant cell growth at very high concentrations
Se	0.02	Toxic to plants at concentrations as low as 0.025 mg/L and toxic to livestock if forage is grown in soils with relatively high levels of added selenium. An essential element to animals but in very low concentrations
Sn		
Ti		Effectively excluded by plants; specific tolerance unknown
W		
V	0.10	Toxic to many plants at relatively low concentrations
Zn	2.0	Toxic to many plants at widely varying concentrations; reduced toxicity at pH >6.0 and in fine textured or organic soils

[a] The maximum concentration is based on a water application rate which is consistent with good irrigation practices (10000 m^3/ha/yr). If the water application rate greatly exceeds this, the maximum concentrations should be adjusted downward accordingly. No adjustment should be made for application rates less than 10000 m^3/ha/yr. The values given are for water used on a continuous basis at one site.

Source: From Food and Agriculture Organization of the United Nations, 1985, Water quality for agriculture, irrigation, and drainage paper no. 29. Kandiah, A., *Water Quality in Food Production, Water Quality Bulletin*, vol. 12, no. 1, Jan. 1987.

Table 8F.78 Australian Agricultural Irrigation Water Long-Term Trigger Value (LTV), Short-Term Trigger Value (STV), and Soil Cumulative Contaminant Loading Limit (CCL) Triggers for Heavy Metals and Metalloids

Element	Suggested Soil CCL[a] (kg/ha)	LTV in Irrigation Water (Long-Term Use—Up to 100 yrs) (mg/L)	STV in Irrigation Water (Short-Term Use—Up to 20 yrs) (mg/L)
Aluminum	ND	5	20
Arsenic	20	0.1	2.0
Beryllium	ND	0.1	0.5
Boron	ND	0.5	Refer to table 9.2.18 (Volume 3)
Cadmium	2	0.01	0.05
Chromium	ND	0.1	1
Cobalt	ND	0.05	0.1
Copper	140	0.2	5
Fluoride	ND	1	2
Iron	ND	0.2	10
Lead	260	2	5
Lithium	ND	2.5 (0.075 Citrus crops)	2.5 (0.075 Citrus crops)
Manganese	ND	0.2	10
Mercury	2	0.002	0.002
Molybdenum	ND	0.01	0.05
Nickel	85	0.2	2

(Continued)

Table 8F.78 (Continued)

Element	Suggested Soil CCL[a] (kg/ha)	LTV in Irrigation Water (Long-Term Use—Up to 100 yrs) (mg/L)	STV in Irrigation Water (Short-Term Use—Up to 20 yrs) (mg/L)
Selenium	10	0.02	0.05
Uranium	ND	0.01	0.1
Vanadium	ND	0.1	0.5
Zinc	300	2	5

Note: Trigger values should only be used in conjunction with information on each individual element and the potential for off-site transport of contaminants.

[a] ND = Not determined; in sufficient background data to calculate CCL.

Source: From Australian and New Zealand Environment and Conservation Council and Agriculture and Resource Management Council of Australia and New Zealand, 2000, *National Water Quality Management Strategy*, Paper No. 4, *Australian and New Zealand Guidelines for Fresh and Marine Water Quality*, vol. I, The Guidelines, October 2000, www.deh.gov.au.

Table 8F.79 Canadian Water Quality Guidelines for the Protection of Agricultural Uses—Irrigation

Parameter[a]	Irrigation Water Quality Guideline (μg/L)	Date [b]
Aldicarb	54.9[c]	1993
Aluminum[d]	5000	1987
Arsenic[e]	100[f]	1997
Atrazine	10[f]	1989
Beryllium[d]	100	1987
Boron[d]	500–6000[h]	1987
Bromacil	0.2[f]	1997
Bromoxynil	0.33[i]	1993
Cadmium	5.1[i,f]	1996
Chloride[d]	100,000–700,000[k]	1987
Chlorothalonil	5.8[f] (other crops)	1984
Chromium		1997
Trivalent chromium (Cr(III))	4.9[f,n]	1997
Hexavalent chromium (Cr(VI))	8[n]	1987
Cobalt[d]	50	1987
Coliforms, fecal [d]	100 per 100 mL	1987
Coliforms, total[d]	1000 per 100 mL	1987
Copper[d]	200–1000[o]	1987
Cyanazine	0.5[f]	1990
Dicamba	0.006	1993
Diclofop-methyl	0.18	1993
Diisopropanolinine	2000[f]	2005
Dinoseb	16[j]	1982
Fluoride[d]	1000	1987
Iron[d]	5000	1987
Lead[d]	200	1987
Linuron	0.071[f]	1995
Lithium[d]	2500	1987
Manganese[d]	200	1987
MCPA (4-chloro-2-methyl phenoxy acetic acid; 2-Methyl-4-chloro phenoxy acetic acid)	0.0025	1995
Metolachlor	28[f]	1991
Metribuzin	0.5[f]	1990
Molybdenum[d]	10–50[r]	1987
Selenium[d]	20–50[s]	1987
Simazine	0.5[f]	1891
Nickel[d]	200	1987

(Continued)

Table 8F.79 (Continued)

Parameter[a]	Irrigation Water Quality Guideline (μg/L)	Date [b]
Sulfolane	500[f]	2005
Tebuthiuron	0.27 [f] (cereals)	1995
Total dissolved solids (salinity)[d]	500,000– 3,500,000[t]	1887
Uranium[d]	10[f]	1987
Vanadium[d]	100	1987
Zinc[d]	1,000–5,000[u]	1987

Note: ug/L—Micrograms per liter.

[a] Unless otherwise indicated, supporting documents are available from the guidelines and Standards Division, Environment Canada.

[b] The guidelines dated 1987 have been carried over from *Canadian Water Quality Guidelines* (CCREM 1987) and no fact sheet was prepared. The guidelines dated 1989 to 1997 were developed and initially published in CCREM 1987 as appendixes on the date indicated. They are published as fact sheets in this document. Other guidelines dated 1997 and those dated 1999 are published for the first time in this document.

[c] Concentration of total aldicarb residues.

[d] No fact sheet created.

[e] The technical document for the guideline is available from the Ontario Ministry of the Environment.

[f] Interim guideline.

[h] Boron guideline
=500 μg L^{-1} for blackberries
=500–1000 μg L^{-1} for peaches, cherries, plums, grapes, cowpeas, onions, garlic, sweet potatoes, wheat, barley, sunflowers, mungbeans, sesame, lupins, strawberries, Jerusalem artichokes, kidney beans, and lima beans
=1000–2000 μg L^{-1} for red peppers, peas, carrots, radishes, potatoes, and cucumbers
=2000–4000 μg L^{-1} for lettuce, cabbage, celery, turnips, Kentucky bluegrass, oats, corn, artichokes, tobacco, mustard, clover, squash, and muskmelons
=4000–6000 μg L^{-1} for sorghum, tomatoes, alfalfa, purple vetch, parsley, red beets, and sugar beets
=6000 μg L^{-1} for asparagus.

[i] Guideline value slightly modified from CCREM 1987 + Appendixes due to re-evaluation of the significant figures.

[j] Guideline is crop-specific (see fact sheet).

[k] Chloride guideline Foliar damage
=100–178 mg L^{-1} for almond, apricots, and plums
=178–355 mg L^{-1} for grapes, peppers, potatoes, and tomatoes
=355–710 mg L^{-1} for alfalfa, barley, corn, and cucumbers
>710 mg L^{-1} for cauliflower, cotton, safflower, sesame, sorghum, sugar beets, and sunflowers
Rootstocks
=180–600 mg L^{-1} for stone fruit (peaches, plums, etc)
=710–900 mg L^{-1} for grapes
Cultivars
=110–180 mg L^{-1} for strawberries
=230–460 mg L^{-1} for grapes
=250 mg L^{-1} for boysenberries, blackberries, and raspberries.

[n] Substance has been re-evaluated since CCREM 1987 + Appendixes. Either a new guideline has been derived or insufficient data existed to derive a new guideline.

[o] Copper guideline
= 200 μg L^{-1} for cereals
= 1000 μg L^{-1} for tolerant crops

[r] Molybdenum guideline = 50 μg L^{-1} for short-terms use on acidic soils

[s] Selenium guideline = 20 μg L^{-1} for continuous use
=50 μg L^{-1} for intermittent use

[t] Total dissolved solids guideline =500 mg L^{-1} for strawberries, raspberries, beans, and carrots
=500–800 mg L^{-1} for boysenberries, currants, blackberries, gooseberries, plums, grapes, apricots, peaches, pears, cherries, apples, onions, parsnips, radishes, peas, pumpkins, lettuce, peppers, muskmelons, sweet potatoes, sweet corn, potatoes, celery, cabbage, kohlrabi, cauliflower, cowpeas, broadbeans, flax, sunflowers, and corn
=800–1500 mg L^{-1} for spinach, cantaloupe, cucumbers, tomatoes, squash, brussels sprouts, broccoli, turnips, smooth brome, alfalfa, big trefoil, beardless wild rye, vetch, timothy, and crested wheat grass
=1500–2500 mg L^{-1} for beets, zucchini, rape, sorghum, oat hay, wheat hay, mountain brome, tall fescue, sweet clover, reed canary grass, birdsfoot trefoil, perennial ryegrass
=3500 mg L^{-1} for asparagus, soybeans, safflower, oats, rye, wheat, sugar beets, barley, barley hay, and tall wheat grass

[u] Zinc guideline =1000 mg L^{-1} when soil pH < 6.5
=5000 mg L^{-1} when soil pH > 6.5

Source: From Canadian Council of Ministers of the Environment, 2005, Canadian water quality guidelines for the protection of agricultural water uses: summary table, Updated October 2005. In: *Canadian Environmental Quality Guidelines, 1999*, Canadian Council of Ministers of the Environment, Winnipeg.

Table 8F.80 Australian Trigger Values for Thermotolerant Coliforms in Irrigation Water Used for Food and Nonfood Crops

Intended use	Level of thermotolerant coliforms[a] (mL)
Raw human food crops in direct contact with irrigation water (e.g. via sprays, irrigation of salad vegetables)	<10 cfu[b] / 100
Raw human food crops not in direct contact with irrigation water (edible product separated from contact with water, e.g. by peel, use of trickle irrigation); or crops sold to consumer cooked or processed	<1000 cfu / 100
Pasture and fodder for dairy animals (without withholding period)	<100 cfu / 100
Pasture and fodder for dairy animals (with withholding period of 5 days)	<1000 cfu / 100
Pasture and fodder (for grazing animals except pigs and dairy animals, i.e. cattle, sheep, and goats)	<1000 cfu / 100
Silviculture, turf, cotton, etc (restricted public access)	<10 000 cfu / 100

[a] Median values.
[b] cfu = colony forming units.

Source: From Australian and New Zealand Environment and Conservation Council and Agriculture and Resource Management Council of Australia and New Zealand, 2000, *National Water Quality Management Strategy*, Paper No. 4, *Australian and New Zealand Guidelines for Fresh and Marine Water Quality*, vol. I, The Guidelines, October 2000, www.deh.gov.au.

Original Source: Adapted from ARMCANZ, ANZECC & NHMRC (1999).

Table 8F.81 Australian Agricultural Irrigation Water Long-Term Trigger Value (LTV) and Short-Term Trigger Value (STV) Guidelines for Nitrogen and Phosphorus

Element	LTV in Irrigation Water (Long-Term—up to 100 yrs) (mg/L)	STV in Irrigation Water (Short-Term—up to 20 yrs) (mg/L)
Nitrogen	5	25–125[a]
Phosphorus	0.05 (To minimise bioclogging of irrigation equipment only)	0.8–12[a]

[a] Requires site-specific assessment.

Source: From Australian and New Zealand Environment and Conservation Council and Agriculture and Resource Management Council of Australia and New Zealand, 2000, *National Water Quality Management Strategy*, Paper No. 4, *Australian and New Zealand Guidelines for Fresh and Marine Water Quality*, vol. I, The Guidelines, October 2000, www.deh.gov.au.

Table 8F.82 Interim Trigger Value Concentrations for a Range of Herbicides Registered in Australia for Use in or Near Waters

Herbicide	Residue Limits in Irrigation Water (mg/L)[a]	Hazard to Crops from Residue in Water[b]	Crop Injury Threshold in Irrigation Water (mg/L)
Acrolein	0.1	+	Flood or furrow: beans 60, corn 60, cotton 80, soybeans 20, sugar-beets 60 Sprinkler: corn 60, soybeans 15, sugar-beets 15,
AF 100		+	beets (rutabaga) 3.5, corn 3.5,
Amitrol	0.002	++	lucerne 1600, beans 1200, carrots 1600, corn 3000, cotton 1600, grains sorghum 800, oats 2400, potatoes 1300, wheat 1200
Aromatic solvents (Xylene)		+	
Asulam		++	
Atrazine		++	
Bromazil		+++	
Chlorthiamid		++	
Copper sulfate		+	Apparently above concentrations used for weed control
2,4-D		++	Field beans 3.5–10, grapes 0.7–1.5, sugar-beets 1.0–10
Dicamba		++	Cotton 0.18

(Continued)

Table 8F.82 (Continued)

Herbicide	Residue Limits in Irrigation Water (mg/L)[a]	Hazard to Crops from Residue in Water[b]	Crop Injury Threshold in Irrigation Water (mg/L)
Dichlobenil		++	Lucerne 10, corn 10, soybeans 1.0, sugar-beets 1.0–10, corn 125, beans 5
Diquat		+	
Diuron	0.002	+++	
2,2-DPA (Dalapon)	0.004	++	Beets 7.0, corn 0.35
Fosamine		+++	
Fluometuron		++	Sugar-beets, alfalfa, tomatoes, squash 2.2
Glyphosate		+	
Hexazinone		+++	
Karbutilate		+++	
Molinate		++	
Paraquat		+	Corn 10, field beans 0.1, sugar-beets 1.0
Picloram		+++	
Propanil		++	Alfalfa 0.15, brome grass (eradicated) 0.15
Simazine		++	
2,4,5-T		++	Potatoes, alfalfa, garden peas, corn sugar-beets, wheat, peaches, grapes, apples, tomatoes 0.5
TCA (Trichloroacetic acid)		+++	
Terbutryne		++	
Triclopyr		++	

Note: These should be regarded as interim trigger values only.

[a] Guidelines have not been set for all herbicides where specific residue limits are not provided, except for a general limit of 0.01 mg/L for herbicides in NSW.

[b] Hazard from residue at maximum concentration likely to be found in irrigation water: +, low; ++, moderate; +++, high.

Source: From Australian and New Zealand Environment and Conservation Council and Agriculture and Resource Management Council of Australia and New Zealand, 2000, *National Water Quality Management Strategy*, Paper No. 4, *Australian and New Zealand Guidelines for Fresh and Marine Water Quality*, vol. I, The Guidelines, October 2000, www.deh.gov.au.

Original Source: From ANZECC (1992).

Table 8F.83 Australian Trigger Values for Radioactive Contaminants for Irrigation Water

Radionuclide	Trigger Concentration (Bq/L)
Radium 226	5
Radium 228	2
Uranium 238	0.2
Gross alpha	0.5
Gross beta (excluding K-40)	0.5

Source: From Australian and New Zealand Environment and Conservation Council and Agriculture and Resource Management Council of Australia and New Zealand, 2000, *National Water Quality Management Strategy*, Paper No. 4, *Australian and New Zealand Guidelines for Fresh and Marine Water Quality*, vol. I, The Guidelines, October 2000, www.deh.gov.au.

Table 8F.84 Corrosion Potential of Waters on Metal Surface and Fouling Potential as Indicated by pH, Hardness, Langelier Index, Ryznar Index, and the Log of Chloride Carbonate Ratio

Parameter[a]	Value	Comments
Corrision Potential		
pH	<5	High corrosion potential
	5–6	Likelihood of corrosion
	>6	Limited corrosion potential
Hardness	<60 mg/L CaCO3	Increased corrosion potential
Langelier index	<−0.5	Increased corrosion potential
	−0.5–0.5	Limited corrosion potential
Ryznar index	<6	Limited corrosion potential
	>7	Increased corrosion potential
Log of chloride to carbonate ratio	>2	Increased corrosion potential
Fouling potential		
pH	<7	Limited fouling potential
	7–8.5	Moderate fouling potential (groundwater)[b]
	>8.5	Increased fouling potential (groundwater)[c]
Hardness	<350 mg/L $CaCO_3$	Increased fouling potential
Langelier index	>0.5	Increased fouling potential
	−0.5–0.5	Limited fouling potential
Ryznar index	<6	Increased fouling potential
	>7	Limited fouling potential
Log of chloride to carbonate ratio	<2	Increased fouling potential

[a] For further information on these parameters refer to Volume 3, Section 9.2.9.1.
[b] For surface waters, pH range 7 to 9.
[c] For surface waters, pH >9.

Source: From Australian and New Zealand Environment and Conservation Council and Agriculture and Resource Management Council of Australia and New Zealand, 2000, *National Water Quality Management Strategy*, Paper No. 4, *Australian and New Zealand Guidelines for Fresh and Marine Water Quality*, vol. I, The Guidelines, October 2000, www.deh.gov.au.

Table 8F.85 Influence of Water Quality on the Potential for Clogging Problems in Localized (Drip) Irrigation Systems

		Degree of Restriction on Use		
Potential Problem	**Units**	**None**	**Slight to Moderate**	**Severe**
Physical				
Suspended solids	mg/L	<50	50–100	>100
Chemical				
pH		<7.0	7.0–8.0	>8.0
Dissolved solids	mg/L	<500	500–2000	>2000
Manganese[a]	mg/L	<0.1	0.1–1.5	>1.5
Iron[b]	mg/L	<0.1	0.1–1.5	>1.5
Hydrogen sulphide	mg/L	<0.5	0.5–2.0	>2.0
Biological				
Bacterial populations	Maximum number/ML	<10000	10000–50000	>50000

[a] While restrictions in use of localized (drip) irrigation systems may not occur at these manganese concentrations, plant toxicities may occur at lower concentrations.
[b] Iron concentrations >5.0 mg/L may cause nutritional imbalances in certain crops. www.fao.org/icatalog/inter-e.htm

Source: From Food and Agriculture Organization of the United Nations, 1994, Water Quality for Agriculture, Irrigation, and Drainage Paper No 29, Rev. 1, www.fao.org. Reprinted with permission.

Original Source: Adapted from Nakayama (1982).

SECTION 8G WATER QUALITY FOR AQUATIC LIFE

United States National Recommended Water Quality Criteria for Priority Toxic Pollutants

Section 304(a)(1) of the Clean Water Act requires the United States Environmental Protection Agency (EPA) to develop criteria for water quality that accurately reflects the latest scientific knowledge. These criteria are based solely on data and scientific judgments on pollutant concentrations and environmental or human health effects. Section 304(a) also provides guidance to states and tribes in adopting water-quality standards. Criteria are developed for the protection of aquatic life as well as for human health.

The Criteria Maximum Concentration (CMC) is an estimate of the highest concentration of a material in surface water to which an aquatic community can be exposed briefly without resulting in an unacceptable effect. The Criterion Continuous Concentration (CCC) is an estimate of the highest concentration of a material in surface water to which an aquatic community can be exposed indefinitely without resulting in an unacceptable effect. The CMC and CCC are just two of the six parts of an aquatic life criterion; the other four parts are the acute averaging period, chronic averaging period, acute frequency of allowed exceedence, and chronic frequency of allowed exceedence. Because 304(a) aquatic life criteria are national guidance, they are intended to be protective of the vast majority of the aquatic communities in the United States.

The tables below lists all priority toxic pollutants and some non priority toxic pollutants, and both human health effect and organoleptic effect criteria issued pursuant to CWA §304(a). Blank spaces indicate that EPA has no CWA §304(a) criteria recommendations. For a number of nonpriority toxic pollutants not listed, CWA §304(a) "water + organism" human health criteria are not available, but EPA has published MCLs under the SDWA that may be used in establishing water-quality standards to protect water supply designated uses.

The human health criteria for the priority and nonpriority pollutants are based on carcinogenicity of 10^{-6} risk. Alternate risk levels may be obtained by moving the decimal point (e.g., for a risk level of 10^{-5}, move the decimal point in the recommended criterion one place to the right).

The compilation contains 304(a) criteria for pollutants with toxicity-based criteria as well as nontoxicity based criteria. The basis for the nontoxicity based criteria are organoleptic effects (e.g., taste and odor) which would make water and edible aquatic life unpalatable but not toxic to humans. The table includes criteria for organoleptic effects for 23 pollutants. Pollutants with organoleptic effect criteria more stringent than the criteria based on toxicity (e.g., included in both the priority and nonpriority pollutant tables) are footnoted as such.

Table 8G.86 United States National Recommended Water Quality Criteria for Priority Toxic Pollutants

			Freshwater		Saltwater		Human Health for the Consumption of		
	Priority Pollutant	CAS Number	CMC (acute) (μg/L)	CCC (chronic) (μg/L)	CMC (acute) (μg/L)	CCC (chronic) (μg/L)	Water + Organism (μg/L)	Organism Only (μg/L)	FR Cite/Source
1	Antimony	7440360					5.6[A]	640[A]	65FR66443
2	Arsenic	7440382	340[B,C,D]	150[B,C,D]	69[B,C,D,E]	36[B,C,D,E]			65FR31682
							0.018[F,G,H]	0.14[F,G,H]	57FR60848
3	Beryllium	7440417					i		65FR31682
4	Cadmium	7440439	2.0[C,D,E,J]	0.25[C,D,E,J]	40[C,E]	8.8[C,E]	i		EPA-822-R-01–001
									65FR31682
5a	Chromium (III)	16065831	570[C,J,D]	74[C,J,D]			Total[i]		EPA820/B-96–001
									65FR31682
5b	Chromium (VI)	18540299	16[C,D]	11[C,D]	1,100[C,E]	50[C,E]	Total[i]		65FR31682
6	Copper	7440508	13[C,D,J,K]	9.0[C,D,J,K]	4.8[C,K,L]	3.1[C,K,L]	1,300[M]		65FR31682
7	Lead	7439921	65[C,E,J,N]	2.5[C,E,J,N]	210[C,E]	8.1[C,E]			65FR31682
8a	Mercury	7439976	1.4[C,D,O]	0.77[C,D,O]	1.8[C,O,P]	0.94[C,O,P]			62FR42160
8b	Methylmercury	22967926						0.3 mg/kg[Q]	EPA823-R-01–001
9	Nickel	7440020	470[C,J,D]	52[C,J,D]	74[C,E]	8.2[C,E]	610[A]	4,600[A]	65FR31682
10	Selenium	7782492	—[R,S,T]						62FR42160
				5.0[T]	290[C,E,U]	71[C,E,U]			65FR31682
							170[i]	4200	65FR66443
11	Silver	7440224	3.2[C,J,V]		1.9[C,V]				65FR31682
12	Thallium	7440280					0.24	0.47	68FR75510
13	Zinc	7440666	120[C,J,D]	120[C,J,D]	90[C,E]	81[C,E]			65FR31682
							7,400[M]	26,000[M]	65FR66443
14	Cyanide	57125	22[D,W]	5.2[D,W]					EPA820/B-96–001
					1[E,W]	1[E,W]			57FR60848
							140[X]	140[X]	68FR75510
15	Asbestos	1332214					7 million fibers/L[YX]		57FR60848
16	2,3,7,8-TCDD (Dioxin)	1746016					5.0E−9[F]	5.1E−9[F]	65FR66443
17	Acrolein	107028					190	290	65FR66443
18	Acrylonitrile	107131					0.051[A,F]	0.25[A,F]	65FR66443
19	Benzene	71432					2.2[A,F]	51[A,F]	IRIS 01/19/00 &65FR66443
20	Bromoform	75252					4.3[A,F]	140[A,F]	65FR66443
21	Carbon Tetrachloride	56235					0.23[A,F]	1.6[A,F]	65FR66443
22	Chlorobenzene	108907					130[I,M]	1,600[M]	68FR75510
23	Chlorodibromo-methane	124481					0.40[A,F]	13[A,F]	65FR66443
24	Chloroethane	75003							
25	2-Chloroethylvinyl Ether	110758							
26	Chloroform	67663					5.7[F,Z]	470[F,Z]	62FR42160
27	Dichlorobro-momethane	75274					0.55[A,F]	17[A,F]	65FR66443
28	1,1-Dichloroethane	75343							
29	1,2-Dichloroethane	107062					0.38[A,F]	37[A,F]	65FR66443
30	1,1-Dichloroethylene	75354					330	7,100	68FR75510
31	1,2-Dichloropropane	78875					0.50[A,F]	15[A,F]	65FR66443
32	1,3-Dichloropropene	542756					0.34[F]	21[F]	68FR75510
33	Ethylbenzene	100414					530	2,100	68FR75510

34	Methyl Bromide	74839					47[A]	1,500[A]	65FR66443
35	Methyl Chloride	74873							65FR31682
36	Methylene Chloride	75092					4.6[A,F]	590[A,F]	65FR66443
37	1,1,2,2-Tetrachloroethane	79345					0.17[A,F]	4.0[A,F]	65FR66443
38	Tetrachloroethylene	127184					0.69[F]	3.3[F]	65FR66443
39	Toluene	108883					1,300[I]	15,000	68FR75510
40	1,2-Trans-Dichloroethylene	156605					140[I]	10,000	68FR75510
41	1,1,1-Trichloroethane	71556					—[I]		65FR31682
42	1,1,2-Trichloroethane	79005					0.59[A,F]	16[A,F]	65FR66443
43	Trichloroethylene	79016					2.5[F]	30[F]	65FR66443
44	Vinyl Chloride	75014					0.025[F,aa]	2.4[F,aa]	68FR75510
45	2-Chlorophenol	95578					81[A,M]	150[A,M]	65FR66443
46	2,4-Dichlorophenol	120832					77[A,M]	290[A,M]	65FR66443
47	2,4-Dimethylphenol	105679					380[A]	850[A,M]	65FR66443
48	2-Methyl-4,6-Dinitrophenol	534521					13	280	65FR66443
49	2,4-Dinitrophenol	51285					69[A]	5,300[A]	65FR66443
50	2-Nitrophenol	88755							
51	4-Nitrophenol	100027							
52	3-Methyl-4-Chlorophenol	59507					—[M]	—[M]	
53	Pentachlorophenol	87865	19[D,bb]	15[D,bb]	13[E]	7.9[E]	0.27[A,F]	3.0[A,F,cc]	65FR31682 65FR66443
54	Phenol	108952					21,000[A,M]	1,700,000[A,M]	65FR66443
55	2,4,6-Trichlorophenol	88062					1.4[A,F]	2.4[A,F,M]	65FR66443
56	Acenaphthene	83329					670[A,M]	990[A,M]	65FR66443
57	Acenaphthylene	208968							
58	Anthracene	120127					8,300[A]	40,000[A]	65FR66443
59	Benzidine	92875					0.000086[A,F]	0.00020[A,F]	65FR66443
60	Benzo(a) Anthracene	56553					0.0038[A,F]	0.018[A,F]	65FR66443
61	Benzo(a) Pyrene	50328					0.0038[A,F]	0.018[A,F]	65FR66443
62	Benzo(b) Fluoranthene	205992					0.0038[A,F]	0.018[A,F]	65FR66443
63	Benzo(ghi) Perylene	191242							
64	Benzo(k) Fluoranthene	207089					0.0038[A,F]	0.018[A,F]	65FR66443
65	Bis(2-Chloroethoxy) Methane	111911							
66	Bis(2-Chloroethyl) Ether	111444					0.030[A,F]	0.53[A,F]	65FR66443
67	Bis(2-Chloroisopropyl) Ether	108601					1,400[A]	65,000[A]	65FR66443
68	Bis(2-Ethylhexyl) Phthalate[dd]	117817					1.2[A,F]	2.2[A,F]	65FR66443
69	4-Bromophenyl Phenyl Ether	101553							
70	Butylbenzyl Phthalate[ee]	85687					1,500[A]	1,900[A]	65FR66443
71	2-Chloronaphthalene	91587					1,000[A]	1,600[A]	65FR66443
72	4-Chlorophenyl Phenyl Ether	7005723							
73	Chrysene	218019					0.0038[A,F]	0.018[A,F]	65FR66443
74	Dibenzo(a,h) Anthracene	53703					0.0038[A,F]	0.018[A,F]	65FR66443
75	1,2-Dichlorobenzene	95501					420	1,300	68FR75510

(Continued)

Table 8G.86 (Continued)

	Priority Pollutant	CAS Number	Freshwater CMC (acute) (μg/L)	Freshwater CCC (chronic) (μg/L)	Saltwater CMC (acute) (μg/L)	Saltwater CCC (chronic) (μg/L)	Human Health for the Consumption of Water + Organism (μg/L)	Human Health for the Consumption of Organism Only (μg/L)	FR Cite/Source
76	1,3-Dichlorobenzene	541731					320	960	65FR66443
77	1,4-Dichlorobenzene	106467					63	190	68FR75510
78	3,3′-Dichlorobenzidine	91941					$0.021^{A,F}$	$0.028^{A,F}$	65FR66443
79	Diethyl Phthalate[ee]	84662					$17,000^{A}$	$44,000^{A}$	65FR66443
80	Dimethyl Phthalate[ee]	131113					270,000	1,100,000	65FR66443
81	Di-n-Butyl Phthalate[ee]	84742					$2,000^{A}$	$4,500^{A}$	65FR66443
82	2,4-Dinitrotoluene	121142					0.11^{F}	3.4^{F}	65FR66443
83	2,6-Dinitrotoluene	606202							
84	Di-n-Octyl Phthalate	117840							
85	1,2-Diphenyl-hydrazine	122667					$0.036^{A,F}$	$0.20^{A,F}$	65FR66443
86	Fluoranthene	206440					130^{A}	140^{A}	65FR66443
87	Fluorene	86737					$1,100^{A}$	$5,300^{A}$	65FR66443
88	Hexachlorobenzene	118741					$0.00028^{A,F}$	$0.00029^{A,F}$	65FR66443
89	Hexachloro-butadiene	87683					$0.44^{A,F}$	$18^{A,F}$	65FR66443
90	Hexachloro-cyclopentadiene	77474					40^{M}	$1,100^{M}$	68FR75510
91	Hexachloroethane	67721					$1.4^{A,F}$	$3.3^{A,F}$	65FR66443
92	Ideno(1,2,3-cd)Pyrene	193395					$0.0038^{A,F}$	$0.018^{A,F}$	65FR66443
93	Isophorone	78591					$35^{A,F}$	$960^{A,F}$	65FR66443
94	Naphthalene	91203							
95	Nitrobenzene	98953					17^{A}	$690^{A,cc,M}$	65FR66443
96	N-Nitroso-dimethylamine	62759					$0.00069^{A,F}$	$3.0^{A,F}$	65FR66443
97	N-Nitroso-di-n-Propylamine	621647					$0.0050^{A,F}$	$0.51^{A,F}$	65FR66443
98	N-Nitroso-diphenylamine	86306					$3.3^{A,F}$	$6.0^{A,F}$	65FR66443
99	Phenanthrene	85018							
100	Pyrene	129000					830^{A}	$4,000^{A}$	65FR66443
101	1,2,4-Trichlorobenzene	120821					35	70	68FR75510
102	Aldrin	309002	3.0^{V}		1.3^{V}				65FR31682
							$0.000049^{A,F}$	$0.000050^{A,F}$	65FR66443
103	alpha-BHC	319846					$0.0026^{A,F}$	$0.0049^{A,F}$	65FR66443
104	beta-BHC	319857					$0.0091^{A,F}$	$0.017^{A,F}$	65FR66443
105	gamma-BHC (Lindane)	58899	0.95^{D}		0.16^{V}				65FR31682
							0.98	1.8	68FR75510
106	delta-BHC	319868							
107	Chlordane	57749	2.4^{V}	$0.0043^{V,ff}$	0.09^{V}	$0.004^{V,ff}$			65FR31682
							$0.00080^{A,F}$	$0.00081^{A,F}$	65FR66443
108	4,4′-DDT	50293	$1.1^{V,gg}$	$0.001^{V,ff,gg}$	$0.13^{V,gg}$	$0.001^{V,ff,gg}$			65FR31682
							$0.00022^{A,F}$	$0.00022^{A,F}$	65FR66443
109	4,4′-DDE	72559					$0.00022^{A,F}$	$0.00022^{A,F}$	65FR66443
110	4,4′-DDD	72548					$0.00031^{A,F}$	$0.00031^{A,F}$	65FR66443

111	Dieldrin	60571	0.24[D]	0.056[D,hh]	0.71[V]	0.0019[V,ff]	0.000052[A,F]	0.000054[A,F]	65FR31682 65FR66443
112	Alpha-Endosulfan	959988	0.22[V,ii]	0.056[V,ii]	0.034[V,ii]	0.0087[V,ii]	62[A]	89[A]	65FR31682 65FR66443
113	Beta-Endosulfan	33213659	0.22[V,gg]	0.056[V,ii]	0.034[V,ii]	0.0087[V,ii]	62[A]	89[A]	65FR31682 65FR66443
114	Endosulfan Sulfate	1031078					62[A]	89[A]	65FR66443
115	Endrin	72208	0.086[D]	0.036[D,hh]	0.037[V]	0.0023[V,ff]	0.059	0.060	65FR31682 68FR75510
116	Endrin Aldehyde	7421934					0.29[A]	0.30[A,cc]	65FR66443
117	Heptachlor	76448	0.52[V]	0.0038[V,ff]	0.053[V]	0.0036[V,ff]	0.000079[A,F]	0.000079[A,F]	65FR31682 65FR66443
118	Heptachlor Epoxide	1024573	0.52[V,jj]	0.0038[V,ff,jj]	0.053[V,jj]	0.0036[V,ff,jj]	0.000039[A,F]	0.000039[A,F]	65FR31682 65FR66443
119	Polychlorinated Biphenyls PCBs			0.014[ff]		0.03[ff,kk]	0.000064[A,F,ii]	0.000064[A,F,ii]	65FR31682 65FR66443
120	Toxaphene	8001352	0.73	0.0002	0.21[ff]	0.0002[ff]	0.00028[A,F]	0.00028[A,F]	65FR31682 65FR66443

A This criterion has been revised to reflect The Environmental Protection Agency's q1* or RfD, as contained in the Integrated Risk Information System (IRIS) as of May 17, 2002. The fish tissue bioconcentration factor (BCF) from the 1980 Ambient Water Quality Criteria document was retained in each case.

B This recommended water quality criterion was derived from data for arsenic (III), but is applied here to total arsenic, which might imply that arsenic (III) and arsenic (V) are equally toxic to aquatic life and that their toxicities are additive. In the *arsenic criteria document* (EPA 440/5–84–033, January 1985), Species Mean Acute Values are given for both arsenic (III) and arsenic (V) for five species and the ratios of the SMAVs for each species range from 0.6 to 1.7. Chronic values are available for both arsenic (III) and arsenic (V) for one species; for the fathead minnow, the chronic value for arsenic (V) is 0.29 times the chronic value for arsenic (III). No data are known to be available concerning whether the toxicities of the forms of arsenic to aquatic organisms are additive.

C Freshwater and saltwater criteria for metals are expressed in terms of the dissolved metal in the water column. The recommended water quality criteria value was calculated by using the previous 304(a) aquatic life criteria expressed in terms of total recoverable metal, and multiplying it by a conversion factor (CF). The term "Conversion Factor" (CF) represents the recommended conversion factor for converting a metal criterion expressed as the total recoverable fraction in the water column to a criterion expressed as the dissolved fraction in the water column. (Conversion Factors for saltwater CCCs are not currently available. Conversion factors derived for saltwater CMCs have been used for both saltwater CMCs and CCCs). See "*Office of Water Policy and Technical Guidance on Interpretation and Implementation of Aquatic Life Metals Criteria*," October 1, 1993, by Martha G. Prothro, Acting Assistant Administrator for Water, available from the *Water Resource Center* and 40CFR§131.36(b)(1). Conversion Factors applied in the table can be found in Appendix A to the Preamble-Conversion Factors for Dissolved Metals.

D This recommended criterion is based on a 304(a) aquatic life criterion that was issued in the 1995 *Updates*: *Water Quality Criteria Documents for the Protection of Aquatic Life in Ambient Water*, (EPA-820-B-96-001, September 1996). This value was derived using the GLI Guidelines (60FR15393–15399, March 23, 1995; 40CFR132 Appendix A); the difference between the 1985 Guidelines and the GLI Guidelines are explained on page iv of the 1995 Updates. None of the decisions concerning the derivation of this criterion were affected by any considerations that are specific to the Great Lakes.

E This water quality criterion is based on a 304(a) aquatic life criterion that was derived using the *1985 Guidelines* (*Guidelines for Deriving Numerical National Water Quality Criteria for the Protection of Aquatic Organisms and Their Uses*, PB85–227049, January 1985) and was issued in one of the following criteria documents: *Arsenic* (EPA 440/5-84-033), *Cadmium* (EPA-822-R-01-001), *Chromium* (EPA 440/5-84-029), *Copper* (EPA 440/5-84-031), *Cyanide* (EPA 440/5- 84-028), Lead (EPA 440/5-84-027), Nickel (EPA 440/5-86-004), Pentachlorophenol (EPA 440/5-86-009), Toxaphene, (EPA 440/5-86-006), Zinc (EPA 440/5-87- 003).

F This criterion is based on carcinogenicity of 10^{-6} risk. Alternate risk levels may be obtained by moving the decimal point (e.g., for a risk level of 10^{-5}, move the decimal point in the recommended criterion one place to the right).

G EPA is currently reassessing the criteria for arsenic.

H This recommended water quality criterion for arsenic refers to the inorganic form only.

I A more stringent MCL has been issued by EPA. Refer to drinking water regulations (40 CFR 141) or Safe Drinking Water Hotline (1-800-426-4791) for values.

(Continued)

Table 8G.86 (Continued)

[J] The freshwater criterion for this metal is expressed as a function of hardness (mg/L) in the water column. The value given here corresponds to a hardness of 100 mg/L. Criteria values for other hardness may be calculated from the following: CMC (dissolved) = $\exp\{m_A [\ln(\text{hardness})] + b_A\}$ (CF), or CCC (dissolved) = $\exp\{m_C [\ln(\text{hardness})] + b_C\}$ (CF) and the parameters specified in Appendix B- Parameters for Calculating Freshwater Dissolved Metals Criteria That Are Hardness-Dependent.

[K] When the concentration of dissolved organic carbon is elevated, copper is substantially less toxic and use of Water-Effect Ratios might be appropriate.

[L] This recommended water quality criterion was derived in *Ambient Water Quality Criteria Saltwater Copper Addendum* (Draft, April 14, 1995) and was promulgated in the Interim final National Toxics Rule (*60FR22228–222237*, May 4, 1995).

[M] The organoleptic effect criterion is more stringent than the value for priority toxic pollutants.

[N] EPA is actively working on this criterion and so this recommended water quality criterion may change substantially in the near future.

[O] This recommended water quality criterion was derived from data for inorganic mercury (II), but is applied here to total mercury. If a substantial portion of the mercury in the water column is methylmercury, this criterion will probably be under protective. In addition, even though inorganic mercury is converted to methylmercury and methylmercury bioaccumulates to a great extent, this criterion does not account for uptake via the food chain because sufficient data were not available when the criterion was derived.

[P] This recommended water quality criterion was derived on page 43 of the *mercury criteria document* (EPA 440/5-84-026, January 1985). The saltwater CCC of 0.025 ug/L given on page 23 of the criteria document is based on the Final Residue Value procedure in the 1985 Guidelines. Since the publication of the Great Lakes Aquatic Life Criteria Guidelines in 1995 (60FR15393–15399, March 23, 1995), the Agency no longer uses the Final Residue Value procedure for deriving CCCs for new or revised 304(a) aquatic life criteria.

[Q] This fish tissue residue criterion for methylmercury is based on a total fish consumption rate of 0.0175 kg/day.

[R] The CMC = 1/[(f1/CMC1) + (f2/CMC2)] where f1 and f2 are the fractions of total selenium that are treated as selenite and selenate, respectively, and CMC1 and CMC2 are 185.9 g/L and 12.82 g/L, respectively.

[S] This value for selenium was announced (*61FR58444–58449*, November 14, 1996) as a proposed GLI 303(c) aquatic life criterion. EPA is *currently working on this criterion* and so this value might change substantially in the near future.

[T] This recommended water quality criterion for selenium is expressed in terms of total recoverable metal in the water column. It is scientifically acceptable to use the conversion factor (0.996- CMC or 0.922- CCC) that was used in the GLI to convert this to a value that is expressed in terms of dissolved metal.

[U] The selenium criteria document (EPA 440/5-87-006, September 1987) provides that if selenium is as toxic to saltwater fishes in the field as it is to freshwater fishes in the field, the status of the fish community should be monitored whenever the concentration of selenium exceeds 5.0 g/L in salt water because the saltwater CCC does not take into account uptake via the food chain.

[V] This Criterion is based on 304(a) aquatic life criterion issued in 1980, and was issued in one of the following documents: *Aldrin/Dieldrin* (EPA 440/5-80-019), *Chlordane* (EPA 440/5-80-027), *DDT* (EPA 440/5-80-038), *Endosulfan* (EPA 440/5-80-046), *Endrin* (EPA 440/5-80-047), *Heptachlor* (EPA 440/5-80-052), *Hexachlorocyclohexane* (EPA 440/5-80-054), Silver (EPA 440/5-80-071). The Minimum Data Requirements and derivation procedures were different in the 1980 Guidelines than in the *1985 Guidelines*. For example, a "CMC" derived using the 1980 Guidelines was derived to be used as an instantaneous maximum. If assessment is to be done using an averaging period, the values given should be divided by 2 to obtain a value that is more comparable to a CMC derived using the *1985 Guidelines*.

[W] This recommended water quality criterion is expressed as g free cyanide (as CN)/L.

[X] This recommended water quality criterion is expressed as total cyanide, even though the IRIS RFD we used to derive the criterion is based on free cyanide. The multiple forms of cyanide that are present in ambient water have significant differences in toxicity due to their differing abilities to liberate the CN-moiety. Some complex cyanides require even more extreme conditions than refluxing with sulfuric acid to liberate the CN-moiety. Thus, these complex cyanides are expected to have little or no 'bioavailability' to humans. If a substantial fraction of the cyanide present in a water body is present in a complexed form (e.g. $Fe_4[Fe(CN)_6]_3$), this criterion may be over conservative.

[Y] This criterion for asbestos is the Maximum Contaminant Level (MCL) developed under the Safe Drinking Water Act (SDWA).

[Z] Although a new RfD is available in IRIS, the surface water criteria will not be revised until the National Primary Drinking Water Regulations: Stage 2 Disinfectants and Disinfection Byproducts Rule (Stage 2 DBPR) is completed, since public comment on the relative source contribution (RSC) for chloroform is anticipated.

[aa] This recommended water quality criterion was derived using the cancer slope factor of 1.4 (LMS exposure from birth).

[bb] Freshwater aquatic life values for pentachlorophenol are expressed as a function of pH, and are calculated as follows: CMC = $\exp(1.005(\text{pH}) - 4.869)$; CCC = $\exp(1.005(\text{pH}) - 5.134)$. Values displayed in table correspond to a pH of 7.8.

[cc] No criterion for protection of human health from consumption of aquatic organisms excluding water was presented in the 1980 criteria document or in the 1986 *Quality Criteria for Water*. Nevertheless, sufficient information was presented in the 1980 document to allow the calculation of a criterion, even though the results of such a calculation were not shown in the document.

[dd] There is a full set of aquatic life toxicity data that show that DEHP is not toxic to aquatic organisms at or below its solubility limit.

[ee] Although EPA has not published a completed criteria document for butylbenzyl phthalate it is EPA's understanding that sufficient data exist to allow calculation of aquatic criteria. It is anticipated that industry intends to publish in the peer reviewed literature draft aquatic life criteria generated in accordance with EPA Guidelines. EPA will review such criteria for possible issuance as national WQC.

[ff] This criterion is based on a 304(a) aquatic life criterion issued in 1980 or 1986, and was issued in one of the following documents: *Aldrin/Dieldrin* (EPA 440/5-80-019), *Chlordane* (EPA 440/5-80-027), *DDT* (EPA 440/5-80-038), *Endrin* (EPA 440/5-80-047), *Heptachlor* (EPA 440/5-80-052), Polychlorinated biphenyls (EPA 440/5-80-068), Toxaphene (EPA 440/5-86-006). This CCC is currently based on the Final Residue Value (FRV) procedure. Since the publication of the Great Lakes Aquatic Life Criteria Guidelines in 1995 (60FR15393–15399, March 23, 1995), the Agency no longer uses the Final Residue Value procedure for deriving CCCs for new or revised 304(a) aquatic life criteria. Therefore, the Agency anticipates that future revisions of this CCC will not be based on the FRV procedure.

[gg] This criterion applies to DDT and its metabolites (i.e. the total concentration of DDT and its metabolites should not exceed this value).
[hh] The derivation of the CCC for this pollutant (Endrin) did not consider exposure through the diet, which is probably important for aquatic life occupying upper trophic levels.
[ii] This value was derived from data for endosulfan and is most appropriately applied to the sum of alpha-endosulfan and beta-endosulfan.
[jj] This value was derived from data for heptachlor and the criteria document provides insufficient data to estimate relative toxicities of heptachlor and heptachlor epoxide.
[kk] This criterion applies to total pcbs, (e.g. the sum of all congener or all isomer or homolog or Aroclor analyses.)

Source: From United States Environmental Protection Agency, 2005, *National Recommended Water Quality Criteria*, www.epa.gov.

Table 8G.87 United States National Recommented Water Quality Criteria for Nonpriority Pollutants

			Freshwater		Saltwater		Human Health for the Consumption of		
	Nonpriority Pollutant	CAS Number	CMC (acute) (μg/L)	CCC (chronic) (μg/L)	CMC (acute) (μg/L)	CCC (chronic) (μg/L)	Water + Organism (μg/L)	Organism Only (μg/L)	FR Cite/Source
1	Alkalinity	—		20000[a]					Gold Book
2	Aluminum pH 6.5–9.0	7429905	750[b,c]	87[b,c,d]					53FR33178
3	Ammonia	7664417	Freshwater criteria are pH, temperature and life-stage dependent—see document[e] Saltwater criteria are pH and temperature dependent						EPA822-R-99–014
									EPA440/5-88-004
4	Aesthetic Qualities	—	Narrative statement—see document (See table notes)						Gold Book
5	Bacteria	—	For primary recreation and shellfish uses—see document						Gold Book
6	Barium	7440393					1,000[f]		Gold Book
7	Boron	—	Narrative statement—see document						Gold Book
8	Chloride	16887006	860000[b]	230000[b]					53FR19028
9	Chlorine	7782505	19	11	13	q	—[g]		Gold Book
10	Chlorophenoxy Herbicide (2,4,5,-TP)	93721					10[f]		Gold Book
11	Chlorophenoxy Herbicide (2,4-D)	94757					100[f,g]		Gold Book
12	Chloropyrifos	2921882	0.083[b]	0.041[b]	0.011[b]	0.0056[b]			Gold Book
13	Color	—	Narrative statement—see document (See table notes)[a]						Gold Book
14	Demeton	8065483		0.1[a]		0.1[a]			Gold Book
15	Ether, Bis (Chloromethyl)	542881					0.00010[h,i]	0.00029[h,i]	65FR66443
16	Gases, Total Dissolved	—	Narrative statement—see document[a] (See table notes)						Gold Book
17	Guthion	86500		0.01[a]		0.01[a]			Gold Book
18	Hardness	—	Narrative statement—see document						Gold Book
19	Hexachlorocyclo-hexane-Technical	319868					0.0123	0.0414	Gold Book
20	Iron	7439896		1000[a]			300[f]		Gold Book
21	Malathion	121755		0.1[a]		0.1[a]			Gold Book
22	Manganese	7439965					50[f,j]	100[f]	Gold Book
23	Methoxychlor	72435		0.03[a]		0.03[a]	100[f,g]		Gold Book
24	Mirex	2385855		0.001[a]		0.001[a]			Gold Book
25	Nitrates	14797558					10,000[f]		Gold Book
26	Nitrosamines	—					0.0008	1.24	Gold Book
27	Dinitrophenols	25550587					69	5300	65FR66443
28	Nitrosodi-butylamine,[k]	924163					0.0063[f,i]	0.22[f,i]	65FR66443
29	Nitrosodi-ethylamine,[k]	55185					0.0008[f,i]	1.24[f,i]	Gold Book
30	Nitrosopy-rrolidine,[k]	930552					0.016[i]	34[i]	65FR66443
31	Oil and Grease	—	Narrative statement—see document[a] (See table notes)						Gold Book
32	Oxygen, Dissolved Freshwater	7782447	Warmwater and coldwater matrix—see document[k]						Gold Book
	Oxygen, Dissolved Saltwater		Saltwater—see document						EPA-822R-00–012
33	Parathion	56382	0.065[l]	0.013[l]					Gold Book
34	Pentachlo-robenzene	608935					1.4[h]	1.5[h]	65FR66443

35	pH	—		6.5–9[a]		6.5–8.5[a,m]	5–9		Gold Book
36	Phosphorus Elemental	7723140				0.1[a,m]			Gold Book
37	Nutrients	—	See EPA's *Ecoregional criteria* for Total Phosphorus, Total Nitrogen, Chlorophyll *a* and Water Clarity (Secchi depth for lakes; turbidity for streams and rivers) (& Level III Ecoregional criteria)						[n]
38	Solids Dissolved and Salinity	—					250,000[f]		Gold Book
39	Solids Suspended and Turbidity	—	Narrative statement—see document[a] (See table notes)						Gold Book
40	Sulfide-Hydrogen Sulfide	7783064		2.0[a]		2.0[a]			Gold Book
41	Tainting Substances	—	Narrative statement—see document (See table notes)						Gold Book
42	Temperature	—	Species dependent criteria—see document[o]						Gold Book
43	Tetrachloro-benzene,-1,2,4,5-	95943					0.97[h]	1.1[h]	65FR66443
44	Tributyltin (TBT)	—	0.46[p]	0.072[p]	0.42[p]	0.0074[p]			69FR342
45	Trichloro-phenol,2,4,5-	95954					1,800[h,q]	3,600[h,q]	65FR66443

Appendix C — Calculation of Freshwater Ammonia Criterion

1. The one-hour average concentration of total ammonia nitrogen (in mg N/L) does not exceed, more than once every 3 years on the average, the CMC (acute criterion) calculated using the following equations

Where salmonid fish are present:

$$\mathrm{CMC} = \frac{0.275}{1 + 10^{7.204-\mathrm{pH}}} + \frac{39.0}{1 + 10^{\mathrm{pH}-7.204}}$$

Or where salmonid fish are not present:

$$\mathrm{CMC} = \frac{0.411}{1 + 10^{7.204-\mathrm{pH}}} + \frac{58.4}{1 + 10^{\mathrm{pH}-7.204}}$$

2A. The 30 average concentration of total ammonia nitrogen (in mg N/L) does not exceed, more than once every 3 years on the average, the CCC (chronic criterion) calculated using the following equations

When fish early life stages are present:

$$\mathrm{CCC} = \frac{0.0577}{1 + 10^{7.688-\mathrm{pH}}} + \frac{2.487}{1 + 10^{\mathrm{pH}-7.688}} \times \mathrm{MIN}\,(2.85, 1.45 \times 10^{0.028(25-T)})$$

When fish early life stages are absent:

$$\mathrm{CCC} = \frac{0.0577}{1 + 10^{7.688-\mathrm{pH}}} + \frac{2.487}{1 + 10^{\mathrm{pH}-7.688}} \times 1.45 \times 10^{0.028(25-\mathrm{MAX}(T,7))}$$

2B. In addition, the highest 4-day average within the 30-day period should not exceed 2.5 times the CCC.

Notes:

Narrative Statements: National Recommented Water Quality Criteria for Nonpriority Pollutants

Aesthetic Qualities

All waters free from substances attributable to wastewater or other discharges that

(1) settle to form objectionable deposits
(2) float as debris, scum, oil, or other matter to form nuisances
(3) produce objectionable color, odor, taste, or turbidity
(4) injure or are toxic or produce adverse physiological responses in humans, animals, or plants, and

(Continued)

Table 8G.87 (Continued)

(5) produce undesirable or nuisance aquatic life

Color

Waters shall be virtually free from substances producing objectionable color for aesthetic purposes

the source of supply should not exceed 75 color units on the platinum-cobalt scale for domestic water supplies; and

Increased color (in combination with turbidity) should not reduce the depth of the compensation point for photosynthetic activity by more than 10 percent from the seasonally established norm for aquatic life

Gases, Total Dissolved

To protect freshwater and marine aquatic life, the total dissolved gas concentrations in water should not exceed 110 percent of the saturation value for gases at the existing atmospheric and hydrostatic pressures

Oil and Grease

For aquatic life

(1) 0.01 of the lowest continuous flow 96-hour LC50 to several important freshwater and marine species, each having a demonstrated high susceptibility to oils and petrochemicals

(2) Levels of oils or petrochemicals in the sediment which cause deleterious effects to the biota should not be allowed

(3) Surface waters shall be virtually free from floating nonpetroleum oils of vegetable or animal origin, as well as petroleum-derived oils

Solids (Suspended, Settleable) and Turbidity

Freshwater fish and other aquatic life

Settleable and suspended solids should not reduce the depth of the compensation point for photosynthetic activity by more than 10 percent from the seasonally established norm for aquatic life

Tainting Substances

Materials should not be present in concentrations that individually or in combination produce undesirable flavors which are detectable by organoleptic tests performed on the edible portions of aquatic organisms

[a] The derivation of this value is presented in the *Red Book* (EPA 440/9-76-023, July, 1976).

[b] This value is based on a 304(a) aquatic life criterion that was derived using the 1985 Guidelines (*Guidelines for Deriving Numerical National Water Quality Criteria for the Protection of Aquatic Organisms and Their Uses*, PB85-227049, January 1985) and was issued in one of the following criteria documents: Aluminum (EPA 440/5-86-008); Chloride (EPA 440/5-88-001); Chloropyrifos (EPA 440/5-86-005).

[c] This value for aluminum is expressed in terms of total recoverable metal in the water column.

[d] There are three major reasons why the use of Water-Effect Ratios might be appropriate. (1) The value of 87 ug/L is based on a toxicity test with the striped bass in water with pH = 6.5–6.6 and hardness < 10 mg/L. Data in “Aluminum Water-Effect Ratio for the 3M Plant Effluent Discharge, Middleway, West Virginia” (May 1994) indicate that aluminum is substantially less toxic at higher pH and hardness, but the effects of pH and hardness are not well quantified at this time. (2) In tests with the brook trout at low pH and hardness, effects increased with increasing concentrations of total aluminum even though the concentration of dissolved aluminum was constant, indicating that total recoverable is a more appropriate measurement than dissolved, at least when particulate aluminum is primarily aluminum hydroxide particles. In surface waters, however, the total recoverable procedure might measure aluminum associated with clay particles, which might be less toxic than aluminum associated with aluminum hydroxide. (3) EPA is aware of field data indicating that many high quality waters in the U.S. contain more than 87 μg aluminum/L, when either total recoverable or dissolved is measured.

[e] According to the procedures described in the *Guidelines for Deriving Numerical National Water Quality Criteria for the Protection of Aquatic Organisms and Their Uses*, except possibly where a very sensitive species is important at a site, freshwater aquatic life should be protected if both conditions specified in Appendix C to the Preamble- Calculation of Freshwater Ammonia Criterion are satisfied.

[f] This human health criterion is the same as originally published in the Red Book(EPA 440/9-76-023, July, 1976) which predates the 1980 methodology and did not utilize the fish ingestion BCF approach. This same criterion value is now published in the *Gold Book*(Quality Criteria for Water: 1986. EPA 440/5-86-001).

[g] A more stringent Maximum Contaminant Level (MCL) has been issued by EPA under the Safe Drinking Water Act. Refer to drinking water regulations 40CFR141 or Safe Drinking Water Hotline (1-800-426-4791) for values.

[h] This criterion has been revised to reflect EPA’s q1* or RfD, as contained in the Integrated Risk Information System (IRIS) as of May 17, 2002. The fish tissue bioconcentration factor (BCF) used to derive the original criterion was retained in each case.

[i] This criterion is based on carcinogenicity of 10^{-6} risk. Alternate risk levels may be obtained by moving the decimal point (e.g., for a risk level of 10^{-5}, move the decimal point in the recommended criterion one place to the right).

[j] This criterion for manganese is not based on toxic effects, but rather is intended to minimize objectionable qualities such as laundry stains and objectionable tastes in beverages. U.S. EPA. 1973. Water Quality Criteria 1972. EPA-R3-73-033. National Technical Information Service, Springfield, VA.; U.S. EPA. 1977. Temperature Criteria for Freshwater Fish: Protocol and Procedures. EPA-600/3-77-061. National Technical Information Service, Springfield, VA.

[K] U.S. EPA. 1986. *Ambient Water Quality Criteria for Dissolved Oxygen*. EPA 440/5-86-003. National Technical Information Service, Springfield, VA.

[l] This value is based on a 304(a) aquatic life criterion that was issued in the 1995 *Updates*: *Water Quality Criteria Documents for the Protection of Aquatic Life in Ambient Water* (EPA-820-B-96-001). This value was derived using the GLI Guidelines (60FR15393–15399, March 23, 1995; 40CFR132 Appendix A); the differences between the 1985 Guidelines and the GLI Guidelines are explained on page iv of the 1995 Updates. No decision concerning this criterion was affected by any considerations that are specific to the Great Lakes.

[m] According to page 181 of the *Red Book*: For open ocean waters where the depth is substantially greater than the euphotic zone, the pH should not be changed more than 0.2 units from the naturally occurring variation or any case outside the range of 6.5 to 8.5. For shallow, highly productive coastal and estuarine areas where naturally occurring pH variations approach the lethal limits of some species, changes in pH should be avoided but in any case should not exceed the limits established for fresh water, i.e. 6.5–9.0.

[n] *Lakes and Reservoirs in Nutrient Ecoregion*: II EPA 822-B-00-007, III EPA 822-B-01-008, IV EPA 822-B-01-009, V EPA 822-B-01-010, VI EPA 822-B-00-008, VII EPA 822-B-00-009, VIII EPA 822-B-01-015, IX EPA 822-B-00-011, XI EPA 822-B-00-012, XII EPA 822-B-00-013, XIII EPA 822-B-00-014, XIV EPA 822-B-01-011; *Rivers and Streams in Nutrient Ecoregion*: I EPA 822-B-01-012, II EPA 822-B-00-015, III EPA 822-B-00-016, IV EPA 822-B-01-013, V EPA 822-B-01-014, VI EPA 822-B-00-017, VII EPA 822-B-00-018, VIII EPA 822-B-01-015, IX EPA 822-B-00-019, X EPA 822-B-01-016, XI EPA 822-B-00-020, XII EPA 822-B-00-021, XIV EPA 822-B-00-022; and *Wetlands in Nutrient Ecoregion* XIII EPA 822-B-00-023.

[p] EPA announced the availability of a draft updated tributyltin (TBT) document on August 7, 1997 (62FR42554). The Agency *has reevaluated this document* and anticipates releasing an updated document for public comment in the near future.

[q] The organoleptic effect criterion is more stringent than the value presented in the nonpriority pollutants table.

Source: From United States Environmental Protection Agency, 2005, *National Recommended Water Quality Criteria*, www.epa.gov.

Original Source: From United States Environmental Protection Agency Gold Book, *Quality Criteria for Water: 1986*, EPA 44015-86-001.

Table 8G.88 United States National Recommended Water Quality Criteria for Organoleptic Effects

	Pollutant	CAS Number	Organoleptic Effect Criteria (μg/L)	FR Cite/Source
1	Acenaphthene	83329	20	Gold Book[a]
2	Monochlorobenzene	108907	20	Gold Book
3	3-Chlorophenol	—	0.1	Gold Book
4	4-Chlorophenol	106489	0.1	Gold Book
5	2,3-Dichlorophenol	—	0.04	Gold Book
6	2,5-Dichlorophenol	—	0.5	Gold Book
7	2,6-Dichlorophenol	—	0.2	Gold Book
8	3,4-Dichlorophenol	—	0.3	Gold Book
9	2,4,5-Trichlorophenol	95954	1	Gold Book
10	2,4,6-Trichloropehnol	88062	2	Gold Book
11	2,3,4,6-Tetrachlorophenol	—	1	Gold Book
12	2-Methyl-4-Chlorophenol	—	1800	Gold Book
13	3-Methyl-4-Chlorophenol	59507	3000	Gold Book
14	3-Methyl-6-Chlorophenol	—	20	Gold Book
15	2-Chlorophenol	95578	0.1	Gold Book
16	Copper	7440508	1000	Gold Book
17	2,4-Dichlorophenol	120832	0.3	Gold Book
18	2,4-Dimethylpehnol	105679	400	Gold Book
19	Hexachlorocyclopentadiene	77474	1	Gold Book
20	Nitrobenzene	98953	30	Gold Book
21	Pentachlorophenol	87865	30	Gold Book
22	Phenol	108952	300	Gold Book
23	Zinc	7440666	5000	45 FR 79341

Note: These criteria are based on organoleptic (taste and odor) effects. Because of variations in chemical nomenclature systems, this listing of pollutants does not duplicate the listing in Appendix A of 40 CFR Part 423. Also listed are the Chemical Abstracts Service (CAS) registry numbers, which provide a unique identification for each chemical.

[a] The Gold book is Quality Crieteria for Water: 1986. EPA 440/5-86-001.

Source: From United States Environmental Protection Agency, 2005, *Nutrient Recommended Water Quality Criteria*, www.epa.gov.

These tables present the recommended EPA criteria for each of the aggregate nutrient ecoregions for the following parameters: Total Phosphorus (TP), Total Nitrogen (TN), Chlorophyll *a*, and Turbidity or Secchi. Criteria are presented for both Lakes & Reservoirs and Rivers & Streams.

Table 8G.89 United States Recommended Criteria for Each of the Aggregate Nutrients Ecoregions Lakes and Reservoirs

Parameter	Agg Ecor II	Agg Ecor III	Agg Ecor IV	Agg Ecor V	Agg Ecor VI	Agg Ecor VII	Agg Ecor VIII	Agg Ecor IX	Agg Ecor XI	Agg Ecor XII	Agg Ecor XIII	Agg Ecor XIV
TP μg/L	8.75	17.00	20.00	33.00	37.5	14.75	8.00	20.00	8.00	10.00	17.50	8.00
TN μg/L	0.10	0.40	0.44	0.56	0.78	0.66	0.24	0.36	0.46	0.52	1.27	0.32
Chl a μg/L	1.90	3.40	2.00 S	2.30 S	8.59 S	2.63	2.43	4.93	2.79 S	2.60	12.35 T	2.90
Secchi (m)	4.50	2.70	2.00	1.30	1.36	3.33	4.93	1.53	2.86	2.10	0.79	4.50

Note: Chl *a*, Chlorophyll *a* measured by Fluorometric method, unless specified; S is for Spectrophotometric and T is for Trichromatic method.

Source: From United States Environmental Protection Agency, *Nutrient Water Quality Criteria, Ecoregional Criteria*, www.epa.gov/waterscience/criteria/nutrient/ecoregions.

Table 8G.90 United States Recommended Criteria for Each of the Aggregate Nutrient Ecoregions Rivers and Streams

Parameter	Agg Ecor I	Agg Ecor II	Agg Ecor III	Agg Ecor IV	Agg Ecor V	Agg Ecor VI	Agg Ecor VII	Agg Ecor VIII	Agg Ecor IX	Agg Ecor X	Agg Ecor XI	Agg Ecor XII	Agg Ecor XIV
TP μg/L	47.00	10.00	21.88	23.00	67.00	76.25	33.00	10.00	36.56	128[a]	10.00	40.00	31.25
TN μg/L	0.31	0.12	0.38	0.56	0.88	2.18	0.54	0.38	0.69	0.76	0.31	0.90	0.71
Chl a μg/L	1.80	1.08	1.78	2.40	3.00	2.70	1.50	0.63	0.93 S	2.10 S	1.61 S	0.40 S	3.75 S
Turb FTU/NTU	4.25	1.30 N	2.34	4.21	7.83	6.36	1.70 N	1.30	5.70	17.50	2.30 N	1.90 N	3.04

Note: Turb, Turbidity; Chl *a*, Chlorophyll *a* measured by Fluorometric method, unless specified; S is for Spectrophotometric and T is for Trichromatic method; N for NTU. Unit of measurement for Turbidity.

[a] This value appears inordinately high and may either be a statistical anomaly or reflects a unique condition. In any case, further regional investigation is indicated to determine the sources, i.e. measurement error, notational error, statistical anomaly, natural enriched conditions, or cultural impacts. From United States Environmental Protection Agency, Nutrient Water Quality Criteria, Ecoregional Criteria, www.epa.gov/waterscience/criteria/nutrient/ecoregions/.

Source: From United States Environmental Protection Agency, *Nutrient Water Quality Criteria, Ecoregional Criteria*, www.epa.gov/waterscience/criteria/nutrient/ecoregions/.

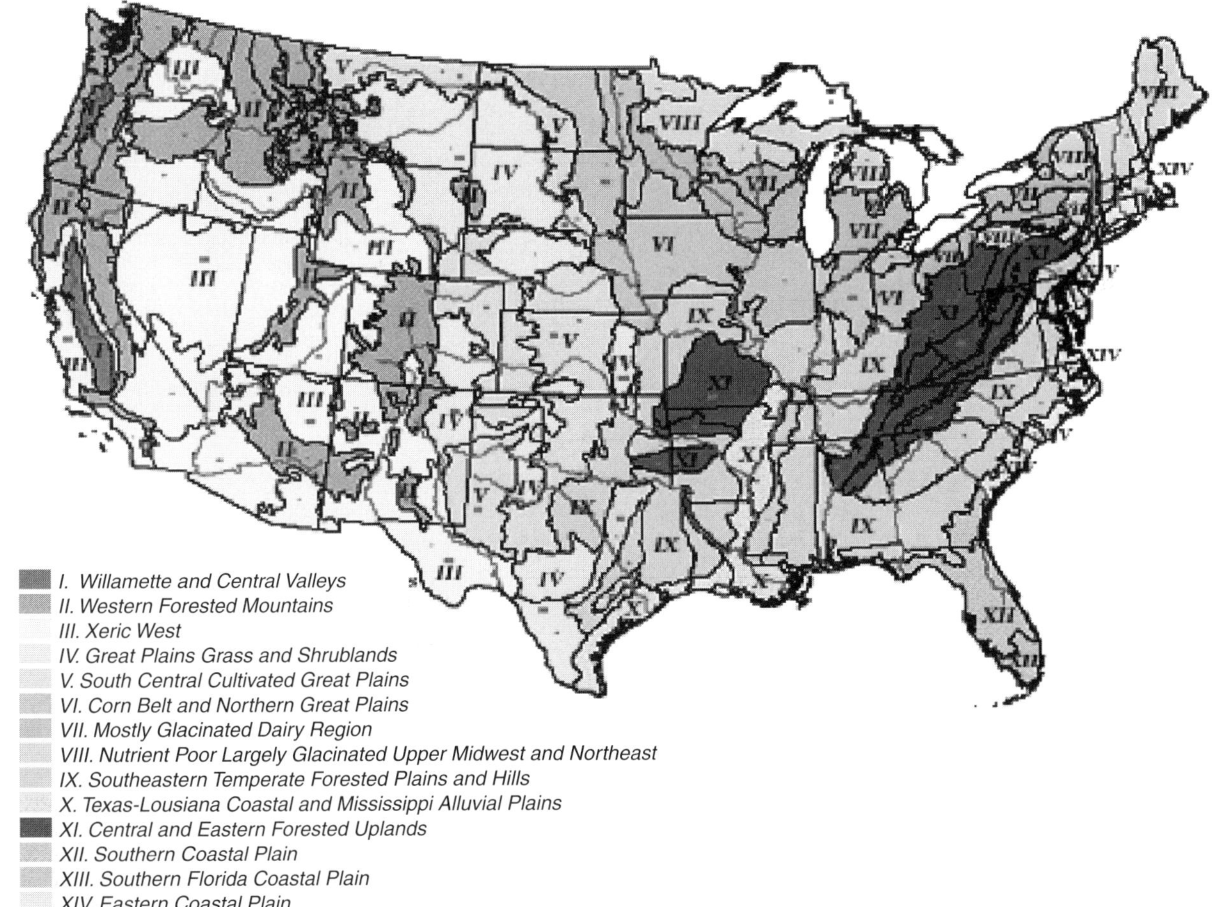

Figure 8G.21 United States draft aggregations of level III ecoregions for the national nutrient strategy. (From United States Environmental Protection Agency, *Nutrient Water Quality Criteria, Where-You-Live*, epa.gov/waterscience/criteria/nutrient/where-you-live.htm).

Table 8G.91 Aquatic Life Criteria for Dissolved Oxygen (Saltwater): Cape Cod to Cape Hatteras

Endpoint	Persistent Exposure (24 hr or Greater Continuous Low DO Conditions)	Episodic and Cyclic Exposure (Less Than 24 hr Duration of Low DO Conditions)
Juvenile and adult survival (minimum allowable conditions)	(1) *a limit for continuous exposure* DO = 2.3 mg/L (criterion minimum concentration, CMC)	(4) *a limit based on the hourly duration of exposure* $DO = 3.70\ n(t) + 1.095$ where: DO, allowable concentration (mg/L); *t*, exposure duration (hrs)
Growth effects (maximum conditions required)	(2) *a limit for continuous exposure* DO = 4.8 mg/L (criterion continuous concentration, CCC)	(5) *a limit based on the intensity and hourly duration of exposure* Cumulative cyclic adjusted percent daily reduction in growth must not exceed 25% $\sum_{1}^{n} \frac{t_i \times 1.56 \times \text{Gredi}}{24} < 25\%$ and $\text{Gredi} = -23.1 \times DO_i + 138.1$ where: Gred_i, growth reduction (%); DO_i, allowable concentration (mg/L); t_i, exposure interval duration (hrs); *i*, exposure interval
Larval recruitment effects[a] (specific allowable conditions)	(3) *a limit based on the number of days a continuous exposure can occur* Cumulative fraction of allowable days above a given daily mean DO must not exceed 1.0 $\sum \frac{t_i(\text{actual})}{t_i(\text{allowed})} < 1.0$ and $DO_i = \frac{13.0}{(2.80 + 1.84e^{-0.10t_i})}$ where: DO_i, allowable concentration (mg/L); t_i, exposure interval duration (d); *i*, exposure interval	(6) *a limit based on the number of days an intensity and hourly duration pattern of exposure can occur* Maximum daily cohort mortality for any hourly duration interval of a DO minimum must not exceed a corresponding allowable days of occurrence where: Allowable number of days is a function of maximum daily cohort mortality (%) Maximum daily cohort mortality (%) is a function of DO minimum for any exposure interval (mg/L) and the duration of the interval (hrs)

[a] Model integrating survival effects to maintain minimally impaired larval populations.

Source: From USEPA, 2000, *Aquatic Life Criteria Document for Dissolved Oxygen (Saltwater): Cape Cod to Cape Hatteras*, EPA-822-R-00-012, November 2000, www.epa.gov.

Table 8G.92 United States Sediment Quality Guidelines for Metals in Freshwater Ecosystems That Reflect Threshold Effect Concentrations (TECs) (i.e., Below Which Harmful Effects are Unlikely to Be Observed)

	Threshold Effect Concentrations						
Substance	**TEL**	**LEL**	**MET**	**ERL**	**TEL-HA28**	**SQAL**	**Consensus-Based TEC**
Metals (mg/kg DW)							
Arsenic	5.9	6	7	33	11	NG	9.79
Cadmium	0.596	0.6	0.9	5	0.58	NG	0.99
Chromium	37.3	26	55	80	36	NG	43.4
Copper	35.7	16	28	70	28	NG	31.6
Lead	35	31	42	35	37	NG	35.8
Mercury	0.174	0.2	0.2	0.15	NG	NG	0.18
Nickel	18	16	35	30	20	NG	22.7
Zinc	123	120	150	120	98	NG	121
Polycyclic aromatic hydrocarbons (μg/kg DW)							
Anthracene	NG	220	NG	85	10	NG	57.2
Fluorene	NG	190	NG	35	10	540	77.4
Naphthalene	NG	NG	400	340	15	470	176
Phenanthrene	41.9	560	400	225	19	1,800	204
Benz[a]anthracene	31.7	320	400	230	16	NG	108
Benzo(a)pyrene	31.9	370	500	400	32	NG	150
Chrysene	57.1	340	600	400	27	NG	166
Dibenz[a,h]anthracene	NG	60	NG	60	10	NG	33.0
Fluoranthene	111	750	600	600	31	6,200	423
Pyrene	53	490	700	350	44	NG	195
Total PAHs	NG	4,000	NG	4,000	260	NG	1,610
Polychlorinated biphenyls (μg/kg DW)							
Total PCBs	34.1	70	200	50	32	NG	59.8
Organochlorine pesticides (μg/kg DW)							
Chlordane	4.5	7	7	0.5	NG	NG	3.24
Dieldrin	2.85	2	2	0.02	NG	110	1.90
Sum DDD	3.54	8	10	2	NG	NG	4.88
Sum DDE	1.42	5	7	2	NG	NG	3.16
Sum DDT	NG	8	9	1	NG	NG	4.16
Total DDTs	7	7	NG	3	NG	NG	5.28
Endrin	2.67	3	8	0.02	NG	42	2.22
Heptachlor epoxide	0.6	5	5	NG	NG	NG	2.47
Lindane (gamma-BHC)	0.94	3	3	NG	NG	3.7	2.37

Note: TEL, threshold effect level; dry weight (Smith et al., 1996); LEL, lowest effect level, dry weight (Persaud et al., 1993); MET, minimal effect threshold; dry weight (EC and MENVIQ, 1992); ERL, effect range low; dry weight (Long and Morgan, 1991); TEL-HA28, threshold effect level for *Hyalella azteca*; 28 day test; dry weight (US EPA 1996a); SQAL, sediment quality advisory levels; dry weight at 1% OC (US EPA 1997a); NG, no guideline.

Source: From MacDonald, D.D, Ingersoll, C.G., and Berger, T.A., 2000, Development and evaluation of consensus-based sediment quality guidelines for freshwater ecostystems, *Archives of Environmental Contamination Toxicology* 39, 20–31, Table 2, © 2000 Springer-Verlag New York, Inc. With kind permission of Springer Science and Business Media.

Table 8G.93 United States Sediment Quality Guidelines for Metals in Freshwater Ecosystems That Reflect Probable Effect Concentrations (PECs) (i.e., Above Which Harmful Effects are Unlikely to Be Observed)

	Probable Effect Concentrations					
Substance	**PEL**	**SEL**	**TET**	**ERM**	**PEL-HA28**	**Consensus-Based TEC**
Metals (mg/kg DW)						
Arsenic	17	33	17	85	48	33.0
Cadmium	3.53	10	3	9	3.2	4.98
Chromium	90	110	100	145	120	111
Copper	197	110	86	390	100	149
Lead	91.3	250	170	110	82	128
Mercury	0.486	2	1	1.3	NG	1.06
Nickel	36	75	61	50	33	48.6
Zinc	315	820	540	270	540	459
Polycyclic aromatic hydrocarbons (μg/kg DW)						
Anthracene	NG	3,700	NG	960	170	845
Fluorene	NG	1,600	NG	640	150	536
Naphthalene	NG	NG	600	2,100	140	561
Phenanthrene	515	9,500	800	1,380	410	1,170
Benz[a]anthracene	385	14,800	500	1,600	280	1,050
Benzo(a)pyrene	782	14,400	700	2,500	320	1,450
Chrysene	862	4,600	800	2,800	410	1,290
Fluoranthene	2,355	10,200	2,000	3,600	320	2,230
Pyrene	875	8,500	1,000	2,200	490	1,520
Total PAHs	NG	100,000	NG	35,000	3,400	22,800
Polychlorinated biphenyls (μg/kg DW)						
Total PCBs	277	5,300	1,000	400	240	676
Organochlorine pesticides (μg/kg DW)						
Chlordane	8.9	60	30	6	NG	17.6
Dieldrin	6.67	910	300	8	NG	61.8
Sum DDD	8.51	60	60	20	NG	28.0
Sum DDE	6.75	190	50	15	NG	31.3
Sum DDT	NG	710	50	7	NG	62.9
Total DDTs	4,450	120	NG	350	NG	572
Endrin	62.4	1,300	500	45	NG	207
Heptachlor epoxide	2.74	50	30	NG	NG	16.0
Lindane (gamma-BHC)	1.38	10	9	NG	NG	4.99

Source: From MacDonald, D.D, Ingersoll, C.G., and Berger, T.A., 2000, Development and evaluation of consensus-based sediment quality guidelines for freshwater ecosystems, *Archives of Environmental Contamination Toxicology* 39, 20–31, Table 3, © 2000 Springer-Verlag New York, Inc. With kind permission of Springer Science and Business Media.

Table 8G.94 United States Sediment Quality Guidelines for Marine Sediment

Chemical	**CAS No.**	**Threshold Effects Level (TEL)**	**Effects Range-Low (ERL)**	**Effects Range-Median (ERM)**	**Probable Effects Level (PEL)**	**Apparent[a] Effects Threshold (AET)**
Aluminum (Al) (%)						1.8% N
Antimony (Sb)						9,300 E
Arsenic (As)		7,240	8,200	70,000	41,600	35,000 B
Barium (Ba)						48,000 A
Cadmium (Cd)		676	1,200	9,600	4,210	3,000 N
Chromium (Cr)		52,300	81,000	370,000	160,400	62,000 N
Cobalt (Co)						10,000 N
Copper (Cu)		18,700	34,000	270,000	108,200	390,000 MO
Iron (Fe) (%)						22% N
Lead (Pb)		30,240	46,700	218,000	112,180	400,000 B

(Continued)

Table 8G.94 (Continued)

Chemical	CAS No.	Threshold Effects Level (TEL)	Effects Range-Low (ERL)	Effects Range-Median (ERM)	Probable Effects Level (PEL)	Apparent[a] Effects Threshold (AET)
Manganese (Mn)						260,000 N
Mercury (Hg)		130	150	710	696	410 M
Nickel (Ni)		15,900	20,000	51,600	42,800	110,000 EL
Selenium (Se)						1,000 A
Silver (Ag)		730	1,000	3,700	1,770	3,100 B
Tin (Sn)						>3,400 N as TBT
Vanadium (V)						57,000 N
Zinc (Zn)		124,000	150,000	410,000	271,000	410,000 I
Sulfides						4,500 MO
Chlorinated dioxins and PCBs						
TCDD 2,3,7,8-	1746016					0.0036 N
Polychlorinated biphenyls	1336363	21.55	22.7	180	188.79	130 M
Semivolatiles						
Benzoic acid	65850					65 O
Benzyl alcohol	100516					52 B
Dibenzofuran	132649	5100 H				110 E
Semivolatile, nitroaromatics						
Nitrobenzene	98953					21 N
N-nitrosodiphenylamine	86306					28 I
Semivolatile, organochlorines						
Aldrin	309002					9.5 AE
Chlordane	57749	2.26	0.5	6	4.79	2.8 A
p,p-DDD (TDE)	72548	1.22	2	20	7.81	16 I
p,p-DDE	72559	2.07	2.2	27	374.17	9 I
p,p-DDT	50293	1.19	1	7	4.77	12 E
DDT, total		3.89	1.58	46.1	51.7	11 B
Dieldrin[b]	60571	0.715	0.02	8	4.3	1.9 E
Heptachlor	76448					0.3 B
Hexachlorobenzene	118741					6 B
Hexachlorobutadiene	87683					1.3 E
Hexachlorocyclohexane (BHC)	608731					
Hexachloroethane	67721					73 BL
Lindane	58899	0.32			0.99	>4.8 N
Semivolatile, PAHs						
Acenaphthene	83329	6.71	16	500	88.9	130 E
Acenaphthylene	208968	5.87	44	640	127.87	71 E
Anthracene	120127	46.85	85.3	1100	245	280 E
Benzo(k)fluoranthene	207089					1800 EI
Benzo(a)pyrene	50328	88.81	430	1600	763.22	1100 E
Benzo(b)fluoranthene	205992					1800 EI
Benzo(ghi)perylene	191242					670 M
Benz(a)anthracene	56553	74.83	261	1600	692.53	960 E
Chrysene	218019	107.77	384	2800	845.98	950 E
Dibenz(a,h)anthracene	53703	6.22	63.4	260	134.61	230 OM
Fluoranthene	206440	112.82	600	5100	1493.54	1300 E
Fluorene	86737	21.17	19	540	144.35	120 E
Indeno(1,2,3-cd)pyrene	193395					600 M
Methylnaphthalene, 2-	91576	20.21	70	670	201.28	64 E
Naphthalene	91203	34.57	160	2100	390.64	230 E
Phenanthrene	85018	86.68	240	1500	543.53	660 E
Pyrene	129000	152.66	665	2600	1397.6	2400 E
LMW PAHs		311.7	552	3160	1442.00	1200 E
HMW PAHs		655.34	1700	9600	6676.14	7900 E
Total PAHs		1684.06	4022	44792	16770.4	
Volatile, aromatic and halogenated						
Dichlorobenzene 1,2-	95501					13 N
Dichlorobenzene 1,4-	106467					110 IM

(Continued)

Table 8G.94 (Continued)

Chemical	CAS No.	Threshold Effects Level (TEL)	Effects Range-Low (ERL)	Effects Range-Median (ERM)	Probable Effects Level (PEL)	Apparent[a] Effects Threshold (AET)
Semivolatile, phenolics						
Chlorophenol 2-	95578					8 A
Dichlorophenol 2,4-	120832					5 A
Dimethylphenol 2,4-	105679					18 N
Methyl phenol 2- [O-cresol]	95487					8 B
Methyl phenol 4- [P-cresol]	106445					100 B
Pentachlorophenol [at pH 7.8][c]	87865					17 B
Phenol	108952					130 E
Trichlorophenol 2,4,5-	95954					3 I
Trichlorophenol 2,4,6-	88062					6 I
Semivolatile, phthalates						
Butyl benzyl phthalate	85687					63 M
Di[2-ethylhexyl] phthalate	117817	182.16			2646.51	1300 I
Diethyl phthalate	84662					6 BL
Dimethyl phthalate	131113					6 B
Di-*N*-octyl phthalate	117840					61 BL
Di-N-butyl phthalate	84742					58 BL
Dichloropropene	542756					4 EL
Tetrachloroethylene	127184					57 I
Trichlorbenzene 1,2,4-	120821					>4.8 E
Trichloroethylene	79016					41 N
Xylene	1330207					4 BL

Note: The Effects Range-Low (ERLs) and Effects Range-Median (ERMs) plus the marine Threshold Effects Levels (TELs) and Probable Effects Levels (PELs) are based upon a similar data compilations, but use different calculations. The ERL is calculated as the lower 10th percentile concentration of the available sediment toxicity data which has been screened for only those samples which were identified as toxic by original investigators. It is *not* an LC_{10}. Since the ERL is at the low end of a range of levels at which effects were observed in the studies compiled, it represents the value at which toxicity may begin to be observed in sensitive species. The ERM is simply the median concentration of the compilation of just toxic samples. It is *not* an LC_{50}. The TEL is calculated as the geometric mean of the 15th percentile concentration of the toxic effects data set and the median of the no-effect data set; as such, it represents the concentration below which adverse effects are expected to occur only rarely. The PEL, as the geometric mean of the 50% of impacted, toxic samples and the 85% of the non-impacted samples, is the level above which adverse effects are frequently expected.

From Apparent Effect Thresholds (AETs) relate chemical concentrations in sediments to synoptic biological indicators of injury (i.e., sediment bioassays or diminished benthic infaunal abundance). Individual AETs are essentially equivalent to the concentration observed in the highest nontoxic sample. As such, they represent the concentration above which adverse biological impacts would *always* be expected by that biological indicator due to exposure to that contaminant alone. Conversely, adverse impacts are known to occur at levels below the AET. Only the lowest of the potential AETs is listed. AET values were developed for use in Puget Sound (Washington) and are not easily compared directly to other benchmarks based on single-chemical models and broader data sources. SquiRT cards have been updated with interim AET values which are subject to change. (All sediment and soil values in ppb dry weight, except as noted.)

[a] Entry is lowest value among AET tests. I, infaunal community impacts; A, amphipod; B, bivalve; M, microtox; O, Oyster larvae; E, Echinoderm larvae; L, $larval_{max}$; or, N-Neanthes bioassays.

[b] EPA proposed criteria, based on equilibrium partitioning, for Dieldrin are 11,000 and 20,000, and for Endrin are 4,200 and 760 μg/kg O.C. in freshwater and marine sediment, respectively.

[c] For PCP, freshwater $CMC = e^{1.005pH - 4.869}$ and $CCC = e^{1.005pH - 5.134}$.

Source: From NOAA, 1999, Screening Quick Reference Tables (SQuiRTs), www.response.restoration.noaa.gov/cpr/sediment/squirt/squirt.html.

Table 8G.95 United States Equilibrium Partitioning Sediment Benchmarks (ESB_{WQC}s) for Dieldrin and Endrin Using the Water Quality Criteria (WQC) FCVs as the Effect Concentration

Type of Water Body	$Log_{10}K_{OW}$ (L/kg)	$Log_{10}K_{OC}$ (L/kg)	FCV (μg/L)	ESG_{OC} (μg/g_{OC})
Dieldrin				
Freshwater	5.37	5.28	0.06589	12[a]
Saltwater	5.37	5.28	0.1469	28[b]
Endrin				
Freshwater	5.06	4.97	0.05805	5.4[a]
Saltwater	5.06	4.97	0.01057	0.99[b]

[a] $ESB_{WQCOC} = (10^{5.28}$ L/$kg_{OC}) \times (10^{-3}$ $kg_{OC}/g_{OC}) \times (0.6589$ μg dieldrin/L$) = 12$ μg dieldrin/g_{OC}.
[b] $ESB_{WQCOC} = (10^{5.28}$ L/$kg_{OC}) \times (10^{-3}$ $kg_{OC}/g_{OC}) \times (0.1469$ μg dieldrin/L$) = 28$ μg dieldrin/g_{OC}.
[c] $ESB_{WQCOC} = (10^{4.97}$ L/$kg_{OC}) \times (10^{-3}$ $kg_{OC}/g_{OC}) \times (0.05805$ μg endrin/L$) = 5.4$ μg endrin/g_{OC}.
[d] $ESB_{WQCOC} = (10^{4.97}$ L/$kg_{OC}) \times (10^{-3}$ $kg_{OC}/g_{OC}) \times (0.01057$ μg endrin/L$) = 0.99$ μg endrin/g_{OC}.

Source: From USEPA, 2003, *Procedures for the Derivation of Equilibrium Partitioning Sediment Benchmarks (ESBs) for the Protection of Benthic Organisms: Dieldrin*, EPA-600-R-02-010, August 2003; USEPA, 2003, *Procedures for the Derivation of Equilibrium Partitioning Sediment Benchmarks (ESBs) for the Protection of Benthic Organisms: Endrin*, EPA-600-R-02-009, August 2003, www.epa.gov.

Table 8G.96 Canadian Water Quality Guidelines for the Protection of Aquatic Life

	Freshwater		Marine	
Parameter[a]	Concentration (μg/L)	Date[b]	Concentration (μg/L)	Date[b]
Acenaphthene [see Polycyclic aromatic hydrocarbons (PAHs)]				
Acridine [see Polycyclic aromatic hydrocarbons (PAHs)]				
Aldicarb	1[c]	1993	0.15[c]	1993
Aldrin + Dieldrin[d]	~~0.004~~[e,f]	1987		
Aluminum[d]	5–100[g]	1987		
Ammonia (total)[h]	See factsheet	2001		
Ammonia (unionized)	19	2001		
Aniline	2.2[i]	1993	Insufficient data	1993
Anthracene [see Polycyclic aromatic hydrocarbons (PAHs)]				
Arsenic[j]	5.0[k]	1997	12.5[c]	1997
Atrazine	1.8[i]	1989		
Benz*(a)*anthracene [see Polycyclic aromatic hydrocarbons (PAHs)]				
Benzene[j]	370[c,k]	1999	110[c]	1999
Benzo*(a)*pyrene [see Polycyclic aromatic hydrocarbons (PAHs)]				
2,2-Bis(*p*-chlorophenyl)-1,1,-trichloroethane [see DDT (total)]				
Bromacil	5.0[c,i]	1997	Insufficient data	1997
Bromoform [see Halogenated methanes, tribromomethane]				
Bromoxynil	5.0[i]	1993	Insufficient data	1993
Cadmium	0.017[c,l]	1996	0.12[i]	1996
Captan	1.3[c]	1991		
Carbaryl	0.20[i]	1997	0.32[c,i]	1997
Carbofuran	1.8[i]	1989		
Carbon tetrachloride [see Halogenated methanes, tetrachloromethanes]				
Chlordane[d]	~~0.006~~[e,f]	1987		
Chlorinated benzenes				
Monochlorobenzene	1.3[c,k]	1997	25[c,k]	1997
1,2-Dichlorobenzene	0.70[c,k]	1997	42[c,k]	1997
1,3-Dichlorobenzene	150[c,k]	1997	Insufficient data[k]	1997
1,4-Dichlorobenzene	26[c,k]	1997	Insufficient data[k]	1997
1,2,3-Trichlorobenzene	8.0[c,k]	1997	Insufficient data[k]	1997
1,2,4-Trichlorobenzene	24[c,k]	1997	5.4[c,k]	1997

(Continued)

Table 8G.96 (Continued)

Parameter[a]	Freshwater Concentration (μg/L)	Freshwater Date[b]	Marine Concentration (μg/L)	Marine Date[b]
1,3,5-Trichlorobenzene[d]	Insufficient data[k]	1997	Insufficient data[k]	1997
1,2,3,4-Tetrachlorobenzene	1.8[c,k]	1997	Insufficient data[k]	1997
1,2,3,5-Tetrachlorobenzene[d]	Insufficient data[k]	1997	Insufficient data	1997
1,2,4,5-Tetrachlorobenzene	Insufficient data[k]	1997	Insufficient data[k]	1987
Pentachlorobenzene	6.0[c,k]	1997	Insufficient data	1997
Hexachlorobenzene[d]	Insufficient data[e,f,k]		Insufficient data	1997
Chlorinated ethanes				
1,2-Dichloroethane	100[c,i]	1991	Insufficient data	1991
1,1,1-Trichloroethane	Insufficient data	1991	Insufficient data	1991
1,1,2,2-Tetrachloroethane	Insufficient data	1991	Insufficient data	1991
Chlorinated ethenes				
1,1,2-Trichloroethene (Tichloroethylene; TCE)	21[c,i]	1991	Insufficient data	1991
1,1,2,2-Tetrachloroethene (Tetrachloroethylene; PCE)	111[c,i]	1993	Insufficient data	1993
Chlorinated methanes [see Halogenated methanes]				
Chlorinated phenols[d]				
Monochlorophenols	7	1987		
Dichlorophenols	0.2	1987		
Trichlorophenols	18	1987		
Tetrachlorophenols	1	1987		
Pentachlorophenol (PCP)	0.5	1987		
Chlorine, reactive [see Reactive chlorine species]				
Chloroform [see Halogenated methanes, trichloromethane]				
4-Chloro-2-methyl phenoxy acetic acid [See MCPA]				
Chlorothalonil	0.18[c]	1994	0.36[c]	1994
Chlorpyrifos	0.0035	1997	0.002[c]	1997
Chromium				
Trivalent chromium (Cr(III))	8.9[c,k]	1997	56[c,k]	1997
Hexavalent chromium (Cr(VI))	1.0[k]	1997	1.5[k]	1997
Chrysene [see Polycyclic aromatic hydrocarbons (PAHs)]				
Color	Narrative	1999	Narrative	1999
Copper[d]	2–4[m]	1987		
Cyanazine	2.0[c,i]	1990		
Cyanide[d]	5 (as free CN)	1987		
DDAC (Didecyl dimethyl ammonium chloride)	1.5	1999		
DDT (total)[d] (2,2-Bis(*p*-chlorophenyl)-1,1,1-trichloroethane; dichloro diphenyl trichloroethane)	~~0.001~~[e,f]	1987		
Debris (litter/settleable matter)			Narrative[c]	1996
Deltamethrin	0.0004	1997	Insufficient data	1997
Deposited bedload sediment [see Total particulate matter]				
Dibromochloromethane [see Halogenated methanes]				
Dicamba	10[c,i]	1993		
Dichlorobenzene [see Chlorinated benzenes]				
Dichlorobromomethane [see Halogenated methanes]				
Dichloro diphenyl trichloroethane [see DDT (total)]				
Dichloroethane [see Chlorinated ethanes]				
Dichloroethylene [see Chlorinated ethanes, 1,2-dichloroethane]				
Dichloromethane [see Halogenated methanes]				
Dichlorophenols [see Chlorinated phenols]				
2,4-Dichlorophenoxyacetic acid [see Phenoxy herbicides]				
Diclofop-methyl	6.1	1993		
Didecyl dimethyl ammonium chloride [see DDAC]				
Diethylene glycol [see Glycols]				
Di(2-ethylhexyl) phthalate [see Phthalate esters]				
Diisopropanolamine	1600[c]	2005	Insufficient data	2005
Dimethoate	6.2[c]	1993	Insufficient data	1993
Di-*n*-butyl phthalate [see Phthalate esters]				

(Continued)

Table 8G.96 (Continued)

	Freshwater		Marine	
Parameter[a]	**Concentration (μg/L)**	**Date**[b]	**Concentration (μg/L)**	**Date**[b]
Di-*n*-octyl phthalate [see Phthalate esters]				
Dinoseb	0.05	1992		
Dissolved gas supersaturation	Narrative	1999	Narrative	1999
Dissolved oxygen	5500–9500[k,n]	1999	>8000 and narrative[c,k]	1996
Endosulfan[d]	0.02	1987		
Endrin[d]	~~0.0023~~[e,f]	1987		
Ethylbenzene[j]	90[c,k]	1996	25[c,k]	1996
Ethylene glycol [see Glycols]				
Fluoranthene [see Polycyclic aromatic hydrocarbons (PAHs)]				
Fluorene [see Polycyclic aromatic hydrocarbons (PAHs)]				
Glycols				
Ethylene glycol	192,000[k]	1997	Insufficient data	1997
Diethylene glycol	Insufficient data[k]	1997	Insufficient data	1997
Propylene glycol	500,000[k]	1997	Insufficient data	1997
Glyphosate	65[c]	1989		
Halogenated methanes				
Monochloromethane (Methyl chloride)[d]	Insufficient data	1992	Insufficient data	1992
Dichloromethane (Methylene chloride)	98.1[c,i]	1992	Insufficient data	1992
Trichloromethane (Chloroform)	1.8[c,i]	1992	Insufficient data	1992
Tetrachloromethane (Carbon tetrachloride)	13.3[c,i]	1992	Insufficient data	1992
Monobromomethane (Methyl bromide)[d]	Insufficient data	1992	Insufficient data	1992
Tribromomethane (Bromoform)[d]	Insufficient data	1992	Insufficient data	1992
Dibromochloromethane[d]	Insufficient data	1992	Insufficient data	1992
Dichlorobromomethane[d]	Insufficient data	1992	Insufficient data	1992
HCBD [see Hexachlorobutadiene (HCBD)]				
Heptachlor (Heptochlor epoxide)[d]	0.01[e,f]	1987		
Hexachlorobenzene [see Chlorinated benzenes]				
Hexachlorobutadiene (HCBD)	1.3[c,k]	1999		
Hexachlorocyclohexane (Lindane)[d]	0.01	1987		
Hypochlorous acid [see Reactive chlorine species]				
Inorganic fluorides	0.12[c]	2002	NRG[o]	2002
3-Iodo-2-propynyl butyl carbamate [see IPBC]				
IPBC (3-Iodo-2-propynyl butyl carbamate)	1.9	1999		
Iron[d]	300	1987		
Lead[d]	1–7[p]	1987		
Lindane [see Hexachlorocyclohexane]				
Linuron	7.0[c]	1995	Insufficient data	1995
MCPA (4-Chloro-2-methyl phenoxy acetic acid (2-methyl-4-chloro phenoxy acetic acid)	2.6[c]	1995	4.2[c]	1995
Mercury[w]				
Inorganic mercury[w]	0.026	2003	0.016[c,x]	2003
Methyl mercury[w]	0.004[c,x]	NRG		
Methyl bromide [see Halogenated methanes, Monobromomethane]				
Methyl chloride [see Halogenated methanes, Monochloromethane]				
2-Methyl-4-chloro phenoxy acetic acid [see MCPA]				
Methylene chloride [see Halogenated methanes, Dichloromethane]				
Metolachlor	7.8[c]	1991		
Metribuzin	1.0[c]	1990		
Molybdenum[j]	73[c]	1999		
Monobromomethane [see Halogenated methanes]				
Monochloramine [see Reactive chlorine species]				
Monochlorobenzene [see Chlorinated benzenes]				
Monochloromethane [see Halogenated methanes]				

(Continued)

Table 8G.96 (Continued)

	Freshwater		Marine	
Parameter[a]	Concentration (μg/L)	Date[b]	Concentration (μg/L)	Date[b]
Monochlorophenols [see Chlorinated phenols]				
MTBE (methyl *tertiary* butyl ether)	10,000	2003	5,000[c]	2003
Naphthalene [see Polycyclic aromatic hydrocarbons (PAHs)]				
Nickel[d]	25–150[q]	1987		
Nitrate[d]	13,000[c,v]	2003	16,000[c,v]	2003
Nitrite[d]	60	1987		
Nonylphenol and its ethoxylates	1.0[r]	2002	0.7[c,r]	2002
Organotins				
Tributyltin	0.008[c]	1992	0.001	1992
Tricyclohexyltin	Insufficient data	1992	Insufficient data	1992
Triphenyltin	0.022[c,l]	1992	Insufficient data	1992
Oxygen, dissolved [see Dissolved oxygen]				
PAHs [see Polycyclic aromatic hydrocarbons (PAHs)]				
PCBs [see Polychlorinated biphenyls (PCBs)(total)]				
PCE [see Chlorinated ethenes, 1,1,2,2-Tetrachloroethene]				
PCP [see Chlorinated phenols, Pentachlorophenol]				
Pentachlorobenzene [see Chlorinated benzenes]				
Pentachlorophenol [see Chlorinated phenols]				
pH	6.5–9[d]	1987	7.0–8.7 and narrative	1996
Phenanthrene [see Polycyclic aromatic hydrocarbons (PAHs)]				
Phenols (mono- & dihydric)	4.0[k]	1999		
Phenoxy herbicides[d,s]	4.0	1987		
Phophorous	narrative[v]			
Phthalate esters				
Di-*n*-butyl phthalate	19[c]	1993	Insufficient data	1993
Di(2-ethylhexyl) phthalate	16[c]	1993	Insufficient data	1993
Di-*n*-octyl phthalate	Insufficient data	1993	Insufficient data	1993
Picloram	29[c]	1990		
Polychlorinated biphenyls (PCBs) (total)[d]	~~0.001~~[e,f]	1987	~~0.01~~[e,f]	1991
Polycyclic aromatic hydrocarbons (PAHs)				
Acenaphthene	5.8[c]	1999	Insufficient data	1999
Acridine	4.4[c]	1999	Insufficient data	1999
Anthracene	0.012[c]	1999	Insufficient data	1999
Benz*(a)*anthracene	0.018[c]	1999	Insufficient data	1999
Benzo*(a)*pyrene	0.015[c]	1999	Insufficient data	1999
Chrysene	Insufficient data	1999	Insufficient data	1999
Fluoranthene	0.04[c]	1999	Insufficient data	1999
Fluorene	3.0[c]	1999	Insufficient data	1999
Naphthalene	1.1[c]	1999	1.4[c]	1999
Phenanthrene	0.4[c]	1999	Insufficient data	1999
Pyrene	0.025[c]	1999	Insufficient data	1999
Quinoline	3.4[c]	1999	Insufficient data	1999
Propylene glycol [see Glycols]				
Pyrene [see Polycyclic aromatic hydrocarbons (PAHs)]				
Quinoline [see Polycyclic aromatic hydrocarbons (PAHs)]				
Reactive chlorine species (hypochlorous acid and monochloramine)	0.5	1999	0.5	1999
Salinity			<10% fluctuation[c]	1996
Selenium[d]	1.0	1987		
Silver[d]	0.1	1987		
Simazine	10	1991		
Streambed substrate [see Total particulate matter]				
Styrene	72[c]	1999		
Sulfolane	50,000[c]	2005	Insufficient data	2005

(Continued)

Table 8G.96 (Continued)

Parameter[a]	Freshwater Concentration (μg/L)	Freshwater Date[b]	Marine Concentration (μg/L)	Marine Date[b]
Suspended sediments [see Total particulate matter]				
TCE [see Chlorinated ethenes, 1,1,2-Trichloroethene]				
Tebuthiuron	1.6[c]	1995	Insufficient data	1995
Temperature	Narrative[t]	1987	Not to exceed ± 1°C[c]	1996
Tetrachlorobenzene [see Chlorinated benzenes]				
Tetrachloroethane [see Chlorinated ethanes]				
Tetrachloroethene [see Chlorinated ethenes]				
Tetrachloroethylene [see Chlorinated ethenes, 1,1,2,2-Tetrachloroethene]				
Tetrachloromethane [see Halogenated methanes]				
Tetrachlorophenols [see Chlorinated phenols]				
Thallium[j]	0.8	1999		
Toluene	2.0[c,j,k]	1996	215[c,k]	1996
Total particulate matter[u]				
Deposited bedload sediment	Insufficient data	1999	Insufficient data	1999
Streambed substrate	Narrative	1999	Narrative	1999
Suspended sediments	Narrative	1999	Narrative	1999
Turbidity	Narrative	1999	Narrative	1999
Toxaphene[d]	~~0.008~~[e,f]	1987		
Triallate	0.24[c]	1992		
Tribromomethane [see Halogenated methanes]				
Tributyltin [see Organotins]				
Trichlorobenzene [see Chlorinated benzenes]				
Trichloroethane [see Chlorinated ethanes]				
Trichloroethene [see Chlorinated ethenes]				
Trichloroethylene [see Chlorinated ethenes, 1,1,2-Trichloroethene]				
Trichloromethane [see Halogenated methanes]				
Trichlorophenols [see Chlorinated phenols]				
Tricyclohexyltin [see Organotins]				
Trifluralin	0.20[i]	1993		
Triphenyltin [see Organotins]				
Turbidity [see Total particulate matter]				
Zinc[d]	30	1987		

[a] Unless otherwise indicated, supporting documents are available from the Guidelines and standards division, environment Canada.
[b] The guidelines dated 1987 have been carried over from *Canadian Water Quality Guidelines* (CCREM 1987) and no fact sheet was prepared. The guidelines dated 1989 to 1997 were developed and initially published in CCREM 1987 as appendixes on the date indicated. They are published as fact sheets in this document. Other guidelines dated 1997 and those dated 1999 are published for the first time in this document.
[c] Interim guideline.
[d] No fact sheet created. For more information on this guideline, please refer to *Canadian Water Quality Guidelines* (CCREM 1987).
[e] This guideline (originally published in *Canadian Water Quality Guidelines* [CCREM 1987 + Appendixes] in 1987 or 1991 [PCBs in marine water]) is no longer recommended and the value is withdrawn. A water quality guideline is not recommended. Environmental exposure is predominantly via sediment, soil, and/or tissue, therefore, the reader is referred to the respective guidelines for these media.
[f] This substance meets the criteria for Track 1 substances under the national CCME Policy for the Management of Toxic Substances (PMTS) (i.e., persistent, bioaccumulative, primarily the result of human activity, and CEPA-toxic or equivalent), and should be subject to virtual elimination strategies. Guidelines can serve as action levels or interim management objectives toward virtual elimination.
[g] Aluminium guideline $=5$ μg/L at pH <6.5; $[Ca^{2+}]<4$ mg/L; DOC <2 mg/L $=100$ μg/L at pH ≥ 6.5; $[Ca^{2+}]\geq 4$ mg/L; DOC ≥ 2 mg/L.
[h] Ammonia guideline: Guideline for total ammonia is temperature and pH dependent, please consult factsheet for more information.
[i] Guideline value slightly modified from CCREM 1987 + Appendixes due to re-evaluation of the significant figures.
[j] The technical document for the guideline is available from the Ontario Ministry of the Environment.
[k] Substance has been re-evaluated since CCREM 1987 + Appendixes. Either a new guideline has been derived or insufficient data existed to derive a new guideline.
[l] Cadmium guideline $=10^{\{0.86[\log(\text{hardness})]-3.2\}}$.
[m] Copper guideline $=2$ μg/L at $[CaCO_3]=0$–120 mg/L; $=3$ μg/L at $[CaCO_3]=120$–180 mg/L; $=4$ μg/L at $[CaCO_3]>180$ mg/L.
[n] Dissolved oxygen for warm-water biota: early life stages $=6{,}000$ μg/L; other life stages $=5{,}500$ μg/L; for cold-water biota: early life stages $=9{,}500$ μg/L; other life stages $=6{,}500$ μg/L
[o] No recommended guideline.

(Continued)

Table 8G.96 (Continued)

[p] Lead guideline = 1 g/L at $[CaCO_3]$ = 0–60 mg/L; = 2 μg/L at $[CaCO_3]$ = 60–120 mg/L; = 4 μg/L at $[CaCO_3]$ = 120–180 mg/L; = 7 μg/L at $[CaCO_3]$ = > 180 mg/L.

[q] Nickel guideline = 25 μg/L at $[CaCO_3]$ = 0–60 mg/L; = 65 μg/L at $[CaCO_3]$ = 60–120 mg/L; = 110 μg/L at $[CaCO_3]$ = 120–180 mg/L; = 150 μg/L at $[CaCO_3]$ = > 180 mg/L.

[r] Expressed on a TEQ basis using NP TEFs, see Table 2 in factsheet.

[s] The guideline of 4.0 μg/L for phenoxy herbicides is based on data for ester formulations of 2,4-dichlorophenoxyacetic acid.

[t] Temperature: (for more information, see CCREM 1987).
Thermal Stratification: Thermal additions to receiving waters should be such that thermal stratification and subsequent turnover dates are not altered from those existing prior to the addition of heat from artificial origins.
Maximum Weekly Average Temperature: Thermal additions to receiving waters should be such that the maximum weekly average temperature is not exceeded.
Short-term Exposure to Extreme Temperature: Thermal additions to receiving waters should be such that the short-term exposures to maximum temperatures are not exceeded. Exposures should not be so lengthy or frequent as to adversely affect the important species.

[u] The technical document for the guideline is available from British Columbia Ministry of Environment, Lands and Parks.

[v] For protection from direct toxic effects; the guidelines do not consider indirect effects due to eutrophication.

[w] May not prevent accumulation of methylmercury in aquatic life, therefore, may not protect wildlife that consume aquatic life; see factsheet for details. Consult also the appropriate Canadian Tissue Residue Guideline for the Protection of Wildlife consumers of Aquatic Biota

[x] May not protect fully higher trophic level fish; see factsheet for details.

[y] Canadian Trigger Ranges (for further narrative see factsheet). Total Phosphorus (ugL^{-1}):
ultra-oligotrophic < 4
oligotrophic 4–10
mesotrophic 10–20
eutrophic 35–100
hyper-eutrophic > 100

Source: From Canadian Council of Ministers of the Environment, 2005. Canadian Water Quality Guidelines for the Protection of Aquatic Life: Summary Table, Updated October 2005. In: *Canadian Environmental Quality Guidelines*, 1999, Canadian Council of Ministers of the Environment, Winnipeg, www.ec.gc.ca/CEQG-RCQE.

Table 8G.97 Canadian Tissue Residue Guidelines for the Protection of Wildlife Consumers of Aquatic Biota

Parameter[a]	Guideline (μg/kg diet ww)[b]	Date
DDT (total)	14.0	1997
Methylmercury	33.0	2001
Polychlorinated biphenyls (PCBs)	Mammalian: 0.79 ng TEQ/kg diet ww[c]	1998
	Avian: 2.4 ng TEQ/kg diet ww[d]	1998
Polychlorinated dibenzo-*p*-dioxins/polychlorinated dibenzofurans	Mammalian: 0.71 ng TEQ/kg diet ww[c]	2001
	Avian: 4.75 ng TEQ/kg diet ww[d]	2001
Toxaphene	6.3	1997

[a] Supporting documents are available from the Guidelines and Standards Division, Environment Canada.

[b] Guideline refers to the total concentration of the contaminant found in an aquatic organism on a wet weight (ww) basis that is not expected to result in adverse effects in predaceous wildlife.

[c] TEQ refers to dioxin toxic equivalents using toxic equivalency factors (TEFs) for PCBs for mammals developed by the World Health Organization in 1998. See fact sheet or supporting document for more details.

[d] TEQ refers to dioxin toxic equivalents using toxic equivalency factors (TEFs) for PCBs for birds developed by the World Health Organization in 1998. See fact sheet or supporting document for more details.

Source: From Canadian Council of Ministers of the Environment, 2001, Canadian Tissue Guidelines for the Protection of Wildlife Consumers of Aquatic Biota: Summary Table. Updated. In: *Canadian Environmental Quality Guidelines*, 1999, Canadian Council of Ministers of the Environment, www.ec.gc.ca.

Table 8G.98 Summary of Water Quality Guidelines for Turbidity, Suspended and Benthic Sediments, British Columbia, Canada

Water Use	Maximum Induced Turbidity—NTU or % of Background	Maximum Induced Suspended Sediments—mg/L or % of Background	Streambed Substrate Composition
Drinking Water—raw untreated	1 NTU when background is less than or equal to 5	No Guideline	No Guideline
Drinking Water—raw treated	5 NTU when background is less than or equal to 50 10% when background is greater than 50	No Guideline	No Guideline
Recreation and Aesthetics	Maximum 50 NTU secchi disc visible at 1.2 m	No Guideline	No Guideline
Aquatic Life Fresh Marine Estuarine	8 NTU in 24 hrs when background is less than or equal to 8 Mean of 2 NTU in 30 days when background is less than or equal to 8	25 mg/L in 24 hrs when background is less than or equal to 25 Mean of 5 mg/L in 30 days when background is less than or equal to 25	Fines not to exceed –10% as less than 2 mm –19% as less than 3 mm –25% as less than 6.35 mm at salmonid spawning sites
Aquatic Life Fresh Marine Estuarine	8 NTU when background is between 8 and 80 10% when background is greater than or equal to 80	25 mg/L when background is between 25 and 250 10% when background is greater than or equal to 250	Geometric mean diameter not less than 12 mm Fredle number not less than 5 mm
Terrestrial Life Wildlife Livestock water Irrigation Industrial	10 NTU when background is less than or equal to 50 20% when background is greater than or equal to 50	20 mg/L when background is less than or equal to 100 20% when background is greater than or equal to 100	No guideline

Source: From British Columbia Approved Water Quality Guidelines (Criteria) 2006 Edition, Updated: August 2006, www.env.gov.bc.ca. Reprinted with Permission.

Table 8G.99 Canadian Sediment Quality Guidelines for the Protection of Aquatic Life

	Freshwater		Marine	
Parameter	ISQG[a](ug/kg)	PEL[b] (ug/kg)	ISQG[a] (ug/kg)	PEL[b] (ug/kg)
Arsenic	5,900	17,000	7,240	41,600
Cadmium	600	3,500	700	4,200
Chlordane	4.50	8.87	2.26	4.79
Chromium	37,300	90,000	52,300	160,000
Copper	35,700	197,000	18,700	108,000
DDD (2,2-Bis (*p*-chlorophenyl)-1,1-dichloroethane; Dichloro diphenyl dichloroethane)[c]	3.54	8.51	1.22	7.81
DDE(1,1-Dichloro-2,2-bis(*p*-chlorophenyl)-ethene, Diphenyl dichloro ethylene[c]	1.42	6.75	2.07	374
DDT(2,2-Bis(*p*-chlorophenyl)-1,1,1-trichloroethane; Dichloro diphenyl trichloroethane)[c]	1.19[d]	4.77[e]	1.19	4.77
Dieldrin	2.85	6.67	0.71	4.3
Endrin	2.67	62.4	2.67[d]	62.4[e]
Heptachlor (Heptachlor epoxide}	0.60	2.74	0.60[d]	2.74[e]
Lead	35,000	91,300	30,200	112,000
Lindane (Hexachlorocyclohexane)	0.94	1.38	0.32	0.99
Mercury	170	486	130	700
Nonylphenol and its ethoxylates	1.4 mg/kg dw[f,g]		1.0 mg/kg dw[f,h]	
Polychlorinated biphenyls (PCBs)				
Aroclor 1254	0.60	340[i]	63.3	709
Total PCBs	34.1	277	21.5	189

(Continued)

Table 8G.99 (Continued)

Parameter	Freshwater		Marine	
	ISQG[a](ug/kg)	PEL[b] (ug/kg)	ISQG[a] (ug/kg)	PEL[b] (ug/kg)
Polychlorinated dibenzo-*p*-dioxins/dibenzo furans (PCDD/Fs)	0.85 ng TEQ/kg dw[k]	21.5 ng TEQ/kg dw[k]	0.85 ng TEQ/kg dw[k]	21.5 ng TEQ/kg dw[k]
Polycyclic aromatic hydrocarbons (PAHs)				
Acenapthene	6.71[d]	88.9[e]	6.71	88.9
Acenaphthylene	5.87[d]	128[e]	5.87[d]	128[e]
Anthracene	46.9[d]	245[e]	46.9	245
Benzo(a)anthracene	31.7	385	74.8	693
Benzo(a)pyrene	31.9	782	88.8	763
Chrysene	57.1	862	108	846
Dibenz(a,h)anthracene	6.22[d]	135[e]	6.22	135
Fluoranthene	111	2,355	113	1,494
Fluorene	21.2[d]	144[e]	21.2	144
2-Methylnaphthalene	20.2[d]	201[e]	20.2	201
Naphthalene	34.6[d]	391[e]	34.6	391
Phenanthrene	41.9	515	86.7	544
Pyrene	53.0	875	153	1,398
Toxaphene	0.1[l]	—[m]	0.1[l]	—[m]
Zinc	123,000	315,000	124,000	271,000

[a] ISQG: Interim sediment quality guideline Canadian Environmental Quality Guidelines Summary Table December 2003.
[b] PEL: Probable effect level.
[c] Sum of *p,p'* and *o,p'* isomers.
[d] Provisional: adoption of marine ISQG.
[e] Provisional: adoption of marine PEL.
[f] Provisional; use of equilibrium partitioning approach.
[g] Note that the incidence of adverse biological effects below the TEL, between the TEL and PEL, and above the PEL were 22%, 24%, and 65%, respectively, prior to the application of a safety factor.
[h] Expressed in a toxic equivalency (TEQ) basis using NP TEFs; assumes 1% TOC.
[i] Provisional: adoption of lowest effect level from Ontario (Persaud et al. 1993).
[j] Provisional: 1% TOC: adoption of severe effect level of 34μg.g-*1*TOC from Ontario (Persaud et al. 1993).
[k] Values are expressed as toxic equivalency (TEQ) units, based on WHO 1998 TEF values for fish.
[l] Provisional; 1% TOC: adoption of the chronic sediment quality criterion of 0.01 ug/gl TOC of the New York State Department of Environmental Conservation (NYSDEC 1994).
[m] No PEL derived.

Source: From Canadian Council of Ministers of the Environment, 2002, Canadian Sediment Quality Guidelines for the Protection of Aquatic Life: Summary Table, Updated 2002. In: *Canadian Environmental Quality Guidelines*, 1999, Canadian Council of Ministers of the Environment, Winnipeg, www.ec.gc.ca/CEQG-RCQE.

Table 8G.100 ECE Standard Statistical Classification of Surface Freshwater Quality for the Maintenance of Aquatic Life

Variables	Class I	Class II	Class III	Class IV	Class V
Oxygen regime					
DO (%)					
Epilimnion (stratified waters)	90–110	70–90 or 110–120	50–70 or 120–130	30–50 or 130–150	<30 or >150
Hypolimnion (stratified waters)	90–70	70–50	50–30	30–10	<10
Unstratified waters	90–70	70–50 or 110–120	50–30 or 120–130	30–10 or 130–150	<10 or >150
DO(mg/L)	>7	7-6	6-4	4-3	<3
COD-Mn (mg O_2 /L)	<3	3-10	10-20	20-30	>30
COD-Cr (mg O_2/L)	—	—	—	—	—
Eutrophication					
Total P (μg/L)[a]	<10 (<15)	10–25 (15–40)	25–50 (40–75)	50–125 (75–190)	>125 (>190)
Total N (μg/L)[a]	<300	300–750	750–1,500	1,500–2,500	>2,500
Chlorophyll a (μg/L)[a]	<2.5 (<4)	2.5–10 (4–15)	10–30 (15–45)	30–110 (45–165)	>110 (>165)
Acidification					
pH[b]	9.0–6.5	6.5–6.3	6.3–6.0	6.0–5.3	<5.3
Alkalinity (mg $CaCO_3$/l)	>200	200–100	100–20	20–10	<10
Metals					
Aluminium (μg/l; pH 6.5)	<1.6	1.6–3.2	3.2–5	May-75	>75
Arsenic (μg/L)[c]	<10	10–100	100–190	190–360	>360
Cadmium (μg/L)[d]	<0.07	0.07–0.53	0.53–1.1	1.1–3.9	>3.9
Chromium (μg/L)[c]	<1	1-6	6-11	11-16	>16
Copper (μg/L)[d]	<2	2-7	7-12	12-18	>18
Leader (μg/L)[d]	<0.1	0.1–1.6	1.6–3.2	3.2–82	>82
Mercury (μg/L)[d]	<0.003	0.003–0.007	0.007–0.012	0.012–2.4	>2.4
Nickel (μg/L)[d]	<15	15–87	87–160	160–1,400	>1,400
Zinc (μg/L)[d]	<45	45–77	77–110	110–120	>120
Chlorinated micropollutants and other hazardous substances					
Dieldrin (μg/L)	na	na	<0.0019	0.0019–2.5	>2.5
DDT and metabolites (μg/L)	na	na	<0.001	0.001–1.1	>1.1
Endrin (μg/L)	na	na	<0.0023	0.0023–0.18	>0.18
Heptachlor (μg/L)	na	na	<0.0038	0.0038–0.52	>0.52
Lindane (μg/L)	na	na	<0.08	0.08–2.0	>2.0
Pentachlorophenol (μg/L)	na	na	<13	13–20	>20
PCBs (μg/L)	na	na	<0.014	0.014–2.0	>2.0
Free ammonia (NH_3)	na	na	—	—	—
Radioactivity					
Gross-alpha activity (mBq/L)	<50	50–100	100–500	500–2,500	>2,500
Gross-beta activity (mBq/L)	<200	200–500	500–1,000	1,000–2,500	>2,500

Note: Measures falling on the boundary between two classes are to be classified in the lower class; na, Not applicable; —, No value set at present.

[a] Data in brackets refer to flowing waters.
[b] Values >9.0 are disregarded in the classification of acidification.
[c] Applicable for hardness from about 0.5 to 8 meq/L. Arsenic V and chromium III to be converted to arsenic III and chromium VI, respectively.
[d] Applicable for hardness from about 0.5 to 8 meq/L.

Source: From Ute S. Enderlein, Rainer E. Enderlein and W. Peter Williams, 1997, Chapter 2—Water Quality Requirements, In: World Health Organization, *Water Pollution Control—A Guide to the Use of Water Quality Management Principles*, Copyright © 1997 WHO/UNEP.

Original Source: From The United Nations is the author of the original material. United Nations Economic Commission for Europe, 1994, Standard Statistical Classification of Surface Freshwater Quality for the Maintenance of Aquatic Life. In: *Readings in International Environment Statistics*, United Nations Economic Commission of Europe, United Nations, New York, and Geneva.

Table 8G.101 Water-Quality Objectives for the River Rhine Related to Organic Substances

Water-Quality Variable	Water-Quality Objective (μg/L)	Basis for Elaboration[a]
Tetrachloromethane	1.0	Drw+aqL
Trichloromethane	0.6	aqL
Aldrin, Dieldrin, Endrin, Isodrin	0.0001 (per substance)	aq+terrL
Endosulfan	0.003	aqL
Hexachlorobenzene	0.0005	aqL
Hexachlorobutadien	0.001	aqL
PCB 28, 52, 101,180, 138, 153	0.001 (per substance)	aqL
1-Chloro-4-nitro-Benzen	1.0	Drw
1-Chloro-2-nitro-Benzen	1.0	Drw+aqL
Trichlorobenzene	0.1	aqL
Pentachlorophenol	0.001	aq+terrL
Trichloroethen	1.0	Drw
Tetrachloroethen	1.0	Drw
3,4-Dichloroanilin	0.1	aqL
2-Chloroanilin	0.1	Drw+aqL
3-Chloroanilin	0.1	Drw
4-Chloroanilin	0.01	aqL
Parathion(-ethyl)	0.0002	aqL
Parathion(-methyl)	0.01	aqL
Benzene	0.1	aqL
1,1,1-Trichloroethane	1.0	Drw
1,2-Dichloroethane	1.0	aqL
Azinphos-methyl	0.001	aqL
Bentazon	0.1	Drw
Simazine	0.1	Drw+aqL
Atrazine	0.1	Drw+aqL
Dichlorvos	0.001	aqL
2-Chlorotoluol	1.0	Drw
4-Chlorotoluol	1.0	Drw
Tributyl tin-substances	0.001	aqL
Triphenyl tin-substances	0.001	aqL
Trifluralin	0.1	aqL
Fenthion	0.01	aqL

[a] Water-quality objectives have been set on the basis of water-quality criteria for drinking-water supply (Drw), drinking-water supply and aquatic life (Drw+aqL) and/or aquatic life (aqL), as well as on the basis of toxicity testing on selected species of aquatic and terrestrial life (aq+terrL).

Source: From Ute S. Enderlein, Rainer E. Enderlein and W. Peter Williams, 1997, Chapter 2 — Water Quality Requirements, In: *Water Pollution Control — A Guide to the Use of Water Quality Management Principles.* Copyright © 1997 WHO/UNEP.

Original Source: From ICPR, *Konzept zur Ausfüllung des Punktes A.2 des APR* über *Zielvorgaben,* 1991. Lenzburg, den 2. Juli (Methodology to implement item A.2 of the Rhine Action Programme related to water quality objectives, prepared at Lenzbourg on 2 July 1991). PLEN 3/91, *International Commission for the Protection of the Rhine against Pollution,* Koblenz, Germany, 1991, www.who.int.

Table 8G.102 Maximum Permissible Concentrations and Negligible Concentrations for Metals, The Netherlands

	Surface Water (ug/L)								SED (mg/kg)				
Metal	**MPA (fresh)**	**NA (fresh)**	***Cb* (fresh)**	**MPC (fresh)**	**NC (fresh)**	***Cb* (marine)**	**MPC (marine)**	**NC (marine)**	**MPA (sed) (d)**	**NA (sed)**	***Cb* (sed)**	**MPC (sed)**	**Nc (sed)**
Antimony(Sb)	6.2[a]	0.062	0.29	6.5	0.35				16	0.16	3.0	19	3.2
Arsenic (As)	24[b]	0.24	0.77	25	1.0				160	1.6	29	190	31
Barium (Ba)	150[a]	1.5	73	220	75				150	1.5	155	300	157
Beryllium (Be)	0.16[b]	0.0016	0.02	0.18	0.022				0.096	0.00096	1.1	1.2	1.1
Cadmium (Cd)	0.34[b]	0.0034	0.08	0.42	0.083	0.025	0.37	0.028	29	0.29	0.8	30	1.1
Chromium (Cr)	8.5[b]	0.085	0.17	8.7	0.26				1,620	16	100	1,720	116
Cobalt (Co)	2.6[b]	0.026	0.20	2.8	0.23				10	0.10	9.0	19	9.1
Copper (Cu)	1.1[b]	0.011	0.44	1.5	0.45	0.25	1.4	0.26	37	0.37	36	73	36
Lead (Pb)	11[b]	0.11	0.15	11	0.26	0.02	11	0.13	4,700	47	85	4,800	132
Mercury (Hg)	0.23[b]	0.0023	0.01	0.24	0.012	0.0025	0.23	0.0048	26	0.26	0.3	26	0.56
Methyl-mercury	0.01[b]	0.0001	0.01	0.02	0.01	0.0025	0.013	0.0026	1.1	0.011	0.3	1.4	0.31
Molybdenum (Mo)	290[a]	2.9	1.4	290	4.3				250	2.5	0.5	250	3.0
Nickel (Ni)	1.8[b]	0.018	3.3	5.1	3.3				9.4	0.094	35	44	35
Selenium (Se)	5.3[b]	0.053	0.04	5.3	0.093				2.2	0.022	0.7	2.9	0.72
Thallium (Tl)	1.6[a]	0.016	0.04	1.6	0.056				1.6	0.016	1.0	2.6	1.0
Tia (Sn)	18[a]	0.18	0.0002	18	0.18				22,000	220	19	22,000	239
Vanadium (V)	3.5[a]	0.035	0.82	4.3	0.86				14	0.14	42	56	42
Zinc (Zn)	6.6[b]	0.066	2.8	9.4	2.9	0.35	7.0	0.42	480	4.8	140	620	145

Note: MPA, maximum permissible addition; NA, Negligible Addition, Cb, background concentration; MPC, maximum permissible concentration; NC, negligible concentration for metals; fresh, freshwater; marine, saltwater.

[a] MPA based on modified EPA-method

[b] MPA based on statistical extrapolation

[c] Values are given as concentrations in mg/kg standard soil/sediment (soil/sediment containing 10% organic matter and 25% clay)

[d] MPA based on equilibrium partitioning.

Source: From Crommentuijn, T., M. Polder, and E. van de Plassche. 1997. Maximum permissible concentrations and negligible concentrations for metals, taking background concentrations into account. *Nat. Inst. Public Health and the Environ.*, Bilthoven, The Netherlands. RIVM Report 601501001, tivm.nl.

Australia and New Zealand
Physico-Chemical Stressor Low-Risk Guideline Trigger Values for Ecosystems

The default approach to deriving trigger values has used the statistical distribution of reference data collected within five geographical regions across Australia and New Zealand. Here, depending on the stressor, a *measureable perturbation* in slightly to moderately disturbed ecosystems has been defined using the 80th and/or 20th percentile of the reference data.

First, New Zealand and Australian state and territory representatives used percentile distributions of available data and professional judgement to derive trigger values for each ecosystem type in their regions.Trigger values were then collated, discussed and agreed for southeast Australia (VIC, NSW, ACT, south-east QLD, and TAS,) southwest Australia (southern WA), tropical Australia (northern WA, NT, northern QLD), south central Australia —— low rainfall area (SA) and New Zealand.

The default trigger values in the present guidelines were derived from ecosystem data for unmodified or slightly-modified ecosystems supplied by state agenicies. However, the choice of these reference systems was not based on any objective biological criteria. This lack of specificity may have resulted in inclusion of reference systems of varying quality, and further emphasises that the default trigger values should only be used until site- or ecosystem-specific values can be generated.

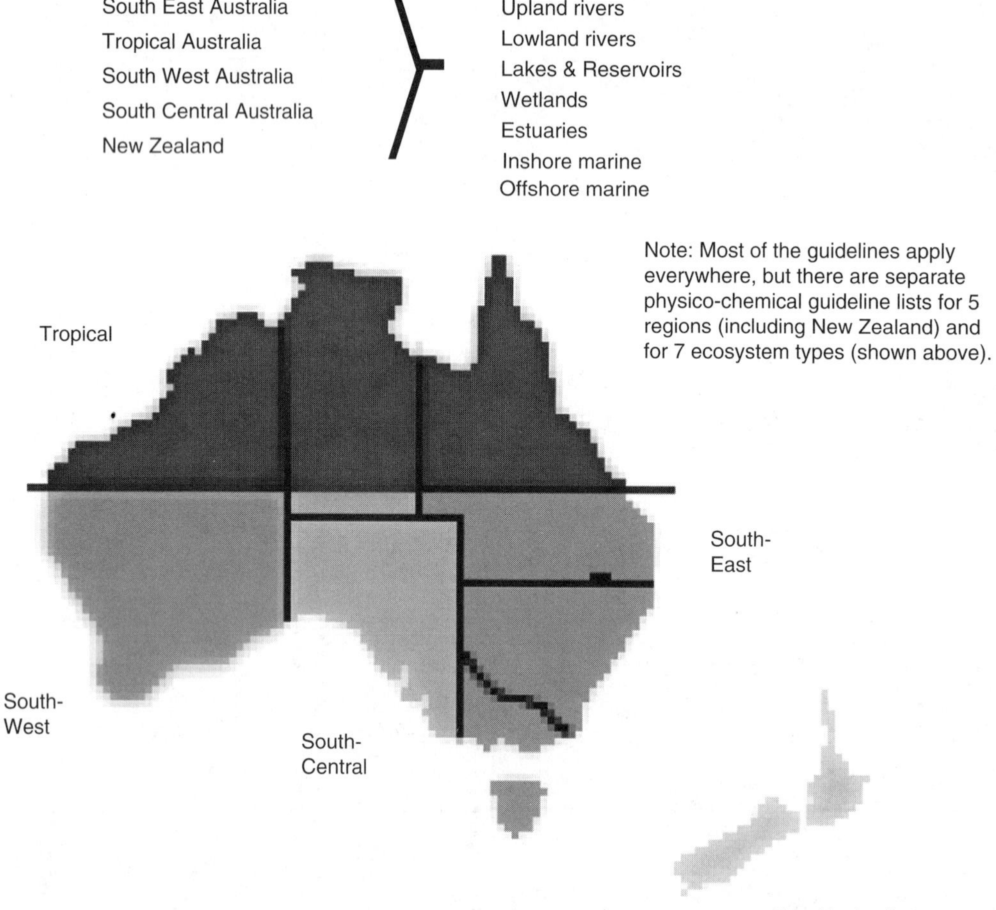

Figure 8G.22 Geographic regions of Australia and New Zealand. (From Australia Water Quality — Introduction, *The Guideline Value Lists*; www.ozh2o.com/ozh3a.html.)

Table 8G.103 Default Trigger Values for Physical and Chemical Stressors for Southeast Australia for Slightly Disturbed Ecosystems

							DO (% saturation)[a]		pH	
Ecosystem Type	**Chl *a* (μg/L)**	**TP (μg P/L)**	**FRP (μg P/L)**	**TN (μg N/L)**	**NO_x (μg N/L)**	**NH_4^+ (μg N/L)**	**Lower Limit**	**Upper Limit**	**Lower Limit**	**Upper Limit**
Upland river	na[b]	20[c]	15[d]	250[e]	15[f]	13[k]	90	110	6.5	7.5[g]
Lowland river[h]	5	50	20	500	40[i]	20	85	110	6.5	8.0
Freshwater lakes & Reservoirs	5[j]	10[k]	5	350	10	10	90	110	6.5	8.0[g]
Wetlands	no data	no data	no data	no data	no data	no data	no data	no data	no data	no data
Estuaries[l]	4[m]	30	5[n]	300	15	15	80	110	7.0	8.5
Marine[l]	1[o]	25[o]	10	120	5[p]	15[p]	90	110	8.0	8.4

Note: Trigger values are used to assess risk of adverse effects due to nutrients, biodegradable organic matter and pH in various ecosystem types. Data derived from trigger values supplied by Australian states and territories. Chl *a*, chlorophyll *a*; TP, total phosphorus; FRP, filterable reactive phosphate; TN, total nitrogen; NO_x, oxides of nitrogen; NH_4^+, ammonium; DO, dissolved oxygen; na, not applicable.

[a] Dissolved oxygen values were derived from daytime measurements. Dissolved oxygen concentrations may very diurnally and with depth. Monitoring programs should assess this potential variability.
[b] Monitoring of periphyton and not phytoplankton biomass is recommended in upland rivers — values for periphyton biomass (mg Chl a m^{-2}) to be developed.
[c] Values are 30 μg/L for Qld rivers, 10 μg/L for Vic. Alpine streams and 13 μg/L for Tas. Rivers.
[d] Values is 5 μg/L for Vic. alpine streams and Tas. rivers;
[e] Values are 100 μg/L for Vic. alpine streams and 480 μg/L for Tas. rivers.
[f] Value is 190 μg/L for Tas. rivers.
[g] Values for NSW upland rivers are 6.5–8.0, for NSW lowland rivers 6.5–8.5, for humic rich Tas. lakes and rivers 4.0–6.5.
[h] Values are 3 μg/L for Chl *a*, 25 μg/L for TP and 350 μg/L for TN for NSW & Vic. east flowing coastal rivers.
[i] value is 60 μg/L for Qld rivers.
[j] Values are 3 μg/L for Tas. lakes.
[k] Values is 10 μg/L for Qld. rivers.
[l] No data available for Tasmanian estuarine and marine waters. A precautionary approach should be adopted when applying default trigger values to these systems.
[m] Values is 5 μg/L for Qld estuaries.
[n] Value is 15 μg/L for Qld. estuaries.
[o] values are 20 μg/L for TP for offshore waters and 1.5 μg/L for Chl a for Old inshore waters.
[p] Values of 25 μg/L for NO_x and 20 μg/L for NH_4^+ for NSW are elevated due to frequent upwelling events.

Source: From Australian and New Zealand Environment and Conservation Council and Agriculture and Resource Management Council of Australia and New Zealand, 2000, *National Water Quality Management Strategy*, Paper No. 4, *Australian and New Zealand Guidelines for Fresh and Marine Water Quality*, vol. I, The Guidelines, October 2000, www.deh.gov.au.

Table 8G.104 Ranges of Default Trigger Values for Conductivity (EC, salinity), Turbidity and Suspended Particulate Matter (SPM) Indicative of Slightly Disturbed Ecosystems in Southeast Australia

Ecosystem type	**Salinity (μS/cm)**	**Turbidity (NTU)**
Upland rivers	30–350	2–25
Lawland rivers	125–2,200	6–50
Lakes & reservoirs	20–30	1–20
Estuarine & marine		0.5–10

Note: Ranges for turbidity and SPM are similar and only turbidity is reported here. Values reflect high site-specific and regional variability. Explanatory notes provide detail on specific variability issues for ecosystem type, can be found in the guidance document.

Source: From Australian and New Zealand Environment and Conservation Council and Agriculture and Resource Management Council of Australia and New Zealand, 2000, *National Water Quality Management Strategy*, Paper No. 4, *Australian and New Zealand Guidelines for Fresh and Marine Water Quality*, Volume I, The Guidelines, October 2000, www.deh.gov.au.

Table 8G.105 Default Trigger Values for Physical and Chemical Stressors for Tropical Australia for Slightly Disturbed Ecosystems

Ecosystem type		Chl *a* (μg/L)	TP (μg P/L)	FRP (μg P/L)	TN (μg N/L)	NO_x (μg N/L)	NH_4^+ (μg N/L)	DO (% saturation)[a] Lower limit	DO (% saturation)[a] Upper limit	pH Lower limit	pH Upper limit
Upland river[b]		na[c]	10	5	150	30	6	90	120	6.0	7.5
Lowland river[b]		5	10	4	200–300[d]	10[e]	10	85	120	6.0	8.0
Freshwater lakes & Reservoirs		3	10	5	350[f]	10[e]	10	90	120	6.0	8.0
Wetlands		10	10–50[g]	5–25[g]	350–1,200[g]	10	10	90[e]	120[e]	6.0	8.0
Estuaries[b]		2	20	5	250	30	15	80	120	7.0	8.5
Marine	Inshore	0.7–1.4[h]	15	5	100	2–8[h]	1–10[h]	90	no data	8.0	8.4
	Offshore	0.5–0.9[h]	10	2–5[h]	100	1–4[h]	1–6[h]	90	no data	8.2	8.2

Note: Trigger values are used to assess risk of adverse effects due to nutrients, biodegradable organic matter and pH in various ecosystem types. Data derived from trigger values supplied by Australian states and territories for the Northern Territory and regions north of Carnarvon in the west and Rockhampton in the east. Chl *a*, chlorophyll *a*; TP, total phosphorus; FRP, filterable reactive phosphate; TN, total nitrogen; NO_x, oxides of nitrogen; NH_4^+, ammonium; DO, dissolved oxygen; na, not applicable.

a Dissolved oxygen values were derived from daytime measurements. Dissolved oxygen concentrations may vary diurnally and with depth. Monitoring programs should assess this potential variability.
b No data available for tropical WA estuaries or rivers. A precautionary approach should be adopted when applying default trigger values to these systems.
c Monitoring of periphyton and not phytoplankton biomass is recommended in upland rivers — values for periphyton biomass (mg Chl a m^{-2}) to be developed.
d Lower values from rivers draining rainforest catchments.
e Northern Territory values for 5 μg/L for No_x, and <80 (lower limit) and >110% saturation (upper limit) for DO.
f This value represents turbid lakes only. Clear lakes have much lower values.
g Higher values are indicative of tropical WA river pools.
h The lower values are typical of clear coral dominated waters (e.g. Great Barrier Reef), while higher values typical of turbid macrotidal systems (e.g. Northwest Shelf of WA).

Source: From Australian and New Zealand Environment and Conservation Council and Agriculture and Resource Management Council of Australia and New Zealand, 2000, *National Water Quality Management Strategy*, Paper No. 4, *Australian and New Zealand Guidelines for Fresh and Marine Water Quality*, vol. I, The Guidelines, October 2000, www.deh.gov.au.

Table 8G.106 Ranges of Default Trigger Values for Conductivity (EC, salinity), Turbidity and Suspended Particulate Matter (SPM) Indicative of Slightly Disturbed Ecosystems in Tropical Australia

Ecosystem Type	Salinity (μS/cm)	Turbidity (NTU)
Upland & lowland rivers	20–250	2–15
Lakes, reservoirs & wetlands	90–900	2–200
Estuarine & marine	—	1–20

Note: Ranges for turbidity and SPM are similar and only turbidity is reported here. Values reflect high site-specific and regional variability. Explanatory notes provide detail on specific variability issues for groupings of ecosystem type can be found in the guidance document.

Source: From Australian and New Zealand Environment and Conservation Council and Agriculture and Resource Management Council of Australia and New Zealand, 2000, *National Water Quality Management Strategy*, Paper No. 4, *Australian and New Zealand Guidelines for Fresh and Marine Water Quality*, Volume I, The Guidelines, October 2000, www.deh.gov.au.

Table 8G.107 Default Trigger Values for Physical and Chemical Stressors for South-West Australia for Slightly Disturbed Ecosystems

Ecosystem type		Chl *a* (μg/L)	TP (μg P/L)	FRP (μg P/L)	TN (μg N/L)	NO_x (μg N/L)	NH_4^+ (μg N/L)	DO (% saturation)[a] Lower Limit	DO (% saturation)[a] Upper Limit	pH Lower Limit	pH Upper Limit
Upland river[b]		na[c]	20	10	450	200	60	90	na	6.5	8.0
Lowland river[b]		3–5	65	40	1,200	150	80	80	120	6.5	8.0
Freshwater lakes & reservoirs		3–5	10	5	350	10	10	90	no data	6.5	8.0
Wetlands[d]		30	60	30	1,500	100	40	90	120	7.0[e]	8.5[e]
Estuaries		3	30	5	750	45	40	90	110	7.5	8.5
Marine[f,g]	Inshore[h]	0.7	20[i]	5[i]	230	5	5	90	na	8.0	8.4
	Offshore	0.3[i]	20[i]	5	230	5	5	90	na	8.2	8.2

Note: Trigger values are used to assess risk of adverse effects due to nutrients, biodegradable organic matter and pH in various ecosystem types. Data derived from trigger values supplied by Western Australia. Chl *a*, chlorophyll *a*; TP, total phosphorus; FRP, filterable reactive phosphate; TN, total nitrogen; NO_x, oxides of nitrogen; NH_4^+, ammonium; DO, dissolved oxygen; na, not applicable.

a Dissolved oxygen values were derived from daytime measurements. Dissolved oxygen concentrations may vary diumally and with depth. Monitoring programs should assess this potential variability.
b All values derived during base river flow conditions not storm events.
c Monitoring of periphyton and not phytoplankton biomass is recommended in upland rivers—values for periphyton biomass (mg Chl a m^{-2}) to be developed.
d Elevated nutrient concentrations in highly coloured wetlands (given >52 g_{440}/m) do not appear to stimulate algal growth.
e In highly coloured wetlands (given >52 g_{440}/m) pH typically ranges 4.5–6.5.
f Nutrient concentrations alone are poor indicators of marine trophic status.
g These trigger values are generic and therefore do not necessarily apply in all circumstances, e.g. for some unprotected coastlines, such as Albany and Geographe Bay, it may be more appropriate to use offshore values for inshore waters.
h Inshore waters defined as coastal lagoons (excluding estuaries) and embayments and waters less than 20 m depth.
i Summer (low rainfall) values, values higher in winter for Chi a (1.0 μg/L), TP (40 μg P/L), FRP (10 μg P/L).

Source: From Australian and New Zealand Environment and Conservation Council and Agriculture and Resource Management Council of Australia and New Zealand, 2000, *National Water Quality Management Strategy*, Paper No. 4, *Australian and New Zealand Guidelines for Fresh and Marine Water Quality*, vol. I, The Guidelines, October 2000, www.deh.gov.au.

Table 8G.108 Ranges of Default Trigger Values for Conductivity (EC, salinity), Turbidity and Suspended Particulate Matter (SPM) Indicative of Slightly Disturbed Ecosystems in Southwest Australia

Ecosystem type	Salinity (μS/cm)	Turbidity (NTU)
Upland & lowland rivers	120–300	10–20
Lakes, reservoirs & wetlands	300–1,500	10–100
Estuarine & marine	—	1–2

Note: Ranges for turbidity and SPM are similar and only turbidity is reported here. Values reflect high site-specific and regional variability. Explanatory notes that provide detail on specific variability issues for ecosystem types can be found in the guidance document.

Source: From Australian and New Zealand Environment and Conservation Council and Agriculture and Resource Management Council of Australia and New Zealand, 2000, *National Water Quality Management Strategy*, Paper No. 4, *Australian and New Zealand Guidelines for Fresh and Marine Water Quality*, vol. I, The Guidelines, October 2000, www.deh.gov.au.

Table 8G.109 Default Trigger Values for Physical and Chemical Stressors for South-Central Australia for Slightly Disturbed Ecosystems

							DO (% Saturation)		pH	
Ecosystem Type	Chl *a* (μg/L)	TP (μg P/L)	FRP (μg P/L)	TN (μg N/L)	NO_x (μg N/L)	NH_4^+ (μg N/L)	Lower Limit	Upper Limit	Lower Limit	Upper Limit
Upland river	No data	No data	No data	No data	No data	No data	No data	No data	No data	No data
Lowland river	No data	100	40	1,000	100	100	90	No data	6.5	9.0
Freshwater lakes & reservoirs	No data	25	10	1,000	100	25	90	No data	6.5	9.0
Wetlands	No data	No data	No data	No data	No data	No data	No data	No data	No data	No data
Estuaries	5	100	10	1,000	100	50	90	No data	6.5	9.0
Marine	1	100	10	1,000	50	50	No data	No data	8.0	8.5

Note: Trigger values are used to assess risk of adverse effects due to nutrients, biodegradable organic matter and pH in various ecosystem types. Data derived from trigger values supplied by South Australia. Chl *a*, chlorophyll *a*; TP, total phosphorus; FRP, filterable reactive phosphate; TN, total nitrogen; NO_x, oxides of nitrogen; NH_4^+, ammonium; DO, dissolved oxygen.

Source: From Australian and New Zealand Environment and Conservation Council and Agriculture and Resource Management Council of Australia and New Zealand, 2000, *National Water Quality Management Strategy*, Paper No. 4, *Australian and New Zealand Guidelines for Fresh and Marine Water Quality*, vol. I, The Guidelines, October 2000, www.deh.gov.au.

Table 8G.110 Ranges of Default Trigger Values for Conductivity (EC, salinity), Turbidity, and Suspended Particulate Matter (SPM) Indicative of Slightly Disturbed Ecosystems in South-Central Australia

Ecosystem Types	Salinity (μS/cm)	Turbidity (NTU)
Lowland rivers	100–5,000	
Upland & lowland rivers		1–50
Lakes, reservoirs & wetlands	300–1,000	1–100
Estuarine & marine		0.5–10

Note: Ranges for turbidity and SPM are similar and only turbidity is reported here. Values reflect high site-specific and regional variability. Explanatory notes provide detail on specific variability issues for groupings of ecosystem types can be found in the guidance document.

Source: From Australian and New Zealand Environment and Conservation Council and Agriculture and Resource Management Council of Australia and New Zealand, 2000, *National Water Quality Management Strategy*, Paper No. 4, *Australian and New Zealand Guidelines for Fresh and Marine Water Quality*, vol. I, The Guidelines, October 2000, www.deh.gov.au.

Table 8G.111 Default Trigger Values for Physical and Chemical Stressors in New Zealand for Slightly Disturbed Ecosystems

							DO[a] (% Saturation)		pH[a]	
Ecosystem Type	Chl *a* (μg/L)	TP (μg P/L)	FRP (μg P/L)	TN (μg N/L)	NO_x (μg N/L)	NH_4^+ (μg N/L)	Lower Limit	Upper Limit	Lower Limit	Upper Limit
Upland river	na[b]	26[c]	9[c]	295[c]	167[c]	10[c]	99	103	7.3	8.0
Lowland river	no data	33[d]	10[d]	614[d]	444[d]	21[d]	98	105	7.2	7.8

Note: Trigger values are used to assess risk of adverse effects due to nutrients, biodegradable organic matter and pH in various ecosystem types. Chl *a*, chlorophyll *a*; TP, total phosphorus; FRP, filterable reactive phosphate; TN[e], total nitrogen; NO_x, oxides of nitrogen; NH_4^+, ammonium nitrogen; DO, dissolved oxygen, na, not applicable.

[a] DO and pH percentiles may not be very useful as trigger values because of diurnal and seasonal variation—values listed are for daytime sampling.

[b] Monitoring of periphyton and not phytoplankton biomass is recommended in upland rivers—values for periphyton biomass (mg Chl a m^{-2}) to be developed. New Zealand is currently making routine observations of periphyton cover.

[c] Values for glacial and lake-fed sites in upland rivers are lower;

[d] Values are lower for Haast River which receives waters from alpine regions;

[e] Commonly referred to dissolved reactive phosphorus in New Zealand.

Source: From Australian and New Zealand Environment and Conservation Council and Agriculture and Resource Management Council of Australia and New Zealand, 2000, *National Water Quality Management Strategy*, Paper No. 4, *Australian and New Zealand Guidelines for Fresh and Marine Water Quality*, Volume I, The Guidelines, October 2000, www.deh.gov.au.

Table 8G.112 Default Trigger Values for Water Clarity (Lower Limit) and Turbidity (Upper Limit) Indicative of Unmodified or Slightly Disturbed Ecosystems in New Zealand

	Upland rivers[a,b]		Lowland rivers	
Ecosystem types	Clarity (1/m)[c,d]	Turbidity (NTU)[c,d]	Clarity (1/m)	Turbidity (NTU)
	0.6	4.1	0.8	5.6

[a] Light availability is generally less of an issue in NZ rivers and streams than is visual clarity because, in contrast to many of Australia's rivers, most NZ rivers are comparatively clear and/or shallow. Davies-Colley et al. (1992) recommend that visual clarity, light penetration and water colour are important optical properties of an ecosystem which need to be protected. Neither turbidity nor visual clarity provide a useful estimate of light penetration—light penetration should be considered separately to turbidity or visual clarity. Clarity relates to the transmission of light through water and is measured by the visual range of a black disk (see NZ Ministry for the Environment (1994)) or a Secchi disk.

[b] Recent work has shown that at least some NZ indigenous fish are sensitive to low levels of turbidity; however, it may also be desirable to protect the naturally high turbidities of alpine glacial lakes to prevent possible ecological impacts, such as change in predator-prey relationships.

[c] Note that turbidity and visual water clarity are closely and inversely related, and the 80th percentile for turbidity is consistent with the 20th percentile for visibility and vice versa.

[d] Clarity and turbidity values for glacial sites in upland rivers are lower and higher, respectively.

Source: From Australian and New Zealand Environment and Conservation Council and Agriculture and Resource Management Council of Australia and New Zealand, 2000, *National Water Quality Management Strategy*, Paper No. 4, *Australian and New Zealand Guidelines for Fresh and Marine Water Quality*, vol. I, The Guidelines, October 2000, www.deh.gov.au.

Table 8G.113 General Framework for Applying Australian Levels of Protection for Toxicants to Different Ecosystem Conditions

Ecosystem Condition	Level of Protection
1 High conservation/ecological value	For anthropogenic toxicants, detection at any concentration could be grounds for source investigation and management intervention; for natural toxicants background concentrations should not be exceeded[a] *Where local biological or chemical data have not yet been gathered,* apply the 99% protection levels (Table 8.114) as default values Any relaxation of these objectives should only occur where comprehensive biological effects and monitoring data clearly show that biodiversity would not be altered In the case of effluent discharges, Direct Toxicity Assessment (DTA) should also be required on the effluent Precautionary approach taken to assessment of post-baseline data through trend analysis or feedback triggers
2 Slightly to moderately disturbed ecosystems	Always preferable to use local biological effects data (including DTA) to derive guidelines *If local biological effects data unavailable*, apply 95% protection levels (Table 8.114) as default, low-risk trigger values.[b] 99% values are recommended for certain chemicals as noted in Table 8.114[c] Precautionary approach may be required for assessment of post-baseline data through trend analysis or feedback triggers In the case of effluent discharges DTA may be required
3 High disturbed ecosystems	Apply the same guidelines as for slightly-moderately disturbed systems. However, the lower protection levels provided in the Guidelines may be accepted by stakeholders DTA could be used as an alternative approach for deriving site-specific guidelines

[a] This means that indicator values at background and test sites should be statistically indistinguishable. It is acknowledged that it may not be strictly possible to meet this criterion in every situation.

[b] For slightly disturbed ecosystems where the management goal is no change in biodiversity, users may prefer to apply a higher protection level.

[c] 99% values recommended for chemicals that bioaccumulate or for which 95% provides inadequate protection for key test species. Jurisdictions may choose 99% values for some ecosystems that are more towards the slightly disturbed end of the continuum.

Source: From Australian and New Zealand Environment and Conservation Council and Agriculture and Resource Management Council of Australia and New Zealand, 2000, *National Water Quality Management Strategy*, Paper No. 4, *Australian and New Zealand Guidelines for Fresh and Marine Water Quality*, vol. I, The Guidelines, October 2000, www.deh.gov.au.

Table 8G.114 Australian Trigger Values for Toxicants at Alternative Levels of Protection

	Trigger Values for Freshwater (μg/L) Level of Protection (% species)				Trigger Values for Marine Water (μg/L) Level of Protection (% Species)			
Chemical	**99%**	**95%**	**90%**	**80%**	**99%**	**95%**	**90%**	**80%**
Metals & metalloids								
Aluminum pH >6.5	27	**55**	80	150	ID	ID	ID	ID
Aluminum pH<6.5	ID	ID	ID	ID	ID	ID	ID	ID
Antimony	ID	ID	ID	ID	ID	ID	ID	ID
Arsenic (As III)	1	**24**	94[a]	360[a]	ID	ID	ID	ID
Arsenic (AsV)	0.8	**13**	42	140[a]	ID	ID	ID	ID
Beryllium	ID	ID	ID	ID	ID	ID	ID	ID
Bismuth	ID	ID	ID	ID	ID	ID	ID	ID
Boron	90	**370[a]**	680[a]	1,300[a]	ID	ID	ID	ID
Cadmium[b]	0.06	**0.2**	0.4	0.8[a]	**0.7[c]**	5.5[a,c]	14[a,c]	36[c,d]
Chromium (Cr III)[b]	ID	ID	ID	ID	7.7	**27.4**	48.6	90.6
Chromium (CrVI)	0.01	**1.0[a]**	6[d]	40[d]	0.14	**4.4**	20[a]	85[a]
Cobalt	ID	ID	ID	ID	0.005	**1**	14	150[a]
Copper[b]	1.0	**1.4**	1.8[a]	2.5[a]	0.3	**1.3**	3[a]	8[d]
Gallium	ID	ID	ID	ID	ID	ID	ID	ID
Iron	ID	ID	ID	ID	ID	ID	ID	ID
Lanthanum	ID	ID	ID	ID	ID	ID	ID	ID
Lead[b]	1.0	**3.4**	5.6	9.4[a]	2.2	**4.4**	6.6[a]	12[a]
Manganese	1,200	**1,900[a]**	2,500[a]	3,600[a]	ID	ID	ID	ID
Mercury (inorganic)[c]	**0.06**	0.6	1.9[a]	5.4[d]	**0.1**	0.4[a]	0.7[a]	1.4[a]
Mercury (methyl)	ID	ID	ID	ID	ID	ID	ID	ID
Molybdenum	ID	ID	ID	ID	ID	ID	ID	ID
Nickel[b]	8	**11**	13	17[a]	**7**	70[a]	200[d]	560[d]
Selenium (Total)[c]	**5**	11	18	34	ID	ID	ID	ID
Selenium (SelV)[c]	ID	ID	ID	ID	ID	ID	ID	ID
Silver	0.2	**0.05**	0.1	0.2[a]	0.8	**1.4**	1.8	2.6[a]
Thallium	ID	ID	ID	ID	ID	ID	ID	ID
Tin (inorganic, SnIV)	ID	ID	ID	ID	ID	ID	ID	ID
Tributyltin (as μg/L Sn)	ID	ID	ID	ID	0.0004	**0.006[a]**	0.02[a]	0.05[a]
Uranium	ID	ID	ID	ID	ID	ID	ID	ID
Vanadium	ID	ID	ID	ID	50	**100**	160	280
Zinc[b]	2.4	**8.0[a]**	15[a]	31[a]	7	**15[a]**	23[a]	43[a]
Nonmetallic Inorganics								
Ammonia[e]	320	**900[a]**	1,430[a]	2,300[d]	500	**910**	1,200	1,700
Chlorine[f]	0.4	**3**	6[d]	13[d]	ID	ID	ID	ID
Cyanide[g]	4	**7**	11	18	2	**4**	7	14
Nitrate[h]	17	**700**	3,400[a]	17,000[d]	ID	ID	ID	ID
Hydrogen sulfide[i]	0.5	**1.0**	1.5	2.8	ID	ID	ID	ID
Organic Alcohols								
Ethanol	400	**1,400**	2,400[a]	4,000[a]	ID	ID	ID	ID
Ethylene glycol	ID	ID	ID	ID	ID	ID	ID	ID
Isopropyl alcohol	ID	ID	ID	ID	ID	ID	ID	ID
Chlorinated Alkanes								
Chloromethanes								
Dichloromethane	ID	ID	ID	ID	ID	ID	ID	ID
Chloroform	ID	ID	ID	ID	ID	ID	ID	ID
Carbon tetrachloride	ID	ID	ID	ID	ID	ID	ID	ID
Chloroethanes								
1,2-dichloroethane	ID	ID	ID	ID	ID	ID	ID	ID
1,1,1-trichloroethane	ID	ID	ID	ID	ID	ID	ID	ID
1,1,2-trichloroethane	5,400	**6,500**	7,300	8,400	140	**1,900**	5,800[a]	18,000[a]
1,1,2,2-tetrachloroethane	ID	ID	ID	ID	ID	ID	ID	ID
Pentachloroethane	ID	ID	ID	ID	ID	ID	ID	ID
Hexachloroethane[c]	**290**	360	420	500	ID	ID	ID	ID
Chloropropanes								
1,1-dichloropropane	ID	ID	ID	ID	ID	ID	ID	ID
1,2-dichloropropane	ID	ID	ID	ID	ID	ID	ID	ID
1,3-dichloropropane	ID	ID	ID	ID	ID	ID	ID	ID

(Continued)

Table 8G.114 (Continued)

	Trigger Values for Freshwater (μg/L) Level of Protection (% species)				Trigger Values for Marine Water (μg/L) Level of Protection (% Species)			
Chemical	**99%**	**95%**	**90%**	**80%**	**99%**	**95%**	**90%**	**80%**
Chlorinated Alkenes								
Chloroethylene	ID	ID	ID	ID	ID	ID	ID	ID
1,1-dichloroethylene	ID	ID	ID	ID	ID	ID	ID	ID
1,1,2-trichloroethylene	ID	ID	ID	ID	ID	ID	ID	ID
1,1,2,2-tetrachloroethylene	ID	ID	ID	ID	ID	ID	ID	ID
3-chloropropene	ID	ID	ID	ID	ID	ID	ID	ID
1,3-dichloropropene	ID	ID	ID	ID	ID	ID	ID	ID
Anilines								
Aniline	**8**	250[d]	1,100[d]	4,800[d]	ID	ID	ID	ID
2,4-dichloroaniline	0.6	**7**	20	60[a]	ID	ID	ID	ID
2,5-dichloroaniline	ID	ID	ID	ID	ID	ID	ID	ID
3,4-dichloroaniline	1.3	**3**	6[a]	13[a]	85	**150**	190	260
3,5-dichloroaniline	ID	ID	ID	ID	ID	ID	ID	ID
Benzidine	ID	ID	ID	ID	ID	ID	ID	ID
Dichlorobenzidine	ID	ID	ID	ID	ID	ID	ID	ID
Aromatic Hydrocarbons								
Benzene	600	**950**	1,300	2,000	**500[a]**	700[a]	900[a]	1,300[a]
Toluene	ID	ID	ID	ID	ID	ID	ID	ID
Ethylbenzene	ID	ID	ID	ID	ID	ID	ID	ID
o-xylene	200	**350**	470	670	ID	ID	ID	ID
m-xylene	ID	ID	ID	ID	ID	ID	ID	ID
p-xylene	140	**200**	250	340	ID	ID	ID	ID
m+*p*-xylene	ID	ID	ID	ID	ID	ID	ID	ID
Cumene	ID	ID	ID	ID	ID	ID	ID	ID
Polycyclic Aromatic Hydrocarbons								
Naphthalene	2.5	**16**	37	85	**50[a]**	70[a]	90[a]	120[a]
Anthracene[c]	ID	ID	ID	ID	ID	ID	ID	ID
Phenanthrene[c]	ID	ID	ID	ID	ID	ID	ID	ID
Fluoranthene[c]	ID	ID	ID	ID	ID	ID	ID	ID
Benzo(a)pyrene[c]	ID	ID	ID	ID	ID	ID	ID	ID
Nitrobenzenes								
Nitrobenzene	230	**550**	820	1,300	ID	ID	ID	ID
1,2-dinitrobenzene	ID	ID	ID	ID	ID	ID	ID	ID
1,3-dinitrobenzene	ID	ID	ID	ID	ID	ID	ID	ID
1,4-dinitrobenzene	ID	ID	ID	ID	ID	ID	ID	ID
1,3,5-trinitrobenzene	ID	ID	ID	ID	ID	ID	ID	ID
1-methoxy-2-nitrobenzene	ID	ID	ID	ID	ID	ID	ID	ID
1-methoxy-4-nitrobenzene	ID	ID	ID	ID	ID	ID	ID	ID
1-chloro-2-nitrobenzene	ID	ID	ID	ID	ID	ID	ID	ID
1-chloro-3-nitrobenzene	ID	ID	ID	ID	ID	ID	ID	ID
1-chloro-4-nitrobenzene	ID	ID	ID	ID	ID	ID	ID	ID
1-chloro-2,4-dinitrobenzene	ID	ID	ID	ID	ID	ID	ID	ID
1,2-dichloro-3-nitrobenzene	ID	ID	ID	ID	ID	ID	ID	ID
1,3-dichloro-5-nitrobenzene	ID	ID	ID	ID	ID	ID	ID	ID
1,4-dichloro-2-nitrobenzene	ID	ID	ID	ID	ID	ID	ID	ID
2,4-dichloro-2-nitrobenzene	ID	ID	ID	ID	ID	ID	ID	ID
1,2,4,5-tetrachloro-3-nitrobenzene	ID	ID	ID	ID	ID	ID	ID	ID
1,5-dichloro-2,4-dinitrobenzene	ID	ID	ID	ID	ID	ID	ID	ID
1,3,5-trichloro-2,4-dinitrobenzene	ID	ID	ID	ID	ID	ID	ID	ID
1-fluoro-4-nitrobenzene	ID	ID	ID	ID	ID	ID	ID	ID
Nitrotoluenes								
2-nitrotoluene	ID	ID	ID	ID	ID	ID	ID	ID
3-nitrotoluene	ID	ID	ID	ID	ID	ID	ID	ID
4-nitrotoluene	ID	ID	ID	ID	ID	ID	ID	ID
2,3-dinitrotoluene	ID	ID	ID	ID	ID	ID	ID	ID

(Continued)

Table 8G.114 (Continued)

Chemical	Trigger Values for Freshwater (μg/L) Level of Protection (% species)				Trigger Values for Marine Water (μg/L) Level of Protection (% Species)			
	99%	95%	90%	80%	99%	95%	90%	80%
2,4-dinitrotoluene	**16**	65[a]	130[a]	250[a]	ID	ID	ID	ID
2,4,6-trinitrotoluene	100	**140**	160	210	ID	ID	ID	ID
1,2-dimethyl-3-nitrobenzene	ID	ID	ID	ID	ID	ID	ID	ID
1,2-dimethyl-4-nitrobenzene	ID	ID	ID	ID	ID	ID	ID	ID
4-chloro-3-nitrotoluene	ID	ID	ID	ID	ID	ID	ID	ID
Chlorobenzenes and Chloronaphthalenes								
Monochlorobenzene	ID	ID	ID	ID	ID	ID	ID	ID
1,2-dichlorobenzene	120	**160**	200	270	ID	ID	ID	ID
1,3-dichlorobenzene	160	**260**	350	520[a]	ID	ID	ID	ID
1,4-dichlorobenzene	40	**60**	75	100	ID	ID	ID	ID
1,2,3-trichlorobenzene[c]	**3**	10	16	30[a]	ID	ID	ID	ID
1,2,4-trichlorobenzene[c]	**85**	170[a]	220[a]	300[a]	**20**	80	140	240
1,3,5-trichlorobenzene[c]	ID	ID	ID	ID	ID	ID	ID	ID
1,2,3,4-tetrachlorobenzene[c]	ID	ID	ID	ID	ID	ID	ID	ID
1,2,3,5-tetrachlorobenzene[c]	ID	ID	ID	ID	ID	ID	ID	ID
1,2,4,5-tetrachlorobenzene[c]	ID	ID	ID	ID	ID	ID	ID	ID
Pentachlorobenzene[c]	ID	ID	ID	ID	ID	ID	ID	ID
Hexachlorobenzene[c]	ID	ID	ID	ID	ID	ID	ID	ID
1-chloronaphthalene	ID	ID	ID	ID	ID	ID	ID	ID
Polychlorinated Biphenyls (PCBs) & Dioxins								
Capacitor 21[c]	ID	ID	ID	ID	ID	ID	ID	ID
Aroclor 1016[c]	ID	ID	ID	ID	ID	ID	ID	ID
Aroclor 1221[c]	ID	ID	ID	ID	ID	ID	ID	ID
Aroclor 1232[c]	ID	ID	ID	ID	ID	ID	ID	ID
Aroclor 1242[c]	**0.3**	0.6	1.0	1.7	ID	ID	ID	ID
Aroclor 1248[c]	ID	ID	ID	ID	ID	ID	ID	ID
Aroclor 1254[c]	**0.01**	0.03	0.7	0.2	ID	ID	ID	ID
Aroclor 1260[c]	ID	ID	ID	ID	ID	ID	ID	ID
Aroclor 1262[c]	ID	ID	ID	ID	ID	ID	ID	ID
Aroclor 1268[c]	ID	ID	ID	ID	ID	ID	ID	ID
2,3,4′-trichlorobipheny[c]	ID	ID	ID	ID	ID	ID	ID	ID
4,4′-dichlorobiphenyl	ID	ID	ID	ID	ID	ID	ID	ID
2,2′,4,5,5′-pentachloro-1,1′-biphenyl[c]	ID	ID	ID	ID	ID	ID	ID	ID
2,4,6,2′4,6′-hexachlorobiphenyl[c]	ID	ID	ID	ID	ID	ID	ID	ID
Total PCBs[c]	ID	ID	ID	ID	ID	ID	ID	ID
2,3,7,8-TCDD[c]	ID	ID	ID	ID	ID	ID	ID	ID
Phenols and Xylenols								
Phenol	85	**320**	600	1,200[a]	270	**400**	520	720
2,4-dimethylphenol	ID	ID	ID	ID	ID	ID	ID	ID
Nonylphenol	ID	ID	ID	ID	ID	ID	ID	ID
2-chlorphenol[j]	**340[a]**	490[a]	630[a]	870[a]	ID	ID	ID	ID
3-chlorophenol[j]	ID	ID	ID	ID	ID	ID	ID	ID
4-chlorophenol[j]	160	**220**	280[a]	360[a]	ID	ID	ID	ID
2,3-dichlorophenol[j]	ID	ID	ID	ID	ID	ID	ID	ID
2,4-dichlorophenol[j]	**120**	160[a]	200[a]	270[a]	ID	ID	ID	ID
2,5-dichlorophenol[j]	ID	ID	ID	ID	ID	ID	ID	ID
2,6-dichlorophenol[j]	ID	ID	ID	ID	ID	ID	ID	ID
3,4-dichlorophenol[j]	ID	ID	ID	ID	ID	ID	ID	ID
3,5-dichlorophenol[j]	ID	ID	ID	ID	ID	ID	ID	ID
2,3,4-trichlorophenol[j]	ID	ID	ID	ID	ID	ID	ID	ID
2,3,5-trichlorophenol[j]	ID	ID	ID	ID	ID	ID	ID	ID
2,3,6-trichlorophenol[j]	ID	ID	ID	ID	ID	ID	ID	ID
2,4,6-trichlorophenol[c,j]	**3**	20	40	95	ID	ID	ID	ID
2,3,4,5-tetrachlorophenol[c,j]	ID	ID	ID	ID	ID	ID	ID	ID
2,3,4,6-tetrachlorophenol[c,j]	**10**	20	25	30	ID	ID	ID	ID
2,3,5,6-tetrachlorophenol[c,j]	ID	ID	ID	ID	ID	ID	ID	ID

(Continued)

Table 8G.114 (Continued)

Chemical	Trigger Values for Freshwater (μg/L) Level of Protection (% species)				Trigger Values for Marine Water (μg/L) Level of Protection (% Species)			
	99%	95%	90%	80%	99%	95%	90%	80%
Pentachlorophenol[c,j]	**3.6**	10	17	27[d]	**11**	22	33	55[d]
Nitrophenols								
2-nitrophenol	ID	ID	ID	ID	ID	ID	ID	ID
3-nitrophenol	ID	ID	ID	ID	ID	ID	ID	ID
4-nitrophenol	ID	ID	ID	ID	ID	ID	ID	ID
2,4-dinitrophenol	13	**45**	80	140	ID	ID	ID	ID
2,4,6-trinitrophenol	ID	ID	ID	ID	ID	ID	ID	ID
Organic Sulfur Compounds								
Carbon disulfide	ID	ID	ID	ID	ID	ID	ID	ID
Isopropyl	ID	ID	ID	ID	ID	ID	ID	ID
n-propyl sulfide	ID	ID	ID	ID	ID	ID	ID	ID
Propyl disulfide	ID	ID	ID	ID	ID	ID	ID	ID
Tert-butyl sulfide	ID	ID	ID	ID	ID	ID	ID	ID
Phenyl disulfide	ID	ID	ID	ID	ID	ID	ID	ID
Bis(dimethylthiocarbamyl) sulfide	ID	ID	ID	ID	ID	ID	ID	ID
Bis(diethylthiocarbamyl) disulfide	ID	ID	ID	ID	ID	ID	ID	ID
2-methoxy-4H-1,3,2-benzodioxaphosphorium-2sulfide	ID	ID	ID	ID	ID	ID	ID	ID
Xanthates								
Potassium amyl xanthate	ID	ID	ID	ID	ID	ID	ID	ID
Potassium ethyl xanthate	ID	ID	ID	ID	ID	ID	ID	ID
Potassium hexyl xanthate	ID	ID	ID	ID	ID	ID	ID	ID
Potassium isopropyl xanthate	ID	ID	ID	ID	ID	ID	ID	ID
Sodium ethyl xanthate	ID	ID	ID	ID	ID	ID	ID	ID
Sodium isobutyl xanthate	ID	ID	ID	ID	ID	ID	ID	ID
Sodium isopropyl xanthate	ID	ID	ID	ID	ID	ID	ID	ID
Sodium sec-butyl xanthate	ID	ID	ID	ID	ID	ID	ID	ID
Phthalates								
Dimethylphthalate	3,000	**3,700**	4,300	5,100	ID	ID	ID	ID
Diethylphthaiate	900	**1,000**	1,100	1,300	ID	ID	ID	ID
Dibutylphthalate[c]	**9.9**	26	40.2	64.6	ID	ID	ID	ID
Di(2-ethylhexyl)phthalate[c]	ID	ID	ID	ID	ID	ID	ID	ID
Miscellaneous Industrial Chemicals								
Acetonitrile	ID	ID	ID	ID	ID	ID	ID	ID
Acrylonitrile	ID	ID	ID	ID	ID	ID	ID	ID
Poly(acrylonitrile-co-butadiene-co-styrene)	200	**530**	800[a]	1,200[a]	200	**250**	280	340
Dimethylformamide	ID	ID	ID	ID	ID	ID	ID	ID
1,2-diphenylhydrazine	ID	ID	ID	ID	ID	ID	ID	ID
Diphenylnitrosamine	ID	ID	ID	ID	ID	ID	ID	ID
Hexachlorobutadiene	ID	ID	ID	ID	ID	ID	ID	ID
Hexachlorocyclopentadiene	ID	ID	ID	ID	ID	ID	ID	ID
Isophorone	ID	ID	ID	ID	ID	ID	ID	ID
Organochlorine Pesticides								
Aldrin[c]	ID	ID	ID	ID	ID	ID	ID	ID
Chlordane[c]	**0.03**	0.08	0.14	0.27[a]	ID	ID	ID	ID
DDE[c]	ID	ID	ID	ID	ID	ID	ID	ID
DDT[c]	**0.006**	0.01	0.02	0.04	ID	ID	ID	ID
Dicofol[c]	ID	ID	ID	ID	ID	ID	ID	ID
Dieldrin[c]	ID	ID	ID	ID	ID	ID	ID	ID
Endosulfan[c]	**0.03**	0.2[d]	0.6[d]	1.8[d]	**0.005**	0.01	0.02	0.05[d]
Endosulfan alpha[c]	ID	ID	ID	ID	ID	ID	ID	ID
Endosulfan beta[c]	ID	ID	ID	ID	ID	ID	ID	ID
Endrin[c]	**0.01**	0.02	0.04[a]	0.06[d]	**0.004**	0.008	0.01	0.02
Heptachlor[c]	**0.01**	0.09	0.25	0.7[d]	ID	ID	ID	ID

(Continued)

Table 8G.114 (Continued)

Chemical	Trigger Values for Freshwater (μg/L) Level of Protection (% species)				Trigger Values for Marine Water (μg/L) Level of Protection (% Species)			
	99%	95%	90%	80%	99%	95%	90%	80%
Lindane	0.07	**0.2**	0.4	1.0[d]	ID	ID	ID	ID
Methoxychlor[c]	ID	ID	ID	ID	ID	ID	ID	ID
Mirex[c]	ID	ID	ID	ID	ID	ID	ID	ID
Toxaphene[c]	**0.1**	0.2	0.3	0.5	ID	ID	ID	ID
Organophosphorus Pesticides								
Azinphos methyl	**0.01**	0.02	0.05	0.11[d]	ID	ID	ID	ID
Chlorpyrifos[c]	0.00004	**0.01**	0.11[d]	1.2[d]	0.0005	**0.009**	0.04[d]	0.3[d]
Demeton	ID	ID	ID	ID	ID	ID	ID	ID
Demeton-S-methyl	ID	ID	ID	ID	ID	ID	ID	ID
Diazinon	0.00003	**0.01**	0.2[d]	2[d]	ID	ID	ID	ID
Dimethoate	0.1	**0.15**	0.2	0.3	ID	ID	ID	ID
Fenitrothion	0.1	**0.2**	0.3	0.4	ID	ID	ID	ID
Malathion	0.002	**0.05**	0.2	1.1[d]	ID	ID	ID	ID
Parathion	0.0007	**0.004**[a]	0.01[a]	0.04[d]	ID	ID	ID	ID
Profenofos[c]	ID	ID	ID	ID	ID	ID	ID	ID
Temephos[c]	ID	ID	ID	ID	0.0004	**0.05**	0.4	3.6[d]
Carbamate & Other Pesticides								
Carbofuran	**0.06**	1.2[d]	4[d]	15[d]	ID	ID	ID	ID
Methomyl	0.5	**3.5**	9.5	23	ID	ID	ID	ID
S-methoprene	ID	ID	ID	ID	ID	ID	ID	ID
Pyrethroids								
Deltamethrin	ID	ID	ID	ID	ID	ID	ID	ID
Esfenvalerate	ID	0.001[k]	ID	ID	ID	ID	ID	ID
Herbicides & Fungicides								
Bypyridilium herbicides								
Diquat	0.01	**1.4**	10	80[d]	ID	ID	ID	ID
Paraquat	ID	ID	ID	ID	ID	ID	ID	ID
Phenoxyacetic acid herbicides								
MCPA	ID	ID	ID	ID	ID	ID	ID	ID
2,4-D	140	**280**	450	830	ID	ID	ID	ID
2,4,5-T	3	**36**	100	290[d]	ID	ID	ID	ID
Sulfonylurea herbicides								
Bensulfuron	ID	ID	ID	ID	ID	ID	ID	ID
Metsulfuron	ID	ID	ID	ID	ID	ID	ID	ID
Thiocarbamate herbicides								
Molinate	0.1	**3.4**	14	57	ID	ID	ID	ID
Thiobencarb	1	**2.8**	4.6	8[a]	ID	ID	ID	ID
Thiram	**0.01**	0.2	0.8[a]	3[d]	ID	ID	ID	ID
Triazine herbicides								
Amitrole	ID	ID	ID	ID	ID	ID	ID	ID
Atrazine	0.7	**13**	45[a]	150[a]	ID	ID	ID	ID
Hexazinone	ID	ID	ID	ID	ID	ID	ID	ID
Simazine	0.2	**3.2**	11	35	ID	ID	ID	ID
Urea herbicides								
Diuron	ID	ID	ID	ID	ID	ID	ID	ID
Tebuthiuron	0.02	**2.2**	20	160[a]	ID	ID	ID	ID
Miscellaneous herbicides								
Acrolein	ID	ID	ID	ID	ID	ID	ID	ID
Bromacil	ID	ID	ID	ID	ID	ID	ID	ID
Glyphosate	**370**	1,200	2,000	3,600[d]	ID	ID	ID	ID
Imazethapyr	ID	ID	ID	ID	ID	ID	ID	ID
Ioxynil	ID	ID	ID	ID	ID	ID	ID	ID
Metolachlor	ID	ID	ID	ID	ID	ID	ID	ID
Sethoxydim	ID	ID	ID	ID	ID	ID	ID	ID
Trifluralin[c]	**2.6**	4.4	6	9[d]	ID	ID	ID	ID

(Continued)

Table 8G.114 (Continued)

	Trigger Values for Freshwater (μg/L) Level of Protection (% species)				Trigger Values for Marine Water (μg/L) Level of Protection (% Species)			
Chemical	**99%**	**95%**	**90%**	**80%**	**99%**	**95%**	**90%**	**80%**
Generic Groups of Chemicals								
Surfactants								
Linear alkylbenzene sulfonates (LAS)	65	**280**	520[a]	1,000[a]	ID	ID	ID	ID
Alcohol ethoxyolated sulfate (AES)	340	**650**	850[a]	1,100[a]	ID	ID	ID	ID
Alcohol ethoxylated surfactants (AE)	50	**140**	220	360[a]	ID	ID	ID	ID
Oils & Petroleum Hydrocarbons	ID	ID	ID	ID	ID	ID	ID	ID
Oil Spill Dispersants								
BP 1100X	ID	ID	ID	ID	ID	ID	ID	ID
Corexit 7664	ID	ID	ID	ID	ID	ID	ID	ID
Corexit 8667		ID	ID	ID	ID	ID	ID	ID
Corexit 9527	ID	ID	ID	ID	230	1100	2,200	4,400[d]
Corexit 9550	ID	ID	ID	ID	ID	ID	ID	ID

Note: ID, Insufficient data to derive a reliable trigger value. Users advised to check if a low reliability value or an ECL is available **Section 8.3.7.** Values in grey shading are the trigger values applying to typical *slightly-moderately disturbed systems*; see Table 8-113. And Section 3.4.2.4 of the guidance document for guidance on applying these levels to different ecosystem conditions. Where the final water quality guideline to be applied to a site is below current analytical practical quantitation limits, see Section 3.4.3.3 of guidance document. Most trigger values listed here for metals and metalloids are *High reliability* figures, derived from field or chronic NOEC data. The exceptions are *Moderate reliability* for freshwater aluminium (pH >6.5), manganese and marine chromium (III). Most trigger values listed here for non-metallic inorganices and organic chemicals are *Moderate reliability* figures, derived for acute LC_{50} data. The exceptions are *High reliability* for freshwater ammonia, 3,4-DCA, endosulfan chlorpyrifos, esfenvalerate, tebuthiuron, three surfactants and marine for 1, 1,2-TCE and chlorpyrifos.

[a] Figure may not protect key test species from chronic toxicity (this refers to experimental chronic figures or geometric mean for species)—check Section 8.3.7 of guidance document for spread of data and its significance. Where grey shading and 'C' coincide, refer to text in Section 8.3.7. of guidance document.

[b] Chemicals for which algorithms have been provided in Table 3.4.3 of guidance document to account for the effects of hardness. The values have been calculated using a hardness of 30 mg/L $CaCO_3$. These should be adjusted to the site-specific hardness (see Section 3.4.3) of guidance document.

[c] Chemicals for which possible bioaccumulation and secondary poisoning effects should be considered.

[d] Figure may not protect key test species from acute toxicity (and chronic)—check Section 8.3.7 for spread of data and its significance. 'A' indicates that trigger value $>$ acute toxicity figure; note that trigger value should be $<1/3$ of acute figure.

[e] Ammonla as TOTAL ammonia as (NH_3-N) at pH 8. For changes in trigger value with ph refer to Section 8.3.7.2 of the guidance document.

[f] Chlorine as total chlorine, as [Cl].

[g] Cyanide as unionised HCN, measured as [CN] of the guidance document.

[h] Figures protect against toxicity and do not relate to eutrophication issues. Refer to Section 3.3 if eutrophication is the issue of concern.

[i] Sulfide as un-lonised H_2S, measured as [S] of the guidance document.

[j] Tainting or flavour impairment of fish flesh may possibly occur at concentrations below the trigger value.

[k] *High reliability* figure for esfenvalerate derived form mesocosm NOEC data (no alternative protection levels available).

Source: From Australian and New Zealand Environment and Conservation Council and Agriculture and Resource Management Council of Australia and New Zealand, 2000, *National Water Quality Management Strategy*, Paper No. 4, *Australian and New Zealand Guidelines for Fresh and Marine Water Quality*, Volume I, The Guidelines, October 2000, www.deh.gov.au.

Table 8G.115 Australian Guidelines for the Protection of Human Consumers of Fish and Other Aquatic Organisms from Bacterial Infection

Toxicant	Guideline *in Shellfishing Water*	Standard *in Edible Tissue*
Faecal (thermotolerant) coliforms	The median faecal coliform bacterial concentration should not exceed 14 MPN/100 mL, with no more than 10% of the samples exceeding 43 MPN/100 mL	Fish destined for human consumption should not exceed a limit of 2.3 MPN *E. coli lg* of flesh with a standard place count of 100,000 organisms/g

Note: MPN: Most probable number, The guideline for faecal (thermotolerant) coliforms should not be used in conjunction with the data from a sanitary survey of the shellfish harvesting areas for the purpose of harvesting area classification.

Source: From Australian and New Zealand Environment and Conservation Council and Agriculture and Resource Management Council of Australia and New Zealand, 2000, *National Water Quality Management Strategy*, Paper No. 4, *Australian and New Zealand Guidelines for Fresh and Marine Water Quality*, vol. I, The Guidelines, October 2000, www.deh.gov.au.

Table 8G.116 Australian Guidelines of Chemical Compounds in Water Found to Cause Tainting of Fish Flesh and Other Aquatic Organisms

Parameter	Estimated Threshold Level in Water (mg/L)
Acenaphthene	0.02
Acetophenone	0.5
Acrylonitrile	18.0
Copper	1.0
m-cresol	0.2
o-cresol	0.4
p-cresol	0.1
Cresylic acids (meta, para)	0.2
Chlorobenzene	0.02
n-butylmercaptan	0.06
o-sec. butylphenol	0.3
p-tert. butylphenol	0.03
o-chlorophenol	0.0001–0.015
p-chlorophenol	0.0001
2,3-dinitrophenol	0.08
2,4,6-trinitrophenol	0.002
2,4-dichlorophenol	0.0001–0.014
2,5-dichlorophenol	0.02
2,6-dichlorophenol	0.03
3,4-dichlorophenol	0.0003
2-methyl-4-chlorophenol	2.0
2-methyl-6-chlorophenol	0.003
3-methyl-4-chlorophenol	0.02–3.0
o-phenylphenol	1.0
Pentachlorophenol	0.03
Phenol	1.0–10.0
Phenols in polluted rivers	0.15–0.02
2,3,4,6-tetrachlorophenol	0.001
2,3,5-trichlorophenol	0.001
2,4,6-trichlorophenol	0.002
2,4-dimethylphenol	0.4
Dimethylamine	7.0
Diphenyloxide	0.05
B,B-dichlorodiethyl ether	0.09–1
o-dichlorobenzene	<0.25
Ethylbenzene	0.25
Ethanethiol	0.2
Ethylacrylate	0.6
Formaldehyde	95.0
Gasoline	0.005
Guaicol	0.08
Kerosene	0.1
Kerosene plus kaolin	1.0
Hexachlorocyclopentadiene	0.001
Isopropylbenzene	<0.25
Naphtha	0.1
Naphthalene	1.0
Naphthol	0.5
2-Naphthol	0.3
Nitrobenzene	0.03
a-methylstyrene	0.25
Oil, emulsifiable	>15.0
Pyridine	5–28
Pyrocatechol	0.8–5
Pyrogallol	20–30
Quinoline	0.5–1
p-quinone	0.5
Styrene	0.25

(Continued)

Table 8G.116 (Continued)

Parameter	Estimated Threshold Level in Water (mg/L)
Toluene	0.25
Outboard motor fuel as exhaust	7.2
Zinc	5.0

Source: From Australian and New Zealand Environment and Conservation Council and Agriculture and Resource Management Council of Australia and New Zealand, 2000, *National Water Quality Management Strategy*, Paper No. 4, *Australian and New Zealand Guidelines for Fresh and Marine Water Quality*, vol. I, The Guidelines, October 2000, www.deh.gov.au.

Original Source: Reproduced from ANZECC (1992), an adaptation of NAS/NAE (1973).

Table 8G.117 Australian Recommended Sediment Quality Guidelines

Contaminant	ISQG-Low (Trigger Value)	ISQG-High
Metals (mg/kg dry wt)		
Antimony	2	25
Cadmium	1.5	10
Chromium	80	370
Copper	65	270
Lead	50	220
Mercury	0.15	1
Nickel	21	52
Silver	1	3.7
Zinc	200	410
Metalloids (mg/kg dry wt)		
Arsenic	20	70
Organometallics		
Tributyltin (μg Sn/kg dry wt)	5	70
Organics (μg/kg dry wt)[a]		
Acenaphthene	16	500
Acenaphthalene	44	640
Anthracene	85	1,100
Fluorene	19	540
Naphthalene	160	2,100
Phenanthrene	240	1,500
Low Molecular Weight PAHs[b]	552	3,160
Benzo(a)anthracene	261	1,600
Benzo(a)pyrene	430	1,600
Dibenzo(a, h)anthracene	63	260
Chrysene	384	2,800
Fluoranthene	600	5,100
Pyrene	665	2,600
High Molecular Weight PAHs[b]	1,700	9,600
Total PAHs	4,000	45,000
Total DDT	1.6	46
p.p′-DDE	2.2	27
o,p′-+p,p′-DDD	2	20
Chlordane	0.5	6
Dieldrin	0.02	8

(Continued)

Table 8G.117 (Continued)

Contaminant	ISQG-Low (Trigger Value)	ISQG-High
Endrin	0.02	8
Lindane	0.32	1
Total PCBs	23	—

[a] Normalised to 1% organic carbon;

[b] Low molecular weight PAHs are the sum of concentrations of acenaphthene, acenaphthalene, anthracene, fluorene, 2-methylnaphthalene, naphthalene and phenanthrene; high molecular weight PAHs are the sum of concentrations of benzo(a)anthracene, benzo(a)pyrene, chrysene, dibenzo(a,h)anthrancene, fluoranthene and pyrene.

Source: From Australian and New Zealand Environment and Conservation Council and Agriculture and Resource Management Council of Australia and New Zealand, 2000, *National Water Quality Management Strategy*, Paper No. 4, *Australian and New Zealand Guidelines for Fresh and Marine Water Quality*, vol. I, The Guidelines, October 2000, www.deh.gov.au.

Original Source: Primarily adapted from Long et al. (1995).

Table 8G.118 Maximum Concentrations of Copper Sulfate Safe for Fish

Fish	Safe Copper Sulfate Concentration	
	ppm	lb/mill gal
Trout	0.14	1.2
Carp	0.30	2.5
Suckers	0.30	2.5
Catfish	0.40	3.5
Pickerel	0.40	3.5
Goldfish	0.50	4.0
Perch	0.75	6.0
Sunfish	1.20	10.0
Black Bass	2.10	17.0

Source: From U.S. Public Health Service.

Table 8G.119 Observed Lethal Concentration of Selected Chemicals in Aquatic Environments

Chemical	Organism Tested	Lethal Concentration (mg/L)	Exposure Time (hr)
ABS (100 percent)	Fathead minnow	3.5–4.5	96
ABS (100 percent)	Bluegills	4.2–4.4	96
Household syndets	Fathead minnow	39–61	96
Alkyl sulfate	Fathead minnow	5.1–5.9	96
LAS (C12)	Bluegill fingerlings	3	96
LAS (C14)	Bluegill fingerlings	0.6	96
Acetic acid	Goldfish	423	20
Alum	Goldfish	100	12–96
Ammonia	Goldfish	2–2.5 NH_3	24–96
Ammonia	Perch, roach, rainbow trout	3N	2–20
Sodium arsenite	Minnow	17.8 As	36
Sodium arsenate	Minnow	234 As	15
Barium chloride	Goldfish	5,000	12–17
Barium chloride	Salmon	158	—
Cadmium chloride	Goldfish	0.017	9–18
Cadmium nitrate	Goldfish	0.3 Cd	190
CO_2	Various species	100–200	—
CO	Various species	1.5	1–10
Chloramine	Brown trout fry	0.06	—

(Continued)

Table 8G.119 (Continued)

Chemical	Organism Tested	Lethal Concentration (mg/L)	Exposure Time (hr)
Chlorine	Rainbow trout	0.03–0.08	—
Chromic acid	Goldfish	200	60–84
Copper sulfate	Stickleback	0.03 Cu	160
Copper nitrate	Stickleback	0.02 Cu	192
Cyanogen chloride	Goldfish	1	6–48
H_2S	Goldfish	10	96
HCl	Stickleback	pH 4.8	240
HCl	Goldfish	pH 4.0	4–6
Lead nitrate	Minnow, stickleback, brown trout	0.33 Pb	—
Mercuric chloride	Stickleback	0.01 Hg	204
Nickel nitrate	Stickleback	1 Ni	156
Nitric acid	Minnow	pH 5.0	—
Oxygen	Rainbow trout	3 cc/L	—
Phenol	Rainbow trout	6	3
Phenol	Perch	9	1
Potassium chromate	Rainbow trout	75	60
Potassium cyanide	Rainbow trout	0.13 Cn	2
Sodium cyanide	Stickleback	1.04 Cn	2
Silver nitrate	Stickleback	70 K	154
Sodium fluoride	Goldfish	1,000	60–102
Sodium sulfide	Brown trout	15	—
Zinc sulfate	Stickleback	0.3 Zn	120
Zinc sulfate	Rainbow trout	0.5	64
Pesticides			
1. Chlorinated hydrocarbons			
Aldrin	Goldfish	0.028	96
DDT	Goldfish	0.027	96
DDT	Rainbow trout	0.5–0.32	24–36
DDT	Salmon	0.08	36
DDT	Brook trout	0.032	36
DDT	Minnow, guppy	0.75 ppb	29
DDT	Stoneflies (species)	0.32–1.8	96
BHC	Goldfish	2.3	96
BHC	Rainbow trout	3	96
Chlordane	Goldfish	0.082	96
Chlordane	Rainbow trout	0.5	24
Dieldrin	Goldfish	0.037	96
Dieldrin	Bluegill	0.008	96
Dieldrin	Rainbow trout	0.05	24
Endrin	Goldfish	0.0019	96
Endrin	Carp	0.14	48
Endrin	Fathead minnow	0.001	96
Endrin	Various species	0.03–0.05 ppb	—
Endrin	Stoneflies (species)	0.32–2.4 ppb	96
Heptachlor	Rainbow trout	0.25	24
Heptachlor	Goldfish	0.23	96
Heptachlor	Bluegill	0.019	96
Heptachlor	Redear sunfish	0.017	96
Methoxychlor	Rainbow trout	0.05	24
Methoxychlor	Goldfish	0.056	96
Toxaphene	Rainbow trout	0.05	24
Toxaphene	Goldfish	0.0056	96
Toxaphene	Carp	0.1	—
Toxaphene	Goldfish	0.2	24
Toxaphene	Goldfish	0.04	170
Toxaphene	Minnows	0.2	24
2. Organic phosphates			
Chlorothion	Fathead minnow	3.2	96
Dipterex	Fathead minnow	180	96
EPN	Fathead minnow	0.2	96

(Continued)

Table 8G.119 (Continued)

Chemical	Organism Tested	Lethal Concentration (mg/L)	Exposure Time (hr)
Guthion	Fathead minnow	0.093	96
Guthion	Bluegill	0.005	96
Malathion	Fathead minnow	12.5	96
Parathion	Fathead minnow	1.4–2.7	96
TEPP	Fathead minnow	1.7	96
3. Herbicides			
Weedex	Young roach and trench	40–80	1 m
Weeda Zol		15–30	1 m
Weeda Zol T.L.		20–40	1 m
Simazine (no plants present)	Minnow	0.5	<3 d
Atrazine (A361) (plants present)	Minnow	5.0	24
Atrazine in Gesaprime	Minnow	3.75	24
4. Bactericides			
Algibiol	Minnow	20	24
Soricide tetraminol	Minnow	8	48

Source: From McGauhey, *Engineering Management of Water Quality*, McGraw-Hill, copyright 1968.

SECTION 8H RECREATIONAL WATER QUALITY

Table 8H.120 Water Quality Characteristics Relevant to Recreational Use

Characteristics	Primary Contact (e.g. swimming)	Secondary Contact (e.g. boating)	Visual Use (no contact)
Microbiological guidelines	X	X	
Nulsance organisms (e.g., algae)	X	X	X
Physical and chemical guidelines:			
Aesthetics	X	X	X
Clarity	X	X	X
Color	X	X	X
pH	X		
Temperature	X		
Toxic chemicals	X	X	
Oil, debris	X	X	X

Source: From Australian and New Zealand Environment and Conservation Council and Agriculture and Resource Management Council of Australia and New Zealand, 2000, *National Water Quality Management Strategy*, Paper No. 4, *Australian and New Zealand Guidelines for Fresh and Marine Water Quality*, vol. 1, The Guidelines, October 2000, www.deh.gov.au.

Table 8H.121 Guides for Evaluating Recreational Waters

Determination	Water Contact		Boating and Aesthetic	
	Noticeable Threshold	Limiting Threshold	Noticeable Threshold	Limiting Threshold
Coliforms, MPN per 100 mL	1,000[a]	—[b]		
Visible solids of sewage origin	None	None	None	None
ABS (detergent) (mg/L)	1[a]	2	1[a]	5
Suspended solids (mg/L)	20[a]	100	20[a]	100
Flotable oil and grease, mg/L	0	5	0	10
Emulsified oil and grease (mg/L)	10[a]	20	20[a]	50
Turbidity, silica scale units	10[a]	50	20[a]	—[c]
Color, standard cobalt scale units	15[a]	100	15[a]	100
Threshold odor number	32[a]	256	32[a]	256
Range of pH	6.5–9.0	6.0–10.0	6.5–9.0	6.0–10.0
Temperature, maximum °C	30	50	30	50
Transparency, Secchi disk (ft)	—	—	20[a]	—[c]

Note: Noticeable threshold represents the level at which people begin to notice and perhaps to complain. Limiting threshold in the level at which recreational use of water is prohibited or seriously impaired.

[a] Value not to be exceeded in more than 20 percent of 20 consecutive samples, nor in any 3 consecutive samples.
[b] No limiting concentration can be specified in the basis of epidemiological evidence, provided no fecal pollution is evident.
[c] No concentration likely to be found in surface waters would impede use.

Source: From California State Water Quality Control Board, 1963.

Table 8H.122 United States Criteria for Bathing (Full Body Contact) Recreational Waters

Freshwater

Based on a statistically sufficient number of samples (generally not less than 5 samples equally spaced over a 30-day period), the geometric mean of the indicated bacterial densities should not exceed one or the other of the following[a]:

E. coli	126 per 100 mL; or
Enterococci	33 per 100 mL;

No sample should exceed a one sided confidence limit (C.L.) calculated using the following as guidance:

Designated bathing beach	75% C.L.
Moderate use for bathing	82% C.L.
Light use for bathing	90% C.L.
Infrequent use for bathing	95% C.L.

based on a site-specific log standard deviation, or if site data are insufficient to establish a log standard deviation, then using 0.4 as the log standard deviation for both indicators

Marine Water

Based on a statistically sufficient number of samples (generally not less than 5 samples equally spaced over a 30-day period), the geometric mean of the enterococci densities should not exceed 35 per 100 mL;

No sample should exceed a one sided confidence limit using the following as guidance:

Designated bathing beach	75% C.L.
Moderate use for bathing	82% C.L.
Light use for bathing	90% C.L.
Infrequent use for bathing	95% C.L.

based on a site-specific log standard deviation, or if site data are insufficient to establish a log standard deviation, then using 0.7 as the log standard deviation

[a] Only one indicator should be used. The Regulatory agency should select the appropriate indicator for its conditions.

Source: From United States Environmental Protection Agency, 2003, *Bacterial Water Quality Standards for Recreational Waters (Freshwater and Marine Waters) Status Report*, EPA-823-R-03-008, www.epa.gov.

Table 8H.123 WHO Guidelines for Safe Practice in Managing Recreational Waters

Guidance Level or Situation	How Guidance Level Derived	Health Risks	Typical Actions[a]
Relatively low probability of adverse health effects			
20,000 cyanobacterial cells/mL or 10 μg chlorophyll-a/L with dominance of cyanobacteria	From human bathing epidemiological study	Short-term adverse health outcomes, e.g., skin irritations, gastrointestinal illness	Post on site risk advisory signs Inform relevant authorities
Moderate probability of adverse health effects			
100,000 cyanobacterial cells/ml or 50 μg chlorophyll-a/L with dominance of cyanobacteria	From provisional drinking-water guideline value for microcystin-LR[b] and data concerning other cyanotoxins	Potential for long-term illness with some cyanobactieral species Short-term adverse health outcomes, e.g., skin irritations, gastrointestinal illness	Watch for scums or conditions conducive to scums Discourage swimming and further investigate hazard Post on-site risk advisory signs Inform relevant authorities
High probability of adverse health effects			
Cyanobacterial scum formation in areas where whole-body contact and/or risk of ingestion/aspiration occur	Inference from oral animal lethal poisonings Actual human illness case histories	Potential for acute poisoning Potential for long-term illness with some cyanobacterial species Short-term adverse health outcomes, e.g., skin irritations, gastrointestinal illness	Immediate action to control contact with scums; possible prohibition of swimming and other water contact activities Public health follow-up investigation Inform public and relevant authorities

[a] Actual action taken should be determined in light of extent of use and public health assessment of hazard.
[b] The provisional drinking-water guideline value for microcystin-LR is 1 μg/L (WHO, 1998).

Source: From WHO, 2003, *Guidelines for Safe Recreational Water Environments, Volume 1: Coastal and Fresh Waters*, Copyright © World Health Organization 2003. www.who.int.

Original Source: Derived from Chorus & Bartram, 1999.

Table 8H.124 WHO Guideline Values for Microbial Quality of Recreational Waters

95th Percentile Value of Intestinal Enterococci/100 mL (rounded values)	Basis of Derivation	Estimated Risk Per Exposure
≤40 **A**	This range is below the NOAEL in most epidemiological studies	**<1% GI illness risk** **<0.3% AFRI risk** This upper 95th percentile value of 40/100 mL relates to an average probability of less than one case of gastroenteritis in every 100 exposures. The AFRI burden would be negligible
41–200 **B**	The 200/100 ml value is above the threshold of illness transmission reported in most epidemiological studies that have attempted to define a NOAEL or LOAEL for GI illness and AFRI	**1–5% GI illness risk** **0.3–1.9% AFRI risk** The upper 95th percentile value of 200/100 mL relates to an average probability of one case of gastroenteritis in 20 exposures. The AFRI illness rate at this upper value would be less than 19 per 1,000 exposures, or less than approximately 1 in 50 exposures
201–500 **C**	This range represents a substantial elevation in the probability of all adverse health outcomes for which dose-response data are available	**5–10% GI illness risk** **1.9–3.9% AFRI risk** This range of 95th percentiles represents a probability of 1 in 10 to 1 in 20 of gastroenteritis for a single exposure. Exposures in this category also suggest a risk of AFRI in the range of 19–39 per 1,000 exposures, or a range of approximately 1 in 50 to 1 in 25 exposures
>**500** **D**	Above this level, there may be a significant risk of high levels of minor illness transmission	>**10% GI illness risk** >**3.9% AFRI risk** There is a greater than 10% chance of gastroenteritis per single exposure. The AFRI illness rate of the 95th percentile point of >500/100 mL would be greater than 39 per 1,000 exposures, or greater than approximately 1 in 25 exposures

Note: 1. Abbreviations used: A–D are the corresponding microbial water quality assessment categories used as part of the classification procedure; AFRI = acute febrile respiratory illness; GI = gastrointestinal; LOAEL = lowest-observed-adverse-effect level; NOAEL = no observed-adverse-effect level.
2. The "exposure" in the key studies was a minimum of 10 min of swimming involving three head immersions. It is envisaged that this is equivalent to many immersion activities of similar duration, but it may underestimate risk for longer periods of water contact or for activities involving higher risks of water ingestion.
3. The "estimated risk" refers to the excess risk of illness (relative to a group of non-bathers) among a group of bathers who have been exposed to faecally contaminated recreational water under conditions similar to those in the key studies.
4. The functional form used in the dose-response curve assumes no further illness outside the range of the data (i.e. at concentrations above 158 intestinal enterococci/100 mL). Thus, the estimates of illness rate reported above this value are likely to be underestimates of the actual disease incidence attributable to recreational water exposure.
5. This estimated risks were derived from sewage-impacted marine waters. Different sources of pollution and more or less aggressive environments may modify the risks.
6. This table is derived from risk to healthy adult bathers to marine waters in temperate north European waters.

Source: From WHO, 2003, *Guidelines for Safe Recreational Waters, Volume 1: Coastal and Fresh Waters.* Copyright © World Health Organization 2003, www.who.int.

Table 8H.125 Summary of Guidelines for Canadian Recreational Water Quality

Parameter	Guideline
Microbiological	
Escherichia coli (fecal coliforms)	The geometric mean of at least five samples, taken during a period not to exceed 30 days, should not exceed 2,000 *E. coli*/L. Resampling should be performed when any sample exceeds 4,000 *E. coli*/L
Enterococci	The geometric mean of at least five samples, taken during a period not to exceed 30 days, should not exceed 350 enterococci/L. Resampling should be performed when any sample exceeds 700 enterococci/L
Coliphages	Limits on coliphages can not be specified at this time. See Health and Welfare Canada (1992) for additional information
Waterborne pathogens	The pathogens most frequently responsible for diseases associated with recreational water use are described in Health and Welfare Canada (1992), i.e. *Pseudomonas aeruginosa*, *Staphylococcus aureus*, *Salmonella*, *Shigella*, *Aeromonas*, *Campylobacter jejuni*, *Legionella*, human enteric viruses, *Giardia lamblia*, and *Cryptosporidium*
Cyanobacteria (blue–green algae)	Limits have not been specified. Health Canada is in the process of developing a numerical guideline for microcystin, a cyanobacterial toxin. Water with blue-green surface scum should be avoided because of reduced clarity and possible presence of toxins
Chemical characteristics	Limits for chemicals have not been specified because of lack of data. Decisions for use should be based on an environmental health assessment and the aesthetic quality. Dermal exposures to environmental contaminants has recently been reviewed by Moody and Chu (1995)
Aquatic plants	Bathers should avoid areas with rooted or floating plants; very dense growths could affect other activities such as boating and fishing
Aesthetics	All water should be free from • materials that will settle to form objectionable deposits; • floating debris, oil, scum, and other matter; • substances producing objectionable color, odor, taste, or turbidity; and • substances and conditions or combinations thereof in concentrations that produce undesirable aquatic life
Nuisance organisms	Bathing areas should be as free as possible from nuisance organisms that • endanger the health and physical comfort of users or • render the area unusable Common examples include biting and nonbiting insects and poisonous organisms, for example jellyfish
pH	The pH of the waters used for total body contact recreation should be in the pH range of 6.5 to 8.5. If the water has a very low buffering capacity, pH values from 5.0 to 9.0 should be acceptable
Temperature	The thermal characteristics of waters used for bathing and swimming should not cause an appreciable increase or decrease in the deep body temperature of bathers and swimmers
Turbidity	50 Nephelometric Turbidity Units (NTU)
Clarity—Light penetration	Water should be sufficiently clear that a Secchi disc is visible at a minimum depth of 1.2 m
Color	Color should not be so intense as to impede visibility in areas used for swimming. A maximum limit of 100 platinum-cobalt (Pt-Co) units was proposed by Environment Canada (1972)
Oil and grease	Oil or petrochemicals should not be present in concentrations that • Can be detected as a visible film, sheen, or discoloration on the surface • Can be detected by odor • Can form deposits on shorelines and bottom sediments that are detectable by sight or odor (International Joint Commission 1977)

Note: See guidance document for a complete narrative of recreational water quality guidelines.

Source: From Health and Welfare Canada, 1992, *Guidelines for Canadian Recreational Water Quality*, www.ccme.ca.

Table 8H.126 Summary of Australian Water Quality Guidelines of Recreational Waters

Parameter	Guideline
Microbiological	
Primary contact[a]	The median bacterial content in fresh and marine waters taken over the bathing season should not exceed 150 fecal coliform organisms/100 mL or 35 enterococci organisms/100 mL. Pathogenic free-living protozoans should be absent from bodies of fresh water[b]
Secondary contact[a]	The median value in fresh and marine waters should not exceed 1,000 fecal coliform organisms/100 mL or 230 enterococci organisms/100 mL[b]
Nuisance organisms	Macrophytes, phytoplankton scums, filamentous algal mats, sewage fungus, leeches, etc., should not be present in excessive amounts[a]. Direct contact activities should be discouraged if algal levels of 15,000–20,000 cells/mL are present, depending on the algal species. Large numbers of midges and aquatic worms should also be avoided
Physical and chemical	
Visual clarity & color	To protect the aesthetic quality of a waterbody: • the natural visual clarity should not be reduced by more than 20% • the natural hue of the water should not be changed by more than 10 points on the Munsell Scale • the natural reflectance of the water should not be changed by more than 50% To protect the visual clarity of waters used for swimming, the horizontal sighting of a 200 mm diameter black disc should exceed 1.6 m
pH	The pH of the water should be within the range 5.0–9.0, assuming that the buffering capacity of the water is low near the extremes of the pH limits
Temperature	For prolonged exposure, temperatures should be in the range 15–35°C
Toxic chemicals	Water containing chemicals that are either toxic or irritating to the skin or mucous membranes are unsuitable of recreation. Toxic substances should not exceed values in Table 8.127 and 8.128
Surface films	Oil and petrochemicals should not be noticeable as a visible film on the water nor should they be detectable by odor

[a] Refer to Section 3.3 of this revised Australian Guidelines relating to nutrient concentrations necessary to limit excessive aquatic plant growth.

[b] Sampling frequency and maximum values are given in Section 5.2.3.1 of the guidance document.

Source: From Australian and New Zealand Environment and Conservation Council and Agriculture and Resource Management Council of Australia and New Zealand, 2000, *National Water Quality Management Strategy*, Paper No. 4, *Australian and New Zealand Guidelines for Fresh and Marine Water Quality*, vol. I, The Guidelines, October 2000, www.deh.gov.au.

Table 8H.127 Summary of Australian Water-Quality Guidelines for Recreational Purposes: General Chemicals

Parameter	Guideline Values (μg/L, Unless Otherwise Stated)
Inorganic:	
Arsenic	50
Asbestos	NR
Barium	1,000
Boron	1,000
Cadmium	5
Chromium	50
Cyanide	100
Lead	50
Mercury	1
Nickel	100
Nitrate-N	10,000
Nitrite-N	1,000
Salenium	10
Silver	50
Organic:	
Benzene	10
Benzo(*a*)pyrene	0.01
Carbon tetrachloride	3
1,1-Dichloroethene	0.3
1,2-Dichloroethane	10
Pentachlorophenol	10
Polychlorinated biphenyls	0.1
Tetrachloroethene	10
2,3,4,6-Tetrachlorophenol	1
Trichloroethene	30
2,4,5-Trichlorophenol	1
2,4,6-Trichlorophenol	10
Radiological:	
Gross alpha activity	0.1 Bq/L
Gross beta activity (excluding activity of ^{40}K)	0.1 Bq/L
Other chemicals:	
Aluminum	200
Ammonia (as N)	10
Chloride	400,000
Copper	1,000
Oxygen	>6.5 (>80% saturation)
Hardness (as $CaCO^3$)	500,000
Iron	300
Manganese	100
Organics (CCE & CAE)	200
pH	6.5–8.5
Phenolics	2
Sodium	300,000
Sulfate	400,000
Sulfide	50
Surfactant (MBAS)	200
Total dissolved solids	1,000,000
Zinc	5,000

Note: NR = No guideline recommended at this time; MBAS Methylene blue active substances.

Source: From Australian and New Zealand Environment and Conservation Council and Agriculture and Resource Management Council of Australia and New Zealand, 2000, *National Water Quality Management Strategy*, Paper No. 4, *Australian and New Zealand Guidelines for Fresh and Marine Water Quality*, vol. I, The Guidelines, October 2000, www.deh.gov.au.

Table 8H.128 Summary of Water-Quality Guidelines for Recreational Purposes: Pesticides

Compound	Maximum Concentration (μg/L)	Compound	Maximum Concentration (μg/L)
Acephate	20	Fenvalerate	40
Alachlor	3	Flamprop-methyl	6
Aldrin	1	Fluometuron	100
Amitrol	1	Formothion	100
Asulam	100	Fosamine (ammonium salt)	3,000
Azinphos-methyl	10	Glyphosate	200
Barban	300	Heptachlor	3
Benomyl	200	Hexaflurate	60
Bentazone	400	Hexazinone	600
Bioresmethrin	60	Lindane	10
Bromazil	600	Maidison	100
Bromophos-ethyl	20	Methidathion	60
Bromoxynil	30	Methomyl	60
Carbaryl	60	Metolachlor	800
Carbendazim	200	Metribuzin	5
Carbofuran	30	Mevinphos	6
Carbophenothion	1	Molinate	1
Chlordane	6	Monocrotophos	2
Chlordimeform	20	Nabam	30
Chlorfenvinphos	10	Nitralin	1,000
Chloroxuron	30	Omethoate	0.4
Chlorpyrifos	2	Oryzalin	60
Clopzralid	1,000	Paraquat	40
Cyhexatin	200	Parathion	30
2,4-D	100	Parathion-methyl	6
DDT	3	Pendimethalin	600
Demeton	30	Perfluidone	20
Diazinon	10	Permethrin	300
Dicamba	300	Pioloram	30
Dichlobenil	20	Piperonyl butoxide	200
3,6-Dichloropicolinic acid	1,000	Pirimicarb	100
Dichlorvos	20	Pirimiphos-ethyl	1
Diclofop-methyl	3	Pirimiphos-methyl	60
Dicofol	100	Profenofos	0.6
Dieldrin	1	Promecarb	60
Difenzoquat	200	Propanil	1,000
Dimethoate	100	Propargite	1,000
Diquat	10	Propoxur	1,000
Disulfoton	6	Pyrazophos	1,000
Diuron	40	Quintozene	6
DPA	500	Sulprofos	20
Endosulfan	40	2,4,5-T	2
Endothal	600	Temephos	30
Endrin	1	Thiobencarb	40
EPTC	60	Thiometon	20
Ethion	6	Thiophanate	100
Ethoprophos	1	Thiram	30
Fenchlorphos	60	Trichlorofon	10
Fenitrothion	20	Triclopyr	20
Fenoprop	20	Trifuralin	500
Fensulfothion	20		

Source: From Australian and New Zealand Environment and Conservation Council and Agriculture and Resource Management Council of Australia and New Zealand, 2000, *National Water Quality Management Strategy*, Paper No. 4, *Australian and New Zealand Guidelines for Fresh and Marine Water Quality*, vol. 1, The Guidelines, October 2000, www.deh.gov.au.

Original Source: From NHMRC & AWRC (1987), NHMRC (1989).

SECTION 8I WATER QUALITY FOR LIVESTOCK AND AQUACULTURE

Table 8I.129 Guides for Evaluating the Quality of Water Used by Livestock

Quality Factor (mg/L)	Threshold Concentration[a]	Limiting Concentration[b]
Total dissolved solids (TDS)	2500	5000
Cadmium	5	
Calcium	500	1000
Magnesium	250	500[c]
Sodium	1000	2000[c]
Arsenic	1	
Bicarbonate	500	500
Chloride	1500	3000
Fluoride	1	6
Nitrate	200	400
Nitrite	None	None
Sulfate	500	1000[c]
Range of pH	6.0–8.5	5.6–9.0

[a] Threshold values represent concentrations at which poultry or sensitive animals might show slight effects from prolonged use of such water. Lower concentrations are of little or no concern.
[b] Limiting concentrations based on interim criteria, South Africa. Animals in lactation or production might show definite adverse reactions.
[c] Total magnesium compounds plus sodium sulfate should not exceed 50 percent of the total dissolved solids.

Source: From California State Water Quality Control Board, 1963.

Table 8I.130 Guide to the Use of Saline Waters for Livestock and Poultry

Total Soluble Salts Content of Waters	Uses
Less than 1,000 mg/L (EC<1.5 mmhhos/cm)	Relatively low level of salinity. Excellent for all classes of livestock and poultry
1,000–3,000 mg/L (EC=1.5–5 mmhos/cm)	Very satisfactory for all classes of livestock and poultry. May cause temporary and mild diarrhea in livestock not accustomed to them; may cause watery droppings in poultry
3,000–5,000 mg/L (EC=5–8 mmhos/cm)	Satisfactory for livestock, but may cause temporary diarrhea or be refused at first by animals not accustomed to them. Poor waters for poultry, often causing watery feces, increased mortality, and decreased growth, especially in turkeys
5,000–7,000 mg/L (EC=8–11 mmhos/cm)	Can be used with reasonable safety for dairy and beef cattle, sheep, swine, and horses. Avoid use for pregnant or lactating animals. Not acceptable for poultry
7,000–10,000 mg/L (EC=11–16 mmhos/cm)	Unfit for poultry and probably for swine. Considerable risk in using for pregnant or lactating cows, horses or sheep, or for the young of these species. In general, use should be avoided although older ruminants, horses, poultry, and swine may subsist on them under certain conditions
Over 10,000 mg/L (EC>11–16 mmhos/cm)	Risks with these highly saline waters are so great that they cannot be recommended for use under any condition

Source: From Soltanpour, P. N. and Raley, W. L., Livestock Drinking Water Quality, Colorado State University Cooperative Extension – Agriculture, no. 4.908, www.ext.colostate.edu.

Original Source: From Environmental Studies Board, Nat. Acad. of Sci., Nat. Acad. of Eng., *Water Quality Criteria,* 1972. Ayers, R.S. and D.W. Westcot. *Water Quality for Agriculture.* Food and Agriculture Organization of the United Nations, Rome, 1976. P.N. Soltanpour, Colorado State University professor, soil and crop sciences; and W.L. Raley, former Colorado State University Cooperative Extension western district director. 10/93. Reviewed 3/99.

Table 8I.131 Canadian Water-Quality Guidelines for the Protection of Agricultural Water Uses–Livestock

Parameter[a]	Livestock Water (μg/L)
Aldicarb	11[c]
Aluminum[d]	5,000
Arsenic[e]	25[f]
Atrazine	5[f,g]
Beryllium[d]	100[f]
Blue-green algae(Cyanobacteria)[d]	
Boron[d]	5,000
Bromacil	1,100[f]
Bromoxynil	11[f]
Cadmium	80
Calcium[d]	1,000,000
Captan	13[f,i]
Carbaryl	1,100
Carbofuran	45
Chlordane	7[l,m]
Hexachlorobenzene	0.52[f,n]
1,2-Dichloroethane	5[f]
1.1,2-Trichloroethene (Trichloroethylene, TCE)	50[f]
Chlorothalonil	170[f]
Chlorpyrifos	24[f]
Chromium	
Trivalent chromium (Cr(III))	50[f,n]
Hexavalent chromium (Cr(VI))	50[f,n]
Cobalt[d]	1,000
Color	Narrative
Copper[d]	500–5,000[p]
Cyanazine	10[f]
Cyanobacteria (Blue-green algae)	Narrative
2,4-D [See 2,4-Dichlorophenoxyacetic acid]	See also Phenoxy herbicides
DDT(2,2-Bis(p-chlorophenyl)-1,1,1-trichloroethane; Dichloro diphenyl trichloroethane)[d]	30[l,m]
Deltamethrin	2.5
Dicamba	122
2,4-Dichlorophenoxyacetic acid (2,4-D) [See also Phenoxy herbicides]	See also Phenoxy herbicides
Diclofop-methyl	9[f]
Dimethoate	3[f]
Dinoseb	150
Endrin[d]	0.2[l,m]
Ethylbenzene[d,e]	2.4
Fluoride[d]	1,000–2,000[n]
Glyphosate[d]	280
Dichloromethane[d] (Methylene chloride)	50[f]
Trichloromethane[d] (Chloroform)	100[g]
Tetrachloromethane[d] (Carbon tetrachloride)	5[f]
Tribromomethane (Bromoform)	100[g]
Dichlorobromomethane	100[g]
Dibromochloromethane	100[g]
Heptachlor (Heptachlor epoxide)[d]	3[l,m]
Lead[d]	100
Lindane (Hexachlorocyclohexane)[d]	4
MCPA (4-chloro-2-methyl phenoxy acetic acid; 2-Methyl-4-chloro phenoxy acetic acid)	25[f]
Mercury[d]	3
Metolachlor	50[f]
Metribuzin	80
Molybdenum[d]	500
Nickel[d]	1,000
Nitrate + Nitrite[d]	100,000
Nitrite[d]	10,000
Tributyltin	250
Tricyclohexyltin	250[f]
Triphenyltin	820[f,i]

(Continued)

Table 8I.131 (Continued)

Parameter[a]	Livestock Water (μg/L)
Phenols[d]	2
Phenoxv herbicides[d]	100
Picloram[d]	190
Selenium[d]	50
Simazine	10[f]
Sulphate[d]	1,000,000
Tebuthiuron	130[f]
Toluene[d,e]	24
Total dissolved solids (salinity)[d]	3,000,000
Toxaphene[d]	5[l,m]
Triallate[d]	230[f]
Trifluralin	45[f]
Uranium[d]	200
Vanadium[d]	100
Zinc[d]	50,000

Note: μg/L is microgram per liter.

[a] Unless otherwise indicated, supporting documents are available from the Guidelines and Standards Division, Environment Canada.

[b] The guidelines dated 1987 have been carried over from *Canadian Water Qualify Guidelines* (CCREM 1987) and no fact sheet was prepared. The guidelines dated 1989 to 1997 were developed and initially published in CCREM 1987 as appendixes on the date indicated. They are published as fact sheets in this document. Other guidelines dated 1997 and those dated 1999 are published for the first time in this document.

[c] Concentration of total aldicarb residue.

[d] No fact sheet created.

[e] The technical document for the guideline is available from the Ontario Ministry of the Environment.

[f] Interim guideline.

[g] During the initial development of this guideline, insufficient data were available to derive a livestock watering guideline value. Therefore, the Canadian drinking water quality guideline (Health and Welfare Canada 1987) was adopted. Since then, this value has been revised by Health Canada (1996). This revised drinking water quality guideline in now adopted as the guideline for livestock water.

[h] Guideline value slightly modified from CCREM 1987 + Appendixes due to re-evaluation of the significant figures.

[i] Guideline is crop-specific (see fact sheet).

[j] This guideline (originally published in *Canadian Water Quality Guidelines* [CCREM 1987]) is no longer recommended and the value is withdrawn. A water quality guideline is not recommended. Environmental exposure is predominantly via sediment, soil, and/or tissue, therefore, the reader is referred to the respective guidelines for these media.

[k] This substance meets the criteria for Track substances under (the national CCME Policy for the Management of Toxic Substances (PMTS) (i.e., persistent, bioaccumulative, primarily result of human activity, and CEPA-toxic or equivalent) and should be subject to virtual elimination strategies. Guidelines can serve as action levels or interim management objectives towards virtual elimination.

[l] Substance has been re-evaluated since CCREM 1987 + Appendixes. Either a new guideline has been derived or insufficient data existed to derive a new guideline.

[m] Copper guideline = 500 $\mu g L^{-1}$ for sheep, 1000 $\mu g\ L^{-1}$ for cattle, 5000 $\mu g\ L^{-1}$ for swine and poultry.

[n] Fluoride guideline = 1000 $\mu g\ L^{-1}$ if feed contains fluoride.

[o] Molybdenum guideline = 50 $\mu g\ L^{-1}$ for short-term use on acidic soils.

Source: From Canadian Council of Ministers of the Environment, 2005, Canadian water quality guidelines for the protection of agricultural water uses: summary table, Updated October 2005. In: *Canadian Environmental Quality Guidelines*, 1999, Canadian Council of Ministers of the Environment, Winnipeg, ec.gc.ca/CEQC-RCQE.

Table 8I.132 Tolerances of Livestock to Total Dissolved Solids (Salinity) in Drinking Water

	Total Dissolved Solids (mg/L)		
Livestock	**No Adverse Effects on Animals Expected**	**Animals May Have Initial Reluctance to Drink or There May Be Some Scouring, but Stock Should Adapt without Loss of Production**	**Loss of Production and a Decline in Animal Condition and Health Would Be Expected. Stock may Tolerate These Levels for Short Periods if Introduced Gradually**
Beef cattle	0–4,000	4,000–5,000	5,000–10,000
Dairy cattle	0–2,500	2,500–4,000	4,000–7,000
Sheep	0–5,000	5,000–10,000	10,000–13,000[a]
Horses	0–4,000	4,000–6,000	6,000–7,000
Pigs	0–4,000	4,000–6,000	6,000–8,000
Poultry	0–2,000	2,000–3,000	3,000–4,000

[a] Sheep on lush green feed may tolerate up to 13,000 mg/L TDS without loss of condition or production.

Source: From Australian and New Zealand Environment and Conservation Council and Agriculture and Resource Management Council of Australia and New Zealand, 2000, *National Water Quality Management Strategy*, Paper No. 4, *Australian and New Zealand Guidelines for Fresh and Marine Water Quality*, vol. 1. The Guidelines, October 2000, www.deh.gov.au.

Original Source: From ANZECC (1992), adapted to incorporate more recent information.

Table 8I.133 Recommended Water Quality Trigger Values (Low Risk) for Heavy Metals and Metalloids in Livestock Drinking Water

Metal or Metalloid	Trigger Value (low risk)[a] (mg/L)
Aluminum	5
Arsenic	0.5
	up to 5[c]
Beryllium	ND[b]
Boron	5
Cadmium	0.01
Chromium	1
Cobalt	1
Copper	0.4 (sheep)
	1 (cattle)
	5 (pigs)
	5 (poultry)
Fluoride	2
Iron	Not sufficiently toxic
Lead	0.1
Manganese	Not sufficiently toxic
Mercury	0.002
Molybdenum	0.15
Nickel	1
Selenium	0.02
Uranium	0.2
Vanadium	ND
Zinc	20

[a] Higher concentrations may be tolerated in some situations (details provided in Volume 3, Section 9.3.5).
[b] ND = not determined, insufficient background data to calculate.
[c] May be tolerated if not provided as a food additive and natural levels in the diet are low.

Source: From Australian and New Zealand Environment and Conservation Council and Agriculture and Resource Management Council of Australia and New Zealand, 2000, *National Water Quality Management Strategy*, Paper No. 4, *Australian and New Zealand Guidelines for Fresh and Marine Water Quality*, vol. I. The Guidelines, October 2000, www.deh.gov.au.

Table 8I.134 Australian Trigger Values for Radioactive Contaminants in Livestock Drinking Water

Radionuclide	Trigger Value (Bq/L)
Radium 226	5
Radium 228	2
Uranium 238	0.2
Gross alpha	0.5
Gross beta (excluding K-40)	0.5

Source: From Australian and New Zealand Environment and Conservation Council and Agriculture and Resource Management Council of Australia and New Zealand, 2000, *National Water Quality Management Strategy*, Paper No. 4, *Australian and New Zealand Guidelines for Fresh and Marine Water Quality*, vol. 1. The Guidelines, October 2000, www. deh.gov.au.

Table 8I.135 Australian Physico-Chemical Stressor Guidelines for the Protection of Aquaculture Species

	Recommended Guideline (mg/L)	
Measured Parameter	**Freshwater Production**	**Saltwater Production**
Alkalinity	≥20[e]	>20[c]
Biochemical oxygen demand (BODs)	<15[a]	ND
Chemical oxygen demand (COD)	<40[a]	ND
Carbon dioxide	<10	<15
Color and appearance of water	30–40[b] (Pt-Co units)	30–40[b] (Pt-Co units)
Dissolved oxygen	>5[c]	>5[c]
Gas supersaturation	<100%[f]	<100%[f]
Hardness ($CaCO_3$)	20–100[e]	NC[f]
pH	5.0–9.0	6.0–9.0
Salinity (total dissolved solids)	<3,000[f]	33,000–37,000[f] (3,000–35,000 Brackish)[f]
Suspended solids	<40	<10 (<75 Brackish)
Temperature	<2.0°C change over 1 hr[d]	<2.0°C change over 1 hr[d]

Note: Unless noted, guidelines are based on professional judgements.

[a] Schlotfeldt & Aldeman (1995).
[b] O'Connor pers. comm.
[c] Meade (1989).
[d] ANZECC (1992).
[e] DWAF (1996).
[f] Lawson (1995).

Source: From Australian and New Zealand Environment and Conservation Council and Agriculture and Resource Management Council of Australia and New Zealand, 2000, *National Water Quality Management Strategy*, Paper No. 4, *Australian and New Zealand Guidelines for Fresh and Marine Water Quality*, vol. 1. The Guidelines, October 2000, www.deh.gov.au.

Table 8I.136 Australian Toxicant Guidelines for the Protection of Acquaculture Species

	Guideline (μg/L)	
Measured Parameter	**Freshwater Production**	**Saltwater Production**
Inorganic Toxicants (Heavy metals and others)		
Aluminum	<30 (pH>6.5)[a] <10 (pH<6.5)	<10[a]
Ammonia (unionized)	<20 (pH>8.0) coldwater[b] <30 warmwater[b]	<100
Arsenic	<50[a,b]	<30[a,b]
Cadmium (varies with hardness)	<0.2–1.8[b]	<0.5–5[a]
Chlorine	<3[a]	<3[a]
Chromium	<20[b]	<20
Copper (varies with hardness)	<5[b]	<5[c]
Cyanide	<5[a]	<5[a]

(Continued)

Table 8I.136 (Continued)

	Guideline (μg/L)	
Measured Parameter	**Freshwater Production**	**Saltwater Production**
Fluorides	<20[d]	ND
Hydrogen sulfide	<1[b]	<2
Iron	<10[a]	<10[a]
Lead (varies with hardness)	<1–7[d]	<1–7[d]
Magnesium	<15,000[a]	ND
Manganese	<10[a,e]	<10[a,e]
Mercury	<1	<1
Nickel	<100[a]	<100[a]
Nitrate (NO_3^-)	<50,000[f]	<100,000[c,g]
Nitrite (NO_2^-)	<100[a,g]	<100[a,g]
Phosphates	<100[b]	<50
Selenium	<10[a]	<10[a]
Silver	<3[a]	<3[a]
Tributyltin (TBT)	<0.026[a]	<0.01[a]
Total available nitrogen (TAN)	<1,000[a]	<1,000[a]
Vanadium	<100[a]	<100[a]
Zinc	<5[a]	<5[a]
Organic Toxicants (nonpesticides)		
Detergents and surfactants	<0.1[h]	ND
Methane	<65,000[i,j]	<65,000[i,j]
Oils and greases (including petrochemicals)	<300[f]	ND
Phenols and chlorinated phenols	<0.6–1.7[f]	ND
Polychlorinated biphenyls (PCBs)	<2[a]	<2[a]
Pesticides		
2,4-dichlorophenol	<4.0[b]	ND
Aldrin	<0.01[b,c,h]	ND
Azinphos-methyl	<0.01[b]	ND
Chlordane	<0.01[k]	0.004[k]
Chlorpyrifos	<0.001[b]	ND
DDT (including DDD & DDE)	<0.0015[b]	ND
Demton	<0.01[k]	ND
Dieldrin	<0.005[b]	ND
Endosulfan	<0.003[b,k]	0.001[k]
Endrin	<0.002[b]	ND
Gunthion (see also Azinphos-methyl)	<0.01[k]	ND
Hexachlorobenzole	<0.00001[f]	ND
Heptachlor	<0.005[b]	ND
Lindane	<0.01[k]	0.004[k]
Malathion	<0.1[e,k]	ND
Methoxychlor	<0.03[k]	ND
Mirex	<0.001[b,k]	ND
Paraquat	ND	<0.01
Parathion	0.04[k]	ND
Toxaphene	<0.002[b]	ND

Note: ND, Not determined—insufficient information, NC, Not of concern; Unless noted, guidelines are based on professional judgements.

[a] Meade (1989).
[b] DWAF (1996).
[c] Piliay (1990).
[d] Tebbutt (1972).
[e] Zweig et al. (1999).
[f] Schlotfeldt & Alderman (1995).
[g] Coche (1981).
[h] Langdon (1988).
[i] McKee & Wolf (1963).
[j] Boyd (1990).
[k] Lannan et al. (1986).

Source: From Australian and New Zealand Environment and Conservation Council and Agriculture and Resource Management Council of Australia and New Zealand, 2000, *National Water Quality Management Strategy*, Paper No. 4, *Australian and New Zealand Guidelines for Fresh and Marine Water Quality*, vol. 1. The Guidelines, October 2000, www.deh.gov.au.

SECTION 8J WATER TREATMENT PROCESSES

Table 8J.137 Common Water Quality Problems, Effects, and Treatment

Probable Cause	General Effect	Probable Remedy
Hardness (Calcium and Magnesium)	Scales in pipes and water heaters; causes "soap curd" on fixtures, tiles, dishes and laundry; low sudsing characteristics	Removal by ion exchange softener
Iron, Manganese, Copper, Zinc	Causes discolored water; red, brown, orange or black stains on fixtures, appliances and laundry; dark scale in pipes and water heaters	Low level (2 ppm) removal by ion exchange softener when hardness is also present; best removed by oxidizing iron filter, aeration and/or chlorination followed by filtration in some cases
Iron, Manganese, Sulfur Bacteria	Same general effects as above plus slimey deposits that form in pumps, pipes, softeners and toilet tanks	Low level removal possible by oxidizing iron filter; best removed by disinfection followed by filtration
Hydrogen Sulfide Gas	Foul rotten-egg odor; corrosion to plumbing; tarnishes silver and stains fixtures and laundry; ruins the taste of foods and beverages	Best removed by aeration, scrubbing and filtration; also removed by oxidizing filters or chlorination followed by filtration
Turbidity	Suspended matter in water; examples include mud, clay, silt and sand; can ruin seals and moving parts in appliances	Removal by backwashing sediment filters; extra fine treatment utilizing sediment cartridge elements
Acid Water (low pH)	Corrosive water attacks piping and other metals; red and/or green staining of fixtures and laundry	Best corrected by neutralizing filters or soda ash feeding
Taste, Odor, Color (organic matter)	Makes water unpalatable; can cause staining	Depending on the nature of contaminant, aeration followed by filtration; carbon filtration; oxidation followed by filtration
Tannins, Humic Acid	Can impart an "iced-tea" color to water; causes light staining; can affect the taste of foods and beverages	Removal by special ion exchange or oxidizing agents and filtration
Coliform Bacteria, Cryptosporidium, Giardia Lamblia, Viruses	Can cause serious disease and intestinal disorders	Disinfection and filtration is most widely practiced
Organic Halides (e.g., herbicides and pesticides)	Can cause serious disease and/or poisoning	Most are readily removed by absorption with carbon filters; some can also be removed by hydrolysis and oxidation
Nitrates, Chlorides and Sulfates	Can cause health-related problems if quantities are high	Removal by special ion exchange, deionization process or reverse osmosis
Sodium Salts	Imparts an alkaline or soda taste to water	Removal by deionization process or reverse osmosis, distillation can be used
Arsenic[a]	Can cause health-related problems; Known carcinogen	Removal using conventional coagulation with iron or aluminum salts followed by filtration; may also be removed by adsorption onto activated alumina or by ion exchange
Radionuclides[b]	Can cause health-related problems	Best Available Technology (BAT) identified for removal include ion exchange, reverse osmosis, lime softening, and conventional coagulation followed by filtration
Synthetic Organic Compounds[a]	Can cause health-related problems	Removal using aeration, air stripping for volatile organic compounds; removal using granular activated carbon adsorption or chemical oxidation for volatile or non-volatile compounds

[a] J.M. Montgomery Consulting Engineers, Inc., 1985, *Water Treatment Principles and Design*, John Wiley & Sons, Inc.

[b] USEPA, 2005, A Regulator's Guide to the Management of Radioactive Residuals from Drinking Water Treatment Technologies, EPA 816-R-05-004, July 2005

Source: From Chandler, J., A comprehensive look at water treatment, *Water Well Journal*, May 1988. Copyright Water Well Publ. Co. Reprinted with permission. Amended.

Table 8J.138 Summary of Conventional Processes and Systems for Water Quality Control

Type or Process	Common Application	Approximate Limit of Quality Input	Principal Change in Quality Factors (Approximate)
		Gravity Separation	
Plain sedimentation	Reduction in suspended solids in raw water to be pumped	No theoretical limit: 3,000–5,000 mg/L typical maximum in flood waters	Removes larger and heavier suspended solids
	Primary sewage treatment	Unspecified	50% reduction in suspended solids 35–40% reduction in BOD 50% reduction in turbidity
	Secondary sewage treatment	Unspecified	Unreported
	Concentrating return activated sludge (secondary treatment)	Unspecified	Thickens sludge to 20–25% original volume
	Concentrating or reducing suspended solids in industrial wastes, organic and inorganic	Unspecified	Highly dependent upon nature of waste treated
	Grit removal-raw sewage	Unspecified	Removes heavy suspended solids not transported at velocity of 1 ft/sec
Plain sedimentation plus skimming	Primary sewage treatment	Unspecified	25–40% reduction in BOD 40–70% reduction in suspended solids 25–75% reduction in bacteria 2% reduction in detergents
	Various industrial wastes	Unspecified	Dependent upon nature of waste
Trickling filter plus plain sedimentation	Secondary sewage treatment	0.25-3.0 lb BOD/cu yd/filter	80–95% reduction in BOD 70–92% reduction in suspended solids 90–95% reduction in bacteria 30–35% reduction in ABS 80–90% reduction in LAS
	Organic industrial wastes (e.g., milk process)	Dependent upon waste treated	Dependent upon nature of waste
Activated sludge plus plain sedimentation	Secondary sewage treatment	Unspecified	80–95% reduction in BOD 85–95% reduction in suspended solids 95–98% reduction in bacteria 50% reduction in BAS 90–99% reduction in LAS
Sedimentation after mechanical flocculation	Raw sewage (experimentally)	Unspecified	64% reduction in turbidity 40% reduction in suspended solids 60% reduction in BOD
	Industrial wastes	Unspecified	Variable, depending upon nature of wastes treated
Sedimentation after chemical coagulation	Municipal and industrial water supply water softening	Unspecified	Seldom evaluated separate from filtration
	Raw sewage (not common)	Unspecified	50–85% reduction in BOD 70–90% reduction in suspended solids 40–80% reduction in bacteria
	Industrial wastes	Dependent upon waste	Variable, dependent upon nature of waste

Chemical coagulation plus sedimentation	Municipal water supply	Unspecified	Coalesces and precipitates dispersed clay colloids Reduces turbidity Reduces color
	Phosphate removal from waste waters	Unspecified	Reduces soluble phosphates to trace amounts
	Lime-soda softening of water supplies	Applicable to waters containing Ca and Mg sulfates and bicarbonates; Iron and Mg in natural waters (e.g. maximum from 10 mg/L; minimum, 3 mg/L)	Reduces hardness to approximately 75 mg/L; by excess lime to 30–50 mg/L; by hot process to <10 mg/L as $CaCO_3$ Reduces Fe to 0.1 mg/L ($\pm$) Removes CO_2—requiring restabilization 80–100% reduction in bacteria by excess lime
		Filtration	
Slow sand (gravity)	Tertiary treatment of sewage effluent Water reclamation systems	Relatively low turbidity	90–95% reduction in BOD 85–95% reduction in suspended solids 95–98% reduction in bacteria 90–99% reduction in surfactants
	Municipal water supply	Turbidity 40 mg/L	99% reduction in bacteria 95–100% reduction in turbidity 30% reduction in color Odors and tastes removed 60% reduction in iron
	Industrial wastes	Unspecified	Varies with nature of waste
Rapid sand (gravity)	Municipal and industrial water supply (little used without coagulation)	Low turbidity, e.g., 50 mg/L, maximum coliform MPN 5,000/100 mL	95% reduction in bacteria 90% reduction in turbidity
Rapid sand plus chemical coagulation (gravity)	Municipal and industrial water supply	No limit specified for maximum turbidity Maximum coliform MPN 5,000–20,000/100 mL	90–99% reduction in bacteria 100% ($-$) reduction in turbidity Color reduction to less than 5 mg/L Alkali increased 7.7 mg/L/gr. alum CO_2 increased 6.8 mg/L/gr. alum Slight reduction in iron Odor and taste partially removed
Rapid sand plus chemical coagulation, chlorination, and activated carbon	Municipal and industrial water supply	No limit specified for maximum turbidity Maximum coliform MPN 5,000–20,000/100 mL	Approximately 100% reduction in bacteria 100% reduction in turbidity Color reduced to near zero Iron and Mn reduced Taste and odor removed
Rapid sand (pressure) (precoat with chemical floc)	Small municipal supplies Institutional water supply Swimming pools Industrial supply and process Emergency and military use	Generally unspecified Low turbidity desirable	Similar to rapid sand filter but more variable in performance

(Continued)

Table 8J.138 (Continued)

Type or Process	Common Application	Approximate Limit of Quality Input	Principal Change in Quality Factors (Approximate)
		Filtration (Continued)	
Diatomaceous earth (pressure and vacuum)	Small municipal supplies Institutional water supply Swimming pools Industrial supply and process Emergency and military use	None specified, but operation depends upon nature of water	Capable of good clarification of water; efficiency, however, not well documented 40–90% reduction in suspended solids 50% reduction in color
Contact filters	Manganese removal Iron removal	None specified	Reduces to USPHS Standards 88% reduction in iron
Bag filters	Swimming pools	Unspecified	Strains out hair and coarser suspended solids, reduces bacteria to level controllable by chlorination practice
Microstraining	Primary clarification of water prior to filtration Clarification of sewage effluents Treatment of industrial wastes	Size of particles to be removed greater than screen size Material suitable for microstraining	87–96% reduction in microscopic organisms 60–90% reduction in microscopic particulates 50–60% reduction in suspended solids trickling filter effluent 30–40% reduction on turbidity
Fine screening	Raw sewage	None specified	5–10% reduction in BOD 2–20% reduction on suspended solids 10–20% reduction in bacteria
	Industrial wastes (e.g., cannery, pulp mill, etc.)	None specified	Varies with nature of waste
Carbon filters	Special municipal and industrial water applications	Very low turbidity, other not specified	Adsorbs exotic organic chemicals, including surfactants Removes tastes and odors Adsorbs miscellaneous gases
		Aeration	
Spray or cascade	Municipal and industrial water supply Industrial waste treatment	Unspecified	Releases gases producing taste and odor Reduces CO_2 in groundwaters to normal surface water levels Partial removal of H_2S Partial removal of gases of decomposition Oxidation and removal of soluble iron in groundwaters; 80–97% reduction observed
Pressure aerators	Treatment of sewage and industrial wastes	Limits variable or unspecified	Grit precipitated Grease concentrated at surface Separates various solids by flotation Maintains aerobic conditions in biological systems, e.g., activated sludge, aerated ponds Reduces ABS or LAS 1–2 mg/L Reduces septicity of sewage

Aeration (Continued)			
Oxidation ponds	Treatment of domestic sewage and organic industrial wastes	No toxic substances	75–96% reduction in BOD 90–99% reduction in suspended solids 98–99.9% reduction in bacteria 56–93% reduction in LAS
Deminerlization			
Ion exchange (natural or synthetic zeolite)	Softening of groundwater supplies for municipal or industrial use	Hardness (Ca and Mg sulfates and bicarbonates) of natural waters >850–1,000 mg/L $CaCO_3$ Iron <1.5–2 mg/L Low in silica $CO_2 < 15$ mg/L	Increases sodium content by exchange with removed Ca and Mg
Ion exchange (greensand or styrene base gels)	Iron or Mn removal from groundwater	Iron less than approximately 2.0 mg/L	90–100% removal of iron Mn partially removed
Ion exchange (organic cation exchangers)	Special water conditioning for industry and commerce	Unspecified	Removes all cations (Na, K, Mg, Fe, Cu, Mn)
Ion exchange (anion exchangers)	Special water conditioning for industry and commerce	Unspecified	Removes SO_4, Cl, NO_3, etc.
Ion exchange (fluoride exchangers)	Defluoridation of public water supply	More than 1.5 mg/L F in water supply	Approximately 100% removal possible Normally reduced to <1.5 mg/L
Electrochemical desalting	Reclaiming water from saline sources, public and industrial supplies Demineralizing municipal waste effluents	Applicable to highly saline or brackish waters	Removes anions and cations
Reverse osmosis	Reclamation of water from brackish natural or waste waters (experimental)	Brackish waters, upper limit not specified	Reduces ions depending upon concentration difference across membrane 97–98% reduction in TDS, ABS, and COD
Distillation	Reclamation of water from saline sources Specialty industrial and commercial supplies	No limit	Produces distilled water (may be contamination with NH_3, volatile organics, etc.)
Freezing	Reclamation of water from saline sources Specialty industrial and commercial supplies	No limit	(Experimental)
Disinfection			
Liquid Cl_2 and Cl_2 Compounds such as chlorine dioxide and chloramines[a]	Public water supply Industrial water supply	Turbidity low for waters to be sterilized by Cl_2 Total organic carbon (TOC) concentration <2 mg/L to minimize disinfection byproduct (DBP) formation	Reduces bacterial load on filters Oxidizes organic matter Reduces odor Assists in color removal 100% (—) reduction in bacteria Controls plankton growth in reservoirs Reduces Mn concentration in breakpoint

(Continued)

Table 8J.138 (Continued)

Type or Process	Common Application	Approximate Limit of Quality Input	Principal Change in Quality Factors (Approximate)
	Disinfection (Continued)		
	Municipal and industrial waste-water treatment and management.	Unspecified	Assists in grease removal Controls filter fly nuisance Cleans air stones in aeration systems Removes H_2S Removes NH_3 Controls slime formation in sewers and cooling towers Assists in control of digester foaming Disinfects effluent; 98–99% reduction in bacteria
Ozone[a]	Public water supply	Bromide concentrations less than 0.1 mg/L to minimize potential for bromate formation	Oxidizes organics Microbial inactivation including coliform, giardia, cryptosporidium, and viruses Reduces odor Removes color
Utraviolet Radiation[a]	Small public water supply	Low turbidity	Microbial inactivation including coliform and viruses Some organic oxidation
	Digestion		
Anaerobic digestion	Stabilization of sewage solids Stabilization of organic industrial wastes	pH above 6.8 Acids limited No toxic substances in significant amounts Minimum of grit	Reduces organic sludges to humus and relatively stable chemical compounds Produces offensive supernatant

[a] USEPA, 1999, Alternative Disinfectants and Oxidants Guidance Manual, EPA 815-R-99-014, April 1999.

Source: From McGauhey, Engineeering Management of Water Quality, McGraw-Hill, Copyright 1968. Amended.

Table 8J.139 Potential Water Treatment Efficiencies

	Conventional Processes					Special Processes						
						Activated Carbon Absorption						
Parameter	Aeration	Chemical Oxidation (Chlorination, etc.)	Coagulation Flocculation	Lime Softening	Filtration	PAC	GAC	Air Stripping	Demineralizing (Reverse Osmosis, etc.)	Ion Exchange	Ozone	Comments
Aldrin	P		P			G	VG				VG	
Antimony			X		A		X					
Arsenic		A	L–G	G–VG	A	P			G–VG	VG		Valencies important
Asbestos			G–VG		G							
Barium			P	G–VG	A	P	P		VG	VG		
Boron			X				G–VG		X	G–VG		
Cadmium			L–G	VG	A		P–L					pH important
Chlordane		P	L	L		VG	VG					
Chloride									VG	VG		
Chromium			G	G	A	P	P		X	X		Valencies important
Color			VG		A						VG	
Copper	A		F–G		A							
Cyanide		VG									VG	
2,4-D		P	P		A	VG	X					
DDT		P	L–VG	F		VG	X				P	
Diazinon							X(L)					
Dieldrin			P–L		A		G–VG	L				
Endrin			L			G–VG	VG					
Fluoride									G	G–VG		
Heptachlor						VG	X(VG)					
Heptachlor Epoxide						VG	X					
Iron	A	A		A	VG					VG		
Lead			G–VG	VG	A		X		G–VG	X		
Lindane		P	P		P	G	G–VG					
Manganese		A	L–G	G	A					VG		
Mercury			G	F–G	A	VG	VG					Form important
Methoxychlor			G	G–VG	A	VG	VG					
Methyl Parathion						X	X				X	
Nitrate									F	F–VG		
NTA		P									G–VG	
Odor	A	VG				VG	VG				VG	
Parathion		P–VG	P	P	A	VG	L–VG				G–VG	
Ph	A		A	A								
Phenol		G	P			G–VG	X				G–VG	
Radionuclides												
^{226}Ra			P	G–VG	A				VG	G–VG		
^{90}Sr			P	G–VG	A					G–VG		
^{137}Cs				P	P–F				VG	VG		
^{131}I			P	L					VG			
Selenium			P–G	P–F	A				F–G	X		Valencies important
Silver			F–G	G–VG	A		P		X	X		
Sulphate									G–VG	G–VG		
Sulphide	F–VG	F–VG									F–VG	pH important
2,4,5-TP		P	X(F)			X(G)	X(G–VG)					

(Continued)

Table 8J.139 (Continued)

	Conventional Processes					Special Processes						
						Activated Carbon Absorption						
Parameter	Aeration	Chemical Oxidation (Chlorination, etc.)	Coagulation Flocculation	Lime Softening	Filtration	PAC	GAC	Air Stripping	Demineralizing (Reverse Osmosis, etc.)	Ion Exchange	Ozone	Comments
T. Dissolved Solids									G–VG	G–VG		
Toxaphene		P	P			VG	X(VG)	X				
Trihalomethanes							F–G	F–G				Process generated
Turbidity			G–VG		A							
Uranium			L–G	F–G	A		P			VG		
Zinc			P	F–G	A							

Note: VG, 90–100% removal; G, 70–90% removal; F, 50–70% removal; L, 25–50% removal; P, 0.25% removal; A, auxiliary process; X, possible candidate process (data lacking); PAC, Powdered activated carbon; GAC, Granular activated carbon. Treatment based on available full-scale, pilot or bench studies and should only be used as potential indicators. Treatabilty studies and/or site experience should be assessed for specific applications.

Source: From Canadian Council of Resource and Environment Ministers, March 1987. *Canadian Water Quality Guidelines.* Data provided by McDonald & Associates Consulting Engineers, Regina, Saskatchewan.

Table 8J.140 Treatment Technology Removal Effectiveness Reported for Organic Contaminants (Percent)

Contaminant	Coagulation/ Filtration	GAC	PCA	PAC	Dilfused Aeration	Oxidation[a]	Reverse Osmosis
Acrylamide	5	NA	0–29	13	NA	NA	0–97
Aidicarb	NA	NA	0–29	NA	NA	NA	94–99
Alactior	0–49	70–100	70–100	36–100	NA	70–100	70–100
Benzene	0–29	70–100	70–100	NA	NA	70–100	0–29
Carbofuren	54–79	70–100	0–29	45–75	11–20	70–100	70–100
Carbon tetrachloride	0–29	70–100	70–100	0–25	NA	0–29	70–100
Chlordane	NA	70–100	0–29	NA	NA	NA	NA
Chlorobenzene	0–29	70–100	70–100	NA	NA	30–69	70–100
2,4-D	0–29	70–100	70–100	69–100	NA	W	0–65
1,2-Dichloroethane	0–29	70–100	70–100	NA	42–77	0–29	15–70
1,2-Dichloropropane	0–29	70–100	70–100	NA	12–79	0–29	10–100
Dibroinochloropropane	0–29	70–100	30–69	NA	NA	0–29	NA
Dichlorobenzene	NA	70–100	NA	NA	NA	NA	NA
o-Dichlorobenzene	0–29	70–100	70–100	38–95	14–72	30–88	30–69
p-Dichlorobenzene	0–29	70–100	70–100	NA	NA	30–69	0–10
1,1-Dichloroethylene	0–29	70–100	70–100	NA	97	70–100	NA
cis-1,2-Dichloroethylene	0–29	70–100	70–100	NA	32–85	70–100	0–30
Trans-1,2-Dichloroelhylene	0–29	70–100	70–100	NA	37–96	70–100	0–30
Epichlorohydrin	NA	NA	0–29	NA	NA	0–29	NA
Ethylbenzene	0–29	70–100	70–100	33–99	24–89	70–100	0–30
Ethylene dibromide	0–29	70–100	70–100	NA	NA	0–29	37–100
Heptachlor	64	70–100	70–100	53–97	NA	70–100	NA
Heptachlor epoxide	NA	NA	NA	NA	NA	25	NA
High molecular weight Hydrocarbons (gasoline, dyes, amines, humics)	NA	W	NA	NA	NA	NA	NA
Lindane	0–29	70–100	0–29	82–97	NA	0–100	50–75
Methoxychlor	NA	70–100	NA	NA	NA	NA	>90
Monochlorobenzene	NA	NA	NA	14–99	14–85	86–98	50–100
Natural organic material	P	P	NA	P	NA	W	P
PCBs	NA	70–100	70–100	NA	NA	NA	95
Phenol and chlorophenol	NA	W	NA	NA	NA	W	NA
Pentachlorophenol	NA	70–100	0	NA	NA	70–100	NA
Styrene	0–29	NA	NA	NA	NA	70–100	NA
Tetrachloroethylene	NA	70–100	NA	NA	73–95	W	70–90
Trichloroethane	0–29	70–100	70–100	NA	53–95	30–69	0–100
Trichlorethane	NA	70–100	NA	NA	NA	NA	NA
1,1,1-Trichloroethane	0–29	70–100	70–100	40–65	58–90	0–29	15–100
Toluene	0–29	70–100	70–100	0–67	22–89	70–100	NA
2,4,5-TP	63	70–100	NA	82–99	NA	30–69	NA
Toxaphene	0–29	70–100	70–100	40–99	NA	NA	NA
Vinyl chloride	0–29	70–100	70–100	NA	NA	70–100	NA
Xylenes	0–29	70–100	70–100	60–99	18–89	70–100	10–85

Note: W, well removed; P, poorly removed; NA, not available. Little or no specific performance data were available for: 1. Multiple Tray Aeration; 2. Catenary Aeration; 3. Higee Aeration; 4. Resins; 5. Ultrafiltration; 6. Mechanical Aeration.

[a] The specifics of the oxidation processes effective in removing each contaminant are provided in Chapter 8 of the USEPA report.

Source: From United States Environmental Protection Agency, 1990, *Technologies for Upgrading Existing or Designing New Drinking Water Treatment Facilities*, EPA/625/4-89/023, March 1990, http://nepis.epa.gov.

Table 8J.141 Removal Effectiveness for Inorganic Contaminant

Treatment	Contaminant																				
	Ag	As	As^{III}	AS^{V}	Ba	Cd	Cr	Cr^{III}	Cr^{VI}	F	Hg	Hg(0)	Hg^{III}	NO_3	Pb	Ra	Rn	Se	$Se^{(V)}$	$Se^{(II)}$	U
Conventional treatment	H	—	M	H	L	H	—	H	H	L	—	M	M	L	H	L	—	—	M	L	M
Coagulation-aluminum	H	—	—	H	—	M	—	H	—	—	M	—	—	—	H	—	—	—	—	—	—
Coagulation-Iron	M	—	—	H	—	—	—	H	H	—	—	—	—	—	—	—	—	—	—	—	—
Lime softening	—	—	M	H	H	H	—	H	L	M	—	L	M	L	H	H	—	—	M	L	H
Reverse osmosis and electrodialysis	H	—	M	H	H	H	H	—	—	H	H	—	—	M	H	H	—	H	—	—	H
Cation exchange	—	L	—	—	H	H	—	H	L	L	—	—	—	L	H	H	—	L	—	—	H
Anion exchange	—	—	—	—	M	M	—	M	H	—	—	—	—	H	M	M	—	H	—	—	H
Activated alumina	—	—	H	—	L	L	—	—	—	H	—	—	—	—	—	L	—	H	—	—	—
Powdered activated carbon	L	—	—	—	L	M	—	L	—	L	—	M	M	L	—	L	—	—	—	—	—
Granular activated carbon	—	—	—	—	L	M	—	L	—	L	—	H	H	L	—	L	H	—	—	—	—

Note: H, High $\geq$80% removal; M, Medium 20–80% removal; L, Low $\leq$20% removal; "—", indicate no data were provided.

Source: From United States Environmental Protection Agency, 1990, *Technologies for Upgrading Existing or Designing New Drinking Water Treatment Facilities*, EPA/625/4-89/023, March 1990, http://nepis.epa.gov.

Table 8J.142 Arsenic Treatment Technologies Summary Comparison

Factors	Sorption Processes: Ion Exchange IX	Sorption Processes: Activated Alumina[a] AA	Sorption Processes: Iron Based Sorbents IBS	Membrane Processes: Reverse Osmosis RO
USEPA BAT[b]	Yes	Yes	No[c]	Yes
USEPA SSCT[b]	Yes	Yes	No[c]	Yes
System Size[b,d]	25–10,000	25–10,000	25–10,000	501–10,000
SSCT for POU[b]	No	Yes	No[c]	Yes
POU System Size[b,d]	—	25–10,000	25–10,000	25–10,000
Removal Efficiency	95%[e]	95%[e]	up to 98%[e]	>95%[e]
Total Water Loss	1–2%	1–2%	1–2%	15–75%
Pre-Oxidation Required[f]	Yes	Yes	Yes[g]	Likely[h]
Optimal Water Quality Conditions	pH 6.5–9 [e] <5 mg/L NO_3^- [i] <50 mg/L SO_4^{2} [j] <500 mg/L TDS [k] <0.3 NTU Turbidity	pH 5.5–6 [i] pH 6–8.3 [l] <250 mg/L Cl^- [i] <2 mg/L F^- [i] <360 mg/L SO_4^{2-} [k] <30 mg/L Silica[m] <0.5 mg/L Fe^{+3} [j] <0.05 mg/L Mn^{+2} [l] $<1{,}000$ mg/L TDS[k] <4 mg/L TOC[k] <0.3 NTU Turbidity	pH 6–8.5 <1 mg/L PO_4^{-3} [n] <0.3 NTU Turbidity	No Particulates
Operator Skill Required	High	Low[a]	Low	Medium
Waste Generated	Spent Resin, Spent Brine, Backwash Water	Spent Media, Backwash Water	Spent Media, Backwash Water	Reject Water
Other Considerations	Possible pre & post pH adjustment. Pre-filtration required Potentially hazardous brine waste, Nitrate peaking Carbonate peaking affects pH	Possible pre & post pH adjustment Pre-filtration may be required Modified AA available	Media may be very expensive[o] Pre-filtration may be required	High water loss (15–75% of feed water)
Centralized Cost	Medium	Medium	Medium	High
POU Cost	—	Medium	Medium	Medium

Factors	Precipitative Processes: Enhanced Lime Softening LS	Enhanced (Conventional) Coagulation Filtration CF	Coagulation Assisted Micro-Filtration CMF	Coagulation-Assisted Direct Filtration CADF	Oxidation Filtration OxFilt
USEPA BAT[b]	Yes	Yes	No	Yes	Yes
USEPA SSCT[b]	No	No	Yes	Yes	Yes
System Size[b,d]	25–10,000	25–10,000	500–10,000	500–10,000	25–10,000
SSCT for POU[b]	No	No	No	No	No
POU System Size[b,d]	—	—	—	—	—
Removal Efficiency	90% [p]	95% (w/$FeCl_3$)[p] <90%(w/Alum)[p]	90% [p]	90%[p]	50–90%[p]
Total Water Loss	0%	0%	5%	1–2%	1–2%
Pre-Oxidation Required[f]	Yes	Yes	Yes	Yes	Yes
Optimal Water Quality Conditions	pH 10.5–11 [i] >5 mg/L Fe^{+3} [i]	pH 5.5–8.5 [q]	pH 5.5–8.5 [q]	pH 5.5–8.5 [q]	pH 5.5–8.5 <0.3 mg/L Fe Fe:As Ratio >20:1
Operator Skill Required	High	High	High	High	Medium

(Continued)

Table 8J.142 (Continued)

	Precipitative Processes				
Factors	**Enhanced Lime Softening LS**	**Enhanced (Conventional) Coagulation Filtration CF**	**Coagulation Assisted Micro-Filtration CMF**	**Coagulation-Assisted Direct Filtration CADF**	**Oxidation Filtration OxFilt**
Waste Generated	Backwash Water, Sludge (high volume)	Backwash Water, Sludge	Backwash Water, Sludge	Backwash Water, Sludge	Backwash Water, Sludge
Other Considerations	Treated water requires pH adjustment	Possible pre & post pH adjustment	Possible pre & post pH adjustment	Possible pre & post pH adjustment	None
Centralized Cost	Low[r]	Low[r]	High	Medium	Medium
POU Cost	N/A	N/A	N/A	N/A	N/A

[a] Activated alumina is assumed to operate in a nonregenerated mode.
[b] USEPA 2002a.
[c] IBS's track record in U.S. was not established enough to be considered as Best Available Technology (BAT) or Small System Compliance Technology (SSCT) at the time the rule was promulgated.
[d] Affordable for systems with the given number of people served.
[e] USEPA, 2000.
[f] Pre-oxidation only required for As(III).
[g] Some iron based sorbents may catalyze the As(III) to As(V) oxidation and therefore would not require a pre-oxidation step.
[h] RO will remove As(III), but its efficiency is not consistent and pre-oxidation will increase removal efficiency.
[i] AwwaRF, 2002.
[j] Kempic, 2002.
[k] Wang, 2000.
[l] AA can be used economically at higher pHs, but with a significant decrease in the capacity of the media.
[m] Clifford, 2001.
[n] Tumalo, 2002.
[o] With increased domestic use, IBS cost will significantly decrease.
[p] Depends on arsenic and iron concentrations.
[q] Fields, et al., 2002a.
[r] Cost for enhanced LS and enhanced CF are based on modification of an existing technology. Most small systems will not have this technology in place.

Source: From United States Environmental Protection Agency, 2003, *Arsenic Treatment Technology Evaluation Handbook for Small Systems*, EPA-816-R-03-014, July 2003, www.epa.gov.

Table 8J.143 Treatment Utilized by Major Water Utilites in the United States in 1996

	Water Source (Percent)						Predisinfection/ Oxidation								Rapid Mix				Flocculation/ Coagulation					
Utility Name	Groundwater	Surface Water	Purchased Water	Lime/Soda Ash Softening	Lime/Soda Ash Softening	Aeration (conventional use)	Chlorine	Chlorine Dioxide	Chloramines	Ozone	Potassium Permanganate	UV Radiation	Lime/Soda Ash Softening	Recarbonation with CO_2	In-Line Hydraulic	In-Line Mechanical	In-Line Static	Basin Mechanical	Aluminum Salts	Iron Salts	pH Adjustment	Alkalinity Adjustment	Activated Silica	Clays
ALASKA																								
Anchorage Water/WW Util	22	78	0		y										y				y	y	y	y		
ALABAMA																								
Mobile Area Water & Sewer Service System	0	100	0		y			y								y	y		y		y	y		
ARIZONA																								
Glendale, (City of)	28	72	0		y		y	y			y						y	y	y					
Phoenix, (City of)	7	94	0		y		y									y			y	y	y			
Scottsdale, (City of)	40	41	19		y	y	y				y					y		y	y					
Tempe, (City of) Water Mngmt Div	0	100	0		y		y									y		y	y					
Tucson Water	100	0	0				y																	
CALIFORNIA																								
Anaheim, (City of)	75	0	25		y		y			y								y	y					
Coachella Valley Water Dist	100	0	0				y																	
Contra Costa Water Dist	0	100	0		y					y	y				y				y					
Corona, (City of) Util Svcs	42	58	0				y									y	y		y					
Corte Madera/Marin Muni Water Dist	0	81	19				y								y			y	y					
Cucamonga County Water Dist	51	18	31		y		y								y	y	y	y	y		y			
Downey/Park Water Co	10	0	90																					
El Monte/San Gabriel Valley Water Co	100	0	0				y																	
Fremont/Alameda County Water Dist	26	51	24				y			y	y				y			y	y	y				
Fresno, (City of)	100	0	0																					
Fullerton, (City of)	75	25	0				y																	
Glendale Public Service Dept	6	0	94																					
Granite Bay/San Juan Suburban Water Dist	0	100	0				y						y			y		y	y					
Irvine Ranch Water Dist	53	0	47				y																	
La Mesa/Helix Water Dist	0	28	72				y									y			y		y			
Long Beach/Dominguez Services Inc	100	0	0			y	y		y															
Long Beach Water Dept	46	0	54		y	y	y				y							y	y					
Los Angeles Water & Power	17	44	39		y					y								y		y				
Los Angeles/Metro Water Dist of So Calif	0	100	0				y								y	y			y					
Modesto Public Works	49	51	0							y				y				y	y					
Monterey Division—CA-American Water Co	48	9	43				y			y	y				y					y	y	y		
Oakland/East Bay Muni Util Dist	0	100	0			y	y			y	y		y		y			y	y					
Riverside Water Div	100	0	0																					
Sacramento/Citizens Util Co of California	100	0	0		y		y				y													
Sacramento, (City of) Util Dept	20	80	0				y								y				y					
San Bernardino Water Dept	100	0	0		y	y	y																	
San Diego, (City of)	0	100	0		y		y				y					y				y	y			
San Francisco Water Dept	0	100	0			y	y			y					y				y					
San Jacinto/Eastern Muni Water Dist	18	0	82			y																		
San Jose Water Co	45	11	45															y	y					
San Jose/Santa Clara Valley Water Dist	0	100	0		y		y		y							y	y	y	y					
Santa Barbara, (City of)	0	100	0			y	y									y			y					
Santa Clarita/Castaic Lake Water Agency	0	100	0				y			y					y			y	y	y				
Stockton East Water Dist	0	100	0				y				y				y				y					
Sunnyvale, (City of) Public Works	9	0	91	y																				
Thousand Oaks/Calleguas Muni Water Dist	0	0	100		y		y			y						y			y					
Torrance Muni Water Dept	5	0	95		y	y	y																	
Vacaville/Suisun Solano Water Auth	5	95	0				y												y					
Vallejo, (City of)	0	100	0		y		y			y					y	y	y	y	y		y	y		
Vista Irrigation Dist	0	61	39																					
COLORADO																								
Arvada, (City of)	0	100	0		y		y				y			y		y		y	y		y	y		
Colorado Springs Util	1	99	0		y	y	y									y	y	y	y					
Denver Water Dept	0	100	0		y		y		y									y	y		y			
Fort Collins Water Util	0	100	0		y		y				y							y	y					
Pueblo Board of Waterworks	0	100	0				y		y		y							y	y					
CONNECTICUT																								
Bridgeport Hydraulic Co - Main System	13	87	0			y	y											y	y		y			
Hartford Metropolitan Dist	0	100	0			y										y			y					
New Haven/Regional Water Auth	13	88	0				y				y						y	y		y	y			
WASHINGTON DC																								
Washington Aqueduct	0	100	0		y										y				y					

(Continued)

Table 8J.143 (Continued)

	Water Source (Percent)						Predisinfection/ Oxidation								Rapid Mix				Flocculation/ Coagulation					
Utility Name	Groundwater	Surface Water	Purchased Water	Lime/Soda Ash Softening	Lime/Soda Ash Softening	Aeration (conventional use)	Chlorine	Chlorine Dioxide	Chloramines	Ozone	Potassium Permanganate	UV Radiation	Lime/Soda Ash Softening	Recarbonation with CO_2	In-Line Hydraulic	In-Line Mechanical	In-Line Static	Basin Mechanical	Aluminum Salts	Iron Salts	pH Adjustment	Alkalinity Adjustment	Activated Silica	Clays
DELAWARE																								
Newark/Artesian Water Company Inc	72	0	28			y	y				y						y				y			
Wilmington/United Water Delaware	0	96	4		y		y				y				y			y		y	y			
FLORIDA																								
Clearwater Water Div	26	0	74			y	y																	
Cocoa, (City of) Util Dept	100	0	0		y	y	y		y				y	y										
Coral Gables/Miami-Dade Water & Sewer Dept	100	0	0										y	y				y			y	y	y	
Jacksonville, (City of) Water Div	100	0	0		y	y																		
Jacksonville/United Water Florida	97	0	3			y	y																	
Kissimmee, (City of)	100	0	0		y	y																		
Lakeland Electric & Water Util	100	0	0				y						y											
Orlando/Orange County Util	100	0	0			y	y																	
Orlando Util Comm	100	0	0			y	y																	
Pensacola/Escambia County Util Auth	100	0	0				y																	
Pompano Beach/Broward County Enviro Services	88	0	12				y		y				y			y				y				
Sarasota County Util	39	0	61			y	y																	
Tallahassee Water Util	100	0	0																					
Tampa Water Dept	5	95	0				y				y		y	y				y		y	y	y		
W Coast Regional Water Supply	100	0	0		y	y	y																	
West Palm Beach/Palm Beach County Water Util	100	0	0			y			y	y				y				y			y	y		
GEORGIA																								
Atlanta-Fulton County Water	0	100	0		y	y	y				y					y			y			y		
Cobb County-Marietta Wtr Auth	0	100	0				y	y								y			y					
Columbus Water Works	0	100	0					y								y			y					
Morrow/Clayton County Water Auth	0	100	0		y		y				y					y		y	y		y	y		
Stone Mountain/DeKalb County Public Works	0	100	0				y				y				y			y	y					
HAWAII																								
Honolulu/Board of Water Supply	100	0	0				y																	
Pearl Harbor/Navy Public Wrks Util Dept	99	0	1																					
IOWA																								
Cedar Rapids Water Dept	100	0	0			y							y	y				y						
Des Moines Water Works	59	41	0		y						y		y	y				y	y	y				
ILLINOIS																								
Champaign Division - Northern Illinois Water Inc	100	0	0										y	y				y		y				
Chicago, (City of) Dept of Water	0	100	0				y									y		y	y					
Decatur, (City of)	0	100	0								y		y	y	y				y					
Elmhurst/DuPage Water Comm	0	0	100																					
Evanston Water Dept	0	100	0				y									y			y					
Interurban District - IL-American Water Co.	0	100	0				y				y						y		y					
Lake Bluff/Central Lake Cnty Water Agency	0	100	0							y	y							y	y					
Peoria District - IL-American Water Co.	69	31	0				y				y								y					
Rockford City Water Div	100	0	0				y																	
Springfield/City Water, Light and Power	0	100	0										y	y	y					y				
Wilmette Water Plant	0	100	0				y											y	y					
INDIANA																								
Evansville Water & Sewer Dept	0	100	0					y									y		y					
Gary/Northwest Indiana Water Co.	0	100	0				y				y							y	y					
Indianapolis Water Co.	6	94	0			y	y									y	y		y		y	y		
KANSAS																								
Kansas City Public Util Board	0	100	0		y		y	y							y	y			y		y		y	
KENTUCKY																								
Covington/Kenton County Water Dist No.1	0	100	0		y		y				y							y		y	y			
Lexington/KY-American Water Co.	0	100	0			y	y				y							y	y	y	y	y		
Louisville Water Company	0	100	0		y		y								y	y	y	y		y				
LOUISIANA																								
The Baton Rouge Water Co.	100	0	0																					
Harahan/Jefferson Parish Water Dept	0	100	0													y	y							
Lake Charles, (City of) Water Div	100	0	0			y	y				y													
New Orleans Water/Sewer Board	0	100	0						y				y					y		y	y			
MASSACHUSETTS																								
Fall River Water Dept	0	100	0				y											y	y		y			
Lowell Water Dept	0	100	0				y										y		y		y	y		
New Bedford, (City of) Water Dept	0	100	0													y			y					
Springfield Water Dept	0	100	0		y													y						

(Continued)

Table 8J.143 (Continued)

	Water Source (Percent)						Predisinfection/ Oxidation								Rapid Mix				Flocculation/ Coagulation					
Utility Name	Groundwater	Surface Water	Purchased Water	Lime/Soda Ash Softening	Lime/Soda Ash Softening	Aeration (conventional use)	Chlorine	Chlorine Dioxide	Chloramines	Ozone	Potassium Permanganate	UV Radiation	Lime/Soda Ash Softening	Recarbonation with CO_2	In-Line Hydraulic	In-Line Mechanical	In-Line Static	Basin Mechanical	Aluminum Salts	Iron Salts	pH Adjustment	Alkalinity Adjustment	Activated Silica	Clays
MARYLAND																								
Annapolis/Anne Arundel County DPW	78	0	22			y	y						y				y	y			y			
Laurel/Washington Suburban Sanit Comm	0	100	0				y				y						y		y	y	y			
MAINE																								
Portland Water District	4	96	0							y														
MICHIGAN																								
Ann Arbor Util Dept	0	100	0		y								y	y		y								
Grand Rapids, (City of)	0	100	0				y				y							y	y	y				
Holland/Wyoming Util Dept	0	100	0				y				y					y			y					
Kalamazoo Dept Public Util	100	0	0																					
MINNESOTA																								
St. Paul Water Util	0	100	0				y		y		y		y	y	y				y	y	y			
MISSOURI																								
Independence Water Dept	100	0	0			y	y														y	y		
Kansas City Water Svcs Dept	10	90	0		y	y					y		y	y	y				y	y	y	y		
Springfield City Util	24	76	0		y		y				y					y			y	y				
St. Louis County Water Co	0	100	0		y		y		y				y			y				y				
NORTH CAROLINA																								
Asheville, (City of)	0	100	0				y								y									
Charlotte-Mecklenburg Util	0	100	0		y		y										y		y					
Durham, (City of) Dept Water Res	0	100	0		y	y	y				y							y	y		y	y		
Fayetteville Public Works Comm	0	100	0													y		y	y		y	y		
Greensboro, (City of) Util Dept	0	100	0		y						y							y	y					
NEBRASKA																								
Lincoln Water System	100	0	0			y	y			y														
Omaha Metro Util Dist	47	53	0		y		y						y					y	y		y			
NEW HAMPSHIRE																								
Manchester Water Works	0	100	0				y									y			y		y			
NEW JERSEY																								
Brick Township Muni Util Auth	35	65	0			y		y					y					y	y		y			
Clifton/Passaic Valley Water Comm	0	54	46				y	y							y				y		y			
Haddon System - NJ-American Water Co.	93	0	7		y	y	y			y							y	y		y	y	y		
Harrington Park/United Water New Jersey	1	97	2							y					y				y					
Iselin/Middlesex Water Company	28	65	6				y												y		y			
Little Falls/Newark Div of Water/Sewer Util	0	100	0			y	y												y					
Short Hills System - NJ-American Water Co.	15	30	54		y		y												y					
Shrewsbury/NJ-American Water Co.- Monmouth System	0	80	20		y	y	y	y							y			y	y					
Westfield/Elizabethtown Water Co.	11	88	0				y				y				y	y			y		y	y		
NEW MEXICO																								
Albuquerque Public Works Dept	100	0	0																					
NEVADA																								
Boulder City/Southern Nevada Water System	0	100	0			y	y									y				y				
Las Vegas Valley Wtr Dist	16	84	0																					
North Las Vegas, (City of)	9	0	91				y																	
NEW YORK																								
Buffalo/Erie County Water Auth	0	100	0				y				y					y		y	y					
Lynbrook/Long Island Water Inc	100	0	0			y	y																	
Oakdale/Suffolk County Water Auth	100	0	0				y				y													
Rochester, (City of) Water Bureau	0	86	14				y				y							y	y	y				
Syracuse/Onondaga County Water Auth	0	40	60				y				y						y		y					
Syracuse Dept of Water	0	98	2				y																	
West Nyack/United Water New York	77	23	0			y					y				y				y					
OHIO																								
Akron Public Util Bureau	0	100	0				y	y										y	y					
Cincinnati Water Works	12	88	0		y		y				y		y		y		y	y	y					
Cleveland Div of Water	0	100	0		y	y	y				y				y	y		y	y					
Columbus Water Div	16	84	0				y				y		y	y		y			y		y	y		
Dayton Water Supply/Treatment	100	0	0										y	y		y		y			y	y		
Fairfield/Hamilton Util	100	0	0			y		y					y	y		y								
Toledo, (City of) Water Div	0	100	0										y	y	y				y					
OKLAHOMA																								
Lawton Water/Wastewater Plants	0	100	0			y	y		y						y				y		y	y		y
Oklahoma City, (City of)	0	100	0		y		y		y	y	y		y	y		y		y		y	y	y		
Tulsa Public Works Dept	0	100	0		y		y				y		y			y			y		y			
OREGON																								
Eugene Water & Electric Board	0	100	0				y										y		y		y			
Portland Water Bureau	4	96	0																					

(Continued)

Table 8J.143 (Continued)

	Water Source (Percent)						Predisinfection/ Oxidation								Rapid Mix				Flocculation/ Coagulation					
Utility Name	Groundwater	Surface Water	Purchased Water	Lime/Soda Ash Softening	Lime/Soda Ash Softening	Aeration (conventional use)	Chlorine	Chlorine Dioxide	Chloramines	Ozone	Potassium Permanganate	UV Radiation	Lime/Soda Ash Softening	Recarbonation with CO_2	In-Line Hydraulic	In-Line Mechanical	In-Line Static	Basin Mechanical	Aluminum Salts	Iron Salts	pH Adjustment	Alkalinity Adjustment	Activated Silica	Clays
PENNSYLVANIA																								
Allentown Muni Water System	50	50	0		y		y											y	y					
Bethlehem, (City of)	0	100	0		y		y										y		y		y			
Bryn Mawr/Philadelphia Suburban Water Co.	14	80	6				y	y			y					y			y		y			
Harrisburg/United Water PA	11	88	1				y	y			y								y					
Lancaster, (City of) Water Bureau	0	100	0				y				y		y					y	y		y			
Philadelphia Water Dept	0	100	0		y		y	y			y								y	y	y			
Pittsburgh/West View Water Auth	0	100	0				y				y					y				y	y	y		
Reading Area Water Auth	0	100	0				y				y		y		y				y					
The York Water Co.	0	100	0		y		y				y							y	y					
RHODE ISLAND																								
Pawtucket Water Supply Board	15	85	0		y	y	y									y			y		y			
SOUTH CAROLINA																								
Charleston Comm Public Works	0	100	0						y									y	y		y			
Spartanburg Water System	0	100	0				y								y				y		y			
SOUTH DAKOTA																								
Sioux Falls, (City of)	0	100	0								y		y	y			y	y		y				
TENNESSEE																								
Chattanooga/TN-American Water Co	0	100	0				y										y		y					
Memphis Light Gas & Water	100	0	0			y	y												y					
Nashville/Metro Water Services	0	100	0		y	y	y				y				y				y		y			
TEXAS																								
Abilene, (City of)	0	100	0		y	y		y	y				y		y	y		y	y		y	y		
Amarillo Muni Water System	45	55	0		y		y								y	y			y					
Arlington/Trinity River Auth	0	100	0				y	y	y		y		y			y			y	y	y			
Austin, (City of) Water/WW	0	100	0				y						y	y	y			y		y	y			
Dallas Water Util	0	99	1		y		y		y	y			y	y	y			y		y	y	y		
El Paso Water Util	57	43	0		y		y	y		y			y			y				y				
Gavelston, (City of) Dept Pub Wks	10	90	0				y		y															
Houston, (City of)	41	59	0		y		y		y						y	y	y		y					
Killeen/Bell County WCID #1	0	100	0				y		y		y					y			y		y			
Lubbock, (City of) Water Util	15	85	0				y		y		y				y				y	y				
San Antonio Water System	100	0	0				y																	
Wichita Falls, (City of)	0	100	0		y			y	y				y	y	y					y	y	y		
UTAH																								
Layton/Weber Basin Water Cons Dist	22	78	0		y		y				y				y			y	y					
Orem/Central Utah Water Cons Dist	0	100	0								y				y				y					
Salt Lake City/Metropolitan Water Dist	0	100	0			y	y				y					y			y		y	y		
Salt Lake City Public Util	14	43	43				y				y				y	y	y		y	y		y		
West Jordan/Salt Lake Cnty Water Cons Dist	15	79	6		y		y				y	y			y				y		y			
VIRGINIA																								
Chesapeake, (City of)	9	46	45					y										y	y		y			
Chesterfield County Util Dept	0	36	64								y					y			y		y			
Merrifield/Fairfax County Water Auth	0	99	0			y					y					y			y		y	y		
Newport News Waterworks	0	100	0								y					y			y					
Petersburg/Appomattox River Water Auth	0	100	0				y				y						y		y		y	y		
Richmond/Henrico Co Public Util	13	0	87																					
Richmond Dept of Public Util	0	100	0		y	y	y									y			y					
Woodbridge/Prince William Co. Service Auth	13	0	87	y																				
WASHINGTON																								
Everett Water Util	0	100	0		y		y										y		y					
Seattle Water Dept	6	94	0																					
Tacoma Public Util	10	90	0				y																	
Vancouver, (City of)	100	0	0																					
WISCONSIN																								
Green Bay Water Util	1	99	0				y				y					y			y					
Kenosha Water Util	0	100	0				y				y					y			y					
Madison Water Util	100	0	0		y		y																	
Milwaukee Water Works	0	100	0				y									y		y	y					
Sheboygan Water Util	0	100	0				y				y					y			y					
WEST VIRGINIA																								
Charleston/WV-American Water Co. - Kanawha Valley District	0	100	0				y									y								

(Continued)

Table 8J.143 (Continued)

	Polymers			Settling/ Sedimentation					Filtration							Filter Backwash Practices			
Utility Name	Anionic Polymers	Cationic Polymers	Nonionic Polymers	Conventional Sedimentation/ Clarification	Upflow Clarifiers	Tube Settlers	Lamella Plates	Dissolved Air Flotation	Direct Filtration	Microstrainer	Slow Sand	Rapid Sand	Dual/Multi-Media	Diatomaceous Earth	Pressure Filtration	Air Scouring	Surface Wash	Recycle Filter Backwash	Filter-to-Waste
ALASKA																			
Anchorage Water/WW Util	y		y	y					y			y	y			y	y	y	y
ALABAMA																			
Mobile Area Water & Sewer Service System				y								y	y				y	y	y
ARIZONA																			
Glendale, (City of)		y	y	y									y			y	y	y	
Phoenix, (City of)	y	y		y					y				y			y	y	y	y
Scottsdale, (City of)		y		y									y			y		y	
Tempe, (City of) Water Mngmt Div		y	y	y									y				y		y
Tucson Water																			
CALIFORNIA																			
Anaheim, (City of)		y				y										y		y	
Coachella Valley Water Dist																			
Contra Costa Water Dist		y	y	y					y				y			y	y	y	y
Corona, (City of) Util Svcs	y	y		y		y			y				y			y	y	y	
Corte Madera/Marin Muni Water Dist	y	y	y		y								y				y	y	
Cucamonga County Water Dist	y	y		y									y			y	y	y	y
Downey/Park Water Co.																			
El Monte/San Gabriel Valley Water Co.																			
Fremont/Alameda County Water Dist	y	y	y	y	y								y			y	y	y	y
Fresno, (City of)																			
Fullerton, (City of)																			
Glendale Public Service Dept														y					
Granite Bay/San Juan Suburban Water Dist		y	y	y		y			y			y	y				y	y	
Irvine Ranch Water Dist																			
La Mesa/Helix Water Dist	y	y		y									y				y		
Long Beach/Dominguez Services Inc																			
Long Beach Water Dept		y		y									y				y	y	
Los Angeles Water & Power		y							y							y		y	
Los Angeles/Metro Water Dist of So Calif	y	y	y	y					y				y				y	y	
Modesto Public Works	y	y		y					y							y			
Monterey Division - CA-American Water Co.		y										y	y		y	y	y		y
Oakland/East Bay Muni Util Dist		y	y	y					y			y	y				y	y	y
Riverside Water Div																			
Sacramento/Citizens Util Co of California			y						y						y		y	y	
Sacramento, (City of) Util Dept			y	y									y				y	y	
San Bernardino Water Dept																			
San Diego, (City of)		y		y									y				y		
San Francisco Water Dept		y	y	y					y				y			y	y	y	y
San Jacinto/Eastern Muni Water Dist				y															
San Jose Water Co.		y							y			y				y		y	
San Jose/Santa Clara Valley Water Dist	y	y	y	y	y	y							y			y		y	y
Santa Barbara, (City of)		y		y									y				y	y	
Santa Clarita/Castaic Lake Water Agency	y	y	y	y		y							y			y		y	y
Stockton East Water Dist		y		y									y				y	y	y
Sunnyvale, (City of) Public Works																			
Thousand Oaks/Calleguas Muni Water Dist	y	y	y						y							y		y	
Torrance Muni Water Dept																			
Vacaville/Suisun Solano Water Auth		y			y										y		y	y	y
Vallejo, (City of)	y	y	y	y								y	y				y	y	y
Vista Irrigation Dist	y	y		y									y				y	y	
COLORADO																			
Arvada, (City of)			y	y					y				y				y	y	
Colorado Springs Util	y	y	y	y			y					y	y			y	y	y	y
Denver Water Dept	y	y	y	y								y	y				y	y	
Fort Collins Water Util	y	y	y	y	y	y	y						y			y	y	y	
Pueblo Board of Waterworks		y		y									y					y	
CONNECTICUT																			
Bridgeport Hydraulic Co. - Main System	y						y						y			y		y	y
Hartford Metropolitan Dist				y							y	y					y		
New Haven/Regional Water Auth		y	y	y					y				y				y	y	
WASHINGTON DC																			
Washington Aqueduct		y	y	y									y				y	y	y

(Continued)

Table 8J.143 (Continued)

	Polymers			Settling/ Sedimentation					Filtration							Filter Backwash Practices			
Utility Name	Anionic Polymers	Cationic Polymers	Nonionic Polymers	Conventional Sedimentation/ Clarification	Upflow Clarifiers	Tube Settlers	Lamella Plates	Dissolved Air Flotation	Direct Filtration	Microstrainer	Slow Sand	Rapid Sand	Dual/Multi-Media	Diatomaceous Earth	Pressure Filtration	Air Scouring	Surface Wash	Recycle Filter Backwash	Filter-to-Waste
DELAWARE																			
Newark/Artesian Water Company Inc	y														y			y	y
Wilmington/United Water Delaware	y	y		y	y		y						y		y		y	y	y
FLORIDA																			
Clearwater Water Div																			
Cocoa, (City of) Util Dept					y								y				y	y	
Coral Gables/Miami-Dade Water & Sewer Dept				y	y								y				y	y	
Jacksonville, (City of) Water Div																			
Jacksonville/United Water Florida																			
Kissimmee, (City of)																			
Lakeland Electric & Water Util	y				y								y				y		
Orlando/Orange County Util																			
Orlando Util Comm																			
Pensacola/Escambia County Util Auth																			
Pompano Beach/Broward County Enviro Services		y	y		y								y				y	y	
Sarasota County Util															y			y	
Tallahassee Water Util																			
Tampa Water Dept		y	y	y									y				y	y	
W Coast Regional Water Supply																			
West Palm Beach/Palm Beach County Water Util	y		y		y								y			y	y	y	y
GEORGIA																			
Atlanta-Fulton County Water							y						y			y		y	y
Cobb County-Marietta Wtr Auth	y			y								y	y				y	y	y
Columbus Water Works				y									y				y	y	
Morrow/Clayton County Water Auth				y									y				y	y	y
Stone Mountain/DeKalb County Public Works		y		y									y				y	y	y
HAWAII																			
Honolulu/Board of Water Supply																			
Pearl Harbor/Navy Public Wrks Util Dept																			
IOWA																			
Cedar Rapids Water Dept	y				y							y	y			y	y	y	y
Des Moines Water Works				y								y				y		y	
ILLINOIS																			
Champaign Division - Northern Illinois Water Inc					y								y				y	y	
Chicago, (City of) Dept of Water		y		y								y					y	y	
Decatur, (City of)		y			y								y			y		y	
Elmhurst/DuPage Water Comm																			
Evanston Water Dept		y		y								y	y				y	y	
Interurban District - IL-American Water Co.		y	y	y			y					y	y				y	y	y
Lake Bluff/Central Lake Cnty Water Agency							y						y			y		y	y
Peoria District - IL-American Water Co.		y	y	y								y	y		y		y		
Rockford City Water Div																			
Springfield/City Water, Light and Power		y			y								y				y		y
Wilmette Water Plant				y									y				y	y	
INDIANA																			
Evansville Water & Sewer Dept				y									y				y		
Gary/Northwest Indiana Water Co.		y		y									y				y		
Indianapolis Water Co.		y		y								y			y		y	y	
KANSAS																			
Kansas City Public Util Board		y		y									y				y		y
KENTUCKY																			
Covington/Kenton County Water Dist No.1			y	y	y	y						y	y				y		
Lexington/KY-American Water Co.		y		y	y								y			y	y	y	
Louisville Water Company		y		y									y			y	y		
LOUISIANA																			
The Baton Rouge Water Co.																			
Harahan/Jefferson Parish Water Dept		y		y	y							y	y				y		y
Lake Charles, (City of) Water Div				y								y	y		y	y	y	y	y
New Orleans Water/Sewer Board		y		y	y							y	y			y	y	y	y
MASSACHUSETTS																			
Fall River Water Dept				y								y							y
Lowell Water Dept				y									y						y
New Bedford, (City of) Water Dept				y									y			y			
Springfield Water Dept		y							y		y	y	y			y	y		

(Continued)

Table 8J.143 (Continued)

Utility Name	Polymers			Settling/ Sedimentation					Filtration							Filter Backwash Practices			
	Anionic Polymers	Cationic Polymers	Nonionic Polymers	Conventional Sedimentation/ Clarification	Upflow Clarifiers	Tube Settlers	Lamella Plates	Dissolved Air Flotation	Direct Filtration	Microstrainer	Slow Sand	Rapid Sand	Dual/Multi-Media	Diatomaceous Earth	Pressure Filtration	Air Scouring	Surface Wash	Recycle Filter Backwash	Filter-to-Waste
MARYLAND																			
Annapolis/Anne Arundel County DPW				y	y	y					y	y			y	y	y	y	y
Laurel/Washington Suburban Sanit Comm		y		y		y							y				y	y	y
MAINE																			
Portland Water District																			
MICHIGAN																			
Ann Arbor Util Dept		y		y									y				y	y	
Grand Rapids, (City of)				y	y							y					y		y
Holland/Wyoming Util Dept	y	y		y	y							y	y				y	y	
Kalamazoo Dept Public Util																			
MINNESOTA																			
St. Paul Water Util				y								y	y			y		y	
MISSOURI																			
Independence Water Dept				y								y					y	y	
Kansas City Water Svcs Dept		y		y	y							y					y		y
Springfield City Util		y		y								y	y				y	y	y
St. Louis County Water Co.	y	y		y	y							y	y				y		y
NORTH CAROLINA																			
Asheville, (City of)		y							y				y			y			y
Charlotte-Mecklenburg Util				y									y				y		y
Durham, (City of) Dept Water Res			y	y									y			y	y	y	y
Fayetteville Public Works Comm				y		y							y				y		y
Greensboro, (City of) Util Dept				y		y							y				y		y
NEBRASKA																			
Lincoln Water System		y										y	y			y	y		
Omaha Metro Util Dist		y			y								y			y			
NEW HAMPSHIRE																			
Manchester Water Works			y	y								y						y	
NEW JERSEY																			
Brick Township Muni Util Auth	y			y		y						y	y			y	y	y	y
Clifton/Passaic Valley Water Comm			y	y		y							y					y	
Haddon System - NJ-American Water Co.		y			y										y	y	y	y	y
Harrington Park/United Water New Jersey		y							y				y			y			
Iselin/Middlesex Water Company	y	y	y	y		y							y				y	y	
Little Falls/Newark Div of Water/Sewer Util		y	y						y								y	y	
Short Hills System - NJ-American Water Co.	y	y	y	y	y								y				y	y	
Shrewsbury/NJ-American Water Co. - Monmouth System	y	y		y								y	y			y	y	y	y
Westfield/Elizabethtown Water Co.		y		y		y							y				y	y	
NEW MEXICO																			
Albuquerque Public Works Dept																			
NEVADA																			
Boulder City/Southern Nevada Water System		y							y				y			y		y	
Las Vegas Valley Wtr Dist																			
North Las Vegas, (City of)																			
NEW YORK																			
Buffalo/Erie County Water Auth				y		y							y				y	y	
Lynbrook/Long Island Water Inc			y									y	y		y	y			y
Oakdale/Suffolk County Water Auth																			
Rochester, (City of) Water Bureau		y							y				y			y		y	y
Syracuse/Onondaga County Water Auth									y				y				y	y	y
Syracuse Dept of Water																			
West Nyack/United Water New York		y		y									y				y		
OHIO																			
Akron Public Util Bureau				y								y	y				y		
Cincinnati Water Works		y		y	y		y					y	y				y	y	
Cleveland Div of Water		y		y								y	y				y	y	y
Columbus Water Div				y	y								y					y	
Dayton Water Supply/Treatment				y	y							y	y				y	y	
Fairfield/Hamilton Util				y	y							y	y				y	y	
Toledo, (City of) Water Div				y								y					y		
OKLAHOMA																			
Lawton Water/Wastewater Plants		y			y							y					y		y
Oklahoma City, (City of)		y		y	y								y			y	y	y	y
Tulsa Public Works Dept	y	y	y	y									y			y	y	y	y
OREGON																			
Eugene Water & Electric Board				y									y				y		
Portland Water Bureau																			

(Continued)

Table 8J.143 (Continued)

	Polymers			Settling/ Sedimentation					Filtration							Filter Backwash Practices			
Utility Name	Anionic Polymers	Cationic Polymers	Nonionic Polymers	Conventional Sedimentation/ Clarification	Upflow Clarifiers	Tube Settlers	Lamella Plates	Dissolved Air Flotation	Direct Filtration	Microstrainer	Slow Sand	Rapid Sand	Dual/Multi-Media	Diatomaceous Earth	Pressure Filtration	Air Scouring	Surface Wash	Recycle Filter Backwash	Filter-to-Waste
PENNSYLVANIA																			
Allentown Muni Water System				y								y	y				y		y
Bethlehem, (City of)									y				y			y		y	y
Bryn Mawr/Philadelphia Suburban Water Co.		y	y	y					y				y			y	y	y	y
Harrisburg/United Water PA		y	y	y									y				y		y
Lancaster, (City of) Water Bureau	y		y	y	y							y					y	y	y
Philadelphia Water Dept			y	y								y	y				y	y	
Pittsburgh/West View Water Auth					y								y				y	y	y
Reading Area Water Auth				y								y	y			y			y
The York Water Co.				y									y				y	y	y
RHODE ISLAND																			
Pawtucket Water Supply Board		y		y															
SOUTH CAROLINA																			
Charleston Comm Public Works		y		y								y				y			
Spartanburg Water System		y		y									y				y		y
SOUTH DAKOTA																			
Sioux Falls, (City of)	y											y				y	y	y	
TENNESSEE																			
Chattanooga/TN-American Water Co.				y									y				y	y	
Memphis Light Gas & Water												y	y					y	y
Nashville/Metro Water Services			y		y	y						y	y				y	y	y
TEXAS																			
Abilene, (City of)		y			y								y				y	y	y
Amarillo Muni Water System				y								y					y		
Arlington/Trinity River Auth	y	y		y									y				y	y	
Austin, (City of) Water/WW				y	y								y				y	y	
Dallas Water Util		y		y									y			y	y	y	
El Paso Water Util	y	y		y									y			y		y	y
Gavelston, (City of) Dept Pub Wks																			
Houston, (City of)	y	y	y	y									y				y	y	
Killeen/Bell County WCID #1		y		y	y							y	y				y	y	
Lubbock, (City of) Water Util				y								y					y		
San Antonio Water System																			
Wichita Falls, (City of)		y		y	y							y	y			y	y	y	
UTAH																			
Layton/Weber Basin Water Cons Dist				y								y	y			y	y	y	y
Orem/Central Utah Water Cons Dist		y							y								y	y	
Salt Lake City/Metropolitan Water Dist	y	y		y									y				y	y	y
Salt Lake City Public Util	y	y		y		y					y		y			y	y	y	y
West Jordan/Salt Lake Cnty Water Cons Dist	y	y	y	y					y				y			y	y	y	y
VIRGINIA																			
Chesapeake, (City of)			y	y								y	y						y
Chesterfield County Util Dept		y		y									y				y		y
Merrifield/Fairfax County Water Auth	y	y		y	y	y						y	y				y	y	y
Newport News Waterworks		y		y	y		y						y			y	y		y
Petersburg/Appomattox River Water Auth			y	y									y				y		y
Richmond/Henrico Co Public Util																			
Richmond Dept of Public Util			y	y		y							y			y	y	y	
Woodbridge/Prince William Co. Service Auth																			
WASHINGTON																			
Everett Water Util		y	y						y							y	y	y	y
Seattle Water Dept																			
Tacoma Public Util																			
Vancouver, (City of)																y		y	y
WISCONSIN																			
Green Bay Water Util				y								y	y				y	y	
Kenosha Water Util				y								y					y		
Madison Water Util																			
Milwaukee Water Works				y								y	y				y		
Sheboygan Water Util				y									y				y	y	y
WEST VIRGINIA																			
Charleston/WV-American Water Co. - Kanawha Valley District		y			y								y				y	y	y

(Continued)

Table 8J.143 (Continued)

	Post-Disinfection							Corrosion Control			Membrane Processes												
Utility Name	Chlorine	Chlorine Dioxide	Chloramines	Ozone	UV Radiation	Fluoridation	Defluoridation	pH Adjustment	Alkalinity Adjustment	Corrosion Inhibitors	Reverse Osmosis	Nanofiltration	Ultrafiltration	Microfiltration	Electrodialysis	Ion Exchange	Iron/Manganese Removal	Manganese Green Sand	GranularActivated Carbon	Powdered Activated Carbon	Resin Adsorption	Air Stripping	Off Gas Treatment
ALASKA																							
Anchorage Water/WW Util	y					y																	
ALABAMA																							
Mobile Area Water & Sewer Service System	y					y		y		y													
ARIZONA																							
Glendale, (City of)	y					y		y												y			
Phoenix, (City of)	y					y		y															
Scottsdale, (City of)	y																y			y		y	
Tempe, (City of) Water Mngmt Div	y					y		y	y											y			
Tucson Water																							
CALIFORNIA																							
Anaheim, (City of)	y																						
Coachella Valley Water Dist	y																						
Contra Costa Water Dist			y	y		y		y											y	y			
Corona, (City of) Util Svcs	y																						
Corte Madera/Marin Muni Water Dist			y			y		y		y													
Cucamonga County Water Dist	y							y															
Downey/Park Water Co	y																						
El Monte/San Gabriel Valley Water Co.																	y						
Fremont/Alameda County Water Dist	y		y			y		y															
Fresno, (City of)	y					y																	
Fullerton, (City of)																							
Glendale Public Service Dept	y																						
Granite Bay/San Juan Suburban Water Dist	y							y	y														
Irvine Ranch Water Dist	y																						
La Mesa/Helix Water Dist			y					y												y			
Long Beach/Dominguez Services Inc											y												
Long Beach Water Dept	y		y			y		y		y													
Los Angeles Water & Power	y									y												y	y
Los Angeles/Metro Water Dist of So Calif	y		y					y															
Modesto Public Works	y							y	y										y				
Monterey Division - CA-American Water Co.	y							y		y							y		y				y
Oakland/East Bay Muni Util Dist	y					y		y											y				
Riverside Water Div	y																						
Sacramento/Citizens Util Co of California	y					y											y	y				y	
Sacramento, (City of) Util Dept	y							y	y														
San Bernardino Water Dept																			y			y	
San Diego, (City of)			y					y															
San Francisco Water Dept	y					y		y									y						
San Jacinto/Eastern Muni Water Dist	y									y													
San Jose Water Co.														y									
San Jose/Santa Clara Valley Water Dist	y		y					y		y										y			
Santa Barbara, (City of)	y																			y			
Santa Clarita/Castaic Lake Water Agency	y							y		y										y			
Stockton East Water Dist	y							y											y				
Sunnyvale, (City of) Public Works																							
Thousand Oaks/Calleguas Muni Water Dist	y									y													
Torrance Muni Water Dept																							
Vacaville/Suisun Solano Water Auth	y																						
Vallejo, (City of)	y					y		y											y				y
Vista Irrigation Dist																							
COLORADO																							
Arvada, (City of)	y					y		y									y			y			
Colorado Springs Util	y							y									y						
Denver Water Dept	y		y			y		y															
Fort Collins Water Util						y		y	y								y			y			
Pueblo Board of Waterworks	y		y			y														y			
CONNECTICUT																							
Bridgeport Hydraulic Co - Main System	y					y		y		y							y						
Hartford Metropolitan Dist	y					y		y												y			
New Haven/Regional Water Auth	y					y		y		y							y					y	
WASHINGTON DC																							
Washington Aqueduct	y					y		y												y			

(Continued)

Table 8J.143 (Continued)

	Post-Disinfection							Corrosion Control			Membrane Processes											
Utility Name	Chlorine	Chlorine Dioxide	Chloramines	Ozone	UV Radiation	Fluoridation	Defluoridation	pH Adjustment	Alkalinity Adjustment	Corrosion Inhibitors	Reverse Osmosis	Nanofiltration	Ultrafiltration	Microfiltration	Electrodialysis	Ion Exchange	Iron/Manganese Removal	Manganese Green Sand	GranularActivated Carbon	Powdered Activated Carbon	Resin Adsorption	Air Stripping	Off Gas Treatment
DELAWARE																							
Newark/Artesian Water Company Inc	y					y		y		y							y	y				y	
Wilmington/United Water Delaware	y					y		y												y			
FLORIDA																							
Clearwater Water Div	y																						
Cocoa, (City of) Util Dept			y			y		y														y	
Coral Gables/Miami-Dade Water & Sewer Dept			y			y		y		y												y	
Jacksonville, (City of) Water Div	y																						
Jacksonville/United Water Florida	y									y													
Kissimmee, (City of)	y					y		y															
Lakeland Electric & Water Util						y		y															
Orlando/Orange County Util								y														y	
Orlando Util Comm	y					y		y															
Pensacola/Escambia County Util Auth								y											y				
Pompano Beach/Broward County Enviro Services	y		y			y																	
Sarasota County Util			y					y		y					y								
Tallahassee Water Util	y																						
Tampa Water Dept	y	y	y			y		y												y			
W Coast Regional Water Supply	y							y															
West Palm Beach/Palm Beach County Water Util			y						y	y													
GEORGIA																							
Atlanta-Fulton County Water	y					y				y							y			y			
Cobb County-Marietta Wtr Auth	y					y		y															
Columbus Water Works	y					y		y												y			
Morrow/Clayton County Water Auth	y					y				y							y			y			
Stone Mountain/DeKalb County Public Works	y					y		y									y						
HAWAII																							
Honolulu/Board of Water Supply	y																		y				
Pearl Harbor/Navy Public Wrks Util Dept	y					y																	
IOWA																							
Cedar Rapids Water Dept	y					y		y		y													
Des Moines Water Works	y					y				y						y				y			
ILLINOIS																							
Northern Illinois Water Inc - Champaign Division	y					y		y	y														
Chicago, (City of) Dept of Water	y					y				y										y			
Decatur, (City of)	y	y				y		y	y	y										y			
Elmhurst/DuPage Water Comm	y																						
Evanston Water Dept	y					y				y										y			
Interurban District - IL-American Water Co.	y		y			y		y		y									y	y			
Lake Bluff/Central Lake Cnty Water Agency						y				y									y				
Peoria District - IL-American Water Co.	y					y		y		y							y		y	y			
Rockford City Water Div						y				y									y				
Springfield/City Water, Light and Power	y		y			y														y			
Wilmette Water Plant	y					y				y										y			
INDIANA																							
Evansville Water & Sewer Dept	y					y		y															
Gary/Northwest Indiana Water Co.			y			y				y										y			
Indianapolis Water Co.			y			y																	
KANSAS																							
Kansas City Public Util Board	y		y			y		y												y			
KENTUCKY																							
Covington/Kenton County Water Dist No.1	y					y		y		y										y			
Lexington/KY-American Water Co.			y			y		y	y										y	y			
Louisville Water Company			y			y		y	y											y			
LOUISIANA																							
Baton Rouge Water Co.			y																				
Harahan/Jefferson Parish Water Dept	y		y			y				y										y			
Lake Charles, (City of) Water Div			y			y		y		y							y	y					
New Orleans Water/Sewer Board						y		y												y			
MASSACHUSETTS																							
Fall River Water Dept	y					y		y		y									y				
Lowell Water Dept	y					y													y				
New Bedford, (City of) Water Dept	y							y															
Springfield Water Dept	y							y		y													

(Continued)

Table 8J.143 (Continued)

	Post-Disinfection							Corrosion Control			Membrane Processes											
Utility Name	Chlorine	Chlorine Dioxide	Chloramines	Ozone	UV Radiation	Fluoridation	Defluoridation	pH Adjustment	Alkalinity Adjustment	Corrosion Inhibitors	Reverse Osmosis	Nanofiltration	Ultrafiltration	Microfiltration	Electrodialysis	Ion Exchange	Iron/Manganese Removal	Manganese Green Sand	GranularActivated Carbon	Powdered Activated Carbon	Resin Adsorption	Air Stripping	Off Gas Treatment
MARYLAND																							
Annapolis/Anne Arundel County DPW	y					y		y									y						
Laurel/Washington Suburban Sanit Comm	y					y		y												y			
MAINE																							
Portland Water District	y		y					y		y												y	
MICHIGAN																							
Ann Arbor Util Dept			y	y		y		y											y				
Grand Rapids, (City of)	y					y				y										y			
Holland/Wyoming Util Dept	y					y														y			
Kalamazoo Dept Public Util	y					y				y												y	
MINNESOTA																							
St. Paul Water Util	y					y		y												y			
MISSOURI																							
Independence Water Dept			y							y													
Kansas City Water Svcs Dept	y		y			y		y	y	y										y			
Springfield City Util	y					y		y	y											y			
St. Louis County Water Co.			y			y														y			
NORTH CAROLINA																							
Asheville, (City of)	y					y		y		y													
Charlotte-Mecklenburg Util	y					y		y	y											y			
Durham, (City of) Dept Water Res						y		y		y							y						
Fayetteville Public Works Comm	y					y		y		y										y			
Greensboro, (City of) Util Dept						y		y	y	y													
NEBRASKA																							
Lincoln Water System			y			y											y						
Omaha Metro Util Dist	y					y		y	y											y			
NEW HAMPSHIRE																							
Manchester Water Works	y							y		y									y				
NEW JERSEY																							
Brick Township Muni Util Auth	y							y															
Clifton/Passaic Valley Water Comm	y																						
Haddon System - NJ-American Water Co.	y							y		y							y		y				y
Harrington Park/United Water New Jersey			y					y															
Iselin/Middlesex Water Company	y							y		y										y		y	
Little Falls/Newark Div of Water/Sewer Util	y							y	y														
Short Hills System - NJ-American Water Co.	y							y		y									y	y		y	
Shrewsbury/NJ-American Water Co.- Monmouth System	y					y	y	y		y							y		y	y		y	
Westfield/Elizabethtown Water Co.	y		y			y		y	y	y							y	y		y		y	
NEW MEXICO																							
Albuquerque Public Works Dept	y					y																	
NEVADA																							
Boulder City/Southern Nevada Water System	y									y										y			
Las Vegas Valley Wtr Dist	y																						
North Las Vegas, (City of)																							
NEW YORK																							
Buffalo/Erie County Water Auth	y					y		y	y											y			
Lynbrook/Long Island Water Inc								y									y					y	
Oakdale/Suffolk County Water Auth	y							y		y							y	y	y				
Rochester, (City of) Water Bureau	y					y		y		y													
Syracuse/Onondaga County Water Auth	y					y														y			
Syracuse Dept of Water						y																	
West Nyack/United Water New York										y													
OHIO																							
Akron Public Util Bureau	y					y		y		y										y			
Cincinnati Water Works	y					y		y											y	y			
Cleveland Div of Water	y					y				y										y			
Columbus Water Div	y					y		y	y	y										y			
Dayton Water Supply/Treatment	y					y																	
Fairfield/Hamilton Util		y				y																	
Toledo, (City of) Water Div	y	y				y		y	y											y			
OKLAHOMA																							
Lawton Water/Wastewater Plants	y		y			y		y	y											y			
Oklahoma City, (City of)	y		y			y		y		y									y	y			
Tulsa Public Works Dept	y					y		y	y										y	y			
OREGON																							
Eugene Water & Electric Board								y															
Portland Water Bureau			y																				

(Continued)

Table 8J.143 (Continued)

	Post-Disinfection							Corrosion Control			Membrane Processes												
Utility Name	**Chlorine**	**Chlorine Dioxide**	**Chloramines**	**Ozone**	**UV Radiation**	**Fluoridation**	**Defluoridation**	**pH Adjustment**	**Alkalinity Adjustment**	**Corrosion Inhibitors**	**Reverse Osmosis**	**Nanofiltration**	**Ultrafiltration**	**Microfiltration**	**Electrodialysis**	**Ion Exchange**	**Iron/Manganese Removal**	**Manganese Green Sand**	**GranularActivated Carbon**	**Powdered Activated Carbon**	**Resin Adsorption**	**Air Stripping**	**Off Gas Treatment**
PENNSYLVANIA																							
Allentown Muni Water System	y																			y			
Bethlehem, (City of)	y					y		y	y	y													
Bryn Mawr/Philadelphia Suburban Water Co.			y			y		y	y	y										y			
Harrisburg/United Water PA	y							y		y										y			
Lancaster, (City of) Water Bureau	y					y		y												y			
Philadelphia Water Dept			y			y				y										y			
Pittsburgh/West View Water Auth	y					y		y	y										y	y			
Reading Area Water Auth			y			y		y		y										y			
The York Water Co.			y					y									y			y			
RHODE ISLAND																							
Pawtucket Water Supply Board	y					y		y		y													
SOUTH CAROLINA																							
Charleston Comm Public Works	y	y	y			y		y		y										y			
Spartanburg Water System	y					y		y		y							y			y			
SOUTH DAKOTA																							
Sioux Falls, (City of)	y					y		y												y			
TENNESSEE																							
Chattanooga/TN-American Water Co.	y					y		y		y									y	y			
Memphis Light Gas & Water	y					y				y													
Nashville/Metro Water Services	y					y				y										y			
TEXAS																							
Abilene, (City of)			y					y	y	y													
Amarillo Muni Water System	y																			y			
Arlington/Trinity River Auth			y			y		y															
Austin, (City of) Water/WW			y			y														y			
Dallas Water Util			y			y		y	y											y			
El Paso Water Util	y									y									y	y			y
Gavelston, (City of) Dept Pub Wks	y		y					y															
Houston, (City of)	y		y			y		y												y			
Killeen/Bell County WCID #1			y			y																	
Lubbock, (City of) Water Util																				y			
San Antonio Water System																							
Wichita Falls, (City of)			y			y		y	y											y			
UTAH																							
Layton/Weber Basin Water Cons Dist	y																		y	y			
Orem/Central Utah Water Cons Dist	y																			y			
Salt Lake City/Metropolitan Water Dist	y							y	y														
Salt Lake City Public Util	y								y														
West Jordan/Salt Lake Cnty Water Cons Dist	y							y												y			
VIRGINIA																							
Chesapeake, (City of)	y		y			y				y										y		y	
Chesterfield County Util Dept	y					y				y							y						
Merrifield/Fairfax County Water Auth	y		y			y		y	y										y				
Newport News Waterworks	y					y		y	y	y							y			y			
Petersburg/Appomattox River Water Auth	y					y		y		y							y		y				
Richmond/Henrico Co Public Util										y													
Richmond Dept of Public Util			y			y				y										y			
Woodbridge/Prince William Co. Service Auth																							
WASHINGTON																							
Everett Water Util	y					y		y	y														
Seattle Water Dept																							
Tacoma Public Util						y		y														y	
Vancouver, (City of)	y					y											y	y				y	y
WISCONSIN																							
Green Bay Water Util	y					y											y			y			
Kenosha Water Util	y					y																	
Madison Water Util						y																	
Milwaukee Water Works	y		y			y														y			
Sheboygan Water Util	y					y				y										y			
WEST VIRGINIA																							
Charleston/WV-American Water Co. - Kanawha Valley District	y					y				y										y			

Note: Selected utilities serving a population of 100,000 or more.

Source: Adapted from water:\Stats: 1996 Survey, by permission. Copyright © 2000, American Water Works Associations. (Updated information available electronically from AWWA at 800-926-7337), www.awwa.org.

Table 8J.144 Water Treatment Processes Considered for Best Available Technology

Conventional Processes	Advanced Processes
Coagulation, sedimentation, filtration	Activated alumina
Direct filtration	Adsorption
Diatomaceous earth filtration	GAC
Slow sand filtration	Powdered activated carbon
Lime softening	Resins
Ion exchange	Aeration
Oxidation-disinfection	Packed column
Chlorination	Diffused air
Chorine dioxide	Spray
Chloramines	Slat tray
Ozone	Mechanical
Bromine	Cartridge filtration
Others	Electrodialysis
	Reverse osmosis
	Ultrafiltration
	Ultraviolet light (UV)
	UV with other oxidants

Source: From Dyksen, J.E., Hiltebrand, D.J., and Raczko. R.F., 1988, SDWA amendments: effects on the water industry. *J. Am. Water Works Assoc.*, vol. 80, no. 1. Copyright AWWA. Reprinted with permission.

Table 8J.145 Chemicals Used for Treatment by Public Water-Supply Systems in the United States and Canada

Chemical	Total Use Tons
Quick lime	330,988
Aluminum sulfate	188,986
Chlorine	79,034
Hydrated lime	44,679
Caustic soda	39,030
Carbon dioxide	18,111
Soda ash	13,750
Ferrous sulfate	10,590
Powdered activated carbon	9,016
Ferric sulfate	7,956
Sodium silicofluoride	7,903
Polyelectrolytes	5,915
Ammonia	5,232
Phosphate	3,970
Copper sulfate	2,825
Granular activated carbon	2,587
Potassium permanganate	1,231
Sodium aluminate	1,129
Hypochlorites	1,112
Sodium chloride	828
Clays	133

Note: Based on data from 430 of the largest U.S. utilities and 24 of the 75 largest Canadian utilities.

Source: From American Water Works Association, 1988. Grigg, N.S., 1984 Water Utility Operating Data, Summary Report. Reprinted with permission.

Table 8J.146 Costs of Some Water Treatment Technologies

Population Range	Type of Treatment	Cost per Family per Year
501–1,000	conventional coagulation	$125
50,001–75,000	filtration and disinfection to	$ 50
>1,000,000	control microbial contaminants	$ 25
501–1,000	corrosion control (stabilization	$ 60
50,001–75,000	with lime) to control lead and	$ 15
>1,000,000	other corrosion products	<$ 10
501–1,000	packed tower aeration to	$ 55
50,000–75,000	control organic chemicals	$ 28
>1,000,000		$ 20
501–1,000	granular activated carbon to	$190
50,001–75,000	control synthetic organic	$130
>1,000,000	chemicals	$ 40

Note: Basic Assumptions: 3.2 persons per household; each person using 180 gallons per day; total cost per household including operation, maintenance and amortization of capital at 10 percent per year for 20 years.

Source: From U.S. Environmental Protection Agency; League of Women Voters Education Fund, 1987, Safety on Tap. Reprinted with permission.

Table 8J.147 Cost of Treating Contaminated Groundwater (Portable Treatment Systems, 1987)

Unit type	Cost (dollars/unit[a])
In situ biological treatment (suspended growth reactor)	$15–40/cu yd, treated
Rotating biodisks	0.20–1.10
Trickling filter	0.08–0.15
Activated sludge	0.10–0.30
Packed towers	0.02–0.10
Aeration basins	0.02–0.08
Carbon adsorption	0.20–0.90
Ultraviolet/hydrogen peroxide	0.04–0.18
Belt press	0.01–0.05
Mixing tangs (including chemicals)	0.03–0.29
Equalization tanks	0.005–0.01
Clarifiers	0.008–0.06
Solidification of solids in situ	0.20–1.00

Note: Portable treatment systems — 1987.

[a] All costs per 1,000 gal except as noted.

Source: From Estimated by Geraghty & Miller Inc Oak Ridge, TN.

Table 8J.148 Treatment Costs for Removal of Trichloroethylene from Drinking Water [US Dollars; Cost Data as of 1984]

	Treatment Technique					
	GAC absorption[a]			Packed-Tower Aeration[b]		
	System Size					
Cost Component[c]	0.037 mgd	0.95 mgd	36.8 mgd	0.037 mgd	0.95 mgd	36.8 mgd
Capital[d] (*Thousand$*)	24	240	9,000	69	264	4,789
Annual O&M (*Thousand $*)	4.5	86	710	1.4	18	617
Total (¢ per 1,000 gal)	57.0	34.0	14.0	79.0	15.5	9.4

Note: Raw water concentration 500 ug/L; assuming 99-percent removal.

[a] Based on 10-min empty bad contact time.

[b] Does not include air pollution controls.

[c] USEPA estimates in August 1983 dollars.

[d] Includes site work, engineering, contractor overhead and profit, and contingencies.

Source: From American Water Works Association, 1988, *New Dimensions in Safe Drinking Water.* Copyright AWWA. Reprinted with permission.

Table 8J.149 Cost of Removal of Volatile Organic Chemicals in Drinking Water

	Capacity[a]			Cost (Dollars per Thousand Gallons)[b]		
Contaminant (1)	Millions of Gallons per Day (2)	μg/L (3)	Percent Removal (4)	Tower (5)	Aeration Basin (6)	Carbon Adsorption (7)
Trichloroethylene	0.5	100	90	0.273	0.546	0.868
		10	99	0.287	0.793	0.918
		1	99.9	0.296	1.032	1.010
		0.1	99.99	0.303	1.270	1.124
	1	100	90	0.182	0.383	0.637
		10	99	0.191	0.611	0.679
		1	99.9	0.196	0.850	0.765
		0.1	99.99	0.202	1.088	0.867
	10	100	90	0.083	0.207	0.356
		10	99	0.088	0.403	0.390
		1	99.9	0.093	0.587	0.458
		0.1	99.99	0.099	0.755	0.543
Tetrachloroethylene	0.5	100	90	0.279	0.637	0.610
		10	99	0.293	0.935	0.660
		1	99.9	0.302	1.228	0.705
		0.1	99.99	0.308	1.486	0.805
	1	100	90	0.186	0.460	0.453
		10	99	0.194	1.752	0.502
		1	99.9	0.201	1.046	0.548
		0.1	99.99	0.206	1.296	0.651
	10	100	90	0.085	0.277	0.197
		10	99	0.091	0.514	0.224
		1	99.9	0.098	0.726	0.251
		0.1	99.99	0.103	0.905	0.313
1.1.1-Trichloroethane	0.5	100	90	0.270	0.502	1.445
		10	99	0.289	0.825	1.651
		1	99.9	0.307	1.421	1.945
		0.1	99.99	0.332	2.572	2.605
	1	100	90	0.180	0.348	1.396
		10	99	0.192	0.644	1.500
		1	99.9	0.205	1.234	1.801
		0.1	99.99	0.230	2.313	2.402
	10	100	90	0.082	0.176	0.802
		10	99	0.089	0.430	0.973
		1	99.9	0.102	0.860	1.229
		0.1	99.99	0.122	1.821	1.818
Carbon tetrachloride	0.5	100	90	0.264	0.428	0.942
		10	99	0.287	0.531	1.021
		1	99.9	0.272	0.600	1.132
		0.1	99.99	0.280	0.648	1.340
	1	100	90	0.176	0.292	0.703
		10	99	0.181	0.371	0.775
		1	99.9	0.184	0.427	0.940
		0.1	99.99	0.186	0.470	1.063
	10	100	90	0.081	0.133	0.408
		10	99	0.683	0.196	0.467
		1	99.9	0.084	0.247	0.550
		0.1	99.99	0.085	0.286	0.719
Cis-1,2-Dichloroethylene	0.5	100	90	0.284	0.727	2.513
		10	99	0.296	1.010	2.791
		1	99.9	0.304	1.281	3.153
		0.1	99.99	0.310	1.572	3.511
	1	100	90	0.189	0.547	2.156
		10	99	0.196	0.828	2.417
		1	99.9	0.202	1.098	2.760
		0.1	99.99	0.206	1.379	3.099

(Continued)

Table 8J.149 (Continued)

	Capacity[a]			Cost (Dollars per Thousand Gallons)[b]		
Contaminant (1)	**Millions of Gallons per Day (2)**	**μg/L (3)**	**Percent Removal (4)**	**Tower (5)**	**Aeration Basin (6)**	**Carbon Adsorption (7)**
	10	100	90	0.087	0.350	1.735
		10	99	0.093	0.571	1.989
		1	99.9	0.099	0.763	2.327
		0.1	99.99	0.104	0.966	2.660
1,2-Dichloroethane	0.5	100	90	0.276	0.587	1.286
		10	99	0.285	0.749	1.465
		1	99.9	0.292	0.901	1.748
		0.1	99.99	0.297	1.054	2.322
	1	100	90	0.184	0.415	1.015
		10	99	0.190	0.568	1.177
		1	99.9	0.194	0.720	1.437
		0.1	99.99	0.197	0.871	2.980
	10	100	90	0.084	0.237	0.675
		10	99	0.087	0.368	0.820
		1	99.9	0.090	0.489	1.057
		0.1	99.99	0.094	0.603	1.566
1,1-Dichlorethylene	0.5	100	90	0.262	0.406	0.880
		10	99	0.265	0.448	0.963
		1	99.9	0.270	0.500	1.066
		0.1	99.99	0.272	0.531	1.243
	1	100	90	0.174	0.274	0.647
		10	99	0.177	0.307	0.721
		1	99.9	0.180	0.348	0.814
		0.1	99.99	0.181	0.371	0.977
	10	100	90	0.080	0.121	0.364
		10	99	0.081	0.144	0.423
		1	99.9	0.082	0.176	0.499
		0.1	99.99	0.083	0.196	0.640

Note: US dollars; cost data as of 1984.

[a] To convert from mgd to m^3/day, multiply by 3,785.
[b] To convert from dollars/1,000 gal to dollars/m^3 multiply by 0.26412.

Source: From Clark, R.M., Eilers, R.G., and Goodrich, J.A., 1984, *VOC's in Drinking Water: Cost of Removal*, U.S. Environmental Protection Agency, Cincinnati, OH 45268; PB85-166429.

Table 8J.150 Cost of Removal of MTBE from Drinking Water for Various Treatment Systems

Flow (gpm)	Influent (μg/L)	Effluent (μg/L)	Removal (%)	Air Stripping			AOPs				
				Packed Tower	Low Profile	Packed Tower w/OGT	H_2O_2/MP-UV	O_3/H_2O_2	GAC	Resin Sorption	Lowest Unit Cost Amongst the Technologies Evaluated
60	20	5	75.00	**$1.66**	NE	NR	$2.18	$2.63	NE	$2.50	Packed Tower
	20	0.5	97.50	**$1.75**	$1.86	NR	$2.50	$2.68	$2.30	$2.81	Packed Tower
	200	20	90.00	**$1.66**	$1.70	NR	$2.32	$2.65	NE	$4.16	Packed Tower
	200	5	97.50	**$1.75**	$1.80	NR	$2.50	$2.68	NE	$4.16	Packed Tower
	200	0.5	99.75	**$1.82**	$1.89	NR	$2.72	$2.98	$3.10	$4.16	Packed Tower
	2,000	20	99.00	$1.79[a]	$1.90	**$3.08**	$3.07	$3.29	NE	$4.56	Packed Tower with OGT
	2,000	5	99.75	$1.82[a]	$2.02	**$3.20**	$3.47	$3.31	NE	$4.57	Packed Tower with OGT
	2,000	0.5	99.98	NE	NE	NE	$4.11	**$3.62**	$4.61	$4.57	O_3/H_2O_2
600	20	5	75.00	$0.30	$0.78	NR	$0.57	$0.82	NE	$1.01	Packed Tower
	20	0.5	97.50	$0.34	$0.92	NR	$0.91	$0.90	$0.77	$1.01	Packed Tower
	200	20	90.00	$0.32[a]	$0.85	**$0.57**	$0.71	$0.84	NE	$1.16	Packed Tower with OGT
	200	5	97.50	$0.34[a]	$0.96	**$0.59**	$0.96	$0.90	NE	$1.17	Packed Tower with OGT
	200	0.5	99.75	$0.37[a]	$1.09	**$0.62**	$1.27	$0.95	$1.15	$1.17	Packed Tower with OGT
	2,000	20	99.00	$0.36[a]	$0.96	**$0.90**	$1.52	$1.07	NE	$1.32	Packed Tower with OGT
	2,000	5	99.75	$0.37[a]	$1.09	**$0.91**	$1.75	$1.13	NE	$1.36	Packed Tower with OGT
	2,000	0.5	99.98	NE	NE	NE	$2.08	**$1.19**	$2.37	$1.38	O_3/H_2O_2
6,000	20	5	75.00	$0.13[a]	$0.34	$0.36	$0.32	$0.35	NE	**$0.30**	Resin Sorption
	20	0.5	97.50	$0.16[a]	$0.48	$0.39	$0.52	$0.43	$0.50	**$0.36**	Resin Sorption
	200	20	90.00	$0.15[a]	$0.41	$0.38	$0.42	**$0.37**	NE	$0.39	O_3/H_2O_2
	200	5	97.50	$0.16[a]	$0.48	**$0.39**	$0.60	$0.43	NE	$0.41	Packed Tower with OGT
	200	0.5	99.75	$0.17[a]	$0.64	**$0.40**	$0.74	$0.48	$0.97	$0.41	Packed Tower with OGT
	2,000	20	99.00	$0.17[a]	NE	NE	$0.65	$0.56	NE	**$0.53**	Resin Sorption
	2,000	5	99.75	$0.18[a]	NE	NE	$1.24	$0.59	NE	**$0.54**	Resin Sorption
	2,000	0.5	99.98	NE	NE	NE	$1.59	$0.68	$2.22	**$0.58**	Resin Sorption

Note: Costs are in 1999 dollars, total amortized costs 1,000 gallons; NE, not evaluated due to lack of data; NR, off-gas treatment not required; Bold numbers indicate the lowest unit cost amongst the technologies evaluated; OGT, off-gas treatment; AOP, Treatment Costs for byproduct and residual oxidant removal not included.

[a] Off-gas treatment is expected to be required based on 1 lb/day emission standards.

Source: From Stocking, A. et al., 2000, Appendix 24, MTBE Treatability, Executive Summary, Treatment Technologies for Removal of Methyl Tertiary Butyl Ether (MTBE) from *Drinking Water: Air Stripping, Advanced Oxidation Processes, Granular Activated Carbon, Synthetic Resin Sorbents*, Second Edition, February 2000, Written for the California Agencies MTBE Research Partnership, Center for Groundwater Restoration and Protection, National Water Research Institute, NWRI-99-08, www.nwri-usa.org.

SECTION 8K WATER TREATMENT FACILITIES

Table 8K.151 Data on Selected Large Rapid-Sand Filter Plants in the United States

	Chemical Feed and Mix					Sedimentation				Filtration							
City	Capacity (mgd)	Type Feed	Alum Used (Grains/gal)	Lime Used (Grains/gal)	Mixing Time (min)	Type of Mix	Time (hr)	Flow V (ft/min)	Basin Depth (ft)	Filter Rate (mgd/acre)	Unit Size (mgd)	Sand Depth (in.)	Sand Size (mm)	Gravel Depth (in.)	Wash Rate (in./min)	Wash Water (percent)
Detroit, MI	320	Dry	0.74	0	3	Baffles	2.0	4.2	16	160	4	30	0.45	17	26–30	2.0
Milwaukee, WI	200	Dry	0.5	0.11	64	Mechanical	4.0	2.32	27	125	6.25	27	0.51	24	24	2.2
St. Louis, MO	160	Solution	1.2	4.7	45	Baffles	36.0	1.5	16–23	125	4	30	0.4	12	24	1.67
Louisville, KY	120	Dry	0.65	0.31	—	None	2.0	5.7	17.5	125	6 and 3	26–30	0.4–0.5	14–24	22–36	2.0
Toledo, OH	80	Dry	1.12	0.5	40	Mechanical	2.8	1.5	15	94	2	22	0.42	18	20	2.43
Denver, CO	64	Dry	0.1–1.0	0.05–0.5	20	Baffles	1.0	12.0	13.5	156	4.5	48[a]	0.62	15	25	1.4
Atlanta, GA	54	Dry	0.59	0.26	23	Baffles	9.3	0.7	14–24	125	3 and 5	24–27	0.4–0.5	18	30	1.3
Dallas, TX	48	Dry	1.0	5.2	12	Baffles	8.0	1.1	18	120	2	30	0.4–0.45	18	20	1.5
New Orleans, LA	40	Solution	0.75	5.0	60	Baffles	18.0	0.55	13.75	122	4	30	0.33	9	24	0.3
Albany, NY	32	Dry	1.5	0	20	Baffles	2.25	1.5	10–18	125	4	30	0.33	18	24	2.5
Richmond, VA	30	Dry	2.0	0	10	Combined	10.0	6.0	10	120	3	26	0.43	16	24	1.0

[a] Coal.

Source: From Cosens, *J. Am. Water Works Assoc*, 1956.

Table 8K.152 Community Water System Treatment Schemes (Percentage of Plants Using Each Treatment Scheme)

	System Service Population Category								
Water Source	**100 or Less**	**101–500**	**501–3,300**	**3,301–10,000**	**10,001–50,000**	**50,001–100,000**	**100,001–500,000**	**Over 500,000**	**All Sizes**
Groundwater Plants									
Disinfection with no additional treatment	68.7	55.5	52.1	37.4	43.5	54.2	60.3	84.1	54.5
Other chemical addition	7.7	21.7	11.4	30.5	15.8	15.0	2.7	0.6	16.1
Ion exchange, activated alumina, aeration	0.0	9.8	22.0	18.7	23.1	25.3	14.1	14.9	13.7
Other filtration (not direct or conventional)	12.3	9.5	3.9	4.6	9.0	0.6	7.0	0.2	7.6
Direct filtration	0.1	0.0	0.1	0.2	0.3	0.6	0.2	0.0	0.1
Conventional filtration	0.0	0.0	0.0	2.6	0.0	0.0	0.0	0.0	0.3
Membranes	1.1	0.0	0.1	0.5	0.0	0.6	0.6	0.0	0.3
Softening	10.0	3.6	6.9	5.5	6.9	1.4	3.7	0.2	6.1
Observations	83	98	127	125	168	191	394	469	1,655
Surface Water Plants									
Disinfection with no additional treatment	50.0	14.3	3.6	1.0	0.3	0.0	2.4	2.6	10.7
Other chemical addition	0.0	0.0	4.3	1.0	0.0	1.0	4.2	6.0	1.5
Ion exchange, activated alumina, aeration	0.0	2.3	2.1	7.1	6.5	5.3	4.6	10.2	4.0
Other filtration (not direct or conventional)	12.8	28.4	14.5	6.0	0.0	11.1	2.3	0.0	11.8
Direct filtration	7.2	18.8	9.2	16.9	15.5	8.9	13.5	15.4	13.5
Conventional filtration	11.9	8.9	37.4	35.7	59.2	60.7	64.5	45.4	34.7
Membranes	6.4	5.8	1.3	1.0	0.0	0.0	0.6	0.9	2.5
Softening	11.7	20.8	27.6	30.2	16.3	12.9	6.7	12.0	20.6
Observations	50	58	76	82	85	81	169	115	716
Mixed Plants									
Disinfection with no additional treatment	0.0	89.8	100.0	46.7	14.2	0.0	3.2	0.0	51.5
Other chemical addition	0.0	0.0	0.0	0.0	2.7	0.0	14.2	0.0	1.8
Ion exchange, activated alumina, aeration	0.0	0.0	0.0	14.6	0.0	28.5	16.3	0.0	5.9
Other filtration (not direct or conventional)	0.0	3.4	0.0	0.0	0.0	0.0	3.2	0.0	1.6
Direct filtration	0.0	6.8	0.0	7.3	9.6	0.0	3.2	0.0	6.8
Conventional filtration	100.0	0.0	0.0	16.8	31.5	57.1	40.5	100.0	17.9
Membranes	0.0	0.0	0.0	0.0	0.0	0.0	0.0	0.0	0.0
Softening	0.0	0.0	0.0	14.6	42.0	14.4	16.3	0.0	14.2
Observations	1	3	1	9	11	4	29	6	64
All Plants									
Disinfection with no additional treatment	67.0	52.9	47.5	31.8	33.5	37.3	43.9	68.5	49.4
Other chemical addition	7.0	19.8	10.7	25.1	12.1	10.6	3.6	1.5	14.3
Ion exchange, activated alumina, aeration	0.0	9.1	20.1	16.7	19.0	19.3	11.9	13.9	12.5
Other filtration (not direct or conventional)	12.4	10.8	4.9	4.7	6.9	3.7	5.7	0.2	8.0
Direct filtration	0.7	1.5	1.0	3.1	3.8	3.1	3.6	2.8	1.7
Conventional filtration	1.1	0.7	3.5	8.2	13.5	18.9	17.3	9.1	4.4
Membranes	1.6	0.4	0.2	0.6	0.0	0.4	0.6	0.2	0.5
Softening	10.2	4.8	8.9	9.6	9.7	5.0	4.9	2.3	7.8
Observations	134	159	204	216	264	276	592	590	2,435

Note: Excludes plants that treat purchased water. The tabulations presented in the Community Water-System Survey 2000 are based on data collected from a sample of U.S. Water Systems, not from a census of every water system in the United States.

Source: From USEPA, 2002, *Community Water System Survey 2000*, EPA 815-R-02-005A.

Table 8K.153 Community Water System Surface Water Treatment Practices (Percentage of Plants Performing Each Treatment)

Surface Water Treatment Practice	System Service Population Category								
	100 or Less	101–500	501–3,300	3,301–10,000	10,001–50,000	50,001–100,000	100,001–500,000	Over 500,000	All Sizes
Chlorination only	49.5	32.4	7.9	6.0	0.3	1.0	6.8	10.2	16.2
Raw water storage/ presedimentation basin	12.5	20.5	11.5	16.3	20.5	29.8	19.1	26.5	17.3
Predisinfection/oxidation prior to sedimentation									
Chlorine	18.6	2.9	37.6	50.9	51.7	45.8	52.2	45.4	34.5
Chlorine dioxide	0.0	0.0	0.0	3.0	3.9	11.4	7.4	4.3	2.2
Chloramines	0.0	0.0	0.0	1.0	8.3	0.0	10.0	11.1	2.4
Ozone	0.5	0.9	0.0	1.0	0.0	5.3	3.9	6.0	1.0
Potassium permanganate	0.0	1.1	24.9	29.4	37.1	29.2	27.8	26.6	20.3
Other predisinfection	0.0	0.0	4.8	1.0	0.0	0.0	1.7	0.0	1.3
Predisinfection/oxidation prior to filtration									
Chlorine	7.4	11.9	24.2	24.3	35.4	38.3	31.5	37.7	23.0
Chlorine dioxide	0.0	2.0	0.0	0.0	1.1	2.0	2.8	0.0	0.8
Chloramines	0.0	0.0	0.0	0.0	4.6	1.3	6.9	7.7	1.4
Ozone	0.0	15.3	0.0	0.0	1.1	2.3	4.1	6.0	3.5
Potassium permanganate	0.0	1.2	6.9	10.1	5.5	7.8	3.9	2.6	5.0
Other predisinfection	0.0	0.0	4.4	0.0	0.0	2.0	0.0	0.0	1.0
Rapid mix	7.8	13.1	43.3	67.5	90.9	75.6	77.8	62.4	48.9
Coagulation/flocculation	11.0	22.6	63.1	87.9	96.4	82.0	85.2	77.0	60.8
Polymers	15.5	21.5	54.4	63.7	62.1	53.4	58.4	53.9	46.0
Setting/sedimentation	11.9	15.9	60.8	67.9	79.8	74.7	71.7	63.3	51.5
Softening									
Lime/soda ash	0.0	20.8	27.6	34.2	20.7	12.9	7.2	12.0	20.5
Recarbonation	0.0	0.0	0.0	6.1	7.7	5.0	4.4	3.4	2.9
Ion exchange	11.7	0.0	0.0	0.0	0.0	0.0	0.0	0.0	1.6
Filtration									
Direct filtration	2.6	2.3	0.8	8.1	5.3	4.0	9.6	6.8	4.1
Micro strainer	0.0	0.0	0.0	0.0	0.0	0.0	1.7	0.9	0.1
Slow sand	4.9	10.8	4.8	3.1	0.0	0.0	1.7	0.0	4.3
Bag and cartridge	18.9	18.6	0.0	0.0	1.0	0.0	0.6	0.0	6.3
Rapid sand	0.0	9.4	24.5	32.5	17.6	7.9	16.2	23.2	17.3
Green sand	0.7	0.0	3.0	1.0	0.0	0.0	0.6	0.0	0.9
Diatomaceous earth	2.6	1.0	2.5	2.0	0.0	13.1	2.9	0.0	2.2
Dual/multi media	13.7	14.0	51.8	46.6	73.2	68.7	59.3	52.1	43.5
Pressure filtration	9.5	20.7	2.2	2.6	0.0	2.0	0.0	0.0	6.2
Other filtration	0.6	0.8	4.1	0.0	3.9	2.3	3.0	2.6	2.1
Post-disinfection after filters									
Chlorine	27.8	58.3	77.4	85.3	83.7	75.3	74.5	53.9	68.7
Chlorine dioxide	0.0	0.0	0.0	1.0	5.0	1.5	0.6	0.0	1.1
Chloramines	0.0	0.0	0.0	4.1	22.5	18.9	23.6	21.3	7.1
Ozone	0.0	0.7	0.0	0.0	0.0	5.3	0.0	1.7	0.4
UV	11.7	0.0	0.0	0.0	0.0	0.0	0.6	0.0	1.6
Other post disinfection	0.0	0.0	7.9	0.0	2.1	1.0	1.1	0.0	2.1
Clearwell	11.0	37.4	80.0	81.0	82.7	85.4	82.0	64.1	63.2
Membranes									
Reverse osmosis	2.0	0.0	0.0	0.0	0.0	0.0	0.0	0.0	0.3
Micro filtration	6.4	5.8	1.3	1.0	0.0	0.0	0.6	0.9	2.5
Ultrafiltration	0.0	0.0	0.0	0.0	0.0	0.0	0.0	0.0	0.0
Nanofiltration	0.0	0.0	0.0	0.0	0.0	0.0	0.0	0.0	0.0
Corrosion control	1.5	14.1	40.8	48.9	70.6	62.9	69.7	72.6	40.2
Miscellaneous									
Ion exchange	0.0	0.0	0.0	1.0	0.0	0.0	0.0	0.0	0.2
Granular activated carbon	2.2	2.6	0.8	21.1	14.9	15.9	11.0	6.0	8.6
Activated alumina	0.0	2.3	0.0	1.0	0.0	1.3	1.1	0.9	0.7
Aeration	0.0	0.0	2.1	6.1	7.6	5.0	5.7	10.2	3.6
Other									
Flouride	0.0	2.5	37.4	57.3	69.4	60.4	64.4	71.0	38.0
PAC	0.0	1.1	5.7	6.1	18.7	12.3	18.1	20.5	7.6

(Continued)

Table 8K.153 (Continued)

Surface Water Treatment Practice	System Service Population Category								
	100 or Less	101–500	501–3,300	3,301–10,000	10,001–50,000	50,001–100,000	100,001–500,000	Over 500,000	All Sizes
pH adjust	0.0	0.0	6.1	1.0	1.1	1.0	3.3	2.5	1.9
Iron/mag. removal/seq.	0.0	2.0	0.0	0.0	0.0	0.0	0.0	0.0	0.4
Taste/odor	0.0	0.0	0.0	1.0	0.3	0.0	1.2	0.0	0.3
Filter aid	0.0	0.0	7.7	5.9	2.2	1.0	0.0	0.0	3.1
Clarify	0.0	0.0	0.0	0.0	0.0	0.0	0.0	0.0	0.0
Blending	0.0	0.0	0.0	0.0	0.0	0.0	0.0	0.0	0.0

Note: Represents treatment practices for plants treating water that comes entirely or partly from surface sources. Chlorination only is indicated when a plant chlorinated but did not filter. It includes plants that only chlorinated and plants that chlorinated and used other nonfiltration practices. The tabulations presented in the Community Water System Survey 2000 are based on data collected from a sample of U.S. Water Systems, not from a census of every water system in the United States. Percentages may not sum to 100 percent because systems may perform more than one treatment.

Source: From USEPA, 2002, *Community Water System Survery 2000*, EPA 815-R-02-005A.

Table 8K.154 Community Water System Groundwater Treatment Practices

	System Service Population Category								
Groundwater Treatment Practice	**100 or Less**	**101–500**	**501–3,300**	**3,301–10,000**	**10,001–50,000**	**50,001–100,000**	**100,001–500,000**	**Over 500,000**	**All Sizes**
Chlorination only	81.4	73.2	72.7	72.5	67.8	71.1	68.6	96.9	74.3
Raw water storage/ Presedimentation basin	1.5	0.0	0.4	1.6	1.4	1.2	5.8	0.0	0.8
Predisinfection/oxidation prior to sedimentation									
Chlorine	0.0	7.5	9.4	10.6	11.3	5.9	9.5	0.0	7.2
Chlorine dioxide	0.0	0.0	0.0	0.0	0.7	10.5	0.4	0.0	0.2
Chloramines	0.0	0.0	0.0	0.0	0.0	0.6	4.5	0.0	0.1
Ozone	0.5	0.0	0.0	0.0	0.0	0.0	0.9	0.0	0.1
Potassium permanganate	0.0	3.7	3.2	4.4	0.7	1.2	0.4	0.0	2.6
Other Predisinfection	1.3	0.0	0.0	0.0	0.7	0.0	4.9	0.0	0.4
Predisinfection/oxidation prior to filtration									
Chlorine	1.5	7.1	10.0	7.5	10.8	1.2	3.0	2.5	6.9
Chlorine dioxide	0.0	0.0	0.0	0.0	0.0	0.4	0.0	0.0	0.0
Chloramines	0.0	0.0	0.0	0.0	0.0	1.2	0.4	0.0	0.0
Ozone	0.0	0.0	0.0	0.0	0.0	0.0	0.0	0.0	0.0
Potassium permanganate	1.0	1.4	5.0	2.4	4.0	0.0	0.4	0.0	2.6
Other predisinfection	0.0	1.2	0.0	0.0	0.0	0.0	0.0	0.0	0.4
Rapid mix	0.0	0.0	2.4	5.0	0.0	3.0	2.7	0.0	1.3
Coagulation/flocculation	0.0	0.0	4.3	6.0	2.6	4.7	1.7	0.0	2.1
Polymers	0.1	0.0	8.4	3.5	2.2	2.2	1.0	0.0	2.9
Settling/sedimentation	0.0	0.0	4.3	8.1	5.4	3.6	3.0	0.2	2.6
Softening									
Lime/soda ash	3.4	2.5	10.3	8.2	9.1	4.7	12.7	0.4	6.1
Recarbonation	0.0	0.0	1.0	5.1	2.2	1.8	7.1	0.2	1.2
Ion exchange	6.6	2.2	6.0	2.3	3.6	0.6	0.0	0.0	4.1
Filtration									
Direct filtration	1.1	0.0	0.0	2.5	0.0	1.2	0.2	0.0	0.5
Micro strainer	0.0	0.0	0.0	0.0	0.0	0.0	0.0	0.0	0.0
Slow sand	0.0	0.0	1.0	3.4	0.0	0.0	0.2	0.0	0.7
Bag and cartridge	8.9	0.1	0.4	0.0	0.0	0.0	0.2	0.0	1.8
Rapid sand	0.0	0.9	2.7	7.2	4.4	1.2	2.0	0.0	2.2
Green sand	1.6	7.5	4.2	4.4	2.9	0.0	0.4	0.0	4.5
Diatomaceous earth	0.0	0.0	0.0	0.0	0.3	0.0	0.0	0.0	0.0
Dual/multi media	0.1	0.0	4.6	3.4	5.7	3.2	10.5	2.5	2.4
Pressure filtration	0.0	3.7	8.1	2.6	7.2	1.2	1.0	0.0	4.2
Other filtration	0.8	2.1	0.0	0.3	0.3	0.8	0.0	0.0	0.8
Post-disinfection after filters									
Chlorine	8.5	7.8	15.3	21.0	19.9	3.3	11.5	2.7	12.3
Chlorine dioxide	0.0	0.0	0.0	0.0	0.0	0.0	6.6	0.0	0.1
Chloramines	0.0	0.0	0.0	0.8	1.4	1.8	3.4	0.0	0.3
Ozone	0.0	0.0	0.0	0.0	0.0	0.0	0.0	0.0	0.0
UV	0.0	0.0	0.0	0.0	0.0	0.0	0.0	0.0	0.0
Other post disinfection	0.0	0.0	1.4	0.0	0.0	5.8	0.0	0.0	0.5
Clearwell	11.4	15.4	24.7	23.6	16.0	6.9	13.3	0.4	17.7
Membranes									
Reverse osmosis	1.1	0.0	0.1	0.5	0.0	0.6	0.6	0.0	0.3
Micro filtration	0.0	0.0	0.0	0.3	0.0	0.0	0.0	0.0	0.0
Ultrafiltration	0.0	0.0	0.0	0.0	0.0	0.0	0.0	0.0	0.0
Nanofiltration	0.0	0.0	0.0	0.0	0.0	0.0	0.0	0.0	0.0
Corrosion control	8.0	17.6	17.4	28.9	21.2	19.0	7.3	0.8	16.9
Miscellaneous									
Ion exchange	0.0	0.0	0.0	0.0	1.4	0.0	0.0	0.0	0.1
Granular activated carbon	0.0	0.0	0.0	0.8	1.9	0.6	8.6	0.2	0.4
Activated alumina	0.0	0.0	0.0	0.0	0.0	18.1	0.0	0.0	0.3
Aeration	0.0	9.8	22.0	19.0	21.7	7.9	17.4	14.9	13.4
Other									
Flouride	0.0	4.4	18.3	27.6	20.5	9.4	12.1	0.6	11.4

(Continued)

Table 8K.154 (Continued)

Groundwater Treatment Practice	System Service Population Category								
	100 or Less	101–500	501–3,300	3,301–10,000	10,001–50,000	50,001–100,000	100,001–500,000	Over 500,000	All Sizes
PAC	0.0	0.0	0.0	0.0	0.0	0.0	0.4	0.0	0.0
pH adjust	0.0	1.2	4.7	0.0	2.2	0.0	0.0	0.0	1.8
Iron/mag. removal/seq.	2.3	3.8	3.0	4.7	1.4	0.6	0.0	0.0	3.1
Taste/odor	0.0	0.0	0.0	1.6	0.0	0.0	0.0	0.0	0.2
Filter aid	0.0	0.0	3.5	0.0	0.0	0.0	5.8	0.0	1.0
Clarify	0.0	0.0	0.0	0.0	0.0	0.0	0.0	0.0	0.0
Blending	0.0	2.0	0.0	6.0	0.7	1.2	0.2	0.0	1.4

Note: Represents treatment practices for plants treating water that comes entirely or partly from ground sources. Percentages may not sum to 100 percent because systems may perform more than one treatment. Chlorination only is indicated when a plant chlorinated but did not filter. It includes plant that only chlorinated and plants that chlorinated and used other nonfiltration practices. The tabulations presented in the Commonly Water System Survery 2000 are based on data collected from a sample of U.S. Water Systems, not from a census of every water system in the United States.

Source: From USEPA, 2002, *Community Water System Survery 2000*, EPA 815-R-02-005A.

CHAPTER 9

Wastewater

William H. Lynch

CONTENTS

SECTION 9A WASTEWATER CHARACTERISTICS

Ratio of Q peak hourly/Q design ave
10.0
8.0
6.0
5.0
4.0
3.0
2.0
1.5
1.0
0.1
0.2
0.3
0.4
0.5
0.7
1.0
2
3
4
5
7
10
20
30
40
50
70
100
Population in thousands

Q peak hourly: Maximum rate of wastewater flow (Peak hourly flow)
Q design ave: Design average daily wastewater flow

Source: Q peak hourly/Q design ave = $\frac{18 + \sqrt{P}}{4 + \sqrt{P}}$ - - - (P = population in thousands)

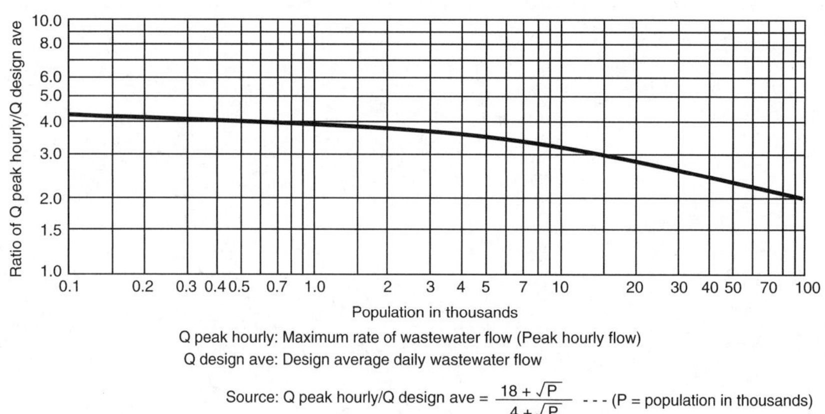

Figure 9A.1 Ratio of peak hourly flow to design average flow. (From Board of State and Provincial Public Health and Environmental Managers, Health Education Services Division, Recommended Standards for Wastewater Facilities, Figure 1, p. 10.5, 2004 Edition. www.hes.org.)

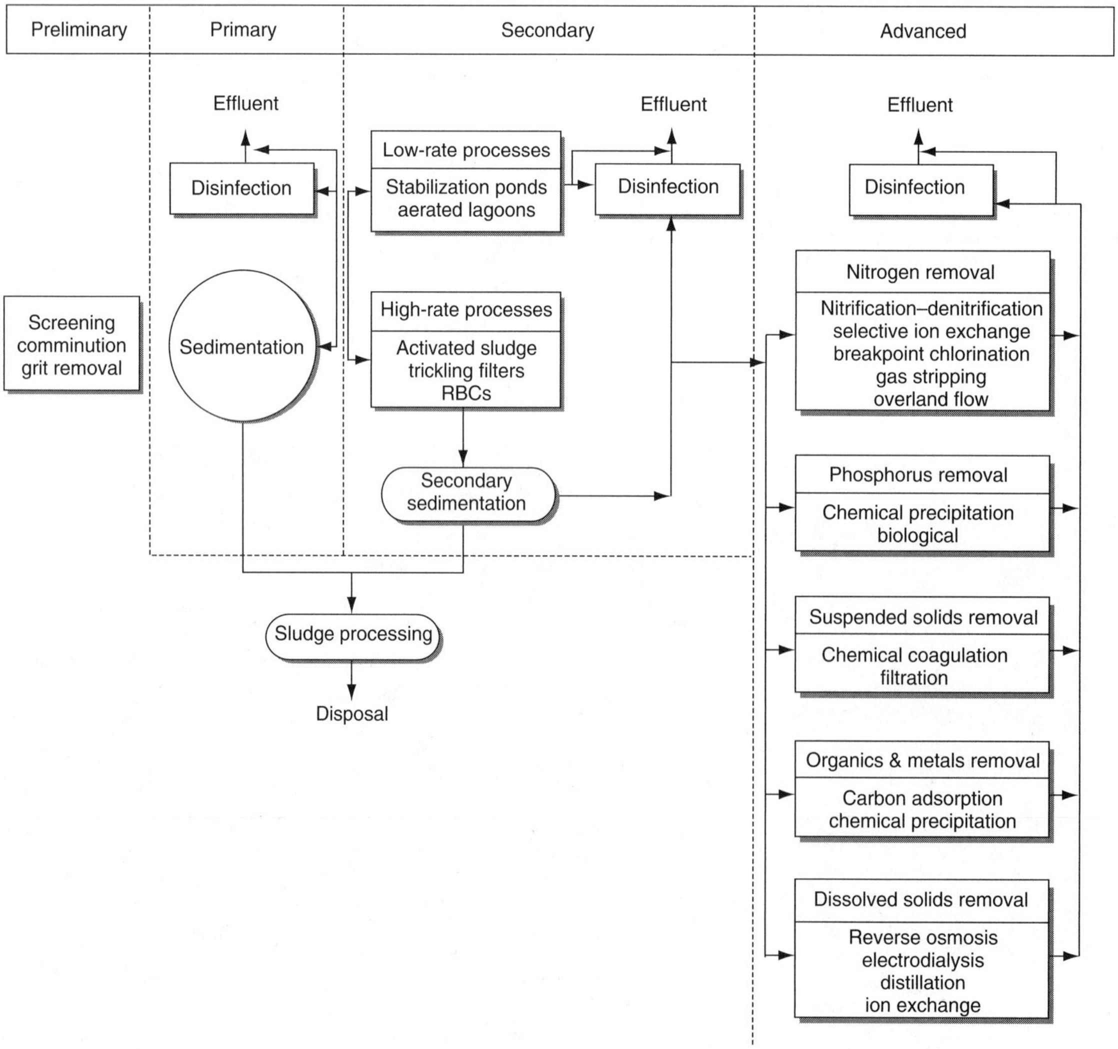

Preliminary
Primary
Secondary
Advanced
Effluent
Effluent
Effluent
Disinfection
Low-rate processes
Stabilization ponds aerated lagoons
Disinfection
Disinfection
Screening comminution grit removal
Sedimentation
High-rate processes
Activated sludge trickling filters RBCs
Secondary sedimentation
Nitrogen removal
Nitrification–denitrification selective ion exchange breakpoint chlorination gas stripping overland flow
Phosphorus removal
Chemical precipitation biological
Suspended solids removal
Chemical coagulation filtration
Organics & metals removal
Carbon adsorption chemical precipitation
Dissolved solids removal
Reverse osmosis electrodialysis distillation ion exchange
Sludge processing
Disposal

Figure 9A.2 Generalized flow sheet for wastewater treatment. (From USEPA, *Manual Guidelines Water Reuse*, Office of Water, Figure 12 (EPA/625/R-92/004), September 1992.)

Table 9A.1 Typical Wastewater Flowrates from Urban Residential Sources in the United States

Household Size, No. of Persons	Flowrate, gal/capita d		Flowrate, l/capita d	
	Range	Typical	Range	Typical
1	75–130	97	285–490	365
2	63–81	76	225–385	288
3	54–70	66	194–335	250
4	41–71	53	155–268	200
5	40–68	51	150–260	193
6	39–67	50	147–253	189
7	37–64	48	140–244	182
8	36–62	46	135–233	174

Source: From Metcalf & Eddy, Inc., McGraw-Hill, *Wastewater Engineering Treatment and Reuse, Fourth Edition*, 2003, Table 3.1, p. 156. With permission. Adapted in part from AWWARF (1999).

Table 9A.2 Typical Wastewater Flowrates from Recreational Facilities in the United States

Facility	Unit	Flowrate, gal/unit d		Flowrate, l/unit d	
		Range	Typical	Range	Typical
Apartment, resort	Person	50–70	60	190–260	230
Cabin, resort	Person	8–50	40	30–190	150
Cafeteria	Customer	2–4	3	8–15	10
	Employee	8–12	10	30–45	40
Camp					
With toilets only	Person	15–30	25	55–110	95
With central toilet and bath facilities	Person	35–50	45	130–190	170
Day	Person	15–20	15	55–76	60
Cottages, (seasonal with private bath)	Person	40–60	50	150–230	190
Country club	Member present	20–40	25	75–150	100
	Employee	10–15	13	38–57	50
Dining hall	Meal served	4–10	7	15–40	25
Dormitory, bunkhouse	Person	20–50	40	75–190	150
Fairground	Visitor	1–3	2	4–12	8
Picnic park with flush toilets	Visitor	5–10	5	19–38	19
Recreational vehicle park					
With individual connection	Vehicle	75–150	100	280–570	380
With comfort station	Vehicle	40–50	45	150–190	170
Roadside rest areas	Person	3–5	4	10–19	15
Swimming pool	Customer	5–12	10	19–45	40
	Employee	8–12	10	30–45	40
Vacation home	Person	25–60	50	90–230	190
Visitor center	Visitor	3–5	4	10–19	15

Source: From Metcalf & Eddy, Inc., McGraw-Hill, *Wastewater Engineering Treatment and Reuse, Fourth Edition*, 2003, Table 3.4, p. 159. With permission. Adapted from Metcalf & Eddy (1991), Salvato (1992), and Crites and Tchobanoglous (1998).

Table 9A.3 Typical Wastewater Flowrates from Commercial Sources in the United States

		Flowrate, gal/unit d		Flowrate, l/unit d	
Source	**Unit**	**Range**	**Typical**	**Range**	**Typical**
Airport	Passenger	3–5	4	11–19	15
Apartment	Bedroom	100–150	120	380–570	450
Automobile service station	Vehicle served	8–15	10	30–57	40
	Employee	9–15	13	34–57	50
Bar/cocktail lounge	Seat	12–25	20	45–95	80
	Employee	10–16	13	38–60	50
Boarding house	Person	25–65	45	95–250	170
Conference center	Person	6–10	8	40–60	30
Department store	Toilet room	350–600	400	1,300–2,300	1,500
	Employee	8–15	10	30–57	40
Hotel	Guest	65–75	70	150–230	190
	Employee	8–15	10	30–57	40
Industrial building (sanitary waste only)	Employee	15–35	20	57–130	75
Laundry (self-service)	Machine	400–550	450	1,500–2,100	1,700
	Customer	45–55	50	170–210	190
Mobile home park	Unit	125–150	140	470–570	530
Motel (with kitchen)	Guest	55–90	60	210–340	230
Motel (without kitchen)	Guest	50–75	55	190–290	210
Office	Employee	7–16	13	26–60	50
Public lavatory	User	3–5	4	11–19	15
Restaurant:					
Conventional	Customer	7–10	8	26–40	35
With bar/ cocktail lounge	Customer	9–12	10	34–45	40
Shopping center	Employee	7–13	10	26–50	40
	Parking space	1–3	2	4–11	8
Theater (Indoor)	Seat	2–4	3	8–15	10

Source: From Metcalf & Eddy, Inc., McGraw-Hill, *Wastewater Engineering Treatment and Reuse, Fourth Edition*, 2003, Table 3.2, p. 157. With permission. Adapted from Metcalf & Eddy (1991), Salvato (1992), and Crites and Tchobanoglous (1998).

Table 9A.4 Typical Wastewater Flowrates from Institutional Sources in the United States

		Flowrate, gal/unit d		Flowrate, l/unit d	
Source	**Unit**	**Range**	**Typical**	**Range**	**Typical**
Assembly hall	Guest	3–5	4	11–19	15
Hospital	Bed	175–400	250	660–1,500	1,000
	Employee	5–15	10	20–60	40
Institutions other than hospitals	Bed	75–125	100	280–470	380
	Employee	5–15	10	20–60	40
Prison	Inmate	80–150	120	300–570	450
	Employee	5–15	10	20–60	40
School, day:					
With cafeteria, gym, and showers	Student	15–30	25	60–120	100
With cafeteria only	Student	10–20	15	40–80	60
School, boarding	Student	75–100	85	280–380	320

Source: From Metcalf & Eddy, Inc., McGraw-Hill, *Wastewater Engineering Treatment and Reuse, Fourth Edition*, 2003, Table 3.3, p. 158. With permission. Adapted from Metcalf & Eddy (1991), Salvato (1992), and Crites and Tchobanoglous (1998).

Table 9A.5 Terminology Used to Quantify Observed Variations in Flowrate and Constituent Concentrations

Item	Description
Average dry-weather flow (ADWF)	The average of the daily flows sustained during dry-weather periods with limited infiltration
Average wet-weather flow (AWWF)	The average of the daily flows sustained during wet-weather periods when infiltration is a factor
Average annual daily flow	The average flowrate occurring over a 24-h period based on annual flowrate data
Instantaneous peak	Highest record flowrate occurring for a period consistent with the recording equipment. In many situations the recorded peak flow may be considerably below the actual peak flow because of metering and recording equipment limitations
Peak hour	The average of the peak flows sustained for a period of 1 h in the record examined (usually based on 10-min increments)
Maximum day	The average of the peak flows sustained for a period of 1 day in the record examined (the duration of the peak flows may vary)
Maximum month	The average of the maximum daily flows sustained for a period of 1 month in the record examined
Minimum hour	The average of the minimum flows sustained for a period of 1 h in the record examined (usually based on 10-min increments)
Minimum day	The average of the minimum flows sustained for a period of 1 day in the record examined (usually for the period from 2 a.m. to 6 a.m.)
Minimum month	The average of the minimum daily flows sustained for a period of 1 month in the record examined
Sustained flow (and load)	The value (flowrate or mass loading) sustained or exceeded for a given period of time (e.g., 1 h, 1 day, or 1 month)

Source: From Metcalf & Eddy, Inc., McGraw-Hill, *Wastewater Engineering Treatment and Reuse, Fourth Edition*, 2003, Table 3.11, p. 179. With permission. Adapted in part from Crites and Tchobanoglous (1998).

Table 9A.6 Terminology Commonly Used in the Field of Wastewater Engineering

Term	Definition
Biosolids	Primarily an organic, semisolid wastewater product that remains after solids are stabilized biologically or chemically and are suitable for beneficial use
Class A biosolids[a]	Biosolids in which the pathogens (including enteric viruses, pathogenic bacteria, and viable helminth ova) are reduced below current detectable levels
Class B biosolids[a]	Biosolids in which the pathogens are reduced to levels that are unlikely to pose a threat to public health and the environment under specific use conditions. Class B biosolids cannot be sold or given away in bags on other containers or applied on lawns or home gardens
Characteristics (wastewater)	General classes of wastewater constituents such as physical, chemical, biological, and biochemical
Composition	The makeup of wastewater, including the physical, chemical, and biological constituents
Constituents[b]	Individual components, elements, or biological entities such as suspended solids or ammonia nitrogen
Contaminants	Constituents added to the water supply through use
Disinfection	Reduction of disease-causing microorganisms by physical or chemical means
Effluent	The liquid discharged from a processing step
Impurities	Constituents added to the water supply through use
Nonpoint sources	Sources of pollution that originate from multiple sources over a relatively large area
Nutrient	An element that is essential for the growth of plants and animals. Nutrients in wastewater, usually nitrogen and phosphorus, may cause unwanted algal and plant growths in lakes and streams
Parameter	A measurable factor such as temperature
Point sources	Pollutional loads discharged at a specific location from pipes, outfalls, and conveyance methods from either municipal wastewater treatment plants or industrial waste treatment facilities
Pollutants	Constituents added to the water supply through use
Reclamation	Treatment of wastewater for subsequent reuse application or the act of reusing treated wastewater

(Continued)

Table 9A.6 (Continued)

Term	Definition
Recycling	The reuse of treated wastewater and biosolids for beneficial purposes
Repurification	Treatment of wastewater to a level suitable for a variety of applications including indirect or direct potable reuse
Reuse	Beneficial use of reclaimed or repurified wastewater or stabilized biosolids
Sludge	Solids removed from wastewater during treatment. Solids that are treated further are termed biosolids
Solids	Material removed from wastewater by gravity separation (by clarifiers, thickeners, and logoons) and is the solid residue from dewatering operations

[a] U.S. EPA (1997b).
[b] To avoid confusion, the term “constituents” is used in this text in place of contaminants, impurities, and pollutants.

Source: From Metcalf & Eddy, Inc., McGraw-Hill, *Wastewater Engineering Treatment and Reuse, Fourth Edition*, 2003, Table 1.1, p. 4. With permission. Adapted in part from Crites and Tchobanoglous (1998).

Table 9A.7 Levels of Wastewater Treatment

Treatment Level	Description
Preliminary	Removal of wastewater constituents such as rags, sticks, floatables, grit, and grease that may cause maintenance or operational problems with the treatment operations, processes, and ancillary systems
Primary	Removal of a portion of the suspended solids and organic matter from the wastewater
Advanced primary	Enhanced removal of suspended solids and organic matter from the wastewater. Typically accomplished by chemical addition or filtration
Secondary	Removal of biodegradable organic matter (in solution or suspension) and suspended solids. Disinfection is also typically included in the definition of conventional secondary treatment
Secondary with nutrient removal	Removal of biodegradable organics, suspended solids, and nutrients (nitrogen, phosphorus, or both nitrogen and phosphorus)
Tertiary	Removal of residual suspended solids (after secondary treatment), usually by granular medium filtration or microscreens. Disinfection is also typically a part of tertiary treatment. Nutrient removal is often included in this definition
Advanced	Removal of dissolved and suspended materials remaining after normal biological treatment when required for various water reuse applications

Source: From Metcalf & Eddy, Inc., McGraw-Hill, *Wastewater Engineering Treatment and Reuse, Fourth Edition*, 2003, Table 1.4, p. 11. Adapted in part from Crites and Tchobanoglous (1998).

Table 9A.8 Commonly Used Treatment Processes and Optional Treatment Methods

Treatment Objective	Treatment Process	Treatment Methods
Suspended solids removal	Sedimentation	Septic tank Free water surface constructed wetland Vegetated submerged bed
	Filtration	Septic tank effluent screens Packed-bed media filters (incl. dosed systems) Granular (sand, gravel, glass, bottom ash) Peat, textile Mechanical disk filters Soil infiltration
Soluble carbonaceous BOD and ammonium removal	Aerobic, suspended-growth reactors	Extended aeration Fixed-film activated sludge Sequencing batch reactors (SBRs)
	Fixed-film aerobic bioreactor	Soil infiltration Packed-bed media filters (incl. dosed systems) Granular (sand, gravel, glass) Peat, textile, foam Trickling filter Fixed-film activated sludge Rotating biological contactors
	Lagoons	Facultative and aerobic lagoons Free water surface constructed wetlands
Nitrogen transformation	Biological Nitrification (N) Denitrification (D)	Activated sludge (N) Sequencing batch reactors (N) Fixed film bio-reactor (N) Recirculating media filter (N, D) Fixed-film activated sludge (N) Anaerobic upflow filter (N) Anaerobic submerged media reactor (D) Submerged vegetated bed (D) Free water surface constructed wetland (N, D)
	Ion exchange	Cation exchange (ammonium removal) Anion exchange (nitrate removal)
Phosphorus removal	Physical/Chemical	Infiltration by soil and other media Chemical flocculation and settling Iron-rich packed-bed media filter
	Biological	Sequencing batch reactors
Pathogen removal (bacteria, viruses, parasites)	Filtration/Predation/Inactivation	Soil infiltration Packed-bed media filters Granular (sand, gravel, glass bottom, ash) Peat, textile
	Disinfection	Hypochlorite feed Ultraviolet light
Grease removal	Flotation	Grease trap Septic tank
	Adsorption	Mechanical skimmer
	Aerobic biological treatment (incidential removal will occur; overloading is possible)	Aerobic biological systems

Source: From USEPA, On-Site *Wastewater Treatment Systems Manual*, Office of Water, Office of Research and Development, EPA, (EPA/625/R-00/0008). www.epa.gov/ord/NRMRL/Pubs/625R00008/625R00008totaldocument.pdf.

Table 9A.9 Number of Operational Treatment Facilities and Collection Systems in 2000

State	Treatment Facilities	Collection Systems
Alabama	272	275
Alaska	45	46
Arizona	118	132
Arkansas	335	367
California[a]	586	797
Colorado[a]	311	391
Connecticut	91	137
Delaware	18	42
District of Columbia	1	1
Florida	277	317
Georgia	352	403
Hawaii	21	21
Idaho	168	207
Illinois	721	1,018
Indiana	404	482
Iowa	726	756
Kansas	634	673
Kentucky	224	255
Louisiana	355	382
Maine	137	171
Maryland	156	201
Massachusetts	126	230
Michigan	396	663
Minnesota	514	655
Mississippi	303	352
Missouri	678	751
Montana	194	204
Nebraska	464	469
Nevada[b]	85	117
New Hampshire	85	117
New Jersey	156	575
New Mexico	55	64
New York[a]	588	1,048
North Carolina	491	617
North Dakota	282	284
Ohio	765	1,008
Oklahoma	489	495
Oregon	207	254
Pennsylvania	779	1,553
Rhode Island	21	34
South Carolina	186	206
South Dakota[a]	271	274
Tennessee	246	281
Texas	1,363	1,675
Utah	97	164
Vermont	81	97
Virginia	227	290
Washington	235	331
West Virginia	212	289
Wisconsin	592	823
Wyoming[b]	96	121
American Samoa[b]	2	2
Guam[b]	7	7
N. Mariana Islands[b]	2	2
Puerto Rico[b]	30	30
Virgin Islands[b]	12	12
Total	16,255	21,107

[a] California, Colorado, New York, and South Dakota did not have the resources to complete the updating of these data.

[b] Results presented in this table for American Samoa, Guam, Northern Mariana Islands, Nevada, Puerto Rico, Virgin Islands, and Wyoming are from the 1996 survey because these States and Territories did not participate in the CWNS 2000.

Source: From 2000 Clean Watersheds Needs Survey Report to Congress, Published 2003, Appendix C, Table C.1, p. C.2. epa.gov/owm/mtb/cwns/2000rtc/cwns2000-appendix-c.pdf, epa.gov/owm/mtb/cwns/2000rtc/toc.htm.

Table 9A.10 Number of Operational Treatment Facilities and Collection Systems if All Documented Needs Are Met

State	Treatment Facilities	Collection Systems
Alabama	279	285
Alaska	50	51
Arizona	232	258
Arkansas	360	406
California[a]	579	799
Colorado[a]	331	430
Connecticut	99	159
Delaware	18	49
District of Columbia	1	1
Florida	302	346
Georgia	345	405
Hawaii	27	27
Idaho	177	219
Illinois	754	1,056
Indiana	424	510
Iowa	744	775
Kansas	665	712
Kentucky	301	369
Louisiana	371	405
Maine	145	184
Maryland	180	303
Massachusetts	141	267
Michigan	403	673
Minnesota	518	661
Mississippi	372	475
Missouri	729	848
Montana	208	218
Nebraska	475	483
Nevada[b]	52	56
New Hampshire	85	120
New Jersey	164	600
New Mexico	58	68
New York[a]	657	1,175
North Carolina	518	702
North Dakota	282	286
Ohio	837	1,213
Oklahoma	487	496
Oregon	219	270
Pennsylvania	1,013	1,936
Rhode Island	20	36
South Carolina	187	222
South Dakota[a]	273	276
Tennessee	251	286
Texas	1,469	1,850
Utah	114	188
Vermont	84	100
Virginia	254	383
Washington	240	337
West Virginia	404	626
Wisconsin	628	974
Wyoming[b]	96	121
American Samoa[b]	2	2
Guam[b]	6	7
N. Mariana Islands[b]	2	2
Puerto Rico[b]	30	30
Virgin Islands[b]	12	12
Total	17,674	23,748

[a] California, Colorado, New York, and South Dakota did not have the resources to complete the updating of these data.

[b] Results presented in this table for American Samoa, Guam, Northern Mariana Islands, Nevada, Puerto Rico, Virgin Islands, and Wyoming are from the 1996 survey because these States and Territories did not participate in the CWNS 2000.

Source: From 2000 Clean Watersheds Needs Survey Report to Congress, Published 2003, Appendix C, Table C.2, p. C.3. www.epa.gov/owm/mtb/cwns/2000rtc/cwns2000-appendix-c.pdf, www.epa.gov/owm/mtb/cwns/2000rtc/toc.htm.

Table 9A.11 Number of Treatment Facilities by Flow Range

Treatment Facilities in Operation in 2000[a,b]		
Existing Flow Range (mgd)	**Number of Facilities**	**Total Existing Flow (mgd)**
0.001–0.100	6,583	290
0.101–1.000	6,462	2,339
1.001–10.000	2,665	8,328
10.001–100.000	487	12,741
100.001 and greater	46	11,201
Other[c]	12	—
Total	16,255	34,899
Treatment Facilities in Operation in 2000 if All Documented Needs Are Met[a,b]		
Design Flow Range (mgd)	**Number of Facilities**	**Total Future Design Flow Capacity (mgd)**
0.001–0.100	6,112	298
0.101–1.000	7,223	2,750
1.001–10.000	3,525	12,081
10.001–100.000	748	19,873
100.001 and greater	64	15,040
Other[c]	2	—
Total	17,674	50,042

[a] California, Colorado, New York, and South Dakota did not have the resources to complete the updating of these data.

[b] Results presented in this table for American Samoa, Guam, Nevada, Northern Mariana Islands, Puerto Rico, Virgin Islands, and Wyoming are from the 1996 survey because these States and Territories did not participate in the CWNS 2000.

[c] Flow data for these facilities were unavailable.

Source: From 2000 Clean Watersheds Needs Survey Report to Congress, Published 2003, Appendix C, Table C.3, p. C.4. www.epa.gov/owm/mtb/cwns/2000rtc/cwns2000-appendix-c.pdf, www.epa.gov/owm/mtb/cwns/2000rtc/toc.htm.

Table 9A.12 Improvements in Treatment Level of the Nation's Municipal Wastewater Treatment Facilities

Level of Treatment	1992 Number of Facilities	1996 Number of Facilities	Change 1992–1996 (%)	2000 Number of Facilities	Change 1992–2000 (%)	Change 1996–2000 (%)
No discharge[a]	1,981	2,032	2.6	1,938	−2.2	−4.6
Less than secondary[b]	868	176	−79.7	47	−94.5	−73.3
Secondary	9,086	9,388	3.3	9,156	0.8	−2.5
Greater than secondary	3,678	4,428	20.4	4,892	33.0	10.5
Total facilities	15,613	16,024	2.6	16,255[c]	4.1	1.4

Note: A secondary treatment level is defined as meeting an effluent quality of 30 mg/L for biochemical oxygen demand (BOD) and suspended solids.

[a] *No discharge* refers to facilities that do not discharge effluent to surface waters (e.g., spray irrigation, groundwater recharge).

[b] Includes facilities granted section 301(h) waivers from secondary treatment for discharges to marine waters. As of January 1, 2000, waivers for 34 facilities in the CWNS 2000 database had been granted or were pending.

[c] The number of facilities includes 222 facilities that provide partial treatment and whose flow goes to another facility for further treatment.

Source: From 2000 Clean Watersheds Needs Survey Report to Congress, Published 2003, Table 3.2, p. 3–4. www.epa.gov/owm/mtb/cwns/2000rtc/toc.htm, www.epa.gov/owm/mtb/cwns/2000rtc/cwns2000-chapter-3.pdf.

Table 9A.13 Comparison of Total Needs for the 1992 Needs Survey, 1996 Clean Water Needs Survey, and CWNS 2000 (January 2000 Dollars in Billions)

Needs Category		1992[a]	1996[a]	2000
	Publicly Owned Wastewater Treatment and Collection Systems and Storm Water Management Programs			
I	Secondary wastewater treatment	39.3	29.4	36.8
II	Advanced wastewater treatment	19.4	19.4	20.4
III-A	Infiltration/inflow correction	3.4	3.7	8.2
III-B	Sewer replacement/rehabilitation	4.6	7.7	16.8
IV-A	New collector sewers and appurtenances	22.5	12.0	14.3
IV-B	New interceptor sewers and appurtenances	18.4	11.9	14.8
V	Combined sewer overflow correction	51.7[b]	49.6	50.6
VI	Storm water management programs	0.1[b]	8.2[b]	5.5
	Nonpoint Source Pollution Control Projects			
VII-A	Agriculture (cropland)	4.7[b]	4.2[b]	0.5
VII-B	Agriculture (animals)	3.4[b]	2.3[b]	0.7
VII-C	Silviculture	3.0[b]	3.9[b]	0.04
VII-D	Urban	—	1.1	4.4
VII-E	Groundwater protection: unknown source	1.4	1.1	0.9
	Estuaries[c]	0.01	0.04	—
	Wetlands[c]	0.04	0.01	—
VII-F	Marinas	—	—	0.002
VII-G	Resource extraction	—	—	0.04
VII-H	Brownfields	—	—	0.4
VII-I	Storage tanks	—	—	1.0
VII-J	Sanitary landfills	—	—	1.8
VII-K	Hydromodification	—	—	4.1
	Total needs	172.0	154.6	181.2
	Treatment categories I and II only	58.7	48.8	57.2
	Collection and conveyance categories III and IV only	48.9	35.3	54.1
	Category I to V subtotal	159.3	133.7	161.9

[a] The needs from 1992 and 1996 were inflated to January 2000 dollars for comparison with CWNS 2000 data.
[b] Modeled needs.
[c] Documented needs for estuaries and wetlands were provided by States during the 1992 and 1996 surveys, but they are no longer reported as individual categories.

Source: From 2000 Clean Watersheds Needs Survey Report to Congress, Published 2003, Table 3.4, p. 3–6. www.epa.gov/owm/mtb/cwns/2000rtc/toc.htm, www.epa.gov/owm/mtb/cwns/2000rtc/cwns2000-chapter-3.pdf.

Table 9A.14 Number of Treatment Facilities by Level of Treatment

Treatment Facilities in Operation in 2000[a,b]

Level of Treatment	Number of Facilities	Present Design Capacity (mgd)	Number of People Served	Percent of U.S. Population
Less than secondary[c]	47	1,023	6,426,062	2.3
Secondary	9,156	19,268	88,221,896	32.0
Greater than secondary	4,892	22,165	100,882,207	36.6
No discharge[d]	1,938	2,039	12,283,047	4.5
Partial treatment[e]	222	563	—	—
Total	16,255	45,058	207,813,212[f]	75.4

Treatment Facilities in Operation in 2000 if All Documented Needs Are Met[a,b]

Level of Treatment	Number of Facilities	Future Design Capacity (mgd)	Number of People Served	Percent of U.S. Population
Less than secondary[c]	27	481	3,851,000	1.2
Secondary	9,463	20,008	103,716,058	31.9
Greater than secondary	5,739	26,239	140,251,554	43.2
No discharge[d]	2,221	2,579	21,224,596	6.5
Partial treatment[e]	224	734	—	—
Total	17,674	50,041	269,043,208[f]	82.8

[a] California, Colorado, New York, and South Dakota did not have the resources to complete the updating of these data.

[b] Results presented in this table for American Samoa, Guam, Nevada, Northern Mariana Islands, Puerto Rico, Virgin Islands, and Wyoming are from the 1996 survey because these States and Territories did not participate in the CWNS 2000.

[c] Less-than-secondary facilities include facilities granted or pending section 301(h) waivers from secondary treatment for discharges to marine waters.

[d] No-discharge facilities do not discharge treated wastewater to the Nation's waterways. These facilities dispose of wastewater via methods such as industrial reuse, irrigation, or evaporation.

[e] These facilities provide some treatment to wastewater and discharge their effluents to wastewater facilities for further treatment and discharge.

[f] This table does not include the results for approximately 3.3 million people (present) and 3.5 million people (future) that are receiving centralized collection because the data related to flow and effluent levels were not complete for the CWNS 2000.

Source: From 2000 Clean Watersheds Needs Survey Report to Congress, Published 2003, Appendix C, Table C.4, p. C.5. www.epa.gov/owm/mtb/cwns/2000rtc/cwns2000-appendix-c.pdf, www.epa.gov/owm/mtb/cwns/2000rtc/toc.htm.

Table 9A.15 Clean Watersheds Needs Survey 2000 Total Needs (January 2000 Dollars in Millions)

	Category of Need										
State	Total	I	II	III-A	III-B	IV-A	IV-B	V	VI	VII	Total (I–V)
Alabama	2,720	14	951	135	1,168	386	66	0	0	0	2,720
Alaska	560	306	7	7	65	163	7	5	0	0	560
Arizona	6,199	726	2,368	126	240	319	1,081	0	1,251	88	4,860
Arkansas	500	37	117	22	24	41	71	0	0	188	312
California	14,402	3,916	3,748	111	3,114	82	1,853	426	352	800	13,250
Colorado	1,340	183	812	5	179	16	37	9	48	51	1,241
Connecticut	2,349	399	923	85	16	170	161	500	0	95	2,254
Delaware	288	33	23	0	68	58	4	102	0	0	288
District of Columbia	1,478	305	37	14	64	0	0	1,019	37	2	1,439
Florida	9,966	299	2,853	129	562	1,191	1,012	0	680	3,240	6,046
Georgia	2,336	114	205	1,004	25	9	61	918	0	0	2,336
Hawaii	1,743	575	19	471	441	88	149	0	0	0	1,743
Idaho	207	119	29	3	18	18	20	0	0	0	207
Illinois	11,888	795	103	27	1,204	95	169	9,450	0	45	11,843
Indiana	7,222	626	171	65	419	291	176	5,468	0	6	7,216
Iowa	1,954	240	22	23	79	36	19	1,534	1	0	1,953
Kansas	1,419	373	100	213	2	65	270	396	0	0	1,419
Kentucky	2,797	654	101	193	280	756	592	217	3	1	2,793
Louisiana	2,370	410	146	1,167	216	240	189	0	0	2	2,368
Maine	1,102	176	7	3	31	88	16	653	0	128	974
Maryland	4,779	1,239	837	94	739	407	369	396	456	242	4,081
Massachusetts	4,675	874	249	59	92	662	406	2,324	0	9	4,666
Michigan	4,092	837	73	107	307	301	30	2,437	0	0	4,092
Minnesota	2,319	660	101	42	281	45	104	6	120	960	1,239
Mississippi	856	92	129	156	152	184	143	0	0	0	856
Missouri	4,998	725	22	720	297	301	193	1,180	0	1,560	3,438
Montana	516	170	70	14	55	100	60	0	0	47	469
Nebraska	1,194	149	56	7	11	11	75	861	24	0	1,170
Nevada	NR	NR	NR	NR	NR	NR	NR	NR	NR	NR	NR
New Hampshire	906	127	47	7	33	6	135	485	0	66	840
New Jersey	12,827	2,818	368	339	610	1,007	411	4,385	89	2,800	9,938
New Mexico	206	94	15	9	42	18	21	0	0	7	199
New York	20,422	9,853	776	75	2,072	538	173	5,497	16	1,422	18,984
North Carolina	5,927	423	1,737	291	205	1,725	1,535	3	1	7	5,919
North Dakota	52	27	0[a]	2	17	0	1	0	4	1	47
Ohio	8,722	1,219	391	1,493	112	725	533	3,623	0	626	8,096

(Continued)

Table 9A.15 (Continued)

State	Total	Category of Need									
		I	II	III-A	III-B	IV-A	IV-B	V	VI	VII	Total (I–V)
Oklahoma	586	85	25	1	207	33	45	0	190	0	396
Oregon	1,477	540	155	4	654	16	34	74	0	0	1,477
Pennsylvania	8,060	845	204	121	119	963	197	5,431	17	163	7,880
Rhode Island	1,415	109	113	12	52	345	119	633	0	32	1,383
South Carolina	1,309	551	334	1	13	283	125	0	0	2	1,307
South Dakota	142	16	29	0	44	13	6	2	14	18	110
Tennessee	604	66	45	48	107	58	36	244	0	0	604
Texas	9,152	2,009	813	235	1,323	616	1,890	0	2,225	41	6,886
Utah	848	347	74	0[a]	97	98	217	0	5	10	833
Vermont	144	45	32	0[a]	0[a]	33	2	31	0	1	143
Virginia	3,519	727	777	111	358	516	570	460	0	0	3,519
Washington	2,744	1,000	52	226	136	198	521	608	0	3	2,741
West Virginia	2,529	298	12	134	47	691	478	869	0	0	2,529
Wisconsin	3,338	588	141	54	365	260	462	342	16	1,110	2,212
Wyoming	NR	NR	NR	NR	NR	NR	NR	NR	NR	NR	NR
American Samoa	NR	NR	NR	NR	NR	NR	NR	NR	NR	NR	NR
Guam	NR	NR	NR	NR	NR	NR	NR	NR	NR	NR	NR
N. Mariana Islands	NR	NR	NR	NR	NR	NR	NR	NR	NR	NR	NR
Puerto Rico	NR	NR	NR	NR	NR	NR	NR	NR	NR	NR	NR
Virgin Islands	NR	NR	NR	NR	NR	NR	NR	NR	NR	NR	NR
Total	181,198	36,833	20,419	8,165	16,762	14,265	14,844	50,588	5,549	13,773	161,876

Categories

I Secondary wastewater treatment
II Advanced wastewater treatment
III-A Infiltration/inflow correction
III-B Sewer replacement/rehabilitation
IV-A New collector sewers and appurtenances
IV-B New interceptor sewers and appurtenances
V Combined sewer overflow correction
VI Storm water management programs
VII NPS pollution control (see Table A.2 for totals by subcategory)

Note: NR, not reported. American Samoa, Guam, Nevada, Northern Mariana Islands, Puerto Rico, Virgin Islands, and Wyoming did not participate in the CWNS 2000.

[a] Estimate is less than $0.5 million.

Source: From 2000 Clean Watersheds Needs Survey Report to Congress, Published 2003, Appendix A, Table A.1, p. A.2 and A.3. www.epa.gov/owm/mtb/cwns/2000rtc/toc.htm, www.epa.gov/owm/mtb/cwns/2000rtc/cwns2000-appendix-a.pdf.

Table 9A.16 Number of Treatment Facilities and Population Served Per State by Level of Treatment for Year 2000

	Number of Facilities Providing Listed Effluent Level				Population Served by Listed Effluent Level			
State	Less than Secondary[a]	Secondary	Greater than Secondary	No Discharge[b]	Less than Secondary[a]	Secondary	Greater than Secondary	No Discharge[b]
Alabama	0	130	129	8	0	732,009	1,994,219	7,593
Alaska	5	30	0	9	207,994	108,879	0	21,920
Arizona	0	17	18	81	0	111,767	2,215,703	1,378,004
Arkansas	0	118	207	9	0	726,471	803,753	12,155
California[c]	5	182	77	309	4,198,270	12,159,009	7,919,130	3,577,181
Colorado[c]	0	246	38	22	0	1,556,854	2,142,434	7,788
Connecticut	0	49	38	4	0	1,266,574	813,536	1,210
Delaware	0	3	11	4	0	10,476	728,997	13,070
District of Columbia[d]	0	0	1	0	0	0	1,298,601	0
Florida	0	17	84	175	0	238,764	6,155,714	4,931,819
Georgia	0	227	80	35	0	1,721,572	2,594,389	89,249
Hawaii	2	5	2	12	532,378	139,609	20,286	89,512
Idaho	0	107	5	55	0	562,008	265,812	60,303
Illinois	0	415	301	1	0	683,543	9,811,768	572
Indiana	0	125	274	0	0	410,940	3,416,852	0
Iowa	0	707	9	3	0	1,925,926	181,763	1,393
Kansas	0	355	79	197	0	694,512	1,277,425	101,964
Kentucky	0	123	94	0	0	1,242,187	921,134	0
Louisiana	1	184	163	1	3,000	2,268,451	878,478	207
Maine	12	116	2	7	9,303	624,604	16,038	5,956
Maryland	0	75	75	6	0	949,367	2,045,325	3,920
Massachusetts	1	77	35	7	20,074	4,235,095	822,135	17,043
Michigan	0	204	120	68	0	1,254,599	6,161,491	108,121
Minnesota	1	411	100	0	42	967,813	2,073,977	0
Mississippi	0	195	75	1	0	1,139,734	507,809	524
Missouri	0	578	77	21	0	3,757,717	451,630	2,663
Montana	0	107	5	80	0	397,988	89,635	63,564
Nebraska	0	298	19	146	0	977,825	155,078	64,166
Nevada[e]	0	44	3	4	0	139,996	252,229	237,442
New Hampshire	1	70	2	10	25,409	555,435	17,890	7,984
New Jersey	0	94	55	1	0	6,762,536	1,090,502	34,307
New Mexico	0	57	1	17	0	898,530	7,150	135,338
New York[c]	0	360	178	27	0	11,273,282	3,748,413	116,814
North Carolina	0	313	134	33	0	1,056,606	2,576,092	112,989
North Dakota	0	254	1	27	0	468,946	21,531	5,909
Ohio	0	169	593	2	0	1,401,922	7,404,543	956
Oklahoma	0	249	39	199	0	1,716,478	712,679	151,004
Oregon	1	101	67	37	625	1,333,432	1,219,279	33,050
Pennsylvania	2	360	397	2	1,476	6,237,683	4,157,929	2,314
Rhode Island	0	19	2	0	0	687,805	10,184	0
South Carolina	0	123	53	7	0	1,769,072	549,626	30,628
South Dakota[c]	0	234	8	29	0	268,874	164,144	14,467
Tennessee	0	110	130	5	0	1,459,559	1,700,862	4,193
Texas	2	524	661	160	1,070	2,538,924	14,025,086	640,857
Utah	0	49	4	44	0	1,636,148	190,027	134,011
Vermont	0	48	31	2	0	90,497	193,684	722
Virginia	0	157	60	2	0	2,166,150	2,318,144	1,373
Washington	0	201	7	27	0	2,847,237	894,801	31,127
West Virginia	3	142	63	0	2,205	581,527	374,677	0
Wisconsin	0	283	279	26	0	573,346	3,250,360	20,360
Wyoming[e]	0	78	3	14	0	244,075	87,923	3,030
American Samoa[e]	2	0	0	0	5,511	0	0	0
Guam[e]	2	2	0	2	62,639	9,236	0	4,275
N. Mariana Islands[e]	0	2	0	0	0	1,118	0	0
Puerto Rico[e]	6	22	2	0	1,336,535	581,405	151,290	0

(Continued)

Table 9A.16 (Continued)

	Number of Facilities Providing Listed Effluent Level				Population Served by Listed Effluent Level			
State	Less than Secondary[a]	Secondary	Greater than Secondary	No Discharge[b]	Less than Secondary[a]	Secondary	Greater than Secondary	No Discharge[b]
Virgin Islands[e]	1	10	1	0	19,531	58,294	50	0
Total	47	9,156	4,892	1,938	6,426,062	88,221,896	100,882,207	12,283,047

[a] Less-than-secondary facilities include facilities granted or pending section 301(h) waivers from secondary treatment for discharges to marine waters.
[b] No-discharge facilities do not discharge treated wastewater to the Nation's waterways. These facilities dispose of wastewater via methods such as industrial reuse, irrigation, or evaporation.
[c] California, Colorado, New York, and South Dakota did not have the resources to complete updating of these data.
[d] The reported population served for the District of Columbia includes populations from Maryland and Virginia that receive wastewater treatment at the Blue Plains facility in the District of Columbia.
[e] Results presented in this table for American Samoa, Guam, Northern Mariana Islands, Nevada, Puerto Rico, Virgin Islands, and Wyoming are from the 1996 survey because these States and Territories did not participate in the CWNS 2000.

Source: From 2000 Clean Watersheds Needs Survey Report to Congress, Published 2003, Appendix C, Table C.7, p. C.10 and C.11. www.epa.gov/owm/mtb/cwns/2000rtc/cwns2000-appendix-c.pdf, www.epa.gov/owm/mtb/cwns/2000rtc/toc.htm.

Table 9A.17 Typical Wastewater Pollutants of Concern

Pollutant	Reason for Concern
Total suspended solids (TSS) and turbidity (NTU)	In surface waters, suspended solids can result in the development of sludge deposits that smother benthic macroinvertebrates and fish eggs and can contribute to benthic enrichment, toxicity, and sediment oxygen demand. Excessive turbidity (colloidal solids that interfere with light penetration) can block sunlight, harm aquatic life (e.g., by blocking sunlight needed by plants), and lower the ability of aquatic plants to increase dissolved oxygen in the water column. In drinking water, turbidity is aesthetically displeasing and interferes with disinfection
Biodegradable organics (BOD)	Biological stabilization of organics in the water column can deplete dissolved oxygen in surface waters, creating anoxic conditions harmful to aquatic life. Oxygen-reducing conditions can also result in taste and odor problems in drinking water
Pathogens	Parasites, bacteria, and viruses can cause communicable diseases through direct/indirect body contact or ingestion of contaminated water or shellfish. A particular threat occurs when partially treated sewage pools on ground surfaces or migrates to recreational waters. Transport distances of some pathogens (e.g., viruses and bacteria) in groundwater or surface waters can be significant
Nitrogen	Nitrogen is an aquatic plant nutrient that can contribute to eutrophication and dissolved oxygen loss in surface waters, especially in lakes, estuaries, and coastal embayments. Algae and aquatic weeds can contribute trihalomethane (THM) precursors to the water column that may generate carcinogenic THMs in chlorinated drinking water. Excessive nitrate-nitrogen in drinking water can cause methemoglobinemia in infants and pregnancy complications for women. Livestock can also suffer health impacts from drinking water high in nitrogen
Phosphorus	Phosphorus is an aquatic plant nutrient that can contribute to eutrophication of inland and coastal surface waters and reduction of dissolved oxygen
Toxic organics	Toxic organic compounds present in household chemicals and cleaning agents can interfere with certain biological processes in alternative OWTSs. They can be persistent in groundwater and contaminate downgradient sources of drinking water. They can also cause damage to surface water ecosystems and human health through ingestion of contaminated aquatic organisms (e.g., fish, shellfish)
Heavy metals	Heavy metals like lead and mercury in drinking water can cause human health problems. In the aquatic ecosystem, they can also be toxic to aquatic life and accumulate in fish and shellfish that might be consumed by humans
Dissolved inorganics	Chloride and sulfide can cause taste and odor problems in drinking water. Boron, sodium, chlorides, sulfate, and other solutes may limit treated wastewater reuse options (e.g., irrigation). Sodium and to a lesser extent potassium can be deleterious to soil structure and SWIS performance

Source: From USEPA, *On-Site Wastewater Treatment Systems Manual*, Office of Water, Office of Research and Development, EPA, (EPA/625/R-00/0008), Table 3.16, p. 3–23. www.epa.gov/ord/NRMRL/Pubs/625R00008/625R00008totaldocument.pdf; Adapted in part from Tchobanoglous and Burton, 1991.

Table 9A.18 Wastewater Constituents of Concern and Representative Concentrations in the Effluent of Various Treatment Units

Constituents of Concern	Example Direct or Indirect Measures (Units)	Tank-Based Treatment Unit Effluent Concentrations					SWIS Percolate into Groundwater at 3 to 5 ft Depth (% Removal)
		Domestic STE[a]	Domestic STE with N-Removal Recycle[b]	Aerobic Unit Effluent	Sand Filter Effluent	Foam or Textile Filter Effluent	
Oxygen demand	BOD_5 (mg/L)	140–200	80–120	5–50	2–15	5–15	>90
Particulate solids	TSS (mg/L)	50–100	50–80	5–100	5–20	5–10	>90
Nitrogen	Total N (mg N/L)	40–100	10–30	25–60	10–50	30–60	10–20
Phosphorus	Total P (mg P/L)	5–15	5–15	4–10	<1–10^4	5–15^4	0–100
Bacteria (e.g., *Clostridium perfringens, Salmonella Shigella*)	Fecal coliform (organisms per 100 mL)	10^6–10^8	10^6–10^8	10^3–10^4	10^1–10^3	10^1–10^3	>99.99
Virus (e.g., hepatitis, polio, echo, coxsackie, coliphage)	Specific virus (pfu/mL)	0–10^5 (episodically present at high levels)	0–10^5 (episodically present at high levels)	0–10^5 (episodically present at high levels)	0–10^5 (episodically present at high levels)	0–10^5 (episodically present at high levels)	>99.9%
Organic chemicals (e.g., solvents, petrochemicals, pesticides)	Specific organics or totals (μg/L)	0 to trace levels (?)	0 to trace levels (?)	0 to trace levels (?)	0 to trace levels (?)	0 to trace levels (?)	>99%
Heavy metals (e.g., Pb, Cu, Ag, Hg)	Individual metals (μg/L)	0 to trace levels	0 to trace levels	0 to trace levels	0 to trace levels	0 to trace levels	>99%

[a] Septic tank effluent (STE) concentrations given are for domestic wastewater. However, restaurant STE is markedly higher particularly in BOD, COD, and suspended solids while concentrations in graywater STE are noticeably lower in total nitrogen.
[b] N-removal accomplished by recycling STE through a packed bed for nitrification with discharge into the influent end of the septic tank for denitrification.

Source: From Van Cuyk, S.M., R.L. Siegrist, and A.L. Logan. 2001. Evaluation of virus and microbiological purification in wastewater soil absorption systems using multicomponent surrogate and tracer additions. *On-Site Wastewater Treatment: Proceedings of the Ninth National Symposium on Individual and Small Community Sewage Systems.* American Society of Agricultural Engineers, St. Joseph, MI; USEPA, *On-Site Wastewater Treatment Systems Manual*, Office of Water, Office of Research and Development, EPA (EPA/625/R-00/0008), Table 3.19, p. 3–29. www.epa.gov/ord/NRMRL/Pubs/625R00008/625R00008totaldocument.pdf.

Table 9A.19 Proposed On-Site System Treatment Performance Standards in Various Control Zones

Standard	BOD (mg/L)	TSS (mg/L)	PO.-P (mg/L)	NH.-N (mg/L)	N03-N (mg/L)	Total N (% Removed)[a]	Fecal Coliforms (CFU/1,000 mL)[b]
T81—primary treatment							
T81u—unfiltered	300	300	15	80	NA	NA	10,000,000
T811—filtered	200	80	15	80	NA	NA	10,000,000
T82—secondary treatment	30	30	15	10	NA	NA	50,000
T83—tertiary treatment	10	10	15	10	NA	NA	10,000
T84—nutrient reduction							
T84n—nitrogen reduction	10	10	15	5	NA	50	10,000
T84p—phosphorus reduction	10	10	2	10	NA	25	10,000
T84np—N & P reduction	10	10	2	5	NA	50	10,000
T85—bodily contact disinfection	10	10	15	10	NA	25	200
T86—wastewater reuse	5	5	15	5	NA	50	14
T87—near drinking water	5	5	1	5	10	75	<1b

Note: NA, not available.

[a] Minimum percentage reduction of total nitrogen (as nitrate-nitrogen plus ammonium nitrogen) concentration in the raw, untreated wastewater.
[b] Total coliform colony densities <50 per 100 mL of effluent.

Source: From Hoover, M.T., A. Arenovski, D. Daly, and D. Lindbo. 1998. A risk-based approach to on-Site system siting, design and management. In *On-Site Wastewater Treatment. Proceedings of the Eighth National Symposium on Individual and Small Community Sewage Systems.* American Society of Agricultural Engineers, St. Joseph, MI; USEPA, *On-Site Wastewater Treatment Systems Manual*, Office of Water, Office of Research and Development, EPA, (EPA/625/R-00/0008), Table 3.27, p. 3–48. www.epa.gov/ord/NRMRL/Pubs/625R00008/625R00008totaldocument.pdf.

Table 9A.20 Typical Wastewater Constituent Data for Various Countries

Country/ Constituent	BOD, g/capita d	TSS, g/capita d	TKN, g/capita d	NH_3-N, g/capita d	Total P, g/capita d
Brazil	55–68	55–68	8–14	ND	0.6–1
Denmark	55–68	82–96	14–19	ND	1.5–2
Egypt	27–41	41–68	8–14	ND	0.4–0.6
Germany	55–68	82–96	11–16	ND	1.2–1.6
Greece	55–60	ND	ND	8–10	1.2–1.5
India	27–41	ND	ND	ND	ND
Italy	49–60	55–82	8–14	ND	0.6–1
Japan	40–45	ND	1–3	ND	0.15–0.4
Palestine[a]	32–68	52–72	4–7	3–5	0.4–0.7
Sweden	68–82	82–96	11–16	ND	0.8–1.2
Turkey	27–50	41–68	8–14	9–11	0.4–2
Uganda	55–68	41–55	8–14	ND	0.4–0.6
United States[b]	50–120	60–150	9–22	5–12	2.7–4.5

[a] West Bank and Gaza Strip.
[b] From Table 3.11.

Source: From Metcalf & Eddy, Inc., McGraw-Hill, *Wastewater Engineering Treatment and Reuse, Fourth Edition*, 2003, Table 3.14, p. 184. With permission. Adapted from Henze et al. (1997), Ozturk et al. (1992), Andreadakis (1992), and Nashashibi and van Duijl (1995).

SECTION 9B CENTRALIZED WASTEWATER TREATMENT

Table 9B.21 Gravity Sewer Average Design Flows for Development Types

Type of Development	Design Flow (GPD)
Residential	
General	100/person
Single family	370/residence
Townhouse unit	300/unit
Apartment unit	300/unit
Commercial	
General	2,000/acre
Motel	130/unit
Office	20/employee
	0.20/net sq.ft
Industrial (varies with type of industry)	
General	10,000/acre
Warehouse	600/acre
School site (general)	16/student

Source: From Darby, 1995; USEPA, *Collection Systems Technology Fact Sheet, Sewers, Conventional Gravity*, Office of Water, Municipal Technology Branch, Table 1, (EPA/823/F-02/007), September 2002. epa.gov/owm/mtb/congrasew.pdf.

Table 9B.22 Minimum Slope for Gravity Sewers

Nominal Sewer Size	Minimum Slope (in ft per 100 ft Im/100 m)
8 in. (200 mm)	0.40
10 in. (250 mm)	0.28
12 in. (300 mm)	0.22
14 in. (350 mm)	0.17
15 in. (375 mm)	0.15
16 in. (400 mm)	0.14
18 in. (450 mm)	0.12
21 in. (525 mm)	0.10
24 in. (600 mm)	0.08
27 in. (675 mm)	0.067
30 in. (750 mm)	0.058
33 in. (825 mm)	0.052
36 in. (900 mm)	0.046
39 in. (975 mm)	0.041
42 in. (1,050 mm)	0.037

Source: From Board of State and Provincial Public Health and Environmental Managers, Health Education Services Division, Recommended Standards for Wastewater Facilities, 2004 Edition. hes.org.

Table 9B.23 Force Main Capacity

	Velocity = 2 fps		Velocity = 4 fps		Velocity = 6 fps	
Diameter (in.)	**gpm**	**lps**	**gpm**	**lps**	**gpm**	**lps**
6	176	11	362	22	528	33
8	313	20	626	40	626	60
10	490	31	980	62	1,470	93
18	1,585	100	3,170	200	4,755	300
24	2,819	178	5,638	356	8,457	534
36	6,342	400	12,684	800	19,026	1,200

Source: From Metcalf and Eddy, 1981; USEPA, *Wastewater Technology Fact Sheet Sewers, Force Main*, Office of Water, Municipal Technology Branch, Table 2 (EPA/823/f-00/071), September 2000. epa.gov/own/mtb/force_main_sewers.pdf.

Table 9B.24 Common Sewer Cleaning Methods

Technology	Uses and Applications
Mechanical	
Rodding	Uses an engine and a drive unit with continuous rods or sectional rods. As blades rotate they break up grease deposits, cut roots, and loosen debris Rodders also help thread the cables used for TV inspections and bucket machines Most effective in lines up to 300 mm (12 in.) in diameter
Bucket machine	Cylindrical device, closed on one end with 2 opposing hinged jaws at the other Jaws open and scrape off the material and deposit it in the bucket Partially removes large deposits of silt, sand, gravel, and some types of solid waste
Hydraulic	
Balling	A threaded rubber cleaning ball that spins and scrubs the pipe interior as flow increases in the sewer line Removes deposits of settled inorganic material and grease build-up Most effective in sewers ranging in size from 13 to 60 cm (5–24 in.)
Flushing	Introduces a heavy flow of water into the line at a manhole. Removes floatables and some sand and grit Most effective when used in combination with other mechanical operations, such as rodding or bucket machine cleaning
Jetting	Directs high velocities of water against pipe walls. Removes debris and grease build-up, clears blockages, and cuts roots within small diameter pipes Efficient for routine cleaning of small diameter, low flow sewers

Technology	Applications
Scooter	Round, rubber-rimmed, hinged metal shield that is mounted on a steel framework on small wheels. The shield works as a plug to build a head of water Scours the inner walls of the pipe lines Effective in removing heavy debris and cleaning grease from line
Kites, bags, and poly pigs	Similar in function to the ball Rigid rims on bag and kite induce a scouring action Effective in moving accumulations of decayed debris and grease downstream
Silt traps	Collect sediments at convenient locations Must be emptied on a regular basis as part of the maintenance program
Grease traps and sand/oil interceptors	The ultimate solution to grease build-up is to trap and remove it These devices are required by some uniform building codes and/or sewer-use ordinances. Typically sand/oil interceptors are required for automotive business discharge Need to be thoroughly cleaned to function properly Cleaning frequency varies from twice a month to once every 6 months, depending on the amount of grease in the discharge Need to educate restaurant and automobile businesses about the need to maintain these traps
Chemicals Before using these chemicals review the material safety data sheets (MSDS) and consult the local authorities on the proper use of chemicals as per local ordinance and the proper disposal of the chemicals used in the operation. If assistance or guidance is needed regarding the application of certain chemicals, contact the U.S. EPA or state water pollution control agency	Used to control roots, grease, odors (H_2S gas), concrete corrosion, rodents and insects Root control — Longer lasting effects than power rodder (approximately 2–5 years) H_2S gas — Some common chemicals used are chlorine (Cl_2), hydrogen peroxide (H_2O_2), pure oxygen (O_2), air, lime ($Ca(OH_2)$), sodium hydroxide (NaOH), and iron salts Grease and soap problems — Some common chemicals used are bioacids, digester, enzymes, bacteria cultures, catalysts, caustics, hydroxides, and neutralizers

Source: From information provided by Arbour and Kerri, 1997 and Sharon, 1989; USEPA, *Collection Systems, O&M Fact Sheet, Sewer Cleaning and Inspection*, Office of Water, Municipal Technology Branch, Table 1 (EPA/823/f-99/031), September 1999. www.epa.gov/owm/mtb/sewcl.pdf.

Table 9B.25 Frequency of Maintenance Activities for Sewer Lines

Activity	Average (% of system/year)
Cleaning	29.9
Root removal	2.9
Manhole inspection	19.8
CCTV inspection	6.8
Smoke testing	7.8

Source: From ASCE, 1998; USEPA, *Collection Systems, O&M Fact Sheet, Sewer Cleaning and Inspection*, Office of Water, Municipal Technology Branch, Table 2 (EPA/823/f-99/031), September 1999. www.epa.gov/owm/mtb/sewcl.pdf.

Table 9B.26 Limitations of Standard Inspection Techniques for Sewer Lines

Inspection Technique	Limitations
Visual inspection	In smaller sewers, the scope of problems detected is minimal because the only portion of the sewer that can be seen in detail is near the manhole. Therefore, any definitive information on cracks or other structural problems is unlikely. However, this method does provide information needed to make decisions on rehabilitation
Camera inspection	When performing a camera inspection in a large diameter sewer, the inspection crew is essentially taking photographs haphazardly, and as a result, the photographs tend to be less comprehensive
Closed circuit television (CCTV)	This method requires late night inspection and as a result the TV operators are vulnerable to lapses in concentration. CCTV inspections are also quite expensive and time-consuming
Lamping inspection	The video camera does not fit into the pipe and during the inspection it remains only in the maintenance hole. As a result, only the first 10 ft of the pipe can be viewed or inspected using this method

Source: From Water Pollution Control Federation, 1989; USEPA, *Collection Systems, O&M Fact Sheet, Sewer Cleaning and Inspection*, Office of Water, Municipal Technology Branch, Table 3 (EPA/823/F-99/031), September 1999. www.epa.gov/owm/mtb/-sewcl.pdf.

Table 9B.27 Limitations of Cleaning Methods for Sewer Lines

Cleaning Method	Limitations
Balling, jetting, scooter	In general, these methods are only successful when necessary water pressure or head is maintained without flooding basements or houses at low elevations. Jetting—The main limitation of this technique is that caution needs to be used in areas with basement fixtures and in steep-grade hill areas. Balling—Balling cannot be used effectively in pipes with bad offset joints or protruding service connections because the ball can become distorted
	Scooter—When cleaning larger lines, the manholes need to be designed to a larger size in order to receive and retrieve the equipment. Otherwise, the scooter needs to be assembled in the manhole. Caution also needs to be used in areas with basement fixtures and in steep-grade hill areas
Bucket, machine	This device has been known to damage sewers. The bucket machine cannot be used when the line is completely plugged because this prevents the cable from being threaded from one manhole to the next. Set-up of this equipment is time-consuming
Flushing	This method is not very effective in removing heavy solids. Flushing does not remedy this problem because it only achieves temporary movement of debris from one section to another in the system
High velocity cleaner	The efficiency and effectiveness of removing debris by this method decreases as the cross-sectional areas of the pipe increase. Backups into residences have been known to occur when this method has been used by inexperienced operators. Even experienced operators require extra time to clear pipes of roots and grease
Kite or bag	When using this method, use caution in locations with basement fixtures and steep-grade hill areas
Rodding	Continuous rods are harder to retrieve and repair if broken and they are not useful in lines with a diameter of greater than 300 mm (0.984 ft) because the rods have a tendency to coil and bend. This device also does not effectively remove sand or grit, but may only loosen the material to be flushed out at a later time

Source: From USEPA, 1993; USEPA, *Collection Systems, O&M Fact Sheet, Sewer Cleaning and Inspection*, Office of Water, Municipal Technology Branch, Table 4 (EPA/823/F-99/031), September 1999. www.epa.gov/owm/mtb/sewcl.pdf.

Table 9B.28 Comparison of Various Sewer Rehabilitation Techniques

Method		Diameter Range (mm)	Maximum Installation (m)	Liner Material
In-line expansion	Pipe bursting	100–600 (4–24 in.)	230 (750 ft)	PE, PP, PVC, GRP
Sliplining	Segmental	100–4,000 (4–158 in.)	300 (1,000 ft)	PE, PP, PVC, GRP (−EP & −UP)
	Continuous	100–1,600 (4–63 in.)	300 (1,000 ft)	PE, PP, PE/EPDM, PVC
	Spiral wound	150–2,500 (6–100 in.)	300 (1,000 ft)	PE, PVC, PP, PVDF
Cured-in-place Product linings	Inverted-in-place	100–2,700 (4–108 in.)	900 (3,000 ft)	Thermoset Resin/fabric composite
	Winched-in-place	100–1,400 (4–54 in.)	150 (500 ft)	Thermoset resin/fabric composite
	Spray-on-linings	76–4,500 (3–180 in.)	150 (500 ft)	Epoxy resins/cement mortar
Modified cross-sectional methods	Fold and form	100–400 (4–15 in.)	210 (700 ft)	PVC
	Deformed/reformed	100–400 (4–15 in.)	800 (2,500 ft)	(Thermoplastics) HDPE (thermoplastics)
	Drawdown	62–600 (3–24 in.)	300 (1,000 ft)	HDPE, MDPE
	Rolldown	62–600 (3–24 in.)	300 (1,000 ft)	HDPE, MDPE
	Thin-walled lining	500–1,100 (20–46 in.)	960 (3,000 ft)	HDPE
Internal point Repair	Robotic repair	200–760 (8–30 in.)	N/A	Exopy resins Cement mortar
	Grouting/sealing & spray-on	N/A	N/A	Chemical grouting
	Link seal	100–600 (4–24 in.)	N/A	Special sleeves
	Point CIPP	100–600 (4–24 in.)	15 (50 ft)	Fiberglass/polyester, etc

Note: Spiral wound sliplining, robotic repair, and point CIPP can only be used only with gravity pipeline. All other methods can be used with both gravity and pressure pipeline. EPDM, ethylene polypelene diene monomer; GRP, glassfiber reinforced polyester; HDPE, high density polyethylene; MDPE, medium density polyethylene; PE, polyethylene; PP, polypropylene; PVC, poly vinyl chloride; PVDF, poly vinylidene chloride.

Source: From Iseley and Najafi (1995); USEPA, *Collection Systems, O&M Fact Sheet, Trenchless Sewer Rehabilitation*, Office of Water, Municipal Technology Branch, Table 1 (EPA/823/F-99/0032), September 1999. www.epa.gov/owm/mtb/rehabl.pdf.

Table 9B.29 Limitations of Trenchless Sewer Rehabilitation Techniques

Method	Limitations
Pipe bursting	Bypass or diversion of flow required Insertion pit required Percussive action can cause significant ground movement may not be suitable for all materials
Sliplining	Insertion pit required Reduces pipe diameter Not well suited for small diameter pipes
CIPP	Bypass or diversion of flow required Curing can be difficult for long pipe segments Must allow adequate curing time Defective installation may be difficult to rectify Resin may clump together on bottom of pipe Reduces pipe diameter
Modified cross section	Bypass or diversion of flow required The cross section may shrink or unfold after expansion Reduces pipe diameter Infiltration may occur between liner and host pipe unless sealed Liner may not provide adequate structural support

Source: From USEPA, *Collection Systems, O&M Fact Sheet, Trenchless Sewer Rehabilitation*, Office of Water, Municipal Technology Branch, Table 2 (EPA/823/F-99/0032), September 1999. www.epa.gov/owm/mtb/rehabl.pdf.

Table 9B.30 Characteristics of Common Force Main Pipe Materials

Material	Application	Key Advantages	Key Disadvantages
Cast or ductile iron, cement lined	High pressure available sizes of 4–54 in.	Good resistance to pressure surges	More expensive than concrete and fiberglass
Steel, cement lined	High pressure all pipe sizes	Excellent resistance to pressure surges	More expensive than concrete and fiberglass
Asbestos cement	Moderate pressure for 36-in. + pipe sizes	No corrosion slow grease buildup	Relatively brittle
Fiberglass reinforced epoxy pipe	Moderate pressure for up to 36-in. pipe sizes	No corrosion slow grease buildup	350 psi max pressure
Plastic	Low pressure for up to 36-in. pipe sizes	No corrosion slow grease buildup	Suitable for small pipe sizes and low pressure only

Source: From Sanks, 1998; USEPA, *Wastewater Technology Fact Sheet Sewers, Force Main*, Office of Water, Municipal Technology Branch, Table 1 (EPA/823/F-99/040), September 1999. www.epa.gov/owm/mtb/force_main_sewers.pdf.

Table 9.31 Design Parameters for Static Screens

Hydraulic loading, gal/min/ft of width	100–180
Incline of screens, degrees from vertical[a]	35
Slot space, μm	250–1,600
Automatic controls	None

Note: gal/min/ft × 0.207= l/m/s.

[a] Bauer Hydrasieves™ have 3-stage slopes on each screen: 25°, 35°, 45°.

Source: From USEPA, *Combined Sewer Overflow, Technology Fact Sheet, Screens Office of Water*, Municipal Technology Brach, Table 1 (EPA/823/F/F-99/040), September 1999. www.epa.gov/own/mtb/screens.pdf.

Table 9B.32 Design Parameters for Drum Screens and Rotary Screen

Parameter	Drum/Band Screen	Rotary Screen
Screen spacing, μm	100–420	74–167 105 recommended
Screen material	Stainless steel or plastic	Stainless steel or plastic
Drum speed, r/min		
Speed range	2–7	30–65
Recommended speed	5	55
Peripheral speed, ft/s		14–16
Submergence of drum, %	60–70	
Flux density, gal/ft^2/min of submergence screen	20–50	70–150
Hydraulic efficiency, % of inflow		75–90
Headloss, in.	6–24	
Backwash		
Volume, % of inflow	0.5–3	0.02–2.5
Pressure, lb/in.^{2}	30–50	50

Note: gal/ft^2/min × 2.44= m^3/h/m^2 in. × 2.54= cm ft × 0.305= cm; lb/in.^{2} × 0.0703= kg/cm^2.

Source: From USEPA, *Combined Sewer Overflow, Technology Fact Sheet, Screens Office of Water*, Municipal Technology Brach, Table 2 (EPA/823/F/F-99/040), September 1999. www.epa.gov/owm/mtb/screens.pdf.

Table 9B.33 Typical Design Parameters for Package Plant

	Extended Aeration	SBR	Oxidation Ditch
BOD_5 loading (F:M) (lb BOD_5/lb MLVSS)	0.05–0.15	0.05–0.30	0.05–0.30
Oxygen required avg. at 20 °C (lb/lb BOD_5 applied)	2–3	2–3	2–3
Oxygen required peak at 20 °C (value × avg. flow)	1.5–2.0	1.25–2.0	1.5–2.0
MLSS (mg/L)	3,000–6,000	1,500–5,000	3,000–6,000
Detention time (hours)	18–36	16–36	18–36
Volumetric Loading (lb BOD_5/d/10^3 cu ft)	10–25	5–15	5–30

Source: Adapted from Metcalf and Eddy, 1991 and WEF, 1998; USEPA, *Wastewater Technology Fact Sheet, Package Plants*, Office of Water, Municipal Technology Branch, Table 1 (EPA/823/F-00/016) September 2000. www.epa.gov/owm/mtb/package_plant.pdf.

Table 9B.34 Extended Aeration Performance

	Typical Effluent Quality	Aldie WWTP (Monthly Average)
BOD (mg/L)	<30 or <10	5
TSS (mg/L)	<30 or <10	17
TP (mg/L)	<2[a]	[b]
NH_3-N (mg/L)	<2	[b]

[a] May require chemicals to achieve.
[b] DEQ does not require monitoring of these parameters.

Source: From Sloan, 1999 and Broderick, 1999; USEPA, *Wastewater Technology Fact Sheet, Package Plants*, Office of Water, Municipal Technology Brach, Table 1 (EPA/823/F-00/016) September 2000. www.epa.gov/owm/mtb/package_plant.pdf.

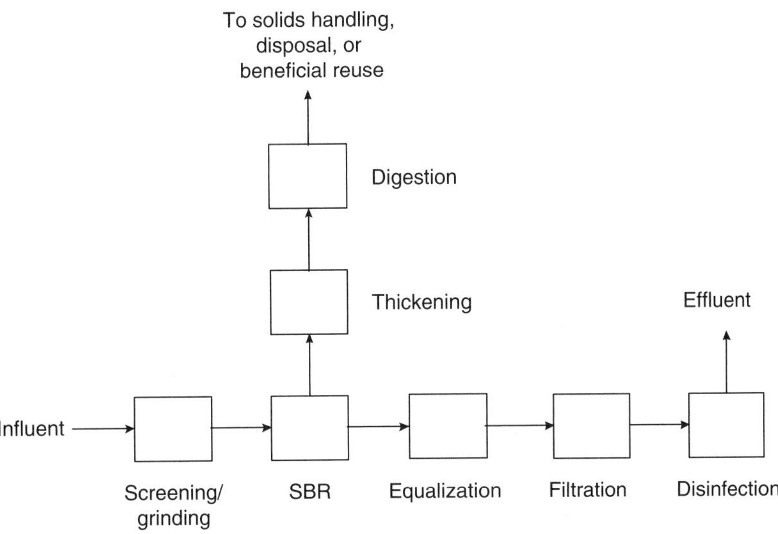

Figure 9B.3 Sequencing batch reactors key design parameters for a conventional load. (From USEPA, *Wastewater Technology Fact Sheet, Sequencing Batch Reactors*, Office of Water, Municipal Technology Branch, Figure 1 (EPA/823/F-99/073) September 1999. www.epa.gov/owm/mtb/sbr_new.pdf.)

Table 9B.35 Sequencing Batch Reactors Key Design Parameters for A Conventional Load

	Municipal	Industrial
Food to mass (F:M)	0.15–0.4/day	0.15–0.6/day
Treatment cycle duration	4.0 h	4.0–24 h
Typically low water level mixed liquor suspended solids	2,000–2,500 mg/L	2,000–4,000 mg/L
Hydraulic retention time	6–14 h	Varies

Source: From USEPA, *Wastewater Technology Fact Sheet, Sequencing Batch Reactors*, Office of Water, Municipal Technology Brach, Table 1 (EPA/823/F-00/073) September 1999. www.epa.gov/owm/mtb/sbr_new pdf.

Table 9B.36 Sequencing Batch Reactors Performance

		Harrah WWTP	
	Typical Effluent	% Removal	Effluent
BOD (mg/L)	10	98	3
TSS (mg/L)	10	98	3
NH_3 (mg/L)	<1	97	0.6

Source: From Sloan, 1999 and Reynolds, 1999; USEPA, *Wastewater Technology Fact Sheet, Sequencing Batch Reactors*, Office of Water, Municipal Technology Brach, Table 2 (EPA/823/F-99/073) September 1999. www.epa.gov/owm/mtb/sbr_new pdf.

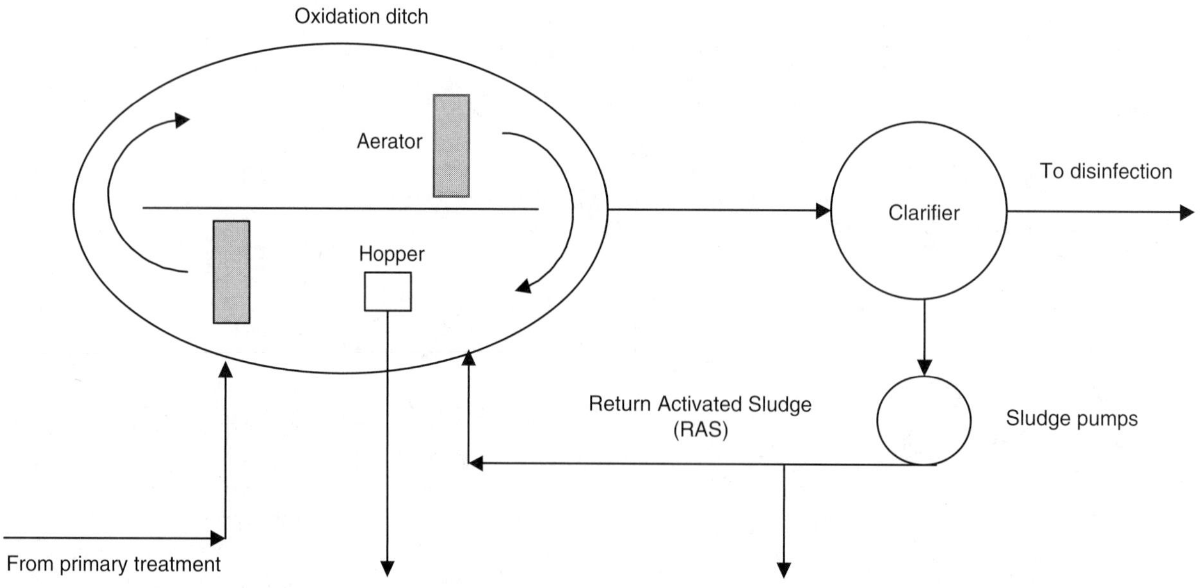

Figure 9B.4 Typical oxidation ditch activated sludge system. (From USEPA, *Wastewater Technology Fact Sheet, Oxidation Ditches*, Office of Water, Municipal Technology Branch, Figure 1 (EPA /823/F-00/013) September 2000. www.epa.gov/owm/mtb/oxidation_ditch.pdf.)

Table 9B.37 Oxidation Ditch Performance

	Typical Effluent Quality		Ocoee WWTP	
	With 2° Clarifier	With Filter	% Removal	Effluent
CBOD (mg/L)	10	5	>97	4.8
TSS (mg/L)	10	5	>97	0.32
TP (mg/L)	2	1	NA	NA
N–NO_3 (mg/L)	NA	NA	>95	0.25

Note: 2°, secondary; NA, not available.

Source: From Kruger, 1999 and Holland, 1999; USEPA, *Wastewater Technology Fact Sheet, Package Plants*, Office of Water, Municipal Technology Branch, Table 4 (EPA/823/F-00/016) September 2000. www.epa.gov/owm/mtb/package_plant.pdf.

Table 9B.38 Trickling Filters Operational Parameters

Parameter	Low-Rate	High-Rate
Hydraulic loading gpd/ft	25–100	200–1000
Organic loading lb BOD_5/1000 ft^3	5–25	25–300
Depth, ft	6–8	15–40
Filter media	Rock	Plastic

Source: From USEPA, *Technology Transfer, Summary Report, Small Community Water and Wastewater Treatment*, Table 1, p. 20 (EPA/625/R-92/010) September 1992.

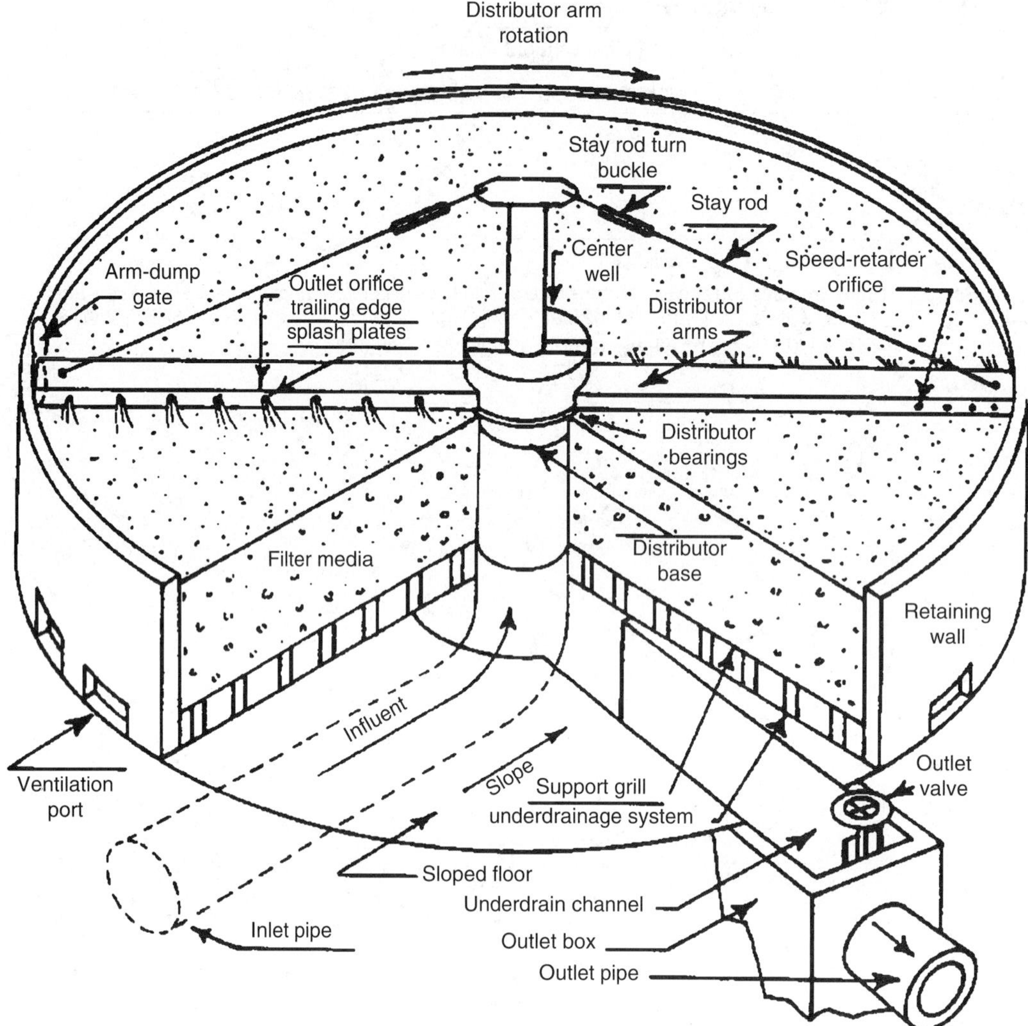

Figure 9B.5 Typical trickling filter. (From USEPA, *Wastewater Technology Fact Sheet, Trickling Filter*, Office of Water, Municipal Technology Branch, Figure 1 (EPA/823/F-00/014) September 2000. www.epa.gov/owm/mtb/trickling_filter.pdf.)

Table 9B.39 Typical Loading Rates for Single-Stage Nitrification in Trickling Filters

TF Media	% Nitrification	Loading Rate lb BOD/1,000 ft^3/d (g BOD/m^3/d)
Rock	75–85	10–6 (160–96)
	85–95	6–3 (96–48)
Plastic	75–85	181–12 (288–192)
Tower TF	85–95	12–6 (192–96)

Source: From Metcalf & Eddy, Inc. with permission from the McGraw-Hill Companies, 1991; USEPA, *Wastewater Technology Fact Sheet, Trickling Filter Nitrification*, Office of Water, Municipal Technology Branch, Table 1 (EPA/823/F-00/015) September 2000. www.epa.gov/owm/mtb/trickling_filt_nitrification.pdf.

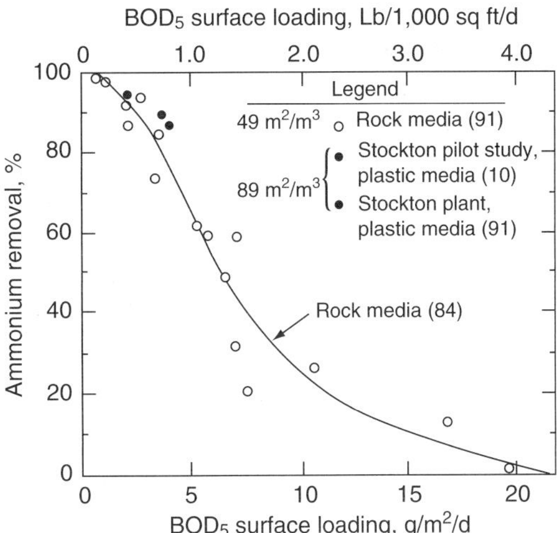

Figure 9B.6 Effect of BOD_5 Surface loading on nitrification efficiency of rock and plastic media trickling filters. (From USEPA, *Wastewater Technology Fact Sheet, Trickling Filter Nitrification*, Office of Water, Municipal Technology Branch, Figure 2 (EPA/823/F-00/015) September 2000. www.epa.gov/owm/mtb/trickling_filt_nitrification.pdf.)

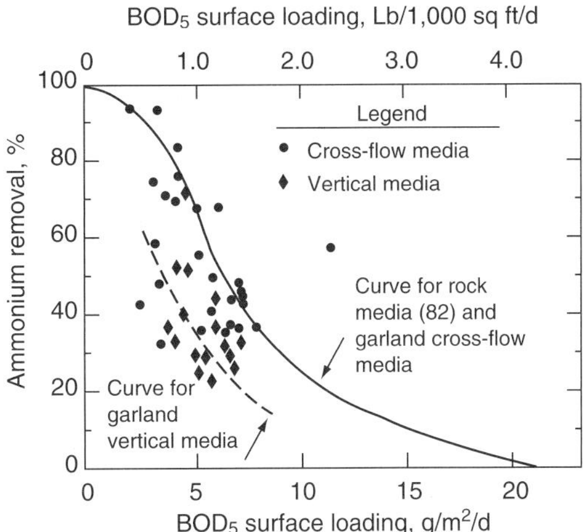

Figure 9B.7 Performance comparison of various trickling filter media. (From USEPA, *Wastewater Technology Fact Sheet, Trickling Filter Nitrification*, Office of Water, Municipal Technology Branch, Figure 2 (EPA/823/F-00/015) September 2000. www.epa.gov/owm/mtb/trickling_filt_nitrification.pdf.)

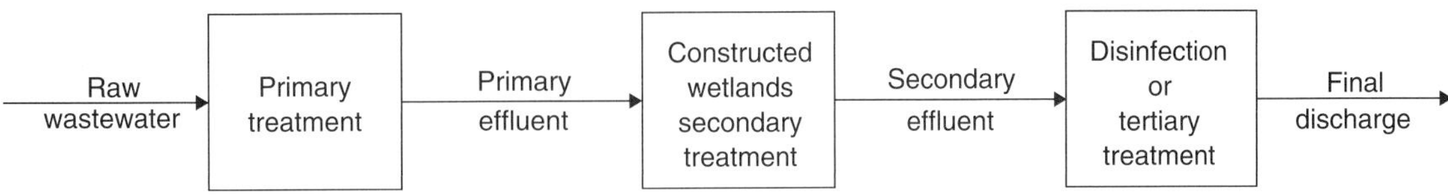

Figure 9B.8 Constructed wetland in wastewater treatment train. (From USEPA, *Manual, Constructed Wetlands Treatment of Municipal Wastewaters*, Office of Research and Development, Figure 2.1, p. 22 (EPA/625/R-99/010) September 2000. www.epa.gov/owow/wetlands/pdf/Design_Manual2000.pdf.)

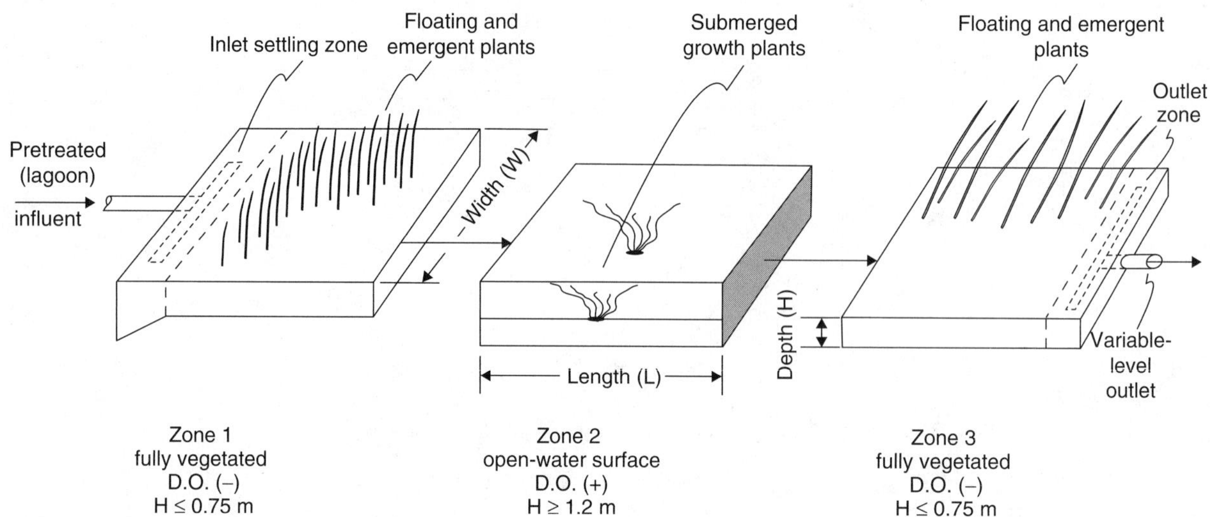

Figure 9B.9 Elements of a free water surface constructed wetland. (From USEPA, *Manual, Constructed Wetlands Treatment of Municipal Wastewaters*, Office of Research and Development, Figure 2.2, p. 22 (EPA/625/R-99/010) September 2000. www.epa.gov/owow/wetlands/pdf/Design_Manual2000.pdf.)

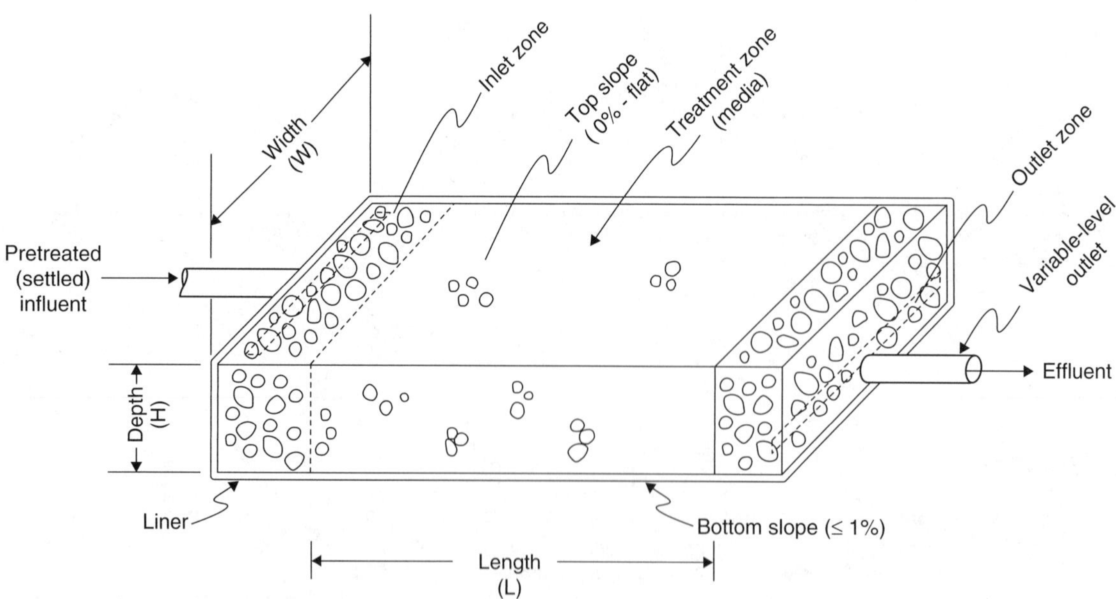

Figure 9B.10 Elements of a vegetated submerged bed system. (From USEPA, *Manual, Constructed Wetlands Treatment of Municipal Wastewaters*, Office of Research and Development, Figure 2.3, p. 22 (EPA/625/R-99/010) September 2000. www.epa.gov/owow/wetlands/pdf/Design_Manual2000.pdf.)

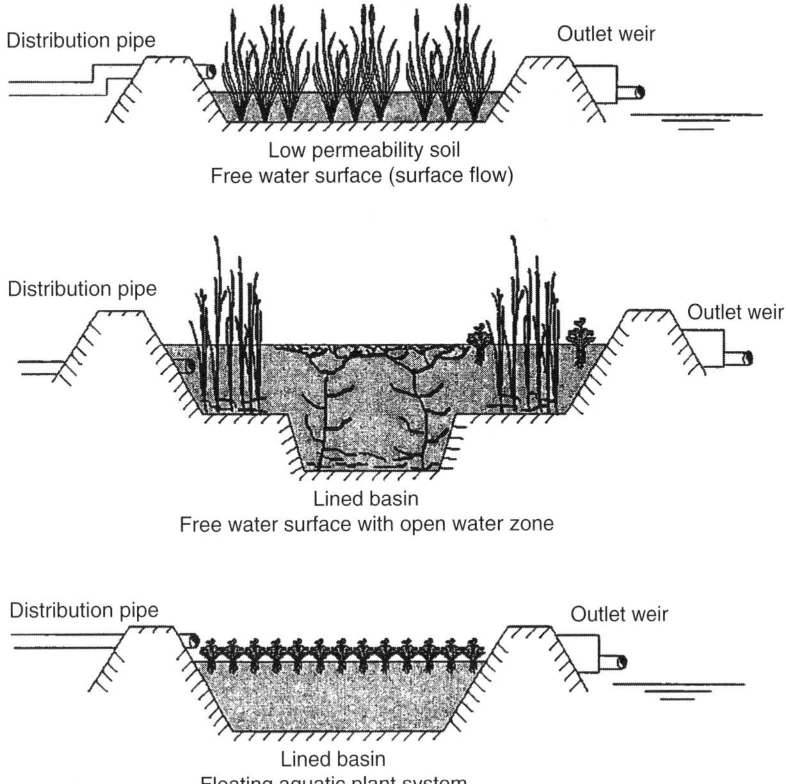

Figure 9B.11 Floating aquatic plant system. (From USEPA, *Free Water Surface Wetlands for Wastewater Treatment, A Technologic Assessment*, Office of Water, Figure 1.1, p. 1–4, (EPA/832/S-99/002) June 1999. www.epa.gov/owow/wetlands/pdf/FW_Surface_wetlands.pdf.)

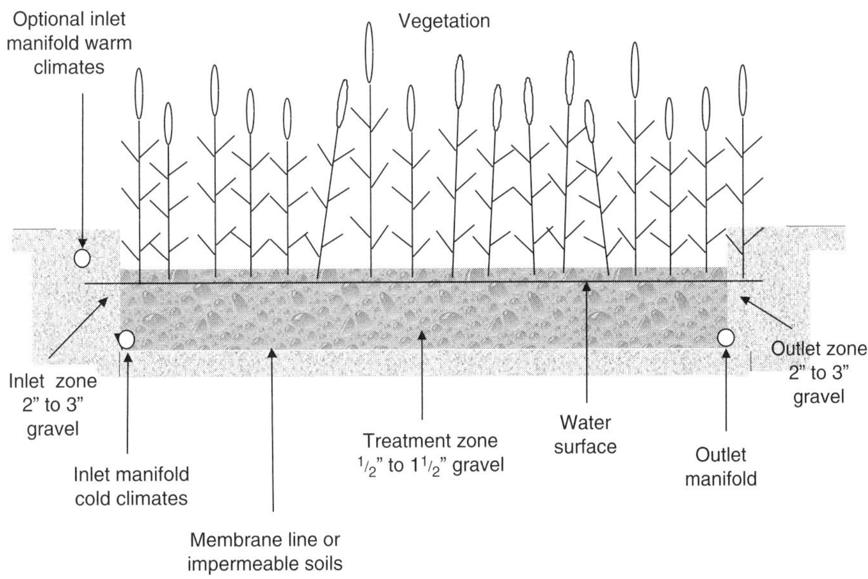

Figure 9B.12 Subsurface flow constructed wetlands. (From USEPA, *Wastewater Technology Fact Sheet, Wetlands: Subsurface Flow*, Office of Water, Municipal Technology Branch, Figure 1 (EPA/823/F-00/023) September 2000. www.epa.gov/owm/mtb/wetlands-subsurface_flow.pdf.)

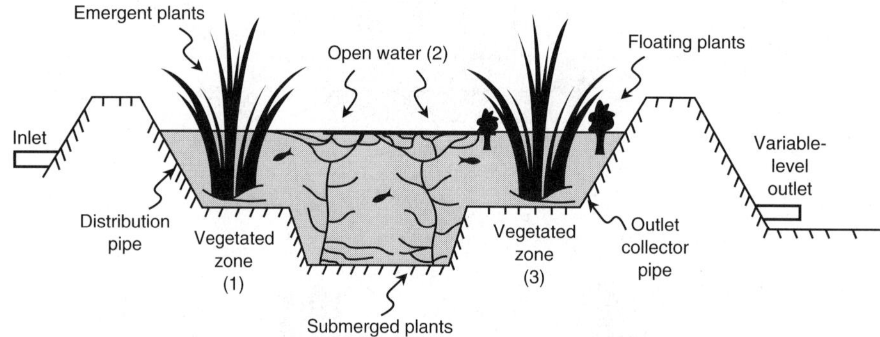

Figure 9B.13 Profile of a three-zone free water surface constructed wetland cell. (From USEPA, *Manual, Constructed Wetlands Treatment of Municipal Wastewaters*, Office of Research and Development, Figure 2.3, p. 22 (EPA/625/R-99/010) September 2000. www.epa.gov/owow/wetlands/pdf/Design_Manual2000.pdf.)

Table 9B.40 Typical Constructed Wetland Influents

Constituent (mg/L)	Septic Tank Effluent[a]	Primary Effluent[b]	Pond Effluent[c]
BOD	129–147	40–200	11–35
Sol. BOD	100–118	35–160	7–17
COD	310–344	90–400	60–100
TSS	44–54	55–230	20–80
VSS	32–39	45–180	25–65
TN	41–49	20–85	8–22
NH_3	28–34	15–40	0.6–16
NO_3	0–0.9	0	0.1–0.8
TP	12–14	4–15	3–4
OrthoP	10–12	3–10	2–3
Fecal coli (log/100 mL)	5.4–6.0	5.0–7.0	0.8–5.6

[a] EPA (1978), 95% confidence interval. Prior to major detergent reformulations which reduce P species by ~50%.
[b] Adapted from Metcalf and Eddy, (1991) assuming typical removal by primary sedimentation-soluble BOD = 35–45% total.
[c] EPA (1980).

Source: From USEPA, *Manual, Constructed Wetlands Treatment of Municipal Wastewaters*, Office of Research and Development, Table 3.1, p. 41 (EPA/625/R-99/010) September 2000. www.epa.gov/owow/wetlands/pdf/Design_Manual2000.pdf.

Table 9B.41 Loading and Performance Data for Systems Analyzed in This Document (DMDB)

	Pollutant Loading Rate (kg/ha-day)			Influent (mg/L)			Effluent (mg/L)		
Constituent	Min	Mean	Max	Min	Mean	Max	Min	Mean	Max
BOD_5	2.3	51	183	6.2	113	438	5.8	22	70
TSS	5	41	180	12.7	112	587	5.3	20	39
NH_4–N	0.3	5.8	16	3.2	13.4	30	0.7	12	23
TKN	1.0	9.5	20	8.7	28.3	51	3.9	19	32
TP	—	—	—	0.56	1.39	2.41	0.68	2.42	3.60
FC	—	—	—	42000	73000	250000	112	403	713

BOD, Biochemical Oxygen Demand (5 day); TSS, Total Suspended Solids; NH_4-N, Ammonia Nitrogen; TKN, Total Kjeldahl Nitrogen; TP, Total Phosphorus; FC, Fecal Coliform, cfu/100 mL.

Source: From USEPA, Manual, *Constructed Wetlands Treatment of Municipal Wastewaters*, Office of Research and Development, Table 4.1, p. 67 (EPA/625/R-99/010) September 2000. www.epa.gov/owow/wetlands/pdf/Design_Manual2000.pdf.

Table 9B.42 Typical Areal Loading Rates for SF Constructed Wetlands

Constituent	Typical Influent Concentration mg/L	Target Effluent Concentration mg/L	Mass Loading Rate lb/ac/d[a]
Hydraulic Load (in./d)	3–12[b]		
BOD	30–175	10–30	60–140
TSS	30–150	10–30	40–150
NH_3/NH_4 as N	2–35	1–10	1–10
NO_3 as N	2–10	1–10	3–12
TN	2–40	1–10	3–11
TP	1–10	0.5–3	1–4

Note: Wetland water temperature >> 20 °C.

Source: From USEPA, *Waster Technology Fact Sheet, Wetlands: Subsurface Flow*, Office of Water, Municipal Technology Branch, Table 1 (EPA/823/F-00/023) September 2000. www.epa.gov/owm/mtb/wetlands-subsurface_flow.pdf.

Table 9B.43 Typical Media Characteristics for Subsurface Wetlands

Media Type	Effective Size D_{10} (mm)[a]	Porosity, n (%)	Hydraulic Conductivity k_s ($ft^3/ft^2/d$)[b]
Coarse sand	2	28–32	300–3,000
Gravelly sand	8	30–35	1,600–16,000
Fine gravel	16	35–38	3,000–32,000
Medium gravel	32	36–40	32,000–160,000
Coarse rock	128	38–45	16×10^4–82×10^4

[a] mm $\times$ 0.03937 = inches.
[b] $ft^3/ft^2/d \times 0.3047 = m^3/m^2/d$, or $\times$ 7.48 = $gal/ft^2/d$.

Source: From Reed, S.C., R.W. Crites, E.J. Middlebrooks (1995) *Natural Systems for Waste Management and Treatment - Second Edition*, McGraw-'Hill Co, New York; USEPA, *Waste Water Technology Fact Sheet, Wetlands: Subsurface Flow, Office of Water*, Municipal Technology Branch, Table 3 (EPA/823/F-00/023) September 2000. www.epa.gov/owm/mtb/wetlands-subsurface_flow.pdf.

Table 9B.44 Summary of Performance for 14 Subsurface Flow Wetland Systems

Constituent	Mean Influent (mg/L)	Mean Effluent (mg/L)
BOD_5	28[a] (5–51)[b]	8[a] (1–15)[b]
TSS	60 (23–118)	10 (3–23)
TKN as N	15 (5–22)	9 (2–18)
NH_3/NH_4 as N	5 (1–10)	5 (2–10)
NO_3 as N	9 (1–18)	3 (0.1–13)
TN	20 (9–48)	9 (7–12)
TP	4 (2–6)	2 (0.2–3)
Fecal coliforms (#/100 ml)	270,000 (1,200–1,380,000)	57,000 (10–330,000)

Mean detention time 3 d (range 1–5 d).

[a] Mean value.
[b] Range of values.

Source: From USEPA, 1993; USEPA, *Wastewater Technology Fact Sheet, Wetlands: Subsurface Flow*, Office of Water, Municipal Technology Branch, Table 4 (EPA/823/F-00/023) September 2000. www.epa.gov/owm/mtb/wetlands-subsurface_flow.pdf.

Table 9B.45 Typical Areal Loading Rates for Free Water Surface Wetlands

Constituent	Typical Influent Conc. (mg/L)	Target Effluent Conc. (mg/L)	Mass Loading Rate (lb/ac/d)
Hydraulic Load (in./d)	0.4–4		
BOD	5–100	5–30	9–89
TSS	5–100	5–30	9–100
NH_3/NH_4 as N	2–20	1–10	1–4
NO_3 as N	2–10	1–10	2–9
TN	2–20	1–10	2–9
TP	1–10	0.5–3	1–4

Source: From USEPA, *Wastewater Technology Fact Sheet, Free Water Surface Wetlands*, Office of Water, Municipal Technology Branch, Table 1 (EPA/823/F-00/024) September 2000. www.epa.gov/owm/mtb/free_water_surface_wetlands.pdf.

Table 9B.46 Summary of Performance for 27 Free Water Surface Wetlands

Constituent	Mean Influent (mg/L)	Mean Effluent (mg/L)
BOD_5	70	15
TSS	69	15
TKN as N	18	11
NH_3/NH_4 as N	9	7
NO_3 as N	3	1
TN	12	4
TP	4	2
Dissolved P	3	2
Fecal coliforms (#/100 mL)	73,000	1,320

Source: USEPA, 2000; USEPA, *Wastewaster Technology Fact Sheet, Free Water Surface Wetlands*, Office of Water, Municipal Technology Branch, Table 1 (EPA/823/F-00/024) September 2000. www.epa.gov/owm/mtb/free_water_surface_wetlands.pdf.

Table 9B.47 Wastewater Characteristics Affecting Chlorination Performance

Wastewater Characteristic	Effects on Chlorine Disinfection
Ammonia	Forms chloramines when combined with chlorine
Biochemical oxygen demand (BOD)	The degree of interference depends on their functional groups and chemical structures
Hardness, iron, nitrate	Minor effect, if any
Nitrite	Reduces effectiveness of chlorine and results in THMs
pH	Affects distribution between hypochlorous acid and hypochlorite ions and among the various chloramine species
Total suspended solids	Shielding of embedded bacteria and chlorine demand

Source: From Darby et al., with permission from the water environment research foundation, 1995; USEPA, *Wastewater Technology Fact Sheet, Chlorine Disinfection*, Office of Water, Municipal Technology Branch, Table 2 (EPA/823/F-99/063) September 1999. www.epa.gov/owm/mtb/chlo.pdf.

Table 9B.48 Wastewater Characteristics Affecting Ultraviolet Disinfection Performance

Wastewater Characteristic	Effects on UV Disinfection
Ammonia	Minor effect, if any
Nitrite	Minor effect, if any
Nitrate	Minor effect, if any
Biochemical oxygen demand (BOD)	Minor effect, if any. Although, if a large portion of the BOD is humic and/or unsaturated (or conjugated) compounds, then UV transmittance may be diminished
Hardness	Affects solubility of metals that can absorb UV light. Can lead to the precipitation of carbonates on quartz tubes
Humic materials, iron	High absorbency of UV radiation
pH	Affects solubility of metals and carbonates
TSS	Absorbs UV radiation and shields embedded bacteria

Source: From USEPA, *Wastewater Technology Fact Sheet, Ultraviolet Disinfection*, Office of Water, Municipal Technology Branch, Table 2 (EPA/823/F-99/064) September 1999. www.epa.gov/owm/mtb/uv.pdf.

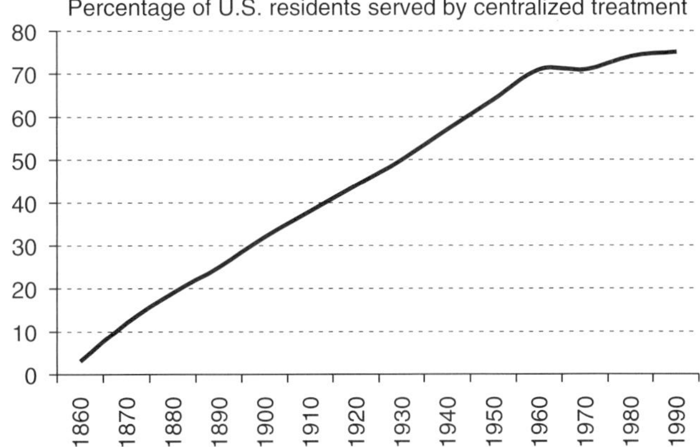

Figure 9B.14 Percentage of US residents severed by centralized treatment. (From USEPA, *Draft Handbook for Management of On-Site and Clustered (Decentralized) Wastewater Treatment Systems*, Office of Water, Table 1.1, p. 12 (EPA /PA 823/P-03/001), February 2003.)

Table 9B.49 Comparison of Design and Operating Parameters, Land Treatment Systems

Parameters	Irrigation	Overland Flow	Rapid Infiltration
Weekly application rate (in.)	0.5–4.0	2.4–6.0	4–96
Annual application rate (ft)	2–18	8–40	20–410
Estimated land required for 100,000 gpd (acres)	20–25	5–10	1–7
Minimum preapplication treatment requirements	Lagoons	Screening and grit removal	Lagoons
Climate restrictions	Storage needed for cold and wet climates	Storage needed for cold and wet climates	Cold weather may reduce hydraulic loading cycles
Slopes	<20%	Smooth sloes of 2–8%	Not critical
Soil permeability	Slow to moderate	Impermeable (clays, silts, soils with impermeable barriers)	Rapid

Source: From USEPA, *Wastewater Technology Transfer, Summary Report*, Small Community Water and Wastewater Treatment, Table 2, p. 34 (EPA/525/R-92/010) September 1992.

Table 9B.50 Design Criteria for Rapid Land Infiltration Land Treatment

Item	Range
Basin infiltration area	0.3–5.5 ha/$10_3 m_3$/d (3–56 acres/MGD)
Hydraulic loading rate	6–90 m/yr (20–300 ft/yr) [6–92 m_3/m_2/yr (150–2250 gal/ft_2/yr)]
BOD loading	22–112 kg/ha/d (20–100 lb/acre/d)
Soil depth	At least 3–4.5 m (10–15 ft)
Soil permeability	At least 1.5 cm/h (0.6 in./h)
Wastewater application period	4 h to 2 wks
Drying period	8 h to 4 wks
Soil texture	Coarse sands, sandy gravels
Individual basin size (at least 2 basins in parallel)	0.4–4 ha (1–10 acres)
Height of dikes	0.15 m (0.5 ft) above maximum expected water level
Application method	flooding or sprinkling
Pretreatment required	primary or secondary

Source: From Crites, et al., 2000; USEPA, *Wastewater Technology Fact Sheet, Rapid Land Infiltration Land Treatment*, Office of Water, Municipal Technology Branch, Table 1 (EPA/823/F-03/025) September 2003. www.epa.gov/owm/mtb/final_rapidinfiltration.pdf.

Table 9B.51 Effluent Quality for Rapid Land Infiltration Land Treatment

Parameter	% Removal
BOD_5	95–99
TSS	95–99
TN	25–90
TP	0–90
Fecal coliform	99.9–99.99 + percent

Source: From USEPA, *Wastewater Technology Fact Sheet, Rapid Land Infiltration Land Treatment*, Office of Water, Municipal Technology Branch, Table 2 (EPA/823/F-03/025) September 2003. www.epa.gov/owm/mtb/final_rapidinfiltration.pdf.

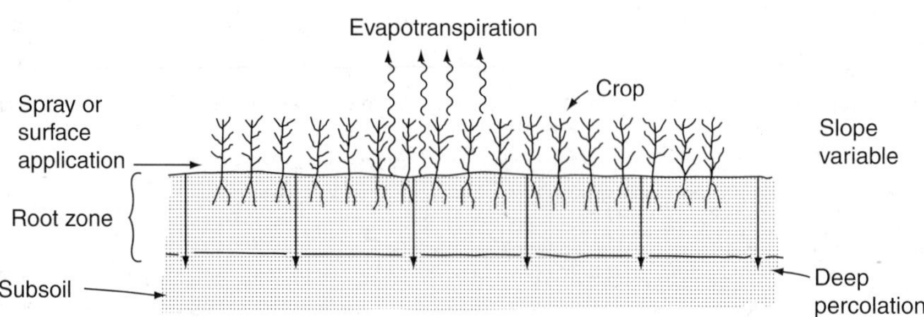

Figure 9B.15 Schematic of spray irrigation. (From USEPA, *Technology Transfer, Summary Report, Small Community Water and Wastewater Treatment*, Figure 2, p. 33 (EPA/625/R-92/010) September 1992.)

Table 9B.52 Design Criteria for Slow Rate Land Treatment

Item	Range
Field area	56–560 acres/MGD
Application rate	2–20 ft/yr (0.5–4 in./wk)
BOD Loading	0.2–5 lb/acre/d
Soil depth	At least 2–5 ft
Soil permeability	0.06–2.0 in./h
Lower temperature limit	25°F
Application method	Sprinkler or surface
Pretreatment required	Preliminary & secondary
Particle size (for sprinkler applications)	Solids less than 1/3 sprinkler nozzle

Source: From Crites, et al., 2000; USEPA, *Wastewater Technology Fact Sheet, Slow Rate Land Treatment*, Office of Water, Municipal Technology Branch, Table 1 (EPA/823/F-02/012) September 2002. www.epa.gov/owm/mtb/sloratre.pdf.

Table 9B.53 Maximum Metal Concentrations for Land Application

Metal	Ceiling Concentration (mg/kg)	Cumulative Pollutant Loading Rates (kg/hectare)	Pollutant Concentrations (mg/kg)
Arsenic	75	41	41
Cadmium	85	39	39
Copper	4,300	1,500	1,500
Lead	840	300	300
Mercury	57	17	17
Molybdenum	75	NL	NL
Nickel	420	420	420
Selenium	100	100	100
Zinc	7,500	2,800	2,800

NL, No limit.

Source: From USEPA, 1993 and 1994; USEPA, *Wastewater Technology Fact Sheet, Land Application*, Office of Water, Municipal Technology Branch, Table 1 (EPA/823/F-00/064) September 2000. www.epa.gov/owm/mtb/land_application.pdf.

Table 9B.54 Range of Expected Centrifuge Performance

Type of Wastewater Solids	Feed (%TS)	Polymer (lb/DTS)	Cake (%TS)
Primary undigested	4–8	5–30	25–40
WAS undigested	1–4	15–30	16–25
Primary+WAS undigested	2–4	5–16	25–35
Primary+WAS aerobic digested	1.5–3	15–30	16–25
Primary+WAS anaerobic digested	2–4	15–30	22–32
Primary anaerobic digested	2–4	8–12	25–35
WAS aerobic digested	1–4	20	18–21
Hi-Temp aerobic	4–6	20–40	20–25
Hi-Temp anaerobic	3–6	20–30	22–28
Lime stabilized	4–6	15–25	20–28

Source: From various centrifuge manufactures; Ireland and Balchunas, 1998; Henderson and Schultz, 1999; Leber and Garvey, 2000; USEPA, *Biosolids Technology Fact Sheet, Centrifuge Thickening and Dewatering*, Office of Water, Municipal Technology Branch, Table 1 (EPA/823/F-00/053) September 2000. www.epa.gov/owm/mtb/centrifuge_thickening.pdf.

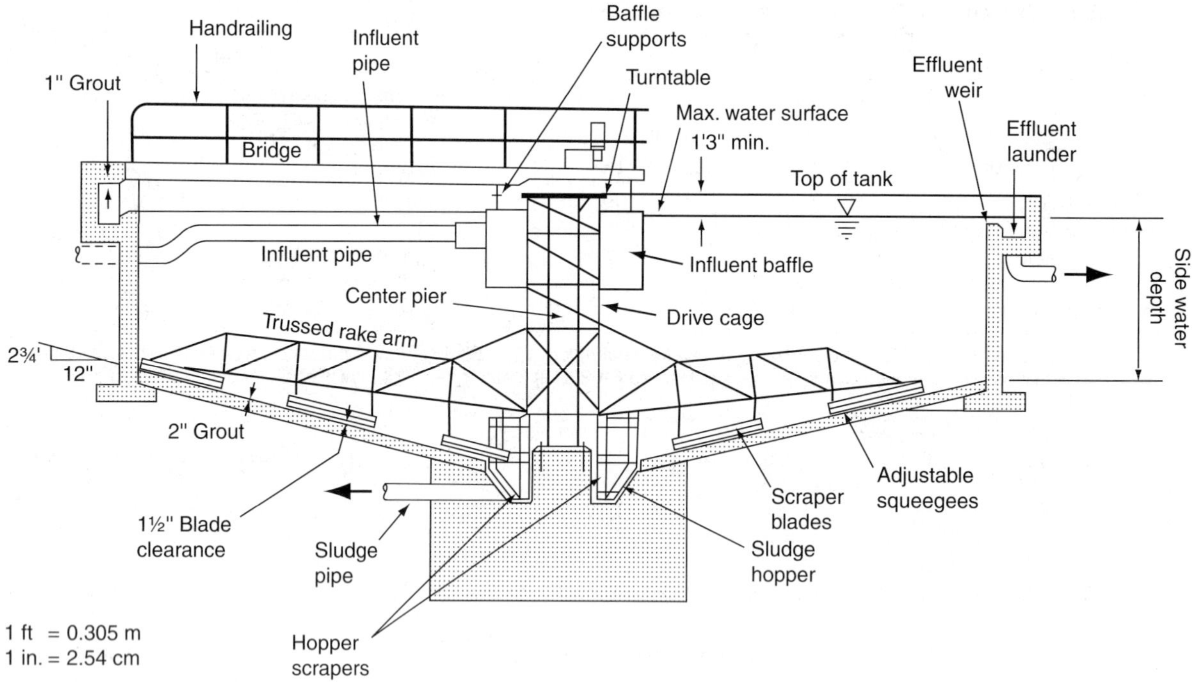

Figure 9B.16 Gravity thickener. (From USEPA, *Biosolids Technology Fact Sheet, Gravity Thickening and Dewatering*, Office of Water, Municipal Technology Branch, Figure 2 (EPA /823/F-03/022) September 2002. www.epa.gov/owm/mtb/final_gravitythickening.pdf.)

Table 9B.55 Factors Affecting Gravity Thickening Performance

Factor	Effect
Nature of the solids feed	Affects the thickening process because some solids thicken more easily than others
Freshness of feed solids	High solids age can result in septic conditions
High volatile solids concentrations	Hamper gravity settling due to reduced particle specific gravity
High hydraulic loading rates	Increase velocity and cause turbulence that will disrupt settling and carry the lighter solids past the weirs
Solids loading rate	If rates are high, there will be insufficient detention time for settling. If rates are too low, septic conditions may arise
Temperature and variation in temperature of thickener contents	High temperatures will result in septic conditions. Extremely low temperatures will result in lower settling velocities. If temperature varies, settling decreases due to stratification
High solids blanket depth	Increases the performance of the settling by causing compaction of the lower layers, but it may result in solids being carried over the weir
Solids residence time	An increase may result in septic conditions. A decrease may result in only partial settling
Mechanism and rate of solids withdrawal	Must be maintained to produce a smooth and continuous flow. Otherwise, turbulence, septic conditions, altered settling, and other anomalies may occur
Chemical treatment	Chemicals—such as potassium permanganate, polymers, or ferric chloride—may improve settling and/or supernatant quality
Presence of bacteriostatic agents or oxidizing agents	Allows for longer detention times before anaerobic conditions create gas bubbles and floating solids
Cationic polymer addition	Helps thicken waste-activated solids and clarify the supernatant
Use of metal salt coagulants	Improves overflow clarity but may have little impact on underflow concentration

Source: From Parsons, 2003; USEPA, *Biosolids Technology Fact Sheet, Gravity Thickening and Dewatering*, Office of Water, Municipal Technology Branch, Table 2 (EPA/823/F-03/022) September 2002. www.epa.gov/owm/mtb/final_gravitythickening.pdf.

Table 9B.56 Performance of Conventional Gravity Thickening

Type of Solids	Feed (%TS)	Thickened solids (%TS)
Primary (PRI)	2–7	5–10
Trickling filter (TF)	1–4	3–6
Rotating biological contactor (RBC)	1–3.5	2–5
Waste activated solids (WAS)	0.2–1	2–3
PRI + WAS	0.5–4	4–7
PRI + TF	2–6	5–9
PRI + RBC	2–6	5–8
PRI + Lime	3–4.5	10–15
PRI + (WAS + iron)	1.5	3
PRI + (WAS + aluminum salts)	0.2–0.4	4.5–6.5
Anaerobically digested PRI + WAS	4	8

Source: From WEF, 1987; USEPA, *Biosolids Technology Fact Sheet, Gravity Thickening and Dewatering*, Office of Water, Municipal Technology Branch, Table 1 (EPA/823/F-03/022) September 2002. www.epa.gov/owm/mtb/final_gravitythickening.pdf.

Table 9B.57 Typical Data for Various Types of Sludges Dewatered on Belt Filter Presses

Type of Wastewater Sludge	Total Feed Solids (%)	Polymer (g/kg)	Total Cake Solids (%)
Raw primary	3–10	1–5	28–44
Raw WAS	0.5–4	1–10	20–35
Raw primary + WAS	3–6	1–10	20–35
Anaerobically digested primary	3–10	1–5	25–36
Anaerobically digested WAS	3–4	2–10	12–22
Anaerobically digested primary + WAS	3–9	2–8	18–44
Aerobically digested primary + WAS	1–3	2–8	12–20
Oxygen activated WAS	1–3	4–10	15–23
Thermally conditioned primary + WAS	4–8	0	25–50

Source: From USEPA, 1987; USEPA, *Biosolids Technology Fact Sheet, Belt Filter Press*, Office of Water, Municipal Technology Branch, Table 1 (EPA/823/F-03/057) September 2000. www.epa.gov/owm/mtb/belt_filter.pdf.

Table 9B.58 Typical Biosolids Application Scenarios

Type of Site/Vegetation	Schedule	Application Frequency	Application Rate
Agricultural land			
Corn	April, May, after harvest	Annually	5–10 dry tons per acre
Small grains	March–June, August, fall	Up to 3 times per year	2–5 dry tons per acre
Soybeans	April–June, fall	Annually	5–20 dry tons per acre
Hay	After each cutting	Up to 3 times per year	2–5 dry tons per acre
Forest land	Year round	Once every 2–5 years	5–100 dry tons per acre
Range land	Year round	Once every 1–2 years	2–60 dry tons per acre
Reclamation sites	Year round	Once	60–100 dry tons per acre

Source: From USEPA, 1994; USEPA, *Wastewater Technology Fact Sheet, Land Application*, Office of Water, Municipal Technology Branch, Table 2 (EPA/823/F-00/064) September 2000. www.epa.gov/owm/mtb/land_application.pdf.

Table 9B.59 Pipe Paint Color Scheme Recommendations

Raw sludge line—brown with black bands

Sludge recirculation suction line—brown with yellow bands Sludge draw off line—brown with orange bands Sludge recirculation discharge line—brown Sludge gas line—orange (or red)

Natural gas line—orange (or red) with black bands Nonpotable water line—blue with black bands Potable water line—blue Chlorine line—yellow

Sulfur dioxide—yellow with red bands Sewage (wastewater) line—gray Compressed air line—green

Water lines for heating digesters or buildings—blue with a 6-inch (150 mm) red band spaced 30 in. (760 mm) apart Fuel oil/diesel—red

Plumbing drains and vents—black Polymer-purple

The contents and direction of flow shall be stenciled on the piping in a contrasting color. The use of paints containing lead or mercury should be avoided. In order to facilitate identification of piping, particularly in the large plants, it is suggested that the different lines be color-coded.

Source: From Board of State and Provincial Public Health and Environmental Manager, Health Education Services Division, Recommended Standards for Wastewater Facilities, 2004 edition, p. 50–5. www.hes.org.

SECTION 9C DECENTRALIZED WASTEWATER TREATMENT

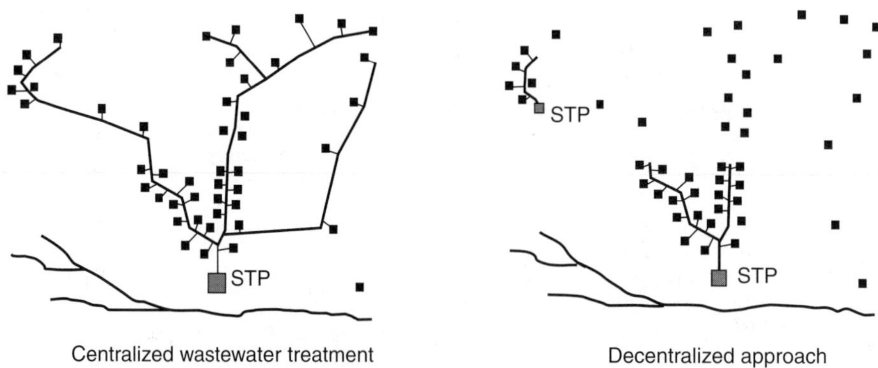

Figure 9C.17 Centralized wastewater treatment vs. decentralized approach. (From USEPA, *Draft Handbook for Management of Onsite and Clustered (Decentralized) Wastewater Treatment Systems*, Office of Water, Figure 1.4 (EPA/PA 823/P-03/001), February 2003. www.epa.gov/ord/NRMRL/Pubs/625R00008/625R00008totaldocument.pdf.)

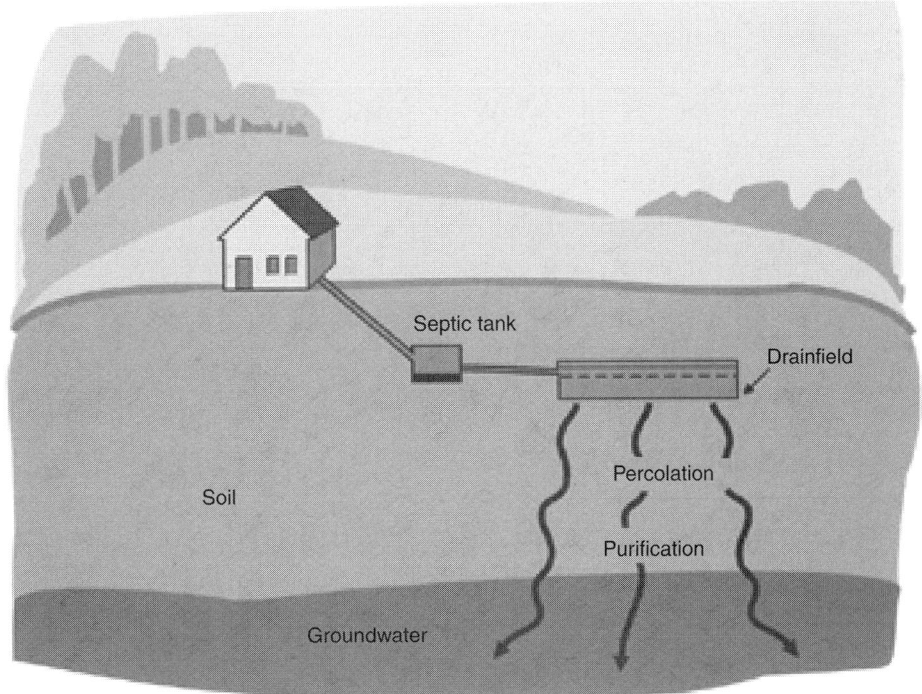

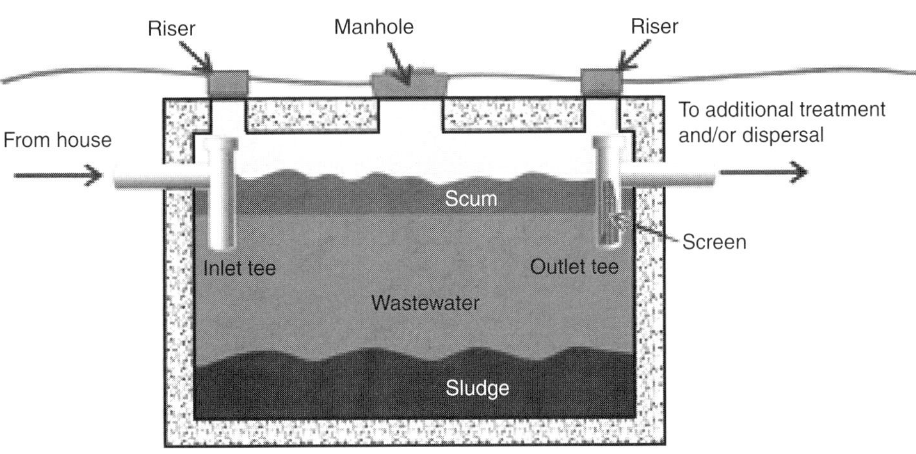

Figure 9C.18 Conventional system. (From USEPA website. www.epa.gov.)

Table 9C.60 Types of Decentralized Wastewater Treatment Systems

Type of System	Description
Individual on-site systems	Systems that serve an individual residence and can range from conventional septic tank/drainfield systems to systems composed of complex mechanical treatment trains
Cluster systems	Wastewater collection and treatment systems that serve two or more dwellings or buildings, but less than an entire community, on a suitable site near the served structures
Commercial, residential, institutional, and recreational facilities	Systems designed to treat larger and sometimes more complex wastewater sources from commercial buildings (e.g. restaurants), apartments, or institutional or recreational facilities

Source: From USEPA, *Draft Handbook for Management of On-Site and Clustered (Decentralized) Wastewater Treatment Systems*, Office of Water, Table 1.1, p. 12 (EPA/PA 823/P-03/001), February 2003.

Table 9C.61 Characteristics of Septage Conventional Parameters

Concentration Parameter	Minimum	Maximum
Total solids	1,132	130,475
Total volatile solids	353	71,402
Total suspended solids	310	93,378
Volatile suspended	95	51,500
Biochemical oxygen demand	440	78,600
Chemical oxygen demand	1,500	703,000
Total kjeldahl nitrogen	66	1,060
Ammonia nitrogen	3	116
Total phosphorus	20	760
Alkalinity	522	4,190
Grease	208	23,368
P^H	1.5	12.6
Total coliform	10^7/100 mL	10^9/100 mL
Fecal coliform	10^6/100 mL	10^8/100 mL

Note: The measurements above are in mg/L unless otherwise indicted.

Source: From USEPA, 1994; USEPA, *Decentralized Systems, Technology Fact Sheet, Septage Treatment/Disposal*, Office of Water, Table 1 (EPA/823/F-99/068), September 1999. www.epa.gov/owm/mtb/septage.pdf.

Table 9C.62 Sources of Septage

Description Rate	Removal Pump-out	Characteristics
Septic tank	2–6 years, but can vary with location local ordinances	Concentrated BOD, solids, nutrients, variable toxins (such as metals), inorganics (sand), odor, pathogens, oil, and grease
Cesspool	2–10 years	Concentrated BOD, solids, nutrients, variable toxins, inorganics, sometimes high grit, odor, pathogens, oil, and grease
Privies/portable toilets	1 week to months	Variable BOD, solids, inorganics, odor, pathogens, and some chemicals
Aerobic tanks	Months to 1 year	Variable BOD, inorganics, odor, pathogens, and concentrated solids
Holding tanks (septic tank with no drain-field, typically a local requirement	Days to weeks	Variable BOD, solids, inorganics, odor, and pathogens, similar to raw wastewater solids
Dry pits (associated with septic fields) Miscellaneous may exhibit characteristics of septage	2–6 years	Variable BOD, solids, inorganics, and odor
Private wastewater treatment plants	Variable	Septic tank
Boat pump-out station	Variable	Portable toilets
Grit traps	Variable	Oil, grease, solids, inorganics, odor, and variable BOD
Grease traps	Weeks to months	Oil, grease, BOD, viscous solids, and odor

Source: From Septage Handling Task Force (1997), copyright Water Environment Federation, used with permission. USEPA, *Decentralized Systems, Technology Fact Sheet, Septage Treatment/Disposal*, Office of Water, Table 2 (EPA/823/F-99/068), September 1999. www.epa.gov/owm/mtb/septage.pdf.

Table 9C.63 Physical and Chemical Characteristics of Septage, as Found in the Literature, with Suggested Design Values

	United States (5) (9–19)				Europe/Canada (4) (20)					
Parameter	**Average**	**Minimum**	**Maximum**	**Variance**	**Average**	**Minimum**	**Maximum**	**Variance**	**EPA Mean**	**Suggested Design Value**
TS	34,106	1,132	130,475	115	33,800	200	123,860	619	38,800	40,000
TVS	23,100	353	71,402	202	31,600	160	67,570	422	25,260	25,000
TSS	12,862	310	93,378	301	45,000	5,000	70,920	14	13,000	15,000
VSS	9,027	95	51,500	542	29,900	4,000	52,370	13	8,720	10,000
BODs	6,480	440	78,600	179	8,343	700	25,000	36	5,000	7,000
COD	31,900	1,500	703,000	469	28,975	1,300	114,870	88	42,850	15,000
TKN	588	66	1,060	16	1,067	150	2,570	17	677	700
NH_3–N	97	3	116	39	—	—	—	—	157	150
Total P	210	20	760	38	155	20	636	32	253	250
Alkalinity	970	522	4,190	8	—	—	—	—	—	1,000
Grease	5,600	208	23,368	112	—	—	—	—	9,090	8,000
pH	—	1.5	12.6	8.0	—	5.2	9.0		6.9	6.0
LAS	—	110	200	2	—	—	—		157	150

Values expressed as mg/L, except for pH.

Source: From Board of State and Provincial Public Health and Environmental Managers, Health Education Services Division, Recommended Standards for Wastewater Facilities, 2004 Edition, p. A4. www.hes.org.

Table 9C.64 Comparison of Septage and Municipal Wastewater[a]

Parameter	Septage[b]	Wastewater[c]	Ratio of Septage to Wastewater
TS	40,000	720	55:1
TVS	25,000	365	68:1
TSS	15,000	220	68:1
VSS	10,000	165	61:1
BODs	7,000	220	32:1
COD	15,000	500	30:1
TKN	700	40	17:1
NH_3–N	150	25	6:1
Total P	250	8	31:1
Alkalinity	1,000	100	10:1
Grease	8,000	100	80:1
pH	6.0		
LAS	150		

Table including footnotes is a taken from the USEPA Handbook entitled "Septage Treatment and Disposal," 1984, EPA-625/6-84-009 and is designated in that document as "Table 3.8."

[a] Values expressed as mg/L, except for pH.

[b] Based on suggested design values in Appendix-Table No.1 (USEPA Table 3.4).

[c] From Metcalf and Eddy, 4th Edition, "Medium strength sewage."

Source: From Board of State and Provincial Public Health and Environmental Managers, Health Education Services Division, Recommended Standards for Wastewater Facilities, 2004 Edition, p. A5. www.hes.org.

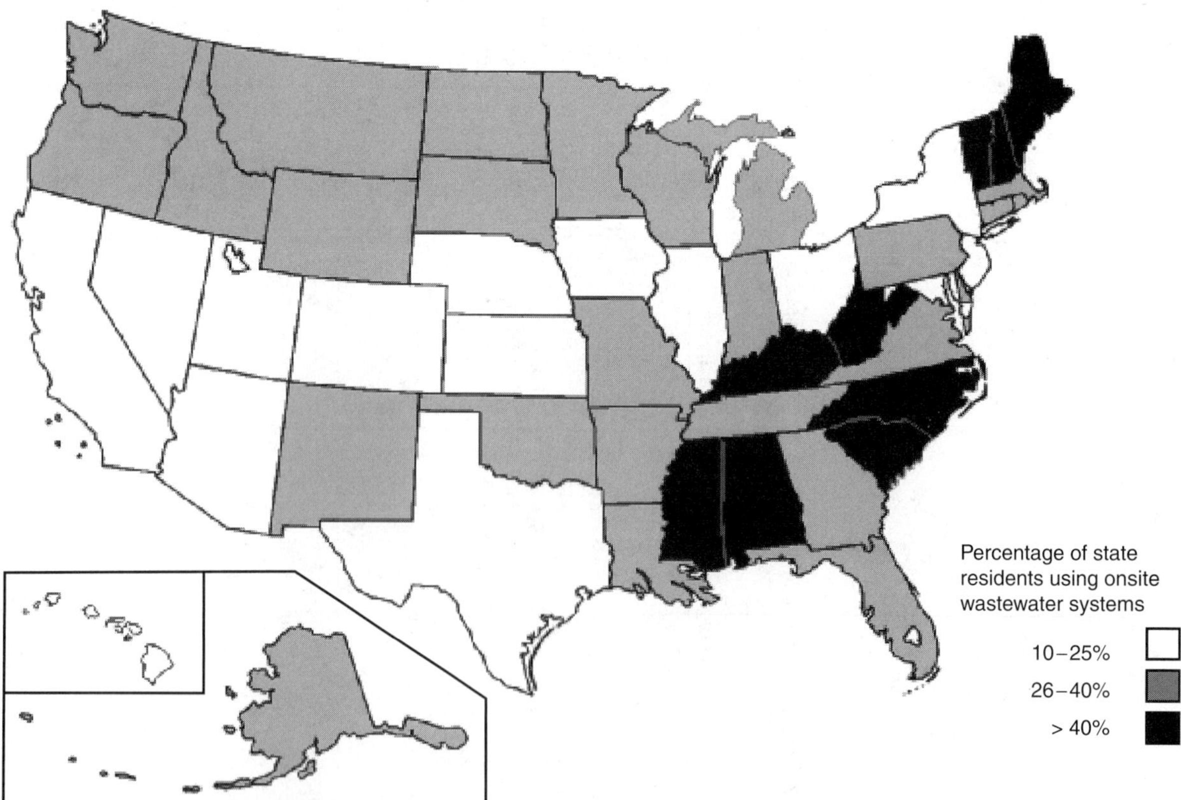

Figure 9C.19 On-site treatment system distribution in the United States. (From USEPA, *Draft Handbook for Management of On-site and Clustered (Decentralized) Wastewater Treatment Systems*, Office of Water, Figure 1.1 (EPA/PA 823/P-03/001), February 2003. www.epa.gov/ord/NRMRL/Pubs/625R00008/625R00008totaldocument.pdf.)

Table 9C.65 Septic Tank Capacities for One- and Two-Family Dwellings

Number of Bedrooms	Septic Tank Volume (gallons)
1	750[a]
2	750[a]
3	1,000
4	1,200
5	1,425
6	1,650
7	1,875
8	2,100

[a] Many States have established 1,000 gallons or more as the minimum size.

Source: From USEPA, *Draft Handbook for Management of On-site and Clustered (Decentralized) Wastewater Treatment Systems*, Office of Water, Table 1.1, p. 12 (EPA/PA 823/ 823/P-03/001), February 2003.

Table 9C.66 Treatment Performance of On-Site Septic Tank and Sand Filter

Parameter	Raw Waste	Septic Tank Effluent	Intermittent Sand Filter Effluent
BOD, mg/L	210–530	140–200	<10
SS, mg/L	237–600	50–90	<10
Total nitrogen, mg/L	35–80	25–60	—
Ammonia-nitrogen, mg/L	7–40	20–60	<0.5
Nitrate-nitrogen, mg/L	<1	<1	25
Total phosphorus, mg/L	10–27	10–30	—
Fecal coliforms (#/100 mL)	10^6–10^{10}	10^3–10^6	10^2–10^4
Viruses (#/100 mL)	Unknown	10^5–10^7	—

Source: Adapted from Tchobanoglous and Burton, 1991; USEPA, *Decentralized Systems, Technology Fact Sheet, Septic Tank Polishing*, Office of Water, Table 1 (EPA/823/F-02/021), September 2002. www.epa.gov/owm/mtb/septn_pol.pdf.

Table 9C.67 Design Parameters for Continuous-Flow, Suspended Growth Systems Extended Aeration Package Plant

Parameter	Extended Aeration
Pretreatment (if needed)	Septic tank or equivalent
Mixed liquor suspended solids (mg/L)a	2,000–6,000
F/M Load (lb BOD/d/MLVSS),	0.05–0.15
Hydraulic retention time (h)	24–120
Solids retention time (days)	20–40
Mixing power inpuf	0.2–3.0 hp/1,000 ft^3
Clarifier overflow rate (gpd/ft^2)	200–400 avg., 800 peak
Clarifier solids loading (lb/d/ft^2)	30 avg., 50 peak
Dissolved oxygen (mg/L)	>2.0
Residuals generated	0.6–0.9 lb TSS/Lb BOD removed
Sludge removal	3–6 months as needed

Source: From USEPA, *On-site Wastewater Treatment Systems Manual Technology Fact Sheet 1, Table 1.1, Continuous-flow, Suspended growth systems (CFSGAS)*, Office of Water, Office of Research and Development, EPA, (EPA/625/ R-00/0008). www.epa.gov/ord/NRMRL/Pubs/625R00008/6250R0008totaldocument.pdf.

Table 9C.68 Common Operational Problems of Extended Aeration Package Plants

Observation	Cause	Remedy
Excessive local turbulence	Diffuser plugging	Remove and clean
In aeration tank	Pipe breakage	Replace as required
	Excessive aeration	Throttle blower
White, thick, billowy foam on aeration tank	Insufficient MLSS	Avoid wasting solids
Thick, scummy, dark tan foam on aeration tank	High MLSS	Waste solids
Dark brown/black foam and mixed liquor in aeration tank	Anaerobic conditions Aerator failure	Check aeration system, aeration tank DO
Billowing sludge washout in clarifier	Hydraulic or solids overload	Waste sludge; check flow to unit
	Bulking sludge	See EPA, 1977
Clumps of rising sludge in clarifier	Denitrification	Increase sludge return rate to decrease sludge retention time in clarifier
	Septic conditions in clarifier	Increase return rate
Fine dispersed floc, turbid effluent	Turbulence in aeration tank	Reduce power input
	Siudae age too high	Waste sludae
Poor TSS and/or BOD removal	Excess flow and strength variations	Install flow smoothing system
Poor nitrification	Low temperatures	Insulate, upgrade to high biomass, etc
	Excessive biocide use	Reduce biocide loading

Source: From USEPA, *On-site Wastewater Treatment Systems Manual Technology Fact Sheet 1, Table 1.2, Continuous-flow, suspended growth systems (CFGAS)*, Office of Water, Office of Research and Development, EPA, (EPA/625/R-00/0008). www.epa.gov/ord/NRMRL/Pubs/625R00008/625R00008totaldocument.pdf.

Table 9C.69 Design Parameters for Fixed Film Systems

Parameter	Trickling Filter	RBC
Pretreatment	Septic tank (primary clarifier)	Septic tank (primary clarifier)
Surface hydraulic loading	10–25 gal/d-ft^2	N/A
Organic loading[a]	5–20 lb BOD/d-ft^2 (3–10 Lb BOD/d-ft^2 to nitrify)	2.5 lb SBOD/d-1,000 ff (6.4 lb BOD/d-1,000 ft^2)
Clarifier overflow rate		
Average flow	600–800 gal/d-ft^2	600–800 gal/d-ft^2
Peak flow	1,000–1,200 gal/d-ft^2	1,000–1,200 gal/d-ft'
Clarifier TSS loading rate		
Average flow	0.8–1.2 LbTSS/d-ft'	0.8–1.2 Lb TSS *Id-ft*^{2}
Peak flow	2.0 Lb TSS *Id-W*	2.0 Lb TSS *Id-ff*
Recirculation	Optional	Optional
Sludge generated[b]	0.6–1.1 Lb TSS lib BOD removed	0.6–1.1 Lb TSS lib BOD removed

[a] Loading rates for RBC are expressed per 1,000 ft^2 of total disk surface.
[b] Sludge generated is in addition to solids removed in septic tank.

Source: From USEPA, *On-site Wastewater Treatment Systems Manual Technology Fact Sheet 2, Table 1, Fixed Film Processes, Office of Water*, Office of Research and Development, EPA, (EPA/625/R-00/0008). www.epa.gov/ord/NRMRL/Pubs/625R00008/625R00008total document.pdf.

Table 9C.70 Design Parameters for IF-Type Sequencing Batch Reactors Systems Treatment Systems

Parameter	SBR Systems
Pretreatment	Septic tank or equivalent
Mixed liquor suspended solids (mg/L)	2,000–6,500
F/M load (Lb BOD/d/MLVSS)	0.04–0.20
Hydraulic retention time (h)	9–30
Total cycle times (h)[a]	4–12
Solids retention time (days)	20–40
Decanter overflow ratea (gpm/ff)	$<$100
Sludge wasting	As needed to maintain performance

[a] Cycle times should be tuned to effluent quality requirements, wastewater flow, and other site constraints.

Source: From USEPA, *On-site Wastewater Treatment Systems Manual Technology Fact Sheet 3, Sequencing Batch Reactors Systems*, Office of Water, Office of Research and Development, EPA, (EPA/625/R-00/0008). www.epa.gov/ord/NRMRL/Pubs/625R00008/625R00008total document.pdf.

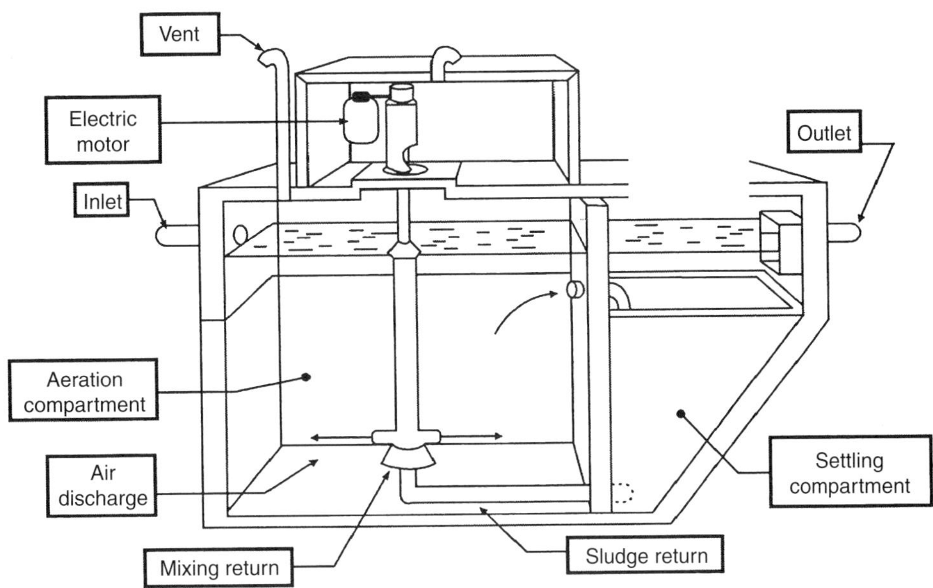

Figure 9C.20 Components of a typical aerobic treatment system. (From USEPA, *Draft Handbook for Management of On-site and Clustered (Decentralized) Wastewater Treatment Systems*, Office of Water, Figure 3 (EPA/PA 823/P-03/001), February 2003. www.epa.gov/ord/NRMRL/Pubs/625R00008/625R00008totaldocument.pdf.)

Table 9C.71 Chlorine Disinfection Dose (in mg/L) Design Guidelines for On-Site Applications

Calcium Hypochlorite	Septic Tank Effluent	Biological Treatment Effluent	Sand Filter Effluent
pH 6	35–50	15–30	2–10
pH 7	40–55	20–35	10–20
pH 8	50–65	30–45	20–35

Note: Contact time = 1 h at average flow and temperature 20 °C. Increase contact time t 02 h at 10 °C and 8 h at 5 °C for comparable efficiency. Dose = mg/L as Cl. Doses assume typical chlorine demand and are conservative estimates based on fecal coliform data.

Source: From USEPA, *On-site Wastewater Treatment Systems Manual Technology Fact Sheet 4, Table 1, Effluent Disinfection Processes*, Office of Water, Office of Research and Development, EPA, (EPA/625/R-00/0008). www.epa.gov/ord/NRMRL/Pubs/625R00008/625R00008total document.pdf.

Table 9C.72 Typical Design Guidance for Aquatic System

Parameter	Facultative Lagoon	Aerated Lagoon	FWS Constructed Wetland
HRT (days)	30–180	3 (max)	6 (min)
Power (hp/1 06gal)	0	30	0
Depth (ft)	3–5	10	2–5
Minimum no. of celis	3	2	3
BOD loading (Lb/acre-day)	20–60	200–600	40–53
TSS loading (Lb/acre-day)	N/A	N/A	27–45

Source: From USEPA, *On-site Wastewater Treatment Systems Manual Technology Fact Sheet 7, Table 1, Stabilization Ponds, FWS Constructed Wetlands, and Other Aquatic Systems*, Office of Water, Office of Research and Development, EPA, (EPA/625/R-00/0008). www.epa.gov/ord/NRMRL/Pubs/625R00008/625R00008total document.pdf.

Table 9C.73 Typical N-Removal Ranges for Managed Systems

Process	Percent TN Removal
RSF	40–50
RSF (with recycle to ST or AUF)	70–80
ST-FFS (with recycle to ST or AUF)[a]	65–75
SBR[a]	50–80
SS and removal	60–80
(SS-TT R)[a]	40–60
ISF-AUF	55–75

Note: RSF, recirculating sand filters; AUF, anaerobic upflow filter; ST, septic tank; FFS, fixed film system; SBR, sequencing batch reactor; SS, source separation; TT, treatment applied to both systems; R, recombined; ISF, intermittent sand filter.

[a] Commercially available systems.

Source: From USEPA, *On-site Wastewater Treatment Systems Manual Technology Fact Sheet 10, Table 1, Enhanced Nutrient Removal-Nitrogen*, Office of Water, Office of Research and Development, EPA, (EPA/625/R-00/0008). www.epa.gov/ord/NRMRL/Pubs/625R00008/625R00008total document.pdf.

Table 9C.74 Types of Mass Loadings to Subsurface Wastewater Infiltrations Systems

Mass Loading Type	Units	Typical Loading Rates
Hydraulic		
Daily	Volume per day per unit area of boundary surface	*Septic tank effluent*: 0.15–1.0 gpd/ft^2 (0.6–4.0 cm/d) *Secondary effluent*: 0.15→2.0 gpd/ft^2 (0.6→8.0 cm/d)
Instantaneous	Volume per dose per unit area of boundary surface	1/24–1/8 of the average daily wastewater volume
Contour (Linear)	Volume per day per unit length of boundary surface contour (which can be a critical design parameter in areas with high water tables)	Depends on soil K_{sai}[a], maximum allowable thickness of saturated zone, and slope of the boundary surface (see section 5.3)
Constituent		
Organic	Mass of BOD per day unit area of boundary surface	0.2–5.0 Lb BOD/1,000 ft^2 (1.0–29.4.4 kg BOD/1,000 m^2)
Other pollutants	Mass of specific wastewater pollutant of concern per unit area of boundary surface (e.g., number of fecal coliforms, mass of nitratre nitrogen, etc	Variable with the constituent, its fate and transport and the considered risk it imposes

[a] K_{sai} is the saturated conductivity of the soil.

Source: From Otis, 2001; USEPA, *On-site Wastewater Treatment Systems Manual, Table 5.1, Office of Water*, Office of Research and Development, EPA, (EPA/625/R-00/0008). www.epa.gov/ord/NRMRL/Pubs/625R00008/625R00008total document.pdf.

SECTION 9D INDUSTRIAL WASTEWATER TREATMENT

Table 9D.75 Toxic Chemical Releases by Industry: 2001

Industry	1987 SIC[a] Code	Total Facilities (Number)	Total On and Off-Site Releases	On-Site Release				Off-Site Releases/ Transfers to Disposal
				Total	Air Emissions	Surface Water Discharges	Other[b]	
Metal mining	10	89	2,782.6	2,782.0	2.9	0.4	2,778.7	0.5
Coal mining	12	88	16.1	16.1	0.8	0.8	14.6	—
Food and kindred products	20	1,688	125.1	118.9	56.1	55.2	7.6	6.2
Tobacco products	21	31	3.6	3.2	2.5	0.5	0.2	0.3
Textile mill products	22	289	7.0	6.2	5.7	0.2	0.3	0.7
Apparel and other textile products	23	16	0.4	0.3	0.3	—	—	0.1
Lumber and wood products	24	1,006	31.4	30.9	30.5	—	0.4	0.5
Furniture and fixtures	25	282	8.0	7.8	7.8	—	—	0.2
Paper and allied products	26	507	195.7	189.9	157.2	16.5	16.2	5.8
Printing and publishing	27	231	19.7	19.3	19.3	—	—	0.4
Chemical and allied products	28	3,618	582.6	501.3	227.8	57.6	215.9	81.3
Petroleum and coal products	29	542	71.4	68.1	48.2	17.1	2.8	3.3
Rubber and misc. plastic products	30	1,822	88.5	78.1	77.1	0.1	0.9	10.5
Leather and leather products	31	60	2.6	1.3	1.2	0.1	—	1.3
Stone, clay, glass products	32	1,027	40.5	35.4	31.3	0.2	4.0	5.1
Primary metal industries	33	1,941	558.6	286.8	57.6	44.7	184.5	271.8
Fabricated metals products	34	2,959	64.0	42.8	40.4	1.7	0.6	21.2
Industrial machinery and equipment	35	1,143	15.4	10.7	8.3	—	2.5	4.6
Electronic, electric equipment	36	1,831	23.9	16.4	12.7	2.9	0.7	7.6
Transportation equipment	37	1,348	80.6	67.7	66.7	0.2	0.8	13.0
Instruments and related products	38	375	9.4	8.6	7.2	1.4	—	0.8
Miscellaneous	39	312	8.4	6.8	6.8	—	—	1.6
Electric utilities	49	732	1,062.2	989.2	717.6	3.5	268.1	73.1
Chemical wholesalers	5169	475	1.5	1.3	1.3	—	—	0.2
Petroleum bulk terminals	5171	596	21.3	21.2	21.2	—	—	0.2
RCRA/solvent recovery	4953/7369	223	219.9	168.4	1.0	—	167.4	51.4
Total[c]	(X)	24,896	6,158.0	5,580.3	1,679.4	220.8	3,680.1	577.7

[In millions of pounds (6,158.0 represents 6,158,000,000), except as indicated. "Original industries" include owners and operators covers facilities that are classified within standard classification code groups 20 through 39,10,12,49, 5169, 5171, and 4953/7169 that have 10 or more full-time employees, and that manufacture, process, or otherwise uses any listed toxic chemical in quantities greater than the established threshold in the course of a calendar year are covered and required to report.

— Represents or rounds to zero, X, Not applicable.

[a] Standard industrial classification, see text, section 12.

[b] Includes underground injection for class I and class II to V wells and land releases.

[c] Includes industries with no specific industry identified, not shown separately.

Source: From Table 378 and Table 379 U.S. Environmental Protection Agency, *Toxics Release Inventory*, Annual. U.S. Census Bureau, Statistical Abstract of the U.S., 2003. www.census.gov/compendia/statab/.

Table 9D.76 Overview of Facility, Impoundment, and Wastewater Quantity Estimates

Characteristic	Direct Dischargers	Zero Dischargers	Total Population
Estimated number of facilities	3,944	512	4,457
Estimated number of impoundments	10,987	876	11,863
Total quantity of wastewaters managed (metric tons)[a]	627,218,336	27,250,309	654,468,645

[a] The estimate of the wastewater quantity for the total population differs from the estimates shown in Table 2.2 and Table 2.15. This is due to missing data associated with this variable. Refer to Appendix A on missing data and Appendix B for the standard error associated with this variable.

Source: From USEPA, *Industrial Surface Impoundments in the United States*, Table 2.1 (530R01005). www.epa.gov/epaoswer/hazwaste/ldr/icr/ldr-impd.htm.

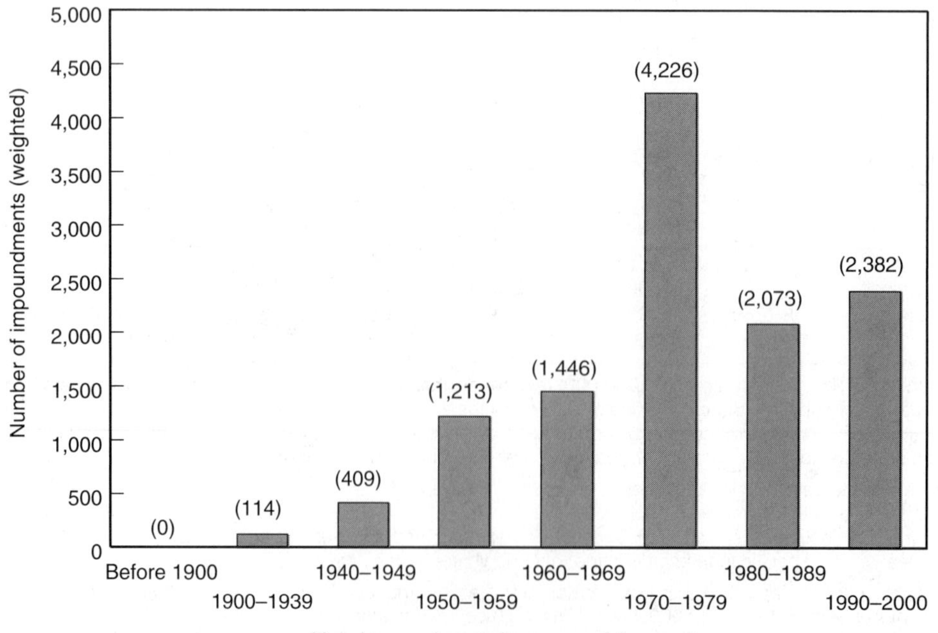

Figure 9D.21 Distribution of 11,863 impoundments by year unit began receiving waste. (From USEPA, *Industrial Surface Impoundments in the United States*, Figure 2.1, (530R01005). www.epa.gov/epaoswer/hazwaste/ldr/icr/ldr-impd.htm.)

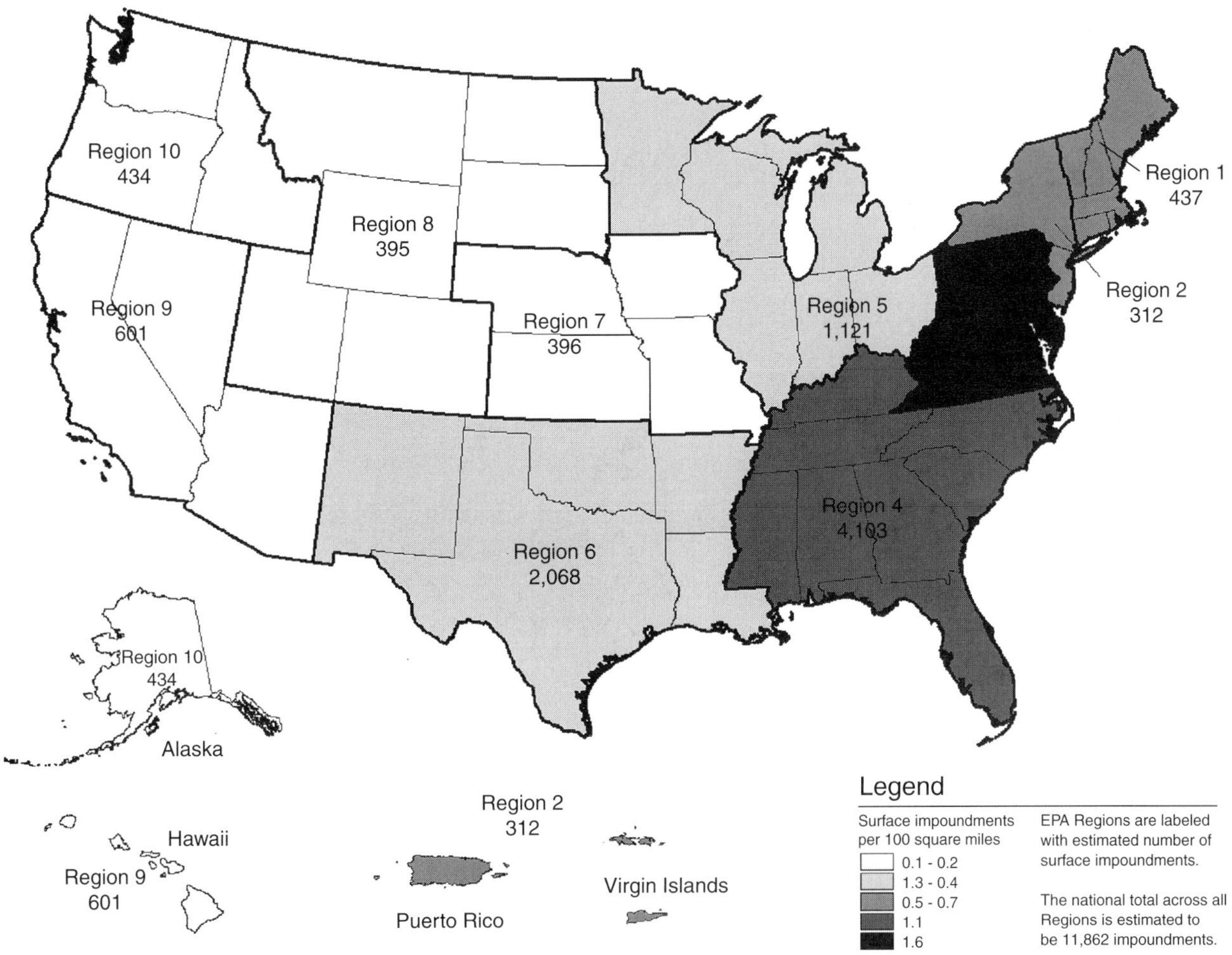

Figure 9D.22 Regional distribution of surface impoundments. (From USEPA, *Industrial Surface Impoundments in the United States*, Figure 2.2, (530R01005). www.epa.gov/epaoswer/hazwaste/ldr/icr/ldr-impd.htm.)

Table 9D.77 Breakdown by 2-Digit SIC Code of Surface Impoundments That Manage Chemicals/pH of Concern and of Quantities of Wastewater Managed

SIC Code Descriptor	Percent of 4,457 Facilities	Percent of 11,863 Impoundments	Percent of 653,314,426[a] Metric Tons Wastewater
Chemical and allied products (SIC 28)	19	23	9
Stone, clay, glass, concrete products (SIC 32)	15	13	1
Wholesale trade-nondurable goods (SIC 51)	12	10	4
Primary metals industry (SIC 33)	10	8	7
Food and kindred products (SIC 20)	8	8	5
Petroleum and coal products (SIC 29)	7	11	6
Paper and allied products (SIC 26)	6	12	66
All other SIC codes	23	15	2

SIC, standard industrial classification.

[a] The estimate of the wastewater quantity for the total population differs from the estimates shown in Table 2.1 and Table 2.15. This is due to missing data associated with this variable. Refer to Appendix A on missing data and Appendix B for the standard error associated with this variable.

Source: From USEPA, *Industrial Surface Impoundments in the United States*, Table 2.2 (530R01005). www.epa.gov/epaoswer/hazwaste/ldr/icr/ldr-impd.htm.

Table 9D.78 Breakdown of Chemical Categories for Wastewater and Sludge (at Different Sampling Points) on Impoundment and Volume Basis

	Wastewater						Sludge					
	Influent		In Impoundment		Effluent		Influent		In Impoundment		Effluent	
Chemical Categories	# Imps	% Vol	# Imps	% Vol	# Imps	% Vol	# Imps	% Vol	# Imps	% Vol	# Imps	% Vol
VOCs	5,866	76	5,412	76	4,815	72	1,690	4	2,006	21	1,311	14
SVOCs	3,824	75	3,786	75	3,508	69	863	7	1,261	24	605	3
Metals	9,966	84	9,982	83	7,762	85	3,925	42	5,551	98	3,078	88
Dioxin-like compounds	291	24	218	21	346	22	247	10	861	35	412	41
Mercury	2,483	27	2,479	30	2,235	31	1,061	0.9	1,745	66	826	6
Any chemicals	10,745	96	10,766	97	8,187	92	4,101	45	5,759	100	3,230	89

Imps, number of impoundments; % Vol, percent of total volume; SVOCs , Semivolatile organic compounds; VOCs, Volatile organic compounds.

Source: From USEPA, *Industrial Surface Impoundments in the United States*, Table 2.5 (530R01005). www.epa.gov/epaoswer/hazwaste/ldr/icr/ldr-impd.htm.

Table 9D.79 Number and Percentage of Impoundments by Liner Status

Liner Status	Number of Impoundments	Percentage of Impoundments
Compacted clay	1,680	14
Flexible membrane (FML)	1,584	13
Composite (FML and clay)	536	5
Concrete	629	5
Asphalt	55	<1
Other	363	3
Unlined[a]	7,017	59
Total	11,863	100

[a] This estimate differs from the estimate of outlined impoundments shown in Table 2.12. This is due to missing data associated with this variable. Refer to Appendix A on missing data and Appendix B for the standard error associated with this variable.

Source: From USEPA, *Industrial Surface Impoundments in the United States*, Table 2.10, (530R01005). www.epa.gov/epaoswer/hazwaste/ldr/icr/ldr-impd.htm.

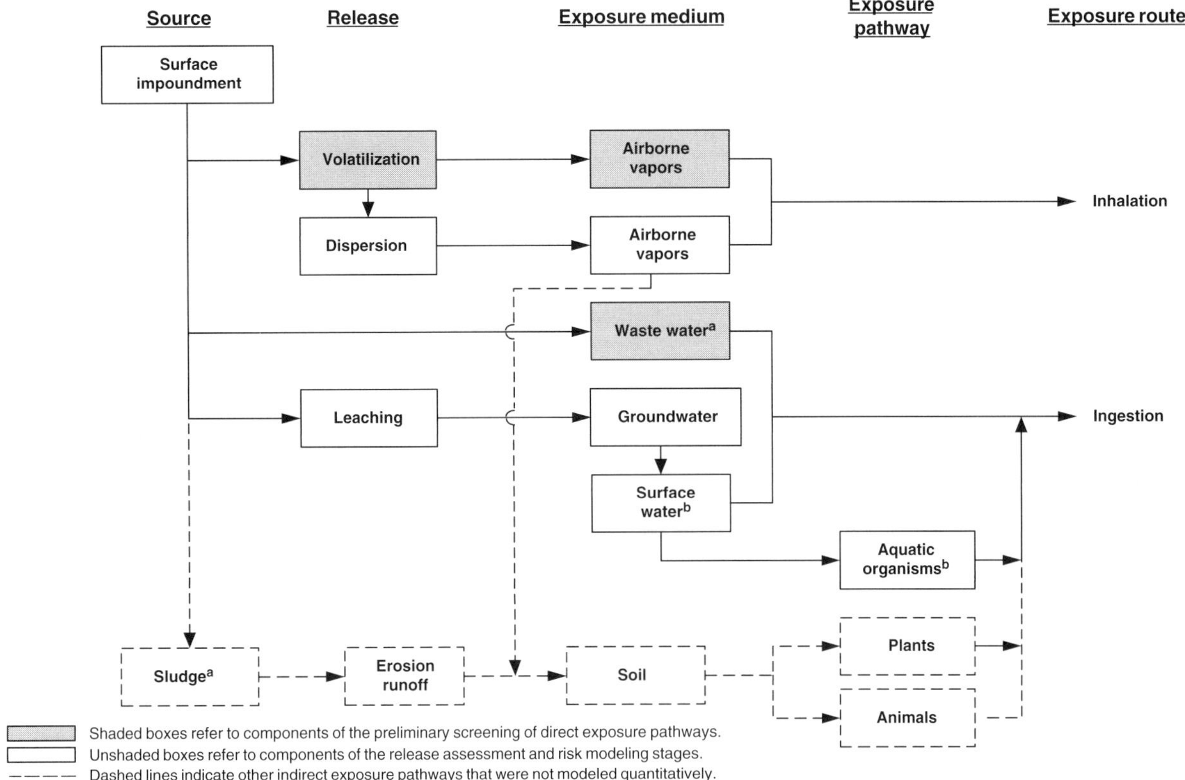

Figure 9D.23 Exposure pathways for active surface impoundments considered for humans and ecological receptors. (From USEPA, *Industrial Surface Impoundments in the United States*, Figure 3.1, (530R01005). www.epa.gov/epaoswer/hazwaste/ldr/icr/ldr-impd.htm.)

CHAPTER 10

Environmental Problems

Katherine L. Thalman

CONTENTS

SECTION 10A POLLUTION SOURCES AND PATHWAYS

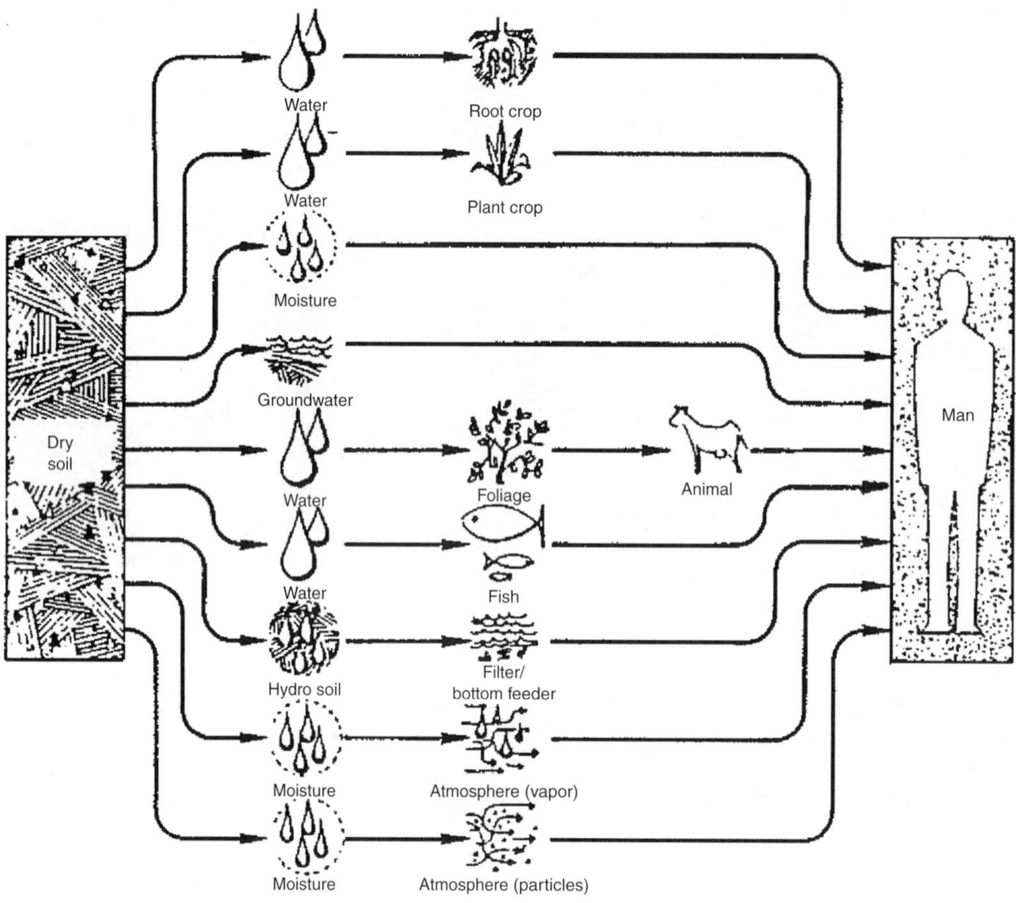

Figure 10A.1 Pollutant pathways from soil to man. (From Dacre, I.C., Rosenblatt, D.H., and Cogley, D.R., 1980, *Preliminary Pollutant Limit Values for Human Health Effects, Environmental Technology* 14: 778–783, Copyright American Chemical Society, Washington, DC.)

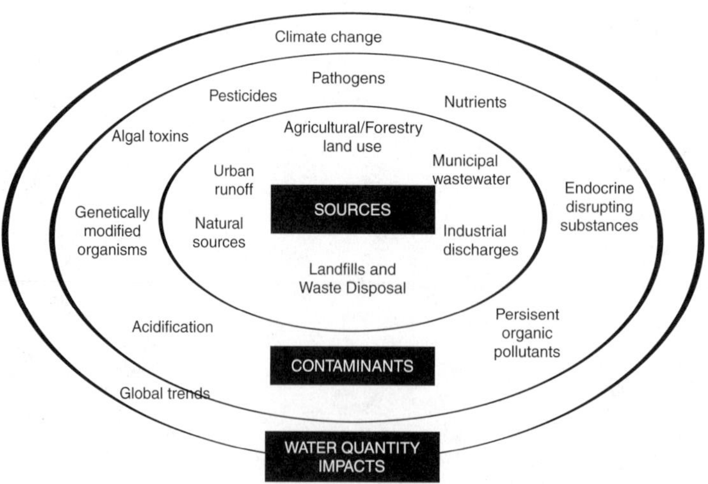

Figure 10A.2 Threats to water sources. (From Threats to Sources of Drinking Water and Aquatic Ecosystem Health in Canada, page x, Environment Canada, 2001. Reproduced with permission from the National Water Research Institute, Environment Canada, 2006.)

Table 10A.1 Causes of Damage to the Quality of Water Resources

Types of Waste	Wastewater Sources	Water-Quality Measures	Effects on Water Quality	Effects on Aquatic Life	Effects on Recreation
Disease-carrying agents — human feces, warm-blooded animal feces	Municipal discharges, watercraft discharges, urban runoff, agricultural runoff, feedlot wastes, combined sewer overflows, industrial discharges	Fecal coliform, fecal streptococcus, other microbes	Health hazard for human consumption and contact	Inedibility of shellfish for humans	Reduced contact recreation
Oxygen-demanding wastes — high concentrations of biodegradable organic matter	Municipal discharges, industrial discharges, combined sewer overflows, watercraft discharges, urban runoff, agricultural runoff, feedlot wastes, natural sources	Biochemical oxygen demand, dissolved oxygen, volatile solids sulfides	Deoxygenation, potential for septic conditions	Fish kills	If severe, eliminated recreation
Suspended organic and inorganic material	Mining discharges, municipal discharges, industrial discharges, construction runoff, agricultural runoff, urban runoff, silvicultural runoff, natural sources, combined sewer overflows	Suspended solids, turbidity, biochemical oxygen demand, sulfides	Reduced light penetration, deposition on bottom, benthic deoxygenation	Reduced photosynthesis, changed bottom organism population, reduced fish production, reduced sport fish population, increased nonsport fish population	Reduced game fishing, aesthetic appreciation
Inorganic materials, mineral substances — metal, salts, acids, solid matter, other chemicals, oil	Mining discharges, acid mine drainage, industrial discharges, municipal discharges, combined sewer overflows, urban runoff, oil fields, agricultural runoff, irrigation return flow, natural sources, cooling tower blowdown, transportation spills, coal gasification	pH, acidity, alkalinity, dissolved solids, chlorides, sulfates, sodium, specific metals, toxicity bioassay, visual (oil spills)	Acidity, salination, toxicity of heavy metals, floating oils	Reduced biological productivity, reduced flow, fish kills, reduced production, tainted fish	Reduced recreational use, fishing, aesthetic appreciation
Synthetic organic chemicals — dissolved organic material, e.g., detergents, household aids, pesticides	Industrial discharges, urban runoff, municipal discharges, combined sewer overflow, agricultural runoff, silvicultural runoff, transportation spills, mining discharges	Cyanides, phenols, toxicity bioassay	Toxicity of natural organics, biodegradable or persistent synthetic organics	Fish kills, tainted fish, reduced reproduction, skeletal development	Reduced fishing, inedible fish for humans
Nutrients — nitrogen, phosphorus	Municipal discharges, agricultural runoff, combined sewer overflows, industrial discharges, urban runoff, natural sources	Nitrogen, phosphorus	Increased algal growth, dissolved oxygen reduction	Increased production, reduced sport fish population, increased nonsport fish population	Tainted drinking water, reduced fishing and aesthetic appreciation
Radioactive materials	Industrial discharges, mining	Radioactivity	Increased radioactivity	Altered natural rate of genetic mutation	Reduced opportunities
Heat	Cooling water discharges, industrial discharges, municipal discharges, cooling tower blowdown	Temperature	Increased temperature, reduced capacity to absorb oxygen	Fish kills, altered species composition	Possible increased sport fishing by extended season for fish, which might otherwise migrate

Source: From Council of Environmental Quality, 1981, *Environmental Trends*.

Table 10A.2 Point- and Nonpoint Sources of Water Pollution

Sources	Common Pollutant Categories
POINT SOURCES	
Municipal sewage treatment plants	BOD; bacteria; nutrients; ammonia; toxics
Industrial facilities	Toxics; BOD
Combined sewer overflows	BOD; bacteria; nutrients; turbidity; total dissolved solids; ammonia; toxics; bacteria
NONPOINT SOURCES	
Agricultural runoff	Nutrients; turbidity; total dissolved solids; toxics; bacteria
Urban runoff	Turbidity; bacteria; nutrients; total dissolved solids; toxics
Construction runoff	Turbidity; nutrients; toxics
Mining runoff	Turbidity; acids; toxics; total dissolved solids
Septic systems	Bacteria; nutrients
Landfills/spills	Toxics; miscellaneous substances
Silvicultural runoff	Nutrients; turbidity; toxics

Source: From U.S. Environmental Protection Agency, *National Water Quality Inventory*, 1986 Report to Congress.

Table 10A.3 Classes of Nonpoint Source Pollution

Class		Description	Common Pollutants
Agriculture			
	Animal feedlots Irrigation Cultivation Pastures Dairy farming Orchards Aquaculture	Runoff from all categories of agriculture leading to surface and groundwater pollution. In northern climates, runoff from frozen ground is a major problem, especially where manure is spread during the winter. Vegetable handling, especially washing in polluted surface waters in many developing countries, leads to contamination of food supply. Growth of aquaculture is becoming a major polluting activity in many countries. Irrigation return flows carry salts, nutrients, and pesticides. Tile drainage rapidly carries leachates such as nitrogen to surface waters	Phosphorus, nitrogen, metals, pathogens, sediment, pesticides, salt, BOD[a], trace elements (e.g., selenium)
Forestry		Increased runoff from disturbed land. Most damaging is forest clearing for urbanization	Sediment, pesticides
Liquid waste disposal		Disposal of liquid wastes from municipal wastewater effluents, sewage sludge, industrial effluents and sludges, wastewater from home septic systems; especially disposal on agricultural land, and legal or illegal dumping in watercourses	Pathogens, metals, organic compounds
Urban Areas			
	Residential Commercial Industrial	Urban runoff from roofs, streets, parking lots, etc. leading to overloading of sewage plants from combined sewers, or polluted runoff routed directly to receiving waters; local industries and businesses may discharge wastes to street gutters and storm drains; street cleaning; road salting contributes to surface and groundwater pollution	Fertilizers, greases and oils, fecal matter and pathogens, organic contaminants (e.g., PAHs[b] and PCBs[c]), heavy metals, pesticides, nutrients, sediment, salts, BOD, COD[d], etc.
Rural sewage systems		Overloading and malfunction of septic systems leading to surface runoff and/or direct infiltration to groundwater	Phosphorus, nitrogen, pathogens (fecal matter)
Transportation		Roads, railways, pipelines, hydroelectric corridors, etc.	Nutrients, sediment, metals, organic contaminants, pesticides (especially herbicides)
Mineral extraction		Runoff from mines and mine wastes, quarries, well sites	Sediment, acids, metals, oils, organic contaminants, salts (brine)
Recreational land use		Large variety of recreational land uses, including ski resorts, boating and marinas, campgrounds, parks; waste and "grey" water from recreational boats is a major pollutant, especially in small lakes and rivers. Hunting (lead pollution in waterfowl)	Nutrients, pesticides, sediment, pathogens, heavy metals
Solid waste disposal		Contamination of surface and groundwater by leachates and gases. Hazardous wastes may be disposed of through underground disposal	Nutrients, metals, pathogens, organic contaminants
Dredging		Dispersion of contaminated sediments, leakage from containment areas	Metals, organic contaminants
Deep well disposal		Contamination of groundwater by deep well injection of liquid wastes, especially oilfield brines and liquid industrial wastes	Salts, heavy metals, organic contaminants
Atmospheric deposition		Long-range transport of atmospheric pollutants (LRTAP) and deposition of land and water surfaces. Regarded as a significant source of pesticides (from agriculture, etc.), nutrients, metals, etc. especially in pristine environments	Nutrients, metals, organic contaminants

[a] BOD, Biological Oxygen Demand.
[b] PAH, Polycyclic Aromatic Hydrocarbons.
[c] PCB, Polycyclic Chlorinated Bi-Phenyls.
[d] COD, Chemical Oxygen Demand.

Source: From Ongley, E.D., 1996, Control of water pollution from agriculture-FAO irrigation and drainage paper 55, *Food and Agriculture Organization of the United Nations*, Rome. Reprinted with permission.

Table 10A.4 Contamination Sources Reported by Public Water-Supply Systems in the United States

Type of Contamination	Water-Supply Source		
	Groundwater	**River/Stream**	**Lake/Reservoir**
Industrial/commercial discharges	62	97	38
Leaking underground tanks	81	33	23
Urban runoff	35	91	24
Landfills	67	49	22
Synthetic or volatile organics	83	56	18
Hazardous waste site(s)	37	31	8
Land development	36	76	32
Underground waste injection	27	5	3
Agricultural runoff (pesticides, fertilizers, etc.)	49	126	86
Algae/bacteria	15	117	124
Overdraft	40	7	4
Water rights disputes	16	22	12
Natural contamination (radionuclides, salinity, etc.)	52	56	35

Note: Number of utilities reporting in each category.

Source: From American Water Works Association, 1984 Water Utility Operating Data; Copyright AWWA.

Table 10A.5 Anthropogenic Sources of Pollutants in the Aquatic Environment

Source	Bacteria	Nutrients	Trace Elements	Pesticides/ Herbicides	Industrial Organic Micro Pollutants	Oils and Greases
Atmosphere		x	xxxG	xxxG	xxxG	
Point sources						
Sewage	xxx	xxx	xxx	x	xxx	
Industrial effluents		x	xxxG		xxxG	xx
Diffuse sources						
Agriculture	xx	xxx	x	xxxG		
Dredging		x	xxx	xx	xxx	x
Navigation and harbors	x	x	xx		x	xxx
Mixed sources						
Urban runoff and waste disposal	xx	xxx	xxx	xx	xx	xx
Industrial waste disposal sites		x	xxx	x	xxx	x

Note: x, Low local significance; xx, Moderate local/regional significance; xxx, High local/regional significance; G, Globally significant.

Source: From Chapman, D. (ed.), 1996, Water quality assessments-A guide to use of biota, *Sediments and Water in Environmental Monitoring—Second Edition*, Copyright © UNESO/WHO/UNEP, 1996, www.who.int. Reprinted with permission.

SECTION 10B SURFACE WATER POLLUTION

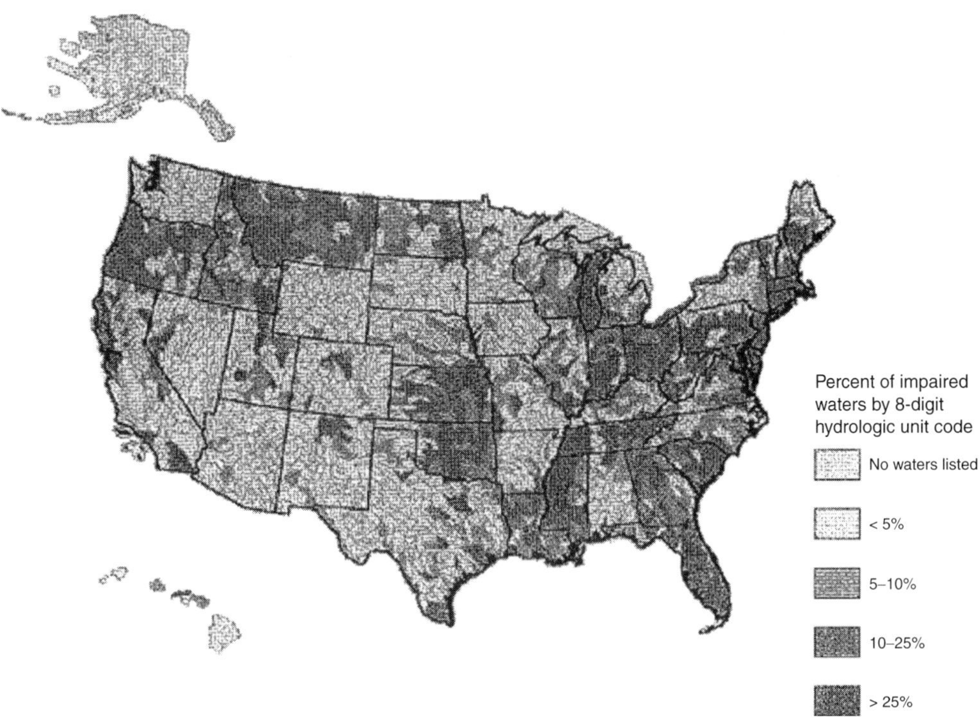

Figure 10B.3 Percentage of impaired waters in the United States by 8-digit hydrologic unit code. (From United States Environmental Protection Agency, 2000, *Atlas of America's Polluted Waters*, EPA-840-B-00-002, May 2000, www.epa.gov.)

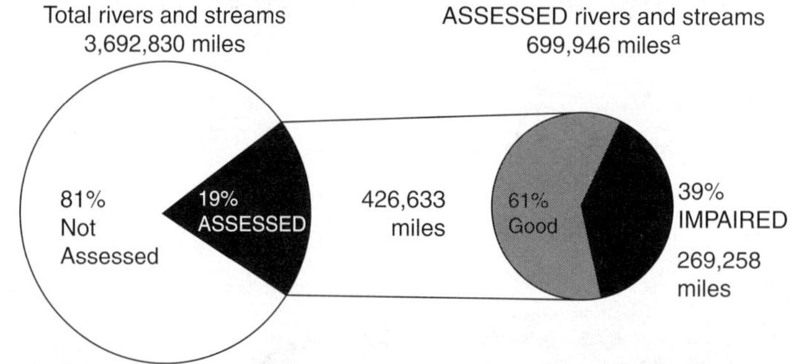

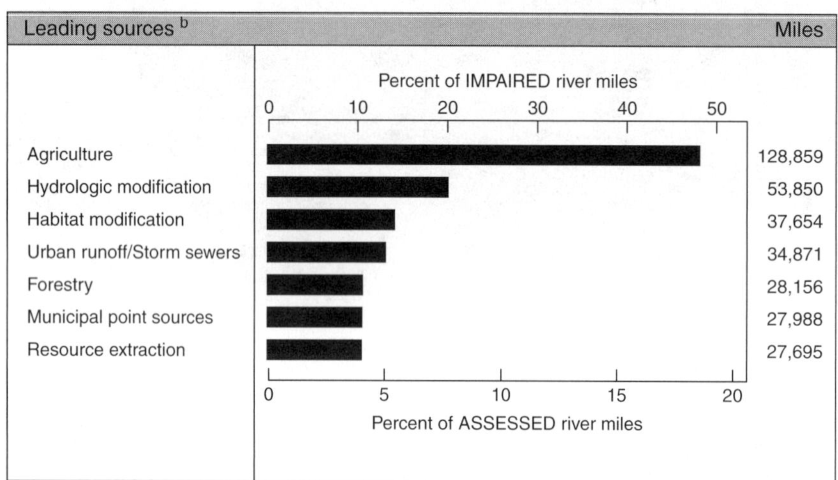

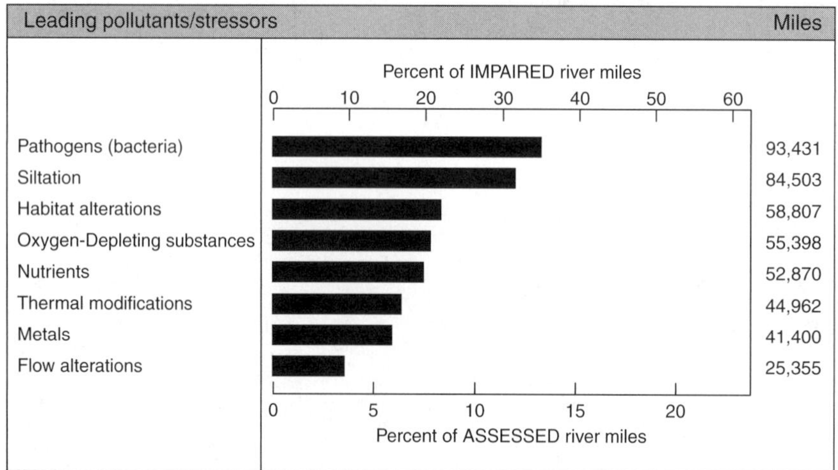

Figure 10B.4 Leading pollutants and sources of river and stream impairment in the United States.[a] Excluding unknown and natural sources.[b] Includes miles assessed as not attainable. Percentages do not add up to 100% because more than one pollutant or source may impair a river segment. (From United States Environmental Protection Agency, 2002, *National Water Quality Inventory 2000 Report*, EPA-841-R-02-001, www.epa.gov.)

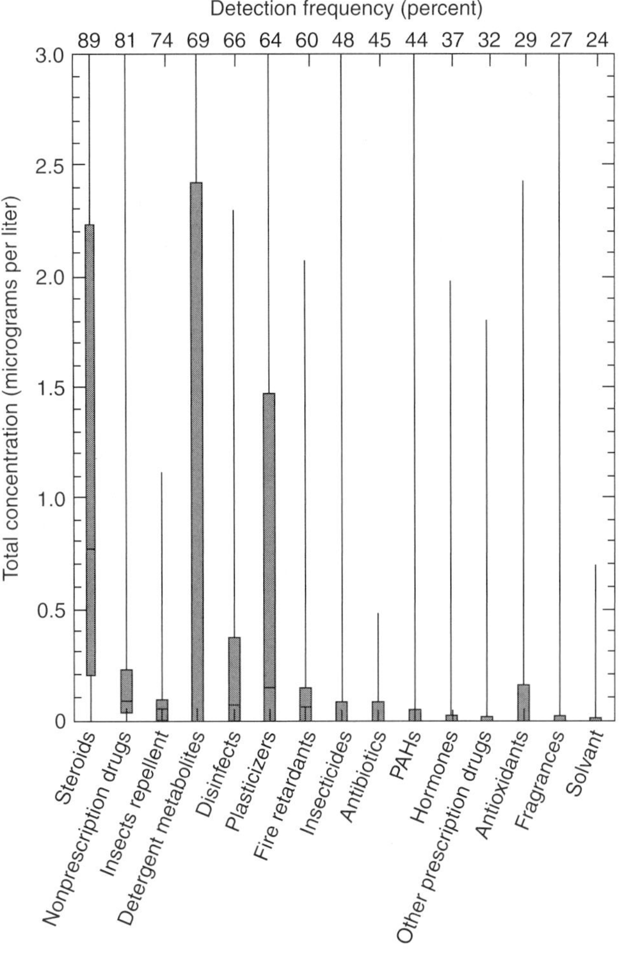

EXPLANATION

Maximum value	"Maximum values not shown:
75thpencentile	Steroids: 18.3
Median	Non prescription drugs: 17.4
25thpencentile	Detergent metabolies: 55.6
Minimum value	Plasticizers: 17.4
	Antibiotics: 3.6
	Fragrances: 4.3

Steroids, nonprescription drugs, and an insects repellent were the three chemical groups most commonly detected in susceptible streams. Detergent and metabolites, steroids, and plasticizers generally were found at the highest concentrations.

Figure 10B.5 Pharmaceuticals, hormones, and other organic wastewater contaminants in United States streams. (From Buxton, H.T. and Kolpin, D.W., 2002, *Pharmaceuticals, Hormones, and Other Organic Wastewater Contaminants in U.S. Streams*, USGS Fact Sheet FS-027-02, June 2002, www.usgs.gov.)

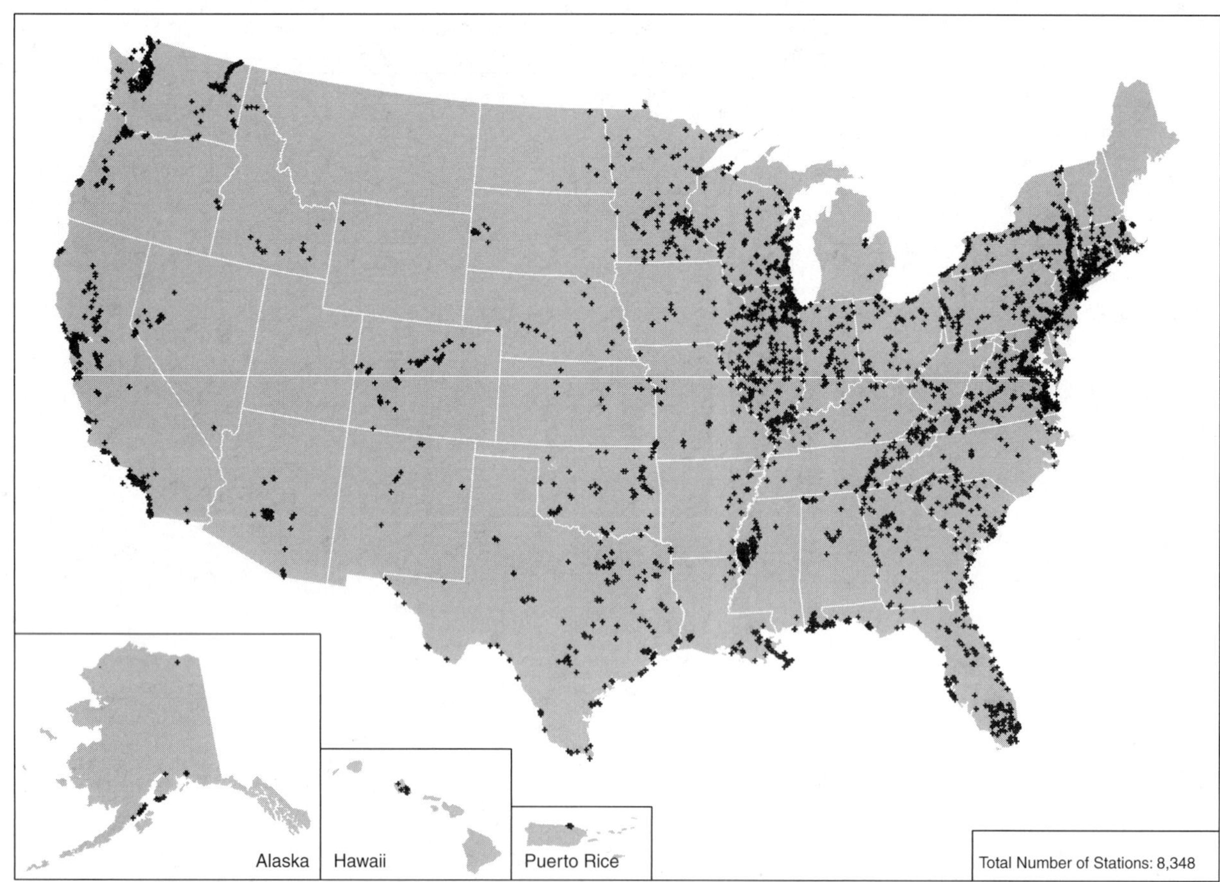

Figure 10B.6 Sampling stations classified as tier 1 (associated adverse effects are probable). (From USEPA, 2004, The incidence and severity of sediment contamination in surface waters of the United States, *National Sediment Quality Survey: Second Edition*, EPA 823-R-04-007, www.epa.gov.)

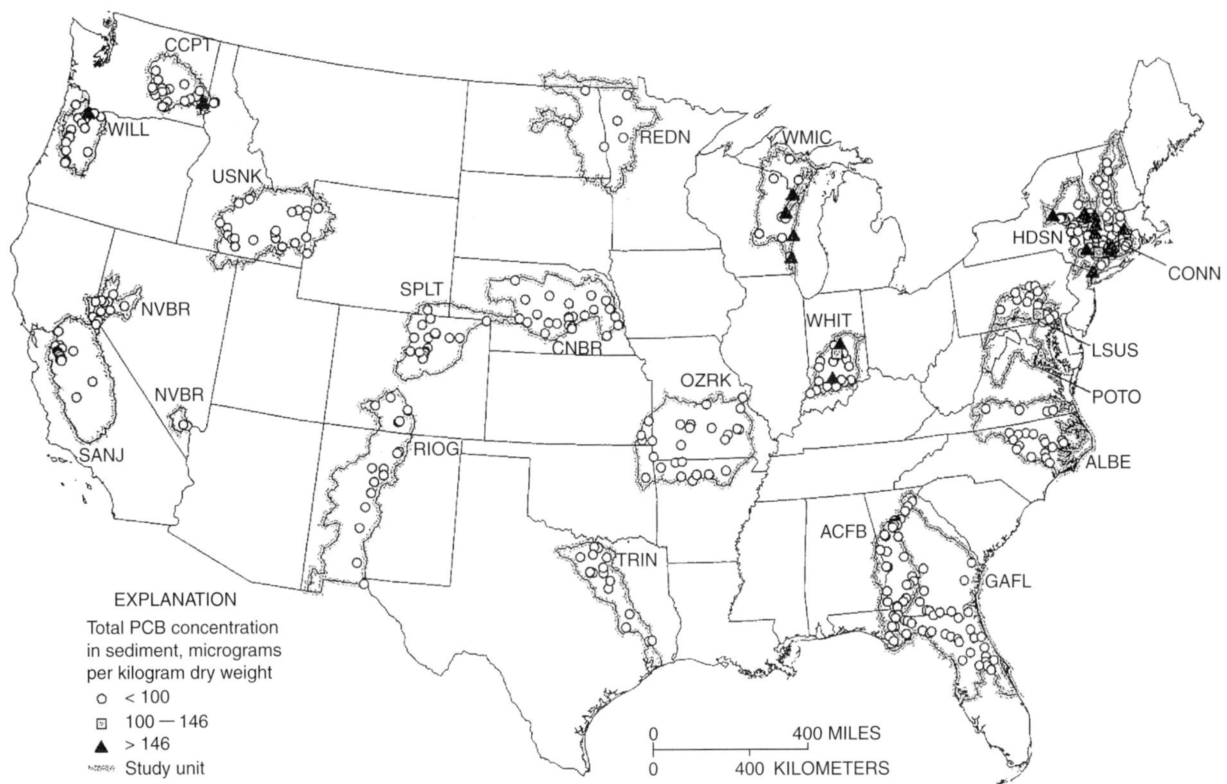

Figure 10B.7 Geographic distribution of total polychlorinated biphenyls in sediment samples in the United States. (From Wong, C.S., Capel, P.D., and Nowell, L.H., 2000, *Organochlorine Pesticides and PCBs in Stream Sediment and Aquatic Biota—Initial Results from the National Water-Quality Assessment Program, 1992–1995*, Water-Resources Investigations Report 00-4053, www.usgs.gov.)

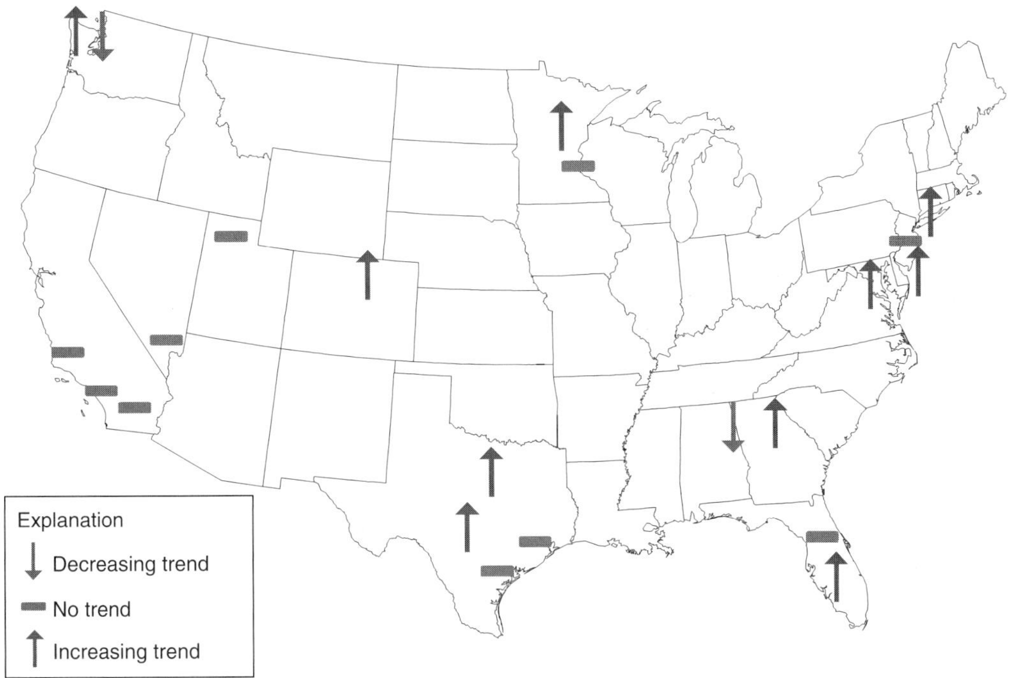

Figure 10B.8 PAH trends throughout the United States using sediment core data from 1970 to top of core. (From USEPA, 2004, The incidence and severity of sediment contamination in surface waters of the United States, *National Sediment Quality Survey: Second Edition*, EPA 823-R-04-007, www.epa.gov.)

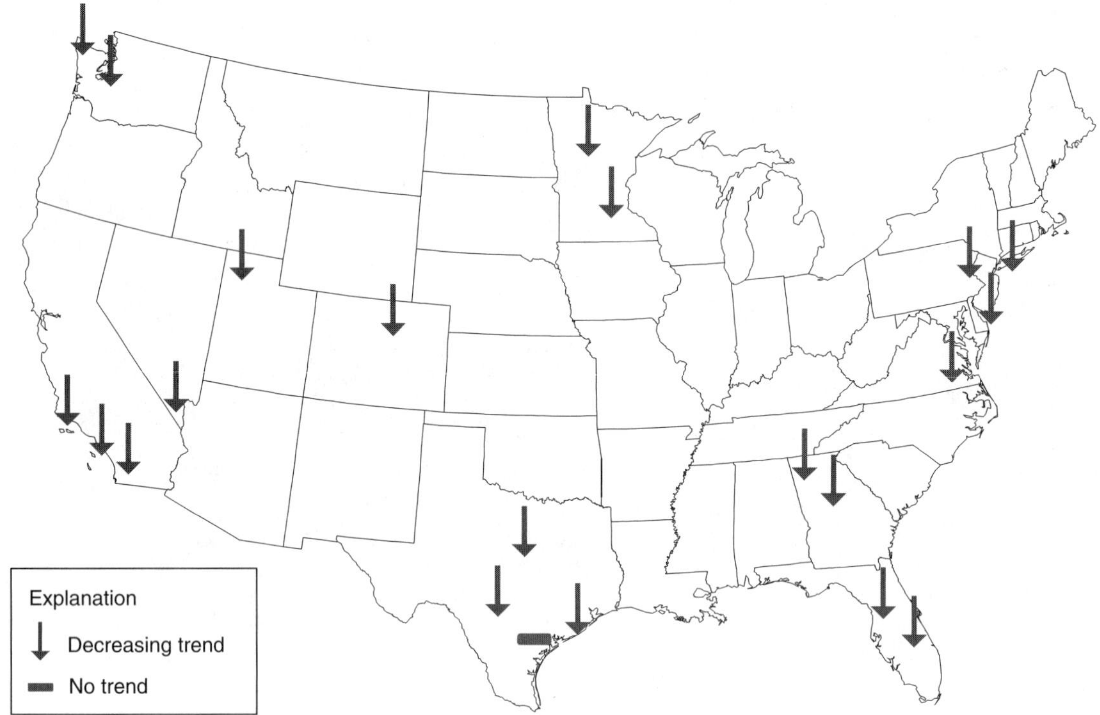

Figure 10B.9 Lead trends throughout the United States using sediment core data from 1975 to top of core. (From USEPA, 2004, The incidence and severity of sediment contamination in surface waters of the United States, *National Sediment Quality Survey: Second Edition*, EPA 823-R-04-007, www.epa.gov.)

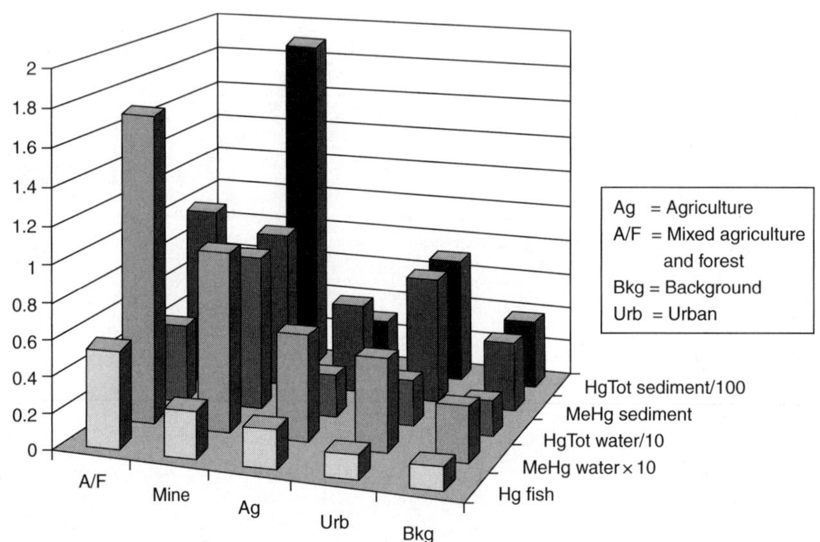

Figure 10B.10 Geometric mean of mercury and methylmercury in fish (μg/g wet), water (μg/L), and sediment (hg/g dry) for land use categories: mixed agriculture and forest, mine, agriculture, urban or industrial activity, and background. (Number of observation = 13, 42, 23, 15, and 34 for A/F, Mine, Ag, Bkg and Urb, respectively. Excludes South Florida Basin.) (From Brumbaugh, W.G. et al., 2001, A national pilot study mercury contamination of aquatic ecosystems along multiple gradients: Bioaccumulation in Fish, *USGS, Biological Science Report USGS/BRD/BSR-2001-0009*, www.usgs.gov.)

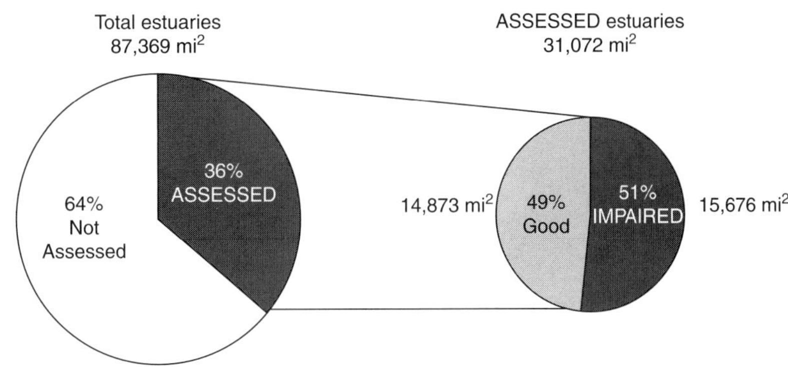

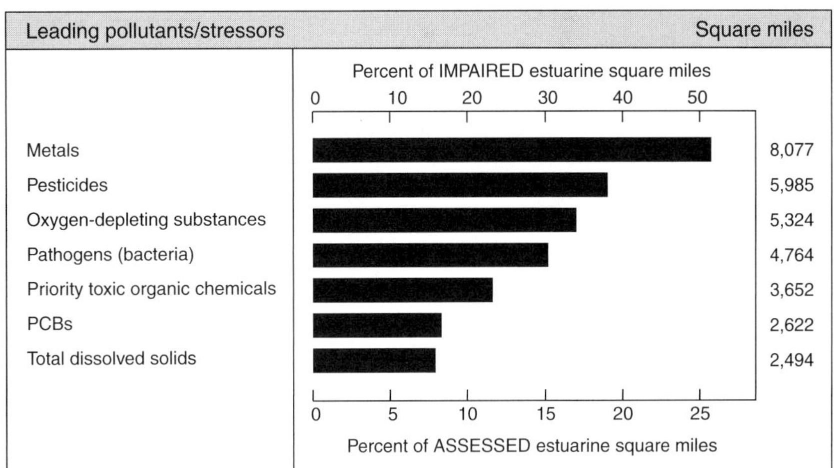

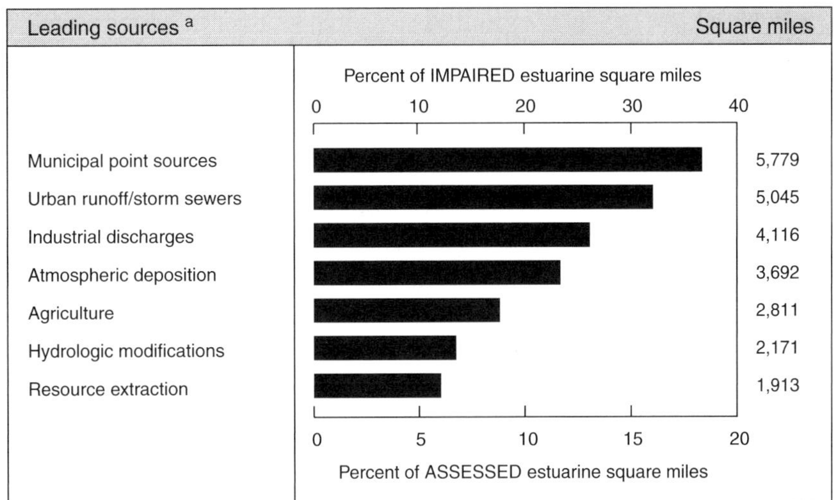

Figure 10B.11 Leading pollutants and sources of estuary impairment in the United States.[a] Excludes unknown, natural, and "other sources." Percentages do not add up to 100% because more than one pollutant or source may impair an estuary. (From United States Environmental Protection Agency, 2002, *National Water Quality Inventory 2000 Report*, EPA-841-R-02-001, www.epa.gov.)

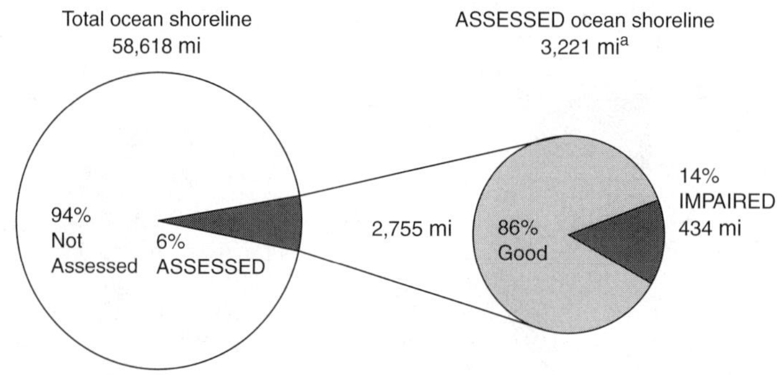

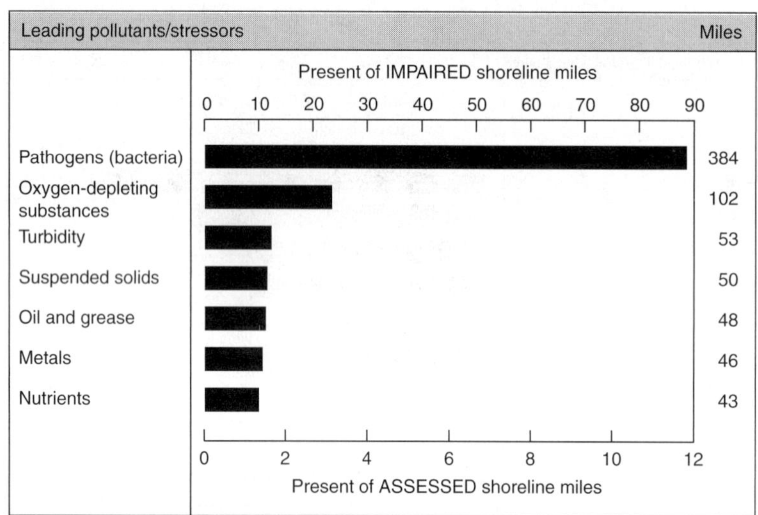

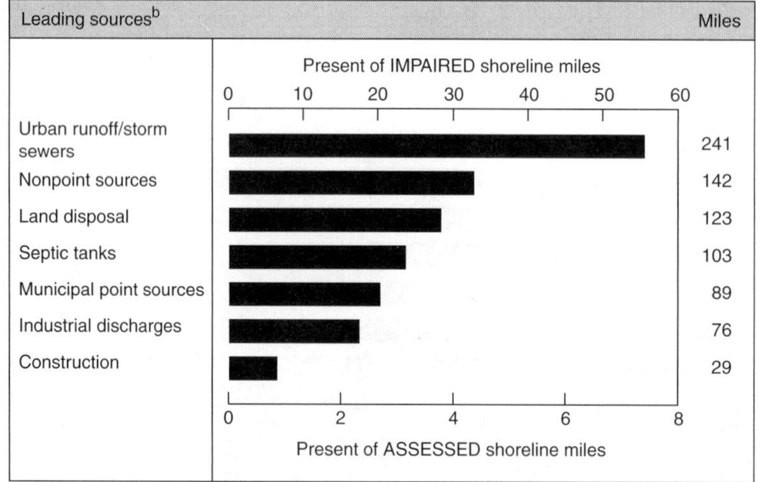

Figure 10B.12 Leading pollutants and sources of ocean shoreline water impairment in the United States.[a] Includes miles assessed as not attainable.[b] Excludes natural sources. Percentages do not add up to 100% because more than one pollutant or source may impair a segment of ocean shoreline. (From United States Environmental Protection Agency, 2002, *National Water Quality Inventory 2000 Report*, EPA-841-R-02-001, www.epa.gov.)

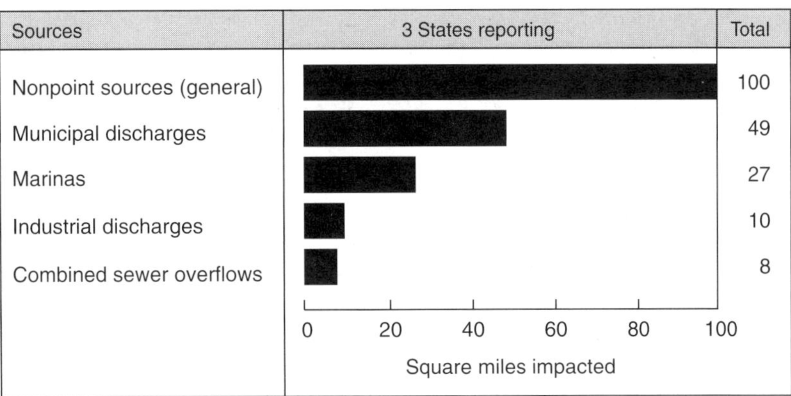

Figure 10B.13 Sources associated with shellfish harvesting restrictions. (From United States Environmental Protection Agency, 2002, *National Water Quality Inventory 2000 Report*, EPA-841-R-02-001, www.epa.gov.)

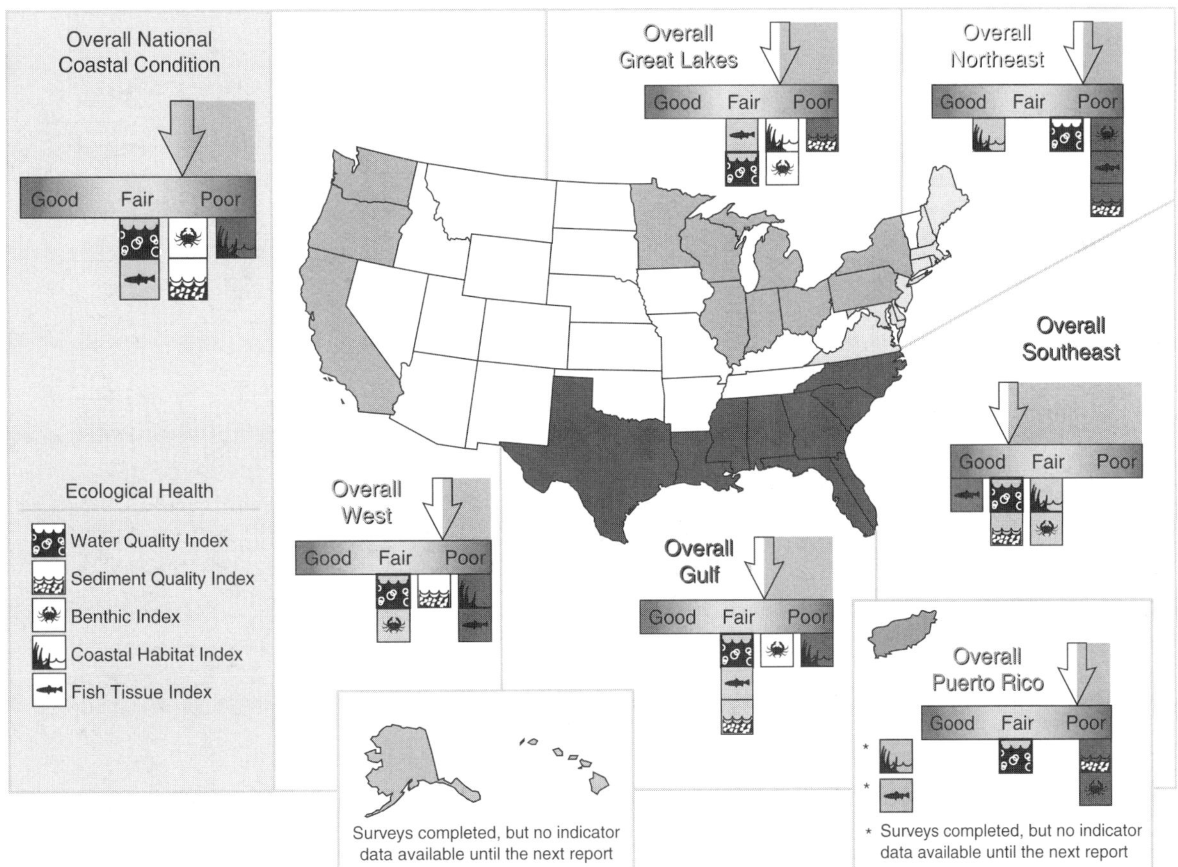

Figure 10B.14 Overall United States national coastal condition. (From United States Environmental Protection Agency, 2004, *National Coastal Condition Report II*, EPA-620/R-03/002, December 2004, www.epa.gov.)

The area and duration of hypoxia are tracked in the Gulf of Mexico and Long Island Sound as indicators of the natural variability in those waterbodies to determine whether actions to control nutrients are having the desired effect and how local species are affected.

The largest of oxygen-depleted coastal waters in the U.S. is in the northern Gulf of Mexico on the Louisiana/Texas continental shelf. Hypoxic waters are most prevalent from late spring through late summer and are more widespread and persistent in some years that in others, depending on river flow, winds, and other environmental variables. Hypoxia occurs mostly in the lower water column, but can encompass as much as the lower half to two-thirds of the entire column.

The midsummer bottom area extent of hypoxic water in the Gulf of Mexico increased from 3,500 mi^2 (9,000 km^2) in 1985 to 8,500 mi^2 (22,000 km^2) in July 2002 (Exhibit 2-3). The primary cause of the hypoxic conditions is probably the eutrophication of those waters from nutrient enrichment delivered to the Gulf by the Mississippi River and its drainage basin.[13,14]

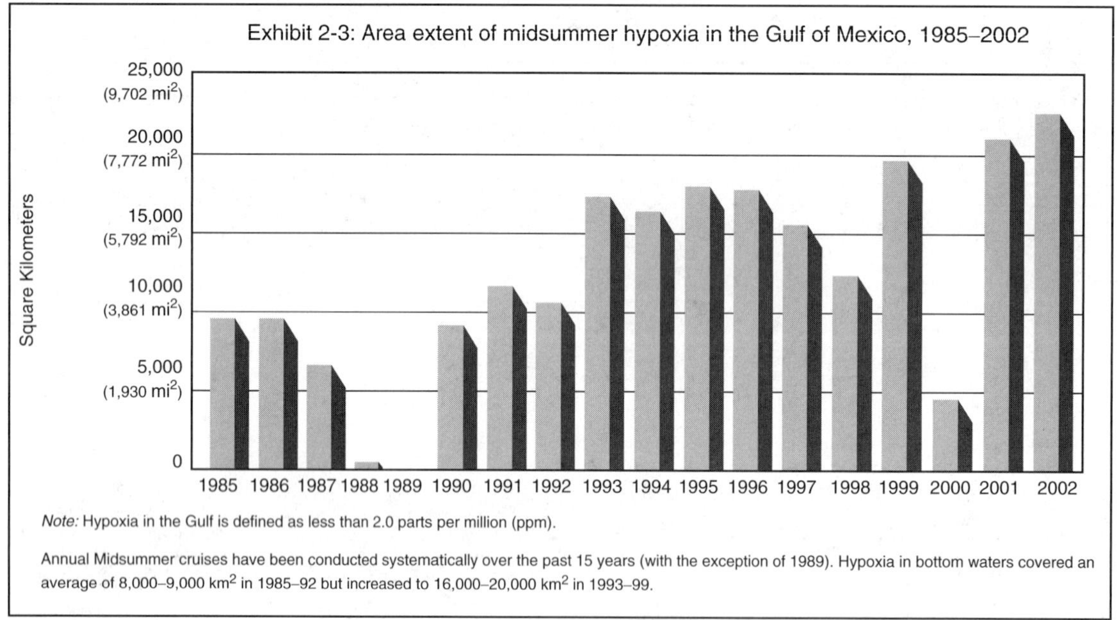

Note: Hypoxia in the Gulf is defined as less than 2.0 parts per million (ppm).

Annual Midsummer cruises have been conducted systematically over the past 15 years (with the exception of 1989). Hypoxia in bottom waters covered an average of 8,000–9,000 km^2 in 1985–92 but increased to 16,000–20,000 km^2 in 1993–99.

The maximum area of hypoxia in Long Island Sound averaged 201 mi^2 *(521* km^2*) from 1987 through 2001. The largest area was 395* mi^2 *(1,023* km^2*) in 1994, and the smallest was 30* mi^2 *(78* km^2*) in 1997 (Exhibit 2-4). The duration of hypoxia averaged 56 days during the same period, with a low of 34 days in 1996 and a high of 82 days in 1989. Hypoxia is typically more severe in the western portions of the sound, where the nitrogen load is higher and mixing of fresh and salt water is more restricted.*[15]

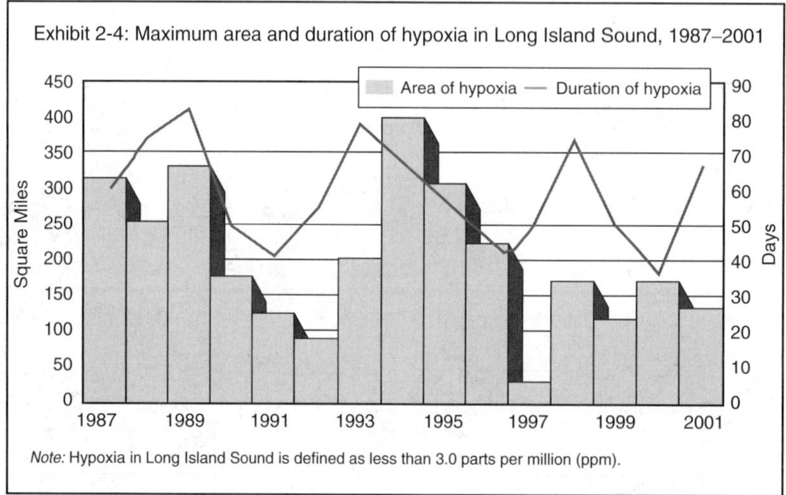

Note: Hypoxia in Long Island Sound is defined as less than 3.0 parts per million (ppm).

Figure 10B.15 Hypoxia in the Gulf of Mexico and Long Island Sound. (From United States Environmental Protection Agency, 2003, *EPA's Draft Report on the Environment*, 2003, EPA 600-R-03-050, www.epa.gov.)

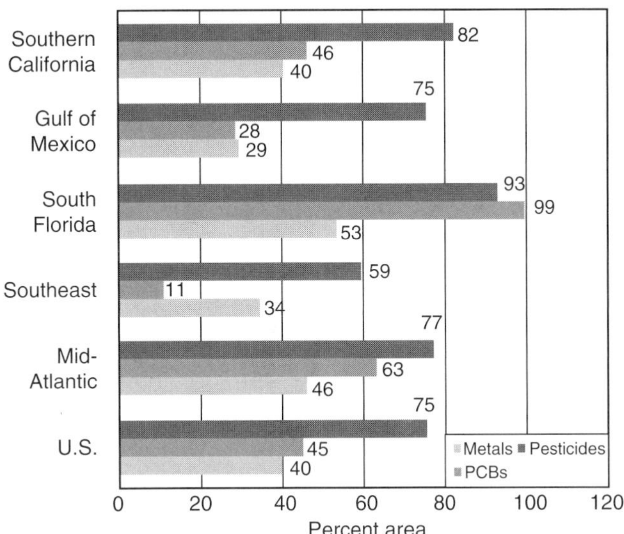

Figure 10B.16 Regional sediment enrichment (1990–1997) in United States coastal waters due to human sources. (From United States Enviornmental Protection Agency, 2003, *EPA's Draft Report on the Environment, 2003*, EPA 600-R-03-050, www.epa.gov.) Original Source: USEPA, *National Coastal Condition Report*, September 2001.

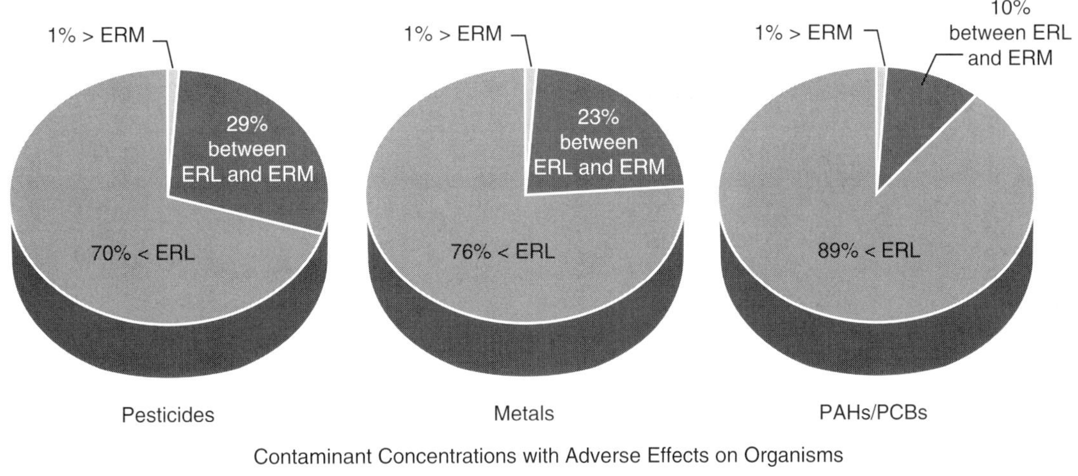

Figure 10B.17 Distribution of sediment contaminant concentrations in sampled estuarine sites, 1990–1997. ERL, NOAA Effects Range Low; ERM, NOAA Effects Range median. (From United States Enviornmental Protection Agency, 2003, *EPA's Draft Report on the Environment, 2003*, EPA 600-R-03-050, www.epa.gov.) Original Source: USEPA, *National Coastal Condition Report*, September 2001.

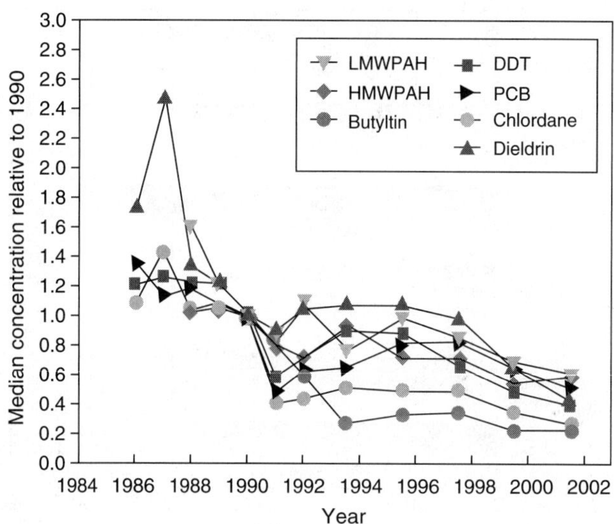

Figure 10B.18 Trends in contaminant concentrations measured in NOAA's mussel water project since 1986. (From United States Environmental Protection Agency, 2004, *National Coastal Condition Report II*, EPA-620/R-03/002, December 2004, www.epa.gov.)

Leading POLLUTANTS in impaired lakes*

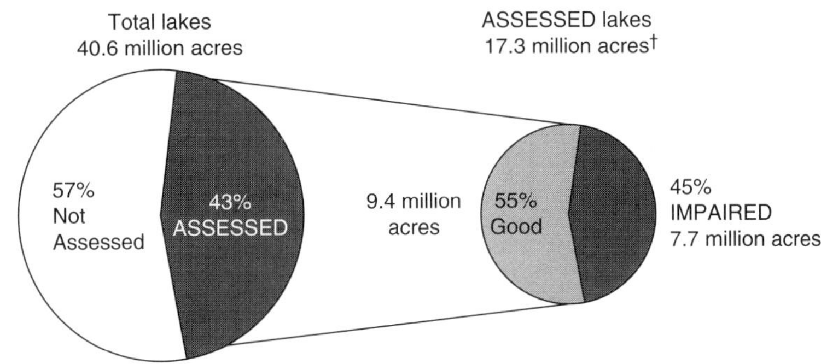

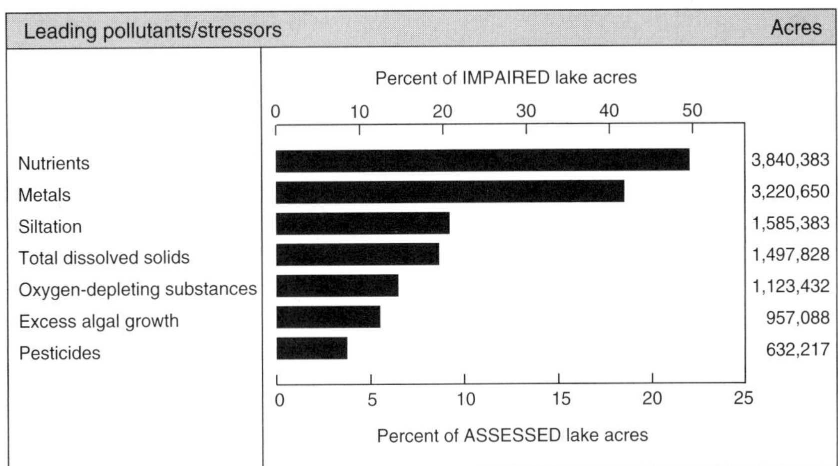

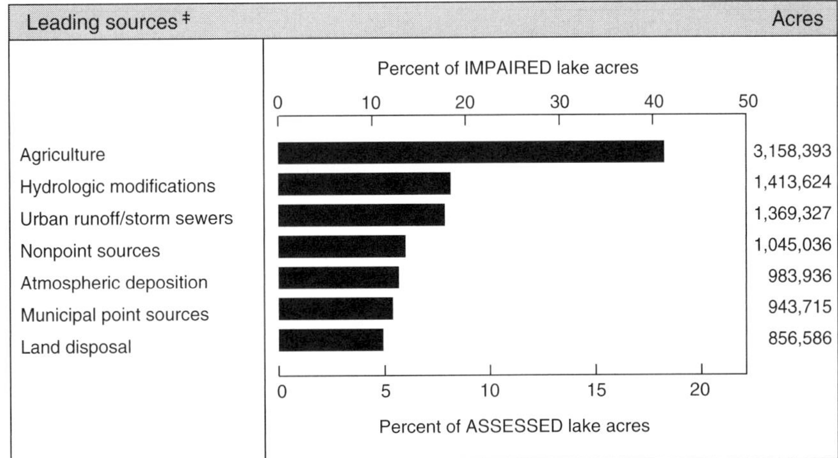

* Eleven states did not include the effects of statewide fish consumption advisories when reporting the pollutants and sources responsible for impairment. Therefore, certain pollutants and sources, such as metals and atmospheric deposition, may be under represented.

‡ Excluding unknown, natural, and "other" sources.

† Includes acres assessed as not attainable.

Note: Percentages do not add up to 100% because more than one pollutant or source may impair a lake.

Figure 10B.19 Leading pollutants and sources in impaired lakes in the United States. (From United States Environmental Protection Agency, 2002, *National Water Quality Inventory 2000 Report*, EPA-841-R-02-001, www.epa.gov.)

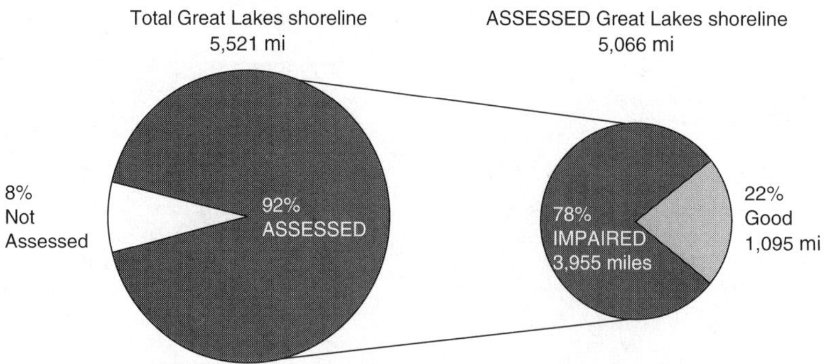

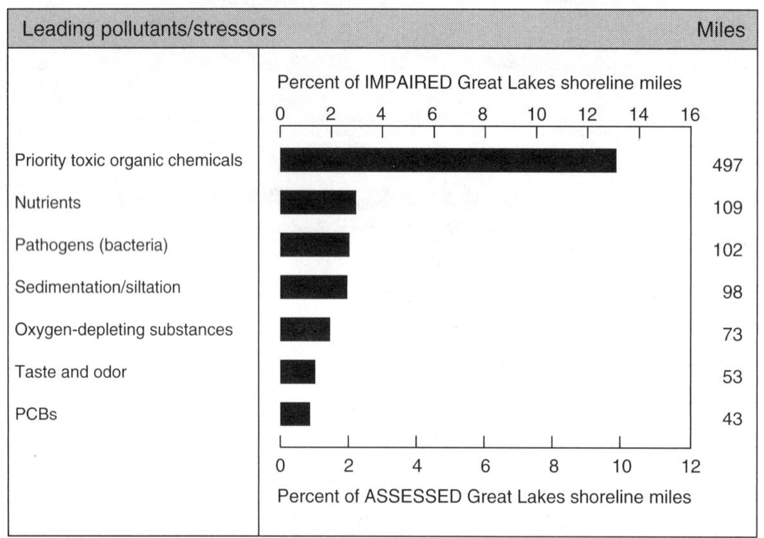

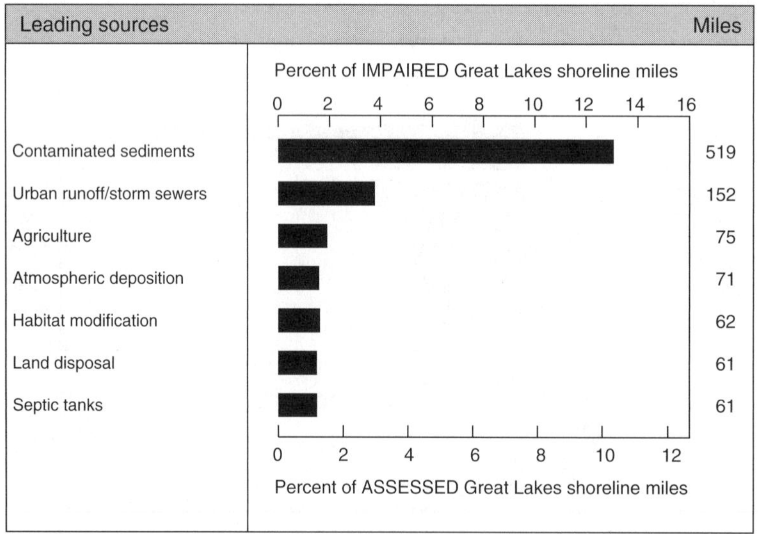

Note: Percentages do not add up to 100% because more than one pollutant or source may impair a segment of great lakes shoreline.

Figure 10B.20 Leading pollutants and sources in impaired Great Lakes shoreline waters in the United States. (From United States Environmental Protection Agency, 2002, *National Water Quality Inventory 2000 Report*, EPA-841-R-02-001, www.epa.gov.)

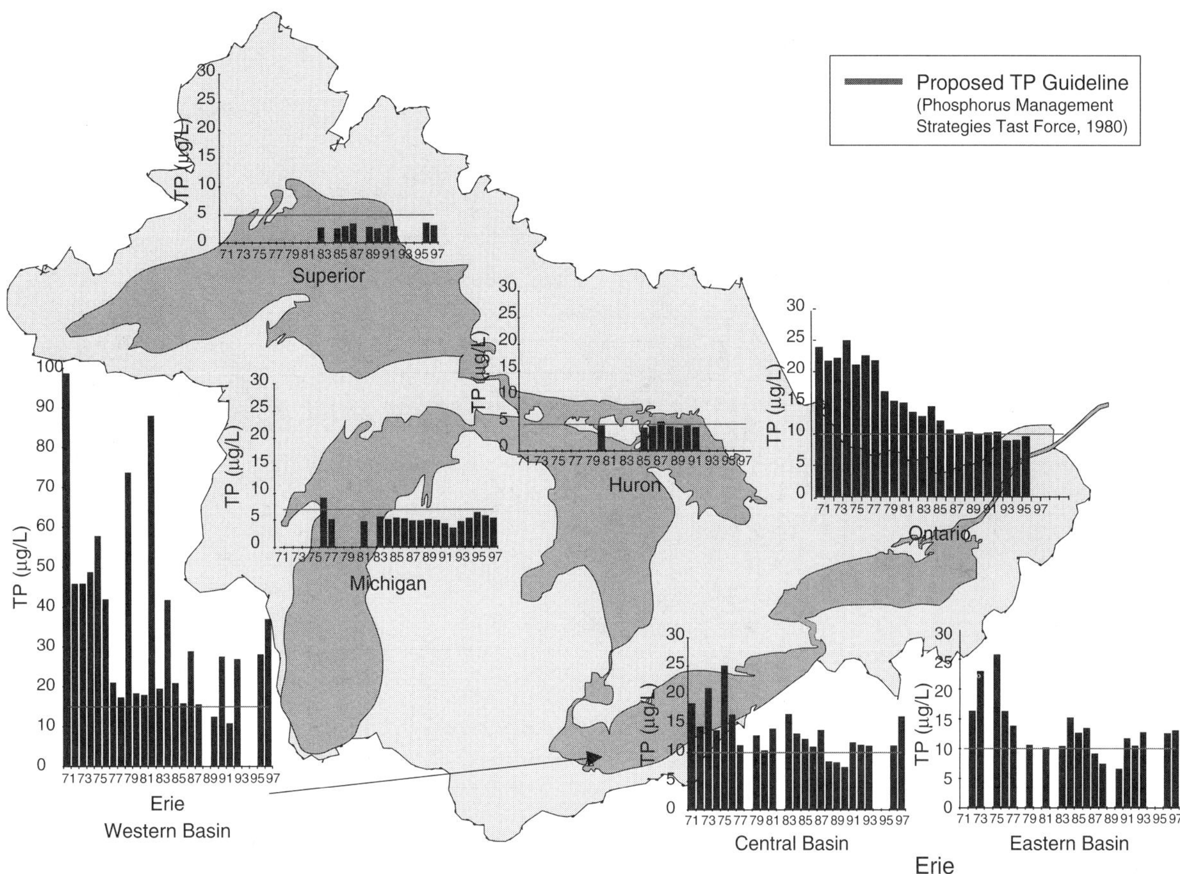

Figure 10B.21 Total phosphorous trends in the Great Lakes from 1971 to 1997 (spring, open lake surface). (From International Joint Commission, 2004, *12th Biennial Report on Great Lakes Water Quality, September 2004*, www.ijc.org.)

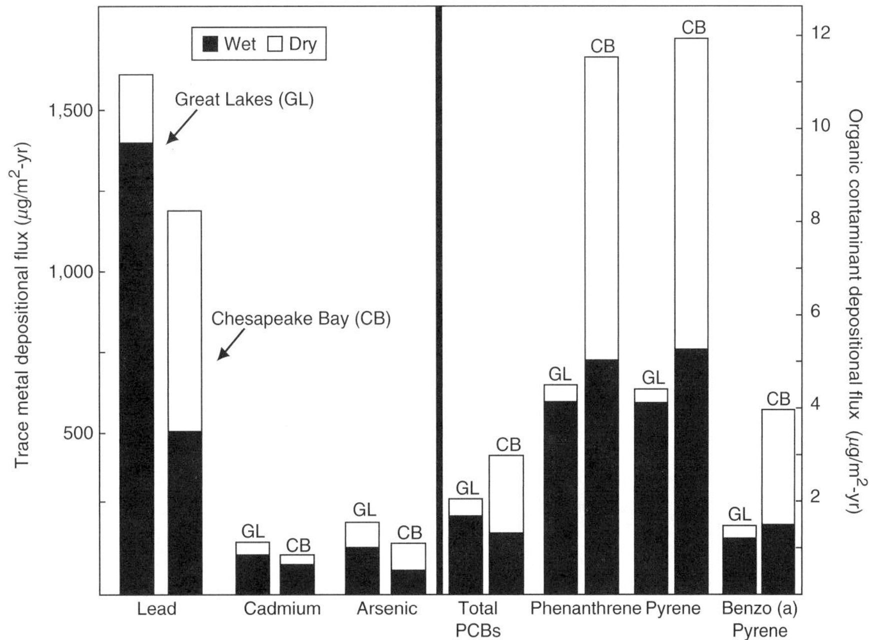

Figure 10B.22 Comparison of Chesapeake Bay and Great Lakes atmospheric depositional fluxes. (From United States Environmental Protection Agency, 1997, *Deposition of Air Pollutants to the Great Waters Second Report to Congress*, EPA-453/R-97-011, www.epa.gov.) Original Source: Baker et al., 1996 (Chesapeake Bay) and Eisenreich and Strachen 1992 (Great Lakes).

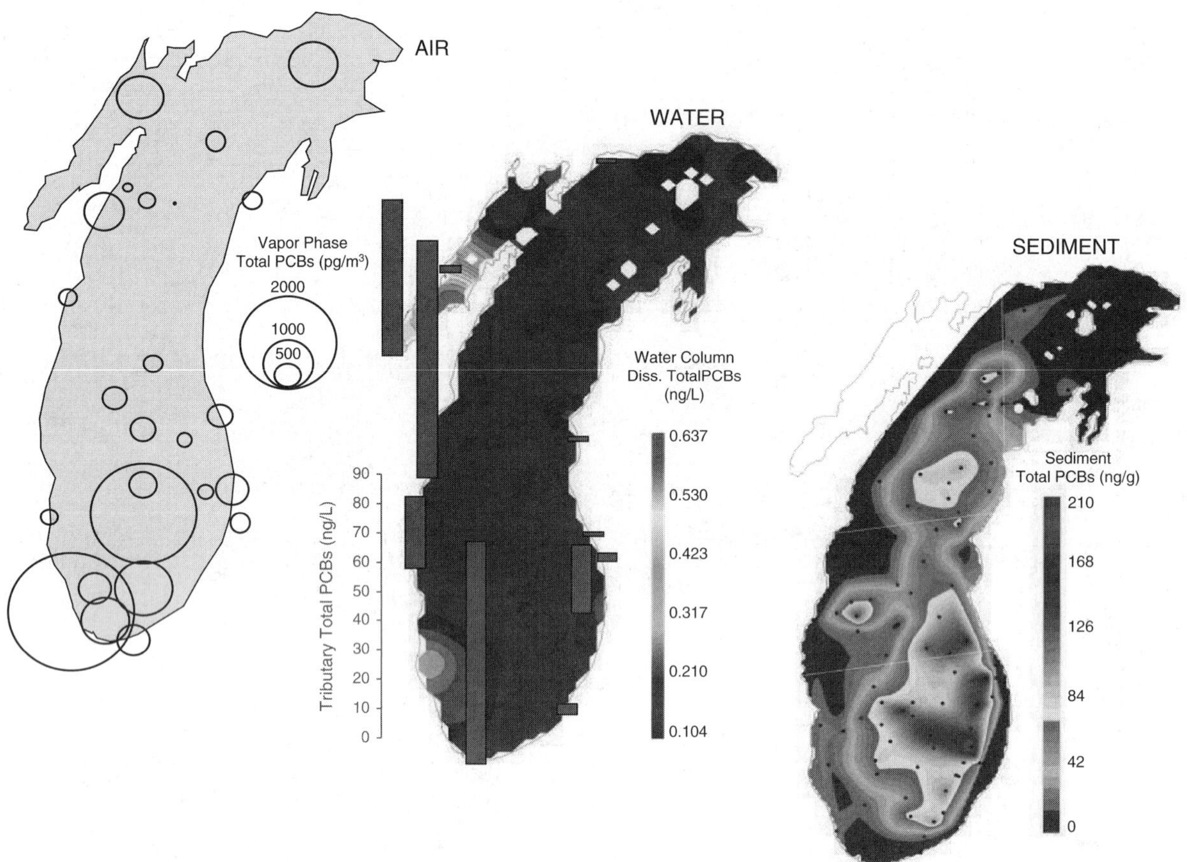

Figure 10B.23 Concentrations of total PCBs in the atmosphere, tributaries, water column and sediments of Lake Michigan. (From McCarty, H.B. et al., United States Environmental Protection Agency, 2004, *Results of the Lake Michigan Mass Balance Study: Polychlorinated Biphenyls and trans-Nonachlor Data Report*, EPA 905 R-01-011, April 2004, www.epa.gov.)

Lake Michigan polychlorinated biphenyls (PCBs) sources, 1970 and 1995 values in kilograms per year

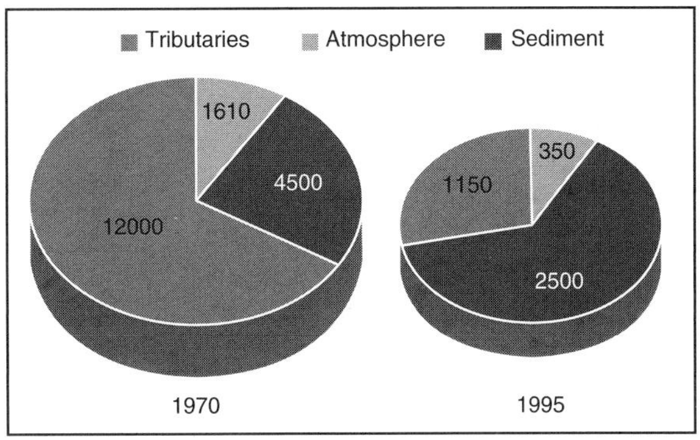

Note: This graphic was created for this report by the EPA Great Lakes National Program Office and the EPA, Office of Research and Developments Large Lakes Research Station using MICHTOX a mass balance and bio accumulation model and air, water, and sediment data drawn from the Great Lakes Environmental Monitoring Database (GLENDA). The 1970 model run was based on available data and extrapolations. The 1995 model run was based on data collected during the Lake Michigan Mass Balance Study that collected over 25,000 samples at 200 locations in 1994–1995.

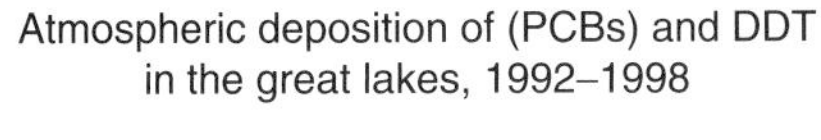

Atmospheric deposition of (PCBs) and DDT
in the great lakes, 1992–1998

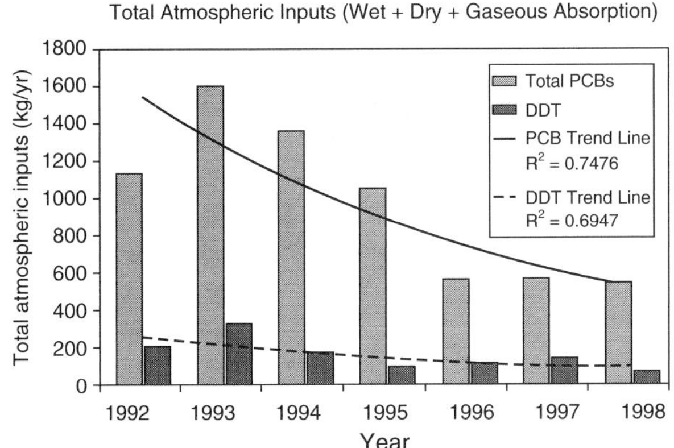

Note: R^2 is the coefficient of determination. It gives a measure of the strength of the correlation.

Polychlorinated biphenyls (PCBs) trends in great lakes fish tissue,* 1972–2000

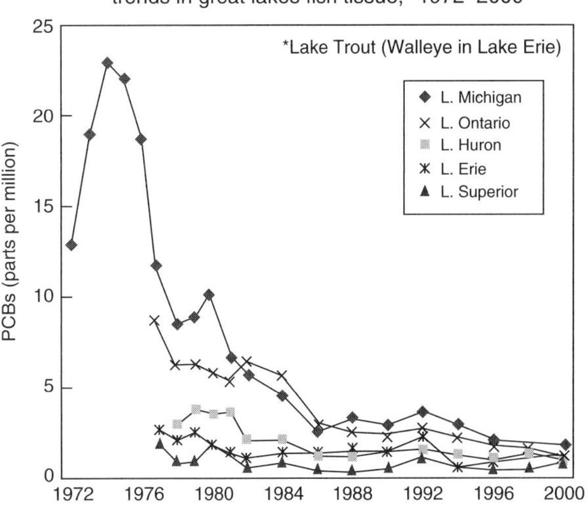

Figure 10B.24 Bioaccumulation of PCBs and DDT in the Great Lakes. (From United States Environmental Protection Agency, 2003, *EPA's Draft Report on the Environment, 2003*, EPA 600-R-03-050, www.epa.gov.)

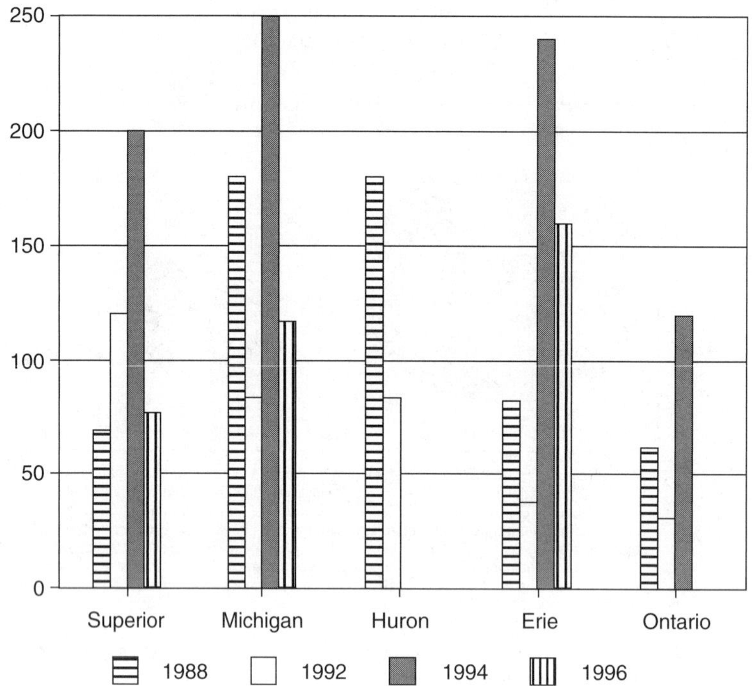

Figure 10B.25 Loading estimates of benzo(a)pyrene to the Great Lakes (kg/yr). (From United States Environmental Protection Agency, 2000, *Deposition of Air Pollutants to the Great Waters Third Report to Congress*, EPA-453/R-00-005, www.epa.gov.)

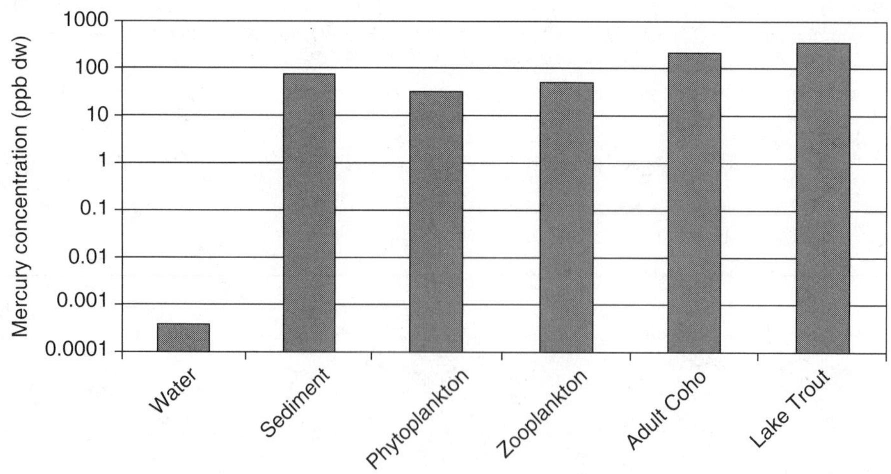

Figure 10B.26 Mercury concentrations in various components of the Lake Michigan ecosystem. (From McCarty, H.B., Brent, R.N., Schofield, J., and Rossmann, R., 2004, *Results of the Lake Michigan Mass Balance Study: Mercury Data Report*, EPA 905 R-01-012, www.epa.gov.)

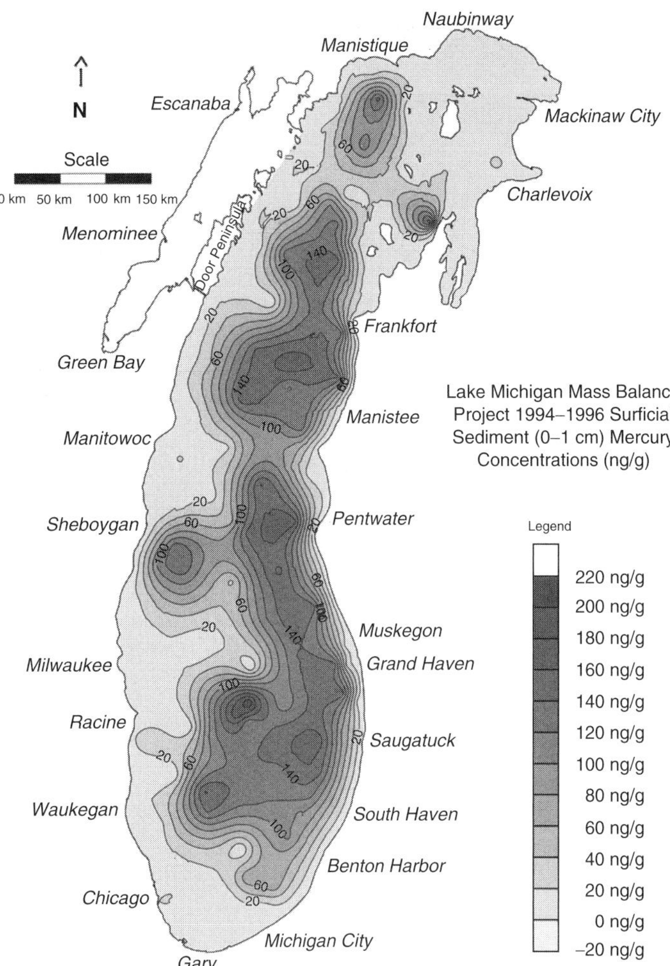

Figure 10B.27 Mercury concentrations (mg/kg) in Lake Michigan surficial sediments (1994–1996). (From McCarty, H.B. et al., 2004, *Results of the Lake Michigan Mass Balance Study: Mercury Data Report*, EPA 905 R-01-012, www.epa.gov.)

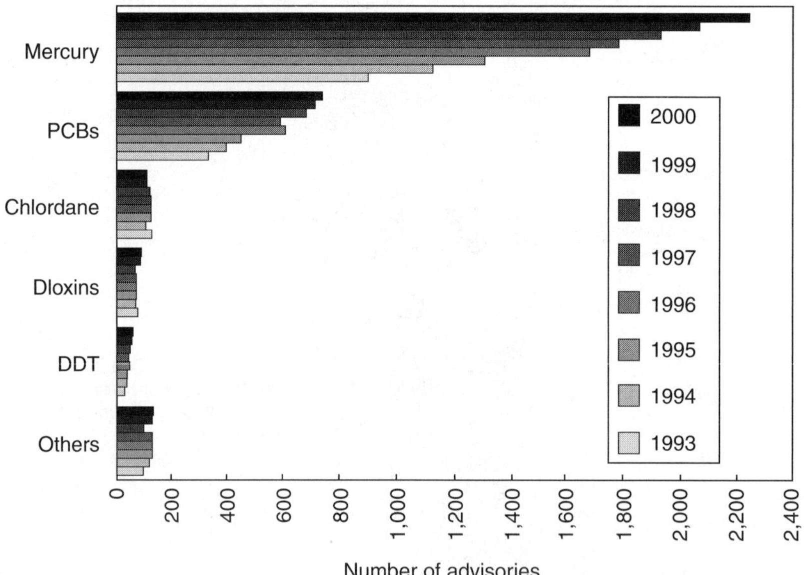

Figure 10B.28 Trends in the number of fish consumption advisories issued for various pollutants. (From United States Environmental Protection Agency, 2001, *Fact Sheet Update: National Listing of Fish and Wildlife Advisories*, EPA-823-F-01-010, April 2001, www.epa.gov.)

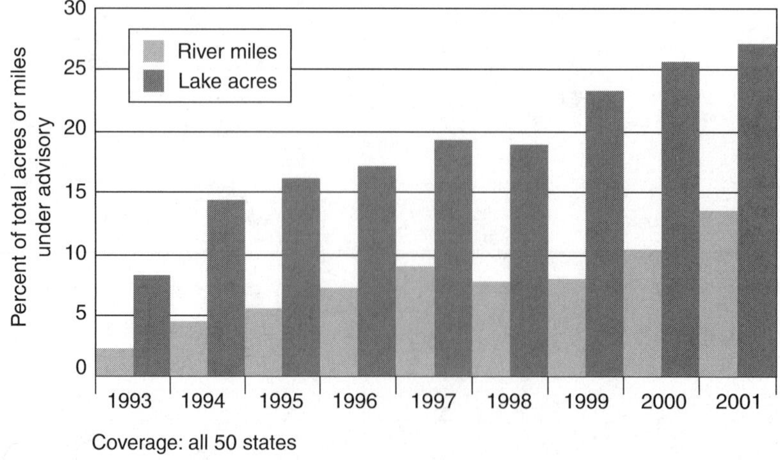

Figure 10B.29 Trends in percentage of river miles and lake acres under fish consumption advisory, 1993–2001. (From United States Environmental Protection Agency, 2003, *EPA's Draft Report on the Environment, 2003*, EPA 600-R-03-050, www.epa.gov.)

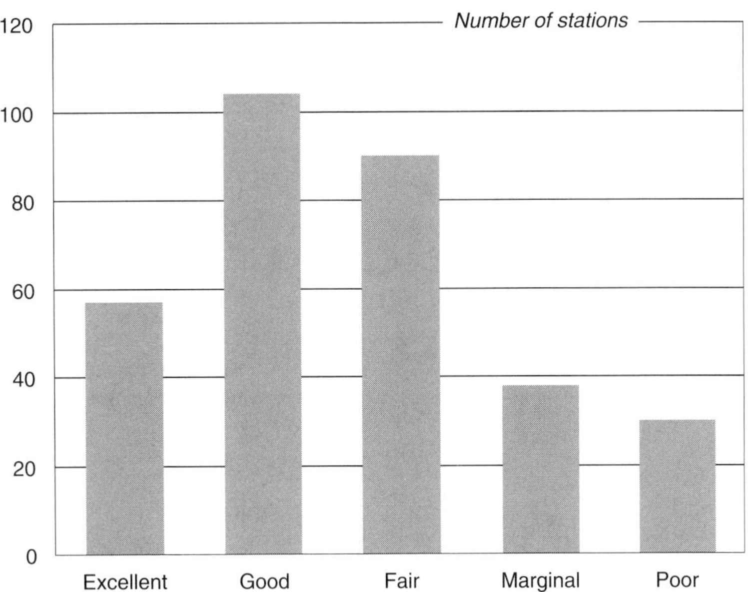

Note: Data were taken over the period 1990 to 2002.

These Water Quality Index (WQI) results are preliminary and should not be regarded as a benchmark or starting point for future trends. Rather, this pilot study provides a first approximation for a national picture of ambient fresh water quality in Canada. Improvements in consistency of application and representation will be sought in the near future.

The WQI values have been calculated by each province and territory (except Quebec) using the methodology developed and endorsed by the Canadian Council of Ministers of the Environment (CCME) in September 2001. According to the CCME user's manual,[1] the specific variables, objectives and time periods used in the index are not specified by the methodology and, because of differences in local conditions, monitoring programs and water quality issues, they vary from one jurisdiction to another. In this regard, it is expected that the variables and objectives chosen to calculate the index provide relevant information about a particular site.

In Quebec, water quality was evaluated using an index other than the CCME WQI: *L'indice de la qualité bactériologique et physico-chimique*. The results between the two indexes have a reasonable degree of comparability. The premise is that the evaluation of water quality in one jurisdiction by water quality experts familiar with the local conditions should be comparable with a similar evaluation by experts in another jurisdiction, even though the index tools may have some variation.

The national portrayal of the WQI results includes information from all provinces and territories except Nunavut and the Yukon, for which suitable data were unavailable at this time. The water bodies included in the WQI calculations do not provide uniform coverage across Canada, but rather tend to be concentrated in the more populated areas of the country where the potential threats to water quality are generally greatest. The coverage and the density of sites are also higher in some provinces than in others.

1. Canadian Council of Ministers of the Environment, 2001, Canadian Water Quality Guidelines for the Protection of Aquatic Life; CCME Water Quality Index 1.0, User's Manual.

Figure 10B.30 Canadian freshwater quality indicator by quality class. Data were taken over the period 1990–2002. (From Produced by Environment Canada based on the Index values or water quality data supplied by the provinces and territories under the auspices of the Water Quality Task Group of the Canadian Council of Ministers of the Environment. Environment Canada, National Round Table on the Environment and the Economy, 2003, *Environment and Sustainable Development Indicators for Canada*, Ottawa.)

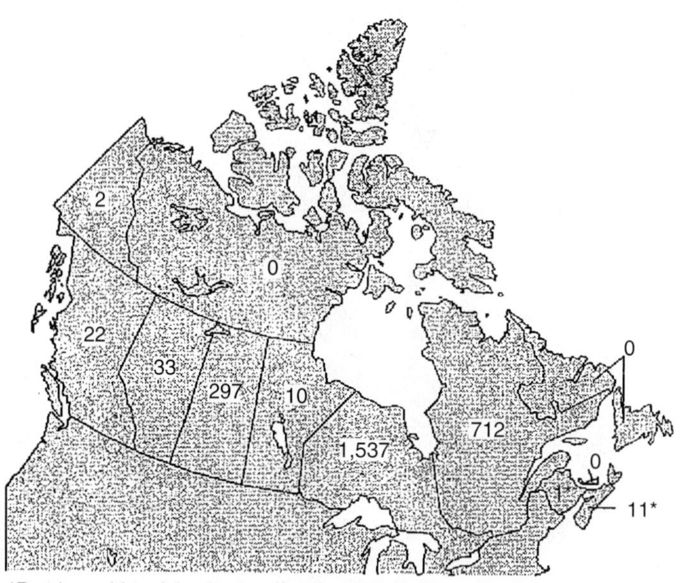

Figure 10B.31 Total number of fish advisories in effect in Canada. (From United States Environmental Protection Agency, 2001, *Fact Sheet Update: National Listing of Fish and Wildlife Advisories*, EPA-823-F-01-010, April 2001, www.epa.gov.)

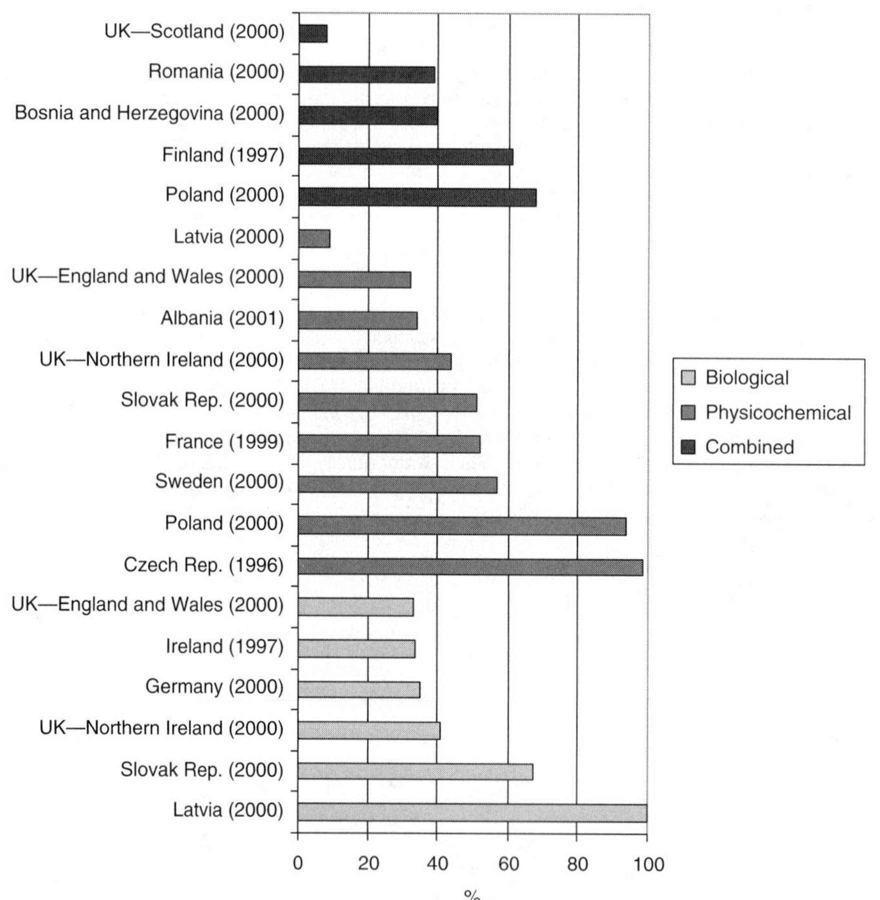

Figure 10B.32 Percentage of European rivers classified as less than good, by country. (From Trent, Z, European Environment Agency, *Indicator Fact Sheet, National River Classification System (WEC04e), Version 13.10.03*, eeaeuropa.eu. Reprinted with permission © EEA.)

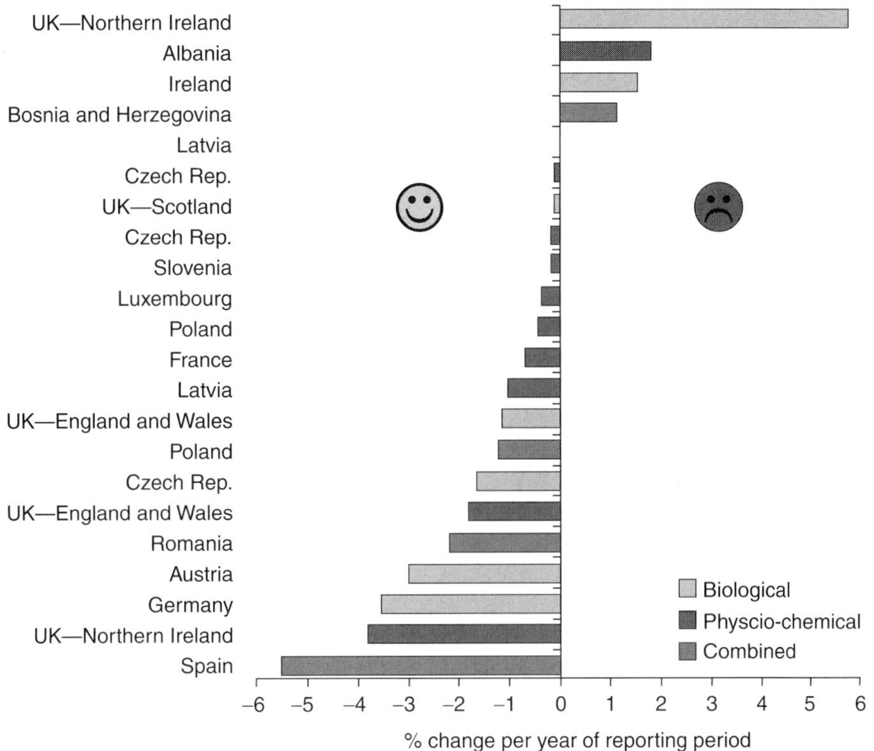

Figure 10B.33 Rate of change in rivers classified as less than good and good as a percentage of the total river classified. (From European Environment Agency, 2003, *Europe's Water: An Indicator-Based Assessment Summary*, EEA, Copenhagen, www.eea.europa.eu. Reprinted with permission © EEA.)

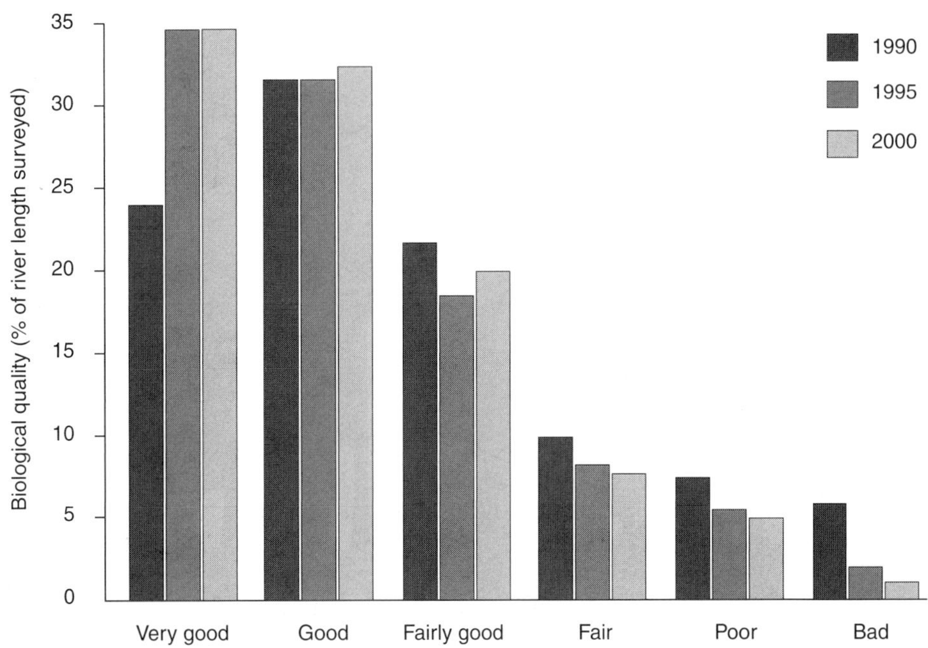

Figure 10B.34 Biological quality of United Kingdom rivers, 1990–2000. (From United Nations Educational, Scientific and Cultural Organization, 2003, *Water for People Water for Life*, The United Nations World Water Development, United Nations Educational, Scientific and Cultural Organization (UNESCO) and Berghahn Books, www.unesco.org. Reprinted with permission.) Original Source: Adapted from Environmental Agency, UK, 2002.

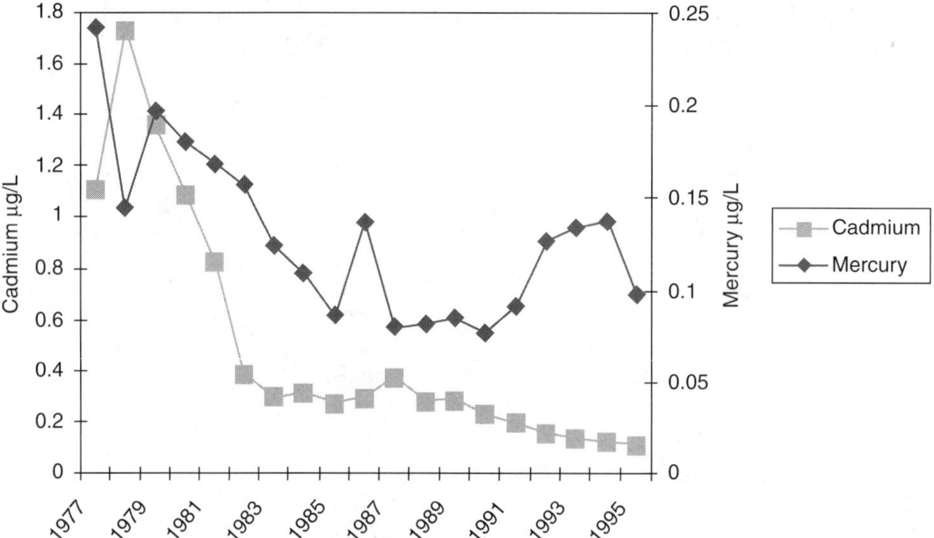

Figure 10B.35 Trends in concentrations of cadmium and mercury at river stations included in the European Union exchange of information decision. The EU environmental quality standards for cadmium and mercury in inland waters are 5μg/L and 1μg/L as annual averages, respectively. In less polluted areas in e.g. Nordic countries concentrations of cadmium and mercury are only 10% and 1% of these values. Average of country annual average concentrations. Cadmium data from Belgium, Germany, Ireland, Luxembourg, Netherlands, UK. Mercury data from Belgium, France Germany, Ireland, Netherlands, UK. (From EEA, Indicator fact sheet, *Hazardous Substances in River Water (WHS02)*, www.eea.europa.eu. Reprinted with permission © EEA.)

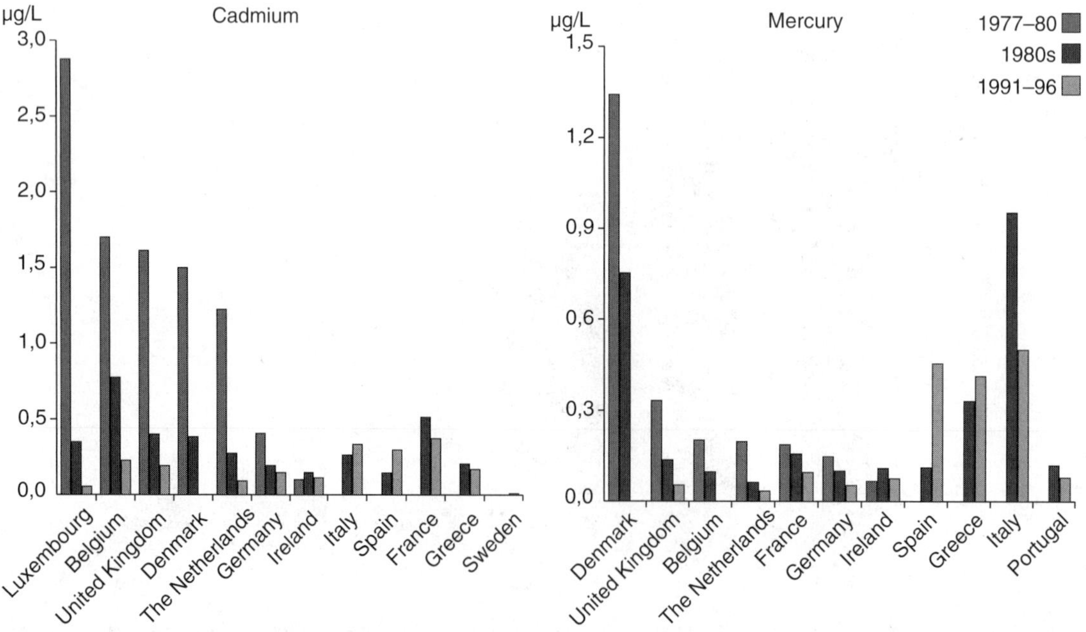

Figure 10B.36 Annual average concentrations of cadmium and mercury in European Union rivers between late 1970s and 1996. (From EEA, 2003, *Europe's Environment, The Third Assessment, Environmental Assessment Report No. 10*, EE1, Copenhagen, eea. Reprinted with permission © EEA.)

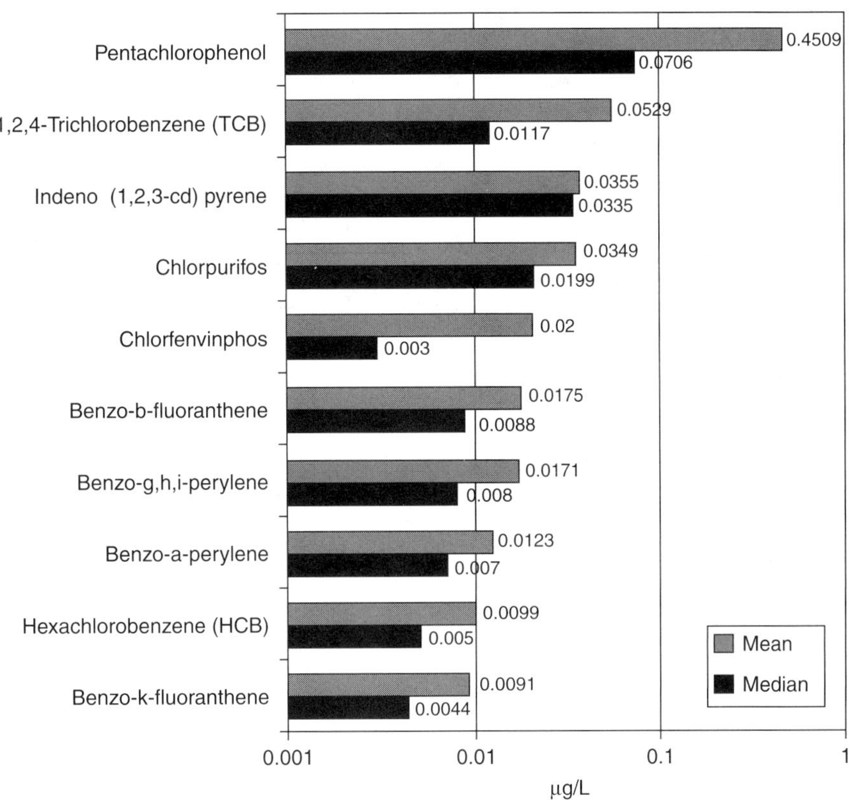

Figure 10B.37 Median and mean concentrations of the 10 most highly ranked substances in the water framework directive priority list in European rivers. (From EEA, Indicator fact sheet, *Hazardous Substances in River Water (WHS02)*, www.eea.europa.eu. Reprinted with permission © EEA.)

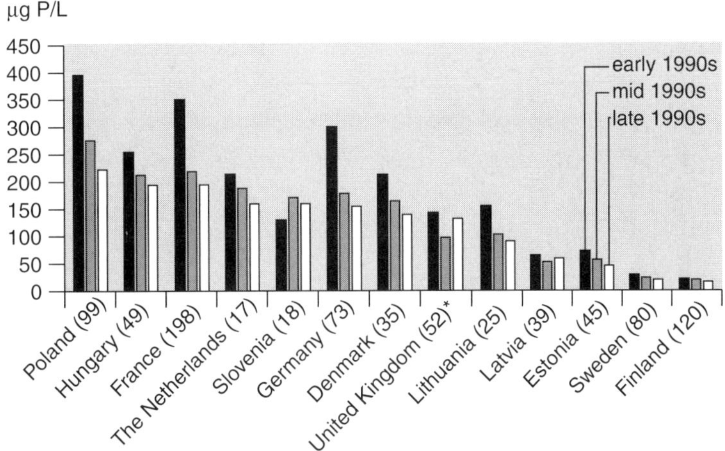

Note: Average of annual median concentrations. Number of stations in brackets:
*UK figures for orthophosphate-p.

Figure 10B.38 Total phosphorus concentrations in rivers, selected European Union and accession countries. (From European Environmental Agency (EEA), 2002, *Environmental Signals 2002—Benchmarking the Millennium, Environmental Assessment Report No. 9*, www.eea.europa.eu. Reprinted with permission © EEA.)

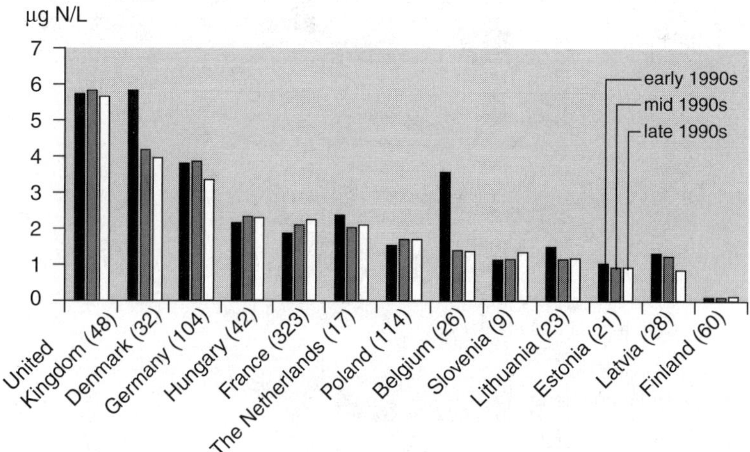

Note: Average of annual median concentrations. Number of stations in brackets.

Figure 10B.39 Nitrate concentrations in rivers, selected European Union and accession countries. (From European Environmental Agency (EEA), 2002, *Environmental Signals 2002—Benchmarking the Millennium, Environmental Assessment Report No. 9*, www.eea.europa.eu. Reprinted with permission © EEA.)

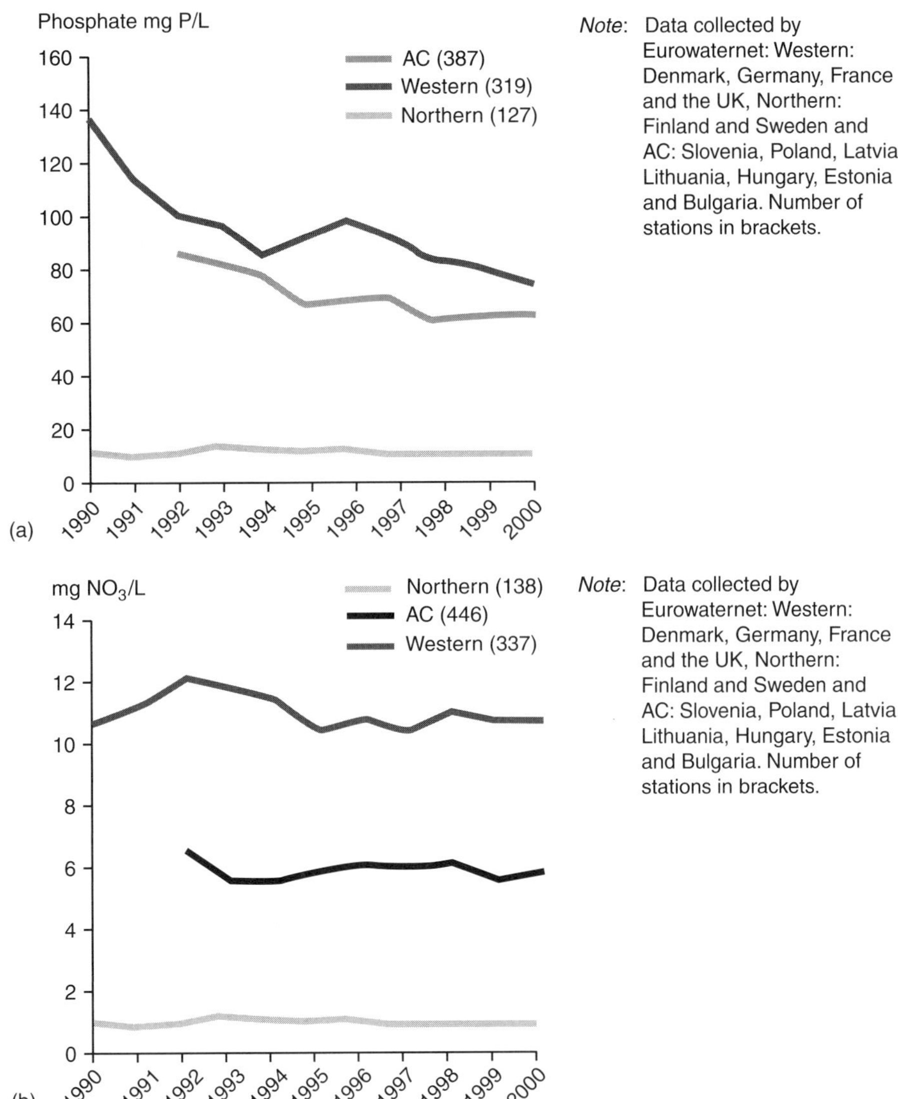

Figure 10B.40 Phosphate and nitrate in European rivers. (From European Environment Agency, 2003, *Europe's Water: An Indicator-Based Assessment Summary*, EEA. Copenhagen. © EEA, www.eea.europa.eu.)

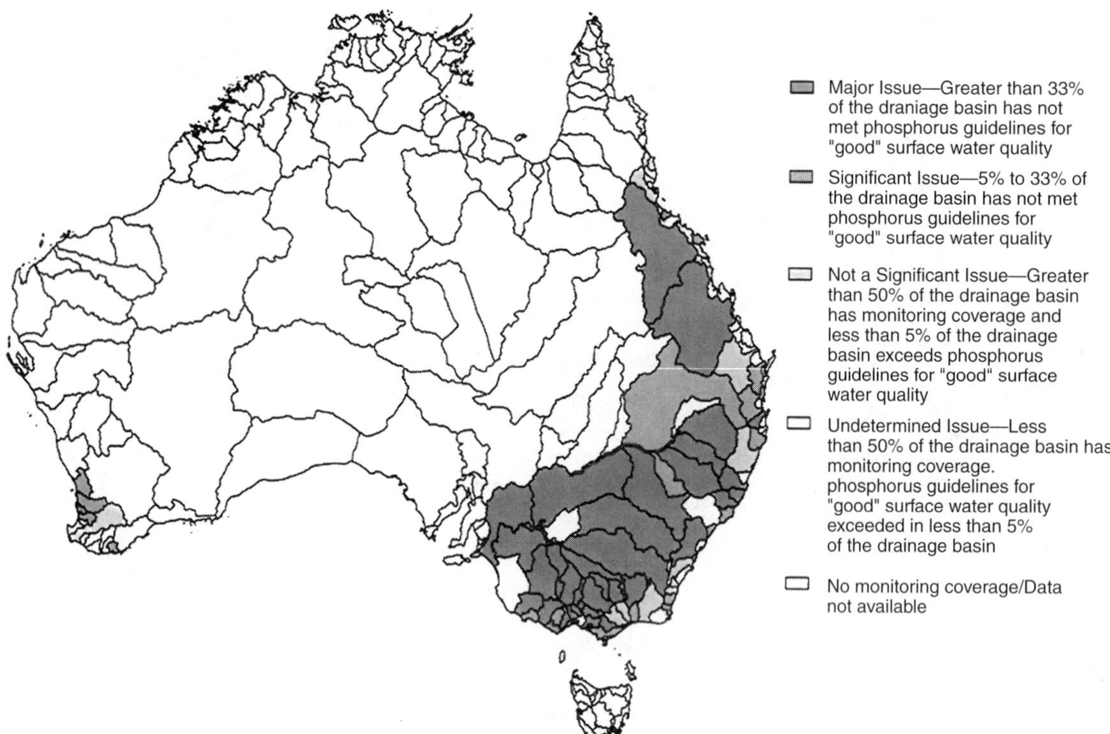
Major Issue—Greater than 33% of the draniage basin has not met phosphorus guidelines for "good" surface water quality
Significant Issue—5% to 33% of the drainage basin has not met phosphorus guidelines for "good" surface water quality
Not a Significant Issue—Greater than 50% of the drainage basin has monitoring coverage and less than 5% of the drainage basin exceeds phosphorus guidelines for "good" surface water quality
Undetermined Issue—Less than 50% of the drainage basin has monitoring coverage. phosphorus guidelines for "good" surface water quality exceeded in less than 5% of the drainage basin
No monitoring coverage/Data not available

Figure 10B.41 Australian river systems where phosphorous levels exceed state or territory guidelines for the protection of ecosystems. (From Ball, J. et al., 2001, *Inland Waters, Australia State of Environment Report 2001 (Theme Report)*, CSIRO Publishing on behalf of the Department of the Environment and Heritage, Canberra, www.deh.gov.au. Reprinted with permission.) Original Source: National Land and Water Resources Audit, 2001a.

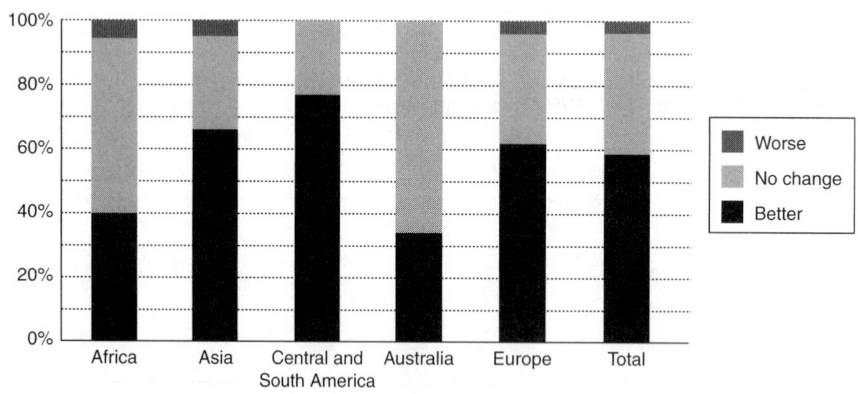
100%
80%
60%
40%
20%
0%
Africa
Asia
Central and South America
Australia
Europe
Total
Worse
No change
Better

Note: This figure is based on a sample of 93 lakes.
Although there has been improvement of lake water condition in some areas of all regions, the overwhelming trend illustrated here is deterioration in quality, most notably in Central and South America where close to 80 percent of sampled lakes deteriorated in the studied period.

Figure 10B.42 Changes in world lake conditions, 1960–1990. (From United Nations Educational, Scientific and Cultural Organization, 2003, *Water for People Water for Life*, The United Nations World Water Development, United Nations Educational, Scientific and Cultural Organization (UNESCO) and Berghahn Books, www.unesco.org.) Original Source: Data collated for Loh et al., 1998.

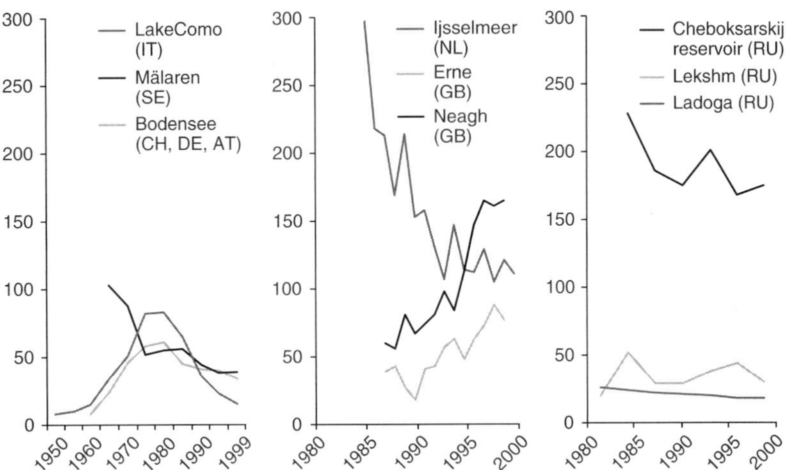

Figure 10B.43 Trends in total phosphorous concentrations in some large European lakes. (From EEA, 2003, Europe's Environment, *The Third Assessment, Environmental Assessment Report No. 10*, EE1, Copenhagen. Reprinted with permission © EEA, www.eea.europa.eu.)

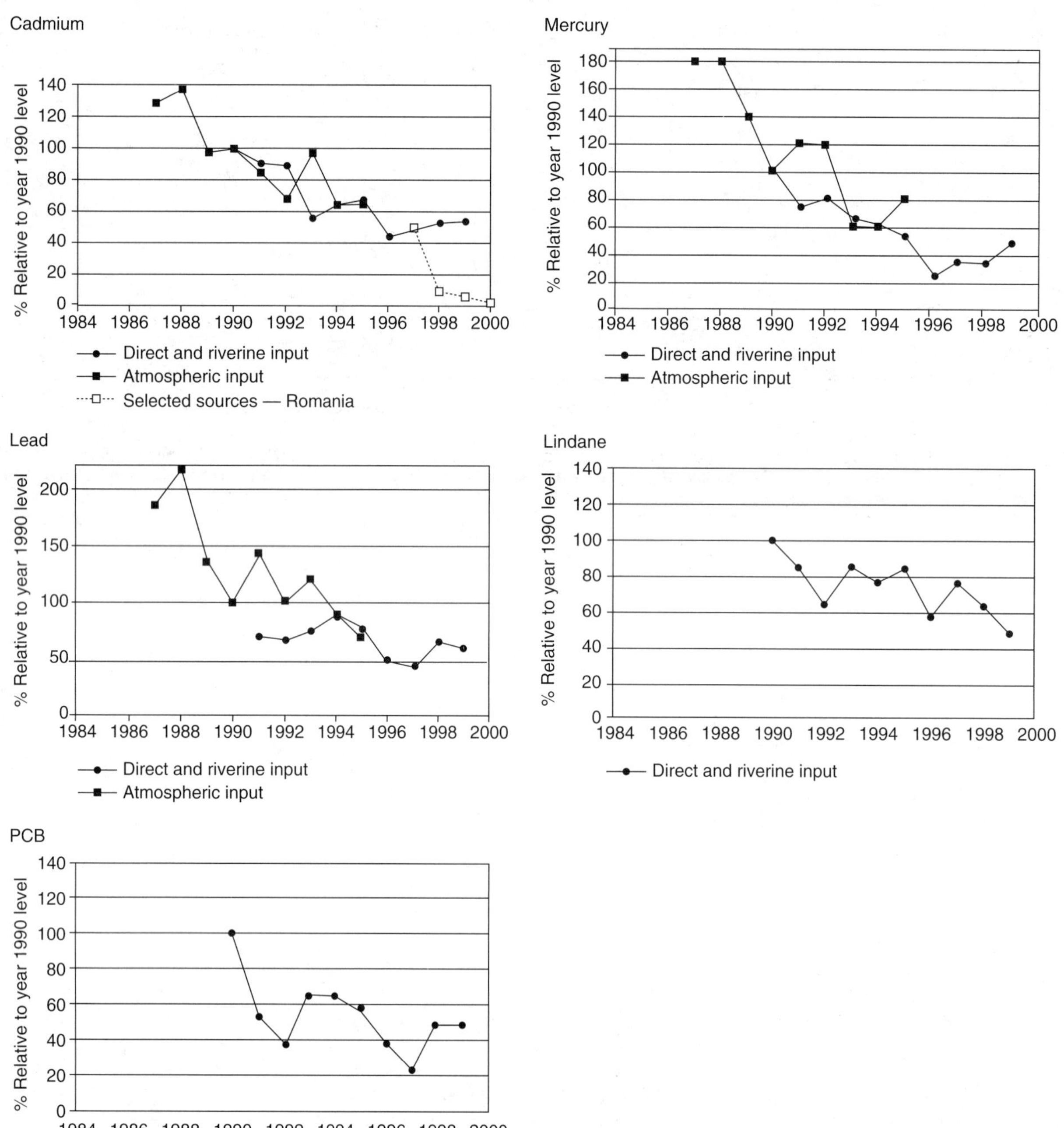

Figure 10B.44 Change (%) in direct riverine and atmospheric inputs of cadmium, mercury, lead, lindane, and PCB in the Northeast Atlantic. (From Green, N. et al., 2003, *Hazardous Substances in the European Marine Environment: Trends in Metals and Persistent Organic Pollutants, European Environment Agency, Topic Report 2/2003*, www.eea.europa.eu. Reprinted with permission © EEA.)

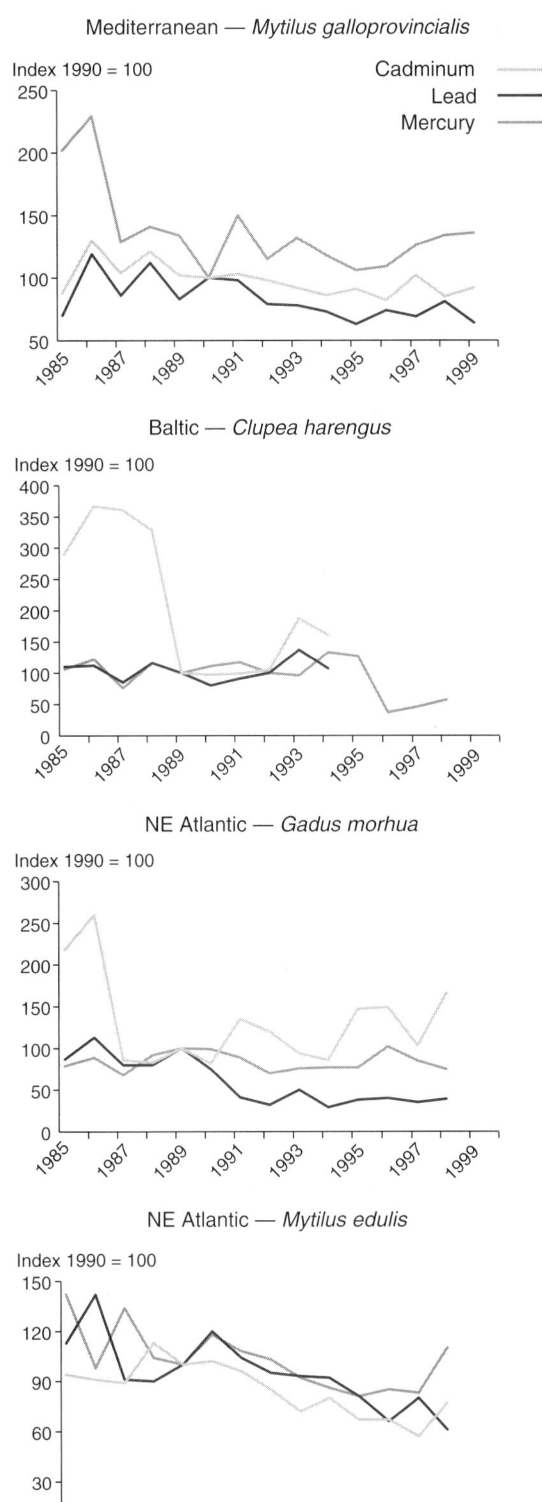

Figure 10B.45 Concentrations of selected metals and synthetic organic substances in marine organisms in the Mediterranean and Baltic Sea, and in the North East Atlantic Ocean. (From European Environmental Agency (EEA), 2003, *Europe's Environment, The Third Assessment, Environmental Assessment Report No. 10*, EE1, Copenhagen, www.eea.europa.eu. Reprinted with permission © EEA.)

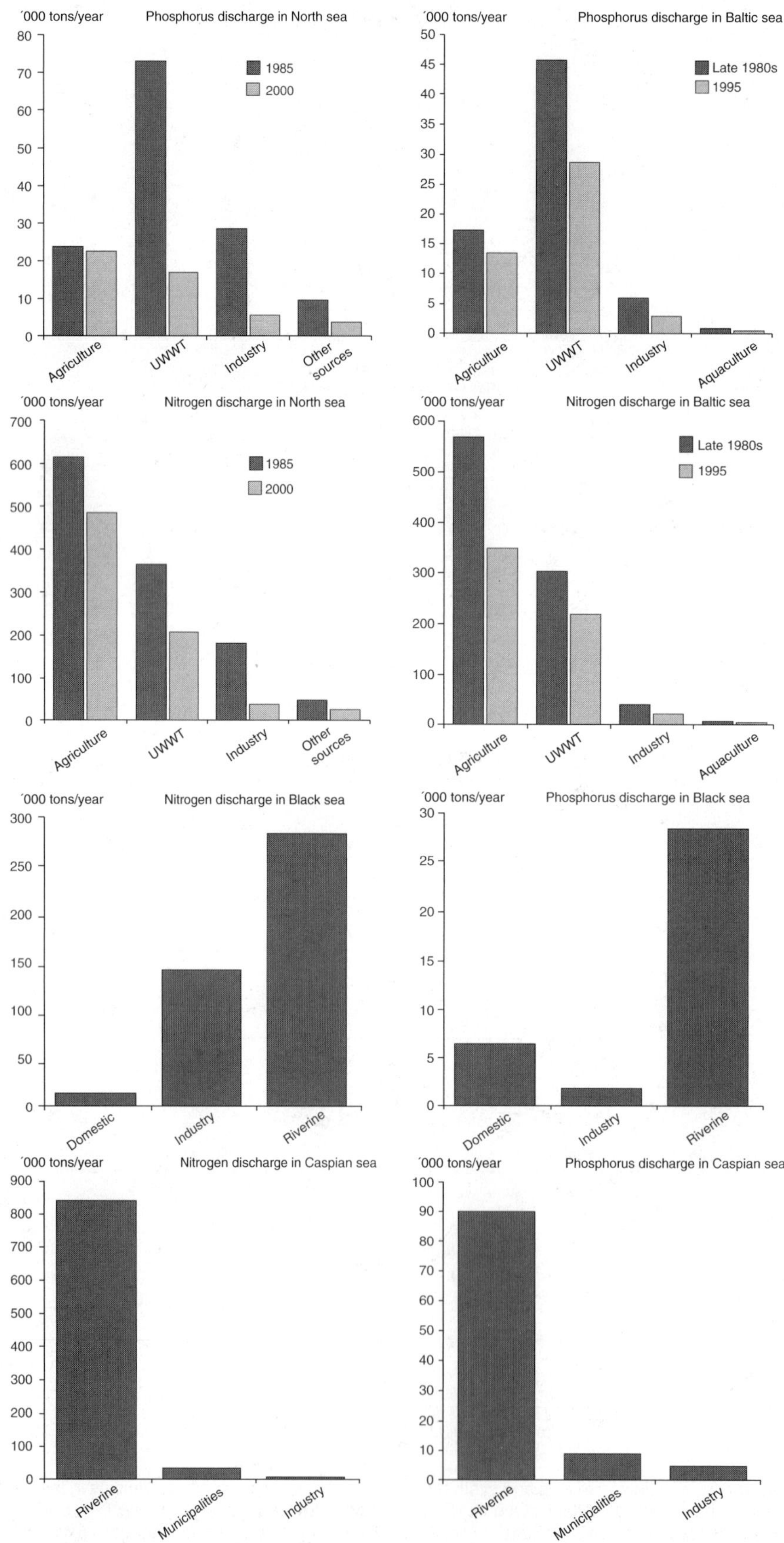

Figure 10B.46 Source apportionment of nitrogen and phosphorus discharges in Europe's seas and percentage reductions. (From European Environmental Agency (EEA), 2003, *Europe's Environment, The Third Assessment, Environmental Assessment Report No. 10*, EE1, Copenhagen, www.eea.europa.eu. Reprinted with permission.) Original Source: North Sea progress report 2002; Finnish Environmental Insitute. 2002, Black Sea Commission, 2002; Caspian Environmental Program, no date. © EEA.

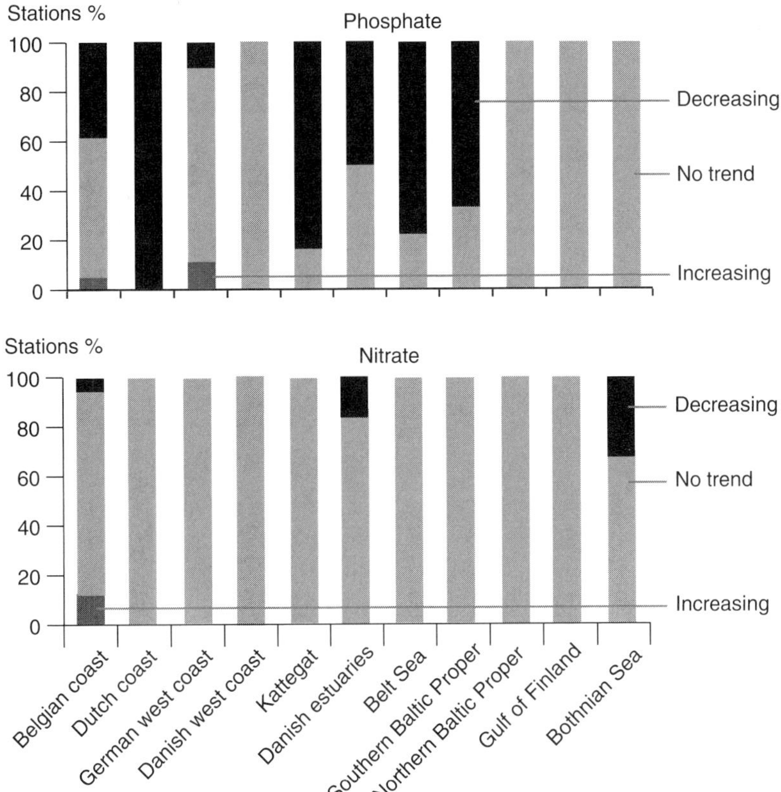

Note: For each station or sampling point in the subregions of the Baltic and North Seas, a trend analysis of winter nutrient concentrations in water from 1985 to 1997/2000 was carried out.The bars in the graph show, at how many sampling points (as %) a decrease or an increase in nutrient concentrations at the 5 % significance level is observed.

Figure 10B.47 Trends in nutrients in the Baltic Sea and coastal North Sea waters, 1985–1997/2000. (From European Environmental Agency (EEA), 2002, *Environmental Signals 2002—Benchmarking the Millenium, Environmental Assessment Report No. 9*, www.eea.europa.eu. Reprinted with permission © EEA.)

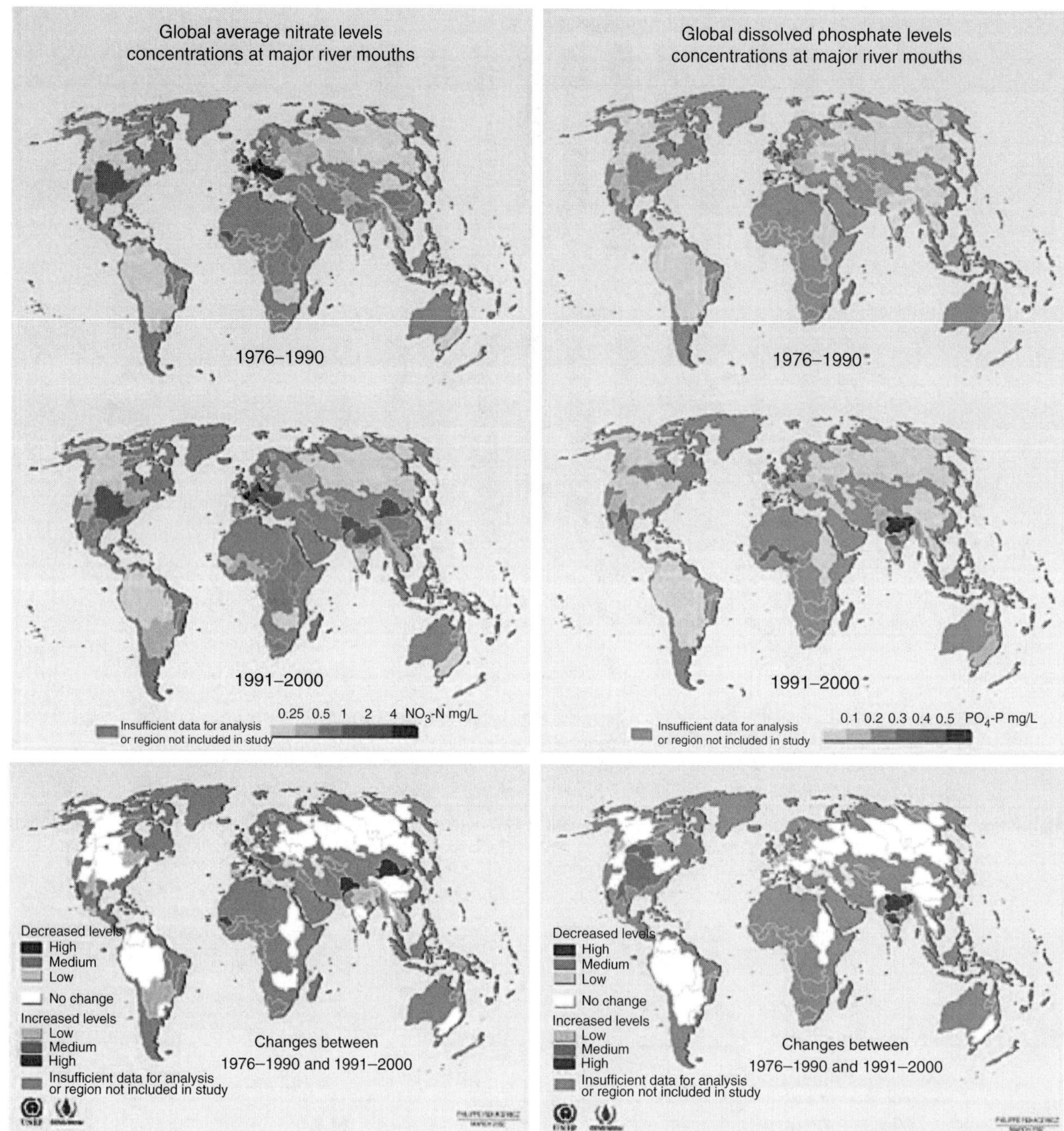

Figure 10B.48 Global average nitrate and dissolved phosphate levels. (From United Nations Environment Programme (UNEP) *Vital Water Graphics, Global Average Nitrate Levels and Global Dissolved Phosphate Levels*, Downloaded 9/22/05, www.unep.org.)

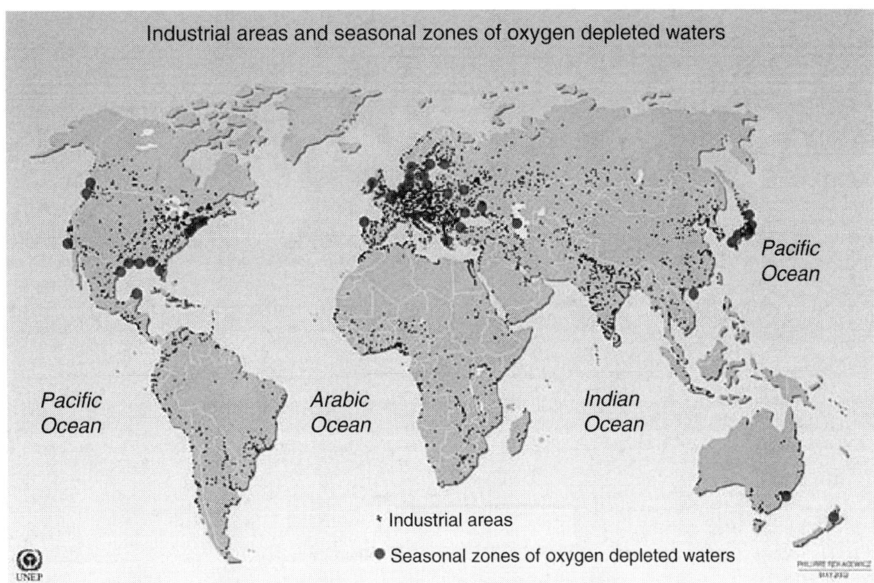

Figure 10B.49 World industrial areas and seasonal zones of oxygen depleted waters. (From United Nations Environment Programme (UNEP), *Vital Water Graphics, Industrial Areas and Seasonal of Oxygen Depleted Waters*, Downloaded 9/22/05, www.unep.org.) Original Source: Malakoff, D., 1998, after Diaz, R.J., and Rosenberg, R., 1995, ESRI, 1990.

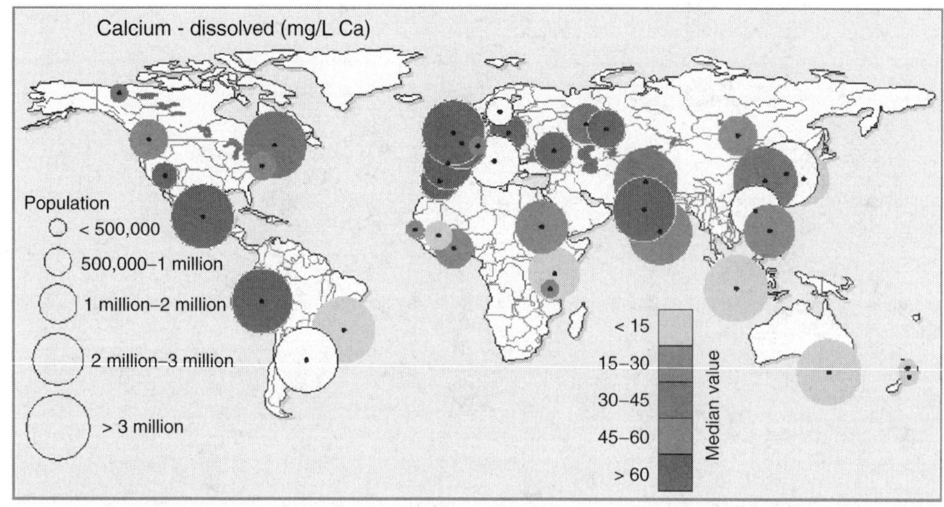

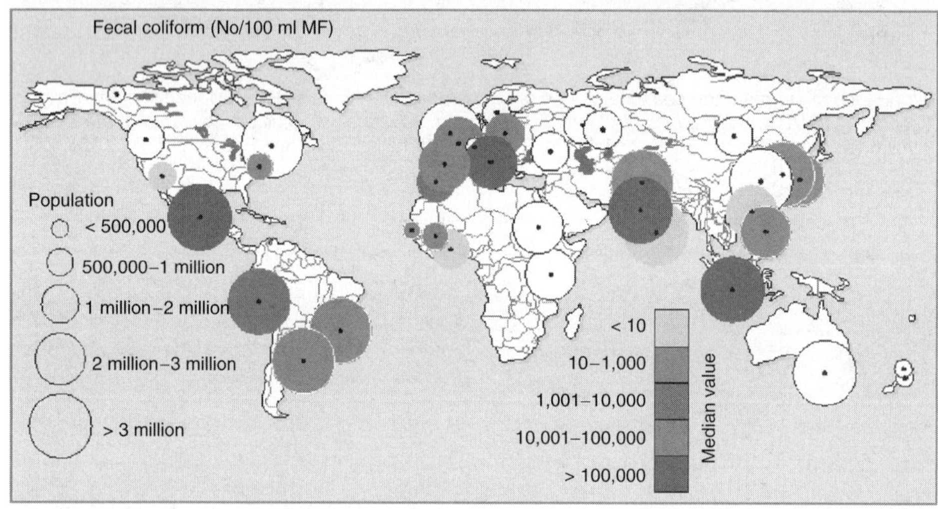

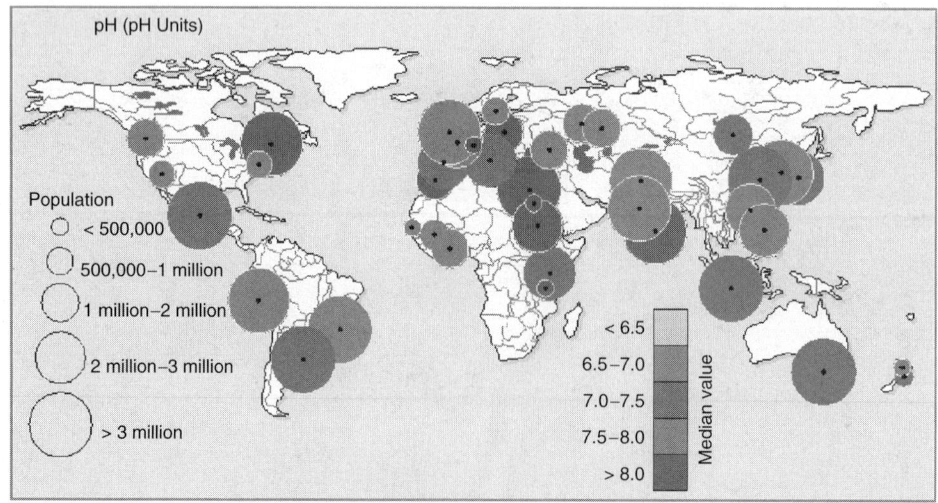

Figure 10B.50 Fecal coliform concentrations and pH values measured in global environment monitoring system stations. (From United Nations Environment Programe, *Global Environment Monitoring System (GEMS) Water Programs, 2004 State of the UNEP GEMS/Water Global Calcium Network and Annual Report*, www.unep.org.)

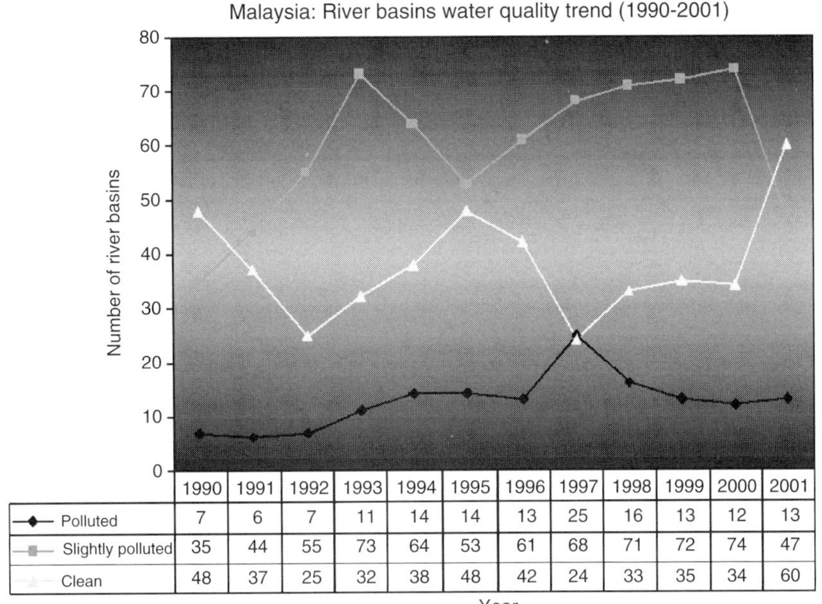

	1990	1991	1992	1993	1994	1995	1996	1997	1998	1999	2000	2001
Polluted	7	6	7	11	14	14	13	25	16	13	12	13
Slightly polluted	35	44	55	73	64	53	61	68	71	72	74	47
Clean	48	37	25	32	38	48	42	24	33	35	34	60

Figure 10B.51 Malaysia: river basins water quality trend (1990–2001). (From Leong, K.H. and Mustafa, A.M., 2003, National Water Quality Standards and Status in Malaysia in United Nations University, *Capacity Development Training for Monitoring of POPS in the East Asian Hydrosphere*, 1–2 September, 2003, UNU Centre, Tokyo, www.unv.org.) Original Source: DOE (1991–2002).

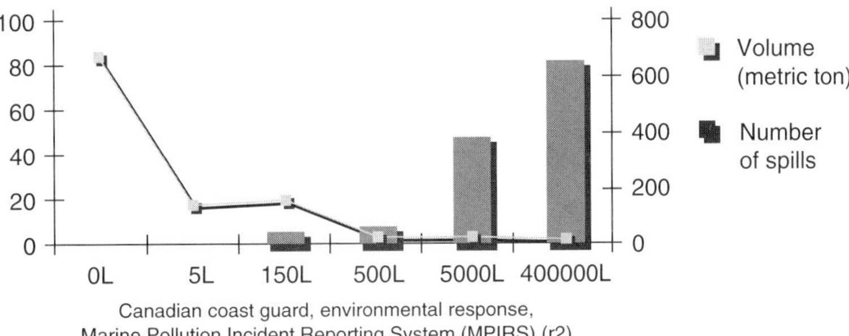

Figure 10B.52 Canadian petrol and chemical spills distribution by volume (metric tonne) and number of spills, 2003, (From *Marine Programs National Performance Report for 2003–2004*, Environmental Response, Fisheries and Oceans Canada, Canadian Coast Guard, 2003. Reproduced with the permission of Her Majesty the Queen in Right of Canada, 2006.)

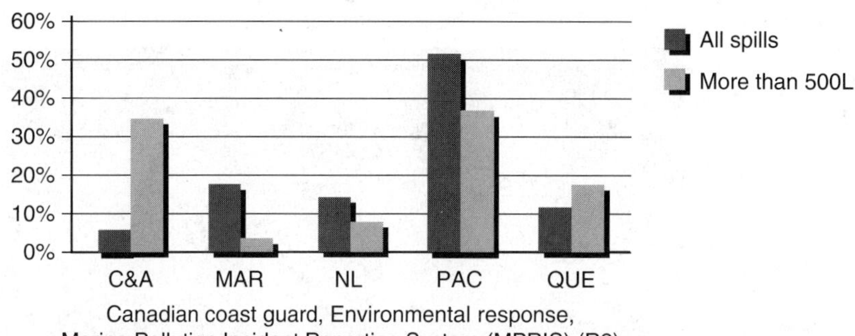

Figure 10B.53 Regional distribution of Canadian spills in percent. (From *Marine Programs National Performance Report for 2003–2004*, Environmental Response, Fisheries and Oceans Canada, Canadian Coast Guard, 2003. Reproduced with the permission of Her Majesty the Queen in Right of Canada, 2006.)

Table 10B.6 Total Assimilative Capacity of Streams of Different Orders

Stream Order	Average Discharge (cfs)	Average Depth (ft)	Average Velocity (ft/sec)	Coefficient of Reaeration (day^{-1})	Total Length of Streams (miles)	Total Assimilative Capacity (tons per day per unit deficiency in dissolved oxygen)	U.S. Rivers Representative of Each Order
1	0.6	—	—	—	1,570,000	—	—
2	2.8	—	—	—	810,000	—	—
3	14	0.55	1.2	9.3	420,000	16,300	—
4	65	0.95	1.6	5.5	220,000	19,000	—
5	310	1.8	1.8	2.6	116,000	20,000	Pecos
6	1,500	2.7	2.0	1.8	61,000	30,000	Shenandoah, Raritan
7	7,000	5	2.5	1.0	30,000	31,000	Allegheny, Kansas, Rio Grande
8	33,000	12	3.0	0.37	14,000	21,000	Tennessee, Wabash
9	160,000	25	4.0	0.19	6,200	18,000	Columbia, Ohio
10	700,000	45	5.0	0.10	1,800	9,400	Mississippi

Source: From U.S. Geological Survey, 1967.

Table 10B.7 Surface Waters Impacted in the United States, 2000 (Assessed Waters Only)

Status	Rivers (thousands of miles)	Lakes (thousands of acres)	Estuaries (thousands of square miles)	Ocean (thousands of shoreline miles)
Full Supporting	367.1	8,027.0	13.8	2.5
Threatened	59.5	1,348.9	1.0	0.2
Impaired	269.3	7,702.4	15.7	0.4
Not Attainable	0.3	0.7	0.0	0.0
Total Assessed	696.2	17,079.0	30.5	3.2

Note: Fully Supporting, Fully supporting of all uses; Threatened, Fully supporting all uses but threatened for one or more uses; Impaired, Partially or not supporting one or more uses; Not Attainable, Not able to support one or more uses.

Source: Abstracted from United States Environmental Protection Agency, 2002, *National Water Quality Inventory 2000 Report*, EPA-841-R-02-001, www.epa.gov.

Table 10B.8 Status of the United States Surface-Water Quality, 1990–2000

Item	Rivers						Lakes[a]						Estuaries					
	1990	1992	1994	1996	1998[b]	2000[c]	1990	1992	1994	1996	1998[b]	2000[c]	1990	1992	1994	1996	1998[b]	2000[c]
	Percent of total water[d]																	
Water systems assessed	36	18	17	19	23	19	47	46	42	40	42	43	75	74	78	72	32	36
	Percent of assessed waters																	
Meeting designated uses[e]																		
Supporting	69	62	64	64	55	53	60	56	63	61	46	47	67	68	63	62	47	45
Partially supporting[f]	21	25	22	36	10	8	19	35	28		9	8	25	23	27		10	<4
Not supporting	10	13	14		35	39	21	9	9	39	45	45	8	9	9	36	44	51
Clean Water Act goals: Fishable																		
Meeting	80	66	69	68	87	61	70	69	69	69	54	61	77	78	70	69	63	47
Not meeting	19	34	31	31	13	39	30	31	31	31	46	39	23	22	30	30	37	53
Not attainable	1	—	—	—	0	—	0	—	—	—	—	—	—	0	0	0	0	—
Clean Water Act goals: Swimmable																		
Meeting	75	71	77	79	69	68	82	77	81	75	69	70	88	83	85	84	88	80
Not meeting	15	20	23	20	27	32	18	22	19	25	31	30	12	17	15	16	12	20
Not attainable	10	9	—	—	5	—	—	—	—	—	—	—	—	0	—	—	0	—

Note: —, less than 1 percent of assessed waters.

[a] Excluding Great Lakes.

[b] United States Environmental Protection Agency, 2000, National Water Quality Inventory: 1998 Report to Congress.

[c] United States Environmental Protection Agency, 2002, National Water Quality Inventory 2000 Report, EPA-841-R-02-001.

[d] Miles of rivers, acres of lakes, square miles of estuaries.

[e] Supporting—water quality meets designated use criteria; partially supporting—water quality fails to meet designated use criteria at times; not supporting—water quality frequently fails to meet designated use criteria.

[f] In 1996, the categories "Partially supporting" and "Not supporting" were combined.

Source: From Ribaudo, M.O., Horan, R.D., and Smith, M.E., 1999, *Economics of Water Quality Protection from Nonpoint Sources: Theory and Practice, Resource Economic Division, Economic Research Service*, U.S. Department of Agriculture, Agricultural Economic Report No. 782 amended with 1998 and 2000 data, www.ers.usda.gov.

Table 10B.9 Ambient Water Quality in United States Rivers and Streams: Violation Rates, 1975–1997

Year	Fecal Coliform Bacteria	Dissolved Oxygen	Total Phosphorus	Total Cadmium, Dissolved	Total Lead, Dissolved
	Percent of all measurements exceeding national water quality criteria				
1975	36	5	5	*	*
1976	32	6	5	*	*
1977	34	11	5	*	*
1978	35	5	5	*	*
1979	34	4	3	4	13
1980	31	5	4	1	5
1981	30	4	4	1	3
1982	33	5	3	1	2
1983	34	4	3	1	5
1984	30	3	4	<1	<1
1985	28	3	3	<1	<1
1986	24	3	3	<1	<1
1987	23	2	3	<1	<1
1988	22	2	4	<1	<1
1989	30	3	2	<1	<1
1990	26	2	3	<1	<1
1991	15	2	2	<1	<1
1992	28	2	2	<1	<1
1993	31	<1	2	na	na
1994	28	2	2	na	na
1995	35	1	4	na	na
1996	na	1	1	<1	<1
1997	na	1	2	<1	<1

Notes: *Base figure too small to meet statistical standards for reliability of derived figures. na, not available. Violation levels are based on the following U.S. Environmental Protection Agency water quality criteria: fecal coliform bacteria — above 200 cells per 100 ml; dissolved oxygen — below 5 milligrams per liter; total phosphorus — above 1.0 milligrams per liter; cadmium, dissolved — above 10 micrograms per liter; and total lead, dissolved — above 50 micrograms per liter.

Source: From The 1997 Annual Report of the Council of Environmental Quality, www.whitehouse.gov/CEQ.

Original Source: From U.S. Geological survey, national-level data, unpublished, Reston, VA, 1999.

Table 10B.10 Major Pollutants Causing Impairment in 1998 in the United States by State

State	Impaired	Sedimentation	Nutrients	Pathogens	Toxics/ Metals/ Inorganics	Toxics/ Organics	Mercury	Pesticides	Other
Alabama	miles	576	385	587	280	281	70	98	1,642
	acres	4,084	70,606	3,628	529	68,083	559	10,147	140,041
Arkansas	miles	910	213	218	45	79	555	0	252
	acres	3,045	4,924	35	4,413	0	25,853	0	1,816
Arizona	miles	959	80	194	304	0	0	244	346
	acres	3,973	241	19,336	3,026	0	3,053	212	5,877
California	miles	5,768	1,086	755	2,464	306	303	2,773	5,717
	acres	151,677	470,153	18,027	709,129	332,963	862,749	818,002	1,450,270
Colorado	miles	233	42	90	1,355	0	29	0	442
	acres	160	0	119	9,100	0	1,561	144	7,928
Connecticut	miles	101	525	1,237	625	615	53	380	2,777
	acres	772	7,577	20,554	6,723	18,532	1,019	4,921	29,010
District of Columbia	miles	0	0	83	59	98	0	0	43
	acres	0	0	2,998	1,250	2,998	0	0	2,125
Delaware	miles	5	775	760	28	76	0	0	1,234
	acres	0	2,334	3,190	0	1,056	0	0	871
Florida	miles	1,219	4,007	2,826	1,045	16	1,445	0	5,616
	acres	33,248	332,502	82,704	179,598	135	226,025	0	333,798
Georgia	miles	149	16	2,649	1,918	1	250	76	5,936
	acres	132	0	45,411	123,012	0	161,96	225	229,830
Hawaii	miles	212	183	10	10	0	0	0	13
	acres	7	7	0	0	0	0	0	0
Iowa	miles	175	71	565	1,447	161	61	94	3,204
	acres	18,535	10,858	27,113	22,228	97,042	5	9,072	79,113
Idaho	miles	6,228	2,653	1,539	503	105	0	134	7,554
	acres	84,145	110,854	5,184	28,942	0	0	4,850	229,183
Illinois	miles	4,001	4,061	204	1,514	582	62	36	4,017
	acres	269,323	270,352	7,825	101,836	95,491	5	12,881	375,681
Indiana	miles	0	78	1,251	673	3,194	2,431	93	1,115
	acres	0	23	47,604	4,695	89,953	75,516	442	4,511
Kansas	miles	317	1,165	9,994	3,635	0	0	435	8,944
	acres	38,291	64,255	19,739	15,589	0	0	6,446	39,851
Kentucky	miles	469	521	1,289	83	656	6	17	835
	acres	1,236	6,885	20,964	536	134,087	0	0	3,276
Louisiana	miles	1,022	1,566	3,390	3,242	592	4,911	697	4,862
	acres	94,235	541,140	590,269	581,092	30,201	804,207	74,589	730,729
Maine	miles	6	147	682	15	185	0	0	987
	acres	0	33,690	52,291	0	8,487	0	0	85,533
Maryland	miles	2,610	5,053	1,138	530	7	26	32	467
	acres	442,854	904,112	121,256	86,927	3,333	13,134	16,916	21,958
Massachusetts	miles	208	332	945	376	363	0	4	1,908
	acres	4,253	4,874	138	897	934	0	150	29,454

Michigan	miles	17	253	612	39	604	254	195	1,818
	acres	0	2,043	19,799	27	333,258	362,042	314,084	792
Minnesota	miles	603	0	703	135	0	160	0	513
	acres	6,808	0	4,528	2,380	0	10,589	0	3,714
Mississippi	miles	72,256	72,241	27,189	1,595	777	358	67,633	1,818
	acres	452,878	447,838	80,919	3,934	4,576	9,396	436,763	792
Missouri	miles	1,120	228	90	479	0	0	67	1,535
	acres	17	2,249	0	47,396	0	0	32,167	336,693
Montana	miles	15,015	6,703	895	6,798	139	0	0	31,731
	acres	550,491	501,586	76,292	338,591	3,342	0	0	1,396,433
Nebraska	miles	54	93	1,987	1,190	0	0	894	263
	acres	3,266	17,979	83,978	26,612	0	0	29,692	1,677
Nevada	miles	587	702	0	924	0	384	0	589
	acres	15,617	51,693	0	27,527	0	14,504	0	14,504
New Hampshire	miles	0	19	678	795	445	0	0	503
	acres	0	750	8,421	26,189	25,166	0	0	5,947
New Jersey	miles	8	518	749	952	106	928	20	7,882
	acres	4,154	7,773	4,007	10,292	1,941	37,219	74	7,550
New Mexico	miles	638	703	154	628	13	185	12	3,675
	acres	995	172	1,460	53	0	63,658	0	5,589
New York	miles	25	99	602	5	920	0	0	733
	acres	1,147	31,369	19,291	24,332	405,342	7,337	13,747	111,143
North Carolina	miles	1,161	64	746	117	126	378	14	1,813
	acres	5,329	19,890	7,499	8,386	25,151	16,685	0	18,035
North Dakota	miles	2,604	2,616	2,691	356	0	0	0	4,327
	acres	11,904	58,964	6,595	6,277	0	0	0	18,714
Ohio	miles	3,563	1,347	321	2,020	885	0	441	11,456
	acres	44,234	50,907	3,250	21,697	14,358	0	11,154	105,140
Oklahoma	miles	4,791	4,066	524	1,531	306	0	4,295	6,614
	acres	445,672	269,714	77,059	150,373	47,244	0	293,999	500,257
Oregon	miles	1,446	598	2,565	308	199	602	317	19,611
	acres	8,207	82,975	8,190	0	0	17,453	0	149,335
Pennsylvania	miles	2,191	920	49	3,190	572	0	462	2,278
	acres	4,353	3,118	0	38,139	48,213	1	33,403	5,677
Puerto Rico	miles	0	1	0	2	0	0	0	210
	acres	0	71	0	0	0	0	0	873
Rhode Island	miles	14	150	192	153	3	3	0	146
	acres	358	1,390	1,857	1,342	46	46	0	1,145
South Carolina	miles	0	10	2,674	230	307	28,527	10	4,908
	acres	0	30,799	122,128	22,000	71,102	209,663	0	88,157
South Dakota	miles	1,828	336	1,049	521	0	0	0	723
	acres	29,647	66,074	9,561	304	0	0	0	43,966
Tennessee	miles	12,309	3,285	4,149	1,608	2,042	152	845	10,072
	acres	212,709	29,289	9,317	7,437	254,155	23,505	179,553	49,513
Texas	miles	0	132	5,463	952	575	927	433	3,648

(Continued)

Table 10B.10 (Continued)

State	Impaired	Sedimentation	Nutrients	Pathogens	Toxics/ Metals/ Inorganics	Toxics/ Organics	Mercury	Pesticides	Other
	acres	0	131	201,824	27,823	9,961	374,434	67,337	399,673
Utah	miles	1,194	708	402	823	414	17	0	5,417
	acres	4,879	109,589	5,215	7,465	5,215	0	0	152,529
Vermont	miles	102	133	237	42	36	72	0	314
	acres	907	13,755	349	228	6,120	10,689	0	6,248
Virginia	miles	0	45	1,691	2,069	101	147	104	4,171
	acres	0	18	16,125	3,086	2,434	21	4,495	5,547
Virgin Islands	miles	3	0	1	0	0	0	0	13
	acres	0	0	0	0	0	0	0	0
Washington	miles	18	1	393	63	28	14	29	812
	acres	0	5,632	1,004	1,357	149	28	6,736	148
West Virginia	miles	0	0	174	5,276	384	0	278	2,633
	acres	2,849	1,899	0	82,124	50,907	0	44,534	2,947
Wisconsin	miles	787	231	49	35	585	661	0	2,050
	acres	41,935	4,460	0	0	149,226	239,092	0	53,827
Wyoming	miles	238	47	517	580	27	0	0	250
	acres	16	215	3,862	5,600	0	0	0	12

Note: Miles include river and shoreline miles, including the Great Lakes and estuarine shorelines. Acres include lake acres, excluding the Great Lakes.

Source: From United States Environmental Protection Agency, *Total Maximum Daily Loads, National Overview, Major Pollutants Causing Impairment by State*, Last updated on Wednesday, February 16th, 2005, www.epa.gov/owow/tmdl/status.html.

Table 10B.11 Top 100 Impairments of Waters in the United States Reported in 1998 for Section 303(d) of the Clean Water Act

General Impairment Name	Impairments Reported	Percent of Reported
Metals	11,538	19.22
Pathogens	7,902	13.16
Nutrients	5,586	9.31
Sediment/siltation	5,047	8.41
Organic enrichment/low DO	4,405	7.34
Fish consumption advis.	3,178	5.29
pH	2,906	4.84
Other habitat alterations	2,389	3.98
Thermal modifications	2,200	3.67
Biological criteria	2,116	3.53
Flow alteration	1,576	2.63
Pesticides	1,468	2.45
Turbidity	1,140	1.90
Salinity/TDS/chlorides	994	1.66
Suspended solids	969	1.61
Cause unknown	896	1.49
PCBs	884	1.47
Unionized ammonia	797	1.33
Priority organics	681	1.13
Sulfates	621	1.03
Algal growth/chlorophylla	476	0.79
Noxious aquatic plants	351	0.58
Oil and grease	286	0.48
Unknown toxicity	282	0.47
Other cause	221	0.37
General WQS (benthic)	195	0.32
Dioxins	184	0.31
Other inorganics	144	0.24
Chlorine	98	0.16
Nonpriority organics	97	0.16
Taste and odor	85	0.14
Stream bottom deposits	75	0.12
Total toxics	48	0.08
Cyanide	47	0.08
Biodiversity impacts	47	0.08
Exotic species	42	0.07
Radiation	22	0.04
Natural limits (wetlands)	13	0.02
Fish kill(s)	12	0.02
Hydrogen sulfide	4	0.01
Ambient water quality criteria	2	0.00
Low nutrients	1	0.00
RDX—Hexahydro-1,3,5-trinitro-1,3,5-triazine	1	0.00
1,2-Diphenylhdrazine	1	0.00

Note: Total impairments reported nationwide: 60,027.

Source: From United States Environmental Protection Agency, *National Section 303(d) List Fact Sheet*, Downloaded 8/28/05, www.epa.gov/owow/tmdl/status.html.

Table 10B.12 Probable Sources of Water-Quality Problems in the Nation's Streams in 1982

Probable Source	Stream Miles	Percentage
Total nonpoint source contribution	367,244	38.4
Agricultural sources	281,241	29.5
Natural sources	212,389	22.2
Total point source contribution	117,684	12.3
Silviculture/logging	71,736	7.5
Municipal point sources	63,816	6.7
Feed lots	59,947	6.3
Individual sewage disposal	47,823	5.0
Industrial point sources	47,097	4.9
Urban runoff	40,376	4.2
Mining (nonpoint)	31,847	3.3
Combined sewers	29,246	3.1
Construction activity	29,110	3.1
Mining (point)	28,686	3.0
Grazing	21,970	2.3
Other	19,445	2.0
Dam releases	19,314	2.0
Landfill leachate	5,504	0.6
Bedload movement	5,299	0.6
Roads	3,569	0.4

Note: Expressed in total stream miles and as percentages of total miles.

Source: From Judy, R.D., and others, 1984, *1982 National Fisheries Survey*, U.S. Fish and Wildlife Service, FWS/OBS-84/06.

Table 10B.13 Sources of Water-Quantity Problems Adversely Affecting the Nation's Streams in 1982

Source	Stream Miles	Percentage
Natural conditions	477,791	50.1
Diversions (agricultural)	130,223	13.6
Dam(s) (water storage)	32,901	3.5
Dam(s) (flood control)	28,002	2.9
Dam(s) (power)	24,821	2.6
Other	18,851	2.0
Diversions (municipal)	10,694	1.1
Channelization	10,629	1.1
Floods/low flows	10,527	1.1
Irrigation	8,897	0.9
Logging	6,271	0.7
Ditches	5,335	0.6
Diversions (industrial)	3,292	0.3

Note: Expressed in total stream miles and as percentages of total miles.

Source: From Judy, R.D., and others, 1984. *1982 National Fisheries Survey*, U.S. Fish and Wildlife Service, FWS/OBS-84/06.

Table 10B.14 Leading Sources Impairing Assessed Rivers and Streams in the United States in 2000

Sources	Stream Miles	Percent of Impaired	Percent of Assessed
Agriculture	128,859	47.9	18.4
Hydromodification	53,850	20.0	7.7
Crop-related sources	53,067	19.7	7.6
Grazing related sources	43,469	16.1	6.2
Source unknown	39,056	14.5	5.6
Habitat modification (other than hydromodification)	37,654	14.0	5.4
Urban runoff/storm sewers	34,871	13.0	5.0
Natural sources	31,033	11.5	4.4
Silviculture	28,156	10.5	4.0
Municipal point sources	27,988	10.4	4.0
Resource extraction	27,695	10.3	4.0
Nonirrigated crop production	26,830	10.0	3.8
Intensive animal feeding operations	24,616	9.1	3.5
Channelization	23,795	8.8	3.4
Bank or shoreline modification/destabilization	18,040	6.7	2.6
Removal of riparian vegetation	17,912	6.7	2.6
Land disposal	17,821	6.6	2.5
Irrigated crop production	17,667	6.6	2.5
Erosion and sedimentation	16,137	6.0	2.3
Unspecified nonpoint source	15,988	5.9	2.3

Source: From United States Environmental Protection Agency, 2002, *National Water Quality Inventory 2000 Report*, EPA-841-R-02-001/, www.epa.gov.

Table 10B.15 Heat Generated and Discharged to the Nation's Fresh and Saline Surface Waters, 1975–2000

	1975		1985		2000	
	Btus $\times 10^{15}$	Percent of Total Discharged	Btus $\times 10^{15}$	Percent of Total Discharged	Btus $\times 10^{15}$	Percent of Total Discharged
Electric power generation						
Heat generated	11.0	—	24.3	—	57.1	—
Heat discharged to water	5.7	66	7.8	98	7.4	98
(Fresh)	(3.9)	(45)	(3.9)	(49)	(2.8)	(37)
(Saline)	(1.8)	(21)	(3.9)	(49)	(4.6)	(61)
Manufacturing						
Heat discharged to water	3.0	34	0.2	2	0.2	2
(Fresh)	(2.2)	(73)	(0.2)	(2)	(0.2)	(2)
(Saline)	(0.8)	(27)	(0)	(0)	(0)	(0)
Total heat discharged	8.7	100	8.0	100	7.6	100
(Fresh)	(6.1)	(70)	(4.1)	(51)	(3.0)	(39)
(Saline)	(2.6)	(30)	(3.9)	(49)	(4.6)	(61)

Source: From U.S. Water Resources Council, 1978, *The Nation's Water Resources 1975–2000, Second National Water Assessment.*

Table 10B.16 Point Source Loadings to Receiving Waters in the United States, mid 1980s

Industry	TSS	BOD	Phosphorus	Metals
Minerals & metals	0.355	0.006	0.255	4.931
Chemical & manufacturing	0.086	0.125	0.267	2.919
Agriculture & fisheries	0.277	0.404	92.800	0.000
POTWs	1.594	1.830	49.555	2.838
Total	2.312	2.365	142.877	10.688

Note: Millions of tons per year for TSS, BOD, and phosphorus and millions of pounds per year for metals; mid1980s. Minerals & metals includes aluminum forming, coal mining, copper forming, foundries, iron and steel, metal finishing, nonferrous metals mining, nonferrous metals forming, ore mining and petroleum refining industries. Chemical & manufacturing includes battery manufacturing, coil coating, electrical and electronic components, organic and inorganic chemicals, plastics, synthetic fibers, pesticide manufacturing, pharmaceuticals manufacturing, plastics molding and forming, porcelain enameling, leather tanning, pulp and paper, and textile industries. Agriculture & fisheries includes animal feedlots, fish hatcheries, food and beverages, fruits and vegetables, meat packing, and seafood industries. POTWs, publicly owned treatment works. TSS, total suspended solids, BOD, biochemical oxygen demand. Metals, cadmium, copper, lead, mercury and zinc. Industrial loadings are direct discharges based on long-term average concentrations and total industry flows at Best Available Technology (BAT) presented in U.S. EPA (1983), except as follows: loadings for electrical and electronic components reflect current level of treatment (U.S. EPA, 1983); conventional loadings for agriculture & fisheries industries represent post-BAT treatment levels; and conventional and toxic pollutant loadings for POTWs include indirect industrial and residential/commercial loadings not removed by the POTWs. Phosphorus loadings for POTWs represent effluent levels of 5 mg/L.

Source: From U.S. Environmental Protection Agency, *Effluent Technology Division*, unpublished data. 1979. Washington, DC; U.S. Geological Survey, National Water Summary 1986.

Table 10B.17 Point Source Discharges to Water in the United States, 1977 (by Sector)

Total Suspended Solids		Total Dissolved Solids		Biochemical Oxygen Demand	
Sector	**Million Pounds Per Year**	**Sector**	**Million Pounds Per Year**	**Sector**	**Million Pounds Per Year**
Municipal sewage plants	3,850.0	Organic chemicals	36,540.4	Municipal sewage plants	3,800.0
Powerplants	1,165.7	Municipal sewage plants	30,255.2	Pulp & paper mills	530.2
Pulp & paper mills	781.8	Powerplants	18,418.1	Organic chemicals	107.6
Feedlots	422.0	Pulp & paper mills	16,825.8	Feedlots	95.9
Iron & steel mills	254.3	Misc. chemicals	8,176.4	Seafoods	86.9
Organic chemicals	144.0	Misc. food & beverages	7,420.2	Misc. food & beverages	54.8
Misc. food & beverages	91.9	Oil & gas extraction	6,077.0	Cane sugar mills	50.4
Textiles	61.7	Petroleum refining	2,389.8	Iron & steel mills	37.8
Mineral mining	52.7	Coal mining	1,328.7	Misc. chemicals	35.2
Seafoods	50.0	Iron & steel mills	1,324.0	Textiles	24.8
Total, top 10 sectors	**6,874.1**	**Total, top 10 sectors**	**128,755.6**	**Total, top 10 sectors**	**4,823.6**
Total, all sectors	**13,746.0**	**Total, all sectors**	**170,759.0**	**Total, all sectors**	**6,944.0**
Top 10 sectors as percent of all sectors	**50%**	**Top 10 sectors as percent of all sectors**	**75%**	**Top 10 sectors as percent of all sectors**	69%

Nitrogen		Phosphorus		Dissolved Heavy Metals	
Sector	**Million Pounds Per Year**	**Sector**	**Million Pounds Per Year**	**Sector**	**Million Pounds Per Year**
Municipal sewage plants	813.5	Municipal sewage plants	73.9	Powerplants	24.4
Pharmaceuticals	87.6	Feedlots	21.8	Municipal sewage plants	9.3
Organic chemicals	41.1	Misc. food & beverages	4.7	Iron & steel mills	7.6
Feedlots	39.9	Meat packing	3.4	Petroleum refining	6.0
Meat packing	36.0	Laundries	3.3	Organic chemicals	3.6
Petroleum refining	15.5	Fertilizers	2.6	Ore mining	2.5
Misc. food & beverages	12.3	Petroleum refining	1.5	Electroplating	0.5
Seafoods	9.5	Seafoods	1.4	Machinery	0.5
Pesticides	8.9	Organic chemicals	1.4	Oil & gas extraction	0.4
Leather tanning	7.1	Poultry	1.2	Foundries	0.1
Total, top 10 sectors	**1,071.4**	**Total, top 10 sectors**	**115.2**	**Total, top 10 sectors**	**54.9**
Total, all sectors	**1,237.0**	**Total, all sectors**	**191.0**	**Total, all sectors**	**59.0**
Top 10 sectors as percent of all sectors	**87%**	**Top 10 sectors as percent of all sectors**	**60%**	**Top 10 sectors as percent of all sectors**	**93%**

Source: From Council on Environmental Quality, 1981, *Environmental Trends.*

Table 10B.18 Nonpoint Source Contributions to Receiving Waters in the United States

Source	TSS	BOD	Nitrogen	Phosphorus
Cropland	1,870.00	9.00	4.30	1.56
Pasture and rangeland	1,220.00	5.00	2.50	1.08
Forest land	256.00	0.80	0.39	0.09
Construction sites	197	NA	NA	NA
Mining sites	59	NA	NA	NA
Urban runoff	20	0.5	0.15	0.019
Rural roadways	2	0.004	0.0005	0.001
Small feedlots	2	0.05	0.17	0.032
Landfills	NA	0.3	0.026	NA
Background	1,260.00	5.00	2.50	1.10
Total	**4,886.00**	**20.35**	**10.04**	**3.88**

Note: By source; millions of tons per year; 1980. TSS, total suspended solids. BOD, biochemical oxygen demand. Excluded from the survey area are 207 million acres in public land (14% of the contiguous United States), mostly in the Rocky Mountains, because of inadequacy of information. Urban runoff includes storm water sewers only.

Source: From U.S. Environmental Protection Agency, Office of Water Regulations and Standards, Nonpoint Source Control Section. 1986. Estimated pollutant contributions to source waters from selected nonpoint sources in the contiguous 48 states (1980). Washington, DC; U.S. Geological survey, National Water Summary 1986.

Table 10B.19 United States Water Quality Conditions by Type of Waterbody: 2000

Item	Rivers and Streams (mi)	Lakes, Reservoirs, and Ponds (acres)	Estuaries (mi^2)	Great Lakes Shoreline (mi)	Ocean Shoreline (mi)
Total size	3,692,830	40,603,893	87,369	5,521	58,618
Amount accessed[a]	699,946	17,339,080	31,072	5,066	3,221
Percent of total size	19	43	36	92	6
Amount accessed as—					
Good[b]	463,441	8,026,988	13,850	—	2,176
Good but threatened[c]	85,544	1,343,903	1,023	1,095	193
Polluted[d]	291,264	7,702,370	15,676	3,955	434
Percent of accessed as—					
Good[b]	53	47	45	—	79
Good but threatened[c]	8	8	4	22	7
Polluted[d]	39	45	51	78	14
Amount impaired by leading sources of pollution[e]					
Agriculture	128,859	3,158,393	2,811	75	(NA)
Atmospheric deposition	(NA)	983,936	3,692	71	(NA)
Construction	(NA)	(NA)	(NA)	(NA)	29
Contaminated sediments	(NA)	(NA)	(NA)	519	(NA)
Forestry	28,156	(NA)	(NA)	(NA)	(NA)
Habitat modification	37,654	(NA)	(NA)	62	(NA)
Hydrologic modification	53,850	1,413,624	2,171	(NA)	(NA)
Industrial discharges/point sources	(NA)	(NA)	4,116	(NA)	76
Land disposal of wastes	(NA)	856,586	(NA)	61	123
Municipal point sources	27,988	943,715	5,779	(NA)	89
Nonpoint sources	(NA)	1,045,036	(NA)	(NA)	142
Resource extraction	27,695	(NA)	1,913	(NA)	(NA)
Septic tanks	(NA)	(NA)	(NA)	61	103
Urban runoff and storm sewers	34,871	13,699,327	5,045	152	241

Note: Section 305(b) of the Clean Water Act requires states and other jurisdiction to assess the health of their waters and the extent to which their waters support quality standards. Section 305(b) requires that states submit reports describing water quality conditions to the Environment Protection Agency every two years. Water quality standard have three elements (designated uses, criteria developed to protect each use, and an antidegradation policy). For information on survey methodology and assessment criteria, see report. —, Representation zero. NA, Not available.

[a] Includes waterbodies accessed as not attainable for one or more uses. Most states do not assess all their waterbodies during the 2-year reporting cycle, but use a "rotating basin approach" whereby all waters are monitored over a set period of time.

[b] Based on accessment of available data, water quality supports all designated uses. Water quality meets narrative and/or numberic criteria adopted to protect and support a designated use.

[c] Although all assessed uses are currently met, data show a declining trend in water quality. Projections based on this trend indicate water quality will be impaired in the future, unless action is taken to prevent further degradation.

[d] Impaired or not attainable. The reporting state or jurisdiction has performed a "use-attainability analysis" and demonstrated that support of one or more designated beneficial uses is not attainable due to specific biological, chemical, physical, or economic/social conditions.

[e] Excludes unknown and natural sources.

Source: From U.S. Census Bureau, *Statistical Abstract of the United States: 2004–2005.*

Original Source: From U.S. Environmental Protection Agency, *National Water Quality Inventory: 2000 Report,* EPA-841-R-02-001, August 2002, www.epa.gov/305b/2000report.

Table 10B.20 Summary of Individual Use Support for Rivers, Lakes, Estuaries, Ocean Shoreline Waters and Great Lakes in the United States in 2000

		Good		Impaired		
	Area Assessed	Full Support (%)	Threatened (%)	Partial Support (%)	Not Supporting (%)	Not Attainable
Rivers and Streams	**Miles**					
Aquatic life support	616,860	55.4	10.1	18.7	15.5	0.3
Fish consumption	205,153	60.9	1.2	15.2	22.8	0.0
Primary contact-swimming	313,832	67.8	3.4	11.9	16.3	0.5
Secondary contact	219,776	73.5	2.3	11.1	13.1	0.0
Drinking water supply	153,155	83.5	2.8	7.6	6.1	0.1
Agriculture	274,736	90.7	0.9	3.8	4.6	0.0
Lakes, Reservoirs and Ponds	**Acres**					
Aquatic life support	11,224,279	60.9	10.3	19.0	9.8	0.0
Fish consumption	8,566,710	60.6	4.4	21.4	13.6	0.0
Primary contact-swimming	12,662,298	70.2	6.5	16.7	6.6	0.0
Secondary contact	5,855,176	73.5	6.9	13.9	5.6	0.0
Drinking water supply	7,244,575	78.5	4.9	10.4	6.2	0.0
Agriculture	4,653,670	84.1	3.0	5.3	7.6	0.0
Estuaries	**Square miles**					
Aquatic life support	22,047	43.7	4.7	37.1	14.5	0.0
Fish consumption	12,940	46.6	5.0	45.3	3.1	0.0
Shellfishing	20,967	75.7	0.6	10.5	13.2	0.0
Primary contact-swimming	21,169	80.0	4.7	14.3	1.0	0.0
Secondary contact	9,524	72.8	4.2	22.2	0.9	0.0
Ocean Shoreline Waters	**Shoreline Miles**					
Aquatic life support	2,079	86.1	7.7	2.2	4.1	0.0
Fish consumption	1,136	87.3	3.9	5.5	3.3	0.0
Shellfishing	1,420	86.4	0.0	11.6	2.0	0.0
Primary contact-swimming	2,521	76.8	8.6	6.6	8.0	0.0
Secondary contact	1,925	87.3	4.2	4.4	4.2	0.0
Great Lakes	**Shoreline Miles**					
Aquatic life support	1,343	3.2	78.6	16.4	1.8	0.0
Fish consumption	4,976	0.0	0.0	33.4	66.6	0.0
Primary contact-swimming	3,663	89.9	7.4	2.5	0.2	0.0
Secondary contact	3,256	99.8	0.0	0.2	0.0	0.0
Drinking water supply	3,313	97.6	0.0	0.0	2.4	0.0
Agriculture	3,250	100.0	0.0	0.0	0.0	0.0

Note: Water-quality conditions: Good, Fully supporting of all of their uses or fully supporting but threatened for one or more uses; Impaired, Partially or not supporting one or more uses; Not Attainable, Not able to support one or more uses. Use Support Levels: Supporting, Fully supporting of all uses; Threatened, Fully supporting all uses but threatened for one or more uses; Partially Supporting, Partially supporting one or more uses; Not Supporting, Not able to support one or more uses.

Source: Abstracted from United States Environmental Protection Agency, 2002, *National Water Quality Inventory 2000 Report*, EPA-841-R-02-001, www.epa.gov.

Table 10B.21 Leading Sources Impairing Assessed Rivers in the United States in 2000 (Expressed in Stream Miles)

	State by EPA Region	Assessed Miles	Miles Impaired	Miles Alloted to Each Source								
				Agriculture	Hydro-modification	Habitat Modification (Other than Hydro-modification)	Urban Runoff/Storm Sewers	Natural Sources	Forestry/ Silviculture	Municipal Point Sources	Resource Extraction	Land Disposal
1	Connecticut	1,207	389	120.5	177.45	9	188.6	7.5	—	194.3	27.3	40.6
	Maine	31,752	729	174.8	—	23.6	101.5	—	1	93.5	1	35.5
	Massachusetts	1,496	969	39.6	146.1	29.9	361.1	—	—	221.35	1.7	83.1
	New Hampshire	2,677	444	59	11.1	11.5	13.4	8	0	11.9	0	7.4
	Rhode Island	649	217	35.07	27.62	7.99	151.1	29.32		35.99	6.53	60.28
	Vermont	5,462	1,169	688.8	418.3	771.6	367.5	373.6	45.4	202.9	67.1	335.4
2	New Jersey	330	209	—	—	—	—	—	—	—	—	—
	New York	2,914	1,081	1,485.6	589.1	1,340.2	944	—	100	545.9	268	1,467.3
3	Delaware	2,506	1,765	984.03	—	—	304.07	423.51	—	118.42	—	213.44
	Delaware river basin	206	206	206	—	—	206	2	—	206	—	—
	District of Columbia	38	38	—	8.8	17.6	38.4	8.3	—	0.9	—	9.8
	Maryland	8,617	3,212	351.82	1,136.33	1,006.02	650.68	1,104.17	—	181.23	56.69	9.29
	Pennsylvania	35,496	7,261	2,426.95	186.17	590.22	1,526.45	43.28	3.43	258.54	2,728.57	117.02
	Virginia	9,190	4,466	1,105.35	18.27	44.8	766.49	1,054.72	—	162.85	140.95	46.35
	West Virginia	11,550	5,313	1,760.29	346.75	392.74	1,157.03	14.87	1,311.83	1,016.42	2,706.79	1,208.95
4	Alabama	2,628	1,930	111.2	—	—	282.6	12	—	151.3	—	—
	Florida	10,159	3,147	1,833.4	1,189.6	—	1,126.6	—	411.1	205.8	465.7	916.3
	Georgia	9,996	5,986	0	65	—	1,925	433	0	203	0	0
	Kentucky	9,923	3,688	1,133.2	172.2	235.3	1,053.7	21.3	100.9	609.9	705.7	1,308.5
	Mississippi	14,972	10,824	10,471.5	224.3	177.82	634.07	235.4	707.32	566.7	151.91	415.15
	North Carolina	37,662	2,143	1,201.71	166	8.7	900.15	37	126.6	398.58	11.88	38.4
	South Carolina	15,405	4,011	1,462.35	35.55	—	2,862.6	—	—	679.09	23.49	219.4
	Tennessee	24,326	7,538	3,886.6	2,672.6	425	1,030.8	—	14.9	451.9	602.7	332.5
5	Illinois	15,587	7,844	4,395.06	2,613.12	795.31	1,020.47	137.19	—	1,640.98	1,047.79	37.51
	Indiana	17,541	4,230	124.84	81.91	38.35	113.25	0.3	—	71.34	—	71.05
	Michigan	13,117	2,456	1,059	248	108	344	30	—	423	24	9
	Minnesota	11,403	7,900	6,601.2	3,889.7	—	2,889	35.6	279.5	524.3	591.4	3,630
	Ohio	8,232	3,743	1,820.49	2,887.33	47.22	556.96	301.41	11.5	978.19	720.22	450.74
	Ohio river valley	981	—	981	—	—	—	—	—	981	—	—
	Wisconsin	23,530	10,029	3,539.9	2,693.25	1,483.05	989.7	1,082.5	89.3	1,169	153.9	51.3
6	Arkansas	8,112	1,177	705.5	—	—	13.5	—	—	97.9	24	—
	Louisiana	7,359	6,575	2,021	810	121	839	2,364	286	1,798	383	1,501
	New Mexico	4,284	2,675	2,531.85	376.7	2,103.35	97.1	422.7	196	262.8	596.2	149.4
	Oklahoma	14,071	7,647	4,481.18	867.38	1,423.18	779.58	86.26	197.5	190.49	1,157.31	847.63
	Texas	15,101	4,548	545	—	—	796	212.1	—	1,398.2	44	33.3
7	Iowa	6,390	1,903	1,018.04	790.99	399.29	45.05	98.63	—	105.43	17.14	24
	Kansas	18,236	14,819	13,128.31	1,171.84	5,911.66	1,295.94	6,148.69	—	4,882.77	2,107.05	575.59

(Continued)

Table 10B.21 (Continued)

				Miles Alloted to Each Source								
	State by EPA Region	Assessed Miles	Miles Impaired	Agriculture	Hydro-modification	Habitat Modification (Other than Hydro-modification)	Urban Runoff/Storm Sewers	Natural Sources	Forestry/ Silviculture	Municipal Point Sources	Resource Extraction	Land Disposal
	Missouri	21,615	10,321	7,624.4	3,758.9	21	44.5	154.5	—	92.8	179.7	0.3
	Nebraska	6,500	3,759	—	—	—	—	—	—	—	—	—
8	Colorado	41,837	1,244	123.3	4.7	—	244.16	474.05	10.52	145.1	757.46	—
	Montana	11,443	8,576	5,833.6	3,620.4	2,093.7	159	656.3	810.7	371.9	2,534.3	172.7
	North Dakota	14,965	7,224	6,982.77	2,621.39	2,458.94	501.38	509.75	—	556.7	489.06	97.25
	South Dakota	3,564	1,778	1,623.5	—	200	48.3	1,172.1	26.2	—	2.1	—
	Utah	10,519	2,825	2,298.25	887.8	960.41	85.79	1,377.24	—	125.39	205.16	—
	Wyoming	2,955	452	101.1	—	—	20.88	45.18	—	11.7	17.45	—
9	Arizona	4,052	986	538.85		30.29	120.22	804.87	—	51.27	301	54.19
	California	25,269	20,949	17,064.33	9,217.46	11,662.3	2,204.68	4,906.15	14,140.2	3,266.88	6,838.16	667.86
	Hawaii	3,904	2,737	1,553.45	958.84	573.78	1,566.25	1,820.74	29.8	18.7	20	1
	Nevada	1,564	953	593.7	247.8	—	61.75	589.33	—	—	—	—
10	Alaska	1,421	518	—	13	21	109.8	—	16.5	—	309.3	108.6
	Idaho	17,333	8,230	—	—	—	—	—	—	—	—	—
	Oregon	53,735	13,937	—	1,624	2,103	505	—	7,707	—	—	—
	Washington	70,439	37,722	11,316.51	6,789.906	—	2,263.3	3,772.17	1,508.86	2,263.3	1,131.65	377.217
	Jurisdictions											
	American Samoa	17	16	—	—	6.7	—	—	—	—	—	—
	Big Sandy Rancheria	0	—	—	—	—	—	—	—	—	—	—
	Guam	167	63	—	—	—	—	—	—	1.67		6.51
	Hoopa Valley Tribe	90	—	—	—	—	—	—	—	—	—	—
	La Posta Band	0	—	—	—	—	—	—	—	—	—	—
	N Mariana Islands	0	0	—	—	—	—	—	—	—	—	—
	Pauma Band	23	0	0	—	0	0	0	0	0	0	0
	Puerto Rico	5,394	4,653	699.6	70.1	0	556.4	0.5	0	43	63.5	2,079.9
	Round Valley Tribe	35	35	15.5	14	—	8	23	24	—	14.5	10
	Total	**699,946**	**269,258**	**128,859**	**53,850**	**37,654**	**34,871**	**31,033**	**28,156**	**27,988**	**27,695**	**17,821**
	Percent of impaired			**47.9%**	**20.0%**	**14.0%**	**13.0%**	**11.5%**	**10.5%**	**10.4%**	**10.3%**	**6.6%**
	Percent of assessed			**18.4%**	**7.7%**	**5.4%**	**5.0%**	**4.4%**	**4.0%**	**4.0%**	**4.0%**	**2.5%**

Note: Includes leading sources of River and Stream Impairment Shown on Fig. 10B.4 plus natural sources and land disposal; —, no data.

Source: Abstracted from United States Environmental Protection Agency, 2002, *National Water Quality Inventory 2000 Report*, EPA-841-R-02-001, www.epa.gov/305b/2000report.

Table 10B.22 Relative Impact of Pollution Sources in Rivers and Streams with Impaired Uses in the United States in 1986 (Percent)

State	Industrial	Municipal	Combined Sewers	Nonpoint Sources	Natural	Other/ Unknown
Alaska	85	1	0	12	0	2
Arizona	26	10	0	20	0	44
California	0	16	0	64	0	20
Connecticut	0	40	20	9	0	31
Delaware	8	6	8	59	19	0
Florida	25	29	0	40	2	4
Georgia	1	95	0	4	0	0
Idaho	2	3	0	78	17	0
Indiana	2	56	30	10	0	2
Iowa	0	3	0	97	0	0
Kansas	7	36	0	25	28	4
Kentucky	26	20	0	54	0	0
Louisiana	7	26	0	46	17	4
Maine	0	100	0	0	0	0
Maryland	5	30	0	50	15	0
Massachusetts	6	26	16	26	14	12
Minnesota	0	42	0	51	0	7
Mississippi	5	23	0	72	0	0
Missouri	0	1	0	99	0	0
Montana	2	3	0	95	0	0
Nebraska	1	7	0	92	0	0
New Hampshire	12	64	6	18	0	0
New Jersey	25	35	0	35	0	5
New Mexico	1	5	0	81	2	11
New York	20	40	13	11	0	16
North Carolina	12	17	0	71	0	0
Ohio	16	36	11	30	0	7
Oregon	3	10	0	57	30	0
Pennsylvania	7	13	1	71	3	5
Rhode Island	42	24	0	19	0	15
South Carolina	12	60	0	26	0	2
South Dakota	4	9	0	34	49	4
Tennessee	5	8	0	76	0	11
Texas	4	71	0	14	11	0
Vermont	11	22	0	50	11	6
Virginia	4	34	1	51	10	0
West Virginia	4	26	0	64	6	0
Wisconsin	1	1	0	98	0	0
Wyoming	10	4	0	43	43	0
Puerto Rico	11	21	0	63	0	5
Guam	5	10	15	50	20	0
Average (weighted)	9	17	1	65	6	2

Source: From U.S. Environmental Protection Agency, *National Water Quality Inventory*, 1986 Report to Congress.

Table 10B.23 River Miles Meeting the Fishable and Swimmable Goals of the Clean Water Act in the United States, by Jurisdiction, in 2000

	Fish Consumption			Swimming		
Jurisdiction	**Total Assessed**	**Full Support**	**Threatened**	**Total Assessed**	**Full Support**	**Threatened**
Alabama	0	—	—	0	—	—
Alaska	0	—	—	0	—	—
American Samoa	—	—	—	0	—	—
Arizona	3,798	3,632	—	3,714	3,457	—
Arkansas	8,112	7,739	—	7,629	7,596	—
Big Sandy Rancheria	0	—	—	0	—	—
California	11,842	1,638	715	17,896	2,151	1,343
Colorado	0	—	—	14,600	14,572	0
Connecticut	1,285	1,175	0	1,080	619	192
Delaware	67	54	0	2,506	99	0
Delaware River Basin	206	0	0	206	194	12
District of Columbia	24	0	0	38	2	0
Florida	3,771	1,002	0	9,524	5,998	552
Georgia	0	—	—	0	—	—
Guam	—	—	—	32	1	—
Hawaii	3,887	3,874	0	3,893	3,892	0
Hoopa Valley Tribe	90	—	90	90	90	0
Idaho	0	0	0	0	0	0
Illinois	3,969	2,920	0	2,944	742	0
Indiana	3,030	0	0	7,300	4,510	0
Iowa	1,892	1,452	441	836	254	149
Kansas	271	92	0	1,697	—	—
Kentucky	2,361	1,574	0	2,810	696	71
La Posta Band	0	—	—	0	—	—
Louisiana	0	—	—	7,063	4,030	74
Maine	31,752	31,325	—	31,752	31,576	—
Maryland	8,617	8,617	—	8,617	8,617	—
Massachusetts	644	240	0	933	457	9
Michigan	1,402	—	—	555	—	—
Minnesota	0	—	—	6,584	1,727	1
Mississippi	1,744	1,005	322	442	37	16
Missouri	21,837	21,671	—	5,405	5,355	—
Montana	9,080	1,510	8	7,066	3,620	0
N. Mariana Islands	—	—	—	0	—	—
Nebraska	2,056	2,056	—	3,239	365	—
Nevada	5	5	—	1,393	1,388	—
New Hampshire	279	0	—	2,769	2,657	—
New Jersey	124	0	30	176	30	0
New Mexico	93	0	—	4,134	4,103	—
New York	427	0	86	157	0	82
North Carolina	0	—	—	0	—	—
North Dakota	147	—	—	9,707	3,484	1,938
Ohio	451	0	—	1,536	805	—
Ohio River Valley	981	0	—	376	21	—
Oklahoma	10	0	0	5,096	1,112	2,016
Oregon	984	84	103	5,062	2,777	48
Pauma Band	23	23	—	23	23	—
Pennsylvania	0	—	—	0	—	—
Puerto Rico	0	—	—	5,394	193	1,008
Rhode Island	6	0	0	574	434	0
Round Valley Tribes	—	—	—	0	—	—
South Carolina	0	—	—	14,726	7,672	—
South Dakota	170	170	—	1,043	342	—
Tennessee	0	—	—	9,182	6,117	0
Texas	3,158	2,841	0	9,598	7,084	0
Utah	16	0	0	518	508	0
Vermont	5,201	4,956	13	5,310	4,115	650
Virginia	9,183	8,716	124	6,510	3,456	5

(Continued)

Table 10B.23 (Continued)

Jurisdiction	Fish Consumption			Swimming		
	Total Assessed	Full Support	Threatened	Total Assessed	Full Support	Threatened
Washington	58,990	15,294	—	70,439	58,892	—
West Virginia	870	201	0	11,408	6,790	2,615
Wisconsin	2,300	1,077	455	0	—	—
Wyoming	0	—	—	251	0	1
Total	**205,153**	**124,941**	**2,387**	**313,832**	**212,659**	**10,779**
Percent of assessed for use		**60.9%**	**1.2%**		**67.8%**	**3.4%**

Note: Fully Supporting, Fully supporting of all uses; Threatened, Fully supporting all uses but threatened for one or more uses; —, no data.

Source: Abstracted from United States Environmental Protection Agency, 2002, *National Water Quality Inventory 2000 Report*, EPA-841-R-02-001, www.epa.gov.

Table 10B.24 Water-Quality Factors Affecting the Nation's Fisheries

Factor	Stream Miles	Percentage
Turbidity	328,261	34.4
High water temperature	250,187	26.2
Nutrient surplus	119,519	12.5
Toxic substances	93,602	9.8
Dissolved oxygen problem	91,022	9.5
Nutrient deficiency	40,603	4.3
Low water temperature	29,877	3.1
Other	26,685	2.8
pH too acidic	24,793	2.6
Low flow	24,364	2.6
Salinity	17,217	1.8
Sedimentation	14,378	1.5
Siltation	9,644	1.0
Gas supersaturation	5,500	0.6
Intermittent water	4,839	0.5
Herbicides and pesticides	4,356	0.5
pH too basic	3,998	0.4
Channelization	2,937	0.3

Note: Expressed in total stream miles and as percentage of total miles.

Source: From Judy, R.D., and others, 1984. *1982 National Fisheries Survey*, U.S. Fish and Wildlife Service, FWS/OBS-84/06.

Table 10B.25 Limiting Factors Adversely Affecting the Nation's Fish Communities

Factor	Stream Miles	Percentage
Fish kills	115,432	12.1
Contamination	81,927	8.6
Overharvest	35,566	3.7
Poaching	28,145	2.9
Diseases/parasites	21,873	2.3
Fish stocking	19,350	2.0
Other	18,063	1.9
Habitat	14,213	1.5
Underharvest	12,714	1.3
Competition	10,836	1.1
Water quality	5,879	0.6
Tumors/lesions	5,101	0.5
Low flow	3,194	0.3
Small channel capacity	1,657	0.2

Note: Expressed in total stream miles and as percentage of total miles.

Source: From Judy, R.D., and others, 1984. *1982 National Fisheries Survey*, U.S. Fish and Wildlife Service, FWS/OBS-84/06.

Table 10B.26 Designated Use Support in Rivers and Streams of the United States by Jurisdiction, in 2000

Jurisdiction	Total River Miles	River Miles Assessed	Miles Fully Supporting	Miles Threatened	Miles Impaired	Miles Not Attainable
Alabama	77,242.0	2,627.9	698.4	0	1,929.5	0
Alaska	365,000.0	1,420.5	902.6	0	517.9	0
American Samoa	169.0	16.9	0.7	0	16.2	0
Arizona	127,505.0	4,052.0	3,066	0	986.0	0
Arkansas	87,617.4	8,113.0	6,935.7	0	1,177.3	0
Big Sandy Rancheria	—	0	—	—	—	—
California	211,513.0	25,274.0	2,463	1,862.0	20,949	0
Colorado	107,403.0	41,470.0	40,226	0	1,244.0	0
Connecticut	5,830.0	1,207.2	479.4	334.2	389.4	4.2
Delaware	2,506.0	2,505.8	740.84	0	1,765.0	0
Delaware River Basin	206.0	206.0	0	0	206.0	0
District of Columbia	39.0	38.4	0	0	38.4	0
Florida	51,858.0	10,159.0	6,460	552.0	3,147.0	0
Georgia	70,150.0	9,999.0	4,013	0	5,986.0	0
Guam	228.0	167.1	23.1	81	63.0	0
Hawaii	3,904.7	3,993.8	1,254.71	0	2,737.1	2
Hoopa Valley Tribe	319.6	0.0	—	—	—	—
Idaho	115,595.0	17,332.6	8,434	669.0	8,229.6	0
Illinois	87,110.0	15,587.2	7,674.06	69.2	7,843.9	0
Indiana	35,673.0	17,541.0	13,310	0	4,230.0	1
Iowa	71,665.0	6,389.7	1,702.57	2,784.0	1,903.1	0
Kansas	134,338.0	18,236.0	3,417	0	14,819.0	0
Kentucky	49,105.0	9,922.7	5,954.75	280.1	3,687.8	0
La Posta Band	—	0.0	—	—	—	—
Louisiana	66,294.0	7,359.0	723	61	6,575.0	0
Maine	31,752.0	31,752.0	31,023	0	729.0	0
Maryland	8,789.0	8,616.8	5,405.3	0	3,211.5	0
Massachusetts	8,229.0	1,496.2	479.9	47	969.3	0
Michigan	51,438.0	10,309.0	7,829	24	2,456.0	0
Minnesota	91,944.0	11,403.4	1,149.6	2,181.7	7,899.6	172.48
Mississippi	84,003.0	14,972.2	3,263.2	855.5	10,824.2	29.2
Missouri	51,977.8	21,615.1	11,129.5	164.2	10,321.4	0

(Continued)

Table 10B.26 (Continued)

Jurisdiction	Total River Miles	River Miles Assessed	Miles Fully Supporting	Miles Threatened	Miles Impaired	Miles Not Attainable
Montana	176,750.0	11,442.7	2,858.2	8.1	8,576.4	0
N. Mariana Islands	59.3	0.0	0	0	0	0
Nebraska	83,258.0	6,500.0	2,741.0	0	3,759.0	0
Nevada	143,578.0	1,564.3	611.5	0	952.8	0
New Hampshire	10,881.2	2,677.4	2,233.1	0	444.3	0
New Jersey	8,050.0	330.0	121.0	0	209.0	0
New Mexico	110,741.0	4,284.0	1,608.8	0	2,675.2	0
New York	52,337.0	2,914.0	0	1,833.0	1,081.0	0
North Carolina	37,662.0	32,072.0	29,929.0	0	2,143.0	0
North Dakota	54,427.4	14,964.5	1,656.9	6,083.5	7,224.2	0
Ohio	29,113.0	8,231.6	3,857.2	631.8	3,742.6	0
Oklahoma	78,778.0	14,070.8	1,558.3	4,764.8	7,647.4	100.3
Oregon	115,472.0	59,735.0	22,292.0	23,506.0	13,937.0	0
Pauma Band	22.9	*22.9*	22.9	0	0	0
Pennsylvania	83,161.0	35,496.0	28,235.0	0	7,261.0	0
Puerto Rico	5,394.2	5,394.2	150.4	590.5	4,653.3	0
Rhode Island	1,383.0	648.8	431.6	0	217.2	0
Round Valley Tribes	384.0	*34.5*	0	0	34.5	0
South Carolina	29,794.0	15,404.6	11,394.0	0	4,010.6	0
South Dakota	9,937.0	3,564.0	1,786.0	0	1,778.0	0
Tennessee	61,075.0	24,326.4	16,755.2	33.6	7,537.6	0
Texas	191,228.0	15,101.4	10,449.6	104.0	4,547.8	0
Utah	85,916.0	10,518.7	7,693.9	0	2,824.8	0
Vermont	7,099.0	5,462.2	3,105.3	1,188.4	1,168.5	0
Virginia	49,460.2	9,190.0	4,088.0	636.0	4,466.0	0
Washington	70,439.4	70,439.5	32,717.8	0	37,721.7	0
West Virginia	32,278.0	11,549.6	3,091.8	3,145.0	5,312.9	0
Wisconsin	55,000.0	23,530.3	6,858.0	6,634.8	10,029.3	8.2
Wyoming	108,767.0	2,954.8	2,124.1	379.2	451.5	0
Total	**3,692,830.0**	**696,207.5**	**367,128.7**	**59,503.6**	**269,257.9**	**317.4**
Percent of assessed for summary of use support			**53%**	**9%**	**39%**	**0%**

Note: Supporting, Fully supporting of all uses; Threatened, Fully supporting all uses but threatened for one or more uses; Impaired, Partially or not supporting one or more uses; Not Attainable, Not able to support one or more uses; —, no data.

Source: Abstracted from United States Environmental Protection Agency, 2002, *National Water Quality Inventory 2000 Report*, EPA-841-R-02-001, www.epa.gov.

Table 10B.27 River and Stream Miles Supporting Uses in the United States, 1972–1982

Status	1972 Miles	Percent	1982 Miles	Percent
Supporting uses	272	36	488	64
Partially supporting uses	46	6	167	22
Not supporting uses	30	4	35	5
Unknown or not reported	410	54	68	9

Note: Thousand of miles and percentage of waters assessed. Forty-nine (49) states reported on water-quality conditions between 1972 and 1982 for 758,000 river and stream miles. Some proportion of the 1972 data unknown or not reported fell into each of the levels of use support.

Source: From The Association of State and Interstate Water Pollution Control Administrators, in cooperation with the U.S. Environmental Protection Agency, 1984, *America's Clean Water: The States' Evaluation of Progress, 1972–1982*. Washington, DC; U.S. Geological Survey, National Water Summary 1986.

Table 10B.28 Condition of Perennial Streams Related to Their Ability to Support Fish in the United States, 1977–1982

Condition (Worst to Best)	1977	1982
0	29.87	29.87
1	48.79	49.31
2	170.07	166.31
3	222.02	228.66
4	155.57	156.24
5	38.20	36.13

Note: Thousands of miles.

Source: From U.S. Department of the Interior, Fish and Wildlife Service. 1984. *1982 National Fisheries Survey*, vol. 1. FWS/OBS-84/06. Washington, DC; U.S. Geological Survey, National Water Summary 1986.

Table 10B.29 Sources of Drinking Water Use Impairment in the United States

Contaminant Group	Specific Contaminant	
Pesticides	Atrazine	Molinate
	Metolachlor	Ethylene dibromide
	Triazine	
Volatile organic chemicals	Trichloroethylene	Dichloromethane
	Tetrachloroethylene	1,1-Dichloroethane
	1,1,1-Trichloroethane	1,1-Dichloroethylene
	cis-1,2-Dichloroethylene	Toluene
	Trihalomethanes	Benzene
	Carbon tetrachloride	Dichlorobenzene
	Ethylbenzene	Methyl tertiary butyl ether
	1,1,2,2-Tetrachloroethane	Xylene
Inorganic chemicals	Arsenic	Fluoride
	Nitrates	Manganese
	Iron	Lead
	Copper	Sodium
	Chloride	
Microbiological contaminants	Exceedance of total coliform rule	Exceedance of fecal coliform rule

Source: From United States Environmental Protection Agency, 2002, *National Water Quality Inventory 2000 Report*, EPA-841-R-02-001, www.epa.gov.

Table 10B.30 Frequency of Detection and Concentrations of Volatile Organic Compounds in Surface Water Source Samples Collected from United States Drinking Water Supplies between May 3, 1999 and October 23, 2000

	River					Reservoir				
Compound	Number of Samples	Number of Detects	Detection Frequency	Minimum Concentration	Maximum Concentration	Number of Samples	Number of Detects	Detection Frequency	Minimum Concentration	Maximum Concentration
Gasoline Oxygenates										
tert-Amyl methyl ether	171	0	0	<0.2	<0.2	204	0	0	<0.2	<0.2
Diisopropyl ether	171	0	0	<0.2	<0.2	204	0	0	<0.2	<0.2
Ethyl *tert*-butyl ether	171	0	0	<0.2	<0.2	204	0	0	<0.2	<0.2
Methyl *tert*-butyl ether	165	24	14	0.2	1.2	198	26	13	0.2	20
Other Gasoline Compounds										
Benzene	171	0	0	<0.2	<0.2	204	1	0.49	0.22	0.22
n-Butylbenzene	171	0	0	<0.2	<0.2	204	0	0	<0.2	<0.2
sec-Butylbenzene	171	0	0	<0.2	<0.2	204	0	0	<0.2	<0.2
tert-Butylbenzene	171	0	0	<0.2	<0.2	204	0	0	<0.2	<0.2
Ethylbenzene	170	0	0	<0.2	<0.2	203	2	0.98	0.26	1.0
Naphthalene	171	1	0.58	1.2	1.2	204	0	0	<0.2	<0.2
Toluene	169	5	3	0.2	0.45	200	2	1.0	0.52	1.0
1,3,5-Trimethylbenzene	171	0	0	<0.2	<0.2	204	0	0	<0.2	<0.2
o-Xylene	170	0	0	<0.2	<0.2	202	2	0.99	0.45	1.4
m- *p*-Xylene	168	1	0.60	0.35	0.35	202	2	0.99	0.71	2.5
Trihalomethanes										
Bromodichloromethane	171	13	7.6	0.39	26	204	14	6.9	0.25	24
Bromoform	171	2	1.2	0.53	3.5	204	3	1.5	0.23	0,62
Chlorodibromomethane	171	8	4.7	0.25	20	204	9	4.4	0.22	8.9
Chloroform	168	27	16	0.20	34	203	25	12	0.21	85
Organic Syntheses										
Acrylonitrile	171	0	0	<0.2	<0.2	204	0	0	<0.2	<0.2
Bromobenzene	171	0	0	<0.2	<0.2	204	0	0	<0.2	<0.2
Bromochloromethane	170	0	0	<0.2	<0.2	204	0	0	<0.2	<0.2
Carbon tetrachloride	171	1	0.58	0.43	0.43	204	1	0.49	0.25	0.25
Chlorobenzene	171	0	0	<0.2	<0.2	204	0	0	<0.2	<0.2
Chloroethane	171	0	0	<0.2	<0.2	204	0	0	<0.2	<0.2
2-Chlorotoluene	171	0	0	<0.2	<0.2	204	0	0	<0.2	<0.2
4-Chlorotoluene	171	0	0	<0.2	<0.2	204	0	0	<0.2	<0.2
Dibromomethane	163	0	0	<0.2	<0.2	195	0	0	<0.2	<0.2
1,2-Dichlorobenzene	171	0	0	<0.2	<0.2	204	0	0	<0.2	<0.2
1,3-Dichlorobenzene	171	0	0	<0.2	<0.2	204	0	0	<0.2	<0.2
1,1-Dichloroethane	171	0	0	<0.2	<0.2	204	0	0	<0.2	<0.2
1,2-Dichloroethane	171	0	0	<0.2	<0.2	204	0	0	<0.2	<0.2
1,1-Dichloroethene	171	0	0	<0.2	<0.2	204	0	0	<0.2	<0.2
cis-1,2-Dichloroethene	171	1	0.58	0.93	0.93	204	0	0	<0.2	<0.2
trans-1,2-Dichloroethene	171	0	0	<0.2	<0.2	204	0	0	<0.2	<0.2
1,2-Dichloropropane	163	0	0	<0.2	<0.2	195	0	0	<0.2	<0.2
1,3-Dichloropropane	170	0	0	<0.2	<0.2	203	0	0	<0.2	<0.2
2,2-Dichloropropane	171	0	0	<0.2	<0.2	204	0	0	<0.2	<0.2
1,1-Dichloropropene	171	0	0	<0.2	<0.2	204	0	0	<0.2	<0.2

(Continued)

Table 10B.30 (Continued)

	River					Reservoir				
Compound	Number of Samples	Number of Detects	Detection Frequency	Minimum Concentration	Maximum Concentration	Number of Samples	Number of Detects	Detection Frequency	Minimum Concentration	Maximum Concentration
Hexachlorobutadiene	171	0	0	<0.2	<0.2	204	0	0	<0.2	<0.2
Hexachloroethane	171	0	0	<0.2	<0.2	204	0	0	<0.2	<0.2
Isopropylbenzene	170	0	0	<0.2	<0.2	204	1.0	0.49	0.37	0.37
p-Isopropyltoluene	171	0	0	<0.2	<0.2	204	0	0	<0.2	<0.2
Methylene chloride	171	1	0.58	2.6	2.6	204	0	0	<0.2	<0.2
Methyl ethyl ketone	171	1	0.98	120	120	204	2	0.58	2.3	9.1
n-Propylbenzene	170	0	0	<0.2	<0.2	204	1.0	0.58	0.21	0.21
Styrene	169	1	0.59	2.3	2.3	202	2	0.99	0.41	0.97
1,1,1,2-Tetrachloroethane	171	0	0	<0.2	<0.2	204	0	0	<0.2	<0.2
1,1,2,2-Tetrachloroethane	171	0	0	<0.2	<0.2	204	0	0	<0.2	<0.2
Tetrachloroethene	171	3	1.8	0.29	5.5	204	0	0	<0.2	<0.2
1,1,1-Trichloroethane	171	0	0	<0.2	<0.2	204	0	0	<0.2	<0.2
1,1,2-Trichloroethane	171	0	0	<0.2	<0.2	204	0	0	<0.2	<0.2
Trichloroethene	171	3	1.8	0.31	2	204	0	0	<0.2	<0.2
1,2,3-Trichloropropane	171	0	0	<0.2	<0.2	204	0	0	<0.2	<0.2
1,2,4-Trimethylbenzene	170	2	1.2	0.43	0.77	204	1	0.49	0.36	0.36
Vinyl bromide	171	0	0	<0.2	<0.2	204	0	0	<0.2	<0.2
Vinyl chloride	171	0	0	<0.2	<0.2	204	0	0	<0.2	<0.2
1,2,3-Trichlorobenzene	171	0	0	<0.2	<0.2	204	0	0	<0.2	<0.2
1,2,4-Trichlorobenzene	171	0	0	<0.2	<0.2	204	0	0	<0.2	<0.2
Fumigants										
Bromomethane	171	1	0.58	0.22	0.22	204	0	0	<0.2	<0.2
1,4-Dichlorobenzene	171	1	0.58	0.22	0.22	203	1	0.49	0.36	0.36
cis-1,3-Dichloropropene	171	0	0	<0.2	<0.2	204	0	0	<0.2	<0.2
trans−1,3-Dichloropropene	171	0	0	<0.2	<0.2	204	0	0	<0.2	<0.2
Refrigerants										
Chloromethane	171	0	0	<0.2	<0.2	204	0	0	<0.2	<0.2
Dichlorodifluoro-methane	171	0	0	<0.2	<0.2	204	0	0	<0.2	<0.2
Trichlorofluoromethane	171	0	0	<0.2	<0.2	204	0	0	<0.2	<0.2
l,l,2-Trichloro-l,2,2-trifluoroethane	171	0	0	<0.2	<0.2	204	0	0	<0.2	<0.2

Note: Concentrations are in micrograms per liter.

Source: From Grady, S., 2003, *A National Survey of Methyl tert-Butyl Ether and Other Volatile Organic Compounds in Drinking Water Sources: Results of a Random Survey*, U.S. Geological Survey Water-Resource Investigations Report 02-4079, www.usgs.gov.

Table 10B.31 Summary of Analytical Results of United States Streams Sampled for 95 Pharmaceuticals, Hormones and Other Organic Wastewater Contaminants, 1999–2000

Chemical (Method)	CASRN	N	RL (μg/L)	freq (%)	max (μg/L)	med (μg/L)	Use	MCL or HAL (23) (μg/L)	Lowest LC_{50} for the Most Sensitive Indicator Species (μg/L)/ No. of Aquatic Studies Identified (24)
				Veterinary and Human Antibiotics					
Carbodox (1)	6804-07-5	104	0.10	0	ND	ND	Antibiotic	—	-/1
Chlortetracycline (1)	57-62-5	115	0.05	0	ND	ND	Antibiotic	—	88,000[a]/3
Chlortetracycline (2)	57-62-5	84	0.10	2.4	0.69	0.42	Antibiotic	—	88,000[a]/3
Ciprofloxacin (1)	85721-33-1	115	0.02	2.6	0.03	0.02	Antibiotic	—	-/0
Doxycycline (1)	564-25-0	115	0.1	0	ND	ND	Antibiotic	—	-/0
Enrofloxacin (1)	93106-60-6	115	0.02	0	ND	ND	Antibiotic	—	40[b]/29
Erythromycin-H_2O (1)	114-07-8	104	0.05	21.5	1.7	0.1	Erythromycin metabolite	—	665,000[b]/35
Lincomycin (1)	154-21-2	104	0.05	19.2	0.73	0.06	Antibiotic	—	-/0
Norfloxacin (1)	70458-96-7	115	0.02	0.9	0.12	0.12	Antibiotic	—	-/6
Oxytetracycline (1)	79-57-2	115	0.1	0	ND	ND	Antibiotic	—	102,000[a]/46
Oxytetracycline (2)	79-57-2	84	0.10	1.2	0.34	0.34	Antibiotic	—	102,000[a]/46
Roxithromycin (1)	80214-83-1	104	0.03	4.8	0.18	0.05	Antibiotic	—	-/0
Sarafloxacin (1)	98105-99-8	115	0.02	0	ND	ND	Antibiotic	—	-/0
Sulfachloropyridazine (2)	80-32-0	84	0.05	0	ND	ND	Antibiotic	—	-/0
Sulfadimethoxine (1)	122-11-2	104	0.05	0	ND	ND	Antibiotic	—	-/5
Sulfadimethoxine (2)	122-11-2	84	0.05	1.2	0.06	0.06	Antibiotic	—	-/5
Sulfamerazine (1)	127-79-7	104	0.05	0	ND	ND	Antibiotic	—	100,000[c]/17
Sulfamerazine (2)	127-79-7	84	0.05	0	ND	ND	Antibiotic	—	100,000[c]/17
Sulfamethazine (1)	57-68-1	104	0.05	4.8	0.12	0.02	Antibiotic	—	100,000[c]/17
Sulfamethazine (2)	57-68-1	84	0.05	1.2	0.22	0.22	Antibiotic	—	100,000[c]/17
Sulfamethizole (1)	144-82-1	104	0.05	1.0	0.13	0.13	Antibiotic	—	-/0
Sulfamethoxazole (1)	723-46-6	104	0.05	12.5	1.9	0.15	Antibiotic	—	-/0
Sulfamethoxazole (3)	723-46-6	84	0.023	19.0	0.52	0.066	Antibiotic	—	-/0
Sulfathiazole (1)	72-14-0	104	0.10	0	ND	ND	Antibiotic	—	-/0
Sulfathiazole (2)	72-14-0	84	0.05	0	ND	ND	Antibiotic	—	-/0
Tetracycline (1)	60-54-8	115	0.05	0	ND	ND	Antibiotic	—	550,000[b]/3
Tetracycline (2)	60-54-8	84	0.10	1.2	0.11	0.11	Antibiotic	—	550,000[b]/3
Trimethoprim (1)	738-70-5	104	0.03	12.5	0.71	0.15	Antibiotic	—	3,000[c]/4
Trimethoprim (3)	738-70-5	84	0.014	27.4	0.30	0.013	Antibiotic	—	3,000[c]/4
Tylosin (1)	1401-69-0	104	0.05	13.5	0.28	0.04	Antibiotic	—	-/0
Virginiamycin (1)	21411-53-0	104	0.10	0	ND	ND	Antibiotic	—	-/0
				Prescription Drugs					
Albuterol (salbutamol) (3)	18559-94-9	84	0.029	0	ND	ND	Antiasthmatic	—	-/0
Cimetidine (3)	51481-61-9	84	0.007	9.5	0.58[d]	0.074[d]	Antacid	—	-/0
Codeine (3)	76-57-3	46	0.24	6.5	0.019	0.012	Analgesic	—	-/0
Codeine (4)	76-57-3	85	0.1	10.6	1.0[d]	0.2[d]	Analgesic	—	-/0
Dehydronifedipine (3)	67035-22-7	84	0.01	14.3	0.03	0.012	Antianginal	—	-/0
Digoxin (3)	20830-75-5	46	0.26	0	ND[d]	ND[d]	Cardiac stimulant	—	10,000,000[a]/24
Digoxigenin (3)	1672-46-4	84	0.008	0	ND	ND	Digoxin metabolite	—	-/0
Diltiazem (3)	42399-41-7	84	0.012	13.1	0.049	0.021	Antihypertensive	—	-/0

(Continued)

Table 10B.31 (Continued)

Chemical (Method)	CASRN	N	RL (μg/L)	freq (%)	max (μg/L)	med (μg/L)	Use	MCL or HAL (23) (μg/L)	Lowest LC_{50} for the Most Sensitive Indicator Species (μg/L)/ No. of Aquatic Studies Identified (24)
Enalaprilat (3)	76420-72-9	84	0.15	1.2	0.046[d]	0.046[d]	Enalapril maleate (antihypertensive) metabolite	—	-/0
Fluoxetine[h] (3)	54910-89-3	84	0.018	1.2	0.012[d]	0.012[d]	Antidepressant	—	-/0
Gemfibrozil (3)	25812-30-0	84	0.015	3.6	0.79	0.048	Antihyperlipidemic	—	-/0
Metformin (3)	657-24-9	84	0.003	4.8	0.15[d]	0.11[d]	Antidiabetic	—	-/0
Paroxetine metabolite (3)	—	84	0.26	0	ND[d]	ND[d]	Paroxetine (antidepressant) metabolite	—	-/0
Ranitidine (3)	66357-35-5	84	0.01	1.2	0.01[d]	0.01[d]	Antacid	—	-/0
Warfarin (3)	81-82-2	84	0.001	0	ND	ND	Anticoagulant	—	16,000[c]/33
				Nonprescription Drugs					
Acetaminophen (3)	103-90-2	84	0.009	23.8	10	0.11	Antipyretic	—	6,000[a]/14
Caffeine (3)	58-08-2	84	0.014	61.9	6.0	0.081	Stimulant	—	40,000[i]/77
Caffeine (4)	58-08-2	85	0.08	70.6	5.7	0.1	Stimulant	—	40,000[i]/77
Cotinine (3)	486-56-6	84	0.023	38.1	0.90	0.024	Nicotine metabolite	—	-/0
Cotinine (4)	486-56-6	54	0.04	31.5	0.57	0.05	Nicotine metabolite	—	-/0
1,7-dimethylxanthine (3)	611-59-6	84	0.018	28.6	3.1[d]	0.11[d]	Caffeine metabolite	—	-/0
Ibuprofen (3)	15687-27-1	84	0.018	9.5	1.0	0.20	Antiinflammatory	—	-/0
				Other Wastewater-Related Compounds					
1,4-dichlorobenzene (4)	106-46-7	85	0.03	25.9	4.3	0.09	Deodorizer	75	1,100[c]/190
2,6-di-*tert*-butylphenol[h] (4)	128-39-2	85	0.08	3.5	0.11[d]	0.06[d]	Antioxidant	—	-/2
2,6-di*tert*-butyl-1,4-benzoquinone[h] (4)	719-22-2	85	0.10	9.4	0.46	0.13	Antioxidant	—	-/0
5-methyl-1H-benzotriazole (4)	136-85-6	54	0.10	31.5	2.4	0.39	Antiocorrosive	—	-/0
Acetophenone (4)	98-86-2	85	0.15	9.4	0.41	0.15	Fragrance	—	155,000[i]/21
Anthracene (4)	120-12-7	85	0.05	4.7	0.11	0.07	Pah	—	5.4[i]/188
Benzo[a]pyrene[h] (4)	50-32-8	85	0.05	9.4	0.24	0.04	Pah	0.2	1.5[a]/428
3-*tert*-butyl-4-hydroxy anisole[h] (4)	25013-16-5	85	0.12	2.4	0.2[d]	0.1[d]	Antioxidant	—	870[c]/14
Butylated hydroxyl toluene (4)	128-37-0	85	0.08	2.4	0.1[d]	0.1[d]	Antioxidant	—	1,440[a]/15
Bis(2-ethylhexyl) adipate (4)	103-23-1	85	2.0	3.5	10 [g]	3 [g]	Plasticizer	400	480[a]/9
Bis(2-ethylhexyl) phthalate[h] (4)	117-81-7	85	2.5	10.6	20 [g]	7 [g]	Plasticizer	6	7,500[a]/309
				Other Wastewater-Related Compounds					
Bisphenol A[h] (4)	80-05-7	85	0.09	41.2	12	0.14	Plasticizer	—	3,600[i]/26
Carbaryl[h] (4)	63-25-2	85	0.06	16.5	0.1[d]	0.04[d]	Insecticide	700	0.4[a]/1,541
cis-Chlordane[h] (4)	5103-71-9	85	0.04	4.7	0.1	0.02	Insecticide	2	7.4[b]/28
Chlorpyrifos[h] (4)	2921-88-2	85	0.02	15.3	0.31	0.06	Insecticide	20	0.1[a]/1,794
Diazinon[h] (4)	333-41-5	85	0.03	25.9	0.35	0.07	Insecticide	0.6	0.56[a]/1,040
Dieldrin[h] (4)	60-57-1	85	0.08	4.7	0.21	0.18	Insecticide	0.2	2.6[c]/1,540
Diethylphthalate[h] (4)	84-66-2	54	0.25	11.1	0.42	0.2	Plasticizer	—	12,000[c]/129

Ethanol,2-butoxy-phosphate (4)	78-51-3	85	0.2	45.9	6.7	0.51	Plasticizer	—	10,400[i]/7
Fluoranthene (4)	206-44-0	85	0.03	29.4	1.2	0.04	PAH	—	74[i]/216
Lindane[h] (4)	58-89-9	85	0.05	5.9	0.11	0.02	Insecticide	0.2	30[c]/1,979
Methyl parathion[h] (4)	298-00-0	85	0.06	1.2	0.01	0.01	Insecticide	2	12[a]/888
4-Methyl phenol (4)	106-44-5	85	0.04	24.7	0.54	0.05	Disinfectant	—	1,400[a]/74
Naphthalene (4)	91-20-3	85	0.02	16.5	0.08	0.02	PAH	20	910[c]/519
N,N-diethyltoluamide (4)	134-62-3	54	0.04	74.1	1.1	0.06	Insect repellant	—	71,250[c]/9
4-Nonylphenol[h] (4)	251-545-23	85	0.50	50.6	40[f]	0.8[f]	Nonionic detergent metabolite	—	130[i]/135
4-Nonylphenol Monoethoxylate[h] (4)	—	85	1.0	45.9	20[f]	1[f]	Nonionic detergent metabolite		14,450[a]/4
4-Nonylphenol diethoxylate[h] (4)	—	85	1.1	36.5	9[f]	1[f]	Nonionic detergent metabolite	—	5,500[a]/6
4-Octylphenol Monoethoxylate[h] (4)	—	85	0.1	43.5	2[f]	0.2[f]	Nonionic detergent metabolite	—	-/0
4-Octylphenol diethoxylate[h] (4)	—	85	0.2	23.5	1[f]	0.1[f]	Nonionic detergent metabolite	—	-/0
Phenanthrene (4)	85-01-8	85	0.06	11.8	0.53	0.04	PAH	—	590[a]/192
Phenol (4)	108-95-2	85	0.25	8.2	1.3[e]	0.7[e]	Disinfectant	400	4,000[c]/2,085
Phthalic anhydride (4)	85-44-9	85	0.25	17.6	1[e]	0.7[e]	Plastic manufacturing	—	40,400[c]/5
Pyrene (4)	129-00-0	85	0.03	28.2	0.84	0.05	PAH	—	90.9[a]/112
Tetrachloroethylene (4)	127-18-4	85	0.03	23.5	0.70[d]	0.07[d]	Solvent, degreaser	5	4,680[c]/147
Triclosan[h] (4)	3380-34-5	85	0.05	57.6	2.3	0.14	Antimicrobial disinfectant	—	180[i]/3
Tri(2-chloroethyl) phosphate (4)	115-96-8	85	0.04	57.6	0.54	0.1	Fire retardant	—	66,000[b]/8
Tri(dichlorisopropyl) Phosphate (4)	13674-87-8	85	0.1	12.9	0.16	0.1	Fire retardant	—	3,600[b]/9
Triphenyl phosphate (4)	115-86-6	85	0.1	14.1	0.22	0.04	Plasticizer	—	280[c]/66
				Steroids and Hormones					
cis-Androsterone[h] (5)	53-41-8	70	0.005	14.3	0.214	0.017	Urinary steroid	—	-/0
Cholesterol (4)	57-88-5	85	1.5	55.3	10[d]	1[d]	Plant/animal steroid	—	-/0
Cholesterol (5)	57-88-5	70	0.005	84.3	60[g]	0.83	Plant/animal steroid	—	-/0
Coprostanol (4)	360-68-9	85	0.6	35.3	9.8[d]	0.70[d]	Fecal steroid	—	-/0
Coprostanol (5)	360-68-9	70	0.005	85.7	150[g]	0.088	Fecal steroid	—	-/0
Equilenin[h] (5)	517-09-9	70	0.005	2.8	0.278	0.14	Estrogen replacement	—	-/0
Equilin[h] (5)	474-86-2	70	0.005	1.4	0.147	0.147	Estrogen replacement	—	-/0
17α-ethynyl estradiol[h] (5)	57-63-6	70	0.005	15.7	0.831	0.073	Ovulation inhibitor	—	-/22
17α-estradiol[h] (5)	57-91-0	70	0.005	5.7	0.074	0.03	Reproductive hormone	—	-/0
17β-estradiol[h] (4)	50-28-2	85	0.5	10.6	0.2[d]	0.16[d]	Reproductive hormone	—	-/0
17β-estradiol[h] (5)	50-28-2	70	0.005	10.0	0.093	0.009	Reproductive hormone	—	-/0
Estriol[h] (5)	50-27-1	70	0.005	21.4	0.051	0.019	Reproductive hormone	—	-/0

(Continued)

Table 10B.31 (Continued)

Chemical (Method)	CASRN	N	RL (μg/L)	freq (%)	max (μg/L)	med (μg/L)	Use	MCL or HAL (23) (μg/L)	Lowest LC_{50} for the Most Sensitive Indicator Species (μg/L)/ No. of Aquatic Studies Identified (24)
Estrone[h] (5)	53-16-7	70	0.005	7.1	0.112	0.027	Reproductive hormone	—	-/11
Mestranol[h] (5)	72-33-3	70	0.005	10.0	0.407	0.074	Ovulation inhibitor	—	-/0
19-Norethisterone[h] (5)	68-22-4	70	0.005	12.8	0.872	0.048	Ovulation inhibitor	—	-/0
Progesterone[h] (5)	57-83-0	70	0.005	4.3	0.199	0.11	Reproductive hormone	—	-/0
Stigmastanol (4)	19466-47-8	54	2.0	5.6	4[d]	2[d]	Plant steroid	—	-/0
Testosterone[h] (5)	58-22-0	70	0.005	2.8	0.214	0.116	Reproductive hormone	—	-/4

Note: CASRN, Chemical Abstracts service Registry Number; *N*, number of samples; RL, reporting level; freq, frequency of detection; max, maximum concentration; med, median detectable concentration; MCL, maximum contaminant level; HAL, health advisory level; LC_{50}, lethal concentration with 50% mortality; Nd, not detected; —, not available; PAH, polycyclic aromatic hydrocarbon.

[a] Daphnia magna (water flea), 48 h exposure LC_{50}.
[b] Other species and variable conditions.
[c] Oncorhynchus mykiss (rainbow trout), 96 h exposure LC_{50}.
[d] Concentration estimated, average recovery <60%.
[e] Concentration estimated, compound routinely detected in laboratory blanks.
[f] Concentration estimated, reference standard prepared from a technical mixture.
[g] Concentration estimated, value greater than highest point on calibration curve.
[h] Compounds suspected of being hormonally active.
[i] Pimephales promelas (fathead minnow), 96 h exposure LC_{50}.

Source: From Kolpin, D.W., Furlong, E.T., Meyer, M.T., Thurman, E. M., Zaugg, S. D., Barber, L.B., and Buxton, H.T., 2002, Pharmaceuticals, hormones, and other organic wastewater contaminants in U.S. streams, 1999–2000. *A National Reconnaissance, Environmental Science & Technology*, vol 36, no. 6, Web Release Date: March 15, 2002, 10.1021/es011055j S0013-936X(01)01055-0, www.usgs.gov.

Table 10B.32 Regions 1–10: River Reach and Watershed Evaluation Summary, United States National Sediment Quality Survey

EPA Region (State)	River Reach Evaluation[a]					Watershed Evaluation					
	Total Number of River Reaches	River Reaches with at Least One Tier 1 Station	River Reaches with at Least One Tier 2 Station and Zero Tier 1 Stations	River Reaches with All Tier 3 Stations	River Reaches with No Data	Total Number of Watersheds	Watersheds Containing APCs	Watersheds with at Least One Tier 1 Station	Watersheds with at Least One Tier 2 Station and Zero Tier 1 Stations	Watersheds with All Tier 3 Stations	Watersheds with No Data
Region 1 (CT, ME, MA, NH, RI, VT)	2,764	97 (3.5%)	23 (0.8%)	5 (0.2%)	2,639 (95.5%)	62	9 (14.5%)	13 (21.0%)	6 (9.7%)	0 (0.0%)	34 (54.8%)
Region 2 (NJ, NY, PR)	1,845	217 (11.8%)	102 (5.5%)	45 (2.4%)	1,481 (80.3%)	71	17 (23.9%)	35 (49.3%)	3 (4.2%)	3 (4.2%)	13 (18.3%)
Region 3 (DE, DC, MD, PA, VA, WV)	3,388	385 (11.4%)	313 (9.2%)	301 (8.9%)	2,389 (70.5%)	126	7 (5.6%)	96 (76.2%)	11 (8.7%)	4 (3.2%)	8 (6.3%)
Region 4 (AL, FL, GA, KY, MS, NC, SC, TN)	10,078	444 (4.4%)	461 (4.6%)	301 (3.0%)	8,872 (88.0%)	307	13 (4.2%)	142 (46.3%)	57 (18.6%)	25 (8.1%)	70 (22.8%)
Region 5 (IL, IN, MI, MN, OH, WI)	6,151	532 (8.6%)	401 (6.5%)	316 (5.1%)	4,902 (79.7%)	278	25 (9.0%)	144 (51.8%)	31 (11.2%)	19 (6.8%)	59 (21.2%)
Region 6 (AR, LA, NM, OK, TX)	7,577	226 (3.0%)	222 (2.9%)	289 (3.8%)	6,840 (90.3%)	402	4 (1.0%)	117 (29.1%)	69 (17.2%)	44 (10.9%)	168 (41.8%)
Region 7 (IA, KS, MO, NE)	4.915	94 (1.9%)	161 (3.3%)	136 (2.8%)	4,524 (92.0%)	238	1 (0.4%)	60 (25.2%)	72 (30.3%)	29 (12.2%)	76 (31.9%)
Region 8 (CO, MT, ND, SD, UT, WY)	13,860	59 (0.4%)	77 (0.6%)	68 (0.5%)	13,656 (98.5%)	385	1 (0.3%)	34 (8.8%)	41 (10.6%)	31 (8.1%)	278 (72.2%)
Region 9 (AZ, CA, HI, NV)	4,686	156 (3.3%)	63 (1.3%)	40 (0.9%)	4,427 (94.5%)	288	19 (6.6%)	41 (14.2%)	19 (6.6%)	15 (5.2%)	194 (67.4%)
Region 10 (AK, ID, OR, WA)	10,462	177 (1.7%)	121 (1.2%)	49 (0.5%)	10,115 (96.7%)	355	10 (2.8%)	48 (13.5%)	29 (8.2%)	21 (5.9%)	247 (69.6%)
Total for United States[b]	**64,591**	**2,298 (3.6%)**	**1,891 (2.9%)**	**1,506 (2.3%)**	**58,896 (91.2%)**	**2,264**	**96 (4.2%)**	**658 (29.1%)**	**302 (13.3%)**	**168 (7.4%)**	**1,040 (45.9%)**

Tier 1, Associated adverse effects on aquatic life or human health are probable.
Tier 2, Associated adverse effects on aquatic life or human health are possible.
Tier 3, No indication of adverse effects.

[a] River reaches based on EPA River Reach File (RF1). RF1 does not include data outside the contiguous United States.

[b] Because some reaches and watersheds occur in more than one region, the total number of reaches and watersheds in each category or the country might not equal the sum of reaches or watersheds in the regions.

Source: From USEPA, 2004, The incidence and severity of sediment contamination in surface waters of the United States, National sediment contamination in surface waters of the United States, *National Sediment Quality Survey: Sectond Edition*, EPA 823-R-04-007, www.epa.gov.

Table 10B.33 Statistical Summary of Semivolatile Organic Compounds in Streambed Sediment That Exceeded Sediment Quality Screening Values in the United States 1992–1995

			Concentration, in μg/kg Dry Weight at the Given Percentile and Maximum Value				Lower Screening Value		Upper Screening Value	
Compound Name	Number of Sites	Percent Detection	75th	90th	95th	Maximum	μg/kg	N	μg/kg	N
Polycyclic aromatic hydrocarbons										
Acenaphthene[a]	530	10.0	<50	51	94	1,500	6.71	53	500	3
Acenaphthylene[a]	535	12.9	<50	77	170	1,500	5.87	69	640	4
Anthracene[b]	533	22	<50	150	520	4,100	57.2	106	845	14
Benz[a]anthracene[b]	518	30.7	79	430	1,100	12,000	108	108	1,050	28
Benzo[b]fluoranthene[a]	533	36.8	92	460	1,400	12,000	3,600	5	—	—
Benzo[k]fluoranthene[a]	532	33.1	84	400	1,100	10,000	3,600	5	—	—
Benzo[a]pyrene[b]	535	32.9	80	350	1,000	9,900	150	101	1,450	17
Benzo[ghi]perylene[a]	526	23.4	<50	200	480	6,700	720	15	—	—
Chrysene[b]	521	35.7	120	570	1,400	16,000	166	111	1,290	31
Dibenz [*a, h*] anthracene[c]	524	16.8	<50	130	280	4,400	33	101	260	26
Fluoranthene[b]	498	39.8	170	1,000	2,700	26,000	423	78	2,230	27
Fluorene[b]	518	10.4	<50	53	140	6,700	77.4	40	536	3
Indeno[1,2,3-cd]pyrene[a]	500	27.8	63	370	710	8,400	690	25	—	—
Naphthalene[b]	496	7.7	<50	<50	78	4,900	176	9	561	1
Phenanthrene[b]	506	30.8	78	500	1,400	15,000	204	82	1,170	31
Pyrene[b]	496	37.9	140	780	2,000	21,000	195	104	1,520	33
Total PAHs[b]							1,610	105	22,800	14
Phthalates										
Bis(2-Ethylhexyl) Phthalate[a]	536	30.0	98	540	1,000	17,000	182	101	2,650	6
Butylbenzylphthalate[a]	536	5.6	<50	<50	76	2,240	900	1	11,000[d]	0
Phenols										
p-Cresol[a]	505	37.8	99	430	870	4,800	670	31	—	—
Halo- and Nitroso-Compounds										
1,2-Dichlorobenzene[a]	516	0.6	<50	<50	<50	86	50	3	340[d]	0
1,4-Dichlorobenzene[a]	518	1.2	<50	<50	<50	140	110	2	350[d]	0
N-Nitrosodiphenylamine[a]	487	0.6	<50	<50	<50	79	28	7	—	—
1,2,4-Trichlorobenzene[a]	517	0.2	<50	<50	<50	68	51	1	9,200[d]	0

Note: N, number of sites exceeding screening value; —, does not exist or apply. Data was collected from the United States Geological Survey National Water-Quality Assessment (NAWQA) Program river basins.

[a] Marine sediment quality guidelines.
[b] Freshwater sediment quality guidelines.
[c] Lower screening value is freshwater and upper screening value is marine sediment quality guideline.
[d] Screening value assuming 1% organic carbon.

Source: Abstracted from Lopes, T.J. and Furlong, E.T., 2001, Occurrence and potential adverse effects of semivolatile organic compounds in streambed sediment, United States, 1992–1995; *Environmental Toxicology and Chemistry*, vol. 20, no. 4, p. 727–737, www.usgs.gov. Printed with permission.

Table 10B.34 Pollutant Discharges into Coastal Waters of the United States, 1980–1985

Coastal Region	BOD	TSS	TN	TP
Northeast	0.57	5.33	0.14	0.05
Mid-Atlantic	0.54	5.42	0.20	0.05
Southeast	0.25	3.18	0.14	0.04
Gulf of Mexico	2.03	149.00	0.88	0.22
West Coast	1.52	101.12	0.64	0.76

Note: (million tons per year) BOD, biochemical oxygen demand; TSS, total suspended solids; TN, total nitrogen; TP, total phosphorus.

Source: From U.S. Department of Commerce, National Oceanic and Atmospheric Administration, National Ocean Survey, Ocean Assessments Division, Strategic Assessment Branch. 1986. Pollutant discharges from East Coast and Gulf of Mexico coastal counties, circa 1980–1985 and unpublished data compiled from the National Coastal Pollutant Discharge Inventory database. Rockville, MD; U.S. Geological Survey, National Water Summary 1986.

Table 10B.35 Point Sources of Pollutants to Coastal Waters of the United States, 1980–1985

Coastal Region	Municipal Wastewater Treatment Facilities	Industrial Wastewater Treatment Facilities
Northeast	540	1,025
Mid-Atlantic	666	1,518
Southeast	1,053	458
Gulf of Mexico	1,651	830
West Coast	521	920

Note: Number of wastewater discharge facilities by region.

Source: From U.S. Department of Commerce, National Oceanic and Atmospheric Administration, National Ocean survey, Ocean Assessments Division, Strategic Assessment Branch. 1986. Pollutant discharges from East Coast and Gulf of Mexico coastal counties, circa 1980–1985 and unpublished data compiled from the National Coastal Pollutant Discharge Inventory database. Rockville, MD; U.S. Geological Survey, National Water Summary 1986.

Table 10B.36 United States Shellfish Growing Waters, 1966–1995

Year	1966	1971	1974	1980	1985	1990	1995
				1000 acres			
Approved for harvest	8,100	10,362	10,560	10,685	11,402	12,304	14,853
Harvested limited	2,090	3,738	4,232	3,533	5,435	6,398	6,721
Conditionally approved	88	410	387	587	1,463	1,571	1,695
Restricted	na	30	34	55	637	463	2,106
Conditionally restricted	na	na	na	na	na	0	119
Prohibited	2,002	3,298	3,811	2,891	3,335	4,364	2,801
Total	10,190	14,100	14,792	14,218	16,837	18,702	21,574

Note: Based on National Shellfish Registers published only in years indicated. Data do not include Alaska, Hawaii, or waters designated as unclassified. The total acreage of classified shellfish growing waters varies with each register. There may be several reasons why shellfish harvest is prohibited, including water-quality problems, lack of funding for complete surveying and monitoring, conservation measures, and other management/administrative actions.

Source: From The 1996 Annual Report of the Council on Environmental Quality, www.whitehouse.gov/CEQ.

Orginal Source: From U.S. Department of Commerce, National Oceanic and Atmospheric Administration, National Ocean Survey, Office of Ocean Resources Conservation and Assessment, Strategic Environmental Assessments Division, *The 1995 National Shellfish Register of Classified Growing Waters (DOC, NOAA, ORCA, Silver Spring, MD, 1997)*, www.nos.noaa.gov.

Table 10B.37 Assessed Estuaries and Ocean Shoreline Waters Supporting Shellfishing Use Requirements in the United States in 2000

	Estuaries (mi^2)			Ocean Shoreline Waters (mi)		
State	**Total Assessed**	**Full Support**	**Threatened**	**Total Assessed**	**Full Support**	**Threatened**
Alabama	0	—	—	0	—	—
Alaska	0	—	—	0	—	—
California	758	43	19	760	694	0
Connecticut	611	273	0	0	—	—
Delaware	15	1	0	15	1	0
Delaware River	679	579	0	—	—	—
District of Columbia	0	—	—	—	—	—
Florida	1,765	1,398	111	0	—	—
Georgia	0	—	—	0	—	—
Hawaii	39	33	0	425	422	0
Louisiana	1,153	1,078	0	0	—	—
Maine	2,852	2,542	0	0	—	—
Maryland	1,839	1,672	—	32	32	—
Massachusetts	2,526	2,254	—	0	—	—
Mississippi	579	550	2	89	0	0
New Hampshire	21	0	—	18	0	—
New Jersey	614	456	0	0	—	—
New York	163	—	0	3	—	0
North Carolina	0	—	—	0	—	—
Oregon	72	4	0	0	—	—
Puerto Rico	0	0	0	0	0	0
Rhode Island	128	96	0	79	79	—
South Carolina	891	613	—	0	—	—
Texas	1,625	1,037	0	0	—	—
Virgin Islands	0	—	—	0	—	—
Virginia	1,735	1,642	2	0	—	—
Washington	2,904	1,274	—	0	—	—
Total	**20,967**	**15,545**	**134**	**1,420**	**1,227**	**0**
Percent of assessed for use		**75.7%**	**0.6%**		**86.4%**	**0.0%**

Note: Supporting, Fully supporting of all uses; Threatened, Fully supporting all uses but threatened for one or more uses; —, no data.

Source: From United States Environmental Protection Agency, 2002, *National Water Quality Inventory 2000 Report*, EPA-841-R-02-001, www.epa.gov.

Table 10B.38 Shellfish Harvesting Restrictions Due to Pathogens as Reported by States, Territories, and Commissions in 2000

			(Sq. Miles)				
Jurisdiction	**Number of Waterbodies with Restrictions**	**Approved (mi^2)**	**Conditionally Approved[a]**	**Restricted[b]**	**Prohibited[c]**	**Management Closures[d]**	**Total Area Affected[e]**
Alabama	—	—	—	—	—	—	—
Alaska	—	—	—	—	—	—	—
American Samoa	—	—	—	—	—	—	—
California	—	—	—	—	—	—	—
Connecticut	—	—	—	—	—	—	—
Delaware	—	—	—	—	—	—	—
District of Columbia	—	—	—	—	—	—	—
Florida	—	—	—	—	—	—	—
Georgia	—	—	—	—	—	—	—
Guam	—	—	—	—	—	—	—
Hawaii	—	—	—	—	—	—	—
Louisiana	26	—	—	—	—	—	—
Maine	—	—	—	—	—	—	—
Maryland	36	1,672.19	58.26	108	0	—	1,838.45
Massachusetts	—	2,254.07	—	40.85	224.2	201.1	2,720.22
Mississippi	—	—	—	—	—	—	—
N. Mariana Islands	—	—	—	—	—	—	—
New Hampshire	11	7.18	0.61	1.57	11.26	0.81	21.43
New Jersey	—	808	—	115	130	—	1,053
New York	—	1,562.5	—	—	312.5	—	1,875
North Carolina	—	—	—	—	—	—	—
Oregon	8	15.84	16.54	—	—	—	32.38
Puerto Rico	—	—	—	—	—	—	—
Rhode Island	19	96	22	—	10	—	128
South Carolina	—	613.06	7.517	151.28	119.233	—	891.09
Texas	—	—	—	—	—	346.1	346.1
Virgin Islands	—	—	—	—	—	—	—
Virginia	—	—	—	3	148	—	151
Washington	—	341.8	46.5	—	104	—	492.3
Total	**100**	**7,370.64**	**151.427**	**419.7**	**1,059.193**	**548.01**	**9,548.97**

Note: —, no data.

[a] Conditionally approved waters do not always meet criteria for harvesting shellfish, but may be harvested when criteria are met.
[b] Restricted water may be harvested if the shellfish are purified with clean water following harvest.
[c] Shellfish may not be harvested in prohibited waters.
[d] Preventative closures due to a lack of data or proximity to point sources or marinas.
[e] Includes water that are classified as conditionally approved, restricted, prohibited, and management closures.

Source: From United States Environmental Protection Agency, 2002, *National Water Quality Inventory 2000 Report*, EPA-841-R-02-001, www.epa.gov.

Table 10B.39 Summary of Fully Supporting, Threatened, and Impaired Waters Assessed in Estuaries and Ocean Shoreline Waters in the United States in 2000

	Estuaries (mi^2)						Ocean Shoreline Waters (mi)					
Jurisdiction	Total Sq. Miles	Assessed	Full Support	Threatened	Impaired	Not Attainable	Total Miles	Assessed	Full Support	Threatened	Impaired	Not Attainable
Alabama	610	541	0	0	541	0	337	0	0	0	0	0
Alaska	33,204	28	3	0	25	0	36,000	25	16	0	9	0
American Samoa	184	0	0	0	0	0	116	53	7	30	16	0
California	2,139	2,033	35	1	1,997	0	1,609	997	775	0	222	0
Connecticut	612	611	139	14	458	0	380	0	0	0	0	0
Delaware	449	0	0	0	0	0	25	0	0	0	0	0
Delaware River	866	866	0	0	866	0	—	—	—	—	—	—
District of Columbia	6	6	0	0	6	0	—	—	—	—	—	—
Florida	4,437	4,037	3,055	121	861	0	8,460	0	0	0	0	0
Georgia	854	858	509	0	349	0	100	0	0	0	0	0
Guam	1	11	0	0	11	0	117	17	1	6	10	0
Hawaii	55	54	23	0	31	0	1,052	871	834	8	29	0
Louisiana	7,656	4,036	318	7	3,711	0	397	0	0	0	0	0
Maine	2,852	2,783	2,473	0	310	0	5,296	0	0	0	0	0
Maryland	2,522	2,478	918	0	1,560	0	32	32	32	0	0	0
Massachusetts	223	128	46	0	82	0	1,519	0	0	0	0	0
Mississippi	760	613	0	62	551	0	245	94	53	41	0	0
N. Mariana Islands	15,989	1	0	0	1	0	52	0	0	0	0	0
New Hampshire	21	21	0	0	21	0	18	18	0	0	18	0
New Jersey	725	614	456	0	158	0	127	127	127	0	0	0
New York	1,530	402	0	11	391	0	120	3	0	0	3	0
North Carolina	3,121	3,115	3,006	0	109	0	320	0	0	0	0	0
Oregon	206	54	8	11	35	0	362	0	0	0	0	0
Puerto Rico		0	0	0	0	0	550	550	302	131	117	0
Rhode Island	151	151	103	0	47	0	79	79	79	0	0	0
South Carolina	401	221	136	0	85	0	190	0	0	0	0	0
Texas	2,394	1,993	1,236	0	758	0	624	0	0	0	0	0
Virgin Islands	3	0	0	0	0	0	209	202	173	21	9	0
Virginia	2,494	1,991	773	796	422	0	120	120	120	0	0	0
Washington	2,904	2,904	611	0	2,293	0	163	—	—	—	—	—
Total	**87,369**	**30,548**	**13,850**	**1,023**	**15,676**	**0**	**58,618**	**3,189**	**2,518**	**237**	**434**	**0**
Percent of assessed			**45.3%**	**3.3%**	**51.3%**	**0.0%**			**79.0%**	**7.4%**	**13.6%**	**0.0%**

Note: Supporting, Fully supporting of all uses; Threatened, Fully supporting all uses but threatened for one or more uses; Impaired, Partially or not supporting one or more uses; Not Attainable, Not able to support one or more uses; —, no data.

Source: Abstracted from United States Environmental Protection Agency, 2002, *National Water Quality Inventory 2000 Report*, EPA-841-R-02-001, www.epa.gov.

Table 10B.40 United States National Coastal Condition Rating Scores by Indicator and Region

Indicator	Northeast Coast	Southeast Coast	Gulf Coast	West Coast	Great Lakes	Puerto Rico	United States[a]
Water-quality index	2	4	3[b]	3	3	3	3.0
Sediment quality index	1	4	3	2	1	1	2.1
Benthic index	1	3	2	3	2	1	2.0
Coastal habitat index	4	3	1	1	2	—[c]	1.7
Fish tissue contaminants index	1	5	3	1	3	—[c]	2.7
Overall condition	1.8	3.8	2.4	2.0	2.2	1.7	2.3

Note: Rating scores are based on a 5-point system, where 1 is poor and 5 is good.

[a] The U.S. score is based on an aerially weighted mean of regional scores.
[b] This rating score does not include the impact of the hypoxic zone in offshore Gulf Coast waters.
[c] No coastal habitat index loss or fish tissue contaminants index results were available for Puerto Rico.

Source: From United States Environmental Protection Agency, 2004, *National Coastal Condition Report II*, EPA-620/R-03/002, December 2004, www.epa.gov.

Table 10B.41 Percent of United States National Coastal Area in Poor Condition by Indicator (except Coastal Habitat Index) and Region

Indicator	Northeast Coast	Southeast Coast	Gulf Coast	West Coast	Great Lakes	Puerto Rico	United States
Water-quality index[a]	19	5	9[b]	3	—	9	11
Sediment quality index[c]	16	8	12	14	—	61	13
Benthic index	22	11	17	13	—	35	17
Coastal habitat index[d]	1.00	1.06	1.30	1.90	—	—	1.26
Fish tissue contaminants index[e]	31	5	14	27	—	—	22
Overall condition[f]	40[g]	23	40	23	—	77	35

Note: The percent area of poor condition is the percentage of total estuarine surface area in the region or the nation (proportional area information is not available for the Great Lakes).

[a] The water-quality index is based on a combination of water quality measurements (dissolved oxygen, chlorophyll *a*, nitrogen, phosphorus, and water clarity).
[b] The area of poor condition does not include the hypoxic zone in offshore Gulf Coast waters.
[c] The sediment quality index is based on a combination of sediment quality measurements (sediment toxicity, sediment contaminants, and sediment TOC).
[d] The coastal habitat index is based on the average of the mean long-term, decadal wetland loss (1780–1990) and the present decadal wetland loss rate (1990–2000).
[e] The fish tissue contaminants index is based on analyses of whole fish (not fillets).
[f] The overall percentage is based on the overlap of the five indicators and includes estuarine area for all of the conterminous 48 states (by region and total) and Puerto Rico.
[g] In Northeast Coast estuaries, at least one of the five indicators is rated poor at sites representing 40% of total estuarine area.

Source: From United States Environmental Protection Agency, 2004, *National Coastal Condition Report II*, EPA-620/R-03/002, December 2004, www.epa.gov.

Table 10B.42 Average PCB and PAH Concentrations in Sediments from Selected Estuaries in the United States, 1984–1992 (Milligrams per Kilogram Dry Weight)

Estuary	Total PCBs	Total PAHs
Machias Bay, ME-Hog Island	24.07	132.20
Machias Bay, ME-Chance Island	25.40	221.03
Frenchman Bay, ME-Long Porcupine Island	28.90	234.99
Penobscot Bay, ME-Colt Head Island	59.31	834.41
Johns Bay, ME–Pemaquid	7.46	95.90
Casco Bay, ME-Great Chebeague Island	98.28	3,138.86
Casco Bay, ME-Cousins Island	161.75	3,097.25
Cape Elizabeth, ME-Richmond Island	13.88	257.17
Merrimack River, MA-Plum Island	33.19	927.98
Salem Harbor, MA-Folger Point	295.81	8,100.99
Boston Harbor, MA-President Roads	5961.75	16,069.24
Boston Harbor, MA-Deer Island	2671.75	8,267.36
Boston Harbor, MA-Quincy Bay	599.39	15,541.18
Boston Harbor, MA-Hull Bay	128.67	2,941.00
Boston Harbor, MA-Mystic River	1436.33	51,263.33
Massachusetts Bay, MA	2.63	28.97
Buzzards Bay, MA-West Island	171.64	653.93
New Bedford Harbor, MA-Clarks Point	2915.50	2,438.17
Narragansett Bay, RI-Prudence Island	265.00	2,899.00
Narragansett Bay, RI-Conanicut Island	309.11	2,152.35
Niantic Bay, CON-Black Point	41.72	704.85
Long Island Sound, NY-New Haven	237.33	3,885.00
Long Island Sound, NY-Norwalk	139.70	2,693.75
Long Island Sound, NY-Long Island Shoal	7.92	93.02
Long Island Sound, NY-Rock Point	9.13	42.26
Long Island Sound, NY-Lloyd Point	8.78	51.81
Long Island Sound, NY-Oak Neck Point	177.36	5,839.20
Hudson River, NY-Englewood Cliffs	947.33	7,843.33
Raritan Bay, NY-Upper Bay	162.37	5,891.03
Raritan Bay, NY-Gravesand Bay	337.67	7,697.50
Raritan Bay, NY-West Reach	386.30	7,385.70
Raritan Bay, NJ-East Reach	468.64	6,711.00
Raritan Bay, NJ-Lower Bay	546.87	6,607.25
Great Bay, NJ-Wells Island	146.00	1,086.67
Great Bay, NJ-Seven Island	85.00	895.12
Great Bay, NJ-Intercoastal Waterway	62.87	444.97
Delaware Bay, DE-Cherry Island Range	485.12	3,117.17
Delaware Bay, DE-Brandywine Shoal	99.05	491.28
Delaware Bay, DE-The Shears	61.10	339.67
Baltimore Harbor, MD, Fort McHenry Channel	616.08	12,282.12
Baltimore Harbor, MD-Brewerton Channel	299.23	13,528.33
Chesapeake Bay, MD-Gibson Island	144.45	3,691.02
Chesapeake Bay, MD-Chester River	75.43	2,116.03
Chesapeake Bay, MD-Kent Island	228.75	5,296.25
Chesapeake Bay, MD-Patuxent River	22.40	489.00
Chesapeake Bay, MD-Smith Island	17.36	355.51
Chesapeake Bay, VA-James River	84.10	2,215.33
Chesapeake Bay, VA-York River	37.14	258.50
Chesapeake Bay, VA-Elizabeth River	165.41	13,649.91
Pamlico Sound, NC-Jones Bay	21.89	443.59
Cape Fear River, NC-Horseshoe Shoal	6.63	21.00
Charleston Harbor, SC-Coastal	246.28	14,135.00
Charleston Harbor, SC-South Channel	29.05	2,082.26
Savannah River, GA-Elba Island	5.97	274.00
Sapelo Sound, GA-South Newport River	1.05	26.23
Sapelo Sound, GA-High Point	0.00	466.92
St. Johns River, FL-Trout River	—	23,228.20
St. Johns River, FL-West Mill Cove	98.35	3,368.49
St. Johns River, FL-Ortega River	—	4,863.00

(Continued)

Table 10B.42 (Continued)

Estuary	Total PCBs	Total PAHs
St. Johns River, FL-Piney Point	—	7,933.67
St. Johns River, FL-Orange Point	—	8,094.77
St. Lucie River, FL-Stuart	46.67	711.00
Biscayne Bay, FL-North Bay	61.20	270.00
Biscayne Bay, FL-Chicken Key	28.33	9.45
Charlotte Harbor, FL-Cape Haze	8.86	44.89
Tampa Bay, FL-Northern Tampa Bay	5.54	69.50
Apalachicola Bay, FL-St. George Island	10.77	218.07
St. Andrew Bay, FL-Military Point	110.33	1,266.43
Choctawhatchee Bay, FL	20.43	34.03
Pensacola Bay, FL	58.19	1,828.31
Mobile Bay, AL-North Point	19.33	134.27
Pascagoula River, MS-Escatawpa River	129.13	1,318.33
Pascagoula River, MS	71.50	525.56
Round Island, MS-Round Island	1.10	83.64
Heron Bay, MS-Heron Bay	11.72	129.29
Mississippi River Delta, LA-Southeast Pass	23.80	922.73
Mississippi River Delta, LA-Head of Passes	581.80	103.80
Barataria Bay, LA-Barataria Pass	3.18	169.77
Calcasieu River, LA-Prien Lake	901.67	133.20
Calcasieu River, LA-West Cove	41.67	5.27
Galveston Bay, TX-East Bay	—	197.00
Galveston Bay, TX-Trinity Bay	—	624.67
Galveston Bay, TX-Greens Bayou	785.07	4,594.17
Galveston Bay, TX-Goat Islands	233.60	1,698.20
Galveston Bay, TX-Morgans Point	32.07	413.93
Galveston Bay, TX-Eagle Point	45.03	194.12
Galveston Bay, TX-Texas City	—	281.90
Lavaca Bay, TX	50.70	263.17
San Antonio Bay, TX-Mosquito Point	3.12	9.93
San Antonio Bay, TX-San Antonio Bay	11.16	132.08
Corpus Christi Bay, TX-Long Reef	5.70	354.34
Lower Laguna Madre, TX-Laguna Heights	5.07	83.25
Lower Laguna Madre, TX-Long Island	6.33	1.00
San Diego Harbor, CA-Outside	10.30	107.88
San Diego Bay, CA-National City	167.84	2,156.23
San Diego Bay, CA-28th Street	393.08	5,096.15
San Diego Bay, CA-North	311.36	4,165.71
San Diego Bay, CA-Harbor Island	103.88	1,178.17
San Diego Bay, CA-Shelter Island	50.70	773.32
Mission Bay, CA-Outside	12.00	3.00
Dana Point Harbor, CA-Outside	8.59	10.64
San Pedro Bay, CA-Seal Beach	60.47	329.75
San Pedro Bay, CA-Long Beach	180.06	899.96
San Pedro Bay, CA-Outer Harbor	279.14	2,391.78
San Pedro Bay, CA-Cerritos Channel	800.20	8,724.80
Santa Monica Bay, CA-Southeast	89.43	212.33
Santa Monica Bay, CA-South	166.00	350.50
Santa Monica Bay, CA-Manhattan Beach	13.58	28.08
Santa Monica Bay, CA-West	118.07	354.69
Santa Monica Bay, CA-Deep	20.08	104.00
Santa Monica Bay, CA-North	75.27	34.33
San Luis Obispo, CA-San Luis Obispo	15.60	32.80
Estero Bay, CA-Estero Bay	1.20	3.40
Monterrey Bay, CA-Indian Head Beach	8.56	13.33
San Francisco Bay, CA-Redwood City	104.50	3,021.50
San Francisco Bay, CA-Hunters Point	92.46	8,388.11
San Francisco Bay, CA-Oakland Estuary	226.50	6,543.30
San Francisco Bay, CA-Southampton Shoal	37.96	1,357.39
San Francisco Bay, CA-Oakland Entrance	61.43	1,701.33

(Continued)

Table 10B.42 (Continued)

Estuary	Total PCBs	Total PAHs
San Francisco Bay, CA-Castro Creek	50.18	1,137.05
San Francisco Bay, CA-San Pablo Bay	24.25	470.12
Bodega Bay, CA-North	10.62	59.13
Coos Bay, OR-North Bend	20.95	831.94
Columbia River Mouth, WA-Desdemona Sands	5.15	53.08
Puget Sound, Nisqually Reach, WA	4.77	11.92
Puget Sound, WA-Commencement Bay	70.07	1,334.87
Puget Sound, WA-Elliott Bay	420.59	6,393.67
Boca de Quadra, AK-Bacrian Point	18.00	115.00
Lutak Inlet, AK-Chilkoot River Mouth	10.33	2.17
Skagway, AK-Skagway River	13.57	630.67
Nahku Bay, AK-East Side	7.02	98.33
Prince William Sound, AK-Port Valdez	13.57	630.67
Gulf of Alaska, AK-Kamishak Bay	16.67	16.33
Bering Sea, AK,-Dutch Harbor	58.83	715.00
Bering Sea, AK,-Port Moller	43.50	0.00
Chukchi Sea, AK-Red Dog Mine	24.80	4.80
Beaufort Sea, AK-Olitok Point	44.28	730.33
Beaufort Sea, AK-Prudhoe Bay	19.18	274.42

Note: PCBs, Polychlorinated biphenyls; PAHs, Polycyclic aromatic hydrocarbons; ND, Not detected.

Source: From Marmon, M.R., Gottholm, W., and Robertson, *A., 1998, A Summary of Chemical Contaminant Levels at Benthic Surveillance Project Sites (1984–1992)—NOAA Technical Memorandum NOS ORCA 124*, www.nos.noaa.gov.

Table 10B.43 Sewage Indicators In Sediments from Selected Estuaries in the Untied States, 1984

Estuary	Clostridium perfringens (cells/g)	Coprostanol (ng/g)
Casco Bay, ME	710.00	221.27
Merrimack River, MA	670.06	206.47
Salem Harbor, MA	57,000.00	7,040.23
Boston Harbor, MA	79,000.00	9,000.00
Buzzards' Bay, MA	413.00	1,376.60
Narragansett Bay, RI	220,00	647.80
East Long Island Sound, NY	290.00	17.00
West Long Island Sound, NY	2,090.00	957.11
Raritan Bay, NJ	24,373.52	5,402.00
Delaware Bay, DE	91.00	148.00
Lower Chesapeake Bay, VA	4.00	781.56
Pamlico Sound, NC	120.00	1,100.00
Charleston Harbor, SC	1,600.00	1,253.23
Sapelo Sound, GA	270.00	510.00
St. Johns River, FL	1,400.00	790.07
Charlotte Harbor, FL	33.00	692.47
Tampa Bay, FL	3.00	350.92
Apalachicola Bay, FL	74.00	687.56
Mobile Bay, AL	100.43	304.90
Round Island, MS	75.00	256.87
Mississippi River Delta, LA	1,200.00	523.98
Barataria bay, LA	45.00	395.68
Galveston Bay, TX	34.00	278.57
San Antonio Bay, TX	6.00	109.78
Corpus Christi Bay, TX	1.00	248.84
Lower Laguna Madre, TX	27.00	240.00
San Diego Harbor, CA	2,600.00	600.00
San Diego Bay, CA	121.79	33.00
Dana Point, CA	103.27	98.00
Seal Beach, CA	832.72	180.00
San Pedro Canyon, CA	6,596.78	780.00
Santa Monica Bay, CA	471.53	230.00
San Francisco Bay, CA	5,093.47	1,860.00
Bodega Bay, CA	29.00	120.00
Coos Bay, OR	30.96	230.00
Columbia River Mouth, OR/WA	261.05	580.00
Nisqually Reach, WA	35.49	5.33
Commencement Bay, WA	5,300.00	1,900.00
Elliott Bay, WA	7,725.95	370.00
Lutak Inlet, AK	91.78	83.33
Nahku Bay, AK	212.13	310.00

Source: From U.S. Department of Commerce, National Oceanic and Atmospheric Administration, National Ocean Survey, Ocean Assessments Division. 1987. *National Status and Trends Program for Marine Environmental Quality. Progress Report on Preliminary Assessment of Findings of the Benthic Surveillance Project, 1984*. Rockville, MD; U.S. Geological Survey, National Water Summary 1986.

Table 10B.44 Average Trace Metal Concentrations Detected in Sediments from Selected Estuaries in the United States, 1984–1992 (Milligrams per Kilogram Dry Weight)

Estuary	Silver	Arsenic	Cadmium	Chromium	Copper	Mercury	Nickel	Lead	Selenium	Tin	Zinc
Machias Bay, ME-Hog Island	0.13	5.90	—	122.64	9.02	—	15.23	21.37	0.25	3.03	50.67
Machias Bay, ME-Chance Island	0.06	7.89	0.08	66.71	11.13	0.05	21.26	19.27	0.14	2.08	55.52
Frenchman Bay, ME-Long Porcupine Island	0.09	10.34	0.18	82.97	17.13	0.05	40.23	27.90	0.42	3.75	102.45
Penobscot Bay, ME-Job Island	0.24	11.29	—	329.93	17.70	0.07	24.80	28.43	0.86	4.40	109.53
Penobscot Bay, ME-Colt Head Island	0.11	11.12	0.12	100.33	19.52	0.21	36.68	31.08	0.48	3.79	115.83
Casco Bay, ME-Great Chebeague Island	0.19	11.91	0.33	92.20	22.62	0.18	30.08	37.18	0.39	4.72	102.53
Casco Bay, ME-Cousins Island	0.25	10.20	0.20	183.94	14.79	0.09	20.70	39.26	0.63	4.01	83.17
Merrimac River, MA-Plum Island	0.03	4.25	0.11	27.70	5.02	0.06	5.02	21.46	0.04	3.94	28.53
Salem Harbor, MA-Folger Point	1.49	13.89	4.87	1,671.81	67.29	0.98	27.41	186.32	0.73	24.34	198.87
Boston Harbor, MA-President Roads	6.12	10.23	1.86	232.02	143.81	1.00	30.28	110.25	0.64	24.86	238.86
Boston Harbor, MA-Deer Island	2.87	10.12	1.30	121.72	88.66	0.67	29.90	95.51	0.30	11.35	156.18
Boston Harbor, MA-Quincy Bay	5.52	12.34	1.20	213.50	130.42	1.39	30.40	120.56	0.39	22.28	194.69
Boston Harbor, MA-Hull Bay	4.02	14.25	0.66	136.13	78.63	0.65	30.35	101.87	0.40	10.75	143.60
Boston Harbor, MA-Mystic River	2.41	26.20	1.81	120.53	126.08	0.74	35.57	158.93	0.50	9.66	223.97
Massachusetts Bay, MA-Plymouth Entrance	0.04	0.95	0.02	2.88	0.72	0.02	1.90	5.47	0.08	0.21	7.57
Buzzards Bay, MA-West Island	0.64	9.55	0.18	67.26	20.49	0.11	20.95	31.15	0.28	2.89	84.79
New Bedford Harbor, MA-Clarks Point	1.79	7.46	0.97	79.80	64.68	0.24	17.52	58.00	0.38	6.01	116.74
Narragansett Bay, RI-Prudence Island	3.24	9.65	0.85	119.72	147.99	0.60	29.86	112.69	0.51	19.64	218.37
Narragansett Bay, RI-Conanicut Island	1.18	8.29	0.41	95.58	76.22	0.34	23.84	61.77	0.35	7.74	143.13
Niantic Bay, CON-Black Point	0.17	3.58	0.15	50.23	9.63	0.03	15.04	20.05	0.07	2.72	114.13
Long Island Sound, NY-Long Island Shoal	0.08	3.62	0.10	42.31	7.92	0.08	10.26	18.50	0.04	2.19	56.76
Long Island Sound, NY-Rocky Point	0.05	3.34	0.06	30.32	4.40	0.04	9.17	18.68	0.27	1.12	40.76
Long Island Sound, NY-Lloyd Point	0.92	7.13	0.47	67.84	57.31	0.25	23.78	52.94	0.21	3.57	215.42
Long Island Sound, NY-Oak Neck Point	1.58	8.01	0.87	123.04	114.56	0.49	32.65	77.12	0.44	9.23	232.30
Raritan Bay, NY-Upper Bay	1.69	7.96	0.84	59.57	44.15	0.55	19.23	74.26	0.19	6.25	122.31
Raritan Bay, NY-Gravesend Bay	3.62	14.94	1.73	120.76	100.44	1.41	34.10	124.64	0.41	13.08	305.15
Raritan Bay, NY-West Reach	3.60	21.67	1.61	138.33	111.69	1.94	36.07	160.04	0.54	15.94	308.87
Raritan Bay, NJ-East Reach	4.68	20.25	1.51	154.68	142.86	2.43	36.86	158.84	0.68	21.10	270.45
Raritan Bay, NJ-Lower Bay	3.85	20.07	2.63	230.89	160.92	2.12	37.61	157.68	0.98	15.86	413.19
Great Bay, NJ-Wells Island	0.25	8.68	0.38	98.42	21.93	0.09	27.87	47.03	0.78	3.36	137.00
Great Bay, NJ-Seven Island	0.69	11.69	0.53	107.65	29.62	0.40	29.53	43.93	0.46	5.12	156.00
Great Bay, NJ-Intercoastal Waterway	0.36	6.85	0.28	45.63	12.05	0.20	14.99	26.80	0.32	3.57	75.96
Delaware Bay, DE-Brandywine Shoal	0.19	7.57	0.34	46.00	10.45	0.10	13.89	21.35	0.12	2.57	81.24
Delaware Bay, DE-The Shears	0.05	5.66	0.13	226.33	12.21	0.07	15.83	27.47	0.34	2.97	90.20
Baltimore Harbor, MD, Fort McHenry Channel	1.94	30.43	3.11	515.67	246.33	0.73	69.80	172.00	1.59	74.17	634.33
Baltimore Harbor, MD-Brewerton Channel	1.12	30.23	0.68	196.78	74.92	0.33	60.07	123.78	1.02	18.39	451.02
Chesapeake Bay, MD-Gibson Island	1.12	30.23	0.68	196.78	74.92	0.33	60.07	123.78	1.02	18.39	451.02
Chesapeake Bay, MD-Chester River	0.40	12.68	0.46	81.42	32.06	0.13	46.49	53.41	0.63	3.24	202.27
Chesapeake Bay, MD-Kent Island	0.18	13.59	0.77	83.87	46.13	0.19	61.67	61.63	1.60	4.38	316.33
Chesapeake Bay, MD-Smith Island	0.08	7.80	0.24	52.25	12.32	0.06	18.34	17.19	0.39	1.65	81.17
Chesapeake Bay, MD-Patuxent River	0.33	9.48	1.03	—	26.59	0.30	—	38.24	—	3.20	170.75
Chesapeake Bay, VA-James River	0.36	8.13	0.33	—	27.95	0.32	—	44.74	—	3.07	174.93

Chesapeake Bay, VA-York River	0.08	5.61	0.24	55.24	9.82	0.09	17.17	15.49	0.18	2.08	64.08
Chesapeake Bay, VA-Elizabeth River	0.51	10.58	1.20	71.97	104.60	0.64	22.87	106.23	0.29	7.37	406.67
Pamlico Sound, NC-Jones Bay	0.08	8.22	0.28	67.70	11.01	0.11	18.77	24.96	0.73	2.82	72.04
Charleston Harboar, SC-South Channel	0.11	12.79	0.16	67.21	15.12	0.09	18.10	23.11	0.44	2.34	68.96
Sapelo Sound, GA-South Newport River	0.02	3.66	0.05	34.47	3.36	—	4.62	9.48	0.15	1.05	26.23
Sapelo Sound, GA-Barbour Island River	0.03	4.39	0.07	35.89	3.44	—	6.79	9.11	0.15	1.07	27.91
Sapelo Sound, GA-Sapelo Sound Inlet	0.02	3.34	0.06	27.02	3.45	—	6.01	8.84	0.09	1.03	20.71
Sapelo Sound, GA-High Point	0.03	8.08	0.11	45.81	6.69	0.05	10.70	15.18	0.29	1.44	42.40
Sapelo Sound, GA-Dog Hammock	0.03	4.95	0.15	38.38	3.83	—	7.53	9.09	0.14	1.06	29.38
St. Johns River, FL-Trout River	0.22	3.41	0.33	58.96	22.13	—	14.14	40.84	0.52	2.85	96.65
St. Johns River, FL-West Mill Cove	0.30	5.33	0.40	50.72	18.39	0.16	12.78	38.30	0.57	2.64	107.32
St. Johns River, FL-Ortega River	0.61	3.03	0.97	66.23	45.00	—	15.65	92.30	1.28	4.81	194.38
St. Johns River, FL-Piney Point	0.25	4.03	0.33	78.63	16.77	—	18.62	35.37	1.51	2.57	86.39
St. Johns River, FL-Orange Point	0.06	1.74	0.14	49.66	6.46	—	12.96	13.89	1.46	2.11	33.20
Charlotte Harbor, FL-Cape Haze	0.01	1.35	0.10	19.34	1.62	0.04	2.86	4.21	0.30	0.76	7.82
Tampa Bay, FL-Northern Tampa Bay	0.11	1.20	0.27	21.52	4.37	0.07	4.10	7.65	0.28	0.86	15.02
Apalachicola Bay, FL-St. George Island	0.06	18.45	0.07	78.75	18.54	0.10	25.65	29.48	0.55	3.57	95.65
St. Andrew Bay, FL-Military Point	0.30	13.76	0.28	82.28	21.04	0.21	21.43	54.15	0.65	1.80	93.23
Choctawhatchee Bay, FL	0.11	38.31	3.22	127.75	22.02	0.27	39.90	41.48	0.70	3.95	161.09
Choctawhatchee Bay, FL-Destin Harbor	0.16	19.97	0.73	39.78	85.14	0.19	10.01	57.37	0.55	2.19	182.78
Pensacola Bay, FL	0.17	20.38	0.21	122.26	21.63	0.17	28.55	38.75	0.81	3.21	129.54
Mobile Bay, AL-North Point	0.10	17.09	0.11	101.34	19.11	0.12	34.91	32.59	0.53	3.59	153.77
Pascagoula River, MS	0.14	6.72	0.21	43.43	15.05	0.11	12.97	18.22	0.28	1.78	82.48
Round Island, MS-Round Island	0.07	7.71	0.08	48.97	8.22	0.08	15.48	17.01	0.26	2.01	64.14
Herron Bay, MS-Heron Bay	0.10	6.84	0.19	41.18	10.79	0.08	14.89	17.05	0.31	1.52	53.36
Mississippi River Delta, LA-Southeast Pass	0.16	8.35	0.42	58.66	18.57	0.07	27.24	20.02	0.33	1.98	86.90
Barataria Bay, LA-Barataria Pass	0.09	5.96	0.17	39.89	9.12	0.05	16.32	14.83	0.21	1.44	55.21
Lake Pontchartrain, LA-North Shore	0.12	5.70	0.20	—	13.08	0.11	—	17.57	—	1.55	73.13
Lake Pontchartrain, LA-South Shore	0.22	6.53	0.37	—	20.89	0.14	—	28.11	—	2.38	125.84
Galveston Bay, TX-East Bay	0.09	3.23	0.06	28.63	5.83	—	11.68	11.23	0.11	0.89	38.83
Galveston Bay, TX-Trinity Bay	0.12	5.89	0.15	53.39	11.15	—	20.37	18.06	0.25	2.34	61.49
Galveston Bay, TX-Cedar Bayou	0.11	38.31	3.22	127.75	22.02	0.27	39.90	41.48	0.70	3.95	161.09
Galveston Bay, TX-Greens Bayou	0.62	7.11	0.96	71.69	41.40	0.41	23.58	63.65	0.36	3.92	183.80
Galveston Bay, TX-Goat Islands	0.18	6.51	0.26	52.09	16.93	0.17	20.04	33.53	0.41	2.25	103.24
Galveston Bay, TX-Morgans Point	0.16	6.36	0.19	55.31	16.04	0.07	20.22	30.82	0.29	2.19	81.70
Galveston Bay, TX-Clear Lake	0.18	4.98	0.16	35.56	23.55	0.07	15.47	23.47	0.27	2.05	72.90
Galveston Bay, TX-Eagle Point	0.09	3.90	0.07	30.73	8.70	0.04	10.31	14.24	0.19	1.52	42.06
Galveston Bay, TX-Texas City	0.22	6.94	0.09	64.79	15.69	—	22.21	25.43	0.15	3.79	79.68
Lavaca Bay, TX	0.11	6.29	0.14	36.05	11.11	0.25	14.04	21.38	0.22	1.60	58.25
San Antonio Bay, TX-San Antonio Bay	0.07	4.41	0.12	31.92	6.92	0.05	10.37	12.96	0.15	1.24	38.09
Corpus Christi Bay, TX-Long Reef	0.09	6.04	0.35	47.73	12.47	0.07	15.32	20.41	0.25	2.35	104.25
Arroyo Colorado, TX-Arroyo City	0.09	2.98	0.15	—	7.86	0.21	—	11.04	—	1.06	72.12
Lower Laguna Madre, TX-Laguna Heights	0.07	6.90	0.08	17.37	5.35	0.04	5.48	12.35	0.12	1.63	31.35
San Diego Bay, CA-Outside	0.19	6.60	0.37	43.30	8.18	0.05	6.03	11.35	0.12	1.85	54.72
Machias Bay, ME-Hog Island	0.13	5.90	—	122.64	9.02	—	15.23	21.37	0.25	3.03	50.67

(Continued)

Table 10B.44 (Continued)

Estuary	Silver	Arsenic	Cadmium	Chromium	Copper	Mercury	Nickel	Lead	Selenium	Tin	Zinc
San Diego Bay, CA-National City	1.43	9.44	0.36	60.13	117.15	0.43	22.87	44.35	0.13	4.81	206.52
San Diego Bay, CA-28th Street	1.09	10.16	0.65	84.98	176.49	0.71	21.38	79.14	0.18	10.19	279.33
San Diego Bay, CA-North	0.78	7.96	0.53	58.71	96.83	0.60	17.00	52.93	0.26	10.92	226.33
San Diego Bay, CA-Harbor Island	0.28	5.47	0.32	55.63	37.57	0.24	10.33	24.82	0.22	6.74	116.00
San Diego Bay, CA-Shelter Island	0.20	5.09	0.25	35.49	30.83	0.15	14.64	24.83	0.29	11.49	87.64
Mission Bay, CA-Outside	0.34	2.28	0.12	40.03	1.62	0.06	16.42	7.80	0.13	2.67	42.29
Ocean Side Harbor, CA-Outside	0.04	3.03	0.10	73.48	4.71	0.02	20.85	8.57	0.07	2.54	70.87
Dana Point Harbor, CA-Outside	0.17	5.75	0.31	35.87	6.39	0.24	14.22	14.18	0.23	2.36	48.94
Dana Point Harbor, CA-Inside Harbor	0.11	8.34	1.20	59.00	40.00	0.04	29.50	26.20	0.53	3.06	118.00
San Pedro Bay, CA-Seal Beach	0.35	6.70	0.24	82.88	24.60	0.31	36.70	32.23	0.13	1.97	111.17
San Pedro Bay, CA-Long Beach	0.30	9.36	0.72	80.13	58.22	0.24	30.24	75.74	0.45	6.12	184.61
San Pedro Bay, CA-Mid Harbor	—	15.87	1.31	151.67	201.00	0.22	53.37	49.83	2.21	—	205.67
San Pedro Bay, CA-Outer Harbor	0.49	8.31	1.00	85.70	106.27	0.32	37.54	38.05	1.39	6.57	173.93
San Pedro Bay, CA-Cerritos Channel	0.13	16.00	0.86	77.42	125.67	0.68	45.97	129.28	0.62	18.73	307.83
Santa Monica Bay, CA-Southeast	2.24	7.34	1.24	90.83	36.85	0.26	26.15	17.62	0.17	2.30	85.33
Santa Monica Bay, CA-South	0.91	5.75	1.19	121.17	36.45	0.29	29.92	25.17	0.30	2.97	91.73
Santa Monica Bay, CA-Manhattan Beach	0.36	8.95	0.29	60.23	10.05	—	12.59	25.73	0.15	1.12	35.48
Santa Monica Bay, CA-West	4.17	4.10	1.65	81.78	43.15	0.13	16.56	22.48	0.33	4.33	87.70
San Luis Obispo, CA-San Luis Obispo	0.60	1.48	0.39	130.00	7.49	0.08	30.40	4.90	—	0.80	43.60
Estero Bay, CA-Estero Bay	0.50	5.01	0.08	5,770.00	5.72	0.08	49.20	—	—	—	38.50
Monterrey Bay, CA-Indian Head Beach	0.12	5.93	0.19	72.80	4.11	0.06	20.45	11.81	0.14	1.53	24.18
Monterrey Bay, CA-Moss Landing	1.77	4.39	0.14	233.33	12.20	—	16.10	14.77	0.15	1.81	54.67
Farallon Islands, CA	—	6.39	3.33	210.00	8.00	0.09	41.90	10.30	0.29	2.27	48.00
San Francisco Bay, CA-Redwood City	0.34	9.90	0.16	167.17	55.10	0.29	92.15	30.15	0.56	6.68	165.50
San Francisco Bay, CA-Hunters Point	0.45	12.16	0.29	229.43	54.13	0.35	76.75	29.53	0.38	3.07	137.71
San Francisco Bay, CA-Islais Creek Channel	0.59	9.40	0.51	145.00	73.00	0.32	108.10	47.20	0.26	2.99	174.50
San Francisco Bay, CA-Oakland Sanctuary	0.42	12.88	0.73	233.84	105.06	1.13	112.38	128.98	0.90	16.32	287.50
San Francisco Bay, CA-Oakland Entrance	0.34	13.00	0.16	196.00	71.67	0.50	104.00	43.50	0.37	1.87	171.67
San Francisco Bay, CA-Southampton Shoal	0.32	9.09	0.24	202.67	30.93	0.16	64.73	17.15	0.21	2.23	102.14
San Francisco Bay, CA-Castro Creek	0.16	8.50	0.18	199.33	46.27	0.16	83.12	30.23	0.19	4.43	128.53
Bodega Bay, CA-North	0.19	4.55	0.16	409.52	7.44	0.14	48.49	4.85	0.13	2.31	44.86
Humbolt Bay, CA-Indian Island	0.08	8.54	0.23	453.67	6.95	0.06	60.07	—	0.08	—	45.60
Coos Bay, OR-North Bend	0.13	6.67	0.35	82.82	14.14	0.15	27.93	13.48	0.30	2.45	68.16
Columbia River Mouth, OR-Youngs Bay	0.62	3.08	0.13	50.33	35.67	0.03	17.87	17.80	0.15	2.71	132.33
Columbia River Mouth, WA-Desdemona Sands	0.12	3.18	0.45	35.73	16.45	0.10	23.94	7.79	0.25	2.60	90.79
Puget Sound, Nisqually Reach, WA	0.23	1.48	0.45	95.33	16.02	0.29	35.77	13.13	0.13	2.54	104.09
Puget Sound, Commencement Bay, WA	0.25	5.20	0.45	57.95	49.60	0.17	30.65	20.07	0.48	2.29	91.45
Puget Sound, WA-Elliott Bay	0.23	8.08	0.54	90.13	101.63	0.61	38.24	42.78	0.34	2.35	168.03
Boca de Quadra, AK-Bacrian Point	0.29	1.90	0.44	52.67	22.00	0.04	7.49	17.87	0.74	3.21	104.33
Lutak Inlet, AK-Chilkoot River Mouth	0.27	1.54	0.64	72.13	28.83	0.14	17.12	18.27	0.43	1.62	180.17
Skagway, AK-Skagway River	0.42	3.45	0.40	43.00	16.90	0.08	4.28	52.20	0.15	2.76	169.33
Nahku Bay, AK-East Side	0.30	1.51	1.09	23.27	9.80	0.23	11.57	43.30	0.87	0.21	191.33
Prince William Sound, AK-Port Valdez	0.36	9.74	0.21	148.33	59.67	0.08	16.60	23.13	0.41	2.79	150.00

Machias Bay, ME-Hog Island	0.13	5.90	—	122.64	9.02	—	15.23	21.37	0.25	3.03	50.67
Gulf of Alaska, AK-Kamishak Bay	0.26	1.75	0.19	82.33	26.67	0.06	9.32	14.67	0.19	2.70	89.67
Bering Sea, AK-Dutch Harbor	0.11	2.22	0.70	30.67	49.67	0.24	5.13	12.93	0.70	2.77	85.33
Bering Sea, AK-Port Moller	0.06	1.67	0.43	63.00	13.33	0.06	5.82	9.84	0.06	3.37	100.67
Bering Sea, AK-Kvichak Bay	0.26	1.75	0.19	82.33	26.67	0.06	9.32	14.67	0.19	2.70	89.67
Chukchi Sea, AK-Red Dog Mine	0.04	5.69	—	213.33	10.00	0.06	32.63	8.49	0.17	2.10	51.00
Beaufort Sea, AK-Olitok Point	0.13	2.33	0.32	70.85	18.29	0.20	23.11	14.02	0.41	1.34	80.98
Beaufort Sea, AK-Prudhoe Bay	0.17	1.34	0.26	58.53	11.83	0.09	17.79	7.88	0.32	2.31	72.23

Source: Abstracted from Marmon, M.R., Gottholm, W., and Robertson, *A., 1998, A Summary of Chemical Contaminant Levels at Benthic Surveillance Project Sites (1984–1992)—NOAA Technical Memorandum NOS ORCA 124*, www.nos.noaa.gov.

Table 10B.45 Cadmium and Lead in Sediments of the Chesapeake Bay Mainstem

	Concentration		Aquatic Life Benchmark			
Contaminant	**Median (ppm)**	**Maximum (ppm)**	**NOEL[a] (ppm)**	**PEL[b] (ppm)**	**Location of Maximum Concentrations**	**Trends**
Cadmium	0.4	2.9	1.0	7.5	Baltimore region south to the Little Choptank River; mouth of Potomac River	Concentrations are declining
Lead	35	86	21	160	Baltimore region	Concentrations are declining in some areas

[a] No Observed Effect Level. Level above which impacts are considered "possible."
[b] Probable Effect Level. Level above which impacts are considered "probable."

Source: From United States Environmental Protection Agency, 2000, *Deposition of Air Pollutants to the Great Waters, Third Report to Congress*, EPA-453/R-00-005.

Original Source: From Eskin et al., 1996.

Table 10B.46 Temporal Trends in Chemical Concentrations Measured Nationally at 206 Mussel Watch Project Sites and at 30 Sites in the NERRS for Which Data Exist for Six Years During 1986–1999

	Trend				Trend		
Organics*	**I**	**D**	**NT**	**Element**	**I**	**D**	
ΣCdane	1	85(8)	120	As	9	15(3)	182
ΣDDTs	1(1)	54(5)	151	Cd	7(4)	20(2)	179
ΣDiel	4(2)	32(4)	170	Cu	9	10(1)	187
ΣPCBs	5(1)	30	171	Hg	13(1)	14(2)	179
ΣPAHs	18(3)	26(1)	162	Ni	11(2)	6(1)	189
ΣBTs	0	100(10)	106	Pb	11(2)	12(3)	183
HCB	16	7	183	Se	14(2)	4	188
Lindane	3	31	172	Zn	7(1)	15(4)	184
Mirex	17	6	183				

Note: I, Increasing; D, Decreasing; NT, No trend. Increasing and decreasing trends for NERRs are given in parentheses.*Individual organic compound concentrations have been aggregated into these groups:

ΣBTs, the sum of the concentrations of tributyltin and its breakdown products dibutyltin and monobutyltin;
ΣCdance, the sum of *cis*-chlordane, *trans*-nonachlor, heptachlor and heptachlorepoxide;
ΣDDTs, the sum of concentrations of DDTs and its metabolities, DDEs and DDDs;
ΣDield, the sum of concentrations of aldrin and dieldrin;
ΣPAHs, the sum of concentrations of the 18 PAH compounds;
ΣPCBs, the sum of the concentrations of homologs, which is approximately twice the sum of the 18 congeners.

Source: From Lauenstein, G.G. and Cantillo, A.Y., 2002, Contaminant Trends in US National Estuaries Research Reserves, Nation Status and Trends Program for Marine Environmental Quality, *NOAA NAS Technical Memorandum NCCOS 156*, October 2002, www.noaa.gov.

Table 10B.47 Trophic Status of Lakes in the United States in 1998

Jurisdiction	Assessed		Oligotrophic		Mesotrophic		Eutrophic		HyperEutrophic		Dystrophic	
	Number of Significant Public Lakes	Acreage of Significant Public Lakes	Number of Significant Public Lakes	Acreage of Significant Public Lakes	Number of Significant Public Lakes	Acreage of Significant Public Lakes	Number of Significant Public Lakes	Acreage of Significant Public Lakes	Number of Significant Public Lakes	Acreage of Significant Public Lakes	Number of Significant Public Lakes	Acreage of Significant Public Lakes
Alabama	33	276,436	3	585	8	54,077	21	210,539	1	11,235	0	0
Alaska	—	—	—	—	—	—	—	—	—	—	—	—
Arizona	11	—	0	—	3	—	8	—	—	—	—	—
Arkansas	—	—	—	—	—	—	—	—	—	—	—	—
California	—	—	—	—	—	—	—	—	—	—	—	—
Colorado	38	47,530	7	5,272	14	15,722	13	15,957	4	10,579	—	—
Connecticut	—	—	—	—	—	—	—	—	—	—	—	—
Cortina Rancheria	—	—	—	—	—	—	—	—	—	—	—	—
Delaware	—	—	—	—	—	—	—	—	—	—	—	—
District of Columbia	—	—	—	—	—	—	—	—	—	—	—	—
Florida	262	1,571	191	604	51	802	13	71	7	94	—	—
Georgia	—	—	—	—	—	—	—	—	—	—	—	—
Hawaii	—	—	—	—	—	—	—	—	—	—	—	—
Idaho	—	—	—	—	—	—	—	—	—	—	—	—
Illinois	329	157,408	6	180	30	3,912	159	79,398	134	73,918	—	—
Indiana	164	54,153	42	4,761	62	37,389	41	10,205	19	1,798	0	0
Iowa	115	41,190	—	—	—	—	115	41,190	—	—		
Kansas	240	123,632	3	140	36	22,052	129	98,521	64	2,919	0	0
Kentucky	105	217,480	15	72,143	33	42,972	54	102,237	3	128	0	0
Louisiana	—	—	—	—	—	—	—	—	—	—	—	—
Maine	—	—	—	—	—	—	—	—	—	—	—	—
Manzanita Band	—	—	—	—	—	—	—	—	—	—	—	—
Maryland	58	21,010	0	0	16	15,172	42	5,838	0	0	0	0
Massachusetts	593	64,688	8	25,790	150	17,057	380	18,912	54	2,892	1	37
Michigan	730	491,931	115	172,591	375	175,307	207	124,881	33	19,152	—	—
Minnesota	1,984	2,131,026	309	210,108	723	1,099,929	667	645,241	285	175,748	0	0
Mississippi	—	—	—	—	—	—	—	—	—	—	—	—
Missouri	145	—	8	—	37	—	89	—	11	—	—	—
Montana	177	797,184	49	289,569	71	425,599	46	81,495	1	500	10	22
Nebraska	81	121,610	2	1,601	3	3,023	29	94,393	47	22,593	—	—
Nevada	17	319,946	3	133,230	12	133,116	2	53,600	—	—	—	—
New Hampshire	671	155,773	199	115,924	315	31,672	157	8,177	—	—	—	—
New Jersey	116	10,462	—	—	3	111	113	10,351	—	—	—	—
New Mexico	—	—	—	—	—	—	—	—	—	—	—	—
New York	—	—	—	—	—	—	—	—	—	—	—	—
North Carolina	161	311,236	44	103,130	29	75,898	70	112,820	4	404	14	18,984
North Dakota	124	617,330	0	0	20	503,386	49	19,152	55	94,792	—	—

(Continued)

Table 10B.47 (Continued)

Jurisdiction	Assessed		Oligotrophic		Mesotrophic		Eutrophic		HyperEutrophic		Dystrophic	
	Number of Significant Public Lakes	Acreage of Significant Public Lakes	Number of Significant Public Lakes	Acreage of Significant Public Lakes	Number of Significant Public Lakes	Acreage of Significant Public Lakes	Number of Significant Public Lakes	Acreage of Significant Public Lakes	Number of Significant Public Lakes	Acreage of Significant Public Lakes	Number of Significant Public Lakes	Acreage of Significant Public Lakes
Ohio	—	—	—	—	—	—	—	—	—	—	—	—
Oklahoma	199	624,343	14	10,568	69	105,325	77	342,706	39	165,744	0	0
Oregon	201	491,255	58	35,280	72	75,212	60	191,310	11	189,453	0	0
Pennsylvania	66	76,122	—	—	13	6,268	39	44,630	14	25,224	—	—
Puerto Rico	18	—	3	—	3	—	12	—	—	—	—	—
Rhode Island	62	7,307	21	1,900	28	4,089	10	1,199	2	99	1	20
South Carolina	27	452,654	—	—	14	247,414	13	205,240	—	—	—	—
South Dakota	112	132,159	2	1,199	10	23,205	37	30,526	63	77,229	0	0
Tennessee	122	538,438	21	100,346	38	320,408	39	73,338	24	44,346	—	—
Texas	—	—	—	—	—	—	—	—	—	—	—	—
Torres—Martinez Desert Band	—	—	—	—	—	—	—	—	—	—	—	—
Utah	129	460,561	47	285,154	57	59,191	24	116,166	1	50	—	—
Vermont	202	42,299	33	9,817	121	25,404	30	6,205	2	473	16	400
Virginia	—	—	—	—	—	—	—	—	—	—	—	—
Washington	—	—	—	—	—	—	—	—	—	—	—	—
West Virginia	81	21,423	17	7,724	31	5,335	33	8,365	—	—	—	—
Wisconsin	—	—	—	—	—	—	—	—	—	—	—	—
Wyoming	—	—	—	—	—	—	—	—	—	—	—	—
Total	7,373	8,808,157	1,220	1,587,615	2,447	3,529,046	2,778	2,752,663	878	919,371	42	19,463

Note: Oligotrophic, Clear waters with little organic matter or sediment and minimum biological activity; Mesotrophic, Waters with more nutrients, and therefore, more biological productivity; Eutrophic, Waters extremely rich in nutrients, with high biological productivity. Some species may be choked out; Hypereutrophic, Murky, highly productive waters, closest to the wetland status. Many clearwater species cannot survive; Dystrophic, Low in nutrients, highly colored with dissolved humic organic material. (Not necessarily a part of the natural trophic progression.); —, no data.

Source: From United States Environmental Protection Agency, 2000, *National Water Quality Inventory: 1998 Report to Congress*, www.epa.gov.

Table 10B.48 Lake Acres Meeting Fishable and Swimmable Goals of the Clean Water Act in the United States, by Jurisdiction, in 2000

	Fishing			Swimming		
Jurisdiction	**Total Assessed**	**Full Support**	**Threatened**	**Total Assessed**	**Full Support**	**Threatened**
Alabama	464,815	391,952	12,650	418,703	318,593	82,955
Alaska	0	—	—	0	—	—
American Samoa	0	—	—	0	—	—
Arizona	135,451	134,896	—	135,379	134,320	—
Arkansas	355,954	339,004	—	339,004	339,004	—
California	484,834	104,952	64,678	634,251	202,876	77,465
Colorado	5,975	0	0	56,658	56,650	0
Connecticut	27,601	26,809.30	0	26,049	16,767	9,011
Delaware	0	—	—	2,954	911	0
District of Columbia	238	0	0	238	0	0
Florida	883,840	654,720	0	1,260,800	760,960	110,080
Georgia	0	—	—	0	—	—
Guam	0	—	—	0	—	—
Hawaii	0	—	—	0	—	—
Idaho	0	—	—	0	—	—
Illinois	123,702	100,646	0	152,628	22,129	0
Indiana	45,540	0	0	0	0	0
Iowa (flood control reservoirs)	29,850	29,850	0	40,850	40,850	0
Iowa (lakes)	21,067	20,844	59	22,924	16,091	1,041
Kansas	13,684	—	13,683	188,506	—	47,903
Kentucky	205,712	197,502	0	215,646	215,427	0
Louisiana	0	—	—	492,913	453,343	0
Maine	987,283	987,283	0	987,283	879,314	71,105
Maryland	21,010	20,910	—	5,069	5,069	—
Massachusetts	10,674	193	0	21,015	2,022	4,190
Michigan	889,600	0	—	3,770	—	—
Minnesota	0	—	—	2,591,796	1,769,686	557
Mississippi	275,720	134,638	112,579	19,821	19,821	0
Missouri	293,305	292,365		262,372	261,757	—
Montana	283,747	215,435	7,550	508,922	205,107	0
N. Mariana Islands	0	—	—	0	—	—
Nebraska	114,734	114,734	—	4,083	4,083	—
Nevada	0	—	—	168,354	168,354	—
New Hampshire	168,002	168,002	0	160,406	158,034	1,085
New Jersey	14,245	0	114	16,820	11,343	0
New Mexico	109,909	410	—	0	—	—
New York	151,557	—	0	125,387	—	32,371
North Carolina	310,727	275,547.00	—	209,819	204,626	—
North Dakota	518,175	—	—	687,315	511,376	28,881
Ohio	103,867	28,682	62,385	78,175	641	51,921
Oklahoma	0	—	—	530,124	61,635	271,024
Oregon	31,489	0	13,008	507,536	479,174	9,428
Pennsylvania	0	—	—	0	—	—
Puerto Rico	0	—	—	12,146	7,741	1,878
Rhode Island	175	0	0	14,493	13,792	0
South Carolina	0	—	—	313,306.0	310,027.9	—
South Dakota	31,438	31,438	—	48,468	48,468	—
Tennessee	0	—	—	494,479	395,923	0
Texas	620,092	265,599	0	480,467	480,067	0
Utah	460,642	460,642	0	162,760	161,760	—
Vermont	51,739	30,781	0	52,943	34,256	10,712
Virginia	116,565	45,487	71,078	109,574	109,469	0
Washington	0	—	—	0	—	—
West Virginia	48	48	0	4,430	—	4,430
Wisconsin	203,704	116,474	22,504	93,663	3,623	1,367
Wyoming	0	—	—	0	—	—
Total	**8,566,710**	**5,189,844**	**380,288**	**12,662,298**	**8,885,090**	**817,405**
Percent of assessed		**60.6%**	**4.4%**		**70.2%**	**6.5%**

Note: Supporting, Fully supporting of all uses; Threatened, Fully supporting all uses but threatened for one or more uses; —, no data

Source: From Abstracted from United States Environmental Protection Agency, 2002, *National Water Quality Inventory 2000 Report*, EPA-841-R-02-001, www.epa.gov.

Table 10B.49 Summary of Acres of Fully Supporting, Threatened, Imparied Waters in Assessed Lakes in the United States by Jurisdiction, in 2000

Jurisdiction	Total Lake Acres	Assessed	Full Support	Threatened	Impaired	Not Attainable
Alabama	490,472	464,811	217,431	131,587	115,793	0
Alaska	12,787,200	16,376	11,438	0	4,938	0
American Samoa	—	0	0	0	0	0
Arizona	400,720	135,451	118,361	0	17,090	0
Arkansas	514,245	355,954	339,004	0	16,950	0
California	1,672,684	754,737	175,282	64,636	514,819	0
Colorado	164,029	62,920	56,669	0	6,251	0
Connecticut	64,973	27,669	19,145	6,984	1,540	0
Delaware	2,954	0	0	0	0	0
District of Columbia	238	238	0	0	238	0
Florida	2,085,120	1,683,000	771,840	110,080	801,080	0
Georgia	425,382	402,849	65,166	0	337,683	0
Guam	169	0	0	0	0	0
Hawaii	2,168	0	0	0	0	0
Idaho	700,000	0	0	0	0	0
Illinois	309,340	154,795	7,855	0	146,940	0
Indiana	142,871	71,120	25,580	0	45,540	0
Iowa (flood control reservoirs)	—	40,850	19,000	16,950	4,900	0
Iowa (lakes)	161,366	43,268	10,336	18,695	14,237	0
Kansas	188,506	188,506	0	26,884	161,622	0
Kentucky	228,385	217,422	100,447	94,839	22,136	0
Louisiana	1,078,031	518,176	40,259	1,926	475,991	0
Maine	987,283	987,283	758,081	80,134	149,068	0
Maryland	77,965	21,010	8,922	0	12,087	0
Massachusetts	151,173	67,749	26,965	652	39,425	706
Michigan	889,600	891,225	0	1,625	889,600	0
Minnesota	3,290,101	2,591,796	1,769,686	557	821,553	0
Mississippi	500,000	291,721	190,239	92,655	8,827	0
Missouri	293,305	293,305	110,189	122,241	60,875	0
Montana	844,802	547,929	64,146	7,550	476,233	0
N. Mariana Islands	—	0	0	0	0	0
Nebraska	280,000	127,926	116,958	0	10,968	0
Nevada	533,239	168,446	168,446	0	0	0
New Hampshire	168,017	160,583	153,191	1,123	6,269	0
New Jersey	72,235	18,359	5,550	12,409	400	0
New Mexico	997,467	154,550	30,410	0	124,140	0
New York	790,782	402,486	0	90,944	311,542	0
North Carolina	311,071	310,513	305,247	0	5,266	0
North Dakota	714,910	0	0	0	0	0
Ohio	118,461	78,175	641	51,921	25,613	0
Oklahoma	1,041,884	592,147	76,188	75,677	440,282	0
Oregon	618,934	507,536	296,173	88,786	122,577	0
Pennsylvania	161,445	42,421	16,157	0	26,264	0
Puerto Rico	12,146	12,146	6,404	1,878	3,864	0
Rhode Island	21,796	16,554	13,742	5	2,808	0
South Carolina	407,505	313,865	74,044	0	239,821	0
South Dakota	750,000	138,857	22,831	0	116,026	0
Tennessee	538,060	530,619	412,538	0	118,081	0
Texas	1,994,600	1,547,955	858,967	97,522	591,466	0
Utah	481,638	460,642	321,453	0	139,189	0
Vermont	228,920	53,350	10,452	12,488	30,410	0
Virginia	149,982	116,399	25,265	87,254	3,880	0
Washington	466,296	243,749	151,763	0	91,986	0
West Virginia	22,373	21,523	2,426	6,295	12,801	0
Wisconsin	944,000	230,006	52,100	44,606	133,300	0
Wyoming	325,048	0	0	0	0	0
Total	**40,603,893**	**17,078,967**	**8,026,988**	**1,348,903**	**7,702,370**	**706**
Percent of assessed			**47.0%**	**7.9%**	**45.1%**	**0.0%**

Note: Supporting, Fully supporting of all uses; Threatened, Fully supporting all uses but threatened for one or more uses; Impaired, Partially or not supporting one or more uses; Not Attainable, Not able to support one or more uses; —, no data.

Source: From United States Environmental Protection Agency, 2002, *National Water Quality Inventory 2000 Report*, EPA-841-R-02-001, www.epa.gov.

Table 10B.50 Estimated Phosphorus Loadings to the Great Lakes, 1976–1991

Year	Lake Superior	Lake Michigan	Lake Huron	Lake Erie	Lake Ontario
		Metric Tons			
1976	3,550	6,656	4,802	18,480	12,695
1977	3,661	4,666	3,763	14,576	8,935
1978	5,990	6,245	5,255	19,431	9,547
1979	6,619	7,659	4,881	11,941	8,988
1980	6,412	6,574	5,307	14,855	8,579
1981	3,412	4,091	3,481	10,452	7,437
1982	3,160	4,084	4,689	12,349	8,891
1983	3,407	4,515	3,978	9,880	6,779
1984	3,642	3,611	3,452	12,874	7,948
1985	2,864	3,956	5,758	11,216	7,083
1986	3,059	4,981	4,210	11,118	9,561
1987	1,949	3,298	2,909	8,381	7,640
1988	2,067	2,907	3,165	7,841	6,521
1989	2,323	4,360	3,277	8,568	6,728
1990	1,750	3,006	2,639	12,899	8,542
1991	2,709	3,478	4,460	11,113	10,475

Note: The 1978 Great Lakes Water Quality Agreement set target loadings for each lake (in metric tons per year): Lake Superior, 3,400; Lake Michigan, 5,600; Lake Huron, 4,360; Lake Erie, 11,000; and Lake Ontario, 7,000. Data do not include loadings to the St. Lawrence River. Data analysis was discontinued after 1991.

Source: From The 1997 Annual Report of the Council of Environmental Quality. Great Lakes Water Quality Board, *Great Lakes Water Quality Surveillance Subcommittee Report to the International Joint Commission,* United States and Canada, (International Joint Commission, Windsor, ON, Canada, Biennial).

Orignal Source: From www.whitehouse.gov/CEQ.

Table 10B.51 Summary of Fully Supporting, Threatened, and Impaired Waters in Assessed Great Lakes Shoreline in the United States in 2000

Jurisdiction	Total Miles	Assessed	Full Support	Threatened	Impaired	Not Attainable
Illinois	63	63	0	63	0	0
Indiana	43	43	0	0	43	0
Michigan	3,250	3,250	0	0	3,250	0
Minnesota	272	0	0	0	0	0
New York	577	457	0	40	417	0
Ohio	236	220	0	185	35	0
Pennsylvania	63	0	0	0	0	0
Wisconsin	1,017	1,017	0	807	210	0
Total	**5,521**	**5,050**	**0**	**1,095**	**3,955**	**0**
Percent of assessed			**0.0%**	**21.7%**	**78.3%**	**0.0%**

Note: Supporting, fully supporting of all uses; Threatened, fully supporting all uses but threatened for one or more uses; Impaired, partially or not supporting one or more uses; Not Attainable, not able to support one or more uses.

Source: From Abstracted from United States Environmental Protection Agency, 2002, *National Water Quality Inventory 2000 Report,* EPA-841-R-02-001, www.epa.gov.

Table 10B.52 Atmospheric Input of Some Organic Contaminants to the Great Lakes

Compound	Lakes Superior	Lake Michigan	Lake Huron	Lake Erie	Lake Ontario
Total PCB	9.8	6.9	7.2	3.1	2.3
Dieldrin	0.5	0.4	0.6	0.2	0.1
Total PAH	163	114	118	51	38
Total DOT	0.6	0.4	0.4	0.2	0.1
p.p′-Methoxychlor	8.3	5.9	6.1	2.6	1.9

Note: Metric tons per Year.

Source: From Great Lakes Water Quality Board, 1985 Report on Great Lakes Water Quality.

Original Source: Eisenreich et al., 1981.

Table 10B.53 Atmospheric Loading Estimates for Selected Pollutants (kg/yr) in the Great Lakes

Pollutant of Concern	Superior	Michigan	Huron	Erie	Ontario
PCBs[a] (wet and dry)					
1988	550	400	400	180	140
1992	160	110	110	53	42
1994	85	69	180	37	64
1996	50	42	N/A	34	N/A
PCBs[a] (net gas transfer)[b]					
1988	−1900	−5140	−2560	−1100	−708
1994	−1,700	−2,700	—	−420	−440
DDT (wet and dry)					
1988	90	64	65	33	26
1992	34	25	25	12	10
1994	17	32	37	46	16
1996	4	12	N/A	2	N/A
DDT (net gas transfer)					
1988	−681	−480	−495	−213	−162
1994	30	67	—	34	13
Benzo(a)pyrene (wet and dry)					
1988	69	180	180	81	62
1992[c]	120	84	84	39	31
1994	200	250	—	240	120

Note: —, Not determined or reported.

[a] Data presented for PCB congeners 18, 44, 52, and 101 (each with 3–5 chlorines in chemical structure).

[b] The convention is to assign a negative number to loss of pollutant from the lake (i.e., volatilization). Thus, the resulting number expresses the mass of a pollutant going into or coming out of the lake per year (i.e., a positive net gas transfer indicates a net input of the pollutant to the lake and a negative net gas transfer indicates a net loss or output from the lake).

[c] Data from 1992 may represent an underestimation in the measurement of benzo(a)pyrene.

Source: Modified From United States Environmental Protection Agency, 1997, *Deposition of Air Pollutants to the Great Waters Second Report to Congress*, EPA-453/R-97-011. 1996; data source: United States Environmental Protection Agency, 2000, *Deposition of Air Pollutants to the Great Waters Third Report to Congress*, EPA-453/R-00-005.

Original Source: From Eisenreich and Strachan 1992; Hillery et al., 1996; Hoff et al., 1996; Strachan and Eisenreich 1988.

Table 10B.54 Sources of PCBs to Lake Superior

Source	kg/yr	Percent of Total
Atmosphere	6,600–8,300	82–86
Tributary	1,300	13–16
Municipal discharges	66	1
Industrial discharges	2	1
Total	8,000–9,000	

Source: From Great Lakes Water Quality Board, 1985 Report on Great Lakes Water Quality.

Original Source: From Eisenreich et al., 1981.

Table 10B.55 Concentration of Total PCBs in Lake Superior Water Column

Year	Total PCB Concentration (μg/L)	Total Concentration of 25 PCB Congeners (μg/L)
1978	0.00173±0.00065	NA
1979	0.00404±0.00056	NA
1980	0.00113±0.00011	0.00099±0.00010
1983	0.0008±0.00007	0.00073±0.00006
1986	0.00056±0.00016	0.00055±0.00015
1988	0.00033±0.00004	0.00020±0.00001
1990	0.00032±0.00003	0.00021±0.00001
1992	0.00018±0.00002	0.00009±0.00001

Note: NA, not applicable.

Source: From United States Environmental Protection Agency, 1997, *Deposition of Air Pollutants to the Great Waters*, Second Report to Congress, EPA-453/R-97-01.

Original Source: From Jeremiason et al., 1994.

Table 10B.56 Comparison of Water-Quality Criteria to Pollutant Concentrations in the Great Lakes (μg/L)

					Total Water Column Concentration[e]				
Pollutant	**National AWQC: Fresh Water Aquatic Life[a]**	**National AWQC: Human Health[b]**	**Great Lakes Water-Quality Agreement Objective[c]**	**Great Lakes Water-Quality Criterion[d]**	**Lake Superior**	**Lake Michigan**	**Lake Huron**	**Lake Erie**	**Lake Ontario**
DDT/DD[f]	0.001	0.00024	0.003	0.000011	<0.00006	NA	<0.00006	<0.00006	<0.00006
Dieldrin	0.0019	0.00071	0.001[g]	0.0000065	**0.00026**	NA	**0.00032–0.00035**	**0.00038**	**0.00028–0.00032**
HCB	—	0.0072	—	0.00045	<0.00004	NA	0.000072	0.000047	0.000036
α-HCH	—	0.092	—	—	0.0011	0.0016	0.0015	0.0011	0.0008–0.0009
Lindane	0.08	0.186	0.01	0.47	0.0004	0.00034	0.00038	0.00049	0.00036
Total PCBs	0.014	0.00079	—	0.0000039	**0.00018**	**0.00020–0.00036**	**0.0007–0.0009**	***0.00122***	***0.0012***
POM[h]	—	0.028	—	—	<0.00046	NA	<0.00046	<0.00046	<0.00046

Note: NA, No data available; Bold texts indicate exceedances of GLWQC; bold-italic texts indicate exceedances of AW/QC for human health.

[a] Values are for freshwater chronic criteria (U.S. EPA 1986).

[b] Values are for human chronic exposure through both fish consumption and drinking water (U.S. EPA 1986).

[c] Values are for protection of the most sensitive user of the water among humans, aquatic life and wildlife (IJC 1978).

[d] Values are the most stringent (i.e., lowest) among those for protection of human health, aquatic life, or wildlife (U.S. EPA 1995a).

[e] Concentrations are taken from De Vault et al. (1995) and L'Italian (1993). Concentrations of dieldrin and PCBs that are reported as ranges represent two different concentrations reported in two different studies. For α-HCH, the range of concentrations in Lake Ontario represents the range reported in a single study.

[f] Sampling data are for *p,p'*-DDE.

[g] Value for aldrin and dieldrin combined.

[h] AWQC for human health is for polycyclic aromatic hydrocarbons (PAHs), a subset of POM, sampling data are for benzo(a)pyrene, a PAH.

Source: From United States Environmental Protection Agency, 1997, *Deposition of Air Pollutants to the Great Waters, Second Report to Congress*, EPA-453/R-97-011, www.epa.gov.

Table 10B.57 Modeled Air Deposition, Depositional Flux, and Waterborne Inputs of Dioxins and Furans to the Great Lakes

Dioxins and Furans	Superior	Huron	Michigan	Erie	Ontario	Total
Atmospheric deposition (g TEQ/yr)	5.6	8.6	13.7	7.3	6.4	42
(range)	(2–17)	(3–25)	(5–43)	(2–21)	(2–18)	(13–124)
Depositional flux (μg/km²/yr)	69	145	238	284	337	172
Waterborne inputs (g TEQ/yr)	1.4	1.4	1.9	11	>3.9	>19.6
Percent contribution from atmospheric sources	80	86	88	40	~62	~68

Source: From United States Environmental Protection Agency, 2000, *Deposition of Air Pollutants to the Great Waters. Third Report to Congress*, EPA-453/R-00-005, www.epa.gov.

Orginal Source: From Cohen et al., 1995.

Table 10B.58 Current Rates of Dioxins and Furans Accumulation in Great Lakes Sediments

Location	Number of Sediment Cores Analyzed	Accumulation Rates in Sediment (pg/cm²/yr)		Percent Contribution from Atmospheric Sources	
		Dioxins[a]	Furans[a]	Dioxins[a]	Furans[a]
Lake Superior	2	7.4–8.0	0.8–0.9	100	100
Northern Lake Michigan	2	44–49	22–25	33–55	5–35
Southern Lake Michigan near Chicago Urban Area	1	17	17	100	5–35
Lake Ontario	3	120–220	130–230	5–35	<5

[a] Includes the total of all dioxin homologs. Includes the total of all furan homologs.

Source: From United States Environmental Protection Agency, 2000, *Deposition of Air Pollutants to the Great Waters, Third Report to Congress*, EPA-453/R-00-005, www.epa.gov.

Orginal Source: From Pearson et al., 1998.

Table 10B.59 Modeled Air Deposition, Depositional Flux, and Waterborne inputs of Hexachlorobenzene to the Great Lakes

HCB	Superior	Michigan	Huron	Erie	Ontario	Total
Total deposition (kg/yr)	11	15	16	15	23	79
(range)	(4–49)	(5–73)	(6–74)	(6–65)	(9–101)	(30–362)
Depositional flux (g/km²/yr)	0.13	0.26	0.27	0.58	1.19	0.32
Waterborne inputs (kg/yr)	0.1	0.8	0.6	<72	35	<108.5

Source: From United States Environmental Protection Agency, 2000, *Deposition of Air Pollutants to the Great Waters, Third Report to Congress*, EPA-453/R-00-005, www.epa.gov.

Orginal Source: From Cohen et al., 1995.

Table 10B.60 Mercury Sources Identified for Lake Michigan, Chesapeake Bay, and Long Island Sound in Mass Balance Studies (by Percent Contribution)

Location	Atmospheric Deposition (%)	Urban Area (%)	Riverine	Groundwater	Direct Discharge	Comments
Lake Michigan[a]	~80	~30 (Chicago)	~17	<1	Not reported	Air-water exchange and sediment burial account for the major loss pathways from the lake, but atmospheric deposition dominates inputs
Chesapeake Bay[b]	>50 (Wet 47–53) (Dry 9–11)	Baltimore (percent not known)	~33–49	Not reported	Not reported	Over 90 percent of the mercury entering the watershed is retained in the terrestrial system and does not reach the aquatic system
Long Island Sound[c]	~10	New York New Jersey (percent not known)	~52	Not reported	~36	Tidal exchange, sediment burial, and air-water exchange are major loss pathways, 45 percent of the mercury entering the sound is re-emitted

Source: From United States Environmental Protection Agency, 2000, *Deposition of Air Pollutants to the Great Waters, Third Report to Congress*, EPA-453/R-00-005, www.epa.gov.
Original Source: From [a] Mason and Sullivan (1997); [b] Mason et al., (1997); [c] Fitzgerald (1998).

Table 10B.61 Preliminary Estimates of Total Atmospheric Mercury Deposition to Lake Michigan

Deposition	Annual Total (kg)	Annual Mean (μg/m^2)
Wet	614±186	10.6±3.2
Aerosol Dry	69±38	1.2±0.7
Reactive Gaseous Mercury[a]	506	8.8
Dissolved Gaseous Mercury[b]	−460	−8.0
Total	729	12.6

[a] Reactive gaseous mercury (RGM) deposition values do not include error bars because they reflect a sensitivity analysis performed on a single measurement in place of direct measurements since no measurement method was available for RGM at the time of the study.

[b] Dissolved gaseous mercury (DGM) deposition values do not include error bars because the values reflect a single measurement taken to establish modeling parameters.

Source: From United states Environmental Protection Agency, 2000, *Deposition of Air Pollutants to the Great Waters, Third Report to Congress*, EPA-453/R-00-005, www.epa.gov.

Orginal Source: From Landis, 1998.

Table 10B.62 Number of Fish Consumption Advisories from the National Listing of Fish and Wildlife Advisories, 2000

Jurisdiction	Rivers	Lakes, Reservoirs, Ponds	Great Lakes	Estuaries	Bayous	Coastal	Canal	Wetland	Multi-class Waters	Regional	Statewide	Total Advisories
Alabama	11	2	—	—	—	1	—	—	—	1	—	15
Alaska	—	—	—	—	—	—	—	—	—	—	—	0
American Samoa	—	—	—	1	—	—	—	—	—	—	—	1
Arizona	3	2	—	—	—	—	—	—	—	—	—	5
Arkansas	8	10	—	—	3	—	—	—	1	—	—	22
California	4	12	—	2	—	11	—	—	1	—	—	30
Colorado	—	11	—	—	—	—	—	—	—	—	—	11
Connecticut	5	6	—	1	—	—	—	—	—		1	13
Delaware	6	5	—	8	—	—	1	—	—	—	—	20
District of Columbia	—	—	—	—	—	—	—	—	—	—	1	1
Florida	27	52	—	—	—	8	1	—	10	—	—	98
Georgia	68	33		3		1		1	—	—	—	106
Hawaii	2	—	—	1	—	—	—	—	—	—	—	3
Idaho	—	1		—	—	—	—	—	—	—	—	1
Illinois	19	11	1	—	—	—	—	—	—	—	—	31
Indiana	127	69	1	—	—	—	—	—	—	—	1	198
Iowa	—	1	—	—	—	—	—	—	—	—	—	1
Kansas	10	1	—	—	—	—	—	—	—	—	—	11
Kentucky	7	1	—	—	—	—	—	—	1	—	1	10
Louisiana	8	12	—	1	6	1	—	—	1	—	—	29
Maine	15	—	—	—	—	1	—	—	—	—	2	18
Maryland	1	1	—	2	—		—	—	—	—	—	4
Massachusetts	17	83	—	4	—	1	—	—	1	—	1	107
Michigan[a]	62	64	13	—	—	—	—	1	—	—	1	141
Minnesota[a]	89	850	1	—	—	—	—	—	—	—	1	941
Mississippi	8	3	—	—	—	1	—	—	—	1	—	13
Missouri	5	—	—	—	—	—	—	—	1	1	1	8
Montana	3	23	—	—	—	—	—	—	—	—	—	26
Nebraska	15	20	—	—	—	—	2	—	—	—	—	37
Nevada	1	1	—	—	—	—	—	—	—	—	—	2
New Hampshire	1	3	—	—	—	1	—	—	—	—	1	6
New Jersey	14	25	—	6	—	2	—	—	1	—	1	49
New Mexico	3	23	—		—		—	—	—	—	—	26
New York	27	45	8	3	—	3	—	—	—	—	1	87
North Carolina	6	7	—	1	—	1	—	—	—	1	1	17
North Dakota	2	18	—	—	—	—	1	—	—	—	—	21
Ohio	53	9	1	—	—	—	—	—	—	—	1	64
Oklahoma	1	—	—	—	—	—	—	—	—	—	—	1
Oregon	6	7	—	—	—	—	—	—	—	—	—	13
Pennsylvania	28	2	2	1	—	—	—	—	—	—	—	33
Rhode Island	1	1	—	—	—	1	—	—	—	—	—	3
South Carolina	39	17	—	—	—	1	2	—	1	—	—	60
South Dakota	—	1	—	—	—	—	—	—	—	—	—	1
Tennessee	7	9	—	—	—	—	—	—	1	—	—	17

(Continued)

Table 10B.62 (Continued)

Jurisdiction	Rivers	Lakes, Reservoirs, Ponds	Great Lakes	Estuaries	Bayous	Coastal	Canal	Wetland	Multi-class Waters	Regional	Statewide	Total Advisories
Texas	4	15	—	2	—	1	—	—	—	—	—	22
Utah	1	1	—	—	—	—	—	—	—	—	—	2
Vermont	1	9	—	—	—	—	—	—	—	—	1	11
Virginia	10	—	—	—	—	—	—	—	—	—	—	10
Washington	1	1	—	8	—	2	—	—	—	—	—	12
West Virginia	10	—	—	—	—	—	—	—	—	—	—	10
Wisconsin[a]	101	364	4	—	—	—	—	—	—	—	1	470
Totals	**837**	**1,831**	**31**	**44**	**9**	**37**	**7**	**2**	**19**	**—**	**17**	**2,838**

Note: Data from the National Listing of Fish and Wildlife Advisories.

[a] Includes Tribal and joint State/Tribal advisories; Alabama, Connecticut, Florida, Georgia, Louisiana, Maine, Mississippi, New Hampshire, New Jersey, New York, North Carolina, Rhode Island, South Carolina, and Texas the coastal advisory extends statewide.

Source: From United States Environmental Protection Agency, 2002, *National Water Quality Inventory 2000 Report*, EPA-841-R-02-001, www.epa.gov.

Table 10B.63 Fish Advisories Issued for the Great Lakes

Great Lakes	PCBs	Dioxins	Mercury	Chlordane	Mirex	DDT
Lake Superior	•	•	•	•		
Lake Michigan	•	•	•	•		•
Lake-Huron	•	•	•	•		
Lake Erie	•	•	•			
Lake Ontario	•	•			•	

Source: From United States Environmental Protection Agency, 2004, *Fact Sheet Update: National Listing of Fish Advisories*, EPA-823-F-04 016, August 2004, www.epa.gov.

Table 10B.64 Pollution Discharges in Navigable Waters of the United States in 2001 (by Size)

	Petroleum			Chemical			Other			
Spill Size (Gallons)	Number of Spills	% of Spill Incidents	Spill Volume (Gallons)	Number of Spills	% of Spill Incidents	Spill Volume (Gallons)	Number of Spills	% of Spill Incidents	Spill Volume (Gallons)	% of Spill Volume
1–100	7,256	96.00	33,276	86	81.90	624	50	90.90	259	4.70
101–1,000	216	2.90	86,955	10	9.50	4,680	3	5.50	775	14.00
1,001–3,000	45	0.60	77,447	1	1.00	2,000	2	3.60	4,500	81.30
3,001–5,000	16	0.20	67,241	1	1.00	4,006				
5,001–10,000	11	0.10	89,224	5	4.80	38,390				
10,001–50,000	14	0.20	376,057	1	1.00	21,680				
100,001–1,000,000	1	0.00	124,320	1	1.00	200,049				
Year-End Statistics	**7,559**	**100.00**	**854,520**	**105**	**100.00**	**271,429**	**55**	**100.00**	**5,534**	**100.00**

Source: From U.S. Coast Guard, Pollution incidents in and around U.S. waters a spill/release compendium: 1969–2001, *Annual Data and Graphics for Oil Spills (1969–2001)*, www.uscg.mil.

Table 10B.65 Pollution Discharges in Navigable Waters of the United States in 2001 (by Type of Source)

	Oil				Chemical				Other			
Detailed Source	Number of Spills	% of Spill Incidents	Spill Volume (Gallons)	% of Spill Volume	Number of Spills	% of Spill Incidents	Spill Volume (Gallons)	% of Spill Volume	Number of Spills	% of Spill Incidents	Spill Volume (Gallons)	% of Spill Volume
Tankship	95	1.30	125,217	14.70	2	1.90	11	0.00	1	1.80	120	2.20
Tankbarge	246	3.30	212,298	24.80	8	7.60	200,059	73.70	6	10.90	11	0.20
Commercial vessel	46	0.60	474	0.10								
Fishing boat	562	7.40	122,454	14.30								
Freight barge	27	0.40	86	0.00								
Freight ship	139	1.80	22,007	2.60								
Industrial vessel	42	0.60	1,284	0.20	1	1.00	1	0.00	1	1.80	1	0.00
Oil recovery	4	0.10	6	0.00								
Passenger	179	2.40	1,163	0.10								
Public vessel	197	2.60	10,192	1.20					1	1.80	3	0.10
Unclassified												
Recreational	613	8.10	8,907	1.00								
Research vessel	36	0.50	162	0.00								
Towboat/Tugboat	382	5.10	12,980	1.50					4	7.30	4	0.10
Unclassified vessel	2,299	30.40	47,512	5.60	6	5.70	6	0.00	8	14.50	78	1.40
Modu	38	0.50	1,775	0.20	5	4.80	23	0.00	1	1.80	1	0.00
OSV	108	1.40	3,314	0.40	4	3.80	4,016	1.50	1	1.80	1	0.00
Public tank ship/barge	4	0.10	12	0.00								
Public freight	4	0.10	13	0.00								
Designated waterfront facility	266	3.50	112,648	13.20	25	23.80	36,515	13.50	3	5.50	591	10.70
Land facility nonmarine	61	0.80	23,099	2.70	3	2.90	2,051	0.80	1	1.80	1	0.00
Other onshore marine facility	3	0.00	10	0.00								
Fixed platform	620	8.20	63,470	7.40	14	13.30	7,195	2.70	4	7.30	118	2.10
Mobile facility	27	0.40	459	0.10	2	1.90	402	0.10				
Municipal facility	18	0.20	1,339	0.20	1	1.00	50	0.00	1	1.80	1,500	27.10
Offshore pipeline	13	0.20	1,241	0.10								
Onshore pipeline	21	0.30	12,336	1.40	1	1.00	1	0.00				
Aircraft	10	0.10	861	0.10								
Other land vehicle	40	0.50	13,860	1.60								
Other railroad equipment	1	0.00	500	0.10	1	1.00	250	0.10	1	1.80	3,000	54.20

Tank truck	1	0.00	5	0.00								
Bridge	1	0.00	40	0.00								
Factory	12	0.20	95	0.00								
Fleeting area	2	0.00	24	0.00								
Industrial facility	28	0.40	7,859	0.90	6	5.70	1,731	0.60	1	1.80	31	0.60
Locks	1	0.00	5	0.00								
Marina	50	0.70	2,207	0.30								
Marpol reception	6	0.10	3,042	0.40								
Non-vessel common carrier	1	0.00	5	0.00								
Outfall/sewer/drain	7	0.10	20	0.00								
Permanently moored	4	0.10	115	0.00								
Shipyard/repair facility	20	0.30	4,134	0.50								
Shoreline	252	3.30	23,149	2.70	10	9.50	10.635	3.90	1	1.80	50	0.90
Unknown or other	1,073	14.20	14,141	1.70	16	15.20	8,483	3.10	20	36.40	24	0.40
Year-end statistics	**7,559**	**100.00**	**854,520**	**100.00**	**105**	**100.00**	**271,429**	**100.00**	**55**	**100.00**	**5,534**	**100.00**

Source: From U.S. Coast Guard, Pollution incidents in and around U.S. waters a spill/release compendium: 1969–2001, *Annual Data and Graphics for Oil Spills (1969–2001)*, www.uscg.mil.

Table 10B.66 Pollution Discharges in Navigable Water in the United States (by Type of Location and Water Body)

	Petroleum				Chemical				Other			
	Number of Spills	% of Spill Incidents	Spill Volume (Gallons)	% of Spill Volume	Number of Spills	% of Spill Incidents	Spill Volume (Gallons)	% of Spill Volume	Number of Spills	% of Spill Incidents	Spill Volume (Gallons)	% of Spill Volume
Waterbody												
Atlantic Ocean	83	1.10	7,168	0.80	1	1.00	5	0.00				
Pacific Ocean	493	6.50	53,295	6.20	7	6.70	16	0.00				
Gulf of Mexico	1,728	22.90	133,872	15.70	28	26.70	19,634	7.20	17	30.90	136	2.50
Great Lakes	109	1.40	1,600	0.20								
Lakes	35	0.50	244	0.00								
Rivers & canals	1,682	22.30	237,980	27.80	20	19.00	20,492	7.50	13	23.60	3,075	55.60
Bays & sounds	1,140	15.10	139,300	16.30	12	11.40	1,070	0.40	2	3.60	2	0.00
Harbors	693	11.80	158,667	18.60	17	16.20	206,869	76.20	8	14.50	213	3.80
Other	1,396	18.50	122,394	14.30	20	19.00	23,343	8.60	15	27.30	2,108	38.10
Year-end statistics	7,559	100.00	854,520	100.00	105	100.00	271,429	100.00	55	100.00	5,534	100.00
Location												
Internal/headlands	3,677	48.60	529,482	62.00	50	47.60	228,432	84.20	22	40.00	3,289	59.40
Coastal (0–3 MI)	796	10.50	121,432	14.20	10	9.50	7,173	2.60	6	10.90	11	0.20
Contiguous zone (3–12 MI)	118	1.60	287	0.00	1	1.00	5	0.00	1	1.80	1	0.00
Ocean (12–200 MI)	1,005	13.30	52,784	6.20	22	21.00	12,465	4.60	10	18.20	124	2.20
Ocean general	503	6.70	26,891	3.10	2	1.90	11	0.00	1	1.80	1	0.00
Other	1,460	19.30	123,664	14.50	20	19.00	23,343	8.60	15	27.30	2,108	38.10
Year-end statistics	7,559	100.00	854,520	100.00	105	100.00	271,429	100.00	55	100.00	5,534	100.00

Source: From U.S. Coast Guard, Pollution incidents in and around U.S. waters a spill/release compendium: 1969–2001, *Annual Data and Graphics for Oil Spills (1969–2001)*, www.uscg.mil.

Table 10B.67 Pollution Discharges in Navigable Waters of the United States in 2001 by State

	Oil				Chemical				Other			
State	Number of Spills	% of Spill Incidents	Spill Volume (Gallons)	% of Spill Volume	Number of Spills	% of Spill Incidents	Spill Volume (Gallons)	% of Spill Volume	Number of Spills	% of Spill Incidents	Spill Volume (Gallons)	% of Spill Volume
Alaska	350	4.60	47,643	5.60	6	5.70	15	0.00				
Alabama	152	2.00	2,450	0.30	2	1.90	2	0.00				
American Samoa	65	0.90	2,423	0.30								
California	641	8.50	24,794	2.90	3	2.90	403	0.10	1	1.80	7	0.10
Connecticut	21	0.30	2,665	0.30								
District of Columbia	11	0.10	5,048	0.60	1	1.00	10,000	3.70				
Delaware	36	0.50	1,621	0.20	1	1.00	1,000	0.40				
Florida	554	7.30	11,508	1.30	1	1.00	2	0.00	1	1.80	3	0.10
Georgia	20	0.30	47	0.00	1	1.00	500	0.20				
Guam	44	0.60	694	0.10								
Hawaii	115	1.50	5,854	0.70								
Iowa	2	0.00	3	0.00								
Illinois	49	0.60	2,034	0.20								
Indiana	10	0.10	39	0.00								
Iowa	2	0.00	65	0.00								
Kentucky	58	0.80	125,917	14.70								
Louisiana	1,487	19.70	147,740	17.30	10	9.50	15,640	5.80	21	38.20	565	10.20
Massachusetts	136	1.80	7,794	0.90								
Maryland	154	2.00	4,095	0.50	3	2.90	256	0.10	1	1.80	3,000	54.20
Maine	53	0.70	1,607	0.20	3	2.90	52	0.00				
Michigan	62	0.80	353	0.00	2	1.90	2,020	0.70				
Minnesota	25	0.30	309	0.00	1	1.00	4	0.00	1	1.80	1	0.00
Missouri	8	0.10	75	0.00	1	1.00	7,600	2.80				
N. Mariana Islands	2	0.00	2	0.00								
Mississippi	119	1.60	2,465	0.30								
North Carolina	126	1.70	2,399	0.30	2	1.90	17	0.00				
New Hampshire	22	0.30	42	0.00								
New Jersey	114	1.50	10,854	1.30	4	3.80	109	0.00	1	1.80	50	0.90
New York	156	2.10	32,114	3.80	5	4.80	1,554	0.60	2	3.60	2	0.00
Ohio	43	0.60	2,123	0.20								
Oregon	67	0.90	18,765	2.20								
Pennsylvania	42	0.60	9,003	1.10	1	1.00	1	0.00	1	1.80	1	0.00
Puerto Rico	52	0.70	779	0.10								
Rhode Island	78	1.00	1,497	0.20					1	1.80	1	0.00
South Carolina	58	0.80	1,227	0.10								
Tennessee	9	0.10	334	0.00					1	1.80	1	0.00
Texas	1,072	14.20	213,653	25.00	25	23.80	206,411	76.00	11	20.00	277	5.00

(Continued)

Table 10B.67 (Continued)

State	Oil				Chemical				Other			
	Number of Spills	% of Spill Incidents	Spill Volume (Gallons)	% of Spill Volume	Number of Spills	% of Spill Incidents	Spill Volume (Gallons)	% of Spill Volume	Number of Spills	% of Spill Incidents	Spill Volume (Gallons)	% of Spill Volume
Virginia	132	1.70	21,383	2.50	3	2.90	21,687	8.00	1	1.80	1	0.00
Virgin Islands	30	0.40	1,364	0.20								
Washington	465	6.20	24,800	2.90	5	4.80	73	0.00	1	1.80	1	0.00
Wisconsin	21	0.30	10,186	1.20	2	1.90	7	0.00	1	1.80	1,500	27.10
West Virginia	14	0.20	543	0.10								
Unknown or beyond state waters	882	11.70	106,209	12.40	23	21.90	4,076	1.50	10	18.20	124	2.20
Year-end statistics	**7,559**	**100.00**	**854,520**	**100.00**	105	**100.00**	**271,429**	**100.00**	**55**	**100.00**	**5,534**	**100.00**

Source: From U.S. Coast Guard, Pollution incidents in and around U.S. waters a spill/release compendium: 1969–2001, *Annual Data and Graphics for Oil Spills (1969–2001)*, www.uscg.mil.

Table 10B.68 Type of Oil Discharged to Navigable Waters of the United States in 2001

Oil Type	Number of Spills	% of Spill Incidents	Spill Volume (Gallons)	% of Spill Volume
Crude oils	1,173	15.50	182,999	21.40
Heavy fuel oils	121	1.60	82,168	9.60
Intermediate fuel oils	2,181	28.90	228,320	26.70
Gasoline products	260	3.40	27,233	3.20
Other petroleum oils	3,788	50.10	333,643	39.00
Nonpetroleum oils	36	0.50	157	0.00
Year-end statistics	**7,559**	**100.00**	**854,520**	**100.00**

Source: From U.S. Coast Guard, Pollution incidents in and around U.S. waters a spill/release compendium: 1969–2001, *Annual Data and Graphics for Oil Spills (1969–2001)*, www.uscg.mil.

Table 10B.69 Oil Spills in U.S. Water — Number and Volume: 1998–2001

	Number of Spills				Spill Volume (Gallons)			
Spill Characteristics	**1998**	**1999**	**2000**	**2001**	**1998**	**1999**	**2000**	**2001**
Total	8,315	8,539	8,354	7,539	885,303	1,172,449	1,431,370	854,520
Size of spill (gallons)								
1–100	7,962	8,212	8,058	7,258	38,093	38,119	39,355	33,276
101–1,000	259	240	219	216	86,606	86,530	78,779	86,955
1,001–3,000	54	42	37	45	96,743	74,582	67,529	77,447
3,001–5,000	15	18	12	16	64,609	73,798	45,512	67,241
5,001–10,000	15	10	16	11	108,148	66,274	112,415	89,224
10,001–50,000	8	12	6	14	216,335	301,510	108,400	376,057
50,001–100,000	—	4	4	—	—	245,406	266,380	—
100,001–1,00,000	2	1	2	1	274,769	285,230	713,000	124,320
1,000,000 and over	—	—	—	—	—	—	—	—
Waterbody								
Atlantic ocean	109	148	150	83	6,674	29,440	135,010	7,168
Pacific ocean	644	758	623	493	192,775	150,694	36,301	53,295
Gulf of Mexico	2,190	1,756	1,838	1,728	181,372	45,786	112,069	133,872
Great Lakes	119	129	95	109	3,006	906	4,535	1,600
Lakes	25	31	32	35	63	624	349	244
Rivers and canals	1,944	1,924	1,816	1,682	280,651	504,254	663,404	237,980
Bays and sounds	891	1,299	1,248	1,140	24,234	136,650	49,783	139,300
Harbors	790	907	801	893	97,223	105,213	273,095	158,667
Other	1,603	1,587	1,750	1,396	99,305	198,872	156,824	122,394
Source								
Tankship	104	82	111	95	56,673	8,414	608,176	125,217
Tankbarge	220	227	229	246	248,089	158,977	133,540	212,298
All other vessels	4,848	5,361	5,220	4,680	316,473	409,084	291,927	232,341
Facilities	937	1,019	1,054	995	166,269	367,537	311,604	201,025
Piperlines	45	25	25	34	47,863	36,140	17,021	13,577
All other non vessels	571	571	566	436	32,584	147,704	45,136	55,921
Unknown	1,590	1,244	1,149	1,073	17,352	44,593	23,966	14,141

Note: Based on reported discharges into U.S. navigable waters, including territorial waters (extending 3 to 12 mi from the coastline), tributaries, the contiguous zone, onto shoreline, or into other waters that threaten the marine environment. Data found in Marine Safety Management System; — Represents or rounds to zero.

Source: From U.S. Census Bureau, *Statistical Abstract of the United States 2004–2005*, www.census.gov.

Orginal Source: From U.S. Coast Guard, www.uscg.mil/hq/g-m/nmc/response/stats/Summary.htm and uscg.mil/hq/g-m/nmc/response/stats/chp2001.pdf (released August 2003).

Table 10B.70 Number and Volume of Spills by Type of Oil in the United States, 1973 to 2001

	Crude Oils		Heavy Fuel Oils		Intermediate Fuel Oils		Gasoline Products		Other Petroleum Oils		Nonpetroleum Oils	
Year	Number	Volume	Number	Volume	Number	Volume	Number	Volume	Number	Volume	Number	Volume
73	4,807	7,219,648	915	2,068,102	1,423	2,271,442	535	1,070,803	985	2,476,361	349	147,224
74	5,243	10,102,280	932	1,858,340	1,712	1,189,120	684	1,763,230	1,051	664,618	377	121,145
75	4,643	7,138,446	826	7,342,076	1,654	1,091,283	684	2,068,473	1,075	3,763,013	410	116,792
76	4,521	5,631,995	809	9,811,141	1,867	1,033,888	735	1,127,213	1,156	571,926	334	341,786
77	4,347	3,331,533	863	1,151,495	1,944	1,528,183	762	993,647	1,285	1,041,758	258	142,518
78	4,481	3,518,825	938	925,233	2,498	1,653,321	854	3,780,524	1,509	759,190	364	227,015
79	4,036	15,411,430	873	950,344	2,396	1,658,465	771	996,134	1,414	1,340,479	344	536,710
80	3,358	7,762,112	773	275,089	2,083	1,328,861	604	1,601,836	1,259	1,512,769	306	116,303
81	3,216	2,440,608	627	4,114,826	1,850	850,052	591	841,469	1,158	427,372	369	246,666
82	3,070	4,761,261	454	2,906,973	1,885	911,026	620	815,359	1,100	434,293	355	515,884
83	3,188	3,763,206	448	409,906	1,873	2,136,142	603	854,318	1,355	1,109,342	449	106,934
84	2,919	4,281,504	400	2,159,878	1,927	7,224,260	664	891,746	1,594	3,360,784	754	87,706
85	2,153	2,012,551	291	964,056	1,596	1,997,320	582	721,325	1,152	2,617,914	395	123,083
86	875	1,720,405	379	840,040	1,772	1,025,190	439	471,692	1,469	139,051	59	85,600
87	555	1,768,240	325	489,686	1,867	858,323	452	272,265	1,585	158,184	57	62,187
88	660	1,856,136	314	448,387	1,941	3,452,480	447	597,773	1,572	203,555	64	27,672
89	1,183	10,997,750	371	957,043	2,578	1,173,326	429	258,328	1,998	84,992	54	7,254
90	1,992	4,730,679	362	499,347	2,829	1,322,094	479	756,680	2,457	603,189	58	3,017
91	1,712	896,683	316	205,063	2,622	366,201	448	98,483	3,423	307,422	48	2,099
92	1,728	803,160	350	192,638	2,659	564,314	542	124,806	4,161	184,873	51	5,876
93	1,598	319,467	234	743,447	2,595	662,415	512	99,164	3,958	217,856	75	25,039
94	1,549	626,821	206	824,254	2,710	345,726	468	219,804	3,929	414,767	98	57,901
95	1,467	195,857	223	213,949	2,963	290,186	469	46,416	3,829	344,664	86	707,157
96	1,783	222,976	175	266,034	2,706	2,206,213	422	267,992	4,177	147,849	72	6,767
97	1,642	412,747	163	19,710	2,477	193,390	419	124,976	3,849	182,535	74	9,216
98	1,593	304,424	122	63,885	2,576	299,337	423	82,499	3,531	70,145	70	65,013
99	1,352	148,654	116	108,408	2,751	305,096	381	77,992	3,852	520,088	87	12,211
2000	1,283	704,259	94	156,135	2,479	274,545	314	32,921	4,121	262,373	63	1,137
2001	1,173	182,999	121	82,168	2,181	228,320	260	27,233	3,788	333,643	36	157
Totals	**72,127**	**103,266,654**	**13,020**	**41,047,653**	**64,414**	**38,440,519**	**15,593**	**21,085,101**	**67,793**	**25,095,005**	**6,116**	**3,908,069**

Source: From U.S. Coast Guard, Pollution incidents in and around U.S. waters a spill/release compendium: 1969–2001. *Cumulative date for oil spills 1973–2001*, www.uscg.mil.

Table 10B.71 Number and Volume of Oil Spills by Source in the United States, 1973–2001

	Tankship		Tankbarge		All Other Vessels		Facilities		Pipelines		All Other Nonvessels		Unknown	
Year	Number	Gallons	Number	Gallons	Number	Gallons	Number	Gallons	Number	Gallons	Number	Gallons	Number	Gallons
73	694	3,153,070	603	1,251,320	1,527	1,049,748	3,317	5,250,092	511	2,353,744	301	424,055	2,061	1,771,550
74	846	1,177,851	754	2,331,302	1,566	317,939	3,844	3,834,292	582	6,833,402	376	588,091	2,031	615,854
75	595	8,723,153	767	2,572,118	1,499	1,415,466	3,139	4,663,214	667	2,769,165	390	1,149,716	2,235	227,251
76	526	9,315,761	894	1,702,772	1,514	281,032	2,978	2,046,062	627	4,283,495	398	647,656	2,485	241,170
77	533	202,590	993	1,566,631	1,760	275,255	2,671	2,353,360	461	2,528,165	426	593,305	2,615	669,827
78	678	329,699	980	3,239,284	2,057	474,151	2,534	4,391,595	406	1,220,486	530	501,074	3,459	707,819
79	647	13,077,598	862	1,162,569	1,833	394,951	2,358	1,824,738	583	3,351,156	506	608,740	3,045	473,806
80	547	1,597,088	799	1,738,003	1,698	290,976	2,011	2,926,797	552	3,067,276	377	382,505	2,399	2,594,326
81	419	1,074,621	718	4,294,542	1,584	341,595	2,007	1,126,966	561	1,338,116	324	313,718	2,198	431,436
82	279	1,219,922	547	2,146,576	1,383	412,484	2,244	1,660,560	598	4,213,862	392	368,696	2,041	322,696
83	258	145,822	523	1,807,897	1,444	378,537	2,443	1,385,766	582	3,036,906	444	323,750	2,222	1,301,170
84	238	4,663,952	499	2,484,481	1,530	1,863,435	2,408	1,193,770	557	1,212,702	565	381,704	2,461	6,205,834
85	164	732,397	385	3,683,548	1,113	446,966	2,032	2,237,558	385	777,017	385	235,654	1,705	323,108
86	196	1,164,962	516	1,510,064	900	160,890	1,382	902,917	91	230,785	158	28,596	1,750	283,764
87	158	1,547,462	413	550,108	1,208	848,200	1,160	317,437	95	196,852	142	36,522	1,665	112,303
88	222	852,287	486	3,164,017	1,300	369,985	1,038	1,368,898	120	704,719	142	39,383	1,690	86,715
89	200	11,272,324	504	746,833	1,564	674,660	1,688	448,792	110	214,920	138	33,030	2,409	88,137
90	249	4,977,251	457	992,025	1,779	417,882	2,287	1,059,302	149	316,928	148	32,242	3,108	119,377
91	220	92,334	428	241,346	1,780	362,809	2,389	445,986	105	49,382	117	10,068	3,530	674,027
92	193	118,075	322	149,212	4,795	398,145	2,045	504,600	36	200,396	815	235,839	1,285	269,400
93	172	69,541	314	697,653	4,944	409,963	2,320	350,141	35	362,399	826	145,796	361	31,895
94	172	69,694	393	955,582	4,681	308,343	2,258	677,016	55	62,340	796	348,577	605	77,721
95	148	125,491	353	1,101,938	4,977	396,724	586	868,900	30	11,894	500	77,428	2,444	55,854
96	122	219,311	313	1,163,258	5,151	298,451	509	406,384	17	978,392	552	23,527	2,671	28,508
97	124	22,429	252	165,649	4,971	192,801	838	204,935	32	224,122	486	72,208	1,921	60,430
98	104	56,673	220	248,089	4,848	316,473	937	166,269	45	47,863	571	32,584	1,590	17,352
99	92	8,414	228	210,383	5,360	357,678	1,019	367,537	25	36,140	571	147,704	1,244	44,593
2000	111	608,176	229	133,540	5,220	291,927	1,054	311,604	25	17,021	566	45,136	1,149	23,966
2001	95	125,217	246	212,298	4,680	232,341	995	201,025	34	13,577	436	55,921	1,073	14,141
Totals	**9,002**	**66,743,165**	**14,998**	**42,223,038**	**78,666**	**13,979,807**	**56,491**	**43,486,513**	**8,076**	**40,653,222**	**12,378**	**7,883,225**	**59,452**	**17,874,030**

Source: From U.S. Coast Guard, Pollution incidents in and around U.S. waters a spill/release compendium: 1969–2001. *Cumulative data for oil spills 1973–2001*, www.uscg.mil.

Table 10B.72 Number and Volume of Oil Spills by Waterbody in the United States, 1973–2001

	Atlantic Ocean		Pacific Ocean		Gulf of Mexico		Great Lakes		Lakes		Rivers & Canals		Bays & Sounds		Harbors		Other	
Year	Number	Gallons	Number	Gallons	Number	Gallons	Number	Gallons	Number	Gallons	Number	Gallons	Number	Gallons	Number	Gallons	Number	Gallons
73	117	2,459,968	455	281,546	38	8,553	62	72,971	0	0	3,566	8,250,726	1,116	387,285	3,503	3,224,849	157	567,682
74	203	185,682	622	218,073	218	157,926	72	131,059	0	0	2,383	5,613,582	1,941	454,526	4,015	4,714,829	545	4,223,054
75	156	6,107,710	477	253,517	727	1,418,791	72	60,463	0	0	2,544	6,486,011	934	3,184,703	3,789	2,937,545	593	1,071,342
76	175	7,569,064	297	436,868	936	850,660	194	179,912	0	0	2,457	3,763,548	967	512,399	3,614	2,655,553	782	2,549,944
77	283		162		731		209		0		3,763		1,125		2,634		552	
78	270	119,135	614	88,314	656	402,392	113	40,706	0	0	4,573	4,924,452	726	2,803,394	3,079	701,866	613	1,783,850
79	225	307,823	487	556,054	867	386,281	129	72,982	0	0	4,223	5,804,497	479	11,060,800	2,654	563,290	770	2,141,835
80	173	242,993	362	210,428	743	437,069	80	155,358	0	0	3,294	7,090,268	366	692,490	2,650	1,590,711	715	2,177,654
81	149	350,511	351	49,965	857	99,120	45	29,508	0	0	2,826	6,144,670	471	152,695	2,388	590,160	724	1,504,366
82	119	31,305	276	127,377	1,061	121,889	18	10,027	0	0	2,593	5,388,921	512	366,107	1,896	1,417,451	1,009	2,881,720
83	135	57,083	332	129,264	1,290	295,736	2	11	0	0	2,666	2,814,518	768	1,601,887	1,784	1,212,231	939	2,269,117
84	157	58,201	468	1,555,741	1,583	2,897,179	37	18,612	0	0	2,453	3,928,947	647	192,080	1,848	7,604,388	1,065	1,750,729
85	124	56,026	361	493,419	951	116,969	23	1,583	2	9	1,808	4,169,267	581	201,348	1,143	722,800	1,176	2,674,827
86	59	4,401	229	171,520	431	97,221	130	6,854	14	1,445	1,379	1,410,325	639	335,795	723	1,172,891	1,389	1,081,527
87	63	8,250	348	1,342,548	218	91,524	176	4,951	10	235	1,273	391,057	817	390,047	692	302,665	1,244	1,077,607
88	68	9,961	366	2,191,448	372	1,076,986	155	32,898	6	98	1,166	1,878,896	932	312,885	724	126,498	1,209	956,333
89	45	30,955	524	480,644	1,063	108,519	179	4,875	10	1,752	1,373	322,326	1,145	11,062,200	881	766,281	1,393	701,139
90	92	13,400	480	624,494	1,834	4,115,264	194	129,131	11	383	1,749	1,775,142	988	263,436	940	455,108	1,889	538,649
91	109	9,009	446	199,306	1,977	100,702	191	5,103	14	1,256	2,010	430,905	938	143,723	916	687,563	1,968	298,385
92	129		594		1,974		229		21		1,999		969		1,241		2,335	
93	132	14,713	649	262,292	1,763	53,265	256	10,602	19	2,300	1,744	942,114	1,004	418,137	1,095	51,842	2,310	312,123
94	206	799,549	666	128,752	1,350	205,151	240	15,984	16	318	1,814	383,171	1,062	72,022	1,016	346,649	2,590	537,677
95	267	48,313	648	69,053	1,485	253,040	282	3,103	26	92	1,849	1,156,002	1,109	41,004	1,176	148,229	2,196	919,393
96	119	27,980	491	29,209	2,403	45,145	228	3,507	19	52	1,984	475,550	793	1,092,207	992	288,252	2,306	1,155,929
97	87	40,875	505	32,841	2,341	105,462	156	4,311	29	210,270	1,821	182,676	811	46,450	858	45,932	2,016	273,775
98	109	6,674	644	192,775	2,190	181,372	119	3,006	25	63	1,944	280,651	891	24,234	790	97,223	1,603	99,305
99	148	29,440	758	150,694	1,756	45,786	129	906	31	624	1,924	504,264	1,299	136,650	907	105,213	1,587	198,872
2000	150	135,010	623	36,301	1,838	112,069	96	4,535	32	349	1,816	663,404	1,248	49,783	801	273,095	1,750	156,824
2001	83	7,168	493	53,295	1,728	133,872	109	1,600	35	244	1,682	237,980	1,140	139,300	893	158,667	1,396	122,394
Total	4,152	19,054,638	13,728	10,653,402	35,381	15,214,328	3,925	1,192,525	320	219,867	66,676	79,244,143	26,418	36,765,442	49,642	34,240,859	38,821	36,257,796

Source: From U.S. Coast Guard, Pollution incidents in and around U.S. waters a spill/release compendium: 1969–2001. *Cumulative data for oil spills 1973–2001*, www.uscg.mil.

Table 10B.73 The World's Major Water-Quality Issues

Issue Scale	Water Bodies Polluted	Sector Affected	Time Lag between Cause and Effect	Effects Extent
Organic pollution	Rivers[a] Lakes[a] Groundwater[b]	Aquatic environment	<1 yr	Local to district
Pathogens	Rivers[a] Lakes[b] Groundwater[b]	Health[a]	<1 yr	Local
Salinization	Groundwater[a] Rivers[b]	Most uses Aquatic environment Health	1–10 yrs	District to region
Nitrate	Rivers[b] Lakes[b] Groundwater[a]	Health	>10 yrs	District to region
Heavy metals	All bodies	Health Aquatic environment Ocean fluxes	<1 to >10 yrs	Local to global
Organic	All bodies	Health Aquatic environment Ocean fluxes	1 to 10 yrs	Local to global
Acidification	Rivers[a] Lakes[a] Groundwater[b]	Health Aquatic environment	>10 yrs	District to region
Eutrophication	Lakes[a] Rivers[b]	Aquatic environment Most uses Ocean fluxes	> 10 yrs	Local
Sediment load (increase and decrease)	Rivers[b] Lakes	Aquatic environment Most uses Ocean fluxes	1–10 years	Regional
Diversions, dams	Rivers[a] Lakes[b] Groundwater[a]	Aquatic environment Most uses	1–10 yrs	District to region

Note: Pollutants of many kinds eventually find their way into water bodies at all levels. Although it may take some years for problems to become evident, poor water quality affects both human health and ecosystem health.

[a] Very serious issue on a global scale.
[b] Series issue on a global scale.

Source: From *Water for People Water for Life*, The United Nations World Water Development, Copyright © United Nations Educational, Scientific and Cultural Organization (UNESCO) - World Water Assessment Programme (UNESC0-WWAP), 2003. Reproduced by permission of UNESCO. www.unesco.org.

Original Source: From WHO/UNEP (World Health Orgauization/United Nations Environment Programme), 1991. *Water Quality: Progress in the Implementation of the Mar del Plata Action Plan and a Strategy for the 1990s*, Nairobi, Earthwatch Global Monitoring System, World Health Organization, United Nations Environment Programme.

Table 10B.74 Median Concentrations of Acidic Drugs in Raw and Finished Drinking Water (ug/L) in Ontario, Canada, Separated According to the Source of the Water

	Row Water			Finished water		
	Wells	Lakes	Rivers	Wells	Lakes	Rivers
Clofibric acid	Nd	0.3	0.5	Nd	0.1	0.4
Ibuprofen	Nd	0.6	54.7	Nd	0.5	13
Gemfibrozil	Nd	1.4	8.6	Nd	Nd	Nd
Fenoprofen	Nd	Nd	0.5	Nd	Nd	Nd
Naproxen	Nd	1.3	69.3	Nd	Nd	Nd
Ketoprofen	Nd	0.3	Nd	Nd	Nd	Nd
Diclofenac	Nd	Nd	5.7	Nd	Nd	Nd
Indomethacin	Nd	Nd	2.5	Nd	Nd	Nd

Note: Nd, not detected.

Source: From Kummerer, K., (eds.), 2004, *Pharmaceuticals in Environment*, Sources, Fate, Effects and Risks, Second Edition, © Springer-Verlag, Berline Heidelberg 2004. Table 6.5. With kind permission of Springer Science and Business Media.

Original Source: From Servos M.R. et al., 2004, Presence and removal of acidic drugs in drinking water treatment plants in Ontario, Canada. Water Qual Res J Can (to be published).

Table 10B.75 Theoretical Environmental Loads of Selected Pharmaceuticals to Italy and Concentrations Measured in the River Po

	Theoretical Environmental Load 1997 (tons yr^{-1})	Theoretical Environmental Load 2001 (tons yr^{-1})	Measured Concentration in 1997 (range; ng 1^{-1})	Measured concentration in 2001 (range; ng 1^{-1})
Pharmaceuticals measured in 1997 and 2001				
Amoxycillin	59.64	125.75	Nd	Nd
Atenolol	7.54	19.86	49.5–84.32	3.4–39.4
Bezafibrate	—	3.80	15.1–22.4	0.8–2.7
Ceftriaxone	11.42	5.93	Nd	Nd
Cyclophosphamide	—	—	Nd	Nd
Diazepam	—	—	0.5–0.7	Nd-2.1
Erythromycin	1.00	0.39	0.7–0.9	1.4–15.9
Furosemide	3.49	5.76	Nd	1.7–67.2
Ibuprofen	1.00	0.19	Nd-4.0	Nd-9.7
Lincomycin	5.11	3.60	1.2–4.6	3.1–248.9
Ranitidine	10.46	10.67	Nd	Nd-4.0
Salbutamol	0.034	0.126	Nd-4.6	Nd-1.7
Spiramycin	—	1.02	Nd	Nd-43.8
Pharmaceuticals measured only in 2001				
Ciprofloxacin	—	2.96	—	Nd-26.1
Clarythromycin	—	8.47	—	0.5–20.3
Enalapril	—	1.47	—	Nd-0.1
Hydrochlorothiazide	—	13.93	—	Nd-24.4
Omeprazole	—	0.67	—	Nd

Source: From Kummerer, K., (eds.), 2004, *Pharmaceuticals in Environment, Sources, Fate, Effects and Risks, Second Edition*, © Springer-Verlag Berline Heidelberg 2004. Table 4.2. With kind permission of Springer Science and Business Media.

Table 10B.76 Concentrations of Total Phosphorus and Total Nitrogen in Selected Lakes of the World

		Total Phosphorus (mgP/L)					Total Nitrogen (mgN/L)				
		1996	1997	1998	1999	Average Last 3 Yrs[a]	1996	1997	1998	1999	Average Last 3 yrs[a]
Canada	Ontario	—	—	—	—	—	—	—	—	—	—
	Huron	—	—	0.004	—	—	—	—	0.11	—	—
	Superior	0.004	—	0.004	—	—	0.11	0.11	—	—	0.11
Mexico	Chapala	0.430	0.380	0.380	—	0.397	0.17	0.12	0.12	—	0.14
	Patzcuaro	—	—	—	—	—	—	—	—	—	—
	Catemaco	—	—	—	—	—	0.02	0.01	—	—	0.02
	Chairel	0.050	0.020	—	—	0.030	0.16	0.13	—	—	0.13
	Cantenano	—	—	—	—	—	0.24	0.16	—	—	0.21
U.S.A.	Twin-Portage (Ohio)	—	—	—	—	—	—	—	—	—	—
Japan	Biwa (North)	0.009	0.009	0.010	0.008	0.009	0.31	0.32	0.33	0.32	0.32
	Biwa (South)	0.018	0.021	0.018	0.020	0.020	0.40	0.42	0.40	0.40	0.41
	Kasumigaura	0.120	0.100	0.100	0.091	0.097	0.91	0.89	1.30	0.93	1.04
Korea	Chunchonho	0.015	0.025	0.016	0.014	0.018	1.51	1.52	1.44	1.60	1.52
	Chungjiuho	0.022	0.027	0.023	0.019	0.023	1.77	3.06	3.38	2.59	3.01
	Paldang lake	0.032	0.141	0.040	0.036	0.039	2.14	2.39	2.24	2.24	2.29
Austria	Mondsee	0.009	0.008	—	—	0.008	—	—	—	—	—
	Ossiachersee	0.009	—	0.011	—	0.010	—	—	—	—	—
	Wallersee	0.014	0.018	—	—	0.016	—	—	—	—	—
	Zellersee	0.008	0.008	—	—	0.008	—	—	—	—	—
Denmark	Dons Norreso	—	—	—	—	—	—	—	—	—	—
	Arreso	0.206	0.230	0.200	—	0.212	2.25	2.64	2.80	—	2.56
	Fureso	0.122	0.109	0.090	—	0.107	0.81	0.80	0.74	—	0.78
	Sobygard	0.543	0.341	0.270	—	0.385	2.18	1.63	1.87	—	1.89
Finland	Pääjärvi	0.012	0.013	—	—	0.012	1.03	1.01	—	—	1.03
	Päijänne	0.006	0.006	—	—	0.006	0.51	0.48	—	—	0.49
	Yli-Kitka	0.007	0.005	—	—	0.006	0.20	0.23	—	—	0.22
France	Parentis-Biscarrosse	0.039	0.050	0.069	0.076	0.065	1.00	0.89	1.07	0.88	0.95
	Cazaux-Sanguinet	—	0.013	—	—	—	—	0.44	—	—	—
	Réservoir Marne	—	—	—	—	—	0.85	—	—	—	—
	Réservoir-Seine	—	—	—	—	—	0.55	—	—	—	—
	Lac d'Annecy	—	—	—	—	—	—	—	—	—	—
Germany	Bodensee-Fischbach-Uttwil	0.019	0.017	—	—	0.019	0.97	0.98	—	—	0.99
Hungary	Ferto	0.130	0.087	0.089	0.049	0.075	—	—	—	—	—
	Balaton	0.075	0.110	0.082	0.056	0.083	0.74	1.06	0.98	0.94	0.99
	Velencei	0.071	0.069	0.059	0.053	0.060	2.53	2.35	2.46	2.12	2.31
Ireland	Ennel	0.017	0.017	—	—	0.018	0.52	0.48	0.48	0.29	0.41
	Owel	0.011	0.011	—	—	0.011	0.09	0.05	0.08	0.05	0.06
	Derg	0.042	0.048	—	0.032	0.041	1.24	1.23	—	—	0.99
	Sheelin	0.023	0.025	0.021	0.019	0.022	—	—	—	—	—
Italy	Maggiore	0.010	0.010	0.012	—	0.011	0.83	0.84	0.84	—	0.84
	Como	0.042	0.046	0.038	—	0.042	0.88	0.89	0.87	—	0.88
	Garda	0.017	0.015	0.022	—	0.018	0.35	0.35	0.36	—	0.35
	Orta	0.001	0.002	0.001	—	0.001	2.50	2.30	2.30	—	2.37
Luxembourg	EschSure	—	—	—	—	—	2.73	5.15	3.31	3.42	3.96

(Continued)

Table 10B.76 (Continued)

		Total Phosphorus (mgP/L)					Total Nitrogen (mgN/L)				
		1996	1997	1998	1999	Average Last 3 Yrs[a]	1996	1997	1998	1999	Average Last 3 yrs[a]
	Weiswarnpach	0.100	0.500	0.340	0.119	0.286	2.71	2.48	2.48	2.48	2.48
Norway	Mjoesa	0.007	0.007	0.005	0.152	0.021	0.45	0.44	0.46	0.44	0.44
	Randsfjorden	0.006	0.007	0.004	0.003	0.005	0.53	0.53	0.55	0.51	0.53
	Tyrifjorden	0.004	0.004	—	—	—	0.47	0.42	—	—	—
Poland	Sriardwy	—	0.043	—	—	—	—	0.90	—	—	—
	Hancza	0.063	0.025	—	—	—	0.94	0.68	—	—	—
	Niegocin	—	—	—	—	—	—	—	—	—	—
	Northern Mamry	—	—	0.052	—	—	—	—	—	1.06	—
Sweden	Mälaren	0.023	0.018	0.019	0031	0.023	0.55	0.67	0.66	0.87	0.73
	Vänern	0.006	0.006	0.006	0.008	0.007	0.80	0.88	0.85	0.83	0.85
	Vättern	0.005	0.004	0.003	0.005	0.004	0.71	0.83	0.79	0.78	0.80
	Hjälmaren	0.033	0.033	0.043	0.039	0.038	0.57	0.67	0.79	0.59	0.68
Switzerland	Léman	0.041	0.038	0.040	0.039	0.039	0.68	0.67	0.66	0.66	0.66
	Constance	0.019	0.018	0.015	0.014	0.016	1.11	1.21	1.18	1.19	1.19
Turkey	Kurtbogazi	—	—	—	—	—	—	—	—	—	—
	Sapanca	—	—	—	—	—	—	—	—	—	—
	Gala	—	—	—	—	—	1.10	1.29	0.65	0.93	0.96
	Altinapa	—	—	—	—	—	1.27	1.84	2.74	1.64	2.07
UK	Lough Neagh	0.095	—	0.158	—	0.124	0.83	—	0.61	—	0.62
	Lomond	—	—	0.000	0.000	0.003	—	—	0.30	0.37	0.35
	Bewl Water	0.077	0.105	0.065	0.032	0.067	2.12	3.08	2.60	1.29	2.32

Total Phosphorus Note: CAN, Ontario, Haron and Superior: data represent annual mean surface values from several hundred open water samples for each lake (mainly spring and summer); MEX, Orthophosphates; U.S.A., W. Twin(Ohio), samples obtained from the deepest point in each lake, generally weekly from late spring-eady fall and less frequently the rest of the year at 0.1, 2, 4, 7 and 10 m; FIN, data refer to surface measurements; DEU, Bodensee, Lac Constance (Switzerland); IRL, Derg, 1996 and 1997 data refer to mean values recorded in samples taken in Apr, Jul, Sep and Oct each year; LUX, Weiswarnpach, 96 and 98, upper limits; POL, Data refer to spring and summer surveys, and to surface measurements; CHE, Lac Constance, Bodensee (Germany); UKD, Lomond and Bewl Water: annual averages were calculated using the limit of detection values; actual average may therefore be lower. Bowl Water: 1994–1999 data refer to Total inorganic Phosphate.

Total Nitrogen Note: CAN, Ontario, Huron and Superior data represent annual mean surface values from several handed open water samples for each lake (mainly spring and summer); MEX, Data refer to nitrates only; U.S.A., W: Twin(Ohio): total inorganic nitrogen ($NH_4+NO_3+NO_2$). Samples obtained from the deepest point in each lake, generally weekly from late-spring early fall, and less frequently the rest of the year at 0.1, 2, 4, 7 and 10 m; AUT, data refer to total inorganic nitrogen ($NH_4+NO_3+NO_2$) measured in the epilimnion; FIN, Data refer to surface measurements; FRA, Réservoirs Marne and Seine, and Lac d'Annecy: Kjeldhai nitrogen. Réservoirs Marme and Seine 1980: 1981 data; DEU, Total inorganic nitrogen ($NH_4+NO_3+NO_2$). Bodensee: Lac Constance (Switzerland); IRL, Nitrates and nitrites only. Ennel and Owel: 1990–99 data refer to nitrate only and to limited surveys (March–October). Derg: 1996 and 1997 data refer to mean values recorded in samples taken in Apr, Jul, Sep and Oct each year; ITA, Maggiore: total inorganic nitrogen ($NH_4+NO_3+NO_2$). Average of monthly samples taken at the deepest point of the lake; LUX, Nitrates only; POL, Data refer to spring and summer surveys; CHE, Lac Constance: Bodensee (Germany); TUR, Total inorganic nitrogen ($NH_4+NO_3+NO_2$); UKD, Neath and Lomond; nitrates (NO_3) only. Lomond and Bewel Water: annual averages were calculated using the limit of detection values; actual averages may therefore be lower.

[a] Average over the latest 3 years available: data prior to 1953 have not been taken into account.

Source: Table 3.5a and 3.5b (data from 1996, 1997, 1998, 1999, and average last 3 years), *OECD Environmental Data Compendium 2002*, © OECD 2002, www.oecd.org.

Table 10B.77 Country Contributions to Sum of Direct and Riverine Inputs in Tonnes Per Year of Cadmium, Mercury, Lead, Lindane and Pcb to the Northeast Atlantic

Substance/Year	B	DK	F	D	IRL	NL	NO	P	E	S	UK	SUM	%
Cadmium													
1990	4.30	0.57	4.70	8.54	3.39	10.30	16.21	9.00	n.a.	0.67	51.30	108.97	100.00
1991	3.90	n.a.	3.30	7.24	3.39	7.00	8.31	15.00	n.a.	0.34	49.35	97.83	89.80
1992	6.70	n.a.	4.50	10.55	2.31	5.70	7.90	24.00	n.a.	0.54	34.69	96.89	88.90
1993	3.42	n.a.	3.20	9.20	1.76	4.70	7.18	1.70	0.06	0.11	29.30	60.62	55.60
1994	3.44	n.a.	3.20	8.14	2.44	12.76	9.23	3.93	0.10	0.42	25.87	69.52	63.80
1995	3.81	n.a.	2.70	7.54	2.17	26.60	8.99	0.20	0.10	0.24	21.31	73.67	67.60
1996	2.90	n.a.	2.20	5.65	3.94	8.40	5.85	n.a.	n.a.	n.a.	19.11	48.05	44.10
1997	2.65	n.a.	n.a.	6.35	2.55	4.05	5.20	1.35	14.50	0.47	15.35	52.47	48.10
1998	2.45	n.a.	n.a.	6.25	3.45	9.70	7.75	0.60	6.80	0.56	19.75	57.31	52.60
1999	2.30	n.a.	n.a.	4.99	2.91	11.90	4.60	0.40	12.80	0.90	18.00	58.80	54.00
Mercury													
1990	4.05	0.13	5.70	10.54	2.45	3.18	0.85	4.90	n.a.	0.11	8.81	40.71	100.00
1991	1.15	n.a.	2.70	10.01	2.45	3.30	0.50	1.75	n.a.	0.13	7.85	29.83	73.30
1992	0.68	n.a.	8.70	11.03	2.40	3.26	0.53	0.04	n.a.	0.13	6.00	32.78	80.50
1993	2.62	n.a.	0.80	10.05	2.40	3.46	0.49	0.26	n.a.	0.11	6.55	26.73	65.70
1994	3.10	n.a.	1.10	5.65	2.40	6.41	0.38	0.85	n.a.	0.10	4.83	24.82	61.00
1995	0.60	n.a.	0.80	4.64	2.40	8.06	0.54	0.35	n.a.	0.15	4.13	21.67	53.20
1996	0.03	n.a.	0.50	2.94	n.a.	3.28	0.47	n.a.	n.a.	n.a.	3.26	10.48	25.70
1997	0.25	n.a.	n.a.	1.95	n.a.	2.70	0.40	1.15	2.60	0.09	4.41	13.54	33.30
1998	0.61	n.a.	n.a.	2.15	n.a.	2.15	2.86	0.75	0.40	0.13	4.68	13.72	33.70
1999	0.72	n.a.	n.a.	2.39	n.a.	2.33	1.46	0.56	8.60	0.20	3.31	19.60	48.10
Lead													
1990	27.5	7.2	150.0	212.8	63.4	346.5	120.9	680.0	n.a.	8.3	591.1	2,207.6	100.0
1991	36.6	n.a.	150.0	297.5	63.4	230.0	102.8	80.0	7.7	5.3	575.6	1,548.8	70.2
1992	32.5	n.a.	110.0	314.9	82.1	221.8	70.5	120.0	2.6	4.3	501.6	1,460.2	66.1
1993	36.6	n.a.	100.0	378.1	55.5	393.8	58.5	7.3	0.9	3.2	631.4	1,665.2	75.4
1994	48.2	n.a.	170.0	247.7	79.6	701.8	82.6	6.1	2.2	10.5	528.3	1,877.1	85.0
1995	43.2	n.a.	55.0	185.5	63.3	871.8	82.5	2.8	1.3	8.2	384.8	1,698.3	76.9
1996	44.5	n.a.	66.0	126.5	115.6	382.9	62.6	n.a.	n.a.	n.a.	311.5	1,109.5	50.3
1997	40.0	n.a.	n.a.	151.5	120.5	233.0	63.7	43.0	17.0	8.7	303.3	980.7	44.4
1998	51.5	n.a.	n.a.	181.5	143.0	257.5	150.0	4.9	54.0	14.3	601.5	1,458.2	66.1
1999	56.5	n.a.	n.a.	125.4	145.0	326.0	53.6	2.1	58.0	17.3	546.0	1,330.0	60.2
Lindane													
1990	102.0	26.0	175.0	342.0	n.a.	16.7	536.0	n.a.	n.a.	n.a.	588.4	1,786.1	100.0
1991	111.0	26.0	175.0	187.8	n.a.	7.2	261.3	n.a.	n.a.	n.a.	759.5	1,527.8	85.5
1992	82.5	26.0	175.0	208.0	n.a.	10.5	97.9	0.7	n.a.	n.a.	543.0	1,143.6	64.0
1993	80.0	26.0	175.0	191.5	n.a.	360.0	90.9	3.4	n.a.	n.a.	601.8	1,528.6	85.6
1994	74.5	26.0	175.0	261.0	n.a.	230.0	87.7	15.0	n.a.	n.a.	501.8	1,371.0	76.8
1995	55.5	26.0	175.0	291.4	n.a.	370.0	107.4	2.8	n.a.	n.a.	484.7	1,512.8	84.7
1996	n.a.	n.a.	100.0	239.0	n.a.	300.0	75.9	n.a.	n.a.	n.a.	312.4	1,027.3	57.5
1997	66.0	n.a.	n.a.	370.0	n.a.	300.0	78.7	n.a.	203.5	n.a.	354.5	1,372.7	76.9
1998	110.5	n.a.	n.a.	261.0	n.a.	259.5	84.8	n.a.	15.0	n.a.	403.8	1,134.6	63.5

(Continued)

Table 10B.77 (Continued)

Substance/Year	B	DK	F	D	IRL	NL	NO	P	E	S	UK	SUM	%
1999	77.0	n.a.	n.a.	68.0	n.a.	279.0	9.4	n.a.	121.0	n.a.	324.0	877.0	49.1
PCB_7													
1990	30.5	30.3	100.0	143.5	n.a.	150.0	484.5	n.a.	n.a.	n.a.	2,158.1	3,096.8	100.0
1991	24.5	30.3	160.0	89.4	n.a.	130.0	38.0	n.a.	n.a.	n.a.	1,183.5	1,655.7	53.5
1992	26.5	30.3	130.0	43.6	n.a.	100.0	40.5	4.9	n.a.	n.a.	764.7	1,140.5	36.8
1993	211.0	30.3	130.0	54.1	n.a.	130.0	21.6	25.0	n.a.	n.a.	1,417.1	2,019.0	65.2
1994	223.5	30.3	160.0	95.5	n.a.	300.0	55.0	84.0	n.a.	n.a.	1,066.9	2,015.2	65.1
1995	45.9	30.3	130.0	144.0	n.a.	470.0	27.8	26.0	n.a.	n.a.	938.7	1,812.6	58.5
1996	n.a.	n.a.	100.0	127.5	n.a.	200.0	16.6	n.a.	n.a.	n.a.	739.6	1,183.6	38.2
1997	66.0	n.a.	n.a.	109.0	n.a.	170.0	22.6	n.a.	n.a.	n.a.	360.0	727.6	23.5
1998	135.0	n.a.	n.a.	88.0	n.a.	181.0	22.4	n.a.	n.a.	n.a.	1,089.4	1,515.8	48.9
1999	164.0	n.a.	n.a.	123.0	n.a.	236.0	n.a.	n.a.	158.0	n.a.	835.0	1,516.0	49.0

Note: Belgium (B), Denmark (DK), France (F), Germany (D), Ireland (IRL), the Netherlands (NL), Norway (NO), Portugal (P), Spain (E), Sweden (S), United Kingdom (UK), n.a. indicates where data was not available. Rounded values from EEA Fact Sheet YIR01HS01, zero or near zero (<1 kg) submissions assumed as data not available; %, percent as of 1990.

Source: From Green, N. et al., 2003, *Hazardous Substances in the European Marine Environment: Trends in Metals and Persistent Organic Pollutants, European Environment Agency, Topic Report 2/2003*, www.eea.europa.eu. Reprinted with permission © EEA.

Table 10B.78 PCBs Concentrations in Freshwater and Seawater in China (ng/L)

Water Body			Sampling Date	Concentration	
				Range	Mean
Freshwater	Pearl River, Guangzhou reach		May 24–26, 2000	$0.70\times10^{-3}\sim3.96\times10^{-3}$	2.3×10^{-3}
	Donghu Lake Wuhan			2.7×10^{-3}	
	Yangtse River, Nanjing section		May 16–17, 1998	$1.74\times10^{-3}\sim2.0\times10^{-3}$	1.91×10^{-3}
Seawater	Pearl River Delta	Humen	May–June, Oct–Nov, 2000	$2.08\times10^{-3}\sim3.92\times10^{-3}$	2.70×10^{-3}
		Hengmen		$0.475\times10^{-3}\sim1.54\times10^{-3}$	0.999×10^{-3}
		Jiaomen		$1.73\times10^{-3}\sim3.26\times10^{-3}$	2.62×10^{-3}
		Dourmen		$1.73\times10^{-3}\sim3.26\times10^{-3}$	1.16×10^{-3}
	Daya Bay		Aug 4, 1999	$0.091\sim1.355$	0.314
	Minjing River Estuary	Surface water	Nov, 1999	$0.204\sim2.473$	0.985
		Pore water		$3.19\sim10.86$	6.37
	Xiamen Harbor		July 20, 1998	$0.12\times10^{-3}\sim1.69\times10^{-3}$	0.74×10^{-3}
	Jiulong River Estuary		Early than 1999	$0.032\sim0.169$	
	Pearl River Estuary		Early than 1999	$0.033\sim1.064$	

Source: From Yeru, H. et al., 2003, Water Quality Standards and POPs Pollution in China in United Nations University, *Capacity Development Training for Monitoring of POPS in the East Asian Hydrosphere*, 1–2 September, 2003, UNU Centre, Tokyo.

Table 10B.79 Polluting Incidents from Vessels in Canadian Waters, 1974–1983

	Transfer Accident		Collision, Ground, Sinking		Other		Total	
Year	**Events**	**Tons**	**Events**	**Tons**	**Events**	**Tons**	**Events**	**Tons**
1974	60	371	21	4,277	60	248	141	4,896
1975	52	116	13	613	28	886	93	1,615
1976	53	206	13	1,613	19	160	85	1,979
1977	47	249	11	931	38	294	96	1,474
1978	51	154	16	1,343	33	73	99	1,570
1979	49	108	6	948	33	8,186	88	9,242
1980	68	145	12	121	68	213	148	479
1981	75	97	13	2,296	33	931	121	3,324
1982	58	199	16	2,106	27	989	101	3,294
1983	28	73	7	504	22	404	57	981
Total	541	1,718	127	14,752	361	12,384	1,029	28,854
Percent	53	6	12	51	35	43	100	100

Note: Tankers, bulk carriers and other vessels; metric tons.

Source: From Environment Canada, *Summary of Spill Events in Canada*, 1974–1983, EPS 5/SP/1, www.ec.gc.ca.

Table 10B.80 Top Releases of Chemicals to Water in Canada, 2001

Chemical	**Release Tons**
Ammonia (total)[a]	26,106
Nitrate ion in solution at pH $\geq$6.0	22,450
Manganese (and its compounds)	1,157
Methanol	697
Zinc (and its compounds)	308

[a] Refers to the total of both ammonia (NH_3) and ammonium ion (NH_4^+) in solution.

Source: From Environment Canada, National Pollutant Release Inventory Database, www.ec.gc.ca/pdb/npri, 2001. Reproduced with the permission of the Minister of Public Works and Government Services, 2006.

Table 10B.81 Water Bodies Receiving over 500 Tons of Pollutants in Canada, 2001

Water Body	**Total Release (tons)**	**Dominant Release**	**Share of Total Release (percent)**
Fraser River	9,168	Ammonia[a]	49.2
Lake Ontario	8,877	Ammonia[a]	41.6
Bow River	8,264	Nitrate Ion	90.8
Ottawa River	3,066	Ammonia[a]	76.6
North Saskatchewan River	2,953	Nitrate ion	61.3
Red River	2,766	Ammonia[a]	72.7
Hamilton Harbour	1,516	Ammonia[a]	70.6
South Saskatchewan River	1,275	Nitrate Ion	62.4
St. Lawrence River	1,088	Nitrate Ion	43.6
Saint John River	984	Methanol	28.6
Frank Lake	818	Nitrate ion	70.3
Detroit River	679	Ammonia[a]	83.8
Kelly Lake	619	Nickel (and its compounds)	29.6
Neroutsos Inlet	526	Nitrate ion	64.4

Note: The information in this table is not intended to be an assessment of environmental impact or water quality. The totals do not include releases to tributaries of the named rivers.

[a] Refers to the total of both ammonia (NH_3) and ammonium ion (NH_4^+) in solution.

Source: From Environment Canada, Pollution Data Branch, National Pollutant Release Inventory Database, www.ec.gc.ca/pdb/npri/npri_dat_rep_e_cfm, 2001. Reproduced with the permission of the Minister of Public Works and Government Services, 2006.

Original Source: From Environment Canada, Pollution Data Branch, National Pollutant Release inventory Database, www.ec.gc.ca/pdb/nprl/npri_dat_rep_e_cfm (accessed March 25, 2003).

Table 10B.82 Accidental Oil Spills in the World from Tankers, 1975–2000 (a)

Year[a]	Date	Name of Ship	Flag	Country Affected	Quality Spilled (tons)	Indemnity Million/s USD (estimates)
1975	6.1	Showa Manu	Japan	Singapore	3 800	10.9
	10.1	British Ambassador	UK	Japan (Pacif.)	45 000	—
	29.1	Jakob Maersk	Denmark	Portugal	84 000	2.8
	31.1	Corinthos/E.M. Queeny	U.S.A.	U.S.A. (Delaware)	40 000	5.9
	4.4	Spartan Lady	Liberia	U.S.A.	25 000	—
	4.1	Shell Barge No 2	U.S.A.	U.S.A.	—	5.7
	17.4	Mitsu Maru 3	Japan	Japan	500	5.7
	13.5	Epic Colocoltroni	Greece	Dominican Rep.	57 000	—
1976	6.2	Saint Peter	Liberia	Colombia	33 000	0.9
	12.5	Urquiola	Spain	Spain	101 000	19.7
	23.6	Nepco 140	U.S.A.	Canada—U.S.A.	1 200	11.1
	12.7	Crelan Star	Cyprus	Indian Ocean	28 600	—
	14.10	Boehlen	GDR	France	11 000	20.3
	15.12	Argo Merchant	Liberia	U.S.A. (Massach.)	28 000	2.5
1977	18.1	Irenes Challenge	Liberia	Pacific	34 000	—
	7.1	Borag	Liberia	East China	4 000	15.6
	25.2	Hawaiian Patriot	Liberia	Honolulu (Pacif.)	99 000	—
	27.5		Panama	Nicaragua	30 000	—
	16.12	Venoil/Venpet	Liberia	South Africa	26 000	5.4
	30.12	Grand Zenith	Panama	U.S.A. (Massach.)	29 000	—
1978	16.3	Amoco Cadiz	Liberia	France	228 000	—
	6.5	Eleni V	Greece	UK	3 000	10.6
	7.7	Cabo Tamar	Chile	Chile	60 000	4.2
	12.10	Christos Bitas	Greece	UK	5 000	13.1
	30.12	Esso Bernica	UK	UK (Shetland)	1 160	9.8
	31.12	Andrus Patria	Greece	Spaint	47 000	6.4
1979	8.1	Betelgeuse	France	Ireland	27 000	36.2
	28.2	Antonio Gransci	U.S.S.R.	Sweden, Finland, U.S.S.R.	6 000	54.1
	2.3	Messlaniki Frontis	Liberia	Greece	6 000	11.5
	15.3	Kurdistan	UK	Canada	7 000	5.1
	28.4	Gino	Liberia	France	42 000	0.8
	28.6	Aviles	Liberia	Arabian Sea	25 000	—
	29.7	Atlantic Express	Greece	Tobago	276 000	1.5
	16.8	Ionnis Angelicoussis	Greece	Angola	30 000	—
	1.9	Chevron Hawaii	U.S.A.	U.S.A.	2 000	12.0
	1.11	Burmah Agate	Liberia	U.S.A. (Texas)	40 000	11.5
	15.11	Independenta	Romania	Turkey	94 600	17.2
1980	28.1	Princess Anne Marie	Greece	Cuba	6 000	51.7
	24.2	Irenes Serenade	Greece	Greece	102 000	12.5
	7.3	Tanio	Madagascar	France, UK	13 500	40.0
	29.12	Juan A. Lavalleja	Uruguay	Algeria	40 000	—

(Continued)

Table 10B.82 (Continued)

Year[a]	Date	Name of Ship	Flag	Country Affected	Quality Spilled (tons)	Indemnity Million/s USD (estimates)
1981	7.1	Jose Marti	U.S.S.R.	Sweden	6 000	6.7
	3.3	Ondina	Dubai	Germany	5 000	7.0
1983	7.1	Assimi	Greece	Oman	51,431	—
	6.8	Castillo de Bellver	Spain	South Africa	255 525	1.0
	27.9	Sivand	Iran	UK	6 000	5.0
	25.11	Feoso Ambassador	China	China	4 000	10.0
	10.12	Pericles GC	Greece	Qatar	46 631	—
1985	14.2	Neptunia	Liberia	Iran	60 000	—
	6.12	Nova	Liberia	Iran	71 120	—
1989	24.3	Exxon Valdez	U.S.A.	U.S.A. (Alaska)	35 000	2000
	19.12	Kharg 5	..	Morocco	70 000	37
	29.12	Aragon	Spain	Madeira	25 000	—
1991	9/11.4	Le Haven	Cyprus	Italy	30 000	—
		ABT Summer	..	..	260 000	—
1992	21.1	Maersk Navigator	Singapore	Sumatra	>25 000	—
1993	5.1	Braer	Liberia	UK	84 000	—
1994	6.1	Red Star	..	Portgual	—	14.0
	8.2	Albinoni	Bahamas	Caribbean	—	50.0
1995	21.1	Maersk Navigator	Denmark	Indonesia	—	33.0
1996	15.2	Sea Empress	..	UK	72 000	30.0
1997	2.1	Nakhodka	Russia	Japan	37,000	1,233
	15.1	Orapin Global & Evoikos	Singapore	Thailand & Cypress	25,000 (Evoikos)	>100
1999	12.12	Erika	Malta	France	19 800	155

Note: Over 25,000 tons or over USD 5 million of indemnity, world.

[a] 1986–1988: no major oil spills over 25,000 tons.

Source: Table 6.1, *OECD Environmental Data Compendium 2002*, © OECD 2002, www.oecd.org.

Original Source: From IMO, IOPC-Fund, ACOPS, IFP, TAC, TOVALOP, SIGMA.

SECTION 10C GROUNDWATER CONTAMINATION

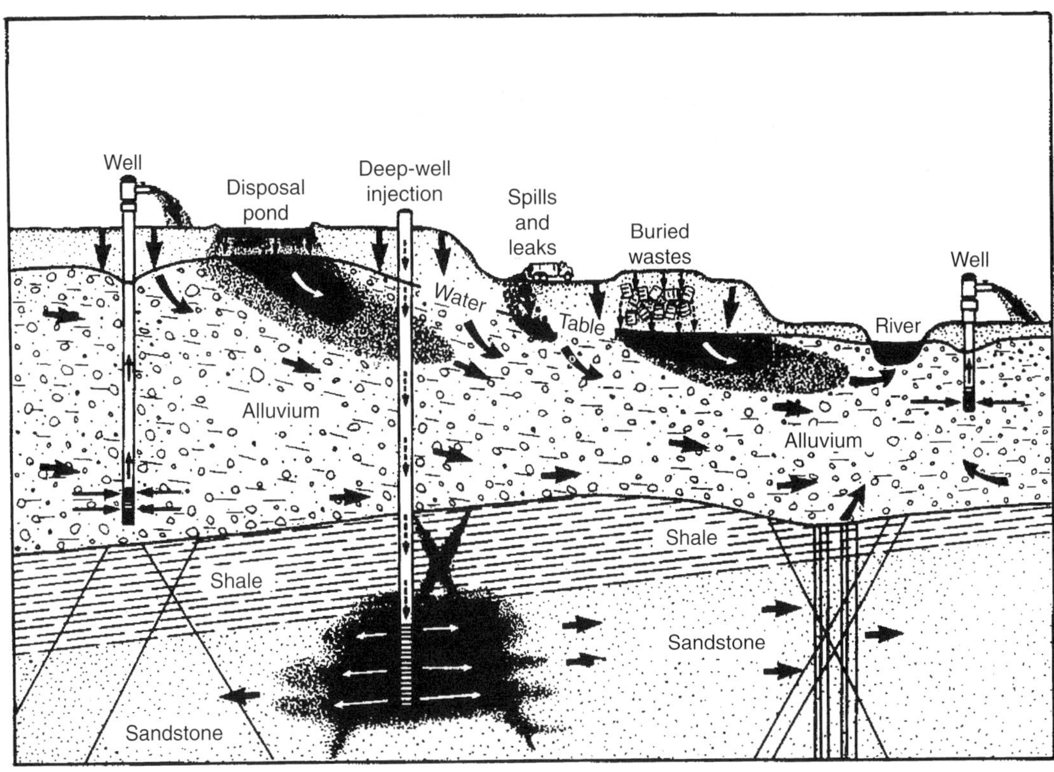

Figure 10C.54 Waste disposal practices and contamination of groundwater. Movement of contaminants in unsaturated zone, alluvial aquifer, and bedrock shown by dark shading. (From U.S. Geological Survey Water Fact Sheet. *Toxic Waste, Groundwater Contamination*, 1983.)

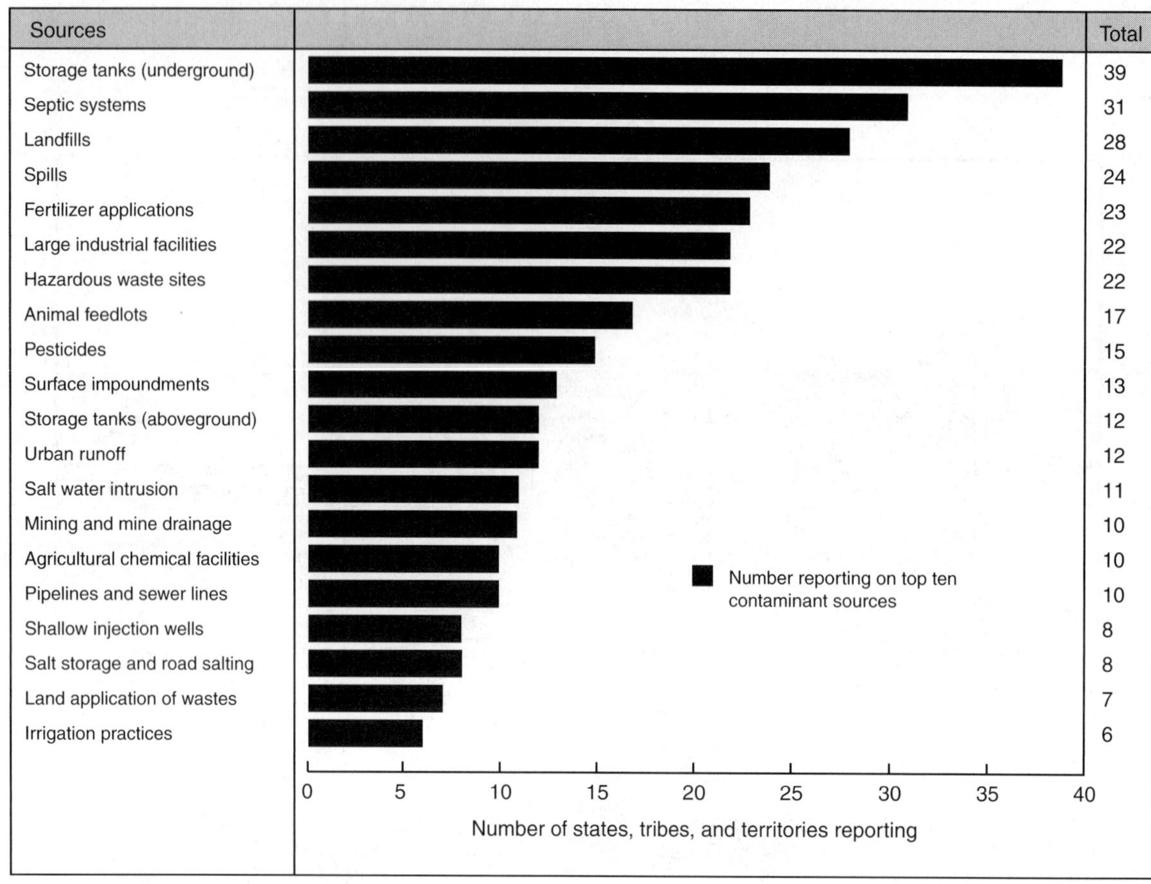

Figure 10C.55 Major source of groundwater contamination. (From United States Environmental Protection Agency, 2002, *National Water Quality Inventory 2000 Report*, EPA-841-R-02-001, www.epa.gov.)

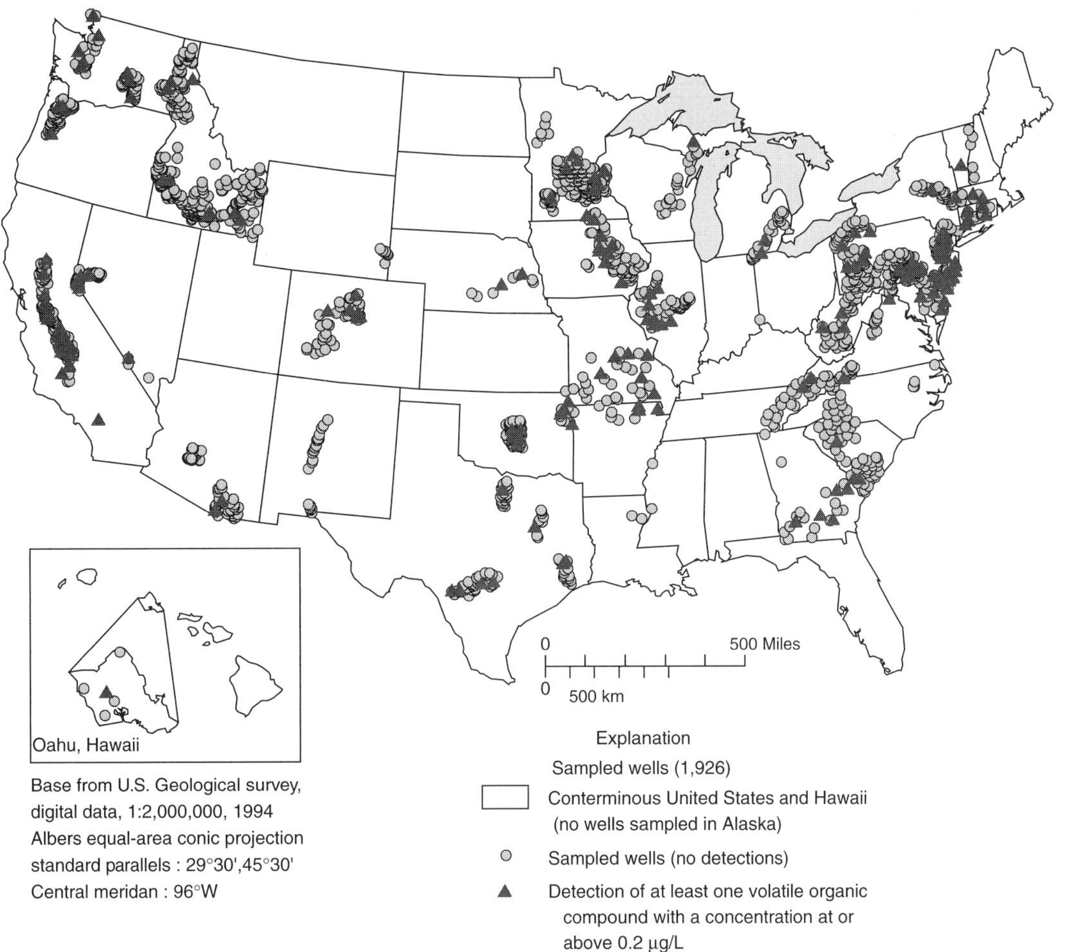

Figure 10C.56 Location of rural, untreated, self-supplied domestic wells in the United States where at least one volatile organic compound was detected at an assessment level of 0.2 μ/L, 1986–1999. (From Moran, M.J., Lapham, W.W., Rowe, B.L., and Zogorski, J.S., 2002, *Occurrence and Status of Volatile Organic Compounds in Groundwater from Rural, Untreated, Self-Supplied Domestic Wells in the Untied States, 1986–1999*, United States Geological Survey, Water Resources Investigation Report 02-4085, www.usgs.gov.)

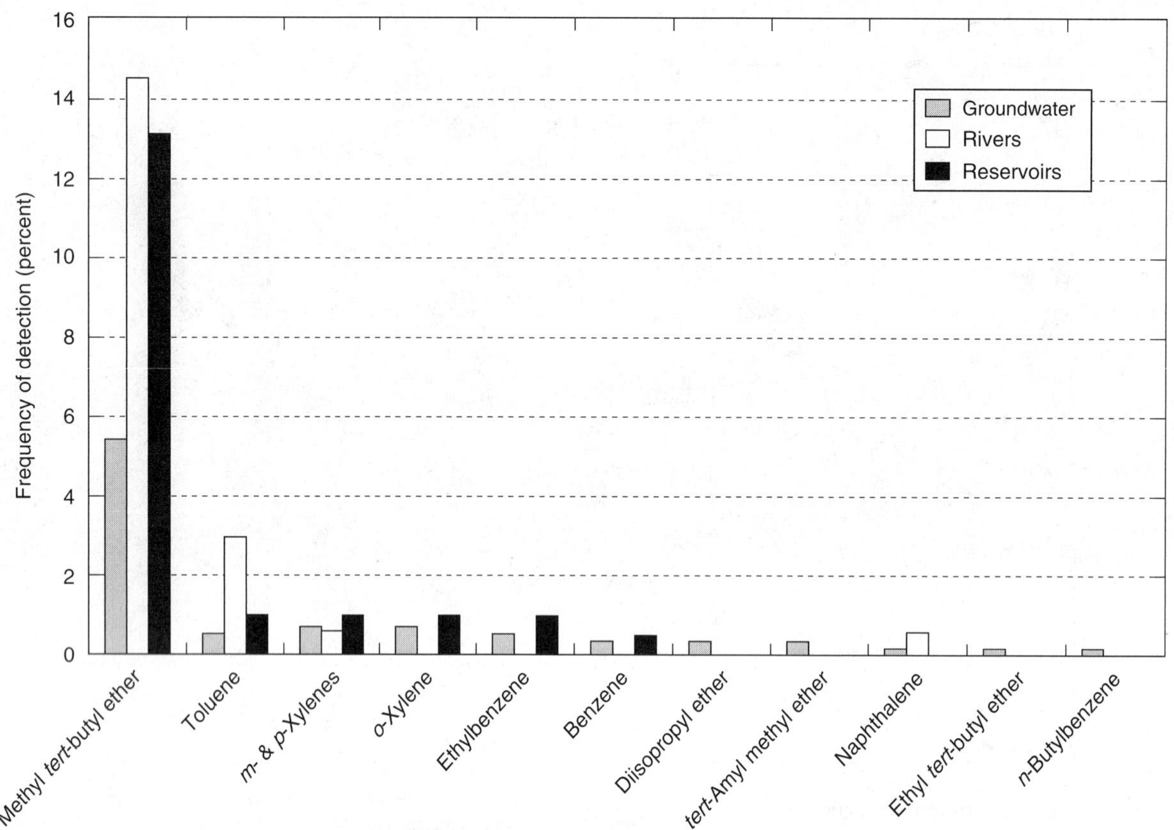

Figure 10C.57 Frequency of detection of gasoline compounds in source-water samples from groundwater, rivers, and reservoirs in the United States from May 3, 1999 through October 23, 2000. (From Grady, S., 2003, *A National Survey of Methyl tert-Butyl Ether and Other Volatile Organic Compounds in Drinking Water Sources: Results of a Random Survey*, United States Geological Survey, Water-Resource Investigations Report 02-4079, www.usgs.gov.)

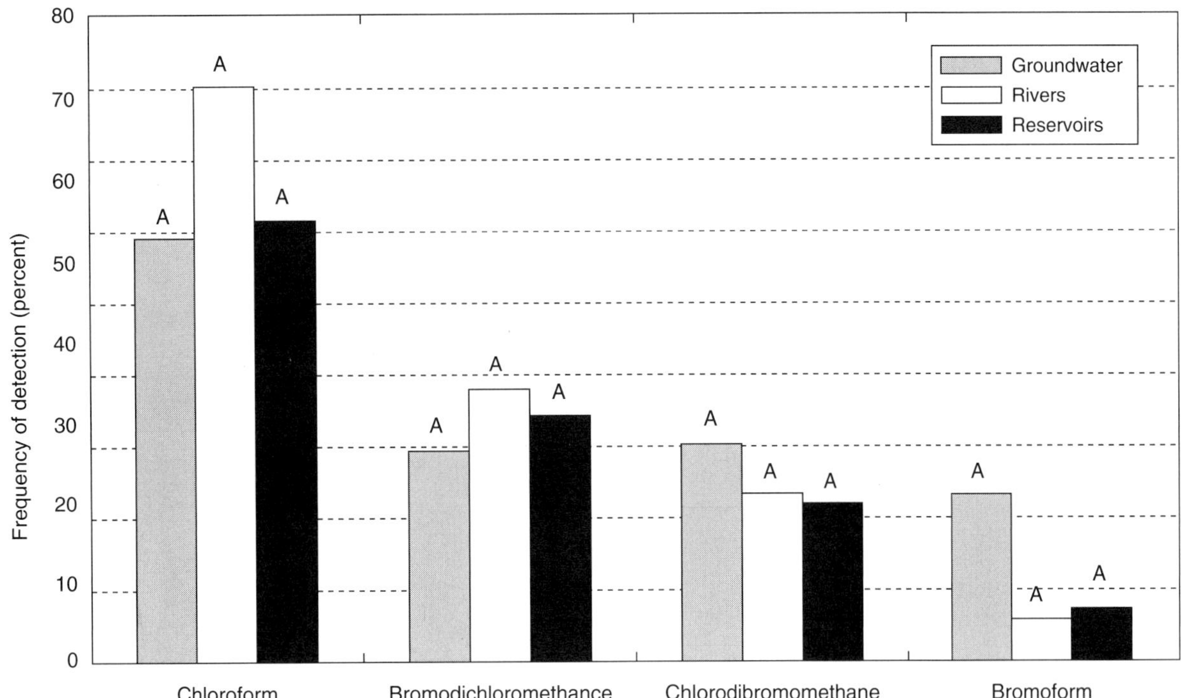

Figure 10C.58 Frequency of detection of trihalomethanes in source-water samples from groundwater, rivers, and reservoirs in the United States from May 3, 1999 through October 23, 2000 (populations that share the same letter symbol are not significantly different at the 95-perecent confidence). (From Grady, S., 2003, *A National Survey of Methyl tert-Butyl Ether and Other Volatile Organic Compounds in Drinking Water Sources: Results of a Random Survey*, United States Geological Survey, Water-Resource Investigations Report 02-4079, www.usgs.gov.)

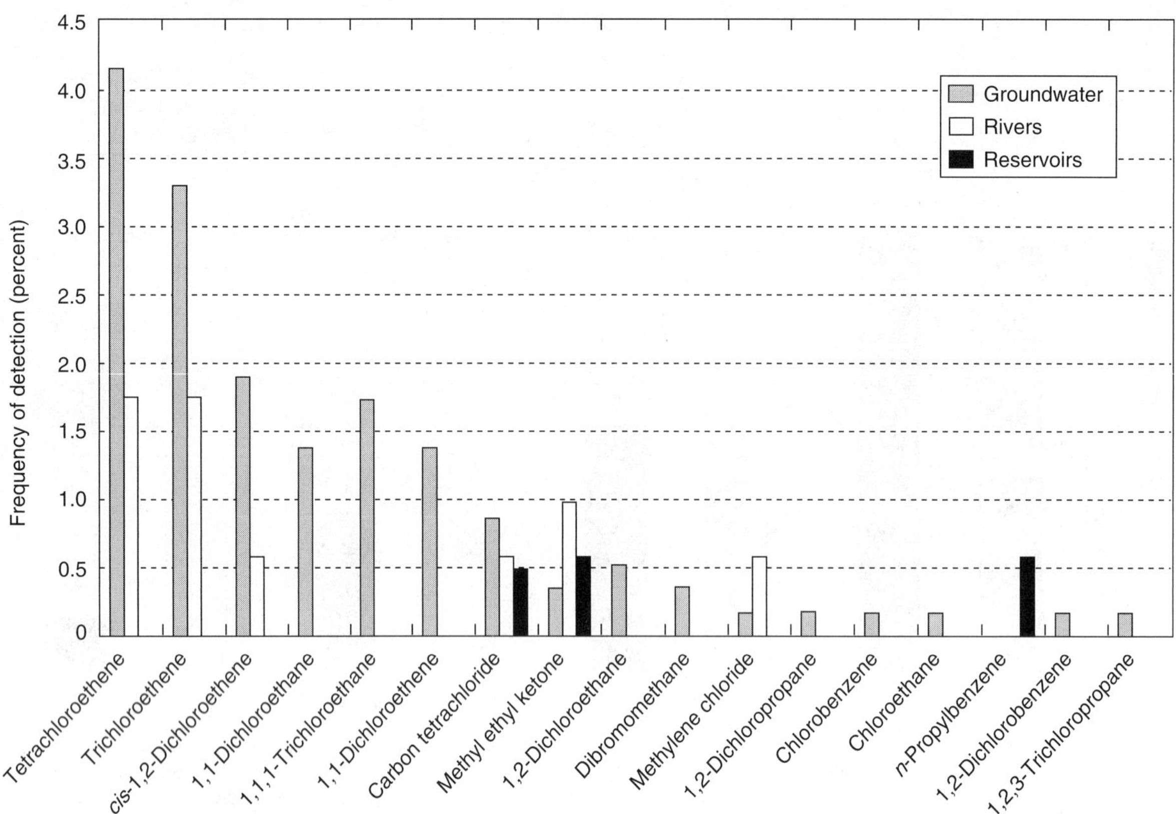

Figure 10C.59 Frequency of detection of solvents in source-water samples from groundwater, rivers, and reservoirs in the United States from May 3, 1999 through October 23, 2000. (From Grady, S., 2003, *A National Survey of Methyl tert-Butyl Ether and Other Volatile Organic Compounds in Drinking Water Sources: Results of a Random Survey, United States Geological Survey*, Water-Resource Investigations Report 02-4079, www.usgs.gov.)

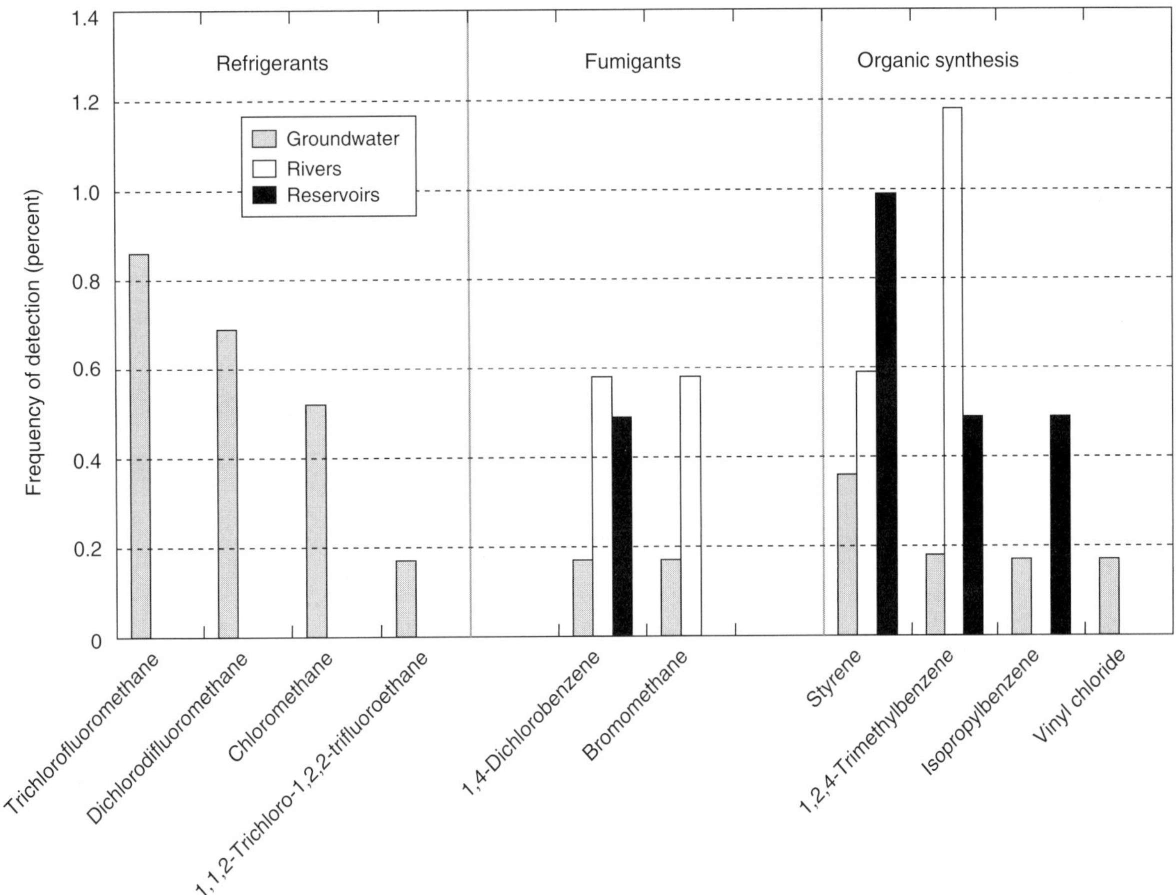

Figure 10C.60 Frequency of detection of refrigerants, fumigants, and organic synthesis compounds in source-water samples from groundwater, rivers, and reservoirs in the United States from May 3, 1999 through October 23, 2000. (From Grady, S., 2003, *A National Survey of Methyl tert-Butyl Ether and Other Volatile Organic Compounds in Drinking Water Sources: Results of a Random Survey*, United States Geological Survey, Water-Resource Investigations Report 02-4079, www.usgs.gov.)

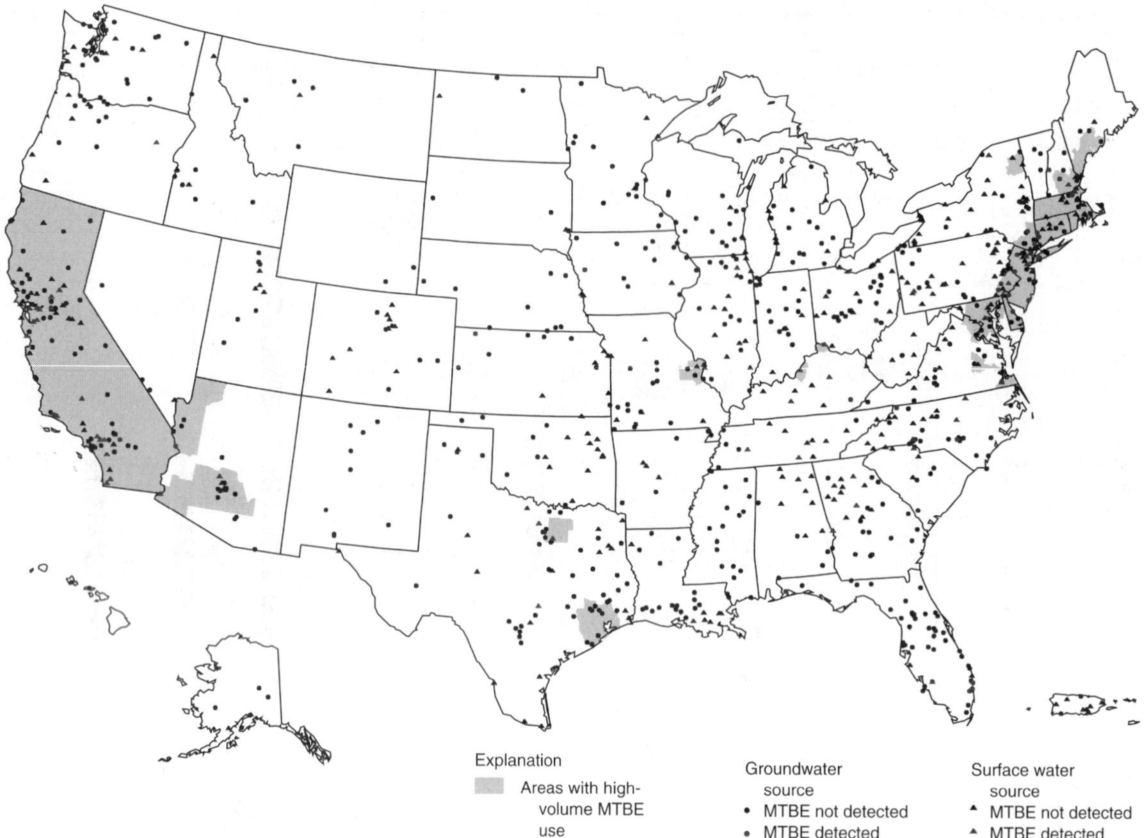

Figure 10C.61 Distribution of methyl *tert*-butyl ether (MTBE) in ground- and surface-water sources in the United States and Puerto Rico in relation to high MTBE-use areas. (From Grady, S., 2003, *A National Survey of Methyl tert-Butyl Ether and Other Volatile Organic Compounds in Drinking Water Sources: Results of a Random Survey*, United States Geological Survey, Water-Resource Investigations Report 02-4079, www.usgs.gov.)

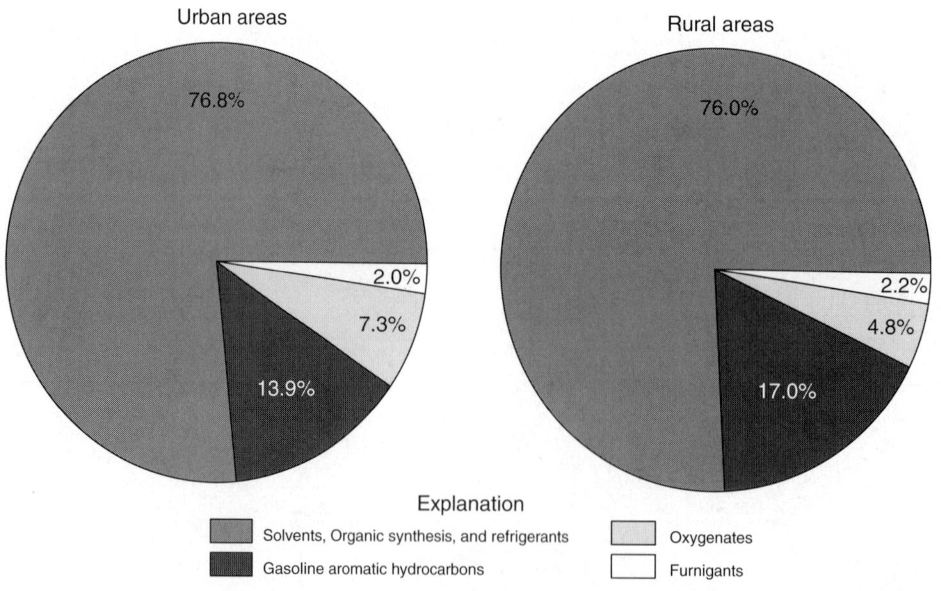

Figure 10C.62 Percent detection of selected classes of VOCs in the United States in urban and rural areas, 1985–1995. (From Squillace, P.J. et al., 1999, volatile organic compounds in untreated ambient groundwater of the United States, 1985–1995, *Environ. Sci. Technol.*, 33, 4167–4187.)

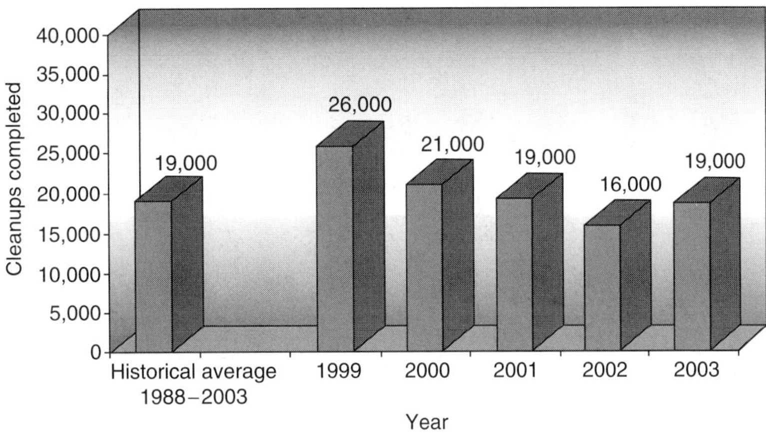

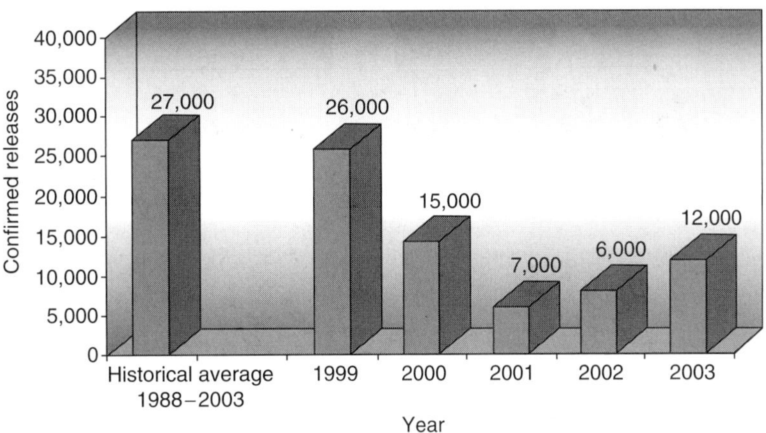

Figure 10C.63 Confirmed underground storage tank releases and cleanups complete in the United States: historical average, 1999–2003. (From United States Environmental Protection Agency, 2004, *Underground Storage Tanks: Building On the Past to Protect the Future*, EPA 510-R-04-001, March 2004, www.epa.gov.)

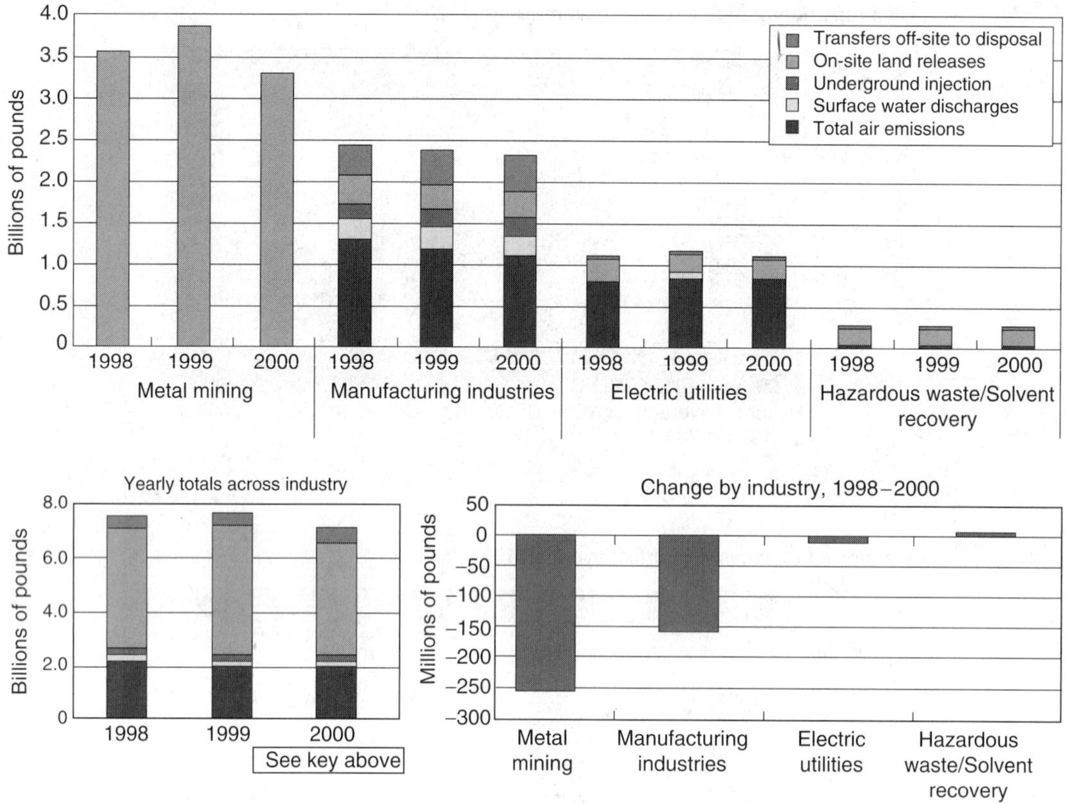

Figure 10C.64 United Sates toxics release inventory (TRI) total releases and change by industry, 1998–2000. (From United States Environmental Protection Agency, 2003, *EPA's Draft Report on the Environment, 2003*, EPA 600-R-03-050, www.epa.gov.)

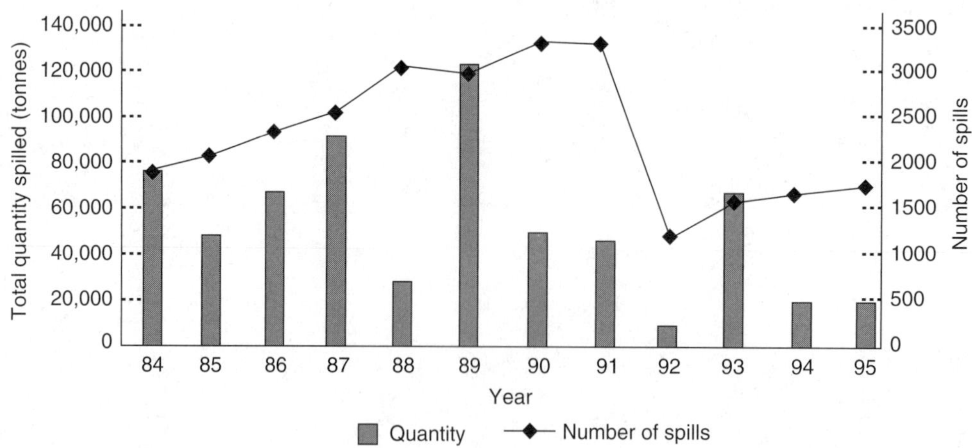

Figure 10C.65 Petroleum sector—number of reported spills in Canada and total quantity, by year. (From *Summary of Spill Events in Canada, 1984–1995*, EPS 5/SP/3, Environment Canada, Canada Environmental Emergencies Program, 1998. Reproduced with the permission of the Minister of Public Works and Government Services, 2006.)

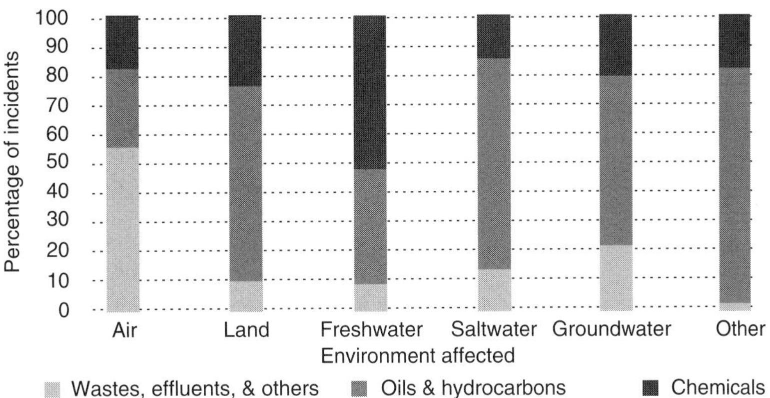

Figure 10C.66 Percent distribution of environment affected by spill category in Canada. (From *Summary of Spill Events in Canada, 1984–1995*, EPS 5/SP/3, Environment Canada, Canada Environmental Emergencies Program, 1998. Reproduced with the permission of the Minister of Public Works and Government Services, 2006.)

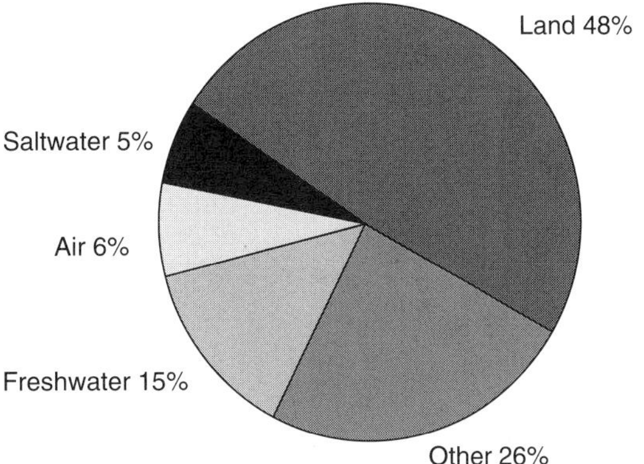

Figure 10C.67 Percent distribution of environment affected by reported spills in Canada. (From *Summary of Spill Events in Canada, 1984–1995*, EPS 5/SP/3, Environment Canada, Canada Environmental Emergencies Program, 1998. Reproduced with the permission of the Minister of Public Works and Government Services, 2006.)

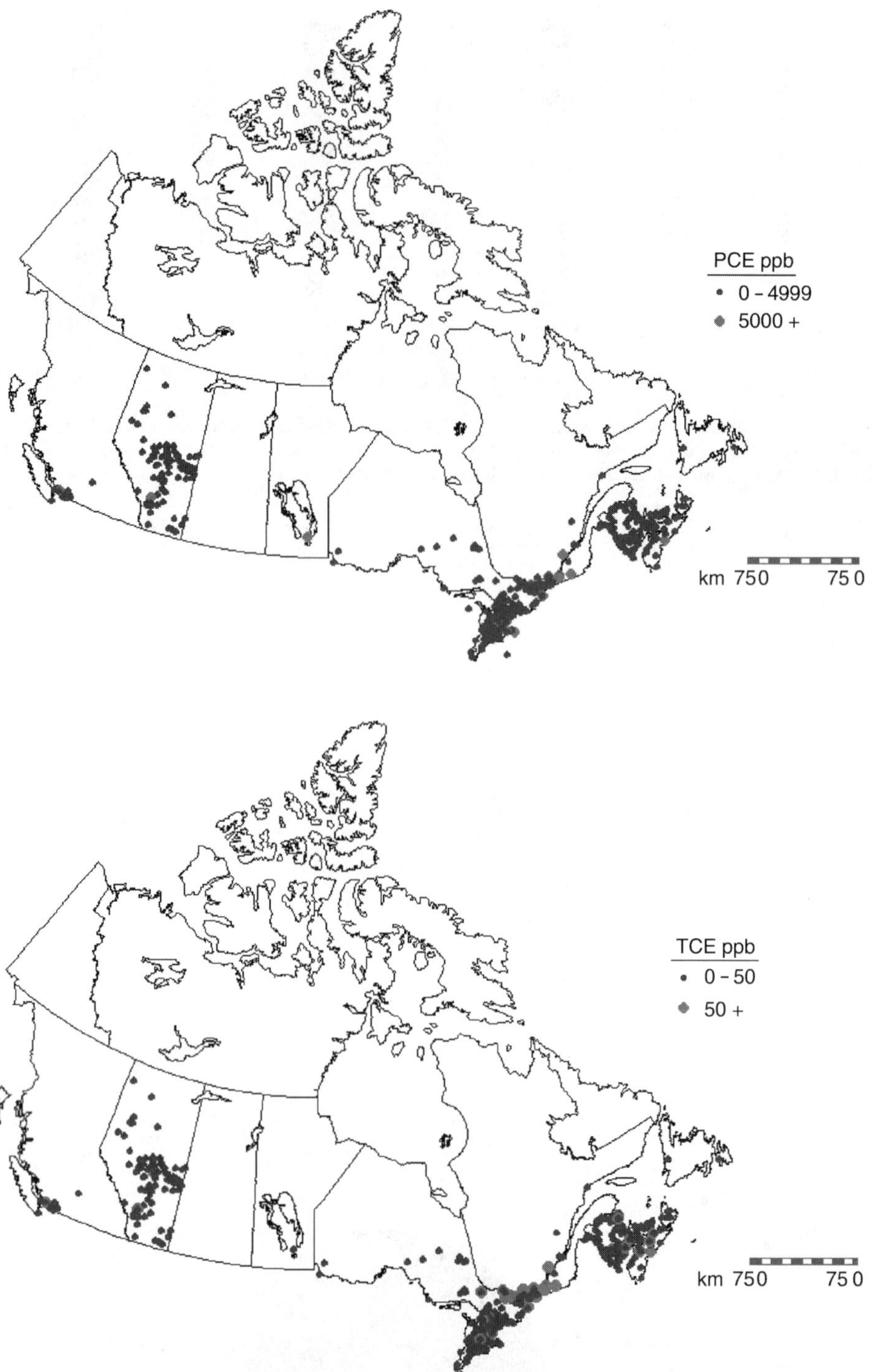

Figure 10C.68 Distribution of tetrachloroethylene and trichloroethylene in groundwater throughout Canada. (From Bajjali, W., 2003, *Study of the Distribution of Tetrachloroetheylene and Trichloroethylene in Groundwater Throughout Canada Using SPANS-GIS*, frontpage.uwsuper.edu/bajjali/proj/can/c1.htm. With permission. Last updated on June 16, 2003.)

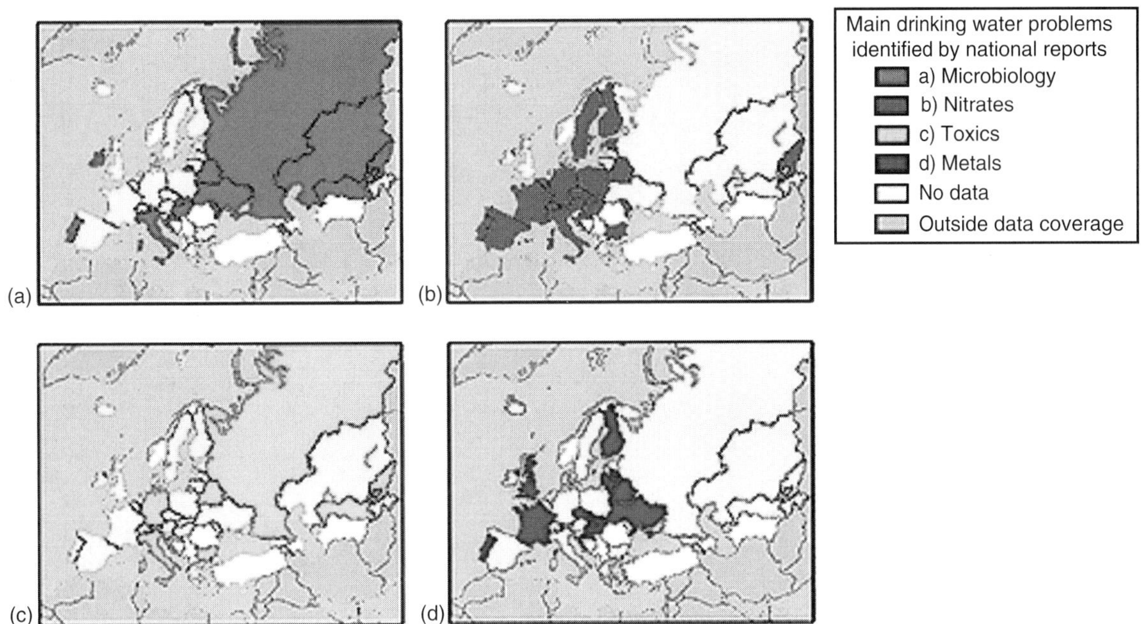

Figure 10C.69 Main drinking water problems identified in the European Union. (From European Environmental Agency, *Indicator Fact Sheet Drinking Water Quality (WEU10) Verslon 13.10.03*, www.eea.europa.eu. Reprinted with permission © EEA.)

Table 10C.83 Major Groundwater Contaminants Reported by States

	Reported as a Major Contaminant	
Contaminant	**No. of States**[a]	**% of States**
Sewage	46	89
Inorganic chemicals		
Nitrates	42	75
Brine/Salinity	36	69
Arsenic	19	37
Fluorides	18	35
Sulfur compounds	7	14
Organic chemicals		
Synthetic	37	71
Volatile	36	69
Metals	34	65
Pesticides	31	60
Petroleum	21	40
Radioactive materials	12	23

[a] Based on a total of 52 States and territories which cited groundwater contaminants in their 305(b) submittals.

Source: From U.S. Environmental Protection Agency, *National Water Quality Inventory-1986 Report to Congress*, EPA-440/4-87-008.

Table 10C.84 Activities Contributing to Groundwater Contamination in the United States

Activity	States Citing	Estimated Sites[a]	Contaminants Frequently Cited as Result of Activity	Remarks
Waste disposal				
Septic systems	41	22 million	Bacteria, viruses, nitrate, phosphate, chloride, and organic compounds such as trichloroethylene	Between 820 and 1,460 billion gallons per year discharged to shallowest aquifers (Office of Technology Assessment, 1984)
Landfills (active)	51	16,400	Dissolved solids, iron, manganese, trace metals, acids, organic compounds, and pesticides	Traditional disposal method for municipal and industrial solid waste. Unknown number of abandoned landfills
Surface impoundments	32	191,800	Brines, acidic mine wastes, feedlot wastes, trace metals, and organic compounds	Used to store oil/gas brines (125,100 sites), mine wastes (19,800), agricultural wastes (17,200), industrial liquid wastes (16,200), municipal sewage sludges (2,400), other wastes (11,100) (U.S. Environmental Protection Agency, 1987)
Injection wells	10	280,800	Dissolved solids, bacteria, sodium, chloride, nitrate, phosphate, organic compounds, pesticides, and acids.	Wells used for injecting waste below drinking-water sources (550), oil/gas brine disposal (161,400), solution mining (22,700), injecting waste into or above drinking-water sources (14), and storm-water disposal, agricultural drainage, heat pumps (69,100) (U.S. Environmental Protection Agency, 1987)
Land application of wastes	12	19,000 land application units	Bacteria, nitrate, phosphate, trace metals, and organic compounds	Waste disposal from municipal waste-treatment plants (11,900), industry (5,600), oil/gas production (730), petroleum and wood-preserving wastes (250), others (620) (U.S. Environmental Protection Agency, 1987)
Storage and handling of materials				
Underground storage tanks	39	2.4–4.8 million	Benzene, toluene, xylene, and petroleum products	Useful life of steel tanks, 15–20 yrs. About 25–30 percent of petroleum tanks may leak (Conservation Foundation, 1987)
Above-ground storage tanks	16	Unknown	Organic compounds, acids, metals, and petroleum products	Spills/overflows may contaminate groundwater
Material handling and transfers	29	10,000–16,000 spills per year	Petroleum products, aluminum, iron, sulfate, and trace metals	Includes coal storage piles, bulk chemical storage, containers, and accidental spills
Mining activities				
Mining and spoil disposal—coal mines	23	15,000 active; 67,000 inactive	Acids, iron, manganese, sulfate, uranium, thorium, radium, molybdenum, selenium, and trace metals	Leachates from spoil piles of coal, metal, and nonmetallic mineral mining contain a variety of contaminants. Coal mines are sources of acid drainage

Oil and gas activities				
Wells	20	550,000 production; 1.2 million abandoned	Brines	Contamination from improperly plugged wells and oil brine stored in ponds or injected underground
Agricultural activities				
Fertilizer and pesticide applications	44	363 million acres[b]	Nitrate, phosphate, and pesticides	Fertilizer applied 1982–83, 42.3 million tons per year (U.S. Bureau of the Census, 1984); active ingredients of pesticides applied 1982, 660 million pounds (Gianessi, 1987)
Irrigation practices	22	376,000 wells; 49 million acres irrigated[c]	Dissolved solids, nitrate, phosphate, and pesticides	Salts, fertilizers, pesticides can concentrate in groundwater. Improperly plugged abandoned wells contamination source
Animal feedlots	17	1,900	Nitrate, phosphate, and bacteria	Primarily in the Corn Belt and High Plains States (Office of Technology Assessment, 1984)
Urban activities				
Runoff	15	47.3 million acres urban land[d]	Bacteria, hydrocarbons, dissolved solids, lead, cadmium, and trace metals	Infiltration from detention basins, drainage wells, pits, shafts can reach groundwater. Karst areas particularly vulnerable
Deicing chemical storage and use	14	Not reported	Sodium chloride, ferric ferrocyanide, sodium ferrocyanide, phosphate, and chromate	Winter 1983, 9.35 million tons dry salts/abrasives, 7.78 million gallons of liquid salts applied (Office of Technology Assessment 1984)
Other				
Saline intrusion or upconing	29	Not reported	Dissolved solids and brines	Present in coastal areas and in many inland areas

[a] Estimated number of sites from U.S. Environmental Protection Agency (1987) unless otherwise indicated.
[b] U.S. Bureau of the Census, 1984, p. 658, 1982 data.
[c] U.S. Bureau of the Census, 1984, p. 639, 1982 data.
[d] U.S. Bureau of the Census, 1984, p. 195, 1980 data.

Source: From U.S. Geological Survey, National Water Survey 1986, Water-Supply Paper 2325.

Table 10C.85 Profile of Sources of Groundwater Contamination in the United States

Source	Potential Contaminants	Number/Volume	Geographic Distribution
Subsurface percolation systems	Organics, metals, nitrates, phosphates, microorganisms	22 million domestic systems 25,000 industrial systems	Highest concentration in eastern third of country and portions of west coast
Injection wells	Organics, metals, inorganic acids, microorganisms, radionuclides	280,752 active wells	Varies by well type Class I (hazardous waste)—Gulf Coast and Great Lakes Class II (oil/gas)—throughout the U.S. Class III (mining)—Southwest Class V—agricultural wells in IA, ID, TX, CA; industrial wells in NY and NJ
Land application	Nitrogen, phosphorous, metals organics, microorganisms	2,463 POTWs—sludge application 1,000 POTWs—land treatment 250 hazardous waste land treatment units 18,889 nonhazardous units	Unknown
Landfills	Organics, inorganics, microorganisms, radionuclides	16,416 landfills 9,284 municipal 3,155 industrial	Urban locations nationwide
Open dumps	Organics, inorganics, microorganisms	1,856–2,396 dumps	55 states and territories
Residential disposal	Organics, metals, other inorganics, microorganisms	Unknown	Nationwide
Surface impoundments	Organics, metals and other inorganics, microorganisms, radionuclides	191,822 surface impoundments 16,232 industrial 2,426 municipal 17,159 agricultural 19,813 mining 125,074 oil and gas 11,118 other	70% in hydrogeologically vulnerable areas 37% over current groundwater sources of drinking water Highest number of non-hazardous are in AR, KS, LA, MN, OH, OK, PA, TX, WV
Waster tailings and piles	Arsenic, sulfuric acid, copper, selenium, molybdenum, uranium, thorium, radium, lead, manganese, vanadium	Total mining—2.3 billion tons/yr Metal—250 million tons/yr Uranium—215 million tons/yr. Hazardous waste—0.39 billion tons	Unknown
Material stockpiles	Metals, inorganics, radionuclides	Annual materials production—3.4 billion tons/yr Stockpiles—700 million tons/yr	Nationwide
Graveyards	Metals, nonmetals, microorganisms	Unknown	Nationwide
Animal burial		Unknown	Unknown
Aboveground storage	Organics, inorganics, microorganisms, radionuclides	Unknown	Nationwide
Underground storage	Organics, inorganics, microorganisms, radionuclides	Steel—2.4–4.8 million tanks Fiberglass—0.1 million tanks Total capacity—25 billion gallons Hazardous storage—2,032 tanks	Nationwide

Pipelines	Microorganisms, organics, inorganics	175,000 miles of petroleum product pipelines (1976) carrying 9.63 billion bbls 700,000 miles of sewer pipeline (1976) carrying 5.6 trillion gallons	Nationwide
Materials transport	Organics, inorganics, microorganisms	10,000–16,000 spills per year; spills account for approximately 0.35 percent of 4 billion tons shipped annually (1984)	Nationwide
Mining/mine drainage	Acids, metals, radionuclides	15,000 active coal mines (1986) 67,000 inactive coal mines phospate mines; metalic ore mines	Varies by mining type
Production wells	Organics, inorganics, microorganisms	548,000 oil wells produced approximately 3.1 billion bbls crude oil (1980) 376,000 irrigation wells for 126,000 farms Up to 1.2 million abandoned wells	Oil Wells—nationwide Geothermal wells—primarily CA, NV, ID Water wells—mostly in the Southwest, Central Plains, Idaho, and Florida
Other wells (monitoring and exploration)	Organics, inorganics, microorganisms radionuclides	Unknown	Unknown
Pesticide application	Organics—1,200–1,400 active ingredients Approximately 280 million acre-treatments annually	552 million pounds of active ingredients applied to crops in 1982	17 pesticides confirmed in 23 states (1986) due to normal agricultural application
Fertilizer applications	Nitrates, phosphates	Fertilizer use has declined from 54 million tons to 42.3 million tons (1980–1983); fertilizers in 1981–1982 contained 11 million tons of nitrogen, 4.8 million tons of phosphates, 5.6 million tons of potash	Highest fertilizer use in 1981–1982: CA, IL, IN, IO, TX
Deicing	Salts	9.35 million tons dry salts, and abrasives; 1.78 million gallons liquid salts applied to U.S. highways (1982–1983)	Northeast, Mid-Atlantic, Midwest
Irrigation practices	Fertilizers, pesticides, naturally occurring contaminants (e.g., selenium), sediment	14 percent of cropland is irrigated	Water, Central, and South Plains, Arkansas, Florida
Percolation of atmospheric pollutants	Sulfhur and nitrogen compounds, asbestos, heavy metals	Unknown	Acid rain around Great Lakes, Northeast Distribution of other pollutants varies
Groundwater/surface water interaction	Organics, inorganics, microorganisms, radionuclides	Unknown	Unknown
Natural leaching	Inorganics, radionuclides	Unknown	Unknown, very localized
Salt water intrusion	Inorganics, radionuclides	Unknown	Predominantly coastal areas—CA, TX, LA, FL, NY, Southwest, Central Plains

Source: From U.S. Environmental Protection Agency, Office of Groundwater Protection, 1987, *EPA Activities Related to Groundwater Contamination.*

Table 10C.86 Detections of Individual Volatile Organic Compounds in Groundwater from Rural, Untreated Self-Supplied Domestic Wells in the United States, 1986–1999

			Detected Concentrations (μg/L)			Type of Drinking-Water Standard, Health Criterion, or Taste/Odor Threshold			
Volatile Organic Compound	Predominant Use	Number of Detections/Samples	Min	Median	Max	MCL, HA, 10^{-4} CR (μg/L)	Number of Concentrations Exceeding Standard of Criterion	Taste/Odor Threshold[a] (μg/L)	Number of Concentrations Exceeding Taste/Odor Threshold
VOCs Not Detected									
Bromomethane	Fumigant	0/1,677	—	—	—	10 (HA)	0	—	0
cis−1,3-Dichloropropene	Fumigant	0/1,685	—	—	—	40 (10^{-4} CR)	0	—	0
trans-1,3-Dichloropropene	Fumigant	0/1,592	—	—	—	40 (10^{-4} CR)	0	—	0
n-Butylbenzene	Gasoline hydrocarbon	0/1,454	—	—	—	—	0	—	0
1,2-Dimethylbenzene	Gasoline hydrocarbon	0/805	—	—	—	10,000 (MCL)	0	—	0
1,3/1,4-Dimethylbenzene	Gasoline hydrocarbon	0/799	—	—	—	10,000 (MCL)	0	17[b]	0
Bromoethene	Organic synthesis	0/688	—	—	—	—	0	—	0
Hexachlorobutadiene	Organic synthesis	0/1,347	—	—	—	1 (HA)	0	6	0
2-Propenal	Organic synthesis	0/472	—	—	—	—	0	110	0
2-Propenenitrile	Organic synthesis	0/693[c]	—	—	—	6 (10^{-4} CR)	0	9,100	0
Ethenylbenzene	Organic synthesis	0/1,915	—	—	—	100 (MCL and HA)	0	11	0
1,2,3-Trichlorobenzene	Organic synthesis	0/1,455	—	—	—	—	0	10	0
Ethyl *tert*-butyl ether	Oxygenate	0/688	—	—	—	—	0	—	0
1,2-Dichlorobenzene	Solvent	0/1,911	—	—	—	600 (MCL)	0	24	0
1,3-Dichlorobenzene	Solvent	0/1,340	—	—	—	600 (HA)	0	77	0
Hexachloroethane	Solvent	0/698	—	—	—	1 (HA)	0	10	0
1,2,4-Trichlorobenzene	Solvent	0/1,464	—	—	—	70 (MCL)	0	5	0
1,1,2-Trichloroethane	Solvent	0/1,686	—	—	—	5 (MCL)	0	50,000	0
Detection Frequency Greater Than 0 and Less Than or Equal to 1 Percent									
1,2-Dibromoethane	Fumigant	[c]3/1,614	0.3	0.6	1.1	0.05 (MCL)	3	—	—
1,4-Dichlorobenzene	Fumigant	1/1,925	1.2	1.2	1.2	75 (MCL)	0	4.5	0
Benzene	Gasoline hydrocarbon	6/1,892	0.2	0.5	4.4	5 (MCL)	0	170	0
Ethylbenzene	Gasoline hydrocarbon	2/1,926	0.2	2.8	5.4	700 (MCL)	0	29	0
Naphthalene	Gasoline hydrocarbon	2/1,464	1.8	2.1	2.5	20 (HA)	0	2.5	0
1,2,4-Trimethylbenzene	Gasoline hydrocarbon	6/1,415	0.2	0.3	12	—	—	500	0
Chloroethene	Organic synthesis	1/1,917	0.4	0.4	0.4	2 (MCL)	0	3,400	0
1,1-Dichloroethene	Organic synthesis	2/1,926	0.5	19.8	39	7 (MCL)	1	1,500	0
(1-Methylethyl) benzene	Organic synthesis	2/1,371	0.4	0.6	0.8	233 (10^{-4} CR)	0	0.8	0
tert-Amyl methyl ether	Oxygenate	1/688	0.8	0.8	0.8	—	—	—	—
Diisopropyl ether	Oxygenate	1/581	22	22	22	—	—	—	—
Chloromethane	Refrigerant	2/1,565	1.1	1.8	2.5	3 (HA)	0	—	—
Trichlorofluoromethane	Refrigerant	7/1,925	0.2	0.3	1.9	2,000 (HA)	0	28	0
1,1,2-Trichloro-1,2,2-trifluoroethane	Refrigerant	1/1,515	0.3	0.3	0.3	2,100,000 (10^{-4} CR)	0	—	—
Chlorobenzene	Solvent	1/1,926	3.5	3.5	3.5	100 (MCL)	0	50	0
Chloroethane	Solvent	2/1,677	0.3	0.5	0.7	—	—	19	0
1,1-Dichloroethane	Solvent	5/1,926	0.2	0.7	1.7	—	—	—	—
1,2-Dichloroethane	Solvent	4/1,910	0.2	1.5	2.9	5 (MCL)	0	7,000	0
cis−1,2-Dichloroethene	Solvent	2/1,705	3	3.7	4.4	70 (MCL)	0	—	—
trans-1,2-Dichloroethene	Solvent	1/1,767	0.4	0.4	0.4	100 (MCL)	0	4.3	0
Dichloromethane	Solvent	12/1,923	0.2	0.6	4	5 (MCL)	0	910	0
1,2-Dichloropropane	Solvent	15/1,926	0.2	0.5	19.4	5 (MCL)	2	10	1
n-Propylbenzene	Solvent	1/1,454	0.6	0.6	0.6	—	—	—	—

Tetrachloromethane	Solvent	5/1,925	0.2	0.4	0.6	5 (MCL)	0	520	0
Trichloroethene	Solvent	16/1,926	0.2	0.7	25	5 (MCL)	3	310	0
1,2,3-Trichloropropane	Solvent	10/1,615	0.2	0.4	2.1	40 (HA)	0	—	—
Bromodichloromethane[d]	Trihalomethane	7/1,926	0.2	1.5	7	80 (MCL)	0	—	—
Chlorodibromomethane[d]	Trihalomethane	7/1,926	0.3	0.6	11	80 (MCL)	0	—	—
Tribromomethane[d]	Trihalomethane	4/1,925	0.3	0.7	8.2	80 (MCL)	0	300	0
Detection Frequency Greater Than 1 and Less Than or Equal to 5 Percent									
1,2-Dibromo-3-chloropropane	Fumigant	16/1,459[e]	0.2	1.3	3.2	0.2 (MCL)	16	—	—
Methylbenzene	Gasoline hydrocarbon	21/1,882	0.2	0.3	12	1,000 (MCL)	0	42	0
Methyl *tert*-butyl ether	Oxygenate	30/1,335	0.2	0.7	30.2	—	—	20[f]	1
Dichlorodifluoromethane	Refrigerant	23/1,916	0.2	0.3	2	1,000 (HA)	0	—	—
Tetrachloroethene	Solvent	32/1,897	0.2	0.3	29	5 (MCL)	3	190	0
1,1,1-Trichloroethane	Solvent	22/1,926	0.2	0.4	120	200 (MCL)	0	970	0
Trichloromethane[d]	Trihalomethane	83/1,926	0.2	0.5	74	80 (MCL)	0	2,400	0

Note: VOC, volatile organic compound; Min, minimum; Max, maximum; MCL, Maximum Contaminant Level; HA, Health-Advisory Level, lifetime 70-km adult consuming 2 L of water per day; 10^{-4} CR, risk of one additional 70-km adult in ten thousand (1×10^{-4}) contracting cancer over a lifetime of exposure consuming 2 L of water per day; μg/L, micrograms per liter; —, not applicable.

[a] Taste/odor thresholds from bender and others, 1999.
[b] 1,3-Dimethylbenzene.
[c] No additional detections exist below 0.2 μg/L.
[d] Bromodichloromethane, chlorodibromomethane, tribromomethane, and trichloromethane can also be classified as solvents or VOCs used in organic synthesis.
[e] When uncensored, the number of detections equals 19.
[f] Lower limit of U.S. Environmental Protection Agency Drinking Water Advisory.

Source: From Moran, M.J. et al., Occurrence and status of volatile organic comounds in groundwater from rural, untreated, self-supplied domestic wells in the United States, 1986–1999, United States Geological Survey, *Water Resources Investigation Report 02-4085*, www.usgs.gov.

Table 10C.87 Frequency of Detection and Concentrations of Volatile Organic Compounds in Groundwater Source Samples Collected from United States Drinking Water Supplies between May 3, 1999 and October 23, 2000

	Groundwater Statistics				
Compound	**Number of Samples**	**Number of Detects**	**Detection Frequency**	**Min Concentration**	**Max Concentration**
Gasoline Oxygenates					
tert-Amyl methyl ether	579	2	0.34	0.21	0.31
Diisopropyl ether	579	2	0.34	0.23	1.7
Ethyl *tert*-butyl ether	579	1	0.17	0.25	0.25
Methyl *tert*-butyl ether	571	31	5.4	0.20	6.3
Other Gasoline Compounds					
Benzene	578	2	0.34	0.75	3
n-Butylbenzene	579	1	0.17	0.21	0.21
sec-Butylbenzene	579	0	0	<0.2	<0.2
tert-Butylbenzene	579	0	0	<0.2	<0.2
Ethylbenzene	566	3	0.53	0.23	0.63
Naphthalene	579	1	0.17	0.22	0.22
Toluene	562	3	0.53	1.1	4.2
1,3,5-Trimethylbenzene	579	0	0	<0.2	<0.2
o-Xylene	566	4	0.71	0.26	0.91
m- *p*-Xylene	564	4	0.71	0.33	1.6
Trihalomethanes					
Bromodichloromethane	578	34	5.9	0.2	7.4
Bromoform	579	27	4.7	0.21	49
Chlorodibromomethane	578	35	6.1	0.21	9.4
Chloroform	575	68	12	0.21	22
Organic Syntheses					
Acrylonitrile	579	0	0	<0.2	<0.2
Bromobenzene	579	0	0	<0.2	<0.2
Bromochloromethane	579	0	0	<0.2	<0.2
Carbon tetrachloride	579	5	0.86	0.38	1.8
Chlorobenzene	579	1	0.17	1.6	1.6
Chloroethane	579	1	0.17	2.6	2.6
2-Chlorotoluene	579	0	0	<0.2	<0.2
4-Chlorotoluene	579	0	0	<0.2	<0.2
Dibromomethane	562	2	0.36	0.33	0.75
1,2-Dichlorobenzene	579	1	0.17	0.37	0.37
1,3-Dichlorobenzene	579	0	0	<0.2	<0.2
1,1 –Dichloroethane	579	11	1.9	0.21	10
1,2-Dichloroethane	579	3	0.52	0.27	0.65
1,1 –Dichloroethene	579	8	1.4	0.22	23
cis-1,2-Dichloroethene	579	11	1.9	0.33	14
trans-1,2-Dichloroethene	579	0	0	<0.2	<0.2
1,2-Dichloropropane	561	1	0.18	0.47	0.47
1,3-Dichloropropane	569	0	0	<0.2	<0.2
2,2-Dichloropropane	579	0	0	<0.2	<0.2
1,1-Dichloropropene	579	0	0	<0.2	<0.2
Hexachlorobutadiene	579	0	0	<0.2	<0.2
Hexachloroethane	579	0	0	<0.2	<0.2
Isopropylbenzene	573	1	0.17	0.38	0.38
p-Isopropyltoluene	579	0	0	<0.2	<0.2
Methylene chloride	577	1	0.17	1.6	1.6
Methyl ethyl ketone	577	2	0.35	3.4	5.8
n-Propylbenzene	579	0	0	<0.2	<0.2
Styrene	562	2	0.36	0.21	0.83
1,1,1,2-Tetrachloroethane	579	0	0	<0.2	<0.2
1,1,2,2-Tetrachloroethane	579	0	0	<0.2	<0.2
Tetrachloroethene	577	24	4.2	0.2	36
1,1,1 -Trichloroethane	578	10	1.7	0.21	13
1,1,2-Trichloroethane	579	0	0	<0.2	<0.2
Trichloroethene	576	19	3.3	0.23	170
1,2,3-Trichloropropane	579	1	0.17	0.31	0.31

(Continued)

Table 10C.87 (Continued)

	Groundwater Statistics				
Compound	**Number of Samples**	**Number of Detects**	**Detection Frequency**	**Min Concentration**	**Max Concentration**
1,2,4-Trimethylbenzene	571	1	0.18	0.46	0.46
Vinyl bromide	579	0	0	<0.2	<0.2
Vinyl chloride	579	1	0.17	3.2	3.2
1,2,3-Trichlorobenzene	579	0	0	<0.2	<0.2
1,2,4-Trichlorobenzene	579	0	0	<0.2	<0.2
Fumigants					
Bromomethane	579	1	0.17	6.4	6.4
1,4-Dichlorobenzene	579	1	0.17	0.25	0.25
cis-1,3-Dichloropropene	579	0	0	<0.2	<0.2
trans-1,3 –Dichloropropene	579	0	0	<0.2	<0.2
Refrigerants					
Chloromethane	579	3	0.52	0.2	0.69
Dichlorodifluoromethane	579	4	0.69	1.1	18
Trichlorofluoromethane	579	5	0.86	0.24	1.3
l,l,2-Trichloro-l,2,2-Trifluoroethane	579	1	0.17	0.91	0.91

Note: Concentrations are in micrograms per liter.

Source: From Grady, S., 2003, *A National Survey of Methyl tert-Butyl Ether and Other Volatile Organic Compounds in Drinking Water Sources: Results of a Random Survey*, United States Geological Survey Water-Resource Investigations Report 02-4079, www.usgs.gov.

Table 10C.88 Summary Comparison of Occurrence of Phase II/V Contaminants In United States Water Systems Using Surface Water vs. Groundwater, from National Cross-Section States

	Percent > MRL		Percent > ½ MCL		Percent > MCL[a]			
Contaminant	Surface Water (%)	Groundwater (%)	Surface Water (%)	Groundwater (%)	Surface Water (%)	Groundwater (%)	National MCL Viol.-SW (%)	National MCL Viol.-GW (%)
Inorganic Chemicals (IOCs)								
IOCs								
Antimony (total)	4.2	3.2	0.8	1.2	0.2	0.4	0.08	0.06
Arsenic	13.0	19.3	0.6	1.6	0.5	0.9	0.00	0.06
Asbestos (MCL in fibers per liter)	8.9	8.5	0.7	0.9	0.7	0.4	0.00	0.01
Barium	49.1	47.3	0.6	0.8	0.5	0.2	0.00	0.06
Beryllium (total)	2.5	2.1	0.4	0.5	0.0	0.2	0.02	0.02
Cadmium	5.1	4.9	1.3	1.2	0.2	0.6	0.05	0.06
Chromium	10.5	13.2	0.3	0.5	0.2	0.2	0.00	0.01
Cyanide	5.1	2.0	0.4	0.5	0.0	0.2	0.02	0.01
Fluoride	77.8	72.5	0.8	3.4	0.5	1.3	0.02	0.18
Mercury	9.0	4.5	1.3	0.7	0.5	0.4	0.03	0.04
Nickel	11.8	10.6	1.2	0.9	0.4	0.4	0.02	0.01
Selenium	11.2	8.6	0.2	0.3	0.0	0.2	0.03	0.07
Thallium (total)	2.5	3.6	0.8	1.3	0.0	0.4	0.12	0.06
Synthetic Organic Chemicals (SOCs)								
2,3,7,8-TCDD (Dioxin)	0.0	1.3	0.0	1.3	0.0	1.3	0.00	0.00
2,4,5-TP (Silvex)	2.0	0.7	0.0	0.0	0.0	0.0	0.00	0.00
2,4-D	11.2	1.2	0.2	0.0	0.0	0.0	0.00	<0.01
Alachlor (Lasso)	7.3	0.3	1.5	0.1	0.2	0.0	0.00	0.00
Atrazine	21.1	2.0	13.2	0.3	10.7	0.1	0.83	0.01
Benzo[a]pyrene	0.5	0.5	0.0	0.1	0.0	0.1	0.00	0.00
bis(2-ethylhexyl) adipate	4.9	6.8	0.5	0.4	0.5	0.3	0.02	0.00
bis(2-ethylhexyl) phthalate	28.9	14.9	3.2	2.7	2.8	1.7	0.03	0.01
Carbofuran (Furadan)	0.8	0.1	0.0	0.0	0.0	0.0	0.00	0.00
Chlordane	0.0	0.1	0.0	0.0	0.0	0.0	0.00	0.00
Dalapon	9.4	0.8	0.2	0.0	0.2	0.0	0.00	0.00
Dibromo-chloropropane (DBCP)	4.8	2.6	1.1	2.3	1.1	2.0	0.00	<0.01
Dinoseb	2.5	0.4	0.0	0.0	0.0	0.0	0.00	0.00
Diquat	3.5	0.8	0.0	0.0	0.0	0.0	0.00	0.00
Endothall	0.3	0.2	0.0	0.0	0.0	0.0	0.00	0.00
Endrin	2.1	0.2	0.2	0.1	0.2	0.0	0.02	0.00
Ethylene Dibromide (EDB)	4.2	1.0	3.8	1.0	3.7	0.7	0.07	0.04
Glyphosate (Roundup)	0.0	0.1	0.0	0.0	0.0	0.0	0.00	0.00
Heptachlor	0.4	0.2	0.0	0.1	0.0	0.0	0.00	0.00
Heptachlor Epoxide	0.4	0.2	0.0	0.0	0.0	0.0	0.00	0.00
Hexachloro-benzene	0.4	0.0	0.0	0.0	0.0	0.0	0.00	0.00
Hexachloro-cyclopentadiene	9.6	0.1	1.0	0.0	0.6	0.0	0.00	0.00
Lindane	1.2	0.3	0.3	0.1	0.3	0.1	0.00	0.00
Methoxychlor	1.0	0.2	0.0	0.0	0.0	0.0	0.00	0.00
Oxamyl (Vydate)	0.0	0.1	0.0	0.0	0.0	0.0	0.00	0.00
Penta-chlorophenol	3.1	0.7	0.4	0.1	0.2	0.0	0.00	<0.01
Picloram (Tordon)	3.7	0.5	0.0	0.0	0.0	0.0	0.00	<0.01
Simazine	15.9	1.4	2.5	0.0	1.0	0.0	0.00	0.00
Total PCBs	0.2	0.2	0.2	0.1	0.2	0.1	0.00	<0.01
Aroclor 1016	0.0	0.1	0.0	0.0	0.0	0.0	0.00	0.00
Aroclor 1221	0.0	0.1	0.0	0.0	0.0	0.0	0.00	0.00
Aroclor 1232	0.0	0.1	0.0	0.0	0.0	0.0	0.00	0.00
Aroclor 1242	0.0	0.1	0.0	0.0	0.0	0.0	0.00	0.00
Aroclor 1248	0.0	0.1	0.0	0.0	0.0	0.0	0.00	0.00
Aroclor 1254	0.0	0.2	0.0	0.0	0.0	0.0	0.00	0.00
Aroclor 1260	0.0	0.0	0.0	0.0	0.0	0.0	0.00	0.00
Toxaphene	0.3	0.1	0.0	0.0	0.0	0.0	0.00	<0.01
Volatile Organic Chemicals (VOCs)								
1,1,1-Trichloroethane	7.3	3.3	0.9	1.3	0.9	1.3	0.00	0.01
1,1,2-Trichloroethane	5.7	0.7	0.6	0.4	0.3	0.3	0.00	0.01
1,1-Dichloroethene	2.9	1.5	0.3	1.0	0.3	0.9	0.02	0.04

(Continued)

Table 10C.88 (Continued)

Contaminant	Percent > MRL		Percent > ½ MCL		Percent > MCL[a]			
	Surface Water (%)	Groundwater (%)	Surface Water (%)	Groundwater (%)	Surface Water (%)	Groundwater (%)	National MCL Viol.-SW (%)	National MCL Viol.-GW (%)
1,2,4-Trichlorobenzene	3.1	1.0	0.0	0.4	0.0	0.4	0.00	0.00
1,2-Dichloroethane	3.1	1.4	0.3	0.6	0.3	0.4	0.00	0.02
1,2-Dichloropropane	3.2	1.0	0.6	0.4	0.5	0.3	0.00	0.01
Benzene	3.9	1.2	0.5	0.5	0.3	0.4	0.02	0.04
Carbon tetrachloride	9.0	1.7	1.6	0.6	1.1	0.4	0.00	0.02
Chlorobenzene	8.1	1.0	0.2	0.3	0.2	0.3	0.00	<0.01
cis-1,2-Dichloroethylene	3.3	1.9	0.3	0.7	0.3	0.6	0.00	0.01
Ethyl benzene	7.3	2.2	0.3	0.3	0.3	0.1	0.00	<0.01
Methylene chloride (Dichloromethane)	25.6	11.1	10.4	3.3	4.7	2.3	0.03	0.05
o-Dichlorobenzene	3.7	1.2	0.2	0.6	0.0	0.5	0.00	0.00
p-Dichlorobenzene	6.2	2.0	0.0	0.9	0.0	0.8	0.00	0.00
Styrene	4.1	2.1	0.0	0.2	0.0	0.2	0.00	0.00
Tetrachloroethylene (PCE)	7.1	4.3	2.5	2.3	1.7	1.8	0.13	0.14
Toluene	11.9	3.8	1.0	0.7	0.5	0.4	0.00	0.00
trans−1,2-Dichloroethylene	2.7	0.7	0.0	0.2	0.0	0.2	0.00	<0.01
Trichloroethene (Trichloroethylene, TCE)	5.6	3.1	1.9	1.8	1.2	1.5	0.05	0.12
Vinyl chloride	3.1	0.5	0.3	0.2	0.3	0.2	0.00	0.02
Xylenes (Total)	12.3	3.9	0.2	0.2	0.0	0.1	0.00	0.00
Group Summaries								
IOCs								
IOCs-All Regulated	83.7	83.5	6.8	9.2	2.5	4.2	0.3	0.6
SOCs							1.0	0.1
SOCs-Group 1	21.9	2.4	13.2	0.3	10.7	0.1		
SOCs-Group 2	20.4	13.4	1.2	1.9	0.9	1.0		
VOCs								
VOCs-All Regulated	41.1	19.9	15.4	7.9	8.2	6.1	0.2	0.4
VOCs-Group 1	19.5	6.6	1.7	1.3	0.9	0.9		
VOCs-Group 2	11.1	6.4	4.5	3.9	2.9	3.2		

Note: IOC, Regulated: includes all the regulated IOCs; SOCs, Group 1: includes alachlor, atrazine, and simazine; SOCs, Group 2: includes bis(2-ethylhexyl)phthalate, bis(2-ethylhexyl)adipate, and benzo(a)pyrene; VOCs, Regulated: includes all the regulated VOCs; VOCs, Group 1: includes benzene, ethyl benzene, toluene, and total xylenes (LNAPLEs); VOCs, Group 2: includes cis-1, 2-dichloroethylene, trans-1,2-dichloroethylene, 1,1-dichloroethene, tetrachloroethylene, trichloroethene, and vinyl chloride (DNAPLEs). Percent MCL Violations Derived from SDWIS Information for 1/1/93-3/31/1998.

[a] % >MCL indicates the proportion of systems with any analytical results exceeding the concentration value of the MCL; it does not necessarily indicate an MCL violation. An MCL violation occurs when the MCL is exceeded by the average results from four quarterly samples or confirmation samples as required by the primacy State.

Source: From United States Environmental Protection Agency, 1999, *A Review of Contaminant Occurrence in Public Water Systems*, EPA 816-R-99-006, www.epa.gov.

Table 10C.89 United States Public Water System Unregulated Contaminant Monitoring Information System (UCRIS) (Round 1) 24-State Cross-Section Summary of Occurrence (1988–1992)

Chemical Name (Threshold in μg/L)	Total # PWS	# GW PWS	# SW PWS	% PWS > MRL	% GW PWS > MRL	% SW PWS > MRL	% PWS > Threshold	% GW PWS > Threshold	% SW PWS > Threshold	99% Value (μg/L)
Synthetic Organic Chemicals (SOCs)										
Dibromochloropropane (MCL = 0.2)	12,827	11,446	1,511	2.49	2.51	2.32	1.32	1.35	0.99	1.03
Ethylene dibromide[a] (MCL = 0.05)	11,450	10,274	1,284	1.14	1.01	2.10	0.16	0.12	0.47	0.01
Volatile Organic Chemicals (VOCs)										
Benzene (MCL = 5)	14,910	13,919	1,119	1.14	1.11	5.18	0.25	0.25	0.27	<2.0
Bromobenzene (N/A)	16,450	14,862	1,726	0.19	0.14	0.64	N/A	N/A	N/A	<2.0
Bromochloromethane (MCL = 10)	12,881	11,576	1,386	0.50	0.44	1.08	0.03	0.03	0.07	<1.0
Bromodichloromethane (HRL = 60)	20,024	17,917	2,324	22.09	14.84	79.69	0.13	0.04	0.86	22.00
Bromoform (HRL = 400)	19,582	17,773	1,979	9.01	7.56	22.13	0.01	0.01	0.00	7.32
Bromomethane (MCL = 10)	20,198	18,472	1,886	0.77	0.71	1.22	0.09	0.08	0.16	<4.0
Carbon tetrachloride (MCL = 5)	15,266	14,176	1,214	1.32	1.09	3.95	0.16	0.15	0.25	1.60
Chlorobenzene (MCL = 100)	20,038	18,337	1,859	0.53	0.26	3.17	0.00	0.00	0.00	<1.0
Chloroethane (N/A)	20,236	18,507	1,882	0.39	0.29	1.33	N/A	N/A	N/A	<2.0
Chloroform (HRL = 600)	20,039	17,874	2,385	28.84	21.69	84.40	0.02	0.01	0.17	87.00
Chloromethane (MCL = 3)	20,246	18,513	1,894	1.22	1.11	2.27	0.45	0.41	0.84	<4.0
cis-1,2-Dichloroethene (MCL = 70)	16,705	15,026	1,832	1.47	1.45	1.53	0.03	0.03	0.00	2.18
cis-1,2-Dichloropropene (N/A)	9,211	8,438	836	0.61	0.52	1.44	N/A	N/A	N/A	<1.0
Dibromochloromethane (HRL = 60)	19,750	17,785	2,158	18.01	12.41	64.55	0.06	0.02	0.32	12.70
Dibromomethane (N/A)	16,549	14,953	1,720	0.36	0.21	1.69	N/A	N/A	N/A	<2.0
Dichlorodifluoromethane (MCL = 1,000)	16,076	14,617	1,588	1.37	1.38	1.39	0.00	0.00	0.00	0.50
1,1-Dichloroethane (MCL = 5)	20,483	18,758	1,876	1.14	1.09	1.55	0.18	0.16	0.37	0.10
1,2-Dichloroethane (MCL = 5)	15,282	14,192	1,215	1.16	1.10	1.73	0.19	0.17	0.41	<5.0
Dichloroethene (MCL = 7)	15,430	14,180	1,380	1.17	1.06	1.45	0.20	0.20	0.22	1.80
Dichloromethane (MCL = 5)	19,287	17,602	1,836	4.05	3.31	11.06	0.77	0.52	3.27	1.30
1,2-Dichloropropane (MCL = 5)	19,591	17,908	1,820	0.67	0.66	0.77	0.08	0.09	0.00	<4.0
1,3-Dichloropropane (N/A)	16,947	15,338	1,748	0.12	0.12	0.11	N/A	N/A	N/A	<1.0
2,2-Dichloropropane (N/A)	16,757	15,138	1,754	0.15	0.14	0.23	N/A	N/A	N/A	<2.0
1,1-Dichloropropene (N/A)	16,947	15,332	1,749	0.13	0.10	0.40	N/A	N/A	N/A	<1.0
1,3-Dichloropropene (HRL = 40)	9,164	8,303	898	0.16	0.12	0.56	0.00	0.00	0.00	<1.0
Ethyl benzene (MCL = 700)	20,081	18,355	1,884	1.62	1.40	3.66	0.00	0.00	0.00	<5.0
Hexachlorobutadiene (HRL = 0.9)	12,284	10,980	1,385	0.35	0.30	0.72	0.11	0.06	0.51	<5.0
Isopropylbenzene (N/A)	12,771	11,480	1,359	0.27	0.28	0.22	N/A	N/A	N/A	<2.0
m-Dichlorobenzene (HAL = 600)	20,429	18,752	1,819	0.25	0.20	0.77	0.00	0.00	0.00	<5.0
m-Xylene (N/A)	11,329	10,145	1,276	1.55	1.47	2.12	N/A	N/A	N/A	<4.0
n-Butylbenzene (N/A)	12,763	11,471	1,371	0.35	0.29	0.88	N/A	N/A	N/A	<2.0
n-Propylbenzene (N/A)	12,724	11,440	1,363	0.33	0.34	0.22	N/A	N/A	N/A	<2.0
Naphthalene (HRL = 140)	13,452	12,034	1,502	1.18	1.08	1.93	0.01	0.02	0.00	<5.0
o-Chlorotoluene (MCL = 100)	15,721	14,154	1,702	0.20	0.16	0.53	0.00	0.00	0.00	<1.0
o-Dichlorobenzene (MCL = 600)	19,953	18,300	1,795	0.28	0.20	1.00	0.00	0.00	0.00	<5.0
o-Xylene (N/A)	13,987	12,638	1,450	1.76	1.69	2.41	N/A	N/A	N/A	<5.0
p-Chlorotoluene (MCL = 100)	15,612	14,057	1,689	0.17	0.15	0.36	0.00	0.00	0.00	<1.0
p-Dichlorobenzene (MCL = 750)	15,494	14,284	1,334	1.25	1.11	2.70	0.00	0.00	0.00	<4.4
p-Isopropyltoluene (N/A)	12,167	10,953	1,282	0.25	0.26	0.08	N/A	N/A	N/A	<2.0
p-Xylene (N/A)	10,127	8,956	1,230	1.58	1.49	2.36	N/A	N/A	N/A	<5.0
sec-Butylbenzene (N/A)	12,343	11,071	1,337	0.23	0.23	0.22	N/A	N/A	N/A	<2.0
Styrene (MCL = 100)	16,623	14,938	1,832	0.57	0.45	1.53	0.00	0.00	0.00	<2.0

tert-Butylbenzene (N/A)	12,353	11,081	1,337	0.19	0.19	0.22	N/A	N/A	N/A	<2.0
1,1,1,2-Tetrachloroethane (HAL=70)	16,956	15,338	1,753	0.18	0.13	0.63	0.00	0.00	0.00	<1.0
1,1,2,2-Tetrachloroethane (HAL=2)	20,407	18,693	1,867	0.45	0.39	1.02	0.05	0.05	0.11	<1.0
Tetrachloroethylene (MCL=5)	19,814	18,298	1,652	3.33	3.38	2.66	0.91	0.93	0.67	13.2
Toluene (MCL=100)	20,089	18,364	1,887	3.50	3.10	7.31	0.00	0.00	0.00	0.7
trans-1,2-Dichloroethene (MCL=100)	19,945	18,267	1,825	0.64	0.59	1.10	0.01	0.01	0.00	<1.0
trans-1,3-Dichloropropene (N/A)	9,883	9,017	959	0.25	0.13	1.36	N/A	N/A	N/A	<1.0
1,2,3-Trichlorobenzene (N/A)	12,876	11,567	1,389	0.49	0.46	0.72	N/A	N/A	N/A	<5.0
1,2,4-Trichlorobenzene (MCL=70)	13,449	11,996	1,589	0.49	0.45	0.78	0.00	0.00	0.00	<5.0
1,1,1-Trichloroethane (MCL=200)	15,279	14,191	1,213	3.66	3.57	4.62	0.03	0.03	0.00	3.7
1,1,2-Trichloroethane (MCL=5)	19,964	18,253	1,853	0.43	0.29	1.78	0.04	0.02	0.16	<1.0
Trichloroethylene (MCL=5)	15,290	14,198	1,220	3.54	3.37	5.66	0.98	1.00	0.66	20.8
Trichlorofluoromethane (HAL=175)	16,851	15,347	1,637	1.48	1.39	2.32	0.01	0.01	0.00	0.6
1,2,3-Trichloropropane (MCL=40)	17,392	15,771	1,758	0.25	0.25	0.23	0.01	0.01	0.00	<2.0
1,2,4-Trimethylbenzene (N/A)	12,755	11,462	1,372	0.83	0.76	1.38	N/A	N/A	N/A	<2.0
1,3,5-Trimethylbenzene (N/A)	12,671	11,379	1,370	0.61	0.59	0.66	N/A	N/A	N/A	<2.0
Vinyl chloride (MCL=2)	15,184	14,099	1,209	0.50	0.44	1.24	0.28	0.23	0.83	<2.0
Xylenes (Total) (MCL=10,000)	9,463	8,841	670	3.04	2.51	10.75	0.00	0.00	0.00	0.6

Note: MCL, maximum contaminant level; HAL, health advisory level (as of December 2000); HRL, health reference level (concentration values used only as reference levels for analyses in this report); MRL, minimum reporting level.

The MCL, HAL, HRL, and MRL values are used in this report only as reference levels to facilitate occurrence assessments.

"% PWS > Threshold" indicates the proportion of systems with any analytical results exceeding the concentration value of the HRL/MCL/HAL. (Note that results for % PWSs greater than an MCL value does not indicate a MCL violation. A formal MCL violation occurs when the MCL is exceeded by the average of four consecutive quarterly samples or confirmation samples as required by the primacy States.)

N/A, there is no HRL/MCL/HAL available.

[a] The high occurrence of ethylene dibromide are, in part, considered false positives related to analytical methods problems.

Source: From United States Environmental Protection Agency, 2001, *Occurrence of Unregulated Contaminants in Public Water Systems - A National Summary*, EPA 815-P-00-002, www.epa.gov.

Table 10C.90 United States Public Water System Safe Drinking Water Information System/Federal Version (SDWIS/FED) (Round 2) Data—20-State Cross-Section Summary of Occurrence (1993 to 1997)

Chemical Name (Threshold in μg/L)	Total PWS	# GW PWS	# SW PWS	% PWS> MRL	% GW PWS	% SW PWS	% PWS> Threshold	% GW PWS>	% SW PWS>	99% Value
Inorganic chemicals (IOCs)										
Sulfate (HRL = 500,000)	16,495	15,009	1,486	88.11	87.76	91.66	1.79	1.83	1.41	560,000
Synthetic Organic Chemicals (SOCs)										
Aldicarb[a] (HRL = 7)	11,972	10,509	1,463	0.01	0.00	0.07	0.00	0.00	0.00	<3.0
Aldicarb sulfone[a] (HRL = 7)	11,968	10,512	1,456	0.08	0.04	0.41	0.00	0.00	0.00	<2.0
Aldicarb sulfoxide[a] (HRL = 7)	11,954	10,500	1,454	0.08	0.03	0.48	0.01	0.01	0.00	<4.0
Aldrin[a] (HRL = 0.002)	11,745	10,420	1,325	0.01	0.01	0.00	0.01	0.01	0.00	<2.0
Butachlor[a] (N/A)	11,940	10,482	1,458	0.04	0.01	0.27	N/A	N/A	N/A	<10.0
Carbaryl[a] (MCL = 700)	12,623	11,086	1,537	0.03	0.02	0.13	0.00	0.00	0.00	<10.0
Dicamba[a] (MCL = 200)	14,034	12,220	1,814	0.34	0.21	1.21	0.00	0.00	0.00	<10.0
Dieldrin[a] (HRL = 0.002)	11,788	10,329	1,459	0.09	0.09	0.14	0.09	0.09	0.14	<1.0
3-Hydroxycarbofuran[a] (N/A)	12,644	11,088	1,556	0.07	0.02	0.45	N/A	N/A	N/A	<10.0
Methomyl[a] (MCL = 200)	12,604	11,068	1,536	0.07	0.05	0.20	0.00	0.00	0.00	<50.0
Metolachlor[a] (HRL = 70)	12,953	11,503	1,450	0.83	0.11	6.55	0.00	0.00	0.00	<5.0
Metribuzin[a] (HRL = 91)	13,512	11,833	1,679	0.01	0.01	0.00	0.00	0.00	0.00	<2.0
Propachlor[a] (MCL = 90)	12,050	10,600	1,450	0.05	0.02	0.28	0.00	0.00	0.00	<5.0
Volatile organic chemicals (VOCs)										
Bromobenzene (N/A)	24,125	21,461	2,664	0.13	0.12	0.23	N/A	N/A	N/A	<1.0
Bromochloromethane[b] (MCL = 10)	22,974	20,507	2,467	0.46	0.32	1.62	0.03	0.02	0.08	<1.0
Bromodichloromethane (HRL = 60)	23,858	21,152	2,706	21.97	16.14	67.52	0.08	0.05	0.30	18.8
Bromoform (HRL = 400)	18,461	16,348	2,113	12.12	11.08	20.11	0.01	0.00	0.05	6.5
Bromomethane (MCL = 10)	23,328	20,872	2,456	0.75	0.74	0.86	0.06	0.05	0.08	<9.0
Chloroethane (N/A)	24,433	21,925	2,508	0.34	0.32	0.56	N/A	N/A	N/A	<2.5
Chloroform (HRL = 600)	23,737	21,021	2,716	27.42	21.84	70.54	0.04	0.01	0.26	110.0
Chloromethane (MCL = 3)	23,478	21,030	2,448	2.25	2.04	4.08	0.58	0.55	0.78	<2.5
Dibromochloromethane (HRL = 60)	23,750	21,059	2,691	18.37	14.55	48.23	0.08	0.05	0.30	9.7
Dibromomethane (N/A)	23,006	20,454	2,552	0.46	0.32	1.53	N/A	N/A	N/A	<1.0
Dichlorodifluoromethane[b] (MCL = 1,000)	22,141	19,836	2,305	1.27	1.23	1.65	0.00	0.00	0.00	<20.0
1,1-Dichloroethane (MCL = 5)	24,808	22,114	2,694	0.74	0.67	1.34	0.08	0.07	0.11	<1.0
1,3-Dichloropropane (N/A)	24,065	21,430	2,635	0.06	0.05	0.11	N/A	N/A	N/A	<2.0
2,2-Dichloropropane (N/A)	24,096	21,445	2,651	0.09	0.07	0.26	N/A	N/A	N/A	<1.0
1,1-Dichloropropene (N/A)	24,069	21,438	2,631	0.07	0.06	0.15	N/A	N/A	N/A	<1.0
1,3-Dichloropropene (HRL = 40)	16,787	15,178	1,609	0.35	0.32	0.62	0.00	0.00	0.00	<0.5
Hexachlorobutadiene[b] (HRL = 0.9)	22,736	20,380	2,356	0.18	0.13	0.59	0.02	0.00	0.13	<1.0
Isopropylbenzene[b] (N/A)	22,995	20,524	2,471	0.24	0.23	0.32	N/A	N/A	N/A	<2.0
m-Dichlorobenzene (HAL = 600)	24,119	21,457	2,662	0.26	0.22	0.53	0.00	0.00	0.00	<1.0
n-Butylbenzene[b] (N/A)	22,972	20,509	2,463	0.13	0.12	0.20	N/A	N/A	N/A	<2.0
n-Propylbenzene[b] (N/A)	22,969	20,501	2,468	0.23	0.19	0.57	N/A	N/A	N/A	<2.0
Naphthalene[b] (HRL = 140)	22,923	20,524	2,399	0.75	0.62	1.92	0.00	0.00	0.00	<2.0
o-Chlorotoluene (MCL = 100)	24,118	21,457	2,661	0.14	0.11	0.38	0.00	0.00	0.00	<2.0
p-Chlorotoluene (MCL = 100)	21,378	18,808	2,570	0.12	0.10	0.27	0.00	0.00	0.00	<2.0
p-Isopropyltoluene[b] (N/A)	22,617	20,320	2,297	0.16	0.15	0.26	N/A	N/A	N/A	<2.0
sec-Butylbenzene[b] (N/A)	22,973	20,509	2,464	0.14	0.14	0.20	N/A	N/A	N/A	<2.0

tert-Butylbenzene[b] (N/A)	22,973	20,508	2,465	0.11	0.10	0.16	N/A	N/A	N/A	<2.0
1,1,1,2-Tetrachloroethane (HAL=70)	24,127	21,462	2,665	0.21	0.16	0.64	0.00	0.00	0.00	<1.0
1,1,2,2-Tetrachloroethane (HAL=2)	24,800	22,106	2,694	0.08	0.05	0.30	0.00	0.00	0.00	<1.0
1,2,3-Trichlorobenzene[b] (N/A)	22,532	20,144	2,388	0.19	0.15	0.50	N/A	N/A	N/A	<2.0
Trichlorofluoromethane[b] (HAL=175)	22,659	20,329	2,330	1.17	0.93	3.22	0.00	0.00	0.00	<2.5
1,2,3-Trichloropropane (MCL=40)	24,088	21,441	2,647	0.08	0.06	0.23	0.00	0.00	0.00	<1.0
1,2,4-Trimethylbenzene[b] (N/A)	22,965	20,504	2,461	0.76	0.63	1.79	N/A	N/A	N/A	<1.0
1,3,5-Trimethylbenzene[b] (N/A)	22,974	20,513	2,461	0.43	0.35	1.10	N/A	N/A	N/A	<2.0

Note: MCL, maximum contaminant level; HAL, health advisory level (as of December 2000); HRL, health reference level (concentration values used only as reference levels for analyses in this report); MRL, minimum reporting level.

The MCL, HAL, HRL, and MRL values are used in this report only as reference levels to facilitate occurrence assessments.

"% PWS>Threshold" indicates the proportion of systems with any analytical results exceeding the concentration value of the HRL/MCL/HAL. (Note that results for % PWSs greater than an MCL value does not indicate a MCL violation. A formal MCL violation occurs when the MCL is exceeded by the average of four consecutive quarterly samples or confirmation samples as required by the primacy States.)

N/A, there is no HRL/MCL/HAL available.

[a] Massachusetts data not included in summary statistics for this contaminant.

[b] New Hampshire data not included in summary statistics for this contaminant.

Source: From United States Environmental Protection Agency, 2001, *Occurrence of Unregulated Contaminants in Public Water Systems - A National Summary*, EPA 815-P-00-002, www.epa.gov.

Table 10C.91 Public Water-Supply Wells in the United States Closed Because of Contamination as of 1984

	Water-Supply Wells					
	In Service or Standby		Closed by Man-Made or Chemical Contamination		Closed by Natural Contamination	
State and Utility	Shallow (<100′)	Deep (>100′)	Shallow (<100′)	Deep (>100′)	Shallow (<100′)	Deep (>100′)
Arizona						
Mesa		24		2		0
Phoenix Wtr & Swr Dept.		115		8		5
Tempe		6		2		0
Tucson		272		7		0
California						
Alhambra		14		3		0
Anaheim		31		4		0
Burbank		4		6		0
Fresno		100		1		0
Fullerton		11		0		1
Garden Grove		14		0		17
Glendale		8		4		0
La Puente/SW Suburban		40		10		2
Long Beach/Dominguez Wtr Corp.		15		0		2
Los Angeles Wtr & Power		180		25		0
Modesto	1	43	0	2	0	0
Pomona		31		2		0
Riverside	4	70	4	28	0	0
Sacremento/Arcade Cnty Wtr		60		1		0
San Bernardino	2	34	0	8	0	0
San Jose/Great Oaks Wtr Co.		10		3		0
Santa Barbara/Goleta Wtr Dist.		9		0		1
Santa Barbara/Santa Barbara Pub.		7		0		2
Colorado						
Colorado Springs	8		0		1	
Connecticut						
Clinton/Conn. Wtr Co.	24	17	6	2	0	0
Delaware						
Newark/Artesian Wtr Co.	7	35	0	3	0	0
Florida						
Boca Raton		50		0		1
Daytona Beach		21		0		5
Englewood	47	9	0	0	7	0
Hollywood	17	6	0	1	0	0
Miami	73		6		0	
Naples	52	7	0	0	2	0
Ocala/Gen. Dev. Util.		4		1		0
Palm Bay/Gen. Dev. Util.		21		1		0
Palm Beach Gdns/Seacoast Util.	39		2		0	
Tallahassee		24		0		2
Hawaii						
Honolulu	60	73	0	9	0	0

(Continued)

Table 10C.91 (Continued)

	Water-Supply Wells					
	In Service or Standby		Closed by Man-Made or Chemical Contamination		Closed by Natural Contamination	
State and Utility	Shallow (<100′)	Deep (>100′)	Shallow (<100′)	Deep (>100′)	Shallow (<100′)	Deep (>100′)
Illinois						
Elgin		11		0		2
Indiana						
Anderson	4	9	2	0	0	2
Richmond/Ind. Am. Wtr. Co.	5		2		0	
South Bend		23		7		0
Iowa						
Sioux City	5	15	0	0	0	1
Louisiana						
Baton Rouge/B.R. Wtr Co.		50		0		2
Michigan						
Kalamazoo	10	108	0	11	0	0
Lansing	3	124	0	1	0	0
Minnesota						
Rochester		19		3		0
Nebraska						
Grand Island	9	30	0	2	0	0
Nevada						
Reno/Sierra Pac. Power Co.	0	16	0	0	3	0
New Jersey						
Brick	7	4	1	0	0	0
Harrington Park/Hackensack Wtr Co.	19	68	0	1	1	2
Merchantville		15		1		1
New Mexico						
Albuquerque		86		1		0
Santa Fe/Sangre de Cristo Wtr Co.		12		1		1
New York						
East Meadow/ Hempstead Wtr Dept.		35		3		3
Elmira	4	0	0	1	0	0
Farmingdale		11		0		1
Lake Success/ Jamaica Wtr Co.	30	55	14	3	0	0
Oakdale/Suffolk Wtr Auth.	31	357	14	20	0	2
West Nyack/Spring Valley Wtr Co.	19	68	0	1	1	2
Ohio						
Mansfield	3	6	0	0	0	1
Oklahoma						
Oklahoma City		4		0		2
Pennsylvania						
Pittsburgh/West View M.A.	24		12		0	
Rhode Island						
Bristol/Bristol Cnty Wtr Co.	2		0		1	
West Warwick/Kent Cnty W.A.	1	4	0	0	1	0

(Continued)

Table 10C.91 (Continued)

	Water-Supply Wells					
	In Service or Standby		Closed by Man-Made or Chemical Contamination		Closed by Natural Contamination	
State and Utility	Shallow (<100′)	Deep (>100′)	Shallow (<100′)	Deep (>100′)	Shallow (<100′)	Deep (>100′)
Texas						
El Paso		140		1		0
Lubbock		330		3		3
Utah						
Ogden		8		0		1
Provo		10		0		2
Salt Lake City/Salt Lake Wtr Dept.	7	14	0	0	0	1
Virginia						
Woodbridge/Prince William Cnty		26		1		0
Wyoming						
Cheyenne		37		1		0

Source: From Compiled from 1984 Water Utility Operating Data issued by the American Water Works Association. Copyright 1986 AWWA.

Table 10C.92 Summary of Contaminant Source Type and Number Reported in the United States in 1998

Source Type	Number of States Reporting Information	Number of Aquifers or Hydrogeologic Settings for Which Information Was Reported	Total Sites	Number of Sites with Confirmed Releases		Number of Sites with Confirmed Groundwater Contamination		Number of Sites with Active Remediation		Number of Sites with Cleanup Completed	
				Number	Percent of Total	Number	Percent of Total	Number	Percent of Total	Number	Percent of Total
LUST	22	72	85,067	48,320	57	15,436	18	3,044	4	21,438	25
Underground injection	17	72	31,480	1,313	4	172	<1	61	<1	452	<2
State sites	17	34	12,202	6,199	51	3,139	26	753	6	3,242	27
DOD/DOE	17	54	8,705	4,470	51	286	3	1,717	20	1,937	22
CERCLA (non-NPL)	19	59	3,506	1,381	39	802	23	229	7	316	9
RCRA corrective action	19	50	2,696	538	20	267	10	95	4	67	3
Nonpoint sources	8	29	2,030	44	2	31	<2	5	<1	3	<1
Landfills	6	26	1,356	110	8	110	8	2	<1	—	—
NPL	22	66	307	275	90	249	81	83	27	33	11

Note: CERCLA, Comprehensive Environmental Response, Compensation, and Liability Act; DOD/DOE = Department of Defense/Department of Energy; LUST, leaking underground storage tank; NPL, national priority list; RCRA, Resource Conversation and Recovery Act; —, not available.

Source: From United States Environmental Protection Agency, 2000, *National Water Quality Inventory 1998. Report to Congress, Groundwater and Drinking Water Chapters*, www.epa.gov.

Table 10C.93 Hazardous Waste Sites on the National Priority List by State and Outlying Area in the United States 2003

State and Outlying Area	Total Sites	Rank	Percent Distribution	Federal	Nonfederal
Alabama	15	24	1.2	3	12
Alaska	6	44	0.5	5	1
Arizona	9	40	0.7	2	7
Arkansas	11	37	0.9	—	11
California	98	2	7.6	24	74
Colorado	18	22	1.4	3	15
Connecticut	16	23	1.2	1	15
Delaware	15	24	1.2	1	14
District of Columbia	1	(X)	0.1	1	—
Florida	52	6	4.1	6	46
Georgia	15	24	1.2	2	13
Hawaii	3	46	0.2	2	1
Idaho	9	41	0.7	2	7
Illinois	45	8	3.5	5	40
Indiana	29	14	2.3	—	29
Iowa	14	29	1.1	1	13
Kansas	12	32	0.9	2	10
Kentucky	14	30	1.1	1	13
Louisiana	15	24	1.2	1	14
Maine	12	33	0.9	3	9
Maryland	19	20	1.5	9	10
Massachusetts	32	12	2.5	7	25
Michigan	69	5	5.4	1	68
Minnesota	24	18	1.9	2	22
Mississippi	4	45	0.3	—	4
Missouri	25	16	1.9	3	22
Montana	15	24	1.2	—	15
Nebraska	11	38	0.9	1	10
Nevada	1	49	0.1	—	1
New Hampshire	20	19	1.6	1	19
New Jersey	116	1	9.0	8	108
New Mexico	12	34	0.9	1	11
New York	91	4	7.1	4	87
North Carolina	29	15	2.3	2	27
North Dakota	—	50	—	—	—
Ohio	35	11	2.7	5	30
Oklahoma	11	39	0.9	1	10
Oregon	12	35	0.9	2	10
Pennsylvania	95	3	7.4	6	89
Rhode Island	12	36	0.9	2	10
South Carolina	25	17	1.9	2	23
South Dakota	2	47	0.2	1	1
Tennessee	13	31	1.0	4	9
Texas	45	9	3.5	4	41
Utah	19	21	1.5	4	15
Vermont	9	42	0.7	—	9
Virginia	30	13	2.3	11	19
Washington	47	7	3.7	14	33
West Virginia	9	43	0.7	2	7
Wisconsin	40	10	3.1	—	40
Wyoming	2	47	0.2	1	1
Guam	2	(X)	(X)	1	1
Puerto Rico	9	(X)	(X)	—	9
Virgin Islands	2	(X)	(X)	—	2
United States	**1,283**	**(X)**	**100.0**	**163**	**1,120**
Total	**1,296**	**(X)**	**(X)**	**164**	**1,132**

Note: As of December 31. Includes both proposed and final sites listed on the National Priorities List for the Superfund program as authorized by the Comprehensive Environmental Response, Compensation, and Liability Act of 1980 and the Superfund Amendments and Reauthorization Act of 1986. —, represents zero.

Source: From U.S. Census Bureau, *Statistical Abstract of the United States: 2004–2005*, www.census.gov.

Original Source: From U.S. Environmental Protection Agency, *Supplementary Materials: National Priorities List, Proposed Rule*, December 2003.

Table 10C.94 Number of Injection Wells in the United States in 1998

Well Class	Number of Wells (Rounded to Nearest 100)	Description of Injection Practice
Class I	500	Inject fluids into deep, confined geologic formations Associated with municipal or industrial waste disposal, hazardous or radioactive waste sites
Class II	164,300	Inject fluids used in oil and gas production into deep, confined geologic formations
Class III	29,600	Inject fluids into shallower formations for mineral extraction
Class IV	Banned by all states and EPA under the Safe Drinking Water Act unless authorized for groundwater cleanup	Inject hazardous or radioactive wastes directly or indirectly into drinking water sources
Class V	Actual numbers unavailable	Includes all injection methods not included in other four categories

Source: From United States Environmental Protection Agency, 2000, *National Water Quality Inventory 1998. Report to Congress, Groundwater and Drinking Water Chapters*, www.epa.gov.

Original Source: From U.S. EPA Office of Groundwater and Drinking Water, 1999.

Table 10C.95 United States Toxic Release Inventory On-Site and Off-Site Disposed or Other Releases, by State, 2003

	On-site Disposal or Other Releases																
	On-site Disposal to Class I Underground Injection Wells, RCRA Subtitle C Landfills, and Other Landfills				Other On-site Disposal or Other Releases												
State	Class I Wells Pounds	RCRA Subtitle C Landfills Pounds	Other On-site Landfills Pounds	Subtotal Pounds	Fugitive Air Emissions Pounds	Point Source Air Emissions Pounds	Surface Water Discharges Pounds	Class II–V Wells Pounds	Land Treatment Pounds	RCRA Subtitle C Surface Impoundments Pounds	Other Surface Impoundments Pounds	Other Land Disposal Pounds	Subtotal Pounds	Total On-Site Disposal or Other Releases Pounds	Total Off-Site Disposal or Other Releases Pounds	Total On-Site and Off-Site Disposal or Other Releases Pounds
Alabama	0	7,739,591	10,578,722	18,318,313	10,050,670	48,028,693	7,805,767	8,305	147,954	0	14,642,205	464,575	81,148,169	99,466,482	18,982,157	118,448,639
Alaska	0	0	310	310	395,715	1,689,402	541,992	21,374,380	0	0	270,833,014	244,609,756	539,444,259	539,444,569	199,327	539,643,896
American Samoa	0	0	0	0	8,460	6	0	0	0	0	0	0	8,466	8,466	0	8,466
Arizona	0	1	5,854,841	5,854,842	1,063,942	3,605,620	6,965	0	167,809	0	6,854,241	30,027,815	41,726,391	47,581,233	632,299	48,213,532
Arkansas	3,303,538	295,473	2,439,167	6,038,178	6,279,944	14,018,639	5,419,186	93,481	501,024	1	2,039,402	43,324	28,395,002	34,433,180	6,168,633	40,801,813
California	0	25,491,811	608,973	26,100,784	3,946,761	14,238,606	4,617,780	30,086	675,350	5,134	130,552	444,069	24,088,337	50,189,121	7,682,581	57,871,702
Colorado	0	97	6,086,523	6,086,620	753,351	2,204,320	2,955,073	0	341	0	2,726,225	3,317,078	11,956,389	18,043,008	4,474,503	22,517,511
Connecticut	0	0	393	393	855,300	2,154,851	722,325	0	0	0	78	778	3,733,332	3,733,725	1,650,391	5,384,116
Delaware	0	0	1,252,178	1,252,178	290,367	7,003,978	918,650	0	9	0	8,178	48	8,221,230	9,473,408	4,139,175	13,612,583
District of Columbia	0	0	0	0	3,338	1	8,062	0	0	0	2,082	0	13,482	13,482	306	13,788
Florida	20,875,927	84	10,441,600	31,317,611	5,640,681	71,659,217	2,507,602	154,144	2,067,467	405,334	9,159,351	123,809	91,717,605	123,035,216	3,417,197	126,452,413
Georgia	0	90	522,944	523,034	9,084,590	90,203,030	9,573,187	0	431,299	35,472	13,660,727	748,427	123,736,732	124,259,766	2,391,094	126,650,860
Guam	0	12	10	22	1,926	162,669	71,627	0	0	0	7	3	236,233	236,255	2,048	238,303
Hawaii	0	0	0	0	180,874	1,900,899	364,067	2,670	19,681	0	0	229,586	2,697,777	2,697,777	419,791	3,117,568
Idaho	0	30,517,289	42,662	30,559,951	995,037	3,038,312	4,642,166	0	669,554	2,239,333	6,571,414	12,027,196	30,183,012	60,742,962	584,846	61,327,809
Illinois	794	23,219,451	4,464,352	27,684,597	8,628,864	47,405,033	7,221,378	0	477,550	29,315	8,987,454	447,772	73,197,365	100,881,962	31,519,211	132,401,173
Indiana	1,075,521	8,094,913	17,522,055	26,692,489	8,722,806	64,952,587	23,296,297	0	1,850,171	1,500	8,571,928	777,844	108,173,133	134,865,622	99,900,349	234,765,971
Iowa	0	0	1,889,630	1,889,630	3,791,907	18,311,363	3,274,619	0	750	0	2,611,789	39,658	28,030,086	29,919,715	7,482,328	37,402,043
Kansas	470,746	33,000	3,535,497	4,039,243	2,641,059	10,557,093	4,021,144	500	419,554	0	3,795,286	127,585	21,562,221	25,601,464	3,262,448	28,863,912
Kentucky	0	0	15,143,069	15,143,069	6,024,718	51,653,888	2,986,815	2,971	333,583	334,515	5,381,917	921,961	67,640,366	82,783,436	7,798,069	90,581,505
Louisiana	35,904,030	7,050,318	7,348,228	50,302,576	12,230,196	42,697,721	11,303,522	0	46,587	289,599	4,062,499	148,267	70,778,392	121,080,968	5,764,045	126,845,013
Maine	0	0	838,723	838,723	686,531	3,611,216	3,334,311	0	0	0	0	9,557	7,641,615	8,480,338	824,733	9,305,071
Maryland	0	0	1,827,936	1,827,936	596,281	35,266,724	2,704,113	50,375	159,930	0	34,954	119,632	38,932,009	40,759,945	4,739,560	45,499,505
Massachusetts	0	0	41,739	41,739	741,565	5,297,813	68,806	0	39	0	80,638	670,421	6,859,282	6,901,021	2,112,307	9,013,328
Michigan	1,907,236	1,046,569	3,757,600	6,711,405	3,915,701	46,250,106	1,218,195	0	123,547	0	5,389,756	1,062,639	57,959,944	64,671,349	36,933,939	101,605,288
Minnesota	0	251	5,528,048	5,526,299	3,110,524	10,522,357	1,246,533	0	71,993	0	5,284,791	63,705	20,299,903	25,826,201	5,622,420	31,448,621
Mississippi	12,654,379	65	1,985,150	14,639,593	5,810,502	25,988,162	7,751,711	0	393,645	15,268	6,767,118	221,233	46,947,640	61,587,233	1,489,728	63,076,960
Missouri	0	199,911	2,187,364	2,387,276	2,811,105	24,870,138	2,620,282	0	12,617	544	40,355,980	21,326,736	91,997,402	94,384,677	8,114,819	102,499,496
Montana	0	81	22,971,757	22,971,839	597,835	3,486,141	49,172	209,716	933	0	10,445,012	6,475,019	21,263,828	44,235,667	970,916	45,206,583
Nebraska	0	0	7,740,974	7,740,974	886,910	5,985,779	18,177,388	0	678,512	1,000	755	277,302	26,007,647	3,748,621	17,722,548	51,471,169
Nevada	0	29,771,405	2,606,310	32,377,715	588,282	1,253,025	88,601	3	43	0	188,495,103	185,492,867	375,917,924	408,295,639	758,761	409,054,400
New Hampshire	0	0	0	0	157,081	5,217,148	86,194	0	0	0	22,805	165	5,483,393	5,483,393	460,749	5,944,142
New Jersey	0	104,607	33,841	138,448	1,840,951	10,484,189	4,148,642	5	6,583	0	6,551	188,018	16,674,938	16,813,386	6,310,629	23,124,015
New Mexico	18,605	0	5,296,008	5,314,613	300,082	647,660	62,237	0	326,660	0	1,270,605	9,831,265	12,438,509	17,753,122	146,129	17,899,251
New York	0	2,919,297	1,137,787	4,057,084	2,855,774	24,254,524	7,937,611	0	23,565	0	257	398,503	35,470,235	39,527,319	4,499,897	44,027,216
North Carolina	0	2,023	3,265,168	3,267,191	6,936,819	93,358,632	8,592,474	0	245,329	0	6,096,625	609,420	115,839,298	119,106,489	10,039,033	129,145,522
North Dakota	0	0	2,263,167	2,263,167	251,193	4,621,269	248,831	0	17,123	0	7,020,805	2,975	12,162,196	14,425,362	9,213,436	23,638,798
Northern Marianas	0	0	0	0	324	5,701	0	0	0	0	2	0	6,027	6,027	0	6,027
Ohio	29,289,527	7,261,217	16,604,064	53,154,807	9,765,195	123,297,398	6,716,124	0	499,432	307,465	12,485,002	178,440	153,249,057	206,403,864	45,193,295	251,597,159
Oklahoma	1,307,526	1,757,603	1,814,396	4,879,525	2,467,942	13,460,673	3,451,751	0	13,387	1	547,749	524,632	20,466,134	25,345,659	4,622,081	29,967,740
Oregon	0	13,208,609	11,729,305	24,937,914	2,538,300	10,784,137	2,452,054	0	187,648	12,337	503	2,350	15,977,329	40,915,243	1,213,679	42,128,922
Pennsylvania	0	11,433	5,711,204	5,722,637	5,871,513	85,113,791	9,684,378	0	12,327	1,790,272	309,090	5,508,710	108,290,080	114,012,717	52,905,725	166,918,443
Puerto Rico	0	13,533	250	13,783	942,381	7,081,569	24,365	0	0	0	0	3,500	8,051,815	8,065,598	734,652	8,800,250
Rhode Island	0	250	0	250	236,638	390,168	6,947	0	0	0	0	0	633,753	634,003	258,275	892,279
South Carolina	0	62,318	2,387,601	2,449,919	6,720,209	45,438,879	3,473,022	0	37,514	85	2,944,266	327,757	58,941,732	61,391,651	22,316,357	83,708,008
South Dakota	0	0	5,533,657	5,533,657	656,431	807,101	3,199,143	101	685	0	23,556	5,350	4,692,347	10,226,004	94,109	10,320,113
Tennessee	0	54,792	18,926,296	18,981,088	11,030,628	77,689,173	2,410,104	0	26,760	0	22,765,611	2,457,322	116,379,598	135,360,685	7,178,908	142,539,594
Texas	85,633,991	3,619,830	24,058,727	113,312,547	35,163,397	56,502,383	21,670,283	42,000	3,090,383	11	3,684,426	1,953,398	122,106,261	235,418,808	26,485,939	261,904,747
Utah	0	7,672,552	12,978,810	20,651,362	1,857,825	7,268,894	56,978	0	292,689	0	130,057,726	78,732,740	218,266,853	238,918,214	3,077,073	241,995,287
Vermont	0	0	4	4	19,824	45,025	136,856	0	0	0	7	5	201,716	201,721	144,981	346,702
Virgin Islands	0	0	0	0	124,893	831,616	355,871	0	597	0	10,124	0	1,323,101	1,323,101	7,916	1,331,017
Virginia	0	344	3,317,037	3,317,381	5,233,866	45,714,310	8,198,926	0	203,048	0	1,887,312	244,389	61,481,850	64,799,231	9,431,206	74,230,438
Washington	0	633,020	191,652	824,673	1,560,909	11,772,038	1,397,187	0	29,989	0	5,035,903	517,707	20,313,733	21,138,405	1,741,529	22,879,934
West Virginia	10	7,844	7,700,777	7,708,631	3,840,246	75,083,656	4,194,526	87	597,291	21,088	4,937,887	342,718	89,017,499	96,726,130	5,443,463	102,169,593

Wisconsin	0	1,677	655,225	656,902	2,873,281	21,768,989	4,595,302	0	821,747	49,651	21,242	302,494	30,423,707	31,080,609	19,717,618	50,798,227
Wyoming	7,960,400	2,909	7,064,110	15,027,419	513,877	1,638,924	10,990	0	1,564	4,341	1,019,873	12,290	3,201,858	18,229,277	1,034,080	19,263,357
Total	**200,402,228**	**170,794,270**	**267,883,840**	**639,080,339**	**205,095,324**	**1,381,295,231**	**222,628,110**	**21,968,824**	**15,675,243**	**5,542,266**	**817,040,382**	**612,362,811**	**3,281,608,191**	**3,920,688,530**	**518,031,287**	**4,438,719,817**

Note: This information does not indicate whether (or to what degree) the public has been exposed to toxic chemicals. Therefore, no conclusions on the potential risks can be made based solely on this information (including any ranking information). For more detailed information on this subject refer to *The Toxics Release Inventory (TRI) and Factors to Consider When Using TRI Data* document at www.epa.gov/trl/tridata.

On-site Disposal or Other Releases include Underground injection to Class I Wells (Section 5.4.1), RCRA Subtitle C Landfills (5.5.1A), Other Landfills (5.5.1B), Fugitive or Non-point Air Emissions (5.1), Stack or Point Source Air Emissions (5.2), Surface Water Discharges (5.3), Class II–V Wells (5.4.2), Land Treatment (5.5.2), RCRA Subtitle C Surface impoundments (5.5.3A), Other Surface Impoundments (5.5.3B) and Other Land Disposal (5.5.4). Off-site Disposal or Other Releases include from Section 6.2 Underground Injection to Class I Wells (M81), RCRA Subtitle C Landfills (M65), Other Landfills (M64, M72), Storage Only (M10), Solidification/Stabilization—Metals and Metal Category Compounds only (M41 or M40), Wastewater Treatment (excluding POTWs)—Metals and Metal Category Compounds only (M62 or M61), RCRA Subtitle C Surface Impoundments (M66), Other Surface Impoundments (M67, M63), Land Treatment (M73), Other Land Disposal (M79), Underground Injection to Class II–V Wells (M82, M71), Other Off-site Management (M90), Transfers to Waste Broker—Disposal (M94, M91), and Unknown (M99) and, from Section 6.1, Transfers to POTWs (metals and metal category compounds only).

Does not include Off-site Disposal or Other Releases transferred to other TRI facilities that reported the amounts as On-site Disposal or Other Releases.

Source: From United States Environmental Protection Agency, 2005, *2003 TRI Public Data Release eReport May 2005*, www.epa.gov.

Table 10C.96 United States Toxic Release Inventory On-Site and Off-Site Disposal or Other Releases, by Industry, 2003

SIC Code Industry	On-site Disposal or Other Releases: On-site Disposal to Class I Underground Injection Wells, RCRA Subtitle C Landfills, and Other Landfills				On-site Disposal or Other Releases: Other On-site Disposal or Other Releases											
	Class I Wells Pounds	RCRA Subtitle C Landfills Pounds	Other On-site Landfills Pounds	Subtotal Pounds	Fugitive Air Emissions Pounds	Point Source Air Emissions Pounds	Surface Water Discharges Pounds	Class II–V Wells Pounds	Land Treatment Pounds	RCRA Subtitle C Surface Impound-ments Pounds	Other Surface Impound-ments Pounds	Other Land Disposal Pounds	Subtotal Pounds	Total On-Site Disposal or Other Releases Pounds	Total Off-Site Disposal or Other Releases Pounds	Total On-Site and Off-Site Disposal or Other Releases Pounds
10 Metal mining	0	0	17,124,349	17,124,349	1,285,236	1,753,800	679,446	21,584,184	13,150	2,236,052	637,974,487	562,019,783	1,227,546,138	1,244,670,487	1,037,942	1,245,708,429
12 Coal mining	0	0	7,952,585	7,952,585	842,399	64,775	199,797	58,768	1,154,725	21,088	1,817,948	795,570	4,955,069	12,907,654	4,925	12,912,580
20 Food	63,205	546	41,361	105,112	16,998,226	35,055,551	83,136,183	17,044	9,873,967	89,218	140,913	428,500	145,739,602	145,844,714	7,340,634	153,185,349
21 Tobacco	0	0	2,376	2,376	58,494	2,416,426	130,052	0	149,836	0	0	0	2,754,807	2,757,183	421,402	3,178,586
22 Textiles	0	0	0	0	1,130,155	4,826,038	262,178	0	123,650	0	160,937	53	6,503,010	6,503,010	894,688	7,397,699
23 Apparel	0	0	0	0	112,094	366,780	0	0	0	0	0	0	478,875	478,875	200,481	679,355
24 Lumber	0	2,273	330,292	332,566	3,370,921	27,040,230	108,277	0	115,179	250	1,108	16,190	30,652,154	30,964,719	2,021,849	33,006,568
25 Furniture	0	0	1	1	691,650	5,410,353	35	0	0	0	0	35,961	6,137,999	6,138,000	71,580	6,209,580
26 Paper	0	42	12,545,504	12,545,546	27,019,698	146,219,717	18,715,017	0	1,064,035	128,025	3,635,346	307,503	197,089,341	209,634,887	5,327,919	214,962,805
27 Printing	0	87	0	87	7,253,937	7,437,745	549	0	0	0	0	4,951	14,697,183	14,697,270	267,911	14,985,181
28 Chemicals	177,818,769	3,804,019	24,987,711	206,610,499	62,053,840	168,589,153	44,537,842	273,096	557,537	23,290	12,922,181	4,731,869	293,688,809	500,299,307	44,440,173	544,739,481
29 Petroleum	2,487,806	56	724,902	3,212,764	16,706,339	34,608,196	17,134,723	32,752	54,496	12	61,682	98,744	68,696,945	71,909,709	3,059,186	74,968,895
30 Plastics	0	15,272	165,188	180,480	14,095,493	51,339,039	125,302	0	0	11	4,997	81,715	65,646,558	65,827,018	9,461,313	75,288,331
31 Leather	0	0	6,725	6,725	269,382	655,731	27,908	0	60	0	0	5	953,086	959,811	1,139,481	2,099,292
32 Stone/ clay/glass	0	94,933	3,332,388	3,427,320	1,556,565	38,084,142	2,133,677	2,971	195	169	230,674	337,690	42,346,084	45,773,404	5,469,229	51,242,633
33 Primary metals	945,916	9,728,105	36,433,952	47,107,973	13,604,646	35,879,537	39,443,391	5	15,115	0	32,623,268	29,540,600	151,006,562	198,114,535	279,358,340	477,472,875
34 Fabricated metals	0	14,789	48,262	63,051	12,327,229	23,708,658	2,331,208	0	17,556	125	8,368	340,049	38,733,193	38,796,244	19,827,879	58,624,123
35 Machinery	0	11	3,688,102	3,688,112	2,666,073	4,127,143	209,418	0	210	0	8	20,986	7,023,837	10,711,949	3,627,929	14,339,878
36 Electrical Equip.	0	239,206	198,932	438,140	3,281,807	6,476,571	3,628,323	0	0	750	2,466	5,939	13,395,857	13,833,997	6,444,050	20,278,047
37 Transportation Equip.	6,354	2,676	236,433	245,463	11,936,992	51,064,653	207,966	0	0	0	34	92,279	63,301,924	63,547,387	11,228,710	74,776,098
38 Measure/Photo.	0	5,765	168	5,933	1,694,548	5,123,745	1,018,609	0	341	0	5	82,747	7,919,995	7,925,928	788,458	8,714,386
39 Miscellaneous	0	755	1,450	2,205	926,391	3,908,177	62,644	0	0	0	5	504	4,897,720	4,899,925	2,155,240	7,055,165
491/493 Electric utilities	0	51,833	144,198,732	144,250,565	284,136	721,277,416	3,340,491	4	1,993,907	1,238,158	126,316,380	5,397,157	859,847,648	1,004,098,213	78,665,493	1,082,763,707
5,169 Chemical wholesale distributors	0	0	0	0	634,593	839,584	1,218	0	0	5	5	6,067	1,281,472	1,281,472	117,925	1,399,396
5,171 Petroleum terminals/bulk storage	0	12	0	12	974,868	1,820,524	12,614	0	0	0	14,333	7,393	2,829,733	2,829,745	349,073	3,178,818
7,389/4,953 Hazardous waste/solvent recovery	19,080,178	156,200,572	15,801,796	191,082,546	287,991	498,873	300,944	0	10	1,803,701	2	127,109	3,018,630	194,101,176	33,007,422	227,108,598
No codes	0	633,316	62,631	695,947	3,131,619	2,902,674	4,880,299	0	541,275	1,412	1,125,238	7,883,444	20,465,982	21,161,909	1,282,055	22,443,964
Total	**200,402,228**	**170,794,270**	**267,883,840**	**639,080,339**	**205,095,324**	**1,361,295,231**	**222,628,110**	**21,968,824**	**15,875,243**	**5,542,266**	**817,040,382**	**612,362,811**	**3,281,608,191**	**3,920,688,530**	**518,031,287**	**4,438,719,817**

Note: This information does not indicate whether (or to what degree) the public has been exposed to toxic chemicals. Therefore, no conclusions on the potential risks can be made based solely on this information (including any ranking information). For more detailed information on this subject refer *Toxics Release Inventory (TRI) and Factors to Consider When Using TRI Data* document at www.epa.gov/tri/tridata.

On-site Disposal or Other Releases include Underground Injection to Class I Wells (Section 5.4.1), RCRA Subtitle C Landfills (5.5.1A), Other Landfills (5.5.1B), Fugitive or Non-point Air Emissions (5.1), Stack or Point Source Air Emissions (5.2), Surface Water Discharges (5.3), Class II–V Wells (5.4.2). Land Treatment (5.5.2), RCRA Subtitle C Surface Impoundments (5.5.3A), Other Surface Impoundments (5.5.3B) and Other Land Disposal (5.5.4). Off-site Disposal or Other Releases include from Section 6.2 Underground Injection to Class I Wells (M81), RCRA Subtitle C Landfills (M65), Other Landfills (M64, M72), Storage Only (M10), Solidification/Stabilization—Metals and Metal Category Compounds only (M41 or M40), Wastewater Treatment (excluding POTWs)—Metals and Metal Category Compounds only (M62 or M61), RCRA Subtitle C Surface Impoundments (M66), Other Surface Impoundments (M67, M63), Land Treatment (M73), Other Land Disposal (M79), Underground Injection to Class II–V Wells (M82, M71), Other Off-site Management (M90), Transfers to Waste Broker—Disposal (M94, M91), and Unknown (M99) and, from Section 6.1, Transfers to POTWs (metals and metal category compounds only).

Does not include Off-site Disposal or Other-Releases transferred to other TRI facilities that reported the amounts as On-site Disposal or Other Releases.

Source: From United States Environmental Protection Agency, 2005, *2003 TRI Public Data Release eReport*, May 2005, www.epa.gov.

Table 10C.97 Number of Reported Spills in Canada in Seven Sectors, 1984–1995

Year	Chemical	Government	Metallurgy	Mining	Petroleum	Pulp & Paper	Service Industry
1984	70	223	31	153	1,831	38	94
1985	130	200	58	83	2,053	44	104
1986	206	206	181	118	2,398	73	157
1987	179	228	139	124	2,512	63	208
1988	405	981	360	172	3,021	148	281
1989	582	1,080	392	172	2,971	224	346
1990	588	1,320	361	191	3,157	312	408
1991	552	1,487	508	195	3,139	291	434
1992	667	1,991	703	194	1,144	340	427
1993	754	1,957	618	186	1,531	371	456
1994	784	2,165	599	199	1,577	458	464
1995	534	2,204	431	184	1,642	353	484
Total	**5,451**	**14,042**	**4,381**	**1,971**	**26,976**	**2,715**	**3,863**

Source: From *Summary of Spill Events in Canada, 1984–1995*, EPS 5/SP/3, Environment Canada, Canada Environmental Emergencies Program, 1998. Reproduced with the permission of the Minister of Public Works and Government Services, 2006.

Table 10C.98 Total Quantity of Reported Spills in Canada in Seven Sectors (Tons), 1984–1995

Year	Chemical	Government	Metallurgy	Mining	Petroleum	Pulp & Paper	Service Industry
1984	1,783	142,556	4,860	113,078	72,121	2,948	433
1985	12,399	140,820	314	16,105	46,029	35,447	211
1986	16,160	11,267	23,923	29,972	62,232	28,138	431,886
1987	17,128	133,863	87,665	126,939	89,773	90,608	616,308
1988	5,498	58,480	23,497	6,752	29,444	26,933	1,115
1989	7,194	189,169	51,266	42,899	120,765	16,322	228
1990	6,629	84,194	79,178	35,247	50,284	35,845	310
1991	1,619	184,449	32,449	26,172	43,963	46,491	5,106
1992	827	1,386,991	193,435	58,667	11,164	25,494	5,625
1993	1,519	677,529	1,425,753	12,094	62,725	35,612	190
1994	178	678,622	27,489	7,262	18,174	19,751	197
1995	325	1,576,576	11,791	4,783	18,176	49,224	763
Total	**71,259**	**5,265,518**	**1,961,620**	**479,969**	**624,852**	**412,814**	**1,062,374**

Source: From *Summary of Spill Events in Canada, 1984–1995*, EPS 5/SP/3, Environment Canada, Canada Environmental Emergencies Program, 1998. Reproduced with the permission of the Minister of Public Works and Government Services, 2006.

Table 10C.99 Hydrocarbon Spills Reported in Canada, 1974–1983

Year	Condensates and Gases	Crude Oil	No. 2 Fuel	No. 6 Fuel	No. 4 & 5 Fuel	Gasoline	Other Oils	Waste Oil	Asphalt	Total
1974	3,623	14,823	1,046	1,106	5,594	810	222	631	82	27,935
1975	600	11,530	836	2,499	2,844	3,095	256	321	266	22,250
1976	7,429	10,901	1,650	2,084	2,231	2,156	220	38	372	27,085
1977	20,865	10,699	3,753	804	1,587	1,638	551	85	103	40,085
1978	845	12,067	2,801	3,288	1,932	1,237	454	72	476	23,170
1979	3,306	12,540	4,724	9,054	1,801	1,567	339	319	392	34,045
1980	705	15,274	3,517	585	649	918	278	108	479	22,510
1981	1,160	8,041	12,589	763	1,184	1,767	1,013	1,886	287	28,690
1982	281,181	10,658	4,602	915	1,067	847	609	46	147	300,070
1983	768	8,553	1,467	803	422	460	300	69	337	13,180
Total	320,482	115,086	36,985	21,901	19,311	14,495	4,242	3,575	2,941	539,020
%	59	21	7	4	4	3	1	1	1	100

Note: By type; volume in metric tons.

Source: From Environment Canada, 1987, Summary of Spill Events, 1974–1983, www.ec.gc.ca. With permission.

Table 10C.100 Quantity Spilled Annually for the Top Five MIACC List 1 Substances in Canada

Tonnes

Year	Anhydrous Ammonia	Chlorine	Gasoline	Hydrochloric Acid	Propane	Total
1984	27	2.9	5,632	36	19	5,716
1985	25	0.2	1,746	57	1,591	3,418
1986	33	409.1	909	53	25	1,430
1987	7	0.3	837	189	1	1,035
1988	17	9.2	1,096	51	1	1,174
1989	27	1.1	746	250	11	1,035
1990	86	0.1	675	106	64	930
1991	4	0.2	508	55	137	704
1992	28	0.5	6,439	346	15	6,829
1993	70	0.4	689	37	57	853
1994	13	8.2	206	72	43	343
1995	18	16.3	247	25	2	308
Total	**355**	**448.4**	**19,730**	**1,276**	**1,965**	**23,775**

Source: From *Summary of Spill Events in Canada, 1984–1995*, EPS 5/SP/3, Environment Canada, Canada Environmental Emergencies Program, 1998. Reproduced with the permission of the Minister of Public Works and Government Services, 2006.

Table 10C.101 Top Five On-Site Releases to Land in Canada, 1996 and 2001

Substance	Releases (Tons)	Share of Total (%)
1996		
Zinc (and its compounds)	4,989.7	35.9
Ethylene glycol	3,209.8	23.1
Manganese (and its compounds)	1,910.2	13.8
Lead (and its compounds)	894.3	6.4
Asbestos (friable form)	848.2	6.1
2001		
Calcium fluoride	10,211.0	31.0
Zinc (and its compounds)	8,143.8	24.8
Manganese (and its compounds)	3,637.2	11.1
Ethylene glycol	2,044.5	6.2
Lead (and its compounds)	1,641.0	5.0

Source: From Environment Canada, Pollution Data Branch, National Pollutant Release Inventory Database, www.ec.gc.ca/pdb/npri, 2001. Reproduced with the permission of the Minister of Public Works and Government Services, 2006.

Original Source: From Environment Canada, *Pollution Data Branch, National Pollutant Release Inventory database*, www.gc.ca/pdb/npri/ (accessed April 1, 2003).

Table 10C.102 Trichloroethene Concentrations Detected in Municipal/Communal and Private/Domestic Water Supplies in Canada

Supply	Number of Supplies	Percentage of Supplies Where TCE Was Detected	Average Maximum Concentration
Municipal/communal	481	8.3	25 ug/L
Private/domestic	215	3.3	1,680 ug/L

TCE Concentration Range	Percentage of Sites within TCE Range	Percent of Population Supplied with Groundwater within TCE Range[a]
Nondetectable levels (<0.01–10 ug/L)	93	49
<1 ug/L	3.6	
1–10 ug/L	1.4	48
10–100 ug/L	0.43	2.1
>100 ug/L	1.3	0.8

[a] 1.67 Million of the 7.1 million Canadians who relied on groundwater for household use in 1995 were covered this study. Most of sites were from Ontario and New Brunswick.

Source: Abstracted from Health Canada, 2005, *Guidelines for Canadian Drinking Water Quality: Supporting Documentation—Trichloroethene*. Water Quality and Health Bureau, Healthy Environments and Consumer Safety Branch, Health Canada, Ottawa, Ontario, www.hc-sc.gc.ca. With permission.

Table 10C.103 Sources of Groundwater Contamination Reported by European Countries

Country/ Pollutant	EEA18							Phare							T	R
	AT	DK	ES	FR	DE	SE	UK	BG	EE	HU	LT	RO	SK	SI	MD	CY
Heavy metals		•	•	•		•	•	•	•	•		•	•	•	•	
Chlorinated hydrocarbons	•	•	•	•	•		•			•		•	•	•		
Hydrocarbons				•	•		•		•	•	•	•	•		•	
Sulphate				•			•		•						•	•
Metals		•								•						
Phosphate												•				
Bacteria				•								•				

Note: T, Tacis; R Others, AT, Austria; DK, Denmark; ES, Spain; FR, France; CY, Cyprus; DE, Germany; SE, Sweden; UK, UK; BG, Bulgaria; EE, Estonia; HU, Hungary; LT, Lithuania; RO, Romania; SK, Slovak Rep; MD, Rep. of Moldova.

Source: From European Environmental Agency, 1999, *Groundwater Quality and Quantity in Europe*, Printed with permission, www.eca.europa.eu. Reprinted with permission © EEA.

Table 10C.104 Known Occurrences of Hydrocarbon Contamination of Groundwater in Australia

State or Territory	Area	Source
Victoria	9 sites	Industrial/manufacturing/storage facilities
	2 sites in Melbourne–Geelong region	Landfill
	1 site	Fuel station
Western Australia	8 sites including 2 sites on the Swan Coastal Plain[a]	Fuel stations
NSW	5 sites, including Anna Bay, Botany and Matraville	Industrial/manufacturing/storage facilities
	1 site	Landfill
ACT	3 sites	Fuel stations
Queensland	1 site in Cairns	Fuel station
South Australia	6 Sites Mt Gambler (2 sites), Bordertown, Jamestown, Fregon and Minlayton	Fuel stations
Northern Territory	1 site Croker Island	Fuel station
Tasmania	Several (number and locations unknown)	Fuel stations

[a] The two sites on the Swan Coastal Plain are documented in Davis et al. (1993).

Source: From Ball, J. et al., 2001, *Inland Waters, Australia State of Environment Report 2001 (Theme Report)*, CSIRO Publishing on behalf of the Department of the Environment and Heritage, Canberra, www.deh.gov.au.

Original Source: Adapted from Knight 1993.

SECTION 10D SOLID WASTE

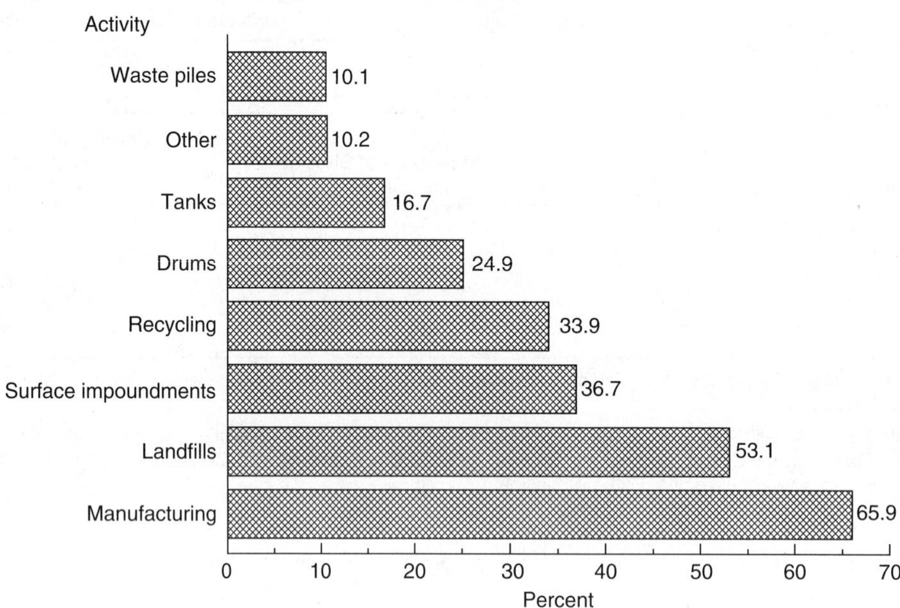

Figure 10D.70 Types of activities at hazardous waste sites in the United States (percent of 1177 final and proposed sites on the National Priorities list as of June 1988; a site may have more than one type of activity). (From U.S. EPA, Office of Emergency Response, Washington, DC 20460.)

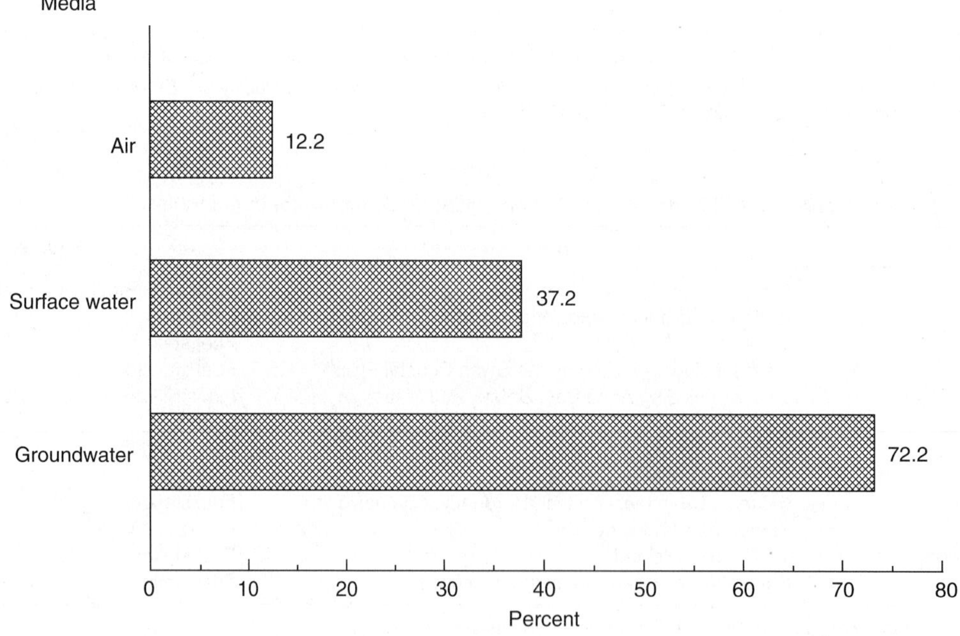

Figure 10D.71 Observed contamination at hazardous waste sites in the United States (percent of 1177 final and proposed sites on the National Priorities list as of June 1988; a site may have more than one type of contamination). (From U.S. EPA, Office of Emergency Response, Washington, DC 20460.)

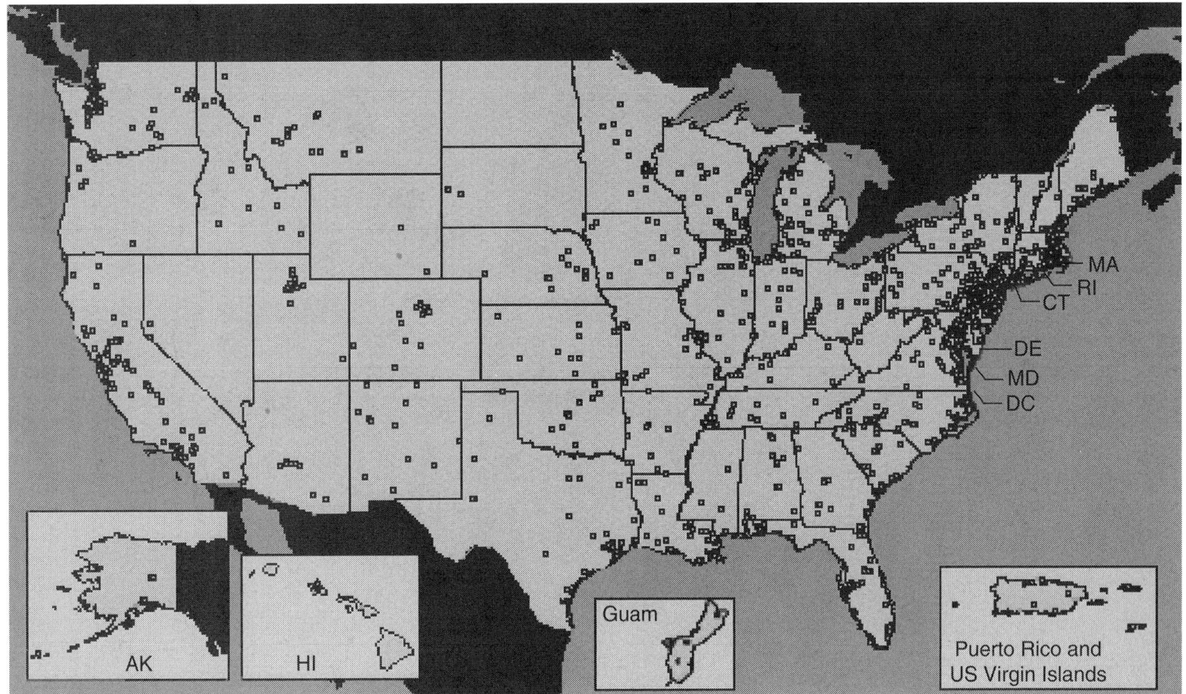

Figure 10D.72 Superfund sites in the United States. (From www.images.google.com.)

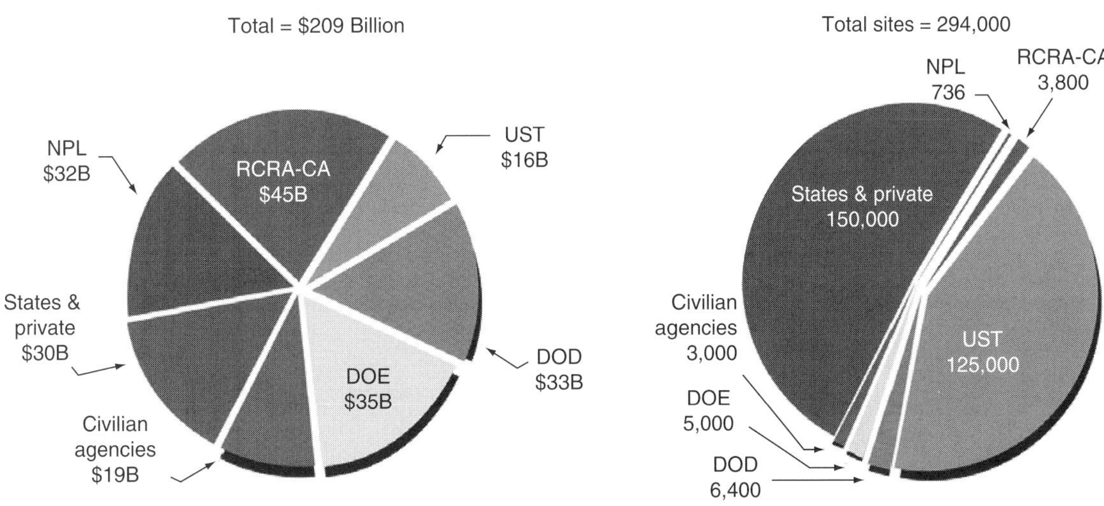

These estimates are derived from judgements regarding the most likely scenarios within a range of estimates. The estimates described in the report, include a number of assumptions such as the average cleanup cost per site, number of new site discoveries, and future additons to the NPL.

NPL: National Priorites List, or Superfund; RCRA-CA: Resource Conservation and Recovery Act Corrective Action program; UST: Underground Storage Tanks; DOD; Department of Defense; DOE: Department of Energy; Civilian agencies: non-DOD and non-DOE federal agencies; and State & private: state mandatory, voluntary, and brownfields sites, and private sites.

Totals may not add due to rounding.

Figure 10D.73 Estimated number of hazardous waste sites and cleanup costs: 2004–2033. (From clu-in.org/download/market/2004market.)

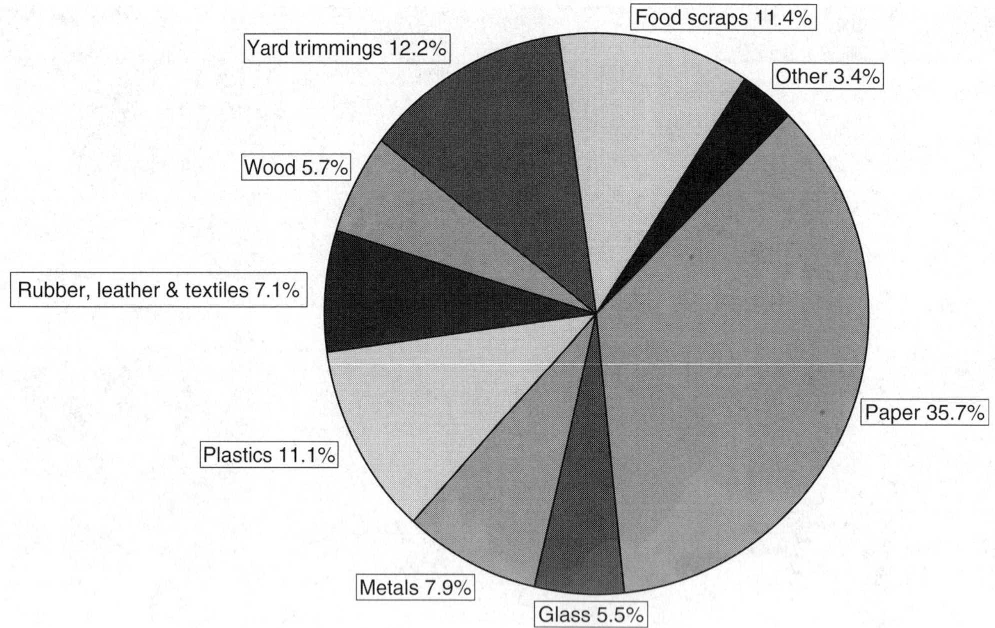

Figure 10D.74 2001 total MSW generation—229 million tons (before recycling). (From www.epa.gov.)

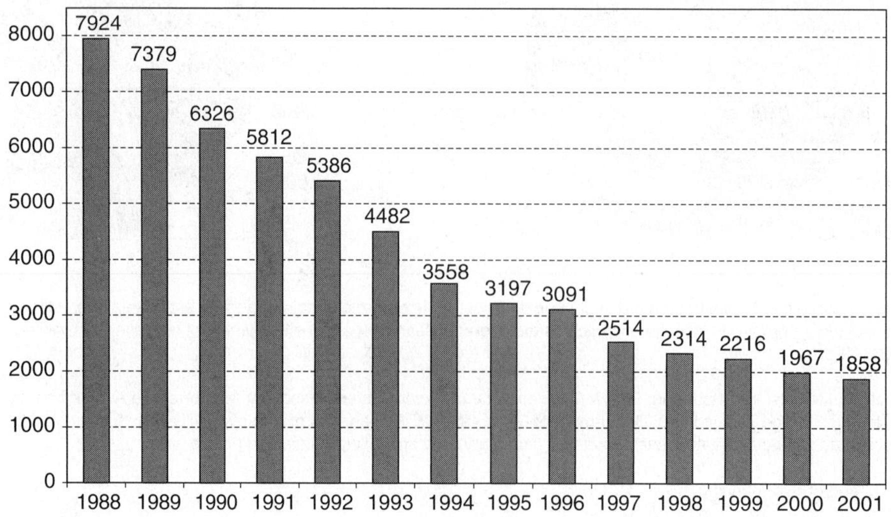

Figure 10D.75 Number of landfills in the U.S. (From www.epa.gov.)

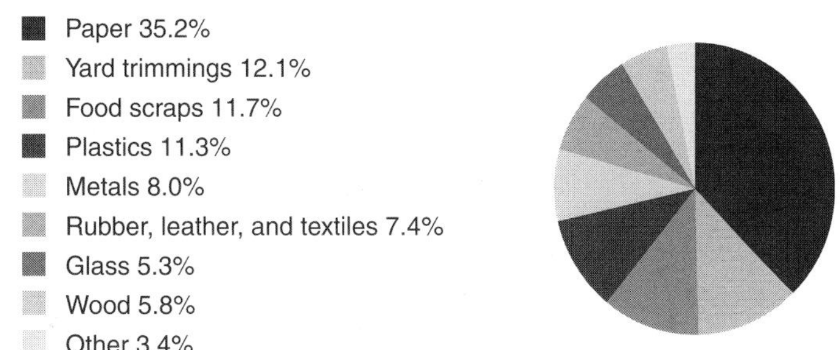

Figure 10D.76 2003 total waste generation—236 million tons (before recycling). (From www.epa.gov.)

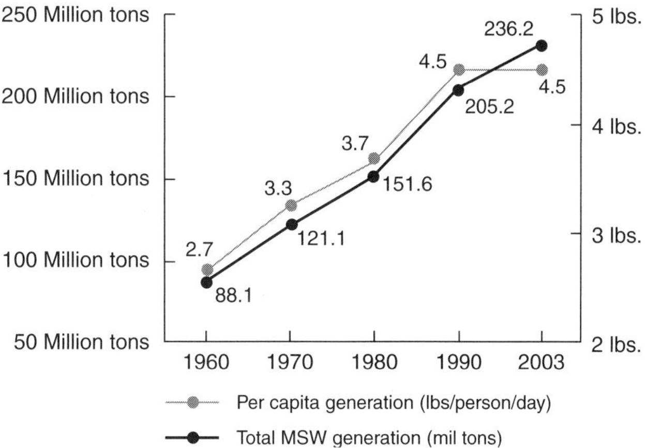

Figure 10D.77 Trends in MSW generation 1960–2003. (From www.epa.gov.)

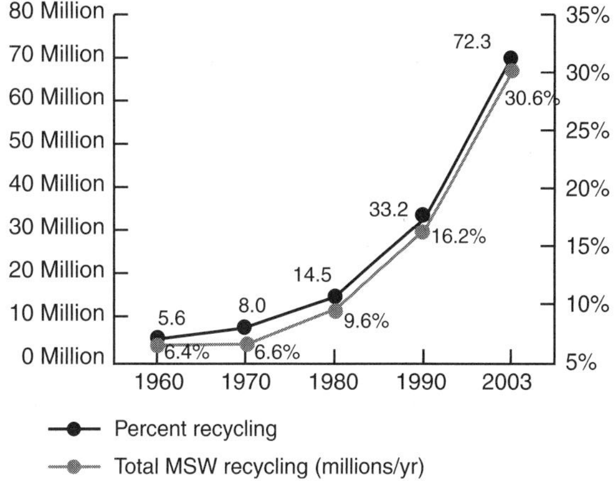

Figure 10D.78 MSW recycling rates 1960–2003. (From www.epa.gov.)

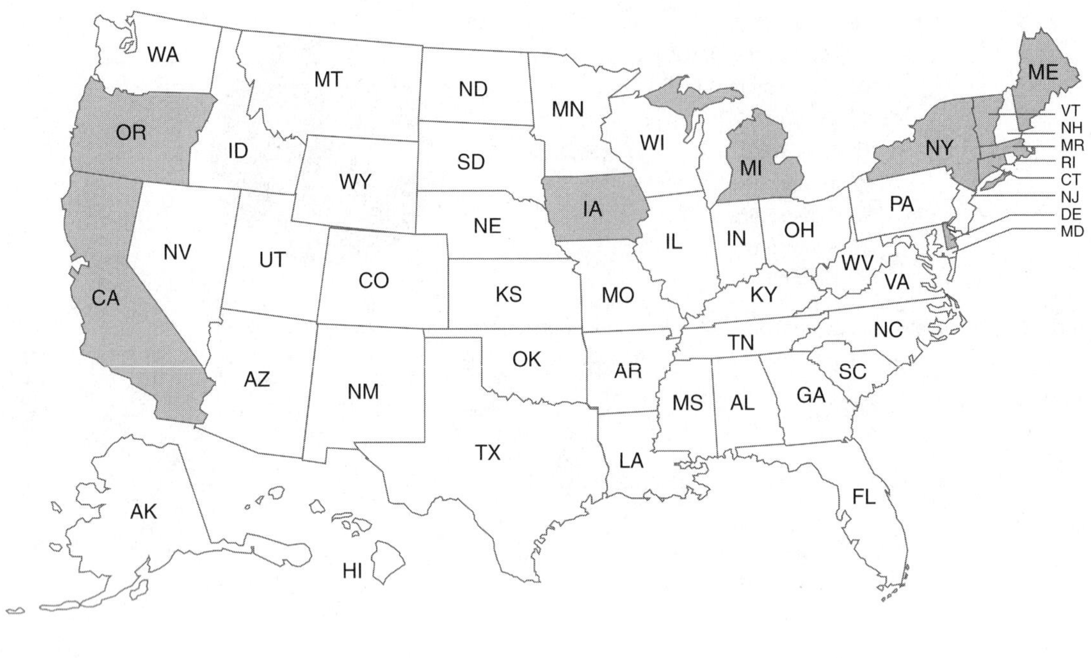

Figure 10D.79 States with bottle deposit rules. (From The Container Recycling Institute 1999, www.epa.gov.)

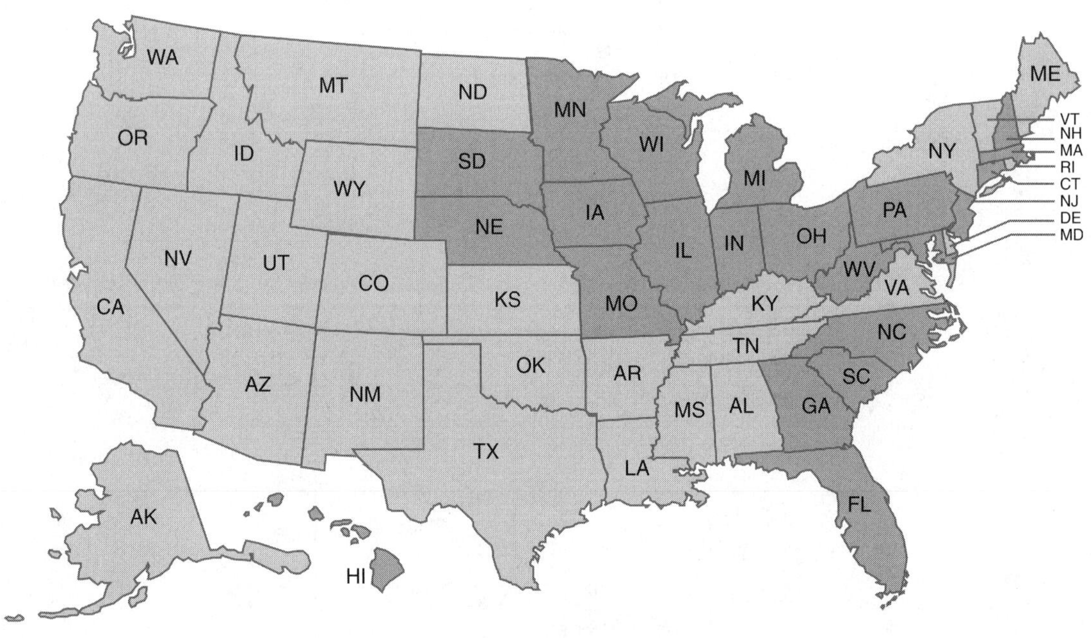

Figure 10D.80 States with yard waste bans. (From *BioCycle Magazine*, May 1998, www.epa.gov.)

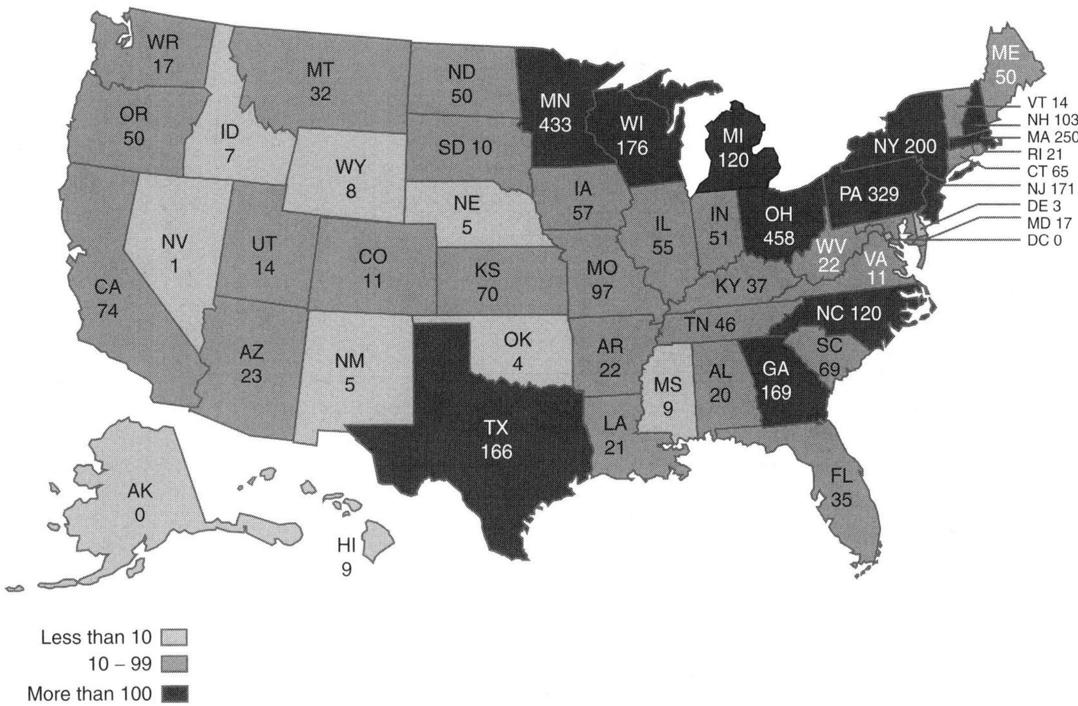

Figure 10D.81 Number of yard waste composting programs. (From *BioCycle Magazine*, April 1999, www.epa.gov.)

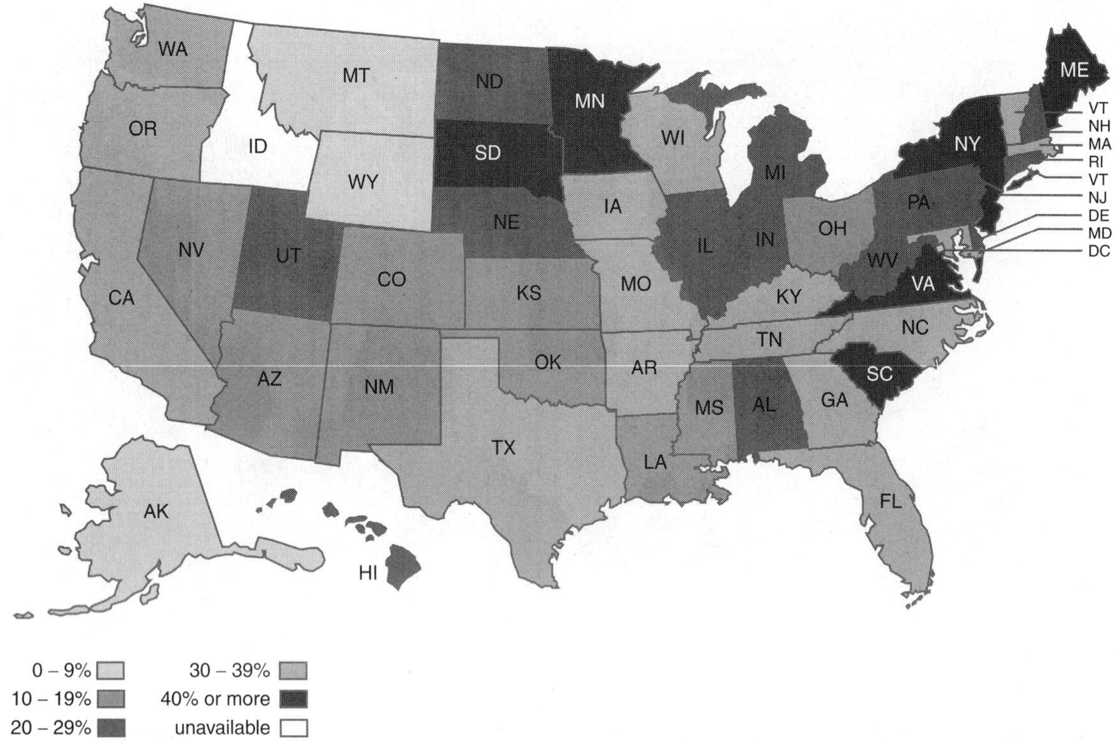

Figure 10D.82 State recycling rates. (From *BioCycle Magazine*, April 1999, www.epa.gov.)

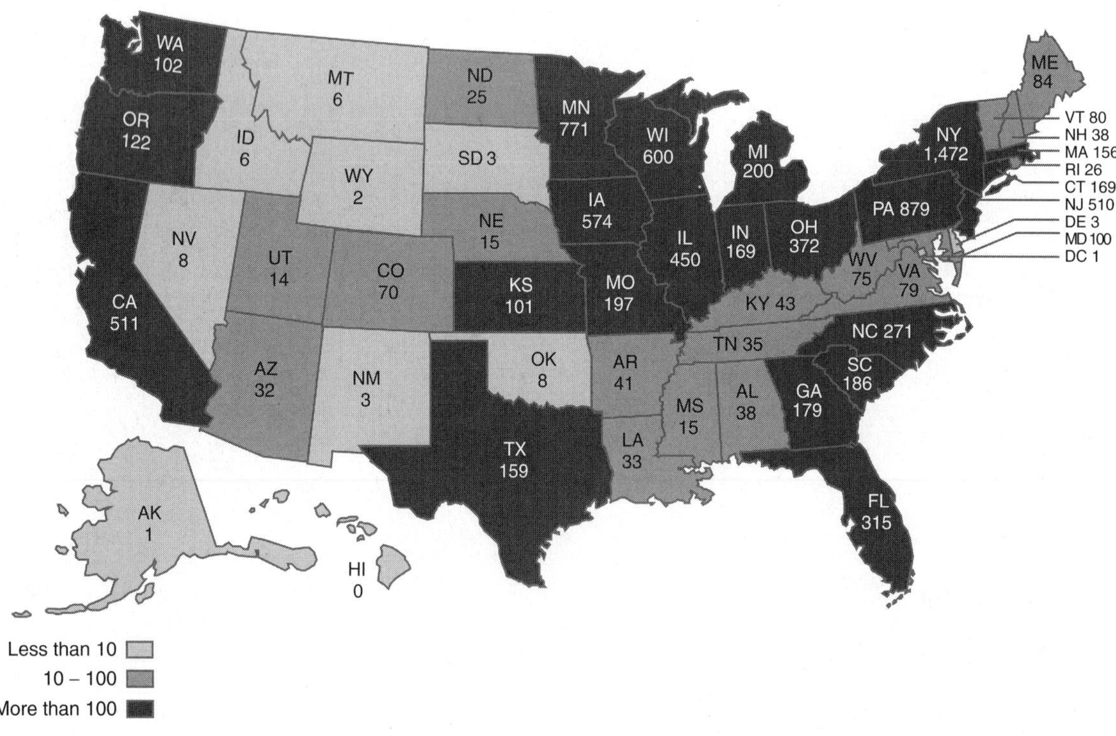

Figure 10D.83 Number of curbside recycling programs. (From *BioCycle Magazine*, April 1999, www.epa.gov.)

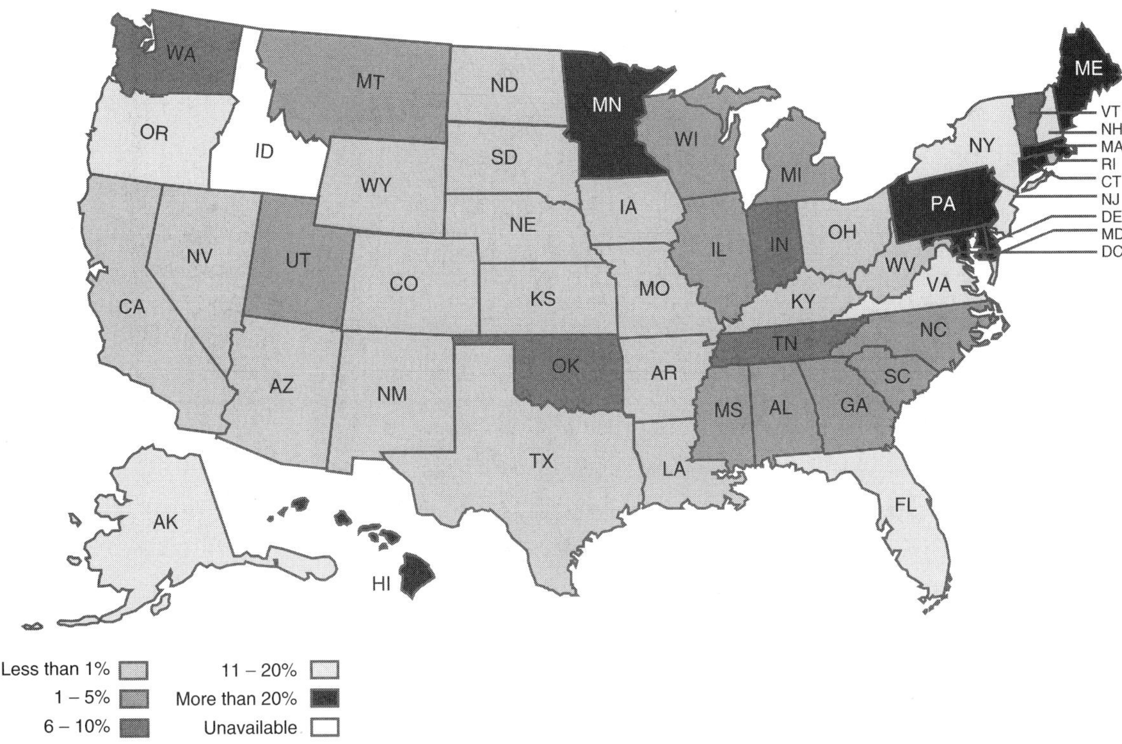

Figure 10D.84 State combustion rates. (From *BioCycle Magazine*, April 1999, www.epa.gov.)

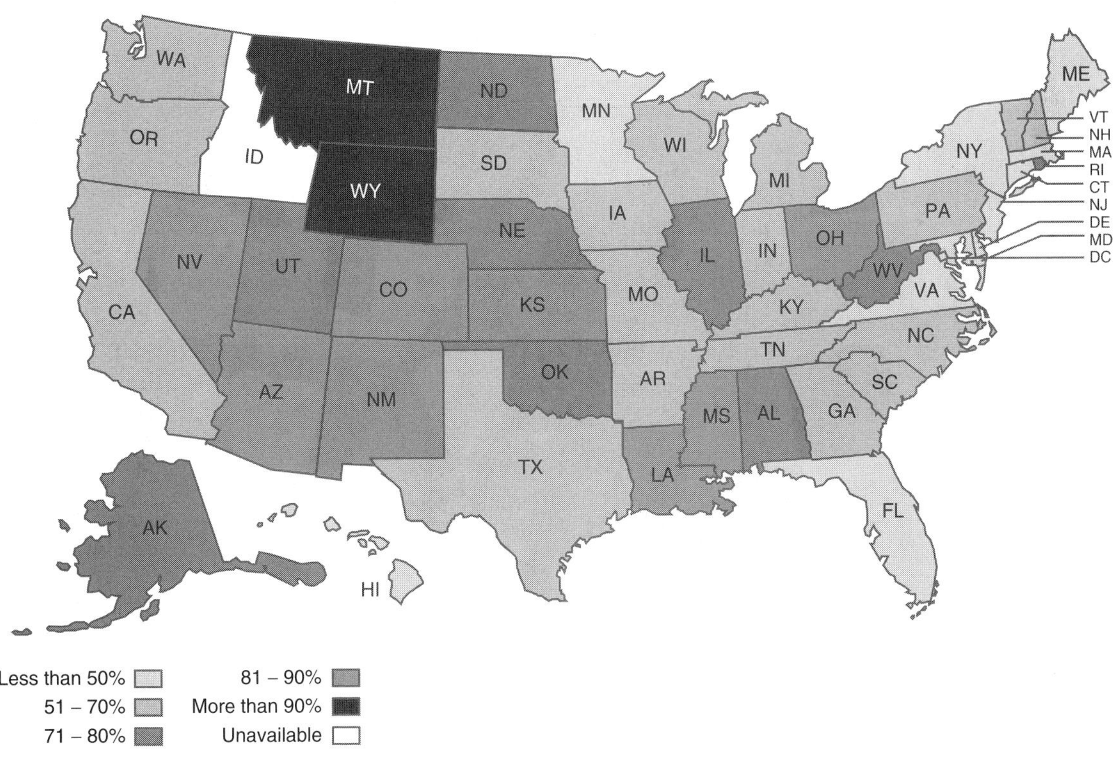

Figure 10D.85 State land disposal rates. (From *BioCycle Magazine*, April 1999, www.epa.gov.)

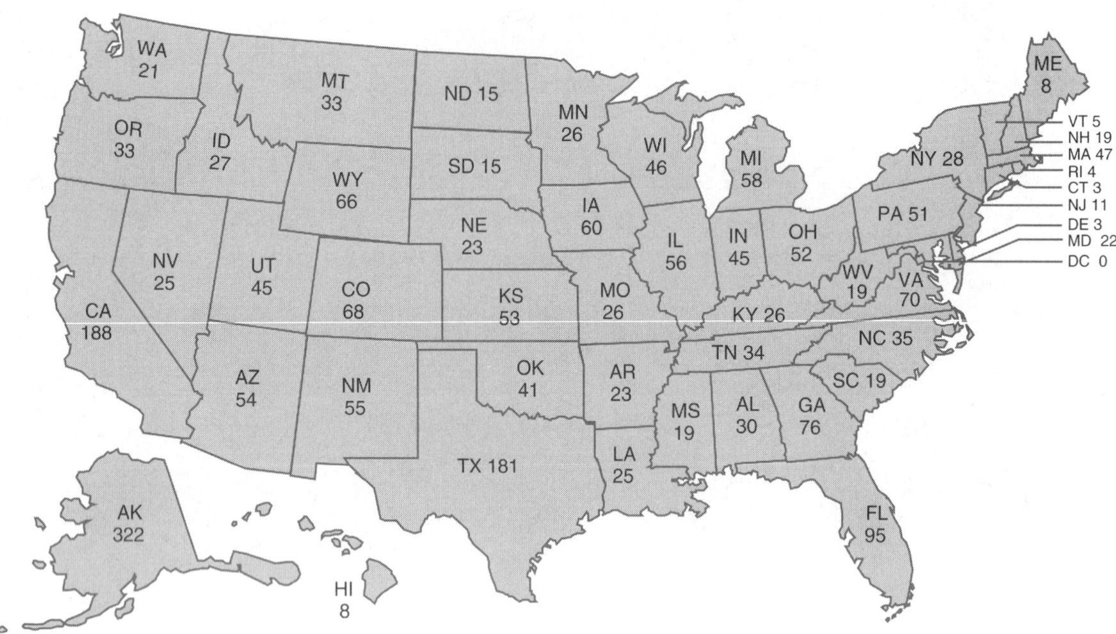

Figure 10D.86 Number of landfills. (From *BioCycle Magazine*, April 1999, www.epa.gov.)

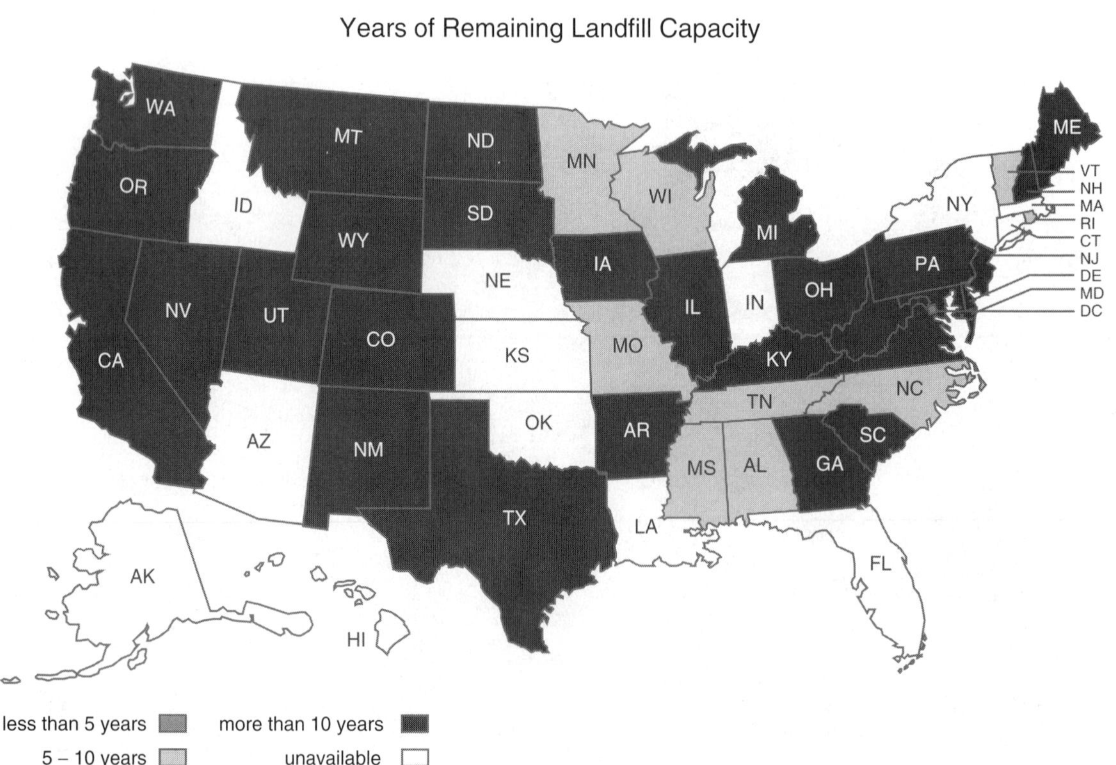

Figure 10D.87 Years of remaining landfill capacity. (From *BioCycle Magazine*, April 1999, www.epa.gov.)

Table 10D.105 Composition of Municipal Solid Waste Discards by Organic and Inorganic Fractions in the United States, 1960–2000

Year	Organics	Inorganics
1960	77.8	22.3
1965	78.3	21.7
1970	75.2	24.8
1975	75.5	24.5
1980	77.1	22.9
1981	77.5	22.5
1982	77.8	22.2
1983	78.7	21.3
1984	79.6	20.4
1985	80.4	19.6
1986	80.8	19.2
1990[a]	80.8	19.2
1995[a]	81.7	18.3
2000[a]	82.5	17.5

Note: In percent of total net discards; discards after materials recovery has taken place, and before energy recovery.

[a] Estimate.

Source: From U.S. Environmental Protection Agency, 1988, *Characterization of Municipal Solid Waste in the United States 1960 to 2000 (Update 1988)*, PB88-232780. Based on study by Franklin Associates, Ltd.

Table 10D.106 Composition of Municipal Solid Waste in the United States, 1960–2000

Materials	1960	1965	1970	1975	1980	1981	1982	1983	1984	1985	1986	1990	1995	2000
Paper and paperboard	24.5	32.2	36.5	34.4	42.0	43.6	41.4	45.8	49.4	48.7	50.1	54.9	60.2	66.0
Glass	6.4	8.5	12.5	13.2	14.2	14.3	11.8	13.3	12.8	12.2	11.8	12.3	12.2	12.0
Metals														
Ferrous	9.9	10.0	12.4	12.0	11.2	11.1	11.0	11.1	11.0	10.4	10.6	11.1	11.3	11.3
Aluminum	0.4	0.5	0.8	1.0	1.4	1.4	1.3	1.5	1.5	1.6	1.7	2.0	2.4	2.7
Other nonferrous	0.2	0.2	0.3	0.3	0.4	0.4	0.3	0.3	0.3	0.3	0.3	0.3	0.3	0.4
Plastics	0.4	1.4	3.0	4.4	7.6	7.8	8.4	9.1	9.6	9.7	10.3	11.8	13.7	15.6
Rubber and leather	1.7	2.2	3.0	3.7	4.1	4.1	3.8	3.4	3.3	3.4	3.9	3.5	3.6	3.8
Textiles	1.7	1.9	2.0	2.2	2.6	3.4	2.8	2.8	2.8	2.8	2.8	3.0	3.1	3.3
Wood	3.0	3.5	4.0	4.4	4.9	4.4	5.0	5.2	5.1	5.4	5.8	5.3	5.7	6.1
Other	0.0	0.0	0.1	0.1	0.1	0.1	0.1	0.1	0.1	0.1	0.1	0.1	0.1	0.1
Total nonfood product wastes	48.2	60.5	74.7	75.6	88.6	90.5	87.8	92.6	95.9	94.5	97.4	104.2	112.5	121.3
Food wastes	12.2	12.4	12.8	13.4	11.9	12.1	12.0	12.0	12.2	12.3	12.5	12.5	12.4	12.3
Yard wastes	20.0	21.6	23.2	25.2	26.5	26.7	27.0	27.5	27.8	28.0	28.3	29.5	31.0	32.0
Miscellaneous inorganic wastes	1.3	1.6	1.8	2.0	2.2	2.3	2.4	2.4	2.4	2.5	2.6	2.8	3.0	3.2
Total wastes discarded[a]	81.7	96.1	112.5	116.2	129.2	131.6	129.1	134.5	138.3	137.3	140.8	149.0	158.9	168.8
Energy recovery[b]	0.0	0.2	0.4	0.7	2.7	2.3	3.5	5.0	6.5	7.6	9.6	13.3	22.5	32.0
Net wastes discarded	81.7	95.9	112.1	115.5	126.5	129.3	125.6	129.5	131.8	129.7	131.2	135.7	136.4	136.8

Note: In millions of tons.

[a] Wastes discarded after materials recovery has taken place.

[b] Municipal solid waste consumed for energy recovery. Does not include residues. Details may not add to totals due to rounding.

Source: From U.S. Environmental Protection Agency, 1988, *Characterization of Municipal Solid Waste in the United States 1960 to 2000 (Update 1988)*, PB88-232780. Based on a study prepared by Franklin Associates, Ltd.

Table 10D.107 Summary Data on Solid Waste Facilities in the United States

Percent of uncontrolled sites that are solid waste facilities	
Of 1,389 sites with actual or presumed problems of releases of hazardous substances	18%
Of 550 sites on National Priority List	20%
Two most prevalent effects at problem solid waste sites	
Leachate migration, groundwater pollution: at 89% of sites	
Drinking water contamination: at 49% of sites	
Mean size of problem solid wastes sites	67.4 acres
Median hazard ranking score[a]	
Solid waste sites on the NPL	40.8
All NPL sites	42.2
Estimates for national number of solid waste sites	
Operating sanitary, municipal landfills	14,000
Closed sanitary, municipal landfills	42,000
Operating industrial landfills	75,000
Closed industrial landfills	150,000
Operating surface impoundments	170,000
Closed surface impoundments	170,000
Total	621,000
Estimate of need for future cleanup	
Low: 5% landfills, 1% impoundments likely to release toxic substances	17,400
High: 10% landfills, 2% impoundments likely to release toxic substances	34,800
Conservative figure used for cleanup by superfund	5,000

[a] 28.5 required for placement on National Priorities List; current highest site score is 75.6.

Source: From Office of Technology Assessment, 1985.

Table 10D.108 Solid Waste Disposal by Selected Industries in the United States, 1975–1983

									1983		
Industry Group	**1975**	**1976**	**1977**	**1978**	**1979**	**1980**	**1981**	**1982**	**Total**	**Hazardous Waste[a]**	**Non-hazardous Waste**
All industries[b]	139.1	156.8	160.0	160.8	163.7	149.9	145.8	99.4	89.0	8.0	81.0
Food	12.6	15.0	13.1	13.4	14.0	14.4	13.2	9.8	9.6	0.2	9.4
Lumber and wood	8.1	9.3	6.3	6.7	6.5	5.9	6.4	3.7	4.0	(Z)	4.0
Paper	9.1	10.1	10.6	10.9	13.3	12.3	11.3	11.5	13.7	0.1	13.5
Chemicals	38.7	50.3	55.7	48.8	45.4	43.4	43.7	36.1	18.8	3.6	15.2
Petroleum	2.0	2.6	2.9	3.6	3.1	4.9	4.7	4.4	3.6	1.5	2.1
Stone, clay, glass	11.3	11.1	12.6	12.7	14.1	13.3	12.1	5.8	6.2	0.2	5.9
Primary metal	42.7	42.4	41.7	46.1	47.8	37.5	36.0	16.7	17.7	1.0	16.7
Fabricated metals	1.9	2.1	2.0	2.0	2.0	1.9	1.8	1.4	2.0	0.2	1.7
Machinery exc. electrical	2.7	3.1	3.6	3.4	3.5	3.0	2.8	1.6	1.8	0.2	1.7
Electric equipment	1.5	1.5	1.5	1.8	2.3	2.1	1.7	1.3	1.7	0.3	1.4
Transportation equipment	3.8	4.3	4.7	5.2	4.3	4.2	4.0	2.9	3.0	0.3	2.7

Note: In millions of short tons. Excludes recovered materials. Data included both wet and dry weight figures. Excludes apparel and other textiles, and, beginning 1978, establishments with less than 20 employees. Z, Less than 50,000 short tons.

[a] Covers waste, which because of its quantity, concentration, or physical, chemical, or infectious characteristics, may cause, or significantly contribute to an increase in serious irreversible, or incapacitating reversible illness; or pose a substantial present or potential hazard to human health or the environment when improperly treated, stored, transported, or disposed of or managed. See Resource Conservation and Recovery Act 1976, Public Law 94-580, for listing of hazardous wastes.

[b] Includes industries not shown separately.

Source: From U.S. Department of Commerce, Statistical Abstract of the United States 1987.

Table 10D.109 Future Use of Containment Technologies for Cleanup of Hazardous Waste Sites in the United States

Technique	Applicability	Effectiveness	Confidence	Capital Cost	Cap/O&M	Projected Level of Use
Barriers						
Slurry wall	2	1	2	2	1	Extensive
Grout curtain	2–3	1	2	2–3	1	Limited
Vibrating beam	2	1	2–3	2–3	1	Moderate
Sheet pile	3	1–2	2	2–3	1	Nil-Limited
Block displacement	3	1	4	3	1	Nil
Hydraulic controls (wells)	2	1,3	1	1	3	Extensive
Subsurface drains	2	1	2	1	2	Moderate
Runon/runoff controls	1	3	1	1	2	Extensive
Surface seals and caps	1	2,3	2	1	1	Extensive
Solidification, etc.	2	1,3	3–4	2	1	Moderate-Limited

Key:
Applicability: (1) Very broadly applicable; little or no site dependency. (2) Broadly applicable; some sites unfavorable. (3) Limited to sites of specific characteristics. *Effectiveness*: (1) Can produce "leak-tight" containment. (2) Can reduce migration—some leakage likely. (3) Used as supporting technique in conjunction with other elements. *Confidence*: (1) Well proven—long-term effectiveness—high. (2) Well proven—long-term effectiveness—unknown. (3) Limited experience; used in other applications. (4) Developmental; little data. *Capital cost for function provided*: (1) Low. (2) Normal. (3) High. *Capital to operation and maintenance (O&M) cost ratio*: (1) Capital higher than O&M. (2) Capital about same as O&M. (3) Capital lower than O&M.

Source: From U.S. Congress, Office of Technology Assessment, 1985, *Superfund Strategy*.

Original source: Little, A.D., *Evaluation of Available Cleanup Technologies for Uncontrolled Sites*, contractor report prepared for the Office of Technology Assessment, Nov. 15, 1984.

Table 10D.110 Future Use of Treatment Technologies for Cleanup of Hazardous Waste Sites in the United States

Technique	Applicability	Effectiveness	Confidence	Capital Cost	Cap/O&M	Secondary Disposal	Projected Level Use of
Biological treatment	Or, 1–2	2	1	1	1–2	3	Moderate
Chemical treatment							
Neutralization/precipitation	In, 1	1	1	1	2	4	Moderate–Extensive
Wet air oxidation	Or, 2	2	2	3	1–2	1	Limited
Chlorination	In, 3	1	2	2	2	1	Limited
Ozonation	Or, 3	2	3	3	2–3	2	Nil
Reduction (Cr)	In, 3	1	2	2	2	3	Limited
Physical treatment							
Carbon adsorption	Or, In, 1	1	1	2	2–3	2–3	Moderate–Extensive
Sedimentation/filtration	Or, In, 1	1	1	1	2–3	4	Moderate–Extensive
Stripping	Or, 2	1	1	1	2	4	Moderate
Flotation	Or, 2	2	1	1	1	4	Limited
Ion exchange	In, 3	1–3	3	3	3	4	Nil
Reverse osmosis	Or, In, 3	1–2	3	3	3	4	Nil
Gas stream controls							
Thermal oxidation	Or, 1	1	1	3	3	1	Limited–Moderate
Carbon adsorption	Or, 1	1	1	3	2–3	2–3	Limited–Moderate

(Continued)

Table 10D.110 (Continued)

Technique	Applicability	Effectiveness	Confidence	Capital Cost	Cap/O&M	Secondary Disposal	Projected Level Use of
Incineration							
Onsite	Or, 1	1	2	3	1	3[a]	Limited
Offsite	Or, 1	1	1	3	NA	3[b]	Moderate
In situ biodegradation	Or, 3	2	3	2	3	1	Limited

Key:
Class: Or, Organic compounds. In, Inorganic compounds. *Range*: (1) Broadly applicable to compounds in indicated class. (2) Moderated applicable: depends on waste composition concentration. (3) Limited to special situations. *Effectiveness*: (1) Highest levels available. (2) Output may need further treatment; may have pockets untreated (in situ). *Confidence*: (1) Well proven—easily transferable to site cleanup. (2) Well proven—but not in clean-up settings. (3) Limited experience. (4) Developmental; little data. *Capital cost for function provided*: (1) Low. (2) Normal. (3) High. *Capital to operations and maintenance (O&M) cost basis*: (1) Capital higher than O&M. (2) Capital about the same. (3) Capital lower than O&M. *Secondary treatment or disposal*: (1) None. (2) Minor. (3) Major, but does not require hazardous waste techniques. (4) Basically a separation process; must be used with subsequent hazardous waste treatment or secure disposal step.

[a] Must dispose solid residues.
[b] Depends on reactive material used.

Source: From U.S. Congress, Office of Technology Assessment, 1985, Superfund Strategy.

Original Source: Little, A.D., *Evaluation of Available Cleanup Technologies for Uncontrolled Sites*, contractor report prepared for the Office of Technology Assessment, Nov. 15, 1984.

Table 10D.111 Generation, Materials Recovery, Composting, and Discards of Municipal Solid Waste, 1960–2001

	Millions of Tons							
	1960	1970	1980	1990	1995	1999	2000	2001
Generation	88.1	121.1	151.6	205.2	213.7	231.4	232.0	229.2
Recovery for recycling	5.6	8.0	14.5	29.0	46.2	50.8	51.2	51.4
Recovery for composting[a]	Neg.	Neg.	Neg.	4.2	9.6	14.7	16.5	16.6
Total materials recovery	5.6	8.0	14.5	33.2	55.8	65.5	67.7	68.0
Discarded after recovery	82.5	113.0	137.1	172.0	158.0	165.9	164.3	161.2

Note: In millions of tons. Details may not add to totals due to rounding.

[a] Composting of yard trimmings, food scraps and other MSW organic material. Does not include backyard composting.

Source: From Franklin Associates, Ltd., www.epa.gov/epaoswer.

Table 10D.112 Generation, Materials Recovery, Composting, and Discard of Municipal Solid Waste, 1960–2001

	Percent of Total Generation							
	1960	1970	1980	1990	1995	1999	2000	2001
Generation (%)	100.0	100.0	100.0	100.0	100.0	100.0	100.0	100.0
Recovery for recycling (%)	6.4	6.6	9.6	14.2	21.6	22.0	22.1	22.4
Recovery for composting (%)[a]	Neg.	Neg.	Neg.	2.0	4.5	6.4	7.1	7.2
Total materials recovery (%)	6.4	6.6	9.6	16.2	26.1	28.4	29.2	29.7
Discarded after recovery (%)	93.6	93.4	90.4	83.8	73.9	71.6	70.8	70.3

Note: In percent of total generation. Details may not add to totals due to rounding.

[a] Composting of yard trimmings, food scraps and other MSW organic material. Does not include backyard composting.

Source: From Franklin Associates, Ltd., www.epa.gov/epaoswer.

Table 10D.113 Generation and Recovery of Materials in MSW, 2001

	Weight Generated	Weight Recovered	Recovery as a Percent of Generation (%)
Paper and paperboard	81.9	36.7	44.9
Glass	12.6	2.4	19.1
Metals			
Steel	13.5	4.6	33.8
Aluminum	3.2	0.8	24.5
Other nonferrous metals[a]	1.4	0.9	64.8
Total metals	18.1	6.3	34.5
Plastics	25.4	1.4	5.5
Rubber and leather	6.5	1.1	17.4
Textiles	9.8	1.4	14.6
Wood	13.2	1.3	9.5
Other materials	4.2	0.9	20.7
Total materials in products	171.5	51.4	30.0
Other wastes			
Food, other[b]	26.2	0.7	2.8
Yard trimmings	28.0	15.8	56.5
Miscellaneous inorganic wastes	3.5	Neg.	Neg.
Total other wastes	57.7	16.6	28.7
Total municipal solid waste	229.2	68.0	29.7

Note: In millions of tons and percent of generation of each material. Includes waste from residential, commercial, and institutional sources. Neg., less than 5,000 tons or 0.05 percent.

[a] Includes lead from lead–acid batteries.
[b] Includes recovery of other MSW organics for composting.

Source: From Franklin Associates, Ltd., www.epa.gov/epaoswer.

Table 10D.114 Generation and Recovery by Products in MSW by Material, 2001

	Weight Generated	Weight Recovered	Recovery as a Percent of Generation (%)
Durable goods			
Steel	10.9	3.0	27.8
Aluminum	1.0	Neg.	Neg.
Other nonferrous metals[a]	1.4	0.9	64.8
Total metals	13.3	4.0	29.6
Glass	1.7	Neg.	Neg.
Plastics	8.0	0.3	3.9
Rubber and leather	5.6	1.1	20.1
Wood	5.0	Neg.	Neg.
Textiles	2.9	0.3	118.8
Other materials	1.2	0.9	73.7
Total durable goods	37.6	3.6	17.5
Nondurable goods			
Paper and paperboard	43.5	15.6	35.9
Plastics	6.1	Neg.	Neg.
Rubber and leather	0.9	Neg.	Neg.
Textiles	6.7	1.1	16.1
Other materials	3.2	Neg.	Neg.
Total nondurable goods	60.4	16.7	27.7
Containers and packaging			
Steel	2.6	1.5	58.8
Aluminum	2.0	0.8	40.0
Total metals	4.6	2.3	50.8

(Continued)

Table 10D.114 (Continued)

	Weight Generated	Weight Recovered	Recovery as a Percent of Generation (%)
Glass	10.9	2.4	22.0
Paper and paperboard	38.4	21.1	55.0
Plastics	11.2	1.1	9.6
Wood	8.2	1.3	15.2
Other materials	0.2	Neg.	Neg.
Total containers and packaging	73.5	28.1	38.3
Other wastes			
Food, other[b]	26.2	0.7	2.8
Yard trimmings	28.0	15.8	56.5
Miscellaneous inorganic wastes	3.5	Neg.	Neg.
Total other wastes	57.7	16.5	28.7
Total municipal solid waste	229.2	68.0	29.7

Note: In millions of tons and percent of generation of each product. Includes waste from residential, commercial, and institutional sources. Details may not add to totals due to rounding. Neg., less than 5,000 tons or 0.05 percent.

[a] Includes lead from lead–acid batteries.
[b] Includes recovery of other MSW organics for composting.

Source: From Franklin Associates, Ltd., www.epa.gov/epaoswer.

Table 10D.115 Projections of Materials Generated in the Municipal Waste Stream: 2000 and 2005

	Million Tons		% Of Total	
Materials	**2000**	**2005**	**2000**	**2005**
Paper and paperboard	87.7	94.8	39.3	39.6
Glass	11.9	11.2	5.3	4.7
Metals	17.6	18.7	7.9	7.8
Plastics	23.4	26.7	10.5	11.2
Wood	14.0	15.8	6.3	6.6
Others	19.7	22.2	8.8	9.3
Total materials in products	174.3	189.4	78.1	79.1
Other wastes				
Food wastes	22.5	23.5	10.1	9.8
Yard trimmings	23.0	23.0	10.3	9.6
Miscellaneous inorganic wastes	3.4	3.6	1.5	1.5
Total other wastes	48.9	50.1	21.9	20.9
Total MSW generated	223.2	239.5	100.0	100.0

Note: In thousands of tons and percent of total generation. Generation before materials recovery or combustion. Details may not add to totals due to rounding.

Source: From Franklin Associates, www.epa.gov/epaoswer.

Table 10D.116 Median Concentrations of Substances Found in MSW Landfill Leachate, in Comparison with Existing Exposure Standards

		Exposure Standards	
Substance[a]	Median Concentration (ppm)	Type[f]	Value (ppm)
Inorganics			
Animony (11)	**4.52**	T	0.01
Arsenic (72)	0.042	N	0.05
Barium (60)	0.853	N	1.0
Beryllium (6)	0.006	T	0.2
Cadmium (46)	**0.022**	N	0.01
Chromium (total) (97)	**0.175**	N	0.05
Copper (68)	**0.168**		0.012
		W	0.018
Cyanide (21)	0.063		0.7
Iron (120)	221	W	1,000
Lead (73)	**0.162**	N	0.05
Manganese (103)	**9.59**	W	0.05
Mercury (19)	0.002	N	0.002
Nickel (98)	**0.326**	T	0.07
Nitrate (38)	1.88	W	10
Selenium (18)	**0.012**	N	0.01
Silver (19)	0.021	N	0.05
Thallium (11)	**0.175**	W	0.04
Zinc (114)	**8.32**	W	0.110
Organics			
Acrolein (1)		W	21
Benzene (35)		T	5
Bromomethane (1)	**170**	T	10
Carbon tetrachloride (2)	**202**	T	5
Chlorobenzene (12)	128	T	1,000
Chloroform (8)	**195**	C	5.7
Bis(chloromethyl) ether (1)	**250**	C	0.0037
p-Cresol (10)	**2,394**	T	2,000
2,4-D (7)	**129**	T	100
4,4-DDT (16)	**0.103**	C	0.1
Di-*n*-butyl phthalate (5)	70.2	T	4,000
1,2-Dichlorobenzene (8)	11.8	W	763
1,4-Dichlorobenzene (12)	13.2		
Dichlorodifluoromethane (6)	237	T	7,000
1,1-Dichloroethane (34)	**1,715**		7
		N	0.58
1,2-Dichloroethane (6)	**1,841**	T	5
1,2-Dichloropropane (12)	66.7	W	5,700
1,3-Dichloropropane (2)		C	0.19
Diethyl phthalate (27)	118		30,000
2,4-Dimethyl phenol (2)		W	2,120
Dimethyl phthalate (2)	42.5	W	313,000
Endrin (3)	**16.8**	T	0.2
Ethyl benzene (41)	274	W	1,400
Bis(2-ethylhexyl) phthalate (10)	184	T	
Isophorone (19)	1,168	W	5,200
Lindane (2)	0.020	T	4
Methylene chloride (68)	**5,352**	C	4.7
Methyl ethyl ketone (24)	**4,151**	W	2,000
Naphthalene (23)	32.4	W	620
Nitrobenzene (3)	**54.7**	T	20
4-Nitrophenol (1)	17	W	150
Pentachlorophenol (3)	173	T	1,000
Phenol (45)	**2,456**	T	1,000
1,1,2,2-Tetrachloroethane (1)	**210**	C	1.75

(Continued)

Table 10D.116 (Continued)

Substance[a]	Median Concentration (ppm)	Exposure Standards	
		Type[f]	Value (ppm)
Tetrachloroethylene (18)	**132**	C	6.9
Toluene (69)	1,016	T	10,000
Toxaphene (1)	1	N	5
1,1,1-Trichloroethane (20)	**887**	N	200
		T	3,000
1,1,2-Trichloroethane (4)	**378**	C	6.1
Trichloroethylene (28)	**187**	N	5
		T	3.2
Trichlorofluoromethane (10)	56.1	T	10,000
1,2,3-Trichloropropane (1)	**230**	T	20
Vinyl chloride (10)	**36.1**	N	2

Note: Types of exposure standards: C, EPA Human Health Criteria, based on carcenogenicity; N, National Interim Primary or Secondary Drinking Water Standard; T, EPA Human Health Criteria, based on systemic toxicity; W, Water-Quality Criteria.

[a] Number of samples in parentheses.

Source: From After U.S. Environmental Protection Agency, *Office of Solid Waste, Summary of Data on Municipal Solid Waste Landfill Leachate Characteristics, Criteria for Municipal Solid Waste Landfills (40 CFR Pert 258)*, EPA/530-SW-88-038 (Washington, DC: July 1988), princeton.edu/cgi-bin.

Table 10D.117 NPL Status (June 2004)

Number of Sites on Final NPL	
Total	1,245
General superfund section	1,087
Federal facilities section	158
Number of sites remaining on proposed NPL	
Total	56
General superfund section	50
Federal facilities section	6
Total number of final and proposed sites	1,301
Number of sites on the construction completion list	899
Number of sites deleted from final NPL	282
Number of sites with partial deletions	45 partial deletions at 37 sites

Note: These numbers reflect the status of sites as of June 29, 2004. Site status changes occurring after this date may affect these numbers at time of rule publication in the **Federal Register**.

Source: From www.epa.gov/superfund/sites.

Table 10D.118 Number of NPL Site Actions and Milestones by Fiscal Year

	1992	1993	1994	1995	1996	1997	1998	1999	2000	2001	2002	2003	2004	2005
Action														
Sites proposed to the NPL	30	52	36	9	27	20	34	37	40	45	9	14	26	7
Sites finalized on the NPL	0	33	43	31	13	18	17	43	39	29	19	20	11	11
Sites deleted from the NPL	2	12	13	25	34	32	20	23	19	30	17	9	16	6
Milestone														
Partial deletions[a]	—	—	—	—	0	6	7	3	5	4	7	7	7	4
Construction completion	88	68	61	68	64	88	87	85	87	48	42	40	40	7

Note: A fiscal year is October 1 through September 30. Partial deletion totals are not applicable until fiscal year 1996, when the policy was first implemented.

[a] These totals represent the total number of partial deletions by fiscal year and may include multiple partial deletions at a site. Currently, there are 50 partial deletions at 42 sites.

Source: From www.epa.gov/superfund/sites.

Table 10D.119 NPL Site Totals by Status and Milestone as of May 19, 2005

	Nonfederal (General)	Federal	Total Sites
Status			
Proposed sites	58	6	64
Final sites	1,085	159	1,244
Deleted sites	283	13	296
Milestone			
Partial deletions	32	10	42[a]
Construction completions	890	43	933

Note: Sites that have achieved these milestones are included in one of the three NPL status categories.

[a] 50 partial deletions have occurred at these 42 sites.

Source: From www.epa.gov/superfund.

Table 10D.120 Municipal Landfills in the United States and Protectorates

State	Number of Active Municipal Landills
Alabama	28
Alaska	217
Arizona	59
Arkansas	67
California	278
Colorado	72
Connecticut	11
Delaware	3
Florida	67
Georgia	159
Hawaii	10
Idaho	37
Illinois	61
Indiana	32
Iowa	77
Kansas	58
Kentucky	12
Louisiana	29
Maine	27
Maryland	25
Massachusetts	106
Michigan	54
Minnesota	26
Mississippi	14
Missouri	30
Montana	82
Nebraska	21
Nevada	56
New Hampshire	33
New Jersey	14
New Mexico	79
New York	42
North Carolina	114
North Dakota	12
Ohio	63
Oklahoma	94
Oregon	88
Pennsylvania	47
Rhode Island	4
South Carolina	37
South Dakota	13
Tennesee	81
Texas	678
Utah	54

(Continued)

Table 10D.120 (Continued)

State	Number of Active Municipal Landills
Vermont	61
Virginia	152
Washington	25
West Virginia	22
Wisconsin	46
Wyoming	59
Subtotal	3,536
Protectorates	
American Samoa	4
Guam	3
Northern Mariana Islands	3
Puerto Rico	33
U.S. Virgin Islands	2
Total	3,581

Source: From ERG Estimates. 20-Mar-96, www.epa.gov/epaoswer/non-hw.

Table 10D.121 Number and Population Served by Curbside Recyclables Collection Programs, 2001

			Population Served	
Region	Number of Programs	Population (in 1000)	(in 1000)	Percent (%)[a]
Northeast	3,421	53,805	43,981	82
South	1,677	101,833	26,496	26
Midwest	3,572	64,687	25,851	40
West	1,034	62,612	43,038	69
Total	9,704	233,931	139,366	60
Percent of total U.S. population				49

[a] Percent of population served by curbside programs was calculated using population of states reporting data.

Source: From U.S. Census Bureau 2002, *BioCycle* December 2001, www.epa.gov/epaoswer/non-hw.

SECTION 10E AGRICULTURAL ACTIVITIES

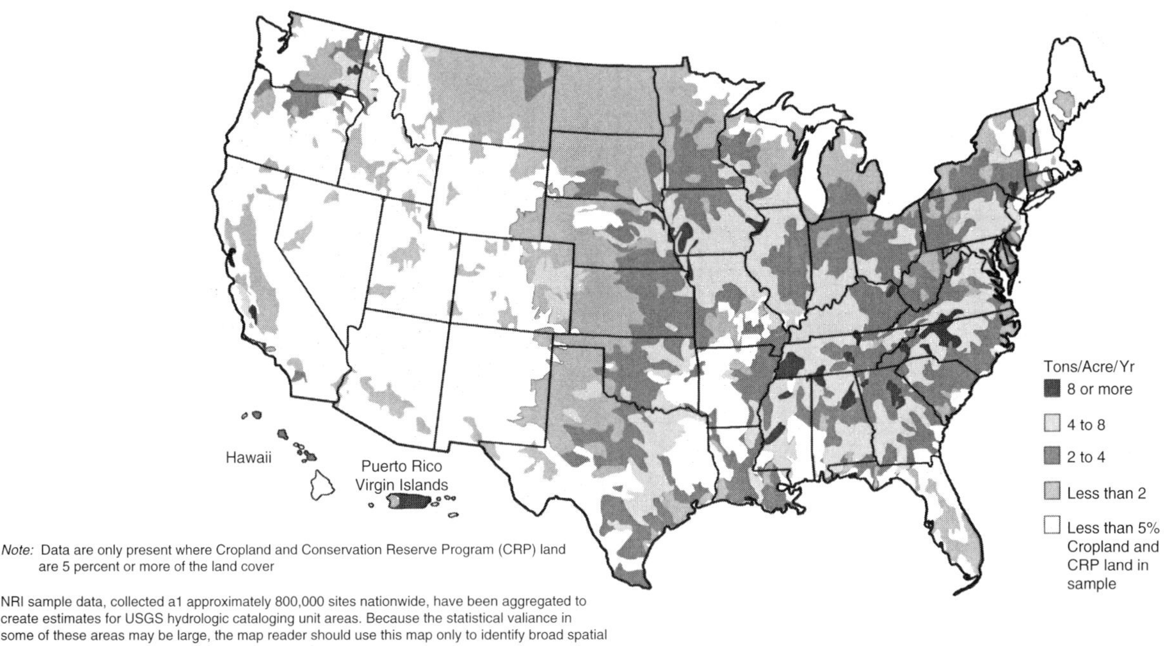

Figure 10E.88 Average annual soil erosion by water on cropland and conservation reserve program land in the United States, 1992. (From United States Department of Agriculture, Natural Resources Conservation Services, 1997, *Water Quality and Agriculture, Status, Conditions, and Trends*, www.nrcs.usda.gov.)

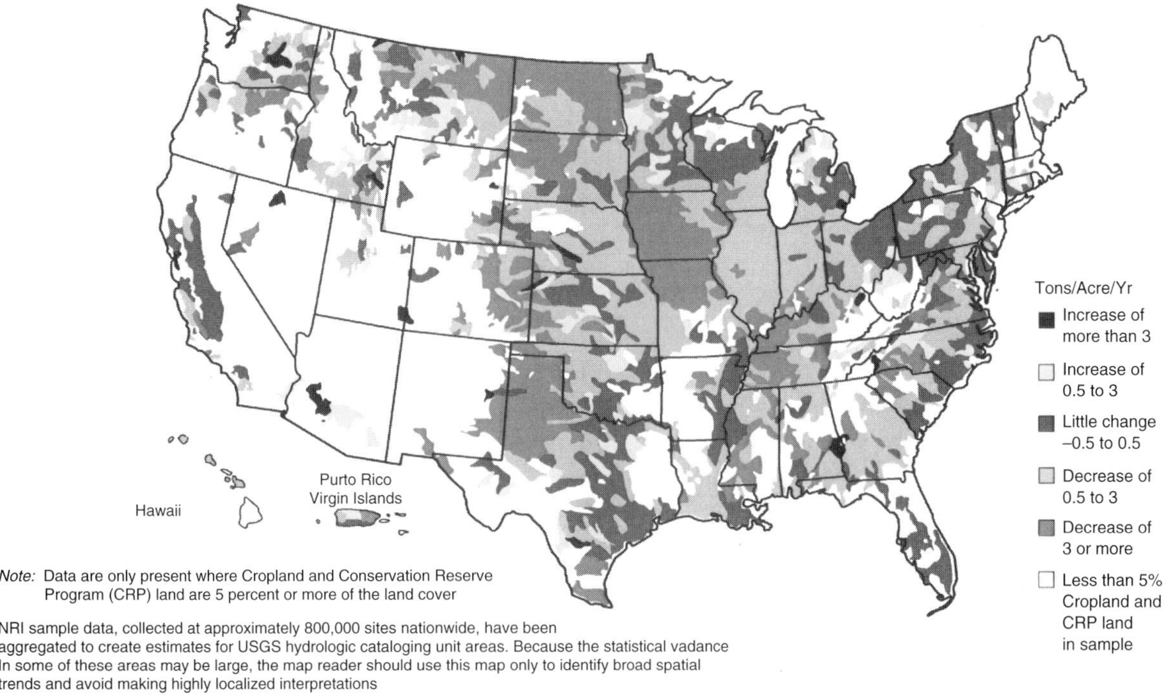

Figure 10E.89 Change in average annual soil erosion by wind and water on cropland and conservation reserve program land in the United States, 1982–1992. (From United States Department of Agriculture, Natural Resources Conservation Services, 1997, *Water Quality and Agriculture, Status, Conditions, and Trends*, www.nrcs.usda.gov.)

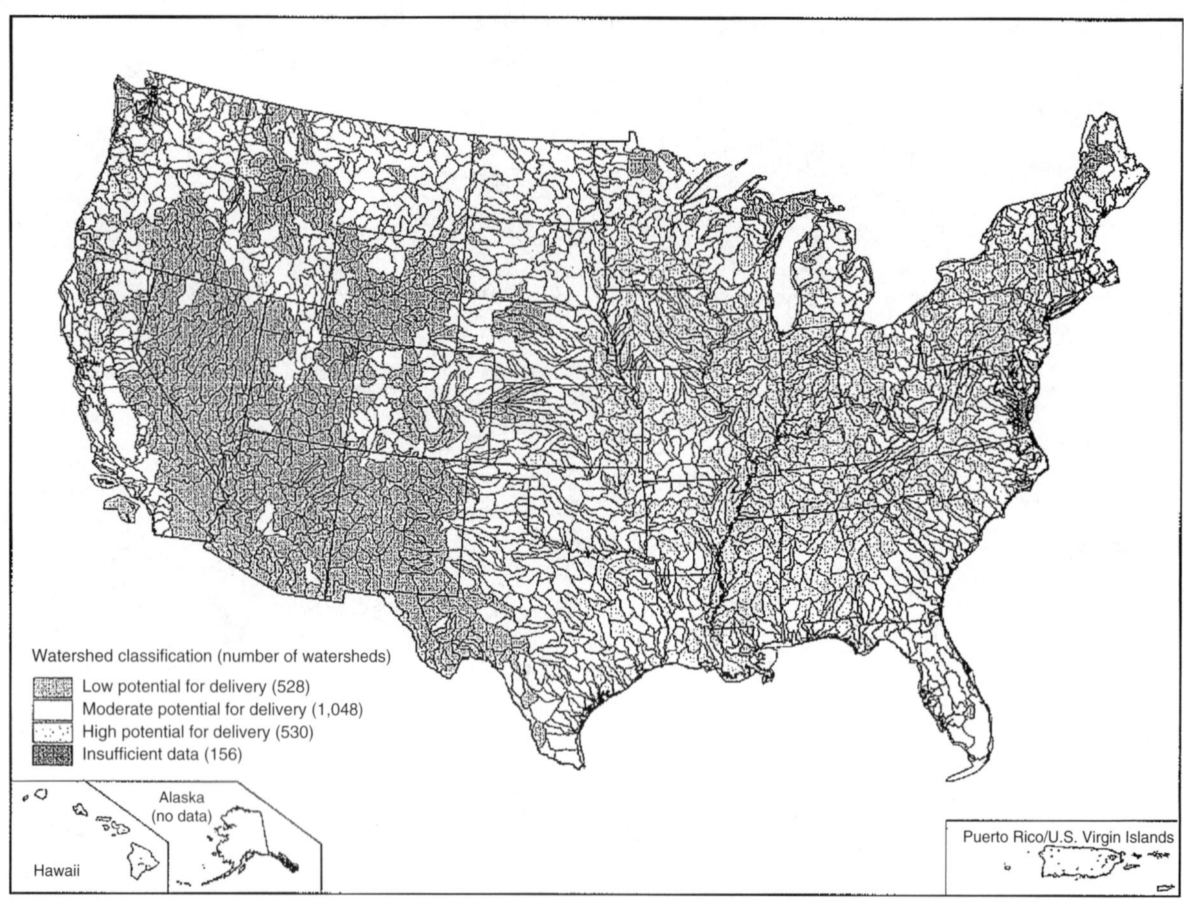

Figure 10E.90 Sediment runoff potential from croplands and pasture lands, 1990–1995. (From United States Environmental Protection Agency, 2003, *EPA's Draft Report on the Environment, 2003*, EPA 600-R-03-050, www.epa.gov in the United States.)

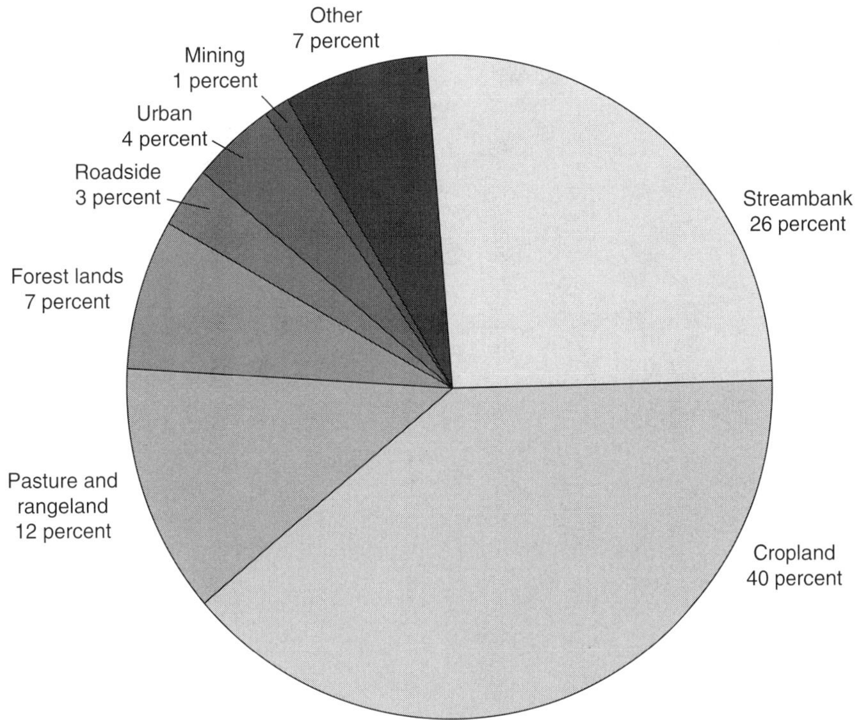

Figure 10E.91 Sources of sediment discharge to surface waters in the United States, 1977. (From United States Department of Agriculture, Soil Conservation Service, 1978, *Environmental Impact Statement: Rural Clean Water Program.*)

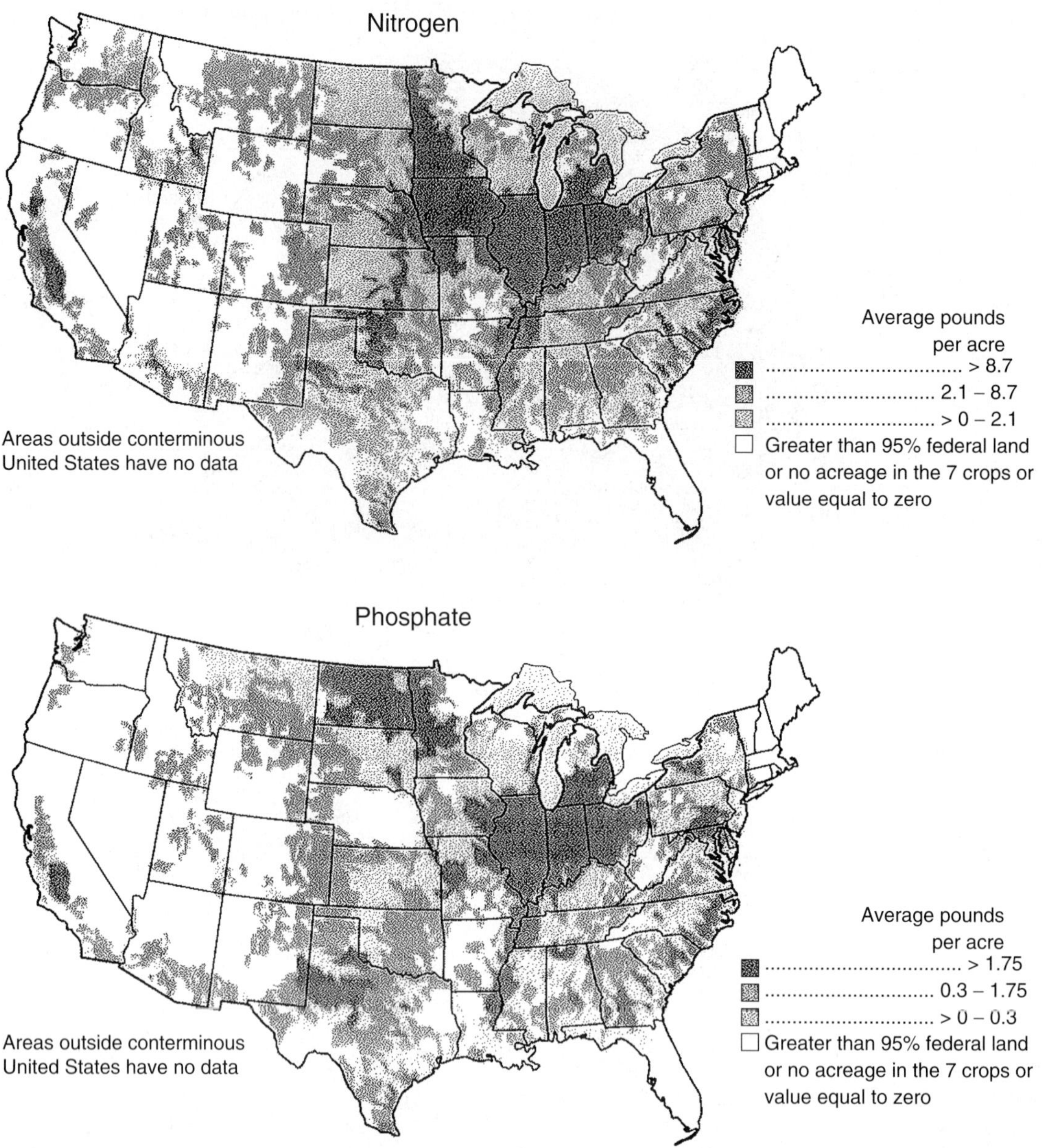

Figure 10E.92 Potential nitrogen and phosphate fertilizer loss from farm fields. (From United States Department of Agriculture, Natural Resources Conservation Services, 1997, *Water Quality and Agriculture, Status, Conditions, and Trends, Working Paper No. 16*, www.nrcs.usda.gov.)

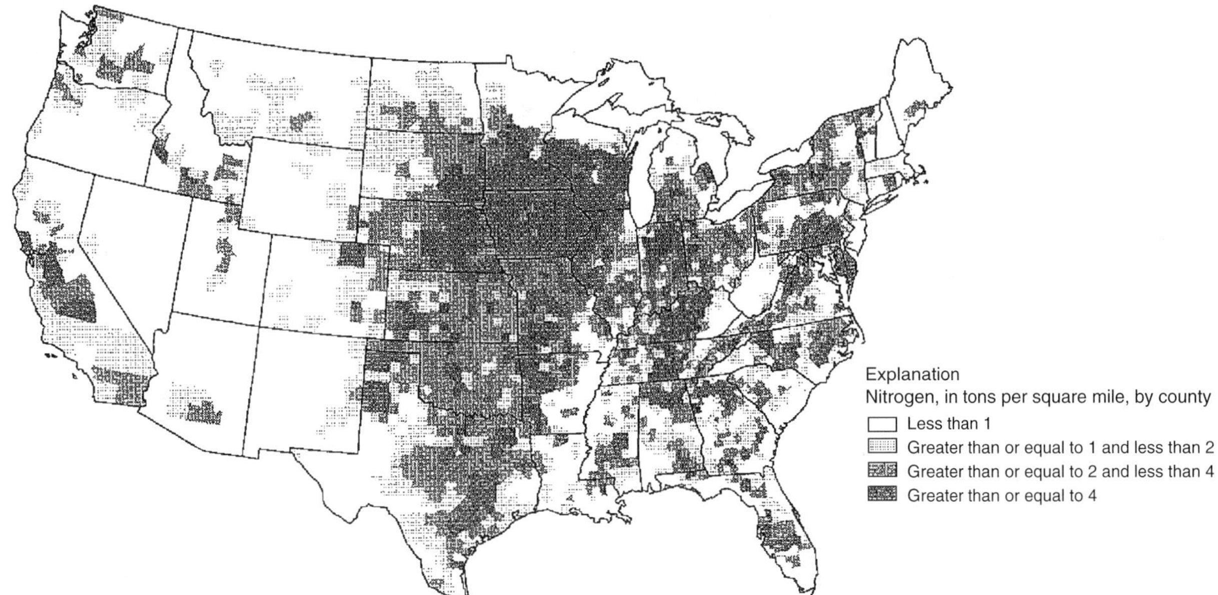

Figure 10E.93 Estimated nonpoint-source inputs of nitrogen applied in animal manure, 1987. (From Nolan, B.T. and Stoner, J.D., Nutrients in groundwaters of the conterminous United States, 1992–1995, *Environmental Science and Technology*, vol. 34, no. 7, 2000, p. 1156–1165, www.usgs.gov.)

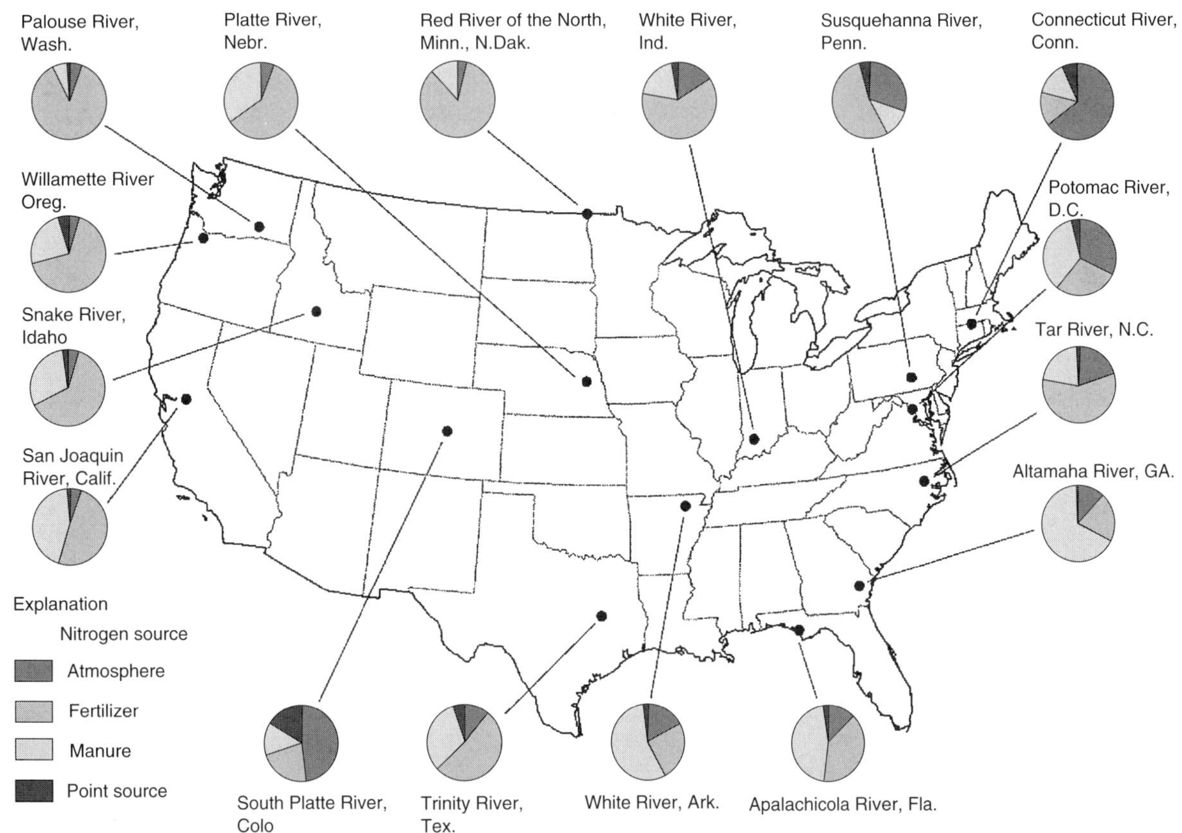

Figure 10E.94 Proportions of nonpoint and point sources of nitrogen in selected United States national water-quality assessment program watersheds. (From Puckett, L.J., 1994, *Nonpoint and Point Sources of Nitrogen in Major Watersheds of the United States, U.S. Geological Survey Water-Resources Investigations Report 94-4001*, www.usgs.gov.)

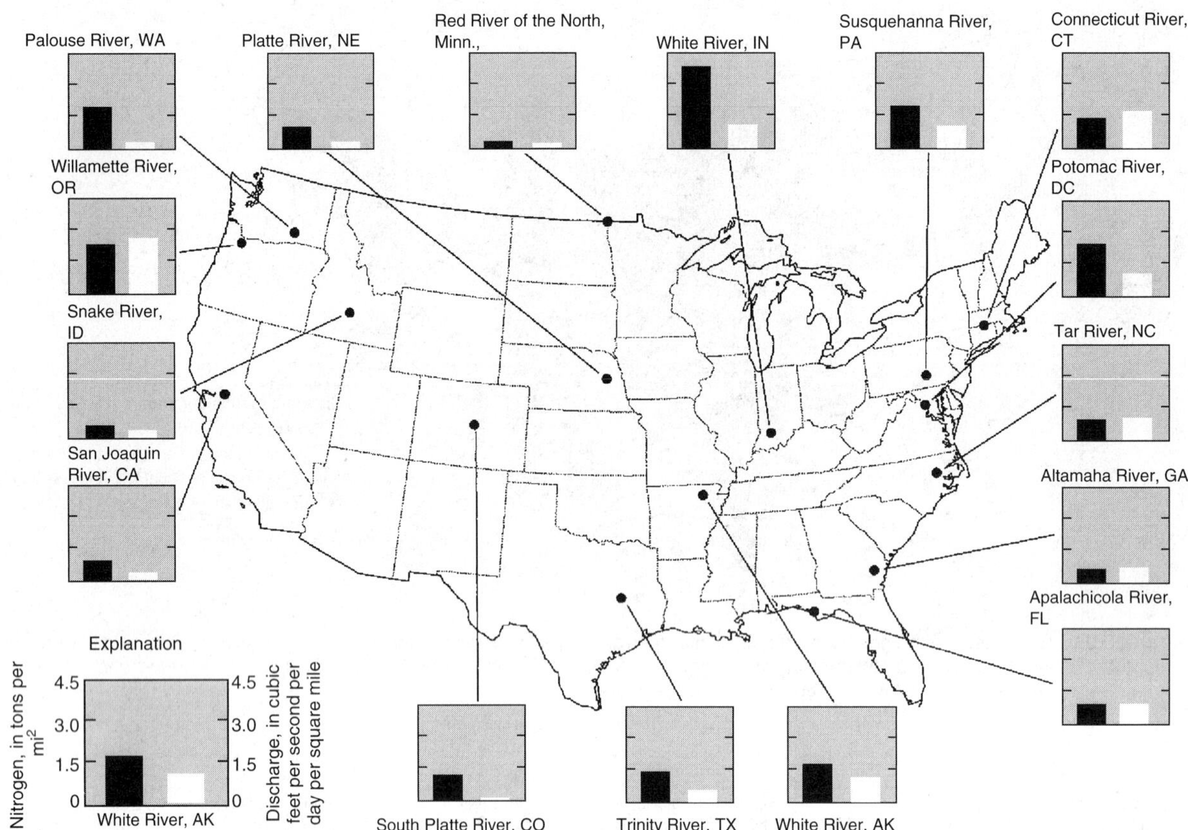

Figure 10E.95 Annual amounts of nitrogen transported in streams and stream discharges in selected United States national water quality assessment program watersheds. (From Puckett, L.J., 1994, *Nonpoint and Point Sources of Nitrogen in Major Watersheds of the United States, U.S. Geological Survey Water-Resources Investigations Report 94-4001*, www.usgs.gov.)

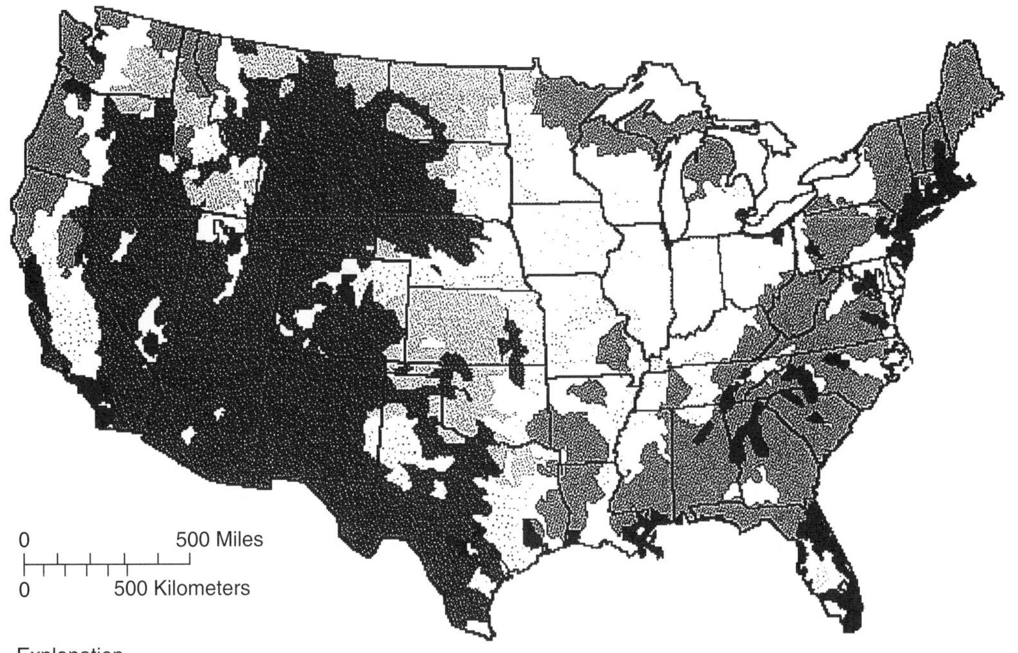

Explanation

Land use shown on map	Nitrate 1980–89		Total phosphorus 1982–89		Suspended sediment 1980–89	
	Yield, in tons per square mile per year	Percentage change per year	Yield, in tons per square mile per year	Percentage change per year	Yield, in tons per square mile per year	Percentage change per year
Agriculture wheat	0.032	*	0.010	–2.8	10	+0.8
Corn and soybeans	0.932	*	0.163	–2.1	100	–1.0
Mixed	0.304	*	0.066	–1.6	79	–0.7
Urban	0.547	+0.2	0.119	–0.6	23	–0.6
Forest	0.255	*	0.063	–0.8	31	–0.3
Range	0.031	*	0.017	–1.9	33	–0.2

* Between –0.1 and +0.1

Figure 10E.96 Yield and percentage change in yield of nitrate, total phosphorous, and suspended sediment in hydrologic cataloging units in the conterminous United States. That are classified as having agricultural (wheat, corn and soybeans, and mixed), urban, forest, and range land use, 1980–1989. (From Smith, R.A., Alexander, R.B., and Lanfear, K.J., 1993, Stream water quality in the conterminous United States—status and trends of selected indicators during the 1980s in national water summary 1990–1991—stream water quality, *U.S. Geological Survey Water-Supply Paper 2400*, www.usgs.gov.)

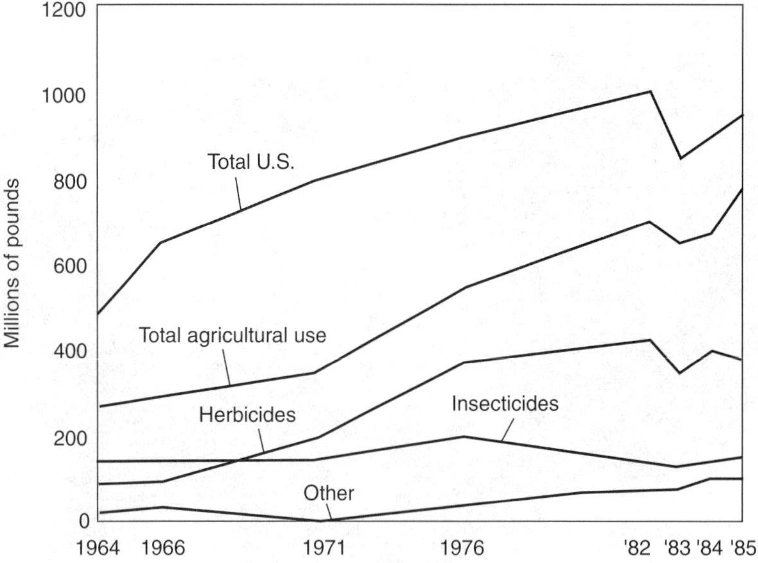

Figure 10E.97 Use of pesticides in the United States, 1964–1984. (From U.S. Environmental Protection Agency, Office of Pesticide and Toxic Substances, 1987, *Agricultural Chemicals in Groundwater: Proposed Pesticide Strategy.*)

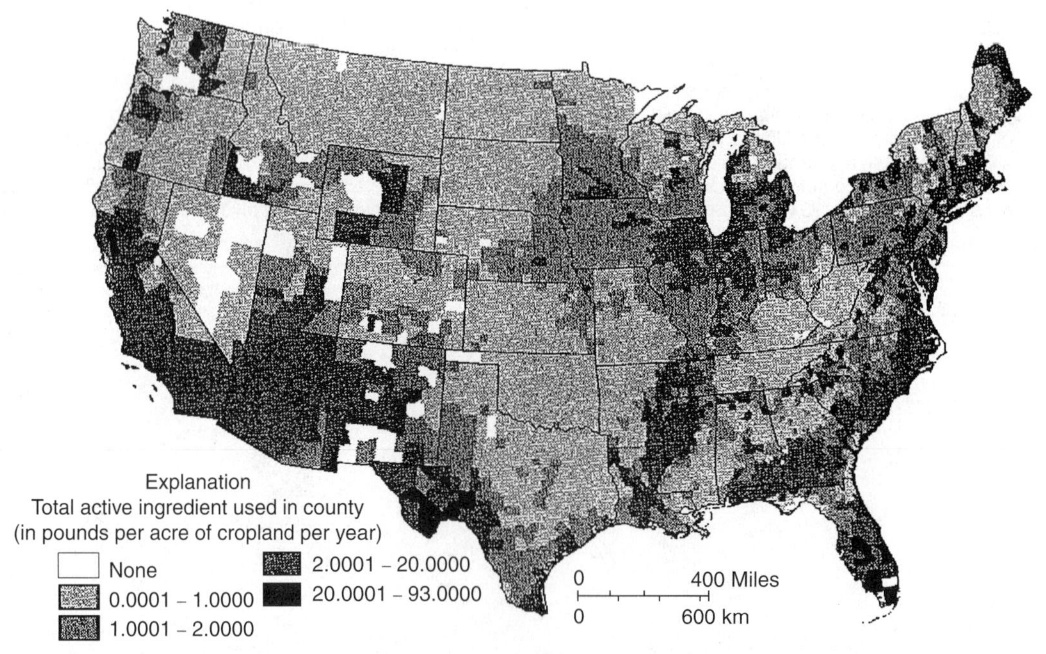

Figure 10E.98 Annual estimated pesticide use per acre of agricultural land in each county in the conterminous United States, based on the index years from 1987 to 1991. (From Barbash, J.E. and Resek, E.A., 1996, *Pesticides in Groundwater Distribution, Trends and Governing Factors*, Ann Arbor Press, Chelsea, Michigan. Printed with permission.)

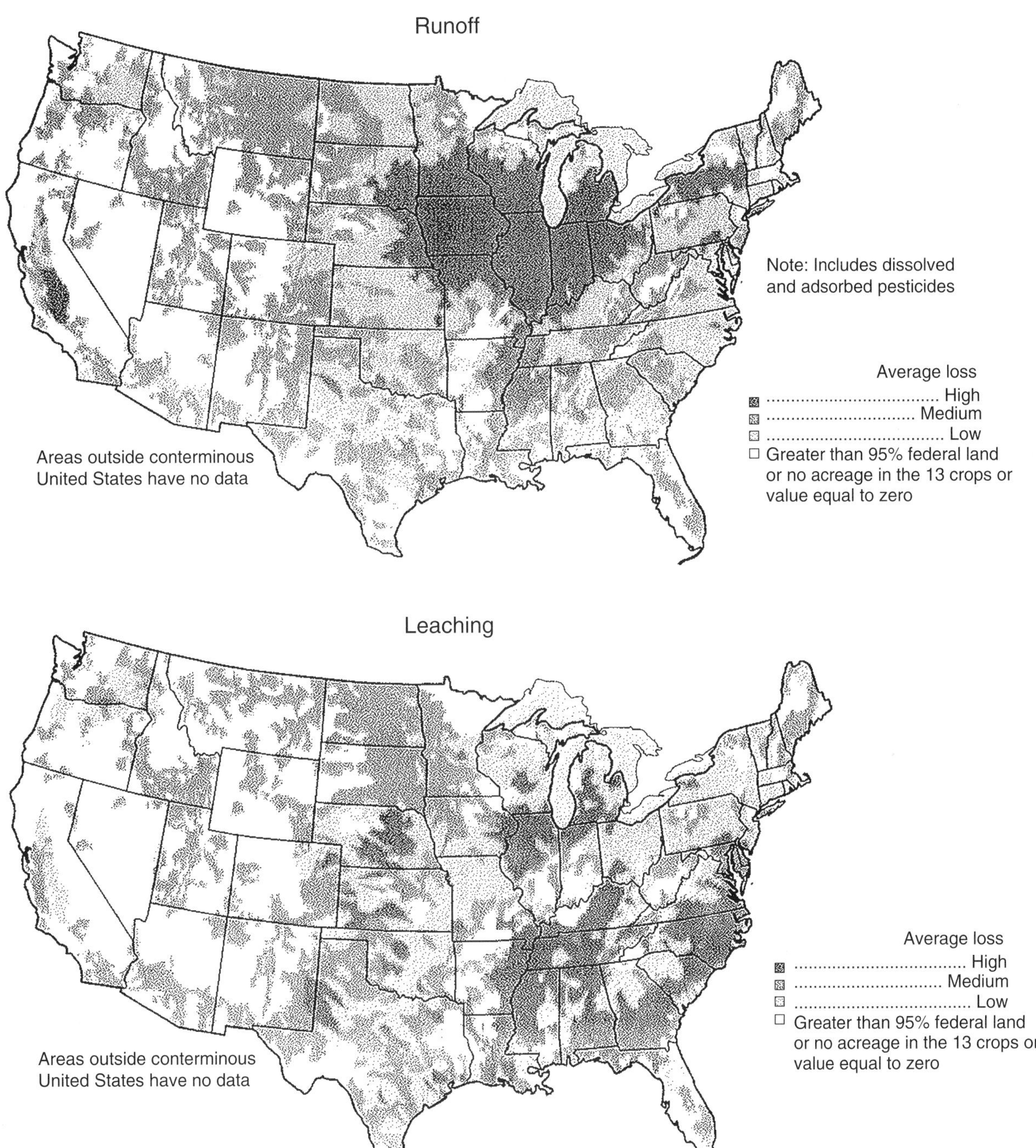

Figure 10E.99 Pesticide runoff and leaching potential for field crop production. (From United States Department of Agriculture, Natural Resources Conservation Services, 1997, *Water Quality and Agriculture, Status, Conditions, and Trends, Working Paper No. 16*, www.nrcs.usda.gov.)

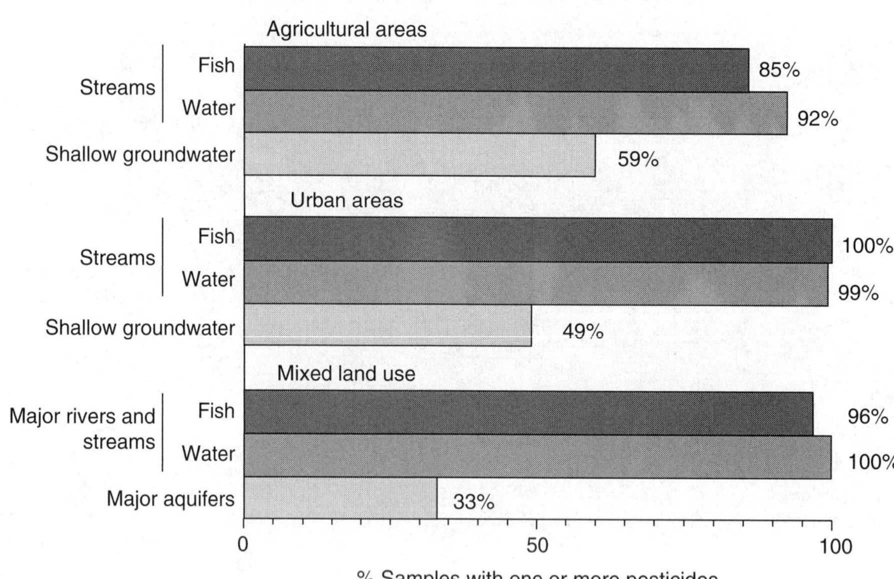

Figure 10E.100 Percent samples collected from national water quality assessment (NAWQA). Pesticide National Synthesis Project with one or more Pesticides. (From United States Geological Survey, 1999, *Pesticides in the Nations Resources, National Water Quality Assessment* (NAWQA), Pesticide National Synthesis Project, April 1999, Powerpoint presentation, ca.water.usgs.gov/pnsp/present/water.)

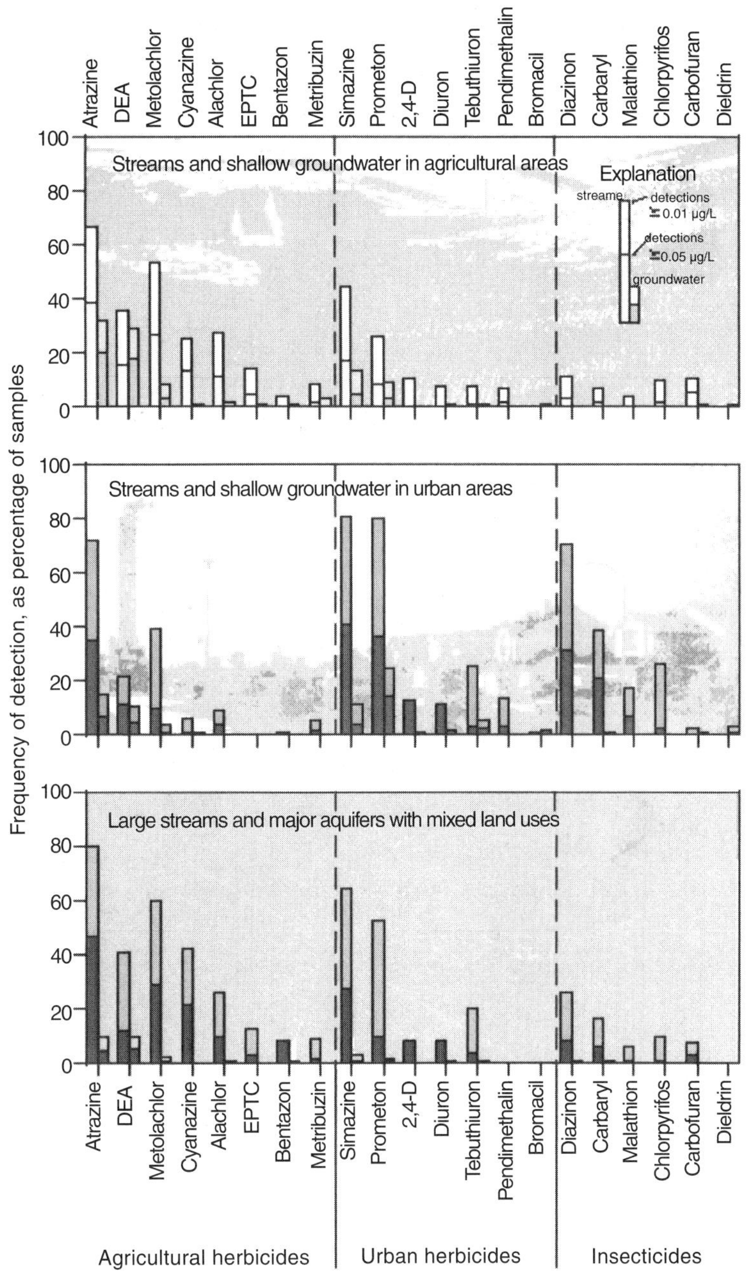

Figure 10E.101 Patterns of occurrence to the 21 most detected compounds in surface and groundwater samples collected from areas included in the United States Geological Survey National Water Quality Assessment Program, 1992–1996. (From United States Geological Survey, 1998, *Pesticides in Surface and Ground Water of the United States: Summary of Results of the National Water Quality Assessment Program* (NAWQA), July 22, 1998, www.usgs.gov.)

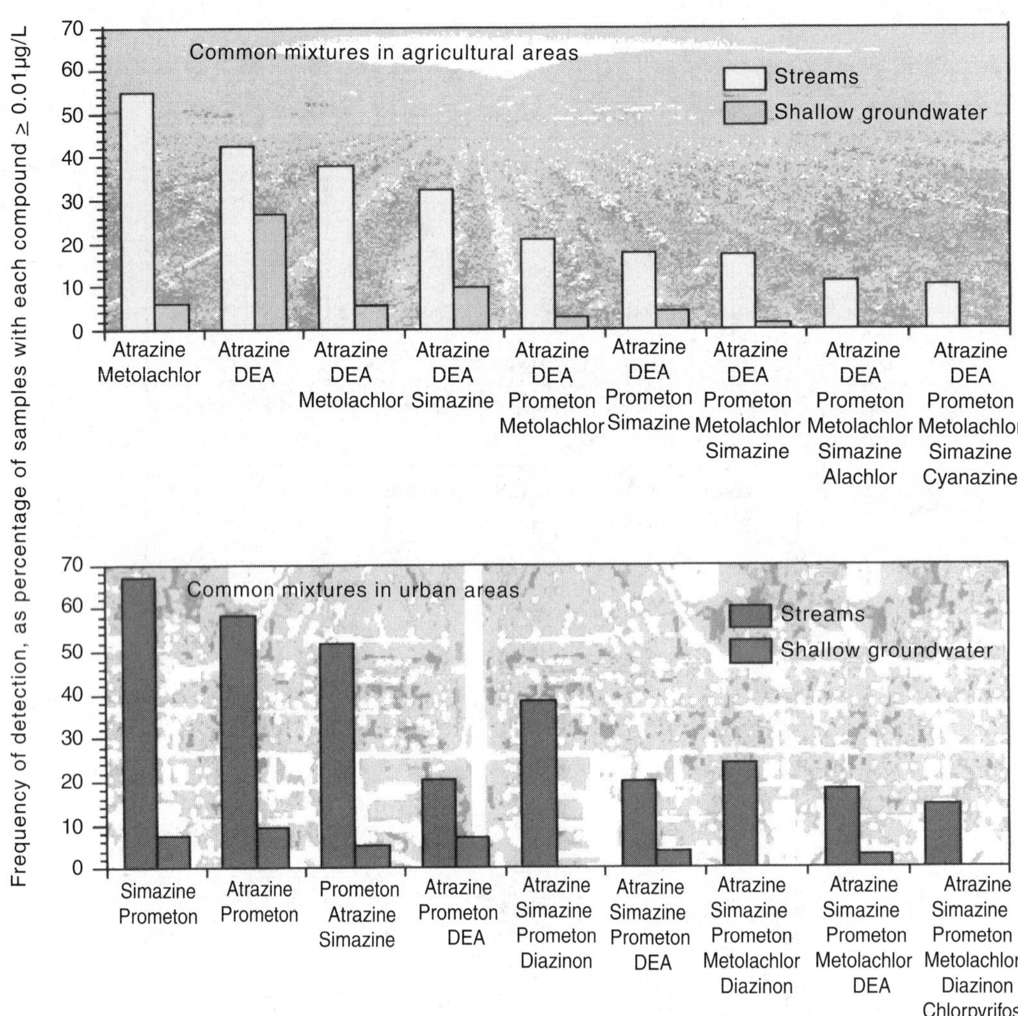

Figure 10E.102 Frequency of composition of common mixtures in surface water and shallow groundwater samples with detections collected from areas included in the United States Geological Survey National Water Quality Assessment Program, 1992–1996. (From United States Geological Survey, 1998, *Pesticides in Surface and Ground Water of the United States: Summary of Results of the National Water Quality Assessment Program* (NAWQA), July 22, 1998, www.usgs.gov.)

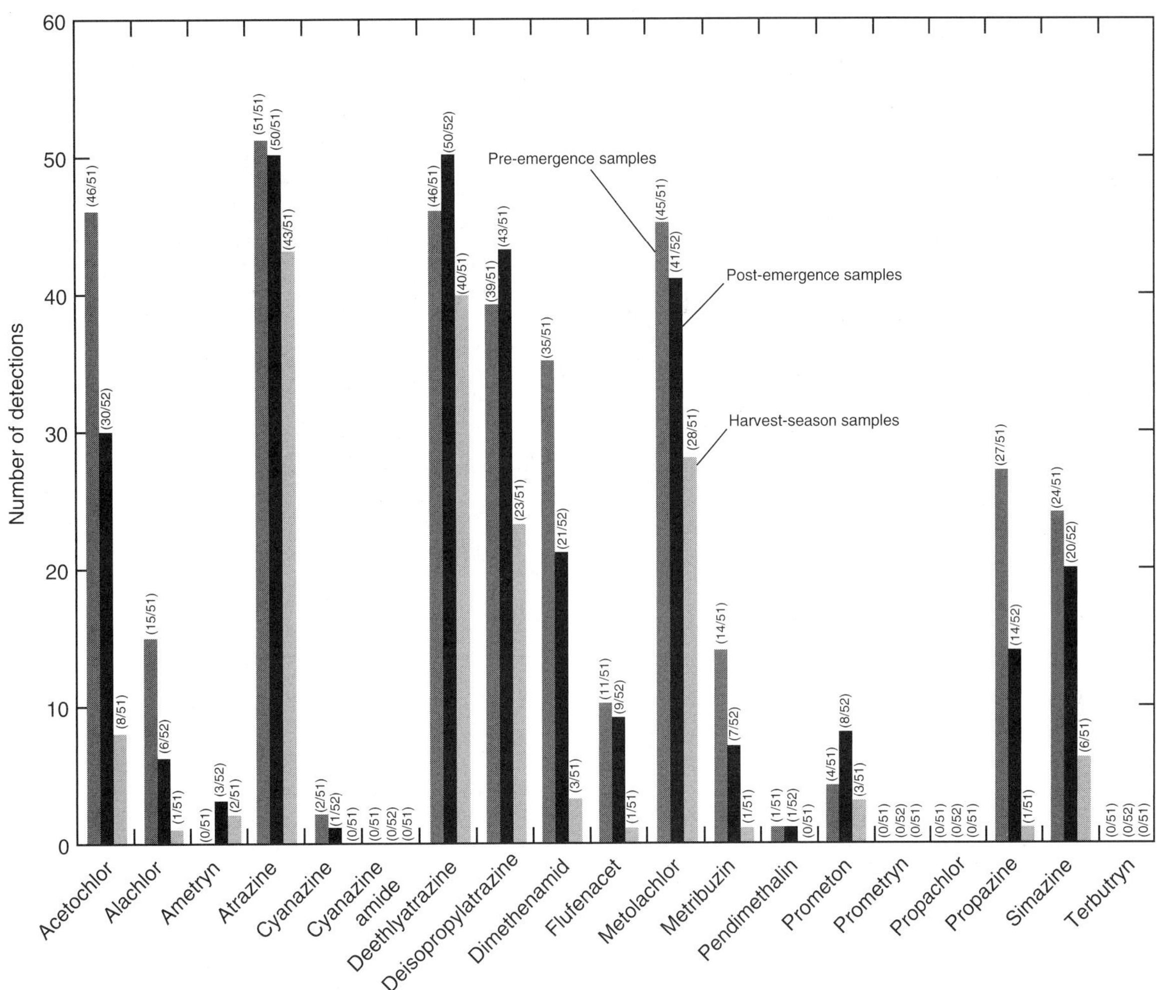

Figure 10E.103 Number of detections of selected herbicides and degradation products for pre-emergence, post-emergence, and Harvest-season runoff samples collected from 51 streams in the Midwestern States, 2002. (From Scribner, E.A., Battaglin, W.A., Dietze, J.E., and Thurman, E.M., 2003, Reconnaissance Data for Glyphosate, Other Selected Herbicides, Their Degradation Products, and Antibiotics in 51 Streams in Nine Midwestern States, 2002, *U.S. Geological Survey Open-File Report 03-217*, www.usgs.gov.)

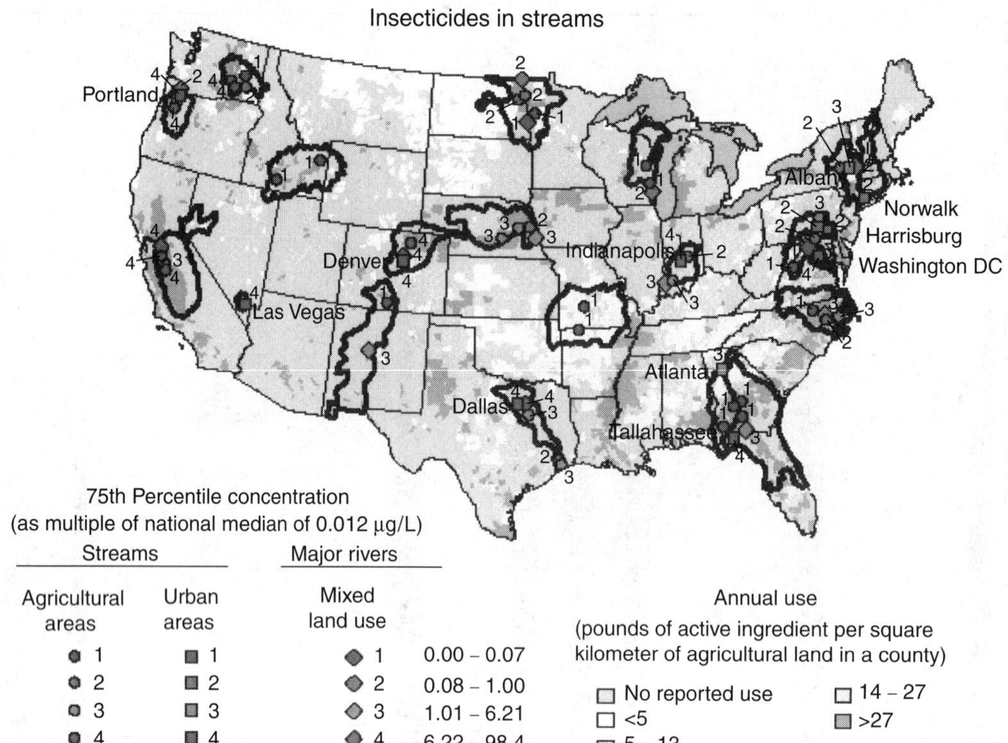

Figure 10E.104 Geographic distribution of insecticides in surface water for streams included in the United States Geological Survey National Water Quality Assessment Program, 1992–1996. (From United States Geological Survey, 1998, *Pesticides in Surface and Ground Water of the United States: Summary of Results of the National Water Quality Assessment Program* (NAWQA), July 22, 1998, www.usgs.gov.)

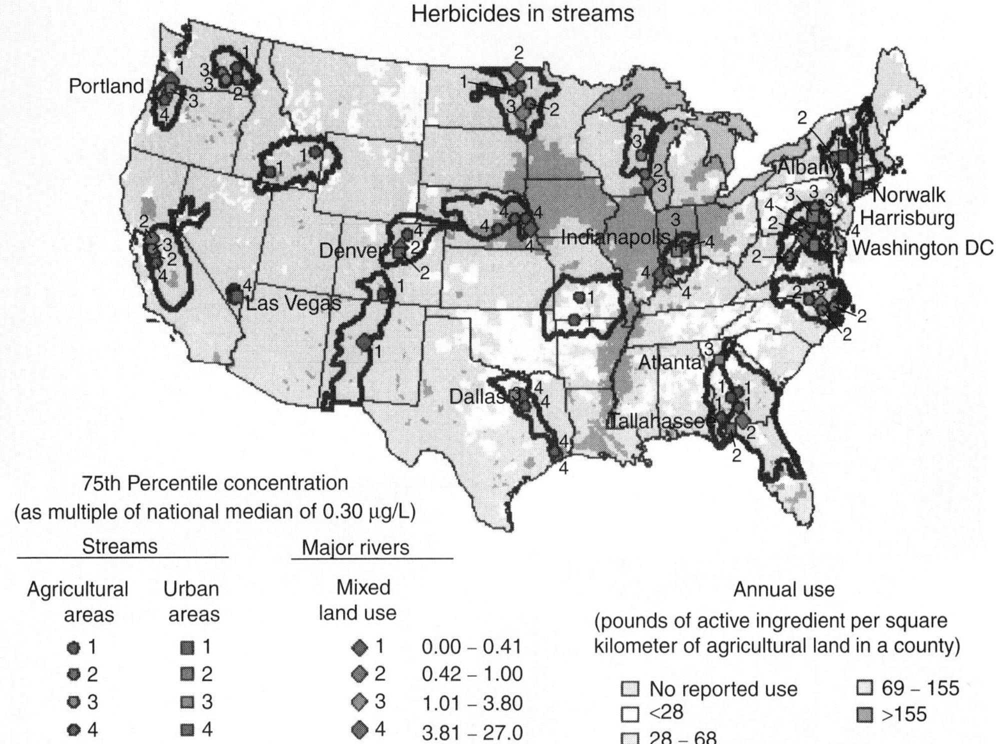

Figure 10E.105 Geographic distribution of herbicides in surface water for streams included in the United States Geological Survey Pesticide National Water Quality Assessment Program, 1992–1996. (From United States Geological Survey, 1998, *Pesticides in Surface and Ground Water of the United States: Summary of Results of the National Water Quality Assessment Program* (NAWQA), July 22, 1998, www.usgs.gov.)

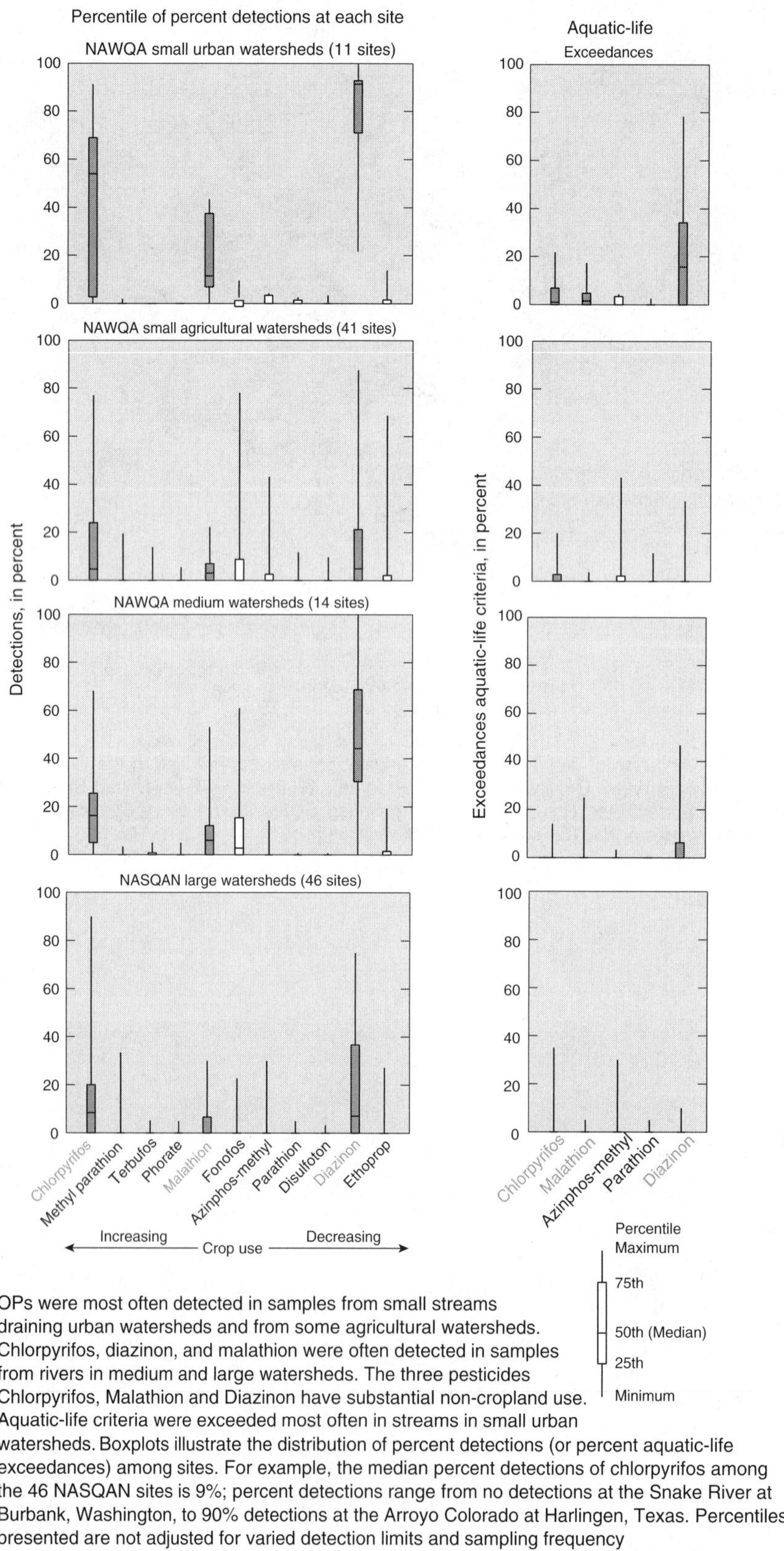

Figure 10E.106 Organophosphorous pesticide occurrence in surface water for streams included in the United States Geological Survey National Water Quality Assessment Program, 1992–1997. (From Hopkins, E. H. et al., 2000, Organophosphorous Pesticide Occurrence and Distribution in Surface and Ground Water of the United States, 1992–1997, *U.S. Geological Survey Open-File Report 00-187*, www.usgs.gov.)

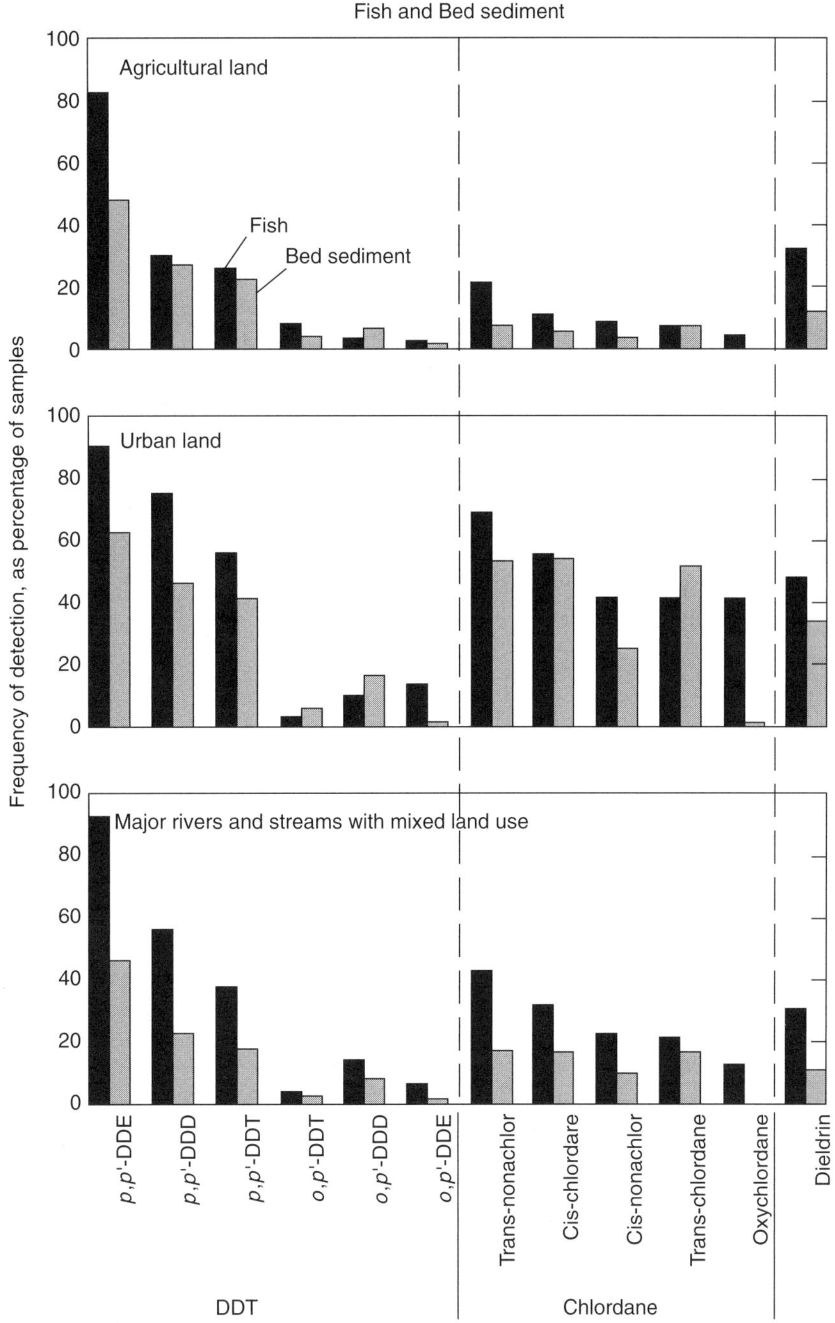

Note: The pesticide compounds found most often in fish and bed sediment are related to three major groups of insecticides that were heavily used in the 1960s. Organochlorine compounds related to DDT and dieldrin were widely used in both agricultural and urban areas, and chlordane was mainly used in urban areas.

Figure 10E.107 The most frequently detected pesticides in fish and bed sediment in the United States, 1991–1996. (From United States Geological Survey, 1999, The Quality of Our Nation's Waters, Nutrients and Pesticides, *U.S. Geological Survey Circular. 1225*, www.usgs.gov.)

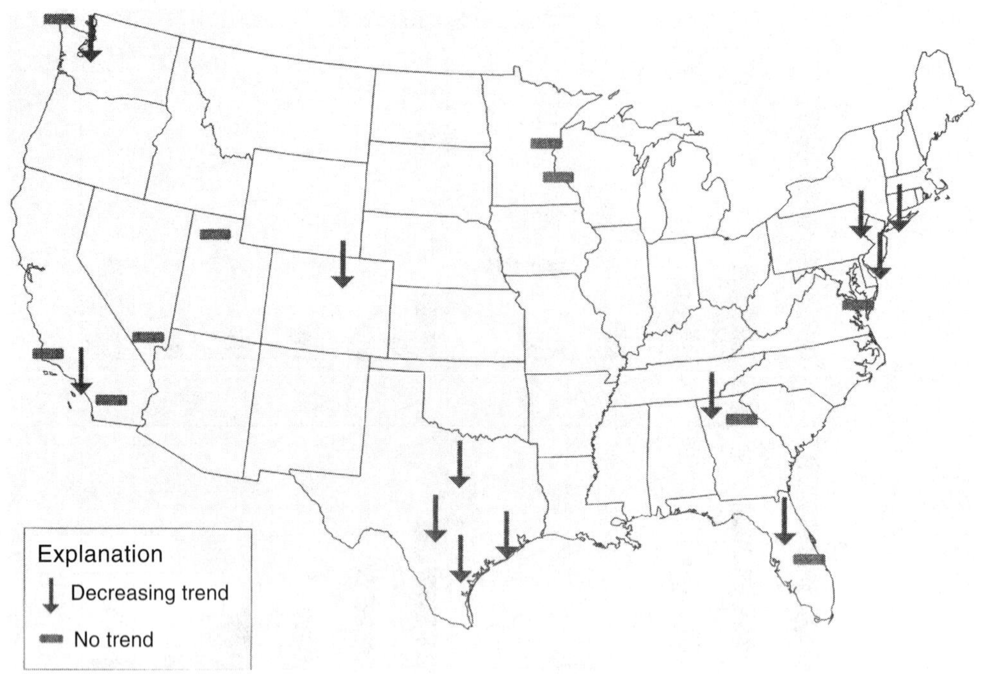

Figure 10E.108 DDT trends throughout the United States using sediment core data form 1970 to top of core. (From United States Environmental Protection Agency, The Incidence and Severity of Sediment Contamination in Surface Waters of the United States, *National Sediment Quality Survey*, Second Edition, www.epa.gov.)

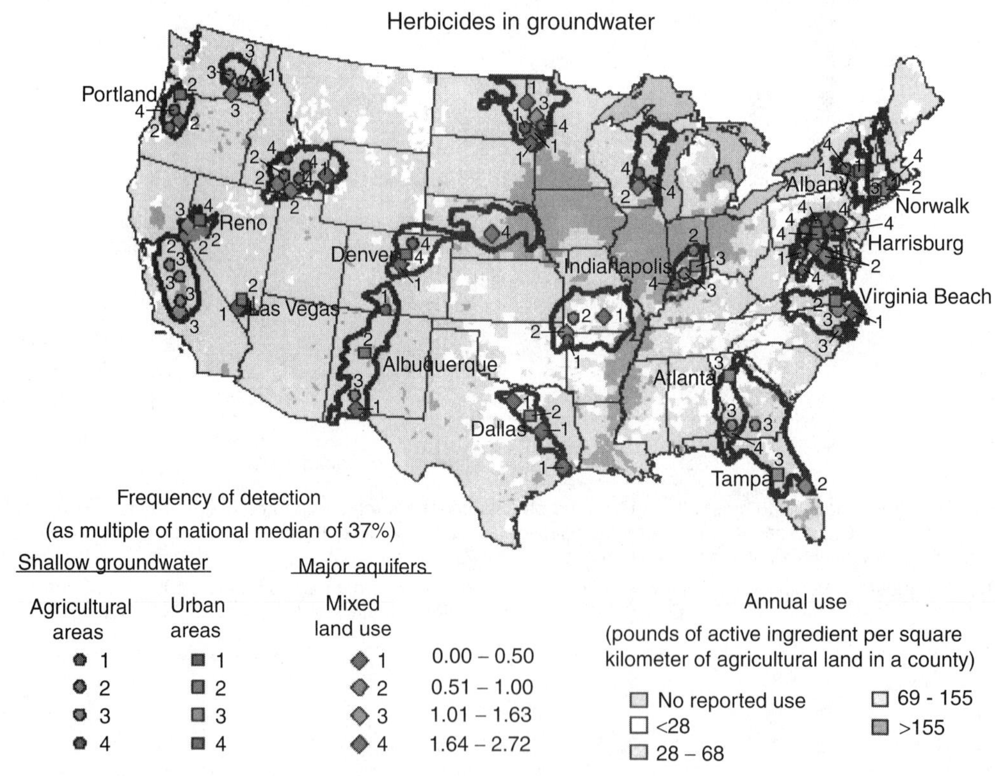

Figure 10E.109 Geographic distribution of insecticides in groundwater in areas included in the United States Geological Survey National Water Quality Assessment Program, 1992–1996. (From United States Geological Survey, *Pesticides in Surface and Ground Water of the United State: Summary of Results of the National Water Quality Assessment Program* (NAWQA), www.usgs.gov.)

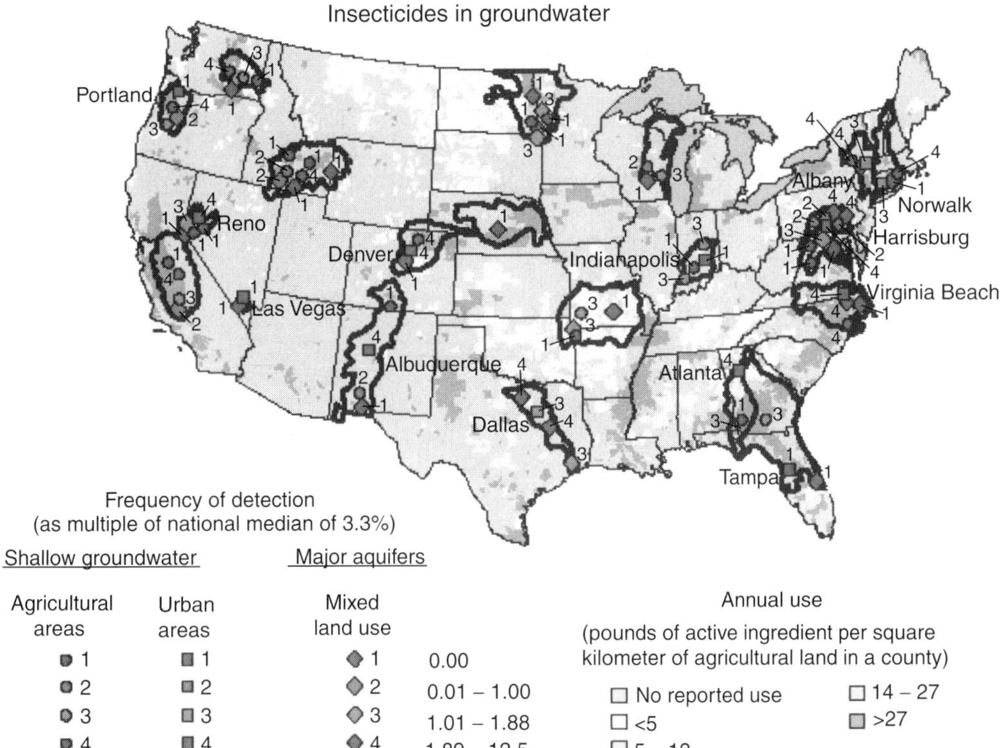

Figure 10E.110 Geographic distribution of herbicides in groundwater in areas included in the United States Geological Survey Pesticide National Water Quality Assessment Program, 1992–1996. (From United States Geological Survey, *Pesticides in Surface and Ground Water of the United State: Summary of Results of the National Water Quality Assessment Program* (NAWQA), www.usgs.gov.)

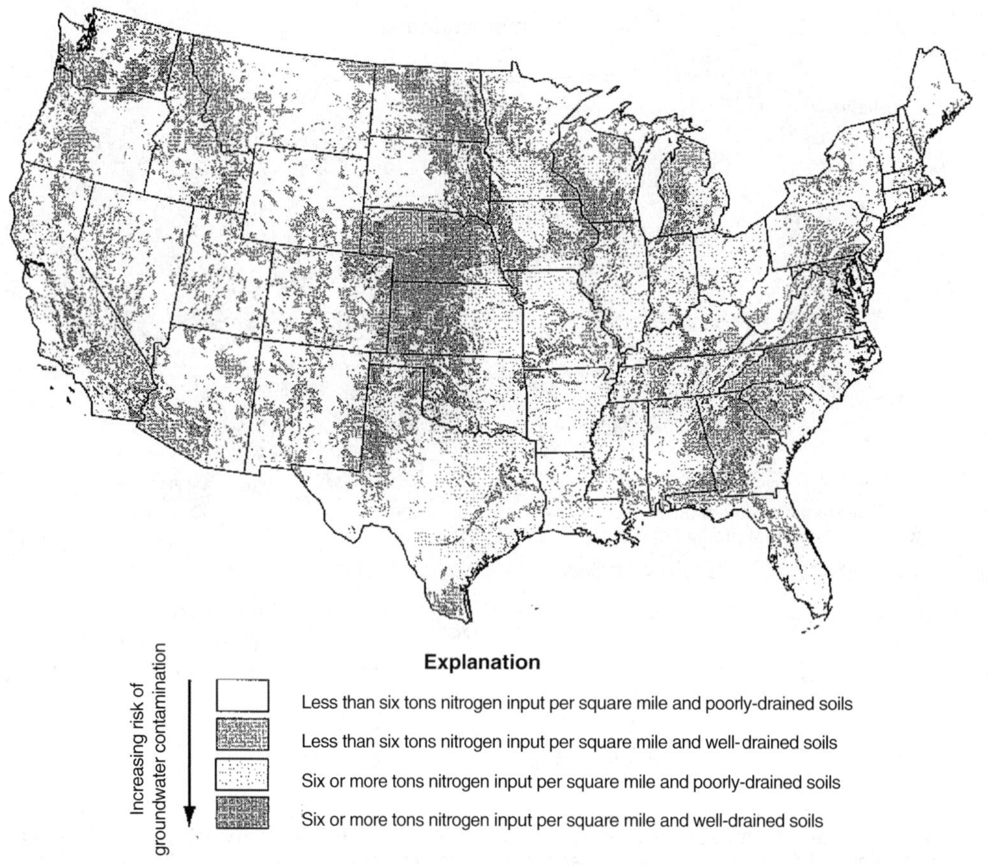

Figure 10E.111 Areas in the United States most vulnerable to nitrate contamination of groundwater. (From Nolan, B. T. and Ruddy, B. C., Nitrate in Ground Waters of the United States-Assessing the Risk, *United States Geological Survey Fact Sheet FS-092-96*, www.usgs.gov.)

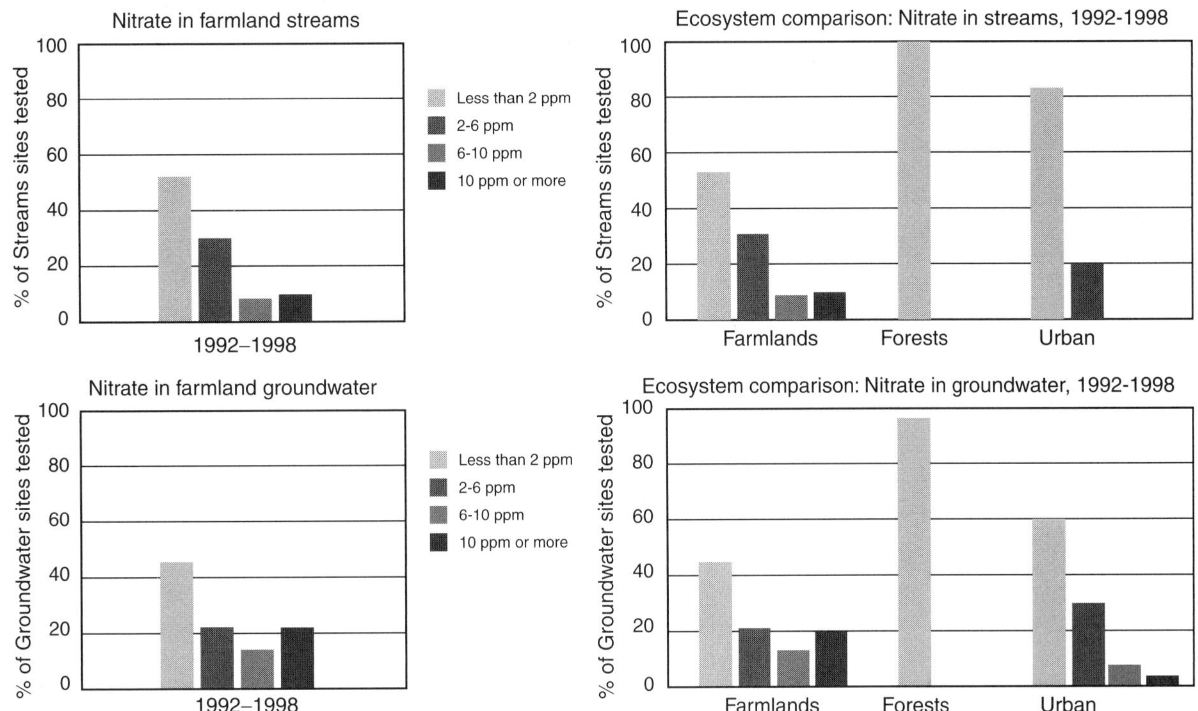

Figure 10E.112 Nitrates in farmland streams and groundwater in the 36 major United States river basins and aquifers sampled by the National Water Quality Assessment Program, 1992–1998. (From United States Environmental Protection Agency, 2003, *EPA's Draft Report on the Environment*, 2003, EPA 600-R-03-050, www.epa.gov. The Heintz Cent, The State of the Nation's Ecosystem 2002.)

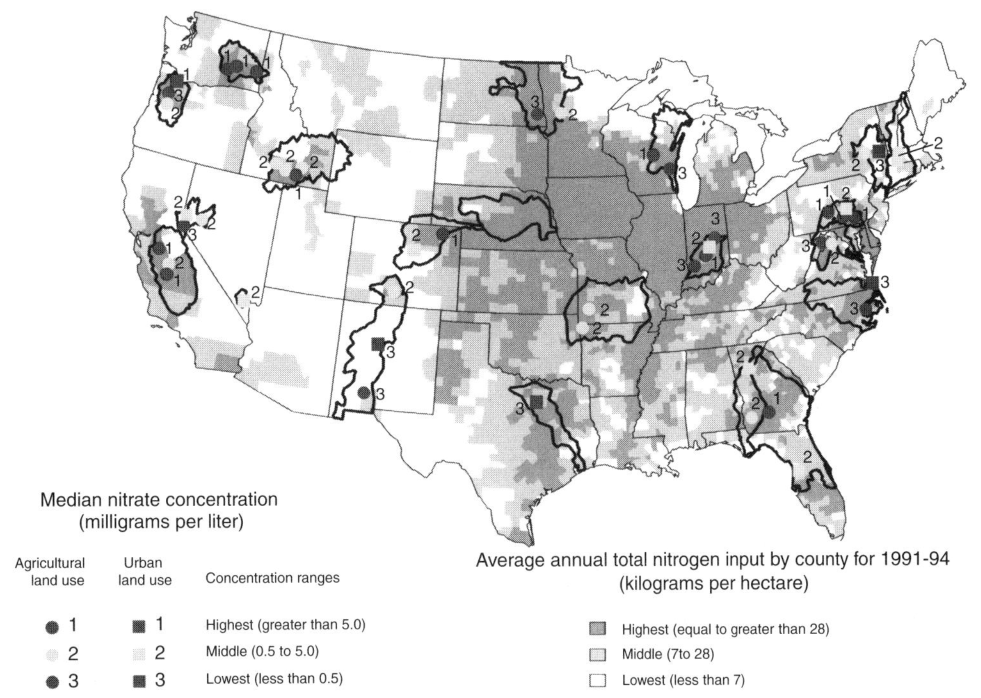

Figure 10E.113 Median Nitrate Concentration in Shallow Groundwater Sampled by the NAWQA Program During 1992–1995. (From Nolan, B.T. and Stoner, J.D., Nutrients in groundwaters of the conterminous United States, 1992–1995, *Environmental Science and Technology*, vol. 34, no. 7, 2000, p. 1156–1165, www.usgs.gov.)

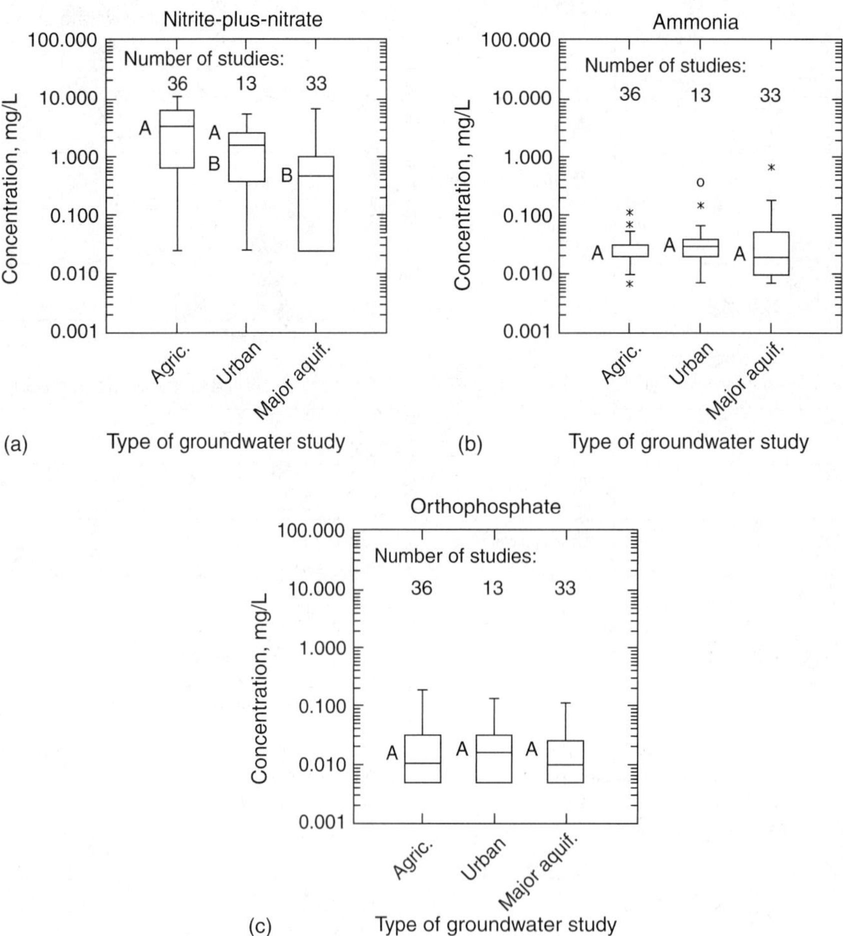

Figure 10E.114 Distribution of nitrate-plus-nitrite, ammonia, and othrophosphate in groundwater samples from land-use studies and major aquifers. (From Nolan, B.T. and Stoner, J.D., Nutrients in groundwaters of the conterminous United States, 1992–1995, *Environmental Science and Technology*, vol. 34, no. 7, 2000, p. 1156–1165, www.usgs.gov.)

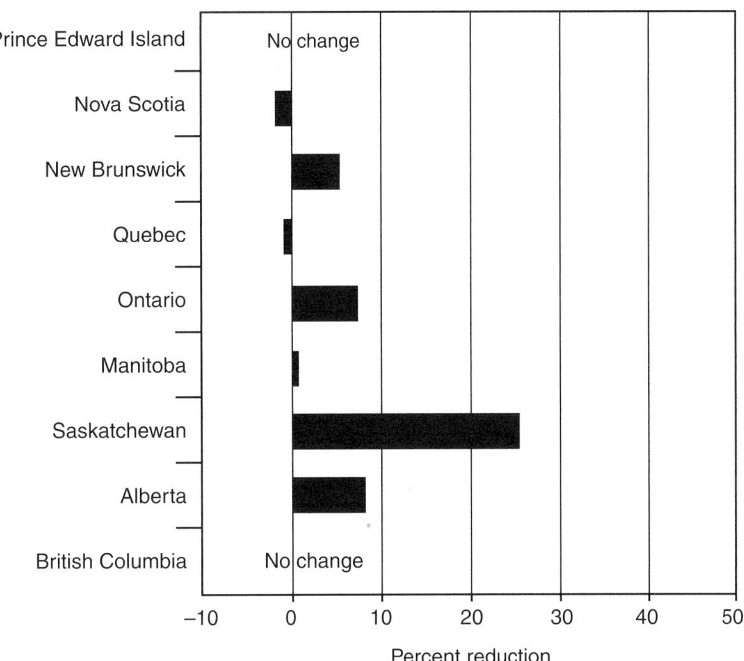

Figure 10E.115 Changes in the area of Canadian cropland at risk of exceeding a tolerable level of water erosion between 1981 and 1996. (From *The Health of Our Water, Toward Sustainable Agriculture in Canada*, Publication 2020/E, Coote, D.R. and Gregorich, L.J. (eds.), Research Branch Agriculture and Agri-Food Canada, 2000. Original Source: McRae, T., Smith, C.A.S., and Gregorich, L.J., (eds.), 2000, Environmental Sustainability of Canadian Agriculture: Agri-Environmental Indicator Project. Agriculture and Agri-Food Canada, Ottawa, Ont. Reproduced with the permission of the Minister of Public Works and Government Services Canada, 2006.)

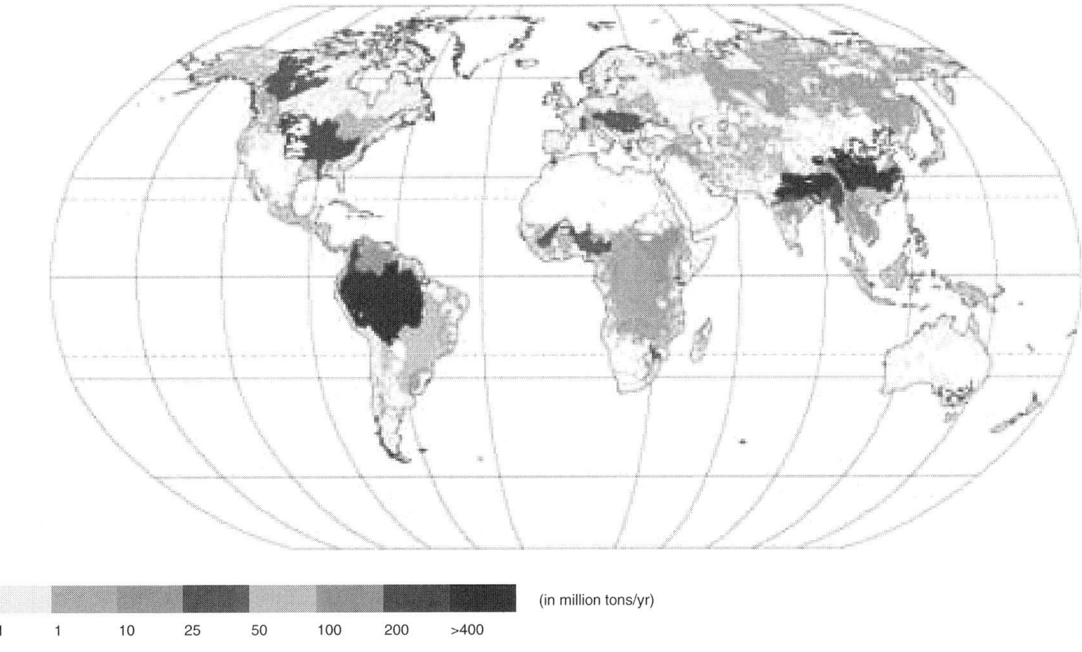

Changes in sediment yield reflect changes in basin conditions, including climate, solis, erosion rates, vegetation, topography and land use. It is influenced strongly by human actions, such as in the construction of dams and levees (see high sediment load in China and the Amason basin, where large dams have been implemeted), forest harvesting and farming in drainage basins

Figure 10E.116 World sediment load by basin. (From *Water for People Water for Life*, The United Nations World Water Development, Copyright © United Nations Educational, Scientific and Cultural Organization — World Water Assessment Programme (UNESCO-WWAP), 2003. Reproduced by permission of UNESCO. www.unesco.org. Original Source: Reprinted from Marine Geology, vol. 154, Syvitski, J.P. and Morehead, M.D., Estimating River-Sediment Discharge to the Ocean: Application to the Eel Margin, Northern California, pp 13–28, 1999. With permission from Elsevier.)

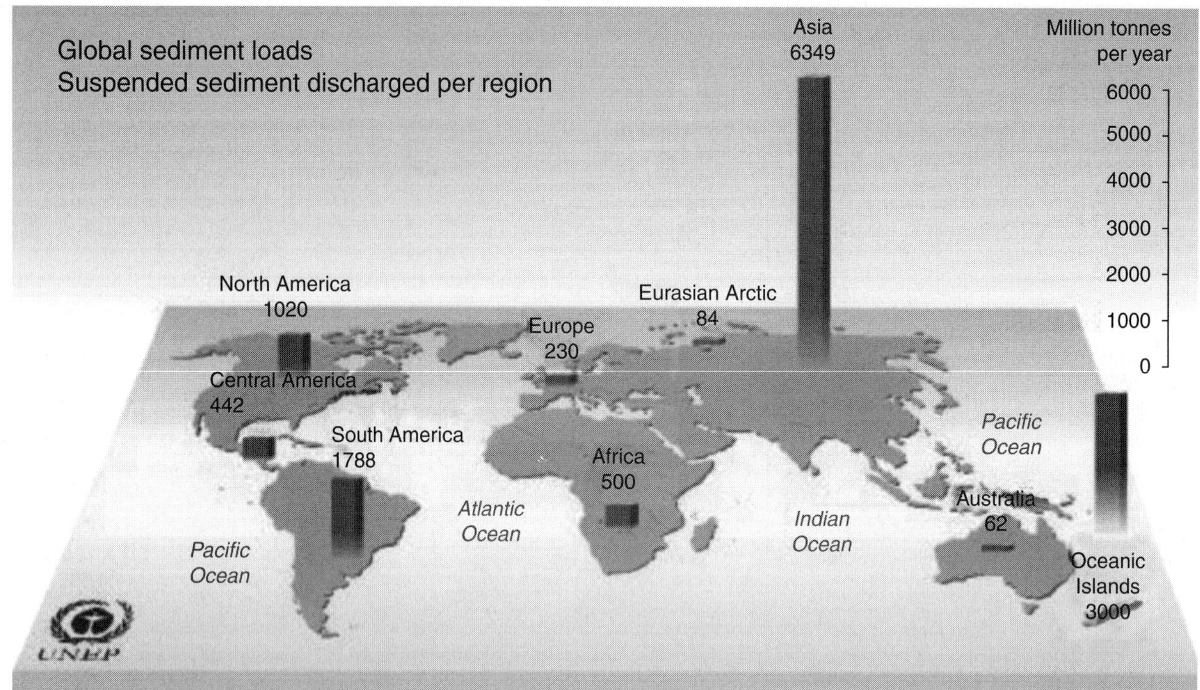

Figure 10E.117 Global sediment loads, suspended sediment discharged by region. (From United Nations Environmental-Programme, UNEP/GRID-Arendal, *Vital Water Graphics, Global Sediment Loads Suspended Sediment Discharged per Region*, Downloaded 9/22/05, www.unep.org/vitalwater/freshwater.htm Data source for chart Gleick. P.H., 1993, Witter in Crisis: A Guide to the World's Fresh Wafer Resources, Oxford University Press, NY. Reprinted with permission.)

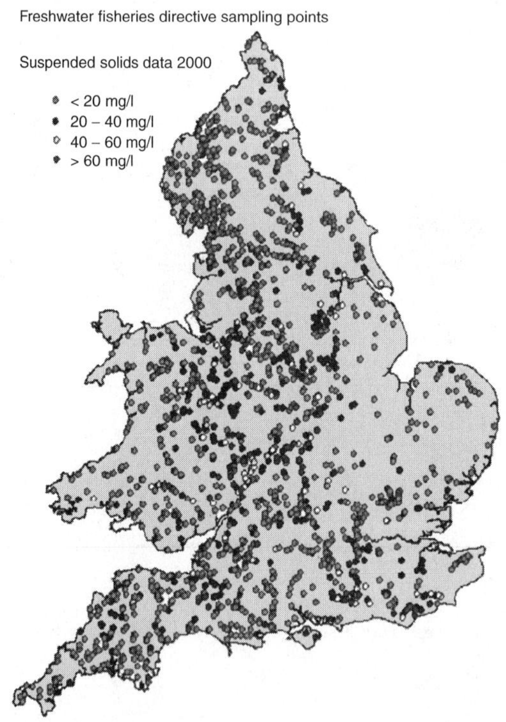

Figure 10E.118 Suspended solids loads in sampling points in England and Wales monitored under the freshwater fish directive. (From Department of Environment, Food and Rural Affairs, 2002, *The Government's Strategic Review of Diffuse Water Pollution from Agriculture in England: Agriculture and Water: A Diffuse Pollution Review*, www.defra.gov.uk. Reproduced under the terms of the Click-Use Licence.)

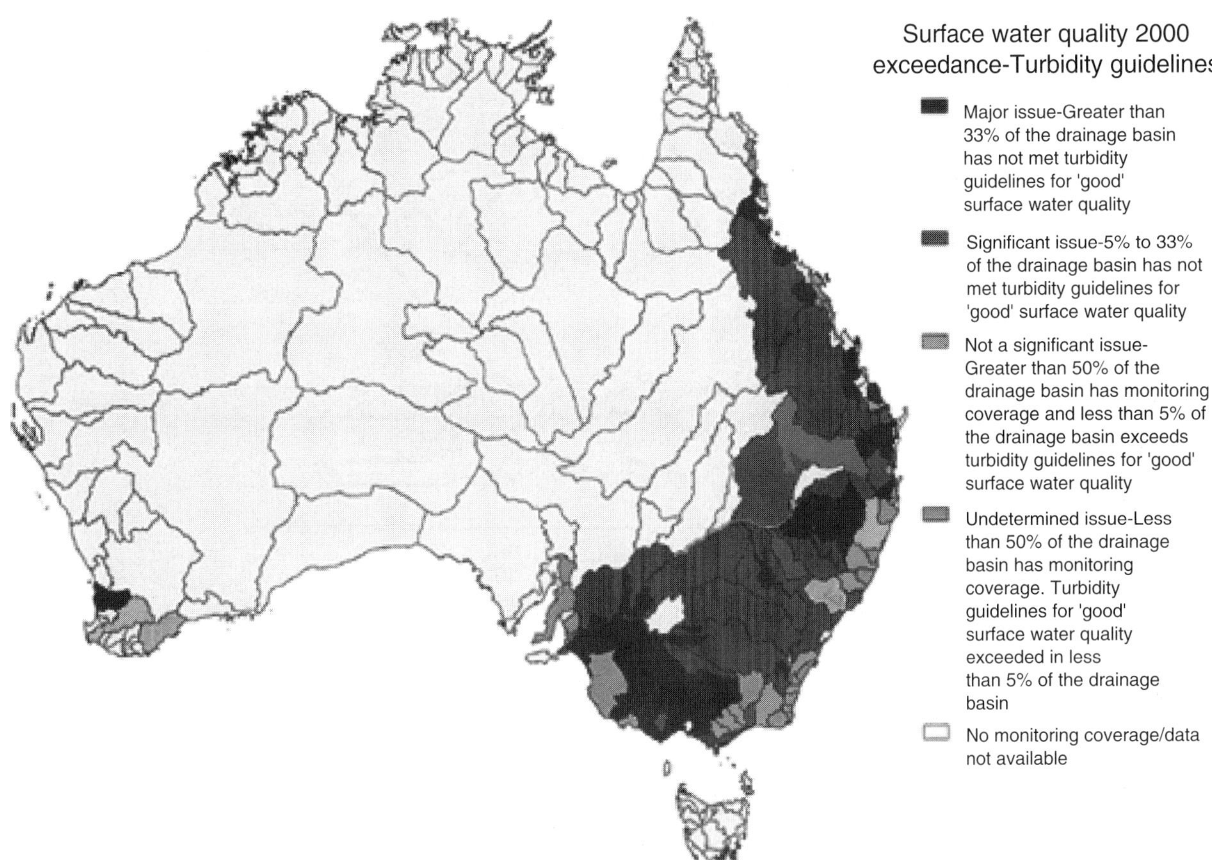
Surface water quality 2000
exceedance-Turbidity guidelines
Major issue-Greater than 33% of the drainage basin has not met turbidity guidelines for 'good' surface water quality
Significant issue-5% to 33% of the drainage basin has not met turbidity guidelines for 'good' surface water quality
Not a significant issue-Greater than 50% of the drainage basin has monitoring coverage and less than 5% of the drainage basin exceeds turbidity guidelines for 'good' surface water quality
Undetermined issue-Less than 50% of the drainage basin has monitoring coverage. Turbidity guidelines for 'good' surface water quality exceeded in less than 5% of the drainage basin
No monitoring coverage/data not available

Figure 10E.119 Australian catchments where turbidity is considered an environmental issue. (From Ball, J., Donnelley, L., Erlanger, P., Evans, R., Kollmorgen, A., Neal, B., and Shirley, M., 2001, *Inland Waters, Australia State of Environment Report 2001* (Theme Report), CSIRO Publishing on behalf of the Department of the Environment and Heritage, Canberra. Original Source: National Land and Water Resources Audit, 2001a, Australian Water. Resources Assessment 2000, www.deh.gov.au. Copyright Commonwealth of Australia reproduced by permission.)

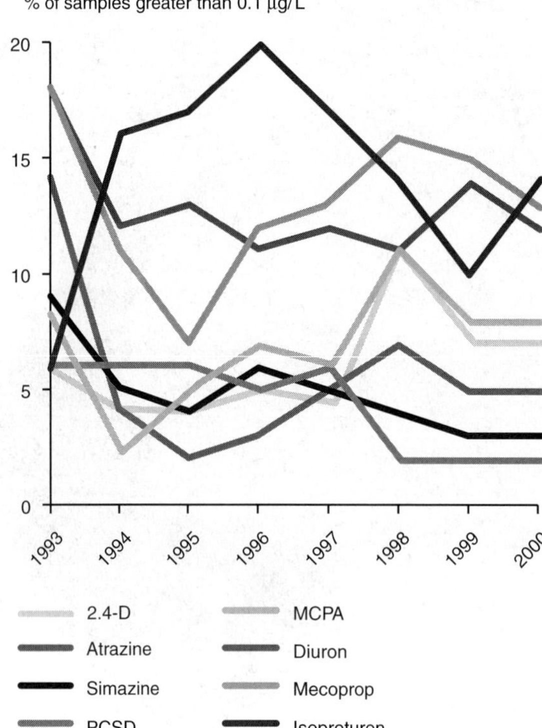

Figure 10E.120 Occurrences of some commonly found pesticides in surface freshwaters in England and Wales, 1993–2000. (From European Environmental Agency (EEA), 2003, *Europe's Environment, The Third Assessment, Environmental Assessment Report No. 10*, EE1, Copenhagen, www.eea.europa.eu. Reprinted with permission © EEA.)

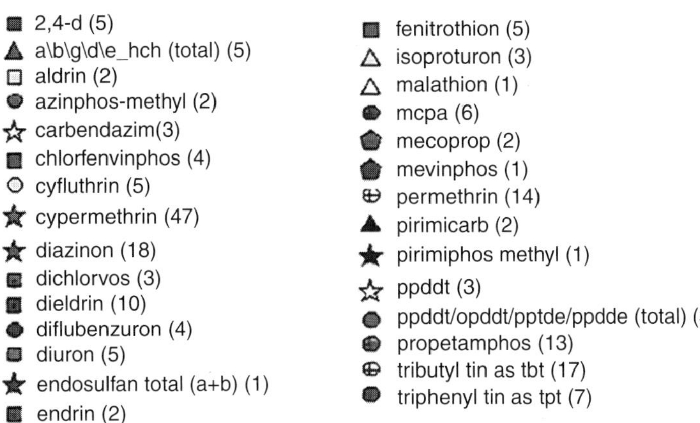

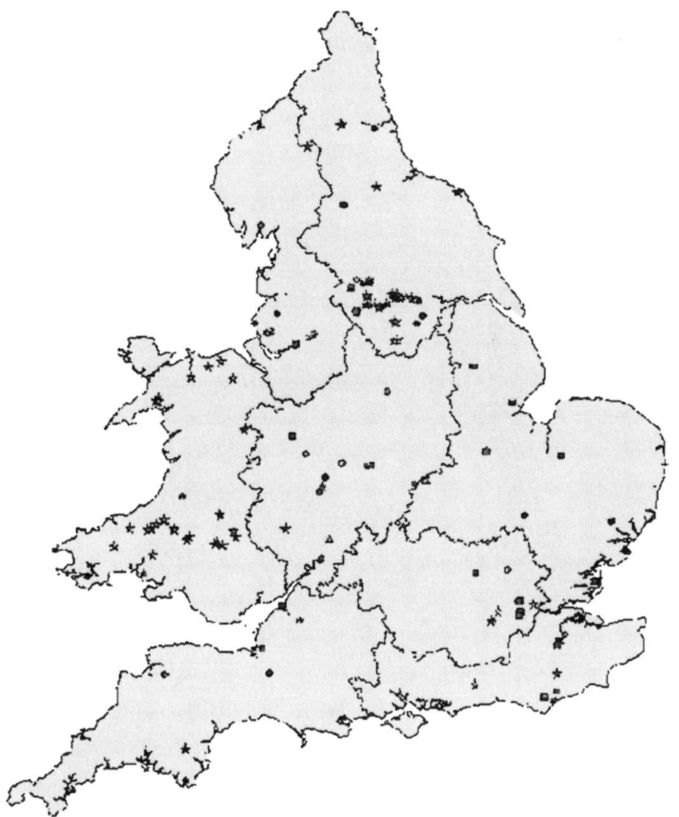

Figure 10E.121 Surface freshwater sites exceeding pesticide environmental quality standards (EQS) For England and Wales, 2000. (From Department of Environment, Food and Rural Affairs, 2002, *The Government's Strategic Review of Diffuse Water Pollution from Agriculture in England: Agriculture and Water: A Diffuse Pollution Review*, www.defra.gov.uk. Reproduced under the terms of the Click-Use Licence.)

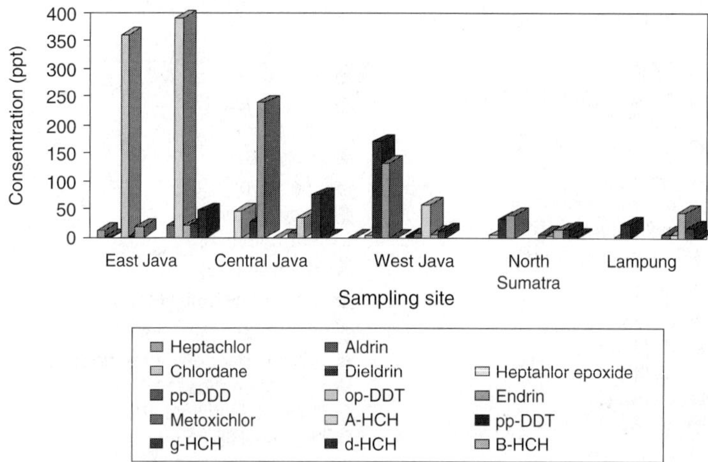

Figure 10E.122 Highest concentrations of POPs and OCs compounds in Indonesian surface water in 2002. (From Syafrul, H., 2003, Environmental Standards Related to POPs in Indonesia in United Nations University, *Capacity Development Training for Monitoring of POPS in the East Asian Hydrosphere*, 1–2 September, 2003, UNUM Centre, Tokyo, www.unu.edu. Printed with permission.)

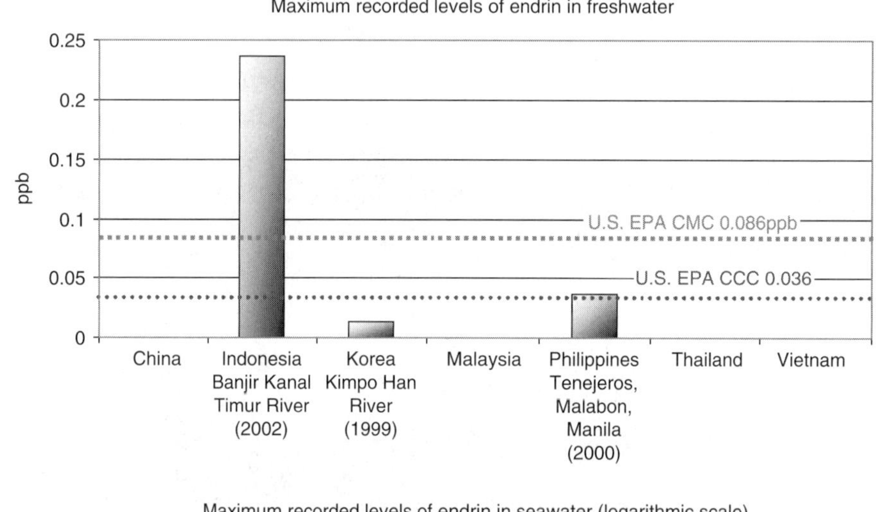

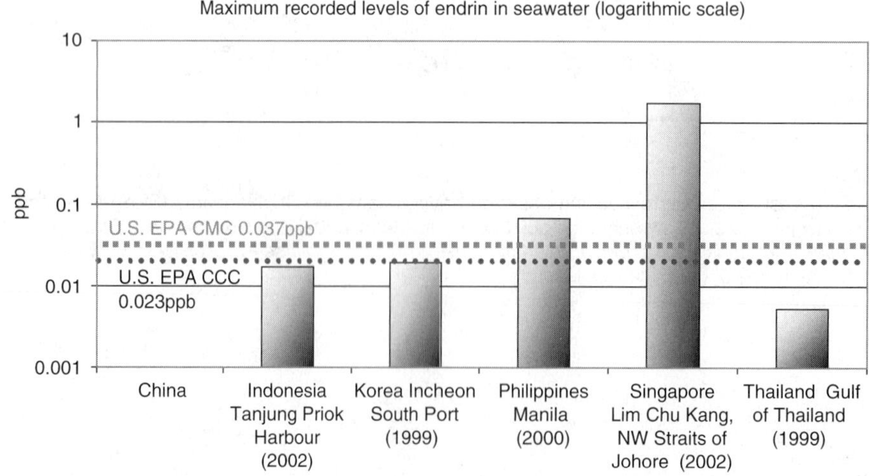

Figure 10E.123 Maximum recorded levels of endrin in freshwater and seawater in East Asia. (From King, C., 2003, Capacity Development for Monitoring Major Persistent Organic Pollutants (POPs) in East Asian Waters: Examples of UNU Monitoring Activities in East Asia in United Nations University, *Capacity Development Training for Monitoring of POPS in the East Asian Hydrosphere*, 1–2 September, 2003, UNU Centre, Tokyo, www.unu.edu. Printed with permission.)

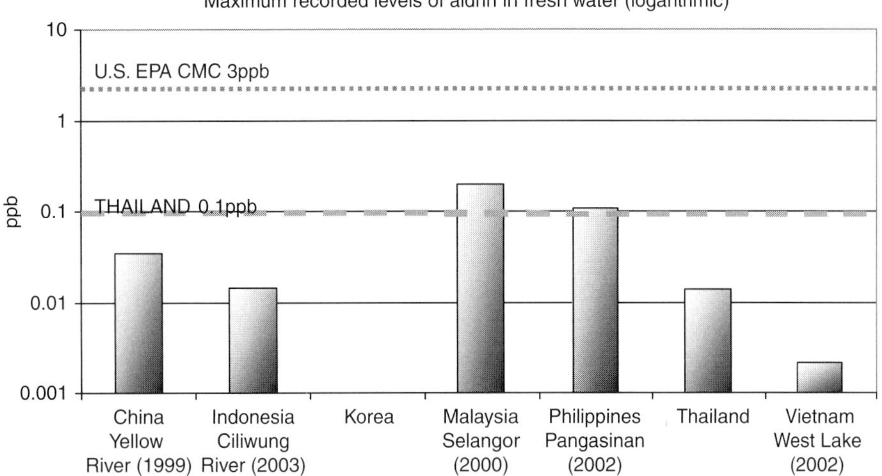

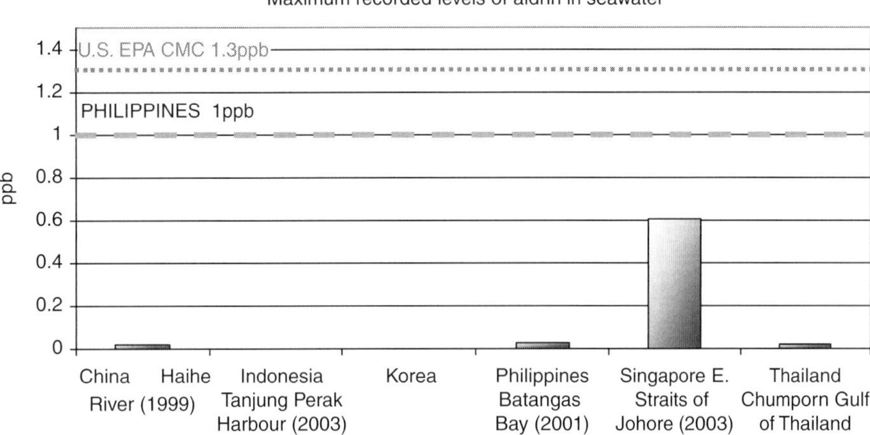

Figure 10E.124 Maximum recorded levels of aldrin in freshwater and seawater in East Asia. (From King, C., 2003, Capacity Development for Monitoring Major Persistent Organic Pollutants (POPs) in East Asian Waters: Examples of UNU Monitoring Activities in United Nations University, *Capacity Development Training for Monitoring of POPS in the East Asian Hydrosphere*, 1–2 September, 2003, UNU Centre, Tokyo, www.unu.edu. Printed with permission.)

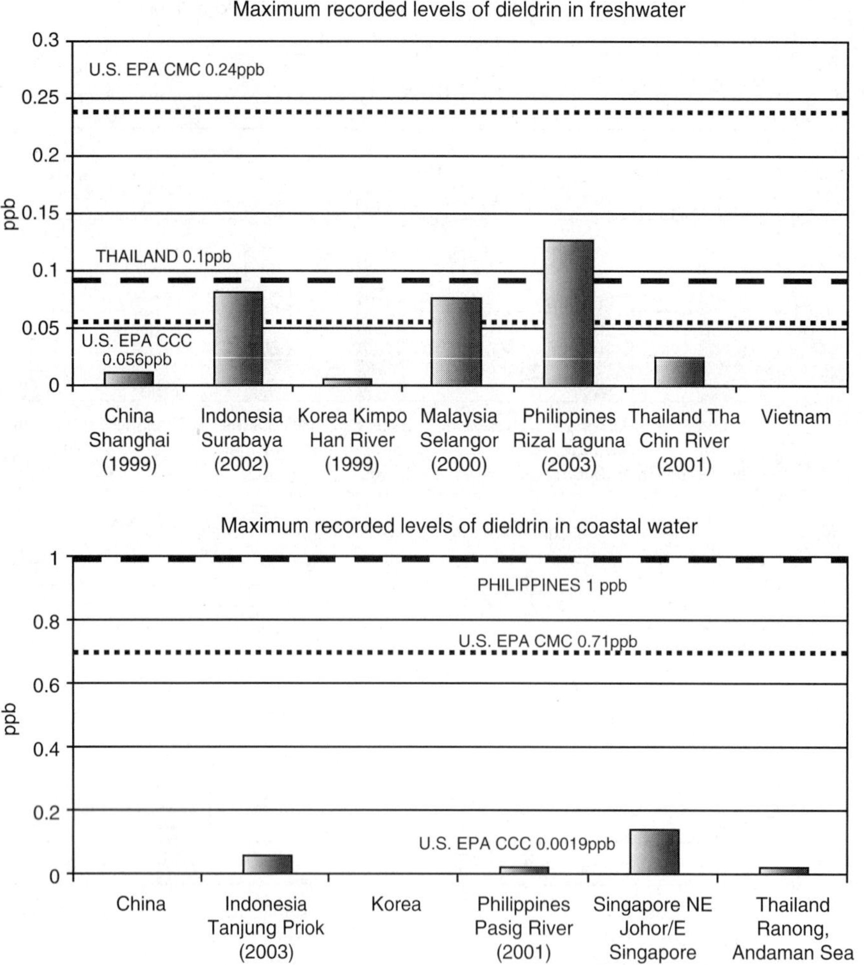

Figure 10E.125 Maximum recorded levels of dieldrin in freshwater and seawater in East Asia. (From King, C., 2003, Capacity Development for Monitoring Major Persistent Organic Pollutants (POPs) in East Asian Waters: Examples of UNU Monitoring Activities in East Asia in United Nations University, *Capacity Development Training for Monitoring of POPS in the East Asian Hydrosphere*, 1–2 September, 2003, UNU Centre, Tokyo, www.unu.edu. Printed with permission.)

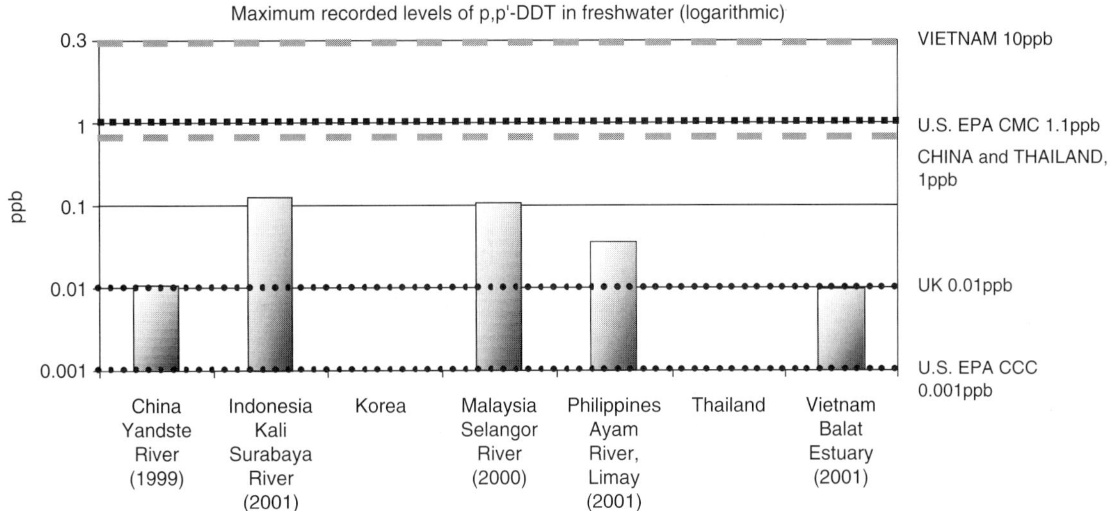

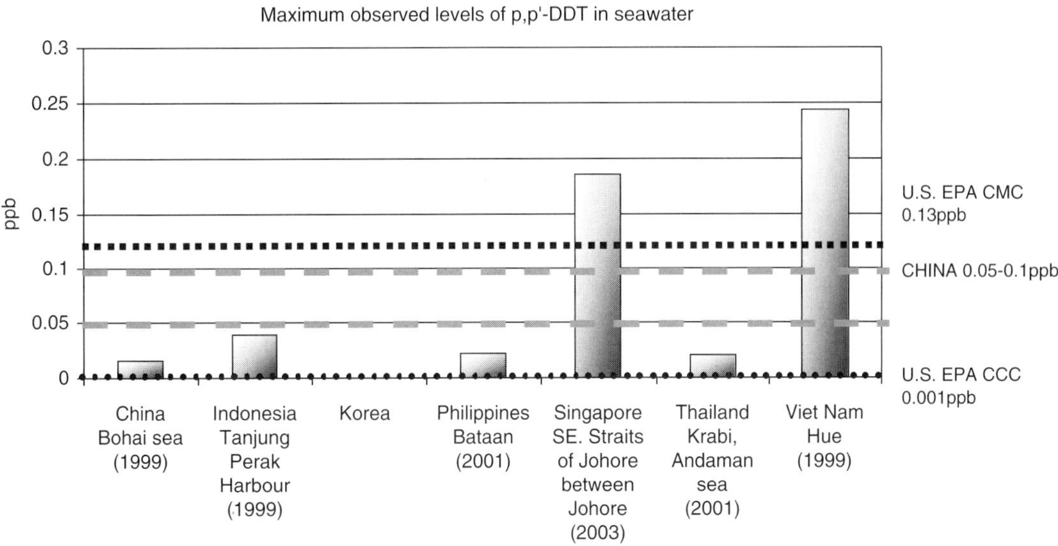

Figure 10E.126 Maximum recorded levels of *p,p′*-DDT in freshwater and seawater 1999–2003. (From King, C., 2003, Capacity Development for Monitoring Major Persistent Organic Pollutants (POPs) in East Asian Waters: Examples of UNU Monitoring Activities in East Asia in United Nations University, *Capacity Development Training for Monitoring of POPS in the East Asian Hydrosphere*, 1–2 September, 2003, UNU Centre, Tokyo, www.unu.edu. Printed with permission.)

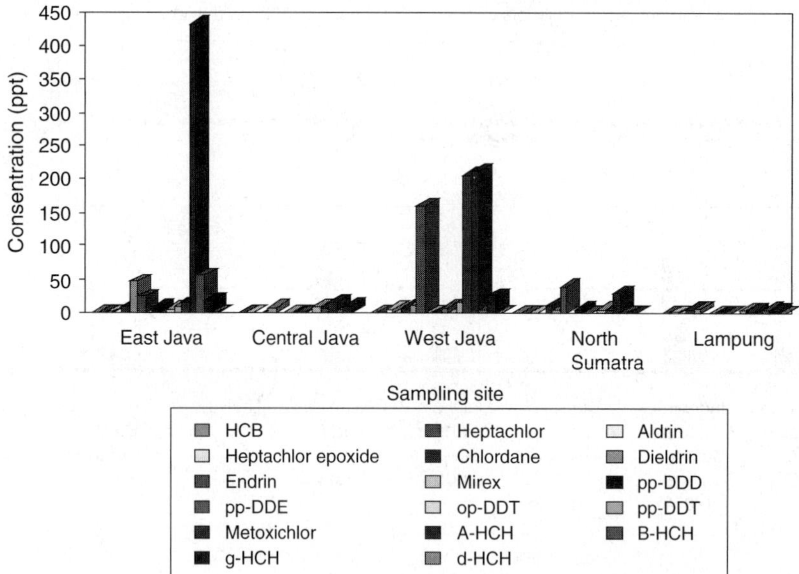

Figure 10E.127 Highest concentrations of POPs and OCs compounds in Indonesian river sediment in 2002. (From Syafrul, H., 2003, Environmental Standards Related to POPs in Indonesia in United Nations University, *Capacity Development Training for Monitoring of POPS in the East Asian Hydrosphere*, 1–2 September, 2003, UNU Centre, Tokyo, www.unu.edu. With permission.)

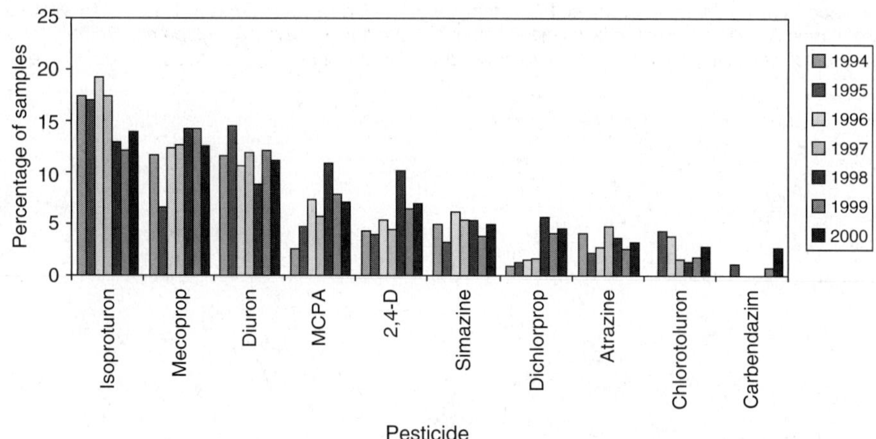

Figure 10E.128 Trends in exceedance of 0.1 ug/L drinking water directive standard for England and Wales, 1994–2000. (From Department of Environment, Food and Rural Affairs, 2002, *The Government's Strategic Review of Diffuse Water Pollution from Agriculture in England: Agriculture and Water: A Diffuse Pollution Review*, www.defra.gov.uk. Reproduced under the terms of the Click-Use Licence.)

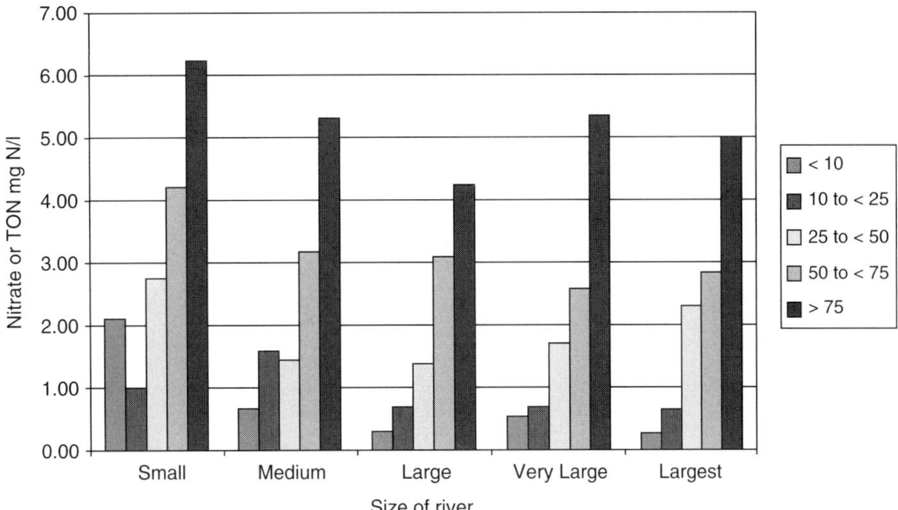

Figure 10E.129 Median annual average nitrate or total oxidized nitrogen concentrations (mg N/L) at stations in different sized rivers in relation to the percentage total agricultural land use in upstream catchment. (From European Environment Agency, *YIR01WQ1 Nitrogen and Phosphorous in River Stations by River Size and Catchment Type*, www.eea.europa.eu. Reprinted with permission © EEA.)

Nitrate concentrations in rivers

Nitrate pollution in rivers is higher in the EU-15 than in the New -10 (but lowest of all in the Nordic countries). This reflects differences in agricultural intensity and practices. In 2000/2001, rivers in 14 European countries (out of 24 with available information) exceeded the EU drinking water directive's guide concentration for nitrate; five also exceeded the maximum allowable concentration. In general nitrate concentrations in rivers are declining: 25% of monitoring stations on Europe's rivers recorded a decrease between 1992 and 2001. However, around 15% of river monitoring stations showed an increasing trend in nitrate concentrations over the same period.

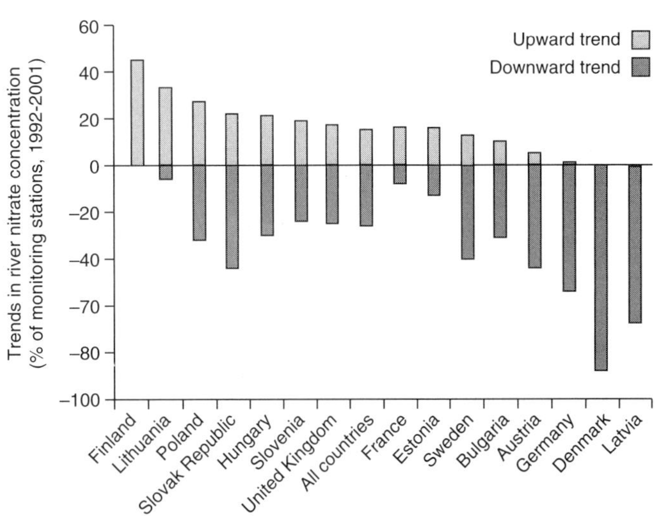

Figure 10E.130 Nitrate concentration trends in European rivers. (From European Environmental Agency, 2004, EEA Signals 2004, *A European Environmental Update on Selected Issues*, www.eea.europa.eu. Reprinted with permission © EEA.)

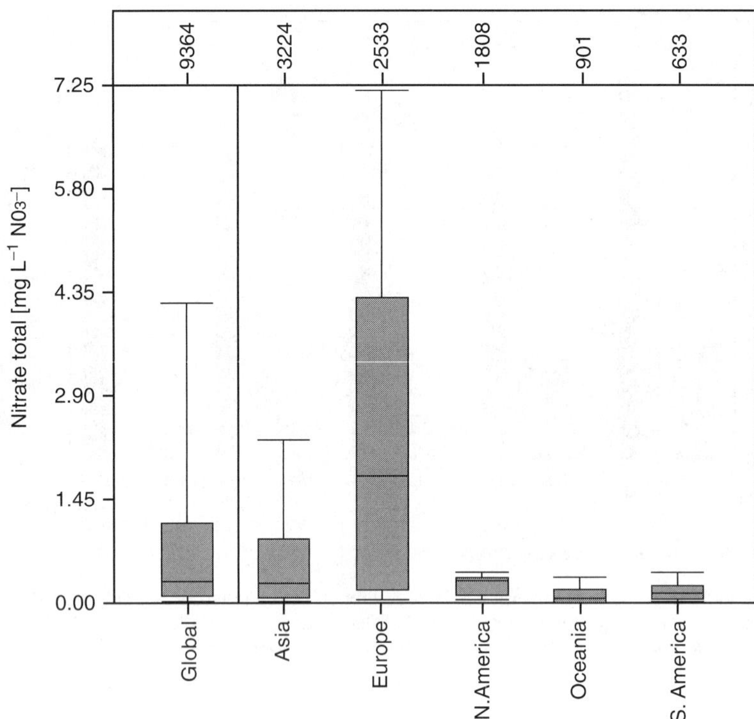

Figure 10E.131 Statistical distribution of nitrate for major global watersheds, 1976–1990. (From United Nations Environmental Programme Global Environment Monitoring System, Water Programme (GEMS/WATER), *The Annotated Digital Atlas of Global Water Quality*, www.gemswater.org. Reprinted with permission.)

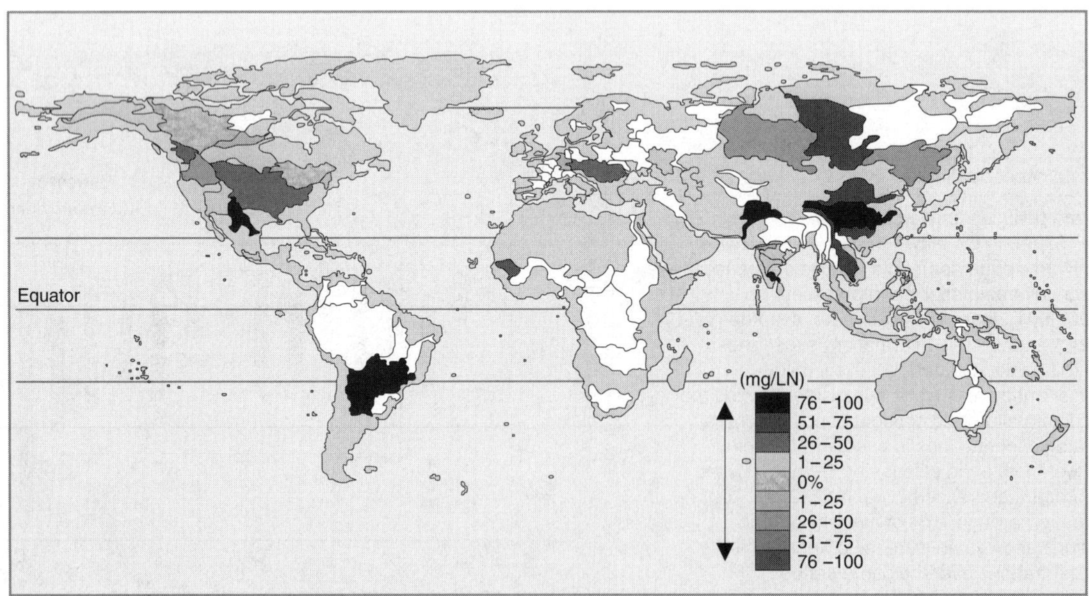

Figure 10E.132 Percent change of nitrogen, nitrate + nitrite in selected 82 watersheds. (From United Nations Environment Program, *Global Environment Monitoring System (GEMS) Water Programs, 2005, 2004 State of the UNEP GEMS/Water Global Network and Annual Report*, 2005, www.gemswater.org. Printed with permission.)

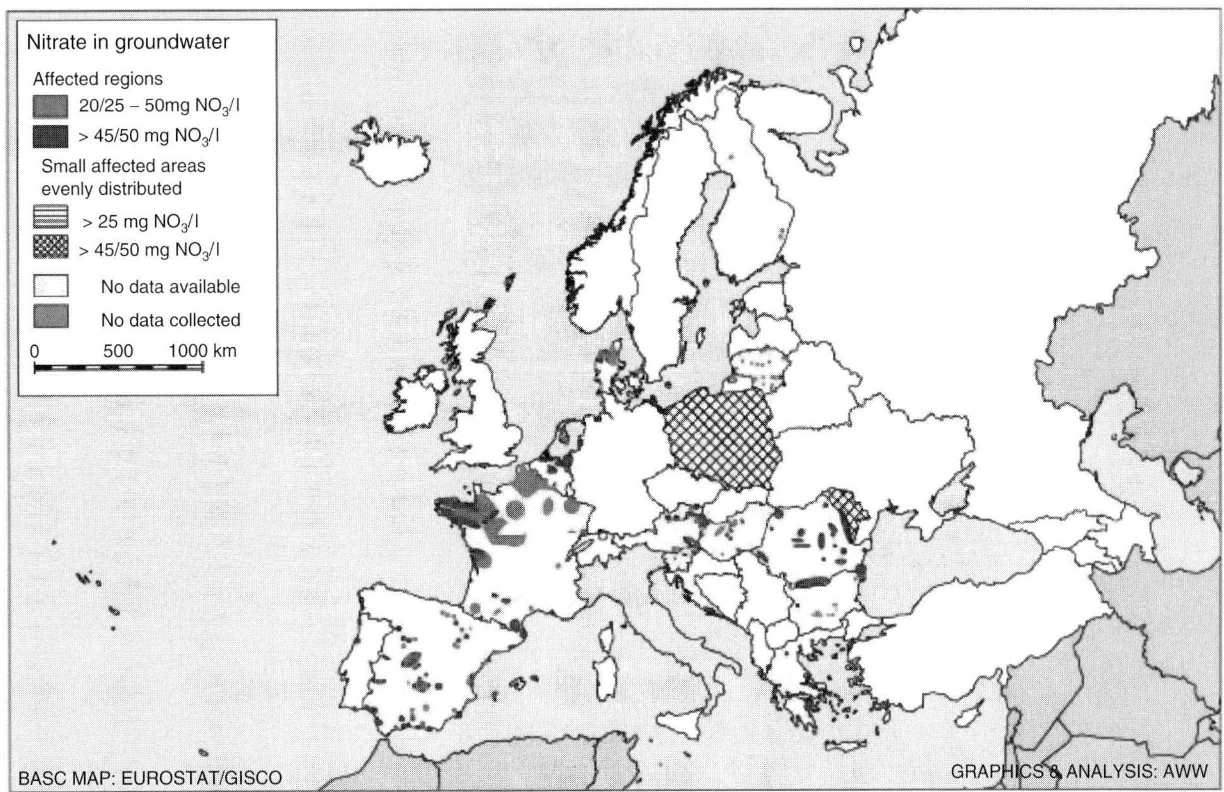

Figure 10E.133 Nitrate in groundwater in Europe. (From European Environmental Agency (EEA), 1999, *Groundwater Quality and Quantity in Europe*, June 1999, www.eea.europa.eu. Reprinted with permission © EEA.)

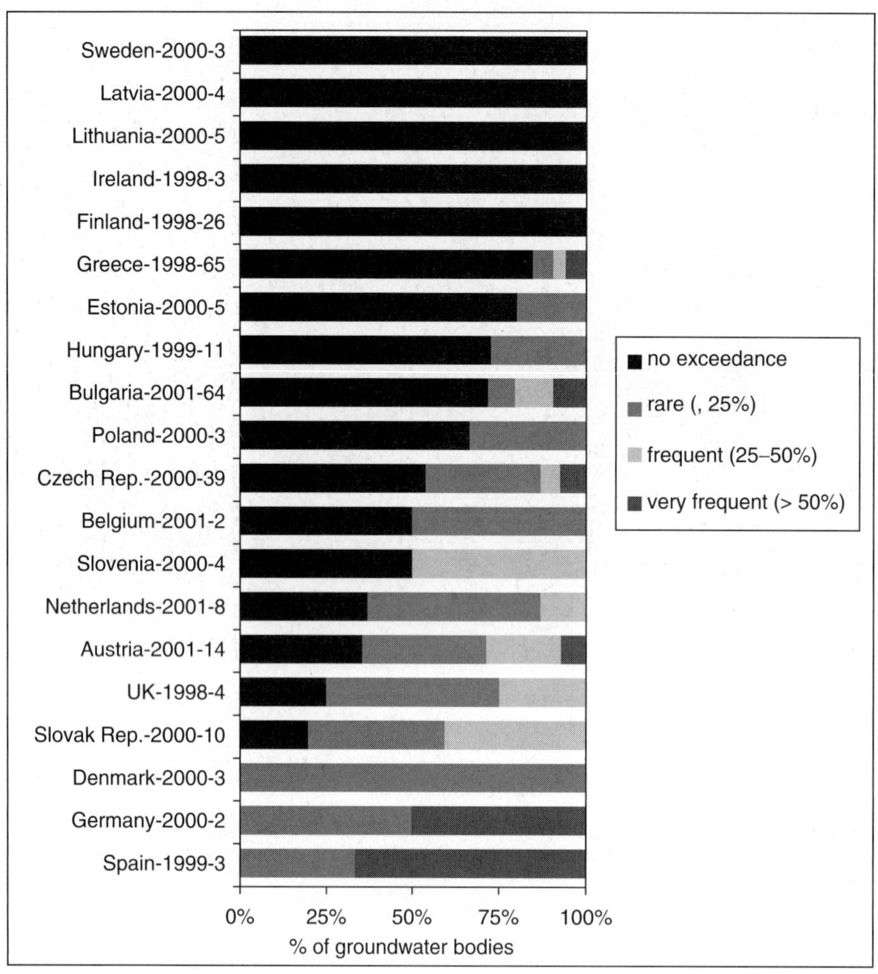

Note: The figure is based on the data for the latest year available (given after the country name). The numbers of groundwater bodies per country indluded in the presentation are given after the year. The four classes represent the percentage of sampling sites within each groundwater body where annual mean nitrate values exceed 50 mg No_3/litre.

Figure 10E.134 Percentage of sampling sites in groundwater bodies where annual mean values exceed 50 mg/L. Nitrate (From European Environment Agency, *Nitrate in Groundwater, Indicator Fact Sheet WEU01*, www.eea.europa.eu. Reprinted with permission © EEA.)

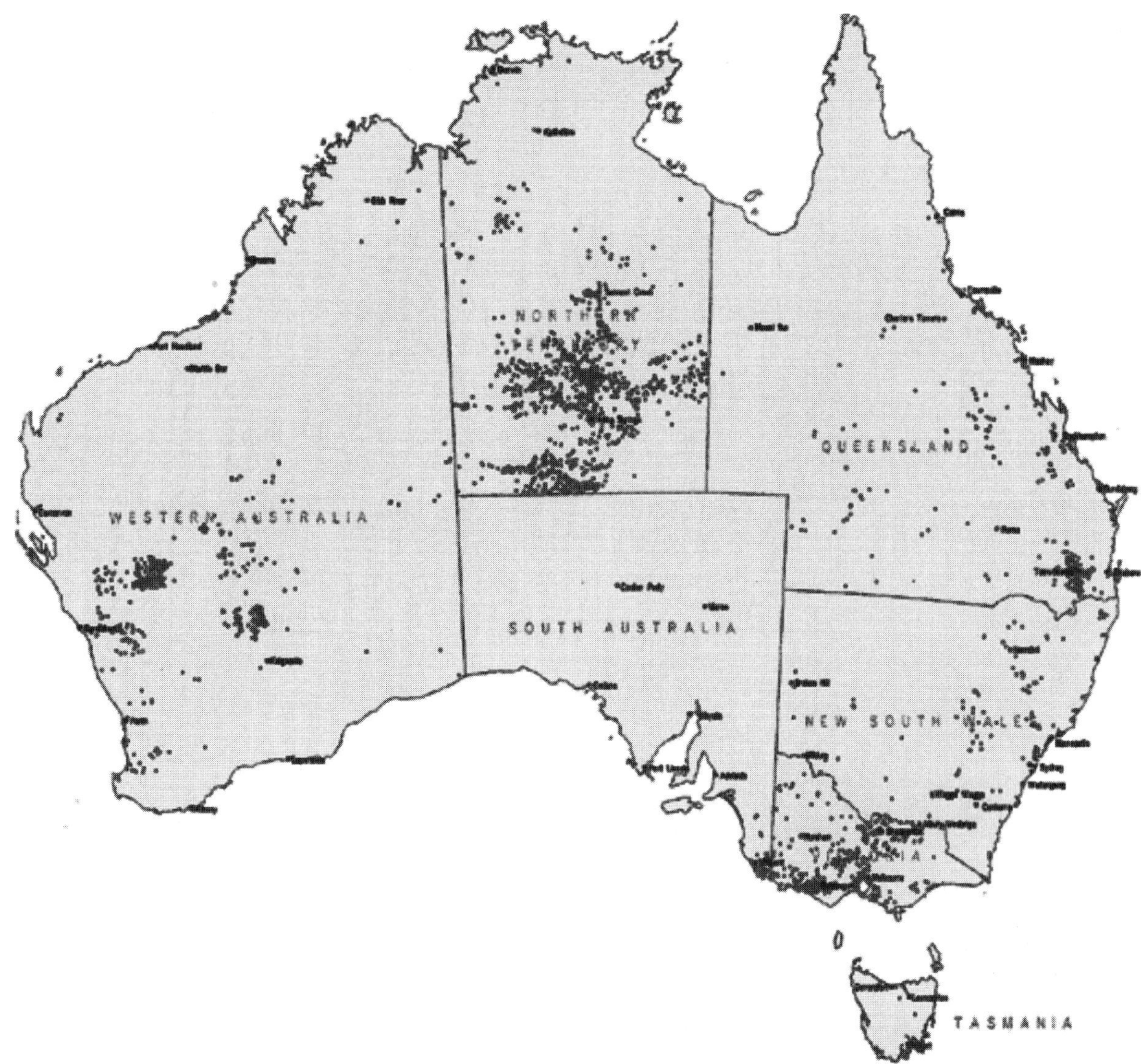

Figure 10E.135 Distribution of bores across Australia, with nitrate levels greater than 10 mg/L. (From Ball, J., Donnelley, L., Erlanger, P., Evans, R., Kollmorgen, A., Neal, B., and Shirley, M., 2001, Inland Waters, *Australia State of Environment Report 2001* (Theme Report), CSIRO Publishing on behalf of the Department of the Environment and Heritage, Canberra www.deh.gov.au. Original Source: LWRRDC, 1999, Contamination of Australian Groundwater Systems with Nitrate, Occasional Paper No. 03/99, Land and Water Resources Research and Development Corporation, Canberra, www.lwrrdc.gov.au. Copyright Commonwealth of Australia reproduced by permission.)

Table 10E.122 Trends in Regional Agricultural Activity and Soil Loss, 1975 and 2000 (High Growth Scenario)

	Acres in Crop Production				Soil Loss				
	1975		2000		1975		2000		
Region	Quantity (10^6 Acres)	Percent of National Total	Percent of 1975 Value	Percent of National Total	Quantity (10^6 Tons)	Percent of National Total	Percent of 1975 Value	Percent of National Total	Annual Average Soil Loss Per Acre (Tons)
I. New England	[a]	[b]	129	[b]	2	[b]	126	[b]	13
II. New York-New Jersey	1	1	119	1	17	[b]	120	[b]	14
III. Middle Atlantic	5	2	127	2	180	5	127	5	36
IV. Southeast	27	12	121	12	900	24	119	24	33
V. Great Lakes	65	29	132	33	880	24	136	27	14
VI. South Central	31	14	108	13	590	16	114	15	20
VII. Central	58	26	107	24	960	26	113	24	18
VIII. Mountain	26	12	105	10	72	1	100	2	3
IX. West	4	2	140	2	10	[b]	144	[b]	3
X. Northwest	7	3	129	3	86	2	129	2	13
Total[c]	**223**	**100**	**118**	**100**	**3,700**	**100**	**121**	**100**	**17**

[a] Less than 0.5 million acres.
[b] Less than 0.5 percent.
[c] Rounding may creat inconsistencies in addition.

Source: From U.S. Environmental Protection Agency, 1980, *Environmental Outlook* 1980.

Table 10E.123 Estimated Acreage and Erosion in the Contiguous United States, Selected Years, 1938–1997

Item	1938	1967	1977	1982	1987	1992	1997
Acreage				**Million acres**			
Cropland and CRP combined	398.8[a]	438.2	413.3	421	406.6	382.3	377
CRP land	—	—	—	—	3.8	34	32.7
Pasture	na	na	na	131.9	127.6	125.9	120
Range	na	na	na	408.9	402.8	398.9	406
Total erosion				**Billion tons/year**			
Cropland and CRP combined							
Sheet and rill	na	2.60[b]	1.93	1.69	1.52	1.21	1.06
Wind	na	na	na	1.38	1.4	0.95	0.84
Pasture							
Sheet and rill[c]	na	na	na	1.45	1.28	1.26	1.08
Wind[d]	na	na	na	0.13	0.13	0.13	0.12
Range							
Sheet and rill[c]	na	na	na	0.49	0.48	0.48	na
Wind[c]	na	na	na	1.92	1.77	1.76	na
Total cropland, pasture, range	na	na	na	7.12	6.46	5.76	na
Erosion per acre				**Tons/acre/year**			
Cropland							
Sheet and rill	na	5.9	4.7	4	3.7	3.1	2.8
Wind	na	na	5.3	3.3	3.2	2.4	2.2
Subtotal	8.9[d]	na	na	7.3	6.9	5.5	5
CRP							
Sheet and rill	—	—	—	—	2	0.6	0.4
Wind	—	—	—	—	6.8	0.7	0.3

(Continued)

Table 10E.123 (Continued)

Item	1938	1967	1977	1982	1987	1992	1997
Subtotal	—	—	—	—	8.8	1.3	0.7
Pasture							
Sheet and rill	na	na	na	1.1	1	1	0.9
Wind	na	na	na	0.1	0.1	0.1	0.1
Range							
Sheet and rill	na	na	na	1.2	1.2	1.2	na
Wind	na	na	na	4.7	4.4	4.4	na

Note: na, not available; CRP, Conservation Reserve Program.

[a] Based on 1939 census estimate of cropland.
[b] Kimberlin (1976), based on 1967 Conservation Needs Inventory.
[c] Based on multiplying published per acre erosion estimates times acreage.
[d] Based on dividing sum of sheet, rill, and wind erosion by total U.S. cropland acres.

Source: From Heimlich, R., 2003, Agricultural Resources and Environmental Indicators 2003, *Agriculture Handbook No. (AH722)*, www.ers.usda.gov; based on data from USDA, ERS, NRCS National Resources Investigations of 1977, 1982, 1987, 1992, and 1997, except where noted.

Table 10E.124 Representative Values for Nutrient Export and Input Rates for Various Land Uses in the United States

Land Use	Total Phosphorus	Total Nitrogen
A. Export rates (kg/ha/yr)[a,b]		
Forest	0.2	2.5
Nonrow crops	0.7	6.0
Pasture	0.8	14.5
Mixed agriculture	1.1	5.0
Row crops	2.2	9.0
Feedlot, manure storage	255.0	2,920.0
B. Total atmospheric input rates (kg/ha/yr)[a,b]		
Forest	0.26	6.5
Agricultural/rural	0.28	13.1
Urban industrial	1.01	21.4
C. Wastewater input rates (kg/capita/yr)[b]		
Septic tank input[c]	1.45	4.65

Note: All values are medians and are only approximations owing to the highly variable nature of data on these rates.

[a] Value in this table are all in kg/ha/yr, which is the standard for such measurements. To convert to pounds per acre per year, multiple by 0.892.
[b] *Source*: Reckhow et al., 1980.
[c] This is prior to absorption to soil during infiltration; generally, soils will absorb 80 percent or more of this phosphorus.

Source: From U.S. Environmental Protection Agency, 1988, *The Lake and Reservoir Restoration Guidance Manual*, EPA 440/5-88-002.

Table 10E.125 Contribution of Total Nitrogen and Phosphorous from Variable Nonpoint Sources (in Pounds per Acre per Year)

	Areal Loading Rates	
Source	**Nitrogen (N)**	**Phosphorus (P)**
Precipitation	5.0–9.8	0.04–0.05
Forest land	2.7–12	0.03–0.8
Range land	—	0.07–0.08
Crop land	0.10–12	0.05–2.7
Land receiving manure	3.6–12	0.7–2.7
Irrigation return flows		
Surface	3.0–27	1.0–4.1
Subsurface	4.0–18	2.9–10
Urban land drainage	6.4–8.9	1.1–5.4
Animal feedlot runoff	90–1,400	8.9–630

Source: From U.S. Geological Survey, 1984, National Water Summary 1983-Hydrologic Events and Issues, *Water Supply Paper 2250*; based on data from Loehr, R.C., 1974.

Table 10E.126 Constituents of Livestock Waste

	Per Day (lb.)	104 Days (lb.)	360 Head/Acre/yr (Tons)
Wet manure and urine	64	8,960	4,200
Dry mineral matter	2.1	294	144
Dry organic matter	8.2	1,148	540
Water	53.7	7,518	30.7 in.
Total nitrogen	0.380	55.0	24.9
Total phosphorus	0.048	6.7	3.2
Total potassium	0.260	36.4	16.8

Note: Some constituents of waste of a 1,000-pound bovine on a daily and feeding period basis, and of 360 head per acre on an annual basis.

Source: From Hansen, R.W., 1971, Livestock Waste Disposal and Water Pollution Control, *Colorado State University Cooperative Extension Service Bulletin* 480a.

Table 10E.127 Fertilizer Elements of Animal Excrements

	Dairy Cattle (lb./day)	Beef Cattle (lb./day)	Hens (lb./day)	Hogs (lb./day)	Sheep (lb./day)
Wet manure	88	64	59	50	37
Total mineral matter	1.80	2.1	4.5	1.3	1.5
Organic matter	7.20	8.2	12.9	5.9	6.9
Nitrogen (N)	0.36	0.38	2	0.4	
Phosphorus (P_2O_5)	0.10	0.048	0.69	0.18	
Potassium (K_2O)	0.15	0.26	0.34	0.1	

Note: Fertilizer elements of various complete animal excrements per 1,000 pounds of liveweight.

Source: From Hansen, R.W., 1971, Livestock Waste Disposal and Water Pollution Control, *Colorado State University Cooperative Extension Service Bulletin* 480a.

Table 10E.128 Trends in Gross Nutrient Discharges in Agricultural Runoff in the United States, 1975 and 2000

		2000 (Tons)	
Pollutant	1975 (Tons)	High Growth	Low Growth
Nitrogen	5,700	8,800	7,600
Phosphorus	3,000	5,700	4,900
Potassium	1,400	2,600	2,200
Total[a]	10,000	17,000	15,000

[a] Rounding may create inconsistencies in addition.

Source: From U.S. Environmental Protection Agency, *Environment Outlook* 1980.

Table 10E.129 Trends in Gross Discharges of Sediment in Agricultural Runoff in the United States, 1975 and 2000

		2000 (10^3 Tons)	
Pollutant	1975 (10^3 Tons)	High Growth	Low Growth
Total suspended solids	94,000	110,000	97,000
Total dissolved solids	40,000	49,000	42,000
Biochemical oxygen demand	630	760	650

Note: Conservative estimates; discharges may uniformly be too low by a factor of four of five.

Source: From U.S. Environmental Protection Agency, *Environmental Outlook* 1980.

Table 10E.130 Yield and Percentage Change in Yield of Nitrate, Total Phosphorous, and Suspended Sediment in Water-Resource Regions of the Conterminous United States, 1980–1989

	Nitrate 1980–1989		Total Phosphorus 1982–1989		Suspended Sediment 1980–1989	
Water-Resources Region	Yield, in Tons Per Square Mile Per Year	Percentage Change Per Year	Yield, in Tons Per Square Mile Per Year	Percentage Change Per Year	Yield, in Tons Per Square Mile Per Year	Percentage Change Per Year
North Atlantic	0.558	[a]	0.077	−1.4	32	−0.4
South Atlantic-Gulf	0.226	[a]	0.092	+0.1	20	+0.2
Great Lakes	0.647	[a]	0.067	−3.3	36	+0.5
Ohio-Tennessee	0.847	[a]	0.125	−1.0	85	−1.3
Upper Mississippi	0.989	−0.4	0.157	−1.2	102	−1.3
Lower Mississippi	0.333	−1.6	0.103	−3.8	111	−1.2
Souris-Red-Rainy	0.011	[a]	0.008	−0.8	4	+1.2
Missouri	0.060	[a]	0.028	−1.7	45	−0.2
Arkansas-White-Red	0.056	[a]	0.039	−3.1	31	−0.7
Texas-Gulf-Rio Grande	0.012	[a]	0.014	−0.9	15	−0.6
Colorado	0.057	[a]	0.036	−2.4	92	−0.8
Great Basin	0.049	[a]	0.018	−2.7	21	−0.2
Pacific Northwest	0.225	[a]	0.063	−1.7	40	−0.1
California	0.047	[a]	0.060	−1.4	21	−0.6

[a] Between −0.1 and +0.1.

Source: From Smith, R. A., Alexander, R. B., and Lanfear, K. J., 1993, Stream water quality in the conterminous United States – status and trends of selected indicators during the 1980's in national water summary 1990–91 — stream water quality, *U.S. Geological Survey Water-Supply Paper 2400*, www.usgs.gov.

Table 10E.131 Nitrogen Runoff Estimates in Coastal Regions by Major Source Category, Selected Years 1982–1987

	Total Nitrogen (1,000 lbs/yr)							
EDA Group	WWTPs	Industries	Urban Runoff	Cropland Runoff	Pature/Range Runoff	Forest Runoff	Upstream Sources	Total
North Atlantic	23,475	1,335	8,291	2,471	108	556	69,930	106,166
Middle Atlantic	178,810	22,237	46,367	36,277	780	1,166	314,539	600,177
South Atlantic	23,265	13,802	15,184	39,115	34	130	82,929	174,460
Gulf of Mexico East	27,516	11,154	17,131	42,310	29,362	51	3,057,405	3,184,927
Gulf of Mexico West	22,917	13,789	7,261	17,021	34,070	46	88,626	183,730
Pacific-South	83,163	1,530	9,437	32,472	33,811	5,589	58,784	224,786
Pacific-North	18,301	3,877	5,630	5,432	89,307	89,307	173,968	303,771

Source: From United States Department of Agriculture, Natural Resources Conservation Services, 1997, *Water Quality and Agriculture, Status, Conditions, and Trends*, www.nrcs.usda.gov.

Original Source: From NOAA, 1994.

Table 10E.132 Phosphorus Runoff Estimates in Coastal Regions by Major Source Category, Selected Years 1982–1987

	Total Phosphorus (1,000 lbs/yr)							
EDA Group	WWTPs	Industries	Urban Runoff	Cropland Runoff	Pasture/Range Runoff	Forest Runoff	Upstream Sources	Total
North Atlantic	14,826	312	1,368	132	1	6	4,114	20,760
Middle Atlantic	112,998	3,823	7,512	2,001	8	12	20,579	146,931
South Atlantic	14,580	10,872	2,311	2,664	0	1	17,159	47,588
Gulf of Mexico East	9,202	27,248	2,607	433	294	1	434,073	473,857
Gulf of Mexico West	13,990	6,580	1,105	173	341	0	20,458	42,646
Pacific-South	59,211	186	1,456	723	338	56	3,611	65,581
Pacific-North	13,749	304	886	113	73	893	27,452	43,470

Source: From United States Department of Agriculture, Natural Resources Conservation Services, 1997, *Water Quality and Agriculture, Status, Conditions, and Trends*, www.nrcs.usda.gov.

Original Source: From NOAA, 1994.

Table 10E.133 Load Yield, and Percentage Change in Load and Yield of Nitrate, Total Phosphorous, and Suspended Sediment in Six Coastal Segments of the Conterminous United States, 1980–1989

	Nitrate 1980–1988			Total Phosphorus 1982–1988			Suspended Sediment 1980–1988		
Coastal Segment	Load (tons/yr)	Yield in Tons Per Square Mile Per Year	Percentage Change Per Year	Load (tons/yr)	Yield, in Tons Per Square Mile Per Year	Percentage Change Per Year	Load (tons/yr)	Yield, in Tons Per Square Mile Per Year	Percentage Change Per Year
North Atlantic	171	0.972	+1.8	18	0.101	[a]	1,177	67	+8.0
South Atlantic	37	0.257	[a]	10	0.068	[a]	257	18	−1.1
Gulf of Mexico	1,178	0.687	−3.4	136	0.079	−3.5	16,607	97	−3.8
Great Lakes	205	1.570	[a]	14	0.108	−2.6	1,484	114	[a]
Pacific Northwest	74	0.267	[a]	15	0.055	−2.2	2,519	91	[a]
California	26	0.232	+2.1	15	0.132	[a]	4,684	502	−21.0
Total load	1,691	—	—	207	—	—	26,728	—	—
Area-weighted average yield	—	0.638	—	—	0.079	—	—	104	—
Change in Total Load	—	—	−2.1	—	—	−3.0	—	—	−6.4

[a] Not statistically significant.

Source: From Smith, R.A., Alexander, R.B., and Lanfear, K.J., 1993, Stream water quality in the conterminous United States – status and trends of selected indicators during the 1980's in national water summary 1990–91 — stream water quality, *U.S. Geological Survey Water-Supply Paper 2400*, www.usgs.gov.

Table 10E.134 Farm Fertilizer Use in the United States, 1939–1998 (Millions of Tons of Primary Nutrients)

Year	Fertilizer Applied	Year	Fertilizer Applied	Year	Fertilizer Applied	Year	Fertilizer Applied
1939	1.6	1954	5.9	1969	15.5	1984	21.8
1940	1.8	1955	6.1	1970	16.1	1985	21.7
1941	1.9	1956	6.1	1971	17.2	1986	19.7
1942	2.1	1957	6.4	1972	17.2	1987	19.1
1943	2.4	1958	6.5	1973	18	1988	19.6
1944	2.6	1959	7.4	1974	19.3	1989	19.6
1945	2.7	1960	7.5	1975	17.6	1990	20.6
1946	3.1	1961	7.8	1976	20.8	1991	20.5
1947	3.3	1962	8.4	1977	22.1	1992	20.7
1948	3.6	1963	9.5	1978	20.6	1993	20.9
1949	3.9	1964	10.5	1979	22.6	1994	22.4
1950	4.1	1965	11	1980	23.1	1995	21.3
1951	4.7	1966	12.4	1981	23.7	1996	22.1
1952	5.2	1967	14	1982	21.4	1997	22.4
1953	5.6	1968	15	1983	18.1	1998	22.3

Source: From U.S. Department of Commerce, Bureau of the Census, 1976. *Historical Statistics of the United States: Colonial Times to 1970*, Series K 193. Washington, DC; U.S. Department of Commerce, Bureau of the Census. 1985. *Statistical Abstracts of the United States*: 1986, no. 1161, p. 654. Washington, DC; U.S. Department of Agriculture, Economic Research Service. 1987. *Inputs Situation and Outlook Report*. AR-5. Washington, DC; U.S. Geological Survey National Water Summary 1986; Heimlich, 2003; Agricultural Resources and Environmental Indicators 2003, *Agriculture Handbook No. (AH722)*, February 2003.

Table 10E.135 United States Commercial Fertilizer Use, 1960–1998

Year	Total Materials[a]	Primary Nutrient Use			
		Nitrogen (N)	Phosphate (P_2O_5)	Potash (K_2O)	Total[b]
		Million Ions			
1960	24.9	2.7	2.6	2.2	7.5
1961	25.6	3.0	2.6	1.2	7.8
1962	26.6	3.4	2.8	2.3	8.4
1963	28.8	3.9	3.1	2.5	9.5
1964	30.7	4.4	3.4	2.7	10.5
1965	31.8	4.6	3.5	2.8	10.9
1966	34.5	5.3	3.9	3.2	12.4
1967	37.1	6.0	4.3	3.6	14.0
1968	38.7	6.8	4.4	3.8	15.0
1969	38.9	6.9	4.7	3.9	15.5
1970	39.6	7.5	4.6	4.0	16.1
1971	41.1	8.1	4.8	4.2	17.2
1972	41.2	8.0	4.9	4.3	17.2
1973	43.3	8.3	5.1	4.6	18.0
1974	47.1	9.2	5.1	5.1	19.3
1975	42.5	8.6	4.5	4.4	17.6
1976	49.2	10.4	5.2	5.2	20.8
1977	51.6	10.6	5.6	5.8	22.1
1978	47.5	10.0	5.1	5.5	20.6
1979	51.5	10.7	5.6	6.2	22.6
1980	52.8	11.4	5.4	6.2	23.1
1981	54.0	11.9	5.4	6.3	23.7
1982	48.7	11.0	4.8	5.6	21.4
1983	41.8	9.1	4.1	4.8	18.1
1984	50.1	11.1	4.9	5.8	21.8
1985	49.1	11.5	4.7	5.6	21.7
1986	44.1	10.4	4.2	5.1	19.7
1987	43.0	10.2	4.0	4.8	19.1
1988	44.5	10.5	4.1	5.0	19.6
1989	44.9	10.6	4.1	4.8	19.6
1990	47.7	11.1	4.3	5.2	20.6
1991	47.3	11.3	4.2	5.0	20.5
1992	48.8	11.4	4.2	5.0	20.7
1993	49.2	11.4	4.4	5.1	20.9
1994	52.3	12.6	4.5	5.3	22.4
1995	50.7	11.7	4.4	5.1	21.3
1996	53.6	12.3	4.5	5.3	22.1
1997	55.0	12.4	4.6	5.4	22.4
1998	55.0	12.3	4.6	5.3	22.3

Note: Includes Puerto Rico. Detailed State data shown in (USDA, 1997). Fertilizer statistics used in this table include commercial fertilizers purchased for use on farms such as chemical fertilizers and natural processed and dried organic materials. Purchased natural processed and dried organic materials historically have represented about 1 percent of total nutrient use.

[a] Includes secondary and micronutrients. Most of the difference between total primary nutrient tons and total fertilizer materials is carrier or filler materials.

[b] Totals may not add due to rounding.

Source: From Heimlich, R., 2003, Agricultural Resources and Environmental Indicators 2003, *Agriculture Handbook No. (AH722)*, February 2003, www.ers.usda.gov.

Original Source: From Fertilizer use estimates for 1960–1984 are from USDA; *Commercial Fertilizers.* 1985 and earlier issues; those for 1985–1994 are from Tennessee Valley Authority (TVA), *Commercial Fertilizers,* 1994 and earlier issues; and those for 1995–1997 are from The Association of American Plant Food Control Officials, *Commercial Fertilizers,* 1995–1998.

Table 10E.136 United States Regional Commercial Nutrient Use for Year Ending June 30, 1989–1998

Region	1989	1990	1991	1992	1993	1994	1995	1996	1997	1998
					1,000 tons					
Nitrogen										
Northeast	313	306	299	328	350	376	349	334	354	351
Lake States	1,011	1,134	1,128	1,119	1,073	1,186	1,108	1,108	1,236	1,172
Corn Belt	3,041	3,215	3,280	3,279	3,003	3,562	3,228	3,354	3,243	3,220
Northern Plains	1,680	1,751	1,978	1,954	2,090	2,319	2,133	2,219	2,373	2,317
Appalachia	613	667	662	718	705	720	694	752	741	763
Southeast	643	670	627	65	682	701	640	694	660	635
Delta States	560	643	609	674	615	663	630	718	607	752
Southern Plains	1,217	1,117	1,223	1,192	1,235	1,377	1,208	1,186	1,258	1,282
Mountain	626	642	628	666	744	775	765	806	862	867
Pacific	916	921	838	849	886	953	953	1,122	1,010	938
U.S. total[a]	10,619	11,065	11,273	11,432	11,382	12,633	11,709	12,294	12,344	12,297
Phosphate										
Northeast	188	197	188	208	211	232	203	183	184	171
Lake States	477	508	479	468	474	465	461	474	537	487
Corn Belt	1,254	1,334	1,262	1,269	1,312	1,317	1,257	1,340	1,304	1,275
Northern Plains	522	550	583	577	646	649	617	626	713	772
Appalachia	361	381	384	409	410	412	399	396	413	409
Southeast	297	308	281	295	314	297	313	332	318	328
Delta States	154	177	154	180	172	192	197	213	198	221
Southern Plains	342	315	334	288	340	363	341	313	319	330
Mountain	253	279	255	270	296	298	300	312	306	338
Pacific	270	289	274	248	257	291	326	335	316	290
U.S. total[a]	4,119	4,339	4,195	4,212	4,431	4,517	4,412	4,523	4,609	4,621
Potash										
Northeast	232	261	262	267	262	299	280	230	234	228
Lake States	852	941	832	809	779	781	760	776	866	848
Corn Belt	1,974	2,132	2,044	1,987	2,034	2,133	1,996	2,098	2,153	2,074
Northern Plains	129	133	134	123	134	123	124	123	147	168
Appalachia	506	538	539	584	575	576	574	592	624	634
Southeast	558	559	517	556	581	535	563	587	563	564
Delta States	212	240	229	280	288	302	336	351	323	316
Southern Plains	149	143	150	146	168	191	168	172	178	172
Mountain	53	65	80	55	80	68	79	79	80	91
Pacific	155	179	200	220	230	252	231	240	249	242
U.S. total[a]	4,820	5,192	4,988	5,026	5,131	5,259	5,112	5,248	5,416	5,335

Note: Totals may not add due to rounding. Northeast, ME, NH, VT, MA, RI, CT, NY, NJ, PA, DE, and MD; Lake States, MI, WI, and MN; Corn Belt, OH, IN, IL, IA, and MO; Northern Plains, ND, SD, NE, and KS; Appalachia, VA, WV, NC, KY, and TN; Southeast, SC, GA, FL, and AL; Delta States, MS, AR, and LA; Southern Plains, OK and TX; Mountain, MT, ID, WY, CO, NM, AZ, UT, and NV; and Pacific, WA, OR, CA, AK, and HA.

[a] Excludes Puerto Rico. Detailed state data shown in (USDA, 1995).

Source: From Heimlich, R., 2003, Agricultural Resources and Environmental Indicators 2003, *Agriculture Handbook No. (AH722)*, February 2003, www.ers.usda.gov.

Original Source: From USDA, ERS, based on Tennessee Valley Authority, *Commercial Fertilizers*, 1994 and earlier issues; The Association of American Plant Food Control Officials, *Commercial Fertilizers*, 1995–1998.

Table 10E.137 Production and Sales of Synthetic Organic Pesticides in the United States, 1960–1984

Item	Unit	1960	1965	1970	1975	1976	1977	1978	1979	1980	1981	1982	1983	1984
Production, total	Mil. lb	648	877	1,034	1,603	1,364	1,388	1,416	1,429	1,468	1,430	1,113	1,017	1,189
Herbicides	Mil. lb	102	263	404	788	656	674	664	657	806	839	623	570	716
Insecticides	Mil. lb	366	490	490	660	566	570	605	617	506	448	379	324	350
Fungicides	Mil. lb	179	124	140	155	142	143	147	155	156	143	111	123	123
Production value[a]	Mil. dol	307	577	1,058	2,900	2,880	3,116	3,342	3,685	4,269	5,136	4,331	3,993	5,056
Sales, total	Mil. lb	570	764	881	1,317	1,193	1,263	1,300	1,369	1,406	1,291	1,147	1,017	1,108
Sales value	Mil. dol	262	497	870	2,359	2,410	2,808	3,041	3,631	4,078	4,652	4,432	4,054	4,730

Note: Includes a small quantity of soil conditioners.

[a] Manufacturers unit value multiplied by production.

Source: From U.S. Department of Commerce, *Statistical Abstract of the United States*, 1987.

Table 10E.138 United States Production and Sales of Conventional Pesticides, 1994–2001

	1994/1995	1996/1997	1998/1999	2000/2001
Production-active ingredient (billions of pounds)	1.3	1.3	1.6	1.6
Sales Value (billons of dollars)	7.0	7.9	9.6	9.3

Source: From Kiely, T., Donaldson, D., and Grube, A., 2004, *Pesticides Industry Sales and Usage 2000 and 2001 Market Estimates*, United States Environmental Agency, Office of Prevention, Pesticides and Toxic Substances, EPA-733-R-04-001, May 2004. Donaldson, D., Kiely, T., and Grube, A., 2002, *Pesticides Industry Sales and Usage 1998 and 1999 Market Estimates*, United States Environmental Agency, Office of Prevention, Pesticides and Toxic Substances, EPA-733-R-02-001, August 2002. Aspelin, A. and Grube, A., 1999, *Pesticides Industry Sales and Usage 1996 and 1997 Estimates*, United States Environmental Agency, Office of Prevention, Pesticides and Toxic, 733-R-99-001, November 1999. Aspelin, A., 1997, *Pesticides Industry Sales and Usage 1994 and 1995 Market Estimates*, United States Environmental Agency, Office of Prevention, Pesticides and Toxic Substances, EPA-733-R-97-002, August 1997, www.epa.gov.

Table 10E.139 Annual Amount of Pesticide Active Ingredient Used in the United States by Pesticide Type, 1982–2001 Estimates All Market Sectors

	Million Pounds of Active Ingredient					
Year	Herbicides/PGR	Insecticides	Fungicides	Other Conv[a]	Other[b]	Total
1982	620	198	117	149	298	1,382
1983	573	185	115	148	287	1,308
1984	634	173	109	145	284	1,345
1985	611	161	110	138	284	1,304
1986	590	151	109	138	278	1,266
1987	532	141	100	133	269	1,175
1988	557	132	99	137	266	1,191
1989	567	123	98	154	251	1,193
1990	564	121	91	173	252	1,201
1991	546	114	86	182	226	1,154
1992	554	116	81	189	246	1,186
1993	527	115	80	192	248	1,162
1994	583	124	79	199	244	1,229
1995	556	125	77	203	249	1,210
1996	578	116	79	222	234	1,229
1997	568	112	81	197	270	1,228
1998	555	103	86	168	294	1,206
1999	534	126	79	173	332	1,244
2000	542	122	74	188	308	1,234
2001	553	105	73	157	315	1,203

Note: Excludes wood preservatives, specialty biocides, and chlorine/hypochlorites.

[a] Other conventional pesticides include nematicides, fumigants, and other conventional pesticides.

[b] "Other" includes sulfur, petroleum, and other chemicals used as pesticides (e.g., sulfuric acid and insect repellents).

Source: From Kiely, T., Donaldson, D., and Grube, A., 2004, *Pesticide Industry Sales and Usage, 2000 and 2001 Market Estimates*, United States Environmental Protection Agency, www.epa.gov, based on Croplife America annual surveys, USDA/NASS (www.usda.gov/nass), and EPA proprietary data.

Table 10E.140 Amount of Conventional Pesticide Active Ingredient Used in the United States by Pesticide Type and Market Sector, 2000 and 2001 Estimates

Year	Herbicides/Plant Growth Regulators		Insecticides/Miticides		Fungicides		Nematicide/Fumigant		Other Conventional[a]		Total	
Sector	Mil. lbs of a.i.	%	Mil. lbs of a.i.	%	Mil. lbs of a.i.	%	Mil. lbs of a.i.	%	Mil. lbs of a.i.	%	Mil. lbs of a.i.	%
2000												
Agriculture	432	80	90	74	44	59	131	84	25	78	722	78
Ind/Comm/Gov	48	9	17	14	19	26	24	15	6	19	114	12
Home & Garden	62	11	15	12	11	15	1	1	1	3	90	10
Total	542	100	122	100	74	100	156	100	32	100	926	100
2001												
Agriculture	433	78	73	70	42	58	102	80	25	83	675	76
Ind/Comm/Gov	49	9	15	14	19	26	24	19	4	13	111	13
Home & Garden	71	13	17	16	12	16	1	1	1	3	102	11
Total	553	100	105	100	73	100	127	100	30	100	888	100

Note: Totals may not add due to rounding. Table does not cover industrial wood preservatives, specialty biocides, chlorine/hypochlorites, and other chemicals used as pesticides (e.g., sulfur and petroleum oil). The abbreviation "a.i." stands for active ingredient.

[a] "Other Conventional" pesticides include rodenticides, molluscicides, aquatic and fish/bird pesticides, and other miscellaneous conventional pesticides.

Source: From Kiely, T., Donaldson, D., and Grube, A., 2004, *Pesticide Industry Sales and Usage, 2000 and 2001 Market Estimates*, United States Environmental Protection Agency, www.epa.gov, EPA estimates based on Croplife America annual surveys, USDA/NASS (www.usda.gov/nass), and EPA proprietary data.

Table 10E.141 Conventional Pesticide Active Ingredient Used in the United States Agricultural and Nonagricultural Market Sector Shares, 1964–2001

Year	Total U.S. Million Pounds of Active Ingredient	Agricultural Sector		Nonagricultural Sector Million Pounds of Active Ingredient
		Million Pounds of Active Ingredient	% of Total U.S.	
1964	617	366	59	251
1965	658	396	60	262
1966	682	414	61	268
1967	712	429	60	283
1968	742	457	62	285
1969	763	491	64	272
1970	760	499	66	261
1971	793	528	67	265
1972	843	575	68	268
1973	882	607	69	275
1974	964	688	71	276
1975	1013	729	72	284
1976	1041	753	72	288
1977	1084	794	73	290
1978	1106	813	74	293
1979	1144	843	74	301
1980	1121	826	74	295
1981	1118	831	74	287
1982	1084	804	74	280
1983	1021	745	73	276
1984	1061	794	75	267
1985	1020	767	75	253
1986	988	739	75	249
1987	906	666	74	240
1988	925	690	75	235
1989	942	712	76	230
1990	949	720	76	229
1991	928	708	76	220
1992	940	723	77	217
1993	914	698	76	216
1994	984	776	79	208
1995	961	765	80	196
1996	996	803	81	193
1997	958	767	80	191
1998	912	724	79	188
1999	912	706	77	206
2000	926	722	78	204
2001	888	675	76	213

Note: Conventional pesticides only, excluding sulfur, petroleum oil, and other chemicals used as pesticides (e.g., sulfuric acid and insect repellants), wood preservatives, specialty biocides, and chlorine/hypochlorites.

Source: From Kiely, T., Donaldson, D., and Grube, A., 2004, *Pesticide Industry Sales and Usage, 2000 and 2001 Market Estimates*, United States Environmental Protection Agency, www.epa.gov, EPA estimates based on Croplife America annual surveys, USDA/NASS (www.usda.gov/nass), and EPA proprietary data.

Table 10E.142 Most Commonly Used Conventional Pesticide Active Ingredients United States, Agricultural Market Sector, 2001, 1999, 1997, and 1987 Estimates

		2001		1999		1997		1987	
Active Ingredient	Type	Rank	Range	Rank	Range	Rank	Range	Rank	Range
Glyphosate	H	1	85–90	2	67–73	5	34–38	17	6–8
Atrazine	H	2	74–80	1	74–80	1	75–82	1	71–76
Metam sodium	Fum	3	57–62	3	60–64	3	53–58	15	5–8
Acetochlor	H	4	30–35	4	30–35	7	31–36	NA	NA
2,4-D	H	5	28–33	6	28–33	8	29–33	5	29–33
Malathion	I	6	20–25	7	28–32	NA	NA	NA	NA
Methyl bromide	Fum	7	20–25	5	28–33	4	38–45	NA	NA
Dichloropropene	Fum	8	20–25	11	17–20	6	32–37	4	30–35
Metolachlor-s	H	9	20–24	12	16–19	NA	NA	NA	NA
Metolachlor	H	10	15–22	8	26–30	2	63–69	3	45–50
Pendimethalin	H	11	15–19	10	17–22	9	24–28	10	10–13
Trifluralin	H	12	12–16	9	18–23	10	21–25	6	25–30
Chlorothalonil	F	13	8–11	13	9–11	15	7–10	19	5–7
Copper hydroxide	F	14	8–10	15	8–10	13	10–13	19	5–7
Chlorpyrifos	I	15	8–10	16	8–10	14	9–13	14	6–9
Alachlor	H	16	6–9	17	7–10	12	13–16	2	55–60
Propanil	H	17	6–9	18	7–10	22	6–8	13	7–10
Chloropicrin	Fum	18	5–9	14	8–10	25	5–6	NA	NA
Dimethenamid	H	19	6–8	20	6–8	20	6–9	NA	NA
Mancozeb	F	20	6–8	21	6–8	17	7–10	21	4–6
Ethephon	PGR	21	5–8	24	5–6	NA	NA	NA	NA
EPTC	H	22	5–8	19	7–9	18	7–10	8	17–21
Simazine	H	23	5–7	NA	NA	NA	NA	NA	NA
Dicamba	H	24	5–7	22	6–8	16	7–10	23	4–6
Sulfosate	H	25	3–7	NA	NA	NA	NA	NA	NA

Note: List is limited to conventional pesticides and does not include sulfur and petroleum oil usage. Ranked by Range in Millions of Pounds of Active Ingredient. H indicates herbicide; I, insecticide; Fum, fumigant; F, fungicide; and PGR, plant growth regulator. NA indicates that an estimate is not available.

Source: From Kiely, T., Donaldson, D., and Grube, A., 2004, *Pesticide Industry Sales and Usage, 2000 and 2001 Market Estimates*, United States Environmental Protection Agency, www.epa.gov, EPA estimates based on USDA/NASS (www.usda.gov/nass) and EPA proprietary data.

Table 10E.143 Quantity of Pesticides Applied to Selected Crops in the United States, 1990–2001

Type of Pesticide and Commodity	1990	1995	1996	1997	1998	1999	2000	2001
Total	497.7	543.3	575.8	579.3	544.4	553.7	539.4	511.1
Herbicides	344.6	324.9	365.7	362.6	340.3	316.8	308.6	307.5
Insecticides	57.4	69.9	59.2	60.2	52.0	75.4	77.4	62.0
Fungicides	27.8	47.5	46.8	48.5	45.7	42.3	36.6	33.2
Other	67.9	101.0	104.0	108.0	106.4	119.1	116.8	108.3
Corn	240.7	201.3	227.7	227.3	212.4	186.0	176.1	187.3
Cotton	50.9	83.7	65.6	68.4	55.4	90.6	94.5	72.8
Wheat	17.8	21.5	32.9	25.5	23.9	21.4	19.2	18.3
Soybeans	74.4	68.7	78.1	83.5	78.8	77.3	79.1	72.2
Potatoes	43.8	53.1	49.5	59.4	63.6	64.6	61.8	60.5
Other vegetables	39.8	78.0	82.8	73.3	67.8	70.5	70.1	66.5
Citrus fruit	11.0	14.0	14.5	15.0	14.1	13.3	13.0	12.8
Apples	8.3	9.0	9.7	10.6	9.3	7.9	7.6	7.6
Other deciduous fruit	10.9	14.1	14.9	16.4	19.2	22.2	18.1	13.1
Pounds of Active Ingredient Per Planted Acre								
Total	2.2	2.4	2.4	2.4	2.3	2.3	2.2	2.2
Herbicides	1.5	1.4	1.5	1.5	1.4	1.3	1.3	1.3
Insecticides	0.3	0.3	0.2	0.2	0.2	0.3	0.3	0.3
Fungicides	0.1	0.2	0.2	0.2	0.2	0.2	0.5	0.1
Other	0.3	0.4	0.4	0.4	0.4	0.5	0.5	0.5

Note: In million pounds of active ingredients, except as indicated (497.7 represents 497,700,000).

Source: From U.S. Census Bureau, Statistical Abstract of the United States, 2002, www.census.gov.

Original Source: From U.S. Dept. of Agriculture, Economic Research Service, *Production Practices for Major Crops in U.S. Agriculture, 1990–1997*, Statistical Bulletin No. 969, August 2000, and unpublished data.

Table 10E.144 Selected Characteristics and Uses of Pesticides Monitored by the U.S. Geological Survey-U.S. Environmental Protection Agency Pesticide Monitoring Network, 1975–1980

	Characteristics						Uses				
		Water-Quality Criteria[b] (μg/L)					National Use on Farms[e] (million lb/yr)				
Chemical	Detection Limit[a] (μg/L)	Human Health	Aquatic Life	Solubility[c] (μg/L)	Relative Persistence within Pesticide Group[d]	Principal Uses and Sources	1966	1971	1976	1982	Total Use, 1981[f] (million lb/yr)
					Organochlorine insecticides						
Aldrin	0.01	0.0007	0.002	13	Low	Corn	15	7.9	0.9	nr	0.8
									(Most farm uses cancelled 1974)		
Dieldrin	0.03	0.0007	0.002	22	Medium	Termite control, degradation product of aldrin	0.7	0.3	nr	nr	0
									(Most farm uses cancelled 1974)		
Chlordane	0.15	0.005	0.004	56	High	Corn, termites, general purpose	0.5	1.9	nr	nr	9.6
									(Most farm uses cancelled 1974)		
DDD	0.05	0.0002	0.001	5	High	Fruits and vegetables, degradation product of DDT	2.9	0.2	nr	nr	0
									(Cancelled 1972)		
DDE	0.03	0.0002	0.001	10	High	Degradation product of DDT and DDD	Nr	nr	nr	nr	0
DDT	0.05	0.0002	0.001	17	High	Cotton, fruits, vegetables, general purpose	27	0.1	nr	nr	0
									(Cancelled 1972)		
Endrin	0.05	*1	0.002	14	nd	Cotton, wheat	0.6	1.4	0.8	nr	0.3
Heptachlor epoxide	0.01	0.003	0.004	30	Low	Degradation product of heptachlor which is used on corn, and termite control	1.5	1.2	0.6	nrt	2.0
Lindane	0.01	*4	0.08	150	Medium	Livestock, seed treatment, general purpose	0.7	0.7	0.2	nr	0.8
Methoxychlor	0.10	*100	*.03	3	nd	Livestock, alfalfa, general purpose	2.6	3.0	3.8	0.6	5.0
Toxaphene	0.25	0.007	0.013	400	nd	Cotton, livestock	35	37	33	5.9	16
					Organophosphate insecticides						
Diazinon	0.10	nd	nd	40,000	High	Corn, general purpose	5.6	3.2	1.6	0.3	9.0
Ethion	0.25	nd	nd	2,000	nd	Citrus fruits	2.0	2.3	nr	nr	2.0
Malathion	0.25	nd	0.1	145,000	Low	General purpose	5.2	3.6	2.8	1.6	28
Methyl parathion	0.25	nd	nd	57,000	Low	Cotton and wheat	8.0	28	23	11	20
Methyl trithion	0.50	nd	nd	nd	nd	Not identified	nr	nr	nr	nr	0.1
Parathion	0.25	nd	0.04	24,000	Low	Wheat, corn, sorghum	8.5	9.5		4.0	5.0
Trithion	0.50	nd	nd	340	nd	General purpose	nr	nr	nr	nr	0.1
					Chlorophenoxy and triazine herbicides						
Atrazine	0.5	nd	nd	33,000	High	Corn	24	54	90	76	92
2,4-D	0.5	*100	nd	900,000	Low	Wheat, rangeland, general purpose	4	31	38	23	60
2,4,5-T	0.5	*10	nd	240,000	Medium	Rice, rangeland, general purpose	0.8	nr	nr	0.2	2.2
Silvex	0.5	nd	nd	140,000	nd	Sugarcane, rice, rangeland	nr	nr	nr	nr	0.4

Note: μg/L, microgram per liter; lb/yr = pounds per year; nd, no available data; nr, none reported. *See footnote[b].

[a] Detection limits shown are for water samples. Bed-sediment reporting limits are 10 times greater and are expressed in units micrograms per kilogram (Lucas and others, 1980).

[b] All criteria are from U.S. Environmental Protection Agency (1980), except for values marked by asterisks, which are from U.S. Environmental Protection Agency (1976). The human-health criteria for all pesticides except endrin, lindane, methoxychlor, 2,4-D, and 2,4,5-T represent the estimated average concentrations associated with an incremental increase in cancer risk of 10^{-8} (one additional cancer per 100,000 people over a lifetime of exposure). The aquatic-life criteria are for freshwater and are 24-hour average concentrations.

[c] Data from Kenaga and Goring (1980).

[d] Relative persistence within each pesticide group as estimated from Hiltbold (1974) and Wauchope (1978).

[e] Data for 1966, from Eichers and others (1970); for 1971, Andrilenas (1974); for 1976, Eichers and others (1978); for 1982, U.S. Department of Agriculture (1983). Data for 1982 do not include use on livestock or use in California, Colorado, Connecticut, Maine, Massachusetts, Nevada, New Hampshire, New Jersey, New Mexico, Oregon, Rhode Island, Utah, Vermont, West Virginia, and Wyoming.

[f] Data from Mark H. Glaze (U.S. Environmental Protection Agency, written communication, 1983).

Source: From U.S. Geological Survey, 1985. *National Water Summary 1984*. Water-Supply Paper 2275.

Table 10E.145 Detection Frequency of Targeted Pesticides in Surface Waters in the United States, 1957–1992

	Sampling Sites			Samples		
Pesticide	**Total Sites**	**Number of Sites with Detections**	**Percent of Sites with Detections**	**Total Samples**	**Number of Samples with Detections**	**Percent of Samples with Detections**
			Insecticides			
Organochlorine Compounds						
Aldin	951	65	7	3,910	224	6
Chlordane	838	154	18	3,366	948	28
DDT[a]	1,185	258	22	5,569	945	17
DDT-total (sum of DDT, DDD, DDE)	75	56	75	77	42	55
Dieldrin	1,016	459	45	4,995	1,412	28
Endosulfan	469	9	2	1,614	42	3
Endrin	944	136	14	4,255	359	8
HCH (all isomers)[b]	1,498	462	31	7,144	2,087	29
Heptachlor	948	102	11	3,877	287	7
Kepone	75	nr	nr	750	nr	nr
Methoxychlor	268	33	12	772	33	4
Mirex	212	2	1	512	13	3
Perthane	81	0	0	285	0	0
Toxaphene	215	16	7	1,490	84	6
Organophosphorus Compounds						
Azinphos-methyl	79	0	0	402	0	0
Chlorpyrifos	108	7	6	987	13	1
Crufomate	33	0	0	33	0	0
DEF	4	2	50	4	2	50
Diazinon	193	36	18	1,836	256	14
Dichlorvos (DDVP)	2	0	0	30	0	0
Dimethoate	33	0	0	33	0	0
Disulfoton	40	0	0	349	0	0
Disyston	4	0	0	4	0	0
Ethion	326	0	0	1,046	0	0
Ethoprop	33	0	0	33	0	0
Fenitrothion	42	0	0	42	0	0
Fensulfothion	9	0	0	9	0	0
Fenthion	232	0	0	538	0	0
Fonofos	94	14	15	945	63	7
Imidan	33	0	0	33	0	0
Malathion	426	16	4	2,415	104	4
Methamidophos	10	0	0	100	0	0
Methidathion	2	2	100	nr	nr	nr
Methyl parathion	387	13	3	2,215	14	1
Methyl trithion	80	0	0	185	0	0
Parathion	326	4	1	1,493	5	0
Phorate	121	0	0	1,008	0	0
Phosphamidon	33	0	0	33	0	0
Ronnel	35	0	0	63	0	0
Sulprofos	33	0	0	33	0	0
Terbufos	94	10	11	945	2	0
Trithion	314	1	0	805	2	0
Other Insecticides[c]						
Aldicarb	4	0	0	4	0	0
Carbaryl	24	6	25	333	32	10
Carbofuran	84	25	30	396	119	30
Deet	26	22	85	nr	nr	nr
Dibutyltin (DBT)	10	1	10	22	4	18
Fenvalerate	4	4	100	nr	nr	nr
Methomyl	8	0	0	8	0	0
Oxamyl	4	0	0	4	0	0
Permethrin	11	4	36	316	3	1
Propargite	7	3	43	316	3	1

(Continued)

Table 10E.145 (Continued)

Pesticide	Sampling Sites: Total Sites	Sampling Sites: Number of Sites with Detections	Sampling Sites: Percent of Sites with Detections	Samples: Total Samples	Samples: Number of Samples with Detections	Samples: Percent of Samples with Detections
Tributyltin (TBT)	10	8	80	22	15	68
Herbicides						
Triazines and Acetanilides						
Acrolein	17	nr	nr	121	nr	nr
Alachlor	372	272	73	1,549	802	52
Ametryn	123	11	9	947	212	22
Atratone	15	0	0	27	0	0
Atrazine	497	440	89	4,650	3,928	84
Cyanazine	366	242	66	1,473	755	51
Cyprazine	15	0	0	27	0	0
Hexazinone	26	5	19	nr	nr	nr
Metolachlor	362	280	77	1,452	827	57
Metribuzin	349	147	42	1,469	245	17
Prometon	270	74	27	828	140	17
Prometryn	62	9	15	523	4	1
Propachlor	12	7	58	450	48	11
Propazine	244	70	29	827	28	3
Simazine	209	119	57	632	312	49
Simetone	15	0	0	27	0	0
Simetryn	22	0	0	36	0	0
Terbutryn	4	0	0	16	0	0
Phenoxy Acids						
2,4-D	215	110	51	1,721	359	21
2,4-D (methyl ester)	6	0	0	84	0	0
2,4-DP	50	2	4	141	4	3
2,4,5-T	166	70	42	1,347	214	16
2,4,5-TP (silvex)	196	27	14	1,576	79	5
Other Herbicides						
Bensulfuron-methyl	3	2	67	54	16	30
Butylate	94	8	9	945	49	5
Chloramben	30	0	0	30	0	0
Dacthal	119	nr	nr	1,074	nr	nr
Dicamba	68	17	25	181	17	9
Dinoseb	4	0	0	16	0	0
Diquat	9	0	0	9	0	0
EPTC	15	7	47	316	63	20
Fluometuron	26	7	27	nr	nr	nr
Linuron	37	9	24	395	2	1
Molinate	27	7	26	16	16	100
Norflurazon	26	5	19	nr	nr	nr
Paraquat	9	0	0	9	0	0
Pendimethalin	15	14	93	316	25	8
Picloram	38	15	39	71	18	25
Propham	8	0	0	8	0	0
Thiobencarb	27	2	7	16	16	100
Trifluralin	104	24	23	1,087	113	10
Fungicides						
Captan	30	0	0	580	0	0
Chlorothalonil	4	0	0	16	0	0
HCB	50	43	86	255	216	85
PCNB	4	0	0	16	0	0
PCP	11	8	73	11	8	73
Transformation products						
Azinhos-methyl oxon	6	0	0	20	0	0
Carbofuran phenol	1	1	100	9	9	100
2-Chloro-2′,2′-diethylacetanilide	26	8	31	nr	nr	nr

(Continued)

Table 10E.145 (Continued)

	Sampling Sites			Samples		
Pesticide	**Total Sites**	**Number of Sites with Detections**	**Percent of Sites with Detections**	**Total Samples**	**Number of Samples with Detections**	**Percent of Samples with Detections**
Cyanazine amide	26	16	62	nr	nr	nr
DDD	876	139	16	3,941	543	14
DDE	1,128	219	19	4,869	939	19
Deethylatrazine	291	254	87	685	559	82
Deisopropylatrazine	242	154	64	685	249	36
Desmethyl norflurazon	26	2	8	nr	nr	nr
Endosulfan sulfate	50	0	0	154	0	0
Endrin aldehyde	50	0	0	154	0	0
ESA (alachlor metabolite)	76	60	79	304	222	73
Heptachlor epoxide	922	181	20	3,714	552	15
2-Hydroxy-2′6′-diethylacetanilide	26	19	73	nr	nr	nr
2-Ketomolinate	1	1	100	nr	nr	nr
4-Ketomolinate	1	1	100	nr	nr	nr
Oxychlordane	14	14	100	14	14	100
Paranitrophenol	1	1	100	9	9	100
Photomirex	14	0	0	14	0	0
Terbufos sulfone	33	0	0	33	0	0

Note: α, alpha; β, beta; γ, gamma; δ, delta. nr, not reported.

[a] Detection frequencies for DDT, DDD, and DDE include both *p,p′*-, and *o,p′*-isomers, as many studies did not report which isomer was targeted.

[b] HCH data for all isomers, including, α, β, γ (lindane), and δ.

[c] Includes compounds used as acaricides, miticides, and nematocides.

Source: From Larson, S.J., Capel, P.D., and Majewski, M.S., 1997, *Pesticides in Surface Waters Distribution, Trends, and Governing Factors, Volume Three of the Series Pesticides in the Hydrologic System*, Ann Arbor Press, Inc., Chelsea, Michigan.

Table 10E.146 Statistical Summary of Concentrations of Glyphosate, Its Degradation Product, AMPA, and Glufosinate Determined for Water Samples Collected form 51 Streams in Nine Midwestern States, 2002

Herbicide	Number of Samples	Number at or Above MRL	25th Percentile	Median	75th Percentile	95th Percentile	Maximum
Pre-emergence runoff samples							
Glyphosate	51	18	<0.10	<0.10	0.20	0.58	1.00
AMPA	51	27	<0.10	0.10	0.28	0.55	1.8
Glufosinate	51	0	<0.10	<0.10	<0.10	<0.10	<0.10
Post-emergence runoff samples							
Glyphosate	52	21	<0.10	<0.10	0.32	1.5	4.5
AMPA	52	43	0.18	0.27	0.42	0.94	2.0
Glufosinate	52	2	<0.10	<0.10	<0.10	<0.10	0.26
Harvest-season runoff samples							
Glyphosate	51	16	<0.10	<0.10	0.14	0.45	8.7
AMPA	51	37	<0.10	0.21	0.51	1.3	3.6
Glufosinate	51	0	<0.10	<0.10	<0.10	<0.10	<0.10

Note: All concentrations in micrograms per liter. MRL, method reporting limit; <, less than; AMPA, aminomethylphosphonic acid.

Source: From Scribner, E.A., Battaglin, W.A., Dietze, J.E., and Thurman, E.M., 2003, Reconnaissance Data for Glyphosate, Other Selected Herbicides, Their Degradation Products, and Antibiotics in 51 Streams in Nine Midwestern States, 2002, *U.S. Geological Survey Open-File Report 03-217*, www.usgs.gov.

Table 10E.147 Average Pesticide Concentrations Detected in Sediments from Selected Estuaries in the United States, 1984–1992 (milligrams per kilogram dry weight)

Estuary	tDDT	TDieldrin	tCdane	Hexachl	Lindane	Mirex
Machias Bay, ME-Hog Island	0.00	0.00	0.13	0.53	0.00	0.00
Machias Bay, ME-Chance Island	0.52	0.26	0.65	0.36	0.51	0.00
Frenchman Bay, ME-Long Porcupine Island	1.14	1.93	0.86	0.76	0.25	0.00
Penobscot Bay, ME-Colt Head Island	4.88	1.08	1.83	1.46	0.99	0.20
Johns Bay, ME-Pemaquid Neck	0.08	0.18	0.02	0.00	0.00	0.00
Casco Bay, ME-Great Chebeague Island	6.53	3.71	3.61	1.05	0.00	0.00
Casco Bay, ME-Cousins Island	6.60	0.23	1.63	1.43	0.25	1.00
Cape Elizabeth, ME-Richmond Island	0.32	0.15	0.07	0.07	0.07	0.00
Merrimac River, MA-Plum Island	0.00	0.17	0.89	0.70	0.21	0.00
Salem Harbor, MA-Folger Point	24.49	1.64	6.12	1.77	2.02	0.31
Boston Harbor, MA-President Roads	87.01	1.33	7.14	1.12	0.52	1.10
Boston Harbor, MA-Deer Island	7.21	1.03	7.04	1.23	0.08	0.25
Boston Harbor, MA-Quincy Bay	32.65	5.16	11.53	0.67	0.44	1.40
Boston Harbor, MA-Hull Bay	6.27	1.67	3.23	0.53	0.17	0.00
Boston Harbor, MA-Mystic RIver	43.67	8.33	22.43	0.67	0.00	0.00
Massachusetts Bay, MA-Plymouth Entrance	0.00	0.00	0.00	0.00	0.00	0.00
Buzzards Bay, MA-West Island	1.77	0.70	0.78	0.26	0.68	0.03
New Bedford Harbor, MA-Clarks Point	14.38	0.00	3.90	1.88	0.07	0.33
Narragansett Bay, RI-Prudence Island	11.00	2.67	4.93	2.33	0.00	0.43
Narragansett Bay, RI-Conanicut Island	8.12	6.14	4.03	0.98	2.56	0.52
Niantic Bay, CON-Black Point	1.85	0.83	0.40	0.37	0.05	0.05
Long Island Sound, NY-New Haven	7.33	0.00	4.00	2.00	0.00	0.33
Long Island Sound, NY – Norwalk	7.50	2.00	3.33	0.95	0.00	0.33
Long Island Sound, NY-Long Island Shoal	0.10	0.02	0.04	0.03	0.06	0.00
Long Island Sound, NY-Rocky Point	0.00	0.00	0.00	0.36	0.00	0.00
Long Island Sound, NY-Lloyd Point	0.00	0.00	0.00	0.29	0.00	2.67
Long Island Sound, NY-Oak Neck Point	8.02	0.49	3.15	0.44	1.41	1.31
Hudson River, NY	26.63	3.50	5.00	0.77	0.00	0.67
Raritan Bay, NY-Upper Bay	9.10	1.25	2.08	0.29	0.16	0.22
Raritan Bay, NY-Gravesend Bay	17.48	1.83	4.35	0.97	0.52	0.27
Raritan Bay, NY-West Reach	31.10	3.50	7.04	0.83	0.71	0.58
Raritan Bay, NJ-East Reach	34.74	2.41	6.22	1.18	0.64	0.41
Raritan Bay, NJ-Lower Bay	35.75	2.05	7.79	1.89	1.23	35.40
Great Bay, NJ-Wells Island	6.97	0.67	1.93	1.30	0.67	0.00
Great Bay, NJ-Seven Island	8.36	0.75	2.29	0.41	0.16	0.29
Great Bay, NJ-Intercoastal Waterway	1.45	0.17	0.13	0.48	0.08	0.00
Delaware Bay, DE-Cherry Island Range	47.60	3.65	4.38	0.70	0.03	1.20
Delaware Bay, DE-Brandywine Shoal	4.05	0.75	6.15	1.03	0.89	0.00
Delaware Bay, DE -The Shears	4.63	0.00	0.53	0.60	0.17	0.00
Baltimore Harbor, MD-Fort McHenry Channel	30.13	2.18	8.97	6.18	0.00	0.42
Baltimore Harbor, MD-Brewerton Channel	15.83	3.75	5.69	5.90	1.70	0.31
Chesapeake Bay, MD-Gibson Island	10.52	0.80	2.45	1.06	0.79	0.00
Chesapeake Bay, MD-Chester River	5.32	1.65	1.55	0.35	0.70	0.58
Chesapeake Bay, MD-Kent Island	7.70	2.25	2.40	0.50	1.50	1.20
Chesapeake Bay, MD-Patuxent River	0.53	0.10	0.07	0.00	0.00	0.00
Chesapeake Bay, MD-Smith Island	1.11	0.19	0.99	0.14	0.50	0.00
Chesapeake Bay, VA-James RIver	3.33	0.90	0.67	0.20	0.00	0.00
Chesapeake Bay, VA-York RIver	2.11	0.19	1.02	0.36	0.59	0.00
Chesapeake Bay, VA-Elizabeth RIver	17.23	0.38	3.18	0.51	0.14	0.00
Pamlico Sound, NC-Jones Bay	1.43	0.00	0.00	0.00	0.00	0.00
Cape Fear River, NC-Horseshoe Shoal	0.10	0.00	0.00	0.00	0.00	0.00
Charleston Harboar, SC – Coastal	47.58	8.00	14.60	0.00	0.00	6.62
Charleston Harboar, SC-South Channel	1.77	0.00	0.03	0.08	0.00	0.14
Savannah River, GA-Elba Island	0.00	0.00	0.00	0.57	0.00	0.23
Sapelo Sound, GA-High Point	1.57	0.00	0.00	0.00	0.00	0.00
St. Johns River, FL -Trout River	5.03	0.47	3.93	0.00	0.00	0.00
St. Johns River, FL-West Mill Cove	6.86	0.28	1.09	0.00	0.00	0.00
St. Johns River, FL-Ortega River	11.70	0.00	4.20	0.00	0.00	0.00
St. Johns River, FL-Piney Point	1.07	0.00	1.60	0.00	0.00	0.00

(Continued)

Table 10E.147 (Continued)

Estuary	tDDT	TDieldrin	tCdane	Hexachl	Lindane	Mirex
St. Johns River, FL-Orange Point	0.00	0.00	0.49	0.00	0.00	0.00
St. Lucie River, FL-Stuart	6.93	0.40	1.17	0.00	0.00	0.00
Biscayne Bay, FL-North Bay	0.90	0.00	0.00	0.57	0.00	0.00
Biscayne Bay, FL-Chicken Key	0.53	0.00	0.00	0.00	0.00	0.17
Charlotte Harbor, FL-Cape Haze	0.14	0.00	0.06	0.10	0.00	0.00
Tampa Bay, FL-Northern Tampa Bay	0.27	0.06	0.02	0.06	0.00	0.00
Apalachicola Bay, FL -St George Island	2.05	0.78	0.30	0.19	0.00	0.00
St. Andrew Bay, FL-Military Point	718.50	0.00	0.00	0.00	0.67	0.00
Choctawhatchee Bay, FL	2.50	0.43	0.47	0.00	1.00	0.07
Pensacola Bay, FL	0.39	0.20	0.39	0.00	0.00	0.00
Mobile Bay, AL-North Point	17.21	0.13	0.15	0.05	0.00	0.00
Pascagoula River, MS-Escatawpa River	5.30	1.20	1.67	0.00	0.33	0.00
Pascagoula River, MS	3.62	0.14	1.06	0.44	0.20	0.54
Round Island, MS-Round Island	0.38	0.00	0.12	0.00	0.00	0.00
Herron Bay, MS-Heron Bay	0.17	0.00	0.00	0.02	0.00	0.00
Mississippi River Delta, LA-Southeast Pass	3.47	0.93	0.44	7.50	0.00	0.09
Mississippi River Delta, LA-Head of Passes	390.00	2.20	6.58	3.00	0.60	1.80
Barataria Bay, LA-Barataria Pass	0.04	0.11	0.01	0.02	0.03	0.01
Calcasieu River, LA-Prien Lake	473.00	2.00	6.07	3.67	0.70	3.33
Calcasieu River, LA-West Cove	25.67	0.83	0.00	0.00	0.00	0.23
Galveston Bay, TX-East Bay	0.00	0.00	0.00	0.00	0.00	0.00
Galveston Bay, TX-Trinity Bay	0.00	0.00	0.00	0.00	0.43	0.00
Galveston Bay, TX-Greens Bayou	123.67	7.17	71.00	32.83	0.22	0.17
Galveston Bay, TX-Goat Islands	6.38	0.94	10.72	10.20	0.12	0.00
Galveston Bay, TX-Morgans Point	0.90	0.00	0.17	1.35	0.00	0.00
Galveston Bay, TX-Eagle Point	1.67	0.20	1.39	1.67	0.04	0.04
Galveston Bay, TX-Texas City	0.00	0.00	0.00	0.00	0.00	0.00
Lavaca Bay, TX	0.52	0.00	0.00	0.37	0.00	0.00
San Antonio Bay, TX-Mosquito Point	0.20	0.00	0.00	0.00	0.00	0.00
San Antonio Bay, TX-San Antonio Bay	0.65	0.09	0.03	0.02	0.03	0.00
Corpus Christi Bay, TX-Long Reef	0.42	0.00	0.00	2.03	0.00	0.00
Lower Laguna Madre, TX -Laguna Heights	0.00	0.00	0.00	0.05	0.03	0.00
Lower Laguna Madre, TX -Long Island	0.00	0.00	0.00	0.00	0.00	0.00
San Diego Bay, CA-Outside	14.71	0.00	0.00	0.00	0.00	0.00
San Diego Bay, CA-National City	8.14	0.10	1.68	0.00	0.00	0.00
San Diego Bay, CA-28th Street	16.36	0.49	7.91	0.00	0.00	0.66
San Diego Bay, CA-North	7.33	0.18	2.01	0.03	0.00	0.12
San Diego Bay, CA-Harbor Island	37.93	0.00	1.12	0.00	0.00	0.00
San Diego Bay, CA-Shelter Island	3.89	0.00	0.86	0.27	0.04	0.00
Mission Bay, CA-Outside	0.33	0.00	0.17	0.23	0.00	0.00
Dana Point Harbor, CA-Outside	2.20	0.00	0.10	0.10	0.04	0.04
San Pedro Bay, CA-Seal Beach	31.26	0.00	1.60	0.00	0.02	0.09
San Pedro Bay, CA-LongBeach	97.60	0.75	8.19	0.18	0.03	0.17
San Pedro Bay, CA-Outer Harbor	498.66	0.94	0.78	0.09	0.17	0.02
San Pedro Bay, CA-Cerritos Channel	221.80	1.18	14.40	0.90	0.00	0.10
Santa Monica Bay, CA-Southeast	149.67	0.27	0.83	0.00	0.00	0.00
Santa Monica Bay, CA-South	833.67	0.32	1.12	0.03	0.12	0.05
Santa Monica Bay, CA-Manhattan Beach	2.51	0.00	0.31	0.00	0.01	0.00
Santa Monica Bay, CA-West	57.09	0.25	1.69	0.26	0.00	0.06
Santa Monica Bay, CA-Deep	24.85	0.00	0.00	0.00	0.00	0.00
Santa Monica Bay, CA-North	98.33	0.77	0.73	0.10	0.33	0.00
San Luis Obispo, CA-San Luis Obispo	6.90	0.00	0.00	0.00	0.00	0.00
Estero Bay, CA-Estero Bay	0.30	0.00	0.00	0.00	0.00	0.00
Monterrey Bay, CA-Indian Head Beach	0.74	0.00	0.01	0.00	0.00	0.00
San Francisco Bay, CA-Redwood City	11.42	2.00	2.27	0.68	0.20	0.12
San Francisco Bay, CA-Hunters Point	6.48	0.62	0.34	0.29	0.03	0.35
San Francisco Bay, CA-Oakland Estuary	27.33	2.10	5.56	0.79	0.18	0.29
San Francisco Bay, CA-Southampton Shoal	1.00	0.00	0.00	0.47	0.53	0.00
San Francisco Bay, CA-Oakland Entrance	5.3	0.00	0.00	0.57	0.00	0.00
San Francisco Bay, CA-Castro Creek	6.6	0.78	1.08	0.75	0.17	0.12

(Continued)

Table 10E.147 (Continued)

Estuary	tDDT	TDieldrin	tCdane	Hexachl	Lindane	Mirex
San Francisco Bay, CA-San Pablo Bay	3.7	0.28	0.33	0.28	0.12	0
Bodega Bay, CA-North	0.4	0.06	0.14	0.01	0.03	0.04
Coos Bay, OR-North Bend	0.83	0.15	0.08	0.14	0.03	0
Columbia River Mouth, WA-Desdemona Sands	0.15	0.08	0.04	0.03	0.00	0.04
Puget Sound, Nisqually Reach, WA	0.00	0.00	0.00	0.00	0.00	0.00
Puget Sound, Commencement Bay, WA	3.48	0.64	1.75	3.18	0.06	0.17
Puget Sound, WA-Elliott Bay	13.25	1.7	1.05	0.27	0.03	0.03
Boca de Quadra, AK-Bacrian Point	0.00	0.00	0.00	0.67	0.00	0.00
Lutak Inlet, AK-Chilkoot River Mouth	2.17	0.00	0.00	0.30	0.00	0.00
Skagway, AK-Skagway River	1.00	0.00	0.00	0.47	0.53	0.00
Nahku Bay, AK-East Side	0.00	0.00	0.00	0.00	0.00	0.00
Prince William Sound, AK-Port Valdez	1.00	0.00	0.00	0.47	0.53	0.00
Gulf of Alaska, AK-Kamishak Bay	0.00	0.00	0.00	0.73	0.70	0.00
Bering Sea, AK,-Dutch Harbor	1.83	0.00	0.00	0.15	0.00	0.00
Bering Sea, AK,-Port Moller	4.67	0.00	0.00	0.00	1.00	0.00
Chukchi Sea, AK-Red Dog Mine	1.53	0.00	0.00	0.00	0.27	0.00
Beaufort Sea, AK-Olitok Point	0	0.00	0.00	0.33	0.00	0.00
Beaufort Sea, AK-Prudhoe Bay	0.17	0.17	0.08	0.22	0.2	0.06

Note: tDDT is the sum of 2,4′-DDD, 4,4′DDD, 2,4′-DDE, 4,4′-DDE, 2,4′DDT, 4,4′-DDT. TDieldrin is the sum of aldrin and dieldrin. tCdane is the sum of cis-chlordane and *trans*-nonachlor, heptachlor, and heptachlorepoxide.

Source: From Abstracted from Marmon, M.R., Gottholm, W., and Robertson, A., 1998, *A Summary of Chemical Contaminant Levels at Benthic Surveillance Project Sites (1984–1992)—NOAA Technical Memorandum NOS ORCA 124*, noa.gov.

Table 10E.148 Selected Results of Studies that Monitored Pesticides in Bed Sediment in the United States, 1960–1994

	All Monitoring Studies (Published 1960–1994)				Recent Monitoring Studies (Published 1984–1994)			
	Concentration Range (μg/kg dry weight)		Total Number of Studies that Reported Data	Percentage of Studies with Detectable Residues in at Least One Sample	Concentration Range (μg/kg dry weight)		Total Number of Studies That Reported Data	Percentage of Studies with Detectable Residues in at Least One Sample
Target Analytes	In Detection Limits	In Maximum Concentrations			In Detection Limits	In Maximum Concentrations		
Aldrin	nr, 0.05–480	nd-1,065	119	28	nr, 0.1–480	nd-1,065	70	24
Chlordane[a]	nr, 0.01–4,800	nd-1,000	111	74	nr, 0.1–4,800	nd-510	62	69
Chlordane, *cis-*	nr, 0.5–100	nd-293	25	24	nr, 0.5–100	nd-90	22	14
Chlordane, *trans-*	nr, 0.5–50	nd-149	28	32	nr, 0.5–50	nd-149	25	24
DDD[a]	nr, 0.01–38	nd-5,100	95	91	nr, 0.01–10	nd-260	46	91
DDD, *o,p′-*	nr, 0.1–50	nd-1,312	29	28	nr, 0.1–50	nd-1,312	27	22
DDD, *p,p′-*	nr, 0.05–960	nd-5,820	50	52	nr, 0.1–960	nd-5,820	44	48
DDE[a]	nr, 0.01–24	nd-10,000	101	88	nr, 0.01–20	nd-430	48	90
DDE, *o,p′-*	nr, 0.1–50	nd-292	29	21	nr, 0.1–50	nd-292	27	15
DDE, *p,p′-*	nr, 0.1–960	nd-1,870	52	60	nr, 0.1–960	nd-1,870	45	53
DDT[b]	nr, 0.01–10	nd-6,480	96	80	nr, 0.01–10	nd-2,280	45	76
DDT, *o,p′-*	nr, 0.1–50	nd-807	35	37	nr, 0.1–50	nd-807	29	28
DDT, *p,p′-*	nr, 0.1–960	nd-3,752	51	53	nr, 0.1–960	nd-3,752	43	47
DDT, total	nr, 0.01–50	nd-30,200,000	44	80	nr, 0.1–50	nd-4,443,500	25	68
Diazinon	nr, 0.1–100	nd-13	42	19	nr, 0.1–50	nd-0.2	16	13
Dieldrin	nr, 0.01–100	nd-440	163	65	nr, 0.01–100	nd-440	97	54
Endosulfan[a]	nr, 0.1–100	nd-4,530	50	26	nr, 0.1–10	nd-339	34	21
Endosulfan I	nr, 0.1–480	nd-96	14	50	nr, 0.1–480	nd-96	13	46
Endosulfan II	nr, 0.1–960	nd-140	14	50	nr, 0.1–960	nd-140	13	46
Endrin	nr, 0.01–100	nd-120	131	25	nr, 0.1–100	nd-43	80	20
HCH[a]	nr, 0.1–100	nd-20	10	30	nr, 0.1–100	nd-20	9	22
HCH, α-	0.03–480	nd-110	44	20	nr, 0.1–480	nd-65	39	13
HCH, β-	nr, 0.1–480	nd-2,800	33	15	nr, 0.1–480	nd-2,800	30	17
HCH, δ-	nr, 0.1–480	nd-0.8	25	4	nr, 0.1–480	nd-0.8	25	4
Heptachlor	nr, 0.01–480	nd-16.6	119	21	nr, 0.1–480	nd-12	72	22
Heptchlor epoxide	nr, 0.01–480	nd-106	123	32	nr, 0.01–480	nd-106	78	28
Hexachlorobenzene	nr, 0.03–400	nd-7,500,000	47	32	nr, 0.1–400	nd-7,500,000	41	24
Lindane	nr, 0.05–480	nd-221	128	19	nr, 0.1–480	nd-65	75	16
Malathion	nr, 0.1–100	0	41	0	nr, 0.1–50	0	14	0
Methoxychlor[c]	nr, 0.1–4,800	nd-366	70	20	nr, 0.1–4,800	nd-366	50	20
Mirex	nr, 0.01–100	nd-1,834	85	13	nr, 0.01–50	nd-310	64	9
Nonachlor, *cis-*	nr, 0.5–100	nd-5.5	22	9	nr, 0.5–100	nd-5.5	22	9
Nonachlor, *trans-*	nr, 0.5–100	nd-16	21	10	nr, 0.5–100	nd-16	21	10
Pentachlorophenol	nr, 0.5–1,200	nd-770	13	31	nr, 0.1–1,200	nd-770	8	25
Toxaphene	nr, 0.05–9,600	nd-1,858,000	100	12	nr, 0.1–9,600	nd-2,800	65	11

Note: Results are presented for all monitoring studies (published 1960–1994), and for recent monitoring studies (published 1984–1994) that are listed in Tables 2.1 and 2.2 of the publication. Concentration range in detection limits: "nr" indicates that one or more studies did not report detection limits. Abbreviations: nd, not detected; nr, not reported; μg/kg, microgram per kilogram.

[a] Total or unspecified.
[b] (*o,p′* + *p,p′*)- or unspecified.
[c] *p,p′*- or unspecified.

Source: From Nowell, L.H., Capel, P.D., and Dileanis, P.D., 1999, *Pesticides in Stream Sediment and Aquatic Biota, Distribution, Trends, and Governing Factors, Volume Four of the Series Pesticides in the Hydrologic System*, Lewis Publishers, Boca Raton.

Table 10E.149 Statistical Summary of Organochlorine Concentrations in Sediment, in the United States, 1992–1995

			Concentration (μg/kg dry weight) at Given Percentile								
Target Analyte	**Number of Samples**	**RL**	**5**	**10**	**25**	**50**	**75**	**90**	**95**	**100**	**Frequency of Detection (percent)**
p,p′-DDE	421	1	<1	<1	<1	<1	2.2	7.28	12.9	220	39.4
Total PCBs[a]	428	100	<100	<100	<100	<100	<100	<100	145.5	13,000	5.8
Total PCBs[b]	207	50	<50	<50	<50	<50	<50	150	336	13,000	18.8
p,p′-DDD	350	1	<1	<1	<1	<1	<1	4.19	9.24	130	24.9
trans-Nonachlor	411	1	<1	<1	<1	<1	<1	2.08	3	18	15.8
p,p′-DDT	358	2	<2	<2	<2	<2	<2	4.03	12.05	180	18.7
Dieldrin	413	1	<1	<1	<1	<1	<1	1.5	3.23	18	14.3
cis-Chlordane	411	1	<1	<1	<1	<1	<1	1.8	3.3	17	15.8
cis-Nonachlor	412	1	<1	<1	<1	<1	<1	<1	1.6	10	9.2
trans-Chlordane	416	1	<1	<1	<1	<1	<1	2.2	3.775	20	17.1
Oxychlordane	405	1	<1	<1	<1	<1	<1	<1	<1	1.3	0.3
o,p′-DDD	350	1	<1	<1	<1	<1	<1	<1	2.09	150	10.9
Pentachloro-anisole	430	50	<50	<50	<50	<50	<50	<50	<50	<50	0
Hexachloro-benzene	442	50	<50	<50	<50	<50	<50	<50	<50	<50	0
o,p′-DDT	354	2	<2	<2	<2	<2	<2	<2	<2	30	3.1
Heptachlor epoxide	412	1	<1	<1	<1	<1	<1	<1	<1	4.6	1.2
o,p′-DDE	405	1	<1	<1	<1	<1	<1	<1	<1	22	2
γ-HCH	413	1	<1	<1	<1	<1	<1	<1	<1	5.2	1
Dacthal	414	5	<5	<5	<5	<5	<5	<5	<5	25	1.5
Endrin	412	2	<2	<2	<2	<2	<2	<2	<2	<2	0
α-HCH	415	1	<1	<1	<1	<1	<1	<1	<1	<1	0
Toxaphene	419	200	<200	<200	<200	<200	<200	<200	<200	240	0.2
Aldrin	418	1	<1	<1	<1	<1	<1	<1	<1	3	0.5
Mirex	418	1	<1	<1	<1	<1	<1	<1	<1	4.4	1.9
o,p′-Methoxychlor	382	5	<5	<5	<5	<5	<5	<5	<5	<5	0
p,p′-Methoxychlor	378	5	<5	<5	<5	<5	<5	<5	<5	71	0.8
β-HCH	411	1	<1	<1	<1	<1	<1	<1	<1	1.2	0.5
Heptachlor	419	1	<1	<1	<1	<1	<1	<1	<1	<1	0.2
Endosulfan	409	1	<1	<1	<1	<1	<1	<1	<1	8.8	2.7
cis-Permethrin	344	5	<5	<5	<5	<5	<5	<5	<5	26	1.2
trans-Permethrin	340	5	<5	<5	<5	<5	<5	<5	<5	15	0.9
Chloroneb	396	5	<5	<5	<5	<5	<5	<5	<5	<5	0
Isodrin	409	1	<1	<1	<1	<1	<1	<1	<1	<1	0

Note: All statistics apply to samples in the national data set. 5. Frequency of detection: percentage of samples with concentrations at or above the reporting limit. PCB, polychlorinated biphenyl; RL, reporting limit in microgram per kilogram dry weight; μg/kg, microgram per kilogram; <, less than.

[a] Total PCBs censored at a reporting limit of 100 μg/kg dry weight.

[b] Total PCBs censored at a reporting limit of 50 μg/kg dry weight.

Source: From Wong, C.S., Capel, P.D., and Nowell, L.H., 2000, Organochlorine Pesticides and PCBs in Stream Sediment and Aquatic Biota — Initial Results from the National Water-Quality Assessment Program, 1992–1995, *United States Geological Survey, Water-Resources Investigations Report 00-4053*, www.usgs.gov.

Table 10E.150 Comparison of Organochlorine Concentrations in Sediment in the United States with Sediment-Quality Guidelines, 1992–1995

	EPA Sediment-Quality Criteria		Boundary Values and Tier Assignments				
					NAWQA Sites in Each Tier (percent)		
Target Analytes	Criterion (μg/kg-oc)	Tier 1 Percent of Sites Exceeding Criterion	Tier 2–3 Boundary Value[a] (μg/kg dry weight)	Tier 1–2 Boundary Value[b] (mg/kg dry weight)	Tier 1	Tier 2	Tier 3
Total chlordane	—	—	0.5	6	10.4	8.3	81.3
Total DDT	—	—	1.58	46.1	4.7	32.1	63.2
Dieldrin	11,000	0	0.02	6.67[c]	0	14.3	85.7
Endosulfan I	—	—	2.9	i	—	1.2	98.8
Endrin	4,200	0	0.02	45[c]	0	0	100
Total HCH	—	—	0.32	3.7	0.2	1.2	98.5
Heptachlor epoxide	—	—	0.6	i	—	1.2	98.8
Hexachlorobenzene	—	—	22	i	—	0	100
Lindane	—	—	0.32	1.38	0.7	0.2	99
Total methoxychlor	—	—	19	i	—	0.5	99.5
Total PCBs	—	—	21.6	189	3.5	2.3	94.2
Toxaphene	—	—	1.5	100	0.2[d]	0[d]	99.8[d]
Probability of adverse effects		High			High	Intermediate	No indication

Note: Percentage of NAWQA sites in Tiers 1, 2, and 3 based on exceedance of sediment-quality guidelines. EPA sediment-quality criteria, and the percent exceedance, are shown because exceedance of this criterion triggers classification in Tier 1 for sites with sediment organic carbon data. Tier 1-2 and Tier 2-3 boundary values are also shown (see text for explanation). The probability of adverse effects on aquatic life is shown for each tier. I, insufficient guidelines to determine a Tier 1-2 boundary value; EPA, U.S. Environmental Protection Agency; PCB, polychlorinated biphenyl; wt., weight; μg/kg, microgram per kilogram; μg/kg-oc, microgram per kilogram of sediment organic carbon; —, no guideline available.

[a] Lowest of the lower screening values.
[b] Second lowest of the upper screening values.
[c] Values are appropriate only for site with no sediment organic carbon data; if sediment organic carbon data are available, the EPA sediment-quality criterion (column 1) should be used instead.
[d] Because the boundary values (1.5 and 100 μg/kg) for this compound are well below its detection limit in the present study (200 μg/kg), the percentage of sites in Tiers 1 and 2 must be considered as underestimates.

Source: From Wong, C.S., Capel, P.D., and Nowell, L.H., 2000, Organochlorine Pesticides and PCBs in Stream Sediment and Aquatic Biota — Initial Results from the National Water-Quality Assessment Program, 1992–1995, *United States Geological Survey, Water-Resources Investigations Report 00-4053*, www.usgs.gov.

Table 10E.151 Current and Maximum Rates of Toxaphene Accumulation in Great Lakes Sediments

Location	Number of Sediment Cores Analyzed	Accumulation Rates in Sediment ($ng/cm^2/yr$) Present	Maximum Historical
Lake Superior	3	0.097–0.14	0.25
Northern Lake Michigan	3	0.52–1.01	1.07
Southern Lake Michigan near Chicago Urban Area	1	0.24	0.32
Lake Ontario	3	0.39–0.60	1.4

Note: All accumulations are focus-corrected.

[a] Total or unspecified.
[b] (*O,p′* + *p,p′*)- or unspecified.
[c] *p,p′*- or total.

Source: From United States Environmental Protection Agency, 2000, *Deposition of Air Pollutants to the Great Waters Third Report to Congress*, EPA-453/R-00-005.

Original Source: From Pearson et al., 1997b, www.epa.gov.

Table 10E.152 Selected Results of Studies That Monitored Pesticides in Whole Fish and the Percentage of Those Studies That Exceeded Guidelines for the Protection of Fish-Eating Wildlife, 1960–1994

	Whole Fish: All Monitoring Studies (Published 1960–1994)					
	Concentration Range (μg/kg wet weight)					
Target Analytes	In Detection Limits	In Maximum Concentrations (C_{max})	Total Number of Studies	Percentage of Studies with Detectable Residues	Percentage of Studies with C_{max} that Exceeded NAS/NAE Guideline[a]	Percentage of Studies with C_{max} that Exceeded NYSDEC Guideline[b]
Aldrin	nr, 0.01–50	nd-12	32	19	0	0
Acephate	50	nd	1	0	—	—
Alachlor	nr, 10–100	nd	2	0	—	—
Aldicarb	50	nd	1	0	—	—
Atrazine	10–8,000	nd	3	0	—	—
Azinphos-methyl	50–300	nd	2	0	—	—
Carbaryl	5–50	nd-50	1	0	—	—
Carbofuran	nr, 20–200	nd-560	1	100	—	—
Carbofuran, 3-hydroxy-	20–200	1,490	1	100	—	—
Carbophenothion	50	nd	1	0	—	—
Chlordane[a]	nr, 0.1–100	nd-870	27	63	33	4
Chlordane, *cis-*	nr, 0.01–100	nd-1,090	44	61	20	5
Chlordane, *trans-*	nr, 0.01–50	nd-970	42	62	12	7
Chlorpyrifos	nr, 10	nd	2	0	—	—
Coumaphos	nr	nd	1	0	—	—
Cyanazine	10	nd	1	0	—	—
D, 2,4-	1	6	1	100	—	—
Dacthal (DCPA)	nr, 2–10	nd-13,400	14	79	—	—
DDD[a]	nr, 0.1–50	nd-12,500	19	84	26	26
DDD, *o,p′*-	nr, 0.01–50	nd-420	34	29	0	9
DDD, *p,p′*-	nr, 0.01–100	nd-31,000	61	77	10	31
DDE[a]	nr, 0.1–100	nd-31,500	21	90	33	62
DDE, *o,p′*-	nr, 0.01–50	nd-360	35	23	0	3
DDE, *p,p′*-	nr, 0.01–100	nd-140,000	61	89	28	43
DDT[a]	nr, 0.1–50	nd-6,750	22	77	27	50
DDT, *o,p′*-	nr, 0.01–50	nd-720	37	35	0	5
DDT, *p,p′*-	nr, 0.01–100	nd-4,600	55	58	7	20
DDT, total	nr, 0.1–100	nd-28,880	34	100	50	82
Demeton	50	nd	1	0	—	—
Diazinon	nr, 100	nd	2	0	—	—
Dichlorvos	nr	nd	1	0	—	—
Dicofol	nr, 1–10	nd-560	10	40	—	—
Dieldrin	nr, 0.01–100	nd-12,500	87	74	31	30
Dimethoate	41–50	nd	1	0	—	—
Disulfoton	nr	nd	1	0	—	—
Endosulfan[a]	nr, 0.1–20	nd-170	6	67	17	—
Endosulfan I	nr, 0.02–10	nd-285	8	38	13	—
Endosulfan II	nr, 2–10	nd-40	7	29	0	—
Endosulfan sulfate	nr, 2–200	nd-20	6	17	—	—
Endrin	nr, 0.01–100	nd-2,060	65	28	5	9
Endrin aldehyde	nr, 200	nd	2	0	—	—
Endrin ketone	nr	nd	1	0	—	—
EPN	200	nd	1	0	—	—
Ethoprop	nr	nd	1	0	—	—
Famphur	nr	nd	1	0	—	—
Fensulfothion	nr	nd	1	0	—	—
Fenthion	nr	nd	1	0	—	—
Fenvalerate	nr	nd-11	2	50	—	—
HCH[a]	nr, 0.1–100	nd-170	12	50	17	17
HCH, α-	nr, 0.0.1–100	nd-610	34	35	—	—
HCH, β-	nr, 0.01–100	nd-900	31	23	—	—
HCH, δ-	nr, 0.01–100	nd-41	25	12	—	—
Heptachlor	nr, 0.01–50	nd-600	34	35	6	6

(Continued)

Table 10E.152 (Continued)

	Whole Fish: All Monitoring Studies (Published 1960–1994)					
	Concentration Range (μg/kg wet weight)					
Target Analytes	**In Detection Limits**	**In Maximum Concentrations (C_{max})**	**Total Number of Studies**	**Percentage of Studies with Detectable Residues**	**Percentage of Studies with C_{max} that Exceeded NAS/NAE Guideline[a]**	**Percentage of Studies with C_{max} that Exceeded NYSDEC Guideline[b]**
Heptachlor epoxide	nr, 0.01–104	nd-480	64	50	5	2
Hexachlorobenzene	nr, 0.01–100	nd-27,000	51	45	—	14
Kepone	nr, 10–50	nd-2,800	6	50	—	—
Lindane	nr, 0.01–100	nd-120	42	24	2	2
Malathion	nr, 1–50	nd	3	0	—	—
Methamidophos	nr	nd	1	0	—	—
Methiocarb	nr	nd	1	0	—	—
Methomyl	nr	nd	1	0	—	—
Methoxychlor[c]	nr, 0.1–100	nd-130	15	47	—	—
Methyl parathion	nr, 50	nd-60	3	33	—	—
Metolachlor	nr	nd	1	0	—	—
Mevinphos	nr	nd	1	0	—	—
Mirex	nr, 0.01–100	nd-1,810	46	28	—	11
Monocrotophos	nr	nd	1	0	—	—
Nonachlor, *cis-*	nr, 0.01–100	nd-156	38	29	—	—
Nonachlor, *trans-*	nr, 0.01–107	nd-1,550	45	64	—	—
Oxadiazon	nr	2,200	1	100	—	—
Oxamyl	nr	nd	1	0	—	—
Oxychlordane	nr, 0.01–105	nd-640	50	34	—	—
Parathion	30–50	nd	2	0	—	—
Pentachloroanisole	nr	33–160	2	100	—	—
Pentachlorophenol	nr, 0.1–3,000	nd-4,520	2	50	—	—
Permethrin	nr	0.53	1	100	—	—
Perthane	0.01–1	nd	3	0	—	—
Phorate	50	nd	1	0	—	—
Phosdrin	20	nd	1	0	—	—
Photomirex	nr	196–400	2	100	—	—
T,2,4,5-	0.2	nd	1	0	—	—
Terbufos	50	nd	1	0	—	—
Tetrachlorvinphos	20	nd	1	0	—	—
Tetradifon	nr, 1–10	nd-2	5	20	—	—
Toxaphene	nr, 0.1–2,000	nd-280,330	51	37	37	—
Trichlorfon	50–80	nd	2	0	—	—
Trifluralin	nr, 2	7–126	2	100	—	—

Note: All concentrations are wet weight, NAS/NAE guideline, maximum recommended concentration for protection of fish-eating wildlife. NYSDEC guideline: New York fish flesh criterion for protection of piscivorous wildlife. Abbreviations and symbols: C_{max}, maximum concentration in the study; NAS/NAE, National Academy of Sciences and National Academy of Engineering; NYSDEC, New York State Department of Environmental Conservation; na, data not available; nd, not detected; nr, one or more studies did not report this information; μg/kg, microgram per kilogram; —, no guideline available.

[a] Total or unspecified.
[b] (*o,p′* + *p,p′*)- or unspecified.
[c] *p,p′*- or total.

Source: From Nowell, L.H., Capel, P.D., and Dileanis, P.D., 1999, *Pesticides in Stream Sediment and Aquatic Biota, Distribution, Trends, and Governing Factors, Volume Four of the Series Pesticides in the Hydrologic System*, Lewis Publishers, Boca Raton.

Table 10E.153 Statistical Summary of Organochlorine Concentrations in Fish in the United States, 1992–1995

Target Analyte	Number of Samples	Concentration (μg/kg wet weight) at Given Percentile									Frequency of Detection (percent)
		RL	5	10	25	50	75	90	95	100	
p,p′-DDE	233	5	<5	<5	6.9	27	80.5	186	326	2,400	79.8
Total PCBs	233	50	<50	<50	<50	<50	150	646	1,400	72,000	44.6
p,p′-DDD	224	5	<5	<5	<5	<5	14	33.5	46	1,200	42.4
trans-Nonachlor	231	5	<5	<5	<5	<5	9.2	21.8	29.4	120	33.8
p,p′-DDT	232	5	<5	<5	<5	<5	7.08	21.7	30.35	430	32.3
Dieldrin	232	5	<5	<5	<5	<5	6.15	21	36.05	260	28.9
cis-Chlordane	231	5	<5	<5	<5	<5	<5	15.8	28.8	150	24.2
cis-Nonachlor	230	5	<5	<5	<5	<5	<5	8.59	11.45	53	18.7
trans-Chlordane	232	5	<5	<5	<5	<5	<5	8.61	15.35	56	16.8
Oxychlordane	232	5	<5	<5	<5	<5	<5	5.77	11.35	30	12.1
o,p′-DDD	231	5	<5	<5	<5	<5	<5	<5	8.82	360	9.5
Pentachloro-anisole	232	5	<5	<5	<5	<5	<5	<5	7.56	87	8.2
Hexachloro-benzene	234	5	<5	<5	<5	<5	<5	<5	8.875	33	7.3
o,p′-DDT	228	5	<5	<5	<5	<5	<5	<5	5.2	140	5.7
Heptachlor epoxide	232	5	<5	<5	<5	<5	<5	<5	7.335	24	5.6
o,p′-DDE	227	5	<5	<5	<5	<5	<5	<5	5.06	130	4.9
γ-HCH	231	5	<5	<5	<5	<5	<5	<5	<5	30	4.3
Dacthal	231	5	<5	<5	<5	<5	<5	<5	<5	67	3
Endrin	232	5	<5	<5	<5	<5	<5	<5	<5	16	1.7
δ-HCH	230	5	<5	<5	<5	<5	<5	<5	<5	5.5	0.4
α-HCH	232	5	<5	<5	<5	<5	<5	<5	<5	5.4	0.4
Toxaphene	233	200	<200	<200	<200	<200	<200	<200	<200	210	0.4
Aldrin	233	5	<5	<5	<5	<5	<5	<5	<5	<5	0
Mirex	233	5	<5	<5	<5	<5	<5	<5	<5	<5	0
p,p′-Methoxy-chlor	233	5	<5	<5	<5	<5	<5	<5	<5	<5	0
o,p′-Methoxy-chlor	231	5	<5	<5	<5	<5	<5	<5	<5	<5	0
β-HCH	232	5	<5	<5	<5	<5	<5	<5	<5	<5	0
Heptachlor	233	5	<5	<5	<5	<5	<5	<5	<5	<5	0

Note: All Statistics apply to samples in the national data set. Compounds are listed in order of detection frequency. Frequency of detection: percentage of samples with concentrations at or above the reporting limit. PCB, polychlorinated biphenyl; RL, reporting limit in microgram per kilogram wet weight; μg/kg, microgram per kilogram; <, less than.

Source: From Wong, C.S., Capel, P.D., and Nowell, L.H., 2000, Organochlorine Pesticides and PCBs in Stream Sediment and Aquatic Biota - Initial Results from the National Water-Quality Assessment Program, 1992–1995, *United States Geological Survey, Water-Resources Investigations Report 00-4053*, www.usgs.gov.

Table 10E.154 Comparison of Organochlorine Concentrations in Whole Fish Samples Collected in the United States with Edible-Fish Guidelines for Protection of Human Health, 1992–1995

		FDA Action Level[b]		EPA Guidance for Use in Fish Advisories: Recommended Screening Value[a,c]	
Target Analytes	**EPA Cancer Group[a]**	**Standard (μg/kg)**	**Percent of Sites Exceeding Standard**	**Guideline (μg/kg)**	**Percent of Sites Exceeding Guideline**
Total chlordane	B2	300	0.4	80[d]	5.7
Total DDT	B2	5,000	0	300[d]	7.8
Total dieldrin	B2	300	0	7[d]	23
Endrin	D/E	300	0	3,000	0
Total heptachlor	B2	300	0	10[d]	3
Hexachlorobenzene	B2	300	0	10[d]	0
Lindane	B2/C	—	—	70[d]	0
Mirex	—[e]	100	0	2,000	0
Total PCBs	B2	2,000[f]	2.6	10[d]	45
Toxaphene	B2	5,000	0	100[d]	0.4

Note: Because chemical concentrations are in whole fish, whereas standards and guidelines apply to edible fish tissues, exceedance indicates that additional sampling of game fish fillets may be warranted. All concentrations are wet weight. EPA cancer group: B2, probable human carcinogen; C, possible human carcinogen; D, not classified; E, no evidence of carcinogenicity. FDA, Food and Drug Administration; EPA, U.S. Environmental Protection Agency; μg/kg, microgram per kilogram; —, no standard or guideline available.

[a] From U.S. Environmental Protection Agency (1995).
[b] From Food and Drug Administration (1990), unless otherwise specified.
[c] Based on chronic toxicity, unless otherwise specified.
[d] Based on 1 in 100,000 cancer risk.
[e] Not classified by EPA; however, classified as a probable human carcinogen by International Agency for Research on Cancer.
[f] FDA tolerance level (from Food and Drug Administration, 1984).

Source: From Wong, C.S., Capel, P.D., and Nowell, L.H., 2000, Organochlorine Pesticides and PCBs in Stream Sediment and Aquatic Biota - Initial Results from the National Water-Quality Assessment Program, 1992–1995, *United States Geological Survey, Water-Resources Investigations Report 00-4053*, www.usgs.gov.

Table 10E.155 Potential Sources of Pesticide Contamination of Groundwater

	Manufacturer/ Formulator	Dealer	Industrial User	Land Application
SPILLS AND LEAKS				
Storage areas	x	x	x	x
Storage tanks/pipelines	x	x	x	x
Loading/unloading	x	x	x	x
Transport accidents	x	x	x	x
DISPOSAL				
Process Waste	x		x	
Off-specification material	x			
Cancelled products	x	x	x	x
Containers	x	x	x	x
Rinsate			x	x
LAND APPLICATION				x
Leaching				x
Backflow to irrigation well				x
Run-in to wells, sinkholes				x
Mixing/loading areas				x
Feed lots				x

Source: From U.S. Environmental Protection Agency, office of Pesticides and Toxic Substances, 1987, *Agricultural Chemicals in Ground Water: Proposed Pesticide Strategy.*

Table 10E.156 Pesticide Concentrations Measured in Groundwater during Multistate Studies and Studies from the Pesticides in Groundwater Database in Relation to Drinking-Water-Quality Criteria in the United States, 1971–1993

Pesticide or Transformation Product	USEPA Water-Quality Criteria for Drinking Water (μg/L)				Ranges of Observed Concentrations (μg/L) in Groundwater						Frequencies of Exceedance of MCL or HA (if MCL not available) Among Sampled Wells (percent)					
	MCL	HA	10-kg Child		MCPS	NPS		NAWWS	MMS	PGWDB	MCPS	NPS		NAWWS	MMS	PGWDB
			Acute (1 d)	Chronic (7 yrs)		CWS	RD					CWS	RD			
Acifluorfen	nsr	nsr	2,000	100	ND	Q[a]			0.003–0.025	ND	NA	NA			NA	
Alachlor	2	nsr	100	NR	0.003–4.270	ND	4.2	0.0385–6.185		0.006–3,000	0.91[b]	ND	0.13	0.02		0.38
Alachlor ESA	nsr	nsr	nsr	nsr	0.100–8.630						NA					
2,6-Diethylaniline	nsr	nsr	nsr	nsr	0.002–0.022					ND	NA					ND
Hydroxyalachlor	nsr	nsr	nsr	nsr						0.910						NA
Aldicarb	3	1	1	1		ND	ND			0.08–1,264.00		ND	ND			4.6
Aldicarb sulfone	2	42	60	60		ND	ND			0.01–153.00		ND	ND			12
Aldicarb sulfoxide	4	9	10	10		ND	ND			0.01–1030.00		ND	ND			9.2
Aldrin	nsr	nsr	0.3	0.3		ND	ND			0.0052–21		ND	ND			0.033[c]
Ametryn	nsr	60	9,000	900	ND	ND	ND			0.01–0.200	ND	ND	ND			0
Arsenic	50[d]									1.6–680.0						18
Atrazine	3	3	100	50	0.003–2.090	Q-0.92	Q-7.0	0.03–6.719		0.001–1,500	0	0	0.13	0.1		0.64
Deethyl atrazine	nsr	nsr	nsr	nsr	0.002–2.320	ND	ND			0.05–2.860	NA	ND	ND			NA
Deisopropyl atrazine	nsr	nsr	nsr	nsr	0.050–1.170					0.100–3.540	NA					NA
Baygon (propoxur)	nsr	3	40	40		ND	ND			2.0–35.0		ND	ND			0.019
Bentazon	nsr	20	300	300	ND	ND	2.9			0.10–41.89	ND	ND	0			0.28
α-BHC	nsr	nsr	50	50	ND	ND	ND			0.0014–0.16[e]	ND	ND	ND			NA
β-BHC	nsr	nsr	nsr	nsr		ND	0.04			0.0014–0.16[a]		ND	ND			
δ-BHC	Nsr	nsr	nsr	nsr		Q[a]	Q[a]				NA	NA				
γ-BHC (lindane)	0.2	0.2	1,200	33		ND	Q-0.42			0.0006–180.000	0	ND	0.13			0.045
Bromacil	nsr	90	5,000	3,000		ND	ND			0.03–951.6		ND	ND			0.12
Butylate	nsr	350	2,000	1,000	ND	ND	ND			0.87–2.23	ND	ND	ND			0
Carbaryl	nsr	700	1,000	1,000	ND	ND	ND			0.03–610.00	ND	ND	ND			0
Carbofuran	40	36	50	50	ND	ND	ND			0.01–176.00	ND	ND	ND			0.26
Carboxin	nsr	700	1,000	1,000		ND	ND			ND		ND	ND			ND
Chloramben	nsr	100	3,000	200		Q[a]	Q[a]			1.00		0	0			0
α-Chlordane	2[f]	nsr	60[f]	0.5[f]		Q-0.01	Q-0.01			0.01–20.000[f]		0[f]	0[f]			0.20[f]
γ-Chlordane		nsr				ND	Q-0.01									
Chlorothalonil	nsr	nsr	200	200						0.140–1.100						0.18[c]
Chlorpyrifos	nsr	20	30	30	0.005–0.024					0.05–0.654	0					0
Cyanazine	nsr	1	100	20	0.010–0.880	ND	ND	0.1205–0.1485		0.002–29.0	0	ND	ND	0		0.29
Cyanazine amide	nsr	nsr	nsr	nsr	0.050–0.550						NA					
2,4-D	70	70	1,000	100	0.100–0.890	ND	ND			0.0079–57.1	0	ND	ND			0.33[c]
Dacthal (DCPA)	nsr	4,000	80,000	5,000	ND	ND	ND			0.010–300.0	ND	ND	ND			0
DCPA hydrolysis products	nsr	nsr	nsr	nsr	0.010–2.220	Q-7.2	Q-2.4			0.21–431.0	NA	NA	NA			NA
Dalapon	200	200	3,000	300		Q[a]	Q[a]			ND		0	0			ND
Diazinon	nsr	0.6	20	5	ND	Q[a]	Q[a]			0.01–3.2		0	ND			0.051
1,2-Dibromo-3-chloropropane (DBCP)	0.2	nsr	200	NR		Q	0.48–0.71			0.001–8000.00		0	0.27			5.5
Dicamba	nsr	200	300	300	0.100	ND	ND			0.006–44.0	0	ND	ND			0
1,3-Dichloropropene	nsr	nsr	30[f]	30[f]		ND	ND			0.279–140[f]		ND	ND			NA

Dieldrin	nsr	nsr	0.5	0.5	ND	ND	ND			0.001–2.600	ND	ND	ND			0.095[c]
Dinoseb	7	7	300	10		3.5	ND			0.008–47.00		0	ND			0.59
Diphenamid	nsr	200	300	300		ND	ND			ND		ND	ND			ND
Diquat	20	20	nsr	nsr						ND						ND
Disulfoton	nsr	0.3	10	3	ND	Q[a]	Q[a]			0.04–100.00		0	0			0.61
Diuron	nsr	10	1,000	300		ND	ND			0.01–5.37		ND	ND			0
Endothall	100	140	800	200						ND						ND
Endrin	2	2	20	3		ND	ND			0.001–3.5		ND	ND			0.024
Ethylene dibromide (EDB)	0.05	nsr	8	NR		ND	0.29			0.001–15,772.4		ND	0.13			10.6
Ethylene thiourea (ETU)[7]	nsr	nsr	300	100		ND	Q-16			0.725		ND	NA			NA
Fenamiphos	nsr	2	9	5		ND	ND			ND		ND	ND			ND
Fluometuron	nsr	90	2,000	2,000		ND	ND			0.8–5.000		ND	ND			0
Fonofos	nsr	10	20	20	ND					0.007–0.90	ND					0
Glyphosate	700	700	20,000	1,000						0.004–150.0						0
Heptachlor	0.4	nsr	10	5.0		ND	ND			0.001–0.8		ND	ND			0.12
Heptachlor epoxide	0.2	nsr	nsr	0.1		ND	ND			0.01–0.22		ND	ND			0.032
Hexachlorobenzene	1	nsr	50	50		Q-0.17	ND			0.0039–0.0056		0	ND			0
Hexazinone	nsr	200	3,000	3,000		ND	ND			0.060–0.720		ND	ND			0
Malathion	nsr	200	200	200	ND					0.007–6.17	ND					0
MCPA	nsr	10	100	100						0.13–5.5						0
Methomyl	nsr	200	300	300		ND	ND			1.0–20.00		ND	ND			0
Methoxychlor	40	40	50	50		ND	ND			0.01–0.312		ND	ND			0
Methyl parathion	nsr	2	300	30	ND					0.01–0.256	ND					0
Metolachlor	nsr	100	2,000	2,000	0.003–1.460	ND	ND	0.0375–3.805	([h])	0.02–157.00	0	ND	ND	0	0	0.013
Metribuzin	nsr	200	5,000	300	0.050–0.220	ND	ND			0.001–25.10	0	ND	ND			0
Metribuzin DA	nsr	nsr	nsr	nsr		ND	ND			ND		ND	ND			ND
Metribuzin DADK	nsr	nsr	nsr	nsr		Q[a]	Q[a]			ND		NA	NA			ND
Metribuzin DK	nsr	nsr	nsr	nsr		Q[a]	Q[a]			ND		NA	NA			ND
Oxamyl	200	200	200	200		ND	ND			0.01–395.00		ND	ND			0.013
Paraquat	nsr	30	100	50						0.01–100.0						0.21
Pentachlorophenol	1	NR	1,000	300		ND	ND			0.001–0.64		ND	ND			0
Picloram	500	500	20,000	700	0.010–0.030	ND	ND			0.01–30.0	0	ND	ND			0
Prometon	nsr	100	200	200	0.050–1.350	Q	Q-0.57			0.05–29.6	0	0	0			0
Pronamide	nsr	50	800	800	ND	Q[a]	Q[a]			ND	ND	0	0			ND
Propachlor	nsr	90	500	100	0.002	ND	ND			0.02–3.5	0	ND	ND			0
Propazine	nsr	10	1,000	500	ND	ND	ND			0.01–0.20	ND	ND	ND			0
Propham	nsr	100	5,000	5,000		ND	ND			6.000		ND	ND			0
Simazine	4	4	500	50	0.002–0.270	Q-0.76	Q	0.043–8.359		0.001–67.0	0	0	0	<0.01		0.40[i]
2,4,5-T	nsr	70	800	300	0.020	ND	ND			0.01–2.99	0	ND	ND			0
Tebuthiuron	nsr	500	3,000	700	0.050	ND	ND			20.700–380.0	0	ND	ND			0
Terbacil	nsr	90	300	300	ND	ND	ND			0.3–8.9	ND	ND	ND			0
Terbufos	nsr	0.9	5	1	ND	Q[a]	Q[a]			0.02–20.0	ND	0	0			0.19
Toxaphene	3	nsr	500	nsr						1.15–18.000						0.070
2,4,5-TP (silvex)	50	50	200	70	ND	ND	ND			0.002–1.4	ND	ND	ND			0
Trifluralin	nsr	5	80	80	0.008–0.016	ND	ND			0.0018–14.890	0	ND	ND			0.018

Table 10E.156 (Continued)

Note: Pesticide or Transformation Product: Compounds listed are those for which one or more of the multistate studies or the studies included in the Pesticides In Ground Water Database conducted analyses, and for which one or more of the indicated criteria were reported by Nowell and Resek (1994); Transformation products are indented. Criteria listed do not account for the protection of aquatic organisms. Ranges of Observed Concentrations: Numbers of significant figures given for concentrations in the original references have been retained; Concentration data for the Pesticides In Ground Water Database exclude values given in original reference (U.S. Environmental Protection Agency, 1992b) as "0," "trace" or "$<$X," where X is some maximum value. CWS, Community water system wells; HA, lifetime health advisory level for a 70-kg adult; MCL, maximum contaminant level (MCL values given are the most recent ones for which, according to Nowell and Resek [1994], a published rationale is available); MCPS, Midcontinent Pesticide Study (Kolpin and others, 1995; Kolpin, 1995b; data shown are based on all three years of sampling reported to date [1991–1993], unless otherwise indicated); MMS, Metolachlor Monitoring Study (Roux and others, 1991a); NA, not applicable; NAS, National Academy of Sciences: NAWWS, National Alachlor Well Water Survey (Holden and others, 1992; Klein, 1993); ND, not detected; NPS, National Pesticide Survey (U.S. Environmental Protection Agency, 1990a); NR, health advisories for exposures over the longer term are not recommended due to the carcinogenic risk associated with this compound; nsr, no standard or guideline reported by Nowell and Resek (1994) for this compound, unless states otherwise; PGWDB, Pesticides In Ground Water Database (U.S.Environmental Protection Agency, 1992b); Q, Compound may have been present, be "[could not] be quantified or reliable detected" (U.S.Environmental Protection Agency, 199a); RD, Rural domestic wells; USEPA, U.S. Environmental Protection Agency; Blank cells indicate no information applicable or available; $<$, less than; kg, kilograms; µg/L, micrograms per liter.

[a] Detection reported in the overall summary of analytical results (U.S. Environmental Protection Agency, 1990a), but not in the data reported for the individual wells sampled for the NPS (U.S. Environmental Protection Agency, 1994b). No distinction made between detections in CWS and RD wells.

[b] Value for frequency of exceedance of MCL for alachlor pertains to 1993 sampling only (Kolpin and Thurman, 1995). No samples exceeded the MCL for alachlor in 1991 or 1992 (Kolpin and others, 1995).

[c] Basis for reported MCL or HA exceedance not provided in original reference (U.S. Environmental Protection Agency, 1992b).

[d] Source: U.S. Environmental Protection Agency (1992b).

[e] Original reference did not differentiate between α, β and δ isomers.

[f] No isomers specified.

[g] Ethylenebis(dithiocarbamate) fungicide transformation product.

[h] Compound detected, but full concentration range not reported.

[i] Exceedance frequency based on an MCL (1 µg/L) that has been superceded by a higher value of 4 mg/L (Nowell and Resek, 1994) since the publication of the Pesticides In Ground Water Database (U.S. Environmental Protection Agency, 1992b.)

Source: From Barbash, J.E. and Resek, E.A., 1996, *Pesticides in Ground Water Distribution, Trends and Governing Factors*, Ann Arbor Press, Chelsea, Michigan. With permission.

Table 10E.157 Upper 90th-Percentile Concentrations of Seven Herbicides Measured in Groundwater by the National Water Quality Assessment Program (1993–1995) and Midwest Pesticide Study Investigations (1991–1994)

Study Name	Sampling Phase	Upper 90th-Percentile Herbicide Concentrations Measured in Groundwater (μg/L)						
		Atrazine	Cyanazine	Prometon	Simazine	Acetochlor	Alachlor	Metolachlor
NAWQA	SGW (agricultural areas)	0.21	‡	0.008	0.013	‡	‡	0.006
	SGW (urban areas)	0.017	‡	0.078	0.010	‡	‡	‡
	SGW (mixed land-use areas)	0.010	‡	‡	‡	‡	‡	0.001
	Deeper aquifers	0.008	‡	‡	‡	‡	‡	‡
	DWA	0.056	‡	0.003	0.009	‡	‡	0.002
MWPS	7–8/92 (Random selection, postplanting)	0.086	‡	‡	0.002	NA	‡	0.003

Note: Study Name: NAWQA, National Water-Quality Assessment; MWPS, Midwest Pesticide Study. Sampling Phase: DWA, drinking-water aquifers; SGW, shallow ground water. NA, not analyzed. ‡, compound detected at fewer than 10 percent of the sties sampled.

Source: From Barbash, J.E., Thelin, G.P., Kolpin, D.W., and Gillion, R.J., 1999, *Distribution of Major Herbicides in Ground Water of the United States*, USGS Water-Resources Investigations Report 98-4245, www.usgs.gov.

Table 10E.158 Frequencies of Detection of Herbicide Degradates in Comparison with Those for the Corresponding Parent Compounds Measured in Groundwater by the National Water Quality Assessment Program (1993–1995) and the Midwest Pesticide Study Investigations (1991–1994)

Pesticide Compound (Degradates Indented)	Frequency of Detection During Different Study Phases (Percent of Sites)								
	NAWQA					MWPS		CGAS	
	RL (μg/L)	SGW (ag)	SGW (urban)	SGW (mixed)	Deeper aquifers	RL (μg/L)	1991–1994	RL (μg/L)	All wells
Alachlor	0.01	1.4	0	0.98	0.85	—	—	—	—
	0.05	0.5	0	0.98	0.42	0.05	3.3	—	—
2,6-Diethylaniline	0.003	1.0	0	0	0	0.003	16.0	—	—
	0.01	0.4	0	0	0	—	—	—	—
Alachlor ESA	—	—	—	—	—	0.10	45.8	—	—
Atrazine	0.01	31.1	14.5	10.3	7.8	—	—	—	—
	0.05	19.5	6.3	3.9	2.8	0.05	22.4	—	—
	—	—	—	—	—	—	—	0.10	23.9
DEA	0.01	28.2	10.4	17.2	6.1	—	—	—	—
	0.05	17.5	1.3	2.5	1.5	0.05	22.8	—	—
	—	—	—	—	—	—	—	0.10	28.8
DIA[a]	—	—	—	—	—	0.05	10.2	—	—
	—	—	—	—	—	—	—	0.10	14.9
Didealky-latrazine	—	—	—	—	—	—	—	0.10	24.1
Hydroxyatrazine	—	—	—	—	—	—	—	0.10	4.5
Deethyl hydroxyatrazine	—	—	—	—	—	—	—	0.10	2.8
Deisopropyl hydroxyatrazine	—	—	—	—	—	—	—	0.10	0.3
Didealkyl hydroxyatrazine	—	—	—	—	—	—	—	0.10	0.5
Cyanazine	0.01	1.2	0.9	0	0.1	—	—	—	—
	0.05	0.5	0	0	0	0.05	2.3	—	—
Cyanazine amide	—	—	—	—	—	0.05	11.0	—	—

(Continued)

Table 10E.158 (Continued)

Pesticide Compound (Degradates Indented)	Frequency of Detection During Different Study Phases (Percent of Sites)								
	NAWQA					MWPS		CGAS	
	RL (μg/L)	SGW (ag)	SGW (urban)	SGW (mixed)	Deeper aquifers	RL (μg/L)	1991–1994	RL (μg/L)	All wells
Deethylcyanazine	—	—	—	—	—	0.05	0	—	—
Deethylcyanazine amide	—	—	—	—	—	0.05	0	—	—

Note: CGAS, Ciba-Geigy atrazine study; MWPS, Midwest Pesticide Study; NAWQA, National Water-Quality Assessment. MWPS data from Kolpin and others (1996b). CGAS data from Balu and others (1998).ag, agricultures; DEA, deethylatrazine; DIA, deisopropylatrazine; ESA, ethanesulfonic acid; RL, reporting limit; SGW, shallow groundwater, μg/L, microgram per liter. —, no data available from sources consulted.

[a] DIA may be produced from the transformation of either atrazine, cyanazine, or simazine.

Source: From Barbash, J.E., Thelin, G.P., Kolpin, D.W.,and Gillion, R.J., 1999, *Distribution of Major Herbicides in Ground Water of the United States*, USGS Water-Resources Investigations Report 98-4245, www.usgs.gov.

Table 10E.159 Nitrate Concentrations in United States Groundwater by Land Use

Land Use Type	Median Concentration (mg/L)	Median Concentrations Exceeding Drinking-Water MCL (percent)
Forest land	0.1	3.0
Range land	1.5	8.5
Agricultural land	3.4	21.1
Urban land	1.9	7.0

Note: Maximum contaminant level for drinking water = 10 mg/L

Source: From United States Department of Agriculture Natural Resources Conservation Services, 1997, *Water Quality and Agriculture, Status, Conditions, and Trends*, www.nrcs.usda.gov.

Original Source: From Mueller et al., 1995.

Table 10E.160 Summary of Nitrate Concentrations in Groundwater Below Agricultural Land, by Region

Region[a,b]	Median Nitrate Concentration (mg/L)	Samples with Concentrations >3 mg/L[c] (percent)	Samples with Concentrations > 10 mg/L[d] (percent)
Northeastern states	4.3	58.2	19.6
Appalachian and Southeastern states	0.2	16.3	2.0
Corn Belt states	0.2	22.6	1.5
Lake states	0.1	25.5	14.6
Northern Plains states	6.0	60.6	35.2
Southern Plains states		(numbers requested)	
Mountain states	0.7	24.9	8.1
Pacific states	5.5	75.0	26.9

[a] Only includes wells less than 100 ft below the earth's surface.
[b] Only 17 of the 20 study units fit easily into the regional classification. Regions are based on R.F. Spalding and Exner, 1991. *Nitrate contamination in the contiguous United States*. Berlin, Springer-Verlag, NATO ASI Series, vol. G 30, P. 12–48. The eight regions correspond to the 10 USDA Farm Production Regions, except that three USDA regions (Delta, Appalachian, and Southeast) are combined.
[c] May indicate elevated concentration resulting from human activities.
[d] Maximum contaminate level for drinking water.

Source: From United States Department of Agriculture Natural Resources Conservation Services, 1997, *Water Quality and Agriculture, Status, Conditions, and Trends*, www.nrcs.usda.gov.

Original Source: From Mueller et al., 1995.

Table 10E.161 Median Concentrations of Nutrients in Samples from NAWQA Groundwater Studies, and Median Depth to Groundwater and to Sampled Zone within Aquifer

Water-Quality Variable or Well Parameter	MDL[a] (mg/L)	Agric. Land Use	Urban Land Use	Major Aquifer
Ammonia as N (mg/L)	0.01, 0.015[b]	0.02	0.03	0.02
Nitrite as N (mg/L)	0.01	<0.01	<0.01	<0.01
Ammonia-plus-organic nitrogen as N (mg/L)	0.20	<0.2	<0.2	<0.2
Nitrite-plus-nitrate as N (mg/L)	0.05	3.4	1.6	0.48
Orthophosphate as P (mg/L)	0.01	0.01	0.02	0.01
Depth to groundwater (m)	—	5.1	3.3	6.8
Depth of top of open interval below water level (m)	—	3.5	1.7	6.8

[a] Method detection limit.
[b] Ammonia method has two detection limits.

Source: From Nolan, B.T. and Stoner, J.D., Nutrients in ground waters of the conterminous United States, 1992–1995, *Environmental Science and Technology*, vol. 34, no. 7, 2000, p. 1156–1165.

Table 10E.162 The Largest Rivers in the World by Mean Annual Discharge with Their Loads

River	Basin Area (km^2)	Mean Annual Discharge (m^3/s)	Maximum Discharge (m^3/s)	Maximum Discharge (m^3/s)	Runoff (mm/yr)	Volume (km^3)	Suspended Solids (million tons/yr)	Dissolved Solids (million tons/yr)
Amazon (South America)	4,640.300	155,432	176,067	133,267	3,653	4,901	275	1,200
Congo (Central Africa)	3,475,000	40,250	54,963	32,873	1,056	1,296	41	43
Orinoco (Venezuela)	836,000	31,061	37,593	21,540	1,172	980	32	150
Yangtze (China)	1,705,383	25,032	28,882	21,377	463	789	247	478
Brahmaputra (India)	636,130	19,674	21,753	18,147	975	620	61	540
Yenisei (Russian Federation)	2,440,000	17,847	20,966	15,543	231	563	68	13
Lean (Russian Federation)	2,430,000	16,622	19,978	13,234	216	524	49	18
Parana (Argentina)	1,950,000	16,595	54,500	4,092	265	516		
Mississippi (United States)	3,923,799	14,703	20,420	10,202	118	464	125	210
Ob (Russian Federation)	2,949,998	12,504	17,812	8,791	134	394		

Note: The world's largest river, the Amazon, contributes by itself some 16 percent of the global total annual stream water flow, and the Amazon and the other four largest river system (Congo, Orinoco, Yangtze, Brahmaputra) combined account for 27 percent.

Source: From *Water for People Water for Life*, The United Nations World Water Development, Copyright © United Nations Educational, Scientific, and Cultural Organization (UNESCO) - World Water Assessment Programme (UNESCO-WWAP), 2003. Reproduced by permission of UNESCO. www.unesco.org.

Original Source: From GRDC, 1996, Berner and Berner, 1987.

Table 10E.163 Comparison of Phosphorus and Nitrogen Loading to Canadian Surface and Groundwaters from Various Sources, 1996

Nutrient Source	Nitrogen (10^3 t/yr)	Phosphorus (10^3 t/yr)
Municipality		
Municipal wastewater treatment plants	80.3	5.6
Sewers	11.8	2.3
Septic systems	15.4	1.9
Industry[a]	11.8	2.0
Agriculture[b] (residual in the field after crop harvest)	293	55
Aquaculture	2.3	0.5
Atmospheric deposition to water	182 (NO_3- and NH_4+ only)	N/A

[a] Industrial N loads are based on NO_3+NH_3 and are thus DIN not TN; industrial loads do not include New Brunswick, Prince Edward Island or Nova Scotia. Quebec data are only for industries discharging to the St. Lawrence River.

[b] Agricultural residual is the difference between the amount of N or P available to the growing crop and the amount removed in the harvested crop; data are not available as to the portion of this residual that moves to surface or groundwaters.

Source: From *Threats to Sources of Drinking Water and Aquatic Ecosystem Health in Canada*, page 24, Environment Canada, 2001. Reproduced with permission from the National Water Research Institute, Environment Canada, 2006.

Table 10E.164 Aspects of Water Quality of Some Alberta Canada Surface Waters

Water Body	Intensity of Farming	Nutrient	No. of Samples	Share That Exceeded Provincial Guidelines for Aquatic life[a] (%)
Streams (1995–1996)	High	Total nitrogen	214	87
		Total phosphorus	220	99
		Ammonia	70	0
	Moderate	Total nitrogen	343	65
		Total phosphorus	341	88
		Ammonia	126	0
	Low	Total nitrogen	163	32
		Total phosphorus	164	89
		Ammonia	162	0
Lakes (1995–1996)	High	Total phosphorus	69	96
	Low	Total phosphorus	23	38
Irrigation canals (1977–1996)		Total phosphorus		
Supply source			183	16
Return flow			1,034	61

[a] The limits for aquatic life are the Alberta Environment Protection guidelines of 1 mg/L total nitrogen, 0.05 mg/L total phosphorous, and the *Canadian Water Quality Guidelines for the Protection of Aquatic Life* maximum acceptable concentration of 1.13–1.81 mg/L ammonia-nitrogen (depending on temperature and pH).

Source: From *The Health of Our Water, Toward Sustainable Agriculture in Canada*, Publication 2020/E, Coote, D.R. and Gregorich, L.J., (eds.), Research Branch Agriculture and Agri-Food Canada, 2000.

Original Source: Canadian-Alberta Environmentally Sustainable Agriculture Water Quality Committee. 1998. Agricultural Impacts on Water Quality in Alberta: An Initial Assessment. CAESA Agreement. Lethbridge, Alta. Reproduced with the permission of the Minister of Public Works and Government Services Canada, 2006.

Table 10E.165 Average Nitrogen Surplus for Selected OECD Countries (Kilograms Per Hectare Per Year)

Country	1985–1987	1995–1997	Percent Change
Canada	7	14	113
Denmark	154	119	−23
France	59	53	−10
Japan	98	89	−9
New Zealand	5	6	27
Untied States	25	31	24

Source: From *The Health of Our Water, Toward Sustainable Agriculture in Canada*, Pubiication 2020/E, Coote, D.R. and Gregorich, L.J., (eds.), Research Branch Agriculture and Agri-Food Canada, 2000.

Original Source: From McRae, T., Smith, C.A.S., and Gregorich, L.J., (eds.), 2000. Environmental Sustainability of Canadian Agriculture: Agri-Environmental Indicator Project. Agriculture and Agri-Food Canada, Ottawa, Ont. Reproduced with the permission of the Minister of Public Works and Government Services Canada, 2006.

Table 10E.166 Total Nutrient Emission for Selected Coastal Regions of Australia (1,000 kg)

Coastal Region	Land Use	Phosphorus		Nitrogen	
		Point[a]	Diffuse[b]	Point[a]	Diffuse[b]
Peel-Harvey (WA)	Agriculture	na	260	na	1,800
Esk/Tamar (Tas)	Agriculture	150	360	390	1,600
Dawson River (Qld)	Agriculture	20	1,700	28	6,400
Latrobe-Thomson (Vic)	Agriculture	18	410	44	3,800
Richmond River (NSW)	Agriculture	5.8	250	na	1,700
Darwin Harbour (NT)	Mix	68	47	270	590
South-east Queensland (Qld)	Mix	1,300	1,500	3,300	7,000
Water Catchment Adelaide (SA)	Urban/Industrial	210	64	920	550
Port Philip Bay (Vic)	Urban/Industrial	2,500	190	8,500	2,300
Botany Bay (NSW)	Urban/Industrial	1,300	48	7,600	280
Lake Illawarra (NSW)	Urban/Industrial	190	25	970	170

Note: na, none reported.

[a] Emissions of Total Phosphorus and nitrogen from Reporting Facilities 1990–2000.

[b] Emissions of Total Phosphorus and Nitrogen to Water from Aggregate Sources 1999–2000. Diffuse source pollution represents aggregated data for which there may be significant error of estimation, ranging up to +/−50% for data for WA and 3–13 times for diffuse source estimates in Tasmania. There are significant qualifications to the National Pollutant Inventory estimates and information on the website should be consulted before quoting and/or interpreting these figures.

Source: From Australian State of the Environment Committee, 2001, *Coasts and Oceans, Australia State of Environment Report 2001* (Theme Report), CSIRO Publishing on behalf of the Department of the Environment and Heritage, Canberra, www.deh.gov.au. With permission.

Table 10E.167 Annual Estimates (Tons) of Total United Kingdom Inputs (Direct Plus Riverine) of Pesticides from Great Britain to Marine Waters

	1991		1992		1993		1994		1995		1996		1997		1998		1999	
	Lower	Upper	Lower	Upper	Lower	Upper	Lower	Upper	Lower	Upper	Lower	Upper	Lower	Upper	Lower	Upper	Lower	Upper
Atrazine	7.36	9.11	4.41	11.65	2.33	7.68	0.85	1.28	0.41	1.20	0.45	1.01	0.53	1.14	0.39	1.64	0.45	1.69
Azinphos-ethyl	0.00	0.79	0.00	0.78	0.00	0.89	0.00	0.36	0.00	0.49	0.00	0.32	0.00	0.36	0.00	0.58	0.05	0.50
Azinphos-methyl	0.00	0.77	0.00	0.89	0.08	1.70	0.09	0.74	0.00	0.74	0.03	0.47	0.01	0.44	0.00	0.70	0.00	1.03
DDT	0.00	0.31	0.03	0.15	0.01	0.26	0.00	0.21	0.00	0.18	0.00	0.10	0.01	0.13	0.01	0.15	0.00	0.29
Dichlorvos	0.00	1.55	0.12	0.18	0.31	0.35	0.32	0.36	0.32	0.38	0.19	0.25	0.22	0.26	0.37	0.40	0.54	0.56
Drins	0.01	0.95	0.02	0.50	0.03	0.68	0.02	0.73	0.00	0.58	0.05	0.34	0.00	0.36	0.01	0.63	0.00	0.82
Endosulphan	0.04	0.37	0.09	0.36	0.01	0.27	0.00	0.35	0.00	0.21	0.00	0.14	0.07	0.16	0.00	0.26	0.01	0.45
Fenitrothion	0.04	0.80	0.00	0.40	0.05	0.89	0.01	0.45	0.00	0.59	0.00	0.29	0.01	0.33	0.01	0.59	0.00	0.63
Fenthion	0.01	0.36	0.06	0.38	0.00	0.47	0.00	0.31	0.00	0.42	0.00	0.25	0.01	0.33	0.00	0.55	0.00	0.45
Lindane (gHCH)	0.61	0.91	0.37	0.69	0.48	0.73	0.36	0.64	0.36	0.63	0.26	0.37	0.29	0.42	0.21	0.59	0.15	0.49
Malathion	0.01	0.93	0.01	1.20	0.00	0.93	0.00	0.45	0.00	0.51	0.01	0.25	0.15	0.41	0.25	0.77	0.01	0.57
Parathion	0.01	0.47	0.01	1.24	0.00	0.67	0.00	0.41	0.00	0.34	0.00	0.33	0.00	0.23	0.00	0.44	0.00	0.35
Parathion-methyl	0.00	0.36	0.04	1.46	0.00	1.05	0.00	0.45	0.00	0.52	.00	0.17	0.00	0.12	0.00	0.29	0.00	0.34
Simazine	3.03	5.86	2.67	10.42	1.86	6.76	1.16	1.59	0.67	1.32	0.53	1.06	0.46	1.01	0.84	1.84	1.18	2.17
Trifluralin	0.01	0.37	0.01	0.14	0.07	0.33	0.11	0.33	0.02	0.35	0.00	0.20	0.07	0.23	0.03	0.38	0.05	0.56
Triphenyltin	0.03	0.03	0.00	0.01	0.00	0.06	0.02	0.09	0.02	0.12	0.02	0.23	0.00	0.23	0.01	0.47	0.04	0.25
Flow Rate (Ml/d)[a]	251,480		289,500		278,620		305,470		261,240		223,270		237,010		314,480		308,810	

Note: Data abstracted from UK submission of input estimates fro Annex 1a List substances to the fifth North Sea Conference.

[a] The flow rates are those provide in the OSPAR RID data report for 2000.

Source: From Department of Environment, Food and Rural Affairs, 2002, *The Government's Strategic Review of Diffuse Water Pollution from Agriculture in England: Agriculture and Water: A Diffuse Pollution Review.* www.defra.gov.uk. Reproduced under the terms of the Click-Use Licence.

Table 10E.168 Apparent Consumption of Commercial Fertilizers World 1980–2000

	1,000 tons												
	1980	1985	1990	1991	1992	1993	1994	1995	1996	1997	1998	1999	2000
Canada	1,939	2,325	2,074	2,173	2,261	2,359	2,395	2,568	2,696	2,726	2,649	2,689	2,477
Mexico	1,238	1,764	1,798	1,619	1,616	1,592	1,648	1,286	1,636	1,644	1,804	1,776	1,832
USA	21,480	17,831	18,587	18,784	18,991	20,350	19,297	20,038	20,310	20,165	19,774	19,563	18,507
Japan	1,816	2,024	1,838	1,752	1,784	1,817	1,764	1,641	1,563	1,510	1,419	1,439	1,454
Korea	803	837	958	928	964	974	960	979	909	992	867	809	782
Australia	1,162	1,155	1,164	1,285	1,418	1,512	1,726	1,867	2,016	2,184	2,250	2,341	2,301
N. Zealand	464	427	343	439	535	626	647	683	660	648	662	713	838
Austria	407	388	303	289	267	262	250	237	265	248	247	230	223
Belgium	447	425	384	354	330	321	319	309	314	307	309	304	287
Czech R.	1,091	1,088	797	278	290	306	337	348	347	314	300	269	300
Denmark	627	634	633	581	507	483	466	438	449	437	412	380	366
Finland	489	498	443	339	345	341	355	341	313	314	309	310	308
France	5,609	5,695	5,683	5565	4,530	4,604	4,712	4,914	5,064	4,989	4,837	4,753	4,145
Germany	5,169	4,823	3,351	2,969	2,844	2,672	2,906	2,821	2,819	2,857	2,938	3,054	2,743
Greece	527	710	696	652	628	509	527	505	560	504	472	469	458
Hungary	1,399	1,338	680	326	189	293	314	368	455	427	371	433	472
Iceland	29	25	23	23	20	23	20	20	19	19	19	22	21
Ireland	601	630	692	665	668	716	752	748	682	660	712	671	626
Italy	2,111	2,150	1,944	1,987	1,922	1,904	1,892	1,822	1,865	1,758	1,748	1,784	1,726
Netherlands	679	701	561	561	546	519	534	535	527	508	485	477	425
Norway	259	234	210	208	206	206	207	210	209	205	201	199	196
Poland	3,468	3,315	2,144	1,175	1,192	1,282	1,428	1,512	1,596	1,700	1,557	1,526	1,518
Portugal	259	242	278	258	242	250	248	244	258	236	252	239	228
Slovak R.	639	646	506	147	100	96	99	108	119	107	120	89	101
Spain	1,662	1,734	1,976	1,882	1,575	1,818	1,927	1,869	2,192	2,108	2,367	2,319	2,149
Sweden	484	419	328	292	314	329	323	294	311	310	294	290	280
Switzerland	181	180	168	159	152	151	145	135	125	121	120	102	102
Turkey	1,456	1,427	1,888	1,771	1,928	2,207	1,507	1,700	1,799	1,826	2,181	2,204	2,089
UK	2,054	2,516	2,388	2,171	2,003	2,075	2,219	2,191	2,376	2,270	2,081	1,996	1,694
Russian Fed.	7,480	9,790	9,923	7,899	5,510	3,851	1,510	1,750	1,580	1,550	1,264	1,361	1,420
N. America	24,656	21,919	22,459	22,576	22,868	24,301	23,340	23,892	24,643	24,536	24,227	24,029	22,816
Australia	1,627	1,582	1,507	1,723	1,952	2,137	2,373	2,550	2,676	2,832	2,911	3,054	3,139
OECD	29,649	29,817	26,076	22,650	20,797	21,367	21,488	21,668	22,663	22,225	22,332	22,119	20,457
EU	21,126	21,565	19,660	18,563	16,720	16,803	17,431	17,268	17,995	17,506	17,463	17,275	15,658
OECD	58,551	56,179	52,838	49,629	48,365	50,596	49,924	50,730	52,454	52,094	51,756	51,450	48,648
World	116,720	129,499	137,829	134,606	125,339	120,480	122,046	129,681	134,579	137,224	138,159	140,537	136,435

Note: a), Includes estimates and provisional data, MEX, Fertilizer year: calendar year, U.S.A., Includes data for Puerto rico, KOR, Fertilizer year: calendar year, BEL, Data for Belgium include Luxembourg, CZE, Data before 1986 are Secretariat estimates, DNK, Fertilizer year: August-July, FRA, Data for phosphate and potash refer to the fertilizer year May-April, GRC, Fertilizer year: calendar year, HUN, Fertilizer year: calendar year; ISL, Fertilizer year: calendar year, SLO, Data before 1993 are Secretariat estimates, ESP, Fertilizer year: calendar year, JWE, Fertilizer year: June-May, TUR, Fertilizer year: calendar year, UKD, Fertilizer year: June-May.

Source: Table10.5c, *OECD Environmental Data Compendium 2002*, © OECD 2002, www.oecd.org.

Table 10E.169 World Fertilizer Consumption by Major Crops

	Share (%) in Total 1997/99	Nutrients, Million Tons 1997/99	2015	2030	p.a. (%) 1997/99–2030
Wheat	18.4	25.3	30.4	34.9	1.0
Rice	17.3	23.8	26.5	28.1	0.5
Maize	16.3	22.5	29.0	34.5	1.3
Fodder	6.2	8.5	9.3	10.0	0.5
Seed cotton	3.5	4.9	6.2	7.1	1.2
Soybeans	3.4	4.6	7.6	11.5	2.9
Vegetables	3.3	4.6	5.3	6.1	0.9
Sugar cane	3.2	4.4	5.5	6.6	1.3
Fruit	2.9	4.1	4.3	7.5	1.9
Barley	2.9	4.0	4.4	4.8	0.6
Other cereals	2.9	3.9	9.2	8.3	2.3
Potato	2.0	2.7	3.3	3.8	1.1
Rapeseed	1.5	2.1	3.5	5.1	2.8
Sweet potato	1.3	1.8	2.0	2.1	0.5
Sugar beet	1.0	1.4	1.6	1.7	0.6
All cereals		79.5	99.5	110.6	1.0
% of total	57.7	57.7	64.8	58.8	
All crops above		118.5	148.2	172.1	1.2
% of total	86.0	86.0	89.8	91.5	
World total		137.7	165.1	188.0	1.0

Note: Crops with a 1997/99 share of at least 1 percent, ordered according to their 1997/99 share in fertilizer use; p.a., per annum

Source: From Food and Agriculture Organization of the United Nations (FAO), World Agriculture: Towards 2015/2030, An FAO Perspective, www.fao.org. Reprinted with permission.

Table 10E.170 World Fertilizer Consumption: Past and Projected

	Nutrients, Million Tons					% p.a.		
Total	**1961/63**	**1979/81**	**1997/99**	**2015**	**2030**	**1961–1999**	**1989–1999**	**1997/99–2030**
Sub-Saharan Africa	0.2	0.9	1.1	1.8	2.6	5.3	−1.8	2.7
Latin America and the Caribbean	1.1	6.8	11.3	13.1	16.3	6.1	4.4	1.2
Near East/North Africa	0.5	3.5	6.1	7.5	9.1	7.3	0.8	1.3
South Asia	0.6	7.3	21.3	24.1	28.9	9.6	4.5	1.0
excl. India	0.2	1.6	4.2	5.4	6.9	9.2	4.6	1.5
East Asia	1.7	18.2	45.0	56.9	63.0	9.3	3.8	1.1
excl. China	0.9	4.1	9.4	13.8	10.3	7.0	3.2	0.3
All above	4.1	36.7	84.8	103.5	119.9	8.5	3.7	1.1
excl. China	3.3	22.6	49.2	60.4	67.3	7.6	3.5	1.0
excl. China and India	2.9	16.9	32.1	41.6	45.3	6.9	3.1	1.1
Industrial countries	24.3	49.1	45.2	52.3	58.0	1.4	0.1	0.8
Transition countries	5.6	28.4	7.6	9.3	10.1	0.7	−14.9	0.9
World	34.1	114.2	137.7	165.1	188.0	3.6	0.2	1.0
Per Hectare	kg/ha (Arable Land)					% p.a		
Sub-Saharan Africa	1	7	5	7	9	4.5	−2.4	1.9
Latin America and the Caribbean	11	50	56	59	67	6.0	0.0	0.6
Near East/North Africa	6	38	71	84	99	5.7	3.9	1.0
South Asia	6	36	103	115	134	9.5	4.5	0.8
excl. India	6	48	113	142	178	8.8	4.3	1.4
East Asia	10	100	194	244	266	8.3	3.6	1.0
excl. China	12	50	96	131	92	6.1	3.3	−0.1
All above	6	49	89	102	111	7.7	3.3	0.7
excl. China	6	35	60	68	71	6.9	3.2	0.5
excl. China and India	7	35	49	58	58	6.0	2.6	0.5
Industrial countries	64	124	117			1.3	0.3	
Transition countries	19	101	29			0.9	−14.4	
World	25	80	92			3.3	0.1	

Note: Kg/ha for 1997/99 are for developing countries calculated on the basis of "adjusted" arable land data. For industrial and transition countries no projections of arable land were made; p.a., per annum.

Source: From Food and Agriculture Organization of the United Nations (FAO), *World Agriculture: Towards 2015/2030, An FAO Perspective*, www.fao.org. Reprinted with permission.

Table 10E.171 World Fertilizer Production and Consumption

	Production (Million MT)			Consumption (Million MT)		
Fertilizer	**1980/1981**	**1990/1991**	**2001/2002**	**1980/1981**	**1990/1991**	**2001/2002**
Nitrogen	63	82	86	61	77	82
Phosphate	35	39	34	32	36	33
Potash	27	27	26	24	25	23
Total	125	148	146	117	138	138

Source: From Food and Agriculture Organization of the United Nations (FAO) 2003, *Summary of Food and Agricultural Statistics*, www.fao.org. Reprinted with permission.

Table 10E.172 Leading Fertilizer Producing and Consuming Countries

	2001/2002			2001/2002	
Producers	Production (Million MT)	% of World Production	Consumers	Consumption (Million MT)	% Word of Consumption
China	30	21	China	35	25
United States of America	19	13	United States of America	20	14
India	15	10	India	17	12
Russian Federation	12	8	Brazil	7	5
Canada	12	8	France	4	3
Rest of the Word	58	40	Rest of the World	55	41

Source: From Food and Agriculture Organization of the United Nations (FAO), 2003, Summary of Food and Agricultural Statistics, www.fao.org. Reprinted with permission.

Table 10E.173 Canadian Agricultural Pesticide Expenditures and Application Rates by Ecozone, 1970 and 1995

	Total Agricultural Pesticide Expenditures			Agricultural Pesticide Applied per km^2 of Cultivated Land		
	1970	1995	Change 1970–1995	1970	1995	Change 1970–1995
Ecozone[a]	1990 Dollars		Percent	1990 Dollars		Percent
Boreal Shield	2,607,889	7,660,443	193.7	298	1,098	268.4
Atlantic Maritime	13,100,429	33,109,343	152.7	1,080	3,545	228.3
Mixed Wood Plains	88,433,803	211,800,054	139.5	1,692	4,408	160.5
Boreal Plains	12,700,961	130,895,084	930.6	174	1,512	768.6
Prairie	47,033,763	540,946,447	1,050.1	169	1,807	966.3
Montane Cordillera	5,639,076	8,236,737	46.1	2,190	6,581	200.5
Pacific Maritime	3,489,107	8,639,126	147.6	2,076	2,290	10.3
Canada	173,005,028	941,287,234	444.1	404	2,067	411.3

Note: Figures may not add up to totals due to rounding. Farm input price indices were used to obtain 1990 constant dollar expenditures.

[a] Limited to those with agricultural activity.

Source: From Statistics Canada, Human Activity and the Environment Annual Statistics 2003, Catalogue 16-201-XPE, released December 3, 2003, page 20. *Statistics Canada information is used with the permission of Statistics Canada. Users are forbidden to copy this material and/or redisseminate the data, in an original or modified form, for commercial purposes, without the expressed permission of Statistics Canada. Information on the availability of the wide range of data from Statistics Canada can be obtained from Statistics Canada's Regional Offices, its World Wide Web site at www.statcan.ca, and its toll-free access number 1-800-263-1136.*

Table 10E.174 World Production of Pesticides (Insecticides, Fungicides, Disinfectants, etc.)

Country	1990	1995 (Thousand MT)	1999	Country	1990 (Thousand MT)	1995	1999
Total	**2,318**	**2,045**	**2,254**	**Europe**	**1,377**	**1,051**	**1,006**
				Albania	9	0	0
Africa	**60**	**75**	**85**	Austria	17	12	12
Algeria	8	—	—	Belgium[a]	100	3	6
Burundi	4	—	—	Bulgaria	10	8	8
Egypt	21	25	—	Croatia	—	7	—
Kenya	3	1	—	Czech Republic	—	22	19
South Africa[b]	23	36	—	Czechoslovakia (former)	17	—	—
Tanzania United Rep	1	—	—	Denmark[c]	—	18	19
North America	**127**	**169**	**228**	Finland	14	14	4
Mexico	127	169	228	France	326	—	—
South America	**88**	**85**	**88**	Germany[d]	342	128	—
Brazil	64	—	—	Greece[e]	7	11	—
Colombia	—	21	—	Hungary	56	19	15
Ecuador	—	2	—	Lithuania	—	1	0
Asia	**461**	**665**	**847**	Poland	20	24	30
China	228	417	625	Portugal	25	15	21

(Continued)

Table 10E.174 (Continued)

Country	1990	1995 (Thousand MT)	1999	Country	1990	1995 (Thousand MT)	1999
Cyprus	—	0	—	Romania	24	15	4
Indonesia	19	18	—	Russian Federation	—	16	10
Korea, Rep. of	182	166	173	Serbia and Montenegro	—	7	6
Turkey	23	32	26	Slovakia	—	—	3
Uzbekistan	—	15	—	Slovenia	—	6	1
Vietnam[f]	9	16	19	Spain[g]	83	91	99
USSR (former)	205			Sweden	—	9	2
				T.F.Yug.Rep. Macedonia	—	0	0
				Ukraine	—	4	2
				United Kingdom	—	306.9	—

Note: Global and regional totals refer to the countries shown only and include estimates for mission data.

[a] Incomplete coverage.
[b] Excluding products usually measured in units of volume.
[c] Sales.
[d] Excluding disinfectants.
[e] Insecticides for household use in 1990.
[f] Insecticides only.
[g] Including insecticides for household use in 1990.

Source: From Food and Agriculture Organization of the United Nations (FAO), 2003, Summary of Food and Agricultural Statistics, www.fao.org. Reprinted with permission.

Table 10E.175 World Consumption of Pesticides[a]

	Year	Total Pesticides	Insecticides	Fungicides	Herbicdies	Other Pesticides
		Tons (Active Ingredients)				
Canada	1994	29,206	3,426	3,780	21,910	90
Mexico	1993	36,000	—	—	—	—
USA	1997	353,802	41,730	24,040	213,188	74,843
Japan	1993	645,000	—	—	—	—
Korea	1997	24,814	9,161	7,332	6,043	2,278
Australia	1999	34,200	8,700	3,000	22,000	500
N.Zealand	1998	3,368	718	702	1,847	102
Austria	2000	3,563	334	1,598	1,609	22
Belgium	2000	9,973	918	3057	5,217	373
Czech Rep.	2000	4,303	158	1,007	2,599	539
Denmark	2000	2,841	41	614	1,982	204
Finland	2000	1,150	55	178	862	55
France	2000	94,693	3103	52,834	30,845	7,911
Germany	2000	31,850	2,357	9,641	16,610	3,242
Greece	1998	11,479	2,505	4,731	2,303	1,940
Hungary	2000	5,472	771	1,590	2,682	429
Icland	2000	4	1	0	3	0
Ireland	1997	2,325	73	679	1,261	312
Italy	1996	48,050	8,992	25,074	9,888	4,096
Luxembourg	1999	421	19	186	198	18
The Netherlands	2000	9,644	1,008	4,460	2,605	1,571
Norway	2000	380	11	53	283	33
Poland	2001	8,855	549	2,815	4,748	743
Portugal	1998	14,382	1,079	10,475	1,914	914
Slovak Rep.	2001	3,444	178	537	2,144	585
Spain	1998	35,070	10,173	11,984	9,413	3,500

(Continued)

Table 10E.175 (Continued)

		Tons (Active Ingredients)				
	Year	**Total Pesticides**	**Insecticides**	**Fungicides**	**Herbicdies**	**Other Pesticides**
Sweden	1998	1,629	27	300	1,269	33
Switzerland	1999	1,528	188	708	613	19
Turkey	1995	33,243	14,850	4,937	7,583	5,873
UK	2000	32,989	667	5,282	22,702	4,339
Russian Fed.	1997	25,961	1,686	12,110	12,152	13

Note: CAN, Data refer to agriculture uses only. Insecticides: data exclude Bacillus thuringiensis. Other pesticides; include growth regulators, animal repellents, rodenticides and fumigants; USA, Agricultural pesticides only. Other pesticides include fumigants, nematocides, rodenticides, molluscicides, insect regulators and other pesticides; JPN, Data refer to national production of pesticides; NZL, Data refer to use in agriculture; AUT, Data refer to sales; BEL, Data refer to sales; CZE, Data refer to agricultural pesticides and sales of chemical pesticides. Other pesticides: include growth regulators, rodenticides, animal repellents, additives, adhesives and other pesticides; DNK, Data refer to sales for use in plant production in open agriculture; FIN, Data refer to sales: other pesticides include growth regulators and forest pesticides; FRA, Data refer to quantities sold to agriculture. Fungicides: include copper and sulphur compounds but not elemental sulphur; DEU, Data refer to sales; GRC, Data refer to sales; IRL, Data based on pesticide imports, which gives a reasonable estimation of the quantities used as the country imports most of its pesticides; NLD, Data refer to sales of chemical pesticides; Other pesticides: include soil disinfectants which correspond, for the years presented, about the half of the total consumption; NOR, Data refer to sales to the agricultural sector; POL, Other pesticides: include growth regulators, rodenticides, animal repellents and other pesticides; PRT, Data refer to sales; ESP, Data refer to sales. Data is provisional; SWE, Data refer to sales; CHE, Data refer to sales and have been estimated to represent 95 percent of the total market volume; Liechtenstein included; TUR, Formulation weight. Powdered sulphur and copper sulphate excluded; UKD, Great Britain only. Insecticides: include acarides, insecticides nematocides and tar oil/acid. Fungicides: include fungicides and sulfur; Herbicides; include herbicides and dessicants.

[a] Unless otherwise specified, data refer to active ingredients. Insecticides: acaricides, molluscicides, nematocides, and mineral oils. Fungicides: bactericides and seed treatments. Herbicides; defoliants and dessicants. Other pesticides: plant growth regulators and rodenticides.

Source: Table 10.6a, *OECD Environmental Data Compendium 2002*, © OECD 2002, www.oecd.org.

Table 10E.176 Trends in the Consumption of Pesticides in the World[a], 1980–2000

		Tons		Index 1985 = 100[a]																
			Base Year	1980	1985	1986	1987	1988	1989	1990	1991	1992	1993	1994	1995	1996	1997	1998	1999	2000
Canada	Insecticides	3,172		—	100	90	94	86	—	57	—	—	—	108	—	—	—	—	—	—
	Fungicides	2,823		—	100	120	97	92	—	90	—	—	—	134	—	—	—	—	—	—
	Herbicides	30,181		—	100	82	83	92	—	90	—	—	—	73	—	—	—	—	—	—
	Other	3,083		—	100	66	99	79	—	82	—	—	—	3	—	—	—	—	—	—
	Total pesticides	39,259		—	100	84	86	90	—	87	—	—	—	74	—	—	—	—	—	—
Mexico	Insecticides	—		—	—	—	—	—	—	—	—	—	—	—	—	—	—	—	—	—
	Fungicides	—		—	—	—	—	—	—	—	—	—	—	—	—	—	—	—	—	—
	Herbicides	—		—	—	—	—	—	—	—	—	—	—	—	—	—	—	—	—	—
	Other	—		—	—	—	—	—	—	—	—	—	—	—	—	—	—	—	—	—
	Total pesticides	36,000	(1993)	—	—	—	—	—	—	—	—	—	100	—	—	—	—	—	—	—
U.S.A.	Insecticides	58,513		126	100	98	94	70	74	70	66	70	62	70	71	65	71	—	—	—
	Fungicides	26,762		100	100	100	88	92	92	85	80	76	80	81	83	86	90	—	—	—
	Herbicides	227,250		101	100	96	85	90	92	91	88	90	85	97	92	96	94	—	—	—
	Other	42,638		106	100	100	97	101	120	141	153	160	164	173	181	202	176	—	—	—
	Total pesticides	355,163		105	100	97	88	88	92	93	91	94	90	100	98	103	100	—	—	—
Japan	Insecticides	45,018		100	100	—	—	—	—	—	—	—	—	—	—	—	—	—	—	—
	Fungicides	18,622		103	100	—	—	—	—	—	—	—	—	—	—	—	—	—	—	—
	Herbicides	19,416		133	100	—	—	—	—	—	—	—	—	—	—	—	—	—	—	—
	Other	—		—	—	—	—	—	—	—	—	—	—	—	—	—	—	—	—	—
	Total pesticides	83,056		108	100	95	92	84	83	82	79	78	78	—	—	—	—	—	—	—
Korea	Insecticides	7,834	(1986)	—	—	100	103	93	101	119	134	126	98	102	114	107	117	—	—	—
	Fungicides	7,054	(1986)	—	—	100	119	115	114	110	120	120	144	143	112	100	104	—	—	—
	Herbicides	4,454	(1986)	—	—	100	105	103	110	124	126	121	118	124	131	134	136	—	—	—
	Other	1,980	(1986)	—	—	100	107	100	122	124	145	153	149	134	162	157	115	—	—	—
	Total pesticides	21,322	(1986)	—	—	100	109	103	109	118	129	125	122	123	121	115	116	—	—	—
Australia	Insecticides	7,430	(1992)	—	—	—	—	—	—	—	—	100	—	—	—	—	—	—	117	—
	Fungicides	94,193	(1992)	—	—	—	—	—	—	—	—	100	—	—	—	—	—	—	3	—
	Herbicides	18,031	(1992)	—	—	—	—	—	—	—	—	100	—	—	—	—	—	—	122	—
	Other	—		—	—	—	—	—	—	—	—	—	—	—	—	—	—	—	—	—
	Total pesticides	119,654	(1992)	—	—	—	—	—	—	—	—	100	—	—	—	—	—	—	29	—
N.Zealand	Insecticides	473	(1994)	—	—	—	—	—	—	—	—	—	—	100	109	103	—	152	—	—
	Fungicides	910	(1994)	—	—	—	—	—	—	—	—	—	—	100	102	110	—	77	—	—
	Herbicides	2,039	(1994)	—	—	—	—	—	—	—	—	—	—	100	115	92	—	91	—	—
	Other	93	(1994)	—	—	—	—	—	—	—	—	—	—	100	132	135	—	109	—	—
	Total pesticides	3,515	(1994)	—	91	—	91	—	—	—	—	—	99	100	111	100	—	96	—	—

Austria	Insecticides	428		80	100	117	—	—	117	132	101	111	109	109	86	74	91	58	83	78
	Fungicides	2,072		81	100	117	—	—	72	81	89	72	76	75	68	82	81	79	67	77
	Herbicides	2,648		83	100	115	—	—	97	73	82	70	71	58	61	58	60	61	63	61
	Other	122		88	100	74	—	—	33	33	37	66	53	36	15	13	9	11	6	18
	Total pesticides	5,270		82	100	115	—	—	88	80	85	74	76	69	65	68	70	66	65	68
Belgium	Insecticides	1,669		94	100	91	82	83	82	77	70	76	67	61	67	71	59	60	58	55
	Fungicides	2,123		51	100	103	105	122	124	129	134	155	131	108	125	113	122	125	141	144
	Herbicides	4,617		99	100	100	105	112	114	113	110	111	121	125	135	129	98	108	97	113
	Other	339		101	100	128	131	135	169	201	160	104	113	128	154	131	120	130	116	110
	Total pesticides	8,748		85	100	100	102	109	113	117	114	119	118	113	125	119	102	108	105	114
Czech Rep.	Insecticides	665	(1989)	—	—	—	—	—	100	78	54	49	28	21	17	16	15	19	18	24
	Fungicides	1,809	(1989)	—	—	—	—	—	100	98	84	58	52	44	54	62	50	49	50	56
	Herbicides	7,816	(1989)	—	—	—	—	—	100	80	52	40	29	32	32	31	33	34	33	33
	Other	927	(1989)	—	—	—	—	—	100	44	44	30	28	24	24	29	36	49	68	58
	Total pesticides	11,217	(1989)	—	—	—	—	—	100	80	57	43	32	33	34	35	35	37	37	38
Denmark	Insecticides	262		—	100	89	60	57	86	99	56	49	41	36	62	14	19	21	18	16
	Fungicides	2,199		—	100	76	51	49	58	63	65	61	47	41	48	29	36	35	33	28
	Herbicides	4,079		—	100	93	96	92	97	77	70	69	65	66	80	71	67	64	46	49
	Other	323		—	100	111	94	80	102	268	59	87	102	76	96	27	32	54	68	63
	Total pesticides	6,863		—	100	89	80	77	84	82	67	67	60	57	70	53	54	53	42	41
Finland	Insecticides	144		130	100	97	89	148	136	67	44	64	86	51	40	38	33	32	47	38
	Fungicides	110		90	100	101	92	131	162	148	133	176	191	190	104	105	140	190	200	162
	Herbicides	1,566		134	100	98	102	91	109	101	88	64	54	59	51	43	47	54	50	55
	Other	144		120	100	104	111	95	128	138	103	81	58	59	64	60	56	53	44	38
	Total pesticides	1,964		130	100	98	101	98	115	104	88	72	64	66	54	48	52	60	58	59
France	Insecticides	6,258		75	100	114	104	106	114	123	113	98	86	73	113	86	97	75	58	50
	Fungicides	48,569		82	100	101	93	102	95	85	114	92	112	102	88	100	132	123	130	109
	Herbicides	36,320		89	100	99	94	99	100	103	93	75	72	82	75	99	92	100	117	85
	Other	6,880		76	100	110	102	97	159	160	103	95	92	82	101	114	89	114	166	115
	Total pesticides	98,027		84	100	102	95	101	102	100	106	86	94	91	86	100	112	111	123	97
Germany	Insecticides	4,094	(1992)	—	—	—	—	—	—	—	—	100	106	98	120	93	115	153	150	58
	Fungicides	9,368	(1992)	—	—	—	—	—	—	—	—	100	82	82	103	111	100	112	104	103
	Herbicides	15,622	(1992)	—	—	—	—	—	—	—	—	100	81	95	103	106	106	111	101	106
	Other	4,401	(1992)	—	—	—	—	—	—	—	—	100	97	73	88	99	92	109	85	74
	Total pesticides	33,485	(1992)	—	—	—	—	—	—	—	—	100	86	89	103	105	103	116	106	95
W.Germany	Insecticides	1,566		149	100	93	86	82	85	97	—	—	—	—	—	—	—	—	—	—
	Fungicides	8,491		77	100	102	110	135	127	129	—	—	—	—	—	—	—	—	—	—

(Continued)

Table 10E.176 (Continued)

		Tons		Index 1985=100[a]																
			Base Year	1980	1985	1986	1987	1988	1989	1990	1991	1992	1993	1994	1995	1996	1997	1998	1999	2000
	Herbicides	17,390		120	100	107	97	99	109	98	—	—	—	—	—	—	—	—	—	—
	Other	2,606		122	100	100	91	96	138	141	—	—	—	—	—	—	—	—	—	—
	Total pesticides	30,053		110	100	105	100	108	115	110	—	—	—	—	—	—	—	—	—	—
Greece	Insecticides	3,119	(1986)	—	—	100	88	93	102	—	69	73	76	99	81	78	78	80	—	—
	Fungicides	2,056	(1986)	—	—	100	94	80	94	—	123	133	120	152	149	158	151	230	—	—
	Herbicides	2,158	(1986)	—	—	100	85	103	140	—	96	99	107	110	99	126	98	107	—	—
	Other	13	(1986)	—	—	100	62	108	92	—	8,462	10,815	11,092	10,615	6,100	11,269	10,600	14,923	—	—
	Total pesticides	7,346	(1986)	—	—	100	89	92	111	—	107	117	117	136	116	134	123	156	—	—
Hungary	Insecticides	5,577		57	100	131	104	100	132	86	59	38	27	24	19	19	12	16	14	14
	Fungicides	9,157		164	100	104	88	94	119	78	45	35	28	31	23	22	17	21	18	17
	Herbicides	10,761		127	100	97	102	101	135	110	91	67	52	43	35	30	23	27	26	25
	Other	821		219	100	63	82	106	141	116	78	64	62	91	101	72	69	67	57	52
	Total pesticides	26,316		128	100	106	97	99	129	94	68	50	39	36	29	26	20	24	22	21
Iceland	Insecticides	—		—	—	—	—	—	—	—	—	—	—	—	—	—	—	—	—	—
	Fungicides	—		—	—	—	—	—	—	—	—	—	—	—	—	—	—	—	—	—
	Herbicides	—		—	—	—	—	—	—	—	—	—	—	—	—	—	—	—	—	—
	Other	—		—	—	—	—	—	—	—	—	—	—	—	—	—	—	—	—	—
	Total pesticides	—		—	—	—	—	—	—	—	—	—	—	—	—	—	—	—	—	—
Ireland	Insecticides	40	(1988)	461	—	—	—	100	—	373	405	360	—	275	285	163	183	200	—	—
	Fungicides	544	(1988)	39	—	—	—	100	99	85	98	122	119	121	99	97	125	—	—	—
	Herbicides	1,089	(1988)	96	—	—	—	100	103	90	101	92	110	99	125	81	116	130	—	—
	Other	139	(1988)	18	—	—	—	100	—	109	87	96	—	222	178	193	224	—	—	—
	Total pesticides	1,812	(1988)	81	—	—	—	100	105	96	106	107	120	119	125	96	128	—	—	—
Italy	Insecticides	34,401		95	100	97	96	107	104	101	96	95	98	92	102	98	114	—	—	—
	Fungicides	85,126		187	100	110	135	136	124	125	105	111	111	98	99	97	99	—	—	—
	Herbicides	28,525		78	100	104	110	109	101	94	91	79	86	87	95	101	101	—	—	—
	Other	18,787		90	100	126	111	135	133	134	126	109	99	97	102	102	78	—	—	—
	Total pesticides	166,839		139	100	108	120	125	117	115	103	102	10.3	95	99	99	100	—	—	—
Luxembourg	Insecticides	10	(1991)	—	—	—	—	—	—	—	100	—	90	120	120	100	90	110	190	—
	Fungicides	113	(1991)	—	—	—	—	—	—	—	100	—	142	140	136	160	161	198	165	—
	Herbicides	121	(1991)	—	—	—	—	—	—	—	100	—	95	115	136	122	100	151	164	—
	Other	9	(1991)	—	—	—	—	—	—	—	100	—	189	211	300	200	222	133	200	—
	Total pesticides	253	(1991)	—	—	—	—	—	—	—	100	—	119	130	141	141	131	170	166	—

The Netherlands	Insecticides	634		—	100	88	79	236	255	310	280	278	335	288	235	279	204	203	182	159
	Fungicides	4,363		—	100	82	93	95	93	95	98	96	92	89	91	83	100	117	105	102
	Herbicides	3,977		—	100	95	98	92	84	87	81	75	70	67	77	76	75	73	71	66
	Other	12,028		—	100	114	80	74	84	77	66	58	24	23	20	16	15	12	14	13
	Total pesticides	21,002		—	100	103	86	87	91	90	82	76	56	53	52	49	50	51	49	46
Norway	Insecticides	39		93	100	122	83	98	71	49	48	70	44	57	53	41	50	59	64	28
	Fungicides	138		69	100	104	80	78	86	111	104	107	130	113	121	101	127	190	158	38
	Herbicides	1,236		97	100	96	86	74	69	78	46	45	41	51	56	41	41	44	36	23
	Other	116		33	100	116	106	111	26	40	39	38	50	49	47	41	48	107	90	28
	Total pesticides	1,529		90	100	99	87	78	68	77	50	51	50	56	61	46	49	62	52	25
Poland	Insecticides	1,305	(1986)	—	—	100	97	97	82	—	26	28	31	23	34	33	45	50	31	44
	Fungicides	3,622	(1986)	—	—	100	147	190	199	—	38	54	51	64	63	82	84	80	71	69
	Herbicides	9,035	(1986)	—	—	100	126	154	131	—	33	46	48	47	44	61	57	49	50	53
	Other	517	(1986)	—	—	100	101	255	92	—	111	51	47	85	59	90	134	143	180	189
	Total pesticides	14,479	(1986)	64	86	100	127	161	142	52	36	47	47	51	48	65	66	60	58	61
Portugal	Insecticides	763	(1984)	90	—	—	—	—	—	—	109	99	102	95	87	95	120	141	—	—
	Fungicides	14,021	(1984)	148	—	—	—	—	—	—	46	28	48	51	65	70	67	75	—	—
	Herbicides	1,074	(1984)	104	—	—	—	—	—	—	168	111	122	146	155	147	165	178	—	—
	Other	108	(1984)	101	—	—	—	—	—	—	196	221	180	137	382	370	619	846	—	—
	Total pesticides	15,966	(1984)	142	—	—	—	—	—	—	59	38	56	60	74	78	80	90	—	—
Slovak Rep.	Insecticides	495	(1991)	—	—	—	—	—	—	—	100	25	56	54	125	45	32	35	28	27
	Fungicides	1,102	(1991)	—	—	—	—	—	—	—	100	51	58	51	188	20	46	40	36	43
	Herbicides	3,115	(1991)	—	—	—	—	—	—	—	100	57	90	106	107	88	81	75	62	68
	Other	191	(1991)	—	—	—	—	—	—	—	—	—	100	139	210	363	170	400	248	326
	Total pesticides	4,713	(1991)	—	—	—	—	—	—	—	100	52	83	93	137	83	75	78	63	71
Spain	Insecticides	13,148	(1986)	—	—	100	121	112	102	70	70	51	45	81	73	74	76	77	—	—
	Fungicides	8,464	(1986)	—	—	100	130	152	153	145	137	121	110	122	107	120	133	142	—	—
	Herbicides	10,822	(1986)	—	—	100	93	107	106	124	127	111	98	67	58	80	85	87	—	—
	Other	6,700	(1986)	—	—	100	106	128	129	70	69	42	54	44	44	70	54	52	—	—
	Total pesticides	39,134	(1986)	—	—	100	113	122	119	101	100	81	75	80	71	85	87	90	—	—
Sweden	Insecticides	140		122	100	108	37	74	28	19	14	21	11	29	12	14	17	19	—	—
	Fungicides	639		78	100	156	80	106	78	101	113	81	50	58	31	38	40	47	—	—
	Herbicides	2,752		136	100	152	64	73	67	59	38	34	40	55	35	45	47	46	—	—
	Other	129		30	100	187	64	57	26	34	32	22	29	34	25	20	22	26	—	—
	Total pesticides	3,660		121	100	153	66	78	66	64	50	41	40	54	33	42	44	45	46	—
Switzerland	Insecticides	385	(1988)	31	—	—	—	100	93	101	75	70	62	64	48	54	48	47	49	—
	Fungicides	1,120	(1988)	100	—	—	—	100	99	88	81	85	88	87	85	79	75	68	63	—

(Continued)

Table 10E.176 (Continued)

		Tons		Index 1985 = 100[a]																
			Base Year	1980	1985	1986	1987	1988	1989	1990	1991	1992	1993	1994	1995	1996	1997	1998	1999	2000
	Herbicides	885	(1988)	93	—	—	—	100	105	93	88	85	76	75	74	71	68	68	69	—
	Other	66	(1988)	—	—	—	—	100	103	123	111	76	61	52	53	35	32	27	29	—
	Total pesticides	2,456	(1988)	84	—	—	—	100	100	93	84	82	79	78	74	71	67	64	62	—
Turkey	Insecticides	20,336		67	100	113	73	74	92	87	51	65	60	55	73	—	—	—	—	—
	Fungicides	5,804		76	100	102	105	110	101	95	96	102	101	84	85	—	—	—	—	—
	Herbicides	6,839		62	100	87	109	115	90	93	105	86	134	124	111	—	—	—	—	—
	Other	3,683		122	100	119	125	130	127	124	136	129	138	118	159	—	—	—	—	—
	Total pesticides	36,662		73	100	107	90	93	96	93	77	81	88	79	91	99	92	—	—	—
UK	Insecticides	1,348		52	100	93	103	103	103	104	104	82	83	91	87	68	69	76	71	49
	Fungicides	5,125		35	100	101	98	146	146	143	143	134	137	126	125	131	129	126	130	103
	Herbicides	30,715		56	100	99	99	68	67	76	76	66	68	74	74	78	78	78	78	74
	Other	3,612		9	100	104	105	88	88	104	93	94	94	98	95	107	107	107	107	120
	Total pesticides	40,801		49	100	100	100	81	80	88	87	78	79	83	83	87	87	87	87	81
Russian Federtion	Insecticides	45,900	(1989)	—	—	—	—	—	100	52	51	23	—	2	2	2	4	—	—	—
	Fungicides	33,700	(1989)	—	—	—	—	—	100	77	67	29	—	1	3	2	36	—	—	—
	Herbicides	69,300	(1989)	—	—	—	—	—	100	51	65	48	—	30	31	21	18	—	—	—
	Other	1,500	(1989)	—	—	—	—	—	100	67	87	0	—	53	87	53	1	—	—	—
	Total pesticides	150,400	(1989)	—	—	—	—	—	100	57	61	36	29	15	17	11	17	—	—	—

Note: See next page.
CAN, Survey coverage has varied greatly (different active ingredients, registrants and products); survey trends therefore may not reflect actual trends but simply changes in the survey coverage. Data include non-agricultural uses, such as home and garden plants, golf course, etc; they represent, however, a small part of the total use (between 1 and 6% of the total depending on category). 1994: data refer to agriculture uses only (non-agricultural uses excluded). Insecticides: data exclude Bacillus thuringiensis. Total: includes animal repellents and fumigants; U.S.A., Agricultural pesticides only. Other pesticides include fumigants, nematocides, rodenticides, molluscicides, insect regulators and other pesticides; JPN, Data refer to national production of pesticides; NZL, Data refer to use in agriculture; AUT, Data refer to sales; BEL, Data refer to sales; CZE, Data refer to agricultural pesticides and sales of chemical pesticides. Other pesticides: include growth regulators, rodenticides, animal repellents, additives, adhesives and other pesticides; DNK, Data refer to sales for use in plant production in open agriculture; FIN, Data refer to sales; other pesticides include growth regulators and forest pesticides; FRA, Data refer to quantities sold to agriculture. Fungicides: include copper and sulphur compounds but not elemental sulphur; DEU, Data refer to sales; GRC, Data refer to sales; IRL, 1994–1997: data based on pesticide imports, which gives a reasonable estimation of the quantities used as the country imports most of its pesticides; ITA, From 1985, data include only agricultural pesticides. Data refer to formulation weight. 1996 total in active ingredients: 48 050 t; NLD, Sales of chemical pesticides. Total: includes soil disinfectants which correspond, for the years presented, to about the half of the total consumption. Insecticides: from 1988 onwards include mineral oils; NOR, Data refer to sales to the agricultural sector; POL, Other pesticides: growth regulators, rodenticides, animal repellents and other pesticides; PRT, Data refer to sales. 1991–1994 data do not include all sales (some enterprises trading in these substances are excluded); since 1995 data refer to all sales (all enterprises included). The increase in sales for 1995 corresponding to the increase in the number of enterprises covered was about 12%. The decrease of sales in 1992 results from the 1991–1992 years drought, which led to a decrease in the agricultural activity; ESP, Data refer to sales. Data is provisional; SWE, A special sales tax has been applied to pesticides since 1987. Another tax was applied in 1995. Data refer to sales; CHE, Data refer to sales and have been estimated to represent 95% of the total market volume; Liechtenstein included; TUR, Formulation weight. Powdered sulphur and copper sulphate excluded; UKD, Great Britain only. Insecticides: include acarides, insecticides nematocides and tar oil/acid. Fungicides: include fungicides and sulfur. Herbicides: include herbicides and dessicants.

[a] Unless otherwise specified, data refer to active ingredients and the reference year is 1985. Insecticides: acaricides, molluscicides, nematocides and mineral oils. Fungicides: bactericides and seed treatments. Herbicides: defoliants and dessicants. Total pesticides may include other pesticides such as plant growth regulators and rodenticides.

Source: Table 10.6b, *OECD Environmental Data Compendium 2002*, © OECD 2002, www.oecd.org.

Table 10E.177 Water Quality in Relation to Pesticides in Four Rivers Located in Quebec's Intensive Corn – Cropping Areas, 1992–1995

		Frequency of Exceeding Canadian Water Quality Guildlines (%)							
		Aquatic Life Guidelines				Drinking Water Guidelines			
Pesticide	**Frequency of Detection, 1992–1995 (%)**	**1992**	**1993**	**1994**	**1995**	**1992**	**1993**	**1994**	**1995**
Herbicides: Atrazine	100	43	36	23	16	22	12	10	6
Metolachlor	94	0	2.5	2.8	1.7	0	0	0	0
Cyanazine	67	0	0.8	2.1	1.4	0	0	0.7	0
Insecticides: Carbaryl	6	0	0	8.5	5.6	0	0	0	0
Diazinon	3	0	0	4.3	6.3	0	0	0	0
Malathion	2	0	0	0	0.7	0	0	0	0
Azinphos-methyl	2	0	0.8	4.3	1.4	0	0	0	0
Chloropyrifos	1	0	0	0	3.5	0	0	0	0

Notes: The total number of samples was 441. Seven other herbicides (phenoxy and benzoic compounds) were detected although their concentrations did not exceed *Canadian Water Quality Guidelines*. Among them, the frequency of detection was 66.4% for dicamba, 59.1% for 2,4-D, 44.3% for mecoprop and 31.2% for MCPA. Six other pesticides were detected in 3% to 67% of the samples and showed concentrations in water lower than the *Canadian Water Quality Guidelines*. Four additional pesticides, for which no water quality guidelines exist, were detected in trace amounts (from less than 0.2% to 20% of the samples).

Source: From *The Health of Our Water, Toward Sustainable Agriculture in Canada*, Publication 2020/E, Coote, D.R. and Gregorich, L.J., (eds.), Research Branch Agriculture and Agri-Food Canada, 2000.

Original Source: From Giroux, I., Duchemin, M., and Roy, M., 1997. Contamination de l'Eau par les Pesticides dans les Régions de Culture Intensive du Maís au Québec. Campagnes d'échantillonnage de 1994–1995. Envirodoq EN970099 PES-8. Direction des écosystèmes aquatiques, Ministère de l'Environnement et de la Faune du Québec, Ste-Foy, Que. Reproduced with the permission of the Minister of Public Works and Government Services Canada, 2006.

Table 10E.178 Summary of Median Flows and Concentrations in Six Lake Ontario Tributaries During 1987 and 1998

STN	Wet/Dry	N	Flow (CMS)	Susp Sol (μg/L)	TKN (μg/L)	TP (μg/L)	Al (μg/L)	Cd (μg/L)	Cr (μg/L)	Cu (μg/L)	Fe (μg/L)	Mn (μg/L)	Ni (μg/L)	Pb (μg/L)
Credit	Dry	8	3.59	6.00	0.43	0.014	59.30	0.00	4.74	2.56	93.45	10.24	0.00	0.00
	Wet	6	19.50	184.00	1.60	**0.370**	720.50	0.00	2.81	**8.45**	**1,003.00**	200.5	24.36	**5.63**
Humber	Dry	5	1.43	11.50	0.38	0.030	126.00	0.00	7.18	2.29	184.00	25.3	0.00	0.00
	Wet	7	15.23	125.00	1.16	**0.250**	737.00	**0.22**	5.91	**11.30**	**1,200.00**	181.00	3.85	**8.71**
Ganaraska	Dry	3	1.77	5.00	0.22	0.012	39.20	**0.26**	5.55	0.51	64.40	14.80	0.00	0.00
	Wet	9	7.15	60.50	0.96	**0.110**	218.00	0.00	3.28	1.30	**411.00**	68.60	0.00	0.00
Trent	Dry	4	45.90	2.00	0.45	0.024	12.60	**0.53**	3.43	0.65	27.80	11.25	0.00	0.00
	Wet	8	175.35	4.25	0.46	0.017	19.40	0.13	1.54	0.77	59.00	10.28	0.00	0.00
20 Mile Cr.	Dry	7	0.08	4.00	0.72	**0.086**	77.30	0.00	3.53	1.44	92.90	52.60	2.90	0.00
	Wet	6	30.65	194.00	2.35	**0.760**	1,415.00	0.00	1.97	**5.10**	**1,105.00**	123.70	4.43	2.87
12 Mile Cr.	Dry	6	213.80	7.75	0.26	0.025	88.85	0.00	2.68	1.35	98.10	8.55	0.00	0.00
	Wet	6	215.51	13.00	0.34	**0.061**	216.50	0.00	0.78	1.87	217.00	13.95	3.09	0.00

STN	Wet/Dry	N	Flow (CMS)	Zn (μg/L)	PCB (μg/L)	BaP (μg/L)	TPAH (μg/L)	-BHC (μg/L)	LINDANE (μg/L)	HPTCHLR (μg/L)	ALDRIN (μg/L)	HPTCHLR EPOXIDE (μg/L)	CHLRDNE (μg/L)	ENDOSUL FAN (μg/L)
Credit	Dry	8	3.59	5.72	**4.2**	0.66	22.25	0.06	0.15	0.00	0.00	0.03	0.00	0.03
	Wet	6	19.50	**31.10**	**2.9**	5.36	119.20	0.28	0.36	0.00	0.00	0.01	0.00	0.22
Humber	Dry	5	1.43	4.54	**4.2**	1.03	32.70	0.13	0.10	0.00	0.00	0.00	0.00	0.04
	Wet	7	15.23	**43.80**	**5.4**	22.15	271.06	0.39	0.29	0.00	0.00	0.02	0.02	0.17
Ganaraska	Dry	3	1.77	1.32	**3.5**	0.09	8.17	0.00	0.00	0.00	0.00	0.00	0.00	0.00
	Wet	9	7.15	5.06	**3.2**	0.15	18.05	0.17	0.09	0.00	0.00	0.00	0.00	0.03
Trent	Dry	4	45.90	2.75	**4.1**	0.05	5.46	0.09	0.10	0.00	0.00	0.00	0.00	0.01
	Wet	8	175.35	2.72	**4.4**	0.00	11.88	0.17	0.13	0.00	0.00	0.00	0.00	0.01
20 Mile Cr.	Dry	7	0.08	4.28	**4.2**	0.00	11.97	0.07	0.09	0.00	0.00	0.00	0.00	0.02
	Wet	6	30.65	**28.30**	**2.4**	0.34	25.69	0.32	0.36	0.00	0.00	0.02	0.00	0.12
12 Mile Cr.	Dry	6	213.80	1.63	**6.4**	0.23	13.17	0.27	0.28	0.00	0.00	0.03	0.00	0.03
	Wet	6	215.51	4.26	**5.6**	0.37	19.56	0.34	0.26	0.00	0.00	0.05	0.00	0.03

STN	Wet/Dry	N	Flow (CMS)	CHLRDN (μg/L)	DIELDRIN (μg/L)	*p,p*-DDE (μg/L)	ENDRIN (μg/L)	ENDSLFN (μg/L)	*p,p*-DDD (μg/L)	*o,p*-DDT (μg/L)	*p,p*-DDT (μg/L)	MTHXY CHLR (μg/L)	MIREX (μg/L)	HCB (μg/L)
Credit	Dry	8	3.59	0.03	0.10	0.00	0.00	0.00	0.02	0.00	0.00	0.00	0.06	0.02
	Wet	6	19.50	0.02	0.06	0.00	0.00	0.20	0.02	0.00	0.07	0.00	0.02	0.03
Humber	Dry	5	1.43	0.04	0.10	0.00	0.00	0.04	0.00	0.00	0.00	0.00	0.00	0.01
	Wet	7	15.23	0.04	0.07	0.00	0.00	0.12	0.03	0.00	0.00	0.00	0.00	0.03
Ganaraska	Dry	3	1.77	0.00	0.15	0.00	0.00	0.00	0.00	0.00	0.21	0.00	0.00	0.01
	Wet	9	7.15	0.00	0.20	0.00	0.00	0.03	0.05	0.00	0.14	0.00	0.00	0.02
Trent	Dry	4	45.90	0.01	0.03	0.00	0.00	0.02	0.00	0.00	0.01	0.00	0.00	0.02
	Wet	8	175.35	0.00	0.03	0.00	0.00	0.00	0.00	0.00	0.00	0.00	0.00	0.02
20 Mile Cr.	Dry	7	0.08	0.01	0.06	0.00	0.00	0.00	0.00	0.00	0.00	0.00	0.00	0.02
	Wet	6	30.65	0.00	0.06	0.00	0.00	0.12	0.01	0.00	0.10	0.00	0.00	0.03
12 Mile Cr.	Dry	6	213.80	0.02	0.11	0.00	0.00	0.00	0.00	0.00	0.00	0.00	0.00	0.02
	Wet	6	215.51	0.01	0.13	0.00	0.00	0.06	0.04	0.00	0.12	0.00	0.00	0.02

Note: Bold Value Indicates Concentration Greater Than PWQO; "0.00" Indicates "Not Detected."

Source: From *Large Volume Sampling at Six Lake Ontario Tributaries During 1997 and 1998: Project Synopsis and Summary of Selected Results*, Boyd, D., Ontario Ministry of the Environment, and Biberhofer, J., Environment Canada, 1999, © Queen's Printer for Ontario, 1999. Reproduced with the permission.

Table 10E.179 Chlordane, Heptachlor and Hexachlorobenzene Concentrations Monitored by the United Nations University Project in River Water in China, 2002 (ng/L)

River or Reservoir		Chlordane	Heptachlor	Hexachlorobenzene
Haihe River	Caihong Bridge	N.D.	N.D.	N.D.~1.37
	Haimen Brideg	N.D.	N.D.	N.D.~6.64
	Erdaozha Sluice Gate	N.D.	N.D.	N.D.~0.26
	Jingang Bridge	N.D.	N.D.	N.D.~0.45
	Bridge Belan	N.D.	N.D.	N.D.~0.36
	Ziya River	N.D.	N.D.	N.D.~0.32
Yongding River	Mentougou Sluice	N.D.	N.D.	N.D.~0.15
	Mentougou	N.D.	N.D.	N.D.
	Xifeng Villa	N.D.	N.D.	N.D.
	Zhenzhu Lake	N.D.	N.D.	N.D.~0.17
	Yanchi	N.D.	N.D.	N.D.~0.12
Chaobai River	Niumutun Village	N.D.	N.D.	0.38~9.80
	Shamian Sands	N.D.	N.D.	N.D.~1.67
	Baimaozhuang	N.D.	N.D.	N.D.~9.11
	Chaobai River Left Embankment	N.D.	N.D.	N.D.~6.07
	Leshan Bridge	N.D.	N.D.	N.D.~0.16
	Yujialing Bridge	N.D.	N.D.	N.D.~0.26
Guanting Reservoir	Guanting Reservoir	N.D.	N.D.	N.D.
	Guanting Bridge	N.D.	N.D.	N.D.
	Youzhou	N.D.	N.D.	N.D.
Miyun Reservoir	Miyun Rubber Dam	N.D.	N.D.	N.D.~1.59
	Chaohe river Dam	N.D.	N.D.	N.D.
	Xinzhuang Bridge	N.D.	N.D.	N.D.
	Qikong Bridge	N.D.	N.D.	N.D.

Source: From Yeru, H. et al., 2003, Water Quality Standards and POPs Pollution in China in United Nations University, *Capacity Development Training for Monitoring of POPS in the East Asian Hydrosphere*, 1-2 September, 2003, UNU Centre, Tokyo, www.unu.edu. With permission.

Table 10E.180 DDT Concentrations Monitored by the United Nations University Project in Fresh Water in China (ng/L)

River	Year	DDE	DDD	DDT
Yellow River (Lanzhou)	1999	N.D.~1.06	N.D.~39.8	N.D.
	2000	N.D.~0.10	N.D.	N.D.~0.77
Jialing River (Chongging)	1999	1.19~31.1	N.D.~ 1.41	N.D.~1.02
Songhua River (Harbin)	1999	34.8	N.D.	0.73
Haihe River	1999	1.64~1.65	N.D.~1.37	1.29~5.16
	2002	N.D.	N.D.~12.2	N.D.~75.2
Huangpu River (Shanghai)	1999	2.08~22.1	N.D.~0.34	0.77~1.75
Chongging	1999	1.08~28.3	N.D.~1.49	1.11~13.8
Shanghai	2000	0.10	N.D.	0.79
Yangtse River				
Yongding River	2002	N.D.	N.D.	N.D.
Chaobai River	2002	N.D.	N.D.	N.D.

Source: From Yeru, H. et al., 2003, Water Quality Standards and POPs Pollution in China in United Nations University, *Capacity Development Training for Monitoring of POPS in the East Asian Hydrosphere*, 1-2 September, 2003, UNU Centre, Tokyo, www.unu.edu. With permission.

Table 10E.181 DDT Concentrations Monitored by United Nations University Project in Seawater in China (ng/L)

Sea	Year	DDE	DDD	DDT
Bohai Sea				
8	1999	0.96 ~ 2.36	N.D. ~ 5.32	N.D. ~ 16.5
9	1999	N.D. ~ 0.63	N.D. ~ 3.85	1.04 ~ 2.34
12	2001	N.D.	N.D.	0.313
13	1999	N.D. ~ 25.5	N.D. ~ 5.28	0.83 ~ 2.71
16	2000	0.32	N.D.	3.86
	2001	0.369	N.D.	0.973
19	2001	0.420	N.D.	1.29
Dalian Bay	2001	0.310 ~ 0.450	N.D.	0.793 ~ 2.50
Yellow Sea	2000	N.D. ~ 1.7	N.D. ~ 2.2	N.D. ~ 3.1
	2001	N.D. ~ 0.414	N.D. ~ 1.01	N.D. ~ 1.30
East Sea	1999	1.65 ~ 2.45	0.53 ~ 2.15	0.88 ~ 3.34
	2000	N.D.	N.D.	N.D.
	2001	N.D.	N.D.	N.D.
South Sea				
Shenzhen	1999	N.D.	N.D.	0.84
Daya Bay	2000	N.D.	N.D.	N.D.
Shantou	2000	N.D.	N.D.	N.D.
Xiamen Bay	2001	0.607 ~ 1.05	1.32 ~ 2.39	0.97 ~ 1.35

Source: From Yeru, H. et al., 2003, Water Quality Standards and POPs Pollution in China in United Nations University, *Capacity Development Training for Monitoring of POPS in the East Asian Hydrosphere*, 1-2 September, 2003, UNU Centre, Tokyo, www.unu.edu. With permission.

Table 10E.182 Concentrations of Chlorinated Pesticides in Water (ng/L) in Babe and West Lake, Vietnam

	Babe Lake (n=20)		West Lake (n=22)	
Compounds	**Rainy Season**	**Dry Season**	**Rainy Season**	**Dry Season**
α-HCH	1.2	0.52	<0.25	0.63
	(<0.25–1.2)	(<0.25–0.52)	(<0.25)	(<0.25–0.63)
β-HCH	<0.25	1.58	<0.25	<0.25
	(<0.25)	(<0.25–2.33)	(<0.25)	(<0.25)
γ-HCH	<0.25	1.55	<0.25	<0.25
	(<0.25)	(<0.25–1.76)	(<0.25)	(<0.25)
δ-HCH	<0.25	<0.25	<0.25	<0.25
	(<0.25)	(<0.25–4.61)	(<0.25)	(<0.25)
HCHs	21.1	3.39	<0.25	0.63
	(<0.25–19.9)	(<0.25–1.76)	(<0.25)	(<0.25–0.63)
HCB	<0.2	<0.2	<0.2	<0.2
	(<0.2)	(<0.2)	(<0.2)	(<0.2)
Heptachlor	<0.2	1.86	0.35	9.25
	(<0.2)	(<0.2–1.86)	(<0.2–0.35)	(<0.2–13.91)
Aldrin	<0.4	2.19	<0.4	<0.4
	(<0.4)	(<0.4–3.31)	(<0.4)	(<0.4)
cis-Chlordane	<0.45	<0.45	<0.45	<0.45
	(<0.45)	(<0.45)	(<0.45)	(<0.45)
Dieldrin	<1.5	<1.5	<1.5	<1.5
	(<1.5)	(<1.5)	(<1.5)	(<1.5)
Endrin	<2	<2	<2	<2
	(<2)	(<2)	(<2)	(<2)
p,p′-DDE	<0.25	0.47	<0.25	1.43
	(<0.25)	(<0.25–0.64)	(<0.25)	(<0.25–2.06)
p,p′-DDD	<0.5	<0.5	<0.5	<0.5
	(<0.5)	(<0.5)	(<0.5)	(<0.5)
p,p′-DDT	<0.1	0.3	<0.1	<0.1
	(<0.1)	(<0.1–0.32)	(<0.1)	(<0.1)
DDTs	<0.1	0.77	<0.1	1.43
	(<0.1)	(<0.1–0.32)	(<0.1)	(<0.1–2.06)

Note: HCHs, sum of α,β and γ-HCH; DDTs, sum of *p,p′*-DDE, *p,p′*-DDD and *p,p′*-DDT.

Source: From Viet, P.H. et al., 2003, Occurrence of Persistent Chlorinated Pesticides in Lakes and Rivers in Vietnam: Levels, Fate, Trends and Environmental Implications in United Nations University, *Capacity Development Training for Monitoring of POPS in the East Asian Hydrosphere*, 1-2 September, 2003, UNU Centre, Tokyo, www.unu.edu. With permission.

Table 10E.183 A Summary of Pesticides Detected in Irrigation Districts of New South Wales, Australia

Catchment/Area	Pesticides Detected	Source of Contamination	Reference
Coleambally Irrigation Area (1991–1993)	**Endosulfan**, **Atrazine**, Diuron, Metolachlor, Simazine, **Diazinon**, **Malathion**, Thlobencarb & ***Molinate***	*Where found:* Drainage water	Buchan (1994)
Coleambally Irrigation Area (1994–1995)	**Endosulfan**, Atrazine, Diuron, Simazine, MCPA, 2,4-D **Thiobencarb** & ***Molinate***	*Where found:* Drainage & supply water	Bowmer et al. (1998)
Murrumbidgee Region NSW	**Molinate**, **Malathion, Chlorpyrifos & Atrazine**	*Crops:* Rice & maize *Where found:* Drainage water	Bowmer et al. (1998)
Murray Irrigation Area (1991–1993)	**Molinate**, **Malathion, Diuron, Atrazine &** Bromacil	*Where found:* Drainage water	Shepheard (1994)
Murray Irrigation Area (1994–1995)	**Endosulfan**, Atrazine, Diuron, Metolachlor, Simazine, Diazinon, Bromacil, MCPA, **2,4-D**, **Malathion**, **Thiobencarb** & **Molinate**	*Where found:* Drainage water, supply water, lakes, streams & rivers	Bowmer et al. (1998)
Gwydir River Basin	**Endosulfan**, Atrazine, Diuron, Prometryne & **Chlorpyriphos**	*Crops:* Irrigated cotton, sorphum, maize *Where found:* Rivers & streams	Cooper (1996)
Gwydir River Basin	**Endosulfan**, Atrazine, Amitraz, **Chlorpyrifos**, Profenofos, Diuron, Fluometuron, Metolachlor, Pendimethalin, Prometryn, Simazine & Trifluralin	*Crops:* Irrigated cotton, sorghum, maize *Where found:* Rivers & streams	Muschal (1998)
Border River Basins	**Endosulfan**, Atrazine, diuron, fluometuron, Metolachlor & Prometryn	*Crops:* Irrigated cotton, sorghum, maize *Where found:* Rivers & streams	Muschal 1998
Border River Basins	**Endosulfan**, Atrazine, Diuron & Prometryne	*Crops:* Irrigated cotton, sorghum, maize *Where found:* Rivers & Streams	Cooper (1996)
Namoi River Basin	**Endosulfan**, Atrazine, **Chlorpyrifos**, **Dimethoate**, Diuron, Fluometuron, Metolachlor, Pendimethalin & Prometryn	*Crops:* Irrigated cotton, sorghum, maize *Where found:* Rivers & streams	Muschal 1998
Namoi River Basin	**Endosulfan**, Atrazine, Prometryn, Diuron & **Chlorpyrifos**	*Crops:* Irrigated cotton, sorghum, maize *Where found:* Rivers & streams	Cooper (1996)
Macquire River Basin	**Endosulfan** & Diuron	*Crops:* Cotton, sorghum *Where found:* Rivers & streams	Cooper (1996)
Macquire River Basin	**Endosulfan**	*Crops:* Cotton, sorghum *Where found:* Rivers & streams	Muschal (1998)
Darling River	**Endosulfan**, Atrazine, Diuron, Fluometuron, Metolachlor & Prometryn	*Crops:* Irrigated cotton, sorghum, maize *Where found:* Rivers & streams	Muschal (1998)

Note: A pesticide in bold indicates that the concentration of that pesticide exceeded guidelines for the protection of aquatic ecosystems. A pesticide in italics indicates that the concentration of that pesticide exceeded guidelines for drinking water.

Source: From Ball, J., Donnelley, L., Erlanger, P., Evans, R., Kollmorgen, A., Neal, B., and Shirley, M., 2001, *Inland Waters, Australia State of Environment Report 2001* (Theme Report), CSIRO Publishing on behalf of the Department of the Environment and Heritage, Canberra, www.deh.gov.au.

Table 10E.184 Concentrations of Chlorinated Pesticides in Sediment (ng/g dry wt) in Babe and West Lake, Vietnam

	Babe Lake (n=20)		West Lake (n=22)
Compounds	**Rainy Season**	**Dry Season**	**Dry Season**
α-HCH	0.31 (<0.20–1.2)	3.46 (<0.20–0.52)	6.1 (<0.20–15.1)
β-HCH	0.5 (<0.2–0.5)	<0.2 (<0.2)	2.55 (<0.2–2.55)
γ-HCH	<0.2 (<0.2)	<0.2 (<0.2)	<0.2 (<0.2–2.36)
δ-HCH	<0.20 (<0.20–1.13)	<0.20 (<0.20)	2.36 (<0.20–2.36)
HCHs	0.59 (<0.2–1.13)	3.46 (<0.2–5.26)	6.92 (<0.2–15.1)
HCB	<0.2 (<0.2)	<0.2 (<0.2)	<0.2 (<0.2)
Heptachlor	0.79 (<0.15–1.02)	1.06 (<0.15–1.06)	<0.15 −0.15
Aldrin	<0.25 (<0.25)	0.78 (<0.25–1.4)	1.04 (<0.25–1.04)
cis-Chlordane	<0.5 (<0.5)	<0.5 (<0.5)	<0.5 (<0.5)
Dieldrin	<1.5 (<1.5)	<1.5 (<1.5)	<1.5 (<1.5)
Endrin	<2 (<2)	<2 (<2)	<2 (<2)
p,p′-DDE	2.47 (<0.35–8.92)	0.84 (<0.35–1.85)	29.64 (<0.35–80.45)
p,p′-DDD	2.8 (<0.35–18.53)	0.56 (<0.35–0.82)	5.37 (<0.35–11.26)
p,p′-DDT	1.03 (<0.15–1.93)	1.99 (<0.15–3.35)	0.73 (<0.15–1.87)
DDTs	2.3 (<0.15–18.53)	1.99 (<0.15–3.35)	39.31 (<0.15–80.45)

Note: HCHs, sum of α,β and γ-HCH; DDTs, sum of *p,p′*-DDE and *p,p′*-DDD and *p,p′*-DDT.

Source: From Viet, P.H. et al., 2003, Occurrence of Persistent Chlorinated Pesticides in Lakes and Rivers in Vietnam: Levels, Fate, Trends and Environmental Implications in United Nations University, *Capacity Development Training for Monitoring of POPS in the East Asian Hydrosphere*, 1-2 September, 2003, UNU Centre, Tokyo, www.unu.edu. With permission.

Table 10E.185 Results of Ontario Canada Well Water Surveys

		% of Wells		
Survey Years	**Number of Wells**	**Nitrate-Nitrogen >10 mg N/L**	**Coliform Bacteria >10/100 mL**	**Pesticide Detections**
1950–1954	484	14	15[a]	—
1980	37	5	43	—
1954–1985	63	21	—	—
c.1985	49	5	—	—
1979–1984	359	—	—	37
1981–1984	102	—	—	14
1984	91	—	—	13
1986	103	15	—	10
1987	76	7	—	5
1990	566	12	37	—
1991–1992	142	7	44[b]	—
1991	301	15	34[b]	10
1991–1992	1292	14	34, 25[a]	12

[a] Data for *E. coll.*
[b] Data for fecal coliform.

Source: From *The Health of Our Water, Toward Sustainable Agriculture in Canada*, Publication 2020/E, Coote, D.R. and Gregorich, L.J., (eds.), Research Branch Agriculture and Agri-Food Canada, 2000. Reproduced with the permission of the Minister of Public Works and Government Services Canada, 2006

Original Source: From Reprinted from J. Contam, Hydrol., Volume 32, Goss, M.J., Barry, D.A.J., and Rudolph, D.L. Contamination in Ontario Farmstead Domestic Wells and its Association with agriculture. 1. Results from Drinking Water, pages 267–293, Copyright 1998, with permission from Elsevier.

Table 10E.186 Pesticide Concentrations in Groundwater in Atlantic Canada

Location	Date	Pesticide	Canadian Guideline for Drinking Water (μg/L)	Wells with Pesticide Detected	Mean or Median Pesticide Concentration (μg/L)
Kings Co., N.S. 102 wells	1989	Atrazine	5	33	0.24
		Simazine	10	5	0.52
		Metribuzin	80	4	0.02
		Alachlor	No guideline	3	0.05
		Metolachlor	50	2	0.86[a]
		Captan	No guideline	1	0.05[a]
		Chlorothalonil	No guideline	1	0.065[a]
		Dimethoate	20	1	0.40[a]
N.B., P.E.I. On-farm tile drainage plots various locations	1987–1990	Dinoseb	10	NA	0.2
		Metribuzin	80	NA	0.22
		Atrazine	5	NA	0.4–2.5
		Desethyl atrazine	5	NA	less than 1.0
N.B. Wells	1990	Chlorothalonil	No guideline	0	Not detectable
N.B. Research multi-level wells	1996–1998	Hexazinone (Velpar)	No Canadian guideline (USEPA=200)	6	1.5
N.B. Drainage water research wells	1996	Metalyxl (Ridomil)	No guideline	10	1
N.B. Research multi-level wells	1992–1993	16 Pesticides	Various	NA	Generally not detectable or less than 1.0
P.E.I. 60 wells	1996	10 Pesticides	Various	0	Generally not detectable

Note: NA, not applicable; Compiled from various published sources by Milburn, P.H., Agriculture and Agri-Food Canada, and by Fairchild, G.L., Eastern Canada Soil and Water Conservation Centre.

[a] Denotes limited number of samples.

Source: From *The Health of Our Water, Toward Sustainable Agriculture in Canada*, Publication 2020/E, Coote, D.R. and Gregorich, L.J., (eds.), Research Branch Agriculture and Agri-Food Canada, 2000. Reproduced with the permission of the Minister of Public Works and Government Services Canada, 2006.

Table 10E.187 Percentage of Sampling Sites with Greater Than 0.1μg/L in European Union Countries (Total Number of Sampling Site in Brackets)

	EEA18								PHARE					TACIS	
	AT	DK	FR	DE	ES	LU	NO	UK	CZ	HU	RO	SK	SI	MD	sum
Atrazine	16.3 (1666)	0 (625)	a	4.1 (11690)		a		a		15.5 (174)			32.1 (84)		8
Simazine	0.2 (1248)	0.3 (625)	a	0.9 (11630)		a				6.8 (177)			4.8 (84)		7
Lindane			a	a	a				0 (215)			25 (8)			5
Atrazine-Desethyl	24.5 (1666)			7.1 (11690)									47.6 (84)		3
Heptachlor			a		a							0 (12)			3
Metolachlor	1.1 (1248)					a							4.8 (84)		3
Bentazone						a	80 (5)								2
DDT									0 (215)			0 (12)			2
Dichlorprop		0.6 (623)					83.3 (6)								2
MCPA							100 (2)			4.8 (168)					2
Methoxychlor									0 (206)			8.3 (12)			2
2,4-D										3.6 (168)					1
Atrazine-Desisopropyl	1.3 (1666)														1
Bromacil				3.5 (6650)											1
DDE, DDD, DDT														•	1
DDD (p,p′), DDT (p, p′)						a									1
Chlortoluron								a							1
Dichlorbenzamid		13.7 (102)													1
Dieldrin		a													1
Diuron								a							11
Endosulfan 1						a									1
Endosulfan sulphate						a									1
GCCG-a, b														a	1
HCH, a, b, d						a									1
Hexachlorobenzene												0 (10)			1
Hexazinon				a											1
Isoproturon								a							1
Linuron								a							1
Mecoprop (MCPP)		0.2 (625)													1
Metalaxyl														a	1
Metazachlor							a								1
Parathion-methyl						a									1
Pentachlorophenol									0 (207)						1
Phosphamid														a	1
Phozalon														a	1
Prometryn													2.4 (84)		1
Propzine			0.6 (10890)												1
sum(HCH)											a				1
sum(HCH + DDT)											a				1

Percentage of sampling sites with pesticide concentrations >0.1μg/L. Total number of sampling sites in brackets.

[a] Data available at the regional level only.

Source: From European Environmental Agency (EEA), 1999, *Groundwater Quality and Quantity in Europe*, June 1999, www.eea.europa.eu. Reprinted with permission © EEA.

Table 10E.188 Documented Cases of (Diffuse) Pesticide Contamination of Groundwater in Australia

State or Territory	Area	Principal Contaminant	Percentage of Groundwater Samples Contaminated
Victoria	Shepparton East[a]	Atrazine and Simazine	50
	Goulburn-Broken catchment[b]	Atrazine and Simazine	
	Strathmerton-Cobram	Atrazine and Simazine	48
	Shepparton East	Atrazine and Simazine	49
	Nagambie-Mangalore	Atrazine and Simazine	5
	Tongala-Kyabram		22
	Ardmona[c]	Dieldrin, Chlorpyriphos and Amitrol	n/s[d]
	Girgarre[c]	No pesticides above ANZECC criteria	—
	Ky Valley[c]	No pesticides above ANZECC criteria	—
Western Australia	Kwinana industrial area near Perth[e]	Herbicides 2,4-D and 2,4,5-T	n/s
NSW	Lower Namoi Valley[e]	Atrazine	4
	Murray-Riverina catchment: Berriquin Denimein area[b]	Atrazine and Simazine	16
	Liverpool Plains[g]	Atrazine	40
		Isolated detections of Simazine, Metolachlor, Trifluraline and Diuron	n/s
Queensland	Burdekin Delta[a]	Atrazine	76
		To a lesser extent the Atrazine Degradation Product (DEA)	n/s
	Logan-Albert catchment (SE Qtd)[h]	No pesticides detected	—
	Border Rivers catchment: Western and Macintyre sectors[b]	Atrazine and Simazine	16
South Australia	Padthaway[a]	Atrazine and Simazine	60
		to a lesser extent the DEA	n/s
	Piccadilly Valley, near Adelaide[i]	Atrazine	5
	Lower south-east of SA[j]	Dieldrin, Lindane, Chlorpyrifos and Alachlor	15
Northern Territory	Darwin rural region[k]	No pesticides detected	—

[a] Bauld, 1994.
[b] Bauld et al., 1998.
[c] Wenig & Lawrence 1998.
[d] n/s, not stated.
[e] Appleyard 1993.
[f] Jiwan and Gates, 1994.
[g] Timms & Cooper 1998.
[h] Please et al., 1996.
[i] Ivkovic et al., 1998.
[j] Schmidt et al., 1996.
[k] Radke et al., 1998.

Source: From Ball, J., Donnelley, L., Erlanger, P., Evans, R., Kollmorgen, A., Neal, B., and Shirley, M., 2001, *Inland Waters, Australia State of Environment Report 2001* (Theme Report), CSIRO Publishing on behalf of the Department of the Environment and Heritage, Canberra, www.deh.gov.au. Copyright Commonwealth of Australia reproduced by permission.

Table 10E.189 Average Nitrate Levels in Groundwater in the Maritime Provinces of Canada

Location and Type of Study	Date	Average Concentration of Nitrate-Nitrogen (mg/L)	Wells with Nitrogen Concentration Exceeding the Guideline (%)
Carleton County, N.B., 300 farm wells	1984–1985	2–30	14–22
3 Farming regions in N.B., 47 farm wells	1973–1976, 1988	9.5	20
4 Watersheds of Kings Co., N.S., 237 wells	1989	4.6	13
P.E.I., 2216 Drinking water analyses	1991–1994	2.7	1
1 Watershed of Bedeque Bay, P.E.I., 283 wells	1995	6	7
P.E.I., 146 Dairy farm wells	1997	9.9	44

Compiled from various published sources by Milburn, P.H., Agriculture and Agri-Food Canada, and by Fairchild, G.L., Eastern Canada Soil and Water Conservation Centre.

Source: From *The Health of Our Water, Toward Sustainable Agriculture in Canada*, Publication 2020/E, Coote, D.R. and Gregorich, L.J., (eds.), Research Branch Agriculture and Agri-Food Canada, 2000. Reproduced with the permission of the Minister of Public Works and Government Services Canada, 2006.

Table 10E.190 European Groundwater Quality—Nitrate, Frequency Distribution for the Latest Year Available

	Number of GW-Bodies	≤10	>10–≤25	>25–≤50	>50
AC 3	62	20	19	18	5
Bulgaria (2003)	62	20	19	18	5
EFTA 4	2	2			
Liechtenstein (2003)	1	1			
Norway (2003)	1	1			
EU 25	629	284	162	137	46
Austria (2003)	14	3	2	8	1
Belgium (2002)	1			1	
Czech Republic (2003)	39	20	10	7	2
Denmark (2003)	3		1	2	
Estonia (2003)	5	5			
Finland (2003)	34	33	1		
France (2003)	308	138	80	69	21
Germany (2003)	9	1	3	4	1
Greece (1998)	65	27	26	8	4
Hungary (2002)	16	12	4		
Ireland (2002)	3	2	1		
Italy (2001)	43	21	6	10	6
Latvia (2003)	1	1			
Lithuania (2003)	5	5			
Luxembourg (2003)	5	1	2	2	
Malta (2003)	2			1	1
The Netherlands (2003)	9	3	2	2	2
Poland (2003)	3	1	2		
Portugal (2003)	8	2	3	2	1
Slovakia (2003)	10	4	6		
Slovenia (2003)	5	1	2	2	
Spain (1999)	3			1	2
Sweden (2003)	3	3			
United Kingdom (2003)	35	1	11	18	5
Western Balkans	9	7	2		
Serbia and Montenegro (2003)	9	7	2		

Note: Disclaimer: The data in Waterbase are sub-samples of national data assembled for the purpose of providing comparable indicators of pressures, state and impact of waters on a Europe-wide scale and the data sets are not intended for assessing compliance with any European Directive or any other legal instrument. Information on the national and sub-national scales should be sought from other sources.

Values below the limit of quantification and values below the limit of detection were replaced by half the value. Interval/Individual classes: classes for nitrate rely on Council Directive 98/83/EC on the quality of water intended for human consumption and on 91/676/EEC the Nitrate Directive: 10 mg/L NO_3—Background level, 25 mg/L NO_3—Drinking water Guide Level (related to replaced 80/778/EC—Drinking water Directive), 50 mg/L NO_3—91/676/EEC and parametric value (98/83/EC).

Unit: The table gives the number of GW-bodies according to concentration classes for nitrate. Annual mean values of nitrate are calculated for each GW-body for the latest year available (based on the annual mean values of the sampling sites) and GW-bodies are assigned to frequency classes accordingly.

Source: From European Environment Agency, *The European Topic Center on Water, EEA Data Service, Waterbase-Groundwater*, Downloaded November 27, 2005, www.eea.europa.eu. Reprinted with permission © EEA.

Table 10E.191 Information on the Occurrence of Excessive Nitrate in Groundwaters in India

Name of State/Union Territory	Name of District	Number of Samples	Number of Samples with NO_3 >45 mg/l	Number of Samples with NO_3 >45 mg/L (%)
Andhra Pradesh	Kurnool	143	26	18.2
Bihar	Rohtas	209	55	26.3
Bihar	Palamu	191	40	20.9
Gujarat	Mehsana	200	37	18.5
Gujarat	Amreli	49	0	0
Haryana	Faridabad	200	45	22.5
Haryana	Gurgaon	415	104	25.1
Karnataka	Gulbarga	529	261	49.3
Lakshadweep		135	0	0
Madhya Pradesh	Jhabua	55	11	20.0
Maharashtra	Nagpur (villages)	114	47	41.2
Maharashtra	Nagpur (city)	100	82	82.0
Maharashtra	Satara	1,001	163	16.3
Maharashtra	Latur	220	92	41.8
Orissa	Phulbani	225	29	12.9
Orissa	Koraput	503	71	14.1
Rajasthan	Barmer	351	220	62.7
Tamil Nadu	Ramanathapuram	66	7	10.6
	Total	4,696	1,290	27.5

Source: From World Health Organization, 2004, Rolling Revision of the WHO Guidelines for Drinking Water Quality, Nitrates and Nitrites in Drinking Water, Draft for Review and Comments. Copyright © World Health Organization 2004, www.who.int.

Original Source: From National Environmental Engineering Research Institute (NEERI), 1991 Water Mission Report on Participation of NEERI. NEERI, Nagpur, India.

SECTION 10F URBAN RUNOFF/DEICING MATERIALS

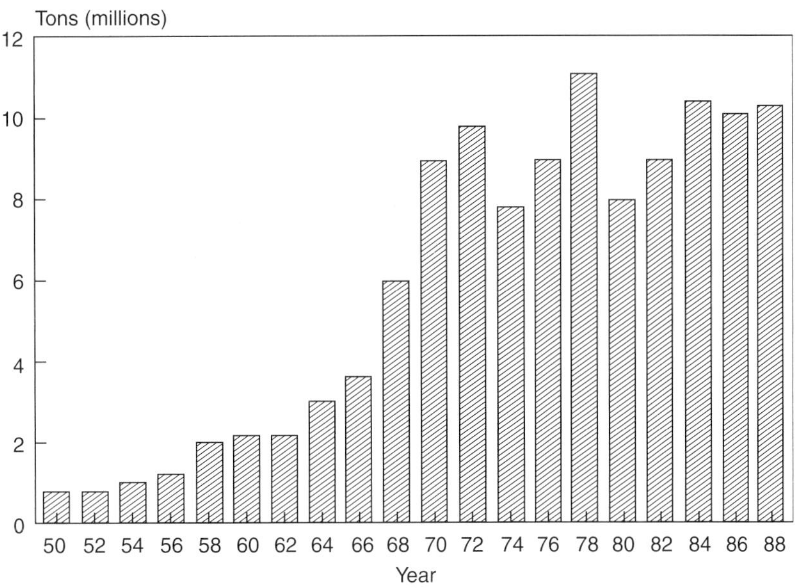

Figure 10F.136 Trends in highway salt use, 1950–1988. (From Salt Institute gulliver.trb.org/publications/sr/sr235.html. With permission.)

Table 10F.192 Trends in Gross Discharges of Major Pollutants in Urban Runoff in the United States (Thousands of Tons)

	High Growth				Low Growth		
Pollutant	**1975**	**1985**	**1990**	**2000**	**1985**	**1990**	**2000**
Biochemical oxygen demand	370	400	420	450	390	400	420
Chemical oxygen demand	3,400	3,400	3,900	4,200	3,600	3,800	3,900
Suspended solids	6,300	6,800	7,100	7,600	6,500	6,900	7,100
Dissolved solids	4,000	4,400	4,600	4,900	4,300	4,400	4,600
Phosphorus	5	5	6	6	5	5	6
Nitrogen	58	63	65	70	61	63	65
Oil and grease	86	94	98	110	92	95	98

Note: Conservative estimates; discharges may uniformly be too low by a factor of 2–5.

Source: From U.S. Environmental Protection Agency, Environmental Outlook 1980.

Table 10F.193 Average Concentrations of Heavy Metals in Street Sweepings and Urban Runoff

Heavy Metal	Street Sweepings (mg/kg)	Urban Runoff (μg/L)
Cadmium (Cd)	3.4	18
Chromium (Cr)	211	33
Copper (Cu)	104	45
Iron (Fe)	22,000	—
Lead (Pb)	1,810	235
Manganese (Mn)	418	—
Nickel (Ni)	35	24
Zinc (Zn)	371	236

Source: From U.S. Geological Survey, 1984, National Water Summary 1983-Hydrologic Events and Issues, Water-Supply Paper 2250. Based on data from W.L. Bradford, 1977, and the U.S. Environmental Protection Agency.

Table 10F.194 Snow and Ice Control Materials Use in the United States, Winter 1982–1983

				Control Materials Used			
					Calcium Chlorido		
State	Population 1983 (thousands)	Total Lane (mil)[a]	Bare Pavement (mil)[a]	Salt (metric tons)	Dry (metric tons)[b]	Liquid (l)[c]	Abrasives (metric tons)[b]
Alaska	100	1,020	900	322	227	757,000	16,534
Arizona	2,178	16,650	0	375	23	0	50,803
Arkansas	2,178	34,852	0	777	223	0	8,305
California	22,000	53,000	12,900	—	—	—	—
Colorado	2,500	32,000	32,000	9,885	0	0	394,484
Connecticut	3,100	10,160	0	47,115	544	0	153,859
Delaware	611	9,971	2,543	6,400	0	0	6,998
Florida	9,740	—	—	—	—	—	—
Georgia	5,463	41,809	13,900	7,439	54	0	7,076
Idaho	944	11,512	0	9,979	0	0	174,182
Illinois	11,114	38,515	—	186,883	472	435,843	—
Indiana	5,300	31,036	14,559	105,825	40	341	98,539
Iowa	2,900	24,300	0	54,795	2,019	35,958	105,235
Kansas	2,364	22,371	5,688	28,695	0	0	58,968
Kentucky	3,661	53,846[e]	4,966[e]	29,905	97	0	0
Louisiana	5,000	38,191	0	21	0	0	0
Maine	1,125	7,877	1,178	44,636	485	—	428,652
Maryland	4,500	14,600	0	74,843	887	—	27,995
Massachusetts	5,737	12,000	—	161,935	2,578	0	86,184
Michigan	9,258	13,667	4,695	207,749	242	0	9,072
Minnesota	4,077	28,724	4,162	116,083	254	0	262,084
Mississippi	2,500	23,391	0	259	—	—	—
Missouri	4,917	69,664	—	68,141	3,060	0	—
Montana	700	18,790[a]	18,790	2,944	25	0	181,440
Nebraska	1,600	22,000	—	22,588	504	98,312	78,385
Nevada	799	12,608	11,340	8,919	0	0	46,267
New Hampshire	921	8,630	8,406	85,107	207	0	137,279
New Jersey	7,364	10,366	10,366	32,387	526	862,980	3,810
New Mexico	1,400	27,450	20,000	20,866	0	0	72,576
New York	17,557	29,780	27,780	272,160	272	0	417,312
North Carolina	5,847	112,573	23,342	33,179	145	0	—
North Dakota	600	15,800	—	7,910	197,588[d]	0	34,474
Ohio	10,797	42,192	0	167,234	177	428,806	117,007
Oklahoma	3,025	25,935	0	17,028	0	—	65,318
Oregon	2,656	17,895	—	414	—	1,608,580	—
Pennsylvania	11,867	77,000	—	209,563	4,536	0	594,216
Rhode Island	947	3,015	3,015	26,578	120	0	40,824
South Carolina	3,122	84,450	3,450	814	259	0	4,309
South Dakota	688	18,216	—	3,354	5	0	40,826
Tennessee	4,591	25,087	25,087	—	—	0	—
Utah	2,000	22,000	—	72,322	0	0	113,400
Vermont	511	6,079	6,079	59,555	0	0	96,757
Virginia	5,347	112,814	17,350	86,184	907	0	231,336
Washington	4,130	16,778	0	6,804	0	0	184,643
West Virginia	1,950	70,000	21,000	47,818	285	0	162,064
Wisconsin	4,705	25,774	25,774	208,477	588	151,400	23,508
Wyoming	430	15,743	0	5,752	0	0	115,214
Total	205,361	1,410,131	321,270	2,560,019 (2,821,141 tons)	217,349 (239,519 tons)	4,379,220 (1,156,990 gallons)	4,650,235 (5,124,559 tons)

Note: As reported by state departments of transportation/highways.

[a] To convert to kilometers, multiply amount by 1.609.
[b] To convert to short tons, multiply amount by 1.102.
[c] To convert to gallons, multiply amount by 0.2642.
[d] Sodium chloride liquid used.
[e] Includes parkways.

Source: From Salt Institute, 1983, Survey of Salt, Calcium Chloride and Abrasives in the United States and Canada for 81/82 ad 82/83. With permission.

Table 10F.195 Average Annual Salt Loadings on State Highways Where Salt Is Normally Applied

Region and State	Average Annual Loading (tons/lane-mile)
New England	
Maine	8.0
Massachusetts	19.4
New Hampshire	16.4
Vermont	17.1
Middle Atlantic	
Delaware	9.0
Maryland	7.1
New Jersey	6.7
New York	16.6
Virginia	3.0
West Virginia	6.3
Great Lakes	
Illinois	6.6
Indiana	9.0
Michigan	12.9
Ohio	9.1
Wisconsin	9.2
Plains	
Iowa	3.8
Minnesota	5.0
Missouri	1.0
Nebraska	1.5
Oklahoma	1.5
South Dakota	1.0
Mountain and West	
Alaska	1.2
California	3.0
Idaho	0.3
Nevada	1.9
New Mexico	0.5

Note: Data are from only those states that responded to relevant questions in survey.

Source: From TRB survey of state highway agencies, gulliver.trb.org/publications/sr/sr235.html. With permission.

Table 10F.196 State Spending on Prevention and Remediation of Drinking Water Contamination by Road Salt, 1989

State	State Expenditure[a] ($)	Amount of Salt Spread Annually by State[b] (tons)
Illinois	200,000	275,000
Indiana	175,000	240,000
Maine	75,000	60,000
Massachusetts	500,000	225,000
Michigan	200,000	350,000
New Hampshire	195,000	125,000
New York	175,000	450,000
Pennsylvania	200,000	425,000
Vermont		35,000
Total	1,755,000	2,250,000

[a] From survey of state highway departments.
[b] From survey of state highway departments and Salt Institute data.

Source: From gulliver.trb.org/publications/sr/sr235.html. With permission.

SECTION 10G AIR EMISSIONS/ACID RAIN/SEA LEVEL RISE

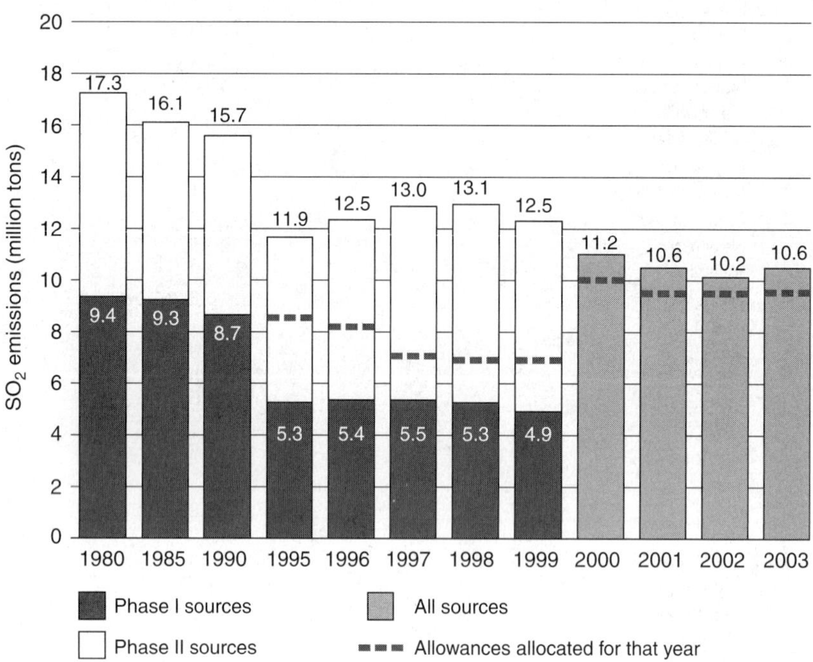

Figure 10G.137 Trends in SO_2 emissions since 1980 for all Title IV affected sources, www.epa.gov/airmarkets.

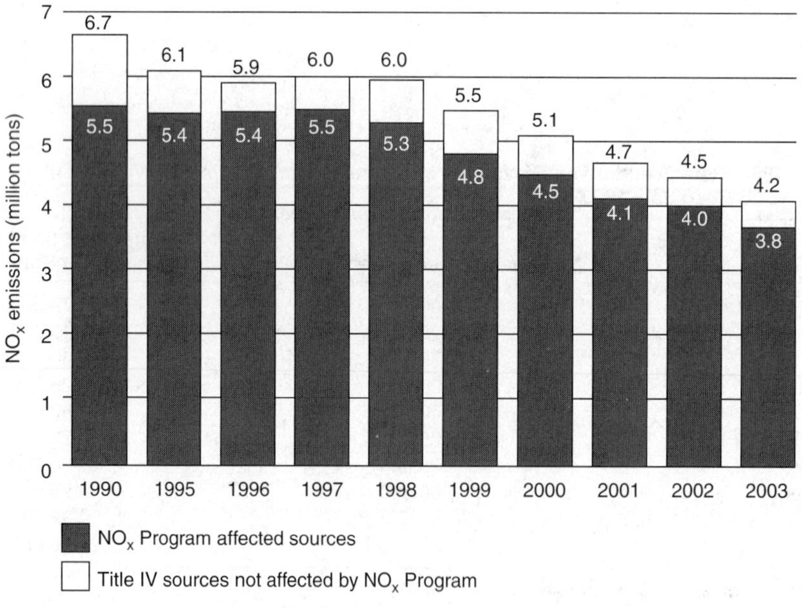

Figure 10G.138 Trends in NO_x emissions under the acid rain program, www.epa.gov/airmarkets.

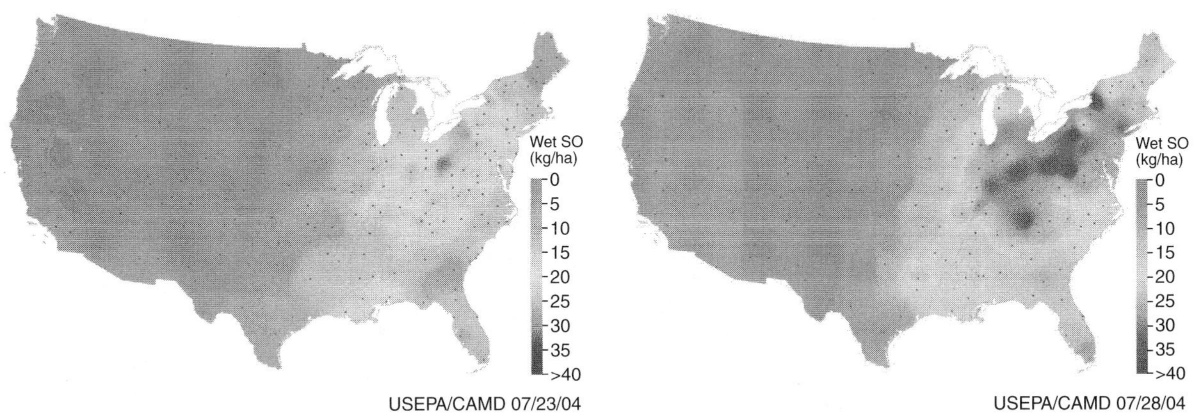

Figure 10G.139 Wet sulfate deposition decreased throughout the early 1990s in much of the Ohio River Valley and Northeastern U.S. Other less dramatic reductions were observed across much of New England, portions of the Southern Appalachian Mountains and in the Midwest. Average decreases in wet deposition of sulfate range from 39 percent in the Northeast to 17 percent in the Southeast, www.epa.gov/airmarkets.

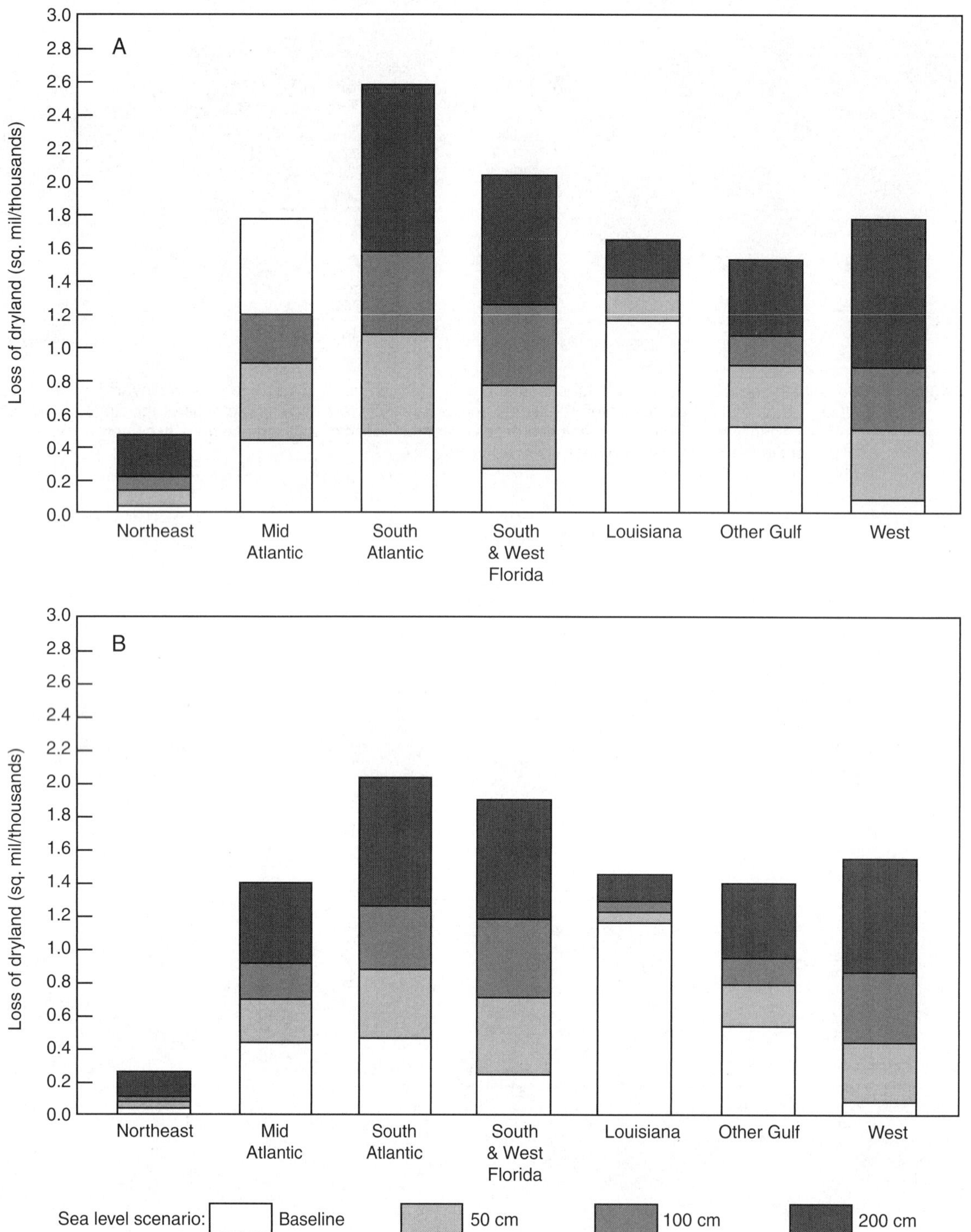

Figure 10G.140 Loss of dry land in the United States by 2100 (A) if no shores are protected and (B) if developed areas are protected for sea level rise. (From U.S.EPA 1988. The Potential Effects of Global Climate Change on the U.S. Draft report to Congress. Prepared by Titus and Greene, adapted from Park and others.)

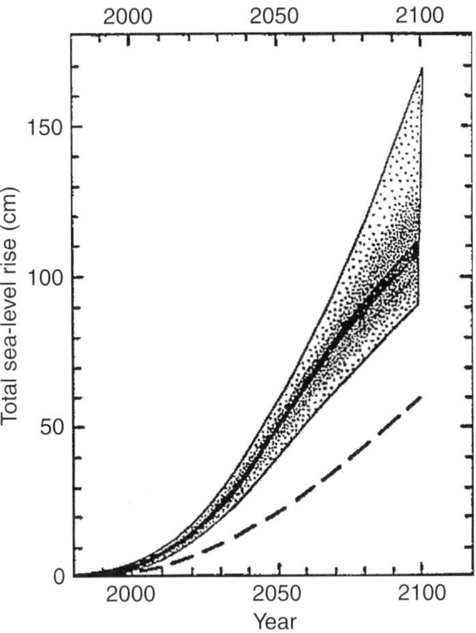

The dark shading indicates the most probable response to the climate scenario shown in Figure 10G-142. The broken line depicts the response to a warming trend delayed 100 years by thermal inertia of the ocean. A global warming of 6°C by 2100, which represents an extreme upper limit, would result in a sea level rise of about 2.3 m, but errors on this estimate are very large.

Figure 10G.141 Total estimated sea-level rise, 1980–2100. (From Thomas, Robert 1986, Future sea-level rise and its early detection by satellite remote sensing, in *Effects of Change in Stratospheric Ozone and Global Climate*, vol. 4.)

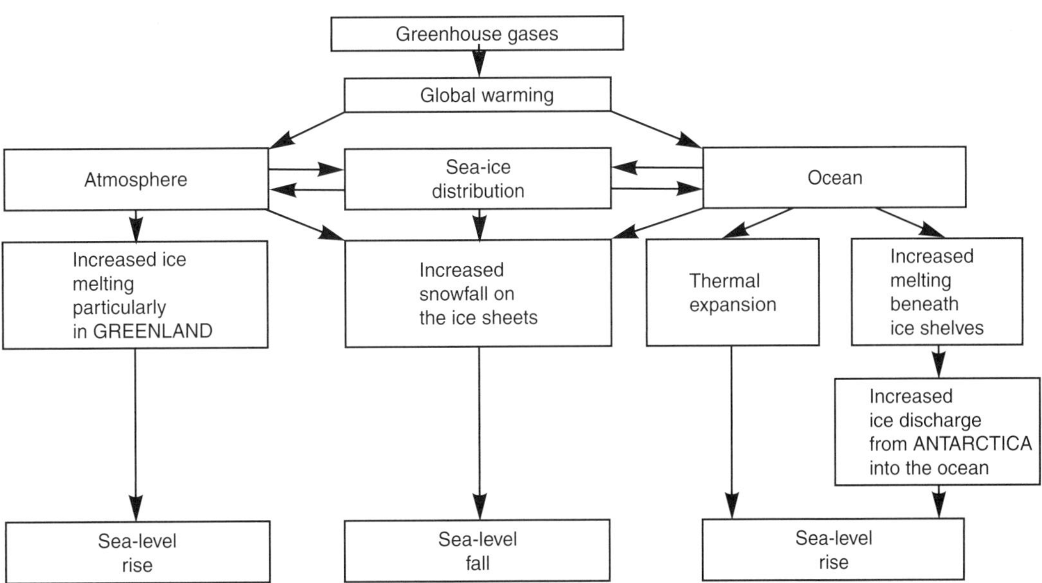

Heat trapped by greenhouse gases raises the temperature of the atmosphere and the ocean. The response of sea level to this warming is strongly determined by the partition of available heat between these two processes. If most of the heat remains in the atmosphere, air temperatures rise rapidly and sea level is affected most by increased melting of ice. Alternatively, rapid transfer of heat into the sea would increase ocean temperatures, and sea level would rise because of thermal expansion and by accelerated Antarctic ice discharge associated with increased melting from beneath the floating ice shelves. Moreover, sea-ice distribution both influences, and is affected by, thermal interactions between atmosphere and ocean.

Figure 10G.142 Major processes relating greenhouse warming to average worldwide sea level. (From Thomas Robert 1986, Future sea-level rise and its early detection by satellite remote sensing, in *Effects of Changes in Stratospheric Ozone and Global Climate*, vol. 4.)

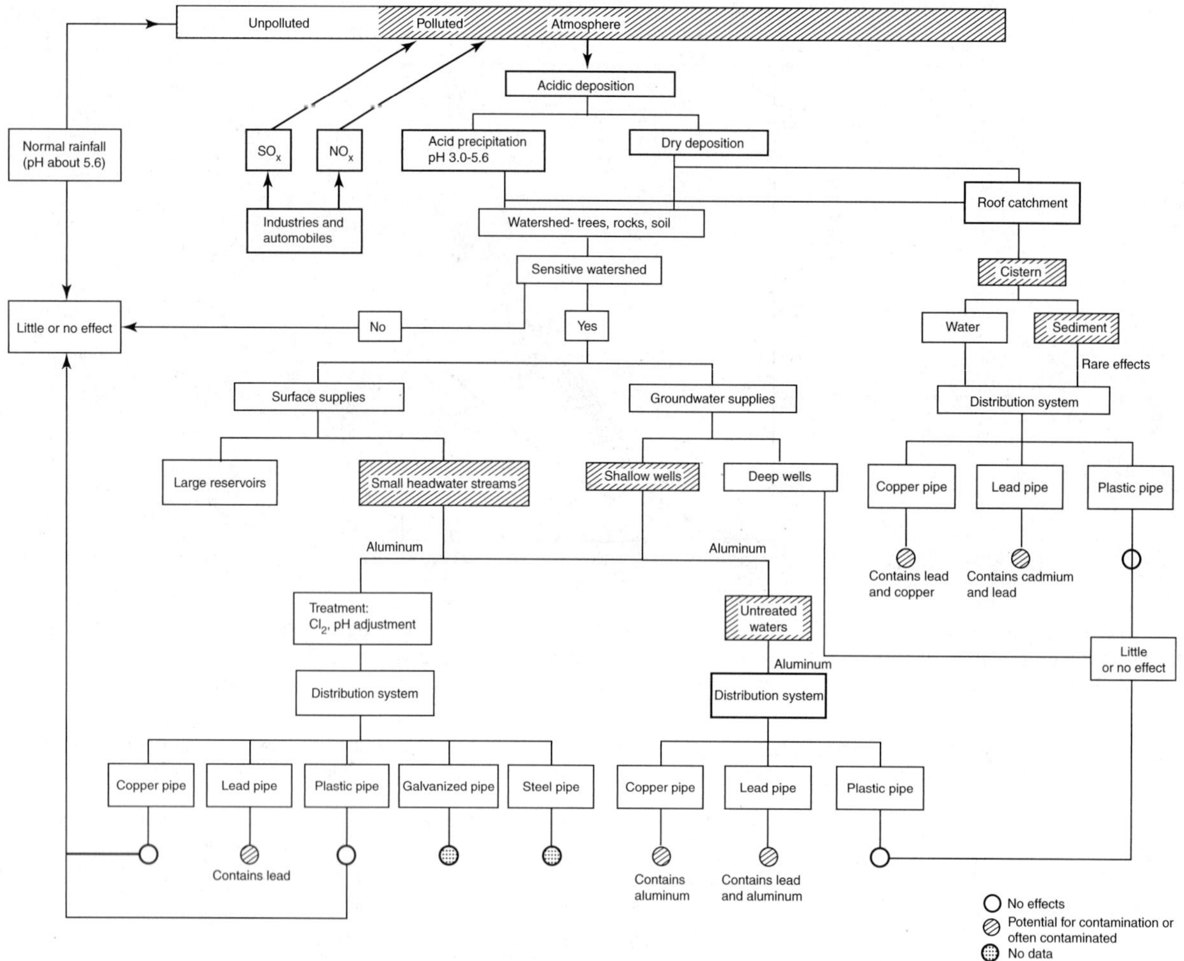

Figure 10G.143 Probable effects of acid deposition on water supplies. (From Perry, J.A., 1984, Current research on the effects of acid deposition, *J. Am. Water Works Assoc.*, vol. 76, no. 3 Copyright. With permission.)

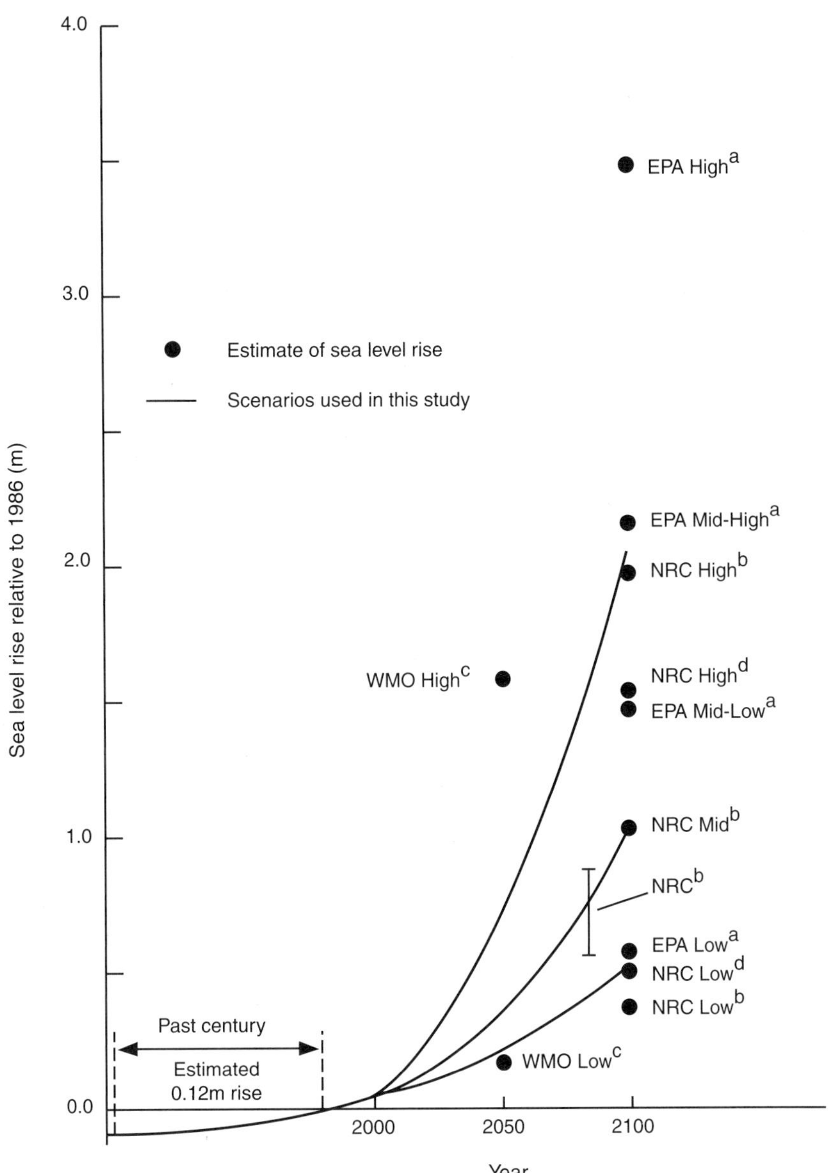

[a] Environmental Protection Agency, reported in JS Hoffman, D Keyes and JG Titus, *Projection Future Sea Level Rise*, US GPO, 1983. [b] Glacial volume estimate of National Research Council, reported in MF Meier et al, *Glaciers, Ice Sheets, and Sea Level*, National Academy Press, 1985, augmented with thermal expansion estimates of the NRC, reported in R Revelle, Probable future changes in sea level resulting from increased atmospheric carbon dioxide, in *Changing Climate*, NAP, 1983. [c] WMO International assessment of the role of carbon dioxide and other greenhouse gases in *Climate Variations and Associated Impacts*, WMO, 1985. [d] NRC, *Responding to Changes in Sea Level: Engineering Implications*, NAP, 1987.

Figure 10G.144 Estimates of future sea level rise. (From yosemite.epa.gov.)

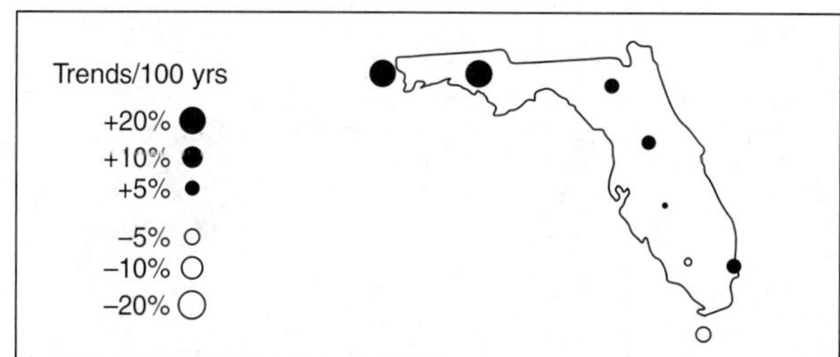

Figure 10G.145 Precipitation trends from 1900 to present. (From Karl et al. (1996), yosemite.epa.gov.)

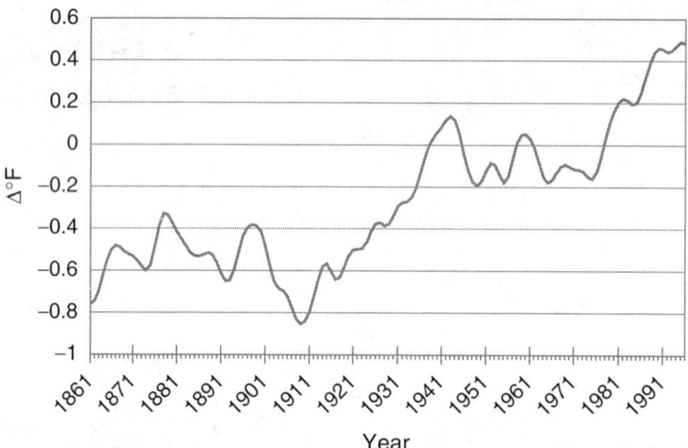

Figure 10G.146 Global temperature changes (1861–1996). (From IPCC (1995), updated yosemite.epa.gov.)

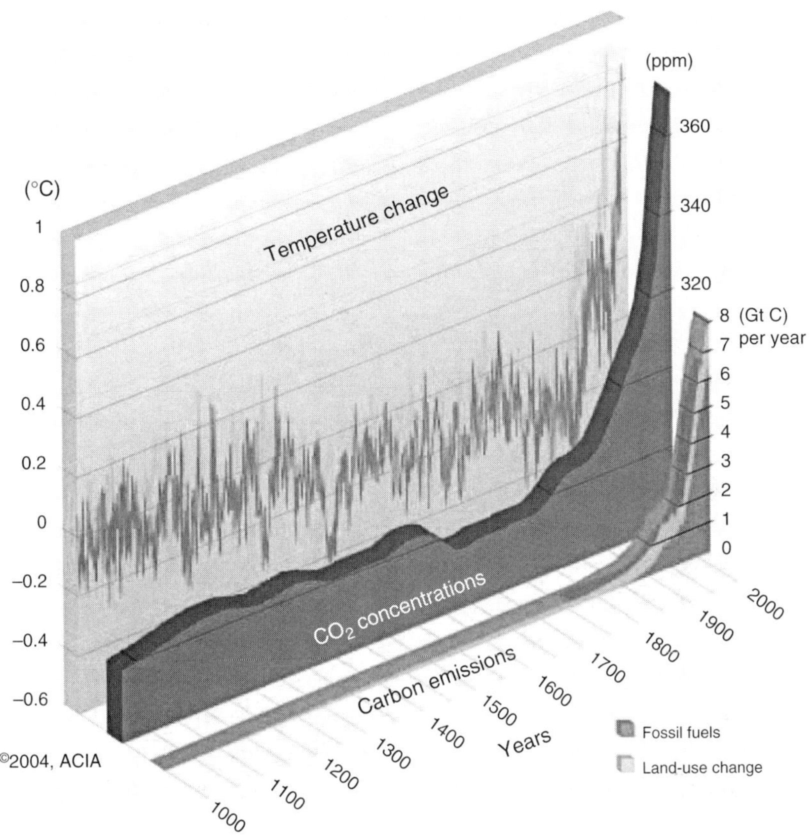

This 1000-year record tracks the rise in carbon emissions due to human activities (fossil fuel burning and land clearing) and the subsequent increase in atmospheric carbon dioxide concentrations, and air temperatures. The earlier parts of this Northern Hemisphere temperature reconstruction are derived from historical data, tree rings, and corals, while the later parts were directly measured. Measurements of carbon dioxide (CO_2) in air bubbles trapped in ice cores form the earlier part of the CO_2 record; direct atmospheric measurements of CO_2 concentration began in 1957.

Figure 10G.147 1000 years of changes in carbon emissions, CO_2 concentrations and temperature, www.amap.no/acia.

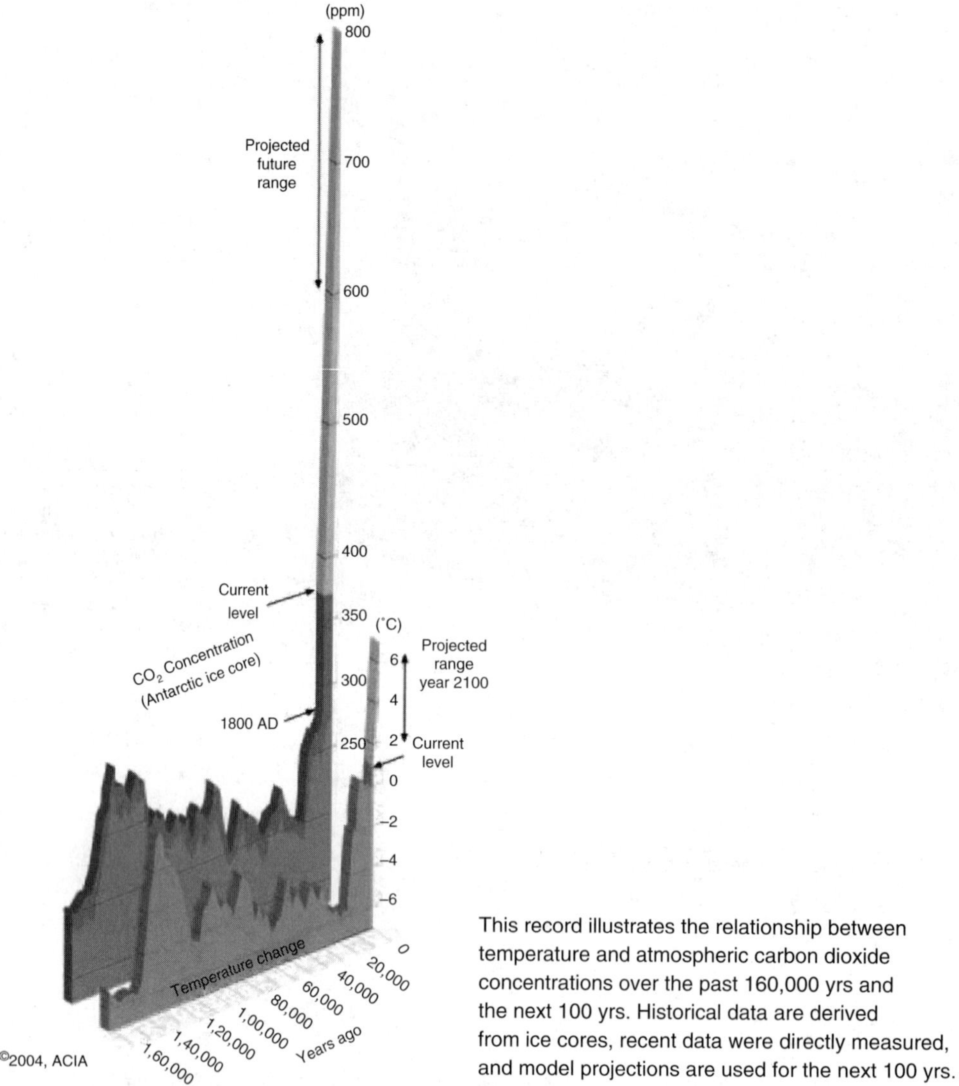

Figure 10G.148 Atmospheric carbon dioxide concentration and temperature change, www.amap.no/acia.

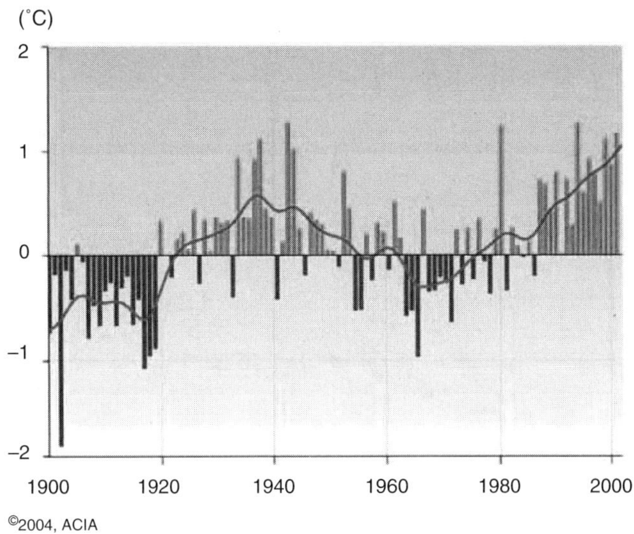

Annual average change in near surface air temperature from stations on land relative to the average for 1961–1990, for the region from 60 to 90°N.

Figure 10G.149 Observed arctic temperature, 1900 to present, www.amap.no/acia.

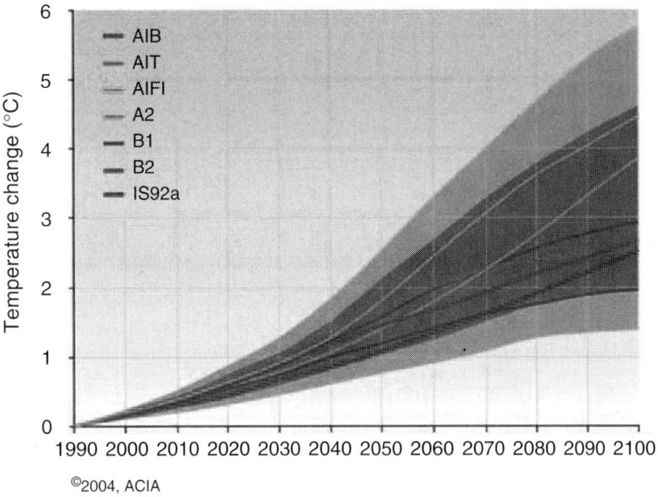

Projections of global temperature change (shown as departures from the 1990 temperature) from 1990 to 2100 for seven illustrative emissions scenarios. The brown line shows the projection of the B2 emissions scenario, the primary scenario used in this assessment and this scenario on which the maps in this report showing projected climate changes are based. The pink line shows the A2 emissions scenario, used to a lesser degree in this assessment. The dark gray band shows the range of results for all the SRES emissions scenarios with one everage model while the light gray band shows the full range of scenarios using climate models.

Figure 10G.150 Projected global temperature rise, www.amap.no/acia.

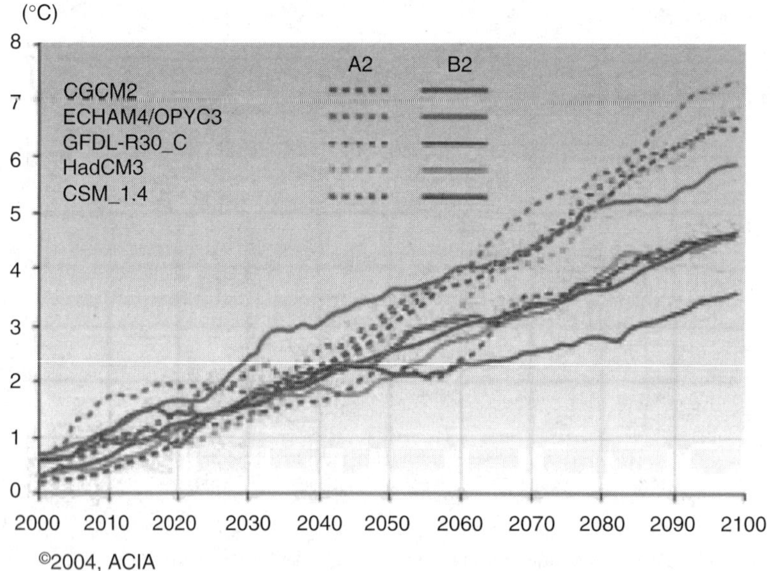

The ten lines show air temperatures for the region from 60°N to the pole as projected by each of the five ACIA global climate models using two different emissions scenarios. The projections remain similar through about 2040, showing about a 2°C temperature rise, but then diverge, showing increases from around 4° to over 7°C by 2100. The full range of models and scenarios reviewed by the IPCC cover a wider range of possible futures. Those used in this assessment fall roughly in the middle of this range, and thus represent neither best- nor worst-case scenarios.

Figure 10G.151 Projected arctic surface air temperatures 2000–2100 60°N—Pole: change from 1981–2000 average, www.amap.no/acia.

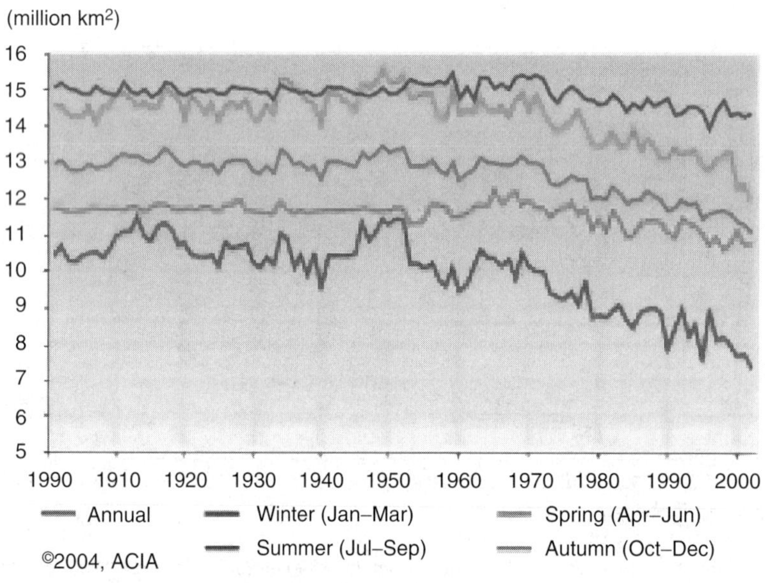

Annual average extent of arctic sea ice from 1900 to 2003. A decline in sea-ice extent began about 50 years ago and this decline sharpened in recent decades, corresponding with the arctic warming trend. The decrease in sea-ice extent during summer is the most dramatic of the trends.

Figure 10G.152 Observed seasonal Arctic sea-ice extent (1900–2003), www.amap.no/acia.

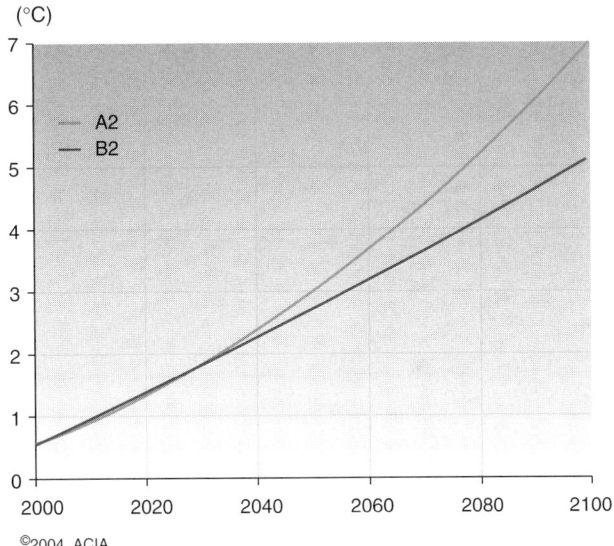

Increases in arctic temperature (for 60°–90°N) projected by an average of ACIA models for the A2 and B2 emissions scenarios, relative to 1981–2000.

Figure 10G.153 Projected arctic temperature rise, www.amap.no/acia.

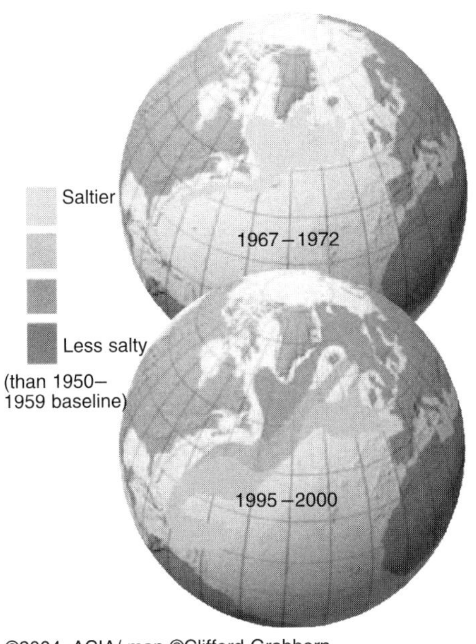

©2004, ACIA/ map ©Clifford Grabhorn

Figure 10G.154 Reduced salinity of North Atlantic waters, www.amap.no/acia.

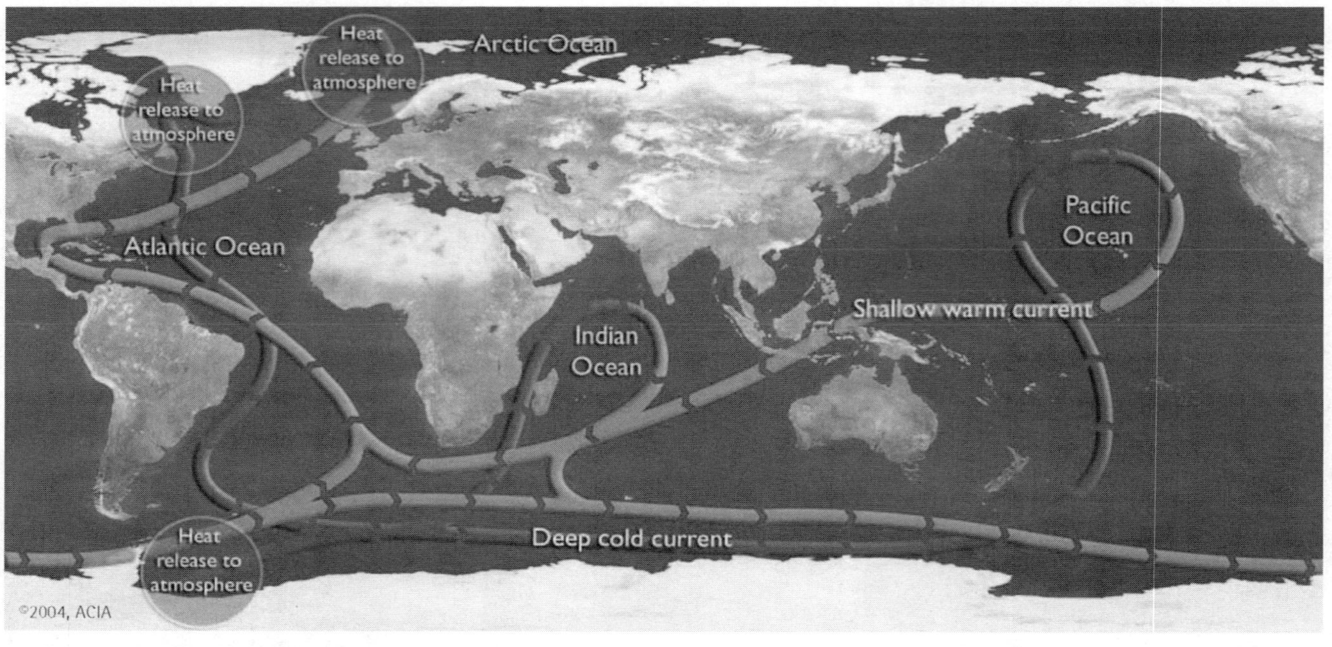

Changes in global ocean circulation can lead to abrupt climate change. Such change can be initiated by increases in arctic precipitation and river runoff, and the melting of arctic snow and ice, because these lead to reduced salinity of ocean waters in the North Atlantic.

Figure 10G.155 Global ocean circulation, www.amap.no/acia.

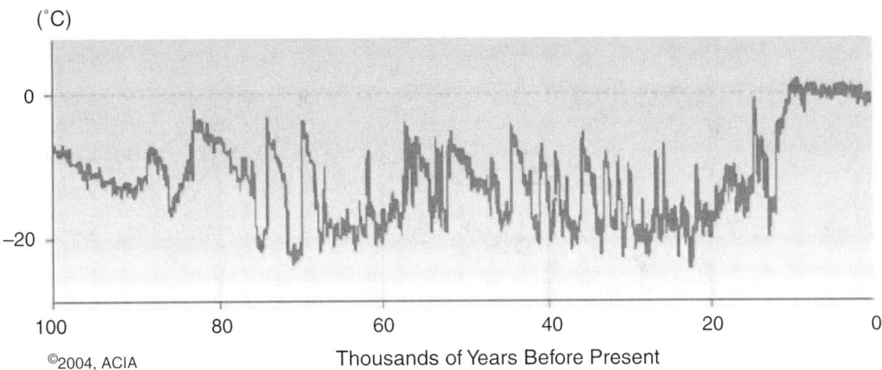

This record of temperature change (departures from present conditions) has been reconstructed from a Greenland ice core. The record demonstrates the high variability of the climate over the past 100,000 yrs. It also suggests that the climate of the past 10000 years or so, which was the time during which human civilization developed, has been unusually stable. There is concern that the rapid warming caused by the increasing concentrations of greenhouse gases due to human activities could destabilize this state.

Figure 10G.156 1000,000 yrs of temperature variation in Greenland, www.amap.no/acia.

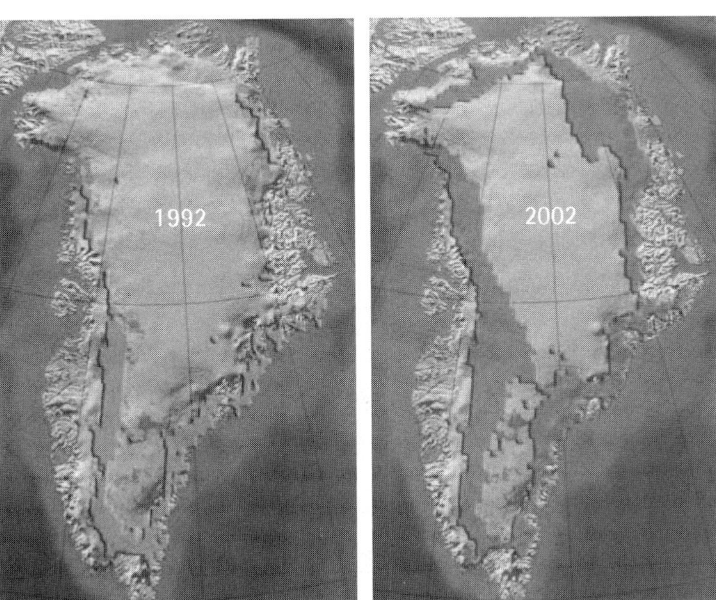

Greenland ice sheet melt extent
(Maximum melt extent 1979 – 2002)
(10^5 km^2)
7 6 5 4 3 2
1980 1985 1990 1995 2000
©2004, ACIA

Seasonal surface melt extent on the Greenland Ice Sheet has been observed by satellite since 1979 and shows an increasing trend. The melt zone, where summer warmth turns snow and ice around the edges of the ice sheet into slush and ponds of meltwater, has been expanding inland and to record high elevations in recent years. When the meltwater seeps down through cracks in the ice sheet, it may accelerate melting and, in some areas, allow the ice to slide more easily over the bedrock below, speeding its movement to the sea. In addition to contributing to global sea-level rise, this process adds fresh water to the ocean, with potential impacts on ocean circulation and thus regional climate.

Figure 10G.157 Greenland ice sheet melt extent, www.amap.no/acia.

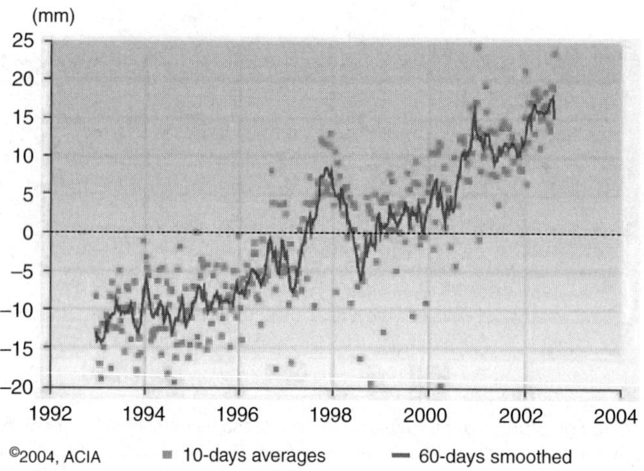

These data, from a satellite launched in 1992, show the rise in global average sea level over the past decade.

Figure 10G.158 Observed global sea level rise, www.amap.no/acia.

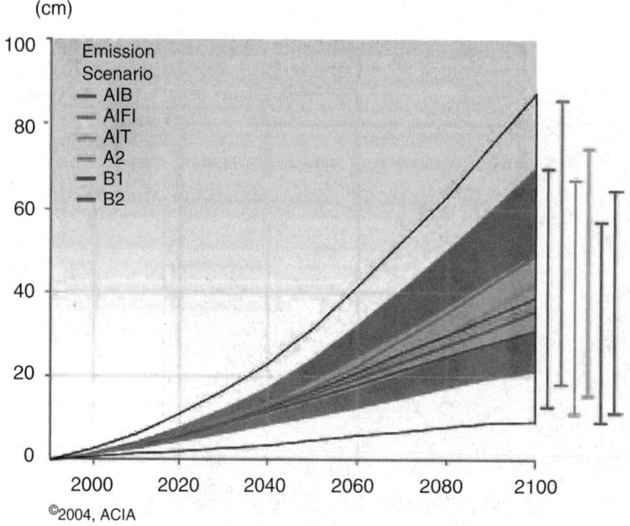

The graph shows future increases in global average sea level in centimeters as projected by a suite of climate models using six IPCC emissions scenarios. The bars at right show the range projected by a group of models for the designated emissions scenarios.

Figure 10G.159 Projected global sea level rise, www.amap.no/acia.

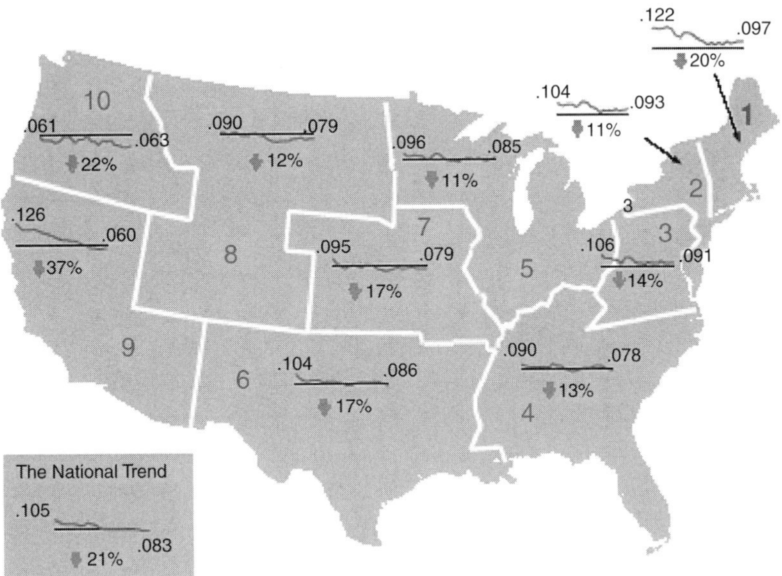

Figure 10G.160 Trend in fourth highest daily maximum 8-hour ozone concentration (ppm) by EPA region, 1980–2003, www.epa.gov/airtrends.

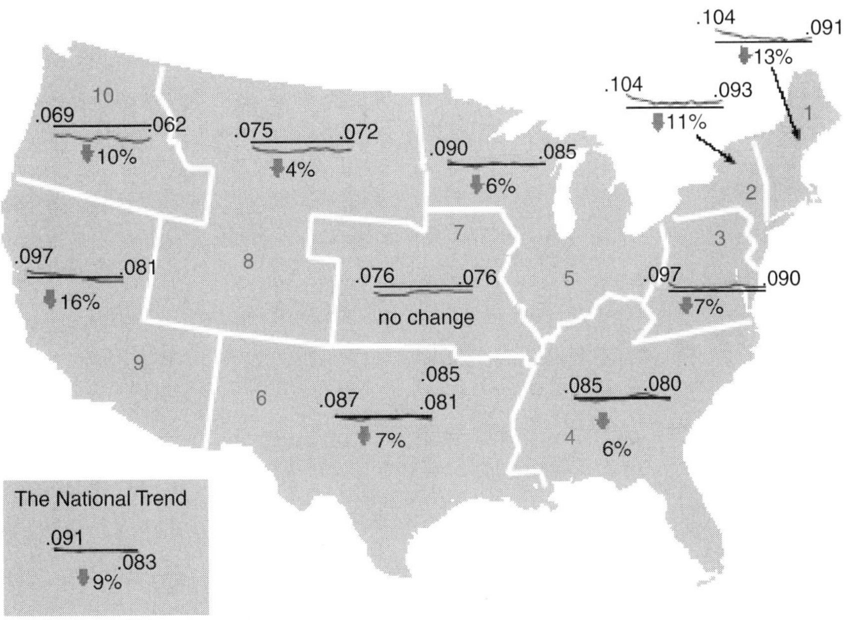

Figure 10G.161 Trend in fourth highest daily maximum 8-hour ozone concentration (ppm) by EPA region 1990–2003, www.epa.gov/airtrends.

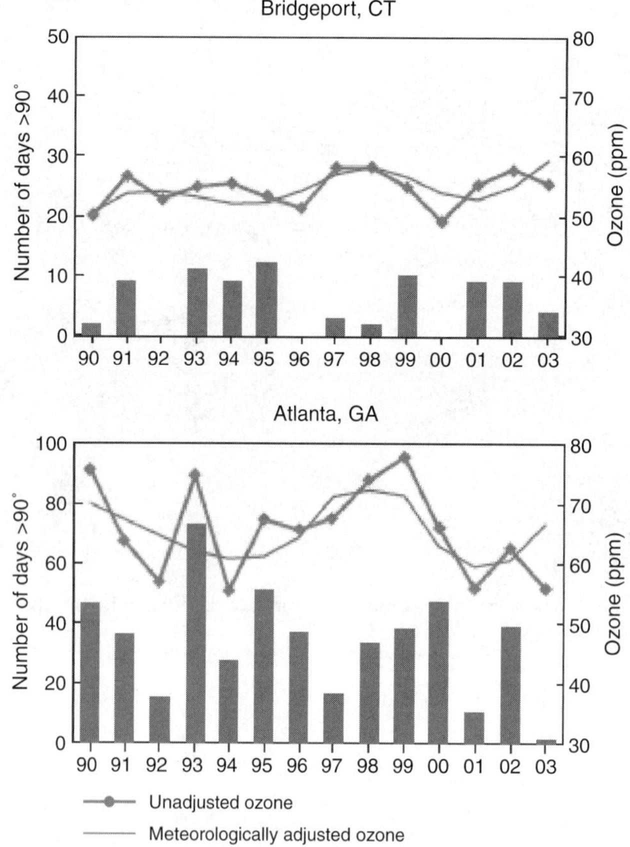

Ozone concentrations are Annual Average Daily Maximum 8-hr values between June and August.

Figure 10G.162 Number of days daily maximum temperatures exceed 90° (bar) compared to unadjusted ozone (red line) and meteorologically adjusted ozone (blue line) for Bridgeport and Atlanta, 1990–2003. Ozone concentrations are annual average daily maximum 8-hr values between June and August, www.epa.gov/airtrends.

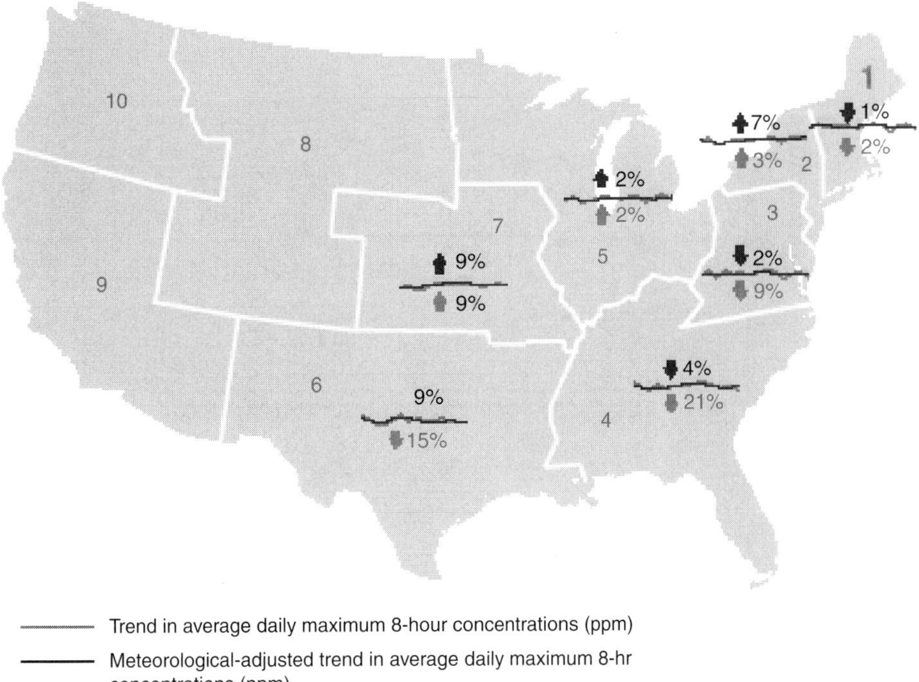

Figure 10G.163 Trends in unadjusted and meteorologically adjusted ozone levels by EPA region, 1990–2003, www.epa.gov/airtrends.

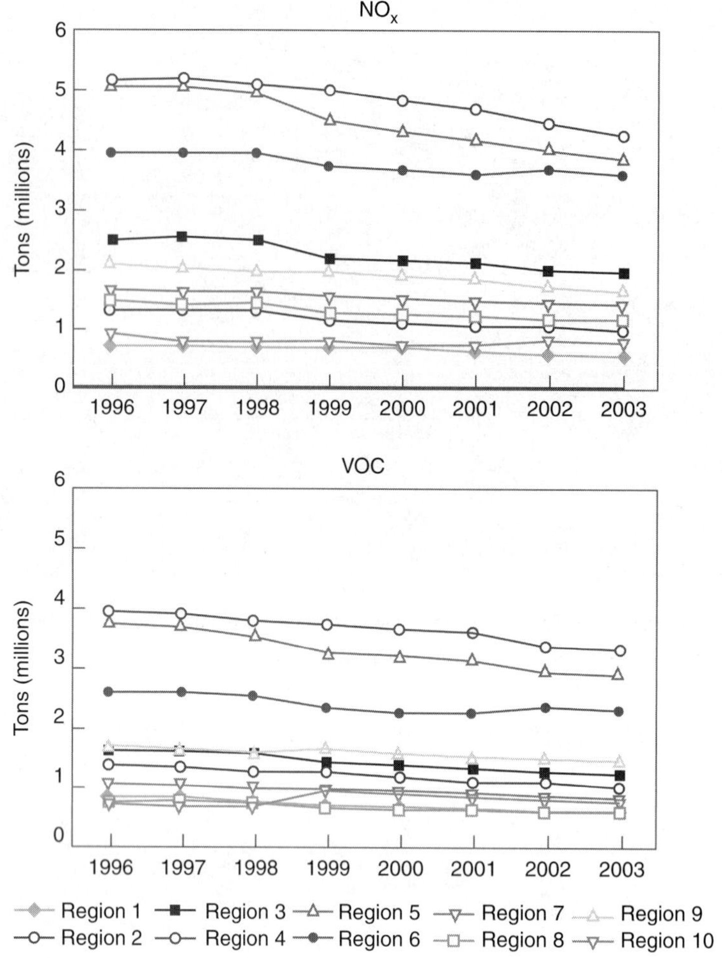

Figure 10G.164 NO_x and VOC emissions in Region 10 from 1998 to 1999 is due to a change in methodology rather than a true emission increase, www.epa.gov/airtrends.

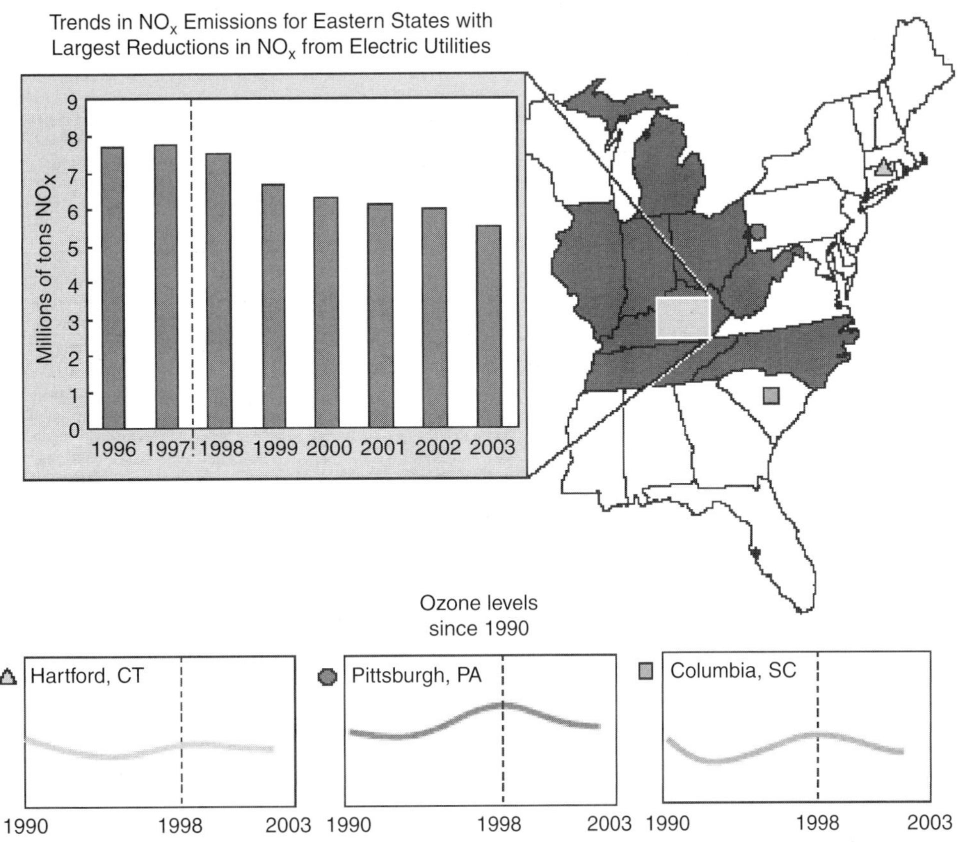

Figure 10G.165 Ozone trends for selected urban areas and corresponding regional emission trends, www.epa.gov/airtrends.

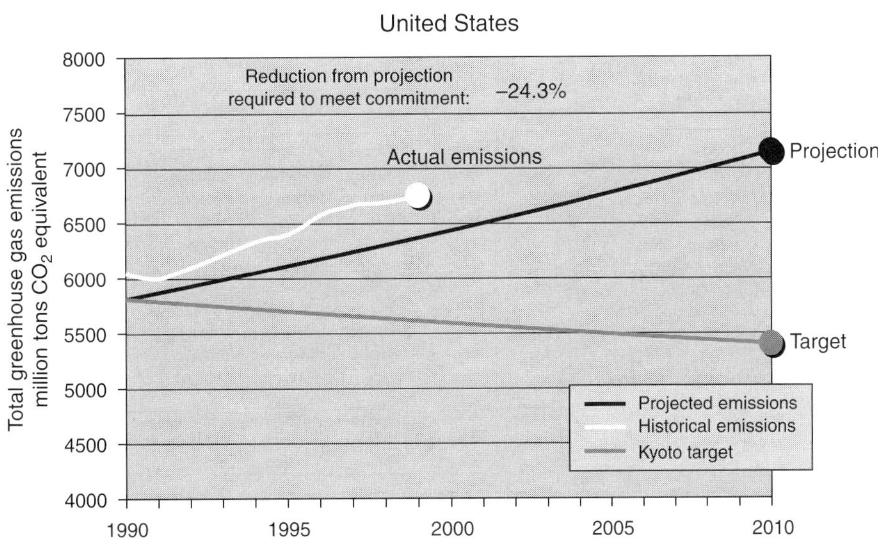

Figure 10G.166 Actual and projected emission of six greenhouse gases (CO_2, CH_4, N_2O, HFCs, PFCs, SF_6). (From Actual emissions UNFCCC/SB12000/11 Table B.1 Projected emissions UNFCCCM998/Add 2 Table C.6, www.epa.gov.)

I. Alaska, Yukon, and Coastal British Columbia
Lightly settled/water-abundant region; potential ecological, hydropower, and flood impacts:
- Increased spring flood risks
- Glacial retreat/disappearance in south, advance in north; impacts on flows, stream ecology
- Increased stress on salmon, other fish species
- Flooding of coastal wetlands
- Changes in estuary salinity/ecology

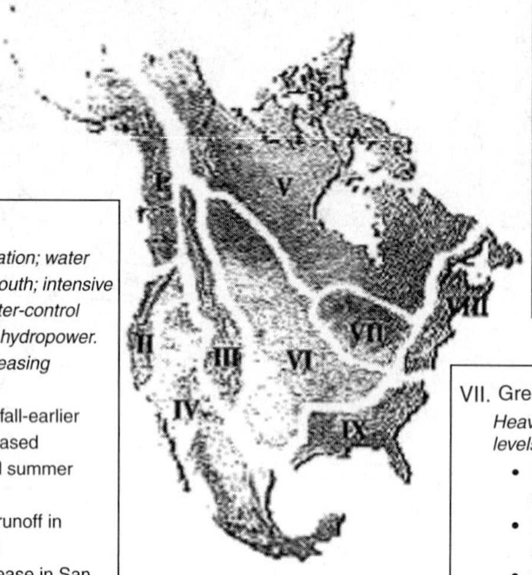

V. Sub-Arctic and Arctic
Sparse population (many dependent on natural systems); winter ice cover important feature of hydrologic cycle:
- Thinner ice cover, 1- to 3-month increase in ice-free season, increased extent of open water
- Increased lake-level variability, possible complete drying of some delta lakes
- Changes in aquatic ecology and species distribution as a result of warmer temperatures and longer growing season

VI. Midwest U.S.A. and Canadian Prairies
Agricultural heartland–mostly rainfed, with some areas relying heavily on irrigation:
- Annual streamflow decreasing/increasing; possible large declines in summer streamflow
- Increasing likelihood of severe droughts
- Possible increasing aridity in semi-arid zones
- Increases or decreases in irrigation demand and water availability–uncertain impacts on farm-sector income, groundwater levels, streamflows, and water quality

II. Pacific Coast States (U.S.A.)
Large and rapidly growing population; water abundance decreases north to south; intensive irrigated agriculture; massive water-control infrastructure; heavy reliance on hydropower. endangered species issues; increasing competition for water:
- More winter rainfall/less snowfall-earlier seasonal peak in runoff, increased fall/winter flooding, decreased summer water supply
- Possible increases in annual runoff in Sierra Nevada and Cascades
- Possible summer salinity increase in San Francisco Bay and Sacramento/San Joaquin Delta
- Changes in lake and stream ecology–warmwater species benefiting; damage to coldwater species (e.g., trout and salmon)

VII. Great Lakes
Heavily populated and industrialized region; variations in lake levels/flows now affect hydropower, shipping, shoreline structures:
- Possible precipitation increases coupled with reduced runoff and lake-level declines
- Reduced hydropower production; reduced channel depths for shipping
- Decreases in lake ice extent–some years w/out ice cover
- Changes in phytoplankton/zooplankton biomass, northward migration of fish species, possible extirpations of coldwater species

VIII. Northeast U.S.A. and Eastern Canada
Large, mostly urban population–generally adequate water supplies, large number of small dams, but limited total reservoir capacity; heavily populated floodplains:
- Decreased snow cover amount and duration
- Possible large reductions in streamflow
- Accelerated coastal erosion, saline intrusion into coastal aquifers
- Changes in magnitude, timing of ice freeze-up/break-up, with impacts on spring flooding
- Possible elimination of bog ecosystems
- Shifts in fish species distributions, migration patterns

III. Rocky Mountains (U.S.A. and Canada)
Lightly populated in north, rapid population growth in south; irrigated agriculture, recreation, urban expansion increasingly competing for water; headwaters area for other regions:
- Rise in snow line in winter-spring, possible increases in snowfall, earlier snowmelt, more frequent rain on snow–changes in seasonal streamflow, possible reductions in summer streamflow, reduced summer soil moisture
- Stream temperature changes affecting species composition; increased isolation of coldwater stream fish

IX. Southeast, Gulf, and Mid-Atlantic U.S.A.
Increasing population–especially in coastal areas, water quality/non-point source pollution problems, stress on aquatic ecosystems:
- Heavily populated coastal floodplains at risk to flooding from extreme precipitation events, hurricanes
- Possible lower base flows, larger peak flows, longer droughts
- Possible precipitation increase–possible increases or decreases in runoff/river discharge, increased flow variability
- Major expansion of northern Gulf of Mexico hypoxic zone possible–other impacts on coastal systems related to changes in precipitation/non-point source pollutant loading
- Changes in estuary systems and wetland extent, biotic processes, species distribution

IV. Southwest
Rapid population growth, dependence on limited groundwater and surface water supplies, water quality concerns in border region, endangered species concerns, vulnerability to flash flooding:
- Possible changes in snowpacks and runoff
- Possible declines in groundwater recharge–reduced water supplies
- Increased water temperatures–further stress on aquatic species
- Increased frequency of intense precipitation events–increased risk of flash floods

Figure 10G.167 Possible water resources impacts in North America, www.epa.gov/oar.

Table 10G.197 Percent Change in Air Quality and Emissions

	Percent Change in Air Quality	
	1983–2002	1993–2002
NO_2	−21	−11
O_3		
1-h	−22	−2[a]
8-h	−14	+4[a]
SO_2	−54	−39
PM_{10}	—	−13
$PM_{2.5}$	—	−8[b]
CO	−65	−42
Pb	−94	−57
	Percent Change in Emissions	
NO_2	−15	−12
VOC	−40	−25
SO_2	−33	−31
PM_{10}[c]	−34[d]	−22
$PM_{2.5}$[c]	—	−17
CO	−41	−21
Pb[e]	−93	−5

Note: Trend data not available. Negative numbers indicate improvements in air quality or reductions in emission. Positive numbers show where emissions have increased or air quality has gotten worse.

[a] Not statistically significant.
[b] Based on percentage change from 1999.
[c] Includes only directly emitted particles.
[d] Based on percentage change from 1985. Emission estimates prior to 1985 are uncertain.
[e] Lead emissions are included in the toxic air pollutant emissions inventory and are presented for 1982–2001.

Source: From www.epa.gov/airtrends/images/enlarge/sixpoll-1lg.gif.

Table 10G.198 Lead (Pb) National Totals (Thousands of Tons)

	Lead (Pb) National Totals (Thousands of Tons)													
Source Category	**1970**	**1975**	**1980**	**1985**	**1990**	**1991**	**1992**	**1993**	**1994**	**1995**	**1996**	**1997**	**1998**	**1999**
Fuel Comb. Elec. Util.	0.327293	0.229773	0.128825	0.063955	0.064244	0.061487	0.058603	0.061661	0.062	0.057	0.061	0.063925	0.068881	0.532
Fuel Comb. Industrial	0.236916	0.075254	0.059612	0.03017	0.017612	0.017531	0.017872	0.018884	0.019	0.018	0.016	0.01581	0.014985	0.414
Fuel Comb. Other	10.05174	10.0423	4.110847	0.421131	0.417995	0.416037	0.414095	0.415973	0.415	0.415	0.415	0.413444	0.410383	0.014
Chemical & Allied Product Mfg	0.102811	0.120227	0.104387	0.118276	0.135801	0.132297	0.093402	0.092	0.096	0.163	0.167	0.187825	0.194212	0.027
Metals Processing	24.22351	9.923236	3.025672	2.096969	2.169625	1.974318	1.77442	1.899989	2.027	2.049	2.055	2.080743	1.991153	1.002
Petroleum & Related Industries	0	0	0	0	0	0	0	0	0	0	0	0	0	0.013
Other Industrial Processes	2.028097	1.337357	0.807826	0.315972	0.168558	0.166802	0.056379	0.055	0.054	0.059	0.051	0.054315	0.053518	0.321
Solvent Utilization	0	0	0	0	0	0	0	0	0	0	0	0	0	0.273
Storage & Transport	0	0	0	0	0	0	0	0	0	0	0	0	0	0.0007
Waste Disposal & Recycling	2.2	1.59515	1.21002	0.870866	0.804311	0.808	0.812	0.825	0.83	0.604	0.787782	0.798363	0.805779	0.178
Highway Vehicles	171.9611	130.2061	60.50133	18.05194	0.420736	0.017849	0.018332	0.018796	0.01903	0.019355	0.01919	0.019996	0.02116	0
Off-highway	9.737102	6.129554	4.204813	0.921121	0.776437	0.574333	0.565182	0.528556	0.525034	0.544384	0.505382	0.502712	0.497295	0.552
Miscellaneous	0	0	0	0	0	0	0	0	0	0	0	0	0	0.029
Total	220.869	159.659	74.153	22.890	4.975	4.169	3.810	3.916	4.047	3.929	4.077	4.137	4.057	3.356
Fires	0	0	0	0	0.000	0.000	0.000	0.000	0.000	0.000	0.000	0.000	0.000	0.000
Total without Fires	220.869	159.659	74.153	22.890	4.975	4.169	3.810	3.916	4.047	3.929	4.077	4.137	4.057	3.356

Source: From www.epa.gov/airtrends/pdfs/leadnational.pdf.

Table 10G.199 Volatile Organic Compounds (VOC) National Totals (Thousands of Tons)

Source Category	Volatile Organic Compounds (VOC) National Totals (Thousands of Tons)																	
	1970	1975	1980	1985	1990	1991	1992	1993	1994	1995	1996	1997	1998	1999	2000	2001	2002	2003
Fuel Comb. Elec. Util.	30	40	45	32	47	44	44	45	45	44	50	52	56	54	62	61	52	56
Fuel Comb. Industrial	150	150	157	134	182	196	187	186	196	206	179	175	174	172	173	176	170	170
Fuel Comb. Other	541	470	848	1,403	776	835	884	762	748	823	893	893	889	919	949	950	790	878
Chemical & Allied Product Mfg	1,341	1,351	1,595	881	634	710	715	701	691	660	388	388	394	251	254	262	214	218
Metals Processing	394	336	273	76	122	123	124	124	126	125	73	78	78	66	67	71	69	72
Petroleum & Related Industries	1,194	1,342	1,440	703	611	640	632	649	647	642	477	487	485	457	428	441	375	380
Other Industrial Processes	270	235	237	390	401	391	414	442	438	450	435	438	443	438	454	420	406	412
Solvent Utilization	7,174	5,651	6,584	5,699	5,750	5,782	5,901	6,016	6,162	6,183	5,477	5,621	5,149	5,036	4,831	5,012	4,692	4,562
Storage & Transport	1,954	2,181	1,975	1,747	1,490	1,532	1,583	1,600	1,629	1,652	1,294	1,328	1,327	1,237	1,176	1,192	1,205	1,178
Waste Disposal & Recycling	1,984	984	758	979	986	999	1,010	1,046	1,046	1,067	509	518	535	487	415	420	457	427
Highway Vehicles	16,910	15,392	13,869	12,354	9,388	8,860	8,332	7,804	7,277	6,749	6,221	5,985	5,859	5,681	5,325	4,952	4,543	4,428
Off-highway	1,616	1,917	2,192	2,439	2,662	2,709	2,754	2,799	2,845	2,890	2,935	2,752	2,673	2,682	2,644	2,622	2,688	2,572
Miscellaneous	1,101	716	1,134	566	1,059	756	486	556	720	551	1,940	816	718	791	733	532	883	704
Miscellaneous	NA	NA	NA	NA	NA	NA	NA	NA	NA	NA	0	0	0	0	0	0	0	0
Total	34,659	30,765	31,106	27,404	24,108	23,577	23,066	22,730	22,569	22,041	20,871	19,530	18,782	18,270	17,512	17,111	16,544	16,056
Fires	917	587	1,024	465	983	678	407	478	638	464	1,870	744	645	667	615	412	785	627
Total without Fires	33,742	30,178	30,082	26,939	23,125	22,899	22,659	22,252	21,931	21,577	19,001	18,786	18,136	17,603	16,898	16,699	15,759	15,429

Source: From www.epa.gov/airtrends/pdfs/vocnational.

Table 10G.200 Sulfur Dioxide (SO_2) National Totals (Thousands of Tons)

	Sulfur Dioxide (SO_2) National Totals (Thousands of Tons)																	
Source Category	**1970**	**1975**	**1980**	**1985**	**1990**	**1991**	**1992**	**1993**	**1994**	**1995**	**1996**	**1997**	**1998**	**1999**	**2000**	**2001**	**2002**	**2003**
Fuel Comb. Elec. Util.	17,398	18,268	17,469	16,272	15,909	15,784	15,416	15,189	14,889	12,080	12,767	13,195	13,416	12,583	11,396	10,850	10,293	10,929
Fuel Comb. Industrial	4,568	3,310	2,951	3,169	3,550	3,256	3,292	3,284	3,218	3,357	2,849	2,805	2,740	2,135	2,139	2,243	2,299	2,227
Fuel Comb. Other	1,490	1,082	971	579	831	755	784	772	780	793	636	648	586	620	628	642	575	596
Chemical & Allied Product Mfg	591	367	280	456	297	280	279	269	275	286	255	259	261	325	338	342	328	329
Metals Processing	4,775	2,849	1,842	1,042	726	612	615	603	562	530	389	407	405	304	313	332	271	285
Petroleum & Related Industries	881	727	734	505	430	378	416	383	379	369	335	344	342	312	316	319	348	323
Other Industrial Processes	846	740	918	425	399	396	396	392	398	403	386	409	415	382	410	429	416	426
Solvent Utilization	NA	NA	NA	1	0	0	1	1	1	1	1	1	1	1	1	1	2	2
Storage & Transport	NA	NA	NA	4	7	10	9	5	2	2	5	5	5	6	6	7	5	6
Waste Disposal & Recycling	8	46	33	34	42	44	44	71	59	47	32	33	34	34	34	35	28	32
Highway Vehicles	273	334	394	455	503	469	436	402	369	335	302	304	300	300	260	248	275	256
Off-highway	278	301	323	354	371	379	385	392	399	406	413	422	432	475	437	440	420	443
Miscellaneous	110	20	11	11	12	11	10	10	15	10	15	7	6	67	70	44	91	88
Miscellaneous	NA	NA	NA	NA	NA	NA	NA	NA	NA	NA	0	0	0	0	0	0	0	0
Total	31,218	28,043	25,925	23,307	23,076	22,375	22,082	21,772	21,346	18,619	18,385	18,840	18,944	17,545	16,347	15,932	15,353	15,943
Fires	NA	NA	NA	NA	12	12	9	9	14	10	15	6	6	67	69	44	91	95
Total without Fires	31,218	28,043	25,925	23,307	23,064	22,363	22,073	21,763	21,332	18,609	18,370	18,834	18,939	17,478	16,278	15,888	15,263	15,848

Source: From www.epa.gov/airtrends/2005.

Table 10G.201 Nitrogen Oxides (Nox) National Emissions Totals (Thousands of Tons)

	Nitrogen Oxides (NO_x) National Emissions Totals (Thousands of Tons)																	
Source Category	1970	1975	1980	1985	1990	1991	1992	1993	1994	1995	1996	1997	1998	1999	2000	2001	2002	2003
Fuel comb. Elec. Util.	4,900	5,694	7,024	6,127	6,663	6,519	6,504	6,651	6,565	6,384	6,164	6,276	6,232	5,721	5,330	4,917	4,699	4,458
Fuel Comb. Industrial	4,325	4,007	3,555	3,209	3,035	2,979	3,071	3,151	3,147	3,144	3,151	3,101	3,050	2,709	2,723	2,757	2,870	2,775
Fuel Comb. Other	836	785	741	712	1,196	1,281	1,353	1,308	1,303	1,298	1,197	1,177	1,101	768	766	779	725	729
Chemical & Allied Product Mfg	271	221	213	262	168	165	163	155	160	158	125	127	129	102	105	107	105	102
Metals Processing	77	73	65	87	97	76	81	83	91	98	83	89	89	86	89	94	84	94
Petroleum & Related Industries	240	63	72	124	153	121	148	123	117	110	139	143	143	120	122	124	149	137
Other Industrial Processes	187	182	205	327	378	352	361	370	389	399	433	460	467	451	479	504	487	504
Solvent Utilization	NA	NA	NA	2	1	2	3	3	3	3	2	3	3	4	4	4	8	7
Storage & Transport	NA	NA	NA	2	3	6	5	5	5	6	15	16	16	14	15	16	16	16
Waste Disposal & Recycling	440	159	111	87	91	95	96	123	114	99	153	157	163	162	129	130	152	137
Highway Vehicles	12,624	12,061	11,493	10,932	9,592	9,449	9,306	9,162	9,019	8,876	8,733	8,792	8,619	8,371	8,394	7,774	7,365	7,381
Off-highway	2,652	2,968	3,353	3,576	3,781	3,849	3,915	3,981	4,047	4,113	4,179	4,178	4,156	4,084	4,167	4,156	4,086	4,103
Miscellaneous	330	165	248	310	369	286	255	241	390	267	412	187	179	251	276	184	356	289
Miscellaneous	NA	NA	NA	NA	NA	NA	NA	NA	NA	NA	0	0	0	0	0	0	0	0
Total	26,883	26,377	27,079	25,757	25,529	25,179	25,260	25,357	25,349	24,956	24,787	24,705	24,348	22,845	22,598	21,549	21,102	20,728
Fires	NA	NA	NA	NA	362	247	234	234	382	258	405	179	172	236	263	171	341	236
Total without Fires	26,883	26,377	27,079	25,757	25,167	24,932	25,026	25,123	24,967	24,698	24,382	24,526	24,176	22,609	22,335	21,378	20,761	20,492

Source: From www.epa.gov/airtrends/pdfs/noxnational.pdf.

Table 10G.202 National Air Pollutant Emissions Estimates (Fires and Dust Excluded) for Major Pollutants

	Millions of Tons Per Year							
	1970	1975	1980	1985[a]	1990	1995	2000[a]	2004[b]
Carbon monoxide (CO)	197.3	184.0	177.8	169.6	143.6	120.0	102.4	87.2
Nitrogen oxides (NO_x)[c]	26.9	26.4	27.1	25.8	25.2	24.7	22.3	18.8
Particulate matter (PM)[d]								
PM10	12.2[a]	7.0	6.2	3.6	3.2	3.1	2.3	2.5
PM2.5[e]	NA	NA	NA	NA	2.3	2.2	1.8	1.9
Sulfur dioxide (SO_2)	31.2	28.0	25.9	23.3	23.1	18.6	16.3	15.2
Volatile organic compounds (VOC)	33.7	30.2	30.1	26.9	23.1	21.6	16.9	15.0
Lead[f]	0.221	0.16	0.074	0.022	0.005	0.004	0.003	0.003
Totals[g]	301.5	275.8	267.2	249.2	218.2	188.0	160.2	138.7

[a] In 1985 and 1996 EPA refined its methods for estimating emissions. Between 1970 and 1975, EPA revised its methods for estimating particulate matter emissions.
[b] The estimates for 2004 are preliminary.
[c] NO_x estimates prior to 1990 include emissions from fires. Fires would represent a small percentage of the NO_x emissions.
[d] PM estimates do not include condensable PM, or the majority of PM2.5 that is formed in the atmosphere from "precursor" gases such as SO_2 and NO_x.
[e] EPA has not estimated PM2.5 emissions prior to 1990.
[f] The 1999 estimate for lead is used to represent 2000 and 2003 because lead estimates do not exist for these years.
[g] PM2.5 emissions are not added when calculating the total because they are included in the PM10 estimate.

Source: From www.epa.gov/airtrends/2005/econ-emission.html.

Table 10G.203 Carbon Monoxide (CO) National Emissions Totals (Thousands of Tons)

	Carbon Monoxide (CO) National Emissions Totals (Thousands of Tons)																	
Source Category	**1970**	**1975**	**1980**	**1985**	**1990**	**1991**	**1992**	**1993**	**1994**	**1995**	**1996**	**1997**	**1998**	**1999**	**2000**	**2001**	**2002**	**2003**
Fuel Comb. Elec. Util.	237	276	322	291	363	349	350	363	370	372	408	423	451	496	484	485	499	530
Fuel Comb. Industrial	770	763	750	670	879	920	955	1,043	1,041	1,056	1,188	1,162	1,151	1,213	1,219	1,253	1,436	1,377
Fuel Comb. Other	3,625	3,441	6,230	7,525	4,269	4,587	4,849	4,181	4,108	4,506	2,741	2,742	2,727	3,829	3,081	3,088	2,498	3,003
Chemical & Allied Product Mfg	3,397	2,204	2,151	1,845	1,183	1,127	1,112	1,093	1,171	1,223	1,053	1,071	1,081	350	361	372	337	329
Metals Processing	3,644	2,496	2,246	2,223	2,640	2,571	2,496	2,536	2,475	2,380	1,599	1,710	1,702	1,255	1,295	1,380	1,294	1,422
Petroleum & Related Industries	2,179	2,211	1,723	462	333	345	371	371	338	348	354	367	366	159	161	162	128	138
Other Industrial Processes	620	630	830	694	537	548	544	594	600	624	561	582	590	571	592	615	635	634
Solvent Utilization	NA	NA	NA	2	5	5	5	5	5	6	1	2	2	52	51	50	51	73
Storage & Transport	NA	NA	NA	49	76	28	17	51	24	25	70	71	72	163	169	178	215	241
Waste Disposal & Recycling	7,059	3,230	2,300	1,941	1,079	1,116	1,138	1,248	1,225	1,185	2,904	2,948	3,121	3,019	1,849	1,851	1,852	1,854
Highway Vehicles	163,231	153,555	143,827	134,187	110,255	104,980	99,705	94,431	89,156	83,881	78,606	75,849	73,244	68,708	68,061	63,476	62,161	58,807
Off-highway	11,371	14,329	16,685	19,029	21,447	21,934	22,419	22,904	23,389	23,874	24,358	23,668	23,689	23,316	24,178	24,677	24,450	24,446
Miscellaneous	7,909	5,263	8,344	7,927	11,122	8,618	6,934	7,082	9,658	7,298	15,016	7,316	7,184	11,410	12,964	8,676	16,498	14,033
Miscellaneous	NA	NA	NA	NA	NA	NA	NA	NA	NA	NA	0	0	0	0	0	0	0	0
Total	204,043	188,398	185,407	176,844	154,186	147,128	140,896	135,901	133,559	126,777	128,858	117,910	115,380	114,541	114,467	106,262	112,054	106,886
Fires	6,766	4,433	7,622	7,289	10,583	10,583	6,389	6,537	9,089	6,705	14,502	6,793	6,654	10,508	12,049	7,744	15,654	13,180
Total without Fires	197,277	183,965	177,785	169,555	143,603	136,545	134,507	129,364	124,470	120,072	114,356	111,117	108,726	104,033	102,418	98,518	96,401	93,706

Source: From www.epa.gov/airtrends/2005/pdfs/conational.pdf.

Table 10G.204 Estimated Sea-Level Change by Year 2100, as a Result of Ice Wastage in a Carbon Dioxide-Enhanced Environment

Ice Mass Contributing to Sea-Level Change	Estimated Sea-Level Change (Range, ft)
Glaciers and small ice caps	+0.3 to 1.0
Greenland ice sheet	+0.3 to 1.0
Antarctic ice sheet	−0.3 to 3[a]

[a] Most likely the change will range from 0 to 0.7 foot.

Source: From National Academy of Sciences, Committee on Glaciology, 1985.

Table 10G.205 Nationwide Impacts of Sea-Level Rise in the United States

	Sea-Level Rise		
	50 cm	100 cm	200 cm
If densely developed areas are protected			
Shore protection costs ($ billions)	32–43	73–111	169–309
Dryland lost (mi^2)	2,200–6,100	4,100–9,200	6,400–13,500
Wetlands lost (%)	20–45	29–69	33–80
If no shores are protected			
Dryland lost (mi^2)	3,300–7,300	5,100–10,300	8,200–15,400
Wetlands lost (%)	17–43	26–66	29–76
If all shores are protected			
Wetlands lost (%)	38–61	50–82	66–90

Source: From U.S. Environmental Protection Agency, 1988, *The potential effects of Global climate change on the United States*, Draft Report to Congress, Data assembled by Titus and Greene.

SECTION 10H OFFSHORE WASTE DISPOSAL

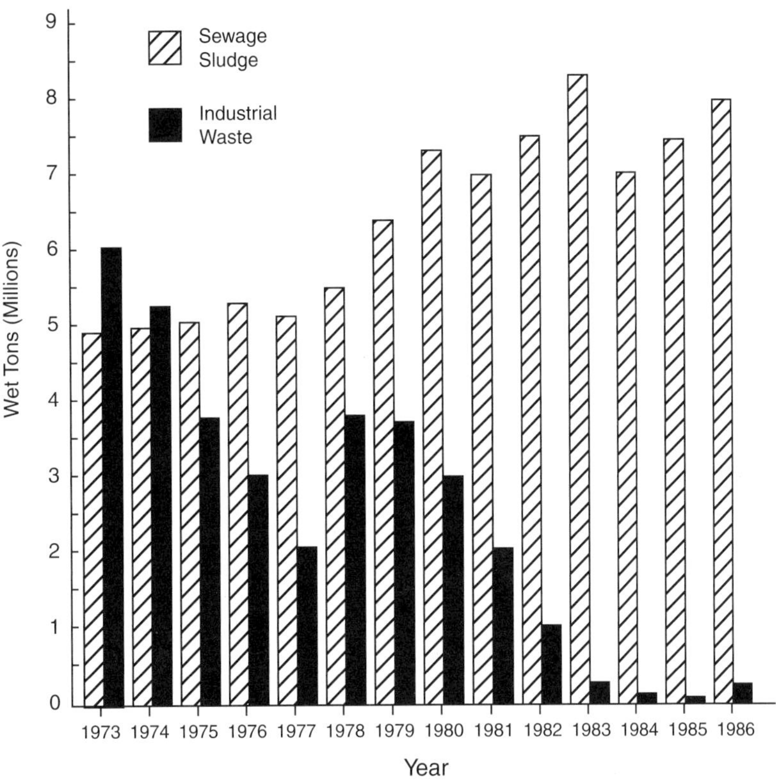

Figure 10H.168 Sewage sludge and industrial waste dumped in U.S. ocean waters from 1973 to 1986. (From U.S. EPA 1988, *Report to Congress on Administration of the Marine Protection Research, and Sanctuaries Act of 1972*, as Amended, 1984–1986, EPA-503/8-88/002.)

Table 10H.206 Offshore Waste Disposal in the United States, 1973–1983

	Offshore Waste Disposal in the United States, 1973–1983										
	1973	1974	1975	1976	1977	1978	1979	1980	1981	1982	1983
Atlantic (A)											
Industrial waste	3,643	3,642	3,322	2,633	1,784	2,548	2,577	2,928	2,271	1,063	283
Sewage sludge	4,898	5,010	5,040	5,271	5,134	5,535	6,442	7,309	6,703	7,670	8,312
Construction debris	974	770	396	315	379	241	107	89	0	0	0
Solid waste/chemicals incinerated	0	0	0	0	0	0	0	0	0	0	0
Explosives	0	0	0	0	0	0	0	0	0.0003	0	0
Wood incinerated	11	16	6	9	15	18	45	11	15	13	31
Gulf of Mexico (B)											
Industrial waste	1,408	938	120	100	60	0.17	0	0	0	0	0
Sewage sludge/construction debris solid waste/explosives/wood incinerated	0	0	0	0	0	0	0	0	0	0	0
Chemicals incinerated	0	12.3	4.1	0	17.6	0	0	0	700[a]	800[a]	0
Pacific (C)											
Industrial waste	0	0	0	0	0	0	0	0.26	23.3	18.8	21.5
Sewage sludge/construction debris solid waste/explosives/wood incinerated	0	0	0	0	0	0	0	0	0	0	0
Solid waste	240	200	0	0	0	0	0	0	0	0	0
Wood incinerated	0	0	0	0	12.1	0	0	0	0	0	0
Totals of (A), (B), (C)											
Industrial waste	5,051	4,580	3,452	2,733	1,844	2,548.17	2,577	2,928.26	2,294.3	1,081.8	304.5
Sewage sludge	4,890	5,010	5,040	5,271	5,134	5,535	6,442	7,309	6,703	7,670	8,312
Construction debris	974	770	396	315	379	241	107	89	0	0	0
Solid waste	240	200	0	0	0	0	0	0	0	0	0
Explosives	0	0	0	0	0	0	0	0	0.0003	0	0
Wood incinerated	11	16	6	9	15	18	45	11	15	13	31
Chemical incinerated Atlantic (A)	0	12.3	4.1	0	17.6	0	0	0	700[a]	800[a]	0

Note: Thousands of tons.

[a] Thousand gallons (prior to incineration).

Source: From U.S. Environmental Protection Agency, Report to Congress January 1981–December 1983.

Table 10H.207 Quantities of Sludge Dumped by Sewage Authorities in United States Ocean Waters, 1984–1986

	Quantities in Thousand Wet Tons		
	1984	**1985**	**1986**
Sewage Authorities			
Bergen County Utilities Authority NJ	255	309	353
Joint Meeting of Essex and Union Counties NJ	335	341	238
Linden Roselle Sewerage Authority NJ	235	95	93
Middlesex County Utilities Authority NJ	966	1,039	1,018
Nassau County Dept. of Public Works NY	520	576	709
New York City Dept. of Environmental Protection NY	3,085	3,345	3,591
Passaic Valley Sewerage Commission NJ	854	884	1,317
Rahway Valley Sewerage Authority NJ	160	187	98
Westchester County Dept. of Environmental Facilities NY	539	470	506
	6,999	7,246	7,923

Note: Nine municipal sewage authorities which had previously held interim permits are dumping sewage sludge pursuant to court orders issued by United States district courts in New York and New Jersey. These authorities have been required to submit permit applications to the USEPA, and currently are shifting their dumping from the 12-Mile Site to the Deepwater Municipal Sludge Dump Site, also known as the 106-Mile Site.

Source: From U.S. Environmental Protection Agency, 1988, *Report to Congress on Administration of the Marine Protection, Research, and Sanctuaries Act of 1972*, as Amended, 1984–1986, EPA-503/8-88/002.

Table 10H.208 Quantities of Industrial Waste Materials Dumped in United States Ocean Waters by Special Permit, 1984–1986

	Quantities in Thousand Wet Tons		
	1984	**1985**	**1986**
USEPA Region II			
Acid waste site (NY Bight Apex) Allied Chemical Corp.[a] NY	40	40	34
Deepwater Industrial Waste Site			
DuPont—Edge Moor[b] DE	19	0	140
DuPont — Grasselli[c] NJ	146	100	73
USEPA Region IX			
Fish Wastes Site			
Samoa Packing, American Samoa	8	4.6	21.4
Star Kist, American Samoa	7.9	20.3	24.1
Oil Drilling Muds and Cuttings THUMS Long Beach, CA	—[d]	2.7	13.6
	220.9	167.6	306.1

[a] Hydrochloric acid waste.
[b] Aqueous iron and miscellaneous chlorides and hydrochloric acid wastes.
[c] Solution of alkaline sodium wastes.
[d] No permit issued.

Source: From U.S. Environmantal Protection Agency, 1988, *Report to Congress on Administration of the Marine Protection, Research, and Sanctuaries Act of 1972*, as Amended, 1984–1986, EPA-503/8-88/002.

SECTION 10I ENERGY DEVELOPMENT

Table 10I.209 Water-Quality Impacts of Various Energy Processes

Process	Water-Quality Impacts	Frequency and Areal Scale		Time Frame	Severity	Effectiveness of Control
		Regional	Local			
Extraction and on-site processing						
Coal mining						
Underground mining	Most damaging problems are acid mines drainage and disruption of aquifers, which affect pH, dissolved solids and specific ion content, and thus impair utility of streams and groundwaters for other uses.					
	a. Surface waters	H		L	3	Poor
	b. Groundwaters		M	L	3	Ineffective
Surface mining	Surface disturbance results in high sediment transport potential. Discharge from mines may impair water quality through increase in dissolved solids and specific ions.					
	a. Surface waters	H		M	3	Fair
	b. Groundwaters		L	S	1	Good
Beneficiation	Release of chemical and physical treatment materials to streams can impair water quality. Leaching of solid wastes results in pollution similar to acid drainage.					
	a. Surface waters	H		M	2	Fair
	b. Groundwaters		L	M	1	Fair
Oil and gas extraction						
Primary recovery	Principal problems are handling of saline waste waters. Leaks in casings, pipes, and storage ponds can release brines to groundwaters and streams.					
	a. Surface waters	M		S	2	Good
	b. Groundwaters		M	M	3	Fair
Secondary and tertiary recovery	Principal concerns are escape of oil and formation waters through casing, pipe, and storage tank leaks releasing organic and inorganic contaminants to the environment.					
	a. Surface waters		L	S	2	Good
	b. Groundwaters		L	L	3	Good
Offshore operation	Blowouts with resulting massive oil contamination are a rare but catastrophic problem.	L		S	3	Good
Oil-shale extraction and processing	Most significant concerns relate to potential for escape of noxious organic and inorganic contaminants to streams. Disruption of aquifers likely, low hazard due to limited occurrence of oil shale.					
Undergoing mining	Concerns center on disruption of aquifers and disposal of sometimes saline dewatering by injection.					
	a. Surface waters		L	L	1	Good
	b. Groundwaters		L	L	1	Good

(Continued)

Table 10I.209 (Continued)

Process	Water-Quality Impacts	Frequency and Areal Scale		Time Frame	Severity	Effectiveness of Control
		Regional	Local			
Surface mining	Surface disturbance results in high sediment transport potential.					
	a. Surface waters		L	S	2	Excellent
	b. Groundwaters		L	M	1	Excellent
Surface retorting	Concerns center on potential for escape of organic and inorganic contaminants from plant site to accidental leaks and spills. A more significant concern is escape of contaminants from waste piles through leaching.					
	a. Surface waters		M	L	2	Good to fair
	b. Groundwater		L	L	1	Good
In-situ recovery	Underground effects mainly involve contamination of groundwaters by organic and inorganic compounds produced in combustion; where applicable surface effects are similar to those of surface retorting.					
	a. Surface waters		L	S	2	Excellent
	b. Groundwaters		M	L	3	Untested
Tar-sands extraction and processing						
Surface mining and processing	Main concerns are accidental release of organic contaminants to streams and potential for failure of waste impoundment structures leading to massive downstream damage from fine waste.					
	a. Surface waters		M	S	2	Good to fair
	b. Groundwaters		L	L	1	Untested
In-situ recovery	Concerns center on potential for escape of noxious organic and inorganic chemicals to groundwaters.					
	a. Surface waters		L	S	2	Untested
	b. Groundwaters		M	L	3	Untested
Geothermal extraction						
Vapor-dominated systems	Main concerns are escape of noxious inorganic contaminants to groundwater from waste disposal, blowouts, and leaks in casings and pipes.					
	a. Surface waters		L	S	2	Good
	b. Groundwaters		M	L	1	Fair
Water-dominated systems	Main problems involve escape of noxious and toxic constituents of thermal waters to surface and groundwaters from production operations, waste disposal, blowouts, and leaks in casings and pipes.					
	a. Surface waters	H		M	4	Fair
	b. Groundwaters		H	L	3	Fair

(Continued)

Table 10I.209 (Continued)

Process	Water-Quality Impacts	Frequency and Areal Scale		Time Frame	Severity	Effectiveness of Control
		Regional	Local			
Uranium mining and milling						
Underground mining	Escape of radioactive and other inorganic contaminants to the environment through disposal of dewatering waste and escape from tailings ponds can seriously impair downstream water uses.					
	a. Surface waters	M		S	4	Good
	b. Groundwaters		M	L	4	Poor
Surface mining	Some concern about high sediment transport potential, but main source of concern is potential for radioactive contamination of streams and groundwaters through leakage from tailings disposal ponds.					
	a. Surface waters	M		S	4	Good
	b. Groundwaters		M	L	4	Poor
Solution mining	Main concern centers on escape of radioactive and inorganic process chemicals to off-site groundwaters.					
	a. Surface waters		L	S	1	Excellent
	b. Groundwater		L	L	4	Excellent
Transportation						
Coal slurry lines	Main concern centers on pipeline breaks and the potential for contamination of streams.					
	a. Surface waters		L	S	2	Excellent
	b. Groundwaters		L	L	1	Excellent
Oil pipelines	Most significant problems are pipeline breaks and resulting oil pollution of streams.					
	a. Surface waters		M	S	2	Fair
	b. Groundwaters		L	L	1	Good
Oil tankers	Escape of oil to marine environment as result of shipwrecks can be catastrophic to marine life over wide areas.	L		M	3	Poor
Refining						
Oil refining	Controlled release of waste water and accidental releases of organic and inorganic contaminants are most significant issues; concerns center on impairment of water supplies of other water users.					
	a. Surface waters		M	M	3	Good
	b. Groundwaters		L	L	3	Good
Nuclear fuel cycle	Accidental releases of radioactive materials to surface and groundwaters from processing and reprocessing plants are main concern; both high- and low-level waste disposal also have potential for escape of radioactivity to the water environment. Controlled release of nonradioactive inorganic chemicals adds to chemical load of receiving waters.					
	a. Surface waters		M	S	4	Good
	b. Groundwaters		M	L	4	Fair to poor

(Continued)

Table 10I.209 (Continued)

Process	Water-Quality Impacts	Frequency and Areal Scale		Time Frame	Severity	Effectiveness of Control
		Regional	Local			
Conversion						
Fossil-fueled steam electric generation	Controlled release of cooling-system blowdown to streams and/or leakage from cooling ponds add dissolved solids and treatment chemicals to stream loads. Once-through cooling contributes to thermal pollution.					
	a. Surface waters	M		M	2	Good
	b. Groundwater		L	L	1	Good
Nuclear steam-electric generation	Small controlled releases of radioactive materials and discharge of cooling-system blowdown add radioactivity, dissolved solids, and treatment chemicals to stream loads. Accidental release of radioactivity through reactor containment failure could endanger human life over wide area.					
	a. Surface waters	M		M	5	Good
	b. Groundwaters		L	L	4	Good
Geothermal electric generation	Disposal of waste and condensate containing noxious inorganic compounds and thermal load to streams impair downstream uses and damages aquatic life.					
	a. Surface waters	H		M	4	Fair
	b. Groundwaters		H	L	3	Fair
Hydroelectric generation	Changes in stream temperature and dissolved gases due to storage and reservoir releases seriously alter the aquatic environment.					
	a. Surface waters	H		L	2	Fair
	b. Groundwaters		L	S	1	Good
Coal conversion processes	Controlled release of cooling system blowdown and accidental releases of organic and inorganic contaminants, as in oil refining and with similar concern about impairment of other water uses.					
	a. Surface waters		M	M	3	Good
	b. Groundwaters		L	L	3	Good

Note: Frequency, H, high, M, medium, L, low. Time frame, L, longer than 10 yrs, M, 1 to 10 yrs, S, less than 1 yr, Severity, 5, direct threat to human life, 4, hazardous to human health, 3, severe economic damage, 2, damage to biota, 1, aesthetic or other Intangible harm.

Source: From Davis, G.H., 1985, *Water and Energy: Demand and Effects, Unesco Studies and Reports in Hydrology 42*. Copyright Unesco. Reprinted with permission.

SECTION 10J WATERBORNE DISEASES/HEALTH HAZARDS

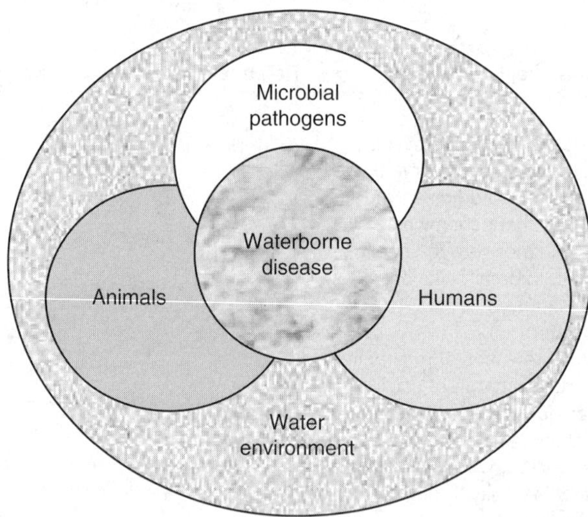

Figure 10J.169 Waterborne disease interactions in the water environment, (From Cotruvo, J.A., et al., 2004, *Waterborne Zoonoses, Identification, Causes, and Control*, Published on Behalf of the World Health Organization by IWA Publishing, www.who.int.)

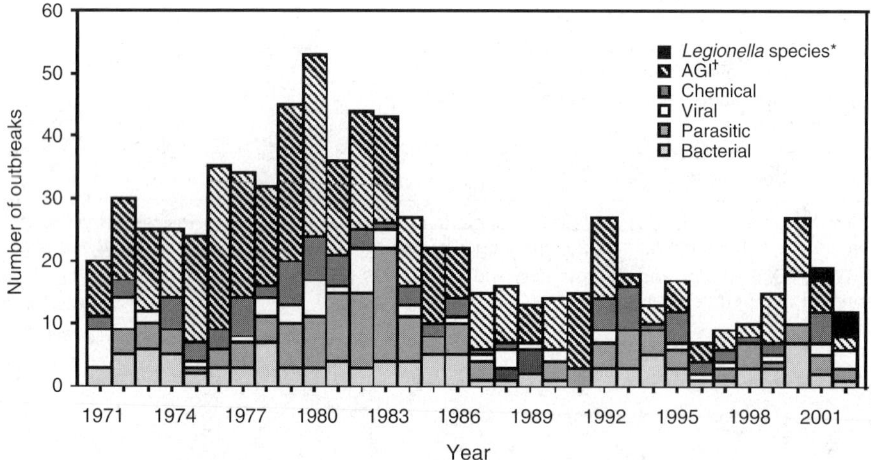

* Beginning in 2001, Legionnaires disease was added to the surveillance system, and *Legionella* species were classified separately.

† Acute gastrointestinal illness of unknown etiology.

Figure 10J.170 Number of waterborne-disease outbreaks ($n = 764$) associated with drinking water, by year and etiologic agent in the United States, 1971–2002. (From Blackburn, B.G. et al., 2004, Surveillance of waterborne-disease outbreaks associated with drinking water — United States, 2001–2002, *MMWR Surveillance Summaries*, vol. 53, no. SS-08, pp. 23–45, www.cdc.gov.)

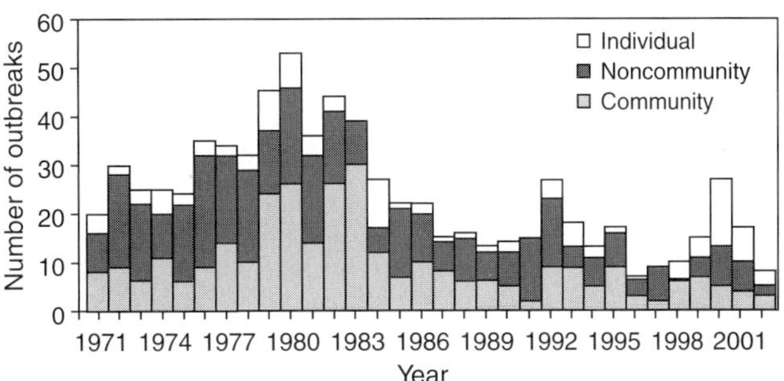

Figure 10J.171 Number of waterborne disease outbreaks ($n=758$)* associated with drinking water, by year and type of water system — United States, 1971–2002. (From Cotruvo, J.A. et al., (ed.), 2004, Waterborne Zoonoses, Identification, Causes, and Control, Published on Behalf of the World Health Organization by IWA Publishing, Copyright © World Health Organization 2004, wlio.int.)

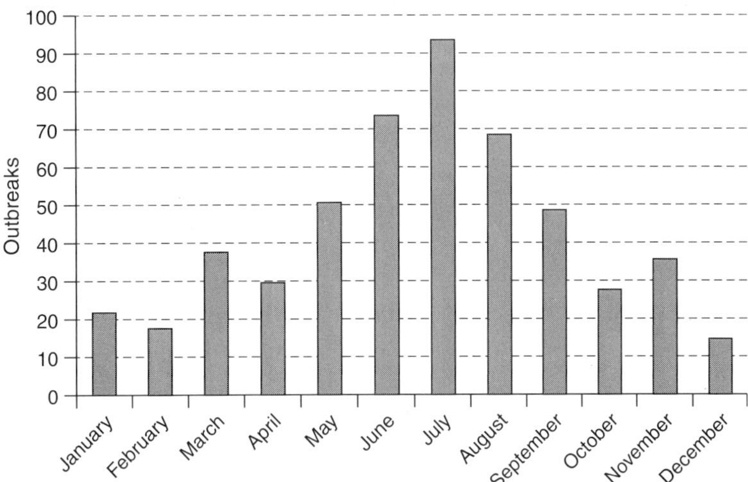

Figure 10J.172 Outbreaks of waterborne disease by month, 1973–1998. (From Gleick, P.H. et al., 2004, The World's Water 2004–2005, *The Biennial Report on Freshwater Resources*, Island Press, Washington, www.worldwater.org.)

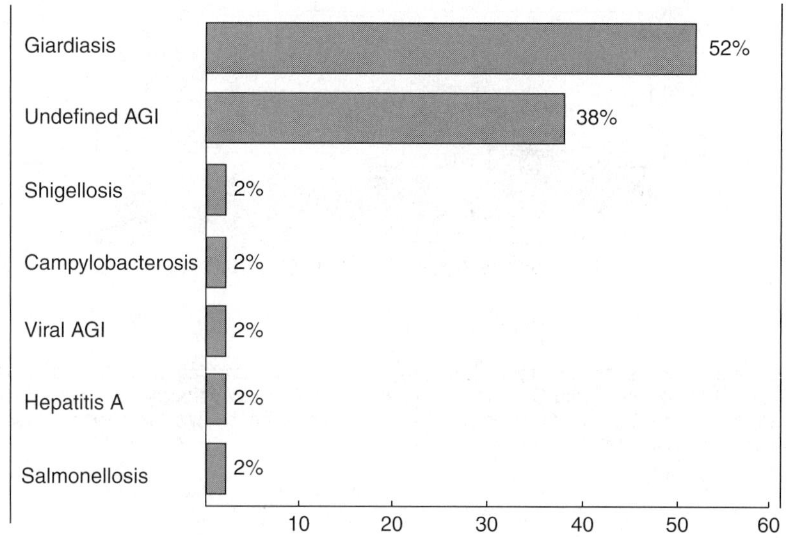

Figure 10J.173 Etiology of waterborne disease outbreaks in untreated, disinfected-only, and filtered surface water systems in the United States, 1971–1985. (From Craun, G.F., 1988, Surface water supplies and health, *J. American Water Works Association*, vol. 80, no. 2. Copyright AWWA. Reproduced with permission.)

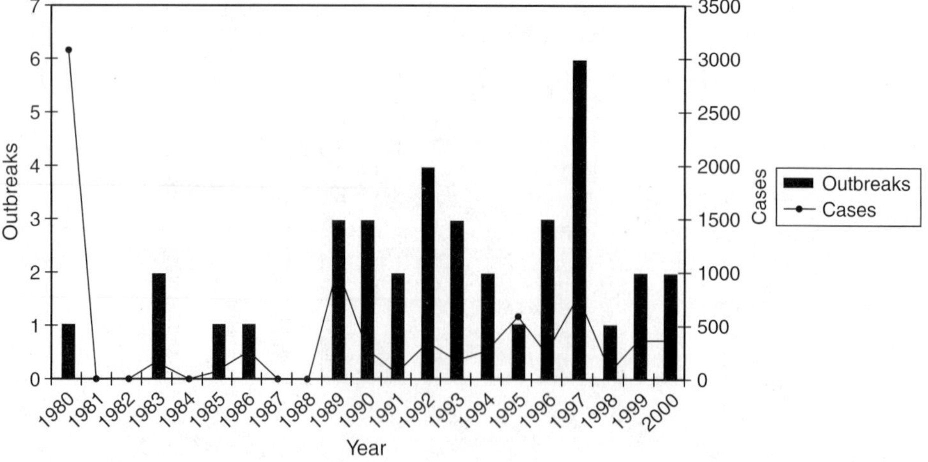

Figure 10J.174 Outbreaks of Waterborne Disease by Month, 1973–1998 (From *World's Water 2004–2005*, by Peter H. Gleick. Copyright © 2004 Island Press. Reproduced by permission of Island Press, Washington, DC.)

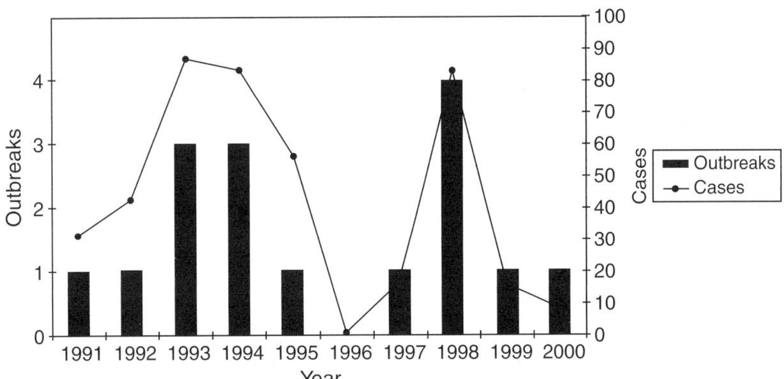

Figure 10J.175 Outbreaks of waterborne disease associated with private water supplies in England and Wales from 1980–2000. (From Stanwell-Smith, R., Anderson, Y., and Levy, D., 2003, National surveillance systems in Hunter, P.R., Waite, M., and Ronchi, El., (eds.), *Drinking Water and Infectious Disease Establishing the Links*, CRC Press LLC, Boca Raton, Florida. Reproduced with permission.)

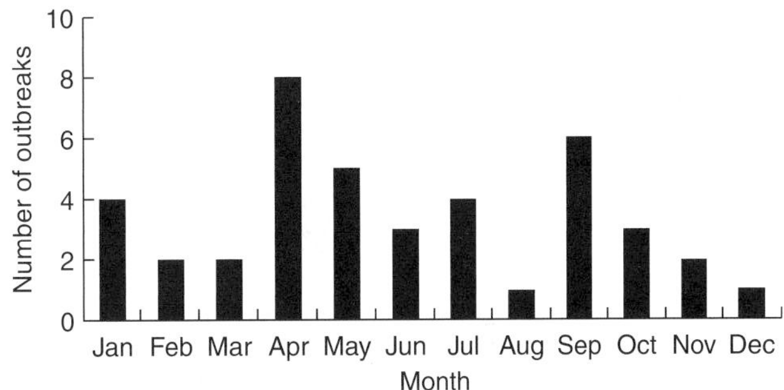

Figure 10J.176 Seasonal distribution of outbreaks associated with both private and public drinking water supplies in England and Wales from 1991 to 2000. (From Stanwell-Smith, R., Anderson Y., and Levy, D., 2006, National Surveillance Systems in Hunter, P.R., Waite, M., and Ronchi, El. (eds.), Drinking Water and Infectious Disease Establishing the Links, CRC Press LLC, Boca Raton, Florida.)

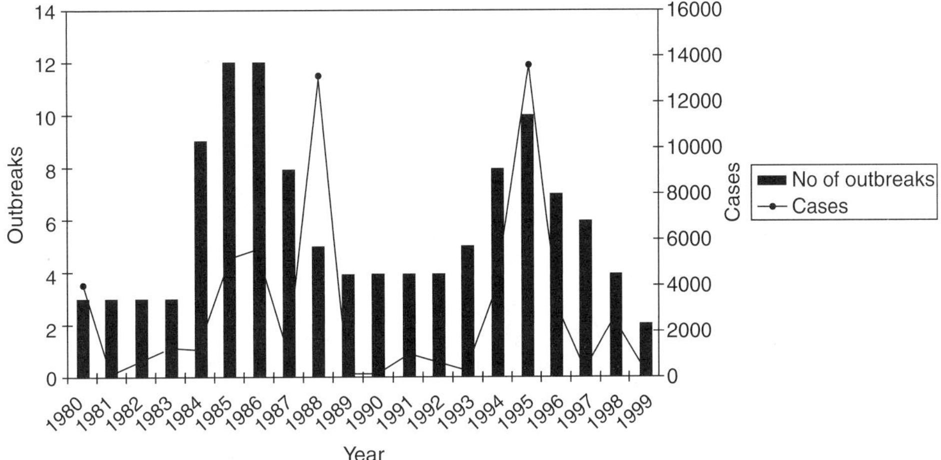

Figure 10J.177 Outbreaks and cases of waterborne disease in Sweden, 1980–1999. (From Stanwell-Smith, R., Anderson Y., and Levy, D., 2006, National Surveillance Systems in Hunter, P.R., Waite, M., and Ronchi, El. (eds.), Drinking Water and Infectious Disease Establishing the Links, CRC Press LLC, Boca Raton, Florida.)

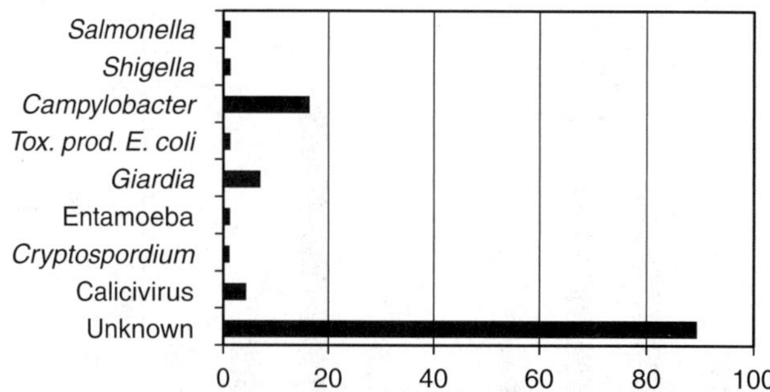

Figure 10J.178 Microbial agents associated with waterborne outbreaks in Sweden, 1980–1999. (From Stanwell-Smith, R., Anderson, Y., and Levy, D., 2003, National surveillance systems in Hunter, P.R., Waite, M., and Ronchi, El., (eds.), *Drinking Water and Infectious Disease Establishing the Links*, CRC Press LLC, Boca Raton, Florida. Reproduced with permission.)

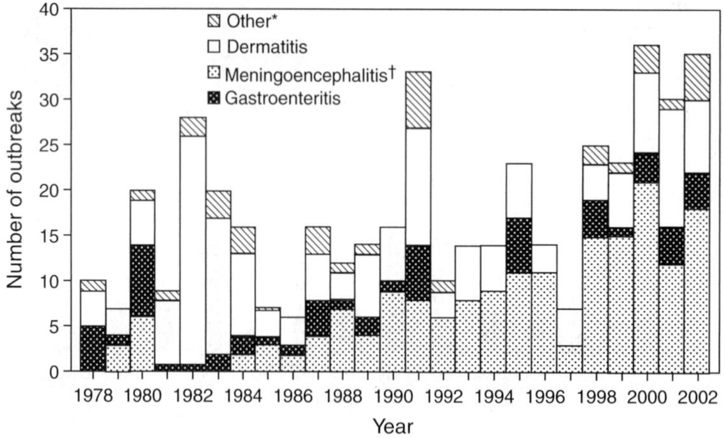

* Includes keratitis, conjunctivitis, otitis, bronchitis, meningitis, hepatitis, leptospirosis, Pontiac fever, and acute respiratory illness.

† Also includes data from report of ameba infections.

Figure 10J.179 Number of waterborne-disease related outbreaks ($n=445$) associated with recreational water by year and illness in the United States, 1978–2002. (From Yoder, J.S. et al., 2004. Surveillance for waterborne-disease outbreaks associated with recreation water — United States, 2001–2002, *MMWR Surveillance Summary*, vol. 53, no. SS08, pp 1–22, October 22, 2004, www.cdc.gov.)

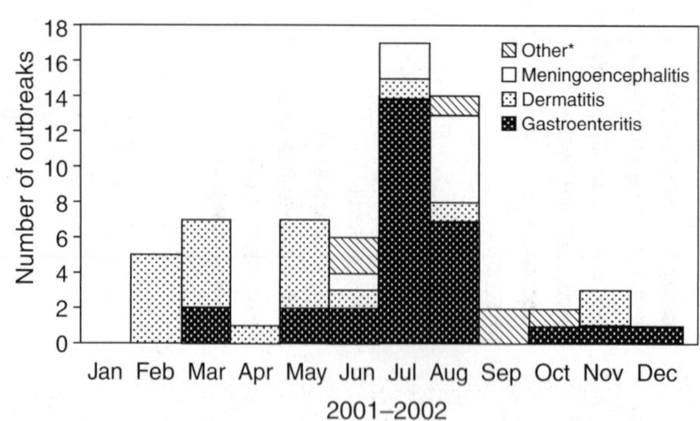

* Acute respiratory illness, Pontiac fever, or chemical exposure.

Figure 10J.180 Number of waterborne-disease related outbreaks ($n=65$) associated with recreational water by illness and month in the United States, 2001–2002. (From Yoder, J.S. et al., 2004, Surveillance for waterborne-disease outbreaks associated with recreation water — United States, 2001–2002, *MMWR Surveillance Summary*, vol. 53, no. SS08, pp 1–22, October 22, 2004, www.cdc.gov.)

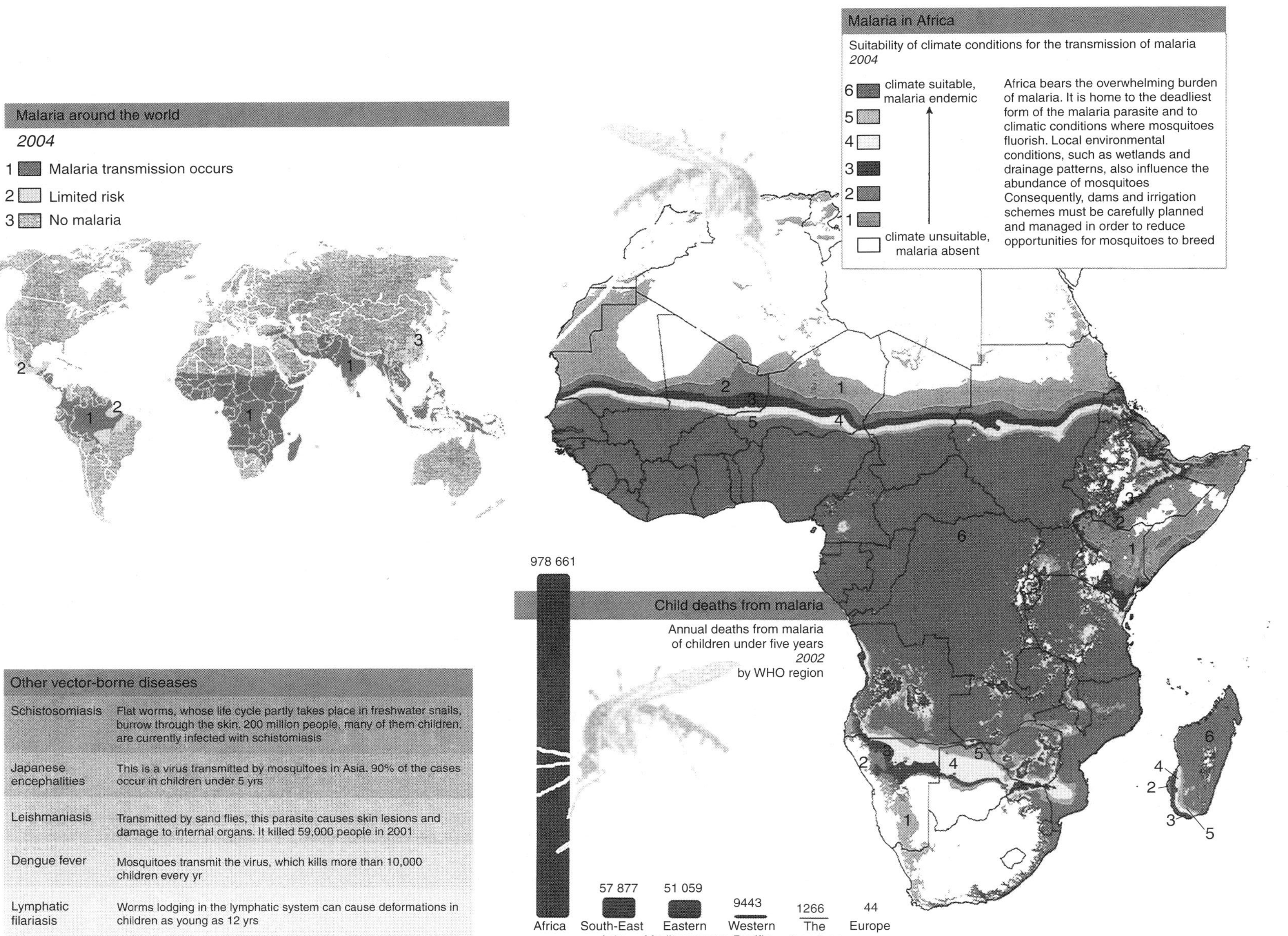

Malaria around the world
2004
1 Malaria transmission occurs
2 Limited risk
3 No malaria
Malaria in Africa
Suitability of climate conditions for the transmission of malaria
2004
6 climate suitable, malaria endemic
5
4
3
2
1
climate unsuitable, malaria absent
Africa bears the overwhelming burden of malaria. It is home to the deadliest form of the malaria parasite and to climatic conditions where mosquitoes fluorish. Local environmental conditions, such as wetlands and drainage patterns, also influence the abundance of mosquitoes Consequently, dams and irrigation schemes must be carefully planned and managed in order to reduce opportunities for mosquitoes to breed
Child deaths from malaria
Annual deaths from malaria of children under five years
2002
by WHO region
978 661 Africa
57 877 South-East Asia
51 059 Eastern Mediterranean
9443 Western Pacific
1266 The Americas
44 Europe
Other vector-borne diseases
Schistosomiasis
Flat worms, whose life cycle partly takes place in freshwater snails, burrow through the skin. 200 million people, many of them children, are currently infected with schistomiasis
Japanese encephalities
This is a virus transmitted by mosquitoes in Asia. 90% of the cases occur in children under 5 yrs
Leishmaniasis
Transmitted by sand flies, this parasite causes skin lesions and damage to internal organs. It killed 59,000 people in 2001
Dengue fever
Mosquitoes transmit the virus, which kills more than 10,000 children every yr
Lymphatic filariasis
Worms lodging in the lymphatic system can cause deformations in children as young as 12 yrs

Figure 10J.181 Malaria. (From Gordon, B., Mackay, R., and Rehfuess, E., 2004, *Inheriting the world: The Atlas of Children's Health and the Environment*, World Health Organization, www.who.int.)

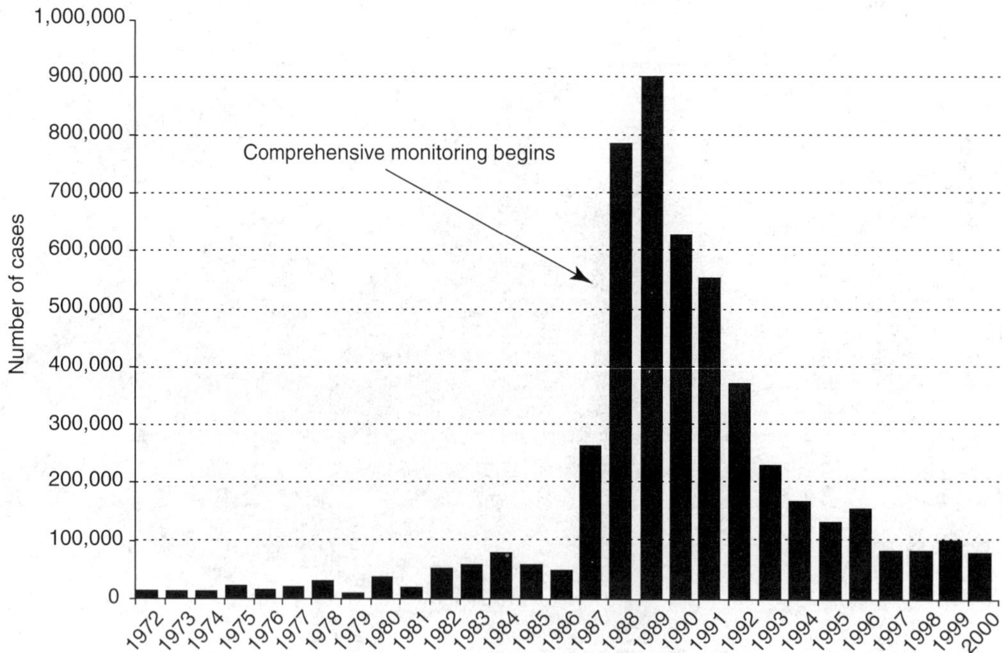

Figure 10J.182 Dracunculiasis (guinea worm) cases worldwide, 1972–2000. (From Gleick. et al., 2002. The world's water, *The Biennial Report on Freshwater Resources*, 2002–2003, Island Press, Washington, www.worldwater.org. Reproduced with permission.)

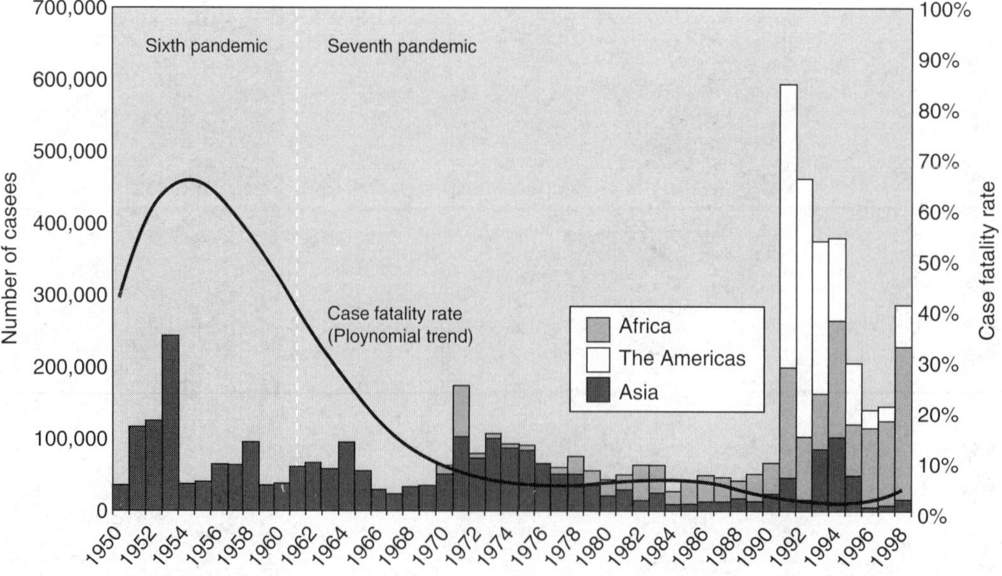

Figure 10J.183 Cholera, reported number of cases and case fatality rates, 1950–1998. (From World Health Organization, WHO Report on Global Surveillance of Epidemic-Prone Infectious Diseases-Cholera, www.who.int.)

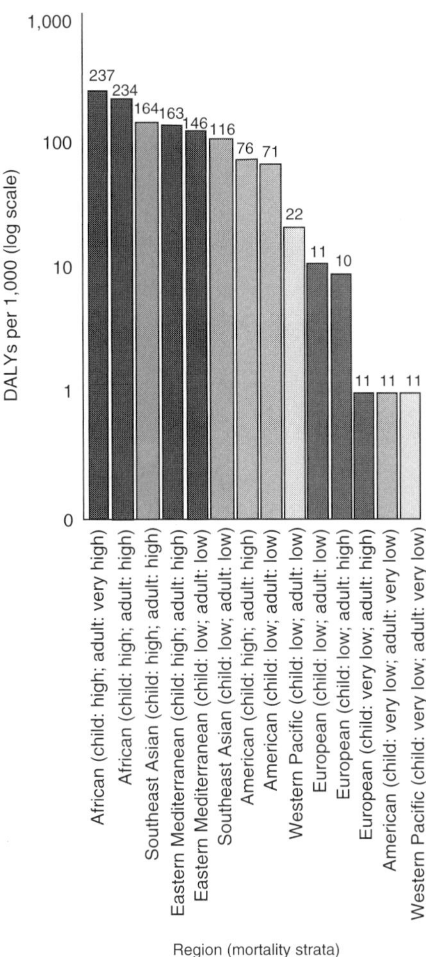

Figure 10J.184 Diarrheal disease from water, sanitation, and hygiene: DALYs per 1,000 children (under 5 years old) by region. (From Prüss, A., Kay, D., Fewtrell, L., and Bartram, J., 2002, Estimating the burden of disease from water, sanitation, and hygiene at a global level, *Environmental Health Prerspectives*, vol. 110, no. 5, May 2002, www.ehponline.org.)

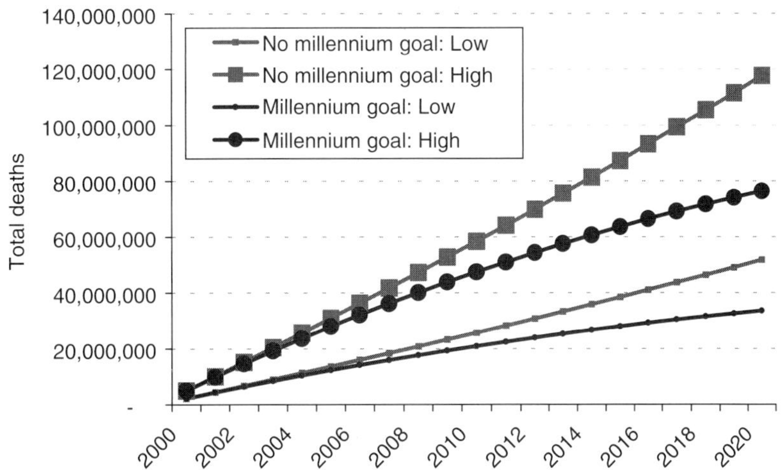

Figure 10J.185 Total water-related deaths between 2000 and 2020. (From Gleick, P. H., 2002, Dirty water: estimated deaths from water-related diseases 2000–2020, *Pacific Institute Research Report*, © 2002 Pacific Institute for Studies in Development, Environment, and Security, on-line at www.pacinst.org.)

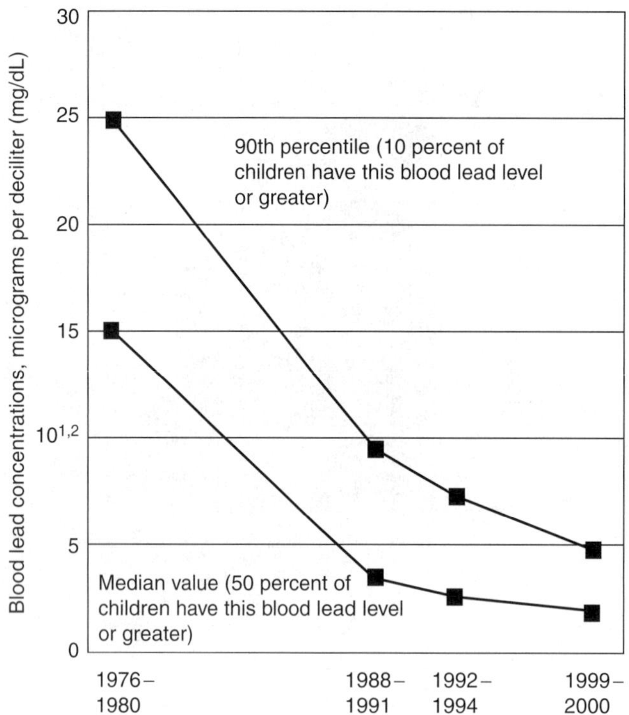

Figure 10J.186 Concentration of lead in blood of children age 5 and under, 1976–1980, 1988–1991, 1992–1994, 1999–2000. (From United States Environmental Protection Agency, 2003, EPA's Draft Report on the Environment, 2003, EPA 600-R-03-050. U.S. Environmental Protection Agency. *America's Children and the Environment Measures of Contaminants, Body Burdens, and Illnesses, Second Edition*, February 2003. Data from CDC National Center for Health Statistics, National Health and Nutrition Examination Survey, 1976–2000, www.epa.gov.

[1] 10 μg/dL of blood lead has been identified by CDC as elevated, which indicates the need for interventions. (CDC *Preventing Lead Poisoning in Young Children*, 1991.)

[2] Recent research suggests that blood levels less than 10 μg/dL may still produce subtle, subclinical health effects in children. (Schmidt, C.W. *Poisoning Young Minds*, 1999.)

Table 10J.210 Magnitude of Waterborne Disease Outbreaks in the United States, 1920–1980

	Frequency of Occurrence (Number of Outbreaks)			
Size of Outbreak (Cases of Illness)	**Community Systems**	**Noncommunity Systems**	**Individual Systems**	**All Systems**
<2	3	0	3	6
2–5	26	35	95	156
6–10	71	50	81	202
11–25	145	119	63	327
26–50	94	124	34	252
51–100	68	82	16	166
101–200	63	50	6	119
201–300	28	14	2	44
301–500	29	14	1	44
501–1,000	29	9	1	39
1,001–3,000	28	3	0	31
3,001–5,000	9	0	0	9
5,001–10,000	5	0	0	5
>10,000	5	0	0	5
Total	603	500	302	1,405

Source: From Craun, G.F., 1986, *Waterborne Diseases in the United States.* Copyright CRC Press, Inc., Boca Raton, FL. Reprinted with permission.

Table 10J.211 Etiology of Waterborne Disease Outbreaks in the United States, 1920–1984

Time Period	Disease	Outbreaks	Cases	Deaths
1920–1925	Typhoid fever	127	7,294	435
	Gastroenteritis	11	27,756	0
1926–1930	Typhoid fever	100	3,072	234
	Gastroenteritis	17	63,902	0
1931–1935	Typhoid fever	85	2,114	140
	Gastroenteritis	25	7,664	0
	Amebiasis	1	1,412	98
	Hepatitis A	1	28	0
1936–1940	Gastroenteritis	91	77,403	2
	Typhoid fever	60	1,281	80
	Shigellosis	10	3,308	0
	Chemical poisoning	1	92	0
	Amebiasis	1	4	0
1947–1945	Gastroenteritis	126	36,118	3
	Thyhoid fever	56	1,450	46
	Shigellosis	10	2,817	6
	Salmonellosis	1	12	0
	Paratyphoid fever	2	14	0
	Chemical poisoning	1	30	0
1946–1950	Gastroenteritis	87	10,718	0
	Typhoid fever	18	264	5
	Hepatitis A	5	173	0
	Shigellosis	4	2,321	1
	Paratyphoid fever	1	5	0
	Leptospirosis	1	9	0
	Tularemia	1	4	0
1951–1955	Gastroenteritis	31	5,297	0
	Typhoid fever	7	103	0
	Hepatitis A	7	340	0
	Shigellosis	4	732	1
	Amebiasis	1	31	2
	Salmonellosis	1	2	0
	Poliomyelitis	1	16	0
1956–1960	Gastroenteritis	21	2,306	0
	Typhoid fever	13	128	3
	Hepatitis A	11	417	0
	Shigellosis	7	3,081	0
	Chemical poisoning	3	14	4
	Salmonellosis	2	17	0
	Amebiasis	1	5	0
	Tularemia	1	2	0
1961–1965	Gastroenteritis	18	20,627	0
	Typhoid fever	11	63	0
	Hepatitis A	10	334	0
	Shigellosis	7	520	4
	Chemical poisoning	5	30	6
	Salmonellosis	3	16,425	3
	Giardiasis	1	123	0
	Paratyphoid fever	1	5	0
1966–1970	Gastroenteritis	21	5,922	0
	Hepatitis A	19	562	1
	Shigellosis	14	1,215	0
	Typhoid fever	4	45	0
	Salmonellosis	4	226	0
	Toxigenic E. coil AGI	4	188	4
	Chemical poisoning	4	15	0
	Amebiasis	3	39	2
	Giardiasis	2	53	0
1971–1975	Gastroenteritis	63	17,752	0
	Shigellosis	14	2,803	0
	Hepatitis A	14	368	0
	Giardiasis	13	5,136	0
	Chemical poisoning	13	513	0
	Typhoid fever	4	222	0
	Salmonellosis	2	37	0
	Toxigenic E. coil AGI	1	1,000	0
1976–1980	Gastroenteritis	114	22,093	0
	Giardiasis	26	14,416	0
	Chemical poisoning	25	3,081	1
	Shigellosis	10	2,392	0
	Viral gastroenteritis	10	3,147	0
	Salmonellosis	6	1,113	0
	Campylobacteriosis	3	3,821	0
	Hepatitis A	2	95	0
1981–1984	Gastroenteritis, undetermined etiology	59	20,772	0
	Giardiasis	48	4,048	0
	Chemical poisoning	11	179	0
	Shigellosis	7	532	0
	Hepatitis A	7	274	0
	Viral gastroenteritis, Norwalk agent	7	1,077	0
	Salmonellosis	2	1,150	0
	Campylobacterosis	6	993	0
	Viral gastroenteritis, rotavirus	1	1,761	0
	Cholera	1	17	0
	Yersiniosis	1	16	0
	Cryptosporidium	1	117	0
	Entamoeba	1	4	0

Source: From Craun, G.F., 1986, *Waterborne Diseases in the United States*, Copyright CRC Press, Inc., Boca Raton, FL. Reprinted with permission; amended with statistics from Center for Disease Control Annual Summaries, 1981–1984.

Table 10J.212 Etiologic Agents Most Frequently Identified in Waterborne Outbreaks of Infectious Diseases in the United States, 1971–1992

Etiologic Agent	Outbreaks	Cases of Illness
Giardia lamblia	118	26,733
Shigella	57	9,967
Norwalk-like virus	24	10,908
Hepatitis A	29	807
Campylobacter	13	5,257
Salmonella	12	2,370
Cryptosporidium parvum	7	17,194
All others[a]	23	4,243
Total	283	77,479

[a] Toxigenic *E. coli*, Yersinia, rotavirus, S. typhi, V. cholera and others.

Source: From Chlorine Chemistry Council, 1997, Drinking Water Chrlorination White Paper, *A Review of Disinfection Practices and Issues*, June 12, 1997, www.c3.org.

Table 10J.213 Etiology of Waterborne Disease Outbreaks in the United States, by Type of Water System, 1991–2000

	Community Water Systems[a]		Non-community Water Systems[b]		Individual Water Systems[c]		All Systems	
Etiological Agent	Outbreaks	Cases	Outbreaks	Cases	Outbreaks	Cases	Outbreaks	Cases
Giardia	11	2,073	5	167	6	16	22	2,256
Cryptosporidium[d]	7	407,642	2	578	2	39	11	408,259
Campylobacter jejuni	1	172	3	66	1	102	5	340
Salmonellae, nontyphoid	2	749	0	0	1	84	3	833
E. coli	3	208	3	39	3	12	9	259
E. coli O157:H7/C. jeuni	0	0	1	781	0	0	1	781
Shigella	1	83	5	484	2	38	8	605
Plesiomonas shigelloides	0	0	1	60	0	0	1	60
Non-01 V.cholerae	1	11	0	0	0	0	1	11
Hepatitis A virus	0	0	1	46	1	10	2	56
Norwalk-like viruses	1	594	4	1,806	0	0	3	2,400
Small, round-structured virus	1	148	1	70	0	0	2	218
Chemical	18	522	0	0	7	9	25	531
Undetermined	11	10,162	38	4,837	11	238	60	15,237
Total	57	422,364	64	8,934	34	548	155	431,846

Note: Data are compiled from CDC Morbidity and Mortality Weekly Report Surveillance Summaries for 1991–1992, 1993–1994, 1995–1996, 1997–1998 and 1999–2000. Figures include adjustments to numbers of outbreaks and illness cases originally reported, based on more recent CDC data.

[a] Community water systems are those that serve communities of an average of at least 25 year-round residents and have at least 15 service connections.

[b] Non-community water systems are those that serve an average of at least 25 residents and have at least 15 service connections and are used at least 60 days y^{-1}.

[c] Individual water systems are those serving less than 25 residents and have less than 15 service connections.

[d] There were 403,000 cases of illness reported in Milwaukee in 1993.

Source: From Chlorine Chemistry Council, 2003, Drinking Water Chlorination, *A Review of Disinfection Practices and Issues*, February 2003, www.c3.org.

Table 10J.214 Waterborne-Disease Outbreaks ($n=25$) Associated with Drinking Water, by Etiologic Agent and Type of Water System (Excluding Outbreaks Caused by *Legionella* Species) — United States, 2001–2002

	Type of Water System[a]							
	Community		Noncommunity		Individual		Total	
Etiologic Agent	Outbreaks	Cases	Outbreaks	Cases	Outbreaks	Cases	Outbreaks	Cases
Unknown	0	0	2	98	5	19	7	117
AGI[b]	0	0	2	98	5	19	7	117
Viruses	1	71	4	656	0	0	5	727
Norovirus	1	71	4	656	0	0	5	727
Parasitic	3	14	0	0	2	16	5	30
Giardia intestinalis	2	12	0	0	1	6	3	18
Cryptosporidium species	0	0	0	0	1	10	1	10
Naegleria fowleri	1	2	0	0	0	0	1	2
Chemical	3	33	1	4	1	2	5	39
Copper	2	30	0	0	0	0	2	30
Copper and other minerals	0	0	1	4	0	0	1	4
Ethyl benzene, toluene, xylene	0	0	0	0	1	2	1	2
Ethylene glycol	1	3	0	0	0	0	1	3
Bacterial (other than *Legionella* species)	0	0	1	12	2	15	3	27
Campylobacterjejuni	0	0	0	0	1	13	1	13
C. jejuni and Yersinia enterocolitica	0	0	1	12	0	0	1	12
Escherichia coli O157:H7	0	0	0	0	1	2	1	2
Total	7	118	8	770	10	52	25	940
Percentage	(28.0)	(12.6)	(32.0)	(81.9)	(40.0)	(5.5)	(100.0)	(100.0)

[a] Com, community; Ncom, noncommunity; Ind, individual. Community and noncommunity water systems are public water systems that serve >15 serv connections or an average of ≥ 25 residents for ≥ 60 days/year. A community water system serves year-round residents of a community, subdivision, mobile home park with ≥ 15 service connections or an average of ≥ 25 residents. A noncommunity water system can be nontransient or transient. Nontransient systems serve ≥ 25 of the same persons for >6 months of the year, but not year-round (e.g., factories or schools), whereas transient systems provide water to places in which persons do not remain for long periods of time (e.g., restaurants, highway rest stations, or parks). Individual water systems are small systems not owned or operated by a water utility that serve <15 connections or <25 persons. Outbreaks associated with water not intended for drinking (e.g., lakes, springs and creeks used by campers and boaters, irrigation water, and other nonpotable sources with or without taps) are also classified individual systems.

[b] Acute gastrointestinal illness of unknown etiology.

Table 10J.215 Total Outbreaks of Drinking-Water Related Disease, United States, 1973–2000

Year	Heinz Center[a] Outbreaks	Schneider[b] Outbreaks	Schneider[b] Cases	CDC Data[c] Outbreaks	CDC Data[c] Cases
1973	21				
1974	20				
1975	16				
1976	29	22	3,860		
1977	26	17	1,911		
1978	22	33	11,435		
1979	26	41	6,761		
1980	29	49	20,005		
1981	24	32	4,430		
1982	30	40	3,456		
1983	35	40	20,905		
1984	20	26	1,755		
1985	16	25	2,117		
1986	18	22	1,569	22	25,846 (1986–1988)
1987	11	15	22,149	15	
1988	15	15	2,159	13	
1989	12	12	2,540	13	
1990	10	14	1,748	14	
1991	11	15	12,960	15	17,464 (1991–1992)
1992	18	27	4,724	27	
1993	8	17	404,183	18	405,366 (1993–1994)
1994	11	13	1,178	12	
1995	11	16	2,375	16	2,567 (1995–1996)
1996	3	6	192	6	
1997	4	7	304	7	2,038 (1997–1998)
1998	7	10	1,734	10	
1999				15	2,068 (1999–2000)
2000				24	
Total	453	514	534,450	227	455,349

[a] Heinz Center *State of the Nations*; *Ecosystem* report, heinzctr.org/ecosystems/fr_water/datasets/freshwater_waterborne_ disease_ outbreaks.shtm.

[b] Data compiled by Dr. Orren D. Schneider and used by permission, water.sesep.drexel.edu/outbreaks/US_summaryto1998.htm.

[c] Data part of the Center for Disease Control. Surveillance for Waterborne-Disease Outbreaks program. *Waterborne Disease Outbreaks, 1986–1988*, www.cdc.gov/epo.mmwr/preview/mmwrhtml/00001596.htm; *Waterborne-Disease Outbreaks, 1991–1992*, www.cdc.gov/epo.mmwr/preview/mmwrhtml/000025893.htm; *Waterborne-Disease Outbreaks, 1993–1994*, www.cdc.gov/epo.mmwr/preview/mmwrhtml/0004088.htm; *Waterborne-Disease Outbreaks, 1995–1996*, www.cdc.gov/epo.mmwr/preview/mmwrhtml/00055820.htm; *Waterborne-Disease Outbreaks, 1997–1998*, www.cdc.gov/epo.mmwr/preview/mmwrhtml/ss4904a1.htm; *Waterborne-Disease Outbreaks, 1999–2000*, www.cdc.gov/epo.mmwr/preview/mmwrhtml/ss5108a1.htm.

Source: From *World's Water 2004–2005*, by Peter Gleick. H. Copyright © 2004 Island Press. Reproduced by permission of Island Press, Washington, DC.

Table 10J.216 Number of Waterborne Outbreaks by Type of Water System and Etiology in the United States, 1971–2000

Water System Type	Unidentified Agents	Protozoa	Viruses	Bacteria	Chemicals
Non-community	228	31	27	43	11
Community	98	96	20	40	54
Treated and untreated recreational water[a]	40	98	18	97	5
Individual	39	16	9	18	21
All water systems	405	241	74	198	91

[a] An outbreak attributed to algal toxins is not included. An outbreak of both *Shigella* and *Cryptosporidium* is included in the protozoa category.

Source: From Cotruvo, J.A. et al., (ed.), 2004, Waterborne Zoonoses, Identification, Causes, and Control, Published on Behalf of the World Health Organization by IWA Publishing, Copyright © World Health Organization 2004, www.who.int.

Table 10J.217 Drinking-Waterborne Outbreaks of Zoonotic Agents in the United States, 1971–2000

		Type of Water System[a]			Water Source[b]		
Etiologic Agent	**Total**	**C**	**NC**	**I**	**GW**	**SW**	**M/U**
Giardia	126	83	29	14	31	90	5
Campylobacter	19	9	7	3	12	3	4
Cryptosporidium	15	11	2	2	8	5	2
Salmonella	15	11	2	2	11	2	2
E. coli O157:H7	11	4	4	3	8	2	1
Yersinia	2	—	1	1	2	—	—
E. coli O6:H16	1	—	1	—	1	—	—
E. coli O0157:H7 and *Campylobacter*	1	—	1	—	1	—	—
Total	190	118	47	25	74	102	14

[a] C, community; NC, noncommunity; I, individual.
[b] GW, groundwater; SW, surface water; M/U, mixed or unknown.

Source: From Cotruvo, J.A. et al., (ed.), 2004, Waterborne Zoonoses, Identification, Causes, and Control, Published on Behalf of the World Health Organization by IWA Publishing, Copyright © World Health Organization 2004, www.who.int.

Table 10J.218 Waterborne Outbreaks Reported in United States Drinking Water Systems by Type of System and Water Source, 1991–1998

	Number of Waterborne Outbreaks			
Water Source	**Community Systems**	**Non-Community Systems**	**Individual Systems**	**All Water Systems**
Groundwater[a]	22	52	11	85
Surface water[b]	22	2	1	25
Unknown	3	8	5	16
Totals	47	62	17	126

[a] Surfacewater, lakes, reservoirs, rivers, streams.
[b] Groundwater, wells and springs.

Source: From Craun, G.F., Calderon, R.L., and Nwachuku, N., 2003, Causes of waterborne outbreaks reported in the United States, 1991-1998 in Hunter, P.R., Waite, M., and Ronchi, El., (eds.), *Drinking Water and Infectious Disease Establishing the Links*, CRC Press LLC, Boca Raton, FL.

Table 10J.219 Etiology of Waterborne Outbreaks in United States Drinking Water Systems, 1991–1998; Number of Outbreaks by Type of Water System and Water Source

	Community Water Systems			Non-Community Water Systems		
Etiological Agent	**Surface-Water**	**Ground-Water**	**Unknown Source**	**Surface-Water**	**Ground-Water**	**Unknown Source**
Undetermined	5	1	1	1	36	5
Chemical	7	9				1
Giardia	6	4			3	
Cryptosporidium	3	3	1	1	1	
Norwalk-like virus	1	1				
Campylobacter		1			2	1
Salmonella, non-typhoid		1				
Escherichia coli O157:H7		1			3	1
Shigella		1			5	
Vibrio cholerae			1			
Hepatitis A virus					1	
Plesiomonas shigelloides					1	
Total	22	22	3	2	52	8

Source: From Craun, G.F., Calderon, R.L., and Nwachuku, N., 2003, Causes of waterborne outbreaks reported in the United States, 1991–1998 in Hunter, P.R., Waite, M., and Ronchi, El., (eds.), *Drinking Water and Infectious Disease Establishing the Links*, CRC Press LLC, Boca Raton, FL.

Table 10J.220 Etiology of Waterborne Outbreaks Reporting in United States Drinking Water Systems; Cases of Illness by Type of Water System and Water Source, 1991–1998

	Community Water Systems			Noncommunity Water Systems		
Etiological Agent	**Surface-Water**	**Ground-Water**	**Unknown Source**	**Surface-Water**	**Ground-Water**	**Unknown Source**
Undetermined	10,210	18	67	250	4,789	101
Chemical	104	409				2
Giardia	1,937	49			128	
Cryptosporidium	403,343	4,294	77	27	551	
Norwalk-like virus	148	594				
Campylobacter		172			51	7
Salmonella, non-typhoid		625				
E. coli O157:H7		157			39	27
Shigella		83			484	
Vibrio cholerae			11			
Hepatitis A virus					46	
Plesiomonas shigelloides					60	
Total	415,742	6,401	155	277	6,148	137

Source: From Craun, G.F., Calderon, R.L., and Nwachuku, N., 2003, Causes of waterborne outbreaks reported in the United States, 1991–1998 in Hunter, P.R., Waite, M., and Ronchi, El., (eds.), *Drinking Water and Infectious Disease Establishing the Links*, CRC Press LLC, Boca Raton, FL.

Table 10J.221 Etiology of Waterborne Outbreaks in Individual Water Systems in the United States Outbreaks and Cases of Illness by Water Source, 1991–1998

	Outbreaks			Cases of Illness		
Etiological Agent	**Surface-Water**	**Ground-Water**	**Unknown Source**	**Surface-Water**	**Ground-Water**	**Unknown Source**
Undetermined	0	2	1	0	43	8
Chemical	0	3	3	0	3	5
Giardia	1	1	0	2	10	0
Cryptosporidium	0	2	0	0	39	0
E. coli	0	1	0	0	3	0
Shigella	0	1	1	0	33	5
Hepatitis A virus	0	1	0	0	10	0
Total	1	11	5	2	141	18

Source: From Craun, G.F., Calderon, R.L., and Nwachuku, N., 2003, Causes of waterborne outbreaks reported in the United States, 1991–1998 in Hunter, P.R., Waite, M., and Ronchi, El., (eds.), *Drinking Water and Infectious Disease Establishing the Links*, CRC Press LLC, Boca Raton, FL.

Table 10J.222 Waterborne Disease Outbreaks Caused by Use of Contaminated, Untreated Surface Water in the United States, 1920–1980

	Type of Water System							
	Community		Noncommunity		Individual		All	
Deficiency	OB[a]	Cases	OB[a]	Cases	OB[a]	Cases	OB[a]	Cases
Contamination on watershed	26	3,498	3	57	12	257	41	3,812
Use of surface water for supplemental source	7	3,613	7	245	2	115	16	3,973
Overflow of sewage or outfall near water intake	3	103	3	39	5	87	11	229
Flooding, heavy rains	2	125	1	93	1	77	4	295
Dead animals in reservoir	—	—	1	100	—	—	1	100
Insufficient data	27	1,228	24	726	28	436	79	2,390
Total	65	8,567	39	1,260	48	972	152	10,799

[a] Number of outbreaks.

Source: From Craun, G.F., 1986, *Waterborne Diseases in the United States*. Copyright CRC Press, Inc., Boca Raton, FL. Reprinted with permission.

Table 10J.223 Waterborne Disease Outbreaks Caused by Use of Contaminated, Untreated Groundwater (Springs) in the United States, 1920–1980

	Type of Water System							
	Community		Non-Community		Individual		All	
Deficiency	OB[a]	Cases	OB[a]	Cases	OB[a]	Cases	OB[a]	Cases
Overflow or seepage of sewage	8	238	3	35	5	39	16	312
Surface runoff	11	265	5	162	7	75	23	502
Flooding	2	76	2	123	—	—	4	199
Creviced limestone	1	200	3	213	—	—	4	413
Contamination of raw water transmission line	2	284	1	7	—	—	3	291
Improper construction	—	—	1	26	1	9	2	35
Insufficient data	12	508	18	1,961	20	415	50	2,884
Total	36	1,571	33	2,527	33	538	102	4,636

[a] Number of outbreaks.

Source: From Craun, G.F., 1986, *Waterborne Diseases in the United States*. Copyright CRC Press, Inc., Boca Raton, FL. Reprinted with permission.

Table 10J.224 Waterborne Disease Outbreaks Caused by Use of Contaminated, Untreated Groundwater (Wells) in the United States, 1920–1980

	Type of Water System							
	Community		Noncommunity		Individual		All	
Deficiency	OB[a]	Cases	OB[a]	Cases	OB[a]	Cases	OB[a]	Cases
Overflow or seepage of sewage	28	14,915	104	10,236	52	675	184	25,826
Surface runoff, heavy rains	25	2,492	26	947	34	824	85	4,263
Creviced limestone, fissured rock	9	1,404	19	2,044	12	660	40	4,108
Improper construction, faulty well casing	8	342	10	414	9	141	27	897
Flooding	9	5,883	3	107	5	211	17	6,201
Chemical contamination	3	77	2	16	10	68	15	161
Contamination by stream or river	3	445	6	392	3	48	12	885
Contamination of raw water transmission line	8	10,481	—	—	—	—	8	10,481
Seepage from abandoned well	3	144	1	50	—	—	4	194
Animal in well	1	34	1	238	2	19	4	291
Insufficient data	19	18,480	67	3,309	40	413	126	22,202
Total	116	54,697	239	17,753	167	3,059	522	75,509

[a] Number of outbreaks.

Source: From Craun, G.F., 1986, *Waterborne Diseases in the Untied States*, Copyright CRC Press, Inc. Boca Raton, FL. Reprinted with permission.

Table 10J.225 Waterborne Disease Outbreak and Disease Rates Attributed to Source Contamination and Treatment Inadequacies in Community Systems in the United States Using Surface Water Sources, 1971–1985

Type of Community Water System	Waterborne Disease Outbreaks per 1,000 Water Systems	Waterborne Illnesses per Million-Person Years
Untreated	32.5	370.9
Disinfected only	40.5	66.3
Filtered and disinfected water	5.0	4.7

Source: From U.S. Environmental Protection Agency, 1987; Craun, G.F., 1987.

Table 10J.226 Water Supply Deficiencies Responsible for Waterborne Outbreaks in the United States, 1971–1985.

Source of Deficiency	Outbreaks	Reported Illnesses
Surface water source		
No treatment	31	1,647
Disinfection only, or inadequate disinfection	67	23,028
Disinfection with other treatment (but no filtration)	5	969
Filtration and disinfection	20	9,852
Totals	123	35,496
Groundwater source		
No treatment	154	11,266
Inadequate disinfection	90	40,893
Disinfection with other treatment	1	22
Totals	245	52,181
Distribution system		
Cross-connection	44	8,124
Contamination of mains/plumbing	14	3,413
Contamination of storage	11	6,244
Corrosive water	10	147
Totals	79	17,928
Grand Total (reported)		
Outbreaks		447
Illnesses		105,605

Source: From U.S. Environmental Protection Agency, 1987; Craun, G.F., 1987.

Table 10J.227 Waterborne Outbreaks and Deficiencies in United States Public Water Systems Surfacewater Sources, 1991–1998

	Community Systems Surfacewater Source		Noncommunity Systems Surfacewater Source	
Type of Contamination	**Outbreaks**	**Percent**	**Outbreaks**	**Percent**
Untreated surface water	0	0	0	0
Inadequate or interrupted disinfection; disinfection only treatment	4	18	1	50
Inadequate or interrupted filtration	4	18	0	0
Distribution system contamination	9	41	0	0
Inadequate control of chemical feed	2	9	0	0
Miscellaneous/unknown	3	14	1	50
Total	22	100	2	100

Source: From Craun, G.F., Calderon, R.L., and Nwachuku, N., 2003, Causes of waterborne outbreaks reported in the United States, 1991–1998 in Hunter, P.R., Waite, M., and Ronchi, El., (eds.), *Drinking Water and Infectious Disease Establishing the Links*, CRC Press LLC, Boca Raton, Florida.

Table 10J.228 Waterborne Outbreaks and Deficiencies in United States Public Water Systems Groundwater Sources, 1991–1998

	Community Systems Groundwater Source		Noncommunity Systems Groundwater Source	
Type of Contamination	**Outbreaks**	**Percent**	**Outbreaks**	**Percent**
Untreated groundwater	5	23	18	35
Inadequate or interrupted disinfection; disinfection only treatment	3	14	21	40
Inadequate or interrupted filtration	1	4	0	
Distribution system contamination	8	36	8	15
Inadequate control of chemical feed	3	14	0	
Miscellaneous/unknown	2	9	5	10
Total	22	100	52	100

Source: From Craun, G.F., Calderon, R.L., and Nwachuku, N., 2003, Causes of waterborne outbreaks reported in the United States, 1991–1998 in Hunter, P.R., Waite, M., and Ronchi, El., (eds.), *Drinking Water and Infectious Disease Establishing the Links*, CRC Press LLC, Boca Raton, Florida.

Table 10J.229 Causes of Waterborne Disease Outbreaks in Community Water Systems in the United States — 1971–1998

		Outbreaks[a] — %				
Time Period	**Number of Outbreaks**	**Distribution System Deficiencies**	**Untreated Groundwater**	**Inadequate, Interrupted Disinfection of Groundwater**	**Inadequate Disinfection of Unfiltered Surface Water**	**Filtered Surface Water**
1971–74	34	32.4	14.7	17.6	20.6	2.9
1975–78	39	41.0	2.6	17.9	17.9	7.7
1979–82	90	24.4	8.0	16.7	23.3	11.1
1983–86	60	31.7	6.7	11.7	28.3	13.3
1987–90	24	16.7	12.5	16.7	29.2	12.5
1991–94	27	29.6	11.1	7.4	11.1	14.8
1995–98	20	45.0	10.0	10	5	10
1971–98	294	30.3	8.8	14.6	21.4	10.5

[a] Rows do not total 100% because miscellaneous and unknown causes of outbreaks are not tabulated.

Source: Reprinted from *Journal AWWA*, vol. 93, no. 9 (September 2001), by permission. Copyright © 2001, American Water Works Association, www.awwa.org.

Table 10J.230 Causes of Waterborne Disease Outbreaks in Noncommunity Water Systems in the United States — 1971–98

		Outbreaks[a] — %				
Time Period	**Number of Outbreaks**	**Distribution System Deficiencies**	**Untreated Groundwater**	**Inadequate, Interrupted Disinfection of Groundwater**	**Inadequate Disinfection of Unfiltered Surface Water**	**Filtered Surface Water**
1971–74	51	0	40.4	28.8	11.5	1.9
1975–78	77	14.3	40.3	26	7.8	0
1979–82	67	4.5	49.3	26.9	4.5	3.0
1983–86	38	2.6	50.0	39.5	7.9	0
1987–90	29	3.4	34.5	41.4	5.1	0
1991–94	41	12.2	34.1	34.1	2.4	0
1995–98	21	14.3	19.0	33.3	0	0
1971–98	325	7.4	40.6	31.1	6.8	1.0

[a] Rows do not total 100% because miscellaneous and unknown causes of outbreaks are not tabulated.

Source: Reprinted from *Journal AWWA*, vol. 93, no. 9 (September 2001), by permission. Copyright © 2001, American Water Works Association, www.awwa.org.

Table 10J.231 Etiology of Outbreaks Caused by Distribution System Contamination in the United States — 1971–1998

	CWS[a]		NCWS[b]	
Etiology	**Outbreaks**	**%**	**Outbreaks**	**%**
Chemical	35	39.3	3	12.5
Unidentified pathogen	29	32.6	11	45.8
Giardia	8	9.0	4	16.7
Salmonella	4	4.5	1	4.2
Norwalk-like virus	3	3.4	1	4.2
Shigella	3	3.4	1	4.2
Campylobacter	3	3.4	1	4.2
Hepatitis A	1	1.1	1	4.1
Salmonella typhimurium	1	1.1		
Cyclospora	1	1.1		
Escherichia coil 0157:H7	1	1.1		
Vibrio cholerae			1	4.1
Total	89	100	24	100

[a] CWS — community water system.
[b] NCWS — noncommunity water system.

Source: Reprinted from *Journal AWWA*, vol. 93, no. 9 (September 2001), by permission. Copyright © 2001, American Water Works Association, www.awwa.org.

Table 10J.232 Number of Waterborne Outbreaks by Deficiencies in Drinking Systems, in the United States 1971–2000

Type of Contamination	*Giardia, Cryptospridium*	*Campylobacter, E. coli, Salmonella, Yersinia*
Distribution system contamination	16	11
Inadequate disinfection; only treatment, surface water[a]	52	3
Inadequate, interrupted, or bypass of filtration; surface water	22	—
Untreated groundwater	14	14
Untreated surface water	14	2
Inadequate or interrupted disinfection; groundwater[b]	13	11
Water not intended for drinking; contaminated faucet or ice; unknown	10	8
Total	141	49

[a] Includes two outbreaks with surface water and groundwater sources.
[b] Includes three outbreaks where groundwater was filtered.

Source: From Cotruvo. J.A. et al., (ed.), 2004. Waterborne Zoonoses, Identification, Causes, and Control, Published on Behalf of the World Health Organization by IWA Publishing, Copyright © World Health Organization 2004, www.who.int.

Table 10J.233 Epidemiological Characteristics of the Principal Pathogenic Agents in Wastewater

Agents	Quantity Excreted per g/feces	Latency[a]	Survival[b]	Multiplication in the Environment	Infecting Dose ID 50[c]
Virus					
Enterovirus (including polio, echo and coxsackie)	10^7	0	3 mo	no	100
Hepatitis A	10^6?	0	?	no	?
Rotavirus	10^6?	0	?	no	?
Bacteria					
Colibacilli	10^8	0	3 mo	yes	$\pm 10^9$
Salmonella thyphi	10^8	0	2 mo	yes	10^7
Other salmonellas	10^8	0	2–3 mo	yes	10^6
Shigella	10^7	0	1 mo	yes	10^4
Campylobacter	10^7	0	7 d	yes	10^6
Cholera	10^7	0	1 mo	yes	10^8
Yersinia enterocolitica	10^5	0	3 mo	yes	10^9
Leptospira	urine	0	7 d	no	low
Parasites					
Dysentery amoeba	10^7	0	25 d	no	10–100
Glardia	10^5	0	25 d	no	25–100
Balantidium coli	?	0	20 d?	no	25–100
Ascaris	10^4	10 d	1 y	no	Several units
Ancyclostoma	10^2	7 d	3 mo	no	1
Anguillula	10	3 d	3 wks	yes	1
Trichocephalus	10^3	20 d	9 mo	no	Several units
Hymenolepis	?	0	10 d	no	1
Taenia	10^4	2 mo	9 mo	no	1
Fasciola hepatica	?	2 mo	4 mo	yes	Several units
Other flukes	10^2	6–8 wks	Life of Host	yes	Several units

[a] Period necessary for excreted pathogenic agent to become infectious to receiving or susceptible individual; (0, is immediate).
[b] In environment, outside final host (man or animal).
[c] Dose sufficient to provoke the appearance of clinical symptoms in 50% of individuals tested.

Source: From Prost, A., 1987, Heath risks stemming from wastewater reutilization, *Water Quality Bulletin*, vol. 12, no. 2.

Table 10J.234 Microbial Pathogens Linked to Drinking Water or Recreational Water Contact

Organism	Disease	Transmission	Clinical Features
Helminths			
Schistosoma spp.	Schistosomaisis	Contact with surface water infected with free swimming cercariae	Urinary and intestinal damage. Bladder cancer
Dracunculus medinentis	Dracunculiasis	Drinking water	Painful ulcers on lower limbs and feet
Protozoa			
Giardia duodenalis	Giardiasis	Faecal oral spread through drinking water or recreational water	Diarrhoea and abdominal pain, weight loss and failure to thrive
Cryptosporidium parvum	Cryptosporidiosis	Faecal oral spread through drinking water or recreational water	Diarrhoea often prolonged
Cyclospora cayetanensis	Cyclosporiasis	Faecal oral spread through drinking water	Diarrhoea and abdominal pain, weight loss and failure to thrive
Entamoeba histolytica	Amebiasis	Faecal and spread through drinking water	Diarrhoea, may be severe dysentery
Toxoplasma gondii	Toxoplasmosis	Drinking water contaminated by feline animals	Glandular fever, foetal damage in pregnant women
Free-living amoebae	Amoebic meningoencephalitis	Aspiration of infected surface water into nose	Fatal encephalitis
Algae			
Cyanobacteria	Various	Toxins in drinking water or direct contact with surface water blooms	Dermatitis, hepatitis, respiratory symptoms, potentially fatal
Pfesteria piscicida	Estuary-associated syndrome	Toxins in water	Respiratory and eye irritation, deficiencies in learning and memory and acute confusional states
Bacteria			
Vibrio cholerae	Cholera	Drinking water	Watery diarrhoea, may be severe
Salmonella spp.	Salmonellosis	Occasional outbreaks with drinking water	Diarrhoea, colicky abdominal pain and fever
Salmonella typhi	Typhoid	Drinking water	Fever, malaise and abdominal pain with high mortality
Shigella spp.	Shigellosis (bacillary dysentery)	Both drinking and recreational water	Diarrhoea frequently with blood loss
Campylobacter spp.	Campylobacteriois	Both drinking and recreational water	Diarrhoea frequently with blood loss
Enterotoxigenic *E. coli*		Drinking water	Watery diarrhoea
Enterohaemorrhagic *E. coli*		Drinking water and recreational water contact	Bloody diarrhoea and haemolyic uraemic syndrome in children
Yersinia spp.	Yersiniosis	Drinking water	Fever, diarrhoea and abdominal pain
Francisella tularensis	Tularaemia	Drinking water	Typhoid-like or mucocutaneous with suppurative skin lesions
Helicobacter pylori		Drinking water	Gastritis that can progress to gastric cancer
Mycobacteria spp. not *M. tuberculosis*	Varies	Potable water systems in hospitals, some recreation	Varies, includes respiratory disease, wound infections, skin disease
Viruses			
Hepatitis A and Hepatitis E viruses	Viral hepatitis	Drinking and recreational water contact	Hepatitis
Various, esp. Norwalk-like viruses	Viral gastroenteritis	Drinking and recreational water contact	Vomiting and diarrhoea
Enteroviruses	Various, including poliomyelitis	Drinking and recreational water contact	Various

Source: From Hunter, P.R., 2003, Climate change and waterborne and vector-borne disease in Sartory, D., Jones, K., Semple, K., and Godfree, A., (eds.), The Society for Applied Microbiology Symposim Series no. 32, *Pathogens in the Environment and Changing Ecosystems*, Blackwell Publishing, Oxford, UK.

Table 10J.235 Occurrence of Cryptosporidium Oocysts in Various Waters throughout the Western United States

Water Sampled	Number of Samples	Number of Samples Positive	Percent Positive	Oocysts/L[a]
Raw sewage	11	10	91	28.4
Treated sewage[b]	22	20	91	17
Reservoir, lake	32	24	75	0.91
Stream, river	58	45	77	0.94
Filtered drinking water	10	2	20	0.001
Nonfiltered drinking water	4	2	50	0.006

[a] Geometric means.
[b] Activated sludge.

Source: From Craun, G.F., 1988. Surface water supplies and health, *J. Am. Water Works Assoc.*, vol. 80, no.2. Copyright AWWA. Reprinted with permission.

Table 10J.236 Cryptosporidiosis Case Reports by State/Area — United States, 1999–2002

	1999				2000				2001				2002			
State/Area	No.	(%)	Rate[a]	No. of Outbreak Cases	No.	(%)	Rate	No. of Outbreak Cases	No.	(%)	Rate	No. of Outbreak Cases	No.	(%)	Rate	No. of Outbreak Cases
Alabama	16	(0.6)	0.4		16	(0.6)	0.4		18	(0.5)	0.4		47	(1.6)	1.0	
Alaska	NR[b]				NR	(0.3)			1	(<0.1)	0.2		1	(<0.1)	0.2	
Arizona	16	(0.6)	0.3		10	(0.8)	0.2		11	(0.3)	0.2		19	(0.6)	0.3	
Arkansas	2	(0.1)	0.1		16	(0.5)	0.6		10	(0.3)	0.4		8	(0.3)	0.3	
California	279	(10.1)	0.8		285	(7.5)	0.7		229	(6.0)	0.7		200	(6.6)	0.6	
Colorado	14	(0.5)	0.3		72	(2.3)	1.7	14	44	(1.2)	1.0		57	(1.9)	1.3	
Connecticut	22	(0.8)	0.6		29	(0.9)	0.9		17	(0.4)	0.5		19	(0.6)	0.5	
Delaware	1	(<0.1)	0.1		9	(0.3)	1.1		6	(0.2)	0.8		4	(0.1)	0.5	
District of Columbia	7	(0.3)	1.2		18	(0.6)	3.1		14	(0.4)	2.4		5	(0.2)	0.9	
Florida	189	(6.8)	1.2	186	240	(7.7)	1.5	233	91	(2.4)	0.6		106	(3.5)	0.6	97
Georgia	170	(6.1)	2.1		191	(6.1)	2.3		162	(4.3)	1.9		123	(4.1)	1.4	
Hawaii	NR				NR				3	(0.1)	0.2	3	2	(0.1)	0.2	2
Idaho	8	(0.3)	0.6		28	(0.9)	2.2		23	(0.6)	1.7		29	(1.0)	2.2	
Illinois	90	(3.3)	0.7		126	(4.0)	1.0		488	(12.8)	3.9	341	121	(4.0)	1.0	
Indiana	47	(1.7)	0.6	1	72	(2.3)	1.2		90	(2.4)	1.5		70	(2.3)	1.1	
Iowa	56	(2.0)	1.9		77	(2.5)	2.6		82	(2.2)	2.8		49	(1.6)	1.7	
Kansas	2	(0.1)	0.1		9	(0.3)	0.3		7	(0.1)	0.1		16	(0.5)	0.6	
Kentucky	7	(0.3)	0.2	7	7	(0.2)	0.2	7	5	(0.1)	0.1	5	10	(0.3)	0.2	10
Louisiana	24	(0.9)	0.5		14	(0.4)	0.3		8	(0.2)	0.2		10	(0.3)	0.2	
Maine	81	(1.1)	2.4		20	(0.6)	1.6		19	(0.5)	1.5		12	(0.4)	0.9	
Maryland	17	(0.6)	0.3		14	(0.4)	0.3		40	(1.1)	0.7		19	(0.6)	0.3	
Massachusetts	71	(2.6)	1.1	55	87	(1.2)	0.6	2	55	(1.5)	0.9		77	(2.6)	1.2	2
Michigan	52	(1.9)	0.5		97	(3.1)	1.0		187	(4.9)	1.9		135	(4.5)	1.3	
Minnesota	91	(3.3)	1.9	2	190	(6.1)	3.9	4	197	(5.2)	4.0	11	206	(6.8)	4.1	32
Mississippi	12	(0.4)	0.4		16	(0.5)	0.6		15	(0.4)	0.5		10	(0.3)	0.3	
Missouri	26	(0.9)	0.5		31	(1.0)	0.6		55	(1.5)	1.0		41	(1.4)	0.7	
Montana	13	(0.5)	1.4		10	(0.3)	1.1		37	(1.0)	4.1		6	(0.2)	0.7	
Nebraska	15	(0.5)	0.9		82	(2.6)	4.8		185	(4.9)	10.8		52	(1.7)	3.0	
Nevada	9	(0.3)	0.5	4	4	(0.1)	0.2	8	7	(0.2)	0.3		4	(0.1)	0.2	
New Hampshire	20	(0.7)	1.6	20	25	(0.8)	2.0	25	17	(0.4)	1.4	17	31	(1.0)	2.4	31
New Jersey	54	(2.0)	0.6		19	(0.6)	0.2		24	(0.6)	0.3		17	(0.6)	0.2	
New Mexico	44	(1.6)	2.4		26	(0.8)	1.4		80	(0.8)	1.6		20	(0.7)	1.1	
New York[c]	452	(16.3)	2.4	4	910	(9.9)	1.6		248	(6.5)	1.3	4	300	(9.9)	1.6	3

(Continued)

Table 10J.236 (Continued)

State/Area	1999				2000				2001				2002			
	No.	(%)	Rate[a]	No. of Outbreak Cases	No.	(%)	Rate	No. of Outbreak Cases	No.	(%)	Rate	No. of Outbreak Cases	No.	(%)	Rate	No. of Outbreak Cases
New York City	260	(9.4)	3.5		171	(5.5)	2.1		123	(3.2)	1.5		147	(4.9)	1.8	
North Carolina	85	(1.3)	0.4	4	28	(0.9)	0.3		31	(0.8)	0.4		40	(1.3)	0.5	1
North Dakota	20	(0.7)	3.1		18	(0.6)	2.8		15	(0.4)	2.4		41	(1.4)	6.5	
Ohio	67	(2.4)	0.6		260	(8.3)	2.3	134	185	(4.9)	1.6		119	(3.9)	1.0	
Oklahoma	14	(0.5)	0.4		30	(1.0)	0.9		16	(0.4)	0.5		16	(0.5)	0.5	
Oregon	98	(3.5)	2.9	61	20	(0.6)	0.6		58	(1.5)	1.7		40	(1.3)	1.1	
Pennsylvania	123	(4.4)	1.0		64	(2.0)	0.5		102	(2.7)	0.8		111	(3.7)	0.9	
Rhode Island	6	(0.2)	0.6		4	(0.1)	0.4		10	(0.3)	0.9		21	(0.7)	2.0	
South Carolina	NR				NR				7	(0.2)	0.2		8	(0.3)	0.2	
South Dakota	7	(0.3)	0.9		15	(0.5)	2.0		8	(0.2)	1.1		42	(1.4)	5.5	22
Tennessee	13	(0.6)	0.2		12	(0.4)	0.2		24	(0.6)	0.4		61	(2.0)	1.1	
Texas	69	(2.5)	0.3		115	(3.7)	0.6		96	(2.5)	0.4		34	(1.1)	0.2	
Utah	4	(0.1)	0.2		26	(0.9)	1.3	1	84	(2.2)	3.7		16	(0.5)	0.7	
Vermont	36	(1.9)	6.0	7	28	(0.9)	4.6		34	(0.9)	5.5		33	(1.1)	5.4	
Virginia	30	(1.1)	0.4		21	(0.7)	0.3		27	(0.7)	0.4		35	(1.2)	0.5	
Washington	NR				NR				NR				46	(1.5)	0.8	
West Virginia	3	(0.1)	0.2		3	(0.1)	0.2		2	(0.1)	0.1		3	(0.1)	0.2	
Wisconsin	386	(13.9)	7.2	2	428	(13.7)	8.0	5	664	(17.5)	12.3	6	515	(17.1)	9.5	5
Wyoming	1	(<0.1)	0.2		5	(0.2)	1.0		7	(0.2)	1.4	2	9	(0.3)	1.8	2
Total	2,769	(100.0)[d]	1.0	353	3,128	(100.0)	1.1	428	3,787	(100.0)	1.3	389	3,016	(100.0)	1.0	207

Note: Population estimates are from the Population Division, U.S. Census Bureau. Estimates of the population of states: ST-99-4 State Rankings of Population Change and Demographic Components of Population Change for the Period July 1, 1998 to July 1, 1999, available at census.goalpopest/archives/1990s/ST-99-01.txt, and Table 1: Annual estimates of the population for the United States and States and Puerto Rico: April 1, 2001, to July 1, 2003 (NST-EST 2003 01), available at www.census.gov/popset/estates/tables/NST-EST2003-01.xls. Estimates of the New York City population: (SU-99-7) Population Estimates for Places (Sorted Alphabetical Within State): Annual Time Series, July 1, 1990 to July 1, 1999 (Includes April 1, 1990 Population Estimates Base), available at www.census.gov/popset/archives/1990a/su-99-07/SU-99-7_NY.txt, and Table 1: Annual Estimates of the Population for incorporated Places over 100,000, Ranked by July 1, 2003 Population; April 1, 2001, to July 1, 2003 (SUB-EST 2003-01) available at www.census.gov/popest/cities/tables/SUB-EST2003-01.xls.

[a] Per 100,000 population on the basis of U.S. Census Bureau population estimates.
[b] No cases reported.
[c] New York State case counts include New York City cases.
[d] Percentages might not total 100% because of rounding.

Source: From Hlavsa, M.C., Watson, J.C. and Beach, M.J., 2005, Cryptosporidiosis surveillance – United States 1999–2002, *MMWR Surveillance Summaries*, vol. 54, no. SS-01, January 28, 2005, www.cdc.gov.

Table 10J.237 Outbreaks and Cases of Cryptosporidiosis in England and Wales, 1983–1997

Year	Number of Outbreaks	Drinking Water Outbreaks[a]	Total Outbreak Cases	Total Drinking Water Cases[a]	Total Cases in England and Wales	Outbreak Cases as a Percentage of the Total	Drinking Water Outbreak Cases as a Percentage of the Total[a]
1983	1	0	16	0	61	26	0
1984	1	0	19	0	876	2	0
1985	3	0	60	0	1875	3	0
1986	4	0	98	0	3560	3	0
1987	1	0	69	0	3277	2	0
1988	2	0	102	0	2750	4	0
1989	6	3	1,090	1,042	7,768	14	13
1990	4	2	92	49	4,682	2	1
1991	9	2	93	46	5,165	2	1
1992	12	4	549	343	5,211	11	7
1993	9	3	358	164	4,832	7	3
1994	7	2	373	257	4,432	8	6
1995	5	1	612	575	5,691	11	10
1996	4	3	244	236	3,660	7	6
1997	12	5	874	743	4,321	20	17
Total	80	25	4,649	3,455	58,161	8	6

[a] Cases include those with a strong probable or possible association with public drinking water.

Source: From Nichos, G., 2003, Using existing surveillance-based data in Hunter, P.R., Waite, M., and Ronchi, El., (eds.), *Drinking Water and Infectious Disease Establishing the Links*, CRC Press LLC, Boca Raton, FL.

Table 10J.238 Some Published Cases of Cryptosporidiosis Outbreaks in Recreational Waters, 1986–2001

Year	Location	Facility	Disinfectant	No. of Cases Estimated (Confirmed)
1986	New Mexico, United States	Lake	None	56
1988	Doncaster, England	Pool	Chlorine	(79)
1988	Los Angeles county, United States	Pool	Chlorine	44 (5)
1990	British Columbia, Canada	Pool	Chlorine	66 (23)
1992	Gloucestershire, England	Pool	Chlorine	(13)
1992	Idaho, United States	Water Slide	Ozone/Chlorine	500
1992	Oregon, United States	Pool (wave)	Chlorine	(52)
1993	Wisconsin, United States	Pool (motel)	Chlorine	51 (22)
1993	Wisconsin, United States	Pool (motel)	Chlorine	64
1993	Wisconsin, United States	Pool	Chlorine	5
1993	Wisconsin, United States	Pool	Chlorine	54
1994	Missouri, United States	Pool (motel)	Chlorine	101 (26)
1994	New Jersey, United States	Lake	None	2,070 (46)
1994	South west England	Pool	Chlorine	14 (8)
1994	Sutherland Australia	Pool	Chlorine	(70)
1995	Kansas, United States	Pool	[a]	101 (2)
1995	Georgia, United States	Water Park	Chlorine	2,470 (6)
1995	Nebraska, United States	Water Park	[a]	(14)
1996	Florida, United States	Pool	[a]	22 (16)
1996	California, United States	Water Park	Chlorine	3,000 (29)
1996	Andover, England	Pool	Chlorine	8
1996	Indiana, United States	Lake	None	3
1997	England and Wales	River	None	27 (7)
1997	England and Wales	Pool	Ozone/Chlorine	(9)
1997	Minnesota, United States	Fountain	(Sand Filter)	369 (73)
1997	Queensland, Australia	Pools	[a]	129
1998	Canberra, Australia	3 Pools	[a]	(210)
1998	Oregon, United States	Pool	[a]	51 (8)

(Continued)

Table 10J.238 (Continued)

Year	Location	Facility	Disinfectant	No. of Cases Estimated (Confirmed)
1998	New South Wales, Australia	Pools	[a]	370
1998	Hutt Valley, New Zealand	Pools	[a]	(171)
1998	Minnesota, United States	Pool	[a]	(26)
1999	Florida, United States	Interactive water fountain	Chlorine	38 (2)
2000	Ohio, United States	Pool	[a]	700 (186)
2000	Nebraska, United States	Pool	[a]	225 (65)
2000	Trent region, England	Pool	Chlorine	41 (41)
2000	London, England	Pool	Chlorine	3 (3)
2001	South west England	Pool	Chlorine	14 (8)
2001	South west England	Stream onto Beach	None	14 (6)

[a] Data not available.

Source: From Pond, K., 2005. *Water Recreation and Disease Plausibility and Associated Infections: Acute Effects, Sequelae, and Mortality*, Published on behalf of the World Health Organization by IWA Publishing, London Copyright © World Health Organization 2005, www.who.int.

Original Source: From Fayer et al., 2002; CDR Weekly website: hpa.org.uk.

Table 10J.239 Total Waterborne Outbreaks of Giardiasis in United States Water Systems, 1965–1995

Time Period	Number of Outbreaks
1965–1970	3
1971–1975	12
1976–1980	26
1981–1985	49
1986–1990	16
1991–1995	11

Source: From United States Environmental Protection Agency, 1998, *Giardia: Human Health Criteria Document*, EPA-823-R-002, www.epa.gov.

Table 10J.240 Waterborne Outbreaks of Giardiasis by Type of Water System in the United States, 1965–1996

	Community Water Systems		Noncommunity Water Systems		Individual Systems & Nonpotable Water		Total	
Year	Outbreaks	Cases	Outbreaks	Cases	Outbreaks	Cases	Outbreaks	Cases
1965	1	123	0	0	0	0	1	123
1966–1968	0	0	0	0	0	0	0	0
1969	0	0	1	19	0	0	1	19
1970	1	34	0	0	0	0	1	34
1971	0	0	0	0	0	0	0	0
1972	1	12	3	112	0	0	3	124
1973	2	52	1	16	1	5	4	73
1974	2	4,878	1	18	1	34	4	4,930
1975	0	0	0	0	1	9	1	9
1976	1	600	2	39	0	0	3	639
1977	2	950	2	62	0	0	4	1,012
1978	2	5,130	1	23	1	18	4	5,171
1979	5	3,789	2	2,120	0	0	7	5,909
1980	7	1,724	0	0	1	6	8	1,730
1981	8	265	2	39	1	7	11	311
1982	9	497	2	60	1	4	12	561
1983	17	2,216	0	0	1	4	18	2,220
1984	3	463	1	400	1	3	5	866
1985	1	703	2	38	0	0	3	741
1986	4	251	1	23	0	0	5	274
1987	2	633	0	0	0	0	2	633
1988	2	262	0	0	0	0	2	262
1989	3	380	1	152	0	0	4	532
1990	1	123	2	42	0	0	3	165
1991	0	0	2	28	0	0	2	28
1992	2	95	0	0	0	0	2	95
1993	2	27	0	0	0	0	2	27
1994	3	358	0	0	0	0	3	358
1995	1	1,449	0	0	1	10	2	1,459
1996	0	0	0	0	0	0	0	0
Total	82	25,014	26	3,191	10	100	117	28,305

Source: From United States Environmental Protection Agency, 1998, *Giardia: Human Health Criteria Document*, EPA-823-R-002, www.epa.gov.

Table 10J.241 Giardiasis Case Reports, by State/Area — United States, 1998–2002

	1998				1999				2000				2001				2002			
State/Area	No.	(%)	Rate[a]	No. of Outbreak Cases	No.	(%)	Rate	No. of Outbreak Cases	No.	(%)	Rate	No. of Outbreak Cases	No.	(%)	Rate	No. of Outbreak Cases	No.	(%)	Rate	No. of Outbreak Cases
Alabama	288	(1.2)	6.6		340	(1.5)	7.8		227	(1.0)	5.1		231	(1.2)	5.2		205	(1.0)	4.6	
Alaska	109	(0.4)	17.7		96	(0.4)	15.5		115	(0.5)	18.3		121	(0.6)	19.1		115	(0.5)	17.9	21
Arizona	250	(1.0)	5.4		255	(1.1)	5.3		313	(1.4)	6.1		267	(1.4)	5.0	2	269	(1.3)	4.9	1
Arkansas	168	(0.7)	6.6		152	(0.7)	6.0		203	(0.9)	7.6		160	(0.8)	5.9		175	(0.8)	6.5	
California	NR[b]				NR				NR				NR				2,561	(12.0)	7.3	
Colorado	618	(2.6)	15.6		704	(3.0)	17.4	4	695	(3.2)	16.2	5	632	(3.2)	14.3	2	571	(2.7)	12.7	5
Connecticut	NR				NR				462	(2.1)	13.6		417	(2.1)	12.1		260	(1.2)	7.5	
Delaware	39	(0.2)	5.2		52	(0.2)	6.9		92	(0.4)	11.7		59	(0.3)	7.4		54	(0.3)	6.7	
District of Columbia	34	(0.1)	6.5		37	(0.2)	7.1		38	(0.2)	6.6		70	(0.4)	12.2		47	(0.2)	8.3	
Florida	1,676	(6.9)	11.2	1,636	1,360	(5.8)	9.0	1,334	1,521	(7.0)	9.5	1,459	1,155	(5.9)	7.1		1,318	(6.2)	7.9	1,226
Georgia	1,215	(5.0)	15.9		1,355	(5.8)	17.4		1,201	(5.5)	14.7		963	(4.9)	11.5		926	(4.3)	10.8	
Hawaii	123	(0.5)	10.3	99	117	(0.5)	9.9	117	105	(0.5)	8.7	105	118	(0.6)	9.6	118	91	(0.4)	7.3	91
Idaho	177	(0.7)	14.4	1	134	(0.6)	10.7	1	139	0.6	10.7		172	(0.9)	13.0	1	137	(0.6)	10.2	1
Illinois	1,472	(6.1)	12.2		1,458	(6.3)	12.0	11	1,093	(5.0)	8.8		1,108	(5.6)	8.9		1,011	(4.8)	8.0	
Indiana	772	(3.2)	(13.1)		654	(2.8)	11.0	3	517	(2.4)	8.5		3	(<0.1)	<0.1		NR			
Iowa	429	(1.8)	15.0	4	377	(1.8)	13.1		420	(1.9)	14.4		345	(1.8)	11.8		314	(1.5)	10.7	
Kansas	226	(0.9)	8.6	7	220	(0.9)	8.3		205	(0.9)	7.6		178	(0.9)	6.6		192	(0.9)	7.1	
Kentucky	NR				NR				NR				NR				NR			
Louisiana	NR				21	(0.1)	0.5		41	(0.2)	0.9		14	(0.1)	0.3		6	(<0.1)	0.1	
Maine	277	(1.1)	20.2		236	(1.0)	19.0		238	(1.1)	18.7		197	(1.0)	15.3	1	213	(1.0)	16.4	
Maryland	NR				119	(0.5)	2.3		125	(0.6)	2.4	8	NR				118	(0.6)	2.2	5
Massachusetts	833	(3.4)	13.6	637	851	(3.7)	13.8	640	632	(2.9)	10.0	12	906	(4.6)	14.2	27	935	(4.4)	14.6	1
Michigan	1,172	(4.8)	11.9	8	1,166	(5.0)	11.8	2	1,135	(5.2)	11.4	1	1,003	(5.1)	10.0	1	923	(4.3)	9.2	
Minnesota	1,324	(5.5)	28.0	8	1,555	(6.7)	32.6	9	1,227	(5.6)	24.9	22	1,061	(6.4)	21.3	16	982	(4.6)	19.5	
Mississippi	131	(0.5)	4.8	131	145	(0.6)	5.2		116	(0.5)	4.1		NR				NR			
Missouri	790	(3.3)	14.5		807	(3.5)	14.8		839	(3.8)	15.0		715	(3.6)	12.7		512	(2.4)	9.0	
Montana	119	(0.5)	13.5	2	83	(0.4)	9.4	1	91	(0.4)	10.1	1	95	(0.5)	10.5		94	(0.4)	10.3	
Nebraska	249	(1.0)	15.0		288	(1.0)	14.3		300	(1.4)	17.5	1	234	(1.2)	13.6		191	(0.9)	11.1	
Nevada	222	(0.9)	12.7	140	215	(0.9)	11.9	121	211	(1.0)	10.6	138	208	(1.1)	9.9		162	(0.8)	7.5	
New Hampshire	83	(0.3	7.0	83	64	(0.3)	5.3	64	56	(0.3)	4.5	56	98	(0.2)	3.0	38	46	(0.2)	3.6	46
New Jersey	218	(0.9)	2.7		NR				NR				494	(2.5)	5.8		474	(2.2)	5.5	
New Mexico	238	(1.0)	13.7		261	(1.1)	15.0	1	164	(0.8)	9.0		148	(0.8)	8.1		153	(0.7)	8.3	
New York[c]	3,789	(15.4)	20.6	55	3,696	(15.9)	20.3	78	3,346	(15.3)	17.6	83	2,903	(14.7)	15.2	66	2,764	(13.0)	14.4	80

New York City	2,079	(8.6)	28.1		1,894	(8.1)	25.5		1,797	(8.0)	21.7		1,620	(7.7)	18.9		1,417	(6.7)	17.6	
North Carolina	NR				NR				NR				NR				NR			
North Dakota	82	(0.3)	12.9		104	(0.4)	16.4		65	(0.3)	10.1		78	(0.4)	12.3		47	(0.2)	7.4	
Ohio	1,093	(4.5)	9.7	1	1,110	(4.8)	9.9	2	1,058	(4.9)	9.3	1	1,090	(5.5)	9.6	1	972	(4.6)	8.5	6
Oklahoma	148	(0.6)	4.4		162	(0.7)	4.5		96	(0.4)	2.8		NR				85	(0.4)	2.4	
Oregon	900	3.7	27.4		808	(3.5)	24.4	2	654	(3.0)	19.1	13	543	(2.8)	15.6	4	447	(2.1)	12.7	4
Pennsylvania	1,681	(6.0)	12.2	2	1,124	(4.8)	9.4	4	1,083	(5.0)	8.8	10	1,150	(5.8)	9.4	8	1,066	(5.0)	8.6	1
Rhode Island	130	(0.5)	13.2		149	(0.6)	15.0		157	(0.7)	15.0		168	(0.9)	15.9		170	(0.8)	15.9	
South Carolina	NR				NR				NR				NR				149	(0.7)	3.6	
South Dakota	181	(0.7)	24.8		143	(0.6)	19.5	5	108	(0.5)	14.3	16	106	(0.5)	14.0	15	83	(0.4)	10.9	7
Tennessee	220	(0.9)	4.0		187	(0.8)	3.4		187	(0.9)	3.3		169	(1.0)	3.3		191	(0.9)	3.3	
Texas	NR				NR				NR				NR				3	(<0.1)	<0.1	
Utah	291	(1.2)	13.9		256	(1.1)	12.0	19	281	(1.3)	12.6	42	284	(1.4)	12.5	3	335	(1.6)	14.4	
Vermont	326	(1.9)	55.2		345	(1.5)	58.1	30	217	(1.0)	35.6		220	(1.1)	35.9		145	(0.7)	23.5	
Virginia	503	(2.1)	7.4		471	(2.0)	6.9	1	487	(2.0)	6.2		417	(2.1)	5.8		386	(1.8)	5.3	
Washington	740	(3.1)	13.0		560	(2.4)	9.7		622	(2.9)	10.6		512	(2.6)	8.5		510	(2.4)	8.4	
West Virginia	90	(0.4)	5.0	4	93	(0.4)	5.1	1	80	(0.4)	4.4	3	89	(0.4)	4.6		78	(0.4)	4.3	
Wisconsin	1,003	(4.1)	19.2		936	(4.0)	17.8	10	811	(3.7)	15.1	3	765	(3.9)	14.2	12	691	(3.2)	12.7	5
Wyoming	45	(0.2)	9.4		37	(0.2)	7.7		49	(0.2)	9.9	1	87	(0.2)	7.5	1	29	(0.1)	5.8	
Total state	24,204	(99.9)	9.0	2,818	23,245	(99.8)	8.5	2,460	21,772	(99.8)	7.7	1,980	19,659	(99.8)	6.9	316	21,206	(99.6)	7.4	1,500
Guam	9	(<0.1)	6.0		23	(0.1)	15.1		17	(0.1)	10.9		9	(<0.1)	5.7		7	(<0.1)	4.3	
Northern Mariana Islands	NR		—		NR		—		NR		—		NR		—		1	(<0.1)	1.4	
Puerto Rico	13	(0.1)	0.3		13	(0.1)	0.3		24	(0.1)	0.6		40	(0.2)	1.0	3	86	(0.4)	2.2	2
Total	24,226	(100)[d]		2,818	23,281	(100.0)		2,460	21,813	(100.0)		1,980	19,706	(100.0)		319	21,300	(100.0)		1,502

Note: Population estimates are from the Population Division, U.S. Census Bureau, Estimates of the population of states: ST-99-4 State Rankings of Population Change and Demographic Components of Population Change for the Period July 1, 1998 to July 1, 1999, available at www.census.gov/popest/archives/states/1990s/ST-90-01.txt, and Table1: Annual estimates of the population for the United States and States and Puerto Rico: April 1, 2001, to July 2003 (NST-EST 2003 01), available at www.census.gov/popest/states/tables/NST-EST2003-01.xls. Estimates of the New York City population: (SU-99-7) Population Estimates for Places (Sorted Alphabetically With in State): Annual Time Series, July 1, 1990 to July 1, 1999 (includes April 1, 1990 Population Estimates Base), available at www.census.gov/popest/archives/1990s/su-99-07/SU-99-7_NY.txt, and Table 1: Annual Estimates of the Population for Incorporated Places over 100,000. Ranked by July 1, 2003 Population: April 1, 2001, to July 1, 2003 (SUB-EST 2003-01), available at www.census.gov/popest/cities/tables/SUB-EST2003-01.xls. Estimates of the population of Guam, the Northern Marians Islands, and Puerto Rico: International Data Base (IDB) Data access — Spreadsheet, available at www.census.gov/ipc/www/dbsprd.html.

[a] Per 100,000 population on the basis of U.S. Census Bureau population estimates.
[b] No cases reported to CDC.
[c] New York State counts include New York City cases.
[d] Percentages might not total 100% because of rounding.

Source: From Hlavsa, M.C., Watson, J.C. and Beach, M.J., 2005, Giardiasis surveillance – United States 1998-2002, *MMWR Surveillance Summaries*, vol. 54, no. SS-01, January 28, 2005, pp. 9–16, www.cdc.gov.

Table 10J.242 Causes of Waterborne Outbreaks of Giardiasis in the United States, 1971–1994

Water Source, Treatment, or Deficiency	Outbreaks
Surface water source	
Untreated	13
Chlorination only	51
Filtered (includes outbreaks when filtration was bypassed)	17
Groundwater source	
Untreated	8
Chlorination only	7
Filtration	1
Contamination of distribution system or storage	12
Use of water not intended for drinking or ingestion during water recreation or other water activities	14
Insufficient information	4
Total	127

Source: From United States Environmental Protection Agency, 1998, *Giardia: Human Health Criteria Document*, EPA-823-R-002, www.epa.gov.

Table 10J.243 Causes of Waterborne Diseases in England and Wales, 1971–2000

Decade	Cause	No. Outbreaks	No. Cases
Public Supplies			
1971–1980	Gastroenteritis	2	3,114
	Giardia	1	60
	Total	3	322
1981–1990	*Cryptosporidium*	7	1,157
	Campylobacter	3	629
	Gastroenteritis	3	310
	Total	13	2,096
1981–2000	*Cryptosporidium*	23	2,837
	Campylobacter	1	281
	Gastroenteritis	1	229
	Total	25	3,347
Private Supplies			
1971–1980	Paratyphoid	1	6
	Gastroenteritis	1	160
	Total	2	166
1981–1990	*Campylobacter*	3	520
	Streptobacillary fever	1	304
	Gastroenteritis	1	56
	Total	5	962
1991–2000	*Campylobacter*	8	178
	Mixed *Campylobacter* and *Cryptosporidium*	1	43
	Cryptosporidium	3	74
	Gastroenteritis	2	81
	Giardia	1	31
	E. coli O157	1	14
	Total		421

Source: From Stanwell-Smith, R., Anderson, Y., and Levy, D., 2003, National surveillance systems in Hunter, P.R., Waite, M., and Ronchi, El., (eds.), *Drinking Water and Infectious Disease Establishing the Links*, CRC Press LLC, Boca Raton, FL.

Table 10J.244 Etiology of Recreational Waterborne Outbreaks, Outbreaks, and Cases of Illness by Type of Water Source in the United States, 1991–1998

	Outbreaks				Cases of Illness			
	Swim				Swim			
Etiological Agent	Lake	Swim Pool[a]	River	Other[b]	Lake	Swim Pool	River	Other[b]
Cryptosporidium	3	19		1	429	9,477		369
Naegleria	9		4	4	9		4	4
E. coli O157:H7	9	2			293	44		
Shigella	11			1	1,216			9
Undetermined	8	1	1	1	1,016	100	15	61
Giardia	3	5	1		59	187	6	
Schistosoma	6			1	173			30
Pseudomonas		5				162		
Leptospira	2			1	48			55
Norwalk-like virus	2			1	48			55
Chemical		2				29		
Salmonella, on-typhoid		1				3		
Adenovirus	1				595			
Total	54	35	6	9	4,219	10,002	25	528

[a] Includes wading pools and pools and other activities at water parks.
[b] Water slide park, dunking booth, hot spring, canal, fountain, ocean unknown.

Source: From Craun, G.F., Calderon, R.L., and Nwachuku, N., 2003, Causes of waterborne outbreaks reported in the United States, 1991–1998 in Hunter, P.R., Waite, M., and Ronchi, El., (eds.), *Drinking Water and Infectious Disease Establishing the Links*, CRC Press LLC, Boca Raton, FL.

Table 10J.245 Waterborne-Disease Outbreaks (n=65) Associated with Recreational Water, by Etiologic Agent and Type of Water in the United States, 2001–2002

	Type				Total	
	Treated		Fresh			
Etiologic Agent	Outbreaks	Cases	Outbreaks	Cases	Outbreaks	Cases
Bacterial	24	571	3	69	27	640
Pseudomonas aeruginosa	18[a]	393	0	0	18	393
Escherichia coli (O157:H7, O26:NM)	1	9	3[b]	69	4	78
Shigella sonnei	2	78	0	0	2	78
Bacillus species	1	20	0	0	1	20
Legionella species	1	68	0	0	1	68
Staphylococcus species	1	3	0	0	1	3
Parasitic	9	1,469	12	34	21	1,503
Cryptosporidum species	9	1,469	2	5	11	1,474
Naagleria fowleri	0	0	8	8	8	8
Giardia intestinalis	0	0	1	2	1	2
Avian schistosomes	0	0	1[a]	19	1	19
Unknown	5	63	3	82	8	145
AGI[c]	4	59	3	82	7	141
ARI[d]	1	4	0	0	1	4
Viruses	2	51	3[b]	95	5	146
Norovirus	2	51	3[b]	95	5	146
Chemical	4	102	0	0	4	102
Chlorine gas	2	50	0	0	2	50
Chloramines	2[a]	52	0	0	2	52
Total	44	2,256	21	280	65	2,536
Percentage	(67.7)	(89.0)	(32.3)	(11.0)	(100)	(100)

[a] Includes outbreaks of suspected etiology on the basis of clinical syndrome and setting.
[b] Includes one mixed-pathogen outbreak.
[c] Acute gastrointestinal illness of unknown etiology.
[d] Acute respiratory illness of unknown etiology.

Source: From Yoder, J.S. et al., 2004, Surveillance for waterborne-disease outbreaks associated with recreation water – United States, 2001–2002, *MMWR*, vol. 53, no. SS08, pp 1–22, October 22, 2004, www.cdc.gov.

Table 10J.246 Causes of Waterborne Disease Outbreaks Associated with Recreational Water in the United States

Source of Contamination or Deficiency	Number of Outbreaks
Fecal accident or ill bather	13
Children in diapers	8
Poor maintenance, inadequate treatment or operation of swimming or wading pool	9
Bather overload or crowding	3
Floods	1
Livestock	1
Geese	2
Seepage or overflow of sewage	3
Total	40

Source: From Craun, G.F., Calderon, R.L., and Nwachuku, N., 2003, Causes of waterborne outbreaks reported in the United States, 1991–1998 in *Drinking Water and Infectious Disease Establishing the Links*, Hunter, P.R., Waite, M., and Ronchi, El., (eds.), CRC Press LLC, Boca Raton, FL.

Table 10J.247 Swimming-Associated Giardiasis Outbreaks Reported in the United States, 1982–1996

State	Year	Cases	Location	Additional Information
Washington	1982	78	Swimming pool	
Illinois	1985	15	Swimming pool	Fecal contamination
New Jersey	1985	9	Indoor pool	
Maryland	1987	266	Swimming pool	Inadequate chlorination
Maryland	1988	34	Swimming pool	
Georgia	1991	9	Swimming pool	Day-care center
Georgia	1991	7	Swimming pool	Day-care center
Maryland	1991	14	Pool	Park; fecal contamination
Washington	1991	4	Swimming pool	Wild animals near lake
Maryland	1993	12	Swimming pool	
New Jersey	1993	43	Swimming pool	Met water quality limits
Washington	1993	6	River	
Indiana	1994	80	Swimming pool	Filter malfunction
Florida	1996	60	Wading pool	

Source: From United States Environmental Protection Agency, 1998, *Giardia: Human Health Criteria Document*, EPA-823-R-002, www.epa.gov.

Table 10J.248 Number of Cases of Shigellosis Associated with Recreational Waters in the United States, 1991–2000

State	Date	Etiologic Agent	Illness	No. of Cases	Source and Settings
Rhode Island	July 1991	*S. sonnei*	GI	23	Lake, swimming area
Mass.	June 1991	*S. sonnei*	GI	203	Lake, park
Virginia	July 1992	*S. sonnei*	GI	9	Lake, camp
New Jersey	June 1992	*S. sonnei*	GI	54	Lake, Campground
Ohio	July 1993	*S. sonnei*	GI	150	Lake, park
New Jersey	June 1994	*S. sonnei*	GI	242	Lake, park
Minnesota	May 1994	*S. flexneri*	GI	35	Lake, park
Colorado	July 1995	*S. sonnei*	GI	81	Lake, recreational area
Colorado	July 1995	*S. sonnei*	GI	39	Lake, recreational area
Penn.	Aug 1995	*S. sonnei*	GI	70	Lake, beach
Mass.	July 1997	*S. sonnei*	GI	9	Pool/fountain, public park
Florida	1999	*S. sonnei*	GI	38	Interactive fountain, beach park
Missouri	Sept 2000	*S. sonnei*	GI	6	Wading pool, municipal pool
Minnesota	July 2000	*S. sonnei*	GI	15	Lake/pond, beach
Minnesota	Aug 2000	*S. sonnei*	GI	25	Lake, public beach

Notes: GI, gastroenteritis; Mass., Massachusetts; Penn., Pennsylvania.

Source: From Pond, K., 2005. Water Recreation and Disease Plausibility and Associated Infections: Acute Effects, Sequelae, and Mortality, Published on behalf of the World Health Organization by IWA Publishing, London Copyright © World Health Organization 2005, www.who.int.

Original Source: From Minnesota Department of Health 1974; Anonymous 1993; 1995; 1996b; Levy et al., 1998; Barwick et al., 2000; Lee 2002.

Table 10J.249 Water- and Sanitation-Related Infections in Developing Countries and Their Control

	Importance of Alternate Control Methods[a]							
Infections	Water Quality	Water Availability	Excreta Disposal	Excreta Treatment	Personal and Domestic Cleanliness	Drainage and Sullage Disposal	Food Hygiene	Public Health Importance[a]
Diarrhoeal diseases and enteric fevers								
Viral agents	2	3	2	1	3	0	2	3
Bacterial agents	3	3	2	1	3	0	3	3
Protozoal agents	1	3	2	1	3	0	2	2
Poliomyelitis and hepatitis A	1	3	2	1	3	0	1	3
Worms with no intermediate host								
Ascaris and *Trichuris*	0	1	3	2	1	1	2	2
Hookworms	0	1	3	2	1	0	1	3
Beef and pork tapeworms	0	0	3	3	0	0	3	2
Worms with intermediate stages								
Schistosomiasis	1	1	3	2	1	0	0	3
Guinea worm	3	0	0	0	0	0	0	2
Worms with two aquatic intermediate stages	0	0	2	2	0	0	3	1
Skin, eye and louse-borne infections	0	3	0	0	3	0	0	2
Infections spread by water-related insects								
Malaria	0	0	0	0	0	1	0	3
Yellow fever and dengue	0	0	0	0	0	1	0	3
Bancroftian filariasis	0	0	3	0	0	3	0	3

[a] 0, no importance; 1, little importance; 2, moderate importance; 3, great importance.

Source: From Feachem, R.G., 1984, The Health Dimension of the Decade, in world Water '83: the World Problem, Proceedings of July 1983 Conference of Institution of Civil Engineers, London.

Table 10J.250 Some Water-Associated Diseases by Cause and Sex Estimates for 2001

	Deaths (in Thousands)						Burden of Disease (in Thousands)					
	Both Sexes		Males		Females		Both Sexes		Males		Females	
	Number	%	Number	%	Number	%	Number	%	Number	%	Number	%
Total burden of disease (000s of DALYs)							1.467,183	100	768,064	100	699,119	100
Total deaths (000s)	56,552	100	29,626	100	26,926	100						
Communicable diseases, matemal and perinatal conditions and nutritional deficiencies overall	18,374	32.5	9,529	32.2	8,846	32.9	615,737	42.0	304,269	39.6	311.468	44.6
Infectious and parasitic diseases altogether	10,937	19.3	5,875	19.8	5,063	18.8	359,377	24.5	184,997	24.1	174,380	24.9
Diarrhoeal diseases	2,001	3.5	1,035	3.5	966	3.6	62,451	4.3	31,633	4.1	30,818	4.4
Malaria	1,123	2.0	532	1.8	591	2.2	42,280	2.9	20,024	2.6	22,256	3.2
Schistosomiasis	15	0.0	11	0.0	5	0.0	1,760	0.1	1,081	0.1	678	0.1
Lymphatic filariasis	0	0.0	0	0.0	0	0.0	5,644	0.4	4,316	0.6	1,327	0.2
Onchocerciasis	0	0.0	0	0.0	0	0.0	987	0.1	571	0.1	416	0.1
Dengue	21	0.0	10	0.0	11	0.0	653	0.0	287	0.0	365	0.0
Japanese encephalitis	15	0.0	8	0.0	8	0.0	767	0.1	367	0.0	400	0.1
Trachoma	0	0.0	0	0.0	0	0.0	3,997	0.3	1,082	0.1	2,915	0.4
Intestinal nematode infections	12	0.0	6	0.0	5	0.0	4,706	0.3	2,410	0.3	2,296	0.3
Ascariasis	4	0.0	2	0.0	2	0.0	1,181	0.1	604	0.1	577	0.1
Trichuriasis	2	0.0	1	0.0	1	0.0	1,649	0.1	849	0.1	800	0.1
Hookworm infection	4	0.0	2	0.0	2	0.0	1,825	0.1	932	0.1	893	0.1
Unintentional injuries overall	3,508	6.2	2,251	7.6	1,256	4.7	129,853	8.9	82,378	10.7	47,475	6.8
Drowning	402	0.7	276	1.0	126	0.5	11,778	0.8	8,150	1.1	3,628	0.5

Note: The burden of disease is calculated through an indicator of population health the DALY: a DALY represents a lost year of healthy life and is the unit used to estimate the gap between the current health of a population and an idea situation in which everyone in that population in which everyone in that population would live into old age in full health. This table shows the total deaths and burden of disease caused by communicable diseases, maternal and perinatal conditions and nutritional deficiencies noncommunicable diseases and injuries related to water.

Source: From *Water for People Water for Life*, The United Nations World Water Development, Copyright © United Nations Educational, Scientific and Cultural Organization (UNESCO)-World Water Assessment Programme (UNESCO-WWAP), 2003. Reproduced by permission of UNESCO. www.unesco.org.

Original Source: From WHO (World Health Organization), 2002, Report on the Status of the Dracunculiasis Eradication Campaign in 2001. Document WHO/CDS/CPE/CEE/2002.30, Geneva. Reprinted with permission.

Table 10J.251 Reported Dracunculasis Cases by Country, 1972–2000

Region and Country	1972	1973	1974	1975	1976	1977	1978	1979	1980	1981	1982	1983	1984	1985	1986	1987	1988	1989	1990	1991	1992	1993	1994	1995	1996	1997	1998	1999	2000
Africa																													
Benin	1,480		820													400	33,962	7,172	37,414	4,006	4,315	13,887	4,302	2,273	1,472	855	695	492	186
Burkina Faso	5,822	4,404	4,008	6,277	1,557		2,885	2,694	2,565			4,362	1,739	458	2,558	1,957	1,266	45,004	42,187		11,784	8,281	6,861	6,281	3,241	2,477	2,227	2,184	1,956
Cameroon				251									0	168	86		752	871	742	393	127	72	30	15	17	19	23	8	5
Central African Republic																								18	9	5	34	26	35
Chad							172						1,472	9	314						156	1,231	640	149	127	25	3	1	3
Côte d'Ivoire	4,891	4,654	6,283	4,971	4,656	5,207	6,993		6,712	7,978		2,259	2,573	1,889	1,177	1,272	1,370	1,555	1,360	12,690		8,034	5,061	3,801	2,794	1,254	1,414	476	297
Ethiopia													2,882	1,467	3,385	2,302	1,487	3,565	2,233		303	1,120	1,252	514	371	451	366	249	60
Ghana	693	1,606	1,226	4,052	1,421	1,617	1,676		2,703	853	3,413	3,040	4,244	4,501	4,717	18,398	71,767	179,556	123,793	66,697	33,464	17,918	8,432	8,894	4,877	8,921	5,473	9,027	7,402
Kenya																		5	6			35	53	23	0	6	7	1	4
Mali	668	786	737	542	760	1,084			816	777	401	428	5,008	4,072	5,640	435	564	1,111	884	16,024		12,011	5,581	4,218	2,402	1,099	650	410	290
Mauritania						127			651	663	903	1,612	1,241	1,291		227	608	447	8,036		1,557	3,533	5,029	1,762	562	388	379	255	136
Niger					2,600	3,000	5,560		1,906	2,113	1,530			1,373		699		288		32,829	500	21,564	18,562	13,821	2,956	3,030	2,700	1,920	1,166
Nigeria	98			1,007					1,693				8,777	5,234	2,821	216,848	653,492	640,008	394,082	281,937	183,169	75,752	39,774	16,374	12,282	12,590	13,420	13,237	7,869
Senegal		334	208	65	137				161					62	128	132	138		38	1,341	728	630	195	76	19	4	0	0	0
Sudan															822	399	542				2,447	2,984	53,271	64,608	118,578	43,596	47,977	66,097	54,890
Togo				3,261	1,648		2,617	2,673	1,748	951	2,592		1,839	1,456	1,325		178	2,749	3,042	5,118	8,179	10,349	5,044	2,073	1,626	1,762	2,128	1,589	828
Uganda													6,230	4,070			1,960	1,309	4,704		126,852	42,852	10,425	4,810	1,455	1,374	1,061	321	96
Asia and Middle East																													
India						7,052	6,827	2,846	2,729	5,406	42,926	44,818	39,792	30,950	23,070	17,031	12,023	7,881	4,798	2,185	1,081	755	371	60	9	0	0	0	0
Pakistan							250		14,155							2,400	1,110	534	160	106	23	2	0	0	0	0	0	0	0
Yemen		25																					94	82	62	7	0	0	0
Number of Countries Reporting	6	6	6	8	7	6	8	3	11	7	6	6	12	14	12	13	15	15	15	11	15	18	19	20	20	20	20	20	20
Number of Cases	13,652	11,809	13,282	20,426	12,779	18,087	26,980	8,213	35,839	18,741	51,765	56,519	75,797	57,000	46,043	262,136	781,219	892,055	623,844	547,575	374,202	229,773	164,973	129,852	152,814	77,863	78,557	96,293	75,223

Source: From *World's Water 2002–2003*, by Peter H. Gleick. Copyright © 2002 Island Press. Reproduced by permission of Island Press, Washington, DC.

Table 10J.252 Deaths and DALYs from Selected Water-Related Diseases, 2000

	Deaths	DALYs
Diarrheal diseases	2,019,585	63,345,722
Childhood cluster diseases		
Poliomyelitis	1,136	188,543
Diphtheria	5,527	187,838
Tropical-cluster diseases		
Trypanosomiasis	49,129	1,570,242
Schistosomiasis	15,335	1,711,522
Trachoma	72	3,892,326
Intestinal nematode infections		
Ascariasis	4,929	1,204,384
Trichuriasis	2,393	1,661,689
Hookworm disease	3,477	1,785,539
Other Intestinal Infections	1,692	53,222
Total	2,103,274	75,601,028

Notes: DALYs, The DALY is a measure of population health that combines in a single indicator years of life lost from premature death and years of life lived with disabilities. One DALY can be thought of as one lost year of "healthy" life. This table excludes mortality and DALYs associated with water-related insect vectors, such as malaria, onchocerciasis, and dengue fever. Trachoma, while few deaths from trachoma are reported, approximately 5.9 million cases of blindness or severe complications occur annually.

Source: From *World's Water 2004–2005*, by Peter H. Gleick. Copyright © 2004 Island Press. Reproduced by permission of Island Press, Washington, DC.

Table 10J.253 Annual Number of Diarrhoeal Cases Avoided if the Combined Water and Sanitation Millennium Goals Are Achieved by WHO subregion

WHO Sub-Region	Region/ Country	Pop. (m)	Cases of Diarrhoea (million)	Number of Cases Avoided Per Year ('000s), by Intervention				
				1[a]	2[b]	3[c]	4[d]	5[e]
2	AFR-E	481	619	28′548	87′405	127′049	345′132	439′980
5	AMR-D	93	93	3′250	9′307	13′208	48′679	64′106
11	EUR-D	223	43	112	568	1′056	19′816	27′983
13	SEAR-D	1,689	1,491	26′895	146′829	272′361	807′596	1′04′922
15	WPR-B1	1,488	1,193	39′454	131′171	239′104	659′687	844′381
World		7,183	5,388	154′854	545′950	903′004	2′860′951	3′717′971

Notes: Therefore, in the analysis, account is taken of the proportion of populations in each country who did not have access to improved" water and sanitation in 2000.
Level VI, No improved water supply and no basic sanitation in a country which is not extensively covered by those services, and where water supply is not routinely controlled;
Level Vb, Improved water supply and no basic sanitation in a country which is not extensively covered by those services, and where water supply is not routinely controlled;
Level Va, Improved sanitation but no improved water supply in a country which is not extensively covered by those services, and where water supply is not routinely controlled;
Level IV, Improved water supply and improved sanitation in a country which is not extensively covered by those services, and where water supply is not routinely controlled;
Level III, Improved water supply and improved sanitation in a country which is not extensively covered by those services, and where water supply is not routinely controlled, plus household water treatment;
Level II, Regulated water supply and full sanitation coverage, with partial treatment for sewage, corresponding to a situation typically occurring in developed countries;
Level I, Ideal situation, corresponding to the absence of transmission of diarrhoeal disease through water, sanitation and hygiene.

[a] Intervention 1. Millennium targets: halving the proportion of people who do not have access to improved water sources by 2015, with priority given to those already with priority given to those already with improved sanitation. This means: Scenario Va to IV (applied to half the population without improved water supply).

[b] Intervention 2. Millennium targets with sanitation targets: halving the proportion of people who do not have access to improved water sources and improved sanitation facilities, by 2015. This means: Scenario VI to IV, or Scenario Va or Vb to IV (applied to half the population without improved water supply and half the population without improved sanitation).

[c] Intervention 3. Access for all to improved water and improved sanitation. This means: Scenario VI, Va and Vb to IV (applied to the entire population without improved water and the entire population without improved water and the entire population without improved sanitation).

[d] Intervention 4. A minimum of water disinfected at the point of use for all, on top of improved water and sanitation services. This means: Scenarios VI, Va, Vb and IV go to Scenario III.

[e] Intervention 5. Access for all to a regulated piped water supply and sewage connection into their houses. This means: Scenarios VI, Va, Vb, IV and III go to Scenario II. All the interventions were compared to the situation in 2000, which was defined as the baseline year.

Source: From Hutton, G. and Haller, L., 2004, Evaluation of Costs and Benefits of Water and Sanitation Improvements at the Global Level, Water, Sanitation and Health Protection of the Human Environment, Work Health Organization, Geneva. Copyright © World Health Organization 2004, www.who.int.

Table 10J.254 Clinical Syndromes and Incubation Periods of Infectious and Chemical Agents Causing Acute Waterborne Disease in the United States

Agent	Incubation Period	Clinical Syndrome
Bacteria		
Campylobacter jejuni	2–5 d	Gastroenteritis, often with fever
Enterotoxigenic *Escherichia coli*	6–36 hr	Gastroenteritis
Salmonella	6–48 hr	Either gastroenteritis (often with fever), enteric fever, or extraintestinal infection
Salmonella typhi	10–14 d	Enteric fever — fever, anorexia, malaise, transient rash, splenomegaly, and leukopenia
Shigella	12–48 hr	Gastroenteritis, often with fever and bloody diarrhea
Vibrio cholerae 01	1–5 d	Gastroenteritis, often with significant dehydration
Yersinia enterocolitica	3–7 d	Either gastroenteritis, mesenteric lymphadenitis, or acute terminal ileitis; may mimic appendicitis
Viruses		
Hepatitis A	2–6 wks	Hepatitis — nausea, anorexia, jaundice, dark urine
Norwalk virus	24–48 hr	Gastroenteritis, of short duration
Rotavirus	24–72 hr	Gastroenteritis, often with significant dehydration
Parasites		
Entamoeba histolytica	2–4 wks	Varies from mild gastroenteritis to acute fulminating dysentery with fever and bloody diarrhea
Giardia lamblia	1–4 wks	Chronic diarrhea, epigastric pain, bloating, malabsorption, and weight loss
Chemicals		
Fluoride	<1 hr	Nausea, vomiting, and abdominal cramps
Heavy metals		
Antimony		
Cadmium		
Copper		
Lead		
Tin		
Zinc, etc	<1 hr	Nausea, vomiting, and abdominal cramps, often accompanied by a metallic taste
Others		
Pesticides		
Petroleum products, etc	Variable	Variable

Source: From Craun, G.F., 1986, *Waterborne Diseases in the United States*, Copyright CRC Press, Inc. Boca Raton, FL. Reprinted with permission.

Table 10J.255 Pathogens in Drinking Water, Infectious Dose, Estimated Incidence through Consumption of Drinking Water in the United States, Survival in Drinking Water, and Potential Survival Strategies

	Infectious Dose[a]	Estimated Incidence[b]	Survival in Drinking Water (Days)	Survival Strategies[c]
Bacteria				
Vibrio cholera	10^8	(very few)[c]	30	VNC, IC
Salmonella spp.	10^{6-7}	59,000	60–90	VNC, IC
Shigella spp.	10^2	35,000	30	VNC, IC
Toxigenic Escherichia coli	10^{2-9}	150,000	90	VNC, IC
Campylobacter spp.	10^6	320,000	7	VNC, IC
Leptospira spp.	3	?[d]	?	?
Francisella tularensis	10	?	?	?
Yersinia enterocolitica	10^9	?	90	?
Aeromonas spp.	10^8	?	90	?
Helicobacter pylori	?	High	?	?
Legionella pneumophila	>10	13,000[e]	Long	VNC, IC
Mycobacterium avium	?	?	Long	IC
Protozoa				
Giardia lambia	1–10	260,000	25	Cyst
Cryptosporidium parvum	1–30	420,000	?	Oocyst
Naegleria fowleri	?	?	?	Cyst
Acanthamoeba spp.	?	?	?	Cyst
Entamoeba histolica	10–100	?	25	Cyst
Cyclospora cayetanensis	?	?	?	Oocyst
Isospora belli	?	?	?	Oocyst
The microsporidia	?	?	?	Spore, IC[d]
Ballantidium coli	25–100	?	20	Cyst
Toxoplasma gondii	?	?	?	Oocyst
Viruses[f]				
Total estimates	1–10	6,500,000	5–27[g]	Adsorption/ absorption

Note: ?, unknown; IC, intracellular survival and/or growth; VNC, viable but not culturable.

[a] Infectious dose is number of infectious agents that produce symptoms in 50% of teted volunteers. Volunteers are not usually susceptible individuals, and therefore these numbers are not useful for risk estimates.

[b] U.S. point estimates.

[c] Very few outbreaks of cholera occur in the United States, and these are usually attributable to imported foods (14).

[d] Possible IC with microsporidialike organisms (15).

[e] Data form Breiman and Butler (14).

[f] Includes Norwalk virus, poliovirus, coxsachievirus, echovirus, reovirus, adenovirus, HAV, HEV, rotavirus, SRSV, astrovirus, coronavirus, calicivirus, and unknown viruses.

[g] Estimated for HAV, Norwalkvirus, and rotavirus (13).

Source: From Ford, T.E., 1999, Microbiological safety of drinking water: United States and global perspectives, *Environmental Health Perspectives Supplements*, vol. 107, no. S1, February 1999, www.ehponline.org

Original Source: Except where noted, data are compiled from Morris and Levin (2), WHO (10), Hazen and Toranzos (11), and Geldreich (12).

Table 10J.256 Survival of Bacteria In Various Media

Organism	Survival Time	Media
E. coli	63 days	Recharge well
E. coli	3–3.5 months	Groundwater in the field
E. coli	4–4.5 months	Groundwater held in lab
Coliforms	17 hours-50 percent reduction	Well water
Salmonella	44 days	Water infiltrating sand column
S. typhi	2–85 days	Soil
S. typhi	24–27 days	Septic tank
S. typhi	25–41 days	Soil
Shigella	24 days	Water infiltrating sand column
S. flexneri	26.8 hours-50 percent reduction	Well water
Vibrio cholerae	7.2 hours-50 percent reduction	Well water

Source: From McGinnis, J.A. and DeWalle, F.B., 1983, The movement of typhoid organisms in saturated permeable soil, *J. Am. Water Works Assoc.*, Vol. 75, no. 6. Copyright AWWA. Reprinted with permission.

Table 10J.257 Survival Times of Organisms

Organism	Media	Survival Time Days
Ascaris ova	Vegetables	27–35
	Soil	730–2010
Entamoeba histolytica	Vegetables	3
	Soil	6–8
	Water	60+
Mycobacterium tuberculosis	Soil	180+
	Grass	10–40
	Water	30–90
Salmonella (spp.)	Vegetables	3–40+
	Soil	15–280+
	Pasture	200+
	Grass	100+
Salmonella typhi	Vegetables	10–53
	Lettuce	18–21
	Soil	2–120
	Water	87–104
Shigella (spp.)	Vegetables	7
	Grass	42
Shigella sonnel	Tomatoes	2–10
Streptococcus faecalis	Soil	26–77
Vibrio cholerae	Vegetables	5–14
Vibrio comma	Water	32
Poliovirus	Water	20

Source: From Crook, James, 1985, Water reuse in California, *Journal American Water Works Assoc.*, Vol. 77, no. 7. Copyright AWWA. Reprinted with permission.

Table 10J.258 Movement of Bacteria in Soil in Relation to Groundwater Velocity

Pollution	Organism	Length of Travel (m)	Medium	Velocity (m/d)
River water	*E. coli*	0.30	Dune sand	0.11
Sewage	Bacteria	0.31	Dense soil	0.01[a]
Sewage	Bacteria	0.61	Porous soil	0.10[a]
Sewage	*Coliforms*	0.91	Coarse sand	0.08
Sewage	Bacteria	1.50	Fine sand	0.03[a]
Sewage	*E. coli*	2.00	Fine sand	0.128
River water	*E. coli*	3.00	Sand	0.004
Sewage	*B. coli*	3.10	Fine sand	0.50
Sewage	Bacteria	4.50	Fine sand	0.03[a]
Wastewater	Coliforms	6.10	Fine sand	0.03[a]
River water	Bacteria	7.50	Sand	0.08
Sewage	*B. coli*	10.70	Sand gravel	1.50
Sewage	*C. welchii*	15.20	Fine sand	0.50
Polluted water	Bacteria	18.30	Sand gravel	0.53[a]
Sewage	*B. coli*	19.80	Fine sand	0.11
Sewage	*B. coli*	24.40	Sand gravel	4.10
Pure culture	*B. stearothermophilis*	28.70	Crystalline rock	25.50
Sewage	Coliforms	30.50	Sand	21.80
Sewage	Bacteria	61.00	Sand gravel	30.50
Sewage[b]	*S. typhi*	64.00	Gravel loam	42.70
Sewage	Bacteria	70.70	Fine sand	0.11
Sewage	Coliforms	91.00	Sand gravel	0.24
Sewage	Coliforms	457.20	Coarse gravel	30.50
Sewage	Coliforms	830.00	Sand gravel	1.60
Polluted water	Bacteria	850.00	Coarse gravel	5.50[a]
Polluted water	Bacteria	1,000.00	Fractured limestone	5.50[a]

[a] Estiamted from soil description, using a hydraulic gradient of 0.01. Other groundwater velocity values were measured by other authors.
[b] Yakima outbreak.

Source: From McGinnis, J.A. and DeWalle, F.B., 1983, The movement of typhoid organisms in saturated permeable soil, *J. Am. Water Works Assoc.*, Vol. 75, no. 6. Copyright AWWA. Reprinted with permission.

Table 10J.259 Migration of Viruses Beneath Land Treatment Sites

Site Location	Site Type	Maximum Distance of Virus Migration (m)	
		Depth	Horizontal
St. Petersburg, FL	S	6.0	—
Gainesville, FL	S	3.0	7.0
Lubbock, TX	S	30.5	—
Kerrville, TX	S	1.4	—
Muskegon, MI	S	10.0	—
San Angelo, TX	S	27.5	—
East Meadow, NY	R	11.4	3.0
Holbrook, NY	R	6.1	45.7
Sayville, NY	R	2.4	3.0
12 Pines, NY	R	6.4	—
North Massapequa, NY	R	9.1	—
Babylon, NY	R	22.8	408.0
Ft. Devons, MA	R	28.9	183.0
Vineland, NJ	R	16.8	250.0
Lake George, NY	R	45.7	400.0
Phoenix, AZ	R	18.3	3.0
Dan Region, Israel	R	31–67	60–270

Note: R, Rapid infiltration; S, Slow-rate infiltration.

Source: Adapted from Gerba, C.P. and Goyal, S.M., 1985, Pathogen removal from wastewater during groundwater recharge, in Asano, T., *Artificial Recharge of Groundwater*, Copyright Butterworth Publ, Boston. Reprinted with permission.

Table 10J.260 Selected Human Health and Environmental Effects from Toxic Chemicals

Chemical	Human Health Effects[a]			Environmental Effects
	Carcinogen[b]	Teratogen[b]	Others	
Aldrin/dieldrin	○		Tremors, convulsions, kidney damage	Toxic to aquatic organisms, reproductive failure in birds and fish, bioaccumulation in aquatic organisms
Arsenic	●	○	Vomiting, poisoning, liver and kidney damage	Toxic to legume crops
Benzene	●		Anemia, bone marrow damage	Toxic to some fish and aquatic invertebrates
Cadmium		○	Suspected causal factor in many human pathogies: tumors, renal dysfunction, hypertension, arteriosclerosis, Itai-itai disease (weakened bones), emphysema	Toxic to fish, bioaccumulates significantly in bivalve mollusks
Carbon tetrachloride	●		Kidney and liver damage, heart failure	
Chromium			Kidney and gastrointestinal damage, respiratory complications	Toxic to some aquatic organisms
Copper			Gastrointestinal irritant, liver damage	Toxic to juvenile fish & other aquatic organisms
DDT	○	○ (minimal)	Tremors, convulsions, kidney damage	Reproductive failure of birds and fish, bioaccumulates in aquatic organisms, biomagnifies in food chain
Di-*n*-butyl phthalate			Central nervous system damage	Eggshell thinning in birds, toxic to some fish
Dioxin	○	○	Acute skin rashes, systemic damage, mortality	Bioaccumulates, lethal to aquatic organisms birds and mammals
Lead		○	Convulsions, anemia, kidney and brain damage	Toxic to domestic plants and animals, bio-magnifies to some degree in food chain

(Continued)

Table 10J.260 (Continued)

Chemical	Human Health Effects[a]			Environmental Effects
	Carcinogen[b]	Teratogen[b]	Others	
Methyl Mercury		○	Irritability, depression, kidney and liver damage, Minamata disease	Reproductive failure in fish species, inhibits growth and kills fish; biomagnifies
PCBs	○	○	Vomiting, abdominal pain, temporary blindness, liver damage	Liver damage in mammals, kidney damage and eggshell thinning in birds, suspected reproductive failure in fish
Phenols			Effects on central nervous system, death at high doses	Reproductive effective in aquatic organisms toxic to fish
Toxaphene	○	○	Pathological changes in kidney & liver; changes in blood chemistry	Decreased productivity of phytoplankton communities, birth defects in fish and birds, toxic to fish and invertebrates

[a] In many cases human health effects are based upon the results of animal tests.
[b] If a substance is identified as a carcinogen, there is evidence that it has the potential for causing cancer in humans; if it is identified as a teratogen, it has the potential for causing birth defects in humans.

Source: From U.S. Environmental Protection Agency; National Water Quality Inventory, 1984 Report to Congress and The Conservation Foundation, State of the Environment 1982.

Table 10J.261 Chemicals Causing Distribution System Outbreaks in the United States — 1971–98

Chemical	CWS[a]	NCWS[b]
Copper	16	2
Chlordane	3	
Nitrite	2	
Unidentified herbicide	2	
Ethylene glycol	2	
Oil	1	1
Other chemicals[c]	9	
Total	35	3

[a] CWS — community water system.
[b] NCWS — noncommunity water system.
[c] Each of the following was responsible for one outbreak; fluoride, lead, chromium, sodium hydroxide, chlorine, liquid soap, ethyl acrylate, morpholine, and hydroquinone.

Source: From Reprinted from *Journal AWWA*, vol. 93, no. 9 (September 2001), by permission.

Table 10J.262 Potential Exposure to Lead in Tap Water

Age of House	pH of Water[a]	Percent of Samples >20 ug/L First-Flush (%)	Percent of Samples >20 ug/L Fully-Flushed (2 min) (%)
0–2 yrs	<6.4	93	51
	7.0–7.4	83	5
	≥:8.0	72	0
2–5 yrs	≤6.4	84	19
	7.0–7.4	28	7
	≥8.0	18	4
6+yrs	≤6.4	51	4
	7.0–7.4	14	0
	≥8.0	13	3

Note: Percentage of samples taken from kitchen taps exceeding 20 ug/L of lead at different pH levels, by age of house. The United States Environmental Protection Agency (EPA) sets drinking water standards and has determined that lead is a health concern at certain levels of exposure. There is currently an action level of 15 parts per billion (μg/L). The most common cause of lead entering drinking water is corrosion, a reaction between the water and the lead pipes or the lead-based solder. When water stands in the pipes of a residence for several hours without use, there is a potential for lead to leach, or dissolve, into the water if a lead source is present. Soft water (water that makes soap suds easily) can be more corrosive and, therefore, has higher levels of dissolved lead. Some home water treatment devices may also make water more corrosive.

[a] A measure of the concentration of hydrogenions and potential electrochemical corrosion.

Source: From U.S. EPA, 1987, Preliminary Results from "Lead Solder Aging Study" More recent data are available at: U.S. EPA, 2005, Is there Lead in my Drinking Water? www.epa.gov/safewater/lead.

Table 10J.263 Geometric Mean and Selected Percentiles of Lead in Blood in the United States. Population Aged 1 Yr and Older, National Health, and Nutrition Examination Survey, 1999–2002

	Survey Yrs	Geometric Mean (95% Conf. Interval)	Selected Percentiles (95% Confidence Interval) 50th	75th	90th	95th	Sample Size
Total, age 1 and older	99–00	1.66 (1.66–1.72)	1.60 (1.50–1.60)	2.40 (2.30–2.60)	3.80 (3.60–3.90)	4.90 (4.60–5.30)	7,970
	01–02	1.45 (1.39–1.51)	1.40 (1.30–1.40)	2.20 (2.10–2.20)	3.40 (3.10–3.50)	4.40 (4.20–4.70)	8,945
Age group							
1–5 yrs	99–00	2.23 (1.96–2.53)	2.20 (1.90–2.50)	3.30 (2.80–3.80)	4.80 (4.00–6.60)	7.00 (6.10–8.30)	723
	01–02	1.70 (1.55–1.87)	1.50 (1.40–1.70)	2.50 (2.20–2.80)	4.10 (3.40–5.00)	5.80 (4.70–6.90)	898
6–11 yrs	99–00	1.51 (1.36–1.66)	1.30 (1.20–1.50)	2.00 (1.70–2.40)	3.30 (2.70–3.60)	4.50 (3.40–6.20)	905
	01–02	1.25 (1.14–1.36)	1.10 (1.00–1.30)	1.60 (1.50–1.80)	2.70 (2.40–3.00)	3.70 (3.00–4.70)	1,044
12–19 yrs	99–00	1.10 (1.04–1.17)	1.00 (0.900–1.10)	1.40 (1.30–1.60)	2.30 (2.10–2.30)	2.80 (2.60–3.00)	2,135
	01–02	0.942 (0.899–0.986)	0.800 (0.800–0.900)	1.20 (1.20–1.30)	1.90 (1.80–2.00)	2.70 (2.30–2.90)	2,231
20 yrs and older	99–00	1.75 (1.68–1.81)	1.70 (1.60–1.70)	2.50 (2.50–2.60)	3.90 (3.70–4.00)	5.20 (4.80–5.50)	4,207
	01–02	1.56 (1.49–1.62)	1.60 (1.50–1.60)	2.20 (2.20–2.30)	3.60 (3.30–3.70)	4.60 (4.20–4.90)	4,772
Gender							
Males	99–00	2.01 (1.93–2.09)	1.80 (1.80–1.90)	2.90 (2.80–3.00)	4.40 (4.10–4.80)	6.00 (5.40–6.40)	3,913
	01–02	1.78 (1.71–1.86)	1.70 (1.70–1.80)	2.70 (2.50–2.80)	3.90 (3.70–4.10)	5.30 (5.00–5.50)	4,339
Females	99–00	1.37 (1.32–1.43)	1.30 (1.20–1.30)	1.90 (1.90–2.10)	3.00 (2.90–3.20)	4.00 (3.70–4.20)	4,057
	01–02	1.19 (1.14–1.25)	1.10 (1.10–1.20)	1.80 (1.70–1.80)	2.60 (2.40–2.70)	3.60 (3.00–3.80)	4,606
Race/ethnicity							
Mexican Americans	99–00	1.83 (1.75–1.91)	1.80 (1.60–1.80)	2.70 (2.60–2.90)	4.20 (3.90–4.50)	5.80 (5.10–6.60)	2,742
	01–02	1.46 (1.34–1.60)	1.50 (1.30–1.60)	2.20 (2.00–2.60)	3.60 (3.30–4.00)	5.40 (4.40–6.60)	2,268
Nonhispanic blacks	99–00	1.87 (1.75–2.00)	1.70 (1.60–1.90)	2.80 (2.50–2.90)	4.20 (4.00–4.60)	5.70 (5.20–6.10)	1,842
	01–02	1.65 (1.52–1.80)	1.60 (1.40–1.70)	2.50 (2.30–2.80)	4.20 (3.80–4.60)	5.70 (5.30–6.50)	2,219
Nonhispanic whites	99–00	1.62 (1.55–1.69)	1.60 (1.50–1.60)	2.40 (2.30–2.40)	3.60 (3.40–3.70)	5.00 (4.40–5.70)	2,716
	01–02	1.43 (1.37–1.40)	1.40 (1.30–1.40)	2.10 (2.10–2.20)	3.10 (3.00–3.40)	4.10 (3.90–4.50)	3,806

Source: From Department of Health and Human Services, Centers of Disease Control and Prevention, 2005, Third National Report on Human Exposure to Environmental Chemicals, NCEH Pub. No. 05-0570, July 2005, www.cdc.gov.

Table 10J.264 Pesticide Residues In Human Adipose Tissue in the United States

Pesticide Residues	1970	1971	1972	1973	1974	1975	1976	1977	1978	1979	1980	1981	1983
Sample size	1,386	1,560	1,886	1,092	900	779	682	789	827	796	98	384	407
DDT	7.95	8.06	6.97	5.96	5.15	4.76	4.35	3.14	3.52	3.10	2.82	2.24	1.67
Dieldrin	0.16	0.22	0.18	0.17	0.14	0.12	0.09	0.09	0.09	0.08	0.10	0.05	0.06
Oxchlordane	(NA)	(NA)	0.10	0.12	0.12	0.11	0.11	0.10	0.11	0.10	0.12	0.09	0.09
Heptachlor Epoxide	0.09	0.09	0.07	0.09	0.08	0.08	0.08	0.07	0.07	0.07	0.08	0.09	0.09
Trans-Nonachlor	(NA)	(NA)	(NA)	(NA)	(NA)	0.06	0.13	0.10	0.12	0.12	0.14	0.11	0.12
Beta-Benzene Hexachloride	0.37	0.35	0.19	0.25	0.21	0.19	0.18	0.14	0.14	0.15	0.12	0.09	0.10
Hexachlorobenzene	(NA)	(NA)	(NA)	(NA)	0.03	0.04	0.04	0.04	0.04	0.04	0.04	0.04	0.03

Note: Concentration levels, 1970–1983 in parts per million (ppm), Data represent geometric means and are based on a sample of measurements of pesticide residues and associated chemicals found in human tissue collected by medical pathologists from selected cities in the conterminous 48 states as part of the National Human Adipose Tissue Monitoring Program. NA, Not available.

Source: From U.S. Department of Commerce, Statistical Abstract of the United States 1987.

Table 10J.265 Geometric Mean and Selected Percentiles of Blood Serum Concentrations for Organochlorine Pesticides in the United States Population Aged 12 Yrs and Older, National Health and Nutrition Examination Survey, 2001–2002

		Selected Percentiles (95% Confidence Interval)				
	Geometric Mean	50th	75th	90th	95th	Sample Size
Hexachlorobenzene						
Lipid adjusted (ng/g of lipid)	—[a]	<LOD	<LOD	<LOD	<LOD	2,277
Whole weight (ng/g of serum)	—[a]	<LOD	<LOD	<LOD	<LOD	2,277
p,p′-DDT						
Lipid adjusted (ng/g of lipid)	—[a]	<LOD	<LOD	<LOD	26.5 (22.4–32.7)	2,305
Whole weight (ng/g of serum)	—[a]	<LOD	<LOD	<LOD	0.184 (0.161–0.221)	2,305
p,p′-DDE						
Lipid adjusted (ng/g of lipid)	295 (267–327)	250 (227–277)	597 (521–699)	1,400 (1,210–1,500)	2,320 (1,830–2,780)	2,298
Whole weight (ng/g of serum)	1.81 (1.64–2.01)	1.57 (1.37–1.72)	3.97 (3.43–4.59)	8.81 (7.85–10.1)	15.4 (12.9–17.6)	2,298
o,p′-DDT						
Lipid adjusted (ng/g of lipid)	—[a]	<LOD	<LOD	<LOD	<LOD	2,279
Whole weight (ng/g of serum)	—[a]	<LOD	<LOD	<LOD	<LOD	2,279
Oxychlordane						
Lipid adjusted (ng/g of lipid)	11.4 (<LOD–12.5)	11.1 (<LOD–12.5)	21.7 (19.2–24.2)	36.3 (31.4–41.4)	49.7 (42.0–61.2)	2,249
Whole weight (ng/g of serum)	0.070 (0.063–0.077)	0.069 (0.058–0.078)	0.143 (0.126–0.160)	0.248 (0.215–0.297)	0.352 (0.289–0.441)	2,249
trans-Nonachlor						
Lipid adjusted (ng/g of lipid)	17.0 (15.2–18.9)	17.9 (15.5–20.5)	33.7 (30.2–37.2)	56.3 (49.6–65.9)	78.2 (63.6–111)	2,286
Whole weight (ng/g of serum)	0.104 (0.093–0.116)	0.112 (0.097–0.124)	0.217 (0.191–0.243)	0.389 (0.328–0.470)	0.589 (0.432–0.797)	2,286
Hetachlor epoxide						
Lipid adjusted (ng/g of lipid)	—[a]	<LOD	<LOD	14.8 (13.0–17.8)	21.6 (18.1–26.2)	2,259
Whole weight (ng/g of serum)	—[a]	<LOD	<LOD	0.102 (0.089–0.121)	0.153 (0.125–0.179)	2,259
Mirex						
Lipid adjusted (ng/g of lipid)	—[a]	<LOD	<LOD	15.8 (<LOD–73.7)	57.1 (13.2–230)	2,257
Whole weight (ng/g of serum)	—[a]	<LOD	<LOD	0.101 (0.049–0.468)	0.414 (0.080–1.73)	2,257
Aldrin						
Lipid adjusted (ng/g of lipid)	—[a]	<LOD	<LOD	<LOD	<LOD	2,275
Whole weight (ng/g of serum)	—[a]	<LOD	<LOD	<LOD	<LOD	2,275
Dieldrin						
Lipid adjusted (ng/g of lipid)	—[a]	<LOD	<LOD	15.2 (14.3–17.0)	20.3 (18.7–22.4)	2,159
Whole weight (ng/g of serum)	—[a]	<LOD	<LOD	0.109 (0.099–0.121)	0.146 (0.129–0.164)	2,159
Endrin						
Lipid adjusted (ng/g of lipid)	—[a]	<LOD	<LOD	<LOD	5.10 (<LOD–5.20)	2,187
Whole weight (ng/g of serum)	—[a]	<LOD	<LOD	<LOD	0.021 (0.020–0.021)	2,187

Note: <LOD means less than the limit of detection, which may vary for some chemicals by individual sample. Maximum detection limits (ng/g of lipid) for Hexachlorobenzene, *p,p′*-DDT, *o,p′*-DDT, Heptachlor epoxide, Mirex, Aldrin, Dieldrin, Endrin were 31.4, 17.4, 17.4, 10.5, 10.5, 5.94, 10.5, 5.09, respectively.

[a] Not calculated. Proportion of results below limit of detection was to high to provide a valid result.

Source: From Department of Health and Human Services, Centers of Disease Control and Prevention, 2005, Third National Report on Human Exposure to Environmental Chemicals, NCEH Pub. No. 05-0570, July 2005, www.cdc.gov.

Table 10J.266 Selected Synthetic Organic Chemicals Detected in Drinking Water Wells in the United States

Chemical	Cancer Classification
Benzene	A
α-BHC	C
β-BHC	B2
γ-BHC (Lindane)	B2–C
Bis(2-ethylhexyl)phthalate	B2
Bromoform	B2
Butyl benzyl phthalate	C
Carbon tetrachtoride	B2
Chloroform	B2
Chloromethane	C
Cyclohexane	D
Dibromochloropropane (DBCP)	B2
Dibromochloromethane	C
1,1-Dichloroethane	C
1,2-Dichloroethane	B2
1,1-Dichloroethylene	C
1,2-Dichloroethylene	D
Di-n-butyl phthalate	D
Dioxane	B2
Ethylenedibromide(EDB)	B2
Isopropyl benzene	NTA
Methylene chloride	B2
Parathion	C
Tetrachloroethylene	B2–C
Toluene	D
1,1,1-Trichloroethane	D
1,1,2-Trichloroethane	C
Trichloroethylene (TCE)	B2–C
Trifluorotrichloroethane	D
Vinyl chloride	A
Xylene	D

Note: A, Known human carcinogen; B1, Probable human carcinogen — limited evidence in humans; B2, Probable human carcinogen — sufficient evidence in animals and inadequate data in humans; C, Possible human carcinogen — limited evidence in animals; D, Inadequate evidence to classify.

Source: From U.S. Environmental Protection Agency, *Integrated Risk Information System URL:* www.epa.gov/iris.

Table 10J.267 Summary of Toxic Effects of Some Organic and Inorganic Chemicals Known to Occur In Groundwater

Containment	1	2	3	4	5	6	7	8	9	10	11	12	13	14	15	16	17	18	19	20	21	22	23	24
Aromatic Hydrocarbons																								
Alkyl benzene sulfonates	X	X	X											X										
Aniline									X							X								X
Anthracene			X																				X	X
Benzene	X	X	X						X	X	X	X		X										X
Benzidine																								X
Benzyl alcohol		X																						
Chrysene																								X
Dibenz (a.h.) anthracene																								X
4,4-Dinitrosodiphenylamine																								X
Ethylbenzene	X	X		X		X	X	X																
4,4-Methylene-bis-2-chloroaniline (MOCA)	X			X		X	X									X							X	X
Naphthalene	X	X		X							X							X						
o-Nitroaniline																X							X	
Nitrobenzene						X	X		X		X					X			X	X	X			
4-Nitrophenol					X											X								
n-Nitrosodiphenylamine						X																		X
Phenanthrene			X																				X	X
n-Propylbenzene										X														
Pyrene		X			X	X					X			X								X	X	
Styrene (vinyl benzene)	X	X		X		X	X		X					X							X		X	
Toluene	X	X		X		X	X		X				X											
1,2,4-Trimethylbenzene		X			X				X															
Xylenes (*m,o,p*)	X			X		X	X		X				X											
Oxygenated Hydrocarbons																								
Acetic acid	X	X		X	X																			
Acetone	X			X				X																
Benzophenone			X																					
Di-*n*-butyl phthalate	X			X				X																
Diethyl ether	X	X		X				X																
Diethyl phthalate		X		X																				
1,4-Dioxane	X			X		X	X																	X
Formic acid	X	X		X					X															
Methanol	X					X			X									X						
Phenols (e.g., *p*-*tert*-Butyl phenol)		X																						
Tetrahydrofuran		X		X		X	X		X															
Hydrocarbons with Specific Elements																								
Aldicarb (sulfoxide and sulfone; Temik)															X									
Aldrin		X				X	X		X											X				X
Atrazine		X			X				X		X										X		X	
Benzoyl chloride		X		X																				
Bromobenzene	X	X			X	X		X															X	
Bromacil						X			X															
Bromochloromethane					X	X	X		X															

Containment	1	2	3	4	5	6	7	8	9	10	11	12	13	14	15	16	17	18	19	20	21	22	23	24
Bromoform	X			X		X			X															
Carbofuran						X									X									
Carbon tetrachloride						X	X		X											X				X
Chlordane				X		X			X				X											X
Chlorobenzene	X			X	X	X	X		X											X			X	
Chloroform	X	X				X	X		X					X							X			X
Chloromethane					X	X	X		X															
2-Chloronaphthalene		X				X																		
Chlorpyrifos						X									X									
o-Chlorotoluene	X	X							X															
Dibromochloromethane																							X	
Dibromochloropropane (DBCP)	X	X			X	X	X													X			X	X
Dichlofenthion (DCFT)															X									
o-Dichlorobenzene	X	X		X	X	X	X		X		X													
p-Dichlorobenzene	X			X	X	X	X		X		X													
Dichlorobenzidine	X			X	X									X										X
Dichlorodiphenyldichlor-ethane (DDD, TDE)																			X					X
Dichlorodiphenyltrichloro-ethane (DDT)					X	X			X		X									X			X	X
1,1-Dichloroethane						X	X		X												X			
1,2-Dichloroethane	X	X			X	X	X		X				X	X					X	X			X	X
1,1-Dichloroethylene (vinylidiene chloride)	X	X		X	X	X	X		X				X						X		X		X	X
Dichloromethane (methylene chloride)	X	X				X	X		X												X			
2,4-Dichlorophenoxyacetic acid (2,4-D)									X	X												X	X	
1,2-Dichloropropane						X	X																X	
Dieldrin						X			X											X				X
Dimethyl disulfide																						X		
2,4-Dinotrophenol (Dinoseb, DNBP)	X	X					X		X											X				
Dioxins (e.g.,TCDD)						X			X	X							X			X		X	X	X
Dodecyl mercaptan (lauryl mercaptan)	X	X	X						X														X	
Endosulfan						X	X		X												X		X	
Endrin						X			X															
Bis-2-ethylhexylphthalate	X	X				X														X	X	X		X
Fluoroform																							X	
Hexachlorocyclo-pentadiene	X			X	X	X	X						X						X					
Hexachloroethane						X	X		X												X			X
Kepone						X	X													X	X			X
Methoxychlor							X													X				
Parathion															X					X	X			
Pentachlorophenol (PCP)	X	X		X		X			X												X			
Phorate (Disulfoton)															X									

Polybrominated biphenyls (PBBs)						X						X									X	X		
Polychlorinated biphenyls (PCBs)					X	X						X	X	X						X	X	X	X	X
RDX (Cyclonite)																							**X**	
Simazine									X															
Tetrachloroethane (1,1,1,2)						X															X		X	
Tetrachloroethane (1,1,2,2)				X		X	X		X					X							X	X	X	X
Tetrachloroethylene (1,1,2,2) (perchloroethylene, PCE)	X			X	X		X		X												X			X
Toxaphene						X	X		X												X		X	X
Trichloroethanes (1,1,1 & 1,1,2)	X	X		X	X	X	X		X				X										X	X
1,1,2-Trichloroethylene (TCE)		X			X	X	X		X		X		X											X
Trichlorofluoromethane (Freon 11)									X				X											
2,4,6-Trichlorophenol						X																	X	X
2,4,5-Trichlorophenoxyacetic acid (2,4,5-T)						X	X		X													X	X	
2,4,5-Trichlorophenoxypropionic acid (2,4,5-TP, or Silvex)						X	X		X															
Vinyl chloride		X				X	X		X		X												X	X
Other Hydrocarbons																								
Alkyl sulfonates	X	X	X																					
Cyclohexane						X	X		X															
Dicyclopentadiene (DCPD)	X	X		X	X		X		X															
Fuel oil		X			X		X		X					X										
Gasoline		X		X					X					X										
Jet fuels						X	X		X	X														
Propane													X											
Metals and Cations																								
Aluminum														X										
Antimony		X			X	X	X				X		X											
Arsenic					X					X			X				X			X		X	X	X
Barium		X		X	X		X			X									X					
Beryllium	X	X	X	X	X	X																		X
Cadmium				X	X		X		X											X	X	X		X
Chromium		X	X	X	X	X															X	X	X	X
Cobalt	X	X	X		X						X		X	X										
Copper						X	X					X												
Iron					X	X								X										
Lead									X		X									X				
Lithium							X		X				X	X										
Magnesium	X								X															
Manganese					X	X			X															
Mercury							X		X									X				X		
Nickel	X		X	X	X				X					X						X				X
Palladium							X				X													
Selenium	X			X	X	X			X					X										

Containment	1	2	3	4	5	6	7	8	9	10	11	12	13	14	15	16	17	18	19	20	21	22	23	24
Silver					X		X				X			X			X	X						
Thallium						X	X		X				X	X						X		X		
Vanadium			X		X	X	X		X		X													
Zinc	X	X									X								X	X				
Nonmetals and Anions																								
Ammonia	X	X		X	X	X	X		X															
Cyanides	X	X	X	X					X															
Fluorides							X		X															

Numerical key of toxic effects: (1) Eye irritation, (2) Skin irritation, (3) Allergic sensitization, (4) Upper respiratory tract irritation, (5) Lung/respiratory effects, (6) Liver damage, (7) Kidney damage, (8) Pancreatic damage, (9) central nervous system, (CNS) effects, (10) Peripheral nervous system effects, (11) Blood cell disorders, (12) Immunological effects, (13) Cardiovascular effects, (14) Gastrointestinal effects, (15) Cholinesterase inhibition, (16) Methemoglobinemia, (17) Skin damage, (18) Visual damage, (19) Endocrine effects, (20) Reproductive effects, (21) Embryotoxicity, (22) Teratogenicity, (23) Mutagenicity, (24) Carcinogenicity.

Source: From Office of Technology Assessment. Compiled from a partial survey of literature conducted by Environ Corp., 1983.

Table 10J.268 Estimated Drinking Water Concentration Corresponding to a 1×10^{-6} Excess Lifetime Cancer Risk for Selected Known, Probable, or Possible Carcinogens for a Lifetime Consumption of Water

Chemical	Drinking Water Concentration (μg/L)	Sex, Species	Tumor Type
Benzene	1–10	Human	Leukemia
α-BHC	0.006	M mouse	Hepatocellular carcinoma
β-BHC	0.02	M mouse	Hepatocellular carcinoma
Carbon tetrachloride	0.3	Various	Hepatocellular carcinoma
1,2-Dichloroethane	0.4	M rat	Hemangiosarcoma of circulatory system
1,4-Dioxane[a]	3	M rat	Nasal squamous cell carcinoma
Ethylene dibromide (EDB)	0.02	Mrat	Forestomach tumors, hemangiosarcomas, thyroid follicular cell adenomas or carcinomas
DDT	0.1	Mouse/rat	Benign and malignant liver tumors
Heptachlor	0.008	Mouse	Hepatocellular carcinomas
1,1,2,2-Tetrachloroethane	0.2	Mouse	Hepatocellular carcinoma
1,1,2-Trichloroethane	0.5	Mouse	Hepatocellular carcinoma
Vinyl chloride — continuous lifetime exposure from birth	0.024	F rat	Total of liver angiosarcoma, hepatocellular carcinoma, and neoplastic nodules
Vinyl chloride — continuous lifetime exposure during adulthood	0.048	F rat	Total of liver angiosarcoma, hepatocellular carcinoma, and neoplastic nodules

[a] Toxicity values under review; revised assessment due in 2006 or 2007.

Source: From U.S. Environmental Protection Agency, *Integrated Risk Information System (IRIS)*, 2005, www.epa.gov/iris.

CHAPTER 11

Water Resources Management

Gustavõ Suarez

CONTENTS

SECTION 11A DAMS

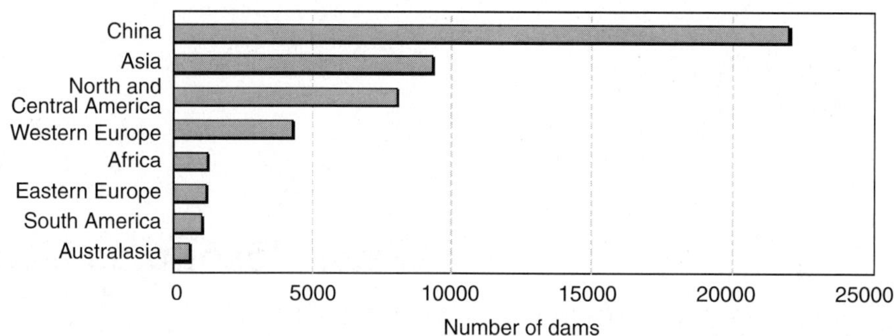

Figure 11A.1 Regional distribution of large dams in 2000. (From WCD compilation of various sources and ICOLD, 1998. damsreport.org. Annex V, Dams, Water and Energy: A Statistical Profile.)

Table 11A.1 World's Largest Dams

Dam	Location	Volume (Thousands)		Year Completed
		m^3	yds^3	
Syncrude Tailings	Canada	540,000	706,320	UC
Chapetón	Argentina	296,200	387,410	UC
Pati	Argentina	238,180	274,026	UC
New Cornelia Tailings	United States	209,500	274,026	1973
Tarbela	Pakistan	121,720	159,210	1976
Kambaratinsk	Kyrgyzstan	112,200	146,758	UC
Fort Peck	Montana	96,049	125,628	1940
Lower Usuma	Nigeria	93,000	121,644	1990
Cipasang	Indonesia	90,000	117,720	UC
Atatürk	Turkey	84,500	110,522	1990
Yacyretá-Apipe	Paraguay/Argentina	18,000	105,944	1998
Guri (Raul Leoni)	Venezuela	78,000	102,014	1986
Rogun	Tajikistan	75,500	98,750	1985
Oahe	South Dakota	70,339	92,000	1963
Mangla	Pakistan	65,651	85,872	1967
Gardiner	Canada	65,440	85,592	1968
Afsluitdijk	Netherlands	63,400	82,927	1932
Oroville	California	59,639	78,008	1968
San Luis	California	59,405	77,700	1967
Nurek	Tajikistan	58,000	75,861	1980
Garrison	North Dakota	50,843	66,500	1956
Cochiti	New Mexico	48,052	62,850	1975
Tabka (Thawra)	Syria	46,000	60,168	1976
Bennett WAC	Canada	43,733	57,201	1967
Tucuruíi	Brazil	43,000	56,242	1984
Boruca	Costa Rica	43,000	56,242	UC
High Aswan (Sadd-el-Aali)	Egypt	43,000	56,242	1970
San Roque	Philippines	43,000	56,242	UC
Kiev	Ukraine	42,841	56,034	1964
Dantiwada Left Embankment	India	41,040	53,680	1965
Saratov	Russia	40,400	52,843	1967
Mission Tailings 2	Arizona	40,088	52,435	1973
Fort Randall	South Dakota	38,227	50,000	1953
Kanev	Ukraine	37,860	49,520	1976
Mosul	Iraq	36,000	47,086	1982
Kakhovka	Ukraine	35,640	46,617	1055

(Continued)

Table 11A.1 (Continued)

Dam	Location	Volume (Thousands) m^3	Volume (Thousands) yds^3	Year Completed
Itumbiara	Brazil	35,600	46,563	1980
Lauwerszee	Netherlands	35,575	46,532	1969
Beas	India	35,418	46,325	1974
Oosterschelde	Netherlands	35,000	45,778	1986

Note: UC, under construction in 2004. China's Three Gorges dam, on the Yangtze River, begun in 1993 and expected to be completed in 2009, will be the world's largest and highest dam.

Source: From Department of the Interior, Bureau of Reclamation and International Water Power and Dam Construction.

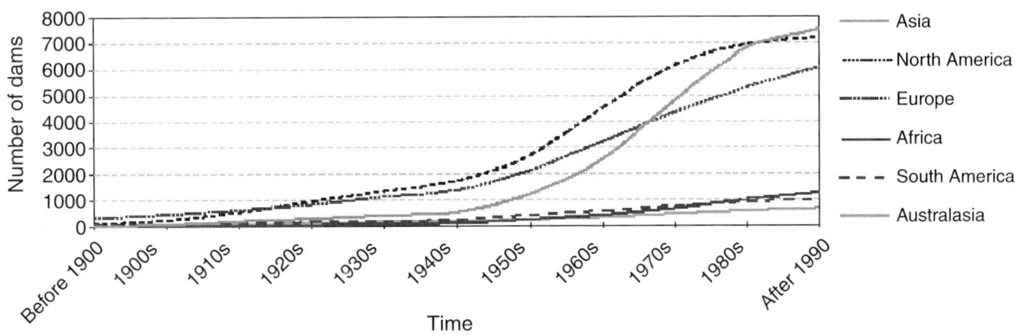

Figure 11A.2 Cumulative commissioning of large dams in the 20th century. (From ICOLD 1998, excluding over 90% of large dams in China. damsreport.org. Annex V, Dams, Water and Energy: A Statistical Profile.)

Table 11A.2 Number of Dams in the World in 1950 and 1982 by Continent

Continent	1950 Dams	1950 %	1982 Dams	1982 %
Europe	1,292	25	3,800	11
Asia	1,541	30	22,701	65
America	2,090	40	7,241	21
Africa	123	2	610	2
Australasia	150	3	446	1
Total	5,196	100	34,798	100
China	8	0.2	18,595	53

Source: From International Commission on Large Dams, 1984, *Register of Dams; percentages rounded.*

Table 11A.3 Dams in 140 Countries

Regions and Countries	Number of Dams	Regions and Countries	Number of Dams	Regions and Countries	Number of Dams
Africa		Finland	55	Nicaragua	4
South Africa	539	Cyprus	52	Trinidad & Tobago	4
Zimbabwe	213	Greece	46	Jamaica	2
Algeria	107	Iceland	20	Antigua	1
Morocco	92	Ireland	16	Haiti	1
Tunisia	72	Belgium	15	**Total**	**8,010**
Nigeria	45	Denmark	10	**Asia**	
Côte d'Ivoire	22	Netherlands	10	China	22,000
Angola	15	Luxembourg	3	India	4,291
Dem. Rep. of Congo	14	**Total**	**4,277**	Japan	2,675
Kenya	14	**South America**		South Korea	765
Namibia	13	Brazil	594	Turkey	625
Libya	12	Argentina	101	Thailand	204
Madagascar	10	Chile	88	Indonesia	96
Cameroon	9	Venezuela	74	Russia	91
Mauritius	9	Colombia	49	Pakistan	71
Burkina Faso	8	Peru	43	North Korea	70
Ethiopia	8	Ecuador	11	Iran	66
Mozambique	8	Bolivia	6	Malaysia	59
Lesotho	7	Uruguay	6	Taipei China	51
Egypt	6	Paraguay	4	Sri Lanka	46
Swaziland	6	Guyana	2	Syria	41
Ghana	5	Suriname	1	Saudi Arabia	38
Sudan	4	**Total**	**979**	Azerbaijan	17
Zambia	4	**Eastern Europe**		Armenia	16
Botswana	3	Albania	306	Philippines	15
Malawi	3	Romania	246	Georgia	14
Benin	2	Bulgaria	180	Uzbekistan	14
Congo	2	Czech Republic	118	Iraq	13
Guinea	2	Poland	69	Kazakstan	12
Mali	2	Yugoslavia	69	Kyrgyzstan	11
Senegal	2	Slovakia	50	Tajikistan	7
Seychelles	2	Slovenia	30	Jordan	5
Sierra Leone	2	Croatia	29	Lebanon	5
Tanzania	2	Bosnia Herzegovina	25	Myanmar	5
Togo	2	Ukraine	21	Nepal	3
Gabon	1	Lithuania	20	Vietnam	3
Liberia	1	Macedonia	18	Singapore	3
Uganda	1	Hungary	15	Afghanistan	2
Total	**1,209**	Latvia	5	Brunei	2
Western Europe		Moldova	2	Cambodia	2
Spain	1,196	**Total**	**1,203**	Bangladesh	1
France	569	**North and Central America**		Laos	1
Italy	524	United States	6,575	**Total**	**31,340**
United Kingdom	517	Canada	793	**Austral-Asia**	
Norway	335	Mexico	537	Australia	436
Germany	311	Cuba	49	New Zealand	86
Sweden	190	Dominican Republic	11	Papua New Guinea	3
Switzerland	156	Costa Rica	9	Fiji	2
Austria	149	Honduras	9	**Total**	**577**
Portugal	103	Panama	6		
		El Salvadaor	5		
		Guatemala	4		

Source: Based on ICOLD, 1998, ID, 1000 and other sources. damsreport.org. Annex V, Dams, Water and Energy: A Statistical Profile. With permission.

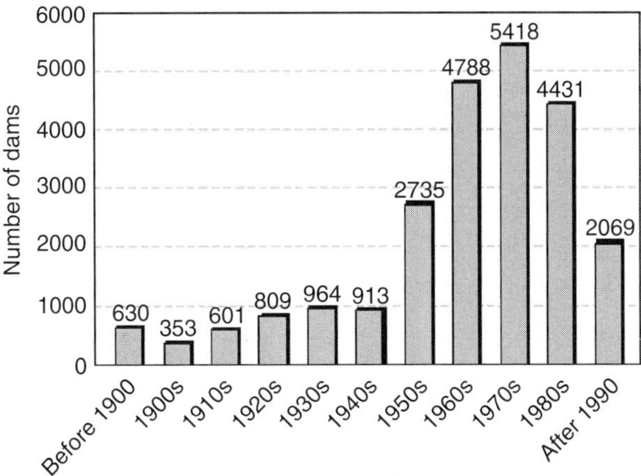

Figure 11A.3 Commissioning of large dams globally by decade in the 20th century. (From ICOLD, 1998, excluding over 90% of large dams in China. damsreport.org. Annex V, Dams, Water and Energy: A Statistical Profile.)

Table 11A.4 Countries with More Than 100 Dams in 1982

China	18,595	(16,500)	Brazil	489	(443)	Germany	184	(158)
U.S.A.	5,338	(5,046)	Mexico	487	(433)	Czechoslovakia	142	(131)
Japan	2,142	(2,006)	France	432	(388)	Sweden	134	(132)
India	1,085	(999)	Italy	408	(403)	Switzerland	130	(128)
Spain	690	(630)	Australia	374	(320)	Yugoslavia	114	(93)
Korea	628	(558)	South Africa	342	(316)	Austria	112	(97)
Canada	580	(494)	Norway	219	(205)	Bulgaria	108	(106)
Great Britain	529	(519)				Romania	106	(81)
	29,587			2,751			1,030	

Note: Numbers in brackets represent number of dams in service at the end of 1977.

Source: From International Commission on Large Dams, 1984, *Register of Dams.*

Table 11A.5 Rate of Dam Construction in the World, 1950–1982

	Outside of China			China			
Period	Number per Period	Total Number	Number per Annum	Number per Period	Total Number	Number per Annum	Ratio China Compared to Outside China
Up to 1950	5,188	5,188			8		
1951 to 1974	8,948	14,136	373				
1975 to 1977	752	14,888	251				
1951 to 1977	9,700	14,888	359	16,492	16,500	611	1.70
1978 to 1982	1,315	16,203	263	2,095	18,595	419	1.60
1975 to 1982	2,067	16,203	258				
1951 to 1982	11,015	16,203	344	18,587	18,595	581	1.69

Source: From International Commission on Large Dams, 1984, *World Register of Dams.*

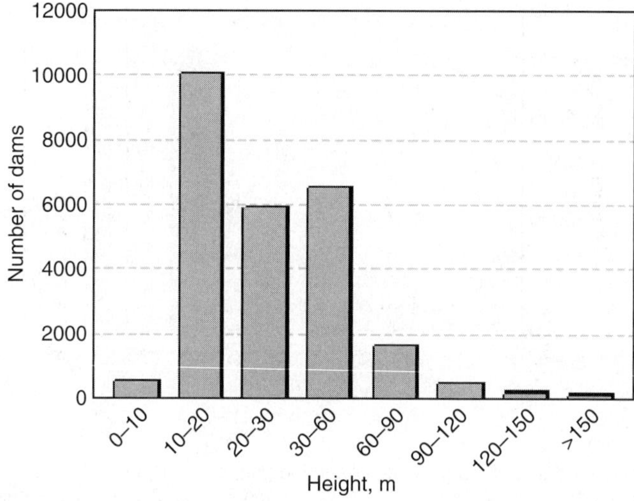

Figure 11A.4 Global distribution of dam heights (m). (From ICOLD, 1998. damsreport.org. Annex V, Dams, Water and Energy: A Statistical Profile.)

Table 11A.6 Classification of Dams in the World by Type and Height

	Number of Dams/Height						
	Total	**15–30 m**	**30–60 m**	**60–100 m**	**100–150 m**	**150–200 m**	**>200 m**
Earth (TE) and Rockfill (ER)	28,844	24,567	3,657	477	116	21	6
Gravity (PG)	3,954	2,222	1,294	361	65	8	4
Arch (VA)	1,527	775	428	204	83	24	13
Buttress (CB)	337	175	110	40	12	—	—
Multi-Arch (MV)	136	74	48	13	—	—	1
Total	34,798	27,813	5,537	1,095	276	53	24

Source: From International Commission on Large Dams, 1984, *World Register of Dams.*

Table 11A.7 World's Highest Dams

No	Height above Lowest Foundation (m)	Type[a]	Name	Country	Year
	335	TE/ER	Rogun	U.S.S.R.	U/C
1	300	TE	Nurek	U.S.S.R.	1980
2	285	PG	Grande Dixence	Switzerland	1961
3	272	VA	Inguri	U.S.S.R.	1980
4	262	VA	Vajont	Italy	1961
5	261	ER	Manuel Moreno Torres (Chicoasén)	Mexico	1980
	261	ER/TE	Tehri	India	U/C
	253	ER/TE	Kishau	India	U/C
	245	VA/PG	Sayano-Shushensk	U.S.S.R.	U/C
	243	ER	Guavio	Colombia	U/C
6	242	TE	Mica	Canada	1972
7	237	ER	Chivor	Colombia	1975
8	237	VA	Mauvoisin	Switzerland	1957
	234	VA	El Cajón	Honduras	U/C (1984)
9	233	VA	Chirkey	U.S.S.R.	1978
10	230	TE	Oroville	U.S.A.	1968
11	226	PG	Bhakra	India	1963
12	221	VA/PG	Hoover	U.S.A.	1936

(Continued)

Table 11A.7 (Continued)

No	Height above Lowest Foundation (m)	Type[a]	Name	Country	Year
13	220	VA	Contra	Switzerland	1965
14	220	VA	Mratinje	Yugoslavia	1976
15	219	PG	Dworshak	U.S.A.	1973
16	216	VA	Glen Canyon	U.S.A.	1966
17	215	PG	Toktogul	U.S.S.R.	1978
18	214	MV	Daniel Johnson	Canada	1968
19	213	VA	Dez	Iran	1962
	210	ER	San Roque	Philippines	U/C
20	208	VA	Luzzone	Switzerland	1963
21	207	ER/PG	Keban	Turkey	1974
22	202	VA	Almendra	Spain	1970
	201	VA	Khudoni	U.S.S.R.	U/C
23	200	VA	Karoun	Iran	1975
24	200	VA	Kolnbrein	Austria	1977
25	196	PG/ER/TE	Itaipu	Brazil	1982
	195	ER	Altinkaya	Turkey	U/C
26	194	VA	New Bullard's Bar	U.S.A.	1970
	192	PG	Lakhwar	India	U/C
27	191	ER	New Melones	U.S.A.	1979
28	186	VA	Kurobe	Japan	1964
29	186	TE	Swift	U.S.A.	1958
	186	VA	Zillergründl	Austria	U/C
30	185	VA	Mossyrock	U.S.A.	1968
	185	VA	Oymapinar	Turkey	U/C
	184	ER	Atatürk	Turkey	U/C
31	183	PG	Shasta	U.S.A.	1945
32	183	TE	WAC Bennett	Canada	1967
33	180	VA	Amir Kabir	Iran	1964
34	180	ER	Dartmouth	Australia	1979
35	180	VA	Emmosson	Switzerland	1974
	180	VA	Tehchi	Taiwan	1974
36	180	VA	Tignes	France	1952
37	176	ER	Takase	Japan	1978
38	175	ER	Ayvacik	Turkey	1981
39	174	PG	Alpe Gera	Italy	1964
40	173	TE	Don Pedro	U.S.A.	1971
	173	VA	Karakaya	Turkey	U/C
41	172	VA	Hungry Horse	U.S.A.	1953
	172	PG	Longyangxia	China	U/C
	171	VA	Cabora Bassa	Mozambique	1974
42	169	VA	Idukki	India	1974
43	168	ER	Charyak	U.S.S.R.	1977
	168	ER	Gura Apelor	Romania	U/C
44	168	ER	La Grande 2	Canada	1978
45	168	PG	Grand Coulée	U.S.A.	1942
46	167	ER	Fierze	Albania	1978
	167	VA/PG	Daniel Palacios	Ecuador	U/C
47	166	VA	Vidraru	Romania	1965
48	165	TE	Kremasta	Greece	1965
49	165	VA	Ross	U.S.A.	1949
50	165	PG	Wujiangdu	China	1981
	164	ER	Thomson	Australia	U/C
51	164	TE	Trinity	U.S.A.	1962
	162	PG/ER	Guri	Venezuela	U/C
52	162	ER	Talbingo	Australia	1971
53	100	ER	Foz de Areia	Brazil	1980
	160	TE/ER	Grand-Maison	France	U/C
	160	ER	Salvajina	Columbia	U/C
	160	ER/TE	Thein Dam Ranjit	India	U/C
54	160	VA	Yellowtail	U.S.A.	1966

(Continued)

Table 11A.7 (Continued)

No	Height above Lowest Foundation (m)	Type[a]	Name	Country	Year
	158	ER	Canales	Spain	U/C
	158	TE	Yacambu	Venezuela	U/C
55	158	ER	Cougar	U.S.A.	1964
56	158	ER	Emborcacão	Brazil	1982
57	158	VA	Gökcekaya	Turkey	1972
	158	ER	Naramata	Japan	U/C
	157	VA	Dongjiang	China	U/C
58	157	PG	Okutadami	Japan	1961
59	157	VA	Speccheri	Italy	1957
60	156	PG	Sakuma	Japan	1956
61	156	VA–TE	Zeuzier	Switzerland	1957
62	155	ER	Goescheneralp	Switzerland	1960
63	155	VA	Monteynard	France	1962
64	155	VA	Nagawado	Japan	1969
65	155	VA	Place Moulin	Italy	1965
	155	PG	Sadar Sarovar	India	U/C
66	154	VA/PG	Bhumibol	Thailand	1964
67	154	ER	Tedorigawa	Japan	1979
68	153	VA	Curnera	Switzerland	1967
69	153	VA	Flaming George	U.S.A.	1964
70	153	ER	Gepatsch	Austria	1965
	153	PG/ER	Revelstoke	Canada	U/C
71	153	VA	Santa Giustina	Italy	1950
	151	PG	Dorna	Spain	U/C
	151	ER	Menzelet	Turkey	U/C
72	151	VA	Zervreila	Switzerland	1957
	150	PG	Baishan	China	U/C
73	150	VA	Canelles	Spain	1960
74	150	ER	Finstertal	Austria	1980
	150	ER	Kenyir	Malaysia	U/C
75	150	VA/CB	Roselend	France	1961
76	150	TE	Big Horn	Canada	1972

Note: U/C, under construction.

[a] TE, Earth; ER, Rockfill; PG, Gravity; CB, Buttress; VA, Arch; MV, Multi-Arch.

Source: From International Commission on Large Dams, 1984, *World Register of Dams.*

Table 11A.8 High Dams of the World Arranged by Country

Name of Dam	Year of Completion	River	Nearest City	State Province or Country	Type	Height above Lowest Foundation (m)	Length of Crest (m)	Volume Content of Dam (103 m^3)	Gross Capacity of Reservoir (10^3 m^3)	Purpose	Maximum Discharge Capacity of Spillways (m^3/sec)
Afghanistan											
Kajakaiv	1952	Helmand	Kandahar		ER	98	274	3,230	2,680,000	I	9,300
Albania											
Fierze	1978	Drin	B. Curri	Tropoje	ER	167	400	8,000	2,700,000	H	
Koman	C	Drin	Shkoder	Shkoder	ER	133	275	4,500	450,000	H	
Algeria											
Bou Hanifia	1948	El Hammam	Bou Hanifia	Mascara	ER	99	464	1,530	73,000	IS	5,500
Bou Roumi	C (1984)	Bou Roumi	Bou Medfa	Blida	TE	100	300	4,200	188,000	I	800
Keddara	C (1986)	Boudouaou	Boudouaou	Blida	ER	108	560	4,380	146,000	S	380
Sly	C (1985)	Sly	El Asnam	El Asnam	TE	87	395	3,565	286,000	I	1,700
Argentina											
Las Pirquitas	1961	Del Valle	Piedra Blanca	Catamarca	TE	83	410	2,900	65,000	ICH	1,400
Gral. M. Belgrano Cabra Corral Res.	1973	Juramento	Coronel Moldes	Salta	TE	112	470	8,000	3,100,000	IH	1,500
Las Maderas (B.L)	1974	Las Maderas	El Carmen	Jujuy	TE	98	460	4,500	300,000	IH	30
Futaleufu AmutuiQuimei Res.	1976	Futaleufú	Trevelin	Chubut	TE	130	600	6,000	5,600,000	H	3,000
Los Reyunos	1980	Diamante	25 de Mayo	Mendoza	TE	131	266	3,220	260,000	IH	2,300
Alicura	C (1984)	Limay	S.C.de Bariloche	Rio Negro/ Neuquén	TE	130	880	13,000	3,215,000	H	3,000
Cerro Pelado	C (1985)	Grande	Amboy	Córdoba	TE	104	410	3,700	370,000	HICR	3,300
Australia											
Upper Yarra	1957	Yarra	Melbourne	Victoria	TE/ER	89	610	5,660	207,200	S	2,165
Eucumbane	1958	Eucumbene	Cooma	NSW	TE	116	579	6,735	4,798,400	H	191
South Para	1958	South Para	Adelaide	Sth Aust.	TE	48	284	581	51,190	S	736
Glenbawn	1958	Hunter	Scone	NSW	ER	78	823	7,650	360,000	IC	1,700
Warragamba	1980	Warragamba	Sydney	NSW	PG	137	351	1,233	2,057,000	SH	12,740
Geehi	1966	Geehi	Cooma	NSW	ER	91	265	1,421	21,093	H	1,534
Blowering	1968	Tumut	Tumut	NSW	ER	112	808	8,563	1,628,000	IH	2,350
Corin	1968	Cotter	Canberra	ACT	ER	76	282	1,376	74,970	S	1,190
Talbingo	1971	Tumut	Tumut	NSW	ER	162	701	14,490	921,400	H	4,290
Cethana	1971	Forth	Devonport	Tasmania	ER	110	213	1,376	109,000	H	1,980
Wyangla	1971	Lachlan	Cowra	NSW	ER	85	1,510	3,580	1,220,000	I	14,700
Ord River Lake Argyle (Res.)	1972	Ord	Wyndham	West Aust.	ER	99	341	1,908	5,797,000	I	3,500
Cardinia	1973	Cardinia Creak	Melbourne	Victoria	TE/ER	86	1,542	5,150	288,905	S	13
Copeton	1976	Gwydir	Inverell	NSW	ER	113	1,484	8,333	1,364,000	I	14,800

(Continued)

Table 11A.8 (Continued)

Name of Dam	Year of Completion	River	Nearest City	State Province or Country	Type	Height above Lowest Foundation (m)	Length of Crest (m)	Volume Content of Dam (103 m^3)	Gross Capacity of Reservoir (10^3 m^3)	Purpose	Maximum Discharge Capacity of Spillways (m^3/sec)
Dartmouth	1979	Mitta Mitta	Mitta Mitta	Victoria	ER	180	670	14,100	4,000,000	IHS	2,584
Split-Yard Creek	1980	Pryde Creek	Ipswich	Queensland	ER	76	1,140	3,371	28,700	H	570
Winneke	1980	Sugarloaf Creek	Melbourne	Victoria	ER	89	1,000	4,700	100,000	S	7
Mangrove Creek	1982	Mangrove Creek	Wyong	NSW	ER	80	380	1,340	170,000	S	570
Blue Rock	C (1984)	Tanjil	Moe	Victoria	TE	75	640	1,530	200,000	S	1,018
Thomson	C (1984)	Thomson	Moe	Victoria	ER	164	1,180	13,300	1,175,000	SI	1,040
Lower Pieman	C (1985)	Pieman	Queenstown	Tasmania	ER	122	374	2,950	641,000	H	4,714
Austria											
Gepatch	1965	Faggenbach	Landeck	Tyrol	ER	153	600	7,100	140,000	H	325
Oschaniksee	1972–1978	tr. Fragant	Obervellach	Carinthia	ER	116	530	2,250	33,000	H	6
Kölnbrein	1977	Malta	Oṁnd	Carinthia	VA	200	626	1,580	205,000	H	188
Rotlech	1977	tr. Lech	Reutte	Tyrol	PG/ER	32	128	14/11	1,260	H	200
Finstertal	1980	tr. Naderbach	Ötz	Tyrol	ER	150	652	4,500	60,500	H	23
Zillergründl	C (1986)	Ziller	Mayrhofen	Tyrol	VA	186	506	1,355	90,000	H	215
Brazil											
Euclides Da Cunha	1960–1977	Pardo	S.Jcsé do Rio Pardo	São Paulo	TE	92	312	2,200	13,400	H	2,340
Tres Marias	1960	São Francisco	Tres Marias	Minas Gerais	TE	75	2,700	14,250	19,790,000	CHI	8,700
Furnas	1963	Grande	Passos	Minas Gerais	ER/PG	127	779	9,697	22,950,000	HC	13,000
Estreito	1969	Grande	Pedregulho	M.Gerais/São Paulo	ER/PG	92	715	4,446	1,418,000	H	13,000
Xavantes	1970	Parana-panema	Xavantes	São Paulo	TE/ER	92	500	6,000	8,750,000	H	3,200
Paraitinga	1975	Paraitinga	Paraibuna	São Paulo	TE	105	586	11,051	2,430,000	HC	600
Paraitinga Dike Paraitinga Res.	1975		Paraibuna	São Paulo	TE	80	530	3,391	2,430,000		—
Itauba	1975	Jacui	Julio de Castilhos	Rio Grande do Sul	ER	90	385	3,410	510,000	H	8,130
Paraibuna	1978	Paraibuna	Paraibuna	São Paulo	TE	94	1,285	7,892	2,463,000	HCI	1,500
São Simão	1978	Paranaiba	São Simão	Minas Gerais (Goiás)	TE/ER/PG	120	3,611	27,378	12,540,000	H	24,100
Foz Do Areia	1980	Iguaçu	Bituruna	Paraná	ER	160	850	13,000	6,100,000	H	11,000
Itumbiara	1980	Paranaiba	Itumbiara	Golás/Minas Gerais	TE/PG	106	6,780	38,820	17,030,000	HC	16,200
Salto Santiago (Main)	1980	Iguaççú	Laranjeiras do Sul	Paraná	ER	80	1,400	9,860	6,750,000	H	24,000
Jequitai	1981	Jequital	Jequital	Minas Gerais	TE	80	580	2,300	1,200,000	I	1,850
Emborcação	1982	Paranaiba	Araguari	Minas Gerais/Goiás	ER	158	1,607	25,000	17,600,000	H	7,800

Itaipu	1982	Paraná	Foz do Iguaçu	Brasil/Paraguai	ER/PG/TE	196	7,900	29,200	29,000,000	H	61,400
Tucuruí	C (1983)	Tocantis	Tucuruí	Pará	TE/ER/PG	93	10, 667	64,300	43,000,000	HN	110,000
Itaparica	C (1986)	São Francisco	Petrolándia	Pernambuco	ER	105	4,150	16,530	10,700,000	IH	28,700
Pedra Do Cavalo	C	Paraguassu	Cachoeira de S. Félix	Bahia	ER	142	510	6,500	5,330,000	HCS	12,000
Segredo	C (1987)	Iguaçu	Pinhão	Paraná	ER	140	700	6,700	3,000,000	H	13,000
Bulgaria											
Belmeken	1976	Kriva	Sestrimo	S	ER	94	760	3,560	145,000	H	
Canada											
Kenney	1952	Nechako	Prince George	British Columbia	ER	104	457	3,071	23,700,000	H	—
Lajoie	1955	Bridge	Goldbridge	British Columbia	ER	87	1,033	2,860	720,728	H	566
WAC Bennett	1967	Peace	Hudson Hope	British Columbia	TE	183	2,042	43,733	70,308,930	H	10,194
Daniel Johnson (Manic 5)	1968	Manicouagan	Baie Comeau	Quebec	MV	214	1,314	2,255	141,851,350	H	
Outardes 4 No. 1	1968	Outardes	Baie Comeau	Quebec	ER	122	649	7,533		H	
Outardes 4 No. 2	1968	Outardes	Baie Comeau	Quebec	ER	108	726	4,688		H	
Lower Notch	1971	Montreal	North Bay	Ontario	TE/PG	132	1,969	1,817	170,960	H	1,425
Big Horn Abraham Lake Res.	1972	North Sask.	Nordeag	Alberta	TE	150	472	4,330	1,768,000	H	
Mica	1972	Columbia	Revelstoke	British Columbia	TE	242	792	32,111	24,699,800	H	4,248
Manicouagan 3	1975	Manicouagan	Baie Comeau	Quebec	TE	108	366	9,175	10,422,991	H	
Revelstoke	C	Columbia	Revelstoke	British Columbia	PG/ER	153	1,620	13,000	5,180,000	H	7,080
Barrage principal	1978	La Grande Riviere	Radisson	Quebec	ER	168	2,826	23,192	61,715,000	H	16,000
Barrage Nord	1981	La Grande	Radisson	Quebec	ER	93	2,156	13,511	60,020,000	H	9,970
Sud		Riviere	Radisson			93	1,689	8,601			
Barrage	1981	La Grande Riviere	Radisson	Quebec	TE/ER	125	3,780	18,800	19,530	H	7,350
Digue QA-1	1981	R. aux Meandres	Radisson	Quebec	TE	95	1,435	3,000	—	H	—
Digue QA-8	1982	Riviere Stephane	Radisson	Quebec	TE	90	1,940	10,100	—	H	—
Chile											
Paloma	1967	Grande	Ovalle	IV Región	TE	96	1,000	7,350	740,000	I	6,500
Digua	1968	Cato	Parral	VII Región	TE	89	420	3,650	220,000	I	300
Colihues A	1981	Cauquenes	Rancagua	VI Región	TE	83	1,200	9,400	170,000	T	220
Los Leones – 1st stage	1981	Los Leones	Los Andes	V Región	TE	128	330	4,360	42,500	T	10
Colbun	C (1985)	Maule	Linares	VII Región	TE	116	530	13,870	1,490,000	I–H	7,570
China											
Shuifeng	1943	Yalu Jiang	Kuandian	Liaoning	PG	106	899	3,400	14,700,000	H	37,500
Fengman	1955	Songhua Jiang	Jilin	Jilin	PG	81	1,080	1,940	10,778	CIH	10,000
Nangudong	1960	Loushui He	Linxian	Henan	ER	78	164	1,750	64,500	IC	3,285
Sanmenxia	1960	Huang He	Sanmenxia	Henan	PG	106	713	1,630	35,400,000	HI	3,843

(Continued)

Table 11A.8 (Continued)

Name of Dam	Year of Completion	River	Nearest City	State Province or Country	Type	Height above Lowest Foundation (m)	Length of Crest (m)	Volume Content of Dam (103 m^3)	Gross Capacity of Reservoir (10^3 m^3)	Purpose	Maximum Discharge Capacity of Spillways (m^3/sec)
Xin'Anjiang	1960	Xin'an Jiang	Hangzhou	Zhejiang	PG	105	462	1,380	21,626	CI	14,000
Songtao	1970	Nandu Jiang	Woxian	Guangdong	TE	80	730	4,915	3,340,000	HIS	6,300
Yunfeng	1970	Yalu Jiang	Jian	Jilin	PG	114	828	2,740	3,911,000	H	24,200
Huairen	1972	Hun Jiang	Huairen	Liaoning	PG	79	593	1,196	3,450,000	HI	23,910
Shanmei	1972	tr. Dong Jiang	Nan'an	Fujian	ER	76	305	3,464	655,000	HIS	8,570
Nanshui	1973	Bel Jiang	Shaoguan	Guangdong	ER	81	215	1,711	1,218,000	CH	400
Zhaikou	1973	Hongnong He	Lingbao	Henan	TE	76	245	4,660	168,000	H	1,950
Danjiangkou	1974	Han Jiang	Xiangfan	Hubei	PG	97	246	2,928	20,900	CIHM	4,700
Bikou	1976	Bailong Jiang	Wenxian	Gansu	ER	101	297	4,241	521,000	IH	2,310
Zhuzhuang	1979	Nanli He	Xingtai	Hebei	VA	95		1,040		I	
Fengtan	1980	Youshui	Yuanling	Hunan	PG	111	488	1,080	1,550,000	IHC	23,300
Wujiangdu	1981	Wu Jiang	Zun'yi	Guizhou	VA	165	368	1,930	2,300,000	H	19,796
Ankang	C	Han Jiang	Ankang	Shaanxi	PG	120	542	2,100	2,580,000	HN	30,900
Baishan	C	Songhua Jiang	Huadian	Jilin	VA	150	670	1,630	6,215,000	H	6,300
Baoquan	C	Luo He	Huixian	Henan	PG	116		2,790	46,000	HIC	
Guxian	C	Luo He	Luoning	Henan	PG	121	315	1,340	1,200,000	HC	
Hunanzhen	C	Wuxi Jiang	Juzhou	Zhejiang	CB	129	440	1,450	2,060,000	HNS	10,600
Longyangxia	C	Huang He	Gonghw	Qinghai	VA	172	342	1,750	24,700,000	HIS	10,500
Lubuge	C	Huangni He	Luoping	Yunnan	ER	97		1,850	110,000	IH	
Shibianyu	C		Changan	Shaanxi	ER	85		2,080	28,000	I	
Shitouhe	C	tr. Wei He	Meixian	Shaanxi	ER	105	600	8,550	147,000	IH	7,150
Wuqiangxi	C	Yuan Shui	Yuanling	Hunan	PG	104		2,670	10,800,000	HI	
Colombia											
Calima I	1965	Calima	Buga	Valle	ER	115	240	2,820	563,000	H	370
Prado	1971	Prado	Ibagué	Tolima	TE	90	260	2,000	1,400,000	HI	1,200
Alto Anchicaya	1974	Anchicayá	Cali	Valle	ER	140	240	2,500	45,000	H	46,000
Chivor(La Esmeralda)	1975	Batá	Guateque	Boyacá	ER	237	280	10,800	815,000	H	10,600
Chuza (Golillas)	1978	Chuza	Bogotá	Cundinamarca	ER	135	106	1,400	257,000	H	545
Punchina	1982	Guatape	Medellin	Antioquía	TE	77	750	5,800	72,000	S	7,500
Betania	C (1984)	Magdalena	Neiva	Huila	ER/PG	90	670	6,300	1,971,000	H	19,000
Salvajina	C (1985)	Cauca	Popayan	Cauca	ER	160	360	3,500	904,000	HICS	3,550
Guavio	C (1987)	Guavio	Gachalá	Cundinamarca	ER	243	390	17,755	1,020,000	H	3,500
Czechoslovakia											
Orlik	1963	Vitava	Pribram	Bohéme C.	PG	91	550	1,030	703,800	HCS	2,555
Dalesice	1979	Jihlava	Trebíc	Moravie S.	ER	100	330	1,800	127,300	HSRI	442
Dominican Republic											
Tavera	1974	Yaque del Norte	Tavera	Santiago	TE	82	405	1,850	170,000	IH	6,900
Sabana Yegua	1978	Yaque del Sur	Los Bancos	San Juan	TE	90	1,200	14,700	677,000	IHC	1,885

Tavera-Bao	C (1980)	Bao	Sabana Iglesia	Santiago	TE	112	425	3,050	280,000	HIS	[a]
Ecuador											
Daniel Palacios (Amaluza Res.)	C (1983)	Paute	Cuenca	Azuay	PG/VA	167	420	1,200	120,000	H	7,700
Daule-Peripa	C (1980)	Daule	Quevedo	Los Ríos	TE/ER	90	250	3,000	6,000,000	M	3,600
Egypt											
Aswan High Dam (Sadd-el-Aali)	1970	Daule	Aswan	Egypt	TE/ER	111	3,830	44,300	168,900,000	GR	11,000
El salvador											
Cerron Granda (Silencio)	1973	Lempa	Hutyapa		ER	80	900	5,100	1,430,000	H	11,000
Fiji											
Monasavu	1982	Nanuku Ck	Suva	Fiji	ER	85	485	1,738	133,000	H	625
Germany (F.R.)											
Rur	1959	Rur	Heimbach	Nordrhein	TE	77	480	2,600	181,800	CNH	450
Frauenau	C (1983)	Kleiner Regen	Zwiesel	Bayern	ER	86	640	2,263	20,300	SNCH	58
France											
Serre-Poncon	1960	Durance	Gap	Htes Alpes	TE	129	600	14,100	1,270,000	HI	3,430
Mont-Cenis	1968	Cenise	Modane	Savoie	TE/ER	120	1,400	14,850	332,200	H	265
Grand'Maison	C (1985)	Eau d'Olle	Grenoble	Isère	TE/ER	160	550	12,500	140,000	H	65
Pla De Soulcem	C (1983)	Mounicou	Tarascon	Ariège	TE	76	275	1,675	29,300	H	134
Ghana											
Akosombo (Main)	1965	Volta	Accra/Tema	Ghana	ER	134	671	7,991	147,960,000	H	14,160
Great Britain											
Scammonden	1970	Black Brook	Huddersfield	Best Yorkshire	TE/ER	76	624	4,304	7,873	S	
Llyn Brianne	1972	Towy	Llandovery	Dyfed	ER	91	274	2,085	60,000	S	850
Greece											
Kremasta	1965	Achelöos	Agrinion	Etolo-Akarnanie	TE	165	460	8,170	4,750,000	H	3,000
Polyphyton	1974	Aliakmon	Kozani	Makedhonia	ER	112	298	3,459	2,244,000	HI	1,375
Mornos	1979	Mornos	Lidhoriki	Phocide	TE	126	815	17,000	780,000	S	1,135
Pournari	1980	Arachthos	Arta	Ipiro	TE	102	574	9,500	730,000	H	6,100
Sfikia	C (1984)	Aliakmon	Veria	Makedhonia	ER	80	230	1,620	99,000	H	1,600
Peghai (Main Dam)	C (1987)	Aöos	Metsovon	Ipiro	TE	78	295	2,800	262,000	H	160
Guatemala											
Pueblo Viejo	C (1983)	Chixoy	San Cristobel	Alta Verapaz	ER	130	230	3,200	460,000	H	3,850
Hong Kong											
High Island East	1977	Kwun Mun	Sai Kung	New Territories	ER	85	457 }	3443	284,375	S	435
High Island West	1977	Kwun Mun	Sai Kung	New Territories	ER	76	762 }	6120			
India											
Koyna (Shivaji Sagar)	1961	Koyna	Karad	Maharashtra	PG	103	808	1,555	2,796,500	H	3,823
Rihand (Gobind Ballabh Pant Sagar)	1962	Rihan	Mirzapur	Uttar Pradesh	PG	93	934	1,680	10,600,000	H	13,339
Bhakra Dam (Gobind Sagar)	1963	Satluj	Nangal Township	Himachal Pradesh	PG	226	518	4,130	9,621,000	IH	8,372

(Continued)

Table 11A.8 (Continued)

Name of Dam	Year of Completion	River	Nearest City	State Province or Country	Type	Height above Lowest Foundation (m)	Length of Crest (m)	Volume Content of Dam (103 m^3)	Gross Capacity of Reservoir (10^3 m^3)	Purpose	Maximum Discharge Capacity of Spillways (m^3/sec)
Sholayar	1972	Sholayar	Coimbatore	Tamil Nadu	TE,PG	105	1,282	2,496	160,660	IH	1,474
Ukai Dam	1972	Tapi	Fort Songadh	Gujarat	TE,PG	81	5,065	25,180	8,511,000	IHC	35,960
Nagarjunasagar Dam	1974	Krishna	Hyderabad	Andhra Pradesh	TE,PG	125	4,865	7,960	11,550,000	IH	53,450
Pong Dam (Beas Project)	1974	Beas	Mukerian	Himachal Pradesh	TE	133	1,950	35,500	8,570,000	IH	12,374
Cheruthoni	1976	Cheruthoni	Idukki	Kerala	PG	138	650	1,700	1,996,000	H	5,012
Balimela Dam	1977	Sileru	Jeypur	Orissa	TE	75	4,363	19,096	3,610,000	IH	10,930
Ramganga	1978	Ramganga	Dhampur	Uttar Pradesh	TE	128	743	11,013	2,442,600	IH	8,467
Bhatsa	C	Bhatsa	Bombay	Maharashtra	PG	89	938	1,679	915,000	SIH	3,775
Chakra	C	Chakra	Shimoga	Karnataka	TE,ER	84	568	1,715	219,000	H	1,416
Dudhganga	C	Dudhganga	Kolhapur	Maharashtra	TE,PG	89	1,264	1,449	708,260	I	2,247
Hasdeo Project	C	Hasdeo	Bilaspur	Madhya Pradesh	TE,PG	86	2,688	3,219	3,417,000	IH	20,530
Karjan (Lower)	C	Karjan	Rajpipla	Gujarat	PG	100	903	1,937	630,000	I	17,275
Lakhwar	C	Yamuna	Dehradun	Uttar Pradesh	PG	192	440	2,000	580,000	IH	8,000
Sardar Sarovar	C	Narmada	Rajpipla	Gujarat	PG	155	1,210	4,100	9,500,000	IHC	62,269
Srisailam H.E. Project	C	Krishna	Hyderabad	Andhra Pradesh	PG	143	512	1,953	8,722,000	H	37,400
Supa (Kalinadi Project)	C	Kali	Dandeli	Karnataka	PG	101	322	1,150	4,418,000	H	2,830
Tehri Dam	C	Bhagirathi	Tehri	Uttar Pradesh	TE,ER	261	570	22,750	3,539,000	IH	13,150
Thein Dam Ranjit Sagar (Res.)	C	Ravi	Pathankot	Punjab	ER,TE	160	565	16,187	3,280,000	IHC	21,890
Warna	C	Warna	Kolhapur	Maharashtra	TE,PG	91	1,580	15,310	964,000	I	3,222
Indonesia											
IR.H.Juanda (Jatiluhur)	1967	Citarum	Purwakarta	W–Java	ER	100	1,225	9,000	3,345,000	IHCSR	3,000
Karangkates	1972	Brantas	Malang	E–Java	ER	100	823	6,156	342,000	IHCR	1,310
Saguling	C	Citarum	Cianjur	W–Java	ER	98	300	2,570	609,000	HCR	2,400
Iran											
Karoon	1975	Karun	Masjed Soliman	Iran	VA	200	380	1,570	3,005,000	IH	16,200
Ghesh-Lagh	1978	Ghesh-Lagh	Sanandaj	Iran	TE	80	300	2,000	224,000	S	2,600
Lar	1982	Lar	Tehran	Iran	TE	105	1,500	1,300	960, 000	IH	1,700
Iraq											
Derbendikhan	1951	Diyala	Sulaymaniya	Sulaymaniya	ER	128	535	7,480	3,000,000	ICH	11,400
Mosul	1983	Tigris	Mosul	Nienava	ER	131	3,500	23,000	12,500,000	ICH	17,000
Italy											
Alpe Gera	1964	Cormor	Sondrio	Adige Lombardia	PG	174	628	1,700	68,088	H	59
Place Moulin	1965	Buthier	Aosta	Valle d'Aosta	VA	155	678	1,510	105,000	H	473
Japan											
Sakuma	1956	Tenryu	Toyohashi	Aichi	PG	156	294	1,120	327,000	H	7,700
Ogochi	1957	Tama	Ome	Tokyo	PG	149	353	1,680	189,000	SH	1,800
Miboro	1960	Sho	Gifu	Gifu	ER	131	405	7,950	370,000	H	1,800

Tagokura	1960	Tadami	Aizuwa-kamatsu	Fukushima	PG	145	462	1,990	494,000	H	2,200
Arimine	1961	Joganji	Toyama	Toyama	PG	140	500	1,570	218,000	H	380
Makio	1961	Kiso	Matsumoto	Nagano	ER	105	260	2,616	75,000	IHS	3,200
Okutadami	1961	Tadami	Koide	Fukushima	PG	157	480	1,640	601,000	H	1,500
Kurobe	1964	Kurobe	Omachi	Toyama	VA	186	489	1,360	199,000	H	1,500
Oshirakawa	1964	Oshirakawa	Gifu	Gifu	ER	95	390	1,700	14,200	H	330
Tsuruta	1965	Sendai	Kagoshima	Kagoshima	PG	118	448	1,124	123,000	Ch	4,965
Yanase	1965	Nabari	Aki	Kochi	ER	115	202	2,842	105,000	H	1,900
Kuzuryu	1968	Kuzuryu	Ono	Fukui	ER	128	355	6,300	320,000	HC	1,560
Shimokubo	1968	Kanna, tr. Tone	Fujioka	Saitama	PG	129	626	1,190	130,000	HIS	1,600
Kisenyama	1969	Samutani, tr. Yodo	Uji	Kyoto	ER	91	255	2,338	7,230	H	6
Koshibu	1969	Koshibu, tr. Tenryu	Iida	Nagano	VA	105	293	311	58,000	IHC	2,160
Misakubo	1969	Misakubo, tr. Tenryu	Tenryu	Shizuoka	ER	105	268	2,410	30,000	H	900
Shimokotori	1973	Kotori, tr. Jintsu	Takayama	Gifu	ER	119	321	3,530	123 037	H	1,920
Aburatani	1974	Aburatani, tr. Kuma	Yatsushiro	Kumamoto	er	82	189	1,277	5,420	h	430
Fukuchi	1974	Fukuchi	Nago	Okinawa	ER	92	260	1,622	51,500	ICS	800
Kajigawa	1974	Kaji	Shibata	Niigata	PG	107	286	433	22,500	C	1,920
Kurokawa	1974	Ichi	Himeji	Hyogo	ER	98	325	3,623	33,390	H	175
Matsukawa	1974	Matsu, tr Tenryu	Iida	Nagano	PG	84	165	263	7,400	CIS	900
Miyama	1974	Naka	Kuroiso	Tochigi	ER	76	334	1,967	25,800	IHS	840
Niikappu	1974	Niikappu	Tomakomai	Hokkaido	ER	103	326	3,071	145,000	H	1,500
Sameura	1974	Yoshino	Nankoku	Kochi	PG	106	400	1,200	316,000	IHCS	6,000
Taisetsu	1974	Ishikari	Asahikawa	Hokkaido	ER	87	440	3,874	66,000	CHIS	1,000
Hirose	1975	Fufuki, tr. Fuji	Enzan	Yamanashi	ER	75	255	1,400	14,300	IHCS	1,380
Nabara	1975	Nabara, tr. Ota	Hiroshima	Hiroshima	ER	86	305	2,213	5,658	H	475
Iwaya	1976	Mase, tr. Kiso	Minokamo	Gifu	ER	128	366	5,780	173,500	IHCS	2,950
Kusaki	1976	Watarase, tr. Tone	Kiryu	Gunma	PG	140	405	1,374	60,500	IHCS	4,320
Myojin	1976	Nabara, tr. Ota	Hiroshima	Hiroshima	ER	89	402	3,268	6,145	H	80
Terauchi	1977	Sada, tr. Chikugo	Amagi	Fukuoka	ER	83	420	3,012	18,000	ICS	1,300
Futai	1978	Kiyotsu	Nagaoka	Niigata	ER	87	280	2,350	18,300	H	1,950
Kassa	1978	Kassa	Nagaoka	Niigata	ER	90	487	4,450	13,500	H	114
Miho	1978	Sakawa	Minamia-shigara	Kanagawa	ER	95	588	5,816	64,900	CHS	3,100
Nanakura	1978	Takase	Omachi	Nagano	ER	125	340	7,380	32,500	H	1,950

(Continued)

Table 11A.8 (Continued)

Name of Dam	Year of Completion	River	Nearest City	State Province or Country	Type	Height above Lowest Foundation (m)	Length of Crest (m)	Volume Content of Dam (103 m^3)	Gross Capacity of Reservoir (10^3 m^3)	Purpose	Maximum Discharge Capacity of Spillways (m^3/sec)
Takase	1978	Takase	Omachi	Nagano	ER	176	362	11,600	76,200	H	1,700
Seto	1978	Setodani	Gojo	Nara	ER	111	343	3,740	16 850	H	230
Tedorigawa	1979	Tedori	Kanazawa	Ishikawa	ER	154	420	10,050	231,000	HCSI	3,500
Terauchi	1980	Sada	Amagi	Fukuoka	ER	83	420	3,000	18,000	CIS	1,300
Urushizawa	1980	Naruse	Furusawa	Miyagi	ER	80	310	2,143	18,000	HCS	1,500
Inamura	1981	Seto	Kochi	Kochi	ER	88	325	3,100	5,800	H	230
Tamahara	1982	Hotchi	Numata	Gunma	ER	116	570	5,435	14,800	H	160
Agigawa	C (1986)	Kiso	Ena	Gifu	ER	102	430	4,400	48,000	CS	2,000
Arakawa	C (1985)	Ara	Kofu	Yamanashi	ER	88	320	3,000	10,800	CS	1,680
Arima	C (1985)	Iruma	Hanno	Saitama	ER	84	260	1,600	7,600	CS	670
Doyo	C (1984)	Doyo	Yonago	Tottori	ER	87	480	2,700	7,680	H	80
Igarashigawa	C (1990)	Igarashi	Sanjo	Niigata	ER	76	360	2,278	21,100	CS	1,420
Jozankei	C (1986)	Ishikari	Sapporo	Hokkaido	PG	113	405	1,150	82,300	CS	675
Kuriyama	C (1985)	Nebesawa	Imaichi	Tochigi	ER	89	340	2,200	6,890	H	52
Kyuragi	C (1985)	Matsuura	Taku	Saga	PG	117	386	1,045	7,400	CSH	1,080
Naramata	C (1986)	Naramata	Numata	Gunma	ER	158	520	12,300	90,000	CIS	1,650
Nitchu	C (1986)	Oshikiri	Kitakata	Fukushima	ER	106	468	5,010	24,600	ICS	1,120
Ogaki	C (1985)	Ukedo	Haramachi	Fukushima	ER	85	262	1,729	19,500	I	1,680
Okawa	C (1984)	Agano	Aizuwa-kamatsu	Fukushima	PG	78	407	1,000	57,500	CSIH	5,230
Ouchi	C (1985)	Ono	Aizuwa-kamatsu	Fukushima	ER	102	340	4,400	18,500	H	176
Sagae	C (1985)	Mogami	Sagae	Yamagata	ER	115	510	9,490	109,000	CSIH	2,600
Sagurigawa	C (1988)	Saguri	Ojiya	Niigata	ER	116	420	7,214	27,500	CS	1,690
Takami	C (1984)	Shizunai	Tomakomai	Hokkaido	ER	120	427	5,120	229,000	CH	2,400
Tamagawa	C (1987)	Omono	Omagari	Akita	PG	100	432	1,105	254,000	CISH	3,500
Shichigashuku	C (1988)	Abukuma	Shiroishi	Miyagi	ER	93	565	5,050	109,000	CSI	2,620
Shintsuruko	C (1985)	Nyu	Obanazawa	Yamagata	ER	93	303	2,645	31,500	I	1,100
Shitoki	C (1984)	Shitoki	Iwaki	Fukushima	ER	84	300	2,580	12,100	CSI	1,540
Tokachi	C (1984)	Tokachi	Obihiro	Hokkaido	ER	84	443	3,658	112,000	CH	2,600
Yasaka	C (1987)	Oze	Iwakuni	Yamaguchi	PG	120	540	1,600	48,000	CS	4,050
Jordan											
King Talal Dam	1977	Zarqa River	Jarash	Salt District	TE/ER	94	330	4,216	52,000	IC	2,950
Wadi Arab Dam	C (1985)	Wadi Arab	North Shuneii	Irbid District	ER	82	482	2,968	20,000	IS	430
Libya											
Ghan	1982	Ghan	Ghrian	NE/Ghrian	ER	80	316	1,650	39,500	IC	1,640
Malaysia											
Temengor	1978	S. Perak	Grik	Perak	ER	115	128	6,980	570,000	HC	2,720
Kenyir	C (1984)	S. Trengganu	Kuala Brang	Trengganu	ER	150	800	16,500	13,600,000	HC	7,000
Batang Ai	C (1985)	Batang Ai	Lubok Antu	Sarawak	ER	85	810	4,000	2,360,000	H	2,175

Mexico											
Lazaro Cardenas (El Palmito)	1947	Nazas	C. Lerdo	Durango	TE	95	330	5,300	3,152,000	IC	6,000
Sanalona	1948	Tamazula	Culiacán	Sinaloa	TE	81	1,031	4,900	845,000	IH	6,300
Alvaro Obregon (Oviachic)	1952	Yaqui	Obregón	Sonora	TE	90	1,457	8,773	3,237,000	IHC	11,100
Adolfo Ruiz Cortines (Mocúzari)	1955	Río Mayo	Navojoa	Sonora	TE	81	780	4,196	1,014,700	IHC	8,000
Presidente Aleman (Temascal)	1955	Río Tonto	C.Alemán	Oaxaca	TE	76	830	4,059	6,515,000	ICH	5,500
Migual Hidalgo (El Mahone)	1956	Río Fuerte	El Fuerte	Sinaloa	TE	86	3,230	10,200	3,290,000	IGH	16,450
Presidente Benito Juarez (El Marques)	1961	Tehuantepec	Tehuantepec	Oaxaca	TE	86	375	3,540	942,000	IC	5,500
El Infiernillo	1963	Balsas	Apatzingan	Michoacán	TE	148	350	5,500	9,340,000	H	10,350
Netzahualcoyoti	1964	Grijalva	Cárdenas	Chiapas	TE	138	478	5,077	8,300,000	HC	21,750
Pte.Adolfo Lopez Mateos (Humaya)	1964	Humaya	Culiacán	Sinaloa	TE	107	820	7,141	3,150,000	IC	5,600
Internacional La Amistad	1968	Bravo	Acuña	Coahuila	PG/TE	87	9, 760	11,620	4,379,000	IHCS	43,690
La Angostura	1974	Grijalva	Tuxtla Gutierrez	Chiapas	TE	146	323	4,030	9,200,000	H	6,900
Manuel Moreno Torres (Chicoasén)	1980	Grijalva	Tuxtla Gutiérrez	Chiapas	TE	261	485	15,370	1,613,000	H	15,000
Jose Lopez Portillo Pte. (Comedero)	1981	San Lorenzo	Cosalá	Sinaloa	TE	136	400	7,090	2,850,000	IHC	5,000
Gustavo Diaz Ordaz Pdte. (Bacurato)	1982	Sinaloa	Guamuchil	Sinaloa	TE	114	800	9,315	1,800,000	IHC	7,000
El Sabinal	C (1985)	Ocoroni	Guasave	Sinaloa	TE	79	400	2,367	300,000	IC	2,450
Carlos Ramirez Ulloa (Caracol)	C (1985)	Balsas	Iguala	Guerrero	TE	126	347	6,327	782,000	H	17,000
Chilatan	C (1986)	Tepalcatepec	Apatzingan	Jalisco	TE	104	1,150	5,898	600,000	IC	7,000
Morocco											
Moulay Youssef	1970	Tessaout	Marrakech	Marrakech	TE	100	725	5,300	200,000	IH	3,000
Hassan Addakhil	1971	Ziz	Er Rachidia	Er Rachidia	TE	85	785	5,800	380,000	I	1,700
Youssf Ben Tachfine	1973	Masa	Tiznit	Tiznit	ER	85	670	3,700	310,000	IS	3,400
Sidi Mohamed Ben Abdellah	1974	Bou Regreg	Rabat	Rabat	ER	100	340	3,000	493,000	SI	5,000
Ait Chouarit	C (1986)	Lakhdar	Demnate	Azilal	TE	145	380	9,500	270,000	IHS	1,820
Nepal											
Kulekhani Dam	1981	Kulekhani	Kathmandu	Nepal	ER	114	406	4,419	85,300	H	2,540
New Zealand											
Benmore	1965	Waitaki	Oamaru	Otago	TE	118	957	12,500	2,200,000	H	3,400
Matahina	1965	Rangataiki	Whakatane	South Auckland	TE	86	345	3,500	25,000	H	1,900
Mangatangi	1977	Mangatangi	Manukau	South Auckland	TE	78	340	2,240	39,000	S	510

(Continued)

Table 11A.8 (Continued)

Name of Dam	Year of Completion	River	Nearest City	State Province or Country	Type	Height above Lowest Foundation (m)	Length of Crest (m)	Volume Content of Dam (103 m^3)	Gross Capacity of Reservoir (10^3 m^3)	Purpose	Maximum Discharge Capacity of Spillways (m^3/sec)
Patea	C	Patea	Wanganui	Taranaki	TE	82	190	1,100	138,000	H	2,800
Nigeria											
Shiroro	C (1984)	Kaduna/ Dinya	Minna	Niger	ER	125	700	3,457	7,000 000	H	7,500
Norway											
Hyttejuvet	1965	Walldaiselv	Haugesund	Hordaland	ER	90	350	1,450		H	375
Digea	1970	Sira-Digea	Flekkefjord	Vest-Agder	ER	90	400	2,700		H	45
Svartevatn	1977	Sira	Stavanger	Vest-Agder	ER	129	420	4,715		H	
Sysenvatn	1979	Leiro	Bergen		ER	84	1,140	3,624		H	250
Oddatjorn Blasjo	C (1987)	Oddeana	Haugesund	Rogaland	ER	140	500	5,400	see Storvatn	H	
Storvatn Reservoir	C (1987)	Brattliana	Haugesund	Rogaland	ER	98	1,460	9,700	3,105,000	H	600
Pakistan											
Mangla	1967	Jehlum	Dehlum	Punjab	TE	138	3,139	65,379	7,251,811	HI	31,144
Tarbela	1976	Indus	Taxila	NWFP	TE/ER	143	2,743	105,570	13,689,644	IH	42,186
Auxiliary-1 Jari	1967	Saddle Dam	Mirpur	A. Kashmir	TE	84	2,073	26,240		I	—
Paraguay											
Itaipu	1982	Paraná	Hernandarias	Brazil/Paraguay	ER/PG/TE	190	7,655	33,690	29,000,000	H	62,000
Peru											
Yauliyacu Arriba	1982	Yauliyacu	Casapalca	Lima	TE	75	220	3,000	7,500	M	
Condorma	C (1985)	Colca	Chivay	Arequipa	ER	92	503	4,300	260,000	I	1,300
Gallito Ciego	C (1987)	Jequete-peque	Pacasmayo	Cajamarca	ER	112	750	14,200	400,000	IH	1,830
Chinchan	C (2007)	Yuracocha	Casapalca	Lima	TE	90	230	5,800	13,500	M	
Philippines											
Ambuklao	1956	Agno	Baguio	Benguet	ER	129	452	6,000	327,170	H	7,300
Binga	1960	Agno	Baguio	Benguet	ER	107	215	2,000	63,000	H	5,200
Angat	1967	Angat	Manila	Bulacan	ER	131	368	7,000	1,099,000	IH	7,500
Pantabangan	1977	Upper Pampanga	Cabanatuan	Neva Ecija	TE	107	1,615	12,300	2,996,000	IHC	4,200
Magat	1982	Magat	Santiago	Isabela	TE/ER	106	2,925	13,200	1,250,000	IH	30,400
San Roque	C	Agno	Dagupan	Pangesinan	ER	210	1,130	43,150	990,000	IHC	12,600
Portugal											
Paradela	1958	Cávado	Chaves	Vila Real	ER	110	540	2,700	164,500	H	720
Alto Rabagão	1964	Rabagão	Chaves	Vila Real	VA/PG	94	1,897	1,117	569,000	H	500
Santa Clara	1968	Mira	Odemira	Beja	TE	86	428	3,966	485,000	I	208
Romania											
Izvorul Muntelui (Bicaz)	1961	Bistrita	P.Nearnt	Nearnt	PG	127	430	1,625	1,230,000	H	2,400
Fintinele Somes	1978	Somesul Cald	Huedin	Cluj	ER	92	400	2,320	225,000	HR	700
Cerna Principal	1979	Cerna	Tg.Jiu	Gorj	ER	110	342	2,550	124,000	SHI	1,080

Oasa	1979	Sebes	Sebes	Alba	ER	91	300	1,600	136,000	HCR	264
Colibita	C (1983)	Bistrita	Bistrita	Bistrita Násâud	ER	92	250	1,600	90,000	SH	650
Gura Apelor	C (1984)	Rîul Mare	Hateg	Hunedoara	ER	168	450	9,020	225,000	H	1,750
Pecineagu	C (1984)	Dîmbovita	Tirgoviste	Dimbovita	TE	105	270	2,400	68,900	SHI	600
Mineciu	C (1985)	Teleajen	Vâlenii de Munte	Prahova	TE	75	720	5,000	60,000	SCHI	1,200
Riusor	C (1985)	R. Tirgului	Cimpulung	Arges	ER	120	380	3,500	60,000	SH	620
Siriu	C (1985)	Buzàu	Nehoiu	Buzâu	ER	122	440	8,800	155,000	CISH	3,400
Vija	C (1986)	Bistrita	Tg.Jiu	Gorj	ER	93	270	1,700	29,400	HS	800
Poiana Marului	C (1987)	Bistra Mârului	Otelul Rosu	Caras Severin	ER	130	400	5,320	96,000	HI	830
South Africa											
P K Le Roux	1977	Orange	Petrusville	Orange Free State	VA	107	853	1,300	3,237,000	IH	20,400
Sterkfontein	1980	Nue Jaar Spruit	Harrismith	Orange Free State	TE	93	3,060	19,800	2,656,000	S	—
Goedertrouw	1982	Mhlatuze	Eshowe	Natal	TE	88	660	5,330	321,000	IS	7,000
South Korea											
So Yang Gang	1973	Han	Chunchon	Kangwondo	ER	123	530	9,591	2,900,000	IHCR	5,500
Sam Rang Jin Upper part Dam	C	Nakdong	Samrangjin	Kyeongsang-namdo	ER	85	250	1,003	6,140	H	
An Dong	1976	Nakdong	Andong	Keongsang-namdo	ER	83	624	4,015	1,248,000	IHCS	5,360
Spain											
Mequinenza	1966	Ebro	Mequinenza	Zaragoza	PG	81	451	1,000	1,533,800	H	12,800
Porto De Mouros	1967	Ulla	Arzua	La Coruña	ER	93	460	2,337	297,000	H	1,550
Grado I	1969	Cinca	El Grado	Huesca	PG	130	958	1,225	399,000	IH	3,420
Iznajar	1969	Genil	Rute	Cordoba	PG	122	407	1,450	980,000	HIS	6,550
Almandra	1970	Tormes	Almendra	Salamanca	VA	202	557	2,186	2,649,000	H	3,000
El Atazar	1972	Lozoya	Atazar	Madrid	VA	134	484	1,200	426,000	S	410
Arenos	1979	Mijares	Montanejos	Castellon	ER	108	428	3,014	132,000	I	1,300
Beninar	1983	Grande De Adra	Beninar	Almeria	ER	87	386	3,800	70,000	IS	232
Limonero	1983	Guadal-medina	Malaga	Malaga	ER	93	410	3,188	27,000	S	283
Canales	C (1983)	Benil	Guejar Sierra	Granada	ER	159	340	1,217	71,000	IS	226
Cuevas De Almanzora	C (1983)	Almanzora	Cuevas De Almanzora	Almeria	ER	113	623	6,510	191,000	IH	2,520
Negratin	C (1983)	Guadiana Menor	Freila	Granada	PG/ER	75	439	1,150	546,000	IH	3,440
Sallente	C (1983)	Flamisell	Torre De Capdella	Lerida	ER	89	398	1,100	6,000	H	63
La Viñuela	C (1983)	Guaro	La Viñuela	Malaga	ER/TE	94	460	3,345	25,000	IS	120
Zahara	C (1983)	Guadalete	Zahara	Cadiz	ER	85	500	2,011	212,000	I	1,000
Sri Lanka											
Kotmale	C (1985)	Kotmale Oya	Gampola	CP	ER	87	600	4,159	175,000	H	5,550
Randenigala	C (1986)	Mahaweli	Mahiyangana	UP	ER	94	495	3,700	860,000	HIC	8,085

(Continued)

Table 11A.8 (Continued)

Name of Dam	Year of Completion	River	Nearest City	State Province or Country	Type	Height above Lowest Foundation (m)	Length of Crest (m)	Volume Content of Dam (103 m^3)	Gross Capacity of Reservoir (10^3 m^3)	Purpose	Maximum Discharge Capacity of Spillways (m^3/sec)
Sweden											
Trangsist	1961	Dalälven	Mora	Kopparberg,M	ER	125	850	7,200	880,000	H	1,000
Holjes	1961	Klarälven	Hagfors	Värmland,M	ER/TE	80	400	1,750	270,000	H	1,280
Messaure	1963	Lule älv	Jokkmokk	Norrbotten,N	TE	101	1,900	10,500	50,000	H	2,300
Letsi	1967	Lule älv	Jokkmokk	Norrbotten,N	ER	85	570	2,300	67,000	H	1,500
Seitevare	1968	Lule älv	Porius	Norrbotten,N	ER	106	1,450	4,900	1,650,000	H	875
Switzerland											
Marmorera (Castilleto)	1954	Julia	Bivio	Grisons	TE	91	400	2,700	62,600	H	200
Mauvolsin	1957	Drance de Bagnes	Fionnay	Valais	VA	237	520	2,030	181,500	H	100
Goescheneralp	1950	Göschen-erreuss	Göschenen	Uri	ER	155	540	9,300	76,000	H	200
Grande Dixence	1961	Dixence	Hérémence	Valais	PG	285	695	6,000	401,000	H	—
Luzzone	1963	Brenno di Luzzona	Olivone	Tessin	VA	208	530	1,330	88,000	H	88
Mattmark	1967	Seaser Vispa	Saas-Fee	Valais	TE	120	780	10,500	101,000	H	150
Emosson	1974	Barberine	Finhaut	Valais	VA	180	555	1,090	227,000	H	60
Taiwan											
Shihmen	1964	Tahan	Chungli	Taiwan	ER	133	360	7,059	309,120	IHCS	13,400
Tsengwen	1973	Tsengwen	Tainan	Taiwan	TE	133	400	9,296	707,530	ICSH	9,470
Thailand											
Sirikit	1972	Nan	Uttaradit	N	TE	114	800	9,800	10,550,000	IHC	3,250
Bang Lang	1981	Pattani	Yala	S	TE/ER	85	422	2,900	1,360,000	IHC	4,500
Srinagarind	1981	Quae Yai	Kanchanaburi	Central Region	ER	140	610	12,100	17,745,000	IHC	2,420
Khao Laem	C(1984)	Quae Noi	Kanchanaburi	Central Region	ER	90	910	8,000	7,450,000	IHC	3,200
Turkey											
Seyhan	1956	Seyhan	Adana	South A	TE	77	1,955	7,500	1,200,000	ICB	2,500
Hirfanli	1959	Kizilirmak	Kirsehir	Inner A	ER	83	364	2,000	5,980,000	ICH	2,300
Demirköprü	1960	Gediz	Man sa	West.A	TE	77	543	4,300	1,320,000	ICH	200+6272
Almus	1965	Yesilirmak	Tokat	NEA	TE	95	371	3,500	950,000	IH	1,550
Kozan	1972	Kilgen	Adana	South.A	TE/ER	83	289	1,195	163,000	I	1,250
Keban	1974	Firat	Elazig	East	PG/ER	207	1,126	15,585	30,600,000	CH	17,000
Ayvacik	1981	Yesilirmak	Samsun	North	ER	175	405	2,327	10,800,000	CH	11,000
Güzelhisar	1982	Güzelhisar	Izmir	West	ER	89	511	3,204	158,000	IS	2,550
Gönen	C	Gönen	Balikesir	West	TE	78	293	2,036	164,000	I	2,785
Doganci	C	Nilüfer	Bursa	West	ER	82	288	2,278	50,000	S	1,978
Çamlidere	C	Bayindir	Ankara	Inner A	TE/ER	106	278	2,487	133,000	IS	662
Aslantas	C	Ceyhan	Adana	South	TE/ER	95	566	8,000	1,150,000	ICH	11,930
Adigüzel	C	B.Menderes	Den zli	West	ER	145	377	5,892	1,188,000	ICH	4,260
Kiliçkaya	C	Kelkit	Sivas	Inner	ER	135	405	6,030	14,000 000	ICH	2,450

Karakaya	C	First	Diyarbakir	SE	VA	173	462	2,000	9,580,000	H	17,000
Çatalan	C	Seyhan	Adana	South	TE	95	309	7,664	1,629,000	ICH	8,900
Karacaoren	C	Asagiaksu	Burdur	West	TE	95	428	3,500	1,340,000	ICH	4,495
Uluborlu	C	Pupa	Isparta	SE	TE	75	315	1,800	24,000	I	295
Gezende	C	Göksu	Mersin	South	TE	75	171	1,110	66,000	H	4,385
Altinkaya	C	Kizilirmak	Samsun	North	ER	195	604	2,600	5,763,000	ICH	11,800
Menzelet	C	Ceyhan	K.Maras		ER	151	425	8,000	19,500	IH	4,850
Ataturk	C	Firat	Diyarbakir	East	ER	184	746	85,000	48,700	ICH	16,800
United States											
Ashokan	1916	Esopus Creek	Olive Bridge	New York	TE	77	1,417	1,950	484,018	S	5,938
Calaveras	1925	Calaveras Creek	Sunol	California	ER	75	366	2,646	123,348	S	702
Dix	1925	Dix	High Bridge	Kentucky	ER	87	311	1,343	222,027	H	1,300
Tieton	1925	Tieton	Naches	Washington	TE	97	280	1,567	244,229	ICR	1,416
Cobble Mountain Reservoir	1931	Little	Westfield	Mass.	TE	80	221	2,294	86,380	S	113
New Exchequer	1926	Merced	Snelling	California	ER	146	378	3,952	1,265,552	H	9,911
Salt Springs	1931	N Fork Mokelumne		California	ER	9	396	2,294	171,947	HS	1,580
El Capitan	1934	San Diego	Lakeside	California	TE/ER	82	357	2,049	88,811	S	4,831
Hoover (Boulder)	1936	Colorado	Boulder	Nevada	VA	221	379	364	34,852,028	IHCN	11,327
Fort Peck	1937	Missouri	Frazer	Montana	TE	76	6,534	96,050	22,118,763	CHIN	6,514
Alcova	1938	North Platte	Casper	Wyoming	TE/ER	81	233	1,250	227,577	IHR	1,557
Mathews	1938	Tr Cajalco Creek	Corona	California	TE	84	1,988	7,309	224,494	S	382
Tygart	1938	Tygart	Grafton	W Virginia	PG	76	586	1,055	135,190	NC	8,948
Quabbin Winsor	1939	Swift	Ware	Mass.	TE	85	805	3,058	1,561,256	S	425
San Gabriel No 1	1939	San Gabriel	Azusa	California	ER	123	463	8,104	54,725	CS	7,524
Friant	1942	San Joaquin	Fresno	California	PG	97	1,063	1,632	642,027	ISCR	2,350
Grand Coulee	1942	Columbia	Coulee Dam	Washington	PG	168	1,272	8,093	11,794,553	ICHN	26,986
Marshall Ford Lake Travis (res)	1942	Colorado	Austin	Texas	TE	85	1,230	1,251	1,446,381	HCSR	16,197
Nantahala Lake	1942	Nantahala	Nantahala	N Carolina	TE	76	318	5,532	142,590	H	2,503
Green Mountain	1942	Blue	Hot Sulphur Springs	Colorado	TE/ER	94	351	3,333	190,696	IHR	708
Fontana	1944	Little Tennessee	Fontana Village	N Carolina	PG	146	721	2,734	603,715	H	4,474
Merriman	1945	Roundout Creek	Lackawack	New York	TE	114	732	4,434	189,956	S	5,097
Shasta	1945	Sacramento	Redding	California	PG	183	1,055	6,660	5,614,809	ISHN	5,239
Mud Mountain	1948	White	Buckley	Washington	ER	130	213	1,758	130,698	C	3,892
Watauga	1948	Watauga	Elizabethton	Tennessee	ER	97	274	2,660	398,415	CHNR	1,756
Anderson Ranch	1950	S Fork Boise	Boise	Idaho	TE	139	411	7,380	620,071	ICRH	566
Leroy Anderson	1950	Coyote Creek	San Jose	California	TE	77	421	2,485	112,617	I	1,195

(Continued)

Table 11A.8 (Continued)

Name of Dam	Year of Completion	River	Nearest City	State Province or Country	Type	Height above Lowest Foundation (m)	Length of Crest (m)	Volume Content of Dam (103 m^3)	Gross Capacity of Reservoir (10^3 m^3)	Purpose	Maximum Discharge Capacity of Spillways (m^3/sec)
South Holston	1950	S Fork Holston	Bluff City	Tennessee	ER	87	488	4,499	402,115	CHNR	1,756
Bull Shoals	1951	White	Cotter	Arkansas	PG	78	688	1,606	3,759,653	CH	14,158
Center Hill	1951	Caney Fork	Lancaster	Tennessee	TE/PG	76	658	2,736	1,033,658	CHR	12,856
Wolf Creek	1951	Cumberland	Burkesville	Kentucky	TE	79	1,748	8,713	4,927,760	HCR	15,659
Bradbury	1953	Santa Ynez	Santa Barbara	California	TE	85	1,021	5,119	252,864	ISR	4,559
Detroit	1953	N Santiam	Mill City	Oregon	PG	141	482	1,147	561,152	HCRI	4,984
Hungry Horse	1953	S Fork Flathead	Kalispell	Montana	VA	172	645	2,359	4,277,715	IHCN	1,501
Lookout Point	1953	Middle Fork Whillamette	Eugene	Oregon	TE	84	968	5,892	562,385	CINH	7,646
Yale	1953	Lewis	Woodland	Washington	TE	98	472	3,211	495,860	HCR	4,899
Lucky Peak	1954	Boise	Boise	Idaho	TE	104	713	4,511	377,445	CRI	2,642
Pine Flat Lake	1954	Kings	Piedra	California	PG	134	561	1,835	1,233,492	CIRH	11,072
Folsom	1956	American	Sacramento	California	PG	104	3,109	6,866	1,245,817	ISHC	16,056
Beardsley	1957	Mid Fk Stanislaus	Melones	California	TE	85	250	2,294	120,264	H	2,027
Brownlee	1958	Snake	Oxbcw Village	Idaho	ER	120	421	4,587	1,759,808	HCR	8,593
Courtwright	1958	Helms Creek	Piedra	California	ER	95	263	1,193	152,088	HS	400
Oahe	1958	Missouri	Pierre	So. Dakota	TE	75	2,890	70,339	27,432,595	CHIN	2,266
Swift	1958	Lewis	Woodland	Washington	TE	186	640	11,774	932,512	HCR	4,106
Wishon	1958	N Fk Kings	Piedra	California	ER	80	1,021	2,829	157,886	HS	1,379
Casitas	1959	Coyote Creek	Santa Barbara	California	TE	102	610	6,967	313,304	ISC	210
Table Rock	1959	White	Branson	Missouri	TE/PG	77	1,958	3,479	3,332,900	CH	15,801
Mammoth Pool	1960	San Joaquin		California	TE	124	250	4,094	151,718	HS	4,757
Arthur R. Bowman (Prineville)	1961	Crooked	Bend	Oregon	TE	75	244	1,089	190,573	IRC	230
Ball Mountain	1961	West	Jamaica	Vermont	TE	83	279	1,767	2,760	CR	4,248
Lewis Smith	1961	Dipsey Fork	Dilworth	Alabama	TE	93	671	3,930	1,714,540	HCR	5,873
Sly Creek	1961	Lost Crook	Oroville	California	TE	83	640	4,000	80,238	H	328
Hills Creek	1962	M Fk Willamette	Oakridge	Oregon	TE	104	703	8,257	439,055	CHIS	4,010
Smith	1962	Smith	Belknap Springs	Oregon	TE	101	351	1,911	18,502	H	255
Trinity	1962	Trinity	Redding	California	TE	164	747	22,486	3,019,563	IHCR	680
Abiquiu	1963	Rio Chama	Abiquiu	New Mexico	TE	99	469	9,017		CR	447
Dillon	1963	Blue	Silverthorne	Colorado	TE	94	1,798	9,140	311,674	S	333
Lermon	1963	Florida	Durango	Colorado	TE	87	415	2,326	49,463	I	272
Navajo	1963	San Juan	Blanco	New Mexico	TE	123	1,112	20,521	2,108,020	IR	963
Union Valley	1963	Silver Crk	Coloma	California	TE	138	549	7,646	334,274	S	1,260
Whiskeytown	1963	Clear Creek	Redding	California	TE	86	1,190	3,412	297,269	IHCR	815
Briones	1964	Bear Creek	El Sobrante	California	TE	87	629	7,578	83,285	S	93

Cougar	1964	S Fork Mckenzie	Springfield	Oregon	ER	158	488	9,939	270,463	HCIR	2,152
Homestake	1964	M Fk Home-stake	Minturn	Colorado	ER	81	608	2,628	53,780	S	99
Round Butte	1964	Deschutes	Warm Springs	Oregon	TE	134	442	7,340	659,913	HR	1,286
Summersville	1965	Gauley	Swiss	W Virginia	ER	119	695	10,371	236,215	CRS	11,667
Blue Mesa	1966	Gunnison	Montrose	Colorado	TE/ER	119	239	2,355	1,160,706	HCR	954
Glen Canyon	1966	Colorado	Less Ferry	Arizona	VA	216	475	3,747	33,304,009	HSCR	7,815
John W. Flannagan	1966	Pound	Elkhorn City	Virginia	TE	79	279	1,824	83,261	CR	1,240
Lost Creek	1966	Lost Creek	Devils Slide	Utah	TE	76	329	1,401	27,753	ISRC	70
Lower Hell Hole	1966	Rubicon	Auburn	California	ER	125	472	6,357	257,058	SH	132
Millwood	1966	Little	Ashdown	Arkansas	ER	89	5,350	6,117	189,000	CS	13,403
Yellowtail	1966	Bighorn	Hardin	Montana	VA	160	451	1,182	1,076,830	ICHR	2,605
San Luis	1967	San Luis Creek	Los Banos	California	TE	116	5,639	59,559	2,517,536	ISHR	29
Alamo (res)	1968	Bill Williams	Parker	Arizona	TE	105	297	2,328	539,350	CRI	1,175
Blue River	1968	Blue	Springfield	Oregon	TE	95	381	3,726	110,380	CR	1,501
Oroville	1968	Feather	Oroville	California	TE	230	2,073	59,635	4,297,451	SCHR	4,248
Ruedi	1968	Fryingpan	Basalt	Colorado	TE/ER	98	318	2,863	126,432	IRC	157
International Amistad	1969	Rio Grande	Del Rio	Texas/Mex.	TE/PG	77	9,754	2,652	4,323,847	CIHR	42,673
Lopez	1969	Arroyo Grande Crk	Arroyo Grande	California	TE	85	341	2,705	64,758	S	1,256
New Bullards Bar	1970	North Yuba	Marysville	California	VA	194	671	1,988	1,195,984	SH	4,219
Coedar Springs	1971	W Fk Mojave	Victorville	California	ER	76	725	6,040	96,212	IRS	913
Don Pedro	1971	Tuolumne	La Grange	California	TE	173	549	12,233	2,503,968	H	13,380
Heron	1971	Willow Creek	Tierra Amarillo	New Mexico	TE/ER	84	372	2,317	495,243	ISR	19
Castaic	1973	Castaic Creek	Castaic	California	TE	125	1,585	33,640	431,719	IRS	2,220
Dworshak	1973	N. Fork Clearwater	Ahsahka	Idaho	PG	219	1,002	4,931	4,259,213	HCR	6,258
Jocassee	1973	Keowee		S Carolina	ER	133	549	8,869	1,431,206	H	1,761
Libby	1973	Kootenai	Libby	Montana	PG	129	881	2,875	7,165,296	HCR	4,060
Pyramid	1973	Piru Creek	Piru	California	ER	122	329	5,315	220,793	IRSH	4,248
Soldier Creek	1973	Strawberry	Duchesne	Utah	TE	77	393	2,440	1,365,464	ICR	
Carters	1974	Coosawattee	Carters	Georgia	ER	141	594	11,468	465,146	CHR	520
Cochiti	1975	Bio Grande and Santa Fe	Cochiti Pueblo	New Mexico	TE	77	8,785	50,228	52,726	CIR	4,280
Lost Creek	1976	Rogue	Shady Cove	Oregon	ER	105	1,097	8,257	573,485	CHSR	4,474
Ririe	1976	Willow Creek	Idaho Falls	Idaho	TE	77	326	2,046	123,348	IC	1,133
Gross	1977	South Boulder Crk	Louisville	Colorado	PG	104	332	7,809	50,557	SH	444

(Continued)

Table 11A.8 (Continued)

Name of Dam	Year of Completion	River	Nearest City	State Province or Country	Type	Height above Lowest Foundation (m)	Length of Crest (m)	Volume Content of Dam (103 m^3)	Gross Capacity of Reservoir (10^3 m^3)	Purpose	Maximum Discharge Capacity of Spillways (m^3/sec)
Little Blue Run	1977	Little Blue Run of Ohio	East Liverpool	Penn.	TE	122	640	9,939	13,568		48
Gathright	1978	Jackson	Cov ngton	Virginia	ER	78	368	1,988	152,582	RC	173
New Melones	1979	Stanislaus	Modesto	California	ER	191	475	12,233	2, 960,356	CIHR	3,171
Bloomington	1981	N Branch of Potomac	Bloomington	Maryland	ER	90	649	7,646	116,793	CSR	5,465
Bath County Upper	C	Little Back Creek	Warm Springs	Virginia	ER/TE	143	731	18,000	43,790	H	509
Warm Springs	C	Dry Creek	Cloverdale	California	TE	97	914	22,920	469,365	CSR	1,056
U.S.S.R.											
Mingechaur	1953	Kura	Mingechaur	Azerb. SSR	TE	80	1,550	15,600	16,000,000	HNIC	3,600
Bukhtarma	1960	Irtysh	Ust-Kamenogorsk	Kazakh, SSR	PG	90	380	1,170	49,800,000	HN	1,000
Sioni	1963	Iori	Tbilisi	Georg. SSR	TE	86	780	6,300	325,000	IH	596
Bratsk	1964	Angara	Bratsk	Irkutsk	PG	125	1,430	4,415	169,000,000	HNS	7,090
					TE	36	2,987	6,547			
					TE	40	723	2,147			
Serebrianka No 1	1970	Voronlya	Murmansk	Murmansk	ER	78	2,625	5,660	4,170,000	H	675
Sarsang	1976	Terter	Yevlakh	Azerb. SSR	TE	125	590	5,820	560,000	IH	800
Charvak	1977	Chirchik	Tashkent	Uzbek. SSR	ER	168	764	21,600	2,000,000	HI	1,200
Medeo	1977	Malaya Almaatinka	Alma-Ata	Kazakh. SSR	ER	144	530	8,500	—	C	—
Ust-Ilim	1977	Angara	Ust-Ilimsk	Irkutsk	PG	102	1,477	3,800	59,300,000	HN	9,700
					TE/ER	47	2,248	5,066			
Chirkey	1978	Sulak	Makhachkala	Daghest. SSR	VA	233	333	1,358	2,780,000	HIS	2,870
Toktogul	1978	Naryn	Naryn	Kirgh. SSR	PG	215	293	3,345	19,500,000	HI	2,340
Zeya	1978	Zeye	Blagoveshchensk	Amur.	CB	115	758	2,160	68,400,000	HCN	6,600
Andizhan	1980	Karadarya	Osh	Kirghiz. SSR	CB	115	920	3,700	1,750,000	HI	2,392
Inguri	1980	Inguri	Zugdidi	Georgian SSR	VA	272	680	3,960	1,100,000	HI	2,500
Nurek	1980	Vakhsh	Nurek	Tadjik SSR	TE	300	74	58,000	10,500,000	HI	4,000
Sayano-Shushensk	C	Yenisei	Minusinsk	Krasnoyarsk	VA/PG	245	1,066	9,075	31,300,000	NH	13,600
Bureya	C	Bureya	Blagoveshchensk	Kharbarovsk	PG	139	810	3,561	20,900,000	HC	19,100
Irganai	C	Avar Koisu	Makhachkala	Daghest. SSR	TE/ER	111	312	5,827	705,000	H	2,760
Khudoni	C	Kolyma	Magadan	Magadan	ER	126	759	12,550	14,600,000	H	17,500
Rogun	C	Vakhsh	Nurek	Tajik SSR	TE/ER	335	660	75,500	13,300,000	HI	3,500
Spandarian	C	Vorotan	Sisian	Armen. SSR	TE	87	317	2,250	277,000	HI	160
Zhinvali	C	Aragvi	Tbilisi	Georg. SSR	TE	102	412	5,200	520,000	HS	2,500
Venezuela											
Onia	1978	Onia	El Vigia	Mórida	TE	301	450	1,300	6,600	C	435
Tucupido	C	Tucupido	Guanare	Portuguesa	TE	92	290		3,300	IHC	—

Yacambú	C	Yacambú	Sanare	Lara	TE	158	107	3,000	427,000	ICS	480
Las Palmas	C	Cojedes	Acarigua	Cojedes	TE	77	750	7,000	810,000	IHC	
Las Cuevas	C		San Cristóbal	Tachira	TE	108		7,500	1,400,000	H	
La Vueltosa	C	Caparo	Buena Vista	Mérida	TE	118		15,000	5,300,000	H	
Borde Seco	C	Camburito	San Cristobal	Táchira	TE	120		6,600		H	
Taguaza	C	Taguaza	Sta. Lucia	Miranda	TE	100	300	2,000	212,000	S	430
La Honda	C	Uribante	San Cristóbal	Táchira	TE	108		7,800	770,000	H	
Yugoslavia											
Kokin Brod	1962	Uvac	Nova Varos	SR Srbija	ER	82	1,220	2,480	250,000	H	1,500
Tikves	1968	Crna Reka	Kavadarci	SR Makedonija	ER	114	338	2,722	475,000	J	2,050
Spilje	1969	Crni Drim	Debar	SR Makedonija	ER	112	330	2,699	520,000	H	2,890
Rama	1969	Rama	Prozor	SR Bosna and Hercegovina	ER	103	230	1,510	487,000	H	400
Turija	1970	Turija	Strumica	SR Makedonija	ER	93	417	1,978	65,000	J	76
Gazivode	1977	Ibar	Titova Mitrovica	SAP Kosovo	ER	108	520	5,000	370,000	H	720
Sjenica	1979	Veliki Uvac	Nova Varos	SR Srbija	ER	106	310	2,430	190,000	H	1,000
Lazici	C (1983)	Beli Rzav	Bajina Basta	SR Srbija	ER	123	120	2,170	150,000	H	180
Zavoj	C (1987)	Visocica	Pirot	SR Srbija	ER	80	250	1,400	16,000	H	1,800
Zambia/Zimbabwe											
Kariba	1959	Zambezi	Lusaka	Zambia/ Zimbabwe	VA	128	579	1,032	160,368	H	9,500

Note: Height—over 75 m; Volume—over 1,000,000 cubic m. Dam types are identified by the following abbreviations: Earth TE; Rockfill ER; Gravity PG; Buttress CB; Arch VA; Multi-Arch MV. Purpose for which reservoir is used is indicated by the following abbreviations: Irrigation I; Hydroelectric H; Flood Control C; Navigation N; Water Supply S; Mine Tailings T; Recreational R. Under year of completion, C is under construction and P is planned.

Source: Compiled from *World Register of Dams*, 1984, published by the International Commission on Large Dams, 151 Boulevard Haussmann, 75008 Paris.

Table 11A.9 World's Highest Dams

Name	River, Location	Structural Height		Gross Reservoir Capacity		Year Completed
		ft	m	Thousands of acre ft	Millions of cum	
Rogun	Vakhsh, Tajikistan	1,099	335	9,404	11,600	1985
Nurek	Vakhsh, Tajikistan	984	300	8,512	10,500	1980
Grande Dixence	Dixence, Switzerland	935	285	324	400	1962
Inguri	Inguri, Georgia	892	272	801	1,100	1984
Vaiont	Vaiont, Italy	859	262	137	169	1961
Manuel M. Torres	Grijalva, Mexico	856	261	1,346	1,660	1981
Tehri	Bhagirathi, India	856	261	2,869	3,540	UC
Alvaro Obregon	Mextiquic, Mexico	853	260	n.a.	n.a.	1926
Mauvoisin	Drance de Bagnes, Switzerland	820	250	146	180	1957
Alberto Lleras	Orinoco, Colombia	797	243	811	1,000	1989
Mica	British Columbia, Canada	797	243	20,000	24,670	1972
Sayano-Shushenskaya	Yenisei, Russia	794	242	25,353	31,300	1980
Ertan	Yangtze/Yalong, China	797	240	4,720	5,800	1999
La Esmeralda	Batá, Colombia	778	237	661	815	1975
Kishau	Tons, India	774	236	1,946	2,400	1985
Oroville	Feather, CA, U.S.A.	770	235	3,538	4,299	1968
El Cajón	Humuya, Honduras	768	234	4,580	5,650	1984
Chirkey	Sulak, Russia	764	233	2,252	2,780	1977
Bhakra	Sutlej, India	741	226	8,002	9,870	1963
Luzzone	Brenno di Luzzone, Switzerland	738	225	71	87	1963
Hoover	Colorado, AZ-NV, U.S.A.	732	223	28,500	35,154	1936
Contra	Verzasca, Switzerland	722	220	70	86	1965
Mratinje	Piva, Herzegovina	722	220	713	880	1973
Dworshak	North Fork Clearwater, ID, U.S.A.	717	219	3,453	4,259	1974
Glen Canyon	Colorado, AZ, U.S.A.	710	216	27,000	33,304	1964
Toktogul	Naryn, Kyrgyzstan	705	215	15,800	19,500	1978
Daniel Johnson	Manicouagan, Canada	703	214	115,000	141,852	1968
Keban	Firat, Turkey	689	210	25,110	31,000	1974
Zimapan	Moctezuma, Mexico	679	207	n.a.	n.a.	1994
Karun	Karun, Iran	673	205	2,351	2,900	1976
Lakhwar	Yamuna, India	669	204	470	580	1985
Dez	Dez, Abi, Iran	666	203	2,707	3,340	1963
Almendra	Tormes, Spain	662	202	2,148	2,649	1970
Berke	Ceyhan, Turkey	659	201	n.a.	n.a.	2000
Khudoni	Inguri, Georgia	659	201	n.a.	n.a.	n.a.
Kölnbrein	Malta, Austria	656	200	166	205	1977
Altinkaya	Kizil Irmak, Turkey	640	195	4,672	5,763	1986
New Bullards Bar	No. Yuba, CA, U.S.A.	637	194	960	1,184	1968
New Melones	Stanislaus, CA, U.S.A.	625	191	2,400	2,960	1979
Itaipu	Paraná, Brazil/Paraguay	623	190	23,510	29,000	1982
Kurobe 4	Kurobe, Japan	610	186	162	199	1964
Swift	Lewis, WA, U.S.A.	610	186	756	932	1958
Mossyrock	Cowlitz, WA, U.S.A.	607	185	1,300	1,603	1968
Oymopinar	Manavgat, Turkey	607	185	251	310	1983
Atatürk	Firat, Turkey	604	184	39,482	48,700	1990
Shasta	Sacramento, CA, U.S.A.	602	183	4,550	5,612	1945
Bennett WAC	Peace, Canada	600	183	57,006	70,309	1967
Karakaya	Firat, Turkey	591	180	7,767	9,580	1986
Tignes	Isère, France	591	180	186	230	1952
Amir Kabir (Karad)	Karadj, Iran	591	180	166	205	1962
Tachien	Tachia, Taiwan	591	180	188	232	1974
Dartmouth	Mitta-Mitta, Australia	591	180	3,243	4,000	1978
Özköy	Gediz, Turkey	591	180	762	940	1983
Emosson	Barberine, Switzerland	590	180	184	225	1974
Zillergrundl	Ziller, Austria	590	180	73	90	1986
Los Leones	Los Leones, Chile	587	1799	86	106	1986
New Don Pedro	Tuolumne, CA, U.S.A.	585	178	2,030	2,504	1971

(Continued)

Table 11A.9 (Continued)

Name	River, Location	Structural Height		Gross Reservoir Capacity		Year Completed
		ft	m	Thousands of acre ft	Millions of cum	
Alpa-Gera	Cormor, Italy	584	178	53	65	1965
Kopperston Tailings 3	Jones Branch, WV, U.S.A.	580	177	—	—	1963
Takase	Takase, Japan	577	176	62	76	1979
Nader Shah	Marun, Iran	574	175	1,313	1,620	1978
Hasan Ugurlu	Yesil Irmak, Turkey	574	175	874	1,078	1980
Revelstoke	Columbia, B.C., Canada	574	175	4,298	5,300	1984
Hungry Horse	S.Fk., Flathead, MT, U.S.A.	564	172	3,470	4,280	1953
Longyangxia	Huanghe, China	564	172	20,025	24,700	1983
Cabora Bassa	Zambezi, Mozambique	561	171	51,075	63,000	1974
Maqarin	Yarmuk, Jordan	561	171	259	320	1987
Amaluza	Paute, Ecuador	558	170	81	100	1982
Idikki	Periyar, India	554	169	1,618	1,996	1974
Charvak	Chirchik, Uzbekistan	552	168	1,620	2,000	1970
Gura Apelor Retezat	Riul Mare, Romania	552	168	182	225	1980
Grand Coulee	Columbia, Washington	550	168	9,390	11,582	1942
Boruca	Terraba, Costa Rica	548	167	12,128	14,960	UC
Vidraru	Arges, Romania	545	166	380	465	1965
Kremasta (King Paul)	Achelöus, Greece	541	165	3,850	4,750	1965
Pauti-Mazar	Mazar, Ecuador	541	165	405	500	1984

Note: UC, under construction in 2004, n.a., not available. China's Three Gorges dam on the Yangtze River, begun in 1993 and expected to be completed in 2009, will be the world's largest and highest dam.

Source: From International Commission on Large Dams, *World Register of Dams 1998*, and other sources.

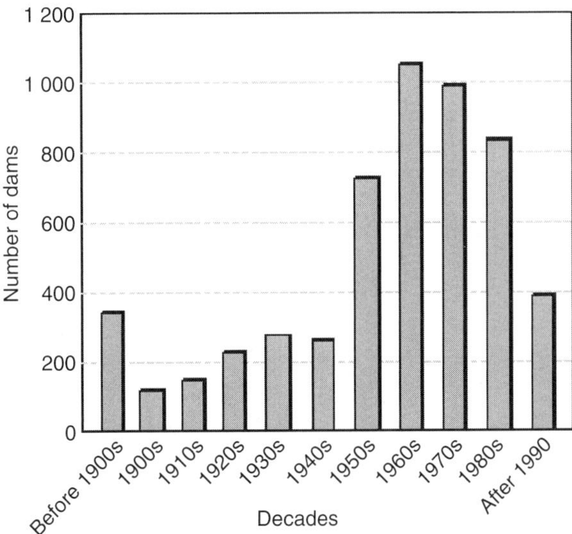

Figure 11A.5 Large dams commissioned per decade in Europe. (From ICOLD, 1998. damsreport.org. Annex V, Dams, Water and Energy: A Statistical Profile.)

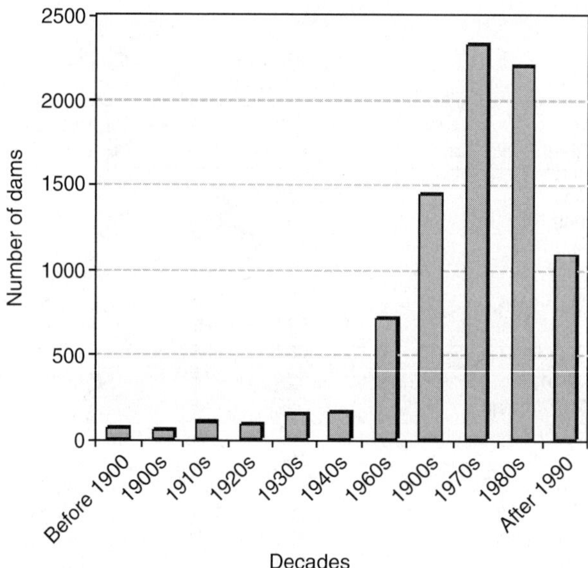

Figure 11A.6 Large dams commissioned per decade in Asia. (From ICOLD, 1998. damsreport.org. Annex V, Dams, Water and Energy: A Statistical Profile.)

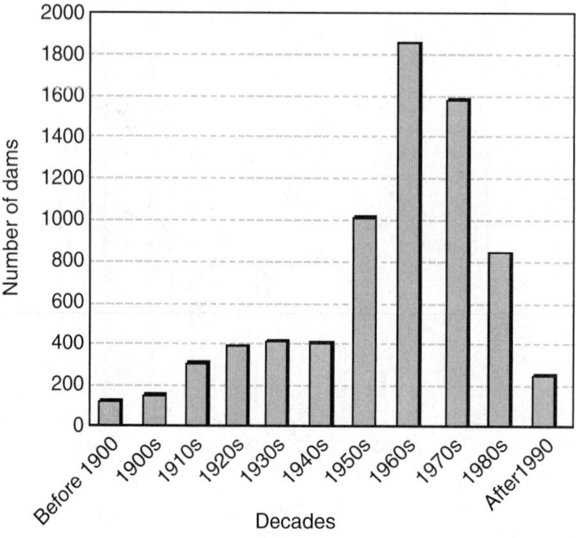

Figure 11A.7 Large dams commissioned per decade in North and Central America. (From ICOLD, 1998. Note: Rates of dam commissioning in the 1990s are underreported. damsreport.org. Annex V, Dams, Water and Energy: A Statistical Profile.)

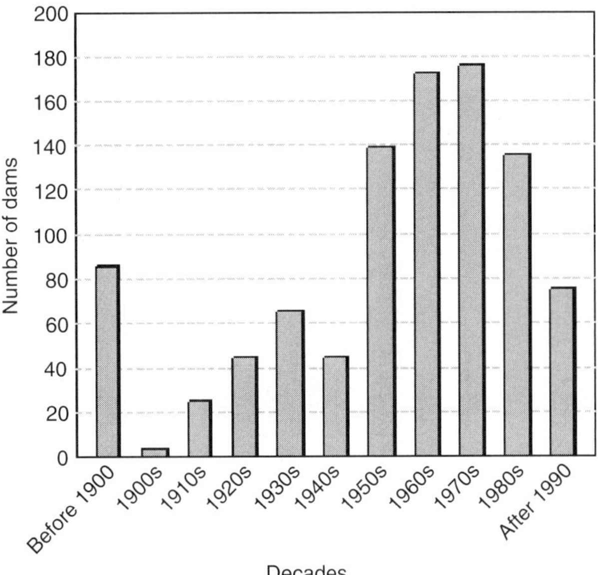

Figure 11A.8 Large dams commissioned per decade in South America. (From ICOLD, 1998. Note: Rates of dam commissioning in the 1990s are underreported.)

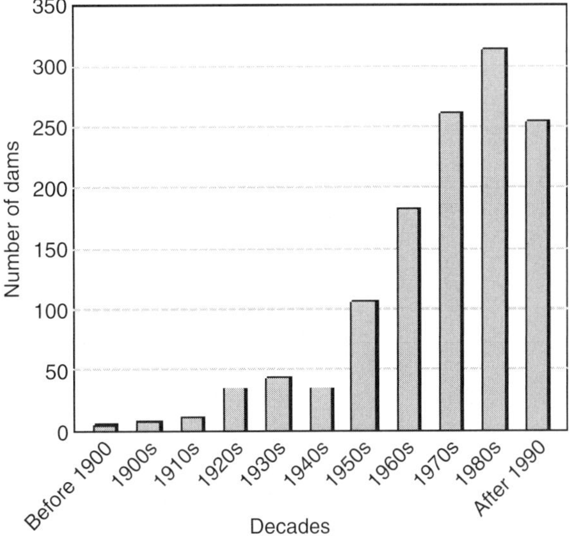

Figure 11A.9 Large dams commissioned per decade in Africa. (From ICOLD, 1998. Note: Rates of dam commissioning in the 1990s are underreported. damsreport.org. Annex V, Dams, Water and Energy: A Statistical Profile.)

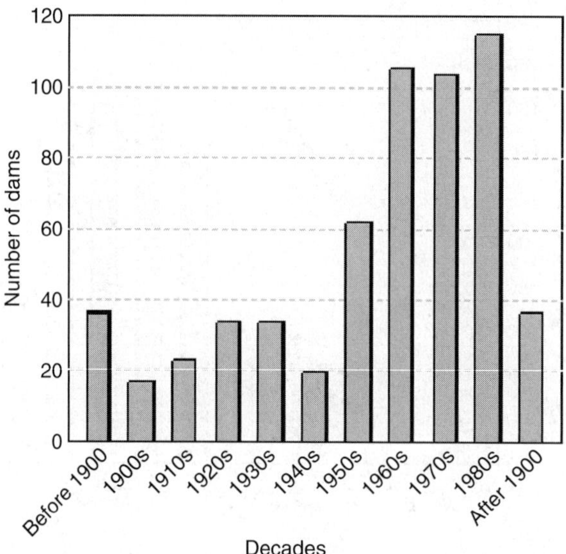

Figure 11A.10 Large dams commissioned per decade in Austral-Asia. (From ICOLD, 1998. Note: Rates of dam commissioning in the 1990s are underreported. damsreport.org. Annex V, Dams, Water and Energy: A Statistical Profile.)

Table 11A.10 Modes and Causes of Earth Dam Failures

Form	General Characteristics	Causes	Preventive or Corrective Measures
	Hydraulic Failures (30% of all failures)		
Overtopping	Flow over embankment, washing out dam	Inadequate spillway capacity	Spillway designed for maximum flood
		Clogging of spillway with debris	Maintenance, trash booms, clean design
		Insufficient freeboard due to settlement, skimpy design	Allowance for freeboard and settlement in design; increase crest height or add flood parapet
Wave erosion	Notching of upstream face by waves, currents	Lack of riprap, too small riprap	Property designed riprap
Toe erosion	Erosion of toe by outlet discharge	Spillway too close to dam	Training walls
		Inadequate riprap	Properly designed riprap
Gullying	Rainfall erosion of dam face	Lack of sod or poor surface drainage	Sod, fine riprap; surface drains
	Seepage Failures (40% of all failures)		
Loss of water	Excessive loss of water from reservoir and/or occasionally increased seepage or increased groundwater levels near reservoir	Previous reservoir rim or bottom	Blanket reservoir with compacted clay or chemical admix; grout seams, cavities
		Previous dam foundation	Use foundation cutoff; grout; upstream blanket
		Pervious dam	Impervious core
		Leaking conduits	Watertight joints; waterstops; grouting
		Settlement cracks in dam	Remove compressible foundation, avoid sharp changes in abutment slope, compact soils at high moisture
		Shrinkage cracks in dam	Use low-plasticity clays for core, adequate compaction

(Continued)

Table 11A.10 (Continued)

Form	General Characteristics	Causes	Preventive or Corrective Measures
Seepage erosion or piping	Progressive internal erosion of soil from downstream side of dam or foundation backward toward the upstream side to form an open conduit or "pipe." Often leads to a washout of a section of the dam	Settlement cracks in dam	Remove compressible foundation, avoid sharp changes, internal drainage with protective filters
		Shrinkage cracks in dam	Low-plasticity soil; adequate compaction; internal drainage with protective filters
		Pervious seams in foundation	Foundation relief drain with filter; cutoff
		Pervious seams, roots, etc. in dam	Construction control; core; internal drainage with protective filter
		Concentration of seepage at face	Toe drain; internal drainage with filter
		Boundary seepage along conduits, walls	Stub cutoff walls, collars; good soil compaction
		Leaking conduits	Watertight joints; waterstops; materials
		Animal burrows	Riprap, wire mesh
Structural Failures (30% of all failures)			
Foundation slide	Sliding of entire dam, one face, or both faces in opposite directions, with bulging of foundation in the direction of movement	Soft or weak foundation	Flatten slope; employ broad berms; remove weak material; stabilize soil
		Excess water pressure in confined sand or silt seams	Drainage by deep drain trenches with protective filters; relief wells
Upstream slope	Slide in upstream face with little or no bulging in foundation below toe	Steep slope	Flatten slope or employ berm at toe
		Weak embankment soil	Increased compaction; better soil
		Sudden drawdown of pond	Flatten slope, rock berms; operating rules
Downstream slope	Slide in downstream face	Steel slope	Flatten slope or employ berm at toe
		Weak soil	Increased compaction; better soil
		Loss of soil strength by seepage pressure or saturation by seepage or rainfall	Core; internal drainage with protective filters; surface drainage
Flow slide	Collapse and flow of soil in either upstream or downstream direction	Loose embankment soil at low cohesion, triggered by shock, vibration, seepage, or foundation movements	Adequate compaction

Source: From National Academy Press, 1983, Safety of existing dams: evaluation and improvement. With permission.

Original Source: From Sowers, G.F., 1961, *The Use and Misuse of Earth Dams, Consulting Engineering*, July.

Table 11A.11 World's Largest Dams According to Spillway Capacity

No	Capacity, (m^3 sec)	Name	Country	Year
	113,000	Gezhouba	China	U/C
	110,000	Tucurui	Brazil	U/C
1	82,300	Dajiangkou	China	1974
2	64,845	The Dalles	U.S.A.	1957
	64,600	Burdekin Falls	Australia	U/C
3	63,713	John Day Lock and Dam	U.S.A.	1968
4	62,297	MacNary Lock and Dam	U.S.A.	1957
	62,296	Sardar Sarovar	India	U/C
5	61,400	Itaipu	Brazil	1982
	60,000	Oosterscheldekering	Netherlands	U/C
6	59,000	Shuifeng	China/Korea	1943
7	58,500	Saratov	U.S.S.R.	1967
8	58,400	Gavins Point	U.S.A.	1958
9	57,000	Salto Grande	Uruguay/Argentina	1979
	54,862	Jhuj	India	U/C
10	54,400	Panjiakou	China	1979
11	53,450	Nagarjuna Sagar	India	1974
	52,000	Porto Primavera	Brazil	U/C
12	50,000	Jupia	Brazil	1968
	49,800	Wanan	China	U/C
13	49,600	Kadana	India	1978
	47,000	Ankang	China	U/C
	46,970	Gohira	India	U/C
14	46,259	Alvin Wirtz	U.S.A.	1950
	45,312	Sri Rama Sagar	India	U/C
15	45,307	Bonneville	U.S.A.	1937
	44,752	Bargi	India	U/C
	43,000	Owen Falls	Uganda	1954
16	43,690	International la Amistad	U.S.A./Mexico	1969
17	42,459	Hirakud	India	1957
18	42,186	Tarbela	Pakistan	1976
19	41,280	Xinanjiang	China	1965
20	40,300	Kuibyshev	U.S.S.R.	1955
21	40,000	Ilha Solteira	Brazil	1973
22	39,644	Priest Rapids	U.S.A.	1959
23	39,644	Wanapum	U.S.A.	1963
24	39,158	Tom Miller	U.S.A.	1939
	37,945	Narayanpur	India	U/C
25	37,500	Shuifeng	China	1943
	37,400	Srisailam HE	India	1943
26	36,200	Fengman	China	1954
27	25,960	Ukai	India	1972
28	35,620	Daheiting	China	1980
	35,000	Guri	Venezuela	U/C
29	33,980	Kentucky	U.S.A.	1944
30	33,800	Fuchunjing	China	1968
31	33,600	Chief Joseph	U.S.A.	1955
32	33,414	Wells	U.S.A.	1967
33	33,131	Conowingo	U.S.A.	1928
34	31,400	Xijin	China	1966
35	31,152	Marala	Pakistan	1968
36	31,144	Mangla	Pakistan	1967

Note: U/C, under construction.

Source: From International Commission on Large Dams, 1984, *World Register of Dams*.

Table 11A.12 Summary of Regional Statistics on Large Dams

	World[a]	Europe	Asia	North and Central America	South America	Africa	Austral-Asia
Total number of large dams	25,420[a]–48,000[a]	5,480	31,340	8,010	979	1,269	577
Average height[b](m)	31	33	33	28	37	28	33
Average reservoir area[b] (km^2)	23	7	44	13	30	43	17
Avg. reservoir capacity[b] (million m^3)	269	70	268	998	1,011	883	205
Technically feasible hydroelectric potential[c] (TWh/yr)	14,370	1,225	6,800	1,660	2,665	1,750	270
Annual hydroelectric production[c] (TWh/yr)	2,643	552	753	700	534	62	42
Exploited technically feasible hydroelectric potential[c] (%)	>18%	>45%	>11%	>42%	>20%	>3.5%	

[a] The primary source of data is ICOLD 1998, but the regional divisions in this Table and in Figure V.16 through Figure V.27 follow those described in Table V.3. The 1998 ICOLD Register has 25,420 dams registered. Reporting depends on the member countries. Table V.I indicates how the global estimate of neatly 48,000 large dams is arrived at, with the main issue being the number of dams in China.
[b] The ICOLD 1998 database was used to calculate the average dam height, reservoir capacity, and surface area by region.
[c] I/HD 2000. Technical Flexibility is based on the conversion of all river lead and flow in the major rivers in region into energy.

Source: Based on ICOLD, 1998. damsreport.org. Annex V, Dams, Water and Energy: A Statistical Profile. With permission.

Table 11A.13 Dam Incident Summary

NPDP ID	Dam Name	Incident Date	Incident Type	Dam Failure
CO00203	Maple Grove	2/1979	Vandalism	Yes
DE00018	Millsboro Pond	2/1979	Piping	Yes
DE00052	Wiggins Mill Pond Dam	2/1979	Internal Erosion	Yes
FL00723	Martin Plant Cooling Water Reservoir	10/30/1979	Piping	Yes
IDS00011	Mud Creek Dam	1/12/1979	Inflow Flood—Hydrologic Event	Yes
KYS00004	Samsel	2/2/1979	Seepage	No
MS00128	Spring Lake	9/1979	Erosion	No
MS02734	Pine Lake Dam	4/1979	Erosion	Unknown
MS01731	Vance Lake	4/13/1979	Piping	Yes
MDS00002	Big Millpond (Route 50)	2/25/1979	Undermining	Yes
IDS00013	Dam on Pierce Park Gulch	2/12/1979	Inflow Flood—Hydrologic Event	Yes
IA01967	Fertile Mill Dam	8/22/1979	Piping	Yes
NJ00235	Gropps Lake Dam	5/27/1979	Sliding	Yes
NM00445	United Nuclear Churchrock	7/16/1979	Cracking and Internal Erosion	Yes
NYS00007	Swimming Poll Dam	1979	Inflow Flood—Hydrologic Event	Unknown
SC01458	Huttos Lake Dam	9/9/1979	Inflow Flood—Hydrologic Event	Yes
UT00379	Goshen	2/14/1979	Inflow Flood—Hydrologic Event	Yes
WI00160	Pulcifer	6/1979	Deterioration	No
WI00053	Little Falls	4/1979	Gate Structural Failure	Unknown
TN15712	Edwards	4/2/1979	Piping	No
NYS00018	Sherman Dam	9/14/1979	Inflow Flood—Hydrologic Event	Yes
NMS00006	Phelps Dodge Corporation Tailings Dam No. 3	10/13/1980	Structural Failure	Yes
SD00921	East Lemmon	6/5/1980	Inflow Flood—Hydrologic Even	Yes
VTS00016	Steward	1980	Piping	Yes
VT00122	Fairfield Swamp Pond	4/13/1980	Piping	Yes
UTS00009	Lower North Eden Reservoir	4/20/1980	Inflow Flood—Hydrologic Event	Yes
UTS00008	South Eden Reservoir	4/20/1980	Inflow Flood—Hydrologic Event	Yes
SCS00004	Tutens Mill Pond	3/14/1980	Inflow Flood—Hydrologic Event	Yes
NY13130	Tannersville Reservoir #1 Dam	3/22/1980	Embankment Erosion	Yes
OR00185	Crump Reservoir	4/19/1980	Piping	Yes
NY00561	Snow Blrd Lake Dam	9/1/1980	Piping	Yes
WIS00005	Kohlsville Dam	4/1980	Seepage	Yes
WI10607	Lepper	3/1980	Gate Structural Failure	Unknown
WI00248	Wyocena	1/1/1980	Abutment Erosion	Unknown

(Continued)

Table 11A.13 (Continued)

NPDP ID	Dam Name	Incident Date	Incident Type	Dam Failure
COS00006	Prospect Reservoir Dam	2/10/1980	Piping	Yes
DE00028	Lake Como Dam	7/29/1980	Inflow Flood—Hydrologic Event	Yes
GAS00214	Rail Road Lake Dam	5/23/1980	Concrete Deterioration	Yes
MN00519	Pickwick	9/21/1980	Inflow Flood—Hydrologic Event	Yes
ID00001	Saint John	5/7/1980	Piping	Yes
CA10268	Cascade	1981	Earthquake	Yes
KYS00006	Eastover Mining Company Dam	12/18/1981	Sabotage—Other	Yes
NHS00007	Mascoma River Dam No. 2	2/12/1981	Structural Failure	Yes
TXS00002	Unnamed Dam (TXS00002)	1981	Inflow Flood—Hydrologic Event	No
UTS00011	Upper North Eden Reservoir	4/20/1981	Spillway Failure	Yes
WI00146	Necedah	9/20/1981	Not Known	Unknown
ORS00005	Mann Creek Dam	11/8/1982	Piping	Yes
WA00240	Alexander Lake Dam	12/3/1982	Inflow Flood—Hydrologic Event; Animal Attack	Yes
WAS00013	Alexander Lake	12/3/1982	Inflow Flood—Hydrologic Event	Yes
UTS00006	Milk Pond Dam	6/23/1982	Inflow Flood—Hydrologic Event	Yes
COS00004	Lawn Lake	7/15/1982	Piping	Yes
CT00157	Gorton Pond	6/4/1982	Inflow Flood—Hydrologic Event	Yes
CT00396	Jennings Pond	6/4/1982	Inflow Flood—Hydrologic Event	Unknown
CT00417	Upper Millpond	6/4/1982	Inflow Flood—Hydrologic Event	Unknown
CT00426	Bushy Hill Pond	6/1982	Inflow Flood—Hydrologic Event	Yes
CT00545	Holbrook Pond	6/1982	Inflow Flood—Hydrologic Event	Yes
CTS00008	Comstock Pond (CT 424	6/4/1982	Inflow Flood—Hydrologic Event	Yes
CTS00007	C. D. Batchelor Pond	6/4/1982	Inflow Flood—Hydrologic Event	Yes
CTS00006	Bushy Pond (CT 391)	6/4/1982	Inflow Flood—Hydrologic Event	Yes
CTS00005	Bronson Company Dam (CT 691)	6/4/1982	Inflow Flood—Hydrologic Event	Yes
CTS00003	Urban Pond	6/4/1982	Inflow Flood—Hydrologic Event	Yes
CTS00002	Hempstead Pond	6/4/1982	Inflow Flood—Hydrologic Event	Yes
CTS00001	Crystal Lake	6/4/1982	Inflow Flood—Hydrologic Event	No
CT00662	Leesville Dam	6/4/1982	Inflow Flood—Hydrologic Event	Yes
CT00427	Pratt Read	6/4/1982	Erosion	Unknown
IDS00009	Cameron Dam	4/24/1982	Inflow Flood—Hydrologic Event	Yes
IDS00008	Howard Dam	4/24/1982	Inflow Flood—Upstream Dam Failure	Yes
CTS00017	Whalebone Creek Pond (CT 1024)	6/4/1982	Inflow Flood—Hydrologic Event	Yes
CTS00016	Upper Pond (CT 433)	6/4/1982	Inflow Flood—Hydrologic Event	Yes
CTS00015	Mansure Pond (CT 517)	6/4/1982	Inflow Flood—Hydrologic Event	Yes
CTS00014	Main Street Pond (CT 880)	6/4/1982	Inflow Flood—Hydrologic Event	Unknown
CTS00013	Lower Mill Pond (CT 1190)	6/4/1982	Inflow Flood—Hydrologic Event	Yes
CTS00012	Lower Pond (CT 1512)	6/4/1982	Inflow Flood—Hydrologic Event	Yes
CTS00011	Ivoryton Pond (CT 882)	6/4/1982	Inflow Flood—Hydrologic Event	Yes
CTS00010	Forman Pond	6/4/1982	Inflow Flood—Hydrologic Event	Yes
CTS00009	Falls River Pond (CT 884)	6/4/1982	Inflow Flood—Hydrologic Event	Yes
CTS00023	Abbott Pond (CT 774)	6/4/1982	Inflow Flood—Hydrologic Event	Unknown
CTS00022	Dennison Road Pond (CT 1504)	6/4/1982	Inflow Flood—Hydrologic Event	Unknown
CTS00021	Dolan Pond (CT 883)	6/4/1982	Inflow Flood—Hydrologic Event	Unknown
CTS00020	Hunts Brook Dam	6/4/1982	Inflow Flood—Hydrologic Event	Yes
CTS00019	Mile Creek Pond (CT 1191)	6/4/1982	Inflow Flood—Hydrologic Event	Unknown
CTS00018	Shady Brook Pond (CT 846)	6/4/1982	Inflow Flood—Hydrologic Event	Yes
CT00423	Mill Pond	6/4/1982	Inflow Flood—Hydrologic Event	Yes
CT00404	Deer Lake	6/1982	Inflow Flood—Hydrologic Event	Yes
CT00339	Johnson Pond	6/4/1982	Inflow Flow—Hydrologic Event	Yes
COS00005	Cascade Lake Dam	7/15/1982	Inflow Flow—Hydrologic Event	Yes
MNS00002	Fishhook River Dam Cofferdam	10/1983	Seepage	No
MS01738	Lakeview Reservoir Dam	11/27/1983	Not Known	Yes
UT00080	Dmad	6/23/1983	Not Known	Unknown
WAS00010	Peters Reservoir No. 2	3/10/1983	Not Known	Yes
WAS00009	Peters Reservoir No. 1	3/10/1983	Not Known	Yes
NJS00062	Unnamed Dam (NJS00062)	1984	Inflow Flood—Hydrologic	No

(Continued)

Table 11A.13 (Continued)

NPDP ID	Dam Name	Incident Date	Incident Type	Dam Failure
WIS00003	Grettum Flowage	6/13/1984	Inflow Flow—Hydrologic Event	Unknown
VTS00015	Halls Lake	5/31/1984	Piping	Yes
VTS00014	Laraway	6/7/1984	Inflow Flow—Hydrologic Event	Yes
VTS00013	Wards Pond	6/7/1984	Inflow Flow—Hydrologic Event; Piping	Yes
VT00081	North Montpelier Pond	6/7/1984	Inflow Flow—Hydrologic Event	Yes
UTS00001	Doty-Tex Johnson (Utah)	4/12/1984	Inflow Flow—Hydrologic Event	Yes
TX05119	Bass Haven Lake Dam	8/17/1984	Embankment Erosion	Yes
TNS00017	Unnamed Dam on Del Rio Creek	5/6/1984	Piping	Yes
TNS00011	Shangri La Lake	9/1984	Sliding	No
SD00695	Dimock	6/20/1984	Inflow Flow—Hydrologic Event	Yes
CTS00004	Haas Pond Dam	5/20/1984	Piping	Yes
NE02138	Merlyn Schrunk Dam	4/1984	Inflow Flow—Hydrologic Event	Yes
NES00003	Atkinson Reservoir	4/8/1984	Spillway Failure	No
MTS00006	Brownes Lake	6/20/1984	Plugged Spillway	Yes
IL00712	Riverview Dam	11/21/1985	Seepage	Yes
MO31374	Richardet Dam	12/1985	Seepage; Embankment Slide	Yes
WI00726	Port Wing	9/1/1985	Inflow Flow—Hydrologic Event	Yes
OK00663	Scs-Upper Red Rock Creek Site–20	10/3/1986	Seepage; Piping	Yes
OK11073	Cedar Lake	11/8/1986	Not Known; Seepage; Piping	Yes
WA00074	Upriver Dam	5/20/1986	Gate Closure	Yes
VTS00012	Noonan	1986	Piping	Yes
UT00301	Trial Lake	6/7/1986	Seepage; Piping	Yes
PAS00005	Unnamed Dam (PAS00005)	1986	Not Known	No
TNS00014	Demery's Lake	4/1986	Inflow Flood—Hydrologic Event	Yes
MI00209	Barryton Dam	9/10/1986	Inflow Flood—Hydrologic Event	Yes
MI00573	Danaher Lake Dam	9/10/1986	Inflow Flood—Hydrologic Event	Yes
MI00616	Rainbow Lake Dam	9/10/1986	Inflow Flood—Hydrologic Event	Yes
MIS00001	Bruce Nordland Dam	9/10/1986	Inflow Flood—Hydrologic Event	Yes
NE00492	Haeker Dam	8/1986	Inflow Flood—Hydrologic Event	Yes
ND00426	Simpson Dam; Alvin	7/17/1986	Inflow Flood—Hydrologic Event; Seepage; Piping	Yes
MIS00004	Childsdale Dam	9/10/1986	Inflow Flood—Hydrologic Event	Yes
MIS00003	Cat Creek Dam	9/10/1986	Inflow Flood—Hydrologic Event	Yes
MIS00002	Carson City Dam	9/10/1986	Inflow Flood—Hydrologic Event	Yes
MI00678	Hesperia Dam	9/10/1986	Inflow Flood—Hydrologic Event	Yes
MI00574	Luther Pond Dam	9/10/1986	Inflow Flood—Hydrologic Event	Yes
MI00526	White Cloud Dam	9/10/1986	Inflow Flood—Hydrologic Event	Yes
MI00281	Hart Lake	9/10/1986	Inflow Flood—Hydrologic Event	Yes
OK20844	Scs—Little Washita River Site–13	9/1987	Piping	Yes
TNS00008	Tomkins Lake	12/25/1987	Inflow Flood—Hydrologic	Yes
TNS00009	Sky Lake No. 1	12/24/1987	Piping	Yes
TN10305	Rebecca Lake	12/25/1987	Inflow Flood—Hydrologic Event	No
WI00638	Bog Brook	5/1988	Inflow Flood—Hydrologic Event; Animal Attack	Yes
WI00330	Bischel	3/1988	Piping	Yes
UT00514	Quail Creek	12/31/1988	Seepage; Piping; Embankment Erosion	Yes
MO31923	Marschke Lake Dam	4/19/1988	Not Known	Yes
MTS00002	Hein Coulee Structure (Lower Birch Creek Watershed)	7/6/1988	Piping	Yes
KYS00005	Unnamed Dam (KYS00005)	1989	Inflow Flood—Hydrologic Event	Yes
MNS00004	Unnamed Dam (MNS00004)	1989	Embankment Failure	No
MNS00005	Unnamed Dam (MNS00005)	1989	Embankment Failure	No
NC02149	Evans Dam	9/15/1989	Not Known	Yes
NC02152	Lockwood Dam	9/15/1989	Not Known	Yes
MNS00006	Unnamed Dam (MNS00006)	1989	Embankment Failure	No
MNS00007	Unnamed Dam (MNS00007)	1989	Embankment Failure	No

(Continued)

Table 11A.13 (Continued)

NPDP ID	Dam Name	Incident Date	Incident Type	Dam Failure
NJS00014	Holmdel Park Dam	7/1989	Inflow Flood—Hydrologic Event; Seepage; Piping	Yes
NYS00015	Unnamed Dam (NYS00015)	1989	Not Known	Unknown
TNS00010	Southern Clay Company Dam No. 2	9/8/1989	Piping	Yes
ORS00004	Avion Water District	1989	Faulty Design and Construction	Yes
TNS00018	Summit Landfill Dam	9/29/1989	Inflow Flood—Hydrologic Event	Yes
TNS00025	Summit Landfill	10/1/1989	Inflow Flood—Hydrologic Event	Yes
TX03546	Nix Lake Dam	3/29/1989	Inflow Flood—Hydrologic Event	Yes
TNS00019	Unnamed Dam (TNS00019)	6/23/1989	Inflow Flood—Hydrologic Event	Yes
OH00575	Acton Lake Dam	1990	Drain Gate Failure	No
VAS00006	Cockram Mill Dam	7/14/1990	Inflow Flood—Hydrologic Event	Yes
VT00217	Beaver Pond	7/23/1990	Inflow Flood—Hydrologic Event	Yes
VTS00011	Eight Trout Club	3/15/1990	Piping	Yes
WAS00015	Unnamed Dam (WAS00015)	11/1990	Inadequate Spillway Capacity	Yes
WI00191	Leland	6/1990	Inflow Flood—Hydrologic Event	Yes
WA01707	Chinook Water District Dam	11/1990	Inflow Flood—Hydrologic Event; Inadequate Spillway Capacity	Yes
VTS00010	Riddel Pond	5/1990	Piping	Yes
SCS00005	Unnamed Dam (SCS00005)	7/18/1990	Piping	Yes
SC02298	Brewer Gold Company Dam 1	10/10/1990	Not Known	Yes
SC00459	Kendall Lake Dam	10/10/1990	Inflow Flood—Hydrologic Event	Yes
AL00539	Caddis Lake Dam	1990	Inflow Flood—Hydrologic Event	No
ALS00006	Campbell	1990	Inflow Flood—Hydrologic Event	Yes
ALS00007	Unnamed Dam (ALS00007)	1990	Inflow Flood—Hydrologic Event	Yes
NE02363	Timperley Wildlife Res	12/1990	Animal Attack	Yes
NC03195	Landrum Lake Dam(Failer)	2/16/1990	Structural Failure	Yes
MOS00004	St. Joe State Park Sediment Impoundment	2/15/1990	Inflow Flood—Hydrologic Event; Inadequate Spillway Capacity	Yes
MO20145	Christiansen Lake Dam	5/1990	Embankment Erosion	Yes
MO12279	Hester Lake Dam	6/27/1990	Not Known	Yes
IL50319	Wilmington Dam (Kankakee River Mill Race Dam)	6/1990	Not Known	Yes
IL50233	Lake Carroll Sedimentation Pond 2 Dam	5/1990	Inflow Flood—Hydrologic Event	Yes
IDS00002	Kirby Dam (Manns Lake; Atlanta Dam)	7/9/1990	Not Known; Seepage; Embankment Erosion	Yes
ALS00008	Unnamed Dam (ALS00008)	1990	Inflow Flood—Hydrologic Event	Yes
NCS00039	Unnamed Dam (NCS00039)	4/29/1991	Not Known	Yes
NHS00006	Unnamed Dam (NHS00006)	8/8/1991	Seepage	Yes
NYS00009	Unnamed Dam (NYS00009)	1991	Not Known	Unknown
TX00309	Lake Center Dam	8/4/1991	Not Known	Yes
SCS00002	Upper Twin Lake Dam	8/2/1991	Inflow Flood—Hydrologic Event	Yes
SCS00001	Lower Twin Lakes Dam	8/2/1991	Inflow Flood—Upstream Dam Failure	Yes
RIS00003	Unnamed Dam (RIS00003)	1991	Not Known	Unknown
RIS00002	Unnamed Dam (RIS00002)	1991	Not Known	Unknown
RI04258	Burton Pond Dam	1991	Concrete Deterioration	Yes
NYS00010	Unnamed Dam (NYS00010)	1991	Not Known	Unknown
NYS00011	Unnamed Dam (NYS00011)	1991	Not Known	Unknown
NYS00012	Unnamed Dam (NYS00012)	1991	Not Known	Unknown
NYS00014	Unnamed Dam (NYS00014)	1991	Not	Unknown
NYS00013	Unnamed Dam (NYS00013)	1991	Not	Unknown
WVS0005	Unnamed Dam (WVS00005)	1991	Inflow Flood—Hydrologic Event	Unknown
WAS00001	Seminary Hill Reservoir, City of Centralia	10/5/1991	Landslide; Seepage; Concrete Deterioration	Yes
VTS00009	Swanson	4/1991	Piping	Yes
NJ00541	Port Republic dam	1992	Inflow Flood—Hydrologic Event	Yes
OK01437	Scs-Upper Black Creek Site–62	1992	Inflow Flood—Hydrologic Event	No
WY00037	Wyoming Hereford Ranch No. 2	1992	Not Known	Yes

(Continued)

Table 11A.13 (Continued)

NPDP ID	Dam Name	Incident Date	Incident Type	Dam Failure
WI00468	La Blonde	7/1992	Inflow Flood—Hydrologic Event; Debris-Reservoir	Yes
NV00024	Bilk Creek Res	1992	Inflow Flood—Hydrologic	Yes
MDS00001	Black Rock Estates Pond	4/21/1992	Inflow Flood—Hydrologic Event; Piping	Yes
MOS00015	Unnamed Dam (MOS00015)	6/5/1992	Erosion	Yes
COS00007	Greeley	1/1993	Animal Burrows	Yes
NE01419	Krone Dam	1993	Inflow Flood—Hydrologic Event	Yes
ND00107	Jund (Zeeland) Dam	7/16/1993	Inflow Flood—Hydrologic Event; Biological Attack (i.e. bush, tree growth); Embankment Erosion; Inadequate Spillway Capacity	Yes
MOS00002	Norman Swinney's Dam	5/26/1993	Inadequate Compaction	Yes
MO32026	Freddies Lake Dam	9/26/1993	Inflow Flood—Hydrologic Event	Yes
MO31996	Boyd Lake Dam	9/25/1993	Embankment Slide	Yes
MO31526	Bockelman Lake Dam	7/1993	Inflow Flood—Hydrologic Event	Yes
MO12370	Harrison County Lake	1/3/1993	Inflow Flood—Hydrologic Event	Yes
MO10107	Stevens Lake Dam	6/1993	Inflow Flood—Hydrologic Event	Yes
MD00330	Annapolis Mall Swm Pond	3/4/1993	Inflow Flood—Hydrologic Event	Yes
OHS00002	Middletown Hydraulic Dam	5/13/1993	Undermining	Yes
WI00112	Rock	6/20/1993	Inflow Flood—Hydrologic Event	Yes
WI00158	Briggsville	3/23/1993	Reconstruction	Yes
WI12788	Hatfield Headrace	7/20/1993	Inflow Flood—Hydrologic Event	Yes
WI10130	Lake Family	4/23/1993	Inflow Flood—Hydrologic Event	Yes
WI01073	Fairchild	6/1993	Inflow Flood—Hydrologic Event	Yes
WI00154	Partridge Lake	1/1993	Piping	Yes
WI00035	Cambria	6/26/1993	Inflow Flood—Hydrologic Event; Gate Misoperation	Yes
OHS00005	Unnamed Dam (OHS00005)	1993	Inflow Flood—Hydrologic Event	Unknown
OHS00004	Unnamed Dam (OHS00004)	1993	Inflow Flood—Hydrologic Event	Unknown
WI00016	Hatfiled	6/18/1993	Inflow Flood—Hydrologic Event; Seepage; Piping; Gate Structural Failure	Yes
WA00408	Iowa Beef Processors Waste Pond No. 1	1/25/1993	Inflow Flood—Hydrologic Event; Animal Attack; Seepage; Inadequate Spillway Capacity	Yes
OHS00003	Unnamed Dam (OHS00003)	1993	Inflow Flood—Hydrologic Event	Unknown
NV10311	Buckskin Tailings	9/12/1994	Earthquake	No
WI00791	Ladysmith	9/15/1994	Inflow Flood—Hydrologic Event	Yes
WI00502	Gomulak And Profit	9/16/1994	Inflow Flood—Hydrologic Event	Yes
WA00063	Sherry Lake Dam	1994	Inflow Flood—Hydrologic Event	Yes
VT00035	Newport No. 11 Diversion Dam	5/1/1994	Inflow Flood—Hydrologic Event; Embankment Erosion	Yes
TX04210	Cade Lake Number 3 Dam	1994	Inadequate Spillway Capacity	Yes
TX03916	Bearfoot Lake Dam	1994	Spillway Failure	No
TX03876	Lake Tinkle Dam	1994	Spillway Failure	Yes
WI10532	Eleva Roller Mill	3/26/1994	Piping; Biological Attack (i.e. bush, tree growth)	Yes
SC00167	Lake Pauline Dam	6/27/1994	Inflow Flood—Upstream Dam Failure	Yes
SC00149	Crystal Lake Dam	6/27/1994	Inflow Flood—Upstream Dam Failure	Yes
SC00142	Saxe-Gotha Millpond Dam	6/27/1994	Inflow Flood—Hydrologic Event; Gate Misoperation	Yes
PA00899	Fishpond	4/28/1994	Concrete Deterioration	Yes
PA00780	Arrowhead Lake	4/26/1994	Piping	Yes
OH00355	Invex of Ohio Upper Lake Dam	8/13/1994	Inflow Flood—Hydrologic Event	Yes
OH00088	Chopper's Lake Dam	4/10/1994	Inflow Flood—Hydrologic Event	Yes

(Continued)

Table 11A.13 (Continued)

NPDP ID	Dam Name	Incident Date	Incident Type	Dam Failure
FLS00001	IMC–AGRICO Hopewell Mine	11/19/1994	Inflow Flood—Hydrologic Event; Seepage; Piping	Yes
GA00211	Gibson—Cary Development Corp Dam	7/5/1994	Inflow Flood—Hydrologic Event	Yes
GA00238	Houston Lake Dam	7/6/1994	Inflow Flood—Hydrologic Event	Yes
GA00239	Mossy Lake Dam	7/6/1994	Inflow Flood—Hydrologic Event	Yes
GA00287	Giles Lake Dam	8/30/1994	Inflow Flood—Hydrologic Event	Yes
GA00831	Grisp County (Warwick)	7/9/1994	Inflow Flood—Hydrologic Event	Yes
GA02902	Holoka Lake Dam	8/17/1994	Inflow Flood—Hydrologic Event	Yes
GA02899	Mulkey Lake Dam	8/17/1994	Inflow Flood—Hydrologic Event	Yes
GA02746	Whatley Lake Dam	7/21/1994	Inflow Flood—Hydrologic Event	Yes
GA02744	Rustins Pond Dam	7/5/1994	Inflow Flood—Hydrologic Event	Yes
GA02743	Hortmans Pond Dam	7/21/1994	Inflow Flood—Hydrologic Event	Yes
GA02677	Goose Lake Dam	7/21/1994	Inflow Flood—Hydrologic Event	Yes
GA01679	McGill Lake Dam	8/17/1994	Inflow Flood—Hydrologic Event	Yes
GA01419	Philema Lake Dam	7/6/1994	Inflow Flood—Hydrologic Event	Yes
GA01070	Phillips Pond Dam	8/17/1994	Inflow Flood—Hydrologic Event	Yes
GA03574	Lies Lake Dam	8/17/1994	Inflow Flood—Hydrologic Event	Yes
GA03572	Coffin Lake Dam	8/17/1994	Inflow Flood—Hydrologic Event	Yes
GA03571	Wolhwender Lake Dam	8/17/1994	Inflow Flood—Hydrologic Event	Yes
GA03568	Shellhouse Lake Dam	8/17/1994	Inflow Flood—Hydrologic Event	Yes
GA03567	Merritt Lake Dam	8/17/1994	Inflow Flood—Hydrologic Event	Yes
GA03540	Lake Yohola Dam	7/7/1994	Inflow Flood—Hydrologic Event	Yes
GA03527	McKemie Lake North Dam	9/12/1994	Inflow Flood—Hydrologic Event	Yes
GA03526	McKemie Lake Dam	9/12/1994	Inflow Flood—Hydrologic Event	Yes
GA03203	Thomas Millpond Dam	7/6/1994	Inflow Flood—Hydrologic Event	Yes
GAS00005	Cordrays Pond Dam	8/17/1994	Inflow Flood—Hydrologic Event	Yes
GAS00004	City of Senoia Dam	7/5/1994	Inflow Flood—Hydrologic Event	Yes
GAS00003	Birch Creek Farms	7/6/1994	Inflow Flood—Hydrologic Event	Yes
GAS00002	Baker/Austin Pond Dam	7/6/1994	Inflow Flood—Hydrologic Event	Yes
GAS00001	6 acre lake on Forrest Road	7/8/1994	Inflow Flood—Hydrologic Event	Yes
GA04987	Suggs Millpond	7/5/1994	Inflow Flood—Hydrologic Event	Yes
GAS00086	Unnamed small dam	8/17/1994	Inflow Flood—Hydrologic Event	Yes
GAS00085	Unnamed small dam	8/17/1994	Inflow Flood—Hydrologic Event	Yes
GAS00084	Unnamed small dam	8/17/1994	Inflow Flood—Hydrologic Event	Yes
GAS00083	Unnamed small dam S of US 27 By-Pass & Webster St.	8/17/1994	Inflow Flood—Hydrologic Event	Yes
GAS00082	Unnamed small dam 2 miles SW of Cuthbert	8/17/1994	Inflow Flood—Hydrologic Event	Yes
GAS00081	Unnamed small dam 3 miles S of Cuthbert, u/s of US	8/17/1994	Inflow Flood—Hydrologic Event	Yes
GAS00080	Unnamed small dam SE of County Rds. #134 & #152	8/17/1994	Inflow Flood—Hydrologic Event	Yes
GAS00079	Unnamed small dam SE of County Rds. #134 ˜	8/17/1994	Inflow Flood—Hydrologic Event	Yes
GAS00078	Unnamed small dam on Collins Mill Creek	8/17/1994	Inflow Flood—Hydrologic Event	Yes
GAS00077	Unnamed small dam north of Bethlehem Church	8/17/1994	Inflow Flood—Hydrologic Event	Yes
GAS00076	Unnamed small dam	9/12/1994	Inflow Flood—Hydrologic Event	Yes
GAS00075	Unnamed small dam	8/30/1994	Inflow Flood—Hydrologic Event	Yes
GAS00074	Unnamed small dam	8/30/1994	Inflow Flood—Hydrologic Event	Yes
GAS00073	Unnamed small dam	8/30/1994	Inflow Flood—Hydrologic Event	Yes
GAS00072	Unnamed small dam	8/30/1994	Inflow Flood—Hydrologic Event	Yes
GAS00071	Unnamed small dam	8/30/1994	Inflow Flood—Hydrologic Event	Yes
GAS00070	Unnamed small dam	8/30/1994	Inflow Flood—Hydrologic Event	Yes
GAS00069	Unnamed dam	8/17/1994	Inflow Flood—Hydrologic Event	Yes
GAS00068	Unnamed small dam 1 mile NW of Buena Vista Lake Dam	8/17/1994	Inflow Flood—Hydrologic Event	Yes
GAS00067	Unnamed small dam	8/17/1994	Inflow Flood—Hydrologic Event	Yes

(Continued)

Table 11A.13 (Continued)

NPDP ID	Dam Name	Incident Date	Incident Type	Dam Failure
GAS00066	Unnamed small dam	8/17/1994	Inflow Flood—Hydrologic Event	Yes
GAS00065	Unnamed dam	7/21/1994	Inflow Flood—Hydrologic Event	Yes
GAS00064	Unnamed dam	7/21/1994	Inflow Flood—Hydrologic Event	Yes
GAS00063	Unnamed small dam west of SR 214, north of SR26	8/17/1994	Inflow Flood—Hydrologic Event	Yes
GAS00062	Unnamed small dam west of SR 128	8/17/1994	Inflow Flood—Hydrologic Event	Yes
GAS00061	Unnamed small dam at RR, SE of Montezuma	8/17/1994	Inflow Flood—Hydrologic Event	Yes
GAS00060	Unnamed small dam on Meadow Creek	8/17/1994	Inflow Flood—Hydrologic Event	Yes
GAS00059	Unnamed small dam 1 mile SW of Marshallville	8/17/1994	Inflow Flood—Hydrologic Event	Yes
GAS00058	Unnamed small dam 1 mile E of Marshallville Lake Dam	8/17/1994	Inflow Flood—Hydrologic Event	Yes
GAS00057	Unnamed small dam on Gin Creek, west of Flint River	8/17/1994	Inflow Flood—Hydrologic Event	Yes
GAS00056	Unnamed dam	7/29/1994	Inflow Flood—Hydrologic Event	Yes
GAS00055	Unnamed dam	7/29/1994	Inflow Flood—Hydrologic Event	Yes
GAS00054	Unnamed dam	7/29/1994	Inflow Flood—Hydrologic Event	Yes
GAS00053	Unnamed small dam	8/30/1994	Inflow Flood—Hydrologic Event	Yes
GAS00052	Unnamed small dam	8/30/1994	Inflow Flood—Hydrologic Event	Yes
GAS00051	Unnamed small dam	8/30/1994	Inflow Flood—Hydrologic Event	Yes
GAS00050	Unnamed small dam	8/30/1994	Inflow Flood—Hydrologic Event	Yes
GAS00049	Unnamed small dam	8/30/1994	Inflow Flood—Hydrologic Event	Yes
GAS00048	Unnamed small dam	8/30/1994	Inflow Flood—Hydrologic Event	Yes
GAS00047	Unnamed small dam	7/14/1994	Inflow Flood—Hydrologic Event	Yes
GAS00046	Unnamed small dam 1 mile east of Felder Lake Dam	9/12/1994	Inflow Flood—Hydrologic Event	Yes
GAS00045	Unnamed small dam 1.5 miles NW of Sutton's Corner	9/12/1994	Inflow Flood—Hydrologic Event	Yes
GAS00044	Unnamed small dam 2 miles SE of Bluffton	9/12/1994	Inflow Flood—Hydrologic Event	Yes
GAS00043	Unnamed dam	8/17/1994	Inflow Flood—Hydrologic Event	Yes
GAS00042	Unnamed dam	8/17/1994	Inflow Flood—Hydrologic Event	Yes
GAS00041	Underwood Millpond Dam	8/17/1994	Inflow Flood—Hydrologic Event	Yes
GAS00040	Thorton Place Pond Dam	7/14/1994	Inflow Flood—Hydrologic Event	Yes
GAS00039	Taylor's Mill Pond Dam	7/5/1994	Inflow Flood—Hydrologic Event	Yes
GAS00038	Swearingen Lake Dam	7/21/1994	Inflow Flood—Hydrologic Event	Yes
GAS00037	Small Rovoli Lake Dam	7/5/1994	Inflow Flood—Hydrologic Event	Yes
GAS00036	Small Lake above Double "O" Ranch	7/5/1994	Inflow Flood—Hydrologic Event	Yes
GAS00035	Shofill Lake Dam	8/30/1994	Inflow Flood—Hydrologic Event	Yes
GAS00034	Scout Lake Dam	8/30/1994	Inflow Flood—Hydrologic Event	Yes
GAS00033	Rockhill Lake Dam	7/5/1994	Inflow Flood—Hydrologic Event	Yes
GAS00032	Parish Lake Dam South	9/12/1994	Inflow Flood—Hydrologic Event	Yes
GAS00031	Owens Lake Dam	8/17/1994	Inflow Flood—Hydrologic Event	Yes
GAS00030	Old Farm Dam	7/5/1994	Inflow Flood—Hydrologic Event	Yes
GAS00029	Minor's Millpond Lake Dam	7/21/1994	Inflow Flood—Hydrologic Event	Yes
GAS00028	McNeil Lake Dam	9/12/1994	Inflow Flood—Hydrologic Event	Yes
GAS00027	McMath Millpond Upper	7/6/1994	Inflow Flood—Hydrologic Event	Yes
GAS00026	McMath Millpond Lower	7/6/1994	Inflow Flood—Hydrologic Event	Yes
GAS00025	Lower Leisure Lake Dam	7/5/1994	Inflow Flood—Hydrologic Event	Yes
GAS00024	Loki Lake Dam	7/6/1994	Inflow Flood—Hydrologic Event	Yes
GAS00023	Levee @ Macon	7/6/1994	Inflow Flood—Hydrologic Event	Yes
GAS00022	Lamar County Reservoir Dam	7/5/1994	Inflow Flood—Hydrologic Event	Yes
GAS00021	Lake Jennifer Upper	7/29/1994	Inflow Flood—Hydrologic Event	Yes
GAS00020	Lake Jennifer Lower	7/29/1994	Inflow Flood—Hydrologic Event	Yes
GAS00019	Lake Corinth Dam	7/6/1994	Inflow Flood—Hydrologic Event	Yes
GAS00018	Iris "B" Lake Dam	7/6/1994	Inflow Flood—Hydrologic Event	Yes
GAS00017	Hutcheson Lake Dam	7/6/1994	Inflow Flood—Hydrologic Event	Yes
GAS00016	Housers Millpond Dam	8/30/1994	Inflow Flood—Hydrologic Event	Yes

(Continued)

Table 11A.13 (Continued)

NPDP ID	Dam Name	Incident Date	Incident Type	Dam Failure
GAS00015	Harell Lake Dam	8/17/1994	Inflow Flood—Hydrologic Event	Yes
GAS00014	Hancock	7/6/1994	Inflow Flood—Hydrologic Event	Yes
NE00419	Waterman Dam	8/1994	Inflow Flood—Hydrologic Event	Yes
NE00212	Morgan Dam	1994	Inflow Flood—Hydrologic Event	Yes
IL50330	East Peoria Dredge Disposal Facility	11/11/1994	Seepage; Piping	Yes
GAS00210	Unnamed Dam (GAS00210)	1994	Flood	Yes
GAS00183	Williamson Downs Lake Dam	7/5/1994	Inflow Flood—Hydrologic Event	Yes
GAS00182	Wells Millpond Lake Dam	7/29/1994	Inflow Flood—Hydrologic Event	Yes
GAS00181	Wainwright Lake Dam (2)	7/21/1994	Inflow Flood—Hydrologic Event	Yes
GAS00180	Wainwright Lake Dam (1)	7/21/1994	Inflow Flood—Hydrologic Event	Yes
GAS00179	Vinny Mill Pond Dam	7/6/1994	Inflow Flood—Hydrologic Event	Yes
GAS00178	Upper & Lower Marimac	7/8/1994	Inflow Flood—Hydrologic Event	Yes
GAS00177	Upper Jackson north of County Road #152	8/17/1994	Inflow Flood—Hydrologic Event	Yes
GAS00176	Upper Iris "A"	7/6/1994	Inflow Flood—Hydrologic Event	Yes
GAS00175	Unnamed small dam	8/17/1994	Inflow Flood—Hydrologic Event	Yes
GAS00174	Unnamed small dam	8/17/1994	Inflow Flood—Hydrologic Event	Yes
GAS00173	Unnamed small dam	8/17/1994	Inflow Flood—Hydrologic Event	Yes
GAS00172	Unnamed small dam	8/17/1994	Inflow Flood—Hydrologic Event	Yes
GAS00171	Unnamed small dam	8/17/1994	Inflow Flood—Hydrologic Event	Yes
GAS00170	Unnamed small dam	8/17/1994	Inflow Flood—Hydrologic Event	Yes
GAS00169	Unnamed small dam	8/17/1994	Inflow Flood—Hydrologic Event	Yes
GAS00168	Unnamed small dam	8/17/1994	Inflow Flood—Hydrologic Event	Yes
GAS00167	Unnamed small dam	8/17/1994	Inflow Flood—Hydrologic Event	Yes
GAS00166	Unnamed small dam	8/17/1994	Inflow Flood—Hydrologic Event	Yes
GAS00165	Unnamed small dam	8/17/1994	Inflow Flood—Hydrologic Event	Yes
GAS00164	Leverett Pond Dam	8/17/1994	Inflow Flood—Hydrologic Event	Yes
GAS00163	Unnamed small dam within city limits of Weston	8/17/1994	Inflow Flood—Hydrologic Event	Yes
GAS00162	Unnamed small dam 1.5 miles NE of Weston	8/17/1994	Inflow Flood—Hydrologic Event	Yes
GAS00161	Unnamed small dam N of County Rd #18, E of PBS tower	8/17/1994	Inflow Flood—Hydrologic Event	Yes
GAS00160	Unnamed small dam N of County Rd #18, E of PBS tower	8/17/1994	Inflow Flood—Hydrologic Event	Yes
GAS00159	Unnamed small dam just upstream of Kennedy	8/17/1994	Inflow Flood—Hydrologic Event	Yes
GAS00158	Unnamed small dam upstream of Holoka Lake Dam	8/17/1994	Inflow Flood—Hydrologic Event	Yes
GAS00157	Unnamed small dam North of Bear Creek	8/17/1994	Inflow Flood—Hydrologic Event	Yes
GAS00156	Unnamed small dam	8/17/1994	Inflow Flood—Hydrologic Event	Yes
GAS00155	Unnamed small dam	8/17/1994	Inflow Flood—Hydrologic Event	Yes
GAS00154	Unnamed small dam	8/17/1994	Inflow Flood—Hydrologic Event	Yes
GAS00153	Unnamed small dam	8/17/1994	Inflow Flood—Hydrologic Event	Yes
GAS00152	Unnamed small dam	8/17/1994	Inflow Flood—Hydrologic Event	Yes
GAS00151	Unnamed small dam	8/17/1994	Inflow Flood—Hydrologic Event	Yes
GAS00150	Unnamed small dam 1/2 mile U/S SR 45 on Mossy Creek	8/17/1994	Inflow Flood—Hydrologic Event	Yes
GAS00149	Unnamed small dam above Taylor Mill Pond Dam	7/5/1994	Inflow Flood—Hydrologic Event	Yes
GAS00148	Unnamed small dam above Taylor Mill Pond Dam	7/5/1994	Inflow Flood—Hydrologic Event	Yes
GAS00147	Unnamed small dam above Taylor Mill Pond Dam	7/5/1994	Inflow Flood—Hydrologic Event	Yes
GAS00146	Unnamed small dam above Taylor Mill Pond Dam	7/5/1994	Inflow Flood—Hydrologic Event	Yes
GAS00145	Unnamed small dam, Bottsford quad in draw, E. of T	7/29/1994	Inflow Flood—Hydrologic Event	Yes

(Continued)

Table 11A.13 (Continued)

NPDP ID	Dam Name	Incident Date	Incident Type	Dam Failure
GAS00144	Unnamed small dam above Tharpe Lake Dam	7/29/1994	Inflow Flood—Hydrologic Event	Yes
GAS00143	Unnamed dam in NW part of Sumeter County	7/29/1994	Inflow Flood—Hydrologic Event	Yes
GAS00142	Unnamed dam in NW part of Sumter County	7/29/1994	Inflow Flood—Hydrologic Event	Yes
GAS00141	Unnamed small dam 1m. SW of Powell Dairy Lake Dam	7/29/1994	Inflow Flood—Hydrologic Event	Yes
GAS00140	Unnamed small dam NW of Plains	7/29/1994	Inflow Flood—Hydrologic Event	Yes
GAS00139	Unnamed small dam 0.75 miles SW of Lake Collins	7/29/1994	Inflow Flood—Hydrologic Event	Yes
GAS00138	Unnamed small dam below Kornonia Lake	7/29/1994	Inflow Flood—Hydrologic Event	Yes
GAS00137	Unnamed small dam below Kornonia Lake	7/29/1994	Inflow Flood—Hydrologic Event	Yes
GAS00136	Unnamed small dam below Kornonia Lake	7/29/1994	Inflow Flood—Hydrologic Event	Yes
GAS00135	Unnamed small dam 1 mile west of Americus	7/29/1994	Inflow Flood—Hydrologic Event	Yes
GAS00134	Unnamed small dam 1 mile north of Americus	7/29/1994	Inflow Flood—Hydrologic Event	Yes
GAS00133	Unnamed small dam 1 mile east of Americus	7/29/1994	Inflow Flood—Hydrologic Event	Yes
GAS00132	Unnamed small dam S. of SR27, 1 mile E. of Americus	7/29/1994	Inflow Flood—Hydrologic Event	Yes
GAS00131	Unnamed dam in Americus	7/9/1994	Inflow Flood—Hydrologic Event	Yes
GAS00130	Unnamed dam	8/17/1994	Inflow Flood—Hydrologic Event	Yes
GAS00129	Unnamed dam	8/17/1994	Inflow Flood—Hydrologic Event	Yes
GAS00128	Unnamed dam	8/17/1994	Inflow Flood—Hydrologic Event	Yes
GAS00127	Unnamed dam	8/17/1994	Inflow Flood—Hydrologic Event	Yes
GAS00126	Unnamed dam	8/17/1994	Inflow Flood—Hydrologic Event	Yes
GAS00125	Unnamed dam	8/17/1994	Inflow Flood—Hydrologic Event	Yes
GAS00124	Unnamed dam	8/17/1994	Inflow Flood—Hydrologic Event	Yes
GAS00123	Unnamed dam	8/17/1994	Inflow Flood—Hydrologic Event	Yes
GAS00122	Unnamed dam	8/17/1994	Inflow Flood—Hydrologic Event	Yes
GAS00121	Unnamed dam	8/17/1994	Inflow Flood—Hydrologic Event	Yes
GAS00120	Unnamed dam	8/17/1994	Inflow Flood—Hydrologic Event	Yes
GAS00119	Unnamed dam	9/12/1994	Inflow Flood—Hydrologic Event	Yes
GAS00118	Unnamed dam	9/12/1994	Inflow Flood—Hydrologic Event	Yes
GAS00117	Unnamed dam	9/12/1994	Inflow Flood—Hydrologic Event	Yes
GAS00116	Unnamed dam	9/12/1994	Inflow Flood—Hydrologic Event	Yes
GAS00115	Unnamed dam	9/12/1994	Inflow Flood—Hydrologic Event	Yes
GAS00114	Unnamed small dam south of Gussie Lake Dam	9/12/1994	Inflow Flood—Hydrologic Event	Yes
GAS00113	Unnamed dam 1 mile north of Union	9/12/1994	Inflow Flood—Hydrologic Event	Yes
GAS00112	Unnamed small dam	8/17/1994	Inflow Flood—Hydrologic Event	Yes
GAS00111	Unnamed small dam	8/17/1994	Inflow Flood—Hydrologic Event	Yes
GAS00110	Unnamed small dam	8/17/1994	Inflow Flood—Hydrologic Event	Yes
GAS00109	Unnamed small dam upstream of Wolhwender	8/17/1994	Inflow Flood—Hydrologic Event	Yes
GAS00108	Unnamed small dam SW of intersect. Of US #19 & SR 2	8/17/1994	Inflow Flood—Hydrologic Event	Yes
GAS00107	Unnamed small dam S of Ebenezer Rd on Little Mucka	8/17/1994	Inflow Flood—Hydrologic Event	Yes
GAS00106	Unnamed small dam	8/17/1994	Inflow Flood—Hydrologic Event	Yes
GAS00105	Unnamed small dam	8/17/1994	Inflow Flood—Hydrologic Event	Yes
GAS00104	Unnamed small dam	8/17/1994	Inflow Flood—Hydrologic Event	Yes
GAS00103	Unnamed small dam	8/17/1994	Inflow Flood—Hydrologic Event	Yes
GAS00102	Unnamed small dam	8/17/1994	Inflow Flood—Hydrologic Event	Yes

(Continued)

Table 11A.13 (Continued)

NPDP ID	Dam Name	Incident Date	Incident Type	Dam Failure
GAS00101	Unnamed small dam	8/17/1994	Inflow Flood—Hydrologic Event	Yes
GAS00100	Unnamed small dam	8/17/1994	Inflow Flood—Hydrologic Event	Yes
GAS00099	Unnamed small dam	8/17/1994	Inflow Flood—Hydrologic Event	Yes
GAS00098	Unnamed small dam	8/17/1994	Inflow Flood—Hydrologic Event	Yes
GAS00097	Unnamed small dam	8/17/1994	Inflow Flood—Hydrologic Event	Yes
GAS00096	Unnamed small dam	8/17/1994	Inflow Flood—Hydrologic Event	Yes
GAS00095	Unnamed small dam	8/17/1994	Inflow Flood—Hydrologic Event	Yes
GAS00094	Unnamed small dam	8/17/1994	Inflow Flood—Hydrologic Event	Yes
GAS00093	Unnamed small dam	8/17/1994	Inflow Flood—Hydrologic Event	Yes
GAS00092	Unnamed small dam	8/17/1994	Inflow Flood—Hydrologic Event	Yes
GAS00091	Unnamed small dam	8/17/1994	Inflow Flood—Hydrologic Event	Yes
GAS00090	Unnamed small dam	8/17/1994	Inflow Flood—Hydrologic Event	Yes
GAS00089	Unnamed small dam	8/17/1994	Inflow Flood—Hydrologic Event	Yes
GAS00088	Unnamed small dam	8/17/1994	Inflow Flood—Hydrologic Event	Yes
GAS00087	Unnamed small dam	8/17/1994	Inflow Flood—Hydrologic Event	Yes
GAS00013	Goodroe Lake Dam	8/30/1994	Inflow Flood—Hydrologic Event	Yes
GAS00012	Goffs Mill Lake Dam	8/17/1994	Inflow Flood—Hydrologic Event	Yes
GAS00011	Free Lake Dam	7/21/1994	Inflow Flood—Hydrologic Event	Yes
GAS00010	Fountain Lake Dam	8/30/1994	Inflow Flood—Hydrologic Event	Yes
GAS00009	Forbes Lake Dam	8/17/1994	Inflow Flood—Hydrologic Event	Yes
GAS00008	Ferguson Lake Dam	8/17/1994	Inflow Flood—Hydrologic Event	Yes
GAS00007	English Lake Dam	8/17/1994	Inflow Flood—Hydrologic Event	Yes
GAS00006	Edgemon	7/29/1994	Inflow Flood—Hydrologic Event	Yes
GA04904	Andrews Lake Dam	8/17/1994	Inflow Flood—Hydrologic Event	Yes
GA04832	Silberman Lake Dam	7/21/1994	Inflow Flood—Hydrologic Event	Yes
GA04765	Harper Lake Dam	7/29/1994	Inflow Flood—Hydrologic Event	Yes
GA04764	Yara Lake Dam	7/29/1994	Inflow Flood—Hydrologic Event	Yes
GA04763	Pace Lake Dam South	7/29/1994	Inflow Flood—Hydrologic Event	Yes
GA04712	Kennedy Lake Dam	8/17/1994	Inflow Flood—Hydrologic Event	Yes
GA04537	West Leisure Lake Dam	7/6/1994	Inflow Flood—Hydrologic Event	Yes
GA03189	Pace Lake Dam	8/17/1994	Inflow Flood—Hydrologic Event	Yes
GA03188	Esperanza Farms Lake Dam	7/29/1994	Inflow Flood—Hydrologic Event	Yes
GA03187	Tharpe Lake Dam	7/29/1994	Inflow Flood—Hydrologic Event	Yes
GA03186	Reeves Lake Dam	7/29/1994	Inflow Flood—Hydrologic Event	Yes
GA03042	Horsehead Creek Lake Dam	7/21/1994	Inflow Flood—Hydrologic Event	Yes
GA03041	Whitewater Creek Lake Dam	7/21/1994	Inflow Flood—Hydrologic Event	Yes
GA03018	Flat Creek Lake Dam	6/5/1994	Inflow Flood—Hydrologic Event	Yes
GA03017	Kersey Lake Dam	8/30/1994	Inflow Flood—Hydrologic Event	Yes
GA01067	Cloud Lake Dam	8/17/1994	Inflow Flood—Hydrologic Event	Yes
GA01050	Barnesville Reservior Dam	7/5/1994	Inflow Flood—Hydrologic Event	Yes
GA01029	Hicks Millpond Dam	7/5/1994	Inflow Flood—Hydrologic Event	Yes
GA01019	Garant Lake Dam	7/21/1994	Inflow Flood—Hydrologic Event	Yes
GA00835	Flint River	7/7/1994	Inflow Flood—Hydrologic Event	Yes
GA01417	Statham Lake Dam	7/6/1994	Inflow Flood—Hydrologic Event	Yes
GA01412	Browns Millpond Lake Dam	7/9/1994	Inflow Flood—Hydrologic Event	Yes
GA01411	Shipp Lake Dam	7/29/1994	Inflow Flood—Hydrologic Event	Yes
GA01410	Able Acres Lake Dam	7/29/1994	Inflow Flood—Hydrologic Event	Yes
GA01364	Tyrone Lake Dam	7/5/1994	Inflow Flood—Hydrologic Event	Yes
GA01235	Kraftsmans Association Lake Dam	7/5/1994	Inflow Flood—Hydrologic Event	Yes
GA01193	McKnight Lake Dam	7/6/1994	Inflow Flood—Hydrologic Event	Yes
GA00289	Lake Clopine Dam	8/30/1994	Inflow Flood—Hydrologic Event	Yes
GA00242	Wilkinson Lake Dam	8/30/1994	Inflow Flood—Hydrologic Event	Yes
CO00390	Frenchman Creek	6/4/1995	Piping	Yes
MI00109	Barnes Dam	5/29/1995	Inflow Flood—Hydrologic Event	Yes
MS03334	Lake Gary Dam	9/1995	Seepage; Piping	Yes
NC01137	Lake Lynn Dam	6/19/1995	Inflow Flood—Hydrologic Event; Piping	Yes
NC04944	Jaycees Pond Dam	6/19/1995	Inflow Flood—Hydrologic Event; Seepage	Yes

(Continued)

Table 11A.13 (Continued)

NPDP ID	Dam Name	Incident Date	Incident Type	Dam Failure
NCS00004	McReady Chicken Waste Lagoon Dike	7/5/1995	Not Known	Yes
NHS00001	Willey House Dam	10/28/1995	Inflow Flood—Hydrologic Event	Unknown
ND00339	Appert	7/15/1995	Inflow Flood—Hydrologic Event; Seepage	Yes
NCS00013	Unnamed small dam	6/19/1995	Inflow Flood—Hydrologic Event	Yes
NCS00011	Southern Pines Country Club Golf Course Dam 3	7/1995	Inflow Flood—Hydrologic Event	Yes
NSC00010	Southern Pines Country Club Golf Course Dam 2	7/1995	Not Known	Yes
NCS00009	Southern Pines Country Club Golf Course Dam 1	7/1995	Inflow Flood—Hydrologic Event	Yes
NCS00007	Reedy Swine Farm Lagoon Dike	7/18/1995	Rupture	Yes
NCS00003	Lancaster Dam	6/26/1995	Inflow Flood—Hydrologic Event	Yes
NC02159	Moose Lodge Dam	6/19/1995	Inflow Flood—Hydrologic Event	Yes
MT03839	Eureka Holding Pond Dike	7/8/1995	Piping	Yes
CO01967	Vincent No. 2	6/9/1995	Inflow Flood—Hydrologic Event	Yes
ID00151	Troy	2/5/1995	Seepage; Piping	Yes
NJ00408	Kenilworth Lake Dam	1/18/1995	Partial Dam Breach	Yes
WY02028	Cottonwood	7/1995	Not Known	Unknown
WI12794	Hazel Lake	12/15/1995	Piping	Yes
WI00092	Mount Morris Dam	8/29/1995	Inflow Flood—Hydrologic Event; Inflow Flood—Upstream Dam Failure; Seepage; Piping	Yes
WAS00002	CSC Orchards, Frost Protection Pond	7/1995	Seepage; Piping	Yes
VT00182	Wolcott Pond	8/5/1995	Inflow Flood—Hydrologic Event	Yes
VA03102	Timber Lake Dam	6/22/1995	Inflow Flood—Hydrologic Event	Yes
NV00051	Boyd Reservoir	1/1995	Piping	Yes
NV00070	Milk Ranch Dam	7/8/1995	Inflow Flood—Hydrologic Event; Embankment Erosion	Yes
NJS00001	Mendham Reservoir Dam	4/1996	Piping	Yes
PA00422	Brookville Waterworks	7/19/1996	Inflow Flood—Hydrologic Event; Embankment Erosion	Yes
TNS00001	Dillard Dam	12/17/1996	Inflow Flood—Hydrologic Event	Yes
WA01045	Olufson Dam	12/11/1996	Seepage	No
WV07719	Bruceton Mills Dam	1/19/1996	Inflow Flood—Hydrologic Event	Yes
WI10386	Hamilton Mill	4/20/1996	Seepage	Yes
WI01102	Vernon Marsh-Ref. Flowage	5/19/1996	Piping	Yes
WI00450	Cranberry Creek	4/21/1996	Not Known; Piping	Yes
WA01782	Boeing Creek North Stormwater Pond	12/31/1996	Inadequate Spillway Capacity	Yes
WA01406	Yelm Diversion	2/7/1996	Inflow Flood—Hydrologic Event	Unknown
VTS00008	Rinse	6/14/1996	Piping	Yes
TX07035	Casa Monte Dam	11/26/1996	Inflow Flood—Hydrologic Event	Yes
TX01961	Roberts Tank Dam	7/14/1996	Inflow Flood—Hydrologic Event	Yes
TN15768	Mallard Lake	11/3/1996	Animal Attack; Piping; Embankment Erosion; Embankment Slide	Yes
CO01122	Henry	4/10/1996	Piping	Yes
KSS00002	Speer Dam	11/14/1996	Inflow Flood—Hydrologic Event	Yes
NH00600	Bergerson Dam	3/13/1996	Piping	Yes
MT01173	Canyon Lake	7/1/1996	Inflow Flood—Hydrologic Event	Yes
MI00876	Hollenbeck Dam	1996	Inflow Flood—Hydrologic Event	Yes
ME00072	Highland Lake Dam	10/20/1996	Inflow Flood—Hydrologic Event	Yes
KSS00001	Decker Dam	1996	Inflow Flood—Hydrologic Event	Yes
IL00918	Aurora-West Dam	7/17/1996	Inflow Flood—Hydrologic Event	Yes
IL01110	Puddle Pond Dam	7/17/1996	Inflow Flood—Hydrologic Event	Yes
IL50394	Channahon Dam	7/17/1996	Inflow Flood—Hydrologic Event	Yes
AZ00187	Udall	8/11/1997	Piping	Yes
AZS00003	Centennial Narrows Dam	9/26/1997	Inflow Flood—Hydrologic Event; Piping	Yes

(Continued)

Table 11A.13 (Continued)

NPDP ID	Dam Name	Incident Date	Incident Type	Dam Failure
AZS00001	Middle Goose Tailings Dam (No. 2 Tailings Impoundment)	10/21/1997	Liquefaction Failure	Yes
NE00695	Scott Dam	4/7/1997	Inflow Flood—Hydrologic Event	Yes
MTS00001	Anita Dam	3/26/1997	Piping	Yes
MO20164	Lake Venita Dam	2/21/1997	Seepage; Piping	Yes
MI00496	Hamilton Dam	6/20/1997	Inflow Flood—Hydrologic Event	Yes
ME00567	Apple Valley Lake Dam	4/16/1997	Debris—Reservoir; Piping	Yes
MA00376	East Head Pond Dam	1997	Not Known; Seepage; Piping	Yes
GA04975	Forsyth Reservoir	12/24/1997	Inflow Flood—Hydrologic Event; Seepage; Piping; Concrete Deterioration	Yes
FLS00003	Ridgewood Avenue Dam (Lake Apopka Dam)	11/27/1997	Not Known; Seepage; Piping	No
FLS00002	RGC Minerals Containment	4/27/1997	Inflow Flood—Hydrologic Event	Yes
NJ00716	Moss Mill Lake Dam	8/20/1997	Inflow Flood—Hydrologic Event	Yes
NJS00009	Stockton College Dam	8/20/1997	Inflow Flood—Hydrologic Event	Yes
NJS00006	Tarkiln Pond Dam (Route 548 Dam)	8/20/1997	Inflow Flood—Hydrologic Event	Yes
WA01377	Horn Rapids Dam	2/1997	Inflow Flood—Hydrologic Event	Yes
WA00133	Wishkah Reservoir No. 2 Dam	3/19/1997	Inflow Flood—Hydrologic Event; Seepage; Piping	Yes
VTS00007	Sibley	1997	Spillway Failure	Yes
TX02021	Holland Dam Site A	1/1/1997	Piping	Yes
TNS00003	Patton	3/1/1997	Inflow Flood—Hydrologic Event	Yes
TNS00002	Mullens Farm Pond	3/5/1997	Inflow Flood—Hydrologic Event	Yes
TN11306	Johnson Ck #4	3/1/1997	Inflow Flood—Hydrologic Event	Yes
SD00025	Woodruff (Breached 1997)	3/22/1997	Inflow Flood—Hydrologic Event	Yes
SC00377	Starnes/Brown Dam	7/24/1997	Inflow Flood—Upstream Dam Failure	Yes
WI10610	Linnie Lac Dam	6/21/1997	Inflow Flood—Hydrologic Event	No
WA01741	Galbreath Sediment Dam	1/18/1997	Piping	Yes
SC00307	Malcolm B. Rawls Dam	7/24/1997	Inflow Flood—Upstream Dam Failure	Yes
OH02964	Green Acres Levee	3/4/1997	Inflow Flood—Hydrologic Event	Yes
OH02900	Thomas Pond Dam	3/1/1997	Inflow Flood—Hydrologic Event	Yes
NY13600	Henry Kaufman Pond Dam	6/14/1997	Inflow Flood—Hydrologic Event	Yes
NV00223	Carson City Wastewater Dam	10/30/1998	Earthquake	No
ORS00002	Lacomb Diversion	5/11/1998	Inflow Flood—Hydrologic Event	Yes
VTS00006	Name Unknown	1/7/1998	Piping	Yes
VTS00005	Sanville	1/16/1998	Inflow Flood—Hydrologic Event	Yes
VTS00004	Name Unknown	6/27/1998	Inflow Flood—Hydrologic Event	Yes
VTS00003	Name Unknown	6/27/1998	Inflow Flood—Hydrologic Event	Yes
VTS00002	Clay Brook Water Supply	6/27/1998	Inflow Flood—Hydrologic Event	Yes
VTS00001	Golf Course Pond	6/27/1998	Inflow Flood—Hydrologic Event	Yes
VT00241	Sunset Lake	8/11/1998	Spillway Failure	Yes
VT00229	Lake Runnemede	7/18/1998	Piping	Yes
TX01580	Jan Land Company Lake No. 1 Dam	10/1998	Inflow Flood—Hydrologic Event	Yes
TN09910	Johnson Lake	7/13/1998	Inflow Flood—Hydrologic Event; Inadequate Spillway Capacity	Yes
TN09902	Bennett Lake	7/13/1998	Inflow Flood—Hydrologic Event	Yes
WA01756	Klickitat Mill Pond Dam	2/8/1998	Fire	Yes
RI03201	Peace Dale Pond Dam	2/18/1998	Inflow Flood—Hydrologic Event	Yes
NY00494	Camp Weona Dam	6/8/1998	Inflow Flood—Hydrologic Event	Yes
NY13643	Peru Water Supply Dam	6/27/1998	Inadequate Spillway Capacity	Yes
OH01978	Bookhamer Lake Dam	7/5/1998	Spillway Failure	Yes
NY01539	Gouldtown-Mill 5 West Channel	1/7/1998	Inflow Flood—Hydrologic Event	Yes
NY12015	Natural Dam	1/7/1998	Inflow Flood—Hydrologic Event	Yes
CO00508	Vertrees	5/29/1998	Piping	Yes
GA00084	Little Ocmulgee Lake Dam	3/11/1998	Inflow Flood—Hydrologic Event	Yes
GAS00185	Bay Meadows	3/7/1998	Inflow Flood—Hydrologic Event	Yes

(Continued)

Table 11A.13 (Continued)

NPDP ID	Dam Name	Incident Date	Incident Type	Dam Failure
GAS00188	Pine Cove Pond Dam	4/5/1998	Inflow Flood—Hydrologic Event	Yes
GAS00191	County Road 15 Dam	3/8/1998	Inflow Flood—Hydrologic Event	Yes
NC00725	Ramseur Lake Dam	2/19/1998	Inflow Flood—Hydrologic Event	Yes
MS01402	Archusa Crk Wtr Park Lk	1/8/1998	Inflow Flood—Hydrologic Event	Yes
JY00174	Hematite	6/11/1998	Not Known; Seepage; Piping	Yes
GAS00189	Boy Scout Camp Lake Dam	2/2/1998	Inflow Flood—Hydrologic Event; Embankment Erosion	Yes
GAS00186	Not Named (Exempt) Dam	3/7/1998	Inflow Flood—Hydrologic Event	Yes
GA05053	Big Sandy Plantation, Inc. Lake Dam	2/2/1998	Inflow Flood—Hydrologic Event	Yes
GA01826	Southern States Lake Dam	2/9/1998	Seepage	Yes
CO00629	Carl Smith	5/2/1998	Embankment Slide	Yes
FLS00004	E.R. Jahna-Independent North Sand Mine Tailings	7/2/1999	Inflow Flood—Hydrologic Event; Piping	Yes
MD00074	Nagels Mill Pond	9/16/1999	Piping	Yes
NCS00031	Colee Naylor Pond Dam	9/20/1999	Inflow Flood—Hydrologic Event	Yes
NCS00029	House Autry Mill Dam	9/21/1999	Inflow Flood—Hydrologic Event	Yes
NCS00022	Hog Waste Lagoon Dike	4/18/1999	Sabotage—Other	Yes
NC03056	Winkler Lake Dam Lower (Flat Rock Lakes)	4/20/1999	Seepage	Yes
NC01191	Dubose Lake Dam	9/21/1999	Inflow Flood—Hydrologic Event	Yes
NC01084	Hall Lake Dam	9/21/1999	Inflow Flood—Hydrologic Event	Yes
NC00946	Kellys Pond Dam	9/15/1999	Inflow Flood—Hydrologic Event	Yes
MD00345	Rolling Green Community Lake	2/11/1999	Piping	Yes
NH00598	Nubble Pond Dam	10/1/1999	Not Known	Yes
NH00270	Cold Brook	10/1/1999	Not Known	Yes
MD00342	Not Known	9/16/1999	Inflow Flood—Hydrologic Event	Yes
MD00319	Stubbs Farm Dam	9/16/1999	Inflow Flood—Hydrologic Event	Yes
MD00190	Riley Mill Pond	9/16/1999	Inflow Flood—Hydrologic Event	Yes
MD00189	Foreman Branch Dam	9/16/1999	Inflow Flood—Hydrologic Event	Yes
MD00170	Jones Lake Dam	9/16/1999	Inflow Flood—Hydrologic Event	Yes
MD00152	Lake Lanahan	5/7/1999	Inflow Flood—Hydrologic Event; Embankment Erosion	Yes
MD00149	Tuckahoe State Park Dam	9/16/1999	Inflow Flood—Hydrologic Event	Yes
MD00098	Frazers Dam	9/15/1999	Inflow Flood—Hydrologic Event	Yes
NJ00039	Bostwicks Pond Dam	9/16/1999	Inflow Flood—Hydrologic Event	Yes
NHS00002	Unregistered Dam (No Name)	9/16/1999	Inflow Flood—Hydrologic Event	Yes
MDS00003	Stubbs	9/16/1999	Inflow Flood—Hydrologic Event	Yes
MD00025	Sassafras Mill Dam	9/16/1999	Inflow Flood—Hydrologic Event	Yes
GAS00194	Lake "Jimmy Carter"	6/28/1999	Inflow Flood—Hydrologic Event	Yes
MA00537	Lake Bray Dam	9/17/1999	Inflow Flood—Hydrologic Event	Yes
MA01258	J.B. Dunnell Dam	9/17/1999	Inflow Flood—Hydrologic Event	Yes
MA02531	Veteran's Memorial State Park Dam	9/17/1999	Inflow Flood—Hydrologic Event	Yes
IL50396	Pittsfield Dredge Disposal Pond Dam	4/29/1999	Seepage; Piping	Yes
FLS00005	Caloosa Sand Mine Reclamation Lake	8/25/1999	Inflow Flood—Hydrologic Event	Yes
NJ00565	Lookover Lake Dam	3/21/1999	Embankment Erosion	Yes
NJ00634	Kirbys Mill Dam	9/16/1999	Inflow Flood—Hydrologic Event	Yes
VAS00005	Unnamed Dam	9/1999	Inflow Flood—Hydrologic Event	Yes
VAS00003	Saddler Dam	9/1999	Inflow Flood—Hydrologic Event	Yes
VAS00002	Unnamed Dam (at Williamsburg Country Club)	9/1999	Inflow Flood—Hydrologic Event	Yes
VA13303	Sydnors Millpond Dam	9/1999	Inflow Flood—Hydrologic Event	Yes
VA12709	Old Forge Pond Dam	9/1999	Inflow Flood—Hydrologic Event	Yes
VA11912	Lower Rosegill Lake Dam	9/1999	Inflow Flood—Hydrologic Event	Yes
VA11911	Town Bridge Pond Dam	9/1999	Inflow Flood—Hydrologic Event	Yes
VA11906	Rosegill Upper Dam	9/1999	Inflow Flood—Hydrologic Event	Yes
VA09704	Allens Mill Dam	9/1999	Inflow Flood—Hydrologic Event	Yes
VA09512	Lake Powell Dam	9/1999	Inflow Flood—Hydrologic Event; Wind; Animal Attack; Biological Attack (i.e., bush, tree growth)	Yes

(Continued)

Table 11A.13 (Continued)

NPDP ID	Dam Name	Incident Date	Incident Type	Dam Failure
VA07308	Cypress Shores	9/1999	Inflow Flood—Hydrologic Event	Yes
VA07305	Cow Creek Dam	9/1999	Inflow Flood—Hydrologic Event	Yes
VA05707	Essex Mill Dam	9/1999	Inflow Flood—Hydrologic Event	Yes
TNS00007	Deer Creek	7/1/1999	Inflow Flood—Hydrologic Event	Yes
TNS00006	Bent Tree Dam	4/22/1999	Piping	Yes
SD00536	Covey Dam	5/9/1999	Inflow Flood—Hydrologic Event	Yes
SD00444	W. Day	9/3/1999	Inflow Flood—Hydrologic Event	Yes
PAS00002	Longo Pond Dam	1/26/1999	Inflow Flood—Hydrologic Event	Yes
OH02867	Beldon Pond Lake Dam	4/1999	Piping	Yes
OH02795	Crown City Mining Pond No. 24 Dam	5/19/1999	Inflow Flood—Hydrologic Event	Yes
NY13278	Lake Hyenga Dam	9/17/1999	Inflow Flood—Hydrologic Event	Yes
NJS00028	Spencer/Estates Detention Basin	9/16/1999	Piping	Yes
NY01345	High Falls	11/27/1999	Concrete Spillway Cap	Yes
NJ00768	Seneca Lake Dam	8/12/2000	Inflow Flood—Hydrologic Event	Yes
RI04389	Mill Pond	8/6/2000	Embankment Erosion	Yes
WIS00002	Chenowith Dam	3/14/2000	Not Known	Yes
TX09244	Camp La Junta Dam	10/23/2000	Inflow Flood—Hydrologic Event	Yes
TX03797	Powell Lake Dam	3/22/2000	Inflow Flood—Hydrologic Event	Yes
NY14821	Murtha Pond Dam	7/5/2000	Inflow Flood—Hydrologic Event	Yes
NJS00046	Furnace Pond Dam	8/12/2000	Inflow Flood—Hydrologic Event	Yes
NJS00050	Edison Pond Dam	8/12/2000	Inflow Flood—Hydrologic Event	Yes
AK00144	City Of Kake Dam	7/24/2000	Inflow Flood—Hydrologic Event	Yes
NJ00010	Tomahawk Lake Dam	8/12/2000	Not Known	Yes
NHS00004	Middle Pond	6/7/2000	Animal Attack; Inadequate Spillway Capacity	Yes
NH01364	Mountain Lake	1/7/2000	Not Known	Yes
ND00540	Grand Forks Co. Com. #1	6/12/2000	Inflow Flood—Hydrologic Event	Yes
MS01687	Ascalmore Creek Str Y-17a-11	4/4/2000	Animal Attack	Yes
GAS00196	Lott Dam	9/18/2000	Deterioration	Yes
ARS00001	Ponca Dam	6/2000	Inflow Flood—Hydrologic Event	Yes
GA00459	Pritchard Lake Dam	3/16/2001	Deterioration	Yes
GAS00203	Ingles Shopping Center	6/13/2001	Inflow Flood—Hydrologic Event	Yes
IL01077	Wardens Pond North Dam	7/19/2001	Not Known	Yes
NY16046	Eagle Lake Dam	7/16/2001	Piping	Yes
TN03510	HIll #1	2001	Inflow Flood—Hydrologic Event	Yes
OR00467	Smith River Lbr. Co. Pond	5/31/2002	Piping	Yes
RIS00006	Sweet's Mill	5/8/2002	Biological Attack (i.e., bush, tree growth); Embankment Erosion)	Yes
WA83006	Swift No 2 Hydroelectric Project	4/21/2002	Seepage; Piping	Yes
GAS00205	Clarke Apple Orchard Lake Dam No. 1	12/24/2002	Piping	Yes
MS01611	Big Sand Creek Str Y-32-32	4/2002	Not Known; Animal Attack; Piping	Yes
MN00486	Wild Rice River	6/2002	Inflow Flood—Hydrologic Event	Yes
CA00724	Las Tablas Cr	3/17/2004	Earthquake	
CA00812	Nacimiento	3/17/2004	Earthquake	
CA00813	San Antonio	3/17/2004	Earthquake	
CA10122	El Piojo	3/17/2004	Earthquake	
CA10123	Hughes	3/17/2004	Earthquake	

Note: Number of Events Found: 680. Time period: 1979 to 2004; Incident: all; Dam type: all; Dam failure: all; State: all.

Source: From npdp.standford.edu. With permission.

Table 11A.14 Causes of Dam Incidents in the United States

	Type of Dam								
	Concrete		Embankment		Other[a]		Totals		
Cause	**F**	**A**	**F**	**A**	**F**	**A**	**F**	**A**	**F&A**
Overtopping	6	3	18	7	3		27	10	37
Flow erosion	3		14	17			17	17	34
Slope protection damage				13				13	13
Embankment leakage, piping			23	14			23	14	37
Foundation leakage, piping	5	6	11	43	1		17	49	66
Sliding	2		5	28			7	28	35
Deformation		2	3	29	3		6	31	37
Deterioration		6	2	3			2	9	11
Earthquake instability				3				3	3
Faulty construction	2			3			2	3	5
Gate failures	1	2	1	3			2	5	7
Total	19	19	77	163	7		103	182	285

Note: F, failure; A, accident (an incident where failure was prevented by remedial work or operating procedures, such as drawing down the pool).

[a] Steel, masonry-wood, or timber crib.

Source: From National Academy Press, 1983, Safety of Existing Dams: Evaluation and Improvement. Based on Schnitter, 1979, Lessons from Dam Incidents U.S.A., ASCE/USCOLD, and supplementary date supplied by U.S. Committee on Large Dams for period to 1979.

SECTION 11B RESERVOIRS

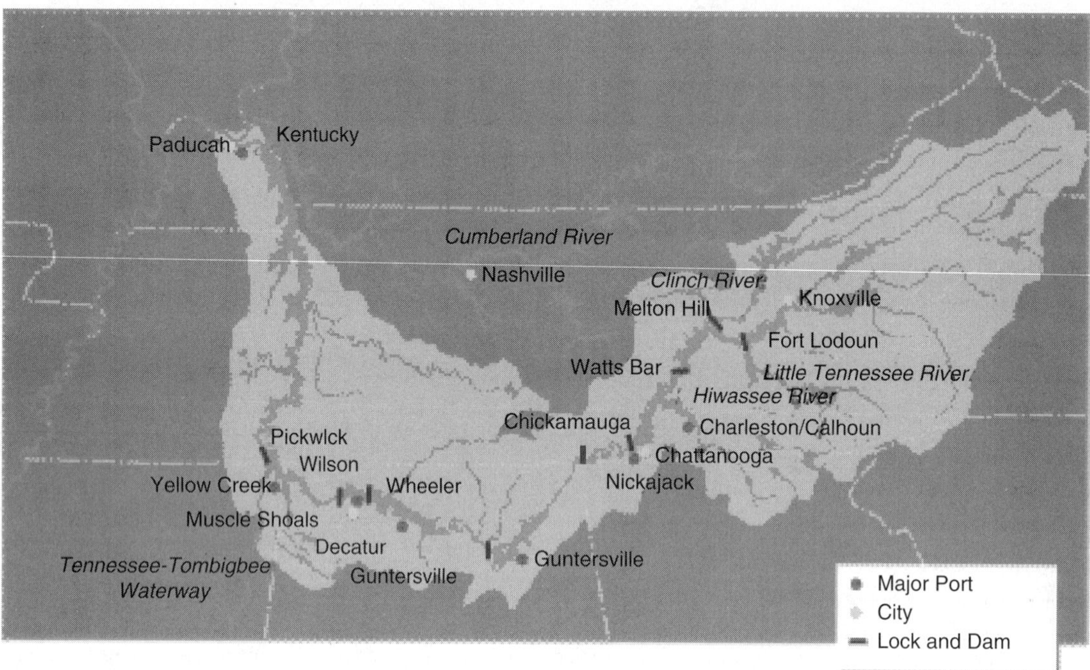

Figure 11B.11 Route of the river. The Tennessee River's main navigable channel is 652 miles long. It officially begins a mile above Knoxville, Tennessee, and eventually empties into the Ohio River at Paducah, Kentucky. Commercial navigation also extends into three major tributaries: 61 miles up the Clinch River, 29 miles up the Little Tennessee River, and 21 miles up the Hiwassee River. Another 150 miles of channel—too shallow for commercial traffic—is marked for recreational use. (From www.tva.gov.)

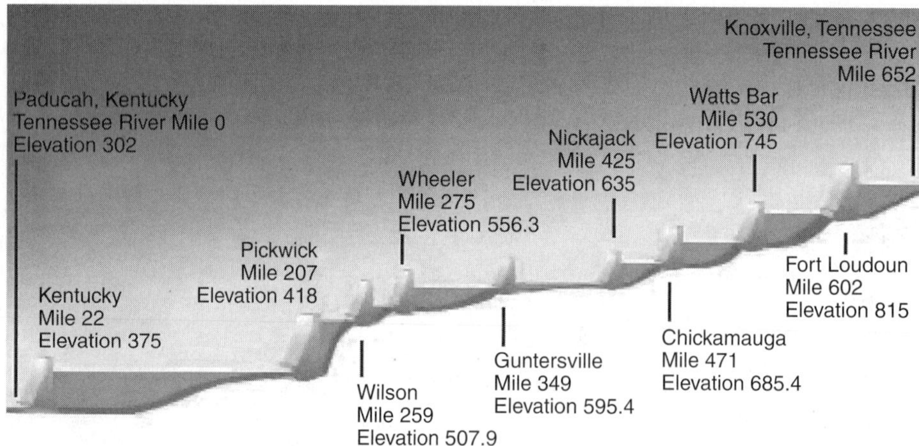

Figure 11B.12 Tennessee river system. Nine main-river dams form a "staircase" of quiet, pooled water, and controlled current—a continuous series of reservoirs that stretches along the entire length of the Tennessee River. From its beginning just above Knoxville, the Tennessee drops a total of 513 ft in elevation before it empties into the Ohio River. (From www.tva.gov.)

Table 11B.15 The Tennessee Valley Authority Multipurpose Reservoir System

The TVA reservoir system was designed to provide for navigation, flood control, and the production of hydroelectricity. Today the reservoir system is also operated for such other purposes as providing municipal and industrial water supplies, regulating flows to minimize the effects of effluents including thermal discharges, fluctuating water levels for control of mosquitos and troublesome aquatic vegetation, and controlling flows and levels for various recreational uses

This reservoir system encompasses more than 11,000 miles of shoreline and 600,000 acres of water. The overall benefit of this system to the Nation and the region are incalculable, but some indication is provided by a consideration of the more quantifiable benefits: flood damages prevented, transportation savings, hydroelectric power production, and recreation visits

There are 39 dams operating in the Tennessee Valley's integrated water control system. They are:

35 TVA dams — Apalachia, Blue Ridge, Boone, Chatuge, Cherokee, Chickamauga, Douglas, Fontana, Fort Loudoun, Fort Patrick Henry, Guntersville, Hiwassee, Kentucky, Melton Hill, Nickajack, Normandy, Norris, Nottely, Ocoee No. 1, Ocoee No. 2, Ocoee No. 3, Pickwick Landing, South Holston, Tellico, Tims Ford, Watauga, Watts Bar, Wheeler, Wilson, Bear Creek, Little Bear Creek, Nolichucky, Upper Bear Creek, Cedar Creek, and Wilbur

4 Alcoa dams — Calderwood, Cheoah, Chilhowee, and Santeetlah

Raccoon Mountain Pumped Storage Project stores energy (generated elsewhere) to meet peak power demands.

There are 9 dams in the Cumberland River basin for which TVA distributes power generation. They include one TVA dam, Great Falls, and 8 Corps of Engineer dams — Barkley, Center Hill, Cheatham, Cordell Hull, Dale Hollow, J. Percy Priest, Old Hickory, and Wolf Creek

The 4 Bear Creek projects (Bear, Little Bear, and Upper Bear and Cedar), the 2 Duck River projects (Normandy and Columbia), and the Tims Ford dam were planned under Tributary Area Development programs but have been counted among those in the Integrated Water Control System because they were all partially justified as having system flood control value. Tims Ford also contributes power to the system

In the 1960's TVA built two systems of small dams in tributary watersheds:

Beech River Project — Beech, Cedar, Dogwood, Lost Creek, Pine, Pin Oak, Redbud, and Sycamore

Bristol Project — Beaver Creek and Clear

There are 12 dams owned by Alcoa in The Little Tennessee River Valley which are not included in TVA's integrated water control system

The Elk River Dam operated by the Air Force, the Burnett Dam operated by the city of Asheville, and the Walters Dam of the Carolina Power and Light Company are other structures in the Tennessee Valley, but are not controlled by TVA

Table 11B.16 Tennessee Valley Authority Reservoirs

				Lake Elevation (Feet above Sea Level)			
Main River Projects	Length of Lake (Miles)	Length of Lake Shoreline (Miles)	Area of Lake (Acres)	Normal Minimum	Top of Gates	Lake Volume (Acre-Feet) at Top of Gates	Useful Controlled Storage in Reservoir (Acre-Feet)
Kentucky	184.3	2,380	160,300	354	375	6,129,000	4,008,000
Pickwick Landing	52.7	496	43,100	408	418	1,105,000	417,700
Wilson	15.5	154	15,500	504.5	507.88	640,200	53,600
Wheeler	74.1	1,063	67,100	550	556.28	1,069,000	349,000
Guntersville	75.7	949	67,900	593	595.44	1,049,000	162,400
Nickajack	46.3	192	10,370	632	635	251,600	31,500
Chickamauga	58.9	810	35,400	675	685.44	737,300	345,300
Watts Bar	95.5	771	39,000	735	745	1,175,000	379,000
Fort Loudoun	60.8	360	14,600	807	815	393,000	111,000
Raccoon Mtn. (Pumped Storage Project)	—	—	528	1,530	—	37,310	35,110
Tributary Projects							
Columbia	54	236	12,600	603	635	363,000	283,000
Normandy	17	73	3,160	859	880	126,100	60,500
Tims Ford	34.2	246	10,600	865	895	608,000	282,600
Apalachia	9.8	31	1,100	1,272	1,280	57,800	8,650
Hiwassee	22.2	163	6,090	1,450	1,526.5	434,000	306,000
Chatuge	13	132	7,050	1,905	1,928	240,500	123,000
Ocoee No. 1	7.5	47	1,890	811	830.76	83,300	31,030
Ocoee No. 2	—	—	—	—	1,115.2	—	Silted

(Continued)

Table 11B.16 (Continued)

				Lake Elevation (Feet above Sea Level)			
Main River Projects	Length of Lake (Miles)	Length of Lake Shoreline (Miles)	Area of Lake (Acres)	Normal Minimum	Top of Gates	Lake Volume (Acre-Feet) at Top of Gates	Useful Controlled Storage in Reservoir (Acre-Feet)
Ocoee No. 3	7	24	480	1,413	1,435	4,180	3,629
Blue Ridge	11	65	3,290	1,590	1,691	195,900	183,900
Nottely	20.2	106	4,180	1,735	1,780	174,300	117,140
Melton Hill	44	173	5,690	790	796	126,000	31,900
Norris	129	800	34,200	960	1,034	2,552,000	1,922,000
Tellico	33.2	373	15,860	807	815	424,000	120,000
Fontana	29	248	10,640	1,580	1,710	1,443,000	946,000
Douglas	43.1	555	30,400	940	1,002	1,461,000	1,251,000
Cherokee	54	393	30,300	1,020	1,075	1,541,000	1,148,000
Ft. Patrick Henry	10.4	37	872	1,258	1,263	26,900	4,250
Boone	32.7	130	4,310	1,330	1,385	193,400	148,400
S. Holston	23.7	168	7,580	1,675	1,742	764,000	438,300
Watauga	16.3	106	6,430	1,915	1,975	677,000	353,000
Wilbur	1.8	3.6	72	1,645	1,650	715	327
Great Falls	22	120	3,080	780	805.3	50,200	35,700
Nolichucky	—	26	383	1,238.9	1,240.9	2,003	496
Totals[a]		11,195	641,455			23,771,708	13,408,432

[a] Does not include the Columbia Dam Project.

Source: From Tennessee Valley Authority, 1988.

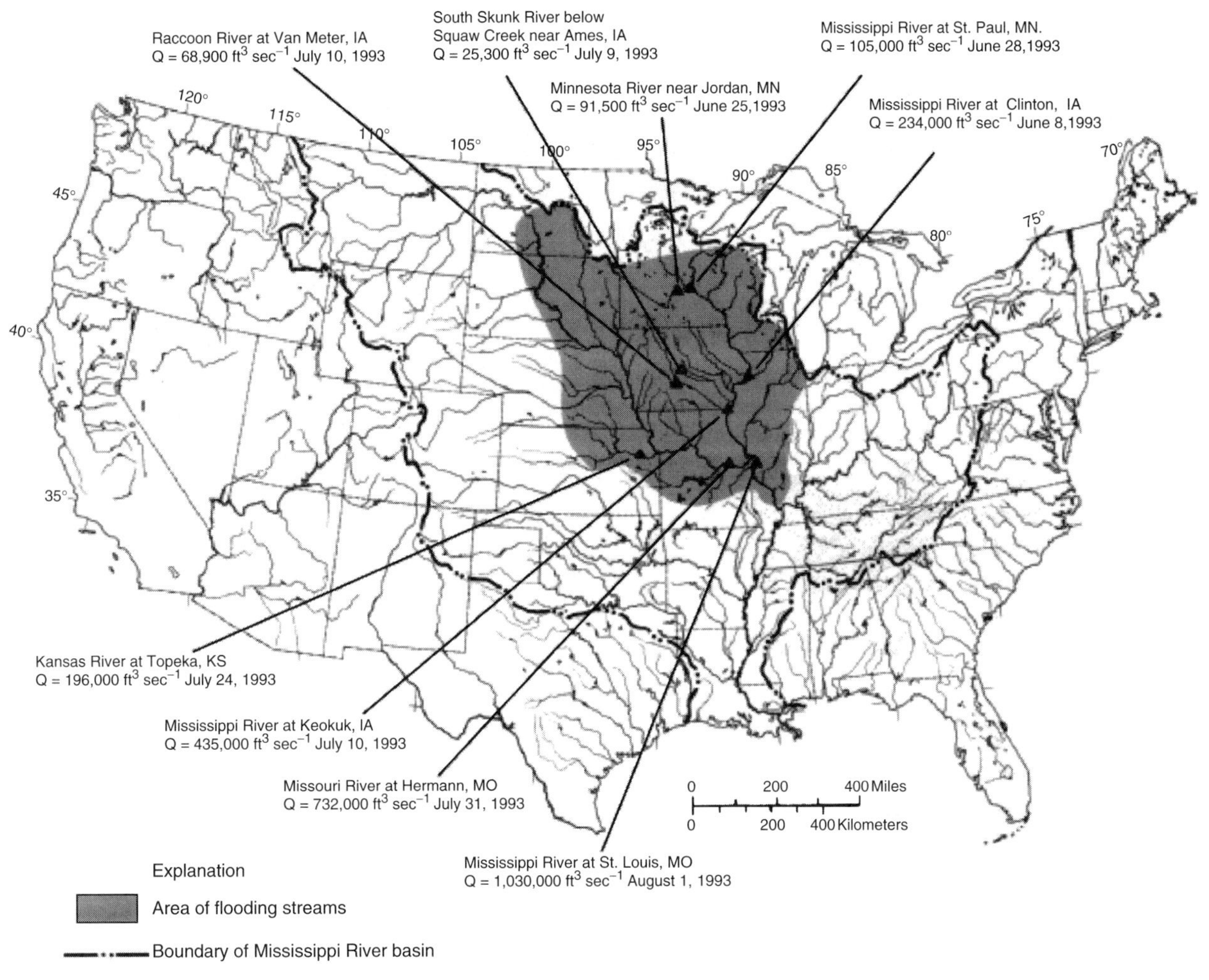

Figure 11B.13 Peak discharges (Q) and dates of occurrence for the 1993 flood at selected streamflow-gaging stations in the upper Mississippi River basin. (From www.geo.mtu.edu.)

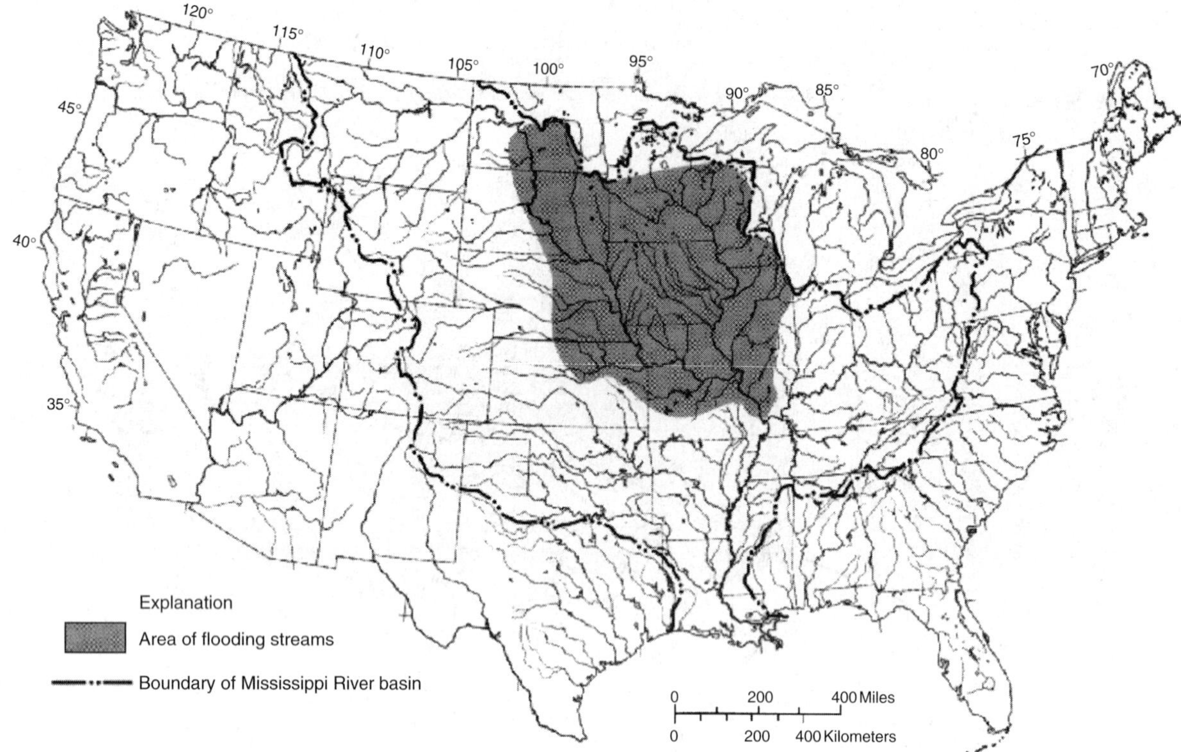

Figure 11B.14 The Mississippi River basin and general area of flooding streams, June to August 1993. (From www.geo.mtu.edu.)

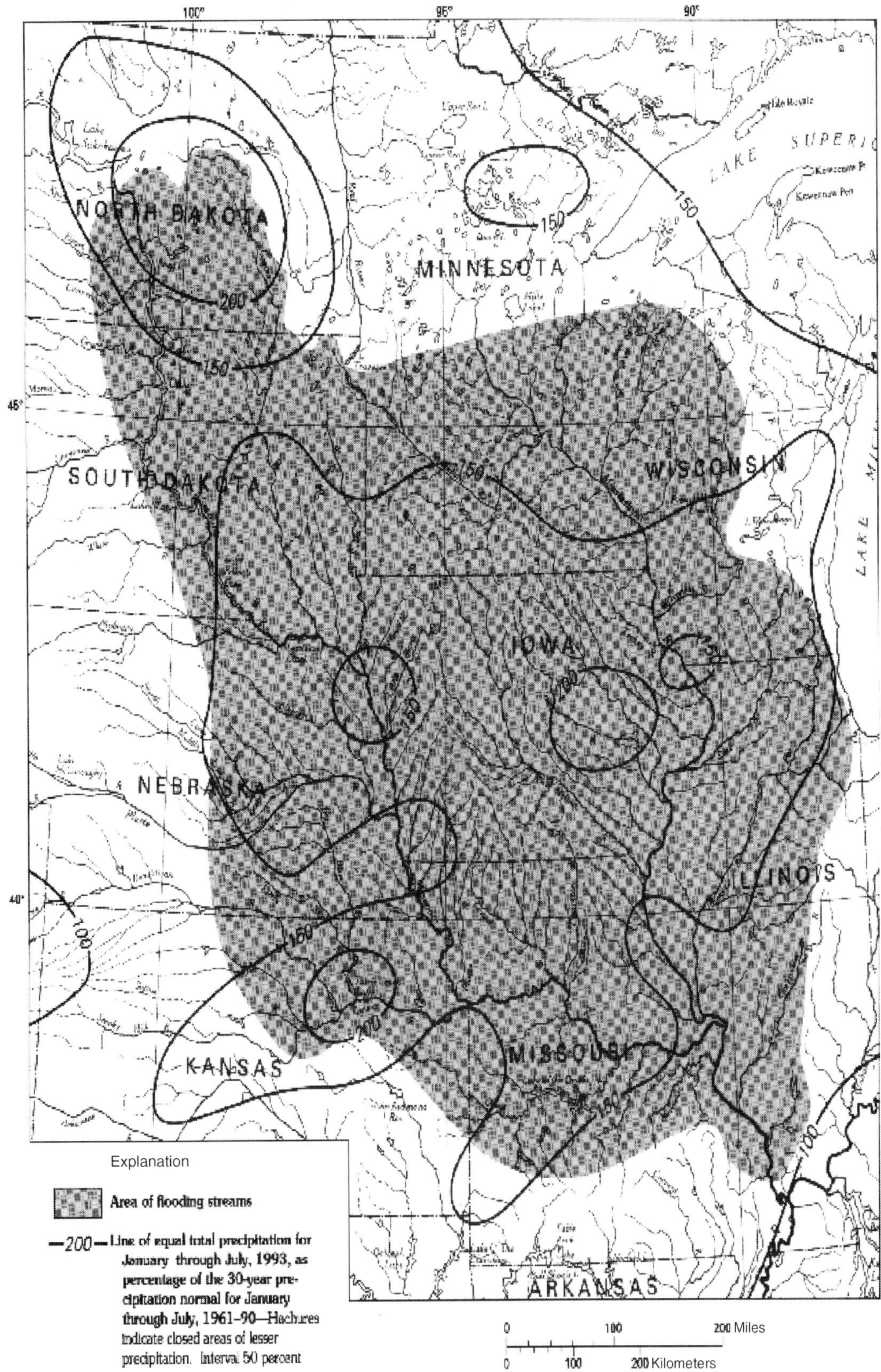

Figure 11B.15 Areal distribution of total precipitation in the area of flooding in the upper Mississippi River basin, January to July 1993. (From www.geo.mtu.edu.)

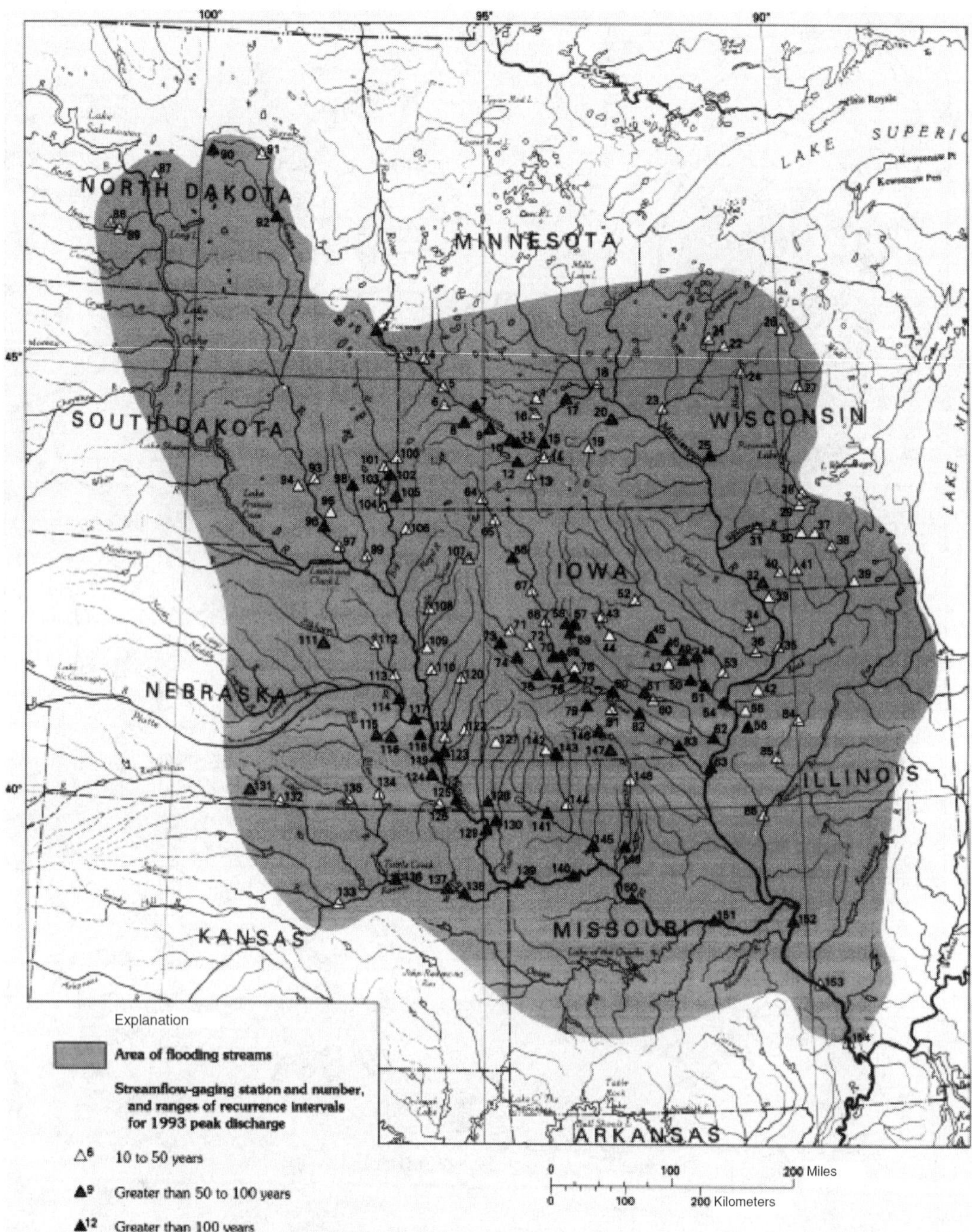

Figure 11B.16 Location of selected streamflow-gaging stations and ranges in recurrence interval for the 1993 peak discharges in the upper Mississippi River basin. (From www.geo.mtu.edu.)

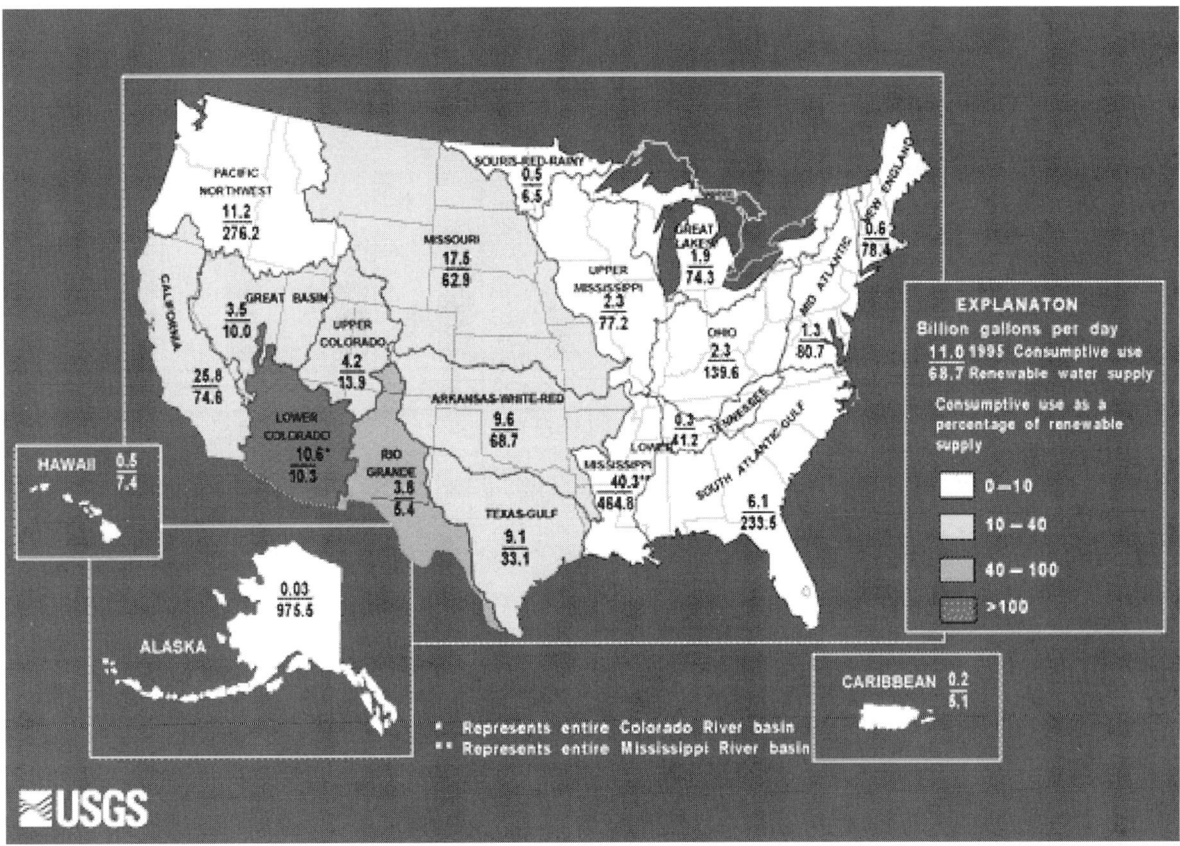

Figure 11B.17 Consumptive use and renewable water supply by water resources region. (From *U.S. Geological Survey, 1984*, updated using 1995 estimates of water use.)

(Data are provisional and subject to change)

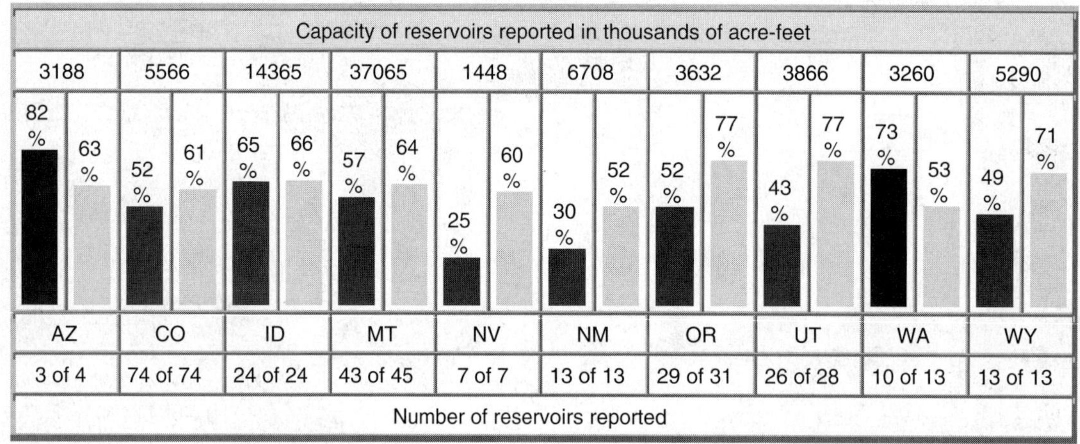

Select here for the reservoir dataset for this period

Storage is below average (% of capacity)

Storage is at or above average (% of capacity)

Average storage as % of capacity

* = Data are not available for this state.

Figure 11B.18 Reservoir storage as percent of capacity for April 1, water year 2005. (From www.wcc.nrcs.usda.gov.)

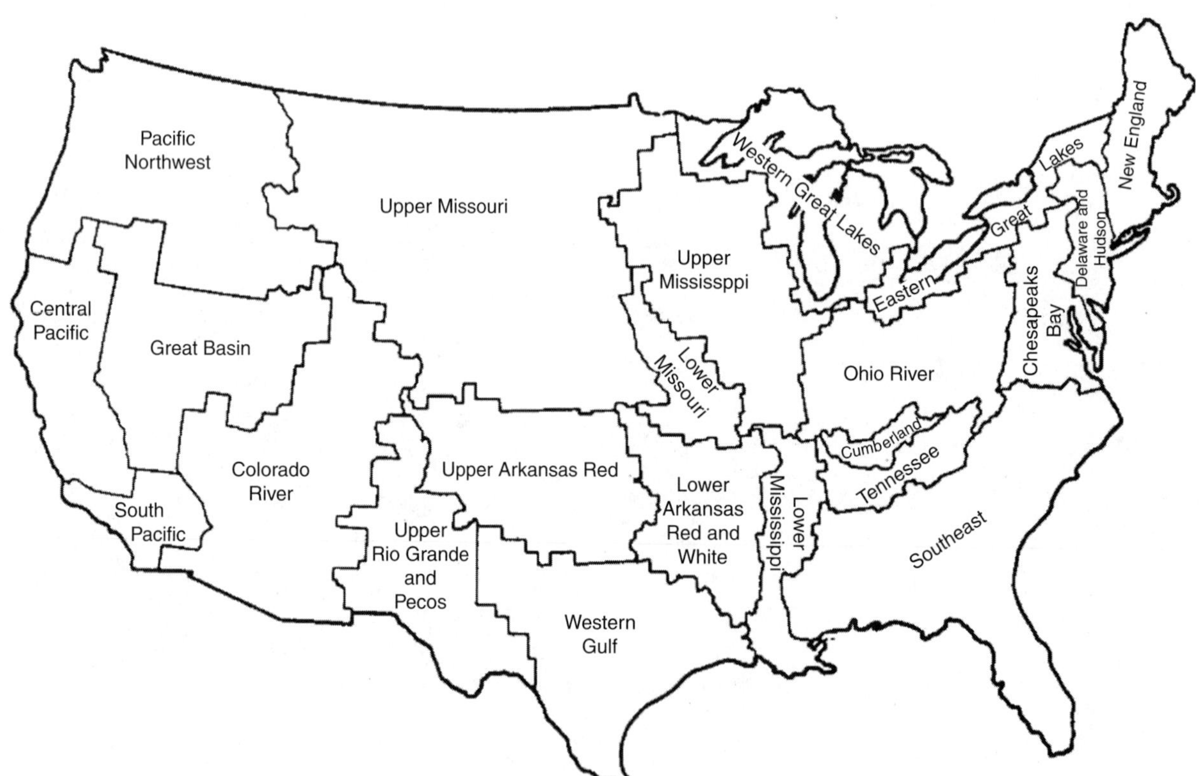

Figure 11B.19 Water resources regions of the United States. (From U.S. Geological Survey.)

Table 11B.17 Normal Surface-Water Reservoir Capacity in the United States

Water Resources Region	Area of Region, in Thousands of Square Miles	Average Renewable Supply, in Billion Gallons per Day[a]	Normal Reservoir Capacity[b]		
			In Million Acre-feet	In Acre-feet of Storage per Square Mile	As a Percentage of Annual Renewable Supply
New England	69	78.4	13.0	188	15
Mid-Atlantic	103	80.7	10.3	100	11
South Atlantic-Gulf	271	233.5	38.7	143	15
Great Lakes	134	74.3	6.9	51	8
Ohio (exclusive of Tennessee Region)	160	139.5	19.6	123	13
Tennessee	43	41.2	11.2	260	24
Upper Mississippi (exclusive of Missouri Region)	181	77.2	12.2	67	14
Mississippi (entire basin)	1241	464.3	164.8	133	32
Souris-Red Rainy	55	6.5	8.0	145	110
Missouri	511	62.5	84.3	165	120
Arkansas-White-Red	244	68.6	31.8	130	41
Texas-Gulf	178	33.1	24.7	139	67
Rio Grande	137	5.1	10.4	76	182
Upper Colorado	103	14.7	37.7	366	229
Colorado (entire basin)	258	15.6	70.4	273	403
Great Basin	139	9.9	3.3	24	30
Pacific Northwest	271	276.2	60.9	225	20
California	165	70.2	38.8	235	49
Alaska	586	975.5	1.5	3	0.1
Hawaii	6	7.4	0.0	2	0.1
Caribbean	4	5.1	0.3	90	5

[a] Adjusted by adding exports and subtracting imports.
[b] About two-thirds of maximum capacity.

Source: From U.S. Geological Survey, 1984, National Water-Summary 1983—Hydrologic Events and Issues, Water—Supply Paper 2250.

Table 11B.18 Summary of Reservoir Storage, Including Controlled Natural Lakes, in the United States and Puerto Rico

Reservoir Storage (Range, in Acre-feet)	Number of Reservoirs	Total Reservoir Storage	
		Acre-feet	Percentage of Total
Greater than 10,000,000	5	107,655,000	22.4
100,000–10,000,000	569	322,852,000	67.3
50,000–100,000	295	20,557,000	4.3
25,000–50,000	374	13,092,000	2.7
5,000–25,000	1,411	15,632,000	3.3
Total[a]	2,654	479,788,000	100.0

Note: Reservoir storage is expressed as normal capacity, which is the total storage space in a reservoir below the normal retention level, including dead storage and inactive storage, and excluding any flood control or surcharge storage.

[a] In addition, there are perhaps at least 50,000 reservoirs with capacities ranging from 50 to 5,000 acre-feet, and about 2 million smaller farm ponds used for storage.

Source: From U.S. Army Corps of Engineers, 1981.

Table 11B.19 Hydraulic Test Data from Aquifer Storage and Recovery Well Systems in Southern Florida

	Production Well Identifier	Open Interval Tested (ft)	Test Type	Monitoring Well	Pumping Rate (gal/min)	Length of Test (hr)	Transmisivity (ft^2/d)	Storage Coefficient (unitless)	Leakance (1/day)	Method of Analysis	Specific Capacity (gallons per minute per foot)	Back-Ground Measurements	Problems and Comments
Broward County													
Deerfield Beach West WTP													
NR	ASR-1	956–1130[a]	S	—	950–2,100	NR	—	—	—	—	42.6–32.0	NR	Step-drawdown test was performed prior to multiwell, but date of test was not given in report
12/10/92	ASR-1	956–1130[a]	M	MW-1	1,200	5.7	24,200	1.33×10^{-6}	6.3×10^{-2}	H-J			
Broward County WTP 2A													
11/21/96	ASR-1	995–1200[a]	S	—	1,050–2,950	8	—	—	—	—	90.6–51.1	Data collected for multiwell test but no information on whether corrections were made	Walton method should give same transmissivity if no leakance as C-J method Curve matches look okay
11/26/96	ASR-1	995–1200[a]	M	MW-1	1,000	24	28,900	1.1×10^{-4}	None	Walton	—		
11/26/96	ASR-1	995–1200[a]	M	MW-1	1,000	24	37,200	5.3×10^{-5}	—	C-J	—		
11/26/96	ASR-1	995–1200[a]	M	ASR-1	1,000	24	44,000	—	—	Theis Rec	NR		
9/19/96	MW-1	990–1200[a]	S	—	48	6	—	—	—	—	NR		
Springtree WTP													
6/6/97	ASR-1	1,110–1,270[a]	S	—	700–1,900	8		—	—	—	18.4–16.5	NR	Acidization of well done after step test and before constant rate test. Interpretation of recovery data favored late time data
7/28/97	ASR-1	1,110–1,270[a]	R	—	2,115	48	5,700	—	—	Theis Rec	22.75		
Fiveash WTP													
1/12/98	FMW-1	998–1,028	P, R	—	160	4	4,700	—	—	Theis Rec	4.7	NR	Third packer test showed apparent failure after about 10 min, which may have affected data. FMW-1 apparently not used in 24-hour test of ASR-1
1/13/98	FMW-1	998–1,042	P, R	—	160	4	8,000	—	—	Theis Rec	5		
1/15/98	FMW-1	1,058–1,175[a]	P	—	600	10 min	23,500	—	—	C-J	46		
3/16/98	FMW-1	1,055–1,175[a]	S	—	100–160	4	—	—	—	—	3.8–3.6		
3/17/98	FMW-1	1,055–1,175[a]	S	—	164	24	NR	—	—	—	~3.5		
3/25/98	ASR-1	1,055–1,200[a]	S	—	968–2,104	4	—	—	—	—	25.5–17.7		
3/30/98	ASR-1	1,055–1,200[a]	R	—	2,100	24	19,500	—	—	Theis Rec	~17.5		
Charlotte County													
Shell Creek WtP													
11/5/97	ASR-1	700–755	P	—	300–550	NR	—	—	—	—	10.0–5.6	NR	
11/17/97	ASR-1	700–764	S	—	231–597	12	—	—	—	—	5.3–4.6		
11/18/97	ASR-1	700–764	R	—	546	54	1,300	—	—	SC	4.4		

6/28/99	ASR-1	764–933[a]	S	—	610–600	8	—	—	—	—	10.6		
Englewood South Regional WWTP													
NR	TPW-1	563–583	P, R	—	10	1.33	450	—	—	Theis Rec	1	NR	Multiwell aquifer test planned in the future
3/7/00	TPW-1	630–807	P	—	131	4	1,300	—	—	SC	5		
3/31/00	TPW-1	507–700[a]	S	—	490–1,050	NR	4,700	—	—	SC	17.34 (avg)		
4/20/00	SZMW-1	510–700[a]	S	—	31–101	NR	3,700	—	—	SC	13.87 (avg)		
4/18/00	IMW-1	280–320	S	—	28.3–59.5	NR	2,300	—	—	SC	5.5–4.0		
4/18/00	SMW-1	170–205	S	—	9.3–17.7	NR	300	—	—	SC	1 (avg)		
Collier County													
Manatee Road													
11–90	MW-A	360–460	P, S	—	NR	NR	2,400	—	—	See comment	NR	NR	All of the step-drawdown tests are indicated to be packer tests; however, some could have been a test of an open interval below casing. For step test, transmissivity was determined at each step by an unspecified method in Walton (1970).
11-90	MW-A	465–530[a]	P, S	—	NR	NR	20,000	—	—	See comment	NR		
11–90	MW-A	680–760	P, S	—	NR	NR	15,000	—	—	See comment	NR		
11–90	MW-A	930–1020	P, S	—	NR	NR	6,700	—	—	See comment	NR		
11–90	MW-A	1,180–1,220	P, S	—	NR	NR	6,300	—	—	See comment	NR		
11–90	MW-A	1,345–1,606	P, S	—	NR	NR	5,700	—	—	See comment	NR		
NR	ASR-1	465–528[a]	M	MW-B	670	17	9,400	1.00×10^{-4}	3.7×10^{-4}	H-J	—		
NR	ASR-1	465–528	M	MW-B	670	17	12,000	—	—	Theis Rec	—		
Macro Lakes													
NR	DZMW	296–399	S	—	220–600	NR	67,000	—	—	Walton	220–170	NR	Estimated transmissivity in lower Hawthorn zone I (550 to 622 ft) was too low for consideration as an aquifer storage and recovery interval. No dates were reported for any tests. Good agreement between tests
NR	DZMW	296–399	R	—	600	3	42,400	—	—	Theis Rec	NR		
NR	DZMW	550–622	P, R	—	5	4	47	—	—	Theis Rec	NR		
NR	DZMW	745–811[a]	R	—	187	4.5	8,200	—	—	Theis Rec	NR		
NR	ASR-1	745–790[a]	M	ASR-1	463	8.3	16,300	—	—	Theis Rec	NR		
NR	ASR-1	745–790[a]	M	DZMW	463	8.3	9,100	6.5×10^{-5}	7×10^{-4}	H-J	—		
NR	ASR-1	745–790[a]	M	DZMW	463	8.3	12,000	—	—	Theis Rec	—		
NR	ASR-2	736–780[a]	S	ASR-1	400–650	NR	—	—	—	Theis Rec	25–24		
NR	ASR-3	736–780[a]	S, M	ASR-2,		ASRZMW	400–820	NR	8,000–8,100	—	—	C-J	17.4–15
Lee County													
Lee County WTP													

(Continued)

Table 11B.19 (Continued)

	Production Well Identifier	Open Interval Tested (ft)	Test Type	Monitoring Well	Pumping Rate (gal/min)	Length of Test (hr)	Transmisivity (ft²/d)	Storage Coefficient (unitless)	Leakance (1/day)	Method of Analysis	Specific Capacity (gallons per minute per foot)	Back-Ground Measurements	Problems and Comments
NR	ASR-1	445–600a	M	NR	350	48	800	1.00×10^{-4}		H-J	NR	NR	Storage zone in ASR-1 is located in the lower Hawthorn producing zone of the Upper Floridan aquifer as defined by Reese (2000)
Corkscrew WTP													
08–94	MW-A	428–515	P	—	NR	NR	500	—	—	SC	2.5	Data collected for multiwell test in September 1995	Test of MW-A, interval 744 to 778 ft, was of upper 34 of 240 ft thick Suwannee Limestone. Second multiwell test of ASR-1 followed back-plugging of MW-A to injection zone. Transmissivity values for step-drawdown test of ASR-2, 3, 4, and 5 were obtained
8/17/94	MW-A	524–578	P	—	39	5 min.	500	—	—	SC	NR		
8/24/94	MW-A	744–778	P, S	—	15–74	4	13,000	—	—	SC	1.3–0.6		
NR	MW-B	452–504	NR	—	NR	NR	100	NR	NR	NR	NR		
09–95	ASR-1	328–397[a]	M	MW-C	400	115.5	3,410	7.70×10^{-5}	1.6×10^{-5}	Hantush	—		
09–95	ASR-1	328–397[a]	M	MW-C	400	115.5	3,380	6.70×10^{-5}	—	C-J	—		
09–95	ASR-1	328–397[a]	M	MW-C	400	115.5	3,460	—	—	Theis Rec	—		
06–96	ASR-1	328–397[a]	M	MW-A	415	120	1,760	2.30×10^{-4}	—	Theis			
06–96	ASR-1	328–397[a]	M	MW-A	415	120	1,900	1.70×10^{-4}	—	C-J	—		
06–96	ASR-1	328–397[a]	M	MW-C	415	120	3,180	5.70×10^{-5}	—	Theis	—		
06–96	ASR-1	328–397[a]	M	MW-C	415	120	3,410	4.90×10^{-5}	—	C-J	—		
7/13/99	ASR-2	337–397[a]	S	—	115–410	NR	2,040	—	—	SC	7.8–6.6		
2/12/99	ASR-3	285–347[a]	S	—	129–497	NR	7,350	—	—	SC	26.9–19.4		
6/25/99	ASR-4	310–368[a]	S	—	153–450	NR	4,020	—	—	SC	15.0–11.7		
7/8/99	ASR-5	253–291[a]	S	—	163–380	NR	—	—	—	—	7.4–5.2		
7/20/99	ASR-5	253–291[a]	S	—	130–490	NR	13,400	—	—	SC	50.0–36.1		
North Reservoir													
12/7/98	MW-1	480–518	S	—	92–430	NR	14,400	—	—	SC	44.4–41.3	Data collected for 3 days prior to the multiwell test	Fit of line to ASR-1 recovery data for multiwell test is poor
12/9/98	MW-1	529–619[a]	P, S	—	73–295	NR	5,200	—	—	SC	9.7–3.5		
12/11/98	MW-1	640–703	P, S	—	79–281	NR	2,040	—	—	SC	6.6–2.8		
12/16/98	MW-1	808–890	P, S	—	55–190	NR	680	—	—	SC	2.8–1.8		
12/18/98	MW-1	904–977	P, S	—	85–322	NR	9,590	—	—	SC	10.5–3.1		
3/3/99	ASR-1	540–642[a]	S	—	162–590	4	2,220	—	—	SC	8.65–7.00		
3/8/99	ASR-1	540–642[a]	M	MW-1	379	72	8,290	3.27×10^{-4}	7.33×10^{-4}	H-J	—		

3/8/99	ASR-1	540–642[a]	M	MW-1	379	72	8,740	4.64×10^{-4}	—	C-J (recovery)	—		
3/8/99	ASR-1	540–642[a]	M	ASR-1	379	72	8,570	—	—	C-J (recovery)	NR		
Winkler Avenue													
NR	ASR-1	455–554[a]	S	—	135	15 min	—	—	—	—	59.7	Ninety hours collected prior to multiwell test, but apparently not used to correct drawdown data	Pumping rate for multiwell test was 1,540 for first 6.5 hr, then changed to 1,400. No attempt was made to analyze multiwell test data for storage coefficient or leakance. Storage zone is located in the lower Hawthorn producing zone of Upper Florida
NR	ASR-1	455–647	S	—	160	15 min	—	—	—	—	86.3		
6/16/99	ASR-1	455–574[a]	P	—	479	70 min	29,100	—	—	C-J (recovery)	NR		
6/17/99	ASR-1	455–575[a]	P	—	483	18 min	26,600	—	—	C-J (recovery)	NR		
10/23/99	ASR-1	455–553[a]	M	ASR-1	1,540–1,400	27	24,700	—	—	C-J	NR		
10/23/99	ASR-1	455–553[a]	M	ASR-1	1,540–1,400	27	25,400	—	—	C-J (recovery)	—		
10/23/99	ASR-1	455–553[a]	M	SZMW-1	1,540–1,400	27	27,400	NR	NR	C-J	—		
10/23/99	ASR-1	455–553[a]	M	SZMW-1	1,540–1,400	27	29,000	—	—	C-J (recovery)	—		
San Carlos Estates													
6/7/99	TPW-1	650–701[a]	S	—	710–1,480	8	—	—	—	—	250–130	Unknown for multiwell test. Test followed 10 days of recharge at 1,955 gal min^{-1} and then 6-day static period	High specific capacity in TPW-1 due to two pilot holes in open interval. Pumping rate for test on 11/10/99 of 985 gallons per minute was natural flow. C-J solution for drawdown in SZMW-1R is suspect. Solution is for very late time only, and background changes due to prior recharge may have affected response
7/29/99	SZMW-1R	659–721[a]	S	—	170–350	8	—	—	—	—	15–9.0		
8/2/99	SMW-1	234–321	S	—	150–220	8	—	—	—	—	8.9–6.5		
11/10/99	TPW-1	650–701[a]	M	SZMW-1R	985	8	39,000	1.00×10^{-2}	—	C-J	—		
11/10/99	TPW-1	650–701[a]	M	SZMW-1R	985	8	70,000	—	—	Theis Rec	—		
Olga WTP													
1/5/99	MW-1	515–605	S	—	110–400	NR	2,500	—	—	SC	NR	Measured for multiwell test, but unknown if used to correct drawdown	H-J results for multiwell test agree better with single well and packer test than C-J results. A second constant rate test was run but is not reported here. Storage zone is about 150 ft below top of Suwannee Limestone
1/7/99	MW-1	612–689	P	—	70–200	NR	1,300	—	—	SC	NR		
2/3/99	MW-1	835–935[a]	P	—	70–355	NR	7,600	—	—	SC	NR		
2/4/99	MW-1	710–935	P	—	70–350	NR	7,600	—	—	SC	NR		
2/4/99	MW-1	835–935[a]	P	—	70–350	NR	7,600	—	—	SC	NR		

(Continued)

Table 11B.19 (Continued)

	Production Well Identifier	Open Interval Tested (ft)	Test Type	Monitoring Well	Pumping Rate (gal/min)	Length of Test (hr)	Transmisivity (ft²/d)	Storage Coefficient (unitless)	Leakance (1/day)	Method of Analysis	Specific Capacity (gallons per minute per foot)	Back-Ground Measurements	Problems and Comments
2/8/99	MW-1	945–1,101	P	—	6–15	NR	33	—	—	SC	NR		
3/17/99	MW-3	740–820	S	—	78–480	NR	1,900	—	—	SC	NR		
3/25/99	MW-3	830–945[a]	P	—	80–340	NR	9,000	—	—	SC	NR		
3/25/99	MW-3	854–945[a]	P	—	75–350	NR	6,400	—	—	SC	NR		
3/26/99	MW-3	857–945[a]	R	—	300	NR	8,700	—	—	Theis Rec	NR		
11/1/99	ASR-1	859–920[a]	S	—	112–545	5	5,000	—	—	SC	14.9–8.5		
11/3/99	ASR-1	859–920[a]	M	MW-1	500	60	7,200	5.10×10^{-5}	5.2×10^{-3}	H-J	—		
11/3/99	ASR-1	859–920[a]	M	MW-1	500	60	12,000	4.10×10^{-5}	—	C-J	—		
11/3/99	ASR-1	859–920[a]	M	MW-3	500	60	9,400	5.50×10^{-5}	6.0×10^{-2}	H-J	—		
11/3/99	ASR-1	859–920[a]	M	MW-3	500	60	11,000	4.20×10^{-4}	—	C-J	—		
Miami-Dade County													
Hialeah													
2/10/75	MW-1	953–1,060[a]	M	ASR-1	250	1.66	11,000	8.4×10^{-5}	—	J-L	—	NR	Transmissivity estimate from Meyer (1989b). Storage coefficient estimated by model simulation of pumping test
West Well Field													
1/26/97	ASR-1	850–1,302[a]	S	—	1,400–4,000	8	—	—	—	—	269–52.1	Measured for Multiwell test. Correction was not done, due to negligibility	All three aquifer storage and recovery wells were heavily acidized prior to all tests. Late time drawdown data problematic because of pump going down several times
2/25/97	ASR-2	845–1,250[a]	S	—	1,500–3,800	8	—	—	—	—	126.6–51.1		
4/8/97	ASR-3	835–1,210[a]	S	—	1,500–3,800	8	—	—	—	—	46.1–38.2		
12/9/97	ASR-1	850–1,302[a]	M	ASR-1	3,500	72	10,300	N/A	N/A	C-J	NR		
12/9/97	ASR-1	850–1,302[a]	M	ASR-2	3,500	72	15,400	3.90×10^{-4}	1.6×10^{-3}	Walton	—		
12/9/97	ASR-1	850–1,302[a]	M	ASR-2	3,500	72	18,200	2.90×10^{-4}	N/A	C-J	—		
12/9/97	ASR-1	850–1,302[a]	M	ASR-3	3,500	72	15,400	4.40×10^{-4}	3.90×10^{-5}	Walton	—		
12/9/97	ASR-1	850–1,302[a]	M	ASR-3	3,500	72	19,700	3.30×10^{-4}	N/A	C-J	—		
Monroe County													
Marathon													
5/3/90	ASR-1	387–432[a]	M	OW-1	105	25	2,290	3.20×10^{-4}	NR	Walton	—	Measured for 3 weeks prior to multiwell test. A regional increasing trend in water level was determined	Leakance using Walton (1962) method not determined
5/3/90	ASR-1	387–432[a]	M	OW-1	105	25	2,510	3.70×10^{-4}	—	C-J	—		
5/3/90	ASR-1	387–432[a]	M	OW-1	105	25	1,760	—	—	Theis-Rec	—		
5/3/90	ASR-1	387–432[a]	M	OW-2	105	25	2,180	5.20×10^{-4}	NR	Walton	—		

5/3/90	ASR-1	387–432[a]	M	OW-2	105	25	4,090	—	—	Theis-Rec	—		
5/6/90	ASR-1	387–432[a]	S	—	95–350	NR	—	—	—	—	3.9–2.7		
Okeechobee County													
Taylor Creek-Nubbin Slough (Lake Okeechobee)													
4/20/98	MW-1	1,175–1,227	P	—	10	6.4	706	NR	—	C-J	NR	Water-level data taken for 5 days prior to constant rate test; corrections made using a long-term increasing trend	Leakance derived by extrapolation; longer pumping period required for more accurate value
4/20/98	MW-1	1,175–1,227	P, R	—	10	6.4	2,940	—	—	Recovery	NR		
8/2/98	ASR-1	1,268–1,710[a]	M	ASR-1	6,500	24	620,000	N/A	N/A	C-j (recovery)	1,600		
8/2/98	ASR-1	1,268–1710[a]	M	MW-1	6,500	24	586,000	1.25×10^{-3}	0.01–0.001	H-J	1,600		
8/2/98	ASR-1	1,268–1,710[a]	M	MW-1	6,500	24	765,000	1.90×10^{-4}	N/A	C-J (recovery)	1,600		
Palm Beach County													
Boynton Beach East WTP													
4/9/92	ASR-1	804–900[a]	S	—	320–2,100	NR	6,800–13,000	NR	—	C-J	18–28	NR	Second step test is with permanent equipment installed in well. Average of estimates for transmissivity was 70,000 gallons per day per foot
10/15/92	ASR-1	804–909[a]	S	—	798–1,723	NR	Not calculated	—	—	C-J	29–27		
Delray Beach North Storage Reservoir													
6/5/96	ASR-1	849–899	P	—	49	NR	—	—	—	—	0.37	NR	Second step test performed after acidization of well. For the second step test, pump malfunctioned after about 10 min of pumping during the last step at 2,550 gal min^{-1}
6/11/96	ASR-1	900–952	P	—	83	NR	—	—	—	—	0.9		
6/14/96	AR-1	974–1,020	P	—	90	NR	—	—	—	—	2.4		
6/18/96	ASR-1	1,020–1,100	P	—	98	NR	—	—	—	—	2		
9/20/96	ASR-1	1,020–1,200[a]	S	—	575–1,100	24	—	—	—	—	10.8–7.8		
2/24/98	ASR-1	1,020–1,200[a]	S	—	760–2,550	13.2	—	—	—	—	17.2–15.7		
West Palm Beach WTP													
8/22/96	FAMW	975–1,091	P, S	—	64–142	NR	—	—	—	—	194–86	NR	Recovery was allowed following step test of ASR-1 before multiwell test was begun. Large deviation from Theis curve during late time, but leaky aquifer solution not used
9/14/96	FAMW	1,304–1,384	P, S	—	55–110	NR	—	—	—	—	220–110		
8/29/96	FAMW	975–1,090	S	—	300–584	NR	—	—	—	—	75–58		
9/1/96	FAMW	975–1,190[a]	S	—	300–584	NR	—	—	—	—	110–101		
9/4/96	FAMW	975–1,290	S	—	550–740	NR	—	—	—	—	116–105		
9/6/96	FAMW	975–1,384	S	—	550–740	NR	—	—	—	—	116–99		

(Continued)

Table 11B.19 (Continued)

	Production Well Identifier	Open Interval Tested (ft)	Test Type	Monitoring Well	Pumping Rate (gal/min)	Length of Test (hr)	Transmissivity (ft^2/d)	Storage Coefficient (unitless)	Leakance (1/day)	Method of Analysis	Specific Capacity (gallons per minute per foot)	Back-Ground Measurements	Problems and Comments
11/19/96	FAMW	975–1,191[a]	S	—	550–732	NR	—	—	—	—	62–42		
1/30/97	ASR-1	985–1,200[a]	S	—	508–704	24	—	—	—	—	390–306		
2/1/97	ASR-1	985–1,200[a]	M	FAMW	700	24	138,000	1.00×10^{-4}	—	C-J	—		
2/1/97	ASR-1	985–1,200[a]	M	FAMW	700	24	108,000	8.00×10^{-4}	—	Theis	—		
Western Hillsboro Canal, Site 1													
4/5/00	EXW-1	1,160–1,225	P	—	95	NR	—	—	—	—	22.6	NR	Acidized EXW-1 with 4,300 gallons of 36 percent HCl on 6-2-00
4/10/00	EXW-1	1,015–1,150	P	—	105	NR	—	—	—	—	10.9		
5/25/00	EXW-1	1,015–1,225[a]	S	—	1,000–3,000	NR	—	—	—	—	31.1–26.2		
St. Lucie County													
8/24/82	MW-1	600–775[a]	M	ASR-1	388	72	5,910	1.64×10^{-4}	4.3×10^{-2}	H-J (drawdown)	—	NR	Also conducted four pump tests of ASR-1 during drilling with total depth ranging from 627 to 1,000 ft and casing at 600 feet. Transmissivity was calculated from these tests based on recovery data from ASR-1
8/24/82	MW-1	600–775[a]	M	ASR-1	388	72	6,430	2.67×10^{-4}	4.7×10^{-4}	H-J (recovery)	—		

Note: Depths are in feet below land surface. Test type: M, multiwell constant rate; P, Packer test; R, single well constant rate recovery; S, step drawdown. Method of analysis; SC, specific capacity; Theis, Theis (1935) confined aquifer; C-J, Cooper and Jacob (1946) confined aquifer; Theis (1935) residual drawdown recovery; H-J, Hantush and Jacob (1955) leaky aquifer; Walton, Walton (1962) leaky aquifer; J-L, Jacob and Lohman (1952). Other annotations: WTP, water treatment plant; WWTP, wastewater treatment plant; —, not applicable; NR, not reported.

[a] Open interval tested is the same (or about the same) as the storage zone.

Source: From damsreport.org.

Table 11B.20 Reservoir Storage Required to Produce Selected Dependable Flows in the United States

Region	Storage Required to Produce Indicated Flow 100% of Time									
	Flow	Storage	Flow	Storage	Flow	Storage	Flow	Storage	Flow	Storage
New England	6,300	1,300	9,700	1,900	16,000	4,200	22,000	7,300	39,000	26,000
Delaware and Hudson Rivers	3,200	590	4,800	900	7,800	2,000	11,000	3,800	19,000	11,000
Chesapeake Bay	5,600	960	8,400	1,400	13,000	3,400	18,000	6,000	32,000	20,000
Southeast	21,000	4,900	31,000	7,800	49,000	14,000	71,000	25,000	126,000	78,000
Western Great Lakes	8,400	780	12,000	1,100	16,000	2,800	21,000	5,100	32,000	20,000
Eastern Great Lakes	2,300	720	3,700	1,100	6,500	2,200	9,700	3,700	19,000	11,000
Ohio River	7,400	4,100	9,400	5,200	15,000	7,200	21,000	11,000	46,000	29,000
Cumberland River	1,500	160	2,100	290	3,300	620	4,500	1,100	7,800	3,100
Tennessee River	9,000	400	11,000	600	15,000	1,200	19,000	2,500	28,000	9,600
Upper Mississippi River	7,800	1,100	12,000	2,300	18,000	5,200	25,000	8,700	41,000	26,000
Upper Missouri River	1,200	340	1,800	660	3,200	1,200	4,500	2,100	9,000	5,700
Lower Missouri River	410	660	550	840	1,200	1,300	2,200	2,300	5,800	5,200
Upper Arkansas-Red Rivers	410	250	700	410	1,300	720	2,100	1,200	4,500	3,000
Lower Arkansas-Red and White Rivers	1,000	1,400	2,100	2,500	4,400	5,000	8,400	9,400	20,000	22,000
Lower Mississippi River	2,100	900	3,600	1800	6,500	3,200	10,000	5,500	21,000	14,000
Upper Rio Grande and Pecos Rivers	[a]	[a]	[a]	[a]	[a]	[a]	[a]	[a]	[a]	[a]
Western Gulf	920	1,300	1,700	1,700	3,400	3,800	5,900	5,500	14,000	12,000
Colorado	210	90	320	120	560	230	830	380	1,700	1,200
Great Basin	300	72	470	100	780	240	1,200	420	2,100	1,000
South Pacific	30	10	40	13	60	26	100	43	180	119
Central Pacific	1,000	880	1,900	1,760	3,800	3,100	6,300	5,300	16,000	11,400
Pacific Northwest	9,700	2,600	21,000	6,600	26,000	9,300	39,000	16,000	76,000	47,000
United States	90,000	24,000	140,000	40,000	210,000	71,000	300,000	120,000	560,000	350,000

Note: Flow is in millions of gallons per day, and storage is in thousands of acre-feet.

[a] Appropriations currently exceed supply.

Source: From Select Committee on National Water Resources, U.S. Senate 1960.

Table 11B.21 Advantages and Disadvantages of Subsurface and Surface Reservoirs

Subsurface Reservoirs	Surface Reservoirs
Advantages	**Disadvantages**
1. Many large-capacity sites available	1. Few new sites available
2. Slight to no evaporation loss	2. High evaporation loss even in humid climate
3. Require little land area	3. Require large land area
4. Slight to no danger of catastrophic structural failure	4. Ever-present danger of catastrophic failure
5. Uniform water temperature	5. Fluctuating water temperature
6. High biological purity	6. Easily contaminated
7. Safe from immediate radioactive fallout	7. Easily contaminated by radioactive material
8. Serve as conveyance systems—canals or pipeline across lands of others unnecessary	8. Water must be conveyed
Disadvantages	**Advantages**
1. Water must be pumped	1. Water may be available by gravity flow
2. Storage and conveyance use only	2. Multiple use
3. Water may be mineralized	3. Water generally of relatively low mineral content
4. Minor flood control value	4. Maximum flood control value
5. Limited flow at any point	5. Large flows
6. Power head usually not available	6. Power head available
7. Difficult and costly to investigate, evaluate, and manage	7. Relatively easy to evaluate, investigate, and manage
8. Recharge opportunity usually dependent on surplus surface flows	8. Recharge dependent on annual precipitation
9. Recharge water may require expensive treatment	9. No treatment required of recharge water
10. Continuous expensive maintenance of recharge areas or wells	10. Little maintenance required of facilities

Source: From U.S. Bureau of Reclamation, 1977, *Groundwater Manual*, U.S. Department of the Interior.

Table 11B.22 Summary of Storage Reservoirs on United States Bureau of Reclamation Projects

	Active Capacity			Total Capacity	
Category	No. of Reservoirs	1000 m^3	Acre-Feet	1000 m^3	Acre-Feet
Constructed and operated by Bureau of Reclamation	110	106,709,244	86,510,251	144,632,109	117,254,697
Rehabilitated and operated by Bureau of Reclamation	10	247,069	200,302	248,694	201,618
Constructed by others, operated by Bureau of Reclamation	11	6,337,601	5,137,957	7,009,548	5,682,711
Under construction by Bureau of Reclamation	8	837,118	678,660	5,469,294	4,434,012
Constructed by Bureau of Reclamation, operated by others	109	18,183,188	14,741,292	19,878,391	16,115,611
Rehabilitated by Bureau of Reclamation, operated by others	14	852 561	691,180	865,130	701,370
Constructed and operated by others	71	78,238,538	63,428,765	100,529,146	81,499,984
Constructed or rehabilitated under loan program	22	444,806	360,609	493,310	399,932
Total	355	211,850,129	171,749,017	279,125,626	226,289,936

Note: As of September 30, 1986.

Source: From U.S. Department of the Interior, Bureau of Reclamation, 1986, *Statistical Compilation of Engineering Features on Bureau of Reclamation Projects*.

Table 11B.23 Results of Bathymetric Surveys of 14 Reservoirs in Puerto Rico

Reservoir	Original Capacity $(Mm)^3$	Const. Year	Study Year	Age	Storage Capacity $(Mm)^3$	Total Vol. Loss $(Mm)^3$	Long-Term Volume Loss (m^3/yr)	Loss in Percent	Long-Term Storage Loss per Year (Percent)	Drainage Area $(km)^2$	Surface Area $(km)^2$	Deposition Rate (cm/yr)	Sediment Yield (m^3/km^2/yr)	Storage Loss (m^3/km^2/yr)
Caonillas	55.66	1948	2000	52	42.27	13.39	257,500	24	0.5	126.65	2.70	9.5	2,186	2,033
Carite	13.95	1913	1999	86	10.74	3.21	37,326	23	0.3	20.51	1.20	3.1	1,938	1,820
Cidra	6.54	1946	1997	51	5.76	0.78	15,294	12	0.2	21.39	1.08	1.4	768	715
Dos Bocas	37.50	1942	1999	57	18.04	19.46	341,404	52	0.9	310.00	1.78	19.2	1,299	1,103
Garzas	5.80	1943	1996	53	5.11	0.69	13,019	12	0.2	15.60	0.40	3.2	878	834
Guajataca	48.46	1928	1999	71	42.28	6.18	87,042	13	0.2	79.77	3.42	2.5	1,188	1,091
Guayo	19.20	1956	1997	41	16.57	2.63	64,146	14	0.3	24.86	1.09	5.9	2,660	2,580
La Plata	40.21	1974	1998	24	35.46	4.75	197,917	12	0.5	469.00	3.09	6.4	483	422
Loiza	26.81	1953	1994	41	14.20	12.61	307,561	47	1.1	538.00	2.67	11.5	750	572
Loco	2.40	1951	2000	49	0.87	1.53	31,224	64	1.3	21.76	0.29	10.8	1,774	1,435
Lucchetti	20.35	1952	2000	48	11.88	8.47	176,458	42	0.9	44.81	1.11	15.9	4,102	3,937
Patillas	17.64	1914	1997	83	13.84	3.80	102,703	22	0.6	65.27	1.35	7.6	1,739	1,617
Prieto	0.76	1955	1997	42	0.22	0.54	12,857	71	1.7	24.80	0.06	21.4	900	518
Yahuecas	1.76	1956	1997	41	0.33	1.43	34,878	81	2.0	45.17	0.22	15.8	1,430	772
Average							119,952	35	0.7			9.6	1,578	1,389

Note: Const., construction; vol., volume; m^3/yr, cubic meter per year; Mm^3, mega cubic meter; km^2, square kilometer; cm/yr, centimeter per year; m^3/km^2/yr, cubic meter per square kilometer per year.

Source: From www.usgs.gov.

Table 11B.24 Percent of Annual Inflow Storage and Water Demand Storage for Lago Caonillas, Lago Cidra, Lago Dos Bocas, Lago Guajataca, Lago Loiza, and Lago La Plata

Reservoir	Average Annual Inflow (Million Cubic Meters)	Percent of Annual Inflow Stored	Annual Demand (Mm^3)	Ability to Store Annual Water Demand (Percent)
Lago Cidra	14.5	40.0	6.9	83
Lago Dos Bocas	400.00	4.0	103.61	17
Lago Guajataca	104.50	40.0	62.16	68
Lago Loiza	363.85	4.0	138.15	10
Lago La Plata	270.98	13.1	96.52	37

Note: Mm^3, mega cubic meter.

Source: From www.usgs.gov.

Table 11B.25 Major Storage Reservoirs on United States Bureau of Reclamation Projects

Project, State, Reservoir Name (Dam Name), Stream, Operator	Category	Purpose	Active Capacity Acre-Feet	Total Capacity Acre-Feet	Reservoir Area Acres	Year Completed
Pacific Northwest Region						
Boise ID-OR						
Cascade Reservoir North Fork Payette River	A	I–P–FC	653,000	703,000	28,300	1948
Columbia Basin, WA						
Banks Lake (North and dry falls) Offstream	A	I	715,000	1,280,000	27,000	1951[b]
Billy Clapp Lake (Pinto) Offstream	A	I–FC	21,200	64,200	1,010	1948
Franklin D. Roosevelt LK (Grand Coulee) Columbia River	A	I–P–FC–RR–N	5,190,000	9,390,000	82,300	1942
Potholes Reservoir (O Sullivan) Lower Crab Creek	A	I–FC	332,000	512,000	27,800	1949
Hungry Horse, MT						
Hungry Horse Reservoir South Fork Flathead River	A	I–P–FC–N	2,980,000	3,470,000	23,800	1953
Minidoka-Palisades, ID-WY						
American Falls Reservoir Snake River	B	I–FC–MI	1,670,000	1,670,000	58,100	1978[c]
Island Park Reservoir Henrys Fork River	A	I–FC	127,500	1,280,000	7,794	1938[c]
Jackson Lake Snake River	A	I–FC	624,000	847,000	25,500	1916[c,e]
Palisades Reservoir South Fork Snake River	A	I–P–FC–FW	1,200,000	1,401,000	16,150	1957
Owyhee, ID-OR						
Lake Owyhee Owyhee River Owyhee Project North Board of Control	D	I–FC	715,000	1,120,000	12,700	1932
Yakima, WA						
Cle Elum Lake Cle Elum River	A	I	437,000	710,000	4,812	1933
Mid-Pacific Region						
Central Valley, CA						
Clair Engle Lake (Trinity) Trinity River	A	I–P	2,140,000	2,450,000	16,500	1962

(Continued)

Table 11B.25 (Continued)

Project, State, Reservoir Name (Dam Name), Stream, Operator	Category	Purpose	Active Capacity Acre-Feet	Total Capacity Acre-Feet	Reservoir Area Acres	Year Completed
Folsom Lake	B	I–P–FC	920,000	1,010,000	11,400	1956
American River						
Lake Isabella	E	I–FC	567,900	570,000	11,400	1953
Kern River						
U.S. Corps of Engineers						
Millerton Lake (Friant)	A	I–FC	434,000	521,000	4,900	1942
San Joaquin River						
New Hogan Lake	E	I–FC	309,000	324,000	4,410	1964
Calaveras River						
U.S. Corps of Engineers						
New Melones Lake	B	I–P–FC	2,100,000	2,400,000	12,500	1979
Stanislaus River						
O Neill Reservoir	D	I–P	20,800	56,400	2,250	1967
San Luis Creek						
California Department of Water Resources						
Pine Flat Lake	E	I–FC	1,000,000	1,000,000	5,970	1954
Kings River						
U.S. Corps of Engineers						
San Luis Reservoir	D	I–P	1,960,000	2,040,000	13,000	1967
San Luis Creek						
California Department of Water Resources						
Shasta Lake	A	I–P–FC–RR–N–MI	3,970,000	4,550,000	29,700	1945
Sacramento River						
Klamath, CA-OR						
Clear Lake Reservoir	A	I–FC	513,000	527,000	25,800	1910
Lost River						
Gerber Reservoir	A	I	94,300	94,300	3,830	1925[c]
Miller Creek						
Upper Klamath Lake	E	I–P	465,000	873,000	90,900	1921[a,c]
Upper Klamath Lake Outlet						
Pacific Power and Light						
Newlands, NV						
Lake Tahoe	D	I–P	732,000	732,000	120,000	1913[c]
Truckee River						
Truckee-Carson Irrigation District						
Solano, CA						
Lake Berryessa (Monticello)	A	I–FC–MI	1,590,000	1,600,000	20,700	1957
Putah Creek						
Lower Colorado Region						
Boulder Canyon, AZ-CA-NV						
Lake Mead (Hoover)	A	I–P–FC–RR–N–MI	17,400,000	2,8500,000	163,000	1936
Colorado River						
Colo R Front Work-Levee System Senator Wash Reservoir	A		12,300	13,800	470	1966[b]

(Continued)

Table 11B.25 (Continued)

Project, State, Reservoir Name (Dam Name), Stream, Operator	Category	Purpose	Active Capacity Acre-Feet	Total Capacity Acre-Feet	Reservoir Area Acres	Year Completed
Senator Wash						
Parker-Davis, AZ-CA-NV						
Lake Havasu (Parker) Colorado River	A	P–FC–MI	180,000	619,400	20,400	1938
Salt River, AZ						
Theodore Roosevelt Lake Salt River Salt River Valley Water Users Assoc.	D	I–P–MI	1,336,700	1,336,700	17,315	1936[e]
Upper Colorado Region						
Central Utah, UT						
Strawberry Reservoir Strawberry River	A	I–MI–FC	951,000	1,106,500	17,000	1974
Colorado River Storage Blue Mesa Reservoir Gunnison River	A	I–P–FC	748,500	940,800	9,180	1966
Flaming Gorge Reservoir Green River	A	I–P	3,515,700	3,788,700	42,020	1964
Lake Powell (Glen Canyon) Colorado River	A	P–R	20,876,000	27,000,000	161,390	1964
Navajo Reservoir San Juan River	A	I–FC–RR	1,036,100	1,708,600	15,610	1963
Southwest Region						
Brantley, NM						
Brantley Reservoir	C	FC–I	42,000	3,485,000	15,320	
Canadian River, TX						
Lake Meredith (Sanford) Canadian River Canadian River Municipal Water Authority	D	FC–MI–FW	1,304,554	1,382,478	16,513	1965
Colorado River, TX						
Marshall Ford Reservoir Colorado River Lower Colorado River Authority	D	I–P–FC–RR–N	1,590,763	1,953,936	18,929	1942
Middle Rio Grande, NM						
Cochiti Lake Rio Grande River U.S. Corps of Engineers	E	I–FC–S–FW	676,217	722,000	10,690	1975
Nueces River, TX						
Choke Canyon Reservoir Frio River	A	MI–FW	689,480	691,130	25,733	1982
Rio Grande, NM-TX						
Elephant Butte Reservoir Rio Grande River	A	I–P	2,060,000	2,110,000	36,521	1916
San Angelo, TX						
Twin Buttes Reservoir Middle South Concho River; Spring Creek San Angelo Water Supply Corporation	D	I–FC–MI–FW	632,214	640,568	9,079	1963
Tucumcari, NM						
Conchas Lake Canadian River	E	I–FC	451,161	528,951	13,664	1940

(Continued)

Table 11B.25 (Continued)

Project, State, Reservoir Name (Dam Name), Stream, Operator	Category	Purpose	Active Capacity Acre-Feet	Total Capacity Acre-Feet	Reservoir Area Acres	Year Completed
U.S. Corps of Engineers						
Missouri Basin Region						
Colorado-Big Thompson, CO						
Lake Granby	A	I–P	466,000	540,000	7,260	1950
Colorado River						
Fort Peck, MT						
Fort Peck Lake	E	I–P–FC–N	14,626,000	18,909,000	249,000	1940
Missouri River						
U.S. Corps of Engineers						
Kendrick, WY						
Seminoe Reservoir	A	I–P	985,603	1,017,273	20,300	1939
North Platte River						
North Platte, NE-WY						
Pathfinder Reservoir	A	I–P	985,102	1,016,507	22,000	1909
North Platte River						
Pick-Sloan Missouri Basin Program						
Lake Francis Case (Fort Randall)	E	P–FC–N	4,033,000	5,603,000	102,100	1953
Missouri River						
U.S. Corps of Engineers						
Lake Oahe	E	I–P–FC	17,886,400	23,337,600	372,800	1962
Missouri River						
U.S. Corps of Engineers						
Lake Sharpe (Big Bend)	E	P	1,882,302	1,884,000	64,000	1964
Missouri River						
U.S. Corps of Engineers						
Lewis and Clark Lake (Gavins Point)	E	P–FC–N	156,000	504,000	32,000	1955
Missouri River						
U.S. Corps of Engineers						
Bostwick Division, KS-NE						
Harlan County Lake	E	I–FC	691,121	825,782	23,064	1952
Republican River						
U.S. Corps of Engineers						
Boysen Division, WY						
Boysen Unit						
Boysen Reservoir	A	I–P–FC	802,000	952,400	22,166	1952
Wind River						
Garrison Division, ND						
Audubon Lake (Snake Creek)	B	I	356,000	500,000	19,400	1965[d]
Offstream						
Lake Sakakawea (Garrison)	E	I–P–FC–N–MI–FW	18,933,500	23,923,000	381,900	1956[a]
Missouri River						
U.S. Corps of Engineers						
Helena-Great Falls Div, MT						
(Canyon Ferry Unit)						

(Continued)

Table 11B.25 (Continued)

Project, State, Reservoir Name (Dam Name), Stream, Operator	Category	Purpose	Active Capacity Acre-Feet	Total Capacity Acre-Feet	Reservoir Area Acres	Year Completed
Canyon Ferry Lake	A	I–FC–P–MI–F–W	1,605,057	2,050,519	35,181	1954
Missouri River						
Lower Bighorn Div, MT-WY						
(Yellowtail Unit)						
Bighorn Lake (Yellowtail Dam)	A	P	834,776	1,328,360	17,298	1966
Bighorn River						
Marias Division, MT						
(Lower Marias Unit)						
Lake Elwell (Tiber)	A	I–FW	790,533	1,368,157	23,152	1956[f]
Marias River						
Oregon Trail Div, NE-WY						
(Glendo Unit)						
Glendo Reservoir	A	I–P–FC	726,254	789,402	12,400	1958
North Platte River						
Solomon Division, KS						
(Glen Elder Unit)						
Waconda Lake (Glen Elder)	A	I–FC–MI	927,104	963,775	12,600	1969
Solomon River						

Note: As of September 30, 1986; total capacity of 500,000 acre-feet or more. Purpose, FW, fish wildlife; FC, flood control; I, irrigation; MI, municipal industrial; N, navigation; P, power; RR, river regulation; S, sediment. Capacity-active, storage available for project, usually the storage above the lowest point of release. Total, storage at highest controlled water surface. Reservoir area, water surface area at active conservation capacity. Year completed, date original construction completed except as indicated in footnote e. Operating agency, agency directly and officially responsible for operation and maintenance of feature. Does not include agencies that may administer recreational or other secondary function as a service to the primary operator. Category, A, constructed and operated by bureau of reclamation; B, constructed by others, operated by bureau of reclamation; C, under construction by bureau of reclamation; D, constructed by bureau of reclamation, operated by others; E, constructed and operated by others.

[a] The bureau of reclamation has responsibility for the proper care and management of five reservoirs by special agreement.
[b] Reregulating reservoir.
[c] Dead storage not evaluated.
[d] Not yet in operation.
[e] Date indicates bureau of reclamation rehabilitation work.
[f] Tiber dam modification completed in 1981.

Source: From U.S. Department of the Interior Bureau of Reclamation, 1986, *Statistical Compilation of Engineering Features on Bureau of Reclamation Projects.*

Table 11B.26 Flood Control Reservoirs in the United States

Name	River Basin	Stream	Community in Vicinity	Year Placed in Operation	Total Storage (acre-ft)	Permanent Pool (Acreage) or No Pool (NPP)	Project Functions[a]	Characteristics of Dam: Type	Height (ft)	Length (ft)
Alaska										
Chena River Lakes	Yukon-Kuskokwim	Chena River	Fairbanks	1981	2,000	NPP	FR	Earth	50	40,200
Arizona										
Adobe	Gila	Skunk Creek	Phoenix	1982	18,350	NPP	FR	Earth	109	2,275
Alamo	Colorado	Bill Williams River	Wenden	1968	1,046,310	560	FRWX	Earth	283	975
Cave Buttes	Gila	Cave Creek	Phoenix	1979	46,600	NPP	FRX	Earth	109	2,275
Dreamy Draw	Gila	Dreamy Draw	Phoenix	1973	320	NPP	FRX	Earth	50	448
New River	Gila	New River	Phoenix	1985	43,520	NPP	F	Earth	104	2,320
Painted Rock	Gila	Gila River	Gila Bend	1959	2,491,700	NPP	FRWX	Earth	181	4,780
Tat Momolikot	Gila	Santa Rosa Wash	Casa Grande	1974	198,550	NPP	FWX	Earth	75.5	12,500
Whitlow Ranch	Gila	Queen Creek	Superior	1960	35,500	NPP	FX	Earth	149	837
Arkansas										
Blakely Mountain Dam	Ouachita	Ouachita	Hot Springs	1955	2,768,500	20,900	FP	Earth	235	1,100
Blue Mountain	Arkansas	Petit Jean River	Paris	1947	257,900	2,910	FRWX	Earth	115	2,800
DeGray	Ouachita	Caddo	Arkadelphia	1971	881,900	6,400	FPQRS	Earth	243	3,400
DeQueen	Red	Rolling Fork River	DeQueen	1977	136,100	1,680	FSQRW	Earth	160	2,360
Dierks	Red	Saline River	Dierks	1975	96,800	1,360	FSRAW	Earth and Rock	153	2,500
Gillham	Red	Cossatot River	Gillham	1975	221,800	1,370	FSQW	Earth and Rock	160	1,750
Millwood	Red	Little River	Ashdown	1966	1,854,930	29,200	FSW	Earth	88	17,554
Narrows Dam	Ouachita	Little Missouri	Murfreesboro	1949	407,900	2,500	FP	Concrete	175	941
Nimrod	Arkansas	Fourche La Fave River	Plainview	1942	336,010	3,550	FSWX	Concrete	97	1,012
California										
Black Butte	Sacramento	Stony Creek	Orland	1963	160,000	770	FIRX	Earth	156	2,970
Brea	Santa Ana	Brea Creek	Fullerton	1942	4,010	NPP	FRX	Earth	87	1,765
Buchanan Dam-H.V. Eastman Lake	San Joaquin	Chowchilla River	Chowchilla	1975	150,000	470	FIRW	Earth and Rock	205.5	1,800
Carbon Canyon	Santa Ana	Carbon Canyon River	Brea	1961	6,610	NPP	FRX	Earth	99	2,610
Coyote Valley	Russian	East Fork Russian River	Ukiah	1959	122,500	1,700	FRX	Earth	160	3,500
Dry Creek (Warm Springs) Lake and Channel	Russian	Dry Creek	Healdsburg	1983	381,000	500	FRSW	Earth	319	3,000
Farmington	San Joaquin	Littlejohn Creek	Farmington	1952	52,000	NPP	F	Earth	60	7,800
Fullerton	Santa Ana	East Fullerton Creek	Fullerton	1941	760	NPP	FRX	Earth	46	575
Hansen	Los Angeles	Big Tujunga Wash	Los Angeles	1940	25,450	120	FRWX	Earth	97	10,475
Harry L. Englebright	Sacramento	Yuba River	Marysville	1941	69,000	400	DR	Concrete	280	1,142
Hidden Dam-Hensley Lake	San Joaquin	Fresno River	Madera	1975	90,000	5,000	FIRW	Earth	163	5,730
Isabella	San Joaquin	Kern River	Bakersfield	1953	570,000	1,850	FIRW	Earth	185	4,952
Lopez	Los Angeles	Pacoima Wash	San Fernando	1954	440	NPP	FX	Earth	50	1,300

(Continued)

Table 11B.26 (Continued)

Name	River Basin	Stream	Community in Vicinity	Year Placed in Operation	Total Storage (acre-ft)	Permanent Pool (Acreage) or No Pool (NPP)	Project Functions[a]	Characteristics of Dam		
								Type	Height (ft)	Length (ft)
Martis Creek	Sacramento	Martis Creek	Reno	1971	20,400	71	FSR	Earth	113	2,670
Merced County Stream Group										
Bear	San Joaquin	Bear Creek	Merced	1954	7,700	NPP	F	Earth	92	1,830
Burns	San Joaquin	Burns Creek	Merced	1950	7,000	NPP	F	Earth	55	4,075
Mariposa	San Joaquin	Mariposa Creek	Merced	1948	15,000	NPP	F	Earth	88	1,330
Owens	San Joaquin	Owens Creek	Merced	1949	3,600	NPP	F	Earth	75	790
Mojave River	Mojave	Mojave	Victorville	1971	89,670	NPP	FRWX	Earth	200	2,200
New Hogan	San Joaquin	Calaveras	Valley Springs	1963	325,000	715	FIRX	Earth and Rock	210	1,960
North Fork	Sacramento	American River	Auburn	1939	14,700	280	DR	Concrete	155	620
Pine Flat	San Joaquin	Kings River	Piedra	1954	1,000,000	NPP	FIRX	Concrete	429	1,820
Prado	Santa Ana	Santa Ana River	Corona	1941	196,240	NPP	FRX	Earth	106	2,280
San Antonio	Santa Ana	San Antonio Creek	Upland	1956	7,700	NPP	FX	Earth	160	3,850
Santa Fe	San Gabriel	San Gabriel River	Duarte	1948	32,110	NPP	FRX	Earth	92	23,800
Sepulveda	Los Angeles	Los Angeles River	Van Nuys	1941	17,430	NPP	FRX	Earth	57	15,444
Success	San Joaquin	Tule River	Porterville	1960	85,000	400	FIRX	Earth	142	3,490
Terminus	San Joaquin	Kaweah River	Visalia	1961	150,000	345	FIRX	Earth	250	2,375
Whittier Narrows	San Gabriel	San Gabriel River and Rio Hondo	El Monte	1957	35,150	NPP	FRWX	Earth	56	16,960
Colorado										
Bear Creek	Missouri	Bear Creek	Denver	1978	30,810	109	FRX	Earth	180	5,300
Chatfield	Missouri	South Platte River	Denver	1974	231,429	1,412	FRX	Earth	148	12,500
Cherry Creek	Missouri	Cherry Creek	Denver	1950	93,920	852	FRX	Earth	141	14,300
John Martin	Arkansas	Arkansas River	Lamar	1943	615,500	1,844	FIR	Concrete and Earth	106	13,274
Trinidad	Arkansas	Purgatoire River	Trinidad	1977	123,500	280	FIRX	Earth	200	6,610
Connecticut										
Black Rock	Housatonic	Branch Brook	Thomaston	1970	8,700	20	FR	Earth	154	933
Colebrook River	Connecticut	West Branch, Farmington River	Riverton	1969	97,700	760	FRSX	Earth	223	1,300
Hancock Brook	Housatonic	Hancock Brook	Plymouth	1960	4,030	40	FRW	Earth	57	630
Hop Brook	Housatonic	Hop Brook	Middlebury	1968	6,970	21	FR	Earth	97	520
Mansfield Hollow	Thames	Natchaug River	Willimantic	1952	52,000	450	FRW	Earth	68	12,420
Northfield Brook	Thames	Northfield Brook	Thomaston	1965	2,430	8	FRW	Earth	118	810
Thomaston	Housatonic	Naugatuck River	Thomaston	1960	42,000	NPP	F	Earth	142	2,000
West Thompson	Thames	Quinebaug	Thompson	1965	26,800	200	FRW	Earth	70	2,550
Idaho										
Lucky Peak	Columbia	Boise River	Boise	1956	306,000	2,850	FIR	Earth	250	1,700
Illinois										
Cariyle	Upper Mississippi	Kaskaskia River	Carlyle	1967	983,000	26,000	FSNRWA	Earth	67	6,570
Farmdale	Upper Mississippi	Farm Creek	East Peoria	1951	15,500	NPP	F	Earth	80	1,275
Fondulac	Upper Mississippi	Fondulac Creek	East Peoria	1951	3,780	NPP	F	Earth	67	1,000
Shelbyville	Upper Mississippi	Kaskaskia River	Shelbyville	1970	684,000	11,100	FSNRW	Earth	108	3,000
Rend Lake	Upper Mississippi	Big Muddy River	Benton	1970	294,000	18,900	FQRSW	Earth	54	10,600

Indiana										
Brookville	Ohio	East Fork of Whitewater River	Brookville	1974	359,600	2,250	FRSW	Earth and Rock	182	3,000
Cagles Mill	Ohio	Mill Creek	Terre Haute	1952	228,120	1,400	FRX	Earth	150	950
Cecil M. Harden	Ohio	Raccoon Creek	Rockville	1960	132,800	1,100	FRX	Earth	117	1,790
Huntington	Ohio	Wabash River	Huntington	1969	153,100	500	FRW	Earth	91	5,332
Mississinewa	Ohio	Mississinewa	Peru	1967	368,400	1,100	FRW	Earth	137	8,100
Monroe	Ohio	Salt Creek	Harrodsburg	1964	441,000	3,280	FARS	Earth	93	1,400
Patoka	Ohio	Patoka River	Ellsworth	1978	301,600	2,010	FRSQW	Earth and Rock	84	1,550
Salamonie	Ohio	Salamonie	Wabash	1966	263,600	976	FRW	Earth	133	6,100
Iowa										
Coralville	Upper Mississippi	Iowa River	Iowa City	1958	492,000	1,820	FARW	Earth	100	1,400
Rathbun	Missouri	Chariton River	Centerville	1969	552,000	11,000	FNRWXQ	Earth	86	10,600
Red Rock	Upper Mississippi	Des Moines River	Des Moines	1969	1,830,000	8,950	FARWQ	Earth	110	5,676
Saylorville	Upper Mississippi	Des Moines River	Des Moines	1975	602,000	74,000	FARWQ	Earth	125	6,750
Kansas										
Clinton	Missouri	Wakarusa River	Lawrence	1977	397,200	7,000	FSWAXR	Earth	114	9,250
Council Grove	Arkansas	Grand (Neosho)	Council Grove	1964	112,265	3,235	FSQR	Earth	96	6,500
El Dorado	Arkansas	Walnut River	El Dorado	1981	236,200	8,000	FSQR	Earth	99	20,930
Elk City	Arkansas	Elk River	Independence	1966	284,300	4,450	FSQ	Earth	107	4,840
Fall River	Arkansas	Fall River	Fall River	1949	256,400	2,350	FSX	Earth	94	6,015
Hillsdale	Missouri	Big Bull Creek	Kansas City	1981	160,000	4,580	FSQR	Earth	75	11,600
John Redmond	Arkansas	Grand (Neosho)	Burlington	1964	630,250	9,280	FSQR	Earth	86.5	21,790
Kanopolis	Missouri	Smoky Hill River	Salina	1948	450,000	3,815	FRWX	Earth	131	15,360
Marion	Arkansas	Cottonwood River	Marion	1968	143,850	6,200	FRQS	Earth	67	8,375
Melvern	Missouri	Marias Des Cygnes	Melvern	1972	363,000	6,930	FRQWX	Earth	98	9,700
Milford	Missouri	Republican River	Junction City	1965	1,160,000	15,600	FRSXWQ	Earth and Rock	126	6,300
Pearson Skubitz Big Hill	Arkansas	Big Hill Creek	Cherryvale	1981	40,600	1,240	FSR	Earth	83	3,870
Perry	Missouri	Delaware River	Perry	1969	770,000	12,200	FRSXW	Earth and Rock	95	7,750
Pomona	Missouri	110 Mile Creek	Pomona	1963	230,000	4,000	FRSWXQ	Earth and Rock	85	7,750
Toronto	Arkansas	Verdigris River	Toronto	1960	200,800	2,660	FX	Earth	90	4,712
Tuttle Creek	Missouri	Big Blue River	Manhattan	1962	2,346,000	15,800	FRWXQAN	Earth and Rock	157	7,500
Wilson	Missouri	Saline River	Wilson	1964	776,000	9,000	FIRWXNA	Earth	160	5,600
Kentucky										
Barren River	Ohio	Barren River	Glasgow	1964	815,200	4,340	FARS	Earth	146	3,970
Buckhorn	Ohio	Middle Fork of Kentucky River	Buckhorn	1960	168,000	550	FR	Earth	162	1,020
Carr Fork	Ohio	Carr Fork	Hazard	1976	47,700	530	FQRW	Earth and Rock	130	720
Cave Run	Ohio	Licking River	Farmers	1974	614,100	6,790	FQRW	Earth and Rock	148	2,740
Dewey	Ohio	Johns Creek	Paintsille	1949	93,000	1,100	FARW	Earth	118	913
Fishtrap	Ohio	Levisa Fork, Big Sandy River	Pikeville	1968	164,360	569	FARW	Rock	195	1,100
Grayson	Ohio	Little Sandy	Grayson	1967	118,990	1,050	FQRW	Earth and Rock	120	1,460
Green River	Ohio	Green River	Campbellsville	1969	723,200	5,070	FRSQW	Earth and Rock	142	2,350
Martins Fork	Cumberland	Martins Fork	Harlan	1978	21,000	578	FQ	Concrete	97	574
Paintsville	Ohio	Paint Creek	Paintsville	1983	73,500	261	FQRW	Earth and Rock	160	1,600
Nolin	Ohio	Nolin River	Kyrock	1963	609,400	2,890	FAR	Earth and Rock	174	990
Rough River	Ohio	Rough River	Leitchfield	1958	334,400	2,180	FRX	Earth and Rock	124	1,530

(Continued)

Table 11B.26 (Continued)

Name	River Basin	Stream	Community in Vicinity	Year Placed in Operation	Total Storage (acre-ft)	Permanent Pool (Acreage) or No Pool (NPP)	Project Functions[a]	Characteristics of Dam		
								Type	Height (ft)	Length (ft)
Taylorsville	Ohio	Salt River	Taylorsville	1983	291,670	1,625	FQRW	Earth and Rock	164	1,280
Louisiana										
Bayou Bodcau	Red	Bayou Bodcau	Shreveport	1949	357,300	NPP	FRW	Earth	76	12,850
Caddo Lake	Red	Cypress Bayou	Shreveport	1971	175,000	32,700	NFRS	Concrete and Earth		3,600
Wallace Lake	Red	Cypress Bayou	Shreveport	1946	88,300	2,300	FR	Earth	48	4,934
Maryland										
Jennings Randolph Lake	Potomac	North Branch	Barnum	1981	130,900	952	FQRS	Earth and Rock	296	2,130
Massachusetts										
Barre Falls	Connecticut	Ware River	Barre	1958	24,000	NPP	FRW	Earth and Rock	62	885
Birch Hill	Connecticut	Millers River	So.Royalston	1941	49,900	NPP	FRW	Earth and Rock	56	1,400
Buffumville	Thames	Little River	Charlton	1958	12,700	200	FRW	Earth and Rock	66	3,255
Charles River Natural Valley Storage	Charles	Charles River	Millis	1983	35,000	NPP	F	Nonstructural	—	—
Conant Brook	Connecticut	Conant Brook	Monson	1966	3,740	NPP	F	Earth and Rock	85	1,060
East Brimfield	Connecticut	Quinebaug River	Fiskdale	1960	30,000	360	FRW	Earth and Rock	55	520
Hodges Village	Connecticut	French River	Oxford	1959	12,800	NPP	FRW	Earth and Rock	55	2,140
Knightville	Connecticut	Westfield River	Huntington	1941	49,000	NPP	FRW	Earth and Rock	160	1,200
Littleville	Connecticut	Middle Branch, Westfield River	Chester	1965	32,400	275	FRWS	Earth and Rock	1,164	1,360
Tully	Connecticut	Tully River	Fryville	1949	22,000	300	FRW	Earth and Rock	62	1,570
West Hill	Blackstone	West River	Uxbridge	1960	12,350	NPP	FRW	Earth and Rock	51	2,400
Westville	Thames	Quinebaug River	Sturbridge	1961	11,100	23	FRW	Earth and Rock	78	560
Minnesota										
Big Stone Lake-Whetstone River	Upper Mississippi	Minnesota River	Ortonville	1973	45,000	12,700	FRW	Earth	25	13,700
Lac Qui Parle Chippewa River	Upper Mississippi	Chippewa River	Montevideo	1950	—	NPP	FRWX	Earth and Rock	23.3	17,975
Lac Qui Parle	Upper Mississippi	Minnesota River	Montevideo	1950	122,800	6,500	FRWX	Earth and Rock	21	4,100
Marsh Lake	Upper Mississippi	Minnesota River	Montevideo	1950	35,000	5,100	FRWX	Earth and Rock	19.5	11,800
Orwell	Red River of the North	Otter Tail River	Fergus Falls	1953	14,100	210	FARS	Earth and Rock	47	1,355
Red Lake	Red River of the North	Red Lake River	Red River	1951	2,680,000	279,000	FARSX	Earth and Rock	15.5	36,500
Mississippi										
Arkabutla Lake	Lower Mississippi	Coldwater River	Arkabutla	1943	525,300	5,100	F	Earth and Rock	81	11,500
Enid Lake	Lower Mississippi	Yocona River	Enid	1951	660,000	6,100	F	Earth and Rock	99	8,400
Grenada Lake	Lower Mississippi	Yalobusha River	Grenada	1954	1,337,400	9,800	F	Earth and Rock	102	13,900
Okatibbee	Pascagoula	Okatibbee Creek	Meridian	1969	142,400	1,280	FQSR	Earth	67	6,543
Sardis Lake	Lower Mississippi	Little Tailahatchie River	Sardis	1940	1,570,000	10,700	F	Earth and Rock	117	15,300
Missouri										
Clearwater	White	Black River	Piedmont	1948	413,700	1,630	FRWX	Earth and Rock	154	4,225
Long Branch	Grande Chariton	Little Chariton	Macon	1980	65,000	2,430	FRSQW	Earth	71	3,800

Longview	Missouri	Little Blue River	Kansas City	1986	46,900	930	FRWQ	Earth and Rock	120	1,900
Pomme de Terre	Missouri	Pomme de Terre River	Hermitage	1961	650,000	7,820	FRWX	Earth and Rock	155	4,630
Smithville	Missouri	Little Platte River	Smithville	1982	246,500	7,190	FSQRW	Earth	95	4,200
Wappapello	Lower Mississippi	St. Francis River	Wappapello	1941	613,200	4,100	FR	Earth and Rock	109	2,700
Nebraska										
Harlan County	Missouri	Republican River	Republican City	1952	850,000	13,600	FIRWX	Earth and Rock	107	11,827
Papillion Creek and Tributaries:										
Glenn Cunningham (Site 11)	Missouri	Knight Creek	Omaha	1975	17,910	392	FQRX	Earth	67	1,940
Standing Bear Lake (Site 16)	Missouri	Trib. of Big Papillion Creek	Omaha	1973	5,220	137	FRX	Earth	70	1,460
Salt Creek and Tributaries										
Olive Creek (Site 2)	Missouri	S. Trib. Olive Br. Creek	Kramer	1964	5,470	174	FR	Earth	45	3,020
Blue Stem (Site 4)	Missouri	N. Trib. Olive Br. Creek	Sprague	1963	10,260	315	FR	Earth	57	2,760
Wagon Train (Site 8)	Missouri	N. Trib. Hickman Creek	Holland	1963	9,280	303	FR	Earth	52	1,650
Stagecoach (Site 9)	Missouri	S. Trib. Hickman Creek	Hickman	1964	6,640	196	FR	Earth	48	2,250
Yankee Hill (Site 10)	Missouri	Cardwell Creek	Denton	1966	7,560	208	FR	Earth	52	3,100
Conestoga (Site 12)	Missouri	Holmes Creek	Denton	1964	10,640	230	FR	Earth	63	3,000
Twin Lake (Site 13)	Missouri	Middle Creek	Pleasantdale	1966	8,080	255	FR	Earth	58	2,075
Pawnee (Site 14)	Missouri	N. Middle Creek	Emerald	1965	29,520	728	FR	Earth	65	5,000
Holmes Park Lake (Site 17)	Missouri	Antelope Creek	Lincoln	1963	6,510	100	FR	Earth	55	7,700
Branched Oak (Site 18)	Missouri	Oak Creek	Raymond	1968	97,560	1,780	FR	Earth	70	5,200
Nevada										
Mathews Canyon	Colorado	Mathews Canyon	Caliente	1957	6,270	NPP	FX	Earth	71	800
Pine Canyon	Colorado	Pine Canyon	Caliente	1957	7,750	NPP	FX	Earth	92	884
New Hampshire										
Blackwater	Merrimack	Blackwater River	Webster	1941	46,000	NPP	FRW	Earth	75	1,150
Edward MacDowell	Merrimack	Nubanusit Brook	West Peterborough	1950	12,800	NPP	FRW	Earth	67	1,030
Franklin Falls	Merrimack	Pemigewasset River	Franklin	1943	154,000	NPP	FRW	Earth	140	1,740
Hopkinton-Everett	Merrimack	Contoocook River	West Hopkinton	1962	71,500	200	FRW	Earth	76	790
	Merrimack	Piscataquog River	East Weare	1962	87,500	120	FRW	Earth	115	2,000
Otter Brook	Connecticut	Otter Brook	Keene	1958	18,300	85	FRW	Earth	133	1,288
Surry Mountain	Connecticut	Ashuelot River	Keene	1941	32,500	265	FRW	Earth	86	1,670
New Mexico										
Abiquiu	Rio Grande	Rio Chama	Abiquiu	1963	1,212,000	NPP	FXS	Earth	325	1,540
Cochiti	Rio Grande	Rio Grande	Pena Blanca	1975	596,300	1,200	FRWX	Earth	251	28,300
Conchas	Arkansas	Canadian River	Tucumcari	1939	529,000	3,000	FI	Concrete and Earth	200	19,400
Galisteo	Rio Grande	Galisteo Creek	Albuquerque	1970	89,000	NPP	FX	Earth	158	2,820
Jemez Canyon	Rio Grande	Jemez River	Bernalillo	1953	102,700	NPP	FX	Earth	135	780
Two Rivers										
Diamond “A” Dam	Rio Grande	Rio Hondo	Roswell	1963	168,000	NPP	FX	Earth	98	4,885

(Continued)

Table 11B.26 (Continued)

Name	River Basin	Stream	Community in Vicinity	Year Placed in Operation	Total Storage (acre-ft)	Permanent Pool (Acreage) or No Pool (NPP)	Project Functions[a]	Characteristics of Dam		
								Type	Height (ft)	Length (ft)
Rocky Dam	Rio Grande	Rocky Arroyo							118	2,940
Santa Rosa Dam Reservoir	Pecos	Pecos	Santa Rosa	1979	447,000	NPP	FIX	Earth	212	1,950
New York										
Almond	Susquehanna	Canacadea Creek	Hornell	1949	14,600	124	FRW	Earth	90	1,260
Arkport	Susquehanna	Canisteo Creek	Arkport	1940	7,900	NPP	F	Earth	113	1,200
East Sidney	Susquehanna	Ouleout Creek	Franklin	1950	33,550	210	FRW	Concrete and Earth	130	2,010
Mount Morris	Genesee	Genesee River	Mount Morris	1952	337,000	170	FR	Concrete	210	1,028
Whitney Point	Susquehanna	Otselic River	Whitney Point	1942	86,440	1,200	FRW	Earth	95	4,900
North Carolina										
B. Everett Jordan	Cape Fear	New Hope	Durham	1982	753,500	14,300	FQRSWX	Earth	112	1,330
Falls	Neuse	Neuse	Raleigh	1983	335,600	11,300	FQRSWX	Earth	92	1,915
W. Kerr Scott	Yadkin-Pee Dee	Yadkin	Wilkesboro	1963	153,000	1,470	FARSX	Earth	148	1,740
North Dakota										
Baldhill	Red River of the North	Sheyenne River	Valley City	1950	70,000	325	FARS	Earth	61	1,650
Bowman-Haley	Missouri	North Fork, Grand River	Haley	1967	92,980	1,750	FSRWX	Earth	79	5,730
Homme	Red River of the North	South Branch of Park River	Park River	1951	3,650	51	FARS	Earth	67	865
Pipestem	James River	Pipestem Creek	Jamestown	1974	146,880	885	FRWX	Earth	108	4,000
Ohio										
Alum Creek	Ohio	Alum Creek	Africa	1975	134,800	348	FRSW	Concrete and Earth	93	10,000
Berlin	Ohio	Mahoning River	Deerfield	1943	91,200	240	FARSWQ	Concrete and Earth	96	5,750
Caesar Creek	Ohio	Caesar Creek	Wilmington	1978	242,200	13,300	FRSQW	Earth and Rock	165	2,750
Clarence J. Brown	Ohio	Buck Creek	Springfield	1974	63,700	1,010	FQRW	Earth and Rock	72	6,620
Deer Creek	Ohio	Deer Creek	New Holland	1968	102,500	727	FRW	Earth	93	3,880
Delaware	Ohio	Olentangy River	Delaware	1951	132,000	950	FARWX	Earth	92	18,600
Dillon	Ohio	Licking River	Zanesville	1961	273,000	1,325	FRWX	Earth	118	1,400
Michael J. Kirwan	Ohio	West Branch, Mahoning River	Newton Falls	1966	78,700	580	FAQRSW	Earth	83	9,900
Mosquito Creek	Ohio	Mosquito Creek	Cortland	1944	104,100	700	FARSWQ	Earth	47	5,650
Muskingum River Reservoirs										
Atwood	Ohio	Indian Fork	New Cumberland	1937	49,700	1,540	FRX	Earth	65	3,700
Beach City	Ohio	Sugar Creek	Beach City	1937	71,700	420	FRX	Earth	64	5,600
Bolivar	Ohio	Sandy Creek	Bolivar	1938	149,600	NPP	FR	Earth	87	6,300
Charles Mill	Ohio	Black Fork	Mufflin	1936	88,000	1,350	FRX	Earth	48	1,390
Clendening	Ohio	Brushy Fork	Tippecanoe	1937	54,000	1,800	FRX	Earth	64	950
Dover	Ohio	Tuscarawas River	Dover	1938	203,000	350	FRX	Concrete	83	824
Leesville	Ohio	McGuire Creek	Leesville	1937	37,400	1,000	FRX	Earth	74	1,694
Mohawk	Ohio	Walhonding River	Nellie	1937	285,000	NPP	FR	Earth	111	2,330
Mohicanville	Ohio	Lake Fork	Mohicanville	1936	102,000	NPP	FR	Earth	46	1,220
Piedmont	Ohio	Stillwater Creek	Piedmont	1937	65,000	2,270	FRX	Earth	56	1,750

Pleasant Hill	Ohio	Clear Fork	Perrysville	1938	87,700	850	FRX	Earth	113	775
Senecaville	Ohio	Seneca Fork	Senecaville	1937	88,500	3,550	FRSX	Earth	45	2,350
Tappan	Ohio	Little Stillwater Creek	Tappan	1936	61,600	2,350	FRX	Earth	52	1,550
Wills Creek	Ohio	Wills Creek	Conesville	1937	196,000	900	FRX	Earth	87	1,950
North Branch, Kokosing River Lake	Ohio	North Branch of Kokosing River	Fredericktown	1973	14,900	98	FRW	Earth	71	1,400
Paint Creek	Ohio	Paint Creek	New Petersburg	1972	145,000	710	FRSQW	Earth and Rock	118	700
Tom Jenkins	Ohio	East Branch, Sunday Creek	Gloucester	1951	26,900	394	FRSWX	Concrete	84	944
West Fork Mill Creek	Ohio	Mill Creek	Mount Healthy	1952	11,380	200	FRX	Earth	100	1,100
William H. Harsha	Ohio	Little Miami River	Williamsburg	1978	284,500	18,760	FRSQW	Earth	200	1,450
Oklahoma										
Birch	Arkansas	Birch Creek	Barnsdall	1977	58,200	1,137	FSQRW	Earth	97	3,193
Canton	Arkansas	North Canadian River	Canton	1948	377,100	7,910	FSI	Earth	73	15,140
Copan	Arkansas	Little Caney River	Copan	1983	227,700	4,850	FSQRW	Earth	70	7,730
Fort Supply	Arkansas	Wolf Creek	Fort Supply	1942	100,700	1,820	FSX	Earth	85	12,225
Great Salt Plains	Arkansas	Salt Fork, Arkansas River	Cherokee	1941	271,400	8,690	FRWX	Earth	68	6,010
Heyburn	Arkansas	Polecat Creek	Sapulpa	1950	55,030	880	FRWXS	Earth	89	2,920
Hugo	Red	Kiamichi River	Hugo	1974	966,700	13,250	FSQRW	Earth	101	10,200
Hulah	Arkansas	Caney River	Caney	1951	289,000	3,570	FSAX	Earth	94	5,200
Kaw	Arkansas	Arkansas River	Ponca City	1976	1,348,000	17,040	FSQRW	Earth	125	9,466
Oologah	Arkansas	Verdigris River	Oologah	1963	1,519,000	29,460	FSN	Earth	137	4,000
Optima	Arkansas	North Canadian River	Hardesty	1978	229,500	5,340	FSRW	Earth	120	15,200
Pine Creek	Red	Little River	Wright City	1969	465,780	3,750	FSQW	Earth	124	7,712
Sardis	Red	Jackfork Creek	Clayton	1983	429,600	14,360	FSRW	Earth	81	14,138
Skiatook	Arkansas	Hominy Creek	Skiatook	1985	500,700	10,190	FSQRW	Earth	143	3,590
Waurika	Red	Beaver Creek	Waurika	1977	343,500	10,100	FISQWR	Earth	106	16,600
Wister	Arkansas	Poteau River	Wister	1949	427,900	5,360	FSAX	Earth	99	5,700
Oregon										
Applegate	Rogue River	Applegate River	Medford	1981	82,000	988	AFIQRSW	Gravel Embankment	242	1,300
Blue River	Columbia	Blue River	Blue River	1968	85,000	975	FINR	Earth	319	1,329
Cottage Grove	Columbia	Coast Fork, Willamette River	Cottage Grove	1942	30,060	1,155	FINR	Concrete and Earth	114	2,110
Dorena	Columbia	Row River	Cottage Grove	1949	70,500	1,885	FINR	Concrete and Earth	145	3,352
Fall Creek	Columbia	Middle Fork, Willamette River	Eugene	1965	115,000	1,865	FINR	Rockfill and Concrete	193	5,100
Fern Ridge	Columbia	Long Tom River	Eugene	1941	110,000	10,305	FINR	Rockfill and Concrete	49	6,624
Willow Creek	Columbia	Willow Creek	Heppner	1983	13,250	96	FRN	Roller Compacted Concrete	160	1,780
Pennsylvania										
Alvin R. Bush	Susquehanna	Kettle Creek	Renovo	1962	75,000	160	FRW	Earth and Rock	165	1,350
Aylesworth Creek	Susquehanna	Aylesworth Creek	Archbald	1970	1,700	NPP	F	Earth and Rock	90	1,270
Beltzville	Delaware	Pohopoco	Lehighton	1971	68,250	947	FQRSW	Earth and Rock	170	4,300
Blue Marsh	Delaware	Tulephocken	Blue Marsh	1978	22,900	963	FAQRS	Earth and Rock	98	1,775

(Continued)

Table 11B.26 (Continued)

Name	River Basin	Stream	Community in Vicinity	Year Placed in Operation	Total Storage (acre-ft)	Permanent Pool (Acreage) or No Pool (NPP)	Project Functions[a]	Characteristics of Dam		
								Type	Height (ft)	Length (ft)
Conemaugh	Ohio	Conemaugh River	Saltsburg	1952	274,000	300	FW	Concrete and Earth	137	1,265
Cowanesque	Susquehanna	Cowanesque River	Lawrenceville	1980	89,000	410	FR	Earth and Rock	151	3,100
Crooked Creek	Ohio	Crooked Creek	Ford City	1940	93,900	350	FRW	Earth	143	1,480
Curwensville	Susquehanna River	West Branch, Susquehanna River	Curwensville	1965	124,200	790	FR	Earth	131	2,850
East Branch, Clarion River	Ohio	East Branch, Clarion River	Wilcox	1952	84,300	90	FARQW	Earth	184	1,725
Foster Joseph Sayers	Susquehanna	Bald Eagle Creek	Blanchard	1969	99,000	1,730	FRW	Earth	100	6,835
Francis E. Walter (Bear Creek)	Delaware	Lehigh River	White Haven	1961	110,000	90	FNRW	Earth and Rock	234	3,000
Gen. Edgar Jadwin	Delaware	Dyberry Creek	Honesdale	1960	24,500	NPP	F	Earth	109	1,225
Indian Rock	Susquehanna	Codorus Creek	York	1942	28,000	NPP	FRW	Earth	83	1,000
Kinsua	Ohio	Allegheny River	Warren	1965	1,180,000	1,900	PFAQRW	Concrete and Earth	177	1,877
Loyalhanna	Ohio	Loyalhanna Creek	Saltsburg	1942	95,300	210	FRW	Concrete and Earth	114	960
Mahoning Creek	Ohio	Mahoning Creek	New Bethlehem	1941	74,200	170	FRW	Concrete	162	926
Prompton	Delaware	Lackawaxen River	Honesdale	1960	52,000	290	FNRW	Earth	140	1,230
Raystown	Susquehanna	Raystown Branch, Juniata River	Huntingdon	1973	762,000	8,300	FRW	Earth and Rock	225	1,700
Shenango	Ohio	Shenanago River	Sharpsville	1966	191,400	1,910	FAQRW	Concrete	68	720
Stillwater	Susquehanna	Lackawanna River	Uniondale	1960	12,000	85	FS	Earth	77	1,700
Tioga-Hammond Lakes	Susquehanna	Tioga River	Tioga	1978	62,000	470	FR	Earth and Rock	140	2,710
Hammond Lakes	Susquehanna	Crooked Creek	Tioga	1978	63,000	680	FR	Earth and Rock	122	6,450
Tionesta	Ohio	Tionesta Creek	Tionesta	1940	133,400	480	FRW	Earth	154	1,050
Union City	Ohio	French Creek	Union City	1970	47,640	NPP	F	Earth	88	1,420
Woodcock Creek	Ohio	French Creek	Meadville	1973	20,000	118	FQRA	Earth	90	4,650
Youghiogheny River	Ohio	Youghiogheny River	Confluence	1943	254,000	450	FARWQ	Earth	184	1,610
South Dakota										
Cold Brook	Missouri	Cold Brook	Hot Springs	1953	7,200	36	FRWX	Earth	127	925
Cottonwood Springs	Missouri	Cottonwood Springs Creek	Hot Springs	1970	8,385	41	FRWX	Earth	123	1,190
Lake Traverse										
Reservation Control Dam	Red River of the North	Bois de Sioux River	Wheaton	1941	164,500	10,925	FRX	Earth	14	9,100
White Rock	Red River of the North	Bois de Sioux River	Wheaton	1941	85,000	6,500	FRX	Earth	16	14,400
Texas										
Addicks	San Jacinto	South Mayde Creek	Addicks	1948	204,500	NPP	FX	Earth	49	61,166
Aquilla	Brazos	Aquilla Creek	Hillsboro	1983	146,000	3,280	FSX	Earth	104.5	11,890
Bardwell	Trinity	Waxahachie Creek	Ennis	1965	140,000	3,570	FRSX	Earth	82	15,400
Barker	San Jacinto	Buffalo Bayou	Barker	1945	207,000	NPP	FX	Earth	37	72,844

Belton	Brazos	Leon River	Belton	1954	1,097,600	12,300	FIRSX	Earth	192	5,524
Benbrook	Trinity	Clear Fork, Trinity River	Fort Worth	1952	258,600	3,770	FNRXA	Earth	130	9,130
Canyon	Guadalupe	Guadalupe	New Braunfeis	1964	740,900	8,240	FRSX	Earth	224	6,830
Ferrells Bridge Dam Lake O' the Pines	Red	Cypress Creek	Jefferson	1959	842,100	18,700	FRS	Earth	97	10,600
Granger Dam and Lake	Brazos	San Gabriel River	Granger	1980	244,200	4,400	FRSWX	Earth	115	16,320
Grapevine	Trinity	Denton Creek	Grapevine	1952	425,500	7,280	FNRSXA	Earth	137	12,850
Hords Creek	Colorado	Hords Creek	Coleman	1948	25,310	510	FARSX	Earth	91	6,800
Lake Kemp	Red	Wichita River	Wichita Falls	1972	502,900	15,590	FX	Earth and Rock	115	8,890
Lavon	Trinity	East Fork, Trinity River	Fort Worth	1953	748,200	21,400	FRSX	Earth	81	19,483
Lewisville	Trinity	Elm Fork, Trinity River	Lewisville	1954	989,700	23,280	FRSX	Earth	125	32,888
Navarro Mills	Trinity	Richland Creek	Corsicana	1962	212,200	5,070	FRSX	Earth	82	7,570
North San Gabriel Dam, Lake Georgetown	Brazos	North Fork, San Gabriel River	Georgetown	1980	130,800	1,310	FRSWX	Rock	164	6,700
O. C. Fisher	Colorado	North Concho River	San Angelo	1952	396,400	5,440	FRSX	Earth	128	40,885
Pat Mayse	Red	Sanders Creek	Paris	1967	189,100	5,993	FRSX	Earth	96	7,080
Proctor	Brazos	Leon River	Comanche	1963	374,200	4,610	FRSX	Earth	86	13,460
Somerville	Brazos	Yegua Creek	Somerville	1967	507,500	11,460	FRSX	Earth	80	26,175
Stillhouse Hollow	Brazos	Lampasas River	Belton	1968	630,400	6,430	FRSX	Earth	200	15,624
Waco	Brazos	Bosque River	Waco	1965	726,400	7,270	FRSX	Concrete and Earth	140	24,618
Wright Patman	Red	Sulphur River	Texarkana	1957	2,654,300	20,300	FRSX	Earth	100	18,500
Vermont										
Ball Mountain	Connecticut	West River	Jamaica	1961	54,600	75	FRW	Concrete and Earth	265	915
North Hartland	Connecticut	Ottauguechee River	North Hartland	1960	71,420	220	FRW	Concrete and Earth	185	1,520
North Springfield	Connecticut	Black River	Springfield	1960	51,067	290	FRW	Concrete and Earth	120	2,940
Townshend	Connecticut	West River	Townshend	1961	33,700	100	FRW	Concrete and Earth	133	1,700
Union Village	Connecticut	Ompompanoosuc River	Union Village	1950	38,000	NPP	FRW	Concrete and Earth	170	1,100
Virginia										
John W. Flannagan	Ohio	Pound River	Haysi	1963	145,700	310	FAQR	Concrete and Earth	250	960
Gathright Dam and Lake Moomaw	James	Jackson	Alleghany	1979	123,739	2,532	FQR	Earth and Rock	257	1,172
North Fork of Pound River	Ohio	North Fork, Pound River	Pound	1966	11,293	106	FR	Rock	122	600
Washington										
Howard A. Hanson	Green	Green River	Kanaskat	1961	106,000	1,600	FAS	Rock	235	675
Mill Creek	Columbia	Mill Creek	Walla Walla	1942	8,300	225	FR	Earth	145	3,200
Mud Mountain	Puyallup	White River	Enumclaw	1953	106,000	NPP	FR	Rock	425	700
Wynoochee	Chehalis	Wynoochee River	Montesano	1972	70,000	1,150	FSARI	Concrete and Earth	177	1,700
West Virginia										
Beech Fork	Ohio	Beech Fork	Lavalette	1977	37,540	450	FRW	Earth	86	1,080
Bluestone	Ohio	New River	Hinton	1952	631,000	1,800	FRWX	Concrete	180	2,048

Note: Operable as of September 30, 1986.

[a] Project Functions: A, Low Flow Augmentation; D, Debris Control; F, Flood Control; I, Irrigation; N, Navigation; P, Power; Q, Water Quality Control; R, Public Recreation (Annual Attendance exceeding 5,000); S, Water Supply; W, Fish and Wildlife (Federal and State); X, Water Conservation and Sedimentation.

Table 11B.27 Largest Manmade Reservoirs in the United States

Dam Name	Reservoir	Location	Owner	Acre-Feet	Completed
Hoover	Lake Mead	Nevada	Bureau of Reclamation	28,255,000	1936
Glen Canyon	Lake Powell	Arizona	Bureau of Reclamation	27,000,000	1964
Oahe	Lake Oahe	South Dakota	Corps of Engineers	19,300,000	1966
Garrison	Lake Sakakawea	North Dakota	Corps of Engineers	18,500,000	1953
Fort Peck	Fort Peck Lake	Montana	Corps of Engineers	15,400,000	1957
Grand Coulee	F D Roosevelt Reservoir	Washington	Bureau of Reclamation	9,562,000	1942
Libby	Lake Koocanusa	Montana	Corps of Engineers	5,809,000	1973
Shasta	Lake Shasta	California	Bureau of Reclamation	4,552,000	1966
Toledo Bend	Toledo Bend Reservoir	Louisiana	Sabine River Authority	4,477,000	1966
Fort Randall	Lake Francis Case	South Dakota	Corps of Engineers	3,800,000	1954
Flaming Gorge	Flaming Gorge Reservoir	Utah	Bureau of Reclamation	3,788,900	1964
Oroville	Lake Oroville	California	California DWR	3,540,000	1968
Hungry Horse	Hungry Horse Reservoir	Montana	Bureau of Reclamation	3,468,000	1953
Dworshak	Dworshak Reservoir	Idaho	Corps of Engineers	3,468,000	1973
Amistad	Amistad International Reservoir	Texas	International Boundary Water Commission	3,384,000	1969
Bull Shoals	Bull Shoals Lake	Arkansas	Corps of Engineers	3,048,000	1951
Sam Rayburn Dam	Sam Rayburn Reservoir	Texas	Corps of Engineers	2,898,200	1965

Source: From USBR Register of Dams, www.usbr.gov.

Table 11B.28 World's Largest Reservoirs in Terms of Capacity

No	Capacity, 10^8 m^3	Name	Country	Year
1	204,800	Owen Falls[a]	Uganda	1954
2	169,270	Bratsk	U.S.S.R.	1964
3	168,900	High Aswan	Egypt	1970
4	160,368	Kariba	Zimbabwe/Zambia	1959
5	147,960	Akosombo	Ghana	1965
6	141,851	Daniel Johnson	Canada	1968
	135,000	Guri	Venezuela	U/C
7	73,300	Krasnoyarsk	U.S.S.R.	1967
8	70,309	W A C Bennett	Canada	1967
9	68,400	Zeya	U.S.S.R.	1978
	63,000	Cabora Bassa	Mozambique	1974
10	61,715	La Grande 2	Canada	1978
11	60,020	La Grande 3	Canada	1981
12	59,300	Ust-Ilim	U.S.S.R.	1977
13	58,000	Kuibyshev	U.S.S.R.	1955
14	53,790	Caniapiscau Barrage KA 3	Canada	1980
	50,700	Upper Wainganga	India	U/C
15	49,800	Bukhtarma	U.S.S.R.	1960
	48,000	Atatürk	Turkey	U/C
16	46,000	Irkutsk	U.S.S.R.	1956
	43,000	Tucurui	Brazil	U/C
17	35,900	Vilyui	U.S.S.R.	1967
18	35,400	Sanmenxia	China	1960
19	34,852	Hoover	U.S.A.	1936
20	34,100	Sobradinho	Brazil	1979

(Continued)

Table 11B.28 (Continued)

No	Capacity, $10^8\,m^3$	Name	Country	Year
21	33,304	Glen Canyon	U.S.A.	1966
22	32,203	Skins Lake No 1	Canada	1953
23	31,790	Jenpeg	Canada	1975
24	31,500	Volgograd	U.S.S.R.	1958
	31,300	Sayano-Shushensk	U.S.S.R.	U/C
25	30,600	Keban	Turkey	1974
26	29,959	Iroquois	Canada	1958
27	29,000	Itaipu	Brazil	1982
28	28,973	Churchill Falls (GR-1)	Canada	1971
29	28,370	Missi Falls Control	Canada	1976
30	28,100	Kapchagay	U.S.S.R.	1970
31	29,000	Loma de la Lata	Argentina	1977
32	27,920	Garrison	U.S.A.	1953
33	27,675	Kossou	Ivory Coast	1972
34	27,433	Oahe	U.S.A.	1958
35	26,000	Razzaza Dyke	Iraq	1970
36	25,400	Rybinsk	U.S.S.R.	1941
	24,700	Longyangxia	China	U/C
37	24,700	Mica	Canada	1972
38	24,700	Tsimlyansk	U.S.S.R.	1952
39	23,700	Kenney	Canada	1952
40	23,550	Ust-Khantaika	U.S.S.R.	1970
41	22,950	Furnas	Brazil	1963
42	22,119	Fort Peck	U.S.A.	1937
43	21,626	Xinanjiang	China	1960
44	21,166	Ilha Solteira	Brazil	1973

Note: U/C, under construction.

[a] This capacity is not fully obtained by a dam; the major part of it is the natural capacity of a lake; Owen Falls is not the greatest manmade lake.

Source: From International Commission on Large Dams, 1984, *World Register of Dams*.

Table 11B.29 Depth of Reservoirs in the United States and Loss by Evaporation

Region	Median Depth (ft)	Loss (ft/acre)
New England	19	0.55
Delaware and Hudson	29	0.96
Chesapeake Bay	25	0.64
Southeast	23	0.53
Eastern Great Lakes	16	0.68
Western Great Lakes	18	0.79
Ohio	24	0.47
Cumberland	44	0.54
Tennessee	40	0.36
Upper Mississippi	7	0.83
Lower Mississippi	20	0.81
Upper Missouri	25	1.97
Lower Missouri	21	1.11
Upper Arkansas-White-Red	9 (15)	3.27
Lower Arkansas-White-Red	10	1.33
Western Gulf	8 (15)	2.75
Upper Rio Grande and Pecos	20	4.26
Colorado	16	3.79
Great Basin	15	3.00
Pacific Northwest	25	1.44
Central Pacific	33	2.58
South Pacific	44	4.81

Source: From U.S. Geological Survey, 1960.

Table 11B.30 Multipurpose Reservoirs in the United States

Project	River	Community in Vicinity	Total Storage Capability (acre-ft)[a]	Flood Control and/or Nav. Feature Placed in Useful Operation	Initial Power In FY	Existing Installation (kW)	Schedu-led Instal-lation (kW)	Ultimate Installation (kW)	Project Functions	Type[b]	Height (ft)	Length (ft)
Albeni Falls, ID	Pend Oreille	Newport, WA	1,153,000	1952	1955	42,600		42,600	NFPR	C	90	1,055
Allatoona Lake, GA	Etowah	Cartersville, GA	670,000	1950	1950	74,000		74,000	FPRW	C	190	1,250
Barkley Dam and Lake Barkley, KY and TN	Cumberland	Grand Rivers, KY	2,082,000	1964	1966	130,000		130,000	NPFR	CE	157	9,959
Beaver Lake, AR	White	Eureka Springs, AR	1,952,000	1963	1965	112,000		112,000	FPSR	CE	228	2,575
Big Bend Dam (Lake Sharpe), SD	Missouri	Chamberlain, SD	1,883,000	1964	1965	468,000		468,000	FPRIW	E	95	10,570
Blakely Mountain Dam-Lake Ouachita, AR	Ouachita	Mt. Pine, AR	2,768,000	1953	1956	75,000		75,000	FPRW	E	235	1,100
Bonneville L&D Lake Bonneville, OR & WA	Columbia	Bonneville, OR	537,000	1938	1938	1,076,600		1,076,600	NPR	C	122	2,690
Broken Bow Lake, OK	Mountain Fork	Broken Bow, OK	1,368,230	1968	1970	100,000		100,000	FPWSR	E	225	2,690
Buford Dam Lanier, GA	Chattahoochee	Buford, GA	2,554,000	1956	1957	86,000		86,000	NFPW	E	192	5,400
Bull Shoals Lake, AR & MO	White	Mountain Home, AR	5,408,000	1952	1953	340,000		340,000	FPR	C	258	2,256
Clarence Cannon Dam	Salt	Perry, MO	1,428,000	1983	1985	58,000		58,000	FNPRSW	CE	138	1,700
Carters Dam, GA	Coosawatte	Carters, GA	472,756	1975	1975	500,000		500,000	FPRW	ER	450	1,950
Center Hill Lake, TN	Caney Fork	Lancaster, TN	2,092,000	1948	1951	135,000		135,000	FPR	CE	250	2,160
Cheatham L&D, TN	Cumberland	Ashland City, TN	104,000	1952	1958	36,000		36,000	NPR	C	75	801
Chief Joseph Dam (Rufus Woods Lake), WA	Columbia	Bridgeport, WA	593,100	1955	1956	2,069,000	204,160[c]	2,273,160	PIR	C	220	4,300
Clarks Hill Lake, GA & SC	Savannah	Augusta, GA	2,900,000	1952	1953	280,000		280,000	NFPRSW	CE	200	5,680
Cooper River, Charleston Harbor, SC	Santee	St. Stephen, SC	2,580,000	N/A	1985	84,000		84,000	PW	CE	86	876
Cordell Hull L&D, TN	Cumberland	Carthage, TN	310,900	1973	1974	100,000		100,000	NPR	CE	445	1,738
Cougar Lake, OR	S. Fork McKenzie	Blue River, OR	219,000	1963	1964	25,000		64,600	NFPRI	ER	445	1,738
Dale Hollow Lake, TN & KY	Obey	Celina, TN	1,706,000	1943	1949	54,000		54,000	FPR	C	200	1,717
Dardanelle L&D, AR	Arkansas	Dardanelle, AR	486,200	1969	1965	124,000		124,000	NPR	C	68	2,683
DeGray Lake, AR	Caddo	Arkadelphia, AR	831,900	1969	1972	68,000		108,000	FNPRS	E	243	3,400
Denison Dam (Lake Texoma), TX & OK	Red	Denison, TX	5,312,300	1944	1945	70,000		175,000	FPSRN	E	165	17,200
Detroit Lake, OR, including Big Cliff Lake, OR	North Santiam	Mill City, OR	461,000	1963	1954	118,000		118,000	NFPRI	C	382	1,528
Dworshak Dam and Reservoir, ID	N. Fork, Clearwater	Orofino, ID	3,453,000	1972	1973	400,000		1,060,000	PNFR	C	717	3,300
Eufaula Lake, OK	Canadian	Eufaula, OK	3,825,400	1964	1964	90,000		90,000	ENPS	E	114	3,200
Fort Gibson Lake, OK	Grand (Neosho)	Ft. Gibson, OK	1,284,400	1950	1953	45,000		67,500	FP	CE	110	2,990
Fort Peck Lake, MT	Missouri	Glasgow, MT	18,909,000	1938	1944	185,250		185,250	NFPRIW	E	251	21,026

(Continued)

Table 11B.30 (Continued)

Project	River	Community in Vicinity	Total Storage Capability (acre-ft)[a]	Flood Control and/or Nav. Feature Placed in Useful Operation	Initial Power In FY	Existing Installation (kW)	Scheduled Installation (kW)	Ultimate Installation (kW)	Project Functions	Type[b]	Height (ft)	Length (ft)
Fort Randal Dam (Lake Francis Case), SD	Missouri	Lake Andes, SD	5,574,000	1953	1954	320,000		320,000	NFPRIW	E	165	10,700
Garrison Dam (Lake Sakakawea), ND	Missouri	Riverdale, ND	24,137,000	1954	1956	430,000		430,000	NFPRIW	E	210	11,300
Gavins Point Dam (Lewis and Clark Lake), SD & NE	Missouri	Yankton, SD	504,000	1956	1957	100,000		100,000	NFPRIW	E	74	8,700
Green Peter Lake, OR, including Foster Lake, OR	Middle Santiam	Sweet Home, OR	491,000	1967	1967	100,000		100,000	PFNIR	C	340	1,380
Greers Ferry Lake, AR	Little Red	Heber Springs, AR	2,844,000	1962	1964	96,000		96,000	FPRW	CE	96	5,000
Harry S. Truman Dam and Res.	Osage	Warsaw, MO	5,202,000	1982	1982[d]	160,000		160,000	FPRW	CE	96	5,000
Hartwell Lake, GA & SC	Sevannah	Hartwell, GA	2,842,700	1961	1962	344,000		344,000	NFPRS	CE	204	17,852
Hills Creek Lake, OR	Middle Fork Willamette	Oakridge, OR	356,000	1961	1962	30,000		30,000	NFPRI	GE	304	2,150
Ice Harbor L&D (Lake Sacajawea), WA	Snake	Pasco, WA	417,000	1962	1962	603,000		603,000	NPRI	CE	130	2,790
Jim Woodruff Dam (Lake Seminole), FL, GA & AL	Appalachicola	Chattahoochee, FL	367,300	1957	1957	30,000		30,000	NPRW	CE	67	6,150
John Day L&D (Lake Umatilla), OR & WA	Columbia	Rufus, OR	2,500,000	1968	1969	2,160,000		2,700,000	NPRFI	CE	161	5,900
John H. Kerr Dam and Reservoir, NC & VA	Roanoke	Boydton, VA	2,750,300	1952	1953	204,000		204,000	FPRW	CE	144	22,285
Robert F. Henry L&D, AL	Alabama	Benton, AL	234,200	1972	1975	68,000		68,000	NPRW	CE	101	14,962
J. Percy Priest Dam and Reservoir, TN	Stones	Nashville, TN	652,000	1967	1970	28,000		28,000	FPRW	CE	147	2,716
Keystone Lake, OK	Arkansas	Tulsa, OK	1,737,600	1964	1968	70,000		70,000	FNPWS	E	121	4,600
Laurel River Lake, KY	Laurel	London, KY	435,600	1973	1978	61,000		61,000	FPRW	R	282	1,420
Libby Dam, Lake Koocanusa, MT	Kootenai	Libby, MT	5,809,000	1972	1975	525,000		840,000	FPR	C	420	3,055
Little Goose L&D (Lake Bryan), WA	Snake	Starbuck, WA	565,000	1970	1970	810,000		810,000	NPRI	CE	160	2,670
Lookout Point Lake, Including Dexter Lake, OR	Willamette Middle Fork, Willamette	Lowell, OR	483,000	1954	1955	135,000		135,000	NFPRI	CE	243	3,381
Lost Creek Lake, OR	Rogue	Trail, OR	465,000	1977	1977	49,000		49,000	DFPISWR	CE	345	3,600
Lower Granite L&D, WA	Snake	Pomeroy, WA	484,000	1975	1975	810,000		810,000	NPRIF	CE	146	3,200
Lower Monumental L&D, WA	Snake	Kahlotus, WA	376,000	1969	1969	810,000		810,000	NPRI	CE	135	3,800
McNary L&D—Lake Wallula, OR & WA	Columbia	Umatilla, CR	1,550,000	1953	1954	980,000		1,625,000	NPRI	CE	183	7,300

Millers Ferry L&D, AL	Alabama	Camden, AL	331,800	1969	1970	75,000		75,000	NPRW	CE	90	11,380
Narrows Dam—Lake Greeson, AR	Little Missouri	Murfreesboro, AR	407,900	1950	1950	25,500		25,500	FPRW	C	183.5	941
New Melones Lake, CA[e]	Stanislaus	Oakdale, CA	2,400,000	1978	1979	300,000		300,000	FIPRW	ER	625	1,560
Norfolk Lake, AR& MO	North Fork	Norfolk, AR	1,983,000	1943	1944	80,550		163,000	FPRS	C	216	2,624
Oahe Dam (Lake Oahe), SD & ND	Missouri	Pierre, SD	23,337,000	1959	1962	595,000		595,000	NFPRIW	E	245	9,300
Old Hickory L&D, TN	Cumberland	Hendersonville, TN	545,000	1954	1957	100,000		100,000	NPR	CE	98	3,605
Ozark-Jeta Taylor L&D, AR	Arkansas	Ozark, AR	148,400	1969	1973	100,000		100,000	NPR	C	58	2,480
Philpott Lake, VA	Roanoke	Bassett, VA	318,500	1951	1954	14,000		14,000	FPR	C	220	892
Robert S. Kerr L&D and Reservoir, OK	Arkansas	Sallisaw, OK	525,700	1970	1971	110,000		110,000	NPR	E	75	7,230
Sam Rayburn Dam and Reservoir, TX	Angelina	Jasper, TX	3,997,600	1965	1966	52,000		52,000	FPWR	CE	120	19,430
St. Mary's River, MI	Great Lakes	Sault Ste. Marie, MI	—	1855	1952	18.400		18,400	NP	Control Gate		
Snettisham, AK[f]	Speel	Juneau, AK	352,400		1973	46,700	27,000[g]	73,700	P	C[h]	18	338
Stockton Lake, MO	Sac	Stockton, MO	1,647,000	1969	1973	45,200		45,200	FPRW	CE	128	5,100
Table Rock Lake, AR & MO	White	Branson, MO	3,462,000	1958	1959	200,000		200,000	FPR	CE	252	6,423
Tenkiller Lake, OK	Illinois	Gore, OK	1,230,800	1952	1953	34,000		34,000	FP	E	197	3,000
The Dalles L&D (Lake Celilo), WA & OR	Columbia	The Dalles, OR	53,000	1957	1957	1,806,800		1,806,800	NPR	CR	300	8,875
Walter F. George L&D, GA & FL	Chattahoochea	Fort Gaines, GA	934,000	1963	1963	130,000		130,000	NPRW	CE	114	13,585
Webbers Falls L&D, OK	Arkansas	Webbers Falls, OK	170,100	1970	1973	60,000		60,000	NP	E	84	4,370
West Point Lake, AL & GA	Chattahoochee	West Point, GA	604,500	1975	1975	73,375		108,375	FPRW	CE	97	7,250
Whitney Lake, TX	Brazos	Whitney, TX	1,999,500	1953	1954	30,000		30,000	FPR	CE	159	17,695
Wolf Creek Dam (Lake Cumberland), KY	Cumberland	Jamestown, KY	6,089,000	1950	1952	270,000		270,000	FPR	CE	258	5,736

Note: In operation September 30, 1986. Nomenclature for project functions: Fy, Fiscal year; D, debris control; F, flood control; I, irrigation; N, navigation; P, power; R, public recreation (annual attendance exceeding 5000); W, fish and wildlife (federal or state).

[a] Total of all storage functions, including inactive and dead storage to normal full pool level.
[b] G, gravel; R, rock; C, concrete; E, earth.
[c] Chief Joseph Additional Units and Operating Units 1–16.
[d] All units are synchronized-to-line and two units have passed the pumpback test. However, due to damaging effects to fish, no further pumping will be done for their testing or operation until a solution to the problem is found.
[e] Being operated for the Department of Interior by the Bureau of Reclamation.
[f] Being operated by the Alaska Power Administration.
[g] Crater Lake Unit.
[h] Weir for Long Lake.

Source: From Department of the Army Annual Report FY86 of the Secretary of the Army on Civil Works Activities, Washington, D.C.

SECTION 11C HYDROELECTRIC POWER

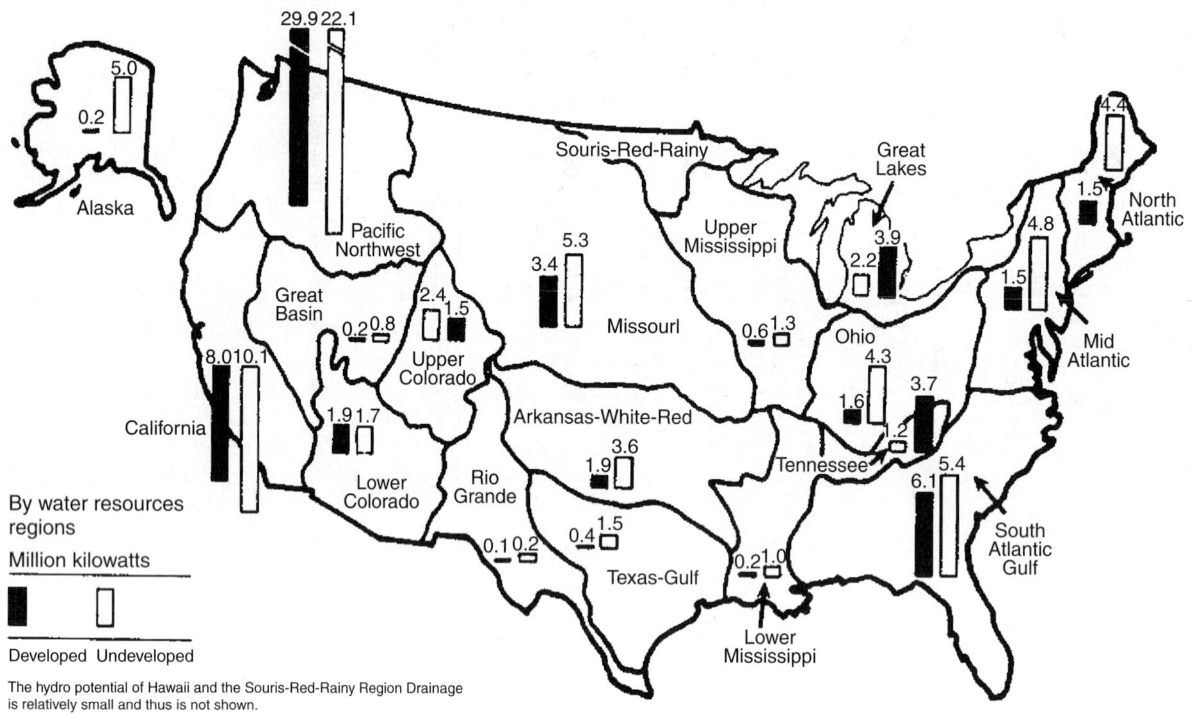

Figure 11C.20 Conventional hydroelectric power developed and undeveloped, January 1, 1984. (From Federal Energy Regulatory Commission, 1984.)

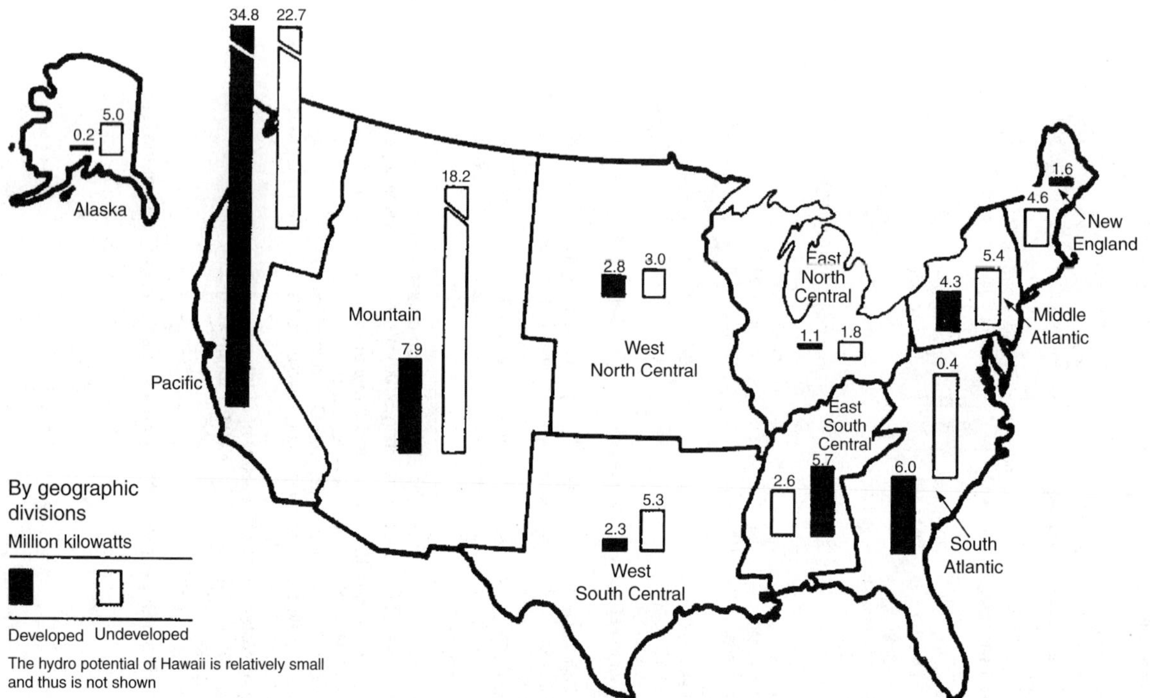

Figure 11C.21 Conventional hydroelectric power developed and underdeveloped, January 1, 1984. (From Federal Energy Regulatory Commission, 1984.)

Table 11C.31 Conventional and Pumped Storage Hydroelectric Plants in the United States

Plant	Stream	State	Year of Initial Oper.	Gross Static Head (ft)	Usable Power Storage 1000 Acre-Feet[a]	Installed Capacity — Kilowatts[b]		
						Developed	Under Construction	Authorized
Corps of Engineers								
Lincoln School	St John R	ME	—	76	32	—	—	70,000
Tocks Island	Delaware R	NJ	—	107	426	—	—	70,000
Gathright	Jackson R	VA	—	194	61	—	—	49,000
John H Kerr	Roanoke R	VA	1952	90	1029	204,000	—	204,000
Philpott	Smith R	VA	1953	164	111	14,000	—	14,000
St. Stephen	Santee & Cooper R	SC	—	70	1420		84,000	84,000
Clark Hill	Savannah R	GA	1953	152	1510	280,000	—	280,000
Richard B Russell	Savannah R	GA	—	162	127	—	3,00,000	300,000
Richard B Russell	Savannah R	GA	—	162	127	—	3,00,000 R	300,000 R
Hartwell	Savannah R	GA	1962	178	1415	344,000	—	344,000
Jim Woodruff L&D	Apalachicola R	FL	1957	31	37	30,000	—	30,000
Walter F George L&D	Chattahoochee R	GA	1963	92	207	130,000	—	130,000
West Point	Chattahoochee R	GA	1975	78	86	73,375	—	108,375
Buford	Chattahoochee R	GA	1957	155	1049	86,000	—	86,000
Lower Auchumpkee	Flint R	GA	—	84	124	—	—	77,000
Lazer Creek	Flint R	GA	—	123	60	—	—	83,000
Spewrell Bluff	Flint R	GA	—	144	242	—	—	100,000
Spewrell Bluff	Flint R	GA	—	144	242	—	—	50,000 R
Millers Ferry L&D	Alabama R	AL	1970	49	17	75,000	—	75,000
Jones Bluff	Alabama R	AL	1975	45	12	68,000	—	68,000
Carters	Coosawattee R	GA	1975	387	135	250,000	—	250,000
Carters	Coosawattee R	GA	1977	387	135	250,000 R	—	250,000
Allatoona	Etowah R	GA	1950	145	284	74,000	—	110,000
St. Marys Falls	St Marys R	MI	1951	21	NA	18,400	—	18,400
Barkley	Cumberland R	KY	1966	46	258	130,000	—	130,000
Cheatham L&D	Cumberland R	TN	1958	22	19	36,000	—	36,000
J Percy Priest	Stones R	TN	1970	85	34	28,000	—	28,000
Old Hickory L&D	Cumberland R	TN	1957	46	63	100,000	—	100,000
Center Hill	Caney Fk, Cumberland R	TN	1950	161	492	135,000	—	135,000
Cordell Hull	Cumberland R	TN	1973	49	54	100,000	—	100,000
Dale Hollow	Obey R	TN	1948	135	496	54,000	—	54,000
Celina	Cumberland R	KY	—	60	15	—	—	108,000
Wolf Creek	Cumberland R	KY	1951	163	2142	270,000	—	270,000
Laurel	Laurel R	KY	1977	251	185	61,000	—	61,000
Booneville	S Fk Kentucky R	KY	—	140	258	—	—	7,300
Bluestone	New R	WV	—	125	238	—	—	180,000
Rowlesburg	Cheat R	WV	—	938	8	—	—	350,000 R
Clarence F Cannon	Salt R	MO	—	107	437	—	27,000	27,000

(Continued)

Table 11C.31 (Continued)

Plant	Stream	State	Year of Initial Oper.	Gross Static Head (ft)	Usable Power Storage 1000 Acre-Feet[a]	Installed Capacity — Kilowatts[b]		
						Developed	Under Construction	Authorized
Clarence F Cannon	Salt R	MO	—	107	437	—	31,000 R	31,000 R
Red Rock	Des Moines R	IA	—	53	400	—	—	17,200
Rock Island L&D 15[c]	Sylvan Slough, Miss R	IL	1919	16	NA	2,752	—	2,752
Narrows	Little Missouri R	AR	1950	145	202	25,500	—	25,500
Degray	Caddo R	AR	1971	188	393	40,000	—	80,000
Degray	Caddo R	AR	1971	188	393	28,000 R	—	28,000 R
Blakely Mountain	Ouachita R	AR	1955	181	1286	75,000	—	75,000
Harry S Truman	Osage R	MO	1979	46	90	160,000 R	—	160,000 R
Pomme de Terre	Pomme de Terre R	MO	—	117	165	—	—	16,800
Stockton	Sac R	MO	1973	86	660	45,200	—	45,200
Gavins Point	Missouri R	NE	1956	44	100	100,000	—	100,000
Fort Randall	Missouri R	SD	1954	131	3500	320,000	—	320,000
Corps of Engineers								
Big Bend	Missouri R	SD	1964	67	260	468,000	—	468,000
Oahe	Missouri R	SD	1962	189	17000	595,000	—	595,000
Garrison	Missouri R	ND	1956	172	17900	430,000	—	430,000
Fort Peck	Missouri R	MT	1943	213	13800	185,300	—	185,300
Broken Bow	Mountain Fk R, Little R	OK	1970	194	9	100,000	—	100,000
Denison	Red R	TX	1944	108	1730	70,000	—	175,000
Dardanelle	Arkansas R	AR	1965	49	65	124,000	—	124,000
Ozark	Arkansas R	AR	1973	23	19	100,000	—	100,000
Robert S Kerr	Arkansas R	OK	1971	42	NA	110,000	—	110,000
Eufula	Canadian R	OK	1964	82	1481	90,000	—	90,000
Tenkiller Ferry	Illinois R	OK	1953	143	345	39,100	—	39,100
Webbers Falls	Arkanasas R	OK	1973	30	NA	60,000	—	60,000
Fort Gibson	Neosho R	OK	1953	60	NA	45,000	—	67,500
Keystone	Arkansas R	OK	1968	76	351	70,000	—	70,000
Greers Ferry	Little Red R	AR	1964	183	763	96,000	—	96,000
Norfork[d]	North Fork R	AR	1944	174	445	80,550	—	165,550
Bull Shoals	White R	AR	1952	194	952	340,000	—	340,000
Table Rock	White R	MO	1959	204	1098	200,000	—	200,000
Beaver	White R	AR	1965	190	956	112,000	—	112,000
Town Bluff	Neches R	TX	—	24	NA	—	—	3,000
Sam Rayburn	Angelina R	TX	1966	71	1400	52,000	—	52,000
Dam A	Neches R	TX	—	13	NA	—	—	2,700
Rockland	Neches R	TX	—	61	1488	—	—	13,500
Belton	Leon R	TX	—	110	399	—	—	19,000
Whitney	Brazos R	TX	1953	89	133	30,000	—	30,000
Big Cliff RRG	N Santiam R	OR	1954	97	243	18,000	—	18,000
Detroit	N Santiam R	OR	1953	357	40	100,000	—	100,000
Foster RRG	S Santiam R	OR	1968	284	28	20,000	—	20,000

Green Peter	M Fk Santiam R	OR	1967	310	63	80,000	—	80,000
Strube RRG	S Fk Mckenzie R	OR	—	64	3	—	—	4,500
Cougar	S Fk Mckenzie R	OR	1964	435	10	25,000	—	60,000
Dexter RRG	M Fk Willamette R	OR	1955	48	4	15,000	—	15,000
Lookout Point	M Fk Willamette R	OR	1954	231	12	120,000	—	120,000
Hills Creek	M Fk Willamette R	OR	1962	318	49	30,000	—	30,000
Bonneville	Columbia R	OR	1938	59	87	518,400	—	544,600
Bonneville 2nd Ph	Columbia R	WA	1981	59	87	558,220	—	558,220
The Dalles	Columbia R	WA	1957	83	53	1,806,800	—	1,806,800
John Day	Columbia R	OR	1968	105	300	1,957,500	—	1,957,500
John Day	Columbia R	WA	1968	105	300	202,500	—	742,500
McNary	Columbia R	OR	1953	74	185	980,000	—	980,000
Ice Harbor	Snake R	WA	1961	98	24	603,000	—	603,000
Lower Monumental	Snake R	WA	1969	100	20	810,000	—	810,000
Little Goose	Snake R	WA	1970	98	49	810,000	—	810,000
Lower Granite	Snake R	WA	1975	100	44	810,000	—	810,000
Dworshak	N Fk Clearwater R	ID	1973	630	2000	400,000	—	1,060,000
Chief Joseph	Columbia R	WA	1955	167	NA	2,069,000	—	2,069,000
Albeni Falls	Pend Oreille R	ID	1955	28	1153	42,600	—	42,600
Libby	Kootenai R	MT	1975	341	4934	420,000	420,000	840,000
Lost Creek	Rogue R	OR	1977	321	315	49,000	—	49,000
Maryville	Yube R	CA	—	210	775	—	—	50,000
Total Conventional						19,011,197	831,000	22,297,897
Total Reversible						438,000 R	331,000 R	1,169,000 R
Bureau of Reclamation								
Prairie Creek	Prairie Cr, Platte R	NE	—	51	NA	—	—	16,800 R
Flatiron 1 and 2	Rattlesnake Cr, Big Thoom	CO	1954	1113	1	74,500[e]	—	74,500
Flatiron 3	Dry Cr, Big Thompson R	CO	1954	292	100	8,500 R	—	8,500 R
Big Thompson	Big Thompson R	CO	1954	835	NA	4,500	—	4,500
Pole Hill	Dry Cr, Big Thompson R	CO	1954	835	NA	33,250	—	33,250
Estes	Big Thompson R	CO	1950	570	NA	45,000	—	45,000
Marys Lake	Fish Cr, Big Thompson R	CO	1951	217	NA	8,100	—	8,100
Guernsey	N Platte R	WY	1927	92	41	4,800	—	4,800
Glendo	N Platte R	WY	1958	133	511	24,000	—	24,000
Alcova	N Platte R	WY	1955	164	1842	36,000	—	36,000
Fremont Canyon	N Platte R	WY	1960	350	1015	48,000	—	48,000
Kortes	N Platte R	WY	1950	204	4	36,000	—	36,000
Seminoe	N Platte R	WY	1939	218	1011	45,000	—	45,000
Yellowtail	Bighorn R	MT	1966	463	613	250,000	—	250,000
Heart Mountain	Shoshone R	WY	1948	275	421	5,000[f]	—	5,000
Boysen	Bighorn R	WY	1952	110	742	15,000	—	15,000
Pilot Butte	Wyoming Cnl (Wind R)	WY	1935	106	NA	1,600	—	1,600
Canyon Ferry	Missouri R	MT	1953	147	1512	50,000	—	50,000
Otero	Arkansas Cnl (Arkansas R)	CO	—	270	93	—	—	11,000

(Continued)

Table 11C.31 (Continued)

Plant	Stream	State	Year of Initial Oper.	Gross Static Head (ft)	Usable Power Storage 1000 Acre-Feet[a]	Installed Capacity — Kilowatts[b]		
						Developed	Under Construction	Authorized
Mt Elbert	Arkansas Cnl (Arkansas R)	CO	1981	464	7	100,000 R	100,000 R	200,000 R
Elephant Butte	Rio Grande	NM	1940	190	1730	24,300	—	24,300
Glen Canyon	Colorado R	AZ	1964	566	20876	1,042,000	—	1,042,000
Flaming Gorge	Green R	UT	1963	435	3516	108,000	—	108,000
Fontenelle	Green R	WY	1968	110	149	10,000	—	10,000
Crystal	Gunnison R	CO	1977	220	18	28,000	—	28,000
Morrolw Point	Gunnison R	CO	1970	403	42	120,000	—	120,000
Blue Mesa	Gunnison R	CO	1967	356	774	60,000	—	60,000
Upper Molina	Cottonwood Cr, Plateau Cr	CO	1962	2663	6	8,640	—	8,640
Lower Molina[g]	Plateau Cr, Colorado R	CO	1962	1600	6	4,860	—	4,860
Green Mountain	Blue R	CO	1943	261	146	26,000	—	26,000
Senator Wash	Senator Wash Cr	CA	—	74	9	—	—	7,210 R
Parker	Colorado R	CA	1942	78	218	120,000	—	120,000
Davis	Colorado R	AZ	1951	131	1809	240,000	—	240,000
Hoover	Colorado R	AZ	1936	530	7227	717,000	—	717,000
Hoover	Colorado R	NV	1936	530	7927	717,000	—	717,000
Stampede	L Truckee R	CA	—	183	NA	—	—	3,000
Watasheamu	E Fk Carson R	NV	—	236	90	—	—	8,000
Deer Creek	Provo R	UT	1958	144	149	4,950	—	4,950
Dyne	Diamond Fk Pipeline	UT	—	848	700[h]	—	—	33,000
Sixth Water	Sixth Water Cr, Diamond	UT	—	823	700[h]	—	—	90,000
Syar	Strawberry Offstream	UT	—	431	700	—	—	10,500
Black Canyon	Payette R	ID	1925	94	NA	8,000	—	8,000
Boise	Boise R	ID	1912	40	NA	1,500	—	1,500
Anderson Ranch	S Fk Boise R	ID	1950	300	423[i]	40,000	—	40,000
Minidoka	Snake R	ID	1909	48	95[j]	13,000	—	13,400
Palisades	Snake R	ID	1957	244	1200	118,750	—	118,750
Chandler	Chandler Pwr Cnl (Yakima R)	WA	1956	121	NA	12,000	—	12,000
Roza(Canal)	Roza Cnl (Yakima R)	WA	1958	160	NA	11,250	—	11,250
Grand Coulee	Columbia R	WA	1941	343	5232[k]	6,180,000[l]	—	6,180,000
Grand Coulee P/G	Columbia R	WA	1974	280	53	150,000 R	150,000 R	300,000 R
Hungry Horse	S Fk Flathead R	MT	1952	477	2982[m]	285,000	—	285,000
Green Springs	Emigrant Cr, Bear Cr	OR	1960	1984	U	16,000	—	16,000
Trinity	Trinity R	CA	1964	469	2285	105,556	—	105,556
Nimbus	American R	CA	1955	43	2	13,500	—	13,500
Folsom	American R	CA	1955	333	920	198,720	—	198,720
Auburn	N Fk American R	CA	—	675	1966	—	—	300,000
Judge Francis Carr	Clear Cr Tnl (Trinity R)	CA	1963	695	2289	160,000	—	160,000

Keswick	Sacramento R	CA	1949	87	23	75,000	—	75,000
Spring Creek	Spring Cr, Sacramento R	CA	1964	625	2517	150,000	—	150,000
Shasta	Sacramento R	CA	1944	482	4050	539,000^{o}	—	539,000
New Melones	Stanislaus R	CA	1979	583	2050	300,000^{n}	—	300,000
O'Neill	Delta Mendota Cnl (San Jose)	CA	1967	56	20	25,200R	—	25,200 R
San Luis	Calif Aqueduct	CA	1968	327	1961	424,000 R^{p}	—	424,000 R
Siphon Drop	Yuma Main Cnl (New R)	CA	—	15	NA	—	—	1,600
Total Conventional					12,139,176		12,596,276	
Total Reversiable					707,700R	250,000R	981,710R	
Tennessee Valley Authority								
Great Falls	Caney Fk, Cumberland R	TN	1916	148	39	31,860	—	31,860
Kentucky	Tennessee R	KY	1944	50	715	175,000	—	175,000
Pickwick landing	Tennessee R	TN	1938	46	236	220,040	—	220,040
Wilson	Tennessee R	AL	1942	88	47	629,840	—	629,840
Wheeler	Tennessee R	AL	1936	48	328	361,800	—	361,800
Tims Ford	Elk R	TN	1972	133	240	45,000	—	45,000
Guntersville	Tennessee R	AL	1939	39	131	115,200	—	115,200
Nickajack	Tennessee R	TN	1968	35	21	103,950	—	103,950
Raccoon MT	Tennessee R	TN	1979	1040	35	1,530,000 R	—	1,530,000 R
Chickamauga	Tennessee R	TN	1940	45	220	120,000	—	120,000
Ocoee 1	Ocoee R	TN	1912	111	33	18,000	—	18,000
Ocoee 2	Ocoee R	TN	1913	245	NA	21,000	—	21,000
Ocoee 3	Ocoee R	TN	1943	308	4	28,800	—	28,800
Blue Ridge	Toccoa R	GA	1931	148	186	20,000	—	20,000
Apalachia	Hiwassee R	TN	1943	440	35	82,800	—	82,800
Hiwassee/	Hiwassee R	NC	1940	246	350	57,600	—	57,600
Hiwassee	Hiwassee R	NC	1956	246	352	59,500 R	—	59,500 R
Nottely	Nottely R	GA	1956	168	171	15,000	—	15,000
Chatuge	Hiwassee R	NC	1954	124	229	10,000	—	10,000
Watts Bar	Tennessee R	TN	1942	54	214	166,500	—	166,500
Melton Hill	Clinch R	TN	1964	58	25	72,000	—	72,000
Norris	Clinch R	TN	1936	193	1761	100,800	—	100,800
Fonatna	Little Tennessee R	NC	1945	430	1136	238,500	—	238,500
Fort Londoun	Tennessee R	TN	1943	70	81	139,140	—	139,140
Douglas	French Broad R	TN	1943	135	1136	120,600	—	120,600
Cherokee	Holston R	TN	1942	147	1411	135,180	—	135,180
Ft Patrick	S Fk Holston R	TN	1953	67	4	36,000	—	36,000
Boonce	S Fk Holston R	TN	19583	119	150	75,000	—	75,000
Wlbur	Watauga R	TN	1912	65	U	10,700	—	10,700
Watauga	Watauga R	TN	1949	312	518	57,600	—	57,600
South Holston	S Fk Holston R	TN	1951	240	519	35,000	—	35,000
Total Conventional					3,242,910		3,242,91	
Total Reversible					1,589,500 R	R	1,589,500 R	

(Continued)

Table 11C.31 (Continued)

Plant	Stream	State	Year of Initial Oper.	Gross Static Head (ft)	Usable Power Storage 1000 Acre-Feet[a]	Installed Capacity — Kilowatts[b]		
						Developed	Under Construction	Authorized
National Park Service								
Cascades	Merced R	CA	1918	356	NA	2,000	—	2,000
Total Conventional					2,000		2,000	
International Boundary & Water Commission								
Falcon	Rio Grande	TX	1954	115	2767	31,500	—	31,500
Amistad	Rio Garnde	TX	1983	213	500	66,000	—	66,000
Total Conventional					97,500	—	97,500	
Bureau of Indian Affairs								
Coolidge	Gila R	AZ	1929	204	1164	10,000	—	10,000
Drop 2	Yakima R (offstream)	WA	1942	30	NA	2,500	—	2,500
Drop 3	Yakima R (offstream)	WA	1932	34	NA	1,700	—	1,700
Big Creek	Big Cr, Flathead R	MT	1916	585	NA	360	—	360
Total Conventional					14,560		14,560	
Alaska Power Administration								
Eklutna[q]	Eklutna R	AK	1955	851	174	30,000	—	30,000
Snettisham[r]	Speel R	AK	1973	823	138	47,160	—	74,160
Total Conventional					77,160		104,160	
Grand Total Conventional					34,584,503	831,000	38,355,303	
Grand Total Reversible					2,735,200 R	581,000 R	3,740,210 R	

Note: Federally owned, developed, under construction, and authorized January 1, 1984. R Capacity shown is in reversible equipment.

a U—Less than 1,000 acre-ft. NA—Data not available.
b Includes main generating units only.
c Operated by Rock island Army Arsenal. Nameplate rating is 3,440 kilowatts which cannot be obtained.
d Storage Capacity for ultimate development is 420,000 acre-ft.
e All water used for generation must be returned by pumping.
f Storage in Buffalo Bill Reservoir is used jointly for irrigation and power.
g Pipeline collects water form Big Creek and Cottonwood Creek, tributaries of Plateau Creek.
h Water to be stored in the enlarged Strawberry Reservoir which is located upstream.
i Used jointly for irrigation, power, and flood control.
j Used jointly for irrigation and power.
k Used jointly for irrigation, power, flood control, and navigation.
l Includes two of three 10,000 kilowatt station service units used to supply commercial power.
m Used jointly for power and flood control.
n Dam and reservoir were constructed by the Corps of Engineers.
o Includes one of two 2,000 kilowatt station service units used to supply commercial power.
p 222,000 kilowatts are allocated to the state of California under contact.
q Constructed by the Bureau of Reclamation.
r Constructed by the Corps of Engineers.

Source: From Federal Energy Regulatory Commission, 1984, *Hydroelectric Power Resources of the United States—Developed and Undeveloped*, FERC-0070, Washington, DC.

Table 11C.32 Trends in Pumped Storage Capacity Development in the United States

	Installed Capacity in Reversible Units (millions of kilowatts)					
	Developed			Under Construction		
Year as of January 1	Pure	Combined	Total	Pure	Combined	Total
1960	0	0.1	0.1	0	0.2	0.2
1964	0.4	0.3	0.7	0.7	0.5	1.2
1968	1.6	0.5	2.1	1.2	1.6	2.8
1972	2.6	1.3	3.9	6.0	1.4	7.4
1976	7.3	2.4	9.7	2.7	1.6	4.3
1980	9.3	3.6	12.9	3.2	1.5	4.7
1984	10.1	3.7	13.8	4.9	0.4	5.3

Note: A pure pumped storage project with a large peaking capacity can be developed at a site with two potential reservoirs of reasonable size in close proximity and with a relatively large difference in elevations. Projects are usually more economically developed at sites with high usable heads; consequently, the more favorable sites are normally located in mountainous terrain. However, consideration has been given to the construction of pumped storage projects in areas of level terrain by placing the lower reservoir in an underground cavern or excavated area. For any development, an assured supply of water at least sufficient to replace evaporation, seepage, and other losses is essential.

Source: From Federal Energy Regulatory Commission, 1984.

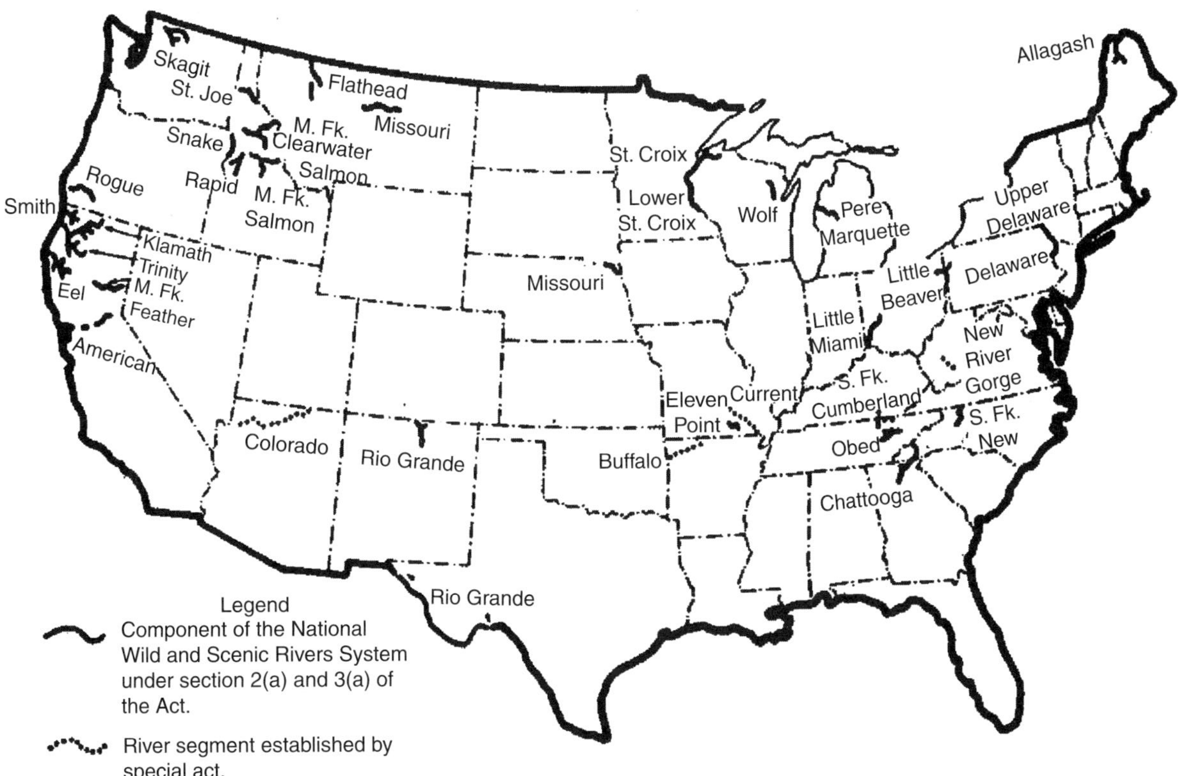

Figure 11C.22 River segments covered by the Wild and Scenic Rivers Act and special acts precluded from hydroelectric development. (From Federal Energy Regulation Commission, 1984.)

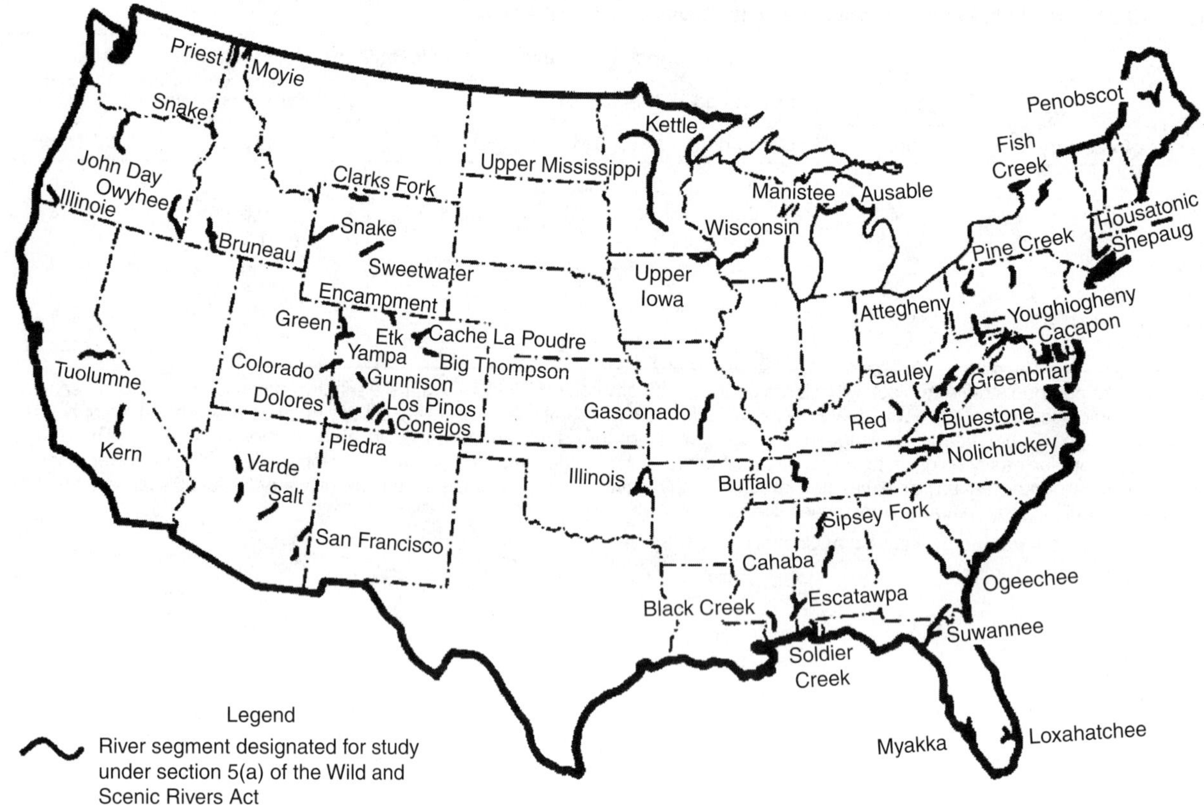

Figure 11C.23 River segments designated for study under section 5(a) of the Wild and Scenic Rivers Act. (From Federal Energy Regulation Commission, 1984.)

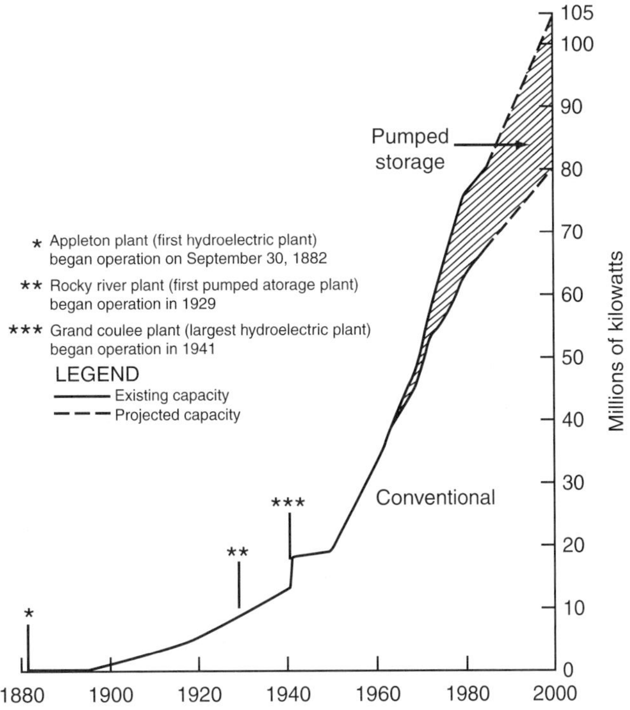

Figure 11C.24 Hydroelectric capacity in the United States 1882–2000. (From Federal Energy Regulatory Commission, 1984.)

Table 11C.33 Small Hydroelectric Capacity in the United States

	5,000 kW and Less		15,000 kW and Less		30,000 kW and Less	
Category	**1980**	**1984**	**1980**	**1984**	**1980**	**1984**
Developed						
Number of sites	751	864	946	1,124	1,071	1,252
Capacity (MW)	1,194.6	1,351.1	3,294.7	3,729.7	5,834.6	6,574.0
Generation (GWh)	4.8	16.8	16.8	18.8	28.0	30.8
Under Construction						
Number of sites	16	137	29	153	33	158
Capacity (MW)	23.5	168.6	135.6	325.8	229.9	444.8
Generation (GWh)	0.1	0.7	1.0	1.4	1.3	2.0
Planned (NERC)[a]						
Number of sites	12	25	23	39	25	50
Capacity (MW)	34.2	54.3	136.2	191.4	182.4	436.8
Generation (GWh)	0.1	0.3	0.5	0.9	0.6	1.9
Projected[b]						
Number of sites	157	1,699	227	2,278	279	2,367
Capacity (MW)	317.5	2,587.1	1,241.8	6,669.4	2,317.5	8,5724.4
Generation (GWh)	1.2	20.0	4.9	29.4	8.9	37.3
Totals						
Number of sites	936	2,725	1,225	3,594	1,408	3,827
Capacity (MW)	1,569.8	4,160.8	4,808.3	10,916.3	8,564.4	16,028.0
Generation (GWh)	6.2	27.6	23.2	50.3	38.8	72.0

Note: Developed, Under Construction, and Projected.

[a] In reports of the Regional Electric Reliability Councils.

[b] Potential developments not under construction or included in NERC reports but which have FERC licensing or exemption status, are authorized or recommended for Federal construction, or have structural provisions for plant additions.

Source: From Federal Energy Regulatory Commission, 1984.

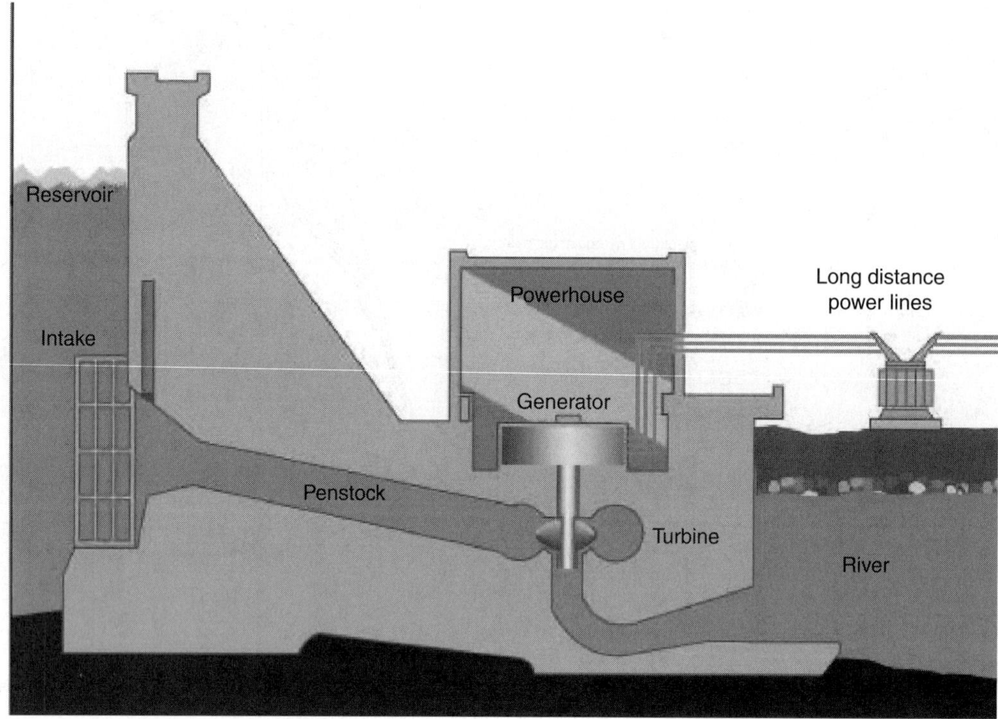

Figure 11C.25 Hydroelectric dam. (From www.tva.gov.)

Table 11C.34 Hydroelectric Plants Having Potential Conventional Capacity over 1,000,000 Kilowatts

			Installed Capacity in Conventional Units Kilowatts		
Plant	**River**	**Owner[a]**	**Developed**	**Under Construction**	**Ultimate Authorized**
Grand Coulee	Columbia	Bureau	6,180,000	0	6,180,000
John Day	Columbia	COE	2,160,000	0	2,700,000
Chief Joseph	Columbia	COE	2,069,000	0	2,069,000
R. Moses Niagara	Niagara	PASNY	1,950,000	0	1,950,000
The Dalles	Columbia	COE	1,806,800	0	1,806,800
Hoover	Colorado	Bureau	1,434,000	0	1,434,000
Rocky Reach	Columbia	CC PUD No. 1	1,213,950	0	1,213,950
Wanapum	Columbia	GC PUD No. 2	831,250	0	1,151,250
Priest Rapids	Columbia	GC PUD No. 2	788,500	0	1,108,500
Bonneville	Columbia	COE	1,076,620	0	1,102,820
Dworshak	N. Fork Clearwater	COE	400,000	0	1,060,000
Glen Canyon	Colorado	Bureau	1,042,000	0	1,042,000
Boundary	Pend Oreille	Seattle	634,600	392,000	1,026,600
Total		21,586,720	392,000	23,834,920	

[a] Bureau, Bureau of Reclamation; COE, Corps of Engineers; PASNY, Power Authority, State of New York; GC, Grant County; CC, Chelan County; and Seattle, Seattle Dept. of Lighting.

Source: From Federal Energy Regulatory Commission, 1984.

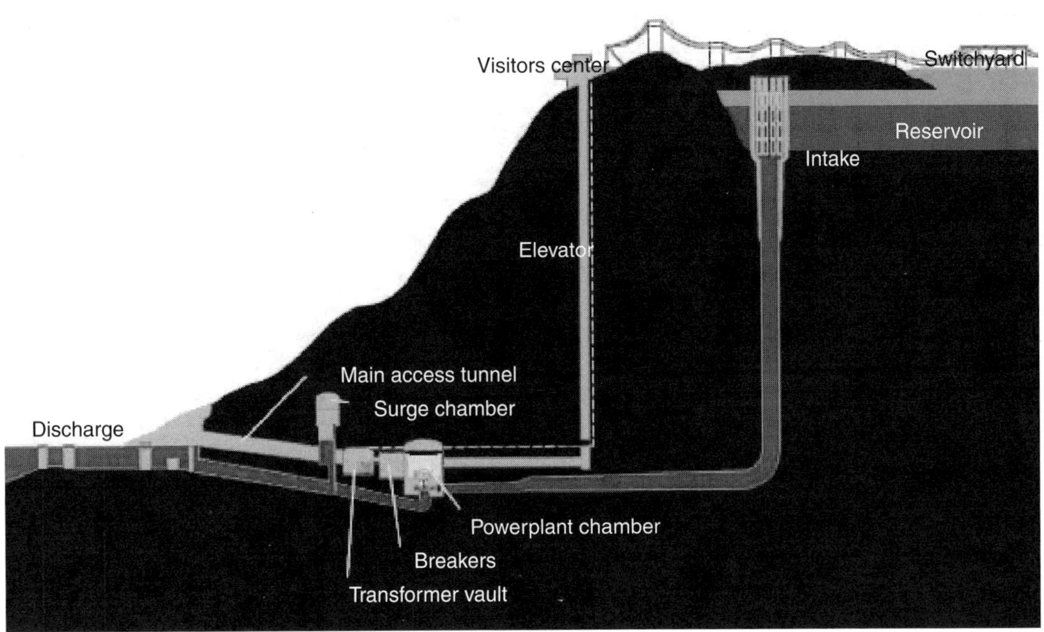

Figure 11C.26 Pumped-storage plant. (From www.tva.gov.)

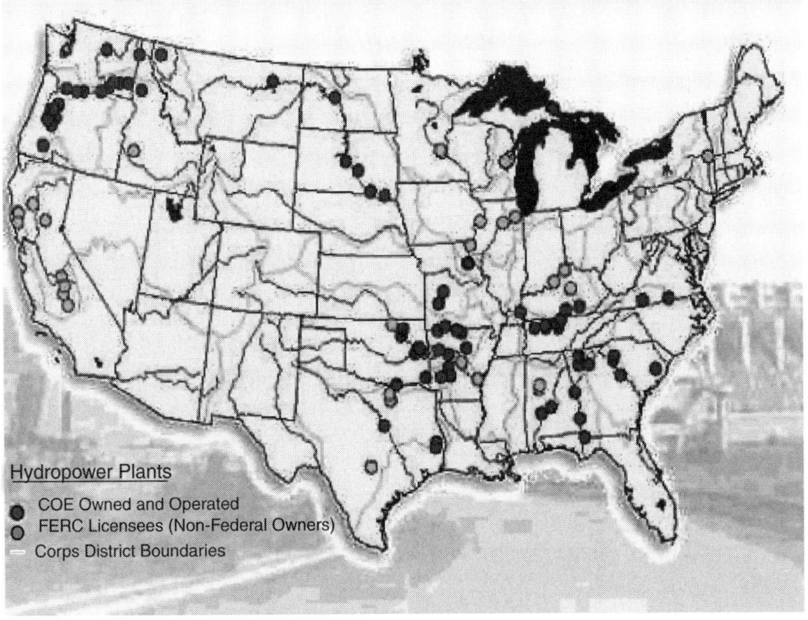

Background. The Corps of Engineers (Corps) is the single largest owner and operator of hydropower in the U.S., with 24% of the nation's hydropower generating capacity. The percentage is 16% for the Bureau of Reclamation and 6% for the Tennessee Valley Authority. Corps dams have a total nameplate capacity of close to 21,000 megawatts (MW) and produce an average of almost 100 million kilowatt-hours (kWh) of energy annually. Nonfederal power plants at Corps facilities add about another 2,000 MW of capacity.

Figure 11C.27 Hydropower plants. (From nwd-wc.usace.army.)

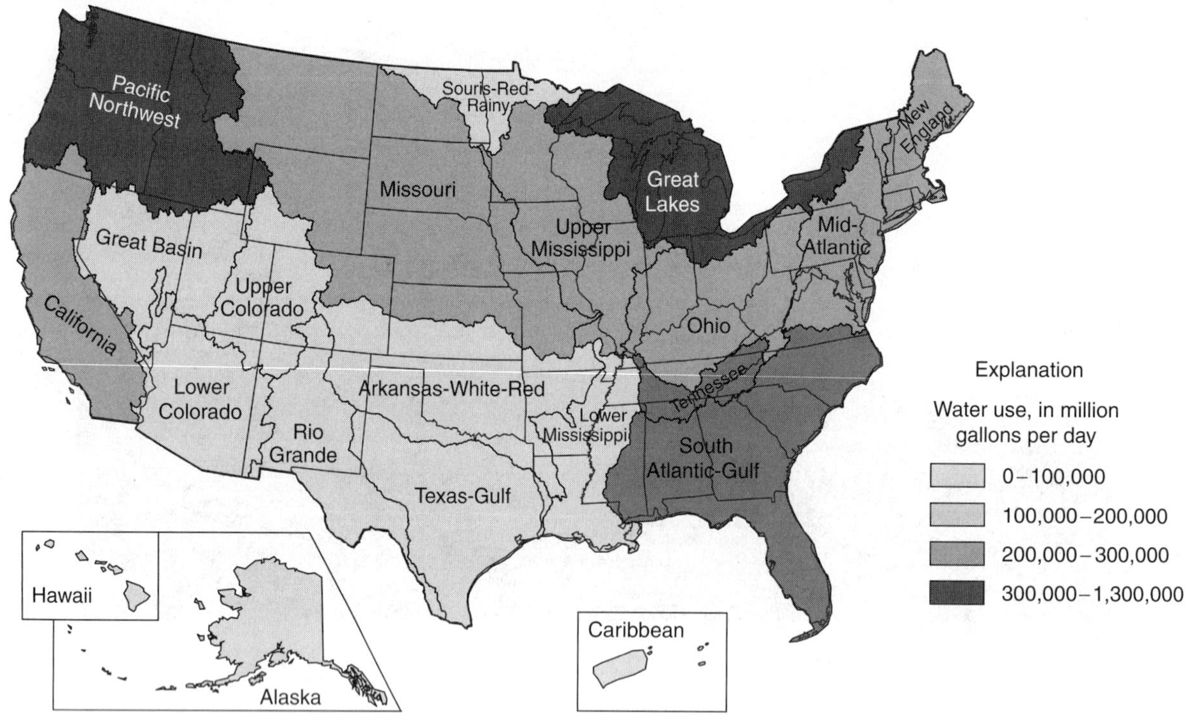

Figure 11C.28 Hydroelectric power water use by water resources region, 1995. (From www.usgs.gov.)

Table 11C.35 Hydroelectric Power Water Use by Water Resources Region, 1995

Region	Water Use: Mgal/d	Water Use: 1000 acre-ft/yr	Power Generated (million kWh)
New England	156,000	175,000	6,720
Mid-Atlantic	144,000	162,000	5,260
South Atlantic-Gulf	229,000	258,000	17,100
Great Lakes	340,000	382,000	24,200
Ohio	172,000	192,000	5,250
Tennessee	209,000	235,000	16,000
Upper Mississippi	119,000	133,000	2,990
Lower Mississippi	78,200	87,700	1,320
Souris-Red-Rainy	3,970	4,450	100
Missouri Basin	141,000	159,000	16,000
Arkansas-White-Red	95,400	107,000	6,740
Texas-Gulf	14,500	16,300	1,050
Rio Grende	3,860	4,320	464
Upper Colorado	17,900	20,000	7,220
Lower Colorado	23,400	26,300	9,740
Great Basin	5,060	5,670	633
Pacific Northwest	1,260,000	1,140,000	140,000
California	140,000	157,000	47,000
Alaska	2,090	2,340	1,440
Hawaii	229	256	148
Caribbean	349	391	101
Total	3,160,000	3,540,000	310,000

Note: Figures may not add to totals because of independent rounding. Mgal/d, million gallons per day; kWh, kilowatt-hour.

Source: From http://npdp.stanford.edu; www.usgs.gov.

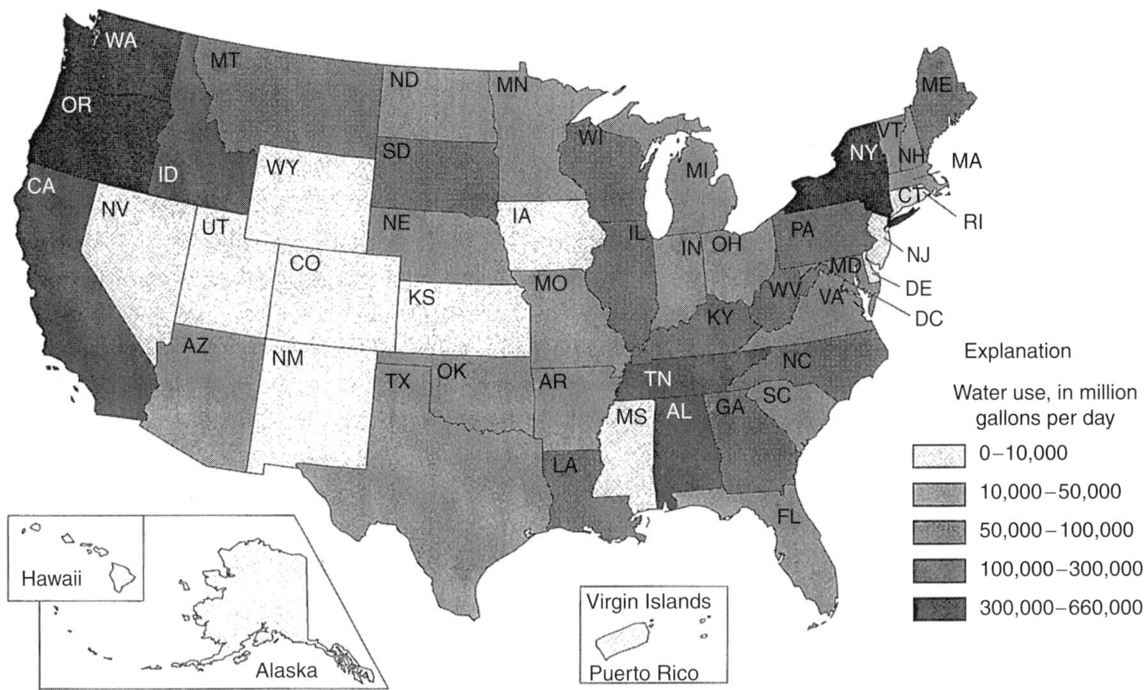

Figure 11C.29 Hydroelectric power water use by state, 1995. (From www.usgs.gov.)

Table 11C.36 Hydroelectric Power Water Use by State, 1995

State	Water Use		Power Generated (million kWh)
	Mgal/d	1000 acre-ft/yr	
Alabama	157,000	177,000	9,510
Alaska	2,090	2,340	1,440
Arizona	21,200	23,700	7,960
Arkonsas	42,700	47,900	2,630
California	146,000	164,000	47,100
Colorado	6,810	7,630	2,140
Connecticut	3,610	4,050	317
Delaware	0	0	0
District of Columbia	0	0	0
Florida	16,900	19,000	443
Georgia	50,900	57,100	4,850
Hawaii	229	256	148
Idaho	115,000	129,000	11,300
Illinois	55,800	62,500	1,010
Indiana	12,300	13,800	467
Iowa	2,350	2,630	21
Kansas	1,250	1,410	11
Kentucky	83,000	93,100	2,880
Louisiana	76,100	85,400	1,110
Maine	85,200	95,500	3,440
Maryland	14,400	16,100	1,450
Massachusetts	24,200	27,100	992
Michigan	39,800	44,600	1,410
Minnesota	19,800	22,200	1,030
Mississippi	0	0	0
Missouri	17,100	19,200	1,920
Montana	66,200	74,200	10,400
Nebraska	15,000	16,800	1,040
Nevada	6,080	6,810	6,320
New Hampshire	33,000	37,000	1,460
New Jersey	309	346	241
New Mexico	2,750	3,090	353
New York	356,000	399,000	24,600
North Carolina	56,400	63,200	5,810
North Dakota	13,900	15,600	2,480
Ohio	14,200	15,900	227
Oklahoma	49,100	55,100	3,300
Oregon	458,000	511,000	40,400
Pennsylvania	55,900	62,600	352
Rhode Island	339	380	6.1
South Carolina	42,200	47,300	3,070
South Dakota	62,400	69,900	6,420
Tennessee	122,000	137,000	9,430
Texas	18,600	20,900	1,520
Utah	3,720	4,170	931
Vermont	17,500	19,600	983
Virginia	14,800	16,600	922
Washington	653,000	733,000	82,300
West Virginia	51,500	57,700	1,210
Wisconsin	50,800	57,000	1,600
Wyoming	5,150	5,770	793
Puerto Rico	349	391	101
Virgin Islands	0	0	0
Total	3,160,000	3,540,000	310,000

Note: Figures may not add to totals because of independent rounding, Mgal/d, million gallons per day; kWh, kilowatt-hour.

Source: From npdp.stanford.edu. 1988.

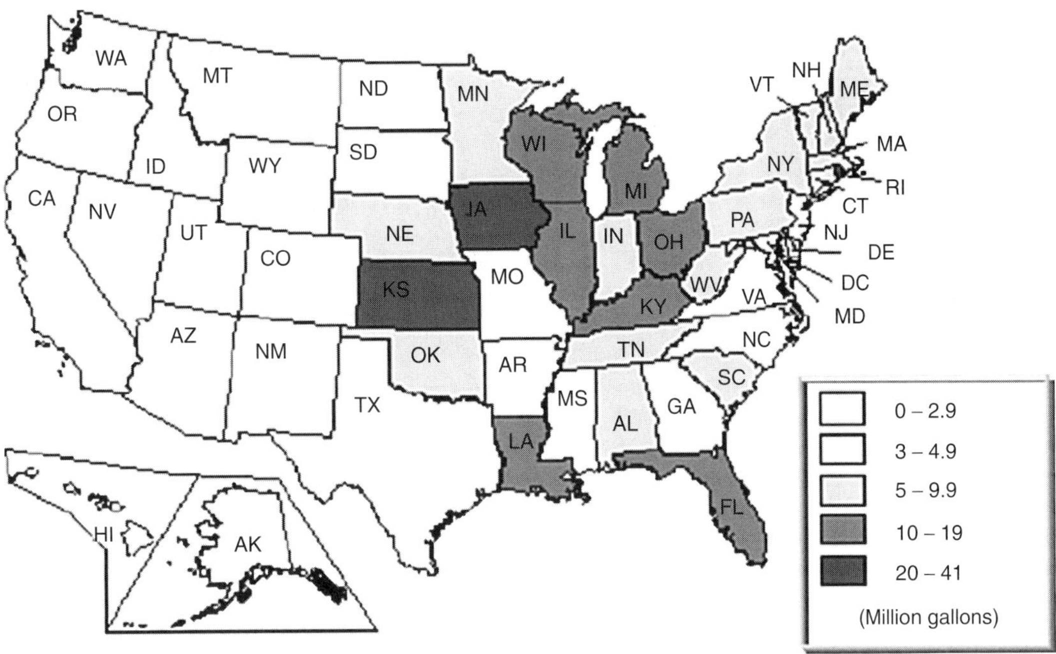

Figure 11C.30 Water used to produce 1 kWh of hydroelectric power in the United States in 1990. (From ga.water.usgs.gov.)

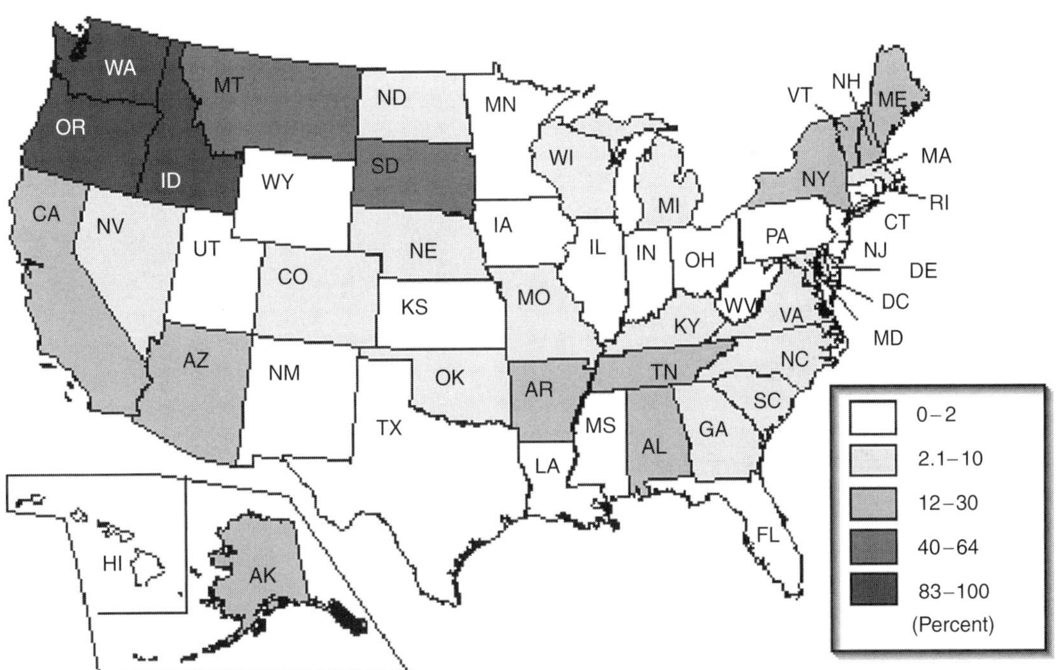

Figure 11C.31 Percent of total power produced coming from hydroelectric sources in the United States in 1990. (From ga.water.usgs.gov.)

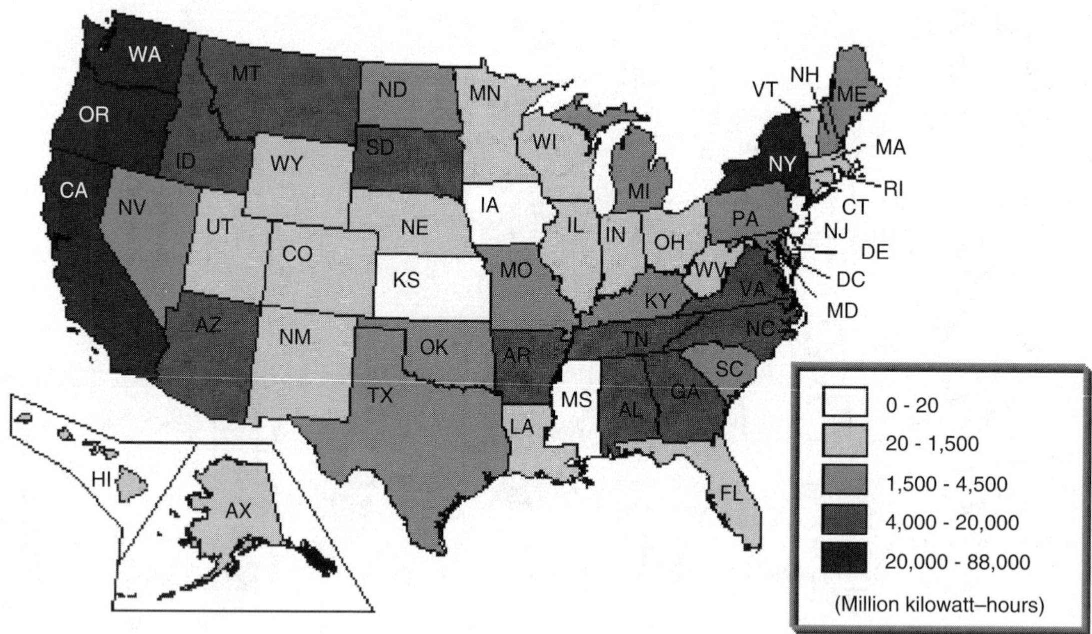

Figure 11C.32 Hydroelectric power production in the United States in 1990. (From ga.water.usgs.gov.)

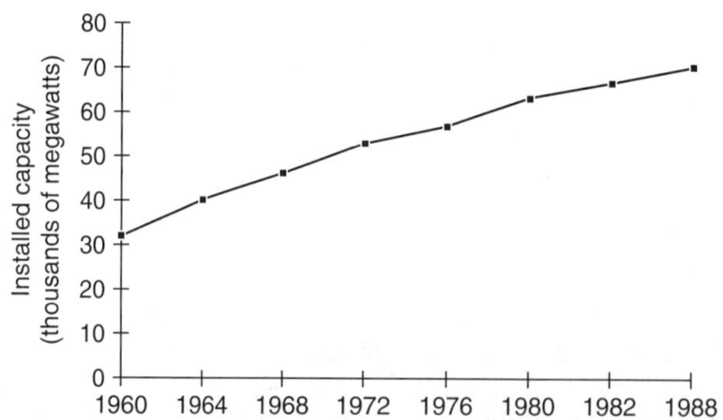

Figure 11C.33 Developed conventional hydroelectric capacity in the United States (From FERC, 1988a, xix. With permission.)

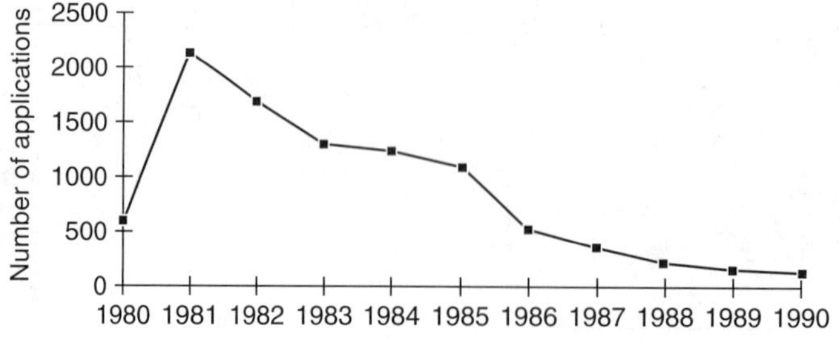

Figure 11C.34 U.S. Hydro applications a 1980–1990. (From FERC, 1988a, xix. With permission.)

Table 11C.37 U.S. Hydro Operations and Potential 1988

	Conventional	Pumped Storage[a]	Total
# Plants in operation	2010	37	2047
Developed capacity	70,800 MW	17,100 MW	87,900 MW
Mean plant capacity	35 MW	462 MW	43 MW
% of developed capacity under FERC jurisdiction	46.2%	81.3%	53.0%
Total potential capacity	146,900 MW	19,100 MW[b]	166,000 MW[b]
% of potential capacity developed	48.2%	89.5%	53.0%

[a] Includes both pure pumped storage and combined conventional and pumped storage facilities.
[b] Because of the enormous number of sites potentially suitable for pumped storage, only developed facilities and those actually under construction are included in the total potential capacity estimates for pumped storage facilities.

Source: From FERC, 1988a, vii–viii.

Table 11C.38 Hydro Development and Potential in the UMRB 1988

	Number of Plants	Average Plant Size	Developed Capacity (MW)	Undeveloped Capacity (MW)	Total Potential Capacity (MW)	Percentage of Total Potential Developed
Illinois	7	4,268	29.9	285.9	315.8	9
Iowa	6	21,939	131.6	388.0	519.6	25
Minnesota	31	6,121	189.7	299.9	489.6	39
Missouri	8	133,000	1,064.0	857.8	1,921.8	55
Wisconsin	116	4,045	469.2	492.4	961.6	49
5 states	168	11,217	1,884.5	2,324.0	4,208.5	45
United States	2047	42,936	87,889.7	78,048.5	165,938.2	53

Source: From FERC, 1988a, pp. 101, 132–41, 264.

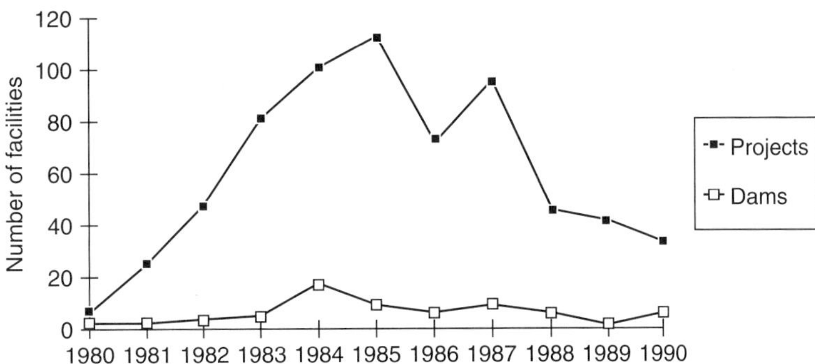

Figure 11C.35 New conventional hydro projects and dams in the United States (From FERC, 1988a, xix. With permission.)

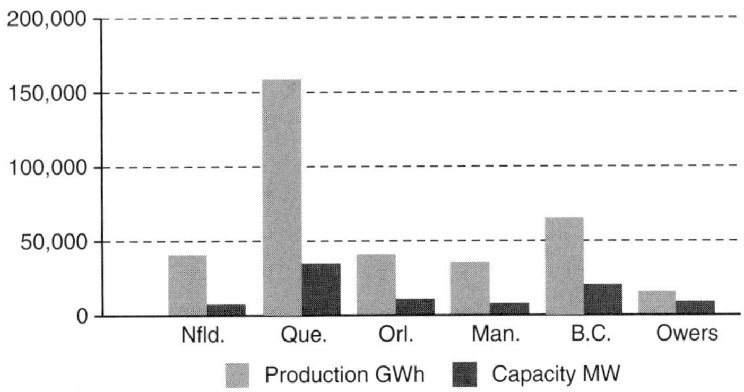

Figure 11C.36 Hydroelectricity by province. (From nrcan.gc.ca.)

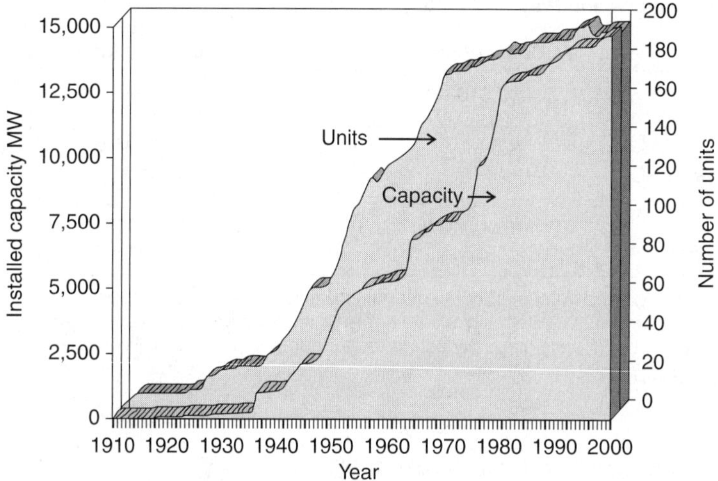

Figure 11C.37 Reclamation hydroelectric development 1909–2000. (From U.S. Department of the Interior, Bureau of Reclamation Power Resources Office—October 2004.)

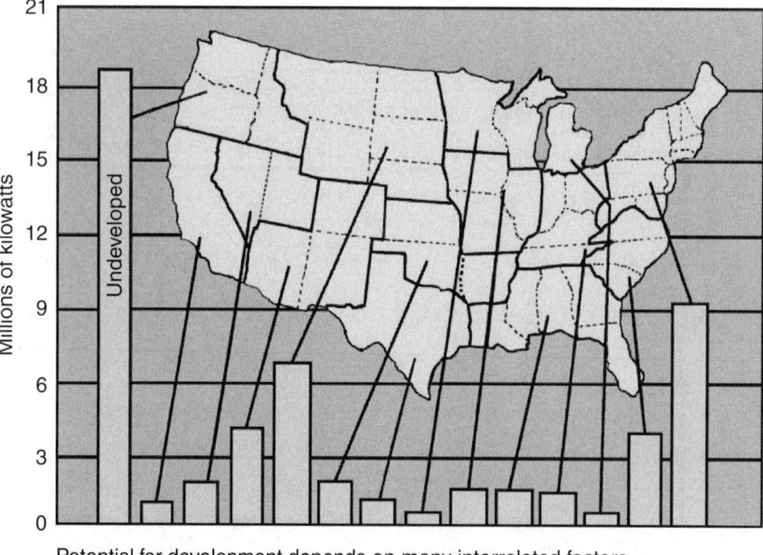

Figure 11C.38 Undeveloped hydroelectric power. (From U.S. Department of the Interior, Bureau of Reclamation Power Resources Office—October 2004.)

Table 11C.39 Conventional Hydroelectric Power in the United States Developed, Undeveloped, and Total Potential — January 1, 1984

Water Resource Region	Developed			Undeveloped			Total Potential		Percent Total Potential Cap. Now Developed
	No. of Plants	Installed Capacity kW	Average Annual Generation 1,000 kWh	No. of Sites	Installed Capacity kW[a]	Average Annual Generation 1,000 kWh	Installed Capacity kW	Average Annual Generation 1,000 kWh	
North Atlantic Region	248	1,538,631	6,359,847	669	4,387,567	13,299,262	5,926,198	19,659,209	25.9
Mid Atlantic Region	122	1,506,299	6,091,861	409	4,786,750	13,732,622	6,293,049	19,824,483	23.9
South Atlantic-Gulf Region	128	6,094,348	16,190,853	261	5,396,194	10,633,130	11,490,542	26,826,983	53.0
Great Lakes Region	209	3,895,123	24,354,228	229	2,243,431	9,852,598	6,138,554	34,206,826	63.4
Ohio Region	36	1,646,672	6,104,170	200	4,313,176	13,330,111	5,959,848	19,443,281	27.6
Tennessee Region	47	3,749,075	16,249,440	33	1,150,985	2,777,781	4,900,060	19,027,221	76.5
Upper Mississippi Region	105	617,237	3,187,639	104	1,267,796	5,459,595	1,885,033	8,47,234	32.7
Lower Mississippi Region	5	205,800	430,400	29	1,000,075	3,106,275	1,205,875	3,536,675	17.0
Souris-Red-Rainy Region	8	13,000	68,000	8	45,880	158,290	58,880	226,290	22.0
Missouri Region	62	3,462,023	15,549,519	234	5,307,568	17,595,761	8,769,591	33,145,280	39.4
Arkansas-White-Red Region	23	1,854,460	5,341,446	128	3,581,693	9,065,790	5,436,153	14,407,236	34.1
Texas-Gulf Region	18	393,970	1,066,359	87	1,481,627	3,032,279	1,875,597	4,098,638	21.0
Rio Grande Region	5	131,408	389,521	25	233,886	632,431	365,294	1,021,952	35.9
Upper Colorado Region	26	1,452,497	5,923,214	137	2,370,246	6,221,787	3,822,743	12,145,001	37.9
Lower Colorado Region	17	1,910,463	6,601,700	33	1,650,983	5,085,441	3,561,446	11,687,141	53.6
Great Basin Region	60	188,422	655,837	112	765,697	1,862,008	954,119	2,517,845	19.7
Pacific Northwest Region	176	29,909,275	143,542,447	1,128	22,104,337	67,221,564	52,013,612	210,764,011	57.5
California Region	211	8,033,509	37,644,560	740	10,099,470	25,983,648	18,132,979	63,628,208	44.3
Alaska Region	26	155,737	634,082	96	5,041,107	22,192,656	5,196,844	22,826,738	2.9
Hawaii Region	14	17,652	104,100	15	62,890	310,950	80,542	415,050	21.9
Total United States	1,546	66,775,601	296,489,223	4,677	77,291,358	231,563,079	144,066,959	528,052,302	46.3

[a] Includes potential capacity additions or subtractions at existing plants.

Source: From Federal Energy Regulatory Commission, 1984.

Table 11C.40 Conventional Hydroelectric Power in the United States Developed, Undeveloped, and Total Potential — January 1, 1984

	Developed			Undeveloped			Total Potential		
Geographic Division	No. of Plants	Installed Capacity kW	Average Annual Generation 1,000 kWh	No. of Sites	Installed Capacity kW[a]	Average Annual Generation 1,000 kWh	Installed Capacity kW	Average Annual Generation 1,000 kWh	Percent Total Potential Cap. Now Developed
New England	279	1,618,739	6,684,066	74	4,555,434	13,887,230	6,174,170	20,571,296	26.2
Middle Atlantic	157	4,297,372	25,703,831	421	5,370,969	18,152,584	9,668,341	43,856,415	44.4
East North Central	210	1,087,054	5,159,366	208	1,789,913	7,158,041	2,876,967	12,317,407	37.7
West North Central	65	2,808,465	12,187,539	157	3,011,876	9,999,018	5,820,341	22,186,557	48.2
South Atlantic	166	6,013,182	17,216,643	355	8,395,423	17,795,208	14,408,605	35,011,851	41.7
East South Central	57	5,734,292	22,024,618	119	2,842,266	9,337,538	8,576,558	31,362,156	66.8
West South Central	43	2,338,020	6,516,705	214	5,328,495	12,734,953	7,666,515	19,251,658	30.4
Mountain	192	7,869,611	32,866,583	876	18,201,259	53,698,630	26,070,870	86,565,213	30.1
Pacific	337	34,835,480	167,391,690	1,472	22,691,726	66,296,271	57,527,206	233,687,961	60.5
Alaska	26	155,737	634,082	96	5,041,107	22,192,656	5,196,844	22,826,738	2.9
Hawaii	14	17,652	104,100	15	62,890	310,950	80,542	415,050	21.9
Total United States	1,546	66,775,601	296,489,223	4,677	77,291,358	231,563,079	144,066,959	528,052,302	46.3

[a] Includes potential capacity additions or subtractions at existing plants.

Source: From Federal Energy Regulatory Commission, 1984.

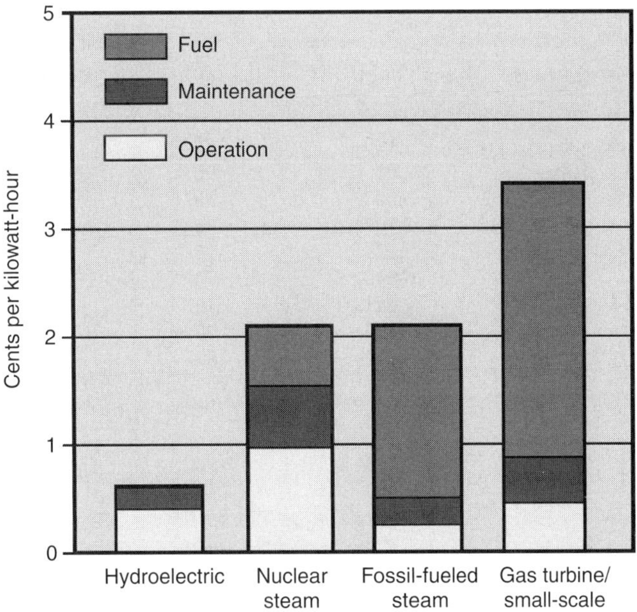

Hydropower is an economical source of electrical energy. It is one type of electricity that is immune to rising fuel costs. Hydropower costs above include pumped-storage.

Figure 11C.39 Average power production expenses per kWh, 1995–1999. (From Energy Information Administration Financial Statistics of Major U.S. Investor-Owned Utilities.)

Table 11C.41 Federal Hydroelectric Capacity by Operating Agency, January 1, 1984

	Installed Capacity Conventional and Reversible (Kilowatts)		
	In Operation	**Under Construction**	**Ultimate Authorized**
Corps of Engineers	19,449,197	1,162,000	23,466,897
Bureau of Reclamation	12,846,876	250,000	13,577,986
Tennessee Valley Authority	4,832,410	—	4,832,410
International Boundary and Water Commission	97,500	—	97,500
Alaska Power Administration	77,160	—	104,160
Bureau of Indian Affairs	14,560	—	14,560
National Park Service	2,000	—	2,000
Total	37,319,703	1,412,000	42,095,513

Source: From Federal Energy Regulatory Commission, 1984.

Table 11C.42 TVA's Hydro Plants

	Dam Height (ft)	Dam Length (ft)	Reservoir Length (mil)	Capacity[a] (MW)	Construction Span
Tennessee River					
Fort Loudeun	125	4,190	61	123	1940–43
Watts Bar	122	2,960	96	175	1939–42
Chickamauga	129	5,800	59	129	1936–40
Nickajack	86	3,767	46	100	1964–67
Guntersville	97	3,979	76	114	1935–39
Wheeler	72	6,342	74	355	1933–36
Wilson	137	4,541	16	611	1918–24
Pickwick Landing	113	7,715	53	193	1934–38
Kentucky	206	8,422	184	199	1938–44
Raccoon Mountain Pumped Storage	230	8,500	1.2	1,532	1970–78
Clinch River					
Norris	265	1,860	129	111	1933–36
Melton Hill	103	1,020	44	72	1960–63
French Broad River					
Douglas	215	1,705	43	90	1942–43
Holston River					
South Holston	285	1,600	24	46	1942–50
Boone	168	1,697	33	83	1950–52
Fort Patrick Henry	95	737	10	32	1951–53
Cherokee	183	6,760	59	126	1940–41
Walauga River					
Walauga	331	925	16	68	1942–48
Wilbur	77	375	2	12	1912
Little Tennessee River					
Fontana	480	2,365	29	241	1942–44
Hiwassee River					
Chaluge	150	3,336	13	10	1941–42
Nottety	199	3,915	20	16	1941–42
Hiwassee	307	1,376	22	121	1936–40
Apalachia	150	1,308	10	74	1941–43
Ocoee River					
Blue Ridge	174	1,553	11	13	1925–30
Ocoee 1	135	840	8	25	1910–11
Ocoee 2	30	450	0.8	19	1912–13
Ocoee 3	110	612	7	28	1941–42
Elk River					
Tims Ford	175	1,580	34	37	1966–70
Caney Fork River					
Great Falls	92	800	22	37	1915–16

Note: To find information on dam releases, reservoir levels, and other river system data, go to http://lakeinfo.tva.com.

[a] Net winter dependable capacity; the amount of power a plant can produce on an average winter day, minus the electricity used by the plant itself.

Source: From www.tva.com.

Table 11C.43 Bureau of Reclamation Power Facilities

Region	Project	Plant	State Location	River	Initial Date in Service	Number of Units	Installed Capacity (kW)	Gross Generation (kWh)
PN	Boise	Anderson Ranch	Idaho	So. Fork, Boise	12–50	2	40,000	117,267,000
		Black Canyon	Idaho	Payette	12–25	2	10,200	66,221,000
		Boise River div.	Idaho	Boise	5–12	3	3,450	7,054,000
	Columbia Basin	Grand Coulee	Washington	Columbia	3–41	33	6,809,000	18,854,034,875
	Hungary Horse	Hungry Horse	Montana	So. Fork, flathead	10–52	4	428,000	851,474,000
	Minidoka	Minidoka	Idaho	Snake	5–09	4	27,700	84,345,000
	Palisades	Palisades	Idaho	So. Fork, Snake	2–57	4	176,564	422,987,000
	Rogue River Basin	Green Springs	Oregon	Trans. Mtn. Div.	5–60	1	17,290	53,649,000
	Yakima	Chandler	Washington	Yakima	2–56	2	12,000	40,885,000
		Roza	Washington	Yakima	8–58	1	12,937	59,568,700
Total						56	7,537,141	20,557,485,575
MP	Central Valley	Judge F. Carr	California	Tunnel, Lewiston	5–63	2	154,400	494,153,000
		Folsom	California	American	6–55	3	198,720	493,687,000
		Keswick	California	Sacramento	10–49	3	117,000	470,903,000
		New Melones	California	Stanislaus	6–79	2	300,000	342,689,000
		Nimbus	California	American	5–55	2	13,500	55,253,000
		O'Neill	California	San Luis Creek	11–67	6	25,200	6,094,000
		San Luis	California	San Luis Creek	3–68	8	[a]202,000	176,083,000
		Shasta	California	Sacramento	6–44	7	646,000	2,209,689,300
		Spring Creek	California	Tunnel, Clear Creek	1–64	2	180,000	582,083,000
		Trinity	California	Trinity	2–64	2	140,000	594,842,000
		Lewiston	California	Trinity	2–64	1	350	3,075,000
	Washoe	Stampede	California	Little Truckee	9–87	2	3,650	13,657,130
Total						40	1,980,820	5,442,208,430
LC	Boulder Canyon	Hoover	Arizona/Nevada	Colorado	9–36	19	2,078,800	4,040,245,040
	Parker-Davis	Davis	Arizona	Colorado	1–51	5	255,000	1,170,088,000
		Parker	California	Colorado	12–42	4	120,000	458,320,999
Total						28	2,453,800	5,668,654,039
UC	Collbran	Lower Molina	Colorado	Pipeline	12–62	1	4,860	14,369,675
		Upper Molina	Colorado	Pipeline	12–62	1	8,640	24,818,100
	Colo. River Storage	Blue Mesa	Colorado	Gunnison	9–67	2	86,400	142,539,000
		Crystal	Colorado	Gunnison	6–78	1	28,000	4,705,000

(Continued)

Table 11C.43 (Continued)

Region	Project	Plant	State Location	River	Initial Date in Service	Number of Units	Installed Capacity (kW)	Gross Generation (kWh)
		Flaming Gorge	Utah	Green	11–63	3	151,950	236,681,000
		Glen Canyon	Arizona	Colorado	9–64	8	1,296,000	3,328,793,000
		Morrow Point	Colorado	Gunnison	12–70	2	173,334	195,118,000
	Provo River	Deer Creek	Utah	Provo	2–58	2	4,950	14,261,154
	Rio Grande	Elephant Butte	New Mexico	Rio Grande	11–40	3	27,945	28,224,800
	Seedskadee	Fontenelle	Wyoming	Green	5–68	1	10,000	45,472,000
	Dolores	McPhee	Colorado	Dolores	12–92	1	1,283	2,655,481
		Towaoc	Colorado	Towaoc Canal	5–93	1	11,495	16,486,900
Total						26	1,804,857	4,054,124,110
GP	Colo.-Big Thompson	Big Thompson	Colorado	Trans. Mtn. Div.	4–59	1	4,500	10,010,000
		Estes	Colorado	Trans. Mtn. Div.	9–50	3	45,000	107,647,000
		Flatiron	Colorado	Trans. Mtn. Div.	1–54	3	94,500	229,031,000
		Green Mountain	Colorado	Blue	5–43	2	26,000	27,690,000
		Marys Lake	Colorado	Trans. Mtn. Div.	5–51	1	8,100	38,686,000
		Pole Hill	Colorado	Trans. Mtn. Div.	1–54	1	38,200	180,065,000
	Frying Pan-Arkansas	Mt. Elbert	Colorado	Trans. Mtn. Div.	6–81	2	200,000	347,142,760
	Kendrick	Alcova	Wyoming	North Platte	7–55	2	41,400	70,301,000
		Seminoe	Wyoming	North Platte	8–39	3	51,750	63,319,000
	North Platte	Guernsey	Wyoming	North Platte	7–27	2	6,400	10,150,000
	Pick-Sloan mo. Basin	Boysen	Wyoming	Wind	8–52	2	15,000	30,139,000
		Canyon Ferry	Montana	Missouri	12–53	3	50,000	239,121,000
		Fremont Canyon	Wyoming	North Platte	12–60	2	66,880	144,776,000
		Glendo	Wyoming	North Platte	12–58	2	38,000	46,405,000
		Kortes	Wyoming	North Platte	6–50	3	36,000	83,793,000
		Yellowtail	Montana	Big Horn	8–66	4	250,000	322,981,000
	Shoshone	Buffalo Bill	Wyoming	Shoshone	7–92	3	18,000	47,604,000
		Heart Mountain	Wyoming	Shoshone	12–48	1	5,000	9,115,000
		Pilot Butte	Wyoming	Wind	1–25	2	1,600	4,416,000
		Shoshone	Wyoming	Shoshone	6–92	1	3,000	18,413,000
		Spirit Mountain	Wyoming	Shoshone	10–94	1	4,500	17,087,000
Total						44	1,003,750	2,047,891,760

Note: Hydroelectric powerplants for which the bureau of reclamation has operating responsibility fiscal year 2004. Grand Total Number of Plants, 58; Grand Total Number of Units, 194; Grand Total Installed Capacity, 14,780,368 kW; Grand Total Gross Generation, 37,770,363,914 kWh. Revised December 16, 2004.

[a] Federal share of 424,000 kW installed capacity-plant operated by the State of California-Navajo stem plant: reclamation share is 546,750 kW installed capacity with 2004 gross generation of 4,713,459,920 kWh.

Source: From www.usbr.gov.

Table 11C.44 Number of Dams Found: 160

NPDP ID	Dam Name	State	Dam Type	Dam Height (ft)	Primary Purpose	Secondary Purpose
VA17702	North Anna Dam	VA	Earth	90	Hydroelectric	Flood Control and Storm Water Management
VA01706	Bath Co. Pumped Storage—Upper Dam	VA	Earth	381	Hydroelectric	Flood Control and Storm Water Management
VA01707	Bath Co. Pumped Storage—Lower Dam	VA	Earth	136	Hydroelectric	Flood Control and Storm Water Management
UT00310	Logan No. 3	UT	Concrete Gravity	31.5	Hydroelectric	Flood Control and Storm Water Management
TN02702	Dale Hollow Dam	TN	Gravity	200	Hydroelectric	Flood Control and Storm Water Management
OR00015	Cougar	OR	Rockfill	519	Hydroelectric	Flood Control and Storm Water Management
OR00004	Detroit	OR	Gravity	463	Hydroelectric	Flood Control and Storm Water Management
OR00011	John Day Dam	OR	Earth	230	Hydroelectric	Flood Control and Storm Water Management
OR00012	Foster	OR	Rockfill	126	Hydroelectric	Flood Control and Storm Water Management
OK83001	Salina Dike	OK	Earth	45	Hydroelectric	Flood Control and Storm Water Management
OK10314	Fort Gibson Lake	OK	Earth Gravity	110	Hydroelectric	Flood Control and Storm Water Management
OK00134	Robert S. Kerr	OK	Gravity Earth	87	Hydroelectric	Flood Control and Storm Water Management
OK00135	Pensacola	OK	Multiple Arch Gravity	151	Hydroelectric	Flood Control and Storm Water Management
NY83112	Bennetts Bridge—Dike A	NY	Earth	5	Hydroelectric	Flood Control and Storm Water Management
NY83028	Stillwater—South Dam	NY	Earth	20	Hydroelectric	Flood Control and Storm Water Management
NY00374	Bennetts Bridge	NY	Concrete Gravity	45	Hydroelectric	Flood Control and Storm Water Management
NY00375	Bennetts Bridge—Dike B	NY	Earth	13	Hydroelectric	Flood Control and Storm Water Management
NY00376	Bennetts Bridge—Dike C	NY	Earth	10	Hydroelectric	Flood Control and Storm Water Management
NY00397	Cranberry Lake	NY	Gravity Concrete	19	Hydroelectric	Flood Control and Storm Water Management
NY00316	Stillwater—North Dam	NY	Earth Concrete Gravity	55	Hydroelectric	Flood Control and Storm Water Management
NY00146	Conklingville	NY	Earth	95	Hydroelectric	Flood Control and Storm Water Management
NY00120	Clark Mills Upper	NY	Concrete Buttress	23	Hydroelectric	Flood Control and Storm Water Management
NH127	Stone Dam	NH	Stone	16.5	Hydroelectric	Flood Control and Storm Water Management
NH00052	Shelburne	NH	Concrete Timber Crib	17	Hydroelectric	Flood Control and Storm Water Management
MT00652	Libby	MT	Gravity	422	Hydroelectric	Flood Control and Storm Water Management
MT00561	Madison Dam	MT	Timber Crib	39	Hydroelectric	Flood Control and Storm Water Management
MT00568	Canyon Ferry	MT	Concrete Gravity	225	Hydroelectric	Flood Control and Storm Water Management
MT00559	Holter Dam	MT	Gravity	124	Hydroelectric	Flood Control and Storm Water Management
MT00560	Hauser Dam	MT	Gravity	125	Hydroelectric	Flood Control and Storm Water Management
MT00134	Hebgen Dam	MT	Earth	88	Hydroelectric	Flood Control and Storm Water Management
MO30014	Osage	MO	Gravity	148	Hydroelectric	Flood Control and Storm Water Management

(Continued)

Table 11C.44 (Continued)

NPDP ID	Dam Name	State	Dam Type	Dam Height (ft)	Primary Purpose	Secondary Purpose
MO20725	Harry S. Truman Dam	MO	Earth Rockfill Gravity	98	Hydroelectric	Flood Control and Storm Water Management
MN83002	International Falls	MN	Gravity Masonry	29	Hydroelectric	Flood Control and Storm Water Management
MI00547	Secord	MI	Gravity	57	Hydroelectric	Flood Control and Storm Water Management
MI00548	Smallwood	MI	Gravity Earth	36	Hydroelectric	Flood Control and Storm Water Management
MI00549	Edenville	MI	Gravity Earth	54	Hydroelectric	Flood Control and Storm Water Management
ME96172	Pleasant Pond Dam	ME	Concrete Rockfill	12	Hydroelectric	Flood Control and Storm Water Management
ME83048	North Twin—Dike 4	ME	Earth	11	Hydroelectric	Flood Control and Storm Water Management
ME83049	North Twin—Dike 5	ME	Earth	11	Hydroelectric	Flood Control and Storm Water Management
ME83050	North Twin—Dike 6	ME	Earth	13	Hydroelectric	Flood Control and Storm Water Management
ME83051	Stone Dam—Dike 8	ME	Earth	9	Hydroelectric	Flood Control and Storm Water Management
ME96079	Davee Brook #1 Dam	ME	Concrete Earth	26	Hydroelectric	Flood Control and Storm Water Management
ME96100	Andres Mill Dam	ME	Concrete Timber Crib	4	Hydroelectric	Flood Control and Storm Water Management
ME83006	Gilman Falls Dam	ME	Concrete Gravity	18	Hydroelectric	Flood Control and Storm Water Management
ME83038	Stone Dam—Dike 1	ME	Earth	10	Hydroelectric	Flood Control and Storm Water Management
ME83039	Stone Dam—Dike 2	ME	Earth	10	Hydroelectric	Flood Control and Storm Water Management
ME83040	Stone Dam—Dike 3	ME	Earth	10	Hydroelectric	Flood Control and Storm Water Management
ME83041	Stone Dam—Dike 4	ME	Earth	9	Hydroelectric	Flood Control and Storm Water Management
ME83042	Stone Dam—Dike 5	ME	Earth	9	Hydroelectric	Flood Control and Storm Water Management
ME83043	Stone Dam—Dike 6	ME	Earth	9	Hydroelectric	Flood Control and Storm Water Management
ME83044	Stone Dam—Dike 7	ME	Earth	9	Hydroelectric	Flood Control and Storm Water Management
ME83045	North Twin—Dike 1	ME	Earth	11	Hydroelectric	Flood Control and Storm Water Management
ME83046	North Twin—Dike 2	ME	Earth	11	Hydroelectric	Flood Control and Storm Water Management
ME83047	North Twin—Dike 3	ME	Earth	15	Hydroelectric	Flood Control and Storm Water Management
ME10103	Automatic Dam Messalonskee #4	ME	Concrete Gravity	40	Hydroelectric	Flood Control and Storm Water Management
ME00534	#1 Fitch's Mill Pond Dam	ME	Concrete Earth Stone	18	Hydroelectric	Flood Control and Storm Water Management
ME00535	Rich Mill Dam	ME	Earth Rockfill Stone	9	Hydroelectric	Flood Control and Storm Water Management
ME00508	Rainbow Lake Dam	ME	Timber Crib Gravity Rockfill	9	Hydroelectric	Flood Control and Storm Water Management
ME00518	Ripley Pond Dam	ME	Earth	6	Hydroelectric	Flood Control and Storm Water Management
ME00468	Long Pond Storage Dam	ME	Timber Crib Rockfill Gravity	7	Hydroelectric	Flood Control and Storm Water Management

ME00483	Mapleton Dam	ME	Timber Crib Rockfill Gravity	11	Hydroelectric	Flood Control and Storm Water Management
ME00430	Parson's Mill Dam	ME	Concrete	10	Hydroelectric	Flood Control and Storm Water Management
ME00398	Long Pond Dam	ME	Stone Earth	8	Hydroelectric	Flood Control and Storm Water Management
ME00352	Morneau's Dam	ME	Timber Crib Gravity	10	Hydroelectric	Flood Control and Storm Water Management
ME00360	Blake Dam	ME	Concrete Earth Rockfill	16	Hydroelectric	Flood Control and Storm Water Management
ME00341	Upper Dam	ME	Concrete	8	Hydroelectric	Flood Control and Storm Water Management
ME00283	Behind The Mill Dam	ME	Concrete Gravity	20	Hydroelectric	Flood Control and Storm Water Management
ME00291	Mainstream Dam, Main Stem	ME	Concrete Timber Crib Stone	21	Hydroelectric	Flood Control and Storm Water Management
ME00305	Sokokis Lake Dam	ME	Concrete Stone Gravity	15	Hydroelectric	Flood Control and Storm Water Management
ME00254	Alford Lake Dam	ME	Concrete Earth Stone	7	Hydroelectric	Flood Control and Storm Water Management
ME00225	Mars Hill Dam	ME	Concrete Earth Gravity	14	Hydroelectric	Flood Control and Storm Water Management
ME00198	Lock Dam	ME	Timber Crib Earth Gravity	14	Hydroelectric	Flood Control and Storm Water Management
ME00200	East Millinocket	ME	Concrete Gravity Earth	28	Hydroelectric	Flood Control and Storm Water Management
ME00201	Dolby	ME	Concrete Gravity Earth	66	Hydroelectric	Flood Control and Storm Water Management
ME00202	Stone Dam	ME	Concrete Gravity	27	Hydroelectric	Flood Control and Storm Water Management
ME00203	North Twin	ME	Earth Concrete Gravity	35	Hydroelectric	Flood Control and Storm Water Management
ME00204	Ripogenus	ME	Concrete Gravity Earth	83	Hydroelectric	Flood Control and Storm Water Management
ME00205	Millinocket Lake	ME	Concrete Gravity Earth	20	Hydroelectric	Flood Control and Storm Water Management
ME00206	Seboomook	ME	Concrete Gravity Earth	60	Hydroelectric	Flood Control and Storm Water Management
ME00209	Ragged Lake	ME	Earth Concrete Gravity	30	Hydroelectric	Flood Control and Storm Water Management
ME00211	Caucomgomoc	ME	Earth Concrete Gravity	16	Hydroelectric	Flood Control and Storm Water Management
ME00215	Canada Falls	ME	Concrete Gravity	50	Hydroelectric	Flood Control and Storm Water Management
ME00166	Branns Mill Dam	ME	Timber Crib Earth Stone	14	Hydroelectric	Flood Control and Storm Water Management

(Continued)

Table 11C.44 (Continued)

NPDP ID	Dam Name	State	Dam Type	Dam Height (ft)	Primary Purpose	Secondary Purpose
ME00187	Bridge Street	ME	Masonry Gravity	8	Hydroelectric	Flood Control and Storm Water Management
ME00128	Spencer Lake Dam	ME	Concrete Timber Crib	14	Hydroelectric	Flood Control and Storm Water Management
ME00133	Brassua	ME	Concrete Buttress Earth	50	Hydroelectric	Flood Control and Storm Water Management
ME00135	First Roach Dam	ME	Concrete Rockfill Gravity	18	Hydroelectric	Flood Control and Storm Water Management
ME00143	Mattaceunk	ME	Concrete Gravity Earth	45	Hydroelectric	Flood Control and Storm Water Management
ME00157	Upper Dam	ME	Concrete Gravity	12	Hydroelectric	Flood Control and Storm Water Management
ME00125	Mill Pond Dam	ME	Concrete Earth Gravity	9	Hydroelectric	Flood Control and Storm Water Management
ME00127	Flagstaff	ME	Earth Concrete Gravity	43	Hydroelectric	Flood Control and Storm Water Management
ME00071	Sebago Lake Dam	ME	Concrete Earth Stone	32	Hydroelectric	Flood Control and Storm Water Management
ME00091	Moosehead—East Outlet	ME	Concrete Gravity Earth	20	Hydroelectric	Flood Control and Storm Water Management
ME00092	Moosehead—West Outlet	ME	Concrete Gravity Earth	17.5	Hydroelectric	Flood Control and Storm Water Management
ME00095	Gardiner Water District Dam	ME	Concrete Timber Crib Masonry	21	Hydroelectric	Flood Control and Storm Water Management
ME00029	Rangeley Lake Dam	ME	Concrete Gravity	17	Hydroelectric	Flood Control and Storm Water Management
ME00058	Lermond Pond Dam	ME	Concrete Earth	12	Hydroelectric	Flood Control and Storm Water Management
ME00062	Great Works Pond Dam	ME	Concrete Masonry Gravity	38	Hydroelectric	Flood Control and Storm Water Management
ME00002	Brunswick Dam/Topsham Dam	ME	Concrete Gravity	45	Hydroelectric	Flood Control and Storm Water Management
MA83013	Collins	MA	Gravity Rockfill Timber Crib	14	Hydroelectric	Flood Control and Storm Water Management
MA00719	Chicopee	MA	Masonry Gravity	10	Hydroelectric	Flood Control and Storm Water Management
MA00501	New Home	MA	Gravity	9	Hydroelectric	Flood Control and Storm Water Management
KY03001	Barkley Dam	KY	Earth Gravity	157	Hydroelectric	Flood Control and Storm Water Management
KY03010	Wolf Creek	KY	Earth Gravity	258	Hydroelectric	Flood Control and Storm Water Management
IL83003	Upper Sterling	IL	Gravity	9	Hydroelectric	Flood Control and Storm Water Management

ID00319	Albeni Falls	ID	Gravity	180	Hydroelectric	Flood Control and Storm Water Management
ID00056	Brownlee	ID	Rockfill Gravity	395	Hydroelectric	Flood Control and Storm Water Management
GA01701	J. Storm Thurmond Dam	GA	Concrete Gravity Earth	200	Hydroelectric	Flood Control and Storm Water Management
GA01702	Hartwell Dam	GA	Concrete Gravity Earth	204	Hydroelectric	Flood Control and Storm Water Management
GA00068	Richard B. Russell Dam	GA	Concrete Gravity Earth	195	Hydroelectric	Flood Control and Storm Water Management
CT01677	Putnam	CT	Other Concrete Gravity	24	Hydroelectric	Flood Control and Storm Water Management
CA01352	Collett Afterbay	CA	Earth	13	Hydroelectric	Flood Control and Storm Water Management
CA00863	New Bullards Bar	CA	Concrete Arch	645	Hydroelectric	Flood Control and Storm Water Management
CA00421	New Drum Afterbay	CA	Arch	95	Hydroelectric	Flood Control and Storm Water Management
CA00200	San Gabriel	CA	Earth Rockfill	381	Hydroelectric	Flood Control and Storm Water Management
CA00095	Upper Gorge Pp	CA	Gravity	57	Hydroelectric	Flood Control and Storm Water Management
AR00150	Blakely Mountain Dam	AR	Earth	240	Hydroelectric	Flood Control and Storm Water Management
AL01420	Lewis Smith	AL	Rockfill	300	Hydroelectric	Flood Control and Storm Water Management
AL01425	Martin	AL	Gravity Earth	175	Hydroelectric	Flood Control and Storm Water Management
AL01413	Point "A"	AL	Gravity Earth	41	Hydroelectric	Flood Control and Storm Water Management
AL01414	Gantt	AL	Buttress Earth	35	Hydroelectric	Flood Control and Storm Water Management
AL01417	Logan Martin	AL	Gravity Earth	142	Hydroelectric	Flood Control and Storm Water Management
WA00004	Chelan	WA	Gravity	40	Hydroelectric	Flood Control and Storm Water Management
WA00084	Rock Island	WA	Gravity	128	Hydroelectric	Flood Control and Storm Water Management
WA00085	Wanapum	WA	Gravity Rockfill Concrete	205.5	Hydroelectric	Flood Control and Storm Water Management
WA00086	Rocky Reach	WA	Gravity	130	Hydroelectric	Flood Control and Storm Water Management
WA00088	Priest Rapids	WA	Gravity Rockfill Concrete	187	Hydroelectric	Flood Control and Storm Water Management
WA00135	Yale Saddle Dam	WA	Earth Arch	40	Hydroelectric	Flood Control and Storm Water Management
WA00147	Swift No. 1	WA	Earth	412	Hydroelectric	Flood Control and Storm Water Management
WA00148	Yale	WA	Earth	323	Hydroelectric	Flood Control and Storm Water Management
WA00151	Mossyrock	WA	Arch Gravity	606	Hydroelectric	Flood Control and Storm Water Management
WI00141	Lower Watertown Dam	WI	Gravity	10	Hydroelectric	Flood Control and Storm Water Management
WI00724	Castle Rock	WI	Gravity Earth Concrete	38	Hydroelectric	Flood Control and Storm Water Management
WI00740	Petenwell	WI	Gravity Earth Concrete	50	Hydroelectric	Flood Control and Storm Water Management
WI00741	Rice	WI	Gravity Earth Concrete	19	Hydroelectric	Flood Control and Storm Water Management
WI00749	Spirit	WI	Gravity Earth Concrete	26	Hydroelectric	Flood Control and Storm Water Management
WI00764	Burnt Rollways	WI	Gravity Earth Concrete	16	Hydroelectric	Flood Control and Storm Water Management
WI00765	Seven Mile	WI	Gravity Earth Concrete	10	Hydroelectric	Flood Control and Storm Water Management

(Continued)

Table 11C.44 (Continued)

NPDP ID	Dam Name	State	Dam Type	Dam Height (ft)	Primary Purpose	Secondary Purpose
WI00766	Sugar Camp	WI	Gravity Earth Concrete	10	Hydroelectric	Flood Control and Storm Water Management
WI00767	Minocqua	WI	Gravity Concrete Earth	9	Hydroelectric	Flood Control and Storm Water Management
WI00772	North Pelican	WI	Gravity Earth Concrete	11	Hydroelectric	Flood Control and Storm Water Management
WI00774	Willow	WI	Gravity Earth Concrete	35	Hydroelectric	Flood Control and Storm Water Management
WI00775	Rainbow	WI	Gravity Earth Concrete	39	Hydroelectric	Flood Control and Storm Water Management
WI00777	Combined Locks	WI	Gravity Concrete	28	Hydroelectric	Flood Control and Storm Water Management
WI00784	Du Bay	WI	Concrete Gravity Earth	40	Hydroelectric	Flood Control and Storm Water Management
WI00804	Buckatahpon	WI	Gravity Earth Concrete	8	Hydroelectric	Flood Control and Storm Water Management
WI00808	Twin Lakes	WI	Gravity Earth Concrete	10	Hydroelectric	Flood Control and Storm Water Management
WI01005	Long-On-Deerskin	WI	Gravity Earth Concrete	10	Hydroelectric	Flood Control and Storm Water Management
WI01017	Little St Germain	WI	Gravity Earth Concrete	9	Hydroelectric	Flood Control and Storm Water Management
WI01138	Lac Vieux Desert	WI	Gravity Earth Concrete	8.5	Hydroelectric	Flood Control and Storm Water Management
WI01146	Big St Germain	WI	Gravity Earth Concrete	7	Hydroelectric	Flood Control and Storm Water Management
WI83003	Squirrel Lake	WI	Gravity Earth Concrete	7	Hydroelectric	Flood Control and Storm Water Management
WI83005	Old Badger	WI	Gravity Concrete	21.5	Hydroelectric	Flood Control and Storm Water Management
WI83006	New Badger	WI	Gravity Concrete	24	Hydroelectric	Flood Control and Storm Water Management
WI83007	Rapide Croche	WI	Gravity Concrete	23.6	Hydroelectric	Flood Control and Storm Water Management
WI83010	Eau Pleine	WI	Gravity Earth Concrete	45	Hydroelectric	Flood Control and Storm Water Management
WI83012	Sawyer Dike	WI	Earth	20	Hydroelectric	Flood Control and Storm Water Management
WI83014	Cth "E" Dike	WI	Earth	24	Hydroelectric	Flood Control and Storm Water Management
WI83015	Jim Hall Dike	WI	Earth	22	Hydroelectric	Flood Control and Storm Water Management
WI83034	Willow Doberstein Dike	WI	Earth	10	Hydroelectric	Flood Control and Storm Water Management

Source: From http://npdp.stanford.edu.

Table 11C.45 Hydro Development in UMRB States 1/1/88 to 9/12/90

	Developed Capacity 1988 (MW)	Developed Capacity 1990 (MW)	Percent Change	Change in Number of Plants
Illinois	29.9	32.6	9.2	+2
Iowa	131.6	131.9	0.2	+1
Minnesota	189.7	209.1	10.2	+3
Missouri	1,064.0	1,062.3	−0.2	+1
Wisconsin	469.2	502.3	7.0	−4
5 states	1,884.5	1,938.2	2.9	+3

Source: From FERC, 1988a, pp. 101, 132–41; FERC, 1990j.

Table 11C.46 Hydro Capacity on the Upper Mississippi River and Illinois Waterway

Mississippi River Site	Owner	Class of Ownership[a]	Developed Capacity (MW)	Potential Capacity (MW)	Status of Potential Capacity[b]
Bemidji	Otter Tail Power Co	P	0.7		
Lake Winnibigoshish	Ball Club Association	R		1.0	
Grand Rapids[c,d]	Blandin Paper Co	I	2.1	3.2	
Brainerd[c,d]	Potlatch Corp. NW Ppr	I	3.3	7.0	
Little Falls 2[c,d]	Minnesota Pwr & Lt Co	P	0.6		
Little Falls 1[c,d]	Minnesota Pwr & Lt Co	P	3.9		
Blanchard[c]	Minnesota Pwr & Lt Co	P	18.0		
Sartell[c]	Champion Intl Corp	I	3.2	6.3	
St. Cloud[c]	St. Cloud, City of	M	8.0		
St. Cloud	St. Cloud Association	R		2.1	
Clearwater				11.2	
Monticello				11.2	
Bailey				11.2	
Coon Rapids	Coon Rapids Hydro Assoc	M		10.5	PO
St. Anthony Falls—Upper[c]	Northern States Power	P	12.4	2.0	
St. Anthony Falls—Lower[c]	Northern States Power Co.	P		16.0	LA
L&D 1[c]	Ford Motor Co.	I	17.9	7.2	
L&D 2[c]	Hastings, City of	I	4.0		
Lake Pepin[e]	Southern Minn Municipal Power Agency	P		500.0	PO
L&D 5	Winona Hydro Partners	R		10.0	
L&D 5A	Mountain City Assoc.	R		6.0	
L&D 6	Trempealeau Assoc.	R		10.8	
L&D 7	Wisc. Public Power Inc.	M	12.7		
L&D 8	Upper Mississippi Hydro Assoc.	R		10.5	PO
L&D 9	United Hydro Partnership	R		10.0	
L&D 10	Guttenburg Partners Ltd.	R		10.0	
L&D 11	Three City Miss. R. Hydro	M		18.4	
L&D 12	Bellevue, City of	M		19.3	
L&D 13	Winnetka, Village of	M		6.8	
L&D 14	LeClaire, City of	M		24.0	MA
L&D 15	Iowa-Illinois G&E Co.	P	3.6	19.7	
L&D 15	Corps of Engineers	F	2.8		
L&D 15	Davenport Hydro Assoc.	R		28.0	MA
L&D 16	Lock 16 Assoc.	R		14.0	
L&D 17	Wapello Assoc.	R		10.5	
L&D 18	Burlington Hydro Assoc.	R		28.0	
L&D 19	Union Electric Co.	P	124.8		
L&D 19	Corps of Engineers	F		103.8	
L&D 20	Canton Assoc.	R		30.0	
L&D 21	Quincy Assoc.	R		10.0	
L&D 22	Hannibal Assoc.	R		10.0	
L&D 24	Clarksville Hydro Assoc.	R		50.0	PO
L&D 25	Winfield Hydro Assoc.	R		50.0	PO

(Continued)

Table 11C.46 (Continued)

Mississippi River Site	Owner	Class of Ownership[a]	Developed Capacity (MW)	Potential Capacity (MW)	Status of Potential Capacity[b]
L&D 26	Missouri Joint Mun. Elec. Uty. Comm.	M		78.0	
L&D 27	Corps of Engineers	F		15.3	
Mississippi Total			205.3	1139.4	

Illinois Waterway Site	Owner	Class of Ownership[a]	Developed Capacity (MW)	Potential Capacity (MW)	Status of Potential Capacity[b]
Calumet/e	Edc Fndtn of Chicago	M		30.0	
Lockport	Sanitary Dist. of Chicago		13.6		
Brandon Road	Rockdale, Village of	M		7.3	MO
Dresden Island	Channahon, Village of	M		7.76	MO
Marseilles	Illinois Power Co.	P	2.0		
Marseilles	Marseilles, City of	M		11.0	PO
Starved Rock	Peru, City of	M		7.6	
Illinois Total			15.6	63.5	
UMRS Total			220.9	1202.9	

[a] F, federally owned; I, industrially owned; M, nonfederal publicly owned; P, private utility owned; R, private non-utility owned.
[b] LA, License amendment pending; PO, Preliminary permit outstanding; MA, Major license applied for (>1.5 MW); MO, Major license outstanding (>1.5 MW).
[c] Existing plant is operating under a FERC license or exemption. Other plants are either federal or are not under FERC's jurisdiction because of their age.
[d] License expires December 31,1993.
[e] Pumped storage facility.

Relicensing: In addition to the development of new projects, decisions regarding the relicensing of existing projects will have an important effect on the character of hydropower in the UMRB. Hydro projects are typically licensed for a period of 30 to 50 years. Many projects under FERC's jurisdiction are due for relicensing in the next few years. Nationwide, the licenses for 227 projects will expire between 1990 and 1999. Twenty-two percent of these projects are in the five UMRB states, which is far greater than the region's proportion of hydro plants. Wisconsin has the highest relicensing workload of any state in the country between 1990 and 1999. (See Appendix C.)

Source: From FERC, 1988a; FERC, 1990j.

Table 11C.47 River Segments Protected from Hydro Development in UMRB States

	State	Exiting Plants	Capacity of Undeveloped Sites (MW)
Wild and Scenic Rivers			
Wolf	WI	0	16.0
Lower St. Croix	MN–WI	1	26.2
Eleven Point	MO	0	14.2
St. Croix	MN–WI	2	47.5
Vermillion, Middle Fork	IL	0	NA
Special Act			
Current and Jacks Fork (Ozark National Scenic Riverway)	MO	0	170.7
			274.6

Source: From FERC, 1988a, pp. 222–23.

Table 11C.48 Existing Plants in Upper Mississippi River Basin States

Project Number[a]	Plant Name	River	Owner	State	County	Project Status[b]	Capacity (kW)	Exp. Date (YYMMDD)
	Waverly (East Hydro)	Cedar R	Waverly, City of	IA	Bremer		495	
	Maquoketa	Maquoketa R	Iowa Elec Lt and Pwr Co.	IA	Jackson		1,200	
	Keokuk L&D 19	Mississippi R	Union Electric Co	IA	Lee		124,800	
	Iowa Falls	Iowa R	Iowa Elec Lt and Pwr Co.	IA	Hardin		540	
925	Ottumwa	Des Moines R	Ottumwa, City of	IA	Wapello	MO	3,000	080430
4344	Five-In-One Dam	Cedar R	Cedar Rapids, City of	IA	Linn	LE	1,600	
8364	Anamusa	Wapsipinicon R	Iowa Elec Lt and Pwr Co.	IA	Jones	LE	288	
	Moline L&D 15	Sylvan Slough, Mississippi R	Iowa IL G & E Co.	IL	Rock Island		3,600	
	Marseilles	Illinois R	Illinois Power Co.	IL	La Salle		2,024	
	Rock Island L&D 15	Sylvan Slough, Mississippi R	Corps of Engineers	IL	Rock Island	FA	2,752	
287	Dayton	Fox R	Hydro-Op One Assoc.	IL	La Salle	MO	3,600	040410
2373	Rockton	Rock R	South Beloit WG & E Co.	IL	Winnebago	NO	1,100	931231
2446	Dixon	Rock R	Commonwealth Edison Co.	IL	Lee	MO	3,200	931231
2866	Lockport	Chicago Sani & Ship Cnl, Des Plaines	Sanitary Dist of Chicago	IL	Will	MO	13,600	011130
2936	Sears	Rock R	White Hydropower Co.	IL	Rock Island	LE	746	
7004	Upper Sterling	Rock R	Rock Falls, City of	IL	Whiteside	LE	2,000	
	Rochester	Zumbro R	Rochester, City of	MN	Wabasha		2,680	
	Lanesboro Hy	S Br Root R	Lanesboro, City of	MN	Fillmore		477	
	Redwood Falls	Redwood R	Redwood Falls, City of	MN	Redwood		500	
	Bemidji	Mississippi R	Otter Tail Power Co.	MN	Beltrami		740	
	Thief River Falls	Red Lake R	Thief River Falls, City of	MN	Pennington		550	
346	Blanchard	Mississippi R	Minnesota Pwr and Lt Co.	MN	Morrison	MO	18,000	030824
362	Twin City L&D 1	Mississippi R	Ford Motor Co.	MN	Hennepin	MO	17,920	030606
469	Winton	Kawishiwi R	Minnesota Pwr and Lt Co.	MN	Lake	MO	4,000	031031

(Continued)

Table 11C.48 (Continued)

Project Number[a]	Plant Name	River	Owner	State	County	Project Status[b]	Capacity (kW)	Exp. Date (YYMMDD)
2056B	Hennepin Island	Mississippi R	Northern States Power Co.	MN	Hennepin	MO	12,400	001231
2056A	Lower Dam	Mississippi R	Northern States Power Co.	MN	Hennepin	MO	8,000	001231
2360C	Scanlon	St. Louis R	Minnesota Pwr and Lt Co.	MN	Carlton	MO	1,600	931231
2360D	Knife Falls	St. Louis R	Minnesota Pwr and Lt Co.	MN	Carlton	MO	2,400	931231
2360B	Thompson	St. Louis R	Minnesota Pwr and Lt Co.	MN	Carlton	MO	68,600	931231
2360A	Fond Du Lac	St. Louis R	Minnesota Pwr and Lt Co.	MN	Carlton	MO	12,000	931231
2361	Prairie River	Prairie R	Minnesota Pwr and Lt Co.	MN	Itasca	NO	1,084	931231
2362	Grand Rapids	Mississippi R	Blandin Paper Co.	MN	Itasca	MO	2,100	931231
2363	Cloquet	St. Louis R	Potlatch Corp.	MN	Carlton	MO	6,514	931231
2454	Sylvan	Crow Wing R	Minnesota Pwr and Lt Co.	MN	Morrison	MO	1,800	931231
2532B	Little Falls 2	Mississippi R	Minnesota Pwr and Lt Co.	MN	Morrison	MO	600	931231
2532A	Little Falls 1	Mississippi R	Minnesota Pwr and Lt Co.	MN	Morrison	MO	3,900	931231
2533	Brainerd	Mississippi R	Potlatch Corp, NW Ppr	MN	Crow Wing	MO	3,342	931231
2663	Pillager	Crow Wing R	Minnesota Pwr and Lt Co.	MN	Morrison	MO	1,600	970511
3071	Rapidan	Blue Earth R	Rapidan Revelopment Ltd.	MN	Blue Earth	LE	5,000	
4108	St. Cloud	Mississippi R	St. Cloud, City of	MN	Stearns	MO	8,000	241130
4306	Mississippi L&D 2	Mississippi R	Hastings, City of	MN	Dakota	MO	4,000	330630
5223	International Falls	Rainy R	International Falls Pwr Co.	MN	Koochiching	MO	10,800	271130
6299	Lake Byllesby	Cannon R	Dakota & Goodhue Counties	MN	Dakota	LE	2,500	
8315	Startell	Mississippi R	Champion Intl Corp.	MN	Benton	MO	3,172	250228
8423	Granite Falls	Minnesota R	Granite Falls, City of	MN	Chippewa	LE	1,367	
10853D	Hoot Lake	Otter Tail R	Otter Tail Power Co.	MN	Otter Tail	MA	1,000	
10853C	Central (Wright)	Otter Tail R	Otter Tail Power Co.	MN	Otter Tail	MA	400	

10853B	Pisgah	Otter Tail R	Otter Tail Power Co.	MN	Otter Tail	MA	520	
10853E	Friberg (Tablin Gorge)	Otter Tail R	Otter Tail Power Co.	MN	Otter Tail	MA	560	
10853A	Dayton Hollow	Otter Tail R	Otter Tail Power Co.	MN	Otter Tail	MA	970	
	Clarence F. Cannon	Salt R	Corps of Engineers	MO	Ralls	FA	27,000	
	Clarence F. Cannon[c]	Salt R	Corps of Engineers	MO	Ralls	FA	31,000	
	Harry S. Truman[c]	Osage R	Corps of Engineers	MO	Benton	FA	160,000[d]	
	Stockton	Sac R.	Corps of Engineers	MO	Cedar	FA	45,200	
	Table Rock	White R	Corps of Engineers	MO	Taney	FA	200,000	
	Riverdale	Finley Cr., James R	Turner, Glenn O. Frances	MO	Christian		112	
459	Osage (Bagnell)	Osage R	Union Electric Co.	MO	Miller	MO	172,000	060228
2221	Ozark Beach	White R	Empire Dist Elec Co.	MO	Taney	MO	16,000	930831
2277	Taum Sauk[c]	E Fk Black R	Union Electric Co.	MO	Reynolds	MO	408,000	100630
2561	Niangua	Niangua R	Sho Me Power Corp.	MO	Camden	MO	3,000	931231
	Figor	Duck Cr, Tagatz Cr	Figor, D.J.	WI	Marquette		38	
	Lawrence	Duck Cr., Tagatz Cr	Pioneer Pwr & Lt Co.	WI	Marquette		200	
	Montello	Montello Lk, Montello R	Montello Granite Co.	WI	Marquette		379	
	Harrisville	Montello R	Miller, Duane J.	WI	Marquette		187	
	Wautoma	Wautoma R	North Amer Hydro Inc.	WI	Waushara		252	
	Pardeeville	Fox R	Pardeeville Elec Comm.	WI	Columbia		50	
	Appleton	Fox R	Wisconsin Elec Pwr Co.	WI	Outagamie		1,940	
	Appleton (Lower)	Fox R	Consolidated Papers Inc.	WI	Outagamie		1,445	
	Neenah Dam	Fox R	Bergstrom Paper Co.	WI	Winnebago		400	
	Nepco Lake	Four Mile Cr., Wisconsin R	Nekoosa Papers Inc.	WI	Wood		2,920	
	Merillan	Halls Cr., Black R	Merillan, Village of	WI	Jackson		100	
	Kilbourn	Wisconsin R	Wisconsin Power & Light Co.	WI	Sauk		8,200	

(Continued)

Table 11C.48 (Continued)

Project Number[a]	Plant Name	River	Owner	State	County	Project Status[b]	Capacity (kW)	Exp. Date (YYMMDD)
	Barron	Yellow R	Barron, City of	WI	Barron		100	
	Merrill	Prairie R	Ward Paper Company	WI	Lincoln		135	
	La Valle	Baraboo R	La Valle El Co.	WI	Sauk		50	
	Pine River	Pine R	Miller, Clair	WI	Richland		100	
	Island Woolen	Baraboo R	Island Woolen Co.	WI	Sauk		412	
	Radisson	Couderay R	North Central Power Co.	WI	Sawyer		393	
	Linen Mill	Baraboo R	McArthur, George and Son	WI	Sauk		100	
	Oak Street	Baraboo R	McArthur, George and Son	WI	Sauk		156	
	Prairie Du Sac	Wisconsin R	Wisconsin Power & Light Co.	WI	Sauk		28,440	
	Nancy	Totagatic R	Dahlberg Lt and Pwr Co.	WI	Washburn		450	
	Clam Falls	Clam R	Northwestern Wis Elec Co.	WI	Burnett		112	
	St. Croix Falls	St. Croix R	Northern States Power Co.	WI	Polk		23,200	
	Gordon	Eau Claire R	Dahlberg Lt and Pwr Co.	WI	Douglas		257	
	Balsam Lake	Balsam R, Apple R	Northerwestern Wis Elec Co.	WI	Polk		68	
	Barton (Gadow Milling)	Milwaukee R	Roy Kleisch	WI	Washington		17	
	Muscoda	Mill Cr, Wisconsin R	Muscoda, Village of	WI	Richland		60	
710	Shawano (Upper)	Wolf R	Wisconsin Power & Light Co.	WI	Shawano	MO	700	770719
1510	Kaukauna (Lower)	Fox R	Kaukauna, City of	WI	Outagamie	MO	4,800	190331
1940	Tomahawk	Wisconsin R	Wisconsin Public Service Corp	WI	Lincoln	MO	2,600	161231
1953	Du Bay	Wisconsin R	Consolidated Water Pwr Co.	WI	Portage	MO	7,200	910630
1957	Otter Rapids	Wisconsin R	Wisconsin Public Service Corp	WI	Vilas	NO	500	300630
1960	Flambeau	Flambeau R	Dairyland Power Coop	WI	Rusk	MO	15,000	010228
1966	Grandfather Falls	Wisconsin R	Wisconsin Public Service Corp	WI	Lincoln	MO	17,240	180331
1968	Hat Rapids	Wisconsin R	Wisconsin Public Service Corp	WI	Oneida	LE	1,700	

1979	Alexander	Wisconsin R.	Wisconsin Public Service Corp	WI	Lincoln	MO	4,200	040630
1981	Stiles	Oconto R.	Oconto Elec Coop	WI	Oconto	MO	1,000	000229
1982	Holcombe	Chippewa R.	Northern States Power Co.	WI	Chippewa	MO	33,900	980630
1984B	Petenwell	Wisconsin R.	Wisconsin River Power Co.	WI	Adams	MO	20,000	980131
1984A	Castle Rock	Wisconsin R.	Wisconsin River Power Co.	WI	Adams	MO	15,000	980131
1989	Merrill	Wisconsin R.	Wisconsin Public Service Corp	WI	Lincoln	LE	2,340	
1999	Wausau	Wisconsin R.	Wisconsin Public Service Corp	WI	Marathon	MO	5,400	950630
2064	Last Fork (Winter)	E Fk Chippewa R.	North Central Power Co.	WI	Sawyer	MO	600	011130
2110	Stevens Point	Wisconsin R.	Consolidated Water Pwr Co.	WI	Portage	MO	4,800	000630
2161	Rhinelander	Wisconsin R.	Rhinelander Paper Co.	WI	Oneida	MO	2,120	000630
2180	Grandmother	Wisconsin R.	Nekoosa Corp.	WI	Lincoln	MO	3,000	030630
2181	Menomonie	Red Cedar R.	Northern States Power Co	WI	Dunn	MO	5,400	050331
2192	Biron 2	Wisconsin R.	Consolidated Water Pwr Co.	WI	Wood	MO	3,800	000630
2207	Mosinee	Wisconsin R.	Mosinee Paper Mills Co.	WI	Marathon	MO	3,050	041231
2212	Rothschild	Wisconsin R.	Weyerhaeuser Co.	WI	Marathon	MO	3,640	930731
2239	Kings	Wisconsin R.	Tomahawk Pwer & Pulp Co.	WI	Lincoln	MO	320	930731
2255	Centralia	Wisconsin R.	Nekoosa Papers Inc.	WI	Wood	MO	3,200	930731
2256	Wis Rpds 1	Wisconsin R.	Consolidated Water Pwr Co.	WI	Wood	MO	4,400	930731
2291	Port Edwards	Wisconsin R.	Nekoosa Papers Inc.	WI	Wood	MO	2,920	930731
2292	Nekoosa	Wisconsin R.	Nekoosa Papers Inc.	WI	Wood	MO	3,800	930731
2347	Central (Janesville)	Rock R.	Wisconsin Power & Light Co.	WI	Rock	NO	500	931231
2348	Blackhawk	Rock R.	Wisconsin Power & Light Co.	WI	Rock	NO	380	931231
2390	Big Falls	Flambeau R.	Northern States Power Co.	WI	Rusk	MO	9,000	931231
2395	Pixley	N Fk Flambeau R.	Flambeau Paper Corp.	WI	Price	NO	960	931231

(Continued)

Table 11C.48 (Continued)

Project Number[a]	Plant Name	River	Owner	State	County	Project Status[b]	Capacity (kW)	Exp. Date (YYMMDD)
2417	Hayward	Namekagon R.	Northern States Power Co.	WI	Sawyer	NO	200	931231
2421	Lower Hydro	N Fk Flambeau R.	Flambeau Paper Corp.	WI	Price	NO	1,200	931231
2430	Ladysmith	Flambeau R.	Lake Superior Dist Pwr Co.	WI	Rusk	LE	3,900	
2440	Chippewa Falls	Chippewa R.	Northern States Power Co.	WI	Chippewa	MO	21,600	931231
2444	White River	White R.	Northern States Power Co.	WI	Ashland	NO	1,000	931231
2464	Weed Dam	Red R.	Gresham, Village of	WI	Shawano	NO	700	150630
2473	Crowley Rapids	N Fk Flambeau R.	Flambeau Paper Corp.	WI	Price	NO	1,500	931231
2475	Thornapple	Flambeau R.	Northern States Power Co.	WI	Rusk	NO	1,400	931231
2476	Jersey	Tomahawk R.	Wisconsin Public Service Corp	WI	Lincoln	NO	512	931231
2484	Gresham	Red R.	Gresham, Village of	WI	Shawano	NO	423	191231
2486	Pine	Pine R.	Wisconsin Elec Pwr Co.	WI	Florence	MO	3,600	931231
2491	Jim Falls	Chippewa R.	Northern States Power Co.	WI	Chippewa	MO	48,600	330930
2522	Johnson Falls	Peshtigo R.	Wisconsin Public Service Corp	WI	Marinette	MO	3,520	931231
2523	Oconto Falls (Upper)	Oconto R.	Wisconsin Elec Pwr Co.	WI	Oconto	NO	1,320	931231
2525	Caldron Falls	Peshtigo R.	Wisconsin Public Service Corp	WI	Marinette	MO	6,400	931231
2536	Little Quinnesec Falls	Menominee R.	Niagara of Wis Paper Corp.	WI	Marinette	MO	8,388	930630
2546	Sandstone Rapids	Peshtigo R.	Wisconsin Public Service Corp	WI	Marinette	MO	3,840	931231
2550	Weyauwega	Waupaca R.	Wisconsin Elec Pwr Co.	WI	Waupaca	NO	400	931231
2560	Potato Rapids	Peshtigo R.	Wisconsin Public Service Corp	WI	Marinette	NO	1,380	931231
2567	Wissota	Chippewa R.	Northern States Power Co.	WI	Chippewa	MO	35,280	000630
2581	Peshtigo	Peshtigo R.	Wisconsin Public Service Corp	WI	Marinette	NO	584	931231
2588	Little Chute	Fox R.	Kaukauna, City of	WI	Outagamie	MO	3,300	000731

2590	Wisconsin R. Div.	Wisconsin R.	Consolidated Water Pwr Co.	WI	Portage	MO	1,800	930630
2595	High Falls	Peshtigo R.	Wisconsin Public Service Corp	WI	Marinette	MO	7,000	931231
2639	Cornell	Chippewa R.	Northern States Power Co.	WI	Chippewa	MO	30,800	231130
2640	Upper Hydro	N Fk Flambeau R.	Flambeau Paper Corp.	WI	Price	NO	900	931231
2670	Dells	Chippewa R.	Northern States Power Co et al	WI	Eau Claire	MO	9,500	000831
2677B	Badger (Old)	Fox R.	Kaukauna, City of	WI	Outagamie	MO	2,000	191231
2677A	Rapid Croche	Fox R.	Kaukauna, City of	WI	Outagamie	MO	2,400	191231
2677C	Badger (New)	Fox R.	Kaukauna, City of	WI	Outagamie	MO	3,600	191231
2684	Arpin	E Fk Chippewa R.	North Central Power Co.	WI	Sawyer	NO	1,450	190430
2689	Oconto Falls	Oconto R.	Scott Paper Co.	WI	Oconto	MO	1,810	931231
2697	Cedar Falls	Red Cedar R.	Northern States Power Co.	WI	Dunn	MO	6,000	010130
2711	Trego	Namekagon R.	Northern States Power Co.	WI	Washburn	NO	1,200	930331
2715	Combined Locks	Fox R.	Kaukauna, City of	WI	Outagamie	MO	7,000	240729
2744B	Park Mill	Menominee R.	Menominee Co.	WI	Marinette	MO	1,724	150228
2894	Black Brook	Apple R.	Northwestern Wis Elec Co.	WI	Polk	NO	650	201231
2970	Argyle	E Pecatonica R.	Argyle, Village of	WI	Lafayette	LE	50	
3052	Black River Falls	Black R.	Black River Falls, City of	WI	Jackson	NO	900	010830
4914	West De Pere	Fox R.	International Paper Co.	WI	Brown	NO	1,078	041130
6476	Neshonoc	La Crosse R.	North Amer Hydro Inc.	WI	La Crosse	LE	500	
7264	Appleton (Middle)	Fox R.	Fox Valley Corp. et al.	WI	Outagamie	NO	1,290	050630
8015	Shawano	Wolf R.	Little Rapids Corp.	WI	Shawano	LE	380	
8286	Chippewa	Chippewa R.	Northern States Power Co. et al.	WI	Sawyer	LE	3,100	
9002	Apple River	Apple R.	Northern States Power Co.	WI	St. Croix	LE	3,700	
9003	Riverdale	Apple R.	Northern States Power Co.	WI	St.Croix	LE	620	
9184	Danbury	Yellow R.	Northwestern Wis Elec Co.	WI	Burnett	NO	1,076	070531
9185	Clam River	Clam R.	Northwestern Wis Elec Co.	WI	Burnett	NO	1,200	070331
10489A	Junction	Kinnickinnic R.	River Falls, City of	WI	Pierce	NO	250	180831
10489B	River Falls	Kinnickinnic R.	River Falls, City of	WI	Pierce	NO	125	180831

(Continued)

Table 11C.48 (Continued)

Project Number[a]	Plant Name	River	Owner	State	County	Project Status[b]	Capacity (kW)	Exp. Date (YYMMDD)
10674	Midtec	Fox R.	Midtec Paper Corp.	WI	Outagamie	MA	2,700	
10805	Hatfield	Black R.	Midwest Hydraulic Co.	WI	Jackson	PO	4,800	920931

[a] Plants without project numbers are not under FERC jurisdiction and thus have no entries under the project status or license expiration columns.
[b] Project Status: FA, Federally authorized; LE, Exempted from licensing; MA, Major license applied for ($>$1.5 MW); MO, Major license outstanding or annual license during major relicensing proceeding ($>$1.5 MW); NO, Minor license outstanding or annual license during minor relicensing proceeding ($\leq$1.5 MW); PO, Preliminary Permit Outstanding.
[c] Pumped storage.
[d] Reversible capacity could be used for conventional generation.

Source: From FERC, 1990j.

Table 11C.49 Small Hydroelectric Capacity in the United States

	5,000 kW and Less		15,000 kW and Less		30,000 kW and Less	
Category	**1980**	**1984**	**1980**	**1984**	**1980**	**1984**
Developed						
Number of sites	751	864	946	1,124	1,071	1,252
Capacity (MW)	1,194.6	1,351.1	3,294.7	3,729.7	5,834.6	6,574.0
Generation (GWh)	4.8	16.8	16.8	18.6	28.0	30.8
Under Construction						
Number of sites	16	137	29	153	33	158
Capacity (MW)	23.5	168.6	135.6	325.8	229.9	444.8
Generation (GWh)	0.1	0.7	1.0	1.4	1.3	2.0
Planned (NERC)[a]						
Number of sites	12	25	23	39	25	50
Capacity (MW)	34.2	54.3	136.2	191.4	182.4	436.8
Generation (GWh)	0.1	0.3	0.5	0.9	0.6	1.9
Projected[b]						
Number of sites	157	1,699	227	2,278	279	2,367
Capacity (MW)	317.5	2,587.1	1,241.8	6,669.4	2,317.5	8,572.4
Generation (GWh)	1.2	20.0	4.9	29.4	8.9	37.3
Totals						
Number of sites	936	2,725	1,225	3,594	1,408	3,827
Capacity (MW)	1,569.8	4,160.8	4,808.3	10,916.3	8,564.4	16,028.0
Generation (GWh)	6.2	27.6	23.2	50.3	38.8	72.0

Note: Developed, Under Construction, and Projected.

[a] In reports of the Regional Electric Reliability Councils.

[b] Potential developments not under construction or included in NERC reports but which have FERC licensing or exemption status, are authorized or recommended for Federal construction, or have structural provisions for plant additions.

Source: From Federal Energy Regulatory Commission, 1984.

Table 11C.50 Trends in Pumped Storage Capacity Development in the United States

	Installed Capacity in Reversible Units (Millions kW)					
	Developed			Under Construction		
Year as of January 1	**Pure**	**Combined**	**Total**	**Pure**	**Combined**	**Total**
1960	0	0.1	0.1	0	0.2	0.2
1964	0.4	0.3	0.7	0.7	0.5	1.2
1968	1.6	0.5	2.1	1.2	1.6	2.8
1972	2.6	1.3	3.9	6.0	1.4	7.4
1976	7.3	2.4	9.7	2.7	1.6	4.3
1980	9.3	3.6	12.9	3.2	1.5	4.7
1984	10.1	3.7	13.8	4.9	0.4	5.3

Note: A pure pumped storage project with a large peaking capacity can be developed at a site with two potential reservoirs of reasonable size in close proximity and with a relatively large difference in elevations. Projects are usually more economically developed at sites with high usable heads; consequently, the more favorable sites are normally located in mountainous terrain. However, consideration has been given to the construction of pumped storage projects in areas of level terrain by placing the lower reservoir in an underground cavern or excavated area. For any development, an assured supply of water at least sufficient to replace evaporation, seepage, and other losses is essential.

Source: From Federal Energy Regulatory Commission, 1984.

Table 11C.51 Hydroelectric Plants Having Potential Conventional Capacity over 1,000,000 kW

			Installed Capacity in Conventional Units Kilowatts		
Plant	**River**	**Owner[a]**	**Developed**	**Under Construction**	**Ultimate Authorized**
Grand Coulee	Columbia	Bureau	6,180,000	0	6,180,000
John Day	Columbia	COE	2,160,000	0	2,700,000
Chief Joseph	Columbia	COE	2,069,000	0	2,069,000
R. Moses Niagara	Niagara	PASNY	1,950,000	0	1,950,000
The Dalles	Columbia	COE	1,806,800	0	1,806,800
Hoover	Colorado	Bureau	1,434,000	0	1,434,000
Rocky Reach	Columbia	CC PUD No. 1	1,213,950	0	1,213,950
Wanapum	Columbia	GC PUD No. 2	831,250	0	1,151,250
Priest Rapids	Columbia	GC PUD No. 2	788,500	0	1,108,500
Bonneville	Columbia	COE	1,076,620	0	1,102,820
Dworshak	N. Fork Clearwater	COE	400,000	0	1,060,000
Glen Canyon	Colorado	Bureau	1,042,000	0	1,042,000
Boundary	Pend Oreille	Seattle	634,600	392,000	1,026,600
	Total		21,586,720	392,000	23,834,920

[a] Bureau, Bureau of Reclamation; COE, Corps of Engineers; PASNY, Power Authority, State of New York; GC, Grant County; CC, Chelan County; and Seattle, Seattle Dept. of Lighting.

Source: From Federal Energy Regulatory Commission, 1984.

Table 11C.52 Federal Hydroelectric Capacity by Operating Agency, January 1, 1984

	Installed Capacity Conventional and Reversible (kW)		
	In Operation	**Under Construction**	**Ultimate Authorized**
Corps of Engineers	19,449,197	1,162,000	23,466,897
Bureau of Reclamation	12,846,876	250,000	13,577,986
Tennessee Valley Authority	4,832,410	—	4,832,410
International Boundary and Water Commission	97,500	—	97,500
Alaska Power Administration	77,160	—	104,160
Bureau of Indian Affairs	14,560	—	14,560
National Park Service	2,000	—	2,000
Total	37,319,703	1,412,000	42,095,513

Source: From Federal Energy Regulatory Commission, 1984.

SECTION 11D COSTS OF WATER PROJECTS

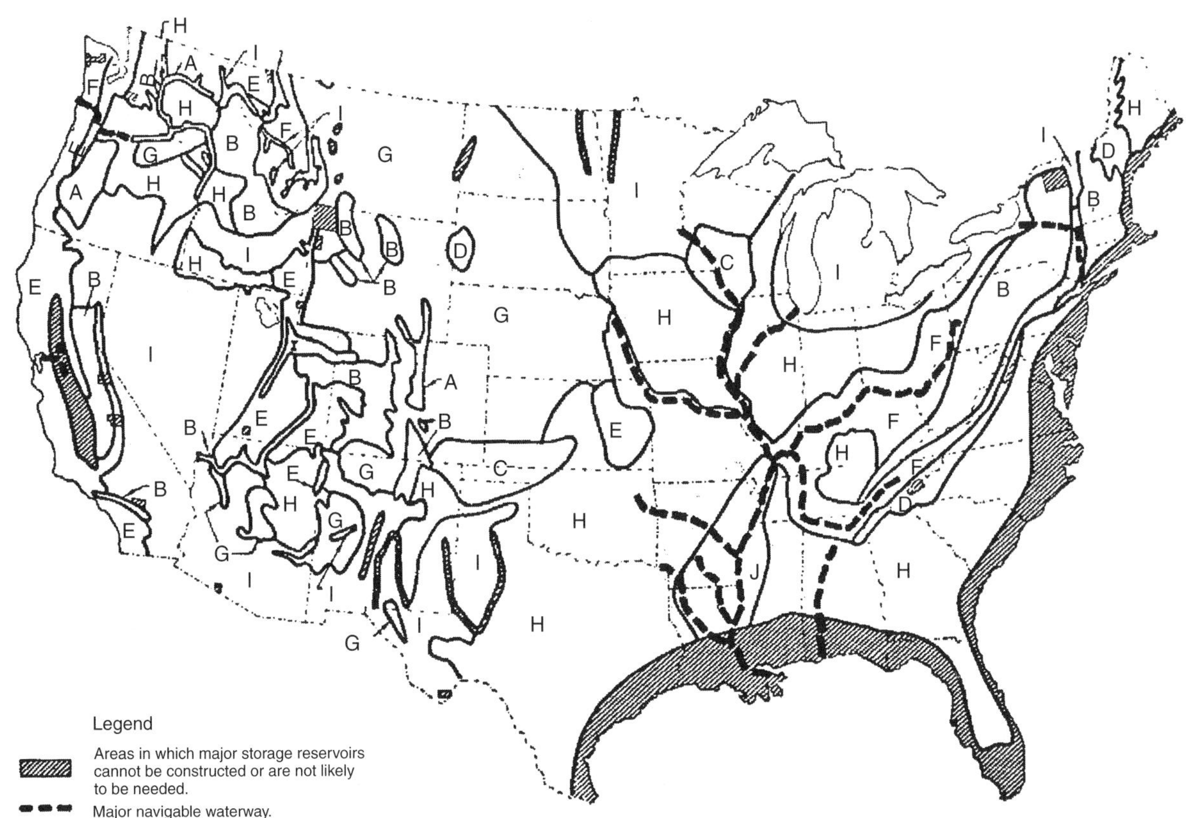

Figure 11D.40 Map of physiographic regions in the United States (for reservoir costs). (From Corps of Engineers, U.S. Army, 1960.)

Table 11D.53 Costs of Reservoirs in the Physiographic Regions of the United States

Physiographic Region	Size Class										
	10	30	50	80	150	300	700	1500	3000	7000	30,000
A	$230	$190	$175	$162	$145	$130	$113	$100	$88	$77	$60
B	200	160	145	132	115	100	85	75	64	55	40
C	180	135	120	110	95	80	66	56	49	40	28
D	160	120	105	93	78	65	52	42	35	28	18
E	155	115	98	85	73	60	47	38	32	25	16
F	145	106	90	80	65	55	43	34	28	21	14
G	140	103	87	75	60	50	38	30	25	19	12
H	120	86	73	62	50	40	30	24	18	14	10
I	95	65	55	46	38	30	22	18	15	10	8
J	65	43	37	32	25	20	15	12	10	8	6

Note: Values are costs in dollars of reservoir storage per acre-feet. Size class in thousand acre-feet.

Source: From Corps of Engineers, U.S. Army, 1960.

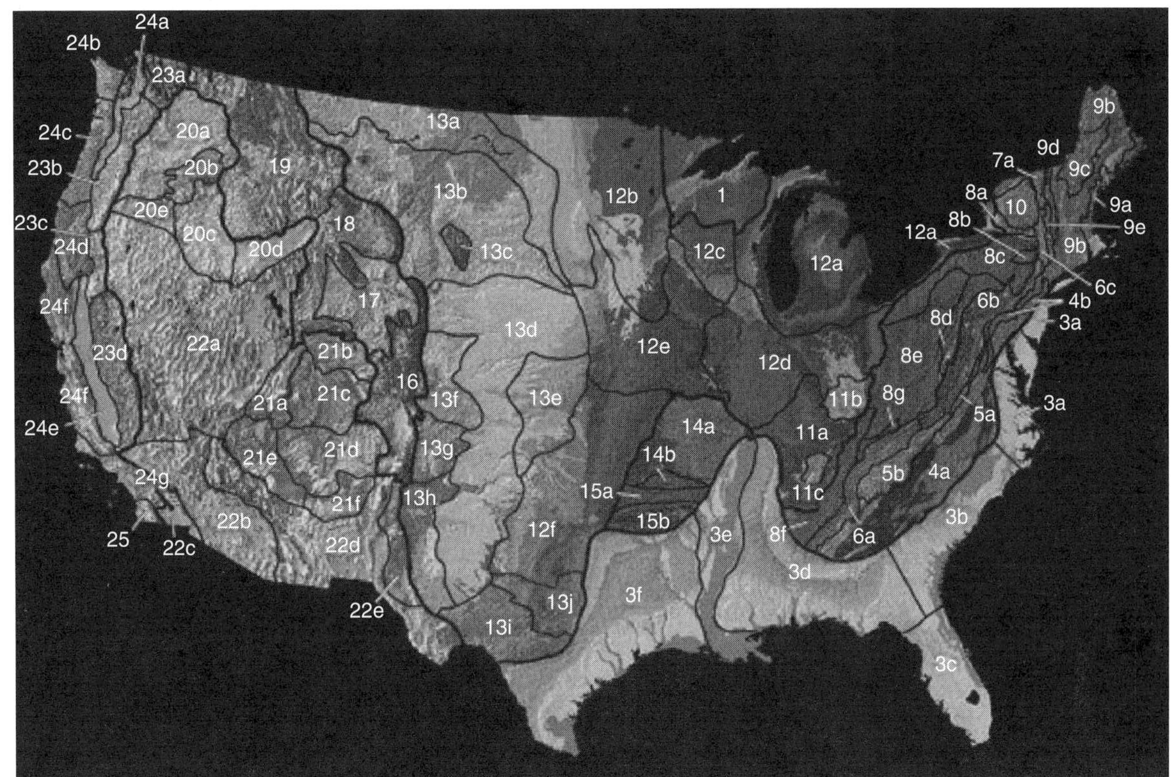

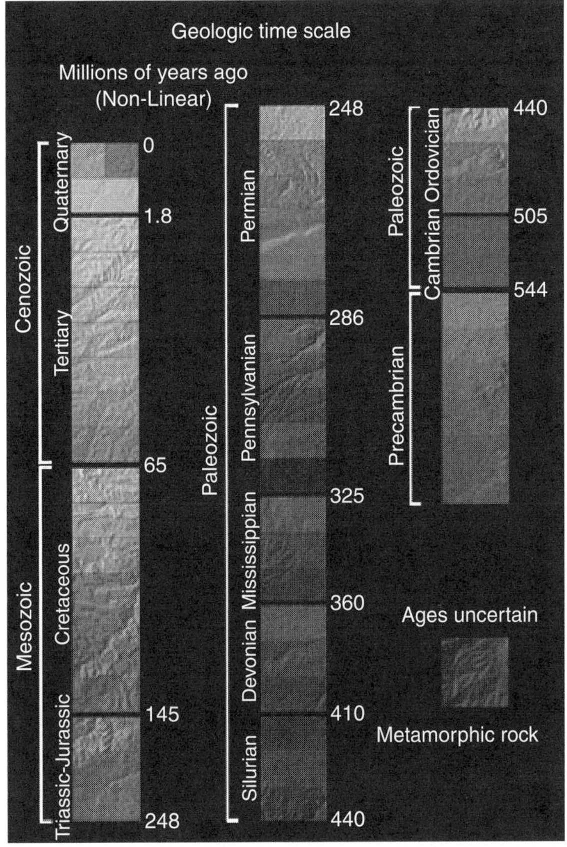

Figure 11D.41 U.S. physiographic region. (From www.tapestry.usgs.gov.)

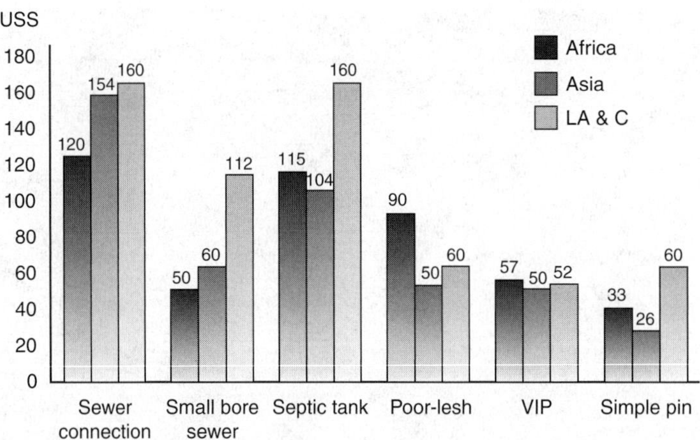

Figure 11D.42 Average construction cost of sanitation facilities for Africa, Asia, and Latin America and the Caribbean, 1990–2000. (From www.who.int.)

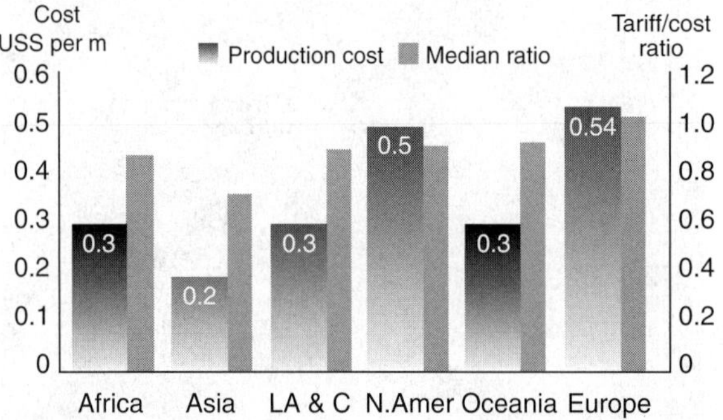

Figure 11D.43 A comparison of the median unit production cost of urban water supply and the median tariff/production cost ratio by region 1990–2000. (From www.who.int.)

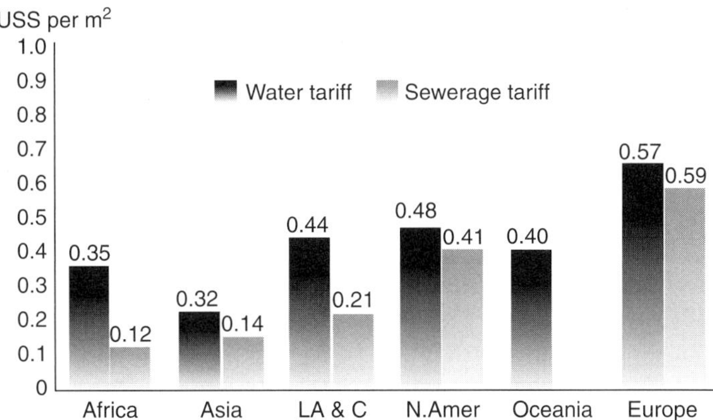

Figure 11D.44 Median water supply and sewerage tariffs by region, 1990–2000. Average construction cost per person served of water supply facilities for Africa, Asia and Latin America and the Caribbean 1990–2000. (From www.who.int.)

Table 11D.54 Typical Costs of Irrigation Development in Latin America, Africa, Asia, and Far East

Region	$/ha	Gravity Schemes Share in Total (%)	Pump and Tubewells ($/ha)	All Schemes Weighted Average Cost ($/ha)
Latin America	6,000	50	3000	4000
Africa (excl. Sudan)	11,000	70	6000	9500
Sudan	5,000	50	4000	4500
Near East	7,000	70	4000	6100
Asia and Far East (excl. South Asia)	4,000	60	2000	3200
South Asia (Bangladesh, India, Pakistan)	2,500	40	1000	1600
Rehabilitation in all regions except South Asia, $1760/ha				
Rehabilitation in South Asia, $800/ha				

Note: 1980 prices; US $ per hectare.

Source: From FA0, 1982, The State of Food and Agriculture.

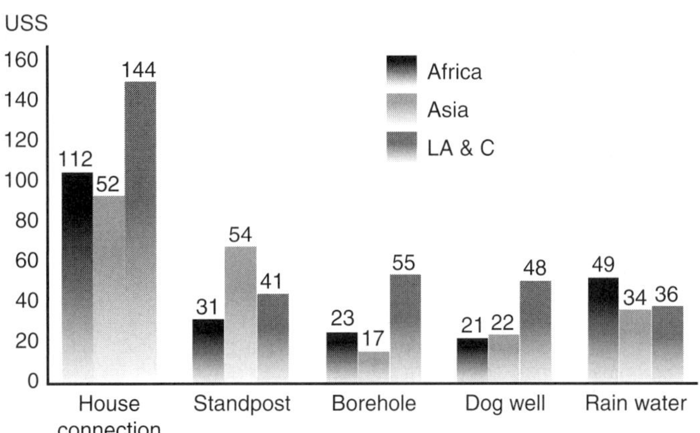

Figure 11D.45 Average construction cost per person served of water supply facilities for Africa, Asia, and Latin America and the Caribbean, 1990–2000. (From www.who.int.)

Table 11D.55 Typical Cost Structures for Water-Supply Systems in the United States

	Small Systems[a]		Large Systems[b]	
	Percentage of Operating Expenses	Percentage of Operating Expenses Excluding Interest Charges	Percentage of Operating Expenses	Percentage of Operating Expenses Excluding Interest Charges
Acquisition	19	22	15	19
Treatment	15	18	10	13
Distribution and Transmission	36	43	31	38
Support Services	14	17	25	30
Interest Charges	16	—	19	—
Total	100	100	100	100

[a] Serving between 300 and 75,000 people.
[b] Serving over 75,000 people.

Source: From U.S. Environmental Protection Agency, 1986, *Guidelines for Ground Water Classification under the EPA Ground Water Protection Strategy*; ACT Systems Inc., 1977, 1979.

Table 11D.56 Typical Water System Costs in the United States (1984 $/million of gallons produced)

	Source	
Population Served by System	Surface Water	Groundwater
1000–3300	1,085	1,493
3300–10,000	1,063	924
10,000–25,000	795	718
25,000–75,000	727	710
75,000–500,000	596	606
Over 5,00,000	457	574

Note: Operating expenses (including depreciation and capital charges), inflated to 1984 dollars.

Source: From U.S. Environmental Protection Agency, 1986, *Guidelines for Ground Water Classification under the EPA Ground Water Protection Strategy*; Survey of Operating and Financial Characteristics of Community Water Systems, Temple, Barker and Sloane, Inc. 1982.

Table 11D.57 Costs of Groundwater Supply Systems in the United States

Population Served by System	Annual Cost
25–1000	4,616
1000–3300	1,493
3300–10,000	924
10,000–25,000	718
25,000–75,000	710
75,000–5,00,000	606
Over 5,00,000	574

Note: By population size category; 1984 $/million gallons produced. Operating expenses (including depreciation and capital charges, inflated to 1984 dollars).

Source: From U.S. Environmental Protection Agency, 1986, *Guidelines for Ground Water Classification under the EPA Ground Water Protection Strategy*; Survey of Operating and Financial Characteristics of Community Water Systems, Temple, Barker and Sloane, Inc. 1982.

Table 11D.58 Costs of Community Water Supply Technology in Developing Countries

	Low			High		
Technology	**Handpumps**	**Standpipes**	**Yardtaps**	**Handpumps**	**Standpipes**	**Yardtaps**
Capital cost (US$)						
Wells[a]	4000	2,000	2,500	10,000	5,000	6,000
Pumps (hand/motor)	1300	4,000	4,500	2,500	8,000	9,000
Distribution[b]	None	4,500	16,000	None	10,000	30,000
Sub-total	5300	10,500	23,000	12,500	23,000	45,000
Cost per capita	13.3	26.3	57.5	31.2	57.5	112.5
Annual cost (US$/year)						
Annualized capital[c]	700	1,500	3,200	1,400	3,000	6,000
Maintenance	200	600	1,000	400	1,200	2,000
Operation (fuel)	None	150	450	None	300	900
Sub-total (cash)	900	2,250	4,650	1,800	4,700	8,900
Haul costs (labor)[d]	1400	1,100	None	3,000	2,200	None
Total (including labor)	2300	3,350	4,650	4,800	6,900	8,900
Total annualized cost per capita						
Cash only	2.3	5.6	11.6	4.5	11.8	22.3
Cash + labor	5.8	8.4	11.6	12.0	17.3	22.3

Note: Capital and recurrent costs for a community of 400 people.

[a] Pumping water level assumed to be 20 m. Two wells assumed for handpump system (200 persons per handpump).

[b] Distribution system includes storage, piping, and taps with soakaway pits.

[c] Capital costs with replacement of mechanical equipment after 10 years annualized at a discount rate of 10% over 20 years.

[d] Labour costs for walking to the water point, queuing, filling the container, and carrying the water back to the house. Time valued at US $ 125/h.

Source: From Arlosoroff, Saul, and others, 1987, *Community Water Supply: The Handpump Option*, The World Bank, Washington, DC.

Table 11D.59 Projection of Irrigation Expansion in Developing Countries and Related Costs (1993–2000)

	Total Irrigated Land-1990 (millions of hectacres)	Projected Increases	Unit Cost (US dollars)	Total Costs (billions of dollars)
Asia (30 countries)	132.11	2.02	400	2.88
Near East (10 countries)	9.50	0.45	800	2.32
Latin America (40 countries)	16.31	0.84	000	7.20
Africa (50 countries)	14.21	0.07	200	7.20
Total (130 countries)	172.11	5.24		5.52

Source: From www.munfw.org.

Table 11D.60 Estimated Targets and Costs for the Modernization of Existing Irrigation Schemes (1990–2000)

	Total Irrigated Area-1990 (millions of hectacres)	Total Area Upgraded	Unit Cost (US dollars)	Total Costs (US dollars)
Asia	132.11	3.21	600	
Near East	95.9	5.1	450	1.4
Latin America	16.3	1.63	1000	1.6
Africa	14.2	1.42	1800	2.52
Developing Countries—130	172.11	7.2	113.1	4

Source: From www.munfw.org.

Table 11D.61 Options for Community Water Supply: A Comparative Study

Step	Type Of Service	Water Source	Quality Protection	Water Use LPC[a]	Energy Source	Operation and Maintenance Needs	Costs	General Remarks
5	House connections	Groundwater Surface water Spring	Good, no treatment May need treatment Good, no treatment	100 to 150	Gravity, Electric, Diesel	Well trained operator; Reliable fuel and chemical supplies; many spare parts: wastewater disposal	High capital and O&M costs, except for gravity schemes	Most desirable service level, but high resources needs
4	Yardtaps	Groundwater Surface water Spring	Good, no treatment May need treatment Good, no treatment	50 to 100	Gravity, Electric, Diesel	Well trained operator; reliable fuel and chemical supplies: many spare parts	High capital and O&M costs, except gravity schemes	Very good access to safe water: fuel and institutional support critical
3	Stand pipes	Groundwater Surface water Spring	Good, no treatment May need treatment Good, no treatment	10 to 40	Gravity, Electric, Diesel, Wind, Solar	Well trained operator; reliable fuel and chemical supplies: many spare parts	Moderate capital and O&M costs, except gravity schemes; collection time	Good access to safe water: cost competitive with handpumps at high pumping lifts
2	Handpumps	Groundwater	Good, no treatment	10 to 40	Manual	Trained repairer: few spare parts	Low capital and O&M costs; collection time	Good access to safe water: sustainable by villagers
1	Improved traditional sources (partially protected)	Groundwater Surface water Spring Rainwater	Variable Poor Variable Good, If protected	10 to 40	Manual	General upkeep	Very low capital and O&M costs: collection time	Improvement if traditional source was badly contaminated
0	Traditional Sources (unprotected)	Surface water Groundwater Spring Rainwater	Poor Poor	10 to 40	Manual	General upkeep	Low O&M costs (buckets, etc.) collection time	Starting point for supply improvement

[a] Liters per capita per day.

Source: From www.skipumps.com.

SECTION 11E PROJECT PLANNING AND ANALYSIS

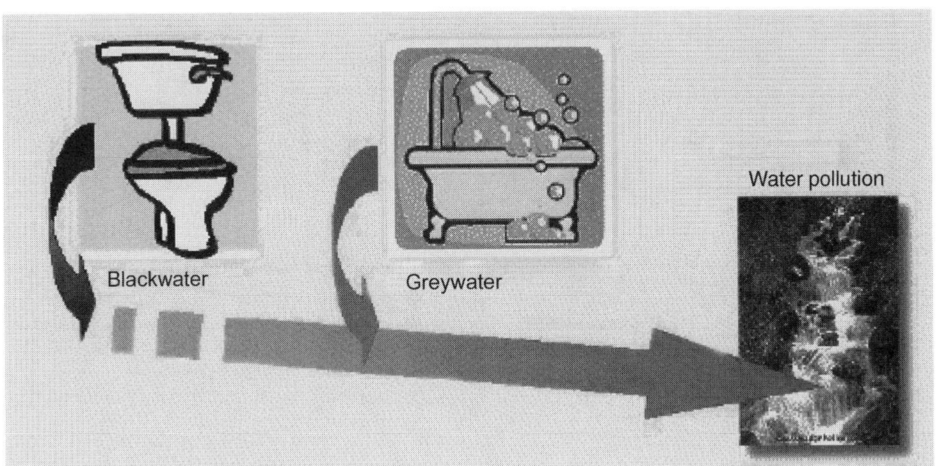

Figure 11E.46 Linear flows. (From The African Water Page, adapted from Esrey et al. 2001, www.thewaterpage.com. With permission.)

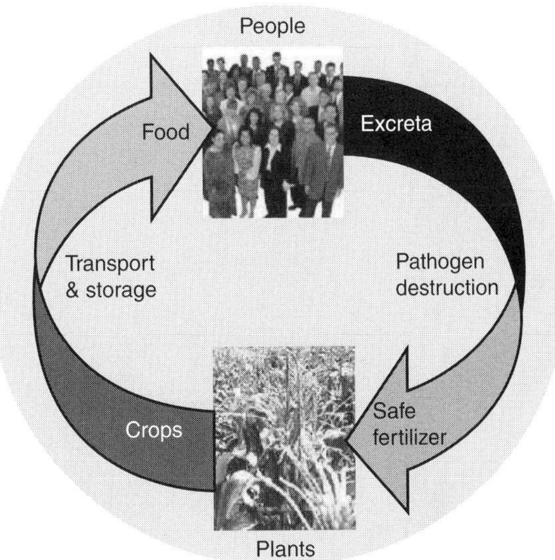

Figure 11E.47 Circular flows. (From The African Water Page, www.thewaterpage.com. With permission.)

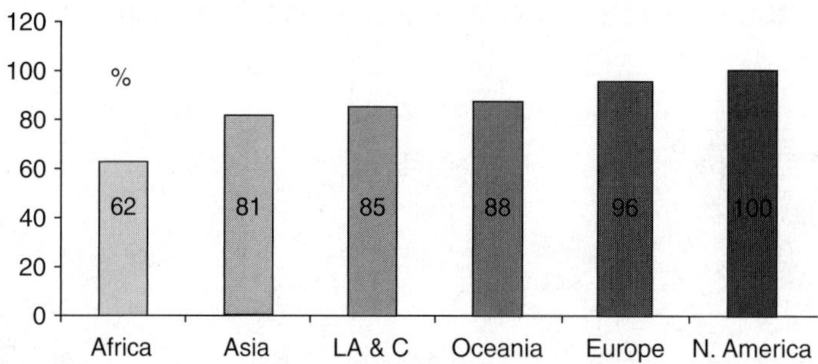

Figure 11E.48 Proportion (%) of population with water supply services. (From The African Water Page, www.thewaterpage.com. WHO/UNICEF, 2000.)

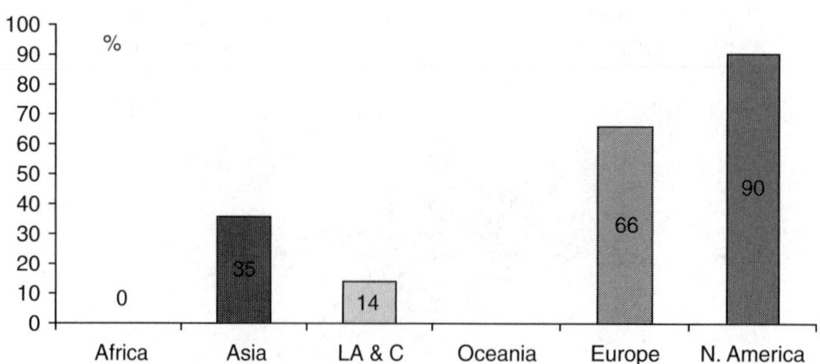

Figure 11E.49 Lack of treatment of waste water. (From The African Water Page, www.thewaterpage.com. WHO/UNICEF, 2000.)

Table 11E.62 Elements Comprising a Water Resources Project Investigation

A. Purposes of a multiple-purpose project
1. Irrigation
2. Drainage
3. Domestic or industrial water supply
4. Flood control
5. Hydro-power generation
6. Navigation
7. Fish and wildlife conservation
8. Recreation
9. Water quality control
10. Salinity control
11. Watershed management
12. National defense
13. International relation

B. Land resources
1. Land classification
2. Land use and capabilities
3. Development
4. Settlement
5. Drainage

C. Water resources
1. Water supply, surface and groundwater and salvage
2. Water quality and treatment
3. Water requirements, all purposes
4. Water rights including international treaties
5. Flood studies
6. Sediment, including transport, erosion and aggradation
7. Project operation studies
8. Forecasting for operation
9. Hydraulic design requirements

D. Engineering and geology
1. Aerial photography, surveying and mapping
2. Geology, foundation and materials
3. Anticipated construction problems
4. Plans and cost estimates, physical plan formulation
5. Anticipated operation, maintenance and replacement problems and estimates of cost

E. Economics
1. Existing economy and resource use
2. Future economy without the project
3. Future economy with the project, and regional and national impact
4. Economic criteria for plan formulation
5. Economic justification

F. Financial considerations
1. Cost allocation to various purposes
2. Repayment of capital investment
3. Payment of annual operation, maintenance and replacement costs

G. Legal considerations
1. Right to use of water
2. International agreements and treaties
3. Land acquisition and rights-of-way

H. Public relations
1. Determination of public interest in contemplated development
2. Dissemination of factual information on progress and objectives of investigation
3. Establishment of government policy and enabling legislation

I. Reports
1. Reconnaissance reports
   (a) Basin plan
   (b) Preliminary project report
2. Special interim or progress reports
3. Feasibility
4. Definite plan

J. Administration
1. Organizational requirements for supervision of construction and operation of proposed projects
2. Program and budget requirements and control

Source: From ECAFE, United Nations, 1964.

Table 11E.63 Purposes of a Water Resources Project

Purpose	Description	Type of Works and Measures
Flood control	Flood-damage prevention or reduction, protection of economic development, conservation storage, river regulation, recharging of groundwater, water supply, development of power, protection of life	Dams, storage reservoirs, levees, floodwalls, channel improvements, floodways, pumping stations, floodplain zoning, flood forecasting
Irrigation	Agricultural production	Dams, reservoirs, wells, canals, pumps and pumping plants, weed-control and desilting works, distribution systems, drainage facilities, farmland grading
Hydroelectric	Provision of power for economic development and improved living standards	Dams, reservoirs, penstocks, power plants, transmission lines
Navigation	Transportation of goods and passengers	Dams, reservoirs, canals, locks, open-channel improvements, harbor improvements
Domestic and industrial water supply	Provision of water for domestic, industrial, commercial, municipal, and other uses	Dams, reservoirs, wells, conduits, pumping plants, treatment plants, saline-water conversion, distribution systems
Watershed management	Conservation and improvement of the soil, sediment abatement, runoff retardation, forests and grassland improvement, and protection of water supply	Soil-conservation practices, forest and range management practices, debris-detention dams, small reservoirs, and farm ponds
Recreational use of water	Increased well-being and health of the people	Reservoirs, facilities for recreational use, works for pollution control, reservations of scenic and wilderness areas
Fish and wildlife	Improvement of habitat for fish and wildlife, reduction or prevention of fish or wildlife losses associated with man's works, enhancement of sports opportunities, provision for expansion of commercial fishing	Wildlife refuges, fish hatcheries, fish ladders and screens, reservoir storage, regulation of streamflows, stocking of streams and reservoirs with fish, pollution control, and land management
Pollution abatement	Protection or improvement of water supplies for municipal, domestic, industrial, and agricultural use and for aquatic life and recreation	Treatment facilities, reservoir storage for augmenting low flows, sewage-collection systems, legal control measures
Insect control	Public health, protection of recreational values, protection of forests and crops	Proper design and operation of reservoirs and associated works, drainage, and extermination measures
Drainage	Agricultural production, urban development, and protection of public health	Ditches, tile drains, levees, pumping stations, soil treatment
Sediment control	Reduction of control of slit load in streams and protection of reservoirs	Soil conservation, sound forest practices, proper highway and railroad construction, desilting works, channel and revetment works, bank stabilization, special dam construction and reservoir operations
Salinity control	Abatement or prevention of salt-water contamination of agricultural, industrial, and municipal water supplies	Reservoirs for augmenting low streamflow, barriers, groundwater recharge, coastal jetties

(Continued)

Table 11E.63 (Continued)

Purpose	Description	Type of Works and Measures
Artificial precipitation	Control of precipitation within meteorological limits	Portable cloud-seeding equipment, ground generators
Employment	Stimulation of employment and sources for increased income in depressed areas of unemployment and underdevelopment	Area Redevelopment Act and Area Redevelopment Administration
Public works acceleration	Acceleration of Federal, state, and local constructions of public works on cost-sharing basis	Public Law 87–658
New water-resources policies	New policies to be used by Federal agencies, according to S. Document no. 97 (27), approved by the President May 15, 1962, affecting the economies of project justification as well as project formulation and composition	Senate Document no. 97

Source: From Dixon, In *Chow, Handbook of Applied Hydrogeology*, McGraw-Hill, Copyright, 1964. With permission.

Table 11E.64 Items to be Considered in Planning a Multiple-Purpose Water Resources Project

A. Physical and Related Items

1. Project area
   - a. Physical geography: location and size; physiography; climate; soils
   - b. Settlement: history; population; cultural background, both rural and urban
   - c. Development: industry; transportation; communication; commerce; power; land uses; water uses; minerals; undeveloped resources
   - d. Economic conditions: general; relief problems; community needs; national needs
   - e. Investigations and reports: previous investigations; history; scope
2. Hydrologic data
   - a. Hydrologic records and networks: gaging and observation stations; data-collecting agencies
   - b. Hydrometeorological data: precipitation; evaporation and evapotranspiration
   - c. Surface water: low flows; normal flows; maximum floods; "design floods;" drought; quality
   - d. Groundwater: aquifers; recharge; quality
3. Supply of water
   - a. Sources of supply: surface-water supply; groundwater supply; reservoirs
   - b. Variation of supply: variability; consumptive use; regulation; diversion requirements; return flow; evapotranspiration losses; seepage losses or gains
   - c. Quality of water: physical, chemical, biological and radioactive qualities; quality requirements; pollution
   - d. Legal rights: water rights; development of project rights; operation rights
4. General considerations for design and planning
   - a. Geology: explorations; geological formations; foundation problems
   - b. Design problems: design criteria; methods of analysis; project operation and maintenance
   - c. Construction problems: accessibility to project site; rights of way and relocation; construction materials; construction period; flow diversion; manpower; equipment, accessibility
   - d. Alternative plans: comparison of alternative plans; supplementary plans; possible alternative plans; relationships to areas to be served
   - e. Estimates of costs
   - f. Intrastate, interstate, and international problems
   - g. Organizations involved: public and/or private; technical, social, and political
5. Flood control
   - a. Flood characteristics of the project area: Historical floods; flood magnitude and frequency
   - b. Design criteria: project design storms and floods; degree of protection
   - c. Damage: survey of flooding areas and things affected by floods, nearby or quite a distance away, including commerce, good will, dates of delivery of goods, etc.
   - d. Measure of control: reduction of peak flow by reservoirs; confinement of flow by levees, floodwalls, or a closed conduit; reduction of peak stage by channel improvement; diversion of floodwater through bypasses or floodways; flood-plain zoning and evacuation; floodproofing and flood insurance of specific properties; reduction of flood runoff by watershed management
   - e. Existing remedial works
6. Agricultural use of water (irrigation and related drainage)
   - a. Factors for land classification: soil texture; depth to sand, gravel, shale, raw soil, or penetrable lime zone; alkalinity; salinity; slopes; surface cover and profile; drainage; water logging
   - b. Present and anticipated development: crops; livestock; financial resources; improvements; organizations; development period
   - c. Water requirements, if any: total crop requirement; irrigation-water demand; farm-delivery losses; diversion amounts
   - d. Available water: sources; quality; quantity; distribution
   - e. Irrigation methods: flooding; furrow irrigation; sprinkling; subirrigation; supplemental irrigation
   - f. Structural works: storage reservoirs; dams; spillways; diversion works; canals and distribution systems
7. Hydroelectric power
   - a. Development: sources; present potential and future capacities
   - b. Alternative sources of power: stream; oil; gas; nuclear power; interties

- c. Types of power plants: run-of-river; storage; pumped storage
- d. Structural components: dams; canals; tunnels; penstock; forebay; powerhouse; tailrace
- e. Power problems: load demand and distribution; interties (interconnections with other power transmission systems)
- f. Markets; revenues; costs

8. Navigation
   - a. Water traffic: present and future needs and savings in shipping costs, if any, on the basis of which the justifications are primarily judged at the present time
   - b. Alternative means of transportation: air; land
   - c. Navigation requirements: depth, width, and alignment of channels; locking time; current velocity; terminal facilities
   - d. Methods of improving navigation: channel improvement by reservoir regulation; contraction works; bank stabilization, straightening, and snag removal; lock-and-dam construction; canalization; dredging
9. Domestic and industrial water supply
   - a. Sources of supply: surface and/or groundwater; location and capacity; desalination
   - b. Water demand: climate; population characteristics; industry and commerce; water rates and metering; size of project area; fluctuation
   - c. Water requirements: quantity; pressure; quality (tastes, odors, color, turbidity, bacteria content, chemicals, temperature, etc.)
   - d. Methods of purification: plain sedimentation; chemical sedimentation or coagulation; filtration; disinfection; aeration; water softening
   - e. Treatment plant: location; design; purpose or purposes
   - f. Distribution systems: reservoirs; pumping stations; elevated storage; layout and size of pipe systems; location of fire hydrants
   - g. Waterworks organizations: maintenance and operation of supply, distribution, and treatment facilities
10. Recreational use of water
    - a. Population tributary (population near enough to the project area to use it for recreational purposes)
    - b. Facilities: boating; fishing; swimming; etc.
    - c. Water requirements: depth of water; area of water surface; sanitation
11. Fish and wildlife
    - a. Biological data: species; habits
    - b. Facilities: reservoirs; fish ladders
    - c. Water requirements: temperature; current velocity; biological qualities
12. Drainage
    - a. Existing projects
    - b. Drainage conditions: rainfall excess; soil condition; topography; disposal of water
    - c. Drainage system: urban; farmland
13. Water-quality control
    - a. Problems involved: sources; nature and degree of pollution; sediment; salinity; temperature; oxygen content; radioactive contamination
    - b. Hydrologic information and measurement
    - c. Methods of control

B. Economic Aspects of Project Formulation

1. Benefits and damages: identification and evaluation
2. Costs: identification and estimation
3. Financial feasibility
4. Allocation of costs
5. Reimbursement requirements and sharing of allocated costs
6. Methods and costs of financing the project, whether federal, state, or local, bringing all benefits and all costs to an annual basis and recognizing interest on the investment not only during construction, but throughout the entire proposed “life of the project”
7. Benefit-cost-ratio analysis: alternative plans

Source: From Dixon, In *Chow, Handbook of Applied Hydrogeology*, McGraw-Hill, Copyright, 1964. With permission.

Table 11E.65 Potential Benefits of Water Quality Improvements

In-Stream use	Commercial fisheries, shell fisheries, and aquaculture; navigation Recreation (e.g., fishing, hunting, boating, swimming) Subsistence fishing Human health risk reductions
Near-stream use	Water-enhanced noncontact recreation (e.g., picnicking, photography, jogging, biking, camping) Nonconsumptive use (e.g., wildlife viewing, hiking near water) Flood control (reduced property loss and risk to health and safety)
Diversionary use	Industry/commercial (process and cooling waters) Agriculture/irrigation Municipal drinking water (treatment cost savings, water storage dredging and construction savings, and human health risk reduction)
Aesthetic use	Residing, working, traveling, and owning property near water, etc.
Passive use	Existence (satisfaction gained from knowing the resources exist and knowing others enjoy the resources; ecologic value, including reduced mortality and morbidity, improved reproductive success, increased diversity of aquatic and piscivorous wildlife, improved habitat for threatened and endangered species, and improved integrity of aquatic and aquatic-dependent ecosystems) Bequest (intergenerational equity)

Note: Previous analyses have included option value as a potential benefit of environmental improvement. For this analysis, EPA adopted Freeman's (1993) conclusion that option value does not exist as a separate benefit category.

Source: From www.epa.gov.

Table 11E.66 Some Examples of Water Management Purposes and the Need for Planning

Function	Primary Responsibility[a]	Capital Investment	Operating Plans
Irrigation	F, S, L, P	Best construction of systems	Best use of water and money
Municipal water supply	L	Cost effectiveness	Contingencies, best use of facilities
Industrial water supply	P	Cost effectiveness	Contingencies, best use of facilities
Energy cooling water	P	Development of supplies	Best use of facilities, meeting standards
Hydropower	F, P	Development of economic power	Maximization of energy production
Wastewater treatment	L	Cost effectiveness	Meeting standards, reducing costs
Navigation	F	National economic efficiency	Operation of facilities
Flood damage reduction	F, S, L	Optimum facilities	Optimum operation
Urban drainage	L	Plans for economical systems	Maintenance, warning, etc
Agricultural drainage	L	Plans for systems	Operation of systems
Recreation	F, S, L	Development of facilities	Effective operation
Fish and wildlife	F, S	Preservation and enhancement of species	Effective operation
Watershed management	L	Best plans	Maintenance and operation
Preservation of ecological systems	F, S	Preservation of systems	NA
Preservation of systems of unique value	F, S	Preservation of systems	NA

[a] F = Federal
S = State
L = Local
P = Private

Source: From Grigg, N.S., 1985, *Water Resources Planning*. Copyright 1985 McGraw-Hill, Inc. Reprinted with permission.

Table 11E.67 Stages of Planning for Water Resources Management

Policy Planning Examples of Activities	Framework Planning Examples of Activities	General Appraisal Planning Examples of Activities	Implementation Planning Examples of Activities
Assess broad national needs	From viewpoint of broad regionwide totals and on "no-project" basis	On basis of local projects or measures, and over regional of watershed areas	For specific projects or measures
Hypothesize national goals and objectives	Inventory and evaluate available data	Estimate present and future water use and environmental needs	Evaluate specific water use and environmental needs
Identify problems and opportunities in achieving goals	Assess present and future water use and environmental needs	Estimate available water and related land resources	Evaluate available water and related land resources
Identify costs and benefits in achieving goals	Assess available water and related land resources	Make preliminary evaluations of alternative water quality management approaches	Evaluate regulation potential for different degrees of storage
Recommend policy choices	Evaluate general regulation potential and identify water quality management approaches	Make preliminary estimates of costs, benefits and consequences of specific alternative projects and measures	Evaluate degree of water quality control with different types of facility
Coordinate national priorities	Inventory present status of development	Compare alternative projects and measures	Prepare conceptual designs and cost estimates
Review programs for achievement of goals	Inventory general means available to satisfy needs	Devise alternative early action and future programs	Make economic analyses of benefits and consequences
	Assess general alternatives to meet different goals	Recommend specific early action and alternative future programs, including selection of projects or measures for implementation study	Make financial analyses to demonstrate payout
	Identify problem areas that need priority attention		Compare alternatives on basis of costs, benefits and payout
	Recommend actions that can be taken at present and those that require further study		Recommend an alternative to be selected
			Recommend concerning authorization

Source: From National Water Commission, 1972, *Water Resources Planning*.

Table 11E.68 Water Body Classification Categories

Description of Desired Use[a]	Guidelines for Desired Use[b]
Category I: Water bodies typically used for direct contact recreational activities, including swimming, scuba diving, snorkeling and waterskiing	Minimum summer Secchi disc depth of at least 1.0 meters; summer average of at least 1.4 meters. Total phosphorus concentrations less than 45 g/L. Chlorophyll-a concentrations less than 20 g/L. Carlson TSI index (Secchi disc based) no greater than 55
Category II: Water bodies typically used for indirect contact recreational activities, including sail boating, motor boating, canoeing, and fishing. These activities involve incidental contact with lake water, but do not generally require the water clarity found in direct contact recreational waters. Algal blooms in mid- to late-summer may limit direct contact recreational activities such as swimming and waterskiing	Mean summer Secchi disc depth of at least 0.9 meters, but less than 1.4 meters Total phosphorus concentration of at least 45 g/L, but less than 75 g/L Chlorophyll-a concentration of at least 20 g/L, but less than 40 g/L Carlson TSI index (Secchi disc based) should be no greater than 60
Category III: Water bodies that typically serve important functions such as wildlife habitat and aesthetics. May also provide opportunities for warm water fishing, provided winterkill does not occur. Generally accessible to the public for education, interpretation, and nature appreciation.	Of primary importance are guidelines related to aesthetic enjoyment and wildlife habitat to maintain/improve desired use of these water bodies — see Table 5.2 for a listing of aesthetic and habitat indicators. Of secondary importance are the following water quality guidelines: Mean summer Secchi disc depth of at least 0.7 meters Total phosphorus concentration of at least 75 g/L, but less than 105 g/L Chlorophyll-a concentration of at least 40 g/L, but less than 60 g/L Carlson TSI index (Secchi disc based) no greater than 65
Category IV — Nutrient Traps: The intended use of these water bodies is to reduce downstream loading of phosphorus and other nutrients that contribute to water pollution. These ponds are generally artificially modified to improve their nutrient trapping capacity. These ponds may become hypereutrophic, and frequent summer algal blooms would be considered normal	Design for phosphorus removal efficiencies of at least 50%. Depth should be managed to prevent or reduce odors associated with algal blooms. No numeric standards for water quality parameters are defined for this category
Category V — Sediment Traps: These water bodies are similar to Category IV water bodies, but too small to effectively remove a significant fraction of nutrients	Generally have phosphorus removal efficiencies less than 50%. No numeric standards for water quality parameters are defined for this category

[a] Categories I–III could also include ecologically or biologically unique resources, or water bodies that directly or indirectly affect such a resource.

[b] The water quality criteria for Categories I–III may not apply in the case of ecologically or biologically unique resource; resource-specific criteria may be required.

Source: From Black Dog Watershed Management Organization, www.dakotacountyswcd.org. With permission.

Table 11E.69 Aesthetic and Habitat Indicators and Recommended City Actions

Indicator	Prevalence	Recommended Actions[a]
Localized shoreline erosion	10% or less of shore	A (note # of sites), B
	10% to 25%	A (note # of sites), B, C
	25% or more	A (note # of sites), B, D
Shoreline erosion resulting from upstream activity	Present	E
Sedimentation/deltas	Present	E
	10 cubic yards or more	F
Shoreline buffer/percent vegetated[b]	>75%	A
	50% to 75%	A, B
	25% to 50%	A, B, Computer
	<25%	A, B, D
Purple loosestrife, non-native invasive species	Present	A (note # of colonies), G, H
	Not present	A
Glossy and common buckthorn, non-native invasive species	Present	A, G, I
	Not present	A
Average density of shallow water aquatic plants[c]	<1.5	A, B (to encourage growth of desirable plants), J
	1.5 to 2.5	A
	>2.5	A, K
Non-native invasive shallow water aquatic plants[d]	Present	A, B (to encourage growth of desirable plants), L
	Not present	A

[a] Key to recommended city actions: A, Monitor annually; B, Targeted education materials regarding the importance of lakescaping and buffers; C, Lakescaping demonstration project; D, Lakescaping incentive program (e.g. small grants to encourage residents to implement lakescaping techniques); E, Evaluate source of problem and develop methods to address problem; F, Excavate sediment; G, Targeted education materials regarding the detrimental impacts of the plants, the importance of removing them, methods to remove them, and replacement plantings; H, Recruit citizens to take part in the DNR's purple loosestrife (beetle) program to eradicate plants; I, Recruit citizens to remove/replace plants; J, Evaluation needed to determine why so few plants are present and if more are needed; K, Evaluate extent of plant growth and develop/implement aquatic plant management plan; L, Evaluate extent of plant growth and develop/implement aquatic plant management plan — may not include eradication of non-native invasive plants, especially if they are the only aquatic plants present.

[b] Percent vegetated — percent of shoreline that is vegetated, exclusive of landscaped lawns, with buffer at least 10 f in width.

[c] Based on density ratings given for numerous locations within the littoral zone of each water body; density of 1, low, 2, moderate, 3, high.

[d] Non-native invasive aquatic plants include Eurasian watermilfoil and curly leaf pondweed.

Source: From Black Dog Watershed Management Organization, www.dakotacountyswcd.org. With permission.

Table 11E.70 Recommended Lake Water Quality Management Actions for Category I and II Lakes

		Type(s) of Management Action Needed		
Most Recent Summer Average Lake Water Quality, as Compared to Management Action Level	**5–10-Year Water Quality Trend**	**Watershed Management**	**Lake Monitoring**	**Runoff Monitoring or Equivalent**
Better than	No trend analysis available	No action	Continue existing water quality monitoring program: Survey Level—Category I Secchi Disc—Category II	None
	Improving	No action	Survey Level—Category I Secchi Disc—Category II	None
	Steady	No action	Survey Level—Category I Secchi Disc—Category II	None
	Degrading	No action	Management Level—Category I Survey Level—Category II	Watershed land use review
Worse than	No trend analysis available	No action	Management Level—Category I Survey Level—Category II	None
	Improving	Implementation of runoff BMPs assumed; no further action required	Management Level—Category I Survey Level—Category II	None
	Steady	Diagnostic study (e.g., P8 modeling)	Management Level—Category I Survey Level—Category II	Watershed land use review; And subsequent, focused runoff water quality monitoring in potential problem subwatersheds
	Degrading	Comprehensive lake/watershed diagnostic feasibility study	Intensive lake monitoring as part of diagnostic feasibility study	Detailed runoff water quality monitoring as part of diagnostic feasibility study

Source: From Black Dog Watershed Management Organization, www.dakotacountyswcd.org. With permission.

Table 11E.71 Detailed Listing of Possible Best Management Practices

Type of Practice	Area of Benefit	Storm Protection Benefit	Pollutants Controlled	Construction Requirements
Institutional Source Controls				
Public education (billing inserts, news releases, radio public service announcements, school programs, and pamphlets)	Not applicable	Reduced pollutant load to storm drain system	Can reduce improper disposal of paints, varnishes, thinners, pesticides, fertilizers, and household cleansers, and chemicals, etc.	None
Litter control	Site dependent	Reduced potential for clogging and discharge	Household and restaurant paper, plastics, and glass	Increase number of trash receptacles and regularly service
Recycling programs	Site dependent	Reduction in potential for clogging and harmful discharge	Household paper, glass, aluminum, and plastics. Oil and grease from auto maintenance	Collection and sorting stations
"No Littering" ordinance	Storm drain system and receiving water	Prohibits littering and prevents litter from entering storm drains	Paper, plastics, glass, food wrappers, and containers	None
"Pooper Scooper" ordinance	Storm drain system and receiving water	Requires animal owners to clean up and properly dispose of animal wastes	Coliform bacteria and nitrogen/urea	None
Develop and enact spill response plan	Site dependent	Prevent pollutants from entering storm drain	Hazardous chemical, harmful chemicals, oil, and grease	None
Clean up vacant lots	Site dependent	Prevent debris from accumulating on lot. Prevent site from appearing as a "dump" for others to use for disposal. Eliminate sources of hazardous waste	Hazardous and/or harmful chemicals, wind blown for water borne debris	None
Prohibit illegal and illicit connections and dumping into storm drain system	Storm drain system and receiving water	Reduces pollutant load entering storm drains	Coliform bacteria, nitrogen, contaminants, and toxic or harmful chemicals	None
Identify, locate, and prohibit illegal or illicit discharge to storm drain system	Area-wide	Halt hazardous and harmful discharges, whether intentional or negligent	Sewage from cross connections, oil, grease, direct disposal of pesticides and fertilizers, contaminated water, paint, varnish, solvents, water from site dewatering, swimming pool and spa water, flushing water from radiators and cooling systems, and hazardous or harmful chemicals	Monitor storm drain system for flows and water quality

(Continued)

Table 11E.71 (Continued)

Type of Practice	Area of Benefit	Storm Protection Benefit	Pollutants Controlled	Construction Requirements
Require proper storage, use, and disposal of fertilizers, pesticides, solvents, paints and varnishes, and other household chemicals (oil, grease, and antifreeze, etc.)	Site dependent (city, state, or county-wide)	Reduce pollutant load to storm system	Household hazardous materials	None
Restrict paving and use of nonporous cover materials in recharge areas	Recharge area site	Promotes infiltration to groundwater and reduces runoff volume and velocity. Filters pollutants		Establishment of vegetation or use of recharge/infiltration materials
Nonstructural Source Controls				
Street sweeping	Street right-of-way	Reduction in potential for clogging storm drains with debris. Some oil and grease control possible	Paper and plastics, leaves and twigs, dust, and oil and grease	Acquire street sweeping equipment
Sidewalk cleaning	Sidewalk right-of-way in areas of heavy foot traffic	Reduction in pollutants entering storm drain	Oil and dirt	None
Clean and maintain storm drain channels annually	Channel capacity and receiving water. Upstream flood control benefits. Includes benefits to channel wildlife habitat and vegetation	Prevent erosion in channel. Improve capacity by removing silt and sedimentation. Remove debris that is habitat destroying or toxic to wildlife	Silt and sediment and the contaminants contained therein. Plastic, glass, paper, and metal thrown or washed in channel	None
Clean and inspect storm inlets and catch basins annually	Site dependent flood control benefits	Allows proper drainage to prevent flooding and continued proper operation of facilities	Silt and sediment and the contaminants contained therein. Plastic, glass, paper, and metal thrown or washed into facilities	None
Clean and inspect debris basins annually	Site dependent flood control benefits	Allows proper drainage to prevent flooding and continued proper operation of facilities	Silt and sediment and the contaminants contained therein. Plastic, glass, paper, and metal thrown or washed into facilities	None
Storm drains cleaned and maintained every 3 to 6 years	Flood control and water quality benefits	Allows proper drainage to prevent flooding and continued proper operation of facilities	Silt and sediment and the contaminants contained therein. Plastic, glass, paper, and metal thrown or washed into facilities	None
Storm system pump stations cleaned and maintained annually	Site dependent flood control and water quality benefits	Prevents flooding and allows continued proper operation of facilities	Silt and sediment and the contaminants contained therein. Plastic, glass, paper, and metal thrown or washed into facilities	None

Inspect and maintain sewer system	Storm drain system and receiving water	Prevents and eliminates sewer system surcharges	Contaminants, toxics, and coliform bacteria	None
Minor Structural Source Controls				
Storm drain inlet protection	Storm drain drainage area	Prevent debris from entering storm drain	Dirt, leaves, twigs, paper, plastic, and other incidentals	Not available
Outlet protection	Storm drain receiving water	Prevent erosion at the outlet of pipes or paved channels and protect downstream water quality	Turbidity and sediment	Structural apron lining at the outlet location. Made of riprap, grouted riprap, concrete, or other structural materials
Slope stabilization and erosion control measures	Site and topography dependent	Reduce silt and sediment load to storm drains	Silt and sediment and the contaminants therein	None
Interceptor swale	Dependent on flow velocity. Max. velocity for earth channel is 6 fps. Max. velocity for vegetated or riprap channel is 8 fps	Shorten length of exposed slopes and intercept and divert storm runoff from erodible areas	Sediments and silt and the contaminants contained therein	Excavation drainageway across disturbed areas or rights-of-way
Improve and maintain natural channels	Channel capacity and receiving water. Upstream flood control benefits. Includes benefits to channel wildlife habitat and vegetation	Prevent erosion in channel. Improve capacity by removing silt and sedimentation. Remove debris that is habitat destroying or toxic to wildlife	Silt and sediment and the contaminants contained therein. Plastic, glass, paper, and metal thrown or washed in channel	None
Diversion channel	Dependent of flow velocity. Maximum velocities: 5 fps for vegetated channel and 8 fps for riprap channel. Not for use on slopes greater than 15%. Drainage area should be 5 acres of less	Intercept and convey runoff to outlets at nonerosive velocity	Sediment and erosion controls	Lined drainageway of trapezoidal cross section
Grass-lined channel	Site dependent but of larger capacity than interceptor or perimeter swales	Intercept runoff and convey runoff from site	Sediment and silt and the contaminants contained therein	Excavation of channel or improvements to natural channel. Stabilization with vegetation
Storm drain drop inlet protection	Areas less than 1 to 2 acres	Filters sediment from runoff before it enters inlet. Provides relatively good protection	Sediment and the contaminants contained therein	Barrier around storm drain inlet. Useful for areas where storm drain is operational before area runoff area is stabilized
Riprap	Site dependent	Provides stabilization and erosion control for stream banks and channels, outlet, and slopes	Erosion and sediment	Placement of rock on area to be stabilized. May also require use of filter fabric liner

(Continued)

Table 11E.71 (Continued)

Type of Practice	Area of Benefit	Storm Protection Benefit	Pollutants Controlled	Construction Requirements
Gabions	Site dependent	Provides stabilization and erosion control for stream banks, outlet, and slopes	Erosion and sediment	Placement of wire cage will with rocks over area to be stabilized. May also require use of filter fabric liner
Vegetative control	Applicable and effective for most sites	Provides stabilization and erosion control for streambanks, swales, channels, outlets, slopes, open disturbed areas. Can be up to 99% effective with established cover. Temporary seeding can be up to 90% effective	Erosion and sediment	Site preparation (can include land leveling and installation of irrigation system), seeding or planting, and netting or mulching to establish seed. Can also include other sodding, ground cover, shrubs, trees, and native plants
Filter strips	Site dependent	Receives overland flow slowing runoff and trapping particulates. Can be 30 to 50% effective for sediment control	Silt, sediment, trash, organic matter, and to an extent, soluble pollutants through infiltration	Grading and vegetative establishment. Should have a minimum width of 15 to 20 feet. Good performance is achieved with a 50- to 75-foot width
Fence open channels	Site dependent	Prevent windblown trash from entering channel. Prevents illegal dumping in channel	Trash and pollutants	Construction of fences
Discharge Elimination Methods				
French drains and subsurface drains	Dependent on site topography and soil permeability	Provides drainage of "wet" soils to allow establishment of vegetation. Can reduce runoff	Sediment	Underground perforated pipe leading to a surface water outlet. Pipe size, bedding and depth is dependent on site conditions
Infiltration trench and dry well	Small drainage areas. Runoff from rooftops, parking lots, residential, etc.	Provides temporary storage of runoff and infiltration to soil. Not for use in areas where groundwater could become contaminated	Prevents 100% of pollutants from entering surface water. Oil, grease, floating organic matter, and settleable solids should be removed before water enters trench	Excavation of a shallow trench 2′ to 10′ deep. Backfilled with coarse stone aggregate
Exfiltration trench	Site dependent	Prevent silting on underlying filter gravel or rock bed. Retain first flush, reduce runoff volume and peak discharge rate and promote water quality improvement	Prevents pollutants from entering surface water. Oil, grease, floating organic matter, and settleable solids should be removed before water enters trench	Uses perforated pipe with suitable membrane filter material. Installed before receiving water outlet or in groundwater recharge area

Porous pavement	Site dependent. Requires relatively flat surface	Allow infiltration of surface runoff. Reduce runoff volume and pollutant loadings from low volume traffic areas	Oil and grease	Install porous pavement. May require twice as much paving materials as standard asphalt to achieve same strength
Retention basin	Best for sites of 5 to 50 acres	Promotes infiltration to groundwater and reduces runoff volume and velocity. Filters pollutants	Sediment, trace metals, nutrients, and oxygen-demanding substances	Excavation of a basin over permeable soils. Size is site dependent. Depth is 3 to 12 feet
Floatables and Oil Removal				
Clarifies and oil and water separators on parking structures	Parking lot structure and receiving water	Collect debris before it can enter storm drain	Oil, grease, and antifreeze from vehicles and foods and food wrappers	Install grit and separators
Oil and grit separators	Site dependent. For heavy traffic areas or areas with high potential for oil spills	Remove pollutants	Sediments and hydrocarbons	Install oil and grit separators on storm drains
Sediment/grease trap	Installed on storm drain inlets	Intercept and trap sediment and grease from runoff	Sediment, oil, and grease	Install sediment and grease traps
Solids Removal				
Detention basin	Four acres of drainage area for each acre/foot of storage provided to retain a permanent pool of water	Temporary storage of storm runoff until release. Can also improve water quality	Sediment, trace metals, hydrocarbons, nutrients, and pesticides	Excavation of a basin over soils which will cause excessive seepage. May require a liner. Can be used aesthetically as a small pond in landscaping
Extended detention basin	Size for a minimum detention time of 24 hours	Temporary storage of runoff for an extended period of time. Can improve water quality	Sediment, trace metals, hydrocarbons, nutrients, and pesticides	Excavation of a basin over soils which will cause excessive seepage. May require a liner. Can be used aesthetically as a small pond in landscaping
Bar screens	Site dependent	Restrict passage of objects which may obstruct pump station suction bays	Large debris	Install bar screens before pump station suction bays
Wetlands	Requires large area, 3% of the watershed area	Remove pollutants. Provide habitat and recreational area	Hydrocarbons, silt and sediment, oxygen-demanding substances, bacteria, and nutrients	Create a new wetlands area or use existing wetlands
Microorganism Removal				
Conversion of wastewater treatment plants to wet weather facilities	Site dependent. Is abandoned treatment facility available?	Treats stormwater flows prior to discharge	Process dependent. Chlorination facilities may be added to remove microorganisms	Treatment conversion

(Continued)

Table 11E.71 (Continued)

Type of Practice	Area of Benefit	Storm Protection Benefit	Pollutants Controlled	Construction Requirements
Install treatment facilities on "Dirty" storm drains	Site and need dependent	Treats stormwater flows. Dry weather flows should be halted or routed to existing wastewater treatment facility if possible	Microorganisms	Site specific
Swirl concentrators and chlorination/dechlorination	Site and need dependent	Treats stormwater flows prior to discharge	Floatables, settleable solids, suspended solids, and coliform bacteria	Install concentrators
Chlorination/dechlorination facilities	Site and need dependent	Treats stormwater flows prior to discharge	Microorganisms	Install chlorination/ dechlorination facilities
Primary clarifiers	Site and need dependent	Treats stormwater flows prior to discharge	Floatables, settleable solids, suspended solids, and coliform bacteria	Install primary clarifiers
Primary clarifiers and filters	Site and need dependent	Treats stormwater flows prior to discharge	Suspended solids, nutrients and coliform bacteria	Construct sedimentation basins and filters
Metals Removal				
Primary clarifiers and lime precipitation	Site and need dependent	Treats stormwater flows prior to discharge	Floatable, settleable solids, suspended solids, coliform bacteria, and metals	Install primary clarifiers and lime precipitation facilities
Detention basin and wetland treatment	Requires large area, 3% of the watershed area	Remove pollutants. Provide habitat and recreational area	Hydrocarbons, silt and sediment, oxygen-demanding substances, bacteria, metals, and nutrients	Create a new wetlands area or use existing wetland

Source: From Black Dog Watershed Management Organization ENR 4900, www.dakotacountyswcd.org. With permission.

Table 11E.72 Recommended Best Management Practices According to Watershed

	Watersheds Tributary to a Ponding Area that Treats Stormwater Runoff (Treated Watershed)				Watersheds Receiving No Treatment	
Best Management Practice	**No Outlet[a]**	**NURP Detention[b]**	**Wet Detention[c]**	**Dry Detention[d]**	**No Detention[e]**	**Direct Discharge[f]**
Retrofit dry ponds to wet detention ponds				X		
Retrofit dry ponds to constructed water quality wetlands				X		
Retrofit dry ponds to extended detention basins				X		
Construct sedimentation (pretreatment) basins, when opportunity arises			X		X	X
Create additional water quality storage to meet NURP standards			X			
Perform stormwater system maintenance in accordance with local plan	X	X	X	X	X	
Install "Stormceptors" or other precast stormwater treatment systems					X	
Prioritize street sweeping	X	X	X	X	X	X
Construct skimmers to prevent downstream discharge of oil and floatables		X	X	X		
Develop/implement lawn fertilizer ordinance	X	X	X	X	X	X
Construct filter strips/grassed swales					X	X
Increase opportunities for infiltration	X	X	X	X	X	X
Stabilize slopes and implement other permanent erosion/sediment controls	X	X	X	X	X	X
Develop and implement education program emphasizing good housekeeping practices[g]	X	X	X	X	X	X
Provide vegetative buffers around ponds and wetlands	X	X	X	X	X	X

[a] No outlet watershed — area tributary to a landlocked basins (i.e., no outlet); receives highest level of treatment (i.e., 100% total phosphorus removal).
[b] NURP watershed — area tributary to a wet detention pond that provides treatment to NURP standards (i.e., 40%–60% total phosphorus removal).
[c] Wet detention watershed — area tributary to a wet detention pond that provides treatment to less than NURP standards (i.e., 5%–40% total phosphorus removal).
[d] Dry detention watershed — area tributary to a dry detention basin that provides for settlement of larger particles and traps floatables, but provides minimal water quality treatment (i.e., 0%–10% total phosphorus removal).
[e] No detention watershed — area tributary to a storm sewer system that does not receive any type of detention storage or treatment.
[f] Direct watershed — area that directly discharges as sheet flow into a major waterbody without any treatment.
[g] "Good housekeeping practices" include: fertilizer and chemical management, lawn and garden care guidelines, litter control, control of illegal dumping/illicit discharges, pet waste management, vacant lot cleanup, recycling programs, etc. See also Table 5.4.

Source: From Black Dog Watershed Management Organization, www.dakotacountyswcd.org. With permission.

Table 11E.73 Current Status the WHO/UNICEF/WSSCC Global Water Supply and Sanitation Assessment 2000 Report Provides Information on the Current Status of Basic Water and Sanitation Services throughout the World

	1990 Population (Millions) (76% of Global Population Represented)				2000 Population (Millions) (89% of Global Population Represented)			
	Total Population	Population Served	Population Unserved	% Served	Total Population	Population Served	Population Unserved	% Served
Global								
Urban water supply	2292	2179	113	95	2845	2672	173	94
Rural water supply	2974	1961	1013	66	3210	2284	926	71
Total water supply	5266	4140	1126	79	6055	4956	1099	82
Urban sanitation	2292	1877	415	82	2845	2442	403	86
Rural sanitation	2974	1028	1946	35	3210	1210	2000	38
Total sanitation	5266	2905	2361	55	6055	3652	2403	60
Africa	(72% of regional population represented)				(96% of regional population represented)			
Urban water supply	197	166	31	84	297	253	44	85
Rural water supply	418	183	235	44	487	231	256	47
Total water supply	615	349	266	57	784	484	300	62
Urban sanitation	197	167	30	85	297	251	46	84
Rural sanitation	418	206	212	49	487	220	267	45
Total sanitation	615	373	242	61	784	471	313	60
Asia	(88% of regional population represented)				(94% of regional population represented)			
Urban water supply	1029	972	57	94	1352	1254	98	93
Rural water supply	2151	1433	718	67	2331	1736	595	75
Total water supply	3180	2405	775	76	3683	2990	693	81
Urban sanitation	1029	690	339	67	1352	1055	297	78
Rural sanitation	2151	496	1655	23	2331	712	1619	31
Total sanitation	3180	1186	1994	37	3683	1767	1916	48
Latin American and the Caribbean	(77% of regional population represented)				(99% of regional population represented)			
Urban water supply	313	287	26	92	391	362	29	93
Rural water supply	128	72	56	56	128	79	49	62
Total water supply	441	359	82	82	519	441	78	85
Urban sanitation	313	267	46	85	391	340	51	87
Rural sanitation	128	50	78	39	128	62	66	49
Total sanitation	441	317	124	72	519	402	117	78
Oceania	(64% of regional population represented)				(85% of regional population represented)			
Urban water supply	18	18	0	100	21	21	0	98
Rural water supply	8	5	3	62	9	6	3	63
Total water supply	26	23	3	88	30	27	3	88
Urban sanitation	18	18	0	99	21	21	0	99
Rural sanitation	8	7	1	89	9	7	2	81
Total sanitation	26	25	1	96	30	28	2	93
Europe	(15% of regional population represented)				(44% of regional population represented)			
Urban water supply	522	522	0	100	545	542	3	100
Rural water supply	200	199	1	100	184	161	23	87
Total water supply	722	721	1	100	729	703	26	96
Urban sanitation	522	522	0	100	545	537	8	99
Rural sanitation	200	199	1	100	184	137	47	74
Total sanitation	722	721	1	100	729	674	55	92
Northern America	(99.9% of regional population represented)				(99.9% of regional population represented)			
Urban water supply	213	213	0	100	239	239	0	100
Rural water supply	69 0	69	0	100	71	71	0	100
Total water supply	282	282	0	100	310	310	0	100
Urban sanitation	213	213	0	100	239	239	0	100
Rural sanitation	69	69	0	100	71	71	0	100
Total sanitation	282	282	0	100	310	310	0	100

Note: The report charts the developments since 1990. The current status is provided in the table above.

Source: From thewaterpage.com. With permission.

Table 11E.74 Table Annual Investments in Water Supply and Sanitation Indicating Proportional Disparity

Region	Water Supply (US$ Billion)	%	Sanitation (US$ Billion)	%
Africa	4.091	88	0.542	12
Asia	6.063	85	1.104	15
LA & C	2.41	62	1.503	38
Total	12.564	80	3.148	20

Note: Investment figures indicate that a higher priority has been given over the 90s for water supply as opposed to sanitation, both through national governments and by the international community.

Source: From WHO/UNICEF, 2000. thewaterpage.com. With permission.

SECTION 11F RESEARCH AND EXPENDITURES

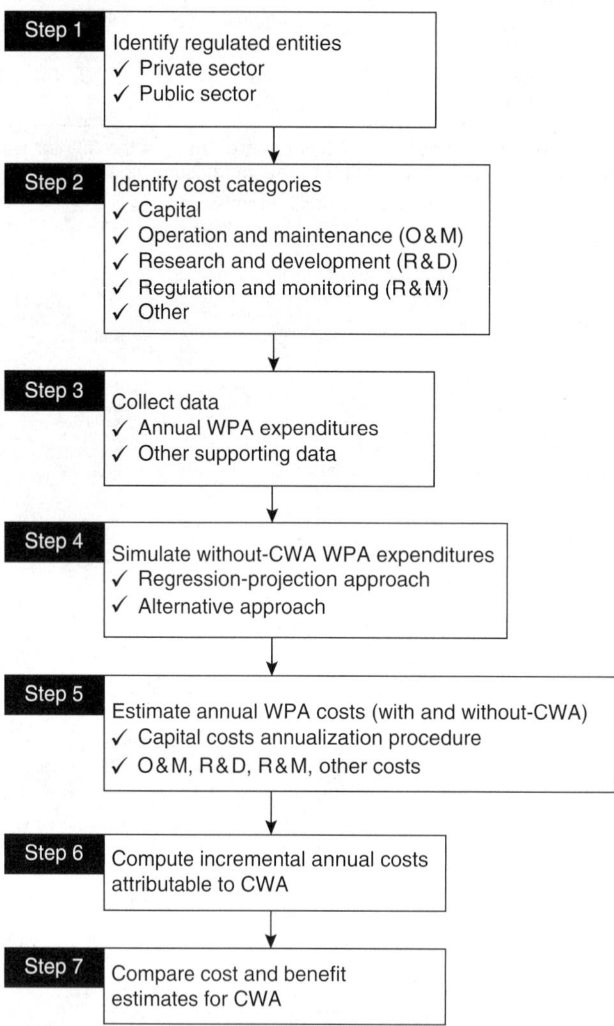

Figure 11F.50 Outline of methodology for assessing CWA costs. (From A Retrospective Assessment of the Costs of the Clean Water Act—1972–1997 Final Report, www.epa.gov.)

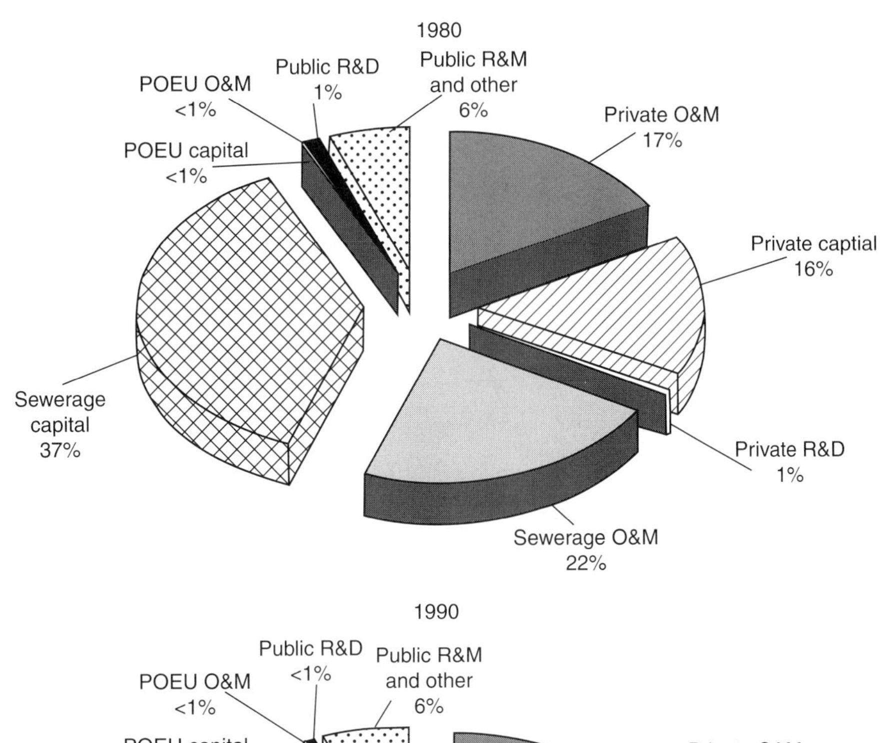

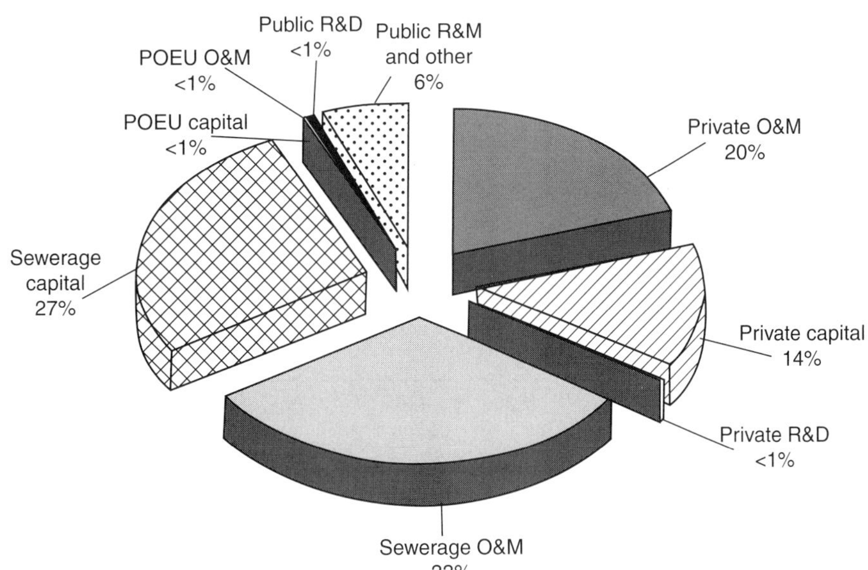

Figure 11F.51 Distribution of estimated water pollution abatement expenditures, 1980 and 1990. (From A Retrospective Assessment of the Costs of the Clean Water Act—1972–1997 Final Report, www.epa.gov.)

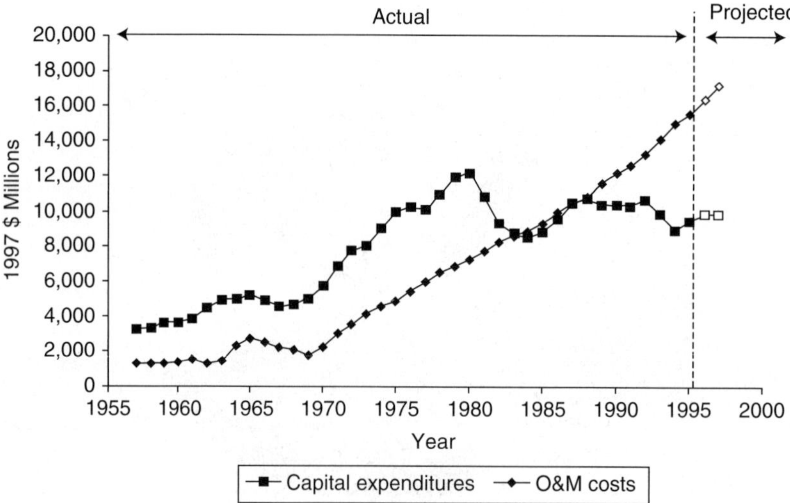

Figure 11F.52 Public sewerage expenditures, 1957–1997 (1997 $ millions). (From A Retrospective Assessment of the Costs of the Clean Water Act—1972–1997 Final Report, www.epa.gov.)

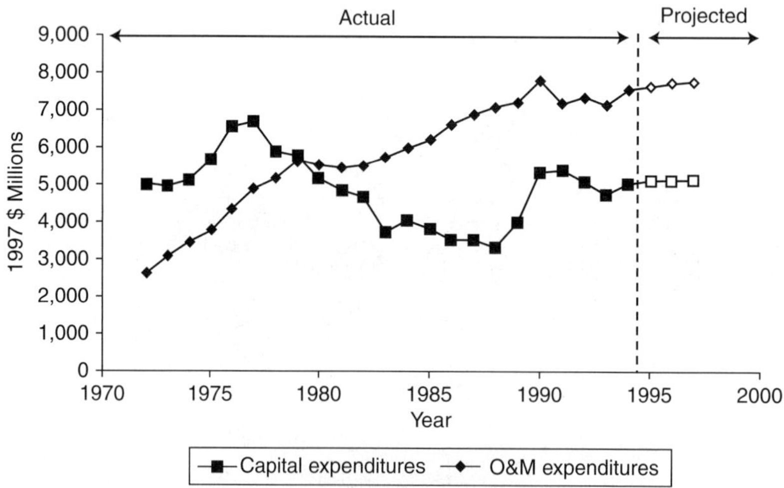

Figure 11F.53 Private water pollution abatement expenditures, 1972–1997 (1997 $ millions). (From A Retrospective Assessment of the Costs of the Clean Water Act—1972–1997 Final Report, www.epa.gov.)

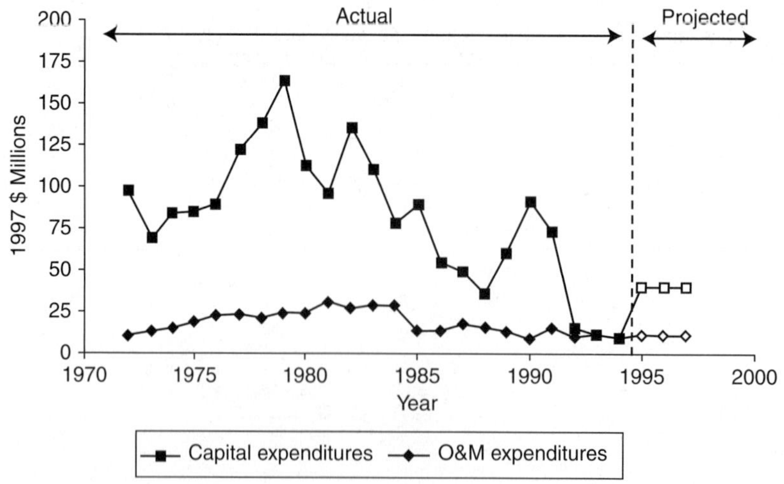

Figure 11F.54 Water pollution abatement expenditures by publicly owned electric utilities, 1972–1997 (1997 $ millions). (From A Retrospective Assessment of the Costs of the Clean Water Act—1972–1997 Final Report, www.epa.gov.)

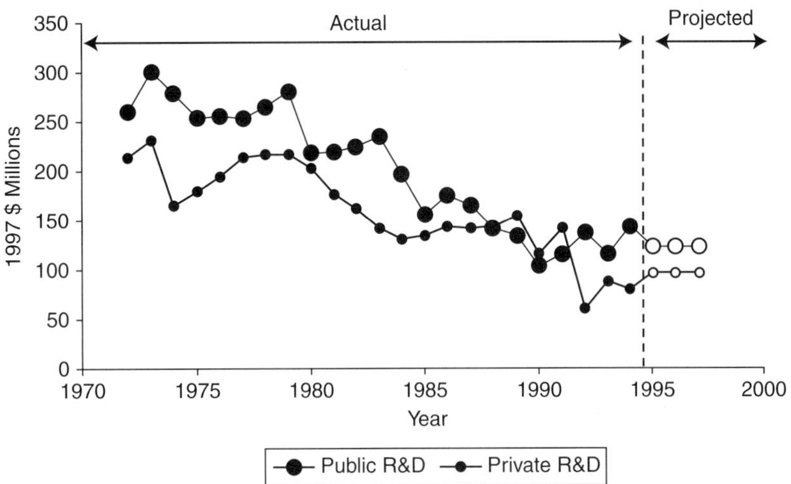

Figure 11F.55 Public and private R&D expenditures for water pollution abatement 1972–1997 (1997 $ millions). (From A Retrospective Assessment of the Costs of the Clean Water Act—1972–1997 Final Report, www.epa.gov.)

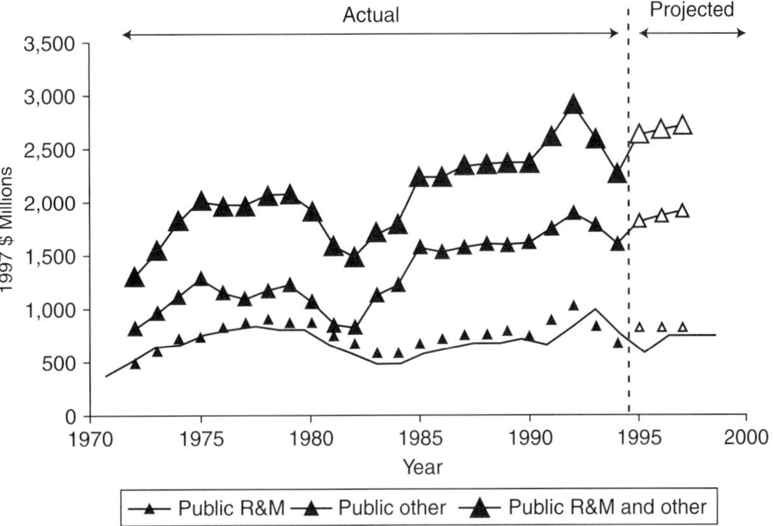

Figure 11F.56 Public water pollution abatement expenditures for R&M and other costs not elsewhere classified, 1972–1997 (1997 $ millions). (From A Retrospective Assessment of the Costs of the Clean Water Act—1972–1997 Final Report, www.epa.gov.)

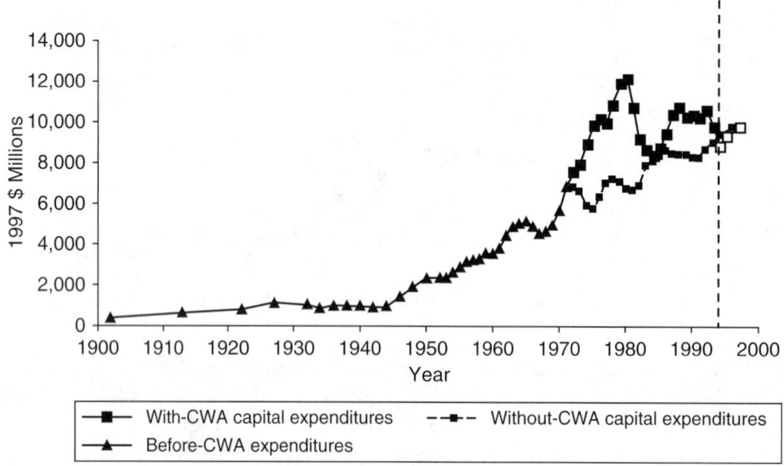

Figure 11F.57 Annual sewerage capital expenditures under the with-CWA and without-CWA scenarios, 1902–1997 (1997 $ millions). (From A Retrospective Assessment of the Costs of the Clean Water Act—1972–1997 Final Report, www.epa.gov.)

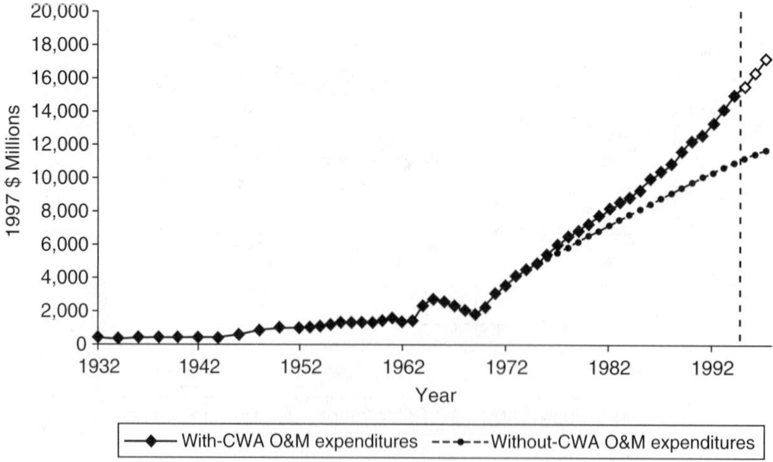

Figure 11F.58 Annual sewerage O&M expenditures under the with-CWA and without-CWA scenarios, 1932–1997 (1997 $ millions). (From A Retrospective Assessment of the Costs of the Clean Water Act—1972–1997 Final Report, www.epa.gov.)

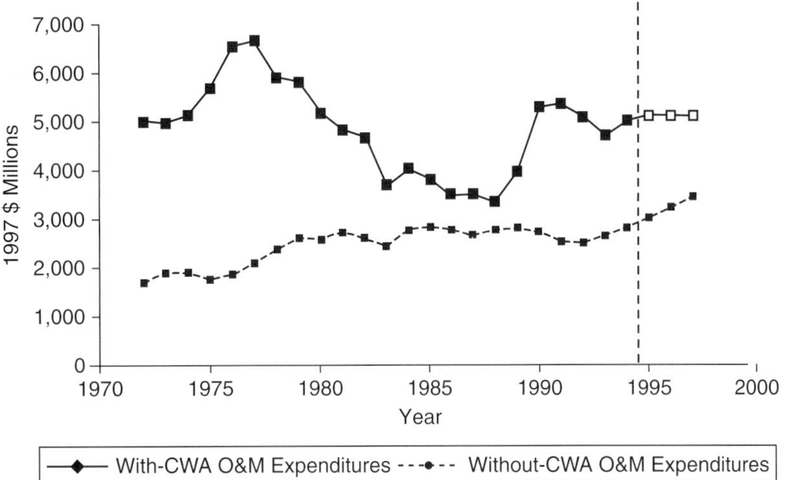

Figure 11F.59 Annual private capital expenditures for water pollution abatement under the with-CWA and without-CWA scenarios, 1972–1997 (1997 $ millions). (From A Retrospective Assessment of the Costs of the Clean Water Act—1972–1997 Final Report, www.epa.gov.)

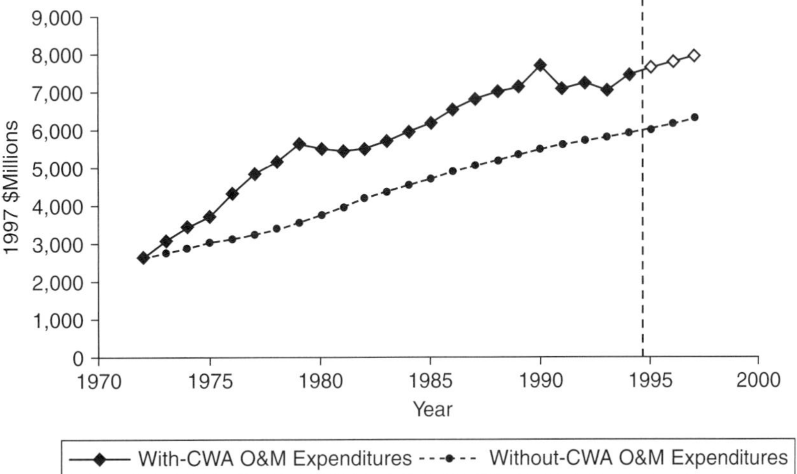

Figure 11F.60 Annual private O&M expenditures for water pollution abatement under the with-CWA and without-CWA scenarios, 1972–1997 (1997 $ millions). (From A Retrospective Assessment of the Costs of the Clean Water Act—1972–1997 Final Report, www.epa.gov.)

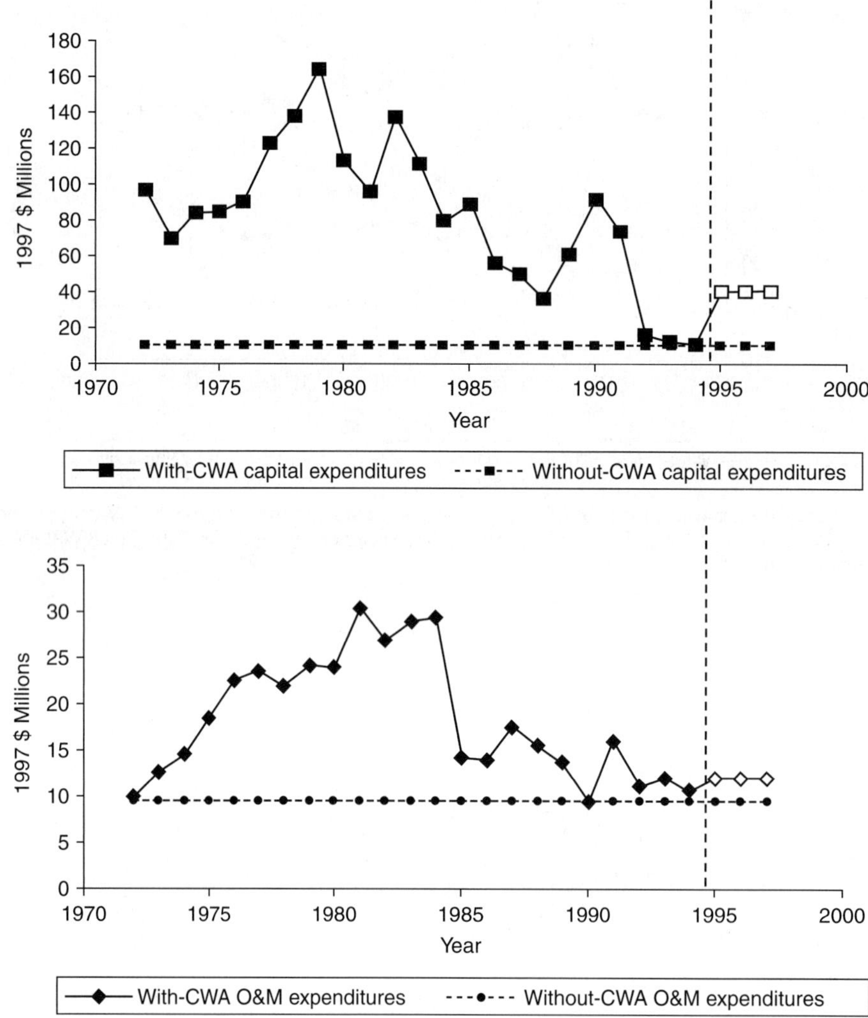

Figure 11F.61 Water pollution abatement expenditures by publicly owned electric utilities, with and without the CWA, 1972–1997 (1997 $ millions). (From A Retrospective Assessment of the Costs of the Clean Water Act—1972–1997 Final Report, www.epa.gov.)

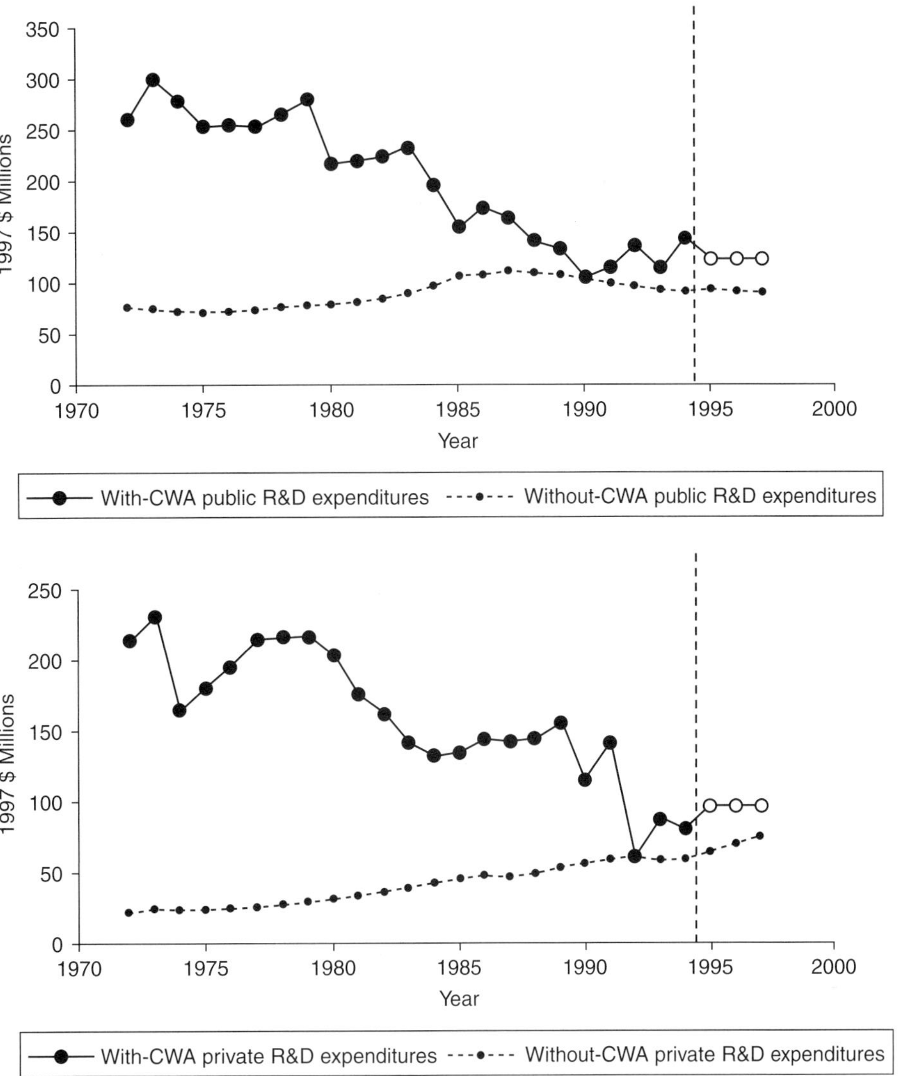

Figure 11F.62 R&D expenditures for water pollution abatement, with and without the CWA, 1972–1997 (1997 $ millions). (From A Retrospective Assessment of the Costs of the Clean Water Act—1972–1997 Final Report, www.epa.gov.)

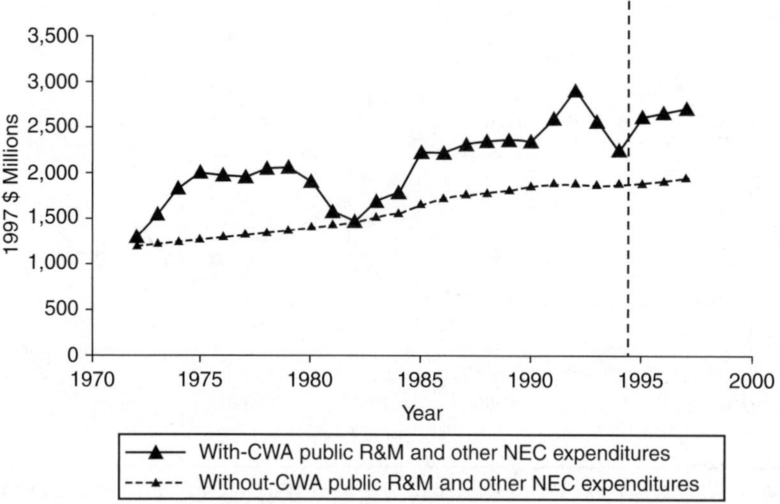

Figure 11F.63 Public R&M and other expenditures related to water pollution abatement, with and without the CWA, 1972–1997 (1997 $ millions). (From A Retrospective Assessment of the Costs of the Clean Water Act—1972–1997 Final Report, www.epa.gov.)

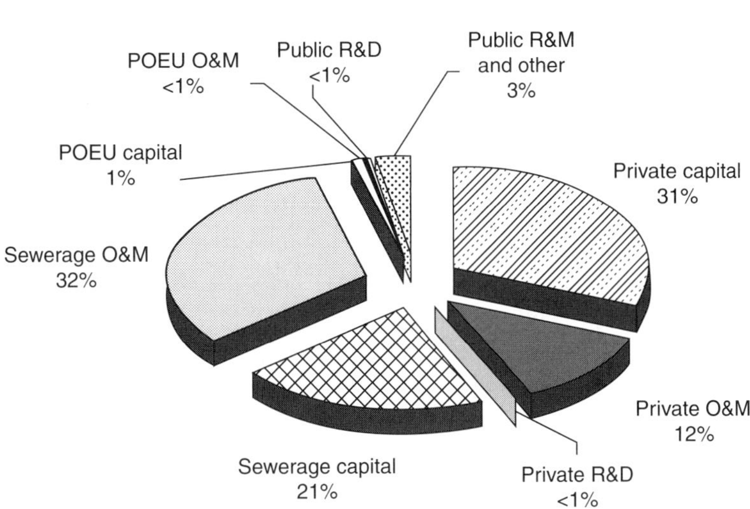

(a) Total incremental costs: $12.6 billion[a]

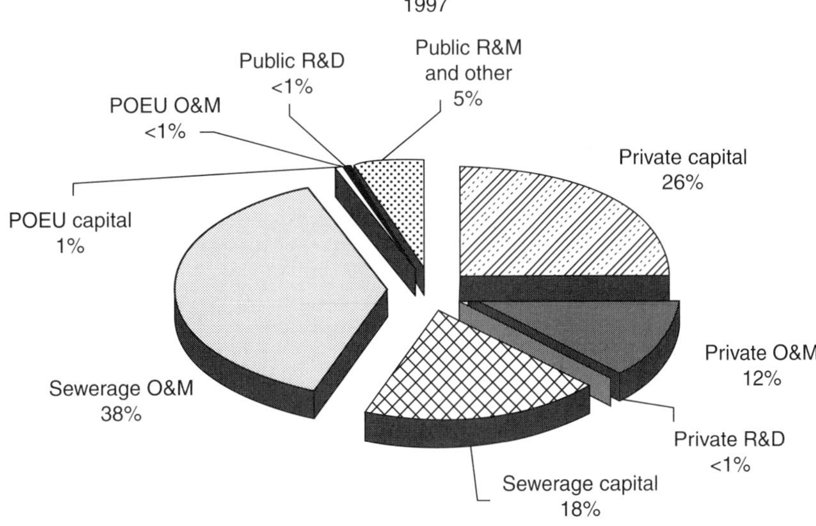

(b) Total incremental costs: $14.1 billion[a]

Figure 11F.64 The distribution of the incremental annual costs attributable to the CWA (3 percent discount rate). (From A Retrospective Assessment of the Costs of the Clean Water Act—1972–1997 Final Report, www.epa.gov.)

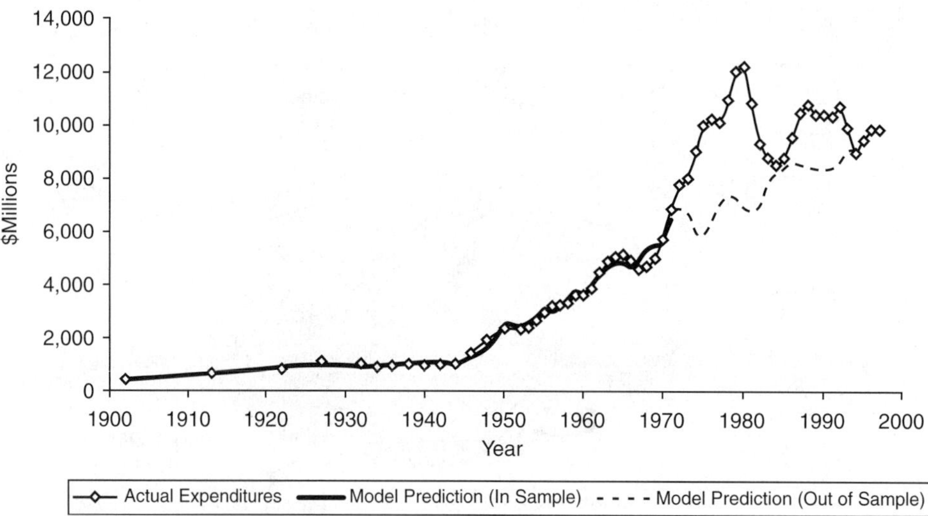

Figure 11F.65 Public sewerage capital expenditures, actual vs. model predictions, 1902–1997 (1997 $). (From A Retrospective Assessment of the Costs of the Clean Water Act—1972–1997 Final Report, www.epa.gov.)

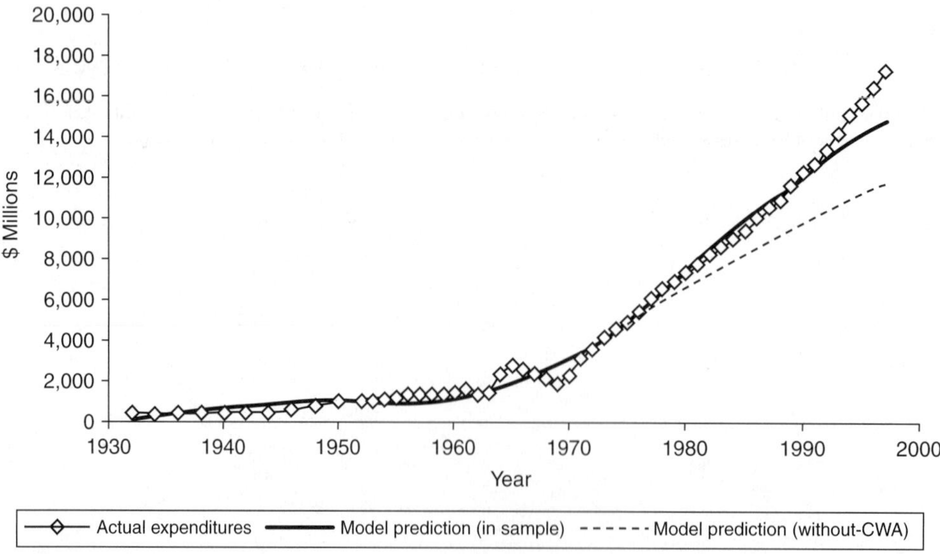

Figure 11F.66 Public sewerage O&M expenditures, actual vs. model predictions, 1932–1997 (1997 $). (From A Retrospective Assessment of the Costs of the Clean Water Act—1972–1997 Final Report, www.epa.gov.)

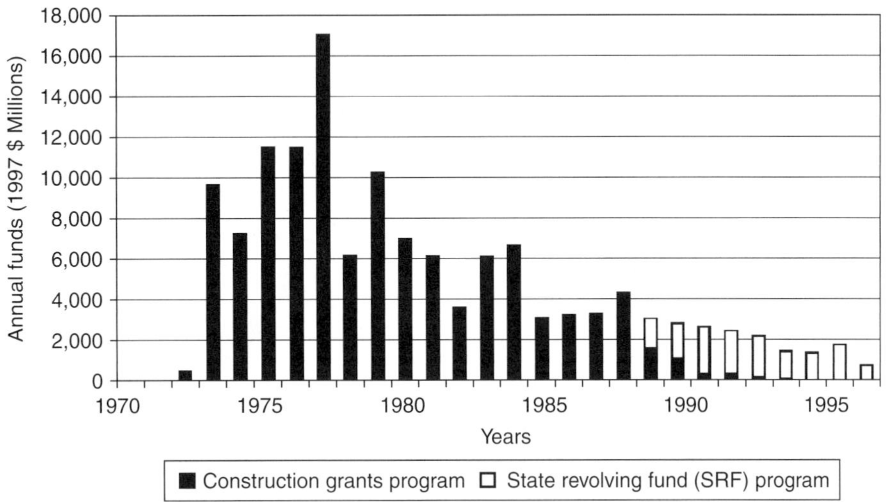

Figure 11F.67 EPA's municipal water pollution control programs annual funds: 1970–1999 (1997 $ millions). (From A Retrospective Assessment of the Costs of the Clean Water Act—1972–1997 Final Report, www.epa.gov.)

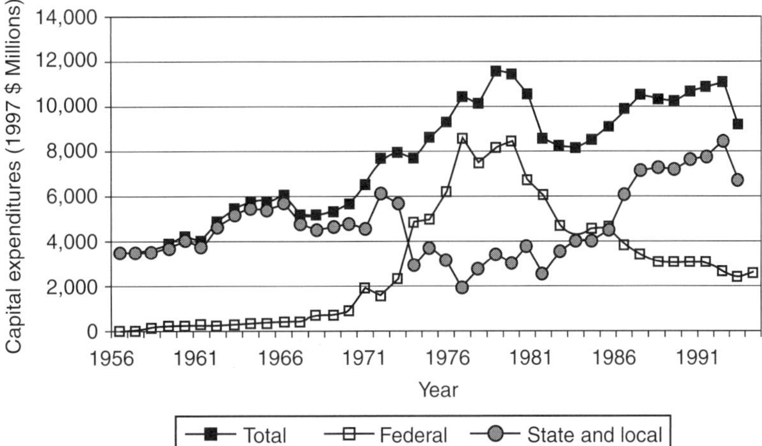

Figure 11F.68 Public wastewater capital expenditures 1956–1995 (1997 $ millions). (From A Retrospective Assessment of the Costs of the Clean Water Act—1972–1997 Final Report, www.epa.gov.)

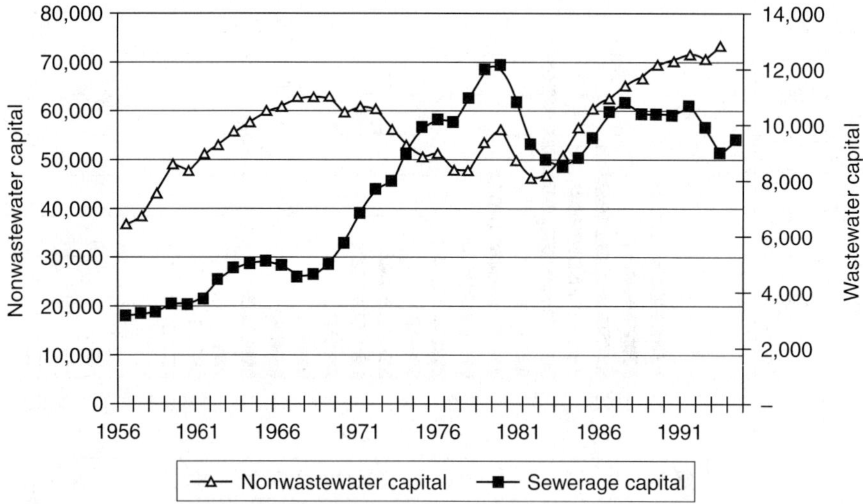

Figure 11F.69 Public infrastructure capital expenditures, 1956–1995 (1997 $ millions). (From A Retrospective Assessment of the Costs of the Clean Water Act—1972–1997 Final Report, www.epa.gov.)

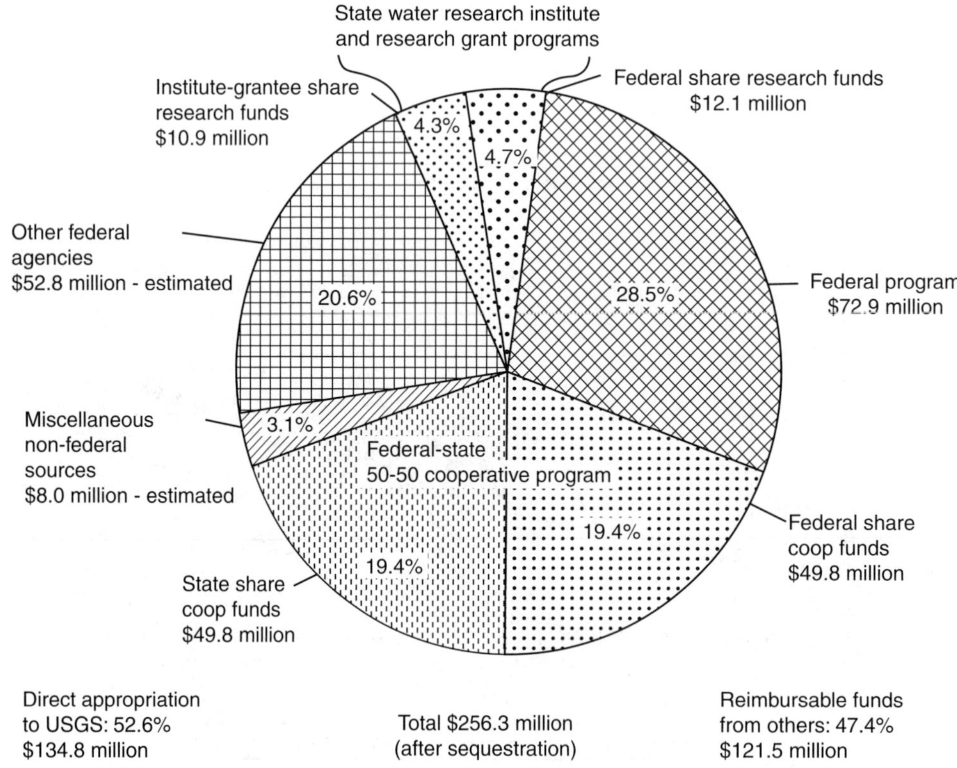

Figure 11F.70 Source of funds for U.S. Geological Survey water resources investigations in 1986. (From Cardin, G.W. et al., 1986, Water Resources Division in the 1980s, *U.S. Geological Survey Circular 1005*.)

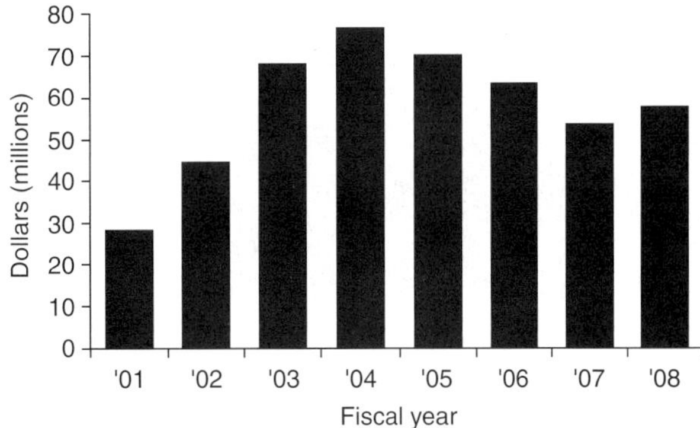

Figure 11F.71 WMD water resource development funding levels. (From dep.state.fl.us.gov.)

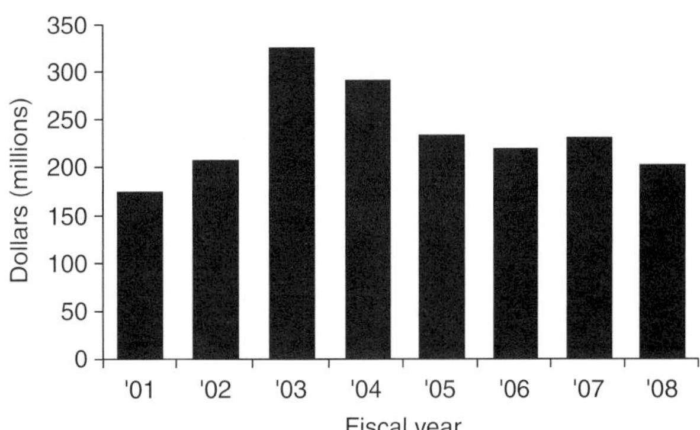

Figure 11F.72 CERP water resource development funding levels. (From dep.state.fl.us.gov.)

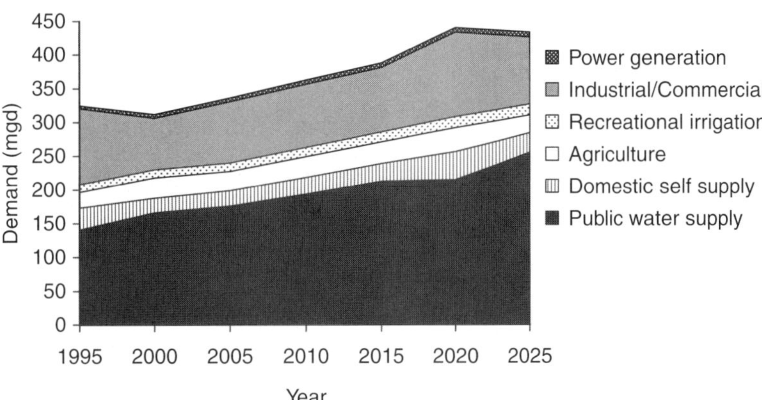

Figure 11F.73 NWFWMD districtwide demand projections. (From dep.state.fl.us.gov.)

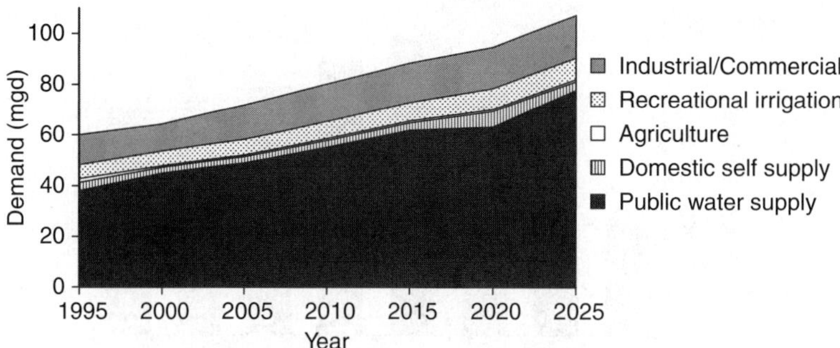

Figure 11F.74 NWFWMD planning region II demand projections. (From www.dep.state.fl.us.gov.)

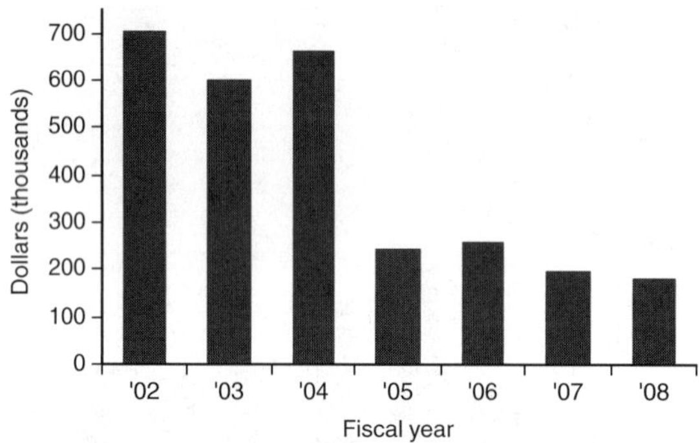

Figure 11F.75 NWFWMD funding for water resource development. (From www.dep.state.fl.us.gov.)

Table 11F.75 Northwest Florida Water Management District

	FY 96–97		FY 96–97		FY 97–98	
	% of Total	Budgeted Amount	% of Total	Actual Amount	% of Total	Budgeted Amount
Ad Valorem Taxes	7.25%	1,538,980	7.13%	1,463,439	3.73%	1,683,660
Ag Privilege Tax	0.00%	0	0.00%	0	0.00%	0
Canyover	36.20%	7,687,168	36.69%	7,530,026	19.01%	8,586,614
Permit and License Fees	1.92%	407,125	2.07%	424,295	0.91%	411,725
Local Revenues	1.04%	221,093	0.84%	172,132	0.25%	111,117
Ecosystem Management Trust Fund	6.66%	1,415,000	3.40%	697,804	3.25%	1,466,951
Water Management Lands Trust Fund	26.79%	5,688,528	26.50%	5,439,483	63.84%	28,832,639
Other State Revenues	16.81%	3,569,188	12.75%	2,617,545	4.83%	2,181,521
Federal Revenues	1.61%	341,886	1.36%	279,957	2.72%	1,230,603
Miscellaneous Revenues	1.71%	363,730	9.26%	1,900,201	1.46%	660,835
Total Revenues	100.00%	21,232,428	100.00%	20,524,882	100.00%	45,165,665
Salaries and Benefits	19.81%	4,206,979	34.14%	3,769,000	10.26%	4,634,713
Other Personal Services	7.72%	1,639,961	7.92%	874,258	5.05%	2,279,983
Expenses	5.55%	1,178,662	8.19%	904,625	2.58%	1,164,141
Operating Capital Outlay	2.40%	503,961	3.65%	402,946	0.65%	293,450
Fixed Capital Outlay	11.89%	2,523,840	22.66%	2,501,328	56.50%	25,519,900
Interagency Expenditures	0.07%	15,000	0.00%	0	1.47%	664,500
Debt	12.18%	2,586,888	23.43%	2,586,888	5.71%	2,578,682
Reserves	40.37%	8,571,137	0.00%	0	4.66%	8,030,296
Transfers	0.00%	0	0.00%	0	0.00%	0
Total Expenditures	100.00%	21,232,428	100.00%	11,039,045	100.00%	45,165,665

(Continued)

Table 11F.75 (Continued)

	FY 96–97		FY 96–97		FY 97–98	
	% of Total	**Budgeted Amount**	**% of Total**	**Actual Amount**	**% of Total**	**Budgeted Amount**
Water Resources Planning and Monitoring	24.61%	5,226,032	20.39%	2,251,404	11.03%	4,982,451
Acquisition, Restoration and Public Works	29.93%	6,355,196	48.71%	5,376,872	65.57%	29,614,245
Operation and Maint. of Lands and Works	16.79%	3,564,120	11.94%	1,317,559	9.74%	4,397,358
Regulation	5.10%	1,083,356	12.46%	1,375,545	3.25%	1,468,615
Outreach	1.84%	391,361	2.70%	297,348	0.92%	416,903
District Management and Administration	21.72%	4,612,363	3.80%	419,717	9.48%	4,286,093
Total Allocations by Program	**100.00%**	**21,232,428**	**100.00%**	**11,039,045**	**100.00%**	**45,165,665**

Note: Budget summary comparison.

Source: From www.myflorida.com. With permission.

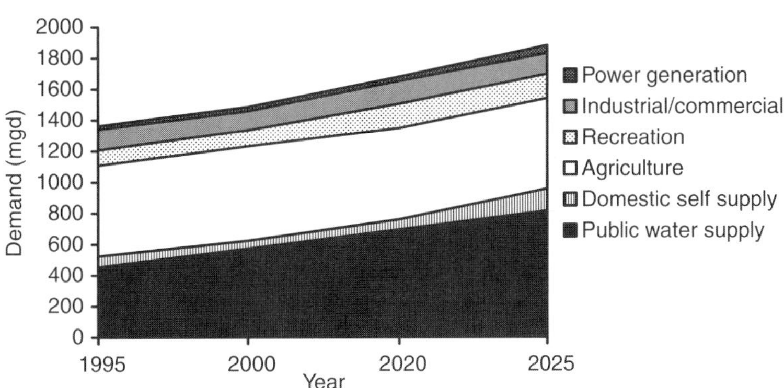

Figure 11F.76 SJRWMD demand projections. (From www.dep.state.fl.us.gov.)

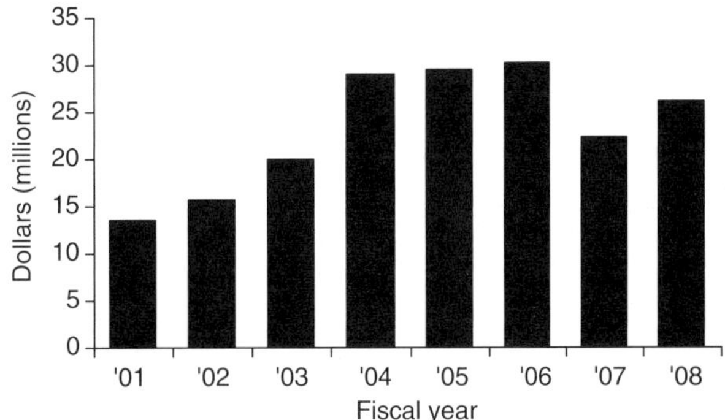

Figure 11F.77 SJRWMD funding for water resource development. (From www.dep.state.fl.us.gov.)

Table 11F.76 St. Johns River Water Management District

	FY 96–97		FY 96–97		FY 97–98	
	% of Total	**Budgeted Amount**	**% of Total**	**Actual Amount**	**% of Total**	**Budgeted Amount**
Ad Valorem Taxes	31.63%	53,154,717	42.335	53,821,960	40.91%	56,165,728
Ag Privilege Tax	0.00%	0	0.00%	0	0.00%	0
Canyover	22.91%	38,494,743	0.00%	0	27.50%	37,752,891
Permit and License Fees	0.85%	1,435,000	1.56%	1,985,787	1.05%	1,435,000
Local Revenues	2.74%	4,596,176	5.02%	6,382,236	2.44%	3,351,826
Ecosystem Management Trust Fund	1.18%	1,975,000	0.00%	0	0.29%	398,334
Water Management Lands Trust Fund	36.88%	61,971,000	38.87%	49,433,409	21.64%	29,700,000
Other State Revenues	2.85%	4,789,693	1.38%	1,753,092	3.83%	5,256,395
Federal Revenues	0.81%	1,367,657	10.71%	13,623,113	2.08%	2,851,000
Miscellaneous Revenues	0.15%	250,350	0.13%	161,352	0.26%	363,475
Total Revenues	100.00%	168,034,336	100.00%	127,160,949	100.00%	137,274,649
Salaries and Benefits	19.28%	32,391,440	27.93%	32,023,319	25.62%	35,175,920
Other Personal Services	8.36%	14,053,875	7.96%	9,125,788	12.40%	17,023,288
Expenses	17.32%	29,101,707	15.09%	17,301,201	2.58%	28,874,227
Operating Capital Outlay	2.44%	4,092,823	2.65%	3,035,139	2.58%	3,547,346
Fixed Capital Outlay	47.48%	79,778,928	40.26%	46,158,710	31.29%	42,958,743
Interagency Expenditures	0.00%	0	0.00%	0	0.00%	0
Debt	4.16%	6,995,563	6.10%	6,995,563	5.10%	6,995,313
Reserves	0.96%	1,620,000	0.02%	21,000	1.97%	2,699,812
Transfers	0.00%	0	0.00%	0	0.00%	0
Total Expenditures	100.00%	168,034,336	100.00%	114,660,720	100.00%	137,274,649
Water Resources Planning and Monitoring	8.99%	15,100,215	10.37%	11,889,183	13.36%	18,333,253
Acquisition, Restoration and Public Works	63.68%	106,997,469	56.06%	64,275,456	51.43%	70,597,860
Operation and Maint. of Lands and Works	9.85%	16,544,807	11.55%	13,247,852	14.34%	19,687,942
Regulation	5.23%	8,786,502	7.03%	8,055,333	6.64%	9,116,784
Outreach	2.79%	4,691,488	2.42%	2,771,517	2.53%	3,479,199
District Management and Administration	9.47%	15,913,855	12.58%	14,421,379	11.70%	16,059,611
Total Allocations by Program	**100.00%**	**168,034,336**	**100.00%**	**114,660,720**	**100.00%**	**137,274,649**

Note: Budget summary comparison.

Source: From www.myflorida.com. With permission.

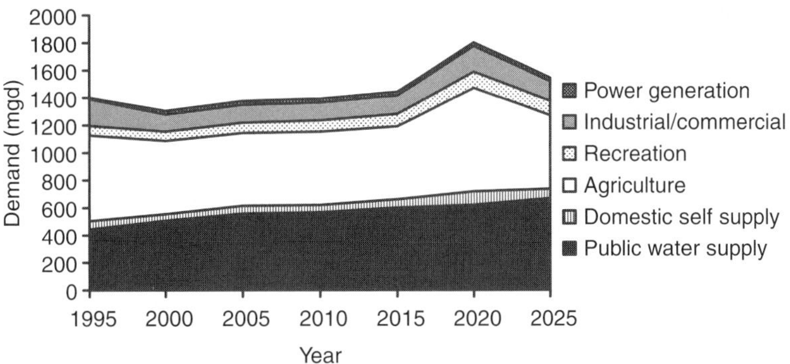

Figure 11F.78 SWFWMD districtwide demand projections. (From www.dep.state.fl.us.gov.)

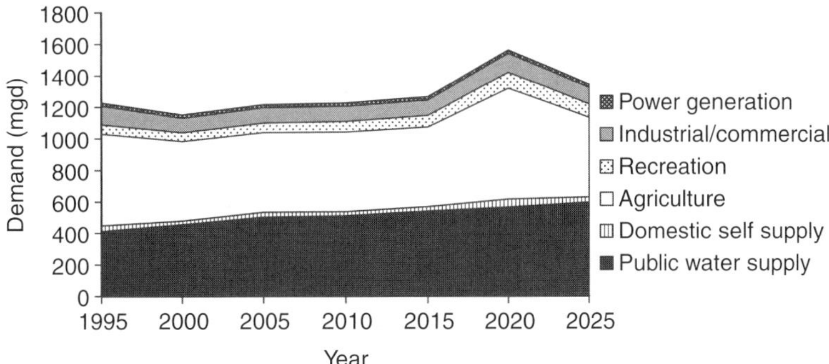

Figure 11F.79 SWFWMD demand projections for RWSP planning region. (From www.dep.state.fl.us.gov.)

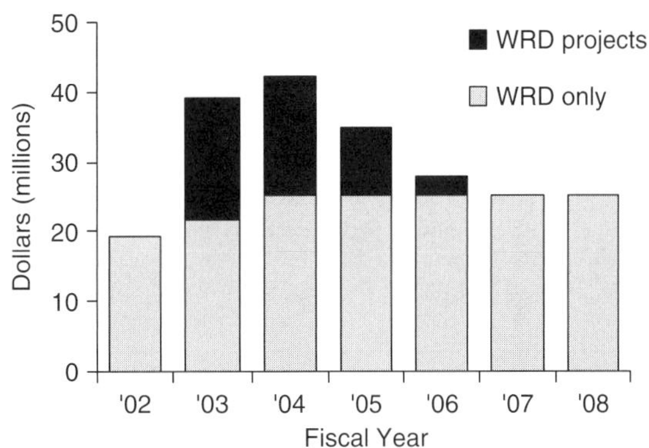

Figure 11F.80 SWFWMD funding for water resource development. (From www.dep.state.fl.us.gov.)

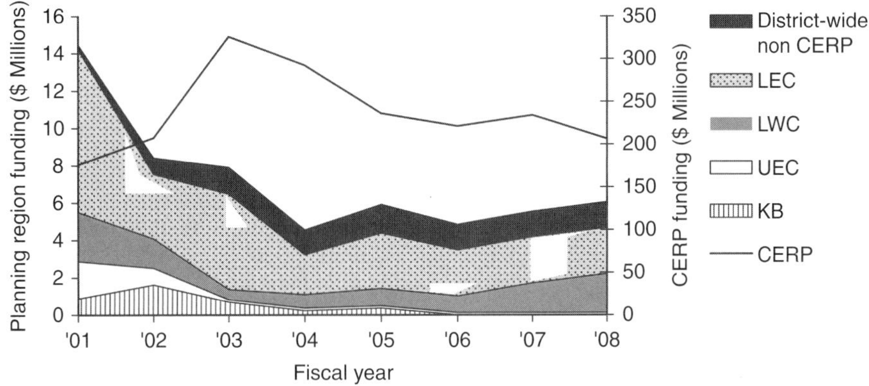

Figure 11F.81 SFWMD funding for water resource development. (From www.dep.state.fl.us.gov.)

Table 11F.77 Southwest Florida Water Management District

	FY 96–97		FY 96–97		FY 97–98	
	% of Total	Budgeted Amount	% of Total	Actual Amount	% of Total	Budgeted Amount
Ad Valorem Taxes	56.08%	84,214,617	65.85%	84,602,369	57.61%	88,033,728
Ag Privilege Tax	0.00%	0	0.00%	0	0.00%	0
Canyover	11.85%	17,790,898	13.85%	17,790,898	8.12%	12,408,347
Permit and License Fees	1.00%	1,500,000	1.49%	1,914,065	1.10%	1,680,000
Local Revenues	1.49%	2,241,900	0.35%	455,762	1.80%	2,747,000
Ecosystem Management Trust Fund	1.32%	1,975,000	0.92%	1,176,059	0.26%	398,333
Water Management Lands Trust Fund	22.90%	34,390,755	7.33%	9,423,549	23.25%	35,522,479
Other State Revenues	0.65%	982,628	1.34%	1,717,221	1.85%	2,833,327
Federal Revenues	0.07%	100,000	0.00%	0	0.00%	0
Miscellaneous Revenues	4.65%	6,979,138	8.88%	11,403,192	6.01%	9,183,821
Total Revenues	100.00%	15,174,936	100.00%	128,483,105	100.00%	152,807,035
Salaries and Benefits	23.53%	35,342,618	37.85%	35,343,410	24.13%	36,879,607
Other Personal Services	10.90%	16,366,679	13.12%	12,252,750	11.02%	16,843,524
Expenses	8.05%	12,092,405	12.55%	11,719,061	8.17%	12,479,555
Operating Capital Outlay	1.23%	1,845,113	2.93%	2,739,702	1.48%	2,255,571
Fixed Capital Outlay	20.61%	30,947,900	9.05%	8,449,110	20.84%	31,837,414
Interagency Expenditures	26.21%	39,365,754	23.58%	22,017,359	27.91%	42,648,355
Debt	0.00%	0	0.00%	0	0.00%	0
Reserves	8.34%	12,530,329	0.00%	0	4.66%	7,123,188
Transfers	1.12%	1,684,138	0.91%	847,700	1.79%	2,739,821
Total Expenditures	100.00%	150,174,936	100.00%	93,369,092	100.00%	152,807,035
Water Resources Planning and Monitoring	5.38%	8,077,256	9.69%	9,050,500	6.63%	10,130,046
Acquisition, Restoration and Public Works	27.92%	41,933,068	17.00%	15,870,244	28.31%	43,263,249
Operation and Maint. of Lands and Works	7.07%	10,623,684	10.47%	9,771,258	7.00%	10,701,419
Regulation	8.11%	12,178,886	11.94%	11,147,497	8.19%	12,520,095
Outreach	35.61%	53,483,156	28.54%	26,646,697	33.42%	51,064,962
District Management and Administration	15.90%	23,878,886	22.37%	20,882,896	16.44%	25,127,264
Total Allocations by Program	**100.00%**	**150,174,936**	**100.00%**	**93,369,092**	**100.00%**	**152,807,035**

Note: Budget summary comparison.

Source: From www.myflorida.com. With permission.

Table 11F.78 South Florida Water Management District

	FY 96–97		FY 96–97		FY 97–98	
	% of Total	Budgeted Amount	% of Total	Actual Amount	% of Total	Budgeted Amount
Ad Valorem Taxes	37.53%	177,099,989	56.61%	179,297,173	45.42%	195,363,519
Ag Privilege Tax	2.73%	12,880,532	3.86%	12,230,695	2.87%	12,357,026
Canyover	22.85%	107,813,178	0.00%	0	21.37%	91,904,398
Permit and License Fees	0.49%	2,328,250	0.65%	2,062,164	0.45%	1,941,075
Local Revenues	0.03%	156,140	0.14%	441,006	0.11%	477,000
Ecosystem Management Trust Fund	0.45%	2,100,000	0.06%	183,791	0.09%	398,000
Water Management Lands Trust Fund	2.47%	11,640,000	3.93%	12,459,257	2.99%	12,875,000
Other State Revenues	15.88%	74,939,224	15.51%	49,133,150	8.51%	36,610,955
Federal Revenues	14.44%	68,118,314	11.51%	36,455,517	14.37%	61,824,950
Miscellaneous Revenues	3.13%	14,751,946	7.73%	24,475,208	3.81%	16,401,080
Total Revenues	100.00%	471,827,573	100.00%	316,737,961	100.00%	439,152,903
Salaries and Benefits	20.71%	97,718,919	31.82%	91,204,979	17.14%	73,712,088
Other Personal Services	7.08%	33,426,318	6.76%	19,368,300	14.32%	61,606,721
Expenses	10.44%	49,257,512	12.84%	36,796,249	10.63%	45,744,725
Operating Capital Outlay	3.81%	17,957,747	2.59%	7,434,809	3.59%	15,430,762
Fixed Capital Outlay	45.59%	215,111,342	39.37%	112,841,705	40.85%	175,702,565
Interagency Expenditures	5.13%	24,203,548	4.46%	12,784,047	9.98%	42,935,634
Debt	1.31%	6,172,033	2.15%	6,172,033	1.83%	7,866,097
Reserves	5.93%	27,980,154	0.00%	0	1.66%	7,154,311
Transfers	0.00%	0	0.00%	0	0.00%	0
Total Expenditures	100.00%	471,827,573	100.00%	286,602,122	100.00%	430,152,903
Water Resources Planning and Monitoring	9.71%	45,811,978	12.19%	34,941,956	14.41%	61,969,108
Acquisition, Restoration and Public Works	61.28%	289,154,312	48.75%	139,727,913	51.64%	222,144,554
Operation and Maint. of Lands and Works	12.93%	61,020,642	19.07%	54,649,903	14.58%	62,711,481
Regulation	2.37%	11,162,461	3.67%	10,518,609	2.79%	12,009,635
Outreach	2.76%	13,034,200	2.60%	7,441,626	5.25%	22,571,014
District Management and Administration	10.95%	51,643,980	13.72%	39,322,115	11.33%	48,747,111
Total Allocations by Program	**100.00%**	**471,827,573**	**100.00%**	**286,602,122**	**100.00%**	**430,152,903**

Note: Budget summary comparison.

Source: From www.myflorida.com. With permission.

Table 11F.79 Suwannee River Water Management District

	FY 96–97		FY 96–97		FY 97–98	
	% of Total	Budgeted Amount	% of Total	Actual Amount	% of Total	Budgeted Amount
Ad Valorem Taxes	12.99%	2,500,000	12.85%	2,504,734	12.49%	2,675,000
Ag Privilege Tax	0.00%	0	0.00%	0	0.00%	0
Canyover	12.30%	2,367,000	3.40%	663,536	9.80%	2,099,000
Permit and License Fees	1.12%	2,15,000	1.02%	198,318	1.00%	215,000
Local Revenues	0.00%	0	0.00%	0	0.00%	0
Ecosystem Management Trust Fund	9.74%	1,875,000	0.92%	179,841	5.51%	1,179,700
Water Management Lands Trust Fund	53.30%	1,0,259,200	68.76%	3,401,803	49.03%	10,500,000
Other State Revenues	7.75%	1,492,000	7.77%	1,515,299	18.48%	3,957,700
Federal Revenues	0.78%	1,49,500	0.44%	86,606	1.84%	395,000
Miscellaneous Revenues	2.03%	390,000	4.83%	941,828	1.84%	394,400
Total Revenues	100.00%	1,9,247,700	100.00%	19,491,965	100.00%	21,415,800
Salaries and Benefits	20.41%	3,929,000	27.65%	3,544,152	17.95%	3,843,100
Other Personal Services	26.54%	5,108,300	22.45%	2,877,909	29.47%	6,310,300
Expenses	5.82%	1,120,000	5.77%	739,199	5.89%	1,261,200
Operating Capital Outlay	2.09%	402,300	2.67%	341,992	1.23%	263,200
Fixed Capital Outlay	38.54%	7,418,000	31.55%	4,044,309	39.34%	8,426,000
Interagency Expenditures	0.00%	0	0.00%	0	0.00%	0
Debt	6.60%	1,270,100	9.91%	1,269,591	5.89%	1,262,000
Reserves	0.00%	0	0.00%	0	0.23%	50,000
Transfers	0.00%	0	0.00%	0	0.00%	0
Total Expenditures	100.00%	19,247,700	100.00%	12,817,152	100.00%	21,415,800
Water Resources Planning and Monitoring	23.17%	4,459,500	21.08%	2,702,432	19.23%	4,117,900
Acquisition, Restoration and Public Works	48.50%	9,334,300	46.95%	6,017,836	43.88%	9,398,200
Operation and Maint. of Lands and Works	6.70%	1,289,900	7.48%	958,889	7.26%	1,554,200
Regulation	7.28%	1,401,500	7.75%	992,874	16.09%	3,445,300
Outreach	1.92%	369,900	2.60%	333,180	2.03%	435,500
District Management and Administration	12.43%	2,392,600	14.14%	1,811,941	11.51%	2,464,700
Total Allocations by Program	**100.00%**	**19,247,700**	**100.00%**	**12,817,152**	**100.00%**	**21,415,800**

Note: Budget summary comparison.

Source: From www.myflorida.com. With permission.

Table 11F.80 Water Resources Research Categories

I. Nature of Water

Category I deals with fundamental research on the water substance

A. *Properties of water*—Study of the physical and chemical properties of pure water and its thermodynamic behavior in its various states

B. *Aqueous solutions and suspensions*—Study of the effects of various solutes on the properties of water; surface interactions; colloidal suspensions

II Water Cycle

Category II covers generally research on the natural processes involving water. It is an essential supporting effort to applied problems in later categories

A. *General*—Studies involving two or more phases of the water cycle such as hydrologic models; rainfall-runoff relations; surface and groundwater relationships; watershed studies, etc.

B. *Precipitation*—Investigation of spatial and temporal variations of precipitation; physiographic effects; time trends; extremes; probable maximum precipitation; structure of storms, etc.

C. *Snow, ice, and frost*—Studies of the occurrence and thermodynamics of water in the solid state in nature; spatial variations of snow and frost; formation of ice and frost; breakup of river and lake ice; glaciers, permafrost, etc.

D. *Evaporation and transpiration*—Investigation of the process of evaporation from lakes, soil, and snow and of the transpiration process in plants; methods of estimating actual evapotranspiration; energy balance; etc.

E. *Streamflow*—Mechanics of flow in streams; flood routing; bank storage; space and time variations (includes high and low-flow frequency); droughts; floods, etc.

F. *Groundwater*—Study of the mechanics of groundwater movements; multiphase systems; sources of natural recharge; mechanics of flow to wells and drains; subsidence; properties of aquifers; etc.

G. *Water in soils*—Infiltration, movement and storage of water in the zone of aeration, including soil

H. *Lakes*—Hydrologic, hydrochemical, and thermal regimes of lakes; water level fluctuations; currents and waves

I. *Water and plants*—Role of plants in hydrologic cycle; water requirements of plants; interception

J. *Erosion and sedimentation*—Studies of the erosion process; prediction of sediment yield; sedimentation in lakes and reservoirs; stream erosion; sediment transport, etc.

K. *Chemical processes*—Chemical interactions between water and its natural environment; chemistry of precipitation

L. *Estuarine problems*—Special problems of the estuarine environment; effect of tides on flow and stage; deposition of sediments; sea water intrusion in estuaries

III Water Supply Augmentation and Conservation

As water use increases we must pay increasing attention to methods for augmenting and conserving available supplies. Research in Category III is largely applied research devoted to this problem area

A. *Saline water conversion*—Research and development related to methods of desalting sea water and brackish water

B. *Water yield improvement*—Increasing streamflow or improving its distribution through land management; water harvesting from impervious areas; phreatophyte control; reservoir evaporation suppression

C. *Use of water of impaired quality*—Research on methods of agricultural use of water of high salinity; use of poor quality water in industry; crop tolerance to salinity

D. *Conservation in domestic use*—Methods for reducing domestic water needs without impairment of service

E. *Conservation in industry*—Reduction in both consumption and diversion requirements for industry

F. *Conservation in agriculture*—More efficient irrigation practices. Chemical control of evaporation and transpiration; lower water use plants, etc.

IV Water Quantity Management and Control

Category IV includes research directed to the management of water, exclusive of conservation, and the effects of related activities on water

A. *Control of water on the land*—Effects of land management on runoff; land drainage, potholes; etc.

B. *Groundwater management*—Artificial recharge; conjunctive operation; relation to irrigation

C. *Effects of man's related activities on water*—Impact of urbanization, highways, logging, etc. on water yields and flow rates

D. *Watershed protection*—Methods of controlling erosion to reduce sediment load of streams and conserve soil

V Water Quality Management and Protection

An increasing population increases the wastes and other pollutants entering our water supplies. Category V deals with methods of identifying, describing and controlling this pollution

A. *Identification of pollutants*—Techniques of identification of physical, chemical and biological pollutants; rational measures of character and strength of wastes

B. *Sources and fate of pollution*—Determination of the sources of pollutants in water; the nature of the pollution from various sources; path of pollutant from source to stream of groundwater

C. *Effects of pollution*—Definition of the effect of pollutants, singly and in combination, on man, aquatic life, agriculture and industry under conditions of sustained use; eutrophication

D. *Waste treatment processes*—Research to improve conventional treatment methods to gain efficiency or reduce cost; processes to treat new types of waste; advanced treatment methods ofr more complete removal of pollutants including purification for direct reuse

E. *Ultimate disposal of wastes*—Disposal of residual material removed from water and sewage during the treatment process; disposal of waste brines

(Continued)

Table 11F.80 (Continued)

	F. *Water treatment*—Development of more efficient and economical methods of making water suitable for domestic or industrial use
	G. *Water quality control*—Research on methods to control stream and reservoir water quality such as flow augmentation; stream and reservoir aeration; control of natural pollution; control of pollution from pesticides and agricultural chemicals; control of acid mine drainage; etc.
VI	Water Resources Planning
	The problems of achieving an optimal plan of water development are becoming increasingly complex. Category VI covers research devoted to determining the best way to plan, the appropriate criteria for planning and the nature of the economic, legal and institutional aspects of the planning process
	A. *Techniques of planning*—Application of systems analysis to project planning; treatment of uncertainty; probability studies
	B. *Evaluation process*—Development of methods, concepts and criteria for evaluating project benefits; discount rate; project life; methods for economic, social and technological projections; reliability of projections; research on the value of water in various uses
	C. *Cost allocation, cost sharing, pricing/repayment*—Research on methods of calculating repayment and establishing prices for vendible products; techniques of cost allocation; cost sharing, pricing and repayment policy
	D. *Water demand*—Research on the water quantity and quality requirements of various uses, both diversion and consumption
	E. *Water law and institutions*—A study of state and Federal water law looking to changes and additions which will encourage greater efficiency in use; investigation of institutional structures and constraints which influence decisions on water at all levels of government
	F. *Nonstructural alternatives*—Exploration of methods to achieve water development aims by nonstructural methods such as flood plain zoning
	G. *Ecologic impact of water development*—Effects of water management operations on overall ecology of the area. Excludes effect of pollution under V-C
VII	Resources Data
	Planning and management of our water resources require information. Category VII includes research oriented to data needs and the most efficiency method of meeting these needs
	A. *Network design*—Studies of data requirements and of the most effective methods of collecting the data
	B. *Data acquisition*—Research on new and improved instruments and techniques for collection of water resources data; telemetering equipment
	C. *Evaluation, processing and publication*—Studies of effective methods of processing data, form and nature of published data; maps of data
VIII	Engineering Works
	To implement water development plans requires engineering works. Category VIII describes research on design, materials and construction which is generally useful to all aspects of water management. Works relevant to a single specific goal, such as water treatment or desalination, are included elsewhere if an appropriate category exists
	A. *Design*—Research leading to improved design of dams, canals, pipelines, locks, fishways, and other works required for water resource development
	B. *Materials*—Research to improve existing structural materials and to develop new materials; subsurface exploration of foundations; corrosion; etc.
	C. *Construction and operation*—Research on efficient construction methods, operating systems, and maintenance procedures
IX	Manpower, Grants and Facilities
	Trained manpower is an essential ingredient of research on water resources and the planning and design of water development projects. Category IX describes plans for support of education and training. It also includes grant and contract programs for which advance distribution to specific categories is impossible
	A. *Education*—extramural—Support of education in water resources at universities (not including reaerch support under other categories)
	B. *Education*—in-house—Government employee training programs
	C. *Research facilities*—Laboratories, field stations, etc.
	D. *Grants, contracts and research and allotments*—Allotments to University Water Resources Research Institutes under P.L. 88–379; OWRR, HEW, NSF, CSRS, and other grants which cannot be distributed to categories in advance

Source: From Committee on Water Resources Research, *Federal Council for Science and Technology*, 1966.

Table 11F.81 Water Resources Research in the United States — Areas of Interest in 1989

General Areas of Interest

1. Aspects of the hydrologic cycle
2. Supply and demand for water
3. Demineralization of saline and other impaired waters
4. Conservation and best use of available supplies of water and methods of increasing such supplies
5. Water reuse
6. Depletion and degradation of groundwater supplies
7. Improvements in the productivity of water when used for agricultural, municipal, and commercial purposes
8. The economic, legal, engineering, social, recreational, biological, geographic, ecological, and other aspects of water problems
9. Scientific information-dissemination activities, including identifying, assembling and interpreting the result of scientific and engineering research on water resources problems
10. Providing means for improved communications of research results, having due regard for the varying conditions and needs for the respective States and regions

Specific Areas of Interest

A. Groundwater quality

1. An improved understanding of the transport and fate of contaminants in groundwater and in the unsaturated zone, from both field and modeling studies. This will require attention to the physical, microbial, and chemical interactions encountered by dissolved, miscible, and immiscible substances in groundwater transport
2. Better equipment and methodology for monitoring and characterizing the chemical, physical, and biological conditions of the subsurface environment, including sampling strategies that lead to an approved understanding of groundwater contamination at scales larger than single contamination sources

B. Water quality management

1. An improved understanding of hazardous substances in water, including interactions of synthetic and natural organics, interactions of waterborne toxic materials with organisms, the binding of chemical species to waterborne sediments, and microbial alteration of organic compounds in aqueous environments
2. Improved methods for treatment or control of contaminants in natural water systems, emphasizing reduction, removal, or alteration to prevent further contamination or to increase potable supplies
3. Applications of biotechnology to water resources, including detection of specific pollutants using DNA probes and genetic engineering of microorganisms for water quality management
4. Use of waters of impaired quality, including improved physical and chemical methods for treatment of impaired waters and the cost-effective utilization of brackish waters

C. Institutional change in water resource management

1. Analysis of the social aspects and the distributive consequences of alternative institutional arrangements
2. Description and evaluation of institutions for consideration/compensation of third-party interests; i.e., those individuals or groups affected by, but external to, reallocation decisions
3. A description of the market structure that has evolved or is likely to evolve if institutional change involves development of water markets
4. Representative case studies providing estimates of economic efficiency gains and the distributive consequences from institutional changes

Source: From *U.S. Geological Survey*, 1988, Announcement 7442, Request for Applications Under the Water Resources Research Act of 1984 (Section 105).

Table 11F.82 Demand and Funding Summary of Planning Regions

	Dates for RWSPs		Current and Projected Water Demands (mgd)			Funding and Water Made Available	
Planning Region	Completed	Update	2000 Demand	Previously Projected 2020 Demand	Current Projected 2025 Demand	WRD for Next 5 Years ($ millions)	Additional Water to Be Made Available by WRD (mgd)
NWFWMD (Region II)	2001	Oct. 2005	64	95	107	1.5	81 by 2020
SJRWMD (Entire WMD)	2000	Oct. 2005	1,485	1,679	1,880	136.8	751 total potential
SWFWMD (Central and Southern part of WMD)	2001	Dec. 2005	1,151	1,562	1,432	155.5	42 by 2004, 678 total potential
SFWMD							
Kissimmee Basin	2000	July 2005	264	663	398	0.61	71 by 2008
Upper East Coast	2000	June 2004	292	660	338	0.63	63 by 2008
Lower West Coast	2000	Oct. 2005	Not available	1,099	Not available	6.2	190 by 2008
Lower East Coast	2000	Dec. 2005	Not available	2,521	Not available	12.5	155 by 2008
Districtwide Projects						6.95	

Source: From USGS Water Use Data 2000.

Table 11F.83 Funding Source Water Protection Activities

Funding Sources	Polluted Runoff Control						Resource Protection and Restoration						Wastewater	
	Agriculture	Forestry	Mining	Brownfields	Storage Tanks (above ground & underground)	Landfills	Hydromodification/ Habitat Modification	Wetlands/Riparian Zone	Land Acquisition/ Conservation Easements	Groundwater Protection Measures	Land Use Controls	Public Education	Stormwater	On-Site Sewage
Brownfields Cleanup Revolving Loan Fund Pilots FY '03—Grants to $1M		•	•									•		
Chesapeake Bay Program Grants FY '02—$750,000 FY '03—$2M	•	•	•	•	•	•	•	•	•	•	•	•	•	•
Clean Water State Revolving Fund Loans FY '02—$4.4B FY '03—$4.0B (est.)	•	•	•	•	•	•	•	•	•	•	•	•	•	•
Drinking Water State Revolving Fund Set-Asides Up to 31% Cap. Grant	•	•			•		•	•	•	•	•	•	•	•
Great Lakes Program FY '02—$5.7M FY '03—$5.7M (est.)							•	•		•		•		
Nonpoint (319) Source Implementation Grants FY '02—$237.5M FY '03—$238.4M	•	•	•	•	•	•	•	•	•	•	•	•	•	•
Pollution Prevention Incentives for States FY '02—$5M FY '03—$5M										•	•	•		
Tribal DW Capacity Building/SWP Grants FY '02—$1.9M FY '03—$1.9M	•	•					•	•	•	•				

EPA	Tribal Grants for Surface/ Grandwater Protection FY '02—$447,700 FY '03—$445,500	•						•	•	•	•		•		
EPA	Underground Injection Control Grants FY '02—$10M FY '03—$10M										•		•		
EPA	Water Pollution Control (106) Program Support FY '02—$192.5M FY '03—$180.4M	•									•	•	•	•	
EPA	Water Quality Cooperative Agreements FY '02—$18.96M FY '03—$18.84M	•	•	•	•	•	•	•	•		•		•	•	•

Funding Source Details and Contact Information

Brownfields Cleanup Revolving Loan Fund Pilots: The objective of the Brownfields Cleanup Loan Fund grants is to capitalize loan funds that can make loans or grants to facilitate cleanup of brownfield sites contaminated with hazardous substances or petroleum products, as well as mine scarred land and 'drug labs.' Eligible organizations include businesses, nonprofit groups, local governments, state/territorial agencies, or tribal agencies. For more information see: www.epa.gov/brownfields

Chesapeake Bay Program Grants: Provides grant awards to reduce and prevent pollution and to improve the living resources of the Chesapeake Bay Watershed. Grants may be provided to state and local governments and nonprofit organizations for implementation projects, as well as monitoring and other related activities in the Chesapeake Bay Basin (Maryland, Pennsylvania, Virginia, and District of Columbia). For more information see: www.chesapeakebay.net

Clean Water State Revolving Fund Loans: Funds are used to make low interest loans to communities, individuals and others for water-quality improvement activities. Traditionally the funds have been used for wastewater treatment facilities, however loans are used increasingly for other water quality management activities including nonpoint source and estuary projects. The CWSRF is currently funding approximately $4 billion annually. For more information see: www.epa.gov/owm/cwfinance/cwsrf

Drinking Water State Revolving Fund Set-Asides: Up to 31% of the DWSRF capitalization grant may be used for set-aside activities including loans for the acquisition of land or easements for source water protection or for implementation of source water protection measures, or direct assistance for wellhead protection programs. For more information see; www.epa.gov/safewater/dwsrf.html

Great Lakes Program: EPA's Great Lakes Program issues awards to monitor Great Lakes ecosystem indicators; provide public access to Great Lakes data; help communities address contaminated sediments; support local protection and restoration activities; promote pollution prevention; and provide assistance to implement community-based Remedial Action Plans and for Lakewide Management Plans. For more information see: www.epa.gov/glnpo/fund/glf.html

Nonpoint (319) Source Implementation Grants: Provides grants to states and tribes to implement nonpoint source projects and programs. These include Best Management Practices (BMP) installation for animal wastes and sediment, pesticide and fertilizer control, stream bank restoration, lake protection/restoration, septic system restoration and management, etc. Al beneficiaries, except for tribes, are required to provide 40% of the total project or program costs. For more information see: www.epa.gov/owow/nps/cwact.html

Pollution Prevention Incentives for States: Provides grants focused on institutionalizing multimedia (air, water, land) pollution prevention techniques. Eligible entities include state and local agencies, universities, nonprofit organizations, and private business. Projects include technical assistance, data collection, education and outreach, training, environmental auditing, demonstration projects, and the integration of pollution prevention into state regulatory programs. For more information see: www.epa.gov/p2/grants/

Tribal Drinking Water Capacity Building/Source Water Protection Grants: Grants are intended to increase tribal capability to provide safe drinking water to consumers, and to prevent tribal sources of drinking water from being contaminated. Eligible projects might include source water assessments and the development and implementation of a source water protection program, and projects to improve a drinking water system's technical, financial and managerial capacity. For more information see: www.epa.gov/safewater/protect.html

Tribal Grants for Surface and Groundwater Protection, Pesticide Management Planning: Funds are available for tribes to develop Pesticide Management Plans (PMPs) and to address other pesticide-related groundwater concerns. The funding is intended for short term projects, stand alone components of larger projects, or projects that are expected to be self-sustaining once funding is used. For more information see: www.epa.gov/oppfead1/rstfield/

Underground Injection Control Grants: Provides grant funds to states for such purposes as state regulation review, program plan developments, data management, inventory of injection facilities, identification of aquifers, public participation, technical assistance and review, permit approval and enforcement, and surveillance and investigation. For more information see: epa.gov/opwdw000/uic.html

Water Pollution Control (106) Program Support: Provides grants to states, tribes and interstate water pollution control agencies to support the prevention and abatement of surface and groundwater pollution from point and nonpoint sources. Eligible activities include water quality planning, monitoring, permitting, surveillance, enforcement, advice and assistance to local agencies, etc. for the purpose of establishing and maintaining water pollution control programs. For more information see: epa.gov/cwfinance/pollutioncontrol.htm

Water Quality Cooperative Agreements (104(b)(3) Grants): Provides grants to support innovative demonstration projects for addressing stormwater, combined sewer overflows, sludge, pretreatment, mining, animal feeding operations, and other sources relating to the National Pollutant Discharge Elimination System (NPDES) program. This includes research, investigations, experiments, training, surveys, and studies related to the causes, effects, and prevention of pollution. For more information see: epa.gov/owm/cwfinance/waterquality.htm

		Polluted Runoff Control						Resource Protection and Restoration						Wastewater	
	Funding Sources	Agriculture	Forestry	Mining	Brownfields	Storage Tanks (above ground & underground)	Landfills	Hydromodification/ Habitat Modification	Wetlands/Riparian Zone	Land Acquisition/ Conservation Easements	Groundwater Protection Measures	Land Use Controls	Public Education	Stormwater	On-Site Sewage
EPA	Watershed Assistance Grants FY '02—$200,000 FY '03—$200,000										•		•		
EPA	Wetlands Program Development Grants FY '02—$15M FY '03—$14.9M		•					•	•				•	•	
USDA	Conservation Reserve Program FY '02—Not Available FY '03—Not Available	•						•	•						
USDA	Conservation Security Program FY '03—$2B (authorized)	•	•												
USDA	Environmental Quality Incentives Program FY '02—$387M FY '03—$695M	•						•	•	•	•		•		
USDA	Farmland Protection Program FY '02—$50M FY '03—$100M	•	•						•	•					

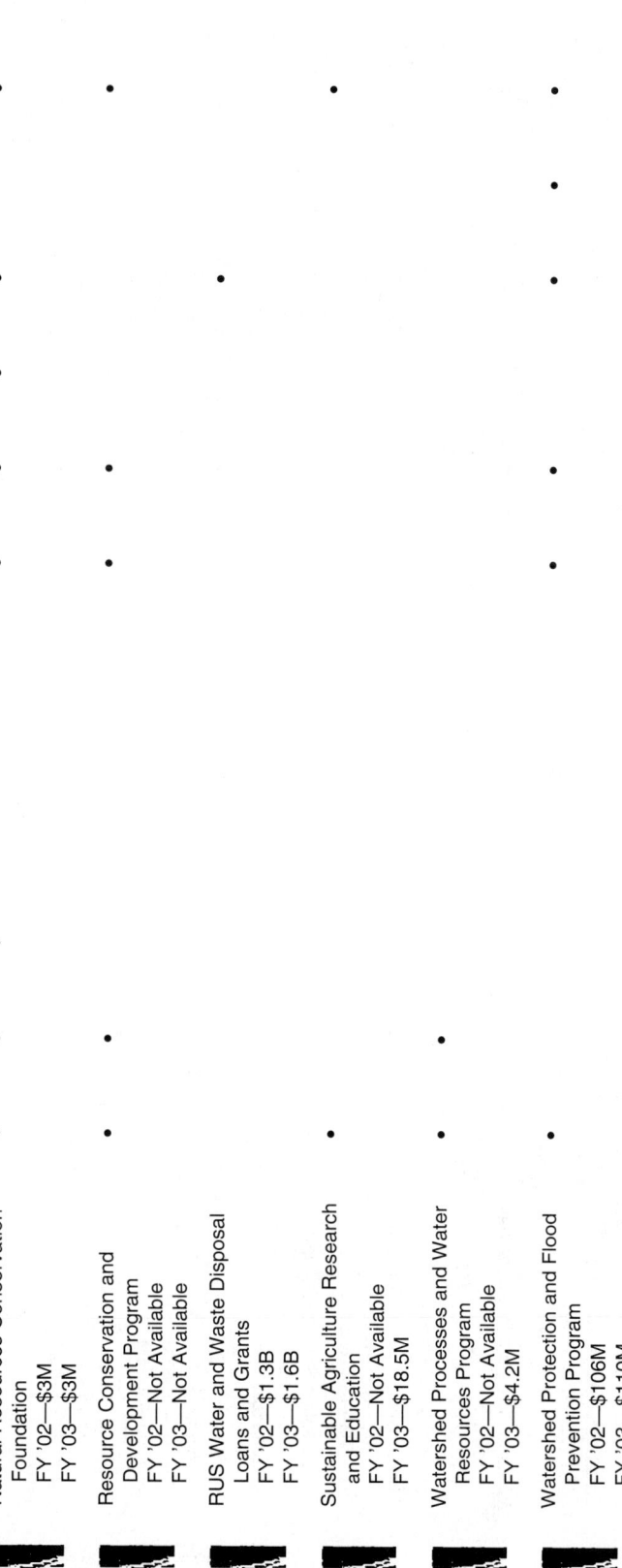

Natural Resources Conservation Foundation FY '02—$3M FY '03—$3M	•	•	•				•	•	•	•		•		
Resource Conservation and Development Program FY '02—Not Available FY '03—Not Available	•	•					•	•				•	•	•
RUS Water and Waste Disposal Loans and Grants FY '02—$1.3B FY '03—$1.6B										•			•	•
Sustainable Agriculture Research and Education FY '02—Not Available FY '03—$18.5M	•											•		
Watershed Processes and Water Resources Program FY '02—Not Available FY '03—$4.2M	•	•												
Watershed Protection and Flood Prevention Program FY '02—$106M FY '03—$110M	•						•	•		•	•	•		

Funding Source Details and Contact Information

Watershed Assistance Grants (WAG): Builds cooperative agreements between nonprofits and other eligible to support watershed partnerships and long-term effectiveness. Funding then supports organizational development and capacity building for watershed partnerships with a diverse membership. Grants will be distributed to a pool of applicants, which are diverse in terms of geography, watershed issues, the type of partnership, and approaches. For more information see: rivernetwork.org/howwecanhelp/index.cfm?doc_id=94#wag

Wetlands Program Development Grants: Provides financial assistance to states, tribes and local governments to support development or enhancement of wetland protection, management or restoration programs. Projects must demonstrate a direct link to an increase in the states, tribes, or local governments ability to protect wetland resources. Funding may only be used to enhance and develop new and existing state wetlands programs, not for their operational support. For more information see: www.epa.gov/owow/wetlands/grantguidelines

Conservation Reserve Program: Voluntary program that offers long-term rental payments and cost-sharing assistance to establish protective covering on cropland and marginal pasture land. Protective covering reduces soil erosion, improves water quality, and enhances wildlife habitat. The land must be owned or operated by the applicant for at least 12 months, enrolled in the Water Bank Program (WBP), or contain other environmentally sensitive land. For more information see: www.nrcs.usda.gov/programs/crp

Conservation Security Program (CSP): Managed through the NRCS, the CSP will provide payments to farmers/producers who already are practicing good stewardship on agricultural lands and incentives for those who want to do more. Land enrolled in the Conservation Reserve Program, Wetlands Reserve Program, or Grassland Reserve Program are not eligible. Neither are those converted to cropland after 2002. For more information see: usda.gov/farmbill

Environmental Quality Incentives Program (NRCS): Voluntary locally-led conservation program that provides technical, education and financial assistance to farmers and ranchers who establish conservation practices and system that will address soil, water, and related natural resource problems. Cost-sharing and incentive payments are provided through five to ten year contracts to help producers in complying with environmental laws and regulations, including clean water. For more information see: www.nrcs.usda.gov/programs/eqip

Farmland Protection Program: Provides matching funds to existing farmland protection programs for the purchase of conservation easements. Eligible property includes farm or ranch lands that have prime, unique, statewide, or locally important soil and includes all cropland, rangeland, grassland, pasture land, incidental forest land, or wetlands. For more information see: usda.gov/farmbill

Natural Resources Conservation Foundation: Funding for the Foundation will come from private donations or grants from individuals, corporations, businesses, and nonprofit organizations and agencies. The Foundation will have the authority to enter into cooperative agreements and contracts with federal, state, tribal, and local agencies and organizations and to grant monies for conservation activities to conserve natural resources on private lands. For more information see: usda.gov/farmbill/conservation_fb.html

Resource Conservation and Development Program: Provides technical assistance for planning and installation of approved projects in RC and D area plans, for land conservation, water management, community development, and environmental enhancement. Not a grant program, individuals work with local RC and D Councils to find funding. For more information see: www.nrcs.usda.gov/programs/rcd

Rural Utilities Service Water and Waste Disposal Loans and Grants: Provides assistance for water and waste disposal facilities to low income rural communities whose residents face significant health risks. Project grants and direct loans are available for local governments, Indian tribes, US Territories, nonprofit associations, state governments, and cooperations.For more information see: usda.gov/rus/water.programs.htm

Sustainable Agriculture Research and Education Program (SARE): Provides grants to advance farming systems that are more profitable and environmentally sound. SARE funds scientific investigation and education to reduce the use of chemical pesticides, to improve management of on-farm resources; to optimize conservation practices; and to promote partnership activities. Research and Education Projects are conducted by interdisciplinary research teams to include farmers as participants. For more information see:www.sare.org/htdocs/sare/cfp.html

Watershed Processes and Water Resources Program: Sponsors research that address two areas: 1) understanding fundamental processes controlling source areas, the flow pathways of water, and the fate of water, sediment, and organisms within forest, rangeland, and agricultural environments as influenced by watershed characteristics; and 2) developing appropriate technology and management practices for improving the effective sue of water and water quality for agricultural and forestry production. For more information see: reeusda.gov/

Watershed Protection and Flood Prevention Program: The 'Watershed,' or 'PL 566,' program provides technical and financial assistance for water resource challenges on a watershed basis. Projects related to flood mitigation, water supply, water quality, erosion and sediment control, wetland creation and restoration, fish and wildlife habitat enhancement, and public recreation are eligible for assistance. Technical and financial assistance is also available for planning new watershed surveys. For more information see: www.nrcs.usda.gov/programs/watershed/

UNITED STATES • ENVIRONMENTAL PROTECTION AGENCY	Polluted Runoff Control						Resource Protection and Restoration						Wastewater	
Funding Sources	Agriculture	Forestry	Mining	Brownfields	Storage Tanks (above ground & underground)	Landfills	Hydromodification/ Habitat Modification	Wetlands/Riparian Zone	Land Acquisition/ Conservation Easements	Groundwater Protection Measures	Land Use Controls	Public Education	Stormwater	On-Site Sewage
USDA Wetlands Reserve Program FY '02—250,000 acres FY '03—Not Available	•	•					•	•	•				•	
USDA Wildlife Habitat Incentives Program FY '02—$15M FY '03—$30M (est.)	•	•					•	•				•		
Abandoned Mine Land Reclamation Program FY '02—$159.6M FY '03—$144M			•							•				
Acid Mine Drainage FY '02—$2.75M FY '03—$2.75M (est.)			•							•				
Environmental Mgmt on Indian Lands					•	•						•	•	•

FY '02—Not Available FY '03—Not Available														
Land and Water Conservation Fund FY '02—$140M FY '03—$94.4M	•	•	•				•	•	•		•			
Landowner Incentive Program (non-tribal) FY '02—$40M	•	•					•	•	•	•	•	•	•	
Partners for Fish and Wildlife Program FY '02—$20.5M FY '03—$19M	•	•						•	•			•		
Water Resources on Indian Lands FY '02—Not Available FY '03—Not Available										•				
NOAA's Community-Based Restoration Program FY '02—$10M FY '03—$10M							•	•						
HUD Community Development Block Grants FY '02—$4.4B FY '03—$4.4B				•					•		•	•	•	•
Transportation Equity Act for the 21st Century FY '02—Not Available FY '04—$10.96B							•	•	•	•	•	•	•	

Funding Source Details and Contact Information

Wetlands Reserve Program: Voluntary program to restore wetlands on private property. The program provides landowners with financial incentives and technical assistance to enhance, restore and protect wetlands in exchange for retiring marginal agricultural land. Participants must have owned the land for at least 12 months and it must be restorable and he suitable to provide wildlife benefits. For more information see: nrcs.usda/gov./programs/wrp

Wildlife Habitat Incentives Program (WHIP): WHIP is a voluntary program for people who want to develop and improve wildlife habitat primarily on private land. This program provides both technical assistance and up to 75% cost-share assistance to establish and improve fish and wildlife habitat. For more information see: www.nrcs.usda.gov/programs/whip

Abandoned Mine Land Reclamation Program: Provides grants to states and tribes to correct environmental damage caused primarily by coal mining that occurred prior to August 3, 1977. The program provides for the restoration of eligible lands and waters mined and abandoned or left inadequately restored after mining. The Appalachian Clean Streams Initiative (a subprogram) also provides funding as seed money to accelerate the clean up of water pollution related to acidity, metals, and toxicity. For more information see: www.osmre.gov/osmaml.htm

Acid Mine Drainage (AMD) Reclamation Program: Designed to support for efforts of local not-for-profit organizations, especially watershed groups, to complete construction projects designed to clean streams impacted by AMD. Eligible organizations include community watershed groups, nonprofit groups, and conservation districts. For more information see: www.osmre.gov/acsifunding.htm

Environmental Management on Indian Lands: The program provides funds to improve environmental management in Indian Country and at Bureau of Indian Affairs facilities, under all environmental statutes, including hazardous waste handling, drinking and wastewater systems, solid waste management and open dump closures, fuel storage in underground and above storage tanks, and management of PCBs, lead-based paint, and asbestos in schools. For more information see: www.doi.gov/bureau-indian-affairs.html

Land and Water Conservation Fund (LWCF): LwCF uses offshore oil leasing revenues to support grants to States, and through States, local units of government for the acquisition and development of state and local park and recreation areas that guarantee public use in perpetuity. All funded projects must be available for public recreational use. For more information see: www.nps.gov/lwef/

Landowner Incentive Program (non-tribal): Provides matching grants to states, territories, and DC to establish or supplement landowner incentive programs. Programs can provide technical and financial assistance to landowners for projects that protect or restore habitats. Projects can involve the removal of exotic plants, changes in grazing practices, instream structural improvements, road closures, and conservation easements. Although not directly eligible, nonprofits may work directly with their states. For more information see: fws.gov

Partners for Fish and Wildlife Program: Since 1987, the program has partnered with more than 28,725 landowners to restore over 639,000 acres of wetlands; 1,070,000 acres of prairie, native grassland, and other upland habitats; and 4,740 miles of in-stream aquatic and riparian habitat. In addition, the program has reopened more than 300 miles of stream habitat for fish and other aquatic species by removing barriers to passage. For more information see: partners.fws.gov/

Water Resources on Indian Lands: This program assists Indian tribes with the management, planning, protection, and development of their water resources and related land resources. Previously funded projects have included geographic and hydrologic quantitative and qualitative analysis of water, groundwater and surface water quality and quantity monitoring, aquifer classification, and stream gauging. For more information see: www.doi.gov/bureau-indian-affairs.html

NOAA's Community-Based Restoration Program (CRP): Provides funds for small-scale, locally driven habitat restoration projects that foster natural resource stewardship within communities. The program seeks to bring together diverse partners to implement habitat restoration projects to benefit living marine resources. Partnerships are sought at the national and local level to contribute funding, land, technical assistance, workforce support, or other in-kind services. For more information see: www.nmfs.noaa.gov/habitat/restoration

HUD Community Development Block Grants: Provides grants to develop viable urban communities by providing housing and a suitable living environment. Activities include water, sewer and other facilities. Entitlement communities must be either a central city in a Metropolitan Statistical Area (MSA), a city with population above 50,000 in the MSA or an urban county of at least 200,000 people. Funds are also awarded to states for distribution to smaller (non-entitlement) communities. For more information see: www.hud.gov/progdesc/cdbg-st.cfm

Transportation Equity Act for the 21st Century Funding Programs (TEA-21): TEA-21 funds numerous programs such as the Surface Transportation Program (STP) ($5.9B) and the National Highway System ($5.06B). States may spend up to 20% of their STP dollars for restoration and pollution abatement projects. Each state may also set aside 10 percent of STP funds for transportation enhancement projects, including conservation easements, wetland mitigation, and pollution abatement. For more information see: thwa.dot.gov/tea21/

Source: From www.epa.gov, www.nalusda.gov, and cfda.gov.

Table 11F.84 Expenditures for Water Management Projects by the United States Corps of Engineers, 1977–1986

Item	1986	1985	1984	1983	1982	1981	1980	1979	1978	1977
Appropriations ($ Millions)										
Navigation	299	304	257	577	552	607	485	395	357	261
Flood control total	673	676	647	647	882	774	772	873	802	859
Flood control										
Mississippi River and tributaries[a]	(227)	(225)	(203)	(291)	(177)	(145)	(211)	(158)	(188)	(173)
Multipurpose, including power	93	112	171	221	247	275	538	358	155	464
Beach erosion control	17	12	25	5	7	12	9	12	12	16
Total new work[b]	1,082	1,104	1,127	1,685	1,580	1,666	1,905	1,567	1,719	1,600
Other work[c]	1,658	1,797	1,561	1,734	1,417	1,331	1,296	1,223	1,070	887
Total	2,740	2,901	2,688	3,419	2,997	2,997	3,201	2,790	2,789	2,487

[a] Included in flood control total.
[b] Advance engineering and design, and construction (including major rehabilitation projects). Savings and slippage applied to projects.
[c] Operation and maintenance, surveys, administration and miscellaneous programs and activities.

Source: From Department of the Army Annual Report FY88 of the Secretary of the Army on Civil Works Activities, Washingon, DC.

Table 11F.85 United States Geological Survey Budget for Water-Resources Investigations, 1981–1986

Budget Activity	1981	1982	1983	1984	1985	1986
Water Resources Investigations	194,016	190,096	199,697	220,390	238,131	248,598
Direct program	115,458	108,637	1,15,096	1,29,441	1,33,408	135,152
Reimbursable program	78,558	81,459	84,601	90,949	104,723	113,446
States, counties, and municipalities	45,138	46,938	48,782	52,113	56,500	56,650
Miscellaneous non-Federal sources	2,088	2,679	3,914	3,600	3,327	2,161
Other Federal agencies	31,332	31,842	31,905	35,236	44,896	54,635

Note: By sources of funds; dollars in thousands.

Source: From U.S. Geological Survey Yearbook Fiscal Year 1986.

Table 11F.86 United States Geological Survey Budget for Water-Resources Investigations for Fiscal Year 1986

Activity/Subactivity/Program Element	Fiscal Year 1986[a] Enacted
Water Resources Investigations	$1,34,802
National Water Resources Research and Information System—Federal Program	72,875
Data Collection and Analysis	19,492
National Water Data and Information Access Program	2,090
Coordination of National Water Data Activities	903
Regional Aquifer System Analysis	13,474
Core Program Hydrologic Research	8,082
Improved Instrumentation	1,888
Water Resources Assessment	1,287
Toxic Substances Hydrology	11,182
Nuclear Waste Hydrology	7,020
Acid Rain	2,998
Scientific and Technical Publications	2,081
National Water Resources Research and Information System—Federal–State Cooperative Program	49,774
Data Collection and Analysis, Areal Appraisals, and Special Hydrological Studies	41,956
Water Use	3,670
Coal Hydrology	4,148
National Water Resources Research and Information System—State Research Institute and Information Grants Program	12,153
State Water Resources Research Institutes	6,159
National Water Resources Research Grants Program	4,767
Program Administration	1,227

Note: By activity and program element; dollars in thousands.

[a] Funding shown represents appropriated dollars and does not include reimbursable funding from federal, state, and other non-federal sources.

Source: From U.S. Geological Survey Yearbook Fiscal Year 1986.

Table 11F.87 Water Resources Investigations

Program	FY 2000 Estimate	Uncontrol and Related Changes	Program Changes[a]	FY 2001 Budget Request	Change from FY 2000
Water Resources Assessments and Research	91,037	+2,012	−2,694	90,355	−682
Water Data Collection and Management	29,167	+949	+9,159	39,275	+10,108
Federal-State Cooperative Water Program	60,553	+2,326	0	62,879	+2,326
Water Resources Research Act Program	5,062	+5	0	5,067	+5
Total Requirements $000	185,819	+5,292	+6,465	197,576	+11,757

[a] See Program Change section for details.

Source: From U.S. Geological Survey Activity Summary.

Table 11F.88 Water Pollution Abatement Expenditures in the United States, 1975–1984

Year	Total[b]	Industrial		Public Sewer Systems[c]	
		Facilities	Operations[a]	Facilities	Operations[a]
1975	22,840	4200	2950	8,997	3428
1976	24,445	4625	3375	9,370	3713
1977	24,652	4415	3706	9,409	4055
1978	26,631	4277	3934	10,090	4392
1979	26,470	4013	4222	9,758	4583
1980	24,647	3725	4081	8,942	4694
1981	21,984	3259	4180	6,882	4880
1982	21,199	3080	4022	6,148	5156
1983	21,461	2813	4402	5,551	5475
1984, preliminary	23,192	2902	4651	6,353	5773

Note: In millions of dollars.

[a] Operation of facilities.

[b] Includes nonpoint sources not shown separately.

[c] Includes expenditures for private connectors to sewer systems, by owners of feedlots, and by government enterprises.

Source: From U.S. Department of Commerce, 1987 Statistical Abstract of the United States.

Table 11F.89 Proposed Cost Sharing for New Water Projects after 1983

Project Purpose	Up-front Nonfederal Share of Costs, %
Hydropower	100
Municipal and industrial supply	100
Flood control	35
Recreation	50[a]
Commercial navigation	75[b]
Irrigation	35
Beach erosion	50

[a] Could be repayment instead of up-front payment.

[b] Twenty-five percent of federal financing is reimbursable; the rest must be an up-front cash contribution.

Source: From Grigg, N.S., 1985, *Water Resources Planning.* Copyright 1985 McGraw-Hill, Inc. Reprinted with permission.

Original Source: S, 1031.

Table 11F.90 Federal Groundwater Quality Research and Development in the United States

Federal Organization	Categories of Groundwater Quality R and D[a]									
	1	2	3	4	5	6	7	8	9	10
National Science Foundation			X		X				X	
Department of Agriculture										
Agricultural Research Services			X			X				
Forest Service			X							
Soil Conservation Service					X	X			X	
Department of Commerce										
National Bureau of Standards	X									
Department of Defense										
Army Corps of Engineers		X			X	X			X	X
Army Medical Bioengineering R and D Laboratory	X								X	
Army Toxic and Hazardous Materials Agency	X									
Department of Energy		X								
Department of Interior										
Bureau of Indian Affairs					X					
Bureau of Land Management					X					
Bureau of Reclamation					X	X				
Fish and Wildlife Service					X					
Geological Survey		X	X	X	X	X	X			
National Park Service				X	X					
Office of Surface Mining					X		X			
Office of Water Policy		X	X					X	X	X
Environmental Protection Agency										
Environmental Monitoring Systems Laboratory	X		X							
R.S. Kerr Environmental Research Laboratory			X							
Environmental Research Laboratory			X							
Office of Pesticide Programs			X							
Office of Radiation Programs	X				X					
Office of Research and Development		X	X		X				X	X
Office of Solid Waste					X					
Office of Water								X		
Nuclear Regulatory Commission		X	X							

The listing is not exhaustive but covers principal programs and activities related to groundwater quality R and D. Examples of other Federal R and D activities omitted here address quantity estimates, use patterns, source inventories, recharge information exchange, socioeconomic effects of alternative supplies, and environmental effect of contamination

[a] Key for categories of groundwater research and development: 1, Standards certification, quality assurance, and water quality criteria; 2, Hydrogeologic investigations and dynamics of groundwater flow; 3, Subsurface fate and transport of contaminants; 4, Background monitoring of groundwater quality; 5, Detection of groundwater contamination from various sources; 6, Salt-water intrusion and salinity problems; 7, Surface water–groundwater interactions; 8, Control of groundwater contamination from various sources; 9, Treatment technologies; 10, Evaluation of alternatives.

Source: From Office of Technology Assessment, 1984, Protecting the Nation's Groundwater from Contamination.

Table 11F.91 Federal Funds Authorized for Implementation of the Safe Drinking Water Act (as Amended)

Area Funded	Fiscal Year 87	88	89	90	91
	Millions of Dollars				
Sole source aquifer demonstration program	$10	$15	$17.5	$17.5	$17.5
Wellhead protection areas	20	20	35	35	35
Grants to alleviate emergency situations	7.65	7.65	8.05	8.05	8.05
Technical assistance and training	35.6	35.6	38.02	38.02	38.02
Technical assistance to small public water systems	10	10	10	10	10
State program grants					
PWS program	37.2	37.2	40.15	40.15	40.15
UIC program	19.7	19.7	20.85	20.85	20.85
Assistance to small systems for monitoring unregulated contaminants	30	—	—	—	—

Source: From American Water Resources Assoc., 1988, *New Dimensions in Safe Drinking Water*. Copyright AWWA. Reprinted with permission.

Table 11F.92 Annual Expenditures of the Water Industry in the United States in 1984

	Annual Expenditures, U.S. $ Billions
Local water supply operations	12.0
Local wastewater operations	14.0
Local water management, other than above	2.5
Hydropower and thermoelectric cooling	8.5
Industrial pollution control	3.0
Industrial water supply	7.5
Regional water management	8.0
Federal agency operations	3.0
Dams, ports, harbors, and waterways	7.5
Home water treatment	1.0
Bottled water and sparkling water	0.8
Well drilling	3.0
Irrigation	6.0
State government	0.2
Education, research, publishing	0.2
	77.2

Source: From Grigg, N.S. 1985, *Water Resources Planning*. Copyright 1985 McGraw-Hill, Inc. Reprinted with permission.

Table 11F.93 Incremental Annual Costs of the CWA Assuming a 3 Percent Discount Rate (1997 $million)

	1994 Costs			1997 Costs		
	Actual (A)	Without-CWA (B)	Incremental (C)=(A)+(B)	Actual (D)	Without-CWA (E)	Incremental (F)=(D)−(E)
Industry						
Capital	6,490.3	2,520.9	3,969.4	6,345.0	2,694.6	3,650.4
O&M	7,492.1	5,946.1	1,546.1	8,005.2	6,356.0	1,649.2
R&D	80.7	60.2	20.5	97.6	76.1	21.5
Offsets	−312.1	−121.2	−190.9	−267.5	−113.6	−153.9
Public						
Sewerage, capital	13,308.0	10,676.4	2,631.6	13,979.9	11,375.9	2,604.0
Sewerage, O&M	15,035.5	10,912.8	4,122.8	17,171.4	11,704.8	5,466.6
Electric, capital	114.6	14.3	100.3	103.3	14.3	89.0
Electric, O&M	10.6	9.5	1.1	11.9	9.5	2.3
R&D	144.4	93.7	50.7	124.0	91.5	32.5
R&M and other	2,259.3	1,891.4	367.9	2,707.5	1,962.0	745.5
Total	44,623.3	32,003.9	12,619.4	48,278.1	34,171.0	14,107.1

Source: From "A Retrospective Assessment of the Costs of the Clean Water Act: 1972 to 1997. Final Report—October 2000." Prepared for USEPA, Office of Water, Office of Policy, Economic and Innovation.

Table 11F.94 Federal Water Pollution Control Laws

Title	Year of Enactment	Key Provisions
Water Pollution Control Act	1948	Authorized federal research
Water Pollution Control Act Amendments	1956	State responsibility to establish water quality criteria Discretionary federal responsibility to convene enforcement conferences
Water Quality Act	1965	Ambient standards State implementation and enforcement responsibility
Federal Water Pollution Control Act (Clean Water Act)	1972	Technology-based standards for conventional pollutants Federal share of municipal treatment plant construction costs set at 75 percent Federal implementation and enforcement responsibility
Coastal Zone Management Act	1972	Directed towards coastal water pollution problems
Clean Water Act Amendments	1977	Increased coverage to toxic pollutants
Municipal Wastewater Treatment Construction Grant Amendments	1981	Reduced federal share of municipal treatment plant financing to 55 percent
Clean Water Act Reauthorization	1987	Increased focus on nonpoint source pollution Fedral support for municipal treatment plant construction converted from grant assistance to loans

Source: From Adapted from Freeman, A. Myrick. 1990. "Water Pollution Policy." In *Public Policies for Environmental Protection*, Washington, DC: Resources for the Future.

Table 11F.95 Regulated Entities under the CWA

Group	SIC Codes
Manufacturing Industries	
Food	20
Paper	26
Chemical	28
Petroleum and coal products	29
Other nondurables	21, 22, 23, 27, 30, 31
Primary metals	33
Machinery	35
Motor vehicles	37
Other durables	24, 25, 32, 34, 36, 38, 39
Nonmanufacturing Industries	
Mining	10, 12, 13, 14
Public and cooperative electric utilities	Part of 491, 493
Other nonmanufacturing	
Construction	15, 16, 17
Transportation	40–47
Services	492, 494–497, 70–89
Wholesale and retail	50–59
Finance etc.	60–67
Public Enterprises	
Publicly owned sewerage treatment plants	Part of 4952
Publicly owned electric utilities	Part of 491, 493

Source: From usepa.gov.

Table 11F.96 Annual Private Water Pollution Abatement Capital Costs under the with CWA and without CWA Scenarios, 1992–1997 (1997 $ millions)

	Discount Rate = 3%		Discount Rate = 7%	
Year	With CWA	Without CWA	With CWA	Without CWA
1992	6504.0	2443.4	9133.8	3431.4
1993	6507.1	2484.0	9138.1	3488.4
1994	6490.3	2520.9	9114.5	3450.1
1995	6482.9	2565.4	9104.2	3602.7
1996	6442.3	2627.4	9047.1	3689.7
1997	6345.0	2694.6	8910.5	3784.2

Source: From www.usepa.gov.

Table 11F.97 Annual Public Sewerage Capital Costs under the with CWA and without CWA Scenarios, 1992–1997 (1997 $ millions)

	Discount Rate = 3%		Discount Rate = 7%	
Year	With CWA	Without CWA	With CWA	Without CWA
1992	12,733.2	10,244.6	20,112.5	16,181.7
1993	13,051.6	10,463.0	20,615.4	16,526.6
1994	13,308.0	10,676.4	21,020.3	16,863.6
1995	13,509.2	10,903.5	21,338.1	17,222.3
1996	13,728.1	11,125.8	21,683.9	17,573.4
1997	13,979.9	11,375.9	22,081.7	17,968.5

Source: From www.usepa.gov.

Table 11F.98 Annual Water Pollution Abatement Capital Costs for Publicly Owned Electric Utilities under the with CWA and without CWA Scenarios, 1992–1997 (1997 $ millions)

	Discount Rate = 3%		Discount Rate = 7%	
Year	With CWA	Without CWA	With CWA	Without CWA
1992	123.9	14.3	174.0	20.0
1993	118.4	14.3	166.3	20.0
1994	114.6	14.3	160.9	20.0
1995	109.6	14.3	153.9	20.0
1996	106.6	14.3	149.8	20.0
1997	103.3	14.3	145.1	20.0

Source: From www.usepa.gov.

Table 11F.99 Private Cost Offsets with and without the CWA, 1992–1997 (1997 $ millions)

Year	With CWA	Without CWA
1992	379	143
1993	305	117
1994	312	121
1995	296	117
1996	282	115
1997	268	114

Source: From www.usepa.gov.

Table 11F.100 Incremental Annual Costs of the CWA (Assuming a 3 Percent Discount Rate) (1997 $ millions)

	1992 Costs			1993 Costs		
	Actual	Without CWA	Incremental	Actual	Without CWA	Incremental
Industry						
Capital	6,504.0	2,443.4	4,060.6	6,507.1	2,484.0	4,023.0
O&M	7,276.6	5,763.4	1,513.2	7,079.7	5,859.3	1,220.4
R&D	61.4	61.4	0.0	88.0	60.0	28.1
Offsets	−379.3	−142.5	−236.8	−305.4	−116.6	−188.8
Public						
Sewerage, capital	12,733.2	10,244.6	2,488.6	13,051.6	10,463.0	2,588.6
Sewerage, O&M	13,309.2	10,405.1	2,904.1	14,171.6	10,668.6	3,503.0
Electric, capital	123.9	14.3	109.6	118.4	14.3	104.2
Electric, O&M	11.2	9.5	1.6	12.0	9.5	2.4
R&D	137.2	98.3	38.9	116.3	95.3	21.0
R&D and other	2,913.1	1,893.7	1,019.4	2,581.5	1,887.2	694.3
Total	**42,690.4**	**30,791.2**	**11,899.1**	**43,420.7**	**31,424.6**	**11,996.1**
	1994 Costs			**1995 Costs**		
	Actual	**Without CWA**	**Incremental**	**Actual**	**Without CWA**	**Incremental**
Industry						
Capital	6,490.3	2,520.9	3,969.4	6,482.9	2,565.4	3,917.5
O&M	7,492.1	5,946.1	1,546.1	7,694.0	6,051.2	1,642.8
R&D	80.7	60.2	20.5	97.6	65.7	31.8
Offsets	−312.1	−121.2	−190.9	−296.5	−117.3	−179.2
Public						
Sewerage, capital	13,308.0	10,676.4	2,631.6	13,590.2	10,903.5	2,605.7
Sewerage, O&M	15,035.5	10,912.8	4,122.8	15,581.6	11,166.5	4,415.2
Electric, capital	114.6	14.3	100.3	109.6	14.3	95.3
Electric, O&M	10.6	9.5	1.1	11.9	9.5	2.3
R&D	144.4	93.7	50.7	124.0	95.3	28.6
R&D and other	2,259.3	1,891.4	367.9	2,614.6	1,900.7	713.9
Total	**44,623.3**	**32,003.9**	**12,619.4**	**45,928.8**	**32,654.7**	**13,274.1**
	1996 Costs			**1997 Costs**		
	Actual	**Without CWA**	**Incremental**	**Actual**	**Without CWA**	**Incremental**
Industry						
Capital	6,442.3	2,627.4	3,814.9	6,345.0	2,694.6	3,650.4
O&M	7,852.5	6,197.3	1,655.2	8,005.2	6,356.0	1,649.2
R&D	97.6	70.4	27.2	97.6	76.1	21.5
Offsets	−281.6	−114.9	−166.8	−267.5	−113.6	−153.9
Public						
Sewerage, capital	13,728.1	11,125.8	2,602.3	13,979.9	11,375.9	2,604.0
Sewerage, O&M	16,357.2	11,417.0	4,940.2	17,171.4	11,704.8	5,466.6
Electric, capital	106.6	14.3	92.4	103.3	14.3	89.0
Electric, O&M	11.9	9.5	2.3	11.9	9.5	2.3
R&D	124.0	93.1	30.8	124.0	91.5	32.5
R&D and other	2,661.0	1,926.8	734.3	2,707.5	1,962.0	745.5
Total	**47,099.5**	**33,366.7**	**13,732.8**	**48,252.6**	**34,171.0**	**14,107.1**

Source: From www.usepa.gov.

Table 11F.101 Incremental Annual Costs of the CWA (Assuming a 7 Percent Discount Rate) (1997 $ millions)

	1992 Costs			1993 Costs		
	Actual	**Without CWA**	**Incremental**	**Actual**	**Without CWA**	**Incremental**
Industry						
Capital	9,133.8	3,431.4	5,702.4	9,138.1	3,488.4	5,649.6
O&M	7,276.6	5,763.4	1,513.2	7,079.7	5,859.3	1,220.4
R&D	61.4	61.4	0.0	88.0	60.0	28.1
Offsets	−379.3	−142.5	−236.8	−305.4	−116.6	−188.8
Public						
Sewerage, capital	20,112.5	16,181.7	3,930.8	20,615.4	16,526.6	4,088.8
Sewerage, O&M	13,309.2	10,405.1	2,904.1	14,171.6	10,668.6	3,503.0
Electric, capital	174.0	20.0	153.9	166.3	20.0	146.3
Electric, O&M	11.2	9.5	1.6	12.0	9.5	2.4
R&D	137.2	98.3	38.9	116.3	95.3	21.0
R&D and other	2,913.1	1,893.7	1,019.4	2,581.5	1,887.2	694.3
Total	**52,749.5**	**37,722.0**	**15,027.5**	**53,663.4**	**38,498.3**	**15,165.1**
	1994 Costs			**1995 Costs**		
	Actual	**Without CWA**	**Incremental**	**Actual**	**Without CWA**	**Incremental**
Industry						
Capital	9,114.5	3,540.1	5,574.3	9,104.2	3,602.7	5,501.5
O&M	7,492.1	5,946.1	1,546.1	7,694.0	6,051.2	1,642.8
R&D	80.7	60.2	20.5	97.6	65.7	31.8
Offsets	−312.1	−121.2	−190.9	−296.5	−117.3	−179.2
Public						
Sewerage, capital	21,020.3	16,863.6	4,156.7	21,338.1	17,222.3	4,115.8
Sewerage, O&M	15,035.5	10,912.8	4,122.8	15,581.6	11,166.5	4,415.2
Electric, capital	160.9	20.0	140.8	153.9	20.0	133.9
Electric, O&M	10.6	9.5	1.1	11.9	9.5	2.3
R&D	144.4	93.7	50.7	124.0	95.3	28.6
R&D and other	2,259.3	1,891.4	367.9	2,614.6	1,900.7	713.9
Total	**55,006.2**	**39,216.2**	**15,790.0**	**56,423.4**	**40,016.6**	**16,406.7**
	1996 Costs			**1997 Costs**		
	Actual	**Without CWA**	**Incremental**	**Actual**	**Without CWA**	**Incremental**
Industry						
Capital	9,047.1	3,689.7	5,357.4	8,910.5	3,784.2	5,126.3
O&M	7,852.5	6,197.3	1,655.2	8,005.2	6,356.0	1,649.2
R&D	97.6	70.4	27.2	97.6	76.1	21.5
Offsets	−281.6	−114.9	−166.8	−267.5	−113.6	−153.9
Public						
Sewerage, capital	21,683.9	17,573.4	4,110.5	22,081.7	17,968.5	4,113.2
Sewerage, O&M	16,357.2	11,417.0	4,940.2	17,171.4	11,704.8	5,466.6
Electric, capital	149.8	20.0	129.7	145.1	20.0	125.1
Electric, O&M	11.9	9.5	2.3	11.9	9.5	2.3
R&D	124.0	93.1	30.8	124.0	91.5	32.5
R&D and other	2,661.0	1,926.8	734.3	2,707.5	1,962.0	745.5
Total	**57,703.3**	**40,882.5**	**16,820.8**	**58,946.8**	**41,859.0**	**17,128.2**

Source: From www.usepa.gov.

Table 11F.102 Private Capital Expenditures for Water Pollution Abatement: 1972–1994 (Current $ millions)

	Manufacturing									
Year	Primary Metals	Transportation Equipment	Machinery	Other Durables	Chem-ical	Paper	Petroleum and Coal	Food	Other Non-Durables	All Manu-facturing*
1972	NA	NA	NA	NA	NA	NA	NA	NA	NA	NA
1973	84.7[a]	41.7[a]	15.6[a]	81.4[a]	214.6[a]	161.0[a]	96.1[a]	104.8[a]	28.0[a]	827.8[a]
1974	132.7[a]	41.5[a]	17.6[a]	91.2[a]	264.4[a]	193.2[a]	119.7[a]	111.7[a]	36.7[a]	1,008.8[a]
1975	187.5[a]	36.4[a]	20.8[a]	99.0[a]	385.7[a]	266.0[a]	155.7[a]	93.9[a]	34.9[a]	1,280.1[a]
1976	197.8[a]	53.6[a]	21.8[a]	109.3[a]	577.4[a]	278.6[a]	199.8[a]	97.6[a]	54.5[a]	1,599.2[a]
1977	250.7[a]	39.4[a]	42.4[a]	150.5[a]	603.8[a]	261.9[a]	196.0[a]	109.3[a]	41.1[a]	1,695.1[a]
1978	219.5[a]	57.9[a]	27.8[a]	129.6[a]	392.9[a]	189.2[a]	100.7[a]	99.6[a]	24.5[a]	1,262.9[a]
1979	227.7[a]	59.5[a]	38.2[a]	117.0[a]	367.2[a]	180.8[a]	119.6[a]	117.6[a]	26.4[a]	1,262.2[a]
1980	180.7[b]	60.7[b]	34.9[b]	113.7[b]	350.0[b]	111.2[b]	114.2[b]	133.0[b]	32.2[b]	1,146.5[b]
1981	144.1[b]	60.0[b]	28.0[b]	125.4[b]	322.2[b]	86.5[b]	131.7[b]	104.8[b]	24.0[b]	1,028.4[b]
1982	133.7[b]	36.5[b]	42.2[b]	122.6[b]	256.5[b]	93.7[b]	165.7[b]	110.9[b]	15.3[b]	977.4[b]
1983	100.2[b]	55.0[b]	19.0[b]	105.2[b]	187.4[b]	65.9[b]	164.7[b]	105.1[b]	13.3[b]	819.0[b]
1984	72.9[b]	116.9[b]	22.6[b]	185.7[b]	212.4[b]	68.2[b]	96.8[b]	91.8[b]	16.5[b]	887.8[b]
1985	84.3[c]	165.1[c]	35.1[c]	159.6[c]	271.5[c]	106.0[c]	88.4[c]	77.4[c]	19.4[c]	1,017.9[c]
1986	74.6[c]	81.8[c]	25.7[c]	173.0[c]	325.5[c]	96.9[c]	121.5[c]	108.2[c]	25.6[c]	1,038.7[c]
1987	NA	NA	NA	NA	NA	NA	NA	NA	NA	NA
1988	100.6[c]	80.4[c]	33.2[c]	169.5[c]	487.8[c]	97.2[c]	203.7[c]	91.0[c]	25.8[c]	1,289.4[c]
1989	138.7[c]	84.6[c]	54.8[c]	233.6[c]	598.6[c]	261.0[c]	230.4[c]	183.6[c]	39.2[c]	1,824.5[c]
1990	166.8[d]	142.6[d]	41.3[d]	184.5[d]	995.0[d]	509.6[d]	400.8[d]	163.3[d]	47.5[d]	2,651.4[d]
1991	131.9[d]	94.7[d]	27.6[d]	276.0[d]	942.3[d]	552.7[d]	373.3[d]	359.5[d]	56.5[d]	2,814.6[d]
1992	123.5[d]	69.2[d]	31.7[d]	146.7[d]	1017.3[d]	373.4[d]	492.6[d]	202.6[d]	48.1[d]	2,509.8[d]
1993	92.0[d]	67.1[d]	20.2[d]	157.4[d]	937.9[d]	289.2[d]	567.2[d]	113.6[d]	46.8[d]	2,294.9[d]
1994	98.5[d]	60.8[d]	152.1[d]	239.6[d]	1005.6[d]	195.9[d]	466.9[d]	152.8[d]	56.7[d]	2,428.9[d]

	Non-manufacturing					
Year	Mining	Electric	Petroleum and Coal	All Non-manufacturing	PACE Aggregate[i]	BEA Total
1972	NA	NA	NA	NA	NA	1501
1973	70[e]	410[e]		620[e]	1,447.8	1,570[g]
1974	60[e]	430[e]		630[e]	1,638.8	1,765[g]
1975	60[e]	400[e]		650[e]	1,930.1	2,145[g]
1976	100[e]	540[e]		850[e]	2,449.2	2,607[g]
1977	100[e]	650[e]		920[e]	2,651.1	2,827[g]
1978	230[e]	900[e]		1,310[e]	2572.9	2,683[g]
1979	180[e]	1,010[e]		3,350[e]	2612.2	2,873[g]
1980	110[e]	890[e]		1,140[e]	2,286.5	2,795[g]
1981	90[e]	810[e]		1,030[e]	2,058.4	2,848[g]
1982	110[e]	900[e]		1,170[e]	2,147.4	2,937[g]
1983	100[e]	750[e]		980[e]	1,799.0	2,422[g]
1984	180[e]	900[e]		1,230[e]	2,117.8	2,730[g]
1985	120[f]	870[f]		1,150[f]	2,167.9	2,670[g]
1986	130[f]	650[f]		930[f]	1,968.7	2,534[g]
1987	56[h]	599[h]	1186.3[h]		NA	2,614[g]
1988	84[h]	526[h]	998.9[h]		2,694.6	2,581[g]
1989	113[h]	482[h]	1,174.5[h]		3,363.4	3196[g]
1990	138[i]	673[i]	1,643.6[i]		4,705.4	4,430[g]
1991	120[i]	573[i]	1,503.3[i]		4,637.3	4,666[g]
1992	145[i]	567[i]	1,380.5[i]		4,109.4	4532[g]
1993	161[i]	621[i]	1,115.1[i]		3,624.4	4335[g]
1994	191[j]	606[i]	1,113.6[i]		3,872.0	4,720[g]

Note: NA, not available.

* The manufacturing category is complied separately by the PACE reports; it does not always equal the sum of the individual columns. It does not include SIC code group 23, Apparel and Other Textile Products.

[a] U.S. Department of Commerce. 1981. *Current Industrial Reports*. "Pollution Abatement Costs and Expenditures, 1979." Table 1A. Pollution Abatement Capital Expenditure and Operating Costs, by Form of Abatement and Major Industry Group: 1973–1979. Washington, DC: Government Printing Office.

(Continued)

Table 11F.102 (Continued)

[b] U.S. Department of Commerce. 1986. *Current Industrial Reports*. "Pollution Abatement Costs and Expenditures, 1984." Table 1. Pollution Abatement Capital Expenditures and Operating Costs, by Form of Abatement by Major Group: 1980–1984. Washington, DC: Government Printing Office.
[c] U.S. Department of Commerce. 1991. *Current Industrial Reports*. "Pollution Abatement Costs and Expenditures, 1989." Table 1. Pollution Abatement Capital Expenditures and Operating Costs, by Form of Abatement for Major Groups: 1985–1989. Washington, DC: Government Printing Office.
[d] U.S. Department of Commerce. 1996. *Current Industrial Reports*. "Pollution Abatement Costs and Expenditures, 1994." Table 1.Summary Capital Expenditures and Operating Costs, by Major Groups: 1990–1994. Washington, DC: Government Printing Office.
[e] Environmental Economic Division. February 1986. *Survey of Current Business*. "Plant and Equipment Expenditures by Business for Pollution Abatement." Table 1. New Plant and Equipment Expenditures by U.S. Nonfarm Business: Total and for Pollution Abatement.
[f] Rutledge, Gary L. and Nikolaos A. Stergioulas. November 1988. *Survey of Current Business*. "Plant and Equipment Expenditures by Business for Pollution Abatement, 1987 and 1988." Table 1. New Plant and Equipment Expenditures by U.S. Nonfarm Business: Total and for Pollution Abatement.
[g] Vogan, Christine R. September 1996. *Survey of Current Business*. "Pollution Abatement and Control Expenditures, 1972–1994." Table 12. Business and Government Expenditures for Air and Water Pollution Abatement in Current Dollars and Chain-Type Quantity and Price Indexes, 1972–1994.
[h] U.S. Department of Commerce. 1992. *Current Industrial Reports*. "Pollution Abatement Costs and Expenditures, 1990." Table A. New Capital Expenditures for Companies in Mining, Petroleum, and Electric Utilities: Total and for Pollution Abatement 1987–1990. Washington, DC: Government Printing Office.
[i] U.S. Department of Commerce. 1996. *Current Industrial Reports*. "Pollution Abatement Costs and Expenditures, 1994." Table 14. Summary New Capital Expenditures by Companies in Mining, Petroleum, and Electric Utilities for all Media: 1990–1994. Washington, DC: Government Printing Office.
[j] Calculated based on all available data from *Current Industrial Reports* for manufacturing and nonmanufacturing categories. For years 1973 through 1986, data for nonmanufacturing sector are from the BEA source (Vogan, 1996). For 1988 through 1994, total includes PACE nonmanufacturing data for the mining, electric utilities, and petroleum and coal sectors and PACE manufacturing data (minus the petroleum and coal sectors to avoid double counting).

Source: From www.usepa.gov.

Table 11F.103 Real Private Water Pollution Abatement Expenditures, 1972–1997 (1997 $ millions)

Year	Capital Expenditures[a]	Operation and Maintenance Expenditures[b]
1972	5,011.0	2,634.0
1973	4,962.2	3,072.1
1974	5,120.2	3,446.3
1975	5,685.9	3,734.9
1976	5,528.9	4,322.6
1977	6,651.4	4,856.2
1978	5,883.3	5,168.4
1979	5,804.8	5,633.1
1980	5,168.0	5,519.3
1981	4,813.7	5,425.5
1982	4,669.2	5,510.1
1983	3,693.6	5,723.4
1984	4,011.9	5,954.7
1985	3,793.4	6,180.2
1986	3,508.5	6,564.3
1987	3,511.2	6,834.4
1988	3,344.5	7,032.4
1989	3,974.3	7,171.5
1990	5,278.2	7,735.1
1991	5,349.2	7,134.2
1992	5,056.4	7,276.6
1993	4,712.2	7,079.7
1994	5,011.0	7,492.1
1995	5,081.4	7,570.2
1996	5,081.4	7,649.0
1997	5,081.4	7,728.6

[a] Values for 1995 through 1997 are assumed to be equal to the average observed value for 1990 through 1994.
[b] Values for 1995 through 1997 are projected based on observed trend from 1972 through 1994.

Source: From Vogan, Christine R. September 1996. *Survey of Current Business*, "Pollution Abatement and Control Expenditures, 1972–1994" Table 12. Business and Government Expenditures for Air and Water Pollution Abatement in Current Dollars and Chain-Type Quantity and Price Indexes, 1972–94. With permission.

Table 11F.104 Water Pollution Abatement Expenditures under the with and without CWA Scenarios, 1972–1997

		With CWA WPA Capital Expenditures[b]		Without CWA WPA Capital Expenditures[b] (1997 $ millions)	
Year	Total Industrial Capital Expenditures[a] (1997 $ millions)	1997 $ millions	Percent of Total Capital Expenditures	0.3% of Total Capital Expenditures	0.5% of Total Capital Expenditures
1972	420,974	5,011.0	1.19%	1,262.9	2,104.9
1973	474,093	4,962.2	1.05%	1,422.3	2,370.5
1974	480,395	5,120.2	1.07%	1,441.2	2,402.0
1975	447,976	5,685.9	1.27%	1,343.9	2,239.9
1976	468,819	6,528.9	1.39%	1,406.5	2,344.1
1977	525,146	6,651.4	1.27%	1,575.4	2,625.7
1978	596,443	5,883.3	0.99%	1,789.3	2,982.2
1979	652,609	5,804.8	0.89%	1,957.8	3,263.0
1980	647,712	5,168.0	0.80%	1,943.1	3,238.6
1981	685,206	4,813.7	0.70%	2,055.6	3,426.0
1982	651,646	4,669.2	0.72%	1,954.9	3,258.2
1983	609,090	3,693.6	0.61%	1,827.3	3,045.5
1984	688,201	4,011.9	0.58%	2,064.6	3,441.0
1985	713,206	3,793.4	0.53%	2,139.6	3,566.0
1986	685,093	3,508.5	0.51%	2,055.3	3,425.5
1987	665,444	3,511.2	0.53%	1,996.3	3,327.2
1988	687,561	3,344.5	0.49%	2,062.7	3,437.8
1989	704,089	3,974.3	0.56%	2,112.3	3,520.4
1990	686,172	5,278.2	0.77%	2,058.5	3,430.9
1991	627,437	5,349.2	0.85%	1,882.3	3,137.2
1992	622,449	5,056.4	0.81%	1,867.3	3,112.2
1993	656,658	4,712.2	0.72%	1,970.0	3,283.3
1994	701,333	5,011.0	0.71%	2,104.0	3,506.7
1995	755,180	4,707.9	0.62%	2,265.5	3,775.9
1996	802,501	4,620.7	0.58%	2,407.5	4,012.5
1997	860,700	4,536.6	0.53%	2,582.1	4,303.5

[a] As estimated by private gross nonresidential fixed investment in NIPA (CEA, 2000).

[b] Vogan, Christine R. September 1996. *Survey of Current Business*. "Pollution Abatement and Control Expenditures, 1972–1994." Table 12. Business and Government Expenditures for Air and Water Pollution Abatement in Current Dollars and Chain-Type Quantity and Price Indexes, 1972–1994. 1995 through 1997 are average of 1990 through 1994.

Source: From Vogan, Christine R. September 1996. *Survey of Current Business*, "Pollution Abatement and Control Expenditures, 1972–1994" Table 12. Business and Government Expenditures for Air and Water Pollution Abatement in Current Dollars and Chain-Type Quantity and Price Indexes, 1972–94. 1995 through 1997 are average of 1990 through 1994. With permission.

Table 11F.105 Annualized Private Water Pollution Abatement Capital Costs under the with CWA and without CWA Scenarios, 1972–1997 (1997 $ millions)

	With CWA			Without CWA		
Year	WPA Capital Expenditures[a]	Estimated Gross WPA Capital Stock	Annualized WPA Capital Cost	WPA Capital Expenditures[b]	Estimated Gross WPA Capital Stock	Annualized WPA Capital Cost
1972	5,011.0			1,262.9		
1973	4,962.2			1,422.3		
1974	5,120.2			1,441.2		
1975	5,685.9			1,343.9		
1976	6,528.9			1,406.5		
1977	6,651.4			1,575.4		
1978	5,883.3			1,789.3		
1979	5,804.8			1,957.8		
1980	5,168.0			1,943.1		
1981	4,813.7			2,055.6		
1982	4,669.2			1,954.9		
1983	3,693.6			1,827.3		
1984	4,011.9			2,064.6		
1985	3,793.4			2,139.6		
1986	3,508.5			2,055.3		
1987	3,511.2			1,996.2		
1988	3,344.5			2,062.7		
1989	3,974.3			2,112.3		
1990	5,278.2			2,058.5		
1991	5,349.2			1,882.3		
1992	5,056.4	96,763.4	6,504.0	1,867.3	36,352.0	2,443.4
1993	4,712.2	96,808.7	6,507.1	1,970.0	36,956.4	2,484.0
1994	5,011.0	96,558.7	6,490.3	2,104.0	37,504.1	2,520.9
1995	5,081.4	96,449.6	6,482.9	2,265.5	38,166.9	2,565.4
1996	5,081.4	95,845.2	6,442.3	2,407.5	39,088.5	2,627.4
1997	5,081.4	94,397.7	6,345.0	2,582.1	40,089.6	2,694.6

Note: Capital Life: 20 Years Discount Rate: 3 percent.

[a] 1995 through 1997 expenditures are average of 1990 through 1994 expenditures.

[b] Without-CWA based on 0.3% of total private capital.

Source: From Vogan, Christine R. September 1996. *Survey of Current Business*, "Pollution Abatement and Control Expenditures, 1972–1994" Table 12. Business and Government Expenditures for Air and Water Pollution Abatement in Current Dollars and Chain-Type Quantity and Price Indexes, 1972–94. U.S. Bureau of Economic Analysis. Summary National Income and Product Time Series, 1929–1997. (Computer file). With permission.

Table 11F.106 Approximated Total Gross Private Capital Stock, 1972–1997 (1997 $ millions)

Year	Total Industrial Capital Expenditures[a]	Approximated Gross Industrial Capital Stock[b]	Annual Growth Rate (%)
1952	178,669		
1953	194,059		
1954	189,685		
1955	210,103		
1956	231,570		
1957	238,935		
1958	208,703		
1959	226,057		
1960	235,894		
1961	230,344		
1962	247,101		
1963	257,185		
1964	284,394		
1965	331,356		
1966	366,972		
1967	358,979		
1968	371,765		
1969	396,702		
1970	390,568		
1971	388,841		
1972	420,975	1,247,400	
1973	474,093	1,341,600	7.6
1974	480,395	1,456,500	8.6
1975	447,976	1,587,400	9.0
1976	468,819	1,717,400	8.2
1977	525,146	1,860,100	8.3
1978	596,443	2,035,800	9.4
1979	652,610	2,265,300	11.3
1980	647,712	2,541,800	12.2
1981	685,206	2,842,900	11.8
1982	651,646	3,199,700	12.6
1983	609,090	3,556,800	11.2
1984	688,201	3,900,600	9.7
1985	713,207	4,306,500	10.4
1986	685,094	4,734,400	9.9
1987	665,444	5,144,800	8.7
1988	687,561	5,555,000	8.0
1989	704,090	5,993,500	7.9
1990	686,172	6,456,800	7.7
1991	627,438	6,926,000	7.3
1992	622,449	7,361,600	6.3
1993	656,659	7,793,400	5.9
1994	701,334	8,247,500	5.8
1995	755,181	8,742,500	6.0
1996	802,501	9,301,200	6.4
1997	860,700	9,901,900	6.5

[a] U.S. Bureau of Economic Analysis. Summary National Income and Product Time Series, 1929–1997. [Computer file].

[b] Assumes 20-year capital life.

Source: From www.usepa.gov.

Table 11F.107 Private O&M Expenditures for Water Pollution Abatement under with CWA and without CWA Scenarios, 1972–1997 (1997 $ millions)

Year	With CWA O&M Expenditures[a,b]	Without CWA O&M Expenditures
1972	2634.0	2634.0
1973	3072.1	2749.3
1974	3446.3	2882.5
1975	3734.9	3020.7
1976	4322.6	3133.9
1977	4856.2	3246.7
1978	5168.4	3382.8
1979	5633.1	3567.3
1980	5519.3	3770.2
1981	5425.5	3966.0
1982	5510.1	4182.4
1983	5723.4	4374.8
1984	5954.7	4542.2
1985	6180.2	4734.2
1986	6564.3	4915.9
1987	6834.4	5067.2
1988	7032.4	5212.9
1989	7171.5	5363.1
1990	7735.1	5509.3
1991	7134.2	5649.9
1992	7276.6	5763.4
1993	7079.7	5859.3
1994	7492.1	5946.1
1995	7694.0	6051.2
1996	7852.5	6197.3
1997	8005.2	6356.0

[a] Vogan, Christine R. September 1996. *Survey of Current Business.* "Pollution Abatement and Control Expenditures, 1972–1994." Table 12. Business and Government Expenditures for Air and Water Pollution Abatement in Current Dollars and Chain-Type Quantity and Price Indexes, 1972–1994. 1995 through 1997 estimated with regression on time.

[b] 1995 through 1997 are projected based on the trend from 1972 through 1994.

Source: From www.usepa.gov.

Table 11F.108 Public Sewerage and Other Sanitation Expenditures, 1902–1996 Fiscal Years (current $ millions)

	Sewerage			
Fiscal Year	Capital	O&M	Total	Total Sanitation
1902				51[e]
1913				97[e]
1922				189[e]
1927				312[e]
1932				223[e]
1934				177[e]
1936				204[e]
1938				226[e]
1940				207[e]
1942				229[e]
1944				245[e]
1946				370[e]
1948				670[e]
1950				834[e]
1952				992[e]
1953				908[e]
1954				1,058[e]
1955				1,142[e]
1956				1,326[e]

(Continued)

Table 11F.108 (Continued)

Fiscal Year	Sewerage			Total Sanitation
	Capital	O&M	Total	
1957			906[d]	1,443[e]
1958	649[a]	254[a]	903[a]	1,505[e]
1959	708[a]	273[a]	981[a]	1,609[e]
1960	767[a]	266[a]	1,033[a]	1,727[e]
1961	726[a]	304[a]	1,030[a]	1,774[e]
1962	886[a]	344[a]	1,230[a]	1,958[e]
1963	1,019[a]	204[a]	1,223[a]	1,996[e]
1964	1,095[a]	408[a]	1,503[a]	2,267[e]
1965	1,107[a]	605[a]	1,712[a]	2,360[e]
1966	1,202[a]	614[a]	1,816[a]	2,571[e]
1967	1,069[a]	555[a]	1,624[a]	2,523[e]
1968	1,107[a]	524[a]	1,631[a]	2,707[e]
1969	1,207[a]	510[a]	1,717[a]	2,969[e]
1970	1,385[a]	423[a]	1,808[a]	3,413[e]
1971	1,744[a]	781[a]	2,525[a]	
1972	2,202[a]	962[a]	3,164[a]	
1973	2,428[a]	1,176[a]	3,604[a]	
1974	2,640[a]	1,440[a]	4,080[a]	
1975	3,569[b]	1,693[b]	5,262[b]	
1976	3,955[b]	1,982[b]	5,937[b]	
1977	4,208[b]	2,329[b]	6,537[b]	
1978	4,365[b]	2,777[b]	7,142[b]	
1979	5,619[b]	3,176[b]	8,795[b]	
1980	6,271[b]	3,621[b]	9,892[b]	
1981	6,912[b]	4,209[b]	11,121[b]	
1982	5,895[b]	4,902[b]	10,797[b]	
1983	5,806[b]	5,433[b]	11,239[b]	
1984	5,663[b]	5,853[b]	11,516[a]	
1985	5,925[b]	6,260[b]	12,185[b]	
1986	6,461[b]	6,847[b]	13,308[b]	
1987	7,306[b]	7,555[b]	14,861[b]	
1988	8,300[c]	8,029[c]	16,329[c]	
1989	8,343[c]	8,696[c]	17,039[c]	
1990	8,356[c]	9,953[c]	18,309[c]	
1991	9,104[b]	10,571[b]	19,675[b]	
1992	8,926[c]	11,418[c]	20,344[c]	
1993	10,252[c]	12,440[c]	22,692[c]	
1994	7,989[c]	13,635[c]	21,624[c]	
1995	8,894[c]	14,690[c]	23,583[c]	
1996	9,326[c]	15,339[c]	24,665[c]	

[a] U.S. Bureau of the Census, *Government Finance.* Annual, as reported by Clark, Edwin. Council on Environmental Quality. "Estimating Baseline Pollution Abatement Expenditures." Table 7. Expenditures for Municipal Wastewater Treatment Facilities 1958–1974.
[b] U.S. Bureau of the Census. *Government Finances.*
[c] U.S. Bureau of the Census. State and Local Government Finance Estimates, by State. census.gov/govs/www/estimate.html Last revised April 26, 1999.
[d] Government Finance Data (BOC, Census of Governments, Compendium of Government Finance, Fiscal Years 1957, 1962, 1967. Table 8. Direct General Expenditure by Function, by Type of Government.)
[e] U.S. Department of Commerce, Bureau of the Census. 1975. *Historical Statistics of the United States, Colonial Times to 1970.* Series Y 533–566. Federal, State, and Local Government Expenditure, By Function: 1902 to 1970. Washington, DC: Government Printing Office.

Source: From usepa.gov.

Table 11F.109 Estimated Public Sewerage Expenditures Before the CWA: FY1902-FY1971

		Total Sewerage (S)[b]		Sewerage Capital[c]		Sewerage O&M[c]
Fiscal Year	Total Sanitation (TS)[a] (Current $ million)	(Current $ million)	(As Percentage of TS) (%)	(Current $ million)	(As Percentage of S) (%)	(Current $ million)
1902	51			23		9
1913	97			43		17
1922	189			84		34
1927	312			138		56
1932	223			99		40
1934	177			78		32
1936	204			90		36
1938	226			100		40
1940	207			92		37
1942	229			101		41
1944	245			109		44
1946	370			164		66
1948	670			297		120
1950	834			370		149
1952	992			440		177
1953	908			402		162
1954	1,058			469		189
1955	1,142			506		204
1956	1,326			588		237
1957	1,443	906	63	646	71	260
1958	1,505	903	60	649	72	254
1959	1,609	981	61	708	72	273
1960	1,727	1,033	60	767	74	266
1961	1,774	1,030	58	726	70	304
1962	1,958	1,230	63	886	72	344
1963	1,996	1,223	61	1,019	83	204
1964	2,267	1,503	66	1,095	73	408
1965	2,360	1,712	73	1,107	65	605
1966	2,571	1,816	71	1,202	66	614
1967	2,523	1,624	64	1,069	66	555
1968	2,707	1,631	60	1,107	68	524
1969	2,969	1,717	58	1,207	70	510
1970	3,413	1,808	53	1,385	77	423
1971		2,524		1,744	69	781

[a] U.S. Department of Commerce, Bureau of the Census. 1975. *Historical Statistics of the United States, Colonial Times to 1970.* Series Y 533–566. Federal, State, and Local Government Expenditure, By Function: 1902 to 1970. Washington, DC: Government Printing Office.

[b] Government Finance Data (BOC, Census of Governments, Compendium of Government Finance, Fiscal Years 1957, 1962, 1967. Table 8. Direct General Expenditure by Function, by Type of Government.)

[c] Estimated for 1902 through 1957 assuming that the ratio of capital to total sewerage expenditures is constant and equal to the average ratio observed in 1958 through 1971 (71 percent). For 1902 through 1956, total sewerage expenditures are assumed to be 62 percent of total sanitation expenditures, based on data for 1957 through 1970.

Source: From www.usepa.gov.

Table 11F.110 Comparison of Total U.S. Population, Housing Starts, and Public Sewerage Capital Expenditures (Actual and Projected), 1902–1997

Calendar Year	U.S. Population (millions)[a]	Housing Starts (thousands)[a]	Sewerage Capital Expenditures (1997 $ millions) Actual before CWA[b]
1902	79	157	427
1913	97	307	665
1922	110	574	773
1927	119	643	1,123
1932	125	64	1,040
1934	126	49	884
1936	128	211	979
1938	130	262	1,017
1940	132	397	979
1942	135	227	933
1944	138	96	1,021
1946	141	738	1,433
1948	147	977	1,944
1950	152	1,387	2,365
1952	158	1,051	2,358
1953	160	1,019	2,408
1954	163	1,113	2,664
1955	166	1,182	2,946
1956	169	963	3,208
1957	172	854	3,256
1958	176	955	3,332
1959	178	1,076	3,585
1960	181	888	3,579
1961	184	946	3,820
1962	187	1,053	4,458
1963	189	1,149	4,889
1964	192	1,118	5,018
1965	194	1,012	5,163
1966	197	788	4,937
1967	199	903	4,584
1968	201	1,096	4,670
1969	203	1,079	4,996
1970	205	1,018	5,727
1971	208	1,502	6,868

Calendar Year	U.S. Population (millions)[a]	Housing Starts (thousands)[a]	Sewerage Capital Expenditures (1997 $ millions)	
			Actual with CWA[b]	Projected without CWA[c]
1972	210	1,720	7,728	6,897
1973	212	1,495	8,009	6,695
1974	214	923	9,006	5,968
1975	216	760	9,972	5,867
1976	218	1,044	10,222	6,429
1977	220	1,377	10,085	7,076
1978	223	1,432	10,946	7,315
1979	225	1,241	12,012	7,194
1980	227	914	12,188	6,851
1981	229	760	10,823	6,771
1982	232	785	9,301	6,955
1983	234	1,351	8,745	7,945
1984	236	1,415	8,515	8,176
1985	238	1,494	8,799	8,435
1986	240	1,546	9,531	8,662
1987	242	1,372	10,481	8,546
1988	244	1,243	10,783	8,501

(Continued)

Table 11F.110 (Continued)

Calendar Year	U.S. Population (millions)[a]	Housing Starts (thousands)[a]	Sewerage Capital Expenditures (1997 $ millions)	
			Actual with CWA[b]	Projected without CWA[c]
1989	247	1,128	10,383	8,484
1990	249	947	10,402	8,388
1991	252	789	10,335	8,333
1992	255	932	10,698	8,738
1993	258	1,032	9,914	9,072
1994	260	1,183	8,962	9,469
1995	263	1,106	9,454	9,520
			Projected with CWA	**Projected without CWA**
1996	265	1,211	9,873	9,839
1997	268	1,221	9,873	10,025

[a] U.S. Bureau of the Census. Historical Housing Starts. Inside Standard Metropolitan Areas. Received Fax November 8, 1999. 1946 through 1958 based on average share that SMSA housing starts were of all private housing starts from 1902 through 1998.
[b] Values for 1996 through 1997 are an average of values for 1990 through 1994.
[c] Values for 1972 through 1997 estimated based on the estimated historical relationship (1902 through 1971) between capital expenditures, and housing starts (see Eq. [6.1]).

Source: From www.usepa.gov.

Table 11F.111 Comparison of Total Public Sewerage Operation and Maintenance Expenditures (Actual and Projected), 1902–1997

Calendar Year	Sewerage O&M Expenditures (1997 $ millions) Actual before CWA[a]
1902	172
1913	268
1922	312
1927	453
1932	420
1934	357
1936	395
1938	411
1940	395
1942	377
1944	412
1946	578
1948	784
1950	954
1952	951
1953	972
1954	1,075
1955	1,189
1956	1,295
1957	1,294
1958	1,294
1959	1,310
1960	1,366
1961	1,536
1962	1,282
1963	1,415
1964	2,308
1965	2,726
1966	2,541
1967	2,273
1968	2,087
1969	1,798
1970	2,204
1971	3,034

Calendar Year	Sewerage O&M Expenditures (1997 $ millions)	
	Actual with CWA[a]	Projected without CWA[b]
1972	3,569	3,705
1973	4,134	4,069
1974	4,544	4,433
1975	4,871	4,796
1976	5,398	5,154
1977	6,007	5,497
1978	6,527	5,866
1979	6,867	6,197
1980	7,239	6,551
1981	7,700	6,848
1982	8,215	7,170
1983	8,606	7,487
1984	8,900	7,834
1985	9,311	8,174
1986	9,970	8,505
1987	10,467	8,828
1988	10,836	9,154
1989	11,595	9,480

(Continued)

Table 11F.111 (Continued)

Calendar Year	Sewerage O&M Expenditures (1997 $ millions)	
	Actual with CWA[a]	**Projected without CWA[b]**
1990	12,227	9,791
1991	12,604	10,106
1992	13,309	10,405
1993	14,172	10,669
1994	15,036	10,913
1995	15,582	11,166
	Projected with CWA	**Projected without CWA**
1996	16,357	11,417
1997	17,171	11,705

[a] Values for 1996 and 1997 are projected based on the observed time trend between 1972 and 1995.
[b] Values for 1972 through 1997 estimated based on the estimated historical relationship (1932 through 1971) between O&M expenditures and the estimated size and average age of the sewerage capital stock (see Eq. [6.2]).

Source: From www.usepa.gov.

Table 11F.112 Public Sewerage Capital Costs under the with CWA and without CWA Scenarios (1997 $ millions)

	With CWA			Without CWA		
Year	Capital Expenditures	Estimated Capital Stock[a]	Annualized Capital Costs	Capital Expenditures	Estimated Capital Stock[a]	Annualized Capital Costs
1952	2,358			2,358		
1953	2,408			2,408		
1954	2,664			2,664		
1955	2,946			2,946		
1956	3,208			3,208		
1957	3,256			3,256		
1958	3,332			3,332		
1959	3,585			3,585		
1960	3,579			3,579		
1961	3,820			3,820		
1962	4,458			4,458		
1963	4,889			4,889		
1964	5,018			5,018		
1965	5,163			5,163		
1966	4,937			4,937		
1967	4,584			4,584		
1968	4,670			4,670		
1969	4,996			4,996		
1970	5,727			5,727		
1971	6,868			6,868		
1972	7,728			6,897		
1973	8,009			6,695		
1974	9,006			5,968		
1975	9,972			5,867		
1976	10,222			6,429		
1977	10,085			7,076		
1978	10,946			7,315		
1979	12,012			7,194		
1980	12,188			6,851		
1981	10,823			6,771		
1982	9,301	183,458	9,360	6,955	149,530	7,629
1983	8,745	190,402	9,714	7,945	154,128	7,863
1984	8,515	196,739	10,037	8,176	159,665	8,146
1985	8,799	202,589	10,336	8,435	165,176	8,427
1986	9,531	208,442	10,635	8,662	170,666	8,707
1987	10,481	214,764	10,957	8,546	176,120	8,986
1988	10,783	221,990	11,326	8,501	181,410	9,255
1989	10,383	229,441	11,706	8,484	186,578	9,519
1990	10,402	236,239	12,053	8,388	191,477	9,769
1991	10,335	243,061	12,401	8,333	196,286	10,014
1992	10,698	249,576	12,733	8,738	200,799	10,245
1993	9,914	255,817	13,052	9,072	205,079	10,463
1994	8,962	260,842	13,308	9,469	209,262	10,676
1995	9,454	264,786	13,509	9,520	213,712	10,903
1996	9,873	269,077	13,728	9,839	218,070	11,126
1997	9,873	274,012	13,980	10,025	222,972	11,376

[a] Based on 30-year life of capital and a 3 percent discount rate.

Source: From www.usepa.gov.

Table 11F.113 Water Pollution Abatement Capital and O&M Expenditures by Publicly Owned Electric Utilities, 1972–1997

	Expenditures (current $ millions)[a]		Expenditures (1997 $ millions)[b]	
Year	**Capital**	**O&M**	**Capital**	**O&M**
1972	29	3	97	10
1973	22	4	70	13
1974	29	5	84	15
1975	32	7	85	19
1976	36	9	90	23
1977	52	10	122	24
1978	63	10	138	22
1979	81	12	164	24
1980	61	13	113	24
1981	57	18	96	30
1982	86	17	137	27
1983	73	19	111	29
1984	54	20	79	29
1985	63	10	90	14
1986	40	10	55	14
1987	37	13	50	17
1988	28	12	36	16
1989	49	11	61	14
1990	77	8	92	10
1991	64	14	73	16
1992	14	10	16	11
1993	11	11	12	12
1994	10	10	11	11
1995			41	12
1996			41	12
1997			41	12

[a] Vogan, Christine R. September 1996. *Survey of Current Business*. "Pollution Abatement and Control Expenditures, 1972–1994." Table 12. Business and Government Expenditures for Air and Water Pollution Abatement in Current Dollars and Chain-Type Quantity and Price Indexes, 1972–1994.

[b] Values for 1995 through 1997 are assumed to be equal to the average observed value for 1990 through 1994.

Source: From www.usepa.gov.

Table 11F.114 Water Pollution Abatement O&M Expenditures by Publicly Owned Electric Utilities without the CWA, 1972–1997 (1997 $ millions)

Year	O&M Expenditures[a]	Without CWA O&M Expenditures[b]
1972	10	10
1973	13	10
1974	15	10
1975	19	10
1976	23	10
1977	24	10
1978	22	10
1979	24	10
1980	24	10
1981	30	10
1982	27	10
1983	29	10
1984	29	10
1985	14	10
1986	14	10
1987	17	10
1988	16	10
1989	14	10
1990	10	10
1991	16	10
1992	11	10
1993	12	10
1994	11	10
1995	12	10
1996	12	10
1997	12	10

[a] Values for 1995 through 1997 are assumed to be equal to the average observed value for 1990 through 1994.

[b] Assumed to be equal to the minimum observed value for WPA O&M expenditures with the CWA (1972 through 1997).

Source: From www.usepa.gov.

Table 11F.115 Water Pollution Abatement Capital Expenditures and Costs by Publicly Owned Electric Utilities without the CWA, 1972–1997 (1997 $ millions)

	With CWA			Without CWA		
Year	Capital Expenditures[a]	Estimated Gross Capital Stock	Capital Cost	Capital Expenditures[b]	Estimated Gross Capital Stock	Capital Cost
1972	97			11		
1973	70			11		
1974	84			11		
1975	85			11		
1976	90			11		
1977	122			11		
1978	138			11		
1979	164			11		
1980	113			11		
1981	96			11		
1982	137			11		
1983	111			11		
1984	79			11		
1985	90			11		
1986	55			11		
1987	50			11		
1988	36			11		
1989	61			11		
1990	92			11		
1991	73			11		
1992	16	1,843		11	212	
1993	12	1,762		11	212	
1994	11	1,704		11	212	
1995	41	1,631		11	212	
1996	41	1,587		11	212	
1997	41	1,537		11	212	

[a] Based on a 20-year life of capital and a 3 percent discount rate. Vogan, Christine R. September 1996. *Survey of Current Business.* "Pollution Abatement and Control Expenditures, 1972–1994." Table 12. Business and Government Expenditures for Air and Water Pollution Abatement in Current Dollars and Chain-Type Quantity and Price Indexes, 1972–1994. Values for 1995 through 1997 are assumed to be equal to the average observed value for 1990 through 1994.

[b] Assumed to be equal to the minimum observed value for WPA capital expenditures with the CWA (1972 through 1997).

Source: From www.usepa.gov.

Table 11F.116 Public and Private R&D, R&M, and Other Not-Elsewhere-Classified Expenditures for Water Pollution Abatement, 1972–1994 (current $ millions)

	Public			Private
Year	R&D	R&M	Other	R&D
1972	78	145	171	64
1973	95	190	171	73
1974	96	247	189	57
1975	96	279	210	68
1976	102	328	204	78
1977	108	370	188	91
1978	121	405	218	99
1979	139	425	257	107
1980	118	465	299	110
1981	130	434	286	104
1982	141	416	276	102
1983	153	381	290	93
1984	134	388	337	90
1985	110	464	391	95
1986	126	510	426	104
1987	123	560	542	106
1988	110	580	484	112
1989	108	626	489	125
1990	89	620	514	97
1991	101	761	537	124
1992	123	922	538	55
1993	107	750	575	81
1994	136	622	472	76

Source: From www.usepa.gov.

Table 11F.117 Comparison of R&D Expenditures for Water Pollution Abatement and Total R&D Expenditures, 1972–1997 (1997 $ millions)

	Public			Private		
	Total R&D[a]	WPA R&D		Total R&D[a]	WPA R&D	
Year	1997 $millions	1997 $millions	% of Total	1997 $millions	1997 $millions	% of Total
1972	54,264	260	0.48%	39,109	214	0.55%
1973	53,290	300	0.56%	42,023	231	0.55%
1974	50,789	279	0.55%	43,174	165	0.38%
1975	49,797	254	0.51%	41,937	180	0.43%
1976	51,455	255	0.50%	44,323	195	0.44%
1977	52,640	254	0.48%	46,205	214	0.46%
1978	54,184	265	0.49%	49,246	217	0.44%
1979	55,722	281	0.50%	52,722	216	0.41%
1980	56,143	218	0.39%	57,186	203	0.36%
1981	57,860	220	0.38%	60,778	176	0.29%
1982	59,987	224	0.37%	64,714	162	0.25%
1983	64,060	233	0.36%	69,047	142	0.21%
1984	69,104	197	0.28%	76,745	132	0.17%
1985	75,701	156	0.21%	82,417	135	0.16%
1986	76,734	174	0.23%	84,574	144	0.17%
1987	79,648	165	0.21%	84,176	142	0.17%
1988	79,130	143	0.18%	88,216	145	0.16%
1989	76,534	134	0.18%	93,374	155	0.17%
1990	74,845	106	0.14%	99,338	116	0.12%
1991	71,123	116	0.16%	106,023	142	0.13%
1992	69,398	137	0.20%	107,558	61	0.06%
1993	67,295	116	0.17%	105,113	88	0.08%
1994	66,122	144	0.22%	105,459	81	0.08%
1995	67,265	124	0.18%	115,191	98	0.08%
1996	65,743	124	0.19%	123,402	98	0.08%
1997	64,566	124	0.19%	133,308	98	0.07%

[a] U.S. Bureau of the Census. October 2, 1998. *Statistical Abstract of the United States*: 1998. "No. 988. R&D Expenditures: 1960–1997."

Source: From www.usepa.gov.

Table 11F.118 Public and Private R&D Expenditures for Water Pollution Abatement, with and without the CWA, 1972–1997 (1997 $ millions)

Year	Public R&D Expenditures		Private R&D Expenditures	
	With CWA	Without CWA[a]	With CWA	Without CWA[a]
1972	260.4	76.9	213.7	22.3
1973	300.3	75.5	230.7	24.0
1974	278.5	72.0	165.4	24.6
1975	254.4	70.5	180.2	23.9
1976	255.4	72.9	195.3	25.3
1977	254.1	74.6	214.1	26.4
1978	265.4	76.8	217.1	28.1
1979	280.9	78.9	216.2	30.1
1980	218.2	79.5	203.4	32.6
1981	219.7	82.0	175.8	34.7
1982	224.2	85.0	162.2	36.9
1983	233.3	90.8	141.8	39.4
1984	196.9	97.9	132.3	43.8
1985	156.3	107.3	135.0	47.0
1986	174.5	108.7	144.0	48.3
1987	165.2	112.8	142.4	48.0
1988	142.5	112.1	145.1	50.3
1989	134.3	108.4	155.4	53.3
1990	106.0	106.0	115.6	56.7
1991	115.8	100.8	142.2	60.5
1992	137.2	98.3	61.4	61.4
1993	116.3	95.3	88.0	60.0
1994	144.4	93.7	80.7	60.2
1995	124.0	95.3	98.0	65.7
1996	124.0	93.1	98.0	70.4
1997	124.0	91.5	98.0	76.1

[a] Values estimated based on observed minimum ratio of public WPA R&D to total public R&D from 1972 through 1994.

Source: From www.usepa.gov.

Table 11F.119 Comparison of Public R&M and Other Not-Elsewhere-Classified Expenditures for Water Pollution Abatement and Total Public Consumption Expenditures, 1972–1997 (1997 $ millions)

Year	Total Public Consumption Expenditures[a] 1997 $ millions	Public R&M and Other NEC WPA Expenditures[b]	
		1997 $millions	% of Total
1972	746,578	1,306	0.17%
1973	756,637	1,561	0.21%
1974	775,167	1,833	0.24%
1975	794,858	2,014	0.25%
1976	804,909	1,976	0.25%
1977	827,059	1,969	0.24%
1978	840,570	2,059	0.24%
1979	852,293	2,079	0.24%
1980	880,917	1,919	0.22%
1981	898,073	1,589	0.18%
1982	918,760	1,479	0.16%
1983	944,334	1,699	0.18%
1984	977,076	1,788	0.18%
1985	1,030,118	2,235	0.22%
1986	1,073,110	2,235	0.21%
1987	1,100,470	2,326	0.21%
1988	1,113,256	2,349	0.21%
1989	1,134,544	2,364	0.21%
1990	1,163,708	2,356	0.20%
1991	1,175,513	2,609	0.22%
1992	1,176,727	2,913	0.25%
1993	1,172,717	2,582	0.22%
1994	1,175,284	2,259	0.19%
1995	1,181,092	2,615	0.22%
1996	1,197,291	2,661	0.22%
1997	1,219,200	2,707	0.22%

[a] Council of Economic Advisers. *Economic Report of the President.* "B-83. Federal and State and Local Government Receipts and Current Expenditures, National Income and Product Accounts, 1959–1998." www.access.gpo.gov/usbudget/fy01/erp.html. As obtained on December 3, 1999.

[b] Values for 1995 through 1997 are projected based on the observed time trend between 1972 and 1994.

Source: From www.usepa.gov.

Table 11F.120 Public Water Pollution Abatement Expenditures for Regulation and Monitoring and Other Expenditures Not Elsewhere Classified, with and without the CWA, 1972–1997 (1997 $millions)

Year	With CWA R&M and Other Expenditures	Without CWA R&M and Other Expenditures[a]
1972	1,305.5	1,201.4
1973	1,561.3	1,217.6
1974	1,833.5	1,247.5
1975	2,014.3	1,279.1
1976	1,976.0	1,295.3
1977	1,969.4	1,331.0
1978	2,059.2	1,352.7
1979	2,079.2	1,371.6
1980	1,919.4	1,417.6
1981	1,588.9	1,445.2
1982	1,478.5	1,478.5
1983	1,698.9	1,519.7
1984	1,788.4	1,572.4
1985	2,234.7	1,657.7
1986	2,234.8	1,726.9
1987	2,326.4	1,771.0
1988	2,349.4	1,791.5
1989	2,363.8	1,825.8
1990	2,355.5	1,872.7
1991	2,609.2	1,891.7
1992	2,913.1	1,893.7
1993	2,581.5	1,887.2
1994	2,259.3	1,891.4
1995	2,614.6	1,900.7
1996	2,661.0	1,926.8
1997	2,707.5	1,962.0

[a] Values estimated based on observed minimum ratio of public WPA R&M and other not-elsewhere-classified expenditures to all government consumption expenditures from 1972 through 1994.

Source: From www.usepa.gov.

Table 11F.121 Canadian Federal Resources Directed to Water-Related Programs by Type of Program, Fiscal Year 1985–1986

		Person Years		Expenditures	
Program	Agencies	Number	% of Total	Millions of $	% of Total
1. Research and Information					
Water research	Environment, Fisheries and Oceans	725	20	60	16
Water data collection	Environment, Fisheries and Oceans	750	21	54	14
2. Water Quality					
Water pollution control	Environment, Fisheries and Oceans	630	18	44	12
Drinking water	Agriculture, Indian Affairs and Northern Development, Health and Welfare	10	—	41	11
3. Navigation and Ports					
Canals and channels	Environment, transport, Public Works	420	12	51	14
Ports and harbors	Fisheries and Oceans, Transport	60	2	36	10
4. Water Supplies					
Irrigation	Agriculture	500	14	38	10
Flood control	Environment, Transport	160	5	18	5
5. International and Interprovincial Agreements					
	Environment, External Affairs, International Joint Commission	230	6	24	6
6. Administration of Northern Waters					
	Indian Affairs and Northern Development	55	2	7	2
Total		3,540	100	373	100

Note: Expenditures in Canadian dollars.

Source: From Pearse, P.H., and others, *Currents of Change*, Final Report on Federal Water Policy, Ottawa, Canada, 1985.

Table 11F.122 Canadian Federal Resources Directed to Water-Related Programs by Department, Fiscal Year 1985–1986

		Expenditures	
Agencies	**Person Years**	**Millions of $**	**% of Total**
Environment	1,880	169	45
Fisheries and Oceans	620	62	17
Indian Affairs and Northern Development	60	39	10
Transport and Public Works	270	37	10
Agriculture	545	50	13
Energy, Mines and Resources	80	8	3
External Affairs	25	4	1
International Joint Commission	45	3	1
Other	15	1	—
Total	3,540	373	100

Note: Expenditures in Canadian dollars.

Source: From Pearse, P.H., and others, *Currents of Change*, Final Report on Federal Water Policy, Ottawa, Canada, 1985. With permission.

Table 11F.123 World Bank Lending for Water Supply and Sewerage Projects, 1985–1987

	1985		1986		1987		1985–1987	
Region	**$**	**%**	**$**	**%**	**$**	**%**	**$**	**%**
Eastern and Southern Africa	49.00	6.3	9.50	1.6	54.80	5.6	113.30	4.8
Western Africa	101.00	12.9	10.00	1.7	7.00	0.7	118.00	5.0
East Asia and Pacific	175.00	22.4	212.30	35.1	—	—	387.30	16.4
South Asia	—	—	78.00	12.9	284.00	29.3	362.00	15.4
Europe, Middle East and N Africa	292.00	37.4	120.00	19.8	559.60	57.7	971.60	41.3
Latin America and Caribbean	163.80	21.0	175.00	28.9	64.00	6.6	402.80	17.1
Total	780.80	100.0	604.80	100.0	969.40	100.00	2,355.00	100.0

Note: Millions of United States dollars.

Source: From World Water, *Mediterranean Region Swamps World Bank's Lending.* Thomas Telford Ltd, Box 124, Liverpool L69 2LQ, England, 1987. With permission.

Table 11F.124 Water Pollution Abatement Expenditures in the United States, 1975–1984

		Industrial		Public Sewer Systems[c]	
Year	**Total[b]**	**Facilities**	**Operations[a]**	**Facilities**	**Operations[a]**
1975	22,840	4,200	2,950	8,997	3,428
1976	24,445	4,625	3,375	9,370	3,713
1977	24,652	4,415	3,706	9,409	4,055
1978	26,631	4,277	3,934	10,090	4,392
1979	26,470	4,013	4,222	9,758	4,583
1980	24,647	3,725	4,081	8,942	4,694
1981	21,984	3,259	4,180	6,882	4,880
1982	21,199	3,080	4,022	6,148	5,156
1983	21,461	2,813	4,402	5,551	5,475
1984, preliminary	23,192	2,902	4,651	6,353	5,773

Note: In millions of dollars.

[a] Operation of facilities.

[b] Includes nonpoint sources not shown separately.

[c] Includes expenditures for private connectors to sewer systems, by owners of animal feedlots, and by government enterprises.

Source: From U.S. Department of Commerce, Statistical Abstract of the United States, 1987.

SECTION 11G DESALINATION

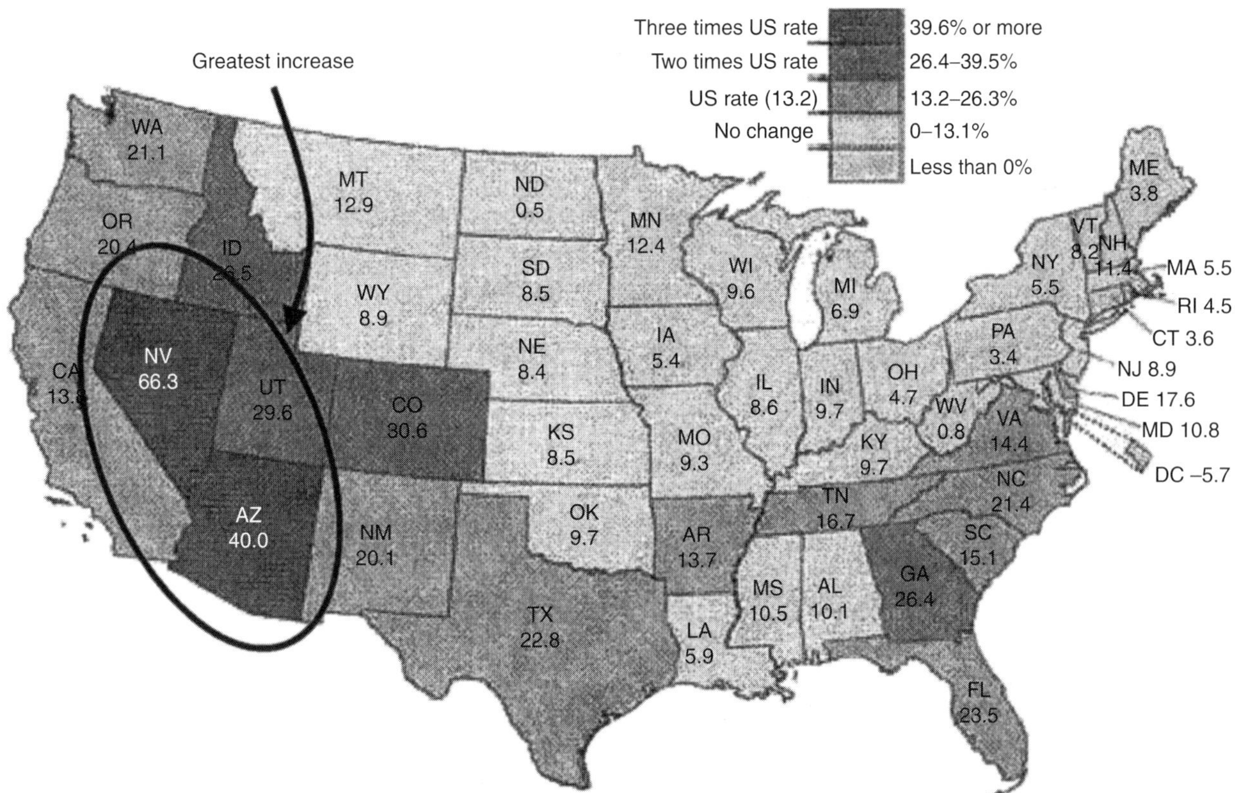

Figure 11G.82 Federal interest in desalination: change in population 1900–2000. (From www.worldbank.org. With permission.)

Average inches of annual precipitation in the United States 1961–1990

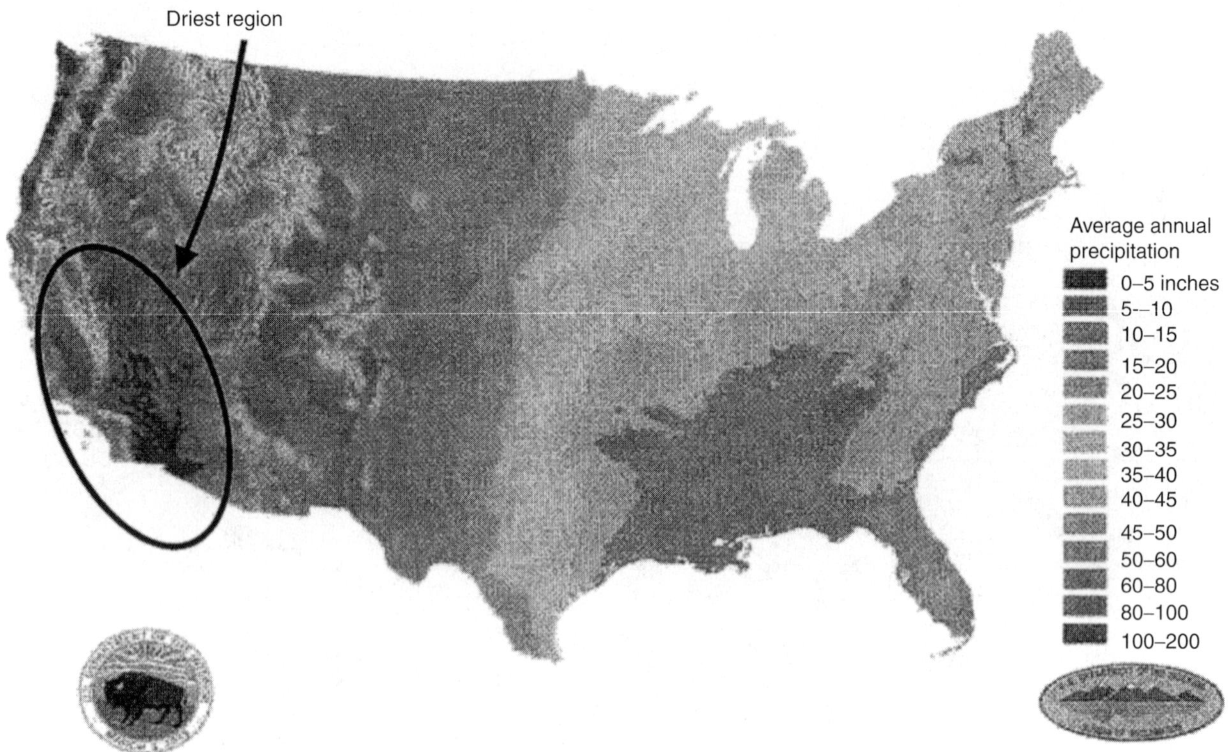

Figure 11G.83 Federal interest in desalination: precipitation. (From www.worldbank.org. With permission.)

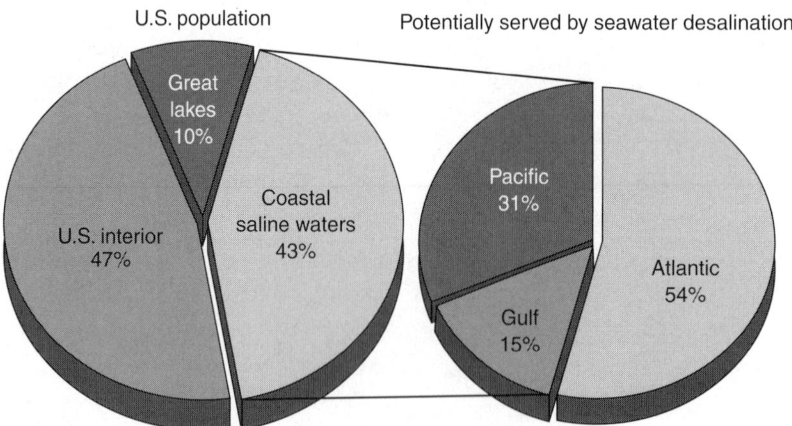

Figure 11G.84 Federal interest in desalination: potential for seawater desalination. (From www.worldbank.org. With permission.)

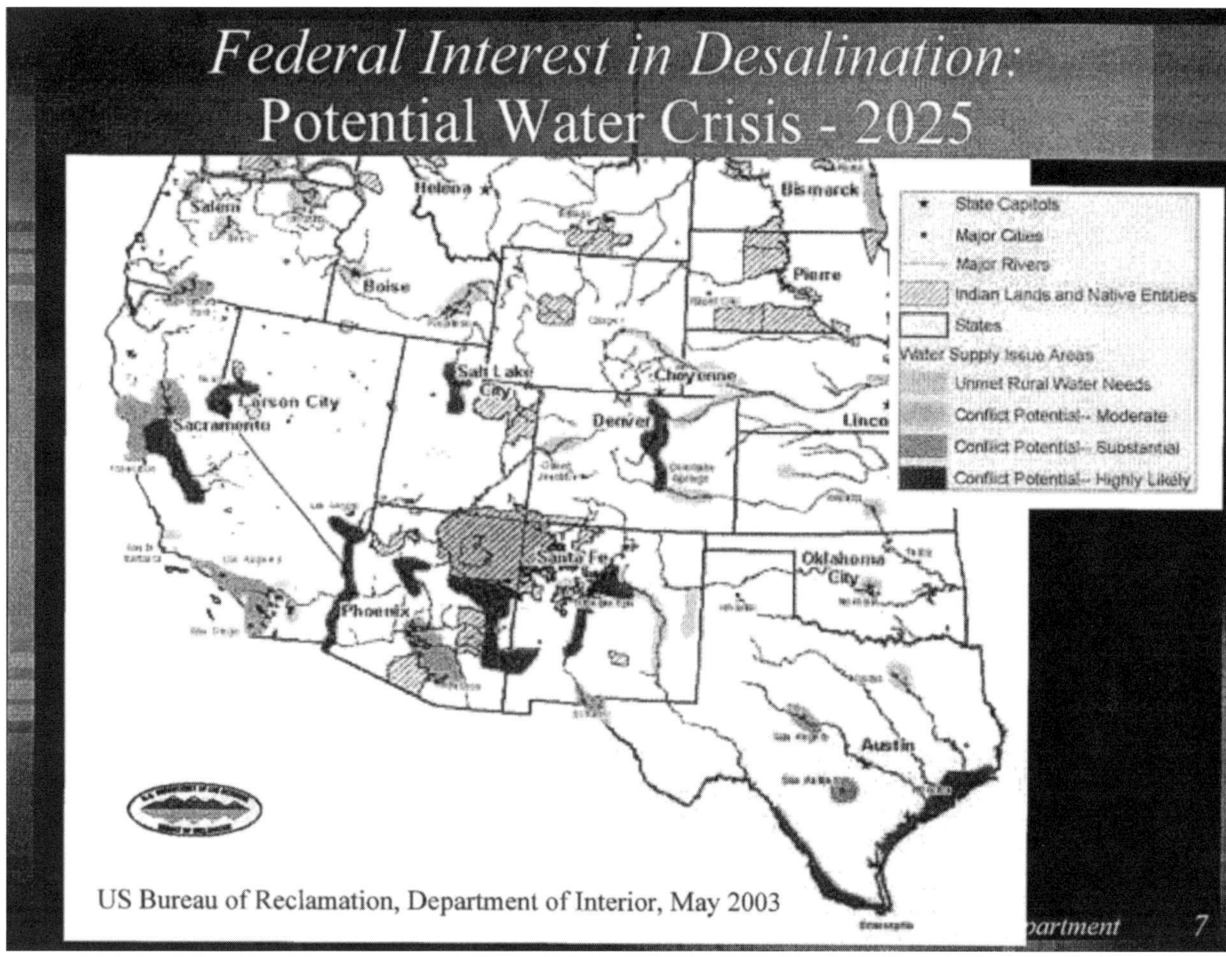

Figure 11G.85 Federal interest in desalination: potential water crisis—2025. (From www.worldbank.org. With permission.)

Table 11G.125 Classification of Desalting Processes Based on Properties of Solids and Liquids

A. Processes dependent on phase changes of water
1. Evaporation
Multiple-effect distillation, in which the latent heat comes from a solid surface
Multiple stage flash distillation, in which the latent heat comes from cooling of the liquid being evaporated
Supercritical distillation, in which all evaporation occurs above the critical temperature of pure water
Solar distillation in which the latent heat is derived from direct solar radiation
Vapor compression distillation, in which the latent heat is obtained regeneratively
2. Crystallization
Freeze-separation, in which the crystals involved are those of pure water
Hydrate-separation, in which the crystals contain molecules of the hydrating agent
B. Processes dependent on the surface properties of membranes in contact with water
1. Electrodialysis, in which the unwanted ions are caused to migrate through membranes due to electrical forces
2. Hyperfiltration or reverse osmosis, in which water is caused to migrate through membranes preferentially to the salt ions, due to pressure
C. Processes dependent on the properties of solids and liquids in contact with water
1. Ion exchange, in which unwanted ions are exchanged for less offensive ions loosely bonded to certain double salts in solid form
2. Solvent extraction, in which certain liquids dissolve water more readily than the salt ions contained in the saline water

Source: From Howe, University of California, Berkeley, 1968. With permission.

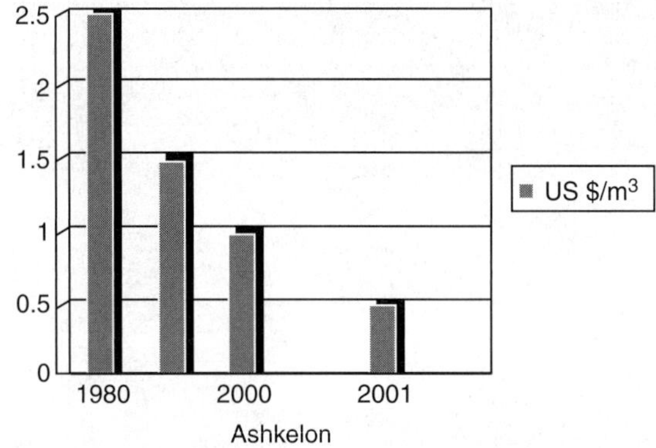

Figure 11G.86 Desalination cost trend. (From www.worldbank.org. With permission.)

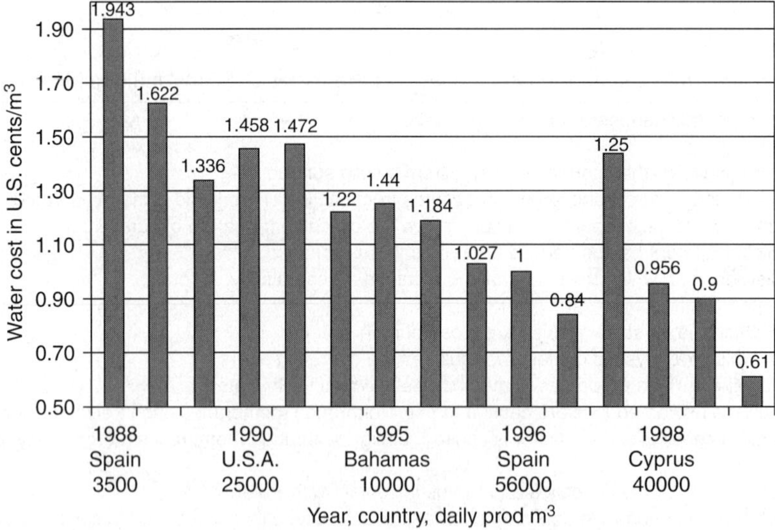

Figure 11G.87 Desalination cost trend: seawater desalination (reverse osmosis) in cents/m^3 from boot schemes from international tenders years 1988–1999. (From www.worldbank.org. With permission.)

Table 11G.126 Desalination Costs in the United States

	Plant Size (mgd)	Overall Cost (1985 dollars/1,000 gal)
Brackish Water		
Reverse osmosis	1	1.67
	3	1.41
	5	1.33
	10	1.23
	25	1.21
Electrodialysis (reversing)	1	1.72
	5	1.47
	10	1.37
	25	1.26
Seawater		
Distillation		
Multi-stage flash	1[a]	9.73
	5[a]	6.78
	10[a]	6.50
	25[a]	6.10[b]
Multiple-effect	1	8.31
	5[a]	5.70
	10[a]	5.36
	25[a]	5.36[b]
Reverse osmosis	0.01	13.42
	0.1	9.88
	1	7.40
	3	6.64
	5[a]	6.36
	10[a]	6.03[c]
	25[a]	5.96[c]

[a] Theoretical costs, since no plants of this size are operating in the United States.
[b] Approximated from Reed.
[c] Extrapolated cost.

Source: From United Nations, "The Use of Nonconventional Water Resources in Developing Countries," adapted from Reed, S.A., "Desalting Seawater and Brackish Water: 1981 Cost Update." With permission.

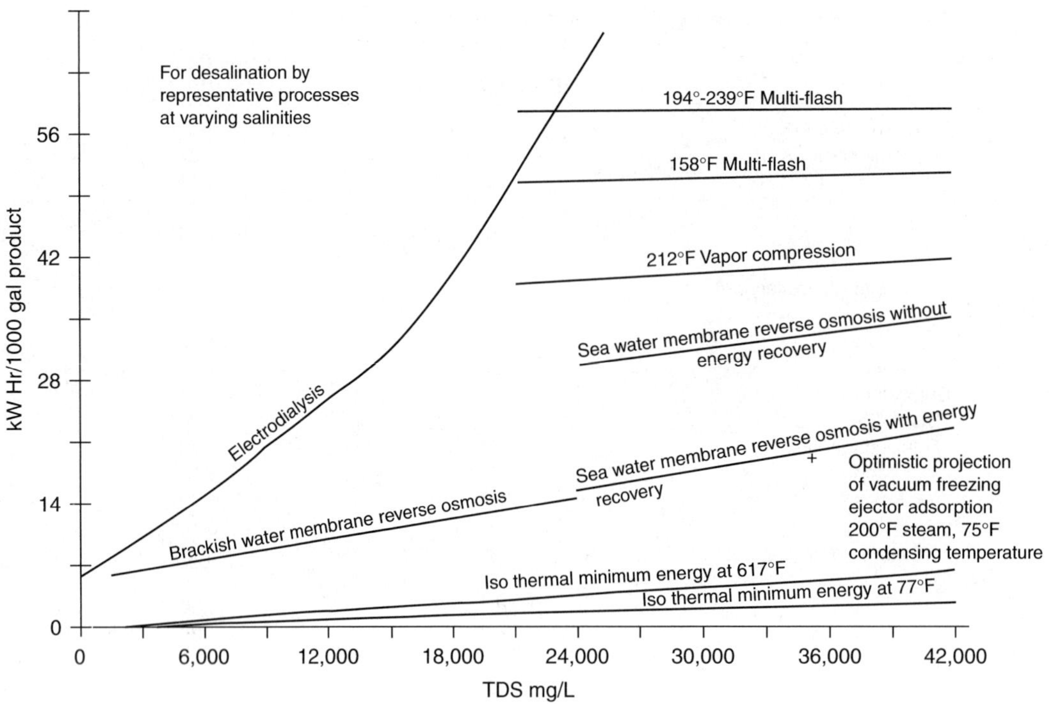

Figure 11G.88 Energy consumption. (From www.usace.army.mil.)

Table 11G.127 Classification of Desalting Processes Based on Type of Energy Required

A. Processes requiring thermal energy	C. Processes requiring electrical energy
Multiple effect distillation	Electrodialysis
Multistage flash distillation	
Solar distillation	
Supercritical distillation	
B. Processes requiring mechanical energy	D. Processes requiring chemical energy
Vapor compression distillation	Ion exchange
Freeze separation	Solvent extraction
Hydrate separation	
Hyperfiltration or reverse osmosis	

Source: From Howe, University of California, Berkeley, 1968. With permission.

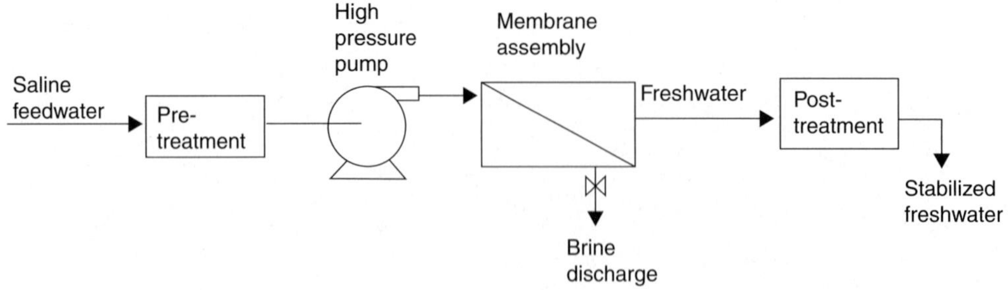

Figure 11G.89 Flow diagram of a reverse osmosis system (courtesy of USAID) (Kahn, 1986). (From www.coastal.ca.gov.)

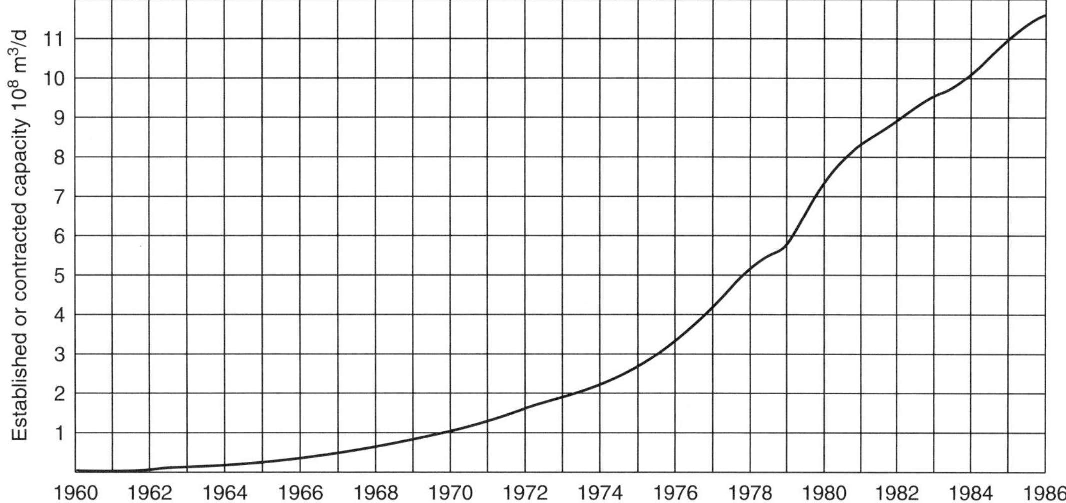

Figure 11G.90 Cumulative worldwide capacity of land-based desalting plants (Plants capable of producing 100 m^3/unit or more of fresh water). (From Wangnick, Klaus, 1987, 1986 World Market of Desalting Plants, *The IDA Magazine*, Vol. 1, No. 4, Copyright International Desalting Assoc. Reproduced with permission.)

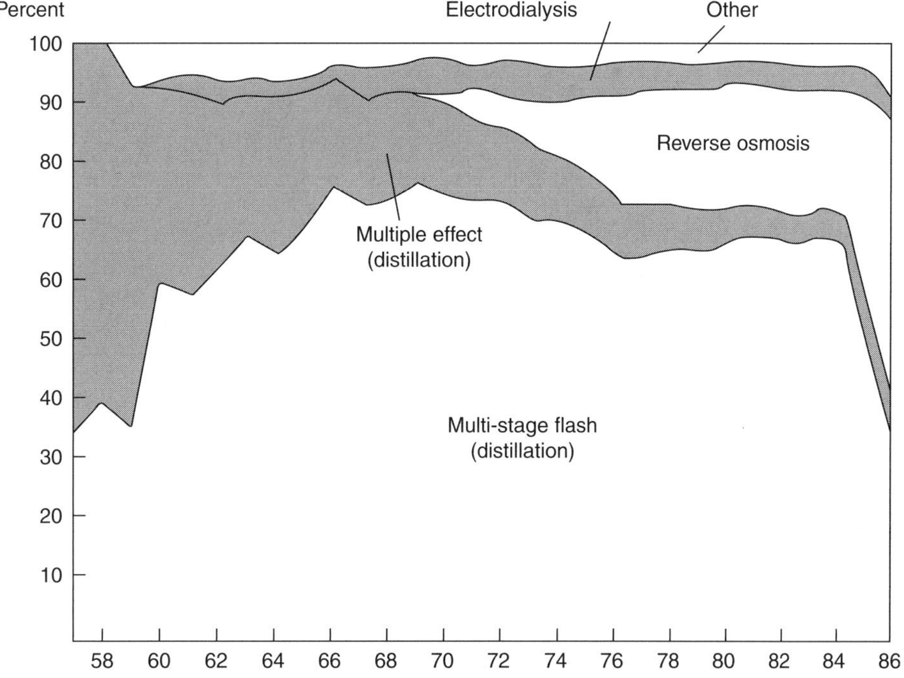

Figure 11G.91 Processes used by land-based desalting plants worldwide. (From Wangnick, Klaus, 1987, 1986 World Market of Desalting Plants, *The IDA Magazine*, Vol. 1, No. 4, Copyright International Desalting Assoc. Reproduced with permission.)

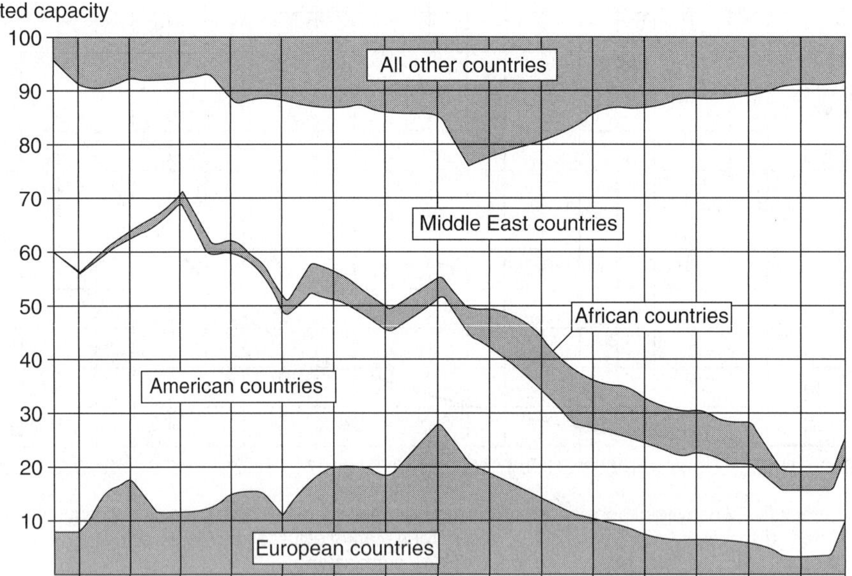

Figure 11G.92 Worldwide distribution of desalting plant capacity. (From Wangnick, Klaus, 1987, 1986 World Market of Desalting Plants, *The IDA Magazine*, Vol. 1, Copyright International Desalting Assoc. Reproduced with permission.)

Table 11G.128 Summary of Worldwide Desalination Capacity to 1998, Split by Plant Type and Process Capacity Range

Desalting Process	Percent	Capacity ($\times 10^6$ m^3 d^{-1})	Capacity (10^6 gal d^{-1})	No. of Plants
Unit capacity				
100–60,000 m^3 d^{-1}				
Multistage flash	44.4	10.02	2204	1244
Reverse osmosis	39.1	8.83	1943	7851
Multiple effect	4.1	0.92	202	682
Electrodialysis	5.6	1.27	279	1470
Vapor compression	4.3	0.97	213	903
Membrane softening	2.0	0.45	99	101
Hybrid	0.2	0.05	11	62
Others	0.3	0.06	13	120
	100.0	22.57	4965	12,433
Unit capacity				
500–60,000 m^3 d^{-1}				
Multistage flash	46.8	10.00	2200	1033
Reverse osmosis	37.9	8.10	1782	3835
Multiple effect	3.8	0.81	178	653
Electrodialysis	4.7	1.00	220	230
Vapor compression	4.2	0.90	198	486
Membrane softening	2.1	0.45	99	64
Hybrid	0.2	0.04	9	27
Others	0.23	0.05	11	11
EDI	0.05	0.01	2	97
	100.0	21.36	4699	6436
Unit capacity				
4000–60,000 m^3 d^{-1}				
Multistage flash	64.0	9.27	2039	496
Reverse osmosis	25.7	3.72	818	613
Multiple effect	3.6	0.52	114	48
Electrodialysis	2.1	0.31	68	60

(Continued)

Table 11G.128 (Continued)

Desalting Process	Percent	Capacity ($\times 10^6$ m^3 d^{-1})	Capacity (10^6 gal d^{-1})	No. of Plants
Vapor compression	1.9	0.28	62	42
Membrane softening	2	0.36	79	50
Hybrid	0	0.02	4	2
Others	0	0.00	0	0
	100.0	14.48	3186	1311

Source: From 1998 *IDA Worldwide Desalting Plants.* Inventory Report No. 15. Wangnick Consulting GmbH. With permission.

Table 11G.129 Summary of Worldwide Desalination Capacity, to 1998 Showing Contribution Made by Arab Nations

Desalting Process	% of Total World Capacity	Capacity (10^6 m^3 d^{-1})	Capacity (10^6 gal d^{-1})
Global			
Multistage flash	44.4	10.02	2204
Reverse osmosis	39.1	8.83	1943
Multiple effect	4.1	0.92	202
Electrodialysis	5.6	1.27	279
Vapor compression	4.3	0.97	213
Membrane softening	2.0	0.45	99
Hybrid	0.2	0.05	11
Others	0.3	0.07	15
	100.0	22.58	4968
Arab States			
Multistage flash	37.6	8.50	1870
Reverse osmosis	12.0	2.70	594
Multiple effect	0.5	0.12	26
Electrodialysis	1.6	0.35	77
Vapor compression	1.6	0.36	79
Membrane softening	0.0	0.00	0
Hybrid	0.0	0.01	2
Others	0	0	0
	53.3	12.04	2649
AGCC States			
Multistage flash	35.0	7.90	1738
Reverse osmosis	9.1	2.05	451
Multiple effect	0.2	0.05	11
Electrodialysis	0.5	0.12	26
Vapor compression	1.0	0.22	48
Membrane softening	0	0	0
Hybrid	0	0	0
Others	0	0	0
	45.8	10.34	2275

Note: AGCC, Arabian Gulf Co-operation Council.

Source: From 1998 *IDA Wordwide Desalting Plants.* Inventory Report No. 15. Wangnick Consulting. With permission.

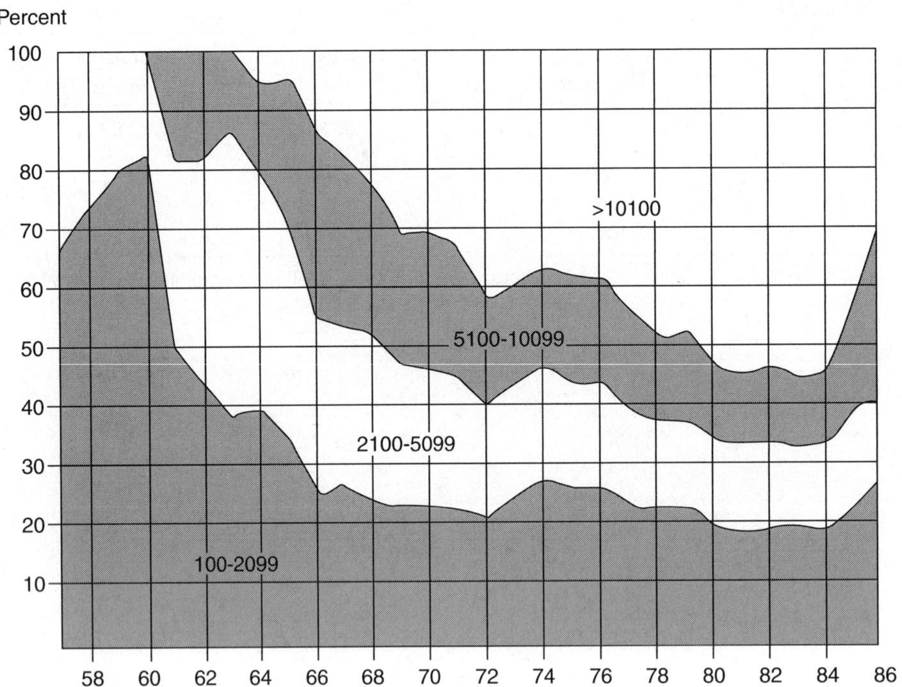

Figure 11G.93 Individual capacity of desalting plants worldwide (in m^3/d). (From Wangnick, Klaus, 1987, 1986 World Market of Desalting Plants, *The IDA Magazine*, Vol. 1. Copyright International Desalting Assoc. Reproduced with permission.)

Table 11G.130 Preliminary Desalination Process Selection

	A	B	C	D	E	F
Rule	**If the Freshest Source of Water Is**	**And if the Desired Output Water Will Be**	**And if Electricity Is to Be Generated**	**And if the Projected Cost Ratio of (264°F Steam)/ (Electricity)**	**Then Investigate the Cost of**	**And Have the Following Tests Performed**
1	More salty than sea water	Potable water			Transportation of fresher water: distillation can be used but at great expense	TDS
2	Sea water	High-pressure boller feed water	By steam turbine		Distillation followed by ion exchange	TDS, Ca^{++}, $SO_4^{=}$, $CO_3^{=}$, pH refer to water testing requirements in Appendix B
3	Sea water	Potable water	By steam turbine	Greater than $(10\times10^6$ BTU)/1 kWhr	Thermal distillation with or without vapor compression	TDS, Ca^{++}, $SO_4^{=}$, $CO_3^{=}$, pH
4	Sea water	Potable water	By internal combustion engine		Vapor compression distillation and waste heat	TDS, bacterial count, turbidity
5	Sea water	Potable water	No	Less than $(10\times10^6$ BTU)/1 kWh	Reverse osmosis	TDS, Ca^{++}, $SO_4^{=}$, $CO_3^{=}$, pH, bacterial count, silt density index, turbidity, oil & grease refer to list for reverse osmosis, Appendix B
6	Brackish water	Potable water			Reverse osmosis	TDS, Ca^{++}, $SO_4^{=}$, $CO_3^{=}$, pH, bacterial count, silt density index, turbidity, oil & grease

(Continued)

Table 11G.130 (Continued)

Rule	A If the Freshest Source of Water Is	B And if the Desired Output Water Will Be	C And if Electricity Is to Be Generated	D And if the Projected Cost Ratio of (264°F Steam)/(Electricity)	E Then Investigate the Cost of	F And Have the Following Tests Performed
7	Slightly saline brackish water	Potable water			Electrodialysis reversal	TDS, full ionic breakdown, bacterial count, turbidity refer to list for electrodialysis reversal, Appendix B

Source: From www.usace.army.mil.

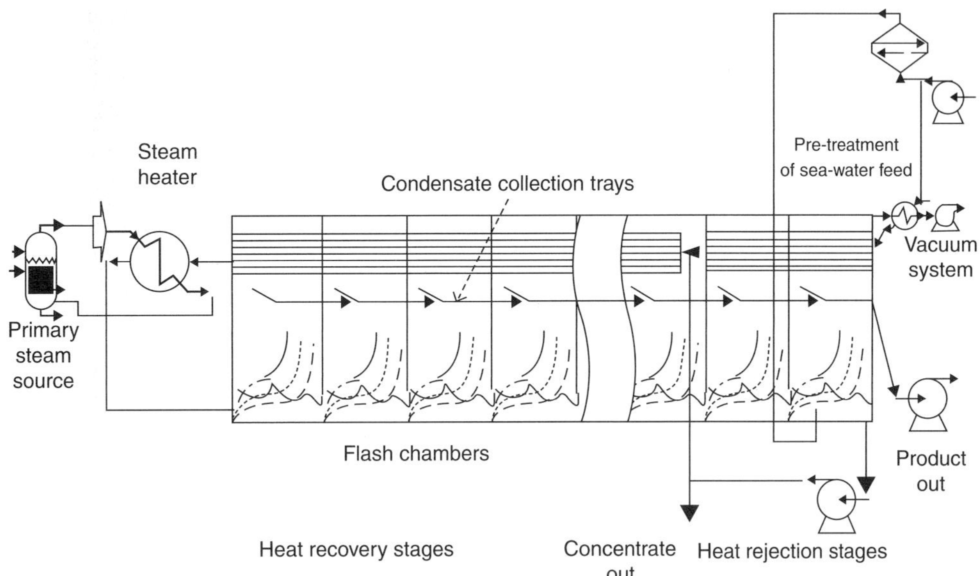

Figure 11G.94 Schematic presentation of a multi-stage flash desalination plant. (From IWRA, *Water International*, Vol. 25, March 2000. With permission.)

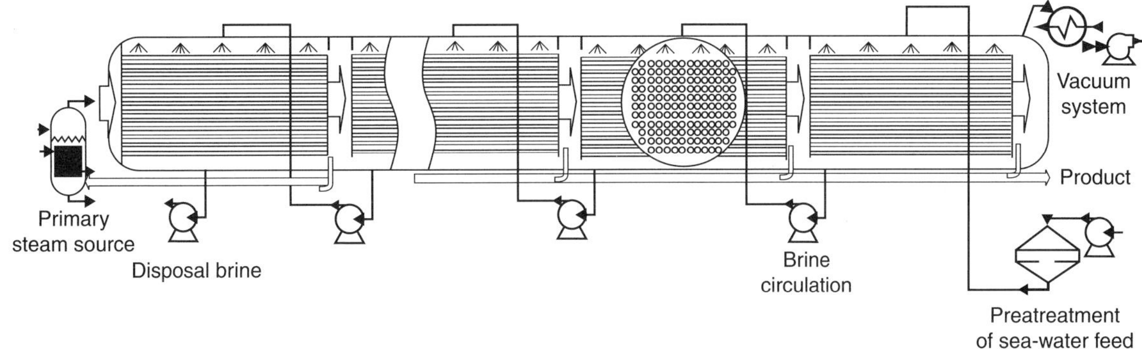

Figure 11G.95 Schematics of a horizontal tube multiple effect distillation plant. (From IWRA *Water International*, Vol. 25, March 2000. With permission.)

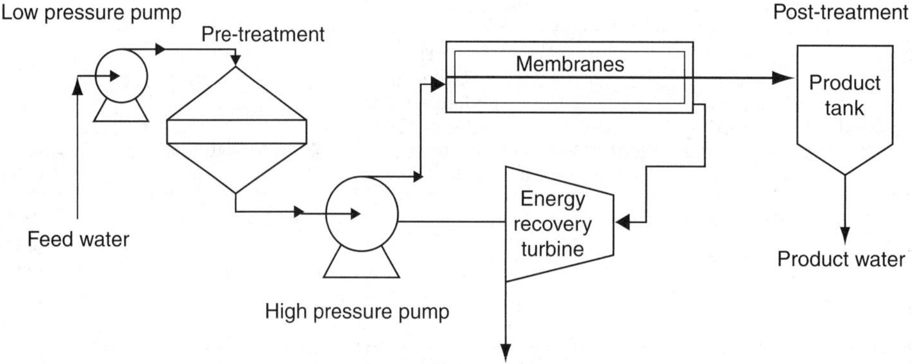

Figure 11G.96 Schematic presentation of a reverse osmosis desalination plant. (From IWRA *Water International*, Vol. 25, March 2000. With permission.)

Table 11G.131 Seawater Desalination Plants Using Reverse Osmosis Permasep Permeators

Location	Country	Capacity		Start-Up
		1000 GPD	m^3/d	
Installed				
Ras Abu-Jarjur[a]	Bahrain	12,000	45,420	84
Ghar Lapsi	Malta	5,300	20,000	83
Ghar Lapsi Expansion	Malta	1,060	4,600	86
Ras Tanajib	Saudi Arabia	3,500	13,248	85
Key West	U.S.A.	3,000	11,355	80
Yanbu	Saudi Arabia	1,300	4,920	80
Cadafe I	Venezuela	1,000	3,788	80
Cadafe II	Venezuela	1,000	3,788	83
Al Birk	Saudi Arabia	600	2,271	83
Jeddah Air Defense	Saudi Arabia	600	2,271	83
KISR	Kuwait	264	1,000	84
Paradise Island	Bahamas	240	908	81
Grand Cayman	U.S.A.	160	600	86
Cape Verde	Cape Verde Islands	150	568	83
PRASA	Puerto Rico	150	568	82
Jeddah Conference Palace	Saudi Arabia	150	568	84
Magna	Saudi Arabia	150	568	84
Mykonos	Greek Island	132	500	81
Ithaki	Greek Island	132	500	83
Puerto Santo	Madeira Island	132	500	80
Formentera	Spain	132	500	85
Raymond Int.	Ras Al Khaimah, UAE	125	473	78
Dura	Saudi Arabia	100	379	84
Al-Wajh (Coast Guard)	Saudi Arabia	100	379	85
Hakato Cho	Japan	79	300	81
Ras Al Mishab	Saudi Arabia	75	284	80
Under Construction				
Al-Dur (SWCC Gift)	Bahrain	12,100	46,000	87/88
Lanzarote	Spain	1,300	5,000	86
Safaniya	Saudi Arabia	1,044	4,000	86
Kutch Lignite[a]	India	1,000	3,788	87
St. Maarten	Netherlands	400	1,514	85/86
Codelco	Chile	160	600	86
Mullet Bay, St. Maarten	Netherlands	150	568	85/86
Shipyard	Italy	135	511	85/86
Misurata	Libya	132	500	86
Maho Bay	U.S.A.	100	379	86
WED Remote Islands	Abu Dhabi, UAE	4,800	18,168	86/87

[a] Highly brackish.

Source: From *The IDA Magazine*, Vol. 1, no. 4, 1987. Copyright International Desalting Association. Reprinted with permission.

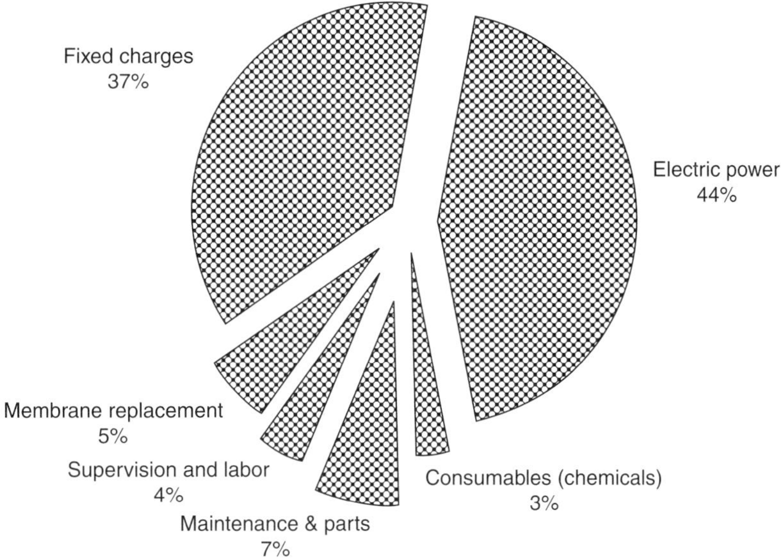

Figure 11G.97 Typical sea water reverse osmosis water cost. (From IWRA, *Water International*, Vol. 25, March 2000. With permission.)

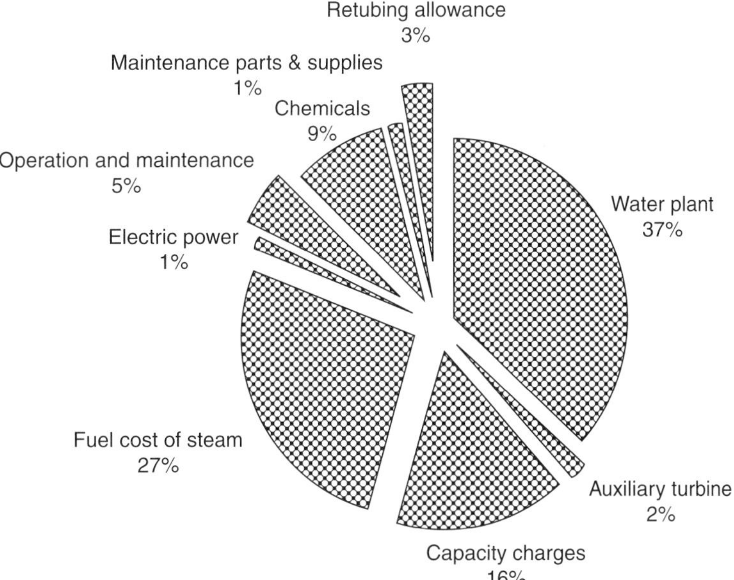

Figure 11G.98 Cost parameters of the MWD Tower MED desalination design. (From IWRA, *Water International*, Vol. 25, March 2000. With permission.)

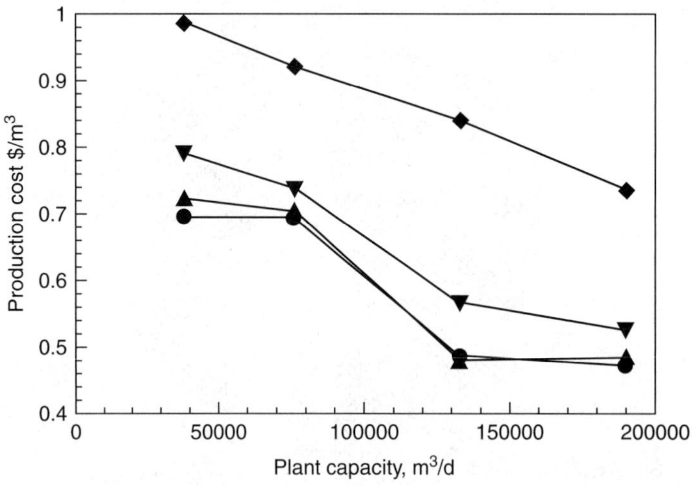

Figure 11G.99 Water cost dependence on plant production. Cost calculation for the four runners-up bidder at the Tampa Bay project. RO Seawater desalination. (From IWRA, *Water International*, Vol. 25, March 2000. With permission.)

Table 11G.132 Developer's Nominal Costs for Desalinated Water, Tampa Bay Project (Using Tax Exempt Financing), 94,625 m^3/d

Developer	First Year Cost $/m^3	30 Years Average Cost $/m^3
Florida seawater desalination company	0.54	0.65
Florida water partners (Parsons & IDE Technologies)	0.53	0.60
Progress energy corp. ionics partnership	0.56	0.67
Stone & Webster — TIC — Citizens Utilities	0.45	0.55

Source: From IWRA, *Water International*, Vol. 25, No. 1, March 2000. With permission.

Table 11G.133 Classification of Water Demineralization Processes Based on the Variation of Energy Requirement with Initial Salinity

Type of Energy	Conversion Process	
Processes in which the energy requirement is essentially independent of initial salinity	Multiple effect distillation	Vacuum flash distillation
	Multistage flash distillation	Solar distillation
	Vapor compression distillation	Freezing
	Supercritical distillation	Reverse osmosis
Processes in which the energy requirement depends on initial salinity	Electrodialysis	
	Ion exchange	
	Chemical precipitation	

Source: From California Department of Water Resources, 1960.

Table 11G.134 Typical Applications of Desalination Techniques

	Typical Applications			
	Brackish Water			
Technique	**0–3,000 ppm**	**3,000–10,000 ppm**	**Seawater 35,000 ppm**	**Higher Salinity Brines**
Distillation	T	S	P	P
Electrodialysis	P	S	T	P
Reverse osmosis	P	P	P	S
Ion exchange	P			

P = Primary application
S = Secondary application
T = Technically possible, but not economic

Source: From Office of Technology Assessment, 1987; Office of Technology Assessment, 1988, *Using Desalination Technologies for Water Treatment*, OTA-BP-0-46. With permission.

Table 11G.135 Relative Distribution of Different Types of Desalination Plants Worldwide

Process	Number of Plants	Percent of Total	Capacity (mgd)	Percent of Total
Distillation				
Multi-Stage Flash (MSF)	532	15.1	1,955	64.5
Multiple Effect (ME)	329	9.3	145	4.8
Vapor Compression (VC)	275	7.8	66	2.2
Membrane				
Reverse Osmosis (RO)	1,742	49.4	709	23.4
Electrodialysis (ED)	564	16.0	139	4.6
Other	85	2.4	18	0.6
Total	3,527	100	3,032	100

Source: From International Desalination Association desalination plant inventory, 1987; Office of Technology Assessment, 1988, *Using Desalination Technologies for Water Treatment*, OTA-BP-0-46. With permission.

Table 11G.136 BOOT Contract Costs

	Tampa Bay	Trinidad	Larnaca	Dheke lia	Singapore	Ashkelon	Algeirs
Design capacity, t/d	95,000	135,000	40,000	40,000	136,000	274,000	200,000
Developer	Poseidon	Ionics	IDE	Caramondani Desalination Plants Ltd	Hyflux	V.I.D. Desalination Company Ltd	Ionics
Feedwater	Power plant condenser discharge	Open water intake	Open water intake	Open water intake		Open water intake	
Seawater salinity, ppm	26,000	38,000	40,000	40,000		40,000	40,000
Energy cost, $/kWh	0.04	0.04	0.057	0.053			
Contract term, yr	30	23	10	10	20	25	25
Contract year			2000	1996	2002	2002	2003
Contracted water price, $/m^3							
Capital recovery	0.21	*	0.37	0.56		0.30	
Non-capital components	0.25	*	0.43	0.53		0.22	
Total first-year water price	0.46	0.71	0.80	1.09		0.52	
Normalized water price for energy cost of $0.04/kWh:				1.02			
Reduction in water price for energy cost of $0.04/kWh	—	—	(0.07)	0.068			
Total first-year water price (US$/m^3)	0.46	0.71	0.73	1.09	0.45	0.52	0.8182

Source: From worldbank.org. With permission.

Table 11G.137 Discharge of Dissolved Solids from Surface and Groundwater into the Oceans

Continent	Surface Water; Dissolved Solids Discharge (million t/yr)	Average Salinity of River Water (g/L)	Groundwater; Dissolved Solids Discharge (million t/yr)	Average Dissolved Solids of Groundwater (g/L)
Europe	240	0.077	30	0.4
Asia	850	0.065	296	0.9
Africa	310	0.072	288	1.0
North America	410	0.069	149	0.4
South America	550	0.053	113	0.3
Australia (including Tasmania, New Guinea, New Zealand)	120	0.060	199	0.5
World's Total	2,480	0.063	1,045	0.6
Percent of surface water; dissolved solids discharge			42	

Note: Data on surface water discharge of dissolved solids and average river water salinity are according to Lvovich, 1974, "World's Water Resources and Their Future"; data on discharge of dissolved solids in groundwater on islands excluded.

Source: From Zektser, I.S. and others, 1984, The Effect of Groundwater on the Salt Balance of Seas and Oceans, *Water Quality Bulletin*, Vol. 9, no. 1.

Table 11G.138 Selecting Desalination Processes after Water Quality Data Are Obtained

Rule	A If the Feed Water TDS Is (Mg/L)	B And if the Raw Feed Water Suspended Solids Are	C And if the Product of the (Ca^{++}) $(SO_4^{=})$ Moles²/Lr² Is in the Reject Brine (See Sample Problem A-3)	D And if the Oil and Grease in the Raw Feed Water Is	E Then Investigate the Cost of	F And Have the Following Pretreatment Processes Investigated for Effectiveness
1	Greater than 50,000				Transportion of fresher water; distillation of this water is extremely expensive	Precipitation of less soluble salts
2	Between 20,000–50,000	Over 20 NTU	Considerably less than 2×10^{-4}	Greater than 10 mg/L	Reverse osmosis or distillation and steam and electricity	Alum jar tests, pH adjustment 10-micron or smaller filter plugging
3	Between 20,000–50,000	Over 1 NTU		Less than 10 mg/L	Reverse osmosis	Alum jar tests 10-micron or smaller filter plugging UV sterilization
4	Between 20,000–50,000	Less than 1 NTU SDI greater than 3		Less than 10 mg/L	Spiral-wound membrane reverse osmosis	PH adjustment, UV sterilization, chlorine disinfection, chlorine residual
5	Between 20,000–50,000	SDI under 3		Less than 10 mg/L	Hollow fine-fiber membrane reverse osmosis	10-micron or smaller filter test, UV sterilization
6	Between 3,000–20,000	Over 1,000 mg/L	Considerably less than 2×10^{-4}	Greater than 10 mg/L	Distillation	PH adjustment, alum jar test
7	Between 3,000–20,000			Less than 10 mg/L	Reverse osmosis	PH adjustment, alum jar test, silt density index, UV sterilization
8	Between 500–4,000				Electrodialysis reversal	PH adjustment, alum jar test, 10-micron filter plugging, chlorine disinfection

Source: From www.usace.army.mil.

Table 11G.139 Final Selection of a Desalination Technique from Treatability Data

Rule	A If the Treated Feed Water Salinity (mg/1) Will Be (See Note)	B And the Cost Ratio of (264(F Steam)/(1 kwh Electricity) Will Be	C Alkaline Earths on the Raw Water Are Such That (See Sample Problem A-3)	D And the Treated Suspended Solids Are	E And the Designated Chlorine Residual Is	F Then Investigate the Cost of	G With the Following Pre- and Post-Treatment Technique Costs
1	Between 20,000–50,000		Within 65% of saturation	Less than 1 NTU but SDI greater than 3	0.0 mg/L	Spiral-wound membrane reverse osmosis	Whatever treatment is necessary to produce D and E
2	Between 20,000–50,000		Within 66% of saturation	SDI less than 3	0.0 mg/L	Hollow fine-fiber membrane reverse osmosis	Whatever treatment is necessary to produce D and E
3	Between 20,000–50,000			Less than 1 NTU	Between 1.0 and 0.0 mg/L	Chlorine-resistant membrane reverse osmosis	Whatever treatment is necessary to produce D and E
4	Between 20,000–50,000	Greater than 10× 106 BTU/1 kWhr	Within 50% of saturation	Greater than 1 NTU	More than 1 mg/L	Some form of distillation under 185°F	Anti-scalent
5	Between 20,000–50,000	Greater than 10× 106 BTU/1 kWhr		Greater than 1 NTU	More than 1 mg/L	Some form of distillation	Acid feed (hydrochloric is best)
6	Between 20,000–50,000	Greater than 10× 106 BTU/1 kWhr	[Ca++] multiplied by [SO_4^2] Well Under 2 (10×5	Greater than 1 NTU	More than 1 mg/K	Some form of distillation	No pretreatment for calcium sulfata scale control
7	Between 3,000–20,000			Less than 1 NTU but SDI greater than 3	Less than 1.0 mg/L**	Brackish water spiral-wound membrane reverse osmosis	Whatever treatment is necessary to produce D and E
8	Between 3,000–20,000			SDI less than 3	Less than 1.0 mg/L**	Brackish water hollow fine-fiber membrane reverse osmosis	Whatever treatment is necessary to produce D and E
9	Between 500–4,000 and especially when expected to vary by more than 15%			Will not plug 10-micron filter	0.0 mg/L**	Electrodialysis reversal	Turbidity removal to 1 NTU and disinfection to less than 1/100 ml.
10	Is stable at some value between 500–1000			SDI less than 3	Less than 1.0 mg/L	Low pressure/high flux membrane reverse osmosis	Whatever treatment is necessary to produce D and E

Source: From www.usace.army.mil.

Table 11G.140 Cost Comparison for Different Desalination Techniques

		MSF	MSF (Singapore)[a]	MED	MED-MWD[a]	VC	RO	RO-Tampa Bay[a]
Installation costs	$\$/m^3/d$	1,200–1,500	2,300	900–1,000	660	950–1,000	700–900	1,000–1,350
Product costs	Cents/m^3	110–125	150	75–85	46	87–95	68–92	45–56

[a] Estimation, based on publications or recent proposals.

Source: From IWRA, *Water International*, Vol. 25, No. 1, MARCH 2000. With permission.

SECTION 11H WATER TRANSFER

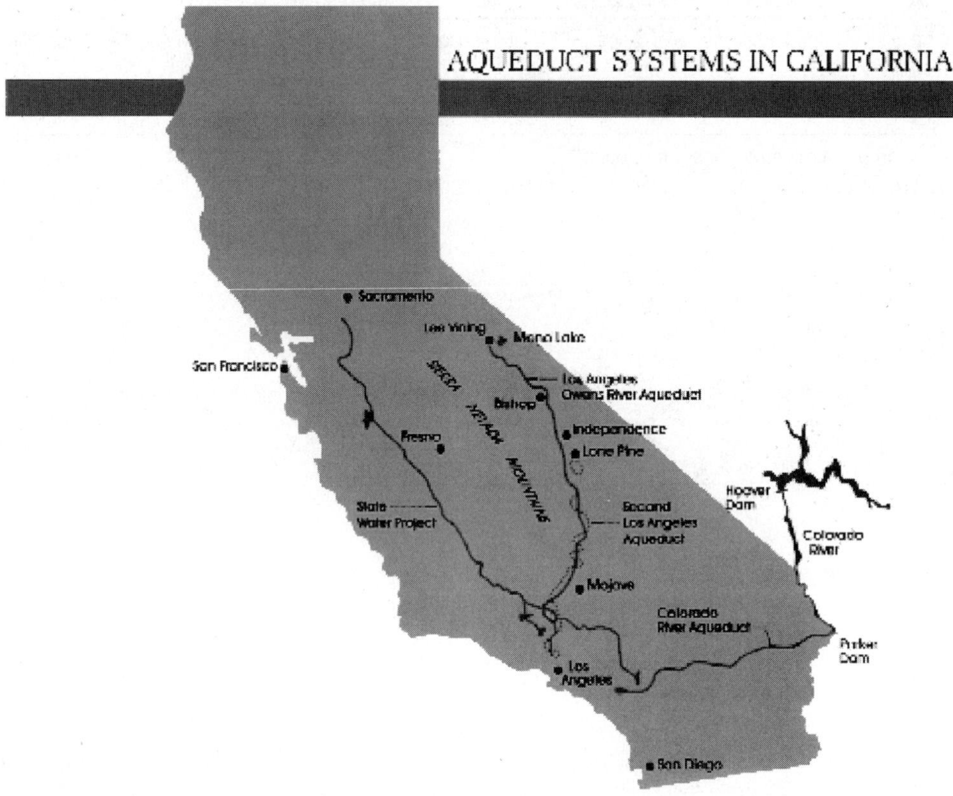

Figure 11H.100 Aqueduct systems in California. (From www.wsu.edu. With permission.)

Table 11H.141 Major Aqueducts in California

	Capacity (ft^3/s)[a]	Length (mi)	Owner	Initial Year of Operation
Los Angeles	710	244	LADWP	1913
Mokelumne River	590	90	EBMUD	1929
Hetch Hetchy	460	152	SF	1934
All American	15,100	80	USBR	1938
Contra Costa	350	48	USBR	1940
Colorado Rive	1,600	242	MWD	1941
Friant-Kern	4,000	152	USBR	1944
Coachella	2,500	123	USBR	1947
San Diego No. 1	200	71	SD	1947
Delta-Mendota	4,600	116	USBR	1951
Madera	1,000	36	USBR	1952
Putah South	960	35	USBR	1957
Santa Rosa-Sonoma	62	31	SCWA	1959
San Diego No. 2	1,000	93	SD	1960
Corning	500	21	USBR	1960
Petaluma	16	26	SCWA	1961
Tehama-Colusa	2,530	113	USBR	1961[b]
South Bay	360	43	DWR	1965
North Bay	46	27	DWR	1968[c]
California	13,100	444	DWR	1972[d]
Folsom South	3,500	27	USBR	1973[e]
Cross Valley	740	20	KCWA	1975

Note: DWR, California Department of Water Resources; EBMUD, East Bay Municipal Utility District; KCWA, Kern County Water Agency; LADWP, Los Angeles Department of Water and Power; MWD, Metropolitan Water District of Southern California; SCWA, Sonoma County Water Agency; SD, City of San Diego; SF, City and County of San Francisco; USBR, U.S. Bureau of Reclamation.

[a] Initial reach only for most irrigation canals.
[b] Tehama and Glenn Counties.
[c] Interim facilities.
[d] To Southern California.
[e] Reaches 1 and 2.

Source: From California Department of Water Resources, 1987. California Water: Looking to the Future, Bulletin 160-87.

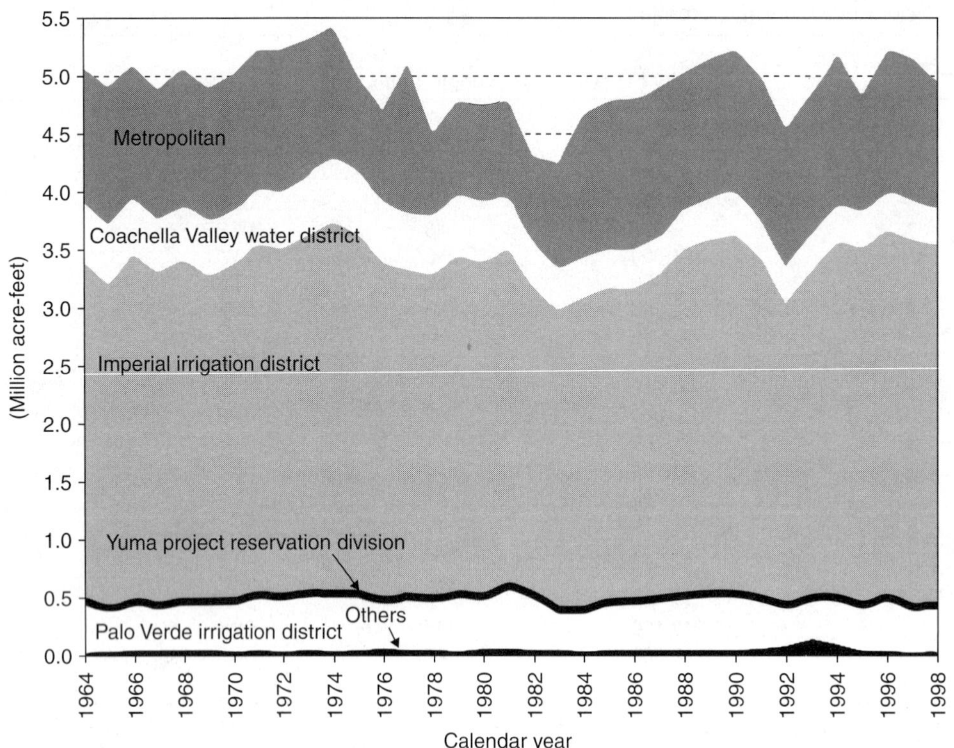

Figure 11H.101 California's net diversions from the Colorado River, includes reductions for unmeasured returns estimated since 1994. (From California's Colorado River Water Use Plan, June 2, 2000, www.crb.ca.gov. With permission.)

Table 11H.142 Apportionment of Colorado River Water Supply in California

Agency and Description of Service Area		Beneficial Consumptive Use in Acre-Feet per Year	
Priority Number		**Per California Seven-Party Agreement**	**After Start of Central Arizona Project**
1	Palo Verde irrigation district		
2	Yuma project, California portion		
3	Imperial irrigation district	3,850,000	3,850,000
	Coachella Valley water district		
	Palo Verde irrigation district (mesa lands)		
4	Metropolitan water district		
5	Metropolitan water district	550,000[a]	550,000[a]
		662,000	0
6	Imperial irrigation district		
	Coachella Valley water district	300,000	0
	Palo Verde irrigation district (mesa lands)		
	Totals	5,362,000[b]	4,400,000[b]

[a] Includes Indian water rights and miscellaneous present perfected rights totalling 58,000 acre-ft that reduce Metropolitan's entitlement to 492,000 acre-ft.

[b] Plus not more than one-half of any excess or surplus water in the lower Colorado River.

Source: From California Department of Water Resources, 1987. California Water: Looking to the Future, Bulletin 160-87.

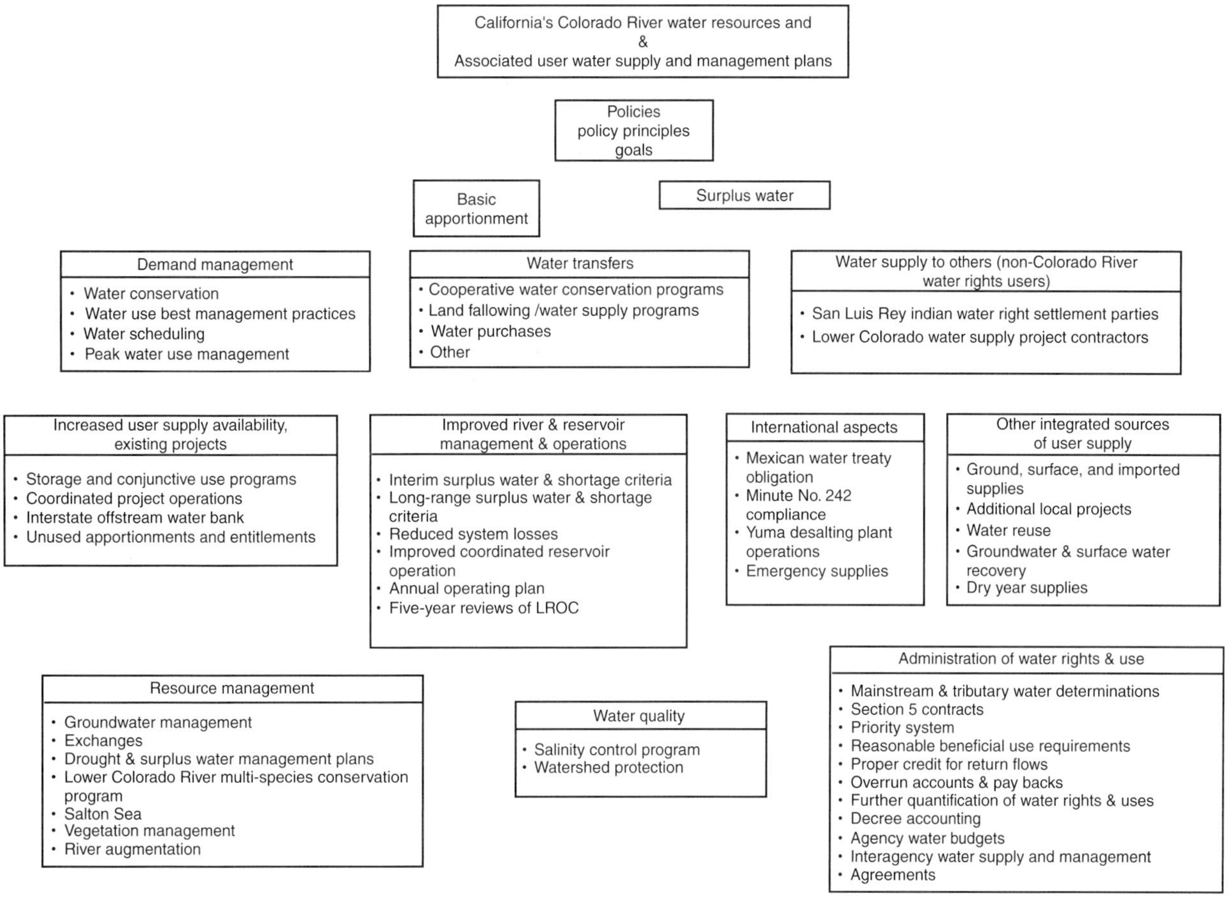

California's Colorado River water resources and
&
Associated user water supply and management plans
Policies
policy principles
goals
Basic
apportionment
Surplus water
Demand management
• Water conservation
• Water use best management practices
• Water scheduling
• Peak water use management
Water transfers
• Cooperative water conservation programs
• Land fallowing /water supply programs
• Water purchases
• Other
Water supply to others (non-Colorado River water rights users)
• San Luis Rey indian water right settlement parties
• Lower Colorado water supply project contractors
Increased user supply availability, existing projects
• Storage and conjunctive use programs
• Coordinated project operations
• Interstate offstream water bank
• Unused apportionments and entitlements
Improved river & reservoir management & operations
• Interim surplus water & shortage criteria
• Long-range surplus water & shortage criteria
• Reduced system losses
• Improved coordinated reservoir operation
• Annual operating plan
• Five-year reviews of LROC
International aspects
• Mexican water treaty obligation
• Minute No. 242 compliance
• Yuma desalting plant operations
• Emergency supplies
Other integrated sources of user supply
• Ground, surface, and imported supplies
• Additional local projects
• Water reuse
• Groundwater & surface water recovery
• Dry year supplies
Administration of water rights & use
• Mainstream & tributary water determinations
• Section 5 contracts
• Priority system
• Reasonable beneficial use requirements
• Proper credit for return flows
• Overrun accounts & pay backs
• Further quantification of water rights & uses
• Decree accounting
• Agency water budgets
• Interagency water supply and management
• Agreements
Resource management
• Groundwater management
• Exchanges
• Drought & surplus water management plans
• Lower Colorado River multi-species conservation program
• Salton Sea
• Vegetation management
• River augmentation
Water quality
• Salinity control program
• Watershed protection

Figure 11H.102 Framework components of California's Colorado River water use plan. (From California's Colorado River Water Use Plan, June 2, 2000, www.crb.ca.gov. With permission.)

600,000
500,000
400,000
300,000
200,000
100,000
0
Acre-feet
IID/CVWD conservation phase II
IID/CVWD conservation phase I
IID/SDCWA transfer
MWD/CVWD transfer and exchange
All American canal lining project
IID/MWD conservation program
Coachella canal lining project
1 2 3 4 5 6 7 8 9 10 11 12 13 14 15 16 17 18 19 20 21 22 23 24 25 26 27 28 29 30
Year

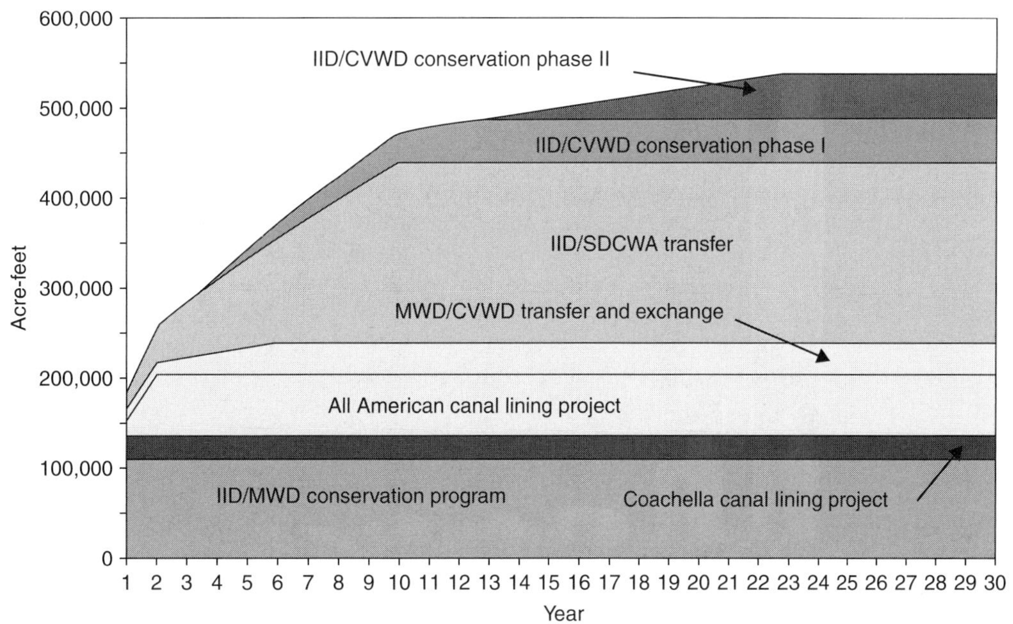

Figure 11H.103 Core water conservation/transfer projects, and exchanges. (From California's Colorado River Water Use Plan, June 2, 2000, www.crb.ca.gov. With permission.)

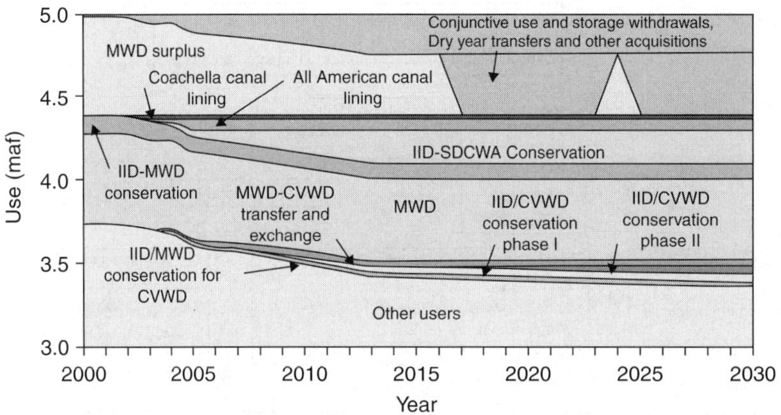

Figure 11H.104 California use of water. (From California's Colorado River Water Use Plan, June 2, 2000, www.crb.ca.gov. With permission.)

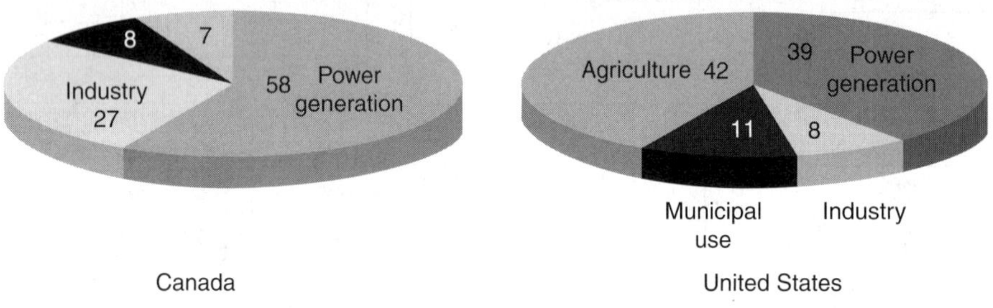

Withdrawals of renewable supplies :

Canada:	1.6%
United States:	18.9%

Figure 11H.105 Freshwater withdrawals by sector. (From www.unep.org. With permission.)

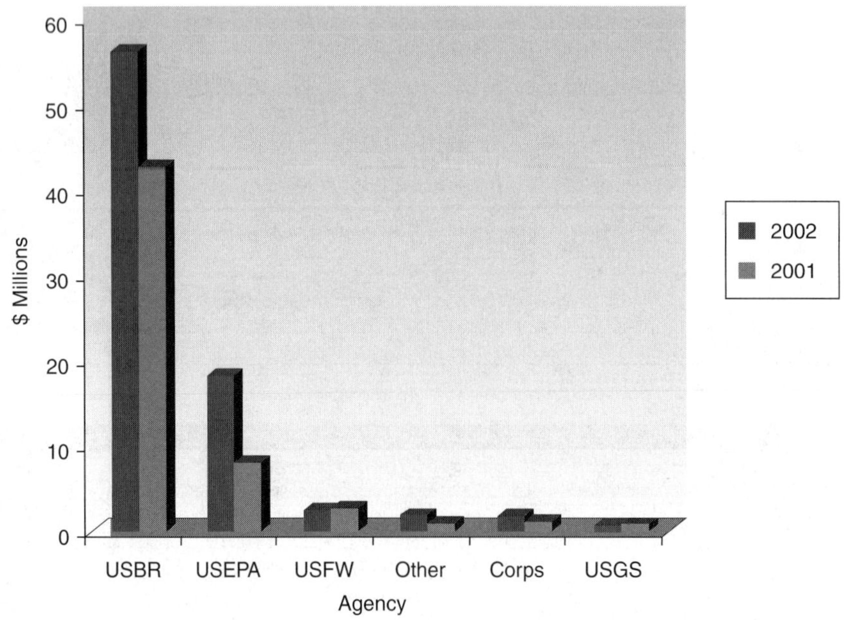

Figure 11H.106 Federal appropriations for Calfed funding by agency (FY 2001 and 2002). (From Calfed Annual Report 2001, www.swf.usace.army.mil. With permission.)

Table 11H.143 Calfed Bay-Delta Program Funding by Source for Stage 1 Program Implementation (California Fiscal Year 2001)

		State					Federal				
Program	Total Year 1 Funding	General Funds	Prop 204	Prop 13	Other	Subtotal	USBR	Corps	Other	Subtotal	Local Subtotal
Ecosystem restoration	$236.0	$3.5	$134.9	$46.2	$6.1	$190.7	$4.3	$0.4	$6.3	$11.0	$34.3
Environmental	$59.1	$59.1	—	—	—	$59.1	—	—	—	—	—
Water use efficiency	$204.1	$17.0	—	$12.3	—	$29.3	$26.0	—	—	$26.0	$148.8
Water conservation	($31.2)	($17.0)	—	($3.2)	—	($20.2)	($1.9)	—	—	($1.9)	($9.1)
Water recycling	($172.9)	—	—	($9.1)	—	($9.1)	($24.1)	—	—	($24.1)	($139.7)
Water transfers	$1.1	$1.1	—	—	—	$1.1	—	—	—	—	—
Watershed	$33.0	$18.9	$1.3	—	$1.0	$21.2	—	—	$2.3	$2.3	$9.5
Drinking water quality	$37.5	$13.5	—	$24.0	—	$37.5	—	—	—	—	—
Levees	$35.4	$0.1	$1.7	$28.5	—	$30.3	—	—	—	—	$5.1
Storage	$95.5	$24.7	—	$69.0	—	$93.7	$1.8	—	—	$1.8	—
Oversight and coordination	($2.9)	($2.9)	—	—	—	($2.9)	—	—	—	—	—
Surface	($13.8)	($13.8)	—	—	—	($12.0)	($1.8)	—	—	($1.8)	—
Groundwater	($78.8)	($78.8)	—	($69.0)	—	($78.8)	—	—	—	—	—
Conveyance	$22.3	$4.2	—	$4.8	—	$9.0	$2.6	—	—	$2.6	$10.7
Science	$28.2	$13.2	—	—	$2.3	$15.5	$4.0	$0.2	$1.9	$6.1	$6.6
Calfed science	($13.8)	($13.8)	—	—	—	($13.0)	—	—	($0.8)	($0.8)	—
Interagency ecological program	($14.4)	($14.4)	—	—	—	($2.5)	($4.0)	($0.2)	($1.1)	($5.3)	($6.6)
Oversight and coordination	$13.8	$13.5	—	—	—	$13.5	—	$0.3	—	$0.3	—
Total	$766.0	$168.8	—	$184.8	$9.4	$500.9	$38.7	$0.9	$10.5	$50.1	$215.0

Note: Local funds include State Water Project Funds and Central Valley Project Improvement Act (CVPIA) that are collected from state water contractors and Central Valley Project water users.

Source: From www.swf.usace.army.mil.

Table 11H.144 Calfed Bay-Delta Program Funding by Source for Stage 1 Program Implementation (California Fiscal Year 2002)

Program	Total Year 2 Funding	State					Federal				Local Subtotal
		General Funds	Prop 204	Prop 13	Other	Subtotal	USBR	Corps	Other	Subtotal	
Ecosystem restoration	$188.2	$2.8	$126.3	$10.0	—	$139.1	$2.2	$1.2	$3.1	$6.5	$42.6
Environmental water account	$48.0	$1.0	$28.2	$6.3	—	$35.5	$12.5	—	—	$12.5	—
Water use efficiency	$333.7	$11.8	—	$43.3	$57.9	$113.0	$19.8	—	$18.2	$38.0	$182.7
Water conservation	($37.1)	($11.8)	—	($18.3)	—	($30.1)	($2.3)	—	—	($2.3)	($4.7)
Water recycling	($296.6)	—	—	($25.3)	($57.9)	($82.9)	($82.9)	—	($18.2)	($35.7)	($178.0)
Water transfers	$1.1	$0.9	—	—	—	$0.9	$0.2	—	—	$0.2	—
Watershed	$17.3	$7.3	—	$10.0	—	$17.3	—	—	—	—	—
Drinking water quality	$16.2	$4.1	—	$12.1	—	$16.2	—	—	—	—	—
Levees	$17.2	$4.9	$8.4	—	—	$13.3	—	$0.3	—	$0.3	$3.6
Storage	$123.2	$14.1	—	$103.0	—	$117.1	$6.2	—	—	$6.2	—
Oversight and coordination	($1.5)	($1.5)	—	—	—	($1.5)	—	—	—	—	—
Surface	($15.0)	($8.8)	—	—	—	($8.8)	($6.2)	—	—	($6.2)	—
Groundwater	($106.8)	($3.8)	—	($103.0)	—	($106.8)	—	—	—	—	—
Conveyance	$72.2	$3.3	—	$37.6		$40.9	$4.0	—	—	$4.0	$27.3
Science	$20.1	$5.1	—	—	$2.3	$7.4	$3.9	$0.2	$2.0	$6.1	$6.6
Calfed science	($6.6)	($4.9)	—	—	—	($4.9)	—	—	($1.7)	($1.7)	—
Interagency ecological program	($13.5)	($0.2)	—	—	($2.3)	($2.5)	($3.9)	($0.2)	($0.3)	($4.4)	($6.6)
Oversight and coordination	$14.9	$7.3	—	—	—	$7.3	$7.5	$0.1	—	$7.6	—
Total	$852.2	$62.6	$162.9	$222.3	$60.2	$508.0	$56.3	$1.8	$23.3	$81.4	$262.8

Note: Local funds include State Water Project Funds and Central Valley Project Improvement Act (CVPIA) that are collected from state water contractors and Central Valley Project water users.

Source: From www.swf.usace.army.mil.

Table 11H.145 Interbasin Water Transfers in the Western United States

West Resources Regions Origin	Exported to	1978	1979	1980	1981	1982
Missouri	Upper Mississippi	267	511	754	1,005	1,257
	Arkansas-White-Red	15,980	15,980	17,269	6,327	11,226
	Upper Colorado	NA	NA	NA	NA	NA
	Total	16,247	16,491	18,023	7,332	12,483
Arkansas-White-Red	Missouri	12,669	12,539	9,429	12,519	2,329
Texas-Gulf	Lower Mississippi	—	—	—	32,570	32,430
	Arkansas-White-Red				460	8,940
	Total				33,030	41,370
Rio Grande	Arkansas-White-Red	517	1,460	900	181	1,290
Upper Colorado	Missouri	511,409	354,254	272,346	437,214	437,222
	Arkansas-White-Red	126,114	156,661	128,461	102,357	173,600
	Rio Grande	108,351	167,852	147,439	57,711	132,388
	Great Basin	98,440	116,820	83,330	111,080	77,340
	Total	844,314	795,587	631,576	708,362	820,550
Lower Colorado	Great Basin	3,060	3,420	3,780	4,170	3,420
	California	4,107,730	4,060,690	4,174,130	4,234,260	3,808,170
	Mexico	274,720	253,326	284,510	241,830	255,800
	Total	4,385,510	4,317,436	4,462,420	4,480,260	4,067,390
Great Basin	California	3,130	3,940	3,950	3,740	5,010
Pacific Northwest	Great Basin	900	1,000	1,000	1,300	1,400
California	Pacific Northwest	25,540	21,000	20,900	21,400	23,300
Grand total		5,288,827	5,169,453	5,148,198	5,268,124	4,975,122

Note: Between Water Resources Regions, water years 1978–1982; in acre-ft.

Source: Compiled from Petsch, H. E., Jr., 1985, Inventory of interbasin transfers of water in the conterminous United States, *U.S. Geological Survey Open-File Report 85–166*.

Table 11H.146 Cooperative Water Conservation/Transfer Projects

Cooperative Water Conservation/Transfer Projects	Annual Yield (af)	Estimated Start Date
MWD/IID 1988 Water Conservation Program	100,000–110,000[a]	Completed
SDCWA/IID Transfer and SDCWA/MWD exchange	130,000–200,000[b]	2002
MWD/CVWD SWP Water Transfer/Colorado River water exchange	35,000	2003
Coachella Canal lining MWD/SLR[c]	26,000	2005[d]
All American Canal lining MWD/SLR[c]	67,700	2006[d]
IID/CVWD/MWD Conservation program	100,000[e]	2007

[a] Yield to MWD, except for 20,000 acre-ft yr to be made available to CVWD.
[b] Yield to SDCWA.
[c] Yield to MWD and San Luis Rey Indian Water Rights Settlement Parties.
[d] Date by which full conservation benefits will be achieved.
[e] Yield to CVWD, MWD has an option to acquire water CVWD does not need. MWD assumes responsibility for 50,000 acre-ft/yr to CVWD after year 45 of Quantification Agreement.

Source: From California's Colorado River Water Use Plan. www.crb.ca.gov.

Table 11H.147 Major Canals on United States Bureau of Reclamation Projects

Project, State, and Canal Name	Length (km)	Initial Capacity (m^3 s)	Initial Reach Width (m)	Initial Reach Depth (m)	Date Constructed
Boise, ID, Main South					
Side (New York) Canal	64	79.3	21.3	2.4	1906–1908
Columbia Basin, WA					
East Low Canal	132.6	127.4	20.7	4.7	1946–1954
Feeder Canal	2.9	736.3	24.4	7.6	1946–1951
Main Canal	20.8	373.8	15.2	6.3	1946–1951
Potholes East Canal	100.4	110.4	14.6	4.8	1949–1953
West Canal	132.3	144.4	11.6	5.0	1946–1955
Minidoka-Palisades, ID-WY					
Milner-Gooding Canal	112.7	76.5	15.2	3.8	1928–1932
Yakima, WA					
Roza Div. Roza Canal	145.5	62.3	16.8	3.7	—
Central Valley, CA					
Delta Cross Channel	1.9	99.1	64.0	7.9	1950–1951
Folsom South Canal	43.0	99.1	10.4	5.4	1970–1973
Friant-Kern Canal	243.3	141.6	11.0	5.2	1945–1951
O'Neill Forebay Inlet Channel	.8	118.9	24.4	—	1965–1966
San Luis Canal	163.5	371.0	33.5	10.0	1963–1968
Tehama-Colusa Canal	88.8	71.6	7.3	4.8	1965–1980
Klamath, CA-OR					
Lost river diversion Channel	12.6	85.0	18.3	—	1911–1912
Boulder Canyon, AZ-CA-NV					
All American Canal	128.7	429.1	48.8	6.3	1934–1940
Coachella Canal	199.6	70.8	18.3	3.1	1938–1948
Central Arizona, AZ					
Granite Ref Aqueduct	279.4	85.0	7.3	5.0	1974–1985
Salt Gila Aqueduct	92.4	77.9	7.3	4.8	1981
Tucson Aqueduct	60.7	63.7	6.7	4.3	1984
Gila, AZ					
Gila Gravity Main Canal	32.2	62.3	6.7	4.1	1936–1939
Salt River, AZ					
Arizona Canal	62.0	53.8	15.2	2.1	1911–1912
Yuma, AZ-CA					
Yuma Main Canal	7.6	56.6	15.2	2.7	1907–1909
Harlingen Irr. Dist., TX					
Project Canals	47.0	99.1	6.1	1.2	1960
North Platte, NE-WY					
Interstate Canal	152.2	59.5	13.4	3.0	1905–1915
Garrison Division, ND					
McClusky Canal	118.4	55.2	7.6	5.3	1970–1978
Wind division, WY: Riverton unit					
Wyoming Canal	100.4	62.3	19.8	3.0	1920–1951

Note: Canals with capacity of 50 m^3/s or greater.

Source: Adapted from U.S. Department of the Interior Bureau of Reclamation, 1986, Statistical Compilation of Engineering Features on Bureau of Reclamation Projects.

Table 11H.148 Physical Features and Population

		Superior	Michigan	Huron	Erie	Ontario	Totals
Elevation[a]	(ft)[f]	600	577	577	569	243	
	(m)	183	176	176	173	74	
Length	(mil)[e]	350	307	206	241	193	
	(km)	563	494	332	388	311	
Breadth	(mil)[e]	160	118	183	57	53	
	(km)	257	190	245	92	85	
Average depth[a]	(ft)[f]	483	279	195	62	283	
	(m)	147	85	59	19	86	
Maximum depth[a]	(ft)[e]	1,332	925	750	210	802	
	(m)	406	282	229	64	244	
Volume[a]	(mi^3)[e]	2,900	1,180	850	116	393	5,439
	(km^3)	12,100	4,920	3,540	484	1,640	22,684
Water area	(sq. mi)[e]	31,700	22,300	23,000	9,910	7,340	94,250
	(km^2)	82,100	57,800	59,600	25,700	18,960	244,160
Land drainage area[b]	(sq. mi)[e]	49,300	45,600	51,700	30,140	24,720	201,460
	(km^2)	127,700	118,000	134,100	78,000	64,030	521,830
Total area	(sq. mi)[e]	81,000	67,900	74,700	40,050	32,060	295,710
	(km^2)	209,800	175,800	193,700	103,700	82,990	765,990
Shoreline length[c]	(mil)[e]	2,726	1,638	3,827	871	712	10,210[d]
	(km)	4,385	2,633	6,157	1,402	1,146	17,017[d]
Retention time	(ys)[f]	191	99	22	2.6	6	
Population:	U.S. (1990)[g]	425,548	10,057,026	1,502,687	10,017,530	2,704,284	24,707,075
	Canada (1991)	181,573		1,191,467	1,664,639	5,446,611	8,484,290
	Totals	607,121	10,057,026	2,694,154	11,682,169	8,150,895	33,191,365
	Outlet	St. Marys River	Straits of Mackinac	St. Clair River	Niagara River/ Welland Canal	St. Lawrence River	

[a] Measured at low water datum.
[b] Land Drainage Area for Lake Huron includes St. Marys River. Lake Erie includes the St. Clair-Detroit system. Lake Ontario includes the Niagara River.
[c] Including islands.
[d] These totals are greater than the sum of the shoreline length for the lakes because they include the connecting channels (excluding the St. Lawrence River).
[e] Coordinating Committee on Great Lakes Basic Hydraulic and Hydrologic Data, *Coordinated Great Lakes Physical Data* May, 1992.
[f] *Extension Bulletins E-1866–70*, Michigan Sea Grant College Program, Cooperative Extension Service, Michigan State University, E. Lansing, Michigan, 1985.
[g] 1990–1991 population census data were collected on different watershed boundaries and are not directly comparable to previous years.

Source: From www.epa.gov.

Table 11H.149 Water Withdrawals

		Superior	Michigan	Huron	Erie	Ontario	Totals
Municipal	Canada[a,b]	40		120	190	660	1,010
		36		107	170	589	902
	U.S.[a,b]	70	2,940	310	2,820	380	6,520
		62	2,262	277	2,515	339	5,455
	Total[a,b]	110	2,940	430	3,010	1,040	7,530
		98	2,622	384	2,685	927	6,716
Manufacturing	Canada[a,b]	860		1,360	1,900	2,760	6,880
		767		1,213	1,694	2,462	6,136
	U.S.[a,b]	410	9,650	1,060	9,110	530	20,760
		366	8,608	945	8,126	473	18,518
	Total[a,b]	1,270	9,650	2,420	11,010	3,290	27,640
		1,133	8,608	2,158	9,820	2,935	24,652
Power production	Canada[a,b]	70		2,870	1,160	8,370	12,470
		62		2,560	1,035	7,466	11,123
	U.S.[a,b]	760	13,600	2,570	13,180	6,520	36,360
		678	12,131	2,292	11,757	5,816	32,674
	Total[a,b]	830	13,600	5,440	14,340	14,890	49,100
		740	12,131	4,852	12,791	13,282	43,796
Grand totals	[a,b]	2,210	26,190	8,290	28,360	19,220	84,270
		1,971	23,361	7,394	25,296	17,144	75,166

[a] Cubic feet per second.
[b] Millions of cubic meters per year.

Source: From www.epa.gov. *Bulletins E-1866-70*, Sea Grant College Program, Cooperative Extension Service, Michigan State University, E. Lansing, Michigan, 1985.

Table 11H.150 Water Consumed

		Superior	Michigan	Huron	Erie	Ontario	Totals
Municipal	Canada[a,b]	10		20	30	100	160
		9		18	27	89	143
	U.S.[a,b]	10	190	170	280	70	720
		9	169	152	249	62	641
	Total[a,b]	20	190	190	310	170	880
		18	169	170	276	152	783
Manufacturing	Canada[a,b]	20		70	80	100	270
		18		62	71	89	240
	U.S.[a,b]	60	880	30	1,500	40	2,510
		53	785	27	1,338	36	2,239
	Total[a,b]	80	880	100	1,580	140	2,780
		71	785	89	1,409	125	2,479
Power production	Canada[a,b]	0		20	10	60	90
		0		18	9	54	81
	U.S.[a,b]	10	240	50	190	120	610
		9	214	45	169	108	545
	Total[a,b]	10	240	70	200	180	700
		9	214	62	178	174	673
Grand Totals	[a,b]	110	1,310	360	2,090	490	4,360
		98	1,168	321	1,863	450	3,935

[a] Cubic feet per second.
[b] Millions of cubic meters per year.

Source: From www.epa.gov. *Bulletins E-1866*–70, Sea Grant College Program, Cooperative Extension Service, Michigan State University, E. Lansing, Michigan, 1985.

Table 11H.151 Water Diversions in and Affecting Canada, 1985

No.	Jurisdiction	Project	Contributing Basin(s)	Receiving Basin	Average Annual Diversion (cfs)	Uses	Operational Date	Owner
1	British Columbia	Kemano	Nechako (Fraser)	Kemano	4,060	Hydro	1952	Alcan Ltd.
2	British Columbia		Bridge	Seton Lake	3,250	Hydro	(1934) 1959	British Columbia Hydro
3	British Columbia		Cheakamus	Squamish	1,300	Hydro	1957	British Columbia Hydro
4	British Columbia		Coquitam Lake	Buntzen Lake	1,000	Hydro	(1902) 1912	British Columbia Hydro
5	Saskatchewan		Tazin Lake	Charlot (Lake Athabasca)	1,000	Hydro	1958	Eldorado Nuclear
6	Manitoba	Churchill Diversion	Churchill (Southern Indian Lake)	Rat. Burntwood (Nelson)	27,350	Hydro	1976	Manitoba Hydro
7	Ontario		Lake St. Joseph (Albany)	Root (Winnipeg)	3,020	Hydro	1957	Ontario Hydro
8	Ontario		Ogoki (Albany)	Lake Nipigon (Superior)	3,990	Hydro	1943	Ontario Hydro
9	Ontario		Long Lake (Albany)	Lake Superior	1,480	Hydro/logging	1939	Ontario Hydro
10	Ontario		Little Abitibi (Moose)	Abitibi (Moose)	1,410	Hydro	1963	Ontario Hydro
11	Ontario	Welland	Lake Erie	Lake Ontario	8,830	Hydro navigation	(1829) 1951)	Government of Canada
12	Quebec	James Bay	Eastmain-Opinaca	La Grande	28,600	Hydro	1980	J.B. Energy Corp.
13	Quebec	James Bay	Fregate	La Grande	1,090	Hydro	1982	J.B. Energy Corp.
14	Quebec	James Bay	Caniapiscau	La Grande		Hydro	1983	J.B. Energy Corp.
15	Newfoundland	Churchill Falls	Julian-Unknown	Churchill	27,400	Hydro	1971	Newfoundland and Labrador Hydro
16	Newfoundland	Churchill Falls	Naskaupi	Churchill	7,060	Hydro	1971	Newfoundland and Labrador Hydro
17	Newfoundland	Churchill Falls	Kanairktok	Churchill	4,590	Hydro	1971	Newfoundland and Labrador Hydro
18	Newfoundland	Bay d'Espoir	Victoria, White Bear, Grey, and Salmon	Norhtwest Brook (Bay d'Espoir)	6,530	Hydro	1969	Newfoundland and Labrador hydro
19	Illinois	Chicago	Lake Michigan	Illinois (Mississippi)	3,200 sanitation	Municipal	(1848) 1900	Chicago Sanitary Dist.

Source: From Quinn, F., *Interbasin Water Diversions: A Canadian Perspective*. Journal of Soil & Water Conservation, November–December, 1987. Copyright Soil and Water Conservation Society. Reprinted with permission.

Table 11H.152 Proposed Interbasin Water Transfer Schemes from Canada into the United States

Proposal (Author)	Year Proposed	Water Source	Annual Diversion (km^3)	Estimated Construction Costs (billions of $)
Grand Canal Plan (Kierans)	1959–1983	James Bay diked, water diverted to Great Lakes and United States	347	100
Great Lakes-Pacific Waterways Plan (Decker)	1963	Skeena, Nechako & Fraser of B.C., Peace, Athabasca, Saskatchewan of Prairie Provinces	142	Not Available
North America Water & Power Alliance (NAWAPA) (Parsons)	1964	Primarily the Pacific & Yukon and B.C.; also tributaries of James Bay	310	100
Magnum Plan (Magnusson)	1965	Peace, Athabasca & N. Saskatchewan in Alberta	31 at border	Not Available
Kuiper Plan (Kuiper)	1967	Peace, Athabasca & N. Saskatchewan in Alberta, Nelson & Churchill in Manitoba	185	50
Central North American Water Project (CeNAWP) (Tinney)	1967	Mackenzie, Peace, Athabasca, N. Saskatchewan, Nelson & Churchill	185	30–50
Western States Water Augmentation Concept (Smith)	1968	Primarily Liard & Mackenzie drainages	49 at border	90
NAWAPA-MUSHEC or Mexican-United States Hydroelectric Commission (Parsons)	1968	NAWAPA sources + lower Mississippi & Sierra Madre Oriental Rivers of Southern Mexico	195 + 159 Nawapa Muschec	Not Available
North American Waters, A Master Plan (NAWAMP) (Tweed)	1968	Yukon & Mackenzie Rivers, drainage to Hudson Bay	1850	Not Available

Source: From Pearse, P.H., Bertrand, F., and MacLaren, J.W., 1985, *Currents of Change, Final Report on Federal Water Policy*, Ottawa, Canada. With permission.

Table 11H.153 Costs of Some Water Transfer Projects

Country	Project	Capacity km^3/yr	Route Length (km)	Lift Height (m)	Approximate Cost Billion Dollars		Year of Plan
					Total	Per 1 (km^3)	
U.S.A. Canada	NAWAPA	100–300	2,000	—	100	0.33–1.0	Project of 1964
U.S.A.	Beck Plan	12.2	1,000	840	3.5	0.29	Project of 1968
U.S.A.	Grand Canal	20.0	—	—	5.0	0.25	Project of 1959
U.S.A.	Texas Water Plan	20.5	1,800	1,300	9.0	0.44	Project of 1968
U.S.A.	Hudson Institute Plan	41.5	1,800	—	12.2	0.29	Project of 1968
U.S.A.	California Aqueduct	5.2	300	1,100	2.3	0.44	1973
U.S.S.R.	Irtysh-Central Asia	25	2,300	130	19	0.76	Project of 1980
U.S.S.R.	Sukhona-Volga	4.0	480	6	0.43	0.11	Project of 1980
U.S.S.R.	Onega R. (upper)-Volga	2.0	390	9	0.22	0.11	Project of 1980
U.S.S.R.	Pechora-Kama	9.8	300	30	2.3	0.24	Project of 1980
U.S.S.R.	Onega Bay	38	800	130	8.2	0.22	Project of 1981
India	Rajasthan Canal	16.5	180	—	0.56	0.034	Under construction
India	Sarda Sahayak	15.4	260	—	0.23	0.015	Under construction
Australia	Snowy Mountains	2.4	220	—	0.80	0.33	1974
Israel	Yarkon-Negev	0.23	110	220	0.051	0.22	In operation
Israel	Jordan Waterway	0.32	240	350	0.115	0.36	In operation

Source: From Shiklomanov, I.A., 1985, *Large Scale Water Transfers*. Chapter 12 of Facets of Hydrology II, edited by John C. Rodda. Copyright 1985 John Wiley and Sons. Reprinted with permission.

Table 11H.154 Authorizations for Water Supply, Water Quality and Related Infrastructure Projects and Studies

Act Section	Provision	T, Traditional	NT, Non-traditional	New Federal $(000)[b]	State(s)
WRDA 1986, PL 99-662, 17 November 1986					
401(a)	Santa Ana River Mainstem, CA. Study to investigate the feasibility of including water supply and conservation storage at Prado Dam	T		0	CA
707	Capital Investment Needs for Water Resources. Authorized the Assistant Secretary of the Army for Civil Works to estimate long-term capital investment needs for, among other things, municipal and industrial water supply	T		0	Nationwide
729	Study of Water Resources Needs of River Basins and Regions. Requires the Assistant Secretary of the Army for Civil Works, in coordination with the Secretary of the Interior and in consultation with other governmental agencies, to study "water resources needs of river basins and regions of the United States." This section specifically requires consultation with "State, interstate, and local governments"	T		5,000	Nationwide
818	Brazos River Basin, TX. Modifies Section 10 of the 1946 FCA to insert "or water supply" after "irrigation"	T		0	TX
834	Curwensville Lake. Authorization to construct a water line with pumps in order to provide water for municipal use		NT	225	PA
838	Denison Dam (Lake Texoma). Authorization to reallocate 300,000 acre-ft of hydropower storage to municipal and industrial water supply	T		0	OK & TX
843	Beaver Lake. Authorization to study and undertake a project to preserve and enhance water quality of the lake		NT	3,825	AR
931	Interim use of Water Supply for Irrigation. Authorizes the temporary use of unused municipal and industrial water supply storage for irrigation	T		0	Nationwide
1103	Upper Mississippi River Plan. This is a major environmental restoration authorization for the Corps. This authorization recognized the Upper Mississippi River as a nationally significant ecosystem and a nationally significant commercial navigation system. It authorized the Secretary of the Army to enter into agreements with basin states and to transfer funds to the DOI as necessary to carry out the provisions of the section. In consultation with the DOI and the basin states the Secretary of the Army was also authorized to determine the need for river rehabilitation and environmental enhancement and protection based on the conditions of the environmental, project developments, and projected environmental impacts from implementing any proposal resulting from recommendations made under provisions of the section	T		188,000 over 10-years	IL, IA, MN, MO and WI,

(Continued)

Table 11H.154 (Continued)

Act Section	Provision	T, Traditional	NT, Non-traditional	New Federal $(000)[b]	State(s)
1121	Ogallala Aquifer. To establish a comprehensive research and development program to assist those portions of the High Plains region dependent on water from the Ogallala Aquifer		NT	65,000	CO, KS, NE, NM, OK, SD, TX, & WY
1135	Project Modifications for Improvement of Environment. Authorizes the review of water resources projects to determine the need for modifications in the structures and operations of such projects for the purpose improving the quality of the environment in the public interest	T		0	Nationwide
1141	Groundwater Recharge. Authorization to plan, engineer and design a project for recharge of groundwater in the drainage basis of the Tucson and Scottsdale, AZ metropolitan areas		NT	250	AZ
1157	Miami River Water Quality Commision. Authority to make a grant to establish a commission to develop a plan for improving the water quality of the Miami River and tributaries		NT	50	FL
Total WRDA '86	13 Sections (4 of which apply nationwide); 8 sections traditional and 5 non-traditional			262,350	18
WRDA 1988, PL 101-676,17 November 1988					
23	Louisiana Water Supply. Authorized to review the water supply problems related to drought at a water supply reservoir and to respond as appropriate		NT	0	LA
Total WRDA '88	1 Section; non-traditional			0	1
WRDA 1990 PL 101-640, 28 November 1990					
116(d)	Southern California Infrastructure Restoration. Conduct a feasibility study, in consultation with FEMA, on the problems and alternative solutions of the infrastructure of the region		NT	1,500	CA
116(g)	Santa Rosa, CA. Authorize a study to evaluate storage facilities associated with wastewater reclamation and irrigation		NT	0	CA
116(p)	Water Supply, Minnesota and North Dakota. Conduct a study to determine alternate plans to augment flows in the Red River of the North including plans to supplement flows for municipal, industrial, agricultural, and fish and wildlife purposes		NT	0	MN & ND
116(w)	Buffalo, New York. To evaluate a city plan on flooding and associated water quality problems (including those associated with combined sewer over flow, sewer backups and riverside outfalls)		NT	0	NY
116(x)	Caesar's Creek Lake, Ohio. To conduct a study of the water supply needs of Clinton County, Ohio		NT	0	OH

116(y)	Liberty, Ohio. To conduct a study of the water supply needs of the city	NT	0	OH
116(z)	Washingtonville, Ohio. To conduct a study of the water supply needs of the city	NT	0	OH
116(dd)	Radium Removal. To study and provide technical assistance to small communities on methods of mitigating radium contamination in groundwater used as a source of public drinking water	NT	0	Nationwide
Total WRDA '90	8 Sections (including one nationwide); all non-traditional		1,500	5
WRDA 1992, PL 102-580, 31 October 1992				
114	Brockton MA. Study of water supply needs and of water quality and quantity to meet future needs	NT	0	MA
217	Reuse of Wastewater. Authorizes assistance to non-Federal interests for planning and design of reuse systems	NT	5,000	CA
218	Demonstration of Waste Water Technology, Santa Clara Valley Water District and San Jose, CA. Authorized, in cooperation with EPA, to provide design and construction assistance to the Water District for demonstrating and field testing public use innovative processes which advance the technology of waste water reuse and treatment and which promote the use of treated waste water for critical water supply purposes	NT	10,000	CA
219	Environmental Infrastructure. Authorizes technical and planning and design assistance	NT	5,000 *	Nationwide
220	Environmental Infrastructure Assistance. Authorizes assistance in design and construction of water transmission line	NT	5,000	AR
221	Environmental Infrastructure Assistance. Authorizes design and construction assistance for Combined Sewer System and storm water projects	NT	7,000	NY
222	Environmental Infrastructure Assistance. Authorizes design and construction assistance for storm water project	NT	200	PA
304	Broad Top Region of Pennsylvania. Authorizes a Watershed Reclamation and Wetlands Pilot Project along the Juniata River and its tributaries, PA	NT	5,500 *	PA
307	Water Quality Projects. Authorizes design and construction of storm water projects	NT	70,000 *	LA, ME, NY, & RI
313	South Central Pennsylvania Environmental Restoration Infrastructure and Resource Protection Development. Establishes a pilot program for design and construction of wastewater treatment facilities, water supply storage, treatment and distribution facilities and surface water development and protection	NT	17,000	PA

(Continued)

Table 11H.154 (Continued)

Act Section	Provision	T, Traditional	NT, Non-traditional	New Federal $(000)[b]	State(s)
322	Water Supply needs of Mahoning Valley Sanitary District, Ohio. Cooperate with the state in review of water supply needs		NT	0	OH
324	Hackensack Meadowlands Area. Authorizes design and construction assistance for an environmental improvement program		NT	5,000	NJ
340	Southern West Virginia Environmental Restoration and Infrastructure Resources Protection Development Project. Modified by Section 359 of WRDA '96		NT	5,000 *	WV
362	Quonset Point-Davisville, Rhode Island. Authority to construct two water supply towers and relocate sewer lines		NT	1,875	RI
Total WRDA '92	14 sections (including one nationwide); all non-traditional			136,075	11
WRDA 1996. Public Law 104-303, 12 October 1996					
359	Southern West Virginia. Modifies cost sharing and increases funding levels of Section 340 of WRDA '92		NT	15,000 *	VW
401	Rural Sanitation Projects. Authorize a study to report on the advisability and capability of the Corps to implement rural sanitation projects for rural and native villages in Alaska		NT	0	AK
503	Watershed Management, Restoration and Development. Authority to provide technical, planning and design assistance to non-Federal interests for carrying out watershed management, restoration and development projects at specific locations		NT	15,000	AZ, CA, GA, NE, WV
504	Environmental Infrastructure. Amend Section 219 of WRDA '92 by adding specific authorizations of appropriations for construction assistance for six specific environmental infrastructure projects		NT	73,000 *	DC & MD, GA, KY, MI, MS, NH
522	Jackson County, AL. Authority to provide technical, planning and design assistance for wastewater treatment and related facilities, remediation of point and non-point sources of pollution and contaminated riverbed sediments		NT	3,000	AL
531	Southern and Eastern Kentucky. Authority to establish a program to provide environmental assistance to non-Federal interests to design and construct water related environmental infrastructure including wastewater treatment, water supply and surface water protection and development		NT	10,000	KY
552	New York City Watershed. Authority to provide design and construction assistance for environmental infrastructure and resource protection and development projects in the watershed to protect and enhance the quality and quantity of the New York City water supply		NT	22,500	NY

566	Southeastern Pennsylvania. Authority to establish a program to provide environmental assistance to non-Federal interests to design and construct water related environmental infrastructure including wastewater treatment, water supply, and surface water protection and development		NT	25,000	PA
585	Overflow Management Facility. Authority to provide assistance to Narragansett Bay Commission for the construction of a combined river overflow management facility		NT	30,000	RI
Total WRDA '96	9 Sections; all non-traditional			188,500	15
WRDA 1999, Public Law 106-53, 17 August 1999					
101(b)(4)	Success Dam, Tule River Basin, CA. Authorizes a project for flood damage reduction and water supply	T		11,635	CA
101(b)(15)	Howard Hanson Dam, WA. Authorizes a project for water supply and ecosystem restoration		NT	36,900	WA
211	Watershed Management, Restoration and Development. Expands Section 503 of WRDA '96 to extend authorization assistance to sites in six additional states		NT	0	FL, IL, NV, NC, OR
212	Flood Mitigation and Riverine Restoration Program. Authorizes a program for the purpose of conducting projects to reduce flood hazards and restore the natural functions and values of rivers throughout the United States	T		200,000	AZ, CA, KY, MN, ND, NH, NJ, NM, NY, NC, OH, OR, PA, RI, SD, VA, WI
331	Jackson County Mississippi. Modifies Section 219(c)(5) of WRDA '92 to provide cost sharing credit		NT	0	MS
340	New York City Watershed. Modifies Section 552 of WRDA '96 to change local cooperation wording. Increases $22.5 million to $42.5 million		NT	20,000 *	NY
343	Broken Bow Lake. Project modified to require a seasonal adjustment to the top of the conservation pool in the interest of water and related resources	T		0	OK
351	South Central Pennsylvania. Modifies Section 313(g)(3) of WRDA '92 to increase appropriations and cost sharing (increases $80 million to $180 million)		NT	100,000 *	PA
374	White River Basin, AR and MO. Authorizes project operation modifications and storage reallocations in the interest of trout fisheries		NT	0	AR & MO
403	Green Ferry Lake, AR. Conduct a study to determine the feasibility of constructing water intake facilities		NT	0	AR

(Continued)

Table 11H.154 (Continued)

Act Section	Provision	T, Traditional	NT, Non-traditional	New Federal \$(000)[b]	State(s)
502	Environmental Infrastructure. Modifies Section 219(e) of WRDA '92 to increase appropriations and to authorize additional assistance at 43 projects in 20 different states. The assistance to include, but not necessarily limited: to eliminate or control combined sewer overflows, or for water related infrastructure, or groundwater recharge, or wastewater infrastructure projects, or wastewater treatment, or to provide water supply facilities, or for a project for recycled water or for industrial water reuse project		NT	494,000 *	CA, CT, GA, IN, LA, MA, MI, MO, MS, NC, NH, NJ, NY, OR, OK, PA, SC, TN, UT & VA
513	Design and Construction Assistance. Modifies Section 507 of WRDA '96 to include "expansion and improvement of Long Pine Run Dam, PA and associated water infrastructure."		NT	5,000	PA
521	Beaver Lake, AR. Reallocate storage in Beaver Lake at no cost to the local water districts		NT	0	AR
531	Kanopolis Lake, KS. Offers the State of Kansas the right to purchase storage at certain prescribed costs.		NT	0	KS
532	Southeastern and Eastern Kentucky. Modifies Section 531 of WRDA '96 to increase (from \$10 to \$25 million) funding and to expand to include "small stream flooding, local storm water drainage, and related problems		NT	15,000 *	KY
545	Sardis Reservoir, OK. Offers the State of Oklahoma the right to purchase storage a certain prescribed costs.		NT	0	OK
548	Bradford and Sullivan Counties, PA. Authority to provide assistance for water-related infrastructure and resource protection and development projects		NT	0	PA
552	Southeastern Pennsylvania. Modifies Section 566(b) of WRDA '96 to include "environmental restoration" as well as the originally authorized water supply and related facilities.		NT	0	PA
560	Abandoned and Inactive Noncoal Mine Restoration. Authority to provide technical, planning and design assistance to Federal and non-Federal interests for carrying out projects to address water quality problems caused by drainage and related activities from abandoned and inactive noncoal mines		NT	5,000	Nationwide
569	Northeastern Minnesota. Authorizes a pilot program to provide environmental infrastructure assistance to non-Federal interests to include design and construction for wastewater treatment and related facilities, water supply and related facilities, environmental restoration, and surface water resource protection and development		NT	40,000	MN

570	Alaska. Authorizes a pilot program to provide environmental assistance to non-Federal interests to include design and construction for wastewater treatment and related facilities, water supply and related facilities, and surface water resource protection and development	NT	25,000	AK
571	Central West Virginia. Authorizes a pilot program to provide environmental assistance to non-Federal interests to include design and construction for wastewater treatment and related facilities, water supply and related facilities, and surface water resource protection and development	NT	10,000 *	WV
573	Onondaga Lake, NY. Authorization to plan, design and construct projects that are consistent with the Onondaga Lake Management Plan and comply with the amended consent judgment and the project labor agreement for the environmental restoration, conservation and management of Onondaga Lake, NY	NT	1,000 *	NY
592	Mississippi. Authorizes a pilot program to provide environmental assistance to non-Federal interests to include design and construction assistance for projects for wastewater treatment and related facilities, elimination or control of combined sewer overflows, water supply and related facilities environmental restoration, and surface water resource protection and development	NT	25,000	MS
593	Central New Mexico. Authorizes a pilot program to provide environmental assistance to non-Federal interests to include design and construction assistance for projects for wastewater treatment and related facilities, water supply, conservation, and related facilities, storm water resource protection and development	NT	25,000 *	NM
594	Ohio. Authorizes a program to provide environmental assistance in the form of design and construction to non-Federal interests to include wastewater treatment and related facilities; combined sewer overflow, water supply, storage, treatment, and related facilities; mine drainage; environmental restoration; and surface water resources protection and development	NT	60,000	OH
595	Rural Nevada and Montana. Authorizes a program to provide environmental assistance to non-Federal interests in the form of design and construction assistance for projects for wastewater treatment and related facilities, water supply and related facilities, environmental restoration ad surface water resource projection and development	NT	50,000	MT & NV
Total WRDA '99	27 Sections (including one nationwide); 3 traditional and 24 non-traditional		1,123,535	33

(Continued)

Table 11H.154 (Continued)

Act Section	Provision	T, Traditional	NT, Non-traditional	New Federal $(000)[b]	State(s)
337	Buchanan, Dickenson, and Russell Counties, VA. Authority to reallocate storage to water supply in the John Flannagan Reservoir under the authority of Section 322 of WRDA '90		NT	0	VA
447	Fremont, Ohio. To conduct a study to determine the feasibility of carrying out projects for water supply and environmental restoration at the Ballville Dam		NT	0	OH
Total WRDA 00	2 Sections; both non-traditional			0	2

[a] While detailed records are not available, traditional provisions are more than likely part of the Administration's proposal and non-traditional provisions are more than likely Congressional adds.

[b] Dollar values shown are the authorized appropriations in the legislation. Sections with an asterisk (*), (available only for Fiscal Years 1992–2002, indicate funding was received. Actual funding, however, may have been greater than, less than, or as authorized.)

Source: From www.swf.usace.army.mil.

SECTION 11I GROUNDWATER

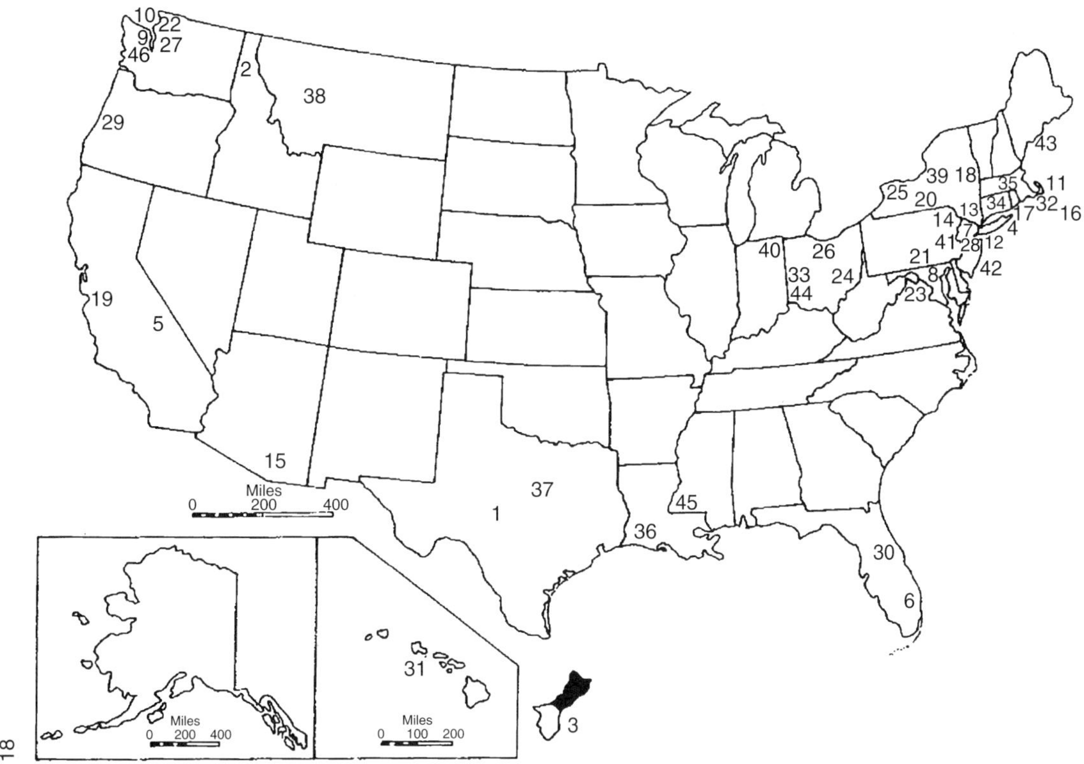

Figure 11I.107 Designated sole-source aquifers in the United States. [For explanation and listing of aquifers see Table 11I.17. (From U.S. Environmental Protection Agency, 1988. With permission.)]

Table 11I.155 Designated Sole-Source Aquifers in the United States

USEPA Region	Map Number	Aquifer and/or Location	State	Date Filed	Federal Register Notice	
					Citation	Publication Date
VI	1	Edwards Aquifers	TX	1/03/75	40 FR 58344	12/16/75
X	2	Spokane Valley Rathdrum Prairie Aquifer	WA-ID	fall of 76	43 FR 5566	02/09/78
IX	3	Northern Guem	Guam	11/20/75	43 FR 17868	04/26/78
II	4	Nassau/Suffolk Counties Long Island	NY	1/21/75	43 FR 26611	06/21/78
IX	5	Fresno County	CA	8/09/76	44 FR 52751	09/10/79
IV	6	Biscayne Aquifer	FL	5/08/78	44 FR 58797	10/11/79
II	7	Buried Valley Aquifer System	NJ	1/15/78	45 FR 30537	05/08/80
III	8	Maryland Piedmont Aquifer Montgomery, Frederick, Howard, Carroll Counties	MD	09/12/75	45 FR 57165	08/27/80
X	9	Camano Island Aquifer	WA	4/13/81	47 FR 14779	04/06/82
X	10	Whidbey Island Aquifer	WA	4/13/81	47 FR 14779	04/06/82
I	11	Cape Cod Aquifer	MA	3/04/81	47 FR 30282	07/13/82
II	12	Kings/Queens Counties	NY	6/18/79	49 Fr 2950	01/24/84
II	13	Ridgewood Area	NY/NJ	7/04/79	49 FR 2943	01/24/84
II	14	Upper Rockaway River Basin Area	NJ	11/30/79	49 FR 2946	01/24/84
IX	15	Upper Santa Cruz & Avra Altar Basin Aquifers	AZ	6/29/81	49 FR 2948	01/24/84
I	16	Nantucket Island Aquifer	MA	12/02/82	49 FR 2952	01/24/84
I	17	Block Island Aquifer	RI	2/18/83	49 FR 2952	01/24/84
II	18	Schenectady/Niskayuna Schenectady, Saratoga and Albany Counties	NY	8/20/82	50 FR 2022	01/14/85
IX	19	Santa Marqarita Aquifer Scotts Valley, Santa Cruz County	CA	9/07/77	50 FR 2023	01/14/85
II	20	Clinton Street-Ballpark Valley, Aquifer System, Broome and Tioga Counties	NY	2/26/81	50 FR 2025	01/14/85
III	21	Seven Valleys Aquifer York County	PA	9/24/81	50 FR 9126	03/06/85
X	22	Cross Valley Aquifer	WA	7/29/83	52 FR 18606	05/18/87
III	23	Prospect Hill Aquifer Clark County	VA	6/27/85	52 FR 21733	06/09/87
V	24	Pleasant City Aquifer	OH	8/27/84	52 FR 32342	08/27/87
II	25	Cattaraugus Creek-Sardinia	NY	2/28/85	52 FR 36100	09/25/87
V	26	Bass Island Aquifer Catawba Island	OH	3/17/86	52 FR 37009	10/02/87
X	27	Newberg Area Aquifer	WA	1/16/84	52 FR 42474	11/5/87
II	28	Highlands Aquifer system	NY/NJ	3/14/85	52 FR 37213	10/05/87
X	29	North Florence Dunal Aquifer	OR	6/02/85	52 FR 37519	10/07/87
IV	30	Volusia-Florida Aquifer	FL	6/18/82	52 FR 44221	11/18/87
IX	31	Southern Oahu Basal Aquifer	HI	5/03/83	52 FR 45496	11/30/87
I	32	Martha's Vineyard Regional Aquifer	MA	6/16/87	53 FR 3451	02/05/88
V	33	Buried Valley Aquifer System (BVAS)	OH	11/25/87	53 FR 15876	05/04/88
I	34	Pawcatuck Basin Aquifer System	RI/CT	11/30/87	53 FR 17108	05/13/88
I	35	Hunt-Annaquatucket-Pettaquamscutt Aquifer	RI	12/30/87	53 FR 19026	05/26/88
VI	36	Chicot Aqufier	LA	12/05/86	53 FR 20893	06/09/88
VI	37	Edwards Aquifer Austin Area	TX	08/29/86	53 FR 20897	06/07/88
VIII	38	Missoula Valley Aquifer	MT	11/23/87	53 FR 20895	06/07/88

(Continued)

Table 11I.155 (Continued)

USEPA Region	Map Number	Aquifer and/or Location	State	Date Filed	Federal Register Notice: Citation	Federal Register Notice: Publication Date
II	39	Cortland-Horner-Preble Aquifer System	NY	9/15/87	53 FR 22045	06/13/88
V	40	St. Joseph Aquifer System (Elkhart Co)	IN	12/11/87	53 FR 23682	06/23/88
II	41	N.J. Fifteen Basin Aquifer Systems	NJ/NY	11/18/85	53 FR 23685	06/23/88
II	42	N.J. Coastal Plain Aquifer System	NJ	12/04/78	53 FR 23791	06/24/88
I	43	Monhegan Island	ME	5/16/88	53 FR 24496	06/29/88
V	44	OKI-Miami Buried Valley Aquifer	OH	03/10/88	53 FR 25670	07/08/88
VI	45	Southern Hills Aquifer System	LA/MS	5/19/80	53 FR 25538	07/07/88
X	46	Cedar Valley Aquifer	WA	1/88		

Note: As of July 1988; the Sole Source Aquifer (SSA) program allows individuals and organizations to petition the Environmental Protection Agency (EPA) to designate aquifers as the "sole or principal" source of drinking water for an area. The program was established under Section 1424(e) of the Safe Drinking Water Act (SDWA) of 1974. The primary purpose of the designation is to provide EPA review of Federal financially assisted projects planned for the area to determine their potential for contaminating the aquifer. Based on this review, no commitment of Federal financial assistance may be made for projects "which the Administrator (of EPA) determines may contaminate such (an) aquifer," although Federal funds may be used to modify projects to ensure that they will not contaminate the aquifer; for location of aquifers see Figure 8.17.

Source: From U.S. Environmental Protection Agency, 1988.

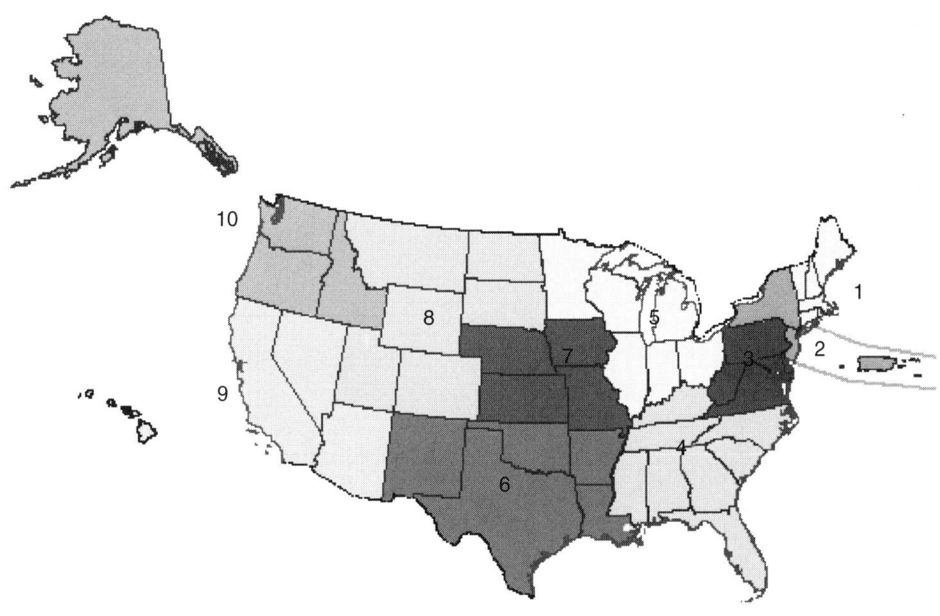

Figure 11I.108 EPA regions. (From www.epa.gov. With permission.)

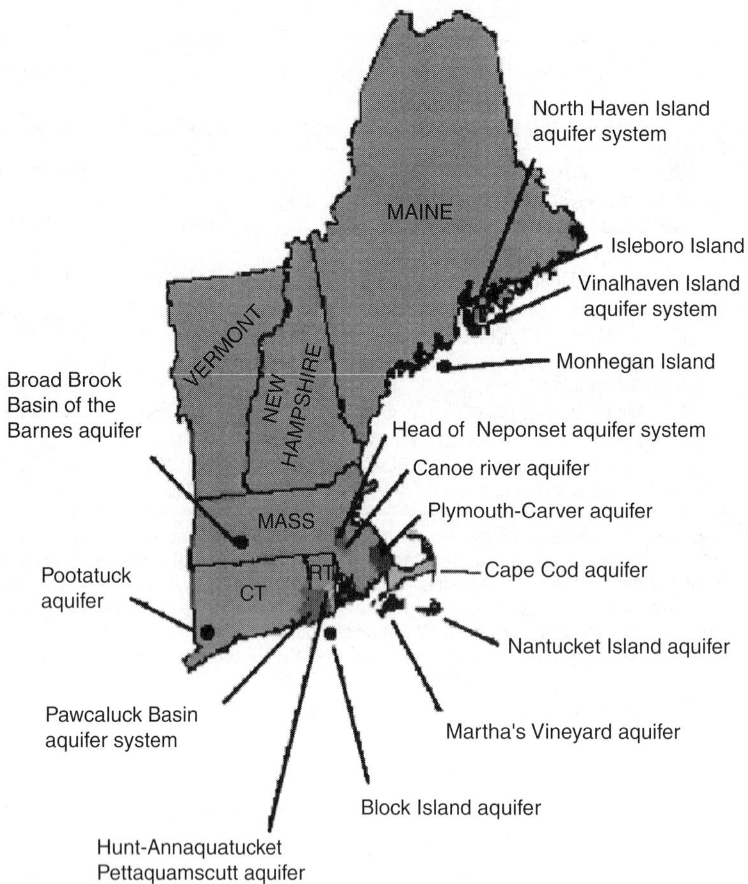

Figure 11I.109 Designated sole source aquifers in EPA region I, Connecticut, Maine, Massachusetts, New Hampshire, Rhode Island, Vermont. (From www.epa.gov. With permission.)

Table 11I.156 Sole Source Aquifer Designations in Region I

State	Sole Source Aquifer Name	Federal Reg. Citation	Public Date
CT	Pootatuck Aquifer	55 FR 11056	03/26/90
MA	Cape Cod Aquifer	47 FR 30282	07/13/82
MA	Nantucket Island Aquifer	49 FR 2952	01/24/84
MA	Martha's Vineyard Aquifer	53 FR 3451	02/05/88
MA	Head of Neponset Aquifer System	53 FR 49920	12/12/88
MA	Plymouth-Carver Aquifer	55 FR 32137	08/07/90
MA	Canoe River Aquifer	58 FR 28402	05/13/93
MA	Broad Brook Basin of the Barnes	60 FR 20989	04/28/95
ME	Monhegan Island	53 FR 24496	06/29/88
ME	Vinalhaven Island Aquifer System	54 FR 29779	07/14/89
ME	North Haven Island Aquifer System	54 FR 29934	07/17/89
ME	Isleboro Island Aquifer System	64 FR 186	09/27/99
RI	Block Island Aquifer	49 FR 2952	01/24/84
RI/CT	Pawcatuck Basin Aquifer System	53 FR 17108	05/13/88
RI	Hunt-Annaquatucket Pettaquamscutt	53 FR 19026	05/26/88

Source: From www.epa.gov.

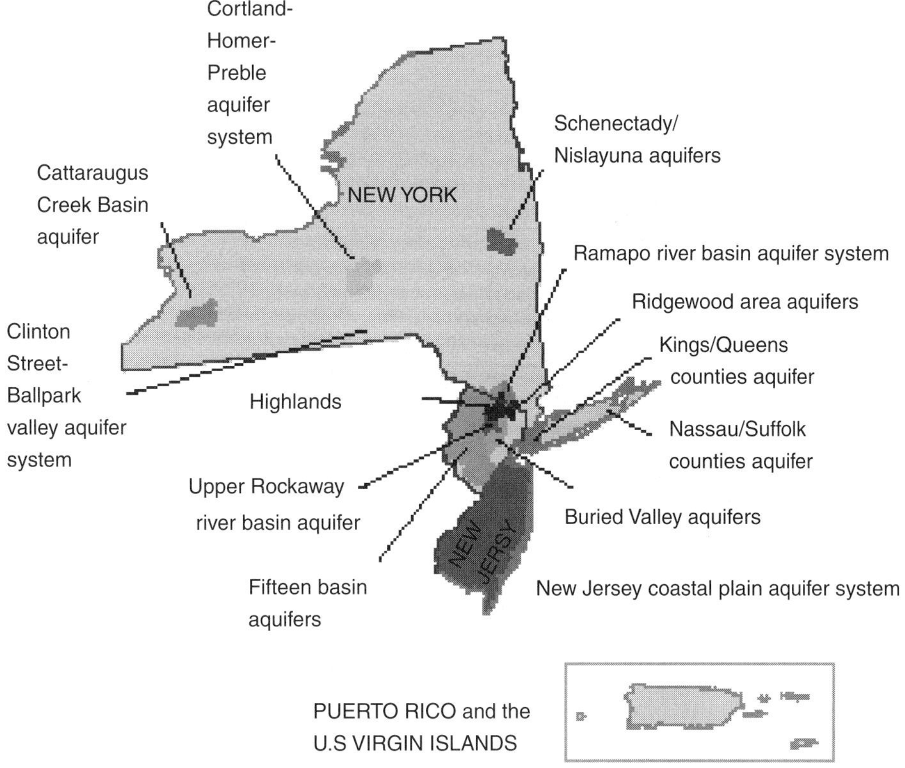

Figure 11I.110 Designated sole-source aquifers in EPA Region II, New York, New Jersey, Puerto Rico, U.S. Virgin Islands. (From www.epa.gov. With permission.)

Table 11I.157 Designated Sole Source Aquifers in Region II

State	Sole Source Aquifer Name	Federal Register Cit.	Public. Date
NJ	Buried Valley Aquifers, Central Basin, Essex and Morris Counties	45 FR 30537	05/08/80
NJ	Upper Rockaway River Basin	49 FR 2946	01/24/84
NJ	Ridgewood Area Aquifers	49 FR 2943	01/24/84
NJ/NY	Highlands Aquifer System Passaic, Morris & Essex Co's NJ; Orange Co. NY	52 FR 37213	10/05/87
[a]NJ/DE/PA	NJ Coastal Plain Aquifer System	53 FR 23791	06/24/88
NJ/NY	NJ Fifteen Basin Aquifers	53 FR 23685	06/23/88
NJ/NY	Ramapo River Basin Aquifer Systems	57 FR 39201	O8/28/92
NY	Nassau/Suffolk Co., Long Island	43 FR 26611	06/21/78
NY	Kings/Queens Counties	49 FR 2950	01/24/84
NY	Schenectady/Niskayuna	50 FR 2022	01/14/85
NY	Clinton Streen-Ballpark Valley Aquifer System, Broome and Tioga Co's.	50 FR 2025	01/14/85
NY	Cattaraugus Creek Basin Aquifer, WY & Allegany Cos.	52 FR 36100	09/25/87
NY	Cortland-Homer-Preble Aquifer System	53 FR 22045	06/13/88

[a] The New Jersey Coastal Plains Aquifer is jointly managed with Region 3. While listed in both regions it is counted only once in the national total of 70.

Source: From www.epa.gov.

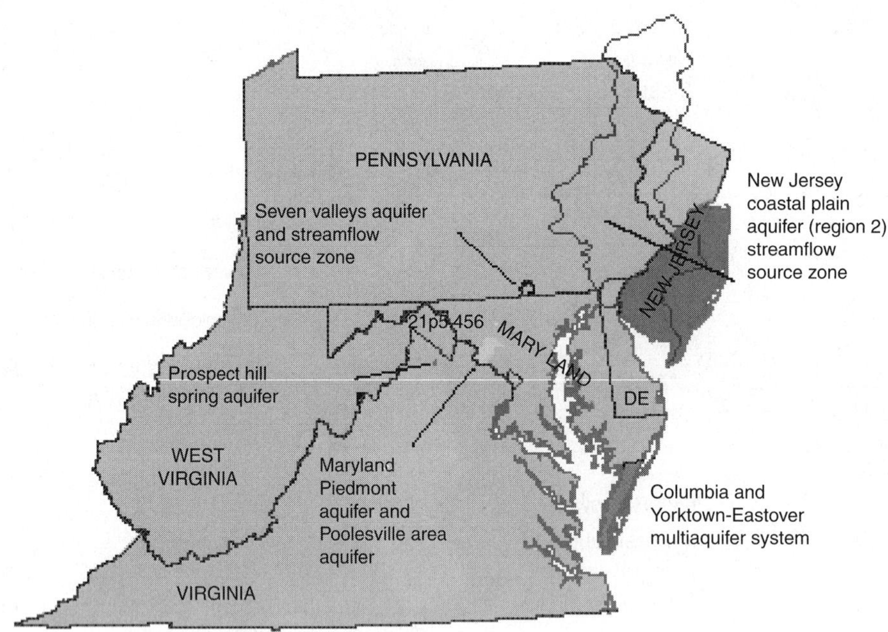

Figure 11I.111 Designated sole source aquifers in EPA Region III, District of Columbia, Delaware, Maryland, Pennsylvania, Virginia, West Virginia. (From www.epa.gov. With permission.)

Table 11I.158 Sole Source Aquifers in Region III

State	Sole Source Aquifer Name	Federal Register Citation	Publication Date
[a]DE/PA/NJ	New Jersey Coastal Plain Aquifer	53 FR 23791	06/24/88
MD	Maryland Piedmont Aquifer Montgomery, Howard, Carroll Counties	45 FR 57165	08/27/80
MD	Poolesville Area Aquifer Extension of the Maryland Piedmont Aquifer	98 FR 3042	02/06/98
PA	Seven Valleys Aquifer, York County	50 FR 9126	03/06/85
VA	Prospect Hill Aquifer, Clark County	2 FR 21733	06/09/87
VA	Columbia and Yorktown, Eastover Multi-aquifer System Accomack and North Hampton Counties	62 FR 17187	04/09/97

[a] The New Jersey Coastal Plains Aquifer is jointly managed with Region II.

Source: From www.epa.gov.

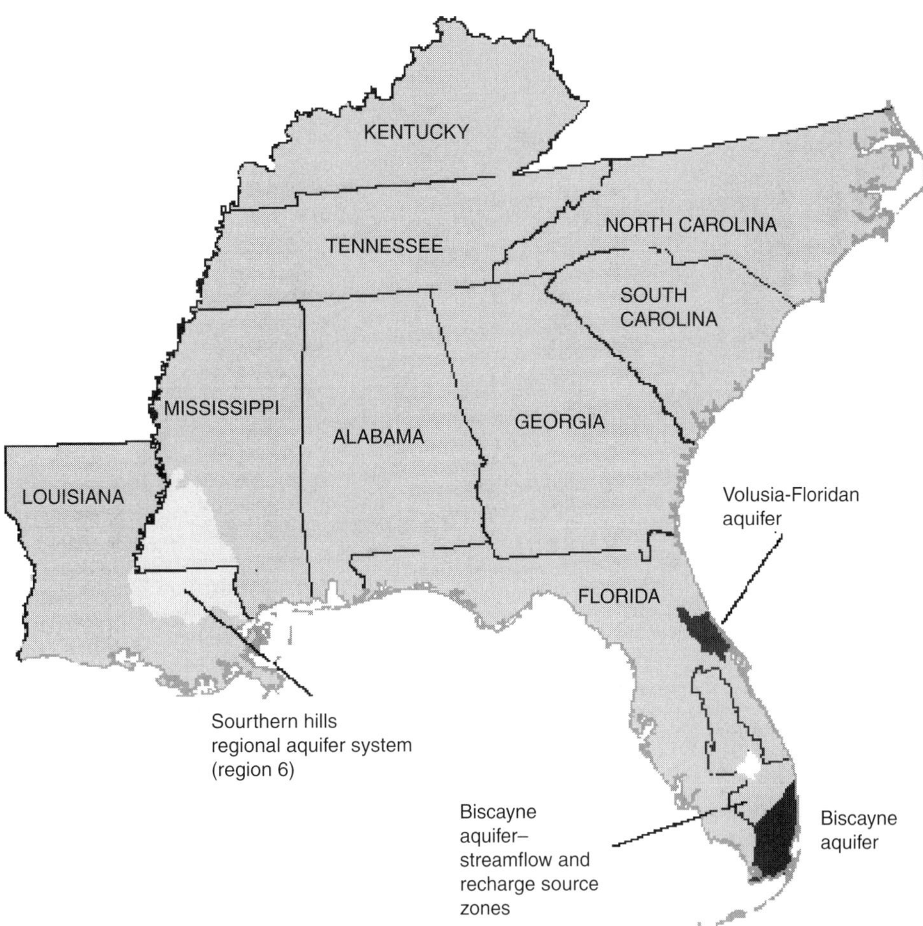

Figure 11I.112 Designated sole source aquifers in EPA Region IV, Alabama, Florida, Georgia, Kentucky, Mississippi, North Carolina, South Carolina, Tennessee. (From www.epa.gov. With permission.)

Table 11I.159 Designated Sole Source Aquifers in Region IV

State	Sole Source Aquifer Name	Federal Register Cit.	Public. Date
FL	Biscayne Aquifer, Broward, Dade, Monroe & Palm Beach Counties	44 FR 58797	10/11/79
FL	Volusia-Floridan Aquifer, Flagler & Putnam Counties	52 FR 44221	11/18/87
[a]LA/MS	Southern Hills Regional Aquifer System	53 FR 25538	07/07/88

[a] The Southern Hills Regional Aquifer system is jointly managed with Region VI. While listed in both regions, it is counted only once in the national total of 70.

Source: From www.epa.gov.

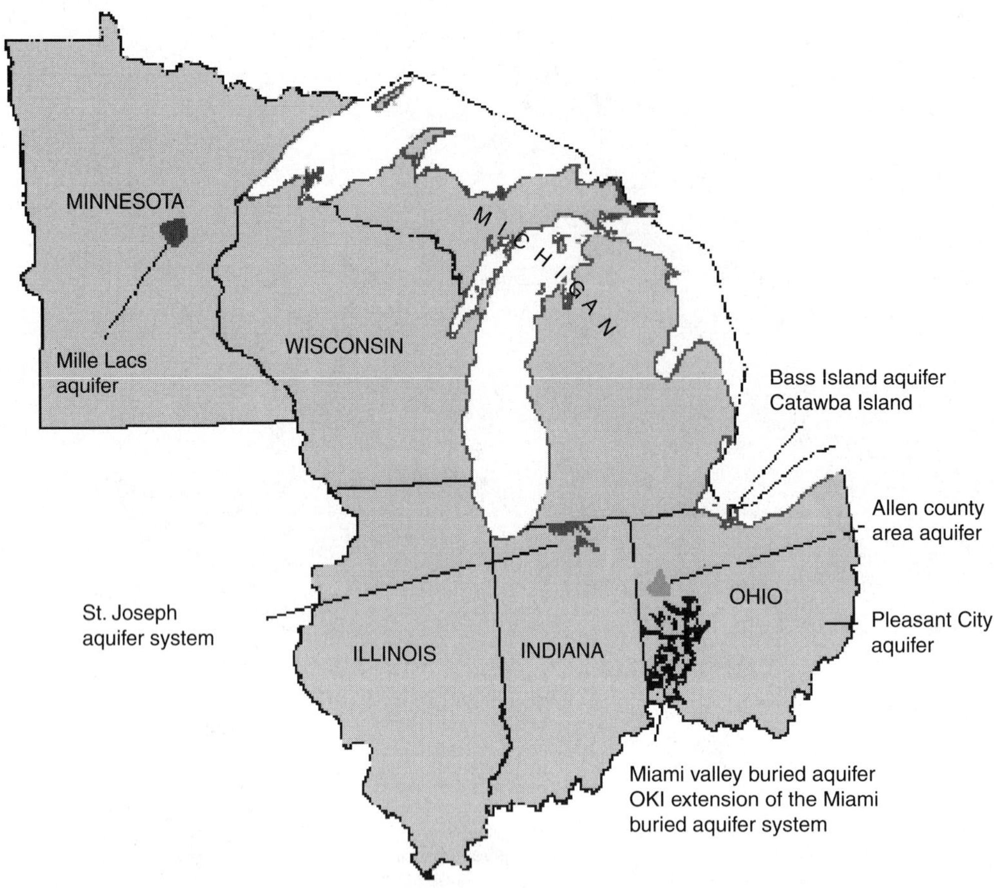

Figure 11I.113 Designated sole source aquifers in EPA Region V, Illinois, Indiana, Michigan, Minnesota, Ohio, Wisconsin. (From www.epa.gov. With permission.)

Table 11I.160 Designated Sole Source Aquifers in Region V

State	Sole Source Aquifer Name	Federal Register Cit.	Public. Date
IN	St. Joseph Aquifer System	53 FR 23682	06/23/88
MN	Mille Lacs Aquifer	55 FR 43407	10/29/90
OH	Pleasant City Aquifer	52 FR 32342	08/27/87
OH	Bass Island Aq., Catawba Island	52 FR 37009	10/02/87
OH	Miami Valley Buried Aquifer	53 FR 15876	05/04/88
OH	OKI extension of the Miami Buried Valley Aquifer	53 FR 25670	07/08/88
OH	Allan County Area Combined Aquifer System	57 FR 53111	11/06/92

Source: From www.epa.gov.

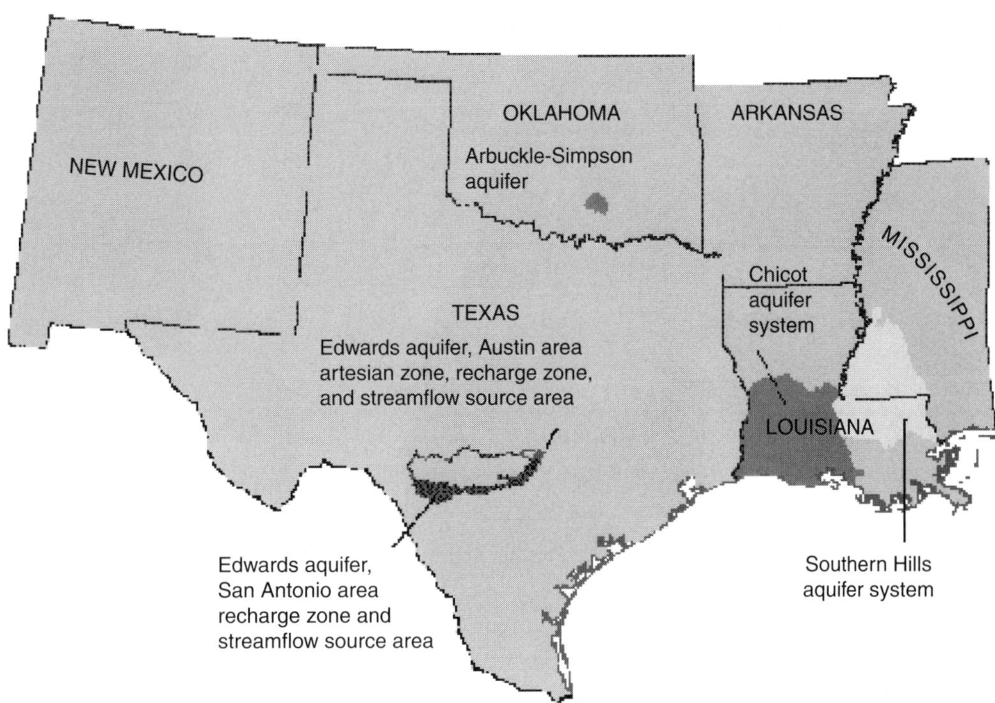

Figure 11I.114 Designated sole source aquifers in EPA region VI, Arkansas, Louisiana, New Mexico, Oklahoma, Texas. (From www.epa.gov. With permission.)

Table 11I.161 Designated Sole Source Aquifers in Region 6

State	Sole Source Aquifer Name	Federal Reg. Cit.	Public. Date
LA	Chicot Aquifer System	53 FR 20893	06/07/88
[a]LA/MS	Southern Hills Aquifer System	53 FR 25538	07/07/88
OK	Arbuckle-Simpson Aquifer, South Central Oklahoma	54 FR 39230	09/25/89
TX	Edwards Aquifer, San Antonio Area	40 FR 58344	12/16/75
TX	Edwards Aquifer, Austin Area	53 FR 20897	06/07/88

[a] The Southern Hills Regional Aquifer system is jointly managed with Region IV. While listed in both regions, it is counted only once in the national total of 70.

Source: From www.epa.gov.

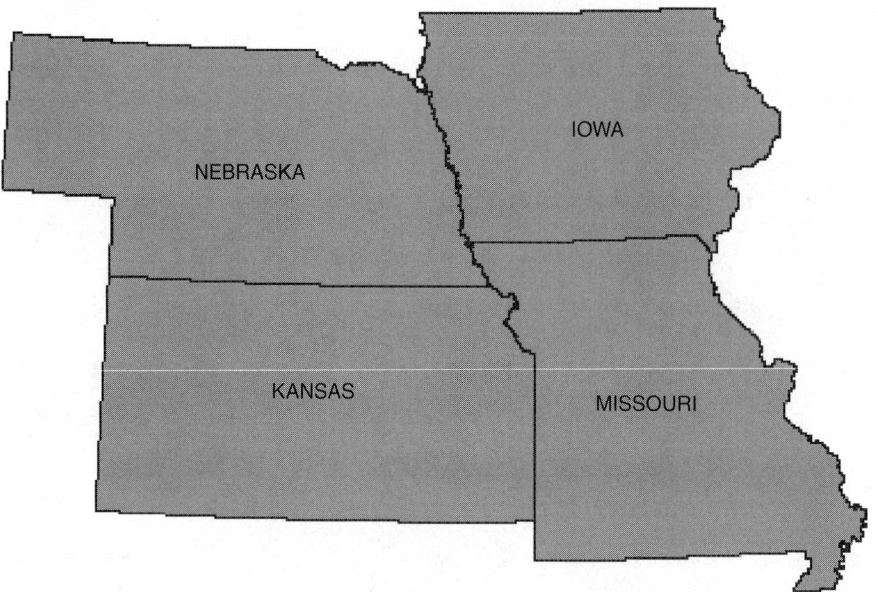

Figure 11I.115 Designated sole source aquifers in EPA region VII. Iowa, Kansas, Missouri, Nebraska. (From www.epa.gov. With permission.)

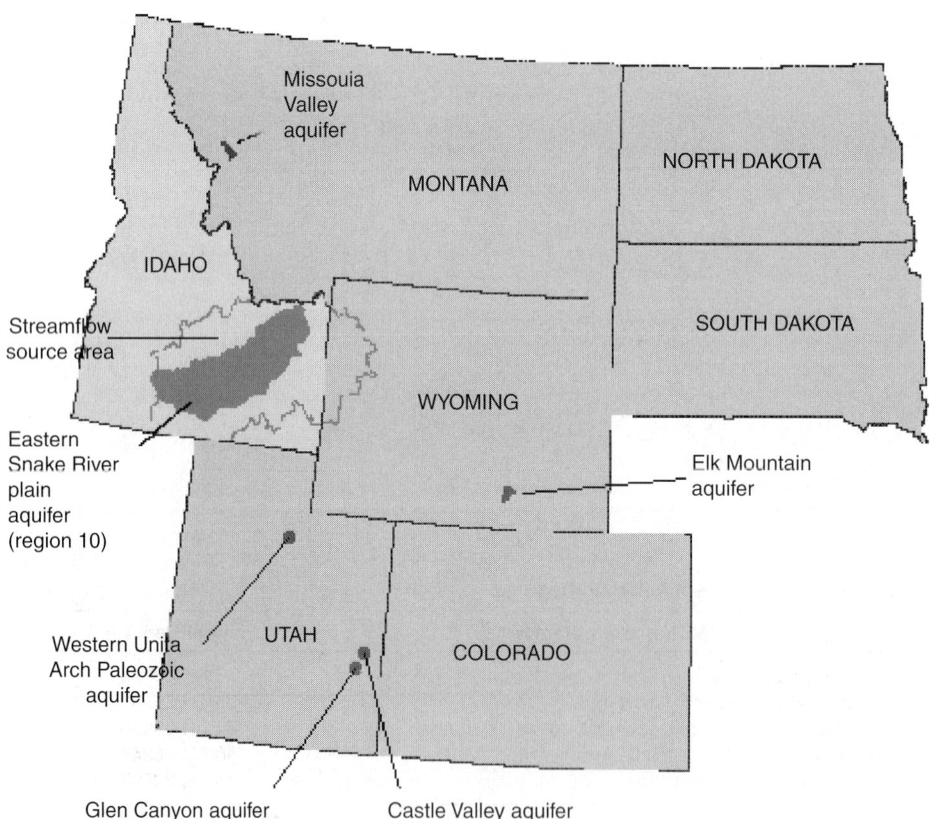

Figure 11I.116 Designated sole source aquifers in EPA region VIII, Colorado, Montana, North Dakota, South Dakota, Utah, Wyoming. (From www.epa.gov. With permission.)

Table 11I.162 Designated Sole Source Aquifers in Region VIII

State	Sole Source Aquifer Name	Federal Reg. Cit.	Publ. Date
MT	Missoula Valley Aquifer	53 FR 20895	06/07/1988
UT	Castle Valley Aquifer System	66 FR 41027	08/06/2001
UT	Western Unita Arch Paleozoic Aquifer System at Oakley, UT	65 FR 232	12/01/2000
UT	Glen Canyon Aquifer System	67 FR 736	01/07/2000
WY[a]	Eastern Snake River Plain Aquifer Stream Flow Source Area	56 FR 50638	10/07/1991
WY	Elk Mountain Aquifer	63 FR 38167	07/15/1998

[a] The Eastern Snake River Plain Aquifer is jointly managed with Region X. While listed in both regions, it is counted only once in the national total of 73.

Source: From www.epa.gov.

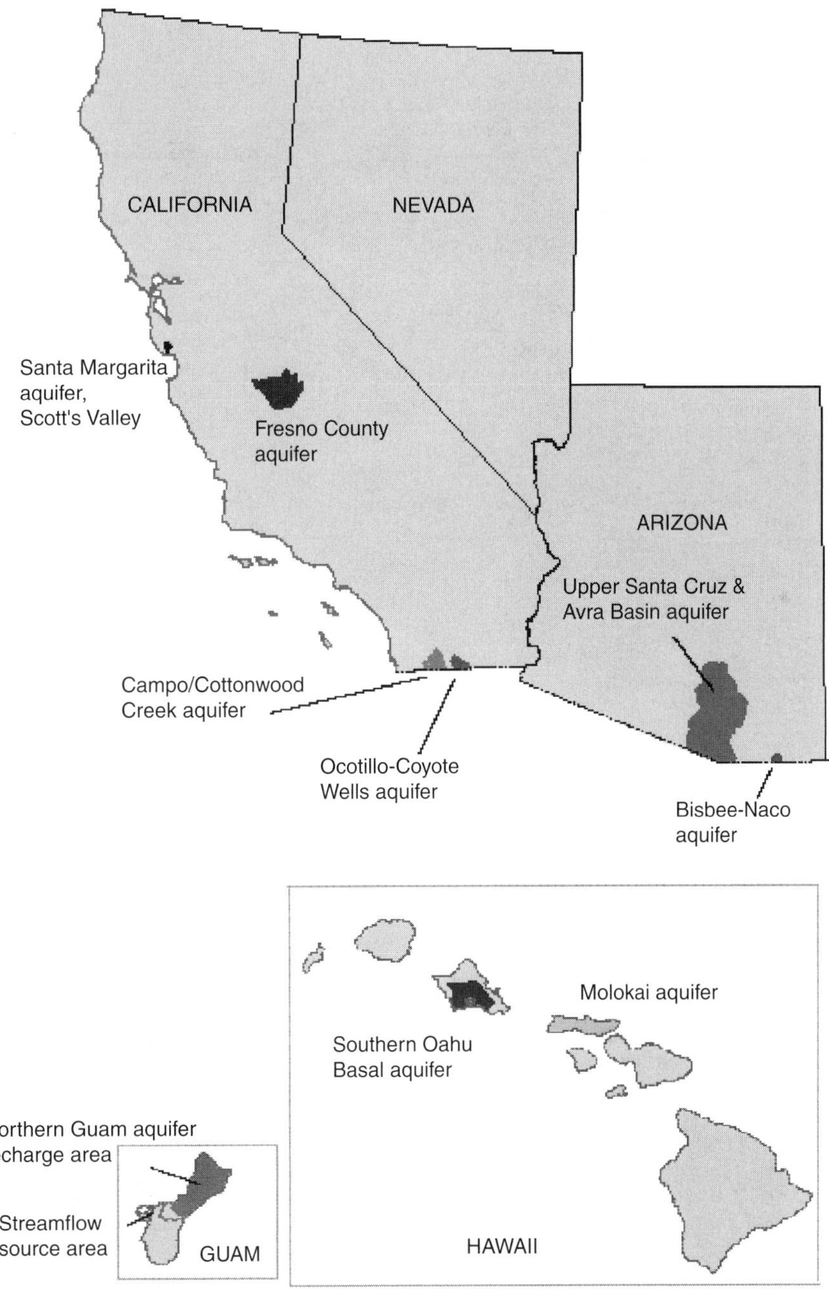

Figure 11I.117 Designated sole source aquifers in EPA region IX, Arizona, California, Hawaii, Nevada, Guam, and American Samoa. (From www.epa.gov. With permission.)

Table 11I.163 Designated Sole Source Aquifers in Region IX

State	Sole Source Aquifer Name	Federal Reg. Cit.	Publ. Date
AZ	Upper Santa Cruz & Avra Basin Aquifer	49 FR 2948	01/24/84
AZ	Bisbee-Naco Aquifer	53 FR 38337	09/30/88
CA	Fresno County Aquifer	44 FR 52751	09/10/79
CA	Santa Margarita Aquifer, Scotts Valley	50 FR 2023	01/14/85
CA	Campo/Cottonwood Creek	58 FR 31024	05/28/93
CA	Ocotillo-Coyote Wells Aquifer	61 FR 47752	09/10/96
GU	Northern Guam Aquifer System	43 FR 17867	04/26/78
HI	Southern Oahu Basal Aquifer	52 FR 45496	11/30/87
HI	Molokai Aquifer	59 FR 23063	04/20/93

Source: From www.epa.gov.

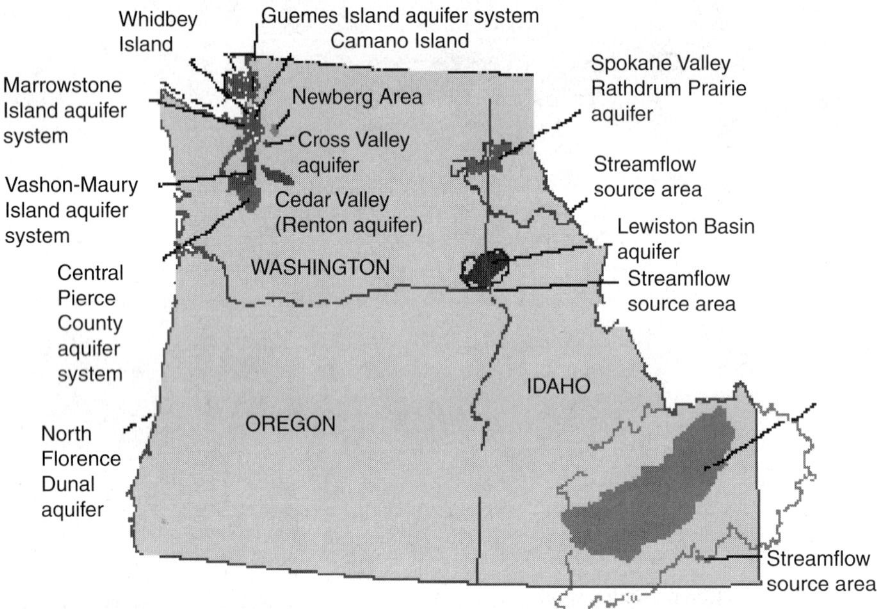

Figure 11I.118 Designated sole source aquifers in EPA region X, Alaska, Idaho, Oregon, Washington. (From www.epa.gov. With permission.)

Table 11I.164 Designated Sole Source Aquifers in Region X

State	Sole Source Aquifer Name	Federal Reg. Cit.	Publ. Date
ID/WY	Eastern Snake River Plain Aquifer	56 FR 50638	10/07/91
OR	North Florence-Dunal Aquifer	52 FR 37519	10/07/87
WA/ID	Spokane Valley Rathdrum Prairie Aquifer	42 FR 5566	02/09/78
WA	Camano Island Aquifer	47 FR 14779	04/06/82
WA	Whidbey Island Aquifer	47 FR 14779	04/06/82
WA	Cross Valley Aquifer	52 FR 18606	05/18/87
WA	Newberg Area Aquifer	52 FR 37215	10/05/87
WA	Cedar Valley (Renton Aquifer)	53 FR 38779	10/03/88
WA/ID	Lewiston Basin Aquifer	53 FR 49920	12/12/88
WA	Central Pierce Cty. Aquifer Syst.	59 FR 224	01/03/94
WA	Marrowstone Island Aquifer Syst.	59 FR 28752	06/02/94
WA	Vashon-Maury Island Aquifer Syst.	59 FR 34468	07/05/94
WA	Guemes Island Aquifer System	62 FR 5928–3	12/01/97

Source: From www.epa.gov.

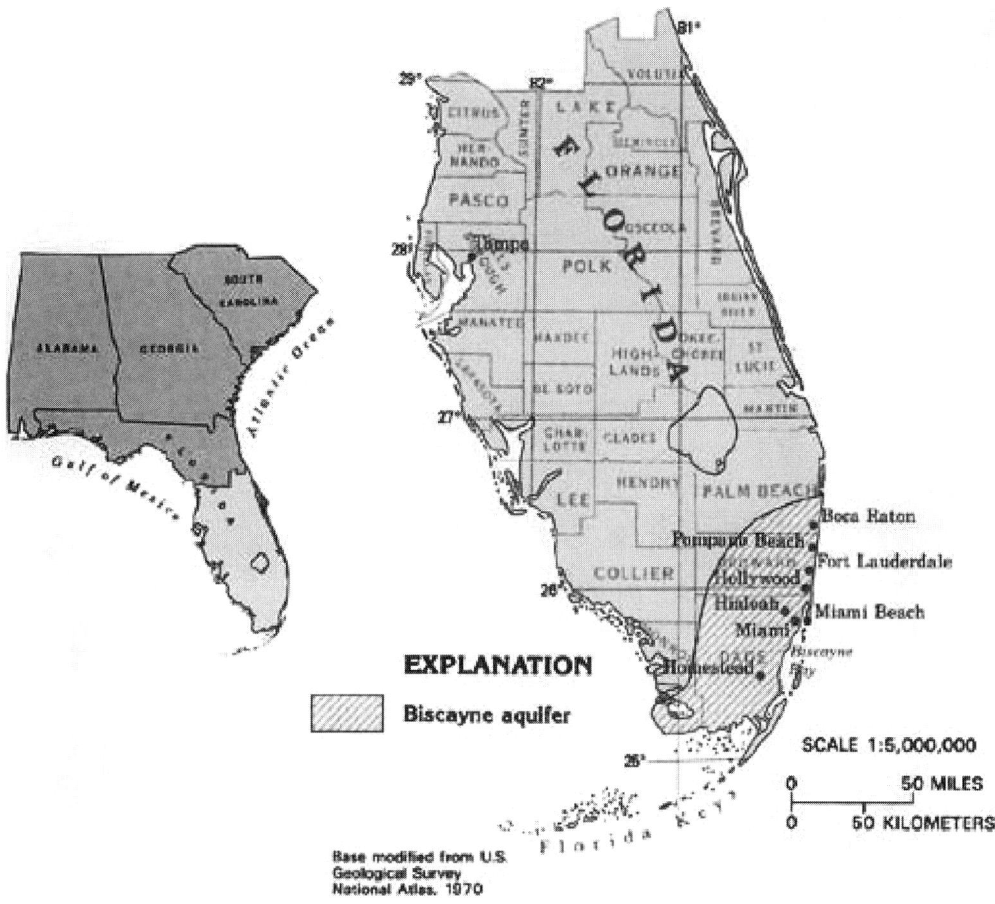

Figure 11I.119 The Biscayne aquifer underlies part of four counties in southeastern Florida, and consists predominantly of limestone. (From www.capp.water.usgs.gov. With permission.)

Table 11I.165 Relative Evaluation of the Elements of the Biscayne Aquifer Protection Plan (Florida)

Category	Element Number	Description	Ease of Implementation	Effectiveness	Personnel Resources[a]	Cost[b]	Time to Implement[c]	Public Acceptability[d]
Regulation	2	Regulate land use within well field protection zones	Difficult	High	Moderate	Moderate	Moderate	Moderate
Regulation	3	Surveillance over small quantity generators	Moderately difficult	Moderate	Moderate	Low	Short	Moderate
Regulation	4	Regulate activities of small quantity generators	Difficult	High	High	Moderate	Long	Moderate
Regulation	6	Regulate storage tanks	Moderately difficult	High	Moderate	High	Moderate	Moderate
Regulation	7	Control handling/disposal by commercial haulers	Moderately difficult	High	Low	Low	Short	High
Regulation	18	Stormwater/ wastewater review	Easy	Moderate	Moderate	Moderate	Short	Moderate
Waste Management	1	Local waste storage/transfer facilities for small generators, individuals	Moderately difficult	High	Low	Moderate	Short	Moderate
Waste Management	5	Public awareness/ education program	Easy	High	Moderate	Moderate	Moderate	High
Waste Management	9	Spill prevention, control, and countermeasure program	Moderately difficult	High	Moderate	Moderate	Moderate	High
Waste Management	12	Adopt liability policy	Moderately difficult	High	Low	Low	Moderate	High
Waste Management	13	Emergency spill cleanup program	Difficult	Moderate	Low	High	Moderate	High
Waste Management	15	Control agricultural groundwater pollution	Difficult	Moderate	Moderate	Moderate	Moderate	Moderate
Waste Management	16	Collect/recycle automobile drain oils	Easy	Moderate	Low	Low	Short	High
Construction/Treatment	8	Construct leak-proof sewers	Difficult	Moderate	Low	High	Moderate	High
Construction/Treatment	10	Pretreat commercial/ industrial waste	Moderately difficult	Moderate	Moderate	Moderate	Short	High

Construction/Treatment	11	Control sewer exfiltration	Difficult	Moderate	Moderate	Moderate	Long	High
Information Needs	14	Public reporting program on improper disposal	Easy	Moderate	Low	Low	Short	High
Information Needs	17	Tri-county coordinating committee	Easy	Low	Low	Low	Short	High
Information Needs	19	Determine safe soil contamination levels	Difficult	Moderate	Moderate	Moderate	Long	Moderate
Information Needs	20	Monitor groundwater	Moderately difficult	Moderate	Moderate	High	Moderate	High

[a] "Personnel Resources" refers to the manpower needed by each county to develop and implement each program. Low? 1 to 4 full-time persons from each county; Moderate? 5 to 10 ful-time persons; High? More than 10 full-time persons.

[b] Includes cost of development and implementation of the program by county agencies as well as costs incurred by the public or the industries.

[c] Short? 0 to 2 years; Moderate? 2 to 5 years; Long? Greater than 5 years.

[d] Public acceptability tended to be lower for a specific program if large numbers of people were directly affected.

Source: From Sing, U.P., Orban, J.E., and Docal, A.L., 1985, *The Biscayne Aquifer Protection Plan*, American Water Resources Association Proceeding Tucson AZ Symposium on Groundwater Contamination and Reclamation. Copyright AWWA. Reprinted with permission.

Series	Stratigraphic and hydrologic units		Lithology and water-yielding characteristics	Thickness (ft)
Holocene	Organic soils Lake Flirt marl	Confining Unit	Peat and muck; water has high color content. Almost impermeable, Lake Flirt is shelly, calcaroous mud	0–18
Pleistocene	Pamlico sand	Biscayne Aquifer	Quartz sand; water high in iron. Small yields to domestic wells	0–40
	Miami oolite		Sandy, oolitic limestone. Large yields	0–40
	Fort Thompson formation		Alternating marine shell beds and freshwater limestone. Generally high permeability. Large yields	0–150
	Anastesia formation		Coquina, sand, sandy limestone, marl. Moderate to large yields.	0–120
	Key Largo limestone		Coralline reef rock. Large yields	0–60
Pliocene	Caloosahatchee marl		Sand, shell, silt, and marl. Moderate yields	0–25
	Tamiami formation	Confining Unit	Limestone, clay, and marl. Occasional moderate yields in upper few feet. Remainder forms upper part of basal confining unit.	25–220

Stratigraphic Limits are equivalent in part. Order does not necessarily reflect relative age.

Modified from Klein and Causaraa, 1982

Figure 11I.120 Several geologic units comprise the Biscayne aquifer. Most of these units are of Pleistocene age. (From www.capp.water.usgs.gov. With permission.)

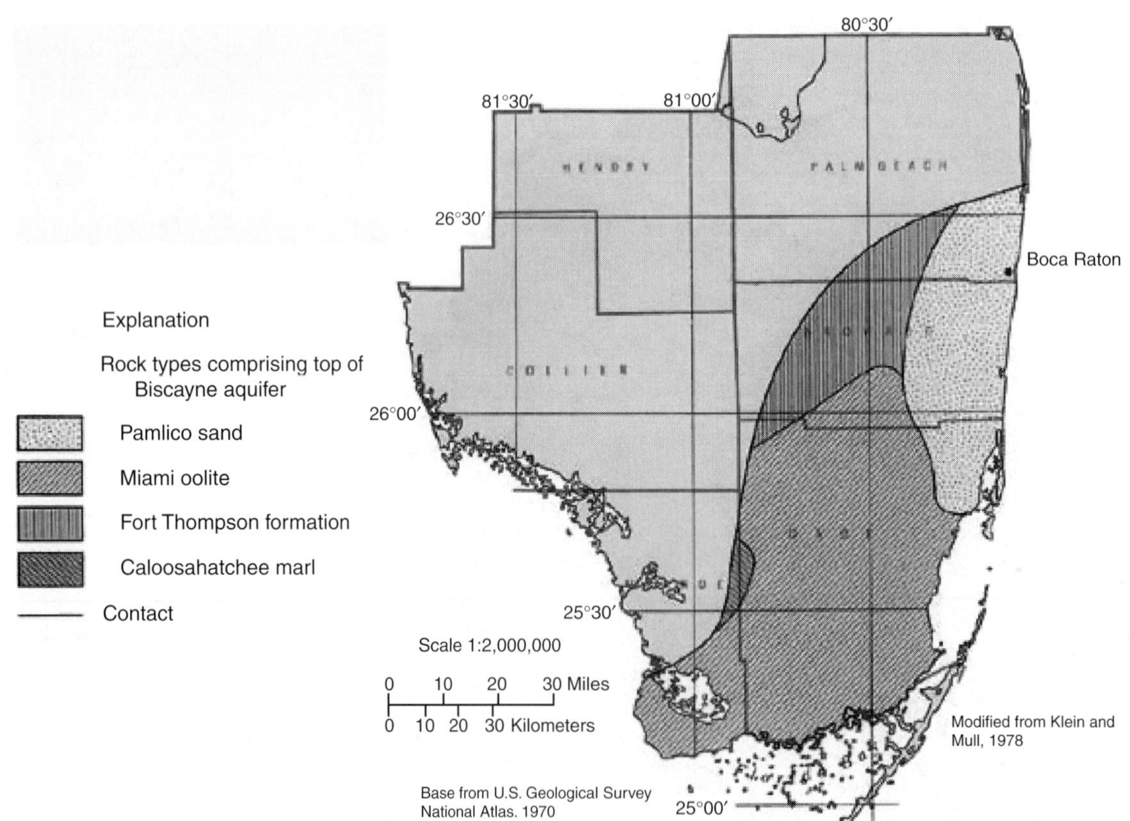

Figure 11I.121 The rocks that comprise the top of the Biscayne aquifer vary in character. They are mostly limestone, but sand marks the top of the aquifer to the northeast. (From www.capp.water.usgs.gov. With permission.)

Table 11I.166 Implementation Phases of the Biscayne Aquifer Protection Plan

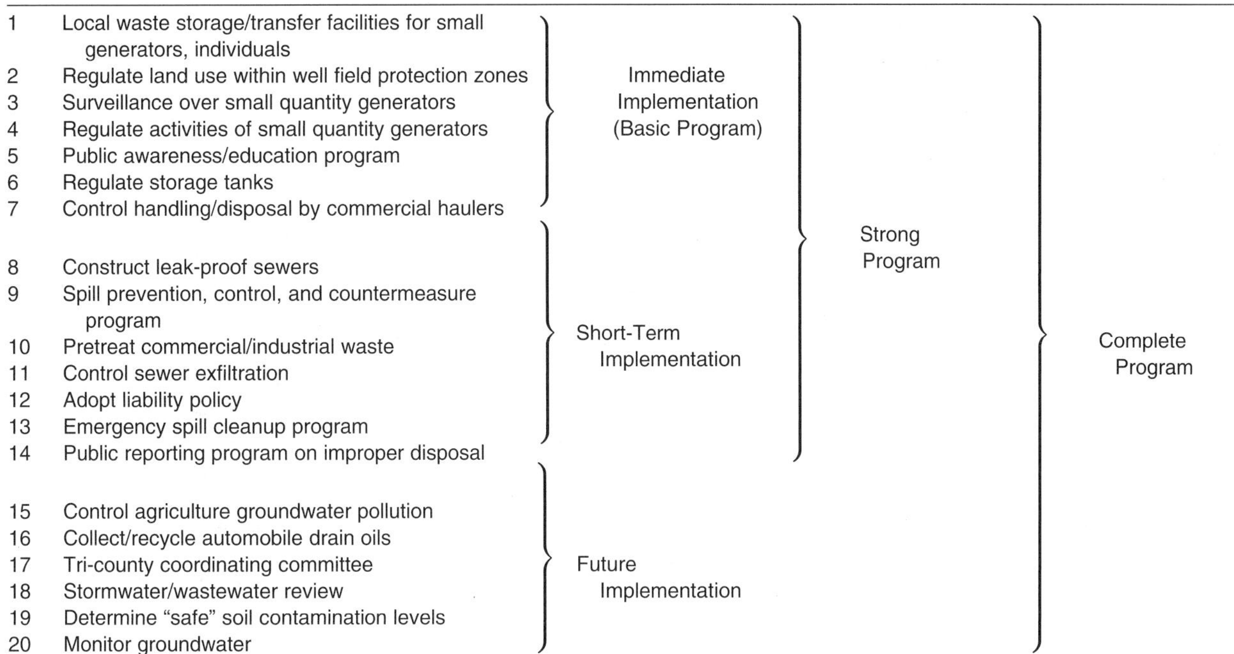

No.	Measure	Implementation Phase		
1	Local waste storage/transfer facilities for small generators, individuals	Immediate Implementation (Basic Program)	Strong Program	Complete Program
2	Regulate land use within well field protection zones			
3	Surveillance over small quantity generators			
4	Regulate activities of small quantity generators			
5	Public awareness/education program			
6	Regulate storage tanks			
7	Control handling/disposal by commercial haulers			
8	Construct leak-proof sewers	Short-Term Implementation		
9	Spill prevention, control, and countermeasure program			
10	Pretreat commercial/industrial waste			
11	Control sewer exfiltration			
12	Adopt liability policy			
13	Emergency spill cleanup program			
14	Public reporting program on improper disposal			
15	Control agriculture groundwater pollution	Future Implementation		
16	Collect/recycle automobile drain oils			
17	Tri-county coordinating committee			
18	Stormwater/wastewater review			
19	Determine “safe” soil contamination levels			
20	Monitor groundwater			

Source: From Sing, U.P., Orban, J.E., and Docal, A.L., 1985, *The Biscayne Aquifer Protection Plan*, American Water Resources Association Proceedings, Tucson AZ Symposium on Groundwater Protection and Reclamation. Copyright AWWA. Reprinted with permission.

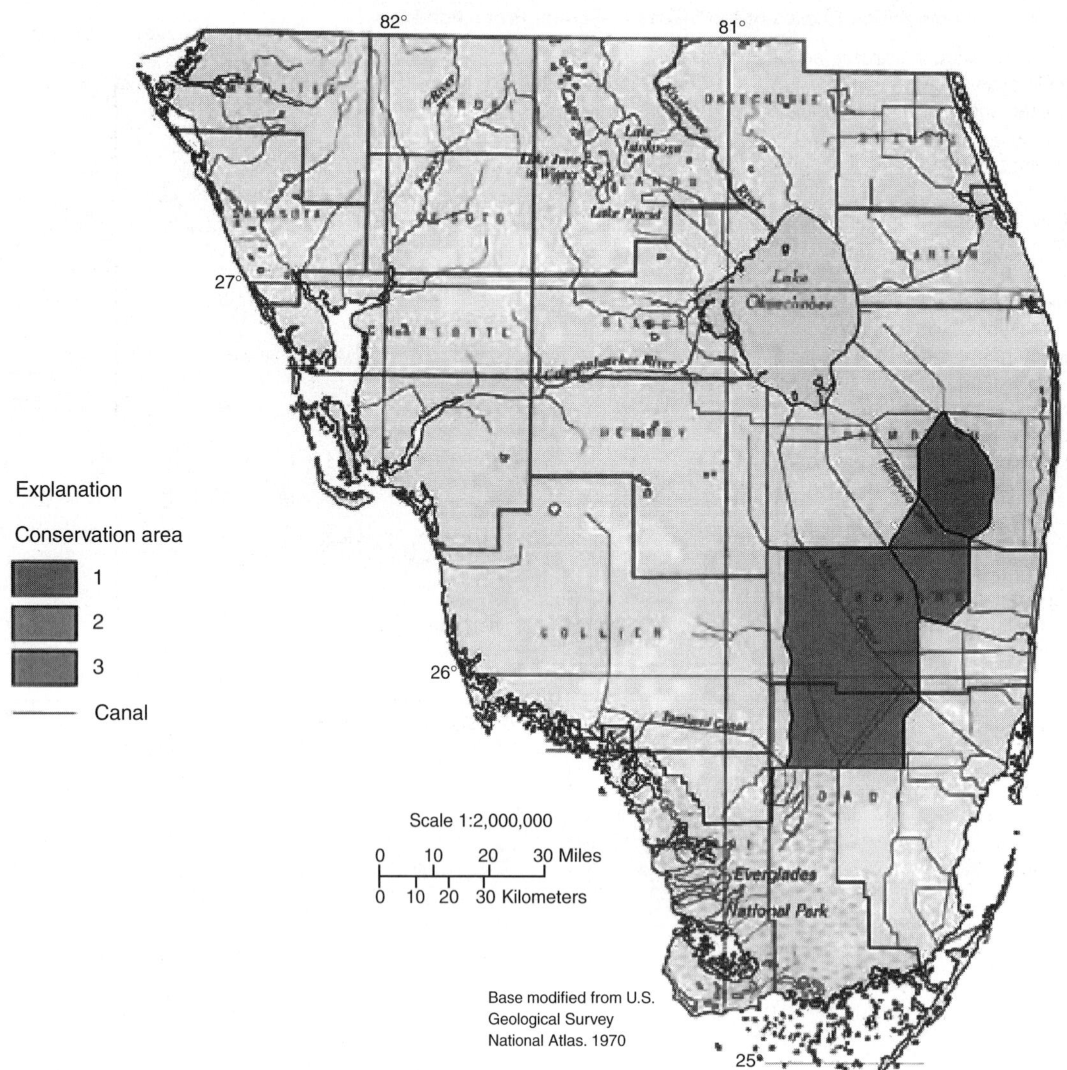

Figure 11I.122 Lake Okeechobee, an extensive canal system, and three water-conservation areas are all linked as parts of a sophisticated water-management system. Dikes, pumps, and control structures on the canals are used to store stormwater and release it when and where it is later needed. (From www.capp.water.usgs.gov. With permission.)

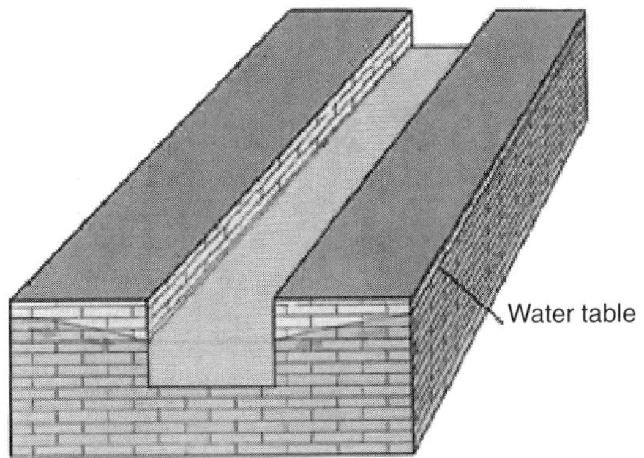

(a) When the water level in an aquifer is higher than that in a canal that penetrates it, water moves toward the canal.

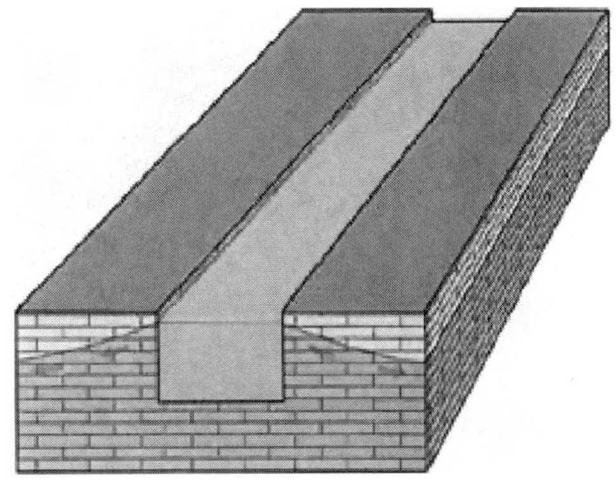

(b) When the water level in a canal is higher than that in the aquifer it penetrates, water moves into the aquifer.

Modified from Klein and others, 1975.

Figure 11I.123 (a) During periods of less than normal precipitation, water passes freely from the aquifer into canals that are dug into it. (b) When storm runoff raises canal water levels, the movement of water is reversed. (From www.capp.water.usgs.gov. With permission.)

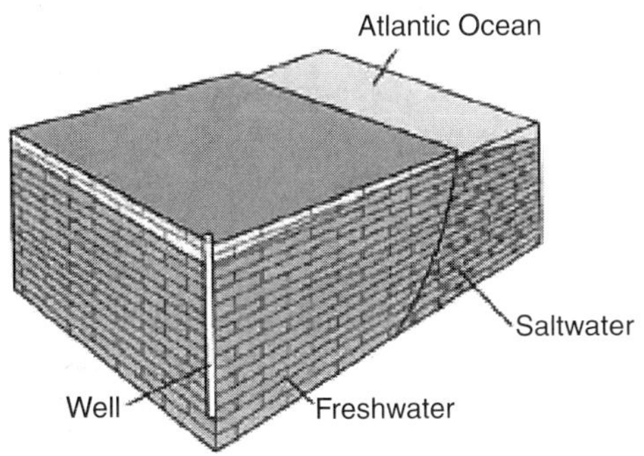

The freshwater-saltwater interface was nearly stable before coastal canals were built.

(a)

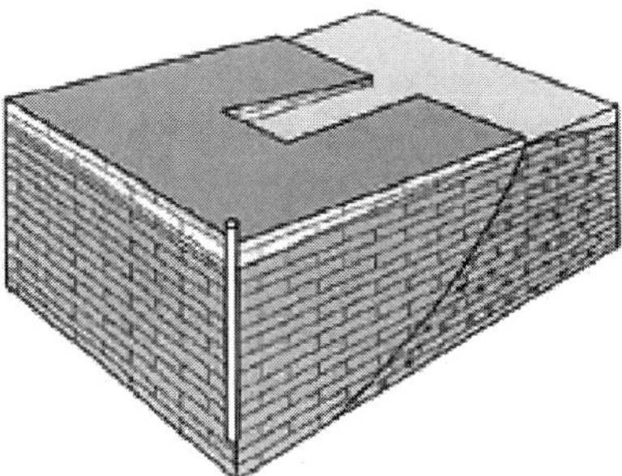

(c) Uncontrolled tidal canals cause saltwater encroachment by lowering freshwater levels and providing open channels to the sea. The interface in the aquifer shifts inland adjacent to the canal.

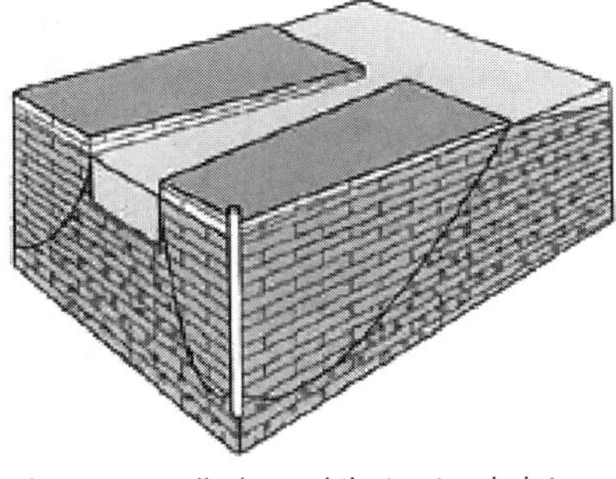

(b) An uncontrolled canal that extends into an area of heavy pumping can convey saltwater inland to contaminate freshwater supplies.

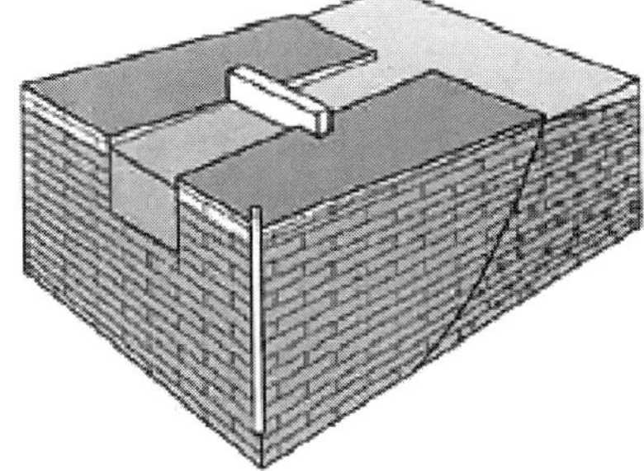

(d) In contrast, a controlled canal can provide a perennial supply of freshwater to prevent saltwater encroachment and to recharge a well field by bringing in freshwater from upgradient areas.

Figure 11I.124 The original, virtually stable freshwater–saltwater interface (a) was disturbed by construction of drainage canals (b). Uncontrolled canals (c) are open conduits, allowing saltwater to migrate inland, contaminating the aquifer, and in places, encroaching into pumped wells. A properly located control structure (d) prohibits the saltwater encroachment. (From www.capp.water.usgs.gov. With permission.)

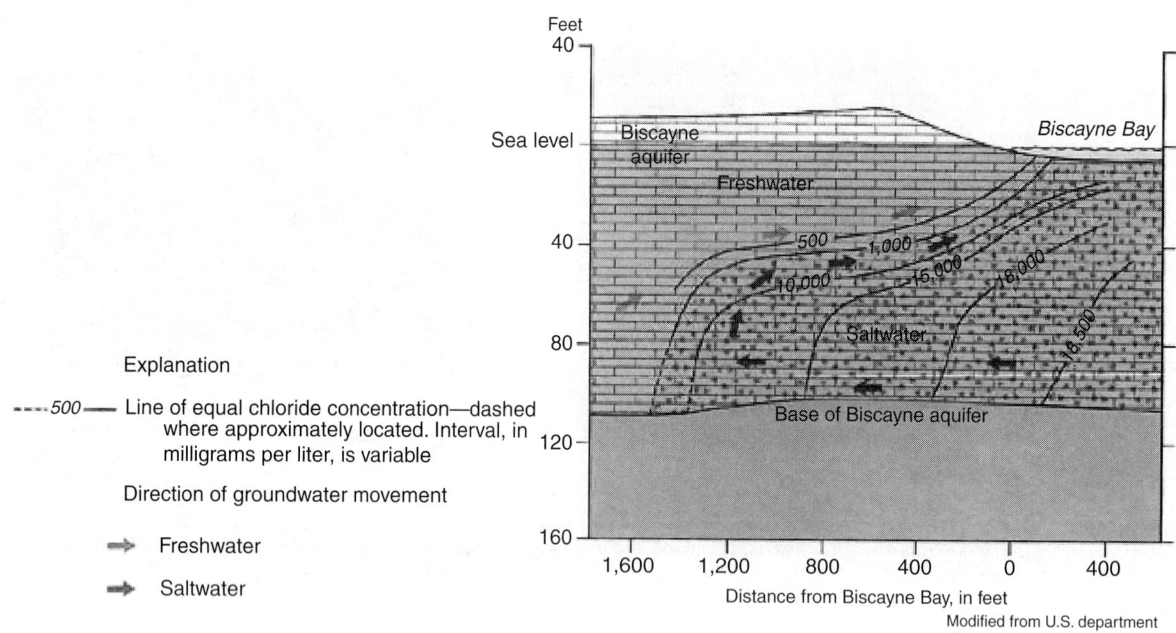

Figure 11I.125 Saltwater extends farther inland near the base of the Biscayne aquifer rather than near the top of it. (From www.capp.water.usgs.gov. With permission.)

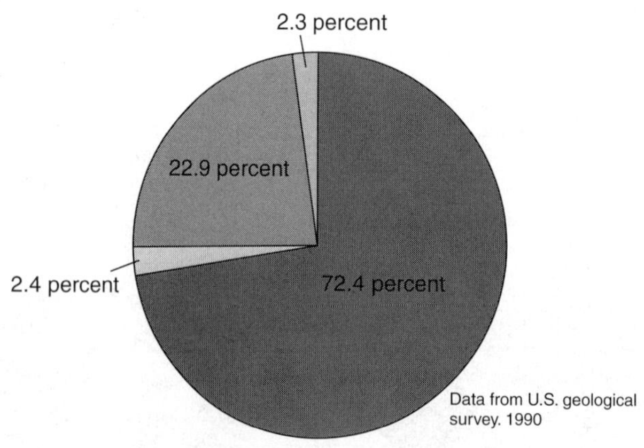

Explanation

Use of fresh groundwater withdrawals during 1985, in percent–total fresh groundwater withdrawls during 1985 were 786 million gallons per day

72.4 Public supply

2.4 Domestic and commercial

22.9 Agricultural

2.3 Industrial, mining, and thermoelectric power

Figure 11I.126 Most of the freshwater withdrawn from the Biscayne aquifer during 1985 was used for public supply and agricultural purposes. (From www.capp.water.usgs.gov. With permission.)

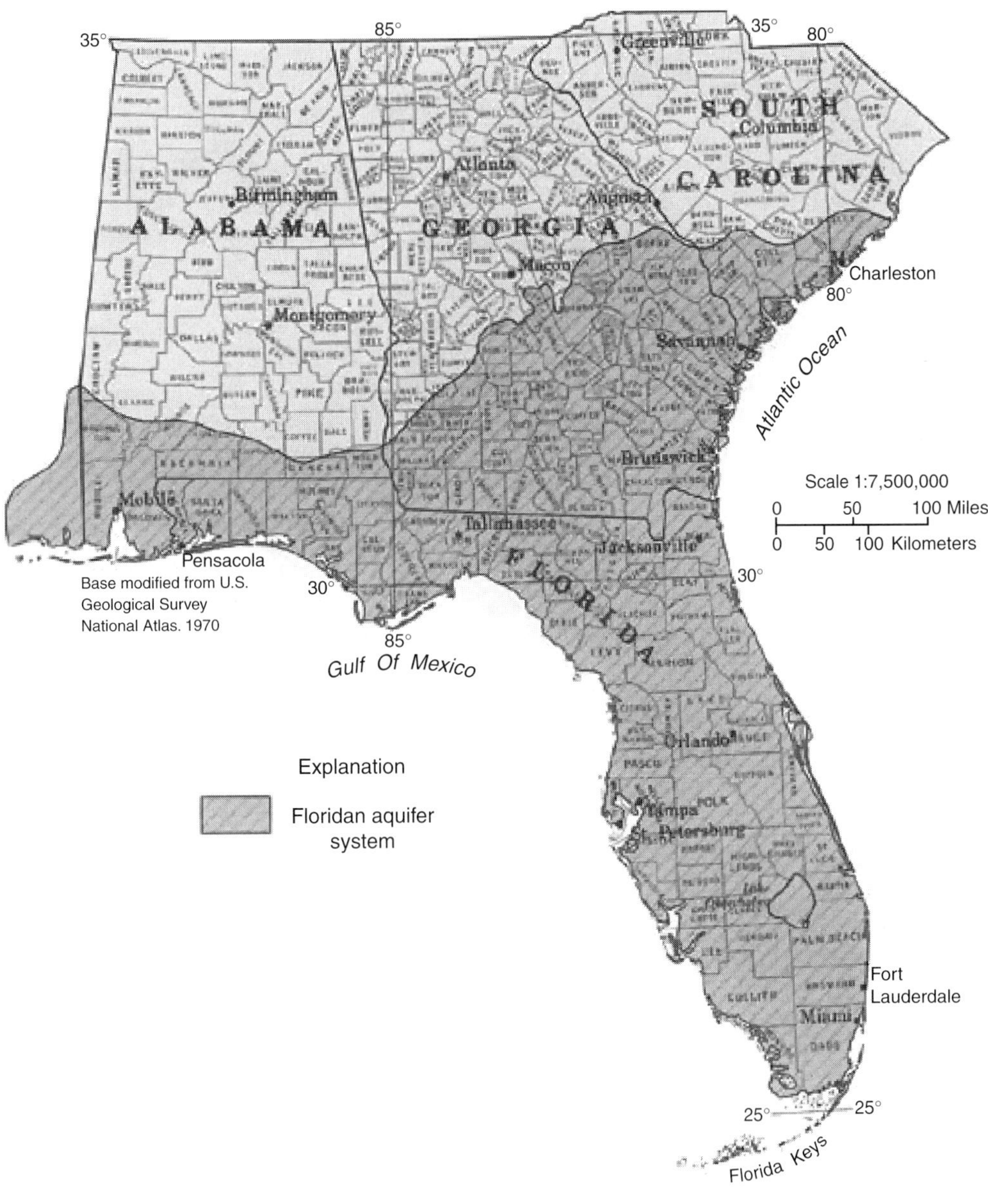

Figure 11I.127 The carbonate rocks of the Floridan aquifer system underlie all of Florida, most of the coastal plain of Georgia, and extend for short distances into Alabama and South Carolina. (From www.capp.water.usgs.gov. With permission.)

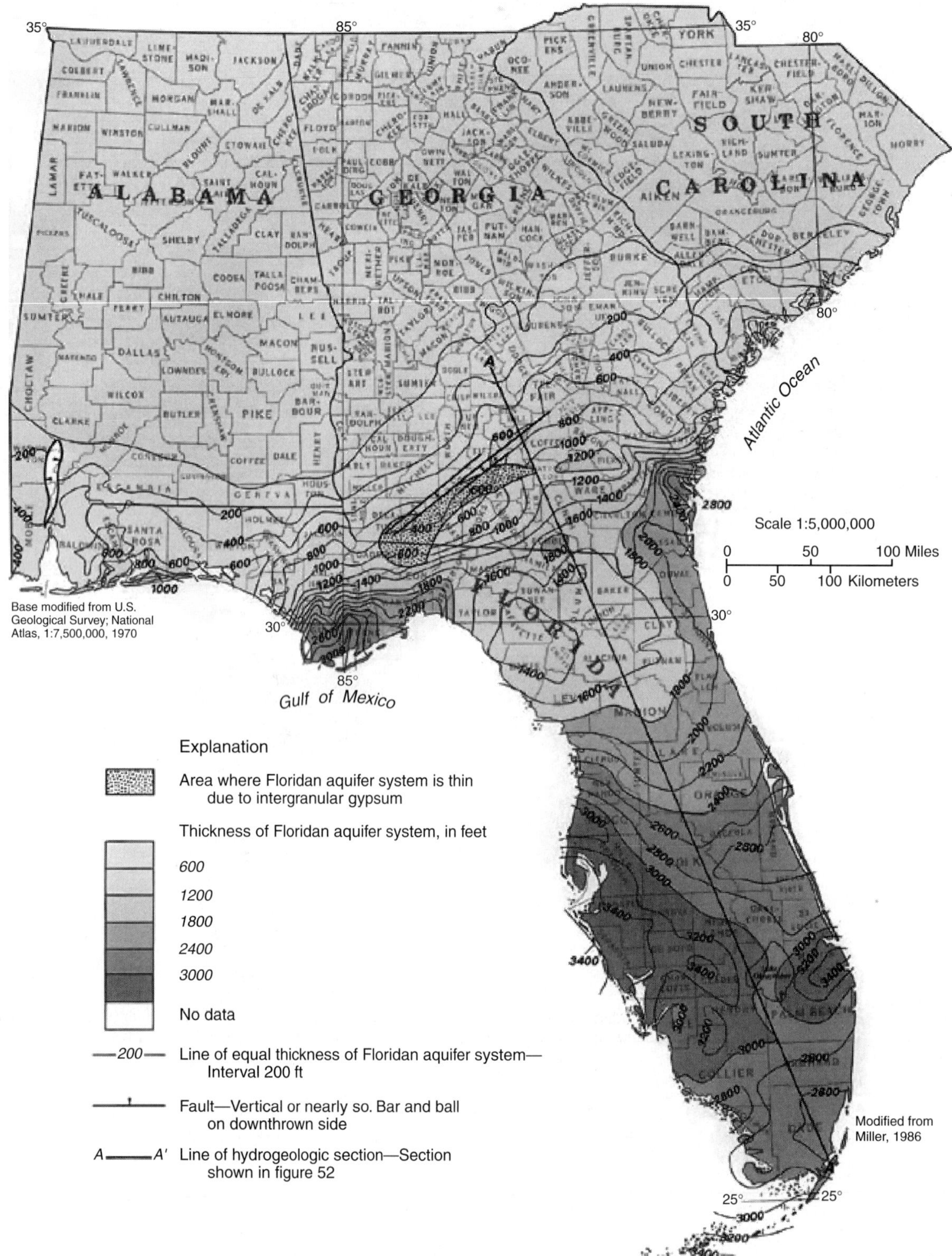

Figure 11I.128 The thickness of the Floridan aquifer system varies considerably and reflects some major warping during deposition and fracturing following deposition. (From www.capp.water.usgs.gov. With permission.)

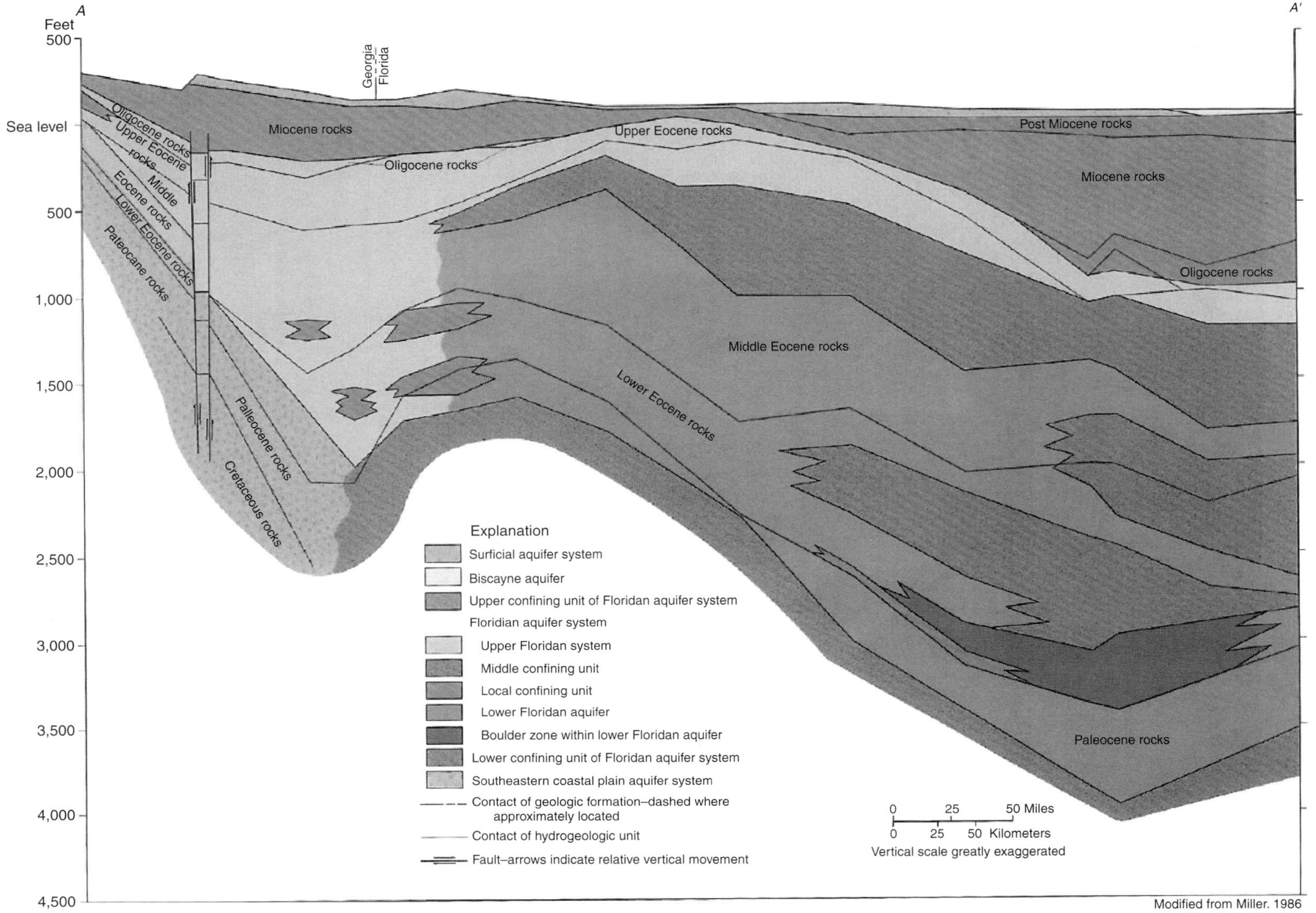

Figure 11I.129 The Floridan aquifer system changes significantly from south-central Georgia to southern Florida. Aquifers and confining units in the system thicken and thin from well to well, and generally resemble complexly interlingering, lens-shaped bodies of rock. (From www.capp.water.usgs.gov. With permission.)

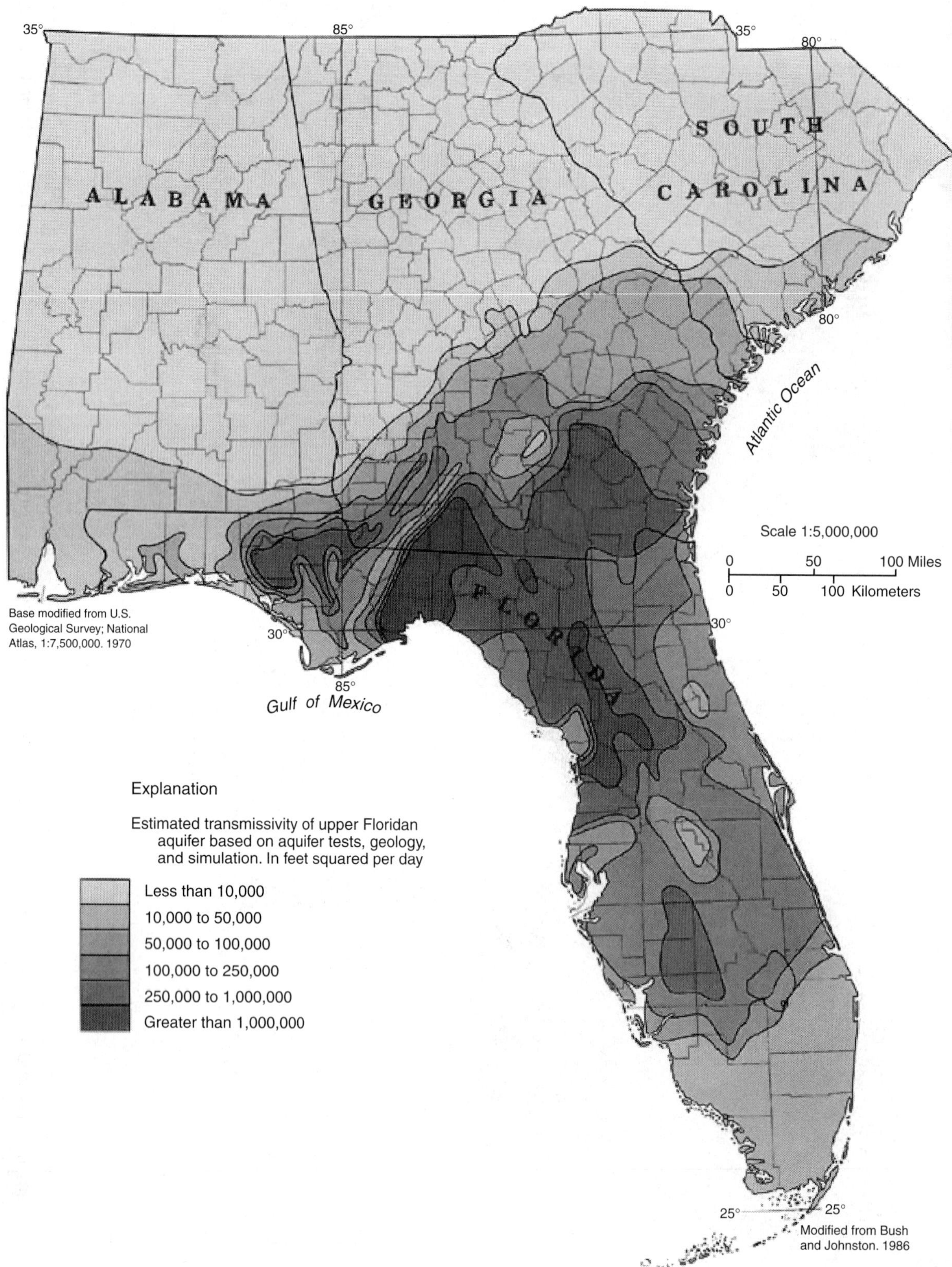

Figure 11I.130 Transmissivity is a measure of the ease with which an aquifer transmits water. The most transmissive parts of the upper Floridan aquifer are located where many solution openings have formed. Such transmissivity not only means that water will move more easily, but also that wells will yield large volumes of water. (From www.capp.water.usgs.gov. With permission.)

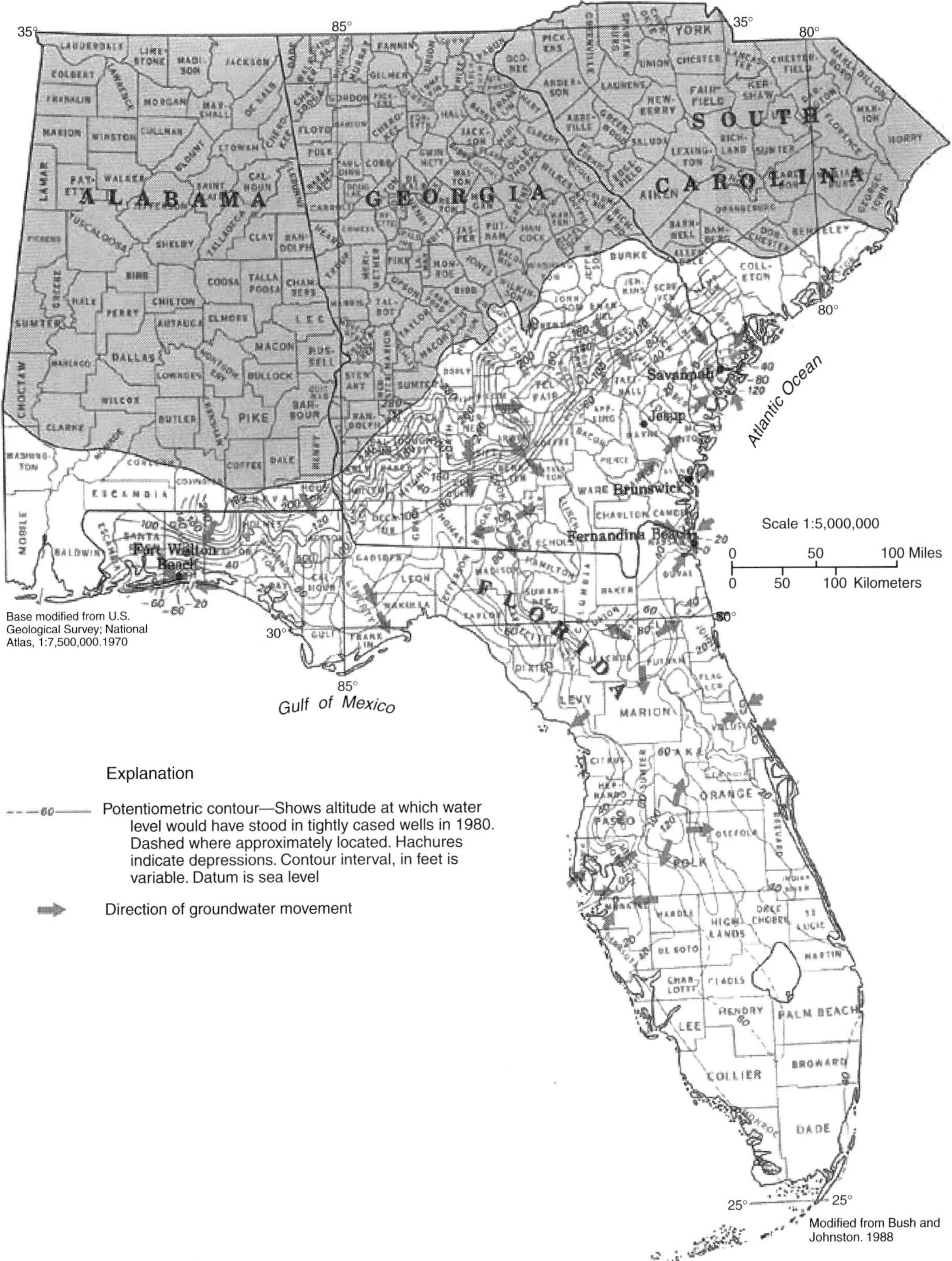

Figure 11I.131 Withdrawals from the Floridan aquifer system had developed deep cones of depression in several places by 1980. The direction of groundwater flow was locally changed, even reversed in places, by the pumpage. (From www.capp.water.usgs.gov. With permission.)

Table 11I.167 Major Components of an Information System Needed for Groundwater Management Decisions

- Hydrogeology
  - Soil and unsaturated zone characteristics
  - Aquifer characteristics
    - Depths involved
    - Flow patterns
    - Recharge characteristics
    - Transmissive and storage properties
    - Amblent water quality
    - Interaction with surface water
    - Boundary conditions
    - Mineralogy, including organic content
- Water extraction (withdrawals) and use patterns
  - Locations
  - Amounts
  - Purpose (domestic, industrial, agricultural)
  - Trends
- Potential contamination sources and characteristics
  - Point sources
    - Industrial and mining waste discharges
    - Commercial waste discharges
    - Hazardous material and waste storage
    - Domestic waste discharges
  - Nonpoint sources
    - Agricultural
    - Septic tanks
    - Land applications of waste
    - Urban runoff
    - Transportation spills (can also be considered a print source)
    - Pipelines (energy and waste) (can also be considered a point source)
- Population patterns
  - Demographic
  - Economic trends
  - Land-use patterns

Source: From U.S. Geological Survey, *National Water Summary 1986*; modified from National Research Council, 1986.

Table 11I.168 Criteria for Effective Groundwater Protection Programs

Criterion	Scope
Goals and objectives	Protection programs should clearly define goals and objectives, reflect understanding of groundwater problems, have adequate legal authority, and have criteria for evaluating program success and the need for modifications
Information	Programs should be based on information that permits resources and issues to be defined and preventive strategies to be evaluated
Technical basis	Effective programs require a sound technical basis with which to link actions to results
Source elimination and control	Long-term program goal should be to eliminate or reduce the sources of groundwater contamination
Intergovernmental and interagency linkages	Comprehensive protection program must link actions at every level of government into coherent, coordinated action
Effective implementation and adequate funding	Programs must have adequate legal authority, resources, and stable institutional structures to be effective
Economic, social, political, and environmental impacts	A preventive program assumes that groundwater protection is the least costly strategy in the long run. Protective actions should be evaluated in terms of their economic, social, political, and environmental effects
Public support and responsiveness	Programs must be responsive and credible to the public

Source: From U.S. Geological Survey, *National Water Summary 1986;* modified from National Research Council, 1986.

Table 11I.169 Common Elements of "MATURE" Groundwater Protection Programs

Setting Goals and Documenting Progress

- • Groundwater protection goals which accounts for present and future uses of the resource
- • Yearly action plan for achieving the goal, which includes a mechanism for evaluating progress toward accomplishing the goal and provides for periodic review

Characterizing the Resource and Setting Priorities for Actions

- • Comprehensive assessment of aquifer systems and their associated recharge and discharge areas
- • Procedure for inventorying and ranking potential sources of contamination that may cause an adverse effect on human health, or ecological systems
- • Process used for setting priorities for actions taken to protect or remediate the resource, such as a use designation/classification scheme that considers use, value, vulnerability, yield, current quality, etc., including wellhead protection and cost benefit analyses

Developing and Implementing Prevention and Control Programs

- • A Coordinated pollution prevention and source reduction program aimed at eliminating and reducing the amount of pollution that could potentially affect groundwater, including wellhead and recharge area protection programs, sitting criteria, improved management practices and technology standards, etc.
- • Enforceable quality standards that are health based for drinking water supplies and ecologically based in areas where groundwater is closely hydrologocally connected to surface water (Note: For actions under State law that are independent of any Federally authorized program, it is the State's prerogitive to determine whether to establish its own standards or to use EPA's standard)
- • Regulatory and nonregulatory authorities to control sources of contamination currently under State or local jurisdictions, e.g., permitting, sitting and zoning authorities on State and local level
- • Monitoring, data collection, and data analysis activities to determine the extent of contamination, update control strategies and assess any needed changes in order to meet the groundwater protection goal
- • Compliance and enforcement authorities given to the appropriate State and local officials through legislative or administrative processes
- • Water well programs, including private drinking water wells, covering areas such as well testing, driller certification, well construction, and plugging abandoned wells
- • Statement of how Federal, State and local resources will be used to adequately fund the program
- • Public participation activities to involve the public in the development and implementation of the program

Defining Roles Within the State and the Relationship to Federal Programs

- • Delineation of State agencies' responsibilities in the groundwater program covering areas such as planning, implementation, enforcement and coordination
- • Statement indicating how the State will or does provide local governments with authorities to address local groundwater protection issues
- • Statement of the State's role under groundwater related EPA statues, including RCRA, CERCLA SDWA, CWA, and FIFRA, e.g., EPA-approved program such as a RCRA authorization should be listed and integrated, as a part of the State's overall groundwater protection program yet continue operating as free-standing programs
- • Mechanisms for dealing with other Federal agencies that affect State groundwater programs (e.g., MOUs or other arrangements with USDA, DOI, DOD)
- • Statement indicating how the State intends to integrate water quantity and quality management
- • Corodination of groundwater programs with other relevant natural resource protection programs, including surface water managment

Source: From Protecting the Nation's groundwater: EPA's strategy for the 1990's—The final report of the EPA groundwater task force. www.epa.gov.

Table 11I.170 EPA's Groundwater Related Grants

Statutory Authority	Match	Eligible Activities	Limitations[a]	FY 91 $ Appropriation
		Clean Water Act		
106	None	General: Prevention & abatement of surface & groundwater pollution	Allotment based on extent of pollution problem, not the quality of the State program. No authorization ceiling in FY91	$81.7 million (Groundwater portion $12.2m)
		Specific: Permitting, pollution control studies, planning, surveillance & enforcement, assistance to locals, training & public information		
104 (b) (3)	None	General: Pollution prevention, reduction, & elimination programs Specific: Research, experiments, training, demonstrations, surveys, studies, investigations	Not for program operation	$16.5 million
205 (g)	None	Delegated administration of construction grants program, 402 or 404 permit program, 208(b) (4) planning program, & construction grants management for small communities		0 (Congress cut off funding)
205 (j) (1) 604 (b)	None	Develop water quality management plans	Not for implementation; 40% to regional comprehensive planning agencies	0 $16 million
205 (j) (5) 201 (g) (1) (b)	None	Develop & implement nonpoint source management programs	201(g)(1)(b): Construction grant deobligations and allotment funds available	0 (Congress cut off funding)
319 (h)	40%	Implement nonpoint source management programs	No more than 15% of total available to any one State. Financial assistance for demonstrations only (cannot be used for cost sharing programs. Limits on administration costs	$51 million
319 (i)	50%	Carry out groundwater protection activities	$150K per State	See 319(h)

(Continued)

Table 11I.170 (Continued)

Statutory Authority	Match	Eligible Activities	Limitations[a]	FY 91 $ Appropriation
		Federal Insecticide, Fungicide and Rodenticide Act		
23(a)(1)	15%	General: Implement pesticide enforcement programs		$26.8 million (Groundwater portion $5m)
		Toxic Substances Control Act		
28	25%	General: Establish & operate toxics control programs	Authorization expired in 1982. Appropriations committees should be notified before funds are used for new groundwater program	$8.1 million
		Specific: Monitoring, analysis, surveillance & general program activities (currently used for asbestos & SARA Title III activities)		
		Resource Conservation and Recovery Act		
3011	25%	General: State hazardous waste management programs Specific: Planning for hazardous waste treatment, storage & disposal facilities		$83 million
		Safe Drinking Water Act		
1443 (a)	25%	Public water system supervision; State drinking water programs	Funds available only to States with primacy	$47.5 million
1443 (b)	25%	General: Underground injection control programs Specific: Program costs, inventories, data management, technical assistance, etc	Funds available only to State with primacy	$10.5 million
		Comprehensive Environmental Response, Compensation & Liability Act		
104 (b)	10%	General: Superfund activities under core program cooperative agreements Specific: Implementation, coordination, enforcement, training, community relations, site inventory and assessment, administration of remedial activities, legal assistance relating to CERCLA implementation	Not for site-specific activities	$14 million

[a] Authorities in this matrix may be used to fund groundwater activities either in separate categorical grants or consolidated grants. Further, the scope of eligible groundwater activities varies among authorities. Regions should consult their Grants Management Office and Regional Counsel regarding these issues.

Source: From Protecting the Nation's groundwater: EPA's strategy for the 1990's—The final report of the EPA groundwater task force. www.epa.gov.

Table 11I.171 Overview of Activities Included in State Groundwater Protection Programs

Activities	Number of States Initiating	Percent of States Initiating
Groundwater Mapping/Resource Assessment	51	91
Groundwater Monitoring Program	46	82
Policy/Strategy Development	38	68
Groundwater Discharge Permit Program	26	46
Groundwater Classification System	23	41
Public Information/Education Program	23	41
Groundwater Standards Development	22	39
Data Management System	14	25
Local Management Plan Development	2	4
Specific Source Control		
Underground Injection Control Program	47	84
Septic Management	38	68
Underground Storage Tank Program	30	54
Hazardous Waste Sites	34	61
Agricultural Contamination	23	41
Landfill Sites	20	36
Surface Impoundments	14	25

Source: From U.S. Environmental Protection Agency, *National Water Quality Inventory—1986 Report to Congress*. EPA-440/4-87-008.

SECTION 11J WATER CONSERVATION

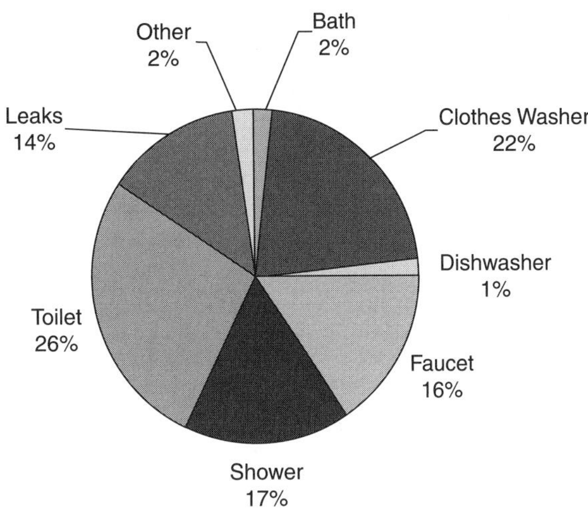

Figure 11J.132 Indoor per capita use by fixture source. (From www.usagreen.org.)

Table 11J.172 Federal Water Efficiency Standards for Plumbing Fixtures and Fixture Fittings Required by the U.S. Energy Policy Act of 1992

Product	Maximum Water Use	Compliance Date
Toilets[a]		
Gravity tank-type	1.6 gal/flush	1/1/94
Flushometer tank	1.6 gal/flush	1/1/94
Electromechanical hydraulic	1.6 gal/flush	1/1/94
Blowout[b]	3.5 gal/flush	1/1/94
Commercial gravity tank-type, white two-piece[d]	3.5 gal/flush	1/1/94 to 12/31/96
Commercial gravity tank-type, white two-piece[d]	1.6 gal/flush	1/1/97
Flushometer valve[c]	1.6 gal/flush	1/1/97
Uritials[b,c]	1.0 gal/flush	1/1/94
Showerheads[d]	2.5 gpm (80 psi)	1/1/94
Faucets[d]		
Lavatory[d]	2.5 gpm (80 psi)	1/1/94
Lavatory replacement aerators	2.5 gpm (80 psi)	1/1/94
Kitchen	2.5 gpm (80 psi)	1/1/94
Kitchen replacement aerators	2.5 gpm (80 psi)	1/1/94
Metering	0.25 gal/cycle (80 psi)	1/1/94

[a] Compliance with ASME–ANSI Standards A 112.19.2M-1990 and A112.19.6–1990.
[b] No data on conversion to lower volume.
[c] Must bear conspicuous label that states "For Commercial Use Only."
[d] Compliance with ASME–ANSI Standard A112.18.1M-1989.

Source: From cepis.ops-oms.org.

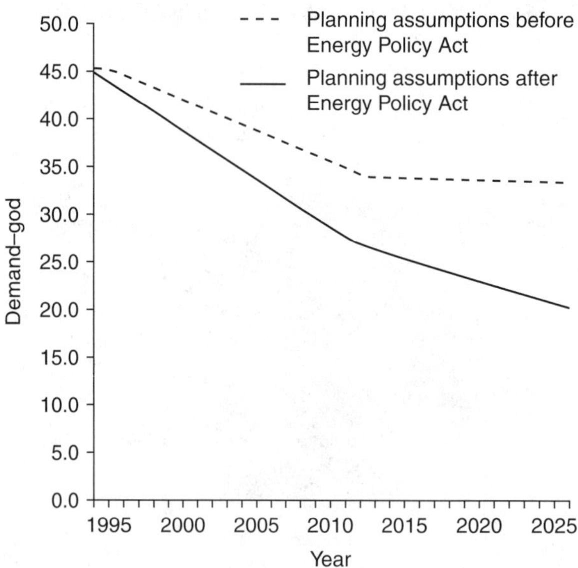

Figure 11J.133 Estimated daily residential water demand by toilets, shower heads, and faucets per person. (From www.cepis.ops-oms.org.)

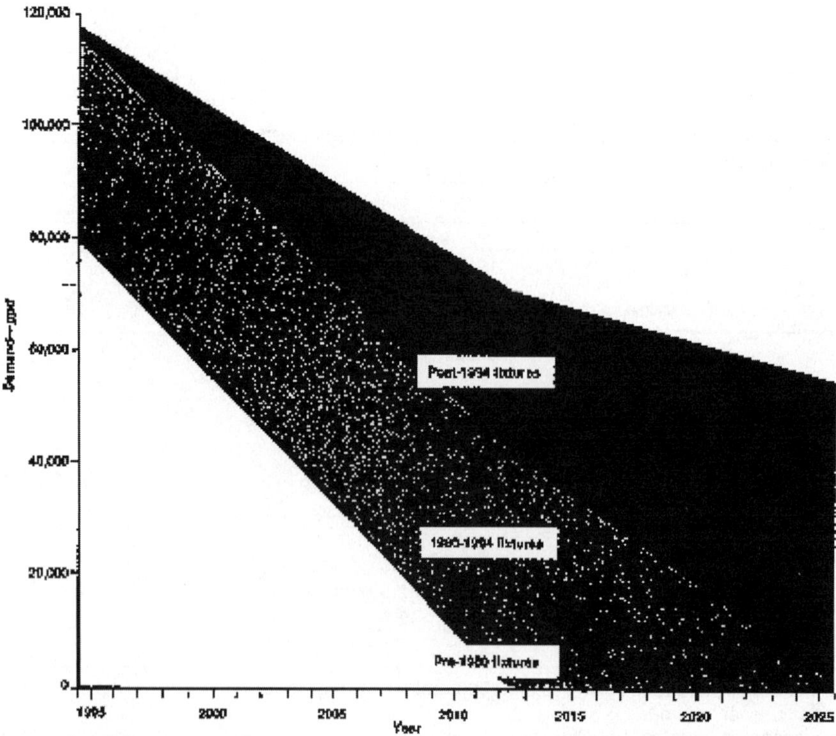

Figure 11J.134 Estimated daily residential water demand by toilets, shower heads, and faucets per 1,000 households (assuming 2.63 persons per household). (From cepis.ops-oms.org.)

Table 11J.173 Estimated Residential Water Saving with 1.6-gal/Flush Toilet, 2.5-gpm Showerhead, and 2.5-gpm Faucet Replacement

	Maximum Water Use — gpd[a]		Water Savings — gpd	
Water Use, All Fixtures	**Per Capita**	**Per 2.63-Person Household[b]**	**Per Capita**	**Per 2.63-Person Household[b]**
Post-1994	21.4	56.3		
1980–1994	33.9	91.5	12.9	34.8
Pre-1980	54.5	143.3	33.1	87.1

[a] Assumes an average of 4.0 toilet flushes, 4.8-minute showering time, and 4.0-minute faucet-use time daily per person, with adjustments for throttleback effect with showerhead and faucet use; factors based on findings derived from 1984 HUD study.
[b] Per 2.63-person household based on 1990 US census.

Source: From cepis.ops-oms.org.

Table 11J.174 Water Conservation Measures

Management	Regulations	Education
Metering	Federal and state laws and policies	Direct mail
Meter maintenance	Local codes and ordinances	Pamphlets
Pressure regulation	Plumbing codes for new structures	Leaflets
Leak detection and repair	Retrofitting resolutions	Posters
System rehabilitation	Sprinkling ordinances	Bill inserts
Pricing (conservation oriented)	Changes in landscape design	Newsletters
Marginal cost pricing	Reduction in lawn sizes	Handbooks
Seasonal peak load pricing	Increases in impervious area	Buttons
Uniform unit pricing	Planting of low-water using plants	Bumper stickers
Demand changes	Water recycling	News media
Summer surcharge	Hook-up moratorium	Radio/tv ads
Excess-use charge	Restrictions	Movie
Increasing unit	Rationing by	Radio announcements
Hook-up fees	Fixed allocations	Newspaper articles
Penalty charges	Variable percentage plan	Personal contact
Economic incentives	Per capita use	Speaker program
Rebates	Prior use bases	Slide show
Tax credits	Determination of priority uses	Booths at fairs
Subsidies	Restrictions on private and public	Customer assistance
Penalties	recreational uses	Special events
Implementing water-saving devices	Restrictions on commercial and	School talks
Devices for new construction	institutional uses	Slogan/poster contests
Shallow trap toilet	Banned wasteful practices	Posters around town
Pressurized tank toiled	Car washing	Billboards
Vacuum toilet	Pool filling	Displays
Incinerator toilet	Landscape irrigation	Reminder items
Pressurized flush toilet	Watering with hand-held hose only	Decals
Wastewater recycling toilet	Scheduled irrigation	Serving water on request in restaurants
Oil flush toilet		County fair exhibit
Composer toilet		
Dual flush toilet		
Micropore toilet		
Premixed water system		
Water recycling system		
Compressed air toilet		
Retrofit devices		
Water closet inserts		
Water dams		
Toilet flush adapters		
Flush valve toilets		
Shower mixing valves		
Shower flow-control devices		
Air-assisted showerheads		
Pressure-reducing valves		

(Continued)

Table 11J.174 (Continued)

Management	Regulations	Education
Toilet inserts		
Faucet aerators		
Faucet flow restrictors		
Spray taps		
Pressure balancing mixing valves		
Hot water pipe insulation		
Swimming pool covers		
Low water-using diswashers		
Low flush toilets		
Thermostatic mixing valves		
Minimum use showers		
Low flow shortheads		
Low water-using clothes washers		
Devices for landscape irrigation		
Moisture sensors		
Hose meters		
Sprinklers timers		
Distribution of water conservation kits		
Free distribution and installation of water-saving devices		
Distribution of leak detection kits		
Reuse of water works facility washwater		

Source: From U.S. Army Corps of Engineers, Engineer Institute for Water Resources, Analytical Bibliography for Water Supply and Conservation Techniques, Contract Report 82-C07, 1982.

Table 11J.175 Water Conservation Techniques

- Inspect the plumbing system for leaks
- Install flow control devices in showers
- Turn off all water during vacations or long periods of absence
- Check the frequency with which home water softening equipment regenerates and backwashes. It can use as much as 100 gallons of water each time it does this
- Insulate hot water pipes to avoid having to clear the "hot" line of cold water during use
- Check all faucets, inside and out, for drips. Make repairs promptly. These problems get worse—never better
- Reduce the volume of water in the toilet flush tank with a quart plastic bottle filled with water (bricks lose particles, which can damage the valve)
- Never use the toilet as a trash basket for facial tissues, etc. Each flush uses 5 to 7 gallons of water. Items carelessly thrown in could clog the sewage disposal system
- Accumulate a full laundry load before washing, or use a lower water level setting
- Take showers instead of baths
- Turn off shower water while soaping body, lathering hair, and massaging scalp
- Bottle and refrigerate water to avoid running excess water from the lines to get cold water for meals. Shake bottle before serving to incorporate air in the water so that it doesnot taste flat
- To get warm water, turn hot water on first; then add cold water as needed. This is quicker this way and saves water, too
- Wash only full loads of dishes. A dishwasher uses about 9 to 13 gallons to water per cycle
- When washing dishes by hand, use one pan of soapy water for washing and a second pan of hot water for rinsing. Rinsing in a pan requires less water than rinsing under a running faucet
- Use rinse water—"gray water"—saved from bathing or clothes washing to water indoor plants. Do not use soapy water on indoor plants. It could damage them
- Vegetables requiring more water should be grouped together in the garden to make maximum use of water applications
- Mulch shrubs and other plants to retain moisture in the soil longer
- Spread leaves, lawn clippings, chopped bark or cobs, or plastic around the plants. Mulching also controls weeds that complete with garden plants for water
- Mulches should permit water to soak into the soil
- Try "trickle" or "drip" irrigation systems in outdoor gardens. These methods use 25 to 50 percent less water than hose or sprinkler methods. The tube for the trickle system has many tiny holes to water closely spaced plants. The drip system tubing contains holes or openings at strategic places for tomatoes and other plants that are more widely spaced
- Less frequent but heavier lawn watering encourages a deeper root system to withstand dry weather better
- Plan landscaping and gardening to minimize watering requirements
- When building or remodeling, consider
  — Installing smaller than standard bath tubs to save water
  — Locating the water heater near where hottest water is needed
  — usually in the kitchen/laundry area

Source: From U.S. Environmental Protection Agency, 1988, *The Lake and Reservoir Restoration Guidance Manual*, EPA 440/5-88-002.

Table 11J.176 Means to Improve Effective Use of Water and Energy for Irrigation

Direct Means	More efficient irrigation
1. Reduce water applied	More uniform irrigation
	Better use of natural rainfall
	Reduction of losses
	Compromise yield (Optimize yield ratio)
2. Reduce operating pressures	Use different irrigation devices
	Use different irrigation practices
Indirect Means	
1. Change irrigation practices to allow	Less tillage
	Fewer secondary operations
	Spraying
	Fertilizing
	Cultivating
2. Change irrigation practices to prevent	Need for ripping
	Need for land preparation
	Loss of fertilizer and nutrients
3. Substitute irrigation for other more energy consumptive operations	Environmental control
	Sprinkling for frost protection vs. wind machines or heaters
4. Improve yields thereby reducing acreage needed	Control amount and timing of water application
	Irrigation scheduling
5. Time irrigation to better use limited energy resource	Peak use avoidance (load management). Irrigation scheduling

Source: From von Bernuth, R.D., 1980, *New Concepts in AG Sprinklers*, Copyright, The Irrigation Assoc., 1980 Technical Conference Proceedings. Reprinted with permission.

Table 11J.177 Drought Tolerant Varieties Recommended for Nebraska

Trees		**Mature Size (ft)**
Acre ginnala	Amur Maple	25
Acer platanoides	Norway Maple	50'
Carya sps.	Hickory	45'
Celtis occidentalis	Common Hackberry	60'
Cercis canadensis	Eastern Redbud	20'
Crataegus crusgalli inermis	Thornless Cockspur Hawthorn	20'
Crataegus phaenapyrum	Washington Hawthorn	25'
Fraxinus pennsylvanica	Green Ash	60'
Ginko Bilboa	Ginko	60'
Gleditsia triacanthos inermis	Thornless Common Honeylocust	60'
Gymnocladus dioicus	Kentucky Coffeetree	60'
Juniperus spp.	Junipers	25–35'
Koelreuteria paniculata	Golden Raintree	30'
Maclura pimifera	Osage-orange (thornless male)	50'
Malus cultivars	Crabapple	15–25'
Picea pungens	Colorado Spruce	50'
Picea glauca, densata	Black Hills Spruce	45'
Pinus banksiana	Jack Pine	35'–40'
Pinus nigra	Austrian Pine	50'
Pinus ponderosa	Ponderosa Pine	60'
Pinus sylvestris	Scotch Pine	60'
Pseudotsuga menziesii	Douglas Fir	50'
Quercus coccinea	Scarlet Oak	65'
Quercus imbricaria	Shingle Oak	40'
Quercus macrocarpa	Bur Oak	50'
Quercus marilandica	Blackjack Oak	25'
Quercus velutina	Black Oak	55'
Quercus bicolor	Swamp White Oak	50'
Robinia speudoacadia	Black Locust	60'
Taxodium distichum	Baldcypress	50'
Shrubs		
Amelancher alnifolia	Saskatoon Serviceberry	6'
Amorpha canescens	Leadplant Amorpha	8'–20'
Aronia meloncarpa	Black Chokeberry	10'

(Continued)

Table 11J.177 (Continued)

Berberis X mentorensis	Mentor Barberry	7'
Berberis thunbergii	Japanese Barberry	7'
Caragana arborescens	Siberian Peashrub	15–20'
Ceanothus americanus	New Jersey Tea	3'
Chaenomeles species	Flowering Quince	3'–6'
Cornus mas	Corneliancherry Dogwood	20'
Cornus racemosa	Gray Dogwood	12'
Cotinus coggygria	Smoketree	15'
Diervilla lanicera	Bush Honeysuckle	5–6'
Hamamelis virginiana	Common Witchhazel	15'
Juniperus chinensis	Juniper	10'–15'
Kolkwitzia amabilis	Beautybush	10'
Ligustrum species	Privet	15'–20'
Lonicera korolkowii	Blueleaf Honeysuckle	15'
Lonicera X, Emerald Mound	Emerald Mound Honeysuckle	2'
Physocarpus opulifolius	Common Ninebark	10'
Pine mugo	Mugo Pine	5'–15'
Prunus tomentosa	Nanking Cherry	15'
Prunus besseyi	Western Sandcherry	4'
Pyracantha coccinea	Scarlet Firethorn	6'
Rhus species	Sumac	6'
Ribes missourense	Missouri Gooseberry	3'
Ribes odoratum	Clove Currant	8'
Rosa rugosa	Rugose Rose	4'
Sambucus candensis	Elderberry	8'
Spiraea X vanhouttei	Vanhoutteis Spirea	5'
Symphoricarpos occidentalis	Western Snowberry	3'–4'
Symphoricarpos orbiculatus	Coralberry	3'
Syringa species	Lilac	10'
Viburnum lantana	Wayfaringtree	10'
Viburnum lentago	Nannyberry	25'
Viburnum prunifolium	Blackhaw Viburnum	15'
Yucca glauca	Soapweed	3'
Yucca filamentosa	Adams-Needle or Yucca	3'
Forbs (wildflowers & perennials)		
Achillea sps.	Yarrow	
Antennaria species	Pussytoes	
Artemisia sps.	Artemisia	
Aster novae-angliae	New England Aster	
Asclepias tuberosa	Butterfly Milkweed	
Baptisia sps.	False Indigo	
Coreopsis sps.	Coreopsis cultivars	
Dalea purpurea	Purple Prairie Clover	
Echinacea purpurea	Purple Coneflower	
Gaillarada species	Blanket Flower	
Hemerocallis sps.	Daylily	
Iris sps.	Iris species	
Liatris sps.	Gayfeather	
Monarda species	Beebalm	
Sedum sps.	Sedum	
Sempervirens sps.	Hens-N-Chickens	
Turfgrasses		
Buchloe dacryloides	Buffalograss	
Festuca arundinacea	Tall Fescue (turf types)	
Ornamental Grasses		
Andropogon gerardia	Big Bluestem	
Bouteloua curtipendula	Sideoats Grama	
Calamagrostis acutiflora	Feather Reedgrass	
Eragrostis tricbodes	Sand Lovegrass	
Miscanthus sps.	Eulaliagrass	
Panicum virgatum	Switchgrass	
Pennisetum sps.	Fountaingrass	
Schizachyrium scoparium	Little Bluestem	
Sorghastrum nutans	Indiangrass	

Source: From ianrpubs.unl.edu. With permission.

Table 11J.178 A Selection of Drought-Tolerant Landscape Plants

Shade Trees	
Red Maple	(Acer rubrum) (wet or dry situations, and depends on seed source)
Green Ash	(Fraxinus pennsylvanica)
Ginkgo	(Ginkgo biloba)
Tupelo	(Nyssa sylvatica)
Red Oak	(Quercus rubra)
Pin Oak	(Quercus palustris)
Zelkova	(Zelkova serrata)
Japanese Pagodatree	(Sophora japonica)
Golden Raintree	(Kolreuteria paniculata)
Japanese Tree Lilac	(Syringa reticulata)
Crapemyrtle	(Lagerstroemia indica)
Tulip Poplar	(Liriodendron tulipfera)
Evergreen Trees	
Atlas Cedar	(Cedrus atlantica)
Colorado Blue Spruce	(Picea pungens glauca)
American Holly	(Ilex opaca)
Red Cedar	(Juniperus virginiana)
Deciduous Shrubs	
Red Chokeberry	(Aronia arbutifolia)
Smoketree	(Cotinus coggyria)
Pinxterbloom Azalea	(Rhododendron nudiflorum)
Flameleaf Sumac	(Rhus copallina)
Burning Bush	(Euonymus alatus)
Fragrant Wintersweet	(Chimonanthus praecox)
Japanese Barberry	(Berberis thunbergii)
Evergreen Shrubs	
Heavenly Bamboo	(Nandina domestica)
False-Holly	(Osmanthus heterophyllus)
Mugo Pine	(Pinus mugo)
Pyracantha	(Pyracantha coccinea)
Glossy Abelia	(Abelia x grandiflora)
Junipers	(Juniperus species)
Herbaceous Plants	
May–June Blooming	
Cushion Spurge	(Euphorbia sp)
Candy tuft	(Iberis sempervirens)
Dwarf Crested Iris	(Iris cristata)
Sea Pink	(Armeria maritima)
Perennial Bachelor's Button	(Centaurea montana)
Gas Plant	(Dictamnus albus)
Day Lily	(Hemerocallis)
Beaded Iris	(Iris sp.)
Iceland Poppy	(Papaver nudicaule)
June–Sept Blooming	
Black Eyed Susan	(Rudbeckia hirta)
Yarrow	(Achillea filipendulina)
Coreopsis	(Corepsis lanceolata)
Coreopsis	(Coreopsis verticillata)
Foxtail Lily	(Eremurus x isabellinus)
Blanket Flower	(Gaillardia x grandiflora)
Perennial Baby's Breath	(Gypsophila paniculata)
Red-Hot Poker	(Kniphofia uvaria)
Moneywort	(Lysimachia nummularia)
Perennial Salvia	(Salvia x superba)
Yucca	(Yucca filamentosa)
Blue Fescue	(Festuca glauca)
Hens and Chicks	(Sempervivum tectorum)
Goldenrod	Spp. (Solidago)
Lamb's Ear	(Stachys byzantina)
Pearly Everlasting	(Anaphalis margaritacea)
Silvermound	(Artemesia schmidtiana)
Liatris	(Liatris spicata)
Sedum	(Sedum spectabile)

Note: The plants listed are some that generally survive well under dry, hot conditions that would normally harm most other plants. It must be kept in mind that even these plants will require regular watering during the first season they are planted. The frequency of watering of established landscapes will vary based on the plant species, its conditions and maturity and soil conditions of the site.

Source: From agnr.umd.edu. With permission.

Table 11J.179 Hydrozones for Urban Landscape Water Conservation

	Area with Highest Intensity of Watering	Area with Lowest Intensity of Watering
Level of human contact and activity	Most direct, active, intense	Least direct and intense; most passive
Level of visibility	Highly visible	Least visible
Maximum compatible water use	Most intense water use	Least intense water use
Priority for watering during drought	First priority	Last priority
Typical types of plants	Turf; exotic ornamentals	Native and drought-tolerant trees and shrubs
Examples on golf course	Greens	Rough or fairway
Examples in commercial area	Urban plaza; building entry area	Parking and service screens
Examples in residential community	Play area; building entry area; central common area	Buffer zones; parking screens
Examples in single-family residence	Active areas in rear yard	Side yard; privacy screens

Note: Allocation of different intensities of irrigation to different areas.

Source: From Ferguson, B.K., 1987, *Water Conservation Methods in Urban Landscape Irrigation: An Exploratory Overview, Water Resources Bull.*, vol. 23, no. 1. Copyright American Water Resources Assoc. Reprinted with permission.

Table 11J.180 Water Conserving Plants, Shrubs, and Trees in the Western United States

Plant	Mature Size	Flower Color	Water Requirements
Flowering Plants			
Callistemon citrinus Lemon bottlebrush	25' high 15' wide	O,R	S
Cassia artemisiodes Feathery cassia	3' to 5' high	Y	V
Cistus Rockrose	2' to 6' high 6' wide	V	V
Coreopsis verticillata	2–1/2' high, 2' wide	Y	C
Cytisus and Spartium Broom	3' to 8' high 4' to 6' wide	V	V
Escallonia	3' to 15'	V	V
Lantana	1' to 6' high	V	V
Nerium oleander Oleander	8' to 12' high 8' to 12' wide	V	S
Ochna serrulata Mickey Mouse plant	4' to 8' high	Y	V
Plumbago auriculata Cape plumbago	To 6' high 8' to 10' wide	B,W	V
Poinciana gilliesii Bird of paradise bush	10' high and nearly as wide	Y	V
Limonium Perezil Sea Lavender	2' to 4' high	L	V
Foliage Plants			
Arbutus unedo Strawberry tree	8' to 35' high and wide	W	V
Dodonaea viscose Hopseed bush	12' high 6' to 8' wide	—	C
Elaeagnus Prunus lyoni P. ilicifolia	Mostiy very big shrubs	—	V
P. caroliniana	6' to 40' high	W	C
Rhus ovata Sugar bush	2–1/2' to 10' high and wide	P	C
Xylosma congestum	8' to 10' high and often wider	—	V
Yucca	2' to 20' high and more	W	V
Vines			
Bougainvillea	Big mound or huge vine	V	V
Campsis trumpet creeper	Can bury a house or yard	O,R	V

(Continued)

Table 11J.180 (Continued)

Plant	Mature Size	Flower Color	Water Requirements
Solanum jasminoides	To 30’	W	S
Potato vine			
Tecomaria capensis	Bush or fence	R	V
Cape honeysuckle	Vine		
Wisteria	100’ or more	B,W	V
Trees			
Acacia	Large shrubs	Y	V
Certain species	Small trees		
Cupressus glabra	20’ to 40’ high	—	C
Eucalyptus	Shrubs or trees	V	C
Geijera parvifolia	To 30’ high, 20’ high	—	C
Heteromeles arbutifolia	Shrub to 10’ or	W	C
Toyon	25’ tree		
Olea europaea	25’ to 30’	—	V
Olive			
Palms	20’ to 100’	—	V
Rhus lancea	To 25’ spreading	—	V
Robinia (tree forms)	40’ to 70’ high	P,W	V
Locust			
Ground Cover			
Aloe	1’ to 18’ high	O,R	V
Baccharis pilularis	8" to 24" mat	—	V
Dwarf coyote bush	with 6’ spread		
Gazania	6" to 8" high	V	S
Pennisetum setaceum	2’ to 4’ clumps	G	C
Fountain grass			
Rosmarinus officinalis	Up to 3’	B	V
Prostrata			
Other			
Ceanothus	Mat, shrubs trees 2’ to 14’	B	C
Juniperus	Ground covers,	—	S
Juniper	shrubs, trees		

Note: Flower Colors: B, Blue; G, Gray; L, Lavender; O, Orange; P, Pink; R, Red; V, Varies; W, White; Y, Yellow. Water Requirements: S, Somewhat unthirsty, more than 6 waterings annually; V, Very unthirsty, 4 to 6 waterings annually; C, Completely unthirsty, annual rainfall sufficient.

Source: From Los Angeles Department of Water and Power, October 1986.

Table 11J.181 Potential Water Savings of Household Fixtures

Water User	Without Conservation	With Conservation
Toilet	5–7 gallons per flush	3–4 gall per flush British dual cycle water closets use 2.5 gall to flush solids and 1.25 gall to flush liquids One-gallon toilets have recently become available for about $150 One-half gallon toilets which are aided by compressed air are also available
Showers	6–10 gal/min	2–3 gal/min
Faucets	5 gal/min	1.5 gal/min
Dishwasher	10–14 gall per load	8.5–10 gall per load
Washing machine	47.5 gall per load	42 gall per load

Source: From New York State Senate Research Service Task Force on Critical Problems, 1986, *Water Conservation*, The Hidden Supply, Albany, NY. With permission.

Table 11J.182 Water Saving Equipment in Place of Conventional Plumbing Fixtures, Fittings, and Appliances

Fixture/Fitting/Appliance	Water Use in Gallons Per
Vintage Toilet[a]	4–6 flush
Conventional Toilet[b]	3.5 flush
Low Consumption Toilet[c]	1.6 flush
Conventional Showerhead[a]	3–10 min
Low-Flow Showerhead	2–2.5 min
Faucet Aerator[a]	3–6 min
Flow Regulating Aerator	0.5–2.5 min
Top-Loading Washer	40–55 load
Front-Loading Washer	22–25 load
Dishwasher	8–12 load

[a] Manufactured before 1978.
[b] Manufactured from 1978 to 1993.
[c] Manufactured since January 1, 1994.

Source: From usangreen.org.

Table 11J.183 Annual Utility Electric Energy Demands Associated with Water Used by Residential Plumbing Fixtures[a]

Fixture	Per Capita kWh[b]	Per Household[c] kWh[b]	Per 1,000 Households kWh[b]
Post-1994 fixtures	22	59	59,463
1980–1994 fixtures	35	94	94,206
Pre-1980 fixtures	57	151	151,776

[a] Toilets, showerheads, and faucets.
[b] Combined average energy use for water treatment (1,500 kWh/mgd) and wastewater treatment (1,400 kWh/mgd)
[c] Per 2.63-person household based on 1990 US census.

Source: From cepis.ops-oms.org.

Table 11J.184 Annual Utility Emissions of Carbon Dioxide, Nitrogen Oxides, and Sulfur Dioxide Associated with Energy Demand Created by Plumbing Fixture Water Use

	Utility		
Fixture	Per Capita lb/kWh[a]	Per Household lb/kWh[a]	Per 1,000 Households lb/kWh[a]
Post-1994 fixtures	43.4	114.1	114,088
1980–1994 fixtures	68.7	180.8	180,747
Pre-1980 fixtures	110.7	291.2	291,203

Note: For water use by toilets, showerheads, and faucets.

[a] Emissions per kWh — 1.89 lb carbon dioxide, 0.00914 lb nitrogen oxides, and 0.0195 lb sulfur dioxide, based on total electric energy demands.

Source: From cepis.ops-oms.org.

CHAPTER 12

Agencies and Organizations

Pedro Fierro, Jr.

CONTENTS

SECTION 12A FEDERAL AGENCIES

Table 12A.1 Federal Agencies Concerned with Water-Related Matters

Federal Citizen Information Center (http://www.firstgov.gov)
Phone: (800) 333-4636
Since 1970, the Federal Citizen Information Center (FCIC) has provided information related to government services. Serves as the official gateway to all government information

Bureau of Land Management (http://www.blm.gov)
Office of Public Affairs, 1849 C Street, Room 406-LS, Washington, DC 20240, Phone: (202) 452-5125

Bureau of Reclamation (http://www.usbr.gov)
Commissioner, Bureau of Reclamation, 1849 C Street NW, Washington DC 20240-0001, Phone: (202) 513-0501

Council on Environmental Quality (http://www.whitehouse.gov/ceq)
722 Jackson Place, N.W., Washington, DC 20503, Phone: (202) 395-5750

Extension Service, Department of Agriculture (reeusda.gov)
1400 Independence Ave., SW, Washington, DC 20250-0003, Phone: (202) 720-7441

Federal Emergency Management Agency (http://www.fema.gov)
500 C Street, SW, Washington, DC 20472, Phone: (202) 566-1600

Federal Highway Administration (FHWA) (http://www.fhwa.dot.gov)
Federal Highway Administration, 400 Seventh Street, SW, Washington, DC 20590, FHWA Personnel Locator (202) 366-0537

Forest Service, U.S. Department of Agriculture (http://www.fs.fed.us/)
USDA Forest Service, 1400 Independence Ave., SW, Washington, DC 20250-0003, Phone: (202) 205-8333

Minerals Management Service (http://www.mms.gov)
Chief of Public Affairs, 1849 C Street, N. W., Washington, DC 20240, Phone: (202) 208-3985

National Geodetic Survey (NGS) (http://www.ngs.noaa.gov)
NGS Information Services, NOAA, N/NGS12, National Geodetic Survey, SSMC-3, #9202, 1315 East-West Highway, Silver Spring, MD 20910-3282, Phone: (301) 713-3242, Fax: (301) 713-4172

National Marine Fisheries Service (http://www.nmfs.noaa.gov)
NOAA Fisheries, 1315 East West Highway, SSMC3, Silver Spring, MD 20910, Phone: (301) 713-2239

National Oceanic and Atmospheric Administration (NOAA) (http://www.noaa.gov)
14th Street & Constitution Avenue, NW, Room 6217, Washington, DC 20230, Phone: (202) 482-6090

National Park Service (http://www.nps.gov)
1849 C Street NW, Washington, DC 20240, Phone: (202) 208-6843

Natural Resources Conservation Service (NRCS) (http://www.nrcs.usda.gov)
National Headquarters, Postal Mail, Natural Resources Conservation Service, Attn: Conservation Communications Staff, P.O. Box 2890, Washington, DC 20013

Street Address
Natural Resources Conservation Service, 14th and Independence Avenue, SW, Washington, DC 20250

National Science Foundation (http://www.nsf.gov)
4201 Wilson Boulevard, Arlington, VA 22230, Phone: (703) 292-5111

National Sea Grant College Program (http://www.nsgo.seagrant.org)
NOAA/Sea Grant, R/SG, 1315 East-West Highway, SSMC-3, Eleventh Floor, Silver Spring, MD 20910, Phone: (301) 713-2448

National Weather Service (http://www.nws.noaa.gov)
1325 East West Highway, Silver Spring, MD 20910, Phone: (301) 713-4000

National Wetlands Inventory (enterprise.nwi.fws.gov)
National Wetlands Inventory Center, 9720 Executive Center Drive North, Monroe Building, Suite 101, Saint Petersburg, FL 33702, Phone: (727) 570-5400

Ocean Pollution Data and Information Network (http://www.nodc.noaa.gov)
NOAA/NESDIS E/OC, SSMC3, 4th Floor, 1315 East-West Highway, Silver Spring, MD 20910-3282, Phone: (301) 713-3277

(Continued)

Table 12A.1 (Continued)

Office of Ocean and Coastal Resource Management (http://www.ocrm.nos.noaa.gov)
N/ORM 10th Floor SSMC4, 1305 East-West Highway, Silver Spring, MD 20910, Phone: (301) 713-3155

Office of Surface Mining (http://www.osmre.gov)
1951 Constitution Ave. N.W., Washington, DC 20240, Phone: (202) 208-2719

Saint Lawrence Seaway Development Corporation (http://www.seaway.dot.gov)
400 7th Street, S.W., Room 5424, Washington, DC 20590, Phone: (202) 366-0091

Soil Conservation Service (http://www.nrcs.usda.gov)
14th and Independence Ave, SW, Room 5006-S, Washington, DC 20250, Phone: (202) 720-4630

Tennessee Valley Authority (http://www.tva.gov)
400 W. Summit Hill Dr., Knoxville, TN 37902-1499, Phone: (865) 632-2101

UNIDATA Integrated Earth Information Server (atm.geo.nsf.gov)
UCAR Office of Programs, Unidata Program Center, P.O. Box 3000, Boulder, CO 80307-3000, Phone: (303) 497-8643

U.S. Army Corps of Engineers (USACE) (see separate listing)

U.S. Bureau of Land Management (USBLM) (http://www.blm.gov)
Bureau of Land Management - Eastern States, State Director's Office; Mail Stop ES-910, 7450 Boston Boulevard, Springfield, VA 22153-3121, Phone: (703) 440-1703, Fax: (703) 440-1701

U.S. Bureau of Reclamation (USBR) (http://www.usbr.gov)
Trudy Harlow, Chief, Public Affairs, Bureau of Reclamation, 1849 C Street NW, Washington DC 20240-0001, Phone: (202) 513-0575

U.S. Coast Guard (http://www.uscg.mil)
Commandant, U.S. Coast Guard, 2100 Second Street, SW, Washington, DC 20593, Phone: (202) 267-1587

U.S. Department of Agriculture — (http:www.ks.nrcs.usda.gov/)
Natural Resources Conservation Service, USDA, NRCS, 760 South Broadway, Salina, KS 67401, Phone: (785) 823-4500, Fax: (785) 823-4540

U.S. Department of Energy (http://www.energy.gov)
1000 Independence Ave., SW, Washington, DC 20585, Phone: (800) 342-5363

U.S. Department of the Interior (USDOI) (http://www.doi.gov)
Department of the Interior, 1849 C Street, N.W., Washington, DC 20240, Phone: Phone: (202) 208-3100

U.S. Environmental Protection Agency (see separate listing)

U.S. Fish and Wildlife Service (http://www.fws.gov)
1849 C Street. N.W., Number 3256, Washington, DC 20240, (202) 208-4717

U.S. Geological Survey Water Resources Information (see separate listing)

U.S. Nuclear Regulatory Agency (USNRC) (http:www.nrc.gov)
U.S. Nuclear Regulatory Commission, Office of Public Affairs (OPA), Washington, DC 20555, Toll-free: (800) 368-5642, Local: (301) 415-8200

U.S. Office of Surface Mining (USOSM) (http://www.osmre.gov)
Office of Surface Mining, 1951 Constitution Ave. N.W., Washington, DC 20240, Phone: (202) 208-2719

Source: From www.epa.gov.

Table 12A.2 U.S. Army Corps of Engineers

The U.S. Army Corps of Engineers is the Federal government's largest water resources development and management agency. The Corps began its water resources (Civil Works) program in 1824 when Congress for the first time appropriated money for improving navigation. Since then, the Corps has been involved in improving river navigation and harbors, reducing flood damage and controlling beach erosion. At projects designed for these missions, the Corps also generates hydropower; supplies water for cities, industry and agriculture; and manages a recreation and natural resources program. The Corps also has a legal mandate to regulate development by others in the Nation's waterways and wet lands, and at sea and lakeshores.

The Civil Works Program is serviced by 40 district offices, along with the divisions that supervise the districts, and the Washington Headquarters (202-761-1878).

Headquarters U.S. Army Corps of Engineers (http://www.hq.usace.army.mil) 441 G St., N.W., Washington, DC 20314-1000, Phone: (202) 761-0000 Fax: (202) 761-1683

U.S. Army Engr. Div., Great Lakes & Ohio River, Great Lakes *Regional HQ* (http://www.usace.army.mil/ncd) 111 N. Canal St., Suite 1200, Chicago IL 60606-7205, Phone: (312) 353-6310 Fax: (312) 353-5233

Ohio River *Regional HQ* (usace.mil/ceord.html) P.O. Box 1159 [*550 Main St., Rm. 10001*] Cincinnati, OH 45201-1159, Phone: (513) 684-3002, Fax: (513) 684-2085

U.S. Army Engineer District, Buffalo (www.ncb.usace.army.mil) 1776 Niagara Street, Buffalo, NY 14207-3199, Phone: (716) 879-4200 Fax: (716) 879-4195

U.S. Army Engineer District, Chicago (http://www.usace.army.mil/ncc) 111 N. Canal Street, Suite 600, Chicago, IL 60606-7206, Phone: (312) 353-6400 Fax: (312) 353-2525

U.S. Army Engineer District, Detroit (sparky.nce.usace.army.mil) P.O. Box 1027 [*477 Michigan Ave.*] Detroit, MI 48231-1027 [*48226*], Phone: (313) 226-6762, Fax: (313) 226-6009

U.S. Army Engineer District, Europe USAEDE, CMR 410, Box 1, APO AE 09096, [*Konrad Adenauer Ring 31, Box 1 65187 Wiesbaden, Germany*], Phone: 011-49-611-816-2700 DSN 703-695-0441, Fax: 011-49-611-816-2705

U.S. Army Engineer District, New England 696 Virginia Road, Concord, MA 01742-2751, Phone: (978) 318-8220, Fax: (978) 318-8821

U.S. Army Engineer District, Huntington (155.80.20.141/) 502 8th Street, Huntington, WV 25701-2070, Phone: (304) 529-5395 Fax: (304) 529-5591

U.S. Army Engineer District, Louisville (http://www.orn.usace.army.mil/) P.O. Bx 59 [*600 Dr. Martin Luther King Pl.*] Louisville, KY 40201-0059 [*40202*], Phone: (502) 315-6102 Fax: (502) 315-6771

U.S. Army Engineer District, Nashville (http://www.orn.usace.army.mil) P.O. Box 1070 [*110 Ninth Ave S.*] Nashville, TN 37202-1070 [*37203-3863*], Phone: (615) 736-5626 Fax: (615) 736-2052

U.S. Army Engineer District, Pittsburgh (orp-wc.usace.army.mil) Wm. S. Moorhead Federal Bldg. 1000 Liberty Avenue, Rm. 1828, Pittsburgh, PA 15222-4186, Phone: (412) 395-7103 Fax: (412) 644-4093

U.S. Army Engr. Division, Mississippi Valley (lmv.usace.army.mil) P.O. Box 80 [*1400 Walnut St.*], Vicksburg, MS 39180-0080 [*39181*], Phone: (601) 634-5750 Fax (601) 634-5666

U.S. Army Engineer District, Memphis (lmv.usace.army.mil/lmm.html) 167 N. Main St., Rm. B-202, Memphis, TN 38103-1894, Phone: (901) 544-3221, Fax: (901) 544-3628

U.S. Army Engineer District, Omaha (mro.usace.army.mil) 215 N.17th Street, Omaha, NE 68102-4978, Phone: (402) 221-3900, Fax: (402) 221-3128

U.S. Army Engineer District, Portland (npd.usace.army.mil/cenpp.html) P.O. Box 2946, [*333 S.W. First Ave.*], Portland, OR 97208-2946 [*97204-3495*], Phone: (503) 808-4500, Fax: (503) 808-4505

U.S. Army Engineer District, New Orleans P.O. Box 60267, [*7400 Leake Ave.*] New Orleans, LA 70160-0267 [*70118*], Phone: (504) 862-2204, Fax: (504) 862-1259

U.S. Army Engineer District, Rock Island (ncr.usace.army.mil) Clock Tower Bldg., P.O. Box 2004, Rock Island, IL 61204-2004, Phone: (309) 794-5224, Fax: (309) 794-5181

U.S. Army Engineer District, St. Louis (lms61.lms.usace.army.mil) 1222 Spruce St., St. Louis, MO 63103-2833, Phone: (314) 331-8010 Fax: (314) 331-8770

U.S. Army Engineer District, St. Paul 190 Fifth St. East, St. Paul, MN 55101-1638, Phone: (651) 290-5300, Fax: (651) 290-5478

U.S. Army Engineer District, Vicksburg 4155 Clay St., Vicksburg, MS 39183-3435, Phone: (601) 631-5010 Fax: (601) 631-5296

U.S. Army Engineer Division, North Atlantic (http://www.usace.mil/cenad.html) 302 General Lee Ave., Fort Hamilton, NY 11252-6700, Phone: (718) 765-7000

U.S. Army Engineer District, Baltimore P.O. Box 1715, [*10 S Howard St Rm 11000*], Baltimore, MD 21201-1715 [*21201*], Phone: (410) 962-4545, Fax: (410) 962-7516

U.S. Army Engineer District, Japan USAED-J, APO AP 96343-0061, [*Bg. 250, Camp Zama, Zama-shi, Kanagawa-ken 228, Japan*], Phone: 011-81-311-763-3025 (DSN 315-263-3025), Fax: 011-81-311-763-4887

(Continued)

Table 12A.2 (Continued)

U.S. Army Engineer District, New York
26 Federal Plaza, New York, NY 10278-0090, Phone: (212) 264-0100, Fax: (212) 264-5490

U.S. Army Engineer District, Norfolk
(155.78.30.111)
Waterfield Bg., 803 Front Street, Norfolk, VA 23510-1096, Phone: (757) 441-7601, Fax: (757) 441-7115

U.S. Army Engineer District, Philadelphia
(nap-wc.usace.army.mil)
Wannamaker Building, 100 Penn Square East, Philadelphia, PA 19107-3390, Phone: (215) 656-6501, Fax: (215) 656-6899

U.S. Army Engineer Division, Northwestern
(mrd.usace.army.mil)
Missouri River Regional HQ, 12565 West Center Rd., Omaha, NE 68144, Phone: (402) 697-2400, Fax: (402) 697-2720

North Pacific Regional HQ
P.O. Box 2870, [*220 N.W. Eighth Ave.*], Portland, OR 97208-2870 [*97209*], Phone: (503) 808-3700, Fax: (503) 808-3706

U.S. Army Engineer District, Kansas City
(mrk-wc.usace.army.mil)
700 Federal Building, 601 E. 12th Street, Kansas City, MO 64106-2896, Phone: (816) 983-3206

U.S. Army Engineer District, Albuquerque
(swa-wc.usace.army.mil)
4101 Jefferson Plaza, NE, Albuquerque, NM 87109-3435, Phone: (505) 342-3432, Fax: (505) 342-3199

U.S. Army Engineer District, Los Angeles P.O. Box 532711, [*911 Wilshire Blvd.*], Los Angeles, CA 90053-2325 [*90017*], Phone: (213) 452-3967, Fax: (213) 452-4214

U.S. Army Engineer District, Sacramento
(http://www.usace.mil/cespk.html)
1325 J Street, Sacramento, CA 95814-2922, Phone: (916) 557-7490, Fax: (916) 557-7859

U.S. Army Engineer District, Seattle
(nps.usace.army.mil)
P.O. Box 3755 [*4735 E. Marginal Way S.*], Seattle, WA 98124-2255, [*98134-2385*], Phone: (206) 764-3690, Fax: (206) 764-6544

U.S. Army Engineer District, Walla Walla
(npw.usace.army.mil)
201 N. Third Avenue, Walla Walla, WA 99362-1876, Phone: (509) 527-7700, Fax: (509) 527-7804

U.S. Army Engineer Division, Pacific Ocean
(http://www.pod.usace.army.mil)
Building 525, Fort Shafter, HI 96858-5440, Phone: (808) 438-1500, Fax: (808) 438-8387

U.S. Army Engineer District, Alaska
(usace.army.mil/alaska)
P.O. Box 898, Anchorage, AK 99506-0898, [*2201 3rd St.; Elmendorf AFB, AK 99506*], Phone: (907) 753-2504, Fax: (907) 753-5610

U.S. Army Engineer District, Far East
[*Korea*]
Far East Unit #15546, APO AP 96205-0610, Phone: 011-82-2-270-7300 (DSN 351-721-7300), Fax: 011-82-2-265-8440

U.S. Army Engineer District, Honolulu
Building 230, Fort Shafter, HI 96858-5440, Phone: (808) 438-1069, Fax: (808) 438-8351

U.S. Army Engineer District, Tulsa
1645 S. 101-East Avenue, Tulsa, OK 74128-4609, Phone: (918) 669-7201, Fax: (918) 669-7207

Centers

U.S. Army Engr. Support Center, Huntsville
(http://www.hnd.usace.army.mil)
P.O. Box 1600 [*4820 University Square*], Huntsville, AL 35807-4301 [*35816-1822*], Phone: (205) 895-1300, Fax: (205) 895-1910

U.S. Army Engr. Transatlantic Programs Center
(usace.army.mil/hypertext)
P.O. Box 2250, [*201 Prince Frederick Dr.*], Winchester, VA 22604-1450 [*22602*], Phone: (540) 665-4073, Fax: (540) 665-4023

U.S. Army Engineer Division, South Atlantic
(usace.army.mil/sad)
60 Forsyth Street, SW, Rm. 9M15, Atlanta, GA 30303-8801, Phone: (404) 562-5003, Fax: (404) 562-5002

U.S. Army Engineer District, Charleston
P.O. Box 919 [*4600 Goer Dr*], Charleston, SC 29402-0919, Phone: (843) 329-8000

U.S. Army Engineer District, Jacksonville
(http://www.saj.usace.army.mil)
P.O. Box 4970, [*400 West Bay St.*], Jacksonville, FL 32232-0019 [*33202*], Phone: (904) 232-2241, Fax: (904) 232-1213

U.S. Army Engineer District, Mobile
(sam.usace.army.mil)
P.O. Box 2288, [*109 Saint Joseph St.*], Mobile, AL 36628-0001, [*36602-3630*], Phone: (334) 690-2511, Fax: (334) 690-2525

U.S. Army Engineer District, Savannah
(http://mac.sas.usace.army.mil)
P.O. Box 889, [*100 W. Oglethorpe Ave.*], Savannah, GA 31402-0889 [*31401-3640*], Phone: (912) 652-5226, Fax: (912) 652-5222

U.S. Army Engineer District, Wilmington
P.O. Box 1890, [*69 Darlington Ave.*], Wilmington, NC 28402-1890 [*28403*], Phone: (910) 251-4501, Fax: (910) 251-4185

U.S. Army Engineer Division, South Pacific
(usace.mil/cespd.html)
333 Market Street, Suite 1101, San Francisco, CA 94105-2195, Phone: (415) 977-8001, Fax: (415) 977-8316

Field Operating Agencies

U.S. Army Corps of Engrs. Finance Center
5720 Integrity Drive, Millington, TN 38054-5005, Phone: (901) 874-8410, Fax: (901) 874-8622

U.S. Army Humphreys
Engr. Ctr. Spt. Activity, 7701 Telegraph Rd., Alexandria, VA 22310-3860, Phone: (703) 428-6169, Fax: (703) 428-6188

(Continued)

Table 12A.2 (Continued)

U.S. Army Engineer District, San Francisco
333 Market Street, Rm. 923, San Francisco, CA 94105, Phone: (415) 977-8600, Fax: (415) 977-8316

U.S. Army Engineer Division, Southwestern
(usace.mil/ceswd.html)

U.S. Army Engineer District, Fort Worth
(usace.army.mil/swf)
P.O. Box 17300 [*819 Taylor St., Rm. 3A24*], Fort Worth, TX 76102-0300, Phone: (817) 978-2300, Fax: (817) 978-3311

U.S. Army Engineer District, Galveston
(usace.army.mil/swg)
P.O. Box 1229, [*2000 Fort Point Road*], Galveston, TX 77553-1229 [*77550*], Phone: (409) 766-3001, Fax: (409) 766-3951

U.S. Army Engineer District, Little Rock
P.O. Box 867 [*700 W. Capitol St.*], Little Rock, AR 72203-0867 [*72201*], Phone: (501) 324-5531, Fax: (501) 324-6968

Laboratories

U.S. Army Engineer Research & Development Center, 3903 Halls Ferry Road, Vicksburg, MS 39180-6199, Phone: (601) 634-2513, Fax: (601) 634-2388

U.S. Army, Cold Regions
Research & Eng. Lab, 72 Lyme Rd., Hanover, NH 03755-1290, Phone: (603) 646-4200, Fax: (603) 646-4448

U.S.Army, Construction
Engr'g Research Lab., P.O. Box 9005, [*2902 Newmark Dr.*], Champaign, IL 61826-9005 [*61821-1076*], Phone: (217) 373-7203, Fax: (217) 373-6776

U.S. Army, Topographic Engineering Center
7701 Telegraph Rd., Bldg. 2592, Rm. L-1A, Alexandria, VA 22315-3864, Phone: (703) 428-6602, Fax: (703) 428-8154

USACE
Installation Support Center, 7701 Telegraph Rd., Alexandria, VA 22315-3862, Phone: (703) 428-6933, Fax (703) 428-2805

U.S. Army
Institute for Water Resources, Casey Bldg., 7701 Telegraph Rd., Alexandria, VA 22315-3868, Phone: (703) 428-8250, Fax: (703) 428-8171

U.S. Army
Marine Design Center, Wannamaker Bldg. Rm. 630 South, 100 Penn Square East, Philadelphia, PA 19107-3390, Phone: (215) 656-6850, Fax: (215) 656-6868

249th Engineer Battalion, (Prime Power), Fort Belvoir, VA 22050, Phone: (703) 805-2656

Source: From www.usace.army.mil.

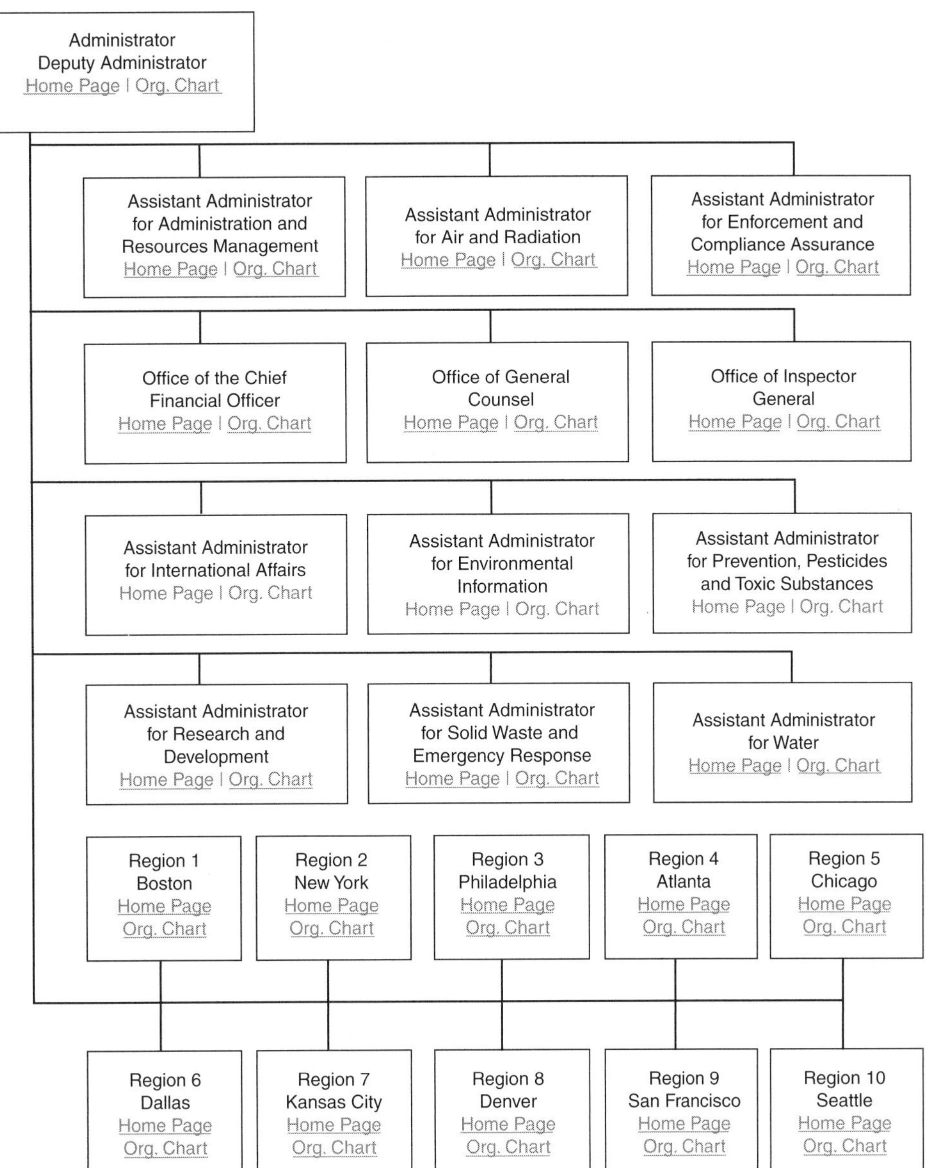

Figure 12A.1 Organization of the U.S. Environmental Protection Agency (www.epa.gov).

Table 12A.3 Organization of the U.S. Environmental Protection Agency Home Page Addresses

Administrator
Deputy Administrator
http://www.epa.gov/adminweb/

Assistant Administrator for Administration and Resource Management
http://www.epa.gov/oarmweb1/index.htm

Assistant Administrator for Air and Radiation
http://www.epa.gov/oar/

Assistant Administrator for Enforcement and Compliance Assurance
http://www.epa.gov/compliance/

Office of the Chief Financial Officer
http://www.epa.gov/ocfopage/

Office of General Counsel
http://www.epa.gov/ogc/

Office of Inspector General
http://www.epa.gov/oigearth/

Assistant Administrator for International Affairs
http://www.epa.gov/oia/

Assistant Administrator for Environmental Information
http://www.epa.gov/oei/

Assistant Administrator for Prevention, Pesticides, and Toxic Substances
http://www.epa.gov/oppts/

Assistant Administrator for Research and Development
http://www.epa.gov/ORD/

Assistant Administrator for Solid Waste and Emergency Response
http://www.epa.gov/swerrims/

Assistant Administrator for Water
http://www.epa.gov/OW/

Region 1 — Boston
http://www.epa.gov/region1/

Region 2 — New York
http://www.epa.gov/region2/

Region 3 — Philadelphia
http://www.epa.gov/region03/index.htm

Region 4 — Atlanta
http://www.epa.gov/region4/

Region 5 — Chicago
http://www.epa.gov/region5/
Region 6 — Dallas
http://www.epa.gov/region6/

Region 7 — Kansas City
http://www.epa.gov/region7/

Region 8 — Denver
http://www.epa.gov/region7/

Region 9 — San Francisco
http://www.epa.gov/region9/

Region 10 — Seattle
http://www.epa.gov/region10/

Source: From www.epa.gov.

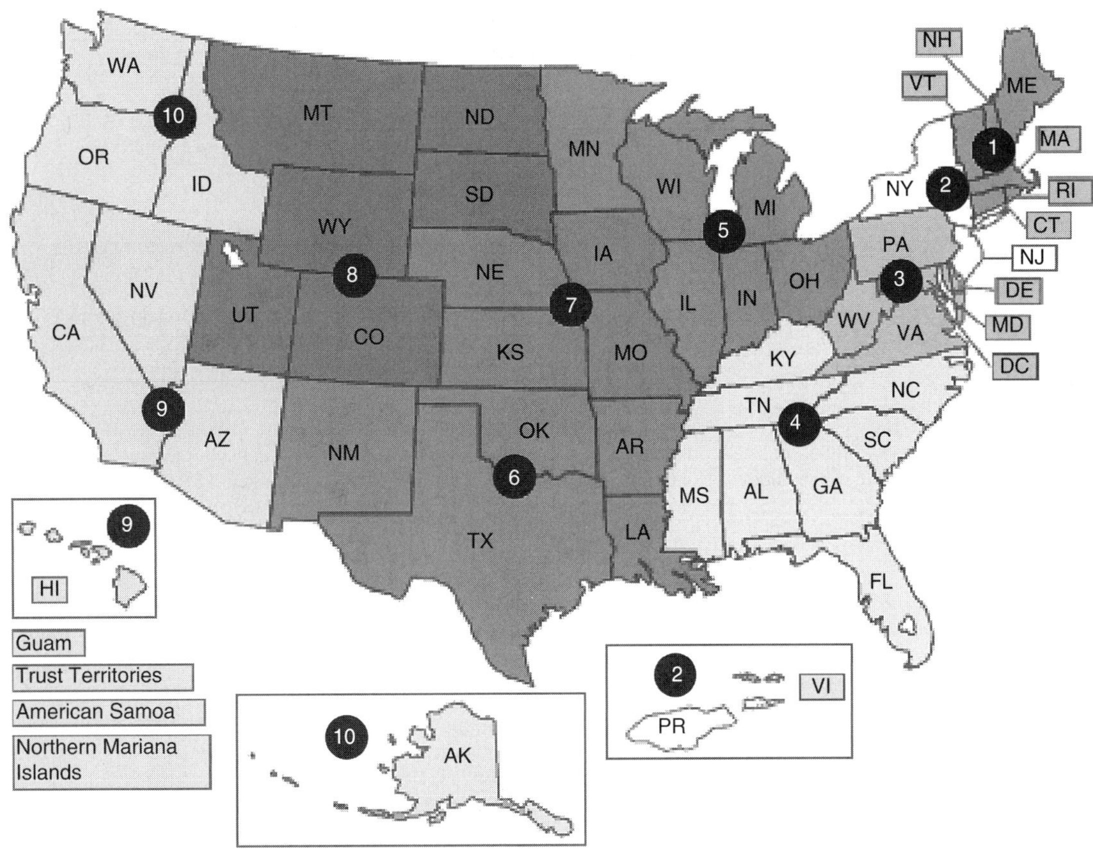

REGIONS		REGIONS		REGIONS	
Alabama	4	Maine	1	Pennsylvania	3
Alaska	10	Maryland	3	Rhode Island	1
Arizona	9	Massachusetts	1	South Carolina	4
Arkansas	6	Michigan	5	South Dakota	8
California	9	Minnesota	5	Tennessee	4
Colorado	8	Mississippi	4	Texas	6
Connecticut	1	Missouri	7	Utah	8
Delaware	3	Montana	8	Vermont	1
District of Columbia	3	Nebraska	7	Virginia	3
Florida	4	Nevada	9	Washington	10
Georgia	4	New Hampshire	1	West Virginia	3
Hawaii	9	New Jersey	2	Wisconsin	5
Idaho	10	New Mexico	6	Wyoming	8
Illinois	5	New York	2	American Samoa	9
Indiana	5	North Carolina	4	Guam	9
Iowa	7	North Dakota	8	Puerto Rico	2
Kansas	7	Ohio	5	Virgin Islands	2
Kentucky	4	Oklahoma	6		
Louisiana	6	Oregon	10		

Figure 12A.2 States covered by U.S. Environmental Protection Agency regional offices (www.epa.gov).

Table 12A.4 Environmental Protection Agency Headquarters, Regional Offices, and Laboratories

EPA Headquarters

Standard Mailing Address

Environmental Protection Agency, Ariel Rios Building, 1200 Pennsylvania Avenue, N.W., Washington, DC 20460, Phone: (202) 272-0167

Overnight Package Delivery Mailing Addresses

Environmental Protection Agency, Ariel Rios Building, 1200 Pennsylvania Avenue N.W., Washington, DC 20004

Environmental Protection Agency, EPA West Building, 1301 Constitution Avenue N.W., Washington, DC 20004

Environmental Protection Agency, EPA East Building, 1201 Constitution Avenue N.W., Washington, DC 20004

Environmental Protection Agency, Ronald Reagan Building, 1300 Pennsylvania Avenue N.W., Washington, DC 20004

EPA Regional Offices

Region 1 (CT, MA, ME, NH, RI, VT)

Environmental Protection Agency, 1 Congress St. Suite 1100, Boston, MA 02114-2023, http://www.epa.gov/region01/, Phone: (617) 918-1111, Fax: (617) 565-3660, Toll free within Region 1: (888) 372-7341

Region 2 (NJ, NY, PR, VI)

Environmental Protection Agency, 290 Broadway, New York, NY 10007-1866, http://www.epa.gov/region02/, Phone: (212) 637-3000, Fax: (212) 637-3526

Region 3 (DC, DE, MD, PA, VA, WV)

Environmental Protection Agency, 1650 Arch Street, Philadelphia, PA 19103-2029, http://www.epa.gov/region03/, Phone: (215) 814-5000, Fax: (215) 814-5103, Toll free: (800) 438-2474, Email: r3public@epa.gov

Region 4 (AL, FL, GA, KY, MS, NC, SC, TN)

Environmental Protection Agency, Atlanta Federal Center, 61 Forsyth Street, SW, Atlanta, GA 30303-3104, http://www.epa.gov/region04/, Phone: (404) 562-9900, Fax: (404) 562-8174, Toll free: (800) 241-1754

Region 5 (IL, IN, MI, MN, OH, WI)

Environmental Protection Agency, 77 West Jackson Boulevard, Chicago, IL 60604-3507, http://www.epa.gov/region5/, Phone: (312) 353-2000, Fax: (312) 353-4135, Toll free within Region 5: (800) 621-8431

Region 6 (AR, LA, NM, OK, TX)

Environmental Protection Agency, Fountain Place 12th Floor, Suite 1200, 1445 Ross Avenue, Dallas, TX 75202-2733, http://www.epa.gov/region06/, Phone: (214) 665-2200, Fax: (214) 665-7113

Region 7 (IA, KS, MO, NE)

Environmental Protection Agency, 901 North 5th Street, Kansas City, KS 66101, http://www.epa.gov/region07/, Phone: (913) 551-7003, Toll free: (800) 223-0425

Region 8 (CO, MT, ND, SD, UT, WY)

Environmental Protection Agency, 999 18th Street Suite 500, Denver, CO 80202-2466, http://www.epa.gov/region08/, Phone: (303) 312-6312, Fax: (303) 312-6339, Toll free: (800) 227-8917, Email: r8eisc@epa.gov

Region 9 (AZ, CA, HI, NV)

Environmental Protection Agency, 75 Hawthorne Street, San Francisco, CA 94105, http://www.epa.gov/region09/, Phone: (415) 947-8000, (866) EPA-WEST (toll free in Region 9), Fax: (415) 947-3553, Email: r9.info@epa.gov

Region 10 (AK, ID, OR, WA)

Environmental Protection Agency, 1200 Sixth Avenue, Seattle, WA 98101, http:www.epa.gov/region10/, Phone: (206) 553-1200, Fax: (206) 553-0149, Toll free: (800) 424-4372

Other Locations

U.S. EPA, 26 Martin Luther King Drive, Cincinnati, OH 45268

U.S. EPA, Research Triangle Park, NC 27711, http://www.epa.gov/rtp/

Satellite Locations and Laboratories

U.S. EPA National Air and Radiation Environmental Laboratory (NAREL), 540 South Morris Avenue, Montgomery, AL 36115-2601, http://www.epa.gov/narel/, Phone: (334) 270-3400, Fax: (334) 270-3454

U.S. EPA National Enforcement Investigations Center Laboratory, Box 25277, Bldg. 53, Denver Federal Center, Denver, CO 80225, http://www.epa.gov/compliance/criminal/forensics/laboratory/index.html, Phone: (303) 236-5132

U.S. EPA National Exposure Research Laboratory (NERL), Mail Code: D305-01, Research Triangle Park, NC 27711, http://www.epa.gov/nerl/, Phone: (919) 541-2106, Fax: (919) 541-0445

U.S. EPA National Health and Environmental Effects Research Laboratory (NHEERL), Mid-Continent Ecology Division, 6201 Congden Boulevard, Duluth, MN 55804, http://www.epa.gov/med/, Fax: (218) 720-5703

U.S. EPA National Health and Environmental Effects Research Laboratory (NHEERL), Western Ecology Division, 200 SW 35th Street, Corvallis, OR 97333, http://www.epa.gov/wed/, Phone: (541) 754-4600, Fax: (541) 754-4799

National Risk Management Research Laboratory, 26 Martin Luther King Drive, Cincinnati, Ohio 45268, http://www.epa.gov/ordntrnt/ORD/NRMRL/, Fax: (513) 569-7680

(Continued)

Table 12A.4 (Continued)

U.S. EPA National Exposure Research Laboratory (NERL), Ecosystems Research Division, 960 College Station Road, Athens, GA 30605-2700, http://www.epa.gov/AthensR/, Phone: (706) 355-8005

U.S. EPA National Exposure Research Laboratory (NERL), Environmental Sciences Division, P.O. Box 93478, Las Vegas, NV 89193-3478, http://www.epa.gov/nerlesd1/, Phone: (702) 798-2100, Fax: (702) 798-2637

U.S. EPA National Health and Environmental Effects Research Laboratory (NHEERL), Mail Code: B305-01, Research Triangle Park, NC 27711, http://www.epa.gov/nheerl/, Phone: (919) 541-2281, Fax: (919) 541-4324

U.S. EPA National Health and Environmental Effects Research Laboratory (NHEERL), Atlantic Ecology Division, 27 Tarzwell Drive, Narragansett, RI 02882, http://www.epa.gov/aed/, Phone: (401) 782-3001, Fax: (401) 782-3030

U.S. EPA National Health and Environmental Effects Research Laboratory (NHEERL) Gulf Ecology Division, Sabine Island Drive, Gulf Breeze, FL 32561, http://www.epa.gov/ged/, Phone: (850) 934-9200, Fax: (850) 934-9201

National Risk Management Research Laboratory (NRMRL), Subsurface Protection and Remediation Division, P.O. Box 1198, Ada, OK 74820, http:www.epa.gov/ada/, Phone: (580) 436-8500

National Risk Management Research Laboratory (NRMRL), Water Supply and Resources Division, Urban Watershed Management Branch, 2890 Woodbridge Avenue (MS-104), Edison, NJ 08837, http://www.epa.gov/ednnrmrl/

National Risk Management Research Laboratory (NRMRL), Air Pollution Prevention and Control Division, Mail Code E343-04, Research Triangle Park, NC 27711, http://www.epa.gov/appcdwww/, Phone: (919) 541-2821, Fax: (919) 541-5227

U.S. EPA National Vehicle and Fuel Emissions Laboratory (NVFEL), 2000 Traverwood Drive, Ann Arbor, MI 48105, http://www.epa.gov/otaq/01-nvfel.htm, Phone: (734) 214-4200

U.S. EPA Radiation and Indoor Environments National Laboratory, P.O. Box 98517, Las Vegas, NV 89193-8517, http://www.epa.gov/radiation/rienl/, Phone: (702) 798-2476

Source: From www.epa.gov.

Table 12A.5 EPA Libraries and Information Centers

The U.S. Environmental Protection Agency {EPA) is comprised of a Headquarters Office in Washington, DC, 10 Regional offices, and 13 specialized, scientific laboratories located throughout the country. There are 28 EPA network libraries located in Headquarters and all Regional offices and laboratories to support this organizational structure.

The libraries contain a combined collection of over 122,422 books, 5,414 journals, 377,217 hard copy reports, 3,075,443 documents on microfilm and microfiche, 9,000 journal article reprints and 2,000 maps. Most of the EPA library network's holdings are cataloged on OCLC, a national cataloging system.

Headquarters Library — Washington, D.C.
Lucy Park, Library Manager, ASRC Aerospace Corp., contractor to U.S. EPA, 1200 Pennsylvania Ave, NW (3404T), Washington, DC 20460, Phone:(202) 566-0556, http://www.epa.gov/natlibra/hqirc/

U.S. EPA Office of Congressional & Intergovernmental Relations
ATTN: Craig Freer, 1200 Pennsylvania Avenue, NW, Mail Code 1301A, Washington, D.C. 20460, Phone: (202) 564-5200, Fax: (202) 501-1519, http://www.epa.gov/ocirpage/leglibrary/index.htm

Office of General Counsel Law Library
U.S. Environmental Protection Agency, Office of General Counsel Law Library, 1200 Pennsylvania Ave., N.W, Ariel Rios North Building, Room 1315, Washington, D.C. 20460, Phone: (202) 564-3971, http://www.epa.gov/natlibra/libraries/law.htm

Prevention, Pesticides & Toxic Substances (OPPTS) Chemical Library
EPA Headquarters, 1301 Constitution Avenue NW, 3231 EPA West, Washington, D.C. 20004, Phone: (202) 566-0800, http://www.epa.gov/opptintr/library/

Water Resource Center (RC-4100) (Mailing Address),
U.S. Environmental Protection Agency, 1200 Pennsylvania Avenue NW, Washington, D.C. 20460, Phone: (202) 566-1729, Phone: (800) 832-7828 (Wetlands Helpline), http://www.epa.gov/safewater/resource/

Water Resource Center (Physical Address),
U.S. Environmental Protection Agency, EPA West Room 1119, 1301 Constitution Avenue NW, Washington, D.C. 20004

Libraries in Research Triangle Park, North Carolina

NERL — Atmospheric Modeling Division Library
Phone: (919) 541-4536, Fax: (919) 541-1379, epa.gov/asmdnerl/library/library.htm

EPA — RTP Library
U.S. Environmental Protection Agency, 109 TW Alexander Drive, Durham, NC 27709, Phone: (919) 541-7645, http://www.epa.gov/rtp/library/index.htm

EPA Region 1 Library, Boston
1 Congress St, Suite 1100, Boston, MA 02114-2023, Phone: (617) 918-1990, Fax: (617) 918-1992 http://www.epa.gov/region01/oarm/index.html

EPA Region 1 RCRA Research Library, Boston
1 Congress St, Suite 1100, Boston, MA 02114-2023, Phone: (617) 918-1990, Fax: (617) 918-1992 http://www.cpa.gov/rcgion01/oarm/index.html

EPA Region 2 Library, New York City
http://www.epa.gov/Region2/library/

MAIN REGIONAL OFFICE
290 Broadway, New York, NY 10007-1866, Phone: (212) 637-5000

(Continued)

Table 12A.5 (Continued)

EDISON LABORATORIES
2890 Woodbridge Ave., Ms100, Edison, NJ 08837-3679, Phone: (732) 321-6754, Fax: (732) 321-4381

CARIBBEAN ENVIRONMENTAL PROTECTION DIVISION
Centro Europa Building, 1492 Ponce Deleon Avenue, Suite 417, San Juan, PR 00907-4127, Phone: (787) 977-5870, Fax: (787) 729-7747

VIRGIN ISLANDS FIELD OFFICE
Tunick Building, Suite 102, 1336 Beltjen Road, St. Thomas, VI 00801, Phone: (340) 714-2333, Fax: (340) 714-2332

HUDSON RIVER FIELD OFFICE
421 Lower Main Street, Hudson Falls, NY 12839, Phone: (518) 747-4389, Fax: (518) 747-8149

NIAGARA FALLS PUBLIC INFORMATION CENTER
345 Third Street, Suite 530, Niagara Falls, NY 14303, Phone: (716) 285-8842, Fax: (716) 285-8788

Region 3 Regional Center for Environmental Information, Philadelphia
US Environmental Protection Agency, Regional Center for Environmental Information, Second Floor (3PM52), 1650 Arch Street, Philadelphia, PA 19103, Phone: (215) 814-5254, Fax: (215) 814-5253, epa.gov/reg3rcei/contacts2.htm

Region 4 Library, Atlanta
United States Environmental Protection Agency, Region 4, Sam Nunn Atlanta Federal Center, 61 Forsyth Street, SW, Atlanta, GA 30303-3104, Phone: (404) 562-9900, Fax: (404) 562-8174, epa.gov/region4/library/index.htm

Region 5 Library, Chicago
77 W. Jackson Blvd., Chicago, IL 60604, 12th floor, Phone: (312) 353-2022, Fax: (312) 353-2001, epa.gov/region5/library/

Region 6 Library, Dallas
EPA Region 6 Main Office, 1445 Ross Avenue *(maps)*, Suite 1200, Dallas, TX 75202, Phone: (214) 665-6444, epa.gov/earth1r6/6md/6lib/index.htm

Houston Laboratory
10625 Fallstone Road, Houston, TX 77099, Phone: (281) 983-2100, Fax: (281) 983-2248

Border Liaison Office
4050 Rio Bravo, Suite 100, El Paso, TX 79902, Phone: (915) 533-7273, Fax: (915) 533-2327

Brownsville Border Outreach Office
3505 Boca Chica, Suite 302, Brownsville, TX 78521, Phone: (956) 548-0898

Underground Injection Control
Pawhuska Section, Osage Nations Federal Programs, 627 Grandview Ave., Pawhuska, OK 74056, Phone: (918) 287-4041, Fax: (918) 287-2322

Water Quality Field Office
707 Florida Street, Room B21, Baton Rouge, LA 70801, Phone: (225) 389-0735, Fax: (225) 389-0704

Region 7 Information Resource Center, Kansas City
U.S. EPA Region 7, 901 N. 5th Street, Kansas City, KS 66101, Phone: (913) 551-7241, Fax (913) 551-8762, r7-library@epa.gov

Region 8 Environmental Information Service Center, Denver
U.S. EPA Region 8 Environmental Information Svc Ctr., 999 18th Street, Suite 300 OC-L, Denver, CO 80202-2466, Phone: (800) 227-8917, Phone: (303) 312-6312, r8eisc@epa.gov

U.S. EPA Region 8 Technical Library
999 18th Street, Suite 300 OC-L, Denver, CO 80202-2466, Phone: (303) 312-6312 or Phone: (800)227-8917 in Region 8 states, Fax: (303) 312-7061, library-reg8@epa.gov

U.S. EPA Region 9
75 Hawthorne Street, San Francisco, CA 94105, Phone: (866) EPA-WEST, Phone: (415) 947-8000, http://www.epa.gov/region09/visitor.html

U.S. EPA Region 10
1200 Sixth Avenue, Seattle, WA 98101, Phone: (800) 424-4EPA, Fax: (206) 553-6346, library-reg10@epa.gov

Laboratory Libraries

Andrew W. Breidenbach Environmental Research Center
U.S. Environmental Protection Agency, 26 W. Martin Luther King Dr., Cincinnati, OH 45268, Phone: (513) 569-7703, Fax: (513) 569-7709, http://www.epa.gov/oarmcinc/library.htm

U.S. Environmental Protection Agency, Environmental Science Center Library
701 Mapes Road, Fort Meade, MD 20755-5350, Phone: (410) 305-2603, Fax: (410) 305-3099, http://www.epa.gov/region3/esc/library/

NERL — Atmospheric Sciences Modeling Division Library
MD-267-02, Environmental Protection Agency, 109 T.W. Alexander Drive, Research Triangle Park, NC 27711, Phone: (919) 541-4536, Fax: (919) 541-1379, epa.gov/asmdnerl/library/general.htm

NERL — Environmental Sciences Division Technical Research Center
U.S. Environmental Protection Agency, National Environmental Research Laboratory, Environmental Sciences Division, 944 East Harmon, Las Vegas, NV 89119, Phone: (702) 798-2100, Fax: (702) 798-8147, epa.gov/nerlesd1/trc/home6.htm

NERL — Ecosystem Research Division Library NERL/ERD Library
U.S. EPA, 960 College Station Road, Athens, GA 30605-2700, Phone: (706) 355-8011, Fax: (706) 355-8440, http://www.epa.gov/natlibra/libraries/athens.htm

NHEERL — Atlantic Ecology Division Library, U.S. EPA AED — NHEERL Library
27 Tarzwell Drive, Narragansett, RI 02882, Phone: (401) 782-3025, Fax: (401) 782-3025, http://www.epa.gov/natlibra/libraries/aed.htm

NHEERL — Gulf Ecology Division Library
1 Sabine Island Drive, Gulf Breeze, FL 32561; Phone: (850) 934-9208, Fax (850) 934-2406, http://www.epa.gov/ged/overview_dw.htm

NHEERL — Midcontinent Ecology Division Library
U.S. EPA — Mid-Continent Ecology Division—Duluth, 6201 Congdon Boulevard, Duluth, MN 55804-2595, Phone: (218) 529-5000, Fax: (218) 529-5003, epa.gov/med/facilities/scientific_library.htm

(Continued)

Table 12A.5 (Continued)

NHEERL — Western Ecology Division Library
U.S. Environmental Protection Agency, NHEERL WED Library, 200 SW 35th Street, Corvallis, OR 97333, Phone: (541) 754-4731, Fax: (541) 754-4799, http://www.epa.gov/natlibra/libraries/wed.htm

NRMRL — Groundwater and Ecosystems Restoration Division Library
919 Kerr Research Drive, P.O. Box 1198, Ada, OK 74821-1198, Phone: (580) 436-8505, Fax: (580) 436-8503, http://www.epa.gov/natlibra/libraries/ada.htm

National Vehicle & Fuel Emissions Laboratory Library
U.S. EPA NVFEL Library, 2000 Traverwood, Ann Arbor, MI 48105, Phone: (734) 214-4311, Fax: (734) 214-4525, http://www.epa.gov/natlibra/libraries/nvfel.htm

RTP Library Services
U.S. Environmental Protection Agency, 109 TW Alexander Drive, Durham, NC 27709, Phone: (919) 541-7645, http://www.epa.gov/rtp/library/index.htm

Environmental Financing Information Network
U.S. Environmental Protection Agency, Office of the Comptroller, Environmental Finance Program (Mail Code) 2731R, Ariel Rios Building 1200 Pennsylvania Ave., NW, Washington, D.C. 20460, Phone: (202) 564-4994, Fax: (202) 565-2587, http://www.epa.gov/efinpage/efp.htm

National Enforcement Investigations, Center Environmental Forensics, Library
U.S. EPA NEIC Library, P.O. Box 25277, Bldg. 53, Denver Federal Center, Denver, CO 80225, Phone: (303) 236-6136, Fax: (303) 236-3218 epa.gov/compliance/criminal/forensics/neiclibrary/index.html

Source: From www.epa.gov.

Table 12A.6 Databases Used by EPA Libraries

Accidental Release Information Program (ARIP)

The *Accidental Release Information Program (ARIP) database (July, 1999)* is contained in a zip file (around 1 MB) that contains the ARIP database file (DBF format) and supporting documentation. http://yosemite.epa.gov/oswer/ceppoweb.nsf/content/ds-epds.htm#arip

Acid Rain Program

Acid Rain Program quarterly reports use Electronic Data Reporting (EDR) format version 2.1 and are due 30 days following every calendar quarter. http://www.epa.gov/airmarkets/reporting/#arp

Agricultural Pollution Prevention

Shows how farmers can save money and reduce pollution by sensible use of fertilizers, pesticides and herbicides. agpp.html. http://www.epa.gov/seahome/agpp.html

Air CHIEF CD-ROM, Version 11 April 2004 — Released June 2004

The U. S. Environmental Protection Agency is working to provide current emissions data to federal, state, and local regulatory agencies, businesses, and the general public. The Air Clearing House For Inventories And Emission Factors (*Air CHIEF*) gives the public and private sector user's access to air emission data specific to estimating the types and quantities of pollutants that may be emitted from a wide variety of sources. Updated annually, *Air CHIEF* offers thousands of pages contained in some of EPA's most widely used and requested documents. Included are the U.S. EPA Emission Inventory Group's most popular emission estimation tools. http://www.epa.gov/ttn/chief/software/airchief/index.html

Air Facility System (AFS)

The Air Facility System (AFS) contains compliance and permit data for stationary sources regulated by the U.S. EPA and state and local air pollution agencies. http://www.epa.gov/Compliance/planning/data/air/afssystem.html

Air Quality Index

The Air Quality Index (AQI) is an index for reporting daily air quality. It tells you how clean or polluted your outdoor air is, and what associated health effects might be a concern for you. http://www.epa.gov/airnow/background.html

AQS Web Home

AQS has been migrated and deployed on the Web to eliminate the installation and updates of the AQS software on users' machines and to eliminate the need for the SecuRemote software. epa.gov/ttn/airs/airsaqs/aqsweb/aqswebhome.html

AirData

The site gives you access to air pollution data for the entire United States. AirData produces reports and maps of air pollution data based on criteria that you specify. http://www.epa.gov/air/data/index.html

AIRS Executive Software

A personal computer software and database that contains air pollution data extracted from the AIRS database. The Windows interface makes it easy to generate many different reports, charts, and maps. AIRS Executive for Windows provides ambient air quality data from monitoring sites across the U.S.A., and pollutant emissions estimates for stationary sources regulated by the EPA. AIRS Executive for Windows also includes ambient air quality data from about 50 nations that voluntarily provided data in the early 1990s to the GEMS/AIR Programme sponsored by the United Nations World Health Organization. http://www.epa.gov/airs/aexec.html

Alternatives for Unsewered Communities

This program is an extensive guide to the facilities planning process for small communities. Topics covered include needs documentation, development of alternative solutions and selection of the best response, as well as treatment management and implementation. An improved and expanded section on sludge is included as well. http://www.epa.gov/seahome/unsewer.html

Applicability Determination Index

EPA is committed to helping entities comply with regulatory requirements and improve environmental performance through compliance assistance (CA). Compliance assistance is defined by EPA to include activities, tools or technical assistance which provide clear and consistent information for helping the regulated community understand and meet its obligations under environmental laws and regulations. http://cfpub.epa.gov/adi/

AQUATOX Model

AQUATOX is a simulation model for aquatic systems that predicts the fate of various pollutants, such as nutrients and organic chemicals, and their effects on the ecosystem, including fish, invertebrates, and aquatic plants. AQUATOX is a valuable tool for ecologists, biologists, water quality modelers, and anyone involved in the performing ecological risk assessments for aquatic ecosystems. http://www.epa.gov/waterscience/models/aquatox/

ArcHydro

Redlands, California — ArcHydro: GIS for Water Resources, the latest title from ESRI Press, describes how a water resources data model, ArcHydro, is being applied within geographic information system (GIS) technology to provide a wide variety of hydrologic solutions. esri.com/news/releases/02_3qtr/archydro.html

(Continued)

Table 12A.6 (Continued)

ASSESS Software

A computer-based tool for statistically assessing measurement errors in the collection of soil samples. http://www.epa.gov/nerlesd1/databases/assess/abstract.htm

Assessment Database Software

The Assessment Database (ADB) is a relational database application for tracking water quality assessment data, including use attainment, and causes and sources of impairment. States need to track this information and many other types of assessment data for thousands of waterbodies, and integrate it into meaningful reports. The ADB is designed to make this process accurate, straightforward and user-friendly for participating States, territories, tribes and basin commissions. http://www.epa.gov/waters/adb/

BEACH Watch

National Health Protection Survey of Beaches for the 2002 Swimming Seas. http://yosemite.epa.gov/water/beach2003.nsf

Benchmark Dose Software (BMDS)

Benchmark dose risk assessment software (BMDS) was designed by EPA to generate dose-response curves and facilitate the analysis, interpretation and synthesis of toxicological data. lhttp://cfpub2.epa.gov/ncea/cfm/recordisplay.cfm

Best Management Practices for Soil Erosion

This program uses graphics and hypertext to provide information about soil erosion. Information is provided about the severity of erosion worldwide, in the U.S. and in the Midwest in particular. The types of water erosion are described and illustrated along with detailed descriptions of management and structural practices to control erosion.erosion.html http://www.epa.gov/seahome/erosion.html

Better Assessment Science Integrating Point & Nonpoint Sources (Basins)

A multi-purpose environmental analysis system that integrates a geographical information system (GIS), national watershed data, and state-of-the-art environmental assessment and modeling tools into one convenient package. http://www.epa.gov/waterscience/basins/

BIOPLUME III Software

BIOPLUME III is a 2D, finite difference model for simulating the natural attenuation of organic contaminants in groundwater due to the processes of advection, dispersion, sorption, and biodegradation. Biotransformation processes are potentially important in the restoration of aquifers contaminated with organic pollutants. http://www.epa.gov/ada/csmos/models/bioplume3.html

Biennial Reporting System (BRS)

The Hazardous Waste Report (Biennial Report) collects data on the generation, management, and minimization of hazardous waste. This provides detailed data on the generation of hazardous waste from large quantity generators and data on waste management practices from treatment, storage, and disposal facilities. http://www.epa.gov/enviro/html/brs/

Biogenic Emissions Inventory System (BEIS)

BEIS estimates volatile organic compound (VOC) emissions from vegetation and nitric oxide (NO) emissions from soils. Non-air-quality model users may wish to consider GloBEIS. http://www.epa.gov/asmdnerl/biogen.html

Building for Environmental and Economic Sustainability (BEES)

BEES software provides a technique for selecting cost-effective, environmentally preferable building products. BEES reduces complex, science-based technical content (e.g. over 400 environmental flows from raw material acquisition through product disposal) to decision-enabling results and delivers them in a graphical format. http://www.epa.gov/oppt/epp/tools/bees.htm

Catalog of Federal Funding Sources for Watershed Protection

A searchable database of financial assistance sources (grants, loans, cost-sharing) available to fund a variety of watershed protection projects. http://cfpub.epa.gov/fedfund/

Center for Exposure Assessment Modeling Software

CEAM provides proven predictive exposure assessment techniques for aquatic, terrestrial, and multimedia pathways for organic chemicals and metals. http://www.epa.gov/ceampubl/

Center for Subsurface Modeling Support (CSMoS)

The Center for Subsurface Modeling Support (CSMoS) provides public domain groundwater and vadose zone modeling software and services to public agencies and private companies throughout the nation. The primary aims of CSMoS are to provide direct technical support to EPA and State decision makers in subsurface model applications and to manage and support the groundwater models and databases resulting from the research at NRMRL. http://www.epa.gov/ada/csmos.html

Central Data Exchange

The Central Data Exchange (CDX) enables fast, efficient, and more accurate environmental data submissions from state and local governments, industry and tribes to the Environmental Protection Agency (EPA) and participating program offices. EPA's CDX is the point of entry on the Environmental Information Exchange Network (Exchange Network) for environmental data submissions to the Agency. http://www.epa.gov/cdx/

(Continued)

Table 12A.6 (Continued)

Certification and Fuel Economy Information System (CFEIS)
Information about the process of entering data into the EPA's Certification and Fuel Economy Information System for light-duty vehicle and truck manufacturers. http://www.epa.gov/otaq/cfeis.htm

CHEMFLO Software
A program that enables users to simulate water movement and chemical transport in unsaturated soils by solving the Richards equation (water) and the convection-dispersion equation (chemicals). http://www.epa.gov/nerlesd1/databases/chemflo/access.htm

Chemical Contamination in Fish
A Windows based update of the original FISH program, it contains video clips illustrating concepts such as bioaccumulation, and risk reduction techniques (including cooking and filleting tips). Users can search a list of chemical contaminants and learn about their effects, or see photos and video of target species. Included are an updated list of contaminants from 1997 guidance, as well as references to an online database of fish advisories and state contacts. http://www.epa.gov/seahome/fish.html

Chemical Hazard Information Profiles (CHIPS)
CHIPS summarize readily available information on health and environmental effects and exposure relating to a specific chemical substance. The database contains information about chemical substances that have been referenced in these profiles. The Existing Chemicals Program currently managed by OPPT, has superseded these profiles. http://oaspub.epa.gov/srs/srs_proc_qry.navigate?P_REG_AUTH_ID=1&P_DATA_ID=11681&P_VERSION=1

Chemical Screening Tool For Exposures & Environmental Releases (ChemSTEER)
Estimates occupational inhalation and dermal exposure to a chemical during industrial and commercial manufacturing, processing, and use operations involving the chemical. Estimates releases of a chemical to air, water, and land that are associated with industrial and commercial manufacturing, processing, and use of the chemical. http://www.epa.gov/opptintr/exposure/docs/chemsteer.htm

Children's Environmental Health & Safety Inventory of Research (CHESIR)
Children's Environmental Health and Safety Inventory of Research is a publicly accessible database created and maintained to ensure that researchers and Federal research agencies have access to information on all research conducted or funded by the Federal Government that is related to adverse health risks in children resulting from exposure to environmental health risks or safety risks. http://oaspub.epa.gov/chehsir/chehsir.page

Chemicals On Reporting Rules (CORR) Database
The CORR database contains information on chemicals which are regulated under specific sections of the Toxic Substances Control Act (TSCA), or section 313 of the Emergency Planning and Community Right-to-Know Act (EPCRA). http://www.epa.gov/oppt/CORR/

Clean Water State Revolving Fund National Information Management System
The Clean Water State Revolving Fund (CWSRF) program is an innovative method of financing a wide range of projects related to water quality. The CWSRF National Information Management System produces annual reports that provide a record of progress and accountability for the program. http://www.epa.gov/r5water/cwsrf/

Clean Watersheds Needs Survey
The Office of Wastewater Management conducts the Clean Watersheds Needs Survey (CWNS) on a periodic basis. The CWNS, a joint effort between States and EPA, is conducted in response to Section 205(a) and 516 of the Clean Water Act. The CWNS has information on publicly-owned wastewater collection and treatment facilities, facilities for control of sanitary sewer overflows (SSOs), combined sewer overflows (CSOs), stormwater control activities, nonpoint sources, and programs designed to protect the nation's estuaries. http://www.epa.gov/owm/mtb/cwns/

COMMUTER Model
Reports and information that relate to the U.S. Department of Transportation's Congestion Mitigation and Air Quality Improvement Program, the environmental impacts of transportation, induced travel/induced demand, intelligent transportation systems, EPA's Commuter Model, and other transportation/air quality issues. http://www.epa.gov/otaq/transp/traqmodl.htm

Comparative Risk Assessment Software
Comparative risk assessment is a methodology which uses sound science, policy, economic analysis and stakeholder participation to identify and address the areas of greatest environmental risks and provide a framework for prioritizing environmental problems. http://www.epa.gov/seahome/comprisk.html

Comply Software
A computerized screening tool for evaluating radiation exposure from atmospheric releases of radionuclides. May be used for demonstrating compliance with some EPA and Nuclear Regulatory Commission regulations. http://www.epa.gov/radiation/assessment/comply.html

Computational Toxicology
Computational toxicology is the integration of modern computing and information technology with molecular biology and chemistry to improve EPA's prioritization of data requirements and risk assessments for toxic chemicals. http://www.epa.gov/comptox/

(Continued)

Table 12A.6 (Continued)

Computer Aided Management of Emergency Operations (CAMEO)
CAMEO® is a system of software applications used widely to plan for and respond to chemical emergencies. CAMEO can be used to access, store, and evaluate information critical for developing emergency plans and support regulatory compliance by helping users meet the chemical inventory reporting requirements of the Emergency Planning and Community Right-to-Know Act (EPCRA, also known as SARA Title III). http://www.epa.gov/ceppo/cameo/

Comprehensive Environmental Response, Compensation, and Liability Information System (CERCLIS)
Superfund is a program administered by the EPA to locate, investigate, and clean up the worst hazardous waste sites throughout the United States. http://www.epa.gov/enviro/html/cerclis/

Confidence Interval Calculation for Source Partitioning Using Stable Isotopes
ISOERROR1_04 is a Microsoft Excel 2000™ spreadsheet which calculates estimates and confidence intervals of source proportional contributions to a mixture, using stable isotope analyses. http://www.epa.gov/wed/pages/models/isotopes/isoerror1_04.htm

Consolidated Human Activity Database (CHAD)
Consolidated Human Activity Database (CHAD) contains data obtained from pre-existing human activity studies that were collected at city, state, and national levels. CHAD is intended to be an input file for exposure/intake dose modeling and/or statistical analysis. http://www.epa.gov/chadnet1/

Cornell Mixing Zone Expert System (CORMIX)
Cornell Mixing Zone Expert System (CORMIX) is no longer distributed or supported by the EPA Center for Exposure Assessment Modeling (CEAM). Information on a Windows version of CORMIX is available from the CORMIX Home Page. http://www.epa.gov/ceampubl/swater/cormix/index.htm

Decentralized Onsite Management for Treatment of Domestic Wastes Software
This program provides operation and maintenance information for on-site wastewater treatment systems, such as septic systems. http://www.epa.gov/seahome/decent.html

DEFT Software
A program that allows users to evaluate the financial feasibility of incorporating selected data quality objective (DQO) constrains into a statistical sampling design before developing a final plan — the program assists with the seven-step DQO process, which is used to develop statistical sampling design plans. http://www.epa.gov/nerlesd1/databases/deft/access.htm

Delisting Risk Assessment Software (DRAS)
The Delisting Risk Assessment Software (DRAS) is a Windows-based computer program that predicts risks to human health and the environment from the disposal of wastes to a landfill or surface impoundment. The DRAS is used to calculate potential chemical releases, fate and transport of the chemicals and to determine risk associated with exposure to the chemicals released. epa.gov/region6/6pd/rcra_c/pd-o/dras/dras.htm

DFLOW Software
DFLOW is a Windows-based tool developed to estimate user selected design stream flows for low flow analysis and water quality standards. http://epa.gov/waterscience/dflow/

Dietary Exposure Potential Model
A Model Using Extant Food Databases to Estimate Dietary Exposure to Chemical Residues. http://www.epa.gov/nerlcwww/depm.htm

Drinking Water Contaminant Candidate List
The drinking water Contaminant Candidate List (CCL) is the primary source of priority contaminants for evaluation by EPA's drinking water program. The Safe Drinking Water Act (SDWA), as amended in 1996, requires EPA to publish a list of contaminants every five years which, at the time of publication, are not subject to any proposed or promulgated national primary drinking water regulations. Contaminants on the CCL are known or anticipated to occur in public water systems and may require regulations under SDWA. http://www.epa.gov/safewater/ccl/cclfs.html

Ecological Structure Activity Relationships (ECOSAR)
ECOSAR (Ecological Structure Activity Relationships) is a personal computer software program that is used to estimate the toxicity of chemicals used in industry and discharged into water. The program predicts the toxicity of industrial chemicals to aquatic organisms such as fish, invertebrates, and algae by using Structure Activity Relationships (SARs). The program estimates a chemical's acute (short-term) toxicity and, when available, chronic (long-term or delayed) toxicity. http://www.epa.gov/opptintr/newchems/21ecosar.htm

ECOTOX Database
The ECOTOXicology database is a source for locating single chemical toxicity data for aquatic life, terrestrial plants and wildlife. ECOTOX integrates three toxicology effects databases: AQUIRE (aquatic life), PHYTOTOX (terrestrial plants), and TERRETOX (terrestrial wildlife). These databases were created by the U.S. EPA, Office of Research and Development (ORD), and the National Health and Environmental Effects Research Laboratory (NHEERL), Mid-Continent Ecology Division. http://www.epa.gov/ecotox/

(Continued)

Table 12A.6 (Continued)

8(e) TRIAGE Chemical Studies Database

The primary purpose of this information product is to make TSCA 8(e) submission information accessible to the general public and organizations whose efforts are associated with protection of health and safety. http://www.epa.gov/opptintr/8e_triag/

Electronic Wetlands Herbarium

A detailed look at the plants typical of 11 different wetland types from bog to wet prairie. Each entry gives detailed botanical information as well as a photo of the plant. The outstanding photos are identification quality. Examples are taken from the U.S. Midwest region. http://www.epa.gov/seahome/wetherb.html

Emergency Response Notification System

The Emergency Response Notification System (ERNS) is a computer database containing information on release notifications of oil and hazardous substances that have occurred throughout the United States and have been reported to the National Response Center (NRC) and/or one of the 10 EPA Regions. http://www.epa.gov/region08/community_resources/ppt/pptemerge.html

Emissions & Generation Resource Integrated Database (eGRID)

eGRID is a database that provides information on the air quality attributes of almost all the electric power generated in the United States. eGRID provides you with many search options, including information for individual power plants, generating companies, states, and regions of the power grid. http://www.epa.gov/cleanenergy/egrid/index.html

Endangered Species Protection Program Databases

Provides the endangered species by county and the species information database. http://www.epa.gov/espp/database.htm

Energy Star Software Tools

Energy star is a government-backed program helping businesses and individuals protect the environment through superior energy efficiency. http://www.energystar.gov

Enforcement & Compliance History Online (ECHO)

Use ECHO to determine whether compliance inspections have been conducted by EPA or State/local governments; violations were detected; and enforcement actions were taken and penalties were assessed in response to environmental law violations. http://www.epa.gov/echo/

Enforcement Economic Models — ABEL, BEN, CASHOUT, PROJECT, INDIPAY, MUNIPAY

The economic enforcement models are used to analyze the financial aspects of enforcement actions. http://www.epa.gov/compliance/civil/programs/econmodels/index.html#models

Enhanced Stream Water Quality Model, Windows (QUAL2E)

The Enhanced Stream Water Quality Model (QUAL2E) is a steady state model for conventional pollutants in branching streams and well mixed lakes. The model can be used to study impact of waste loads on in-stream water quality and identify magnitude and quality characteristics of nonpoint waste loads. http://www.epa.gov/ceampubl/swater/qual2eu/index.htm

Envirofacts

Envirofacts, is a single point of access to select U.S. EPA environmental data. This website provides access to several EPA databases to provide information about environmental activities that may affect air, water, and land anywhere in the United States. http://www.epa.gov/enviro/index.html

Environmental Assessment Case Study

A companion to the 1992 Environmental Assessment Resource Guide, the program utilizes real-life examples of an environmental impact assessment and leads the user through the environmental assessment process. http://www.epa.gov/seahome/eacase.html

Environmental Assessment Resource Guide

EARG is a generic source of information to help assist in the conduct of environmental assessments of virtually any type of project. Topics covered include scoping, generation of alternatives, impact identification and analysis, mitigation, decision-making and post-decision analysis. http://www.epa.gov/seahome/earg.html

Environmental Data Registry (EDR)

The Environmental Data Registry (EDR) is a comprehensive, authoritative reference for information about the definition, source, and uses of environmental data. The EDR supports the creation and implementation of data standards that are designed to promote the efficient sharing of environmental information among EPA, states, tribes, and other information trading partners. The EDR also catalogs data elements in application systems. The EDR does not contain environmental data—it provides descriptive information to make the data more meaningful. http://www.epa.gov/edr/

Environmental Financing Information Network (EFIN)

The Environmental Financing Information Network (EFIN) provides information services needed by state, sub-state, municipal and EPA officials and small business owners involved in funding environmental programs and projects. These information services include maintaining the EFP Web site and the EFIN database, which are the main methods for disseminating information on environmental financing altermatives; an infoline that provides referrals and assistance with locating environmental financing information, and distribution of EPA publications on the topic of environmental finance. http://www.epa.gov/efinpage/efin.htm#desc

(Continued)

Table 12A.6 (Continued)

Environmental Information Management System (EIMS)
EPA's Office of Research and Development (ORD) has developed a scientific environmental information management system (EIMS) that stores, manages, and delivers descriptive information (metadata) for data sets, databases, documents, models, multimedia, projects, and spatial information. The EIMS design also provides a repository for scientific documentation that can be easily accessed with standard Web browsers to place a virtual library on the desktop of EPA staff and others with Internet access. epa.gov/eims/

Environmental Planning for Small Communities (TRILOGY)
This program offers a complete one-stop introduction to a wide range of environmental issues and decisions that affect small to medium-sized communities. It offers communities the chance to judge their own needs and preferences, and to make informed decisions on their own. Major sections cover Environmental laws and regulations, Self-assessment, Planning and comparative risk analysis, Financial tools and financial self-analysis, Case studies, and Contact and information directory. http://www.epa.gov/seahome/trilogy.html

Environmentally Preferable Purchasing (EPP) Database
The Environmentally Preferable Purchasing (EPP) Database — a tool to make it easier to purchase products and services with reduced environmental impacts. Environmental information on over 600 products and services is included in this database. epa.gov/opptintr/epp/database.htm

Environmental Radiation Ambient Monitoring System (ERAMS) Database
The Environmental Radiation Ambient Monitoring System (ERAMS) is a national network of monitoring stations that regularly collect air, precipitation, drinking water, and milk samples for analysis of radioactivity. The ERAMS network has been used to track environmental releases resulting from nuclear emergencies and to provide baseline data during routine conditions. Data generated from ERAMS provides the information base for making decisions necessary to ensure the protection of public health. http://www.epa.gov/enviro/html/erams/

Enviroene Solvent Substitution Data Systems
Enviroene, part of the U.S. EPA's web site, provides a single repository for pollution prevention, compliance assurance, and enforcement information and data bases. The search engine searches multiple web sites (inside and outside the EPA), and offers assistance in preparing a search. http://es.epa.gov/ssds/ssds.html

Enviroene VendInfo
Vendor Information (VendInfo) search page, a repository of more than 1200 listings of pollution prevention equipment, products, or services. This page provides listings using a free text method, or by selecting from the various equipment categories. http://es.epa.gov/vendors/

EPANET Software
EPANET is a Windows 95/98/NT program that performs extended period simulation of hydraulic and water-quality behavior within pressurized pipe networks. A network can consist of pipes, nodes (pipe junctions), pumps, valves and storage tanks or reservoirs. EPANET tracks the flow of water in each pipe, the pressure at each node, the height of water in each tank, and the concentration of a chemical species throughout the network during a simulation period comprised multiple time steps. In addition to chemical species, water age and source tracing can also be simulated. http://www.epa.gov/ORD/NRMRL/wswrd/epanet.html

EPA REACH IT — Remediation and Characterization Innovative Technologies
EPA REACH IT, sponsored by EPA's Office of Superfund Remediation and Technology Innovation (OSRTI), is a system that lets environmental professionals use the Internet to search, view, download, and print information about innovative remediation and characterization technologies. epareachit.org/

Estimation Program Interface (EPI) Suite
The EPI (Estimation Programs Interface) SuiteTM is a Windows® based suite of physical/chemical property and environmental fate estimation models developed by the EPA's Office of Pollution Prevention Toxics and Syracuse Research Corporation (SRC). EPI SuiteTM uses a single input to run the following estimation models: KOWWINTM, AOPWINTM, HENRYWINTM, MPBPWINTM, BIOWINTM, PCKOCWINTM, WSKOWWINTM, BCFWINTM, HYDROWINTM, and STPWINTM, WVOLWINTM, and LEV3EPITM. EPI SuiteTM was previously called EPIWIN. epa.gov/opptintr/exposure/docs/episuite.htm

Exposure Analysis Modeling System (EXAMS)
The Exposure Analysis Modeling System (EXAMS) is an interactive software application for formulating aquatic ecosystem models and rapidly evaluating the fate, transport, and exposure concentrations of synthetic organic chemicals including pesticides, industrial materials, and leachates from disposal sites. http://www.epa.gov/ceampubl/swater/exams/index.htm

Exposure, Fate Assessment Screening Tool
Provides screening-level estimates of the concentrations of chemicals released to air, surface water, landfills, and from consumer products. E-FAST Version 2 is being designed to support both new chemicals and existing chemical programs. Estimates provided are potential inhalation, dermal and ingestion dose rates resulting from these releases. Modeled estimates of concentrations and doses are designed to reasonably overestimate exposures, for use in screening level assessment. http://www.epa.gov/opptintr/exposure/docs/efast.htm

(Continued)

Table 12A.6 (Continued)

Extremely Hazardous Substances (EHS) Chemical Profiles and Emergency First Aid Guides

This is information about each of the 300+ EHS currently listed as part of Section 302 of the Emergency Planning and Community Right-to-Know Act. Each chemical profile includes physical/chemical properties, health hazards, fire and explosion hazards, reactivity data, precautions for safe handling and use, and protective equipment for emergency situations. The first aid guide provides signs and symptoms of poisoning and emergency treatment for first responders. The chemical profiles and first aid guides may be accessed from either the CAS No. or alphabetical list of EHS. http://yosemite.epa.gov/oswer/ceppoehs.nsf/EHS_Profile?openform

Facility Registry System (FRS)

The Facility Registry System (FRS) is a centrally managed database that identifies facilities, sites or places subject to environmental regulations or of environmental interest. FRS creates high-quality, accurate, and authoritative facility identification records through rigorous verification and management procedures that incorporate information from program national systems, state master facility records, data collected from EPA's Central Data Exchange registrations and data management personnel. The FRS provides Internet access to a single integrated source of comprehensive (air, water, and waste) environmental information about facilities, sites or places. http://www.epa.gov/enviro/html/fii/index.html

Factor Information REtrieval (FIRE)

The Factor Information REtrieval (FIRE) Data System is a database containing EPA's emission estimation factors for criteria and hazardous air pollutants in an easy to use Windows program. Users can browse through records in the database or select specific emission factors by source category, source classification code (SCC), pollutant name, CAS number, or control device. FIRE 6.25 contains emission factors from the Compilation of Air Pollutant Emission Factors (AP-42 Fifth Edition) for all AP42 sections posted by September 1, 2004, the Locating and Estimating (L&E) series of documents, and the retired AFSEF and XATEF documents. All EPA Source Classification Codes (SCC) through September 1, 2004 are in the FIRE database. http://www.epa.gov/ttn/chief/software/fire/index.html

Federal Energy Management Program (FEMP) Software

Analytical software tools are intended to help choose conservation measures that are most cost effective and environmentally friendly. Used at the facility evaluation and assessment stage of energy project development, the tools compare potential energy conservation measures by performing complex energy consumption analyses and modeling, as well as comparative life-cycle costing analyses. eere.energy.gov/femp/information/access_tools.cfm

FEMWATER/LEWASTE Model

Three-Dimensional Finite Element Model of Water Flow Through Saturated-Unsaturated Media (3DFEMWATER) and Three-Dimensional Lagrangian–Eulerian Finite Element Model of Waste Transport Through Saturated–Unsaturated Media (3DLEWASTE) are related and can be used together to model flow and transport in three-dimensional, variably-saturated porous media under transient conditions with multiple distributed and point sources/sinks. These models can be used to apply the assimilative capacity criterion to development of wellhead protection areas, as each U.S. state is required to do under the 1986 Amendments to the Safe Drinking Water Act. http://www.epa.gov/ceampubl/gwater/femwater/index.htm

Fertilizer Storage and Handling Software

This HTML program gives a general overview of fertilizer handling and storage, and features a risk assessment section, and recommendations for correcting fertilizer storage problems. http://www.epa.gov/seahome/farmfert.html

FIFRA Section 18 Database

The EPA FIFRA Section 18 Emergency Exemption database provides information about current and recent actions under Section 18. For detailed information about the tolerances associated with a particular action, the Federal Register can be accessed to locate the tolerance document according to the date it was published. This database is updated approximately every 2 weeks. http://www.epa.gov/opprd001/section18/#dbinfo

FLOW-CALC Software

FLOW-CALC is a software tool developed by EPA's Clean Air Markets Division for stack testers who perform flow RATAs for units under the Acid Rain Program, the OTC NOx Budget Program and the SIP Call NOx Budget Trading Program. The software is designed to improve data quality and facilitate the entry of flow RATA data into EDR v2.1 format. http://www.epa.gov/airmarkets/monitoring/mdc/flowcalc.html

Food and Gill Exchange of Toxic Substances (FGETS) Software Model

Food and Gill Exchange of Toxic Substances (FGETS) is a FORTRAN simulation model that predicts temporal dynamics of fish whole body concentration [ug chemical/(g live weight fish)] of non ionic, nonmetabolized, organic chemicals that are bioaccumulated from either: (a) water only — the predominant route of exchange during acute exposures, or (b) water and food jointly — characteristic of chronic exposures. http://www.epa.gov/ceampubl/fchain/fgets/index.htm

Fuel Storage Practices on the Farm

An overview of the importance and techniques of proper fuel storage. The risk assessment portion prompts the user to describe how petroleum products are stored and then evaluates the safety of these practices and provides detailed information about any problems discovered through the questionnaire. http://www.epa.gov/seahome/farmfuel.html

(Continued)

Table 12A.6 (Continued)

Fuels Models

This Web page provides information on models estimating emissions impacts to changes in fuel properties and composition. http://www.epa.gov/otaq/fuelsmodel.htm

GCSOLAR Model

GCSOLAR program is a set of routines that computes direct photolysis rates and half-lives of pollutants in the aquatic environment. The half-lives are calculated as a function of season, latitude, time-of-day, depth in water bodies, and ozone layer thickness. This program operates in an interactive screen mode to facilitate data and program command entry by the user. Input values, with few restrictions, are format free. The user controls program flow by entering program execution commands. http://www.epa.gov/ceampubl/swater/gcsolar/index.htm

Geo-EAS Software

A collection of interactive software tools for performing two-dimensional geostatistical analyses of spatially distributed data. The principal functions of the package are the production of grids and contour maps of interpolated (kriged) estimates from sample data. Geo-EAS can produce data maps, univariate statistics, scatter plots/linear regression, and variogram computation and model fitting. http://www.epa.gov/nerlesd1/databases/geo-eas/access.htm

GEOPACK Software

A comprehensive geostatistical software package that allows both novice and advanced users to undertake geostatistical analyses of spatially correlated data. The program generates graphics (i.e. linear or logarithmic line plots, contour and block diagrams); computes basic statistics (i.e., mean, median, variance, standard deviation, skew, and kurtosis); runs programs for linear regression, polynomial regression, and Kolomogorov-Smirnov tests; calculates linear estimations and nonlinear estimations; and determines sample semivariograms and cross-semivariograms. http://www.epa.gov/nerlesd1/databases/geo-pack/access.htm

Geophysics Advisor Software

This program aids in the sampling and monitoring of hazardous waste sites. Utilizing an initial set of questions, the program develops an initial profile that is further refined by additional question based upon earlier answers. The program considers several geophysical methods: electromagnetic induction; resistivity; ground-penetrating radar; magnetic; seismic; soil gas; gravity; and radiometric. The program recommends the type(s) of geophysics that will most likely fit the site requirements for determining the location of contamination and providing site characterization. various methods is shown on screen, indicating the degree of superiority of one method over another. This version contains a database of the physical and chemical properties of 94 substances selected from EPA's first priority list. http://www.epa.gov/nerlesd1/databases/geophy-adv/geophysi.htm

GEOS Software

A public-domain software package developed by EPA to facilitate the collection and analysis of geoenvironmental data. The term "geoenvironmental" includes soil, geologic, and groundwater data collected to assess or monitor environmental conditions at a site. Major GEOS functions include easy data entry (standard database format allows the user to enter data in a spreadsheet format); simple data management and reporting (such as sorts and queries); preliminary site visualization (contouring of soil and groundwater chemical concentration isopleths, creation of actual and interpolated geologic cross sections, and viewing of well screen intervals in relation to subsurface geology); data exchange (existing GRITS/STAT facility data can be imported into GEOS for contouring of groundwater quality data using the PSV module); access to other geoenvironmental software (GEOS provides a convenient framework for moving between multiple programs). http://www.epa.gov/nerlesd1/databases/geos/abstract.htm

Geospatial Data Clearinghouse

The U.S. Environmental Protection Agency (EPA) Node of the National Geospatial Data Clearinghouse is a component of the *National Spatial Data Infrastructure (NSDI)*. This node provides a pathway to find information about geospatial data available from the EPA. Geospatial data is used in Geographic Information Systems (GIS) to identify the geographic location and characteristics of natural or man-made features and boundaries on the earth. http://www.epa.gov/nsdi/

Global Endocrine Disruptor Research Inventory (GEDRI)

This compilation of on going research projects related to endocrine disruptions was assembled following the recommendation of the Intergovernmental Forum on Chemical Safety (IFCS) and the 1997 Declaration of the Environment Leaders of the Eight on Children's Environmental Health. The initial goal was to bring together and update the existing United States Inventory with those of the Canadian government and the European Union. With this accomplished, the organization is now encouraging and accepting submission of research projects by other countries and private industry. http://oaspub.epa.gov/endocrine/pack_edri.All_Page

EPA Grant Writing Tutorial Software

This interactive software tool walks the user through the grant-writing process and helps them learn to write more competitive grants. The program includes: detailed information and tips on writing a grant proposal; how to complete a grant application package; program-specific sections on three EPA grant programs: (1) Environmental Justice, (2) Environmental Justice through Pollution Prevention; and (3) Environmental Education; examples of good, complete grant packages; references; a glossary of terms; resources and contacts; and a mock grant-writing activity where the user is able to compare their results to a successful grant application epa.gov/seahome/grants.html

(Continued)

Table 12A.6 (Continued)

Grants Information and Control System (GICS)

EPA's management information system for grants programs is the Grants Information and Control System (GICS), which awards, administers, and monitors grants. Grants are regularly awarded to Federal, State, or local government agencies, universities, and other institutions that support EPA's environmental programs. Specific types of agreements include assistance agreements, grants, cooperative agreements, interagency agreements, and other types of program support agreements administered by Headquarters or EPA regions. These program support agreements provide for research, demonstration projects, training, fellowships, investigations, surveys, studies, and other types of program support activities. http://www.epa.gov/enviro/html/gics/

Great Lakes Adventure Software

This program (Teacher and Student versions) is aimed at a youth audience (grades 9–12) and will run on both Wintel and Mac computers. Investigate the natural history of the basin, the water cycle, stratification and turnover of the lakes. Explore six Hot Topics: (1) Habitat loss; (2) Beach closures; (3) Fish communities; (4) Toxics; (5) Exotic Species; and (6) Eutrophication

Great Lakes Environmental Database

The *Great Lakes National Program Office* (GLNPO) collects environmental data on a wide variety of constituents in water, biota, sediment, and air. Central to the data management effort is a computerized relational database system to house *Lake Michigan Mass Balance* and other project results. That system, the Great Lakes Environmental Database (GLENDA), was developed to provide data entry, storage, access and analysis capabilities to meet the needs of mass balance modelers and other potential users of Great Lakes data. http://www.epa.gov/glnpo/monitoring/data_proj/glenda/

Green Chemistry Expert System

Green chemistry is the design of chemical products and processes that reduce or eliminate the use or generation or hazardous waste. The Microsoft Access driven GCES software is a stand-alone version of the program designed to perform custom applications. http://www.epa.gov/greenchemistry/gces.html

GRITS/STA Software

A comprehensive database system for storing, analyzing, and reporting information from groundwater monitoring programs at RCRA, CERCLA, and other regulated facilities and sites. This software program integrates EPA's Groundwater Information Tracking System, a database of groundwater information, with STAT, a statistical analysis system. The package provides a nationally responsive system that incorporates data elements from appropriate EPA program offices. The system supports data entry, report generation, export of data to other software applications (e.g., modeling programs), and statistical analysis. http://www.epa.gov/nerlesd1/databases/grits-stat/abstract.htm

Ground Water Primer

This update of the original groundwater program educates users about the nature of groundwater and the principles of groundwater protection. It contains a detailed introduction to hydrogeology, information on numerous drinking water contaminants, and a section on what you can do to protect your groundwater. http://www.epa.gov/seahome/gwprimer.html

Hazards Analysis for Toxics Analysis (HATS) Software

EPA and NOAA developed this web site to facilitate the use of CAMEO and to offer online technical support. http://www.epa.gov/ceppo/cameo/

Hazardous Waste Management on the Farm Software

This tutorial is intended to serve as a guide towards proper hazardous waste management. Knowing the regulations, different methods of disposal of hazardous waste, and how to provide a safer environment are all part of proper hazardous waste management. A risk assessment included in this program will allow you to identify high risk practices on farms. http://www.epa.gov/seahome/farmhaz.html

Health Effects Notebook for Hazardous Air Pollutants

The fact sheets available on this Web page describe the effects on human health of substances that are *defined as hazardous* by the 1990 amendments of the *Clean Air Act*. These substances include certain volatile organic chemicals, pesticides, herbicides, and radionuclides that present tangible hazard, based on scientific studies of exposure to humans and other mammals. There is uncertainty in the precise degree of hazard, and readers are cautioned that the fact sheets may be revised as additional data become available. http://www.epa.gov/ttnatw01/hlthef/hapindex.html

HELP Software

A quasi-two-dimensional modeling program that simulates water movement into and out of landfills based on a waste management systems particular design. The user can conduct water-balance analyses of solid waste disposal and containment facilities. The program allows comparison of proposed landfill designs by estimating runoff, evapotranspiration, drainage, leachate collection, and liner leakage. Modeling incorporates information on cover soils, waste cells, lateral drain layers, low permeability barrier soils, synthetic geomembrane liners, and weather. Results are expressed as daily, monthly, annual, and long-term water budgets. Although applicable to most landfills, HELP was developed specifically for modeling hazardous and municipal solid waste landfills as required by RCRA. http://www.epa.gov/nerlesd1/databases/help/abstract.htm

High Production Volume (HPV) Voluntary Challenge Chemical List

Searchable database of the 1990 or 1994 list by CAS Number, Chemical Name, Indicator value, Chemical Sponsorship Status, or Sponsor Commitment Status. epa.gov/chemrtk/opptsrch.htm

(Continued)

Table 12A.6 (Continued)

The Hotelling/Williams Test for the Difference between Two Dependent Correlations
The Hotelling/Williams procedure tests the null hypothesis of equality between two dependent product-moment correlations. Specifically, it tests whether the correlation between Z and X differs from the correlation between Z and Y, where X, Y and Z are three variables measured on the same set of observational units. http://www.epa.gov/wed/pages/models/correlations/correlations.htm

Household Waste Management Software
This HTML program teaches the user how to safely and efficiently manage waste, and particularly hazardous waste, in the home. Users can visit a 'virtual house' and choose extended discussions of the many products they might encounter there. Safe homemade alternatives are presented and evaluated. An enormous state-by-state database of mail, telephone, and Internet contacts puts users in touch with a wealth of free downloadable publications. Provides over 30 quizzes and activities for K-12 students, sorted by grade level. http://www.epa.gov/seahome/hwaste.html

Human Exposure Database Software
HEDS is the Human Exposure Database System. HEDS is an integrated database system that contains chemical measurements, questionnaire responses, documents, and other information related to EPA research studies of the exposure of people to Environmental contaminants. http://www.epa.gov/heds/index.htm

Human Exposure Model (HEM)
The Human Exposure Model (HEM) is used primarily for performing risk assessments for major point sources (usually producers or large users of specified chemicals) of air toxics. The HEM only addresses the inhalation pathway of exposure, and is designed to predict risks associated with emitted chemicals in the ambient air (i.e., in the vicinity of an emitting facility but beyond the facility's property boundary). The HEM provides ambient air concentrations, as surrogates for lifetime exposure, for use with unit risk estimates and inhalation reference concentrations to produce estimates of cancer risk and noncancer hazard, respectively, for the air toxics modeled. http://www.epa.gov/ttn/fera/human_hem.html

Hydrodynamic, Sediment, and Contaminant Transport Model (HSCTM2D)
The Hydrodynamic, Sediment, and Contaminant Transport Model (HSCTM2D) is a finite element modeling system for simulating two-dimensional, vertically-integrated, surface water flow (typically riverine or estuarine hydrodynamics), sediment transport, and contaminant transport. The modeling system consists of two modules, one for hydrodynamic modeling (HYDRO2D) and the other for sediment and contaminant transport modeling (CS2D). One example problem is included. The HSCTM2D modeling system may be used to simulate both short term (less than 1 year) and long term scour and/or sedimentation rates and contaminant transport and fate in vertically well mixed bodies of water. http://www.epa.gov/ceampubl/swater/hsctm2d/index.htm

Hydrological Simulation Program — FORTRAN (HSPF)
Hydrological Simulation Program — FORTRAN (HSPF) is a comprehensive package for simulation of watershed hydrology and water quality for both conventional and toxic organic pollutants. HSPF incorporates watershed-scale ARM and NPS models into a basin-scale analysis framework that includes fate and transport in one dimensional stream channels. It is the only comprehensive model of watershed hydrology and water quality that allows the integrated simulation of land and soil contaminant runoff processes with In-stream hydraulic and sediment-chemical interactions. The result of this simulation is a time history of the runoff flow rate, sediment load, and nutrient and pesticide concentrations, along with a time history of water quantity and quality at any point in a watershed. HSPF simulates three sediment types (sand, silt, and clay) in addition to a single organic chemical and transformation products of that chemical. http://www.epa.gov/ceampubl/swater/hspf/

IMES Software
A computer-based tool for matching a site's characteristics with the appropriate model or models. This integrated system for model evaluation has three elements: (1) A selection system for use in choosing an exposure assessment model (currently for air, groundwater, nonpoint source, and surface water media models); (2) A validation database that includes information on models for air, groundwater, nonpoint source, and surface water assessment; and (3) A model uncertainty database (currently only for six surface water models). Each IMES element can be used independently. Once a model(s) has been selected, information on model uncertainty can be accessed. http://www.epa.gov/nerlesd1/databases/imes/abstract.htm

Improving Silage Storage on the Farm Software
To prevent possible well contamination, silos should be located as far away from wells as practical. This program gives a general overview of silage storage, and features a risk assessment section, and recommendations for correcting silage storage problems. http://www.epa.gov/seahome/silage.html

Indoor Air Quality Software
Indoor air quality problems carry a high degree of human health risk. This program explores how the environment of a typical home or office affects indoor air quality. Colorful building cross-sections offer users self-guided home, office and pollutant tours. Sources of various indoor air pollutants are discussed, along with associated health risks for each pollutant, and how to reduce pollutant levels. http://www.epa.gov/seahome/indoor.html

(Continued)

Table 12A.6 (Continued)

Industrial Waste Air Model (IWAIR)

EPA developed IWAIR to assist facility managers and regulatory agency staff in evaluating inhalation risks. Workers and residents in the vicinity of a unit may be exposed to volatile chemicals from the unit in the air they breathe. Exposure to some of these chemicals at sufficient concentrations may cause a variety of cancer and noncancer health effects (such as developmental effects in a fetus or neurological effects in an adult). With a limited amount of site-specific information, IWAIR can estimate whether specific wastes or waste management practices may pose an unacceptable risk to human health. http://www.epa.gov/epaoswer/non-hw/industd/iwair.htm

Industrial Waste Management Evaluation Model (IWEM)

The IWEM software is designed to assist in determining the most appropriate waste management unit design to minimize or avoid adverse groundwater impacts, by evaluating types of liners, the hydrogeologic conditions of the site, and the toxicity and expected leachate concentrations of the anticipated waste constituents. The software compares the groundwater protection afforded by various liner systems with the anticipated leachate concentrations to determine what minimum liner system is needed to be protective of human health and groundwater resources (or in the case of land application units (LAUs), determine whether or not land application is recommended). http://www.epa.gov/epaoswer/non-hw/industd/iwem.htm

Integrated Compliance Information System (ICIS)

The Integrated Compliance Information System (ICIS) supports the information needs of the National Enforcement and Compliance program as well as the unique needs of the National Pollutant Discharge Elimination System (NPDES) program. ICIS will integrate data that is currently located in more than a dozen separate data systems. The Web-based system will eventually enable individuals from states, communities, facilities, and EPA to access integrated enforcement and compliance data from any desktop connected to the Internet. EPA's ability to target the most critical environmental problems will improve as the system integrates data from all media. epa.gov/Compliance/planning/data/modernization/index.html

Integrated Data for Enforcement Analysis System (IDEA)

The Integrated Data for Enforcement Analysis system (IDEA) is a comprehensive single-source of environmental performance data on regulated facilities within the EPA. IDEA provides a comprehensive historical profile of inspections, enforcement actions, penalties assessed, toxic chemicals released, and emergency hazardous spills for any EPA regulated facility. This single point of access provides information from the Agency's Air, Water, Hazardous Waste, Toxic Chemical Release Inventory, and Emergency Response Notification Systems. epa.gov/Compliance/planning/data/multimedia/idea/index.html

Integrated Exposure Uptake Biokinetic Model for Lead in Children (IEUBK)

Lead poisoning presents potentially significant risks to the health and welfare of children all over the world today. The Integrated Exposure Uptake Biokinetic Model for Lead in Children (IEUBK) attempts to predict blood-lead concentrations (PbBs) for children exposed to lead in their environment. http://www.epa.gov/superfund/programs/lead/ieubk.htm

Integrated Risk Information System (IRIS)

The Integrated Risk Information System (IRIS) is prepared and maintained by the U.S. Environmental Protection Agency (U.S. EPA), is an electronic database containing information on human health effects that may result from exposure to various chemicals in the environment. IRIS was initially developed for EPA staff in response to a growing demand for consistent information on chemical substances for use in risk assessments, decision-making and regulatory activities. The information in IRIS is intended for those without extensive training in toxicology, but with some knowledge of health sciences. http://www.epa.gov/iris/intro.htm

Internet Geographical Exposure Modeling System (IGEMS)

The IGEMS is a modernization of OPPT's older Graphical Exposure Modeling System and PCGEMS tools. IGEMS brings together in one system several EPA environmental fate and transport models and some of the environmental data needed to run them. IGEMS includes models and data for ambient air, surface water, soil, and groundwater, and makes the models much easier to use than their stand-alone counterparts. IGEMS will have graphics and Geographical Information System (GIS) capabilities for displaying environmental modeling results. http://www.epa.gov/opptintr/exposure/docs/gems.htm

IsoConc: Concentration-Dependent Stable Isotope Mixing Model

ISOCONC1_01.xls is a Microsoft Excel 2000™ spreadsheet which performs calculations for the concentration-weighted stable isotope mixing model outlined in Phillips DL & Koch PL (2002) Incorporating concentration dependence in stable isotope mixing models, Oecologia 130: 114–125. This dual-isotope model takes into account isotopic element concentration differences among the sources in determining the proportional contributions of sources to a mixture. epa.gov/wed/pages/models/isoconc/isoconc1_01.htm

IsoConc: Confidence Interval Calculation for Source Partitioning Using Stable Isotopes

ISOERROR1_04 is a Microsoft Excel 2000™ spreadsheet which calculates estimates and confidence intervals of source proportional contributions to a mixture, using stable isotope analyses. Examples include the proportions of different food sources in an animal's diet, C3 vs. C4 plant inputs to soil organic carbon, etc. Linear mixing models are used to partition two sources with a single isotopic signature (e.g., d13C) or three sources with a second isotopic signature (e.g., d15N). epa.gov/wed/pages/models/isotopes/isoerror1_04.htm

(Continued)

Table 12A.6 (Continued)

IsoSource: Stable Isotope Mixing Model for Partitioning an Excess Number of Sources
IsoSource is a Microsoft Visual Basic™ software package which calculates ranges of source proportional contributions to a mixture based on stable isotope analyses when the number of sources is too large to permit a unique solution (> number of isotope systems +1). Examples include partitioning of pollutant sources in a waste stream, food sources in an animal's diet, water sources for plant uptake, carbon sources in soil organic matter, and many others. epa.gov/wed/pages/models/isosource/isosource.htm

Landfill Air Emissions Estimation Model
The Landfill Air Emissions Model is PC-based software for estimating emissions of methane, carbon dioxide, nonmethane organic compounds, and hazardous air pollutants from municipal solid waste landfills. These emissions are generated by decomposition of refuse in landfills. The mathematical model used in Landfill is based on a first order decay equation that can be run using site-specific data supplied by the user for the parameters needed to estimate emissions or, if data are not available, using default value sets included in Landfill. epa.gov/oar/oaqps/landfill.html

LandView® 5
LandView 5 contains selected Census 2000 demographic data from Summary File 1 (SF1) and maps based on the Census 2000 TIGER/Line® files for all states, the District of Columbia and Puerto Rico. These maps show both streets and Census 2000 legal and statistical areas (including Census 2000 Urban/Rural delineations). LandView 5 also contains recent EPA and USGS Geographic Names Information System (GNIS) data and maps. http://www.census.gov/geo/landview/lv5/lv5.html

LC50 (Lethal Concentration, 50%) Software Model
LC50 (lethal concentration, 50%) estimates LC50 values using the Trimmed Spearman-Karber Method. This program originated at Montana State University and has been modified by data processing staffs at the Duluth and Athens National Exposure Research Laboratories. http://www.epa.gov/ceampubl/fchain/lc50/index.htm

The Lead Contamination Information System (Lead in Drinking Water)
The dangers of lead in drinking water are explained using graphics and hypertext. The program provides basic facts about this vital topic, answering common questions on the effects of lead in drinking water, and presenting techniques for reducing lead exposure. leaddw.html http://www.epa.gov/seahome/leaddw.html

Lead in the Environment Software
A detailed look at the health problems posed by lead. Learn about testing children for lead poisoning. This hypertext program presents sources of lead, health effects, and techniques for reducing lead exposure in and around the home. A quiz helps users review what they have learned, and the reference section points them to local, regional and national sources of information and help. http://www.epa.gov/seahome/leadenv.html

List of Lists Database — Consolidated list of Chemicals Subject to the Emergency Planning and Community Right-to-Know Act (EPCRA and Section 112® of the Clean Air Act
This is a searchable database of EPA's October 2001 Consolidated List of Chemicals Subject to the Emergency Planning and Community Right-to-Know Act (EPCRA) and Section 112(r) of the Clean Air Act. The entire database is available as a PDF file. epa.gov/ceppo/pubs/title3.pdf

Local Emergency Planning Committee (LEPC) Database
The Emergency Planning and Community Right-to-Know Act (EPCRA) of 1986 required each state's Governor to establish a state emergency response commission (SERC) which in turn resulted in the formation of local emergency planning committees. The database provides information related to data compiled by LEPCs nationwide. epa.gov/ceppo/lepclist.htm

Livestock Manure Handling
Similar in content to Livestock Manure Storage, this program gives an overview of manure types and proper management systems for different animals including cattle, swine, sheep, poultry, and horses. Users can evaluate their own risk through the interactive questionnaire, and receive suggestions for reducing the threat to their water supply. http://www.epa.gov/seahome/manure-handle.html

Mean Similarity Analysis
MEANSIM6 contains software for Mean Similarity Analysis, a method of assessing the strength of a classification of many objects (sites) into a relatively small number of groups. Classification strength is measured by the extent to which sites within the same groups are more similar to each other, on average, than they are to sites in different groups. Users supply a matrix of pairwise similarities (or dissimilarities) for all possible pairs of sites to a program (RNDTST6) that calculates mean between and within-group similarities. http://www.epa.gov/wed/pages/models/dendro/mean_similarity_analysis.htm

Mercury in Medical Facilities
This HTML program teaches health care professionals (and anyone else) about the hazards of mercury as commonly used in medical care facilities, especially hospitals and clinics. Mercury can pose a significant health problem for these professionals and also for patients and the environment in general. The program explains the impact of mercury on human health and the integrity of ecosystems, and a 'virtual hospital' allow users to see sources of mercury and their alternatives. http://www.epa.gov/seahome/mercury.html

Meteorological Data
The meteorological data files contain measurements taken at 237 weather stations located throughout the United States for a period extending from 1961 to 1990. Exact collection dates vary by weather station. http://www.epa.gov/ceampubl/tools/metdata/index.htm

(Continued)

Table 12A.6 (Continued)

Milking Center Wastewater Treatment

Considers the special problems posed by milkhouse wastewater. Explores design of the milking facility for cleaning management, determining strength and volume of effluent flow, and devising facilities to receive, treat, and safely dispose of the wastewater. A risk assessment questionnaire evaluates the soundness of the user's situation and suggests improvements. http://www.epa.gov/seahome/milking.html

MINTEQA2

MINTEQA2 is a equilibrium speciation model that can be used to calculate the equilibrium composition of dilute aqueous solutions in the laboratory or in natural aqueous systems. The model is useful for calculating the equilibrium mass distribution among dissolved species, adsorbed species, and multiple solid phases under a variety of conditions including a gas phase with constant partial pressures. A comprehensive data base is included that is adequate for solving a broad range of problems without need for additional user-supplied equilibrium constants. http://www.epa.gov/ceampubl/mmedia/minteq/index.htm

MOBILE Model (On-Road Vehicles)

This web page contains information on the MOBILE vehicle emission factor model, which is a software tool for predicting gram per mile emissions of hydrocarbons (HC), carbon monoxide (CO), oxides of nitrogen (NOx), carbon dioxide (CO_2), particulate matter (PM) and air toxics from cars, trucks, and motorcycles under various conditions. http://www.epa.gov/otaq/mobile.htm

Models-3 Air Quality Modeling System

The primary goals for the Models-3/Community Multiscale Air Quality (CMAQ) modeling system are to improve 1) the environmental management community's ability to evaluate the impact of air quality management practices for multiple pollutants at multiple scales and 2) the scientist's ability to better probe, understand, and simulate chemical and physical interactions in the atmosphere. epa.gov/asmdnerl/models3/index.html

MOFAT Software

A two-dimensional, finite element model for simulating coupled multiphase flow and multicomponent transport in planar or radically symmetric vertical sections. MOFAT evaluates flow and transport for water, nonaqueous phase liquid (NAPL), and gas. The program also can be used when gas and/or NAPL phases are absent in part or all of the domain. The flow module can analyze either two-phase flow of water and NAPL in a system of constant gas pressure or explicit three-phase flow of water. The transport module can accommodate up to five components partitioning among water, NAPL, gas, and solid phases, assuming either local equilibrium interphase mass transfer or first-order kinetically controlled mass transfer. http://www.epa.gov/nerlesd1/databases/mofat/abstract.htm

Most Probable Number of Microorganisms

This program can be used to calculate Most Probable Number (MPN) values for the quantitation of microorganisms. http://www.epa.gov/nerlcwww/other.htm

AIRMaster+

AIRMaster+ provides comprehensive information on assessing compressed air systems, including modeling, existing and future system upgrades, and evaluating savings and effectiveness of energy efficiency measures. oit.doe.gov/bestpractices/software_tools.shtml

MOVES (Motor Vehicle Emission Simulator)

To keep pace with new analysis needs, modeling approaches, and data, the EPA's Office of Transportation and Air Quality (OTAQ) is developing a modeling system termed the MOtor Vehicle Emission Simulator (MOVES). This system will estimate emissions for on-road and nonroad sources, cover a broad range of pollutants, and allow multiple scale analysis, from fine-scale analysis to national inventory estimation. When fully implemented MOVES will serve as the replacement for MOBILE6 and NONROAD. http://www.epa.gov/otaq/ngm.htm

Multi-Chamber Concentration and Exposure Model (MCCEM) Version 1.2

Estimates average and peak indoor air concentrations of chemicals released from products or materials in houses, apartments, townhouses, or other residences. The data libraries contained in MCCEM are limited to residential settings. However, the model can be used to assess other indoor environments (e.g., schools, offices) if the user can supply the necessary inputs. Estimates inhalation exposures to these chemicals, calculated as single day doses, chronic average daily doses, or lifetime average daily doses. epa.gov/opptintr/exposure/docs/mccem.htm

MULTIMED 2.0 BETA

The Multimedia Exposure Assessment Model (MULTIMED) for exposure assessment simulates the movement of contaminants leaching from a waste disposal facility. The model consists of a number of modules which predict concentrations at a receptor due to transport in the subsurface, surface air, or air. http://www.epa.gov/ceampubl/mmedia/multim2/index.htm

MULTIMED Daughter Process Model

The Multimedia Exposure Assessment Model (MULTIMED) for exposure assessment simulates the movement of contaminants leaching from a waste disposal facility. The MULTIMED model has been modified (MULTIMDP) to simulate the transport and fate of first and second-generation transformation (daughter) products that migrate from a waste source through the unsaturated and saturated zones to a down gradient receptor well. http://www.epa.gov/ceampubl/mmedia/multidp/index.htm

(Continued)

Table 12A.6 (Continued)

The Multimedia Contaminant Fate, Transport, and Exposure Model (MMSOILS)

The Multimedia Contaminant Fate, Transport, and Exposure Model (MMSOILS) estimates the human exposure and health risk associated with releases of contamination from hazardous waste sites. The methodology consists of a multimedia model that addresses the transport of a chemical in groundwater, surface water, soil erosion, the atmosphere, and accumulation in the food chain. The human exposure pathways considered in the methodology include: soil ingestion, air inhalation of volatiles and particulates, dermal contact, ingestion of drinking water, consumption of fish, consumption of plants grown in contaminated soil, and consumption of animals grazing on contaminated pasture. For multimedia exposures, the methodology provides estimates of human exposure through individual pathways and combined exposure through all pathways considered. The risk associated with the total exposure dose is calculated based on chemical-specific toxicity data. http://www.epa.gov/ceampubl/mmedia/mmsoils/index.htm

Multimedia Exposure Assessment Model (MULTIMED)

The Multimedia Exposure Assessment Model (MULTIMED) for exposure assessment simulates the movement of contaminants leaching from a waste disposal facility. The model consists of a number of modules which predict concentrations at a receptor due to transport in the subsurface, surface air, or air. http://www.epa.gov/ceampubl/mmedia/multim1/index.htm

Multimedia Integrated Modeling System

Numerical models are being developed in response to a growing interest in studying and addressing issues that have influences and effects that cross physical media, such as air, water, and soil. Numerical models provide an important tool in studying and addressing such issues. The models are used to test our understanding of ecosystems, diagnose causes of problems, and predict conditions given assumptions about future influences on the environment. EPA's Multimedia Integrated Modeling System (MIMS) project is researching and developing solutions for some of those challenges. epa.gov/AMD/mims/index.html

Multimedia, Multi-Pathway, Multi-Receptor Exposure, and Risk Assessment (3MRA)

The Multimedia, Multi-pathway, Multi-receptor Exposure and Risk Assessment (3MRA) technology provides the ability to conduct screening-level risk-based assessment of potential human and ecological health risks resulting from long term (chronic) exposure to HWIR chemicals released from land-based waste management units (WMUs) containing currently listed waste streams. The 3MRA system consists of a series of components within a system framework. The new modeling system, dubbed 3MRA technology, is envisioned as the foundation for eventually integrating other regulatory support decision tool needs anticipated in the future. http://www.epa.gov/ceampubl/mmedia/3mra/index.htm

Municipal Pollution Prevention Diagnostic Planner Software

This hypertext-based program serves as an early warning system for wastewater treatment plant operators. The user supplies information on influent loadings, discharge monitoring data, pretreatment, financial status, failures and upsets, and other aspects of plant operation and maintenance. http://www.epa.gov/seahome/cmar.html

National Asbestos Registry System

The National Asbestos Registry System (NARS) may be used by building owners to assist in evaluating which asbestos contractors should be hired, or by the general public to determine the compliance history of asbestos contractors working at a specific locale. NARS is also used by delegated agencies to assist in targeting inspections and by EPA managers as a general tool for evaluating the asbestos program. http://cfpub.epa.gov/nars/

National Compliance Data Base System (NCDB)

The National Compliance Data Base System (NCDB) tracks regional compliance and enforcement activity, and manages the Pesticides and Toxic Substances Compliance and Enforcement program at a national level. The system tracks all compliance monitoring and enforcement activities from the time an inspector conducts an inspection until the time the inspector closes the case or settles the enforcement action. epa.gov/Compliance/planning/data/toxics/ncdbsys.html

National Contaminant Occurrence Database (NCOD)

EPA has developed NCOD to satisfy the statutory requirements set by Congress in the 1996 amendments to the *Safe Drinking Water Act (SDWA)* to maintain a national drinking water contaminant occurrence database using occurrence data for both regulated and unregulated contaminants in public water systems. This site provides a library of water sample analytical data (or "samples data") that EPA is currently using and has used in the past for analysis, rulemaking, and rule evaluation. The drinking water sample data, collected at Public Water Systems, are for both regulated and unregulated contaminants. http://www.epa.gov/safewater/data/ncod.html

National Emissions Inventory

The U.S. EPA prepares a national emissions inventory with input from numerous State and local air agencies. These data are used for air dispersion modeling, regional strategy development, regulation setting, air toxics risk assessment, and tracking trends in emissions over time. http://www.epa.gov/ttn/chief/eiinformation.html

National Hydrography Dataset

The National Hydrography Dataset (NHD) is a comprehensive set of digital spatial data that contains information about surface water features such as lakes, ponds, streams, rivers, springs and wells. These linkages enable the analysis and display of these water-related data in upstream and downstream order. http://nhd.usgs.gov/

(Continued)

Table 12A.6 (Continued)

National Listing of Fish Consumption Advisories
Fish may contain chemicals that could pose health risks. When contaminant levels are unsafe, consumption advisories may recommend that people limit or avoid eating certain species of fish caught in certain places. http://www.epa.gov/OST/fishadvice/

National Water Quality Inventory
Water Quality Reports, required by Section 305(b) of the Clean Water Act, contain information on the attainment of water quality standards. States, Tribes, and Territories classify assessed waters as either Fully Supporting, Threatened, or Not Supporting their designated use(s). http://www.epa.gov/waters/305b/

NONROAD Model
The Draft Nonroad Model is used for estimation of air pollution inventories by professional mobile source modelers, such as state air quality officials and consultants. The model is still draft, so some emission rates and activity levels predicted by the model may substantially change. http://www.epa.gov/otaq/nonrdmdl.htm

NPS (Nonpoint Source) Modeling
Nonpoint source assessment procedures and modeling techniques are reviewed and discussed for both urban and nonurban land areas. Detailed reviews of specific methodologies and models are presented, along with overview discussions focusing on urban methods and models, and on non-urban (primarily agricultural) methods and models. http://www.epa.gov/ceampubl/tools/nps/index.htm

Nutrient Criteria Database
EPA has developed a National Nutrient Database to store and analyze nutrient water quality data and to serve as an information resource for states, tribes, and others in establishing scientifically defensible numeric nutrient criteria. It contains ambient data from EPA's Legacy STOrage and RETrieval (STORET) data system, the U.S. Geological Survey's National Stream Quality Accounting Network (NASQAN) data and National Water Quality Assessment (NAWQA) data, and other relevant sources such as universities and states/tribes. The ultimate use of the data is to derive ecoregional waterbody-specific numeric nutrient criteria. http://www.epa.gov/waterscience/criteria/nutrient/database/index.html

Oil Spill Exercise Generator
This database-driven program can construct an oil spill exercise or response drill in a matter of hours, and fill out checklists for use in a final evaluation report. http://www.epa.gov/seahome/oilspill.html

On-Site: The On-line Site Assessment Tool
On-Site was developed to provide modelers and model reviewers with prepackaged tools ("calculators") for performing site assessment calculations. The philosophy behind On-Site is that the convenience of the prepackaged calculators helps provide consistency for simple calculations, and access to methods and data that are not commonly available. The latter include data on fuel composition and models for leaching from fuel lenses. http://www.epa.gov/athens/learn2model/part-two/onsite/index.html

Online Tracking Information System (OTIS)
The OnlineTracking Information System (OTIS) is a collection of search engines which enables EPA staff, state/local/tribal governments, and federal agencies to access a wide range of data relating to enforcement and compliance. OTIS is a web application that sends queries to the Integrated Data for Enforcement Analysis (IDEA) system. Only State and Federal agencies have access to the OTIS data system. http://www.epa.gov/Compliance/planning/data/multimedia/aboutotis.html

Oxygen Requirements of Fishes (OXYREF) Model
The Oxygen Requirements of Fishes (OXYREF) database contains test results reported in the published literature on the respiratory oxygen requirements of fishes. This file is based on results of a computerized literature search, over the time period 1969–1986, of three separate journal referencing services: Biological Sciences Information Service (BIOSIS PREVIEWS), Aquatic Sciences and Fisheries Abstracts (ASFA), and Canadian Water Resources References (AQUAREF). http://www.epa.gov/ceampubl/swater/oxyref/index.htm

Ozone Depletion Software
This program includes information and graphics depicting the ozone layer, how ozone is created, causes and effects of ozone depletion, the ozone hole, the greenhouse effect, and how an individual can help prevent ozone depletion. The program also examines applicable government legislation, including portions of the Clean Air Act Amendments of 1990. http://www.epa.gov/seahome/ozone.html

Polychlorinated Biphenyls (PCBs)
EPA provides various paths for the public to access information about PCBs. This website provides information on the health effects of PCBs, laws and regulations that govern PCBs, lists of approved companies that can handle PCB wastes including storage and disposal companies, alternative decontamination procedures and scrap metal recovery ovens. EPA also maintains two databases-PCB Activity Database (PADS) and Transformer Registration and applicable forms. http://www.epa.gov/opptintr/pcb/index.html

Permit Application Software System (PASS) for NPDES
The Permit Application Software System (PASS) is a software solution provided by the United States Environmental Protection Agency (U.S. EPA) to facilitate applications for National Pollutant Discharge Elimination System (NPDES) permits. http://cfpub2.epa.gov/npdes/permitissuance/pass.cfm

Permit Compliance System (PCS)
The Permit Compliance System (PCS) provides information on companies which have been issued permits to discharge waste water into rivers, when a permit was issued and expires, how much the company is permitted to discharge, and the actual monitoring data showing what the company has discharged. http://www.epa.gov/enviro/html/pcs/index.html

(Continued)

Table 12A.6 (Continued)

Personal Greenhouse Gas Calculator

EPA has developed tools to help individuals reduce household greenhouse gas emissions. These tools are designed for individuals who want to take action and for businesses and organizations interested in reaching their employees and members about what they can do at home to help protect our climate. http://yosemite.epa.gov/oar/globalwarming.nsf/content/ResourceCenterToolsGHGCalculator.html

PESTAN Software

A model used to estimate the vertical migration of dissolved organic solutes through the vadose zone to groundwater. Estimates are based on a closed-form analytical solution of the advective–dispersive–reactive transport equation. The model is intended for use in conducting initial screening assessments of the potential for contamination of groundwater from currently registered pesticides and those submitted for registration. http://www.epa.gov/nerlesd1/databases/pestan/abstract.htm

Pesticide Assessment Tool for Rating Investigations of Transport (PATRIOT)

Pesticide Assessment Tool for Rating Investigations of Transport (PATRIOT) provides rapid analyses of groundwater vulnerability to pesticides on a regional, state, or local level. PATRIOT assesses groundwater vulnerability by quantifying pesticide leaching potential in terms of pesticide mass transported to the water table. PATRIOT is composed of: (1) a pesticide fate and transport model (PRZM2), (2) a comprehensive database, (3) an interface that facilitates database exploration, (4) a directed sequence of interactions that guide the user in providing necessary information to perform alternative model analyses, and (5) user-selected methods for summarizing and visualizing results. http://www.epa.gov/ceampubl/gwater/patriot/index.htm

Pesticide Data Submitters List (PDSL)

The Pesticide Data Submitters List is a compilation of names and addresses of registrants who wish to be notified and offered compensation for use of their data. It was developed to assist pesticide applicants in fulfilling their obligation as required by sections 3(c)(1)(f) and 3(c)(2)(D) of the Federal Insecticide, Fungicide, and Rodenticide Act (FIFRA) and 40 CFR Part 152 sub part E regarding ownership of data used to support registration. http://www.epa.gov/opppmsd1/DataSubmittersList/

Pesticide Database Descriptions

The Office of Pesticide Programs has five environmental databases that it uses to assess hazards to the environment and to wildlife, aquatic organisms, and plants. Some of these databases are in the initial phases of development, while others are fully operational. http://www.epa.gov/oppefed1/general/databasesdescription.htm

Pesticide Product Information System (PPIS)

The Pesticide Product Information System (PPIS) contains information concerning all pesticide products registered in the United States. http://www.epa.gov/opppmsd1/PPISdata/

Pesticide Product Label System (PPLS)

The Pesticide Product Label System is a collection of images of pesticide labels which have been approved by the Office of Pesticide Programs (OPP) under Section 3 of the Federal Insecticide, Fungicide, and Rodenticide Act. The collection contains the initially approved label for pesticide products registered under *FIFRA Section 3* as well as subsequent versions of labels which have changed via amendment or notification. http://www.epa.gov/pesticides/pestlabels/

Pesticide Reregistration Status

EPA is reviewing older pesticides (those initially registered prior to November 1984) under the Federal Insecticide, Fungicide, and Rodenticide Act (FIFRA) to ensure that they meet current scientific and regulatory standards. This process, called reregistration, considers the human health and ecological effects of pesticides and results in actions to reduce risks that are of concern. EPA also is reassessing tolerances (pesticide residue limits in food) to ensure that they met the safety standard established by the Food Quality Protection Act (FQPA) of 1996. http://cfpub.epa.gov/oppref/rereg/status.cfm?show=rereg

Pesticide-Restricted Use Products Report (RUP)

The Restricted Use Products Report is a compilation of both active and cancelled pesticide products classified as "Restricted Use". The "Restricted Use" classification restricts a product, or its uses, to use by a certificated pesticide applicator or under the direct supervision of a certified applicator. http://www.epa.gov/opprd001/rup/index.htm

Pesticide Storage and Handling Practices on the Farm

Explains proper handling and storage of pesticides on farmsteads. Provides a general overview of pesticide handling and storage, a risk assessment section, and recommendations for correcting problems related to pesticide storage. http://www.epa.gov/seahome/farmpest.html

Pesticide Water Models

Scientists in the Office of Pesticide Programs (OPP) in the Environmental Protection Agency (EPA) frequently use simulation models to predict pesticide concentrations in surface and groundwater for use in both human health and aquatic ecological exposure assessments. http://www.epa.gov/oppefed1/models/water/index.htm

(Continued)

Table 12A.6 (Continued)

PLUMES Models

PLUMES includes two initial dilution plume models (RSB and UM) and a model interface manager for preparing common input and running the models. Two farfield algorithms are automatically initiated beyond the zone of initial dilution. PLUMES also incorporates the flow classification scheme of the Cornell Mixing Zone Models (CORMIX) with recommendations for model usage, thereby providing a linkage between the systems. PLUMES models are intended for use with plumes discharged to marine and some freshwater bodies. Both buoyant and dense plumes, single sources, and many diffuser outfall configurations can be modeled. http://www.epa.gov/ceampubl/swater/plumes/index.htm

PM (Particulate Matter) Calculator

EPA has replaced the PM-10 Calculator program with new software that will estimate filterable PM2.5 emissions. PM Calculator is applicable to point sources only and requires the user to input uncontrolled emissions (either total filterable particulate or filterable PM10) for each source, the source category classification (SCC) and the control device, if any. The program will then calculate controlled emissions for filterable PM2.5 and filterable PM10 for each point source. http://www.epa.gov/ttn/chief/software/pmcalc/index.html

Pollution Prevention Software

The Pollution Prevention Act of 1990 established source reduction as the preferred approach to environmental protection. P2 can save money, reduce liability, and improve efficiency, worker safety, and competitiveness. This program outlines the relevant environmental legislation and addresses P2 in industry, agriculture, energy, government, and consumer sectors. Examples include more efficient manufacturing processes, alternative products and procedures that generate less or more benign end products, and ways to reduce the amount of raw materials used. http://www.epa.gov/seahome/prevent.html

PRESTO Software

PRESTO (Prediction of Radiological Effects Due to Shallow Trench Operations) is a computer model for evaluating radiation exposure from contaminated soil layers, including waste disposal, soil cleanup, agricultural land application, and land reclamation. http://www.epa.gov/radiation/assessment/presto.html

Principles and Design of On-site Waste Disposal with Septic Systems Software

The software teaches the basics of on-site treatment using diagrams and animations: septic tanks, drainfields, mounds, water conservation. It covers principles of treatment, design, and siting, as well as soil basics and water conservation. http://www.epa.gov/seahome/onsite.html

Private Pesticide Applicator Training Software

This is an interactive guide to the Federal regulatory requirements and user health concerns for pesticide handling and application. It includes an extended tutorial on pesticide laws and reglulations, a guide to safe handling and mixing, and a short course in reading and interpreting the labels and health warnings. http://www.epa.gov/seahome/privpest.html

Private Water Systems

A complete minicourse in design and construction of private drinking water systems (wells and piping) includes water quantities required, water pumps, systems controls, design considerations and piping. http://www.epa.gov/seahome/private.html

Program to Assist in Tracking Critical Habitat (PATCH)

PATCH is a spatially explicit, individual-based, life history simulator designed to project populations of territorial terrestrial vertebrate species through time. PATCH is ideal for investigations involving wildlife species that are mobile habitat specialists. PATCH's data requirements are minimal: provided with habitat maps, specifications for habitat use (territory size and habitat affinity), vital rates (survival and reproduction), and parameters for species' movement behavior, a simulation can be generated.PATCH's outputs fall into two general categories: pattern-based metrics and demographic analyses. http://www.epa.gov/wed/pages/models/patch/patchmain.htm

Protecting Children's Health in Schools Software

Intended for the decision-maker in a school system, the software aids in identification of potential hazards to assist in protecting the environment and the health of those who use the school. The guide will discuss what those hazards might be, how to test for them, and how to remedy problems found. http://www.epa.gov/seahome/child.html

PRZM3 Model

PRZM3 is the most recent version of a modeling system that links two subordinate models, PRZM and VADOFT, in order to predict pesticide transport and transformation down through the crop root and unsaturated zone. PRZM is a one-dimensional, finite-difference model that accounts for pesticide and nitrogen fate in the crop root zone. PRZM3 includes modeling capabilities for such phenomena as soil temperature simulation, volatilization and vapor phase transport in soils, irrigation simulation, microbial transformation, and a method of characteristics (MOC) algorithm to eliminate numerical dispersion. PRZM is capable of simulating transport and transformation of the parent compound and as many as two daughter species. VADOFT is a one-dimensional, finite-element code that solves Richard's equation for flow in the unsaturated zone. The user may make use of constitutive relationships between pressure, water content, and hydraulic conductivity to solve the flow equations. VADOFT may also simulate the fate of two parent and two daughter products. http://www.epa.gov/ceampubl/gwater/przm3/index.htm

(Continued)

Table 12A.6 (Continued)

Radiation Risk Assessment Software: CAP88 PC
CAP88-PC is a personal computer software system used for calculating dose and risk from annual average releases of radionuclides to the air. http://www.epa.gov/radiation/assessment/CAP88/index.html

Radionuclide Carcinogenicity Slope Factors Software
The source provides the latest information and guidance on using radionuclide slope factors from the Health Effects Assessment Summary Tables (HEAST) — Radionuclides Table (formerly Table 12A.4). EPA, other federal agencies, states and contractors who are responsible for the identification, characterization and remediation of sites contaminated with radioactive materials use radionuclide slope factors in risk assessments to calculate potential risks to the general public. http://www.epa.gov/radiation/heast/download.htm

RCRA Boiler & Industrial Furnace Regulations
In February 1991, the USEPA promulgated a final rule that expands controls on hazardous waste combustion by regulating the burning of hazardous waste in boilers and industrial furnaces (BIFs). The BIF rule sets levels of control for emissions of toxic organic compounds, toxic metals, HCl, chlorine gas and particulate matter. Owners and operators of BIFs are subject to permitting and other standards applicable to hazardous waste treatment, storage and disposal facilities. The site helps to understand the applicability, terms, definitions, technology and requirements associated with the BIF rule. http://www.epa.gov/seahome/bif.html

RCRA Corrective Action Processs
The Corrective Action Process aims to identify, develop, and implement appropriate corrective measures to protect human health and the environment at facilities where hazardous wastes or pollutants are handled or managed. The purpose of CAP is to guide facilities through the many governing authorities and laws while providing an easy to follow path to proper and legal corrective action. Menu items cover interim and stabilization measures, a detailed breakdown of the quality assurance plan, health and ecological assessment, remediation technologies, public participation, an extensive guide to statutory authorities, and a glossary and a bibliography of guidance documents. http://www.epa.gov/seahome/cap.html

RCRAInfo
RCRAInfo is EPA's comprehensive information system, providing access to data supporting the Resource Conservation and Recovery Act (RCRA) of 1976 and the Hazardous and Solid Waste Amendments (HSWA) of 1984. RCRAInfo replaces the data recording and reporting abilities of the Resource Conservation and Recovery Information System (RCRIS) and the Biennial Reporting System (BRS). RCRAInfo characterizes facility status, regulated activities, and compliance histories and captures detailed data on the generation of hazardous waste from large quantity generators and on waste management practices from treatment, storage, and disposal facilities. http://www.epa.gov/epaoswer/hazwaste/data/index.htm

ReduceIt Companion Software to Source Reduction Program Potential Manual
This companion software is designed to be used in connection with the *Source Reduction Program Potential Manual* to help solid waste managers plan for a source reduction program and determine the impact of various source reduction options on their community. The *Source Reduction Program Potential Manual and Software* estimate the program potential for six source reduction options: three residential (grasscycling, home composting, and clothing reuse) and three commercial and industrial (office paper reduction, converting to multi-use pallets, and paper towel reduction). http://www.epa.gov/epaoswer/non-hw/reduce/reduceit/index.htm

Regional Air Pollutant Inventory Development System (RAPIDS)
As the principle component of the Great Lakes Regional Air Toxic Emissions Inventory project, the Regional Air Pollutant Inventory Development System (RAPIDS) is the first ever multi-jurisdictional pollutant emissions inventory software that has been developed. The focus of the study was on the Commission defined list of 49 toxic pollutants, in addition to several other important nontoxic compounds. Now, RAPIDS is used by each Great Lakes state and Province of Ontario. The latest regional inventory for 1997 contains over 80 toxic pollutants for point, area, and mobile (onroad and nonroad source) emissions. http://www.glc.org/air/rapids/rpdsover.html

Regional Vulnerability Assessment (ReVA) Program
EPA's Regional Vulnerability Assessment (ReVA) program is an approach to regional scale, priority-setting assessment being developed by EPA's Office of Research and Development (ORD). ReVA will expand cooperation among the laboratories and centers of ORD, by integrating research on human and environmental health, ecorestoration, landscape analysis, regional exposure and process modeling, problem formulation, and ecological risk guidelines. http://www.epa.gov/reva/#

Registry of EPA Applications and Databases
The Registry of EPA Applications and Databases (READ) is an authoritative registry that uniquely identifies the Environmental Protection Agency's (EPA) diverse information resources including computer application systems, databases, and models. As a catalog of available EPA information, READ provides a single point of entry to identify, locate, and access Agency information resources. READ supports EPA's metadata vision to: create and store metadata in as few places as possible; improve the quality and consistency of metadata; and provide a central point of access to the complete collection of all Agency information resources metadata. http://www.epa.gov/read/

Regulatory Air Models
This website is a source of information on atmospheric dispersion (air quality) models that support regulatory programs required by the Clean Air Act. Documentation and guidance for these computerized models are a major feature of this website. The computer code, data and technical documents offered deal with mathematical modeling for the dispersion of air pollutants. http://www.epa.gov/scram001/

(Continued)

Table 12A.6 (Continued)

Remediation Annual Status Report

The Annual Status Report (ASR) Remediation Database documents the status and achievements of 1,811 technology applications identified for Superfund remedial action sites, including 1,760 treatment applications for soil, other solid wastes, and groundwater and 51 applications of groundwater containment using vertical engineered barriers (VEB). http://cfpub.epa.gov/asr/index.cfm

Reporting on Municipal Solid Waste: A Local Issue

The "Reporter's Guide" is a software application that presents background information to assist print and broadcast media in understanding municipal solid waste (MSW) issues. Examines the role of federal, state, and local governments in MSW management; options for solid waste management (source reduction, recycling, incineration, and landfilling); and regulations for solid waste landfills. http://www.epa.gov/epaoswer/non-hw/muncpl/reporting.htm

Residential Energy Efficiency

This program shows effective ways to reduce home energy consumption. Topics include insulation, windows, doors, weather-stripping, and caulking. The expert system feature allows users to calculate how much they will save by making their homes more energy efficient. http://www.epa.gov/seahome/energy.html

Residential Water Conservation Techniques Software

With more than 100 color graphic screens and animation, this program shows effective ways to save water inside and outside the house. Topics include efficient toilets, showerheads, faucets; leak detection, water efficient lawn care and gardening; car washing and pool operation. http://www.epa.gov/grtlakes/seahome/watcon.html

RETC Software

A program for analyzing the hydraulic conductivity properties of unsaturated soils. The parametric models of Brooks-Corey and van Genuchten are used to represent the soil water retention curve, and the theoretical pore-size distribution models of Mualem and Burdine predict the unsaturated hydraulic conductivity function. http://www.epa.gov/nerlesd1/databases/retc/abstract.htm

Risk Management Plans (via RTK–NET)

These provides a searchable database of relevant information to assist in citizen participation during the preparation of risk management plans. http://www.rtknet.org/rmphelp.html

RMP*Comp

RMP*Comp is a free program you can use to complete the offsite consequence analyses (both worst case scenarios and alternative scenarios) required under the Risk Management Planning Rule. http://yosemite.epa.gov/oswer/ceppoweb.nsf/content/rmp-comp.htm

RMP*Review

RMP*Review is a free software program designed for reviewing and analyzing Risk Management Plans (RMPs) submitted under the Clean Air Act, Section 112(r). http://yosemite.epa.gov/oswer/ceppoweb.nsf/content/rmp_review.htm

RMP*Submit

RMP*Submit 2004 is the free, official EPA, personal computer software for facilities to use in submitting Risk Management Plans (RMP) required under the Risk Management Program. http://yosemite.epa.gov/oswer/ceppoweb.nsf/content/ap-rmsb.htm#rmpsub

Safe Drinking Water Information System (SDWIS)

The Safe Drinking Water Information System (SDWIS) contains information about public water systems and their violations of EPA's drinking water regulations, as reported to EPA by the states. These regulations establish maximum contaminant levels, treatment techniques, and monitoring and reporting requirements to ensure that water systems provide safe water to their customers. http://www.epa.gov/enviro/html/sdwis/sdwis_query.html

Science Inventory

The Science Inventory (SI) is a searchable database of EPA science activities and scientific/technical work products. The Science Inventory provides information about current or recently completed activities, providing a snapshot of what science EPA is conducting in its research laboratories and centers, program and regional offices, and through grants and other assistance agreements to universities and other institutions. The Science Inventory currently contains 5,646 records. http://cfpub.epa.gov/si/

Scout Software

A univariate and multivariate data analysis tool with several classical and robust procedures (e.g., outlier testing, interactive 2D/3D graphics), making it a useful package for environmental and ecological applications. Scout can transform data, assess variable normality, produce histograms and Q–Q plots of raw data and principal component scores (PCSs), and produce scatter plots of raw data, PCSs, and discriminant scores. http://www.epa.gov/nerlesd1/databases/scout/abstract.htm

Section Seven Tracking System (SSTS)

SSTS is one of the major systems that supports the Pesticide Program at EPA. SSTS is the only automated system that EPA uses to track pesticide producing establishments and the amount of pesticides they produce. SSTS records the registration of new establishments and records pesticide production at each establishment. http://www.epa.gov/Compliance/planning/data/toxics/sstsys.html

(Continued)

Table 12A.6 (Continued)

Sector Facility Indexing Project (SFIP)

The Sector Facility Indexing Project focused on compiling information on facilities within five industry categories (automobile assembly, pulp manufacturing, petroleum refining, iron and steel production, and the primary smelting and refining of aluminum, copper, lead, and zinc (nonferrous metals), and selected major federal facilities). SFIP contained records for 591 facilities from the five industry sectors, as well as 273 federal facilities. SFIP information related to compliance and inspection history, chemical releases and spills, and demographics of the surrounding population. http://www.epa.gov/sfipmtn1/access.html

Simplified Method Program — Variable Complexity Stream Toxics Model (SMPTOX3)

U.S. EPA regulatory programs have sponsored development of an interactive computer program for performing waste load allocations for toxics — Simplified Method Program — Variable Complexity Stream Toxics Model (SMPTOX3). SMPTOX3 provides user-friendly access to a technique for calculating water column and stream bed toxic substance concentrations resulting from point source discharges into streams and rivers. It predicts pollutant concentrations in dissolved and particulate phases for water column and bed sediments and total suspended solid. http://www.epa.gov/ceampubl/swater/smptox3/index.htm

Smart Travel Resource Center

EPA makes more than 100 information campaigns that focus on encouraging individuals to make travel decisions which have positive impacts on air quality, congestion, and quality of life. The goal being that the users can learn from one another in designing, implementing, and evaluating these innovative programs. http://yosemite.epa.gov/aa/strc.nsf

Software for Environmental Awareness

Resulting from a partnership between EPA Region 5 and the students and faculty at Purdue University, SEA has produced a number of software programs from simple linear tutorials to expert systems requiring the user to input data in order to receive a computer-calculated solution. The collection includes water-related topics, solid waste, air pollution, and environmental assessment. http://www.epa.gov/glnpo/seahome/

Soil and Geologic Site Evaluation Software

This software provides information about the roles that soils, geology and depth to groundwater play in the protection of groundwater. A risk assessment portion helps determine the potential for contamination at a site based on information supplied by the user. http://www.epa.gov/seahome/farmsite.html

Source Ranking Database

Performs a systematic screening-level review of over 12,000 potential indoor pollution sources to identify high-priority product and material categories for further evaluation. Can also identify the products that contain a specific chemical. epa.gov/opptintr/exposure/docs/srd.htm

Spatial Grid Networks Software

Software and documentation for generating regular grids of hexagons and other shaped cells for the earth as a whole, or for parts of it, as well as the existing grid for the USGS-BRD GAP program. These materials are for the geometric and cartographic components of such grids. http://www.epa.gov/wed/pages/staff/white/getgrid.htm

SPECIATE Software

SPECIATE is EPA's repository of Total Organic Compound (TOC) and Particulate Matter (PM) speciated profiles for a variety of sources for use in source apportionment studies. This release of SPECIATE no longer contains profiles assigned to SCC's. Instead, searching for profiles by keyword has been improved and of course the user can still browse through the profiles. http://www.epa.gov/ttn/chief/software/speciate/index.html

STF Software

A computer-based tool for selecting data on chemicals in the environment and for simulating their fate and transport in site-specific conditions is available. The software system consists of three components: (1) STF 2.0 (Soil Transport and Fate Database), which provides information on the behavior of chemicals in soil environments for use as input data on, for example, degradation rates and partition coefficients. (2) RITZ (Regulatory and Investigative Treatment Zone) and VIP 3.0 models. RITZ simulates hazardous chemical movement and fate during land treatment of oily wastes. VIP evaluates data using six different output options. (3) RITZ and VIP model editors that are directly interfaced with STF 2.0 and aid in the creation of input files for the RITZ and VIP models. http://www.epa.gov/nerlesd1/databases/stf/abstract.htm

STORET

The U.S. Environmental Protection Agency (EPA) maintains two data management systems containing water quality information for the nation's waters: the Legacy Data Center (LDC), and STORET. The LDC is a static, archived database and STORET is an operational system actively being populated with water quality data. http://www.epa.gov/STORET/about.html

Storm Water Management Model (SWMM)

The Storm Water Management Model (SWMM) is a comprehensive computer model for analysis of quantity and quality problems associated with urban runoff. Both single-event and continuous simulation can be performed on catchments having storm sewers, or combined sewers and natural drainage, for prediction of flows, stages and pollutant concentrations. http://www.epa.gov/ceampubl/swater/swmm/index.htm

(Continued)

Table 12A.6 (Continued)

Substance Registry System (SRS)

The Substance Registry System (SRS) provides information on substances and how they are represented in the Environmental Protection Agency (EPA) regulations and information systems. http://www.epa.gov/srs/index.htm

Subsurface Characterization and Monitoring Techniques

An interactive, multimedia version of the two-volume, EPA reference entitled "Subsurface Characterization and Monitoring Techniques" by R. Boulding. The documents include 1- to 2-page descriptions of more than 280 site characterization and field monitoring methods for detecting groundwater contamination and other aspects of the subsurface at hazardous waste sites. Geological and hydrogeological characterization topics covered include surface and borehole approaches, geophysical methods, and solids sampling; drilling; aquifer tests and groundwater sampling; water-state measurement and monitoring; vadose zone hydraulic conductivity/flux measurement; vadose zone water budget characterization; vadose zone soil-solute sampling and gas monitoring; and field chemical analytical methods. http://www.epa.gov/nerlesd1/databases/subsurface/abstract.htm

Superfund Chemical Data Matrix (SCDM)

The Superfund Chemical Data Matrix (SCDM) is a source for factor values and benchmark values applied when evaluating potential National Priorities List (NPL) sites using the Hazard Ranking System (HRS). Factor values are part of the HRS mathematical equation for determining the relative threat posed by a hazardous waste site and reflect hazardous substance characteristics, such as toxicity and persistence in the environment, substance mobility, and potential for bioaccumulation. Benchmarks are environment- or health-based substance concentration limits developed by or used in other EPA regulatory programs. http://www.epa.gov/superfund/sites/npl/hrsres/tools/scdm.htm

Superfund Information Systems

Site provides links to the various Superfund-related information tracking systems that are maintained by the EPA. Included in the list are CERCLIS (general information about sites), Archived Sites (sites with completed site assessment and no further action required), Records of Decision (sites where remedial options have been selected), Five Year Reviews (sites that have been undergoing remediation for an extended time provide 5-year reviews of activities), Site Assessment Documentation (provides preliminary assessment of sites that have recently been identified), Site Spill IDs (tracks logistical issues at Superfund sites), Data Element Dictionary (lexicon of Superfund related terms), and Site Information Products (collection of Superfund Information reports and products). http://www.epa.gov/superfund/sites/siteinfo.htm

Superfund Innovative Technologies

A searchable database of treatment technology applications for soil, groundwater, and other media at Superfund sites. The information in this database was gathered from Superfund decision documents and project managers at Superfund remedial and removal sites. It contains information about source control technologies (addressing soil, sludge, sediment, and other solid-matrix wastes), innovative (*in situ*) groundwater treatment technologies, and subsurface barrier groundwater containment technologies (vertical engineered barriers). The database is searchable by site name, site identification number (CERCLIS ID), state, technology type, contaminant, and media. http://www.epa.gov/tio/databases/

Surf Your Watershed

Surf Your Watershed is a service to help you locate, use, and share environmental information about your state and watershed. http://www.epa.gov/surf/

Surface Water Education System Software

Topics include a general overview of the importance and quantity of surface water, a detailed look at this resource for Indiana, and a discussion of how surface water can become contaminated. Hypertext and numerous photos and graphics illustrate an extensive discussion of agricultural best management practices to protect surface water resources. http://www.epa.gov/seahome/surfwat.html

TANKS Emission Estimation Software

TANKS is a computer software program that estimates volatile organic compound (VOC) and hazardous air pollutant (HAP) emissions from fixed- and floating-roof storage tanks. *TANKS* is based on the emission estimation procedures from *Chapter 7* of EPA's *Compilation Of Air Pollutant Emission Factors (AP-42).* http://www.epa.gov/ttn/chief/software/tanks/index.html

Terminology Reference System (TRS)

The Terminology Reference System (TRS) was created to provide a single source of Environmental Protection Agency (EPA) terminology. The TRS is part of the centralized System of Registries (SoR), which provides access to the Agency's core registry systems. http://www.epa.gov/trs/index.htm

Three-Dimensional Numerical Model of Hydrodynamics and Sediment Transport in Lakes and Estuaries (SED3D)

The Three-Dimensional Numerical Model of Hydrodynamics and Sediment Transport in Lakes and Estuaries (SED3D) simulates the flow and sediment transport in lakes, estuaries, harbors, and coastal waters. SED3D is a dynamic modeling system that can be used to simulate the flow and sediment transport in various water bodies under the forcing of winds, tides, freshwater inflows, and density gradients with the influence of the Coriolis acceleration, complex bathymetry, and shoreline geometry. http://www.epa.gov/ceampubl/swater/sed3d/index.htm

(Continued)

Table 12A.6 (Continued)

Tier2 Submit Software

EPA has developed Tier2 Submit 2004 to help facilities prepare an electronic chemical inventory report. Twenty-two states are accepting Tier2 Submit from facilities for the 2004 reporting year. http://yosemite.epa.gov/oswer/ceppoweb.nsf/content/tier2.htm#t2forms

The Total Maximum Daily Load (TMDL) Universal Soil Loss Equation (USLE)

The Total Maximum Daily Load (TMDL) Universal Soil Loss Equation (USLE) model is a software application for estimating diffuse sediment source loads within a watershed framework. The user interface is similar to a spreadsheet and allows users to easily select and enter parameters used to estimate sediment loading. http://www.epa.gov/ceampubl/swater/usle/index.htm

Total Maximum Daily Loads (TMDL)

TMDL or Total Maximum Daily Load is a calculation of the maximum amount of a pollutant that a waterbody can receive and still meet water quality standards, and an allocation of that amount to the pollutant's sources. Water quality standards are set by States, Territories, and Tribes. They identify the uses for each waterbody, for example, drinking water supply, contact recreation (swimming), and aquatic life support (fishing), and the scientific criteria to support that use. A TMDL is the sum of the allowable loads of a single pollutant from all contributing point and nonpoint sources. The calculation must include a margin of safety to ensure that the waterbody can be used for the purposes the State has designated. The calculation must also account for seasonal variation in water quality. http://www.epa.gov/owow/tmdl/intro.html#definition

Total Risk Integrated Methodology (TRIM) Models

The *TRIM design* includes three individual modules: The *Environmental Fate, Transport, and Ecological Exposure* module, *TRIM.FaTE*, accounts for movement of a chemical through a comprehensive system of discrete compartments (e.g. media and biota) that represent possible locations of the chemical in the physical and biological environments of the modeled ecosystem and provides an inventory, over time, of a chemical throughout the entire system. In addition to providing exposure estimates relevant to ecological risk assessment. The Exposure-Event module, *TRIM.Expo.* evaluates human exposures by tracking either randomly selected individuals that represent an area's population or population groups referred to as "cohorts" and their inhalation and ingestion through time and space. In the Risk Characterization module, *TRIM.Risk*, estimates of human exposures or doses are characterized with regard to potential risk using the corresponding exposure- or dose-response relationships. The *TRIM.Risk* module is also designed to characterize ecological risks from multimedia exposures. The output from TRIM.Risk is intended to include documentation of the input data, assumptions in the analysis, and measures of uncertainty, as well as the results of risk calculations and exposure analysis. http://www.epa.gov/ttn/fera/trim_gen.html

Toxics Release Inventory (TRI) Program

The Toxics Release Inventory (TRI) is a publicly available EPA database that contains information on toxic chemical releases and other waste management activities reported annually by certain covered industry groups as well as federal facilities. http://www.epa.gov/tri/

Toxic Substances Control Act Chemical Substance Inventory

EPA classifies chemical substances as either "existing" chemicals or "new" chemicals. The only way to determine if the substance is a new chemical is by consulting EPA's Toxic Substances Control Act Chemical Substance Inventory (commonly referred to as the TSCA Inventory or just the Inventory).There are approximately 75,000 chemical substances, as defined in Section 3 of the TSCA, on the Inventory at this time. epa.gov/opptintr/newchems/invntory.htm

TRI-ME Toxics Release Inventory Made Easy Software

The Toxics Release Inventory - Made Easy (TRI-ME) software is a tool to help facilities determine and complete their Emergency Planning and Community Right-to-Know Act (EPCRA) section 313 (TRI) reporting obligations. http://www.epa.gov/triinter/report/software/index.htm

TSCATS Standard Reports (via RTK-NET)

TSCATS (Toxic Substances Control Act Test Submissions) is an online index to unpublished, nonconfidential studies covering chemical testing results and adverse effects of chemicals on health and ecological systems. The studies are submitted by U.S. industry to EPA under several sections of the Toxic Substance Control Act (TSCA). There are four types of documents in the database: Section 4 chemical testing results, Section 8(d) health and safety studies, Section 8(e) substantial risk of injury to health or the environment notices, and voluntary documents submitted to EPA known as a For Your Information (FYI) notice. http://www.rtknet.org/tsc/

UIC Class V Injection Wells Software

USEPA's Underground Injection Control (UIC) program is explained in this hypertext-based software package, with a focus on Class V injection wells in particular. These wells inject wastewater directly into or above underground sources of drinking water, so it is important to understand why they exist, why they may pose a danger to the environment, and how they may be operated and maintained safely. http://www.epa.gov/seahome/inject.html

Urban Airshed Model (UAM)

The Atmospheric Sciences Modeling Division (ASMD) of NOAA's Air Resources Laboratory (ARL) collaborates with the U.S. Environmental Protection Agency (EPA) and its predecessor agencies in developing advanced air quality models that can simulate the transport and fate of pollutants in the atmosphere. http://www.epa.gov/asmdnerl/

Vehicle and Engine Emissions Modeling Software

Modeling is EPA's method for estimating emissions from on-road vehicles, nonroad sources, and fuels. Inventories are calculations of total emissions of a pollutant for a given area at a defined time and set of conditions. http://www.epa.gov/otaq/models.htm

(Continued)

Table 12A.6 (Continued)

Visual Plumes Model
The Visual Plumes model system is a software application for simulating surface water jets and plumes. It also assists in the preparation of mixing zone analyses, Total Maximum Daily Loads (TMDLs), and other water quality applications. http://www.epa.gov/ceampubl/swater/vplume/index.htm

VLEACH Software
A one-dimensional, finite difference model for making preliminary assessments of the effects on groundwater from the leaching of volatile, sorbed contaminants through the vadose zone. The program models four main processes: liquid-phase advection, solid-phase sorption, vapor-phase diffusion, and three-phase equilibration. http://www.epa.gov/nerlesd1/databases/vleach/abstract.htm

Wall Paint Exposure Assessment Model (WPEM)
The Wall Paints Exposure Assessment Model (WPEM) estimates the potential exposure of consumers and workers to the chemicals emitted from wall paint which is applied using a roller or a brush. epa.gov/opptintr/exposure/docs/wpem.htm

WAste Reduction Model
EPA created the WAste Reduction Mode (WARM) to help solid waste planners and organizations track and voluntarily report greenhouse gas emissions reductions from several different waste management practices. WARM calculates and totals GHG emissions of baseline and alternative waste management practices — source reduction, recycling, combustion, composting, and landfilling. The model calculates emissions in metric tons of carbon equivalent (MTCE), metric tons of carbon dioxide equivalent ($MTCO_2E$), and energy units (million BTU) across a wide range of material types commonly found in municipal solid waste (MSW). http://yosemite.epa.gov/oar/globalwarming.nsf/content/ActionsWasteWARM.html

Water-Effect Ratio of Metals
This program is based on the EPA's 1994 "Interim Guidance on the Determination and Use of Water-Effect Ratios (WERs) for Metals." The WER procedure is one of three protocols which may be used to derive site-specific water quality criteria for the protection of aquatic life. http://www.epa.gov/seahome/wer.html

Water Efficient Landscape Planner
The Water Efficient Landscape Planner was developed to explain the advantages and principles of water efficient landscaping. The program covers the basics of landscape planning and provides guidelines and suggestions to help users select the most appropriate plants for their needs. It is intended for use by homeowners who are considering changes to residential landscaping. http://www.epa.gov/seahome/landscp.html

Water Quality Analysis Simulation Program (WASP)
WASP is a generalized framework for modeling contaminant fate and transport in surface waters. WASP has been used to answer questions regarding biochemical oxygen demand, dissolved oxygen dynamics, nutrients and eutrophication, bacterial contamination, and organic chemical and heavy metal contamination. http://www.epa.gov/athens/research/modeling/wasp.html

Water Quality Standards and Criteria
Water Quality Standards (WQS) provide a framework for maintenance and improvement of water quality when adopted by States, U.S. Territories, and Indian Tribes. This software presents the three components of State and Tribal WQS: waterbody uses (e.g., swimming, boating); water quality criteria or limits on chemical concentrations that may be present the waterbody; and antidegradation policy to protect existing water quality http://www.epa.gov/seahome/wqs.html

WQS Database
The WQSDB provides access to EPA and state WQS information in text, tables, and maps. This Web site provides access to several WQS reports with information about designated uses, waterbody names, state numeric water quality criteria, and EPA recommended numeric water quality criteria. http://www.epa.gov/wqsdatabase/

Water Radioactivity Software Development Project (WRSDP)
The EPA — Region 1, Radiation Unit has created a number of computer programs to help make necessary decisions regarding groundwater radioactivity, water radiation regulations, and radon mitigation in New England. These programs are specifically intended for the use of people in the water supply community working with water use, supply, and treatment. http://www.epa.gov/ne/eco/software/index.html

WATERSHEDSS Documentation
An Internet-based decision support and educational software system, WATERSHEDSS (WATER, Soil, and Hydro-Environmental Decision Support System) was developed to assist managers of predominantly agricultural watersheds in defining water-quality problems and selecting appropriate nonpoint source (NPS) pollution control measures. epa.gov/ceampubl/tools/watrshed/

WATER9 Software
WATER9, the wastewater treatment model, is a computer program and consists of analytical expressions for estimating air emissions of individual waste constituents in wastewater collection, storage, treatment, and disposal facilities; a database listing many of the organic compounds; and procedures for obtaining reports of constituent fates, including air emissions and treatment effectiveness. epa.gov/ttn/chief/software/water/index.html

(Continued)

Table 12A.6 (Continued)

Well Water Location and Condition on the Farm Software
Well contamination may occur from farmstead activities near drinking water wells. The condition of your well and its proximity to pollution sources determine how much of a risk each source poses to the water you drink. This program contains information useful for identifying and correcting high-risk management and site conditions, and a water well risk assessment program. http://www.epa.gov/seahome/welloc.html

WhAEM2000
The U.S. EPA's Wellhead Analytic Element Model, WhAEM2000, is a groundwater geohydrology computer program. WhAEM2000 is a public domain, groundwater flow model designed to facilitate capture zone delineation and protection area mapping in support of the State's Wellhead Protection Programs (WHPP) and Source Water Assessment Planning (SWAP) for public water supplies in the United States. WhAEM2000 provides an interactive computer environment for design of protection areas based on radius methods, well in uniform flow solutions, and geohydrologic modeling methods. Protection areas are designed and overlaid upon U.S. Geological Survey Digital Line Graph (DLG) or other electronic base maps. Geohydrologic modeling for steady pumping wells, including the influence of hydrological boundaries, such as rivers, recharge, and no-flow contacts, is accomplished using the analytic element method. http://www.epa.gov/ceampubl/gwater/whaem/index.htm

Wellhead Protection Software
Hypertext and graphics explain protection of public water supplies as presented in the Safe Drinking Water Act Amendments of 1986. General concepts reviewed include building a planning team, delineating a wellhead protection area, conducting a contaminant source inventory, managing potential contaminant sources, and planning for the future. It is intended for local/Tribal officials, water supply managers and interested groups or citizens whose community relies on groundwater for its public water supply. Many of the concepts presented in this software may also be useful for protecting private wells. http://www.epa.gov/seahome/wellhead.html

WhAEM Software
A computer-based tool used in the wellhead protection decision-making process to delineate groundwater capture zones and isochrones of residence time — Unlike similar programs, WhAEM can accommodate fairly realistic boundary conditions, such as streams, lakes, and aquifer recharge due to precipitation. http://www.epa.gov/nerlesd1/databases/whaem/abstract.htm

WHPA Software
A semi-analytical groundwater flow simulation program used for delineating capture zones in a wellhead protection area. The program consists of four computational modules (RESSQC, MWCAP, GPTRAC, MONTEC). WHPA is applicable to homogeneous aquifers exhibiting two-dimensional, steady groundwater flow in an areal plane and appropriate for evaluating multiple aquifer types (i.e. confined, leaky-confined, and unconfined). The model is capable of simulating barrier or stream boundary conditions that exist over the entire depth of the aquifer. WHPA can account for multiple pumping and injection wells and can quantitatively assesses the effects of uncertain input parameters on a delineated capture zone(s). http://www.epa.gov/nerlesd1/databases/whpa/abstract.htm

Window to My Environment
"Window To My Environment" (WME) is a web-based tool that provides a wide range of federal, state, and local information about environmental conditions and features in selected areas. This application is provided by U.S. EPA in partnership with federal, state and local government and other organizations. http://www.epa.gov/enviro/wme/

Source: From www.epa.gov.

Table 12A.7 EPA Public Listservs

AGCENTER
National Agriculture Compliance Assistance Center

AIR_INSPECTOR

BASINSINFO
Private list for basins users

BROWNFIELDS
EPA Brownfields Program news

CAMEO
EPA-NOAA CAMEO Team

CERIPUBS
A periodic announcement of new technology transfer products from U.S. EPA's Center for Environmental Research Information

AGING_INITIATIV
List server of the EPA Aging Initiative

ARCHITECTURES

BOPSAT
BOPSAT group

CALLCENTER_OSWER
RCRA, Superfund & EPCRA Call Center

CEAM-USERS
CEAM Software Product Updates and Releases

CFEIS
CFEIS System status updates

(Continued)

Table 12A.7 (Continued)

CHIEF Emission Factor & Inventory Group	**CLEANNEWS** Cleanup Newsletter
CLIMATE_COMMUNICATORS Climate change communications, outreach, and education	**DECENTRALIZED** Onsite/decentralized wastewater management issues
EJ_Hispanic_Outreach Environmental Justice info (Spanish)	**EMWGLIST** Exposure Modeling Work Group, pesticide exposure modeling
ENVIROTECHNEWS Information for the environmental technology industry	**EPA-AIR** Federal Register AIR documents
EPA-CEPP U.S. EPA Chemical Emergency Preparedness and Prevention Office (CEPPO)	**EPA-ECON-FORUM**
EPA-EJ Environmental Justice Information	**EPA-GENERAL** Federal Register GENERAL Documents
EPA-HUDSON Hudson River PCBs	**EPA-IMPACT** Federal Register IMPACT documents
EPA-MEETINGS Federal Register documents for Meetings	**EPA-MOBILENEWS** Emission factors/emission inventories for mobile sources
EPA-PEST Federal Register PESTICIDE documents	**EPA-R2-PRESS**
EPA_RandD_News Weekly news briefs regarding Science and Research	**EPA-SAB** Federal Register SCIENCE ADVISORY BOARD
EPA-SPECIES Federal Register SPECIES documents	**EPA-TOX** Federal register TOXIC documents
EPA-TRI Federal Register TOXIC RELEASE INVENTORY documents	**EPA-WASTE** Federal Register WASTE documents
EPA-WATER Federal Register WATER documents	**EPA-WTC** World Trade Center monitoring and cleanup
EPAFR-CONTENTS Federal Register CONTENTS documents	**EPANEWS** EPA news releases
EPANEWS-ESPANOL EPA news releases in Spanish	**EPAWEB-OWNERS**
ETVOICE Environmental Verification Technology Program info	**FEDENVIRONEWS** Federal facilities environmental news
FEDGB Limited to Federal Employees Working on green buildings issues	**GHGINVENTORY-L** Greenhouse gas emissions data
GreatLakesNews Funding and project news from Great Lakes National Program Office of USEPA	**GREENBYTES-NE** EPA New England newsletter
GREENVEHICLES Updates and Hot Topics of Green Vehicle Guide	**INNOVATION** National Center for Environmental Innovation/OA — General Distribution
Inthenews New England Environmental News	**JTRNET** Public and non-profit recycling market developers
K12_ENVIRONET K-12 New England educators	**MWRAPERMIT** Proposed Modifications to Ambient Monitoring Plan
NCEE-ANNOUNCE U.S. EPA's National Center for Environmental Economics	**NEI-USERGROUP** National Emission Inventory User Group
NEWS-NOTES Distribution of the publication Nonpoint Source News-Notes	**NPPTAC-EPA** Parties Interested in information related to EPA's NPPTAC

(Continued)

Table 12A.7 (Continued)

NPSINFO NPS Information Exchange	**NRMRL** U.S. EPA National Risk Management Research Laboratory
OCORELINK Office of Enforcement and Compliance Assurance Information	**OEISTAKEHOLDER** OEI Stakeholders and Partners
OIG-News Announcements of new postings to the EPA's Office of the Inspector General's web site	**ONLNCALC** Online Calculators and Internet Modeling Course
OSWER_VISION OSWER Quarterly Newsletter	**OTAQ-ANNOUNCE** EPA Office of Transportation and Air Quality (OTAQ) Announcements
P2NEWS2 Pollution Prevention Activities at EPA	**PBTCHEMFOCUS** Waste minimization
PPDCFORUM Pesticide Program Dialogue Committee	**R3MOUNTAINTOP**
R3NEWSRELEASE EPA Mid-Atlantic Region News Releases	**R5NEWS** Region 5 news releases
RPO-LISTSERV Regional Planning Organizations Activities	**SMOKEIMPACT**
STATEPRETCOORD Discussion forum for State Pretreatment Coordinators	**STORETINFO** STORET news and information
SUNPROTECTION Sun protection, ultraviolet (UV) radiation exposure	**TECHDIRECT** Info for site remediation/site assessment professionals
U.S._MEXBORDER U.S. EPA U.S. Mexican Border Discussions	**USEPAR3NEWS** U.S. EPA Region 3 news
USEPAR9NEWS U.S. EPA Region 9 Press Releases	**VOLMONITOR** Volunteer water monitoring
WASTEWISE All WasteWise Partners and Endorsers	**WATERNEWS** Water quality information
WATERSHED-NEWS Information for watershed groups and others working at the grass roots level to protect and restore watersheds	**WEB-AMBASSADOR** EPA web site updates
WIPP-NEWS Periodic information regarding EPA's role at the WIPP	**WQS-NEWS** Office of water Standards and Applied Science Division

Note: On-line subscription services offered by EPA providing periodic informational updates. Subscriptions may be made at www.lists.epa.gov/cgi-bin/1yris.pl.

Source: From www.epa.gov.

Table 12A.8 EPA Newsletters

Activities Update (http://www.epa.gov/epahome/announce.htm) — a monthly publication of EPA highlights and activities including: summaries of press releases, upcoming public meetings, upcoming congressional hearings, EPA senior management statements and testimonies, reports and publications, enforcement actions, Federal Register notices and highlights of EPA rulemakings

AIEO Update (http://www.epa.gov/indian/1up.htm) — quarterly publication from the American Indian Environmental Office

AMTIC News (http://www.epa.gov/ttnamti1/improv2.html) — a quarterly publication of the ambient monitoring technology information center, which describes events, regulations, and technology relating to the measurement of air pollution

Audit Policy Update (http://www.epa.gov/Compliance/resources/newsletters/incentives/auditupdate/index.html) — published periodically for the assistant administrator for enforcement and compliance assurance by its office of regulatory enforcement. The Update is intended to provide information to the public and regulated communities regarding developments under the EPA Audit Policy

(Continued)

Table 12A.8 (Continued)

Chief Newsletter (epa.gov/ttn/chief/newsletter.html) — quarterly publication of the emission factor and inventory group, which keeps the public informed about new emission estimation tools, conferences, and other information relating to emission factors and inventories

CleanupNews (http://www.epa.gov/compliance/resources/newsletters/cleanup/cleanupnews.html) — a quarterly published by the office of site remediation enforcement that highlights hazardous waste cleanup news, policy updates, court cases, technology, events, publications, and more

Coastlines (http://www.epa.gov/nep/coastlines) — information about estuaries and near coastal waters from the national estuary program

Community Based Environmental Protection (CBEP) News Online Newsletter (epa.gov/ecocommunity/news.htm) — a bi-weekly electronic newsletter about CBEP activities within and beyond EPA

EIIP Update (http://www.epa.gov/ttn/chief/eiip/update/index.html) — a newsletter designed to inform the emissions community about the emission inventory improvement program and its mission, goals, and benefits

ENERGY STAR News (www.energystar.gov/index.cfm?c = newsletter.nr_newsletter) — news concerning energy star labeled products

Enforcement Alert (http://www.epa.gov/compliance/resources/newsletters/civil/enfalert/index.html) — an electronic newsletter from the office of enforcement compliance and assurance intended to inform and educate the public and regulated community of important environmental enforcement issues, recent trends and significant enforcement actions

EPA Journal (epa.gov/history/collection/aid10.htm) — back issues of this agency publication on national and global environmental perspectives last published spring 1995

FedFacs (http://www.epa.gov/compliance/resources/newsletters/civil/fedfac/index.html) — an environmental bulletin for federal facilities from EPA's federal facilities enforcement office

GMPO Newsletter (epa.gov/gmpo/newsletters.html) — current activities related to the economic and environmental health of the gulf of Mexico published by the gulf of Mexico program

Green Lights and Energy Star Buildings Update and Bulletin (http://www.energystar.gov/index.cfm?c = bldrs_lenders_raters.pt_homes_newsletter) — a quarterly publication that provides in-depth information about participant accomplishments and program developments

Ground Water Currents http:// (www.clu-in.org/products/newsltrs/gwc/default.htm) — developments in innovative groundwater treatment from the Technology Innovation Office of the U.S. EPA office of solid waste

Inside the Greenhouse (epa.gov/globalwarming/greenhouse/index.html) — focuses on innovative state and local programs to reduce greenhouse gas emissions; Inside the greenhouse is a quarterly, electronic publication produced by the EPA's state and local climate change program

National Small Flows Clearinghouse (http://www.nesc.wvu.edu/nsfc/nsfc_index.htm) — offers two quarterly newsletters, small flows and pipeline, focusing on wastewater information for small communities

Native American Network Newsletter (epa.gov/epaoswer/non-hw/tribal/fednews.htm#nan) — provides information to help interested parties stay on top of municipal solid waste management issues affecting indian country

NERL Research News (http://www.epa.gov/nerl/publications/bulletins/index.html) — information about recent research from the national exposure research laboratory

Nonpoint Source News-Notes (http://www.epa.gov/OWOW/info/NewsNotes/) — is an occasional bulletin dealing with the condition of the water-related environment, the control of nonpoint sources of water pollution, and the ecosystem-driven management and restoration of watersheds

Oil Spill Program Update (http://www.epa.gov/oilspill/document.htm#update) — quarterly update of the activities of EPA's oil spill program

OPPT Tribal News (http://www.epa.gov/opptintr/tribal/pubs/index.html) — Environmental news for Indian tribes from the office of pollution prevention and toxics

P2 News2 (epa.gov/p2/resources/p2news2.htm) — an electronic newsletter produced by the pollution prevention programs of the U.S. EPA Office of Pollution Prevention and Toxics (OPPT)

PESP Update (http://www.epa.gov/oppbppd1/PESP/publications.htm) — a periodical news brief on the pesticide environmental stewardship program

RCRA/UST, Superfund & EPCRA Hotline Reports http:// (www.epa.gov/epaoswer/hotline/mrqs.htm) — a monthly compilation of frequently asked regulatory questions for which EPA has provided answers and prepares them in a question and answer format

Reusable News Newsletters (http://www.epa.gov/epaoswer/non-hw/recycle/reuse.htm) — a collection of information from the U.S. EPA's office of solid waste

(Continued)

Table 12A.8 (Continued)

Science Advisory Board Newsletter (http://www.epa.gov/science1/news.htm) — news and events pertaining to the activities of the science advisory board

SRF's UP (http://www.epa.gov/safewater/dwsrf.html?Facts) — a quarterly newsletter providing up-to-date information about the clean water and drinking water state revolving fund programs

Tech Trends (http://www.clu-in.org/pub1.cfm) — a newsletter that provides descriptions and performance data for innovative source control technologies that have been applied in the field

Tribal Waste Journal (http://www.epa.gov/epaoswer/non-hw/tribal/fednews.htm?twj) — from the waste management in Indian country program, each issue features a different topic and includes a kids section. Published annually, EPA's tribal waste journal replaces the native American network newsletter

Volunteer Monitor (http://www.epa.gov/OWOW/monitoring/volunteer/vm_index.html) — the national newsletter of volunteer water quality monitoring

Water Quality Criteria and Standards Newsletter (http://www.epa.gov/OST/pc/wqnews/) — a periodic newsletter about issues affecting water quality standards from the office of water's office of science and technology

Water Talk (yosemite.epa.gov/R10/WATER.NSF/webpage/WaterTalk?OpenDocument) — is published quarterly by the U.S. EPA, Region 10 and seeks to be a useful tool for those who protect water resources and ecosystems in communities of the Greater Pacific Northwest

Waternews (http://www.epa.gov/ow/waternews/) — a weekly on-line publication that announces publications, policies, and activities of the U.S. EPA's office of water

Watershed Events (http://www.epa.gov/OWOW/info/WaterEventsNews/index.html) — is intended to update interested parties on the development and use of watershed protection approaches

Source: From www.epa.gov.

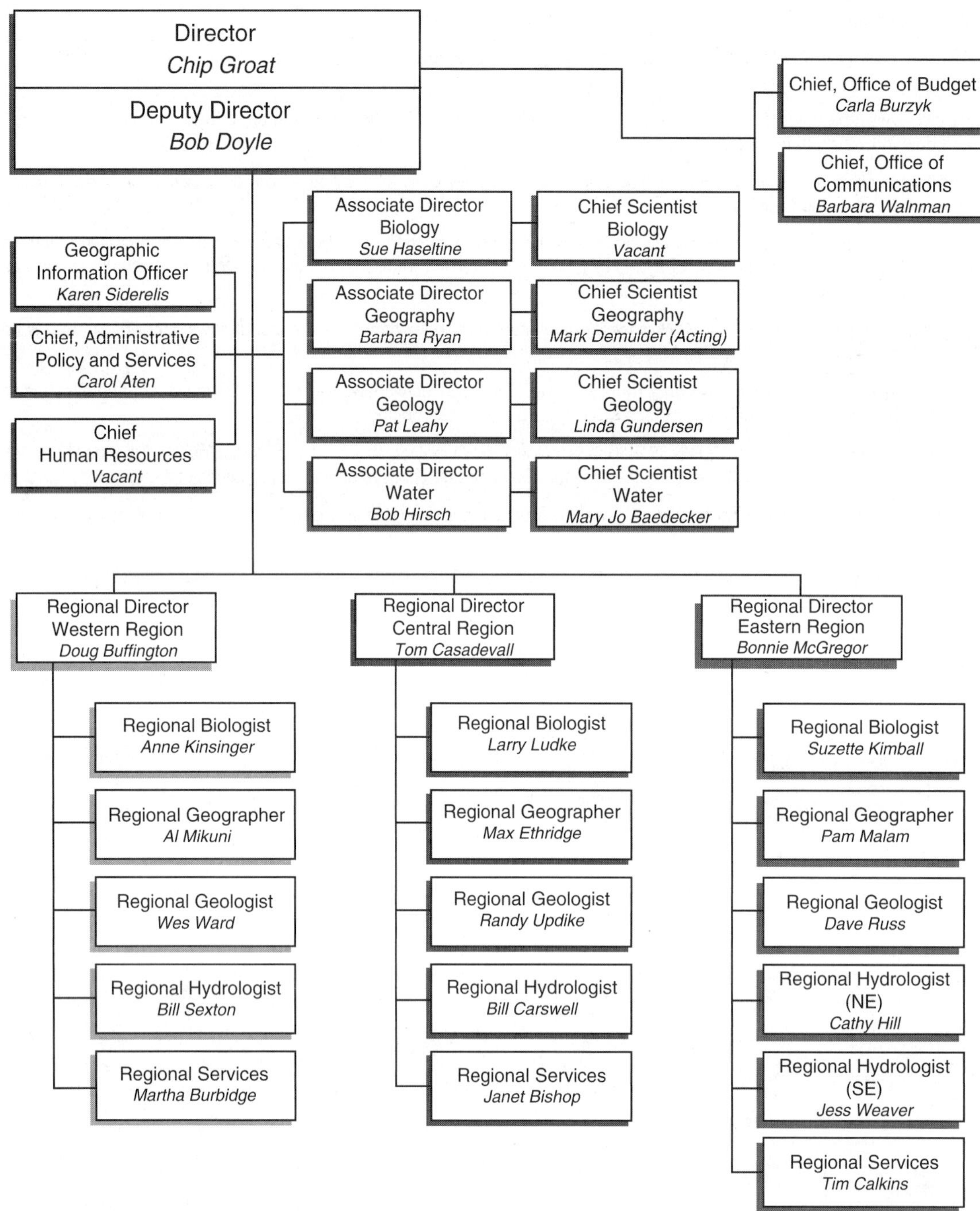

Figure 12A.3 Organization of the U.S. Geological Survey (www.epa.gov).

Table 12A.9 Directory — U.S. Geological Survey Water Resources Division

Office of the Director, U.S. Geological Survey	
Title, Name, Mail Stop	**Phone Number**
Director Charles G Groat MS 100	(703) 648-7411
Staff Assistant Janet N Arneson MS 100	(703) 648-7411
Deputy Director Robert E Doyle MS 101	(703) 648-7412
Staff Assistant Deborah A Riggsbee MS 101	(703) 648-7412
Senior Advisor to the Director Suzanne D Weedman MS 104	(703) 648-4760
Senior Advisor to the Director Amy L Holley MS 104	(703) 648-4411
Education Program Coordinator Robert W Ridky MS 104	(703) 648-4713
Staff Assistant for Special Issues Judy Nowakowski MS 121	(703) 648-4554
Senior Advisor for Science Applications James F Devine MS 106	(703) 648-4423
Secretary Joyce K Copeland MS 106	(703) 648-4191
Associate Director for Water Robert M Hirsch MS 409	(703) 648-5215
Staff Assistant Georgia J Diersing MS 409	(703) 648-5215
Chief Scientist for Hydrology Mary Jo Baedecker MS 436	(703) 648-5041
Secretary Paula A McMahon MS 436	(703) 648-5041
Staff Assistant/Strategic Planning Gail E Mallard MS 409	(401) 322-0902
Staff Assistant/Special Issues, Vacant, MS 409	
Senior Science Advisor for Surface Water Stephen F Blanchard MS 415	(703) 648-5629
Secretary, Maria Calderon, MS 415	(703) 648-5301
Chief, Office of Water Information Katherine Lins, MS 440	(703) 648-5014
Secretary Alice C Dilandro MS 440	(703) 648-5032
Senior Management Officer for Water Gene A Summerhill MS 440	(703) 648-5309
National Coordinator, Cooperative Water Program Glenn G Patterson MS 409	(703) 648-6876
Chief, National Water-Quality Assessment (NAWQA) Program Donna N Myers MS 413	(703) 648-5012
Secretary Sarahnell L Laible MS 413	(703) 648-5715
Chief, Office of Ground Water William M Alley MS 411	(703) 648-5035
Secretary Carolyn L Wakelee MS 411	(703) 648-5001
Chief, Office of Water Quality Timothy L Miller, MS 412	(703) 648-6868
Secretary Nana L Snow MS 412	

Note: U.S. Geological Survey Headquarters, John W. Powell Federal Building, 12201 Sunrise Valley Drive, Reston, Virginia 20192.

Source: From www.usgs.gov.

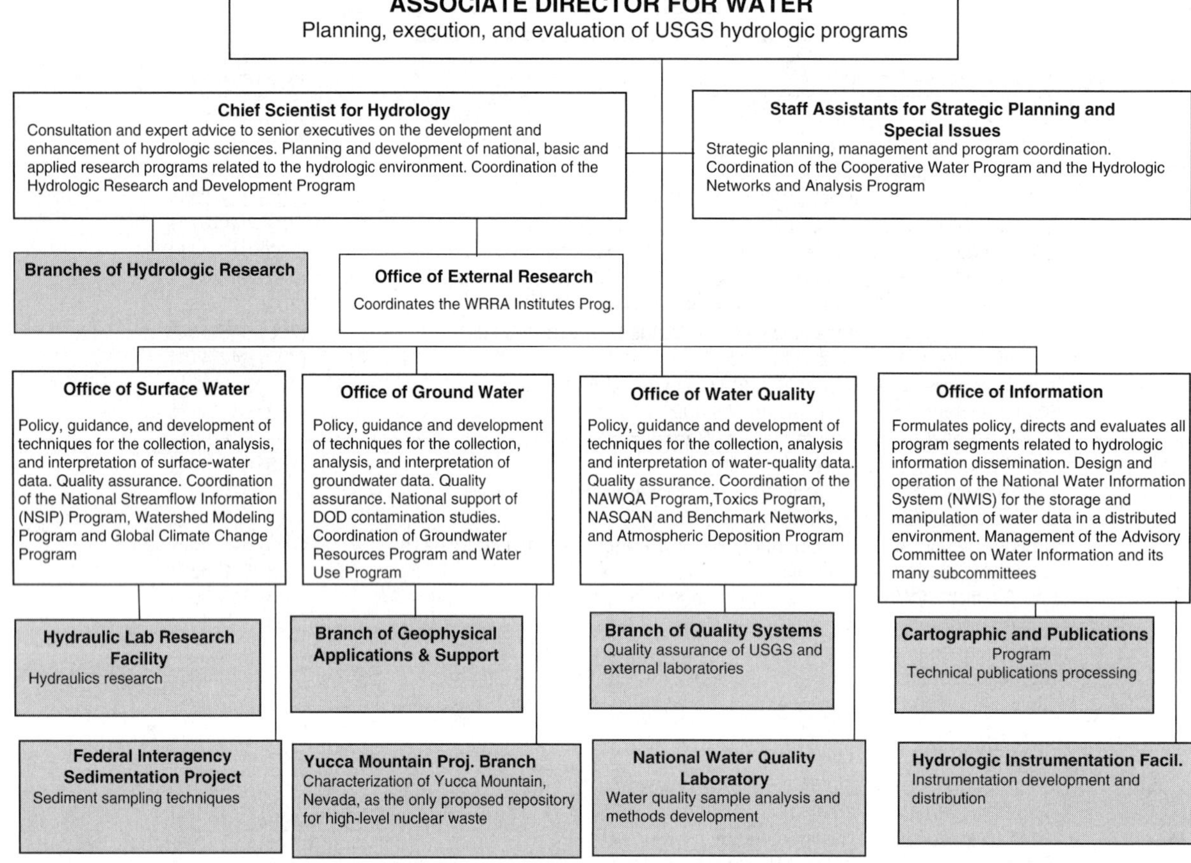

Figure 12A.4 Organization of the water resources division of the U.S. Geological Survey (www.epa.gov).

Table 12A.10 USGS Water Resources District Chiefs

Alabama, Athena P. Clark, dc_al@usgs.gov, 2350 Fairlane Drive, Suite 120, Montgomery, AL 36116, Phone: (334) 213-2332, Fax: (334) 213-2348, Office hours: 7:30 a.m. to 4:00 p.m. Central Time

Arizona, Nick B. Melcher, dc_az@usgs.gov, U.S. Geological Survey Water Resources Division, 520 N. Park Avenue, Suite 221, Tucson, AZ 85719, Phone: (520) 670-6671, ext. 221, Fax: (520) 670-5592, Office hours: 7:30 a.m. to 4:00 p.m. Mountain Time

California, Point of contact: Frederick J. Heimes, Michael V. Shulters, dc_ca@usgs.gov, Placer Hall, 6000 J Street, Sacramento, CA 95819-6129, Phone: (916) 278-3000, Fax: (916) 278-3070, Office hours: 7:30 a.m. to 4:00 p.m. Pacific Time

Colorado, William F. Horak, dc_co@usgs.gov, Bldg. 53, Denver Federal Center, Mail Stop 415, Box 25046, Lakewood, CO 80225-0046, Phone: (303) 236-4882, ext. 258, Fax: (303) 236-4912, Office hours: 8:00 a.m. to 4:30 p.m. Mountain Time

Delaware, William Guertal, dc_de@usgs.gov, 1289 McD Dr., Dover, DE 19901-4907, Phone: (302) 734-2506, Fax: (302) 734-2964, Office hours: 8:00 a.m. to 4:30 p.m. Eastern Time

Alaska, Steven A. Frenzel, dc_ak@usgs.gov, U.S. Geological Survey, 4230 University Dr., Suite 201, Anchorage, AK 99508-4664, Phone: (907) 786-7100, Fax: (907) 786-7150, Office hours: 8:00 a.m. to 4:30 p.m. Alaska Time

Arkansas, John E. Terry JR., dc_ar@usgs.gov, 401 Hardin Rd., Little Rock, AR 72211, Phone: (501) 228-3600, Fax: (501) 228-3601, Office hours: 7:30 a.m. to 4:00 p.m. Central Time

Caribbean, Pedro L. Diaz, dc_pr@usgs.gov, GSA Center, 651 Federal Drive, Suite 400-15, Guaynabo, PR 00965, Phone: (787) 749-4346, Fax: (787) 787 749-4301, Office hours: 7:45 a.m. to 4:30 p.m. Atlantic Time

Connecticut, Virginia A. Delima, dc_ct@usgs.gov, 101 Pitkin Street, East Hartford, CT 06108, Phone: (860) 291-6740, Fax: (860) 291-6799, Office hours: 7:45 a.m. to 4:15 p.m. Eastern Time

District of Columbia, see Maryland

(Continued)

Table 12A.10 (Continued)

Florida, Carl R. Goodwin, dc_fl@usgs.gov, 9100 NW 36th St., Suite 107, Miami, FL 33178, Phone: (305) 717-5845, Fax: (305) 717-5801, St. Pete Offic: (727) 803-8747 (x3070), St. Pete Fax: (727) 803-2031, Office hours: 7:45 a.m. to 4:30 p.m. Eastern Time

Georgia, Edward H. Martin, dc_ga@usgs.gov, Peachtree Business Center, Suite 130, 3039 Amwiler Rd., Atlanta, GA 30360-2824, Phone: (770) 903-9100, Fax: (770) 903-9199, Office hours: 8:00 a.m. to 4:30 p.m. Eastern Time

Guam, see Hawaii

Hawaii, Gordon Tribble, dc_hi@usgs.gov, 677 Ala Moana Blvd., Suite 415, Honolulu, HI 96813, Phone: (808) 587-2405, Fax: (808) 587-2401, Office hours: 8:00 a.m. to 4:30 p.m. Hawaii-Aleutian Time

Idaho, Kathy D. Peter, dc_id@usgs.gov, 230 Collins Rd., Boise, ID 83702-4520, Phone: (208) 387-1300, Fax: (208) 387-1372, Office hours: 7:45 a.m. to 4:15 p.m. Mountain Time

Illinois, Bob Holmes, dc_il@usgs.gov, 221 North Broadway Avenue, Urbana, IL 61801, Phone: (217) 344-0037, ext. 3005, Fax: (217) 344-0082, Office hours: 8:00 a.m. to 4:30 p.m. Central Time

Indiana, James A. Stewart, dc_in@usgs.gov, 5957 Lakeside Blvd., Indianapolis, IN 46278-1996, Phone: (317) 290-3333, Fax: (317) 290-3313, Office hours: 7:30 a.m. to 4:00 p.m. Eastern Time

Iowa, Rob G. Middlemis-Brown, dc_ia@usgs.gov, P.O. Box 1230, Iowa City, IA 52244, Phone: (319) 358-3600, Fax: (319) 358-3606, Office hours: 7:45 a.m. to 4:30 p.m. Central Time

Kansas, Walter R. Aucott, dc_ks@usgs.gov, 4821 Quail Crest Place, Lawrence, KS 66049, Phone: (785) 842-9909, Fax: (785) 832-3500, Office hours: 8:00 a.m. to 4:30 p.m. Central Time

Kentucky, Mark A. Ayers, dc_ky@usgs.gov, 9818 Bluegrass Parkway, Louisville, KY 40299, Phone: (502) 493-1900, Fax: (502) 493-1909, Office hours: 8:00 a.m. to 4:45 p.m. Eastern Time

Louisiana, Charles R. Demas, dc_la@usgs.gov, 3535 S. Sherwood Forest Blvd., Suite 120, Baton Rouge, LA 70816, Phone: (225) 298-5481, Fax: (225) 298-5490, Office hours: 7:45 a.m. to 4:30 p.m. Central Time

Maine, Robert M. Lent, dc_me@usgs.gov, 196 Whitten Road, Augusta, ME 04330, Phone: (207) 622-8202, Fax: (207) 622-8204, Office hours: 7:30 a.m. to 4:15 p.m. Eastern Time

Maryland, James M. Gerhart, dc_md@usgs.gov, U.S. Geological Survey, 8987 Yellow Brick Road, Baltimore, MD 21237, Phone: (410) 238-4200, Fax: (410) 238-4210, Office hours: 8:00 a.m. to 4:30 p.m. Eastern Time

Massachusetts, Wayne H. Sonntag, dc_ma@usgs.gov, 10 Bear Foot Road, Northborough, MA 01532, Phone: (508) 490-5000, Fax: (508) 490-5068, Office hours: 8:00 a.m. to 4:30 p.m. Eastern Time

Michigan, James R. Nicholas, dc_mi@usgs.gov, 6520 Mercantile Way, Suite 5, Lansing, MI 48911, Telephone: (517) 887-8903, Fax: (517) 887-8937, Office hours: 7:45 a.m. to 4:15 p.m. Eastern Time

Minnesota, Jeffrey D. Stoner, dc_mn@usgs.gov, 2280 Woodale Dr., Mounds View, MN 55112, Phone: (763) 783-3100, Fax: (763) 783-3103, Office hours: 8:00 a.m. to 4:30 p.m. Central Time

Mississippi, Michael L. Plunkett, dc_ms@usgs.gov, 308 South Airport Road, Pearl, MS 39208-6649, Phone: (601) 933-2940, Fax: (601) 933-2901, Office hours: 8:00 a.m. to 4:30 p.m. Central Time

Missouri, Michael E. Slifer, dc_mo@usgs.gov, 1400 Independence Rd., Mail Stop 200, Rolla, MO 65401, Phone: (573) 308-3664, Fax: (573) 308-3645, Office hours: 7:30 a.m. to 4:00 p.m. Central Time

Montana, Robert E. Davis, dc_mt@usgs.gov, USGS, WRD, 3162 Bozeman Avenue, Helena, MT 59601, Phone: (406) 457-5900, Fax: (406) 457-5990, Office hours: 8:00 a.m. to 4:30 p.m. Mountain Time

Nebraska, James E. Kircher (Acting), dc_ne@usgs.gov, 100 Centennial Mall North, Lincoln, NE 68508, Phone: (402) 437-5082, Fax: (402) 437-5139, Office hours: 7:45 a.m. to 4:30 p.m. Central Time

Nevada, Kimball Goddard, dc_nv@usgs.gov, 333 West Nye Lane, Rm 203, Carson City, NV 89706, Phone: (775) 887-7600, Fax: (775) 887-7621, Office hours: 8:00 a.m. to 4:30 p.m. Pacific Time

New Hampshire, Brian R. Mrazik, dc_nh@usgs.gov, 361 Commerce Way, Pembroke, NH 03275-3718, Phone: (603) 226-7800, Fax: (603) 226-7894, Office hours: 7:45 a.m. to 4:15 p.m. Eastern Time

New Jersey, Richard H. Kropp, dc_nj@usgs.gov, 810 Bear Tavern Rd., Suite 206, West Trenton, NJ 08628, Phone: (609) 771-3900, Fax: (609) 771-3915, Office hours: 7:45 a.m. to 4:15 p.m. Eastern Time

New Mexico, Linda S. Weiss, dc_nm@usgs.gov, 5338 Montgomery, NE, Suite 400 (for the District Office), Suite 300 (for the Albuquerque Field Office), Albuquerque, NM 87109-1311, Phone: (505) 830-7900, Fax: (505) 830-7998, Office hours: 8:00 a.m. to 4:30 p.m. Mountain Time

New York, Rafael M. Rodriguez, dc_ny@usgs.gov, 425 Jordan Rd. Troy, NY 12180, Phone: (518) 285-5600 Fax: (518) 285-5601 Office hours: 7:30 a.m. to 4:00 p.m. Eastern Time

North Carolina, Gerald L. Ryan, dc_nc@usgs.gov, 3916 Sunset Ridge Road Raleigh, NC 27607, Phone: (919) 571-4000, Fax: (919) 571-4041, Office hours: 8:00 a.m. to 4:45 p.m. Eastern Time

(Continued)

Table 12A.10 (Continued)

North Dakota, Gregg J. Wiche, dc_nd@usgs.gov, 821 E. Interstate Ave., Bismarck, ND 58501-1199, Phone: (701) 250-7401, Fax: (701) 250-7492, Office hours: 8:00 a.m. to 5:00 p.m. Central Time

Ohio, James R. Morris, dc_oh@usgs.gov, 6480 Doubletree Avenue, Columbus, OH 43229-1111, Phone: (614) 430-7702, Fax: (614) 430-7777, Office hours: 7:30 a.m. to 4:30 p.m. Eastern Time

Oklahoma, Kimberly T. Winton, dc_ok@usgs.gov, 202 N.W. 66 St., Building 7 Oklahoma City, OK 73116, Phone: (405) 810-4400, Fax: (405) 843-7712, Office hours: 8:00 a.m. to 4:45 p.m. Central Time

Oregon, Dennis D. Lynch, dc_or@usgs.gov, 10615 S.E. Cherry Blossom Dr., Portland, OR 97216, Phone: (503) 251-3200, Fax: (503) 251-3470, Office hours: 7:30 a.m. to 4:30 p.m. Pacific Time

Pennsylvania, Patricia L. Lietman, dc_pa@usgs.gov, 215 Limekiln Road New Cumberland, PA 17043-1586, Phone: (717) 730-6900, Fax: (717) 730-6997, Office hours: 7:30 a.m. to 4:00 p.m. Eastern Time

Rhode Island, James B. Campbell, dc_ri@usgs.gov, 275 Promenade St., Suite 150 Providence, RI 02908, Phone: (401) 331-9050, Fax: (401) 331-9062, Office hours: 8:00 a.m. to 4:30 p.m. Eastern Time

South Carolina, Marjorie S. Davenport dc_sc@usgs.gov 720 Gracern Rd. Stephenson Center, Suite 129 Columbia, SC 29210, Phone: (803) 750-6100, Fax: (803) 750-6181, Office hours: 7:45 a.m. to 4:30 p.m. Eastern Time

South Dakota, Daniel J. Fitzpatrick, dc_sd@usgs.gov, 1608 Mt. View Rd., Rapid City, SD 57702, Phone: (605) 355-4560, Fax: (605) 355-4523, Office hours: 6:30 a.m. to 4:30 p.m. Mountain Time

Tennessee, W. Scott Gain, dc_tn@usgs.gov, 640 Grassmere Park Drive, Suite 100 Nashville, TN 37211, Phone: (615) 837-4700, Fax: (615) 837-4799, Office hours: 7:45 a.m. to 4:45 p.m. Central Time

Texas, Robert L. Joseph, dc_tx@usgs.gov, USGS WRD 8027 Exchange Drive Austin, TX 78754-4733, Phone: (512) 927-3500, Fax: (512) 927-3590, Office hours: 7:45 a.m. to 4:30 p.m. Central Time

Utah, Patrick M. Lambert, dc_ut@usgs.gov, U.S. Geological Survey 2329 W. Orton Circle West Valley City, UT 84119-2047, Phone: (801) 908-5000, Fax: (801) 908-5001, Office hours: 8:00 a.m. to 4:30 p.m. Mountain Time

Vermont, Brian R. Mrazik, dc_nh@usgs.gov, 361 Commerce Way Pembroke, NH 03275-3718 Phone: (603) 226-7800, Fax: (603) 226-7894, Office hours: 7:45 a.m. to 4:15 p.m. Eastern Time

Virginia, Ward W. Staubitz, dc_va@usgs.gov, 1730 East Parham Road Richmond, VA 23228, Phone: (804) 261-2600, Fax: (804) 261-2659, Office hours: 8:00 a.m. to 4:45 p.m. Eastern Time

Washington, Cynthia Barton, dc_wa@usgs.gov, 1201 Pacific Ave., Suite 600 Tacoma, WA 98402, Phone: (253) 428-3600, ext. 2602, Fax: (253) 428-3614, Office hours: 7:45 a.m. to 4:30 p.m. Pacific Time

West Virginia, Hugh E. Bevans, dc_wv@usgs.gov, 11 Dunbar St. Charleston, WV 25301, Phone: (304) 347-5130, Fax: (304) 347-5133, Office hours: 7:30 a.m. to 4:00 p.m. Eastern Time

Wisconsin, Charles A. Peters, dc_wi@usgs.gov, 8505 Research Way Middleton, WI 53562-3581, Phone: (608) 821-3801, Fax: (608) 821-3817, Office hours: 8:00 a.m. to 4:30 p.m. Central Time

Wyoming, Myron H. Brooks, state_rep_wy@usgs.gov, 2617 E. Lincolnway, Suite B Cheyenne, WY 82001, Phone: (307) 778-2931, ext. 2728, Fax: (307) 778-2764, Office hours: 8:00 a.m. to 4:30 p.m. Mountain Time

Source: From www.usgs.gov.

Table 12A.11 USGS Programs

Biological Informatics — http://biology.usgs.gov/bio
The Biological Informatics Program develops and applies innovative information technologies and practices to the management of biological data, information, and knowledge resulting from research, thereby increasing the value of that research to scientists, planners, policy makers, teachers, students, and private citizens

Coastal and Marine Geology — http://marine.usgs.gov
Provides information relative to marine science for making management decisions that affect all 35 coastal states and the island territories. Monitors changes within the coastal and marine environment, whether naturally occurring or human induced

Contaminant Biology — http://biology.usgs.gov/pub_aff/cont-over.htm
The Contaminants Program conducts research, assessments and monitoring to provide the nation and its natural resource managers with information on the exposure, effects and fate of deleterious substances in the environment

Cooperative Research Units — Biology http://www.coopunits.org
Wildlife and Terrestrial Resources
Terrestrial, Freshwater, and Marine Ecosystems
The Cooperative Research Units Program is a unique collaborative relationship between States, Universities, the Federal government and a non-profit organization. Coop Units conduct research on renewable natural resource questions, participate in the education of graduate students destined to join the natural resource profession, provide technical assistance and consultation to parties who have interests in natural resource issues, and provide various forms of continuing education for natural resource professionals

Cooperative Topographic Mapping — http://nationalmap.usgs.gov
The National Map is a consistent framework for geographic knowledge. It provides public access to high-quality, geospatial data and information from multiple partners to help support decision making by resource managers and the public. The National Map is the product of a consortium of Federal, State, and local partners who provide geospatial data to enhance America's ability to access, integrate, and apply geospatial data at global, national, and local scales

Cooperative Water Program — http://water.usgs.gov/coop
As the primary Federal science agency for water resource information, the USGS Cooperative Water Program provides reliable, impartial, and timely information needed to understand the Nation's water resources through a program of shared efforts and funding with State, Tribal, and local partners to enable decision makers to manage the Nation's water resources

Earth Surface Dynamics — http://geochange.er.usgs.gov
The USGS Global Change Research activities strive to achieve a whole-system understanding of the interrelationships among earth surface processes, ecological systems, and human activities. Activities of the program focus on documenting, analyzing, and modeling the character of past and present environments and the geological, biological, hydrological, and geochemical processes involved in environmental change so that future environmental changes and impacts can be anticipated

Earthquake Hazards — http://earthquake.usgs.gov
Global Seismic Network http://earthquake.usgs.gov/networks/global.html
This web site is provided by the USGS Earthquake Hazards Program as part of the effort to reduce earthquake hazard in the United States

Energy Resources — http://energy.usgs.gov
The USGS Energy Resources Program addresses the challenge of increasing demand for affordable energy from environmentally acceptable energy sources by conducting basic and applied research on geologic energy resources and on the environmental, economic, and human health impacts of their production and use. The Program provides reliable and impartial scientific information and comprehensive analyses of oil, natural gas, and coal resources of the Nation and the World

Fisheries and Aquatic Resources — http://far.nbii.gov/
To aid conservation and restoration efforts, the Fisheries and Aquatic Resources (FAR) node was established to provide an integrated, comprehensive web-based resource that will: 1) serve and access fishery and aquatic databases, 2) link to fishery and aquatic resource information sites, and 3) act as larger scale coordinating site for fisheries and aquatic resources standards

Geographic Analysis and Monitoring — http://gam.usgs.gov
Geographic Analysis and Monitoring Program conducts geographic assessments to improve the understanding of the rates, causes, and consequences of natural and human-induced processes that shape and change the landscape over time

Geomagnetism — http://geomag.usgs.gov
The USGS Geomagnetism Program monitors the field through a network of magnetic observatories and conducts scientific analysis on those data. The Program collects, transports, and can disseminate these data in near-real time, and it also has significant data-processing and management capacities

Groundwater Resources — http://water.usgs.gov/ogw
These pages are designed to provide useful information about groundwater resources of the Nation and groundwater activities of the USGS

(Continued)

Table 12A.11 (Continued)

Hydrologic Networks and Analysis
Hydrologic Research and Development — http://water.usgs.gov/nrp
The National Research Program (NRP) conducts basic and problem oriented hydrologic research in support of the U.S. Geological Survey (USGS)

Invasive Species and Emerging Diseases — http://www.usgs.gov/invasive_species/plw
Provides updates on invasive plants, exotic animals and microbes

Land Remote Sensing — http://remotesensing.usgs.gov
The fundamental goal of the LRS Program is to provide the Federal Government and the public with a primary source of remotely sensed data and applications and to be a leader in defining the future of land remote sensing, nationally, and internationally

Landslide Hazards — http://landslides.usgs.gov/index.html
The Landslides Hazards Group conducts research, gathers information, responds to emergencies and disasters, and produces scientific reports and other products for a broad-based user community. The group provides the results of investigations to private consultants in geology and geotechnical engineering, and to government planners and decision makers

Mineral Resources — http://minerals.usgs.gov
The Mineral Resources Program funds science to provide and communicate current, impartial information on the occurrence, quality, quantity, and availability of mineral resources

National Cooperative Geologic Mapping — http://ncgmp.usgs.gov
The program coordinates an expanded geologic mapping effort by the USGS, the State geological surveys, and universities. The primary goal of the program is to collect, process, analyze, translate, and disseminate earth-science information through geologic maps

National Streamflow Information — http://water.usgs.gov/nsip
Provides information on the quantity and timing of the streamflow in the Nation's rivers. The USGS operates and maintains approximately 7,000 streamgages which provide long-term, accurate, and unbiased information that meets the needs of many diverse users

National Water Quality Assessment — http://water.usgs.gov/nawqa
The NAWQA program has been collecting and analyzing data and information in more than 50 major river basins and aquifers across the Nation. The goal is to develop long-term consistent and comparable information on streams, groundwater, and aquatic ecosystems to support sound management and policy decisions

State Water Resources Research Institute Program — http://water.usgs.gov/wrri
The program cooperates with the USGS in establishing total programmatic direction, reporting on the activities of the Institutes, coordinating and facilitating regional research and information and technology transfer, and in operating the NIWR-USGS Student Internship Program

Toxic Substances Hydrology — http://toxics.usgs.gov
Provides objective scientific information to improve characterization and management of contaminated sites, to protect human and environmental health, and to reduce potential future contamination problems

Volcano Hazards — http://volcanoes.usgs.gov
Provides scientific information on volcanoes including current and historical volcanic activities and reducing associated risks

Source: From www.usgs.gov.

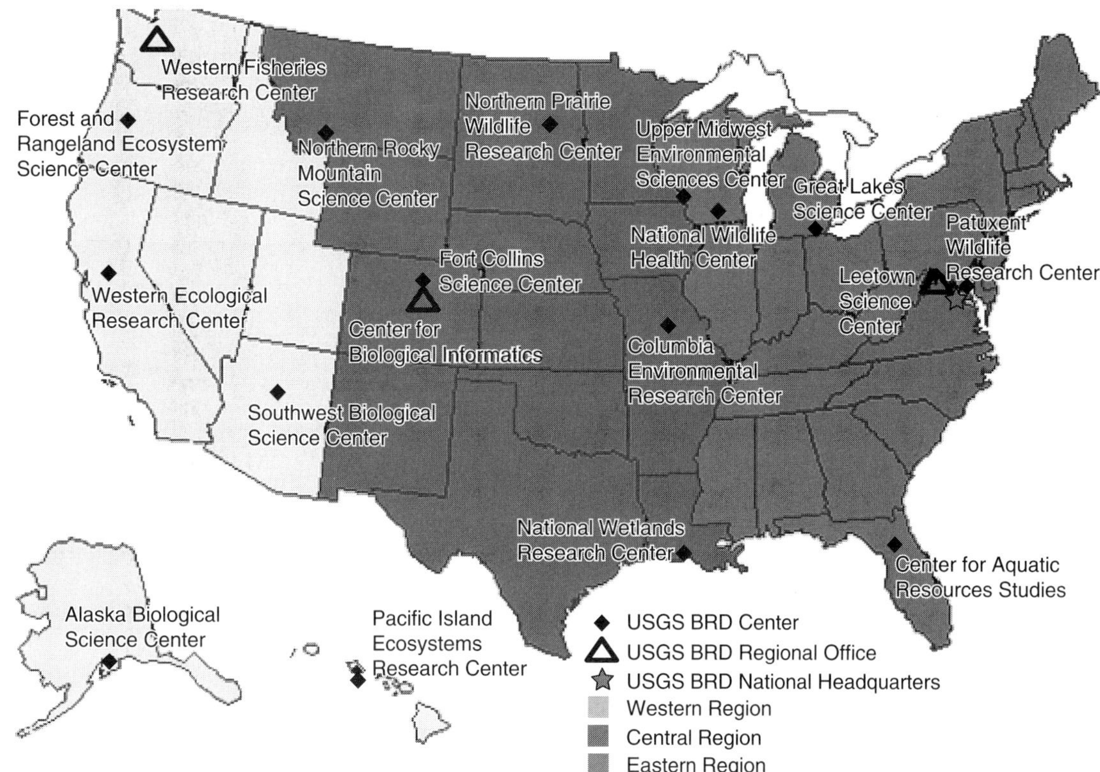

Figure 12A.5 USGS biological resources discipline locations — regional offices (www.usgs.gov).

SECTION 12B STATE WATER AGENCIES

Table 12B.12 State Agency Contacts

Alabama

Alabama Department of Environmental Management, http://www.adem.state.al.us/, 1400 Coliseum Blvd., Montgomery, AL 36110, Phone (334) 271-7700, Fax (334) 271-7950

Alabama Department of Conservation and Natural Resources, http://www.dcnr.state.al.us/, 64 N. Union Street, Suite 468, Montgomery, AL 36130, Phone: (334) 242-3486

Alaska

Department of Environmental Conservation, state.ak.us/dec/, Water Quality Standards Assessment & Reporting Program, Division of Water, 555 Cordova St., Anchorage, AK 99501-2617, Phone: (907) 269-7595, Fax: (907) 269-7649

Arizona

Arizona Department of Environmental Quality, http://www.azdeq.gov/, 1110 W. Washington St., Phoenix, AZ 85007, Phone (602) 771-2300, Fax: (602) 771-4426

Arizona Fish & Game Department, http://www.gf.state.az.us/, Main Office: Phoenix, 2221 W. Greenway Rd., Phoenix, AZ 85023-4399, Phone: (602) 942-3000

Arkansas

Arkansas Department of Environmental Quality, http://www.adeq.state.ar.us/, 8001 National Drive, Little Rock, AR 72209, P.O. Box 8913, Little Rock, AR 72219, Phone: (501) 682-0744

California

California Environmental Protection Agency, http://www.calepa.ca.gov/, 1001 I Street, P.O. Box 2815, Sacramento, CA 95812-2815, Phone: (916) 445-3846

California Department of Conservation, http://www.consrv.ca.gov/index/, DOC Headquarters, 801 K Street, MS 24-01, Sacramento, CA 95814, Phone (916) 322-1080, Fax: (916) 445-0732

California Air Resources Board, http://www.arb.ca.gov/homepage.htm, 1001 "I" Street, P.O. Box 2815, Sacramento, CA 95812, Phone: (916) 322-2990, Fax: (916) 445-5025

Department of Water Resources, http://www.water.ca.gov/, 1416 9th Street, Sacramento, CA 95814, Mailing Address: P. O. Box 942836, Sacramento, CA 94236, Phone: (916) 653-5791

California Integrated Waste Management Board, http://www.ciwmb.ca.gov/, 1001 I Street, P.O. Box 4025, Sacramento, CA 95812-4025, Phone: (916) 341-6000

Colorado

Colorado Revised Statutes, aescon.com/iway/entry.htm, Suite 3175, 1660 Lincoln Street, Denver, CO 80264

Colorado Department of Public Health and Environment, http://www.cdphe.state.co.us/cdphehom.asp, 4300 Cherry Creek Drive South, Denver, CO 80246-1530, Phone: (303) 692-2000

Connecticut

Department of Environmental Protection, http://dep.state.ct.us/, 79 Elm Street, Hartford, CT 06106-5127, Phone (860) 424-3001, Fax (860) 424-4051

Delaware

Delaware Department of Natural Resources and Environmental Control, http://www.dnrec.state.de.us/dnrec2000/, 89 Kings Hwy, Dover, DE 19901, Phone (302) 739-5072

District of Columbia

Environmental Health Administration, http://dchealth.dc.gov/, 51 N Street, NE, Washington, DC 20002, Phone: (202) 535-2500

Florida

Florida Department of Environmental Protection, http://www.dep.state.fl.us/, 3900 Commonwealth Boulevard M.S. 49, Tallahassee, FL 32399, Phone: (850) 245-2118, Fax: (850) 245-2128

Georgia

Georgia Department of Natural Resources, http://www.gadnr.org/, 2 Martin Luther King, Jr. Drive, S. E., Suite 1252 East Tower, Atlanta, GA 30334, Phone: (404) 656-3500

Georgia Environmental Protection Division, http://www.ganet.org/dnr/environ/, 2 Martin Luther King, Jr. Drive, Suite 1152 East Tower, Atlanta, GA 30334, Phone: (404) 657-5947, Fax: (404) 651-5778

Hawaii

State of Hawaii, Department of Land and Natural Resources, http://www.hawaii.gov/dlnr/, Kalanimoku Bldg., 1151 Punchbowl St., Honolulu, HI 96813, Phone: (808) 587-0400, Fax: (808) 587-0390

Idaho

Idaho Bureau of Homeland Security, http://www.bhs.idaho.gov/, 4040 Guard Street, Building 600, Boise, ID 83705-5004, Phone: (208) 334-3460, Fax: (208) 334-2322

(Continued)

Table 12B.12 (Continued)

Idaho Department of Environmental Quality, http://www.deq.state.id.us/index.cfm, 1410 N. Hilton, Boise, ID 83706, Phone: (208) 373-0502, Fax: (208) 373-0417

Idaho Department of Water Resources, http://www.idwr.state.id.us/, 322 E. Front St, P.O. Box 83720, Boise, ID 83720-0098, Phone: (208) 287-4800, Fax: (208) 287-6700

Illinois

Illinois Environmental Protection Agency, http://www.epa.state.il.us/, 1021 North Grand Avenue East, P.O. Box 19276, Springfield, IL 62794-9276, Phone: (217) 782-3397

Illinois Pollution Control Board, http://www.ipcb.state.il.us/, 1021 North Grand Avenue East, P.O. Box 19274, Springfield, IL 62794-9274, Phone: (217) 524-8500

Indiana

Indiana Department of Environmental Management, http://www.ai.org/idem/, Indiana Government Center-North, 100 N. Senate Ave., Indianapolis, IN 46204, Phone: (317) 232-8603

Indiana Department of Natural Resources, http://www.in.gov/dnr/, Department of Natural Resources, 402 West Washington Street, Indianapolis, IN 46204, Phone: (317) 232-4699, Fax: (317) 233-2977

Iowa

Iowa Department of Natural Resources, http://www.iowadnr.com/, 502 E. 9th Street, Des Moines, IA 50319-0034, Phone: (515) 281-5918

Kansas

The Kansas Department of Health and Environment, http://www.kdhe.state.ks.us/, Curtis State Office Building, 1000 SW Jackson, Topeka, KS 66612, Phone: (785) 296-1500, Fax: (785) 368-6368

Kentucky

Kentucky Environmental Quality Commission, eqc.ky.gov/, 14 Reilly Road, Frankfort, KY 40601, Phone: (502) 564-2150, Fax: (502) 564-9729

Department for Natural Resources, naturalresources.ky.gov/, 663 Teton Trail, Frankfort, KY 40601, Phone: (502) 564-2184, Fax: (502) 564-6193

Louisiana

Louisiana Department of Environmental Quality, http://www.deq.state.la.us/, P.O. Box 4301, Baton Rouge, LA 70821-4301, Phone: (225) 219-3953, Fax: (225) 219-3971

Maine

Maine Department of Environmental Protection, http://www.state.me.us/dep/index.shtml, 17 State House Station, Augusta, ME 04333-0017, Phone: (207) 287-7688 • (800) 452-1942

Maryland

Maryland Department of the Environment, http://www.mde.state.md.us/, 1800 Washington Blvd, Baltimore, MD 21230, Phone: (410) 537-3000

Maryland Department of Natural Resources, dnr.state.md.us/, 580 Taylor Avenue, Annapolis, MD 21401, Phone: (877) 620-8367

Massachusetts

Massachusetts Department of Environmental Protection, http://www.mass.gov/dep/dephome.htm, One Winter Street, Boston, MA 02108-4746, Phone: (617) 292-5500

Michigan

Division of Water Supply, Michigan Department of Public Health, P.O. Box 30035, Lansing, MI 48909, Phone: (517) 335-8318

Minnesota

Minnesota Department of Natural Resources, http://www.dnr.state.mn.us/index.html, 500 Lafayette Road, St. Paul, MN 55155-4040, Phone: (651) 296-6157

Minnesota Pollution Control Agency, http://www.pca.state.mn.us/, 520 Lafayette Road, St. Paul, MN 55155-4194, Phone: (651) 296-6300

Mississippi

Mississippi Department of Environmental Quality, http://www.deq.state.ms.us/MDEQ.nsf/, P.O. Box 20305, Jackson, MS 39289-1305, Phone: (601) 961-5171

Missouri

Department of Natural Resources, dnr.state.mo.us/, P.O. Box 176, Jefferson City, MO 65102, Phone: (573) 751-3443

Missouri DNR, Air and Land Protection Division, dnr.state.mo.us/alpd/, Department of Natural Resources, P.O. Box 176, Jefferson City, MO 65102, Phone: (573) 751-3443

Missouri DNR, Water Protection and Soil and Conservation Division, dnr.state.mo.us/wpscd/, Department of Natural Resources, P.O. Box 176, Jefferson City, MO 65102, Phone: (573) 751-3443

Missouri Department of Conservation, http://www.conservation.state.mo.us/, P.O. Box 180 (zip 65102), 2901 W. Truman Blvd., Jefferson City, MO 65109, Phone: (573) 751-4115, Fax: (573) 751-4467

(Continued)

Table 12B.12 (Continued)

Montana

Montana Department of Environmental Quality, http://www.deq.state.mt.us/index.asp, Metcalf Building Office, 1520 E. Sixth Avenue, P.O. Box 200901, Helena, MT 59620-0901, Phone: (406) 444-2544, Fax: (406) 444-4384

Montana Natural Resource Information System, http://www.nris.state.mt.us/, Montana State Library, 1515 East 6th Avenue, P.O. Box 201800, Helena, MT 59620-1800, Phone: (406) 444-3115, Fax: (406) 444-0266

Montana GIS Data Library, http://www.nris.state.mt.us/gis/default.htm, Montana State Library, 1515 East 6th Avenue, P.O. Box 201800, Helena, MT 59620-1800, Phone: (406) 444-3115, Fax: (406) 444-0266

Nebraska

Nebraska Department of Environmental Quality, http://www.deq.state.ne.us/, 1200 "N" Street, Suite 400, P.O. Box 98922, Lincoln, NE 68509, Phone: (402) 471-2186, Fax: (402) 471-2909

Nevada

Nevada Division of Forestry, http://www.forestry.nv.gov/, 2525 South Carson Street, Carson City, NV 89701, Phone: (775) 684-2500, Fax: (775) 687-4244

Nevada Department of Conservation and Natural Resources, http://dcnr.nv.gov/, 123 W. Nye Lane, Room 230, Carson City, NV 89706-0818, Phone: (775) 687-4360, Fax: (775) 687-6122

Nevada Division of Environmental Protection, http://ndep.nv.gov/, 333 W. Nye Lane, Room 138, Carson City, NV 89706-0851, Phone: (775) 687-4670, Fax: (775) 687-5856

New Hampshire

N.H. Department of Environmental Services, des.state.nh.us/, 29 Hazen Drive, P.O. Box 95, Concord, NH 03302-0095, Phone: (603) 271-3503, Fax: (603) 271-2867

New Jersey

New Jersey Department of Environmental Protection, http://www.state.nj.us/dep/, P.O. Box 402, Trenton, NJ 08625-0402, Post Office Box CN-029, Phone: (609) 777-3373

New Mexico

New Mexico Environment Department, http://www.nmenv.state.nm.us/, P.O. Box 26110 - 1190 St. Francis Drive N4050, Santa Fe, NM 87502-0110, Phone: (800) 219-6157, (505) 827-2855

New York

The New York State Department of Environmental Conservation, http://www.dec.state.ny.us/, 625 Broadway, Albany, NY 12233, Phone: (518) 402-8233, Fax: (518) 402-9029

North Carolina

North Carolina Department of Environment & Natural Resources, http://www.enr.state.nc.us/, 1601 Mail Service Center, Raleigh, NC 27699-1601, Phone: (919) 733-4984, Fax: (919) 715-3060

North Carolina Division of Pollution Prevention and Environmental Assistance (DPPEA), http://www.p2pays.org/, 1639 Mail Service Center, Raleigh NC 27699-1639, Phone: (919) 715-6500, (800) 763-0136

North Carolina GIS Database, http://cgia.cgia.state.nc.us/, 301 N. Wilmington St., Suite 700, Raleigh, NC 27601, Phone: (919) 733-2090, Fax: (919) 715-0725

North Dakota

North Dakota State Water Commission, http://www.swc.state.nd.us/, 900 East Boulevard, Bismarck, ND 58505-0850, Phone: (701) 328-2750, Fax: (701) 328-3696

North Dakota Geological Survey, http://www.state.nd.us/ndgs/, 600 East Boulevard Avenue, Bismarck, ND 58505-0840, Phone: (701) 328-8000, Fax: (701) 328-8010

ND Department of Health, Environmental Health Section, http://www.health.state.nd.us/ehs/, 1200 Missouri Ave., P.O. Box 5520, Bismarck, ND 58506-5520, Phone: (701) 328-5150, Fax: (701) 328-5200

Ohio

Ohio Environmental Protection Agency, http://www.epa.state.oh.us/new/divs.html, 122 S Front St., Columbus, OH 43215, Phone: (614) 644-3020

Ohio Air Quality Development Authority, http:www.ohioairquality.org/, 50 W. Broad Street, Suite 1718, Columbus, OH 43215, Phone: (614) 224-3383, Fax: (614) 752-9188

Oklahoma

Oklahoma Conservation Commission, http://www.okcc.state.ok.us/, 2800 N. Lincoln Blvd., Suite 160, Oklahoma City, OK 73105, Phone: (405) 521-2384, Fax: (405) 521-6686

Oklahoma Department of Environmental Quality, http://www.deq.state.ok.us/, P.O. Box 1677, Oklahoma City, OK 73101-1677, Phone: (405) 702-1000, Fax: (405) 702-1001

Oregon

Oregon Department of Environmental Quality, http://www.deq.state.or.us/, 811 SW Sixth Avenue, Portland, OR 97204-1390, Phone: (503) 229-5696, Fax: (503) 229-6124

Oregon Department of Fish and Wildlife, http://www.dfw.state.or.us/, 3406 Cherry Avenue, N.E., Salem, OR 97303, Phone (503) 947-6000

(Continued)

Table 12B.12 (Continued)

Pennsylvania

Pennsylvania Department of Environmental Protection, http://www.dep.state.pa.us/, 16th Floor, Rachel Carson State Office Building, P.O. Box 2063, 400 Market Street, Harrisburg, PA 17105-2063, Phone (717) 787-2814

Pennsylvania Department of Conservation and Natural Resources, http://www.dcnr.state.pa.us/, Rachel Carson State Office Building, P.O. Box 8767, 400 Market Street, Harrisburg, PA 17105-8767, Phone: (717) 787-2869, Fax: (717) 772-9106

Rhode Island

Rhode Island Department of Environmental Management, http://204.139.0.230/, 235 Promenade Street, Providence, RI 02908-5767, Phone: (401) 222-6800

South Carolina

South Carolina Department of Health & Environmental Control, http://www.scdhec.net/, 2600 Bull Street, Columbia, SC 29201, Phone: (803) 898-3432

South Carolina Department of Natural Resources, http://water.dnr.state.sc.us/, P.O. Box 167, Rembert C. Dennis Building, Columbia, SC 29202, Phone: (803) 734-3979, Fax: (803) 734-6310

South Dakota

South Dakota Department of Environment & Natural Resources, http://www.state.sd.us/denr/denr.html, Joe Foss Building, 523 E Capitol, Pierre, SD 57501, Phone: (605) 773-3151, Fax: (605) 773-6035

Tennessee

Tennessee Department of Environment & Conservation, http://www.state.tn.us/environment/, 401 Church Street, L & C Annex, 1st Floor, Nashville, TN 37243-0435, Phone: (615) 532-0109

Texas

Texas Commission on Environmental Quality, http://www.tceq.state.tx.us/, P.O. Box 13087, Austin, TX 78711-3087, 12100 Park 35 Circle, Austin, TX 78753, Phone: (512) 239-1000

City of Dallas Air Pollution Control Program, http://www.dallasair.org/, 320 E. Jefferson St. Room LL13, Dallas, TX 75203, Phone: (214) 948-4435, Fax: (214) 948-4412

Utah

Utah Department of Environmental Quality, http://www.eq.state.ut.us/, 168 North 1950 West, 2nd Floor, P.O. Box 144810, Salt Lake City, UT 84114-4810, Phone: (801) 536-4410, Fax: (801) 536-4273

Utah Automated Geographic Reference Center (AGRC), http://agrc.its.state.ut.us/, 5130 State Office Building, Salt Lake City, UT 84114, Phone: (801) 538-3665, Fax: (801) 538-3317

Vermont

Vermont Agency of Natural Resources, 103 South Main Street, Center Building, Waterbury, VT 05671-0301, Phone: (802) 241-3600, Fax: (802) 244-1102

Virginia

Virginia Department of Environmental Quality, http://www.deq.state.va.us/, 629 East Main Street, P.O. Box 10009, Richmond, VA 23240-0009, Phone: (804) 698-4000

Washington

Washington Department of Ecology, http://www.ecy.wa.gov/, Washington State Department of Ecology, P.O. Box 47600, Olympia, WA 98504-7600, Phone: (360) 407-6000

Washington State Department of Natural Resources, http://www.dnr.wa.gov/, P.O. Box 47001, Olympia, WA 98504-7001, Phone: (360) 902-1004, Fax: (360) 902-1775

Washington State Department of Transportation, http://www.wsdot.wa.gov/environment/default.htm, Transportation Building, Washington State Department of Transportation, 310 Maple Park Avenue SE, P.O. Box 47300, Olympia, WA 98504-7300, Phone: (360) 705-7000

West Virginia

West Virginia Division of Environmental Protection, http://www.dep.state.wv.us/, 601 57th Street, Charleston, WV 25304, Phone: (304) 926-0440

Wisconsin

Wisconsin Department of Natural Resources, http://www.dnr.state.wi.us/, 101 S Webster St, P.O. Box 7921, Madison, WI 53707-7921, Phone: (608) 266-2621, Fax: (608) 261-4380

Wyoming

Department of Environmental Quality, http://deq.state.wy.us/, 122 West 25th St, Herschler Building, Cheyenne, WY 82002, Phone: (307) 777-7937, Fax: (307) 777-7682

Source: From www.epa.gov.

SECTION 12C WATER RESOURCES RESEARCH

Table 12C.13 Universities Council on Water Resources

About 90 universities in the United States and throughout the world comprise the Universities Council on Water Resources (UCOWR) organization. Member institutions engage in education, research, public service, international activities, and information support for policy development related to water resources. Each member university appoints four faculty members as UCOWR lead delegates. Others may join as individual members. See history and goals for more information.

Alabama
Auburn University

Alaska
University of Alaska

Arizona
University of Arizona
Arizona State University

Arkansas
University of Arkansas

California
California State, Sacramento
University of California, Davis
University of California, Riverside
University of Southern California

Colorado
Colorado State University
National Technological University
University of Colorado

Connecticut
Yale University

District of Columbia
The Catholic University of America

Delaware
University of Delaware

Florida
University of Central Florida
University of Florida

Georgia
Georgia Institute of Technology
University of Georgia

Guam
University of Guam

Hawaii
University of Hawaii

Iowa
Iowa State University
University of Iowa

Idaho
University of Idaho

Illinois
Southern Illinois University Carbondale
University of Illinois

(Continued)

Table 12C.13 (Continued)

Indiana
Purdue University

Kansas
University of Kansas

Louisiana
Louisiana State University
University of New Orleans

Massachusetts
Massachusetts Institute of Tech
University of Massachusetts

Maryland
Johns Hopkins University

Maine
Bates College
University of Maine

Michigan
Michigan State University
University of Michigan

Minnesota
University of Minnesota

Missouri
University of Missouri

Mississippi
Mississippi State University

Montana
Montana State University
University of Montana

North Carolina
Duke University
North Carolina State University
University of North Carolina

Nebraska
University of Nebraska

New Hampshire
University of New Hampshire

New Jersey
Rutgers, The State University of New Jersey

New Mexico
New Mexico Institute of Mining/Tech
New Mexico State University
University of New Mexico

Nevada
University of Nevada System

New York
City University of New York
Cornell University
State University of New York, College at Brockport
State University of New York, ES&F, Syracuse
Syracuse University
United States Military Academy

(Continued)

Table 12C.13 (Continued)

Ohio
Central State University
The Ohio State University
University of Cincinnati

Oklahoma
Oklahoma State University
University of Oklahoma

Oregon
Oregon State University

Pennsylvania
Pennsylvania State University
Drexel University

Puerto Rico
University of Puerto Rico

South Carolina
Clemson University

South Dakota
South Dakota State University

Tennessee
Tennessee Tech University
University of Tennessee

Texas
Texas State University
Tarleton State University
Texas A&M University
Texas Tech University
University of Texas at Austin
University of Texas at El Paso
University of Texas at San Antonio

Utah
Utah State University

Virginia
University of Virginia
Virginia Polytechnic Institute

U.S. Virgin Islands
University of the Virgin Islands

Washington
Washington State University

Wisconsin
University of Wisconsin

Wyoming
University of Wyoming

Affiliates
Asian Institute of Technology
University of Calgary
University of New England

Source: From www.ucowr.siu.edu.

Table 12C.14 Water Resources Research Institutes

The Water Resources Research Act authorized by P.L. 101-397 provides for Water Resources Research Institutes in each of the 50 states, the trust territories, and the District of Columbia. Under the Act, these institutes are to:

1. plan, conduct, or otherwise arrange for competent research that fosters (A) the entry of new research scientists into the water resources fields, (B) the training and education of future scientists, engineers, and technicians, (C) the preliminary exploration of new ideas that address water problems or expand understanding of water and water-related phenomena, and (D) the dissemination of research results to water managers and the public.
2. cooperate closely with other colleges and universities in the State that have demonstrated capabilities for research, information dissemination, and graduate training in order to develop a statewide program designed to resolve State and regional water and related land problems.

Each institute shall also cooperate closely with other institutes and other organizations in the region to increase the effectiveness of the institutes and for the purpose of promoting regional coordination.

The state-based program promotes research, training, information dissemination, and other activities meeting the needs of the States and Nation, and encourages regional cooperation among institutes in research into areas of water management, development, and conservation that have a regional or national character.

Alabama

Dr. Upton Hatch, Director, Auburn University Environmental Institute, 101 Comer Hall, Auburn University, Auburn, AL 36849-5431, Phone: (334) 844-4132, Fax: (334) 844-4462, email: hatchlu@auburn.edu

Alaska

Dr. Douglas Kane, Water and Environmental Research Center, University of Alaska-Fairbanks, Fairbanks, AK 99775-5860, Phone: (907) 474-7808, Fax: (907) 474-7979, email: ffdlk@uaf.edu, http://www.uaf.edu/water/

Arizona

Dr. Sharon Megdal, Water Resources Research Center, 350 N. Campbell Ave., The University of Arizona, Tucson, AZ 85721, Phone: (520) 792-9591, Fax: (520) 792-8518, email: wrrc@ag.arizona.edu, http://ag.arizona.edu/AZWATER/

Arkansas

Dr. Ralph Davis, Arkansas Water Resources Center, 112 Ozark Hall, University of Arkansas, Fayetteville, AK 72701, Phone: (501) 575-4515, Fax: (501) 575-3177, email: ralphd@uark.edu, http://www.uark.edu/depts/awrc/

California

Dr. John Letey, Director, Center for Water Resources, Ruibidoux Hall -094, University of California, Riverside, CA 92521-0436, Phone: (909) 787-4327, Fax: (909) 787-5295, email: john.letey@ucr.edu, http://www.waterresources.ucr.edu/

Colorado

Dr. Robert C. Ward, Colorado Water Resources Research Institute, E-102 Engineering Building, Colorado State University, Fort Collins, CO 80523, Phone: (970) 491-6308, Fax: (970) 491-1636, email: robert.ward@colostate.edu, http://cwrri.colostate.edu/

Connecticut

Dr. Glenn Warner, Director, Connecicut Institute of Water Resources, WB Young Building, Room 308, 1376 Storrs Road Unit 4018, The University of Connecticut, Storrs, CT 06269-4018, Phone: (860) 486-2840, Fax: (860) 486-5408, email: gwarner@canrl.cag.uconn.edu, http://www.ctiwr.uconn.edu/

Delaware

Dr. J. Thomas Sims, Delaware Water Resources Center, 113 Townsend Hall, University of Delaware, Newark, DE 19716-2103, Phone: (302) 831-6757, Fax: (302) 831-6758, email: jtsims@udel.edu, http://ag.udel.edu/dwrc/

District of Columbia

Dr. William W. Hare, Director, D.C. Water Resource Research Institute, 4200 Connecticut Ave NW, Building 52, Suite 322D University of the District of Columbia, Washington, DC 20008, Phone: (202) 409-0319, Fax: (202) 274-7016, email: whare@udc.edu, http://www.udc.edu/WRRI/

Florida

Dr. Kirk Hatfield, Interim Director, Water Resources Research Center, Department of Civil Engineering, 124 Yon Hall, P.O. Box 116580, University of Florida, Gainesville, FL 32611-6580, Phone: (352) 392-9537, Fax: (352) 392-3394, email: khatf@ce.ufl.edu, http://www.ce.ufl.edu/~wrrc/

Georgia

Dr. Aris P. Georgakakos, Environmental Resources Center, School of Civil & Environmental Engineering, 790 Atlantic Drive, Georgia Institute of Technology, Atlanta, GA 30332-0335, Phone: (404) 894-2240, Fax: (404) 894-2677, email: ageorgak@ce.gatech.edu, http://www.gwri.org/

Guam

Dr. Leroy F. Heitz, Director, Water and Environmental Research Institute of the Western Pacific, University of Guam, 303 University Drive, Mangilao, Guam 96923, Phone: 011 671 7352685, Fax: 011 671 734-8890, email: lheitz@uog.edu, http://www.uog.edu/weri/index.html

(Continued)

Table 12C.14 (Continued)

Hawaii

Dr. James Moncur, Water Resources Research Center, 2540 Dole Street, Holmes Hall 283, University of Hawaii at Manoa, Honolulu, HI 96822, Phone: (808) 956-7847, Fax: (808) 956-5044, email: jmoncur@hawaii.edu, http://www.wrrc.hawaii.edu/

Idaho

Dr. John C. Tracy, Director, Idaho Water Resources Research Institute, 322 Front Street, University of Idaho, Boise, ID 83702, Phone: (208) 364-9921, Fax: (208) 885-6431, email: iwrri@uidaho.edu, http://www.uidaho.edu/rsrch/iwrri/

Illinois

Dr. Richard Warner, Illinois Water Resources Center, MC-635, University of Illinois at Urbana-Champaign, 1101 West Peabody Drive, Urbana, IL 61801, Phone: (217) 333-0536, Fax: (217) 333-8046, email: dickw@uiuc.edu, http://www.environ.uiuc.edu/iwrc/

Indiana

Dr. Ronald F. Turco, Jr., Indiana Water Resources Research Center, A. A. Potter Engineering Center, Purdue University, West Lafayette, IN 47907-1295, Phone: (765) 494-8041, Fax: (765) 496-3210, email: rturco@purdue.edu, http://ce.ecn.purdue.edu/wrrc.html

Iowa

vacant, Iowa State Water Resources Research Institute, 100 Davidson Hall, Iowa State University, Ames, IA 50011, Phone: (515) 294-1880, Fax: (515) 294-4250, email: http://www.water.iastate.edu/

Kansas

Dr. Jim Koelliker, Interim Director, Kansas Water Resources Research Institute, 44 Waters Hall, Kansas State University, Manhattan, KS 66506, Phone: (913) 532-7419, Fax: (913) 532-6563, email: koellik@ksu.edu

Kentucky

Dr. Lindell E. Ormsbee, Director, Kentucky Water Resources Research Institute, 233 Mining and Mineral Resources Building, University of Kentucky, Lexington, KY 40506-0107, Phone: (606) 257-1299, Fax: (606) 257-6329, email: lormsbee@engr.uky.edu, http://www.uky.edu/WaterResources/

Louisiana

Dr. John Pardue, Director, Louisiana Water Resources Research Institute, 3221 CEBA Building, Louisiana State University, Baton Rouge, LA 70803, Phone: (255) 578-8661, Fax: (255) 578-5043, email: jpardue@lsu.edu, http://www.lwrri.lsu.edu/

Maine

Dr. Steve Kahl, Senator George J. Mitchell Center for Environmental and Watershed Research, 5764 Sawyer Environmental Research Center, University of Maine, Orono, ME 04469-5764, Phone: (207) 581-3286, Fax: (207) 581-3290, email: kahl@maine.edu, http://www.umaine.edu/WaterResearch

Maryland

Prof. Allen P. Davis, Director, Water Resources Research Center, Department of Civil and Environmental Engineering, University of Maryland, College Park, MD 20742-5595, Phone: (301) 405-1958, Fax: (301) 405-2585, email: apdavis@umail.umd.edu, http://www.waterresources.umd.edu/

Massachusetts

Dr. David A. Reckhow, Interim Director, Water Resources Research Center, Dept. of Civil and Environmental Engineering, 18 Marston Hall, University of Massachusetts, Amherst, MA 01003, Phone: (413) 545-5392, Fax: (413) 545-2202, email: reckhow@ecs.umass.edu, http://www.umass.edu/tei/wrrc/

Michigan

Dr. Jon Bartholic, Institute of Water Research, 115 Manly Miles Building, 1405 South Harrison Road, Michigan State University, East Lansing, MI 48823-5243, Phone: (517) 353-3742, Fax: (517) 353-1812, email: bartholi@pilot.msu.edu, http://www.iwr.msu.edu/

Minnesota

Dr. Deborah Swackhamer, Water Resources Center, 173 McNeal Hall, 1985 Buford Avenue, University of Minnesota, St. Paul, MN 55108, Phone: (612) 624-9282, Fax: (612) 625-1263, email: dswack@umn.edu, http://wrc.coafes.umn.edu/

Mississippi

Dr. David Shaw, Director, Mississippi Water Resources Research Institute, GeoResources Institute P.O. Box 9652, Mississippi State University, Mississippi State, MS 39762-9652, Phone: (662) 325-9573, Fax: (662) 325-3621, email: dshaw@gri.msstate.edu, http://www.gri.msstate.edu/

Missouri

Dr. Thomas E. Clevenger, Water Resources Research Center, E1511A Engineering Bldg. East, University of Missouri, Columbia, MO 65211, Phone: (314) 882-3132, Fax: (314) 882-4784, email: Clevengert@missouri.edu, http://web.missouri.edu/~mowrrc/

Montana

Gretchen Rupp, P.E., Director, Montana Water Center, 101 Hoffman Building, Montana State University, Bozeman, MT 59717, Phone: (406) 994-6690, Fax: (406) 994-1774, email: wwwrc@montana.edu, http://water.montana.edu/

(Continued)

Table 12C.14 (Continued)

Nebraska

Dr. Kyle D. Hoagland, Director, Nebraska Water Resources Center, 103 Natural Resources Hall, University of Nebraska, EastCampus, Lincoln, NE 68583-0844, Phone: (402) 472-3305, Fax: (402) 472-3574, email: khoagland@unl.edu, http://watercenter.unl.edu/

Nevada

Dr. John J. Warwick, Director, Water Resources Center, Desert Research Institute, 2215 Raggio Parkway, Reno, NV 89512, Phone: (775) 673-7379, Fax: (775) 673-7363, email: warwick@dri.edu, http://www.dri.edu/

New Hampshire

Dr. William H. McDowell, Director, Water Resources Research Center, c/o Department of Natural Resources, 219 James Hall, University of New Hampshire, Durham, NH 03824-3525, Phone: (603) 862-2249, Fax: (603) 862-4976, email: bill.mcdowell@unh.edu, http://www.wrrc.unh.edu

New Jersey

Dr. Joan G. Ehrenfeld, New Jersey Water Resources Research Institute, Dept. of Ecology, Evolution and Natural Resources, Cook College, Rutgers University, New Brunswick, NJ 08903, Phone: (732) 932-1081, Fax: (732) 932-8746, email: ehrenfel@rci.rutgers.edu, http://njwrri.rutgers.edu/

New Mexico

Dr. Karl Wood, Director, Water Resources Research Institute, Box 30001, Dept. 3167, New Mexico State University, Las Cruces, NM 88003-0001, Phone: (505) 646-4337, Fax: (505) 646-6418, email: kwood@wrri.nmsu.edu, http://wrri.nmsu.edu/

New York

Mr. Keith S. Porter, New York State Water Resources Institute, Rice Hall, Cornell University, Ithaca, NY 14853-5601, Phone: (607) 255-5941, Fax: (607) 255-5945, email: nyswri@cornell.edu, cfe.cornel.edu/wri/

North Carolina

Dr. Greg Jennings, Interim Director, Water Resources Research Institute of the University of North Carolina, Campus Box 7912, North Carolina State University, Raleigh, NC 27695-7912, Phone: (919) 515-2815, Fax: (919) 515-2839, email: water_resources@ncsu.edu, ncsu.edu/ncsu/CIL/WRRI/

North Dakota

Dr. Wei Lin, Interim Director, North Dakota Water Resources Research Institute, Civil/Ind. Building 201A, North Dakota State University, Fargo, ND 58105, Phone: (701) 231-7193, Fax: (701) 231-8831, email: Wei.Lin@ndsu.nodak.edu, http://http://www.ce.ndsu.nodak.edu/wrri/

Ohio

Dr. E. Whitlatch, Water Resources Center, 470 Hitchcock Hall, 2070 Neil Avenue, Ohio State University, Columbus, OH 43210, Phone: (614) 292-6108, Fax: (614) 292-3780, email: whitlatch.1@osu.edu

Oklahoma

Dr. Will Focht, Director, Oklahoma Water Resources Research Institute, 003 Life Sciences East, Oklahoma State University, Stillwater, OK 74078-3011, Phone: (405) 744-9994, Fax: (405) 744-7673, email: wfocht@okstate.edu, http://environ.okstate.edu/

Oregon

Dr. Kenneth Williamson, Center for Water and Environmental Sustainability (CWESt), 210 Strand Agricultural Hall, Oregon State University, Corvallis, OR 97331, Phone: (541) 737-4022, Fax: (541) 737-2735, email: kenneth.williamson@orst.edu, http://cwest.orst.edu/

Pennsylvania

Dr. David R. DeWalle, Penn State Institutes of the Environment, 107 Land and Water Research Building, Pennsylvania State University, University Park, PA 16802, Phone: (814) 863-0291, Fax: (814) 865-3378, email: drdewalle@psu.edu, http://www.pawatercenter.psu.edu/

Puerto Rico

Dr. Jorge Rivera Santos, Water Resources Research Institute, P.O. Box 9040 - College of Engineering, Mayaguez, Puerto Rico 00681-9040, Phone: (787) 265-3826, Fax: (787) 832-0119, email: wrri_rum@rumac.uprm.edu, http://www.ece.uprm.edu/rumhp/prwrri/

Rhode Island

Dr. Leon Thiem, Director, Water Resources Center, Department of Civil & Environmental Engineering, 1 Lippitt Road, University of Rhode Island, Kingston, RI 02881, Phone: (401) 874-2693, Fax: (401) 874-2786, email: thiem@egr.uri.edu, http://www.egr.uri.edu/centers/d.htm#D-3

South Carolina

Dr. Jeffery S. Allen, South Carolina Water Resources Research Institute, Strom Thurmond Institute, Clemson University, Box 345203, Clemson, SC 29634-0125, Phone: (864) 656-4700, Fax: (864) 656-4780, email: jeff@strom.clemson.edu, strom.clemson.edu/teams/water_resources/index.html/

(Continued)

Table 12C.14 (Continued)

South Dakota

Dr. Van Kelley, Acting Director, South Dakota Water Research Institute, Agricultural Experiment Station, AE Bldg., Rm 211, Box 2120, South Dakota State University, Brookings, SD 57007, Phone: (605) 688-4910, Fax: (605) 688-4917, email: van_kelley@sdstate.edu, http://wri.sdstate.edu/

Tennessee

Dr. Tim Gangaware, Associate Director, Tennessee Water Resources Research Center, 600 Henley Street, Suite 311, The University of Tennessee, Knoxville, TN 37996-4134, Phone: (865) 974-2151, Fax: (865) 974-1838, email: gangwrrc@utk.edu, http://eerc.ra.utk.edu/WRRC.html

Texas

Dr. C. Alan Jones, Director, Texas Water Resources Institute, Texas A&M University, 1500 Research Pkwy., College Station, TX 77843-2118, Phone: (979) 845-1851, Fax: (979) 845-8454, email: cajones@tamu.edu, http://twri.tamu.edu

Utah

Dr. Mac McKee, Interim Director, Center for Water Resources Research, Utah State University, 8200 Old Main Hill, Logan, UT 84322-8200, Phone: (435) 797-3188, Fax: (435) 797-3663, email: mmckee@cc.usu.edu, http://www.engineering.usu.edu/uwrl/niwr/

Vermont

Dr. Breck Bowden, Director, Vermont Water Resources and Lake Studies Center, George D. Aiken Center for Natural Resources, The University of Vermont, Burlington, VT 05405, Phone: (802) 656-4057, Fax: (802) 656-8683, email: Breck.Bowden@uvm.edu, http://www.uvm.edu/envnr/vtwater

Virginia

Dr. Tamim Younos, Interim Director, Virginia Water Resources Research Center, 10 Sandy Hall, Virginia Polytechnic Institute and State University, Blacksburg, VA 24060, Phone: (540) 231-5624, Fax: (540) 231-6673, email: tyounos@vt.edu, http://www.vwrrc.vt.edu/

Virgin Islands

Dr. Henry H. Smith, Virgin Islands Water Resources Research Institute, University of the Virgin Islands, St. Thomas, Virgin Islands 00802, Phone: (809) 693-1021, Fax: (809) 693-1025, email: hsmith@uvi.edu, http://rps.uvi.edu/WRRI/wrri.htm

Washington

Dr. Michael E. Barber, Director, State of Washington Water Research Center, P.O. Box 643002, Washington State University, Pullman, WA 99164-3002, Phone: (509) 335-6633, Fax: (509) 335-1590, email: meb@wsc.edu, http://www.swwrc.wsu.edu/

West Virginia

Dr. Paul Ziemkiewicz, West Virginia Water Research Institute, P.O. Box 6064, 617 Spruce Street, West Virginia University, Morgantown, WV 26506-6064, Phone: (304) 293-2867 ext 5441, Fax: (304) 293-7822, email: pziemkie@wvu.edu, nrcce.wvu.edu/wvwrri/index.htm

Wisconsin

Dr. Anders W. Andren, University of Wisconsin - Water Resources Institute, 1975 Willow Drive - 2nd Floor, The University of Wisconsin-Madison, Madison, WI 53706, Phone: (608) 262-0905, Fax: (608) 262-0591, email: awandren@aqua.wisc.edu, http://www.wri.wisc.edu/

Wyoming

Dr. Larry O. Pochop, Director, Wyoming Water Research Program, University of Wyoming, Civil & Arch. Engineering Dept., P.O. Box 3295, Laramie, WY 82071, Phone: (307) 766-3326, Fax: (307) 766-2221, email: pochop@uwyo.edu, http://wwweng.uwyo.edu/civil/research/wwrp/

Source: From niwr.montana.edu.

SECTION 12D PROFESSIONAL, TRADE, AND ENVIRONMENTAL GROUPS

Table 12D.15 Professional, Trade, and Environmental Groups

African Wildlife Foundation — http://www.awf.org/
1400 16th Street, N.W. Suite 120, Washington, DC 20036, Phone: (202) 939-3333, Fax: (202) 939-3332

The Alliance for Justice — http://www.afj.org/
11 Dupont Circle NW, 2nd Floor, Washington, DC 20036, Phone: (202) 822-6070, Fax: (202) 822-6068

American Association for the Advancement of Science — http://www.aaas.org/
AAAS, 1200 New York Avenue NW, Washington, DC 20005, Phone: 202-326-6400

American Association of Pesticide Safety Educators — http://aapse.ext.vt.edu/
495 Borlaug Hall, 41991 Upper Buford Circle, St. Paul, MN 55108, Phone: (612) 624-3477

American Chemical Society (ACS) — http://www.chemistry.org/portal/a/c/s/1/home.html
American Chemical Society, 1155 16th St., NW, Washington, DC 20036, Phone: (800) 227-5558 (U.S. only), Phone: (202) 872-4600 (outside the U.S.), Fax: (202) 776-8258

The American Council for an Energy-Efficient Economy (ACEEE) — http://www.aceee.org/
1001 Connecticut Avenue, NW Suite 801, Washington, DC 20036, Phone: (202) 429-8873, Fax: (202) 429-2248

American Crop Protection Association — http://www.acpa.org/
CropLife America, 1156 15th Street, NW. Suite 400, Washington, DC 20005, Phone: (202) 296-1585, Fax: (202) 463-0474

American Fisheries Society — fisheries.org/html/index.shtml
American Fisheries Society, 5410 Grosvenor Lane, Bethesda, MD 20814, Phone: (301) 897-8616, Fax (301) 897-8096

American Geological Institute — http://www.agiweb.org/
American Geological Institute, 4220 King Street, Alexandria, VA 22302-1502, Phone: (703) 379-2480 (Voice), Fax: (703) 379-7563

American Geophysical Union — http://www.agiweb.org/
2000 Florida Avenue NW, Washington, DC 20009-1277, Phone: (202) 462-6900 (Toll Free in North America: (800) 966-2481), Fax: (202) 328-0566

American Institute of Chemical Engineers — http://www.aiche.org/
American Institute of Chemical Engineers, 3 Park Avenue, New York, NY 10016-5991, Phone (301) 374-2522

American Institute of Hydrology — http://www.aihydro.org/
300 Village Green Circle, Suite 201, Smyrna, GA 30080, Phone: (770) 384-1634, Fax: (770) 438-6172

American Institute of Professional Geologists (AIPG) — aipg.org
1400 W. 122nd Avenue, Suite 250, Westminster, CO 80234, Phone: (303) 412-6205, Fax (303) 253-9220

American National Standards Institute (ANSI) — http://www.ansi.org/
1819 L Street, NW, 6th floor, Washington, DC 20036, Phone: (202) 293.8020, Fax: (202) 293.9287

American Public Health Association — http://www.apha.org/
800 I Street, NW, Washington, DC 20001, Phone: (202) 777-2742, Fax: (202) 777-2534

American Public Works Association (APWA) — http://www.apwa.net/
Washington DC Office, 1401 K Street, NW, 11th Floor, Washington, DC 20005, Phone: (202) 408-9541, Fax: (202) 408-9542

American Society of Agricultural Engineers (ASAE) — http://www.asae.org/
2950 Niles Road, St. Joseph, MI 49085, Phone: (269) 429-0300, Fax: (269) 429-3852

American Society of Agronomy — http://www.agronomy.org/
677 S. Segoe Road, Madison, WI 53711, Phone: (608) 273-8080, Fax: (608) 273-2021

American Society of Civil Engineers — http://www.asce.org/
ASCE World Headquarters, 1801 Alexander Bell Drive, Reston, VA 20191-4400, Phone: (800) 548-2723, Fax: (703) 295-6222

American Society of Ichthyologists and Herpetologists — http://www.asih.org/
Florida International University, Biological Sciences, 11200 SW 8th St. Miami, FL 33199, Phone: (305) 348-1235, Fax: (305) 348-1986

American Society of Limnology and Oceanography — http://www.aslo.org/
ASLO Business Office, 5400 Bosque Boulevard, Suite 680, Waco, TX 76710-4446, Phone: (254) 399-9635, Fax: (254) 776-3767

American Society of Sanitary Engineers — http://www.asse-plumbing.org/
901 Canterbury, Suite A, Westlake, OH 44145, Phone: (440) 835-3040, Fax: (440) 835-3488

(Continued)

Table 12D.15 (Continued)

American Society for Microbiology — http://www.asm.org/
1752 N Street, NW, Washington, DC 20036, Phone: (202) 737-3600

American Society of Testing Materials — http://www.astm.org
100 Barr Harbor Drive, West Conshohocken, PA 19428-2959, Phone: (610) 832-9585, Fax: (610) 832-9555

American Water Resources Association (AWRA) — http://www.awra.org/
American Water Resources Association, 4 West Federal Street, P.O. Box 1626, Middleburg, VA 20118-1626, Phone: (540) 687-8390, Fax: (540) 687-8395

American Water Works Association — http://www.awwa.org/
6666 W. Quincy Ave., Denver, CO 80235, Phone: (303) 794-7711, Fax (303) 347-0804

AOAC International — http://www.aoac.org/
481 North Frederick Avenue Suite 500, Gaithersburg, MD 20877

Association of American State Geologists — http://www.stategeologists.org/
Karl Muessig, State Geologist, New Jersey Geological Survey, Department of Environmental Protection, P.O. Box 427, Trenton, NJ 08625-0427

Association of Engineering Geologists (AEG) — http://www.aegweb.org/
P.O. Box 460518, Denver, CO 80246, Phone: (303) 757-2926, Fax: (303) 757-2969

Association of Metropolitan Sewerage Agencies — http://www.amsa-cleanwater.org/
1816 Jefferson Place, NW, Washington, DC 20036-2505, Phone (202) 833-2672, Fax (202) 833-4657

Association of Metropolitan Water Agencies — http://www.amwa.net/
1620 I Street, NW, Suite 500, Washington, DC 20006, Phone (202) 331-2820, FAX (202) 785-1845

Association of State Drinking Water Administrators — asdwa.org/
1025 Connecticut Avenue, NW — Suite 903, Washington, DC 20036, Phone: (202) 293-7655, Fax: (202) 293-7656

Association of State and Interstate Water Pollution Control Administrators — http://www.asiwpca.org/
750 First Street, NE Suite 1010, Washington, DC 20002

British Hydrological Society (BHS) -http://www.hydrology.org.uk/
Institution of Civil Engineers/Thomas Telford Ltd, 1 Great George Street, London, SW1P 3AA United Kingdom, Phone: +44 207 2227722, Fax: +44 207 2227500

Canada Centre for Inland Waters (CCIW) — cciw.ca
National Water Research Institute, Environment Canada, 867 Lakeshore Road, P.O. Box 5050, Burlington, ON L7R 4A6

Canadian Ground Water Association — http://www.cgwa.org/
1600 Bedford Highway Suite 100–409, Bedford, NS B4A 1E8, Phone: (902) 845-1885, Fax: (902) 845-1886

Center for Watershed Protection — http://www.cwp.org/
8390 Main Street, Second Floor, Ellicott City, MD 21043-4605, Phone: (410) 461-8323, Fax: (410) 461-8324

Clean Water Action Project — http://www.cleanwateraction.org/
4455 Connecticut Avenue, NW Suite A300, Washington, DC 20008, Phone: (202) 895-0420

Clean Water Network — http://www.cwn.org/cwn/
1200 New York Avenue, NW, Suite 400, Washington, DC 20005, Phone: (202) 289-2395, Fax: (202) 289-1060

Coastal Conservation Association — http://www.joincca.org/
6919 Portwest, Suite 100, Houston, TX 77024, Phone: (713) 626-4234

Coastal Society — http://www.thecoastalsociety.org/
P.O. Box 25408, Alexandria, VA 22313-5408, Phone: (703) 933-1599, Fax: (703) 933-1596

CONCERN — http://www.concern.net/
104 East 40th Street, Room 903, New York, NY 10016, Phone: (212) 557-8000, Fax: (212) 557-8004

Conservation International — http://www.conservation.org/
1919 M Street, NW Suite 600, Washington, DC 20036, Phone: (202) 912-1000

Consumer's Union — http://www.consumersunion.org/
1666 Connecticut Avenue, NW, Suite 310, Washington, DC 20009-1039, Phone: (202) 462-6262, Fax: (202) 265-9548

Conservation Foundation — http://www.conservationfoundation.co.uk/
1 Kensington Gore, London SW7 2AR, UK, Phone: +44 207 5913111, Fax: +44 207 5913110

(Continued)

Table 12D.15 (Continued)

Cousteau Society — http://www.cousteau.org/
710 Settlers Landing Road, Hampton, VA 23669, Phone: (757) 722-9300, Fax: (757) 722-8185

CREST — http://www.crest.org/
1612 K Street, NW, Suite 202, Washington, DC 20006, Phone: (202) 293-2898, Fax: (202) 293-5857

Defenders of Wildlife — http://www.defenders.org
National Headquarters, 1130 17th Street, NW, Washington, DC 20036, Phone: (202) 682-9400

Ducks Unlimited — http://www.ducks.org/
One Waterfowl Way, Memphis, TN 38120, Phone (901) 758-3825

Earth First! — http://www.earthfirstjournal.org
P.O. Box 3023, Tucson, AZ 85702, Phone: (520) 620-6900, Fax: (413) 254-0057

Earth Island Institute — http://www.earthisland.org/
300 Broadway, Suite 28, San Francisco, CA 94133, Phone: (415) 788-3666

Earth Share — http://www.earthshare.org/
7735 Old Georgetown Road, Suite 900, Bethesda, MD 20814, Phone: (240) 333-0300

Earthjustice Legal Defense Fund — http://www.earthjustice.org/
426 17th Street, 6th Floor, Oakland, CA 94612-2820, Phone: (510) 550-6700, Fax: (510) 550-6740

Earthwatch — http://www.earthwatch.org/
3 Clock Tower Place, Suite 100, Box 75, Maynard, MA 01754, Phone: (978) 461-0081, Fax (978) 461-2332

Energy & Environmental Research Center (EERC) — http://www.eerc.und.nodak.edu/
University of North Dakota, P.O. Box 9018, 15 North 23rd Street, Grand Forks, ND 58202-9018, Phone: (701) 777-5000, Fax: (701) 777-5181

Environment Canada — http://www.ec.gc.ca/water/index.htm
Sustainable Water Use Branch, Water Policy and Coordination Directorate, Environmental Conservation Service, Environment Canada, 351 St. Joseph Blvd., 7-PVM, Gatineau, Quebec K1A 0H3, Phone: (819) 994-0237

Environmental Action Foundation — http://www.agc.org
333 John Carlyle Street, Suite 200, Alexandria, VA 22314, Phone: (703) 548-3118, Fax: (703) 548-3119

Environmental Defense — http://www.environmentaldefense.org/
257 Park Avenue South, New York, NY 10010, Phone: (212) 505-2100, Fax: (212) 505-2375

Environmental Law Institute — http://www.eli.org
2000 L Street, NW, Suite 620, Washington, DC 20036, Phone: (202) 939-3800, Fax: (202) 939-3868

Environmental Policy Institute — http://www.nepi.org/
1401 K Street, NW Suite M-103, Washington, DC 20005, Phone: (202) 857-4784, Fax: (202) 833-5977

Environmental Working Group — http://www.ewg.org/
1436 U Street, NW Suite 100, Washington, DC 20009, Phone: (202) 667-6982

Estuarine Research Federation — http://erf.org/
P.O. Box 510, Port Republic, MD 20676, Phone: (410) 586-0997, Fax: (410) 586-9226

Food and Agriculture Organization of the United Nations (FAO) — http://www.fao.org/
FAO HEADQUARTERS, Viale delle Terme di Caracalla, 00100 Rome, Italy, Phone: +39 06 57051, Fax: +39 06 57053

Freshwater Foundation — http://www.spiretech.com/~dafoxx/fresh/
2500 Shadywood Road, Box 90, Navarre, MN 55392-0900, Phone: (612) 481-8407

Friends of the Earth U.S.A. — http://www.foe.org/
1717 Massachusetts Avenue, NW, 600, Washington, DC 20036-2002, Phone: (877) 843-8687 — toll free, Fax: (202) 783-0444

Geological Society of America (GSA) — http://www.geosociety.org/index.htm
The Geological Society of America, P.O. Box 9140, Boulder, CO 80301-9140, Phone: (303) 447-2020, Fax: (303) 357-1070

Global Energy and Water Cycle Experiment (GEWEX) — http://www.gewex.org/
1010 Wayne Avenue, Suite 450, Silver Springs, MD 20910-5641, Phone: (301) 565-8345, Fax: (301) 565-8279

Global Environmental Sanitation Initiative (GESI) — http://www.africanwater.org/gesi.htm
Water Web Management Ltd, 1 Dome Hill, Caterham, Surrey CR3 6EE, UK

(Continued)

Table 12D.15 (Continued)

Global Environment Facility (GEF) — http://www.gefweb.org/
1818 H Street, NW, Washington, DC 20433, Call or send a fax to the Secretariat at: Phone: (202) 473-0508, Fax: (202) 522-3240/3245

Global Rivers Environment Education Network (GREEN) — http://www.earthforce.org/section/programs/green/
Earth Force, 1908 Mount Vernon Avenue, Second Floor, Alexandria, VA 22301, Phone: (703) 299-9400, Fax: (703) 299-9485

Global Water — http://www.globalwater.org/
3600 S. Harbor Blvd., #514, Oxnard, CA 93035, Phone: (805) 985-3057, Fax: (805) 985-3688

Global Water Partnership (GWP) — http://www.gwpforum.org
GWP Central America, c/o ASOTEM, 14-5000 Liberia, Guanacaste, Costa Rica, Phone: +506 666 1596, Fax: +506 666 2967

Global Water Unit of the World Bank Group — http://www.gwp.sida.se
The World Bank, 1818 H Street, NW, Washington, DC 20433, Phone: (202) 473-1000, Fax: (202) 477-6391

Greenpeace — http://www.greenpeace.org/international_en/
Ottho Heldringstraat 5, 1066 AZ Amsterdam, The Netherlands, Phone: +31 20 5148150, Fax: +31 20 5148151

Groundwater Protection Council — http://www.gwpc.org/
Ground Water Protection Council, 13308 N. MacArthur, Oklahoma City, OK 73142, Phone: (405) 516-4972, Fax: (405) 516-4973

GSA Hydrogeology Division — http://gsahydrodiv.unl.edu/
The Geological Society of America, P.O. Box 9140, Boulder, CO 80301-9140, Phone: (303) 447-2020, Fax: (303) 357-1070

Groundwater Management Districts Association — http://www.gmdausa.org/
P.O. Box 905, Colby, KS 67701-0905, Phone: (785) 462-3915

Hydrology and Water Resources Programme (HWRP) — http://www.wmo.ch/web/homs/hwrpframes.html
World Meteorological Organization, 7 bis, Avenue de la Paix, CP 2300, CH-1211 Genève 2, Switzerland, Phone: +41 22 73083, Fax: +41 22 730 80 43

Illinois Groundwater Consortium — http://www.siu.edu/worda/igc/
Illinois Groundwater Consortium, Office of Research Development and Administration, Southern Illinois University Carbondale, Carbondale, IL 62901-4709, Phone: (618) 453-4531, Fax: (618) 453-8038

Inter-American Association for Environmental Quality (AIDIS-CANADA) — http://www.rcc.ryerson.ca
AIDIS — Canada, 3750 NW 87th Ave. Suite 300, Miami, FL 33178, Phone: (305) 716-5120, Fax: (305) 716-5155

Inter-American Water Resources Network (IWRN) — http://www.iwrn.net/
1889 F Street, NW, Washington, DC 20006, Phone: (202) 458-3556, Fax: (202) 458-3560

Inter-Islamic Network on Water Resources Development and Management (INWRDAM) — http://www.inwrdam.org/
P.O. Box 1460, Jubieha: Amman, PC 11941, Jordan, Phone: +962 6 5332993, Fax: +962 6 5332969

International Association for Environmental Hydrology (IAEH) — http://www.hydroweb.com/
International Association for Environmental Hydrology, Alexandria, VA 22313, Phone: (703) 683-9768, Fax: (703) 683-6137

International Association for Great Lakes Research — http://www.iaglr.org/index.php
IAGLR Business Office, 2205 Commonwealth Blvd., Ann Arbor, MI 48105, Phone: (734) 665-5303, Fax: (734) 741-2055

International Association for Hydraulic Engineering and Research (IAHR) — http://www.iahr.org
IAHR Secretariat, Paseo Bajo Virgen del Puerto, 3, 28005 Madrid, Spain, Phone: +34 91 3357964, Fax: +34 91 3357935

International Association of Hydrogeology (IAH/AIH) — http://www.iah.org/
IAH Secretariat, P.O. Box 9, Kenilworth, CV8 1JG, United Kingdom, Fax: +44 192 6856561

International Association of Hydrological Sciences (IAHS) — http://www.cig.ensmp.fr/~iahs/
IAHS Press, Centre for Ecology and Hydrology, Wallingford, Oxfordshire OX10 8BB, UK, Phone: 44 1491 692288, Fax: 44 1491 692448/692424

International Association of Theoretical and Applied Limnology (Societas Internationalis Limnologiae) (SIL) — http://www.limnology.org/

International Bottled Water Association — http://www.bottledwater.org/
1700 Diagonal Road, Suite 650, Alexandria, VA 22314, Phone: (703) 683-5213, Fax: (703) 683-4074

International Association on Water Quality (IAWQ) — http://www.iawq.org.uk/
Alliance House, 12 Caxton Street, London SW1H 0QS, UK, Phone: +44 207 6545500, Fax: +44 207 6545555

International Commission on Irrigation and Drainage (ICID) — http://www.icid.org
48 Nyaya Marg, Chanakyapuri, New Delhi 110021, India, Phone: 91-112 6116837, Fax: 91-112 6115962

(Continued)

Table 12D.15 (Continued)

International Commission on Large Dams (ICOLD) — http://www.icold-cigb.net/
Bureau Central de la CIGB, 151 Boulevard Haussmann, 75008 Paris, France, Phone: +33 14 0426824, Fax: +33 14 0426071

International Fund for Agricultural Development (IFAD) — http://www.ifad.org/
International Fund for Agricultural Development, Via del Serafico, 107, 00142 Rome, Italy, Phone: +39 0654591, Fax: +39 06 5043463

International Ground Water Modeling Center (IGWMC) — mines.edu/igwmc/
International Ground Water Modeling Center, Colorado School of Mines, Golden, CO 80401-1887, Phone: (303) 273-3103, Fax: (303) 384-2037

International Groundwater Resources Assessment Centre (IGRAC) — http://www.igrac.nl/
P.O. Box 17, 8200 AA Lelystad, The Netherlands, Phone: +31 32 0298894, Fax: +31 32 0297642

International Hydrological Programme — http://www.unesco.org/water/ihp/
7, Place de Fontenoy, 75352 Paris 07 SP, France, Phone: +33 14 5681000, Fax: +33 14 5671690

International Hydropower Association (IHA) — http://www.hydropower.org/
IHA Central Office, Westmead House (Suite 55), 123 Westmead Road, Sutton, Surrey SM1 4JH, UK, Phone: +44 208 2881918, Fax: +44 208 7701744

International Institute for Land Reclamation and Improvement (ILRI) — http://www.ilri.nl/
Address: Lawickse Allee 11, Wageningen, The Netherlands, Correspondence: Postbus 45, 6700 AA, Wageningen, The Netherlands, Phone: +31 31 7495549, Fax: +31 31 7495590

International Office for Water — http://www.oieau.fr/index.html
International Office for Water, Documentation and Data Department, Rue Edouard Chamberland, 87065 Limoges Cedex, France

International Programme for Technology Research in Irrigation and Drainage (IPTRID) — http://www.hrwallingford.co.uk/projects/IPTRID/
Land and Water Development Division, Food and Agriculture Organization of the United Nations, Viale delle Terme di Caracalla, 00100 Rome, Italy

International Water and Sanitation Centre (IRC) — http://www.irc.nl/
P.O. Box 2869, 2601 CW Delft, The Netherlands, Phone: +31 15 2192939, Fax: +31 15 2190955

International Water Association — http://www.iawq.org.uk/
International Water Association, Alliance House, 12 Caxton Street, London SW1H 0QS, UK, Phone: +44 20 7 654 5500, Fax: +44 20 7 654 5555

International Water Management Institute (IWMI) — http://www.iwmi.cgiar.org/
P.O. Box 2075, Colombo, Sri Lanka, Phone: +94 11 2787404, Fax: +94 11 2786854

International Water Resources Association (IWRA) — http://www.iwra.siu.edu/
IWRA Headquarters, 4535 Faner Hall, Southern Illinois University, Carbondale, IL 62901-4516, Fax: (618) 453-6465

International Water Services Association (IWSA) — http://www.iwahq.org.uk
International Water Association, Alliance House, 12 Caxton Street, London SW1H 0QS, UK, Phone: +44 207 6545500, Fax: +44 (0)207 6545555

Izaak Walton League of America — http://www.iwla.org/
707 Conservation Lane, Gaithersburg, MD 20878, Phone: (301) 548-0150, Fax: (301) 548-0146

LakeNet — http://www.worldlakes.org
P.O. Box 3250, Annapolis, MD 21403, Phone: (410) 268-5155, Fax: (410) 268-8788

League of Conservation Voters — http://www.lcv.org/
1920 L Street, NW, Suite 800, Washington, DC 20036, Phone: (202) 785-8683, Fax: (202) 835-0491

League of Women Voters of the U.S. — http://www.lwv.org/
1730 M Street, NW, Suite 1000, Washington, DC 20036-4508, Phone: (202) 429-1965, Fax: (202) 429-0854

Mekong River Commission (MRC) — http://www.mrcmekong.org/
Mekong River Commission Secretariat, P.O. Box 6101, Unit 18 Ban Sithane Neua, Sikhottabong District, Vientiane 01000, Lao PDR, Phone: +856 2 1263263:, Fax: +856 2 1263264

Mexican Institute of Water Technology (IMTA) — http://www.imta.mx/english/
Paseo Cuauhnáhuac 8532, Jiutepec, Morelos, Mexico, Switchboard: +52 777 3293600

National Academy of Sciences (NAS) — http://www.nas.edu/
500 Fifth St. NW Washington, DC 20001, Phone: (202) 334-2000

(Continued)

Table 12D.15 (Continued)

National Audubon Society — http://www.audubon.org/
National Audubon Society, 700 Broadway, New York, NY 10003, Phone: (212) 979-3000, Fax: (212) 979-3188

National Coalition for Marine Conservation — http://www.savethefish.org/
4 Royal Street SE, Leesburg, VA 20175, Phone: (703) 777-0037, Fax: (703) 777-1107

National Environmental Health Association (NEHA) — http://www.neha.org/
720 S. Colorado Blvd., Suite 970-S, Denver, CO 80246-1925, Phone: (303) 756-9090, Fax: (303) 691-9490

National Environmental Trust — http://www.environet.policy.net/
1200 18th St. NW, Fifth Floor, Washington, DC 20036, Phone: (202) 887-8800, Fax: (202) 887-8877

National Fish and Wildlife Foundation — http://www.nfwf.org/
1120 Connecticut Avenue, NW, Suite 900, Washington, DC 20036, Phone: (202) 857-0166, Fax: (202) 857-0162

National Ground Water Association — http://www.ngwa.org/
601 Dempsey Road, Westerville, OH 43081-8978, Phone: (614) 898-7791, Fax: (614) 898-7786

National Parks and Conservation Association (NPCA) — http://www.eparks.org/
1300 19th Street, NW, Suite 300, Washington, DC 20036, Phone: (800) 628-7275, Fax: (202) 659-0650

National Religious Partnership for the Environment — http://www.nrpe.org/
49 South Pleasant Street Suite 301, Amherst, MA 01002, Phone: (413) 253-1515, Fax: (413) 523-1414

National Rural Water Association — http://www.nrwa.org/
2915 South 13th St. Duncan, OK 73533, Phone: (580) 252-0629, Fax: (580) 255-4476

National Sanitation Foundation — http://www.nsf.org
P.O. Box 130140, 789 N. Dixboro Road, Ann Arbor, MI 48113-0140, Phone: (734) 769-8010, Fax: (734) 769-0109

National Science Foundation (NSF) — http://www.nsf.gov/
4201 Wilson Blvd, Arlington, VA 22230, Phone: (703) 292-5111

National Solid Wastes Management Association — http://www.nswma.org/
4301 Connecticut Avenue, NW, Suite 300, Washington, DC 20008-2304, Phone: (202) 244-4700, Fax: (202) 364-3792

National Water Resources Association — http://www.nwra.org/
3800 North Fairfax Drive, Suite 4, Arlington, VA 22203, Phone: (703) 524-1544, Fax: (703) 524-1548

National Water Resources Institute (NWRI) — http://www.nwri.ca/
70 Crémazie St., Gatineau, Quebec K1A 0H3, Phone: (819) 997-2800 or (800) 668-6767, Fax: (819) 994-1412

National Waterways Conference — http://www.waterways.org/
1130 17th Street, NW Washington, DC 20036-4676, Phone: (202) 296-4415, Fax: (202) 835-3861

National Wildlife Federation — http://www.nwf.org/
National Wildlife Federation, 11100 Wildlife Center Drive, Reston, VA 20190-5362, Phone: (800) 822-9919

Natural Resources Defense Council — http://www.nrdc.org/
Natural Resources Defense Council, 40 West 20th Street, New York, NY 10011, Phone: (212) 727-2700, Fax: (212) 727-1773

The Nature Conservancy — http://nature.org/
The Nature Conservancy, 4245 North Fairfax Drive, Suite 100, Arlington, VA 22203-1606, Phone: (800) 628-6860

North American Lake Management Society — http://www.nalms.org/
P.O. Box 5443, 4513 Vernon Blvd., Suite 100, Madison, WI 53705-0443, Phone: (608) 233-2836, Fax: (608) 233-3186

Northeast Midwest Institute — http://www.nemw.org/
218 D Street, SE, Washington, DC 20003-1900, Phone: (202) 544-5200, Fax: (202) 544-0043

Nuclear Control Institute — http://www.nci.org/
The Nuclear Control Institute, 1000 Connecticut Avenue NW, Suite 410, Washington, DC 20036, Phone: (202) 822-8444, Fax: (202) 452-0892

Oceana — http://www.oceana.org/
2501 M Street, NW, Suite 300, Washington, DC 20037-1311, Phone: (202) 833-3900, Fax: (202) 833-2070

Pacific Institute for Studies in Development, Environment and Security — http://www.pacinst.org/
654 13th Street, Preservation Park, Oakland, CA 94612, Phone: (510) 251-1600, Fax: (510) 251-2203

(Continued)

Table 12D.15 (Continued)

Pacific Water Association — http://pwa.org.fj/index.html
Private Mail Bag, GPO, Suva Fiji Islands, 16 Loa Street, Suva, Fiji Islands, Phone: (679) 337-0402, Fax: (679) 337-0412

Project WET (Water Education for Teachers) — http://www.projectwet.org/
201 Culbertson Hall, Montana State University, P.O. Box 170575, Bozeman, MT 59717-0575, Phone: (406) 994-5392, Fax: (406) 994-1919

Public Agenda — publicagenda.org/issues/frontdoor.cfm?issue_type=environment
Public Agenda, 6 East 39th Street, New York, NY 10016, Phone: (212) 686-6610, Fax: (212) 889-3461

Public Interest Research Group (PIRG) — http://www.pirg.org/
218 D Street, SE, Washington, DC 20003, Phone: (202) 546-9707, Fax: (202) 546-2461

Rainforest Action Network — http://www.ran.org/
221 Pine St., Suite 500, San Francisco, CA 94104, Phone: (415) 398-4404, Fax: (415) 398-2732

Rainforest Alliance — http://www.rainforest-alliance.org/
665 Broadway, Suite 500, New York, NY 10012, Phone: (212) 677-1900

Resources for the Future — http://www.rff.org/
1616 P Street, NW, Washington, DC 20036, Phone: (202) 328-5000

Royal Society of Chemistry (RSC) — http://www.rsc.org/
Royal Society of Chemistry, Burlington House, Piccadilly, London W1J 0BA, UK, Phone: +44 207 4378656, Fax: +44 207 4378883

Scenic America — http://www.scenic.org/
1634 I Street, NW, Suite 510, Washington, DC 20006, Phone: (202) 638-0550, Fax: (202) 638-3171

Sierra Club — http://www.sierraclub.org/
National Headquarters, 85 Second Street, 2nd Floor, San Francisco, CA 94105, Phone: (415) 977-5500, Fax: (415) 977-5799

Society of Environmental Journalists — http://www.sej.org/
P.O. Box 2492, Jenkintown, PA 19046, Phone: (215) 884-8174, Fax: (215) 884-8175

Society of Environmental Toxicology and Chemistry (SETAC) — http://www.setac.org/
SETAC North America, 1010 North 12th Avenue, Pensacola, FL 32501-3367, Phone: (850) 469-1500, Fax: (850) 469-9778

Society of Wetland Scientists — http://www.sws.org/
1313 Dolley Madison Boulevard, Suite 402, McLean, VA 22101, Phone: (703) 790-1745, Fax: (703) 790-2672

Soil Science Society of America (SSSA) — http://www.soils.org/
677 S. Segoe Rd, Madison, WI 53711, Phone: (608) 273-8080, Fax: (608) 273-2021

Soil and Water Conservation Society — http://www.swcs.org/
945 SW Ankeny Road, Ankeny, IA 50021-9764, Phone: (515) 289-2331, Fax: (515) 289-1227

Southern Environmental Law Center — http://www.southernenvironment.org/
201 West Main St., Suite 14, Charlottesville, VA 22902-5065, Phone: (434) 977-4090, Fax: (434) 977-1483

Southern Utah Wilderness Alliance — http://www.suwa.org/
1471 South 1100 East, Salt Lake City, UT 84105, Phone: (801) 486-3161

Stockholm Environment Institute (SEI) — http://www.sei.se/
Lilla Nygatan, 1, Box 2142, S-103 14 Stockholm, Sweden, Phone: +46 8 4121400, Fax: +46 8 7230348

Stockholm International Water Institute (SIWI) — http://www.siwi.org/menu/menu.html
Hantverkargatan 5, 112 21 Stockholm, Sweden, Phone: +46 85 2213960, Fax: +46 85 2213961

Trout Unlimited — http://www.tu.org
1300 N. 17th St., Suite 500, Arlington, VA 22209-2404, Phone: (703) 522-0200, Fax: (703) 284-9400

UNDP/World Bank Water and Sanitation Program — http://www.wsp.org/
Water and Sanitation Program, 1818 H Street, N.W., Washington, DC 20433, Phone: (202) 473-9785, Fax: (202) 522-3313, 522-3228

UNESCO-IHE Institute for Water Education — http://www.unesco-ihe.org/vmp/articles/contentsHomePage.html
P.O. Box 3015, 2601 DA Delft, The Netherlands, Phone: +31 15 2151715, Fax: +31 15 2122921

Union of Concerned Scientists — http://www.ucsusa.org/
National Headquarters, 2 Brattle Square, Cambridge, MA 02238-9105, Phone: (617) 547-5552, Fax: (617) 864-9405

(Continued)

Table 12D.15 (Continued)

United Nations Development Programme (UNDP) — www.undp.org
Presidente Masaryk 29, piso 8, Col. Polanco 11570, México, D.F., Phone: +52 (55) 5 2639600, Fax: +52 (55) 5 2550095

United Nations Educational, Scientific and Cultural Organization (UNESCO) — http://unesco.org/
7, Place de Fontenoy, 75352 Paris 07 SP, France, Phone: +33 14 5681000, Fax: +33 14 5671690

United Nations Environment Programme — Fresh Water Unit — http://www.unep.org/themes/freshwater/
United Nations Avenue, Gigiri, P.O. Box 30552, 00100 Nairobi, Kenya, Phone: +254 2 0621234, Fax: +254 2 0624489

United Nations Environment Programme — Global Programme of Action — http://www.undp.org/
United Nations Avenue, Gigiri, P.O. Box 30552, 00100 Nairobi, Kenya, Phone: +254 2 0621234, Fax: +254 2 0624489

United Nations Environment Programme — Regional Seas Programme — http://www.undp.org/
United Nations Avenue, Gigiri, P.O. Box 30552, 00100 Nairobi, Kenya, Phone: +254 20 621234, Fax: +254 20 624489/90

United Nations Environment Programme — Water Branch — http://www.undp.org/
United Nations Avenue, Gigiri, P.O. Box 30552, 00100 Nairobi, Kenya, Phone: +254 20 621234, Fax: +254 20 624489/90

United Nations World Health Organization (WHO) — http://www.who.int/en/
Avenue Appia 20, 1211 Geneva 27, Switzerland, Phone: +41 22 7912111, Fax: +41 22 7913111

United States Agency for International Development (USAID) — http://www.usaid.gov/
U.S. Agency for International Development, Ronald Reagan Building, Washington, DC 20523-1000, Phone: (202) 712-0000. Fax: (202) 216-3524

United States Public Interest Research Group — http://uspirg.org/
218 D Street, SE, Washington, DC 20003, Phone: (202) 546-9707, Fax: (202) 546-2461

Universities' Water Information Network (UWIN) — http://www.ucowr.siu.edu/
1000 Faner Drive, Room 4543, Southern Illinois University Carbondale, Carbondale, IL 62901, Phone (618) 536-7571, Fax: (618) 453-2671

Wateraid — http://www.wateraid.org.uk/
Prince Consort House, 27-29 Albert Embankment, London SE1 7UB, UK, Phone: +44 20 77934500, Fax: +44 20 77934545

Water Center for the Humid Tropics of Latin America and the Caribbean — http://www.cathalac.org/
TCCC Regional Coordination, Antigua Base de Albrook, P.O. Box 873372, Panama 7, Republic of Panama, Phone: +507 2326851, Fax: +507 2326834

Water Engineering and Development Centre (WEDC) — http://wedc.lboro.ac.uk/
Loughborough University, Leicestershire LE11 3TU, UK, Phone: +44 150 9222885, Fax: +44 150 9211079

Water Environment Federation — http://www.wef.org/index.jhtml
601 Wythe Street, Alexandria, VA, 22314-1994, Phone: (703) 684-2452, Fax: (703) 684-2492

Water For People (WFP) — http://www.water4people.org/
6666 W. Quincy Avenue, Denver, CO 80235, Phone: (303) 734-3490, Fax: (303) 734-3499

Water Online — http://www.wateronline.com/
5340 Fryling Road, Suite 101, Erie, PA 16510

WaterPartners International — http://www.water.org/
P.O. Box 70310, Seattle, WA 98127, Phone: (206) 297-3024

Water Quality Association — http://www.wqa.org/
International Headquarters & Laboratory, 4151 Naperville Road, Lisle, IL 60532-1088, Phone: (630) 505-0160, Fax: (630) 505-9637

Water and Sanitation in Developing Countries (SANDEC) — http://www.sandec.ch/
EAWAG/SANDEC, Ueberlandstrasse 133, P.O. Box 611, CH-8600 Duebendorf, Switzerland, Phone: +41 1 8235286, Fax: +41 1 8235399

Water and Wastewater Equipment Manufacturers Association — http://www.wwema.org/
P.O. Box 17402, Washington, DC 20041, Phone: (703) 444-1777, Fax: (703) 444-1779

Water Supply and Sanitation Collaborative Council (WSSCC) — http://www.wsscc.org/
International Environment House, Chemin des Anémones 9, 1219 Châtelaine, Geneva, Switzerland

WaterWeb — http://www.waterweb.org/
e-mail: webmaster@ces.fau.edu

(Continued)

Table 12D.15 (Continued)

The Wilderness Society — http://www.wilderness.org/
1615 M Street, NW, Washington, DC 20036, Phone: (800) 843-9453

Wildlife Conservation Society — http://www.wcs.org/
North America Program, Wildlife Conservation Society, 2300 Southern Boulevard, Bronx, NY 10460, Phone: (718) 220-1442

Wisconsin Ground Water Association — http://www.wgwa.org
P.O. Box 8593, Madison, WI 53708-8593

The Woods Hole Research Center — http://www.whrc.org/
P.O. Box 296, Woods Hole, MA 02543-0296, Phone: (508) 540-9900, Fax: (508) 540-9700

The World Bank — http://www.worldbank.org/
1818 H Street, NW, Washington, DC 20433, Phone: (202) 473-1000, Fax: (202) 477-6391

World Bank Institute: Water Policy Reform Program — http://www.worldwater.org/links.htm
The World Bank Institute, 1818 H Street, NW, Washington, DC 20433, Phone: (202) 473-2049, Fax: (202) 522-0401

The World Conservation Union (IUCN) — http://www.iucn.org/
IUCN National Committee for Canada, Canadian Museum of Nature; P.O. Box 3443, Station D, Ottawa, Ontario K1P 6P4, Phone: (613) 566-4795, Fax: (613) 364-4022

World Conservation Union (IUCN) Wetlands and Water Resources Programme — http://www.iucn.org/themes/wetlands
Rue Mauverney 28, 1196 Gland, Switzerland, Phone: +41 22 9990001, Fax: +41 22 9990025

World Conservation Union (IUCN) Water & Nature Initiative — http://www.waterandnature.org
Rue Mauverney 28, 1196 Gland, Switzerland, Phone: +41 22 9990000, Fax: +41 22 9990002

World Meteorological Organization: Hydrology and Water Resources Programme — http://www.wmo.ch/index-en.html
7bis, Avenue de la Paix, Case postale No. 2300, CH-1211 Geneva 2, Switzerland, Phone: +41 22 7308111, Fax: +41 22 7308181

World Resources Institute (WRI) — http://www.wri.org/
10 G Street, NE (Suite 800), Washington, DC 20002, Phone: (202) 729-7600, Fax: (202) 729-7610

The World's Water — http://www.worldwater.org/
Pacific Institute, 654 13th Street, Preservation Park, Oakland, CA 94612, Phone: (510) 251-1600, Fax: (510) 251-2203

Worldwatch Institute — http://www.worldwatch.org/
Worldwatch Institute, 1776 Massachusetts Ave., NW, Washington, DC 20036-1904, Phone: (202) 452-1999, Fax: (202) 296-7365

World Water Council (WWC) — http://www.worldwatercouncil.org/
Les Docks Atrium 10.3, 10, place de la Joliette, 13002 Marseille, France, Phone: +33 4 91994100, Fax: +33 4 91994101

World Wide Fund for Nature (WWF) — http://www.panda.org/
WWF-United States, Washington DC, 1250 24th Street NW Washington, Phone: (202) 293-4800, Fax: (202) 293-9211

World Wildlife Fund (U.S.) — http://www.wwf.org/
World Wildlife Fund, 1250 24th Street, NW, Washington, DC 20037, Phone: (202) 293-4800

Note: Selected List of Professional Organizations, Trade Associations and Environmental Groups Concerned with Water Related Matters. For making International calls, please contact your service provider for dialing instructions.

CHAPTER 13

Constants and Conversion Factors

Pedro Fierro, Jr.

CONTENTS

SECTION 13A PHYSICAL PROPERTIES OF WATER

Table 13A.1 Variation of Properties of Pure Water with Temperature

Temperature (°C)	Temperature (°F)	Density at 1 Atmosphere (gm/cm^3)	Dynamic Viscosity, in Centipoises (10^{-2} dyne-sec/cm^2)	Kinematic Viscosity, in Centistokes (10^{-2} cm^2/sec)	Surface Tension Against Air (dyne/cm)	Vapor Pressure (mm Hg)
5	41.0	0.999965	1.5188	1.5189	74.92	6.543
6	42.8	0.999941	1.4726	1.4727	74.78	7.013
7	44.6	0.999902	1.4288	1.4289	74.64	7.513
8	46.4	0.999849	1.3872	1.3874	74.50	8.045
9	48.2	0.999781	1.3476	1.3479	74.36	8.609
10	50.0	0.999700	1.3097	1.3101	74.22	9.209
11	51.8	0.999605	1.2735	1.2740	74.07	9.844
12	53.6	0.999498	1.2390	1.2396	73.93	10.518
13	55.4	0.999377	1.2061	1.2069	73.78	11.231
14	57.2	0.999244	1.1748	1.1757	73.64	11.987
15	59.0	0.999099	1.1447	1.1457	73.49	12.788
16	60.8	0.998943	1.1156	1.1168	73.34	13.634
17	62.6	0.998774	1.0875	1.0889	73.19	14.530
18	64.4	0.998595	1.0603	1.0618	73.05	15.477
19	66.2	0.998405	1.0340	1.0357	72.90	16.477
20	68.0	0.998203	1.0087	1.0105	72.75	17.535
21	69.8	0.997992	0.9843	0.9863	72.59	18.650
22	71.6	0.997770	0.9608	0.9629	72.44	19.827
23	73.4	0.997538	0.9380	0.9403	72.28	21.068
24	75.2	0.997296	0.9161	0.9186	72.13	22.377
25	77.0	0.997044	0.8949	0.8976	71.97	23.756
26	78.8	0.996783	0.8746	0.8774	71.82	25.209
27	80.6	0.996512	0.8551	0.8581	71.66	26.739
28	82.4	0.996232	0.8363	0.8395	71.50	28.349
29	84.2	0.995944	0.8181	0.8214	71.35	30.043
30	86.0	0.995646	0.8004	0.8039	71.18	31.824
31	87.8	0.995340	0.7834	0.7871	71.02[a]	33.695
32	89.6	0.995025	0.7670	0.7708	70.86[a]	35.663
33	91.4	0.994702	0.7511	0.7551	70.70[a]	37.729
34	93.2	0.994371	0.7357	0.7399	70.53[a]	39.898
35	95.0	0.99403	0.7208	0.7251	70.38	42.175
36	96.8	0.99368	0.7064	0.7109	70.21[a]	44.563
37	98.6	0.99333	0.6925	0.6971	70.05[a]	47.067
38	100.4	0.99296	0.6791	0.6839	69.88[a]	49.692

[a] Interpolated.

Source: From U.S. Department of the Interior, Bureau of Reclamation, 1977, *Ground Water Manual.*

Table 13A.2 Density of Water

Temperature (°F)	Density (gm/cm^3)
32	0.99987
35	0.99996
39.2	1.00000
40	0.99999
50	0.99975
60	0.99907
70	0.99802
80	0.99669
90	0.99510
100	0.99318
120	0.98870
140	0.98338
160	0.97729
180	0.97056
200	0.96333
212	0.95865

Note: Density of sea water: approximately 1.025 gm/cm^3 at 15°C.

Table 13A.3 Weight of Water

Temperature	Weight (lb/ft^3)
32	62.416
35	62.421
39.2	62.424
40	62.423
50	62.408
60	62.366
70	62.300
80	62.217
90	62.118
100	61.998
120	61.719
140	61.386
160	61.006
180	60.586
200	60.135
212	59.843

Note: 1 cu in. = 0.0360 lb; 1 cu ft = 62.4 lb; 1 gal = 8.33 lb; 1 Imperial gal = 10.0 lb: Weight of sea water: Approximately 63.93 lb/ft^3 at 15°C.

Table 13A.4 Properties of Water at Various Altitudes

Altitude (Feet)	Barometric Pressure (Inches Hg)	Atmospheric Pressure (psia)	Feet Water
−500	30.5	15.0	34.6
0 (Seasonal)	29.9	14.7	33.9
500	29.4	14.5	33.4
1000	28.9	14.2	32.8
1500	28.3	13.9	32.1
2000	27.8	13.7	31.5
4000	25.8	12.7	29.2
6000	24.0	11.8	27.2
8000	22.2	10.9	25.2

Table 13A.5 Properties of Water at Various Temperatures

Temperature (°F)	Vapor Pressure psfa	Vapor Pressure Feet Water	Specific Weight pcf	Specific Gravity
32.0	12.7	0.20	62.42	0.9999
39.2[a]	16.9	0.27	62.427	1.0000
50.0	25.6	0.41	62.41	0.9997
60.0	36.8	0.59	62.37	0.9990
70.0	52.3	0.84	62.30	0.9980
80.0	73.0	1.17	62.22	0.9966
100.0	136.0	2.19	62.00	0.9931

[a] Temperature when specific gravity = 1.0000.

Source: From Soil Conservation Service, U.S. Department of Agriculture, 1971, Drainage of Agricultural Land.

Table 13A.6 Variation of Boiling Point of Water with Elevation

Elevation, Feet above Mean Sea Level	Boiling Point (°F)
−1,000	213.8
0	212.0
1,000	210.2
2,000	208.4
3,000	206.5
4,000	204.7
5,000	202.9
6,000	201.1
7,000	199.2
8,000	197.4
9,000	195.6
10,000	193.7
11,000	191.9
12,000	190.1
13,000	188.2
14,000	186.4

Note: U.S. standard atmosphere is assumed.

Table 13A.7 Heat of Vaporization, Viscosity, Vapor Pressure, Surface Tension, and Elastic Modulus of Water

Temperature (°F)	Heat Vaporization (Btu/lb)	Viscosity Absolute (lb-sec/ft^2)	Viscosity Kinematic (ft^2/sec)	Vapor Pressure Millibars	Vapor Pressure lb/in.^{2}	Surface Tension (lb/ft)	Elastic Modulus (lb/in.^{2})
32	1073	0.374×10^{-4}	1.93×10^{-5}	6.11	0.09	0.00518	287,000
40	1066	0.323	1.67	8.36	0.12	0.00514	296,000
50	1059	0.273	1.41	12.19	0.18	0.00508	305,000
60	1054	0.235	1.21	17.51	0.26	0.00504	313,000
70	1049	0.205	1.06	24.79	0.36	0.00497	319,000
80	1044	0.180	0.929	34.61	0.51	0.00492	325,000
90	1039	0.160	0.828	47.68	0.70	0.00486	329,000
100	1033	0.143	0.741	64.88	0.95	0.00479	331,000
120	1021	0.117	0.610	—	1.69	0.00466	332,000
140	1010	0.0979	0.513	—	2.89		
160	999	0.0835	0.440	—	4.74		
180	988	0.0726	0.385	—	7.51		
200	977	0.0637	0.341	—	11.52		
212	970	0.0593	0.319	—	14.70		

Note: Heat of fusion at 32°F: 143.5 Btu/lb.

Table 13A.8 Thermal Conductivity of Water

Temperature °C (°F)	Cal/sec/cm^2 under 1°C/cm	Btu/sec/ft^2 under 1°F/in.
0 (32)	0.00139	0.00111
4 (39.2)	0.00138	0.00110
15 (59.0)	0.00144	0.00115
20 (68.0)	0.00143	0.00114

Table 13A.9 Velocity of Sound in Water and Sea Water

Temperature (°C)	Pure Water (m/sec)	Sea Water (m/sec)
0	1,400	1,445
10	1,445	1,485
20	1,480	1,520
30	1,505	1,545

Note: Assuming atmospheric pressure and sea water salinity of 35,000 ppm.

Source: From Berry, Bollay, and Beers, *Handbook of Meteorology*, McGraw-Hill, Copyright 1945. With permission.

SECTION 13B WATER QUALITY

Table 13B.10 Constants and Conversion Factors for Water Quality

To Convert	To	Multiply By
Grains per Imperial gallon	Parts per million	14.3
Grains per gallon	Parts per million	17.12
Parts per million	Grains per gallon	0.05841
Parts per million	Tons per acre-foot	0.00136
Parts per million	Tons per day	Second-feet $\times$ 0.0027
Ca^{+2}	$CaCO_3$	2.497
$CaCl_2$	$CaCO_3$	0.9018
HCO_3^{-1}	$CaCO_3$	0.8202
HCO_3^{-1}	CO_3^{-2}	0.4917
Mg^{+2}	$CaCO_3$	4.116
$MgCl_2$	$CaCO_3$	1.051
Na_2CO_3	$CaCO_3$	0.9442
Fe^{+3}	H_2SO_4	2.634
NO_3^{-1}	N	0.2259
N	NO_3^{-1}	4.4266

Parts Per Million to Equivalent Per Million

Ion	Multiply By	Ion	Multiply By
Aluminum (Al^{+3})	0.11119	Iron (Fe^{+3})	0.05372
Barium (Ba^{+2})	0.01456	Lead (Pb^{+3})	0.00965
Bicarbonate (HCO_3^{-1})	0.01639	Lithium (Li^{+1})	0.14409
Bromide (Br^{-1})	0.01251	Magnesium (Mg^{+2})	0.08224
Calcium (Ca^{+2})	0.04990	Manganese (Mn^{+2})	0.03640
Carbonate (CO_3^{-1})	0.03333	Manganese (Mn^{+4})	0.07281
Chloride (Cl^{-1})	0.02820	Nitrate (NO_3^{-1})	0.01613
Chromium (Cr^{+6})	0.11536	Phosphate (PO_4^{-3})	0.03159
Copper (Cu^{+2})	0.03148	Potassium (K^{+1})	0.02558
Fluoride (F^{-1})	0.05263	Sodium (Na^{+1})	0.04350
Hydrogen (H^{+1})	0.99206	Strontium (Sr^{+2})	0.02282
Hydroxide (OH^{-1})	0.05880	Sulfate (SO_4^{-2})	0.02082
Iodide (I^{-1})	0.00788	Sulfide(S^{-1})	0.06237
Iron (Fe^{+2})	0.03581	Zinc (Zn^{+2})	0.03059

Expressions of Hardness

1 grain per gallon	= 1 grain $CaCO_3$ per U.S. gallon
1 part per million	= 1 part $CaCO_3$ per 1,000,000 parts of water
1 English, or Clark, degree	= 1 grain $CaCO_3$ per Imperial gallon
1 French degree	= 1 part $CaCO_3$ per 100,000 parts of water
1 German degree	= 1 part CaO per 100,000 parts of water

Conversions

1 grain per U.S. gallon	= 17.1 ppm, as $CaCO_3$
1 English degree	= 14.3 ppm, as $CaCO_3$
1 French degree	= 10 ppm, as $CaCO_3$
1 German degree	= 17.9 ppm, as CaCO

For most waters, 1 part per million = 1 mg/l. The U.S. Geological Survey has arbitrarily selected 7,000 ppm as the concentration above which a density correction must be made: the determined concentration of each constituent must be divided by the density to give the correct parts per million. The computation is made before rounding the results.

Parts per million are converted to equivalents per million by multiplying ppm by the reciprocals of combining weights of the appropriate ions (see table). The combining weight is equal to the molecular or ionic weight divided by the ionic charge, or valence.

Specific Conductance Relations (Adapted from USGS Water Supply Paper 1454)

For most natural waters of a mixed type, the specific conductance, in micromhos (1 micromho = 10^{-6} mho. The mho is unit of conductance; thus, a body has a conductance of one mho if one ampere of electric current flows when the potential difference is one volt. The conductance of a body in mhos is the reciprocal of the value of its resistance in ohms.), multiplied by 0.65 approximates the residue on evaporation in parts per million. This does not approach an exact relation, and usually increases as the total dissolved solids exceed 2000–3000 ppm. For waters containing appreciable concentrations of free acid, caustic alkalinity, or sodium chloride, the factor may be much less than 0.65. The relationship for any water becomes indefinite as saturation is approached.

With similar limitations, the specific conductance divided by 100 approximates the equivalents per million of cations or anions.

Table 13B.11 Formula Weights and Equivalent Weights of Ions Found in Water

Ion	Formula Weight	Equivalent Weight
Al^{+3}	27.00	9.00
Ba^{+2}	137.00	68.70
HCO_3^{-1}	61.00	61.00
Br^{-1}	79.90	79.90
Ca^{+2}	40.10	20.00
CO_3^{-2}	60.00	30.00
Cl^{-1}	35.50	35.50
Cr^{+6}	52.00	8.67
Cu^{+2}	63.60	31.80
F^{+1}	19.00	19.00
H^{+1}	1.01	1.01
OH^{-1}	17.00	17.00
I^{-1}	127.00	127.00
Fe^{+2}	55.80	27.90
Fe^{+3}	55.80	18.60
Pb^{+2}	207.00	104.00
Li^{+1}	6.94	6.94
Mn^{+2}	54.90	27.50
Mn^{+4}	54.90	13.70
NO_3^{-1}	62.00	62.00
PO_4^{-3}	95.00	31.70
K^{+1}	39.10	39.10
Na^{+1}	23.00	23.00
Sr^{+2}	87.60	43.80
SO_4^{-2}	96.10	48.00
S^{-2}	32.10	16.00
Zn^{+2}	65.40	32.70

Note: Weights expressed to three significant figures.

SECTION 13C LENGTH

Table 13C.12 Conversion Factors for Length

1 mil	$=1\times10^{-3}$ in.
	=0.0254 mm
1 in.	=0.08333 ft
	=0.0278 yd
	=25.4 mm
	=2.54 cm
	=0.0254 m
1 ft	=12 in.
	=0.3333 yd
	$=1.89\times10^{-4}$ mi
	=30.48 cm
	=0.3048 m
	$=3.048\times10^{-4}$ km
1 yd	=36 in.
	=3 ft
	$=5.68\times10^{-4}$ mi
	=91.44 cm
	=0.9144 m
	$=9.144\times10^{-4}$ km
1 mi	=5280 ft
	=1760 yd
	=1609.3 m
	=1.609 km

(Continued)

Table 13C.12 (Continued)

1 μm	$=1\times10^{-3}$ mm
	$=3.937\times10^{-5}$ in.
1 mm	=0.03937 in.
	=0.1 cm
	$=1\times10^{-3}$ m
1 cm	=0.3937 in.
	=0.03281 ft
	=0.01094 yd
	=10 mm
	$=1\times10^{-5}$ km
1 m	=39.37 in.
	=3.281 ft
	=1.094 yd
	$=6.21\times10^{-4}$ mi
	=1000 mm
	=100 cm
	$=1\times10^{-3}$ km
1 km	=3280.8 ft
	=1093.6 yd
	=0.621 mi
	$=1\times10^{5}$ cm
	=1000 m
1 fathom	=6 ft
1 league (land)	=5280 yd
	=3 mi
	=4.828 km
1 International nautical mile	=6076.1 ft
	=1.151 mi
	=1.852 km
10 mm	=1 cm
10 cm	=1 dm
10 dm	=1 m
10 m	=1 dam
10 dam	=1 hm
10 hm	=1 km

SECTION 13D AREA

Table 13D.13 Conversion Factors for Area

1 sq in.	$=6.944\times10^{-3}$ ft^2
	$=7.716\times10^{4}$ yd^2
	=6.452 cm^2
1 sq ft	=144 in.^{2}
	=0.1111 yd^2
	$=2.296\times10^{-5}$ acre
	$=3.587\times10^{-8}$ mi^2
	=929.0 cm^2
1 sq yd	=1296 in.^{2}
	=9 ft^2
	$=2.066\times10^{4}$ acres
	$=3.228\times10^{-7}$ mi^2
	=0.8361 m^2
	$=8.361\times10^{-5}$ hectare
1 Acre	=43,560 ft^2
	=4840 yd^2
	$=1.562\times10^{-3}$ mi^2
	=4046.9 m^2
	=0.4047 hectare
	$=4.047\times10^{-3}$ km^2

(Continued)

Table 13D.13 (Continued)

1 sq mi	$=2.788\times10^7$ ft^2
	$=3.098\times10^6$ yd^2
	$=640$ acres
	$=2.590\times10^6$ m^2
	$=259.0$ hectares
	$=2.590$ km^2
	$=1$ section (of land)
1 Township	$=36$ mi^2
	$=36$ sections (of land)
1 sq cm	$=0.1550$ in.^{2}
	$=1.076\times10^{-3}$ ft^2
	$=1.196\times10^{-4}$ yd^2
	$=1\times10^{-4}$ yd^2
1 sq m	$=1550.0$ in.^{2}
	$=10.76$ ft^2
	$=1.196$ yd^2
	$=2.471\times10^{-4}$ acre
	$=1\times10^{-4}$ hectare
	$=1\times10^8$ km^2
1 Hectare	$=1.076\times10^5$ ft^2
	$=1.196\times10^4$ yd^2
	$=2.471$ acres
	$=3.861\times10^{-3}$ mi^2
	$=1\times10^4$m^2
	$=0.01$ km^2
1 sq km	$=1.076\times10^7$ ft^2
	$=1.196\times10^6$ yd^2
	$=247.1$ acres
	$=0.3861$ mi^2
	$=1\times10^6$ mi^2
	$=100$ hectares

SECTION 13E VOLUME

Table 13E.14 Conversion Factors for Volume

1 cu in.	$=4.329\times10^{-3}$ gal
	$=5.787\times10^4$ ft^3
	$=16.39$ cm^3
	$=0.01639$ l
1 gal	$=231$ in.^{3}
	$=0.1337$ ft^3
	$=4.951\times10^{-3}$ yd^3
	$=3.785\times10^{-3}$ cm^3
	$=3.785$ l
	$=3.785\times10^{-3}$ m^3
1 cu ft	$=1728$ in.^{3}
	$=7.481$ gal
	$=0.03704$ yd^3
	$=2.832\times10^4$ cm^3
	$=28.32$ l
	$=0.02832$ m^3
1 cu yd	$=4.666\times10^4$ in.^{3}
	$=27$ ft^3
	$=0.7646$ m^3
	$=6.198\times10^{-4}$ acre-ft
1 Acre-in.	$=2.715\times10^4$ gal
	$=3630$ ft^3
1 Acre-ft	$=3.259\times10^5$ gal
	$=43{,}560$ ft^3
	$=1234$ m^3

(Continued)

Table 13E.14 (Continued)

1 cu cm	$=0.06102$ in.^{3}
	$=2.642\times10^4$ gal
	$=3.532\times10^{-5}$ ft^3
	$=1\times10^{-3}$ l
1 l	$=61.02$ in.^{3}
	$=0.2642$ gal
	$=0.03532$ ft^3
	$=1000$ cm^3
	$=1\times10^{-3}$ m^3
1 cu m	$=264.2$ gal
	$=35.32$ ft^3
	$=1.307$ yd^3
	$=1\times10^6$ cm^3
	$=1000$ l
1 Imperial gallon	$=277.4$ in.^{3}
	$=1.201$ gal
	$=0.1604$ ft^3
	$=4.546$ l
1 cfs-day	$=1.98$ acre-ft
	$=0.0372$ in. mi^2
1 in. of rain	$=5.610$ gal/yd^2
	$=2.715\times10^4$gal/acre
1 in.-mi^2.	$=2.323\times10^6$ ft^3
	$=53.3$ acre-ft
	$=26.9$ cfs-days
1 Barrel (oil)	$=42$ gal
1 Million gallons	$=3.069$ acre-ft
1 Pint	$=28.875$ in.^{3}
	$=0.5$ qt
	$=0.473$ l
1 Quart	$=57.75$ in.^{3}
	$=2$ pt
	$=0.25$ gal
	$=0.946$ l

Table 13E.15 Areas and Volumes of Vertical Cylindrical Tanks and Wells

Diameter	Area, sq ft or cu ft per ft of Depth	U.S. gal per ft of Depth	Diameter	Area, sq ft or cu ft per ft of Depth	U.S. gal per ft of Depth	Diameter	Area, sq ft or cu ft per ft of Depth	U.S. gal per ft of Depth
1′	0.785	5.87	6′	28.27	211.5	28′	615.8	4,606
1′1″	0.922	6.89	6′3″	30.68	229.5	28′6″	637.9	4,772
1′2″	1.069	8.00	6′6″	33.18	248.2	29′	660.5	4,941
1′3″	1.227	9.18	6′9″	35.78	267.7	29′6″	683.5	5,113
1′4″	1.396	10.44	7′	38.48	287.9	30′	706.9	5,288
1′5″	1.576	11.79	7′3″	41.28	308.8	31′	754.8	5,646
1′6″	1.767	13.22	7′6″	44.18	330.5	32′	804.3	6,016
1′7″	1.969	14.73	7′9″	47.17	352.9	33′	855.3	6,398
1′8″	2.182	16.32	8′	50.27	376.0	34′	907.9	6,792
1′9″	2.405	17.99	8′3″	53.46	399.9	35′	962.1	7,197
1′10″	2.640	19.75	8′6″	56.75	424.5	36′	1,018	7,616
1′11″	2.885	21.58	8′9″	60.13	449.8	37′	1,075	8,043
2′	3.142	23.50	9′	63.62	475.9	38′	1,134	8,483
2′1″	3.409	25.50	9′3″	67.20	502.7	39′	1,195	8,940
2′2″	3.637	27.58	9′6″	70.88	530.2	40′	1,257	9,404
2′3″	3.976	29.74	9′9″	74.66	558.5	41′	1,320	9,876
2′4″	4.276	31.99	10′	78.54	587.5	42′	1,385	10,360
2′5″	4.587	34.31	10′6″	86.59	647.7	43′	1,452	10,860
2′6″	4.909	36.72	11′	95.03	710.9	44′	1,521	11,370
2′7″	5.241	39.21	11′6″	103.9	777.0	45′	1,590	11,900

(Continued)

Table 13E.15 (Continued)

Diameter	Area, sq ft or cu ft per ft of Depth	U.S. gal per ft of Depth	Diameter	Area, sq ft or cu ft per ft of Depth	U.S. gal per ft of Depth	Diameter	Area, sq ft or cu ft per ft of Depth	U.S. gal per ft of Depth
2′8″	5.585	41.78	12′	113.1	846.0	46′	1,662	12,430
2′9″	5.940	44.43	12′6″	122.7	918.0	47′	1,735	12,980
2′10″	6.305	47.16	13′	132.7	992.9	48′	1,810	13,540
2′11″	6.681	49.98	13′6″	143.1	1,071	49′	1,886	14,110
3′	7.069	52.88	14′	153.9	1,152	50′	1,964	14,690
3′1″	7.467	55.86	14′6″	165.1	1,235	52′	2,124	15,890
3′2″	7.876	58.92	15′	176.7	1,322	54′	2,290	17,130
3′3″	8.296	62.06	15′6″	188.7	1,412	56′	2,463	18,420
3′4″	8.727	65.28	16′	201.1	1,504	58′	2,642	19,760
3′5″	9.168	68.58	16′6″	213.8	1,600	60′	2,827	21,150
3′6″	9.621	71.97	17′	227.0	1,698	62′	3,019	22,580
3′7″	10.08	75.44	17′6″	240.5	1,799	64′	3,217	24,060
3′8″	10.56	78.99	18′	254.5	1,904	66′	3,421	25,590
3′9″	11.04	82.62	18′6″	268.8	2,011	68′	3,632	27,170
3′10″	11.54	86.33	19′	283.5	2,121	70′	3,848	28,790
3′11″	12.05	90.13	19′6″	298.6	2,234	72′	4,072	30,450
4′	12.57	94.00	20′	314.2	2,350	74′	4,301	32,170
4′1″	13.10	97.96	20′6″	330.1	2,469	76′	4,536	33,930
4′2″	13.64	102.0	21′	346.4	2,591	78′	4,778	35,740
4′3″	14.19	106.1	21′6″	363.1	2,716	80′	5,027	37,600
4′4″	14.75	110.3	22′	380.1	2,844	82′	5,281	39,500
4′5″	15.32	114.6	22′6″	397.6	2,974	84′	5,542	41,450
4′6″	15.90	119.0	23′	415.5	3,108	86′	5,809	43,450
4′7″	16.50	123.4	23′6″	433.7	3,245	88′	6,082	45,490
4′8″	17.10	128.0	24′	452.4	3,384	90′	6,362	47,590
4′9″	17.72	132.6	24′6″	471.4	3,527	92′	6,648	49,720
4′10″	18.35	137.3	25′	490.9	3,672	94′	6,940	51,920
4′11″	18.99	142.0	25′6″	510.7	3,820	96′	7,238	54,140
5′	19.63	146.9	26′	530.9	3,972	98′	7,543	56,420
5′3″	21.65	161.9	26′6″	551.5	4,126	100′	7,854	58,750
5′6″	23.76	177.7	27′	572.6	4,283			
5′9″	25.97	194.3	27′6″	594.0	4,443			

Table 13 E.16 Volumes of Horizontal Cylindrical Tanks

Diameter (ft)	Percent Full 10	20	30	40	50	60	70	80	90	100
1	0.3	0.8	1.4	2.1	2.9	3.6	4.3	4.9	5.5	5.87
2	1.2	3.3	5.9	8.8	11.7	14.7	17.5	20.6	22.2	23.50
3	2.7	7.5	13.6	19.8	26.4	33.0	39.4	45.2	50.1	52.88
4	4.9	13.4	23.8	35.0	47.0	59.0	70.2	80.5	89.0	94.00
5	7.6	20.0	37.0	55.0	73.0	92.0	110.0	126.0	139.0	146.9
6	11.0	30.0	53.0	78.0	106.0	133.0	158.0	182.0	201.0	211.5
7	15.0	41.0	73.0	107.0	144.0	181.0	215.0	247.0	272.0	287.9
8	19.0	52.0	96.0	140.0	188.0	235.0	281.0	322.0	356.0	376.0
9	25.0	67.0	112.0	178.0	238.0	298.0	352.0	408.0	450.0	475.9
10	30.0	83.0	149.0	219.0	294.0	368.0	440.0	504.0	556.0	587.5
11	37.0	101.0	179.0	265.0	356.0	445.0	531.0	610.0	672.0	710.9
12	44.0	120.0	214.0	315.0	423.0	530.0	632.0	714.0	800.0	846.0
13	51.0	141.0	250.0	370.0	496.0	621.0	740.0	850.0	940.0	992.9
14	60.0	164.0	291.0	430.0	576.0	722.0	862.0	989.0	1084.0	1152.0
15	68.0	188.0	334.0	494.0	661.0	829.0	988.0	1134.0	1253.0	1322.0

Note: U.S. gallons per foot of length.

SECTION 13F VELOCITY

Table 13F.17 Conversion Factors for Velocity

1 ft/sec	=0.682 mi/h
	=0.3048 m/sec
1 mi/h	=1.467 ft/sec
	=0.869 International knots
	=1.609 km/h
1 International knot	=1.151 mi/h
	=1.852 km/h
1 m/sec	=3.6 km/h
	=3.28 ft/sec
	=2.237 mi/h
	=1.94 knots
1 km/h	=0.621 mi/h
	=0.540 International knot

SECTION 13G FLOW RATE

Table 13G.18 Conversion Factors for Flow Rate

1 gal/min	$=1.44\times10^{-3}$ mgd
	$=2.23\times10^{-3}$ cfs
	$=4.42\times10^{-3}$ AF/day
	=0.0631 l/sec
	$=6.31\times10^{-5}$ m^3/sec
	=5.42 m^3/day
1 Million gallons per day	=694 gpm
	=1.55 cfs
	=3.07 AF/day
	=43.7 l/sec
	=0.0437 m^3/sec
	=3770 m^3/day
1 Cubic foot per second	=449 gpm
	=0.646 mgd
	=1.98 AF/day
	=28.3 l/sec
	=0.0283 m^3/sec
	=2450 m^3/day
1 acre-foot per day	=226 gpm
	=0.326 mgd
	=0.505 cfs
	=14.2 l/sec
	=0.0142 m^3/sec
	=1230 m^3/day
1 l/sec	=15.9 gpm
	=0.0228 mgd
	=0.0353 cfs.
	=0.0703 AF/day
	=0.001 m^3/sec
	=86.4 m^3/day
1 Cubic meter per day second	$=1.58\times10^{4}$ gpm
	=22.8 mgd
	=35.3 cfs
	=70.0 AF/day
	=1000 l/sec
	$=8.64\times10^{4}$ m^3/day
1 Cubic per day	=0.183 gpm
	$=2.64\times10^{-4}$ mgd
	$=4.09\ x\ 10^{-4}$ cfs

(Continued)

Table 13G.18 (Continued)

	$=8.11\times10^{-4}$ AF/day
	$=0.0116$ l/sec
	$=1.16\times10^{-5}$ m^3/sec
1 Miner's inch	=0.025 cubic foot per second (in Arizona, California, Montana, and Oregon)
	=0.02 cubic foot per second (in Idaho, Kansas, Nebraska, New Mexico, North and South Dakota, and Utah)
	=0.026 cubic foot per second (in Colorado)
	=0.028 cubic foot per second (in British Columbia)

Table 13G.19 Conversion of Daily Consumptive Use Rates to Equivalent Continuous Flow Rates per Unit Area

Consumptive Use of Water by Crop		Equivalent Continuous Flow Rates			
		per Hectare		per Acre	
Millimetres per Day	Inches per Day	Litres per Second	Cubic Metres per Day	Cubic Feet per Second	U.S. Gallons per Minute[a]
2	0.08	0.23	20	0.0033	1.5
3	0.12	0.35	30	0.0050	2.2
4	0.16	0.46	40	0.0067	3.0
5	0.20	0.58	50	0.0083	3.7
6	0.24	0.69	60	0.0100	4.5
7	0.28	0.81	70	0.0117	5.2
8	0.32	0.92	80	0.0133	6.0
9	0.36	1.04	90	0.0150	6.7
10	0.40	1.15	100	0.0167	7.5

[a] To convert to imperial gallons per minute, multiply values by 0.833.

Source: From Food and Agriculture Organization of the United Nations, 1974, *Surface Irrigation*. Copyright FAO. Reprinted with permission.

Table 13G.20 Time Required to Supply One Acre with 1 in. of Water

Flow Rate		Time	
gpm	cfs	Hr	Min
100	0.22	4	32
150	0.33	3	01
200	0.45	2	16
250	0.56	1	48
300	0.67	1	30
350	0.78	1	18
400	0.89	1	08
450	1.00	1	00
500	1.11	0	54
600	1.34	0	46
700	1.56	0	39
800	1.78	0	34
900	2.01	0	30
1000	2.23	0	27
1100	2.45	0	24
1200	2.67	0	22
1300	2.90	0	21
1400	3.12	0	20
1500	3.34	0	18

SECTION 13H WEIGHT

Table 13H.21 Conversion Factors for Weight

1 Grain	$=2.286\times10^{-3}$ oz
	$=1.429\times10^{-4}$ lb
1 Ounce	=437.5 gr
	=0.0625 lb
	=28.35 gm
1 Pound	=7000 gr
	=16 oz
	=453.6 gm
	=0.4536 kg
1 Short ton	=2000 lb
	=907.2 kg
	=0.9072 m ton
1 Long ton	=2240 lb
	=1016.0 kg
	=1.016 m ton
1 Metric ton	=2204.6 lb
	=1000 kg
	=1.102 short ton
	=0.9842 long ton
1 g	=15.43 gr
	=0.03527 oz
	$=2.205\times10^{-3}$ lb
	=0.001 kg
1 kilogram	=35.27 oz
	=2.205 lb
	=1000 gm
	=0.001 m ton

SECTION 13I PRESSURE

Table 13I.22 Conversion Factors for Pressure

1 in. of water at 62°F	=0.0361 lb/in.^{2}
	=5.196 lb/ft^2
	=0.0735 in. of mercury at 62°F
1 ft of water at 62°F	=0.433 lb/in.^{2}
	=62.36 lb/ft^2
	=0.833 in. of mercury at 62°F
	$=2.950\times10^{-2}$ atmosphere
1 Atmosphere	=14.7 lb/in.^{2}
	=2116 lb/ft^2
	=33.95 feet of water at 62°F
	=29.92 in. of mercury at 32°F
	=1013.2 millibars
	=760 mm of mercury at 32°F
	=1033 gm/cm^2
	=1.033 kg/cm^2
1 Pound per sq in.	=2.309 feet of water at 62°F
	=2.036 in. of mercury at 32°F
	=0.06804 atmosphere
	=0.07031 kg/cm^2
1 in. of mercury at 32°F	=0.4192 lb/in.^{2}
	=1.133 feet of water at 32°F
1 kg/sq cm	=14.22 lb/in.^{2}
1 Bar	$=1\times10^6$ dynes/cm^2
	=1019.8 gm/cm^2
	=0.9869 atmosphere
	=14.50 lb/in.^{2}
	=1000 millibars

SECTION 13J PERMEABILITY

Table 13J.23 Units and Conversion Factors for Permeability

Permeability Units[a]	
Laboratory (standard) coefficient of permeability	$1\,K_s = \dfrac{1 \text{ gal. of water at } 60°\text{F/day}}{(1 \text{ ft}^2)(1 \text{ ft/ft})}$
Field coefficient of permeability	$1\,K_f = \dfrac{1 \text{ gal. of water at field temperature/day}}{(1 \text{ ft} \times 1 \text{ mi})(1 \text{ ft/mi})}$
Specific (intrinsic) permeability	$1 \text{ darcy} = \dfrac{1 \text{ centipoise} \times 1 \text{ cm}^3\text{/sec})/1 \text{ cm}^2}{1 \text{ atmosphere/1 cm}}$
Conversion Factors	
Permeability	
1 K_s	$= 4.72 \times 10^{-5}$ cm/sec
	$= 0.0408$ m/day
	$= 0.134$ ft/day
	$= 1.43 \times 10^{-6}$ ft/sec
1 Darcy	$= 0.987 \times 10^{-8}$ cm^2
	$= 1.062 \times 10^{-11}$ ft^2
	$= 18.2$ K_s for water at 60°F
	$= 0.966 \times 10^{-3}$ cm/sec for water at 20°C
1 Millidarcy	$= 1.82 \times 10^{-2}$ K_s for water at 60°F
1 cm/sec	$= 1.02 \times 10^{-5}$ cm^2 for water at 20°C
1 m/sec	$= 2.12 \times 10^{6}$ gpd/ft^2
1 m/day	$= 24.5$ gpd/ft^2
Transmissivity	
1 m^2/sec	$= 6.96 \times 10^{6}$ gpd/ft
1 m^2/day	$= 80.5$ gpd/ft

[a] In groundwater investigations, the water-transmitting capability of an aquifer is commonly expressed in terms of a Coefficient of Transmissivity, which is the product of the Coefficient of Permeability of the aquifer material and the thickness of the aquifer.

SECTION 13K POWER

Table 13K.24 Conversion Factors for Power

1 hp	=.746 kW or 746 W
	=33,000 ft lb/min
	=550 ft lb/sec
	=8760 hp-h/yr
	=6535 kWh/yr
H.P. Input	=Horsepower input to motor
	=1.34×kilowatts input to motor
Water H.P.	=H.P. required to lift water at a definite rate to a given distance assuming 100% efficiency
	$=\frac{\text{G.P.M.} \times \text{total head (in. ft)}}{3960}$
Brake H.P.	=H.P. delivered by motor
	=H.P. required by pump
	=H.P. input motor×efficiency
	=1.34×KW input×motor efficiency
	$=\frac{\text{Water H.P.}}{\text{pump efficiency}}$
	$=\frac{\text{G.P.M.} \times \text{total head(ft)}}{3960 \times \text{pump efficiency}}$
	$=\frac{\text{G.P.H.} \times \text{total head(lb/sq in.)}}{103,000 \times \text{pump efficiency}}$
1 kilowatt	=1.3405 Horsepower
	=8760 kWh/yr
Efficiency	$=\frac{\text{Power output}}{\text{Power input}}$
Motor Efficiency	$=\frac{\text{H.P. output}}{\text{KW input} \times 1.34}$
Pump Efficiency	$=\frac{\text{G.P.M.} \times \text{total head(ft)}}{3960 \times \text{B.H.P}}$
Plant Efficiency	$=\frac{\text{G.P.M.} \times \text{total head(ft)}}{5300 \times \text{KW input}}$

Table 13K.25 Horsepower Required to Pump Water

Gallons per Minute	Cubic Feet per Second	Horsepower Required for Elevations of							
		10 ft	20 ft	30 ft	40 ft	50 ft	60 ft	70 ft	80 ft
100	0.22	0.5	1.0	1.5	2.0	2.5	3.0	3.5	4.0
150	0.33	0.8	1.5	2.3	3.0	3.8	4.6	5.3	6.1
200	0.45	1.0	2.0	3.0	4.0	5.0	6.1	7.1	8.1
250	0.56	1.3	2.5	3.8	5.0	6.3	7.6	8.8	10.1
300	0.67	1.5	3.0	4.6	6.1	7.6	9.1	10.6	12.1
350	0.78	1.8	3.5	5.3	7.1	8.8	10.6	12.4	14.1
400	0.89	2.0	4.0	6.1	8.1	10.1	12.1	14.1	16.2
450	1.00	2.3	4.6	6.8	9.1	11.4	13.6	15.9	18.2
500	1.11	2.5	5.0	6.7	10.1	12.6	15.2	17.7	20.2
600	1.34	3.0	6.1	9.1	12.1	15.2	18.2	21.2	24.2
700	1.56	3.5	7.1	10.6	14.1	17.7	21.2	24.8	28.3
800	1.78	4.0	8.1	12.1	16.2	20.2	24.2	28.3	32.3
900	2.01	4.6	9.1	13.6	18.2	22.7	27.3	31.8	36.4
1000	2.23	5.0	10.1	15.2	20.2	25.2	30.3	35.4	40.4
1250	2.78	6.3	12.6	18.9	25.2	31.6	37.9	44.2	50.5
1500	3.34	7.6	15.2	22.7	30.3	37.9	45.4	53.0	60.6

Note: Efficiency of pumping plant 50 percent of theoretical. Use for estimating only.

Table 13K.26 Theoretical Power of American Multiblade Windmills of Different Diameters for Several Wind Velocities

Wind Velocity (mi/h)	Horsepower[a] from Wheel Diameter (in feet) of						
	6	8	10	12	14	16	18
6				0.05	0.06	0.08	0.09
8		0.05	0.07	0.10	0.14	0.18	0.23
10	0.05	0.09	0.14	0.21	0.28	0.36	0.46
12	0.09	0.15	0.24	0.36	0.49	0.63	0.81
15	0.18	0.31	0.48	0.68	0.94	1.21	1.55
20	0.40	0.72	1.12	1.62	2.20	2.88	3.65
25	0.79	1.41	2.21	3.20	4.34	5.67	7.20

[a] Horsepower $= CD^2 V^3$. Where C, the power coefficient for American multiblade type wheel $= 0.45 \times 10^{-6}$, when the tip speed is approximately the same as the velocity of the wind; D = diameter of the wheel in feet; and V = indicated velocity of the wind in feet per second (Robinson anemometer).

Source: From U.S. Dept. of Agriculture.

SECTION 13L TEMPERATURE

Table 13L.27 Conversion Factors for Temperature, Temperature Conversion Fahrenheit ↔ Centigrade

°F to °C (Range 0–279°F)

°F	°C	°F	°C	°F	°C	°F	°C	°F	°C
0	−17.78	56	13.33	112	44.44	168	75.56	224	106.67
1	−17.22	57	13.89	113	45.00	169	76.11	225	107.22
2	−16.67	58	14.44	114	45.56	170	76.67	226	107.78
3	−16.11	59	15.00	115	46.11	171	77.22	227	108.33
4	−15.56	60	15.56	116	46.67	172	77.78	228	108.89
5	−15.00	61	16.11	117	47.22	173	78.33	229	109.44
6	−14.44	62	16.67	118	47.78	174	78.89	230	110.00
7	−13.89	63	17.22	119	48.33	175	79.44	231	110.56
8	−13.33	64	17.78	120	48.89	176	80.00	232	111.11
9	−12.78	65	18.33	121	49.44	177	80.56	233	111.67
10	−12.22	66	18.89	122	50.00	178	81.11	234	112.22
11	−11.67	67	19.44	123	50.56	179	81.67	235	112.78
12	−11.11	68	20.00	124	51.11	180	82.22	236	113.33
13	−10.56	69	20.56	125	51.67	181	82.78	237	113.89
14	−10.00	70	21.11	126	52.22	182	83.33	238	114.44
15	−9.44	71	21.67	127	52.78	183	83.89	239	115.00
16	−8.89	72	22.22	128	53.33	184	84.44	240	115.56
17	−8.33	73	22.78	129	53.89	185	85.00	241	116.11
18	−7.78	74	23.33	130	54.44	186	85.56	242	116.67
19	−7.22	75	23.89	131	55.00	187	86.11	243	117.22
20	−6.67	76	24.44	132	55.56	188	86.67	244	117.78
21	−6.11	77	25.00	133	56.11	189	87.22	245	118.33
22	−5.56	78	25.56	134	56.67	190	87.78	246	118.89
23	−5.00	79	26.11	135	57.22	191	88.33	247	119.44
24	−4.44	80	26.67	136	57.78	192	88.89	248	120.00
25	−3.89	81	27.22	137	58.33	193	89.44	249	120.56
26	−3.33	82	27.78	138	58.89	194	90.00	250	121.11
27	−2.78	83	28.33	139	59.44	195	90.56	251	121.67
28	−2.22	84	28.89	140	60.00	196	91.11	252	122.22
29	−1.67	85	29.44	141	60.56	197	91.67	253	122.78
30	−1.11	86	30.00	142	61.11	198	92.22	254	123.33
31	−0.56	87	30.56	143	61.67	199	92.78	255	123.89
32	0.00	88	31.11	144	62.22	200	93.33	256	124.44
33	0.56	89	31.67	145	62.78	201	93.89	257	125.00

(Continued)

Table 13L.27 (Continued)

°F to °C (Range 0–279°F)

°F	°C	°F	°C	°F	°C	°F	°C	°F	°C
34	1.11	90	32.22	146	63.33	202	94.44	258	125.56
35	1.67	91	32.78	147	63.89	203	95.00	259	126.11
36	2.22	92	33.33	148	64.44	204	95.56	260	126.67
37	2.78	93	33.89	149	65.00	205	96.11	261	127.22
38	3.33	94	34.44	150	65.56	206	96.67	262	127.78
39	3.89	95	35.00	151	66.11	207	97.22	263	128.33
40	4.44	96	35.56	152	66.67	208	97.78	264	128.89
41	5.00	97	36.11	153	67.22	209	98.33	265	129.44
42	5.56	98	36.67	154	67.78	210	98.89	266	130.00
43	6.11	99	37.22	155	68.33	211	99.44	267	130.56
44	6.67	100	37.78	156	68.89	212	100.00	268	131.11
45	7.22	101	38.33	157	69.44	213	100.56	269	131.67
46	7.78	102	38.89	158	70.00	214	101.11	270	132.22
47	8.33	103	39.44	159	70.56	215	101.67	271	132.78
48	8.89	104	40.00	160	71.11	216	102.22	272	133.33
49	9.44	105	40.56	161	71.67	217	102.78	273	133.89
50	10.00	106	41.11	162	72.22	218	103.33	274	134.44
51	10.56	107	41.67	163	72.78	219	103.89	275	135.00
52	11.11	108	42.22	164	73.33	220	104.44	276	135.56
53	11.67	109	42.78	165	73.89	221	105.00	277	136.11
54	12.22	110	43.33	166	74.44	222	105.56	278	136.67
55	12.78	111	43.89	167	75.00	223	106.11	279	137.22

°F to °C (Range 280–300°F)		°C to °F (Range 0–150°C)					
°F	°C	°C	°F	°C	°F	°C	°F
280	137.78	0	32.0	56	132.8	112	233.6
281	138.33	1	33.8	57	134.6	113	235.4
282	138.89	2	35.6	58	136.4	114	237.2
283	139.44	3	37.4	59	138.2	115	239.0
284	140.00	4	39.2	60	140.0	116	240.8
285	140.56	5	41.0	61	141.8	117	242.6
286	141.11	6	42.8	62	143.6	118	244.4
287	141.67	7	44.6	63	145.4	119	246.2
288	142.22	8	46.4	64	147.2	120	248.0
289	142.78	9	48.2	65	149.0	121	249.8
290	143.33	10	50.0	66	150.8	122	251.6
291	143.89	11	51.8	67	152.6	123	253.4
292	144.44	12	53.6	68	154.4	124	255.2
293	145.00	13	55.4	69	156.2	125	257.0
294	145.56	14	57.2	70	158.0	126	258.8
295	146.11	15	59.0	71	159.8	127	260.6
296	146.67	16	60.8	72	161.6	128	262.4
297	147.22	17	62.6	73	163.4	129	264.2
298	147.78	18	64.4	74	165.2	130	266.0
299	148.33	19	66.2	75	167.0	131	267.8
300	148.89	20	68.0	76	168.8	132	269.6
		21	69.8	77	170.6	133	271.4
		22	71.6	78	172.4	134	273.2
		23	73.4	79	174.2	135	275.0
		24	75.2	80	176.0	136	276.8
		25	77.0	81	177.8	137	278.6
		26	78.8	82	179.6	138	280.4
		27	80.6	83	181.4	139	282.2
		28	82.4	84	183.2	140	284.0
		29	84.2	85	185.0	141	285.8
		30	86.0	86	186.8	142	287.6
		31	87.8	87	188.6	143	289.4
		32	89.6	88	190.4	144	291.2

(Continued)

Table 13L.27 (Continued)

°C to °F (Range 0–150°C)					
°C	**°F**	**°C**	**°F**	**°C**	**°F**
33	91.4	89	192.2	145	293.0
34	93.2	90	194.0	146	294.8
35	95.0	91	195.8	147	296.6
36	96.8	92	197.6	148	298.4
37	98.6	93	199.4	149	300.2
38	100.4	94	201.2	150	302.0
39	102.2	95	203.0		
40	104.0	96	204.8		
41	105.8	97	206.6		
42	107.6	98	208.4		
43	109.4	99	210.2		
44	112.2	100	212.0		
45	113.0	101	213.8		
46	114.8	102	215.6		
47	116.6	103	217.4		
48	118.4	104	219.2		
49	120.2	105	221.0		
50	122.0	106	222.8		
51	123.8	107	224.6		
52	125.6	108	226.4		
53	127.4	109	228.2		
54	129.2	110	230.0		
55	131.0	111	231.8		

Note: Temperature Conversions: °F = 9/5 (°C) + 32; °C = 5/9 (°F − 32).

Index

B

E

F

G

H

I

K

L

M

N

O

P

Q

R

S

U

V

W

Y

Z

Related Titles

Field Hydrogeology — A Guide for Site Investigations and Report Preparation
John E Moore
ISBN: 1566705878

Practical Handbook of Environmental Site Characterization and Ground-Water Monitoring, Second Edition
Edited by David M Nielsen
ISBN: 1566705894

Environmental Hydrology, Second Edition
Andrew D Ward and Stanley W Trimble
ISBN: 1566706165

The Handbook of Groundwater Engineering, Second Edition
Edited by Jacques W. Delleur
ISBN: 084934316X

Applied Flow and Solute Transport Modeling in Aquifers: Fundamental Principles and Analytical and Numerical Methods
Vedat Batu
ISBN: 0849335744

Site Assessment and Remediation Handbook, Second Edition
Edited by Martin N Sara
ISBN: 1566705770

In Situ *Remediation Engineering*
Suthan S Suthersan and Fred Payne
ISBN: 156670653X

Watershed Models
Edited by Vijay P Singh and Donald K Frevert
ISBN: 0849336090